中国区域地质志

北京志

北京市地质调查研究所 编著

地质出版社

·北 京·

内容简介

本志是编者在对北京市1986年以来完成的基础性地质调查、综合性与专题性地质科学研究成果等进行系统总结的基础上，编著的一部综合性地质专著。本志以活动论为指导，把三十多年来取得的丰富的地质、地球物理、地球化学资料进行综合研究和集成，使之系统化、规律化和理论化，成为整装的地质资料。主要内容包括：对全市多重地层划分进行了系统整理和厘定，对沉积岩和沉积相进行了系统研究，对第四纪地貌与新构造运动进行了总结研究；对全市侵入岩按断代进行了新的期次划分，以陆块聚散为主线深入分析探讨了各不同时代侵入岩的成因、演化、岩浆作用及其与构造演化的关系，对火山岩与火山作用时空演化进行了总结；按变质单元进行了原岩恢复，分析了变质建造，划分了变质相与变质相系，对变质作用演化进行了总结，探讨了其时空演化及与构造演化的关系；按照板块构造理论，按断代对全市的构造单元进行了新的划分，对构造变形特征、地质构造背景、构造与建造的关系、深部构造与构造动力机制等进行了研究；对不同成因的成矿作用与控矿机制进行了归纳总结。

本志继承了1991年版《北京市区域地质志》以燕山运动为主的山区地质思想，突出了以新构造运动为主的城市地质内容，全面、系统、客观地反映了北京市基础性地质调查研究的最新成果，资料丰富详实，采用新理论新方法，观点论据国内领先，部分研究成果达国际先进的水平。将在未来相当长的时间对北京市内的城市地质调查、地矿勘查规划、矿山生态修复、地质环境调查评价、地质灾害调查与预测、重大工程选址和地质科研等工作具有重要指导意义，为社会经济发展与规划服务。可供广大地质工作者、地质勘查单位、工程勘察单位、科研部门、大专院校、政府管理部门、工矿企业等有关部门和人员参考使用。

图书在版编目（CIP）数据

中国区域地质志·北京志／北京市地质调查研究所编著. — 北京：地质出版社，2024.4
ISBN 978-7-116-10356-6

Ⅰ.①中… Ⅱ.①北… Ⅲ.①区域地质-概况-北京
Ⅳ.①P562

中国版本图书馆CIP数据核字（2017）第116301号

Zhongguo Quyu Dizhizhi · Beijingzhi

责任编辑：吴金键　袁亚琴
责任校对：关风云
出版发行：地质出版社
社址邮编：北京市海淀区学院路31号，100083
电　　话：（010）66554528（邮购部）；（010）66554621（编辑室）
网　　址：https://www.gph.clmpg.com
传　　真：（010）66554621
印　　刷：中煤地西安地图制印有限公司
开　　本：889mm×1194mm　1/16
印　　张：67.25　**附图**：5幅
字　　数：2100千字
版　　次：2024年4月北京第1版
印　　次：2024年4月西安第1次印刷
定　　价：980.00元
审 图 号：京S（2023）022号
书　　号：ISBN 978-7-116-10356-6

《中国区域地质志》编纂指导委员会
（第四届）

主　任：李金发

副主任：徐学义　叶建良　吴登定　侯增谦　肖桂义

成　员（按姓氏笔画排序）：

毛晓长	任收麦	许振强	李　军	李建星
杨志明	汪大明	邱海峻	张海啟	张开军
赵洪伟	郭兴华	姚华军	高延光	夏　鹏
贾伟光	顾晓华	熊自力		

The fourth advisory committee of
Regional Geology of China

Director: Li Jinfa

Associate directors: Xu Xueyi　Ye Jianliang　Wu Dengding　Hou Zengqian　Xiao Guiyi

Committee members (in Chinese-character stroke order of surname):

Mao Xiaochang	Ren Shoumai	Xu Zhenqiang	Li Jun
Li Jianxing	Yang Zhiming	Wang Daming	Qiu Haijun
Zhang Haiqi	Zhang Kaijun	Zhao Hongwei	Guo Xinghua
Yao Huajun	Gao Yanguang	Xia Peng	Jia Weiguang
Gu Xiaohua	Xiong Zili		

《中国区域地质志》编辑委员会
（第一届）

The first editorial board of
Regional Geology of China

《中国区域地质志》编辑委员会
（第二届）

The second editorial board of
Regional Geology of China

《中国区域地质志》编辑委员会
（第三届）

The third editorial board of
Regional Geology of China

项目承担单位

中国地质科学院地质研究所

Project organization institution

Institute of Geology, Chinese Academy of Geological Sciences

项目办公室

主　任： 庞健峰

（耿元生，2010 年—2013 年 8 月；

耿树方，2013 年 8 月—2015 年 12 月）

成　员： 张　聪　闫　斌　张　恒　石成龙　王　军

郑　宁　王振洋　郑洪伟　金　铭　张继彪

Project office

Director: Pang Jianfeng

(Geng Yuansheng from 2010 to August, 2013;

Geng Shufang from August, 2013 to December, 2015)

Staff:

Zhang Cong	Yan Bin	Zhang Heng	Shi Chenglong
Wang Jun	Zheng Ning	Wang Zhenyang	Zheng Hongwei
Jin Ming	Zhang Jibiao		

责任编辑组

郑长胜	祁向雷	孙亚芸	刘亚军	蔡卫东
吴宁魁	柳　青	李　莉	王作勇	袁亚琴
宋永存	肖莹莹	田　野	贺秋梅	张　诚
程　静	关会梅	杨　艺	吴金键	张世刚
孙　琳	顾秋蓓	费详惠		

Responsible editorial group

Zheng Changsheng	Qi Xianglei	Sun Yayun	Liu Yajun
Cai Weidong	Wu Ningkui	Liu Qing	Li Li
Wang Zuoyong	Yuan Yaqin	Song Yongcun	Xiao Yingying
Tian Ye	He Qiumei	Zhang Cheng	Cheng Jing
Guan Huimei	Yang Yi	Wu Jinjian	Zhang Shigang
Sun Lin	Gu Qiubei	Fei Xianghui	

总　序

20世纪80—90年代，我国编纂出版了第一代省级区域地质志，系统总结了20世纪80年代以前各省（区、市）区域地质调查研究成果，为促进我国地质事业和服务经济社会的发展发挥了重要作用。

随着改革开放的不断深入，地质调查和各类科研项目的实施，我国地质事业又获得蓬勃发展，取得了丰硕的调查研究成果。在青藏高原、大兴安岭等地，完成了1：25万区域地质调查和航空磁测，实现了全国陆域中比例尺区域地质调查和航空磁测的全覆盖。开展了各种比例尺的区域地球物理和区域地球化学的调查研究，实施了“深部探测技术与实验研究”专项，拓展了环境地质、农业地质、城市地质等新领域，获得了丰富的基础地质资料，解决了许多基础地质问题，找到了一大批矿产，大幅度地提高了我国的地质研究程度和地质科学技术水平。

《国务院关于加强地质工作的决定》（国发〔2006〕4号）指出，要提高基础地质调查（研究）程度，建立地质图文更新机制。《全国地质勘查规划》（2005—2020）提出，要开展全国基础地质资料综合研究和成果集成工作，编制新一代全国系列基础地质图件，努力创造有宏观影响的大成果。为此，专门组织有关专家，分层次编纂各省（区、市）、跨省（区）成矿区带和重要经济区，以及全国层次的新一代系列地质志书，集中反映我国区域地质调查和地质科学研究工作的重大进展和成果，创造一项有宏观影响的大成果，为地质工作和国家经济社会发展提供一套基础地质资料。

为国家经济建设和社会发展提供矿产资源和地质资料，是地质工作服务经济社会发展和人民生产生活的两项基本任务。通过编纂新一代区域地质志书，把全国各部门多年地质调查研究所积累的丰富地质、地球物理、地球化学等资料进行综合研究和集成，使之系统化、规律化和理论化，不仅为地质工作和经济社会发展提供了一套整装的地质资料，而且为发展地质科学理论奠定了扎实的资料基础。

党的十八届三中全会指出，要紧紧围绕建设美丽中国、深化生态文明体制改革，加快建立生态文明制度，健全国土空间开发、资源节约利用、生态环境保护的体制机制，推动形成人与自然和谐发展的现代化建设新格局。这对地质工作提出了新的更高要求。举凡矿产资源的合理开发利用、国土空间的规划、地质环境的保护、地下水污染的治理、地质公园的建设、地球化学对人体健康的影响，以及土壤性质及农产品的优化等，都需要丰富的基础地质数据和地质信息服务。《中国区域地质志》既可以满足一个省（区、市）、一个成矿区（带）及经济区（带）和全国地质矿产勘查、地质工作规划、部署和管理的需求，又可以在城乡建设、交通、水利、能源和环境保护等重大工程规划、设计、建设和科学管理中发挥重要作用。

新一代《中国区域地质志》是对新时期我国基础地质调查工作的系统总结和高度浓缩，注重在结构和体例上尽量与第一代区域地质志保持有机衔接、汲取其优点和精粹的同时，着力坚持创新，突出反映近30年来取得的新成果和新资料，体现了新的水平和研究高度。

这套志书的编纂和陆续出版是一项浩繁的科学工程，参加单位和人员多、涉及专业领域广、持续时间长，浓缩了几代地质工作者数十年辛勤劳动的成果，是我国地质界的一件大事。参与编纂出版的地质工作者以对历史、对人民负责的高度责任感，孜孜以求，辛勤劳作，力争使志书做到资料翔实、内容精湛、质量上乘、富于创新，打造成高水平的精品著作。在这里，我要向为这项工作付出艰苦努力、做出贡献的地质工作者以及关心、支持这项工作的各有关方面致以崇高敬意和衷心感谢！

中华人民共和国国土资源部副部长

中国地质调查局局长

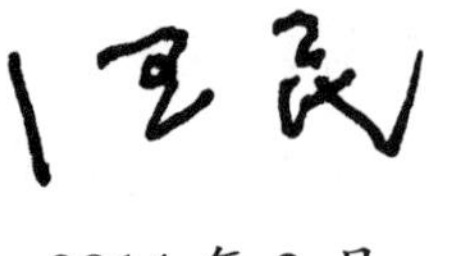

2014年2月

General Preface

Compiled in the 1980s and 1990s, Regional Geology of China of the first generation systematically summarized the research results of regional geology before the 1980s in provinces, autonomous regions or municipalities, and played vital roles in serving social and economic development, and geological advancement in China.

With the deepening of reforms and opening-up, the geological undertakings in China embrace another chance of prosperity. Abundant results of geological survey and research have been achieved along with the implementation of geological survey programs and various scientific research programs. In areas such as the Qinghai-Tibet Plateau and Great Khingan Mountains, regional geological survey and aeromagnetic survey at the scale of 1 : 250 000 have been completed, which means that geological and aeromagnetic survey on medium scales have achieved a full coverage of the whole land area in China. Regional geophysical and geochemical surveys on different scales have been conducted. Sino-Probe, a special program for deep exploration research in China, has been carried out as well. New research fields such as environmental geology, agricultural geology and urban geology have been expanded. All these efforts have brought us with substantial fundamental geological materials, and put a series of geological problems into settlement. Besides, a number of mineral deposits have been found. The intensity of geological survey and the level of geological science and technology have been improved significantly.

As is stated in *Decision of the State Council on Strengthening Geological Work*, the intensity of geological survey and research should be enhanced, and an update mechanism concerning geological materials, including geological maps and reports, should be established. *National Geological Exploration Plan (Program)* (2005—2020) proposes that comprehensive studies of nationwide geological data and integration of research results should be conducted, national series of fundamental geological maps of a new generation should be compiled, and we should strive to create outstanding achievements with macro impact. Under the above mentioned directions, experts in relative disciplines are organized to compile the new series of Regional Geology that successively cover provincial (municipal or autonomous regional) level, cross-provincial (or cross-regional) metallogenic belts or important economic zones level, and national level. The new series of Regional Geology principally represent the significant advancements and achievements of geological survey and research in China. The compilation of the series of Regional Geology constitutes an achievement with remarkable impact, and provides a set of geological data for economic and social development.

Providing mineral resources and supplying geological data for economic construction and social development of China, are two basic tasks of geological work which serves social and economic development and people's production and life. By means of compiling the new series of Regional Geology, abundant geological, geophysical and geochemical data that have been accumulated by different departments of the entire country for many years are integrated systematically, regularly and theoretically. The comprehensive study not only provides an entire set of geological data for both geology itself and economic and social development, but also lays concrete foundation for developing geological theory.

In the Third Plenary Session of the 18th CPC Central Committee, it is proposed that we must deepen ecological environmental management reform by centering on constructing a beautiful China, and that we should accelerate system building to promote ecological progress, improve institutions and mechanisms for developing land space, conserving resources and protecting the ecological environment, and promoting modernization featuring harmonious development between human and nature. This raises greater demand on geological work. The tasks such as rational development and utilization of mineral resources, land space planning, protection of geological environment, groundwater pollution control, geo-park construction, evaluation of geochemical impact on human health and optimization of soil properties and agricultural products, all depend on rich fundamental geological data and geological information services. the *Regional Geology of China* can not only fulfill the requirement for mineral resources exploration and planning, deployment and management of geological work in provincial (municipal or autonomous regional) level, metallogenic zone (belt) and economic zone level and national level, but also hold an important role in planning, design, construction and scientific management of major projects in fields such as urban and rural construction, transportation, water conservancy, energy development and environmental protection.

The *Regional Geology of China* of the new generation systematically summarizes and highly condenses the fundamental geological survey work of the new era. To consistent with the first generation in structure and style as much as possible, the new series follow its advantages. Moreover, the series stick to innovation, and focus on presenting the new advancement and new materials acquired in past 30 years, which stands for the new height and achievement in research.

The compilation and successive publication of the Series is a complex scientific project, in which tremendous time and labor have been input. Scholars of a wide range of professional fields from different institutions were organized for the project. Results of hard work obtained by geologists of several generations in the past decades are condensed in the Series. It is a significant event for the geological community in China. Geologists participating in the project, with strong sense of responsibility for history and people, devoted themselves earnestly and unreservedly to creating excellent works with a high quality. They strive to make the books faithful, informative, sophisticated and innovative. Here, I would like to express my heartfelt thanks to and sincere respect for all the geologists who worked hard and contributed to the project and all concerned who wholeheartedly support this project.

Wang Min,
Vice Minister of the Ministry of Land and Resources, PRC
Director of China Geological Survey

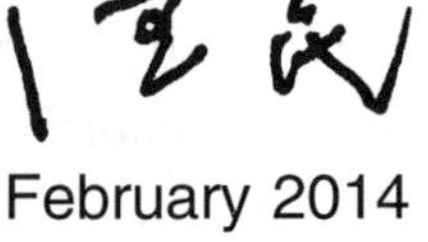

February 2014

总 前 言

2008年，中国地质调查局下达了重编《中国区域地质志》的任务，计划项目名称为“中国地质构造区划综合研究与区域地质调查综合集成”。其总体目标任务是：通过对区域地质调查资料和地学研究成果的综合研究与集成，全面总结30多年来我国区域地质调查、矿产勘查和专题研究所取得的成果，特别是地质大调查实施以来的新资料、新进展、新成果。以活动论和先进的地质科学理论为指导，应用地质编图制图新技术、新方法，按省（区、市）、构造单元（成矿区带、经济区等）和全国3个层次编纂《中国区域地质志》，编制相关地质图件，建立空间数据库；针对综合研究过程中存在的全国性、区域性重大基础地质问题，有重点地开展野外调查和专题研究；进行区域地质调查和地质科学研究资料的系统总结，提高全国地质研究程度和水平，实现地质调查资源共享，提升地质调查和科学研究成果的应用效果和影响，为地质调查、矿产勘查、有关经济和工程建设、地质环境及地质灾害评估，以及地质科研和地质教育等，提供基础地质资料，使地质调查研究成果更好地服务于地质事业和经济社会的可持续发展。

2006年，《国务院关于加强地质工作的决定》（国发〔2006〕4号）指出，要“提高基础地质调查（研究）程度”“系统开展区域地质、地球物理、地球化学和遥感地质等调查，建立地质图文更新机制，为社会提供有效快捷的地质信息服务”。经国务院同意颁布的《全国地质勘查规划》（2005—2020）指出，要“开展重大基础地质问题研究和重点地区基础地质综合研究，力争实现基础地质研究的重大突破。开展全国基础地质资料综合研究和成果集成工作。编制新一代全国系列基础地质图件，开展区域基础地质、矿产、地质环境等系列图件编制与各时代主要地层以及多学科的综合性立典研究”“努力创造有宏观影响的大成果，不断提高地质勘查的精度、深度和广度”。

为经济建设和社会发展提供矿产资源和基础地质资料，一向是地质工作服务经济和社会发展的两项基本任务。通过编纂区域地质志，把各个部门、各种类型地质调查研究取得的丰富地质、地球物理、地球化学、遥感地质等资料进行综合研究和集成，使之系统化、规律化和理论化，可以为国家有关部门和广大民众提供一套整装的地质资料，并为发展地球科学理论开辟通途。因此，国土资源部、中国地质调查局决定重编《中国区域地质志》是一个具有远见卓识的英明决定，是一项带有里程碑意义的战略性综合研究工作，必将为我国地质事业和国家经济社会发展做出重大贡献，在我国地质事业发展史上谱写出光辉的一页。

如果从1916年国民政府农商部改中央地质调查所为地质调查局算起，中国区域地质调查研究已经走过了100多年的时间。在中国地质调查研究的历史上，中国地质界具有编纂地质矿产志的光荣传统，早在20世纪前半叶，就先后编辑出版了《苏浙皖矿志》《中国矿产志》《福建矿务志略》《江苏地质志》《北京西山地质志》《四川西康地质志》等（陈宝国等，2011）。20世纪80—90年代，地质矿产部主持编纂的《中国区域地质志》系统地总结了20世纪80年代末以前我国区域地质调查及专题的研究成果，为我国地质事业和经济社会发展做出了重要贡献，建树了历史的功绩。

近30多年来，我国各有关部门进行了各种类型的地质矿产调查研究，积累了丰富的实际资料，获得大量地质、地球物理、地球化学、遥感地质，特别是高精度同位素年代及地球化学、岩石地球化学数据，解决了一系列长期悬而未决的地质问题，大幅度地提高了我国的地质研究程度，为《中国区域地质志》的编纂创造了良好条件。

为适应各方面的需求，提供更多的地质信息，项目任务要求开展3方面的调查研究，提交3方面的成果：①32个省（区、市）地质志（广东省地质志含香港、澳门特别行政区），15个左右重要地区（构造单元、成矿区带、重要经济区、行政大区等）地质志，全国性地质志（地质、地球物理、地球化学系列专著）。②系列基础地质图件及空间数据库。地质图件包括地质图、地质构造图、岩浆岩地质图、第四纪地质与地貌图、航磁异常图、重力异常图（内部使用），各省（区、市）及各地区可根据地区特点与需要编制其他地质图件，如基岩地质图、岩相古地理图、变质地质图、断裂系统图等。③结合志书编写和系列图件编制开展必要的专题研究，解决一些重要的疑难地质问题，旨在统一地质划分对比和深化对这些疑难地质问题的认识，提高志书的质量。

编纂《中国区域地质志》涉及的范围大、学科领域广，参加的单位和人员多。这是一项对百万地质大军几十年乃至上百年地质调查研究成果的一次大综合、大集成。这样大规模的综合地质研究可以而且应该创造出有宏观影响的大成果。因此，我们提出以下5方面的要求，作为提交有宏观影响大成果的主要标志。

第一，在地质成果上要大综合和大集成。在充分搜集、分析各个部门、各种类型地质调查研究成果的基础上，进行地质、地球物理、地球化学相结合的综合研究和成果集成，使志书内容丰富、资料翔实、数据准确、特色鲜明，如实反映当前地质调查研究的程度，反映志书地域内各种地质现象的认知水平。

第二，在地质研究程度上有大幅度的提高。不断提高全国和一个地区的地质研究程度，是地质工作永恒追求的目标，也是地质工作服务于经济社会发展的必然要求。地质研究程度的提高，是靠不断地进行地质调查研究和持续地开展综合研究和成果集成而实现的。因此，通过编纂地质志，从地质、地球物理、地球化学、遥感地质以及地表与深部地质结合上开展精细的综合研究，大幅度地提高编志地区的地质研究程度和水平，为地质工作和经济社会发展提供高质量的地质资料。

第三，在地质工作和经济社会发展中发挥大的作用。随着人类社会的进步，地质工作的社会功能和服务领域日益宽广，地质工作几乎是兴利于各个部门和行业，造福于人类的方方面面。作为一套完整而翔实的基础地质资料，既可以满足一个省（区、市）、一个成矿区（带）及经济区（带）和全国地质矿产勘查、地质工作规划、部署和管理的需求，又可以在城乡建设、交通、水利、资源和环境保护等重大工程规划、设计、建设和科学管理中发挥重要作用；既可以为地质科研和地质教育开辟捷径，又可以为丰富地质信息库和普及地质科学知识提供翔实的地质资料，充分体现地质工作服务于经济社会发展的强度和广度。

第四，在地质规律的概括和地质理论创新上有大的提升。地质工作是一项科学技术密集的调查研究工作，地质科学基本上属于认识性科学。矿产资源勘查能否取得重大突破，地质环境及地质灾害的评估和预测是否准确，对诸多工程建设规划、设计和实施能否发挥更大作用，归根结底取决于对地质规律的认识水平和地质理论及技术方法的创新程度。地质调查研究的目的，在很大程度上是总结地质规律、认识地质规律和利用地质规律。抓住了地质规律，地质理论的创新也就应运而生了。地质志质量的好坏、水平的高低、贡献的大小，在很大程度上取决于对编志地区地质规律和地质理论概括、认识的深度和高度。因此，在地质志编纂过程中，在详细记录地质事实的基础上，通过实际资料的归纳、分析和总结，概括出地质规律，凝练出新发现、新认识，大幅度地提高对编志地区地质规律和地质总体特征的认知水平。

第五，在国内外地质界得到广泛应用和产生大的影响。一项科学成果，其在社会上应用的广度和影响的大小，主要取决于其实用价值和科学价值，其实用价值和科学价值越大，它的影响力就越大。中国地质构造十分复杂而又十分特殊，有许多地质现象为世界罕见，如高大的青藏高原、北方新生代的黄土堆积、南方岩溶（喀斯特）地质及地貌、中部延展数千千米的高压-超高压变质带、某些标准的地层剖面和罕见的古生物群落、某些特殊类型的矿床等，既是中国地质的特色和区位优势，又是诞生地质科学新理论的摇篮，引起国内外地质科学家的关注和竞相探索，以期地质科学理论上的突破。通过编纂区域地质志，认真总结中国地质，特别是上述特色地质的总体特征及其演化规律，给予理论的概括和解释，不但会对地质勘查实践和经济社会发展发挥更大作用，而且会对发展地质科学理论奠定基础和做出重要贡献，也必将引起国内外地质界的广泛兴趣，产生重大影响。

编纂《中国区域地质志》是地质工作中一项最基础的综合研究工作，其成果将在自然资源部门及其他有关部门和外国各有关地质单位长期使用。因此，确保质量是对这套志书的必然要求。为了确保志书质量，要求做到“高起点、高要求、高水平”。

所谓“高起点”，就是在充分掌握编志地区地质研究程度、地质特点和地质规律的高点上着手进行地质志的编写。为此，首先，要广泛收集编志地区地质资料和成果，包括报告、专著、图件和论文，特别是近些年来的地质调查研究的成果。其次，对收集到的资料进行地质、地球物理、地球化学相结合的综合研究，力求洞悉本区地质特点、地质规律和存在的主要地质问题。最后，利用先进的地质理论和技术方法进行志书编写。

所谓“高要求”，就是在编志过程中，对各项工作，包括资料的收集、整理、样品分析测试、志书编写和图件编制等，都要高标准、严要求，严格按“项目总体设计”“工作指南”“项目管理办法”和自然资源部、中国地质调查局的有关规程、规范执行，切实保证成果质量。

所谓“高水平”，就是力争把这套志书打造成高水平的精品著作，在科学内容上精益求精，忠实记录地质事实，科学总结地质特征和地质规律，使志书做到资料翔实，内容精湛，质量上乘，富于创新。在表现形式上也要精益求精，文字描述要言简意赅，文图并茂；附图、附表要主题鲜明，清晰美观；计量单位和名词术语使用等要符合国标、行标和国际标准及惯例。

为了丰富志书内容，提高志书水平，为地矿行业、社会和公众提供更多的地质信息，充分发挥志书的功能，要求：一要实行地质、地球物理、地球化学、遥感等探测研究资料相结合，从宏观和微观尺度鉴别、认识地质现象，解决地质问题；二要实行地表地质与深部地质相结合，从三维角度揭示各种地质现象、成矿作用和地质灾害的深部根源；三要实行大陆地质与海洋地质相结合，把海陆作为一个整体加以综合研究；四要实行志书、图件和数据库建设相结合，更全面、形象地显示编志地区地质、地貌特点，更好地发挥志书功能；五是实行综合研究与专题研究相结合，在开展综合研究的同时，就一些重大的、有争议的地质问题进行专题研究，统一认识，统一划分对比，解决争论问题，提高志书质量。在指导思想上坚持继承和创新相结合，在继承基础上创新，在创新指导下继承，既要继承所有好的成果，又要大胆创新，敢于改变传统的认识和结论。同时，通过编志，以老带新，互教互学，培养一批学科性和综合性地质科技人才。

地质工作和国民经济发展规划、部署，地质资源勘查及地质环境与地质灾害评估、预测，城乡建设及重大工程规划、设计，普及地质科学知识，提高全民族科学文化素质和水平，开展国内外地质科学技术合作与交流，以及发展地球科学理论等，都需要丰富的地质信息。《中国区域地质志》可以为上述诸方面提供厚实的基础地质资料。

这次地质志的编纂规模大、要求高，参加的单位和人员多，持续的时间长，特别需要加强领导和组织协调，加强项目科学管理，全国、各省（区、市）及各地质区地质志的编写，都要求建立领导机构和学术指导组织。

在项目编纂指导委员会和编辑委员会的领导下，设立了《中国区域地质志》综合研究项目（即“中国地质构造区划综合研究与区域地质调查综合集成”“全国陆域及海区地质图件更新与共享”），由中国地质科学院地质研究所承担实施，负责《中国区域地质志》编纂的科学技术指导和组织协调工作，参加人员及分工如下。

一、项目负责人

项目负责人：丁孝忠　李廷栋　刘　勇　庞健峰

二、特聘专家

组长：肖庆辉

成员（按姓氏笔画排序）：邓晋福　陆松年　张克信　潘桂棠

三、项目专业组主要人员

区域地质与洋板块地质组：李廷栋　丁孝忠　刘　勇　石成龙

区域地层组：张克信　杨崇辉　高林志　万渝生　张　恒　董春燕

区域岩浆岩组：王　涛　薛怀民　刘　翠　张建军　童　英　王晓霞　张　磊　寇彩化

区域构造组：游国庆　王　军　张传恒

第四纪地质与地貌组：王　永　迟振卿　董　进　袁路朋　田　飞　叶梦旎　凌　媛

沉积与岩相古地理组：刘燕学　旷红伟　柳永清　郑　宁　彭　阳

变质地质与前寒武纪地质组：任留东　杨崇辉　杜利林　王彦斌　宋会侠

蛇绿岩组：肖序常　翟庆国　胡培远　高　俊　宋述光

同位素年代组：万渝生　董春燕　颉颃强　刘守偈　朱文斌

地球物理及深部地质组：管　烨　郑洪伟　金　胜　张贵宾

地理底图、数据库与图件设计组：韩坤英　庞健峰　剧远景　张庆合　王丽亚　王振洋　姜　兰　金　铭　丁伟翠

六大区与构造单元区域地质综合研究组：计文化　朱小辉　尹福光　任　飞　杨祝良　余明刚　牛志军　谭满堂　赵凤清　李承东

杨晓平　杨雅军　邢德和　李　敏　邢光福

肖国强　刘燕学

海域区域地质综合研究组：温珍河　李学杰　王中波　姚永坚　侯方辉　杨楚鹏

王保军　高红芳　徐子英　祁江豪

《中国区域地质志》的编纂获得了自然资源部（国土资源部）、财政部、科技部及有关司局、中国地质调查局及有关部（室）和中国地质科学院及有关处（室）的指导和大力支持；获得了中国地质调查局6个地质调查中心、中国地质科学院有关研究所和地质研究所及有关处（室），各省（区、市）自然资源部门（国土资源厅）、地质矿产勘查局、地质调查院（所）和有关地质院校的积极参与和大力支持；地矿行业各部门、全国及各省（区、市）地质资料、地质图书保管等单位为志书编写提供了丰硕的地质资料和信息；参与志书编写的专家、学者和管理人员孜孜以求、精益求精，付出了艰辛的劳动和智慧。在这里，向所有关怀、指导志书编写的部门和领导，向所有关心、帮助志书编写的部门和人士，向所有参与志书编写的专家、学者和管理者，致以衷心的感谢和敬意。

地质工作是一项永恒的事业，只要有人类存在，就有地质工作；哪里有人类存在，哪里就有地质工作。这是因为地质工作服务的领域很广，而且越来越广。地质工作为经济社会发展服务的主要途径之一就是不断地为社会提供有效快捷的地质信息，这也是编纂《中国区域地质志》的初衷和主要宗旨之一。

我们希望这套志书能够集我国地质调查研究成果之大成，走向志书的深处、高处和远处，打造出志书的精品，为我国地质事业的发展做出更大贡献！为优化国土生态环境、建设美丽中国做出更大贡献！为我国经济社会的可持续发展和实现中华民族伟大复兴的中国梦做出更大贡献！为世界地球科学的繁荣昌盛做出更大贡献！

李廷栋

2020年5月

General Foreword

In 2008, a project concerning recompilation of the *Regional Geology of China* entitled "Comprehensive Study of Geological and Tectonic Division and Synthesized Integration of Regional Geological Survey in China", was assigned by China Geological Survey (CGS). The overall objective of the project is to fully summarize achievements acquired in regional geological survey, mineral resources exploration and monographic studies, especially the new materials, new advancement and new achievement since the implementation of the Great Geological Survey in China, by means of comprehensive study and integration of regional geological survey data and geoscience research results. Under the guidance of the Mobilism and advanced geoscience theories, the *Regional Geology of China* and relative geological maps are compiled at three levels respectively: provincial (municipal or autonomous regional) level, tectonic unit (metallogenic belts or economic zones) level and national level, while the spatial database are established, with new technologies and methods of geological mapping. Field surveys and monographic studies are conducted with priority to solve the national or regional fundamental geological problems encountered in comprehensive study. Systematical summaries of regional geological survey data and geoscience research materials are done so as to improve intensity and level of geological survey and research, to realize the sharing of geological survey resources and to raise the applied effectiveness and impact of geological survey and research achievements. The *Regional Geology of China* can provide fundamental geological materials for geological survey, mineral exploration, relative economic development and engineering construction, assessment of geological environment and hazard, as well as geological research and education. That enable achievements of geological survey to better serve geological cause and sustainable development of economy and society.

In 2006, *Decision of the State Council on Strengthening Geological Work* put forward that "the intensity of geological survey and research should be enhanced", and that "regional geological, geophysical, geochemical and geological remote sensing surveys should be conducted systematically and an update mechanism concerning geological maps and texts should be established, to provide effective and efficient geological information service". According to *National Geological Exploration Program* (2005—2020) promulgated with the State Council's consent, "researches on major geological problems and comprehensive study of fundamental geology in key areas should be carried out so as to achieve crucial breakthrough in fundamental geological researches. Comprehensive studies of nationwide geological data and integration of research results should be conducted. The new series of fundamental geological maps should be compiled; compilation of map series of regional fundamental geology, mineral resources, geological environment and multi-disciplinary comprehensive typical studies of various major strata should be conducted", and we should "strive to create outstanding achievements with great impact, and to improve accuracy, depth and breadth of geological exploration".

Providing mineral resources and supplying geological data for economic construction and social development of the country, have always been two basic tasks of geological work that serves social and economic development. Through the compilation of the *Regional Geology of China* abundant geological, geophysical, geochemical and remote sensing data collected in all kinds of geological surveys from various institutions are used for comprehensive research and integration. Geological materials that have been systematized, regularized and theorized, can serve relevant government departments and the general public with a whole set of data, and beat a path for developing theories of earth science. Therefore, to recompile the *Regional Geology of China* is a wise decision with visionary transcendence by China Geological Survey, former Ministry of Land and Resources (MLR) of China; it is a strategic synthetic study with monumental significance, which is bound to contribute greatly to geological cause and development of economy and society, and to write a glorious page in the history of geological development in China.

Dating back to 1916 when the Ministry of Commerce and Agriculture renamed the Central Geological Institute as the Geological Survey, the geological survey and research of China have gone through a history of more than 100 years. Compiling chronicles or records of regional geology has always been a glorious tradition upheld by geological community in the history of geological survey of China. Early in the first half of the 20th century, *Minerals Geology of Jiangsu, Zhejiang and Anhui*, *Minerals Geology of China*, *Mining Records of Fujian*, *Geology of Jiangsu*, *Xishan Geology of Beijing*, *Xikang Geology of Sichuan* were published successively (Chen Baoguo et al., 2011). The *Regional Geology of China* compiled in the 1980s and 1990s, hosted by the former Ministry of Geology and Mineral Resources, systematically summarized results of regional geological survey and monographic studies before the late 1980s in China, which made great contributions to geological cause and development of economy and society, and acquired historic achievements in development of China.

In the past three decades, geological mineral survey and research conducted by relevant departments have accumulated a wealth of practical materials, and have acquired generous data of geology, geophysics, geochemistry, remote sensing geology, and in particular data of high-precision isotopic chronology and rock geochemistry. Besides, a series of pending geological problems have been solved and the intensity of geological survey and research have been enhanced tremendously. All these have created favorable conditions for compiling the new *Regional Geology of China*.

For the purpose of providing more geological data and meeting requirements of various aspects, surveys and researches in 3 aspects should be conducted and 3 kinds of achievements should be submitted in the project: ①Regional geology of 32 provinces (municipalities and autonomous regions) (the Regional Geology of Guangdong Province covers the Hong Kong Special Administrative Region and Macao Special Administrative Region), regional geology of 15 major regions (tectonic units, metallogenic belts, key economic zones, and major administrative regions), and national geology (series of geological, geophysical and geochemical monographs). ②The series of fundamental geological maps and spatial database. The geological map series include geological map, tectonic map, magmatic rock map, Quaternary geology and geomorphology map, aeromagnetic anomaly map and gravity anomaly map (restricted for internal use only). Each province (municipality or autonomous region) or region can adjust to compile other types of maps according to its features and demands, such as geological map of bed rocks, lithofacies and paleogeographic map, metamorphic map and map of fault systems. ③To address major problems of geology, necessary monographic studies should be carried out in the course of compiling regional geology and maps. It is intended to unify recognition on geological division and correlation, to deepen recognition on difficult geological problems, and finally to improve the quality of the series of regional geology.

This compilation work involves a wide range of regions and scholars of broad discipline fields from many departments. The project synthetically integrates achievements of geological survey and researches carried out by a million geologists in decades of years or even a hundred years. It is believed that a comprehensive geological study on such a large scale might and should create achievements of significant influence. Therefore, we propose the following 5 requirements, as major indicators of achievements with macro-influence.

Firstly, geological achievements should be broadly synthesized and integrated. Comprehensive study and integration of geological, geophysical and geochemical data should be conducted on the basis of fully collecting and analyzing various results of geological survey and research from different departments, to make the books informative, faithful, accurate and featured, and to present truthfully the level of geological survey as well as research and cognition of all geological phenomena in target areas.

Secondly, the level of geological survey and research should be highly raised. To continuously raise the level of geological survey and research of the whole country and a particular area is the eternal pursuit of geological work and an inevitable requirement of economic and social development. The raise of geological research level lies in continuous geological survey and comprehensive study as well as synthetic integration. By way of compiling the series of regional geology, sophisticated researches are conducted with integration of geological data, geophysical data, geochemical data, remote sensing data and combination of surface and deep geological data, which will significantly improve the intensity and level of geological researches and provide geological work and social and economic development with geological data of high quality.

Thirdly, the *Regional Geology of China* should play a vital role in geological development and economic and social development. With the progress of human society, the social functions and service sectors of geological work have been expanded. Geology facilitates almost all departments and industries and benefits all aspects of humankind. As a complete set of detailed and faithful fundamental geological materials, the series of the *Regional Geology of China* can not only meet the requirements of mineral exploration, and planning, arrangement and management of geological work in provincial (municipal or autonomous regional) level, in level of metallogenic zone (belt) and economic zone (belt), or in national level, but also play important roles in planning, design, construction and scientific management of major projects in such fields as urban and rural construction, transportation, water conservancy, energy development and environmental protection. Meanwhile, it will not only beat a shortcut for geological research and education, but also provide informative data for enrichment of geological repository and popularization of geological knowledge. All these embody the service intensity and breadth of geology to social and economic development.

Fourthly, there should be great improvement in summarization of geological regularity and innovation of geological theory. Geological work is an investigative work with intensive feature of science and technique. Geoscience essentially belongs to cognitive science. Whether a significant breakthrough in mineral resources exploration will be made or not, whether evaluation and prediction of geological environment and hazards will be accurate or not, and whether geology will play a greater role in planning, design and implementation of project construction, all these questions, in the last analysis, depend on the cognitive level of geological regularity and innovative extent of geological theory and technical method. The purpose of geological survey is to summarize, to recognize and to utilize geological regularity to a large degree. Once geological regularity is obtained, innovation in geological theory will come into being naturally. The quality of the series of the *Regional Geology of China* and their achievements and contribution, largely depend on the depth and breadth of summarization and cognition of geological regularity and theory for the study areas. Therefore, in the course of compiling the series of Regional Geology, practical materials are concluded and analyzed on the basis of detailed records of geological facts so as to summarize the geological regularity and to extract new discoveries and new recognition, resulting in significant enhancement of cognition on geological regularity and general geological features in target areas of the regional geology.

Fifthly, the achievements should be widely used by the geological community both in China and abroad, and great impact should be achieved. The applied breadth and impact of a scientific achievement mainly depend on its practical value and scientific value. The higher its practical value and scientific value are, the greater its influence will be. Completely complex and quite special, geological structures in China hold lots of geological phenomena rare in the world, including the lofty Qinghai-Tibet Plateau, Cenozoic loess accumulation in North China, karst geology and landform in South China, high- or ultrahigh-pressure metamorphic belt that extend thousands of kilometers in Central China, standard stratigraphic sections and rare paleo-biocenosis, and deposits of special types. These geological phenomena not only embody features and regional advantages of geology in China, but also serve as a cradle where new theories of geoscience are incubating. They attract attention of geoscientists both of China and abroad who race to the quest for theoretical breakthroughs in geoscience. General features and evolution regularities of geology of China, especially the geological phenomena mentioned above, are concluded and explained theoretically by means of compiling this series. Such practices will not only play a greater role in geological exploration and economic and social development, but also lay concrete theoretical foundation for development of geoscience and have important contribution to it. It is believed that interests of geoscientists both at home and abroad will be attracted and enormous influence will be extended.

Compiling the *Regional Geology of China* is the most fundamental synthetic study in geological work; its achievements will be used for a long time by departments in relation to natural resources as well as other relevant fields, or even geological departments of other countries. So, to ensure the quality of this series is an inevitable requirement. For this purpose, requirements of high starting point, high requirement and high level are imposed.

The so called "high starting point" means that the compilation of the books and maps should be commenced from the starting point that the intensity of geological researches, geological features and regularities of the target areas have been quite grasped. Hence geological materials and achievements of the target areas, including reports, monographs, geological maps, papers and especially achievements of geological survey in recent years, are collected first of all. And then in the compilation of the books and maps, the researchers should conduct synthetic studies integrated with geological, geophysical and geochemical data that have been gathered and seek to thoroughly understand the geological features, regularities and major geological problems that existed in the target areas. Finally, the books are written and the maps are compiled with advanced geological theories and technical methods.

The so called " high requirement" means that all procedures, including data collecting and sorting, sample analyzing and testing, book and map compiling, should be conducted strictly according to the "General Design of the Project", "Work Guide", "Measures for Project Administration" and relevant regulation issued by the Ministry of Natural Resources (MNR) and CGS, to ensure the quality of the results.

The so called "high level" means that we should strive to create an excellent series of books with high level, to make sure that the books are excellent, informative in scientific content and faithful in geological facts, and should scientifically and creatively summarize geological features and regularities. We pursuit excellence in manifestation of the books as well. The descriptive text should be concise and illustrated with proper figures. The figures and tables should be outstanding in theme, and clear and beautiful in form. All the units of measure and terminology used in the books should conform to national, sectoral and international standards and convention.

In order to enrich the contents and improve the quality of the books, and to fully perform their functions so as to provide more geological information for the geological and mineral industry as well as the public, the following demands are proposed: firstly, integrated researches of geological, geophysical, geochemical and remote sensing data should be done so that geological phenomena can be distinguished and recognized on both macroscopic and microscopic scales and geological problems can be solved; secondly, surface geology and deep geology should be combined with each other so that the deep roots of geological phenomena, mineralization and geohazards can be uncovered from three-dimensional perspective; thirdly, continental geology and marine geology should be associated as a whole for comprehensive study; fourthly, book and map compiling should be connected with construction of database, so that geological and geomorphological features of the target areas are manifested completely and vividly and the functions of this series are better performed; fifthly, comprehensive studies and monographic studies should be integrated; while conducting comprehensive studies, we should carry out monographic studies concerning the major controversial geological problems as well, to unify cognition and to solve disputed problems, and to improve the quality of the books. Adhering to the ideology that inheritance combines with innovation, we seek to innovate on the basis of inheriting and to inherit with the guidance of innovation. We will not only inherit all the great achievements, but also innovate boldly to change traditional understanding and conclusions. Meanwhile, through the compilation of the books and maps, a number of comprehensive geoscientists in various disciplines have been cultivated by means that the senior instruct the junior and they learn from each other.

Aspects such as planning and deployment of geological work and national economic development, geological mineral exploration, geological environment evaluation and hazards forecast, planning and design of urban and rural construction and major engineering projects, popularization of geoscience, improvement of scientific and cultural quality and level of the people, international cooperation and communication of geoscience and technology, and development of theories of earth science, all require abundant geological information. The *Regional Geology of China* can provide substantial fundamental geological data for the aspects mentioned above.

The project of compiling the series of the books, because of its large scale and strict requirements, claims tremendous time and labor input. Hence it is necessary to strengthen leadership and coordination, and to reinforce scientific management of the project. It is required that leading organs and academic guiding organization should be established at the national level, provincial (municipal or autonomous regional) level and the level of geological areas.

Under the leadership of Advisory Committee and Editorial Board, a team responsible for Comprehensive Study Project of the *Regional Geology of China* (known as "Comprehensive Study of Tectonic Division and Synthesized Integration of Regional Geological Survey in China" and "Updating and sharing of national lands and seas geological maps" as well) was set up and by the Institute of Geology, Chinese Academy of Geological Sciences (CAGS) to implement. Participants of the project and members of the team who are responsible for academic guiding, coordinating and organizing in the compilation of the *Regional Geology of China* are as follows:

a. Project Leaders

Chief Project Leaders: Ding Xiaozhong, Li Tingdong, Liu Yong and Pang Jianfeng

b. Distinguished Expert Group

Group Leader: Xiao Qinghui

Members (by surname stroke order): Deng Jinfu, Lu Songnian, Zhang Kexin and Pan Guitang

c. Project Professional Groups

Group of Regional Geology and Ocean Plate Geology: Li Tingdong, Ding Xiaozhong, Liu Yong and Shi Chenglong

Group of Regional Stratigraphy: Zhang Kexin, Yang Chonghui, Gao Linzhi, Wan Yusheng, Zhang Heng and Dong Chunyan

Group of Regional Magmatic Rocks: Wang Tao, Xue Huaimin, Liu Cui, Zhang Jianjun, Tong Ying, Wang Xiaoxia, Zhang Lei, Kou Caihua

Group of Regional Tectonics: You Guoqing, Wang Jun and Zhang Chuanheng

Group of Quaternary Geology and Geomorphology: Wang Yong, Chi Zhenqing, Dong Jin, Yuan Lupeng, Tian Fei, Ye Mengni, Ling Yuan

Group of Sedimentology and Lithofacies Palaeogeography: Liu Yanxue, Kuang Hongwei, Liu Yongqing, Zheng Ning and Peng Yang

Group of Metamorphic Geology and Precambrian Geology: Ren Liudong, Yang Chonghui, Du Lilin, Wang Yanbin and Song Huixia

Group of Ophiolite: Xiao Xuchang, Zhai Qingguo, Hu Peiyuan, Gao Jun and Song Shuguang

Group of Isotopic Chronology: Wan Yusheng, Dong Chunyan, Xie Hangqiang, Liu Shoujie and Zhu Wenbin

Group of Geophysics and Deep Geology: Guan Ye, Zheng Hongwei, Jin Sheng and Zhang Guibin

Group of Geographic Basemap, Database and Map Design: Han Kunying, Pang Jianfeng, Ju Yuanjing, Zhang Qinghe, Wang Liya, Wang Zhenyang, Jiang Lan, Jin Ming and Ding Weicui

Comprehensive Study Group of Regional Geology for the Six Administrative Regions and Tectonic Units: Ji Wenhua, Zhu Xiaohui, Yin Fuguang, Ren Fei, Yang Zhuliang, Yu Minggang, Niu Zhijun, Tan Mantang, Zhao Fengqing, Li Chengdong, Yang Xiaoping, Yang Yajun, Xing Dehe, Li Min, Xing Guangfu, Xiao Guoqiang and Liu Yanxue

Comprehensive Study Group of Marine Geology: Wen Zhenhe, Li Xuejie, Wang Zhongbo, Yao Yongjian, Hou Fanghui, Yang Chupeng, Wang Baojun, Gao Hongfang, Xu Ziying and Qi Jianghao

The compilation work has gained full support and energetic assistance from the MNR, former MLR, MOF, MOST and its relevant departments and bureaus, from the CGS and its relevant departments, and from the CAGS and its relevant departments. Six Centers of the CGS, relevant institutes of the CAGS, relevant departments of the Institute of Geology, CAGS, provincial (municipal or autonomous regional) Departments of Natural and Resources, provincial (municipal or autonomous regional) Bureau of Geology and Mineral Exploration, provincial (municipal or autonomous regional) Bureau (Institute) of Geological Survey and relevant universities or colleges have actively participated in the job. Various departments of the geological and mining industry, National Geological Library of China, provincial (municipal or autonomous regional) Geological Libraries, National Geological Archives of China, and provincial (municipal or autonomous regional) Geological Archives have provided substantial geological materials and information. All experts, scholars and management personnel who participated in the compilation have devoted their wisdom and hard work in pursuit of excellence. Hereby, I would like to express my sincere gratitude to and respect for departments and leaders who exercise care and guide the compilation of the series, to departments and persons who concern and help in the compilation, and to all experts, scholars and management personnel who have been involved in the compilation of the series.

Geological work is an eternal cause, and where there are human beings, there is geological work. The reason of this lies in that the fields which geological work serve are wide and will be wider and wider. One of the main paths by means of which geological work serve social and economic development is to provide effective and efficient geological information. It is one of the major reasons why we strive to compile the series of books.

It is our hope that the series of the *Regional Geology of China* can integrate all great achievements of geological survey in China and mark the culmination of years of hard work. It is our pursuit to create a series of books with excellent quality, to make greater contribution to the development of geological cause in China, to make a great contribution to the optimization of ecological environment and the construction of beautiful China, to make greater contribution to sustainable development of economy and society and realization of the Chinese dream of Chinese Nation's great rejuvenation, and to make greater contribution to the prosperity of Earth Science in the world.

Li Tingdong

李廷栋

May 2020

目　　录

第二篇　沉积岩和沉积作用

第三篇　火山岩及火山作用

第四篇　侵入岩与深成作用

第五篇　变质地质

第六篇　岩石圈结构构造

第七篇　地质构造

第八篇　第四纪地质与地貌

第九篇　平原区三维地质结构

第十篇　城市地质资源与环境

第十一篇　区域地质发展史

Contents

Part Ⅲ Volcanic rocks and volcanism

Part Ⅳ Intrusive rocks and plutonic processes

Part Ⅴ Metamorphic geology

Part Ⅵ Lithosphere structure

Part Ⅶ Geological structure

Part Ⅷ Quaternary geology and geomorphology

Part Ⅸ Three-dimensional geological structure of plain area

Part Ⅹ Urban geological resources and environment

Part XI History of regional geological development

绪　　言

北京市位于华北平原的西北端，地理坐标是北纬39°28′~41°05′，东经115°25′~117°30′，南北横跨纬度1°37′，东西经度相间2°05′。全市面积16410km^2，其中平原区面积6338km^2，占38.6%；山区面积10072km^2，占61.4%。全市共辖16个区，即东城区、西城区、朝阳区、丰台区、石景山区、海淀区、房山区、通州区、顺义区、昌平区、大兴区、门头沟区、怀柔区、平谷区、密云区、延庆区。

北京地势西北高、东南低。西部是属太行山山脉的西山，北部及东北部是属燕山山脉的军都山。延庆盆地位于西北部山区之中。山区多为中低山地形，一般海拔1000~1500m，门头沟区与河北省交界的东灵山海拔2303m，为北京市最高峰。东南部平原区海拔一般在10~60m，通州区柴厂屯一带海拔仅8m，为北京市最低处。

本区属海河流域，分属五大水系：大清河水系、永定河水系、北运河水系、潮白河水系以及蓟运河水系，基本流向是由西北向东南。五大水系中，除北运河水系发源于本区，其余均发源于境外。永定河官厅水库上游三大支流桑干河、洋河和妫水河，出库后称永定河，在三家店地区进入平原区，对北京平原的形成起着十分重要的作用。

本区地域面积虽小，但所处地质构造环境复杂，保存了从新太古代以来2.5Ga的地质演化记录。地质历史发展经历了3个大的阶段，即太古宙—古元古代的克拉通结晶基底形成发展阶段；中元古代—早中生代中三叠世的拗拉槽与稳定克拉通盖层发育阶段；中生代晚三叠世—新生代陆内（板内）造山复杂构造变形阶段。

根据全国地层区划，本区新太古代—新元古代地层属中朝地层大区之华北地层区，寒武纪—中三叠世地层属柴达木-中朝地层大区之华北地层区，晚三叠世—古新世地层属东北-华北地层大区之华北东部地层区，始新世—全新世地层属东北-华北地层大区之华北地层区。除普遍缺失古元古界、新元古界震旦系、古生界上奥陶统至下石炭统、上白垩统、古近系古新统外，其余均有出露，总厚度约33000m。

地质历史上，本区的岩浆侵入活动较为强烈：新太古代以超镁铁质岩浆为代表；中元古代以小规模双峰式富钾碱性岩浆侵入活动为主；晚古生代有零星的岩浆侵入活动发生；中生代是本区岩浆活动最强烈的时期，岩浆侵入活动延续了约100Ma，形成了大小侵入体230余个；新生代的岩浆活动以小规模超基性—基性岩浆活动为主。火山活动主要有5期，即中元古代大红峪期、中生代的南大岭期、髫髻山期、张家口期和新生代前门期。它们分别构成了大红峪组、南大岭组、髫髻山组、张家口组的主要岩层。前门期玄武岩只发育在平原区，被巨厚的新生代松散沉积物覆盖。

本区的地质工作开始于19世纪后半叶。1863年美国学者庞培莱（R. Pumpelly）提出“震旦方向”的概念，表述太行山脉北东向展布的特征。1869—1871年德国学者李希霍芬（F. V. Richthofen）在玉泉山、南口及斋堂等地进行调查，提出了“震旦层系 ”地层名称，1922年葛利普（A. W. Grabau ）明确了震旦系的范围。1910—1912年德国学者梭尔格（F. Solger）绘制1：20万北京西山地质图。1918年瑞典学者安特生（J. G. Andersson）发现周口店北京猿人遗址，1923年奥地利学者师丹斯基（O. Zdansky）发现猿人牙齿化石，1927年加拿大学者步达生（D. Black）和中国学者李捷命名猿人牙齿化石为*Sinanthropus pekinensis*，拉开周口店北京猿人遗址研究的序幕。

20世纪初，我国地质工作者开始在西山地区开展地质调查和研究。1916—1918年，叶良辅等对西山地质作了系统的调查研究，于1920年编写出版了《北京西山地质志》。该书是我国地质学家编写的第一部区域地质专著，不仅是对北京西山地质工作里程碑式的总结，而且代表和体现了当时我国地质研究的水平。

1926年翁文灏根据北京及辽宁北票等地区的调查资料，提出“燕山运动”。1929年裴文中发现第1个完整的猿人头盖骨，确立北京猿人为人类发展的一个阶段，使周口店古人类学的研究领先世界；1936年贾兰坡又发现3个“北京猿人”头盖骨。1930年李善邦等在西山鹫峰寺建立中国第1个地震观测台，开启地震地质学研究。

中华人民共和国成立后，本区的地质工作发展迅速，先后开展了两轮1∶5万区调工作。第一轮是20世纪50年代末至60年代初，由原北京市地质局、原北京地质学院及原北京矿业学院组成联合区调队，进行了山区23个图幅1∶5万区调工作。北京地质学院师生将本轮1∶5万区调成果进行总结并编写出版了《北京的地质》。

第二轮1∶5万区调于1964年开始，大规模工作在20世纪80年代中期至90年代开展。本轮区调采用新理论、新观点、新方法，取得了丰富的资料和新认识。如针对变质岩区采用“构造-岩石-事件法”，在变质火山-沉积岩系（表壳岩）中识别出一批古老变质侵入体；针对沉积地层以现代地层学、沉积学为理论依据，更加注重层序地层学的研究和地层多重划分对比；针对侵入岩按岩石谱系单位建立花岗岩类等级体制进行填图；对平原区第四纪地层尝试性采用“成因类型+微地貌单元”的填图方法。

除上述两轮1∶5万区调，1990年完成了1∶20万怀柔县幅区调、2003年完成了1∶25万北京市幅和延庆县幅区调，并开展了“北京八达岭地区燕山期花岗岩类侵入体成岩地质环境及其演化规律的研究”“华北陆台北部金矿的基本成矿地质条件及找矿方向”和“京南平原区基岩地质研究”等专题研究，提高了区域地质研究水平。

1984年原北京市地质矿产局组织编写了《北京市区域地质志》，于1991年出版。之后陆续出版的主要专著有：

1990年张吉顺、单文琅主编出版了《北京西山地质研究》，对周口店幅1∶5万区调所取得的科研成果进行了总结，通过深入研究西山地区中元古界—下三叠统变质盖层岩系中大规模的流变构造，提出了“褶叠层”这一构造地层的新概念。

1991年白志民等编著出版了《八达岭花岗杂岩》，对八达岭地区燕山旋回120余个侵入体按构造-岩浆建造及演化序列进行侵入岩序列三级划分（超单元-单元-侵入体），深入探讨了成岩地质环境、岩浆演化规律、岩浆作用的物理化学条件以及岩浆活动与成矿作用的关系等问题。

1992年郁建华等编著出版了《北京地区岩浆深成作用》，把北京地区的岩浆岩按“岩系-岩套-岩组-岩体”体系进行划分，详细分析了各岩系的时空分布、岩石类型、岩石学及岩石地球化学特征、共生组合及成因机理，讨论了不同时代侵入活动与火山活动的联系及其演化方向。

1993年周绍林等编著出版了《北京早前寒武纪地质及铁矿》，对北京密云和怀柔地区早前寒武纪地层层序、变质岩区构造特征、变质岩石学和矿物学、区域变质作用与混合岩化作用、含铁变质岩系特征、变质铁矿分布规律和成矿条件、找矿方向以及早前寒武纪地壳演化等作了系统总结。在研究了密云-怀柔地区变质岩系中的变形构造后，提出本区多期变形变质作用叠加的认识，指出本区经历3期主要变形作用的看法。

1994年萧宗正等编著出版了《北京中生代地层及生物群》，遵循当代综合地层学研究的新进展，开展了岩石地层、生物地层、年代地层和构造地层学等多方面研究；对中生代地层中各门类化石进行采集、鉴定和研究，探讨生物群组合、生物演替关系、时代确定和区域对比；应用构造地层方法对印支、燕山构造运动的特征进行了综合分析。

1995年李华章等编著出版了《北京地区第四纪古地理研究》，运用综合自然地理观点，对本区第四纪以来新构造演化、古地貌发育、平原的形成与发展、生物群演化与古气候变迁以及人类活动对自然环境的影响等进行了综合性和区域性古地理研究，按不同时段对当时古地理环境进行复原。

1995年鲍亦冈等编著出版了《北京燕山期火山地质及火山岩》，按照“有联系的成生发展观点、有序的发展演化观点和分级别总结归类对比的观点”学术思路，系统总结了本区中生代燕山期火山地质和火山岩的主要特征，在丰富翔实的资料基础上分析探讨了区域构造-岩浆活动特点和规律。

1996年郁建华等编著出版了《华北地台北部非造山环斑花岗岩及有关岩石》，将北京北部元古代深成岩体和火山岩划分为4条岩浆岩带（体），即大庙斜长岩杂岩体、赤城-古北口斜长岩-富钾花岗岩带、密云沙厂环斑花岗岩杂岩体及平谷-蓟县钾质碱性火山岩带，认为是华北克拉通首次拉张破裂所诱发的典型非造山岩浆活动产物。

1996年鲍亦冈等主编出版了《北京市岩石地层》，对各地层单位的命名由来、划分变化沿革作了详细而严格的考证、订正，规范了岩石地层单位，明确了层型和划分标准，阐明了各地层单位的区域延伸展布与变化，理清了有关岩石地层单位的穿时性，初步建立起各断代地层的沉积格架。上述工作为地层研究的科学化奠定了很好的基础，使地学研究领域中最为基础性的地层工作有了统一的认识。

2000年北京市地质调查所鲍亦冈负责组织完成了“北京市1：5万区调片区总结”项目，对北京地区近百年来的区域地质调查研究所取的成果进行了系统总结，并于2001年编著出版了《北京地质百年研究》。该书以丰富的区调填图资料和科研成果为基础，以现代地学理论为指导，对区域地层、侵入岩、火山岩及地质构造进行了高水平的综合研究，反映了20世纪几代地质工作者在北京地区从事基础地质调查研究取得的丰硕成果。

2000年金文山等编著出版了《北京地区早前寒武纪结晶基底》，将北京地区早前寒武纪岩石划分为混合岩化表壳岩系和非层状岩系两类地质填图单元，对各个单元的构造、岩石、岩石地球化学、变质作用、锆石成因及同位素年龄作了系统论述，提出了本区早前寒武纪地壳演化新模式。

2004年邓晋福等完成的“中国花岗岩重大地质问题研究”项目子课题“燕山造山带花岗岩与大陆地壳改造和生长”，以及国土资源部“中国东部中生代大陆岩石圈剧变过程动力学机制与矿产资源响应”项目，取得了一系列锆石U-Pb测年数据，系统总结燕山地区的火成岩岩石学、岩石化学特征及其形成的地质背景，探讨了岩浆活动与成矿作用的联系。

2007年郭旭东编著出版了《北京第四纪地质导论》，系统论述了本区第四纪冰川与气候问题、冰缘作用与环境、第四纪地层划分、新构造与地震预报、环境地质等，探讨了北京地区第四纪地质的基本特征和演化规律。

2007年邓晋福等编著出版了《中国华北地区岩石圈三维结构及演化》，系统总结了华北地区岩石圈地震、电性、重磁、热能学等三维结构特征，建立了该地区岩石圈构造和演化框架。

除上述出版的专著外，中国地质科学院地质研究所、中国科学院地质与地球物理研究所、中国地质大学（北京）、天津地质调查中心等多家科研及教学单位的众多研究人员在本区开展了多项研究工作，其成果主要是以期刊和硕（博）士毕业论文的形式发表。部分主要成果有：万渝生等（2003）在十三陵长城系常州沟组取得碎屑锆石年龄分布模式，研究了碎屑沉积物来源；高林志等（2008）首次获得西山下马岭组斑脱岩（凝灰岩）高精度测年数据（1368±12）Ma（锆石U-Pb）；和政军等（2011）、李怀坤等（2011）等在密云分别取得了常州沟组之下的环斑花岗岩古风化壳和花岗斑岩脉的锆石U-Pb高精度测年数据，分别是（1682±20）Ma和（1673±10）Ma，限定了常州沟组的底界时限；2010年李怀坤等在延庆高于庄组二段凝灰岩夹层中获得约1560Ma（锆石LA-MC-ICPMS U-Pb）高精度测年数据；陆松年（2010）总结中元古代地层和热-构造事件同位素地质年代学研究工作所取得的主要进展，认为华北克拉通极好地保持了1.8~1.6Ga的一系列非造山裂解事件的岩浆记录，并按时间顺序划分为3期，北京密云环斑花岗岩、平谷大红峪组超钾质火山岩分别是中期和晚期的典型代表。

关于本区及华北地区的燕山运动，任纪舜等（1999）将燕山运动分为早、中、晚3个旋回，时限包括整个侏罗纪和白垩纪，认为同太平洋板块低角度俯冲有关；戴维斯（2001）提出燕山运动是蒙古-鄂霍次克海关闭形成的南北向挤压与太平洋板块向东亚大陆俯冲形成的向西北方向挤压的交替作用的结果；崔盛芹等（2002）将燕山地区的燕山运动分为两个时期，即早—中侏罗世的早燕山期和晚侏罗世—白垩纪的晚燕山期；赵越等（2004）对燕山地区中生代板内变形的研究进展进行了总结，确定翁文灏识别和提出的燕山运动A幕、B幕和中间幕发生在侏罗纪；邓晋福等（2005）将燕山期构造演化划分为初始造山（J_1）、早期造山（J_2）、峰期造山（J_3）、晚期造山（K_1）、后造山（K_2）5个构造幕；董树文等（2007）认为燕山运动起始于约（165±5）Ma之前，结束于83Ma，经历强挤压-陆内造山期（165~136Ma）、主伸展-岩石圈减薄期（135~100Ma）和弱挤压变形期（100~83Ma）三个演化阶段，指示了多向挤压汇聚的动力学机制；张宏仁（2013）认为燕山运动的起始时间应为中侏罗世含煤碎屑岩沉积之后、同造山砾岩形成之前，认为燕山运动是中国东部遭受来自外部的快速打击的结果，并定义为燕山运动的“绪动”或“燕山事件”。

20世纪80—90年代，郑亚东、王玉芳及戴维斯等对云蒙山变质核杂岩进行了研究。白志民（1991）、许继峰（1994）、张建新（1997）及朱大岗（2000）皆主张云蒙山岩基的侵位隆升导致了其南缘大水峪剪切带的伸展运动；张家声等（2007）认为伸展活动与云蒙山岩基的侵位共同导致了该核杂岩的形成，并将四合堂剪切带与大水峪剪切带解释为同期伸展剪切带。关于云蒙山变质核杂岩的发育时间限定，王玉芳等（1989）从大水峪剪切带糜棱岩中获得113~106Ma的黑云母K-Ar年龄，戴维斯等（1996）获得大水峪剪切带内钾长石$^{40}Ar-^{39}Ar$年龄为118Ma；陈印（2014）根据岩脉和岩体锆石U-Pb测年及角闪石、黑云母$^{40}Ar-^{39}Ar$测年资料，认为该核杂岩起始于135Ma，并在125~114Ma期间快速隆升。

21世纪以来，本区的地质调查工作转入平原地区，工作的重点向城市地质领域转变，相继完成了“首都地下水资源调查与评价”“北京市多参数立体地质调查”“顺义等11个规划新城前期区域工程地质勘查”“北京平原区活动断裂监测专项地质调查”“北京市地下空间资源调查评价及关键技术研究”等一批重大项目。建设了地面沉降、地下水污染、浅层地温能、突发性地质灾害监测预警系统，为城市规划和建设提供了基础地质数据。

2008年北京市地质矿产勘查开发局及所属单位完成的“北京市多参数立体地质调查”项目，在充分利用、综合和集成已有地质成果数据的基础上，综合运用现代勘查技术，开展了平原区三维地质结构调查，建立了基岩、新生界和工程建设层三维地质结构模型，对地下水资源、地热资源以及土壤环境质量进行了调查评价，为城市规划、建设与管理提供了科学依据。2008年对项目成果进行总结出版了《北京城市地质丛书》，包括《北京城市地质》《北京城市地质图集》《北京地下水》《北京地质灾害》和《北京浅层地温能资源》。

2013年北京市地质调查研究院完成的“北京平原区活动断裂监测专项地质调查”项目，在全面总结前人工作成果基础上，通过地质专项调查、物探、钻探对北京平原地区隐伏的主要活动断裂（黄庄-高丽营断裂、南口-孙河断裂、顺义断裂和夏垫断裂）进行了精确定位，利用岩石地层、旋回地层、磁性地层的分析方法并结合浅部地层测年数据，对4条断裂第四纪活动特征及活动规律进行了分析。

近年来，北京市地质矿产勘查开发局为满足北京城市发展的需要，以城市地质工作为中心，资源和环境调查工作并重，创建了以重大地质问题战略研究为先导，科学部署北京城市地质工作的理念发展型工作模式，陆续开展了浅层地温能资源调查与开发战略研究、地面沉降对城市安全影响战略研究、突发性地质灾害对城市安全影响战略研究、地下水对城市安全影响战略研究、城乡一体化对地质工作需求程度的战略研究、南水北调进京后地下水蓄养战略研究、矿山地质环境对北京山区生态建设影响的战略研究、城市地质安全预警预报信息系统战略研究、世界城市建设对地质工作需求战略研究、城市地下空间开发利用地质环境战略研究等10项战略研究工作。

1991年《北京市区域地质志》出版以来，北京地区开展了大量的地质工作，积累了丰富的资料、取得了丰富成果，本志是对这些成果的梳理和集成。本志章节安排参照中国地质调查局基础调查部和中国地质科学院地质研究所编写的《中国区域地质志工作指南》的统一要求，内容上以梳理前人的成果为主。本次志书修编工作是在鲍亦冈主持的“北京市1：5万区调片区总结”项目成果的基础上完成的，引用了其主编的《北京地质百年研究》一书的主要内容。北京地区基础地质研究成果丰富，本志的地层篇主要是在《北京市岩石地层》（鲍亦冈等，1996）一书的基础上补充了一些新资料。有些内容如古近系和新近系以及断裂和褶皱构造描述等方面，新资料和新成果较少，为保持志书的完整性，沿用了第一版志书的相关内容。本志的构造分区、地层分区及地貌分区等涉及大的区域方面的内容，参考或沿用了《中国区域地质志·河北志》的成果。

参与本次志书文字编写和编图工作的多是没有参与过北京地区区调工作的年轻同志，在老同志的指导下，他们兢兢业业、刻苦钻研，完成了各自课题的编写及编图工作，实属难能可贵。

本项目出版的成果：

1. 中国区域地质志·北京志；
2. 北京市地质图（1∶25万）；
3. 北京市地质构造图（1∶25万）；
4. 北京市基岩地质图（1∶25万）；
5. 北京市第四纪地质与地貌图（1∶25万）；
6. 北京市航磁 ΔT 等值线平面图（1∶25万）。

参加本志编制工作的人员具体分工如下：

前言由孙永华编写；第一篇第一章由方同明、王世发、吴虎峻编写，第二章至第五章由李莉、孙永华编写，第六章由李瑞杰、吕金波、何晗晗编写；第二篇由何付兵编写；第三篇由何付兵、曹萌萌编写；第四篇由何付兵、徐吉祥、何振军编写；第五篇由方同明、王世发、吴虎峻编写；第六篇由尤世娜、王志辉编写；第七篇由尤世娜、王继明、王增护编写；第八篇第三十五至三十九章及第四十一章由赵勇、张磊、杨誉博、刘予编写，第四十章由吕金波、程金华、刘振华编写；第九篇第四十二、四十三章由王海波、刘鸿编写，第四十四章由刘立岩、曹胜利、王响贵编写；第十篇第四十五章由何静、周圆心编写，第四十六至四十九章、五十一章由刘鸿、何付兵、方同明编写，第五十章由程新彬、郭莉编写，第五十二章由王纯君、高莉编写，第五十三章由吕金波编写；第十一篇由尤世娜、何付兵、赵勇编写；结束语由孙永华、何付兵、尤世娜编写。

全文由孙永华、何付兵、尤世娜、刘鸿负责统稿。

北京市区域地质志系列图件（1∶25万）编制：

地质图由何付兵、方同明、尤世娜、马强编制，王纯君、赵艳英、池京云制图；地质构造图由尤世娜、何付兵、马强编制，王纯君、赵艳英、徐旸制图；第四纪地质与地貌图由程金华、吕金波、赵勇、刘振华编制，王纯君、赵艳英制图；基岩地质图由孙颖、王海波、刘凯编制，许苗娟、赵艳英制图；航磁 ΔT 等值线平面图由王志辉、张晓亮编制，王纯君制图。

地理底图及空间数据库由王纯君、赵艳英、徐旸、尤世娜、何付兵等完成。

参与本志插图清绘的有牛文治、王凯、刘晓勇、刘光泓、倪敬波、周瑜琨、张喜、董静、梁亚南、白凌燕等。

英文翻译由高莉、董静、郭鑫、何付兵完成。

资料归档由尤世娜、徐旸、李莉等完成。

审查后修定由孙永华、何付兵、尤世娜、刘鸿、赵勇、方同明、李莉、王海波、王纯君、李瑞杰、程金华、赵艳英、高莉等完成。

在项目实施过程中，得到了中国地质调查局、中国地质调查局基础调查部、中国地质科学院地质研究所、中国区域地质志项目办公室、中国区域地质志编辑委员会、中国区域地质志特聘专家委员会、天津地质调查中心、北京市地质矿产勘查开发局及所属单位各级领导和专家的精心指导，在此表示诚挚的谢意！

本志编制工作得到了河北省区域地质调查院、天津市地质调查研究院、北京市地质环境监测所、北京市地质灾害防治研究所、北京市工程地质研究所、北京市矿产地质研究所、北京市生态地质研究所、北京市地热调查研究所、北京市地质矿产勘查开发集团有限公司等单位领导和专家的大力支持，在此表示诚挚的谢意！

李廷栋院士、莫宣学院士、陆松年研究员、肖庆辉研究员、潘桂棠研究员、耿树方研究员、丁孝忠研究员、耿元生研究员、卢民杰研究员、刘训研究员、王惠初研究员、尹福光研究员、赵小明研究员、姚建新研究员、万渝生研究员、刘燕学研究员、游国庆教授级高工、王永研究员、任留东研究员、管烨研究员、李瑞敏教授级高工、韩坤英教授级高工等领导与专家，对本志的评审验收工作给予了极大的帮助和多次莅临精心指导，在此表示衷心的感谢！

在志书编写工作中苏德辰、陈广峰、郭高轩、黄骁、曹颖、孙小华、刘宗明、郑佳、雷晓东、贺道兵、张巍、魏波等给予了热情帮助和大力支持，在此表示衷心的感谢！

本志是对北京地区近三十年来各类相关区域地质资料的综合梳理和集成，引用了大量资料和成果，对原创单位和著作者表示衷心的感谢！特别是鲍亦冈、萧宗正、姜守玉、郁建华、陈正邦、周绍林、吴梦源、董得茂、李树祥、白志民等老一辈地质工作者为北京地区基础地质研究工作打下了坚实的基础，在此表示衷心的感谢！若有遗漏标注和不当之处，敬请谅解，在此致以真诚的歉意。

在此，向所有参加过本区地质工作先辈们、同事们表示由衷的感谢！

《中国区域地质志·北京志》编写组

2023 年 7 月

第一篇

区域地层

概　　述

北京区域地层从新太古代地层至新生代地层（除缺失古元古代地层、新元古代地层的南华系和震旦系、古生代地层的志留系和泥盆系以外）基本齐全（表Ⅰ-1）。新太古代—古元古代地层属中朝地层大区之华北地层区；中元古代—新元古代青白口纪地层属中朝地层大区之华北地层区；南华纪—三叠纪地层属华北—柴达木地层大区之华北地层区；侏罗纪—古近纪地层分属东北-阿尔泰地层大区之内蒙-松辽地层区、华北地层大区之华北东部地层区；新近纪地层属东北-华北地层大区之东北地层区和华北地层区；第四纪地层分属东北地层大区之辽吉黑地层区、华北地层大区之华北西部地层区和华北东部地层区。

新太古代表壳岩主要分布于北京市东北部的密云区、怀柔区和平谷区，少量分布在昌平区德胜口—南口、延庆区红旗甸及房山区周口店一带。经区域变质与变形作用改造，表壳岩地层层序严重破坏，多为小面积分散产出在古老变质侵入体中。

中—新元古代地层分布广泛，发育齐全，厚度约6000m，约占北京山区面积的三分之一，此外，也见于平原新生代沉积物之下的基岩中。岩石类型为滨海相—浅海相富镁碳酸盐岩、碎屑岩和黏土岩，夹少量碱性玄武岩系列火山岩。产叠层石、微体及微古植物化石等。

寒武纪—奥陶纪地层属稳定地台型沉积，主要分布于西山地区，其次为北山和京东地区。为陆表浅海碳酸盐岩及少量泥砂质岩，总厚度1300m左右，沉积稳定，横向变化小。三叶虫、头足类、腹足类、牙形石等动物昌盛，其中三叶虫为划阶的标准化石。

晚石炭世—三叠纪地层亦为地台型沉积，主要分布于京西门头沟区和房山区，向东延续，被掩埋在平原地区新生代沉积物之下，为一套海陆交互相—陆相含煤碎屑岩与红色碎屑岩沉积，产蜓类、腕足类、双壳类等海相生物化石及典型华夏植物群化石，厚度约700m。

侏罗纪—白垩纪地层为产于陆相火山-沉积盆地内的一套复杂多变的火山沉积岩系，广泛分布于北京西山、北山的向斜构造及断陷盆地内。侏罗纪从早期的基性火山岩、含煤碎屑岩发展为中、晚期的中性火山岩-火山碎屑沉积岩，含有著名的门头沟植物群及门头沟双壳动物群化石；早白垩世早期发育中性、酸性火山岩-火山碎屑岩沉积，晚期发育河、湖相砾岩与泥、砂岩沉积，含著名的热河生物群及房山生物群化石。总厚度7700~12000m。

新生代地层主要分布于平原区及各大水系与山麓地带，属华北平原新生代断陷盆地堆积的一部分。古近系和新近系以河流相泥砂质堆积为主，间夹玄武岩层，局部含油、含褐煤层，厚度为1100~3400m。第四纪沉积物类型多样，有河、湖相的冲积、洪积、淤积，以及洞穴堆积，厚度从几米至数百米不等。新生代是高等哺乳动物和高等被子植物发展的时代，北京有古人类文化遗迹，为北京猿人的发现地。

1995年由鲍亦冈主持完成的“北京市地层多重划分对比研究”项目，对北京地区新生代以前的地层进行了全面清理，并进行了系统研究，成果代表了本区地层研究的最新进展。本篇的主要内容来自他主编的《北京市岩石地层》（1996）一书。为避免过多重复，本志内容删繁就简，如关于岩石地层仅列出层型和简单概述，如需了解详细内容，请参阅《北京市岩石地层》。

表 I -1 北京市岩石地层序列总表

Table I -1 Stratigraphic sequence of Beijing

宇	界	系	统	阶	修订前岩石地层单位	修订后岩石地层单位
显生宇 Ph	新生界 Cz	第四系 Q	全新统 Qh			Qh（按不同沉积类型划分）
			更新统 Qp		马兰组 Qp*m* 周口店组 Qp*z* 泥河湾组 Qp*n*	Qp（按不同沉积类型划分）
		新近系 N	上新统 N_2		鱼岭组 N*y* 天竺组 N_2tz	鱼岭组 N_2y 天竺组 N_2tz
			中新统 N_1		天坛组 N_1t	天坛组 N_1t
		古近系 E	渐新统 E_3		前门组 $E_{2-3}q$	前门组 $E_{2-3}q$
			始新统 E_2		长辛店组 E*c*	长辛店组 E_2c
			古新统 E_1			
	中生界 Mz	白垩系 K	上白垩统 K_2			
			下白垩统 K_1		夏庄组 K_1x 坨里组 K_1t 上侏罗统—下白垩统 J_3-K_1：九佛堂组 JK*j*；东狼沟组 JK*d*	夏庄组 K_1x 坨里组 K_1t 九佛堂组 K_1j 东狼沟组 K_1d 张家口组 K_1z
		侏罗系 J	上侏罗统 J_3		张家口组 JK*z* 土城子组 J_3tc 髫髻山组 J_3t	土城子组 J_3tc 髫髻山组 $J_{2-3}t$
			中侏罗统 J_2		九龙山组 J_2j 龙门组 J_2l	九龙山组 J_2j 龙门组 J_2l
			下侏罗统 J_1		窑坡组 J_1y 南大岭组 J_1n	窑坡组 J_1y 南大岭组 J_1n
		三叠系 T	上三叠统 T_3		杏石口组 T_3x	杏石口组 T_3x
			下—中三叠统 T_{1-2}		双泉组 PT*s*	双泉组 P_3T_2s
	古生界 Pz	二叠系 P	中—上二叠统 P_{2-3}		石盒子组 P*s*	石盒子组 $P_{2-3}s$
			下二叠统 P_1		山西组 CP*s*	山西组 C_2P_1s
		石炭系 C	上石炭统 C_2		太原组 C*t*	太原组 C_2t
			下石炭统 C_1			
		泥盆系 D				
		志留系 S				
		奥陶系 O	上奥陶统 O_3			
			中奥陶统 O_2		马家沟组 O_2m	马家沟组 O_2m
			下奥陶统 O_1		亮甲山组 O_1l 冶里组 O_1y	亮甲山组 O_1l 冶里组 O_1y
		寒武系∈	芙蓉统∈$_4$	凤山阶 长山阶 崮山阶	炒米店组∈*cm*	炒米店组 ∈$_4cm$
			第三统∈$_3$	张夏阶 徐庄阶	张夏组∈*z*	张夏组 ∈$_3z$
				龙王庙阶	馒头组∈*m*	馒头组 ∈$_{2-3}m$
			第二统∈$_2$	沧浪铺阶	昌平组∈*c*	昌平组 ∈$_2c$
			纽芬兰统∈$_1$			
元古宇 Pt	新元古界 Pt_3	震旦系 Z				
		南华系 Nh				
		青白口系 Qb			青白口系 Qb：景儿峪组 Qb*j*；龙山组 Qb*l*；下马岭组 Qb*x*	景儿峪组 Pt_3^1j 龙山组 Pt_3^1l
	中元古界 Pt_2	待建系 Pt_2^3			蓟县系 Jx：铁岭组 Jx*t*	下马岭组 Pt_2^3x 铁岭组 Pt_2^2t
		蓟县系 Jx			洪水庄组 Jx*h* 雾迷山组 Jx*w* 杨庄组 Jx*y*	洪水庄组 Pt_2^2h 雾迷山组 Pt_2^2w 杨庄组 Pt_2^2y 高于庄组 Pt_2^2g
		长城系 Ch			长城系 Ch：高于庄组 Ch*g*；大红峪组 Ch*d*；团山子组 Ch*t*；串岭沟组 Ch*cl*；常州沟组 Ch*c*	大红峪组 Pt_2d 团山子组 Pt_2t 串岭沟组 Pt_2cl 常州沟组 Pt_2c
	古元古界 Pt_1					
太古宇 Ar	新太古界 Ar_3				密云岩群 Ar*My.*	四合堂岩群 Ar_3 *Sh.* 密云岩群 Ar_3 *My.*

第一章　太古宇

本区仅发育新太古代变质地层，位于华北陆块燕辽-太行构造岩浆岩带中部的华北陆块地层大区之冀东古陆核。集中出露在密云—怀柔地区，面积约 1400km^2。此外在平谷区镇罗营、昌平区德胜口、房山区关坻、延庆区红石湾及门头沟区碾台等地有零星出露。

本区太古宇的调查研究已有 100 余年。孙健初、王曰伦等（1930）首次将密云一带变质岩划归太古界，称桑干杂岩或桑干系。20 世纪 80 年代以前，主要是按沉积地层对变质岩系进行岩石地层划分，之后开展的区调工作采用构造-地层-事件法填图，识别出多个古老变质侵入体，将原变质地层分解为表壳岩和变质深成侵入体。鲍亦冈等（2001）编著《北京地质百年研究》，以第二轮 1：5 万区调成果和《北京市岩石地层》一书中的划分方案为基础，确定本地区 10 个变质岩石地层单位（广义），指出密云岩群形成的时代最老，也是北京地区太古宙唯一的狭义变质岩石地层单位，其余则是由以原地-半原地重熔型和侵入特征明显的岩浆岩为主组成的广义的变质岩石地层单位。

2000—2013 年期间，随着大陆动力学理论的发展和测年方法的进步，变质岩区的研究取得了众多成果。主要表现在两个方面：一是提出了太古宙 TTG 岩石的概念，进一步将具有侵入岩特征的片麻岩从原表壳岩中剥离出来；二是孙会一（2010）、石玉若（2011）、任荣（2011）、万渝生（2012）等在区内开展了测年工作，数据显示原划为中太古代及以前的大多数表壳岩系都形成于新太古代晚期（2.54~2.51Ga），空间上与表壳岩系共生的花岗质岩石也大都形成于这一时期。本区太古宇划分沿革见表 1-1。

综合分析原岩建造类型、变质程度、混合岩化作用、构造特征及矿物组成等方面的差异，本志将本区变质表壳岩划分为密云岩群和四合堂岩群。

一、密云岩群（Ar_3 *My*.）

该岩群主要出露在密云水库南北两侧以及崎峰茶断裂以西地区。南侧主要分布于蔡家洼—霍各庄—沙厂一线，呈近 EW 向展布；北侧分布范围主要位于大漕、小漕、兵马营、高岭以及怀柔马圈子等地，总体沿 NE 向展布；崎峰茶断裂以西地区主要位于怀柔马圈子及其周边地区，地层受云蒙山岩体侵位和崎峰茶断裂逆冲作用的影响，呈 NE—NNE 向展布。为一套麻粒岩相含硅铁建造的变质表壳岩系，是区内最古老的岩石，受后期岩浆侵位和构造作用影响，表现为似层状无序的构造岩片或褶叠层的拼合或堆叠，难以鉴别层序的新老关系。

岩性主要为辉石麻粒岩、含辉石黑云斜长变粒岩、浅粒岩、石榴斜长辉石岩夹多层辉石磁铁石英岩。

其中沙厂地区变质岩石变形变质作用强烈，表现为变质表壳岩，分布不连续，多以大小不等的包体形式残存于后期英云闪长质片麻岩中。主要岩石类型为二辉麻粒岩、黑云辉石变粒岩（呈韵律式互层），夹浅粒岩、斜长辉石岩、斜长角闪岩、磁铁石英岩、角闪二辉石岩透镜体等。据北京市地质局一〇一队北京市密云县沙厂地区地质测量报告实测剖面资料，代表性岩性如下：

9. 石榴斜长片麻岩夹辉石变粒岩、磁铁石英岩	135.2m
8. 磁铁石英岩夹含磁铁石榴辉石变粒岩	8.4m
7. 黑云角闪辉石变粒岩夹磁铁石英岩	11.4m
6. 石榴角闪斜长片麻岩、石榴透辉变粒岩夹磁铁石英岩	66.6m
5. 混合岩化石榴角闪斜长片麻岩、混合岩化石榴透辉变粒岩夹石榴二辉麻粒岩	87.4m

表 1-1　太古宙地层划分沿革

Table 1-1　Division history of Archean stratigraphy

北京地质学院（1961）

时代	群	组
太古代	张家坟群	宋营子组
		山神庙组
		椴树梁组
		石城组
	密云群	阳坡地组
		大漕组
		沙厂组

周绍林等（1984）

时代	群	组
太古代	四合堂群	山神庙组
		西湾子组
		宋营子组
		阳坡地组
	密云群	大漕组
		苇子峪组
		沙厂组

北京市地质调查所（1994）

时代	群	岩系
新太古代		旗杆顶花岗质片麻岩
	潮河主坝基性岩墙群	
	荆子峪片麻岩套	康各庄奥长花岗质片麻岩
		荆子峪花岗闪长质片麻岩
	放马峪基性岩墙群	
中太古代	苇子峪片麻岩套	穆家峪英云闪长质片麻岩
		苇子峪辉长闪长质片麻岩
		黑卧辉长质片麻岩
		沙厂英云闪长质片麻岩
古太古代	密云表壳岩系	

金文山（1994）

时代	群（岩系）
新太古代	四合堂混合岩化表壳岩系
	阳坡地 TTG-M-Me 杂岩
	马圈子 TTG-M-Me 杂岩
	对角沟门 TTG-M-Me 杂岩
	冯家湾 TTG-M-Me 杂岩
中太古代	苇子峪 TTG（A）-M-Me 杂岩
	大漕-沙厂混合岩化表壳岩系

鲍亦冈等（2001）

时代	群	岩系
新太古代		旗杆顶花岗质片麻岩
	潮河主坝基性岩墙群	
	荆子峪片麻岩套	康各庄奥长花岗质片麻岩
		荆子峪花岗闪长质片麻岩
	放马峪基性岩墙群	
中太古代	苇子峪片麻岩套	宋营子英云闪长质片麻岩
		苇子峪辉长闪长质片麻岩
		黑卧辉长质片麻岩
	沙厂英云闪长质片麻岩	
古太古代	密云表壳岩系	

北京市地质调查研究院（2014）

时代	群	岩系
新太古代	官厅花岗质片麻岩	旗杆顶花岗质片麻岩
	潮河主坝基性岩墙群	
	荆子峪片麻岩套	康各庄奥长花岗质片麻岩
		荆子峪花岗闪长质片麻岩
	放马峪基性岩墙群	
中太古代	苇子峪片麻岩套	宋营子英云闪长质片麻岩
		苇子峪辉长闪长质片麻岩
		黑卧辉长质片麻岩
	沙厂英云闪长质片麻岩	
	密云岩群	

本志

时代	群	岩系
新太古代	旗杆顶花岗质片麻岩	
	荆子峪片麻岩套	康各庄奥长花岗质片麻岩
		荆子峪花岗闪长质片麻岩
	放马峪基性岩墙群	
	苇子峪片麻岩套	宋营子英云闪长质片麻岩
		苇子峪辉长闪长质片麻岩
		黑卧辉长质片麻岩
	沙厂英云闪长质片麻岩	
	四合堂岩群	
	密云岩群	

4. 混合质石榴角闪斜长片麻岩，夹黑云辉石变粒岩、二辉麻粒岩	48.5m
3. 透辉石角闪斜长片麻岩、云辉变粒岩夹辉云斜长片麻岩	32.8m
2. 透辉斜长片麻岩夹榴辉变粒岩夹磁铁石英岩	38.9m
1. 混合质透辉角闪斜长片麻岩夹，底部石榴角闪斜长辉石岩	38.5m

大漕地区变质岩系以黑云变粒岩、榴辉斜长角闪岩、榴辉斜长辉石岩、变粒岩、浅粒岩为主，其次有大量的似层状石榴斜长辉石岩、斜长角闪岩和少量的黑云二辉麻粒岩、黑云紫苏麻粒岩、磁铁石英岩等。据北京市地质局一〇一队放马峪-大漕地区 P1-2 实测剖面，代表性岩性如下：

14. 石榴斜长角闪岩夹角闪斜长片麻岩	19.00m
13. 石榴斜长角闪岩	16.91m
12. 混合质黑云变粒岩	16.98m
11. 石榴斜长角闪岩夹石榴斜长辉石岩	76.00m
10. 混合质条带状黑云变粒岩夹薄层磁铁石英岩	145.00m
9. 混合质石榴斜长辉石岩	9.26m
8. 条带状混合质黑云变粒岩，底部为角闪二辉石岩	51.00m
7. 条带状混合质黑云变粒岩与灰白色石榴斜长混合岩互层	39.00m
6. 条带状混合质黑云变粒岩夹角闪黑云变粒岩及薄层磁铁石英岩	250.00m
5. 混合质条带状黑云变粒岩	88.49m
4. 条带状黑云母混合岩及混合质黑云变粒岩	55.85m
3. 条带状混合质黑云变粒岩夹薄层黑云母化二辉石岩	110.51m
2. 条带状混合质黑云变粒岩夹多层磁铁石英岩	69.48m
1. 二辉石岩	31.00m

该岩群普遍受太古宙强烈的麻粒岩相区域变质作用，后期又叠加了角闪岩相变质作用，退变质作用非常发育。岩石成层性较差，越往上成层性越好、延伸亦稳定，石榴斜长辉石岩、斜长角闪岩和变粒岩可作标志层。混合岩化作用不强烈，以原地重熔型为主，伴有碱交代作用。区内变质岩系和磁铁石英岩受变形变质和混合作用，多呈紧闭褶皱，转折端加厚，翼部变薄被拉伸剥蚀。

二、四合堂岩群（Ar_3 *Sh.*）

该岩群集中分布于密云冯家峪—山神庙—四合堂一线，呈 NW 向狭长带状分布。为一套角闪岩相含硅铁建造的变质表壳岩系，与密云岩群难以确立先后关系，为突出该岩群与密云岩群之间的区别，将其划分为四合堂岩群。

岩性主要为斜长角闪岩、混合质黑云变粒岩、黑云角闪变粒岩、混合岩、磁铁石英岩等，在上峪、石炮沟、山神庙、三棱山等地段，常组成一系列级次不等的紧闭线状复式褶皱。

后期受到云蒙山岩体侵位影响，该岩群发生较强的剪切变形改造，原岩结构、构造受到较大破坏，难以划分岩群内部各岩性层的相对层序。层间发育小型褶皱、韧性剪切带、片理化，斜长角闪岩、磁铁石英岩呈透镜状赋存其中。

该套岩群以密云西湾子-山神庙剖面为代表，该变质岩系经受了角闪岩相区域变质作用和较弱的混合岩化作用。主要变质岩石类型为黑云斜长片岩、糜棱岩、黑云变粒岩、浅粒岩、混合岩化斜长角闪岩、磁铁石英岩等，局部有角闪石岩透镜体。混合岩化作用较弱，局部地段出现条带状、雾迷状混合岩。

岩群变质程度一般为角闪岩相，底部为高角闪岩相。成层性良好，延伸稳定，变余层理、微层理构造非常发育。浅粒岩中可见变余粒级序和变余交错层，含有斜长角闪岩（或片岩）、具层状构造的黑云角闪变粒岩、厚层浅粒岩、磁铁石英岩等。混合岩化作用较弱，局部有钾质混合岩化现象，呈条带状、层状分布。岩石发育多级褶皱、韧性变形、片理化带，斜长角闪岩、石英脉等能干性强岩石被挤压拉断呈透镜状、石香肠状，磁铁石英岩形成紧闭褶皱，转折端加厚，翼部拉伸变薄。以北京市地质调查所 1∶5 万琉璃庙幅区调实测的琉璃庙剖面为例，岩性描述如下：

14. 混合质黑云斜长变粒岩	52.12m
13. 初糜岩	120.45m
12. 糜棱岩夹构造片岩	37.58m
11. 磁铁石英岩	32.38m
10. 片理化糜棱岩	7.96m
9. 眼球状糜棱岩	31.34m
8. 糜棱岩化黑云斜长片麻岩	2.51m
7. 眼球状糜棱岩	12.80m
6. 糜棱岩	60.13m
5. 超糜棱岩	0.08m
4. 黑云变粒岩	2.40m
3. 黑云斜长片麻岩	11.36m
2. 混合岩化斜长角闪岩	107.99m
1. 角闪黑云变粒岩	51.98m

三、地层对比

北京地区太古宙岩系遭受了多期变形变质作用，变质程度较深，变形复杂，多被平行化叶理（或片麻理）叠加和置换，破坏了原有各种接触关系。多期构造的韧性剪切和脆性断裂都很发育，致使变质岩被分割呈若干个条带状地块而发生位移，难以确定地层的先后关系。本区同位素年龄资料不全面、不系统，已有的数据也多代表变质作用和混合岩化作用的时间，使老地层“年轻化”。

（一）密云岩群、四合堂岩群对比

前述表壳岩地层划分及其代表性剖面是在密云地区建立的，与区内其他地区变质岩系对比如下：

1）怀柔区北部马圈子、皮条沟一带的新太古代变质杂岩，沿南口山前断裂带和崎峰茶断裂带呈 NE 向带状分布。这套杂岩是经受麻粒岩相区域变质作用的中基性和中性为主的火山-沉积岩。在南口、锥石口一带，以角闪辉石斜长片麻岩和斜长角闪辉石岩为主，夹麻粒岩、浅粒岩、变粒岩，与沙厂地区英云闪长质片麻岩大致相似；在马圈子、皮条沟一带，以榴辉变粒岩、榴辉黑云变粒岩和石榴斜长辉石岩（角闪岩）为主，相当于大漕地区英云闪长质片麻岩。岩石混合岩化作用强烈，退变质作用明显。构造变形强烈，具有多期叠加和塑性形变特征，片理的置换和平行化造成单斜化地段极为普遍；在马圈子、皮条沟等地的含铁矿层的岩石地层具有变质地层特征，可与密云岩群相比，为同时代产物。

2）怀柔北部八道河、喇嘛沟门一带的变质岩系，呈残块零星出露，为太古宙花岗质侵入杂岩体。岩性为黑云斜长片麻岩、黑云角闪斜长片麻岩、斜长角闪岩、夹浅粒岩。

3）平谷区关上—万庄一带的太古宙变质岩出露于弧形背斜的核部，它与北部结晶基底密云岩群呈断层接触。主要变质岩石有黑云角闪片麻岩、角闪斜长片麻岩、黑云变粒岩，夹少量浅粒岩。下部以黑云角闪斜长片麻岩为主，辉石含量相对较多，出现少量含石榴子石的角闪辉石斜长片麻岩和变粒岩；上部以黑云变粒岩为主。该变质岩中未见含铁建造，原岩为辉长岩和闪长岩。

4）密云岩群与四合堂岩群之间关系：①密云岩群由原密云岩群演变而来，《北京地质百年研究》（鲍亦冈等，2001）进一步将含铁表壳岩与太古宙古老侵入体区分，主要为中基性火山岩建造和火山-沉积建造组成的层状表壳岩，原岩主要为中基性—中酸性钙碱系列的火山岩和火山沉积岩，建造具有近似于岛弧特征。包含了3个大的火山喷发-沉积旋回，其中的每个旋回又含有若干个次级小旋回及沉积韵律；普遍经受了麻粒岩相区域变质作用，以麻粒岩相变质岩石组合为主。由下至上麻粒岩减少，上部出现高角闪岩相岩石组合，而且分布在麻粒岩相岩带的外侧，反映了递进的区域变质作用特点；混合岩化作用普遍，但不太强烈，早期以钠质交代为主，晚期以钾质交代为主，常见混合质变质岩和少量条带状混合岩、均质混合岩。②四合堂岩群由原四合堂岩群演变而来，范围比原范围要小，岩性为一套角闪岩相区域变质岩石组合，原岩以中、基性和中酸性火山沉积岩、杂砂岩为主，夹有陆源碎屑沉积岩。由5个不同火山-沉积旋回构成，火山活动由下至上有由中基性到中酸性转化为偏碱性的趋势，靠上部含霓辉石、钠铁闪石的碱性火山岩层和矿层。岩石成层性良好，延伸较稳定，变余砂状结构和变余微层理构造十分复杂，局部见交错层理、递变粒级韵律，岩石中矿物颗粒比密云岩群矿物颗粒要小。四合堂岩群单相区域变质作用明显，较大范围内变质程度相同，除个别地段（近磁铁石英岩围岩等处）含有少量细小石榴子石或透辉石外，通常只出现黑云母-角闪石变质带共生组合，向底部有向高角闪岩相过渡的趋势。混合岩化作用普遍较弱，具选择性钠质交代，为具条带状和间层状混合岩化带或混合岩。四合堂岩群中超镁铁质岩以角闪石岩为主，并有层状浅色辉长岩-闪长岩-斜长岩侵入体组合及重熔再侵入的英云闪长岩等，均遭受角闪相变质作用。

四合堂岩群早期和密云岩群共同经历了中深层次变质作用，由于处于不同构造位置（其中密云岩群主要位于沙厂、大漕地区，四合堂岩群位于云蒙山岩体北部），受后期云蒙山岩体侵位作用，四合堂岩群沿岩体侵位方向发生角闪岩相退变质作用。密云岩群和四合堂岩群没有直接接触，新的测年数据显示，二者之间年龄相差无几，但经野外观察可知，岩石由深到浅显示出由麻粒岩相—高角闪岩相—角闪岩相渐变，并且一般情况下发生角闪岩相变质作用深度应比麻粒岩相变质作用要浅，因此本志将四合堂岩群置于密云岩群的上部层位，时间上四合堂岩群晚于密云岩群。

（二）与邻区太古宇对比

1. 密云岩群与迁西岩群、桑干岩群对比

（1）变质建造及原岩建造

密云岩群由下至上划分为3个变质建造：①二辉石麻粒岩-角闪辉石斜长片麻岩建造，原岩为中基性火山沉积岩夹含铁硅质建造；②黑云角闪辉石斜长片麻岩-石榴角闪二辉麻粒岩建造；③榴辉黑云变粒岩-石榴斜长辉石（角闪）岩建造，原岩为中基性—中酸性火山碎屑沉积岩夹含铁硅质岩建造。其中，第1、2个建造与冀东迁西岩群大致相当；第2个建造与冀西北桑干群基性火山岩建造可能为同期异相产物；第3个建造可与迁西岩群平林镇岩组中基性—中酸性火山岩建造对比。由于密云地区麻粒岩层的数量少、厚度薄、规模小，以及英云闪长岩、紫苏花岗岩、混合花岗岩等岩石系列的岩石都不及迁西地区发育，基性火山岩建造的相对比率也小，反映了本区在整个区域构造演化史上层位偏高的特点。因此，密云岩群的形成时间应晚于迁西岩群，位置相当于其上部。

（2）变质作用特点

密云岩群、迁西岩群、桑干岩群均经历了麻粒岩相—高角闪岩相区域变质作用、角闪岩相变质作用，并在局部地段发生退变质、不同程度混合岩化作用及局部热变质作用。其下部为麻粒岩相，上部为麻粒岩相—高角闪岩相。

（3）构造特征

各岩群均具有太古宙多期变形变质作用和深部岩层塑性形变特征，一般以同斜褶皱最发育，且成群出现。褶皱形态具有下部呈短轴状或穹状、上部呈紧密线状的变化规律，显示了新太古代早期旋回褶皱构造以卵形穹隆和短轴状构造为主的特征。

（4）变质铁矿类型

密云岩群和迁西岩群的硅铁建造均为含透辉石和紫苏辉石的磁铁石英岩，而其上覆地层的硅铁建造却是含角闪石的磁铁石英岩。

（5）形成时代

迁西岩群的锆石 U-Pb 年龄为 2960Ma（伍家善等，1991）、3047Ma（尹庆柱，1988），Sm-Nd 等时线年龄为 3056Ma（河北区调所，1996），密云黑云斜长片麻岩 SHRIMP 锆石 U-Pb 年龄为 2540Ma；桑干岩群麻粒岩中锆石 U-Pb 年龄为 3068Ma（刘宇光，1989），SHRIMP 锆石 U-Pb 年龄为（2761±13）Ma（耿元生，1997），具有可对比性。陆松年（2015）暂将麻粒岩相—高角闪岩相变质岩的迁西岩群、密云岩群以及相伴的 TTG 片麻岩也归属于新太古代，划为新太古代古陆核。

2. 四合堂岩群与遵化岩群、崇礼岩群、滦县岩群等的对比

四合堂岩群与遵化岩群等岩群既有相似性又有区别，主要分属不同构造环境的两套岩石组合。

（1）产出于不同的构造-岩石区

四合堂岩群位于云蒙山北部的角闪岩相区，遵化岩群产于密云-遵化高角闪相—麻粒岩相区，滦县岩群分布于青龙河花岗岩-绿岩区。

（2）变质建造和原岩建造差别

从宏观层序上考虑，遵化岩群处在一个可对比的位置，自下而上均呈基性火山岩至泥质砂岩夹有硅铁质岩或碳酸盐岩的另一火山-沉积旋回；在变质程度上则属于高角闪岩相至麻粒岩相；崇礼岩群、阜平岩群、赞皇岩群硅铁建造均为下部磁铁石英岩建造和上部大理岩建造。

（3）变质程度有差异

四合堂岩群、遵化岩群主体为角闪岩相，局部达到高角闪岩相；遵化岩群主体为高角闪岩相，局部达到麻粒岩相，其与遵化岩群中马兰峪岩组具有较高相似性。

（4）构造特点

均表现为线性构造，显示属于较新层位且大致可对比。

（5）形成时代相同或相近，但年龄值略有差异

与遵化岩群变火山岩建造相比，滦县一带的变火山沉积建造处于较高层位，时代相对较新（耿元生，1998），而四合堂岩群比遵化岩群更新。

四、时代归属

由于众多因素所限，北京地区太古宙的划分至今未能统一。其划分方案曾有过两分法、三分法和四分法。不同部门、不同研究者、在不同的研究时间内根据自己所掌握的资料和认识，进行自己认为适合其工作区情况的划分。20 世纪 90 年代以前主要沿用了原北京地质学院的两分方案。20 世纪 90 年代初北京市地质调查所完成的 1：5 万密云幅等区调，以三分法将测区太古宙划分为早、中、晚 3 个时代。中国地质大学（武汉）完成的 1：5 万不老屯幅区调、北京市地质研究所完成的 1：5 万汤河口等幅区调及天津地质矿产研究所等后来的工作均采用 1991 年 4 月国际地质科学联合会提出的四分法，将测区太古宙地层划归新太古界，其时代确定为新太古代。本志通过系统归纳前人资料，以国际地质科学联合会前寒武系分会 1991 年 4 月第 9 次分会所提出的四分法对北京地区太古宙岩石进行划分归属。

密云岩群、四合堂岩群是本次划分北京地区的表壳岩单位，从其岩性组合、岩石化学、稀土元素含量、同位素年龄、接触关系等综合特征分析，密云岩群和四合堂岩群，从其岩性组合、岩石化学、稀土元素含量、同位素年龄、接触关系等综合特征分析，其形成时代应早于其他各类变质深成体，为北京地区最古老的变质岩石（地层）单位，通过对比分析认为两个岩群成岩时代为新太古代，密云岩群成岩时代略早于四合堂岩群。

1. 密云岩群（Ar_3*My.*）

20 世纪 60—70 年代，北京地区原密云岩群和原四合堂岩群变质岩的 K-Ar 年龄一般小于 1648Ma，均为后期热事件叠加作用造成^{40}Ar 大量扩散和丢失后所得的偏低年龄。

20 世纪 70 年代末，周绍林（1993）采用了微量锆石 U-Pb 同位素法和磷灰石、锆石 U-Th-Pb 同

位素测年法，获得 3 个样品的锆石 U-Pb 和谐年龄为（2444±12）Ma、（2563±50）Ma（原阳坡地组）和（2516±28）Ma（原宋营子组），限定北京地区太古宙变质岩$^{207}Pb-^{206}Pb$ 表面年龄为 2451～1981Ma，受当时仪器和测试方法等限制，测试获得的年龄跨度大，精度差。

贺高品等（1993）发表的由 5 个变质基性侵入岩样品组成的一条 Sm-Nd 等时线年龄为（2825±189）Ma，5 个样品的 Sm-Nd 模式年龄为 2686.21～2183.58Ma。金文山（2000）取其中模式年龄相近的 3 个样品，获得等时线年龄为（2630±160）Ma，全岩 Sm-Nd 等时线年龄一般解释为成岩年龄，考虑到区内深变质岩区早于这些变质基性侵入岩侵位的早期花岗质岩的单颗粒锆石 U-Pb 年龄的最大值为（2662±24）Ma，推测（2630±160）Ma 最为接近密云岩群的实际成岩年龄。

20 世纪 90 年代，王世发等进行 Sm-Nd 同位素测年分析，并结合密云岩群表壳岩特殊分布方式，认为密云岩群在本区太古宙各岩石单位中形成时代最早，Sm-Nd 同位素模式年龄在 3700～3500Ma 之间。据此将其划归古太古代，形成时间应在 3500Ma 以前。

金文山等（2000）对密云-怀柔地区大漕-沙厂混合岩化表壳岩系进行了锆石测量，其中 5 个样品（斜长二辉麻粒岩 97073 样 1～4 锆石和斜长次透辉石岩 97152 样 6 锆石）均为深变质岩，锆石基本上属于变质重结晶锆石，测得单颗粒锆石 U-Pb 年龄为（2553±74）Ma～（2362±31）Ma。认为古老锆石中铅同位素有所丢失，测得年龄值偏低，因此推断其中（2553±74）Ma 为最接近实际的表壳岩系遭受麻粒岩相变质作用的年龄。

石玉若（2012）在对密云水库西南地区采取样品 080202-4 样（斜长角闪岩）的 20 个锆石 24 组数据分析基础上得出锆石的$^{207}Pb-^{206}Pb$ 年龄（2619±25）Ma 和（2614±27）Ma（1σ）。

耿元生（2014）获得迁西王寺峪黑云斜长片麻岩（原岩为英安质火山岩）的锆石 LA-ICP-MS U-Pb 年龄为（2516±9）Ma。目前不论是 3.0Ga 的数据还是 2.5Ga 左右的数据都比较少，难以确证迁西岩群及其中 BIF 的形成时代，根据区域铁矿特征，将迁西岩群置于新太古代。

本志作者于 2016 年采集密云沙厂铁矿围岩黑云斜长片麻岩（原岩为英安质火山岩）样品，获得锆石 LA-ICP-MS U-Pb 年龄为（2544±916）Ma，限定了区内密云岩群形成时代为新太古代。

综上所述，同位素测年数据反映密云岩群变质地层成岩时代为新太古代。

2. 四合堂岩群（Ar_3*Sh.*）

金文山等（1994）采集的 4 个变质岩样品的单颗粒锆石 U-Pb 年龄为（2535±68）～（2497±55）Ma，其锆石基本也属于变质成因，因此推断（2535±68）Ma 为最接近该表壳岩系角闪岩相变质作用的年龄。金文山（2012）认为四合堂表壳岩系角闪岩相变质作用及英云闪长岩主要成岩年龄在 2883Ma 至（2539±68）Ma 之间，以新太古代早期（2800～2700Ma）可能性最大。

石玉若（2012）等在密云四合堂地区采集样品（黑云斜长片麻岩），进行了 SHRIMP 锆石 U-Pb 定年测试，由于遭受强烈变形，变质原因类型难以确定。万渝生（2012）通过系统分析华北克拉通早前寒武纪条带状铁建造形成时代时提出石玉若（2012）在四合堂地区获得的黑云斜长片麻岩年龄 2.54～2.51Ga 可作为区内表壳岩的限制年龄，认为四合堂岩群表壳岩的年龄为新太古代。

耿元生（2014）分析冀东地区原划分的遵化岩群及其中的 BIF 年龄，获得石人沟铁矿围岩角闪斜长片麻岩形成年龄为（2541±21）Ma，变质时代为（2512±13）Ma，认为其形成于新太古代晚期。

陈印（2014）开展“云蒙山变质核杂岩的演化及其与华北克拉通破坏的关系”研究时，对四合堂地区斜长角闪片麻岩、斜长岩以及变形变质的侵入岩类进行了同位素测年，认为该区变质基底的原岩时代为 2502～2406Ma。

综合以上诸多资料，四合堂岩群成岩时代为新太古代，晚于密云岩群成岩年龄（表 1-2）。

表 1-2　北京地区变质岩系测年数据

Table 1-2　Dating data of metamorphic rock series in Beijing

变质岩系	样号	岩性	测试方法	年龄/Ma	测试人员	时间
四合堂岩群表壳岩	97243	斜长角闪岩	t_{DM}	2680	金文山等	2000
	97226	斜长角闪岩	U-Pb	2507±32	金文山等	2000
	97233	斜长角闪岩	U-Pb	2520±33	金文山等	2000
	97234	富黑云斜长变粒岩	U-Pb	2497±55	金文山等	2000
	MY41-3	变形细粒花岗岩脉	U-Pb	2502±46	陈印	2014
	MY108-1	角闪岩	U-Pb	2436±12	陈印	2014
	MY169-2	混合岩化花岗片麻岩	U-Pb	2449±170	陈印	2014
	MY170-2	角闪片麻岩	U-Pb	2406±51	陈印	2014
	MY175	正长岩	U-Pb	2442±59	陈印	2014
	MY210-3	花岗岩脉	U-Pb	2445±150	陈印	2014
	MY389	变形花岗岩脉	U-Pb	2460±390	陈印	2014
	MY23	混合岩化斜长角闪片麻岩	U-Pb	2502±16	陈印	2014
密云岩群表壳岩	97073	斜长二辉麻粒岩	U-Pb	2478±10	金文山等	2000
	97607	细粒黑云二长质岩	U-Pb	2460±12	金文山等	2000
	98794	英云闪长质岩	U-Pb	2518±6.7	金文山等	2000
	97605	黑云二辉斜长变粒岩	t_{DM}	2533	金文山等	2000
	D0502-4-2	单斜辉石斜长片麻岩	SHRIMP	2500±8	石玉若等	2012
	D0502-5-1	黑云斜长片麻岩	SHRIMP	2542±17	石玉若等	2012
	080202-4	斜长角闪岩	SHRIMP	2619±25	石玉若等	2012
	080202-5	斜长角闪岩	SHRIMP	2508±10	石玉若等	2012
	BD007	黑云斜长片麻岩	U-Pb	2544±16	方同明等	2016

第二章　中—新元古界

北京地区中—新元古代地层出露广泛，主要分布于西山和军都山，出露面积约占山区面积的三分之一。另外在平原区新生代地层之下的基岩中亦有分布。中—新元古代地层为华北地区克拉通变质基底之上的第一套稳定盖层，以一套未变质的（仅局部轻微变质）滨海—浅海相富镁碳酸盐岩为主，碎屑岩和黏土岩次之，局部发育碱玄岩系列火山岩。本套地层叠层石发育，并富含微体古生物化石。本区中—新元古代地层进一步划分为长城系、蓟县系、待建系和青白口系，含 12 个组级岩石地层单位（图 2-1；表 2-1）。

本区中—新元古代地层的研究始于 19 世纪晚期，李希霍芬（F. V. Richthofen）（1877）将其划归为“震旦层系”。

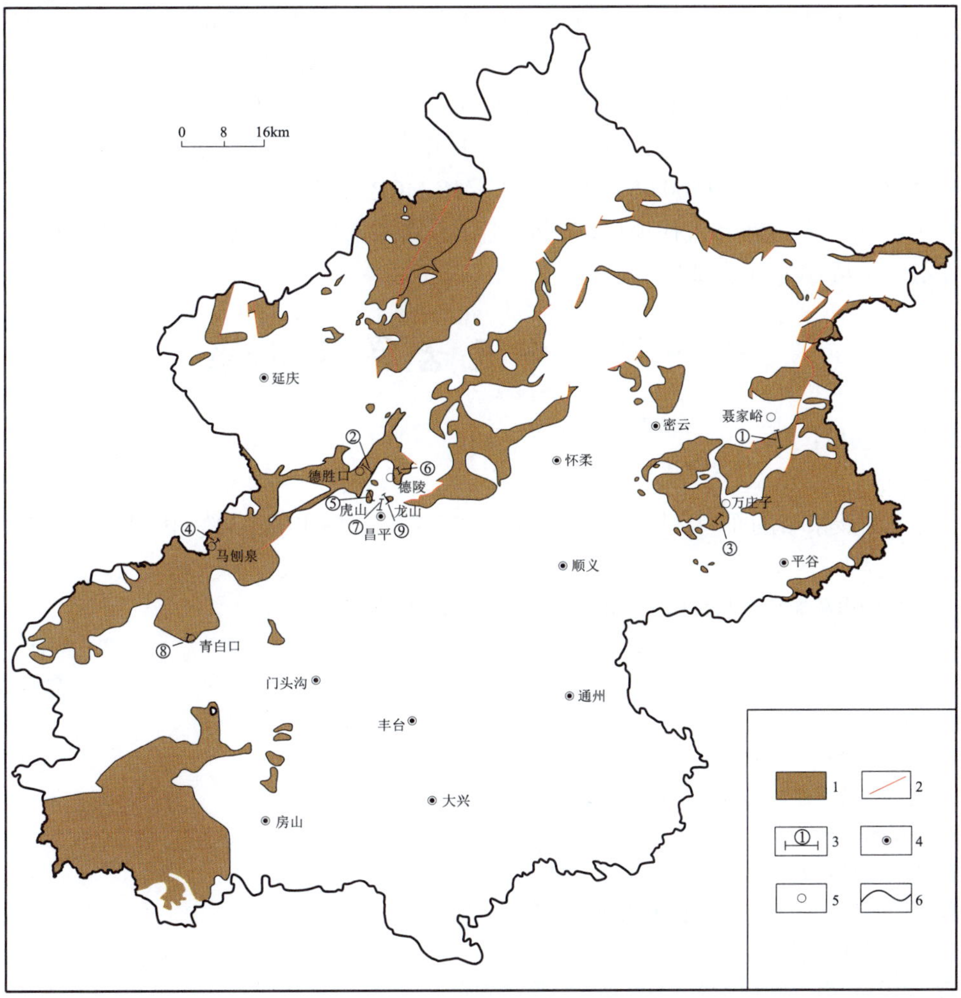

图 2-1　北京市中—新元古代岩石地层及层型剖面分布图

（据《北京市岩石地层》，1996）

Fig. 2-1　Distribution map of Meso-Neoproterozoic lithostratigraphic units and their stratotype section in Beijing

1—中—新元古代岩石地层；2—断层；3—层型剖面：①聂家峪常州沟组剖面，②德胜口串岭沟组、团山子组、高于庄组剖面，③万庄子甲山大红峪组剖面，④马刨泉杨庄组剖面，⑤翠花山-卧虎山雾迷山组剖面，⑥德陵北沟洪水庄组剖面，⑦苏子山铁岭组剖面，⑧青白口下马岭组、景儿峪组剖面，⑨龙山龙山组剖面；4—区政府驻地；5—村庄；6—市界章鸿钊、翁文灏（1916）将覆于片麻岩之上的石英岩及含硅质灰岩称为“南口系”，划属元古宇。《北京西山地质志》（叶良辅等，1920）划分的“矽质石灰岩”至“中灰岩”相当于现今的蓟县系和青白口系。1922 年葛利普（A. W. Grabau）把震旦系限定为前寒武系未变质的沉积岩系。田奇瑪（1923）从层序、化石、地层对比等方面对南口剖面的研究进行了总结。高振西等（1934）在北京地区测制了青白口、北峪沟、下苇甸及横岭-镇边城-大村地质剖面，并与蓟县标准剖面对比对，为华北地区中—新元古代地层的划分和研究奠定了坚实的基础。张文佑（1935）在昌平区龙山的“景儿峪灰岩”上部发现了三叶虫化石，将此层灰岩创名为“昌平灰岩”。孙云铸（1957）认为蓟县“景儿峪灰岩”顶部的角砾岩（相当于昌平灰岩底部的那层底砾岩）代表“震旦系”和寒武系的不整合，其所代表的地壳运动称为蓟县运动。1958—1961 年第一轮 1∶50000 区域地质调查工作，将震旦系的顶界限定在下马岭组之底，并将震旦系二分为长城统和蓟县统。1975 年在北京召开的“中国震旦系讨论会”提出了一个全国中—新元古代地层划分对比方案，将北方“震旦亚界”分为长城系、蓟县系和青白口系。1982 年“晚前寒武纪地层分类、命名会议”决定废除“震旦亚界”一名，改为中、上元古界，在以后开展的北京地区第二轮 1∶50000 区调采用了这一划分方案。1996 年鲍亦冈等对北京地区岩石地层进行清理。

21 世纪以来，随着地质工作的深入展开，一批地质工作者在中—新元古代以碳酸盐岩和碎屑岩为主的地层系统中发现了多层火山岩（斑脱岩）夹层和侵位于其中的岩床，应用新的测年技术，获得了一批新的地质资料和高精度测年数据。高林志等（2007）获得门头沟大村下马岭组中凝灰岩中锆石年龄（1368±12）Ma，李怀坤等（2010）在延庆高于庄组三段上部发现了凝灰岩，并获得锆石（1559±12）Ma（SHRIMP U-Pb）和（1560±5）Ma（LA-MC-ICP-MS U-Pb）年龄，因此，全国地层委员会（2013）对中国北方中—新元古代地层的划分进行了重大调整，下马岭组由原属新元古界调整到中元古界，并暂划归于待建系（表 2-1）。

第一节　中元古界长城系

1. 常州沟组（Pt_2^1c）

（1）代表剖面

以北京市密云区聂家峪剖面为次层型（东经 117°07′27″，北纬 40°22′32″。由肖盈庆等于 1982 年测制）。

上覆地层：串岭沟组灰黑色板状含铁质粉砂岩与条纹状泥质粉砂岩

———— 整　合 ————

常州沟组	总厚度 998. 6m
四段	425. 3m
26. 灰白色厚层状石英岩状砂岩	12. 4m
25. 灰白色薄层石英砂岩与板状泥质粉砂岩互层，具波痕、干裂	3. 0m
24. 白色厚层石英砂岩	19. 2m
23. 紫红色铁质长石石英砂岩，具波痕、干裂	25. 6m
22. 中厚层石英岩状砂岩与紫红色板状铁质长石粉砂岩互层，具干裂	6. 4m
21. 厚层石英岩状砂岩	26. 7m
20. 灰白色中薄层石英砂岩夹紫红色铁质长石质粉砂岩	30. 5m
19. 灰白色石英砂岩	161. 5m
18. 紫红色含细砾长石石英砂岩，具楔状交错层理	18. 5m
17. 灰白色巨厚层石英岩状砂岩，具小型槽状与帚状交错层	121. 5m
三段	240. 2m
16. 石英岩状砂岩夹肉红色长石石英砂岩。具波痕	51. 3m
15. 青灰色薄层石英砂岩与泥质粉砂岩互层，干裂发育	43. 6m

表 2-1 北京市中—新元古代岩石地层划分沿革表

Table 2-1 The classification history of Meso-Neoproterozoic lithostratigraphic units in Beijing

李希霍芬	叶良辅等	田奇瑪	高振西等	北京地质学院	乔秀夫	《华北地区区域地层表·北京市分册》	王长庆等	《北京市区域地质志》	《北京市岩石地层》	本志
1877	1920	1923	1934	1961	1976	1977	1979	1991	1996	
西山	西山	南口	西山	北京	北京	北京	十三陵	北京	北京	北京
震旦系：绿色厚层灰岩	寒武系：中灰岩		震旦系：青白口群：景儿峪灰岩	寒武系：青白口组：景儿峪组	寒武系：昌平组	震旦亚界：寒武系：昌平组	下古生界：寒武系：昌平组	下古生界：寒武系：昌平组	昌平组	下古生界：寒武系：昌平组
灰绿色和黑色砂质页岩与黄色砂岩互层	下灰岩				震旦系：青白口群：景儿峪组	青白口系：景儿峪组	震旦亚界：青白口系：景儿峪组	上元古界：青白口系：景儿峪组	景儿峪组	新元古界：青白口系：景儿峪组
	前寒武系：下马陵层		下马岭页岩	下马岭段	龙山组	龙山组	龙山组	长龙山组	长龙山组	龙山组
					下马岭组	下马岭组	下马岭组	下马岭组	下马岭组	中元古界：待建系：下马岭组
灰白色纯石灰岩	砂质石灰岩		蓟县群：铁岭灰岩	震旦系：蓟县统：铁岭组		蓟县系：铁岭组	蓟县系：铁岭组	中元古界：蓟县系：铁岭组	铁岭组	蓟县系：铁岭组
浅蓝色结晶砂质灰岩；厚层结晶石灰岩；薄层含燧石结晶灰岩；砂质石灰岩			洪水庄页岩	洪水庄组		洪水庄组	洪水庄组	洪水庄组	洪水庄组	洪水庄组
			雾迷山灰岩	雾迷山组		雾迷山组	雾迷山组	雾迷山组	雾迷山组	雾迷山组
			杨庄红页岩	杨庄组		杨庄组	杨庄组	杨庄组	杨庄组	杨庄组
		震旦系：上砂质石灰岩	南口群：高于庄灰岩	高于庄组		高于庄组	南口系：高于庄组	长城系：高于庄组	高于庄组	高于庄组
		上石英岩	大红峪石英岩及安山岩	长城统：大红峪组		大红峪组	大红峪组	大红峪组	大红峪组	长城系：大红峪组
		下砂质石灰岩	串岭沟页岩	串岭沟组		长城系：团山子组	长城系：团山子组	团山子组	团山子组	团山子组
						串岭沟组	串岭沟组	串岭沟组	串岭沟组	串岭沟组
		中石英岩，板岩或页岩下石英岩	长城石英岩	长城组		常州沟组	常州沟组	常州沟组	常州沟组	常州沟组
		五台系	太古界	太古界		太古界	太古界	太古界	太古宇	太古宇

14.	灰白色石英砂岩，波痕发育	76.0m
13.	紫红色石英砂岩	9.1m
12.	黄白、灰白色石英砂岩	41.3m
11.	砖红、紫红色含铁质长石石英砂岩，具干扰菱形波痕	18.9m
二段		243.8m
10.	紫灰色含长石石英砂岩，夹紫红色粉砂岩	107.4m
9.	紫红色含长石石英砂岩	64.7m
8.	紫灰、灰白色石英砂岩与紫灰色砂砾岩。具板状、舟状交错层理	22.5m
7.	上部灰黄白色含长石石英砂岩，下部为细砾岩与含砾不等粒砂岩互层	16.2m
6.	上部黄白色长石石英砂岩，下部含细砾石英长石砂岩，含磷结核	33.0m
一段		89.3m
5.	石英质中砾岩与粗粒长石石英砂岩互层，4个韵律层，叠瓦状构造明显	16.6m
4.	含砾石英长石砂岩	4.1m
3.	灰紫色中—厚层状中粗粒石英长石砂岩	50.0m
2.	石英质中砾岩与粗粒石英长石砂岩互层。3个韵律层，砂砾比1：6，叠瓦状构造明显，发育巨型透镜状层理、大型槽状交错层理	17.8m
1.	黄白色黏土岩（古风化壳）	0.8m

～～～～～ 角度不整合 ～～～～～

下伏地层：太古宙辉石角闪斜长片麻岩

（2）地质特征与区域变化

本组主要分布在北部军都山和平谷北山。岩性以碎屑岩为主，顶部夹有少量黏土岩。可划分为4个岩性段：一段主要为砾岩和砂岩，集中分布于密云-墙子路东西向断裂带以南的火郎峪—安家寨一带和平谷北山地区，由石英质砾岩与粗粒长石石英砂岩组成韵律层序。砾石磨圆度高，具定向排列，有明显的叠瓦状构造。与太古宙片麻岩主要为断层接触。二段为紫灰、紫红色含细砾长石石英砂岩，由细砾岩、砂砾岩和长石石英砂岩组成沉积韵律。三段主要为长石石英砂岩，主要岩性为紫红色和灰白色长石石英砂岩、石英砂岩夹少量泥质粉砂岩。四段主要为灰白色石英岩状砂岩、石英砂岩夹紫红色铁质长石粉砂岩、泥质粉砂岩。

昌平南口—十三陵一带，本组厚度自北东向南西迅速变薄。在十三陵地区厚157m，在王家元地区厚84m。分为3个岩性段，缺少砾岩段。

密云水库周边地区该组厚度变化较大，以密云-墙子路东西向断裂为界，南部常州沟组厚度可达千米以上，以聂家峪剖面为代表，四段发育齐全，最厚处在平谷关上以北至密云张家台一带，厚度可达1200m，密云-墙子路断裂以北，一般仅厚270～330m。在密云大龙门—苍术会一带，缺失底部砾岩段，厚330.57m。崎峰茶阳坡剖面厚338m，底部砾岩仅有0.5m。椴树梁—粘鱼钩一带，由于云蒙山岩体的侵位，地层发生倒转和变质，底部砾岩的砾石被强烈压扁拉长，但砂岩中的斜层理清晰可见。

平谷万庄子—熊耳寨一带，岩性基本同聂家峪剖面，厚200～800m，以大华山为界，东部近800m左右，如花峪及南山剖面厚822m，西部锐减为200～300m。

延庆红旗甸、红石湾、车道沟、营门一带，岩性与聂家峪剖面相比：厚度变薄，缺失一段；二段未有砾岩单层，交错层理不发育；三段砂岩粒度较细，发育较多的细砂岩、粉砂岩，含铁质较高，颜色普遍具灰紫色特征；四段以石英砂岩发育为特征，不发育长石砂岩。其中红旗甸一带厚347m，红石湾与车道沟厚276m左右，营门一带厚177m。

房山磁家务—南观一带，岩性以石英砂岩为主，厚67m，底部被房山花岗岩体侵位。

2. 串岭沟组（$\mathbf{Pt_2^1}\boldsymbol{cl}$）

（1）代表剖面

北京市以十三陵德胜口剖面为次层型（东经116°11′00″，北纬40°17′57″。由汪长庆等1978年测制）。

上覆地层：团山子组黄灰色白云岩与暗灰色白云质泥岩

———— 整　合 ————

串岭沟组	总厚度 46.8m
19. 黑色粉砂质页岩，灰、深灰色厚层泥晶白云岩	7.4m
18. 灰白色中粗粒含长石石英砂岩，含黄铁矿波纹状白云质页岩	6.4m
17. 叠层石白云岩、砂质白云岩夹紫色白云质页岩	9.9m
16. 紫色白云质砂岩、白云质泥岩、条带状粉砂质页岩、白云质砂岩韵律层	7.3m
15. 透镜状砂质白云岩与黑色细条带状硅质板岩韵律互层	9.3m
14. 顶部黄褐色叠层石白云岩和白云质砂岩，中下部含黄铁矿黑色绢云母粉砂质条带状页岩，底部黑色含鲕状黄铁矿页岩	6.5m

———— 整　合 ————

下伏地层：常州沟组灰白色巨厚层石英岩状砂岩

（2）地质特征及区域变化

串岭沟组以黏土岩（泥岩和页岩）为主，地球化学特征富铁富钾。含有微古植物和叠层石，局部地区产有钾矿和铁矿。厚度一般为 50~100m。与上下地层均呈整合接触。

串岭沟组主要分布在军都山和平谷北山一带，出露地点与常州沟组一致。

昌平南口—十三陵一带，该组厚 50~70m。

密云水库周边地区：密云大龙门—古北口—白马关一带，本组以粉砂质页岩和泥质粉砂岩为主，白云岩和砂岩次之，厚度 60~70m，北白岩厚 63m，蔡家沟厚 70~80m，大龙门厚 63m。密云恒河一带厚 114m，主要是灰、灰黑色含钾粉砂质泥质页岩和泥质粉砂岩夹灰色泥质白云岩，底部含十几厘米厚的鲕状黄铁矿。

平谷北山一带以灰、灰黑色粉砂质页岩、页岩和长石粉砂岩为主，夹少量白云岩和砂岩，厚 17~35m。砖瓦窑厚 35m，鱼子山一带厚 32.12m，页岩因含钾量高而成钾矿。

延庆营门、红旗甸及六道河一带，本组以黑色页岩和砂质页岩为主，夹少量砂岩和白云岩，厚 50m 左右，其中营门一带本组底部为肾状、鲕状赤铁矿（宣龙式），厚约 2m。肾状、鲕状构造由叠层石组成。

房山磁家务一带，仅见其残片，主要为灰黑色薄层砂质板岩和细砂岩，出露厚度不大。

3. 团山子组（Pt_2^1t）

（1）代表剖面

以北京市十三陵德胜口剖面为次层型（东经 116°11′03″，北纬 40°17′55″。由汪长庆等 1978 年测制）。

上覆地层：大红峪组灰白色巨厚层沉积石英岩

———— 整　合 ————

团山子组	总厚度 58.9m
二段	36.0m
26. 米黄色细晶灰质白云岩、纹层状泥云岩，砾屑泥云岩，密纹层叠层石白云岩	20.0m
25. 上部米黄色泥云岩，紫灰色白云岩与黑灰色波纹叠层石泥云岩互层；下部灰色锥叠层石泥晶白云岩与纹层状泥云岩韵律互层，具斜层理	10.8m
24. 上部及下部为深灰色白云质泥岩，中部为团块白云岩、紫灰色条带状泥云岩，具斜层理	5.2m
一段	22.9m
23. 黄灰色竹叶状砾屑白云岩、紫灰色厚层条带状白云质泥岩，具波状层理	1.9m
22. 紫黑色白云质泥岩、泥质白云岩，底部砾屑白云岩，见包卷层理	6.7m
21. 黑色薄层白云质泥岩与褐黄色白云岩韵律互层	9.3m
20. 黄灰色白云岩与暗灰色白云质泥岩韵律互层	5.0m

———— 整　合 ————

下伏地层：串岭沟组黑色粉砂质页岩

（2）地质特征及区域变化

团山子组主要为一套含泥质及粉砂质白云岩。可分为两段：一段以紫黑色、黄灰色泥质白云岩为主；二段以灰白色、灰黄色泥质白云岩为主。由于含泥质，本组岩石风化面普遍呈土黄色。上部发育叠层石，与上下地层均为整合接触。

团山子组主要分布在军都山和平谷北山一带。

南口—十三陵一带，岩性主要为含燧石条带（团块）白云岩夹薄板状含粉砂白云质泥岩，地球化学特征富含钾钠。含有微古植物和叠层石。十三陵剖面厚度近 60m，南口一带厚 95m。

密云水库周边地区以大龙门剖面、尖岩剖面和崎峰茶剖面为代表。大龙门剖面本组厚 166m，主要是泥质粉砂质白云岩，含泥质砾屑白云岩，条带、条纹状粉晶白云岩夹白云质页岩。中部见有大型柱状和包心菜状叠层石，顶部为灰绿色玄武岩。尖岩剖面本组厚 280 余米，下部主要为泥质粉晶白云岩夹多层火山碎屑岩和熔岩（安粗岩），反映团山子早期就有火山活动，上部为含硅质条带和团块粉晶白云岩。怀柔崎峰茶一带厚 237m，以灰黄色含粉砂泥质白云岩、条纹状粉晶白云岩为主，发育泥裂、波痕和岩盐假晶，见有燧石条带，中部发育叠层石。

平谷北山主要岩性为灰色厚层—巨厚层条纹条带状（纹层状）含砾屑或砂、粉砂的微—泥晶白云岩，夹灰紫色薄层纹层状粉砂质粉晶白云岩、粉砂质页岩。砖瓦窑厚 115m。

延庆地区主要分布在营门和红旗甸一带。营门一带厚 140m，下部为紫、灰绿色砂质页岩与紫灰色薄层白云岩互层；中部为紫褐、灰白色白云岩及条带状硅质白云岩；上部为灰色中层状白云岩，含有砂岩和叠层石。红旗甸一带厚 339m，下部为砾屑白云岩、含粉砂泥晶白云岩、纹层砾屑泥晶白云岩、叠层石白云岩；上部为厚层藻礁白云岩、硅质团块条带白云岩、薄层粉砂岩。

房山磁家务一带该组厚度 90 余米。底部为灰黑色硅质白云岩与灰绿色钙质页岩互层；下部为薄层含钙细粒长石砂岩；中部为灰黑色硅质白云岩夹页岩；上部为灰白色硅质白云岩与黄白色长石石英砂岩互层。

4. 大红峪组（$\mathbf{Pt}_2^1d$）

（1）代表剖面

选取北京市平谷区万庄子甲山剖面为次层型（东经 116°59′23″，北纬 40°13′27″。由《北京市区域地质志》火山岩编写组 1986 年重测）。

上覆地层：高于庄组灰白色中细粒石英长石砂岩

———— 整　合 ————

大红峪组	总厚度 445.2m
二段	133.4m
35. 上部粉砂质条带白云岩，中下部燧石条带白云岩	32.1m
34. 灰黑色薄层白云岩与硅质页岩互层	4.4m
33. 上部砂质白云岩，下部燧石条带白云岩	29.1m
32. 灰色燧石条带锥叠层石白云岩，锥高 1.5m	8.3m
31. 褐黄色中厚层中粗粒长石石英砂岩，中下部有数层 0.4~0.6m 厚鲕粒重晶石砂岩	19.8m
30. 上部巨厚层燧石团块锥叠层石白云岩，下部燧石条带白云岩，顶底为大波纹状泥晶白云岩	37.6m
29. 灰白色中厚层砂质白云岩，局部为含砾砂岩	2.1m
一段	311.8m
28. 褐红色含火山角砾白云岩	1.8m
27. 褐红色粗面质蚀变熔结角砾岩	12.4m
～～～～ 火山喷发不整合 ～～～～	
26. 灰白色中厚层碧玉条带白云岩	7.6m
25. 灰白色中厚层中粗粒石英砂岩，具有斜层理	1.7m

24.	褐黄色中厚层砂质白云岩	1.7m
23.	暗灰绿色玄武岩与红色粗面岩组成喷发韵律	6.5m
22.	灰绿色玄武岩与浅紫色粗面质角砾岩及凝灰岩组成两个喷发韵律	5.7m
21.	紫色含砾凝灰质砂岩	2.8m
20.	灰绿色玄武岩与浅紫色粗面质角砾岩、凝灰岩构成喷发韵律	5.4m
19.	暗紫色中厚层含砾凝灰质砂岩	3.8m
18.	灰绿色玄武岩、紫红色杏仁状粗面岩与粗面质火山角砾凝灰岩构成两个喷发韵律	66.1m
17.	灰绿色玄武岩与紫红色粗面质角砾凝灰岩构成喷发韵律	30.4m
16.	褐灰色玄武岩与暗灰色粗面质角砾凝灰岩组成两个喷发韵律	12.1m
15.	灰绿色玄武岩（具假枕状）、褐灰色蚀变玄武岩、浅紫红色含角砾玄武岩与紫红色气孔状粗面岩组成喷发韵律	19.3m
	～～～～ 火山喷发不整合 ～～～～	
14.	上部灰白色含砾粗粒石英砂岩，具大型板状斜层理；下部为砾岩，砾石为粗面岩，具叠瓦状构造	2.9m
13.	肉红色粗面质火山角砾岩	1.5m
12.	紫红色凝灰质粉砂岩，具小型交错层理	9.2m
11.	灰绿色玄武岩与紫红色气孔状粗面岩构成两个喷发韵律	49.7m
10.	紫灰色薄层凝灰质粉砂岩	0.8m
9.	暗绿色玄武岩与暗紫色杏仁状粗面岩构成两个喷发韵律，底部为黄白色蚀变玄武岩，具大型喇叭状气孔构造	16.8m
	～～～～ 火山喷发不整合 ～～～～	
8.	灰白色纹层白云岩，大波纹状白云岩与叠层石白云岩	3.4m
7.	黑绿色玄武岩与浅紫红色熔渣状粗面岩构成喷发韵律	15.3m
	～～～～ 火山喷发不整合 ～～～～	
6.	灰色中厚层白云岩	1.7m
5.	底部黄白色气孔状蚀变玄武岩（高岭土化），向上为灰色玄武岩，紫红色气孔状粗面岩	21.3m
	～～～～ 火山喷发不整合 ～～～～	
4.	灰色纹层白云岩，小波纹状白云岩，大波纹状白云岩，叠层石白云岩组成韵律	1.8m
3	浅黄色凝灰质砂岩	3.1m
2.	浅灰色中厚层纹层状白云岩，夹竹叶状砾屑白云岩	4.3m
1.	褐黄色凝灰质砂岩	2.7m
	———— 整　　合 ————	

下伏地层：团山子组浅灰色中厚层燧石条带白云岩

（2）地质特征及区域变化

大红峪组以碎屑岩占优势。一般可分为两个岩性段：一段由石英砂岩与含粉砂泥晶白云岩组成；二段以燧石条带白云岩为主，夹硅质岩。本组主要分布在军都山和平谷北山。在平谷北山地区一段发育火山岩。本组厚82~559m，普遍发育叠层石，与下伏团山子组呈整合接触。

南口—十三陵一带，本组总体呈南厚北薄，南口剖面厚72m，十三陵剖面厚81m，王家元剖面厚128m。岩性下部以石英砂岩及粉砂岩与深灰色硅质条带泥晶白云岩互层。宋天锐（1987）等首次在泰陵发现大红峪组中的硅化岩中含有藻类丝状体及凝灰质成分，杨慧宁（1988）等又在德胜口发现了层凝灰岩，之后宋天锐（2000）研究泰陵一带的大红峪组发现其出露的灰色角砾硅化岩和黑色粉砂凝灰页岩中都含有火山-地震事件的信息。

密云水库周边地区，本组厚度为150~250m。尖岩厚224m，大龙门厚160m，五亩地厚212m，丰各庄厚145m，聂家峪厚164m，怀柔孙胡沟一带厚250m。密云-墙子路断裂以北岩性以白云岩为主，以南该组下部见大量火山岩及火山碎屑沉积岩，沿走向厚度和岩性变化较大。白云岩中产柱状、锥状、包心菜状叠层石。

平谷北山地区为火山岩发育地区，厚度450余米。分为两个段：一段为灰绿色玄武岩、紫红色粗面岩、凝灰质砂岩、粉砂岩、石英砂岩和纹层状白云岩、叠层石白云岩组成火山-沉积岩石组合，火山喷发具有先水下后水上的特点，可分为早、中、晚3期，喷发中心在平谷挂甲峪—熊耳寨一带；二段为硅质条带锥叠层石白云岩、砂质白云岩和少量长石石英砂岩。

延庆地区营门剖面厚170m，下部为杂色石英砂岩；中部为灰白色硅质白云岩，含有硅质条带或团块；上部为乳白色石英岩。河西村剖面厚186m，下部岩性为石英砂岩、硅质条带砂质泥晶白云岩夹含碧玉条带砂质白云岩、叠层石白云岩；上部为碧玉条带砂质白云岩和厚层硅质条带叠层石白云岩。

房山磁家务地区，本组顶部出露不全，厚335m。下部为砖红色长石砂岩夹黄白色长石石英砂岩；中部以灰白色细砂岩，石英砂岩为主；上部为灰白色与砖红色黏土质长石石英砂岩互层。

第二节　中元古界蓟县系

1. 高于庄组（Pt_2^2g）

（1）代表剖面

正层型位于天津市蓟县❶大红峪沟—翟庄村北，1964年由陈晋镳、薛知照重测。北京地区选取十三陵德胜口剖面为次层型（东经116°11′12″，北纬40°17′49″，汪长庆等1978年测制）。

上覆地层：杨庄组紫红色泥云岩，厚层状白云质砂岩

--------------- 平行不整合 ---------------

高于庄组	总厚度1004.9m
四段	520.0m
55. 上部为灰色泥晶白云岩，下部为浅灰色中厚层燧石团块细晶白云岩、燧石条带叠层石白云岩	20.9m
54. 黑色豆状燧石条带白云岩、藻团硅质条带白云岩、小波纹叠层石白云岩	32.7m
53. 浅灰色中厚层含砂屑细晶白云岩、硅质条带白云岩、纹层叠层石白云岩、黑色纹层状燧石条带泥云岩组成沉积韵律	74.8m
52. 硅质条带波纹叠层石白云岩、黑色燧石条带波纹状白云岩、浅灰色白云岩组成沉积韵律	92.5m
51. 硅化核形石白云岩、断续硅质条带波纹叠层石白云岩夹燧石条带泥晶白云岩	58.0m
50. 纹层泥晶白云岩、断续硅质条带白云岩、厚层含砂硅质白云岩、硅质条带锥叠层石白云岩组成沉积韵律	61.3m
49. 浅灰色硅质条带泥晶白云岩、黑色层纹燧石条带白云岩、灰色巨厚层白云岩组成韵律	52.1m
48. 灰白、浅灰色细—粗晶白云岩、硅镁质条带纹层白云岩、灰色泥晶白云岩组成韵律	82.7m
47. 大波纹叠层石白云岩、包心菜叠层石白云岩、硅镁质条带波纹状白云岩、浅灰色纹层叠层石白云岩组成韵律	45.0m
三段	215.9m
46. 上部为深灰色厚层泥晶-细晶白云岩，下部为深灰色含灰白云岩	55.6m
45. 深灰色灰质白云岩与灰黄色含灰白云岩组成密集韵律互层，发育脉状层理、压扁-透镜状层理	35.0m
44. 灰黄色白云质灰岩与深灰色白云岩组成密集韵律互层，发育脉状层理、压扁-透镜状层理	39.2m
43. 灰黑色白云质灰岩与深灰色灰质白云岩组成密集韵律互层，夹赤铁矿锰质透镜体	50.4m
42. 中层板状（砖块状）含灰白云岩与叶片状泥晶白云岩韵律互层，有两层瘤状白云岩，瘤直径约10cm	35.7m
二段	161.2m

❶ 现天津市蓟州区。

41. 灰色厚—巨厚层硅质细条带泥晶白云岩与浅灰色泥质白云岩 20.0m

40. 灰色中—薄层泥质泥晶白云岩、浅灰色泥晶白云岩、黑灰色含锰白云岩、紫红色含锰泥晶白云岩。波痕 $L=18cm$，$H=2cm$ 90.9m

39. 灰色中厚层泥晶白云岩、黑灰、浅灰色泥云岩 34.1m

38. 上部为紫红色含锰白云岩、紫灰色含锰叶片状泥质白云岩，下部为深灰色薄层含锰泥晶白云岩 16.2m

一段 107.8m

37. 浅灰色燧石条带锥叠层石白云岩、深灰色纹层燧石条带泥晶白云岩组成韵律 13.8m

36. 深灰色巨厚层泥晶白云岩、厚层白云岩与硅质条带锥叠层石白云岩组成韵律 48.5m

35. 灰黑色含绢云母泥晶白云岩、含砂屑绢云母白云岩、浅灰色砂屑含泥白云岩、灰黑色含锰泥云岩组成两个韵律层，具斜层理 16.9m

34. 灰色包心菜叠层石白云岩、燧石团块白云岩、黑色燧石条带纹层白云岩、浅灰色砂屑泥晶白云岩、砂质白云岩、竹叶状砾屑砂质白云岩构成沉积韵律，具斜层理 21.9m

33. 灰色巨厚层白云质砂岩、巨厚层白云质粗砂岩、浅灰色厚层中—细粒石英砂岩，具斜层理 6.7m

———— 整　合 ————

下伏地层：大红峪组深灰色硅质条带纹层泥晶白云岩

（2）地质特征及区域变化

高于庄组可分为4个岩性段：

一段底部为厚层白云质砂岩、石英砂岩和粉砂岩，下部为含砂屑白云岩、砂质白云岩、燧石条带白云岩，上部为泥晶白云岩、燧石条带白云岩和锥叠层石白云岩；二段为灰色、紫红色含锰白云岩、泥质泥晶白云岩和燧石条带白云岩；三段为板层状含灰白云岩、灰质白云岩与白云质灰岩不等厚互层，夹页片状泥质白云岩，下部发育多层瘤状灰岩或白云岩；四段为锥叠层石白云岩、角砾状白云岩、燧石条带泥晶白云岩、含砂屑白云岩、核形石白云岩等组合而成，其中上部的叠层石个体较大，高2m以上。

高于庄组主要分布在军都山和平谷北山，在西山也有出露，但仅有中上部地层。与上覆杨庄组呈平行不整合接触，与下伏大红峪组呈整合接触。

南口—十三陵一带厚1005m左右。

密云水库周边地区本组厚度1500m左右，密云北部西石骆驼一带厚1487m，大龙门剖面厚1624m，五亩地剖面厚1372m。其中大龙门剖面一段为粉—泥晶白云岩，上部含少量燧石条带；二段主要为灰白色中—厚层含硅泥晶白云岩，底部为棕、灰色含锰页岩；三段为硅质条带白云岩、灰质白云岩和白云质灰岩，下部和底部发育多层瘤状灰岩；四段以含燧石条带、团块白云岩为主，夹少量砾屑白云岩和藻团纹层状白云岩。

怀柔北台子一带厚1245m，岩性与大龙门一带相似。

平谷北山厚1300~1520m。一段为灰色厚层泥晶白云岩、含燧石条带白云岩夹燧石层。二段为灰色含锰白云岩、粉晶白云岩、硅质条带白云岩，个别地段形成扁豆状锰矿。三段为细晶—泥晶白云岩和白云质灰岩，夹瘤状灰岩。四段为浅灰—灰白色角砾状白云岩、硅质条带白云岩，含粉砂白云岩等。黑豆峪剖面厚1339m。

延庆红旗甸一带厚1096m，一段为石英砂岩、粉砂—泥质泥晶白云岩、硅质条带—燧石条带白云岩组成进积型基本层序；二段由含铁锰白云岩、泥晶白云岩组成退积型基本层序；三段由瘤状灰岩、细晶灰质白云岩、硅质条带白云岩、含砂白云岩组成退积型基本层序；四段由微晶—泥晶白云岩、含硅质条带白云岩组成退积型基本层序。李怀坤等（2010）在延庆红旗甸乡下马鹿沟村附近的高于庄组三段上部发现了凝灰岩，并测得该凝灰岩中锆石SHRIMP U-Pb年龄（1559±12）Ma和LA-MC-ICP-MS U-Pb年龄（1560±5）Ma。

西山地区主要出露在沿河城北和十渡西南靠近河北省边界地区，仅见二段至四段，岩性基本与十三陵地区一致，但三段灰岩不发育。沿河城北一带厚1127m。

2. 杨庄组（Pt_2^2y）

（1）代表剖面

正层型位于天津市蓟县城北罗庄乡三间房，北京市选取昌平区马刨泉剖面为次层型（东经115°51′51″，北纬40°09′53″。由北京市地质调查所区调一队1991年测制）。

上覆地层：雾迷山组深灰色厚层状藻团白云岩

———— 整 合 ————

杨庄组 总厚度64.02m

15.	上部为浅灰色中厚层泥晶白云岩，下部为灰色微红色中厚层状含砂砾屑白云岩	4.59m
14.	灰白色薄层状—中层状砂质砂屑白云岩，平行层理发育。砂屑为泥晶白云岩，底部发育20cm厚的黄灰色薄层状含砾砂质白云岩	3.15m
13.	黄灰色薄层状粉晶白云岩，纹层发育	1.43m
12.	浅灰色中厚层状含硅质条带、团块粉晶白云岩。中部有硅质风暴岩，呈鱼骨状。底部为土黄色薄片状含砂含鲕砂屑粉晶白云岩	5.72m
11.	浅灰色厚层状白云质中粗粒石英砂岩，夹有白云质条带及含砂白云岩，脉状层理发育。底部为土黄色含砂屑泥晶白云岩	5.15m
10.	顶部为灰白色细晶白云岩，中下部为灰色微红色中厚层状粉晶白云岩	8.01m
9.	灰白色中厚层状白云质中粗粒石英砂岩、含云石英砂岩夹泥晶白云岩，斜层理发育	8.58m
8.	灰色微红色中厚层状含硅质团状泥晶白云岩夹少量含砂白云岩	4.03m
7.	浅灰色中厚层状白云质中细粒石英砂岩与泥晶白云岩互层，下部发育斜层理	3.45m
6.	灰白色中薄层状白云质石英砂岩，含砂屑细晶白云岩与泥晶白云岩组成旋回层序，脉状层理发育，中部夹一层灰白色硅质鲕粒白云岩（0.8m）	9.78m
5.	灰白色中厚层状泥晶白云岩，纹层发育，中部发育砂质细条带	1.73m
4.	灰白色巨厚层状含砾含云中粒石英砂岩，平行层理发育	4.20m
3.	肉红色中厚层粉晶白云岩，发育缝合线	1.20m
2.	灰白色厚层状含砾中粗粒石英砂岩，平行层理发育。砾石为硅质，次棱角状，大小2~6cm，底部具小型冲刷面	3.00m

-------- 平行不整合 --------

下伏地层：高于庄组灰色厚层状硅质条带细晶白云岩

（2）地质特征及区域变化

杨庄组在北京地区以富含陆源碎屑为特征，颜色为紫红、粉红—灰白色，厚度一般小于100m，主要由含砾石英砂岩、白云质石英砂岩、砂质白云岩和和泥晶、粉晶白云岩组成。与下伏高于庄组为整合接触或平行不整合接触。

本组主要分布于军都山、平谷北山和西山，岩性和厚度变化非常大。

西山地区本组以（含砾）石英砂岩、白云质砂岩和白云岩为主，局部发育砾岩层。马刨泉至了思台一带厚约64m，沿河城一带厚53m，高庄一带厚约50m。在八宝山附近本组为紫红色泥岩和页岩，厚61m。与下伏高于庄组为平行不整合接触。

南口—十三陵一带厚约78m，分为两段：一段为白云质砂岩、砂质白云岩、燧石团块白云岩和泥晶白云岩、砂砾屑白云岩；二段为波纹状泥质白云岩、硅质条带白云岩、藻团、叠层石白云岩。颜色紫红色、米黄色、灰色。与下伏高于庄组为整合接触。

密云水库周边地区以白云质细砂岩、砂质白云岩、燧石条带粉晶白云岩为主，金叵罗一带厚127m，苍术会一带厚138m，与高于庄组整合接触。

平谷北山地区江水泉—茅山一带最厚，可达684m，分为3个段：一段以砂质白云岩和条带状白云岩为主；二段为紫红色页岩夹灰白色页岩和燧石层；三段为灰白、紫红色泥质白云岩和硅质白云岩。与高于庄组整合接触。

延庆地区主要分布在东北部干沟和大荒地一带。干沟一带厚135m，主要是含砂（砾）白云岩、波纹层燧石条带白云岩和泥粉晶白云岩、内碎屑白云岩，与高于庄组平行不整合接触。水头村厚92m，与高于庄组整合接触。

3. 雾迷山组（Pt_2^2w）

（1）代表剖面

正层型位于天津市蓟县罗庄乡磨盘峪村北山顶至磨盘峪南山，再由二十里铺起沿公路至洪水庄乡洪水庄东南。北京地区选取十三陵翠花山-卧虎山剖面为次层型（东经116°12′13″，北纬40°17′10″。由汪长庆等1978年测制）。

上覆地层：洪水庄组浅灰色薄层泥质白云岩夹黄绿色、黑灰色页岩

———— 整 合 ————

雾迷山组	总厚度 2230.2m
四段	524.0m
20. 深灰色燧石条带纹层状泥晶白云岩，灰色硅质条带细晶白云岩，墙叠层石含粒屑白云岩，深灰色泥晶白云岩，黄色纹层状泥晶白云岩	56.5m
19. 深灰色厚层藻团白云岩，灰色含鲕粒屑燧石条带泥晶白云岩，浅灰色含泥、含屑泥晶白云岩构成约15个韵律，单个韵律厚约16m	247.2m
18. 中层硅质砾砂屑白云岩，灰色藻团（锥叠层石）白云岩。菊花状同生砾岩，波状燧石条带白云岩，深灰色泥晶白云岩构成韵律约14个，单个厚约9m	128.3m
17. 灰色中薄层鲕粒硅质岩及鲕粒白云岩，大波纹状硅质条带纹层状白云岩，泥晶白云岩构成约11个韵律，单个厚约9m	92.0m
三段	854.0m
16. 深灰色厚层含鲕花斑藻团白云岩，深灰色中厚层藻团白云岩（柱叠层石），菊花状同生砾岩，灰色球团粒硅质岩，黄色厚层燧石条带纹层状白云岩，泥晶白云岩组成6个韵律，单个韵律厚约21m	136.0m
15. 灰色燧石条带泥晶白云岩，灰白色厚层砂质白云岩，深灰色藻团白云岩（柱叠层石），灰色鲕粒硅质条带白云岩，棕黄色燧石条带白云岩，含砾砂屑条带泥晶白云岩构成5个韵律，单个韵律厚约22m	115.0m
14. 灰色薄层含泥泥晶白云岩，硅质条带纹层状白云岩，深灰色藻团白云岩（柱叠层石），菊花状同生砾岩，硅质条带含砾屑泥晶白云岩，灰色泥晶白云岩构成韵律，韵律中部藻团含量约80%。具斜层理、波痕，顶部具水平层理，含泥质，具鸟眼、干裂构造	85.0m
13. 浅灰色薄层含砂屑条带泥晶白云岩，灰黑色层纹状燧石条带白云岩（含藻丝体），灰黑色密波纹叠层石白云岩，深灰色厚层藻团白云岩（锥叠层石），灰黑色密纹层—密波纹层叠层石白云岩，灰色燧石条带（团块）白云岩，灰色含泥泥晶白云岩构成5个韵律，单个韵律厚约37m	184.0m
12. 灰色泥晶白云岩，深灰色密硅质条带纹层状白云岩，灰黑色密波纹层叠层石白云岩，深灰色厚层藻团白云岩（柱、锥叠层石），灰色中层硅质砾屑白云岩，灰色波纹状白云岩构成2个韵律，单个韵律厚约44m	88.0m
11. 深灰色薄层含泥泥晶白云岩，灰色燧石条带纹层状白云岩，灰黑色密纹—密波纹叠层石白云岩（小柱叠层石），黑色层纹状硅质条带白云岩（藻丝体），浅灰色大波纹—菊花状同生砾岩，灰色厚层藻团白云岩，柱锥叠层石白云岩，深灰色藻鲕藻团白云岩，灰色硅质条带纹层叠层石白云岩。泥质泥晶白云岩构成韵律	133.0m
10. 黄绿色叶片状含泥粉砂屑泥晶白云岩，灰色菊花状同生砾岩及深灰色含柱叠层石硅质条带白云岩，灰色中厚层藻团白云岩，浅灰色鲕粒（核形石）白云岩，灰色纹层硅质条带白云岩，浅灰色含砂屑条带泥晶白云岩构成韵律	113.0m

二段		402.0m
9.	灰色硅质粒屑白云岩，灰色厚层藻团白云岩，浅灰色燧石条带波纹状白云岩，深灰色含砂砾屑白云岩组成韵律。顶部含粉砂屑	54.3m
8.	米黄色泥质白云岩，黑色层纹状燧石白云岩，灰色厚层藻团白云岩，鲕白云岩，波纹状硅质条带状纹层白云岩，含砂砾屑泥晶白云岩组成韵律	140.9m
7.	灰色中厚层波纹层泥晶白云岩，大波纹状硅质条带白云岩，菊花状同生砾岩，灰黑色豆状藻团白云岩（格锥叠层石）。大波纹状硅质条带纹层状白云岩构成韵律	119.2m
6.	米黄灰色泥晶白云岩，深灰色菊花状硅质条带白云岩，灰色厚层藻团白云岩，黑色层纹状燧石白云岩（藻丝体），格锥叠层石白云岩（藻礁），豆状藻团（鲕）硅质岩组成6个韵律，单个韵律厚约15m。底部为白云质页岩	87.6m
一段		450.2m
5.	浅灰色中薄层含泥泥晶白云岩，灰黑色纹层叠层石白云岩，黑紫色纹层燧石条带白云岩，含鲕硅质条带白云岩，深灰色厚层藻团白云岩，米黄色中层团粒硅质岩，深灰色波纹状燧石条带白云岩构成韵律	79.1m
4.	灰色薄层泥晶白云岩夹密细硅质条带纹层状白云岩，灰色含鲕硅质条带白云岩，深灰色巨厚层藻团白云岩，锥叠层石白云岩组成韵律	133.3m
3.	深灰色中厚层细硅质条带密纹—密波纹层叠层石白云岩，中薄层小杯叠层石硅质岩，深灰色厚层小柱叠层石白云岩，灰色中薄层菊花状同生砾岩，灰色鲕团粒硅质条带含鲕粒屑白云岩组成韵律	81.8m
2.	灰黑色泥晶白云岩，中薄层菊花状同生砾岩，深灰色硅质条带密波纹叠层石白云岩（藻丝体），深灰色厚层藻团白云岩组成韵律	100.4m
1.	灰白色薄层状泥质泥晶白云岩，深灰色燧石条带纹层白云岩，灰色中薄层豆状硅质条带藻团白云岩组成韵律	55.6m

——————— 整　合 ———————

下伏地层：杨庄组灰色硅质小柱叠层石泥晶白云岩

（2）地质特征及区域变化

雾迷山组厚度巨大，厚2000~3500m，为一套韵律性明显，富含有机质的碳酸盐岩地层。根据岩性组合和生物特征，可划分为4个岩性段。一段为灰、深灰、灰黑色薄层泥质泥晶白云岩、燧石条带纹层状白云岩、藻团白云岩、密波纹叠层石白云岩等，夹有少量鲕粒白云岩。叠层石类型以小型假裸枝、小杯、小柱状叠层石为主，富含碳质和沥青质；二段主要由大波纹状硅质条带白云岩、厚层藻团白云岩和大中型柱、锥状叠层石白云岩、藻屑白云岩、密波纹沥青质白云岩、黑色层纹状燧石岩及菊花状同生砾岩组成沉积韵律；三段主要由泥晶白云岩、纹层状白云岩、燧石条带白云岩、藻团白云岩、叠层石白云岩、鲕粒白云岩（或硅质岩）组成沉积韵律，叠层石以大型锥、柱状为主，多由藻团构成；四段主要是硅质条带白云岩、泥晶白云岩、纹层状白云岩。藻团和叠层石则不发育。

本组分布非常广泛，军都山、平谷北山和西山都有出露。

密云溪翁庄一带厚2890m，苍术会一带出露1000m，未见顶，只相当一至三段。北白岩剖面厚2889m，一段底部为薄板—页片状含泥质白云岩，下部以硅质条带纹层状粉晶白云岩为主，夹沥青质白云岩、内碎屑白云岩、藻团白云岩，中部和上部以藻团粉晶白云岩为主；二段下部纹层状粉晶白云岩或含硅质条带纹层状粉晶白云岩与硅质条带细晶或粉晶白云岩互层，中上部主要为含硅质条带细晶白云岩或纹层状白云岩、藻团粉晶白云岩韵律互层，夹叠层石粉晶白云岩；三段下部为燧石条带粉—泥晶白云岩与硅质条带纹层状粉晶白云岩互层，中上部为厚层硅质条带细晶白云岩、硅质条带纹层状细晶白云岩、藻团（叠层石）粉晶白云岩韵律互层；四段以硅质条带细晶白云岩夹含硅质条带纹层状粉晶白云岩为特征。怀柔丁香沟—杨树底下一带厚2195m。

平谷南山甘营—夏各庄一带厚2464m，其岩性特征及古生物组合均与蓟县一致。

延庆虎平沟—大观头一带厚2485m。下德龙湾一带，总厚度大于1374m，一至三段各组厚度基本稳定，厚度为1272m，四段未见顶，仅出露103m，与十三陵剖面相比，各段厚度较薄，在岩石组合上：三段未发育鲕粒白云岩（或硅质岩）、砂质白云岩、白云质砂岩，岩石颜色灰白—灰—深灰色，干裂、鸟眼、透镜状层理等不发育；二段密波纹沥青质白云岩、层纹状燧石岩、菊花状同生砾岩及石英砂岩、砂质白云岩不发育，发育泥质白云岩及泥质条纹泥晶白云岩；一段与十三陵剖面基本一致，但干裂、鸟眼构造不发育。

西山沿河城一带厚1677m。黄土贵-沿河城剖面厚1689m，与下伏杨庄组整合接触，岩性组合特征与十三陵剖面基本一致。房山贾峪口一带出露部分只相当于雾迷山组上部的二至四段，厚714m，一段未见底。

4. 洪水庄组（Pt_2^2h）

（1）代表剖面

正层型位于天津市蓟县洪水庄乡床子岭村西南至老虎顶。北京市选取十三陵德陵北沟剖面为次层型（东经116°15′32″，北纬40°17′13″。由汪长庆等1978年测制）。

上覆地层：铁岭组浅灰色中厚层泥晶白云岩

——————整　合——————

洪水庄组 总厚度100.9m

二段 81.2m

25. 浅灰色纹层状泥晶白云岩。上部夹黑色薄层页岩，下部夹薄层粒屑白云岩 12.7m

24. 浅灰色泥质白云岩与黄绿色泥岩互层，下部夹薄层粒屑白云岩，含钙含菱铁矿结核，大波痕，波长50~60cm 15.7m

23. 黄绿色泥岩与浅灰色页岩互层，底部为黄绿灰黑色板岩，上部夹黑色页岩及块状黄灰色泥岩 52.8m

一段 19.7m

22. 黄灰、浅灰色页岩与白云质泥岩互层 13.5m

21. 浅灰色薄层泥质白云岩夹黄绿色页岩、黑色页岩。上部泥质白云岩，层面见大波痕，$L=15cm$，$H=2cm$ 6.2m

——————整　合——————

下伏地层：雾迷山组黄色薄层纹层状泥晶白云岩

（2）地质特征及区域变化

洪水庄组以黑色和深灰色页岩、粉砂质页岩为主，可分为两段：一段为深灰色页岩、黄绿色粉砂质页岩夹黑色含碳质粉砂质页岩及泥质白云岩；二段为灰、深灰色页岩、黑色页岩夹泥质白云岩及含泥粒屑白云岩。与下伏雾迷山组呈整合接触。

该组主要分布于西山、昌平、延庆、怀柔及密云地区，分布面积小。

昌平地区出露在德陵至东石峪村东北一带，十三陵德陵北沟剖面厚100.9m，十三陵剖面厚82m。

密云北白岩—西智一带较厚，北白岩剖面厚188m，西智剖面厚度可达180m，其他地区较薄，大多50~70m，岩性变化不大。北白岩剖面岩性主要为粉砂质页岩、碳质页岩，底部为白云质泥质页岩夹薄板状泥质泥晶白云岩，顶部为薄板状泥质泥晶白云岩与页片状白云质泥质粉砂岩互层。

平谷地区主要出露在庄果峪、响水寺山一带，呈长条形环带状沿北北西向展布，西八里沟村剖面厚151m，底部为硅质条带白云岩与黑色页岩韵律互层，中部为黄绿色页岩与黑色页岩韵律互层，上部为泥质白云岩与黄绿色页岩韵律互层，并夹有砂岩和硅质结核。

延庆三岔口厚54m，怀柔旧水坑厚82m，岩性相似。

西山雁翅黄土贵一带厚仅36m，主要是灰黑色页片状水云母页岩，夹砂岩，顶部为灰黑色薄层状泥晶白云岩，含铁质结核，水平层理非常发育。沿河城西寺台子一带厚32m。

房山贾峪口—霞云岭—北直河一带，厚20m左右，主要是黑色千枚岩状粉砂质板岩夹白云岩透镜体。贾峪口-山川九道河剖面厚36m，岩性为含砂千枚岩，含白云质，部分含碳质，可见不规则黄铁矿顺层分布，其在本区地层厚度虽较小，但出露稳定，岩性变化不大，厚度变化于15~40m之间。

5. 铁岭组（Pt_2^2t）

（1）代表剖面

正层型位于天津市蓟县洪水庄乡老虎顶。北京地区选取十三陵苏子山剖面为次层型（东经116°13′23″，北纬40°15′06″。由汪长庆等1978年测制）。

上覆地层：下马岭组铁质粉砂岩

--------------- 平行不整合 ---------------

铁岭组	总厚度212.6m
二段	113.1m
32. 浅灰色厚层含燧石团块白云岩，灰色厚层纹层状泥晶白云岩，浅灰色厚层含黑色燧石条带白云岩，浅灰色柱叠层石白云岩，灰色纹层状白云岩	42.1m
31. 浅灰色中—厚层微含锰含泥泥晶白云岩，浅黄、灰色中层含粒屑含泥泥晶白云岩，浅灰色泥质白云岩组成韵律。韵律底部见柱叠层石	38.6m
30. 灰、黄灰色厚层竹叶状砾屑白云岩，灰黄色中—厚层白云质灰岩与灰质白云岩互层，浅灰色中—厚层泥质白云岩组成韵律。韵律底部见分叉柱叠层石	32.4m
一段	99.5m
29. 上部浅棕色中厚层微含锰含泥质白云岩；下部黄灰色中—厚层含锰泥晶白云岩夹浅棕色页岩	17.8m
28. 上部黄灰、灰色中—厚层含泥屑泥晶白云岩夹黄绿色砂质页岩；下部为浅灰色中—厚层纹层状含泥泥晶白云岩夹黑色页岩，灰色中—厚层竹叶状砾屑白云岩组成韵律	32.3m
27. 上部浅灰色纹层状含泥泥晶白云岩与黑色页岩不等厚互层（4∶1），含竹叶状砾屑白云岩组成韵律；下部浅灰色、黄绿色白云质页岩，灰黄黑色燧石条带含泥泥晶白云岩，含粒屑白云岩	21.1m
26. 上部紫红色白云质页岩，浅灰色中—薄层竹叶状砾屑白云岩，含燧石条带纹层波纹层白云岩组成韵律；下部为浅灰色中—厚层粒屑白云岩，浅灰色中—厚层纹层状泥晶白云岩，灰色泥晶白云岩组成韵律。韵律上部见干裂、冲刷面	28.3m

———————— 整　　合 ————————

下伏地层：洪水庄组黑色薄层页岩

（2）地质特征及区域变化

铁岭组分为两个岩性段：一段主要为竹叶状砾屑白云岩、燧石条带白云岩、纹层状泥晶白云岩、微含锰白云岩夹灰绿色薄层页岩；二段主要为竹叶状砾屑白云岩、泥质白云岩、灰质白云岩和叠层石白云岩，含少量燧石团块。与下伏洪水庄组呈整合接触。

铁岭组分布与洪水庄组一致。

昌平地区主要分布在德陵至东石峪北东一带，在龙山西侧也有零星出露。十三陵剖面厚282m，一段下部陆源物质含量较高，石英碎屑和内碎屑发育，并夹有竹叶状砾屑白云岩，具斜层理和波痕，上部含铁、锰质，陆源物质和内碎屑含量减少；二段下部含灰质，上部出现柱叠层石。

密云地区主要分布于东智和西智，厚380m。一段底部以中—厚层含粉砂质粉晶白云岩夹薄层页片状含粉砂白云岩、页岩为主；中、下部为中层状粉晶白云岩、含燧石条带白云岩，上部为含锰白云岩和赤铁矿层。二段下部以浅灰色、深灰色中—薄层云质灰岩、含灰白云岩为主；上部以条带状云质灰岩、灰质白云岩为主，夹叠层石灰岩。

平谷仅在庄果峪以西呈北西向展布，一段出露完整，以含砂白云岩、含灰泥质白云岩、页岩为主；二段出露不全，以白云质灰岩为主，叠层石发育，借用蓟县老虎顶剖面作为代表性剖面，厚295m。

怀柔—延庆地区：主要分布于黄坎、吹风坨、柏木井和转山子一带，以延庆三岔口剖面为代表，厚330m。一段为灰白—浅灰色薄—中层灰质白云岩夹黑色页岩和燧石透镜体，顶部产“四海式铁矿”（铁锰矿层），厚约2m。二段为灰白色白云质灰岩，含燧石团块，底部为含锰灰岩。本段的灰岩可用做水泥原料。产丰富的叠层石。怀柔杨树底下西侧下桃树剖面厚285m。

西山沿河城一带西寺台子厚216m。雁翅黄土贵厚251m，一段为含砂砾白云岩、内碎屑白云岩、硅质条带白云岩夹灰黑色或黄绿色页岩；二段为厚层状硅质条带白云岩、粉—泥晶白云岩，发育叠层石和竹叶状砾屑。

房山区主要分布于房山西部西台子—霞云岭—长操—黄山店一带，厚89~414m。一段下部为深灰色中—厚层硅质白云岩；上部为灰黑、棕黄色页岩或千枚状板岩，含少量锰质。二段为灰白、浅灰色中—厚层白云岩，具少量燧石条带和结核，有时夹黑色千枚状板岩和硅质板岩，含大量叠层石。贾峪口-山川九道河剖面厚215m。

第三节　中元古界待建系

下马岭组（Pt_2^3x）

本组名称源于叶良辅等（1920）称的“下马岭层”。1959年全国地层会议后改称下马岭组。高林志（2007）和李怀坤（2009）分别获得下马岭组斑脱岩和侵位于其中的辉绿岩测年数据，2013年全国地层委员会颁布的中国地层表将下马岭组从青白口系分出，暂列为待建系。

（1）代表剖面

正层型位于门头沟区雁翅镇青白口村西沟（东经115°47′48″，北纬40°01′06″。由汪长庆等1980年重测）。

上覆地层：龙山组灰色巨厚层含砾不等粒石英砂岩

--------------- 平行不整合 ---------------

下马岭组	总厚度 343.9m
三段	177.0m
13. 黄绿、绿灰色页岩，易碎	21.2m
12. 深灰色页岩夹板状粉砂质页岩。具水平纹层	5.4m
11. 灰色粉砂质页岩夹黄绿色页岩。具水平纹层	34.8m
10. 黑色含碳质胶岭石硅质页岩夹黑色碳质页岩	86.7m
9. 黑色胶岭石硅质泥岩。岩石致密坚硬，性较脆，层面平直，可见水平纹层	28.9m
二段	53.7m
8. 绿色伊利石页岩夹灰色铁饼状泥灰岩（灰质白云岩）结核	16.7m
7. 灰色伊利石页岩夹含褐铁矿斑点伊利石页岩	27.1m
6. 灰黑色页岩夹透镜状粉砂岩，页岩具水平纹层与脉状层理	9.9m
一段	113.2m
5. 褐铁矿石英粉砂岩与灰色页岩互层，两者厚度相似，单层厚3~10cm	0.8m
4. 灰绿、灰色伊利石页岩、含粉砂伊利石页岩与粉砂质伊利石页岩呈韵律互层，夹透镜状粉砂岩与含铁粉砂岩，灰色伊利石页岩中含微古植物：*Protoleiosphaeridium solidum*，*Laminarries* cf. *antiquissimus*，*Polyporata* sp.	85.0m
3. 灰绿、绿灰色伊利石页岩。含微古植物：*Protoleiosphaeridium* aff. *faveolatum*，*P. solidum*，*Trachysphaeridum hyalinum*，*T. incrassatum*，*Trematosphaeridium holtedahlii*，*Microconcentrica* cf. *induplicata*，*Laminarites antiquissimus*，*Polyporata obsoleta*	16.9m
2. 灰绿色页岩夹薄层褐铁矿。褐铁矿单层厚2~5cm，主要成分由褐铁矿和黏土质组成	5.8m
1. 褐（赤）铁矿及铁质粉砂岩	4.7m

--------------- 平行不整合 ---------------

下伏地层：铁岭组黄色泥晶白云岩

（2）地质特征及区域变化

下马岭组岩性以页岩占绝对优势，可分为 3 个岩性段：一段主要为灰绿色页岩、伊利石页岩、粉砂质页岩夹粉砂岩和含铁粉砂岩，底部普遍见有铁质风化壳，某些地段形成残积型铁矿；二段主要为灰色、灰绿色伊利石页岩、泥质粉砂岩夹铁饼状泥灰岩（或灰质白云岩）；三段主要是灰黑色硅质页岩、碳质页岩、粉砂质页岩和灰绿色伊利石页岩。与下伏铁岭组为平行不整合接触。

下马岭组主要分布在西山和军都山。

门头沟区沿河城—雁翅青白口一带厚度 340~470m，沿河城西寺台子剖面厚 468m，雁翅黄土贵剖面厚 387m，青白口剖面厚 344m。一段以黑灰色、灰色、黄绿色粉砂质页岩、页岩为主；二段为浅灰绿色粉砂质页岩夹砂岩、页岩及细砂岩，夹 5 层泥灰岩透镜体，泥灰岩呈铁饼状；三段为黑色、灰黑色页岩、粉砂质页岩。

昌平十三陵—怀柔吹风坨一带厚度 320m 左右，十三陵剖面厚 318m，昌平龙山剖面厚 323m，其下部为灰、黄绿色页岩、粉砂质页岩夹含铁粉砂岩和细砂岩；中部为黄绿色和浅灰色页岩；上部为黑色含碳质页岩和黄绿色页岩。

延庆地区仅出露在槽碾沟—黄地沟—马家梁一带，受断裂影响，呈断块或断片产出，地层较薄且发育不全，岩性为灰绿色、灰褐色页岩，粉砂质页岩夹赤铁矿透镜体，顶底部与上下地层呈断层接触，沉积厚度大于 28m。

密云地区主要分布在东智和西智，太子务剖面厚 334m。一段以泥质粉砂岩、细砂岩为主，底部为透镜状铁质角砾岩；二段以绿灰色薄板—页片状泥质粉砂岩为主，夹透镜状细砂岩；三段以硅质泥岩、碳质页岩为主。

房山地区主要分布于西台子—长操—岳各庄一带，出露厚度 90~343m，黄元寺-北将军剖面厚 242m。教军场剖面厚 194m。一段为黑色、暗绿色千枚岩、千枚状页岩、砂质页岩；二段为浅绿、灰绿色千枚状板岩或砂质板岩；三段为黑色硅质页岩及碳质页岩，底部常出现灰色中厚层泥岩，含锰灰岩或泥质白云岩。

第四节　新元古界青白口系

1. 龙山组（Pt_3^1l）

（1）代表剖面

正层型位于昌平区龙山（东经 116°13′32″，北纬 40°14′36″。由乔秀夫 1976 年重测）。

上覆地层：景儿峪组角砾状泥灰岩

———— 整　合 ————

龙山组	总厚度 121.5m
7. 杂色纸片状易碎页岩	25.0m
6. 灰白色中厚—厚层粗粒石英砂岩，石英碎屑的磨圆度和球度均较好，上部局部变为含砾石英砂岩或细砾岩	9.5m
5. 粉砂岩及铁质砂岩，下部以薄层粉砂岩及细砂岩为主，风化面呈棕红色；上部为灰褐—黄褐色中厚层铁质砂岩	31.0m
4. 黄绿色页岩夹粉砂岩	14.0m
3. 灰白色厚层粗粒石英砂岩，具交错层理	6.0m
2. 灰白色薄层中粒石英砂岩及粉砂岩互层，富交错层理，局部夹黄绿色页岩	22.0m
1. 淡黄褐色厚层粗粒石英砂岩，富交错层理；底部为细砾岩，砾石以石英及燧石为主，硅质胶结	14.0m

------ 平行不整合 ------

下伏地层：下马岭组黄色纸片状页岩

（2）地质特征及区域变化

龙山组厚100多米，某些地区可分为两个岩性段。一段主要是（含砾）石英砂岩、长石石英砂岩、粉砂岩及页岩；二段主要是紫红色夹黄绿色易碎页岩，局部地区夹有海绿石细砂岩透镜体或（泥）灰岩透镜体。与下马岭组呈平行不整合接触。

本组分布状况与下马岭组基本一致，主要分布于西山和军都山。

门头沟区主要分布于寺台子、青白口和下苇甸一带，厚度150m左右。寺台子厚167m，底部为底砾岩；下部为含砾砂岩、石英砂岩、海绿石石英砂岩、细砂岩；上部为紫红色、紫灰色泥灰岩、页岩；顶部为紫红色石英砂岩。

昌平-怀柔地区主要分布于昌平龙山和怀柔吹风坨一带。龙山剖面厚121.5m，十三陵剖面厚129m，怀柔河北村厚95m。

密云地区主要分布于西智、东智一带，太子务剖面厚94m，以灰黄、灰绿色薄板状粉砂岩或泥质粉砂岩为主夹石英砂岩、海绿石石英砂岩。底部为含砾粗砂岩，顶部为灰黑色含碳质粉砂质页岩。

房山地区主要分布于长流水—龙门台—黄土台及滴水岩—龙宝峪一带，地层厚度变化大，厚11~225m，下石堡剖面厚60m，北峪村西山梁剖面厚118m。下部为灰白色石英岩，灰、深灰色砂质页岩及（含砾）石英砂岩，中上部为灰黄色厚层含铁质石英砂岩及深灰色砂质页岩。

2. 景儿峪组（Pt_3^1j）

（1）代表剖面

正层型位于天津市蓟县骆驼岭至老鸽顶。北京市选取门头沟区青白口剖面为次层型（东经115°48′35″，北纬40°00′37″。由汪长庆等1980年测制）。

上覆地层：昌平组灰色含角砾粉—泥晶白云岩

--------------- 平行不整合 ---------------

铁岭组		总厚度76.6m
30.	紫色夹黄绿色厚层硅质泥灰岩，上部有1m厚黄褐、灰色硅泥质泥晶灰岩，顶部有18cm厚硅质页岩	19.6m
29.	浅灰色纹层泥晶灰岩，局部含细砾。见少量褐黄色泥灰岩透镜体，长约10cm，厚1~2cm	15.2m
28.	浅灰色中层泥晶灰岩，水平层理。缝合线发育	23.9m
27.	浅棕灰、灰紫色纹层硅质泥灰岩。缝合线发育	6.3m
26.	浅灰色中层具水平纹层含泥泥晶灰岩，偶见细砾，有小型砂纹层理	2.5m
25.	紫红色中层含泥泥晶灰岩，底部夹几层同生角砾岩。含泥泥晶灰岩层面有假角砾和假龟裂构造。角砾岩的角砾成分为蛋青色泥灰岩，个别为下伏地层的钙质页岩，角砾岩层面见干裂，角砾岩与下伏钙质页岩之间呈波状起伏，有微冲刷现象	9.1m

———— 整　合 ————

下伏地层：龙山组紫红、黄绿色易碎页岩

（2）地质特征及区域变化

岩性主要是浅灰、紫红、浅黄及黄绿色泥灰岩、硅质泥晶灰岩等，最顶部有时夹有薄层硅质页岩。

景儿峪组主要分布在西山和军都山，与龙山组一致。

门头沟区主要分布于寺台子、青白口、大村和下苇甸等地。寺台子剖面厚约60m，下苇甸一带厚约154m，黄土贵一带厚85m，青白口厚77m，主要是紫红色、蛋青色、杂色泥灰岩，水平层理较发育。

房山区主要分布在长流水—龙门台—柳林水—黄土台以及斑各庄—龙宝峪一带，出露厚度55~219m。主要岩性为灰白色、黄白色或白色薄层大理岩，普遍发育绿或深灰色泥质条带，顶部有时出现千枚岩及粉红色大理岩。

昌平—怀柔地区主要分布在十三陵、二道关和河北村等地。龙山剖面厚91m，十三陵剖面厚94m，怀柔河北村剖面厚87m。为一套浅紫色、米黄色含泥泥晶灰岩、泥质灰岩和泥灰岩，与下伏龙山组为整合接触。

密云地区主要分布于西智和东智，太子务剖面厚190m，以灰—浅灰色薄—中层状硅泥质条带、纹带状（含）泥质泥晶灰岩为主。

第五节　生物地层

北京地区的中元古代地层中保存有丰富的叠层石和微古植物化石，一直是研究前寒武纪生物特征和演化的重点，但是宏观藻类化石在北京的中元古代地层中至今未见相关报道。新元古代地层中叠层石不甚发育；微古植物较丰富，龙山组和景儿峪组均有产出；宏观藻类化石仅在龙山组有大量发现。目前较可信的遗迹化石在燕山地区仅在青白口系中有所发现，且分布不广，数量有限，总的来说尚在探索阶段。综合前人研究成果，将区内的叠层石、微古植物分别按照岩石地层单位对其进行描述。

（一）叠层石

中元古代是叠层石的繁盛时期，在碳酸盐岩中的分布广泛。叠层石主要产于串岭沟组—铁岭组。萧宗正（1981，1985）划分为5个组合。

组合Ⅰ：产出层位为串岭沟组“宣龙式”铁矿层及白云岩夹层中，以层柱状、柱状为代表，主要分子为：*Kussiella yingmenensis*，*Gruneria yingmenensis*，*G.* cf. *biwabikia*，*G.* cf. *sinensis*，*Eucapsiphora deshengkouensis*，*Cryptozoon* f.，*Parainzeria* f.，*Nordia lanplandica*，*Inzeria nianpanshanensis* 以及核形石 *Osagia* f.。产出具体地点有延庆区营门、昌平区十三陵、怀柔区北湾和密云区大龙门、火郎峪等地。

组合Ⅱ：产出层位为团山子组，以硅化锥、柱层状为代表，叠层石柱体短小，彼此排列紧密，以简单平行分叉为主。主要分子有：*Conophyton* cf. *garganicum*，*C. rugiformis*，*Xiayingella* f.，*Jacutophyton nankouensis*，*Stromacolumna* f.，*Stratifera* f.，*Conophyton garganicum*，*Gaoyuzhuangia* cf. *bulbosa*，*Lochmecolumella* cf. *lufengensis* 等类型。

组合Ⅲ：产于大红峪组和高于庄组，以硅化柱状、锥状叠层石为特征，不分叉或简单平行分叉。主要分子有：*Kusiella* f.，*Conophyton cylindricum*，*C.* cf. *garganicum*，*Gaoyuzhuangia* f. 等。

组合Ⅳ-1：产于杨庄组至雾迷山组二段，为小型丛状叠层石，主要有：*Scyphus parvus*，*Scyphus* f.，*Minglingella wanniangfenensis*，*Pseudogymnosolen mopanyuensis*，*P. shisanlingensis*，*P.* f.，*Conophyton concellosum*，*Conophyton* f.，*Floriscolumella* f. 等。

组合Ⅳ-2：产于雾迷山组三、四段，除假裸枝叠层石外，以个体较大的柱状、锥状叠层石为主，如：*Sinoconophyton gracilium*，*Conophyton lituum*，*C.* f.，*Jacutophyton* f.，*Pseudogymnosolen hushanensis*，*Scyphus minglingensis*，*Scyphus* f.，*Petaliforma* cf. *epicharis*，*Oncocolumna minor*，*Colonnella* f.，*Scopulimorpha irregularis*，*Pseudochihsienella* f.，*Microstylus* f. 等。

组合Ⅴ：产于铁岭组二段，仅有 *Baicalia* f.，*Conicodomeria* f.，*Tielingella* f.，*Chihsienella longotenuia*，*Chihsienella* cf. *chihsienensis* 等。

（二）微古植物

微古植物在整个中—新元古代地层中均有发现，但以大红峪组和下马岭组最为丰富。

1. 常州沟组微古植物特征

十三陵德胜口剖面常州沟组二段泥质岩中发育的微古植物以球形藻群个体为主，有 *Pseudozonosphaera verrucosa*，*Protoleiosphaeridium solidum*，*P. densum*，*Zonosphaeridium* cf. *minutum*，另有具丝纹状的 *Lignum nematoideum*。

2. 串岭沟组微古植物特征

串岭沟组是一套以泥质为主的沉积，因此也是长城系中含微古植物最丰富的一个岩组。十三陵德胜口剖面串岭沟组页岩中发现的微古植物是球形藻群个体 *Protosphaeridium* sp.，*Leiominuscula* sp.。

延庆红旗甸串岭沟组二段黄绿色页岩中发现微古植物：线形类藻丝体 3 属 4 种，即 *Changchengonema densa*，*Changchengonema* sp.，*Eomycetopsis crassiusculum*，*Siphonophycus kestron*；球形个体标本仅见两个，皆为拟蜂巢球形藻的个体，个体直径较大，为巨型和特大型的个体，其种名是 *Pseudofavososphaera gigantea*，*Pseudofavososphaera* sp.；此外尚见有不规则球形或类似船形的个体，由于内部结构不清，均未予定名（朱士兴，1994）。

3. 团山子组微古植物特征

十三陵德胜口剖面团山子组发现的微古植物是具丝纹状的藻类碎片 *Lignum nematoideum* 及船形藻群个体 *Leiofusa digitata*。

延庆红旗甸地区叠层石大部分均由燧石组成，因而都不同程度地保存着微生物化石，其中丝状体类型的化石有 *Gunflintia minuta*，*Gunflintia* cf. *minuta*，*Cyanonema minor*，*Eomycetopsis filiformis*，球状体化石有 *Eosynechococcus* cf. *amadeus*，*Acritarcha* sp.，*Leptoteichos* sp.，*Huroniospora psilata*，其中 *Biocatenoides* sp. 为一种呈链状的化石，细胞呈球形，细小，它们常缠绕呈薄团状产出（朱士兴，1994）。

4. 大红峪组微古植物特征

十三陵德胜口剖面大红峪组发现的微古植物都是球形藻群个体 *Pseudozonosphaera verrucosa*，*Asperatopsophosphaera umishanensis*，*Polyporata obsoleta*。

另外，宋天锐在十三陵地区大红峪组中首次发现具有细胞壁结构的清晰的藻类丝状体，很像蓝藻纲中的颤藻 *Oscillatoriopsis* sp.，其中有些具有分支，像伪枝藻，但又都有些不同（宋天锐，1984）。

5. 高于庄组微古植物特征

十三陵德胜口剖面高于庄组发现的微古植物以球形藻群个体为主，有 *Protoleiosphaeridium* sp.，*Trachysphaeridium* sp.，*Lophosphaeridium* sp.，*Zonosphaeridium* sp.，另有具丝纹状的藻类碎片 *Lignum nematoideum*。

延庆红旗甸高于庄组剖面下部一段至二段厚层及薄层状含燧石条带白云岩中夹有的灰绿及黄绿色粉砂质页岩中发现球形群个体 7 属 10 种，其中较常见的是表面具有细脊和网穴纹饰的，其次是表面光滑和略粗糙的个体，分别为：*Leiosphaeridia infriata*，*L.* aff. *tenuissima*，*L.* cf. *gigantea*，*Trachysphaeridium cultum*，*Pseudozonosphaeridium populosum*，*Pseudofavososphaera gigantea*，*Trematosphaeridium holtedahlii*，*Kildinella miroedichia*，*Stictosphaeridium* aff. *implexum*，*S. hunjiangensis*。线形群藻丝体有 5 属 5 种，*Arctacellularia* aff. *ellipsoidea*，*Rhicnonema* cf. *crassivaginata*，*Gunflintia minuta*，*Siphonophycus kestron*，*Oscillatoriopsis* sp.（朱士兴，1994）。

曹芳在平谷县江水泉—茅山一带的高于庄组剖面上段的含燧石结核白云岩中发现藻类微化石，其中以在数量上占绝对优势的大黏球藻 *Myxococcoides grandis* 及椭球藻 *Archaeoellipsoides grandis* 保存最好，另有球形体化石 *Archaeoellipsoides elongates*，*Asterocapsoides borealis*，丝状体化石 *Oscillatoriopsis regularis*，*Eomycetopsis robusta*，以及呈棒—梗状，末端圆、平或稍尖的化石 *Bactrophycus dolichum*（曹芳，1992）。

6. 杨庄组微古植物特征

杨庄组中发现的微古植物是球形藻群个体 *Protoleiosphaeridium* sp. 以及具丝纹状的藻类碎片 *Lignum nematoideum*。

7. 雾迷山组微古植物特征

雾迷山组微古植物以球形藻群个体为主：*Leiopsophosphaera apertus*，*L.* cf. *pelucidus*，*Asperatopsophosphaera umishanensis*，*Archeofavosina* cf. *simplex*，*Trachysphaeridium hyalinum*，*T. simplex*，*Protoleiosphaeridium solidum*，*Protosphaeridium* sp.，*Trematosphaeridium minutum*，*Pseudozonosphaera verrucosa*，*Pol-*

yporata obsoleta, *Leiominuscula* sp.。另有线形藻群个体 *Taeniatum crassum*，异性藻群个体 *Brocholaminaria* sp.，具孔状的藻类碎片 *Lignum punctulosum*。

曹瑞骥（1984）、宋武（1984）、张昀（1985）、张鹏远（1986）、周小平（1990，1991）等人在十三陵雾迷山组剖面燧石条带中和硅化叠层石中发现了丰富的微体植物化石。综合微体植物群特征：单生球状体有 *Myxococcoides grandis*，*M. kingii*，*Huroniospora* sp.，*Melasmatosphaerae magna*，*M. media*，*Globophycus grandis*，*G. rugosum*，*Gloeodiniopsis lamellosa*，*Micrhystridium* sp.，*Cuihuashania hyndmanii*，*Scissilisphaera regularis*，*Phanerosphaeops parvitas*；群集球状体有 *Eoentophysalis belcherensis*，*Palaeoanacystis vulgaris*，*Coniunctiophycus gaoyuzhuangense*，*Nanococcus vulgaris*，*Eosynechococcus moorei*，*Myxococcoides minor*，*M.* cf. *inornata*；椭球体有 *Archaeoellipsoides grandis*，*A. oblongum*，*A. xishanensis*，*Bactrophycus dolichum*；丝状体有 *Eomycetopsis robusta*，*Siphonophycus inornatum*，*S. kestron*，*S. cuneatus*，*S.* sp.，*Veteronostocale amoeum*，*Archaeonema* cf. *longicellulare*，*Oscillatoriopsis maxima*，*O. connectens*，*O. strictura*，*O. valida*，*O. vulnerata*，*O. planaria*，*O.* sp.，*Cephalophytarion taenia*，*Tylophycus minglingense*，*Palaeolyngbya xishanensis*，*P.* sp.，*Rhicnonema* cf. *antiquum*，*Saccolyngbya pinguis*，*Unnamed filamentous type A*，*Unnamed filamentous type B*，*Palaeoulothrix swbtile*。

组合中最具特色的是形态特殊、丰度很大的 *Archaeoellipsoides grandis* 和 *Bactrophycus dolichum*，前者主要发现于加拿大西北部迪斯马尔湖区群（Horodyski & Donaldson，1980）、河北宣化庞家堡高于庄组（张忠英等，1985）、河北宽城雾迷山组（张鹏远，1987）和北京十三陵雾迷山组（Zhang，1985），后者仅在十三陵雾迷山组及平谷高于庄组中发现；*Myxococcoides grandis* 和 *Archaeoellipsoides grandis* 在本区雾迷山组出现的层位多，数量丰富，并与 *Archaeoellipsoides oblongum* 和 *Bactrophycus dolichum* 相伴生，构成 *Myxococcoides-Archaeoellipsoides-Bactrophycus* 组合，加拿大迪斯马尔湖区群和宽城雾迷山组也有类似组合特征；三段中、上部保存的形态特殊、直径很大的管状体（*Siphonophycus* sp.），在世界其他前寒武纪地层中较少见到；尤其是具有硕大似“结”构造的粗壮丝体（*Unnamed filamentous* type B），形态特殊，以前尚未见过；似螺旋状的丝体（*Unnamed filamentous* type A），从形态和大小上看，相似于 *Obruchevella* 属的分子，如确能归入 *Obruchevella* 属，将大大降低该属出现的最低层位；具刺疑源类 *Micrhystridium* sp. 在雾迷山组的出现表明其并不局限于震旦系—下寒武统；就目前所获资料，从总的面貌和某些特征分子看，十三陵雾迷山组微体植物群比较接近于河北宣化庞家堡高于庄组和加拿大迪斯马尔湖区群的微体植物群，与蓟县雾迷山组微体植物群有一定差异。

8. 洪水庄组微古植物特征

洪水庄组微古植物以球形藻群个体为主，有 *Pseudozonosphaera verrucosa*，*Asperatopsophosphaera partialis*。

9. 铁岭组微古植物特征

铁岭组微古植物以球形藻群个体为主，有 *Trachysphaeridium hyalinum*，*T. simplex*，*T. rugosum*，*T.* aff. *laminaritum*，*Pseudozonosphaera sinica*，*Protoleiosphaeridium* sp.，*Polyporata obsoleta*，*Asperatopsophosphaera bavlensis*。另有线型藻群个体 *Taeniatum crassum*。

10. 下马岭组微古植物特征

下马岭组微古植物以球形藻群个体为主，有 *Trachysphaeridium hyalinum*，*T. incrassatum*，*Protoleiosphaeridium* aff. *faveolatum*，*P. solidum*，*Polyporata obsoleta*，*Trematosphaeridium holtedahlii*，*Microconcentrica* cf. *induplicata*。

另在雁翅下马岭组发现藻类化石古片藻 *Laminarites antiquissimus*，是呈不规则的多角形和宽条带形藻类碎片（孙淑芬，2006）。

11. 龙山组微古植物特征

在雁翅地区龙山组发现的微古植物以球形藻群个体为主，有 *Trachysphaeridium minor*，*Trachysphaeridium cultum*，*Trachysphaeridium rugosum*，*Trachysphaeridium incrassatum*，*Stictosphaeridium pectinale*（孙

淑芬，2006）。

12. 景儿峪组微古植物特征

在昌平龙山景儿峪组剖面紫红色薄皮状含白云质硅质灰岩及含白云质泥灰岩中发现的微古植物是球形藻群个体 *Polyporata microporosa*，*Trematosphaeridium minutum*（齐永安，1999）。

（三）宏观藻类化石

萧宗正（1990）于青白口龙山组剖面中部的黑色页岩内发现宏观藻类化石 *Shouhsienia* cf. *axicula*，*Longfengshania qingbaikouensis*，*L. stipitata*，*Pumilibaxa* cf. *huaiheiana*，*Nephroformia* sp.，这些化石主要为黑色碳质薄膜，形态多样，有椭圆形、卵形、长卵形、肾形及鞋底形等，有的附带有拟茎构造。杜汝霖（1981）在北京昌平龙山组上段中部灰绿色、黄绿色粉砂质页岩中，发现宏观藻类化石 *Chuaria circularis*，*Shouhsienia shouhsienensis*，*S. longa*，*Ovidiscina* sp.，*Pumilibaxa* sp.，*Longfengshania stipitata*。唐烽（1995）在青白口芹峪路口龙山组剖面中发现宏观藻类化石 *Longfengshania stenophylla*，*Paralongfengshania qinyuensis*，*Shouhsienia shouhsienensis*，在十三陵大宫门剖面中发现宏观藻类化石 *Chuaria circularis*，*Tawuia dalensis*。

（四）遗迹化石

宋天锐（1985）首先报道了在北京十三陵地区常州沟组底部的砂—粉砂泥互层中发现了掘穴状充填管状构造，疑是虫孔化石，但迄今尚未能得到公认。罗翠（2010）等认为，这些管状构造不像遗迹化石，反而类似在元古宙地层中独有的所谓“臼齿构造”。

杨式溥（1986）描述了采自北京昌平龙山青白口系龙山组二段下部紫红色砂岩、粉砂岩中的遗迹化石 1 属 1 种（新属种）：昌平龙山迹 *Longshanichnus changpingensis*，在燕山地区首次肯定了动物化石的存在，为寻找动物实体化石提供了重要线索。另外，齐永安（1999）等在昌平龙山龙山组剖面黄褐色中层铁质石英砂岩和粉砂岩中发现痕迹化石 *Helminthopsis* sp. 及 *Cochlichnus* sp.，认为它们为水平分布的觅食潜穴；在昌平龙山景儿峪组剖面紫红色薄皮状含白云质硅质灰岩及含白云质泥灰岩中发现痕迹化石 *Skolithos* sp.，*Monocraterion* sp.，认为它们为垂直居住潜穴。但是，罗翠等（2010）认为北京青白口系的几类遗迹化石，都有重新研究和讨论的必要。

第六节　年代地层

国际地层表中的中—新元古界系指 1600～1000Ma 之间的地层系统，包括盖层系（Calymmian，1600～1400Ma）、延展系（Ectasian，1400～1200Ma）和狭带系（Stenian，1200～1000Ma），而中国区域年代地层表中则将中元古界的底界下延至 1800Ma，包括了国际地层表古元古界的固结系（Statherian，1800～1600Ma）。中元古代在近 2. 0Ga 的元古宙 3 个“代”中是承上启下的一个时代，它的演化历史既有别于古元古代，又有别于新元古代。根据全球演化的阶段性特征，古元古代末哥伦比亚超大陆基本形成，进入中元古代早期时该大陆逐步裂解，经过中元古代的长期演化，到中元古代末期又一个超大陆——罗迪尼亚超大陆逐步汇聚，至新元古代初最终形成。因此，中元古代是超大陆破裂和演化历史占据主导地位的时期。有学者认为，与超大陆破裂有关的盆地和岩浆岩及其共生的巨大的矿产资源潜力与中元古代从 1800～1000Ma 具有行星规模的裂谷事件相联系（planetary-scale rifting events）。

2001 年全国地层委员会根据第三届全国地层大会决议和多年来的传统习惯，以中国北方天津蓟县城北中元古界剖而为标准，将中元古界划分为长城系和蓟县系，时限分别定为 1800～1400Ma 和 1400～1000Ma。标准剖面中的长城系自下而上包括常州沟组、串岭沟组、团山子组、大洪峪组和高于庄组；蓟县系则包括杨庄组、雾迷山组、洪水庄组和铁岭组。蓟县系之上为新元古界青白口系，青白口系底部下马岭组平行不整合覆于蓟县系之上。

20 世纪北京地区可利用的同位素测年资料不多，年代地层学方面多利用紧邻的蓟县层型剖面测年资料。受当时同位素测试技术的限制，多数测年数据的可靠性受到质疑。21 世纪以来，随着地质工作的深入展开，一批地质工作者在中—新元古代以碳酸盐岩和碎屑岩为主的地层系统中发现了多层火山岩（斑脱岩）夹层和侵位于其中的岩床，应用新的测年技术，获得了一批新的地质资料和高精度测年数据，取得了令人瞩目的进展。其中高林志等（2007）取得突破，在北京门头沟大村剖面原划为新元古界下马岭组三段的凝灰岩（斑脱岩）夹层中测得（1368±12）Ma 的数据。由于这一进展，原划中元古界地层时代的归属产生了新问题：第一，从长城系常州沟组至蓟县系的 9 个组的地层时限被限制在 1800～1400Ma 之间的 400Ma 时段内，而不是原先认为的 1800～1000Ma 之间的 800Ma；第二，下马岭组不应划为新元古界，而属中元古界；第三，按传统认识，下马岭组之上原属青白口系的长龙山组和景儿峪组的沉积时限在 900～800Ma 之间，则蓟县剖面缺失从 1320～900Ma 之间长达 420Ma 的地层系统。因而，中元古界年代地层单位不得不重新划分和界定（陆松年，2010）。

2013 年全国地层委员会发布了最新的全国地层表，根据近年来的研究成果，参照蓟县剖面的划分，本志将北京地区中—新元古界年代地层划分进行了修改。

1. 长城系

下部不整合覆盖在太古宙变质岩层之上，上与蓟县系高于庄组整合接触，自下而上包括常州沟组、串岭沟组、团山子组和大红峪组 4 个岩石地层单位。

（1）长城系底界的问题

乔秀夫（2014）认为，依据最新锆石 U-Pb 年龄资料，将燕山地区常州沟组底界年龄定为 1700Ma 是合理的，但它不是长城系或中元古界的底界年龄。长城系常州沟组仅有碎屑锆石 SHRIMP 年龄，最小值为 1805Ma（万渝生，2003），目前在常州沟组内部尚未获得可用年龄值。

北京密云沙厂一带的常州沟组不整合覆盖于环斑花岗岩古风化壳之上。环斑花岗岩侵位于太古宇密云岩群。环斑花岗岩风化壳中获得的锆石 SHRIMP U-Pb 年龄较多：（1685±15）Ma（高维等，2008），（1682±20）Ma，（1708±6）Ma（和政军，2011a，2011b），结合李怀坤等（2011）获得的 LA-MC-ICP-MS 锆石 U-Pb 年龄（1673±10）Ma，密云环斑花岗岩不同岩相的环斑黑云角闪花岗岩、斑状黑云母花岗岩和二云母花岗岩的锆石 TIMS 年龄分别为（1683±4）Ma、（1691±2）Ma 和（1717±13）Ma（Rämö et al.，1995；郁建华等，1996），这些年龄限定了密云地区常州沟组底界年龄值小于 1700Ma。彭澎等（2011）对被长城群底部常州沟组底砾岩不整合覆盖的北京密云水库附近侵入基底岩系的岩墙开展斜锆石的测定，获得 ^{207}Pb-^{206}Pb 年龄为（1731±4）Ma，认为这一年龄限定常州沟组沉积起始时代晚于 1731Ma，并建议将长城系起始时代定为 1730Ma 或者稍晚。

对于上述不同看法，本次修编遵从全国地层委员会的划分建议，将北京地区长城系底界年龄统一为 1800Ma，此年龄并非燕山地区常州沟组底界年龄。

（2）长城系顶界的问题

以往，据有限的同位素数据（主要是河北省兴隆高板河矿区高于庄组中上部的层状方铅矿所测定的铅同位素年龄值），将北京地区长城系的顶界（即高于庄组的上限）推定在 1400Ma。

李怀坤等（2010）在北京延庆高于庄组张家峪亚组（三段）上部发现了凝灰岩，并获得锆石（1559±12）Ma 的 SHRIMP U-Pb 年龄和（1560±5）Ma 的 LA-MC-ICP-MS U-Pb 年龄，为华北北部中元古界年代地层划分研究提供了直接的年代学证据。其结合下伏大红峪组火山岩的锆石 U-Pb 年龄（1625.9±8.9）Ma 和（1625±6）Ma（Gao et al.，2009，2011；Lu et al.，2008），认为可以确切地将高于庄组的底界年龄限定在 1600Ma 左右；结合苏文博（2010）在铁岭组斑脱岩获得的锆石 SHRIMP U-Pb 年龄（1437±21）Ma，建议将华北中元古界蓟县系的底界下移至高于庄组底界。

全国地层委员会发布的《中国地层表》（2013）将蓟县系的底界下移至高于庄组底界，并对应于国际中元古界的盖层系，高于庄组与大红峪组之间的界线则作为蓟县系与长城系的分界标志，长城系顶界年龄统一为 1600Ma。

（3）长城系内部的划分

目前元古宙的年代地层只建立起“系”一级的年代地层单位，系以下没有建立更低一级的年代地层单位。现仅将各岩石地层中获取的同位素年龄数据进行简要总结。

串岭沟组无直接年龄数据，但蓟州地区侵位于串岭沟组的辉绿岩墙的SHRIMP U-Pb锆石年龄为（1638±14）Ma（高林志等，2009），串岭沟组年龄应大于岩墙群年龄，暂定为1640Ma。另外，张拴宏（2013）获得了蓟县青山岭侵位于串岭沟组的闪长玢岩脉中的锆石$^{207}Pb-^{206}Pb$年龄（1634±9）Ma。李怀坤等（1996）采取北京平谷陈庄子村东团山子组富钾粗面安山岩，获得岩浆自生锆石U-Pb同位素年龄（1683±67）Ma。张宏栓（2013）在北京平谷熊耳寨团山子组上部获得的钾质火山岩锆石$^{207}Pb-^{206}Pb$年龄为（1637±15）Ma。北京平谷茅山镇团山子组粗面玄武岩中锆石U-Pb年龄为（1671±5）Ma，平谷熊耳寨大红峪组亚碱性玄武岩中锆石U-Pb年龄为（1645±5）Ma，平谷黄松峪及大华山镇大红峪组粗安岩中锆石U-Pb年龄为（1664±6）Ma和（1652±7）Ma（Wan et al.，2013）。陆松年（1991，2008）获得蓟县大红峪组火山岩中单颗粒锆石年龄（1625.3±6.2）Ma和（1622±23）Ma。高林志等（2009）获得的蓟县剖面大红峪组下部第一套喷发旋回的富钾粗面岩层SHRIMP锆石U-Pb年龄为（1625.9±8.9）Ma。

2. 蓟县系

下与长城系大红峪组整合接触，上被待建系下马岭组平行不整合覆盖，包括5个组级岩石地层单位，自下而上为高于庄组、杨庄组、雾迷山组、洪水庄组和铁岭组。亦没有分统。

蓟县系的底界年龄即长城系的顶界年龄，如前所述推定值为1600Ma；而其顶界年龄，结合高林志（2007）在北京门头沟大村剖面原划为新元古界下马岭组三段的凝灰岩（斑脱岩）夹层中测得（1368±12）Ma的数据，为与国际接轨，全国地层委员会建议蓟县系顶界年龄为1400Ma。

蓟县系各岩石地层单位的同位素年龄数据：张巧大（2002）在北京十三陵高于庄组、杨庄组和雾迷山组中获得碳酸盐岩Pb-Pb年龄（1608±74）Ma、（1488±55）Ma和（1373±92）Ma。另外，张拴宏等获得雾迷山组辉绿岩侵入岩脉的锆石LA-ICP-MS年龄（1353±14）Ma（Zhang et al.，2009）。苏文博等（2010）在冀北—辽西以碳酸盐岩为主的铁岭组内发现了钾质斑脱岩，并获得其锆石SHRIMP U-Pb年龄（1437±21）Ma，这一进展将铁岭组的时代确定为中元古代早期盖层纪（Calymmian Period）末期。

3. 待建系

该系与下伏蓟县系铁岭组为平行不整合接触，与上覆青白口系龙山组呈平行不整合接触。仅有1个岩石地层，即下马岭组。没有分统。

长期以来，华北中—新元古界下马岭组一直被置于新元古界青白口系的第一个组，其时代大致为1000~900Ma。近年来对下马岭组时代的研究取得了突破性进展。如前所述，高林志等（2007）在北京门头沟大村剖面原划为青白口系下马岭组第三段的碳质页岩中识别出凝灰岩（斑脱岩）夹层，并首先测得（1368±12）Ma的数据；李怀坤等（2009）在河北宽城黄家庄获得侵入于下马岭组的辉绿岩锆石和斜锆石的U-Pb同位素年龄（1320±6）Ma；苏文博（2010）又获得了铁岭组中部火山凝灰岩的锆石U-Pb年龄约1440Ma。根据上述年代学研究新资料，下马岭组的下限年龄接近1400Ma，而上限时代应老于（1320±6）Ma。

根据上述同位素测年数据，下马岭组之上原属青白口系的龙山组和景儿峪组的沉积时限在900~800Ma之间，则蓟县剖面缺失1350~900Ma之间长达450Ma的地层系统（迄今长龙山组和景儿峪组的地层时限还缺乏准确可靠的同位素年龄依据）。因此，全国地层委员会建议的蓟县系之上1400~1000Ma时段对应地层为暂定待建系，下马岭组置于该待建系的底部1400~1350Ma时段内。随着研究工作的不断深入，待建系将予以正式界定和命名。乔秀夫（2007，2014）认为下马岭组锆石年龄（1368±12）Ma的样品首次取自北京西山（高林志等，2007；Gao et al.，2007；Su et al.，2008），下马岭组命名地点在北京西山永定河畔的下马岭村（叶良辅，1920），北京西山是中国地质学家的摇篮，在国内外的知名度很高，因此建议将待建系名称定为“西山系”。

4. 青白口系

该系与下伏待建系下马岭组呈平行不整合接触，与上覆地层下寒武统昌平组亦呈平行不整合接触。没有分统。包括 2 个组级岩石地层单位，自下而上为龙山组、景儿峪组，前者为碎屑岩，后者为杂色泥灰岩、硅质泥晶灰岩。

任荣、韩宝福（2011）在十三陵一带龙山组石英砂岩中获得的碎屑锆石 U-Pb 年龄谱峰值分别为 2.50Ga 和 1.80Ga，具有双峰式特征，这些锆石大多为岩浆成因锆石，指示 2.50Ga 和 1.80Ga 是华北克拉通岩浆活动广泛发育的两个时期。结合区域资料，龙山组砂岩沉积应在 1.36Ga 之后。李明荣（1996）在昌平龙山铁塔东和门头沟青白口村的景儿峪组采集样品，获得海绿石 $^{40}Ar-^{39}Ar$ 同位素年龄分别为（810±10）Ma 和（842.2±11.3）Ma。

据全国地层委员会编制的《中国地层表》（2013）提出的参考年龄，青白口系的底界年龄为 1000Ma，顶界年龄为 780Ma。

关于燕山地区中—新元古代地层划分及锆石 U-Pb 年龄数据参见图 2-2。

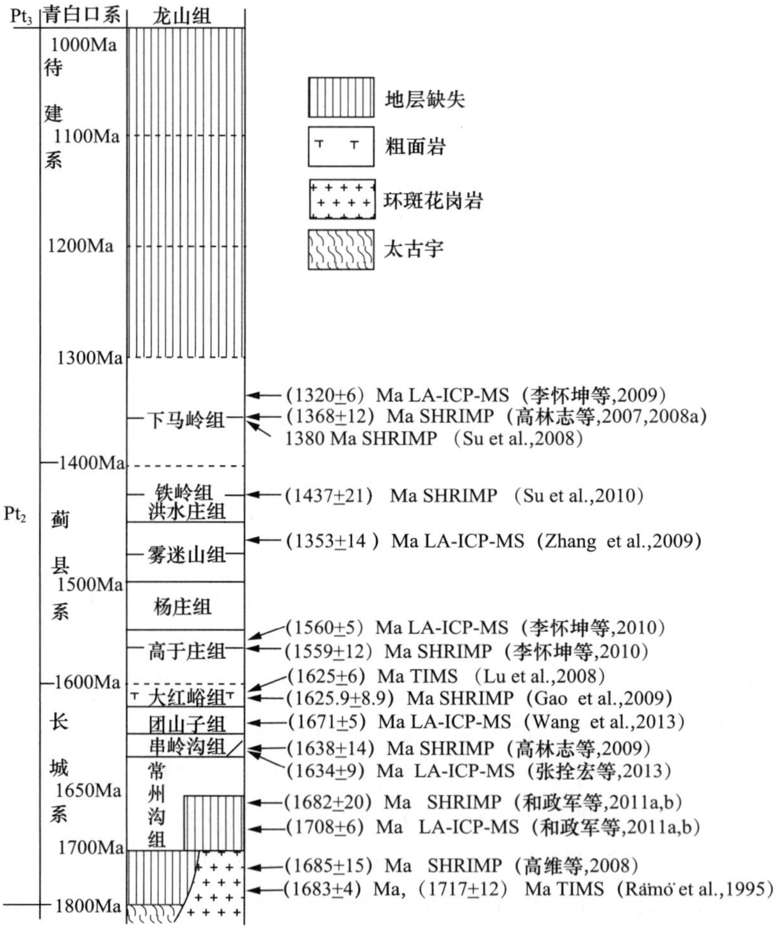

图 2-2 燕山地区中—新元古界地层柱及锆石 U-Pb 年龄

（据乔秀夫等，2014，略修改）

Fig. 2-2 The Meso-Neoproterozoic chronostratigraphic time scale of Yanshan area and zircon U-Pb ages

第七节 事件地层

中—新元古代的地质事件在地层中有所反映。

宋天锐等（1987，1991，2000）在十三陵串岭沟组发现一系列与地质事件有关的现象，其中宏观标志包括地震滑塌、砂体液化、层内错断、管状充填构造（也是砂体液化的一个表现）和角砾化构造等；微观标志包括小型泄水构造和细脉充填，显微层内错动（多数是正断层状），在白云岩中见到火山晶屑、火山玻璃的假象。认为上述现象是由于燕山裂陷槽形成早期的火山活动及由其引发的地震事件形成的。

平谷黄松峪一带的团山子组发育包卷层理、竹叶状砾屑、干裂和岩盐假晶。假晶面有阶梯状“回”形溶蚀漏斗，产出特征有的朝上，有的侧卧或倒转，表明受到强潮流（风暴）的改造而使高盐蒸发环境淡化。巨厚层竹叶状砾屑为异地风暴沉积，可能与古地震、古火山喷发先兆造成的海啸有关。

大红峪组的火山喷是中元古代最引人注目的地质事件。宋天锐（2000）研究十三陵泰陵一带的大红峪组，发现其出露的灰色角砾硅化岩和黑色粉砂凝灰页岩都含有火山-地震事件的信息，其中角砾硅化岩是受火山影响和地震波冲击等事件共同作用的产物，其上部覆盖的凝灰页岩是火山喷出的晶屑经过海浪淘洗而沉积的。

密云岩尖剖面团山子组下部泥质白云岩中夹多层火山碎屑岩和熔岩（安粗岩），反映团山子期早期就有火山活动。

李怀坤等（2010）在延庆高于庄组三段中发现了 1.4m 厚的凝灰岩层，表明该期有较强烈的火山喷发活动。

宋天锐（1988）认为十三陵地区雾迷山组可能存在一套地震-海啸序列。乔秀夫（1996）认为该序列实际代表一个 1200Ma 左右的地震带，在燕山全境均可追踪。苏德辰（2011）在西山地区不同层位的雾迷山组露头中发现了多个期次、不同规模、由古地震引发的软沉积物变形构造，包括液化变形（液化丘、液化混插、底辟、液化脉及液化卷曲层理等）、挤压变形（紧密褶皱、板刺状角砾和丘-槽构造等）、拉伸变形（环状层、拉伸布丁）以及脆性变形（地裂缝、层间断层和震裂岩）。结合该区的构造背景和前人研究资料，认为该区雾迷山组沉积期古地震系沿中元古代燕辽裂陷槽轴部的断裂活动触发的。

洪水庄组以黑色和深灰色页岩、粉砂质页岩为主，沉积物中有机质和硫含量较高，处于缺氧的还原环境，反映沉积环境发生剧烈变化，水深突然增大，是海侵最大时期的产物。

苏文博等（2006）通过对待建系下马岭组硅质岩和斑脱岩的沉积组合岩石学、地球化学和沉积学研究认为，硅质岩的沉积过程受到热水影响，可能与此期火山喷发有关，斑脱岩源岩为中酸性粗安岩-英安岩-流纹岩系列，产生于同碰撞岛弧及板内背景的火山喷发活动，与板块边缘的俯冲碰撞等构造运动密切相关。

第三章　寒武系—奥陶系

北京市寒武纪—奥陶纪地层主要分布于西山，在北山及顺义二十里长山也有少量分布。另外在北京平原区新生代沉积之下也隐伏有寒武纪—奥陶纪地层（图 3-1）。

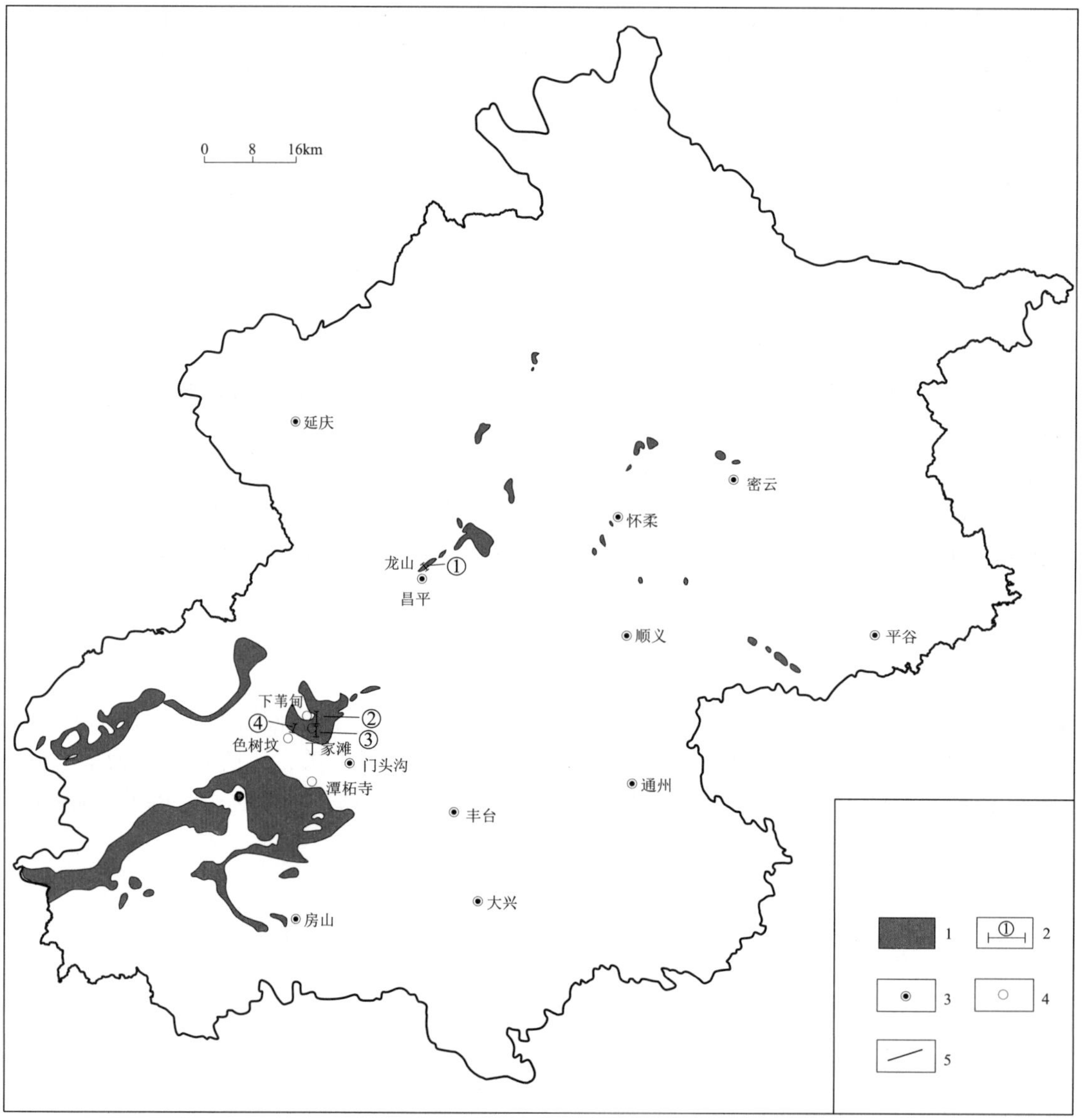

图 3-1　北京市寒武纪—奥陶纪岩石地层及层型分布简图
（据《北京市岩石地层》，1996）
Fig. 3-1　Distribution map of Cambrian-Ordovician lithostratigraphic units and their stratotype section in Beijing
1—寒武纪—奥陶纪岩石地层；2—层型：①龙山昌平组剖面，②下苇甸馒头组—炒米店组剖面，③丁家滩冶里组、亮甲山组剖面，④色树坟马家沟组剖面；3—区政府驻地；4—村庄；5—市界

庞培莱（K. Pumpelly，1867）、李希霍芬（F. V. Richthofen，1869—1871）、叶良辅等（1920）都对西山地区早古生代地层做过粗略的划分。

黄汲清（1927）研究了军庄—南庄一带早古生代地层剖面后，将其划分为馒头页岩、鲕状灰岩系、竹叶状灰岩系、北岭子层、马家沟灰岩。从而奠定了本市寒武系—奥陶系地层划分的基础。张文佑（1935）在昌平龙山原“景儿峪灰岩”上部层位内发现三叶虫化石 *Redlichia chinensis* 等，因而将这段灰岩单独划分出来，创名为“昌平灰岩”，时代定为早寒武统晚期，为在燕山地区确立显生宇寒武系底界提供了确切的古生物依据。

1960—1962 年北京地质学院地层队在郝诒纯的指导下，对门头沟区下苇甸—韭园的寒武系和奥陶系进行研究，划分了 7 阶 5 组，被北京市区域地层表编写组（1977）采用。

20 世纪 70 年代末期开始，又有众多地质学者开展了早古生代碳酸盐岩岩石学及其沉积环境等方面的研究，管守锐等（1979）、张俊明等（1981）、王英华等（1982）、冯增昭（1986）、孟祥化等（1986）先后发表了一批文章和著作。

《北京市区域地质志》（1991）对本区寒武纪—奥陶纪地层进行了岩石地层和年代地层划分。

20 世纪 90 年代以来，王鸿祯（1998）运用层序地层学原则对西山及邻区进行了系统深入的综合研究，识别出了剖面的层序界面，分析了各层序内的沉积相及其演化，划分了副层序和体系域，提出了碳酸盐沉积环境模式，确立了层序地层基本格架，初步建立了西山及邻区寒武系相对海平面变化曲线。苏德辰（1994）对西山地区寒武系第二统—下奥陶统旋回地层格架及相关地层学进行了研究。孟祥化等（1996）对西山丁家滩和青白口一带的寒武系层序地层进行了深入研究，将寒武系分为 10 个三级旋回（$Ⅲ_1$—$Ⅲ_{10}$）。王成述等（1996）将下苇甸寒武纪—奥陶纪地层划分为 12 个层序和 12 个层序界线。近年来，乔秀夫、梅仕龙（1996）、梅冥相（1996，2011）、陈建强等（1997）、彭阳等（2012）等对北京西山地区寒武纪—奥陶纪地层进行了层序地层学及旋回地层学方面的研究。

鲍亦冈等（1996）对本区寒武纪—奥陶纪地层共划分 7 个组级岩石地层单位（表 3-1）。

第一节　岩石地层

1. 昌平组（$€_2c$）

（1）代表剖面

正层型为昌平区龙山剖面（东经 116°13′43″，北纬 40°14′41″）。由张文佑测制，北京市地质局延昌勘探队 1962 年重测。

上覆地层：馒头组紫红色泥质页岩，下部有角砾泥岩

---------------- 平行不整合 ----------------

昌平组	总厚度 52. 48m
10. 浅灰色中厚层细晶白云质灰岩，局部含燧石结核	1. 60m
9. 灰色厚层豹斑状中细晶灰岩，局部含有燧石结核	5. 78m
8. 灰黑色厚层致密块状细晶灰岩	2. 20m
7. 浅灰色中厚层豹斑状中—细晶灰岩，含有不规则燧石结核	13. 56m
6. 灰黑色块状细晶灰岩	1. 20m
5. 深灰色厚层—块状豹斑状细晶灰岩，局部含有燧石结核	5. 70m
4. 灰黑色厚层致密块状细粉晶灰岩	2. 00m
3. 浅灰色厚层豹斑状细晶灰岩	11. 23m
2. 浅灰色细晶白云质灰岩，底部为泥晶白云岩	9. 21m

---------------- 平行不整合 ----------------

下伏地层：景儿峪组紫红色板状泥质灰岩

（2）地质特征及区域变化

昌平组岩性以灰色、深灰色厚层至块状豹斑状灰岩和细晶灰岩为主，与下伏景儿峪组为平行不整合接触。

表 3-1　北京市寒武纪—奥陶纪岩石地层划分沿革表

Table 3-1　The classification history of Cambrian-Ordovician lithostratigraphic units in Beijing

叶良辅等		黄汲清		张文佑		陈恺、熊永先		小幡忠宏		北京地质学院		《华北地区区域地层表·北京市分册》		《北京市区域地质志》		《北京市岩石地层》	本志	
1920		1927		1935		1935		1939		1961		1977		1991		1996		
西山		西山		昌平		西山		军庄		北京		北京		北京		北京	北京	
石炭纪	杨家屯煤系	C—P	杨家屯煤系			C	莫斯科期灰岩	C_2—P_1	杨家屯层	C_2	清水涧组	C_2	本溪组	C_2	清水涧组	太原组	C_2	太原组
奥陶纪（上震旦层）	笋石石灰岩	O_2	马家沟灰岩			O	奥陶纪灰岩	O_2	军庄组	O_2	中奥陶统	O_2	上马家沟组	O_2	上马家沟组	马家沟组	O_2	马家沟组
													下马家沟组		下马家沟组			
		O_1						O_1	西北涧层	O_1	下奥陶统	O_1	亮甲山组	O_1	亮甲山组	亮甲山组	O_1	亮甲山组
			北岭子层										冶里组		冶里组	冶里组		冶里组
奥陶纪（下震旦层）	竹叶状灰岩系	$€_3$—$€_2$	竹叶状灰岩系			$€_3$	炒米店灰岩	$€_3$	十八盘层	$€_3$	炒米店组	$€_3$	凤山组	$€_3$	凤山组	炒米店组	$€_4$	炒米店组
													长山组		长山组			
													崮山组		崮山组			
	鲕状灰岩	$€_2$	鲕状灰岩系			$€_2$	张夏灰岩	$€_2$	南庄层	$€_2$	张夏组	$€_2$	张夏组	$€_2$	张夏组	张夏组	$€_3$	张夏组
													徐庄组		徐庄组			
	上页岩	$€_1$	馒头页岩（未见底）			$€_1$	馒头页岩	$€_1$	桃园层	$€_1$	馒头组	$€_1$	馒头组和毛庄组		毛庄组	馒头组	$€_{2-3}$	馒头组
	中灰岩			$€_1^3$	昌平灰岩	Z	景儿峪灰岩	Z	南口层		青白口组	Qn	昌平组	$€_1$	昌平组	昌平组	$€_2$	昌平组
				Z	景儿峪灰岩								景儿峪组	Qn	景儿峪组	景儿峪组	Pt_3^1	景儿峪组

西山地区本组主要分布在门头沟区的齐家庄—青白口—大村和下苇甸，房山区的龙门台—柳林水—鲁家滩—黄山店。岩性以豹斑状灰岩和白云岩为主。下苇甸剖面厚89m，青白口剖面厚60m，双石头厚62m，鲁家滩厚107m，班各庄一带厚60m，龙门台厚91m。

在北山则零星出露于十三陵、河防口、九渡河、二道关、秤钩湾和西智等地，厚50~141m，岩性主要是细晶灰岩、豹斑状灰岩，底部也有角砾状白云岩。昌平黄岩山厚62m，河防口厚87m，西智厚141m。

2. 馒头组（$\epsilon_{2-3}m$）

（1）代表剖面

选取门头沟区下苇甸剖面为次层型（东经116°01′52″，北纬40°00′21″。由《北京市区域地质志》下古生界编写小组1985年测制）。

上覆地层：张夏组灰紫色鲕粒灰岩，产 *Sunaspis* sp.

———— 整　合 ————

馒头组	总厚度 129m
19. 紫红色粉砂质泥岩夹泥质粉砂岩条带	7. 0m
18. 灰色、灰绿色中薄层含砂泥质条带鲕粒云灰岩，含云母片，产 *Shantungaspis aclis*，*Ptychoparia mantoensis*，*P. theano*	4. 0m
17. 紫红色粉砂质泥岩	2. 0m
16. 灰色、灰绿色薄层含泥质条带砾屑灰岩，含砂，产棘皮动物化石碎片	2. 0m
15. 紫红色薄层粉砂质泥岩夹浅绿色薄层灰岩	3. 0m
14. 灰色、灰绿色中薄层条带状藻团藻屑灰岩，白云质粉砂岩与灰岩互层，含砂质泥晶白云岩条带，夹紫红色粉砂质泥岩，产三叶虫化石碎片	2. 0m
13. 暗紫红色中厚层—薄层粉砂质泥岩	17. 3m
12. 灰白色薄层含砂含棘皮生屑鲕粒灰岩夹紫红色页岩，上部含少量泥晶白云岩，顶部见波痕	2. 8m
11. 紫红色薄层砂质泥岩、页岩	18. 5m
10. 棕黄色薄层条带状砂质泥晶白云岩	4. 2m
9. 灰色厚层泥晶、细晶白云岩	3. 4m
8. 灰黄色中薄层含砂泥晶白云岩、砂屑粉晶白云岩	1. 1m
7. 浅灰色厚层细晶白云岩	1. 5m
6. 灰黄色薄层含砂粉晶云质砾岩	1. 0m
5. 砖红色中厚层粉砂质泥岩	13. 5m
4. 灰白色纹带状泥晶白云岩夹泥质条带	3. 2m
3. 砖红色薄层粉砂质泥岩	13. 6m
2. 土黄色、紫红色泥质角砾岩	23. 3m
1. 土黄色含燧石条带角砾状含砂细粉晶—泥晶白云岩	5. 6m

------- 平行不整合 -------

下伏地层：昌平组深灰色厚层细晶白云岩

（2）地质特征及区域变化

本组以紫红色泥岩（页岩）和白云岩为主，与下伏昌平组为平行不整合接触。

本组主要出露于西山地区，如门头沟区的齐家庄—青白口—大村和九山、下苇甸等地，以及房山区的蒲洼—黄土台—班各庄—龙宝峪一带。下苇甸剖面厚129m，双石头剖面厚174m，九山剖面厚191m，鲁家滩剖面厚135m，龙门台剖面厚147m。

北山地区则零星分布于昌平区文殊峪、十三陵、龙山，怀柔区河防口、九渡河、二道关和密云县西智。昌平黄岩山剖面厚144m，九渡河剖面厚146m，河防口剖面一带厚度>95m，西智剖面厚120m。

3. 张夏组（ϵ_3z）

（1）代表剖面

选取门头沟区下苇甸剖面为次层型（东经 116°02′01″，北纬 40°00′08″。由《北京市区域地质志》下古生界编写小组 1985 年测制）。

上覆地层：炒米店组黄灰色中层泥云质条带泥晶灰岩

——————整　合——————

张夏组　　总厚度 222.1m

25. 深灰色中—厚层亮晶含藻屑鲕粒灰岩夹黄灰色中—薄层粉晶云质灰岩，产三叶虫 *Kolpura* sp.，*Lorenzella* sp.，*Dorypyge*（?）sp.　17.2m
24. 深灰色巨厚层亮晶鲕粒含云灰岩　17.6m
23. 深灰色巨厚层泥亮晶鲕粒灰岩，上部夹泥质条带泥晶灰岩　10.5m
22. 深灰色巨厚层泥亮晶含砂屑、生屑鲕粒灰岩，夹泥晶泥质灰岩条带，含棘皮动物化石碎屑　7.8m
21. 深灰色中层泥亮晶鲕粒灰岩，夹深灰色中—薄层钙质粉砂岩和黄灰色中—薄层条带状泥晶粉砂质白云质灰岩，组成 6 个韵律层。产三叶虫 *Damesella* aff. *blackwelderi*，*D. damesei*，*Anomocarella* sp.，*Poshania* sp.　11.4m
20. 上部为灰色中层竹叶状灰岩，中薄层细晶含鲕粒竹叶状生屑灰岩夹深灰色中—薄层粉砂质钙质页岩，灰质条带粉晶含泥灰岩；下部为深灰色中—薄层泥晶鲕粒灰岩夹深灰色中—薄层粉砂质钙质页岩。砾屑形状似“竹叶”，分选磨圆好，具氧化圈，产三叶虫 *Amphoton* sp.，*Damesella nitida*，*Asaphiscus* sp.，*Aojia spinosa*，*Manchuriella typa*，*Dorypyge* sp.，*Anomocarella* sp.　7.6m
19. 灰色中厚层云化泥晶鲕粒灰岩夹绿灰色薄层钙质粉砂岩。产三叶虫 *Psilaspis* sp.，*Manchuriella* sp.，*Solenoparia* sp.，*Crepicephalus convexus*，*Dorypyge* sp.，*Aojia* sp.，*Anomocarella baucia*，*Lisania* sp.　11.1m
18. 灰绿色薄层钙质粉砂岩，褐灰色中—薄层灰质条带泥晶白云质灰岩夹深灰色中—薄层泥晶鲕粒灰岩，灰色薄层竹叶状鲕粒灰岩，灰色薄层细晶灰岩。底部为灰色中—薄层泥晶灰岩。产三叶虫 *Crepicephalna quadrata*，*Proasaphiscus* sp.，*Asaphiscus* sp.　11.7m
17. 绿灰色薄层含石英粉砂粉晶灰岩夹灰色鲕粒灰岩　10.1m
16. 灰色中厚层泥亮晶鲕粒灰岩，含生物碎屑及粉砂岩砾屑　4.7m
15. 黄灰色中薄层泥云质条带泥晶泥质灰岩夹灰色中薄层泥晶鲕、豆粒灰岩　6.2m
14. 灰色中层鲕粒灰岩与绿灰色薄层钙质粉砂岩不等厚互层夹豆粒灰岩（10cm），其中粉砂岩占 60%　3.8m
13. 绿灰色中层钙质粉砂岩，顶部夹灰岩　4.1m
12. 绿灰色薄层灰质条带含泥粉砂岩，黄灰色中—薄层断续条带状泥晶泥质灰岩与灰色中层颗粒灰岩互层，顶部为灰色薄层细晶灰岩。粉砂岩、泥晶泥质灰岩与颗粒灰岩、泥晶灰岩构成韵律层。产三叶虫 *Crepicephalina* aff. *quadrata*　11.8m
11. 深灰色中层含内碎屑泥亮晶鲕粒灰岩，夹绿灰色中—薄层泥钙质粉砂岩　2.8m
10. 灰绿色薄层含泥钙质粉砂岩，中部夹 30cm 灰色泥晶颗粒灰岩透镜体　6.6m
9. 黄灰色中—薄层泥质条带泥晶灰岩，泥晶灰岩夹灰色中薄层颗粒灰岩，黄灰—浅绿色薄层泥质粉砂岩，见波痕　3.3m
8. 深灰色厚层含竹叶状砾屑鲕粒细晶灰岩。竹叶状砾屑为长条状、蝌蚪状。产三叶虫 *Bailiella lantenoisi*，*Anomocarella* sp.，*Proasaphiscus* sp.；腕足类 *Eoorthis* sp.　11.0m
7. 深灰色中薄层含竹叶状泥、亮晶鲕粒含云灰岩夹薄层灰绿色粉砂岩及深灰色中—薄层含鲕细晶灰岩；顶部为绿灰色、黄色中—薄层粉砂质亮晶竹叶状灰岩，竹叶状砾屑多具红色氧化圈，由下而上陆源成分增多　13.8m
6. 灰色厚层含竹叶状亮晶鲕粒灰岩，顶部为颗粒灰岩，从下向上砾屑增多，砾屑变长，产三叶虫 *Serrodiscus*? sp.，*Solenoparia* sp.，*Manchuriella typa*，*Crepicephalus convexus*；腕足类 *Obolus* sp.　20.5m

5. 浅绿色中—薄层钙泥质粉砂岩，底部夹有厚 20cm 含三叶虫碎屑的细晶灰岩 2. 5m

4. 灰色中—厚层粉砂质鲕粒细晶灰岩，顶部含竹叶状砾屑。产三叶虫 *Proasaphiscus paotaiensis*, *P. suni*, *Anomocarella concava*, *A. chinensis*, *Asaphiscus walcotti*, *Manchuriella typa*, *Crepicephalus convexus*；腕足类 *Obolus* sp. 8. 2m

3. 绿灰色薄层钙质粉砂岩与深灰色中—薄层含竹叶状泥、亮晶鲕粒灰岩不等厚互层 4. 7m

2. 灰色（局部紫色）中厚层泥、亮晶鲕粒灰岩夹绿灰色薄层钙质粉砂岩。鲕粒灰岩含竹叶状砾屑，产三叶虫 *Anomocarella temenus* 8. 3m

1. 灰绿色薄层泥钙质粉砂岩夹灰紫色薄层鲕粒灰岩。产三叶虫 *Sunaspis* sp. 4. 8m

———— 整　合 ————

下伏地层：馒头组紫红色粉砂质泥岩

（2）地质特征及区域变化

整个张夏组由钙质粉砂岩和鲕粒灰岩构成了两个旋回，岩性以鲕粒灰岩为主，与下伏馒头组为整合接触。

张夏组分布于西山和北山，但在北山地区有的出露不全。

在西山地区主要分布于门头沟区的齐家庄—柏峪—青白口—大村一带和九山、下苇甸等地，以及房山区的蒲洼—黄土台—班各庄—龙宝峪和鲁家滩一带。下苇甸剖面厚 222m，青白口剖面厚 199m，九山剖面厚 207m，双石头剖面厚 205m，鲁家滩剖面厚 205m，龙门台剖面厚 111m。

在北山地区，零星分布于昌平区文殊峪、十三陵、怀柔区河防口、二道关、九渡河和密云区西智。文殊峪剖面厚 194m，黄岩山剖面厚 260m，河防口剖面厚 316m，西智太子务剖面未见顶，厚>224m。

4. 炒米店组（€_4cm）

（1）代表剖面

选取门头沟区下苇甸剖面作为本市炒米店组次层型（东经 116°02′02″，北纬 39°59′45″。由《北京市区域地质志》下古生界编写小组 1985 年测制）。

上覆地层：冶里组深灰色厚层含云斑泥晶灰岩

———— 整　合 ————

炒米店组 总厚度 192. 9m

47. 灰色泥晶竹叶状灰岩与灰黄色泥云质条带泥晶灰岩互层。泥云质条带为毫米级，个别可达 1cm 10. 6m

46. 灰色条带泥晶灰岩夹粉晶竹叶状灰岩，云泥质条带较少，夹 9 层竹叶状灰岩，竹叶状砾屑排列杂乱 16. 4m

45. 灰色中—薄层泥晶云泥质灰岩夹泥晶灰岩条带 8. 7m

44. 深灰色厚—中层泥晶灰岩，含云泥质条带，底部为泥晶竹叶状灰岩 5. 9m

43. 深灰色中—薄层云泥质条带泥晶灰岩，含生屑，层理多弯曲呈波状。产三叶虫 *Tellerina chinhsiensis*；腕足类 *Obolus* sp. 9. 0m

42. 灰、黄灰色中—薄层云泥质条带泥晶灰岩夹粉晶竹叶状灰岩 7. 3m

41. 灰、黄褐色中—薄层白云质泥岩和泥晶灰岩夹灰色泥晶竹叶状灰岩，白云质泥岩和泥晶灰岩互成条带，泥晶灰岩条带常断续而形成竹叶状砾屑。产三叶虫 *Calvinella walcotti*；腕足类 *Lingulella marcia* 17. 2m

40. 深灰色厚层云斑、云条粉晶灰岩夹泥晶竹叶状灰岩。云质条带宽 1cm，密集断续平行层理分布，多弯曲变形或断开，有的将灰岩分割成“竹叶”状，有时呈云朵状。泥晶竹叶状灰岩呈透镜状。上部产三叶虫 *Quadraticephalus linyuensis*, *Kaolishania* sp., *Ptychaspis* sp.；腕足类 *Eoorthis* sp.。下部产三叶虫 *Tsinania canens*, *Ptychaspis subglobosa*；腕足类 *Lingulella* sp. 11. 0m

39. 灰紫色薄层泥云质条带泥晶灰岩夹泥亮晶竹叶状白云质灰岩，层间含有厚度变化较大的竹叶状灰岩透镜体，底部为厚 35cm 的竹叶状灰岩。上部产三叶虫 *Mansuyia orientalis*, *Homagnostus hoi*, *Pagodia* sp.；底部产三叶虫 *Kaolishania* sp., *Mansuyia orientalis*, *Anomocarella* sp., *Quadraticephalus*? sp., *Ptychaspis* sp., *Tsinania* sp.；腕足类 *Lingulella* sp., *Obolus* sp. 10. 6m

38. 紫灰色中、薄层泥晶泥云质灰岩夹泥质灰岩条带，底部条带少顶部多，与泥晶泥云质灰岩组成等厚互层，含白云质灰岩角砾 6.3m

37. 上部为灰色薄层云泥质粉砂岩，夹不连续的泥晶灰岩条带，含竹叶状灰岩透镜体。中部为灰色中层竹叶状泥云质灰岩，竹叶状砾屑分选磨圆皆好，具红色氧化圈。下部为浅灰色中、薄层泥质条带泥晶含云灰岩，条带密集。产三叶虫 *Changshania truncata*，*Dikelocephalites* cf. *flabelliformis*，*Pagodia*（?）sp.；腕足类 *Obolus* sp. 3.2m

36. 灰绿色薄层泥云质钙质粉砂岩，夹灰岩透镜体，局部为细晶生屑灰岩 2.5m

35. 紫色泥晶竹叶状灰岩夹泥云质条带泥晶灰岩，竹叶状砾屑具红色氧化圈 3.6m

34. 浅黄绿色薄层钙质粉砂质泥岩，泥云质粉砂岩与泥、亮晶竹叶状灰岩互层。底都为紫色、浅紫色竹叶状灰岩，具氧化圈，竹叶状砾屑定向或不定向排列。粉砂岩中发育扁豆状层理。上部竹叶状灰岩中产三叶虫 *Chuangia* aff. *batia*，*Changshanocephalus reedi* 7.9m

33. 灰色中厚层泥云质条带泥晶颗粒灰岩，底部夹泥晶竹叶状白云质灰岩，竹叶状砾屑多具氧化圈。泥云质条带为深灰色，风化后呈褐黄色 3.1m

32. 浅灰色中—薄层条带状粉晶鲕粒灰岩，条带为紫红色粉砂质泥晶云质灰岩，平行层理密集分布。产三叶虫 *Drepanura* sp.，*D. premesnili*，*Stephanocare* sp.，*Teinistion* sp.，*Blackwelderia* sp.，*Homagnostus hoi*，*H. convexus*，*Pseudagnostus douvillei* 3.9m

31. 深灰色中—薄层泥云质条带泥晶灰岩夹灰绿色薄层钙质粉砂岩，底部夹泥晶鲕粒灰岩及竹叶状灰岩 10.3m

30. 深灰色块状泥晶灰岩，风化面呈鱼鳞状 1.5m

29. 上部为黄灰色中层泥云质条带泥晶灰岩与绿灰色薄层钙质粉砂岩互层，夹3层灰色竹叶状灰岩。中部为2.5m厚的绿灰色薄层钙质粉砂岩夹灰色薄层含生屑钙质泥质白云岩。下部为深灰色中厚层粉晶鲕粒灰岩与粉晶竹叶状灰岩不等厚互层。含生物化石碎片。鲕粒灰岩中夹绿灰色薄层钙质泥岩，部分竹叶状砾屑具氧化圈。产三叶虫 *Blackwelderia sinensis* 18.4m

28. 灰色厚层含颗粒叠层石泥晶灰岩，叠层石为柱状，柱间有鲕粒、生屑、内碎屑充填，底部夹泥晶竹叶状灰岩 22.2m

27. 灰色厚层砂质条带泥晶灰岩，局部含鲕，夹4小层竹叶状灰岩 10.1m

26. 上部为灰色中层含钙质粉砂岩条带细晶竹叶状灰岩，下部为灰、黄、绿等杂色中层含泥云质条带泥晶灰岩，条带断续相连 3.2m

———— 整　合 ————

下伏地层：张夏组深灰色巨厚层亮晶鲕粒灰岩

（2）地质特征及区域变化

本组下部主要是灰色、黄灰色中—薄层状含泥云质条带泥晶灰岩，叠层石灰岩夹竹叶状灰岩和鲕粒灰岩；中部为绿灰色薄层钙质粉砂岩，泥岩和灰色中薄层泥云质条带泥晶灰岩，鲕粒灰岩和竹叶状灰岩，竹叶状砾屑具红色氧化圈；上部为灰色、深灰色中—薄层泥云质条带泥质灰岩，粉晶灰岩夹竹叶状灰岩或与之互层。

炒米店组主要分布于西山门头沟区齐家庄—柏峪—青白口—田庄和九山、下苇甸一带，以及房山区的蒲洼—黄土台—班各庄—龙宝峪一带。下苇甸剖面厚193m，青白口剖面厚244m，双石头剖面厚166m，九山剖面厚147m（未见顶），鲁家滩剖面厚121m，龙门台剖面厚299m。

北山地区出露于怀柔区东口楼和河防口一带以及昌平黄岩山地区。东口楼剖面厚363m（未见顶），河防口剖面厚>141m（未见顶），黄岩山剖面厚201m。

5. 冶里组（O_1y）

（1）代表剖面

正层型位于河北省唐山市赵各庄长山。北京市选取门头沟区丁家滩剖面为次层型（东经116°01′49″，北纬39°59′37″。由《华北地区区域地层表·北京市分册》下古生界编写小组1975年测制）。

上覆地层：亮甲山组灰黑色粉晶含云灰岩

———— 整　合 ————

冶里组 总厚度 93.0m

10. 灰黑色中层含生物碎屑含泥纹细粉晶灰岩夹竹叶状灰岩和页岩，竹叶状灰岩中产三叶虫 *Asaphellus*? sp. 3.0m
9. 灰黑色中层含生物碎屑含云斑细粉晶灰岩夹钙质页岩 5.0m
8. 深灰色中层泥云斑细粉晶、泥晶灰岩夹黄绿色钙质页岩 2.0m
7. 深灰色厚层泥质条带细粉晶、泥晶灰岩夹竹叶状含云灰岩，底部为页岩 6.0m
6. 深灰色中厚层含砂屑泥云纹层泥晶灰岩夹绿色页岩 10.0m
5. 灰黑色中—厚层含泥纹泥、粉晶含云质灰岩夹竹叶状灰岩 9.0m
4. 灰黑色中层含泥纹粉晶、泥晶含云质灰岩夹黄色薄层泥灰岩 3.0m
3. 浅灰色巨厚层细粉晶灰岩，凝块状细晶灰岩夹钙质页岩 33.0m
2. 浅灰色巨厚层含砂屑细粉晶、泥晶含云质灰岩 14.0m
1. 深灰色厚层含云斑泥、粉晶灰岩，下部夹薄层灰岩 8.0m

———— 整　合 ————

下伏地层：炒米店组泥云质条带泥晶灰岩夹竹叶状灰岩

（2）地质特征及区域变化

在丁家滩剖面，下部为深灰—浅灰色厚—巨厚层含云斑和砂屑泥晶—粉晶灰岩，局部为白云岩。中上部为深灰—灰黑色中—厚层含泥质条带、泥质条纹粉晶、泥晶灰岩夹竹叶状灰岩和黄绿色钙质页岩。与下伏炒米店组为整合接触。

冶里组主要分布在西山地区，在顺义二十里长山少量出露。西山丁家滩剖面厚 93m，青白口剖面厚 91m，双石头剖面厚 99m，鲁家滩剖面厚 134m，龙门台剖面厚 304m。顺义二十里长山出露不全，主要是豹斑状灰岩。

6. 亮甲山组（O_1l）

（1）代表剖面

正层型位于河北省唐山市赵各庄长山。北京市选取门头沟区丁家滩剖面为次层型（东经 116°01′45″，北纬 39°59′24″。由《华北地区区域地层表·北京市分册》下古生界编写小组 1975 年测制）。

上覆地层：马家沟组灰黄色含陆源碎屑泥晶白云岩

--------------- 平行不整合 ---------------

亮甲山组 总厚度 246.0m

41. 深灰色中、厚层含砂、砾屑粉晶含灰质白云岩 7.0m
40. 灰色薄层夹中层含燧石粉晶灰质白云岩，局部为角砾状灰质白云岩 6.0m
39. 深灰色中层细晶泥质白云岩，上部夹生屑粉晶灰质白云岩 13.0m
38. 灰色中层粉晶白云岩，含燧石团块，含腹足类 *Ophileta* sp.，*Helicotoma* sp.；双壳类 *Modiolopsis parallelus* 12.0m
37. 灰色中层泥质条带粉晶白云岩夹薄层云质灰岩，含燧石团块，产头足类 *Sactoceras* sp.；腹足类 *Euomphalus* sp. 16.0m
36. 灰黑色中层含燧石团块粉晶泥质白云岩，产腹足类 *Euomphalus* sp. 9.0m
35. 灰黑色中层含燧石粉晶含灰质白云岩 12.0m
34. 灰色中层含燧石细晶泥质白云岩 10.0m
33. 浅灰色中层粉晶白云岩，下部含燧石团块 6.0m
32. 黄灰色中层粉晶白云岩，含燧石团块 23.0m
31. 浅灰色厚层粉晶白云岩，含黑色燧石团块或含不连续燧石条带 15.0m

30. 灰黄色中层粉晶细晶泥质白云岩夹云质灰岩，含燧石团块，产头足类 *Cameroceras* sp.；腹足类 *Ecculiomphalus* sp.，*Ophileta* cf. *plana* 16.0m
29. 浅灰色中层细晶含灰、泥质白云岩 11.0m
28. 灰色中厚层含燧石条带和团块含云斑生屑泥晶灰岩，生屑为三叶虫、棘皮动物化石碎片，约10%，产头足类 *Manchuroceras* aff. *platyventrum*；腹足类 *Ecculiomphalus*? sp.，*Ophileta* sp. 4.0m
27. 灰黑色中层夹巨厚层含燧石团块泥晶灰岩，含头足类 *Piloceras* sp.，*Cameroceras* sp；腹足类 *Ophileta* cf. *plana* 4.0m
26. 灰黑色厚—巨厚层含燧石团块含生屑泥晶灰岩，产头足类 *Piloceras* sp.；腹足类 *Ophileta* cf. *squamosa* 6.0m
25. 灰黑色中层泥质斑纹含生屑粉晶含云质灰岩，底部含燧石 1.0m
24. 灰黑色薄层夹中层粉晶含云质灰岩，底部含燧石 4.0m
23. 深灰色厚层花斑状细粉晶白云质灰岩 9.0m
22. 深灰色中层中晶—粉晶白云质灰岩，含三叶虫化石碎片，产腹足类 *Ophileta* sp. 2.0m
21. 黑色中层含泥纹粉晶含云质灰岩 7.0m
20. 灰黑色中厚层含泥纹含球粒泥晶、细粉晶含云质灰岩，产腹足类 *Ophileta* sp.；头足类 *Piloceras* sp. 7.0m
19. 灰色中层含泥纹泥晶灰岩夹钙质页岩 3.0m
18. 灰黑色中层泥质条带粉晶灰岩夹粉晶含云质灰岩，夹页岩 18.0m
17. 灰黑色中层花斑状泥晶含云质灰岩 2.0m
16. 灰色中层细粉晶含云质泥质灰岩夹似竹叶状灰岩及页岩 5.0m
15. 灰黑色中层泥质条纹泥晶灰岩 2.0m
14. 灰黑色中层含燧石结核含生屑泥晶灰岩，含三叶虫化石碎屑，产腹足类 *Ophileta* sp.，*O.* cf. *plana*，*O.* cf. *squamosa*，*Ecculiomphalus* sp.；头足类 *Cameroceras* sp.，*Proterocameroceras* sp. 2.0m
13. 灰黑色巨厚层含燧石结核细粉晶含云质灰岩，见三叶虫化石碎片 2.0m
12. 灰黑色巨厚层花斑状细粉晶云质灰岩，由下往上云质含量增多 11.0m
11. 灰黑色巨厚层细粉晶含云质灰岩，含燧石团块 1.0m

———— 整　合 ————

下伏地层：冶里组灰黑色中层含泥纹粉晶灰岩夹竹叶状灰岩

（2）地质特征及区域变化

亮甲山组岩性总体为白云质灰岩或灰质白云岩，丁家滩等地与下部发育灰岩。与下伏冶里组为整合接触。

本组主要分布在西山地区，顺义二十里长山也有出露。厚度170~330m，一般在200m左右。岩性变化不大。青白口一带厚188m，双石头一带厚214m，鲁家滩一带受变形变质作用影响，仅厚90余米，黄土岭一带厚141m。昌平仅在西峪村东西两侧发现零星出露，受逆掩断层控制，呈飞来峰状态产出，缺少顶和底，出露最大厚度为37.1m。

7. 马家沟组（O_2m）

（1）代表剖面

选取门头沟区色树坟剖面为次层型（东经116°00′22″，北纬39°58′26″。由《华北地区区域地层表·北京市分册》下古生界编写小组1975年测制）。

上覆地层：太原组底砾岩

---------- 平行不整合 ----------

马家沟组 总厚度404.0m

81. 深灰色中层粉晶灰岩及角砾状灰岩 9.0m
80. 灰色厚—中层含生屑粉晶含云质灰岩，局部呈角砾状，产头足类 *Ormoceras koraiense* 19.0m
79. 灰黑色巨厚层粉晶灰岩，局部浅灰色，花斑状，产头足类 *Discoactinoceras* sp.，*Armenoceras* sp. 6.0m

78. 灰黑色厚—巨厚层粉晶泥质灰岩，夹条纹状含云质灰岩，产头足类 *Actinoceras* cf. *centrale*, *Armenoceras* sp. 6.0m

77. 深灰色巨厚层花斑状细晶含泥灰质白云岩，顶部为灰黑色含生屑粉晶灰岩 8.0m

76. 灰黑色厚层夹中层含泥质球粒灰岩和晶质白云岩，局部薄层状 9.0m

75. 深灰色厚—巨厚层局部夹中层含生屑粉晶含泥质灰岩 32.0m

74. 灰色厚—巨厚层花斑状粉晶白云质灰岩，产头足类 *Stereoplasmoceras*? sp.,? *Ormoceras suanpanoides*,? *Armenoceras tateiwai* 4.0m

73. 灰黑色厚—巨厚层灰岩与粉砂质粉晶灰质白云岩互层，含燧石结核，产头足类 *Armenoceras tateiwai*, *Actinoceras* sp., *Ormoceras suanpanoides* 28.0m

72. 灰黑色厚—巨厚层夹中层含砂屑粉晶含云质灰岩，含燧石结核，产头足类 *Ormoceras suanpanoides* 20.0m

71. 灰黑色巨厚层含燧石结核花斑状含粉屑粉晶灰岩，产头足类 *Ormoceras suanpanoides* 5.0m

70. 灰褐色、灰粉色薄层细晶灰岩、钙质页岩夹泥质白云岩 5.0m

69. 灰白色巨厚层含角砾泥晶含灰质白云岩 10.0m

68. 灰色厚—中层含灰质白云岩及角砾状含灰质白云岩 8.0m

67. 浅灰—浅黄色厚层—巨厚层角砾状粉晶含云质泥质灰岩，夹薄层白云岩 11.0m

66. 浅灰、灰黄色厚层含泥钙质角砾岩 2.0m

65. 深灰色巨厚层泥晶含云质灰岩与灰白色泥质白云岩互层，顶部见波痕 34.0m

64. 灰色巨厚层花斑状粉晶灰质白云岩，含头足类 *Tofangoceras*? sp. 13.0m

63. 浅灰、灰黄色巨厚层粉晶含灰质白云岩 3.0m

62. 灰黑色厚—中层粉晶含云质灰岩 6.0m

61. 深灰色中层亮晶含竹叶砂屑含云质灰岩 1.0m

60. 浅灰色、灰黄色厚—巨厚层细粉晶含云质灰岩 28.0m

59. 灰黑色、深灰色厚—巨厚层粉屑含云质灰岩，含头足类 *Tofangoceras* sp. 3.0m

58. 深灰色中—厚层粉晶含灰质白云岩 3.0m

57. 灰色厚—巨厚层粉晶灰岩，局部含角砾 13.0m

56. 灰黑色夹浅灰色条带巨厚层花斑状粉晶含云质灰岩 5.0m

55. 灰、灰黑色中层花斑状粉晶灰质白云岩，含头足类 *Armenoceras* sp.，腹足类 *Donaldiella* sp. 9.0m

54. 灰黑色巨厚层含海绵骨针泥晶灰岩夹薄层粉晶灰岩 8.0m

53. 灰黑色厚—巨厚层泥晶泥质灰岩，含丰富的头足类化石 *Armenoceras coreanicum*, *A. teicherti*, *A. orientale.*, *Kogenoceras nanpiaoense*, *Tofangoceras* sp., *Wutinoceras aigawaense*, *W. taipingkouense*, *Ormoceras* cf. *nagaoi*, *Actinoceras* sp. 19.0m

52. 黄灰色薄层泥晶泥质白云岩，含泥质条纹，含生屑、粉屑灰岩及含泥晶灰岩条带 3.0m

51. 灰黑色巨厚层含粉砂、球粒含云质灰岩，含头足类 *Tafangoceras* sp. 10.0m

50. 灰黑色厚—巨厚层花斑状泥晶含云质灰岩，含头足类 *Kogenoceras nanpiaoense* 5.0m

49. 深灰色、灰黑色巨厚层夹薄层泥晶灰岩 6.0m

48. 黄灰色、粉灰色薄层夹中层泥晶泥质白云岩 3.0m

47. 杂色厚—巨厚层云泥质条纹泥晶灰岩，底部为泥质白云岩 4.0m

46. 灰黑色厚层角砾状泥晶白云质灰岩和浅灰色细粉晶含云泥质灰岩 9.0m

45. 灰黑色巨厚层含生屑泥晶灰岩，局部呈角砾状 10.0m

44. 浅粉红色薄层泥晶泥质白云岩，含灰质条带 2.0m

43. 灰色角砾状泥晶泥质灰岩，碎屑状细粉晶白云质灰岩，下部为浅黄色、浅灰色角砾状泥晶白云岩 21.0m

42. 灰黄色含陆源碎屑泥晶泥质白云岩 4.0m

--------------- 平行不整合 ---------------

下伏地层：亮甲山组粉晶白云岩

（2）地质特征及区域变化

本组厚度大，岩性主要是灰、深灰及灰黑色厚—巨厚层泥晶灰岩、花斑状、豹斑状灰岩、泥质条带灰岩、白云质灰岩和灰质白云岩，夹少量钙质页岩。在底部以及中部层位，普遍发育角砾状泥晶白云岩和灰岩或角砾岩。与下伏亮甲山组为平行不整合接触。

马家沟组主要分布于西山。在齐家庄—青白口—淤白一线，呈北东向带状展布。在丁家滩—寨口一带构成复式背、向斜翼部。在周口店—龙宝峪—南窖—磁家务呈马蹄形构成北岭向斜翼部。在古迹山庄构成背斜。在芦子水—史家营—白湖呈北东向分布。另外在顺义二十里长山有零星出露。

本组厚度变化较大，蒲洼、柳林水、千军台、色树坟及南辛房厚度较大，向周围减薄。色树坟剖面厚404m，水波涧剖面厚498m，千军台剖面厚489m，南辛房剖面厚482m，鲁家滩剖面厚434m，黄土岭剖面厚493m。在西山西部齐家庄、上清水等地顶部甚至中部遭剥蚀而缺失，厚度显著变薄。双石头剖面厚127m。顺义二十里长山剖面厚300m。

第二节　生物地层

从寒武纪开始海生无脊椎动物大量繁衍，并迅速演化成众多门类。主要有三叶虫、笔石、头足类、小壳类、古杯类、古介形类、苔藓类、珊瑚、牙形石、藻类等，但对年代地层划分有重要意义和研究程度相对较高的主要是三叶虫、头足类、笔石和牙形石。北京地区下古生界所产化石，寒武系以三叶虫为主，奥陶系以头足类和牙形石为主。

三叶虫主要产于北京西山的寒武系，尤其第三统和芙蓉统属种繁多、数量丰富，目前已经建立了16个化石带（表3-2）；头足类主要产于西山地区的奥陶系亮甲山组和马家沟组，但研究程度较低，仅列出12个典型属（或种）；牙形石主要产于北京西山的张夏组至马家沟组，但研究程度很低，仅能列出18个典型分子，还没有条件建立化石带。

表3-2　北京地区寒武纪—奥陶纪生物地层对比简表

Table 3-2　Cambrian-Ordovician biostratigraphic correlation in Beijing

<table>
<tr><th rowspan="2">岩石地层单位</th><th colspan="3">生物地层单位</th><th colspan="2">年代地层单位</th></tr>
<tr><th>三叶虫化石带</th><th>牙形石</th><th>头足类</th><th>统</th><th>系</th></tr>
<tr><td>马家沟组</td><td></td><td>Tangshanodus tangshanensis</td><td>Armenoceras tateiwai
Ormoceras suanpanoides
Discoactinoceras
Tofangoceras
Kogenoceras nanpiaoense
Wutinoceras
Stereoplasmoceras pseudoseptatum</td><td>中奥陶统</td><td rowspan="3">奥陶系</td></tr>
<tr><td>亮甲山组</td><td></td><td>Serratognathus bilobatus</td><td>Manchuroceras
Hopeioceras
Kaipingoceras
Cameroceras
Proterocameroceras</td><td rowspan="2">下奥陶统</td></tr>
<tr><td>冶里组</td><td></td><td>“Acodus” oneotensis
Cordylodus aff. prior
Paltodus nitidus
P. bicostatus
Acontiodus aff. stanfferi
Oneotodus variabilis</td><td></td></tr>
</table>

续表

岩石地层单位	生物地层单位			年代地层单位	
	三叶虫化石带	牙形石	头足类	统	系
炒米店组	*Tellerina-Calvinella* *Quadraticephalus* *Ptychaspis-Tsinania*	*Proconodontus muelleri* *Prooneotodus rotundatus* *P. gallitini*, *P. tenuis* *Rotundoconus jingxiensis*		芙蓉统	寒武系
	Mansuyia-Kaolishania *Changshania* *Chuangia*	*Westergaardodina bicuspidata* *Prooneotodus rotundatus* *P. gallitini*, *P. tenuis*			
	Drepanura *Blackwelderia*	*Westergaardodina matsushitai*			
张夏组	*Damesella* *Amphoton* *Crepicephalina*	*Furnishina latuspa*		第三统	
	Bailiella *Sunaspis*	*Furnishina primitiva*			
馒头组	*Shantungaspis*			第二统	
昌平组	*Redlichia chinensis* *Megapalaeolenus*				

注：此表参考《北京市岩石地层》，略修改。

第四章　石炭系—中三叠统

本区的石炭纪—三叠纪（早、中世）地层主要分布于北京西部的门头沟区、房山区、石景山区及海淀区（图4-1），顺义区和通州区等地新生代沉积物之下也有分布。

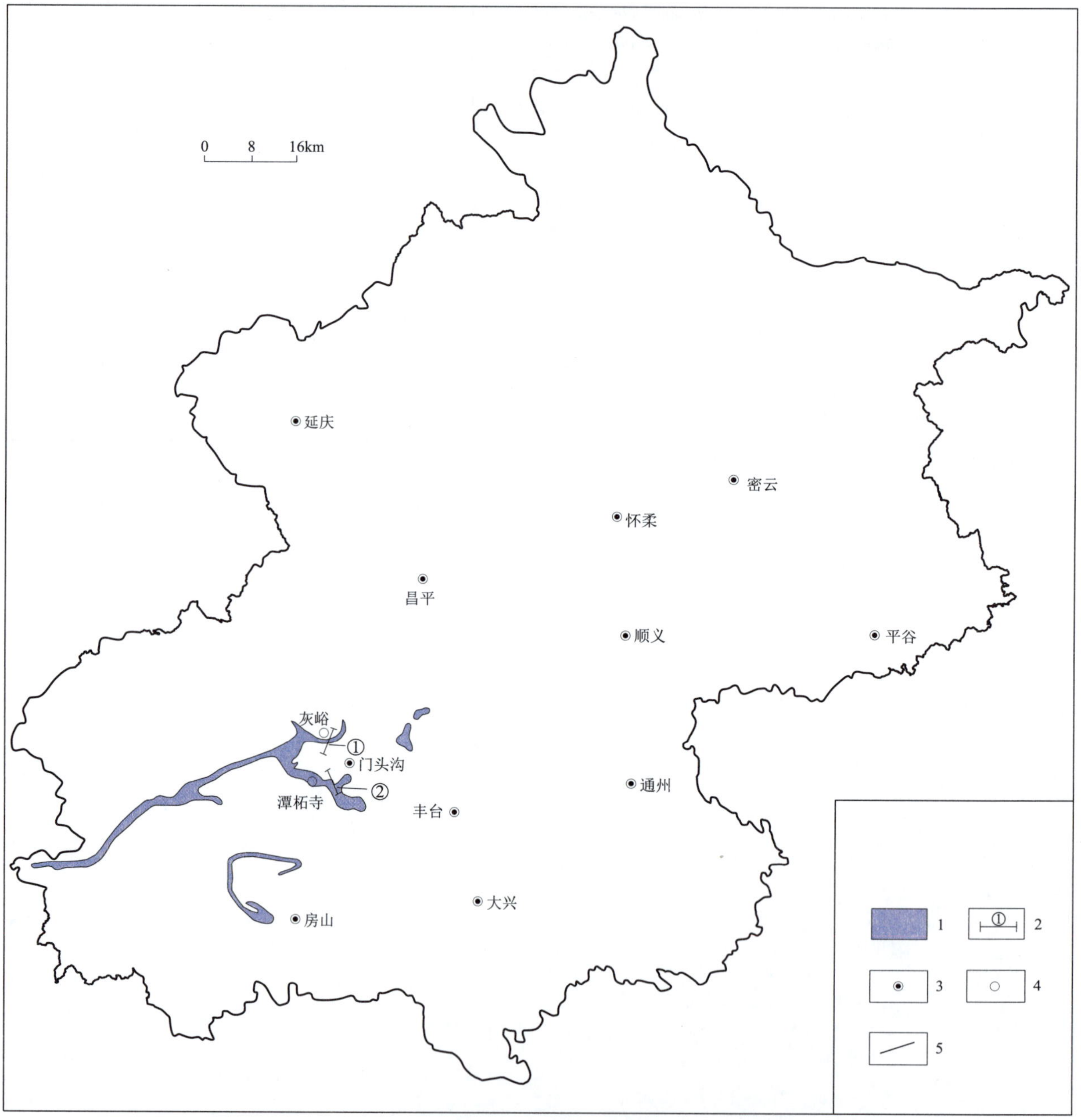

图4-1　北京市石炭纪—中三叠世岩石地层及层型分布简图

（据《北京市岩石地层》，1996）

Fig. 4-1　Distribution map of Carboniferous-Middle Triassic lithostratigraphic units and their stratotype section in Beijing

1—石炭纪—中三叠世岩石地层；2—层型：①灰峪东山梁军庄太原组至石盒子组剖面，②潭柘寺东山梁双泉组剖面；3—区政府驻地；4—村庄；5—市界该套地层底部与下伏的奥陶系马家沟组白云质灰岩之间为一个极明显的平行不整合接触面。此风化剥蚀界面的地质历史延续时期约 140Ma，其间缺失了晚奥陶世至早石炭世地质历史时期的地层沉积。该套地层顶部则由发生于晚三叠世的印支运动构造面所限定，双泉组与上覆地层上三叠统杏石口组之间为不整合接触。该套地层总厚度为 700 余米。

1920 年叶良辅等撰写出版的《北京西山地质志》开始了对北京地区晚古生代—中三叠世沉积地层的划分研究工作。在此基础上，杨曾威等（1928）、计荣森等（1933）、潘钟祥（1933）、熊秉信（1936）、杨杰（1957）、李星学（1963）等老一代地质学家对该断代的地层进行详细研究，基本确定了地层划分方案。1991 年出版的《北京市区域地质志》基本上归纳反映了该断代地层划分与归属方面的各种意见。

1996 年鲍亦冈等对该断代地层进行了全面的清理和研究（表 4-1），尽可能采用了晋冀鲁豫地层区统一地层名称，停止使用以往所惯用的地区性地层单位名称。

第一节　岩石地层

1. 太原组（C_2t）

（1）代表剖面

正层型在山西省太原市西山西铭七里沟。北京市的次层型为门头沟区灰峪村东山梁剖面（东经 160°05′13″，北纬 40°00′29″。1975 年武汉地质学院煤田教研室测制）。

上覆地层：山西组中粗粒杂砂岩

———— 整　合 ————

太原组　　总厚度 58.52m

11. 灰色细砂岩，具波状层理，顶部粉砂岩内含海生动物化石碎片　8.70m
10. 灰色粉砂岩，含植物化石：*Asterophyllile longifolius*，*Neuropteris* cf. *gigantea*，*Neuropteris kaipingiana*　4.00m
9. 灰色细砂岩，钙质胶结，下部水平层理，上部缓波状层理并含植物根化石　10.30m
8. 灰黑色泥质岩，具波状层理，含海相动物化石：腕足类 *Dictyoclostus gruenewaldti*，*Schellwienella* sp.，*Choristites yanghukouensis*；双壳类 *Aviculopecten* sp.　4.85m
7. 灰黑色灰岩，具不明显水平层理，含海相动物化石：有孔虫 *Textularia defrance*，*Ozawainella* sp.　0.40m
6. 深灰色粉砂岩，具水平层理，含铁质结核和海相动物化石：腕足类 *Dictyoclostus* sp.，*Schellwienella* sp.，*Chonetes* sp.；双壳类 *Pseudomonotis* sp.，*Palaeolima* sp.，*Aviculopecten* sp.，*Acanthopecten* cf. *carboniferus*　4.85m
5. 灰色细砂岩，具断续的波状层理，含植物根化石　1.94m
4. 灰色粉砂岩，层理不显，含粉砂质结核，结核大小一般为 4~5cm，中部含有丰富的海相动物化石：腕足类 *Chonetes* sp.，*Choristites yanghukouensis*，*Dictyoclostus gruenewaldti*，*Pugnoides weeler*，*Schellwienella crenistria*；双壳类 *Acanthopecten* cf. *carboniferus*，*Anthraconaia* sp.，*Anthraconauta* sp.，*Aviculopecten* sp.，*Edmondia* sp.，*Pseudomonotis* sp.；其他 *Phillipsia* sp.，*Yangjiatunconularia qingshuijianensis*。顶部含有少量的植物化石碎片　12.03m
3. 浅灰色黏土岩，风化后成粉红色，鲕状结构，鲕粒大小 1mm，鲕粒含铁，呈暗红色　9.20m
2. 灰白色砾岩，砾石成分以石英岩为主（约占 70%），次为燧石、硅质灰岩。砾石直径 1~9cm，一般为 3~5cm，磨圆好，分选中等，孔隙式黏土质胶结。砾岩横向变化大，多成透镜状　2.52m

------------ 平行不整合 ------------

下伏地层：马家沟组灰黑色致密块状灰岩

表 4-1　北京市石炭纪—中三叠世岩石地层单位划分沿革表

Table 4-1 The classification history of Carboniferous-Middle Triassic lithostratigraphic units in Beijing

作者/资料	年代	地区	地层单位（自上而下）
叶良辅等	1920	西山	门头沟煤系；T—P 红庙岭砂岩；C 杨家屯煤系
杨曾威等	1928	西山	门头沟煤系；红庙岭砂岩；上杨家屯煤系；中杨家屯煤系；下杨家屯煤系
计荣森等	1933	西山	门头沟煤系；T_1 双泉组；P_2 红庙岭砂岩
熊秉信	1936	西山	门头沟煤系；双泉层；红庙岭砂岩；上杨家屯煤系；中杨家屯煤系；下杨家屯煤系；清水涧页岩层
《中国区域地层表（草案）》	1956	北京	辉绿岩；双泉组；石河子统：红庙岭砂岩、上杨家屯煤系；山西统：中杨家屯煤系；太原统：下杨家屯煤系；本溪统：清水涧组、房山组
杨杰	1957	西山	门头沟组：辉绿岩、杏石口亚统；八大处统：石景山层、金山口层；红庙岭统：大悲寺层、红山口层
李星学	1963	北京	南岭组；杏石口组；双泉群：P_2^2 上部、P_2^1 下部；P_1^{2-2} 红庙岭组；杨家屯群：P_1^{2-1} 上部、P_1^1 中部、C_3 下部；C_2 清水涧组
《华北地区区域地层表·北京市分册》	1977	北京	J_1 南大岭组、杏石口组；P_2 双泉组；P_1^2 红庙岭组；P_1^1 山西组；C_3 太原组；C_2 本溪组
《中国煤田地质学》	1980	北京	南大岭组；杏石口组；P_2^2—P_2^{1-2} 双泉组；P_2^{1-2} 红庙岭组；杨家屯群：P_1^2 阴山沟组、P_1^1 岔儿沟组、C_3 灰峪组；C_2 水涧组
《北京市区域地质志》	1991	北京	南大岭组；T_3 杏石口组；T_{1-2}—P_2 双泉组；P_2 红庙岭组；P_1 阴山沟组、岔儿沟组；C_3 灰峪组；C_2 清水涧组
萧宗正等	1994	北京	南大岭组；T_3 杏石口组；双泉群：T_2 潭柘寺组、T_1 大悲寺组；红庙岭组
《北京市岩石地层》	1996	北京	南大岭组；杏石口组；双泉组；石盒子组；山西组；太原组
本志		北京	南大岭组；T_3 杏石口组；T_{1-2} 双泉组；P_{2-3} 石盒子组；P_1 山西组；C_2 太原组

（2）地质特征及区域变化

太原组岩性主要由一套海陆交互相或陆相的灰—灰黑色的粉砂岩、细砂岩、泥质岩、泥灰岩（灰岩）组成，夹薄煤层及煤线，底部为透镜状产出底砾岩。与下伏马家沟组呈平行不整合接触。

该组分布在京西的门头沟区、房山区、海淀区，以及京东的顺义区、通州区等地。构造上受九龙山向斜、髫髻山向斜、北岭向斜、大孙各庄向斜所控制，分布于各向斜的翼部。本组的厚度一般在60m左右，灰峪剖面厚58.52m，清水涧剖面厚69.0m，史家营剖面厚53.8m，南车营剖面仅厚31.96m，禅房剖面厚47m。顺义二十里长山新生界之下本组厚度约60m。

九龙山以南的潭柘寺、谷积山庄、大灰厂一带岩石普遍变质为千枚岩、板岩、变质砂岩、叶蜡石化或硬绿泥石化黏土岩等。

2. 山西组（C_2P_1s）

（1）代表剖面

北京市选取次层型为门头沟区灰峪村东山梁剖面（东经116°05′13″，北纬40°00′29″。1975年武汉地质学院煤田教研室测制）。

上覆地层：石盒子组灰白色厚层砾岩

———— 整　合 ————

山西组 总厚度300.28m

45. 深灰色粉砂岩，具有不明显的水平层理，顶部含有丰富而完整的植物化石：*Calamites suckowii*，*Lobatannularia sinensis*，*Taeniopteris* sp.，*Sphenopteris tenuis*，*Plagiozamites tungweiensis*，*Cordaites* sp. 18.10m
44. 细砂岩和粉砂岩互层，砂岩呈灰褐色，常含砾石。粉砂岩多具波状层理，常含植物化石碎片 32.41m
43. 深灰色粉砂岩，具不明显的水平层理，含植物化石：*Calamites suckowii*，*Lobatannularia sinensis*，*Sphenopteris tenuis*，*Cordaites* sp. 5.72m
42. 深灰色粗—中粒砂岩。杂砂岩，颗粒呈棱角状，分选差 4.61m
41. 深色粉砂岩，具波状层理，含植物化石碎片 6.81m
40. 深灰色砾岩，砾石成分以燧石为主。石英次之。砾石直径一般在0.5cm左右，粉砂质胶结。此砾岩厚度不大，但层位较稳定 1.37m
39. 暗绿色粉砂岩，具水平层理 4.51m
38. 灰黑色黏土岩，具滑感和吸水性；上部是“子煤” 1.87m
37. 灰—紫色粉砂岩，具水平层理，下部含铁质结核，上部含植物根部化石 13.73m
36. 灰色砾岩，砾石成分以石英和燧石为主，次为硅质灰岩。磨圆较好，分选差，泥质胶结。与下伏地层呈冲刷接触。此砾岩横向变化大，但层位稳定，是本区重要的标志层之一 21.95m
35. 深灰色泥质岩。从下到上岩性逐渐变为细砂岩。下部含植物化石：*Lobatannularia sinensis*，*Pecopteris* sp.，*Lepidodendron* sp.，*Calamites suckowii*，*Sphenophyllum thonii*，*Emplectopteris triangularis* 11.35m
34. 黏土岩夹煤层，黏土灰黑色，有滑感和吸水性。所夹煤层俗名“黑煤大炭”，为本矿区的重要煤层之一。中部含植物化石：*Lepidodendron* sp.，*Lobatannularia sinensis*，*Pecopteris candolleana*，*Cordaites* sp. 9.40m
33. 深灰色粉砂岩，具水平层理。上部含植物化石：*Sphenophyllum thonii*，*S. oblongifolium*，*S. verticillatum*，*Sphenopteris* sp.，*Calamites suckowii*，*Alethopteris hallei*，*Emplectopteris triangularis*，*Lepidodendron* sp. 7.40m
32. 灰色中粒砂岩，杂砂岩。底部较粗，往上变为细砂岩，具有交错层理和波状层理 24.10m
31. 灰色砾岩。砾石成分主要为黑色燧石和白色石英，其次为硅质灰岩和粉砂岩。砾岩大小一般在2cm左右，次棱角状，分选较好。基底式泥质胶结。与下伏地层呈冲刷接触关系。此层砾岩层位稳定，出露较好 10.90m

30. 深灰色煤层，无烟煤。复杂结构。因含高岭土矿物，俗名“小白煤”。为本矿区重要的可采煤层之一 0.80m

29. 灰黑色黏土岩，具滑感、吸水性和塑性。含植物根化石。此黏土为半软质黏土。是本矿区重要的黏土矿层之一 2.70m

28. 灰黑色粉砂岩。具不明显的水平层理，含有大量保存完好的植物化石：*Sphenophyllum oblongifolium*，*S. laterale*，*Annularia* sp.，*Pecopteris* sp.，*Cordaites* sp.，*Calamites cistii*，*Lobatannularia sinensis*，*Taeniopteris* sp. 3.38m

27. 灰色细—中粒砂岩。碎屑以石英为主，燧石次之，次棱角状，具有大型斜层理。顶部和底部有少量的植物化石：*Lobatannularia sinensis*，*Calamites suckowii*，*Lepidodendron incertum* 22.17m

26. 深灰色粉砂岩，下部含瘤状铁质结核，中上部水平层理发育。是本矿区最丰富的化石层之一，常见的植物化石有：*Sphenophyllum oblongifolium*，*S. laterale*，*S. thonii*，*S. verticillatum*，*Annularia orientalis*，*A.* cf. *steliata*，*Calamites suckowii*，*Bowmanites* sp.，*Lepidodendron oculusfelis*，*L. incertum*，*Bergeria* sp.，*Stigmaria* sp.，*Neuropteris plicata*，*N. otozamioides*，*Alethopteris hallei*，*Pecopteris* sp. 7.50m

25. 深灰色无烟煤。煤层呈复杂结构，厚度稳定，为本矿区的主要煤层之一。因含大量高岭土，俗称“大白煤” 2.10m

24. 黑色黏土岩。具有滑感、吸水性及可塑性，含有瘤状铁质结核及植物根化石。植物化石有：*Sphenophyllum oblongifolium*，*Neuropteris otozamioides*，*Calamites suckowii*。本层属软质耐火黏土，厚度稳定 2.40m

23. 深灰色粉砂岩。具有不明显的水平层理。上部含有丰富的植物化石：*Lepidodendron incertum*，*Stigmaria* sp.，*Neuropteris plicate*，*Neuropteris otozamioides*，*Sphenophyllum thonii*；底部含化石：*Lingulella* sp. 13.60m

22. 深灰色中粒杂砂岩。碎屑半圆状，分选中等，具有不明显的交错层理 3.36m

21. 深灰色粉砂岩。具有不明显的水平层理 13.00m

20. 灰色中粒杂砂岩。碎屑呈次圆状。分选中等，具有波状层理和水平层理 2.15m

19. 深灰色粉砂岩。具有水平层理，风化后呈薄片状 2.70m

18. 灰黑色细砂岩，杂砂岩。具有水平层理和波状层理 6.80m

17. 深灰色粉砂岩。具有不明显的水平层理 9.36m

16. 深灰色无烟煤。煤层呈简单结构，厚度不大，但全区稳定。因煤层中含有大量的黄铁矿，烧时冒黄烟并发出臭味，俗名“黄煤”或“臭煤” 0.40m

15. 深灰色粉砂岩含植物根化石 7.53m

14. 灰色细砂岩、杂砂岩，分选磨圆中等。具水平层理 2.45m

13. 灰色粉砂岩。具波状层理，顶部含煤层。含植物化石：*Sphenophyllum oblongifolium*，*Lepidodendron posthumii*，*Pecopteris feminaeformis* 15.65m

12. 灰色粗—中粒杂砂岩。泥质胶结，碎屑次圆状，分选中等。上部有波状层理，下部具斜层理。本层层位稳定 8.00m

———— 整　合 ————

下伏地层：太原组灰色细砂岩

（2）地质特征及区域变化

山西组在北京地区以深灰、灰黑色砂岩、页岩、黏土岩及砾岩为主，夹有7~13层煤层和煤线，剖面的旋回性强。

分布范围基本上与太原组相同，只是在门头沟区西部髫髻山-百花山向斜的北翼受南大岭组玄武岩喷发超覆等多种因素的影响而变薄甚至缺失。灰峪地区发育最厚，达309.28m，木城涧剖面厚212.8m，史家营剖面厚261m，南车营剖面厚80m，长流水剖面厚145.2m，禅房剖面厚163m。顺义二十里长山新生界之下本组厚度约280m。

3. 石盒子组（$P_{2-3}s$）

（1）代表剖面

门头沟区军庄火车站东侧石盒子组剖面可作为次层型（东经 116°05′46″，北纬 39°59′51″。系 1987 年北京矿务局陈瑞琪、李有纲、姚春青测制）。该剖面下部层位出露不好，1～11 层由处于军庄沟北侧的灰峪东山梁剖面拼接而成（其 1～11 层相当于灰峪东山梁剖面原始剖面原编号的 45～55 层，12～45 层则相当于军庄火车站东侧剖面原始编号的第 6～39 层）。

上覆地层：双泉组黄褐色粉砂岩

——————整　合——————

石盒子组	总厚度 305. 2m
45. 黄绿色含砾粗砂岩	5. 2m
44. 褐紫色粉砂岩	6. 1m
43. 黄绿、黄褐色细砂岩	5. 4m
42. 黄绿色含砾粗砂岩	3. 5m
41. 黄绿色粉砂岩	3. 0m
40. 灰褐、灰绿色细砂岩	2. 3m
39. 黄绿色含砾粗砂岩中夹一层 0. 50m 粉砂岩	4. 4m
38. 褐黄色粉砂岩	1. 9m
37. 灰黄、灰绿色含砾粗砂岩，局部为 0. 1～1. 0cm 砾径的砾岩	4. 8m
36. 灰绿、灰黄色粉砂岩	1. 9m
35. 灰绿色粗砂岩夹灰色薄层细砂岩及含砾粗砂岩透镜体	3. 8m
34. 深灰色粉砂岩，底部 0. 3m 碳质泥岩中含丰富植物化石：*Taeniopteris mucronata*，*Pecopteris chihliensis*，*Yuania striata*，*Asterophyllites* sp.，*Odontopteris*? sp.，*Neuropteris* sp.，*Cordaites principalis*，*Macrostachya*? sp.	3. 1m
33. 灰黄色含砾粗砂岩，含细砾岩透镜体	2. 5m
32. 灰褐色厚层状细砂岩，含铁质结核	1. 6m
31. 黄绿色含砾粗砂岩，具大型斜层理，含铁质结核	9. 6m
30. 灰绿色细砂岩，底为 0. 15m 含碳泥质岩，具斜层理	2. 8m
29. 灰绿色含燧石砾石粗砂岩，砾径 0. 5～1. 5cm，棱角状	2. 3m
28. 灰绿色粉砂岩	1. 2m
27. 灰绿色中粒石英砂岩，底部含砾	8. 8m
26. 深灰色粉砂岩	0. 7m
25. 褐黄色粉砂岩	16. 2m
24. 灰绿色细砂岩	4. 2m
23. 灰绿色粗砂岩，风化后呈灰黄色	5. 5m
22. 灰黄色粉砂岩	6. 5m
21. 灰黄色粗砂岩	6. 6m
20. 灰黄色粉砂岩	4. 9m
19. 灰黄色细砂岩	12. 0m
18. 灰绿色细砂岩夹多层深灰色粉砂岩透镜体	2. 2m
17. 灰绿色含砾粗砂岩	1. 8m
16. 灰黑色碳质泥岩	0. 2m
15. 褐黄色粉砂岩，含铁质结核	11. 7m
14. 灰绿色粗砂岩，风化后呈灰白色	10. 5m
13. 灰黄、灰绿色粉砂岩	3. 1m
12. 灰绿色含砾粗砂岩，顶部夹薄层粉砂岩	6. 9m
11. 暗黄色粉砂岩，具水平层理和缓波状层理	27. 0m

10.	灰黄色粗粒砂岩，碎屑以石英为主，泥质胶结，斜层理较发育	8.0m
9.	暗黄色砂岩粉砂岩互层，砂岩成分以石英为主，泥质胶结，常见斜层理。粉砂岩中可见水平层理和波状层理，含有少量的植物化石：*Calamites* sp.，*Pecopteris* sp.	35.0m
8.	暗黄色含砾砂岩，分选差，砂泥质胶结	6.0m
7.	砾岩：岩性同第2层。横向变化大，有时逐渐过渡为砂岩	7.0m
6.	土黄色粉砂岩，具水平层理。含铁质结核和少量的植物化石：*Calamites suckowii*，*Sphenopteris* sp.，*Walchia* sp.	8.0m
5.	深灰色中粒砂岩，碎屑以石英为主，次为燧石，泥质胶结	10.0m
4.	灰色砾岩，厚层状。砾石成分同第2层	10.0m
3.	灰黄色粗砂岩，碎屑成分以石英和燧石为主，次为长石。磨圆度差，分选性较好，斜层理发育	17.0m
2.	灰白色厚层状砾岩，砾石成分以石英为主，次为燧石，砾石大小一般在0.5cm左右。孔隙式泥质胶结	10.0m

———— 整　合 ————

下伏地层：山西组粉砂岩

（2）地质特征及区域变化

石盒子组在地层颜色上以灰绿色为主，岩性以含砾砂岩、粗砂岩、粉砂岩、粉砂质黏土岩为主，中粒级碎屑岩较少，下部夹煤线。砾岩多分布于该组的下部，单层厚层最大10m，但横向延伸并不稳定，侧向延伸不远即变为含砾砂岩或粗砂岩。

区内石盒子组的分布基本上与山西组相近，分布于京西的门头沟区、海淀区、石景山区及房山区，受九龙山向斜、北岭向斜及髫髻山-百花山向斜的控制，但在该向斜的北翼原始沉积变薄甚至缺失。石盒子组在灰峪—军庄火车站附近厚度最大，达305.2m，向西、向南逐渐变薄，往西史家营剖面厚315m，往南南车营剖面厚285m，而往北则很快变薄缺失，禅房厚102m。顺义二十里长山新生界之下本组厚度约500m。

4. 双泉组（P_3T_2s）

（1）代表剖面

由于天泰山双泉寺剖面上该组顶部岩层出露不全，以北京市门头沟区潭柘寺东山梁剖面为选层型（东经116°01′48″，北纬39°54′19″。由萧宗正等1990年测制）。

上覆地层：杏石口组灰白色石英砾岩

～～～～ 角度不整合 ～～～～

双泉组		总厚度170.5m
26.	土黄色、浅黄色薄层砂质板岩	15.0m
25.	灰白色薄层—中层含岩屑砂岩	1.0m
24.	褐黄色薄—中层板岩。板理发育	9.2m
23.	浅灰绿色厚层含细砂质板岩。上部夹粗砂岩	2.4m
22.	灰褐色薄层含细砂质板岩。上部夹粗—中砂岩	22.8m
21.	灰白色、褐灰色厚层含细砾复成分巨粒砂岩。砾径一般不超过1cm，磨圆度差	13.5m
20.	灰绿色粉砂质板岩。层面可见绢云母，板理发育	8.0m
19.	灰绿色厚层含细砂岩屑粉砂岩	7.8m
18.	灰白色厚层中—粗粒长石砂岩，局部含砾，具斜层理	4.1m
17.	浅灰绿色巨厚层黏土质细砂岩	5.5m
16.	灰绿色黏土质板岩，夹紫红色板岩。板理发育	3.8m
15.	浅灰绿色厚层粉砂质板岩	1.5m
14.	紫色中—厚层含岩屑粉砂岩	0.7m

13. 灰白色、白色厚层—巨厚层含钙中粒砂岩　10.0m
12. 灰绿色、浅绿色厚层—巨厚层细砂质岩屑粉砂岩　12.5m
11. 灰白色中—厚层含钙中粒砂岩　1.6m
10. 灰绿色厚层含细砂岩屑粉砂岩　1.4m
9. 紫色巨厚层含钙细粒岩屑砂岩　14.6m
8. 深灰色、灰色厚层状中粗粒—细粒岩屑粉砂岩。上部含砾及磁铁矿等暗色矿物，具斜层理　5.3m
7. 灰绿色、黄绿色薄—中厚层中—细粒岩屑粉砂岩。顶部颗粒变细，层理发育　2.6m
6. 红色淡红色中厚层中—细粒含岩屑砂岩　2.3m
5. 黄绿色厚层细较岩屑砂岩，具斜层理　1.3m
4. 淡红色、黄绿色厚层硬绿泥石细粒岩屑砂岩，硬绿泥石千枚岩。下部以淡红色为主，上部黄绿色为主，具斜层理　6.7m
3. 紫色、浅紫色厚层细粒岩屑砂岩，硬绿泥石千枚岩。上部黄绿色为主，下部淡红色为主，具斜层理　6.4m
2. 灰色、浅灰色中层中—细粒岩屑砂岩。底部颗粒变粗，局部含砾　2.1m
1. 紫色、紫红色中层细粒岩屑砂岩及叶蜡石化页岩　8.4m

——————　整　　合　——————

下伏地层：石盒子组石英砂岩

（2）地质特征及区域变化

双泉组的显著特征是其具有紫红和灰绿色的醒目色调。下部以紫红色为主，上部多灰绿色、灰白色。在房山北岭向斜一带，岩性多变为变质砂岩、千枚岩、片岩。变质矿物多见绢云母，特征矿物有硬绿泥石、滑石、叶蜡石。

双泉组在北京地区的分布较为局限，出露在海淀、石景山、门头沟和房山等区，在北岭向斜、九龙山-香峪大梁向斜的两翼分布。厚度变化较大，总趋势为东厚西薄，如南车营剖面厚350m，大悲寺剖面厚200m，孟悟剖面厚182m，赵家台剖面厚约100m，潘涧子剖面厚41m，再向西则断续减薄直至尖灭。

第二节　生物地层

本区石炭纪—三叠纪地层中含较丰富生物化石（表4-2），鲍亦冈等（1996）编著的《北京市岩石地层》，对各岩石地层单位所含化石进行了系统清理总结，主要有以下成果：

1）北京地区太原组含有较多的海相动物化石（表4-2）。*Choristites mosquensis* 以及 *Fusulinella* - *Fusulina* 蜓类化石组合是莫斯科期标准化石，时代相当于华南地区晚石炭世的达拉期。植物化石中常见 *Neuropteris gigantea*，*N. kaipingiana* 和 *Linopteris brongniartii*，均为华夏植物群早期的植物组合。本组生物化石资料均表明其时代属于石炭纪二分的晚石炭世中期，或相当于国外的维斯特法期或莫斯科期。

2）山西组所含的植物化石可分为两套组合，以大白煤层的间接顶板灰绿色厚层含砾粗砂岩夹砂砾岩为界。其下地层以 *Neuropteris ovata* 与 *Lepidodendron posthumii* 为主，具典型的早期华夏植物群面貌，其时代为晚石炭世斯蒂芬期，或格热尔期—卡西莫夫期。而大白煤层的间接顶板以上，山西组上部的地层中所含植物化石与华北地区早二叠世早期植物组合面貌基本一致。因此北京地区山西组的地质时代为晚石炭世的斯蒂芬期或格热尔期—卡西莫夫期延续到晚二叠世早期的萨克马尔期。很显然，北京地区开始接受沉积时间要早于层型所在地——山西太原西山，该地区山西组的地质时代为早二叠世早期，两地相比，同一个岩石地层单位，具有明显的穿时性。

3）北京地区的石盒子组下部化石较少，也尚未发现具有地质时代鉴定意义的代表分子。在石盒子组的上部层位内，找到过不少植物化石种属，可归属于华夏植物群晚期，相当于李星学（1979）和盛金章（1982）所划分的二叠纪华夏植物群第3个组合，重要的代表分子如 *Lobatannularia ensifolia*，*Fas-*

cipteris hallei 等都有见及，其时代应为晚二叠世早期。

4）双泉组的植物化石仅发现于下部一段以紫红色调为主的层段中，在其中所夹的浅灰绿色细砂岩和黄绿色泥质岩层中产有植物化石（表 4-2）。在整个植物群中，蕨类植物占 39%（节蕨 7%，真蕨 32%），裸子植物占 54%（种子蕨 3.6%，苏铁 3.6%，科达 3.6%，银杏 7.2%，松柏 14.4%，分类不明的裸子植物 21%），分类不明的种子器官化石占 7%，说明当时蕨类植物与裸子植物都占很重要地位，而后者又略占优势。从植物群面貌来看，这是一个混生植物群，它兼有晚二叠世与早三叠世植物分子的双重面貌。因此推定其地质时代为晚二叠世至早三叠世。双泉组二段迄今未采获任何化石，无法确定其时代，只能依据其上、下地层的时代，以及与邻区相关地层的对比来推断其时代为中三叠世。而整个双泉组的时代大体为晚二叠世晚期至中三叠世。

表 4-2 北京地区石炭纪—中三叠世生物地层对比简表

Table 4-2 Carboniferous-Middle Triassic biostratigraphic correlation in Beijing

岩石地层	生物地层单位					年代地层	
	蜓类	非蜓有孔虫	腕足类	双壳类	植物	统	系
双泉组 P_3T_2s					蕨类植物：*Lobatannularia* cf. *multifolia*，*L. heiaensis*，*Pecopteris* aff. *lativenosa*，*P. orientalis*，*Ptychocarpus* sp.，*Neuropteridium coreanicum*，*N. nervosum*，*Chiropteris* sp.，*Gigantopteris shuangquanensis*，*Danaeopsis hughesi* 裸子植物：*Fascipteris* sp.，*Nilssonia* cf. *simplex*，*Cordaites* sp.，*Psygmophyllum multipartitum*，*Nephropsis* cf. *integerrima*，*Podozamites distans*?，*Ullmannia* sp.，*Albertia* sp.，*Taeniopteris tongshanensis*，*T. denssissima*，*T. norinii*，*T. spatulata* 种子化石：*Hallea pekinensis*	下—中三叠统 T_{1-2}	三叠系 T
石盒子组 $P_{2-3}s$					下部层位： *Pecopteris* cf. *orientalis*，*P.* cf. *hemitelioides*，*Calamites suckowii*，*C. cistii*，*Cordaites principalis*，*C. schenkii*，*Sphenopteris gothenii*，*Sphenophyllum oblongifolium* 上部层位： *Lobatannularia* cf. *lingulata*，*L. ensifolia*，*Pecopteris* sp.，*Fascipteris hallei*，*Taeniopteris szei*，*T.* cf. *multinervis*，*Psygmophyllum ginkgoides*，*P. multipartitum*，*Gigantopteris* sp.，*Neuropteridium polymorphum*，*Sphenophyllum minor*，*Cladophlebis* sp.，*Conchophyllum* sp.，*Radicites* sp.，*Tingia hamaguchii*，*Pelourdea reflexa*，*Sphenopteris* sp.，*Neuropteris* sp.，*Odontopteris* sp.，*Plagiozamites oblongifolius*，*Calamites* sp.，*Cordaites principalis*	中—上二叠统 P_{2-3}	二叠系 P

续表

岩石地层	生物地层单位					年代地层	
	蜓类	非蜓有孔虫	腕足类	双壳类	植物	统	系
山西组 C_2P_1s					下部层位： *Neuropteris ovata*, *N. plicata*, *N. otozamioides*, *N. gigantea*, *Sphenophyllum thonii*, *S. oblongifolium*, *S. laterale*, *S. verticillatum*, *Sphenopteris tenuis*, *Annularia orientalis*, *Cordaites principalis*, *Calamites suckowii*, *Alethopteris* cf. *huiana*, *Lepidodendron posthumii*, *L. incertum*, *Pecopteris feminaeformis* 上部层位： *Sphenophyllum laterale*, *S. oblongifolium*, *S. thonii*, *S. verticillatum*, *Calamites cistii*, *C. suckowii*, *Lobatannularia sinensis*, *Plagiozamites tungweiensis*, *Sphenopteris tenuis*, *S. gothanii*, *Annularia orientalis*, *Alethopteris hallei*, *Lepidodendron* sp., *Emplectopteris triangularis*, *Cordaites principalis*, *Pecopteris* sp.	下二叠统 P_1 / 上石炭统 C_2	石炭系 C
太原组 C_2t	*Fusiella paradoxa*, *Fusulina cylindrica*, *F. quasicylindrica*, *F. konnoi*, *F. schellwieni*, *Profusulinella parva*, *Fusulinella bocki*, *Boultonia rawi*, *Schubertella obscura*	*Bradyina* cf. *nautiliformis*, *Textularia eximia*, *Tetrataxis conica* var. *gibba*, *Schellwienis* sp., *Climacammima longissimoides*, *C. maximum*	*Choristites mosquensis*, *C. yanghukouensis*, *Dictyoclostus taiyuanfuensis*, *D. manchuricus*, *D. semireticulatus*, *D. gruenewaldti*, *Orthotetes regularis*, *Chonetes* sp., *Stenoscisma* sp., *Schellwienella crenistria*, *Chaoiella gruenewaldti*	*Euculopsis anthraconeitoides*, *Parallelodon chihliensis*, *Deltopecten multistriatus*, *Aviculopecten manchuricus*, *Acanthopecten* cf. *carboniferus*, *Allorisma regularis*, *Schizodus shansiensis*, *Pseudomonotis* sp., *Nuculopsis anthraconeiloides*	*Neuropteris gigantea*, *N. kaipingiana*, *N. pseudogigantea*, *Sphenophyllum oblongifolium*, *S. verticillatum*, *S.* cf. *emarginatum*, *Sphenopteris neuropteroides*, *Linopteris brongniartii*, *Pecopteris* cf. *cyathea*, *P. feminaeformis*, *P. acuta*, *Cordaites schenkii*, *C. principalis*, *Annularia* sp., *Lepidodendron galeatum*, *Putrella* sp., *Rhacopteris* cf. *busseana*, *Asterophyllites longifolius*	上石炭统 C_2	石炭系 C

第五章　上三叠统—白垩系

北京市晚三叠世—白垩纪地层分布很广，山区除东部平谷区以外都见有该套地层产出。西山地区集中分布于九龙山-香峪大梁向斜、北岭向斜、百花山向斜、髫髻山-妙峰山向斜及坨里、东岭等断陷盆地中；北部山区则大片分布于岔道、燕羽山、凤坨梁、四海、白河堡、千家店、密北新城子，以及昌平十三陵和怀柔山前等地（图5-1）。

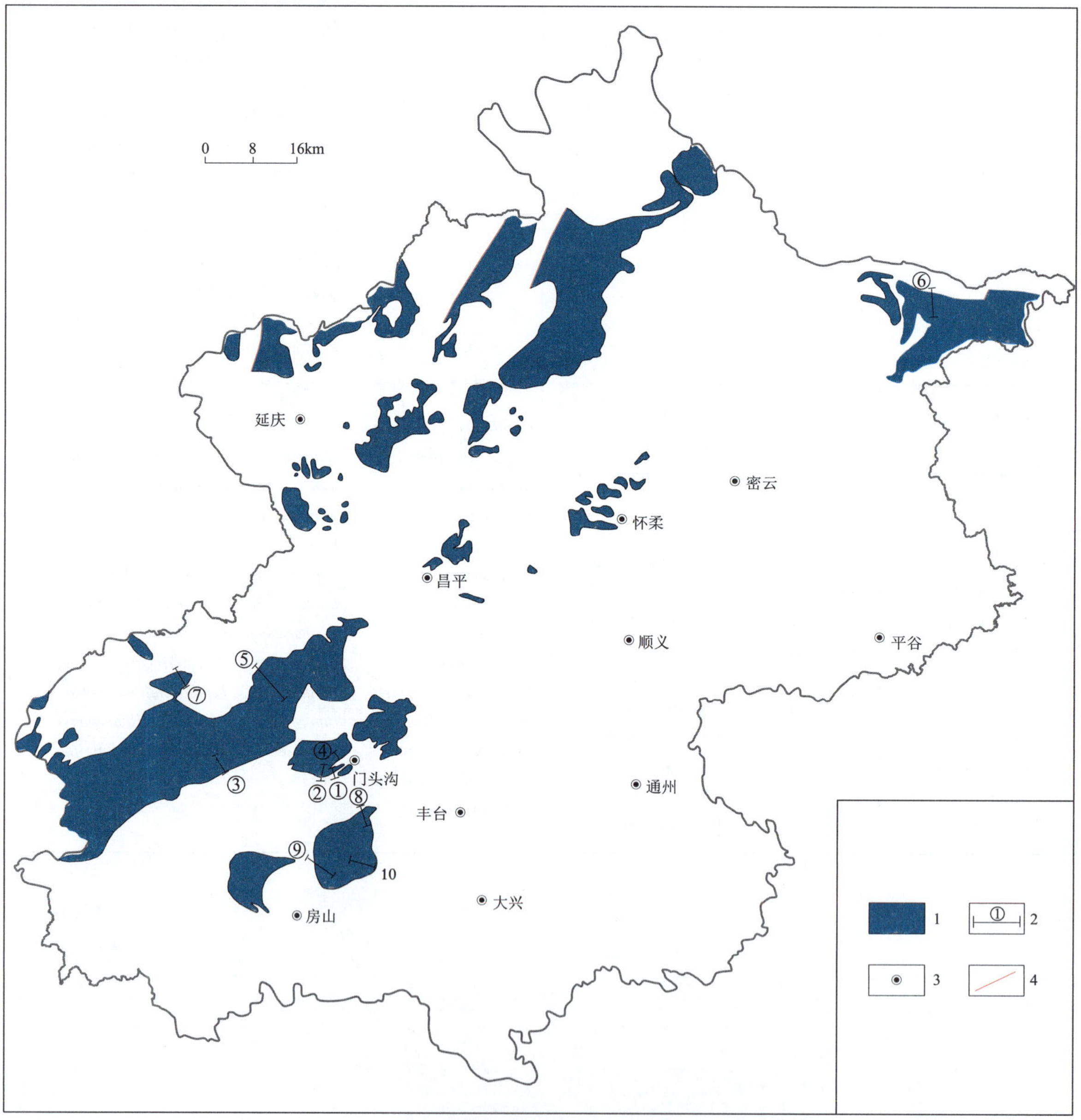

图5-1　北京市晚三叠世—白垩纪岩石地层及层型分布简图

（据《北京市岩石地层》，1996）

Fig. 5-1　Distribution map of Late Triassic-Cretaceous lithostratigraphic units and their stratotype section in Beijing

1—晚三叠世—白垩纪岩石地层；2—层型：①潭柘寺东山梁杏石口组剖面，②官厅-阳坡元南大岭组剖面，③马蹄沟窑坡组剖面，④岳家坡龙门组、九龙山组剖面，⑤田庄村东南髫髻山组剖面，⑥梁西-尚家-大岭土城子组剖面，⑦沿河城村张家口组剖面，⑧大灰厂西南靶场东狼沟组、九佛堂组剖面，⑨辛开口坨里组剖面，⑩芦尚坟-刘太庄夏庄组剖面；3—区级行政中心；4—断层叶良辅等（1920）所著《北京西山地质志》为中生代地层的划分研究奠定了基础。谢家荣（1933）、王竹泉和计荣森（1933）、陈恺和熊永先（1935）等老一代地质学家在西山开展了地层划分和古生物研究。第一轮 1：5 万区调补充完善了该地层系统。20 世纪 60 年代，先后两次对西山八大处虎头山下的杏石口组化石点进行采集，对鱼化石作了详细记述，命名为西山杏石口鳕，并把时代定位早侏罗世。米家榕等（1978，1980）、王自强等（1979）、洪友崇等（1982）、刘本培等（1982）、陈芬等（1984）、刘宪亭（1988）在西山地区开展了生物地层的研究工作，先后报道了地层和古生物的研究成果。萧宗正等（1984，1989）开展了北京中生代地层及生物群的专题研究，1994 年出版了《北京中生代地层及生物群落》。杨进辉（2006）、刘健（2007）、赵越（2006）、赵兴国（2009）等获得一批新的 SHRIMP U-Pb 或同位素测年等年龄数据。

20 世纪 70 年代中期以来，北京地区全面开展第二轮 1：5 万区调工作，解决了许多地层方面的重大问题，如在对坨里盆地详细调查研究之后，提出了取消“辛庄组”、肢解“芦尚坟组”，解决了该区地层问题上历来争论的难点，使岩石地层的划分更趋合理。鲍亦冈等（1996）对本断代地层进行了系统梳理，确立 12 个岩石地层单位。

结合近年来一些新的锆石测年数据，本志将张家口组、东狼沟组、九佛堂组地层分别放置到下白垩统早期、中期和晚期（表 5-1，表 5-2）。

第一节　岩石地层

1. 杏石口组（T_3x）

杨杰于（1957）原称“杏石口亚统”，创名地点在北京八大处杏石口村。1959 年北京矿业学院称杏石口组。以后的地质工作只是在地质时代的归属上有些不同看法，对其地层名称和顶底界线均无改变。

（1）代表剖面

正层型地层出露欠佳，现以门头沟区潭柘寺东山梁剖面作为选层型（东经 116°01′48″，北纬 39°54′19″。由北京市地质研究所地层古生物组 1990 年测制）。

上覆地层：南大岭组玄武岩

--------------- 平行不整合 ---------------

杏石口组　　　　总厚度 29.9m

7. 灰黑色细粒岩屑砂岩及粉砂岩　　1.5m
6. 浅灰色砾岩：砾石以石英为主，次为燧石，砾径最大为 7~8cm，分选性差，磨圆度较好，砂质胶结　　2.6m
5. 灰白色中层含细粒岩屑粉砂岩，层理发育　　4.4m
4. 灰色、灰黄色中—厚层含砾中—粗粒复成分砂岩，粒度变化较大　　3.4m
3. 浅黄色、黄色薄—厚层中粒岩屑砂岩，上部为灰色岩屑粉砂岩含化石：*Cladophlebis ichunensis*，*Czekanowskia* sp.，*Equisetites* sp.，*Podozamites lanceolatus*，*Taeniopteris* sp.，*Carpolithus* sp.　　7.9m
2. 灰色灰白色薄—中层巨粒—粗粒—细粒岩屑砂岩　　7.5m
1. 灰白色石英砾岩，砾石多为石英，燧石次之，砾径 4~8cm，分选性较好　　2.6m

~~~~~~~~ 角度不整合 ~~~~~~~~

下伏地层：双泉组浅黄色砂质板岩

（2）地质特征及区域变化

杏石口组岩性以灰至灰黄色砾岩及岩屑质砂岩为主，底部见有底砾岩（即宝珠洞砾岩），在八大处宝珠洞最发育，向西逐渐变薄。所含砾岩层数亦以东部为多，如大峪南山、冯村西山一带以砾岩、砂岩、粉砂岩组成的旋回性基本层序多而清晰，向西则泥质、碳质增高，至百花山东南的莲花庵一带
~~~~~~~~

表 5-1　北京市侏罗纪地层划分沿革表

Table 5-1　The classification history of Jurassic lithostratigraphic units in Beijing

叶良辅等 1920 西山	王竹泉 计荣森 1933 西山	《中国区域地层表（草案）》 1956 北京	杨杰 1957 西山	北京地质学院 1961 北京	《华北地区区域地层表·北京市分册》 1977 北京	陈芬等 1984 西山	《北京市区域地质志》 1991 北京	萧宗正等 1994 北京	《北京市岩石地层》 1996 北京	本志 北京
	九龙山系	K_1 东岭台组	K_1—J_3 髫髻山组	K_1 东岭台组	J_3 东岭台组	九龙山组	K_1 东岭台组	J_3 西山群 东岭台组	张家口组	K_1 张家口组
J_3 髫髻山组		J_3 髫髻山组		J_3 髫髻山组	J_2 后城组		J_3 后城组	后城组	土城子组	J_3 土城子组
					髫髻山组		髫髻山组	髫髻山组	髫髻山组	J_2 髫髻山组
J_1 九龙山系		J_2 九龙山组	J_3—J_2 九龙山组	J_2 九龙山组	九龙山组		J_2 九龙山组	J_2 门头沟群 九龙山组	九龙山组	九龙山组
门头沟煤系	J_1 龙门系	龙门组	J_1 门头沟统 龙门亚统	J_1 龙门组	J_1 门头沟统 龙门组	J_2 龙门组	龙门组	龙门组	龙门组	龙门组
	上窑坡系	J_1 上窑坡组	窑坡亚统	窑坡组 上段	窑坡组	上窑坡组	J_1 窑坡组	上窑坡组	窑坡组	J_1 窑坡组
	下窑坡系	下窑坡组		窑坡组 下段		J_1 下窑坡组		下窑坡组		
	辉绿岩	辉绿岩		南大岭组	南大岭组	南大岭组	南大岭组	J_1 蔡家岭组	南大岭组	南大岭组
								南大岭组		
		T 双泉组	杏石口亚统	T 双泉组	杏石口组	杏石口组	T_3 杏石口组	T_3 双泉群 杏石口组	杏石口组	T_3 杏石口组

表 5-2　北京市白垩纪地层划分沿革表

Table 5-2　The classification history of Cretaceous lithostratigraphic units in Beijing

谢家荣		陈恺 熊永先		北京矿业学院		北京地质学院		《华北地区区域地层表·北京市分册》		王自强 王濮		洪友崇 王思恩等		郝诒纯等		《北京市区域地质志》		郑桂森等		《北京市岩石地层》		本志	
1933		1935		1959		1961		1977		1979		1982		1986		1991		1994		1996		2018	
西山		西山		北京		北京		北京		西山		西山		西山		北京		西山		北京		北京	
E_2	长辛店砾石层			E_2c	长辛店组	E_2c	长辛店组	E_2c	长辛店组	E_2c	长辛店组	E_2c	长辛店组	E_2c	长辛店组	E_2c	长辛店组	E_2c	长辛店组	E_2c	长辛店组		
K_2	夏庄组			K_2	夏庄组	K_2	夏庄组	K_1	夏庄组	K_1	夏庄组	K_1	夏庄组	K_2	夏庄组	K_1	夏庄组	K_2	夏庄组	K_1	夏庄组	K_1	夏庄组
	芦尚坟页岩与砂岩				芦尚坟组		芦尚坟组		芦尚坟组		芦尚坟组		芦尚坟组		芦尚坟组		芦尚坟组	K_1					
	坨里砾岩				坨里组		坨里组		坨里组		辛庄组		坨里组		坨里组 辛庄组		坨里组		坨里组		坨里组		坨里组
K_1	辛庄紫页岩与砾岩			K_1	辛庄组	K_1	辛庄组	J_3	辛庄组		坨里组			K_1			辛庄组						
	大灰厂系	K_1	大灰厂组		大灰厂组		大灰厂组		大灰厂组	J_3	大灰厂组		大灰厂组		大灰厂组		大灰厂组		大灰厂组	K_1—J_3	九佛堂组		九佛堂组
	髫髻山层		东狼沟集块岩		东岭台组		东岭台组		东岭台组		东狼沟组	J_3	东狼沟组		东岭台组		东狼沟组		东狼沟组		东狼沟组		东狼沟组
			绿色流纹岩	J_3	髫髻山群								东岭台组	J_3			东岭台组				张家口组		张家口组
			东岭台组																				

泥岩增厚，并夹煤线。与下伏双泉组呈角度不整合接触。

该组地层展布于北京西山地区九龙山向斜的两翼和髫髻山向斜的南翼。厚度不大，一般为20~50m。

九龙山向斜两翼本组岩性岩相特征大同小异，横向变化不很显著。潭柘寺东山以砾岩、岩屑质石英砂岩为主，出露较全，上下界线清楚，植物化石丰富，厚30.8m；中门寺沟厚35.01m；大峪坡头附近，夹泥岩较多，厚度减为24.46m；八大处虎头山一带，岩性变化较大，以砂岩页岩为主，未见砾岩，厚度仅12.9m。

髫髻山向斜南翼，本组横向延伸不远，断续分布，一般几百米至千米左右即尖灭。大安山赵亩地剖面厚度最大，达26.55m，由岩屑质石英砂岩及粉砂岩等组成，含大量植物化石及少数叶肢介化石，上下界线清楚；向东至大安山瞧煤涧附近，砾岩增多，夹有“煤线”，亦产植物化石，厚16.83m；再向东的千军台黑阴沟一带，本组由灰黑色岩屑砂岩、黑色薄层状碳质粉砂岩组成，亦夹“煤线”，底部为薄层砾岩，全厚十几米碳质粉砂岩中含植物化石碎片；大台潘涧子沟一带，岩性以浅褐灰色、灰色砾岩及岩屑质石英砂岩为主，夹黑灰色薄层状粉砂岩，含植物化石碎片，厚22.03m。

2. 南大岭组（J_1n）

（1）代表剖面

以北京市门头沟区官厅-阳坡元南大岭组剖面为正层型（东经116°00′22″，北纬39°55′41″。由北京地质学院1959年在开展1∶5万石景山幅、良乡幅区调时测制）。

上覆地层：窑坡组绿泥石板岩

--------------- 平行不整合 ---------------

南大岭组	总厚度520.2m
27. 灰绿色玄武岩，杏仁体很多，直径1~5mm	30.0m
26. 灰绿色玄武岩	56.9m
25. 灰绿色玄武岩，气孔大	14.3m
24. 灰色玄武岩	103.1m
23. 暗紫色玄武岩	77.7m
22. 灰绿色玄武岩	10.4m
21. 灰绿色玄武岩，气孔中充填绿帘石	10.0m
20. 紫色玄武岩	6.7m
19. 灰绿色玄武岩	5.6m
18. 灰绿色玄武岩，气孔发育	23.0m
17. 灰绿色玄武岩	3.8m
16. 灰绿色玄武岩，气孔呈云朵状，被长石和绿帘石充填	5.6m
15. 暗紫色玄武岩	10.5m
14. 灰绿色玄武岩，气孔不发育	3.5m
13. 灰绿色玄武岩	6.4m
12. 暗紫色玄武岩	5.7m
11. 灰绿色玄武岩	8.0m
10. 灰绿色玄武岩夹灰色砂岩	3.8m
9. 灰绿色玄武岩	13.6m
8. 灰白色石英粉砂岩	4.7m
7. 灰绿色玄武岩	24.6m
6. 灰绿色玄武岩，具杏仁构造	4.6m
5. 紫灰色玄武岩	6.0m
4. 灰绿色玄武岩	3.7m
3. 紫褐色玄武岩	55.0m

2. 紫灰色玄武岩 23.0m

--------------- 平行不整合 ---------------

下伏地层：杏石口组黑色页岩

（2）地质特征及区域变化

南大岭组在正层型处厚520.2m，岩石组合简单，以火山溢流相玄武岩为主，夹少量凝灰质砂岩和页岩。整个玄武岩是由多次基性熔浆溢流堆积形成，每一次熔浆连续宁静溢流，形成一个有明显顶底界面可与上下岩层区分开的基本层。

该组主要分布于西山地区，西起门头沟区的清水村和房山区宝儿水村一带，东到颐和园和八宝山附近，范围约为1800km^2。另外在怀柔区南部年丰地区、顺义城区的东北及东南二十里长山地区，新生代沉积物之下亦见有南大岭组。该组岩层厚度以门头沟模式口最大，为676.6m，向西逐渐变薄以至尖灭。桑峪剖面521m，一段为灰绿色、紫灰色玄武岩，仅厚120m，顶部为黑色页岩和泥岩；二段为紫灰色、灰绿色玄武岩，安山岩、安山质熔岩和火山碎屑岩夹部分凝灰岩，厚401m。香峪剖面仅出露一段，厚338m，主要岩性为灰绿色、灰紫色玄武岩。南车营剖面厚92m，主要岩性为变拉斑玄武岩，变玄武岩。横岭西剖面厚110m，一段厚21.2m，主要为玄武岩，顶部为紫色凝灰质粉砂岩；二段厚88.7m，以角砾凝灰岩、集块熔岩及安山岩为主。南大岭组部分地区超覆于古生代不同地层之上。

3. 窑坡组（J_1y）

（1）代表剖面

由于王竹泉和计荣森（1933）于窑坡河谷所测剖面极为粗略，且无确切厚度，故以房山区马蹄沟剖面作为次层型（东经115°45′02″，北纬39°53′29″。由窦亚伟等1963年测制）。

上覆地层：龙门组底部砾岩

--------------- 平行不整合 ---------------

窑坡组	总厚度499.0m
二段	135.6m
63. 黑色粉砂岩	16.9m
62. 深灰色细砂岩	1.2m
61. 黑色粉砂岩，含植物化石 *Equisetites* sp.，*Phoenicopsis* sp.	13.1m
60. 煤层及浮土覆盖	5.8m
59. 灰色含砾中粗粒砂岩	11.1m
58. 黑色粉砂岩，含菱铁质结核	19.7m
57. 煤层及浮土覆盖	7.1m
56. 黑色粉砂岩夹细砂岩。含植物化石 *Coniopteris tatungensis*	11.1m
55. 深灰色细砂岩	3.0m
54. 灰色含砾中粗粒砂岩	11.1m
53. 黑色粉砂岩	1.4m
52. 深灰色细砂岩夹粉砂岩	4.4m
51. 煤层及浮土覆盖	6.2m
50. 深灰色细砂岩夹粉砂岩，含植物化石 *Phoenicopsis* sp.，*Podozamites lanceolatus*，*Pityophyllum* sp.	3.3m
49. 黑色粉砂岩，上部夹煤层，含植物化石 *Equisetites giganteus*，*Coniopteris hymenophylloides*，*Cladophlebis asiatica*，*Elatocladus manchurica*	3.9m
48. 灰色含砾粗砂岩	16.3m

———— 整　合 ————

一段 363.4m

47. 灰黑色细砂岩和粉砂岩互层，含植物化石 *Equisetites beijingensis* 6.6m

46. 黑色粉砂岩，含植物化石 *Phoenicopsis angustifolia* 2.6m

45. 深灰色中细粒砂岩 7.6m

44. 下部灰色中细粒砂岩，其上为黑色粉砂岩夹煤层，含植物化石 *Coniopteris hymenophylloides*, *Phoenicopsis angustifolia*, *Stenorachis lepida*, *Pityophyllum* sp. 2.6m

43. 底部灰黑色细砂岩，其上为黑色粉砂岩，含植物化石 *Pityophyllum* aff. *lindstroemi*, *P.* cf. *nordenskioldi*, *Stenorachis lepida* 3.4m

42. 黑色粉砂岩，含植物化石 *Neocalamites hoerensis*, *Coniopteris hymenophylloides*, *Cladophlebis asiatica*, *Sphenarion latifolia*, *Elatocladus manchurica* 2.5m

41. 灰黑色粉砂岩 2.1m

40. 黑色粉砂岩，含植物化石 *Equisetum laterale*, *Coniopteris hymenophylloides*, *C. tatungensis*, *Cladophlebis* sp., *Ginkgoites sibiricus*, *G. beijingensis*, *Sphenarion latifolia*, *Phoenicopsis angustifolia*, *Pityophyllum longifolium* 7.5m

39. 灰色中粗粒砂岩 20.1m

38. 黑色粉砂岩，含植物化石 *Equisetum laterale*, *Coniopteris hymenophylloides*, *Cladophlebis asiatica*, *Pterophyllum propinquum*, *Baiera concinna*, *B. furcata*, *Czekanowskia rigida*, *Sphenarion latifolia*, *Phoenicopsis angustifolia*, *Elatocladus manchurica*, *Podozamites lanceolatus*, *Taeniopteris* sp. 4.4m

37. 黑色粉砂岩夹煤线 2.6m

36. 深灰色细砂岩夹中砂岩 3.8m

35. 黑色粉砂岩 1.4m

34. 深灰色细砂岩 2.4m

33. 底部深灰色细砂岩，其上为黑色粉砂岩 3.8m

32. 下部深灰色细砂岩，上部黑色粉砂岩夹细砂岩条带 2.9m

31. 底部灰黑色细砂岩，其上为黑色粉砂岩 5.2m

30. 底部灰黑色细砂岩，其上为黑色粉砂岩夹细砂岩条带 2.9m

29. 底部深灰色细砂岩，其上为黑色粉砂岩，含植物化石 *Cladophlebis* sp., *Ctenis* sp., *Phoenicopsis angustifolia*, *Pityophyllum* aff. *lindstroemi*, *P. longifolium*, *Podozamites lanceolatus* 9.0m

28. 底部深灰色细砂岩，其上为黑色粉砂岩 4.6m

27. 底部深灰色细砂岩，其上为黑色粉砂岩夹煤层 6.7m

26. 底部深灰色细砂岩，其上为黑色粉砂岩，含植物化石 *Neocalamites hoerensis*, *Coniopteris hymenophylloides* 7.8m

25. 底部灰黑色细砂岩，其上为黑色粉砂岩，含植物化石 *Neocalamites hoerensis*, *Elatocladus* sp. 11.2m

24. 黑色粉砂岩，含植物化石 *Coniopteris tatungensis*, *Stenopteris* cf. *williamsoni*, *Czekanowskia rigida*, *Podozamites lanceolatus* 4.5m

23. 底部深灰色细砂岩，其上为黑色粉砂岩，含植物化石 *Coniopteris tatungensis*, *Pterophyllum* sp. 1.6m

22. 下部灰黑色细砂岩，上部黑色粉砂岩夹煤层 11.0m

21. 深灰色中细粒砂岩 19.3m

20. 灰黑色粉砂岩，下部夹细砂岩条带，上部夹媒层。含植物化石 *Cladophlebis* sp. 17.1m

19. 深灰色细砂岩，中上部夹粉砂岩。含植物化石 *Sphenopteris* sp. 7.3m

18. 下部灰色细砂岩，中上部黑色粉砂岩夹煤层。含植物化石 *Neocalamites hoerensis*, *Coniopteris hymenophylloides*, *C. tatungensis*, *Cladophlebis* sp., *Podozamites lanceolatus* 0.9m

17. 灰黑色粉砂岩夹细砂岩，含植物化石 *Ginkgoites lepidus*, *Baiera asadai*, *Czekanowskia rigida*, *Phoenicopsis angustifolia*, *Stenorachis lepida* 4.5m

16. 深灰色中粗粒砂岩 8.5m

15. 灰黑色粉砂岩夹细砂岩，顶部含煤 3.5m

14. 深灰色中粒砂岩 13.5m
13. 灰黑色细砂岩 1.8m
12. 黑色粉砂岩，顶部夹煤层，含植物化石 *Baiera furcata*，*Czekanowskia rigida*，*Pityophyllum* sp.，*Podozamites* sp. 1.5m
11. 深灰色含砾粗砂岩 13.7m
10. 灰黑色粉砂岩，顶部夹煤层，含植物化石 *Neocalamites hoerensis*，*Coniopteris hymenophylloides*，*Cladophlebis argutula*，*C.* sp.，*Baiera* sp. 9.9m
9. 浅灰黑色细砂岩 7.5m
8. 灰黑色粉砂岩夹细砂岩 25.0m
7. 浅灰黑色细砂岩 3.2m
6. 黑色粉砂岩 10.8m
5. 黑色含角砾凝灰岩 22.8m
4. 黑色粉砂岩夹煤层 5.8m
3. 灰黑色含角砾凝灰岩 25.8m
2. 浮土覆盖 15.6m
1. 硬绿泥石角岩，残留碎屑结构 4.6m

--------------- 平行不整合 ---------------

下伏地层：南大岭组辉绿岩

（2）地质特征及区域变化

窑坡组为陆相含煤地层，由暗色含砾粗砂岩、砂岩、粉砂岩、泥岩及煤层组成，以粉砂岩为主，总体上粒度较细。岩层沉积旋回结构清晰，每一旋回结构的下部多为深灰—灰黑色中粒或细粒砂岩，上部为灰黑色—黑色粉砂岩、泥岩，并产有煤层。

该组主要分布在西山地区：西起门头沟区清水村西及房山区宝儿水一带，东到石景山、碧云寺以东，北至门头沟牛站、青白口一线，南到房山区猫耳山南麓。另在京东的顺义牛栏山以东与二十里长山西侧之间的龙庭侯地区新生代沉积层之下亦见其分布。整体上表现为近东西向延展。一段一般厚300~360m，在门头沟九龙山南坡大村以东至龙门、河滩一带，厚度最大达460m，向四周厚度逐渐减小。其中，九龙山向斜豹窑沟厚220m；髫髻山向斜南翼，房山马蹄沟厚363m，潘涧子沟厚320m，千军台厚308m，大安山厚304m，大窑厚300m；髫髻山向斜北翼厚度薄，法城厚164m，桑峪厚138m；百花山向斜斋堂厚203m，清水九龙沟厚216m；南部北岭向斜南车营剖面厚166m。二段沉积时，盆地逐渐由南向北掀斜，沉积中心向北迁移，在斋堂、牛角岭、法城口等地均可见二段超覆于南大岭组火山岩之上。厚度一般300~120m，北部髫髻山、杨家峪一带二段岩层厚度超过300m，至门头沟区岳家坡剖面厚195m，而南部北岭地区剖面厚仅100m左右。

4. 龙门组（J_2l）

（1）代表剖面

正层型位于门头沟区岳家坡村北（东经116°02′55″，北纬39°05′22″。由王竹泉、计荣森1933年测制）。本志选取北京市地质调查所于1989年进行1∶5万石景山幅、良乡幅区调时重测岳家坡剖面，描述如下。

上覆地层：九龙山组复成分砾岩及砂岩

--------------- 平行不整合 ---------------

龙门组 总厚度374.74m

26. 土黄色、中厚层状长石砂岩 33.65m
25. 白色薄层状细—粉砂岩，成分为长石和石英 68.53m
24. 底部为土黄色中厚层状岩屑长石砂岩，中部为一层细砂岩。中上部为复成分细砾岩与岩屑长石砂岩及深灰色粉砂岩组成韵律 66.22m

23. 深灰色粉砂岩，顶部见一层 30cm 厚的细砂岩 34.2m

22. 深灰色中厚层状含晶屑凝灰质含细砾细砂岩与灰杂色复成分砾岩互层 18.33m

21. 灰黑色、薄层状粉砂岩。本层中见有铁、锰质结核 11.31m

20. 灰杂色薄层状复成分砾岩、夹中薄层状含砾岩屑粗砂岩 15.58m

19. 黑色薄层状泥质粉砂岩夹中薄层状土黄色泥质细砂岩 2.82m

18. 黄绿色—灰黄色薄层状泥岩、粉砂岩、细砂岩互层 20.59m

17. 黄绿色中薄层状粉砂质泥岩 17.49m

16. 灰白色中薄层状泥质粉砂岩夹薄层状含细砾粗砂岩。本层中见有大量的植物化石碎片，并见有铁、锰质透镜体 33.87m

15. 灰黑色薄层状粉砂岩 5.07m

14. 土灰色中—薄层状复成分砾岩与中层状含砾粗砂岩互层 18.54m

13. 灰黑色薄层状粉砂岩 4.55m

12. 巨厚层状复成分砾岩，砾石含量 80%，成分以砂、粉砂岩为主 23.99m

---------------- 平行不整合 ----------------

下伏地层：窑坡组长石砂岩

（2）地质特征及区域变化

龙门组主要由灰、灰绿、灰黑色复成分砾岩、岩屑砂岩、粉砂岩及粉砂质泥岩组成。在粉砂岩和粉砂质泥岩中常见菱铁矿结核。整体上看龙门组基本继承了窑坡组的沉积特征，发育由砾岩-砂岩-粉砂岩构成的完整沉积旋回结构。

龙门组分布范围大致与窑坡组一致，但在龙门组沉积时盆地西部抬升，因此龙门组分布的西部边界仅达门头沟区青龙涧至房山区秋林铺一带，沉积中心亦随之向东迁移，在门头沟区岳家坡本组最厚达 395m，向四周逐渐减薄，北岭地区为 330m，千军台—大台一带为 150~160m，更往西大安山附近为 75m，斋堂一带厚度仅 50m，厚度变化较大。

5. 九龙山组（J_2j）

（1）代表剖面

正层型为门头沟区岳家坡南山剖面（东经 116°02′55″，北纬 39°55′52″。由北京市地质调查所 1987 年重测），由于顶部剥蚀缺失，发育不全，以门头沟区板桥下港剖面作为次层型（东经 115°53′57″，北纬 39°57′10″。由姚文江 1964 年测制）。

上覆地层：髫髻山组安山岩

~~~~~~~~ 角度不整合 ~~~~~~~~

九龙山组 总厚度 1535.90m

三段 993.00m

25. 紫红色细砾岩，顶部为凝灰岩 15.50m

24. 紫红色凝灰质粉砂岩夹灰色凝灰质细砂岩 118.00m

23. 紫红、灰白色含火山角砾凝灰岩夹火山角砾岩 27.00m

22. 紫红、灰绿色凝灰质粉砂岩夹砾岩，底部为灰绿色凝灰质砾岩 97.00m

21. 紫红色凝灰质粉砂岩夹灰色砂岩 150.00m

20. 上部为灰绿、紫红色凝灰质粉砂岩，下部为灰白色凝灰质砾岩 57.00m

19. 灰绿、灰黑色凝灰质页岩，或与细砂岩互层，有时夹砂岩透镜体 38.00m

18. 凝灰质页岩 12.00m

17. 深灰色凝灰质页岩 32.00m

16. 灰绿色凝灰质粉砂岩夹凝灰质砂岩 107.00m

15. 灰白色凝灰岩及凝灰质细砂岩 8.00m

14. 灰绿色凝灰质粉砂岩 50.20m
~~~~~~~~

13. 灰白色凝灰质砾岩夹凝灰质粉砂岩或两者互层 30. 20m
12. 灰色凝灰质页岩，底部为细砾岩 17. 30m
11. 灰绿色凝灰质砂砾岩、粉砂岩、页岩互层，顶部为页岩。底部为砾岩 115. 00m
10. 灰绿与紫红色凝灰质页岩互层，底部为砂砾岩 49. 80m
9. 紫红色凝灰质页岩、粉砂岩，底部为砾岩 69. 00m

—————— 整 合 ——————

二段 297. 60m

8. 紫红色凝灰质页岩、粉砂岩，含泥灰岩结核，底部为凝灰质砾岩 185. 00m
7. 紫红色凝灰质页岩、粉砂岩、含泥灰岩结核，底部为砾岩 112. 60m

—————— 整 合 ——————

一段 245. 30m

6. 紫红色凝灰质页岩夹粉砂岩、砂砾岩及砂岩透镜休 96. 10m
5. 灰绿色凝灰质砂、页岩互层，顶部为层凝灰岩，底部为凝灰质砾岩 35. 50m
4. 灰绿色凝灰质粉砂岩，底部为凝灰质砾岩 18. 40m
3. 灰绿色凝灰质粉砂岩与砂岩互层，底部为凝灰质砾岩 37. 00m
2. 灰绿色粉砂岩、砂岩与砾岩 27. 90m
1. 灰绿色凝灰质砂岩，底部为灰绿色凝灰质砾岩 30. 40m

--------------- 平行不整合 ---------------

下伏地层：龙门组灰黄色粗砂岩

（2）地质特征及区域变化

九龙山组岩性主要为含有大量火山碎屑物质的各种不同粒级的陆相沉积碎屑岩，以紫红、砖红为主，夹有灰绿、灰黑、灰白等杂色，可划分为3个岩性段：一段由绿灰色砾岩-砂岩-粉砂岩组成沉积韵律；二段由紫红色凝灰质砾岩-砂岩-页岩构成沉积韵律，含泥质、钙质结核；三段由紫红色凝灰质砾岩-砂岩-粉砂岩构成沉积韵律。

该组主要分布于西山地区百花山、髫髻山向斜的两翼、九龙山向斜和北岭向斜的轴部。在昌平十三陵地区及怀柔范各庄附近也见小面积出露。

该组岩性及厚度变化均较大，以髫髻山南麓大台、板桥一带出露完整，厚度最大，可达1500m，往北至桑峪、牛站仅厚300m。昌平十三陵九龙山组厚度仅147m。在岩性变化方面，总的来看表现在剖面上由下往上岩层内所含凝灰物质增多，在平面上则由南而北火山物质亦有增高趋势。在本组沉积过程中，盆地逐渐向东北掀斜，使盆地东北缘青白口—田庄一带，本组超覆于寒武纪或中元古代地层之上。

6. 髫髻山组（$J_{2-3}t$）

（1）代表剖面

北京市地质调查所区调一队（1990）在原命名所在地选择露头好、地层连续的田庄东南剖面作为重测的正层型（东经115°54′05″，北纬40°04′50″）。

上覆地层：土城子组凝灰质粉—细砂岩

—————— 整 合 ——————

髫髻山组 总厚度 2822. 27m

四段 412. 15m

97-96. 灰色、紫色凝灰质砾岩、砂砾岩 82. 96m
95. 灰白色角砾凝灰岩 9. 20m
94-93. 紫灰至灰色凝灰质砾岩 132. 05m
92. 紫灰色粗安质集块角砾熔岩 23. 74m
91-90. 灰色、紫灰色凝灰质砾岩 64. 26m

89.	灰紫色粗安岩	8.54m
88-87.	灰紫色凝灰质砂砾岩	91.40m

—————— 整　合 ——————

三段		1679.97m
86-83.	紫红色粗安岩、粗安质角砾熔岩	203.25m
82.	凝灰质砂砾岩	10.50m
81-77.	紫灰色粗安质角砾熔岩、气孔状粗安岩	338.71m
76-74.	紫灰色岩屑凝灰岩夹粗安岩	38.31m
73.	紫灰色含火山角砾集块岩	60.42m
72-58.	暗灰色至紫色粗安岩、粗安质角砾熔岩，局部含集块或气孔、杏仁体	379.94m
57-56.	紫灰色玄武岩、含角砾玄武岩	68.34m
55-54.	灰紫色粗安质角砾熔岩、含集块粗安岩	10.19m
53-52.	紫色玄武岩、含集块玄武岩	45.01m
51-45.	灰紫色、灰绿色粗安岩、粗安质角砾熔岩	53.74m
44.	紫红色气孔杏仁状玄武岩	4.20m
43-40.	紫灰色粗安岩、粗安质角砾熔岩	72.33m
39.	紫灰色玄武岩	30.74m
38-36.	紫灰色粗安岩、角砾熔岩及黄绿色蚀变粗安岩	32.11m
35.	灰色玄武岩	35.27m
34-28.	灰紫色、灰绿色粗安岩、粗安质角砾熔岩夹凝灰质粉砂岩	296.91m

—————— 整　合 ——————

二段		679.66m
27-25.	灰紫色凝灰质砾岩、砂砾岩与火山角砾岩互层，顶部为凝灰质粉砂岩、晶屑凝灰岩	87.15m
24.	粗安质火山角砾岩	11.93m
23-20.	紫红色、黄绿色复成分砾岩、凝灰质含砾粗砂岩夹晶屑凝灰岩	106.08m
19-17.	灰紫色粗安岩、粗安质角砾熔岩、含集块火山角砾岩	50.09m
16-15.	暗紫色凝灰质砾岩、凝灰质细砂岩	41.98m
14-10.	灰紫色粗安质火山角砾岩、集块角砾岩	198.80m
9-6.	灰绿色、灰色复成分砾岩夹粗安质火山角砾岩	183.63m

—————— 整　合 ——————

一段		50.49m
5.	灰紫色熔结角砾岩	16.39m
4.	暗紫色火山角砾岩夹凝灰质粉砂岩	9.41m
3-1.	暗紫色、灰绿色气孔杏仁状玄武岩夹紫红色凝灰质粉砂岩	24.69m

~~~~~~ 角度不整合 ~~~~~~

下伏地层：九龙山组含砾凝灰质粗砂岩

（2）地质特征及区域变化

根据火山活动特点和岩性特征，本组可划分为 4 个岩性段，一段以粗安岩、粗安质角砾凝灰岩为主，见有少量玄武质、玄武安山质岩石；二段以灰紫、紫红色凝灰质砂岩为主，常夹有薄层熔岩和正常火山碎屑岩；三段以中性粗安质岩石为主；四段以火山-沉积碎屑岩为主，岩石粒度相对较粗，以复成分凝灰质砾岩、沉凝灰角砾岩为主。

髫髻山组火山岩在北京地区分布面积约为 1100km^2，范围内西端起自门头沟区百花山，向东北延至怀柔区长哨营和密云区新城子。髫髻山组厚度以燕羽山地区最大，厚度大于 4971m，十三陵地区厚度为 1134m。凤驼梁厚度为 4943m，四海盆地厚度为 1778m，百花山盆地厚度为 3442m，田庄厚度为 2622m，新城子盆地厚度为 1120m。
~~~~~~

7. 土城子组（J_3tc）

（1）代表剖面

选层型在辽西朝阳市花张营子、项家杖子。以北京市密云梁西-尚家-大岭剖面作为次层型（东经117°11′17″，北纬40°38′32″。由北京市地质调查所区调二队于1993年测制）。

石英正长岩

═══════ 断　　层 ═══════

土城子组	总厚度>1762.27m
三段	>582.46m
63. 砾岩	>187.36m
62. 复成分砾岩	141.04m
61. 砾岩	151.95m
60. 灰质砾岩	88.86m
59. 灰质砾岩夹中粒砂岩	13.25m

——— 整　　合 ———

二段	644.69m
58. 灰绿色粉砂岩	28.37m
57. 灰紫色钙质粉砂岩夹中粒砂岩	57.45m
56. 灰绿色钙质胶结含砾砂岩	29.45m
55. 灰紫色钙泥质胶结细砂岩	35.43m
54. 灰色中粒砂岩	25.31m
53. 灰紫色钙质粉砂岩夹细砂岩	18.33m
52. 绿色含钙质粉砂岩夹少量泥质粉晶灰岩	128.18m
51. 钙质胶结中粒砂岩	3.87m
50. 紫灰色细砂岩	11.60m
49. 灰绿色钙质胶结粉砂岩夹含砾细砂岩（含植物化石碎片）	60.94m
48. 灰紫色钙质粉砂岩	36.43m
47. 灰色钙质胶结含砾杂砂岩	6.36m
46. 灰紫色粉砂岩夹灰色粉砂岩	7.49m
45. 灰绿色粉砂岩	7.49m
44. 灰紫色粉砂岩	33.39m
43. 灰色钙质胶结含砾砂岩	4.65m
42. 紫灰色钙质粉砂岩	10.09m
41. 底部粗砂岩，向上变为细—粉砂岩，夹3层沉凝灰岩	8.20m
40. 粉砂质粉晶灰岩	35.15m
39. 紫灰色细砂岩夹薄层粗砂岩	17.27m
38. 灰紫色粉砂岩夹粉砂质泥质灰岩	79.24m

——— 整　　合 ———

一段	534.85m
37. 灰红色中细粒砂岩	1.60m
36. 粗砂岩、细砂岩、粉砂岩互层	33.14m
35. 底部含砾砂岩，向上渐变为细砂岩—粉砂岩	2.52m
34. 灰色中粒砂岩，上部与粉砂岩韵律性互层	11.06m
33. 灰紫色粉砂岩	4.79m
32. 紫灰色中粒砂岩	2.39m
31. 灰紫色粉砂岩	1.91m
30. 灰色中粒砂岩	2.87m

29. 灰紫色含砾砂岩 2. 39m
28. 含砾砂岩与灰紫色粉砂岩互层 13. 88m
27. 细砂岩与粉砂岩韵律性互层 12. 92m
26. 含砾砂岩夹紫灰色细砂岩（顶部互层） 25. 00m
25. 灰紫色粉砂岩 5. 03m
24. 含砾砂岩 2. 80m
23. 灰紫色粉砂岩 5. 03m
22. 中粒砂岩与灰紫色粉砂岩互层 50. 19m
21. 粗砂岩夹含砾粗砂岩，顶部夹粉砂岩 69. 14m
20. 灰紫色粉砂岩 3. 34m
19. 含砾砂岩夹粗砂岩 16. 01m
18. 灰紫色细砂岩 7. 68m
17. 灰白色细砂岩 6. 38m
16. 细砂岩夹含砾砂岩 14. 58m
15. 砾岩夹细砂岩 11. 42m
14. 细砂岩与含砾砂岩互层 47. 29m
13. 复成分砾岩夹含砾粗砂岩 14. 31m
12. 细砂岩 6. 47m
11. 复成分砾岩 79. 70m
10. 灰紫色凝灰质粉砂岩 4. 83m
9. 灰紫色粗面质岩屑凝灰岩 28. 20m
8. 灰绿色流纹质浆屑凝灰岩 9. 56m
7. 灰紫色凝灰质粉砂岩 24. 10m
6. 砾岩 4. 61m
5. 灰紫色凝灰质含砾砂岩 5. 12m
4. 灰质砾岩 4. 59m

———— 整　合 ————

下伏地层：髫髻山组灰紫色凝灰粉砂岩

（2）地质特征及域区变化

土城子组主要为一套紫红—灰绿色河流相沉积碎屑岩，划分为 3 个岩性段：一段由复成分砾岩、砂砾岩、凝灰质砂岩构成沉积韵律；二段碎屑粒度较细，由凝灰质砂岩、凝灰质粉砂岩、粉砂岩构成沉积韵律；三段以复成分砾岩、含砾粗砂岩为主，由复成分砾岩、含砾粗砂岩、凝灰质砂岩或砂岩构成韵律层。

该组与下伏髫髻山组呈整合或平行不整合接触。

土城子组主要分布于北山白河堡、千家店、新城子、四海-凤驼梁等火山沉积盆地之中，以及西山门头沟区髫髻山盆地的妙峰山一带，各地岩层厚度见表 5-3。

表 5-3　土城子组火山-沉积岩厚度统计表

Table 5-3　The statistical table about the thickness of volcanic-sedimentary rocks of Tuchengzi formation

地区	妙峰山	千家店	白河堡	新城子
第三岩性段	1454. 24	139. 8	135. 71	569. 26
第二岩性段	524. 49	929. 9	822. 56	908. 51
第一岩性段	761. 85	324. 3	285. 46	331. 18
全组统计	2740. 58	1394. 0	1243. 73	1808. 95

注：厚度单位为 m。

8. 张家口组（K_1z）

（1）代表剖面

北京市选取门头沟区沿河城村剖面作为次层型（东经 115°44′00″，北纬 40°04′00″。由北京市地质调查所 1980 年开展沿河城幅 1∶5 万区调时测制）。

上覆地层：东狼沟组紫红、灰绿色复屑凝灰岩

———— 整　合 ————

张家口组	总厚度 1642. 0m
三段	938. 0m
39. 灰紫色流纹质熔结凝灰岩	123. 0m
38. 红、灰绿色晶屑、玻屑凝灰岩	41. 0m
37. 浅灰紫色流纹质玻屑凝灰岩	31. 0m
36. 灰紫色流纹质熔结角砾岩	64. 0m
35. 灰紫色流纹质熔结凝灰岩	80. 0m
34. 紫红色、灰紫色凝灰质粉砂岩，含砾凝灰砂岩夹薄层流纹质熔结凝灰岩	93. 0m
33. 紫红色流纹质熔结凝灰岩	61. 0m
32. 杂色安山质集块角砾熔岩	5. 0m
31. 灰黑色安山岩	69. 0m
30. 浅紫红色玻屑凝灰岩	21. 0m
29. 紫红色凝灰质长石砂岩	12. 0m
28. 浅灰紫色流纹质晶屑、玻屑弱熔结凝灰岩	75. 0m
27. 浅灰紫色流纹质熔结角砾凝灰岩	210. 0m
26. 浅灰紫色流纹质含砾玻屑凝灰岩	21. 0m
25. 紫红色凝灰质粉砂岩	15. 0m
24. 上部流纹质熔结角砾岩，中部紫红色流纹质熔结凝灰岩，下部灰绿色流纹质熔结角砾凝灰岩	17. 0m
———— 整　合 ————	
二段	507. 0m
23. 上部杂色安山质砾岩、砂砾岩、中部紫红色层状凝灰质粉砂岩，下部紫红色凝灰质粉砂岩夹复成分砂砾岩	17. 0m
22. 灰绿色凝灰质砂岩、粉砂岩夹复成分砂砾岩	6. 0m
21. 深灰色紫灰色复成分砾岩夹凝灰质砂岩	85. 0m
20. 浅灰、灰紫色复成分砾岩	67. 0m
19. 紫红色流纹质晶屑、玻屑弱熔结凝灰岩	23. 0m
18. 深灰、灰紫色复成分砾岩夹紫红色英安质玻屑凝灰岩	34. 0m
17. 灰绿色流纹质晶屑、玻屑弱熔结凝灰岩	17. 0m
16. 灰紫色复成分砾岩	200. 0m
15. 灰紫色英安质晶屑、玻屑弱熔结含砾凝灰岩	18. 0m
14. 紫红色英安质熔结凝灰岩	19. 0m
13. 紫褐色复成分砾岩夹红色泥岩、砂砾岩	21. 0m
———— 整　合 ————	
一段	197. 0m
12. 紫红色安山质火山角砾岩	3. 0m
11. 灰黑色玄武安山岩夹火山角砾岩	33. 0m
10. 杂色安山质火山角砾岩	1. 0m
9. 红色粉砂岩	4. 0m
8. 杂色安山质火山角砾岩	2. 0m
7. 灰绿、灰黑色玄武安山岩	14. 0m
6. 灰紫色粗安质角砾熔岩	40. 0m
5. 紫红色凝灰质粉砂岩	2. 0m
4. 灰紫色玄武安山岩	23. 0m
3. 紫红色凝灰质含砾粉砂岩	2. 0m

2. 灰紫色玄武安山岩 48.0m

1. 灰、灰白色砾岩 25.0m

~~~~~~~~ 角度不整合 ~~~~~~~~

下伏地层：洪水庄组黑色页岩

（2）地质特征及区域变化

张家口组划分为3个岩性段。一段以玄武安山岩、粗安岩、粗安质角砾凝灰岩等中性火山喷出岩为主；二段为火山活动间歇期形成的复成分砾岩、砂砾岩，夹多层弱熔结凝灰岩；三段以流纹质熔结凝灰岩、流纹质角砾凝灰岩及流纹岩为主。该组只在七道河盆地见到与土城子组呈不整合接触，其余地区多不整合于髫髻山组、九龙山组、南大岭组以及古生代甚至中—新元古代地层之上。

张家口组分布于东岭、杜家庄—镇厂、上大水、岔道、怀柔、七道河等小型火山盆地或断陷盆地内，昌平棉山、密云五指山也有小面积出露。本组的厚度变化较大（表5-4）。

表5-4 张家口组火山岩厚度统计表

Table 5-4 The statistical table about the thickness of volcanic rocks of Zhangjiakou formation

地区	东岭盆地	上大水盆地	岔道盆地	怀柔盆地	杜家庄镇厂盆地	昌平棉山	七道河盆地
第一段	927.9	>338.0	>529.0	1596.3	2234.2	763	769
第二段	496.0	61.0	338.5	1175.3	29.9		
第三段	197.7	200.0	879.6	1444.5			
全组	1621.6	599.0	1797.1	4216.1	2264.1	763	769

注：厚度单位为m。

9. 东狼沟组（K_1d）

（1）代表剖面

以北京市丰台区大灰厂西南靶场剖面作为选层型（东经116°05′48″，北纬39°51′30″。由北京市地质调查所1989年在开展1∶5万石景山幅、良乡幅区域地质调查时测制）。

上覆地层：九佛堂组黄绿色泥岩

———— 整　合 ————

东狼沟组 总厚度167.85m

二段 117.34m

17. 灰绿色含气孔杏仁状玄武岩 24.82m

16. 黄色凝灰质中厚层含砾岩屑砂岩 0.73m

15. 绿灰色凝灰质泥质粉砂岩及泥岩 2.18m

14. 灰色中层凝灰质粉砂岩与黄色中层凝灰质中粒岩屑砂岩互层 34.70m

13. 灰黄色凝灰质泥质砂砾岩，紫色凝灰质岩屑砂岩 1.58m

12. 紫色凝灰质粉砂岩，灰黄色砂砾岩及砾岩，砂岩层中含化石：腹足类 *Viviparus* sp., *Probaicalia vitimensis*, *Lioplacades* cf. *choluokyi*, *Ammicola* cf. *yunanensis*, *Trochispira donglanggouensis*, *Galba* cf. *tongshanensis*；介形虫 *Cypridea unicostata*, *C. faveolata*, *C. usualis*, *C.*（*C.*）*habros*, *C.*（*C.*）*subscariformis*, *Mongolianella* sp.；双壳类 *Ferganoconcha subcentralis*, *F.* sp.；鱼类 *Sinamia* sp. 等和植物化石碎片 4.56m

11. 灰黄色泥质细砂岩含生物化石碎片 1.01m

10. 灰黄色凝灰质岩屑砂岩，紫色凝灰质粉砂岩 5.57m

9. 灰黄色凝灰质砂砾岩 0.97m

8. 灰绿色致密块状玄武岩，上部为紫色气孔状玄武岩 7.13m

7. 紫色凝灰质砂砾岩 0.65m
~~~~~~~~

6. 灰黄色凝灰质含砾岩屑砂岩 0.65m

5. 灰紫色气孔状玄武岩与灰绿色气孔状玄武岩互层 32.79m

——— 整 合 ———

一段 50.51m

4. 紫灰色凝灰质火山沉角砾岩夹凝灰质岩屑砂岩 28.11m

3. 灰紫色中厚凝灰质砾岩夹凝灰质岩屑砂岩，顶部是一层厚 30cm 的灰紫色玄武岩 7.47m

2. 灰紫色厚层凝灰质砾岩夹紫色岩屑砂岩，砾石中含一定量的白云岩 14.93m

~~~~~~ 角度不整合 ~~~~~~

下伏地层：太原组变质砂岩

（2）地质特征及区域变化

东狼沟组为一套火山-沉积岩系，岩性主要为含集块角砾凝灰岩、凝灰岩、沉凝灰岩、凝灰质砂岩、复成分砾岩，夹有数层玄武岩或粗安岩。在东岭等盆地内为厚层粗安岩。

该组主要出露于坨里火山沉积盆地的北部大灰厂村附近，厚度 420m。在东岭盆地的西北部王龙口村以西，杜家庄-镇厂盆地的西部，以及怀柔凯甲坟附近出露面积都较小，东岭盆地厚度为 507.5m，怀柔盆地厚度为 134.87m，杜家庄-镇厂盆地厚度为 175.22m。

10. 九佛堂（K_1j）

（1）代表剖面

丰台区大灰厂西南靶场剖面作为次层型（东经 116°06′07″，北纬 39°50′55″。由北京市地质调查所区调一队于 1989 年测制）。

上覆地层：坨里组黄色砂砾岩

——— 整 合 ———

九佛堂组 总厚度 28.05m

12. 灰黄色厚层泥质粉砂岩 4.86m

11. 灰黑色碳质页岩 1.94m

10. 灰色碳质页岩夹黑色薄层粉砂岩及黄色页岩，含化石：叶肢介 *Eosestheria* cf. *peipiaoensis*, *E.* cf. *middendorfii*, *E. fengtaiensis*, *E.* cf. *persculpta*, *E.* sp., *Diestheria longinqua*, *D. dahuichangenesis*, *D.* cf. *yixianensis*, *D. suboblonga*, *Mesolimnacliopsis anguangensis*, *Yanjiestheria yanjiensis*，昆虫 *Ephemeropsis trisetalis* 4.24m

9. 土黄色薄层粉砂质泥岩 4.11m

8. 灰色页岩夹黄绿色粉砂质页岩，含动物化石碎片 2.11m

7. 灰黄色粉砂质泥岩，可见化石碎片 4.33m

6. 灰黑色碳质页岩夹灰黄色钙质页岩，含动物化石 0.87m

5. 灰色页岩，含化石碎片 1.91m

4. 黄绿色页岩 1.01m

3. 灰黑色碳质页岩夹薄层粉砂岩 2.11m

2. 灰黑色页岩 0.56m

——— 整 合 ———

下伏地层：东狼沟组凝灰岩

（2）地质特征及区域变化

九佛堂组岩性主要为黑色、灰色、黄绿色页岩、粉砂岩、泥岩，夹岩屑砂岩、细砂岩。页岩含有大量有机质。与上下地层均为整合接触。

九佛堂组仅于丰台区坨里盆地，在大灰厂村附近呈 NE-SW 向带状出露，总面积不足 0.5km^2。
~~~~~~

11. 坨里组（K_1t）

（1）代表剖面

以房山区辛开口剖面为选层型（东经 116°01′20″，北纬 39°47′18″。由北京市地质调查所区调一队于 1989 年测制）。其中坨里组二段利用沙锅村东−岗上剖面，坨里组一段则利用张庄丰浅 2 孔的钻孔资料拼接而成。

上覆地层：夏庄组灰黄色复成分砾岩

———— 整 合 ————

坨里组	总厚度 741.45m
三段	396.67m
44. 紫红色、灰绿色泥质粉砂岩夹薄层黄色砂岩，顶部含砾	7.07m
43. 灰黄色中层复成分砾岩，砾石成分以沉积岩为主，含少量火山砾石	1.86m
42. 灰紫色、黄色含砾岩屑粗砂岩与细砂岩互层	7.22m
41. 灰色中层复成分砾岩与含砾砂岩互层夹白色中薄层泥质粉砂岩	5.44m
40. 中层紫色含砾砂岩夹中薄层泥岩	3.12m
39. 紫红色含砾岩屑砂岩，泥、钙质胶结，夹一层 20~30cm 厚的粉砂岩	0.73m
38. 灰紫色巨厚层复成分砾岩，夹泥质细砂岩和含砾细砂岩	48.05m
37. 深灰紫色中薄层泥质细砂岩夹复成分砾岩及含砾粗砂岩	2.63m
36. 灰绿色复成分砾岩夹深灰紫色中薄层含细砾泥质砂岩	21.54m
35. 黄色中层砂砾岩夹黄绿—灰白色薄层含砾泥质细砂岩	4.21m
34. 灰紫色复成分砾岩夹深灰紫色中薄层含泥质砂岩	29.98m
33. 深灰紫色中薄层含砾泥质中细粒砂岩与灰紫色中层复成分砾岩呈韵律层	14.88m
32. 灰紫色巨厚层复成分砾岩	41.44m
31. 黄色中层砂砾岩	1.39m
30. 灰紫色巨厚层复成分砾岩	9.43m
29. 黄色厚层复成分砾岩与黄色中层复成分砾岩及黄色中层泥质砂岩互层	11.24m
28. 灰紫色巨厚层复成分砾岩夹中薄层含砾砂岩及泥质砂岩	15.58m
27. 黄紫色土黄色厚层复成分砾岩、砂岩互层	55.85m
26. 灰紫色薄层砂砾岩与含砾砂岩互层	6.05m
25. 杂色复成分砾岩	3.81m
24. 灰紫色中层复成分砾岩与紫色砂质泥岩及灰白色中粒含砾砂岩互层	1.59m
23. 灰紫—杂色复成分砾岩	42.86m
22. 黄色中薄层粗砂岩	27.08m
21. 紫色中层复成分砾岩与灰色中层含砾粗砂岩及灰白色中粒砂岩组成韵律层	25.08m
20. 灰紫色巨厚层复成分砾岩	8.54m
———— 整 合 ————	
二段	246.78m
19. 灰紫—灰色中薄层含砾粗砂岩与含砾细砂岩互层	11.07m
18. 灰紫色中薄层含砾粗砂岩与土黄色中薄层细砂岩互层	32.52m
17. 灰—灰紫色中层砂砾岩与灰白色中层含砾砂岩互层	29.99m
16. 黄色中薄层砂岩，含砾砂岩与灰—灰紫色中厚层砂砾岩互层	13.62m
15. 土黄色中薄层粉—细砂岩夹土黄色薄层砂砾岩	9.72m
14. 灰—灰紫色中层砾岩夹土黄色薄层含砾砂岩	9.36m
13. 灰—灰紫色中层含砾粗砂岩与土黄色薄层泥质粉砂岩互层	14.13m
12. 灰—紫色复成分砾岩夹薄层含砾粗砂岩	4.05m
11. 土黄色中层含砾粉砂岩与含砾粗砂岩互层夹少量薄层砂质砾岩	13.99m
10. 土黄色中层含砾细砂岩与粉砂岩互层夹薄层砂质砾岩	26.22m

9.	灰紫色中层复成分砾岩与中薄层泥质粉砂岩互层，泥裂发育	44.93m
8.	灰黄色中层泥质粉砂岩与灰紫色中层复成分砾岩互层	16.16m
7.	土黄色巨厚层复成分砾岩，局部夹砂砾岩和薄层含砾细砂岩	3.16m
6.	灰紫色薄层复成分砾岩夹中薄层含砾泥质砂岩	14.26m
5.	灰紫色中厚—巨厚层复成分砾岩，顶部为一层 50cm 泥质粉砂岩	3.60m

——————— 整　合 ———————

一段 98.00m

4.	黄灰色砾岩。灰—褐色中细砂岩，夹褐棕色泥质粉砂岩、粉砂岩	22.00m
3.	褐棕色粉砂岩，泥质岩，夹褐棕色粉砂岩及褐灰色中砂岩	27.00m
2.	黄灰色砾岩，含砾砂岩，深灰—绿色细砂岩夹黑灰粉砂岩	21.00m
1.	褐灰色粉砂质泥岩夹灰色粗砂岩	28.00m

——————— 整　合 ———————

下伏地层：九佛堂组砂岩及页岩

（2）地质特征及区域变化

坨里组为巨厚层粗碎屑岩系，可分 3 个岩性段。一段主要为黄色至褐色砂砾岩层夹粉砂岩及粉砂质泥岩；二段主要为紫色与黄色互层的砾岩、砂砾岩，即前人所称的“黄色坨里砾岩”；三段主要为杂色复成分砾岩夹各类细粒级的碎屑岩。

该组仅发育于坨里断陷盆地内，见于坨里、南坊、岗上、羊圈头、辛开口一带，出露面积约 30km^2，剖面厚度大于 760m。

12. 夏庄组（K_1x）

（1）代表剖面

以北京市丰台区芦尚坟-刘太庄剖面为夏庄组的选层型（东经 116°06′11″，北纬 39°46′17″。由北京市地质调查所 1989 年在开展 1：5 万石景山幅、良乡幅区调时测制）。

上覆地层：长辛店组砾岩

～～～～～～ 角度不整合 ～～～～～～

夏庄组 总厚度 642.17m

二段 295.99m

65.	黄色厚层含砾砂岩	0.54m
64-60.	底部为黄色含细砾岩屑砂岩，向上为薄层土黄色粉砂岩、泥岩	43.84m
59-56.	底部为土黄色中粒砂岩，向上为中薄层钙质粉砂岩夹薄层泥灰岩	28.70m
55-53.	下部为复成分砾岩，上部由粗砂岩、细砂岩变为杂色页岩	13.19m
52-50.	灰黄色中厚层中粒砂岩夹含砾粗砂岩	13.45m
49-47.	下部为黄色中粒砂岩，中部为杂色页岩，上部为灰黄色泥质粉砂岩	9.13m
46-41.	底部为含砾粗砂岩夹细砂岩，向上为黄褐色粉砂岩、细砂岩、杂色页岩	139.01m
40-37.	复成分砾岩，向上为灰白色泥灰岩与杂色页岩互层夹中薄层细砂岩	48.13m

——————— 整　合 ———————

一段 346.18m

38-34.	下部为含砾砂岩与粉砂岩互层，上部为泥质粉砂岩夹细砂岩	24.24m
33-30.	底部为土黄色厚层含砾粗砂岩，上部为浅黄色细砂岩、泥质粉砂岩	55.67m
29-28.	下部为褐黄色含砾粗砂岩。上部为浅黄色细砂岩、泥质粉砂岩	30.69m
27-23.	底部为灰紫色中层粗砂岩、复成分砾岩，向上为土黄色细砂岩、粉砂岩夹黑色页岩	23.75m
22-19.	底部为紫色含细砾细砂岩，向上为灰黄色中薄层粉砂岩与细砂岩互层	49.48m
18-17.	下部为灰紫色含砾粉砂岩，上部为深灰色泥质粉砂岩夹细砂岩，产 *Ferganoconcha* sp.	33.82m
16-15.	底部为灰紫色粉砂岩夹含砾粗砂岩，中上部为灰黄色砂岩与粉砂岩互层	37.88m
14-13.	下部为土黄色含砾粗砂岩与粉砂岩互层，上部为细砂岩与粉砂岩互层	3.06m

12-8.	底部为褐黄色含砾粗砂岩，往上变为灰黄色、黄绿色中薄层粉砂岩、泥岩、钙质页岩	22.88m
7-6.	下部为复成分砾岩，上部为土黄色薄层砂岩、粉砂岩、泥岩	9.40m
5-3.	底部为黄色复成分砾岩，向上为灰绿、灰黄色砂岩、细砂岩、泥质粉砂岩互层	48.18m
2-1.	下部为灰黄色巨厚层复成分砾岩，上部为灰白色泥质粉砂岩、细砂岩互层	7.13m

———— 整　合 ————

下伏地层：坨里组粉砂岩、砾岩

（2）地质特征及区域变化

夏庄组是一套黄色、黄绿色岩屑砂岩、岩屑粉砂岩、泥岩、页岩，夹有少量复成分砾岩、含砾岩屑砂岩及泥灰岩。可分为两个岩性段：一段由复成分砾岩-含砾粗砂岩-砂岩-粉砂岩-泥岩和含砾砂岩-粉砂岩-泥岩组成旋回性基本层序；二段主要由砂岩-粉砂岩-泥岩（或泥灰岩）组成旋回性基本层序。

该组分布在常乐寺—青龙头水库—大苑村以东，沙锅村—西王佐以西，出露面积约 10km^2。构造上仅限于坨里断陷盆地内。

第二节　生物地层

对于北京地区中生代动植物化石的研究，有不少学者的贡献，积累了不少资料。其中，萧宗正等于 1994 年出版了《北京中生代地层及生物群》，本志在收集、归纳前人研究成果基础上，对北京地区中生代地层中的化石进行了采集、鉴定和研究等工作，在生物群和组合建立、各门类生物演替关系、时代确定和区域对比方面进行了探讨。本部分的主要内容源于此书。

一、侏罗纪古生物群组合

（一）双壳类

北京地区侏罗纪双壳类化石的发现与研究均较晚，已知层位有窑坡组及土城子组，产地已有 10 余处。

产于窑坡组下段的双壳类化石，发现较晚，最初由刘本培（1982）、杨守仁（1982）研究和报道，萧宗正（1994）又在新的地点发现产双壳类化石的层位；萧宗正（1994）还在延庆区花盆乡土城子组二段黑色页岩层内的灰色含细砾砂岩中发现大量双壳类化石。

萧宗正（1994）综合以上各家资料，将北京地区侏罗纪双壳类化石由早到晚划分为下列组合：

1）*Naiadites-Shaanxiconcha-Tutuella* 组合，产于窑坡组下段，其中 *Naiadites* cf. *krasnojarskiensis*，*N. mentougouensis* 为早侏罗世所特有；*Shaanxiconcha* cf. *clinovata*，*S.* aff. *longa*，*S. shijiayingensis* 由晚三叠世延续而来；*Tutuella* 等则常见于早—中侏罗世。故将该组合时代定为早侏罗世。

2）*Unio-Margaritifera* 组合，产于土城子组，在该组合中未见我国中侏罗世的常见分子 *Pseudocardinia*，*Tutuella* 等，也未出现北方晚侏罗世地层中的 *Ferganoconcha*，*Sphaerium* 等属分子，时代为晚侏罗世早期。

（二）叶肢介类

萧宗正（1994）首次在延庆区花盆乡土城子组二段黑色页岩内发现大量叶肢介化石，并将其称为 *Nestoria-Yanshanoleptestheria* 叶肢介群，时代为晚侏罗世早期。其大部分分子见于冀北、辽西的晚侏罗世早期地层中。

（三）昆虫类

萧宗正等（1994）在延庆区花盆村西北土城子组中发现有小型、丰富多样的昆虫化石，经鉴定有：*Rhipidoblattina hebeiensis*，*Triassoblatta fusiformis*，*T. damiaoliangensis*，*Sinonitidulina liugouensis*，*S.*

sp., *Fusoblatta arcuata*, *F.* cf. *arcuata*, *Necrotaulius* sp., *Necropsis paludis*, *N.* sp., 经分析，这个以蜚蠊目、鞘翅目、毛翅目等为主的昆虫组合，其面貌与九龙山昆虫组合（洪友崇，1983）极为相似。萧宗正等还指出，洪氏所指含九龙山昆虫组合的地层应属冀北的“后城组”（土城子组），或属“承德砾岩”，不应为九龙山组。其时代为晚侏罗世。暂以花盆昆虫组合称之，以沟蠊 *Rhipidoblattina* 为代表。

之后洪友崇、萧宗正（1997）又在延庆区花盆村土城子组二段发现了新的昆虫种属，填补了本区土城子组昆虫的空白，充实了土城子组昆虫群的内容，并根据昆虫的特色，建立了一个新的昆虫组合——延庆昆虫组合，并以 *Rhipidoblattina*（*Canaliblatta*）*yangqingensis*-*Samaroblattula houchengensis*-*Protorthophlebia yangqingensis* 组合，反映本区昆虫的特色。将延庆昆虫组合归属后城昆虫群，时代为晚侏罗世早期。

（四）恐龙足迹

张建平（2012）在北京市延庆硅化木国家地质公园核心区土城子组地层发现了大批恐龙足迹，初步研究表明可归属于覆盾甲龙类（cf. *Deltapodus* sp.）、兽脚类、鸟脚类及疑似蜥脚类恐龙足迹。其中大量的覆盾甲龙类足迹表明，早在晚侏罗世，冀北-辽西地区就活动着覆盾甲龙类，它们承接了辽宁朝阳下白垩统义县组发现的甲龙类化石。延庆发现的恐龙足迹群不但是北京地区首次发现恐龙活动的证据，而且极大丰富了土城子组的恐龙类群，为承接其后的热河动物群提供了绝好的演化样本。

（五）古植物

门头沟植物群是一个以 *Coniopteris*-*Phoenicopsis* 为典型代表的著名侏罗纪植物群。斯行健（1931，1933，1956，1962）对门头沟植物群进行过描述。陈芬等（1984）将西山侏罗纪门头沟植物群划为上、下两个组合。下组合为 *Coniopteris hymenophylloides*-*Baiera* 组合，见于窑坡组下段，时代为早侏罗世晚期。上组合为 *Coniopteris hymenophylloides*-*Anomozamites thomasi* 组合，见于窑坡组上段和龙门组，时代为中侏罗世。而萧宗正则将“上、下窑坡组”一起归入中侏罗世。

萧宗正等（1994）将门头沟植物群重新划分为 3 个植物组合及 6 个亚组合：

1. 下组合（南大岭组）

在芦师山、金鸡台等地采到少许 *Podozamites* sp., *Czeknowskia* sp. 等化石，因化石属种较少，植物组合面貌不清。暂借用王自强（1984）划分的 *Otozamites hsiangchieneis* 为此组合名称。

2. 中组合（窑坡组—九龙山组）

此组合化石丰富，为门头沟植物群主体，依化石出现的层位和属种特征，又分为 3 个亚组合。

1）窑坡组下段亚组合，以 *Coniopteris hymenophylloides*-*Baiera* spp. 为代表。据陈芬等（1984）研究与总结，北京西山该亚组合植物化石共有 73 种，以蕨类植物门及银杏纲最繁茂，其次为松柏纲及苏铁纲。蕨类植物有 28 种，银杏植物有 21 中，松柏和苏铁植物各占 10 种左右。

2）窑坡组上段亚组合，以 *Coniopteris hymenophylloides*-*Ctenis kaneharai* 为代表。该亚组合植物较窑坡组下段亚组合贫乏，约 41 种。总体上，蕨类植物门仍占主导地位，共有 14 种；银杏纲大为减少，只有 10 种，*Baiera* 已少见；苏铁 8 种，松柏 6 种，两种的种数相比窑坡组下段亚组合变化不大。

3）龙门组亚组合，以 *Coniopteris hymenophylloides*-*Nilssoniopteris vittata* 为代表。该亚组合植物化石有 27 种，仍以蕨类植物门为主，约有 15 种；银杏纲急速减少，只有 1 种；苏铁纲 6 种；松柏纲也略有减少，有 5 种。

3. 上组合（髫髻山组、土城子组及张家口组）

植物发育程度远逊于下、中组合，但组合分子大体上仍属于 *Coniopteris*-*Phoenicopsis* 植物群范畴，结合其他化石资料及年龄成果，时代应为晚侏罗世。依据层位及化石特点，暂时划分 3 个亚组合。

1）髫髻山组亚组合：暂称为 *Coniopteris hymenophylloides*-*Elatides* sp. 亚组合。

2）土城子组亚组合：植物化石只有零星产出，但硅化木化石丰富。暂称为 *Coniopteris hymenophylloides*-*Xenoxylon* 亚组合。张武等（2000）发现木化石的一个新种——延庆苏格兰木 *Scotoxylon yanqingense*，位于千家店镇下德龙湾村土城子组二段。

3）张家口组亚组合：偶见植物化石碎片，在门头沟区东岭北部采到有节类化石，初步定为 *Neocalamites*？sp.，但保存欠佳，鉴定困难。

二、白垩纪古生物群组合

北京西山白垩系含动植物化石门类较多，数量丰富，演化阶段清楚，时代意义明确。其中，无脊椎动物有双壳类、腹足类、叶肢介、昆虫及介形虫类等；脊椎动物有鱼类；植物有有节类、真蕨类、苏铁类、银杏类、松柏类等裸子植物及少量原始被子植物。

（一）双壳类

白垩系双壳类化石比较丰富，每个层位都有产出，且多与腹足类、介形类等化石共生。王自强等（1979）在坨里组近底部采到双壳类化石，后被萧宗正等重新立定为产于九佛堂组中、上部层位；王璞（1984）研究了夏庄组中的双壳类；萧宗正等（1994）又新采获了许多化石。

萧宗正（1994）将北京地区早白垩世双壳类由早期到晚期划分出 4 个化石组合：

1）*Ferganoconcha* 组合。主要分布在北京西山地区，分布层位是东狼沟组、九佛堂组中上部。其主要分子常见于我国含 *EEL* 生物群的地层中。组合中的主要分子 *Ferganoconcha* 与腹足类的 *Probaicalia* 属化石呈现出密切的共生关系。另见一定数量的球蚬类化石 *Sphaerium* 属，其中的多种在我国其他地区晚侏罗世地层中也常与 *Ferganoconcha* 共生。本组合时代属晚侏罗世晚期—早白垩世早期。

2）*Nakamuranaia* 组合。产于丰台区九佛堂组组上部，并共生有较多的球蚬类化石。小型球蚬类中也有许多南方分子。共生的介形类以 *Cypridea* 占绝对优势，其时代大致属侏罗世晚期—早白垩世早期。

3）*Sphaerium* 组合。产于坨里组，为单调的小型双壳类化石组合。此时为软体动物发展的低潮时期，而以植物和昆虫最繁盛。这一组合面貌同产于山东莱阳组及河南马市坪组的小型单调球蚬类组合比较接近，其时代应属早白垩世早—中期。

（4）*Xishanoconcha* 组合。产于丰台区夏庄组，以西山蚌类占绝对优势。同时共生一些小型球蚬类化石。共生的腹足类呈现出崭新的面貌，叶肢介为 *Yenjiestheria* 群分子，介形类出现 *Candona*，孢粉中出现被子植物花粉，其时代应属早白垩世晚期，可能到晚白垩世早期。

（二）腹足类

萧宗正等（1994）依据有关资料的综合分析，将该图自下而上划分出 4 个腹足类化石组合。

1）*Lioplacodes-Probaicalia vitimensis* 组合：产出层位为东狼沟组，以 *Lioplacodes*，*Amnicola*，*Probaicalia*，*Trochispira*，*Galba* 为主，多为锥形壳体，构成本区晚中生代地层中一个特有的腹足类动物群。时代大致为晚侏罗世中期到早白垩世早期。

2）*Viviparus-Probaicalia gerassimovi* 组合：主要产出层位为九佛堂组，延续至坨里组，其时代为晚侏罗世至早白垩世早期。

3）*Zaptychius* 组合：产于夏庄组的中下部，个体微小，数量较少，以 *Zaptychius* 属的分子占绝对优势，还见少量 *Mesocochliopa*，*Pseudarinia*，*Reesidella* 等属的分子，其时代为早白垩世早期—中期。

4）*Mesocochliopa-Reesidella* 组合：产于夏庄组中上部，属种繁多，数量丰富，个体较大，为本区腹足类最繁盛的时期，在数量上占优势的是 *Mesocochliopa* 属的分子，其次为 *Reesidella* 属的分子，左旋壳种类的 *Physa*，*Pseudarinia*，*Aurisstoma* 等属的分子也相对发育起来，而 *Zaptychius* 属分子相对地减少，并有个体变大的趋势。时代为早白垩世晚期。

（三）叶肢介

九佛堂组及夏庄组产有丰富的叶肢介化石，它们分别代表两个不同的叶肢介动物群：

1）*Eosestheria* 叶肢介群：以 *Eosestheria* 和 *Diestheria* 两属分子为主，种群丰度大，产于九佛堂组黑色页岩中。这一叶肢介群与狼鳍鱼 *Lycoptera* 和三尾拟蜉蝣 *Ephemeropsis trisetalis* 共生，共同构成了著名的热河生物群主体部分。时代为晚侏罗世至早白垩世。

2）*Yanjiestheria* 叶肢介群：产于夏庄组。时代为早白垩世。

（四）昆虫

繁盛于晚侏罗世—早白垩世的热河生物群，广泛分布于我国北方陆相地层中。产于热河生物群中的昆虫以新翅式昆虫占绝对优势，种类繁多，变化复杂。最早 Grabau（1928）曾将 *Ephemeropsis trisetalis* 作为热河生物群的主要成员，洪友崇（1984）重新研究了华北地区热河生物群的昆虫化石，认为"只要发现三尾拟蜉蝣化石，即完全可以肯定是热河生物群的产物。三尾拟蜉蝣的出现与绝灭，在一定程度上反映了热河生物群的发展规律。"可见这个种的重要地位。对这个昆虫群没有再作单独命名，其主要代表种属除 *Ephemeropsis trisetalis* 以外，尚有 *Coptoclava longipoda*，*Mesotendipes gregaria*，*Clypostemma xyphidle* 等。

九佛堂组所产的昆虫化石早有报道，只产单调的三尾拟蜉蝣 *Ephemeropsis trisetalis*，以及少量鞘翅目化石 *Yumenocleus* sp.。时代为晚侏罗世—早白垩世。

洪友崇（1982，1984）报道的坨里组及夏庄组昆虫化石以大量而占统治地位的 *Xishania fusiformis* 为特色，并有与其他生的属种。

萧宗正（1994）将西山白垩世昆虫化石划分为白垩纪房山昆虫群，包括两个组合：热河昆虫组合，以 *Ephemeropsis* 为代表，时代为早白垩世早期；芦尚坟昆虫组合，以 *Coptoclava* 为代表，时代为早白垩世晚期。

（五）介形类

北京地区在东狼沟组、九佛堂组和夏庄组发现介形虫化石。

东狼沟组的介形虫化石是以 *Cypridea faveolata*－*Cypridea unicostata* 为代表的组合，以女星介占优势。东狼沟组未见侏罗纪的介形类化石，其介形类化石具有鲜明的早白垩世早期色彩。

九佛堂组的介形虫化石组合以 *Cypridea amisia*－*Cypridea unicostata* 为代表。其与东狼沟组的介形类化石组合面貌很相似，同属女星介动物群，说明它们的演化具有连续性，时代也应属于早白垩世早期。

夏庄组的介形虫化石，下部为 *Timiriasevia kaitunensis*－*Cyclocypris minipisiformis* 组合，该组合的特点是种属单调，数量丰富，*Timiriasevia* 占绝对优势。上部为 *Cypridea tera*－*Candona orientalis* 组合。其时代应为晚白垩世。

（六）鱼类

北京地区在两个层位中发现鱼化石。一为东狼沟组，产 *Sinamia*；二为九佛堂组，产狼鳍鱼化石 *Lycoptera davidi*，*L. tokunagai*，*L. longicephelus*，*L.* sp. 等。

（七）古植物

萧宗正（1994）提出该植物群以 *Acanthopteris*－*Onychiopsis* 代表比较合适，并分为下、中、上 3 个组合，其中以中、上组合较为重要。

1）下组合，以 *Vitmia*－*Otozamites denticulatus* 为代表，分布于坨里组砾岩内，以真蕨纲、苏铁纲及松柏纲为主。时代相当于早白垩世中期。

2）中组合，以 *Acanthopteris gothani*－*Torreya fangshanensis* 为代表，分布于坨里组顶部至夏庄组底部，以真蕨纲、银杏纲、苏铁纲及松柏纲化石为主，其总体面貌类似阜新植物群，时代相当于早白垩世中期。

3）上组合，以 *Potamogeton? jeholensis*－*Cissites* 为代表，产于夏庄组，此组合以出现原始被子植物为特征，其时代相当于早白垩世晚期，或延续至晚白垩世早期。

（八）孢粉

萧宗正（1994）对北京地区中生代的孢粉化石做了极详细的研究和组合划分：

1. 九佛堂组孢粉组合

裸子植物花粉含量显著超过孢子的含量，前者为90%，后者为10%。

蕨类孢子数量较多的是海金沙科的 *Schizaeoisporites*，*Lygodiumsporites*，*Concavissimisporites* 属的各种孢子，*Cicatricosisporites* 属的各种孢子经常见到（含量为孢子的5%），偶尔见到 *Appendicisporites* 属孢子。出现频率较高的孢子还有 *Densoisporites*，*Reticulatisporites* 属。

裸子植物花粉以 *Classopollis* 属的各种花粉含量最高，达61.2%~80%，在层位中分布稳定。其次是 *Psophosphaera*，*Inaperturopollenites*，*Taxodiaceaepollenites*，*Cycadopites* 及 *Laricoides* 属的各种花粉占有相当数量。组合中经常见到的双囊型花粉，主要为松科的 *Pinuspollenites*，*Abielineaepollenites*，*Piceapollenites*，*Cedripites* 及罗汉松科的 *Podocarpidites* 属的各种花粉。此外尚有一定数量的单囊或周囊型花粉 *Callialasporites*，*Cerebropollenites*，*Quadraculina* 及古老类型的 *Pseudopicea* 属的各种花粉。

2. 坨里组孢粉组合

（1）坨里组下部亚组合

裸子植物花粉减少，但仍占优势，占67.5%，蕨类孢子占32.5%。

蕨类孢子中，海金沙科孢子较九佛堂组明显增多。特别是 *Cicatricosisporites* 属的各种孢子数量及种类均居孢子之首，含量在17.9%以上。其次是 *Lygodiumsporites*，*Concavissimisporites* 属的各种孢子。此外，*Densoisporites velatus*，*D. microrugulatus*，*Hsuisporites multiradiatus*，*H. minor*，*Deltoidospora hallei* 是组合中常见分子。

裸子植物花粉仍以 *Classopollis* 属的各种花粉为多，占孢粉总数的49.7%，但是较九佛堂组组合明显下降。另有 *Monosulcites*，*Cycadopites*，*Inaperturopollenites*，以及少量 *Psophosphaera*。*Taxodiaceaepollenites* 各属的各种花粉也较多。具囊花粉数量较多的是松科 *Pinuspollenites* 属及少量 *Piceapollenites*，*Cedripites*，罗汉松科的 *Podocarpidites* 属等。

（2）坨里组上部亚组合（原“芦尚坟组”孢粉组合的下部亚组合）

其明显特点是蕨类孢子含量明显超过裸子植物花粉的含量，前者约为孢粉总数的63%，后者约为37%。

蕨类孢子中，海金沙科孢子空前繁多，是北京地区早白垩世海金沙科植物发展的鼎盛时期，不仅数量多而且属种多样化，含量高达孢粉总数的48.9%，仅 *Cicatricosisporites*，*Appendicisporites* 两属见到28种之多，大量 *Appendicisporites* 属种孢子的出现是该组合的一个重要特点，与坨里组下部孢粉亚组合相比，出现了大量的新类型。

裸子植物花粉 *Classopollis* 属占14.4%，但与坨里组下部孢粉亚组合相比明显减少。其他如 *Psophosphaera*，*Asaucariacites*，*Inaperturopollenites* 比较常见。具囊花粉中松科的 *Pinuspollenites*，*Piceapollenites*，*Cedripites* 常见。*Podocarpidites*，*Callialasporites*，*Cerebropollenites*，*Quadraculina* 及古老类型的 *Pseudopicea* 属的各种花粉数量不多。另见到 *Ephedripites*，*Welwitschiapites* 属各种花粉及 *Schizosporis reticulatus*，*S. parvus* 等。

3. 夏庄组孢粉组合

该组合的重要特点是被子植物花粉的出现，标志着植被演化进程开始进入一个新的阶段。

（1）夏庄组下部亚组合（原“芦尚坟组”孢粉组合的上部亚组合）

裸子植物花粉含量超过孢子。

蕨类孢子仍以海金沙科孢子数量最多，*Cicatricosisporites* 属较坨里组上部亚组合数量减少，*Appendicisporites*，*Pilosisporites* 属孢子基本消失，而 *Lygodiumsporites*，*Impardecispora* 属孢子明显增多。组合中出现频率较高的孢子还有 *Densoisporites*，*Cyathidites* 属的各种孢子。

裸子植物花粉中 *Classopollis* 属仍占优势，其次是 *Cycadopites*，*Piceapollenites*，*Pinuspollenites*，*Cedripites*，*Podocarpidites*，*Psophosphaera*，*Laricoides*，*Inaperturopollenites*，*Quadraculina*，*Callialasporites*，

Cerebropollenites，*Ephedripites*，*Welwitschiapites* 等属的各种花粉。

（2）夏庄组中部亚组合

蕨类孢子与裸子植物花粉含量大致相当。

蕨类海金沙科孢子有显著含量，为孢粉总数的 35.9%，其中 *Cicatricosisporites* 属的孢子就有 13 种以上。*Appendicisporites* 属孢子只见两个种。

裸子植物花粉 *Classopollis* 属占各种花粉含量 22%以上，*Inaperturopollenites*，*Taxodiaceaepollenites* 属仍保存增长趋势。*Pinuspollenites*，*Piceapollenites*，*Cedripites* 及 *Cycadopites* 属的各种花粉出现频率较高。

被子植物花粉见到 *Tricolpites*，*Retitricolpites* 等少量颗粒。

（3）夏庄组上部亚组合

裸子植物花粉占优势，含量 82%，蕨类孢子贫乏，仅为 15%，被子植物花粉有增多趋势。

海金沙科孢子较中部亚组合数量种类明显减少。

裸子植物花粉 *Classopollis* 属的各种花粉仅为孢粉总数的 5.6%，而 *Inaperturopollenites*，*Taxodiaceae-pollenites* 属的花粉明显增多，分别占 15.7%、9.6%。松科和罗汉科花粉 *Pinuspollenites*，*Piceapollenites*，*Cedripites*，*Podocarpidites*，*Cerebropollenites*，*Pristinosporites* 等属数量和种类均比中部组合多。*Psophosphaera*，*Cycadopites*，*Ephedripites* 等属花粉经常见到。

被子植物花粉仍可见到 *Tricolpites*，*Retitricolpites* 等属，数量有增长趋势。

三、地层时代归属

综合上述各类生物群、生物组合状况，对侏罗纪—白垩纪岩石地层单位的时代归属讨论如下（表 5-5）：

1. 杏石口组

杏石口组产有植物化石：*Danaeopsis*? sp.，*Cladophlebis ichunensis*，*C. gracilis*，*C. asiatica*，*C. kaoiana*，*C.* sp.，*Czekanowskia* sp.，*Equisetites* sp.，*Podozamites lanceolatus*，*Taeniopteris* sp.，*Carpolithus* sp.，*Ctenozamites* cf. *sarrani*，*Glossophyllum shensiense*，*Taeniopteris graenlandica*，*Phoenicopsis* sp. 等；叶肢介化石：Euestheria? sp.（米家榕，1984）；鱼类化石：*Xingshikous xishanensis*。

对于该组的时代归属有很多争论。

晚三叠世的证据：植物化石 *Danaeopsis*? sp.，*Danaeopsis* 是晚三叠世世界性分布的重要属；*Cladophlebis gracilis*，*C. ichunensis*，*Ctenozamites* cf. *sarrani* 为延长植物群和越南鸿基植物群中的分子；*Glossophyllum shensiense*，*Taeniopteris graenlandica* 等常见于我国晚三叠世的化石在杏石口组内均有发现；萧宗正（1994）将杏石口组植物群称为 *Danaeopsis*? -*Glossophyllum*-*Czekanowskia* 组合，用以说明北京地区晚三叠世植物群发育的地方性色彩。

早侏罗世的证据：在杏石口组中也见有 *Czekanowskia*，*Phoenicopsis* 等属于侏罗纪常见分子的植物化石；刘宪亭（1988）定名的鱼类化石 *Xingshikous xishanensis*（西山杏石口鳕）时代确定为早侏罗世；近年来又获得新的杏石口组碎屑锆石测年数据：（205±1）Ma（杨进辉，2006）；（197±6）Ma，（206±2）Ma，（207±2）Ma（刘健等，2007）。

由于目前本地该组地层测年数据较少，未找到其他早侏罗世的证据。因此本志延续前人做法，主要依据植物化石资料及该组底部反映为印支运动的地质界面，将杏石口组时代定位于晚三叠世晚期。杏石口组的时代问题有待对其进一步研究。

2. 南大岭组

赵越（2006）获得北京西山南大岭组同位素年龄（174±8）Ma，根据上下层位推定属早侏罗世。赵兴国（2009）获得的门头沟双石头村和大灰厂靶场北玄武岩锆石 SHRIMP U-Pb 年龄为（211.6±5.8）~（184.1±4.5）Ma，认为南大岭组火山岩的活动时限为早侏罗世。本志将南大岭组时代定为早侏罗世。

表 5-5 中生代地层多重地层划分对比表

Table 5-5 Stratigraphic multiple classification and correlation of the Mesozoic in Beijing

年代地层（系）	年代地层（统）	岩石地层	生物群	化石特征 植物化石：植物化石	化石特征 植物化石：孢粉化石	化石特征 动物化石：双壳类	化石特征 动物化石：叶肢介	化石特征 动物化石：腹足类	化石特征 动物化石：昆虫	化石特征 动物化石：介形类	化石特征 动物化石：脊椎动物类	年龄数据
白垩系	下白垩统	夏庄组	房山生物群	*Acanthopteris gothani* - *Torreya fangshanensis*，早白垩世中期；*Potamogeton? jeholensis* - *Cissites*，出现原始被子植物，时代为早白垩世晚期，或延续至晚白垩世早期	*Classopollis* - *Lygodiumsporites*；*Cicatricosisporites* - *Classopollis* - *Pinuspollenites*，被子植物花粉见到少量颗粒；*Classopollis* - *Cicatricosisporites*，被子植物花粉有增多趋势	*Xishanoconcha* 组合。其时代应属早白垩世晚期，可能到晚白垩世早期	*Yanjiestheria* 叶肢介群。时代为早白垩世	*Zaptychius*，早白垩世早期-中期；*Mesocochliopa* - *Reesidella*，早白垩世晚期	以 *Coptoclava* 属为代表。早白垩世晚期	*Timiriasevia kaitunensis* - *Cyclocypris minipisiformis*；*Cyprideatera* - *Candona orientalis*。晚白垩世		
		坨里组		*Vitmia* - *Otozamites denticulatus* 组合，时代相当于早白垩世中期	*Cicatricosisporites* - *Lygodiumsporites* - *Classopollis*；*Cicatricosisporites* - *Appendicisporites* - *Classopollis*。早白垩世中—晚期	*Sphaerium* 组合，时代应属早白垩世早—中期						
		九佛堂组	热河生物群		*Classopollis* - *Cicatricosisporites* 多数为早白垩世地层的重要分子	*Ferganoconcha* - *Nakamuranaia* 组合。其时代大致属侏罗世晚期—早白垩世早期	*Eosestheria* 叶肢介群。时代为晚侏罗世至早白垩世。	*Viviparus* - *Probaicalia gerassimovi* 组合。其时代为晚侏罗世至早白垩世早期	*Ephemeropsis trisetalis*。时代为晚侏罗世—早白垩世?	*Cypridea amisia* - *Cypridea unicostata* 组合。时代也应属于早白垩世早期。	狼鳍鱼化石 *Lycoptera davidi*，*L. tokunagai* 和 *L. longicephelus*，*L.* sp. 等。晚侏罗世或延续到早白垩世	
		东狼沟组				*Ferganoconcha* 组合。其时代大致属侏罗世晚期—早白垩世早期		*Lioplacodes* - *Probaicalia vitimensis*。晚侏罗世中期到早白垩世早期		*Cypridea faveolata* - *Cypridea unicostata*。具有鲜明的早白垩世早期的特征	*Sinamia*，狼鳍鱼分子之一。晚侏罗世或延续到早白垩世	
		张家口组	门头沟植物群	偶见植物化石碎片，在门头沟区东岭北部采到有节类化石，初步定为 *Neocalamites*? sp.								137.3～118.4Ma
侏罗系	上侏罗统	土城子组		植物化石只有零星产出，暂称为 *Coniopteris hymenophylloides* - *Xenoxylon* 亚组合。另有若干硅化木化石 *Xenoxylon latiporosum* 等。时代为晚侏罗世		*Unio* - *Margaritifera* 组合。时代应为晚侏罗世早期	*Nestoria* - *Yanshanoleptestheria*，时代应为晚侏罗世早期		*Rhipidoblattina* (*Canaliblatta*) *yangqingensis* - *Samaroblattula houchengensis* - *Protorthophlebia yangqingensis*，晚侏罗世早期		恐龙足迹，覆盾甲龙类 cf. *Deltapodusi* sp.	(144.7±4.8) Ma～(130±1) Ma
	中侏罗统	髫髻山组		本组化石稀少，暂称为 *Coniopteris hymenophylloides* - *Elatides* sp. 亚组合。部分为门头沟植物群重要分子，时代应属中晚侏罗世								179 - 137Ma，最大(188.1±19) Ma
		九龙山组										(163.4±1.1) ～138Ma
		龙门组		*Coniopteris hymenophylloides* - *Nilssoniopteris vittata*。其中 *Coniopteris hymenophylloides*，*C. tatungensis*，*Eboracia lobifolia*，*Anomozamites*，*Ginkgoites* 等均是约克郡中侏罗世植物群的分子								
	下侏罗统	窑坡组		*Coniopteris hymenophylloides* - *Baiera*，*Coniopteris hymenophylloides* - *Ctenis kaneharai*。晚三叠世—早侏罗世		*Pseudocardinia* - *Sidireconcha* - *Tutuella*，主要见于下窑坡组。早侏罗世—中侏罗世						
		南大岭组		*Podozamites* sp.，*Czeknowskia*sp.，中国北方早侏罗世地层以不含典型的 *Coniopteris hymenophylloides* 显出早期色彩								(211.6±5.8) Ma～(174±8) Ma
三叠系	上三叠统	杏石口组		*Danaeopsis*? - *Glossophyllum* - *Czekanowskia* 组合。晚三叠世—早侏罗世			*Euestheria*? sp.（米家榕，1984）				*Xingshikous xishanensis*（西山杏石口鳄），早侏罗世	(207±2) Ma～(197±6) Ma

3. 窑坡组和龙门组

窑坡组下段依据其中所含有的植物化石组合 *Coniopteris hymenophylloides-Baiera*，以及双壳类 *Naiadites-Shaanxiconcha-Tutuella* 组合确定为早侏罗世。

窑坡组上段和龙门组出现门头沟组植物群的上组合 *Coniopteris hymenophylloides-Anomozamites thomasi* 组合，因此将其时代定为中侏罗世。

4. 九龙山组和髫髻山组

北京地区未在此两组中发现化石。

20 世纪 90 年代以前有关髫髻山组火山岩同位素年龄测定结果多在 160～152Ma 之间（鲍亦冈，1996）。《中国区域地质志·河北志》（2017）认为髫髻山年龄多在 165.5～146.6Ma 之间，最大为 188Ma，时代定为中—晚侏罗世。近年来所测得的髫髻山组年代数据多在 179～137Ma 之间，最大为（188.1±19）Ma。本志将髫髻山组时代定为中—晚侏罗世。

5. 土城子组

根据在延庆花盆土城子组内所发现的大量动植物化石资料及其组合，将其时代定为晚侏罗世初期。

赵兴国（2009）获取门头沟樱桃沟土城子组流纹质弱熔结凝灰岩中取的锆石 U-Pb SHRIMP 年龄为（144.7±4.8）Ma，为晚侏罗世。

6. 张家口组至九佛堂组

北京地区含热河生物群典型分子狼鳍鱼-东方叶肢介-三尾拟蜉蝣（L-E-E）的层位为东狼沟组和九佛堂组。但是热河动物群的时代归属是侏罗纪或是白垩纪，一直是地质界争论不休的大问题。

本志根据收集的张家口组以往该期火山岩的同位素测年结果（137.3～118.4Ma）（赵兴国，2009；袁洪林等，2005；汪洋等，2003；中国科学院南京地质古生物研究所，1978；地震地质所，1980；地质科学院地质所，1987），结合古生物化石特点，将张家口组划为早白垩世早期火山岩（K_1），东狼沟组与九佛堂组时代也相应定为早白垩世。

7. 坨里组

总的来说，坨里组含化石不丰富，且主要见于下部靠近九佛堂组的一些细碎屑岩夹层内，与九佛堂组热河动物群关系密切，坨里期相当于热河动物群的衰退期，时代可定为早白垩世。

8. 夏庄组

夏庄组化石丰富。根据各类化石组合充分说明夏庄组的时代相当于早白垩世晚期，甚至可延续到晚白垩世早期。

第六章　古近系—新近系

第一节　概　　述

本区古近系—新近系主要隐伏于平原区和延庆盆地第四系之下，西部山麓和延庆盆地边缘地带有零星出露。

本区古近系—新近系的研究始于19世纪20年代，主要是关于地文期的研究，维理士（B. Willis，1904）、安特生（J. G. Andersson，1919）、王竹泉（1926）、巴尔博（G. B. Barbour，1929、1935）等多位学者发表过文章，阐述对北京和整个华北地区地文期的看法。《北京西山地质志》（叶良辅等，1920）把西山的地形时代概括为5个地文期，即北台期、唐县期、汶河期、马兰期、板桥期。

山区古近纪—新近纪的工作主要体现在对长辛店砾岩和周口店猿人遗址的调查研究。

1922年裴文中在发育于长辛店一带的砾石层内找到数块化石，经杨钟健教授鉴定，初步确定其时代至少为第三纪初期。1923年安特生（J. G. Anderson）将这套砾石层定名“长辛店砾岩”，并与山东莱芜一带的始新统进行了对比。1929年，裴文中在长辛店东温庄西小清河东岸采到哺乳动物的前臼齿，经杨钟键鉴定属纯脚类的一种，为始新世化石。计荣森等（1930）、谢家荣（1933）、周明镇（1953）、翟人杰（1977）等对长辛店砾岩开展了调查研究工作。

自1929年在周口店发现中国猿人化石后，在鱼岭（第14地点）浅棕黄色砂层和泥岩洞穴堆积发现很多鱼化石，步达生、德日进、杨钟健、裴文中等定其时代为上新世早期偏晚。刘宪亭（1954）将其分为下鱼层与上鱼层。杨子赓（1985）建鱼岭组（N_2^1y）和东岭子组（N_2^2d）两个地层单位。周口店幅区调以是否含鱼化石为依据将鱼岭组划分为两段，并将东岭子组下部分出建立了新庄组。

平原区的古近纪-新近纪的工作始于20世纪70年代。随着北京平原区地热勘探工作的开展，北京市水文地质工程地质大队对平原区古近系-新近系进行了地层划分。1977年编写的《华北地区区域地层表·北京市分册》（1977）将平原区古近系-新近系由下至上划分为长辛店组、前门组、天坛组和天竺组。《北京市区域地质志》（1991）也沿用了这一划分方案。

20世纪80年代至20世纪末开展的的区调和研究，涉及本区古近纪-新近纪地质方面的有：一是利用平原区的1∶5万青龙桥幅、怀柔幅和通县幅工作中施工的钻孔开展了涉及上新世地层划分、气候环境演化、古地磁等综合性研究；二是在山区1∶5万沿河城幅、周口店幅工作中更加注重了对河谷阶地和洞穴的调查研究；三是对北京地区的第四纪综合研究如《北京猿人遗址综合研究》（1985）《北京地区第四纪古地理研究》（李华章等，1995），对上新世地质环境进行研究。

鲍亦冈等（2001）根据《中国地层指南及中国地层指南说明书》和《国际地层指南》所规定的原则，提出长辛店组、鱼岭组，而前门组、天坛组和天竺组可暂时保留其名作为岩石地层单位使用。

2003年开展的北京市多参数立体地质调查项目，在东坝施工的新5孔揭露了古近系和新近系，对磁性地层进行对比研究，并提出了利用测井资料开展地层划分的思路。2010年以后开展北京平原区活动断裂监测专项调查等项目，施工的部分钻孔基本只揭露了新近系上部，开展了岩性地层和磁性地层的对比研究。

由于没有新的可信的钻孔及其他资料，为保持志书的完整性，本志在古近系—新近系划分等方面，基本沿用《北京市区域地质志》（1991）的相关内容（表6-1），即北京凹陷区内划分为前门组、天坛组和天竺组3个地层单位，暂时作为岩石地层单位使用。大厂凹陷跨入北京地区仅一小角，划分为孔店组、沙河街组、东营组、馆陶组和明化镇组等5个岩石地层单位，详细内容可参阅《中国区域地质志·河北卷》，本志不做详细论述。

表 6-1　北京市古近纪-新近纪地层划分沿革表

Table 6-1　The classification history of Paleogene-Neogene lithostratigraphic units in Beijing

<table>
<tr><td colspan="2" rowspan="3">安特生
袁复礼
(1925)</td><td rowspan="3">裴文中（1930）
张席禔（1936）
刘宪亭（1954）
袁复礼（1955）</td><td colspan="3" rowspan="3">石油工业部
(1972—1978)
渤海湾地区</td><td colspan="3" rowspan="3">李云通
(1978)
华北平
原分区</td><td colspan="3">北京市区域地质志（1991）</td><td colspan="3">鲍亦冈等
(2001)</td><td colspan="4">本志</td></tr>
<tr><td colspan="2" rowspan="2">北京凹陷</td><td rowspan="2">廊固大厂凹陷</td><td colspan="2" rowspan="2">山区</td><td rowspan="2">北京凹陷</td><td colspan="2" rowspan="2">山区</td><td colspan="2">平原</td></tr>
<tr><td>北京凹陷</td><td>廊固大厂凹陷</td></tr>
<tr><td rowspan="3">上新统—中新统</td><td rowspan="3">周口店洞穴上部</td><td rowspan="3">周口店洞穴
石灰华层
第 14 地点</td><td rowspan="3">上新统—中新统</td><td rowspan="2">明化镇组</td><td>上部</td><td rowspan="3">上新统—中新统</td><td rowspan="2">明化镇组</td><td>上部</td><td rowspan="3">上新统—中新统</td><td>天竺组</td><td rowspan="2">明化镇组</td><td rowspan="2">上新统</td><td>东岭子组</td><td>天竺组</td><td rowspan="2">上新统</td><td rowspan="2">鱼岭组</td><td>天竺组</td><td rowspan="2">明化镇组</td></tr>
<tr><td>下部</td><td>下部</td><td rowspan="2">天坛组</td><td>鱼岭组</td><td rowspan="2">天坛组</td><td rowspan="2">天坛组</td></tr>
<tr><td colspan="2">馆陶组</td><td colspan="2">馆陶组</td><td>馆陶组</td><td>中新统</td><td></td><td>中新统</td><td></td><td>馆陶组</td></tr>
<tr><td rowspan="11">渐新统—始新统</td><td rowspan="11"></td><td rowspan="11"></td><td rowspan="5">渐新统</td><td rowspan="3">东营组</td><td>一段</td><td rowspan="7">渐新统—上始新统</td><td rowspan="3">东营组</td><td>上部</td><td rowspan="5">渐新统</td><td rowspan="5">前门组</td><td rowspan="3">东营组</td><td rowspan="5">渐新统</td><td rowspan="5"></td><td rowspan="5">前门组</td><td rowspan="5">渐新统</td><td rowspan="5"></td><td rowspan="5">前门组</td><td rowspan="3">东营组</td></tr>
<tr><td>二段</td><td>中部</td></tr>
<tr><td>三段</td><td>下部</td></tr>
<tr><td rowspan="5">沙河街组</td><td>一段</td><td rowspan="4">沙河街组</td><td>上部</td><td rowspan="4">沙河街组</td><td rowspan="5">沙河街组</td></tr>
<tr><td>二段</td><td>中部</td></tr>
<tr><td rowspan="3">渐始新统</td><td>三段</td><td rowspan="2">下部</td><td rowspan="6">始新统</td><td rowspan="3">长辛店组</td><td rowspan="6">始新统</td><td rowspan="3">长辛店组</td><td rowspan="3">长辛店组</td><td rowspan="6">始新统</td><td rowspan="3">长辛店组</td><td rowspan="3">长辛店组</td></tr>
<tr><td rowspan="2">四段</td></tr>
<tr><td rowspan="3">中始新统</td><td rowspan="4">孔店组</td><td rowspan="2">上部</td><td rowspan="4">孔店组</td></tr>
<tr><td rowspan="3">始新统</td><td rowspan="3">孔店组</td><td>一段</td><td rowspan="3"></td><td rowspan="3"></td><td rowspan="3"></td><td rowspan="3"></td><td rowspan="3"></td><td rowspan="3">孔店组</td></tr>
<tr><td>二段</td><td>中部</td></tr>
<tr><td>三段</td><td>?</td><td>下部</td></tr>
</table>

第二节　岩石地层

一、古近系

（一）山区

山区古近系出露始新统长辛店组（E_2ch），该组由安特生（1923）最早定名“长辛店砾岩”。通过裴文中（1929）、周明镇（1953）、翟人杰（1977）在该组中发现的化石，将时代定为始新世。鲍亦冈（2001）将长辛店组定义为“主要为紫红色砾岩，夹有紫红色黏土质页岩和灰黄色粗砂岩，含有 *Eudinoceras* 及软体动物化石，厚约 50m。与下伏地层为不整合接触”。

1. 代表剖面

本组以丰台区大灰厂铁路东侧剖面为选层型，1977 年陈兆恩测制。

上覆地层：马兰组黄土

～～～～～～ 角度不整合 ～～～～～～

长辛店组		总厚度 52.5m
9.	灰白、紫红色砾岩。砾石成分以安山岩、凝灰岩为主，灰岩及砂岩为辅，钙质胶结，结构疏松。砾石分选差，呈次圆状，最大砾径 25cm，一般约 10cm	>6.0m
8.	灰紫色砂砾岩，上部为砂质泥岩	2.0m
7.	灰白色细砾岩夹砂岩、泥岩透镜体。砾石呈次圆状	8.0m
6.	紫红色泥岩与砂砾岩互层。砾石呈次圆状，砾径 2～5cm，顶面有侵蚀现象	8.0m
5.	紫红色砾岩。砾石成分以火山岩为主，其次为灰岩和砂岩。泥钙质胶结，结构疏松。砾石圆状，但大小混杂，分选性差。砾径一般 10cm 左右，大者达 80cm	4.0m
4.	灰紫色泥岩与砂质泥岩，含钙质结核，风化后呈球状	4.0m
3.	灰白色砾岩夹泥岩、砂岩透镜体。砾石分选较差，次圆状，砾径约 15cm	12.0m
2.	紫红色泥岩，夹 3 层 5～10cm 厚的细砾岩。含钙质结核，直径约 10cm，自下而上增多	4.5m
1.	灰白色砾岩。砾石圆度较好，砾径一般 10～15cm，大者 25cm，钙质胶结，结构疏松	4.0m

～～～～～～ 角度不整合 ～～～～～～

下伏地层：九佛堂组 砂岩、页岩

2. 岩性特征

主要分布于丰台长辛店、辛庄、王佐一带，良乡东南也有小面积出露。近山麓地带地层保存完整，被沟谷切割后呈掌状的台地和缓丘。该组主要为砾岩与砂岩互层或砾岩夹含砾砂岩透镜体。含有 *Eudinoceras* 及软体动物化石，厚约 50m。与下伏地层为不整合接触。西部山麓地带砾岩层较厚，泥岩则相对减少。至东部平原区则相变为河流浅湖相沉积，厚度小于 200m。

（二）平原区

平原区古近系包括长辛店组和前门组，在北京凹陷和固安-大厂凹陷内均有分布。在北京凹陷内，其沉积范围为：西南起自房山琉璃河镇东、过窦店、大南关、良乡一线，到长辛店往西形成一个近乎圆形的较大面积出露区。西北部边界基本沿栗园庄、东辛秤一线，东北则以黄庄-高丽营断裂为界，到奥运村往东折向来广营，直至昌平区回龙观。东部则圈括了东坝环岛，向西凹进为一月牙形，直到永顺镇北。南部则基本以大兴隆起的西北边缘——南苑-通县断裂为界。北京凹陷内，古近系的沉积中心大致宛平、卢沟桥、丰台一带，最厚沉积厚度超过 400m。

在大厂凹陷内，北侧基本以大兴隆起的南侧礼贤断裂带为北界，往东南沉积逐渐加厚，过采育、永乐店、西集一线，沉积厚度超过2000m，到前马坊、军庄、觅子店一线达3000m。

1. 始新统长辛店组（E_2ch）

在北京平原，以中山公园钻孔为代表描述如下：

上覆地层：始—渐新统前门组

---------------- 平行不整合 ----------------

长辛店组	总厚度 131.0m
四段	31.0m
9. 紫棕色粉砂质泥岩，上部夹灰绿色页岩紫棕色泥质砂岩，泥质粉砂岩、灰绿色含砾粗砂岩	15.0m
8. 灰白色含砾、中粗砂硬砂岩、灰白色细粒硬砂岩与灰紫色泥质粉砂岩夹紫褐色粉砂质泥岩	16.0m
三段	13.0m
7. 紫褐色粉砂质泥岩夹紫色泥质粉砂岩	9.0m
6. 紫棕色泥质砂岩，浅紫灰、紫褐色含砾中粒硬砂岩	4.0m
二段	66.0m
5. 紫棕色粉砂质泥岩，泥质粉砂岩，浅紫灰色含粗砂细粒硬砂岩，顶部为紫棕色泥质粉砂岩	29.0m
4. 紫棕色粉砂质泥岩夹泥岩，泥质粉砂岩，紫灰色粗粒硬砂岩	31.0m
3. 灰紫色泥质粉砂岩，浅紫色细粒硬砂岩与含砾中粒硬砂岩	6.0m
一段	21.0m
2. 紫棕色粉砂质泥岩，上部夹紫棕色页岩	9.0m
1. 浅紫灰色含砾硬砂岩夹浅紫灰色泥质粉砂岩，紫棕色砂质泥岩	12.0m

~~~~~~~~ 角度不整合 ~~~~~~~~

下伏地层：下白垩统砂岩、泥岩夹砾岩

在固安凹陷和大厂凹陷内，与长辛店组相当为“沙河街组四段”，其主要岩性上部为深灰、灰黑色泥岩夹灰白色粉砂岩，局部地区夹黑色玄武岩，厚400~1000余米；中部为绿灰、青灰、深灰色泥岩与灰白色粉、细砂岩不等厚互层。局部地区夹数层玄武岩、含膏泥岩和薄层硬石膏等，厚450~782m；下部为灰、深灰色泥岩、钙质泥岩夹钙质页岩、泥灰岩、泥质白云岩及砂质条带，厚0~967m。与下伏孔店组暗色泥岩为平行不整合接触。沙河街组之下的孔店组（下—中始新统）为一套厚约409~1565m河湖相泥岩夹砾、砂岩沉积，局部地区夹石膏、油页岩、暗色玄武岩，生物化石丰富。北京凹陷新5孔内见有黑色玄武岩，岩芯522m和711m处K-Ar年龄分别为（53.46±2.26）Ma和（58.77±2.6）Ma，时代为上古新统—下始新统。

2. 上始新统—渐新统前门组（$E_{2-3}q$）

本组原称“北京组”，1977年编制《华北地层表·北京分册》时改为前门组。在北京平原以中山公园钻孔为代表描述如下：

上覆地层：中—上新统天坛组

~~~~~~~~ 角度不整合 ~~~~~~~~

上始新统—渐新统前门组	总厚度 337.0m
29 深灰绿色含角砾凝灰岩夹绿色玄武岩（2m）	15.0m
28. 深灰绿色玄武岩（两层厚4m与2m）与灰绿色角砾凝灰岩（厚4m）	10.0m
27. 红棕色粉砂质泥岩	9.0m
26. 深灰绿色中—细粒硬砂岩夹灰绿色泥岩	7.0m
25. 黑色碳质页岩与紫色泥岩夹浅灰色泥灰岩	6.0m

24.	灰白色硬砂质细砂岩与浅绿灰色粉砂质泥岩互层夹黑色页岩	24.0m
23.	浅绿灰色硬砂岩、粉砂岩夹绿灰色粉砂质泥岩	16.0m
22.	褐灰色粉砂质泥岩与泥质粉砂岩互层，底部夹灰白色含砾硬砂岩	18.0m
21.	灰褐色玄武质角砾凝灰岩，顶部为深灰绿色玄武岩	19.0m
20.	黑绿色玄武岩夹灰褐色玄武质含砾凝灰岩	18.0m
19.	红褐色玄武质含角砾凝灰岩夹黑绿色玄武岩（2m）	7.0m
18.	黑绿色玄武岩为主夹灰褐色玄武质含角砾凝灰岩（2~4m）	33.0m
17.	灰绿色泥质粉砂岩、粉砂岩夹灰褐色灰质泥岩、粉砂质泥岩	15.0m
16.	灰白色细粒硬砂岩（含砾硬砂岩）夹绿灰色粉砂岩、粉砂质泥岩	30.0m
15.	以绿灰色粉砂质泥岩与灰黑色页岩为主，夹绿灰色粉砂岩，页岩上有炭屑，含植物化石碎片，并有腹足类及介形虫化石，仅保存内膜	15.0m
14.	深绿灰色凝灰岩，底部为深灰绿色杏仁状紫苏辉石玄武岩（2.5m）	12.0m
13.	灰绿色凝灰质含砾砂岩、凝灰质粉砂岩夹灰绿色杏仁状苏辉石玄武岩（2.5m）	10.0m
12.	深灰绿色杏仁状紫苏辉石玄武岩与灰褐色玄武质岩屑凝灰岩、灰褐色玄武质含砾凝灰岩互层	54.0m
11.	以杂色含砾硬砂岩为主，顶部为灰紫色凝灰质粉砂岩	5.0m
10.	上部为灰紫色粉砂质泥岩，下部为绿灰色页岩及泥质粉砂岩。本层孢粉组合以被子植物划分为主，裸子植物花粉次之，蕨类孢子极少。被子植物花粉以葇荑花序类花粉为主，未见草本单子叶植物划分。主要有山核桃属（*Carya*）、桤木属（*Alnus*）、桦属（*Betula*）、栎属（*Quercus*）、化香树属（*Platycarya*）、栗属（*Castanea*）、榛属（*Corylus*）、柳属（*Salix*）等，以及个别的鹰形粉（*Aquilapollenites*）、柳叶菜科（Onagraceae）的 *Jussiena* sp.，山毛榉科（Fagaceae）的 *Nothofagus* sp.，*Gothanipollis* sp.，*Manogemmites pseudosetarins* 等。裸子植物划分有杉科（Taxodiaceae）松属（*Pinus*）南美杉属（*Araucaria*）麻黄属（*Ephedra*）等。蕨类孢子仅见卷柏属（*Selaginella*）	14.0m

--------------- 平行不整合 ---------------

下伏地层：长辛店组

在固安凹陷和大厂凹陷内，该组相当的为沙河街组 1~3 段与东营组，这两组地层在地表均未出露，孢粉组合以被子植物占优势，有胡桃属、山核桃属等。裸子植物次之。至凹陷内部，沙三段为富含有机质的暗色泥岩夹油页岩、泥灰岩、砂岩，厚 538~2300m。沙二段为一套下粗上细的红色砂岩与泥岩沉积夹泥灰岩、生物灰岩、局部夹膏泥岩、碳质泥岩、薄层油页岩。沙一段以褐色泥岩为主，生物灰岩、泥灰岩白云岩及油页岩。东营组在安次县桐柏镇附近的钻孔可见，以棕红色灰绿色泥岩与砂岩为主，夹碳质与砂质泥岩，一般厚 600~800m，与下伏沙河街组为平行不整合接触。

二、新近系

（一）山区新近系（鱼岭组）

1985 年杨子赓建鱼岭组和东岭子组两个地层单位，鲍亦冈（2001）认为其野外分布很零星，且总厚度不大，将其合并统称鱼岭组（N_2y），并将其定义为"下部为棕黄、灰棕等杂色砂砾、粉砂和黏土层，上部多为钙质胶结红色土堆积。多分布于山前丘陵地带，常残留于唐县期夷平面上或与地表水流沟通的洞穴中。鱼类化石较丰富，盛产鲤科魞属。"

1. 代表剖面

以周口店龙骨山剖面为正层型。剖面海拔 143.8m，高出周口店河 58m，由杨子赓等 1985 年测制（图 6-1）。

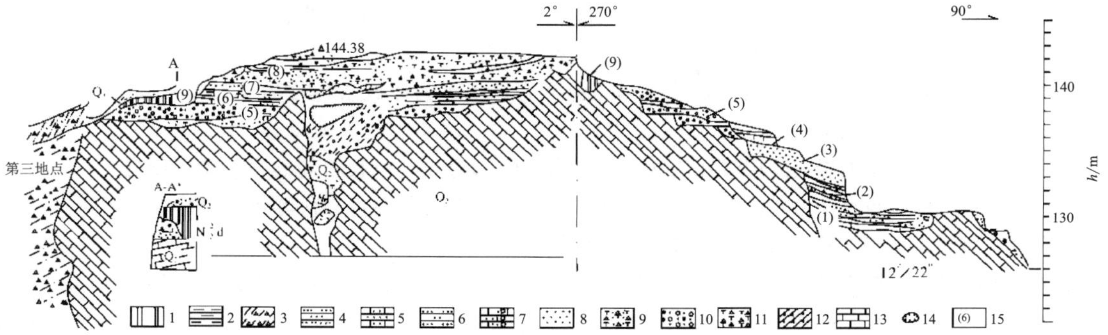

图 6-1　周口店龙骨山鱼岭组剖面图

（据杨子赓，1985）

Fig. 6-1　The profile map of Yuling formation in Zhoukoudian

1—红色黏土；2—薄层粉砂；3—黏土质粉砂角砾；4—薄层细粉砂；5—钙质胶结粉砂；6—中细砂；7—钙质胶结细粉砂；8—粗砂（石灰华层）；9—中粗砂细砾石；10—粗砂细砾石；11—钙质角砾；12—粉砂质黏土角砾；13—灰岩；14—溶洞；15—分层号（注：图中括号内数字为地层序号，详见正文）

上覆地层：第四系砂卵砾石层

～～～～～～ 角度不整合 ～～～～～～

鱼岭组 N_2y	总厚度>17.3m
9.　红色黏土	>2.0m
8.　灰白色钙质胶结含灰岩角砾粗砂层	3.2m
7.　灰白色砂夹砾石透镜体	1.0m
6.　黄棕钙质胶结粉砂层	1.4m
5.　棕红色砂砾层	1.5～2.0m
4.　黄红色钙质胶结的砂—砂砾石层，胶结坚硬	1.0m
3.　黄棕色中—粗砂层	1.7m
2.　黄褐色细砂层，具水平层理，含鲥鱼化石	3.5m
1.　灰黄、灰棕色砂砾层、薄层细粉砂	2.0m

～～～～～～ 角度不整合 ～～～～～～

下伏地层：马家沟组灰岩

2. 地质特征

上新统零星分布于海拔 150m 的唐县期夷平面上，如周口店山前丘陵，苏家坨至西马坊台地、南口山前台地、怀柔柏崖厂台地等地。岩石组合下部为棕黄、灰棕等杂色砂砾、粉砂和黏土层，上部多为钙质胶结红土堆积。柏崖厂一带为一套混杂堆积角砾岩，砾石呈棱角状，无分选，无层理，成分为白云岩，钙质胶结，尚未全部固结成岩，厚度小于 10m。延庆盆地周边及永定河二级阶地也有零星分布，如官厅水库西北的后郝窑一带为红色砂质黏土层（含角砾）与粗砂互层；黄土贵一带发育约 13m 厚的砂砾石层；丰沙线 64 号洞口处永定河二级阶地上为红土和棕红色砂砾石。

（二）平原区新近系

1. 中新统—上新统天坛组（$N_{1-2}t$）

天坛组是 1977 年编制《华北地层表·北京分册》时命名的。以中山公园钻孔为代表描述如下：

上覆地层：第四系砂卵砾石层

～～～～～～ 角度不整合 ～～～～～～

中—上新统天坛组	总厚度 1132.0m
50. 顶部为棕黄色泥质粉砂岩，上部为红棕色粉砂质泥岩，中部为红棕色砂质泥岩与粉砂质泥岩互层，下部为浅红灰色硬砂质长石砂岩与含砾硬砂质长石砂岩互层，底部为浅灰色细砾岩	47.0m

49. 上部为红棕色砂质泥岩与粉砂质泥岩为主夹浅红灰色硬砂岩、长石粉砂岩；中部为红棕色粉砂质泥岩；下部为浅灰红色硬砂质长石砂岩、灰白色含砾长石砂岩夹浅灰色砾岩 35. 0m

48. 上部为棕红色砂质泥岩、粉砂质泥岩夹浅褐棕色泥岩；下部为浅红灰色硬砂质长石砂岩，含砾硬砂岩，底部为浅灰色细砾岩 39. 0m

47. 上部为棕红色粉砂质泥岩夹薄层浅红色含砾硬砂岩；中部为浅红色长石硬砂岩、含砾硬砂岩夹浅灰色细砾岩与红棕色砂质泥岩、泥质砂岩互层；下部为灰白色、灰红色含砾砂岩与红棕色砂质泥岩互层，底部灰色细砾岩夹长石硬砂岩 55. 0m

46. 中上部为红棕色、棕褐色砂质泥岩、粉砂质泥岩与红灰色泥质砂岩互层；下部以灰色细砾岩为主，夹红灰色含砾硬砂岩 63. 0m

45. 上部为红棕色砂质泥岩、粉砂质泥岩夹灰色含砾硬砂岩；中下部为灰红色泥质粉砂岩、细砂岩与含砾硬砂岩互层并夹有灰色细砾岩 30. 0m

44. 上、下部为红棕色砂质泥岩为主，中部以红灰色硬砂岩（或含砾）为主，夹红棕色砂质泥岩 46. 0m

43. 上部为红灰色、灰色含砾硬砂岩与红棕色粉砂质泥岩、砂质泥岩互层；中部为红棕色泥质粉砂岩与棕红色粉砂质泥岩互层；下部以棕红色泥质细砂岩、硬砂岩、含砾硬砂岩为主，夹泥质细砾岩、棕红色粉砂质泥岩 54. 0m

42. 以棕褐色粉砂质泥岩为主，夹红棕色泥质粉砂岩、灰色硬砂岩、含砾硬砂岩 285. 0m

41. 上部为灰色含砾硬砂岩夹灰绿色细砾岩、红棕色砂质泥岩、泥质细砂岩；中部为红棕色粉砂质泥岩夹红灰色泥质粉砂岩；下部为灰绿色细砾岩夹红棕色泥质粉砂岩、粉砂质泥岩 52. 0m

40. 上部为棕褐色粉砂质泥岩夹灰色含砾硬砂岩，下部为灰绿色含砾硬砂岩 21. 0m

39. 上部为红棕色泥质粉砂岩与灰白色含砾硬砂岩互层；中部为红棕色粉砂质泥岩与泥质细砂岩互层夹细砾岩；下部为红棕色泥质细砂岩、灰白色硬砂岩与褐棕色粉砂质泥岩互层；底部以灰白色硬砂岩及含含砾硬砂岩为主，夹灰色细砾岩 52. 0m

38. 上部以红灰色含砾硬砂岩为主，底部为灰色细砾岩；中部为红棕色粉砂质泥岩夹红棕色泥质砂岩；下部为灰白色硬砂岩、含砾硬砂岩夹灰色细砾岩 25. 0m

37. 上部为红棕色粉砂质泥岩为主夹含砾硬砂岩；下部为灰红色泥质粉砂岩与粉砂质泥岩；底部为灰色含砾硬砂岩夹细砾岩 40. 0m

36. 上部为红棕色、褐棕色粉砂质泥岩为主，夹泥质粉砂岩；中、下部为灰红色泥质粉试验，灰色含砾硬砂岩，细砾岩夹红棕色粉砂质泥岩；底部为红棕色粉砂质泥岩褐浅灰色含砾硬砂岩 55. 0m

35. 上部为灰白色硬砂岩、浅灰色含砾硬砂岩夹红棕色粉砂质泥岩；中、下部为红棕色粉砂质泥岩夹浅灰色含砾硬砂岩、红棕色泥质细砂岩；底部为红棕色泥质细砂岩、灰白色硬砂岩、含砾硬砂岩 54. 0m

34. 红棕色粉砂质泥岩为主，夹泥质粉砂岩，上部有灰绿色含砾硬砂岩 53. 0m

33. 上部以灰白色硬砂岩及浅灰色细砾岩为主；中、下部为灰白色硬砂岩、含砾硬砂岩夹红棕色粉砂质泥岩、泥质细砂岩 35. 0m

32. 以浅灰色细砾岩为主，夹灰绿、灰白色含砾硬砂岩及红棕色粉砂质泥岩 19. 0m

31. 中、上部为红棕色粉砂质泥岩夹细砂岩、灰绿色含砾硬砂岩及细砾岩；下部以灰色细砾岩及含砾硬砂岩为主，夹红棕色砂质泥岩 42. 0m

30. 上部为红棕色砂质泥岩与灰白色含砾硬砂岩互层；中部为红棕色砂质泥岩；下部为浅灰色细砾岩夹灰白色含砾硬砂岩与红棕色粉砂质泥岩互层 30. 0m

~~~~~~~~ 角度不整合 ~~~~~~~~

下伏地层：上始新统—渐新统前门组砂质泥岩、页岩和砂岩

天坛组主要分布于北京凹陷和延庆盆地。北京凹陷内该组主要为灰紫色、紫红色、棕红色及班杂色巨厚层泥岩、砂质泥岩，夹灰色、灰紫色薄层砂岩（含砾砂岩）及砾岩。延庆盆地震平 1 孔揭示的该时段岩性为一套红色黏土夹砂砾石层，砾石已风化，属坡洪积相沉积。《北京市区域地质志》(1991) 将廊固凹陷和大厂凹陷内的馆陶组和明化镇组下部对比为天坛组，从岩性特征上看，馆陶组主要分布在武清一带，以灰黑、灰绿色泥岩（砂质泥岩）夹劣质煤，厚 500m；至霸州一带为浅灰、灰
~~~~~~~~

绿色泥岩（砂质泥岩）与灰白、浅灰色砂岩，泥质砂岩不等厚互层，间夹浅灰色含砾砂岩，厚433~956m。明化镇组在凤河营固安—大厂一带分布较广，为棕黄、灰绿、棕红色泥岩、粉砂质泥岩及砾岩，厚400~500m。

2. 上新统天竺组（N_2tz）

天竺组亦是1977年编制《华北地层表·北京分册》时命名的。在北京平原以顺义天竺的顺热1孔为代表描述如下：

上覆地层：下更新统砂、砂砾石夹黏性土

--------------- 平行不整合 ---------------

上新统天竺组	总厚度 336.0m
16. 灰黑色泥砾	18.0m
15. 灰色半胶结粉砂质泥岩	4.0m
14. 深灰色半胶结泥岩	14.0m
13. 灰、深灰色钙质胶结细砂岩，质地坚硬	19.0m
12. 灰黑色半胶结泥岩	7.0m
11. 灰、灰黑色钙质胶结粉砂质泥岩，中部夹泥岩，致密坚硬，含钙质结核	28.0m
10. 浅灰黄色中、细砂，含少量砾石，结构松散	6.0m
9. 灰白色钙质胶结砂砾岩	3.0m
8. 灰黑色半胶结粉砂质砂岩	20.0m
7. 深灰、灰黑色半胶结泥岩为主，中部夹灰白色半胶结砂岩	31.0m
6. 黄色半胶结砂砾岩，以中粗砂为主，砾石多为次圆状，砾径一般约3~4cm	8.0m
5. 棕黄、深灰色半胶结粉砂质泥岩与泥岩互层，泥岩中科见少量结核与铁锈斑	59.0m
4. 灰白、深灰色斑胶结粉砂质泥岩与砂岩、砂砾岩互层，含铁质结核	32.0m
3. 棕黄色斑胶结粉砂质泥岩，含铁锈斑与黑锰膜，具水平层理	23.0m
2. 半胶结泥岩、砂岩及砂砾岩，砾石磨圆较好，砾径一般3~5cm，大者达11cm	42.0m
1. 灰白、棕黄色砂砾岩，其中中粗砂占50%~60%，砾石40%~50%，次棱角—次圆状，泥质胶结，结构疏松	22.0m

--------------- 平行不整合 ---------------

下伏地层：中—上新统天坛组泥岩、砂砾岩

该组分布在北京凹陷东北地区，下部主要为深灰色、灰白色半胶结砂质泥岩与砂岩、砂砾岩互层，上部主要为深灰色、灰黑色钙质胶结粉砂质泥岩、细砂岩。延庆盆地震平1孔1002.2~865m的紫红色黏质砂土和灰绿色黏土，古地磁年龄为3.3~2.48Ma（李华章，1995），相当于该组的沉积时期。在廊固凹陷和大厂凹陷中明化镇组上段与天竺组相当。在凤河营以灰黄、浅棕红、灰绿色泥岩与棕黄色砂质泥岩为主，夹棕黄色、杂色含砾砂岩、细砾岩，厚约129.5m。除北京凹陷外，大兴隆起（新8孔，新11孔，新12孔，新13孔）及京西隆起（新1孔，新3孔，ZK3孔，ZK17孔和ZK18孔）钻孔磁性地层研究揭露均存在上新统。

新近纪，北东向断裂继续活动，沿断裂下盘沉降尤为明显，北京凹陷在原有的基础上沉积范围展宽，西部边界起自韩村河镇、沿石楼、田各庄、阎村一线，直到长辛店东，大宁水库北，东北超过了北西向的南口-孙河断裂，直到顺义石园、荣各庄一带；东部则基本以后桥、芦各庄一线为界；南部则以南苑-通县断裂为界。受顺义断裂继承性活动影响，北京凹陷沉降中心向北东方向迁移，新近系的沉积中心位于房山至海淀北天堂—葆台—四合庄一线，厚度超过1700m。

在大厂凹陷和廊固凹陷内，由于夏垫断裂活动性明显减弱，上下盘呈整体缓慢下沉趋势，因而湖盆在这里主要表现为横向扩展拓宽，其沉积范围北部基本以大兴隆起的东南缘为界，沿北臧村西南的梨园村、东芦垡、马驹桥、张家湾以北一线开始沉积，厚度往东南部增大，最厚处超过400m。

第三节　生物地层

一、古近系

1. 始新统长辛店组（E_2ch）

长辛店组的化石（表6-2）均产于砾石层中的紫红色泥岩夹层内。周明镇（1953）等研究发现了淡水腹足类、龟鳖类、哺乳动物类化石。翟人杰（1977）在高佃村泥岩内发现不等门齿犀、北京高臼兽、麦牙西兽、犬科哺乳动物化石，但未能鉴定其属和种。

表6-2　长辛店组化石一览表

Table 6-2　List of fossils of the Changxindian formation

化石类别	化石名称
淡水腹足类	*Cyclotus* cf. *exaratus* Sandberger，*Planorbina* sp. 1，*Planorbina* sp. 2， *Gastropoda* gen. et sp. indet. ①
龟鳖类	*Chelonia* gen. et sp. indet. ①
哺乳动物类	*Tupaiodon* sp.，*Insectivora* gen. et sp. indet.，? *Eudinaceras* sp.， *Rodentia* gen. et sp. indet. ① 不等门齿犀（*Imequincisoria* sp.），北京高臼兽（*Hypsimylus beijinggensis* Zhai） 麦牙西兽（*Miacis* sp.），犬科（Canidae）②

注：①周明镇，1953；②翟人杰，1977。

与长辛店组时代相当的沙河街组四段，从生物特征分析应划为晚始新世（表6-3）：其介壳类可对比江汉平原上始新统荆沙组化石组合的主要分子；腹足类在 *Sinoplanorbis sinensis* 山西河堤组、辽河下游平原河山东北部平原均有分布，是腹足类化石中主要分子，时代为晚始新世；孢粉可对比渭南地区上始新统白鹿原组，与孔店组孢粉有明显差别；轮藻主要分子在我国分布广泛，同时也普遍见于亚洲，欧洲、北美等地中—晚始新世，尤以晚始新世最丰富，极少数种可延至渐新世。

表6-3　沙河街组四段化石一览表

Table 6-3　List of fossils of the fourth member of Shahejie formation

化石类别	化石名称
介形类	上部：*Eucupris pengzhenensis*；　下部：*Cyprinotusigneus*
腹足类	*Sinoplanorbis sinensis*
孢粉	杉科高含量组合的 *Ephedripites-Ulmoideipites tricostatus-Deltoidospora regularis* 亚组合，其中被子植物花粉占绝对优势：49%~97%，以落叶阔叶的栎粉属（*Quercoidites*），榆粉属（*Ulmipollenites*）和脊榆粉属（*Ulmoideipites*）为主，三孔脊榆粉（*Ulmoideipites tricostatus*）尤为丰富，桦科（Betulaceae）与胡桃科（Juglandaceae）的花粉仅次于栎粉和榆粉属； 裸子植物花粉含量占第2位：3%~49%，由单束松粉属（*Abietineaepollenites*）、双束松粉属（*Pinuspollenites*）和耐旱的麻黄粉属（*Ephedripites*）组成； 蕨类孢子少量：1%~22.5%，以规则三角孢（*Deltoidospora regularis*）最发育
轮藻	*Pseudolatochara parvula-Obtispcjara* 组合中的主要分子 *Gyrogona* 等

长辛店砾岩之下的孔店组根据组内生物资料（表6-4）综合分析其时代以早—中始新世为宜：其介壳类与山东下始新统五图组介形类化石组合相似，而在该组内发现的脊椎动物化石 *Honogalax wutunsis*，*Heptodon niushanensis* 的时代为早始新世；孢粉与抚顺（油页岩下部含植物化石层）的孢粉组合特征基本相似，经宋之琛研究并与国外对比认为其时代为中始新世；轮藻类在山东五图组，湖北新咀组，广东罗佛寨组均有发现，*Gobichara bisulcata*（Peck et Reker）层发现与早始新世地层内。

表 6-4 孔店组化石一览表

Table 6-4 List of fossils of Kongdian formation

化石类别	化石名称
介形类	*Eucypris tuensis-Metacypris changzhoensis* 组合中其他主要分子为 *Cyprois palustris*, *Limnocythere weixianensis*, *L. chagyiensis*, *L. huliachangensis*, *Cycloypris changleensis*, *Labitria chinensis*, *Cypris*, *Candona*, *Candoniella* 等
孢粉	*Aquilapollenites spinulosus - Betulaceage - Deltoidospora* 亚组合的主要属种有 *Betulaceoipollenites prominens*, *Momipites*, *Caryapollernites*, *T. hiatus*, *Aguilapollenites spinulosus*, *Taxodiaceae pollenites*, *Cedripites pacjudemmis*, *C. minufulus*, *Deltoidosporcregularis polupodiacesporites*, *Scabiosapollis* 等
轮藻	以 *Peckichara varians*, *L. grambast*, *P. wutuensis* Sinlun 及 *Gobichara* sp. 为代表的重要分子

2. 上始新统—渐新统前门组（$E_{2-3}q$）

孔昭宸等（1975）对北京凹陷10余个地热钻孔中的前门组黑色页岩的孢粉特征进行了研究，并撰文《北京早白垩纪—第四纪的孢粉组合及其植物群和古气候的初步探讨》，研究表明前门期植物以被子植物为主，裸子植物次之，蕨类植物少（表6-5）。

表 6-5 前门组孢粉化石概况表

Table 6-5 Summary of palynological fossils of Qianmen formation

孢粉类别	孢粉化石名称	
	孔昭宸等，1975	中山公园钻孔剖面
裸子植物	以松科（Pinaceae）为主，包括松属（*Pinus*）、雪松属（*Cedrus*）、原始松科、云杉属（*Picea*）等以及罗汉松（*Podocarpusmacrophyllus*）、银杏（*Ginkgo biloba*）；此外发现古近纪多种麻黄（*Ephedra sinica Stapf*）花粉	杉科（Taxodiaceae）、南美杉属（*Araucaria*）、麻黄属（*Ephedra*）等
被子植物	以葇荑花序植物为主：胡桃科（Juglandaceae）、山毛榉科（壳斗科 Fagaceae）、桦木科（Betulaceae）、榆科（Ulmaceae）占比例大，约占23%~56%	主要有山核桃属（*Carya*）、桤木属（*Alnus*）、桦属（*Betula*）、栎属（*Quercus*）、化香树属（*Platycarya*）、栗属（*Castanea*）、榛属（*Corylus*）、柳属（*Salix*）等，以及个别的鹰形粉（*Aquilapollenites*）、柳叶菜科（Onagraceae）的 *Jussiena* sp.，山毛榉科（Fagaceae）的 *Nothofagus* sp., *Gothanipollis* sp., *Manogemmites pseudosetarins*, *kinthzsin* 等
蕨类植物		仅见卷柏属（*Selaginella*）

被子植物的组合特征说明当时发育着较丰富的喜温落叶阔叶树种，且这些花粉见于晚始新世—早渐新世的苏北、东北的孢粉组合中。

出现热带和亚热带的常绿植物如桃金娘科（Myrtaceae）、山矾科（Symplocaceae）、五加科（Araliaceae）、无患子科（Sapindaceae）、木兰科（Magnoliaceae）、山核桃科、枫香科（Liquidambraraceae），并出现典型的古近纪花粉，如铁青树粉（*Anacolosidites*）、脊榆粉（*Ulmoidepites*）、化香树粉（*Platycarya*）、倒地铃（*Cardiospermum*, cf. *Sporopollis trisulcatus*）。

部分样品中出现反映热带-亚热带气候的海金沙（*Lygodium japonicum*）、膜叶蕨（*Hymenophyllum barbatum*）、莎草科（Cyperaceae）的少量孢子。

河北省内大厂凹陷沙二—沙三段孢粉为栎粉属高含量组合，其中 *Quercoidites microhenrci-Ulmipollenites* 亚组合与东北平原、辽河下游平原沙河街组（二、三段）孢粉组合类同，与吉林延边渐新世孢粉组合基本相同，与日本北海道早渐新世的孢粉组合特征看与沙三、沙三段基本相似。但 *Caryapollenites simplex* 和 *Meliaceoidites rhomboitorus*, *Taxodiaceapollenites hiatus*, *Caryapollenites triangulus* 等曾见于我国江苏晚渐新世的三垛组一段和二段内。但 *Caryapollenites simplex* 在欧洲古近纪晚期较发育，东营组孢粉为榆属高含量组合与沙河街组有更密切关系。

二、新近系

1. 上新统鱼岭组（N_2y）

鱼岭组生物化石较丰富，已发现脊椎动物化石地点有两处。一处在鱼岭，有两个鱼化石层位，刘宪亭将其分为下鱼层与上鱼层（表6-6），这些鱼类属于现代种，已迁徙于长江流域以南；另一处在“上砾石层”上部石灰华层中（表6-6）。这些哺乳动物化石虽然不足以确定地层时代，但在我国动物区划中属于东洋界动物群。

鱼岭组以温带落叶阔叶植物为主混有部分亚热带常绿和落叶阔叶植物群，根据孢粉谱将其分为3段（表6-6），其特点是：禾本植物花粉在多数样品中超过过本草植物花粉；温带落叶阔叶植物花粉占优势（如栎、桦、栗、胡桃、榆等）；含少量亚热带树种花粉（如漆树、五加科、枫杨、紫树、山毛榉、罗汉松、铁杉、枫香、杉科等）；针叶树中普遍含有松花粉；草原类型草本植物花粉有时含量很高；一些样品中有生长于温暖潮湿沙地上立腕藓的孢子，含量较高。该组孢粉特征以温带落叶阔叶植物花粉占优势，含 *Rhus*（漆树）、*Tsuga*（铁杉）、*Fagus*（山毛榉）、*Liquidambar*（枫香）等森林草原孢粉组合，为温暖湿润北亚热带气候。

表6-6　鱼岭组化石一览表

Table 6-6　List of fossils of Yuling formation

化石类别	层位		
	上部	中部	下部
脊椎动物化石	上鱼层①： 四川鲃 *B. szechuanensis* 云南鲃 *B.* cf. *yunnanensis*		下鱼层①： 席褆鲃 *Matsya hsichihi* 短头鲃 *Barbus brevicephalus*
哺乳动物化石	灵猫科 Viveridae 原鼢鼠 *Prosiphneus* sp. 竹鼠 *Rhizomys* sp.，猪 *Sus* sp. 鹿 *Cervus* sp. 蝙蝠 *Myotis* sp. 等，此外尚有零星奇蹄类化石②		
孢粉	*Myotis* sp.，Viverridae，*Prosiphneus* sp.，*Sus* sp.，*Rhizomys* sp.，*Barbus szechyanesis*，*B.* cf. *Yunnanensis*，以栎为主含 *Rhus*，的落叶阔叶林组合	以温带落叶阔叶林为主常 *Araliaceae*，*Nyssa*，*Fagus*，*Rhus*，*Pterocarya*，*Tsuga*，Taxodiaceae，*Podocarpus* 的常绿及落叶阔叶林组合	*Barbus brevicephalus*，*Matsya hsichihi*，以栎为主，含 Liquidambraraceae 或 Araliaceae 的森林草原组合，上部含 *Rhus*，*Fagus*，*Tsuga*

注：①据刘宪亭；②位于上砾石层。

2. 中新统—上新统天坛组（$N_{1-2}t$）

从北京凹陷中收集的天坛组各钻孔孢粉资料表明：从孢粉种类和数量来看，都明显比古近纪少；从孢粉类型上来看，天坛组中下部灰—粉红色页岩（泥岩）中的孢粉（表6-7）以落叶阔叶的葇荑花序植物为主，蕨类植物次之，草本类型的孢粉很少，未见草原类型菊科、禾本科等。天坛组钻孔中上部孢粉葇荑花序植物组合明显减少，以蕨类植物的水龙骨属为主，另有温带落叶阔叶植物、灌木类型植物，并出现草原类型的蒿属和其他菊科、禾本科和车前科花粉。

表 6-7　天坛组孢粉化石概况表

Table 6-7　Summary of palynological fossils of Tiantan formation

层位	孢粉化石名称
天坛组中下部	落叶阔叶的葇荑花序植物为主，栎属（*Quercus*）、胡桃属（*Juglans*）、榆属（*Ulmus*）、桦木属（*Betula*）、鹅耳枥属（*Carpinus*）、榛属（*Corylus*）等 蕨类植物水龙骨（*Rhizoma polypodiodis nipponicae*）、凤尾蕨（*Spider brake*）、紫萁
天坛组中上部	蕨类植物水龙骨属（*Polypodium*）； 温带落叶阔叶植物如栎（*Quercus*）、桤木（*Alnus cremastogyne*）、榛（*Corylus heterophylla*）、鹅耳枥（*Carpinus turczaninowii*）、桦（*Betula*）； 灌木类型的杜鹃（*Rhododendron*）、鼠李（*Rhamnus*）、木犀科（Oleaceae）； 出现草原类型的蒿属（*Artemisia*）； 其他菊科（Asteraceae）、禾本科（Gramineae）和车前科（Plantaginaceae）花粉

3. 上新统天竺组（N_2tz）

天竺组的孢粉很少，其组合特征：蕨类植物的水龙骨草缝孢占2%~6%，凤尾蕨属占4%，以及卷柏属等；裸子植物以松属为主，个别出现了标志寒冷气候的云杉属、冷杉属分子（表6-8）；被子植物以草本为主，其中藜科占28.6%，十字花科占46%，蒿属占26.5%，残留油杉属占1.2%~4%，枫杨属占1.4%；此外含有少量水生植物孢粉，与天坛组上部孢粉已经趋于类似，该二组在区域上往往无直接上下关系，可能为同期不同相的地层。

表 6-8　天竺组孢粉化石概况表

Table 6-8　Summary of palynological fossils of Tianzhu formation

孢粉类别	孢粉化石名称
蕨类植物	水龙骨科（Polypodiaceae）、凤尾蕨属（*Pteris*）、石松属（*Lycopodium*）、卷柏属（*Selaginella*）等
裸子植物	松属（*Pinus*）、杉属（*Piceu*）、冷杉属（*Abies*）、落叶松属（*Larix*）、麻黄属（*Ephedra*）、铁杉属（*Tsuga*）
被子植物	藜科（Chenopodiaceae）、十字花科（Cruciferue）、蒿属（*Aflenigia*）、残留油杉属（*Keteleerria*）、枫杨属（*Pterocarya*）、百合科（Liliseceae）、蓼属（*Polygonium*）、狐尾藻（*Myriophyllum*）
水生植物	水鳖科（Hydrocharitaceae）、莲属（*Nelumbo*）

4. 延庆盆地上新统上部

这一时期孢粉组合中，木本植物含量高达40%~45%，松属（*Pinus*）占35%、云杉属（*Picea*）占10%、桦属（*Betula*）占20%、榆属（*Ulmus*）占22%、椴属（*Tilia*）占5%，此外还有栎（*Quercus*）、鹅耳枥、漆树属（*Rhus*）等；草本植物有藜科（Chenopodiaceae）、蒿属（*Artemisia*）、菊科（Compositae）、麻黄（*Ephedra*）等；水生植物有黑三棱属、狐尾藻属（*Myriophyllum*）等，蕨类孢子有水龙骨科（Polypodiaceae）、凤尾蕨（*Pteris*）、卷柏（*Selaginela*）等。根据本区地貌特点和植被生态特征分析，认为当时山地为针阔混交林带，丘陵及盆地边缘有榆（*Ulmus*）、椴（*Tilia*）等阔叶树，盆地边缘无水地带生长着藜（Chenopodiaceae）、蒿（*Artemisia*）等草丛，湖泊边缘及湖滨浅水地带生长着黑三棱、狐尾藻（*Myriophyllum*）等水生植物。在古湖中生活着布氏土星介（*Ilyocypris bradyi*）、柯氏土星介（*I. cornae*）和纯净小玻璃介（*Candoniella albicans*）等介形类动物，反映出是森林-草原植被类型及温暖的淡水环境。

第四节　沉积环境分析

一、古近系

北京凹陷内沉积了始新世晚期的长辛店组，其沉积范围在长辛店—丰台小井—中心城区附近。长辛店组以紫红色砂砾岩为主，夹泥岩、砂岩。垂向粒序为下粗上细结构。长辛店地区据李容全（1984）资料，砾石的中值粒径等值线在平面上呈扇形分布，表明长辛店组为近源洪积相—洪积扇相。砾石形态特征数值向下游（自辛庄、槐树岭向西峰寺、长辛店方向）有增高的趋势。若按岩性分别统计砾石的平均粒态也可看出其圆度数值有向长辛店方向增高的趋势，其中火山喷出岩的扁平度呈减少的趋势，而灰岩、燧石的扁平度则有增加的趋势。至城区岩性以紫棕色砂砾岩和粉砂质泥岩为主，夹泥岩。泥质粉砂岩与紫灰色硬砂岩，含砾硬砂岩互层构成4个韵律旋回，属洪泛平原沉积，因此由西南向东北长辛店组由山前洪积相过渡到平原的河漫滩相沉积。该时期气候较干燥，地面蒸发强，时有薄层石膏形成。

始新世晚期—渐新世早期，平原区已形成了“二凹一隆”格局（北京凹陷、大兴隆起、大厂凹陷），在北京凹陷内沉积了前门组，在固安-大厂凹陷内沉积了沙河街组和东营组。北京凹陷的前门组以灰绿—灰褐色粉砂质泥岩、粉砂岩为主，夹黑色碳质页岩和绿灰—灰白色细粒硬砂岩及粉砂岩，其中产腹足类、介形类以及脊椎动物的鱼鳖类化石，属河漫滩相—浅湖相沉积。晚始新世—渐新世有数次裂隙式基性火山喷发，同时发育由火山作用形成的各种碎屑物堆积，如含角砾凝灰岩等。固安-大厂凹陷该时期沉降速率远远超过北京凹陷（厚数百米），沉积厚度可达数千米。总体处于河流与浅湖沉积环境，当有机质较丰富，处于还原环境下则形成以湖相为主暗色泥页岩沉积，为油页岩形成的有利时期；当湖沼中植物繁茂，在有利的地质条件下形成褐煤层。

天津、沧州一带渐新统沙河街组一段中见薄球藻，水螅类及喜于含盐度较高的华花介等，说明此期盆地与海有了联系（海泛），形成了过渡相地层，但并未漫及北京广大地区。

始新世—渐新世早期（孔昭宸等，1985）北京地区处于亚热带气候较湿润条件下，发育针阔叶混交林，至后期为含有常绿树种的落叶阔叶林、气候热而略带干燥。组成的主要树种，既有苏铁、银杏、杉、水杉、落叶杉、罗汉松、雪松等针叶树种，也有栎、胡桃、枫杨、山毛榉、椴、桦、榆等落叶阔叶树种和山核桃、枫香、山矾、黄杞、桃金娘、木丝、栎、五加、芸香、无患子等典型亚热带树种，林中有附生寄生植物，草本层中有海金沙、莎草蕨、膜蕨、石松等蕨类植物。上述植被，相当于现今北纬25°~32°中亚热带北部植被，植物丰富、雨量充足，年平均温度较今约高出5℃。

二、新近系

由于受喜马拉雅运动影响，渐新世末地壳抬升，古近系和新近系之间普遍呈角不整合及平行不整合接触关系。在北京凹陷内，发育了中—上新统天坛组与天竺组河、湖相砂砾岩、泥岩沉积，其最大厚度可超过千米，并伴有少量陆相玄武岩喷发。其沉积范围往北扩展遍及顺义地区。而大厂凹陷在新近纪湖盆范围扩大，湖水向西浸漫，甚至淹没了大兴隆起的东南边缘部分，沉积中心却偏移到安次—武清一带，其沉积物以紫色、红棕色泥岩（粉砂质泥岩）为主，其次为砂岩及砂砾岩，反映氧化条件下河流及浅湖相沉积。在北京凹陷的东北顺义地区上部为天竺组，主要为还原—弱还原环境下的河湖相沉积，岩性为灰色、灰黑色及棕黄色半胶结泥岩、粉砂岩及砾岩。中新世，北京平原区森林繁茂以松、胡桃、鹅耳枥、桦、榆、榛等温带针叶落叶阔叶植物占优势，并保留枫香、铁杉、水青冈、紫萁、海金沙、凤尾蕨等，草本类型不多，草原型菊科、藜科、禾本科未见。上新世，葇荑花序植物在组合中明显地减少，而灌木状的杜鹃、鼠李、木樨、蔷薇以及藜科、菊科、禾本科、莎草科等草原型植物明显增多，显示森林草原的特征气候。

北京山区周口店一带在上新世唐县侵蚀面上形成钙质胶结红土堆积，为温热条件下的红色风化壳。

该时地壳相对稳定，为岩溶发育期，地下溶洞数量较多，但规模不大（直径数米至20~30m）。洞穴之间，有的显然相距很近，但彼此孤立，洞内一般少见石钟乳或石笋等。大部分充满水，成为地下湖或水潭，这种充填式的洞穴堆积环境中鱼群随水流而入，并很快地被泥沙所埋，因而形成集中的鲅鱼化石群。在洞外，生长着松、桦、胡桃等温带落叶阔叶树种，其次还有少量亚热带禾木树种。气候温暖，雨量充沛，湖沼广布，水生植物苔藓繁盛。同时频繁出现蒿、藜等旱生植物，适宜偶蹄类如鹿、猪等动物生长。说明当时气候干、湿波动，尤其是上新世晚期，森林覆盖面积减少，草原增加，标志着全球性气候略干燥。

除此而外在山间还出现一些新的小型断陷盆地，如延庆盆地。当时水动力条件很不稳定，盆地边缘以洪冲积相砂，砂质黏土（黏土）夹砾石沉积为主，往中心则颗粒逐渐变细为湖沼相沉积，当时气候旱寒孢粉贫乏，植被以藜科、蒿属和十字花科为主。到晚期盆地渐趋稳定，开始出现少量适应性较广的介形虫。

第五节　区域地层对比

根据地层层序，生物群特征等资料，将北京地区古近系、新近系划分与邻区地层的对比见表6-9。

表6-9　北京地区与邻区古近系、新近系对比表

Table 6-9　Correlation chart of lithostratigraphic units of Paleogene-Neogene in Beijing and its adjacent areas

<table>
<tr><td colspan="2" rowspan="3">年代地层</td><td rowspan="3">年龄/Ma</td><td colspan="6">《中国区域地质志·河北志》(2017)</td><td colspan="3">本志</td></tr>
<tr><td colspan="3" rowspan="2">内蒙高原南缘及太行山-燕山地层分区</td><td colspan="3" rowspan="2">华北平原地层分区</td><td rowspan="2">山区</td><td colspan="2">平原</td></tr>
<tr><td>北京凹陷</td><td>廊固大厂凹陷</td></tr>
<tr><td rowspan="3">新近系</td><td rowspan="2">上新统</td><td rowspan="17">2.588
5.3
23.04
33.8
55.8</td><td colspan="3" rowspan="2">稻地组</td><td rowspan="2">明化镇组</td><td colspan="2">明上段</td><td rowspan="2">鱼岭组</td><td rowspan="2">天竺组</td><td rowspan="2">明化镇组</td></tr>
<tr><td colspan="2">明下段</td></tr>
<tr><td>中新统</td><td colspan="3">石匣组</td><td colspan="3">馆陶组</td><td></td><td>天坛组</td><td>馆陶组</td></tr>
<tr><td rowspan="14">古近系</td><td rowspan="8">渐新统</td><td rowspan="3">雪花山组</td><td rowspan="7">开地坊组</td><td rowspan="7">汉诺坝组</td><td rowspan="3">东营组</td><td colspan="2">一段</td><td rowspan="3"></td><td rowspan="8">前门组</td><td rowspan="6">东营组</td></tr>
<tr><td colspan="2">二段</td></tr>
<tr><td colspan="2">三段</td></tr>
<tr><td rowspan="4">灵山组</td><td rowspan="8">沙河街组</td><td colspan="2">一段</td><td rowspan="5"></td></tr>
<tr><td colspan="2">二段</td></tr>
<tr><td rowspan="3">三段</td><td>上亚段</td></tr>
<tr><td>中亚段</td><td rowspan="5">沙河街组</td></tr>
<tr><td rowspan="7">西坡里组</td><td colspan="2" rowspan="7">蔚县组</td><td>下亚段</td></tr>
<tr><td rowspan="6">始新统</td><td rowspan="3">四段</td><td>上亚段</td><td rowspan="3">长辛店组</td><td rowspan="3">长辛店组</td></tr>
<tr><td>中亚段</td></tr>
<tr><td>下亚段</td></tr>
<tr><td rowspan="3">孔店组</td><td colspan="2">一段</td><td rowspan="3"></td><td rowspan="3"></td><td rowspan="3">孔店组</td></tr>
<tr><td colspan="2">二段</td></tr>
<tr><td colspan="2">三段</td></tr>
</table>

第二篇

沉积岩和沉积作用

第七章 中—新元古代

从中元古代长城纪开始，燕山地区新太古代变质基底发生裂解，延续至新元古代，发育为一个长达 10 亿年演化的盆地。盆地受华北板块北缘构造环境变化的制约，经历了 3 个阶段的演变：裂陷槽阶段、过渡坳陷槽阶段和稳定克拉通阶段（黄学光，2006）。

本区中—新元古代的古地理和古构造格局为燕辽广阔海盆沉积单元的一部分，主要沉积岩为滨海-浅海相富镁碳酸盐岩、碎屑岩，夹少量碱性玄武岩系列火山岩沉积。中元古代长城纪、蓟县纪和待建纪，沉积中心位于天津蓟县至河北兴隆一带，到新元古代青白口纪，海盆向西迁移，沉积中心转移至京西雁翅至河北怀来一带。

第一节 沉积作用及沉积类型

中—新元古代，本区沉积作用以海相为主，海陆过渡相次之，极少陆相沉积作用。海相沉积以障壁岛沉积作用为主，还发育无障壁岛沉积作用。海陆过渡相沉积包括河口湾沉积和三角洲沉积，以前者为主。而陆相沉积仅发育河流相沉积作用。海陆过渡相和陆相沉积范围和强度都远小于海相沉积作用。岩性上则以碳酸盐岩沉积范围最广，厚度最大，碎屑岩沉积次之，还发育少量火山碎屑岩沉积（表 7-1），各岩石地层沉积相特征见表 7-2。

表 7-1 中—新元古代沉积作用与沉积类型统计表

Table 7-1 Sedimentation and sedimentation types of Mesoproterozoic to Neoproterozoic

序号	沉积类型			岩石地层
	沉积相	沉积亚相	沉积微相	
1	海相沉积	内碎屑沉积	潮上（滨岸）、潮间（潮坪）、潟湖、潮下、浅海、深水	景儿峪组、铁岭组、雾迷山组、杨庄组、高于庄组、团山子组、大红峪组
		陆源碎屑沉积	障壁岛陆源碎屑沉积（障壁岛、潟湖、潮坪、滨岸、浅海）	下马岭组、洪水庄组、串岭沟组
			无障壁岛陆源碎屑沉积（海沼沙岭、海滩沙脊）	龙山组上部、常州沟组四段
2	海陆过渡相沉积	河口湾沉积	砂质潮汐水道沉积	常州沟组二、三段
		三角洲沉积	扇三角洲沉积	常州沟组二段
			辫状河三角洲沉积（三角洲前缘、前三角洲）	龙山组
3	陆相沉积	山地河流相		常州沟组一段

表 7-2 中—新元古代岩石地层沉积相标志

Table 7-2 Sign of sedimentary facies for lithostratigraphic of Mesoproterozoic to Neoproterozoic

序号	相类型	相标志									产出层位
1	碳酸盐潮下-浅海	缝合线板状层	舌状波痕								景儿峪组

续表

序号	相类型	相标志									产出层位
2	碎屑海滩-浅海	不对称小波痕	显微层状黄铁矿	氧化铁结核	微波状层理	海绿石					洪水庄组、下马岭组、龙山组
3	碳酸盐潮下带	内层滑塌	假结核	锥状叠层石	波状硅质层	叠层石穿透硅质层	硅质层内层褶曲	显微藻丝细胞构造	石膏假晶	硅质纹层干裂	大红峪组、铁岭组、高于庄组、雾迷山组
		硅质透镜体	核形石岩	沥青质层	地震褶皱角砾	黑色燧石层	白色燧石层	硅质层侵蚀面	硅质斜层理	板刺状硅化岩（片状硅化风暴岩）	
4	碳酸盐潮坪带	片状角砾（砾片风暴岩）	浅水波痕	碳酸盐透镜层	硅质层折断	硅化鱼骨状交错层理	纤维硅质卷缩	显微干裂	双脊波痕	重晶石（天青石）	高于庄组、杨庄组、雾迷山组、铁岭组
		碳酸盐斜层理	断裂构造	显微斜交层	丘状叠层石	显微波痕	墙状叠层石	柱状叠层石	显微藻席层		
5	碳酸盐潮上带	氧化铁质膜	空穴充填	内碎屑角砾	冲刷坑	石膏脉	纤维鸟眼	去白云石化	方解石亮晶充填		串岭沟组
6	潟湖	球状黄铁矿及黄铁矿团块	季节性韵律层状薄纹层构造	显微泄水线	盲脉						串岭沟组
7	砂坝	小型舌状波痕	显微气泡构造								串岭沟组
8	碎屑海滩-潮下	砂-页岩突变界面	次生长石胶结物	风暴层							常州沟组
9	碎屑砂滩 沙坪	透镜状层理	内层错断	流痕	对称波痕	雨痕	鱼骨状交错层理	干裂痕	黄铁矿晶体粒序层		常州沟组、杨庄组、龙山组
	碎屑砂滩 泥坪	胶结构造	掘穴状管状充填构造	爬痕状遗迹化石							常州沟组、下马岭组
10	分流河道	古风化面	碎屑粒序层	交错层理	“花彩弧”层理	层间滑动	长石再生长	长石自形晶	基底式胶结		常州沟组、杨庄组、龙山组

一、陆相沉积作用及沉积类型

陆相沉积作用在北京地区仅发育山地河流沉积作用，表现为地形高差大、坡降大，向源侵蚀作用强烈，河岸陡而河谷深，河道直而支流少，水流急而沉积物粗的网状河沉积特征。

该类沉积作用仅发育于常州沟组底部一段，分布于密云水库周边地区及平谷万庄子—熊耳寨一带，可延伸向河北兴隆、天津蓟县地区。岩石类型属于中—粗砾岩及不等粒长石石英砂岩，缺细砂岩与泥质沉积物。沉积厚度区域变化大，如丰各庄厚 7m、万庄厚 14m、方耳峪厚 79m、聂家峪厚 89.3m、西窝兵湾厚 146m、牛角峪厚 214m。山地河流沉积环境特征表现有（吴梦源，1985）：①具向上变细的正半韵律结构层序（图 7-1），底部发育冲刷面（a），由下向上依次发育石英质中砾岩（b）、粗粒石英

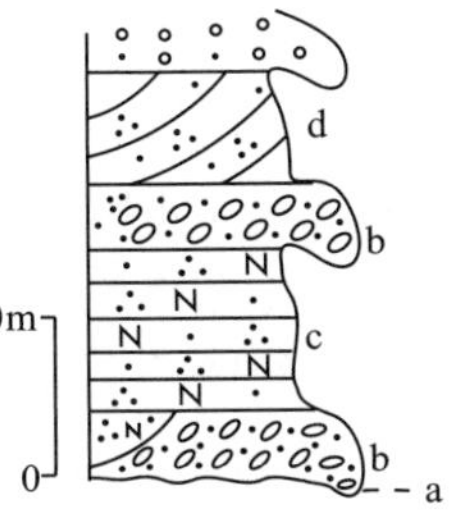

图 7-1　密云区聂家峪常州沟组一段沉积序列图
（鲍亦冈等，2000）
Fig. 7-1　Sedimentary sequence of the first member in Changzhougou formation, Niejiayu of Miyun District

长石砂岩、具斜层理或交错层理砂岩和石英砂岩（c、d），缺失漫流沉积泥质沉积；②岩石沉积厚度大，相变快，相邻剖面不易对比。砂体以透镜体产出，似心滩沉积；③砾石成分较复杂，石英岩砾石占 50%，纹层状砂岩砾石占 17.2%，长石砂岩砾石占 21.8%，砾岩砾石占 1%，磁铁石英岩砾石占 3.7%，脉石英与火山岩砾石占 6.3%，蚀源区为沉积碎屑岩区；④砾石呈叠瓦状排列，颗粒支撑结构，砾石 ab 面定向性显著，砾径 76～97mm，大者 250～350mm，平均砾径 83.7mm，>100mm 砾石占 8%～22%，分选较差，标准偏差 $\sigma=1.32\sim1.42$，统计 ab 面产状表明形成于一种高坡降网状河沉积，古水流方向由 SW 流向 NE，蚀源区位于太行-五台古陆；⑤大中型槽状、板状交错层理发育（图 7-1）。

二、海陆过渡相沉积作用

海陆过渡相沉积环境包括三角洲和河口湾沉积环境，位于海、陆之间的过渡地带（陈建强，2004）。本区中—新元古界这两类沉积环境均有发育。

（一）三角洲沉积

1. 扇三角洲沉积亚相

该类沉积作用表现为水下扇三角洲沉积，发育于常州沟组二段底部层位，分布于密云水库周边及平谷区罗营—关上—西台一带。其沉积厚度不大，为 70～80m，横向上呈透镜状产出，厚度从密怀古陆边缘向东逐渐变薄，总体平面形态呈扇形分布。沉积序列（图 7-2）为底部发育冲刷面（a），向上依次发育为灰紫色石英质细砾岩（b）、含细砾长石石英砂岩（c）、发育板状和帚状交错层理的砂砾岩与石英砂岩（d）、发育透镜状和波状层理及不对称波痕的紫红色铁质含长石石英砂岩与粉砂岩（e），粒度向上变细。砾石成分为石英岩、燧石及片麻岩，次棱角状—次圆状，基质支撑结构，成分、结构成熟度中等—较差，标准偏差 $\sigma=1.37$，分选差，峰度系数 $K_g=0.9\sim1$，偏度系数 $S_k=1$。沉积相标志：下部发育冲刷面、细砾石透镜体、小型槽状和板状交错层理、赤铁磷块岩透镜体、磷结核、砾-砂韵律层理、水流波浪、浅水波浪波痕等；上部发育浅水波浪波痕、人字形交错层理、羽状-帚状-楔状交错层理及海滩冲洗交错层理，砂粒表面具铁质氧化膜，粒径 = 0.13～0.2mm，标准偏差 $\sigma=1$，分选中等，偏度系数 $S_k=-0.1\sim0.3$，$Fe^{3+}/Fe^{2+}>2.5$。

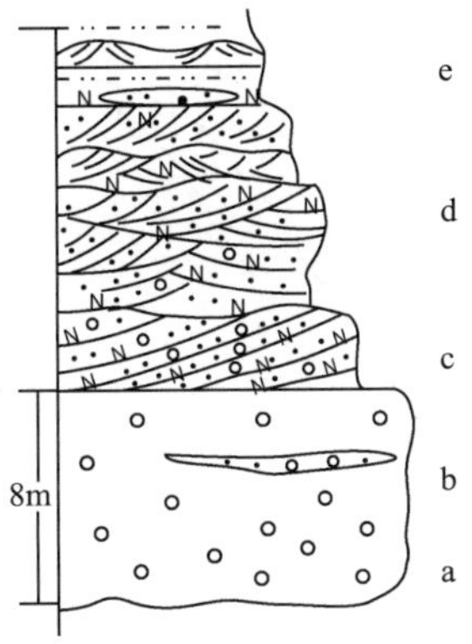

图 7-2　密云区聂家峪常州沟组二段沉积序列图
Fig. 7-2　Sedimentary sequence of the second member in Changzhougou formation, Niejiayu of Miyun District

2. 辫状河三角洲沉积亚相

该类沉积发育于龙山组中下部层位，由三角洲平原、三角洲前缘和前三角洲 3 个亚相单元组成。三角洲平原亚相分布于邻区河北涞源一带，区内仅发育三角洲前缘和前三角洲亚相。根据沉积环境和沉积特征，三角洲前缘亚相细可分为分支河道、河口坝与前缘砂坝及分流间湾 3 类微相。

（1）分支河道沉积微相

分支河道沉积是龙山组早期的第一套碎屑岩沉积。重要沉积相标志：①垂直相序上，由下向上依次发育为含砾含长石不等粒砂岩或细砾岩（a）、交错层理含砾粗砂岩（b）、粗—中砂岩（c）；②由砾-砂岩构成 2～4 个正向粒序韵律（图 7-3），单个韵律厚一般 5～6m，总厚度<15m，向岸方向增厚，

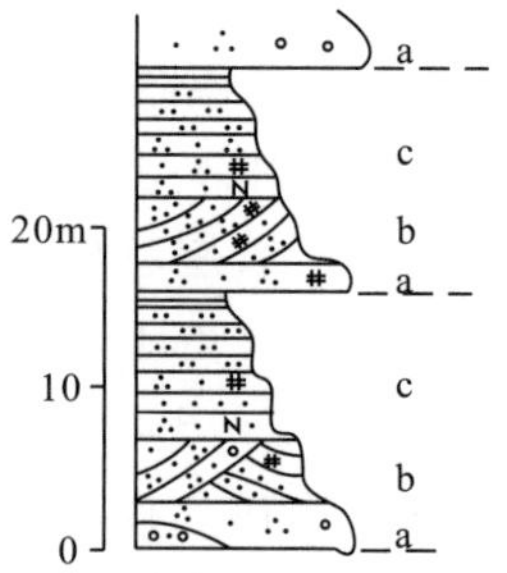

图 7-3　房山区长流水龙山组一段沉积序列图

（鲍亦冈等，2000）

Fig. 7-3　Sedimentary sequence of the first member in Longshan formation, Changliushui of Fangshan District

其中砾岩单层厚<lm，但向岸方向砾岩层数增加，单层厚度增大，砾石的粒径增大，砾石圆度、分选性变差；③砾岩层的粒度分布曲线（图 7-4c）显示出是在推移质、跃移质与悬移质联合沉积作用下的快速沉降，具碎屑重力分选的沉积特点；④发育正粒序层理、双向交错层理和楔状、鱼骨状（人字）、板状交错层理，底面发育印模铸痕，并常见有大、中型冲刷面；⑤古水流玫瑰图（图 7-5）均显示双向态和多向态类型。碎屑粒度曲线均显示为对称弓形（图 7-4a、b）粒度分布。按 Saku（1964）环境判别公式Ⅲ（$Y_{浅海:冲积} = 0.2825M_z - 8.7604\sigma_1^2 - 4.8932Sk_1 + 0.0482K_g$）和Ⅳ（$Y_{冲积:浊流} = 0.7215M_z - 0.403\sigma_1^2 + 6.7322Sk_1 + 5.2927K_g$）计算结果：Ⅲ式均值 $\overline{Y}_x < -7.4190$，波动在 $-8.096 \sim -11.5337$ 间，属冲积三角洲；Ⅳ式均值 $\overline{Y}_x < 9.8433$ 者占 60%，其波动在 $8.0301 \sim 9.3355$ 间，均值 $\overline{Y}_x > 9.8433$ 者占 40%，波动在 $10.2612 \sim 11.0966$ 间，前者属浊流，后者属冲积三角洲。

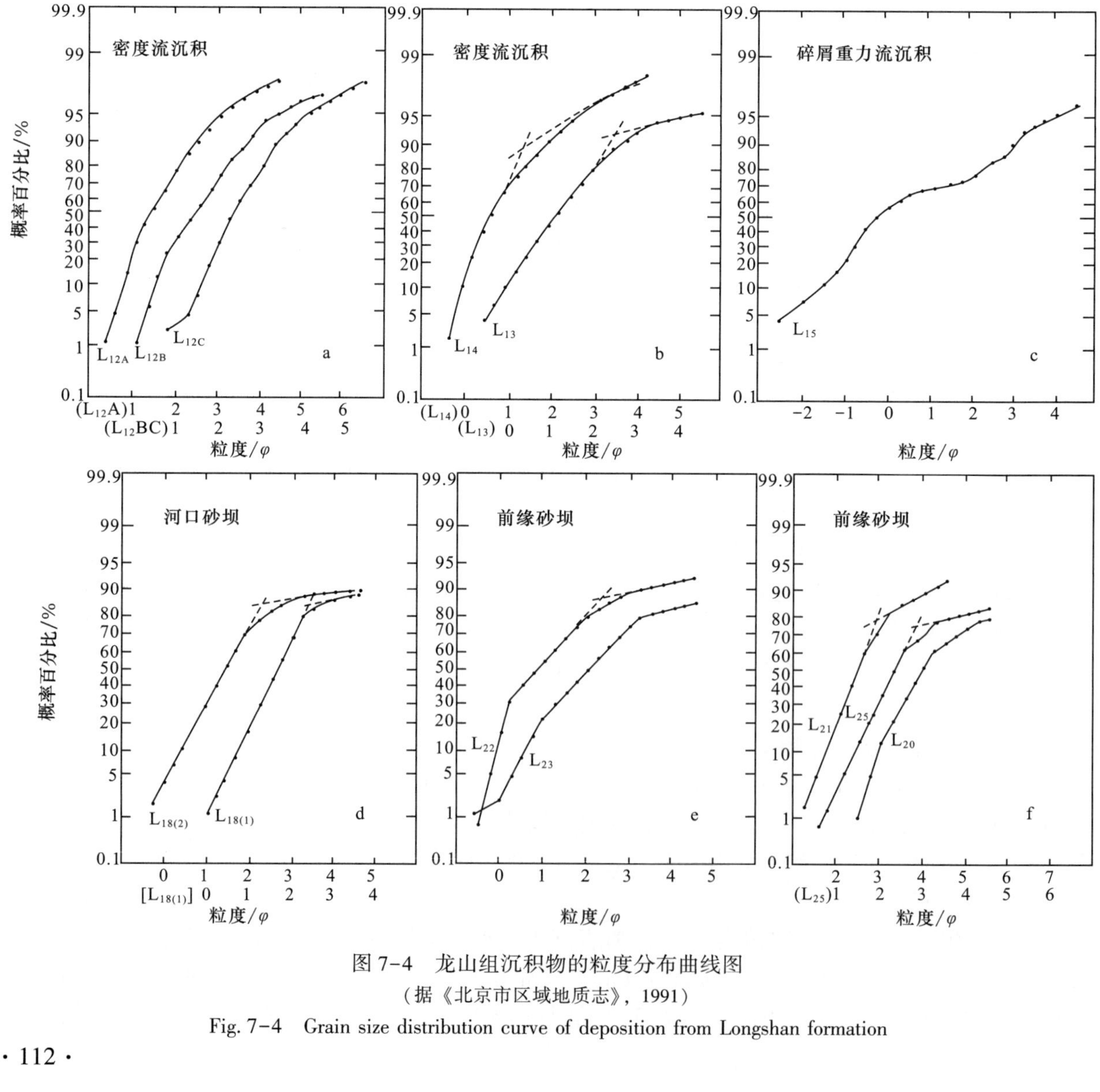

图 7-4　龙山组沉积物的粒度分布曲线图

（据《北京市区域地质志》，1991）

Fig. 7-4　Grain size distribution curve of deposition from Longshan formation

(2) 河口坝与前缘砂坝沉积微相

河口坝沉积发育于龙山组一段中下部层位，其上、下均为分流间湾沉积灰黑色纸片状页岩。垂向相序由下向上依次为极薄层具水平层理的粉砂岩→中薄层具缓斜层理的细粒海绿石石英砂岩→中厚层具波浪交错层理的中细粒含铁石英砂岩→巨厚层具交错层理的中粗粒石英砂岩→中厚层具波浪交错层理的中细粒含铁石英砂岩→巨厚层具交错层理的中粗粒石英砂岩，呈现出典型三角洲沉积层理下薄上厚、粒度下细上粗、岩石成分与结构成熟度下低上高、能量条件下低上高所组成的反向递变层序的沉积特征。碎屑粒度分布曲线为典型两段式，主要由斜率较高（65°）跃移质组成，占80%以上，次要由斜率平缓（<20°）悬移质组成，仅占约10%，通常缺失推移质。上截点呈递变有序悬浮过渡关系（图7-4d）。

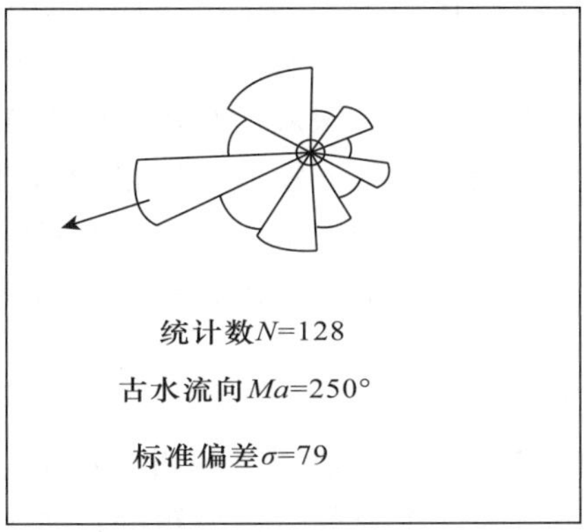

图7-5　龙山组古水流流向统计玫瑰图
（据《北京市区域地质志》，1991）
Fig. 7-5　Rose diagram of paleocurrent direction of Longshan formation

Saku（1964）环境判别式计算结果表明：下部样品Ⅲ式均值$\overline{Y}_x$>-7.4190，为滨浅海，Ⅳ式均值$\overline{Y}_x$<9.8433，为浊流（混水悬移质沉积）；上部样品Ⅲ式均值$\overline{Y}_x$<-7.4190，Ⅳ式均值$\overline{Y}_x$>9.8433，均属冲积三角洲。综上，河口坝曾受波浪作用改造，以跃移质沉积作用为主，发育在以悬移质沉积（混水悬浮沉积）的分流间湾之中，并发生纵向迁移，岩性主要为灰白色块状中粗粒石英砂岩、灰黄色褐色石英砂岩，厚度4~5m。

前缘沙坝沉积发育于龙山组一段顶部层位。其岩性组合在垂向上可分为7层（图7-6），e层之下具有与河口坝相同的反向递变层序特征。e层之上，在层理和粒度演变上属正向递变层序，下部单元沉积特征完全可与河口坝砂体相类比，但上部正向递变层序的出现则有别于河口坝特征，是本区三角洲相利于保存的波浪作用带障积单元。从c层开始波浪冲刷簸选作用加强，到e层达到接近潮下高能带环境，出露巨型对称波浪波痕（龙山脊东侧），L=150~210cm，H=18~20cm，R_I=120，形成圆度极佳、分选好的豆状石英质砾石。顺海进方向，沙坝可能由东向西迁移，叠置在分流间湾砂-泥岩相沉积之上。碎屑粒度曲线分布为此种解释提供了水动力条件方面的旁证。采自b、c、d层样品，其粒度分布曲线均为二段式（图7-4f）。跃移质主总体占75%，斜率高（70°），分选好σ_1=0.7，悬移次总体占10%~20%，斜率30°，上截点均显示有序悬浮递变关系，反映波浪簸选的沉积特点。经用萨胡环境判别式Ⅲ、Ⅳ计算，结果均在冲积（波浪）三角洲范围之内。采于高能带e、f层的样品（图7-4e），反映出明显三段式曲线，主总体分别由两个跃移次总体组成，截点常在0.5~1φ附近截然过渡，说明波浪反复冲刷作用较强；悬移次总体平缓但上截点呈递变有序过渡关系。岩性分选中等—较差，环境判别式计算结果均属冲积（波浪）三角洲环境。

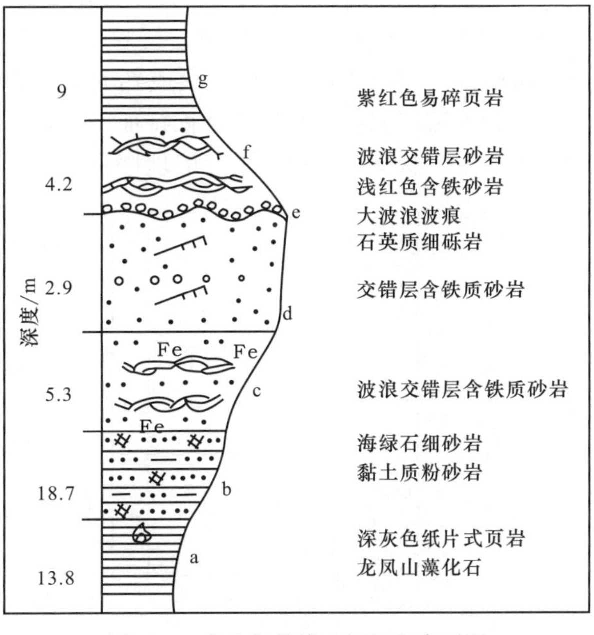

图7-6　龙山组前缘砂坝沉积序列图
（据《北京市区域地质志》，1991）
Fig. 7-6　Sandbar sedimentary sequence of delta front of Longshan formation

(3) 分流间湾沉积微相

分流间湾沉积处于龙山组一段中部，中

间被河口坝相分隔为上下（近岸与远岸）两个单元，属于悬移质沉积作用为主的中性—还原的分流间湾砂泥相沉积环境。平面相图上，因前缘砂坝的障壁作用，使该相处于低能、水体相对滞流的中性—弱还原环境之中。岩石沉积序列：下部为黄绿色粉砂岩夹灰色页岩，上部为深灰、灰色页岩夹黄绿色泥质粉砂岩和海绿石砂岩，常发育水平纹层及纸片状层理。下部单元含长石、铁质，以粉砂屑含量较高为特征，反映近河口水位较浅。上部单元长石含量减少，自生海绿石增加，水云母、伊利石含量增高，水体相应加深。无论上部或下部，指相矿物海绿石均产于泥质沉积之上碎屑沉积位置，富铁、钾海绿石常与细砂—粉砂伴生，并通常在弱还原或近 $E_h=0$ 的环境下生成。

（4）前三角洲沉积亚相

前三角洲沉积处于龙山组二段，沉积为一套厚度比较稳定的红色泥岩相地层，分布于怀来—延庆北—密云北—兴隆一线以北，受后期构造、岩浆破坏，产出零散。岩性主要为紫红、黄绿色页岩夹薄层海绿石细砂岩及钙质泥岩。页岩大部分由鳞片状微晶黏土矿物（水云母）组成，含少量绢云母，粉砂屑含量极少（<2%）。上部有多层海绿石砂岩。本区红色地层与海绿石相伴生，处于富铝（Al_2O_3=15.73%）偏碱性、富钙镁（CaO=5.1%、MgO=1.78%）的泥质中，推测氧化铁被黏土矿物吸附，并一起堆积，从而避免水化而保持红色。元素比值 Ca/Mg=3.9，K/Na=7.74，Na/Ca=0.74，Fe^{2+}/Fe^{3+}=0.24 指示了以悬移质沉积作用为主，水介质盐度较高、氧化条件较弱的前三角洲环境。

（二）河口湾沉积

河口湾沉积仅发育于常州沟组二段上部及三段下部，分布于昌平南口—十三陵、密云水库周边及平谷一带，其沉积厚度较大，一般大于 150m。横向上大致以昌平德胜口—密云河南寨一带为轴线，开口向北东东的喇叭状产出。

河口湾内，潮汐作用和河流径流作用共同控制沉积。下部为砂质潮坪沉积，岩石类型为灰白色中厚层石英砂岩、长石石英砂岩和长石砂岩。潮汐沉积构造十分发育，有小型波浪波痕、大中型改造（菱形、舌形、波曲形）波痕、潮汐与潮汐充填交错层理、人字形交错层理、再作用面、小型板状-槽状交错层理、同沉积断层及准同生变形构造（砂质白云岩枕状体）。粒度分布曲线（图 7-7）为多峰态三段式，涨潮流驱动 0.5φ 以上粗粒级移动，平潮时跃移质随能量衰减依次沉积，跃移质再次分选。十三陵地区底部发育潮道（图 7-7b），据宋天锐（1986，2007）、柳永清（1990）等研究，长石含量达 30%以上，普遍高于其他层位，反映矿物成熟度较低。而结构成熟度其主要结构系数<10.00，胶结物含量较高（多于 10%），属于次成熟的次长石砂岩。根据粒度系数用散点圆和萨胡（Sahu）公式判别，综合河流沉积特征和潮滩沉积特征于一体。沉积相标志上，底部含砾粗砂岩可见大型槽状交错层理、大型鱼骨状交错层理和平行层理序列，向上发育为羽状交错层理、再作用面、沟槽充填交错

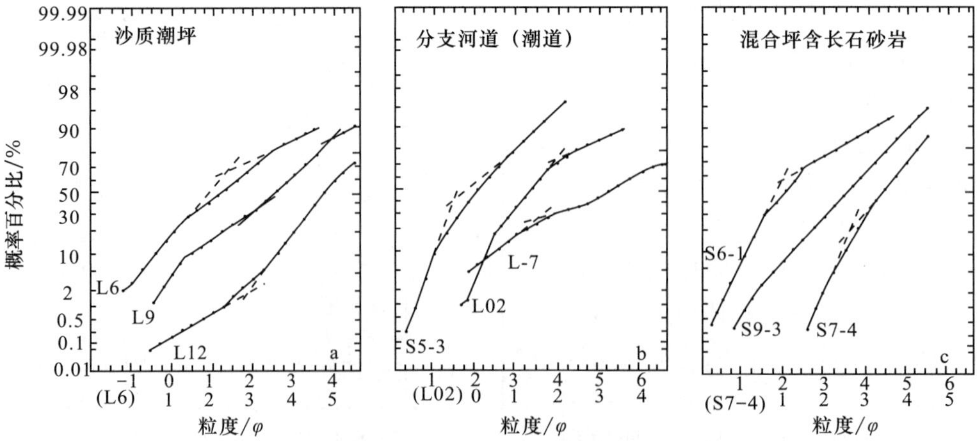

图 7-7 常州沟组二段、三段沉积物粒度分布曲线图

（据北京市区域地质志，1991）

Fig. 7-7 Grain size distribution curve of deposition of Changzhougou formation

层理、小型楔状交错层理、板状交错层理、波痕等。总体反映受潮汐水流影响河道沉积特征，为典型的潮汐水道迁移形成的砂质潮汐水道沉积序列（图 7-8），发育为含砾粗砂岩（a）和不等粒长石石英砂岩与粉砂岩（b）或泥质石英细砂岩-粉砂岩互层（c）。层理统计揭示古水流向 250°，标准偏差 $\sigma=34$，区域海侵方向 280°。

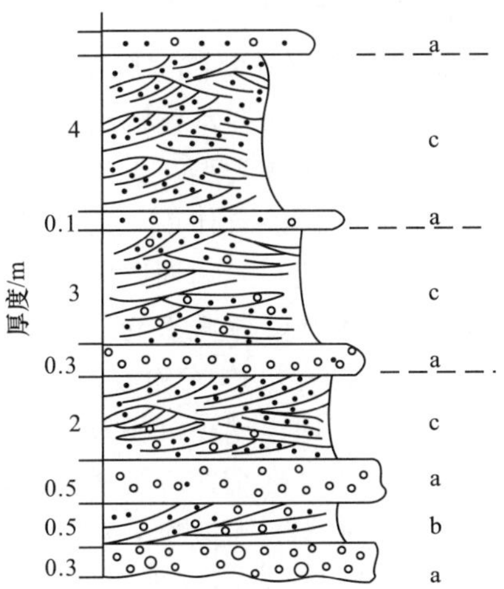

图 7-8　昌平十三陵常州沟组二段河口湾砂质潮汐水道沉积序列图

Fig. 7-8　Estuary sand tidal channel sedimentary sequence of the second member of Changzhougou formation in the Ming Tombs, Changping

十三陵地区上部发育为砂泥混合坪，岩石类型为灰黄色含长石石英砂岩与灰黑色泥质粉砂岩互层组合。富钾、含有机质及黄铁矿，K_2O 含量最高达 16.3%，一般为 8.5%～14.8%，Ca/Mg = 1.44，Na/Ca = 1.45，K/Na = 14.7（图 7-9），砂岩 $Fe^{3+}/Fe^{2+}>1$（1.94～7.31），黑色泥质层 $Fe^{3+}/Fe^{2+}<1$（0.29～0.73），表明在氧化-还原的交替环境下沉积的。潮汐沉积构造十分发育：波状层理、脉状层理、砂-泥韵律层理、泥坡（压扁层理）、砂质透镜层理、干裂、冲流痕、雨痕、双黏土层束状体。粒度分布曲线为二段式，有一个占优势的悬移主总体（65%～85%），斜率 45°，区间 $\varphi=2.5\sim4.5$，跳跃次总体占 15%～30%，斜率高（60°），区间 $\varphi=0\sim1$，两总体截点呈递变有序相接关系，通常缺失推移质，这显然是潮汐沉积作用形成的粒度分布特征（图 7-7c）。

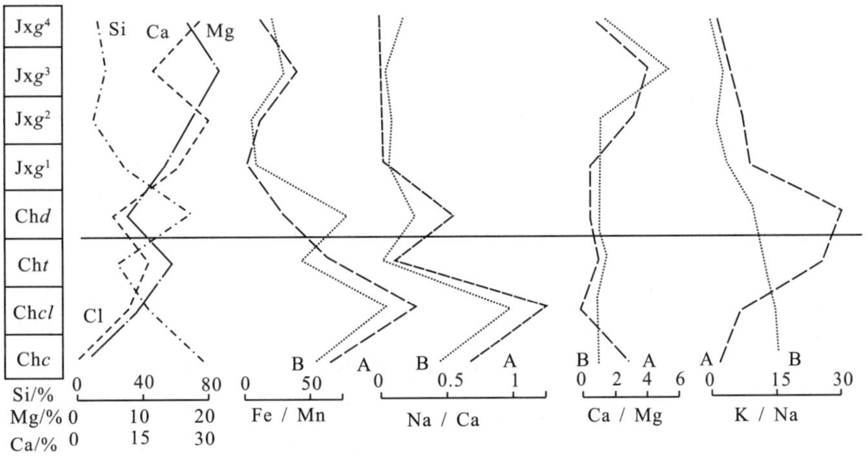

图 7-9　中—新元古代地层元素比值垂向变化曲线图

（据《北京市区域地质志》，1991）

Fig. 7-9　Vertical variability diagram of elements ratios of Mesoproterozoic to Neoproterozoic

A—蓟县剖面（据秦正永，1979）；B—十三陵剖面；Chc—常州沟组；Chcl—串岭沟组；Cht—团山子组；Chd—大红峪组；Jxg—高于庄组

三、海相沉积作用

北京地区中—新元古代的海相沉积作用最为发育，包括陆源碎屑沉积和碳酸盐岩台地沉积两大类。其中，陆源碎屑沉积又根据障壁型地形的有无分为障壁海岸和无障壁海岸沉积两小类。时间序列上，则大致经历了由无障壁岛海岸沉积-障壁岛海岸沉积-碳酸盐岩台地-障壁岛海岸沉积-河流相沉积-碳酸盐岩台地沉积的复杂演变。

（一）陆源碎屑沉积

1. 无障壁岛海岸陆源碎屑沉积

该类沉积主要分布在常州沟组三段上部至四段，龙山组沉积旋回上部也有发育。本区主要表现为砂质无障壁岛海岸沉积类型。无障壁岛海岸与海洋连通性好，海岸带水动力状况复杂，受到风、河流、波浪、潮汐和洋流的综合作用，其中以波浪和潮汐作用为主（陈建强，2004）。该类海岸带沉积物质来源也复杂，以河流、三角洲供应的碎屑物质为主，也包含波浪、潮汐剥蚀的海底物质，以及风从陆地搬来的砂和波浪从陆棚搬来的砂。由海域向陆地方向，破浪带水体扰动大，沉积了砂砾等丰富的粗粒沉积物；碎浪带在沿岸流和波浪两种营力共同作用下，泥砂呈“之”字形沉积。而冲流带水浅流急，垂直于海岸来回运动，沉积较粗颗粒的砂。

常州沟组三段上部至四段沉积为厚层石英砂岩，由下向上相标志组合为浅水小波浪波痕-双脊改造波痕、浅水波曲波痕、水平层理、小型楔状交错层理、波痕层面伴生雨痕（平谷西峪）、泥质层面干裂、脉状-透镜状-压扁状层理（平谷北山）、包卷层理（平谷、密云）、海滩冲洗交错层理等。岩石的结构成熟度和成分成熟度均较高，副矿物特征为锆石-电气石-普通角闪石组合，锆石磨圆好。粒度统计 $\sigma=0.5$ 分选好，平均粒径 $M_d=0.16$mm，峰度系数 $K_g=0.8\sim1$ 接近中等，峰态近对称 $S_k=-0.1\sim0.3$（图 7-10）。地球化学特征（图 7-9）揭示，主量元素：$SiO_2=93\%\sim97\%$，平均 94.4%，$Al_2O_3=1.98\%\sim3.90\%$，$CaO=0.05\%\sim0.06\%$，$Fe_2O_3=0.51\%\sim0.56\%$，$FeO=0.36\%\sim0.39\%$，$K_2O=2.35\%\sim4.88\%$，$Na_2O=0.17\%\sim0.27\%$，$Fe^{3+}/Fe^{2+}=1.83$、$Ca/Mg=1.01\sim1.14$、$Na/Ca=1.14\sim3.98$；微量元素：$Ti=2832\times10^{-6}\sim3608\times10^{-6}$，$V=14\times10^{-6}\sim43\times10^{-6}$，$Cr=72\times10^{-6}\sim116\times10^{-6}$，$Ni=6\times10^{-6}\sim8\times10^{-6}$，$Co=5\times10^{-6}\sim5.4\times10^{-6}$，$Sr=27\times10^{-6}\sim85\times10^{-6}$，$Ba=295\times10^{-6}\sim519\times10^{-6}$，$Mn=142\times10^{-6}\sim285\times10^{-6}$，属海相氧化环境，总体反映无障壁岛滨岸沉积特征。

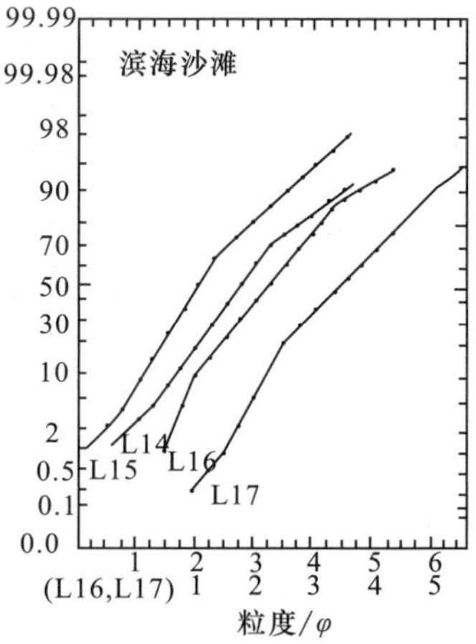

图 7-10 常州沟组三段上部和四段沉积物粒度分布曲线图

（据北京市区域地质志，1991）

Fig. 7-10 Grain size distribution curve of deposition of Changzhougou formation in Changping

龙山组晚期沉积为浅水低能滨浅海沉积（图 7-11），岩性组合：下部为白色石英砂岩（a）与绿色海绿石石英砂岩（b）互层，上部为紫红色泥岩（c）夹海绿石粉砂岩（d），发育水平层理。

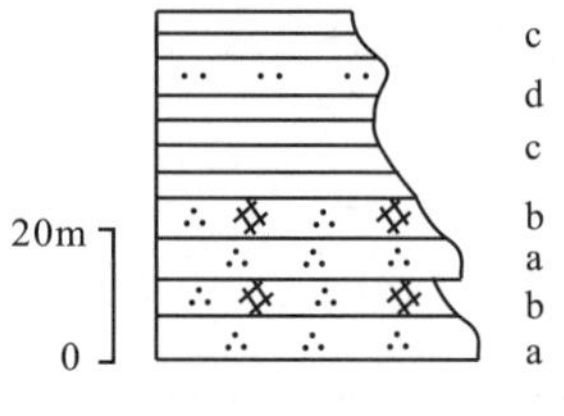

图 7-11 龙山组滨浅海沉积序列图

Fig. 7-11 Littoral neritic sedimentary sequence of Longshan formation

2. 障壁岛海岸陆源碎屑沉积

障壁岛海岸沉积是北京地区较广泛的海相沉积类型之一，串岭沟组、洪水庄组和下马岭组均发育有此类沉积。障壁岛海岸存在砂坝或砂洲、生物礁或水下隆起等障壁地形，使得近岸海域和广海隔绝或部分隔绝。在障壁岛外侧广海地带，水动力条件以波浪为主，而在障壁岛内侧，水动力能量较低，以潮汐作用为主，波浪作用较弱。

障壁岛海岸沉积物源复杂，以陆源碎屑物质为主，有时形成障壁地形的障壁岛也发育为生物礁内源碳酸盐岩沉积。障壁海岸沉积可进一步划分为障壁岛、潮坪、潟湖、滨岸等沉积类型。

（1）障壁岛沉积

本区障壁岛沉积复杂，串岭沟组为藻礁障壁，而洪水庄组和下马岭组则为砂坝障壁。由于洪水庄组和下马岭组砂坝沉积均位于区外，这里不进行论述，仅对串岭沟组藻礁障壁沉积进行总结。

串岭沟组藻礁障壁沉积分布在北西向转换断层和北东向同沉积断层以西的通州—平谷—密云—怀

柔北—延庆北部地区，呈新月形展布。以东（蓟县海槽）沉积厚度537~1167m，以西（冀西海湾）急剧变薄，厚30~100m。藻礁障壁沉积自东向西又可分为二小类沉积，由礁前高能礁滩砂坝和藻礁白云岩构成。

1）礁前高能礁滩砂坝岩石组合为白色石英砂岩（占60%）、紫色含铁质粉砂岩与黑色富钾页岩（占40%），发育波状层理、交错层理、砂泥韵律层理、菱铁矿及褐（黄）铁矿结核等标志组合，潮道中产有磷块岩透镜体（密云盆子峪）。

2）藻礁白云岩由层柱、锥柱状叠层石白云岩组成，以巨厚层大型透镜体直覆于常州沟组沉积石英岩之上，向上为黑、鲜绿色富钾页岩、粉砂岩。平谷—蓟县一带钾含量高（K_2O=8%~12%），钾矿层3~5层，走向延伸达10km。页岩中泥裂、砂质注入构造及波状层理发育。

（2）滨岸沉积

滨岸沉积为临近古陆的海滩上部沉积，相当于平均高潮线之上的潮上带沉积，只有在特大风暴或者异常高潮时才被水淹没，受到波浪和弱水流作用。滨岸沉积物以陆源砂为主，在较浅的地方可能沉积粉砂或泥质。滨岸沉积一般发育干裂、水平层理和小型交错层理，由于沉积物多为砂，沉积构造难以保存下来。串岭沟组和下马岭组发育该类沉积，而洪水庄组由于为燕山海盆第二次海侵期，海域面积大，本区位于海盆内部，不发育此类沉积。

串岭沟组滨岸沉积仅分布在房山磁家务一带，下部以石英砂岩、长石砂岩为主，次为黑色页岩与灰黑色石英砂岩互层，上部以黄褐色长石砂岩为主，次为灰黄色石英砂岩和长石石英砂岩，夹灰色砂质页岩。

下马岭组滨岸沉积分布在密云、怀柔北部，为临近内蒙古陆的边部地带，属下马岭组海进旋回的初期沉积。由于盆地周缘的富镁碳酸盐岩蚀源区曾一度处在温湿古气候环境下，遭受较强的风化作用，使氧化铁、钾、钙、镁等元素淋失迁移、泥粉砂级物质经搬运一起进入沉积盆地，因而造成页岩、粉砂岩（包括赤铁矿岩）密韵律层沉积。页岩与粉砂岩比为1.3∶10，该比值随离岸越远而增大。当粉（细）砂供给不足时形成透镜层理，铁质供给不足则形成铁质透镜体。当粉砂与泥质供给大致相当时，形成毫米级砂泥韵律层理，自下而上可见褐红—灰绿色、灰绿—灰色、灰黑—浅红色韵律交替。颜色交替与有机质及氧化铁含量高低有关。当微古植物或*Chuaria*等藻类相对富集时，页岩碳质增高，颜色发暗，纸片状层理发育。当铁质较多时，则呈浅红褐色，劈开性差，当黏土矿物伊利石、海绿石含量增加时，页岩呈黄绿色。伊利石页岩的化学成分与克拉克统计的页岩平均化学成分大体相仿。元素比值$Fe^{2+}/Fe^{3+}=0.26$，Ca/Mg=0.49，K/Na=16.17，Na/Ca=0.48，属碱性氧化条件下的滨岸粉砂泥岩相沉积。

（3）潟湖沉积

潟湖由障壁岛将其与广海隔开，仅在排水口与外海相连，海水处于封闭或半封闭状态，海水能量低，以潮汐作用为主，沉积物为粉砂和泥质，发育水平层理。根据沉积时气候条件引起的海水含盐量变化情况，分为淡化潟湖和咸化潟湖沉积。

串岭沟组潟湖沉积处于障壁西侧的延庆和昌平地区，古水域呈椭圆形。岩石组合以黑色页岩、泥质粉砂岩为主（75%），含砂泥晶白云岩（铁白云岩）次之（25%）（图7-12）。相标志组合主要是鲕状黄铁矿、黄铁矿粒序韵律、水平纹层构造、白云岩透镜层理、育卡布拉（*Eucapsiphora*）叠层石等。经10个黄铁矿硫同位素测试结果证实，黄铁矿属沉积成因。$\delta^{34}S$=9.84‰~22.57‰，远离CCD（‰）值，均为正值，变化范围小（13.73‰），频率分布直方图显示正偏态塔式结构特征，其中值为18‰。S同位素类型属海相富重硫（^{34}S）型，表明黑色富钾伊利石页岩中黄铁矿是在生物细菌参予下进行强分馏效应

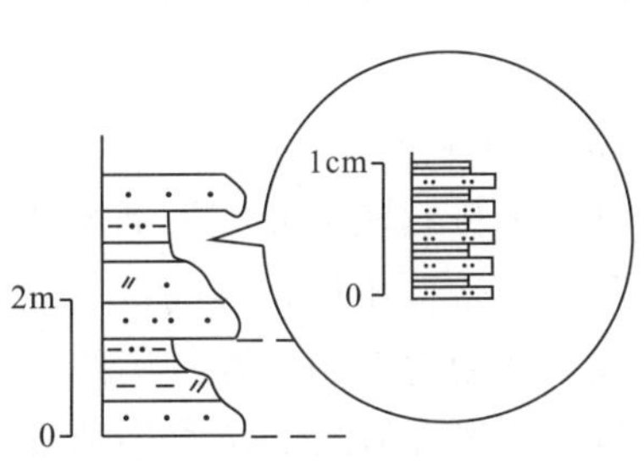

图7-12　串岭沟组潟湖沉积序列图
（据鲍亦冈等，2000）
Fig. 7-12　Lagoon sedimentary sequence of Chuanlinggou formation

作用下形成的。氧化还原电位属强还原相，富有机质，沉积界面远低于氧化还原界面，总体属半封闭潟湖环境。垂向层序上，黄铁矿含量由多变少到无（白云岩）构成韵律层。每个韵律下部为富含黄铁矿（鲕状）泥岩，黄绿色黄钾铁钒发育；中部为含黄铁矿（他形）泥质粉砂岩，黄白色黄钾铁钒发育，上部为白云岩。单个韵律厚2~6m，共3~6个。这种韵律显示潟湖水体由咸化向淡化的演变特征。地球化学标志Ca/Mg=1.8，$Fe^{3+}/Fe^{2+}<1$，Fe/Mn=34，K/Na=8.59，Na/Ca=0.06，属水体不深、碱性还原条件的咸化潟湖环境。

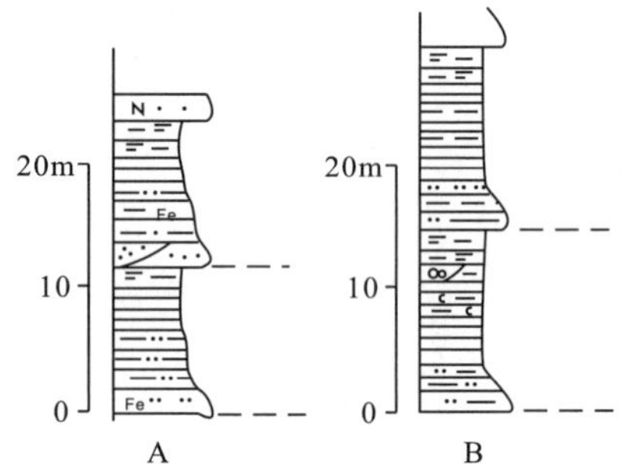

图7-13　下马岭组潟湖沉积序列图
（据鲍亦冈等，2000）
Fig. 7-13　Lagoon sedimentary sequence of Xiamaling formation

下马岭组潟湖沉积广泛发育，区内除怀柔、延庆北部为滨岸沉积外，全区均为此类沉积。根据岩性、沉积构造组合可分为3小类。①含灰-泥岩潟湖相：平面上展布于海盆的周缘地带，垂向上处于下马岭沉积旋回早期（图7-13A）。在沉积中心区则叠置在滨岸含铁粉砂泥岩相之上。岩性为灰—灰黑色页岩夹极薄层粉砂岩透镜体、灰绿色伊利石页岩夹泥灰岩结核。下部出现波状层理、毫米级纸片状层理和粉砂质透镜层理，中上部发育水平层理，多见泥灰岩结核及黄铁矿。下部由于有赤铁质粉砂岩加入，氧化与还原系数波动在-1~+1之间，表现出氧化与还原环境的交替变化，中上部伊利石页岩中分布的泥灰岩结核，其元素比值$Fe^{2+}/Fe^{3+}=1.6$，Na/Ca=0.01，K/Na=1.9，Ca/Mg=2.18，Fe/Mn=7，属水体相对稍深的海相环境，在碱性增加的还原条件下形成的。全组不同层位元素比值Ca/Mg=0.45、K/Na=6.1、Na/Ca=0.35，$Fe^{2+}/Fe^{3+}=1.0$、Fe/Mn=17反映总体属于水体宁静相对稍深、盐度稍高，水体滞流较闭塞的潟湖相沉积环境。②含硅泥岩潟湖沉积（图7-13B），平面上构成本区主体沉积，在沉积中心地区，叠置在含灰-泥岩潟湖相之上。垂相序列上是下马岭组沉积旋回上部的退积序列沉积（相当于中期沉积）。岩性组合上，下部为黑色胶磷石硅质泥岩夹灰色页岩，上部为黑色（碳质）页岩及黄绿色伊利石页岩夹板状粉砂岩。硅质泥（页）岩坚硬性脆，经差热分析黏土矿物为胶磷石、伊利石，含多量有机质与隐晶硅质蛋白石。硅质形成显微浅色层，黏土及有机质形成暗色层，使岩石显现水平纹理。水平纹理是水体宁静缓慢沉积作用的产物。高硅富铝的岩石化学特性可与普通页岩相区别。非晶质氧化硅蛋白石是化学沉积的产物，是由经生物化学作用的氧化硅在沉积时加入到正常页岩内产生的。沉积中心硅质页岩单层厚度大、层数多，向海盆边部有逐渐递减趋势。黑色碳质页岩，普遍发育纸片状层理，含黄铁矿、粉砂量明显增加，有机质含量显著大增加（十三陵为2.47%、夏家沟为5.69%）。钙镁等亲碳酸盐元素在有机质作用下，局部凝聚沉淀，形成泥灰岩结核（怀来等地）。综合上述，该类沉积下部水体滞流宁静，由盐度差形成双层水结构，使底部贫氧条件下有机质腐解，还原作用增强，而上部水体变浅，水介质处于强还原环境，总体属于潟湖相沉积环境。③藻灰岩潟湖沉积，主体分布于河北省内，仅在延庆西部少许发育，这里不做赘述。

（4）潮坪沉积

潮坪为具有明显周期性潮汐活动，但无波浪作用的平缓倾斜的海岸地区。潮坪可分为潮上带、潮间带和潮下带，主体是潮间带。潮坪沉积水动力条件以潮汐作用为主，水流方向具有双向性。沉积物粒度大多数向海变粗，在高潮线以上以泥质为主，在低潮线以下以砂为主，在高潮线和低潮线之间过渡地带沉积砂泥互层。沉积构造复杂，各类沉积层理均有发育。本区串岭沟组发育此类沉积。洪水庄组和下马岭组潮坪沉积主要分布在区外。

串岭沟组发育潮间和潮下沉积。潮间沉积（图7-14A）发育于藻礁障壁岛后壁，为低能混合坪砂-泥岩相沉积，岩石组合为黑色页岩（b）、泥质（c）为主（75%），含砂泥晶白云岩（a）或铁白云岩（d）次之（25%）。其沉积相标志及岩石组合特征基本同潮间低能潟湖相。潮下沉积较为复杂，根据岩性、沉积构造等特征组合划分为两类。①潮下低能泥岩沉积：分布于平谷东、天津蓟县、河北兴隆地区。岩石类型主要为黑色页岩（93%）、砂岩（7%），页岩成分为伊利石，含极少量海绿石，有机碳含量 0.35%~1%。岩性序列下部由细砂岩-粉砂岩-黏土岩组成厘米级韵律层序；中部为黑色碳质伊利石页岩，含黄铁矿结核（褐铁矿化）；上部为黑色伊利石页岩夹深灰色细砂岩及泥晶白云岩透镜体。由下向上相标志组合为：干裂、小型交错层理、波状层理、水下滑动包卷层理、黄铁矿结核、碳质残片、砂质透镜层理等，反映岩类和沉积环境的标型元素主要为钠、钛、铁，含量均高于克拉克值。Fe/Mn=278，K/Na=6.71，$Fe^{3+}/Fe^{2+}=2.8$，Na/Ca=3.96，黏土成熟系数（K+Na）/Al=0.87，是潮间泥坪-潮下泥坪-潮间泥坪的沉积序列。②潮下高能鲕肾状藻铁岩沉积：分布于延庆、怀柔北部，为冀西海湾组成部分。岩石组合为鲕肾状藻铁岩（宣龙式铁矿）、黑—绿色页岩夹铁质砂岩（图 7-14B）。含铁岩系西厚东薄，西部为肾鲕状赤铁岩，东部为菱铁岩。肾状叠层石柱体间伴有鲕粒充填。其组成物质以赤铁矿为主、磁铁矿次之，黏土矿物为高岭土、绿泥石。铁质岩之上为含铁砂岩，发育板状斜层理、波痕和压扁-透镜层理，见海绿石。向上为鲕状铁质岩，单鲕与复鲕并存，磁铁矿与赤铁矿比 2:1。垂向序列上部为黑色碳质页岩、含钾页岩及叠层石白云岩。发育叠层石 *Eucapsiphora* sp.，*Inzeria* sp.，*Linella* sp. 等（杜汝霖，1980）。元素比值 Ca/Mg=1.82，Na/K=5.1，Na/Ca=0.2，上部含碳质、含黄铁矿页岩，$Fe^{3+}/Fe^{2+}<1$，属还原环境，下部藻铁岩 $Fe^{3+}/Fe^{2+}=6.96$ 属氧化环境。铁质来源于内蒙古陆，呈铁的氢氧化物胶溶体运移入海。

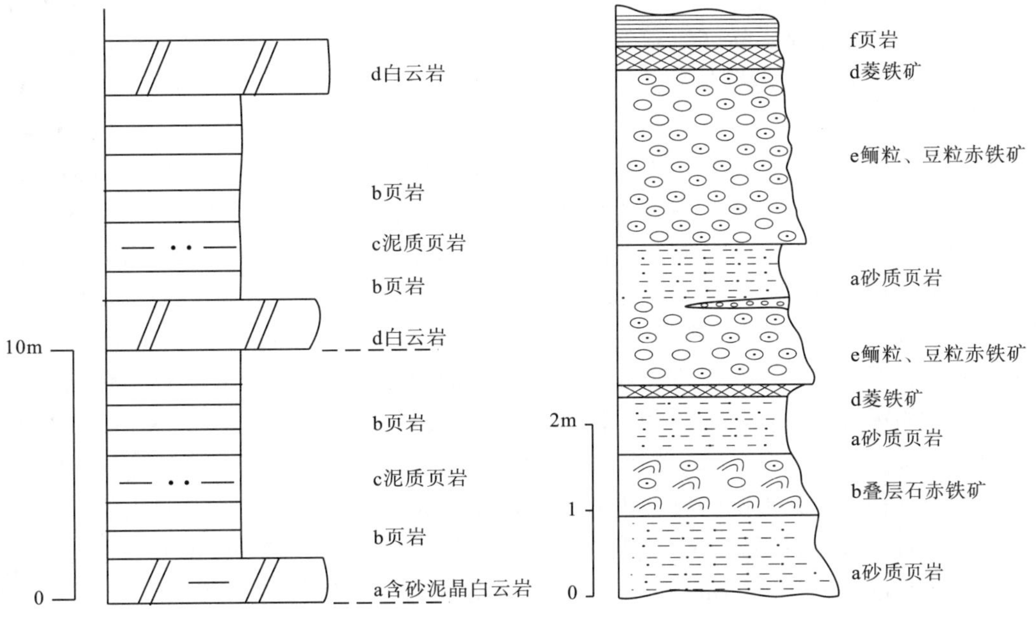

图 7-14 串岭沟组潮间混合砂-泥坪沉积序列图

Fig. 7-14 The intertidal mixed sand and mudflat sedimentary sequence of Chuanlinggou formation

3. 浅海陆棚沉积

浅海陆棚沉积环境一般指正常浪基面以下至水深约 200m 之间的较平坦的广阔浅水海域，平均坡度一般只有几分，最大不超过 4°（陈建强，2004）。浅海水动力作用复杂，受洋流、波浪、潮汐及密度流综合影响，沉积物主要是粉砂质泥及粉砂，发育交错层理、对称或不对称波痕及少量冲刷和沟槽充填构造。

洪水庄组发育此类沉积，根据岩性、沉积构造等特征，可以划分为两小类沉积。①浅海泥岩沉积，分布在海槽中心部位的四海、兴隆一带，黏土岩占 80%~90%，碳酸盐岩占 10%~20%。以黑色碳质页

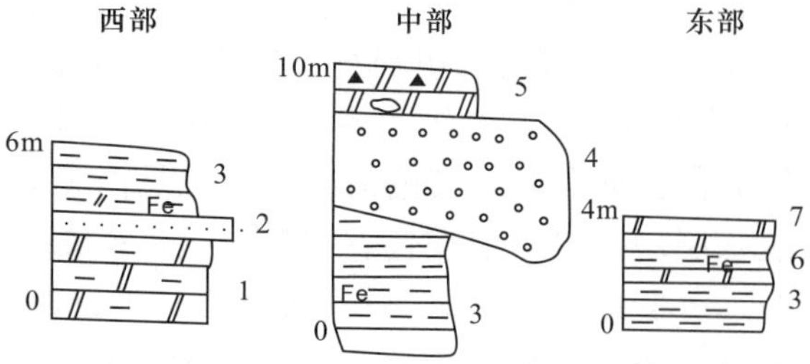

图 7-15 洪水庄组浅海陆棚沉积序列图

Fig. 7-15 Shallow sea shelf sedimentary sequence of Hongshuizhuang formation

1—泥质白云岩；2—石英砂岩；3—泥岩、铁质泥岩；4—滑塌砾岩；5—铁质、沥青质白云岩；6—泥岩与白云岩互层；7—薄层白云岩

岩、黑—黄绿色硅质页岩、灰黑色铁质粉砂质页岩、黄绿色页岩夹泥晶白云岩组合为代表。主要相标志组合：黑色、铁质、水平层理、黄铁矿结核、菱铁矿结核、碳酸盐结核、铁锰质结核、碳质和硅质页岩等。②浅海碳酸盐岩-泥岩混合相沉积，分布于海槽周缘的河北怀来、延庆、怀柔和天津蓟县地区。黏土岩（伊利石）占 50%~80%，碳酸盐岩占 20%~50%，属悬移质混水沉积。岩石以黑色伊利石页岩与泥质白云岩为主（图 7-15），西部（河北怀来、宣化）页岩中含褐铁矿结核和赤铁矿结核，中部（昌平、延庆）含菱铁矿结核，东部（天津蓟县—将军关）含碳酸盐结核。近海盆边缘部位，粉砂、砂质增加、波痕发育。元素比值 Fe/Mn = 16 ~ 75，Ca/Mg = 1.0 ~ 1.5，K/Na = 0.02 ~ 1.5，（K+Na）/Al = 0.4 ~ 0.7，Fe^{2+}/Fe^{3+} = 1.1，水体较深，盐度增高，属碱性弱还原浅海混合沉积。

（二）内碎屑岩（碳酸盐岩）沉积

海相碳酸盐岩沉积是北京地区中—新元古界主要的沉积类型，受生物、水文和自然地理等多种条件的控制，沉积作用因素复杂。按照沉积水动力因素、能量的高低，沉积环境分为潮上、潮间、潮间潟湖、潮下低能、潮下高能（障壁岛）及浅海陆棚沉积。

1. 潮上沉积

潮上沉积位于海平面之上，仅在特大高潮或特大风暴时才被海水淹没，主要岩石类型为白云岩及石膏或硬石膏，陆源碎屑极为普遍，主要为风成及红层沉积，纹理发育，常有泥裂、藻叠层、鸟眼等构造及同沉积构造。团山子组、大红峪组、高于庄组、杨庄组及雾迷山组发育此类沉积。

团山子组潮上沉积为含砂碳酸盐岩沉积，分布范围靠近冀西海湾边部阳原、赤城一带，可能延伸至房山一带，沿着五台-太行古陆呈环带状展布。以陆源碎屑较发育为特征，其垂向序列：下部为灰色中细粒石英砂岩、长石石英砂岩与泥质白云岩互层；中部为深灰色、紫红色粉砂质白云岩，上部为纹层状泥晶白云岩。相标志组合：富陆源砂、长石、菱铁矿、胶磷矿、波痕、泥裂、砂-白云岩韵律层、透镜层理和竹叶状砾屑等，属含砂-碳酸盐潮上-潮间沉积环境。

大红峪组潮上沉积分布在近陆的海盆边部。砂岩占 70%，白云岩占 25%，黏土岩占 5%。垂向上，由含砾长石石英砂岩—长石石英砂岩—石英岩状砂岩夹翠绿色富钾页岩—海绿石石英砂岩—含硅质豆状砂质白云岩构成水进层序。下部富含陆源长石、燧石（玛瑙）质粗砂屑及紫红色细砾石（庞家堡、房山、黄崖关），常见水流波痕，大中型帚状交错层理，中部砂岩成分结构成熟度比下部高，发育浅水干扰波痕；中上部含海绿石及豆状硅质岩、含砂白云岩、显示出潮上沙滩沉积特点。

高于庄组潮上沉积分布于本组内顶、底部（图 7-16），为含硅白云岩沉积，以陆源物质发育为特征。高于庄组底部由巨厚层白云质砂岩—含泥隐晶白云岩—含砾屑白云岩构成潮上—潮间水进韵律，顶部则由球团粒硅质岩—锥状叠层石白云岩—波纹状燧石条带白云岩—砂质白云岩—泥晶白云岩组成水退韵律。相标志组合：斜层理、波浪波痕、近岸浅水干涉波痕等构造。元素比值：Ca/Mg = 1.4，Fe/Mn = 1100，K/Na = 0.6，Na/Ca = 0.01（图 7-9）。

雾迷山组潮上沉积为含砂碳酸盐岩沉积，分布于赤城—古北口一带，其他地区受水退影响，也有分布，为次级沉积相。时间序列上分布于各段的顶、底部。沉积韵律为底部陆源物质（粉砂、泥质）-含砂屑白云岩-泥质白云岩组合。沉积相标志有纹层、断续纹层状、波纹状构造，纹层藻叠层石、浅水波痕、含泥含砂屑、粒屑条带及透镜状层理、沟槽充填结构、干裂、鸟眼构造等。元素比值：Ca/Mg = 1.43，Fe/Mn = 37，K/Na = 0.1。主体反映潮上沉积环境。

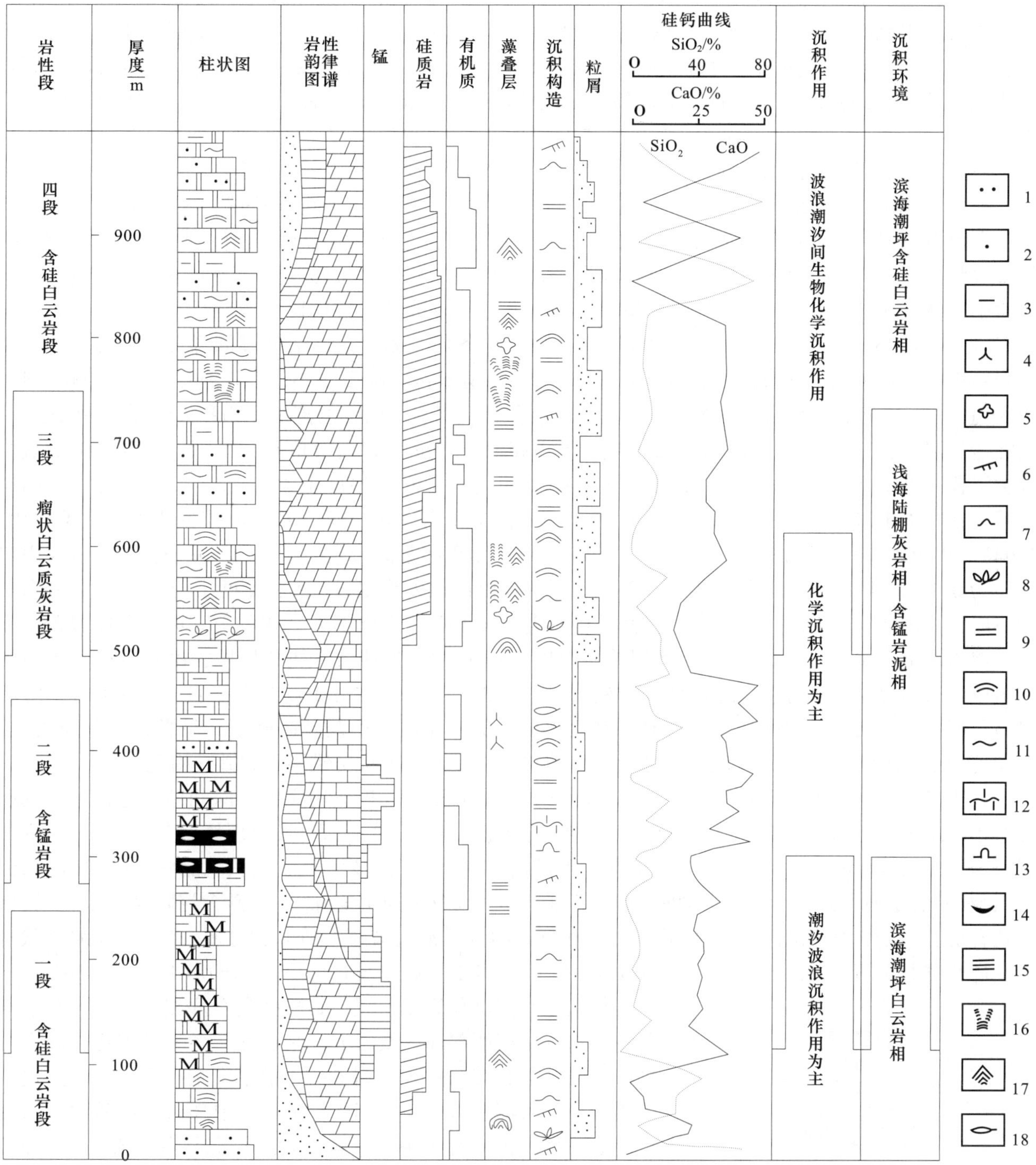

图 7-16　北京十三陵高于庄组沉积相剖面图

（据北京市区域地质志，1991）

Fig. 7-16　The sedimentary facies profile of Gaoyuzhuang formation in Beijing Ming Tombs

1—石英砂；2—粒屑；3—泥；4—藻屑；5—藻团；6—交错层；7—波痕；8—竹叶状砾屑；9—水平层理；10—波状纹层；11—硅质条带；12—硬地；13—瘤状构造；14—泥皮（压扁层理）；15—水平纹层藻叠层；16—柱叠层石；17—锥叠层石；18—透镜状层理

2. 潮间沉积

潮间沉积位于高潮线和低潮线之间，受波浪和潮汐水动力共同作用，但以潮汐作用为主。主要岩石类型为泥晶白云岩，含生物化石，潮汐层理发育，局部可见小型交错层理和干裂构造。团山子组、大红峪组、高于庄组、雾迷山组及铁岭组均有分布，是中—新元古界发育最广泛的沉积类型。

团山子组潮间沉积紧邻潮上沉积，发育为潮间混合坪相。分布于本区北部古子房、古北口一带，由白云岩与粉砂质页岩—含硅砂质白云岩与泥质白云岩（夹玻屑凝灰岩）—细砂岩与砂质页岩构成退积层序，顶部砂岩含长石。标志组合：交错层理、透镜层理、砂-泥-碳酸盐岩韵律层序、干裂等。

大红峪组潮间沉积根据岩性组合分为两小类。①含砂白云岩沉积：分布于延庆、昌平、怀柔、密云、平谷、兴隆、蓟县地区。白云岩占60%，砂岩占35%，黏土岩（硅质岩）占5%。垂向序列：下部由白色石英岩—灰色粉砂质泥云岩构成3~5个海进韵律，上部碳酸盐地层中由纹层状—小波纹状—大波纹状—柱叠层等原生沉积构造组合构成海进韵律层序。总体属于潮间上—潮间下沉积环境。下部韵律中砂体成分结构成熟度高，呈巨厚块状层理，发育浪成波痕。元素比值：Fe/Mn=21.2，Ca/Mg=1.5，K/Na=7.5，Na/Ca=0.03，$Fe^{2+}/Fe^{3+}=1$。属氧化条件下温、浅、清的潮间坪沉积环境。②碳酸盐岩—火山岩相沉积：分布在以平谷为中心的密云—蓟县地区。喷发中心在平谷熊耳寨、挂甲峪一带。距火山源由近及远，大体可划分为火山碎屑基浪沉积相、水下火山碎屑流沉积相和火山硅质—陆源砂-碳酸盐岩混积相三类微相（孟祥化，2004）。火山碎屑流相沉积为火山角砾和沉积岩角砾与碳酸盐岩混合沉积，见于平谷—蓟县大红峪组火山岩-沉积岩相序中，单层厚度1.4~2.2m，仅发育在近火山口地区，在古火山口边缘火山碎屑沉积厚度大于30m，向外侧逐渐减薄，火山角砾明显变小（孟祥化，2004）。火山碎屑基浪相见于平谷古火山口附近，厚度随源区距离增大而呈对数级减小；平谷古火山口直径为0.1~1.8km。火山碎屑基浪沉积距离邻近的火山口距离约2km；沉积厚度小于2m，并且分布范围不广，火山活动以溢流相熔岩为主，表明其形成时火山喷发规模不大；相序上，该相直接覆盖于玄武质熔岩之上，由a、b、c、d四个沉积单元组成；a、b单元为沙波—块状层段，由砂级火山碎屑沉积组成，火山碎屑占碎屑总量的65%~85%，粒度0.03~1.8cm，成分主要为玄武岩屑、晶屑等；陆源碎屑包括长石、石英及碳酸盐岩碎屑等，基质组分为泥、泥质火山灰及碳酸盐，呈颗粒或基质支撑；c、d单元为火山碎屑岩，发育波状层理及小型沙纹交错层理、脉状交错层理和砂丘构造。

高于庄组潮间沉积在空间上广泛分布，是该组一种重要的沉积作用类型。时间序列上，潮间沉积分布于该组上、下部（图7-16）。下部由含硅质结核隐（微）晶白云岩-波纹层燧石条带白云岩-锥叠层石白云岩构成潮间-潮下（上）海进韵律。上部由含泥粒屑白云岩-大波纹状隐晶白云岩-藻团白云岩-锥柱状叠层石白云岩组成多个潮间-潮下海进韵律，以及由同生角砾岩-含砂含泥隐晶白云岩-柱状锥状叠层石白云岩组成潮间-潮上海进韵律。相标志组合：浪成波痕、竹叶状砾屑、波纹-大波纹状构造及硅化锥叠层石。元素比值：Ca/Mg=1.4，Fe/Mn=4.3，K/Na=4.1，Na/Ca=0.01（图7-10）。

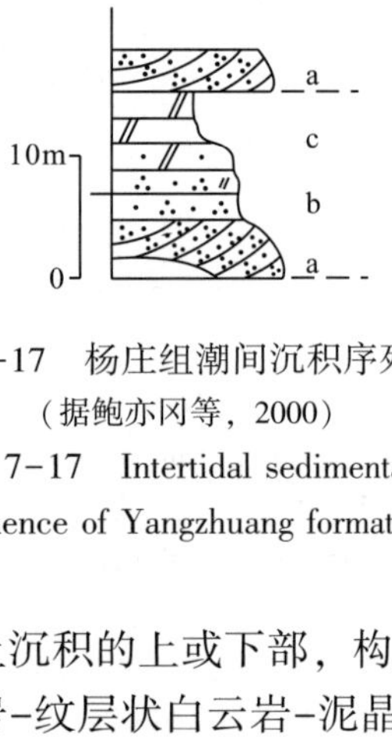

图7-17　杨庄组潮间沉积序列图
（据鲍亦冈等，2000）
Fig. 7-17　Intertidal sedimentary sequence of Yangzhuang formation

杨庄组潮间沉积为砂、泥、碳酸盐混合坪相沉积，分布于宣化—怀来—延庆西部地区及密云、平谷区。各岩类占比：砂岩占17%，泥岩占15%，白云岩占68%。潮汐韵律（图7-17）组成表现为下部石英砂岩（主）、砂岩（a）、中部白云质砂岩、白云质泥岩、紫红色燧石团块白云岩夹紫红色泥晶白云岩（b）和上部红、白两色相间的泥质白云岩或含砂白云岩（c）。韵律底部发育波痕、斜层理，中部发育干裂、砂质注入构造，上部叶片状泥晶白云岩以水平层理、黄白色椭圆形钠质斑点为特征。

雾迷山组潮间沉积空间分布最为广泛，全区内几乎均有分布。时间序列上，各段均有分布，一般同潮上沉积相伴生，多分布在潮上沉积的上或下部，构成水进或水退沉积韵律。岩石组合为大波纹燧石条带白云岩-纹层藻叠层石白云岩-纹层状白云岩-泥晶白云岩。沉积相标志：燧石条带、纹层藻叠层石、纹层状构造、波纹状构造、核形石、收缩裂、泥坡。元素比值：Ca/Mg=1.42，Fe/Mn=5.1，K/Na=0.13，Na/Ca=0.03，（K+Na）/Al=1.08。

铁岭组潮间沉积为混合坪碳酸盐相，处于海盆近边缘部位通州、房山一带。岩性以富含陆源物质的砂岩、页岩为特征。泥晶白云岩富含硅质粒屑与铁质，属混水沉积。

3. 潟湖沉积

潟湖沉积特征同陆源碎屑障壁岛潟湖沉积。海水循环受到限制，一般比较浅，盐度变化较大，淡水、盐水、超盐水均有。潮间、潮下均发育有此类沉积。团山子组、高于庄组、雾迷山组均发育此类沉积。

团山子组潟湖沉积（图7-18）分布于冀西海湾宣化—延庆—昌平一带。下部紫黑色含粉砂质白云岩、白云质泥岩，中部夹2~3层竹叶状砾屑白云岩，上部硅质条带泥晶白云岩。相标志组合：黄铁矿、石膏（新镇西）、岩盐假晶、干裂、水平纹层构造、具氧化晕圈的竹叶状砾屑、包卷层理。元素比值：Fe/Mn = 37.6，Ca/Mg = 1.59，K/Na = 13.21，Na/Ca = 0.02，K_2O=4.46，属富钾碱性还原咸化潟湖相沉积环境。

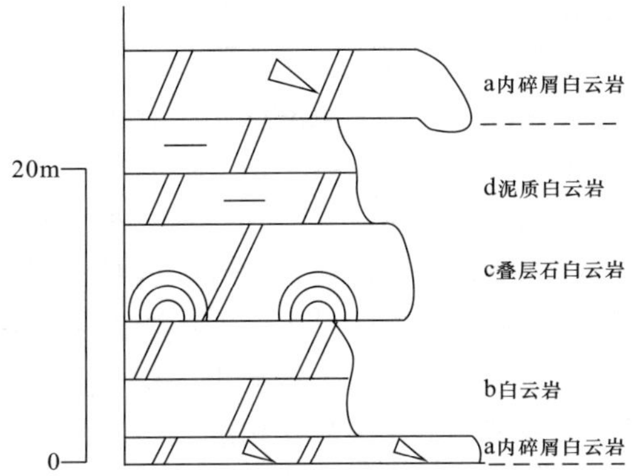

图7-18　团山子组潟湖沉积序列图
（据鲍亦冈等，2000）
Fig. 7-18　Lagoon depositional sequence of Tuanshanzi formation

高于庄组潟湖沉积一般同潮间沉积伴生，空间上分布于延庆—昌平一带。岩性组合以黑色粉砂岩与含锰白云岩组成韵律性互层。沉积相标志：韵律互层、层面发育铁质氧化薄膜、波痕、石膏纤晶、波状纹层（图7-16）。

雾迷山组潟湖沉积分布广泛，区内均有分布。时间序列上，主要分布在一、二、三段，往往与上部潮间带沉积连续过渡。岩石组合为密纹-密波纹层沥青质白云岩、有机质藻白云岩、层纹状燧石岩（产藻丝体）夹风暴岩（菊花状同生角砾岩）。潮下潟湖沉积主要分布于三段，沉积相标志：有机质、沥青质、藻席微结构、密纹层、密波纹层藻叠层石、小型丛状叠层石、微干裂、菊花状构造、沟槽充填结构。潮间潟湖沉积主要分布在一段，沉积相标志：有机质、沥青质、密波纹藻叠层、藻席微结构、微细干裂、鸟眼构造、冲淤构造、菊花状构造、小型丛状叠层石、小杯叠层石。

4. 潮下沉积

潮下沉积位于平均低潮线以下、浪蚀基准面以上的浅水区域，波浪作用频繁，含大量生物。团山子组、高于庄组、杨庄组、雾迷山组及铁岭组、景儿峪组均发育各有特征的此类沉积。

团山子组潮下沉积分布于平谷—蓟县一带，垂向上该相叠覆于潮坪含砂碳酸盐岩相之上。岩类组合：下部为灰白色厚层白云岩、假鲕状硅质条带白云岩、竹叶状砾屑白云岩（0.3~1.9m）；中部灰紫色凝灰质角砾岩、火山碎屑白云岩、碳酸盐化粗面岩（仅平谷地区）；上部为锥柱状叠层石白云岩。相标志组合：波浪波痕、大—中波纹状构造、鲕球粒硅质岩、柱-锥状叠层石等，为潮下高能带沉积环境。

高于庄组潮下沉积为含锰泥岩相（图7-16），空间分布面积最广，是高于庄组主要的沉积作用类型。时间序列上，高于庄组中部（相当于二段）为此类沉积。由含锰页岩、紫红色泥晶白云岩、含锰泥晶白云岩、浅灰色含泥、屑隐晶白云岩组成。发育水平层理、波浪波痕。元素比值：Fe/Mn=2.5，Ca/Mg=1.4，K/Na=1.54，Na/Ca=0.01，属水体较深滞流宁静的潮下低能带沉积环境（图7-9）。

杨庄组潮下沉积为潮下白云岩相，分布于平谷、蓟县、兴隆以东地区。岩石类型以红白两色含泥晶白云岩、紫红色粉砂质泥质白云岩为主。垂向上潮下白云岩相叠覆在潮间混合坪相之上。下部由紫红色含砂白云质灰岩、暗紫红色含泥质白云岩、灰白色微含沥青质白云岩间夹燧石结核白云岩组成潮间-潮下进积层序，下部发育楔状交错层理、波痕；中部发育黄色串珠状结核沿层理断续分布，结核核心呈海蓝色，被蓝、白两色相间的同心环（3~4圈）包围，顶部则为黑白相间的同心圆状和似猫眼状燧石结核，微含沥青质；中上部由紫红色含泥白云质灰岩、灰紫红色粉砂屑白云质灰岩、砖红色燧石团块（结核）白云岩、紫红色钙质页岩及深灰色沥青质细晶白云岩组成潮下-潮间退积层序。沉积构造标志：黑白或红白相间的互层韵律、灰质组分、砂屑构成波痕及小型楔状交错层理，海蓝色燧石结核、黑色豆状燧石发育、干裂、砂质注入构造、岩盐假晶、白色圆形斑点发育，含沥青质、黄铁矿。元素比值：Sr/Ba=0.134，K/Na=7.7、Ca/Mg=1.5，Fe/Mn=29。其中，Ba含量（0.31%）高出克拉克值14倍，远大于正常海水Ba含量值0.5×10^{-6}。且镜下见碎屑状态的重晶石，揭示Ba呈硫酸盐被泥质吸附；Li含量（0.018%）超出克拉克值10倍，Li一般在半封闭咸水潟湖环境中含量较高。

雾迷山组潮下沉积仅分布在三段，分布有限。岩性组合为浅灰色巨厚层状白云岩，贫藻迹。总体反映潮下低能沉积环境。

铁岭组潮下沉积为碳酸盐相，分布于海槽周缘的怀来、昌平及沿河城等地，属波浪沉积作用为主的高能带相区。以白云岩、白云质灰岩、微含锰白云（灰）岩为主，广泛发育竹叶状砾屑柱（锥）叠层石。相标志组合：波状层理、高幅对称波浪波痕（波长=60cm、波高=35cm）、斜层理及交错层理，砾屑粒级递变层序。砾屑最大砾径（长轴）50~80cm，平行或斜交层面排列，其成分为白云质、泥质，系波浪潮汐将上部潮间带未固结成岩的灰泥片携带搬运沉积而成。交错层古水流指向北西向。

景儿峪组潮下沉积为盐泥相沉积。垂向序列上，位于青白口系顶部，属青白口纪末期以化学沉积作用为主的富钙贫镁碳酸盐相沉积。岩性单一，以浅灰、紫红、蛋青、黄绿等杂色泥晶灰岩为主。岩石成分由方解石组成，扫描电镜下，方解石晶体间被球粒状胶磷石、硅质球、石英雏晶和伊利石包围，含量达70%~90%，粒径<0.01mm，含少量棱角状石英粉砂、绢云母鳞片，以自生隐晶硅质为特点。由下向上，泥硅质成分增高。上部岩石硅泥质含量可达30%，成为硅质泥质泥晶灰岩、泥灰岩和硅质泥晶灰岩；顶部为硅质页岩（SiO_2=66.79%，Al_2O_3=9.37%，CaO=6.28%，MgO=2.14%）及白云质硅质泥岩，底部夹多层钙砂质胶结的海绿石泥灰质角砾岩（SiO_2=27.56%，Al_2O_3=2.49%，CaO=5.54%，MgO=0.61%）。发育水平层理、缝合线，含黄铁矿和水下缩裂。Ca/Mg=77~45，Fe^{2+}/Fe^{3+}=1.2~6.67，K/Na=3.6~6.78，Na/Ca=0.006~0.1，属水体较深、宁静低能海相还原环境下的潮下盐泥相沉积。

5. 生物礁沉积

生物礁沉积形成深度从海水表面到水深200m处，有些地方可延伸至水深500m处。生物礁相一般厚度比相邻地区大，具有完整的生物骨架，沉积特征取决于碳酸盐岩陆棚边缘带的坡度。团山子组、雾迷山组发育该类沉积。

团山子组生物礁沉积受燕山裂陷槽盆地同沉积断裂影响，发育为陡坡型沉积，宏观主体分布于盆地断裂西，以密云地区最为典型。垂直层序上，下部为含铁白云岩、泥质白云岩夹页岩，中部为两层竹叶状砾屑白云岩，上部为锥柱状叠层石白云岩、铁白云岩和泥晶白云岩。相标志组合：水平纹层构造、包卷层理、竹叶状砾屑、干裂、岩盐假晶等。元素比值：Fe/Mn=60.58，Ca/Mg=1.55，K/Na=26.01，Na/Ca=0.13，Fe^{2+}/Fe^{3+}=0.72，属碱性富钾高盐蒸发的生物礁相沉积环境。

雾迷山组生物礁沉积广泛分布，尤其在一、二、三段，总体表现为潮间下-潮下上藻礁障壁沉积。岩性组合以鲕白云（硅）岩、巨厚藻团鲕白云岩、竹叶状砾屑风暴岩、大型穹柱、穹锥状叠层石白云岩组成。沉积相标志：凝块、藻团、锥-柱状叠层石（雅库特锥叠层石）、大波浪波痕、交错层、核形石、高能鲕、大波纹状构造、厚层、泥坡、竹叶状砾屑等。元素比值：Fe/Mn=16，Ca/Mg=1.41，K/Na=0.11，Na/Ca=0.012，（K+Na）/Al=1.30。

6. 浅海沉积

浅海沉积位于平均浪蚀基准面以下浅水区域，只有在海洋风暴时候，才受到波浪影响。主体碳酸盐沉积作用始终处于较宁静的水体沉积环境。沉积物源为远洋浮游生物和远洋软泥相互交替。高于庄组、雾迷山组和铁岭组发育该类沉积。

高于庄组浅海沉积为潮下灰岩相，空间分布上位于平谷—蓟县一带，时间序列上相当于三段沉积。由深灰色灰质白云岩→板状含灰白云岩与紫色极薄层泥晶白云岩韵律→瘤状含灰白云岩→灰质白云岩与白云质灰岩韵律→含灰白云岩与泥晶白云岩韵律构成浅海陆棚相退积层序。相标志组合：下部砂屑含量25%，中上部7%，泥质则由下向上增加，上部达30%。板状层理、中层与薄层韵律层序、瘤状构造、铁锰质结核、钙质与镁质条带状层理、碳酸盐透镜-压扁层理（宋天锐，1985）。元素比值：Fe/Mn=30，Ca/Mg=3.1，K/Na=1.8，Na/Ca=0.01，CaO=31.2%，MgO=10.85%（图7-9）。

雾迷山组浅海沉积为浅海陆棚盆地白云岩相。分布于平谷—蓟县一带，时间序列上一段、三段最为发育，往往与潮下带沉积构成连续过渡，总体沉积范围不大、时间不长。由中纹层状白云岩-中厚层燧石条带白云岩构成基本沉积层序（图7-19），无明显的沉积构造。

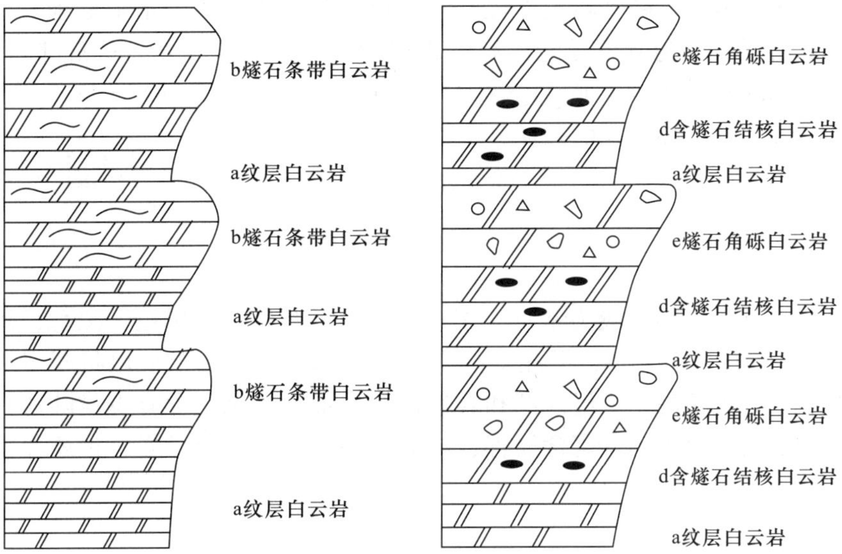

图7-19　雾迷山组浅海陆棚白云岩相沉积序列图

Fig. 7-19　Dolomite facies depositional sequence of shallow sea shelf basin of Wumishan formation

铁岭组浅海沉积也为陆棚铁锰碳酸盆相，空间分布于延庆、四海、密云、兴隆地区。岩石类型主要以巨厚层含锰灰（云）岩夹含锰铁矿层及黑色泥岩。含锰铁矿层以鲕肾状赤铁矿、角砾状扁豆状铁锰矿，似层状及大型透镜状产出为特征。元素比值：Fe/Mn=10.3，Ca/Mg=16.9，Mn=2.86。

第二节　主要沉积岩岩石类型

一、碳酸盐岩类

（一）灰岩类

中—新元古界北京地区以海相富镁沉积为主，实际上并不存在真正的富钙灰岩类沉积。仅在高于庄组三段、铁岭组二段和景儿峪组沉积泥质云质灰岩-泥质白云石灰岩、泥质灰岩-泥灰岩。

1. 泥质白云灰岩和泥质云质灰岩

发育于高于庄组第三段，为潮坪沉积的产物。岩石呈灰色、灰黑色、红色，宏观上由灰岩区至白云岩区构成完整的沉积旋回，每一个旋回从泥质白云石灰岩，向上逐渐转变为泥质云质灰岩及灰质泥质白云岩，层理也由条带状层向透镜状层渐变，下部一般为细粉晶结构，上部为粗粉晶结构。化学成分上MgO含量由3.93%增加到14%。

2. 泥质灰岩-泥灰岩

发育于景儿峪组，为潮下带至浅海海底带相对较深沉积环境。岩石多呈粉红色，含粉砂质或泥质组分。宏观上呈薄板状、厚板状产出，水平层理发育，偶见大型舌状波痕。微观上以微波状纹理相接触，方解石呈泥晶状，并形成0.2~0.5mm波状纹层的夹层，波状纹层上下为间断面，波状纹层实际上是由平行的水云母透镜体聚集组成。岩石化学成分上则以富钙贫镁多硅铝的特征，平均 $CaO=41.18\%$，$MgO=0.73\%$，$Fe_2O_3=0.85\%$，$CO_2=31.51\%$，$FeO=0.48\%$，$S<0.2\%$，$SiO_2=18.16\%$，$Al_2O_3=3.04\%$，$BaO=0.29\%$，$MnO=0.077\%$，$TiO_2=0.025\%$，$SrO=0.14\%$。

景儿峪组泥灰岩在北京西山下马岭一带见层内同生角砾，角砾近定向排列，角度交汇层及斜坡下滑构造发育。

3. 含白云质灰岩、白云质灰岩及叠层石灰岩

仅发育于密云东智、西智一带铁岭组二段，岩石以泥晶结构为主，水平层理发育，硅质高，且发育有大型柱状、墙状叠层石，为潮下低能浅海沉积产物。岩石化学成分上具富硅高钙贫铁锰的特征，平均 $CaO=42.47\%$，$MgO=3.66\%$，$Fe_2O_3=0.56\%$，$P_2O_5=0.02\%$，$SiO_2=13.16\%$，$Al_2O_3=0.85\%$，$K_2O=0.15\%$，$MnO=0.06\%$，$Na_2O=0.12\%$。

（二）白云岩类

1. 原生白云岩

（1）泥晶白云岩

泥晶白云岩在中—新元古界中较为常见，多为灰白色、灰色、灰黑色、黑色，碳酸盐岩矿物颗粒粒径小于0.01mm，白云石少于50%，泥晶支撑，层纹状构造，贝壳至平坦状断口。区内纯度很高的泥晶白云岩较为少见，仅见于高于庄组顶部和雾迷山组下部，常见为含有少量陆源石英粉砂、砂屑及泥屑等物质的泥晶白云岩，还可出现富含沥青、具波纹层理的泥晶白云岩，是还原条件下的富含有机质的产物。

（2）泥晶—细晶白云岩

多发育于中—新元古界雾迷山组，高于庄组、大红峪组及杨庄组也有薄夹层分布。多为灰色、灰白色、黄灰色，碳酸盐矿物颗粒粒径0.01~0.06mm。区内该类白云岩富含灰白色、灰黑色燧石，燧石多呈平行层理分布，多为断续或连续状，按形态可划分为层纹状、条带状、结核状、团块状及放射状等类型。

（3）泥质白云岩

广泛发育于高于庄组、洪水庄组和铁岭组，串岭沟组、团山子组、大红峪组也有分布。新鲜面多为灰色、灰白色、灰黄色，一般都风化为土黄色。

高于庄组的泥质白云岩一般较薄，夹于砂质白云岩之中，有泥质充填于白云岩裂隙之中。微观上表现出不均匀性，其中包含许多亮晶方解石充填的洞穴，反映了潮上淡水的冲洗作用。干缩孔穴充填的泥质占30%（除碳酸盐以外的泥质不溶物），其中一部分呈条线状成层分布，另一部分呈裂隙充填状伸入碳酸盐岩的“V”字形干裂缝中。

洪水庄组和铁岭组的泥质白云岩是在一套灰绿色粉砂质泥岩与黑色页岩之中，其中泥质含量丰富，呈板状平整分布。微观构造呈水平纹层状，白云石与泥质共生。洪水庄组泥质白云岩具由微晶硅质组成之平行纹层，并出现水下胶缩的显微构造，而且白云石多数呈细粉晶自形晶状，反映了潮下带内部封闭-半封闭的富镁条件。

（4）碎屑质白云岩

发育于高于庄组顶部、团山子组、大红峪组、洪水庄组和杨庄组下部。可以进一步划分为以占显著数量碎屑质为主的粉砂质白云岩、砂质白云岩、砂质假团粒白云岩、角砾内碎屑白云岩、假团粒内碎屑白云岩等五种主要类型。前三种主要是粉砂质—砂质白云岩，后二者可归属为内碎屑白云岩。

粉砂质-砂质白云岩：岩石外表风化面可见到砂粒，由于石英类等陆源碎屑抗风化力较强，因此在其具有斜层理或交错层理的砂质白云岩中容易看清这些构造现象。以0.01mm左右的白云石他形晶粒为主，包含占10%~20%的泥岩-粉细砂级石英、长石等陆源碎屑。

内碎屑白云岩：可以分为角砾状内碎屑白云岩和假团粒内碎屑白云岩两大类。角砾状内碎屑白云岩发育于雾迷山组中，剖面上常呈放射状排列，平面上呈波状弯曲，为风暴流的涡动或风暴浪就地掀起堆积或短距离搬运快速堆积的结果，乔秀夫等（2006）认为是古地震作用的结果；假团粒内碎屑白云岩又可根据团粒大小分为球粒状白云岩和鲕粒状白云岩。球粒状白云岩由灰色泥晶球粒状群体相聚集，含量达 70%~80%，球度较好，圆度不一，次棱角状者常见，胶结物多为细晶白云岩，是潮间带最上部沉积环境中水体渐进开始时初始震荡的一种产物。鲕粒白云岩中的鲕粒是一种由核心和包壳组成的粒径<2mm 的球形或椭圆形颗粒，一般认为是产于浅水台地边缘鲕粒滩的产物。孟祥化等（1991）对北京地区鲕粒白云岩微相类型进行过系统的总结（表 7-3）。

（5）叠层石白云岩

本区中—新元古界从串岭沟组的上部起，至铁岭组止，各组段除了杨庄组和洪水庄组以外，叠层石白云岩间断出现，是最发育的一种白云岩类型。

叠层石按形态可分为层纹状、波纹状、锥状、柱状、半球状、穹丘状、球状、墙状等。本区的叠层石白云岩大体归纳为层状（包括波状等）和柱状（包括锥状等）两大类，其他形态都是这两种形态的过渡或组合。一般来说，层状叠层石的生成环境水动力条件较弱，多属潮间带上部的产物；柱状叠层石的生成环境水动力较强，多属潮间带下部及潮下带上部的产物。层纹状叠层石白云岩发育于串岭沟组至高于庄组一、二段，高于庄组三段起至雾迷山组三段则逐渐出现锥状、球状、丘状等叠层石白云岩，雾迷山组四段至铁岭组出现墙状、锥状、丘状叠层石白云岩。

2. 后生白云岩

主要分布在高于庄组顶部及雾迷山组中，尤以雾迷山组最为典型。宏观上，高于庄组顶部往往为藻纹层状，而雾迷山组可出现厚层状。雾迷山组再结晶白云岩由约 0.3mm 半自形—自形白云石组成，晶形长圆似米粒状，并相互交织如窗格状构造。在多数再结晶的晶体内部，可以看到白云石再结晶长大后残留的内部核心，由分散的小点聚集体显示出来。

表 7-3 鲕粒微相类型及特征表

Table 7-3 Oolitic microfacies types and characteristics

动力场环境分带	鲕粒微相类型	特征及成因解释
异常高能带	豆鲕	比正常同心鲕大，直径 2~4mm，具较大的核心，同心外壳圈生长速度较快，搅动能量最强
	复鲕	比正常同心鲕大的包壳包裹有两个以上鲕粒的复杂结构，复杂高能环境多次搅动成鲕作用
高能带	正常同心鲕	包壳厚度大于核心半径，鲕粒直径达 1~2mm，同心包壳由定向方解石和文石组成
间歇高能带	放射同心鲕	包壳由放射状和同心状（切向排列）的方解石和文石交替组成，鲕粒直径一般为 0.5~1.0mm
低能带	薄皮鲕	包壳厚度远小于核心半径，由一层或很少几层同心包壳组成，鲕粒直径为 0.2~0.5mm
	放射鲕	有细小同心微粒，方解石或文石以核心为生长点，呈放射状排列，多形成于盐度不正常的海水
	偏心鲕	鲕核心及包壳纹层不对称，核心向下层面偏移，形成于障壁海岸静水环境，鲕粒直径一般<0.2mm
渗流及潜流带	复鲕	淡水淋滤已沉积的鲕粒，溶蚀形成空心，保留鲕模
	单晶鲕 复晶鲕	淡水淋滤已沉积的鲕粒，溶蚀空心后再被渗流形成方解石充填结晶
其他成岩作用带	变形鲕	压实-压溶鲕粒变形
	云化鲕	成岩白云石化交代部分或全部方解石
	残余鲕	经成岩作用变化，鲕粒原来结构模糊不清，但仍残留鲕粒构造遗迹
	表面硅化鲕	鲕粒形成后被输入至障壁海岸环境，SiO_2 饱和沉淀并交代钙质同心包壳表层而成

（据孟祥化等，1991）

二、碎屑岩类

（一）砾岩类

1. 石英质中砾岩

分布于常州沟组一段。砾状结构，中层状构造，砾石大小不一，砾径大者达30~40cm，一般者3~10cm，磨圆度较高，成分几乎均为石英岩，极少数为脉石英、燧石、火山角砾岩、磁铁石英岩等。砾石表面广泛出现铁质氧化圈及压溶凹坑。其中以底部砾岩层砾径最大，并具明显的叠瓦状特征。底部砾岩以接触式胶结为主，中部和上部以基底式胶结为主。下部胶结物以硅、铁质为主，中部和上部以硅质为主。杂基为砂类，以长石、石英为主，呈不等粒结构，分选极差，含量不等。石英质中砾岩同粗粒长石石英砂岩或含砾长石石英砂岩组合构成典型下粗上细韵律层序，砾砂比为3∶1~6∶1，砂岩以透镜体产出，发育巨型透镜状层理和大型槽状交错层理。

2. 细砾岩、砂砾岩与含砾砂岩

分布于常州沟组二段、杨庄组及龙山组底部。岩石呈灰白色、灰色，（含）砾状结构、中—薄层状构造，砾石分选性较好，砾石组分、杂基成分同石英质中砾岩相似，呈次圆状—次棱角状。砂粒成分以石英颗粒为主，含少量长石砂粒及硅质岩屑。三者组合往往构成正粒序或反粒序沉积韵律，岩石多发育倾角较缓的楔状、板状斜层理。

（二）砂岩类

广泛分布于中—新元古界各组地层中，尤以常州沟组、大红峪组、杨庄组、下马岭组和龙山组最为发育，在其他各组地层中多以夹层或透镜层产出。颜色多取决于胶结物颜色，从粗砂—粉砂级别均有，还常见粉砂岩与泥岩、页岩之间的过渡类型岩石，如钙质粉砂岩、泥质粉砂岩等。在粉砂岩的碎屑物质中，稳定组分较多，成分较单纯，常以石英为主，含少量长石和岩屑，黏土一般较多，碳酸盐胶结物较为常见，铁质和硅质较少，按照砂岩碎屑颗粒组成可细分为以下几类：

1. 含海绿石石英砂岩

分布于龙山组中，岩石以不均匀灰绿色为主，细粒结构，岩石坚硬致密，块状构造，单层厚3~5cm，层面较为平直。砂粒以石英为主，其次为海绿石、硅质岩屑及长石碎斑等，呈圆状—次圆状，分选较好。海绿石呈绿色碎屑球粒，边缘有时不规则，粒度与石英相近，一般略大些，有时以胶结物的形式出现。

2. 长石石英砂岩

主要分布于常州沟组、大红峪组中，灰色、灰白色居多，多呈不等粒砂状结构，板层状构造，粒级范围在0.04~2mm，砂级为主，成分有石英、燧石，占65%~75%，长石以肉红色钾长石、微斜长石为主，斜长石较少，含量占15%~30%。岩石分选较差—中等、碎屑磨圆度较差，多为次棱角状，组分成熟度较低。往往发育双向交错层理，有流水波痕。多同石英岩状砂岩或石英砂岩构成沉积韵律。

3. 杂砂岩

分布于常州沟组上部，根据碎屑长石和石英的含量，细分为石英杂砂岩、长石石英杂砂岩及长石杂砂岩三类。粒级从粗粒—粉砂级别均有，岩石一般分选较差，碎屑磨圆度差，多为次棱角状。多为杂乱式或基底式-孔隙式胶结，发育粒序层理、斜层理。

4. 石英砂岩

广泛分布于中—新元古界各组地层中，尤以常州沟组、大红峪组、杨庄组、下马岭组和龙山组最为发育，在其他各组地层中多以夹层或透镜层产出。主要由石英组成，粒径约1.5mm，分选中等，泥质、硅质、铁质胶结均有，发育纹层状和羽状交错层理、大型板状、槽状交错层理，波痕不明显。

5. 石英岩状砂岩

分布于常州沟组、大红峪组中。岩石颜色以灰黄色、灰白色为主，碎屑以石英为主，含量一般大于95%，分选和圆度中等，具再生长结构，石英颗粒与硅质呈环边复合体，具有加积层理和交错层理。

（三）黏土岩类

1. 泥岩与页岩

在串岭沟组、洪水庄组和铁岭组、下马岭组中广泛发育，其他层位也有零星出现，以泥质页岩为主。颜色多种多样，多为黄绿色、灰绿色、土黄色、灰黑色等。因颜色、混入物及化学成分的不同又可分为钙质泥岩和页岩（雾迷山组、洪水庄组、下马岭组）、铁质泥岩和页岩（杨庄组、串岭沟组）、硅质泥岩和页岩（下马岭组、串岭沟组）、碳质页岩（下马岭组）、锰质页岩（高于庄组）等几种类型。常与薄层灰岩、白云岩互层出现。暗色的页岩和泥岩反映了较深水的还原滞流沉积环境。

2. 粉砂质泥岩与粉砂质页岩

特征与泥岩和页岩基本相似，在中—新元古界中广泛发育，粉砂含量25%~50%，其余则为黏土，黏土矿物以伊利石为主，含高岭石和绿泥石，多发育水平层理和扁平层理、波状层理，对称型微波痕发育，层面上泥裂呈网脉状，不规则，大小不等。部分粉砂质泥岩中可见黄铁矿、菱铁矿结核以及石膏、石盐假晶。

3. 白云质页岩

多发育于雾迷山组中，颜色呈灰、黄褐色，页片状、板状构造，发育水平纹层理，个别层位含海泡石，呈黄绿色，与泥质白云岩共生。

4. 硅质岩（或燧石条带）

主要产出于团山子组、大红峪组、高于庄组、杨庄组、雾迷山组、下马岭组中，多为潮上带硅化作用的产物，是成岩作用或同生阶段的游离硅质交代碳酸盐矿物而成的。根据硅化程度引起的分布形态的不同，可以划分为板刺状、层状、藻丝体状、不均匀状四种类型。

三、火山碎屑岩类

（一）斑脱岩

见于西山青白口、芹峪北、大村等地的下马岭组中。在页岩、泥岩地层序列中呈夹层状产出，其厚度一般为数厘米。风化严重，外观呈土黄—灰白色，质地松软，密度相对较小，其切面常呈蜡状光泽，吸水性较强，并具滑腻的皂状结构。岩石矿物学分析表明，斑脱岩含有较丰富的黏土矿物（多为伊利石-蒙脱石混层物及绿泥石-蒙脱石混层物，少量为高岭石等），并含有较多的钾长石和方解石，同时还含有少量的岩浆型磷灰石、锆石及云母等矿物，是由中酸性岩浆喷发产生的偏碱性火山灰沉降到海盆之后，经水解、成岩而来的典型钾质斑脱岩。

（二）白云石化硅化流纹质玻屑凝灰岩

发育于高于庄组，呈薄层夹层产出，夹在深灰色中厚层状灰质白云岩和薄层板状白云岩之间，呈墨绿色，岩石主要由火山玻屑（65%~70%）、火山尘（30%~35%）和少量晶屑组成，玻屑、晶屑被白云石和硅质（玉髓、微粒石英）顺层不均匀交代。

（三）粗面质晶屑岩屑玻屑凝灰岩

发育于大红峪组和团山子组，前者分布在平谷一带，后者分布在延庆古子房一带，呈透镜状夹层产出。岩石呈灰绿色或紫红色，凝灰结构，微层状构造或块状构造，岩石主要由火山碎屑物与火山灰组成，前者含量常>90%，并主要由钾长石、斜长石晶屑组成，岩屑以粗面岩类岩石为主，晶屑、岩屑之比多为3∶5。元素组成显示具有富钾碱性特征。

（四）粗面质蚀变熔结角砾岩

该类岩石发育于大红峪组和团山子组，呈透镜状夹层产出。岩石呈暗粉红色，具角砾状结构，块状构造。碎屑粗大，呈棱角—次棱角状，直径1~10mm，达角砾级，个别达集块级，分选差。碎屑由大量粗面岩、粗面玄武岩类熔岩及少量大理岩等组成，含量约80%，胶结物主要为碳酸盐，呈充填式水化学胶结。

（五）凝灰质粉砂岩

分布于大红峪组地层中，主要由 0.1~0.01mm 粒级（含量>50%）的碎屑颗粒组成。岩石常见较薄的水平层理、波状层理及透镜状层理，交错层理少见，多为小型的，且斜层倾角比相邻的砂岩小的多。

四、特殊岩类——微亮晶碳酸盐岩

又称臼齿（MT）状碳酸盐岩，是由微亮晶方解石充填的具有褶皱构造的碳酸盐岩，在串岭沟组、团山子组、高于庄组、雾迷山组都有所发现。孟祥化等（2006）、梅冥相（2005）、李现根（2008）、刘自亮（2009）等都对其开展过相关研究，以高于庄组最为典型。

1. 宏观特征

微亮晶碳酸盐岩中臼齿状构造与周围岩石界线清晰，颜色从亮白色到暗灰色。臼齿状构造一般发育在具细纹层理的含黏土碳酸盐岩中，主要发育在泥晶灰岩、粉屑灰岩等细粒碳酸盐岩沉积物中，而在粉砂岩和砂岩中不存在或很少出现，臼齿状构造一般在砂质层位扩散或者消失。

微亮晶碳酸盐岩包括原生和异地成因两大类。原生微亮晶碳酸盐岩臼齿状构造在层面上的典型形态包括丝状、条带状、瘤状、板状和褶皱状五种形态。丝状构造主要发育在泥晶灰岩中，砂屑灰岩中也可出现，长 1~10cm，宽 1~2mm，产状多近垂直于地层，偶尔可见斜交现象；条带状构造分布最为广泛，发育于各类灰岩中，与层面垂直、斜交或平行，长十几到几十厘米，宽 1~5mm；瘤状构造呈不规则复杂瘤状、团块状构造，外型多变，在三维空间上类似球体，长达 10cm 以上，宽 5~6cm 以上；板状构造主要发育在粉屑灰岩中，臼齿状构造个体较大，多呈平直板状，一般沿层理水平方向不连续排列；褶皱构造形态复杂多变，常呈弯弓状、钩状、臼状、肠状等，相互交织，弯带通常向顶层面延伸，穿越薄的纹层，但不穿越层面，臼齿状构造长几厘米至几十厘米，宽 1~5mm。不仅如此，臼齿构造形态一般也随着沉积地层基岩岩性、岩相的变化而发生明显变化。如矿物从细粒到粗粒变化处臼齿条带的宽度和方向常常发生变化。

异地成因微亮晶碳酸盐岩由于搬运过程中受到水动力影响，具有粒序层理，磨圆度较高，常在风暴砾屑灰岩中出现，具有被搬运和磨圆特征，大小不一。

2. 微观特征

臼齿状碳酸盐岩由两部分组成：微亮晶方解石和底基质。在结构上，这两部分有明显区别：底基质表面灰暗，一般由不纯的泥晶方解石、粉屑方解石组成，大小为 3~50μm，通常含有少量黏土物质、有机质，在微层序中，底基质常由下部颜色较浅的物质向上部暗灰色物质逐渐过渡；而微亮晶方解石颗粒则较为纯净，大小均匀、等粒，半透明状，结构比较均一，个体粒度在 5~15μm 之间。也有些微亮晶方解石的粒度差异较大，粒度大小的差异及粒度排列的密集程度不一致显示出类似于流动性的纹理。微亮晶方解石与基质之间界线清晰、截然，边缘常常呈波状，微亮晶条带常穿越基质纹层。

第三节　沉积盆地及沉积建造

一、沉积盆地

本区发育的中—新元古代沉积盆地隶属于燕山-辽西沉积盆地，经历了十亿年演化。在漫长的地质历史时期中，受华北克拉通北缘构造环境变化的制约，该盆地经历了燕山裂陷槽盆地、燕山坳陷槽盆地、克拉通陆表海盆地三个阶段的演变（黄学光，2006）。受盆地的形成、发展、演化和古气候变化的影响，本区沉积建造类型多样。

（一）燕山裂陷槽盆地

中元古代长城纪早期，受区域大规模伸展拉张应力影响，包括华北板块、西伯利亚板块在内的克拉通大陆发生三叉型破裂。其中，三叉型破裂靠南侧的一支，成为夭折的裂谷——燕山裂陷槽（黄学光，2006）。

燕山裂陷槽盆地发育于中元古代长城纪常州沟期至大红峪期。盆地北缘大致以尚义-隆化区域断裂为界。南缘边界则大致位于涿鹿—门头沟—房山—保定—廊坊—蓟县一线，总体呈北缓南陡状。盆地轴部整体呈北东向延伸，表现为密云-兴隆-宽城-凌源和滦县-抚宁-建昌两条盆底同沉积断裂之间的狭长谷地（图7-20）。本区主体位于燕山裂陷槽盆南西部，是华北板块古元古代形成的陆壳，在拉张应力状态下上部地壳拉伸减薄的产物。沉积主要受同盆底断裂斜交近东西走向的密云-兴隆同沉积断裂和赤城-古北口断层控制（同盆底断裂斜交，具有正断走滑性质），沉积物源主要来自盆地北部内蒙古陆和西部五台-太行古陆。沉积物以砂岩、页岩为主，上部有少量白云岩，构成潮汐相潮坪、潟湖砂岩-页岩-碳酸盐岩组合。其中，长城纪早期（常州沟早期）为山间河流相沉积，河流主体流向自南西向北东，长城纪中期后逐渐过渡到滨浅海相碎屑沉积及浅海相碎屑及碳酸盐岩沉积。大红峪期盆地还发育有火山岩、火山碎屑沉积岩和热水二氧化硅沉积。总体沉积物厚度变化大，宏观上从西北向东南逐渐增厚，岩性相变亦较大，沉积物呈楔形分布。

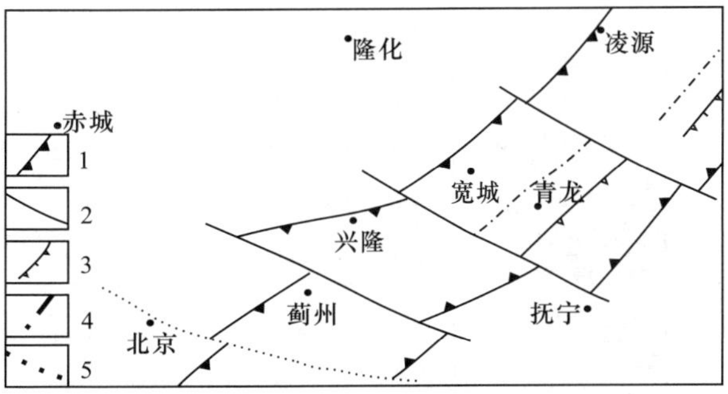

图 7-20　燕山裂陷槽盆地分布图
（据和政军，1994）
Fig. 7-20　Distributional sketch of Yanshan chasmic trough
1—长城纪早期的盆底与盆缘断裂（原文为盆缘断裂）；2—转换断层（原文为传递断层）；3—晚长城纪的盆缘断裂；4—掀斜块段支点连线；5—推测断裂

（二）燕山坳陷槽盆地

燕山坳陷槽盆地是燕山裂陷槽消亡后向克拉通盆地过渡型的盆地。发育于中元古代蓟县纪高于庄期至雾迷山期。该阶段，由于裂陷槽阶段区域地壳应力的释放，以及地幔失热和活化能量衰减，岩石圈上部的构造活动不再是裂陷扩展活动，而表现为挠曲下翘、不均匀沉降。

燕山坳陷槽盆地北缘由于后期构造运动影响，界线不清。南缘边界较裂陷期则大幅度扩大，大致位于山西（阳原—灵丘—阳泉）—河北（赞皇—衡水—沧州）—天津一线。盆地沉积由裂陷槽阶段北东向狭窄的裂陷活动转变为较为宽缓的坳陷作用为主、伴随着频繁震荡运动，沉积作用范围向周边扩展，盆地与外海连通更加广泛，盆地环境系统也趋于开放。不仅如此，盆地活动机制也发生较大变化，转变为以较均衡载荷沉降为主，阵发性断裂活动和局部的沉陷为辅。

本区均位于燕山坳陷槽盆地内部，表现为碳酸盐岩型盆地，沉积物为碳酸盐岩组合，包括高于庄组的潮汐相—浅海相白云岩-灰岩组合，杨庄组的潮汐相—潟湖相含陆源泥砂白云岩组合，雾迷山组潮汐相白云岩组合。该碳酸盐岩岩系分布稳定、相变小，且自盆地边缘向盆地中央沉积相呈规律变化，典型的潮上、潮间、潮下相潮汐韵律十分发育（图7-21），其中叠层石等广泛分布（黄学光，2006）。上述沉积特征基本上具有陆表海性质，标志着期间地壳逐渐刚化，向稳定克拉通转化。黄学光等（2006）研究还认为，该沉积盆地尚存在不稳定性，沉积仍具有一些裂陷槽盆地特征。

（三）克拉通陆表海盆地

克拉通陆表海盆地发育于中元古代蓟县纪铁岭期至青白口纪景儿峪期。铁岭期至下马岭期，承德、兴隆一带逐渐抬起，至青白口纪，盆地北侧与亚洲洋断绝了联系，转而向东与大洋沟通（黄学光等，2006）。

从铁岭期沉积开始及至新元古代，华北克拉通地壳进一步稳定，燕山沉积盆地演化成克拉通盆地，沉积了典型的克拉通盖层沉积。其沉积特征主要表现为：①沉积物构成稳定的潮汐相砂岩-页岩-碳酸盐岩构造-岩石组合；②沉积物为典型的面型分布，沉积物厚度小，区域厚度变化也小，只表现为沉积

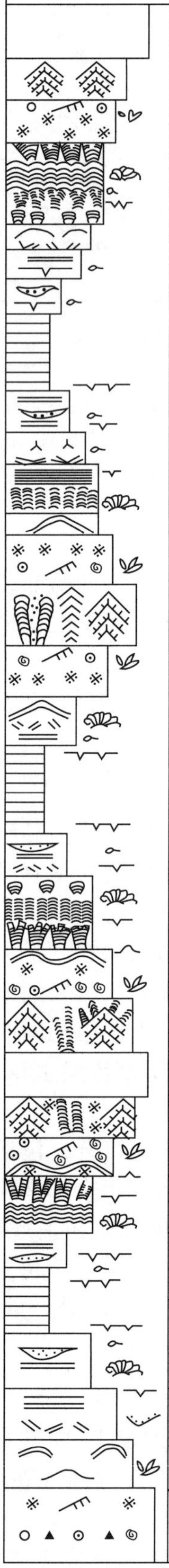

岩相柱状图	相标志及组合	环境 潮上	潮间 上	潮间 下	潮下 上	潮下 下
	灰色，泥晶结构、巨厚层状、水平纹层状结构，贫藻迹					潮下坪
	灰色，柱-锥状叠层石、藻席-藻团微结构、凝块，藻团、鲕粒、交错层理、竹叶状粒屑、大波纹状构造			障壁滩		
	黑色，有机质、沥青质、密波纹藻叠层、藻席微结构、微细干裂、鸟眼、冲淤构造、菊花状构造、小型丛状叠层石、小杯叠层石		潮间潟湖			
	深灰色、灰色、黄绿色，纹层—断续纹层状、波纹状构造、纹层藻叠层石、浅水波痕、含泥含砂屑、粒屑条带与透镜状层理、沟槽充填构造、干裂、鸟眼	高潮坪				
	黄绿色、灰色，水平纹层状构造、砂屑条带与透镜层理、沟槽充填构造、含泥含砂屑、干裂、鸟眼	中高潮坪				
	深灰色、黑色，富有机质、沥青质、密纹层藻叠层石、波纹状断续纹层状构造、菊花状构造、微干裂、鸟眼		潮间潟湖			
	灰色、深灰色，厚—巨厚层、大波纹状硅质条带、波浪波痕、波浪交错层理、竹叶状砾屑、交错层理、凝块、藻团藻鲕、核形石、鲕粒云（硅）岩、柱叠层石与锥叠层石组合、藻礁			藻礁障壁滩		
	黄绿色、灰色，含泥含砂屑，水平纹层状—断续纹层状构造、波纹状硅质条带、砂屑透镜层理、菊花状构造、干裂、鸟眼	中高潮坪				
	蓝灰色、灰色，泥晶结构、水平纹层状构造、含砂屑条带与透镜状层理、沟槽充填构造、断续纹层状构造、干裂、鸟眼	潮坪				
	黑色，含有机质沥青质、藻席微结构、密纹层密波纹层藻叠层石、小型丛状叠层石、微干裂、菊花状构造、沟槽充填构造		潟湖			
	灰色，厚层状、大波纹状燧石条带、藻团、藻鲕、核形石、交错层、对称波浪波痕、波浪交错层理、竹叶状砾屑、大型柱-锥叠层石、藻团凝块微结构、掌状分叉叠层石			藻礁障壁		
	浅灰色，巨厚层状白云岩、贫藻迹					低能
	灰色，厚层状、大波纹状构造、泥披、对称波浪、藻团、藻鲕、核形石、鲕粒、竹叶状砾屑、交错层理、大型分叉柱叠层石、锥穹状叠层石			藻礁障壁		
	黑色，含有机质沥青质、藻席微结构、密纹层密波纹层藻叠层石、小型丛状叠层石、微干裂、菊花状构造		潟湖			
	黄绿色、灰色，含泥含砂屑，水平纹层状构造、粒屑透镜层理、干裂、鸟眼	潮坪				
	灰黄色、灰色，含泥晶白云岩、燧石条带白云岩、含屑条带白云岩，纹层状、叶片状层理、干裂、鸟眼、菊花状构造	高潮坪				
	黄色，燧石条带白云岩、黑色纹层藻叠层白云岩、水平—断续纹层状构造、波纹状燧石条带、收缩裂、泥披灰色大波纹状硅质条带白云岩、硅质砾屑白云岩、核形石、竹叶状砾屑		中低潮坪			
	灰色，巨厚层凝块藻团白云岩、波浪交错层理、底冲刷面、藻团斜层理、墙叠层石			水下浅滩		

图 7-21　北京十三陵雾迷山组沉积相环境模式

（据《北京市区域地质志》，1991）

Fig. 7-21　Sedimentary facies environment model of Wumishan formation in Beijing Ming Tombs

中心的迁移，在长达400Ma的地质历史时期内，沉积的地层厚度不到100m；③沉积物岩性单一，相变小，各组的岩性和岩相变化也很小，洪水庄组页岩、下马岭组页岩、龙山组含海绿石砂岩、粉砂岩、杂色页岩和景儿峪组蛋青色泥灰岩的分布在全区都十分稳定；④地壳进一步稳定的结果，盆地表现出整体同升同降的特点，形成区域性的间断面，如铁岭中期的铁岭上升、铁岭末期的芹峪上升，青白口纪下马岭期末的蔚县上升以及青白口纪末的蓟县运动，在整个燕山地区都可以看到（黄学光等，2006）。

二、沉积建造

本区盆地中发育的沉积建造可分为次稳定型与稳定型两类，各时期主要沉积建造类型综合特征如表7-4所示。

表7-4 中—新元古代各时期沉积建造类型综合特征

Table 7-4 Comprehensive characteristics of sedimentary formation types of Meso-Neoproterozoic

时期	沉积建造系列	盆地类型	建造类型	沉积环境	物源	沉积速率	共生岩石主要类型	典型分布地区
景儿峪期	稳定	克拉通盆地	陆源碳酸盐建造	潮下	混合源	较慢	泥灰岩、硅质条带灰岩、泥质条带灰岩、泥质泥晶灰岩、灰岩、灰质白云岩	北京市全区
龙山期	稳定		单陆屑（含磷）建造	河口、滨岸、潟湖	陆源	较慢	含砾砂岩、含砾粗砂岩、石英砂岩、海绿石石英砂岩、细砂岩、粉砂质页岩、页岩、泥灰岩	北京市全区
下马岭期	稳定		单陆屑（铁质）建造	潟湖、潮坪	陆源	较慢	页岩、粉砂质页岩、粉砂岩、泥灰岩、砂岩、硅质泥岩、硅质岩、鲕状赤铁矿、赤铁矿层	北京市全区
铁岭期	稳定		陆源碳酸盐建造	滨海、浅海	混合源	较慢	灰质白云岩、白云质灰岩、页岩、含锰灰岩、含粉砂白云岩、燧石条带白云岩、硅质白云岩	北京市全区
洪水庄期	稳定		单陆屑（铁质）建造	浅海	陆源	较慢	石英砂岩、页岩、粉砂质页岩、碳质粉砂质页岩、页岩、泥质白云岩、铁质结核、锰质结核、黄铁矿、菱铁矿	北京市全区
雾迷山期	次稳定—稳定	坳陷槽盆地	巨厚沥青藻礁碳酸盐岩建造	潮间、潮下、潟湖	内源	较快	泥晶白云岩、纹层白云岩、叠层石白云岩、白云岩、藻团白云岩、鲕粒白云岩、燧石岩、石英砂岩、砂质白云岩、硅质条带白云岩、核形石白云岩	北京市全区
杨庄期	次稳定—稳定		巨厚沥青碳酸盐建造	潮坪、潟湖	混合源	较慢	白云质石英砂岩、砂质白云岩、泥晶白云岩、粉晶白云岩、白云岩、硅质条带白云岩、砂砾屑白云岩、石英砂岩、白云质细砂岩、页岩	通州—昌平—怀柔—古北口以东和延庆以西
	次稳定—稳定		石英砂砾岩建造（巨厚单碎屑岩建造）	滨海沙滩	陆源	较慢	含砾石英砂岩、砾岩、砂砾岩、石英砂岩、粉砂岩、泥岩	通州—昌平—怀柔—古北口以西、延庆以东
高于庄期	次稳定—稳定		藻礁碳酸盐岩建造	潮坪、浅海	内源	较快	叠层石白云岩、泥晶白云岩、白云岩、粉晶白云岩、瘤状灰岩、白云质灰岩、含锰白云岩、燧石条带白云岩	北京市全区

续表

时期	沉积建造系列	盆地类型	建造类型	沉积环境	物源	沉积速率	共生岩石主要类型	典型分布地区
大红峪期	次稳定	裂陷槽盆地	火山复陆屑建造	潮坪	混合源	较快	玄武岩、粗面岩、凝灰质砂岩、粉砂岩、石英砂岩、泥晶白云岩、粉晶白云岩、白云岩	密云大龙门—平谷熊耳寨、万庄子一带
	次稳定		巨厚单陆屑建造	潮坪	混合源	较缓慢	含长石石英砂岩、石英粗砂岩、石英砂岩、粉砂岩、硅质白云岩、石英岩、粉砂岩、白云岩、泥质白云岩、硅质白云岩、钙质板岩、纹带白云岩、叠层石白云岩、砂质白云岩	延庆—南口—十三陵一带
团山子期	次稳定		沥青碳酸盐岩建造	潮坪、潟湖	混合源	较缓慢	泥质白云岩、泥质粉砂质白云岩、粉晶白云岩、白云质页岩、砂质页岩、硅质白云岩、钙质页岩	北京市全区
串岭沟期	次稳定		灰色复陆屑建造	滨海	陆源	较缓慢	石英砂岩、长石砂岩、长石石英砂岩、砂质页岩、黑色页岩	门头沟—房山—大兴以南一带
	次稳定		巨厚单陆屑建造	潮坪、潟湖、潮下	混合源	较快	页岩、粉砂质页岩、硅质页岩、泥质粉砂岩、白云岩、石英砂岩	门头沟—房山—大兴以北一带
常州沟期	次稳定		巨厚单陆屑建造	潮坪、砂坝、海滩	陆源	较缓慢	石英砂岩、石英粉砂岩、石英岩	北京市全区
	次稳定		灰色复陆屑建造	河流、河口湾、滨海	陆源	较快	砂砾岩、含砾长石石英砂岩、长石石英砂岩、石英长石砂岩、含长石石英砂岩、石英砂岩、粉砂岩	南口—十三陵一带、密云水库周边地区、平谷万庄子—熊耳寨一带

（一）次稳定型沉积建造

次稳定型沉积建造主要产生于裂陷槽-坳陷槽阶段。本区发育有裂谷环境特征的灰色复陆屑建造和火山复陆屑建造，以及裂陷槽环境特征的巨厚单陆屑建造、沥青碳酸盐建造、藻礁碳酸盐建造共 5 类建造类型。前两类建造特点是以双组分陆屑（长石、石英为主）为代表的岩石共生系列，建造厚度较大，建造内部岩相变化或交替迅速，并有清楚的韵律旋回节奏，是裂陷槽构造活动强烈期发育的建造类型。而后三类建造则形成于裂陷构造的中期或结束期等断陷活动相对平静期。主要表现为海水上升或退缩，地形平缓，大区域沉积环境相对单调的滨海、浅海潟湖或潮间、潮下沉积。

1. 灰色复陆屑建造

灰色复陆屑建造属于以陆源碎屑为主的次稳定型沉积建造，主要岩石共生组合为砾岩、含砾砂岩、砂岩、粉砂岩等。本类建造厚度交错，建造内部岩相变化或交替迅速。区内主要发育在两类沉积环境中：一是常州沟组早—中期古元古代已形成的陆壳起始裂解，见于南口—十三陵一带、密云水库周边地区及平谷万庄子—熊耳寨一带山间狭窄断陷盆地或山麓河流型盆地中；二是串岭沟组和大红峪组局地地区，为滨海相沉积。串岭沟期发育于门头沟—房山—大兴以南一带。大红峪期发育于京南房山—大兴一带。

2. 火山复陆屑建造

火山复陆屑建造属于内源与陆源混合源次稳定型沉积建造，主要岩石共生组合为玄武岩、粗面岩、凝灰质砂岩、粉砂岩、石英砂岩、泥晶白云岩、粉晶白云岩、白云岩等。该建造由火山作用产生的 SiO_2 组分、陆源石英砂及内源物质组成。

区内主要形成于大红峪期火山作用过程中的潮坪相沉积，分布于以平谷为中心的密云大龙门—平谷熊耳寨、万庄子一带，近东西向延伸可至蓟县—兴隆一带，而喷发中心在平谷熊耳寨、挂甲峪一带。

3. 巨厚单陆屑建造

巨厚单陆屑建造属于陆源次稳定型建造，主要岩石组合为含砾石英砂岩、砾岩、砂砾岩、石英砂岩、石英粉砂岩、石英岩、页岩、粉砂质页岩、硅质页岩、泥质粉砂岩，包括少许白云岩、硅质白云岩、泥质白云岩。建造沉积物质主要来源于古陆，为稳定性很强、成熟度很高的纯净石英砂或粉砂沉积物。

区内长城纪常州沟期、串岭沟期和大红峪期和蓟县纪杨庄期均有发育。其中，长城纪常州沟期主要发育于该组沉积晚期，为燕辽裂陷槽早期河流三角洲沉积环境下产物；长城纪串岭沟期和大红峪期则仅发育于局部地区，为浅海潮坪、潮下或封闭的潟湖沉积。串岭沟期主要分布于门头沟—房山—大兴以北一带，大红峪期分布于延庆—南口—十三陵一带；而蓟县纪杨庄期仅发育于通州—昌平—怀柔—密云古北口以西、延庆以东地区。与前三组相比，其沉积组合以石英砂砾岩为主，沉积物粒度明显偏粗，为近南北向军都山水下高地滨海沙滩相沉积。

4. 巨厚沥青碳酸盐建造

沥青碳酸盐建造属于以内源物质为主的次稳定型建造，主要岩石组合为泥质白云岩、泥质粉砂质白云岩、粉晶白云岩、白云质页岩、砂质页岩、硅质白云岩、钙质页岩组合。

区内长城纪团山子期和蓟县纪杨庄期均有发育。北京市全区团山子期基本均沉积该类建造，而杨庄期，仅发育于通州—昌平—怀柔—密云古北口以东和延庆以西，即军都山水下高地以东和以西地区，沉积环境较为单一，是坳陷槽阶段构造相对平静期潮坪、潟湖环境下的沉积，沉积速率较为缓慢，建造厚度小。

5. 巨厚沥青藻礁碳酸盐建造

藻礁碳酸盐建造属于内源次稳定型沉积建造，主要岩石共生组合为叠层石白云岩、泥晶白云岩、白云岩、粉晶白云岩、瘤状灰岩、白云质灰岩、含锰白云岩、燧石条带白云岩等。它是由藻礁碳酸盐岩、含藻碳酸盐岩及与其共生的非藻碳酸盐岩所构成的岩石共生组合体（孟祥化，1979）。

区内蓟县纪高于庄期和雾迷山期均有发育，全区均为该类建造。表现为燕辽裂陷槽盆地末期向克拉通盆地转化过程中，盆地以坳陷形式整体沉降或海侵的沉积建造类型。沉积序列交替旋回结构发育，沉积厚度巨大，沉积韵律旋回十分复杂而节奏清晰，形成于气候温暖的潮坪和潟湖环境，形成过程主要受生物化学作用控制，大部分为原地堆积。

（二）稳定型建造

本区稳定型建造主要产生于中元古代后期及新元古代克拉通盆地陆表海沉积阶段，是大陆板块位置经过潮湿型气候带时形成的产物。稳定型建造具有低速率沉积共生特点，在建造层序中既有来自稳定陆源的碎屑岩，也有来自盆地自身的内源物质。

1. 单陆屑建造

包括单陆屑铁质建造和单陆屑磷质建造。

单陆屑铁质建造为最典型的稳定型陆源沉积建造，主要由单一而稳定的陆源碎屑沉积——石英质砾岩、石英砂岩、高岭石黏土岩共生系列构成。其沉积韵律旋回结构简单，厚度薄而横向伸展稳定，相变缓慢，建造体在整个克拉通盆地范围连续分布（孟祥化，1989，1990，1991）。区内蓟县纪洪水庄期、待建纪下马岭期均发育为此类沉积建造，覆盖全区。为浅海潮坪、潟湖及潮下沉积的产物，主要岩石共生组合为石英砂岩、粉砂质页岩、碳质粉砂质页岩、页岩、泥质白云岩或泥灰岩等，共生有铁质结核、锰质结核、黄铁矿、菱铁矿、赤铁矿、鲕状赤铁矿等共生系列。建造中铁矿层及含铁层主要形成在滨海、浅海石英砂浅滩及与其滨外泥质沉积相带之间的交替带内。

单陆屑磷质建造发育于青白口纪龙山期。主要岩石共生组合为含砾砂岩、含砾粗砂岩、石英砂岩、海绿石石英砂岩、细砂岩、粉砂质页岩、页岩、泥灰岩等，是燕辽克拉通盆地整体短期抬升（龙山上升）后再次下沉接受沉积的产物，沉积环境复杂，早期为河口-滨岸-潮间带沉积环境，沉积为砂砾岩、含砂砾岩、石英砂岩、粉砂岩、页岩组合。晚期水体加深，逐渐过渡到浅海沉积环境，形成了一套页岩和泥灰岩组合，颜色呈紫红色，反映当时气候干燥。单陆屑磷质建造厚度不大，其沉积韵律旋回结构同样较为简单，建造横向伸展稳定，相变缓慢，建造体在整个区内甚至于克拉通盆地范围连续分布。

2. 陆源碳酸盐建造

陆源碳酸盐建造属于稳定陆壳板块上克拉通盆地的内源沉积物共生组合体，是大陆板块内的古陆风化剥蚀物质供给停滞、克拉通盆地由于缺乏陆源输入而转变为清水陆表海和内陆棚海环境下以浅水碳酸盐沉积作用为主沉积建造类型。主要岩石共生组合为鲕状灰岩、条带灰岩、泥灰岩及少量页岩、粉砂岩等。区内主要岩石共生组合为泥灰岩、硅质条带灰岩、泥质条带灰岩、泥质泥晶灰岩、灰岩、灰质白云岩、白云质灰岩及少量页岩、含锰灰岩、含粉砂白云岩等。

陆源碳酸盐建造厚度不大，沉积速率较缓慢，能量平衡程度高，杂基组分泥晶微晶组分含量较低。区内蓟县纪铁岭期和青白口纪景儿峪期发育此类建造。建造沉积韵律旋回结构简单，横向伸展稳定，相变缓慢，覆盖整个北京市。

第四节　岩相古地理

由于地壳运动和海平面变化，本区中—新元古代盆地逐渐由裂陷槽盆地演变为克拉通陆表海盆地，后期又有多次抬升和活化。可分为中元古代长城纪常州沟期、串岭沟期、团山子期、大红峪期，蓟县纪高于庄期、杨庄期、雾迷山期、洪水庄期、铁岭期，待建纪下马岭期和新元古代青白口纪龙山期和景儿峪期。

一、常州沟期

常州沟期，区内中—新元古代盆地开始形成，发育陆地河流相至海相沉积。根据朱士兴等（1994）对古藻类的古温度分带和古地磁纬度的研究，本区位于低纬度，属于热带气候。

1. 常州沟期早期

燕辽裂陷槽盆地（简称燕辽盆地）形成。盆地形成初期，分布面积有限，沉积主体位于北京东部—河北—辽宁一带。表现为较封闭条件下河流相沉积为主，局部发育小范围湖相沉积，沉积物为一套成熟度低的砾岩、砂砾岩、石英杂砂岩组合。沉积物来源于盆地周缘的古陆地剥蚀区。和政军（1990）通过大面积的流向统计结果显示河流在平面上自裂陷槽两侧向轴部（谷地）汇聚，轴部两侧为冲积扇、洪积扇，而谷地内部发育为自南西向北东流动的裂陷槽主干河流（图 7-22）。

期间，本区主体地区处于燕辽盆地外围，为剥蚀山地地区。沉积分布有限，仅在密云水库周边地区以及平谷万庄子—熊耳寨一带山间谷地发育小范围山地河流相沉积。受剥蚀山地快速抬升影响，地形高差及坡降大，河流向源侵蚀作用强烈，呈现河岸陡而河谷深、河道直而支流少、水流急而沉积物粗的特征。根据常州沟组一段分布，本区共发育两条古河流，也同断裂密切相关，受控于燕辽盆地控盆密云-墙子路断裂带和盆内具有走滑性质的次级的赤城-古北口断裂带。

2. 常州沟期中期

常州沟期中期是即常州沟组二、三岩性段沉积时期。燕辽盆地控盆断裂及盆内断裂持续强烈活动，盆地面积扩大（图 7-23）。盆地边界由北京东部扩展到北京中西部，海水由东向西入侵。本区除延庆、昌平西部、门头沟和房山西部、怀柔北部为侵蚀陆地外，其余地区由陆地演化为燕辽盆地的一部分。早期发育的两条河流分别向上游退缩，于昌平十三陵—河北后城一带入盆形成河口湾沉积。受海侵影响，东部平谷—兴隆一带（可能包括通州、大兴及北京城区地区），因远离滨岸水流流向表现为近东西单向（图 7-23c），发育潮坪沉积；中部密云西部—怀柔—延庆北部则受北西西向河口湾河流、潮汐、波浪共同影响，水流方向具有双向性（图 7-23b），发育为河口湾沉积；西部的昌平十三陵一带，因密云—墙子路—兴隆一带河流退缩，水流方向表现为近东西（250°）单向，也可能与单向波浪作用有关（图 7-23a），同发育为河口湾沉积；西部靠近古陆的房山东部、石景山一带则发育为滨岸沉积。

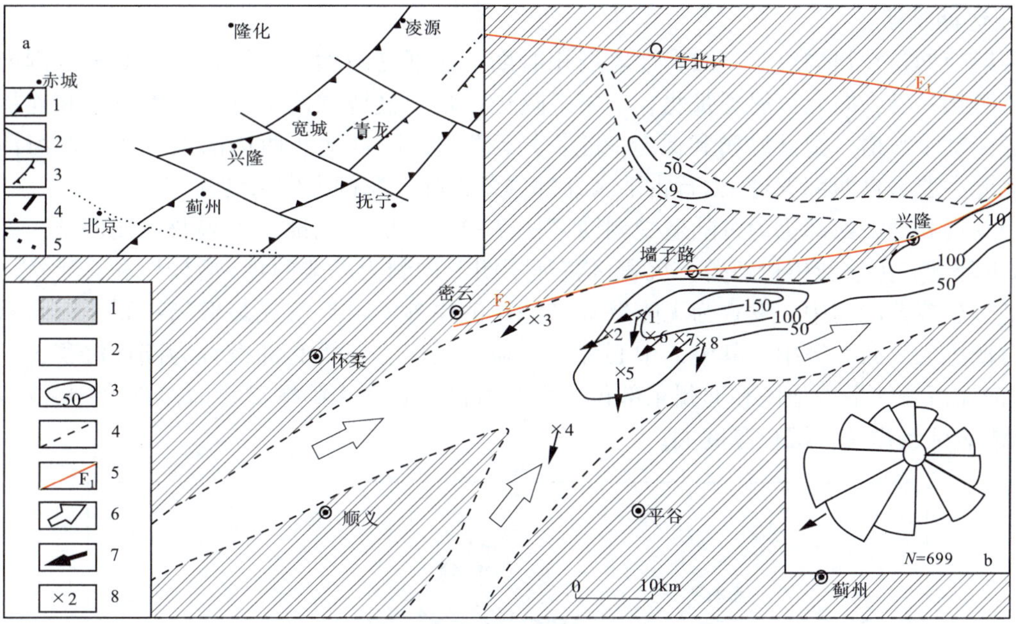

图 7-22　北京地区常州沟期早期沉积古地理图

（据《北京市区域地质志》，1991）

Fig. 7-22　Sedimentary paleogeography map of early-term of Changzhougou period

主图：1—基底剥蚀区；2—河流相堆积区；3—沉积等厚线（m）；4—古河道（推测）边界；5—断裂（F_1—赤城-古北口断裂带；F_2—密云-墙子路-蓟县-兴隆-平泉-凌源断裂带）；6—古流向；7—砾石 ab 面平均倾向；8—剖面及编号，图内数字为砾石测点：①密云聂家峪；②密云前厂；③密云丰各庄；④平谷万庄；⑤平谷上镇南山；⑥平谷镇罗营南山；⑦平谷猪猪峪；⑧平谷史家台

角图 a：1—盆底断层；2—传递断层；3—盆内正断层；4—掀斜块段支点连线；5—推测隐伏传递断层（据和政军，1990，略有改动）

b：N—统计数

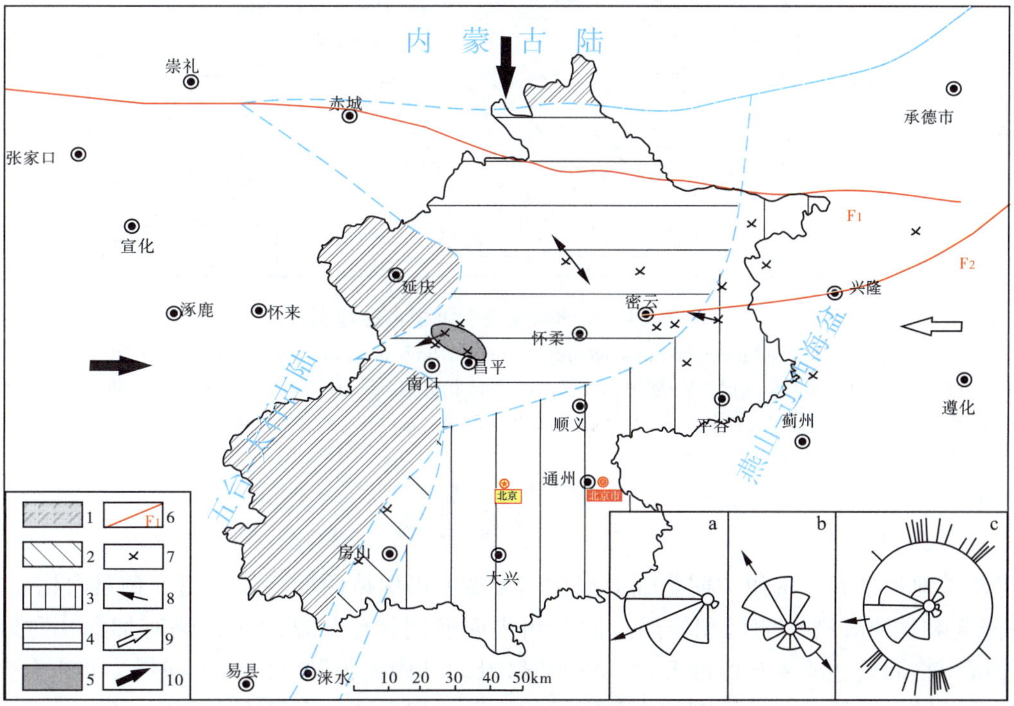

图 7-23　北京地区常州沟期中期沉积古地理图

Fig. 7-23　Sedimentary paleogeography map of mid-term of Changzhougou period

主图：1—基底剥蚀区；2—滨岸相沉积；3—潮坪相沉积；4—河口湾相沉积；5—砂坝相沉积；6—断裂（F_1—赤城-古北口断裂带；F_2—密云-墙子路断裂带）；7—剖面；8—古流向；9—海侵方向；10—物源方向

角图：a—昌平区，Chc^3，$N=87$，$Ma=239°$，$\sigma=34$；b—怀柔区，Chc^2，$N=51$，Ma 总主体 320°，$\sigma=34$，Ma 次主体 143°，$\sigma=103$；c—密平区，Chc^3，$N=132$，$Ma=155°$，$\sigma=46$，外圈为波痕脊线走向频率；N—统计数；Ma—古水流向；σ—标准编号；Chc^3—长城系常州沟组三段

3. 常州沟期晚期

常州沟期晚期是常州沟组三、四岩性段沉积时期。燕辽盆地面积继续扩大，盆地范围向西扩展到河北怀安、蔚县一带，向北扩展到张北、丰宁、隆化一带。盆地轴部发育为两大海槽和连接两大海槽的水下隆起：赞皇海槽、蓟县海槽和保定-北京水下隆起，而盆地北西边部发育为冀西海湾雏形。期间，沉积物源主要来自西部五台-太行古陆及北部内蒙古陆，海侵方向为北东东向。

本区门头沟和房山西部仍为古陆遭受剥蚀（图 7-24）。房山东部、大兴及通州西部、朝阳位于保定-北京水下隆起，由于新生代覆盖，沉积区特征不明，推测为海滩砂坝-后滨沉积；延庆、密云、昌平、怀柔则位于冀西海湾中，发育为无障壁岛型海岸滨岸沙滩、后滨无障壁岛沉积，海水从南西、北东东两大海槽进入冀西海湾中。其中，中心部位赤城、延庆、怀柔北部发育为潮坪沉积，其余发育为砂坝-砂质海滩沉积；平谷地区和通州东部位于蓟县海槽，发育无障壁岛型海岸前滨-临滨沉积。受密云-墙子路断裂活动影响，沉积厚度大。

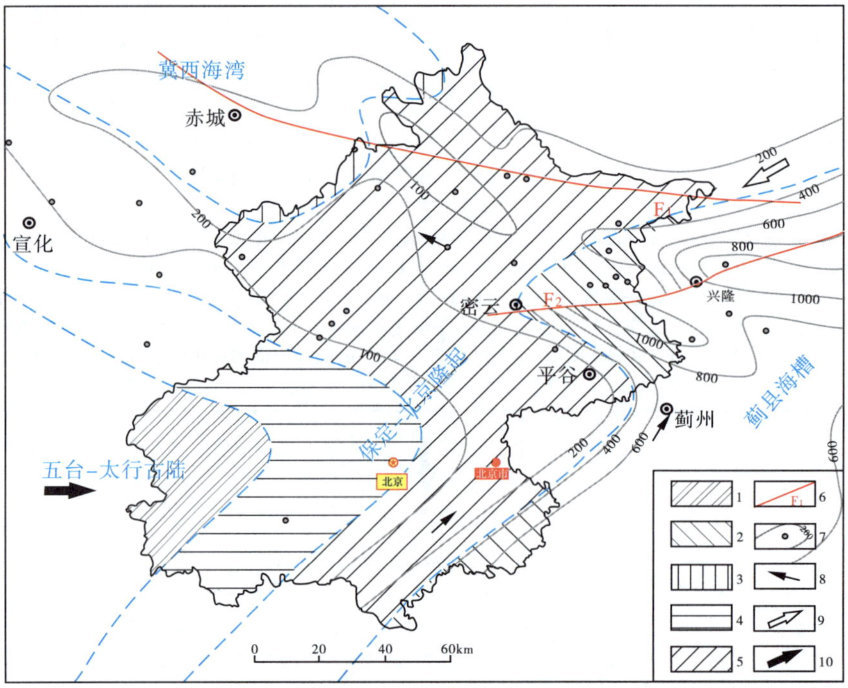

图 7-24 常州沟期晚期沉积古地理图

Fig. 7-24 Sedimentary paleogeography map of late-term of Changzhougou period

1—基底剥蚀区；2—临滨-前滨相沉积；3—潮坪相沉积；4—沙滩相沉积；5—沙滩-后滨相沉积；6—断裂（F_1—赤城-古北口断裂带；F_2—密云-墙子路断裂带）；7—剖面与沉积等厚线（m）；8—古流向；9—海侵方向；10—物源方向

二、串岭沟期

串岭沟期古地理格局在常州沟期基础上演化，宏观上仍保持两槽一隆一湾（蓟县海槽、赞皇海槽、保定-北京隆起和冀西海湾）的沉积格局。同常州沟期晚期相比，盆地水体略有加深，盆地范围略有所扩展。而沉积物源仍主要来自西部五台-太行古陆及北部内蒙古陆，海侵方向仍为北东东向。由于盆地内部断裂活动转弱，加之盆地周围的古陆在常州沟期强烈剥蚀，多已夷平老化，物源供给能力下降，燕辽盆地表现为陆源补偿与盆地沉降处于相对稳定期，沉积一套以细粒黏土岩为主、并发育内源沉积的沉积岩组合系列。该期燕辽盆地古纬度与常州沟期相差不大，仍处于热带条件下，以化学风化作用为主（朱士兴，1994）。

本区雁翅-马兰以西仍为五台-太行古陆剥蚀区，其余则为海相盆地，接受沉积（图 7-25）。沉积受密云-墙子路断裂和赤城-古北口断裂共同影响，沉积模式由常州沟期无障壁岛沉积演化为障壁岛沉积（图7-25），沿着密云-墙子路断裂和赤城-古北口断裂上升盘，发育为环形砂坝及藻礁障壁沉积，且在两条断裂交汇附近古子房—古北口一带演化为连接冀西海湾和蓟县海槽的古水流通道洼地（双古海槽），而障壁岛东部蓟县海槽发育为潮下泥岩沉积，冀西海湾障壁岛北部宣化龙关、营门、青龙杖子及怀柔北部、延庆西部发育为潮下鲕肾状藻铁岩沉积。冀西海湾障壁岛西部延庆、昌平一带受障壁阻挡波浪与潮流能量影响，演化为半封闭滞留还原潟湖泥岩沉积环境，临近五台-太行古陆的门头沟东部、房山东部发育滨岸砂坝、沙滩沉积。

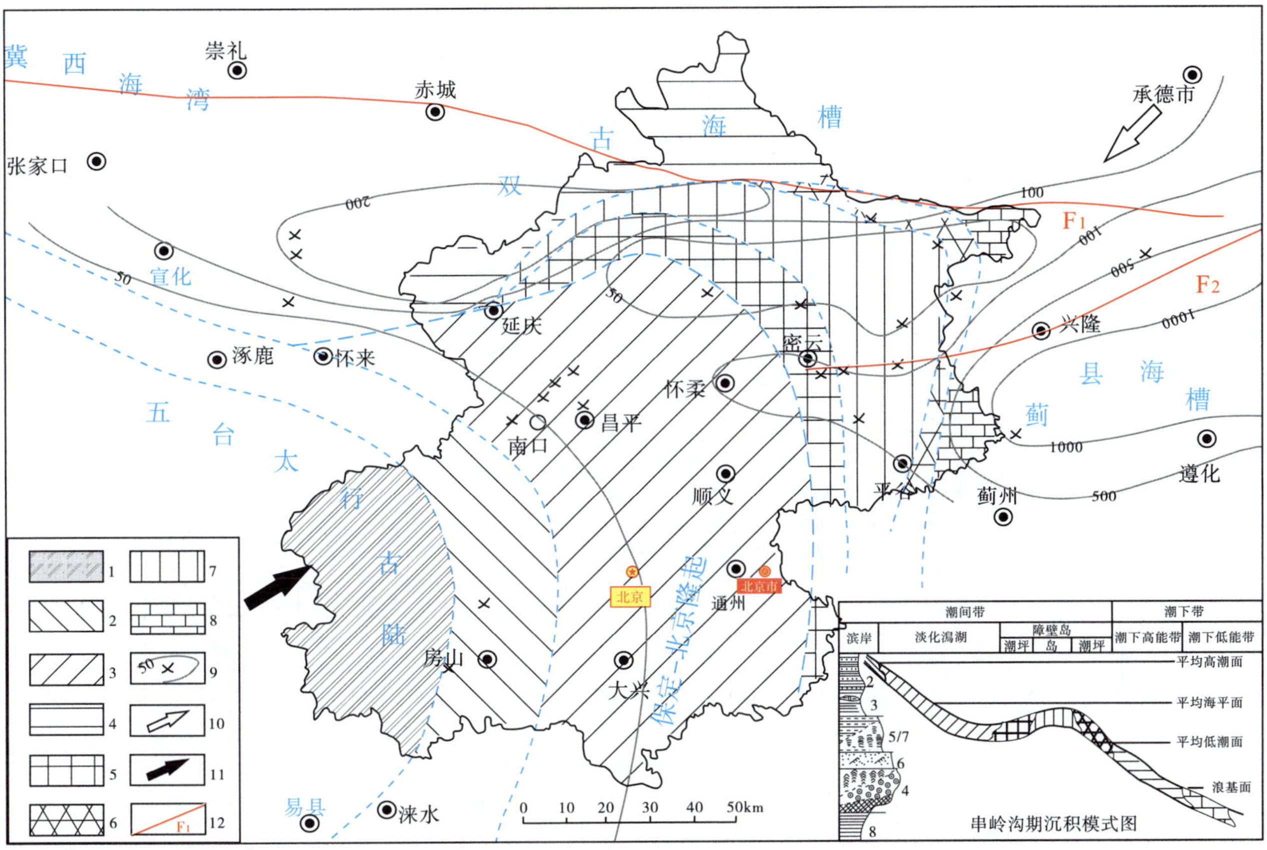

图7-25　串岭沟期沉积古地理图

Fig. 7-25　Sedimentary paleogeography map of Chuanlinggou period

1—基底剥蚀区；2—滨岸沙滩相沉积；3—潮间潟湖相沉积；4—潮下鲕肾状藻铁岩相沉积；5—潮坪混合坪砂泥岩相沉积；6—障壁砂坝相沉积；7—藻礁障壁相沉积；8—潮下泥岩相沉积；9—剖面与沉积等厚线（m）；10—海侵方向；11—物源方向；12—断裂（F_1—赤城-古北口断裂带；F_2—密云-墙子路断裂带）

三、团山子期

团山子期燕辽盆地古地理格局基本继承了串岭沟期的特点。盆地水体总体变浅，岸线总体向盆地中心收缩，但局部发育海水超覆。由于盆地周围的陆地进一步夷平化，陆源碎屑物质供给接近枯竭，沉积岩石组合类型转变为碳酸盐岩为主，总体沉积环境较为宁静。该期盆地沉积仍受盆内断裂控制，北部赤城-古北口断裂和廊坊-蓟县断裂（推断）、密云-墙子路断裂持续活动，形成两个沉积中心：蓟县海槽和宣化-延庆凹地。区域所处古纬度略有增高，处于亚热带气候（朱士兴等，1994），气候较为干燥炎热，海侵方向仍为北东东向。

本区雁翅—马兰以西仍为五台-太行古陆的一部分，遭受剥蚀（图7-26）。冀西海湾内受盆底密云-墙子路断裂同沉积活动影响，密云—顺义一带发育为陡坡生物礁沉积，宣化—延庆一带发育为咸化潟湖相沉积；临近五台-太行古陆的门头沟、房山演化为潮间混合坪含砂碳酸盐岩沉积；密云-墙子路断裂东部平谷、蓟州一带的蓟县海槽为潮下环境；而南部大兴、通州一带为燕辽盆地保定-北京隆起带组成部分，推测为潮坪或生物礁沉积环境。

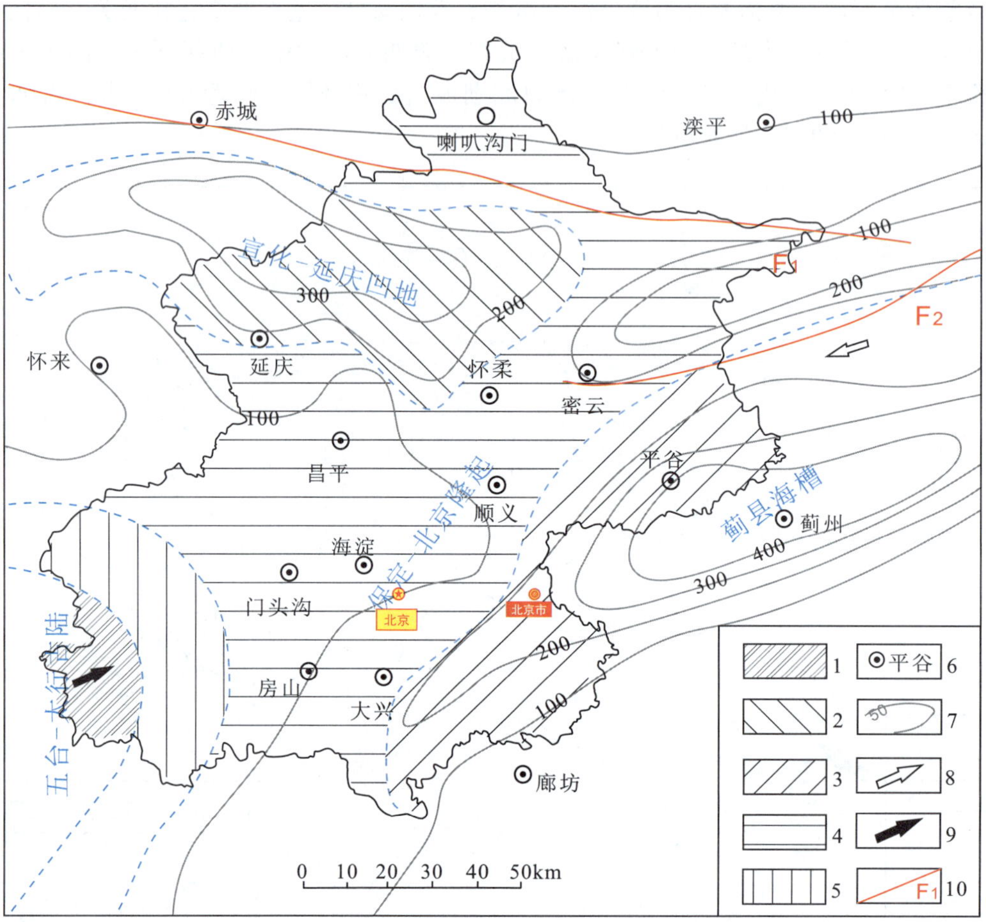

图 7-26　团山子期沉积古地理图

Fig. 7-26　Sedimentary paleogeography map of Tuanshanzi period

1—基底剥蚀区；2—潮间潟湖相沉积；3—潮下白云岩相沉积；4—生物礁相沉积；5—潮坪混合坪含砂碳酸盐岩相沉积；6—地名；7—沉积等厚线（m）；8—海侵方向；9—物源方向；10—断裂（F_1—赤城-古北口断裂带；F_2—密云-墙子路断裂断裂带）

四、大红峪期

大红峪期继承团山子期古地理格局（图 7-27）。燕辽盆地南西地区海水范围大面积缩小，赞皇海槽消失，海水由保定-北京隆起向蓟县海槽收缩，并在蓟县海槽及冀西海湾向盆地北部超覆。该期密云-墙子路断裂及赤城-古北口断裂活动再次增强，严格控制盆地沉积。盆缘赤城-古北口断裂下降盘形成古北口洼地沉积中心，而密云-墙子路断裂下降盘为平谷-兴隆沉积中心。与团山子期相比，由于构造活跃，古陆相对隆起，陆源物质供给明显加大，盆地沉积主要受到波浪作用影响，早期水体很浅，沉积环境主体位于浪基面之上的潮间、潮上及沙滩沉积，岩性为分选良好的具有小型对称波痕和双向交错层理的石英砂岩。晚期水体略有回升，蓟县海槽内发育潮下沉积，岩性为燧石条带白云岩夹硅质岩组合。该期燕辽盆地古纬度进一步提高，为温带气候（朱士兴等，1994）。沉积物来源于西部五台-太行古陆、北部内蒙古陆和南部隆起带，海盆海侵方向转变为近东向。

本区门头沟和房山中、西部仍为五台-太行古陆剥蚀区。由于海退影响，通州南东部、大兴南部也抬升为陆地，遭受剥蚀。西部冀西海湾受海水萎缩、水体变浅影响，南界略有北移，海水面积缩小，沉积环境由潟湖、潮坪沉积转变为环形临滨海沙滩。而东部蓟县海槽也发生较大变化，沉降中心西移，由蓟县海槽演化成为平谷-兴隆海槽，发育为潮坪碳酸盐岩-火山岩相沉积。保定-北京南部水下隆起基本解体，仅在隆起带的最北段大兴、城区、通州一带发育为河流-滨岸沙滩沉积。

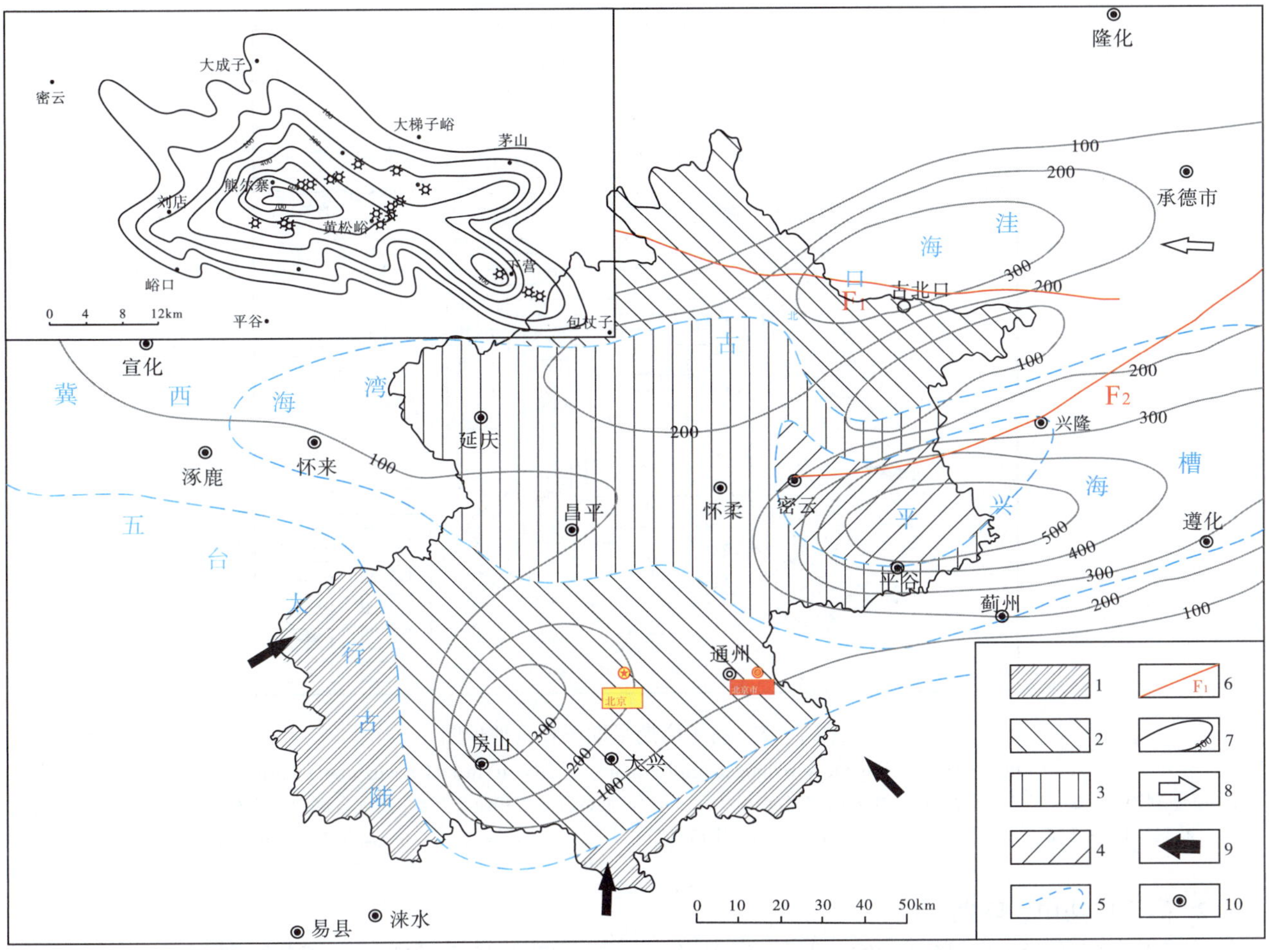

图 7-27　大红峪期沉积古地理图

Fig. 7-27　Sedimentary paleogeography map of Dahongyu period

1—基底剥蚀区；2—河流-滨岸沙滩相沉积；3—潮坪含砂白云岩相沉积；4—潮坪碳酸盐岩-火山岩相沉积；5—岩相界线；6—断裂（F_1—赤城-古北口断裂带；F_2—密云-墙子路断裂带）；7—沉积等厚线（m）；8—海侵方向；9—物源方向；10—地名

五、高于庄期

伴随着燕山裂陷槽盆地演化为坳陷槽盆地，高于庄期燕辽盆地古地理格局发生了重大变化，表现为长城纪以来第一次最广泛海侵、强烈沉降和巨厚碳酸盐岩堆积、菌藻类开始大量繁殖。期间，地壳广泛平稳地下降导致海域面积广域性扩大，向南、西发生广泛海侵超覆，海盆最大海域南界直达任丘、阜平，西界达山西、内蒙古。根据朱士兴等（1994）对古藻类的古温度分带和古地磁纬度的研究，本区古纬度再次降低，属于低纬度亚热带气候。

高于庄期起始，燕辽盆地控盆断裂及盆内断裂对沉积影响转弱，沉积作用更大程度上受垂直升降运动影响，宏观沉积单元更大程度上受古地貌影响。

1. 高于庄期早期

高于庄期早期（图 7-28），即高于庄组一段沉积时期，宏观上基本继承了大红峪期浅水沉积环境特征。房山十渡—霞云岭—门头沟潭柘寺—昌平阳坊—河北东花园一线以西仍为五台-太行古陆，河北张北—丰宁—隆化北部为内蒙古陆。经历常州沟期—大红峪期长期侵蚀夷平作用，这些古陆地势已经非常低缓。而邻近内蒙古陆和五台-太行古陆的密云、怀柔、延庆和昌平为冀西海湾，沿海湾边部的房山东南部—石景山—昌平中部—延庆西南和密云古北口—河北五道营子一带早期为滨海沙滩高能沉积环境，晚期演化为潮间低能沉积环境。而平谷、兴隆所在的平谷-蓟县海槽为潮下低能沉积环境。冀西海湾中部及大兴、北京城区、通州多数地区属于潮坪含硅白云岩相沉积环境。

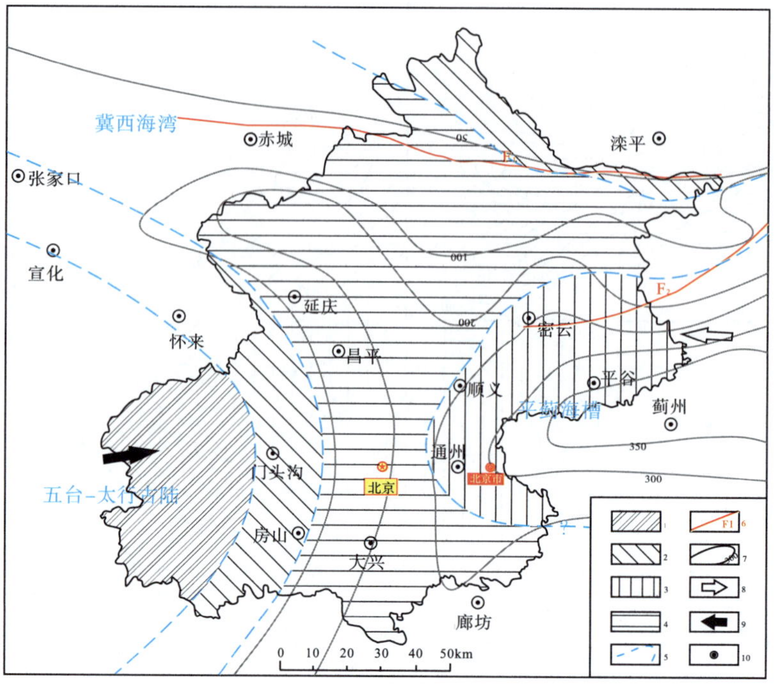

图 7-28 高于庄期早期沉积古地理图

Fig. 7-28 Sedimentary paleogeography map of early-term of Gaoyuzhuang period

1—基底剥蚀区；2—滨岸沙滩相+潮坪含砂白云岩相沉积；3—潮下泥晶白云岩相沉积；4—潮坪含硅白云岩相沉积；5—岩相界线；6—断裂（F_1—赤城-古北口断裂带；F_2—密云-墙子路断裂带）；7—沉积等厚线（m）；8—海侵方向；9—物源方向；10—地名

2. 高于庄期中、晚期

高于庄期中、晚期（图 7-29），即高于庄组二、三、四段沉积时期，沉积物以碳酸盐岩占绝对优势，沉积厚度大，碳酸盐中 $CaCO_3$ 含量显著增高，总体为海泛期沉积。尤其是三段，岩性基本以灰岩为主，是最大海侵时期，沉积水体最深。从二段到三段沉积时期，总体表现为海侵水进期，而从三段到四段沉积期表现为水退期，盆地水体略有变浅。赤城—古北口一线以北及门头沟、房山西部原古陆地区，发育为潮上-潮间潮坪含砂碳酸盐相沉积。延庆、怀柔南部、昌平、海淀西部、门头沟东部发育为潟湖-潮间潮坪碳酸盐岩相沉积。平谷、兴隆及蓟县一带为平谷-兴隆海槽，发育为潮坪-浅海碳酸盐岩沉积。怀柔北部、密云、顺义、朝阳、通州、大兴等地区，处于平谷-兴隆海槽和冀西海湾过渡地带，为潮下-浅海碳酸盐岩相沉积环境。

六、杨庄期

滦县上升结束了高于庄期广泛海浸，太行-五台古陆、内蒙古陆（又称阴山古陆）再次抬升，燕辽盆地分布范围迅速收缩。西南部古陆界线向北东收缩至河北保定东北—蔚县—怀来一带，北部古陆界线则收缩至张家口—赤城—古北口一带，盆地外围原高于庄期盆地的部分地区已经隆起成为陆地。

本期处于热带地区（朱士兴等，1994），古气候变得炎热干燥，蒸发量大于补给量，致使水介质 pH、E_h 升高，高价铁（Fe^{3+}）被泥质吸附从水介质中沉淀出来，演化成为红色的非超补偿的碳酸盐沉积盆地，在横向上产生岩相分异（图 7-30）。总的看来，大致以古北口—密云—高丽营—良乡一线为界，划分为两大沉积单元。该线以西为冀西海湾，以东为平谷-兴隆海槽。冀西海湾内，水体浅，以波浪作用为主，冀西海湾的中心宣化—延庆西部为潮间混合坪沉积，海湾周边临近古陆，发育为潮上滨岸沙滩沉积。平谷-兴隆海槽内，水体略深，潮汐和波浪共同作用，海槽边部密云、平谷、顺义及通州发育为潮间砂、泥、碳酸盐岩混合坪相沉积，而远离海槽边部的兴隆—蓟县一带发育为潮间-潮下白云岩相沉积。

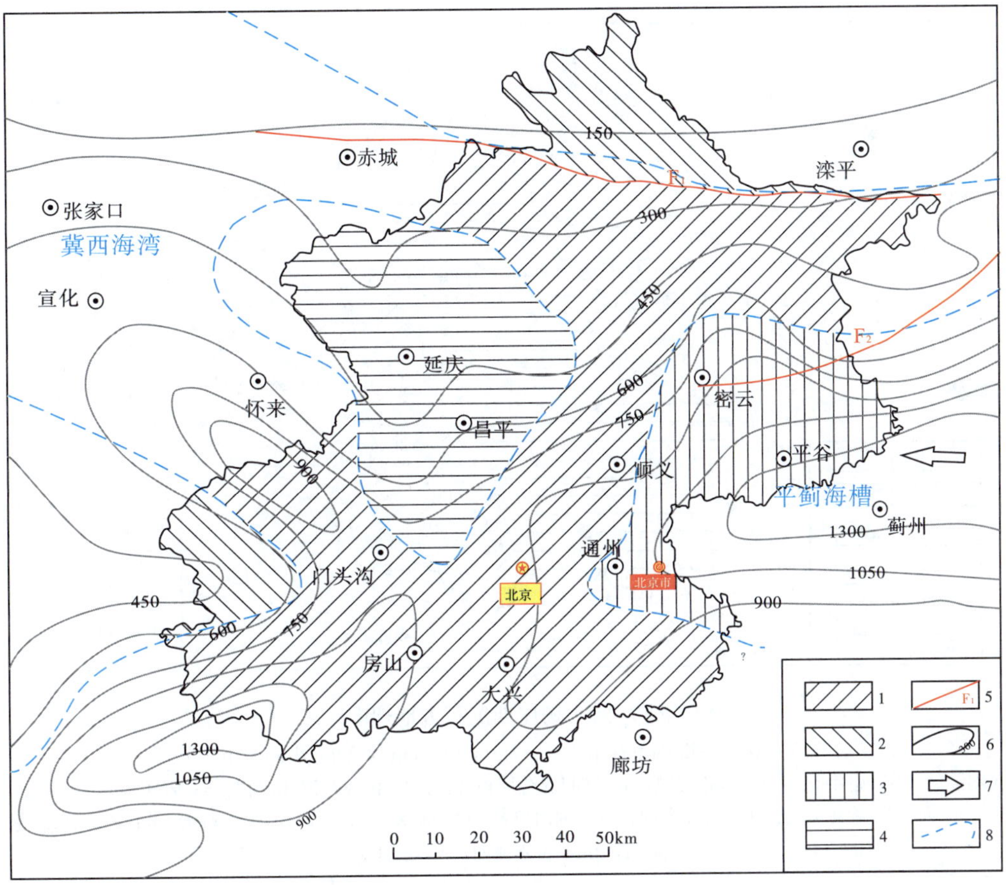

图 7-29 高于庄期中—晚期沉积古地理图

Fig. 7-29 Sedimentary paleogeography map of mid-late term of Gaoyuzhuang period

1—潮坪-浅海碳酸盐岩相沉积；2—潮坪含砂碳酸盐岩相沉积；3—潮下-浅海碳酸盐岩相沉积；4—潮坪-潟湖碳酸盐岩相沉积；5—断裂（F_1—赤城-古北口断裂带；F_2—密云-墙子路断裂带）；6—沉积等厚线（m）；7—海侵方向；8—岩相界线

七、雾迷山期

雾迷山期，燕辽盆地迎来了第二次大规模海侵，海域界线向周边再次急骤扩大，尤其是向南发生广泛海侵超覆。该期燕辽盆地古纬度位于赤道附近、热带地区，气候炎热（杜汝霖等，1980；朱士兴等，1994）。

本区位于盆地内部，全区接受沉积，且沉积岩性岩相比较稳定，总体表现为克拉通开阔稳定型富镁隐藻类碳酸盐岩沉积。由于早期构造运动风化剥蚀及沉积夷平作用，古地形总体比较平坦，沉积大致以平谷—蓟县为沉降中心向周缘厚度小幅度减小，沉积特征明显不同于杨庄期。整个雾迷山期沉积由不同级次的韵律层相互叠置而成（图 7-21），反映了雾迷山期潮汐作用频繁发生，经历了潮间上、潮间、潮下、浅海，继而潮下、潮间、潮间上的海水升降、进退过程，总体沉积环境为深水盆地和潮坪相。

密云南部、顺义东部、平谷为平谷-兴隆海槽（图 7-31），沉积水体深、地层厚度大，主要为潮下和深水盆地沉积环境；临近海槽边部的通州、朝阳、昌平东部，延庆和怀柔南部及密云的中部沉积水体略浅，主要为缓坡浅海和潮下沉积；再向外围发育为潮间-潮下沉积环境；而靠近盆地边部的怀柔北部则发育为混有较多的陆源碎屑沉积物的潮上-潮坪沉积环境。

八、洪水庄期

洪水庄期是雾迷山期海侵的继续。由于燕辽盆地由雾迷山期总体慢速下沉为主转为总体快速抬升，

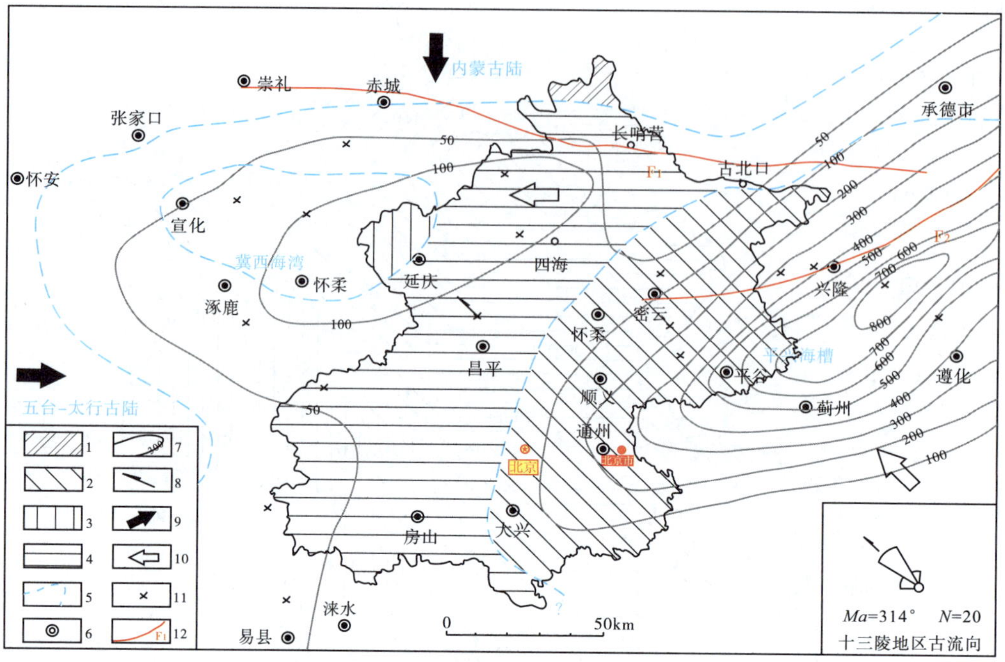

图 7-30　杨庄期沉积古地理图

Fig. 7-30　Sedimentary palaeogeographic map of Yangzhuang period

主图：1—基底剥蚀区；2—潮间-潮下白云岩相沉积；3—潮间混合坪相沉积；4—滨海沙滩相沉积；5—岩相界线；6—地名；7—沉积等厚线（m）；8—古流向；9—物源方向；10—海侵方向；11—剖面位置；12—断裂（F_1—赤城-古北口断裂带；F_2—密云-墙子路断裂带）

角图：*Ma*—古水流向；*N*—统计数

盆地分布面积迅速收缩，但盆内水体深度反而加深。燕辽盆地北界收缩到赤城—古北口一带，南界则到蔚县东—保定东—天津北西一带，宏观上构成一个南北窄、东西长的近东西向海盆。

洪水庄期燕辽盆地周缘陆地地势极度平坦，加上其所处古纬度较雾迷山期略有增大，处于亚热带气候环境（朱士兴等，1994）。周缘陆地风化作用以化学风化作用为主，盆地沉积缺乏物源，总体表现为欠补偿型饥饿沉积。发育一套以细碎屑物质为主的岩石组合，沉积厚度小，最厚也仅有 150m。

赤城—古北口以北地区因盆地收缩演化为陆地，其余地区为燕辽盆地组成部分（图 7-32）。宏观上演化为三大沉积单元：密云-兴隆海槽、冀西海湾和南部海槽周缘地区。密云-兴隆海槽沉积中心可能受东部（凌源一带）抬升影响，由雾迷山期平谷-兴隆海槽向西迁移演化形成，沉积水体深，发育为浅海陆棚泥岩相沉积。围绕密云-兴隆海槽沉积中心，包括冀西海湾和南部海槽边缘地区水体略有变浅，发育为浅海陆棚斜坡型碳酸盐岩-泥岩混合相沉积。

九、铁岭期

铁岭期地壳再次沉降，燕辽盆地古地理格局在洪水庄期基础上进一步演化。和洪水庄期相比，海水深度明显变浅，盆地分布范围向南稍有扩展，达保定南部一带。

本期燕辽盆地古纬度由赤道附近转入湿温气候的中纬度区，演化成为以内源沉积作用为主的，含锰、富铁钙的碳酸盐沉积盆地（图 7-33）。宏观上仍保持密云-兴隆海槽沉积中心，南部直达到涞水、易县一带的浅滩，和西部冀西海湾一起构成一个似“丁”字形的古构造格局和古地理景观。密云-兴隆海槽内（包括延庆、四海、密云、蓟州、兴隆地区）发育为浅海陆棚锰铁碳酸盐岩相沉积；围绕密兴海槽周缘的怀来、昌平及沿河城等地，发育为潮下碳酸盐岩相沉积；而距离海槽较远的燕辽盆地边缘部位，包括南部房山、通州及更南部易县、涞水，北部怀柔中北部一带发育为滨海砂、泥岩-碳酸盐岩混合坪相沉积。

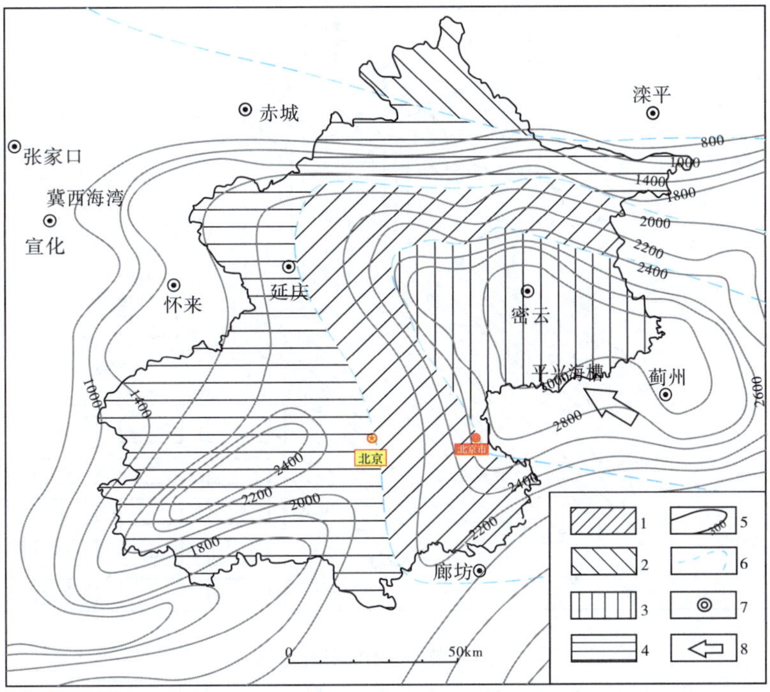

图 7-31　雾迷山期（主期）沉积古地理图

Fig. 7-31　Sedimentary palaeogeographic map of Wumishan period

1—潮下+浅海缓坡碳酸盐岩相沉积；2—潮上+潮坪含砂碳酸盐岩相沉积；3—潮下+深水碳酸盐岩相沉积；4—潮间+潮下碳酸盐岩相沉积；5—沉积等厚线（m）；6—岩相界线；7—地名；8—海侵方向

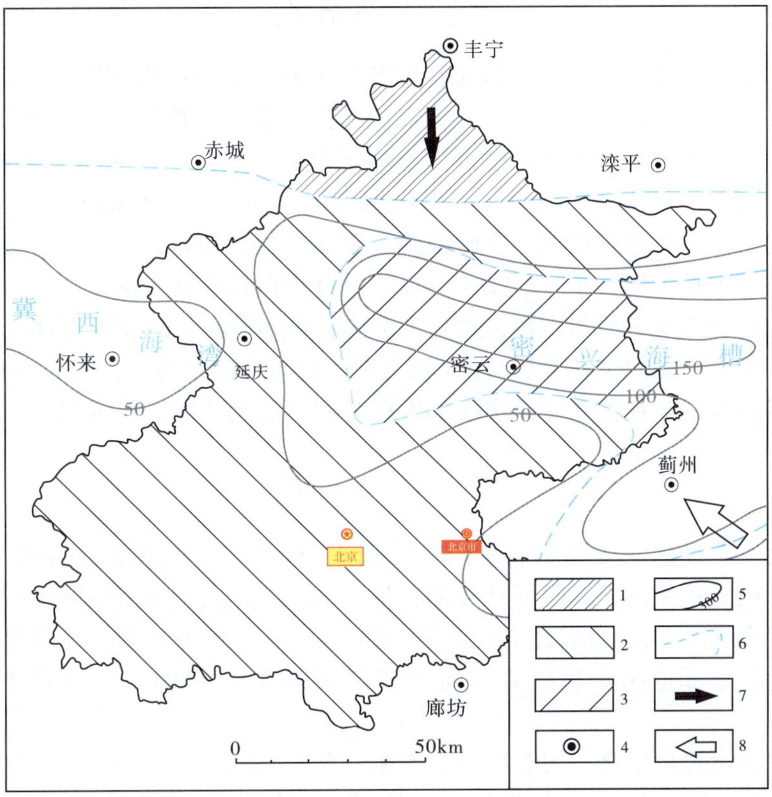

图 7-32　洪水庄期沉积古地理图

Fig. 7-32　Sedimentary palaeogeographic map of Hongshuizhuang period

1—基底剥蚀区；2—浅海陆棚斜坡型碳酸盐岩-泥岩混合相沉积；3—浅海陆棚泥岩相沉积；4—地名；5—沉积等厚线（m）；6—岩相界线；7—物源方向；8—海侵方向

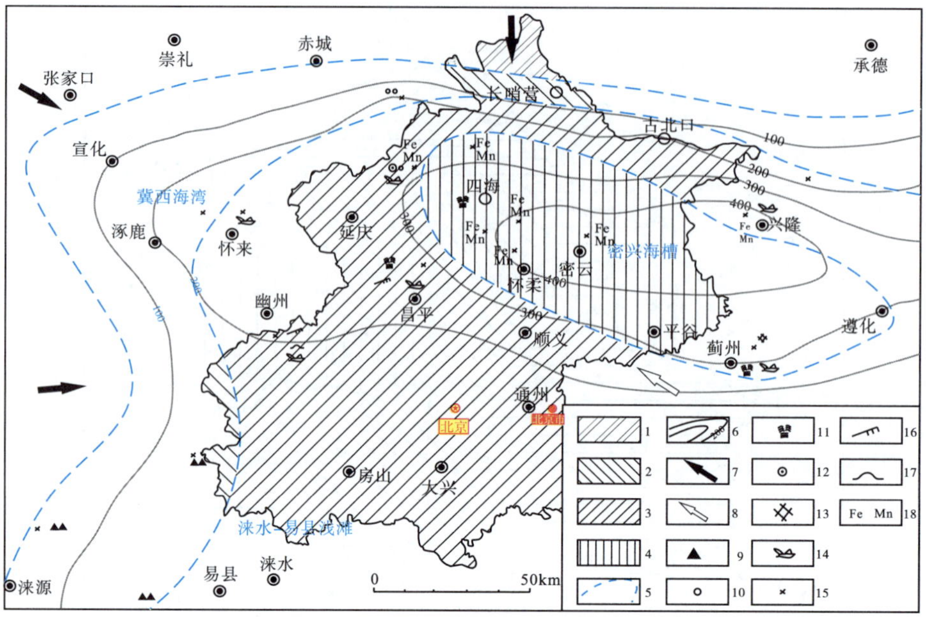

图 7-33　铁岭期沉积古地理图

Fig. 7-33　Sedimentary palaeogeographic map of Tieling period

1—基底剥蚀区；2—潮坪砂、泥-碳酸盐岩混合坪相沉积；3—潮下碳酸盐岩相沉积；4—浅海碳酸盐岩相沉积；5—岩相界线；6—沉积等厚线（m）；7—物源方向；8—海侵方向；9—砾石；10—燧石角砾；11—叠层石；12—鲕粒；13—海绿石；14—竹叶状砾屑；15—剖面位置；16—斜层理；17—波痕；18—铁矿、锰矿

十、下马岭期

在铁岭期之后，盆地整体抬升为陆，燕辽盆地经历了较长时间的沉积间断。区域上，铁岭组和下马岭组之间广泛存在一层薄层风化壳，冀西北怀安—易县一带下马岭组与铁岭组等下伏地层之间还发育底砾岩。至下马岭组沉积时期，燕辽盆地再次整体下沉，接受沉积。期间，盆地分布范围基本同铁岭期一致。沉积中心大致在京西—怀来地区，粗略划分出青白口海盆和密云-兴隆海槽两个古地理单元。下马岭期盆地仍处于温带气候环境，气候温湿（朱士兴等，1994）。

本区是该时期海盆沉积主体分布区（图 7-34），围绕青白口海盆、密云-兴隆海槽的边缘地区，包括密云北部、怀柔北部和大兴—通州一带，发育为滨岸含铁粉砂泥岩沉积；围绕青白口海盆、密云-兴隆海槽的周缘地带，发育为潮间潟湖含灰-泥岩相沉积；而在青白口海盆、密兴海槽中心广大地区发育为潮下潟湖含硅泥岩相沉积，在怀来—延庆西部发育为潮下潟湖藻灰岩相沉积。

十一、龙山期

在下马岭期之后，燕辽盆地又再次经历了较长时间整体隆起，出现沉积间断。至龙山期，盆地再次整体下沉，海水入侵，接受沉积。同下马岭期相比，龙山期燕辽盆地范围有较大的扩展，盆地北部边界位于赤城北—隆化南一线，西部边界位于张家口—蔚县—阜平一线，南部边界位于石家庄—赞皇—衡水—故城之间，东部延拓至渤海湾（河北省区域地质志，2017）。沉积中心位于怀来—延庆一带。

龙山期本区所处纬度再次增高，为亚寒带、寒带气候环境（朱士兴等，1994；陈晋镳等，1980）。古地理格局（图 7-35）在继承下马岭期盆地特征基础上又有所演化，即发育为青白口海盆和怀来-密云海槽两大沉积单元。古海域处于海绿石圈定的边界范围之内，边界以外为冲积搬运沉积作用为主的物源供给区。其中易县—涞水一线以北的京西地区，正处于多条古水流汇合入海的喇叭状河口湾环境，发育分支河道、分流间湾、河口坝等沉积类型。在离河口稍远的京西、怀来、昌平一带，形成横向展

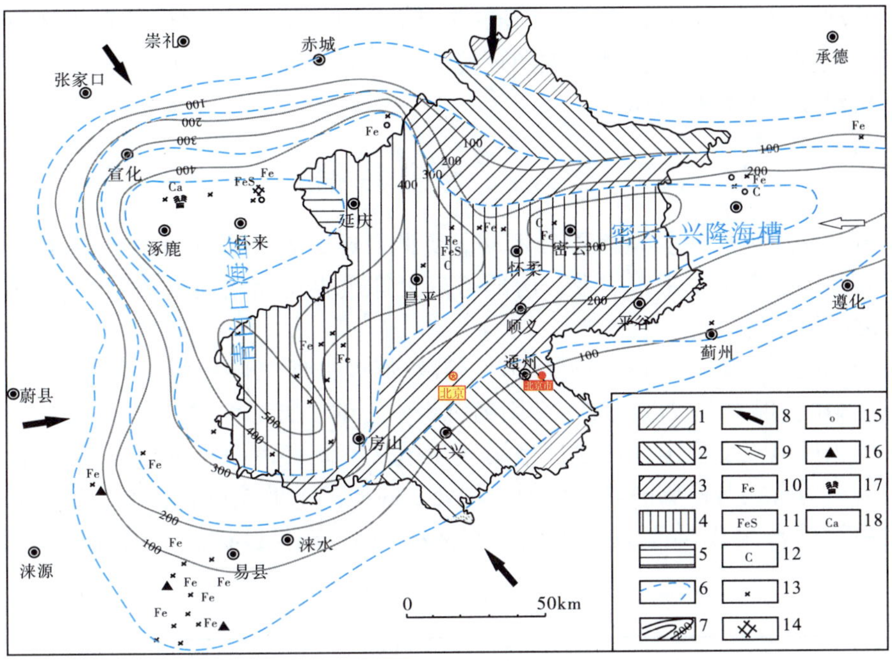

图 7-34　下马岭期沉积古地理图

Fig. 7-34　Sedimentary palaeogeographic map of Xiamaling period

1—基底剥蚀区；2—滨岸含铁砂泥岩相沉积；3—潮间澙湖含灰-泥岩相沉积；4—潮下澙湖含硅泥岩相砾质页岩沉积；5—潮下澙湖藻灰岩相沉积；6—岩相界线；7—沉积等厚线（m）；8—物源方向；9—海侵方向；10—赤铁矿；11—黄铁矿；12—碳质；13—剖面位置；14—海绿石；15—砾石；16—燧石质角砾；17—叠层石；18—钙

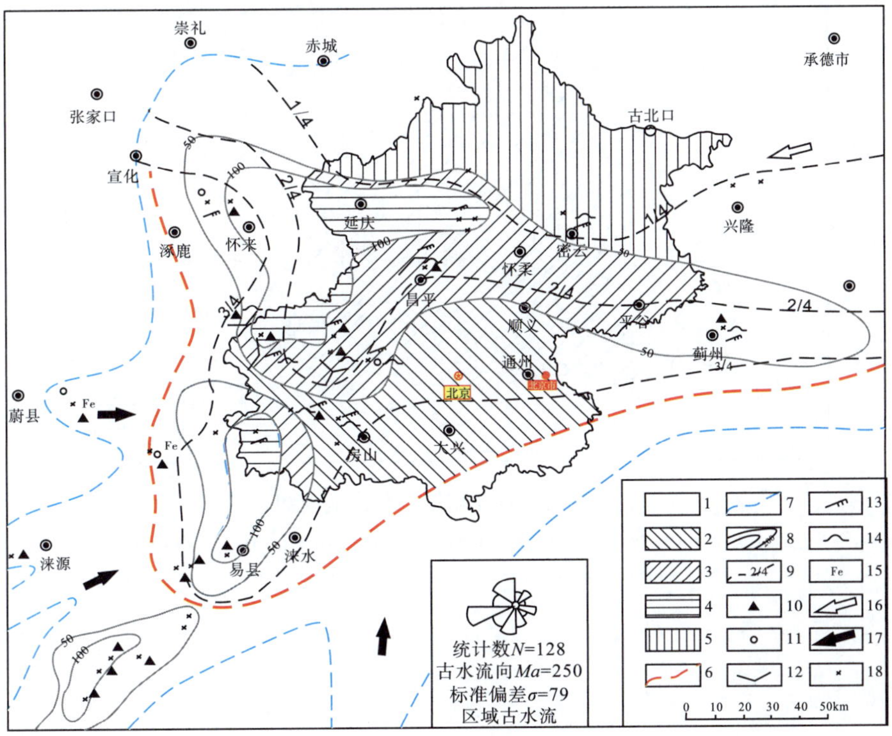

图 7-35　龙山期沉积古地理图

Fig. 7-35　Sedimentary palaeogeographic map of Longshan period

1—山地河流相沉积；2—分流间湾相沉积；3—砂坝相沉积；4—分支河道相沉积；5—前三角洲相沉积；6—海绿石分布界线；7—岩相界线；8—沉积等厚线（m）；9—砂/页岩等比线；10—砾石；11—燧石质角砾；12—交错层古水流；13—交错层；14—波痕；15—铁质；16—海侵方向；17—物源方向；18—剖面位置

布的三角洲前缘沙坝，阻隔削减来自东面的潮流-波浪作用能量，使向岸侧较宽阔区域（房山、大兴、通州、城区）成为较闭塞滞流的类似潮下潟湖环境的分流间湾沉积。沙坝北部密云、怀柔北部、延庆北部则发育前三角洲潮下泥岩相沉积。

十二、景儿峪期

景儿峪期，燕辽盆地再次经受整体抬升运动。受此影响，盆地范围缩小，但水体深度反而增大。盆地西部边界退缩到蔚县东—阜平东一带，南部与东南部边界退缩到新乐—天津北一带，东部边界处于秦皇岛与滦县之间地区。

该期本区所处纬度再次增高，古纬度达76°，为寒带近极地气候环境（朱士兴等，1994；陈晋镳等，1980）。燕辽海盆表现为大致以海淀—朝阳为沉积中心的长方形海盆（图7-36）。由于华北板块的完全克拉通化，沉积环境极度稳定，古地理格局较为单一。北京市内，沉积岩性、相变小，沉积一套以化学沉积作用为主的富钙贫镁碳酸盐岩序列，均处于浅海潮下低能沉积环境，为潮下盐岩泥沉积。

景儿峪期之后至早古生代早寒武世期间，本区一直处于隆起状态，无沉积记录，形成了长达260Ma的沉积间断。

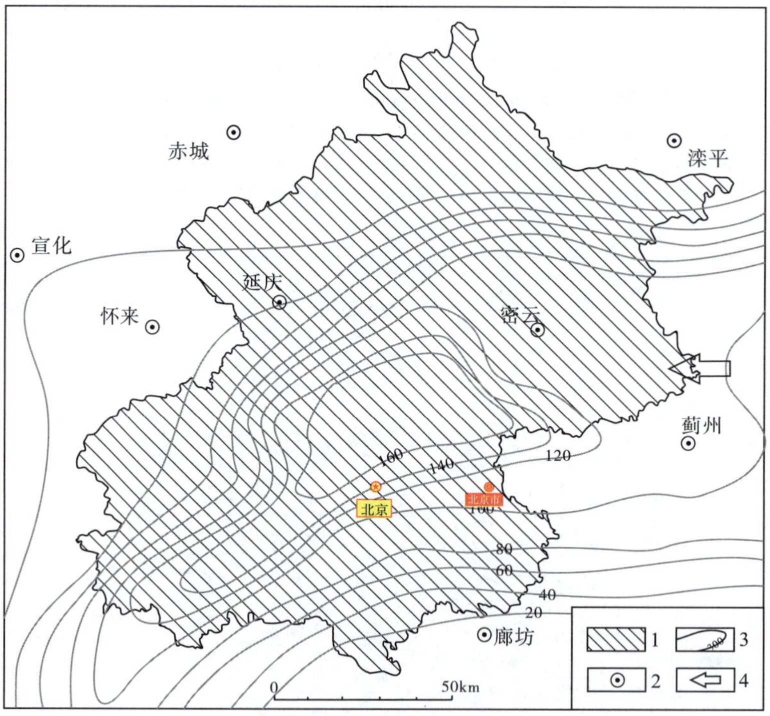

图7-36 景儿峪期沉积古地理图

Fig. 7-36 Sedimentary palaeogeographic map of Jingeryu period

1—浅海潮下盐岩泥相沉积；2—地名；3—沉积等厚线（m）；4—海侵方向

第八章 早古生代

早古生代本区在华北克拉通缓慢升降的大背景下，总体构造条件平稳。沉积自早寒武世昌平期延续至中奥陶世，晚奥陶世至志留纪区内整体抬升，沉积缺失。

第一节 沉积作用及沉积类型

早寒武世沧浪铺晚期，包括本区在内的华北北部燕山地区海水再次入侵，开始了早古生代的新一轮沉积作用。由于克拉通化的华北板块经历了新元古代—早古生代的风化剥蚀接近夷平，致使新一轮海侵近海岸带坡度极缓，平坦的浅水陆棚极度宽广，造成海水局限流通或半局限流通，波浪作用不强，以潮汐作用为主，发育一套以碳酸盐岩为主，间夹碎屑岩的克拉通陆表海沉积（表 8-1；图 8-1，图 8-2，图 8-3）。

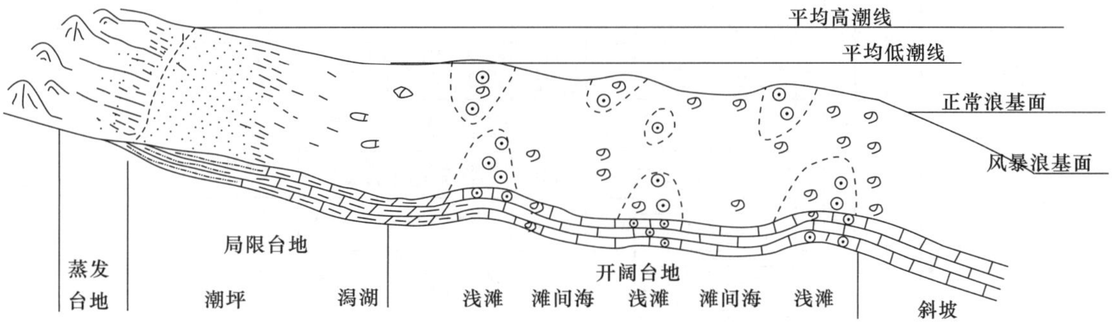

图 8-1 北京地区早古生代碳酸盐沉积相模式

Fig. 8-1 Sketch map showing the carbonate facies model of Early Paleozoic in Beijing

表 8-1 早古生代沉积作用与沉积类型统计表

Table 8-1 Statistical table of sedimentary facies in Early Paleozoic

沉积类型			地层
沉积相	沉积亚相	沉积微相	
海相沉积（内碎屑沉积）	盆地	半深海	冶里组、炒米店组
	开阔台地斜坡	浅缓坡、中缓坡、深缓坡	张夏组、炒米店组、冶里组、亮甲山组
	开阔台地	浅滩、滩间海	张夏组、炒米店组、亮甲山组、马家沟组局部
	局限台地	潮坪、潟湖	昌平组中部、馒头组上部、张夏组底部、炒米店组局部、亮甲山组上部、马家沟组局部
	蒸发台地	潮上、滨岸	昌平组顶部和底部、馒头组下部、马家沟组底部

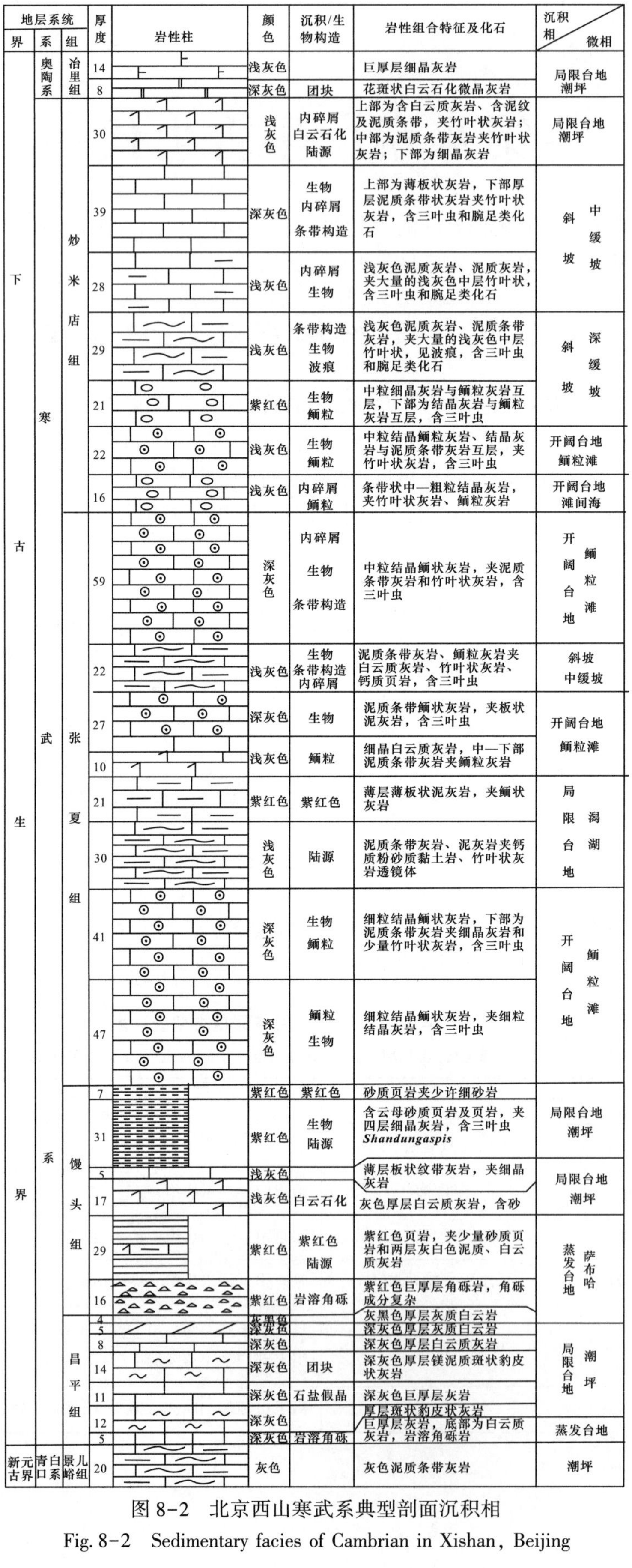

图 8-2　北京西山寒武系典型剖面沉积相

Fig. 8-2　Sedimentary facies of Cambrian in Xishan, Beijing

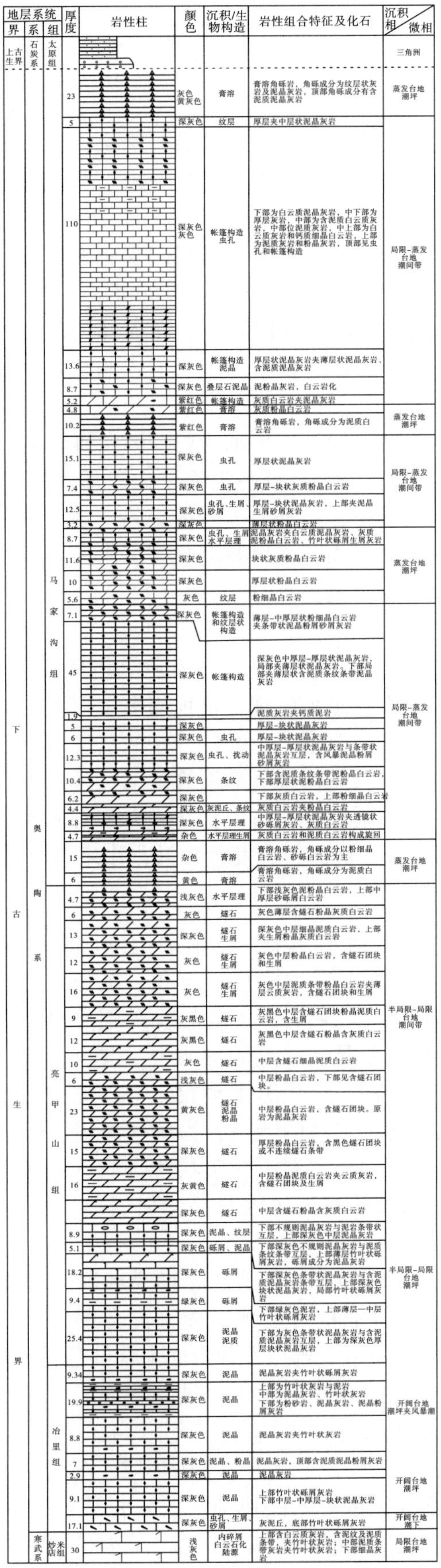

图 8-3 北京西山奥陶系典型剖面沉积相图

Fig. 8-3 Sedimentary facies of Ordovician in Xishan, Beijing

一、正常沉积

（一）蒸发台地沉积

蒸发台地相发育于早古生代3次海侵开始或海退末期阶段，分布于寒武系昌平组底部、馒头组底部、奥陶系马家沟组底部。该相沉积时海水极浅，仅数厘米至数十厘米，蒸发量大，盐度高，不利于生物的繁衍，间歇性水流能量弱。

昌平组底部陆源物质含量较高，含燧石条带，沉积以含砾屑、砂屑粉晶、泥晶白云岩为主，临近西部海岸地带（如延庆秤沟湾）为含砾钙质砂岩及钙质页岩碎屑岩类，其下为角砾状灰岩和风化壳，总体反映了海侵初期水动力条件较强的近岸沉积环境。而顶部为灰色中厚层白云岩，含燧石团块或条带，其上为土黄色泥晶白云岩。顶面有石盐假晶，在下苇店地区发育有紫红色滑塌古喀斯特角砾，反映第一次海退潮上蒸发环境。

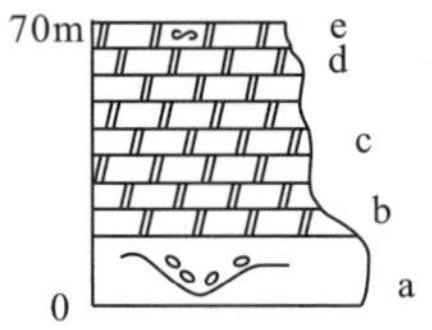

图8-4　馒头组潮上沉积序列

Fig. 8-4　Supratidal deposit sequence of Mantou formation

a—溶塌砾岩；b—灰粉色白云岩；c—灰紫色白云岩；d—灰绿色白云岩；e—灰黄色泥质条带白云岩

馒头组下部为白云质角砾岩和盐类矿物溶解后形成的角砾岩和石英砂岩-紫色泥岩、白云质泥灰岩和泥晶白云岩组合（图8-4）。其中，角砾岩为岩溶塌陷产物，其中发现大量纤维状石膏及其溶蚀残余物，泥岩及泥灰岩、白云岩中发育石盐假晶，总体反映气候干燥炎热的潮上沉积环境，为早古生代第二次海侵开始后形成的沉积。

马家沟组底部为泥晶泥质白云岩，含陆源石英碎屑、白云母，发育水平层理，且顶面多发育泥裂构造。该泥晶白云岩主要形成于准同生期，为海侵开始的蒸发潮上云坪沉积环境产物。由于蒸发泵白云化作用或毛细管浓缩白云化作用形成富镁的超盐度海水，交代潮上坪的文石泥和高镁方解石泥而成。在成岩作用过程中泥晶白云岩局地重结晶变成泥粉晶、细粉晶或者更粗，或发生去云化作用形成晶粒灰岩。

（二）局限台地沉积

局限台地相是本区早古生代主要沉积环境之一，在各组地层中均有分布。

该相在昌平组中期发育，为早古生代首次海侵后沉积，沉积岩厚度及成分变化较大，沿河城发育粉砂质泥岩，青白口发育内碎屑球粒亮晶灰岩，下马岭发育细晶白云岩及灰质白云岩，下苇店发育豹皮灰岩及核形石灰岩，昌平一带发育粉晶灰岩夹豹皮灰岩，含少量三叶虫、棘皮类化石碎片。岩性以厚层灰岩、灰质白云岩为主，颜色较深，单层厚度也较大，以泥、粉晶结构为主，总体反映了介质能量低、物质供应充足稳定、有机质含量较高的特点，为气候温暖、适合生物生长的静水局限海潮下沉积环境。根据上述岩性组合特征，可细分为潮下豹斑状含骨屑球粒灰岩相、潮下球粒泥晶灰岩相、潮下豹斑状粉屑细粉晶灰岩相及潮间含燧石团块细粉晶白云岩相。西山西部见盐壳，为潮间浅水低能环境。

馒头组为区内早古生代第二次海侵沉积，馒头组早期为蒸发台地潮上沉积。伴随着海侵继续，中、晚期发育为潮间潟湖与碳酸盐岩-碎屑岩混合潮坪（简称混合坪）沉积，前者岩性为紫红色薄—中层粉砂质泥岩夹灰色、灰黄色泥晶-粉晶白云岩，后者岩性为紫红色粉砂质泥岩与灰色、灰绿色薄层灰岩互层。灰岩发育经准同生后白云化形成的纹带状水平层理，泥岩中见石盐假晶、泥裂、波痕，顶部灰岩中产数量丰富，含种属单调的三叶虫。

张夏组局限台地相沉积环境发育较少，主体为潮下带环境，沉积岩性主要为灰色、灰黄色泥晶灰岩、白云质灰岩、灰质白云岩、含燧石条带或燧石团块细晶白云岩及杂色角砾状灰岩或角砾状白云质灰岩，另外还含有石膏、方解石等自生矿物。局部发育鸟眼构造、石膏假晶、鸡笼构造、干裂构造、波状层理、水平层理，白云岩中发育角砾状、垂直或近垂直的虫孔以及叠层石等构造，其中叠层石常呈半球状。

炒米店组局限台地相主要发育于顶部，沉积水体能量较低，以泥质条带泥晶灰岩为主，夹叠层石凝块石灰岩、竹叶灰岩，中下部普遍发育叠层石和钙质粉砂岩，为局限台地潮坪-浅滩沉积环境。炒米店组最显著的特征是夹有大量竹叶状砾屑灰岩，其中的竹叶状砾石多为泥晶灰岩，据研究是风暴成因。风暴将半固结的泥晶灰岩破坏，经扰动和磨蚀、溶蚀堆积而成，早中期多为异地堆积，说明水体较浅，风暴经常发生，到晚期竹叶状砾屑明显变小，有时似断似连，说明风暴影响渐弱，海水加深，形成条带状或纹层状薄层灰岩。

冶里组局限台地相发育于底部，岩类以厚层—巨厚层花斑状白云岩、白云岩组合为主，局部夹泥质条带。

亮甲山组局限台地相主要发育于亮甲山组顶部，灰色、灰白色中厚层准同生白云岩可见干裂、波痕构造，局部还可见喀斯特角砾岩，反映沉积环境为局限台地云坪相沉积；岩性为青灰色纹带灰岩及白云岩、钙质泥岩组合，发育多层膏溶角砾，钙质相对减少，镁、钾的含量显著增高，K/Na 值较高，局部达到蒸发岩的数值级别，为局限台地潟湖相沉积。

马家沟组局限台地相沉积最为发育，是主要沉积类型，不同时段局限台地沉积各有特征。早期为杂色角砾状灰岩、白云岩，含黏土与生屑，为水浅而动荡的潮间环境，且叠置在初期（底部）潮上沉积上，而上部发育为开阔台地相沉积；中期发育为云坪间潮间潟湖沉积和潮下静水沉积；晚期沉积由于地壳上升遭剥蚀而保留不全，仅见灰色中层粉晶灰岩，含生屑、云斑，夹角砾状灰岩、薄层泥质灰岩，具水平层理，以潮下低能环境为主。

（三）开阔台地沉积

开阔台地相也是本区早古生代主要沉积环境之一，在各组地层中均有分布。根据沉积地貌特征，开阔台地相可划分为浅滩、滩间海和滩前斜坡三类亚相。

1. 浅滩沉积

浅滩沉积在张夏组、炒米店组和亮甲山组均有发育，岩石微相组合分别属于鲕粒滩、混合滩和生屑滩。

张夏组浅滩沉积为鲕滩沉积，以厚层、巨厚层灰色亮晶鲕粒灰岩为特征，夹叠层石灰岩、泥晶灰岩、藻灰岩、竹叶灰岩。亮晶鲕粒灰岩常伴有一定数量异地生屑及少量砂屑、藻屑，含海绿石。颗粒以鲕粒为主，亮晶胶结，具交错层理、流水波痕，柱状叠层石发育，化石破碎，反映沉积介质能量高，其沉积环境以潮下高能浅滩为主。

炒米店组浅滩沉积发育在下部，岩性为泥亮晶鲕粒灰岩、泥亮晶竹叶灰岩等颗粒灰岩组合，尤以竹叶灰岩为主，呈夹层产出，夹于泥晶灰岩、含颗粒的叠层石灰岩、钙质粉砂岩之中。颗粒灰岩普遍夹砂质条带，颗粒为鲕粒和竹叶，颗粒填隙物主要是泥晶，亮晶胶结居次，为潮下高能浅滩沉积。

亮甲山组浅滩沉积发育于北山，区内沉积地层可能已被剥蚀，仅在邻区河北承德头道杖子一带残留。沉积岩性下部为灰色厚层虫孔泥晶灰岩与灰色厚层含燧石结核泥晶灰岩互层，上部为藻席白云岩、泥晶白云岩和叠层石白云岩组合，反映其由开阔台地斜坡相向台地生物礁滩相转变。

2. 滩间海沉积

本区滩间海相在张夏期海侵达到高潮，水下隆起多成浅滩，滩间海环境较为发育。岩石以深灰色生物泥晶灰岩为主，浅海生物含量较丰富，黏土含量变化较大。

张夏组滩间海相沉积岩性以钙质粉砂岩、泥云质条带泥晶灰岩为主，夹有泥晶鲕粒云质灰岩。具水平层理；为弱氧化色或还原色环境；含三叶虫、腕足类等化石碎片；钙质粉砂岩成层性良好，与灰岩界线清晰且呈不等厚互层，或呈条带夹于灰岩中，当陆源碎屑少时，过渡为粉砂质泥晶灰岩；夹层鲕粒灰岩发生准同生后白云岩化作用。总体反映沉积环境为盐度正常、水体循环程度中等、有一定量氧气供给、生物得以生存的近岸低能局限海沉积环境。

炒米店组滩间海相沉积多发育于中段，岩性为泥晶竹叶灰岩、钙质粉砂岩、云泥质条带泥晶灰岩组合。竹叶灰岩中的竹叶状砾屑多具氧化圈，呈无定向菊花状排列，丁家滩永定河东岸见小型生物丘灰岩，并被碎屑流截切，总体反映沉积环境时有暴露，处于有潮汐流作用并具中低间歇能的潮间环境，间或有潮下低能环境并受风暴作用的影响。

亮甲山组滩间海相（局限海）沉积主要发育于中期，其下部岩石为灰黑色巨厚层含燧石团块含古杯海绵泥晶灰岩及深灰色厚层或中层条带状泥晶灰岩，含泥纹、云斑、砂屑、生屑、球粒、砾屑，夹竹叶灰岩，生物碎屑平行层面分布。上部为灰色中厚层含燧石粉晶白云岩，含丰富的头足类、腹足类等浅海近底生物，为准同生后白云岩。岩石颜色较深，有机质含量高，泥晶结构，生物化石丰富且保存完整，反映当时处于较为稳定，水体深度适当、阳光充足、温度适宜的局限海潮坪-潮下沉积环境。

马家沟组滩间海相沉积特征为浅黄色纹带状含灰泥晶白云岩与深灰色含生屑花斑状泥晶灰岩、球粒灰岩、生物碎屑灰岩、亮晶含云藻粒灰岩、亮晶内碎屑灰岩（竹叶灰岩、砂屑灰岩）构成的韵律层，泥晶白云岩中见虫迹，白云岩与灰岩类之比一般为 3 : 7，有的韵律发育或保存不完整，多具水平层理，顶部见波痕。灰岩为深灰色、灰黑色等还原色，为潮下静水或能量较弱的沉积环境。

3. 滩前斜坡相沉积

滩前斜坡相又称台地前缘斜坡相，本区主要发育于炒米店组、冶里组和亮甲山组，张夏组也局部发育。

张夏组斜坡沉积总体发育较少，为缓坡型边滩斜坡沉积。斜坡带能量不同，形成不同旋回层序（图 8-5）：①深缓坡沉积以泥晶灰岩或含鲕粒泥晶灰岩（上部单元）、灰绿色泥钙质粉砂岩或粉砂质泥岩（下部单元）组合为基本沉积序列，泥岩或粉砂岩层厚，含海绿石，泥晶灰岩有时见小型生物丘；②中缓坡沉积由薄层钙质粉砂质泥岩或泥钙质粉砂岩（下部单元）、薄层泥灰岩（中部单元）和中薄层鲕粒灰岩（上部单元）构成基本沉积序列，下部或中部单元有时缺失；③浅缓坡沉积由中薄层粉砂质泥页岩（下部单元）、中厚层鲕粒灰岩（上部单元）构成基本沉积序列，或由薄层泥晶灰岩（下部单元）、中厚层鲕粒灰岩（上部单元）构成基本沉积序列。

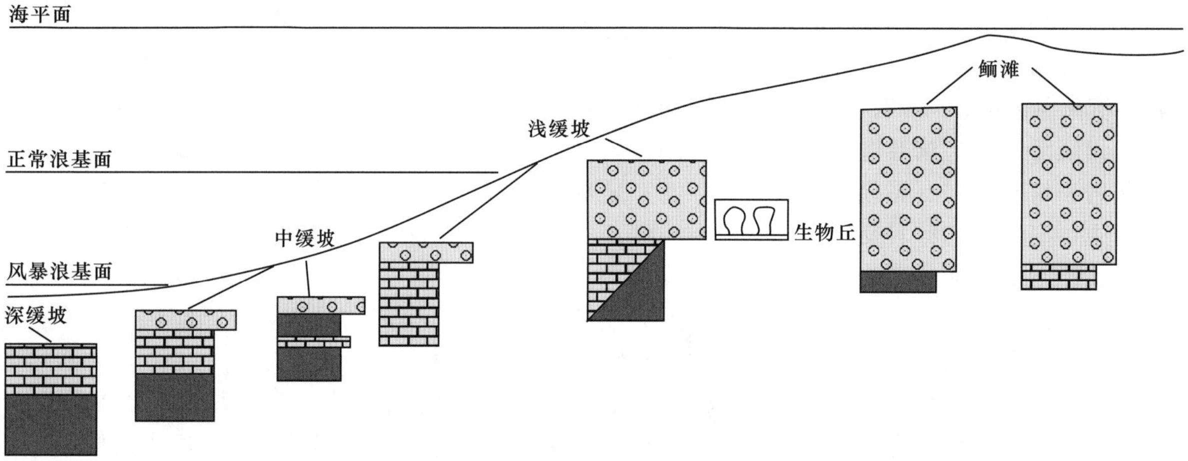

图 8-5　张夏组沉积序列与其生成环境示意图
（据孟祥化等，2004）
Fig. 8-5　Genetic types and their environment feature of meter-scale cycles in Zhangxia formation

炒米店组滩前斜坡沉积主要发育于上部，岩性组合以云泥质条带泥晶灰岩、泥晶灰岩、竹叶灰岩为主，含丰富的三叶虫化石。竹叶砾屑灰岩，竹叶小，不具氧化圈，多为异地风暴成因；泥晶灰岩呈条带产出，为深灰色；云泥质条带泥晶灰岩与竹叶灰岩构成韵律层，云泥质条带泥晶灰岩往往含颗粒，而形成很薄的递变碳酸盐岩层，是一种低密度浊流，具毫米级或厘米级递变纹层。

冶里组滩前斜坡沉积以泥晶灰岩为主，上部夹泥质条带泥晶灰岩和竹叶灰岩，发育水平纹理和波状纹理，含三叶虫、棘皮动物、牙形石、介形虫等，属正常海生物，但种属比寒武纪时减少，深水生物亦减少；经过水体移动、磨蚀的砂级以上的颗粒含量不足 5%，反映沉积环境的水动力条件较弱或很弱，不含准同生白云岩，黏土矿物组合为伊利石和绿泥石组合，未发现浅水盆地边缘相沉积，表明水体较深，反映为开阔浅海斜坡沉积环境。

亮甲山组斜坡沉积分布极少，主要集中于下部。以深色泥晶灰岩为主，发育钙质粉砂岩、泥灰岩、泥岩组合。正常海生物繁盛，含有大量的头足类、腹足类、牙形石以及古杯海绵、双壳类等，所含颗粒为砂屑、生屑、藻屑，砾屑较少。颗粒含量<10%，水动力条件弱。岩层为厚层—巨厚层及中厚层，具水平层理，岩石为还原色（灰黑色、深灰色为主），未见准同生白云岩和浅水盆地边缘相沉积。这些特征表明水体相对较深，反映安静开阔台地斜坡沉积环境。

（四）深水盆地沉积

深水盆地沉积仅发育于第二次海侵峰值期的冶里组，沉积物为粉砂质泥岩、泥灰岩、杂色风暴砾屑灰岩、海绿石生屑灰岩、含海绿石粉砂岩及上部的泥晶灰岩组合。沉积组合含泥质粉砂质岩石和含海绿石泥岩较多，占总岩系的50%以上，且这些泥质粉砂质岩石多为深水沉积，发育凝缩层和硬底构造。

二、风暴沉积

早古生代晚寒武世，包括本区在内的广大燕辽盆地，发生过高强度的风暴沉积。孟祥化（2004）将风暴沉积划分为原地型和异地型。

（一）原地型风暴沉积

原地型风暴沉积特征通过原地型风暴粗颗粒灰岩反映出来。该类岩石是一种扁平的呈放射状、砥柱状排列的碳酸盐砾屑灰岩，以张夏组、炒米店组（尤以原崮山组）最为典型。砾屑成分比较单一，与下伏岩性一致。砾屑表面无氧化圈，无明显搬运磨蚀痕迹。大部分呈颗粒支撑，局部为基质支撑，为风暴潮流就地冲击、掀起、挤压、破碎而原地堆积的产物。

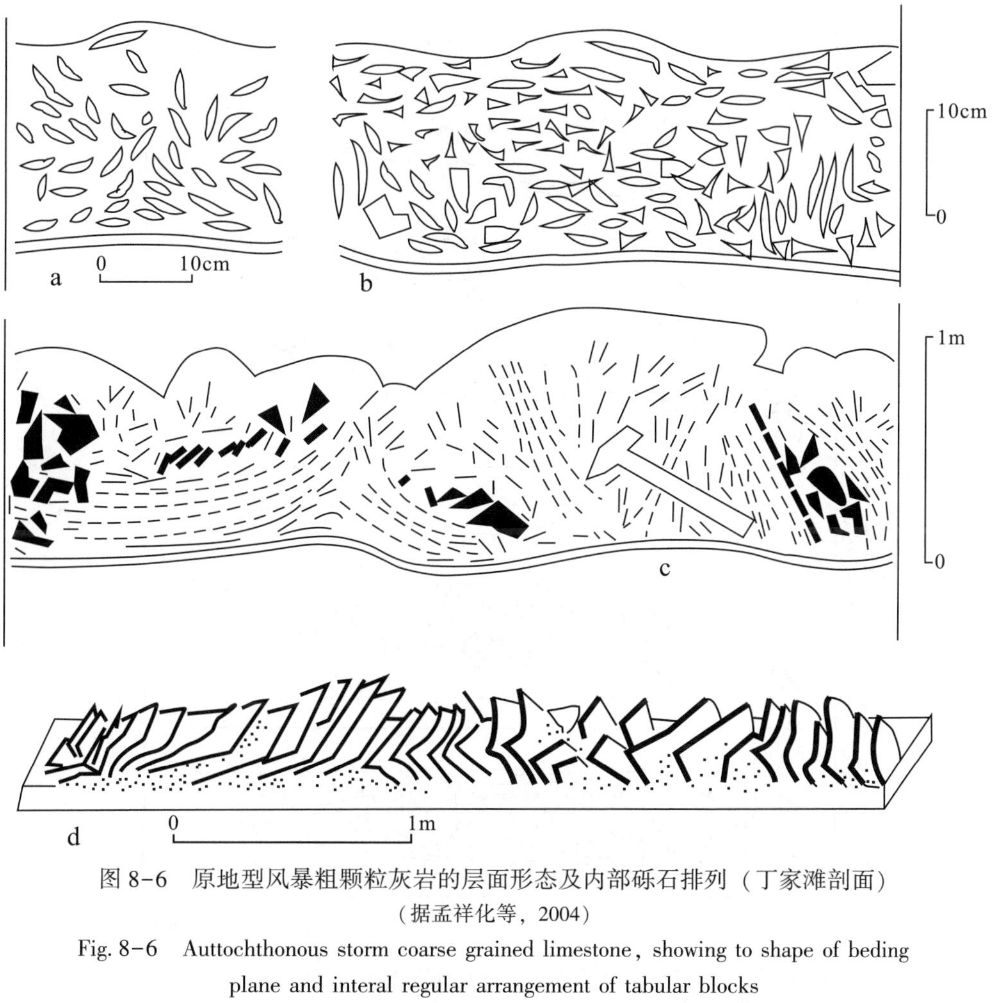

图8-6　原地型风暴粗颗粒灰岩的层面形态及内部砾石排列（丁家滩剖面）
（据孟祥化等，2004）

Fig. 8-6　Auttochthonous storm coarse grained limestone, showing to shape of beding plane and interal regular arrangement of tabular blocks

a—具不平整顶层面，砾石呈放射状排列，沿底部层面砾石片呈协调皱纹排列，绝大多数砾屑为来源于条带状灰岩的板片状碎块（数厘米至数十厘米）；b—具不平整顶层面，砾屑呈倒“小”字形排列；c—具云朵状上层面，颗粒支撑，局部基质支撑；d—板片状砾屑显示出被风暴漩涡流挤压，扭动的特征孟祥化等（2004）总结指示该类沉积的构造类型包括倒“小”字构造（图8-6）、叠瓦构造和巨型云朵-脊椎状构造，并借助现代台风等热带低压气旋理论解释其形成经历前风暴（平静）期、风暴初期、高峰期和衰退期四个过程阶段（图8-7）。

（二）异地型风暴沉积

异地型风暴沉积是在风暴潮流作用下发生异地的沉积作用，通过异地堆积的竹叶灰岩及其沉积构造反映出来，本区内以张夏组、炒米店组（尤以原长山组）最为发育。多形成于陆表海的盆地边缘环境，由成分较复杂的、具菊花状至杂乱状排列的片状灰岩砾屑组成，具有丘状交错层理、粒序层理和截切构造等典型沉积构造。

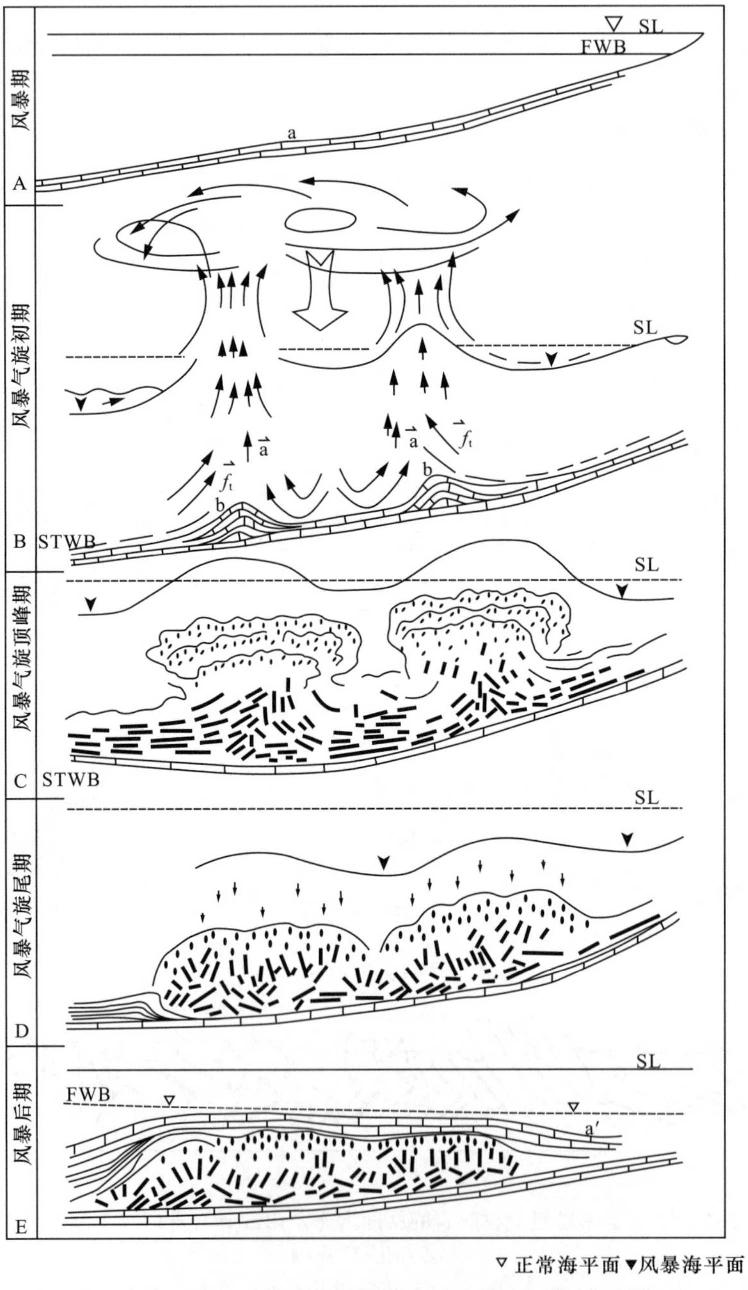

图8-7　原地风暴砾屑灰岩超微层序的成因模型——碳酸盐台地缓坡带上飓风气旋作用的各阶段（A~E）和沉积作用

（据孟祥化等，2004）

Fig. 8-7　Genetic modle of the formation storm sequene

（三）风暴沉积相模式及其典型相序剖面

孟祥化等（1986，1993，2004）对北京西山丁家滩晚寒武世浅海碳酸盐风暴沉积相模式进行了讨论，建立了风暴沉积的典型相序剖面（图8-8），将根据其发生的位置将风暴沉积划分为盆地风暴沉积相和盆地斜坡碎屑流沉积。

1. 盆地风暴沉积相沉积

沉积发生于风暴浪基面附近或以下的浅海盆地，相序由风暴浊积灰岩和盆地相泥质条带灰岩、纹理状泥晶灰岩或泥质页岩组成。

2. 盆地斜坡碎屑流沉积

沉积发生于风暴浪基面以上近源风暴流浅海盆地斜坡部位。沉积层序由异地风暴粗颗粒灰岩、细颗粒灰岩（砾屑碎屑流、粗砂屑碎屑流沉积）与盆地边缘斜坡泥质条带灰岩、细石英粉砂岩层组成。

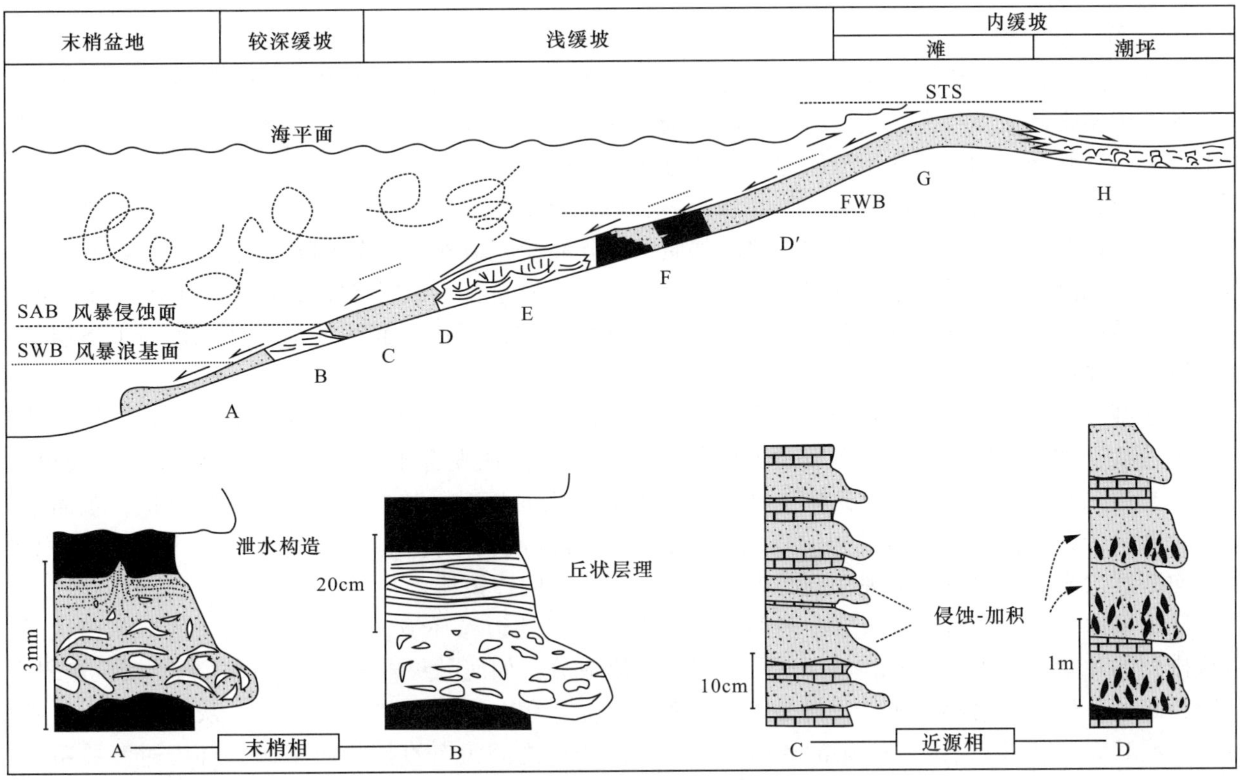

图8-8　华北板块晚寒武世浅海碳酸盐风暴沉积相模式

（据孟祥化等，1986）

Fig. 8-8　Sketch map showing the shallow sea carbonate facies model controlled by storm current

A—末梢盆地的风暴递变沉积相；B—较深缓坡带的含丘状层理的递变沉积相；C—浅缓坡深部风暴砂屑递变层理；D—浅缓坡中深部风暴砾屑递变层理-砾屑块状层理；E—浅缓坡原地风暴砾岩相；F—浅缓坡截断生物丘相；D—浅缓坡上部风暴砾岩相；G—风暴作用改造的鲕滩相（包括土型鲕团块丘状层相）；H—内缓坡风暴作用改造的潮坪相

第二节　主要沉积岩岩石类型

本区早古生代沉积岩以碳酸盐岩为主，占80%~90%，黏土岩占8.2%~17.6%，碎屑岩占0.09%~5.7%，总体表现为贫碎屑、贫泥、富碳酸盐岩（表8-2）。碳酸盐岩成分简单，以灰岩为主，白云岩为次。由于沉积环境多变，成岩作用较复杂，故碳酸盐岩结构复杂，常见的结构组分和岩石类型都有出现。

表 8-2　北京地区下古生界岩石类型统计表

Table 8-2　Statistic table for the Lower Paleozoic rock types in Beijing

岩类	龙凤山	双石头	下苇店	文殊峪	河防口
碎屑岩/%	4.8	5.72	0.09	2.75	4.1
黏土岩/%	12.89	8.23	10	17.55	12.6
碳酸盐岩/%	82.31	86.05	89.91	79.7	83.3

一、碳酸盐岩类

（一）灰岩类

1. 颗粒灰岩

（1）竹叶状灰岩

竹叶状灰岩是早古生代常见岩石类型之一，张夏组、冶里组、马家沟组、炒米店组均有分布。扁平状砾石由条带状微晶—泥晶灰岩破碎而成，为高能沉积环境下的产物。包括潮间坪正常间歇高能环境形成的竹叶状灰岩和风暴作用下原地、异地堆积的竹叶状灰岩。

潮间坪正常间歇高能环境形成的竹叶状灰岩，砾屑多具氧化边，大致顺层排列。竹叶状砾石成分单一，与下伏沉积物成分相同。颗粒支撑的基质部分除亮晶充填外，并有填隙碎屑物，如鲕粒、砂屑、石英砂等，泥晶少，基质颜色为紫红色、红色、灰色，含海绿石。单层厚度较薄，几厘米至几十厘米，为较稳定的似层状体。砾屑分布较均一。上下岩层界面平整，下伏层面往往为一冲刷面。上下层位发育潮坪或浅滩相岩石序列。

原地型风暴沉积的竹叶状灰岩，多形成于陆表海的滩外斜坡环境，其扁片状碳酸盐岩砾屑呈放射状、砥柱状、掌状、倒“小”字或协调皱纹状排列，后者往往发育于岩层底部。砾屑成分比较单一，与下伏岩性一致。大部分呈颗粒支撑，局部基质支撑。底面相对平整，局部有微波状起伏，顶面不规则，轮廓呈云朵状，反映原始堆积的顶面形态，为风暴潮流就地冲击、掀起、挤压，破碎的堆积物。沿走向断续出露，不协调地出现在剖面中，呈厚度不稳定的透镜体。

异地型风暴沉积的竹叶状灰岩，多形成于陆表海的盆地边缘环境，由成分较复杂的、菊花状至杂乱状排列的片状灰岩砾屑组成。包含不同成因的砾屑成分，如泥晶灰岩屑、砂屑灰岩屑、鲕粒灰岩屑、叠层石碎屑、生物碎屑等。泥晶、粉晶基质支撑，砾屑、基质具褐、红、紫、黄、灰、绿等杂色或无色，表明其起源和形成经历了复杂的环境。与下伏沉积层之间没有直接物质来源和环境过渡关系，底面常呈冲刷接触，而与上覆岩层为平行连续过渡，顶面平直，少部分呈云朵状，夹于静水纹层状、条带状泥晶灰岩、页理状泥晶泥质灰岩、透镜状含颗粒泥晶灰岩或页岩中，呈似层状透镜体产出。

（2）角砾灰岩

角砾灰岩最常见于馒头组底部、马家沟组中—上部，冶里组也有少量分布，属于典型的盐溶角砾岩。

馒头组角砾灰岩主要分布在馒头组底部，主要呈紫红色，其次为浅灰、黄灰色。角砾大小混杂，无分选、无定向排列、磨圆度差，一般呈棱角状，显示迅速堆积的特点，成分与同层位的岩石一致，部分角砾与馒头组中段岩性一致，为泥晶白云岩。

马家沟组角砾灰岩呈多层厚层状产出，房山区上万村一带角砾灰岩层数至少是丁家滩一带正常出露层数的两倍。角砾大小混杂，结构组分复杂，含量一般约 70%，多为棱角—次棱角状，主要为碳酸盐岩经过大气淡水或地下水淋滤、渗透作用后，垮塌在原地堆积形成，按照岩石类型可分为灰岩角砾岩和白云岩角砾岩两大类。灰岩角砾较大，白云岩角砾往往与石膏溶解作用有关，局部可见板条状石膏假晶，风化后形成孔洞。

（3）鲕粒灰岩

鲕粒灰岩是寒武系张夏组的主要岩石类型，馒头组和炒米店组也有产出。可进一步划分为泥晶鲕粒灰岩、泥晶豆粒灰岩、亮晶鲕粒灰岩。

泥晶鲕粒灰岩，灰、深灰色，中—薄层，鲕粒多为较低能量环境下形成的放射鲕或藻鲕，主要分布于张夏组下、中部，馒头组中部及炒米店组（原崮山组），含陆源石英砂者分布于馒头组底部。

泥晶豆粒灰岩，灰、深灰色，薄层，豆粒为椭圆形，粒径 2~5mm，同心结构，豆核多为生物碎片，有的为藻屑，豆粒同心层明暗相间，暗层见藻结构，亮层云化程度高。机械作用、生物（藻）作用致使迅速加积形成豆粒，其间常见 0.1~0.2mm 的多晶鲕，有的恰好位于豆粒被溶蚀的沟槽内，基质为泥晶充填，分布于张夏组中下部。

亮晶鲕粒灰岩，灰至深灰色，厚至巨厚层，鲕粒类型以同心鲕、轮辐鲕和复鲕为主，亮晶胶结，高能环境产物，主要分布于馒头组和张夏组中、上部。

（4）生物碎屑灰岩

本类岩石是寒武系、奥陶系的常见岩石类型，但各层位的岩石、生物组合和成熟度都有其各自特征。

昌平组含生屑泥晶灰岩，风化面呈蓝灰色，新鲜面呈深灰色，生物碎屑结构，多具豹斑构造，豹斑成分多为内碎屑，少量为白云岩。生屑中生物种类有三叶虫、腕足类、棘皮动物与软舌螺等。

张夏组生物碎屑灰岩，生物组合复杂，磨圆、分选较好，与鲕粒、砂屑等高能颗粒共生，可见规模不大的生物滩相沉积。

亮甲山组生物碎屑灰岩呈灰黑色，生物碎屑为头足类和腹足类，多含燧石团块，揭示其多为潮下高能沉积环境。

（5）球粒灰岩

昌平组中部产有深色巨厚层球粒泥晶灰岩，亮甲山组下部、马家沟组产有少量球粒灰岩。球粒由泥晶方解石组成，呈圆形或椭圆形，大小均一，无内部结构，属潮下低能环境产物。

（6）藻屑灰岩

分布于寒武系馒头组、张夏组与奥陶系冶里组、亮甲山组各层位中，但量少，产出特点与砂屑灰岩基本相同，藻屑常与其他种类的生屑颗粒共生，形态各异，分选不好，亮晶胶结。

（7）团块灰岩

张夏组亮晶团块灰岩，团块由鲕粒或砂屑黏结而成。昌平组、炒米店组以藻泥晶团块灰岩为主，团块多由藻屑粘结而成，团块灰岩中常有内碎屑、生屑伴生。

2. 泥晶灰岩

寒武系、奥陶系中泥晶灰岩十分发育，各层位都有分布。主要由泥晶方解石组成，但岩性变化大，所含颗粒种类、含量、沉积构造、云化程度差异显著，岩石类型众多。现按沉积构造择常见岩石描述如下。

（1）纹层状泥晶灰岩

灰白、灰色，纹层多由铁泥质、云质顺纹理分布显现出来，通常呈水平或水平波状，为低能环境下形成，多发育于馒头组及马家沟组。

（2）中—薄层泥晶灰岩

灰色，泥晶结构，模糊可见藻的丝状体，藻席发育，明层和暗层清晰。张夏组下部、炒米店组及马家沟组可见。炒米店组中该类岩石发育不规则透镜状层理和压扁层理，含中小型叠层石，发育冲刷沟槽，并向同一方向斜歪，为受风暴潮影响的潮坪沉积。

（3）厚—巨厚层泥晶灰岩

深灰、灰黑色，厚—巨厚层，致密块状，厚度大而稳定，Ca/Mg 值高。分布于寒武系昌平组及炒米店组、奥陶系冶里组下部、亮甲山组、马家沟组局部段。岩石多发生程度不同的重结晶作用，如冶里组下部岩石变成粉晶灰岩并多发生云化，而形成豹斑灰岩，或称云斑灰岩。

（4）豹斑（云斑）泥晶灰岩

豹斑灰岩多为灰、灰黑色，厚层、块状泥晶灰岩，含藻团、球粒、生屑、豆粒、内碎屑等。昌平组中极发育，占66%~80%，层位稳定，为标志岩层。豹斑多由粉晶白云石局部集中而显现出来。斑块主要由0.01~0.05mm的菱形白云石组成，形状不规则，常具环带，核心污浊。白云石沿虫孔、生物扰动构造及裂隙分布，斑块间常由缝合线或细微的砂泥质条带相连。斑块与基质并无明显界线，往往逐渐过渡。风化后浅色的白云质斑块突出于风化面，有的因去云化而淋溶出 Fe^{3+} 使风化面呈黄褐色。氧化镁主要集中于斑块中，灰泥基质中含量低微，具清水沉积特征。

3. 黏结灰岩与骨架灰岩

包括叠层石生物丘灰岩、凝块石生物丘灰岩，在寒武系、奥陶系中都有分布，厚度小，沿走向不连续。

（1）叠层石生物丘灰岩

见于西山下苇店炒米店组底部及张夏组顶部块状颗粒滩相鲕粒灰岩之中。岩石呈深灰色、灰白色，穹隆状，与周围的宿主岩石——块状颗粒鲕粒灰岩之间具有极为明显的凹凸不平突变边界，揭示具有明显的早期石化作用而形成明显的地势起伏面，也具有较为明显的抗浪性。灰岩内部由密集垂直生长的柱状叠层石构成，炒米店组叠层石柱体宽6~10cm不等，高数十厘米乃至1m左右，张夏组柱体较大，顶部生物丘高达2m，直径约3m（图8-9）。叠层石柱体之间充填较多的暗色颗粒（底栖鲕粒）（图8-10）。

叠层石所组成的生物丘，总体上表现为致密泥晶结构，以其内部黏聚着较多三叶虫生物骨屑（主）（图8-10）、腕足类、海绵骨针生屑（偶见）和放射—纤维状底栖鲕粒而明显区别于前寒武纪叠层石。

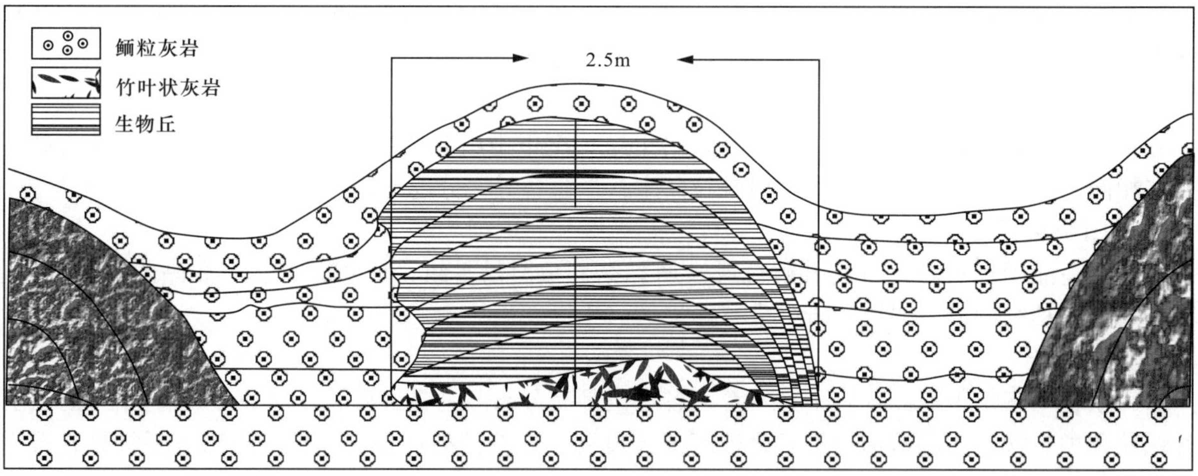

图8-9　下苇店中寒武统张夏组叠层石生物丘灰岩

（据赫云兰，2012）

Fig. 8-9　The stromatolite bioherm of Zhangxia formation in Xiaweidian

（2）凝块石生物丘灰岩

见于西山张夏组底部泥晶灰岩之中。生物丘高约1.8m，直径约3m（图8-11），表面风化为灰褐色，其结构可分为基底和生物丘建造两个部分，基底由风暴成因的竹叶状灰岩构成。生物丘沿基底向上及四周生长，形成了半椭圆形的丘状地质体。生物丘风化面具有席状层理，而新鲜面内部结构则不清晰，风化面纹层厚约5cm，高达几十层。生物丘主要由隐藻凝块石组成，伴有隐藻纹层，藻屑和生物碎屑等。

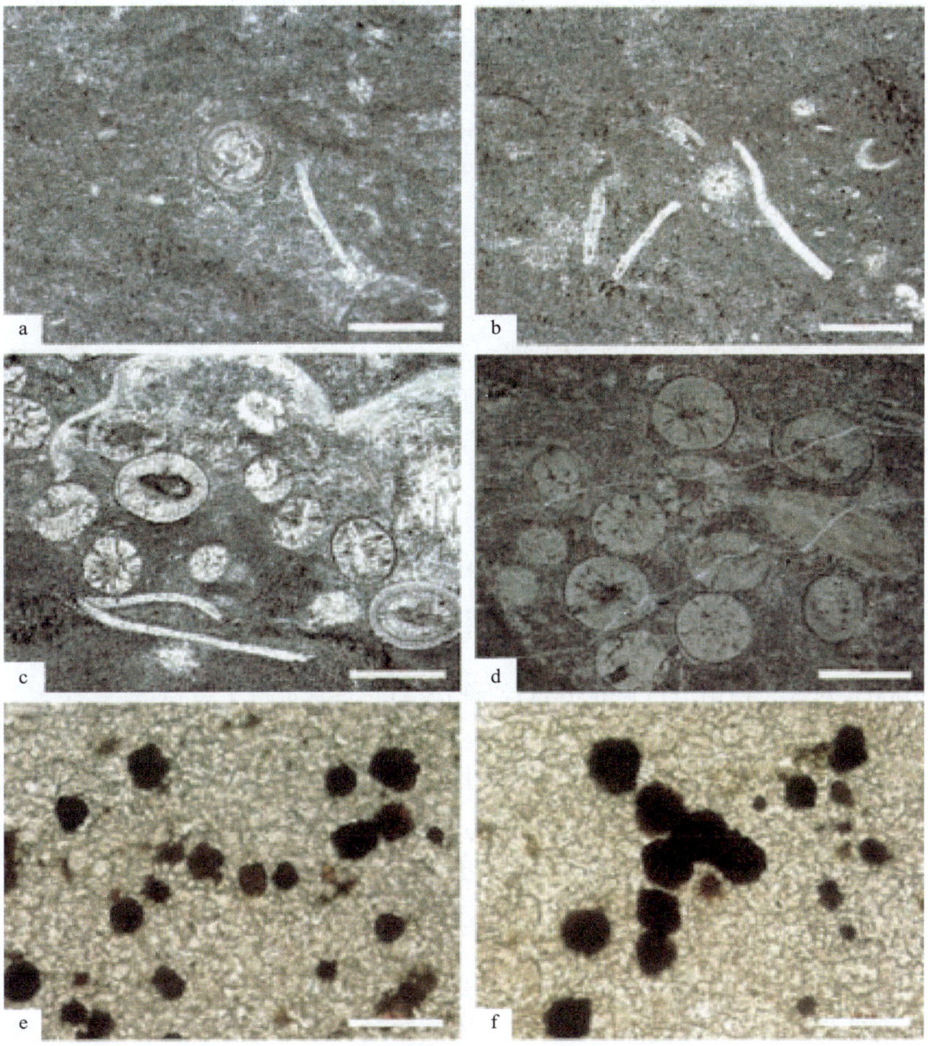

图 8-10　张夏组叠层石生物丘的总体微观组构特征
（据梅冥相，2011）
Fig. 8-10　Images showing microscopically micritic textures for stromatolitic biostrome of Zhangxia formation
a—漂浮有鲕粒和三叶虫骨屑的隐晶质致密泥晶组构；b—见较多三叶虫骨屑和少量鲕粒的隐晶质致密泥晶组构，也见较多黑色小点（为黄铁矿颗粒）；c—含三叶虫骨屑的鲕粒泥晶灰岩组构；d—具有较多鲕粒的颗粒泥晶灰岩组构；e、f—叠层石泥晶组构中异形晶体的黄铁矿

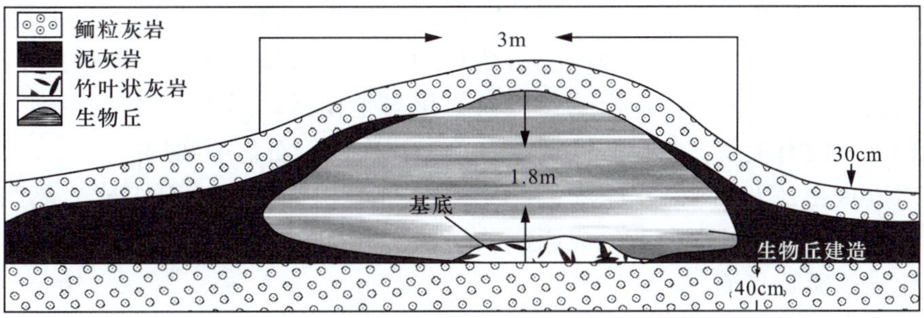

图 8-11　下苇店中寒武统张夏组凝块石生物丘
（据赫云兰，2012）
Fig. 8-11　The clot bioherm of the Middle Cambrian Zhangxia formation in Xiaweidian

凝块石生物丘灰岩造丘生物主要为蓝藻、葛万藻等藻类，伴生生物主要有三叶虫、介形虫等。这些生物群落附着于基底，在生长过程中通过钙化、障积、黏结、捕获等形成了丘状构造。

4. 结晶灰岩

本区结晶灰岩同样广泛发育，基本各组均有分布。主要为细晶灰岩，也见中、粗晶灰岩。颜色多为灰白色、白色、深灰色，细晶、中粗晶结构，厚层—块状构造，为微晶、泥晶发生变质形成，出露层位多发育石英脉或方解石脉。

（二）白云岩类

1. 同生白云岩

主要发育于昌平组底部和顶部、馒头组下部、马家沟组底部和中局部段，以泥晶白云岩为主，亦有部分粉晶白云岩。岩石呈棕黄色、土黄色、浅黄色等，纹层状，泥晶结构，见交代残余泥晶方解石等现象，干裂、鸟眼构造发育，可见石膏或盐类假晶，陆源物含量高达15%～30%，常发育于沉积间断面之上。

2. 后生白云岩

主要分布于昌平组、亮甲山组，以粉晶、粗晶白云岩为主。常具以下特征：①晶粒粗大，等粒或不等粒嵌晶结构，常有环带；②可见残余生物结构；③见残余颗粒结构，如残余绒粒、砾屑、砂屑等；④横向可渐变为灰岩。

彭阳等（2000）对北京西山下古生界不同岩石白云石化过程进行了总结：

（1）灰岩中泥质条带的白云石化

馒头组、张夏组鲕粒灰岩中褐黄色泥质条带及炒米店组灰岩中的泥质条带镜下表现为自形粒状的白云石晶体与铁质细末及残余黏土矿物组成的条带，及竹叶状灰岩中的泥质胶结物也为自形的具铁质镶边的白云石中粒晶体，这是灰岩（如泥质条带灰岩）中的泥质成分在埋藏后的压实过程中，黏土矿物转化时易析出Mg^{2+}，形成沿条带分布的白云石，又称埋藏白云石化作用。

（2）泥晶灰岩的白云石化

寒武系泥晶灰岩中常见两种白云石化。一为零星而均匀的白云石化：准同生期泥晶灰岩中的孔隙水非常有限，因此能量也有限，只能形成漂浮状干净自形的白云石细晶；二为沿缝合线或裂隙发生的白云石化：成岩后期泥晶灰岩受压实作用或构造改造形成各种缝隙，当有黏土矿物及富镁裂隙水参与时即沿缝隙发生埋藏白云石化作用。

（3）颗粒灰岩的白云石化

鲕粒灰岩的白云石化也发生在准同生期，完全白云石化时形成单晶鲕或多晶鲕，弱白云石化时可在原生同心鲕中见到零星的白云石细粒。

竹叶状灰岩的白云石化仅发生在填隙物中，为成岩期白云石化，与黏土矿物的催化作用有关。

（4）豹斑灰岩和生物扰动灰岩的白云石化

豹斑灰岩的原岩为生物扰动砂屑灰岩，原始砂屑泥灰岩受生物扰动改造而呈不均匀状，而后受准同生期白云石化影响，再叠加成岩后生期白云石化作用，白云石呈凝团状、断续条带分布，形成疙瘩状构造。准同生期白云石化受原始构造影响，白云石化与生物活动形成的虫孔有关；成岩后生期白云石化则与原始结构基本没有关系，与顺层微裂隙有关，两次白云石化可叠加。

生物扰动灰岩（或虫迹灰岩），其深灰色“虫孔”部分由细晶白云石组成，晶间可见残余灰泥，浅灰色部分一般为砂屑灰岩。其白云石化成因同豹斑灰岩相同，准同生白云石化往往与虫孔有关，而成岩后生白云石化与微裂隙有关。

（5）砾屑状鲕粒灰岩的白云石化

张夏组巨厚层的砾屑状鲕粒灰岩的砾状构造、断续条带状构造并不全是原生沉积构造，部分砾屑是一种同生或准同生期成岩白云石化的结果。

在野外见到的砾屑为深色原生鲕粒与浅土黄色次生白云石化鲕粒，二者均为不规则团块状，但在镜下深色鲕粒为原生放射同心状鲕粒，浅土黄色鲕粒为白云石化次生鲕粒。次生白云石化鲕粒有 2 种：①单晶白云石鲕粒；②多晶白云石鲕粒。它们在原生鲕粒中集中分布，但与原生鲕粒没有边界，即白云石化鲕粒并没有砾屑特征，因此砾屑并非原始沉积构造，而是白云石化的结果。砾屑状鲕粒灰岩的白云石化发生在同生或准同生期，因为这些白云石化鲕粒的胶结物并没有白云石化，Mg^{2+}的物质来源为原始环境中的孔隙水及原始沉积的文石及高镁方解石向方解石的转变。

二、碎屑岩类

（一）砾岩类

主要分布于昌平组底部，为底砾岩。岩性为褐红色砂砾岩，其砾石成分多为下伏景儿峪组粉砂质泥晶灰岩、黑色硅质岩。砾石大小不一，棱角明显，基质则为碳酸盐泥及砂质。野外产出一般厚度不大，而且变化较大。

（二）砂岩类

1. 石英细砂岩、粉砂岩

石英砂含量 80%~90%，颗粒成熟度高，圆度、分选好，与少量鲕粒共生，具斜层理。分布于馒头组上部，张夏组中下部，为鲕粒灰岩中夹层，为潮间或潮下高能环境沉积物。

2. 钙质、泥质粉砂岩

主要见于张夏组，馒头组、炒米店组。岩石为绿灰色，中—薄层，粉砂结构，水平层理发育。石英砂含量 40%~70%，多为棱角状，粒径大小不一。可区分钙质胶结和泥质胶结，胶结物含量 20%~50%，常与泥晶灰岩条带相间，有时过渡为粉砂质泥晶灰岩。

（三）黏土岩类

以泥岩为主，页岩为次。颜色为紫红、暗紫色和黄绿、灰绿色两种。

紫红、暗紫色岩石集中发育于馒头组，含较多的碎屑混入物，含粉砂鳞片状泥质结构，页片状、条带状—细层状构造，泥质组分为主，含少许粉砂质。泥质多已次生云母化，成分为水云母和绢云母，含少许黑云母，呈细小鳞片状，条带状分布。泥质组分以绿泥石为主，粉砂质组分以石英为主。含极少数细小黄铁矿，见石盐假晶。常见波痕、泥裂，与泥晶白云岩共生，为干燥炎热条件下潮上蒸发泥坪或混合坪沉积。

黄绿色、灰绿色岩石发育于张夏组，陆源碎屑混入物极少，页理发育，含底栖动物化石，常与泥晶灰岩呈互层或为夹层，为斜坡潮下低能带产物。

第三节　沉积盆地及沉积建造

一、沉积盆地

本区位于寒武纪—奥陶纪华北克拉通板内陆表海盆地，属陆源碎屑-碳酸盐岩型及碳酸盐岩型陆表海盆地，盆地地势平坦、地形坡度很小，以均匀缓慢沉降为主，沉积地层厚度达上千米，沉积物岩性组合表现为稳定型陆源建造和稳定型内源建造组合，岩性单一，沉积建造稳定。早古生代寒武纪昌平期末、亮甲山期末虽然也出现盆地短暂隆升，形成沉积间断，但盆地仍表现出整体同升同降的特点，形成区域性的间断面。

二、沉积建造

早古生代沉积建造因盆地克拉通化，在水平方向上延伸稳定，相变缓慢，沉积建造的物源是盆地周围隆起风化剥蚀区的陆源碎屑或陆表海水中直接或间接沉淀下来的内源沉积物。早寒武世晚期至中奥陶世，各建造内部的旋回层序周期性十分明显，建造形成过程表现为一次碳酸盐岩台地形成-成长-成熟-成熟后发展的历史过程。大致分为早寒武世晚期—晚寒武世的台地前缓坡沉积阶段、台地成长成熟阶段、风暴沉积阶段和奥陶世超成熟阶段。具体各时期主要沉积建造类型综合特征如表 8-3 所示。

表 8-3　早古生代各时期沉积建造类型综合特征

Table 8-3　Sedimentary formation types comprehensive characteristics in Early Paleozoic

时期	沉积建造系列	盆地类型	建造类型	沉积环境	物源	沉积速率	共生岩石主要类型	典型分布地区	盆地发育阶段
马家沟期	稳定	克拉通盆地（陆表海）	局限台地（风化壳）型颗粒碳酸盐岩建造	潮坪、潮上	内源	较慢	灰色厚层—巨厚层灰岩夹白云岩、角砾状灰岩、角砾状白云岩	北京市全区	台地建设成熟后阶段
亮甲山期	稳定		局限台地（风化壳）型颗粒碳酸盐岩建造	斜坡、潮坪、潮上	内源	较慢	灰色厚层燧石结核灰岩、中厚层灰岩夹薄板状灰岩及少量竹叶状灰岩	北京市全区	台地建设成熟后阶段
冶里期	稳定		开阔台地（斜坡）型碳酸盐岩建造	斜坡、潮坪、潮上	内源	较慢	以灰色巨厚层豹斑灰岩及灰色厚层泥纹灰岩为主，上部夹灰色薄层泥纹灰岩、竹叶状灰岩及黄绿色页岩	北京市全区	台地建设成熟后阶段
炒米店期	稳定		开阔台地（风暴）型碳酸盐岩建造	斜坡、潮坪、潮上	内源	较慢	灰色和紫灰色亮晶砾屑灰岩，灰色藻层泥晶灰岩，灰色薄层、中厚层泥质条带灰岩，黄绿色和紫红色页岩，灰黄色泥质粉砂岩	北京市全区	台地风暴沉积阶段
			开阔台地（鲕滩）型缓坡碳酸盐岩建造	斜坡、潮坪、潮上	内源	较慢	灰色中厚层状、巨厚层状亮晶鲕粒灰岩为主夹泥质条带（纹）灰岩、生物碎屑灰岩、泥晶灰岩及少量黄绿色页岩、粉砂岩	北京市全区	台地建设成熟阶段
张夏期	稳定		开阔台地（斜坡）型碳酸盐岩建造	缓坡、潮坪、潮上	内源	较慢	灰绿色薄层泥钙质粉砂岩与灰紫色、深灰色中层泥、亮晶鲕粒灰岩、含竹叶状泥、亮晶鲕粒灰岩、条带状灰岩	北京市全区	台地演化成长阶段
馒头期	稳定		蒸发台地（暴露）型陆源碎屑-碳酸盐岩建造	潮坪、潟湖、潮上	混合源	较慢	紫红色和砖红色页岩、泥岩、粉砂岩，灰色和灰白色泥云岩、白云岩、灰岩	北京市全区	台地演化起始阶段
昌平期	稳定		局限台地（风化壳）型颗粒碳酸盐岩建造	潮坪	内源	较慢	细晶灰岩、白云岩、豹斑灰岩、白云质灰岩、角砾状白云岩	青白口—庄户台以东	短期台地，一个完整独立演化时期

1. 蒸发台地（暴露）型陆源碎屑-碳酸盐岩建造

本建造主要形成于馒头期。岩石共生组合包括薄层紫红色、紫色泥岩、页岩、泥晶白云岩。具水平层理，含极少数细小黄铁矿，不发育生物，多发育石盐假晶，常见波痕、泥裂，为干旱蒸发台地潮上萨布哈相沉积产物。

2. 开阔台地（鲕滩、斜坡）型缓坡碳酸盐岩建造

本建造主要形成于张夏期，炒米店期早期也发育此类建造。岩石共生组合包括灰色中厚层状、巨厚层状亮晶鲕粒灰岩为主夹泥质条带（纹）灰岩、生物碎屑灰岩、泥晶灰岩及少量黄绿色页岩。具水平层理、斜层理、交错层理，生物极丰富，主要是三叶虫，其次是藻类，属于开阔台地边缘浅滩相产物。

3. 开阔台地（风暴型）缓坡碳酸盐岩建造

本建造主要形成于炒米店期中晚期，以发育风暴岩-砾屑灰岩为特征。岩石共生组合为灰色和紫灰色亮晶砾屑灰岩，灰色藻层泥晶灰岩，灰色薄层、中厚层泥质条带灰岩，黄绿色和紫红色页岩，灰黄色泥质粉砂岩。具水平层理、斜层理、交错层理，生物丰富，主要是三叶虫，属缓坡相产物。炒米店期风暴型缓坡碳酸盐岩建造是区内乃至整个华北板块发育成为超成熟碳酸盐岩台地后华北板块漂移至低纬度飓风带形成的特殊建造。

孟祥化等（1986）对北京市丁家滩剖面风暴型缓坡碳酸盐岩沉积建造进行了系统的总结，建立了风暴流控制的碳酸盐浅海盆地相模式示意图（图 8-6），并归纳了丁家滩地区晚寒武世浅海碳酸盐风暴环境下各沉积相的岩性组合序列，包括盆地风暴浊流相、盆地斜坡风暴碎屑流相、盆地斜坡原地型风暴沉积相、台缘斜坡风暴流沉积相。

4. 局限台地（风化壳）型碳酸盐岩建造

本建造主要形成于奥陶系冶里组、亮甲山组和马家沟组中，寒武系昌平组也有分布。该建造为海平面周期性升降、碳酸盐生成率接近或达到海平面上升速率时，形成的一套向上变浅-暴露的潮坪和环潮下白云岩层序，建造顶部常表现为白云岩化及形成喀斯特风化壳。区内昌平组主要由厚层和巨厚层豹斑状粉晶—微晶灰岩、灰质白云岩组成；冶里组和亮甲山组主要由厚层—巨厚层豹斑状泥晶灰岩、泥质条纹灰岩、含燧石结构豹斑状泥晶灰岩、含燧石结核白云岩组成；马家沟组由厚层—巨厚层泥晶灰岩夹多层白云岩及角砾状碳酸盐岩组成。其共同特征是顶部均发育风化壳，或具角砾状灰岩、角砾状白云岩（含膏盐的碳酸盐岩暴露后，经膏溶垮塌的产物）或溶蚀漏斗（图 8-12），岩石水平层理发育，含三叶虫、头足类、腕足类及牙形石化石，属于开阔台地相—局限台地相产物。

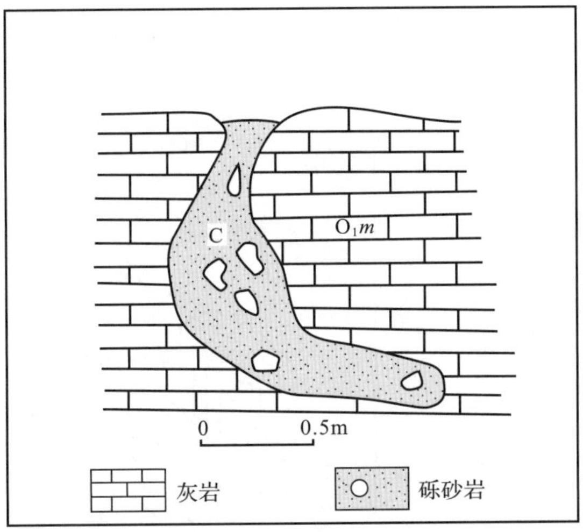

图 8-12　北京门头沟色树坟中奥陶统顶面溶蚀漏斗充填石炭系砂砾岩

Fig. 8-12　Top surface karst funnel in Ordovician filled by Carboniferous glutenite in Seshufen valley, Mentougou District, Beijing

C—石炭系；O_1m—马家沟组

第四节　岩相古地理

新元古代中晚期的抬升造陆运动——蓟县抬升，使包括北京在内的广大燕辽地区整体隆起遭受风化剥蚀。直到早寒武世晚期，华北中南部地区地壳下降，从昌平期开始揭开了早古生代海侵的序幕。华北早古生代逐渐形成的古海洋系，范围辽阔，水体不深。地形虽然复杂，但底部坡度很小，表现为克拉通稳定型内陆浅海沉积海盆。从寒武纪到奥陶纪，共经历了昌平期、馒头期、亮甲山组期和马家沟期 3 个大的沉积旋回。

一、昌平期

昌平期燕山地区发生了显生宙第一次海侵，海水由东向西侵入，形成继新元古代隆起为古陆后再沉降的新燕山海盆。燕山海盆北、西、南三面为古陆，一面通海，形如箕状，表现为半封闭的海湾。

期间，本区主体位于燕山海盆西北边部，大致以新城子—四海—鸡鸣驿一线为界，其北为内蒙古陆，其南为海盆。内蒙古陆受长达 260Ma 的隆起抬升影响，地势平缓，以化学风化侵蚀作用为主，极少提供陆源物源。而燕山海盆沉积受宏观盆底地貌控制，东低西高、北低南高的盆底古地貌总趋势致使沉积物东厚西薄、北厚南薄（图 8-13），由西向东海水逐渐加深。沿着海盆边部太师屯—溪翁庄—东花园—沿河城—清水—九龙一线北西，沉积了灰色、灰黄色中厚层—厚层状白云岩、泥质白云岩、灰岩、泥灰岩、粉砂岩、泥岩组合，为混合云坪沉积；紧邻混合云坪相沉积的盆地边部沉积了豹斑灰岩、白云岩、灰质（白云质）角砾岩组合，为云坪相沉积；庄户台—青白口—大庄科—沙厂一线以东沉积以豹斑灰岩为主，发育粉（细）晶灰岩、核形石灰岩组合，为灰坪相沉积。

昌平期末，燕山地区地壳出现短期上升，海水由西向东退却，结束了早古生代第一次海侵，本区上升成陆，并遭受风化剥蚀，古地形趋于夷平。

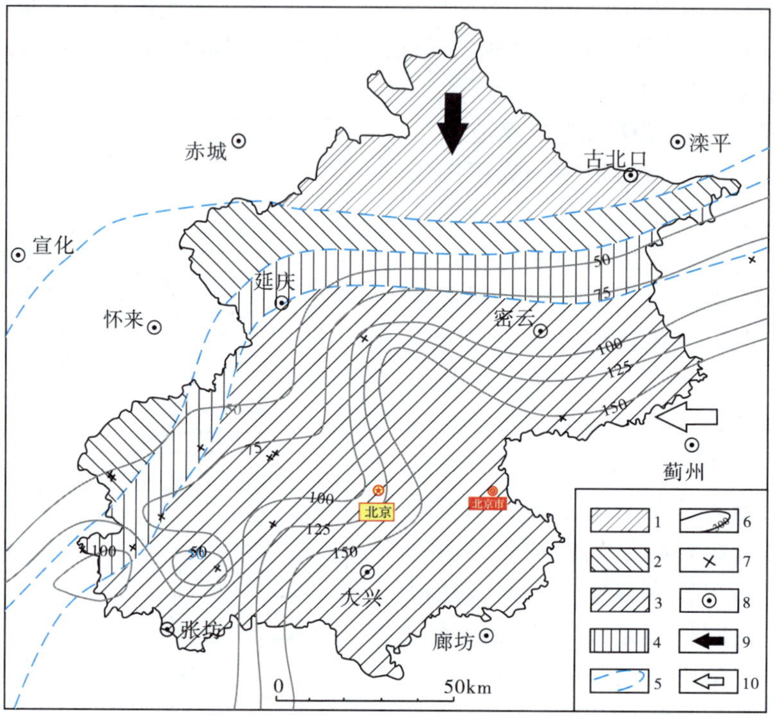

图 8-13　昌平期中期岩相古地理图

Fig. 8-13　Sedimentary paleogeography map of Middle Changping period

1—基底剥蚀区；2—云坪相沉积；3—灰坪相沉积；4—灰云混合坪相沉积；5—岩相界线；6—沉积等厚线（m）；7—剖面位置；8—地名；9—物源方向；10—海侵方向

二、馒头期

馒头期燕山地区再次下沉，开始了早古生代第二次海侵，也是显生宙最广泛、经历时间最长、规模最大的一次海侵，本次海侵一直持续到早奥陶世中期才结束。海水自北东向南西侵入，海域范围极广。

本区全区被海侵覆盖，海水处于半封闭状态，大多数时期水体安静、气候干燥炎热、蒸发作用强烈，海水咸化。整个馒头期，古地理环境分割性很大，岩相在时间、空间上均有变化，形成复杂沉积相类型，形成西山和平谷-兴隆两个沉积中心，围绕这两中心，沉积物厚度向南东、北西方向逐渐减薄（图 8-14）。

馒头期早中期沉积仍受昌平期古地貌面影响，宏观上，区内大致以门头沟南—昌平东—怀柔—密云西为界线，以西发育为潮上萨布哈相沉积，以东发育为潮坪潟湖相。馒头期晚期海侵继续扩大，海水逐渐加深，至末期趋于正常海环境，发育为浅海陆棚稳定环境，并出现生物，发育为潮坪潟湖、潮坪云坪相、潮坪泥坪相和潮坪混合坪相。

三、张夏期

张夏期为馒头期海侵的继续，区内水体逐渐加深，海底地形起伏日趋夷平，水动力增强，碳酸盐

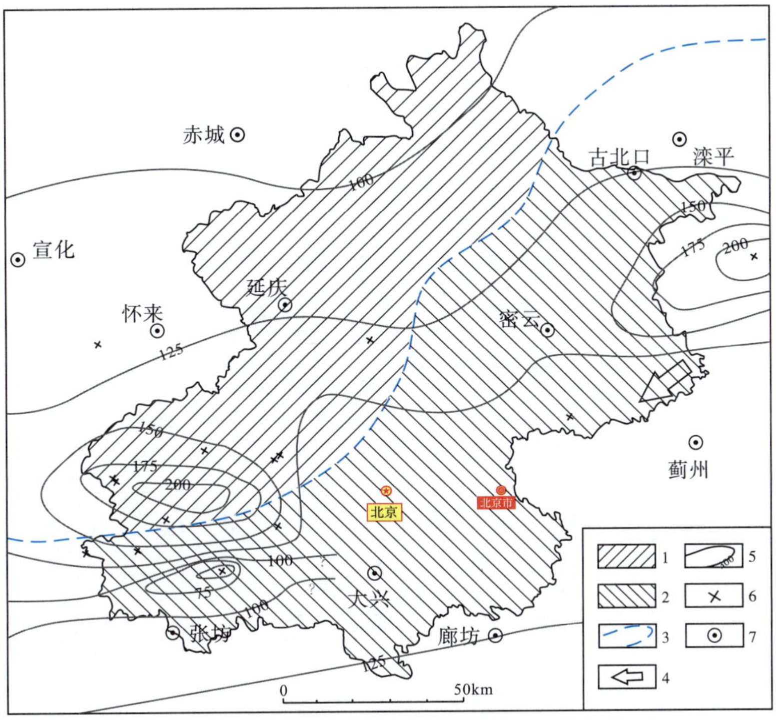

图 8-14　馒头期早期岩相古地理图

Fig. 8-14　Sedimentary paleogeography map of Early Mantou period

1—潮上萨布哈相沉积；2—潮坪潟湖相沉积；3—岩相界线；4—海侵方向；5—沉积等厚线（m）；6—剖面位置；7—地名

岩大量出现，藻类、生物化石丰富。燕山盆地海水由北东向南西方向不断扩展，海域范围不断扩大。本区已演化为温暖潮湿的典型开阔碳酸盐台地，沉降与沉积物堆积速度已达平衡，沉积厚度达 190~230m，全区差异不明显。

至张夏期中晚期，本区延庆—四海—墙子路一线以南区域主体演化为浅滩沉积（图 8-15），发育暴露鲕滩岛、含砂鲕浅滩及高能鲕滩三小类沉积，沉积多种多样的鲕粒以及其他各种内源颗粒，水浅而动荡或宁静而适度的循环，氧气、阳光充足，生物繁盛，三叶虫动物群属种数目发展到高峰阶段，藻类也很发育。而该线以北地区演化为台地滩前斜坡沉积，水体相对较深，发育为浅缓坡边滩相。房山、大兴南部地区沉积厚度相对较薄，遭受剥蚀无沉积地层残留。

四、炒米店期

炒米店期为早古生代第二次海侵继续演化时期。燕山海盆古地形在前期沉积影响下发生变化，宏观上由早期的北高南低、西高东低转变为北低南高，西低东高，致使海水深度由东向西、由北向南递增，但海侵方向保持为北东向南西方向。期间地壳升降频繁，海水动荡不定。同张夏期相比，总体水体明显加深，沉积环境以台地斜坡相沉积（海泛期）为主，时有抬升作用下水体变浅的潮间、潮下沉积（海退期）。宏观沉积厚度南薄北厚，沉积厚度 90~360m，区域差异较大，风暴作用频率高，风暴沉积广泛发育。同时，较深水沉积环境不利于生物繁衍生息，仅初期、末期和抬升时期，即水体较浅情况下发育比较繁盛的三叶虫动物群。

炒米店期海泛期，海水分布范围进一步扩大，水体逐渐加深（图 8-16）。沉积为薄层泥质灰岩、风暴砾屑灰岩、含砂细晶灰岩、含砾屑细晶灰岩、钙质页岩、钙质粉砂岩等沉积组合，沉积环境以台地斜坡沉积为主。根据斜坡坡度及沉积岩性组合序列，由南向北依次为浅缓坡边滩沉积相、中陡坡盆地沉积相和浅缓坡盆地沉积相，最北部可能还发育深水盆地相，具有海绿石凝缩段沉积特征。

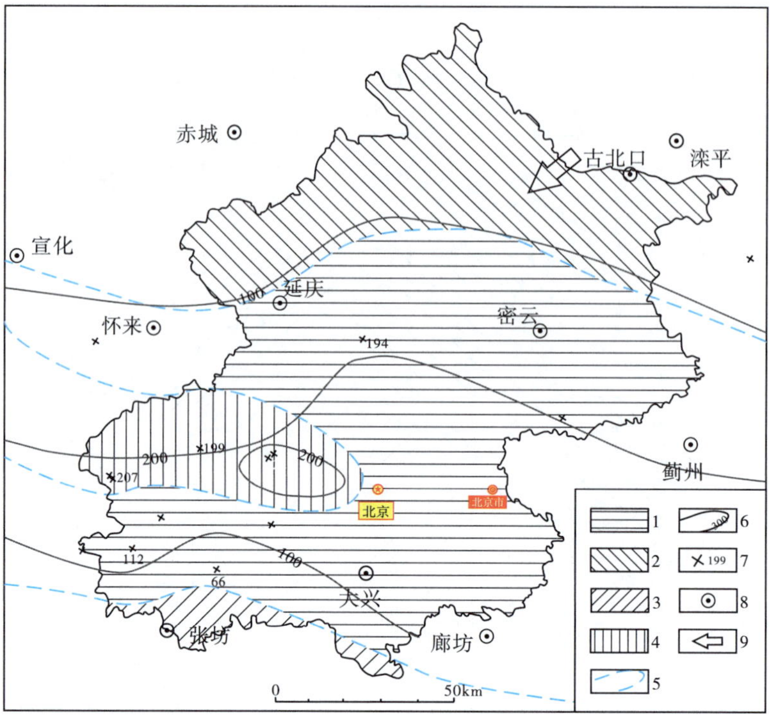

图 8-15　张夏期中晚期岩相古地理图

Fig. 8-15　Sedimentary paleogeography map of middle-late term of Zhangxia period

1—高能鲕滩相沉积；2—滩前中缓坡碳酸盐岩相沉积；3—滩间海相沉积；4—低能暴露鲕滩相沉积；
5—岩相界线；6—沉积等厚线（m）；7—剖面位置及地层厚度（m）；8—地名；9—海侵方向

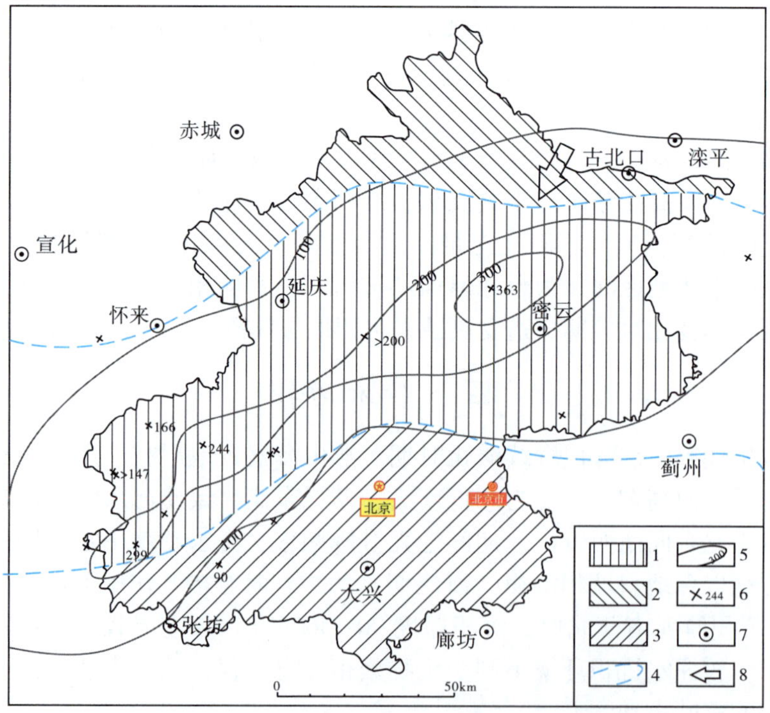

图 8-16　炒米店期海泛期岩相古地理图

Fig. 8-16　Sedimentary paleogeography map of Chaomidian period

1—浅缓坡盆地相；2—中陡坡盆地相；3—浅缓坡边滩沉积相；4—岩相界线；5—沉积等厚线（m）；
6—剖面位置及地层厚度（m）；7—地名；8—海侵方向

炒米店期海退期，海水分布范围萎缩，水体从南向北收缩，形成泥质（晶）灰岩、竹叶砾屑灰岩（具氧化圈）、砂屑灰岩、生物屑灰岩等岩性组合序列，局部甚至经过一定时间的沉积暴露，风化作用强烈，形成风化土壤层，发育为局限台地相或局限海相。

五、冶里期

冶里期古陆轮廓和燕山海盆范围基本继承了炒米店期的古地理格局。同炒米店期相比，冶里期海平面上升，海水加深，同炒米店组共同构成早古生代第二次海侵最大海泛时期。海盆总体沉积不厚，本区为沉积最厚处，约 160m，下陷幅度大，水体较深（图 8-17）。宏观上大致以马坊—杨镇—高丽营—阳坊一线为界，以北发育灰色中—厚层泥质条带（纹）泥晶灰岩、中—薄层泥晶灰岩、中薄层砾屑灰岩，砾屑灰岩层薄且少有发育，为半深水盆地相欠补偿沉积；以南发育灰色巨厚层泥晶灰岩、斑块状灰岩及竹叶状灰岩、灰色泥质条带灰岩、钙质粉砂岩、生屑砂屑泥晶灰岩组合，砾屑灰岩层厚且发育较多，为碳酸盐岩台地斜坡相沉积，沉降速度与沉积物堆积速度基本平衡，沉积厚度差异不明显。

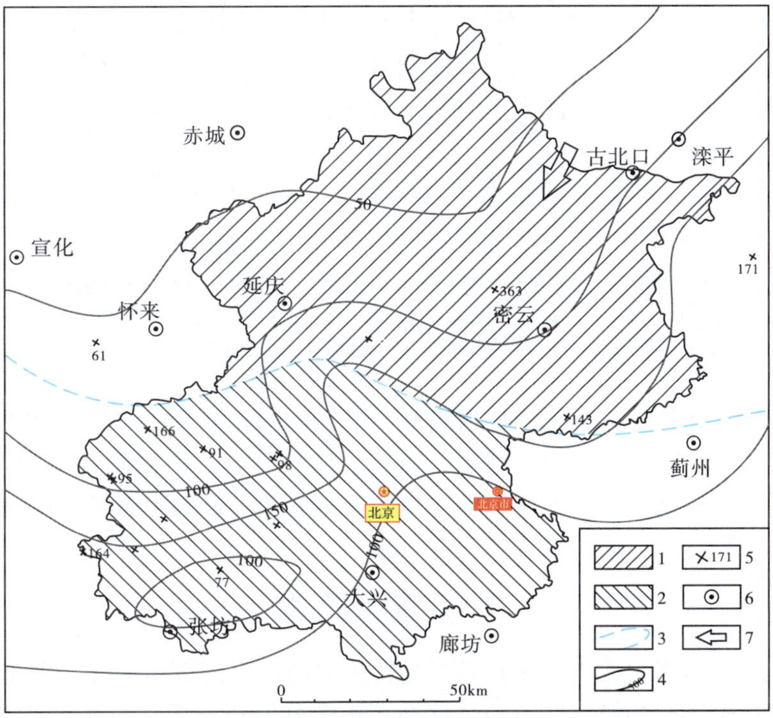

图 8-17 冶里期岩相古地理图

Fig. 8-17 Sedimentary paleogeography map of Yeli period

1—半深水盆地相；2—台地斜坡碳酸盐岩相；3—岩相界线；4—沉积等厚线（m）；5—剖面位置及地层厚度（m）；6—地名；7—海侵方向

六、亮甲山期

亮甲山期是冶里期沉积环境的继续发展时期，海域范围大体相同，水深逐渐变浅。古地形坡度较冶里期加大，沉降幅度增加 2~3 倍，京西沉降幅度最大。亮甲山组南北厚度变化趋势与冶里组相似，南北沉积厚度增长比例基本一致，为冶里组的 2~3 倍（贾振远，1986），本区沉积厚度为 100~300m（图 8-18）。

亮甲山期早中期，大致在门头沟蒲洼乡—潭柘寺—通州漷县一线以南地区，沉积为青灰色纹带灰岩及白云岩、钙质泥岩组合，发育多层膏溶角砾岩，为局限台地潟湖相沉积；该线以北至大约赤城—汤河口—古北口北一线之间地区，沉积为灰色、深灰色中—厚层含燧石团块泥晶白云岩夹灰黑色、黑色含生物碎屑泥晶白云岩组合，为开阔台地局限海潮坪相。

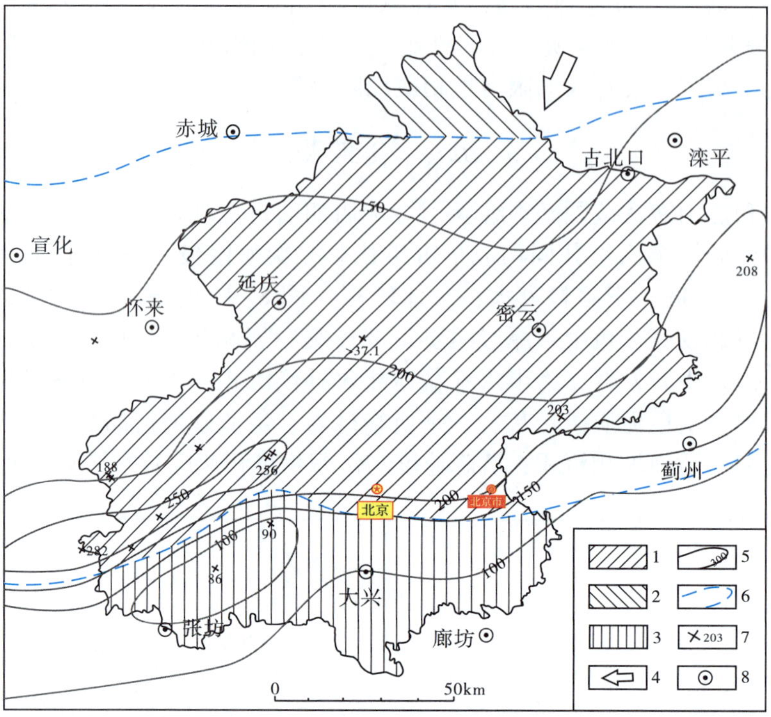

图 8-18　亮甲山期早—中期岩相古地理图

Fig. 8-18　Sedimentary paleogeography map of early to middle term of Liangjiashan period

1—开阔台地局限海潮坪相；2—开阔台地斜坡-生物礁相；3—局限台地潟湖相；4—海侵方向；5—沉积等厚线（m）；6—岩相界线；7—剖面位置及地层厚度（m）；8—地名

亮甲山期晚—末期，受怀远上升运动构造抬升影响，海水由南向北收缩。晚期，本区北部普遍发育准同生后沉积泥晶白云岩，可见干裂、波痕，以及岩溶角砾岩，为局限台地潮坪沉积；南部地层较薄，可能已经抬升为陆地遭受剥蚀。末期，海水退出华北东部，出现短暂沉积间断，也最终结束了早古生代第二次海侵。

七、马家沟期

马家沟期，华北地区地壳再次由整体抬升运动转为波动性下沉运动，海水由北东向南西再次入侵，形成早古生代规摸大、范围广的第三次海侵。该期燕山海盆海域范围广，本区均被海水覆盖，沉积一套厚度超过500m的碳酸盐岩地层，含丰富的头足类化石。全区内该期沉积除底部因海侵初期受古地貌影响不能对比外，其他时间段均能较好对比，沉积区域相变小，主要表现为局限台地和开阔台地交替演化的复杂沉积环境（图8-19）。

海侵初期，受怀远运动的影响，地形不平坦，沉积为土黄色含陆源岩块泥晶泥质白云岩，见泥裂，为海侵初期蒸发台地潮上云坪沉积环境；早期，海水动荡，地形起伏大，常有塌陷作用，沉积为角砾灰岩、白云岩组合，为局限台地潮上—潮间云坪沉积环境；中期持续时间最长，海水频繁升降，沉积为浅灰色与深灰色纹带状含灰泥晶白云岩、深灰色含生屑花斑状泥晶灰岩、球粒灰岩、生物碎屑灰岩、亮晶含云藻粒灰岩、亮晶内碎屑灰岩（竹叶状灰岩、砂屑灰岩）构成的韵律层，以灰岩为主，多具水平层理，为局限台地与开阔台地（局限海与开阔海）交替沉积环境；晚期沉积由于地壳上升遭剥蚀保留不全，沉积以灰色中层粉晶灰岩、粉晶泥质灰岩、条纹状灰岩为主，含生屑、云斑，夹角砾状灰岩、薄层泥质灰岩，为开阔台地潮下低能沉积环境。

马家沟期后，受加里东运动的影响，地壳上升成陆，遭受风化剥蚀，缺失相当于峰峰组以上地层。

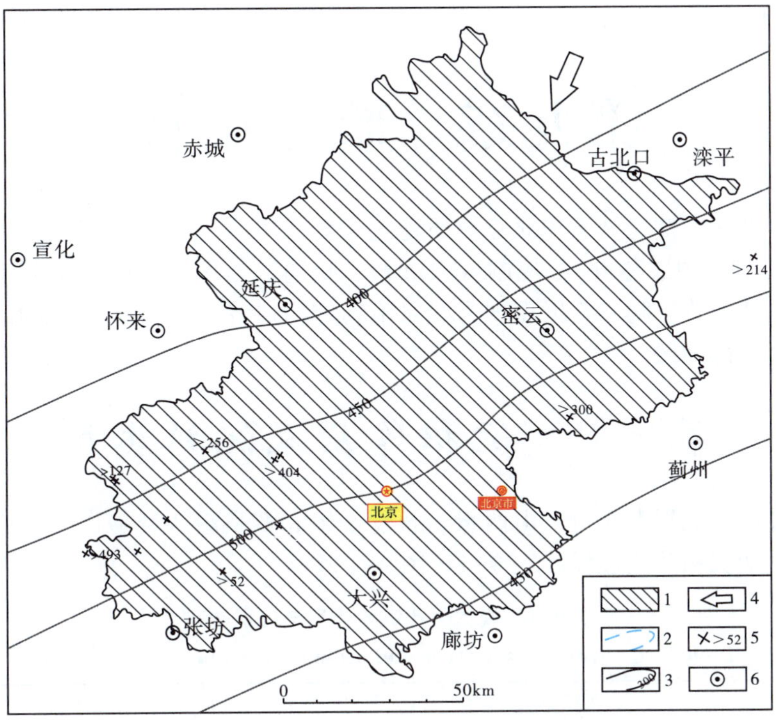

图 8-19　马家沟期中期岩相古地理图

Fig. 8-19　Sedimentary paleogeography map of middle Majiagou period

1—灰岩相；2—沉积等厚线（m）；3—岩相界线；4—海侵方向；5—剖面位置及地层厚度（m）；6—地名

第九章　晚古生代

本区晚古生代早期—中期仍处于华北克拉通稳定大背景下，地壳整体运动，先升（泥盆纪—早石炭世）后降（晚石炭世—中二叠世），控制区内沉积。晚期，受华北克拉通东部活化影响，岩浆开始活动，断裂及褶皱构造发育，晚二叠世表现为相对不稳定构造背景下（挤压）的区域差异性隆起和沉降，沉积发生于沉降局部地区。

第一节　沉积作用及沉积类型

晚石炭世—二叠纪，沉积太原组、山西组、石盒子组和双泉组 4 套地层，构成一个完整的海相—海陆过渡相—陆相复杂沉积演化过程（表 9-1；图 9-1）。

表 9-1　晚古生代沉积作用与沉积类型

Table 9-1　Sedimentary types and sedimentation in Late Paleozoic

序号	沉积类型			地层
	沉积相	沉积亚相	沉积微相	
1	陆相沉积	河流相	决口扇、河道、河漫滩、泛滥平原	石盒子组、双泉组
		湖泊相	滨湖、浅湖、半深湖	
2	海陆过渡相沉积	三角洲沉积	辫状河三角洲沉积（三角洲前缘、三角洲平原、前三角洲）	石盒子组部分层位、山西组、太原组上部
3	海相沉积	障壁岛陆源碎屑沉积	障壁岛、潟湖、潮坪、滨岸、浅海、冲越扇、潮道	太原组下部—中部

一、海相沉积作用及沉积类型

晚古生代海相沉积发育为障壁型陆源碎屑沉积，仅出现于上石炭统太原组下部至中部地层中，可细分为障壁岛、潟湖、潮坪、滨外陆棚沉积亚相（图 9-2）。

（一）障壁砂坝沉积

障壁砂坝沉积主要为上临滨沉积，岩性以细、中、粗砂岩为主，分选中等—好，成分成熟度高，在垂向上常呈向上变粗的逆粒序，发育低角度交错层理、大型交错层理沉积构造。

（二）冲越扇沉积

发育于障壁砂坝后的潟湖之中，由风暴浪将障壁砂坝上的粗粒沉积携带到低能的潟湖中沉积而成，岩性一般为细砂岩，磨圆较好但分选较差，砂体中厚到厚层状，顶底界面突变。砂岩中生物扰动构造比较发育，层理构造因生物扰动变得不清晰，甚至岩石本身成分因生物扰动构造发育变得混杂，垂向上常与潟湖沉积共生。

（三）潟湖沉积

该类沉积岩性主要为黑—灰黑色泥岩粉砂岩，含有较多的菱铁质鲕粒和结核，说明为还原环境，常含有生物化石，但较为单调，比较多见腕足类，软体动物等。另外，见有生物潜穴，层理类型以代表较平静水流环境的水平层理为主，块状层理次之，偶尔发育透镜状层理。

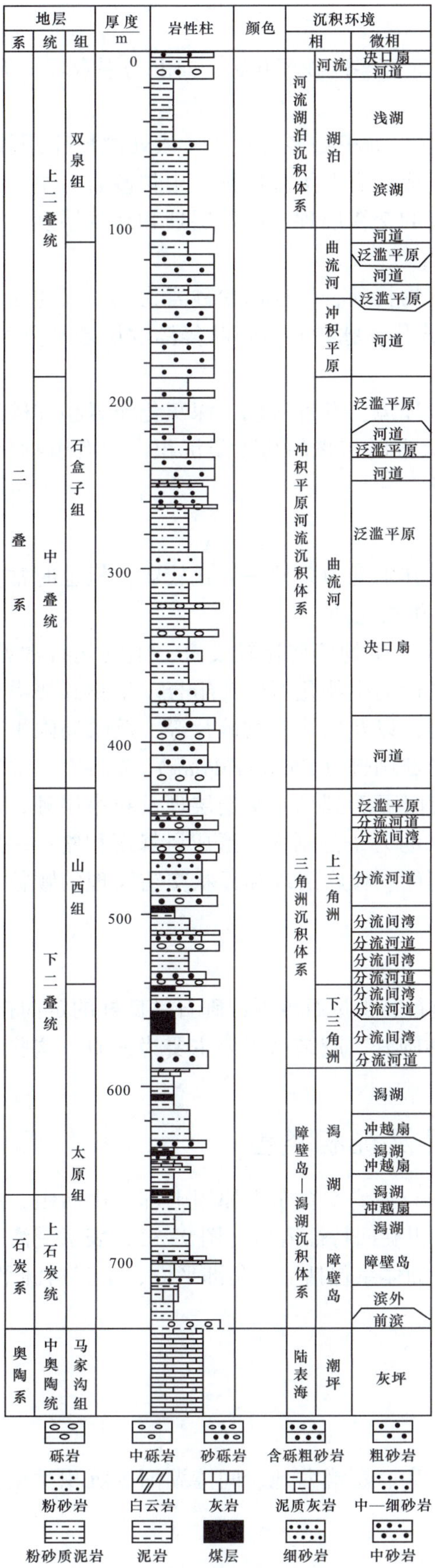

图 9-1　晚古生代典型剖面沉积相

Fig. 9-1　Sedimentary facies model in Late Paleozoic

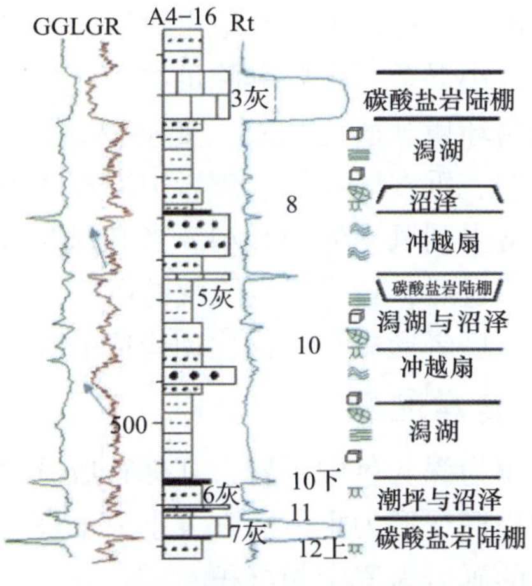

潟湖-冲越扇-滨外碳酸盐岩陆棚沉积环境（太原组）

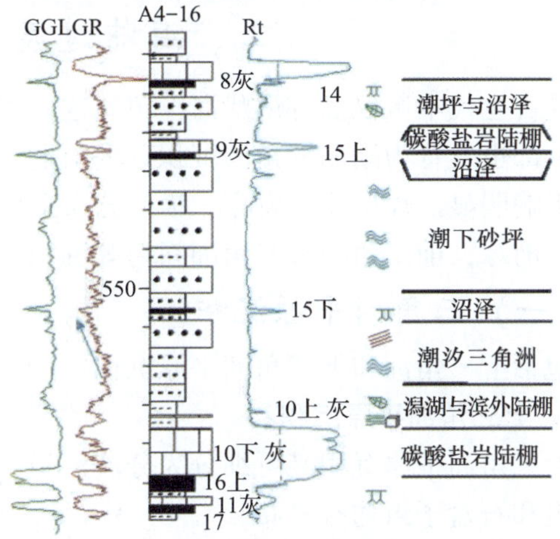

潮坪-潮汐三角洲-滨外陆棚沉积环境（太原组）

图 9-2　晚古生代太原组下部障壁型沉积特征

Fig. 9-2　The sedimentary characteristics of Taiyuan formation in Late Palaeozoic

（四）潮坪沉积

最重要沉积亚相类型，一般具有向上变细的垂直层序，底部多以砂坪开始，向上为混合坪、泥坪，相当于现代海岸沉积的潮滩沉积（王颖等，2003）。

砂坪处在平均高潮线和平均低潮线之间靠近低潮线附近，沉积物多为砂质。该处潮汐作用较强，属较高的能量环境，沉积物受其影响一般由较为纯净的砂组成。岩石类型以细、中粒砂岩为主，夹薄层粉砂岩，砂岩分选中等至好，层理构造较明显，具低角度的交错层理、脉状层理和沙纹层理。局部见少量菱铁矿结核和植物化石，并可见生物扰动构造及潜穴。

混合坪沉积是介于泥坪和砂坪之间的过渡类型，发育生物扰动，岩性以粉砂岩为主，以薄层浅色细砂岩与深色泥岩或砂质泥岩频繁交替的薄互层为特征，波状、脉状和透镜状等潮汐层理十分发育，尤其以薄的砂泥互层层理为典型，常见生物扰动构造。

泥坪在潮坪沉积中位于高潮线附近及潮上地带，属低能环境，发育灰色、深灰色泥岩相和碳质泥岩相，主要由泥岩、碳质泥岩和砂质泥岩组成，偶尔夹有大潮时带来的粉砂质沉积，含有植物化石，并发育生物扰动构造，层理类型多为水平层理和缓坡状层理。

（五）滨外陆棚沉积

该类沉积形成于低潮面以下的滨外浅海陆棚沉积环境，其水动力条件一般比较弱，但也时常受到风暴浪的影响。海水的盐度基本正常，生物较发育，有机质供给充沛。

滨外陆棚沉积发育于太原组，沉积为灰岩层和泥灰岩层，生物碎屑比较发育，生物化石常见有：有孔虫、介形虫、腕足类等原地生物组合，以及一些异地搬运的生物化石碎片组合。生物碎屑类型也十分丰富，大小混杂，主要有腕足类、介形虫、棘屑和藻屑，以及软体动物碎片等，反映局限平静至较为开阔动荡环境。遗迹化石，如动藻迹（*Zoophycos*）、根珊瑚迹（*Rhizocorallium*）常共生，常见水平、倾斜潜穴、垂直潜穴等。灰岩中层理一般不发育，但有时可看到丘状交错层理及粒序层理，反映沉积环境时常受到风暴浪的影响。当有风暴作用发生时，水动力条件剧变，形成风暴沉积物——风暴岩。风暴沉积过后，转为稳定天气的沉积。前者形成具冲刷面、粒序层理和丘状交错层理的颗粒岩和泥质颗粒岩，后者形成具生物扰动构造的灰泥岩。

（六）滨岸沉积

滨岸沉积为深灰色中砾岩，其砾石成分单一，主要为燧石和少量石英砾，砾石呈良好的定向排列，以鹅卵石和扁椭球形为主，砾石大小为1～8cm，分选差，磨圆好，胶结物成分主要为铁质，这些特征指示该时期的砾岩为滨岸相沉积。

二、海陆过渡相沉积作用及沉积类型

晚古生代本区仅发育辫状河三角洲这一海陆过渡相沉积。主要发育于太原组上部和山西组，石盒子组上部也发育为湖相三角洲沉积。垂向上，三角洲下细上粗旋回特征清晰（图9-3），表明受海平面变化影响明显。煤层比较发育，横向分布上比较连续。三角洲体系包括上三角洲平原、下三角洲平原、三角洲前缘、前三角洲及三角洲间湾等沉积。

（一）三角洲平原沉积

包括上三角洲和下三角洲平原沉积。

1. 上三角洲平原沉积

上三角洲平原沉积又可细分为分流河道、天然堤、决口扇、泛滥盆地、间湾湖泊等沉积类型，在山西组和石盒子组均有分布。

（1）分流河道沉积

该相构成上三角洲平原上的主体（图9-1，图9-4），在山西组、石盒子组上部比较发育，岩石类型以中粒砂岩为主，分选中等—较差，具向上变细的正粒序，底部为含砾粗砂岩组成的滞积层和明显

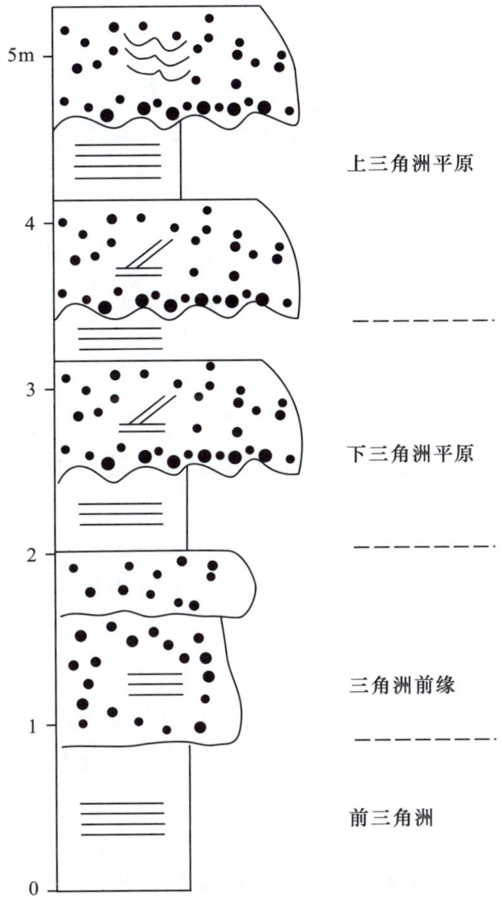

图 9-3　太原组上部—山西组湖相三角洲沉积序列

（据贺聪，2013）

Fig. 9-3　Lake delta sedimentary sequence of the upper Taiyuan formation and Shanxi formation

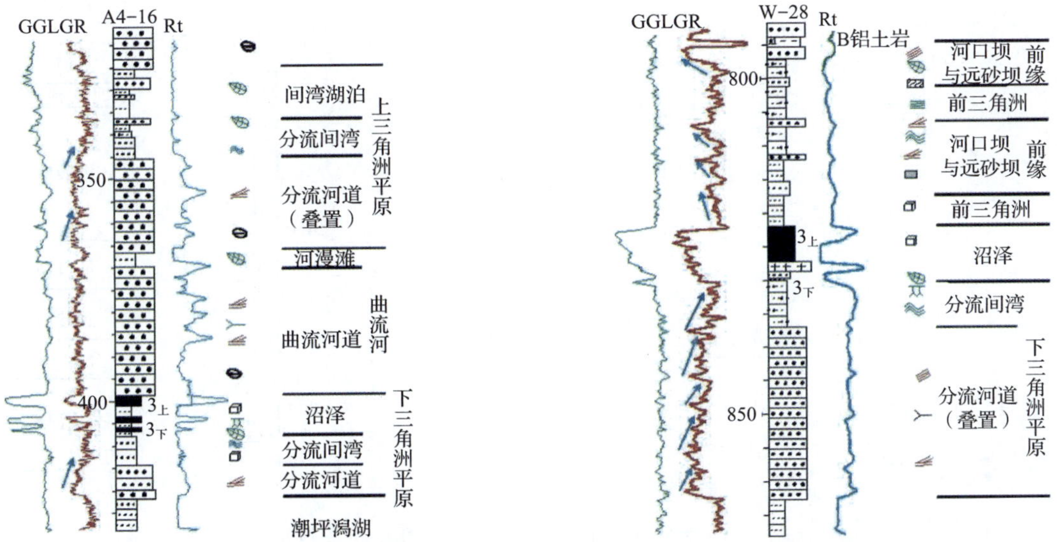

曲流河-三角洲沉积环境（山西组）　　曲流河三角洲沉积环境（山西组）

图 9-4　晚古生代山西组曲流河-三角洲沉积特征

Fig. 9-4　Meandering stream to delta sedimentary characteristics of Shanxi formation of Late Palaeozoic

的冲刷面，发育大型板状、槽状交错层理。垂向上常与天然堤和泛滥平原共生。

（2）天然堤沉积

发育在河道堤岸的两侧，垂向上位于分流河道之上，是在洪水期由洪水中悬浮的较粗物质在河道两岸堆积而成。沉积岩石类型主要为粉砂岩和细砂岩。具小型交错层理、水平层理、沙纹层理，局部可见生物搅动构造。

（3）决口扇沉积

岩性主要为细砂岩及泥质粉砂岩，由切过河岸天然堤的树枝状水系沉积作用形成。因为快速堆积，所以岩石中碎屑分选和磨圆度差，成熟度低，常含较多的杂基，常见小型槽状交错层理及沙纹交错层理，有时具水平层理，平面上呈扇形，剖面上呈小型透镜状。

（4）泛滥平原沉积

发育含植物碎片粉砂岩相、灰—深灰色泥岩相和碳质泥岩相，以湖泊、沼泽沉积为主体，岩石类型由泥岩及粉砂质泥岩组成，含少量的植物化石和碎片，常见有菱铁质结核，沉积构造多水平层理、水平波状层理或块状层理。

2. 下三角洲平原沉积

下三角洲平原沉积又可细分为分流河道、分流间湾等沉积类型，在山西组和石盒子组均有分布，是本区晚古生代主要的沉积相。

（1）分流河道沉积

以冲刷面及大型板状交错层理及大型槽状交错层理的向上变细的中粗砂岩、细砂岩、粉砂岩及泥岩为代表，底部冲刷面规模变小（图9-4，图9-5）。冲刷面上常见树干化石及泥砾，古水流方向以单向为主。

（2）分流间湾沉积

被天然堤或沼泽隔开并与开阔海水之间有一定连通的微咸水体，沉积岩石类型主要为深、暗色的泥岩、含炭泥岩和粉砂质泥岩，常夹有决口形成的砂体透镜体，发育水平层理、水平波状层理及小型沙纹层理，含有少量植物根和叶化石，并含较多的菱铁质结核，微咸水—半咸水环境，垂向上常与分流河道和分流河口坝共生（图9-5）。

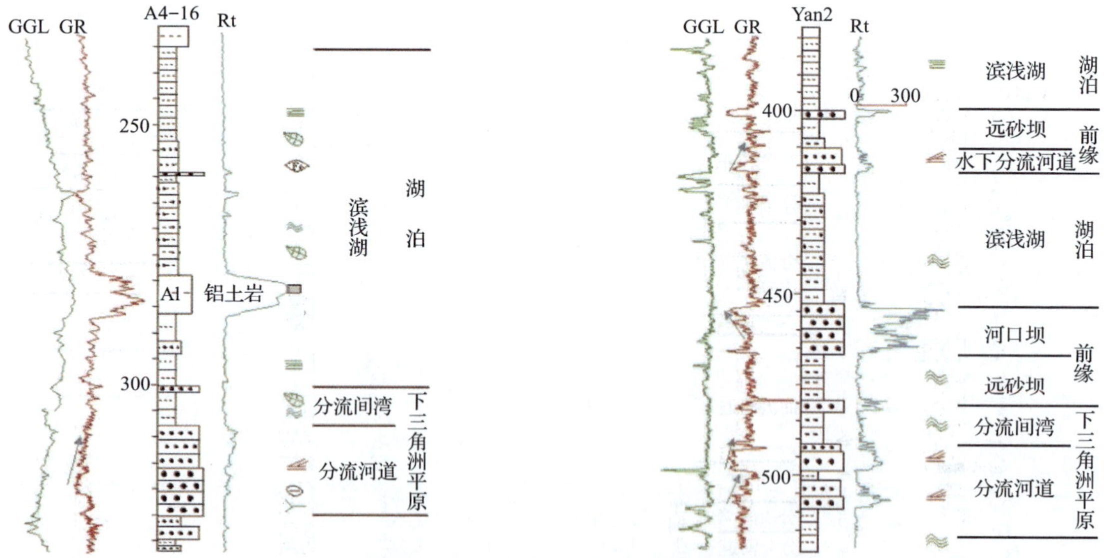

图9-5 晚古生代石盒子组湖泊-三角洲沉积特征

Fig. 9-5 Lake to delta sedimentary characteristics of Shihezi formation of Late Palaeozoic

（二）三角洲前缘沉积

该相又可细分为水下分流河道、河口坝、远砂坝等沉积类型，发育于山西组和石盒子组上部。

1. 水下分流河道沉积

是下三角洲平原分流河道在水下的自然延伸，水下分流河道常呈块状，块状而无粒序，电测曲线形状多呈箱形，有些呈钟形（图9-5）。

2. 河口坝沉积

河口坝沉积主要由比较干净的细砂岩和中砂岩组成，分选中等—较好，泥质杂基含量较低，可能是在河口处受波浪、潮汐作用的反复筛选的结果。层理类型有大型楔状交错层理、低角度交错层理及大型槽状交错层理，有时还看到包卷层理及共生的泥砾。古水流方向受河流和潮汐水流影响较大，常出现向陆方向或多方向变化的分量。砂体在剖面上呈透镜状，粒度由下向上变粗，测井曲线常呈倒钟形，顶部突变或渐变，底部渐变。顶部常被三角洲平原分流河道切割，在垂向上常与分流间湾共生。

3. 远砂坝沉积

远砂坝沉积岩石类型以薄层到中厚层状的中细砂岩与粉砂岩、粉砂质泥岩及泥岩互层组合为主，发育小型沙纹层理，还有薄的砂泥互层的韵律层理、平行层理以及一些生物扰动构造（图9-5）。垂相上与河口坝沉积一起构成粒度向上变粗的序列。

4. 间湾沉积

间湾发育为泥岩、碳质泥岩和菱铁质泥岩组合，水体变浅使得地下水位高于海平面进而发育为泥炭沼泽并形成厚度较大的煤层。

（三）前三角洲沉积

前三角洲相以灰色、深灰色泥岩相为代表，主要由粉砂质泥岩、泥岩及薄层菱铁矿层等细粒沉积物组成，发育水平层理，偶见动物化石。前三角洲与三角洲前缘沉积整体具有向上变粗的特征，在测井曲线上为典型的漏斗形。

三、陆相沉积作用及沉积类型

晚古生代陆相沉积作用包括河流和湖泊沉积作用两大类。

（一）河流沉积作用

按泛滥盆地发育程度可将河流沉积划分为曲流河沉积和辫状河沉积，本区晚古生代地层河流沉积主要发育于石盒子组、双泉组，为曲流河沉积体系。沉积相单元包括河床滞流、边滩（或心滩）、天然堤、决口扇、泛滥平原等，在垂向上表现为向上变细的正粒序结构（图9-6）。

1. 河床滞留沉积

河床滞留沉积岩石类型以细砾岩、含砾粗砂岩为主，分选和磨圆差—极差，成分成熟度低，沉积构造包括块状层理、槽状交错层理、大型板状交错层理。砂体形态多呈小的透镜体状，砾石具明显的定向排列，其最大扁平面倾向河流上游。该沉积层序与下伏岩层呈冲刷接触，上部逐渐过渡为边滩沉积。

2. 边滩沉积

边滩沉积岩石类型主要为中粗粒长石石英砂岩，岩石成熟度较低，杂基含量较高。沉积构造发育有大型板状交错层理及侧向迁移交错层理，也可见到槽状交错层理，反映为单向式水流条件。

3. 天然堤沉积

天然堤沉积岩石类型包括细粒长石石英砂岩和粉砂质泥岩及泥岩，含少量的植物化石碎片，偶见生物潜穴。典型沉积构造为水平层理、沙纹层理及平行层理。

4. 决口扇沉积

决口扇沉积岩石类型为粗粒砂岩、中—细粒砂岩，砂体在横向上表现为延伸不太远的透镜体，具典型的小型沙纹层理，常与下伏岩层冲刷接触。

5. 泛滥平原沉积

泛滥平原沉积以发育含植物碎屑粉砂岩相和紫红色、紫灰色块状泥岩相为特征，泛滥平原为洪水泛滥形成的广阔平坦的河漫沉积区，主要由垂向加积的细粒沉积物组成，一般由互层状的泥岩、粉砂质泥岩和粉砂岩组成，局部发育水平或水平波状层理，含丰富的植物化石碎片。根据沉积物的特征，又可细分为河漫滩、河漫湖泊和岸后沼泽（河漫沼泽）。其中河漫滩以砂质沉积物为主；河漫湖泊以泥质沉积为主；岸后沼泽也以泥质沉积为主，但一般都有泥炭沉积，形成煤线或煤层。

（二）湖泊沉积作用

本区晚古生代湖泊沉积均为碎屑湖相沉积，以双泉组发育最为典型，还包括石盒子组局部层位。双泉组湖泊沉积岩石类型以砂岩为主，粉砂岩次之，地层厚度较大，而含砂率较低，发育水平层理和沙纹层理，表现为以滨湖-浅湖沉积亚相为主的沉积环境（图9-5，图9-7）。滨湖沉积由中细砂或含砾粗砂岩夹粉砂岩、砂质泥岩薄层组成，包括滨湖三角洲沉积和滨湖坝沉积，见水流交错层理等；浅湖沉积由泥岩、粉砂质泥岩夹粉砂岩和细砂岩薄层组成，发育水平层理、平行层理和小型交错层理。滨湖沉积和浅湖沉积在垂相上常交替出现，互层产出。

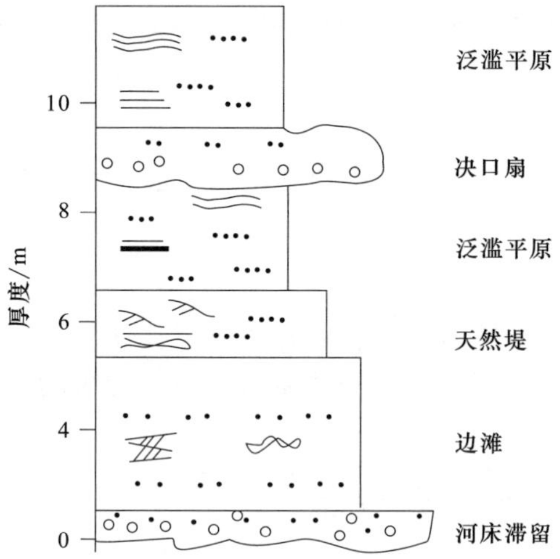

图9-6　石盒子组曲流河相沉积序列

Fig. 9-6　Meandering stream sedimentary sequence of Shihezi formation

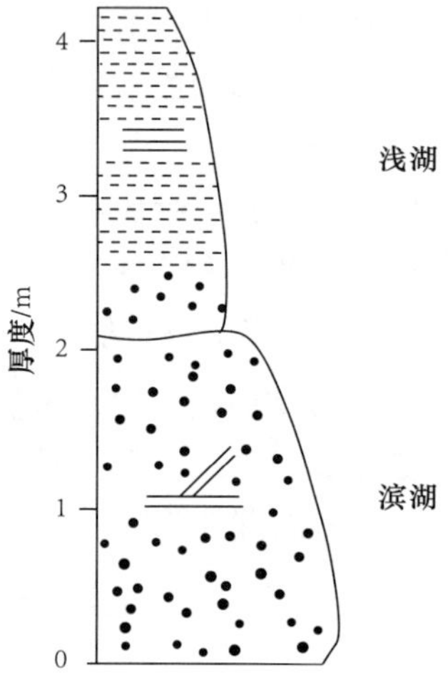

图9-7　双泉组湖相沉积序列

（据贺聪，2013）

Fig. 9-7　Lake sedimentary sequence of Shuangquan formation

第二节　主要沉积岩类型

本区晚古生代沉积岩类型多样，沉积构造、颜色丰富。除在太原组中发育极少海相泥灰岩等碳酸盐岩、在双泉组底部发育极少火山碎屑岩外，大部分为不同粒级的陆源碎屑岩。各组岩石类型和组合各具特征，主要表现有：

1）碎屑物的粒度由下往上有逐渐有变粗的趋势（表9-2）；

表9-2　各组砂岩和砾岩厚度及占比

Table 9-2　Thickness and percentage of sandstone and conglomerate

组别	地层厚度/m	砂岩、砾岩厚度/m	砂岩、砾岩占比/%
双泉组	47.3	37.9	80.08
石盒子组上部	197.9	113.8	57.5
石盒子组下部	130.0	85.5	61.51
山西组上部	183.23	79.13	43.24
山西组下部	126.05	44.93	35.66
太原组	61.9	23.19	37.46

2）岩石的颜色由太原组的灰白色、浅灰色、灰色，变为山西组的深灰色、灰黑色及黑色，最后转变为石盒子组的灰黄色夹紫色、黄褐色、砖红色；

3）岩石的成分比较复杂，碎屑成分除石英外，还有较多的燧石、长石、白云母和硅质灰岩岩屑，几乎全部属硬砂质砂岩。胶结物成分以泥质、粉砂质和钙质为主，随着地层层位由下向上有一定变化规律：太原组多钙质胶结，山西组以泥质胶结为主，钙质消失；石盒子组往上多为粉砂质胶结；

4）碎屑物的分选和磨圆程度由下往上有逐渐变差趋势：太原组的碎屑物分选较好，多呈次磨圆状，山西组下部分选、磨圆中等，往上分选变差，磨圆程度为次棱角状-棱角状；

5）反映水动力条件的层理类型：下部多为水平层理和波状层理，上部以斜层理、交错层理和波状层理为主。而岩层间的接触关系由过渡、渐变接触逐渐发展到以明显、冲刷接触为主。

一、碳酸盐岩类

仅发育于太原组下部层位，岩性为泥灰岩，呈 1~3 层夹层产出。岩石呈灰色、深灰色、灰黑色，生物屑泥晶结构，透镜状产出。基质和颗粒含量之比约为 85∶15，其中基质主要为泥晶方解石。颗粒分布较均匀，分选较差，圆度变化大。产海相动物化石，以腕足类与双壳类较多，偶见三叶虫、珊瑚与海百合茎等。局部呈较纯的灰岩透镜体，基本上由晶粒方解石和生物屑构成。

二、碎屑岩类

（一）砾岩类

整体来看，砾岩总数较少，分布层位和地区也有限，发育于太原组底部、山西组和石盒子组，其颜色、结构、成分、沉积构造差异较大。可细分为 5 种不同的特征类型：

1. 叠瓦状粗砾岩

分布于太原组底部，因其分选、磨圆、胶结都好，又称“三好砾岩”。岩石呈灰白色、浅灰黄色、青灰色及黄褐色，局部紫红色，风化面呈黄褐色，巨厚层状产出。砾石占 80%以上，成分主要为石英和燧石，呈明显定向排列。厚度横向变化大，呈大的透镜体分布。

2. 递变中砾岩

发育于山西组及双泉组底部，呈紫红色、灰白色，中层—巨厚层状。有时含泥砾，呈不明显的大型交错层理，一般向上变细并过渡为砂岩。砾岩砾石成分以白色石英及黑色燧石为主，直径 0.5~4cm，分选、磨圆较好，呈叠瓦状排列。胶结物为深灰色泥质及砂质，悬浮式胶结。

3. 槽状交错层理细砾岩

发育于山西组和石盒子组下部，呈紫灰色、紫红色、浅黄褐色，中层—厚层状。砾石成分以石英为主，直径 3~4mm，分选、磨圆中等，夹泥质条带，钙质胶结，有时含泥砾，发育槽状交错层理。

4. 泥砾岩

仅发育于石盒子组上部，较为少见，呈灰黑色、灰黄色薄层状。由大小不等的泥砾组成，砾石大小为 2~14cm。泥砾岩主要为河流流经泥质沉积为主的区域时，在河床底部形成的薄层泥砾堆积，属于河床滞留沉积。

5. 含砾粗砂岩

广泛分布于太原组、山西组、石盒子组。太原组、山西组为含砾粗粒岩屑砂岩，深灰色厚层状，细砾粗砂结构。碎屑粒度 0.2~20mm。砾石成分主要为石英岩，少量为燧石和粉砂岩，呈次棱角状，分选中等，含量 5%~10%。砂粒主要是不规则的岩屑，石英较少，胶结物多为泥质和碳质。

（二）砂岩类

砂岩是晚古生代地层的主要沉积岩，包含粗砂至粉砂各级粒径的砂岩类型，以粗砂岩和中砂岩为主。

1. 粗砂岩

可根据沉积构造划分为 3 类：

1）大型板状交错层理含砾粗砂岩。发育于石盒子组上部，颜色呈灰黄色、紫色、灰红色、灰色，中粗粒砂状结构，巨厚层—厚层状。砂粒碎屑成熟度低，多为石英、长石，少量岩屑，组成长石石英砂岩或长石石英杂砂岩。所含砾石直径一般为0.5~1cm，含量约5%，分选、磨圆较好，发育大型板状交错层理，有的局部含砾或在底部含泥砾。

2）槽状交错层理粗砂岩。发育于山西组下部和石盒子组，分布较少。岩石呈浅灰红色、灰黄色，巨厚层—中厚层状，见槽状交错层理，分选好、磨圆中等。

3）板状交错层理粗砂岩。发育于石盒子组，岩石呈灰黄色，巨厚层—中厚层状，发育大型板状交错层理及低角度交错层理，底部具冲刷面，有的底部含石英和燧石质砾岩，有的夹薄层细砂岩或含砾砂岩。砂岩以钙质胶结为主，次为铁质胶结，颗粒呈次棱角状，分选差—中等。

2. 中砂岩

可根据沉积构造划分为4小类：

1）平行层理中砂岩。发育于石盒子组下部，紫灰色厚层状，砂体呈透镜状，具平行层理，铁质胶结。

2）低角度交错层理中砂岩。发育于山西组和石盒子组，紫褐色、浅灰色，巨厚层—厚层状，发育大型低角度交错层理，有时夹薄层深灰色泥岩，底部含泥砾和冲刷面。

3）槽状交错层理中砂岩。发育于山西组和石盒子组，浅黄色、黄色、灰白色，厚层状，砂粒碎屑成熟度低，为岩屑长石石英砂岩，局部含砾，发育大型槽状交错层理，底部见冲刷面。

4）板状交错层理中砂岩。发育于山西组和石盒子组，灰黄色、紫灰色，巨厚层状，中粒—中粗粒砂状结构，砂粒成熟度差—中等，局部含砾，分选、磨圆中等—较好，发育大型板状交错层理，底部具冲刷面。

3. 细砂岩

细砂岩在碎屑岩中所占比例虽然小，但在地层中的分布较广，在石炭系山西组至上二叠统双泉组均有不同程度地发育，以山西组和石盒子组最为典型。根据沉积构造可细分为3小类：

1）平行层理细砂岩。褐灰色、灰黄色、灰色，厚层—巨厚层状，通常含大量植物茎化石，平行层面排列，发育平行层理或断续状平行层理，主要为河流天然堤、河口坝或滨海沉积。

2）槽状交错层理细砂岩。灰黄色、灰紫色，巨厚—中厚层状，有时夹粉砂岩，局部菱铁质胶结，发育槽状交错层理，底部见冲刷面，主要为三角洲平原分流河道、前缘水下分流河道等沉积。

3）沙纹层理细砂岩。灰黄色、浅灰色，厚层—中厚层状，发育低角度交错层理及沙纹层理，见菱铁质结核，生物遗迹发育，并含动物化石和植物化石，主要为砂坪、滨岸、决口扇沉积。

4. 粉砂岩

发育于山西组、太原组和石盒子组，深灰色、灰色、灰黄色和紫灰色，中—巨厚层状，发育沙纹层理或水平层理。含生物扰动构造、生物潜穴、生物化石碎片的粉砂岩为潮坪（砂坪）、滨岸、三角洲前缘河口坝沉积，其中，潮坪（砂坪）环境主要分布于平缓海岸的潮间带及潮下带上部；含大量植物碎片或植物根、茎化石的粉砂岩主要为泛滥平原、河漫滩、决口扇沉积。

（三）黏土岩类

1. 泥岩

（1）泥岩

发育于石盒子组和山西组，紫红色、紫灰色、灰色，岩石外表呈致密隐晶质。该类岩石中含植物碎片等化石，部分岩层中含黄铁矿假晶，大小为0.05~0.5mm，或含瘤状铁质结核。该岩性代表了水体低能、还原的沉积环境，主要为河流相泛滥平原沉积或浅湖、潟湖、泥坪沉积。

（2）碳质泥岩

发育于山西组和太原组，形成于各种积水沼泽中，包括河流相牛轭湖、河漫沼泽、三角洲分流间湾泥炭沼泽、潮坪潟湖泥炭沼泽。

2. 页岩

（1）砂质页岩

发育于石盒子组，呈浅灰色，表面有棕色氧化色，且发育明显的揉流痕迹。颗粒以石英为主，约占20%，大小0.25~0.5mm，并且呈条带状分布，基质和胶结物约占70%。镜下水平层理很清楚。其上部多为深色泥岩和薄煤层。岩石多因发生变质作用而板理化，主要为河流泛滥平原沉积。

(2) 铝土质页岩

发育于太原组底部，灰绿色、灰白色，岩石呈层状、似层状、透镜状产出，与铝土矿常呈过渡关系，有时与砂质页岩、碳质页岩共生。隐晶结构，黏土结构，块状、瘤状构造，可见页理。黏土矿物（65%~85%）为高岭石和伊利石。含铝矿物为一水硬铝石（15%~35%），分布不均匀，局部为板柱状晶体组成的瘤状集合体，同生矿物有黄铁矿、菱铁矿、方解石等。

三、火山碎屑岩类

（一）火山集块岩与凝灰质砾岩

仅见于石景山双泉寺一带双泉组中下部地层，分布极少。岩石呈灰色，角砾状或砾状结构，厚层状构造。砾石成分复杂，以石英、燧石为主，还见有少量熔岩角砾，凝灰质、砂质胶结。

（二）凝灰质砂岩

仅见于双泉组，普遍发育，变余凝灰结构、砂状结构，层状构造。砂岩成分复杂，含25%~40%岩屑，成分以安山岩、英安岩为主，为陆源碎屑，次棱角状，晶屑以长板状斜长石为主，斜长石棱角尖锐，破裂不齐，保留有清楚的卡-钠复合双晶，还出现一定数量的黑云母，强退色、暗化，具备火山喷出型黑云母的特点。

第三节　沉积盆地及沉积建造

一、沉积盆地

（一）克拉通陆表海盆地

石炭纪—早二叠纪，相当于太原组—石盒子组下部沉积时期，本区沉积盆地仍为华北克拉通陆表海盆地，盆地特征同寒武纪—奥陶纪陆表海盆地相似，特征是：①盆地沉积的地层厚度较薄（一般厚度为200~300m）；②盆地沉降幅度、速度很小，沉积岩相分布稳定；③地壳构造活动方向差异不大，沉积区内部沉降幅度差异性小；④盆地外围古陆分布范围有限，且具有边隆起边剥蚀使地形保持准平原的特点，而盆地分布面积大，沉降速度慢，且具有边沉降边沉积的沉积平原的特点等。

（二）克拉通板内挤压型挠曲盆地

晚二叠世，相当于石盒子组上部—双泉组沉积时期，由于受华北克拉通活化影响，沉积盆地由稳定型陆表海盆地转化为次稳定陆内挤压型挠曲盆地，盆地性质等同于孟祥化等（2004）、杨明慧等（2012）认为的华北克拉通北缘陆内前陆盆地的隆后挠曲带。盆地为次稳定大陆壳板块区内在挤压环境下缓慢差异性下降作用形成的陆相沉积型盆地。盆地特征是：①盆地沉积厚度明显增大（一般厚度300~500m），且区域差异较大，沉积厚度分布受控于构造应力变形，东西向延伸、南北向迅速减薄或尖灭。石盒子组在灰峪—军庄一带厚达305m，南车营一带厚度仅165m。双泉组在大悲寺厚度达200m，而潘涧沟厚度仅50余米；②古陆扩大，沉积区缩小；③沉降幅度和速度增大；④构造运动差异性增加，早期形成的部分断裂开始重新活动（如赤城-古北口断裂），小规模岩浆侵入上地壳（如杨木栅子一带的具有活动大陆边缘特征的早二叠世、早三叠世岩体）或喷出。构造条件及沉积环境揭示该期盆地性质已经发生变化或者正在处于转变时期。

二、沉积建造

晚古生代，本区除北西部极少数地区一直位于古陆剥蚀区外，主体处于盆地环境接受沉积。其沉积环境复杂，经历了从海相—海陆过渡相—陆相的复杂转变。盆地性质也由稳定型克拉通陆表海盆地转化为次稳定的陆内挤压型挠曲盆地。因此，盆地内部沉积建造复杂，可按照构造-沉积环境划分为稳定型海相单碎屑岩含煤建造、稳定型陆相碎屑岩含煤建造、稳定型河湖相碎屑岩建造、次稳定型陆相火山-碎屑岩建造和次稳定型河湖相碎屑岩建造二大类五小类。具体各时期主要沉积建造类型综合特征如表 9-3 所示。

表 9-3 晚古生代各时期沉积建造类型综合特征

Table 9-3 Comprehensive characteristics of various sedimentary formation types of Late Paleozoic

年代	时期	沉积建造	盆地类型	建造类型	沉积环境	物源	沉积速率	共生岩石主要类型	备注
三叠纪	双泉期	次稳定	板内挤压型坳陷盆地	河湖相碎屑岩建造	陆相河流、湖泊沉积	陆源	较快	杂色岩屑细砂岩、粉砂岩、页岩夹砾岩	
晚二叠世	石盒子期	次稳定		陆相火山-碎屑岩建造	陆相火山盆地	陆源	极快	深灰色石英粗安质角砾熔结凝灰岩、深灰色粗安岩、浅灰色流纹质含角砾熔结凝灰岩及少量灰绿色凝灰质长石岩屑砂岩	北部境外，河北丰宁-隆化断裂以北地区（云雾山组）
		稳定		河湖相碎屑岩建造	陆相山前平原河流、湖泊沉积	陆源	较快	杂色含砾砂岩、粗砂岩、中粒长石石英砂岩、细砂岩、粉砂岩、页岩、泥岩、碳质页岩	境内，河北丰宁-隆化断裂以南地区
早二叠世	山西期	稳定	克拉通盆地（陆表海）	陆相单碎屑岩含煤建造	陆相山前平原湖沼相沉积	陆源	较慢	深灰色、灰黑色砂岩，页岩，泥岩，砾岩及煤层（线）	晚期
石炭纪		稳定			三角洲前缘、前三角洲	陆源	较慢	砾岩、砂岩、页岩和泥岩	早期
	太原期	稳定		海相单碎屑岩含煤建造	障壁岛型海相沉积（潮坪、滨岸、潟湖、障壁岛、冲越扇等）	陆源	较慢	底砾岩；铁质砂岩及扁豆状赤铁矿（山西式铁矿）；浅灰色、灰色铝土质页岩；铝土矿；黄绿色、灰色、黄褐色粉砂岩，细砂岩；灰黑色、黑色砂岩、页岩、碳质页岩及煤层（线）；灰色、灰绿色页岩、泥灰岩	

（一）稳定型陆源建造

1. 海相单碎屑含煤建造

又称海陆交互相含煤建造，发育于晚石炭世，主要为太原组和山西组下部地层。建造以海相地层与陆相地层交替沉积为特征，形成于地势平缓的滨岸淡水与咸水交替作用带。岩石共生组合为底砾岩、铁质砂岩及扁豆状赤铁矿（山西式铁矿）、铝土质页岩、铝土矿、粉砂岩、细砂岩、页岩、碳质页岩及煤层（线）、泥灰岩等，富含珊瑚类、腕足类、蜓类及植物化石。

2. 陆相碎屑岩含煤建造

发育于早二叠世早期的山西组上部，属陆相山前平原湖沼相，为一套近海平原湖沼相含煤沉积岩系。岩石共生组合为砂岩、页岩、泥岩、砾岩及煤层。岩石水平层理、透镜状层理发育，含铁质结核和大量植物化石碎片。

3. 河湖相碎屑岩建造

又称杂色碎屑岩建造，发育于早二叠世晚期的石盒子组，为陆相山前平原河湖相沉积环境，以河流沉积环境为主。岩石共生组合复杂，下部岩石组合为灰白色、灰黑色含砾砂岩、粗砂岩、杂色中粒长石石英砂岩、细砂岩、粉砂岩、页岩、泥岩、碳质页岩、煤线，岩石具小—中型斜层理、平行层理、粒序层理、水平层理，为曲流河相。上部岩石组合为肉红色、紫灰色中粗粒（含砾）长石石英砂岩、粉砂岩、细砂岩、页岩、泥岩，砂岩中发育大中型斜层理及交错层理、粒序层理，为三角洲平原辫状河河流相。

（二）次稳定型陆源建造

1. 陆相火山-碎屑岩建造

发育于晚二叠世的石盒子组。该组地层分布有限，主要分布在丰宁—隆化一线以北，北京境内北部地区可能发育极少部分（已遭剥蚀）。岩石组合为一套中性、酸性火山岩及火山碎屑岩组成。包括深灰色石英粗安质角砾熔结凝灰岩、深灰色粗安岩、浅灰色流纹质含角砾熔结凝灰岩及少量灰绿色凝灰质长石岩屑砂岩。

2. 河湖相碎屑岩建造

又称杂色碎屑岩建造，发育于晚二叠世至三叠纪的双泉组。岩石共生组合略有不同，下部为紫红色岩屑细砂岩、粉砂岩及页岩夹砾岩，上部则为灰绿色、灰白色岩屑碎屑岩组合，凝灰质火山碎屑物增多。

第四节　岩相古地理

中奥陶世以后，华北板块总体上升为陆，燕山地区广泛沉积覆盖的早古生代碳酸盐岩地层以及裸露的铝硅酸盐岩古陆遭受长约140Ma的风化剥蚀，形成了中石炭世时的红土风化壳喀斯特-滨海准平原化地貌。在此地貌基础上，中石炭世地壳再次下沉，海水浸漫，开始了晚古生代盆地沉积。沉积环境从初期的短暂陆表海-障壁岛-潟湖沉积环境，向着海陆交替的过渡陆表海-三角洲沉积环境发展，再向陆表海大陆陆源近海湖泊-河流沉积环境发展，最终演化为陆内河流、湖泊沉积环境。

本区由于古地理位置靠近古陆边缘，因此在整体沉积环境由海相-海陆交互相-陆相发展变化的过程中，较其他地区提早结束（早二叠世）海侵，转为陆相沉积。具体该阶段可分为晚石炭世太原期、晚石炭世—早二叠世山西期、早二叠世—晚二叠世石盒子期和晚二叠世—早三叠世双泉期。

一、太原期

太原期，随着海平面的波动式上升，海水由北东向南西浸漫，直达风化壳带。铁-铝质风化壳物质不断被水流搬运，氧化铝在酸性介质中多半呈胶体状态被搬运，而当进入浅海环境后，由于介质pH的改变，在碱性介质内氧化铝成分发生沉积，在本区发育为高铁高铝黏土岩沉积。根据沉积物的特征反映出当时本区海边陆地地形有一定高差起伏，沉积物源风化作用不够充分，未能像华北克拉通大部分地区一样形成很好的铝土矿床，区内沉积的主要是由水云母、蒙脱石等类黏土矿物为主组成的黏土矿层。这一层位在横向上的岩性变化也很显著，由泥质沉积很快相变为粉砂质、砂质、砂砾质沉积。

初期，燕山地区普遍发育以砾岩、砂岩为主的滨岸沉积。随着海侵继续，中期陆表海-障壁岛-潟湖沉积体系基本形成（图9-8）。大致沿河北谢家堡—昌平居庸关—延寿镇—密云石城—密云古北口一线以北为陆地，主体接受剥蚀；此线以南地区为滨岸砾岩、砂岩沉积；石景山苹果园-大兴安定一带发育为北西向障壁岛砂岩沉积；障壁岛南西部广阔地带发育为潟湖泥岩、泥灰岩沉积或潮坪粉砂岩沉积；而障壁岛北东广阔地带多被后期抬升剥蚀。

太原期末期（图9-9），海侵继续发展，海盆向南、向西扩展，而总体海水变浅。燕山地区发育为陆表海-三角洲-滨海沉积环境。

二、山西期

山西期早期基本继承了太原期末期古地理格局，晚期海水进一步后退，海退方向宏观大致由北西向南东，最终完全退出本区，演化为陆相山前冲积平原沉积环境，发育河湖相沉积（图9-10）。主要有河流冲积相、淡水湖泊相和沼泽相等沉积岩相，这些岩相在剖面上表现一定的序列，每个旋回的起点都是河床相，向上发展渐变为湖泊相，又萎缩衰亡成沼泽相和泥炭沼泽相。上部旋回堆积时都不同程度的冲刷了下部旋回的沉积物，冲刷现象极为清楚，粗碎屑物急剧增多，使得每个旋回都不完整。

整个山西期气候潮湿温暖，是一个有利的成煤期。成煤同湖沼沉积环境密切相关，早期为海陆过渡沉积相海相湖沼，湖沼多且面积大，煤层结构比较简单，层位稳定，且底部多发育与之共生的软质、

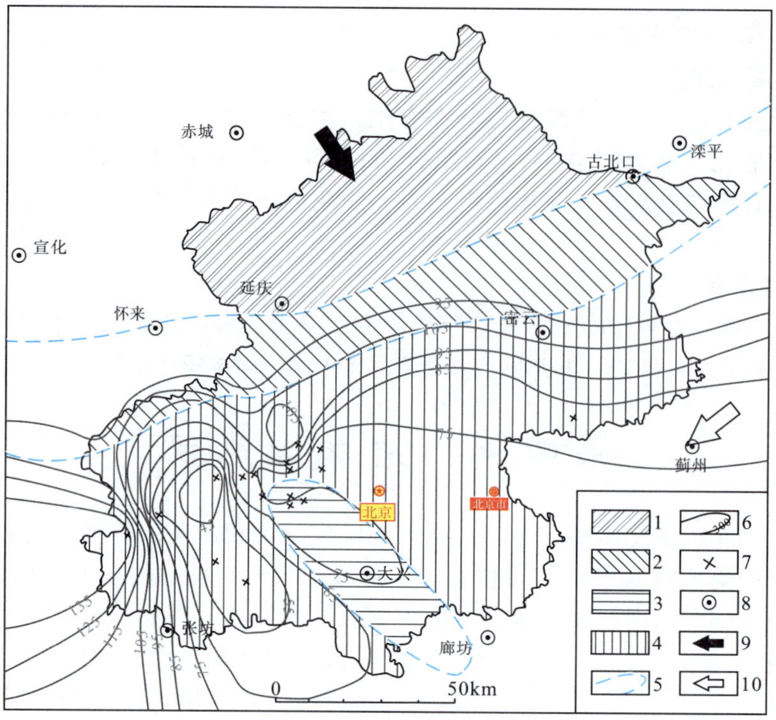

图 9-8　太原期中期岩相古地理图

Fig. 9-8　Lithofacies paleogeographic map of mid-term of Taiyuan period

1—基底剥蚀区；2—滨岸砾砂岩相；3—障壁岛砂岩相；4—潟湖与滨岸相；5—岩相界线；6—地层等厚度线（m）；7-剖面位置；8—地名；9—物源方向；10—海侵方向

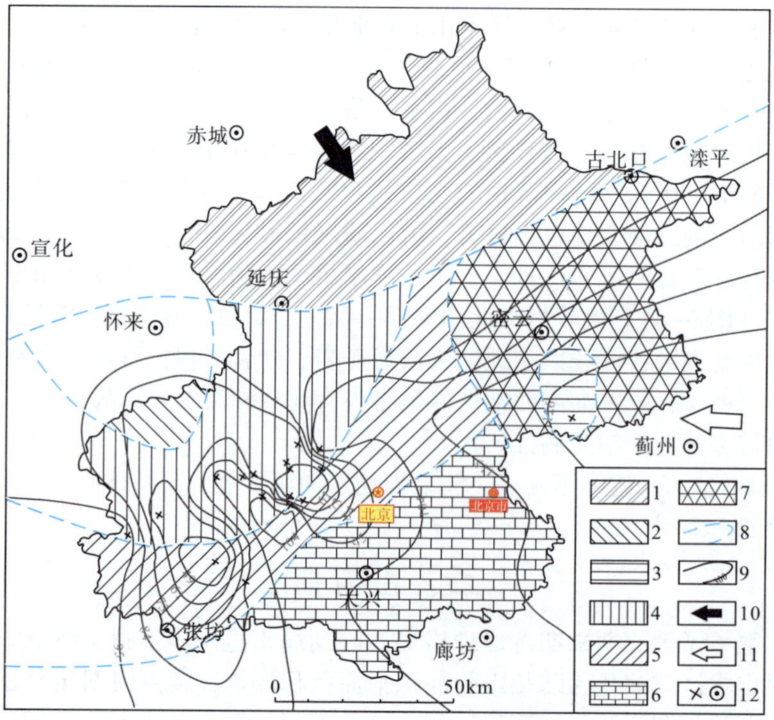

图 9-9　太原期末期岩相古地理图

Fig. 9-9　Lithofacies paleogeographic map of terminal term of Taiyuan period

1—基底剥蚀区；2—上三角洲相；3—障壁岛砂岩相；4—下三角洲前缘相；5—前三角洲泥岩相；6—浅海陆棚相；7—推测潟湖及潮坪相砂岩；8—岩相界线；9—地层等厚度线（m）；10—物源方向；11—海侵方向；12—剖面位置及地名

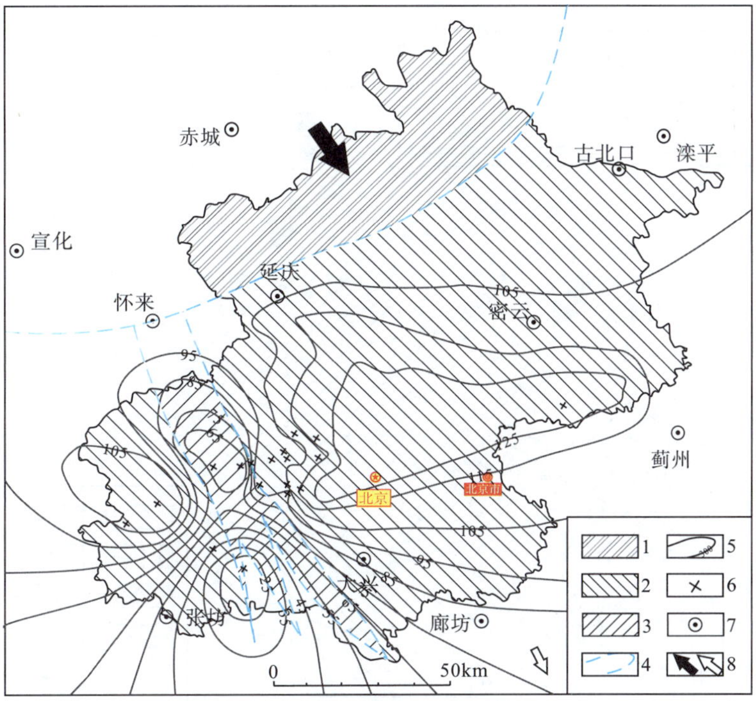

图 9-10　山西期晚期岩相古地理图

Fig. 9-10　Lithofacies paleogeographic map of Late Shanxi period

1—基底剥蚀区；2—山前冲积平原相；3—河道沼泽相；4—岩相界线；5—地层等厚度线（m）；6—剖面位置；7—地名；8—物源方向与海退方向

半软质耐火黏土矿层。而晚期为陆相平原河流湖沼，湖沼多但面积小，煤层厚度变化大，煤层均为复杂结构，夹石层多（多达 10 层以上），很快分叉尖灭，与煤层共生的黏土矿层亦不稳定。

三、石盒子期

石盒子期在山西期末期古地理格局基础上进一步演化。大致沿孙庄子—八达岭—溪翁庄—太师屯一线以南为沉积盆地（图 9-11），盆地性质在晚期开始发生转变，由稳定型陆表海型盆地向板块内部（陆内）挤压型挠曲盆地演化，盆地内部发生次级凹陷与隆起，使各处沉积物厚度差异较大，如灰峪—军庄一带厚度约 305.2m，南车营一带厚仅约 165m。本区沉积特征发生明显变化：沉积相为陆相山前平原河湖相沉积，以河流相（曲流河相）沉积占主导，沉积颗粒的成熟度很低，粗碎屑岩发育，斜层理、交错层理和冲刷现象明显。

石盒子期古气候特征表现为湿-干交替，形成一套以灰绿色为主，下部有灰白色、灰黑色，上部夹有肉红色或紫灰色的碎屑岩建造，下部尚可见到一些煤线，但古气候转变使其不再具备成煤环境。

四、双泉期

双泉期原大盆地可能因盆内岩浆活动（如平谷盘山复合岩体）或断裂活动，演化为若干个大小有别、孤立分布的小盆地，构成宏观走向仍为北东东向盆地群。该期本区的沉积盆地主要发育于西山—大兴一带（图 9-12），从沉积特征来看，与周边的沉积盆地不一定相连，控制沉积物形成和堆积的是内陆河流和湖泊的作用（图 9-12 角图）。盆内早期发育为洪水期河流相砂岩沉积，中期发育为枯水期湖泊相泥岩、粉砂岩及细砂岩互层沉积，而晚期再次发育为洪水期粗杂碎屑沉积。盆地沉积因受近东西向构造挤压影响，发育多个走向近东西的相互平行的沉降中心。门头沟大悲寺一带岩性以砂岩为主，粉砂岩次之，为浅湖-半深湖沉积；周口店一带岩性以中粗砂岩、中细砂岩为主，含砾砂岩、泥砾等次之，为滨湖-河流沉积；大台一带沉积为含砾粗砂岩-含粉砂泥岩或粗砂岩-粉砂岩韵律组合，为河流沉积。沉积物的颜色由紫色到紫灰色再到绿灰色，反映古气候由干燥-半干燥-温湿的转变过程。

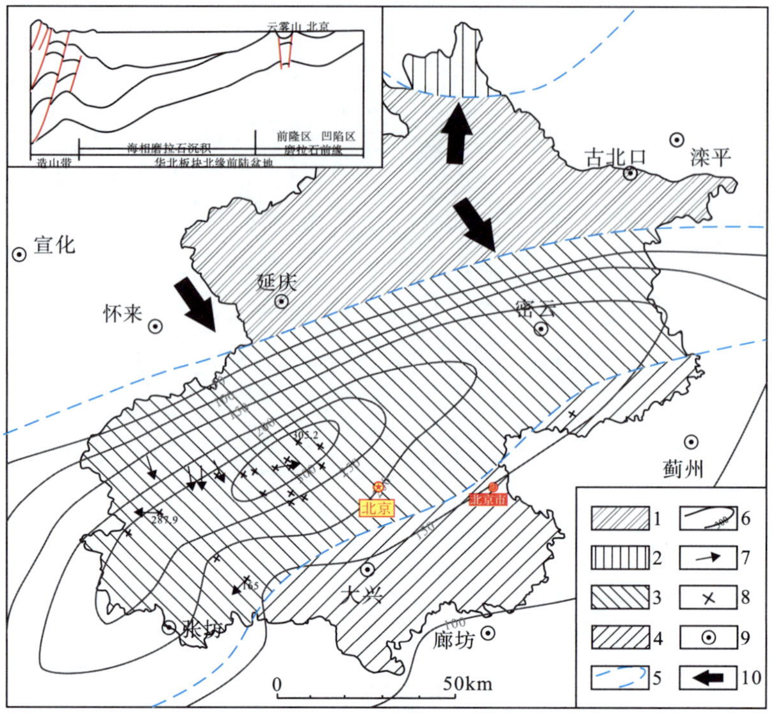

图 9-11　石盒子期岩相古地理图

Fig. 9-11　Lithofacies paleogeographic map of Shihezi period

1—基底剥蚀区；2—火山碎屑相；3—陡坡山前冲积河流相；4—缓坡山前冲积河流相；5—岩相界线；6—地层等厚度线（m）；7—古水流方向；8—剖面位置；9—地名；10—物源方向

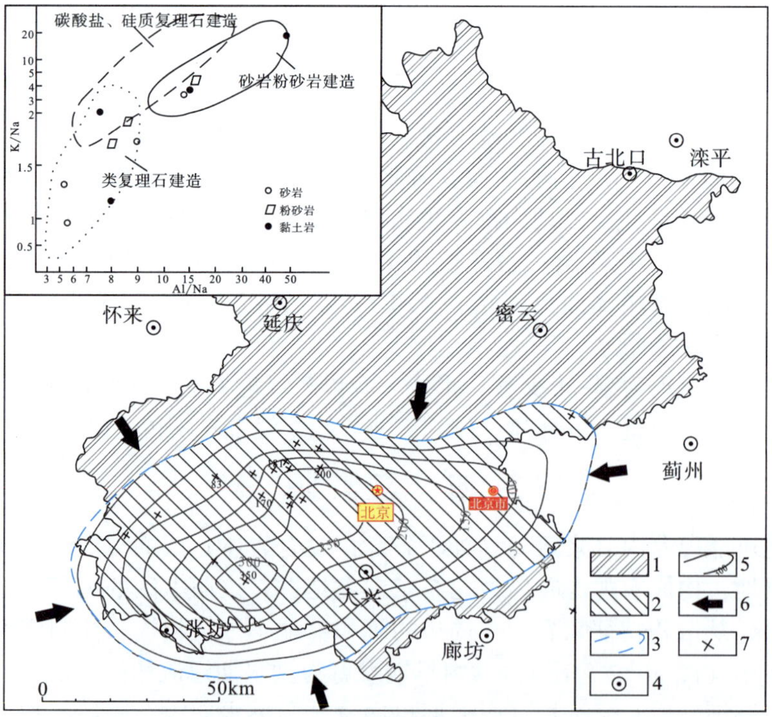

图 9-12　双泉期岩相古地理图

Fig. 9-12　Lithofacies paleogeographic map of Shuangquan period

1—基底剥蚀区；2—陆相河流湖泊相；3—盆地边界；4—地名；5—地层等厚度线（m）；6—物源方向；7—剖面位置

第十章　中生代

中生代本区处于活化的华北克拉通北缘，地壳间歇性运动频繁，发育复杂陆相碎屑、火山碎屑沉积，沉积厚度巨大，总厚度达到9700m。

第一节　沉积作用及沉积类型

通过对中生代地层系统清理，保留13组地层，各组地层沉积类型及特征见表10-1。可按照火山岩是否发育划分为碎屑岩沉积和火山-碎屑岩沉积。

一、碎屑岩沉积

碎屑岩沉积可分为内陆湖盆河流沉积和湖泊沉积，以河流沉积为主，按照河流形式，可划分为曲流河沉积和辫状河沉积两小类。

（一）河流沉积

1. 曲流河沉积

发育于杏石口组、龙门组、坨里组。包括河床滞留、边滩及河漫亚相沉积。由砾岩、含砾粗砂岩、中粒砂岩-粉砂岩基本沉积韵律。砾岩、砂岩的岩石组成以岩屑、长石居多，砂岩局部可见大型板状交错层理，泥岩、粉砂岩局部发育水平层理、波状层理。沉积层序中底部砾岩为河床滞留沉积，中部砂岩、砂砾岩为边滩沉积，上部粉砂岩、泥岩为河漫滩沉积，具有典型下粗上细二元沉积特征，为典型曲流河沉积。

2. 辫状河沉积

发育于窑坡组、九龙山组、坨里组和夏庄组。为湖盆边部山麓冲积、洪积扇、冲积扇上辫状河沉积。其沉积物主要岩性组合往往为砾岩、砂岩组合。砾石大小混杂，分选、磨圆一般均较差，颗粒支撑，钙质胶结，横向上往往岩性变化较大。

（二）湖泊沉积

湖泊沉积主要发育于窑坡组和九佛堂组，九龙山组、夏庄组和龙门组也有发育。

1. 窑坡组

窑坡组主要发育有滨湖三角洲亚相和浅湖亚相。

滨湖三角洲亚相可进一步分为分流河道、河口坝、天然堤、决口扇、分流间湾、沼泽、泥炭沼泽、前三角洲等次级相：①分流河道由中粗粒长石砂岩、岩屑砂岩组成，砂体呈透镜状，底部有冲刷，冲刷面上有泥砾。砂岩中常见树干化石和大型交错层理，其概率曲线呈二段式（图10-1），垂向向上过渡为间湾沉积（图10-2）；②河口坝由粉砂岩、泥岩组成，具向上变粗粒序层理及韵律层理，偶见重荷模及滑塌构造；③天然堤由粉砂岩、泥岩组成，发育有波状层理、水平层理及小型交错层理，平面上分布于分流河道两侧；④决口扇由中细粒长石砂岩、岩屑砂岩组成，中碎屑颗粒的分选、磨圆较差，泥质含量高，见粒序层理；⑤分流间湾由粉砂岩、泥岩组成，颜色深，水平层理、波状层理发育；⑥沼泽和泥炭沼泽主要发育于废弃三角洲朵叶上，其次是在天然堤和分流间湾局部地方，由粉砂岩、泥岩及煤层组成，沉积物呈黑色，含丰富的植物化石；⑦前三角洲主要由泥岩组成，水平层理发育。

浅湖亚相以粉砂岩、泥岩为主，有时夹细砂岩透镜体。由于受波浪作用影响，波状层理、脉状层理及透镜状层理发育，另有对称波痕、水平层理、小型交错层理、菱铁矿结核及双壳类化石。

表 10-1 中生代地层沉积类型及特征

Table 10-1 Stratigraphic sedimentary types and characteristics of Mesozoic

地层单位				岩相	岩性	备注
界	系	统	组			
新生界	古近系	始新统	长辛店组			
中生界	白垩系	下统	夏庄组	河流相	杂色泥、砂岩夹砾岩、泥灰岩	辫状河
				湖相（主）		早期滨湖三角洲
						晚期浅湖-半深湖
			坨里组	洪积扇相	杂色砂泥岩、砾岩，岩屑砂岩、粉砂岩、砾岩	晚期辫状河
				河流相	紫色砂砾岩	早期曲流河
			九佛堂组	湖相	杂色粉砂质页岩、页岩、粉砂岩夹泥灰岩	浅湖相
			东狼沟组	火山-沉积岩相	凝灰岩、角砾凝灰岩、沉凝灰岩、凝灰质砂岩、砾岩夹火山熔岩	
			张家口组	火山-沉积岩相、火山喷发相	酸性熔结凝灰岩，中、酸性火山岩夹砂砾岩、含砾粗砂岩	
	侏罗系	上统	土城子组	火山-沉积岩相	紫红色粉砂质页岩、粉砂岩，杂色复成分砾岩夹砂岩，灰绿色砂岩、凝灰岩夹砾岩	辫状河
			髫髻山组	火山-沉积岩相、火山喷发相	中性火山熔岩、角砾凝灰岩夹凝灰质砂岩、凝灰质砾岩	河流相、湖泊相
		中统	九龙山组	河流相	灰紫色、灰绿色含火山质砾岩、砂岩、粉砂岩及黏土岩	辫状河
				湖泊相（主）		浅湖相
			龙门组	山间河湖相	杂色复成分砾岩、岩屑砂岩、粉砂岩、粉砂质泥岩，局部含煤线	曲流河
		下统	窑坡组	河流相	灰绿色、灰黑色砂岩 粉砂岩、泥岩、煤层	间期辫状河
				湖泊相（主）		早期滨湖三角洲，晚期浅湖相
			南大岭组	火山喷溢相、火山-沉积岩相	中基性火山岩夹少量凝灰质砂岩、页岩	
	三叠系		杏石口组	山间河流相	灰色、灰黑色砾岩、砂岩、粉砂岩、黏土岩，局部夹煤线	曲流河
古生界	二叠系		双泉组	山间河湖相	岩屑砂岩、粉砂岩及页岩夹砾岩	

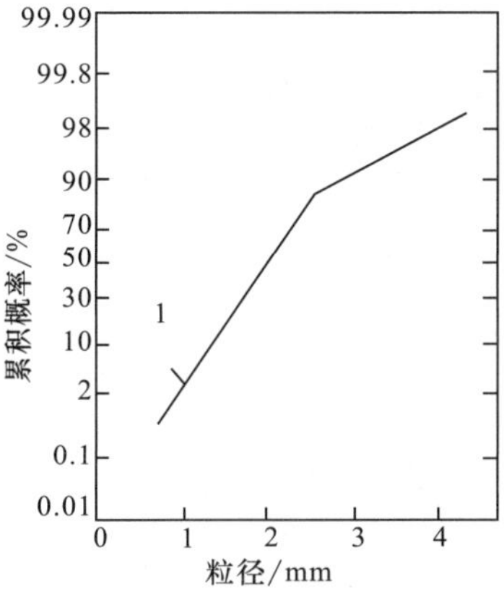

图 10-1 京西潘涧沟窑坡组砂体概率累积曲线
（据杨荣丰等，1992）
Fig. 10-1 Sand body probability cumulative curve of Yaopo formation, Panjiangou, western Beijing

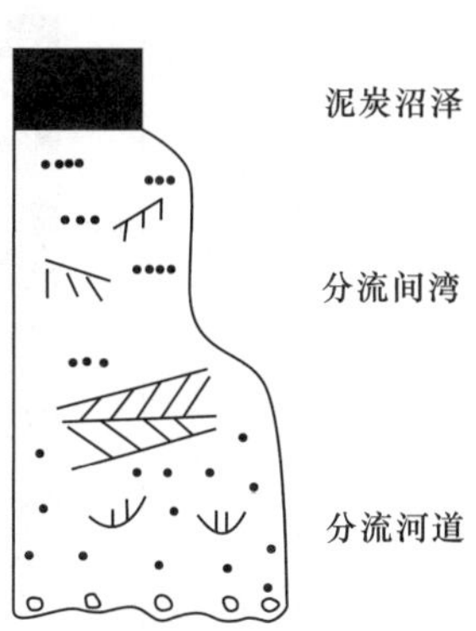

图 10-2 京西潘涧沟剖面向上变细层序
（据杨荣丰等，1992）
Fig. 10-2 Fining-upward sequence of Panjiangou section, western Beijing

2. 龙门组

龙门组砾岩上部的粉砂岩、泥岩细碎屑岩，发育水平层理，且可见菱铁质结核，同窑坡组一致，为浅湖沉积。

3. 九龙山组

九龙山组湖泊沉积岩性组合以紫红色、灰绿色夹灰白色岩屑砂岩、凝灰质细砂岩、粉砂岩及粉砂质泥岩为主，夹砾岩、泥灰岩。沉积韵律变化复杂，旋回结构清楚，表现为河流-湖泊交替沉积特征。其粗颗粒具交错层理，底面一般为冲刷面，为河流相沉积。细碎屑岩颗粒占主体，发育水平纹层，微斜层理，且含大量泥质（泥灰岩）结核，为滨湖-浅湖沉积。

4. 九佛堂组

九佛堂组湖泊沉积岩性组合为杂色钙质（或粉砂质）页岩及粉砂岩，夹油页岩、泥灰岩、砂岩和砾岩。含大量有机质，呈薄层状，具水平层理、微细水平纹层，含泥灰岩夹层及大量腹足类、双壳类、叶肢介、介形虫、昆虫、鱼类及孢粉化石，为浅湖相沉积。

5. 夏庄组

夏庄组湖相沉积发育，主要发育为滨湖三角洲亚相、滨湖亚相和浅湖-半深湖亚相。滨湖三角洲亚相和滨湖亚相为早期沉积，由复成分砾岩-含砾粗砂岩-砂岩-粉砂岩-泥岩和含砾砂岩-粉砂岩-泥岩组成两个旋回性基本层序；浅湖-半深湖亚相发育于晚期，岩性组合以砂岩、粉砂岩、泥岩为主，极少见复成分砾岩，一般由砂岩-粉砂岩-泥岩或泥灰岩组成旋回性基本层序，平行层理和单斜层理发育，且含丰富的植物化石碎片和碳质碎屑。

二、火山-碎屑岩沉积

该类沉积主要发育于土城子组和张家口组，南大岭组、髫髻山组及东狼沟组火山间歇期也有少许发育。岩性组合为复成分砾岩、砂砾岩、凝灰岩、凝灰质砂岩组合。砾岩、砾砂岩粗碎屑颗粒中砾石成分复杂，大小混杂，以火山岩砾石为主，分选极差，无层理，为冲洪积相或火山斜坡泥流相沉积。凝灰质砂岩、凝灰岩等细碎屑岩颗粒多见于粗颗粒之上。碎屑成分复杂，以外来碎屑为主，火山物质较少，分选性和磨圆度较好，岩层层理发育，韵律结构清晰，说明碎屑颗粒经过一定距离的搬运，为河流相沉积。河流通常发育为辫状河，沉积微相可细分为河道、心滩、河间湖泊和溢岸沉积（图 10-3）（郭佳，2011）。

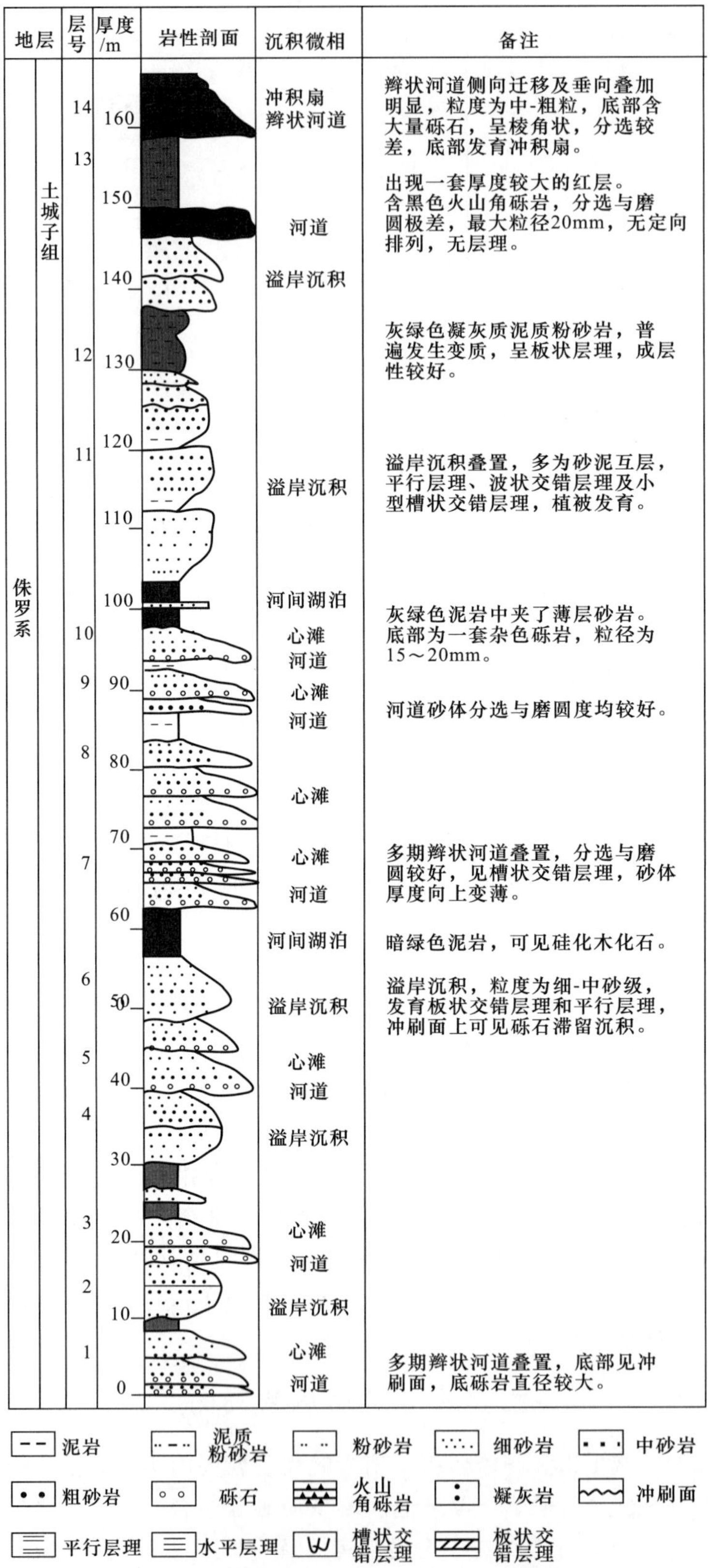

图 10-3　北京延庆地区侏罗系土城子组辫状河沉积序列

（据郭佳，2011）

Fig. 10-3　Braided river sedimentary sequence in Jurassic of Yanqing，Beijing

第二节　主要沉积岩岩石类型

一、碳酸盐岩类

仅发育极少量石灰岩类，岩性为泥灰岩和粉砂质粉晶灰岩、泥质粉晶灰岩。

（一）泥灰岩

见于夏庄组，同杂色页岩互层产出。岩石呈灰白色、灰黄色，薄层状，泥质隐晶结构，层状构造，由细晶方解石组成，含质点状黏土矿物，岩石致密坚硬，单层厚度一般为5~10cm，颜色均一，见有平行层理，含生物化石，且具生物扰动构造，为淡水湖泊沉积物。

（二）粉砂质粉晶灰岩与泥质粉晶灰岩

见于土城子组。粉砂质粉晶灰岩呈紫灰色，粉晶结构，层状构造，发育不对称波痕及缓角度斜层理。泥质粉晶灰岩呈灰绿色，夹于绿色含钙质粉砂岩中含植物化石碎片，见干裂构造，为淡水浅湖相沉积产物。

二、碎屑岩类

（一）砾岩类

1. 复成分砾岩

广泛发育于杏石口组、龙门组、九龙山组、髫髻山组、土城子组、张家口组、东狼沟组、坨里组和夏庄组中。岩石多位于各组底层或层段底部，为地层或层内段的标志层，或为沉积序列的底部，向上同砂砾岩、含砾砂岩、粗砂岩、长石砂岩、杂砂岩等构成韵律性沉积。岩石颜色较杂，多数为厚层—巨厚层状，少数中层状。往往具有较大单向斜层理及交错层理，区域上厚度变化大，数米到几十米（如龙门组厚达20m，髫髻山组厚达46m），均为河流相河道沉积。

杏石口组砾岩呈灰色、浅灰色，中细砾厚层状，粒径1~3cm，成分以硅质岩和粉砂岩为主，少量中酸性火山岩类，充填物为巨粒石英砂和岩屑等，次棱角状-次圆状。

龙门组砾岩呈巨厚层状杂色，砾石含量达80%，成分较复杂，以石英砂岩、粉砂岩、硅质岩为主，其次为灰岩、页岩、斑岩、玄武岩等，分选不好，磨圆度中等，为次棱角状—次圆状，杂乱沉积，内部无层理构造，基质为砂质，胶结物中含有钙质、铁质，胶结程度中等。

九龙山组砾岩砾石成分复杂，有石英岩、石英砂岩、燧石、花岗岩、粉砂岩等，磨圆度较好，而分选性较差，一般砾径2~3cm，最大约10cm。砾石含量高达60%，基质为砂质，胶结物为铁质和硅质。

髫髻山组砾岩多分布于二段、四段底部，砾石含量一般为50%~90%，呈次棱角状—次圆状，一般粒径0.2~1cm，大者可达30cm。组成成分复杂，同其下伏岩性密切相关，具有正粒序变化特征。

土城子组砾岩特征基本同髫髻山组。

东狼沟组砾岩呈紫灰色、土灰色、黄灰色，砾石成分同其下伏岩性密切有关，一般砾径5~6cm，大者可达10cm，次棱角状，少量次圆状，分选性差，砾石总量一般<10%，填隙物为凝灰质、泥砂质。砾岩间夹有凝灰质细砂岩和细砾岩薄层，具有碎屑物经水流搬运作用沉积特征。

坨里组砾岩呈灰色、灰紫色，砾石含量达70%~80%，粗砾成分以火山岩为主，包括安山岩、凝灰岩、细晶岩、流纹岩等砾石，还见有部分灰岩、白云岩、砂岩、粉砂岩等。砾石分选差，磨圆度较好，为次棱角状—次圆状，杂基为砂质、泥质，胶结物中含钙质，胶结程度较低。

夏庄组砾岩仅出现在底部，厚层-巨厚层状，层理极不发育，砾石无分选、无定向，含量均在80%以上，分选不好，粒径0.2~0.5cm，磨圆度中等，为次棱角状-次圆状。砾石成分复杂，火山岩砾石为主，少量砂岩、粉砂岩等碎屑岩。接触式胶结，砾屑支撑，杂基为砂质、泥质，成熟度较差。砾石层局部呈透镜状，上层面常具有冲刷面构造。

2. 砂砾岩

广泛发育于中生代各组地层中，砂砾状结构，层状构造，局部见有斜层理，多为细砾岩。砾石成分复杂，包括玄武岩、安山岩等火山岩，霏细斑岩、花岗岩等侵入岩，以及白云岩、粉砂岩等沉积岩，砾石大小0.2~30cm，以0.5~5cm居多，砾石呈次圆状，含量、分选性在各组中差异较大，含量5%~80%，多分选较差。填隙物为砂质，砂粒主要是0.2~0.005cm的火山灰和岩屑、矿物颗粒，碎屑颗粒分选性一般较差。胶结物多数为钙质，含少量泥质、凝灰质。

（二）砂岩类

1. 含砾砂岩与砾质砂岩

发育于窑坡组、龙门组、九龙山组、土城子组、张家口组、东狼沟组、坨里组和夏庄组。呈杂色，中厚层状，部分岩层中发育斜层理及交错层理。砾石成分有火山岩、侵入岩、沉积岩等，含量5%~30%，磨圆及分选均较好，砾径0.2~5cm，多为1cm。砂质组成粒径差异很大，按照砂屑颗粒大小，可分为含砾粗砂岩、含砾细砂岩、含砾中砂岩、含砾粉砂岩等各种类型。砂粒组成多数为岩屑，成熟度低，石英、长石等矿物含量少，总体体现近源河流沉积特征。岩石胶结物以凝灰质、钙质为主，含少量泥质和铁质。

2. 岩屑砂岩

发育于杏石口组、南大岭组、龙门组、九龙山组、土城子组、东狼沟组。岩石多呈灰色、褐黄色、浅紫灰色，常与长石砂岩、长石石英砂岩等共生，或夹于火山熔岩层中。岩石具砂状结构，孔隙式至接触式胶结，块状—层状构造。砂屑颗粒以岩屑为主（30%~50%），少量为石英、斜长石、云母等矿物碎屑。岩屑成分为安山岩、粗面岩、玄武岩、硅质岩、灰岩、砂岩、黏土岩等。胶结物多为钙质、泥质。

3. 长石砂岩

主要发育于九龙山组、土城子组和九佛堂组。岩石多呈灰色，常与岩屑砂岩共生。砂状结构，块状—层状构造均有发育。碎屑有石英（40%~55%）、长石（30%~55%），含少量硅质岩、黏土岩岩屑及白云母等。碎屑磨圆及分选均较差，其中岩屑磨圆度略高，可呈次圆状，其余颗粒为棱角状—次棱角状。不同粒级的砂岩均有发育，粒径0.06~2mm。胶结物为硅质和钙质，杂基为细碎的长英质、黏土矿物、绢云母等。

4. 长石石英砂岩

主要发育于杏石口组、九龙山组、土城子组和九佛堂组。岩石多呈灰白色、灰色。碎屑以石英（50%~75%）为主，长石（15%~25%）次之，含少量硅质岩、玉髓、黏土岩等岩屑。砂屑弱磨圆，以次圆状—次棱角为主，大小各级粒径均有发育，粒径0.06~1.5mm。胶结物以硅质为主，少许为钙质。杂基多为黏土矿物和与碎屑同成分的细碎屑物质。

5. 粉砂岩

广泛发育于中生代各组地层中，受沉积时古气候、沉积物源影响，颜色较杂，多呈灰色、灰紫色、灰黄色、土黄色、灰黑色、黑色等。通常与页岩、页岩或砂岩呈互层产出。多数粉砂岩中可见生物化石或化石碎片，发育水平层理、斜层理，有些还可见波痕构造。九龙山组粉砂岩中还可见泥灰岩结核。按照碎屑物质组成可分为泥质粉砂岩、钙质粉砂岩、石英粉砂岩、碳质粉砂岩四类。

（1）石英粉砂岩

发育最为广泛，南大岭组、窑坡组、龙门组、九龙山组、土城子组、张家口组、九佛堂组、夏庄组均有出露。具粉砂结构，基底式胶结，层状构造居多，少块状构造。碎屑物有石英，一般含量>70%，其次为长石和极少云母。石英多呈棱角状-次棱角状，少数呈次圆状，粒径0.01~0.06mm。填隙物由硅质、泥质和铁质、钙质等胶结物构成。

（2）泥质粉砂岩

主要发育于九佛堂组、坨里组和夏庄组，其他各组少有发育。岩石粉砂结构、基底式胶结，块状、层状构造。岩石中泥质填隙物含量较高，多为20%～35%。粉砂碎屑以石英（55%～75%）为主，长石（5%～10%）等次之，多呈棱角状-次棱角状。碎屑粒径大多为0.02～0.06mm，少数可达0.25mm。

（3）碳质粉砂岩

主要发育于窑坡组和九佛堂组。粉砂结构，基底式胶结，层状构造。碎屑有石英（55%～60%）、长石（10%～15%）和极少数黑云母（<5%）。石英呈棱角状-次棱角状，粒径0.03～0.06mm。胶结物为碳质和泥质。

（4）钙质粉砂岩

仅发育于九佛堂组。基本特征同碳质页岩，仅胶结物为钙质组分。

（三）黏土岩类

1. 泥岩与粉砂质泥岩

主要发育于南大岭组、九佛堂组和夏庄组，土城子组，九龙山组、杏石口组也有少许发育。岩石呈灰色、灰绿色、灰黄色、黑色、灰黑色等。泥质结构，层理与页理构造，发育水平层理、平行层理，有些还发育干裂、雨痕构造。岩石多与粉砂岩、泥岩、泥质粉砂岩和泥灰岩等共生。同粉砂共生，泥质物和粉砂分别呈条带富集，构成纹层状构造。岩石泥质含量一般超过70%，粒径<0.003mm。九龙山组页岩中往往还含有凝灰物质，夏庄组、杏石口组和土城子组碳质含量较高，局部发育煤线。

2. 页岩与粉砂质页岩

主要发育于南大岭组、九佛堂组和夏庄组，九龙山组也有少许发育。岩石呈灰色、灰绿色、灰黄色、黑色、灰黑色等。泥质结构，层状与块状构造，发育水平层理、平行层理。多与页岩、粉砂岩、泥质粉砂岩共生。岩石泥质含量一般超过70%，粒径<0.003mm，部分岩石含少许石英粉砂。九佛堂组黑色页岩中含有大量有机质，微细水平纹层发育，富含狼鳍鱼等生物化石，为半深湖-深湖相沉积。夏庄组灰色-灰黑色页岩由黏土矿物及有机质组成，微细水平纹层发育，含生物化石，且常含石膏夹层。

三、火山碎屑沉积岩类

北京地区中生代火山活动频繁，各种火山碎屑岩类均有发育。本节仅对同沉积作用密切相关的火山碎屑沉积岩和沉火山碎屑岩进行总结。火山碎屑岩类、火山碎屑熔岩类和熔结火山碎屑岩类在火山岩与火山作用章节叙述。

（一）火山碎屑沉积岩类

1. 凝灰质砾岩与砂砾岩

主要发育于髫髻山组，九龙山组、东狼沟组、张家口组也有少许发育，为陆间河流相沉积。岩石以灰色、灰紫色为主，凝灰砾状结构或凝灰砂砾状结构，厚层状构造。砾径0.2～30cm，多0.5～5cm，组成成分复杂，包括玄武岩、安山岩、粗面岩、霏细斑岩、白云岩、粉砂岩等，外来碎屑所占比例较大，多次圆状-次棱角状，含量40%～60%。胶结物由与砾石同成分的砂级、粉砂级凝灰物质及少量长石、石英晶屑组成，粒度0.05～1mm，岩石胶结物松散，分选性差，层理构造一般不发育。

2. 凝灰质砂岩

主要发育于南大岭组、九龙山组、土城子组、张家口组和东狼沟组。岩石常同凝灰质砂砾岩、凝灰质砾岩共生，构成上细下粗的二元沉积特征。颜色上一般呈紫红色、灰紫色，砂状结构，层状构造，砂粒以粗砂岩为主，分选较差，砂粒组成复杂，火山岩、沉积岩、侵入岩岩屑和石英、长石、云母等矿物碎屑均有，于各组中特征略有所差异。南大岭组为凝灰质含长石石英砂岩（石英碎屑40%～50%，长石10%，其余则为火山灰胶结物）；九龙山组底部发育凝灰质复成分粗砂岩；土城子组多发育含砾砂岩；张家口组发育凝灰质长石砂岩；东狼沟组多发育凝灰质岩屑砂岩或含砾岩屑砂岩。胶结物为火山灰，其次为泥质，接触式胶结。

3. 凝灰质粉砂岩

主要发育于窑坡组、九龙山组、土城子组、髫髻山组、张家口组、东狼沟组。颜色因含有火山物质而以色彩艳丽的灰绿色、灰紫色为主。岩石粉砂结构，薄层-厚层状、块状均有，碎屑组成主要为石英、长石、云母等矿物碎屑和碳酸盐岩类碎屑。碳酸盐岩类，次棱角状-次圆状，粒度约 0.5mm，含量约 5%；晶屑石英，尖棱角状-次棱角状，粒度 0.01~0.2mm，含量 8%；晶屑斜长石，棱角状-次棱角状，聚片双晶，粒度 0.01~0.1mm，含量 5%；晶屑钾长石，棱角状-次棱角状，泥化，粒度 0.01~0.15mm，含量约 3%。胶结物主要为泥质、碳酸盐，少许铁质和细小鳞片状绢云母组成，粒度 0.001~0.01mm，含量达 80%。还包括少量他形粒状磁铁矿、粒状-短柱状磷灰石、他形粒状榍石副矿物，大小 0.01~0.2mm，含量<1%。

4. 凝灰质页岩

仅分布在九龙山组，泥质、凝灰结构，页片状构造。主要由火山灰和泥质组成。一般同凝灰岩、凝灰质粉砂岩共生，为互相沉积产物。

（二）沉火山碎屑岩类

1. 沉火山角砾岩

发育于东狼沟组、髫髻山组。岩石紫色、灰紫色，角砾成分为气孔状玄武岩及致密块状安山岩，占 50%~60%，凝灰岩占 10%，流纹岩占 5%~10%，凝灰质砂岩占 10%，燧石岩、白云岩等占 5%~10%。角砾大者 10~20cm，小者 0.2~0.5cm，一般 5~7cm。砾石呈次棱角状-棱角状，分选性差，砾石总量占 50%~70%。填隙物为泥沙质、凝灰质，基底式胶结，固结程度较低，为火山喷发物及陆屑经水流搬运堆积的产物。

2. 沉角砾凝灰岩

发育于东狼沟组、髫髻山组，呈紫色夹层状分布于沉火山角砾岩中。角砾成分为玄武岩、安山岩、粗面岩，次棱角状，粒径 2~8mm，含量约 30%。凝灰质成分同角砾成分一致，含量约 60%，颗粒支撑，压紧式胶结，为火山喷发物及陆屑经水流作用而成。

3. 沉凝灰岩

发育于土城子组、九龙山组。岩石为灰白色、浅灰色、灰绿色、灰紫色，沉凝灰结构，层状-块状构造，火山碎屑物由<2mm 的凝灰物质及陆源碎屑物质构成，火山碎屑含量 25%~75%，碎屑成分为晶屑。火山尘灰物呈球粒状、长条状，具有塑性特征，含量可达 50%~80%。胶结物为碳酸盐矿物和铁质物，主要是细小方解石。有些岩石中还可见一些火山熔岩角砾，指示由火山作用和沉积作用综合形成。

第三节　沉积盆地及沉积建造

一、沉积盆地

中生代，华北板块内发育了一系列大小不一、活动强度各异的陆相盆地，各类岩石地层均堆积形成于这些盆地中。由于中生代燕山运动发育多期次伸展-挤压转换构造，沉积盆地明显受构造运动的制约，可把盆地划分为三个世代。每个世代由燕山构造旋回三幕次控制，每个构造幕次先伸展后挤压，形成构造体制交替演化过程。因此，每个世代分别形成两种类型盆地：早期为伸展机制下的裂陷（或断陷、坳陷）盆地，晚期则演化为板内造山挤压机制下（继承）挠曲盆地，并最终由于挤压造山使得盆地消失。各个世代的盆地在空间上有一定的分布规律，而且盆地的堆积物随着盆地的生成发展以一定的序列有规律地堆积产出。

（一）裂陷盆地（板内造山幕早期伸展性盆地）

该类盆地发育于燕山运动各幕次早期，燕山运动三幕使裂陷盆地的形成分为三个阶段。期间，盆地总体处于欠补偿沉积状态，具体分述如下：

1. 晚三叠世—早侏罗世早期

相当于杏石口组—窑坡组沉积时期，华北克拉通由北向南（由边部向内部）活化影响波及本区，区域上印支旋回结束进入燕山旋回，构造应力由挤压转变为伸展（伸展机制有认为受中亚造山后伸展影响，也有观念认为与燕山板内前造山伸展影响，本志比较认同总体构造环境表现为次稳定的板内造山前伸展环境）。伸展方向北东东向，发育伸展型坳陷盆地。

本区残存的两个小型山间盆地，分别为西山火山-沉积盆地和昌怀火山-沉积盆地。鲍亦冈等（2001）认为盆地可能受密云-墙子路断裂控制，表现为断陷盆地，但从残存的地层分布特征上看，盆地形成时期基本上表现为整体均匀下沉，各时代地层之间无较大规模的超覆，沉积中心亦无明显迁移现象，可能属坳陷型火山-沉积盆地。由于受后期构造强烈破坏，原始沉积盆地面貌难以恢复，基于盆地周缘地层分布推测上述盆地最初可能为同一盆地。

本区该期火山-沉积盆地沉积速度快但相变不大，总沉积厚度大，邓晋福等（2004）、徐德斌等（2007）总结为类磨拉石沉积特征，其碎屑沉积颗粒相对较细以区别于磨拉石碎屑沉积。在盆地发展的早期阶段（杏石口期-南大岭期），由于强烈的差异性升降作用，在底部往往堆积了山麓粗碎屑洪冲积物，伴随断裂活动导致火山喷发，堆积了大量火山岩；晚期阶段（窑坡期）盆地扩展稳定沉降，形成灰黑色细碎屑内陆河湖相建造，条件适宜时形成煤层或油页岩。

2. 中侏罗世晚期—晚侏罗世早期

相当于髫髻山组和土城子组沉积初期。燕山构造旋回发生第二次构造体制交替，构造背景由挤压再次转变为伸展环境。该期裂陷盆地伸展方向为北东向，使得盆地宏观展布方向 NE50°～60°，受同方向居庸关深断裂、南口-沿河城深断裂及颐和园深断裂（推测）控制。

根据盆地的展布情况，可分为两个带。南带西南起自百花山，往东北方向经髫髻山、妙峰山、十三陵、怀柔山前延至密云新城子一带，分布于南口-沿河城断裂和颐和园断裂之间宽约 20km 的地区，依次发育有百花山、髫髻山、新城子 3 个火山-沉积盆地。东北端的新城子盆地因受到东西向赤城-古北口深断裂的制约，向东西方向偏转延伸。北带处于居庸关深断裂的西北侧，发育有燕羽山、凤驼梁-四海盆地 2 个火山-沉积盆地，总延展方向与南带平行。

3. 早白垩世早期

相当于张家口组晚期、东狼沟组和九佛堂组沉积时期。该期盆地受控于燕山晚期发育的一组截切性极强的区域性 NNE 向深断裂。西界为紫荆关深断裂带，东界为高丽营深断裂带。在这两条深断裂之间发育盆地群，裂陷作用使得盆地总体呈 NE30°方向展布。

该期盆地群南起房山坨里和门头沟区杜家庄一带，北到怀柔七道河地区，两端向外均延出本区，发育有坨里、杜家庄-镇厂、东岭、怀柔、上大水、岔道、七道河等 7 个火山-沉积盆地。该世代断陷盆地均为新生性上叠式火山-沉积盆地。整体看，盆地呈 NNE 向延展，但带内每一个火山盆地的发育和生成则受其具体部位特定盖层断裂构造的控制，盆地的延长方向各异，这是不同于前期盆地的一个重要差异。

（二）板内造山挤压型（继承）挠曲盆地

板内造山挤压型（继承）挠曲盆地发育于燕山运动各幕次后期阶段，为燕山运动挤压背景环境下形成的盆地。盆地性质虽然发生转变，但盆地沉积却基于燕山运动各幕次早期断陷盆地演化，故既有新生性特征，又具有一定继承性特征。

1. 中侏罗世早—中期

相当于龙门组和九龙山组沉积时期。长期孕育的构造变动（燕山运动 A 幕）开始强烈爆发，区内构造环境发生构造反转。在此构造背景下，形成区内第一世代板内造山挤压型（继承）挠曲型盆地。

本区该期盆地仅发育于西山和下花园—延庆一带，分别继承并演化早期杏石口期-窑坡期西山和下花园-延庆断陷盆地演化。继承性表现在总体地层分布上，杏石口组—九龙山组空间上形影相随，一起分布在现今轴向 NE65°~70°的一系列向斜构造内。特征表现为：①盆地内龙门组、九龙山组沉积厚度分异明显，由边部向盆内沉积厚度急剧变薄，并形成两大沉积中心（妙峰山和北岭）；②构造反转为北东东向挤压使得早期近东西向展布的裂陷盆地沉积中心在区域上发生迁移，盆地水流向北西方向超覆，九龙山组在盆地北西部直接超覆于杏石口组等更老地层之上；③上覆的地层同九龙山组角度不整合接触。

2. 晚侏罗世

相当于土城子组中上部沉积时期，盆地发育基本特征同中侏罗世早—中期。该期燕山运动时间短但强度大，形成区域上的最大角度不整合，为燕山期主幕。

该期发育有髫髻山盆地、新城子盆地、后城盆地、千家店盆地和凤驼梁-四海盆地 5 个火山-碎屑岩盆地。髫髻山盆地和新城子盆地为髫髻山期南带继承型盆地，凤驼梁-四海盆地为髫髻山期北带继承型盆地，而后城盆地、千家店盆地为髫髻山期北带新生型盆地。

3. 早白垩世晚期

相当于坨里组、夏庄组沉积时期，盆地发育基本特征同中侏罗世早—中期、晚侏罗世。该期区域上盆地数量少，规模均较小，零星分布，盆地为燕山期向喜马拉雅期过渡过程中相对平静构造背景下，早期断陷盆地残留在山间局地的小盆地，本区仅发育一个坨里盆地。

二、沉积建造

根据中生代盆地演化，结合各组地层沉积时气候环境类型，将中生代沉积建造划分为陆相碎屑岩建造和陆相火山-沉积岩建造两大类、六小类（表 10-2）。

（一）陆相碎屑岩建造

1. 陆相类磨拉石建造

为一套内陆干旱-半干旱气候条件下的山麓河流冲洪积扇相类磨拉石建造，发育于杏石口组，岩性以灰、灰紫色、紫灰色、紫红色复成分砾岩为主，间夹灰绿色砂岩透镜体。

2. 陆相含煤建造

为一套内陆温湿气候条件下的河湖相含煤碎屑岩组合，发育于窑坡组，主要为灰绿色-灰黑色粗砂岩、细砂岩、粉砂岩、页岩及煤层组合，水平层理发育，含丰富植物化石。

3. 陆相河流相复成分砾岩建造

为一套内陆干旱—半干旱气候环境下形成的巨厚河流相或河湖相沉积，发育于龙门组、坨里组，土城子组也有少产出。主要由黄褐色、灰白色、浅灰色、紫红色砾岩、砂砾岩及砂岩等组合，斜层理很发育，砾石成分复杂，分选性差。

4. 陆相湖泊相建造

为一套内陆干旱—半干旱气候环境下形成的湖泊相沉积，主要发育于九佛堂组、夏庄组，九龙山组也有产出。主要为杂色泥、砂岩组合，水平层理发育。

（二）陆相火山-沉积岩建造

1. 陆相碎屑岩夹火山岩建造

为一套河湖相沉积岩系夹少量火山岩系，发育于九龙山组、土城子组中。主要由灰绿色、灰白色、灰色砾岩、砂岩、粉砂岩、页岩、泥岩夹少量中基性火山岩、火山碎屑岩组成，斜层理、平行层理发育，生物化石丰富。

2. 陆相沉积岩-火山岩建造

为中基性火山岩与中酸性火山岩、火山碎屑岩及火山喷发间歇期的沉积岩组合，发育于南大岭组、髫髻山组、张家口组中。沉积岩系主要由黄绿色、黄褐色、紫红色、灰褐色砾岩、砂岩、粉砂岩、泥岩及页岩等组成。

表 10-2 中生代各时期沉积建造类型综合特征

Table 10-2 Comprehensive characteristics of sedimentary formation types of various stages

时期	沉积建造系列	盆地类型	建造类型	沉积环境	物源	沉积速率、沉降幅度	共生岩石主要类型	典型分布地区	构造域	备注
夏庄期	次稳定	挤压型挠曲盆地	陆相河流湖泊碎屑岩相	河流、湖泊沉积	陆源	较快	杂色泥、砂岩夹砾岩、泥灰岩		太平洋构造域（北北东向）	
坨里期	次稳定	伸展向挤压过渡型盆地	陆相河流湖泊碎屑岩相	河流、湖泊沉积	陆源	较快	紫色砂砾岩、杂色砂泥岩、砾岩，岩屑砂岩、粉砂岩、砾岩			
九佛堂期	稳定	伸展型断陷盆地	陆相湖泊碎屑岩相	湖泊沉积	陆源	慢	杂色粉砂质页岩、页岩、粉砂岩夹泥灰岩			
东狼沟期	不稳定	伸展型沉积岩-火山岩盆地	陆相火山-碎屑岩建造	火山间歇河流冲积砾岩沉积	陆源	较快	凝灰岩、角砾凝灰岩、沉凝灰岩、凝灰质砂岩、砾岩夹火山熔岩			
张家口期	不稳定	挤压向伸展过渡性盆地（以伸展为主）	陆相火山-碎屑岩建造	火山间歇河流冲积砾岩沉积	陆源	较慢	紫红色粉砂质页岩、粉砂岩，杂色复成分砾岩夹砂岩，灰绿色砂岩、凝灰岩夹砾岩			
土城子期	不稳定	伸展向挤压过渡性盆地（以挤压为主?）	陆相火山-碎屑岩建造	火山间歇河流冲积砾岩沉积	陆源	快	中性火山熔岩、角砾凝灰岩夹凝灰质砂岩、凝灰质砾岩		古太平洋构造域（北东向）	早期基性火山岩等特征揭示伸展环境，考虑和髫髻山组地层分布具有迁移特征，可能以挤压为主更合适
髫髻山期	不稳定	拉伸型火山-碎屑盆地	陆相沉积岩-火山岩建造	火山间歇河流冲积砾岩沉积	陆源	较慢	酸性熔结凝灰岩，中、酸性火山岩夹砂砾岩、含砾粗砂岩			

续表

时期	沉积建造系列	盆地类型	建造类型	沉积环境	物源	沉积速率、沉降幅度	共生岩石主要类型	典型分布地区	构造域	备注
九龙山期	不稳定	挤压型挠曲盆地	陆相河流湖泊碎屑岩相	河流、湖泊沉积	陆源	较快	砾岩、砂岩、粉砂岩、泥岩		中亚构造域向古太平洋构造域转换时期（多向应力叠加）	晚期
			陆相湖泊碎屑岩相	湖泊沉积	陆源	较快	灰紫色、灰绿色含火山质砾岩、砂岩、粉砂岩及黏土岩			早期
龙门期	次稳定	伸展向挤压转变过渡性盆地	陆相河流湖泊碎屑岩相	河流、湖泊沉积	陆源	较慢	杂色复成分砾岩、岩屑砂岩、粉砂岩、粉砂质泥岩，局部含煤线			
窑坡期	次稳定	继承型山间盆地	陆相含煤建造	河流、湖泊、沼泽沉积	陆源	较快	灰绿色、灰黑色砂岩、粉砂岩、泥岩、煤层			
南大岭期	不稳定	拉伸型火山-碎屑盆地	陆相沉积岩-火山岩建造	火山间歇河流冲积砾岩沉积	陆源	很慢	砾岩、砂岩	门头沟西北部法城口-牛站-清水村	板内前造山伸展，中亚构造域影响（板内构造相对稳定期，北东东向）	上部
				火山间歇短暂湖泊砂岩沉积	陆源	很慢	凝灰质砂岩、页岩			下部
杏石口期	次稳定	继承型山间盆地	陆相碎屑岩建造	山间盆地河流湖泊沉积	陆源	较慢	灰色、灰黑色砾岩、砂岩、粉砂岩、黏土岩，局部夹煤线			北京西山一带，古地貌控制沉积
		拉伸型断陷盆地	陆相碎屑岩建造	山间盆地河流冲积扇沉积	陆源	较快	灰、紫灰色砂砾岩、砾岩、含砾粗砂岩			河北下花园一带
双泉期	次稳定	挤压型挠曲盆地	陆相碎屑岩建造	陆相河流、湖泊沉积	陆源	较快	杂色岩屑细砂岩、粉砂岩、页岩夹砾岩		中亚造山板内远程效应（北东东向）	沉降幅度产生分异

第四节　岩相古地理

本区中生代岩相古地理演化实际反映为盆地演化，可以划分为晚三叠世—中侏罗世早期阶段、中侏罗世晚期—晚侏罗世阶段和早白垩世阶段三大阶段。

一、晚三叠世—中侏罗世早期

在经历了较短时间的印支旋回后，晚三叠世本区进入燕山构造旋回。区域上构造应力场先伸展后挤压，完全进入陆内造山沉积演化阶段，沉积杏石口组-九龙山组一系列地层。

（一）杏石口期

杏石口期，本区处于拉张活动状态，大部分地区上升为剥蚀区，仅在西山和延庆西部（下花园-延庆一带）发育有小型山间裂陷盆地（图 10-4）。

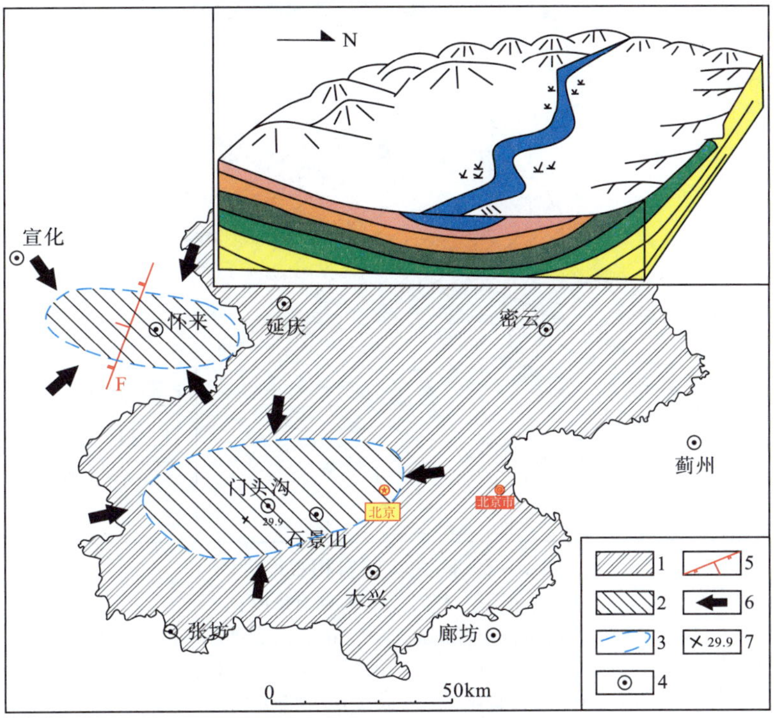

图 10-4　杏石口期岩相古地理图

（角图据赵越，2002，略有修改）

Fig. 10-4　Lithofacies paleogeographic map of Xingshikou period

1—基底剥蚀区；2—陆相山间盆地河流相沉积；3—盆地边界；4—地名；5—断裂带；6—物源方向；7—剖面位置及地层厚度（m）

本区西山盆地表现为一个北东东走向的山间内陆盆地。盆地基底为晚古生代印支运动改造后的宽缓褶皱挠曲地层（图 10-4），沉积主体发育于褶皱挠曲古地貌区。沉积物为一套灰色、灰黑色砾岩、砂岩、粉砂岩及黏土岩组合，沉积厚度不大，一般为 20～50m。底部底砾岩在八大处宝珠洞最发育，向西逐渐变薄。地层中所含砾石层数亦以东部为多，如大峪南山、冯村西山一带以砾岩、砂岩、粉砂岩组成的旋回性基本层序多而清晰，向西则泥质、碳质增高，至百花山东南的莲花庵一带泥岩增厚，并夹有煤线，植物化石亦由东向西逐渐丰富。上述特征表明杏石口期西山盆地为北东东向的开阔谷地曲流河沉积环境（图 10-4），盆地因伸展拉张而处于欠补偿沉积状态，其灰色、灰黑色沉积物及局部煤线发育揭示为潮湿环境。

（二）南大岭期

南大岭期继承了杏石口期古地理格局并进一步演化，区域上构造应力场仍处于拉张状态，沉积仅发育于西山盆地、昌怀盆地（可能为同一西山沉积盆地）和下花园-延庆盆地中，表现为小型山间坳陷盆地沉积。

西山盆地沉积表现为多期次玄武岩和玄武安山岩的溢流、喷发，并在火山活动间歇期发育近源碎屑岩河流相沉积。盆地南北界线基本同杏石口期，略有所扩展，而东西方向明显扩展，使南大岭组地层部分地区超覆于古生代不同岩石地层单位之上。盆内堆积物以灰绿色中、基性玄武岩、安山岩为主，夹少量凝灰质砂岩、页岩组合。溢流玄武岩、玄武安山岩的分布明显受古地形、地貌控制，在喷发中心大台—军庄一带形成新的盆内火山山地古地貌格局（图 10-5）。

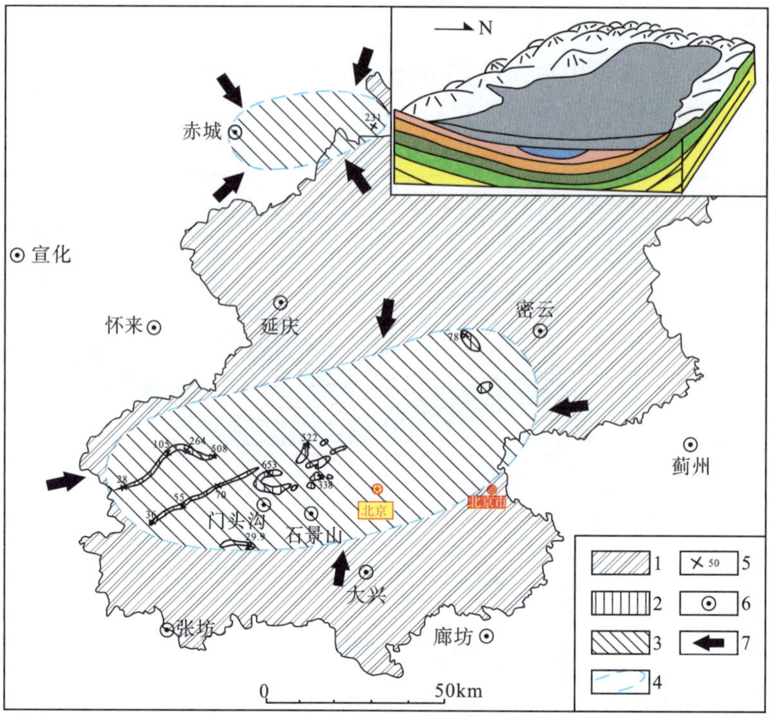

图 10-5 南大岭期岩相古地理图

（角图据赵越，2002，略有修改）

Fig. 10-5 Lithofacies paleogeographic map of Nandaling period

1—基底剥蚀区；2—残留露头地层分布；3—山间裂陷火山-碎屑相沉积；4—推测盆地边界；
5—剖面位置及地层厚度（m）；6—地名；7—物源方向

（三）窑坡期

窑坡期仍继承杏石口期、南大岭期古地理格局并进一步演化，区域上构造应力场仍处于拉张状态，沉积还仅发育于西山盆地和下花园-延庆盆地中，系小型山间裂陷盆地沉积，盆地走向北东东向（图10-6）。

窑坡期（图 10-6）是西山盆地的发展与兴旺期，在区域持续伸展拉张背景环境下快速接受沉积，沉积厚度较大，盆地内部除中心大安山—大台—军庄一带因火山山地古地貌影响厚度较薄（170～300m）外，沉积厚度差异极小，均为 500m 左右。沉积岩性为一套暗色含砾粗砂岩、砂岩、粉砂岩、泥岩及煤层组合，总体上沉积粒度较细，并含大量植物化石及淡水双壳类化石。该期盆地范围基本同南大岭期，宏观展布仍呈北东东向，古气候为温暖潮湿气候。按照盆内沉积演化可划分为早、晚两个阶段。

窑坡期早期，盆内主要发育为陆相湖盆三角洲相-沼泽相环境。窑坡期晚期，随着北京西山盆地周缘山系的缓慢抬升，以及南大岭期形成的盆地中心火山山地地貌被填平，西山盆地逐渐封闭，演化为一个盆底平缓的内陆湖盆。盆地周缘短小河流深入盆内，直通中心，而盆内中心区湖水加深，出现弱还原的沉积环境。受盆地不均匀沉降影响，盆地逐渐由南向北掀斜，沉积中心向北迁移，致使斋堂、牛角峪、法城口等地晚期沉积物直接超覆在南大岭组地层之上。而最厚沉积区由门头沟九龙山南坡大村以东至龙门、河滩一带迁移到门头沟髫髻山、杨家峪一带（厚度大于 300m）。

（四）龙门期

龙门期长期孕育的构造变动开始爆发，区域构造应力场由伸展转化为挤压，发生燕山构造旋回首次构造反转，但构造变动强度不大。正是由于区域构造环境变化，龙门期沉积盆地性质、沉积环境发生巨大变化。北京境内因挤压作用总体呈现区域性的抬升剥蚀，仅在早期两个残留盆地中接受沉积。延庆西部（下花园—延庆一带）山间盆地在此时期也因构造挤压抬升而最终消失。西山盆地仍有残留，但分布面积萎缩。

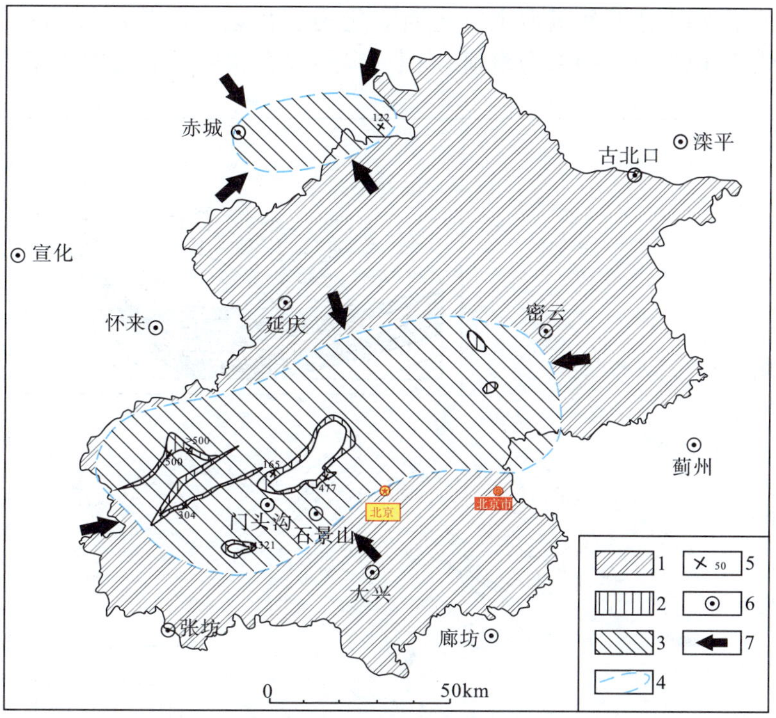

图 10-6　窑坡期岩相古地理图

Fig. 10-6　Lithofacies paleogeographic map of Yaopo period

1—基底剥蚀区；2—残留露头地层分布；3—陆相河湖相沉积；4—推测盆地边界；
5—剖面位置及地层厚度（m）；6—地名；7—物源方向

龙门期西山盆地南、北部边界略有萎缩（图 10-7），但西部边界急剧萎缩至门头沟区青龙涧至房山区秋林铺一带。东部边界因第四纪埋藏而不明确，推测也有萎缩。盆地宏观走向仍保持北东东向，但沉积中心随盆地萎缩向东迁移，由门头沟髫髻山、杨家峪一带迁移至门头沟区岳家坡一带。沉积物岩性为灰绿色（或灰黑色）复成分砾岩、岩屑砂岩、粉砂岩及粉砂质泥岩组合，显示早期为强水流河流相沉积、晚期为湖泊相沉积。沉积物厚度由沉积中心向盆地边部急剧减薄，明显有别于窑坡组盆内沉积厚度变化不大的沉积特征。其沉积中心岳家坡最大厚度约 395m，西北岭地区为 330m，至大安山附近仅厚约 75m。向北部千军台—大台一带急剧减薄至 150~160m，而至雁翅、斋堂沿河城一带仅厚约 50m。向南周口店一带也急剧减薄至约 92m。

（五）九龙山期

九龙山期古地理格局在龙门期基础上演化，表现为更强烈挤压作用。但古气候出现较大变化，气候类型演化为干旱-半干旱气候。受古气候变化和出现较多火山物质综合影响，沉积物颜色发育为灰紫色、灰绿色、紫红色及灰色，岩性为含有大量火山碎屑物质的各种不同粒级的陆相沉积碎屑岩。在沉积物宏观分布上，由盆地沉积中心向盆地边缘厚度急剧变薄，盆地呈现挤压同挠曲充填特征。盆地沉积中心位于大台—板桥一带，沉积厚度可达 1500m，往北至桑峪、牛站一带急剧减薄为约 300m，盆缘昌平十三陵一带厚度仅为 147m；往南至周口店残存约 182m（顶部遭剥蚀）。

由于北部中亚构造叠加东部古太平洋构造影响，西山盆地区域沉积过程中，盆地逐渐向北西掀斜，一方面使得沉积中心由门头沟区岳家坡向北西方向迁移至大台—板桥一带，另一方面使得盆地北西边缘青白口—田庄一带，仅出露九龙山组三段地层，且超覆于寒武纪或中—新元古代地层之上。不仅如此，九龙山期及其后强烈燕山运动 A 幕，使得该期沉积由下往上岩层内凝灰物质增多，在平面分布上则显示由南向北火山物质亦有增高趋势。

根据残留地层展布及沉积相分析，该期盆地西界范围大致在门头沟区清水村和房山区秋林铺一带，东延到怀柔范各庄附近，北至昌平十三陵地区，南到周口店一带，盆地宏观走向仍呈北东东向（图

10-8)。盆地沉积环境早期以湖相沉积为主，晚期演化为冲积扇、辫状河沉积物的快速堆积，北部、东北部多以近端扇顶相沉积为主，中部为扇顶-扇中-扇尾混合相区。向南中段扇中相、远端扇尾相沉积逐渐增多。

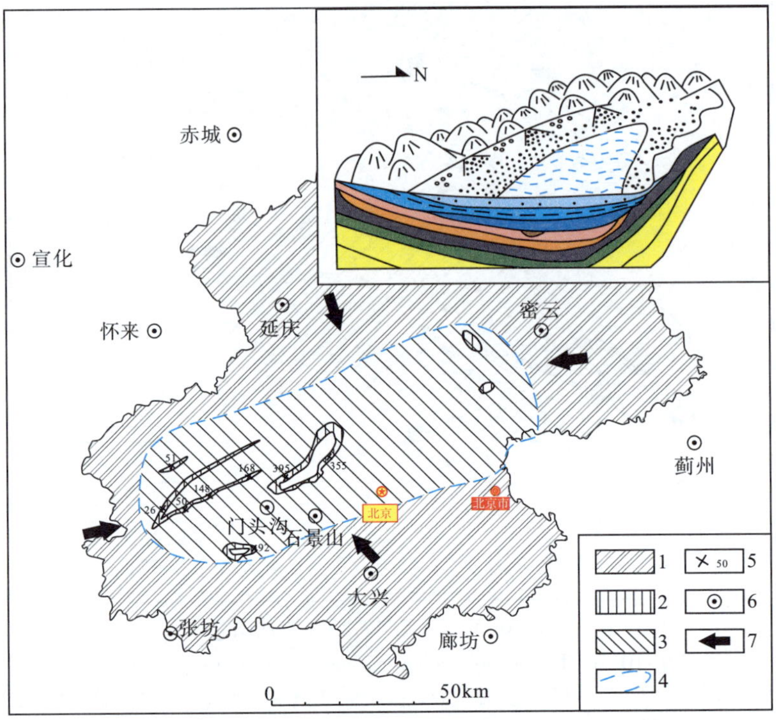

图 10-7　龙门期岩相古地理图

Fig. 10-7　Lithofacies paleogeographic map of Longmen period

1—基底剥蚀区；2—残留露头地层分布；3—陆相河湖相沉积；4—推测盆地边界；5—剖面位置及地层厚度（m）；6—地名；7—物源方向

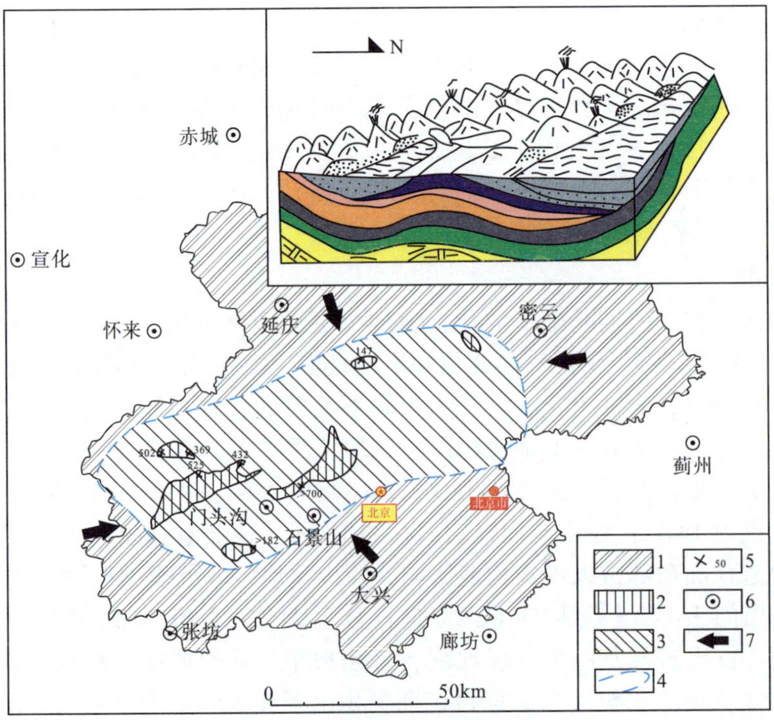

图 10-8　九龙山期岩相古地理图

Fig. 10-8　Lithofacies paleogeographic map of Jiulongshan period

1—基底剥蚀区；2—残留露头地层分布；3—陆相河湖相沉积；4—推测盆地边界；5—剖面位置及地层厚度（m）；6—地名；7—物源方向

二、中侏罗世晚期-晚侏罗世时期

（一）髫髻山期

髫髻山期本区在伸展构造背景下发育两条近平行的北东向火山-碎屑岩断陷盆地（图 10-9），共同构成髫髻山期火山喷发带。火山喷发堆积时期，于盆内发育有两次连续、强烈、大规模的火山喷发活动，形成以粗安质熔岩、正常火山碎屑岩为主的巨厚喷出岩层。这两次火山喷发活动之后均出现较长时间的火山作用间歇期，发育形成巨厚的火山-沉积碎屑岩层。

第一次火山间歇期沉积以灰紫色、紫红色凝灰质砂岩为主，岩石粒度相对较细且比较均匀，成层性好。岩层中常夹有薄层熔岩和正常火山碎屑岩，反映不稳定的小水域湖盆相沉积环境。第二次火山间歇期沉积岩石粒度相对较粗，以复成分凝灰质砾岩、沉凝灰角砾岩为主，砾石粗大，直径多在 2～20cm 之间，且成分复杂，外来物质较多，无层理、分选差，反映冲、洪积特征。密云新城子、昌平十三陵地区砾石成分以火山岩为主，磨圆度较差，属火山泥流相堆积。该期本区为干旱间夹温湿气候环境。

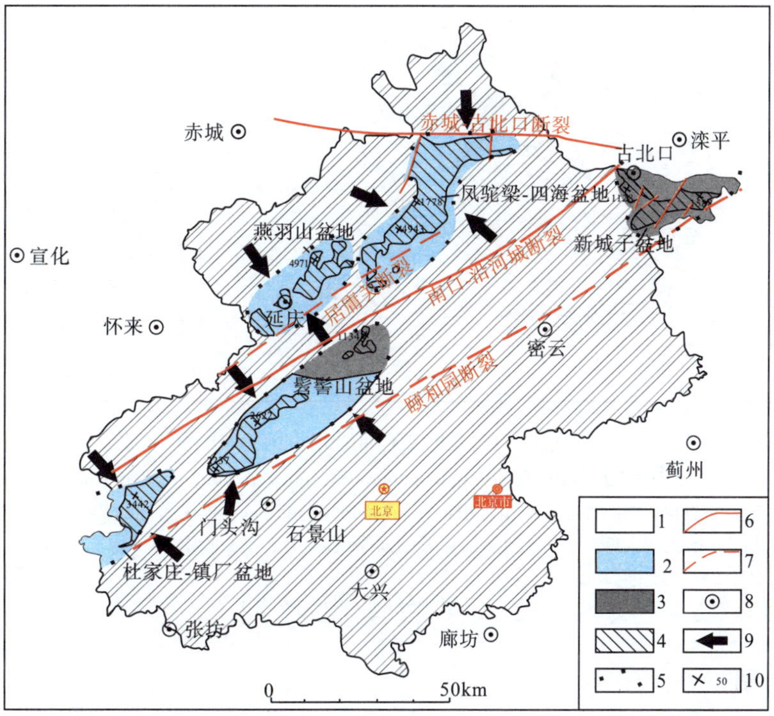

图 10-9　髫髻山期岩相古地理图

Fig. 10-9　Lithofacies paleogeographic map of Tiaojishan period

1—基底剥蚀区；2—火山+湖相+冲洪积相沉积；3—火山+湖相+火山泥流相沉积；4—残留地层分布；5—推测盆地边界；6—断裂；7—推测断裂；8—地名；9—物源方向；10—剖面及剖面厚度（m）

由髫髻山组火山岩构成的北东向火山喷发带，西端起自门头沟百花山，向东北延伸至怀柔长哨营和密云新城子，沿北东 60°方向延伸。带内各盆地中，该组地层厚度差异较大，以雁羽山盆地最厚，达到 4971m，凤驼梁-四海盆地厚度为 1778～4943m，百花山盆地厚度为 3442m，髫髻山盆地厚达 1134～2622m，新城子盆地厚 1120m。

（二）土城子期

受伸展构造背景转化为挤压构造背景影响，本期盆地数量及分布面积较髫髻山期相比急剧减少（图 10-10）。主要残留有早期的髫髻山、新城子盆地及凤驼梁-四海盆地，并新生挠曲型千家店、后城盆地。各盆内在干旱-半干旱气候条件下沉积为一套紫红-灰绿色河流相沉积碎屑岩，以凝灰质砂岩、复成分砾岩为主，其次为紫红色-灰绿色砂岩、粉砂岩及页岩。早期沉积为粗碎屑复成分砾岩、砂砾岩、凝灰质砂岩组合，构成由粗到细的多期次沉积韵律。其中，砾岩为厚层块状，层理不明显，为冲洪积扇沉积；中期沉积为较细碎屑凝灰质砂岩、凝灰质粉砂岩、粉砂岩组合，岩石成层性好，部分韵律层上部还出现页岩，局部地区见有灰岩透镜体。延庆北小川尚见有 3 条煤线，为扇三角洲及滨浅湖相沉积；晚期再次沉积以粗碎屑为主的复成分砾岩、含砾粗砂岩、凝灰质砂岩或砂岩组合并构成由粗到细的沉积韵律，但其中砾岩层相对较薄，为辫状河沉积。

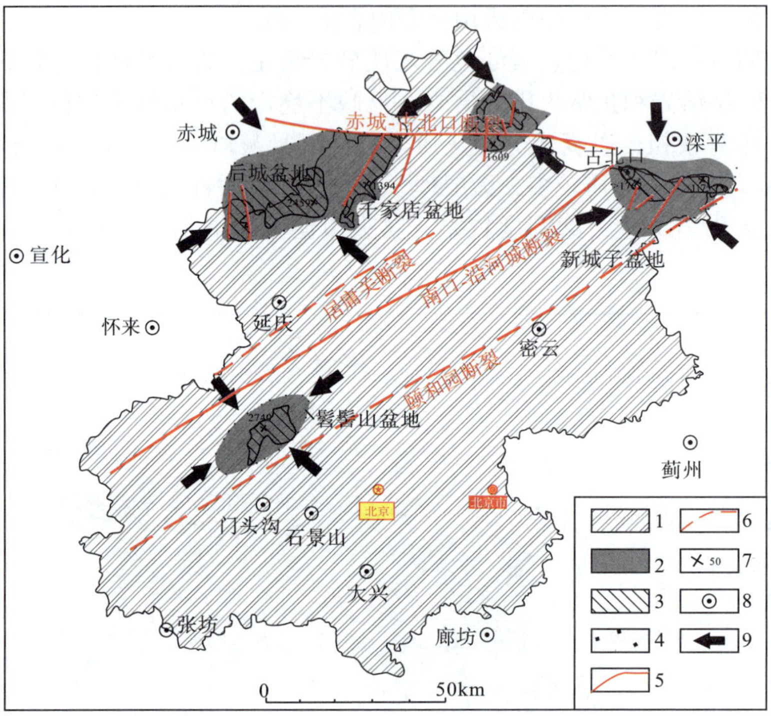

图 10-10　土城子期岩相古地理图

Fig. 10-10　Lithofacies paleogeographic map of Tuchengzi period

1—基底剥蚀区；2—陆相河湖相沉积；3—残留地层分布；4—推测盆地边界；5—断裂；6—推测深大断裂；7—剖面位置及剖面厚度（m）；8—地名；9—物源方向

三、早白垩世时期

（一）张家口期

张家口期本区在第 3 次伸展构造背景下发育北北东走向断裂控制的火山喷发带及火山-碎屑岩断陷盆地（图 10-11）。由于火山活动以中心式喷发为主，致使该期火山-碎屑岩断陷盆地表现为小型火山盆地，发育有东岭盆地、杜家庄-镇厂盆地以及上大水、岔道、怀柔、七道河等断陷盆地。昌平棉山、密云五指山也有小面积本期地层出露，可能也发育有小型盆地。

本期共发育两次强烈的火山喷发活动。第一次以玄武安山岩、粗安岩、粗安质角砾凝灰岩等中性火山喷出岩为主，早期为熔浆喷溢，后期喷溢与爆发作用交替发生，喷发强度逐渐加大。第二次火山活动更加强盛，形成的火山岩层厚度大，分布广泛，各火山盆地内均有产出，以酸性喷出岩占绝对优势，形成了以流纹质熔结凝灰岩、流纹质角砾凝灰岩及流纹岩为主的巨厚堆积。而在这两次强烈火山活动间歇期，形成了巨厚的火山-沉积碎屑岩层。主要岩性为杂色复成分砾岩、砂砾岩，夹多层弱熔结凝灰岩。复成分砾岩中砾石成分复杂，以火山物质为主，大小混杂，分选差，反映沉积环境以河流冲洪积相或火山斜坡泥流相为主。而上部凝灰质砂岩多见于旋回性沉积的上部，为河湖相沉积环境。局部还发育凝灰质砂岩及泥岩组合，属于破火山口湖相沉积环境。

张家口期沉积岩颜色较杂，紫红色、灰绿色、深灰色、暗紫色、灰黑色均有，反映期间气候多变，

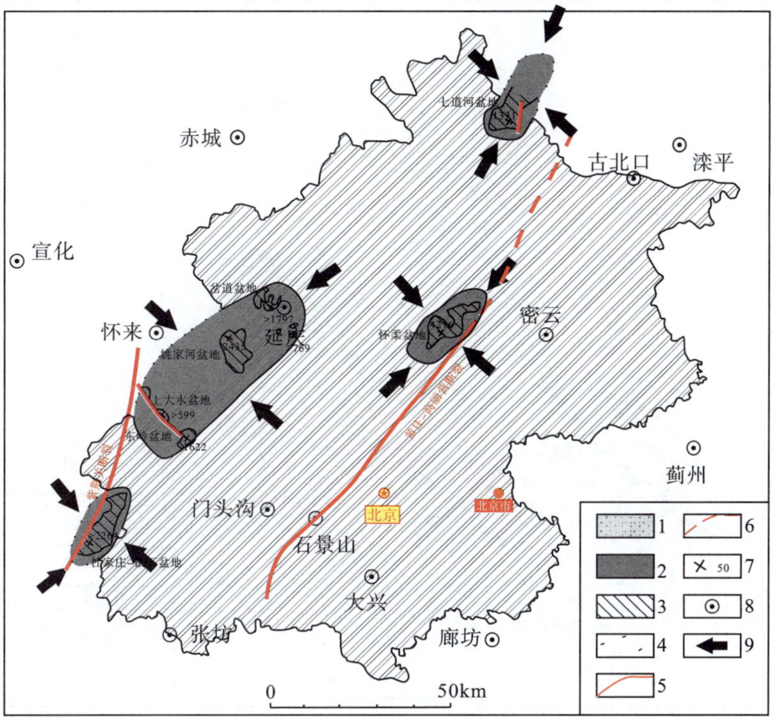

图 10-11　张家口期岩相古地理图

Fig. 10-11　Lithofacies paleogeographic map of Zhangjiakou period

1—基底剥蚀区；2—陆相河湖相沉积；3—残留地层分布；4—推测盆地边界；5—断裂；6—推测深大断裂；7—剖面位置及剖面厚度（m）；8—地名；9—物源方向

干旱、温湿气候频繁交替。

（二）东狼沟期

东狼沟期古地理格局（图 10-12）在张家口期基础上演化，基本特征同张家口期。由于构造、火山活动逐渐转弱，盆地数量及规模减少。北京境内残留有坨里盆地、东岭盆地、怀柔盆地、杜家庄-镇厂盆地。盆内沉积主要为含集块角砾凝灰岩、凝灰岩及沉凝灰岩，夹有玄武岩或粗安岩。沉积碎屑岩厚度不大，碎屑物的粒度由下向上逐渐变细，并夹有间歇流水环境中形成的沉凝灰岩，岩层中见鱼鳞及骨骼化石，并见较多腹足类、双壳类和介形虫化石，为火山间歇期河湖相沉积。

该期盆地的发育仍受盆缘断裂影响。各盆地内沉积厚度存在较大差异，坨里盆地厚度为 420m，东岭盆地厚度为 507rn，怀柔盆地厚度为 134m，杜家庄-镇厂盆地厚度为 175m。沉积环境为温湿间夹干热气候条件下的河湖相沉积。

（三）九佛堂期

九佛堂期本区进入燕山板内造山期中相对稳定的时期。在华北北部地区各类盆地数量和分布总面积逐渐减少。本区仅残存一个坨里盆地，多数地层被新生代覆盖，仅在大灰厂村附近呈带状出露，且厚度不大。

该期演化为温湿间夹凉爽气候环境，构造环境稳定。相对东狼沟期，坨里盆地水体略有加深，盆地演化为浅水湖盆沉积。盆内沉积为黑色、灰色、黄绿色细碎屑岩，主要岩性为页岩、粉砂岩、泥岩，夹岩屑砂岩、细砂岩。沉积相有滨湖相、浅湖相及深湖相。滨湖相由砂岩、泥岩、泥灰岩组成，岩石中含大量生物化石，生物扰动构造明显；深湖相由黑色页岩及少量泥岩组成，页岩含有大量有机质，薄层，具水平层理，发育微细水平纹层。化石丰富，包括腹足类、双壳类、叶肢介、介形虫、昆虫、鱼类及孢粉等。著名的热河生物群典型化石狼鳍鱼-东方叶肢介-三尾拟蜉蝣即产于这个层位，反映沉积环境水动力条件较弱。

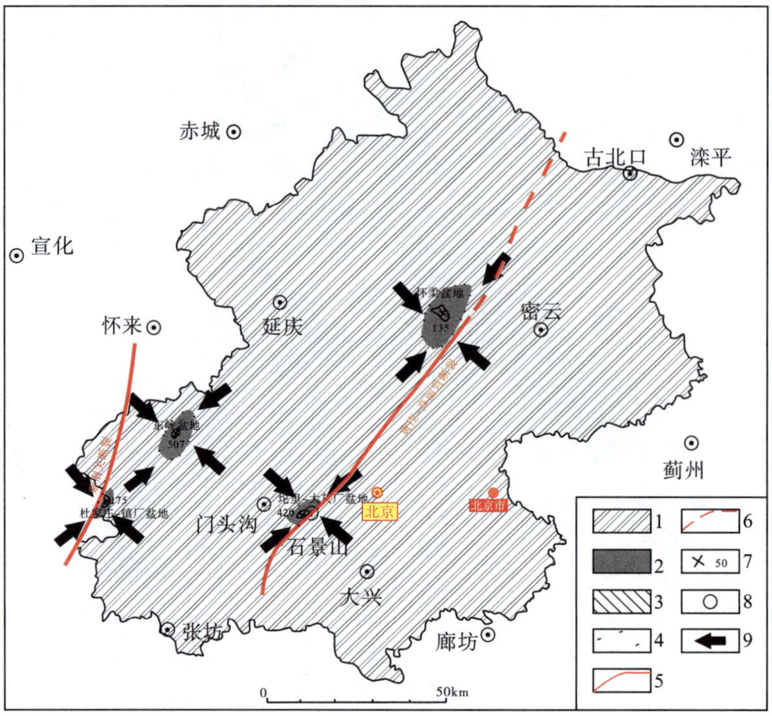

图 10-12　东狼沟期岩相古地理图

Fig. 10-12　Lithofacies paleogeographic map of Donglanggou period

1—基底剥蚀区；2—陆相河湖相沉积；3—残留地层分布；4—推测盆地边界；5—断裂；6—推测深大断裂；7—剖面位置及剖面厚度（m）；8—地名；9—物源方向

（四）坨里期

坨里期，本区因挤压作用总体呈现区域性的抬升剥蚀，仅在早期残留的坨里盆地中接受沉积。盆地分布面积较九佛堂期向盆地中心略有萎缩，盆地总体处于超补偿状态。

坨里期盆地西部边界大致位于房山长沟—晓幼营一带，东部、南部、北部边界因新生代地层埋藏而不明确。根据基岩地质研究推测，南部界线约至房山韩村河一带，北部界线约至朝阳来广营一带，东部界线约为右安门—琉璃河一线。盆地宏观走向保持北东向。盆内沉积一套快速堆积形成的河流相至冲洪积扇相的巨厚层粗碎屑岩系。

早期（一段）沉积为一套黄色至褐色砂砾岩层，沉积厚度约 120m。地表仅在羊圈头有少量出露，表现为河流作用环境下搬运堆积形成的粗粒至细粒碎屑沉积，包括河床滞留沉积、边滩砂岩沉积及洪泛平原沉积。

中期（二段）发育于坨里—公主坟一带，沉积为河流冲积与洪积作用下的紫色岩层与黄色岩层，呈互层产出，向上有砾石变小、砾岩层变薄及紫色岩层增多的趋势，沉积厚度约 300m，恢复古水流流向由北向南，盆地沉积中心位于盆地南侧。

晚期阶段（三段）是坨里盆地大发展的鼎盛时期，沉积总厚度大于 300m。在辛庄、晓幼营至辛开口、土洞、青龙头、岗上、小马庄一带均有出露，沉积表现为以辛开口为扇柄，向岗上及青龙头一带为扇中，再向东和北东方向发展为扇缘的河流洪积扇沉积，在扇中和扇端部位分别发育有扇上网状河沉积，并在晚期局部出现少量岸边沼泽及牛轭湖相沉积。辛开口一带沉积砾石粗、厚度大，颜色呈现为灰紫色、灰黄色，向扇中及扇缘方向沉积渐次变细变薄，颜色也转变为紫红色。因此，该阶段盆地古地貌和沉积物源同早期、中期显著不同，受北东东向构造挤压影响，地貌逐渐演化为西高东低，对坨里盆地沉积产生较大影响。

该期沉积物为砖红色、棕红色、土黄色厚层砾岩、砂岩夹粉砂岩、泥岩，反映古气候环境为干热气候类型。

（五）夏庄期

夏庄期本区因挤压作用总体仍呈现区域性的抬升剥蚀，残留坨里盆地则坳陷加快，水体变深。沉积为杂色细碎屑岩夹泥岩、泥灰岩组合，表现为典型湖相沉积。该期古气候转变为相对潮湿温暖的气候类型。适宜的古气候条件繁殖了大量生物，使得该期为房山生物群的繁盛阶段。

第十一章　古近纪及新近纪

第一节　沉积作用及沉积类型

本区古近系-新近系发育于延庆盆地、北京平原区，最大沉积厚度在凤河营地区，厚度大于3000m，北京凹陷中心区厚700~1600m，北京凹陷东北部顺义地区大于300m，主要为陆相沉积岩，形成于内陆沉积盆地。局部地区由于海侵夹有海陆过渡相的沉积岩。

古新世时期，本区处于隆升状态，缺失沉积建造记录。受到喜马拉雅构造运动的制约，自始新世早期开始，本区发生了不均衡的垂向构造运动，形成规模不等的构造盆地，从而开始了山区与平原的分化，在不同的沉积环境下形成相应的沉积类型。

一、河流沉积作用及沉积类型

河流沉积作用可分为两大类，即河流底层沉积和河流顶层沉积作用。前者主要是推移载荷的侧向加积作用的产物，包括河床底部滞留砾岩沉积、边滩沉积（曲流河为主要沉积）及心滩沉积三类；后者是河流的悬移载荷垂向加积作用的产物，包括堤岸沉积（曲流河顶层沉积）、决口扇沉积（过量洪水冲决天然堤沉积）、串沟填充沉积、河漫湖泊及牛轭湖沉积等。河流沉积类型在区内不同时期地层中均可见。

二、湖泊沉积作用及沉积类型

湖泊沉积物以灰绿色、灰色、棕灰色或棕红色等暗色为主，泥岩呈块状结构，含钙、铁、锰质结核，具有完整旋回，富含生物化石或者炭化植被根系。深湖相以厚层泥岩-砂质泥岩为主，夹有薄层细砂岩；滨湖相以厚层砂岩为主，夹薄层泥岩。常见二者呈交替沉积，如自下而上为厚层砂岩-泥岩的滨湖-深湖沉积；厚层泥岩-砂质泥岩-粉砂岩-砂岩为深湖-滨湖沉积；砂岩与泥岩频繁交互，偶夹巨厚砂岩层，为三角洲相或河口相沉积。沉积物组合主要见于北京凹陷和廊固凹陷内。

三、沼泽沉积作用及沉积类型

平原区该类型主要沉积物为泥炭层或腐殖质黏土、砂质黏土（局部含泥质砂土细条带），厚度较薄。下部具有水平层理和低角度的斜层理，上部层理不发育、多植物根系、虫孔、干裂。该沉积类型多见于晚始新世—渐新世地层。主要分布在北京凹陷和廊固凹陷内，为一套黑色碳质页岩和泥炭。

四、冲洪积扇沉积作用及沉积类型

冲洪积扇主要分布于西山及其山前平原地区。在山区多以单个或数个冲积扇组合为特征；在平原区，尤其是山前地带以冲洪积体系扇或冲洪积裙带体为特征。冲洪积扇的沉积物主要沉积类型是偶遇网状河流沉积的片状砂岩、粉砂岩和砾岩沉积，其次为暂时性水流切入扇体的河床填充沉积物、泥石流沉积物等。

第二节　主要沉积物岩石类型

一、碳酸盐岩类

仅埋藏发育于本区平原区，为淡水灰岩。呈浅灰、紫灰、灰绿、灰褐色，中厚层状，透镜状，包括泥灰岩和泥质白云岩。呈夹层出现，为淡水湖泊浅水环境产物，常与钙质页岩和油页岩伴生。主要产出在北京凹陷和廊固凹陷古近纪地层，如北京凹陷内的前门组，以及廊固凹陷内的孔店组、沙河街组和东营组。

二、碎屑岩类

（一）砾岩类

见于新生代各个时代地层中，其中以始新世长辛店组和洼里砾岩较为典型。厚度由数十米至百米。砾石大小不一，一般为5~30cm，多为次棱角状-次圆状，胶结物多为泥砂质和泥质。砾石成分复杂，与河流流经的物源区及搬运距离长短有关，亦与砾石成熟度和分选性密切相关。形成于沉积序列的底部或者层间，往往与含砾粗砂岩、粗砂岩、长石石英砂岩和杂砂岩等形成韵律沉积。岩石呈紫红、灰、黄褐、灰褐、棕红等色调，厚层—巨厚层，单层厚度小于10m。

（二）砂岩类

1. 中、粗砂岩

分布总体较少，平原区古近纪—新近纪不同年代地层中均有出现，岩石多呈灰白，灰色，砂状结构，部分含细粒，厚层状。碎屑以石英长石为主，含岩屑及少量的黏土矿物。

2. 细砂岩

主要见于平原区古近纪—新近纪地层中，岩石多呈灰白，灰绿色，砂状结构，颗粒支撑，孔隙式、接触式胶结，块状构造。碎屑以石英长石为主，含岩屑及少量的黏土矿物。风化后呈砂粒状，石英大部分呈次圆状—次棱角状，个别为棱角状。岩屑的磨圆度普遍比石英高，多为次圆状—圆状，分选不好，粒径0.05~2mm，填隙物以泥质胶结为主，含少量黏土矿物和长石质杂基。

3. 粉砂岩

主要见于平原区古近纪—新近纪地层夹层中，岩石为灰白色、紫红色，粉砂质结构，纹层状或块状构造。碎屑占60%~80%，除少量云母及长石外，其余多为石英。石英圆度不佳，呈次棱角状-次圆状，粒径0.03~0.05mm。填隙物为黏土矿物和钙质，由于粉砂和填隙物常相间富集，使岩石呈纹层状构造。主要形成于浅湖、河漫滩和三角洲环境，多与砂岩、黏土岩伴生。

（三）黏土岩类

主要由高岭石、伊利石等黏土矿物组成，经成岩后生作用而成。页理不发育的称为泥岩或黏土岩，页理发育的称为页岩，含碳质高者称为碳质泥岩或碳质页岩，含钙质高者称为钙质页岩或钙质泥岩等。常与粉砂岩、粉砂质泥岩、泥灰岩及砂岩互层产出，呈紫红、灰绿、灰、棕黄、黄褐色等。风化后呈小碎块或片状，断面或表面可见绢云母，具微定向构造，块状构造，鳞片结构或黏土结构，常含5%~25%的粉砂，有时粉砂与黏土矿物呈条带状富集，构成纹层构造，形成于湖泊、沼泽、河漫滩等多种环境，是古近纪-新近纪地层中常见的岩石之一。

第三节　沉积盆地及沉积建造

本区古近纪—新近纪盆地共发育有延庆断陷盆地、北京断陷盆地和大厂-廊固断陷盆地3个盆地，按照新生代大地构造位置，前者隶属于太行山山间盆地，后两者隶属于华北平原断陷盆地。

一、沉积盆地

（一）太行山山间盆地区

发育一个断陷盆地——延庆断陷盆地。廷庆断陷盆地属于喜马拉雅旋回晚期山间断陷盆地，其西南与河北省阳原盆地断续相连，由新近纪开始断陷伸展形成，现今仍在继续发展，总体呈北东-南西向的“槽”状（NE50°~60°）展布。

该断陷盆地主要受北东向断裂控制，其边缘和内部皆发育了一系列北东东向张性断裂，使得盆地进一步复杂化，发育为规模更小、沉降幅度各异、长轴为北北东向的菱形凹陷。盆地沉积中心或者深凹部位基本靠近西北侧，南侧逐渐翘起，揭示盆缘北西断裂活动强度强于南东。盆地内震平1号钻孔揭露新近系上新统厚度400.2m，第四系厚度600m。

（二）华北平原断陷（或裂谷）盆地区

在古近纪—新近纪时期，华北平原断陷盆地主要形成两盆夹一隆，即北京断陷盆地、大厂-廊固断陷盆地和大兴隆起（又称大兴凸起）。

1. 北京断陷盆地

始新世由于本区强烈裂陷，黄庄-高丽营断裂、顺义断裂上盘继续下陷，新的南苑-通县断裂发育，使得原有中生代盆地范围扩张，形成由黄庄-高丽营断裂、南苑-通县断裂控制的北京断陷盆地，又接受了近1000m的古近纪—新近纪沉积。由于不同时期的构造作用，断陷内不同地区沉积物有所差异，可进一步细分为琉璃河、丰台及顺义3个凹陷（图11-1）。

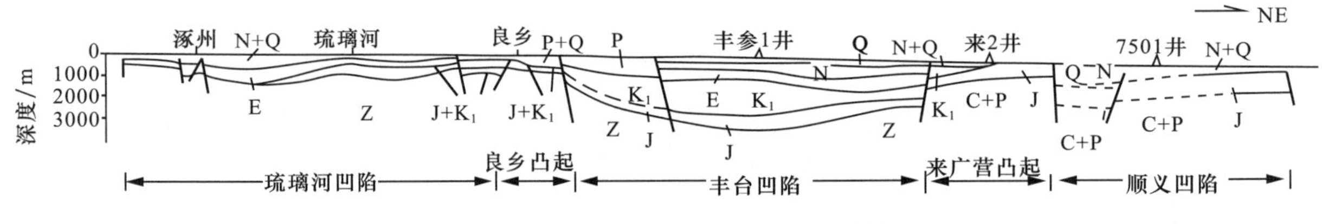

图11-1　北京断陷盆地NE-SW向地质剖面

Fig. 11-1　NE-SW geolgogical profile of Beijing Basin

（1）琉璃河凹陷

位于北京断陷盆地西南部，北西界由黄庄-高丽营断裂带控制，南东界由南苑-通县断裂带控制，北东端与丰台凹陷相邻，南西部伸出区外至河北涿州。布格重力异常图上，沉积盆地形态很不规则，呈现一条大致NNE向弯曲的负异常带。本凹陷长约30km，宽20km，基底为白垩系或蓟县系。内部出现多个沉降中心，可以细分为长辛店、官道-窦店、田各庄-石楼、两间房4个次级的沉降中心。

长辛店沉降中心沉降200~300m的古近纪沉积物，今已出露；官道-窦店沉降中心古近系—新近系厚度为1000m；两间房沉降中心古近系—新近系厚度约430m；田各庄-石楼沉降中心古近系—新近系厚约340m。

（2）丰台凹陷

位于丰台—西四—地坛一带，北东方向展布，长约25km，宽约16km，面积约400km²。西北由黄庄-高丽营断裂控制，东南由南苑-通县断裂控制。西南及东北分别以北东向永定河断裂及东风断裂为界。

盆地内沉积古近系、新近系、第四系（图11-2）。基底南、北冀呈阶梯状，由边部向中心沉积厚度逐渐增大。宏观上，新生界底界西北部和东南部埋藏深度最浅，约为100m；东部和东北部埋藏深度较浅，为300~500m，中部埋藏深度最大，达1500m。沉降中心呈带状分布，位于丰台—前门—北海公园一带，最大厚度达1848m。盆地内古近系沉积厚度60~450m，新近系沉积厚度800~1300m。凹陷内，有新近纪前门期火山活动，沉积3~5层火山岩，累计厚度达300m，以裂隙式岩浆溢流为主，岩性为玄武岩。

（3）顺义凹陷

北界由北东向的黄庄断裂及顺义断裂控制，南界由北东向的南苑-通县断裂控制，南西端与丰台凹陷相邻，北东端由李桥断裂控制。布格重力异常反映明显的负异常。凹陷呈长18km、宽15km的似长方形，基底呈"W"形，中间高，东西低，可细分为酒仙桥、天竺-半壁店两个次一级的沉降中心。

酒仙桥沉降中心范围是孙河断裂以西地区，沉降中心位于东坝西1km，新近系（有一种观点认为在该沉降中心中存在古近系）厚度约750m。基底由北东向的楼梓庄断裂控制，断裂的东南侧为中—新元古界及侏罗系，北西侧为白垩系。

天竺-半壁店沉降中心位置在天竺，南北分别以南苑-通县断裂和顺义断裂控制，东西分别由李桥断裂、南口-孙河断裂控制。第四系埋深约600m，其发育新近系，厚度约340m。基底受断裂影响呈铲形产出，靠近断裂埋深大，往南渐上翘，基底地层为蓟县系。

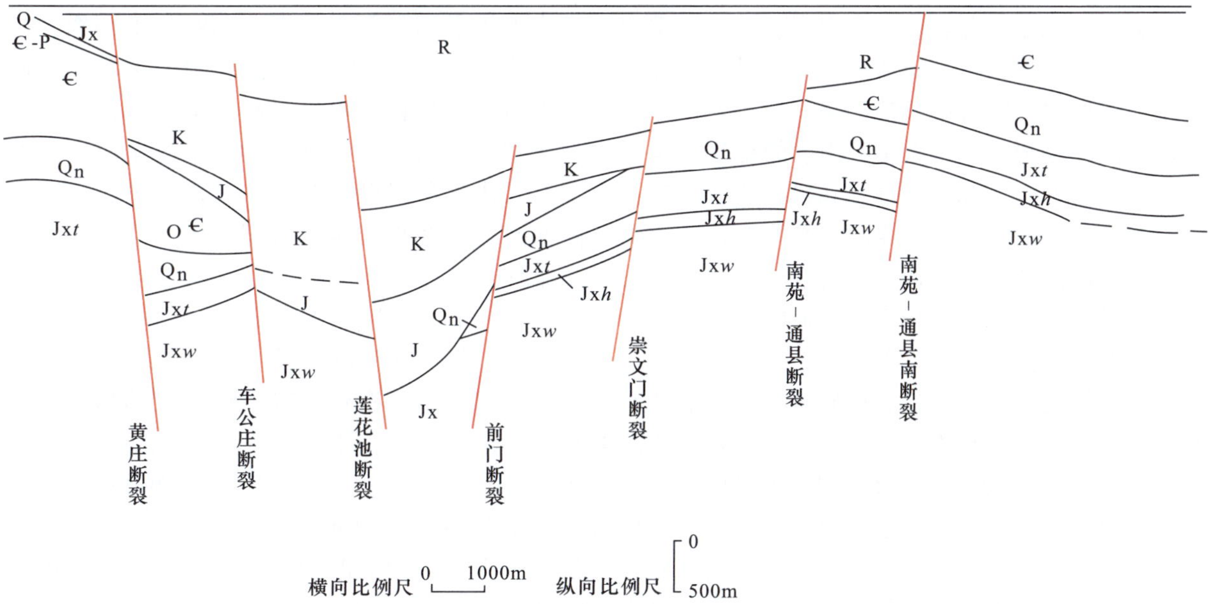

图11-2　丰台凹陷八里庄-南苑地质剖面

Fig. 11-2　Balizhuang-Nanyuan geological profile of Fengtai Depression

2. 大厂-廊固断陷盆地

该盆地位于本区南东及南部通州—大兴一带，主体位于河北、天津境内，以礼贤-夏垫-马坊断裂为界，呈北北东向展布。可进一步细分为大厂凹陷和廊固凹陷。本区仅涉及该断陷盆地边界地带，详细内容请参见《中国区域地质志·河北志》（2017）。

二、沉积建造

本区古近纪—新近纪主体为陆相河湖沉积建造，平原地区局部见有海相沉积，主要沉积建造类型见表11-1。

表 11-1 古近纪—新近纪地层建造特征

Table 11-1 The characteristics of Paleogene-Neogene stratigraphic formation

时期		太行山-燕山地区	华北平原地区
新近纪	中新世—上新世	河湖相棕黄、灰棕等杂色砂砾、粉砂和黏土建造，上部多为钙质胶结红土堆积	河湖相灰绿色泥岩与灰白、浅灰色砂岩，泥质砂岩，夹浅灰色含砾砂岩建造；河湖相灰黄、浅棕红、灰绿色泥岩与棕黄色砂质泥岩，夹杂色含砾砂岩、细砾岩建造
古近纪	始新世—渐新世	河流相紫红色砾岩，夹有紫红色黏土质页岩和灰黄色粗砂岩建造为主，局部为洪积相砂、砾石沉积夹泥岩建造	以湖沼相和河流相为主的深灰、灰黑色泥岩夹灰白色粉砂岩，绿灰、青灰、深灰色泥岩与灰白色粉、细砂岩不等厚互层建造，局部地区夹黑色玄武岩的火山建造

第四节 岩相古地理

从始新世早期以来，大部分断裂复活作张性活动，尤其是太行山-燕山山前区域断裂开始强烈张性活动，致使太行山-燕山山地及华北平原开始分化并逐步形成，新生代沉积在此背景下演化。

一、始新世—渐新世

在北京凹陷内沉积了始新世晚期的长辛店砾岩，其沉积范围在长辛店-丰台小井-城区中心附近。长辛店组以紫红色砂砾岩为主夹泥岩、砂岩，垂向粒序为下粗上细。据李容全（1984）的研究，长辛店地区砾石的中值粒径等值线在平面上呈扇形分布，表明长辛店组的沉积相为近源洪积-冲积扇。砾石形态特征数值向下游有增高的趋势。若按岩性分别统计砾石的平均粒态也可看出其圆度数值有向长辛店方向增高的趋势，其中火山喷出岩的扁平度呈减少的趋势，而石灰岩、燧石的扁平度则有增加的趋势。至城区岩性以紫棕色砂砾岩和粉砂质泥岩为主夹泥岩，由泥质粉砂岩与紫灰色硬砂岩，含砾硬砂岩互层构成4个韵律旋回，属洪泛平原沉积。因此，由西南向东北长辛店组沉积相由山区洪积相过渡到平原的河漫滩相沉积。该时期气候较干燥，地面蒸发强时有薄层石膏形成。

始新世晚期—渐新世早期，平原区已经奠定了二凹一隆（北京凹陷、大兴隆起、大厂-廊固凹陷）格局，在北京凹陷内沉积了前门组，在大厂-廊固凹陷内沉积了沙河街组和东营组。北京凹陷的前门组以灰绿、灰褐色粉砂质泥岩、粉砂岩为主夹黑色碳质页岩和绿灰、灰白色细粒硬砂岩及粉砂岩，其中产腹足类、介形类及脊椎动物的鱼鳖类化石，属河漫滩相-浅湖相沉积。期间，有数次裂隙式基性火山喷发，同时发育由火山作用形成的各种碎屑物堆积，如含角砾凝灰岩等。该时期大厂-廊固凹陷沉积速率远远超过北京凹陷，沉积厚度可达数千米。总体处于河流与浅湖沉积环境，当有机质较丰富、处于还原环境下则形成以湖相为主的暗色泥页岩沉积，为油页岩形成的有利时期，当湖沼中植物繁茂，在有利的地质条件下形成褐煤层。天津、沧县一带渐新世沙河街组一段中见薄球藻、水螅类及喜较高盐度的华花介等，说明此期盆地与海有了联系（海泛），形成了过渡相地层。但并未漫及北京广大地区。

始新世—渐新世早期本区处于亚热带气候较湿润条件下，发育针阔混交林，至后期为含有常绿树种的落叶阔叶林，气候炎热略带干燥（孔昭宸等，1985）。

二、中新世—上新世

渐新世中—末期，区域上经历了一次构造运动后，区内再次演化为准平原，始新世—渐新世早期分化的燕山-太行山地区发育为著名的北台期（又称东灵期）夷平面，而华北平原经历短时间的沉积间断，始新世—渐新世早期沉积地层发生变形，使得中新世—上新世同古近纪地层之间普遍呈角度不整合及平行不整合接触关系。中新世—上新世沉积古地理在北台期准平原背景下演化。

北京凹陷内发育了中新纪—上新统天坛组与天竺组河流、湖泊相砂砾岩、泥岩沉积，其最大厚度可超千米，并伴有少量陆相玄武岩喷发。盆地范围往北扩展遍及顺义地区，发育为还原-弱还原环境下的河湖相沉积，岩性为灰色、灰黑色及棕黄色半胶结泥岩、粉砂岩及砾岩。

大厂-廊固凹陷湖盆范围扩大，湖水向西、北浸漫，大兴隆起仅南部黑庄户—庞各庄一带以残山形式遭受侵蚀，其北部和东南边缘也开始接受沉积，中新世—上新世地层超覆在大兴隆起古生代基底之上。盆地沉积中心位于安次、武清一带。其沉积物以紫色、红棕色泥岩（粉砂质泥岩）为主，其次为砂岩及砂砾岩，反映氧化条件下河流及浅湖相沉积。

这期间沉积不仅发育于分化出来的华北平原区，山区的延怀盆地也开始接受沉积。早期，由于山地与盆地地形差异，盆地四周的入盆河流如大洋河和妫水河将细颗粒物带入盆地，形成湖相沉积。晚期，延怀盆地内老君山地块、军都山地块隆起，延怀盆地分化、演化为涿鹿、官厅、怀来南 3 个独立的小湖（图 11-3），湖泊范围收缩。西部涿鹿古湖的湖水仍主要由大洋河供给，演化为内流湖。延庆湖同怀来南湖贯通，借原大洋河下游经八达岭、南口河流入沙河凹陷，而其原古白河下游废弃（李华章，1994）。

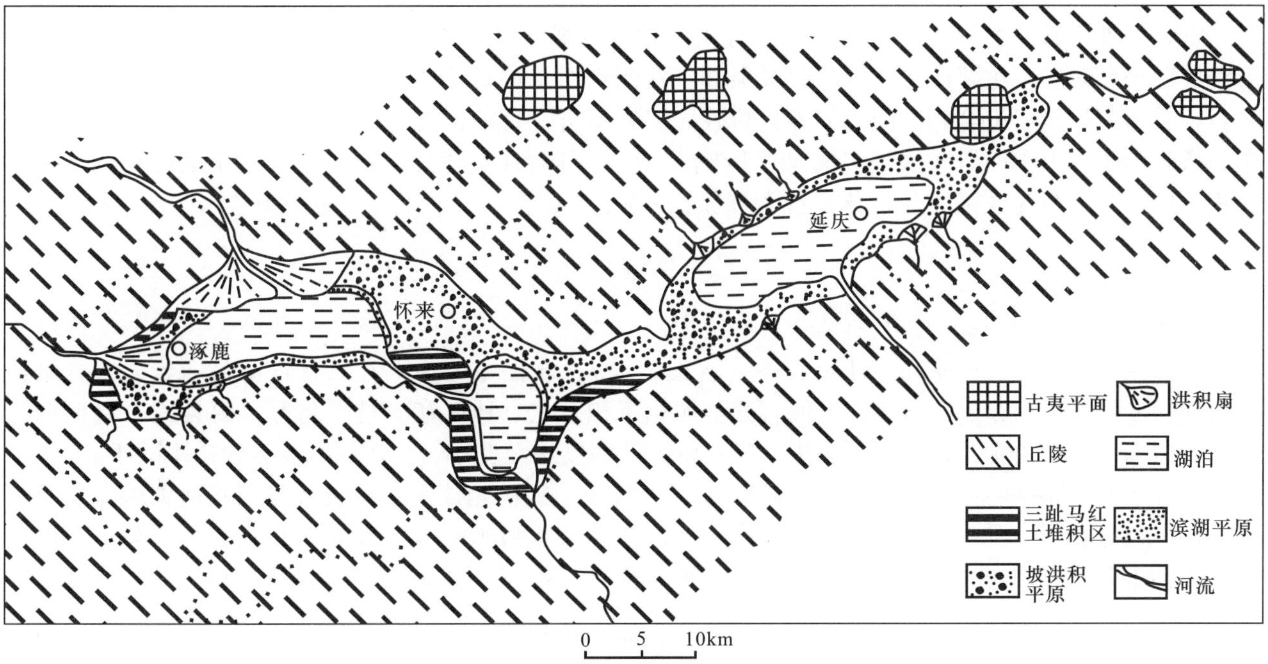

图 11-3 延怀盆地上新世晚期古地理略图

（据李华章，1994）

Fig. 11-3 The paleogeography of Late Pliocene of Yanhuai basin

沉积物黏土矿物及孢粉揭示，中新世—上新世早期古气候呈现干热的氧化环境特征，而上新世晚期则转为暖、湿。同古近纪相比，总体明显偏凉偏干，仍以温暖偏干的大陆性暖温带气候为主要特征，并呈现干湿和暖凉的气候波动。

第十二章　沉积作用与沉积矿产

第一节　中—新元古代沉积作用与沉积矿产

本区与中—新元古代沉积作用有关的沉积矿产主要有以下11种。

一、铁　矿

沉积型外生铁矿主要为中—新元古代海相沉积型铁矿，铁矿类型为赤铁矿，形成于海平面下降、海侵旋回初期阶段，主要产在铁岭组一段顶部，常州沟组、串岭沟组和下马岭组也有分布。主要含矿岩石组合为碳酸盐岩组合和滨浅海的砂岩粉砂岩泥岩组合。目前，北京地区已发现中型矿床1处（延庆四海转山子，习称四海式铁锰矿），小型矿床2处（延庆营门、密云河北庄），属于滨海相胶体化学沉积矿床。大多数为铁锰矿（矿化）或含锰赤铁-菱锰矿矿化点，如平谷上镇南山、永宁水口村等。

二、锰　矿

锰矿为海相沉积型锰矿，共有两个含矿层位。

锰矿主要赋存于长城系高于庄组二段（含锰岩段），岩性主要为页片状含锰泥晶白云岩、含锰白云岩和含砂屑泥晶白云岩，属浅海陆棚含锰盐泥相沉积。北京地区锰矿是河北省“蓟县式”锰矿西延部分。由东到西即由蓟县水厂至平谷中心村一带矿层厚度逐渐变小，含矿层减少且含矿率减小。

赋存于蓟县系铁岭组一段顶部的锰矿，与“四海式”铁矿相伴生，属浅海陆棚铁锰碳酸盐岩相沉积。锰矿物有软锰矿及少量菱锰矿。目前北京地区该矿种无成型矿床，仅有矿化点数处，分布范围西起延庆柏木井，经四海，东至密云玟庄一带，东西长90km、宽13km，其中以延庆四海一带矿化最好。

三、白云岩矿

沉积型白云岩矿为北京市冶金辅助原料矿产的优势矿种之一，矿种质地优良。矿产地主要分布于房山、昌平，其次门头沟、密云、平谷、丰台和延庆亦有零星分布。成矿时代主要为蓟县纪雾迷山期，其次为蓟县纪高于庄期和杨庄期。

蓟县纪高于庄期白云岩矿主要分布于昌平上口—康陵—德胜口一线及延庆佛峪口地区。大、中型矿床各一处，矿点多处。矿体赋存于高于庄组二、三段中，均呈层状、似层状产出，矿体产状与地层围岩一致，顶、底板岩性以燧石条带（团块）白云岩为主。矿石类型简单，均为白云岩，呈中、细粒结构，致密块状构造，矿石矿物成分以白云石为主，少量方解石和石英。矿床成因属中元古代蓟县纪高于庄期浅海相化学沉积矿床。

蓟县纪杨庄期白云岩矿分布于密云金叵罗。北京地区无成型矿床，仅为矿点。矿体赋存于蓟县系杨庄组内，共有三层。矿石类型为白云岩，矿物成分以白云石为主，其次为方解石，含少量燧石及黏土质矿物，隐晶质（泥晶）结构。矿床成因属蓟县纪杨庄期滨海潮间混合坪相化学沉积矿床。

蓟县纪雾迷山期白云岩矿主要分布于房山，其次为昌平、密云，此外在平谷和丰台也有矿点分布。共有产地8处，包括大型矿床2处、中型矿床3处、小型矿床1处，矿点多处。矿体赋存于雾迷山组一、二、三段中，呈层状产出，产状与地层围岩一致，矿层顶、底板以燧石条带白云岩为主。矿石类型均为灰色、灰白色厚层-巨厚层泥晶白云岩。矿石成分主要由白云石组成，其次有少量方解石，偶见石英、绢云母和黄铁矿等。

四、硅石矿

硅石矿发育于常州沟期、大红峪期和杨庄期。

常州沟期硅石矿大型矿床2处，中型矿床2处，矿化点多处。集中分布于延庆佛峪口-营门，昌平锥石口、花园村，怀柔孙胡沟，平谷万庄等地。矿层主要赋存于常州沟组四段（石英岩段），次为二段，矿体产状与地层围岩产状一致。矿石类型主要为白色、乳白色石英岩或石英岩状砂岩，矿石品位稳定，矿石普遍受后期轻微变质作用影响，发生硅化，石英普遍次生加大，对提高矿石质量起有益作用。矿床成因为滨海沙滩相沉积石英岩矿床。

大红峪期硅石矿1处，为佛峪口-营门大型石英岩矿床。矿层赋存于大红峪组石英岩中，产状与地层产状一致，矿石类型为乳白色石英岩，矿石 SiO_2 平均品位约为96.76%。矿床成因属长城纪大红峪期滨海潮间砂坝相沉积石英岩矿床。

杨庄期硅石矿1处，为房山高庄石英砂岩矿点。矿层赋存于杨庄组一段石英砂岩中，矿石自然类型为石英砂岩，矿物成分以石英为主，含量80%~99%，其他为微量的黏土矿物或褐铁矿。矿床成因属杨庄期滨海近岸砂坝相沉积型石英砂岩矿床。

五、钾　矿

钾矿（含钾岩石）成矿时代为长城纪串岭沟期和大红峪期。

串岭沟期为含钾岩石主要成矿期，现有成型产地1处，矿点多处。分布于房山南大寨，怀柔孙胡沟，延庆红旗甸，昌平锥石口—德胜门—南口—古将一带，平谷万庄子、鱼子山，密云大庙峪西沟—孙家沟一带。矿层赋存于富钾黏土岩（伊利石页岩）与陆源碎屑岩（长石砂岩）中。矿石呈层状、似层状产出，矿石自然类型为灰绿色含钾页岩，富钾粉砂岩与富钾长石（石英）砂岩，矿层 K_2O 含量一般为7%~12%，最高达16.5%。

大红峪期含钾岩石主要分布于平谷、密云，延庆新断陷的东缘与房山也有零星分布。矿层主体赋存于大红峪组顶部，一般呈层状、似层状，产状与地层围岩一致。在密云—平谷—蓟县一带的大红峪组下段（火山岩段）中，富钾页岩作为夹层产于火山岩中。矿石类型分为两种：含钾石英砂岩及含钾页岩。一般靠近古陆边部、远离火山喷发中心区的矿石类型以含长石石英砂岩为主，产出层位多；近火山喷发中心区，矿石一般由粉砂质页岩、富钾页岩、凝灰质粉砂岩组成。

六、石灰岩矿

石灰岩矿又可以分为溶剂用石灰岩矿和水泥用石灰岩矿两大类：

1. 溶剂用石灰岩矿

中—新元古代溶剂用石灰岩矿无成型矿床，发育矿点1处，位于密云卸甲山一带，矿层赋存于新元古界景儿峪组中。

2. 水泥用石灰岩矿

中—新元古代水泥用石灰岩矿赋存于蓟县系高于庄组和铁岭组中，矿床成因类型为海相碳酸盐化学沉积矿床。矿产地多处，均为小型矿床，分布于密云地区。铁岭组含矿层位为二段下部，岩性为灰白色中-厚层灰岩与含燧石团块灰岩，矿体形态呈层状，矿石类型为灰白色、深灰色灰岩，隐晶质、微晶砂屑结构，部分矿石呈条带状构造，矿物成分方解石占85%，白云石占10%~15%，燧石、石英、有机质<5%。高于庄组含矿层位为三段，岩性为板层状含灰白云岩、灰质白云岩及瘤状灰岩，矿层呈脉状、压扁-透镜状产出，矿石品位差。

七、磷　矿

未发现成型工业磷矿矿床及矿体，仅发现含磷层位。矿层赋存于长城系常州沟组、团山子组、大红峪组和青白口系龙山组。

常州沟组含磷岩层赋存于该组一段下部的长石石英砂岩及砾岩层中，分布于平谷区上镇南山、老泉口、四道岭、郭家鞍、镇罗营南山、东牛角峪和桃园等地。矿体延伸不远、分散且厚度小，呈透镜状产出，$P_2O_5=2\%\sim5\%$。矿石类型以赤铁磷块岩为主，矿物成分主要由赤铁矿、胶磷矿及微晶至细晶磷灰石组成，含少量长石、石英、锆石和磁铁矿。

团山子组含磷岩层赋存于该组中部含硅质白云岩层中，分布于密云盆子峪。矿层呈透镜状或条带状产出，矿石类型为含磷硅质白云岩及含磷角砾状硅质白云岩。磷矿物为胶磷矿、微晶-细晶磷灰石。矿石品位低。

大红峪组含磷岩层赋存于灰白色石英砂岩及暗绿色石英细砂岩中，分布于密云墙子路、大龙门、杨各庄、汉峪、孙家沟一带。矿层薄且品位低，$P_2O_5=1\%\sim4\%$。

龙山组地层含磷性较差，仅在门头沟区下苇甸发现一含磷层位，为磷块岩透镜体或结核。矿层赋存于龙山组下部的砂质页岩与细粒石英砂岩中。

八、重晶石矿

本区无成型的沉积型重晶石矿矿床，仅发育矿化点，分布于昌平虎峪牧畜后、灰岭口等地。矿化层赋存于长城系高于庄组的灰质白云岩、泥质白云质灰岩中。矿床成因属浅海陆棚碳酸盐坪相沉积。矿化层中重晶石呈团块状、结核状分散于白云质灰岩中，顶板为灰黑色燧石条带灰质白云岩，底板为硅质白云质灰岩。矿化层品位小而分散，平均含矿率约31.5%。

九、大理石矿

中—新元古代大理石矿发育于青白口纪景儿峪期，蓟县纪雾迷山期、杨庄期和高于庄期。

青白口纪景儿峪期大理岩矿赋存于该组杂色泥灰岩、硅质泥晶灰岩及含泥泥晶灰岩中。矿床成因属浅海富钙贫镁碳酸盐化学沉积型矿床，以房山北新村金玉大理石矿点为典型。

蓟县纪雾迷山期大理岩矿赋存于该组地层一段至四段中，以房山大石窝一带大理石矿为典型代表。矿种类型有“黑色”“螺丝转”和“栗色”三类。“黑色”大理石矿一般赋存于潮间潟湖相黑色含有机质密纹层或密波纹层藻叠层石白云岩中，矿体呈中、厚层状产出。“螺丝转”大理石矿一般赋存于一段上部、二段中下部、三段中下部柱叠层石或锥叠层石白云岩中。“栗色”大理石矿一般赋存于该组四段棕红色厚层泥晶白云岩中。

蓟县纪杨庄期大理岩矿赋存于该组下部砂岩段含砂质白云岩中。矿石俗称“京山红”，矿层呈层状，矿石类型为灰白、浅肉红色大理石，肉红色、橘红色大理石，灰紫色大理石。矿物成分以白云石为主，石英、方解石次之。

蓟县纪高于庄期大理岩矿赋存于该组三段板层状泥晶白云岩、瘤状白云岩、灰质白云岩、白云质灰岩、泥晶细晶灰岩中。矿床成因类型属于浅海陆棚灰岩相碳酸盐化学沉积型矿床。大理岩品种包括“黑木纹”“花斑”两大类。“黑木纹”大理岩为黑色、灰黑色硅质白云质灰岩，以平谷樊各庄矿点为典型代表。“花斑”大理石矿为厚层白云质灰岩，以平谷峪口西山矿点为典型代表。

十、玉石矿

玉石矿产地1处，为延庆大观头碧玉（蛋清玛瑙）矿点。矿层赋存于铁岭组二段中部灰白色厚层硅质（碧玉）条带白云质灰岩中。矿体呈小透镜状、条带状分布，矿石自然类型为碧玉，呈暗绿色，风化后呈淡绿色，致密坚硬，贝壳状断口，细粒-隐晶质结构，块状构造。矿石中往往有平行排列的红色细纹层状构造。矿石矿物成分有玉髓、蛋白石、石英、海绿石、云母、锆石、磷灰石等，其中玉髓、蛋白石含量可达90%。

十一、砖用页岩矿

砖用页岩矿产地1处，为北京西山碣石—河西一带矿点。含矿地层为青白口系下马岭组不同色调页岩，主要用于页岩砖烧制。矿石成分主要为黏土矿物水云母（44.3%～66.5%），其他矿物为石英（14.7%~31.3%）、高岭石（6.8%~18.8%）、长石（4.6%~16.4%）、蒙脱石（3.3%~8.3%）。

第二节　早古生代沉积作用与沉积矿产

本区与早古生代沉积作用有关的沉积矿产主要有以下6种。

一、铜　矿

外生型沉积铜矿赋存于馒头组的顶部和底部沉积界面附近，页（泥）岩为含矿有利部位，多出现于灰紫色与黄绿色页岩交互地段的灰绿色页岩中，即氧化还原条件交替变化环境中的还原条件下形成。北京地区此类铜矿均为矿化点，分布于北京西山地区的房山区和门头沟区，含矿层位较稳定，但矿化强度均较弱，达不到边界品位要求。

二、铝土矿

铝土矿赋存于奥陶系马家沟组灰岩侵蚀面之上的石炭系底部，即通常所称“G”层铝土矿。含矿岩系下部为一套海陆交互相的碎屑岩加泥灰岩沉积，含碳量较高，普遍含植物化石，总厚度15～25m。沿走向自东向西碎屑颗粒变细，泥灰岩出露稳定。上部以粗碎屑岩为主，夹细砂岩、页岩及泥灰岩透镜体。铝土矿与铁钒矿、耐火黏土相伴生，为同一成矿地质条件的产物。目前，北京地区仅发现房山芦子水小型矿床1处，且伴生分散元素镓。

三、白云岩矿

沉积型白云岩矿主要形成于寒武纪昌平期，奥陶纪马家沟期也有矿点形成。

寒武纪昌平期白云岩矿产地4处，以小型、中型矿床为主，集中分布于房山区与门头沟区，大多数为溶剂用灰岩矿与电石用灰岩矿伴生矿床。矿体赋存于昌平组顶部，一般呈层状，顶板为馒头组页岩、泥灰岩，底板一般为豹斑状白云质灰岩。矿石类型为灰色、灰白色隐晶质-细晶质白云岩，矿物主要由白云石组成，有少量方解石。

奥陶纪马家沟期白云岩矿无成型矿床，发育矿点多处，分布于房山周口店东山口地区，属水泥用灰岩伴生白云岩矿。矿体赋存于奥陶系马家沟组中，发育于由灰岩、白云质灰岩及白云岩组成的水退韵律顶部。矿石为白云岩，致密块状构造，平均矿石品位：$CaO=34.65\%$，$MgO=16.77\%$，$SiO_2=3.01\%$，$Al_2O_3=1.26\%$，$Fe_2O_3=0.44\%$。矿床成因属滨海潮坪化学沉积白云岩矿床。

四、石灰岩矿

石灰岩矿可以进一步划分为溶剂石灰岩矿，电石灰岩、制碱灰岩矿，水泥用石灰岩，制灰用灰岩四大类。

1. 溶剂石灰岩

主要分布于西山门头沟、房山地区，北山密云、怀柔和昌平等地次之。矿床成因类型均为海相碳酸盐化学沉积矿床，成矿时代为下古生代寒武纪昌平期、张夏期与奥陶纪马家沟期。

寒武纪石灰岩矿矿产地9处，其中大型矿2处，中型矿4处，小型矿3处。集中分布在石景山区“首钢”附近。矿体赋存于寒武系昌平组和张夏组中，均呈层状产出，产状与地层围岩一致。矿石自然类型为黑色石灰岩、豹皮（虎斑）灰岩、鲕状灰岩及混合灰岩，矿石品级大多为Ⅰ、Ⅱ级品。

奥陶纪石灰岩矿矿产地2处，分布在门头沟区龙泉务东山、中灰厂、野溪南山和野溪西山一带，为中型矿产地。矿体赋存于奥陶系马家沟组中，矿石自然类型为石灰岩和白云质灰岩，矿石品位II、III级。

2. 电石灰岩、制碱灰岩矿

主要分布于西山门头沟、房山地区，北山密云、怀柔和昌平等地。其中大型矿床2处，小型矿8处，矿点若干。成因类型均为海相碳酸盐化学沉积型矿床，成矿时代主要是寒武纪、奥陶纪。

寒武纪电石灰岩、制碱灰岩矿赋存于昌平组和张夏组灰色、深灰色厚层石灰岩中。矿体一般呈层状、个别为透镜状，矿石类型单一，主要是黑色、深灰色石灰岩，矿物成分以方解石占绝对优势。矿石品级达I、II级。

奥陶纪电石灰岩、制碱灰岩矿赋存于冶里组底部（第一韵律中上部）和马家沟组顶部（第三韵律中）。冶里组含矿层岩性主要为浅灰色灰岩，丁家滩-千军台一带以外地区则以灰色灰岩为主夹浅灰色灰岩，孟港夹有白云质灰岩，青白口地区因氧化镁含量增高，仅有4m厚石灰岩，双石头及二十里长山不含矿，相变为豹皮灰岩和白云岩。含矿层厚度以丁家滩—千军台一带，最厚达10~30m，向四周显著变薄，含矿性变差。矿层在丁家滩—千军台一带呈层状、似层状产出，其他地区呈似层状、透镜状产出。马家沟组含矿岩性主要为深灰色、黑色石灰岩。矿层厚度相对冶里组要薄，变化比较稳定，矿床规模小，品位为I、II级。

3. 水泥用石灰岩类

绝大多数在西山地区，如房山、门头沟、丰台、海淀等区，其次为北山密云、怀柔、昌平等地，京东顺义也有分布。矿产地多且开采量大。矿床成因类型均为海相碳酸盐化学沉积矿床。成矿时代为寒武纪昌平期、张夏期、炒米店期和奥陶纪冶里期、马家沟期。

寒武纪水泥用石灰岩矿矿产地多处，其中大型矿床2处，中型矿床2处，小型矿床2处，矿点多处。矿层赋存于寒武系昌平组、张夏组和炒米店组。昌平组矿石类型为深灰色灰岩、含豹斑灰岩、白色大理岩等。张夏组矿石类型以鲕状灰岩、方解石细脉灰岩为主，花斑灰岩、泥质条带灰岩次之。炒米店组矿石类型为花斑灰岩、鲕状灰岩、泥质条带灰岩及竹叶状灰岩。

奥陶纪水泥用石灰岩矿矿产地多处，其中以中、小型矿床为主。集中分布在西山房山地区，其次在海淀及京东顺义二十里长山也有分布。矿层赋存于冶里组和马家沟组。冶里组的矿层位于下部，中部、上部一般含矿性较差。矿体呈巨厚层状产出，厚度变化较为稳定，矿石类型由浅灰色灰岩、含泥质斑状灰岩或细泥质条带灰岩为代表，矿石质量达水泥用灰岩I、II级品。马家沟组矿层位于该组中、上部，下部一般不含矿，矿层产出4~5层矿，累积厚度一般为38~60m。西山地区矿石类型比较单一，为深灰色灰岩或暗灰色生物碎屑灰岩，东部二十里长山地区稍复杂，除深灰色灰岩外，尚有豹斑灰岩、含燧石团块灰岩。矿石质量均可达到I、II级品位。

4. 制灰用灰岩

制灰用灰岩矿产地多处，其中大型矿床1处，中型矿床2处，小型及矿点、矿化点多处，主要位于京西丰台后甫营、大灰厂，房山上万、周口店太平山等地，其次在怀柔、延庆也有分布。矿层赋存于奥陶系马家沟组灰岩中，矿石类型一般为石灰岩、角砾状灰岩、白云质灰岩。化学成分含量基本达到制灰用灰岩工业指标要求。

五、大理石矿

沉积型大理石矿赋存于寒武系昌平组和炒米店组地层中，集中分布于昌平、门头沟、房山，顺义也有少许分布。矿石类型有“晚霞”“紫豆瓣”“红色”和“昌平黑”大理石等。

“晚霞”大理石矿赋存于寒武系炒米店组黄白色中厚层泥质灰岩、灰白色薄层泥质灰岩中，以顺义大椴村矿点为典型代表。

“紫豆瓣”大理石矿赋存于寒武系炒米店组紫色粉砂质条带灰岩、鲕状灰岩、结晶灰岩、泥质条带灰岩夹竹叶状灰岩及少量钙质黏土岩、灰色泥质条带灰岩、鲕状灰岩夹竹叶状灰岩中，以门头沟军响塔沟矿点为典型代表。

“红色”大理石矿赋存于寒武系炒米店组黑色结晶灰岩及浅玫瑰色白云岩中，以怀柔龙各庄、昌平文殊峪等矿点为典型代表。

“昌平黑”大理岩矿赋存于寒武系昌平组灰黑色结晶灰岩中，以昌平西石浒矿点为典型代表。

六、页岩矿

早古生代页岩矿矿产地多处，为中小型矿床、矿点，位于门头沟、丰台等地。成矿时代为石炭纪—二叠纪，矿层赋存于石炭系太原组地层中。矿体一般呈层状、似层状产出，并往往在最厚煤层底板附近，有的矿体由于夹石层出现，在纵向、横向上具分叉复合特点。矿石自然类型主要是粗粉砂岩、细砂岩和粉砂质泥岩，一般用于陶粒原料。典型矿床为门头沟区灰峪陶粒页岩矿床。

第三节　晚古生代沉积作用与沉积矿产

本区与晚古生代沉积作用有关的沉积矿产主要有以下5种。

一、铁　矿

晚古生代铁矿可分为硫铁矿和赤铁矿两大类。

硫铁矿矿产地1处，位于房山区堂上。赋存于太原组上部，岩性为泥岩与泥质粉砂岩互层，矿体呈扁豆状、透镜状、团块状或似脉状赋存于含矿层中，多数矿体氧化强烈，地表大部分变为褐铁矿。矿床成因类型为潟湖相沉积型铁矿。

赤铁矿赋存于太原组下部，位于砾岩与其上的硅质岩之间或硅质岩上部的黏土岩中，矿床成因类型为“山西式”风化残积型滨海沉积型铁矿。见于门头沟的清水、王平村、南涧沟、木城涧、淤白地区，房山区羊耳峪北、杏园村东部青灰窑沟北坡一带，分布局限，规模很小，未发现成型矿床，仅发育矿化点和矿点。矿体呈豆荚状、囊状、透镜状产出，矿石类型为致密块状赤铁矿。

二、耐火黏土矿

耐火黏土矿成因类型主要为沉积型，其成矿时代为晚石炭世，早、晚二叠世。晚石炭世黏土矿属滨海-湖沼型沉积矿产，早、晚二叠世黏土矿属平原河流湖沼型沉积矿产。矿产集中分布于门头沟及房山。

石炭纪沉积型耐火黏土矿产地12处。赋存于石炭系太原组底部、中部和上部。底部层位黏土矿同铝土矿层相伴生。中、上部则组成煤层底板，矿体一般呈层状、似层状、透镜状产出。底部矿石可划分为高铝黏土、硬质黏土、铁矾土和高铁黏土四种应用类型。矿石黏土矿物成分主要为一水硬铝石、水白云母（伊利石）及硬绿泥石，次要矿物为石英、赤铁矿、黄铁矿、褐铁矿、绢云母、白云母、磷灰石、绿泥石及碳酸盐等。中、上部矿石则以半软质、软质黏土矿为主，少许硬质黏土。黏土矿物成分由水铝石、蒙脱石和水云母组成。

二叠纪沉积型耐火黏土矿产地5处，其中灰峪与安家滩-王平村耐火属石炭纪—二叠纪的大、中型矿床，其他如军庄、阳坡元-赵家台等处为矿点。矿产赋存于山西组，多为半软质黏土矿，也有硬质黏土矿，矿产似层状或大透镜状产出，绝大多数为煤层底板，有的直接与煤层共生。

三、铁矾土矿

铁矾土矿矿产地多处，其中中型矿床1处，小型矿床2处，主要分布于房山芦子水、门头沟野溪、丰台大灰厂等地，成矿时代为中石炭世太原期。矿层赋存于太原组底部，其层序自下而上为山西式铁矿或高铁黏土矿（铁钒矿）、G层铝土矿或耐火黏土、铁矾土、泥质页岩。多数赋存于耐火黏土下部，为高铝黏土或硬质黏土底板，少数赋存于耐火黏土顶部或夹于耐火黏土中。矿体形态多变，规模大小不一，一般呈层状、似层状、透镜状产出。矿石类型单一，多为铁钒土Ⅱ级品，部分为Ⅰ级品或Ⅲ级品，鲕状、豆状结构，块状构造。矿物由高岭石、水铝石和铁质矿物等组成。

四、页岩矿

页岩矿根据其用途可划分为水泥用页岩和陶瓷用页岩两大类。

水泥用页岩矿矿产地多处，其中大型矿床1处，中型矿床2处。分布于西山地区的丰台辛庄、房山辛庄和门头沟杨坨。矿床成因类型为湖相、河流相淤积型砂、页岩矿床。矿层主要赋存于双泉组下段河流淤积或山间盆地湖积黏土岩中。矿石化学组分含量配比性好，矿石为紫红色、黄绿色黏土岩，多数受后期变质作用影响。矿体品级多数属于Ⅱ级。

陶瓷用页岩矿矿产地多处，其中大型矿床1处，中型矿床2处，位于门头沟灰峪、王家村，石景山秀府村北山等地。矿层赋存于二叠系山西组黑色页岩及二叠系—三叠系双泉组紫色、绿色砂质页岩与风化变质砂岩中。矿床成因类型为湖沼相黏土岩沉积矿床。矿石类型为页岩或泥岩，矿物成分以伊利石和石英为主，少量高岭土、长石。

五、煤　炭

晚古生代煤炭赋存于石炭系—二叠系太原组和山西组深灰、灰黑色砂岩、页岩、泥岩及砾岩中。含煤地层的厚度比较稳定，共计含煤7~13层，可作为区域对比6层，可采3层，可采煤层总厚度为5~9.9m，可采含煤系数为2.1%~4.3%。煤炭产地包括髫髻山-庙安岭煤炭产地、九龙山-香峪大梁煤炭产地、北岭-房山煤炭产地和京东煤炭产地四处。矿床成因为海陆交互潟湖相沉积型矿床为主，陆相湖泊沉积型矿床其次。煤质多为深灰色无烟煤。

第四节　中生代沉积作用与沉积矿产

本区与中生代沉积作用有关的沉积矿产主要有以下2种。

一、煤　炭

中生代煤炭赋存于侏罗系窑坡组灰绿色、灰黑色砂岩、粉砂岩、泥岩等细碎屑岩中。矿床成因为陆相河湖相沉积型矿床。窑坡组地层分为上下两段，下段含煤5~12层，多数可采，为重要含煤层段。上段含煤性差，多不可采，岩层中普遍发育菱铁矿结核。煤质以无烟煤为主，少量贫煤、瘦煤、焦煤，共同特点是低硫、低磷、低灰、高发热量。重要的产地包括髫髻山-庙安岭、九龙山-香峪大梁、北岭-房山、京东和斋堂-百花山。

二、页岩矿

中生代页岩矿形成于二叠纪—三叠纪双泉期和侏罗纪窑坡期。

双泉期页岩矿分布于西山地区的丰台辛庄、房山辛庄和门头沟杨坨。矿产地多处，其中大型矿床1处。矿床成因类型为湖相、河流相淤积型砂、页岩矿床。矿层赋存于双泉组下段河流淤积相与山间盆地湖积相黏土中。矿石为紫红、浅红、黄绿色黏土岩，矿体化学成分含量稳定，且含量配比性好，多为Ⅱ级品。

窑坡期页岩矿主要用于陶粒原料，赋存于侏罗系粉砂岩、页岩及泥质岩中。矿产地多处，以中小型矿及矿点为主，位于西山门头沟、石景山及房山等区县。矿床成因类型为内陆湖沼相沉积型矿床。矿层一般呈层状、似层状产出，产状与地层围岩一致，矿石自然类型以粉砂岩、泥岩为主，次为页岩、粉砂质泥岩等，矿物成分主要为石英、伊利石、绢云母、钾长石及高岭土和碳质。典型矿床为门头沟区木城涧陶粒页岩矿床。

第五节 新生代沉积作用与沉积矿产

本区与新生代沉积作用有关的沉积矿产主要有以下 6 种。

一、铸型用砂矿

北京市铸型用砂的主要物源为西山、北山裸露岩石受物理机械风化作用、由各水系搬运至北京山前平原沉积而成。矿产地多处，主要分布于清河、沙河、潮白河中上游、温榆河、洵河、错河、永定河，大石河的出山口地段，拒马河（北京段）及其上游支流的河床、河漫滩的 I 级或 II 级阶地之上。

铸型用砂矿规模大小不一，上覆层厚度不等，砂体主要赋存于新生代晚更新世以来的地层中，早、中更新世地层也有分布，呈水平层状，一般厚 1～2m，成因类型以河流相冲积、冲洪积型砂矿为主，个别为风成砂型。矿石类型以黏土砂为主，石英砂及石英长石砂次之。矿物成分以石英为主，长石岩屑含量较多，另有暗色矿物即泥质矿物等，粒度以中粗砂、中砂及中细砂为主，SiO_2 含量约为 70%，含泥量稍高。

二、泥　炭

北京地区的泥炭矿矿产地多处，以中、小型矿为主，主要分布在怀柔桃山村，海淀辛庄、冷泉、高里掌、辛力屯、聂各庄、苏家坨，延庆四五里营、大王庄、城关王庄、丁家堡、冯家堡、下营、靳家堡、西羊坊、西卓家营、后黑龙庙，顺义西府南、西府北、相各庄，平谷东高村、韩屯、普贤屯、小辛寨，通州永乐店农场小甸屯—熬哨营一带，大兴区合义分厂-丰台南苑，昌平大汤山阿苏卫、东城、大辛峰，朝阳洼里-仰山等。矿床成因类型为沉积型，按照其产出的地貌部位、沉积相组合等因素可划分为山前平原湖沼型、山前平原河床型和山区沟谷洼地型三类。成矿时代均为第四纪全新世。泥炭造炭植物包括芦苇类（35%～75%）、苔草类（3%～39%）、藓类（3%～30%）和睡莲、棉花莎草、香蒲、木贼（1%～8%）等。主要组分为有机质，平均含量 30%～50%，少数为 50%～70%，矿物质为硅、铝、铁、镁、钙等的氧化物组成的泥炭灰分，总含量 29.02%～67.63%，以及沥青和半纤维素等。

三、高岭土矿

高岭土矿矿产地多处，均为小型矿或矿点。矿床成因类型为风化残积淋滤型矿床，多由燕山期细晶岩脉经受表生风化淋滤而成。矿石类型为高岭土，矿物以高岭土为主，其次为绢云母，少量石英及黄铁矿。呈白色、黄白色，也有淡红色、紫红色，土块状，具可塑性、黏性、吸水性，土质细腻，具滑感。成矿时代为第四纪。

四、建筑用砂、砂砾石

北京市建筑用砂、砂砾石资源丰富，矿产地多处，其中大型矿 1 处，中型矿 8 处，小型矿及矿点多处。矿床集中分布在山前平原的昌平、怀柔、密云、顺义、石景山、丰台、房山和大兴等地。产地的分布受现代水系潮白河及其上游支流、温榆河及其上游支流、永定河与大石河河床严格控制。矿床成因类型为河流相冲积、洪积砂矿床、砂砾石矿床。成矿时代均为新生代第四纪全新世。矿石成分主要受流域面积范围内出露的基岩岩性所控制，如潮白河流经变质岩区，靠近山口地段变质岩岩屑含量较高，永定河的流域范围在西山，尽管有大量的碳酸盐岩出露，但因碳酸盐易于溶解，抗磨蚀的石英、长石成分成为砂矿的主要成分。

五、砖瓦黏土矿

北京市砖瓦黏土矿矿产地多处，其中大型矿 2 处，中型矿 3 处，小型矿及矿点多处，集中位于平原地区，如海淀、丰台、房山、大兴、朝阳、昌平和怀柔等区县。矿床成因类型绝大多数属于冲洪积相亚黏土矿床，少部分为湖相淤积相亚黏土（黏土）矿床。成矿时代主要为第四纪上更新世，个别为全新世。

六、建筑黄土

建筑黄土矿产地主要分布在平原与山区接壤的怀柔、房山及密云等地，延庆盆地及山间沟谷洼地也有少量分布。矿床成因类型以山前冲洪积型黏土、亚黏土矿床为主，少数为平原区湖积型亚黏土矿床。成矿时代均为新生代第四纪上更新世马兰期。产地多处，规模大小不一，以小型矿床及矿点为主。矿层产状平缓，矿石类型为重黏土、亚黏土和亚砂土。矿物成分主要为蒙脱石、水云母。矿石品级 I、II 级。

第三篇

火山岩及火山作用

第十三章 概 述

第一节 发育及分布概况

地质历史上，本区曾出现过多次火山喷发活动，出露的各期火山岩分布总面积约 2000km^2，占全市总面积的 11.9%。太古宙火山岩遭受深度区域变质作用，归入变质地层，本篇只作简单介绍，不再单独论述。中元古代长城系团山子组—大红峪组地层中发育一套钾质碱性玄武岩和钾质响岩类韵律式喷发的火山岩系，受当时燕辽拗拉槽的控制主要呈近东西向分布，从北京市平谷区境东延至天津市蓟州区（鲍亦冈等，2001）；中元古代待建系下马岭组等地层中发育多层凝灰质成分（斑脱岩），是火山喷发远程沉积产物。中生代为火山活动的全盛时期，最强烈的火山活动发生在侏罗世—白垩世。在相当于早、中三叠世双泉组沉积地层中，有较多的凝灰质成分见于沉积碎屑岩内，并见有火山岩砾石（鲍亦冈等，2001），境内未发现火山机构；侏罗世—白垩世火山活动可划分为南大岭期、髫髻山期和张家口期三幕次，分别赋存于早侏罗世南大岭组、中—晚侏罗世髫髻山组—土城子组和早白垩世张家口组—东狼沟组中。这三期火山岩形成于同一燕山构造岩浆旋回内，但受控于不同构造幕，是不同演化阶段的产物（鲍亦冈等，2001）。因此，在分布及产出特点、火山活动方式，岩石类型与演化特征等方面都有极大的共性；但由于不同阶段构造活动强度与岩浆演化阶段的不同，三期火山岩在岩石组合、喷发强度和特征等方面又表现出种种差异（鲍亦冈等，2001）。侏罗世—白垩世火山岩堆积于性质不同的火山-沉积盆地中，现今广泛出露于北京西山和北山地区各主要向斜构造的两翼或向斜轴部（鲍亦冈等，2001）。新生代古近纪巨厚松散沉积物中发育一套基性火山岩系，主要沿深大断裂溢流喷发，埋藏于北京平原区断陷盆地内，地表没有出露，据钻孔控制分布面积约 40km^2。

第二节 研究简史

本区开展火山岩的调查研究工作已有近百年历史。最早比较系统的论述当推叶良辅等（1920）年所著的《北京西山地质志》，将分布于西山地区髫髻山、清水尖、妙峰山一带的火山岩命名为“髫髻山层”（即今髫髻山组），时代归属晚侏罗世，并指出“髫髻山层为阐明西山地质史变迁之一大关键”。此后王恒升（1928）研究了北京妙峰山、髫髻山一带的火山岩，详细介绍了该地喷发岩的分布及性质，论述了喷发幕次及火山岩时代，认为髫髻山火山岩属晚侏罗世，指出“其岩浆之由基性而递转入酸性”。陈恺等（1935）对东岭台地区的火山岩地层进行系统调查后建立“东岭台层”，时代定为白垩世，同时将大灰厂地区火山岩称为“东狼沟集块岩”，时代亦归属白垩世。杨杰（1948）在《北平西山之岩石性质与火山活动》一文中，依据“地壳动，岩浆才动，所以岩浆活动与地壳运动有因果的关系”这一观点，总结出西山地区地壳运动及与之相关的火山活动期次的划分；并对各期深成岩、浅成岩和表成岩（火山岩）作了较为详尽的描述，并首次将辉绿岩（南大岭组）确定为玄武岩，这在当时是一份较为突出的研究成果。

中华人民共和国成立后，对火山地质和火山岩的研究也取得了重要进展。马杏垣（1951）提出平谷区罗峪沟存在震旦纪火山颈构造；孙善平（1961）最早提出北京地区存在熔结凝灰岩，并对中生代凝灰岩类岩石成因类型及岩性特征作了较深入的研究；1958—1961 年开展的北京地区第一轮 1∶5 万区域地质调查，为中生代火山岩的岩石特征、喷发特点和岩石地层划分对比等全面研究奠定了良好的基础。

20世纪80年代至20世纪末，随着火山岩工作方法的改进，研究工作也更加深入。基于第二轮1：5万区调成果，鲍亦冈、白志民等（1995）编写出版了《北京燕山期火山地质及火山岩》，以区域地质学与火山岩石学的理论为指导，采用多学科与多手段综合方法，从岩相、相序列、相组合3个方面对火山岩相做了研究和分析，提出火山活动-堆积序列的5级划分方案，是一部北京地区中生代火山作用与火山研究极为系统而深入的专著。同期，郁建华等（1996）编写出版了《华北地台北部非造山环斑花岗岩及有关岩石》，详细研究探讨了中元古代钾质碱性双峰式火山岩，论述了非造山裂谷作用的拉张构造环境与岩浆活动的关系，火山作用与深成侵入间的内在联系，以及双峰式初始岩浆性质、岩浆混合作用及分离结晶作用过程等科学问题。2000年，鲍亦冈等编写了《北京地质百年研究》，总结概括了北京地区中—新元古代，中生代南大岭期、髫髻山期和张家口期，新生代前门期火山岩的时空分布及变化，各期火山活动各具特色的喷发特点，对主要火山岩类岩石学特征、岩石化学特征做了综合介绍，并探讨了岩浆起源、演化及岩浆成因。上述三部重量级的专著构成本篇总纲，大量成果认识来源于此基础资料。

21世纪以来，随着新技术不断发展，尤其是同位素测试技术的发展，关于岩浆作用与深部壳幔结构演化等科学问题被相继提出，一系列研究论文了提出很多有价值的观点。赵国春博士论文《燕辽地区燕山期火山活动节律与造山-深部过程》（2002）对包括北京在内的燕辽地区燕山期火山活动进行综合研究，总结火山岩的岩石学、矿物学、地球化学和同位素地质学特征，提出了火山活动节律划分，并探讨了不同级别节律火山岩成因、岩浆演化等特征，反演了所对应的深部岩浆房特征及过程，探讨了燕山期火山活动节律与造山-深部过程的耦合关系。邓晋福（1996，1999，2000，2007，2009）基于火山构造组合研究，探索了华北燕山期壳幔相互作用及岩石圈减薄机制与过程，提出了燕山运动及华北克拉通活化的动力学依据。汪洋等（2001，2002，2003，2005，2007，2009）、姬广义等（2002）、李晓勇等（2004）、Wang et al.（2007）对境内中生代火山岩也开展了重点地段岩石学、岩石地球化学等研究，讨论了中生代火山岩岩浆来源及演化和燕山造山运动过程。这些研究成果也为本篇提供了有价值的资料和有益的启迪。

第十四章　火山岩类型及火山岩区带

第一节　火山岩主要类型、火山岩建造

一、火山岩主要类型

火山岩岩石类型划分方式多样。根据岩石结构构造、火山作用方式将火山岩划分为火山熔岩、火山碎屑岩及潜火山岩 3 类。火山碎屑岩类可进一步划分为碎屑熔岩、熔结碎屑岩、普通火山碎屑岩、沉积火山碎屑岩、火山碎屑沉积岩等亚类。本区典型火山岩岩石类型见表 14-1。

根据火山岩形成构造环境将火山岩划分为造山型、非造山型和过渡型 3 类。造山型火山岩由基性-中性-酸性或中性-酸性-碱性火山岩组成，多以中性或酸性火山岩为主，呈单峰式特征；非造山型火山岩由超基性-基性-中酸性或中偏碱性火山岩、基性-偏碱性-碱性火山岩、基性-碱性火山岩组成，个别时代发育不全，呈双峰式特征；过渡型火山岩介于造山型与非造山型之间，由基性-中酸性火山岩组成。本区火山岩成因类型划分见表 14-2。

表 14-1　中元古代至新生代主要火山岩相特征表

Table 14-1　Charateristics of volcanic facies of Mesoproteroic to Cenozoic

时代	岩相类型		主要特征	典型岩石类型	产出或分布
新生代	溢流相	溢流相	岩浆黏度低，平静地从火山口溢出	玄武岩	第四系埋藏
	爆发相	空落相	经空气搬运的细小火山碎屑岩类	玄武质火山碎屑岩	
中生代	溢流相	溢流相	岩浆黏度低，平静地从火山口溢出	安山岩、流纹岩等	广泛
		沸溢相	位于火山机构范围内，介于爆发相与溢流相之间	凝灰熔岩、角砾凝灰熔岩	
	爆发相	火山爆发塌陷相	位于火山口附近，碎屑物的块度大，由火山口向外围碎屑物块度变小	集块岩、角砾岩	广泛
		空落相	经空气搬运的细小火山碎屑岩类	凝灰岩	
		火山碎屑流相	具有密度流搬运、分选的特性；体积差别大，并与特定的火山构造相对应；流动构造发育	熔结凝灰岩、含角砾熔结凝灰岩	广泛
		潜火山岩相	与火山岩同源、同期或稍晚，在空间上有联系	潜流纹岩、流纹斑岩、潜粗安玢岩	石窑岩管等
		隐爆角砾岩相	位于火山机构及其附近，与周围岩石呈“侵入”式接触，形态复杂	隐爆角砾岩	塘石岩管等
		侵出相	火山喷发晚期岩浆沿裂隙、火山通道缓慢侵出地表	碎斑熔岩	土城子组
	火山-沉积相	火山喷发沉积相	火山碎屑物经水动力作用搬运再沉积形成的岩石，层理发育	沉凝灰岩、凝灰质砂岩等	广泛
古生代	火山-沉积相	火山喷发沉积相	火山碎屑物经水动力作用搬运再沉积形成的岩石，层理发育	凝灰质砂岩	双泉组

续表

时代	岩相类型		主要特征	典型岩石类型	产出或分布
中元古代	火山—沉积相	火山喷发沉积相	经空气搬运的细小火山碎屑岩类	斑脱岩、沉凝灰角砾岩	下马岭组、铁岭组、串岭沟组；大红峪组
	溢流相	溢流相	玄武质岩类平静地从火山口溢出	粗玄岩-响岩类	大红峪组
	爆发相	火山爆发塌陷相	位于火山口附近，碎屑物的块度大，由火山口向外围碎屑物块度变小	集块岩、角砾集块岩、火山角砾岩、凝灰岩	大红峪组
		隐爆角砾岩相	位于火山机构及其附近，与周围岩石呈“侵入”式接触，形态复杂	隐爆角砾岩	大红峪组（大华山火山岩管群）
		潜火山岩相	火山喷发晚期岩浆沿裂隙、火山通道缓慢侵出地表	辉绿（玢）岩、正长斑岩	

表 14-2　火山活动旋回及构造阶段综合划分表

Table 14-2　The integrated classification of volcanic activity cycles and its tectonic stage

时代				旋回	亚旋回	火山岩建造	岩石地层	成因类型	构造环境	构造域	构造阶段
新生代	古近纪		E_{2-3}	喜马拉雅	XS2	基性	前门组	双峰组	伸展	太平洋-欧亚大陆构造	非造山
			E_2		XS1	基性	长辛店组				
			E_1						大陆隆起		后造山
中生代	白垩纪	晚期	K_2	燕山	YS3						
		早期	K_1^3				夏庄组		挤压		
							坨里组				
							九佛堂组		伸展		
			K_1^2			基性-中性-酸性	东狼沟组	单峰式			
			K_1^1			中基性-中性-酸性	张家口组		挤压		同造山
	侏罗纪	晚期	J_3		YS2	基性-中性-酸性	土城子组		伸展		
		中晚期	J_{2-3}			基性-中性-酸性（碱性）	髫髻山组				
		中期	J_2		YS1		九龙山组		挤压		前造山
							龙门组				
		早期	J_1				窑坡组		伸展		
						基性-中性	南大岭组	过渡式			
	三叠纪		T_3				杏石口组				非造山
				印支	IS2				大陆隆起	西伯利亚-华北板块构造	后造山
			T_2				双泉组	单峰式（较远）	过渡		
			T_1		IS1				伸展		后碰撞
古生代	二叠纪		P_3						挤压		同碰撞
			P_2	华力西							
			P_1						伸展（岛弧）		前碰撞
	石炭纪		C_2								
			C_1								前造山
	泥盆纪		D_3								
			D_2	加里东					大陆隆起	华北稳定克拉通构造	非造山
			D_1								
	志留纪		S								
	奥陶纪		O								
	寒武纪		€						陆表海		
新元古代	震旦纪		Pt_3^3	扬子					大陆隆起		
	南华纪		Pt_3^2								
	青白口纪		Pt_3^1	晋宁					陆表海		
中元古代	待建纪		Pt_2^3			中酸性-酸性	下马岭组	单峰式（远）			
	蓟县纪		Pt_2^2								
	长城纪		Pt_2^1		JN2	基性-中偏碱性	大红峪组	双峰式	裂谷		

续表

时代			旋回	亚旋回	火山岩建造	岩石地层	成因类型	构造环境	构造域	构造阶段
古元古代		Pt_1	吕梁						Columbia 超大陆	造山-后造山
新太古代	晚期	Ar_3^3	阜平	WT1	基性-中性-酸性	密云岩群、四合堂岩群	单峰式	岛弧-陆缘岩浆弧		前造山

二、主要火山岩相

根据火山岩岩石特征及火山作用方式，本区火山岩相发育有溢流、火山爆发、火山-沉积三大岩相，内含溢流、沸溢、火山爆发崩塌、空落、火山碎屑流、火山爆发沉积、潜火山岩、侵出、火山通道、隐爆角砾岩 10 个亚相。具体火山岩相及特征见表 14-1 及后章节详述。

火山岩相相互组合，形成更高级别的相段、相组和相群，共同反映一定火山活动时期内火山作用发生发展规律及火山作用与其他相关地质事件关系，即相序列（鲍亦冈等，1995）。区内燕山期火山岩四种主要岩相及相序列发育特点对比见图 14-1。

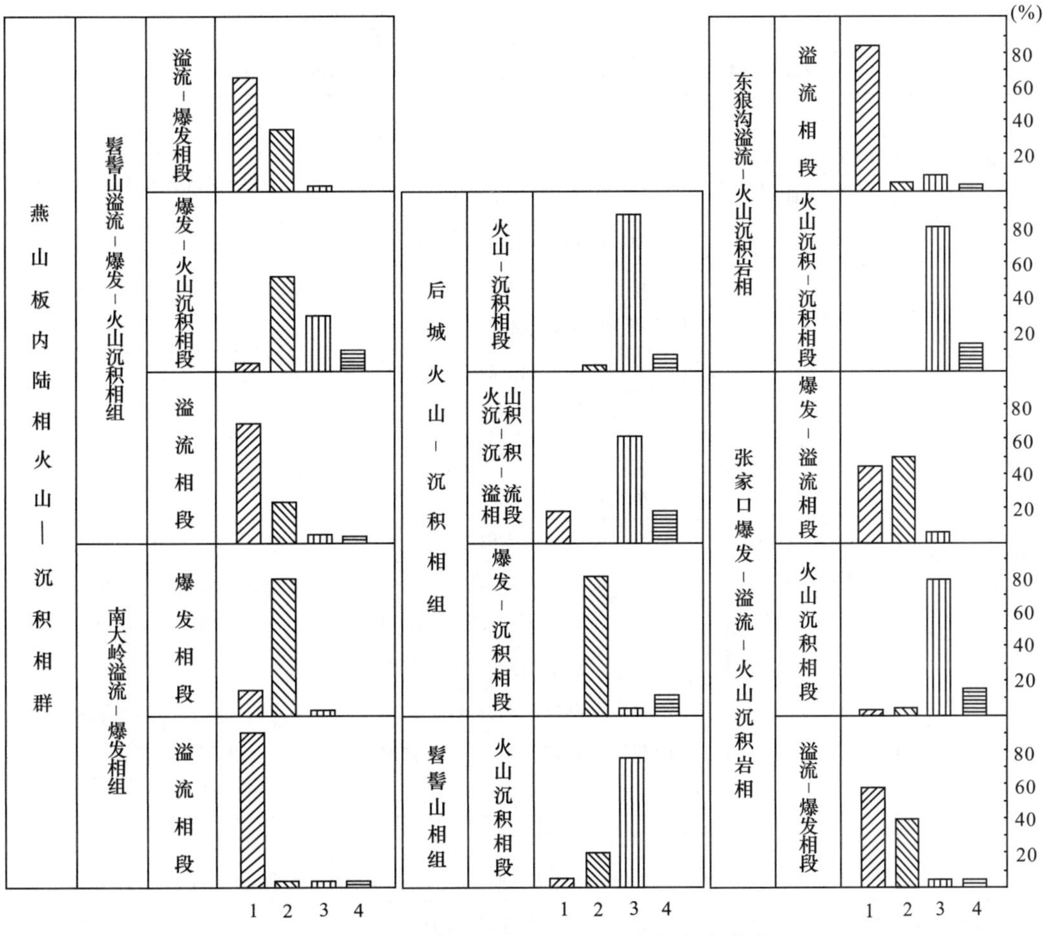

图 14-1 燕山期火山岩四种主要岩相发育特点

（据鲍亦冈等，1995）

Fig. 14-1 Characteristics of four volcanic lithofacies of Yanshanian

1—溢流相；2—爆发相；3—火山沉积相；4—沉积相

三、火山岩建造

本区火山岩建造见表 14-2。根据岩石化学成分将火山岩划分为基性、中性、中酸性、酸性、偏碱性和碱性 6 类火山岩建造，不同的火山岩成因类型对应不同的火山岩建造组合。

第二节　火山活动旋回与构造阶段划分

根据本区不同时代地层中发育的火山岩、火山岩建造及相关研究成果，结合区域构造运动旋回的划分，将本区的火山活动划分为阜平、晋宁、燕山和喜马拉雅等 4 个旋回、7 个亚旋回，并分别对应新太古代—古元古代古陆拼合同造山阶段、中元古代早期燕山裂陷槽阶段、燕山板内造山阶段和新生代陆内裂解阶段（表 14-2）。其中，阜平旋回的火山岩为变质火山岩类，其他 3 个旋回的火山岩为未变质的火山岩类。

第三节　火山构造单元划分

参考《中国区域地质志·河北志》(2017) 火山构造单元划分方案，根据区内火山岩岩性、岩相、分布特征及构造背景等基础资料，将火山构造划分为火山岩区、火山喷发带、火山-沉积盆地和火山机构 4 个级别（表 14-3）。

表 14-3　火山构造单元划分表

Table 14-3　The classification of volcanic tectonic units

<table>
<tr><th>时段</th><th>一级</th><th>二级</th><th colspan="3">三级</th><th>四级</th></tr>
<tr><td rowspan="2">E_2-Q</td><td rowspan="2">华北陆块火山岩区ⅢA_4</td><td rowspan="2">华北平原火山喷发区ⅢA_4^3</td><td colspan="3">廊固火山-沉积盆地</td><td rowspan="2">南苑-通县裂隙式线状火山机构</td></tr>
<tr><td colspan="3">北京火山-沉积盆地</td></tr>
<tr><td rowspan="16">T_3-E_1</td><td rowspan="16">大兴安岭-太行山火山岩带ⅢB_3</td><td rowspan="16">承德-武安火山喷发带ⅢB_3^2</td><td rowspan="7">晚期</td><td colspan="2">坨里-大灰厂火山-沉积盆地</td><td rowspan="7">沿河城卫生院、西台下古火山口；齐家庄南火山穹丘</td></tr>
<tr><td colspan="2">七道河火山-沉积盆地</td></tr>
<tr><td colspan="2">岔道火山-沉积盆地</td></tr>
<tr><td colspan="2">上大水火山-沉积盆地</td></tr>
<tr><td colspan="2">杜家庄-镇厂火山-沉积盆地</td></tr>
<tr><td colspan="2">怀柔火山-沉积盆地</td></tr>
<tr><td colspan="2">东岭火山沉积盆地</td></tr>
<tr><td rowspan="7">中期</td><td rowspan="4">北亚带</td><td>千家店火山-沉积盆地</td><td rowspan="4">窑湾、四司、上王家山古火山口；气不愤、石渣尖、小围场古火山岩管</td></tr>
<tr><td>后城火山-沉积盆地</td></tr>
<tr><td>凤驼梁-四海火山沉积盆地</td></tr>
<tr><td>燕羽山火山-沉积盆地</td></tr>
<tr><td rowspan="3">南亚带</td><td>新城子火山-沉积盆地</td><td rowspan="3">清水河谷裂隙式线状火山机构；黄安坨、大三里、梨元岭古火山；苇子水、田庄、雁翅东、安家庄古火山口</td></tr>
<tr><td>髫髻山火山-沉积盆地</td></tr>
<tr><td>百花山火山-沉积盆地</td></tr>
<tr><td rowspan="2">早期</td><td colspan="2">昌怀火山-沉积盆地</td><td></td></tr>
<tr><td colspan="2">西山火山-沉积盆地</td><td>桑峪古火山口</td></tr>
<tr><td rowspan="2">Pt_2-T_2</td><td rowspan="2">华北陆块火山岩区ⅢA_2</td><td rowspan="2">燕山-辽西火山喷发带ⅢA_2^{2-2}</td><td colspan="3">中二叠世云雾山火山-沉积盆地</td><td></td></tr>
<tr><td colspan="3">中元古界平谷-蓟县裂谷火山盆地</td><td>平谷古火山口</td></tr>
<tr><td rowspan="3">Ar_3^3-Pt_1</td><td rowspan="3">华北陆块火山岩区ⅢA_1</td><td>阴山-冀北火山喷发区ⅢA_1^1</td><td colspan="3">茨营子-大庙沉积-火山盆地</td><td></td></tr>
<tr><td>冀西-冀南火山喷发区ⅢA_1^2</td><td colspan="3">阜平沉积-火山盆地</td><td></td></tr>
<tr><td>冀中-冀东火山喷发区ⅢA_1^3</td><td colspan="3">沧州-迁西沉积-火山盆地</td><td></td></tr>
</table>

一、新太古代—古元古代

本区面积有限，且新太古代至中元古代的变质地层经历了多期次变质变形的叠加改造，其火山构造单元不易划分，仅参考河北省的划分方案进行大致划分：一级单元为华北陆块火山岩区，进一步划分了3个二级单元，每个二级单元各包含1个三级单元，往更大区域则包含8个三级单元（《中国区域地质志·河北志》，2017），以及若干个已经无法识别的四级单元（表14-3）。

二、中元古代—中三叠世

中元古代—中三叠世，本区位于稳定华北克拉通板块内，总体构造背景环境较为稳定，火山活动少，仅中元古代长城纪因发育裂陷槽（后夭折），以及晚古生代中二叠世因中亚造山远程影响发育极少量火山岩。根据火山岩时空分布和构造单元划分，将本区划分为1个一级、1个二级和2个三级构造单元。一级构造单元为华北陆块火山岩，二级构造单元为燕山-辽西火山喷发带，三级构造单元则分别为中元古代平谷-蓟县裂谷火山盆地和早二叠世云雾山火山-沉积盆地（图14-2），四级构造单元为各具体的火山机构。

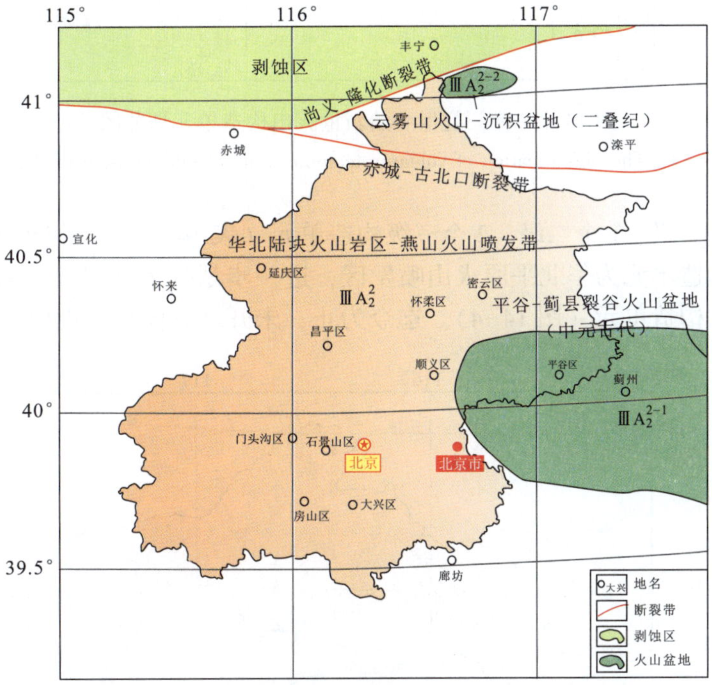

图14-2 中元古代—中三叠世火山构造单元划分图

Fig. 14-2 The classification of volcanic tectonic units of Mesoproterozoic to Middle Triassic

三、晚三叠世—古新世

晚三叠世—古新世，本区火山活动强烈，火山岩广布。根据火山岩时空分布和构造单元划分，将本区划分为1个一级、1个二级、16个三级和若干四级构造单元（表14-3）。一级火山构造单元属大兴安岭-太行山火山岩带，二级火山构造单元属承德-武安火山喷发带，进一步可划分为燕羽山、新城子、髫髻山、百花山等16个三级火山-沉积盆地构造单元（图14-3），包括若干个具体火山机构四级火山构造单元。

四、始新世—第四纪

始新世—第四纪，本区火山岩埋藏分布，研究程度较低。根据火山岩时空分布和区域构造单元划

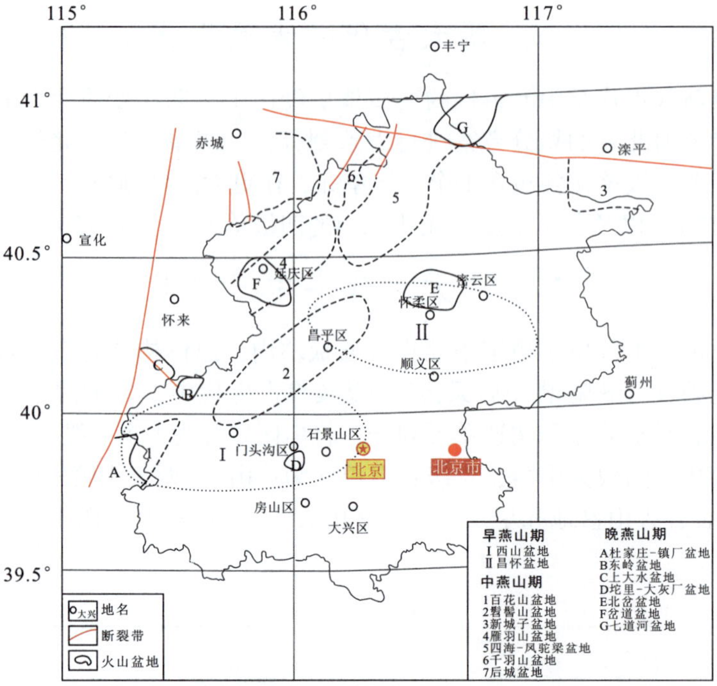

图 14-3 晚三叠世—始新世火山构造单元划分图

Fig. 14-3 The classification of volcanic tectonic units of late Triassic to Paleocene

分，将本区划分为 1 个一级、1 个二级、2 个三级构造单元（表 14-3）。一级火山构造单元属华北陆块火山岩区，二级火山构造单元为华北平原火山喷发区，进一步划分为廊固火山-沉积盆地和北京火山-沉积盆地 2 个三级火山构造单元（图 14-4）。迄今为止，未开展具体火山机构四级火山构造单元研究。

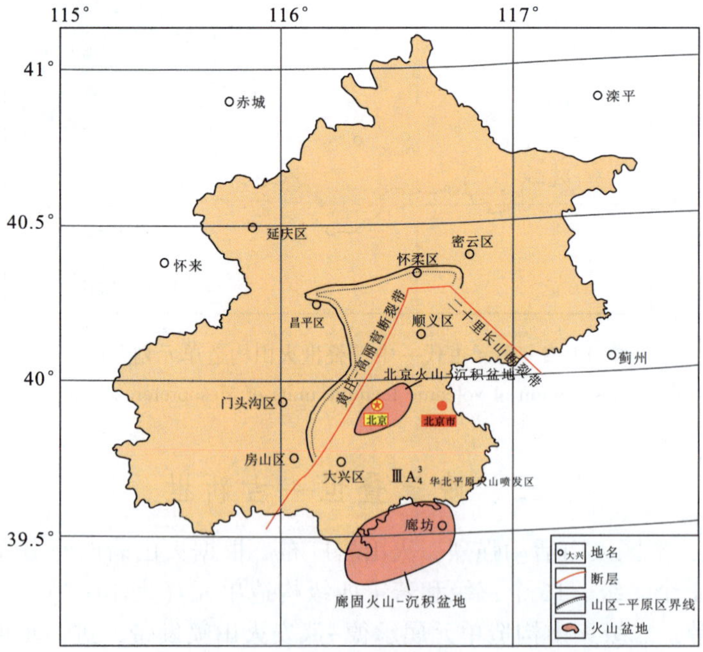

图 14-4 始新世—第四纪火山构造单元划分图

Fig. 14-4 The classification of volcanic tectonic units of Eocene to Quaternary

第十五章 前寒武纪火山岩及火山作用

第一节 太古宙变质火山岩及火山作用概述

本区太古宙变质火山岩的分布较为广泛，而且在时间、空间以及成因上多与岩浆侵入活动关系密切。但由于普遍遭受中深变质作用改造，火山岩的原始厚度和出露面积无法准确统计，火山机构已无法识别，仅能依靠原岩恢复确定太古宙变质火山岩的存在（详见第五篇第一章）。

一、密云岩群火山岩及火山作用

密云岩群中火山岩主要为变质超镁铁质岩类、变质基性（镁铁质）岩类和长英质变质岩类。超镁铁质岩类化学成分、结构特点与变质镁铁质岩相似，稀土元素特征也大体相似。一般呈规模不大的透镜体、夹层或不规则团块状产出。按其产状特征分为两类，一类是超基性顺层小侵入体和岩脉，另一类是成层的火山喷发成因的超基性熔岩，其原岩应为镁铁质火山岩。镁铁质岩类主要为麻粒岩类、斜长辉石岩及斜长角闪岩。长英质变质岩类主要包括（角闪）辉石斜长片麻岩和变粒岩及黑云变粒岩等。

对变质基性和变质超基性岩类进行原岩恢复，并依据 *AFM* 图解将其原岩判定为拉斑玄武岩系列，根据其轻稀土相对富集的特点，其原岩与 K. Condie 划分的 TH_2 型拉斑玄武岩相当。变粒岩具有细粒粒状变晶结构，稀土配分曲线表现为平缓型，原岩恢复为火成岩，可能原岩是变质基性岩，即拉斑玄武岩风化堆积的产物。

密云岩群变质火山岩存在两期，早期以基性火山岩为主的火山活动频繁，晚期中酸性火山活动加强。在火山活动宁静时期，形成了含铁硅质岩及凝灰质半黏土岩、砂质黏土岩，但自始至终缺乏碳酸盐岩。整个岩群构成了一个完整的火山-沉积旋回。因此，综合分析认为其形成于活动大陆边缘海相盆地环境。

二、四合堂岩群火山岩及火山作用

四合堂岩群岩性包括黑云角闪斜长片麻岩、角闪黑云变粒岩、黑云浅粒岩。变质岩石由基性到中酸性有规律地变化，反映出玄武岩-安山岩-英安岩比较完整的火山喷发-沉积旋回。下部变质岩系的原岩反映出火山活动的多旋回性；上部的变质岩石原岩为玄武质层凝灰岩和英安质凝灰岩、凝灰质砂岩互层，反映火山活动总体平静。最上部有少量角斑岩质火山碎屑物和陆源碎屑长石砂岩，反映沉积环境已转向陆棚海。

第二节 中元古代火山岩及火山作用

一、时空分布特征

中元古代火山岩主要发育于长城系大红峪组一段内，在平谷区西北部江米洞地区团山子组二段近顶部也见有同成分的火山熔岩（厚 27.2m）和火山碎屑沉积岩。因此，大红峪期火山活动开始的时间应该在团山子期末期，在大红峪期强烈喷发，故以下简称大红峪期火山活动，是本节重点总结的火山岩。此外，在长城系串岭沟组、蓟县系铁岭组、高于庄组和待建系下马岭组还发育极少量凝灰岩（斑脱岩），具有远程火山灰极细粒沉积特征，火山作用发生在境外。

平谷区大红峪期火山岩呈南东东向带状分布，向东延伸至天津市蓟州区及河北省遵化、滦县，构成东西长约150km，南北最宽处约20km的大型火山-沉积盆地。区内为大红峪期火山活动最为强烈的地段，火山岩分布面积约900km²。火山喷发中心在平谷区熊耳寨一带，该地火山岩厚度最大（624m，占地层总厚68%），向四周逐渐变薄，向北很快减薄至100m，过墙子路东西向断裂带后迅速变薄尖灭。向东在将军关、黄崖关等地火山岩层厚度为200m，再往东延，至天津市蓟州区剖面上实测的火山岩厚度仅84m。往西北在密云金山子村附近火山岩厚数十米，而五亩地一带仅厚10m。该火山岩带的南侧在甲山-鱼子山-黄松峪一带厚度变化较小，甲山剖面大红峪组一段地层总厚度为310.59m，其中火山岩层厚285.64m，占一段总厚度92%（图15-1）。

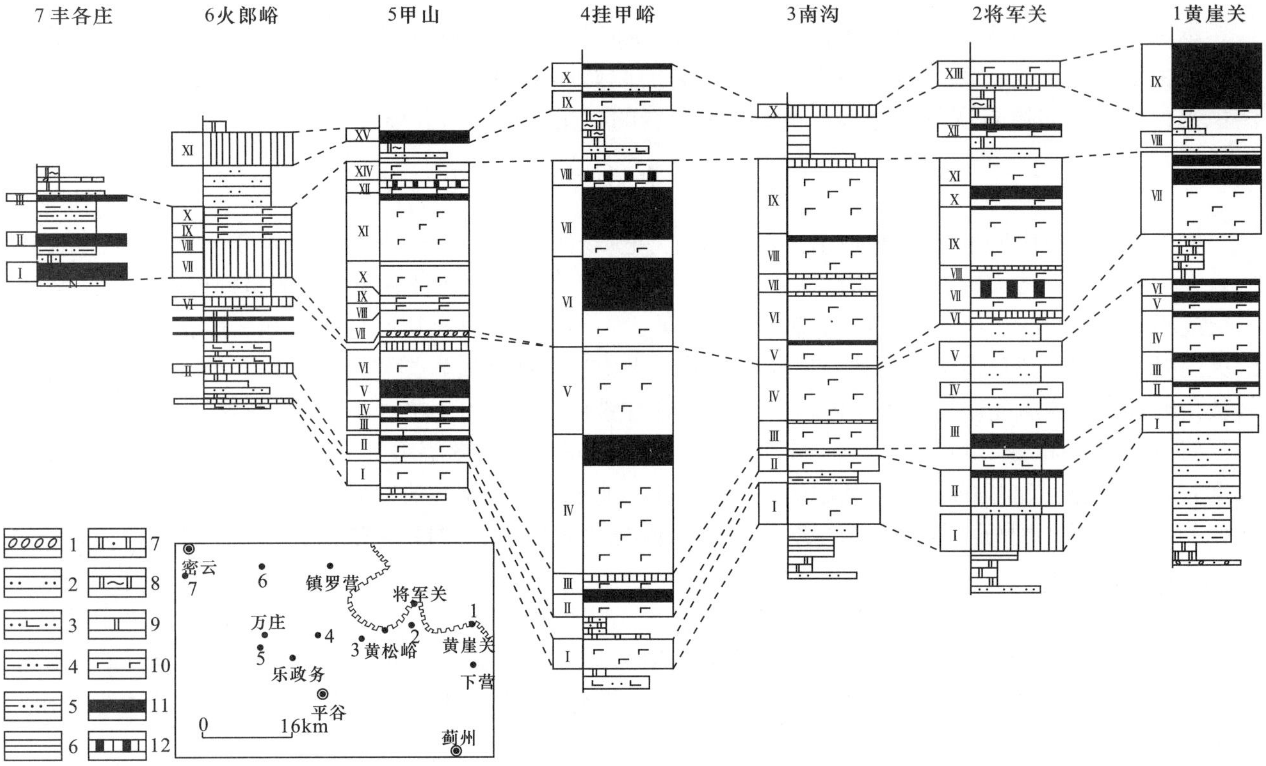

图15-1　北京平谷地区大红峪组火山岩段喷发韵律相序剖面对比图

（据《北京市区域地质志》，1991）

Fig. 15-1　Profile correlation diagram of volcanic eruption lithofacies sequence of Dahongyu formation in Pinggu, Beijing

1—具交错层砂砾岩；2—石英岩状砂岩；3—凝灰质砂岩；4—泥质砂岩；5—砂质页岩；6—绿色富钾页岩；7—含砂白云岩；8—燧石条带白云岩；9—白云岩；10—钾质玄武岩；11—富钾粗面岩；12—火山角砾岩与凝灰岩

大红峪期火山岩形成于燕山裂陷槽持续沉降大背景下，伴随赤城-古北口断裂带两组东西向断裂（长哨营-古北口和密云-兴隆）与密云-墙子路-蓟州-兴隆-平泉-凌源断裂带两组近南北向（黄崖关、程各庄）断裂继承性活动而呈裂隙式喷发，喷发大致经历早、中、晚三期。先兆喷发从团山子晚期开始（李祖湘等，1981），后经历长休止期巨厚潮间带沉积；强烈喷发于大红峪早期，后也经历较长休止期较厚潮间-潮下带沉积；尾声喷发于大红峪中期，其后发育潮下碳酸盐岩沉积。火山活动经历了弱→强→频繁→弱的变化过程。

二、火山岩相及岩石学特征

（一）溢流相

为大红峪期火山活动主要岩相类型，由碱玄质熔岩（厚16～294m，平均169m）和响岩类熔岩（厚度<141m，平均56m）以大致7∶3比例组成的双峰韵律式喷溢，广泛分布于火山岩系中。由于

岩浆混合作用，在两者之间还发育过渡性响岩质碱玄岩和碱玄质响岩。火山熔岩占火山岩厚度>86%，高者（如平谷挂甲峪、黄松峪、黄崖关及蓟州大红峪沟剖面）可达95%~100%，表明火山活动方式以平静的溢流为主。熔岩中气孔杏仁构造十分发育，纵向上呈规律性变化，一般熔岩层底部气孔杏仁体数量较多，多呈扁平状、蝌蚪状，直径0.2~2cm；熔岩层中部气孔杏仁构造最不发育，呈近圆状或不规则状，直径多在1~5cm；熔岩层顶部气孔杏仁体极其发育，呈蝌蚪状、扁平状，直径0.5~10cm。顶部与底部气孔杏仁体具定向排列，可清楚显示熔岩层产状。岩石中有时可见由气孔杏仁体、斜长石碎斑与铁质矿物氧化后形成的同心环状暗色色带组成的涡流构造，说明局部岩浆具快速溢流特征。据杨眉（1990）研究早期碱玄岩类熔岩爆发系数平均为7.6%，相当于夏威夷型；晚期响岩类熔岩爆发系数平均为32.96%，相当于斯通博利型。

按照岩石化学成分组成，溢流相岩石类型发育为碱玄岩、响岩、粗安岩。

1. 碱玄岩

绿灰-灰绿色，风化后呈黄绿色，斑状结构、基质粗玄结构、含长结构，致密块状构造，少数具杏仁状构造，杏仁充填褐铁矿、碳酸盐类物质、绿泥石及硅质。

斑晶矿物含量10%~20%，以斜长石为主，少量辉石及透铁橄榄石。斜长石斑晶半自形板柱状，发育钠长石双晶，偶见环带结构，常边缘熔蚀圆化，粒径1~7mm，类型为拉长石，有序度0.25，属高温无序型。斑晶辉石半自形短柱状，粒径0.5~3mm，常有熔蚀边结构，并在熔蚀边外再生加大辉石边，内嵌斜长石微晶形成含长结构；斑晶辉石成分Wo=38.6~42.4、En=45.1~50.9、Fs=10.6~12.4，类型为含钛普通辉石（TiO_2<1%）；斑晶边部比内部明显富铁而贫镁，显示了岩浆结晶过程向贫镁富铁方向进行。橄榄石斑晶含量少，粒径0.1~0.8mm，常见伊丁石化、蛇纹石化及绿泥石化，类型为透铁橄榄石。

基质由不规则排列的斜长石微晶及充填其间隙中的微粒辉石、橄榄石及隐晶质物质组成。基质斜长石类型为拉长石（An=58.7~69.1），微晶辉石类型为含钛普通辉石及次透辉石，与钛铁矿-钛磁铁矿共生。次生矿物以绿泥石最发育，多呈集合体交代暗色矿物或充填于长石的间隙中。

2. 响岩

紫红色、肉红色、浅褐色，斑状结构、基质粗面结构，气孔状及杏仁状构造，气孔多呈椭圆形或拉长不规则形，定向并分层排列。杏仁体充填物为绿泥石、碳酸盐矿物、硅质及褐铁矿。

斑晶以碱性长石为主，含量约10%，自形板柱状、宽板状，粒径0.5~1.3mm，发育卡式双晶，均匀消光，常见熔蚀边，弱高岭土化，类型为钾透长石。

基质以碱性长石（约80%）为主，自形长柱状，粒径0.05~0.3mm，发育卡式双晶，均匀消光，定向排列，类型为钾透长石。其余基质成分为少量绿泥石、云雾状细晶分散的碳酸盐矿物及铁钛氧化物矿物等。

3. 粗安岩

灰绿色，风化面呈紫红色，斑状结构、基质交织结构，杏仁、气孔构造。主要由钾长石（约45%）、斜长石（约35%）和角闪石（约15%）组成。长石自形—半自形板状，构成岩石主体；角闪石蚀变呈假象填充于长石间，蚀变矿物为绿泥石、黑云母，副矿物为赤铁矿和磁铁矿等。

（二）火山爆发相

在大红峪期火山活动中占比少，但各火山剖面中均有分布，所占比例多<10%。又可划分为火山爆发空落亚相、火山爆发塌陷亚相、隐爆角砾亚相和潜火山岩亚相，依次发育为凝灰岩、火山角砾岩、火山角砾熔岩和辉绿岩-正长斑岩。

1. 凝灰岩

紫红、暗红色，凝灰结构，微层状—块状构造。由晶屑、岩屑和少量玻屑碎屑物质和细晶-隐晶质玻璃质火山灰胶结物组成。晶屑多呈棱角状，大小不一，以碱性长石为主，少量石英和斜长石等。岩屑多为浑圆状响岩和碱玄岩碎屑，粒径0.1~0.2mm。玻屑多为弧面棱角状，脱玻化。岩屑、晶屑、玻屑比例为3：5：1。

2. 火山角砾岩

深褐、浅褐色，角砾组分多为棱角状碱玄岩和响岩，胶结物由凝灰质、碳酸质及少量硅质、铁质物质组成，含石英、长石及不透明矿物等碎屑。

3. 火山角砾熔岩

产于火山喷发角砾岩管，紫红、紫灰色，角砾熔结结构，块状构造。角砾主要由棱角状碱玄岩和响岩组成，少量二者过渡岩石和围岩片麻岩、白云岩等，含量20%~90%，大小2~15cm，还包括少量石英、长石及不透明矿物等碎屑物质。可根据胶结物成分划分为响岩质角砾熔岩和碱玄质角砾熔岩。

4. 辉绿岩与正长斑岩

产于大红峪期火山机构张性裂隙中，火山活动晚期产物，在熊耳寨—小段洼及鱼子山—康石岭一带沿东西向断裂带成群定向分布。

辉绿岩呈灰绿色、墨绿色，斑状结构，基质辉绿结构，致密块状构造。斑晶为斜长石（主）和辉石，含量5%~20%。其中，斜长石自形板状，发育聚片、卡钠复合双晶，粒径3~5mm，类型为中-培长石；基质斜长石为细小板条状，类型为中长石，粒径0.1mm，含量45%~50%；基质中的绿泥石为辉石等暗色矿物蚀变矿物，少部分保留辉石假象，粒径0.05~0.1mm，含量约35%。副矿物为磁铁矿等，含量约15%。

正长斑岩呈肉红色，碎斑结构，似粗面结构，块状构造。斑晶为正长石，碎斑状，发育卡式复合双晶，粒径1~3mm，含量3%~5%。基质正长石细小板条状，杂乱或半定向排列，含量75%~80%；基质绿泥石片状，少部分细小纤维状，绿-淡绿色多色性，由角闪石或黑云母蚀变而成，含量约10%；黑云母片状，浅棕色多色性，部分蚀变绿泥石化，含量约5%；石英他形粒状。

（三）火山沉积相

在串岭沟组、团山子组、高于庄组、下马岭组以及大红峪组中呈夹层产出，发育岩性包括大红峪期火山活动喷发间歇期初沉凝灰角砾岩和其他岩石地层中远源火山灰沉积的斑脱岩。具体岩石特征见第二篇第七章第二节火山碎屑岩类。

统计大红峪期火山活动各岩性（表15-1）组成，火山爆发指数较低，一般为3.2%~17.2%，火山碎屑岩约占大红峪组地层总厚度的7.15%，而熔岩占总厚度的60.88%，并以碱玄岩为主。每个喷发旋回及整个火山岩系中，下部均以碱玄岩为主，向上响岩及同成分火山碎屑岩石不断增多，反映了大红峪期火山岩浆演化及每个旋回的火山喷发都经历了由铁镁质岩浆溢流向碱性硅铝质岩浆喷溢不断转化的过程。在平面上，火山岩带的东部（如黄崖关、蓟县）以碱玄岩所占比例大，有的剖面几乎全由碱玄岩堆积组成；而在火山岩带西北部（如江米洞、东辛庄），响岩增多，江米洞地段形成百米以上巨厚的响岩堆积体。

三、火山韵律与火山旋回

本区喷发中心记录完整火山喷发韵律，由碱玄岩-响岩-火山角砾岩（凝灰岩）-沉积岩构成，火山岩累积厚度大，沉积夹层少且累积厚度小，熔岩累积厚度大，凝灰岩夹层少。离火山喷发中心越远，火山岩累积厚度递减，沉积夹层增多且累积厚度增大，喷发韵律不完整，熔岩累积厚度减小，凝灰岩夹层增加，形成了沉积岩与火山岩交互成层的“掺杂”关系。

火山喷发韵律显示初始水下喷发和中期水上喷发特征（鲍亦冈等，2001）。起始阶段火山边喷溢边冷凝堆积高出水面，可见由单向水流加积形成的以火山岩砾石组成的大型板状斜层理。水下喷发标志有急速冷凝退色边（厚15~180cm）、黄白色蚀变碱玄岩或响岩中具大型喇叭状气孔，气孔中充填碳酸盐物质，火山碎屑岩经水介质重力分选具下粗上细的正粒序韵律（黄松峪）。中期为强烈喷发阶段，曾经历过地表剥蚀留下的大型冲刷面，冲刷面不平整，底部由叠瓦状排列的火山岩（经磨圆）砾石构成大型斜层理，向上为含砾石英质粗砂岩，由单向水流构成斜层系组，显示河流搬运特点。

表 15-1 大红峪期火山岩系区域分布特征表

Table 15-1 Regional distribution characteristics of Dahongyu volcanic rocks

剖面	西-东									南-北			平均值
	甲山 1	东山下 2	挂甲峪 3	鱼子山 4	楼南沟 5	黄松峪 6	将军关 7	黄崖关 8	大红峪沟 9	挂甲峪 3	东辛庄 10	火郎峪 11	
火山岩系地层厚度	311	350	534	555	434	214	453	463	298	534	390	276	401
沉积岩地层厚度	58	76	75	123	111	78	217	197	248	75	114	208	133
火山岩总厚度	253	274	459	432	323	136	236	266	50	459	276	68	268
熔岩总厚度	217	273	434	391	294	132	202	260	50	434	247	18	246
碱玄岩类熔岩厚度	173	117	294	293	285	79	181	165	50	294	153	16	175
响岩类熔岩厚度	45	156	141	98	9. 4	52	21	95	0	141	95	1. 6	71
火山碎屑岩厚度	36	1	25	41	28	4. 4	34	6	0	25	29	50	23
火山岩厚度/火山岩系厚度	81. 5	78. 5	86	77. 9	74. 3	63. 6	52. 2	57. 4	16. 8	86	70. 8	24. 6	63. 5
熔岩厚度/火山岩厚度	85. 5	99. 7	94. 6	90. 5	91	97. 1	85. 6	97. 7	100	94. 6	89. 5	26. 5	85. 9
火山碎屑岩厚度/火山岩厚度	85. 8	99. 7	94. 6	90. 5	91	97. 1	85. 6	97. 7	100	94. 6	89. 5	26. 5	85. 9
火山碎屑岩厚度/火山岩厚度	14. 1	0. 3	5. 4	9. 6	8. 8	3. 2	14. 5	2. 3	0	5. 4	10. 5	73. 5	12. 3
碱玄岩类熔岩厚度/火山岩厚度	68. 2	42. 7	64	67. 8	88. 3	58	76. 8	62	100	64	55. 2	23. 5	64. 2
响岩类熔岩厚度/火山岩厚度	17. 8	56. 9	30. 7	22. 7	2. 9	38. 2	8. 9	35. 7	0	30. 7	34. 4	2. 4	23. 4

注：1、3、4、5、7、8、10、11 剖面据李祖湘等（1981）；6 剖面据杨眉（1990）；9 剖面据蓟州标准剖面与地质大学杨眉实测剖面综合结果；2 剖面据 1∶5 万大华山幅、木林幅、杨镇幅区域地质调查报告（1995）；厚度单位为 m，厚度比值单位为%。

火山喷发韵律组成大红峪期一个完整的火山喷发周期（或称火山-沉积旋回），大致发育有5个喷发-沉积旋回（鲍亦冈等，2001）。第一旋回发育于团山子组顶，经碱玄质和响岩质熔浆溢流后进入较长时期的宁静阶段，是大红峪期大规模火山活动的前奏；第二旋回为大红峪期早期火山喷发，以基性熔浆喷溢为主，早期形成含橄辉石碱性玄武岩，晚期形成薄层响岩或灰紫色响岩质角砾凝灰岩，而后转入短期火山宁静阶段；第三旋回为大红峪期的主要火山活动期，以基性熔岩溢流为主，发育为碱玄岩和响岩交替堆积，而后发育层厚约4m的灰绿色凝灰质砂砾岩沉积；第四旋回为火山活动频繁时期，下部发育为灰绿色碱玄岩与褐红色响岩交替喷发，向上演化为紫红色响岩质凝灰岩层和肉红色响岩夹碱玄岩，最后发育火山间歇期黄色石英砂岩及碧玉条带白云岩、砂质白云岩。第五旋回为火山岩喷发活动的尾声，规模较小，形成褐红色气孔状响岩与灰绿色碱玄岩，夹有灰紫色响岩质含集块角砾凝灰岩与火山角砾岩层。

四、火山构造

本区位于燕山-辽西火山喷发带（二级火山构造单元）的平谷-蓟州沉积-火山盆地（三级火山构造单元）内，是该期火山活动最为强烈的地段。在此火山盆地内，沿密云-墙子路-喜峰口和固安-昌黎深大断裂带发育众多呈线性展布的小型点状火山岩管（图15-2），以将军关、黄松峪和钵尔塘一带最为集中，属四级火山构造单元，仅在北京境内就已发现40多个。

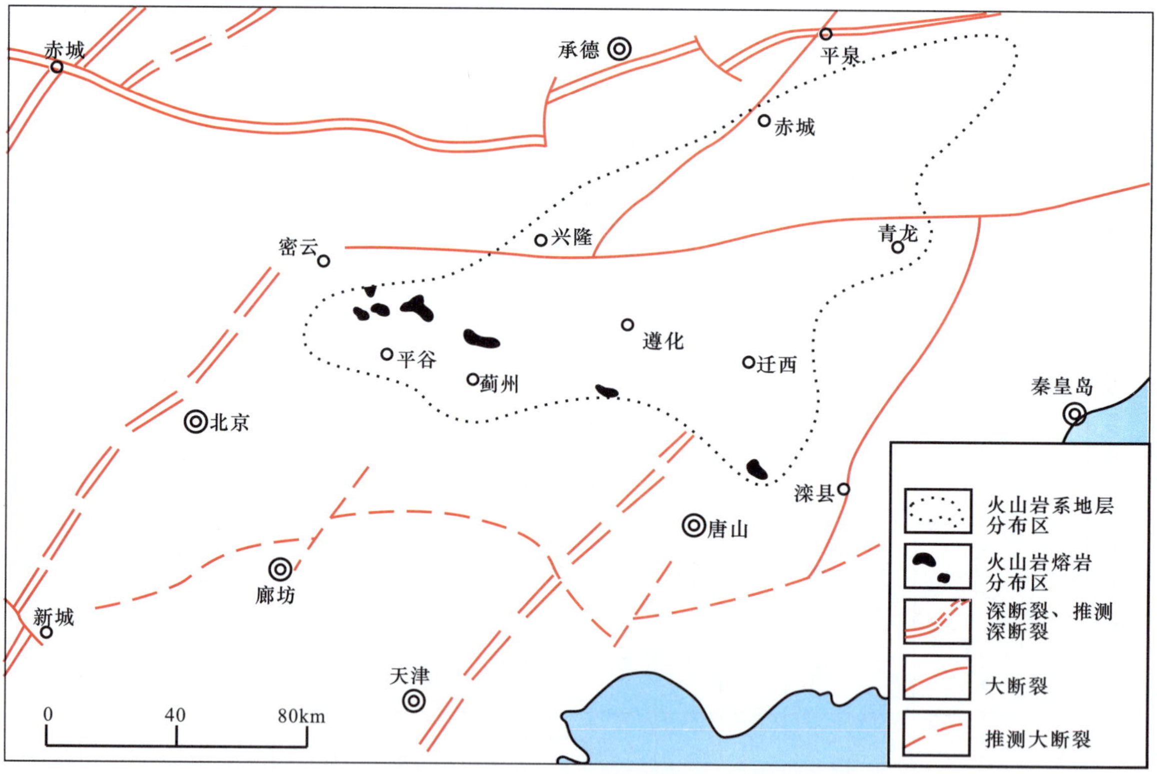

图15-2　大红峪期火山岩分布略图
（据和政军，1990）
Fig. 15-2　The sketch distribution of Dahongyu volcanic rocks

火山岩管平面形态呈近圆形、椭圆形、纺锤形或不规则状，单个出露面积$1.2\times10^4\sim1.8\times10^5m^2$。地貌上，除个别岩管突出地表外，大多数呈钵形、喇叭形凹地。岩管产于新太古代片麻岩—中元古界大红峪组中，与围岩内倾接触，倾角>50°。管内充填物以火山角砾熔岩为主，少数为熔岩、潜火山岩。按充填物成分不同，可将岩管分为角砾凝灰岩（或凝灰角砾岩）、火山角砾岩（火山角砾熔岩）、正长斑岩、辉绿岩4种类型。典型火山机构为程各庄火山岩筒，特征描述如下：

程各庄火山岩筒位于密云程各庄村北，平面呈近东西向钝角三角形（图 15-3），出露面积约 0.9km^2。区内大部分岩管属于此类，剥蚀较浅。火山口切过长城系团山子组、串岭沟组、常州沟组地层，与围岩呈上缓下陡截切突变接触关系。岩管下部岩性为角砾凝灰岩或凝灰角砾岩，向上渐变为凝灰质砂岩，近顶部发育有呈透镜状泥质白云岩和极薄层英安质晶屑凝灰岩。岩性层序及产状特征揭示火山爆发于团山子期潮坪沉积环境，火山活动不很强烈，火山产物分布范围比较狭小，原始火山形态可能为一喇叭状火山口（图 15-3）。

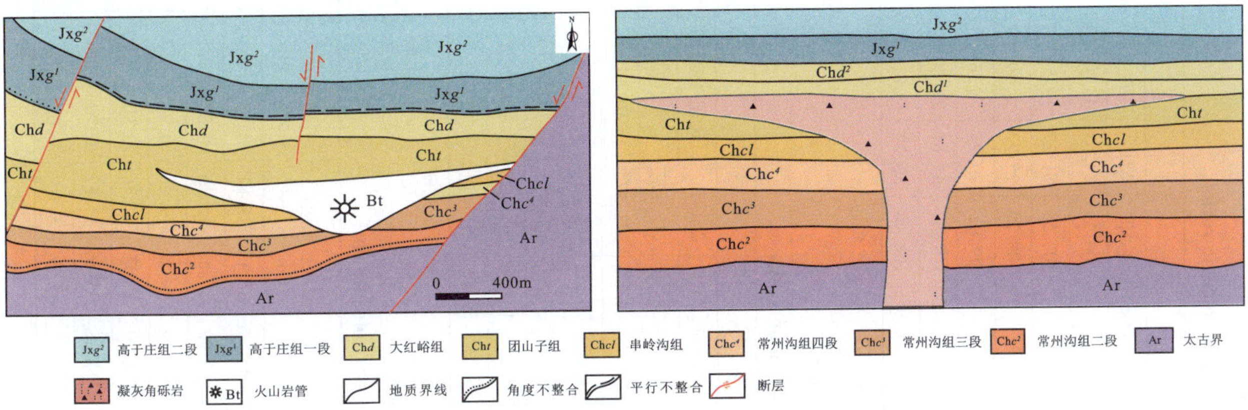

图 15-3 程各庄火山岩筒地质略图（左）与推测火山机构形态图（右）

Fig. 15-3 The Geological sketch of volcanic edifice in Chenggezhuang, Miyun

Ar—太古宇；Chc^2—常州沟组二段；Chc^3—常州沟组三段；Chc^4—常州沟组四段；*Chcl*—串岭沟组；*Cht*—团山子组；*Chd*—大红峪组；Jxg^1—高于庄组一段；Jxg^2—高于庄组二段

五、岩石化学与地球化学特征

大红峪期火山岩化学成分及参数见表 15-2，呈现高 K_2O（K_2O=11.7%~14.9%）、低钠（Na_2O<1.7%）、富铁质（Fe/（Fe+Mg）=0.7~0.9）、贫硅的特征。里特曼指数 σ=10.08~15.42（>4），表明岩石为碱性岩系列；固结指数 *SI*=2.81~31.38，分异指数 *DI*=43.63~86.29，反映岩浆分异程度较低；均位于硅不饱和区，晚期响岩类 SiO_2 含量略有增高；镁铁比值 *m/f*=0.18~1.19，属于铁质-富铁质基性岩；*A/CNK*=0.94~1.19，*A/NK*=1.11~2.18，属于准铝质-过铝质。

在岩石化学 TAS 图解上（图 15-4），成分点位于碱玄岩、响岩及其过渡响岩质碱玄岩、碱玄质响岩区。其中，响岩类 CIPW 标准矿物中大多出现了副长石类矿物，含量介于 0.3%~8.7%。但是，镜下观

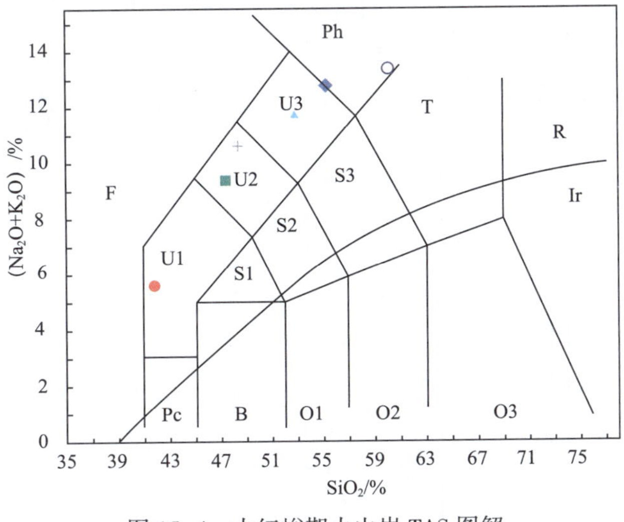

图 15-4 大红峪期火山岩 TAS 图解

Fig. 15-4 TAS diagram of Dahongyu volcanic rocks

表 15-2　大红峪期火山岩岩石化学成分及特征参数一览表

Table 15-2　The chemical composition and main parameters of Dahongyu volcanic rocks

序号	岩性	氧化物含量/%																特征参数					
		SiO_2	TiO_2	Al_2O_3	Fe_2O_3	FeO	MnO	MgO	CaO	Na_2O	K_2O	P_2O_5	H_2O^+	H_2O^-	CO_2	LOS	总和	*DI*	*SI*	σ	*A/CNK*	*A/NK*	*m/f*
1	碱玄岩	41.75	1.55	16.4	6.57	7.41	0.14	8.89	2.85	2.61	3	0.23	5.73	0.91	2.35		100.39	43.63	31.38	13.74	1.29	2.18	1.18
2	响岩质碱玄岩	47.38	2.7	15.75	8.26	4.6	0.06	5.35	1.4	1.2	8.16	0.64	3.83	1.09	0.16		100.58	61.27	19.56	14.34	1.18	1.46	0.79
3	碱玄质响岩	52.81	1.39	16.09	11.16	1.23	0.04	1.67	0.95	0.72	11	0.54	2.33	0.22	0.14		100.29	75.77	6.59	12.81	1.09	1.23	0.26
4	响岩	55.22	0.9	16.61	6.13	2.83	0.14	2.34	0.59	0.3	12.47	0.28	1.82	0.19			99.82	79.25	9.76	12.64	1.10	1.19	0.49
5	辉绿岩	48.34	1.3	16.7	4.34	4.95	0.08	5.97	1.93	0.3	10.3	0.28	2.27	0.41		4.4	101.57	65.61	23.09	15.42	1.10	1.43	1.19
6	正长斑岩	60.08	0.6	16.24	5.08	0.77	0.11	0.55	1.45	0.4	12.9	0.12	4.21	0.55		5.38	108.44	86.29	2.81	10.08	0.94	1.11	0.18

注：数据为均值，引自李祖湘等，1981 和 1∶5 万平谷幅、木林幅、大华山幅区域地质调查报告，1995。

表 15-3　大红峪期火山岩稀土元素成分及特征参数一览表

Table 15-3　REE contents and main parameter of Dahongyu volcanic rocks

序号	岩性	稀土元素含量/10^{-6}																特征参数			
		La	Ce	Pr	Nd	Sm	Eu	Gd	Tb	Dy	Ho	Er	Tm	Yb	Lu	Y	ΣREE	LREE/HREE	$(La/Yb)_N$	δEu	δCe
1	碱玄岩	44.08	83.45	9.12	46.02	8.99	2.71	6.84	1.04	5.72	1.14	2.92	0.44	2.43	0.27	28.94	215.17	9.34	13.01	1.06	1.02
2	响岩质碱玄岩	74.67	149.60	17.72	83.18	15.45	4.04	11.04	1.46	8.77	1.68	4.76	0.59	4.17	0.44	44.84	377.57	10.47	12.84	0.95	1.01
3	碱玄质响岩	102.10	194.50	22.04	101.10	18.30	3.87	13.27	2.10	13.15	2.66	7.70	1.09	6.79	1.05	73.20	489.72	9.24	10.79	0.76	1.01
4	响岩	156.20	283.80	29.99	127.90	20.90	3.37	13.84	2.23	11.77	2.42	7.21	1.02	6.58	0.84	60.56	668.07	13.55	17.03	0.61	1.02
5	辉绿岩	34.15	64.80	7.23	33.30	6.46	1.53	4.85	0.75	4.01	0.82	1.95	0.28	1.77	0.22	19.24	162.12	10.07	13.84	0.84	1.01
6	正长斑岩	123.20	231.50	25.22	110.10	17.69	2.31	11.45	1.74	11.26	2.35	6.70	0.98	6.11	0.80	59.42	551.41	12.32	14.46	0.50	1.02

注：数据为均值，来源同表 15-2。

表 15-4　大红峪期火山岩微量元素成分及特征参数表

Table 15-4　Trace element contents and main parameters of Dahongyu volcanic rocks

序号	岩性	微量元素含量/10^{-6}															特征参数		
		Rb	Ba	Th	U	Ta	Nb	Sr	Zr	Hf	V	Cr	Co	Ni	Cu	Sc	Sc/Cr	Rb/Sr	Nb/Ta
1	碱玄岩	53.00	1763.00	3.89	0.78	1.96	29.00	750.50	160.00	25.00	263.4	34.21	40.42	17.76	38.17	18.81	0.55	0.30	22.52
2	响岩质碱玄岩	92.00	736.30	4.12	0.68	2.66	56.00	476.40	219.00	28.00	69.00	17.21	25.83	4.00	15.83	15.15	0.88	1.14	18.79
3	碱玄质响岩	140.00	341.50	5.62	1.44	2.88	117.00	99.09	773.00	30.00	1.50	12.25	1.00	4.00	5.28	11.45	0.93	0.35	23.85
4	响岩	130.00	337.70	6.00	2.97	3.63	80.00	89.80	492.00	22.00	34.02	16.16	1.20	4.00	9.07	12.9	0.80	1.83	24.29
5	辉绿岩	142.00	1225.40	4.10	0.96	3.51	44.00	211.98	273.50	28.00	64.23	19.30	21.03	4.28	23.86	13.52	0.70	0.43	17.87
6	正长斑岩	86.00	1606.00	5.34	1.20	4.25	96.70	49.56	747.30	30.00	4.19	14.60	1.97	4.00	6.59	9.63	0.66	2.60	16.53

注：数据为均值，来源同表 15-2。

察并未见到副长石类矿物。造成上述矛盾的可能原因是：在含钾较高的岩石中，副长石类矿物以白榴石为主，由于本期火山岩形成年代非常久远，白榴石等副矿物已转化为具白榴石假象的碱性长石等次生矿物，加之含量较少，故镜下难以观察和辨认（鲍亦冈等，2001）。

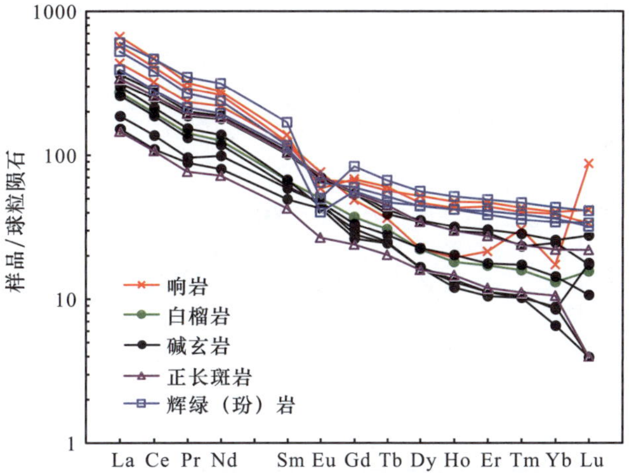

图 15-5　大红峪期火山岩稀土元素配分模式图

Fig. 15-5　Chondrite-normalized REE patterns of Dahongyu volcanic rocks

稀土元素成分及参数见表 15-3。碱玄岩类与响岩类 ΣREE 分别为 162.12 $\times10^{-6}$ ~ 377.57 $\times10^{-6}$ 和 489.72 $\times10^{-6}$ ~ 668.07 $\times10^{-6}$，差异较大；碱玄岩类与响岩类 LREE/HREE 分别为 9.34 ~ 110.47 和 9.24 ~ 13.55，$(La/Yb)_N$ 分别为 112.84 ~ 13.84 和 10.79 ~ 17.03，分馏程度相当，均表现出轻稀土强富集，重稀土相对亏损；δEu 分别为 0.50 ~ 0.76 和 0.84 ~ 1.06，分属于铕负异常和铕弱负异常—无异常；配分曲线特征相似，均为右陡倾光滑曲线（图 15-5），揭示同源岩浆演化特性。

微量元素成分及参数见表 15-4。微量元素蛛网图（图 15-6）中，与球粒陨石相比，过渡族元素 Ti、Sc 呈明显正异常，Cr、Co、Ni 具强烈的负异常；与原始地幔相比，富集大离子亲石元素 K、Rb 和高场强元素 Hf，强烈亏损元素 P，弱亏损 Th、U。碱玄岩类和响岩类元素成分略有所差异，响岩类岩浆演化分异程度增高，元素的富集或亏损呈增强趋势，尤其是不相容元素 Zr、Ti、P、Sr 和过渡族元素 V、Co。微量元素相似特征也揭示了碱玄岩和响岩具有同源岩浆演化分异演化的特点，岩浆可能来源于上地幔岩低度部分熔融，且在分异演化过程中遭受了地壳物质的同化混染。

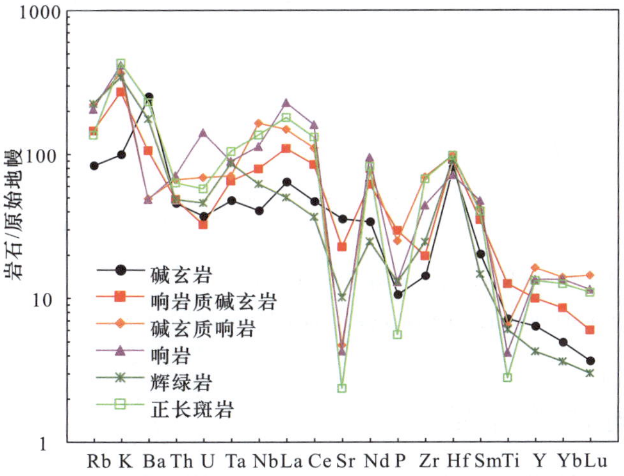

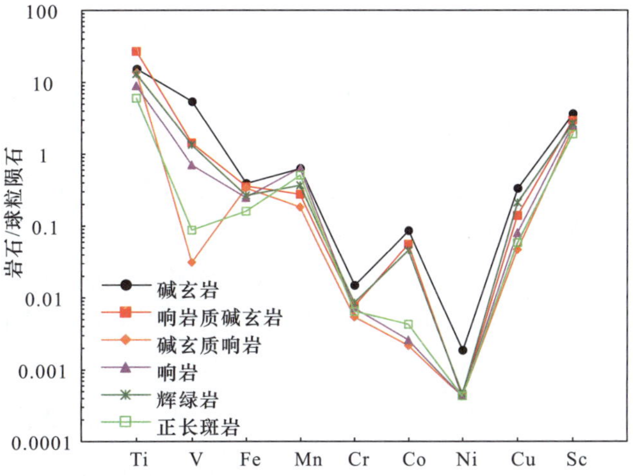

图 15-6　大红峪期火山岩微量元素蛛网图

Fig. 15-6　Primitive mantle-normalized spidergrams for trace elements of Dahongyu volcanic rocks

六、岩石系列划分

该套岩系里特曼指数 σ = 10.08 ~ 15.42，平均值>4，碱度率 AR = 1.82 ~ 7.06，属碱性火山岩。岩石系列图解（图 15-7）上，成分点位于过碱质区，呈伴随 SiO_2 含量升高而碱质不断升高变化趋势，与强地中海型钾质碱性玄武岩（P 线）基本一致。在 Ab′-An-Or 图解和 K_2O-Na_2O-CaO 图解中上（图 15-8），成分点位于钾质岩区，属于钾质碱性岩系。因此，大红峪期火山岩岩石共生组合为大陆拉张条件下裂谷型双模式（双峰式）碱玄岩-响岩组，地球化学为钾质型碱性橄榄玄武岩系列。

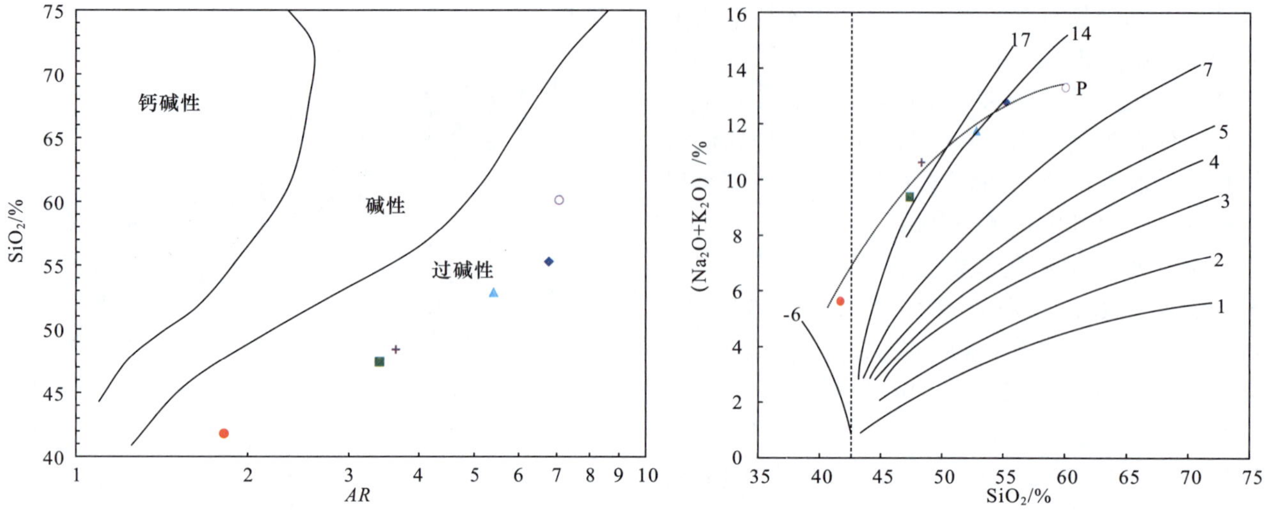

图 15-7 大红峪期火山岩 SiO_2-*AR* 和 SiO_2-（K_2O+Na_2O）图解

Fig. 15-7 The SiO_2-*AR* and SiO_2-（K_2O+Na_2O） diagram of Dahongyu volcanic rocks

（*AR*：碱度率 Alkalinity Ratio）

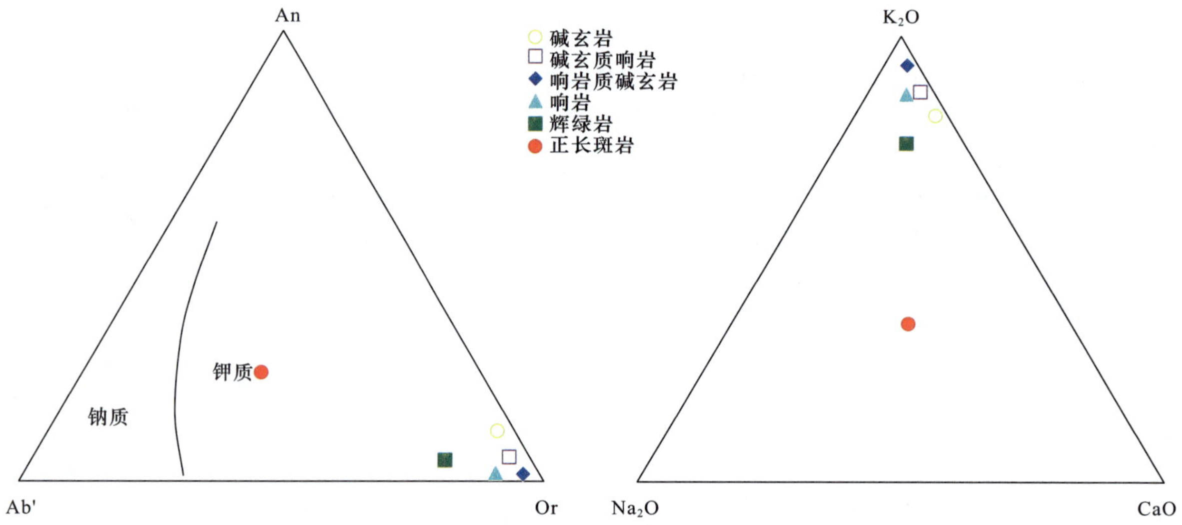

图 15-8 大红峪期火山岩 Ab’-An-Or 和 K_2O-Na_2O-CaO 图解

Fig. 15-8 The Ab’-An-Or and K_2O-Na_2O-CaO diagram of Dahongyu volcanic rocks

七、Sr-Nd 同位素特征

大红峪期火山岩 Sr 同位素略有所差异。火山喷发初期（团山子组顶）碱玄岩（$^{87}Sr/^{86}Sr)_i$ =（0.7047±0.0021）；早—中期（大红峪组下部）碱玄岩（$^{87}Sr/^{86}Sr)_i$ =（0.7045±0.0008）～（0.7049±0.0077）；晚期响岩（大红峪上部）（$^{87}Sr/^{86}Sr)_i$ =（0.7082±0.0012）～（0.7126±0.01）（郁建华等，1996）。碱玄岩初始值明显低于响岩，且变化范围较窄。

大红峪期火山岩 Nd 同位素基本相似，$^{143}Nd/^{144}Nd$ = 0.51189～0.51199，$^{147}Sm/^{144}Nd$ = 0.12003～0.12639；取 t=1650Ma 时，ε_{Nd}（t）=-0.66～+0.63（胡俊良等，2007）。

八、Lu-Hf 同位素特征

大红峪期碱性火山岩$^{176}Hf/^{177}Hf$ = 0.281673～0.281779，取 t = 1650Ma 时，ε_{Hf}（t）= -2.1～+3.8（张健等，2015；Wang et al.，2015a）。

九、岩石成因与岩浆演化

大红峪期钾质碱性火山岩系为双峰式火山岩，钾质碱性玄武岩系与高钾硅铝质的响岩紧密共生韵律式喷发。岩石地球化学特征揭示双峰式铁镁质与硅铝质两种岩浆具有同源岩浆的线性演化成因关系。Sr-Nd 及锆石 Lu-Hf 同位素特征揭示其来源于弱亏损型地幔源区，起源于岩石圈地幔中石榴子石二辉橄榄岩的部分熔融，熔融程度为 2%~3%（张健等，2015；Wang et al.，2015a）。

大红峪期火山岩上部响岩质岩石具有较小的 Sm/Nd 及较高的 Ce/Nd 和 Th/Yb 值；火山岩（$^{87}Sr/^{86}Sr)_i$ 值高，变化范围又宽，明显不同于下部碱玄质岩石。这些特征说明后期岩浆曾受地壳物质的混染，而分离结晶作用不显著，呈现快速上升喷发特性。任富根（1986）认为中元古界长城系中存在大量的富钾质黏土质岩石，在上升过程中有浅部的钾质混染。

大红峪期火山喷发过程中经历较弱岩浆分异作用。由早期碱玄岩类至晚期响岩类，K_2O 成分含量骤增，Na_2O 成分含量略增，CaO 含量减少，*DI* 值明显增大，岩石向高分异、碱性钾质方向演化（图 15-8）。分异指数（*DI*）与各类氧化物关系的图解中，碱玄岩类和响岩类成分大体上分别集中于两段：Al_2O_3 含量变化不大；Fe_2O_3 含量偏高且变化较大；Na_2O 含量偏低，早期碱玄岩类变化不明显，而晚期响岩类则有显著的减少；早期碱玄岩类 K_2O 含量基本正常，晚期响岩类 K_2O 含量迅速增高，高达 7~9（鲍亦冈等，2001）。固结指数 $SI=2.81\sim31.38$，明显低于日本八丈岛火山岩系（$SI=40\sim30$）；MgO、CaO 与 lg*SI* 呈明显线性关系，（K_2O+Na_2O）和 K_2O 与 lg*SI* 呈线性负相关，其他氧化物与 lg*SI* 则不具线性关系，揭示大红峪期火山岩原始岩浆形成后先经历弱分异作用，其后喷发活动过程中遭受地壳物质的混染（鲍亦冈等，2001）。

十、构造环境分析

大红峪期火山岩形成于被动型裂谷大地构造背景环境，深部地球动力学过程可能与地幔柱作用有关。中元古代在地幔隆起和陆壳拉伸作用的影响下，相对固化稳定的克拉通发育三叉裂谷系，本区形成燕辽拗拉槽，受深断裂作用和上地幔底辟隆起的制约，诱发了幔源铁镁质岩浆的上升，在此构造背景下，沿着深断裂火山阵发性喷发活动，韵律式喷溢堆积。与此同时，地幔上隆所提供的热源和深断裂构造作用所产生的热量促使下地壳较低程度部分熔融，形成硅铝质岩浆。两种岩浆在喷发过程中必然互相夹裹出现混合和混熔作用，因而出现响岩质碱玄岩和碱玄质响岩两个端元岩石之间的一系列岩浆混合岩。由于火山作用是快速地喷溢，因此这种混合是极不均匀的，不可能像深成相侵入岩那样进行均匀地混熔，表现在岩石特征上为斑杂状，在玄武质为主的熔岩中有大量紫红色未完全混熔的响岩质条带状、团块状、树枝状包体，岩石中普遍出现钾长石晶块。反之在碱玄质响岩中亦然。而整个喷发过程早期以玄武岩浆为主，中期玄武质响岩质熔浆交替喷发，晚期渐变为以响岩质熔浆为主。

大红峪期火山岩以喷发不整合覆盖于海相沉积物上，说明区内裂谷从常州沟期就开始出现。另外在团山子组白云岩层上发育有大型冲刷面，在其面上常见白云岩砾石呈叠瓦状排列，构成单向斜层理，表明在古风化壳上有陆相的单向水系存在，而这种单向水系是裂谷发育的标志之一。据此也可以推断裂谷早期发育沉积作用，而后地幔上隆和火山作用才开始，属于被动型裂谷。

十一、形成时代

大红峪期火山岩形成于中元古代早期，主要依据包括：①火山-碎屑地层与长城系常州沟组等地层整合接触，共同不整合覆盖于晚太古代变质岩中；②岩石未发生高级变质变形作用；③团山子组粗面安山岩锆石 U-Pb 同位素年龄为 1606~1671Ma（李怀坤等，1996；张拴宏等，2013；Wang et al.，2015），大红峪组粗面岩、玄武岩锆石 U-Pb 同位素年龄为 1622~1671Ma（李怀坤等，1991；高林志等，2008；Lu et al.，2008；Wan et al.，2013；张健等，2015；Wang et al.，2015）；④团山子组粗面安山岩全岩 Rb-Sr 同位素年龄为（1606±19）Ma（郁建华等，1996），大红峪组底部和顶部火山岩全岩 Rb-Sr 同位素年龄分别为（1543.81±69）Ma 和（1371.60±30）Ma（鲍亦冈等，2001）。

第十六章　古生代以来火山岩与火山作用

第一节　古生代火山岩与火山作用

古生代火山岩仅发育于河北省丰宁云雾山一带，出露面积约 40km²。此晚古生代火山活动可能波及北京，表现为同期境内双泉组地层沉积物中发育火山角砾及凝灰物质。

《中国区域地质志·河北志》（2017）将古生代火山-碎屑地层新建立云雾山组，火山岩厚度约 2808m，并划归到额里图小旋回，时代为二叠纪。火山岩以粗安质为主，间夹流纹质，与沉积岩呈互层产出。岩性包括中性粗安质角砾凝灰岩、角砾熔岩、集块角砾岩、粗安岩、粗安质角砾熔结凝灰岩，和酸性流纹质角砾熔结凝灰岩、角砾凝灰岩、流纹岩，岩石普遍发生绿泥石化、绿帘石化蚀变。

该期火山活动以溢流相为主，爆发相、碎屑流相次之。除此以外，地层中还发育潜流纹岩、潜石英粗安岩、潜粗安岩、潜安山岩等潜火山岩相。岩石地球化学反映岩石为钙碱性系列，形成于大陆边缘造山带岩浆弧构造环境。

第二节　中生代火山岩与火山作用

中生代燕山构造旋回地壳活化，火山活动频繁，形成了南大岭、髫髻山、张家口三期典型的陆相火山岩，是区内最重要的火山活动时期。

一、南大岭期火山岩及火山作用

（一）时空分布特征

火山岩分布于南大岭组中，为早燕山期火山喷发阶段产物，可划分为两个喷发周期，相当于火山岩岩石地层的南大岭组第一、第二岩性段。

早期第一岩性段火山岩沿断裂排布构成裂隙喷发带，熔浆沿裂隙喷发带呈脉动式连续多次溢出，形成面积较大的火山岩熔被。火山活动的爆发指数接近于零。该期火山岩呈北东东向延展分布于北京门头沟区清水-房山区抱儿水-海淀区颐和园，分布面积 1800km²。受后期褶皱构造影响，出露于庙安岭、九龙山向斜两翼，北岭向斜的北翼及石景山、八宝山等地区，出露面积 55km²。怀柔南、顺义二十里长山等地新生界之下也有本岩性段分布。该段火山岩厚度以门头沟模式口最大（766.6m），向西逐渐变薄，潘涧沟 357m，木城涧沟 247m，九龙山向斜北翼大窑村 157m、黄岭西 28m，抱儿水 15m（图 16-1）。

晚期第二岩性段火山岩以中心式喷发为主，喷发强度由东向西逐渐减弱。其中，喷发中心位于桑峪—法城一带，粗碎屑的火山角砾集块岩和集块角砾岩占 30%，爆发指数平均为 45.4%，法城口一带高达 96.5%。该期火山岩呈北东向带状分布于百花山-髫髻山向斜的西北翼，即从法城口向西经桑峪、牛站、蔡家岭、双石头、黄岭西至清水村北，走向长约 20km。火山岩带东端被法城断层切断，西段在清水村西被窑坡组地层超覆（鲍亦冈等，1995）。该段火山岩厚度以门头沟区军饷村最大（475m），向西逐渐变薄，牛站为 257m，双石头为 127m，清水村西仅为 6.5m。

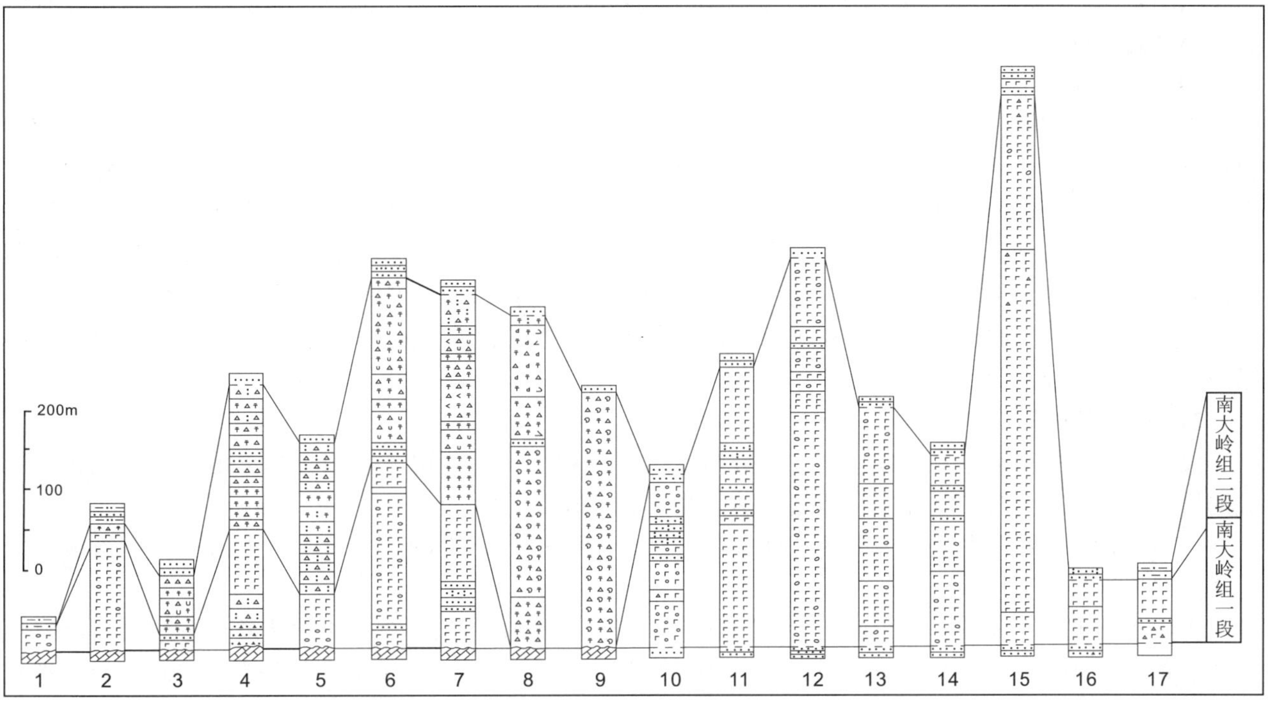

图 16-1　北京地区南大岭组火山岩段喷发韵律相序剖面对比图
（据鲍亦冈等，1995）
Fig. 16-1　Profile correlation diagram of volcanic eruption lithofacies sequence for Nandaling formation in Beijing
1—梯子沟；2—畜牧场；3—黄岭西；4—双石头；5—蔡家岭；6—中点；7—垒峪；8—军响东桥头；9—法城口；10—木城涧；11—潘涧沟；12—官厅-阳坡地；13—东阳坨；14—黑龙潭；15—模式口；16—南东营；17—大窖

（二）火山岩相与岩石学特征

1. 溢流相

为本期火山活动主要火山岩相类型，多发生在火山活动早期。

（1）变玄武岩

广泛发育，呈灰绿、暗紫色，斑状结构，基质拉斑玄武-粗玄结构，块状、气孔-杏仁构造。岩石蚀变钠长石化、碳酸盐化、绢云母化、绿泥石化。杏仁体充填物为绿泥石、方解石，少数石英和绿泥石，顶部形态多呈波浪及云朵状，底部则多呈椭圆形，少数纺锤状。

斑晶由斜长石、橄榄石（假象）、辉石组成，含量8%~15%。斑晶斜长石自形板状，粒径0.5~2mm，发育钠长石双晶，弱绢云母化；斑晶橄榄石铁叶绿泥石及磁铁矿交代而保留矿物外形、解理，多以单晶发育，个别呈聚斑状。

基质由斜长石、辉石、橄榄石（假象）、磁铁矿、赤铁矿等组成。斜长石自形-半自形，粒径0.05~0.6mm，已被钠长石、绿帘石和绢云母所交代而保留微晶外形，类型为钠长石-更长石，有序度0.2~1，变化较大。钠长石成分与岩石化学成分明显不协调，揭示此钠长石可能为钠长石化产物，而非原生岩浆结晶产物；微晶辉石多已绿帘石化，偶见未蚀变者，经成分测定为Ca-Mg-Fe辉石组的透辉石。

（2）玄武岩

仅发育于门头沟双石头，呈灰黑色，斑状结构、基质粗玄结构。块状构造，斑晶主要由斜长石和蚀变橄榄石组成，含量约15%~20%。斑晶斜长石自形-半自形板柱状，粒径2~3.5mm，类型为拉长石，含量约15%；斑晶橄榄石粒径0.5~1mm，发育特征鲜明蚀变：边部伊丁石化环边（宽0.03mm），内核磁铁矿、蛇纹石化。根据蚀变特征推测成岩后可能先遭受低温氧化条件下蚀变作用，形成伊丁石外壳；后遭受中等温度热地质条件下蚀变作用，在已形成伊丁石外壳的基础上发育蛇纹石化内核。

基质矿物以斜长石微晶为主，呈自形短柱状，构成粗玄结构的三角形格架，其钙长石An组分含量为40.02%，属中长石。斑晶斜长石的An含量比微晶高，表明斑晶相结晶早于微晶相。基质中尚有微晶普通辉石，他形粒状，粒径0.2mm。

（3）安山岩

浅灰色、浅绿色，斑状结构，基质玻晶交织结构，块状构造，强烈绢云母化。斑晶主要由角闪石、斜长石、不透明矿物等组成，基质由斜长石微晶和隐晶质物质组成。斑晶角闪石自形，沿解理缝发育不透明矿物，矿物蚀变暗化、绢云母化、绿泥石化，边缘分布一圈不透明矿物，中间为绢云母和绿泥石。较小矿物颗粒蚀变暗化，较大矿物内嵌磷灰石。斑晶斜长石自形-半自形柱状，发育钠长石双晶和卡式双晶，偶见卡钠复合双晶，具溶蚀边缘，强烈绢云母化；基质斜长石微晶自形针柱状，不规则排列，强烈绢云母化、绿泥石化。基质不透明矿物大多为磁铁矿，呈星点状均匀分布；蚀变矿物绢云母、绿泥石不均匀分布，无明显形态，大小不一。

2. 火山爆发相

仅发育于火山活动第二喷发周期，细分为火山空落相、火山碎屑流堆积相和火山崩落堆积相，形成的岩石类型分别为凝灰岩、安山质角砾熔岩和火山角砾凝灰岩。

火山爆发空落相凝灰岩，浅紫色、浅绿色，隐晶质结构，板状构造，主要由火山灰物质组成，多变质为绿泥石板岩。

火山碎屑流堆积相安山质角砾熔岩，灰绿色、紫色，角砾熔岩结构，基质玻晶交织结构，角砾状构造。角砾（少量集块）及岩屑成分均以安山岩碎屑为主，特征与溢流相安山质熔岩相同，少见玄武岩和碳酸盐岩，棱角状-次棱角状，砾径一般<50mm，含量35%～40%；集块含量一般<10%，砾径50～100mm，胶结物以安山质熔浆为主。

火山崩落堆积相火山角砾凝灰岩，紫色、紫灰色，凝灰结构。碎屑物质中角砾含量<10%，岩屑含量约70%，晶屑含量约10%，火山灰含量约10%。岩石蚀变，次生矿物为呈鳞片状集合体产出的绿泥石、绢云母，或交代原生黑云母、角闪石、长石等矿物，或不规则充填裂隙中，分布不均匀。角砾与岩屑棱角状-次棱角状，角砾粒径>3mm，岩屑粒径0.2～2mm，成分以粗安岩、安山岩为主；晶屑尖棱角状或不规则状，粒径0.1～0.3mm，组成为斜长石（主类型为钠长石）、碱性长石、石英、黑云母、角闪石等；火山灰胶结填充于角砾、岩屑、晶屑空隙，隐晶质—显微晶质，略显光性，可能为长英质物质。

（三）火山韵律与火山旋回

南大岭期早期火山活动为基性玄武质熔浆多次连续宁静的溢流，形成一个玄武质熔岩的基本层，发育完好的基本层都由a、b、c、d（从上到下）4个小层组成（图16-2），它们具有不同的颜色、结构、构造和物质成分特点（鲍亦冈等，2001）：

a层多呈紫红色（俗称红顶），位于基本层的最上部，厚度较小。气孔-杏仁构造非常发育，气孔多呈云朵状或不规则长条状，拉长方向与层面近平行。b层玄武岩以灰绿色为主，气孔-杏仁构造的发育明显弱于a层，且多为圆形-椭圆形，直径相对较大，分布较均匀。c层为灰绿-黄绿色，多为致密块状，气孔不发育，厚度相对较大。d层位于基本层最底部，颜色相对较暗（俗称绿底），厚度较小，气孔小而密，多呈椭圆形。

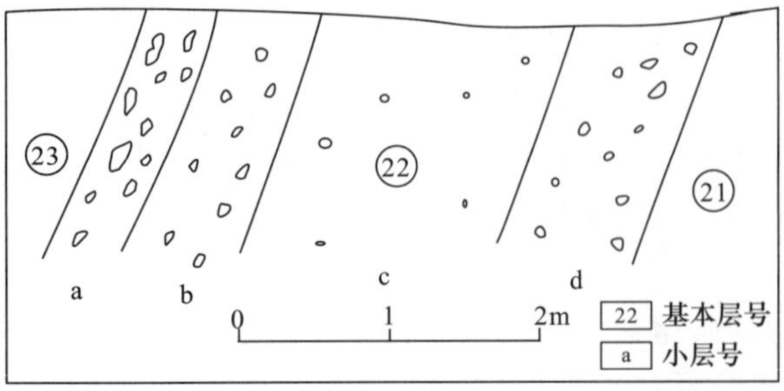

图16-2　潘涧沟剖面变玄武岩第22基本层素描图

（据鲍亦冈等，2001）

Fig. 16-2　Sketch showing the 22nd fundamental layer of the metabasalts in Panjiangou section

每个基本层之间具有突变界面，而基本层内4个小层之间则为渐变关系。有些熔岩层其基本层的剖面结构不完整，常出现缺失某些小层的情况。据对潘涧沟剖面的统计：41个熔岩基本层中结构完整的有27个，有11个基本层缺失d层，两个基本层缺失a层（可能两次喷溢之间间断时间较长，前一熔岩层的顶面已遭剥蚀）。自上而下4个小层的平均厚度之比为1.1∶3.8∶7.3∶1.0。

各小层之间的物质成分变化为：a 层至 d 层，Fe_2O_3 及 TiO_2 呈逐渐降低趋势，而 FeO 和 *OX* 值（OX=FeO/FeO+（Fe_2O_3））呈逐渐增高趋势，表明每一基本层由底到顶氧化程度逐渐增强（图 16-3）。

每一基本层的厚度介于 0.85m～40.69m，一般厚度介于 5～20m，表明每一次熔浆溢流，平均堆积形成 5～20m 厚的玄武岩单层。不同地段其基本层的数量多寡不一，香峪的基本层数最多，达 58 个，潘涧沟 41 个，三家店附近 34 个。这也表明当时喷溢中心在现今门头沟区东部。

在下段的玄武岩层中见有 2～4 层凝灰质砂岩的夹层，有些地方变为页岩夹层。说明在整个喷溢过程中有间歇期，有短时间的小水域湖泊相沉积。南大岭期火山作用的晚期，绝大部分地区都已停止喷溢，但在门头沟区的西北部依然有小面积的中性粗安质熔浆强烈爆发，形成第二喷发周期爆发相火山碎屑岩。

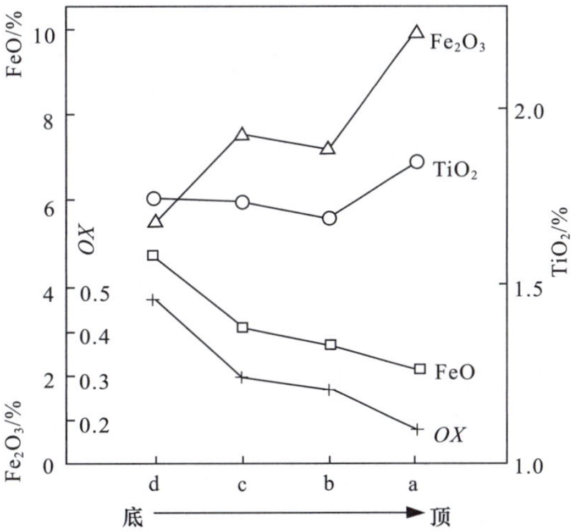

图 16-3　基本层内化学成分变化简图

（据鲍亦冈等，2001）

Fig. 16-3　Schematic map showing chemical composition variation of the fundamental layer

（四）火山构造

参考《中国区域地质志 · 河北志》（2017）火山构造单元划分，本区位于承德-武安火山喷发带（Ⅱ级），发育西山和昌怀两个火山-沉积盆地（Ⅲ级）。

1. Ⅱ级火山构造单元——火山喷发带

南大岭期火山岩分布在密云-墙子路断裂以南，范围基本限定在房山区以北，西起河北蔚县地区，东到平谷及河北省三河市一带，呈近东西向延伸。本市范围内该带东西长约 130km，南北最宽处约 40km。

所处的火山-沉积盆地呈近东西向，明显受到印支期构造的影响和限定，表现为火山岩层等厚线（图 16-4）和紧接南大岭阶段火山喷发后堆积的含煤窑坡组地层等厚线、煤层等厚线均整体上表现为近东西向延展。该火山喷发带的产物受到后期多次褶皱、断裂变形的改造，已彻底改变了其原始面貌（鲍亦冈等，1995）。

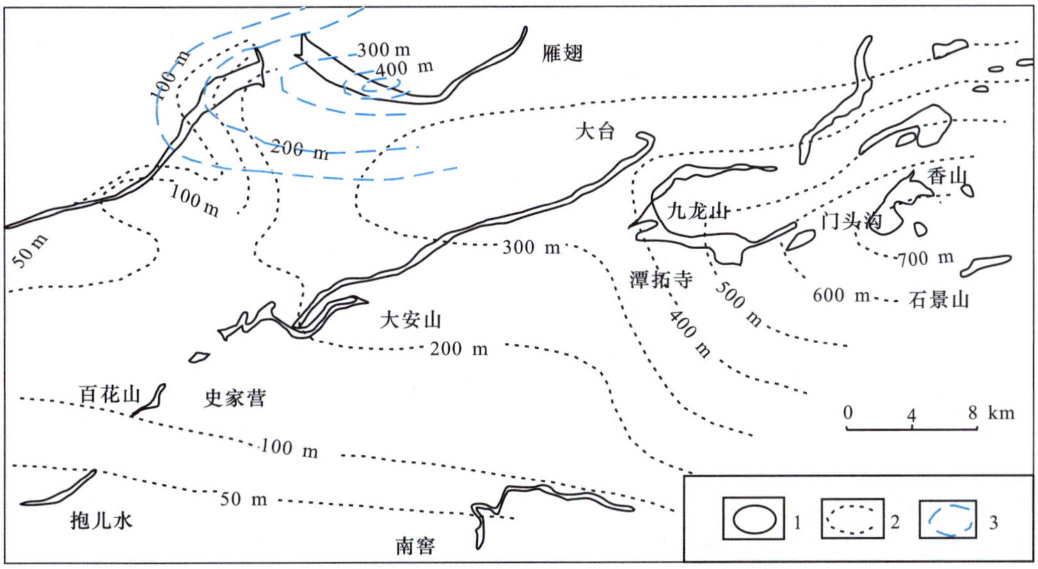

图 16-4　北京西山地区南大岭组火山岩等厚度线图

（据鲍亦冈等，1995）

Fig. 16-4　Isopachous map of volcanic rocks of Nandaling formation in Xishan basin, Beijing

1—J_1n 南大岭组火山岩地表出露区；2—J_1n^1 南大岭组下段火山岩等厚线；3—J_1n^2 南大岭组上段火山岩等厚线

2. Ⅲ级火山构造单元——火山-沉积盆地

（1）西山火山-沉积盆地

位于本区西南部，西自门头沟区上清水村—房山区抱儿水一带，东至丰台区和海淀区颐和园一带；北到海淀区温泉和门头沟青白口一线；南到房山区区政府附近，长约60km，最宽处约35km，盆地呈椭圆形，长轴近东西向，原始分布面积约2000km^2（图16-5）。该盆地形成于前燕山期晚古生代和三叠纪沉积基底上，具有明显的继承性特征，沉积杏石口组、南大岭组、窑坡组、龙门组和九龙山组巨厚层火山喷发物质和陆相碎屑沉积物质，各沉陷阶段的沉积物均呈近东西向分布。

火山-沉积盆地经历两个重要发展演化阶段。早期盆地类型为继承性上叠式伸展型坳陷盆地，表现为整体均匀下沉，各时代地层之间无较大规模的超覆，沉积中心亦无明显迁移现象，也不存在控制整个盆地沉降的盆缘断裂构造。李安宁等（1994）认为，九龙山—清水村一带存在一条隐伏的东西向深断裂，这一断裂印支构造旋回阶段已然存在，并作为热隔障使断裂两侧的印支期构造形变产生差异。早侏罗世早期（南大岭期），在区域北北西-南南东向引张构造背景环境下，该断裂成为南大岭期火山岩浆上升的通道，形成较大面积的裂隙式喷发。大规模基性熔浆溢流结束之后，于盆地西北边缘构造交接部位形成了小面积以中性粗安质岩浆爆发作用为主的第二喷发周期火山活动。早侏罗世晚期（窑坡组），强烈火山活动之后，盆地经历了一段较长时间的火山平静期，接受了一套稳定窑坡组碎屑岩沉积。

由于中侏罗世区域构造应力场由引张转变为挤压，晚期盆地类型转化为挤压挠曲型盆地，表现为沉积地层厚度由中部向两翼减薄，沉积中心发生区域上迁移，九龙山组地层在盆地西、北部超覆在古生代地层之上。该阶段盆内缺乏火山活动，仅沉积了龙门组和九龙山组碎屑岩地层，并最终在过补偿充填状态下盆地逐渐消失。同时，先期盆地内堆积物还发生轴向北东65°~70°纵弯褶皱和剪切滑脱作用，形成了庙安岭、九龙山-香峪和北岭等侏罗山式向斜构造，发育挤压型变质应力矿物——硬绿泥石，并使得南大岭玄武岩矿物定向排列。

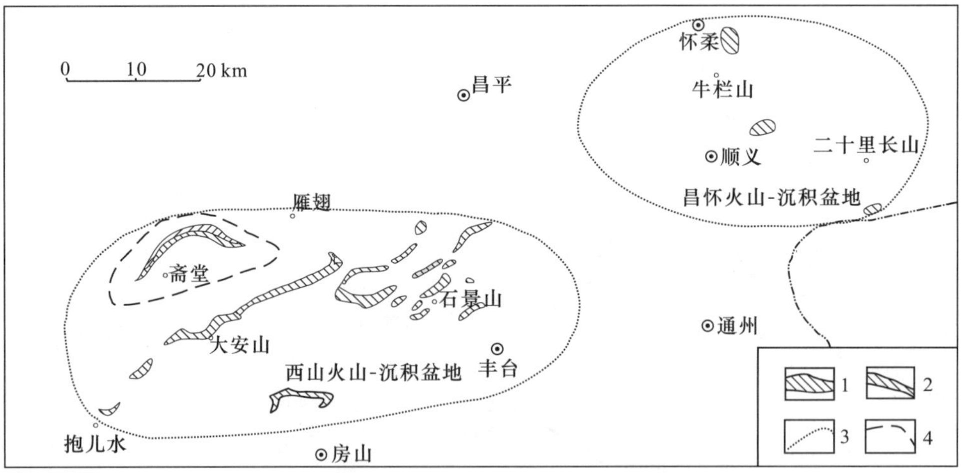

图16-5 北京地区南大岭期火山岩分布略图

（据鲍亦冈等，1995）

Fig. 16-5 The sketch distribution of Nandaling volcanic rocks in Beijing

1—J_1n^1 出露范围；2—J_1n^2 出露范围；3—推测的 J_1n^1 原始分布范围；4—推测的 J_1n^2 原始分布范围

（2）昌怀火山-沉积盆地

位于本区中部，被新生代盆地叠加覆盖。盆地基底为石炭纪—二叠纪地层，盆地内沉积岩性为变玄武岩、凝灰砂岩、凝灰砾岩、砂岩和粉砂岩夹煤层组合。根据众多钻孔资料推测盆地西自昌平南口一带，东至顺义二十里长山一带，南至通州区北部，北至怀柔区南部，长约40km、最宽处约25km，近东西向呈椭圆形（图16-5）。

3. Ⅳ级火山构造单元——火山机构

南大岭期火山活动划分为两个喷发周期。第一喷发周期为裂隙式溢流，形成了大套的面状分布基性熔岩，延伸较好。在桑峪地区，可见基性熔岩裂隙溢出特性，后期有顺裂隙充填线状潜火山岩相辉绿玢岩，与玄武岩呈逐渐过渡的关系（图 16-6 左）。第二喷发周期为中心式喷发，火山机构特征明显：发育多组裂隙；周围分布角砾状熔岩与火山碎屑岩，后又充填潜火山岩相中酸性熔岩，近距离岩相变化大，延伸不远（图 16-6 右）；中酸性熔岩基质隐晶质，边部发育粗安岩包裹体，显示晚期潜火山岩快速冷凝特点。

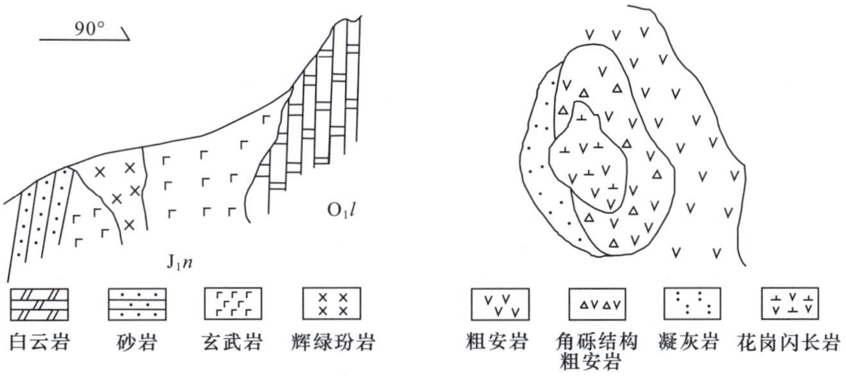

图 16-6　南大岭期火山机构素描图

Fig. 16-6　The sketch map of volcanic apparatus of Nandaling formation

J_1n—南大岭组；O_1l—亮甲山组

（五）岩石化学与地球化学特征

南大岭期火山岩化学成分及参数见表 16-1。与中国玄武岩平均值相比，一段玄武类岩石 SiO_2、Al_2O_3、FeO^*、Na_2O 含量稍高，其余氧化物含量稍低；与世界粗安岩平均值相比，二段粗安类岩石 SiO_2、CaO 含量相近，MgO、FeO 含量低，Al_2O_3、Fe_2O_3 含量高。南大岭期火山岩在 SiO_2 或分异指数（*DI*）变化不大的情况下，K_2O 和 Na_2O 含量有较大波动变化，表明其岩浆可能受到不同程度的同化混染作用或后期蚀变影响。

岩石里特曼指数 σ=1.73~6.12，岩石跨碱性岩和亚碱性岩区；固结指数 *SI*=8.34~23.34，分异指数 *DI*=43.55~69.95，反映岩浆分异程度较低；晚期粗安岩类 SiO_2 含量略有增高，*DI* 值略有增大，分异程度略有提高；镁铁比值 *m/f*=0.35~0.83，属于铁质—富铁质基性岩；*A/CNK*=0.83~1.69，*A/NK*=1.55~2.60，属于准铝质—过铝质。

稀土元素成分及参数见表 16-2。一段、二段 ΣREE 分别为 294.06 $\times10^{-6}$~349.01 $\times10^{-6}$ 和 193.93 $\times10^{-6}$~290.17 $\times10^{-6}$，稀土总量较低，一段略高于二段；LREE/HREE 分别为 9.99~14.46 和 8.75~10.22，$(La/Yb)_N$ 分别为 11.95~21.57 和 8.32~13.28，分馏程度相当，均表现出轻稀土弱富集，重稀土相对亏损；δEu 分别为 0.84~0.96 和 0.76~0.99，均属于铕极弱负异常；一段和二段配分曲线特征相似，均为右缓倾光滑曲线（图 16-7），揭示同源岩浆演化特性。

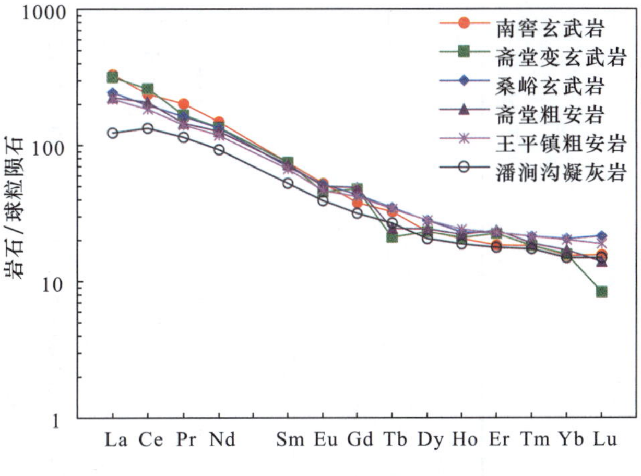

图 16-7　南大岭期火山岩稀土元素配分模式图

Fig. 16-7　Chondrite-normalized REE patterns of Nandaling volcanic rocks

微量元素成分及参数见表 16-3。过渡

表 16-1　南大岭期火山岩岩石化学成分及特征参数一览表

Table 16-1　The chemical composition and main parameters of Nandaling volcanic rocks

序号	取样地点	岩性	氧化物含量/%													特征参数					
			SiO_2	TiO_2	Al_2O_3	Fe_2O_3	FeO	MnO	MgO	CaO	Na_2O	K_2O	P_2O_5	Los	总和	*DI*	*SI*	σ	*A/CNK*	*A/NK*	*m/f*
1	法城-牛站	安粗岩	56. 32	0. 78	18. 31	6. 63	1. 24	0. 19	1. 48	4. 48	4. 06	4. 14	0. 54	1. 54	99. 71	65. 35	8. 58	4. 83	0. 95	1. 64	0. 36
2	斋堂	粗安岩	59. 94	0. 60	17. 16	4. 68	2. 21	0. 18	1. 30	3. 72	5. 17	2. 37	0. 33	2. 13	99. 79	69. 95	8. 34	3. 24	0. 96	1. 55	0. 35
3	潘涧沟	角砾凝灰岩	61. 11	0. 59	18. 03	1. 15	4. 73	0. 05	1. 25	2. 70	3. 97	1. 78	0. 31	1. 64	97. 31	68. 38	9. 70	1. 73	1. 35	2. 13	0. 38
4	潘涧沟	凝灰岩	57. 52	0. 59	19. 87	6. 27	0. 81	0. 16	1. 40	2. 27	2. 89	2. 66	0. 28	1. 48	96. 20	66. 48	10. 22	1. 93	1. 69	2. 60	0. 38
5	斋堂-潘涧沟	玄武粗安岩	55. 17	1. 21	17. 36	6. 94	1. 22	0. 25	2. 92	4. 60	4. 49	2. 38	0. 68	0. 77	97. 99	60. 04	16. 58	3. 61	0. 95	1. 74	0. 67
6	斋堂-潘涧沟	玄武岩	48. 94	2. 03	16. 33	5. 94	5. 11	0. 18	4. 95	6. 38	3. 88	1. 49	0. 77	2. 06	98. 06	43. 55	23. 34	3. 89	0. 83	2. 04	0. 83
7	斋堂	变玄武岩	50. 41	1. 96	17. 82	6. 91	3. 27	0. 12	3. 16	5. 19	3. 95	3. 25	1. 13	0. 98	98. 15	55. 05	15. 58	6. 12	0. 92	1. 78	0. 59

注：数据为均值，1~5 为二段，6~7 为一段；引自 1∶5 万雁翅、阳坊幅区域地质调查报告，1994；1∶5 万周口店幅区域地质调查报告，1988；鲍亦刚等，1995；李晓勇等，2004。

表 16-2　南大岭期火山岩稀土元素成分及特征参数一览表

Table 16-2　REE contents and main parameter of Nandaling volcanic rocks

序号	取样地点	岩性	稀土元素含量/10^{-6}																特征参数			
			La	Ce	Pr	Nd	Sm	Eu	Gd	Tb	Dy	Ho	Er	Tm	Yb	Lu	Y	ΣREE	LREE/HREE	$(La/Yb)_N$	δEu	δCe
1	南窖	玄武岩	78. 53	144. 51	19. 09	68. 91	11. 34	3. 02	7. 75	1. 21	5. 88	1. 16	3. 03	0. 46	2. 61	0. 40	30. 10	347. 89	14. 46	21. 57	0. 99	0. 92
2	斋堂	变玄武岩	74. 34	158. 24	15. 64	62. 37	11. 33	2. 61	9. 72	0. 78	5. 82	1. 17	3. 67	0. 46	2. 65	0. 21	33. 76	349. 01	13. 26	20. 12	0. 76	1. 14
3	桑峪	玄武岩	57. 46	119. 20	15. 30	61. 84	10. 54	2. 97	8. 82	1. 29	7. 07	1. 29	3. 75	0. 54	3. 45	0. 54	34. 30	294. 06	9. 99	11. 95	0. 94	0. 99
4	斋堂	粗安岩	52. 75	125. 60	13. 66	58. 47	10. 92	2. 88	9. 98	0. 91	6. 14	1. 24	3. 91	0. 48	2. 85	0. 35	34. 74	290. 14	10. 22	13. 28	0. 84	1. 15
5	王平镇	粗安岩	51. 30	111. 50	13. 28	54. 67	10. 16	2. 67	8. 58	1. 27	7. 09	1. 36	3. 75	0. 54	3. 39	0. 48	37. 85	270. 03	9. 21	10. 85	0. 87	1. 05
6	潘涧沟	凝灰岩	29. 02	81. 11	10. 77	42. 88	7. 99	2. 26	6. 45	1. 00	5. 17	1. 05	2. 91	0. 44	2. 50	0. 37	25. 80	193. 93	8. 75	8. 32	0. 96	1. 12

注：数据为均值，数据来源同表 15-5。

表 16-3　南大岭期火山岩微量元素成分及特征参数表

Table 16-3　Trace element contents and main parameters of Nandaling volcanic rocks

序号	岩性	微量元素含量/10^{-6}														特征参数	
		Rb	Ba	Th	U	Ta	Nb	Sr	Zr	Hf	V	Cr	Co	Ni	Cu	Rb/Sr	Nb/Ta
1	火山岩	33. 00	1117. 00	2. 63	0. 53	0. 99	19. 86	1636. 75	258. 75	5. 36	166. 75	152. 75	34. 50	81. 00	28. 00	0. 02	20. 16
2	火山岩	24. 58	1238. 79	2. 10	0. 46	0. 83	16. 15	802. 54	258. 50	5. 41	183. 38	172. 75	37. 55	87. 85	35. 00	0. 03	19. 46
3	火山岩	63. 50	3135. 00	1. 67	0. 47	0. 79	13. 97	312. 75	202. 00	4. 58	146. 50	193. 75	36. 25	105. 50	46. 00	0. 20	17. 68
4	火山岩	183. 50	2365. 50	4. 69	1. 55	1. 37	28. 80	113. 50	320. 50	7. 41	155. 50	171. 50	36. 50	73. 50	10. 00	1. 62	21. 02

注：数据为均值，引自汪洋等，1994；李晓勇等，2004；Wang Zhihong et al.，2007。

族元素配分曲线（图 16-8）右呈对称性较差的“W”形，Fe、Ti、Cu 呈明显的尖峰，而 Cr、Co、Ni 成低谷，反映岩浆可能来源于地幔的部分熔融。大离子亲石元素 K、Rb、Sr 等元素含量有所波动，其余元素特征分布形式相似，此与火山岩较强蚀变或受到岩浆混染有关。微量元素蛛网图上（图 16-8）左，富集大离子亲石元素 Ba、K 和 LREE，亏损高场强元素 Nb、Ta、Th、U。微量元素分布特征总体上类似于兴蒙造山带带火山岩，但其亏损 U-Th 的特征则与鲁西基性火山岩相似（李晓勇等，2004）。

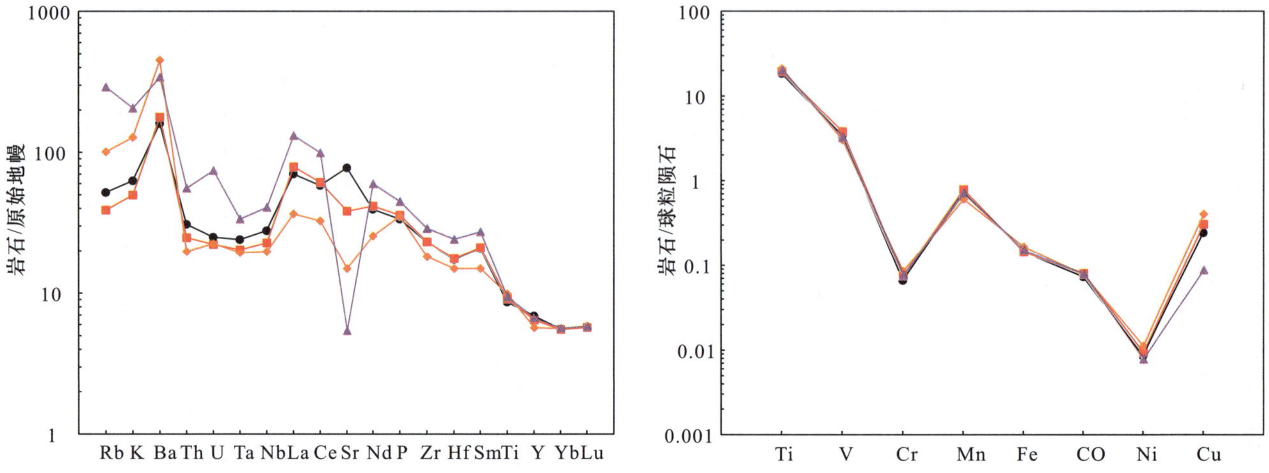

图 16-8　南大岭期火山岩微量元素蛛网图

Fig. 16-8　Primitive mantle-normalized spidergrams for trace elements of Nandaling volcanic rocks

（六）岩石系列划分

该套岩系里特曼指数 $\sigma=1.73\sim6.12$，碱度率 $AR=1.62\sim2.13$，跨碱性岩和亚碱性岩区，与 K_2O-SiO_2 图（图略）中分散于钙碱性-高钾钙碱性-钾玄岩区的成分点特征一致，可能同火山岩发生一定变质作用有关。在 SiO_2-*AR* 和 Ta/Yb-Th/Yb 岩石系列判别图解（图 16-9）上，成分点分别位于碱质-钙碱质区和邻钙碱性区。岩石 K_2O、碱和 K_2O/Na_2O 值均偏低，而 TiO_2 含量较高，不同于典型钾玄岩和碱性岩特性，而属于钙碱性系列。

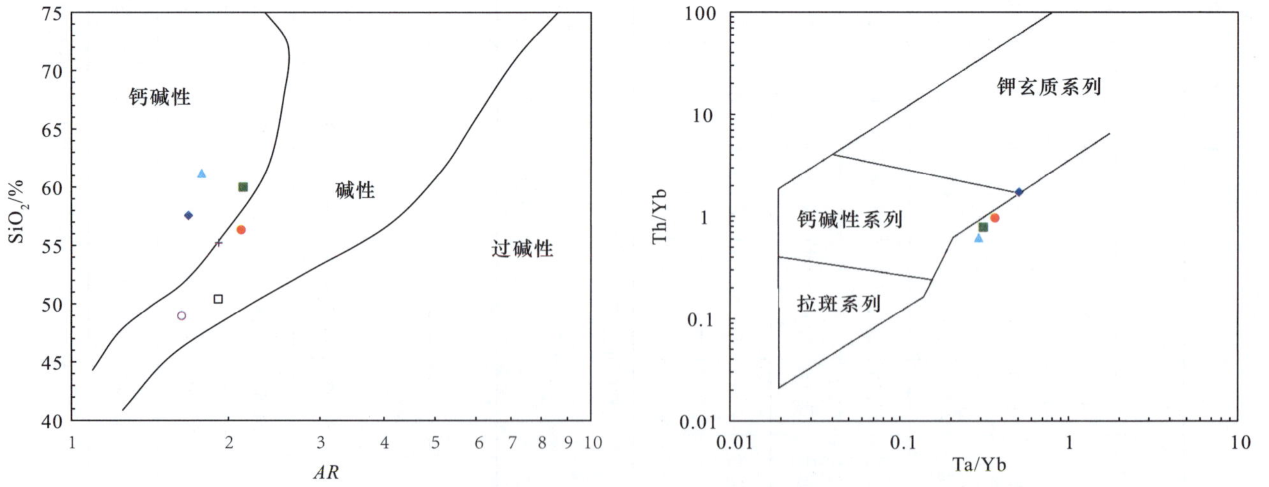

图 16-9　南大岭期火山岩 SiO_2-*AR* 和 Ta/Yb-Th/Yb 图解

Fig. 16-9　The SiO_2-*AR* and Ta/Yb-Th/Yb diagram of Nandaling volcanic rocks

（七）Sr-Nd 同位素特征

南大岭期火山岩 Rb = 4.8 ~ 232.1，Sr = 54.83 ~ 2052.01，$^{87}Rb/^{86}Sr$ = 0.017 ~ 9.756，$^{87}Sr/^{86}Sr$ = 0.704823~0.714215，Sm=5.85~11.56，Nd=28.05~77.61，$^{147}Sm/^{144}Nd$=0.0961~0.1309，$^{143}Nd/^{144}Nd$=0.511813~0.512292；取 t = 190Ma 时，ε_{Nd}（t）= −5.3 ~ −13.8，t_{DM} = 1590 ~ 2020Ma（Wang et al., 2007）。Sr-Nd 同位素特征既不同于鲁西也不同于兴蒙造山带带火山岩，具有较高初始 $^{87}Sr/^{86}Sr$ 比值和较大变化范围，揭示其源区除华北陆块内部古老富集型 EMI 源区贡献，还有壳源放射成因 Sr 的加入（李晓勇等，2004；Wang et al., 2007）。第一段火山岩较第二段火山岩相比，其具有更窄的初始 $^{87}Sr/^{86}Sr$ 比值和更高的 ε_{Nd}（t）值。

（八）Pb 同位素特征

南大岭期火山岩 Pb 同位素变化范围小：$^{206}Pb/^{204}Pb$ = 16.114 ~ 16.884，$^{207}Pb/^{204}Pb$ = 15.176 ~ 15.327，$^{208}Pb/^{204}Pb$=36.029~36.915（Wang et al., 2007），与华北克拉通富集型地幔类似。

（九）岩石成因与岩浆演化

南大岭期火山岩所表现的 LREE 和 LILE 富集和 Sr-Nd-Pb 同位素组成中等富集，反映岩石源于富集岩石圈地幔，表现为在约 70~80km 深处的中生代岩石圈地幔中尖晶石橄榄岩到石榴子石橄榄岩之间过渡型橄榄岩 2%~5%部分熔融（李晓勇等，2004；Wang et al., 2007）。一段火山岩具有较高的 MgO 和 Cr、Ni，且其主微量元素随 SiO_2 含量变化不大，说明岩石可能代表了原始岩浆。二段火山岩具有较低的 MgO 和 Cr、Ni，其含量反映岩石不能代表原始岩浆，经历了较强的岩浆演化过程（李晓勇等，2004；Wang et al., 2007）。两期岩浆均表现出 Th、U 元素的相对亏损，受中上地壳物质混染的可能性较小，而更可能受地幔物质侵入就位过程中下地壳麻粒岩相物质的影响。Wang et al.（2007）认为可能表现为下地壳沉没和再生，形成早侏罗世特有的壳幔相互作用过程，并发生后续强烈的地幔物质上涌。

在 K_2O-Na_2O-CaO 图解（图 16-10）左上，一段和二段火山岩具有明显不同的碱质及 CaO 含量，揭示了岩石由钙碱性向碱性、高钾方向演化。在 ΣREE-La/Sm 图解（图 16-10）右上，不同地区 ΣREE 和 La/Sm 比值差异也较大，宏观表现出由房山向王平镇、色树坟（南向北）逐渐降低的特征。南大岭期火山喷发时间跨度大（见下形成时代），区域上的火山喷发可能并不同时，而呈现南先北后有序变化。南大岭期火山活动还同深大断裂活化密切有关，火山活动与断裂活动时序是否一一对应，有待进一步研究。

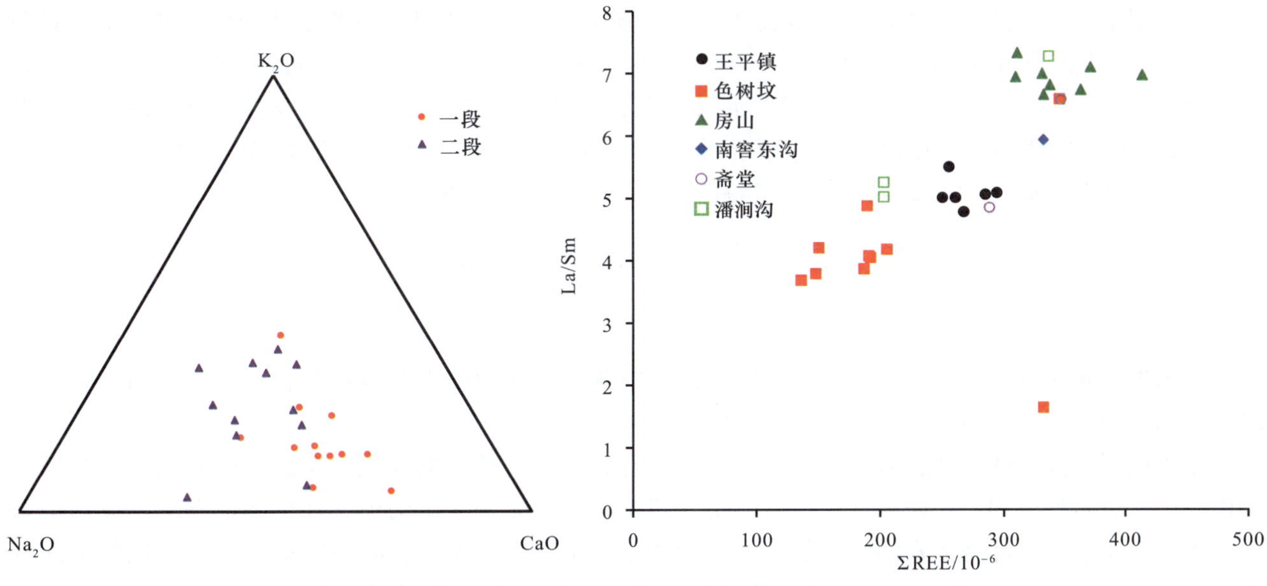

图 16-10　南大岭期火山岩 K_2O-Na_2O-CaO 和 ΣREE-La/Sm 图解

Fig. 16-10　The K_2O-Na_2O-CaO and ΣREE-La/Sm diagram of Nandaling volcanic rocks

（十）构造环境分析

南大岭期火山喷发带呈近东西向，火山喷发强度有限，且与断裂构造密切有关；火山岩岩性较为简单，为钙碱性系列；Ta/Hf-Th/Hf 构造环境判别图解（图 16-11）中，成分点均位于板内玄武岩区。因此，南大岭期火山岩形成于板内伸展构造背景环境，深部地球动力学过程可能同蒙古-鄂霍茨克海的消减、闭合后的伸展作用有关（Davis et al.，2001）。早侏罗世，伴随蒙古-鄂霍茨克海闭合，地温梯度增高，岩石圈地幔物质发生部分熔融形成玄武质岩浆。同时，在构造反转区域应力松弛伸展作用下，先期存在的深大断裂复活构成地壳薄弱带，沿此薄弱带南大岭期火山岩浆上涌溢流或喷发。

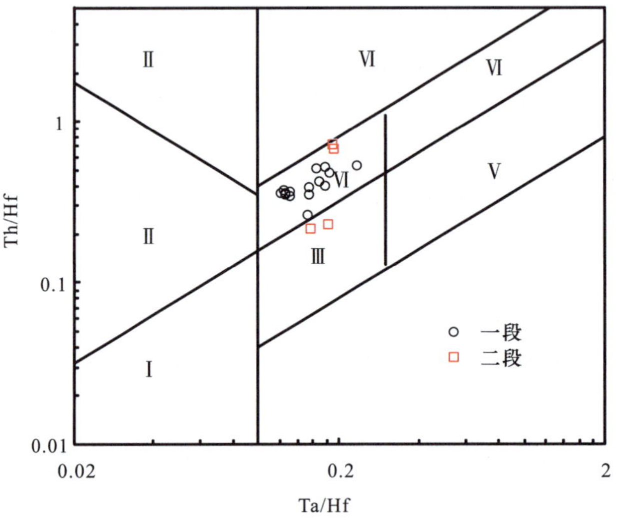

图 16-11　南大岭期火山岩和 Ta/Hf-Th/Hf 构造环境判别图解

Fig. 16-11　The Ta/Hf-Th/Hf diagram of tectonic discrimination of Nandaling volcanic rocks

（十一）形成时代

南大岭期火山岩形成于早侏罗世，主要依据包括：①不整合覆盖于古生界二叠系之上，而与下侏罗统窑坡组地层整合接触；②火山-沉积盆地及火山喷发带走向近东西；③形成于板内伸展构造背景环境；④色树坟地区火山岩锆石 U-Pb 同位素年龄为（174±8）Ma（赵越等，2006），门头沟双石头村、上达摩庄、大灰厂三地火山岩锆石 U-Pb 同位素年龄分别为（211.6±5.8）Ma、（184.1±4.5）Ma 和（185.8±5.1）Ma（赵兴国 2009），彩虹桥、法城口和潘涧沟火山岩锆石 U-Pb 同位素年龄分别为（170±5.81）Ma、（172±2）Ma 和（174±8）Ma（高海龙，2018）。

二、髫髻山期火山岩与火山作用

（一）时空分布特征

火山岩分布于髫髻山组和土城子组地层中，为中燕山期火山喷发阶段产物，可划分为两个喷发阶段，相当于火山岩岩石地层的髫髻山组和土城子组。

据火山活动及岩性特征，髫髻山喷发阶段又可划分出两个喷发周期和两个火山间歇周期，分属髫髻山组一、三岩性段和二、四岩性段。火山喷发周期表现为先爆发后溢流，在剖面上下部形成角砾熔岩或火山角砾岩，上部为正常的熔岩层。多次火山活动则形成爆发-溢流多次重复叠置出现的剖面结构（鲍亦冈等，2001）。第一喷发周期以粗安岩、粗安质角砾凝灰岩为主，见少量玄武质、玄武安山质岩石；第二喷发周期岩石类型复杂，除发育第一喷发周期的岩性外，还发育少量粗面质、流纹质熔岩和火山碎屑岩等。空间上，髫髻山组火山岩呈北东向分布于北京西部及北部数个断陷盆地中，受控于南口-沿河城、颐和园和居庸关断裂，形成南、北两大火山-沉积喷发亚带（见后火山构造）。火山岩厚度以雁羽山地区最大（4971m），新城子盆地最小（1120m）。在火山喷发带的西南段，火山活动强烈，堆积的火山岩层厚度大，向东北则厚度逐渐减薄。据实测剖面资料，南亚带由西南向东北，不同地段髫髻山组厚度为：百花山 3442m、田庄 2622m、十三陵 1134m、新城子 1120m（图 16-12）。总体上看，北亚带各火山盆地髫髻山组的厚度明显大于南亚带（鲍亦冈等，2001）。该套火山岩与下伏九龙山组呈角度或平行不整合接触。区域上，两套岩层褶皱形态及方向明显不同，分属燕山运动期不同构造变动幕形成的褶皱系（鲍亦冈等，1995）。

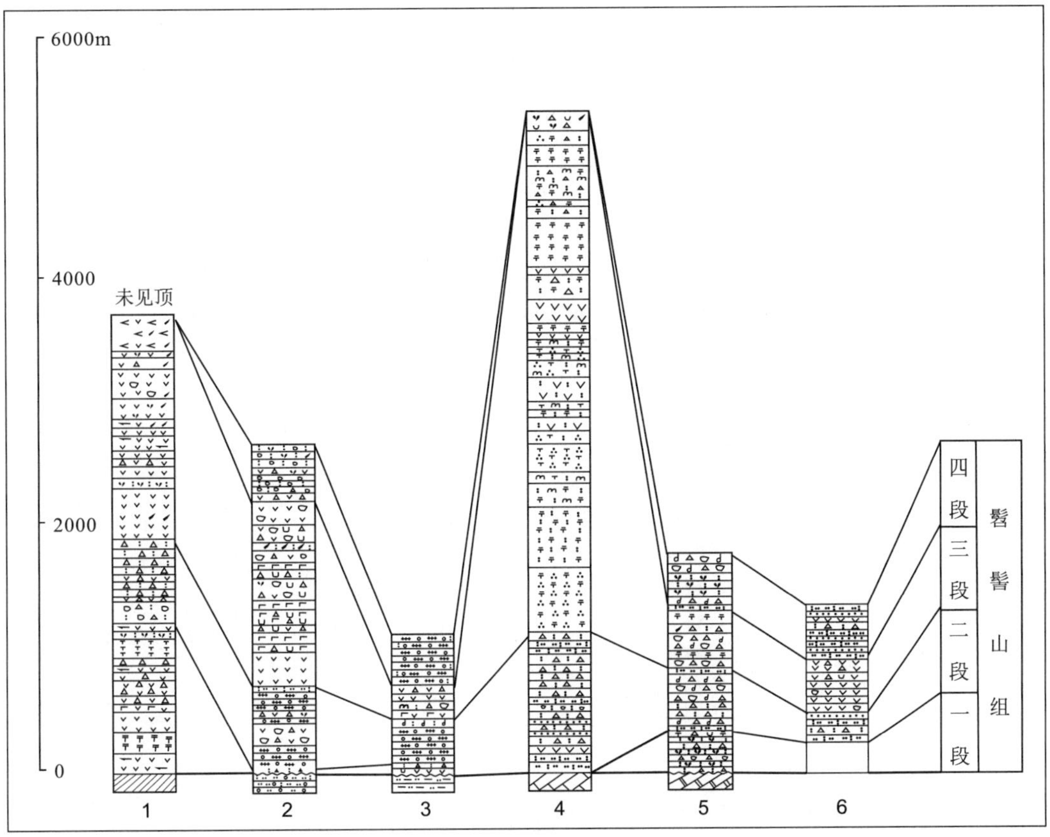

图 16-12　北京地区髫髻山组火山岩段喷发韵律相序剖面对比图
（据鲍亦冈等，1995）
Fig. 16-12　Profile correlation diagram of volcanic eruption lithofacies sequence of Tiaojishan formation in Beijing
1—百花山；2—妙峰山；3—十三陵；4—燕羽山；5—三岔；6—新城子

土城子喷发阶段是在髫髻山强烈喷发阶段之后出现的强弱不等的阵发性火山活动阶段。火山活动强度较弱，形成的火山岩夹于沉积岩地层中，也可划分为两个喷发周期。早期发育为流纹质角砾凝灰岩、流纹质凝灰岩及粗安质凝灰岩组合，晚期则为粗安岩、玄武岩、粗安质角砾凝灰岩、粗面质角砾凝灰岩组合。土城子火山喷发和髫髻山火山喷发同属一个火山喷发旋回，构造形迹特征一致，空间上二者火山岩分布特征基本类似。但该阶段的火山-沉积盆地并不完全继承髫髻山火山-沉积盆地，还新生了白河堡、千家店、四海-凤驼梁多个火山-沉积盆地。

（二）火山岩相与岩石学特征

髫髻山期火山岩广布，相序复杂（鲍亦冈等，1995）。按照火山活动特征，其火山岩相序列划分见表 16-4。同时，各火山-沉积盆地岩相又略有差异（表 16-5）。

表 16-4 北京地区髫髻山期火山岩相序列划分简表

Table 16-4 Charateristics of volcanic facies Series of Tiaojishan volcanic rocks in Beijing

相组	相段	岩相
土城子火山-沉积相组	爆发-沉积相段	火山爆发相（空落、碎屑流堆积、崩落堆积）、火山喷发沉积相
	火山-沉积-溢流相段	火山喷发沉积相、溢流相
	火山-沉积相段	火山喷发沉积相
髫髻山溢流-爆发-火山-沉积相组	溢流相段	溢流相
	爆发-火山-沉积相段	火山爆发相（空落、碎屑流堆积、崩落堆积）、火山喷发沉积相
	溢流-爆发相段	溢流相、火山爆发相（空落、碎屑流堆积、崩落堆积）
	火山-沉积相段	火山喷发沉积相

（据鲍亦冈等，1995）

表 16-5 髫髻山期各火山-沉积盆地火山岩相发育特征简表

Table 16-5 Charateristics of volcanic facies of Tiaojishan volcanic rocks of volcanic basins in Beijing

地区	喷发周期	岩相组合及比例	岩相类型岩浆性质及比例		岩相特征
门头沟百花山	第三喷发周期	溢流相 94.63%	粗安质为主，凝灰岩	粗安岩	
		爆发相 5.37%		5.37%	有些层含晶屑和角砾
	第二间歇周期	爆发相 100%	粗安质为主，层状凝灰岩 100%		含火山角砾和集块
	第一喷发周期	溢流相 79.81%	以玄武粗安质-粗安质为主	粗安岩、玄武粗安岩	
		爆发相 17.14%		层状凝灰岩 3.53%	成层性好，有些层含角砾
				火山角砾岩 2.99%	角砾中可见碳酸盐岩角砾
				凝灰岩 10.62%	晶屑发育，有些层含角砾
		火山-沉积相 3.04%		凝灰质砂岩、砾岩	
门头沟髫髻山	第四间歇周期	溢流相 14.5%			
		火山-沉积相 83.43%			
		沉积相 2.07%			
	第三喷发周期	溢流相 95.91%			
		爆发相 4.09%	粗安质凝灰岩 4.09%		含角砾、集块
	第二间歇周期	溢流相 12.8%			
		爆发相 30.8%	流纹质凝灰岩 30.8%		黄褐色—灰白色，个别层含角砾
		火山-沉积相 5.5%			
		沉积相 50.9%			
	第一喷发周期	溢流相 69.78%	粗安质、流纹质	粗安岩、玄武质粗安岩等	
		爆发相 28.52%		（粗安质）火山集块岩 8%	紫红色
				（粗安质）火山角砾岩 3.8%	紫红色
				（流纹质）凝灰岩 16.72%	含角砾
		沉积相 1.7%	砂岩、砾岩		

续表

地区	喷发周期	岩相组合及比例	岩相类型岩浆性质及比例		岩相特征
延庆燕羽山	第三喷发周期	溢流相 50.54%			
		爆发相 49.46%	粗面质	凝灰岩 3.98%	多数层位含角砾
			粗安质	凝灰岩 27.25%	
			粗面质	熔结凝灰岩 6.08%	
			粗安质	熔结凝灰岩 12.15%	
	第二间歇周期	溢流相 6.34%			
		爆发相 50.48%	粗安质	火山角砾岩 40.83%	凝灰胶结
				凝灰岩 9.65%	含角砾、可见玻屑
		火山-沉积相 43.18%			
昌平十三陵	第四间歇周期	爆发相 12.28%	粗安质凝灰岩，12.28%		含角砾
		火山-沉积相 87.72%			
	第三喷发周期	溢流相 55.52%			
		爆发相 44.48%	玄武粗安质、粗安质	火山角砾岩 39.49%	
				熔结角砾岩 3.45%	含火山凝灰
				凝灰岩 1.54%	含晶屑
	第二间歇周期	爆发相 7.09%	粗安质凝灰岩 7.09%		含角砾
		火山-沉积相 92.91%			
	第一喷发周期	溢流相 54.85%			
		爆发相 45.25%	粗安质	火山角砾岩 41.26%	凝灰胶结
				凝灰岩 3.89%	含角砾
延庆四海	第四间歇周期	爆发相 17.8%	粗安质凝灰岩 17.8%		含集块及角砾
		火山-沉积相 34.8%			
	第三喷发周期	溢流相 34.87%			
		爆发相 30.33%	粗安质-流纹质	火山角砾岩 14.82%	含集块
				凝灰岩 14.88%	含角砾
				层状凝灰岩 0.63%	薄层状，含玻屑
		火山-沉积相 34.8%			
	第二间歇周期	爆发相 2.13%	流纹质熔结凝灰岩 0.63%		
			粗安质凝灰岩 1.5%		含角砾
		火山-沉积相 87.72%			
	第一喷发周期	沉积相 1.31%			
		溢流相 4.91%			
		爆发相 95.09%	粗安质	火山角砾岩 5.51%	含集块
				凝灰岩 89.58%	含集块或角砾

续表

地区	喷发周期	岩相组合及比例	岩相类型岩浆性质及比例		岩相特征
密云高岭	第四间歇周期	溢流相 0.6%			
		爆发相 34.76%	粗安质	火山集块岩 3.14%	含沉积岩角砾
				火山角砾岩 28.5%	含沉积岩角砾
				凝灰岩 3.05%	与砂岩和粉砂岩互层
		沉积相 64.64%	砂岩、砾岩		
	第三喷发周期	溢流相 100%	火山角砾岩 58.49%		
	第二间歇周期	爆发相 66.19%	粗安质	凝灰岩 7.7%	含角砾
				砂岩、砾岩	
		沉积相 33.81%			

（据鲍亦冈等，1995）

1. 溢流相

溢流相广泛分布，尤以髫髻山组第一、第三岩性段最为集中。火山喷发呈北东向多中心多次涌动喷溢特点，形成沿走向（平行火山口排布方向）厚度变化大，沿流向（垂直火山口排布方向）厚度递减快，基本层厚度大（一般为几十至百余米），且构成要素不典型，岩相组合相对复杂的巨厚熔岩；它们常与下伏及上覆地质体呈嵌合接触，沉积或火山-沉积夹层往往呈透镜状产出（鲍亦冈等，1995）。岩性组合复杂，以中性安山岩最为发育，主要岩石特征见表 16-6。

2. 火山爆发相

是髫髻山火山喷发阶段主要岩相，尤以髫髻山喷发周期第二岩性段最为发育。该阶段火山喷发活动具有延续时间长、分布范围广、岩浆性质复杂、产物类型繁多等特点，并且火山喷发作用强度及岩相类型还具有因时、因地而变之特点（鲍亦冈等，1995）。

（1）火山崩落堆积相

发育安山质、石英粗面质和流纹质凝灰岩或角砾凝灰岩，以安山质为主。

安山质凝灰岩：灰绿、绿色，凝灰结构，块状构造。由晶屑、玻屑、岩屑及火山灰等组成。晶屑以斜长石为主，少量黑云母、角闪石、辉石等，含量 10%～25%。岩屑以安山岩为主，棱角状，大小 0.4mm×0.2mm～1.5mm×1mm，强绢云母化、绿帘石化，含量变化大，多数>50%。玻屑为粒度细小的火山玻璃，霏细脱玻结构，含量 10%～40%。胶结物以火山灰为主。此类岩性岩屑较大时发育为角砾凝灰岩。

石英粗面质凝灰岩：灰色、灰紫色，凝灰结构。主要由晶屑、岩屑、玻屑及火山灰组成。晶屑主要为斜长石和碱性长石，碎裂次棱角状，粒径 0.1～2mm，前者发育钠长双晶，强烈绢云母化，后者边缘熔蚀；少量黑云母晶屑，撕裂片状，强烈暗化、帘石化。岩屑以安山岩为主，大小 0.5～2mm，碳酸盐化较强。玻屑呈“弓”状、鸡骨状，霏细脱玻结构，含量 40%～50%，较强绢云母化。胶结物以火山尘为主。

流纹质晶屑凝灰岩：灰绿色，凝灰结构，致密坚硬、块状构造，岩石主要由晶屑和火山灰物质组成。斜长石晶屑次棱角状，半自形，粒径 0.05～1.2mm，发育钠长双晶，中等绢云母化，含量约 15%。正长石晶屑次棱角状，粒径 0.05～0.3mm，发育卡式双晶，中等高岭土化、绢云母化，含量约 6%。黑云母晶屑半自形片状，粒径 0.1～0.4mm，暗化铁质析出仅残留假象，含量约 4%。石英晶屑棱角状，粒径<0.1mm，有被熔蚀现象，含量约 3%。胶结物为火山灰，脱玻化结晶成隐晶质。岩石中可见大量颗粒状、粉末状铁质矿物，大者识别为磁铁矿，含量约 72%。

（2）火山碎屑流堆积相

发育粗安质、石英粗面质、流纹质熔结凝灰岩。

表 16-6 髫髻山期安山岩类岩石主要特征简表

Table 16-6 Major characteristic of Tiaojishan andesites

特征 岩石	颜色	结构	构造	斑晶及含量	基质及含量	交代蚀变
粗面玄武岩	紫色-深紫色	斑状结构，基质间粒结构、粗玄结构	气孔构造、块状构造	橄榄石约 10%；辉石 30% ~ 35%；斜长石少量	斜长石 30% ~ 35%；橄榄石约 10%；辉石约 10%	蛇纹石化、高岭土化
玄武粗安岩	紫灰色	间隐-间粒结构，基质隐晶质结构	气孔构造、杏仁（方解石）构造	辉石约 5%；斜长石约 35%	玻璃质	绢云母化、碳酸盐化
玄武安山岩	紫色-灰紫色	斑状结构，基质具交织结构、玻晶交织结构	块状构造，气孔、杏仁构造	斜长石（拉长石）20% ~ 50%；普通辉石 2% ~ 3%；角闪石约 1%	斜长石（拉长石）50% ~ 60%；普通辉石 3% ~ 5%；橄榄石 2% ~ 3%；副矿物 3% ~ 5%	碳酸盐化、绿泥石化、绢云母化均发育
安山岩	灰色-灰绿色	斑状结构，基质具交织结构、玻晶交织结构	块状构造，气孔、杏仁构造	斜长石（中长石-拉长石）20% ~ 25%；普通辉石 2% ~ 3%；角闪石<3%	斜长石（更长石-中长石）30% ~ 50%；普通辉石 3% ~ 5%；碱性长石少量；副矿物 2% ~ 5%	弱绢云母化、高岭土化、绿泥石化、碳酸盐化
粗安岩	紫红色-灰绿色	斑状结构，基质具交织结构、玻晶交织结构、粗面结构	块状构造，气孔、杏仁构造	斜长石（拉长石、钠长石）2% ~ 3%；碱性长石 1% ~ 11%；黑云母 3%；角闪石 3% ~ 8%；辉石 2% ~ 3%；石英约 3%	斜长石（中长石-钠长石）+碱性长石>50%；普通辉石 2% ~ 3%；副矿物 1 ~ 3%	弱绿泥石化、绢云母化、硅化、碳酸盐化
粗面岩	灰紫色-紫色	斑状结构，基质玻基交织结构	块状构造	斜长石（中长石）5% ~ 7%；辉石 2% ~ 3%	斜长石约 78%；玻璃质约 10%；副矿物约 2%	绢云母化、钠黝帘石化、绿泥石化

（据《北京市区域地质志》，1991）

粗安质熔结凝灰岩：灰-灰紫色，凝灰结构，似流动构造，由塑变岩屑、玻屑、晶屑、岩屑及火山尘组成。塑变岩屑透镜状、边缘撕裂状，定向分布，大小 0.3mm×0.5mm ~ 50mm×70mm，镜下发育黑色肠状和褐-灰褐色火焰状两种类型，含量 10% ~ 30%；塑变玻屑弓状、鸡骨状，少树叉状，形态多变，杂乱分布，具霏细脱玻结构；岩屑等轴状，粒径 0.5 ~ 50mm，成分主要为中性熔岩、中性凝灰岩，少猪肝色硅质岩、粉砂岩、泥灰岩等，含量<5%；晶屑主要为斜长石，An = 44 ~ 45，为中长石，其次为黑云母和碱性长石。

石英粗面质熔结凝灰岩：灰-红色，熔结凝灰结构，块状构造，由晶屑、塑变岩屑、玻屑及火山灰等组成。晶屑主要为乳白色斜长石和红色碱性长石组成，含量 15% ~ 23%，前者类型为更长石（An = 25 ~ 36），后者类型为中-低正微长石；少量黑云母晶屑，弱暗化，含量<1%。塑变岩屑肉红色透镜状-条纹状，风化呈土黄色，定向排列，粒径 1 ~ 10mm，含量 13% ~ 30%；塑变玻屑无色-浅褐色长条状、火焰状，含量约 30%，脱玻化粒径普遍较大，多数呈嵌晶结构。玻屑边部粒径较大，以长石为主，发育梳状结构；中间粒径较小，由自形板条状碱性长石和石英组合，发育球粒或嵌晶结构，大球粒玻屑束状或扇形脱玻结构，核心为晶屑矿物。

流纹质熔结（角砾）凝灰岩：灰绿色、灰紫色，熔结（角砾）凝灰结构，致密块状构造，由浆屑、晶屑、岩屑、玻屑和火山灰组成。塑性浆屑撕裂状、长条状，定向排列；岩屑棱角状，成分为流纹岩，粒径2~30mm，含量各层位不同，最高为70%~80%；晶屑成分为长石和石英。玻屑、浆屑均已脱玻化，在强熔结层中，碎屑边界不清，部分呈塑性似流纹构造。胶结物为火山灰，已部分脱玻化。

（3）侵出相

主要为流纹岩，局部为球泡流纹岩和角砾状流纹岩（近边部发育）。岩石肉红色、浅灰色、灰色，斑状结构，流纹构造。斑晶以条纹长石为主，石英和斜长石次之，少见黑云母、角闪石，粒径0.5~2mm，含量8%~20%。基质霏细、隐晶结构，由长英质物质组成。

侵出相仅分布于太师屯附近。地貌形态上明显高于周围地质体，呈孤峰存在。平面形态呈椭圆形，大小为0.11km×0.08km。它与周围的髫髻山组火山-沉积相地层（凝灰质砂岩、粉砂岩为主）呈侵入接触，接触面产状外倾，倾角极陡（鲍亦冈等，1995）。

（4）潜火山岩相

发育安山玢岩和石英粗面斑岩。

安山玢岩：灰绿色，斑状结构，致密块状构造、气孔-杏仁状构造。气孔圆形或不规则状，大小0.25~2mm，含量约10%，部分内填充沸石、石英杏仁体。斑晶成分：斜长石自形-半自形板状，发育聚片双晶，粒径0.05~3.5mm，含量15%~20%；辉石自形-半自形短柱状，横切面呈八边形，粒径0.5~0.8mm，个别达1.2mm，含量3%~8%。基质玻基交织结构，主要由显微板条状、针状斜长石及隐晶长英质组成，粒度0.01~0.1mm，含量约56%，少量暗色矿物自形-他形柱粒状充填在其他矿物孔隙中，粒径0.05~0.2mm，含量约3%。副矿物为黄铁矿、磷灰石等，含量4%~5%。

石英粗面斑岩：灰褐色，斑状结构，块状构造，弱绢云母化、绿帘石化。斑晶组成：碱性长石自形—半自形板柱状，弱熔蚀，弱高岭土化，类型为条纹长石，粒径0.3~0.5mm，含量约5%；黑云母自形片状，粒径0.3~0.5mm，含量约2%；石英他形粒状，发育熔蚀边，粒径0.3~0.5mm，含量约1%。基质微晶结构，由碱性长石（约80%）、石英（约10%）、黑云母（约2%）、斜长石（约1%）组成。副矿物为磁铁矿、锆石等，粒径0.05~0.1mm。

3. 火山-沉积相

发育于火山喷发间歇期，尤以髫髻山火山喷发周期第四岩性段和土城子火山喷发周期最为发育，发育沉凝灰岩和凝灰质砂岩类。

沉凝灰岩：杂色，沉凝灰结构、块状构造，主要由岩屑、玻屑、晶屑及粉砂级陆源碎屑和胶结物组成。岩屑棱角状-次棱角状，成分为粗面岩、流纹质凝灰岩，大小0.2~2mm，含量2%~3%；玻屑鸡骨头状，脱玻化，带褐红色斑点，大小0.2m~2mm，含量约15%；晶屑主要为自形—半自形长石，极少量棱角状石英，含量约3%；陆源碎屑主要为粉砂级的长石和石英，次棱角—次圆状，粒径0.03~0.2mm，含量15%~20%。胶结物为火山灰和黏土矿物，火山灰多脱玻化，含量约50%。

凝灰质砂岩类：灰紫色、灰绿色，凝灰质砂状结构，致密块状构造。可根据其碎屑物粒级的不同，细分为凝灰质粉砂岩、凝灰质砂岩、凝灰质含砾砂岩等。碎屑组成：陆源粉砂级碎屑，次棱角状，以石英为主，其他成分有云母、帘石、长石类矿物和铁质氧化物等，含量约46%。岩屑棱角状，粒径0.5~1.5mm，少量>2mm，成分为安粗岩、沉凝灰岩、泥岩等；晶屑棱角状，成分为钾长石、斜长石和石英，粒径<0.25mm，含量<2%；火山灰呈极细小的火山碎屑物，脱玻化不明显，含量约8%。胶结物为泥质和火山灰，混合分布，水云母化、绢云母化，另有大量铁质氧化物相伴生，含量约10%。

（三）火山构造

参考《中国区域地质志·河北志》（2017）火山构造单元划分，本区位于承德-武安火山喷发带（Ⅱ级），以深大断裂构造为界，可划分为南、北两个亚带，内含髫髻山等多个火山-沉积盆地（Ⅲ级）。

1. Ⅱ级火山构造单元——火山喷发带

中燕山期火山喷发带呈北东向展布，隶属于承德-武安火山喷发带（Ⅰ级），本区内该期火山岩分布范围南西起自门头沟百花山，北东延至怀柔长哨营和密云新城子，由数个火山-沉积盆地组合而成。

北东向深断裂控制了髫髻山阶段火山喷发及沉积作用的发生与分布，以南口–沿河城及居庸关深断裂为界，可将火山喷发带分成特征不同的南部百花山–新城子火山喷发亚带和北部燕羽山–四海喷发亚带（鲍亦冈等，1995）（图 14–3）。

南亚带位于北东向南口–沿河城、颐和园断裂间，长 185km，宽 12~16km，北东 60°方向展布。范围南西自百花山，向东北经髫髻山、妙峰山、十三陵、怀柔山前、延至密云新城子。发育百花山、髫髻山、新城子三个中—晚侏罗世火山–沉积盆地，分布面积约 560km²。单个盆地主要走向与亚带走向一致呈北东向，但新城子盆地受北东向赤城–古北口深断裂制约，呈近东西向偏转延伸。

北亚带位于居庸关深断裂西北侧，延伸方向平行于南亚带。带内发育燕羽山、凤驼梁–四海两个火山–沉积盆地，分布面积约 548km²。早期髫髻山喷发阶段盆地呈北东向，与喷发带延伸方向一致。至晚期土城子喷发阶段，盆地开始向北迁移，可能受到东西向赤城–古北口深断裂的影响，盆地整体上显示为近东西向排列的趋势。

南北两亚带中的火山喷发物都以中性粗安质岩石为主，并有由中性向偏酸性、碱性方向演化的特点。但两个亚带的火山活动特征有差异：

首先，火山–沉积盆地基底构造特征不一致。南亚带多是在长期坳陷的构造单元基底上发育起来的继承性断陷盆地。而北亚带则是于上隆构造单元的基底上发育起来的新生性上叠式断陷火山–沉积盆地，盆地基底的岩层多是中元古界。

其次，从岩石学、岩石化学特征看，南亚带火山岩比较简单，以粗安质岩石为主，早期见有少量玄武质和玄武安山质岩石。北亚带火山岩相对比较复杂，以粗安质岩石为主，火山活动晚期粗面质、流纹质岩石也有一定比例。燕羽山火山–沉积盆地髫髻山阶段第三喷发周期火山岩中还出现了厚度较大的粗安质、粗面质熔结凝灰岩。岩石的碱度、分异程度明显高于南亚带。

再次，南亚带中火山岩中熔岩所占比例较大，尤其是西南端百花山盆地中更为明显，北亚带火山岩中爆发作用相对较强，喷发成因的火山碎屑岩占优势。

最后，早期南亚带火山活动强烈，特别是西南部的百花山、髫髻山火山盆地中，形成了厚度较大且以中性熔岩为主的第一喷发周期的火山岩地层。同期，北亚带各火山–沉积盆地中，只有少量中性火山熔岩与正常火山碎屑岩的堆积。晚期北亚带火山活动渐强，燕羽山及凤驼梁地区第三喷发周期火山岩层的厚度大大超过南亚带。

2. Ⅲ级构造单元——火山–沉积盆地

中燕山期火山喷发带中发育多个火山–沉积盆地，各盆地基本特征见表 16–7，这里仅对最典型的百花山火山–沉积盆地特征作详细论述。

百花山火山–沉积盆地位于门头沟西部，东自斋堂、马栏，西至张家庄、芦子水，呈长轴北东向椭圆形展布，出露面积 180km²。盆地隶属于北东向中燕山火山喷发带之南亚带，在早燕山期火山–沉积基底上继承演化，基底为九龙山组地层。盆地西上覆晚燕山期杜家庄–镇厂火山–沉积盆地。盆地中地层经褶皱构成了一个北东端扬起、向南西缓慢倾伏的箱形向斜，轴线位于斋堂—田寺一线，走向北东 55°。南东翼岩层走向北东 20°~30°，倾角 40°~50°。北西翼火山岩层走向北东 70°，倾角 35°~45°。盆地中心地层平缓，倾角 15°（图 16–13）。

百花山火山–沉积盆地与髫髻山火山–沉积盆地一致，受控于沿河城–南口与颐和园深断裂。伴随区域引张作用下，沿此二断裂断陷作用火山溢流–喷发，形成中燕山期火山–沉积盆地。盆地发育过程中，火山喷发中心有由北东向南西迁移的事实。

盆地中发育髫髻山火山喷发阶段的火山活动。第一喷发周期形成髫髻山组一段地层，目前呈不规则的半环状分布于盆地的东北端。最初，火山活动以裂隙式喷发为主，中性、中偏基性熔浆沿清水河谷裂隙式线状火山机构强烈喷溢，形成多层似层状粗安质及玄武安山质熔岩流，夹有少量粗安质、安山质角砾熔岩及角砾凝灰岩，堆积厚度达 1479m（鲍亦冈等，1995）。随后，火山活动转为以爆发作用为主，喷发产物形成粗安质角砾岩、集块岩，沿清水河谷两侧分布。第二喷发周期形成髫髻山组三段

表 16-7 北京地区中燕山期火山-沉积盆地的特征简表

Table 16-7 Characteristics of volcannic-sedimentsry basins in middle Yanshanian, Beijing

火山喷发带		火山-沉积盆地名称	盆地类型	出露面积/km^2	原始分布面积/km^2	盆地内部构造形态	盆地长轴方向	盆地火山-沉积岩系的地质时代	主要岩石类型	地层总厚度/m	盆地下覆最新地层	火山活动时期
中燕山期北东向火山喷发带	南亚带	百花山火山-沉积盆地	继承性断陷盆地	180	350	向斜	北东 55°	J_{2-3}	粗安岩、玄武安山岩、粗安质角砾熔岩、凝灰砂岩、凝灰砾岩	>3442	九龙山组	髫髻山阶段
		髫髻山火山-沉积盆地		210	4	单斜	北东 55°	J_{2-3}	粗安岩、角砾集块熔岩、凝灰质砾岩、砂岩、粉砂岩	>5735	九龙山组	髫髻山-土城子阶段
		新城子火山-沉积盆地	新生性上叠式断陷盆地	209		单斜	近 EW 向	J_{2-3}	粗安岩、粗安质凝灰岩、熔结凝灰岩、粗面岩、流纹岩、凝灰砂岩	>2921	蓟县系	髫髻山-土城子阶段
	北亚带	燕羽山火山-沉积盆地		120	330	向斜	北东 55°	J_{2-3}	粗安岩、粗安质角砾凝灰岩、粗面岩、流纹质凝灰岩、凝灰砂岩、砂岩	>4971	蓟县系	髫髻山阶段
		凤驼梁-四海火山-沉积盆地		370	460	向斜	呈“S”型，南段北东 55°，中段 5°，北段北东 60°	J_{2-3}	粉砂岩、凝灰质粉砂岩、砂岩、砾岩夹流纹质凝灰岩、粗面岩、粗安岩	>4943	上寒武统	髫髻山阶段
		后城火山-沉积盆地		>500		向斜	北东 30°	J_3	粉砂岩、砂岩、砾岩夹粗安岩、粗面岩、流纹质凝灰岩	>1244	青白口系	土城子阶段
		千家店火山-沉积盆地		>91		向斜	北东 30°	J_3		>1394	蓟县系	土城子阶段

（据鲍亦冈等，1995）

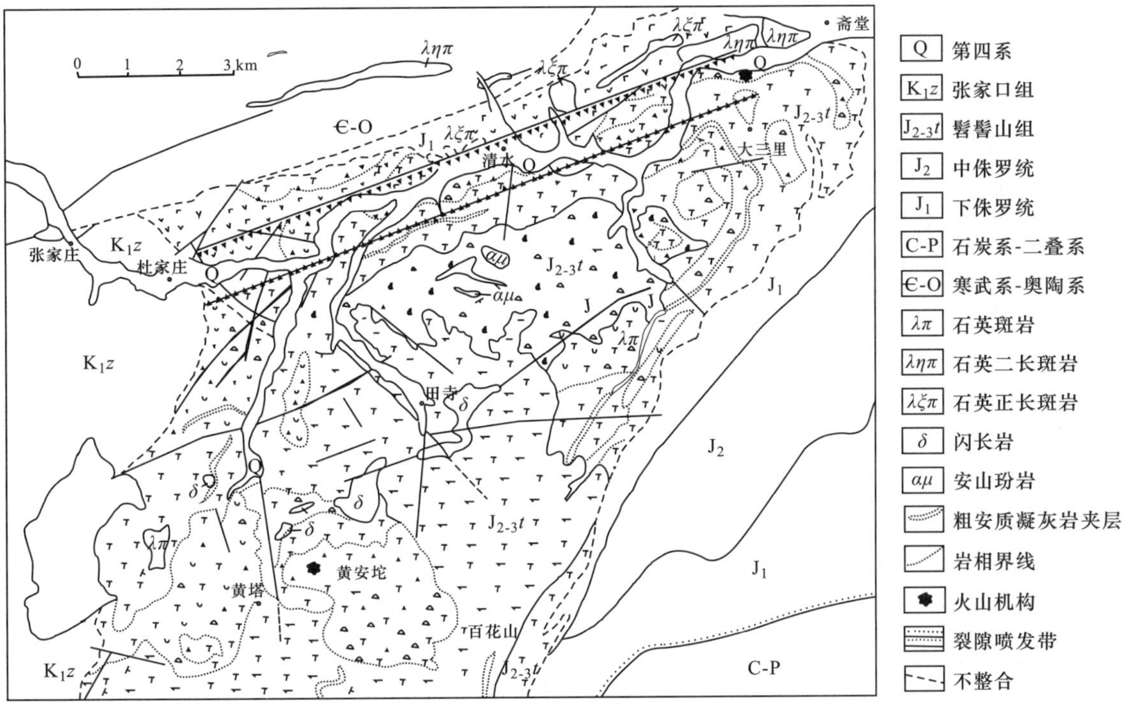

图 16-13 百花山火山-沉积盆地岩性岩相地质图

（据鲍亦冈等，1995）

Fig. 16-13 Lithologic and volcanic lithofacies map of Baihuashan volcanic-sedimentary basin

地层，火山活动转为中心式喷发，出现以黄安坨古火山口为代表的数个中心式火山机构（图 16-13），火山喷发物广泛分布于盆地的中部和南部，以粗安质熔岩及粗安质角砾凝灰岩为主，熔岩及正常火山碎屑岩相间出现，堆积厚度达 1800 余米（鲍亦冈等，1995）。两幕火山喷发周期后为火山活动间歇期，喷发活动基本停止，形成髫髻山组二、四段地层，发育为冲洪积相及小水域浅湖相火山-沉积碎屑岩的堆积，主要岩性为凝灰质砂岩及凝灰质砾岩等。

盆内火山机构潜火山岩高度发育，分布受盖层断裂控制。主要集中出现于两个构造部位：其一分布于清水河谷裂隙式线状火山机构及其北侧，呈北东向串珠状展布，单个面积多介于 $0.2 \sim 1.6km^2$，岩石类型主要有闪长玢岩、石英二长斑岩及石英正长斑岩等，这些潜火山岩体形成于第一喷发期的后期。其二分布于向斜的核部，形成于第三喷发周期的晚期，以安山玢岩、闪长岩为主，受多组断裂控制，呈不规则的岩枝状，面积多介于 $0.2 \sim 0.7km^2$ 之间（鲍亦冈等，1995）。

3. Ⅳ级构造单元——火山机构

髫髻山火山喷发阶段火山机构较为发育，以黄安坨、窖湾、四司、上王家山等古火山口、火山通道为典型代表。对黄安坨和窖湾古火山机构特征总结如下：

（1）黄安坨古火山口

黄安坨古火山口位于百花山主峰西北侧，规模较大，喷发物分布面积较广，是百花山火山-沉积盆地中髫髻山喷发活动阶段第三喷发周期最重要的中心式火山喷发口。火山机构地貌景观明显：坨顶海拔约 1000m，坨上呈一椭圆形平台，东、南侧高山环绕，北、西侧为深谷，山势陡峭，四周水系为放射状。火山喷发物围绕古火山口呈环状分布，岩层向中心倾斜：火山口附近发育集块大小高达 10～20cm 的集块角砾岩、集块角砾熔岩等粗碎屑火山岩，还可见辉石闪长岩、粗安玢岩等潜火山岩体和岩脉；火山口西发育集块大小约 1m 的粗安质集块角砾岩，并见有梨形、椭圆形或不规则形火山弹（图 16-14）。

基于岩石类型及结构构造特征，该火山机构经历 5 次较大规模的喷发活动：第一次为辉石粗安岩浆的喷溢，沿火山口附近 $10km^2$ 范围内均分布有本次喷发物的堆积；第二次为粗安质角砾熔岩的喷发，火山岩主要出露于火山口西，黄塔村以西；第三次为含大量火山弹的粗安质集块角砾岩的强烈喷发，是该火山机构最强烈爆发期，喷发物堆积于黄塔村附近；第四次为角闪粗安岩浆的溢流，喷发物分布面积较小，火山口附近见有少量集块、角砾等粗碎屑物；第五次为粗安质集块角砾熔岩的喷发，火山喷发物仅分布于火山口附近 6km² 范围内，并堵塞了火山通道，岩石中柱状节理发育。

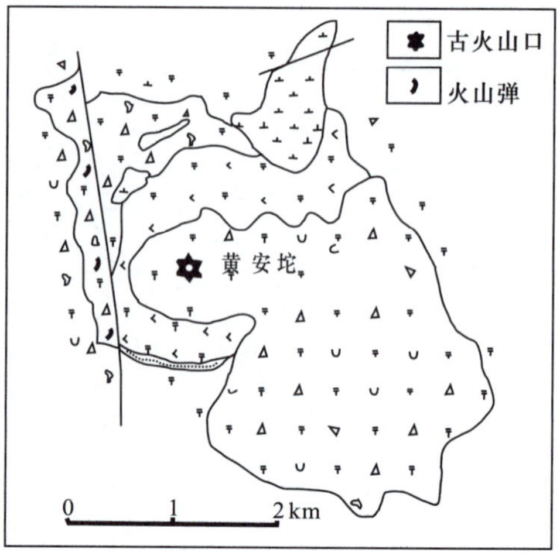

图 16-14　黄安坨火山喷发中心构造岩相图
Fig. 16-14　Structural facies zoning for Huangantuo volcanic edifice

（2）窖湾火山通道

窖湾火山通道位于延庆区窖湾村北东，呈长轴为北西-南东向的椭圆形，出露面积约 7km²（2km×3.5km）。火山通道东被大花岭北西向断层切割破坏，西被晚燕山期碓臼峪花岗岩体侵入破坏，北部被第四系覆盖（图 16-15、图 16-16）。火山机构特征明显：通道中心发育为菱形火山颈相黑云母安山玢岩，出露面积 1km²；向外依次过渡为次火山岩相杏仁状安山玢岩和安山玢岩呈环状或半环状围绕中心分布（图 16-15）。三种安山玢岩均呈侵入接触，生成顺序为黑云母安山玢岩→安山玢岩→杏仁状安山玢岩，侵入髫髻山组二段粗安质熔结凝灰岩、安山质熔结凝灰岩及安山质凝灰岩中（图 16-15）。这些岩层总体上呈北西向展布，与燕羽山向斜构造的特点基本一致，但被火山通道中后期断裂及次火山岩切割成许多小断块，产状杂乱，推测是由于火山通道构造在后期发生了轻微的陷落所致；火山通道构造的南侧，见有呈不规则脉状产出的次火山岩相侵入式粗安质熔结角砾凝灰岩和石英粗面斑岩。前者沿大花岭北西向断层产出，岩石中浆屑产状紊乱，岩石较坚硬，地貌上形成高峰、陡崖。石英粗面斑岩生成较晚，亦呈北西向不规则展布，被后期断裂及碓臼峪花岗岩破坏。

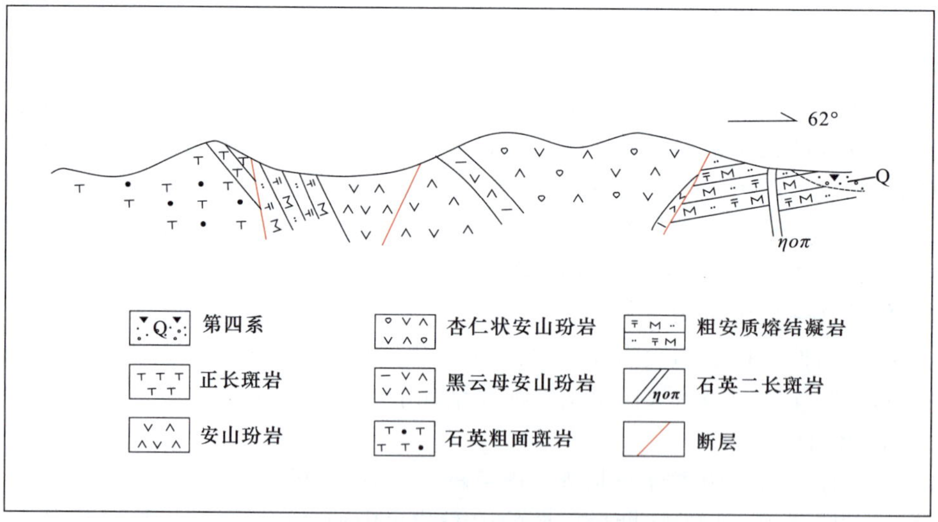

图 16-15　窖湾火山喷发中心构造岩相图
（据鲍亦冈等，1995）
Fig. 16-15　Structural facies zoning for Jiaowan volcanic edifice

（四）岩石化学与地球化学特征

髫髻山期火山岩化学成分及参数见表 16-8，岩石类型和岩石组合复杂，不仅发育玄武粗安岩、粗安岩、粗面岩、流纹岩，还发育少量玄武安山岩、安山岩（图 16-17）。时空上，岩石化学特征也有所

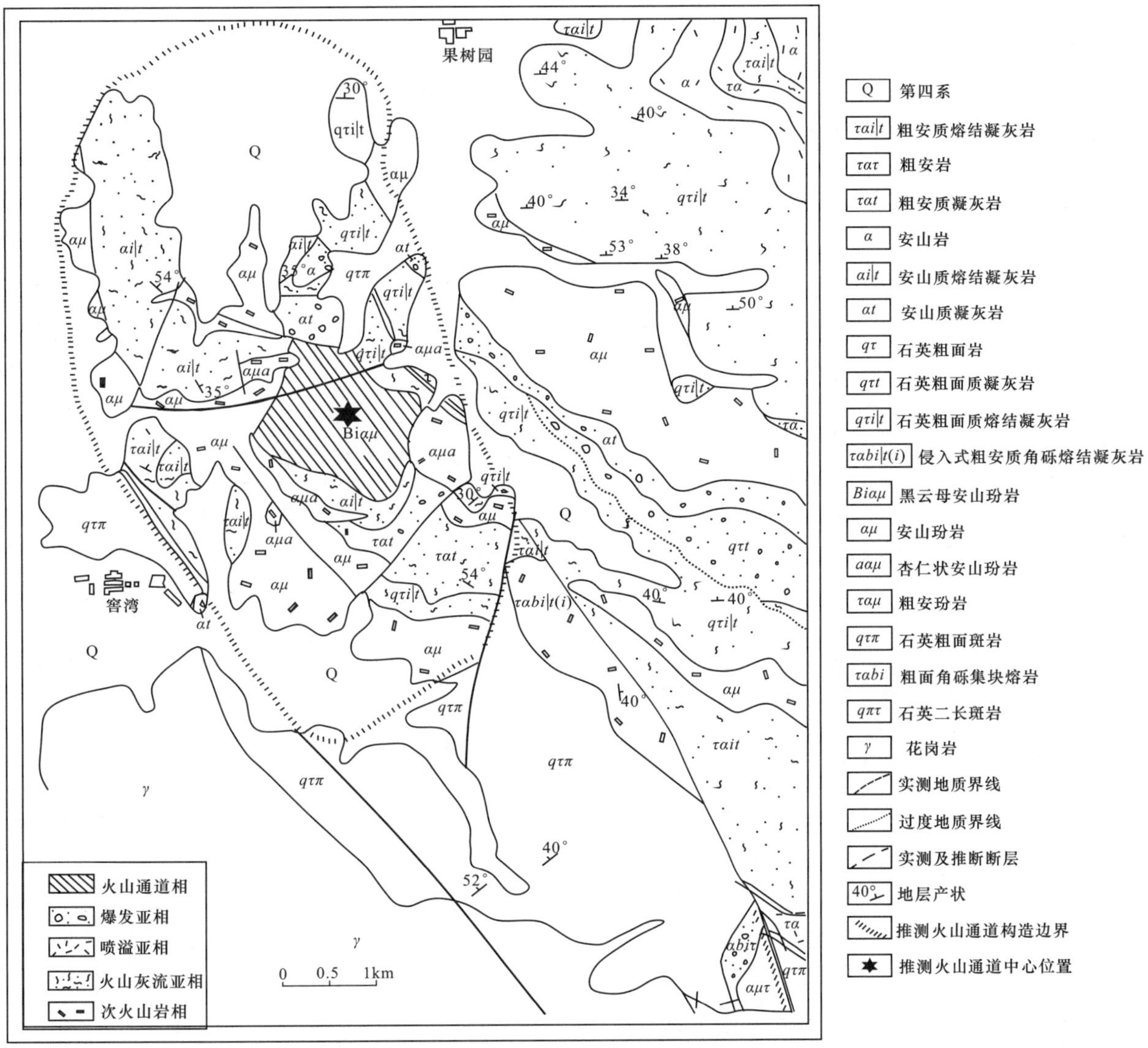

图 16-16 窑湾火山喷发中心构造岩相剖面示意图

（据鲍亦冈等，1995）

Fig. 16-16 Profile section for Jiaowan volcanic edifice

差异（图 16-17）：空间上，北亚带与南亚带相比，CaO 含量稍低，岩石更偏碱性、酸性；时间上，土城子火山喷发阶段与髫髻山火山喷发阶段相比，岩石也更偏碱性、酸性。全岩主要氧化物变化趋势图（图 16-18）上，成分投点比较连续，氧化物含量特点及变化趋势总体与平均粗面玄武岩-粗安岩-碱性流纹岩（系列二）的趋势线一致；Fe_2O_3 的含量明显比对比岩系要高，而 FeO 含量明显偏低，氧化程度较高；岩石呈碱质非过饱和特性，未受明显同化混染作用影响（鲍亦冈等，2001）。同一火山喷发阶段演化趋势表现为：髫髻山阶段由基性、中性向偏酸、偏碱性方向演化，而土城子阶段则从酸性向中性、基性方向变化。

全岩里特曼指数 $\sigma=1.45\sim4.91$，跨碱性岩和亚碱性岩区；固结指数 $SI=1.99\sim26.84$，分异指数 $DI=43.65\sim93.97$，岩浆经历一定分异作用；镁铁比值 $m/f=0.19\sim1.12$，属于铁质—富铁质基性岩；$A/CNK=0.82\sim1.17$，$A/NK=1.10\sim3.00$，属于准铝质—过铝质。

稀土元素成分及参数见表 16-9。髫髻山、土城子火山喷发阶段 ΣREE 分别为 $166.79\times10^{-6}\sim191.84\times10^{-6}$ 和 $232.79\times10^{-6}\sim315.07\times10^{-6}$，稀土总量较低，土城子阶段高于髫髻山阶段；LREE/HREE 分别为 11.12~14.01 和 13.17~18.30，$(La/Yb)_N$ 分别为 16.59~18.44 和 18.55~27.37，分馏程度相当，均表现出轻稀土弱富集，重稀土相对亏损，且轻稀土的分馏程度高于重稀土；δEu 分别为

表 16-8　髫髻山期火山岩岩石化学成分及特征参数一览表

Table 16-8　The chemical composition and main parameters of Tiaojishan volcanic rocks

序号	火山盆地	火山期次	氧化物含量/%													特征参数					
			SiO_2	TiO_2	Al_2O_3	Fe_2O_3	FeO	MnO	MgO	CaO	Na_2O	K_2O	P_2O_5	Los	总和	*DI*	SI	σ	*A/CNK*	*A/NK*	*m/f*
1	风驼梁-四海	髫髻山	64.94	0.70	16.68	2.48	1.27	0.12	0.54	2.14	4.45	3.87	0.24	1.30	98.73	81.90	4.30	3.08	1.08	1.45	0.27
2			70.36	0.14	15.42	1.68	0.65	0.05	0.23	0.66	4.02	5.01	0.07	0.63	98.92	91.17	1.99	2.95	1.17	1.28	0.19
3			57.50	1.13	16.08	4.97	3.10	0.14	2.60	5.10	4.00	3.30	0.44	1.27	99.63	62.47	14.57	3.56	0.83	1.58	0.60
4		土城子	57.73	0.86	18.48	2.25	3.30	0.40	1.90	4.08	4.80	3.42	0.44	1.36	99.02	67.19	12.13	4.40	0.97	1.59	0.59
5			59.42	1.23	15.76	3.83	2.46	0.12	2.33	4.15	3.61	3.42	0.40	1.70	98.43	66.57	14.97	2.86	0.92	1.64	0.69
6			60.46	1.05	16.17	4.62	1.86	0.09	1.67	3.59	4.06	4.04	0.33	1.12	99.06	71.57	10.37	3.64	0.92	1.46	0.49
7			74.02	0.13	13.16	0.53	1.20	0.10	0.52	0.36	4.05	5.20	0.03	0.55	99.85	93.97	4.52	2.75	1.02	1.07	0.52
8	髫髻山	髫髻山	56.44	0.84	17.71	1.66	4.36	0.08	2.72	5.63	3.11	3.34	0.49	1.28	97.66	56.52	17.91	2.88	0.93	2.03	0.82
9			51.42	1.29	17.53	2.63	5.83	0.13	4.59	5.39	3.81	2.35	0.43	0.71	96.11	48.44	23.89	3.83	0.94	1.99	0.98
10			61.91	0.86	16.58	6.36	0.57	0.05	1.13	2.00	6.10	1.35	0.49	2.11	99.51	77.56	7.43	2.84	1.10	1.44	0.32
11			64.53	0.48	17.00	1.56	2.20	0.13	1.38	2.37	4.86	3.50	0.20	2.46	100.67	78.50	10.22	3.19	1.06	1.44	0.66
12			53.34	1.26	16.44	7.94	0.68	0.11	4.47	7.10	3.77	1.83	0.35	0.84	98.13	48.16	24.51	2.77	0.78	2.01	1.00
13	百花山	髫髻山	57.21	1.00	16.80	4.40	2.52	0.12	2.90	6.25	3.50	2.70	0.36	1.28	99.04	56.79	18.26	2.59	0.84	1.94	0.78
14			61.43	0.62	16.85	3.92	1.96	0.11	1.99	2.87	5.26	2.55	0.35	1.94	99.85	73.18	12.77	3.22	1.01	1.48	0.63
15			59.96	0.81	15.83	4.31	2.45	0.09	3.16	5.13	3.27	3.08	0.24	1.65	99.98	60.75	19.57	2.32	0.88	1.82	0.88
16			51.80	1.75	15.82	7.04	3.05	0.42	4.13	6.71	3.10	1.95	1.10	0.68	97.55	46.30	21.78	2.57	0.82	2.19	0.75
17			72.65	0.66	12.39	1.53	0.98	0.06	0.45	1.38	1.63	7.40	0.10	3.96	103.19	90.06	3.76	2.74	0.94	1.16	0.33
18			53.87	1.04	17.88	6.89	1.94	0.15	3.54	7.10	2.58	1.58	0.38	0.92	97.87	43.65	21.89	1.45	0.95	3.00	0.76

续表

序号	火山盆地	火山期次	氧化物含量/%													特征参数					
			SiO_2	TiO_2	Al_2O_3	Fe_2O_3	FeO	MnO	MgO	CaO	Na_2O	K_2O	P_2O_5	Los	总和	*DI*	SI	σ	*A/CNK*	*A/NK*	*m/f*
19	新城子	髫髻山	55.00	0.90	16.86	7.23	0.54	0.12	2.76	5.38	3.30	3.30	0.51	1.14	97.04	57.81	16.49	3.27	0.90	1.87	0.69
20	燕羽山	髫髻山	57.63	0.99	17.09	5.17	2.06	0.15	2.80	6.13	4.24	2.50	0.41	1.23	100.40	58.79	16.89	3.05	0.82	1.77	0.73
21			63.03	0.72	17.04	2.05	1.82	0.20	1.28	2.32	5.31	4.18	0.29	2.18	100.42	80.76	8.74	4.41	0.98	1.29	0.59
22			68.65	0.37	15.89	1.70	1.17	0.15	0.34	0.49	4.27	4.90	0.07	3.31	101.31	90.30	2.75	3.24	1.20	1.29	0.21
23			72.14	0.24	14.09	0.28	1.51	0.08	0.51	0.37	4.26	5.70	0.06	3.85	103.09	93.55	4.16	3.39	1.02	1.07	0.49
24			61.63	0.83	15.95	3.27	2.63	0.11	2.33	4.22	4.22	3.60	0.26	1.85	100.90	69.52	14.55	3.24	0.86	1.47	0.73
25			70.19	0.39	14.58	2.89	0.67	0.08	1.41	1.34	5.63	1.47	0.13	3.53	102.31	84.17	11.79	1.84	1.10	1.34	0.75
26			55.79	1.08	16.66	4.00	3.82	0.10	3.20	5.03	4.50	3.25	0.36	1.08	98.87	61.51	17.08	4.47	0.83	1.53	0.76
27			65.27	0.68	16.00	2.95	1.39	0.07	1.18	2.33	4.85	4.12	0.20	2.51	101.55	81.12	8.18	3.58	0.96	1.29	0.51
28	后城	土城子	71.21	0.18	15.26	0.92	0.30	0.08	0.32	1.13	3.73	5.44	0.03	3.75	102.35	90.78	2.99	2.96	1.08	1.27	0.47
29			61.90	0.79	14.54	4.27	2.60	0.28	0.98	1.50	4.69	5.12	0.31	2.13	99.11	83.31	5.57	4.91	0.91	1.10	0.26
30			73.96	0.24	12.59	1.81	0.44	0.04	0.36	0.58	3.72	5.24	0.04	4.27	103.29	93.96	3.13	2.58	0.98	1.07	0.30
31			50.96	1.66	14.91	4.01	5.84	0.14	6.03	5.04	3.19	3.40	0.83	0.63	96.64	49.46	26.84	4.67	0.82	1.67	1.12
32			64.70	0.49	16.50	3.18	0.85	0.09	1.11	1.50	5.60	3.42	0.26	2.57	100.27	84.10	7.90	3.67	1.06	1.28	0.52
33			55.18	1.17	16.17	4.23	4.05	0.12	4.64	6.25	3.17	1.79	0.27	1.06	98.10	48.23	26.08	1.88	0.87	2.26	1.04

注：1~7引自汪洋等（2003），白志民等（1991）；8~12引自李伍平等（200），白志民等（1991）；13~27引自白志民等（1991）；28-33引自汪洋等（2003）。

表 16-9　髫髻山期火山岩稀土元素成分及特征参数一览表

Table 16-9　REE contents and main parameter of Tiaojishan volcanic rocks

序号	火山盆地	火山期次	稀土元素含量/10^{-6}																特征参数			
			La	Ce	Pr	Nd	Sm	Eu	Gd	Tb	Dy	Ho	Er	Tm	Yb	Lu	Y	ΣREE	LREE/HREE	$(La/Yb)_N$	δEu	δCe
1	髫髻山	髫髻山期	36.00	68.25	8.93	34.49	5.84	1.51	4.21	0.60	3.03	0.58	1.49	0.24	1.40	0.22	15.48	166.79	13.17	18.44	0.93	0.93
2		土城子期	64.98	137.30	15.96	63.02	10.61	2.64	7.81	1.04	5.31	0.97	2.57	0.36	2.20	0.30	18.35	315.07	14.32	21.19	0.89	1.05
3	风驼梁-四海	髫髻山期	48.92	73.99	10.66	37.07	6.68	1.74	4.30	0.63	3.03	0.57	1.84	0.28	1.93	0.20	16.77	191.84	14.01	18.18	0.99	0.79
4		土城子期	60.28	109.50	12.60	42.52	6.88	1.35	4.19	0.49	3.44	0.71	1.81	0.26	1.58	0.26	15.32	245.87	18.30	27.37	0.77	0.97
5	后城	土城子期	51.73	101.50	14.17	40.58	7.30	1.77	4.79	0.72	4.15	0.93	2.40	0.42	2.00	0.33	17.87	232.79	13.79	18.55	0.92	0.92
6	新城子	髫髻山期	34.70	74.30	8.95	35.00	7.15	1.69	5.84	0.55	3.51	0.68	1.97	0.26	1.50	0.24	16.30	176.34	11.12	16.59	0.80	1.03
7		土城子期	52.30	104.00	10.60	45.10	7.68	2.00	6.13	0.58	3.96	0.77	2.16	0.27	1.80	0.26	21.00	237.61	13.92	20.84	0.89	1.08

注：数据为均值，1~2 引自王蕊等，2007 和李伍平等，2001；3~5 引自汪洋等，2003；6~7 引自 1：5 万高岭幅区域地质调查报告，1993。

表 16-10　髫髻山期火山岩微量元素成分及特征参数表

Table 16-10　Trace element contents and main parameters of Tiaojishan volcanic rocks

序号	火山盆地	火山期次	微量元素含量/10^{-6}														特征参数	
			Rb	Ba	Th	U	Ta	Nb	Sr	Zr	Hf	V	Cr	Co	Ni	Cu	Sc	Rb/Sr	Nb/Ta
1	髫髻山	髫髻山	46.46	1403.38	5.31	0.68	0.44	8.60	960.63	183.40	4.52	122.04	33.00	15.98	10.49	57.50	9.29	0.05	19.55
2		土城子	47.95	1627.30	3.68	0.78	0.99	22.86	1507.70	302.90	7.15	226.30	48.09	30.21	25.58	141.00	19.67	0.03	23.16

注：数据为均值，引自王蕊等，2007；李伍平等，2001；汪洋等，2003。

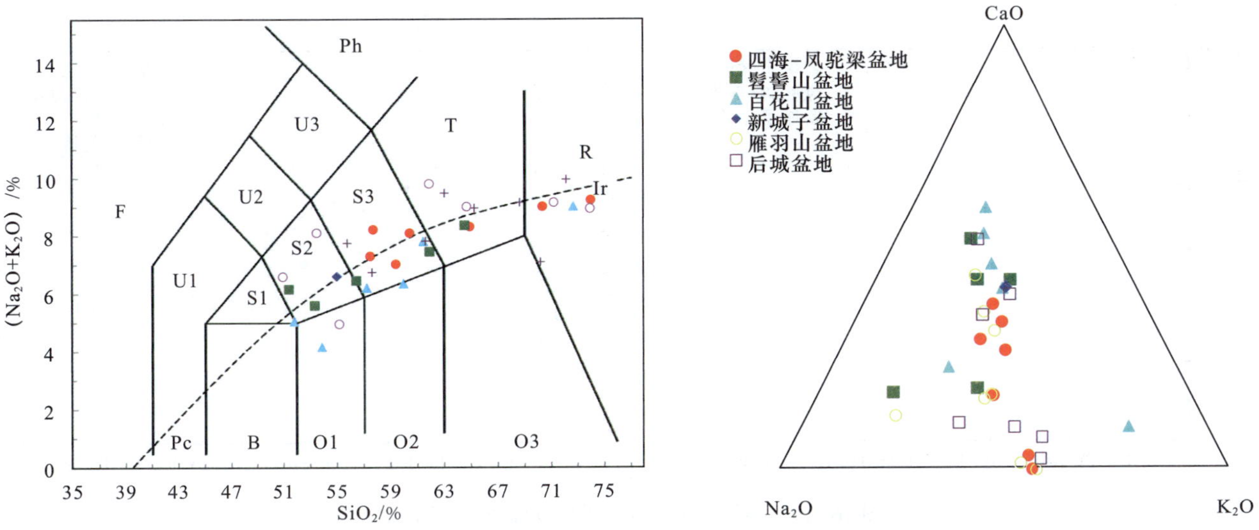

图 16-17　髫髻山期火山岩 TAS 分类图解和 Na_2O-K_2O-CaO 图解

Fig. 16-17　TAS and Na_2O-K_2O-CaO diagram of Tiaojishan volcanic rocks

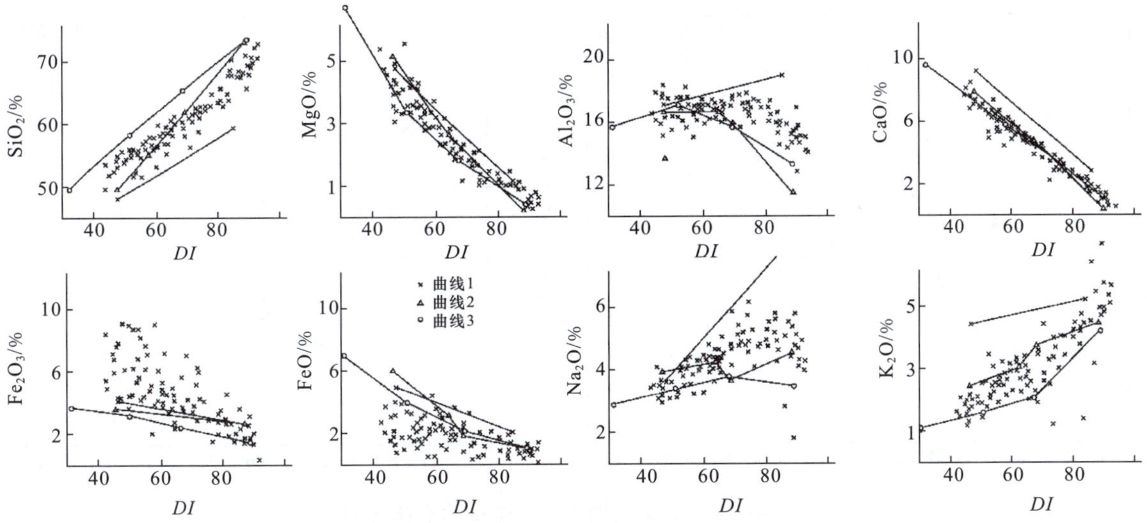

图 16-18　髫髻山期火山岩主要氧化物-*DI* 变化趋势图

（据鲍亦冈等，2000）

Fig. 16-18　Oxides differentiation index plots of Tiaojishan volcanic rocks

曲线 1 为平均碱玄岩-响岩趋势线（系列一）；曲线 2 为平均粗面玄武岩-粗安岩-安粗岩-碱性流纹岩趋势（系列二）；曲线 3 为平均玄武岩-安山岩-英安岩-流纹岩趋势线（系列三）

0.80~0.99 和 0.77~0.92，分属于铕极弱负异常；二者配分曲线特征基本相似，均为右缓倾光滑曲线（图 16-19），揭示同源岩浆演化特性。岩性不同，稀土元素总量有所差异（图略），由玄武安山岩向粗安质岩石、流纹质岩石方向稀土总量递增，且 δEu 由弱正异常向弱负异常转变，也揭示了同源岩浆分异演化的特征。

微量元素成分及参数见表 16-10，土城子阶段火山岩微量元素含量总体略高于髫髻山阶段火成岩。大离子亲石元素 K、Rb、Sr 等元素含量略有所波动，体现了髫髻期火山岩浆形成及演化上的多样性。微量元素蛛网图特征基本相似（图 16-20），富集大离子亲石元素 Ba、K、Sr 和 LREE，强烈亏损高场强元素 Nb、Ta、Th、U 和 HREE；过渡族元素呈对称性雁形，Ti、V、Cu、Sc 呈明显的尖峰，高于原始地幔、球粒陨石，而 Cr、Co、Ni、Fe 成低谷，与南大岭期、张家口期火山岩具有相似性，体现了燕山期同一构造运动对区域地球化学背景条件的制约（鲍亦冈，1995）。

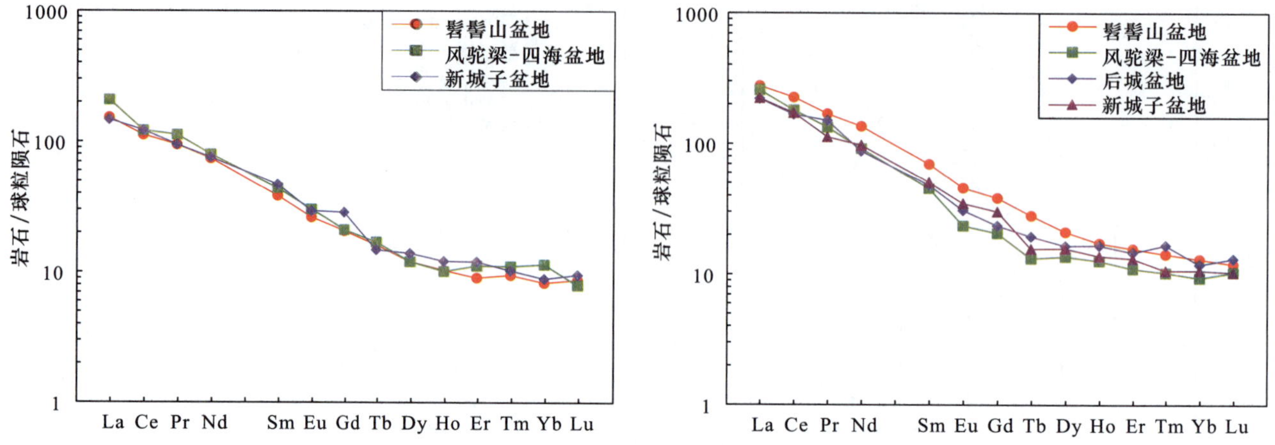

图 16-19 髫髻山期火山岩稀土元素配分模式图

Fig. 16-19 Chondrite-normalized REE patterns of Tiaojishan volcanic rocks

左髫髻山阶段；右土城子阶段

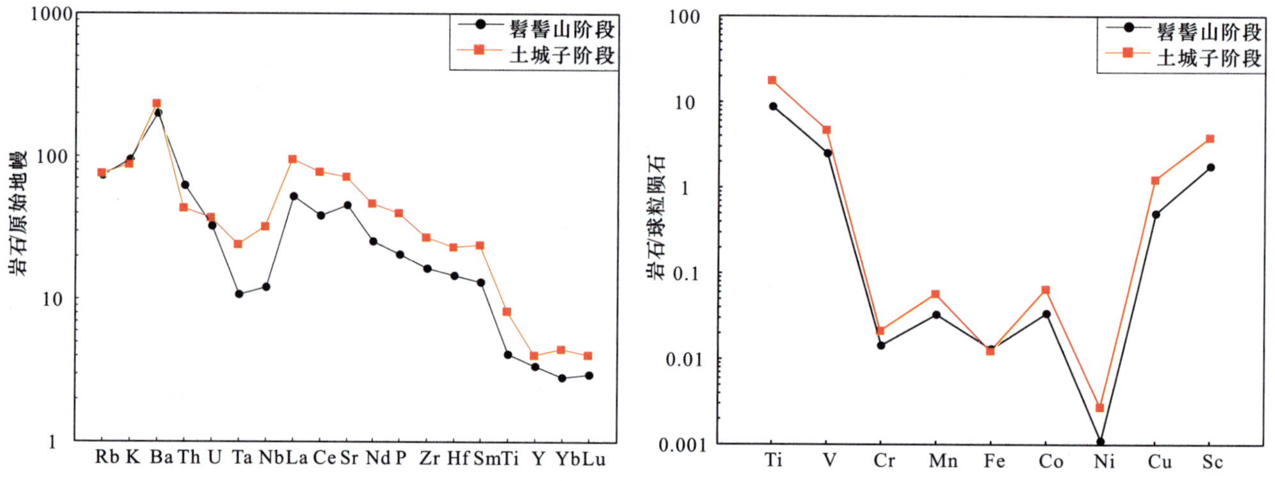

图 16-20 髫髻山期火山岩微量元素蛛网图

Fig. 16-20 Primitive mantle-normalized spidergrams for trace elements of Tiaojishan volcanic rocks

（五）Sr-Nd 同位素特征

髫髻山期火山岩 Sr-Nd 同位素在不同火山喷发阶段和不同盆地略有所差异。髫髻山盆地中，髫髻山阶段火山岩$^{87}Rb/^{86}Sr = 0.0060 \sim 0.1755$，$^{87}Sr/^{86}Sr = 0.705748 \sim 0.706265$，$^{147}Sm/^{144}Nd = 0.0376 \sim 0.1188$，$^{143}Nd/^{144}Nd = 0.511658 \sim 0.511990$，$\varepsilon_{Sr}(t) = 20.13 \sim 24.88$，$\varepsilon_{Nd}(t) = -11.1 \sim -16.7$；土城子阶段火山岩$^{87}Rb/^{86}Sr = 0.0649 \sim 0.1429$，$^{87}Sr/^{86}Sr = 0.70668 \sim 0.70693$，$^{147}Sm/^{144}Nd = 0.0976 \sim 0.1058$，$^{143}Nd/^{144}Nd = 0.51182 \sim 0.51195$，$\varepsilon_{Sr}(t) = 29.8 \sim 32.7$，$\varepsilon_{Nd}(t) = -14.2 \sim -11.8$（李伍平等 2001；汪洋等，2003；王蕊等，2007）。承德盆地中，髫髻山阶段火山岩$^{87}Rb/^{86}Sr = 0.2204 \sim 0.1527$，$^{87}Sr/^{86}Sr = 0.70555 \sim 0.70567$，$^{147}Sm/^{144}Nd = 0.1149 \sim 0.1196$，$^{143}Nd/^{144}Nd = 0.51182 \sim 0.51186$，$\varepsilon_{Sr}(t) = 13.8 \sim 15.5$，$\varepsilon_{Nd}(t) = -14.9 \sim -14.4$（李晓勇等，2003）。

（六）Pb 同位素特征

髫髻山盆地髫髻山阶段火山岩$^{206}Pb/^{204}Pb = 16.37 \sim 17.37$，$^{207}Pb/^{204}Pb = 15.19 \sim 15.38$，$^{208}Pb/^{204}Pb = 36.19 \sim 37.01$，火山岩源区放射成因 Pb 较低，与燕山造山带前寒武纪结晶基底基性麻粒岩和中生代花岗岩 Pb 同位素特征相似（李伍平等 2001）。

（七）Lu-Hf 同位素特征

冀北张家口-宣化地区土城子阶段火山岩锆石颗粒$^{176}Lu/^{177}Hf<0.002$，显示锆石形成以后具有较少的放射性成因 Hf 的积累；英安岩锆石$^{176}Hf/^{177}Hf=0.281976\sim0.282163$，$\varepsilon_{Hf}(t)=-18.8\sim-25.5$（杨进辉等，2006）。

（八）岩石系列划分

该套岩系里特曼指数$\sigma=1.45\sim6.25$，碱度率$AR=1.40\sim3.82$，跨碱性岩和亚碱性岩区，多数属于亚碱性；在 AR-SiO_2 和 K_2O-SiO_2 岩石系列判别图解（图 16-21）上，成分点分散位于碱质-钙碱质区和钙碱性-高钾钙碱性-钾玄岩系列，以高钾钙碱性区较为集中。因此，该套岩系为向碱性岩过渡的钙碱性系列。

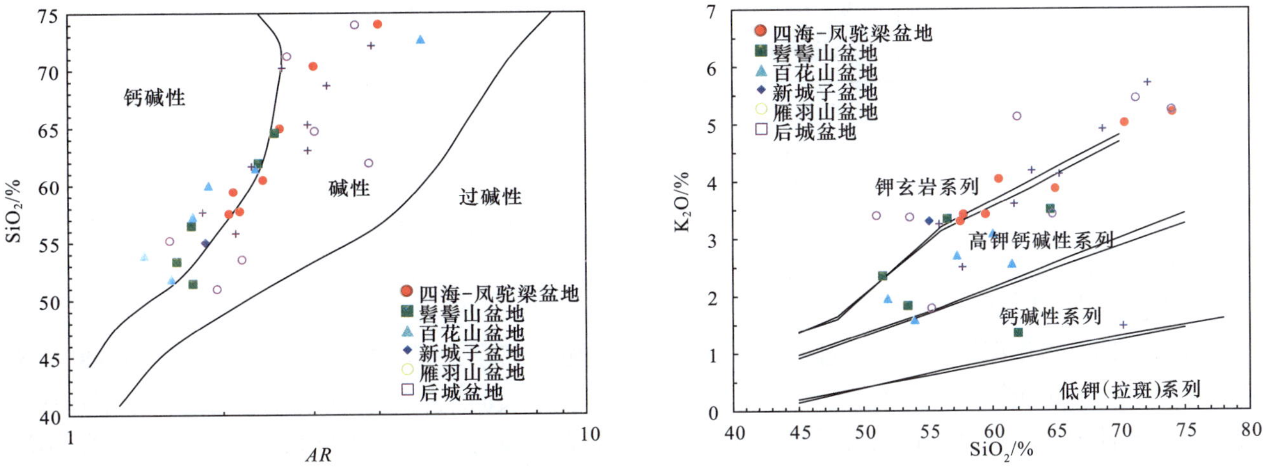

图 16-21 髫髻山期火山岩 SiO_2-AR 和 SiO_2-K_2O 图解

Fig. 16-21 The SiO_2-AR and SiO_2-K_2O diagram of Tiaojishan volcanic rocks

（九）岩石成因与岩浆演化

廖群安等（1993）、邱家骧等（1998）认为该套岩系岩浆是由幔源的钾玄岩浆与壳源的高钾钙碱性酸性岩浆混合形成。鲍亦冈等（1995）、孙善平等（1995）认为其岩浆起源于下地壳的部分熔融作用，可能有少量上地幔熔体加入，并受到地壳物质的混染。李伍平等（2001）研究认为髫髻山组火山岩具有类似埃达克岩的特征，起源于下地壳钙碱性岩石或壳幔过渡带玄武质岩石的部分熔融作用。

该期火山作用表现出以中性岩浆为主的岩浆活动，Sr-Nd-Pb 同位素揭示其岩浆与富集型 EMI 型岩石圈地幔源区具有密切联系。火山岩还具有高 Sr、Ba，高 Sr/Y 比值和相对亏损 HREE 特征，且 Eu 正负异常不明显，绝大多数岩性还具有类似埃达克岩特征（李伍平等，2001）。而 Ba/Nb 与 $\varepsilon_{Nd}(t)$ 表现出较好的相关性，且相关变化于 EMI 型岩石圈地幔和俯冲交代岩石圈地幔混合趋势相一致（李晓勇，2003），暗示源区其还可能来自华北陆块古老富集型 EMI 型岩石圈地幔和俯冲交代地幔两个端元组分混合部分熔融作用（李晓勇，2003），这同 La-La/Sm 图解（图 16-22）揭示的特征基本一致。晚期土城子火山喷发阶段，火山岩发育以酸性岩为主，可能与壳源岩浆有关。

本期火山岩 MgO、FeO 和 Cr、Ni 含量较低，可能暗示深部源区有橄榄石或辉石的分离作用，而哈克图解相关性不明显，固结指数对几种主要氧化物图解呈较好线性排列，Nb/Ta 弱分馏，表明源区可能还存在单斜辉石、斜长石、角闪石结晶作用。因此，结晶分异作用是本期火山岩岩浆演化的重要控制因素，而地壳物质混染作用影响是极小的。分离结晶可以使残余岩浆本来就比较高的 Sr/Y 比值变得更高，同时 Y 的含量变得更低（王蕊等，2007）。

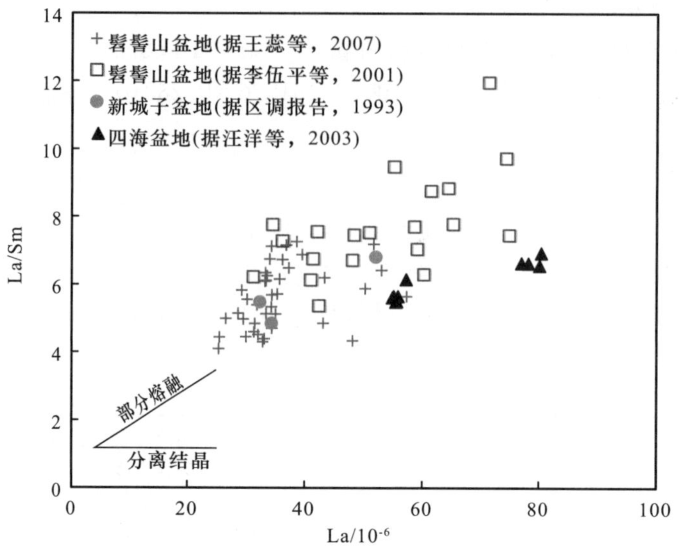

图 16-22 髫髻山期火山岩 La-La/Sm 图解

Fig. 16-22 La-La/Sm diagram diagram of Tiaojishan volcanic rocks

（十） 构造环境分析

髫髻山期火山作用规模巨大，且岩浆来自较深的地幔。邱家骧等（1998）、陈绍海等（1998）、邓晋福等（1998）研究认为中—晚侏罗世燕山地区具有厚度较大的地壳结构，在挤压构造背景下形成大规模中—基性岩浆并喷发至地表几乎不可能。因此，该阶段北京地区主体仍处于伸展拉张构造环境。在此构造背景下，沿着深大断裂发育有髫髻山喷发阶段一系列火山-沉积盆地。

至土城子火山喷发阶段，火山活动强度明显转弱，且岩性发生较大转变，壳源物质成为主要岩浆来源。此时区域构造可能发生应力场反转由北西-南东向伸展转变为挤压作用，主压应力和左旋压扭的构造属性是这一阶段重要特征。目前，造成这一挤压作用的动力来源被认为同中生代太平洋板块俯冲影响（李思田等，1990）或蒙古-鄂霍茨克海的消减、闭合作用有关（Davis et al.，2001）。考虑燕山地区远离俯冲带，其动力原因有待商榷，但此挤压造山作用为主的构造变形在燕山地区广泛分布，如八宝山、新城子等逆冲推覆构造，以及一系列北东向宽缓褶皱，这一构造变形对髫髻山-土城子组以下沉积地层均有一定程度影响。

（十一） 火山岩形成时代

前人对髫髻山期火山岩开展了大量同位素年龄测试（表 16-11），其中 K-Ar 或 Ar-Ar 年龄明显偏年轻，同化石揭示的地层沉积时代相差过远，可能同此类方法测试封闭温度有关。锆石 U-Pb 年龄同化石揭示年代相当，具有参考意义。其中，髫髻山组火山岩年龄集中于 170～145Ma，而上覆的土城子组火山岩年龄约为 145Ma，土城子阶段小规模的火山活动与髫髻山阶段火山活动的时间间隔并不太久，火山活动时间同为中—晚侏罗世。

表 16-11 髫髻山期火山岩同位素年龄一览表

Table 16-11 Isotopic ages of Tiaojishan volcanic rocks

序号	采样地点	岩石名称	测定方法	测试对象	年龄值/Ma	资料来源
1	门头沟九龙台	辉石安山岩	K-Ar	全岩	152.0	北京市地质调查所，1993
2	延庆瓦庙村北	安山岩	K-Ar	黑云母	153.5±2.5	北京市地质调查所，1993

续表

序号	采样地点	岩石名称	测定方法	测试对象	年龄值/Ma	资料来源
3	怀柔上窝铺	石英粗面质熔结凝灰岩	K-Ar	全岩	146.04±2.44	北京市地质研究所，1993
4	延庆瓦庙-四司	粗安岩	Rb-Sr	全岩	160.86±8.37	北京市地质调查所，1993
5	承德	英安岩、流纹岩	U-Pb	锆石	164.4±2.7	张宏等，2008
6	滦平长山峪	安山岩	U-Pb	锆石	153.8±5.2	张宏等，2008
7	滦平小营	安山岩	U-Pb	锆石	162.8±3.2	张宏等，2008
8	西山马兰	安山岩	U-Pb	锆石	157±3	赵越等，2004
9	十三陵	安山岩	Ar-Ar	全岩	161±2	Davis et al.，2001
10	十三陵	安山岩	Ar-Ar	全岩	160.7±0.8	Davis et al.，2001
11	十三陵	安山岩	U-Pb	锆石	153.6±3.8	Davis et al.，2001
12	十三陵	安山岩	Ar-Ar	角闪石	159.7±1.3	Cope et al.，2003
13	十三陵	熔结凝灰岩	Ar-Ar	透长石	152.6±0.3	Cope et al.，2003
14	十三陵	玻屑凝灰岩	U-Pb	锆石	153.3±3.3	Cope et al.，2003
15	西山雁翅	粗安岩	Ar-Ar	全岩	148.91±2.98	李伍平等，2001
16	大台煤矿	英安岩	U-Pb	锆石	137.1±4.5	袁洪林等，2005
17	西山沿河城大台	粗安质熔岩	Sm-Nd	全岩	188.1±19	汪洋等，2003
18	延庆大庄科	辉石安山岩	U-Pb	锆石	169.9±4.3	赵兴国，2009
19	罗家台南	安山岩	U-Pb	锆石	166.7±3.2	赵兴国，2009
20	G109国道54km北	安山岩	U-Pb	锆石	160.7±4.3	赵兴国，2009
21	门头沟沿河口东南	角砾熔结凝灰岩	U-Pb	锆石	179.7±3.5	赵兴国，2009
22	大台	粗面玄武岩	Sm-Nd	全岩	188±19	汪洋等，2003
23	承德	晶屑凝灰岩	U-Pb	锆石	153±1	刘健等，2006
24	承德	粗安岩	U-Pb	锆石	156±3	刘健等，2006
25	延庆羊头山	流纹质晶屑玻屑凝灰岩	K-Ar	全岩	91、100	北京地质学院，1964
26	延庆羊头山	流纹质晶屑玻屑凝灰岩	K-Ar	全岩	143	北京地质学院，1964
27	延庆燕羽山	安山质熔结凝灰岩	K-Ar	全岩	105	北京地质学院，1964
28	延庆燕羽山	安山质熔结凝灰岩	K-Ar	全岩	69	北京地质学院，1964
29	延庆燕羽山	安山岩	K-Ar	全岩	105	北京地质学院，1964
30	延庆燕羽山	粗安质熔结凝灰岩	K-Ar	全岩	145	北京地质学院，1964
31	延庆燕羽山	粗安质熔结凝灰岩	K-Ar	全岩	108	北京地质学院，1964
32	延庆燕羽山	粗安质熔结凝灰岩	K-Ar	全岩	112	北京地质学院，1964
33	张家口-宣化	英安岩	U-Pb	锆石	130±1	杨进辉等*，2006
34	门头沟樱桃沟	流纹质弱熔结凝灰岩	U-Pb	锆石	144.7±4.8	赵兴国*，2009
35	怀柔北部三道石胡	安山玢岩	K-Ar	全岩	144.12±1.70	北京地质调查所*，1993
36	大台	安山岩	SHRIMP	锆石	152.1±1.9、149±2	于海飞等，2016
37	十三陵	安山岩	SHRIMP	锆石	157.2±1.2、155.2±1.7	于海飞等，2016

注：带星号为土城子组，其余为髫髻山组。

三、张家口期火山岩与火山作用

（一）时空分布特征

这一时期形成的火山岩分布于张家口组和东狼沟组中，为晚燕山期火山喷发产物，也可划分为两个喷发阶段，相当于火山岩岩石地层的张家口组和东狼沟组。

张家口期火山岩呈北北东向分布于北京西部及北部数个新生断陷盆地中，主要火山盆地包括东岭、杜家庄-镇厂、上大水、岔道、怀柔、七道河盆地，在昌平棉山、密云五指山等地也小面积出露该期火山岩，总分布面积 550km^2。张家口期火山-沉积盆地组合构成方向性极强的北北东向火山岩带，但带内每一个火山盆地的发育和生成则受其具体部位特定盖层断裂构造的控制，盆地延长方向各异，这是不同于前两期火山盆地的重要特征之一。其中，火山-沉积地层厚度以怀柔火山盆地最大，为 4216m，东岭盆地为 1622m，各盆地火山岩厚度变化较大（图 16-23），无明显规律（鲍亦冈等，2001）。张家口期火山岩只在七道河盆地与土城子组呈角度不整合接触，其余地区多角度不整合于侏罗系髫髻山组、九龙山组、南大岭组及古生界、中—新元古界之上（鲍亦冈等，1995）。

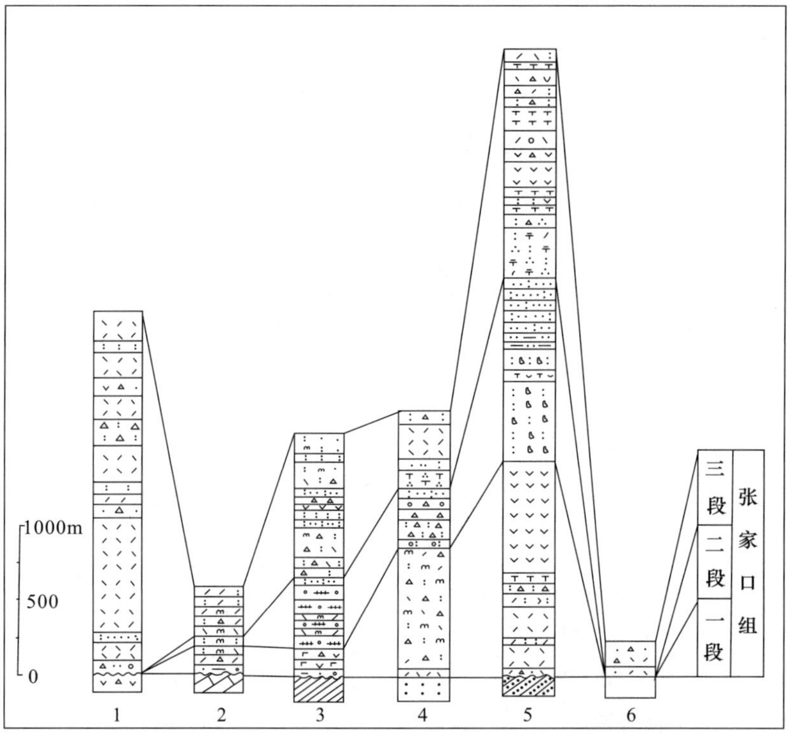

图 16-23　北京地区张家口组火山岩段喷发韵律相序剖面对比图
（据鲍亦冈等，1995）
Fig. 16-23　Profile correlation diagram of volcanic eruption lithofacies sequence of Zhangjiakou formation in Beijing
1—杜家庄；2—上大水村；3—东岭；4—岔道；5—怀柔；6—棉山

据火山活动及岩性特征，张家口火山喷发阶段又可划分出两个喷发周期和一个火山间歇周期，分属张家口组一、三岩性段和二岩性段。第一喷发周期发育岩性复杂，各盆地中演化趋势也有所不同：东岭与上大水盆地发育以玄武安山岩、粗安岩、粗安质角砾凝灰岩等中性火山喷出岩为主，早期为熔浆喷溢，后期喷溢与爆发作用交替发生，喷发强度逐渐增大；上大水盆地在火山活动晚期见有少量流纹质熔结凝灰岩和凝灰岩，火山活动具有由中偏基性向中性及酸性方向演化的特点；怀柔盆地发育为流纹质角砾凝灰岩、粗面岩、粗安岩，火山活动具有由酸性向中性方向演化的特点。第二喷发周期是该阶段火山活动最强盛时期，火山岩层厚度大，分布广泛，各盆地均有出露，以中、酸性火山喷发占绝对优势，发育流纹质熔结凝灰岩、流纹质角砾凝灰岩及流纹岩。怀柔、东岭盆地中见少量粗面质、粗安质熔岩及正常火山碎屑岩。

东狼沟喷发阶段是在张家口强烈喷发阶段之后发育的较弱火山活动阶段，主要发育于东狼沟组第二岩性段，可划分出两个橄榄玄武质熔岩-火山角砾岩-沉玄武质凝灰角砾岩喷发韵律。早期发育为玄武安山质、粗安质岩石，见少量酸性流纹质岩石，晚期火山活动略强，流纹质岩石占主导地位，各火山-沉积盆地略有所差异。由于东狼沟火山喷发和张家口火山喷发同属一个火山喷发旋回，构造形迹特征一致，空间上二者火山岩分布特征基本类似。火山-沉积岩分布于西山坨里-大灰厂、东岭、杜家庄-镇厂盆地，出露面积 10km^2。

（二）火山岩相与岩石学特征

与髫髻山期火山活动特征相似，按照火山活动特征，各火山-沉积盆地火山岩相序列划分见表 16-12。总体上，张家口期火山活动以中心式喷发为主，显示了层火山的特征。喷发活动强烈，爆发作用强，爆发指数高，平均爆发指数为 44%。其中以上大水盆地爆发指数最高，达 88%；东岭盆地和岔道盆地分别为 50%和 52%；而杜家庄盆地却以熔浆溢流为主，爆发指数仅为 10%。

表 16-12 张家口期各火山盆地火山岩相发育特征简表

Table 16-12 Charateristics of volcanic facies of Zhangjiakou volcanic rocks of volcanic basins in Beijing

<table>
<tr><th>地区</th><th>喷发周期</th><th>岩相组合及比例</th><th colspan="2">爆发相岩浆性质、类型及比例</th><th>爆发相特征</th></tr>
<tr><td rowspan="5">门头沟区
杜家庄</td><td rowspan="2">第三喷发周期</td><td>溢流相 76.46%</td><td colspan="2">流纹质凝灰岩 18.4%</td><td>含角砾</td></tr>
<tr><td>爆发相 23.54%</td><td colspan="2">层状凝灰岩 5.14%</td><td>层理发育</td></tr>
<tr><td rowspan="2">第二间歇周期</td><td>溢流相 97.2%</td><td rowspan="2">流纹质为主</td><td>熔结凝灰岩 1.57%</td><td>含 30%~40%玻屑及浆屑</td></tr>
<tr><td>爆发相 2.98%</td><td>凝灰岩 1.41%</td><td>含角砾</td></tr>
<tr><td>第一喷发周期</td><td>溢流相 100%</td><td colspan="2">无爆发相</td><td></td></tr>
<tr><td rowspan="10">门头沟
东岭台</td><td rowspan="3">第三喷发周期</td><td>溢流相 7.93%</td><td rowspan="3">流纹质</td><td>凝灰岩 12.35%</td><td>含玻屑、晶屑</td></tr>
<tr><td>爆发相 79.15%</td><td rowspan="2">熔结凝灰岩 66.8%</td><td rowspan="2">含玻屑、晶屑，个别层含角砾</td></tr>
<tr><td>火山-沉积相 12.92%</td></tr>
<tr><td rowspan="3">第二间歇周期</td><td>爆发相 15.59%</td><td colspan="2" rowspan="3">流纹质为主、熔解凝灰岩 15.59%</td><td rowspan="3">含晶屑，玻屑及角砾，弱熔结</td></tr>
<tr><td>火山-沉积相 15.28%</td></tr>
<tr><td>沉积相 14.62%</td></tr>
<tr><td rowspan="4">第一喷发周期</td><td>溢流相 79.62%</td><td colspan="2" rowspan="4">粗安质、火山角砾岩 3.79%</td><td></td></tr>
<tr><td>爆发相 3.79%</td><td></td></tr>
<tr><td>火山-沉积相 1.97%</td><td></td></tr>
<tr><td>沉积相 14.62%</td><td></td></tr>
<tr><td rowspan="5">延庆
岔道</td><td rowspan="2">第三喷发周期</td><td>溢流相 81.92%</td><td rowspan="2">流纹质</td><td>凝灰岩 15.88%</td><td>含角砾</td></tr>
<tr><td>爆发相 18.08%</td><td>熔结凝灰岩 2.2%</td><td>含角砾</td></tr>
<tr><td>第二间歇周期</td><td>火山-沉积相 100%</td><td colspan="2">无爆发相</td><td></td></tr>
<tr><td rowspan="2">第一喷发周期</td><td>溢流相 5.03%</td><td rowspan="2">流纹质</td><td>凝灰岩 9.06%</td><td>含角砾</td></tr>
<tr><td>爆发相 94.97%</td><td>熔结凝灰岩 85.91%</td><td>含角砾，有些层位弱熔结</td></tr>
</table>

续表

<table>
<tr><th>地区</th><th>喷发周期</th><th>岩相组合及比例</th><th colspan="2">爆发相岩浆性质、类型及比例</th><th>爆发相特征</th></tr>
<tr><td rowspan="10">怀柔北宅</td><td rowspan="5">第三喷发周期</td><td>溢流相 55.96%</td><td colspan="2">流纹质、火山角砾岩 6.53%；</td><td>含集块，凝灰胶结</td></tr>
<tr><td rowspan="4">爆发相 44.04%</td><td colspan="2">流纹质凝灰岩 12.62%</td><td>含角砾</td></tr>
<tr><td colspan="2">粗面质凝灰岩 1.99%</td><td>含角砾</td></tr>
<tr><td colspan="2">粗安质凝灰岩 20.66%</td><td>晶屑发育</td></tr>
<tr><td colspan="2">流纹质熔解凝灰岩 2.24%</td><td>紫红色，含角砾</td></tr>
<tr><td rowspan="3">第二间歇周期</td><td>溢流相 3.23%</td><td colspan="2" rowspan="3">流纹质、凝灰岩 0.68%</td><td rowspan="3">灰紫色含角砾</td></tr>
<tr><td>爆发相 0.68%</td></tr>
<tr><td>火山-沉积相 96.09%</td></tr>
<tr><td rowspan="2">第一喷发周期</td><td>溢流相 87.13%</td><td rowspan="2">流纹质</td><td>凝灰岩 2.28%</td><td>凝灰胶结</td></tr>
<tr><td>爆发相 12.87%</td><td>熔结凝灰岩 10.59%</td><td>晶屑、玻屑较发育</td></tr>
</table>

（据鲍亦冈等，1995）

1. 溢流相

产于张家口组第一、第三喷发周期和东狼沟组第二喷发周期，以中心式涌动喷溢为特点，形成了分布不甚广泛、横向厚度及基本层厚度变化均较大、基本层构成要素不典型、岩相组合比较复杂的相段（鲍亦冈等，1995）。它们多产出在形态较规则圆形或近圆形，但规模较小的火山-沉积盆地中。特别是东狼沟阶段第二喷发周期的喷溢活动，由于接近燕山期构造-岩浆活动的尾声，火山喷溢作用规模较小，溢流相分布局限，仅在丰台大灰厂、门头沟沿河城、怀柔北宅等地局部出露。发育岩性组合复杂，以酸性流纹岩最为发育，主要岩石特征见表 16-13。

表 16-13　张家口期主要火山熔岩特征简表

Table 16-13　Major characteristics of Zhangjiakouan volcanic lava

特征 / 岩石	颜色	结构	构造	斑晶及含量	基质及含量	交代蚀变
玄武安山岩	深灰—灰绿	斑状结构，基质交织结构、拉斑玄武结构	块状构造，杏仁构造	斜长石（拉长石）20%~30%，辉石 3%~5%，橄榄石约 3%	斜长石 50%~60%，橄榄石约 3%，辉石 1%~2%，副矿物 1%~2%	硅化、高岭土化、绢云母化、绿泥石化、碳酸盐化均发育
安山岩	深灰—灰绿	斑状结构，基质交织结构、玻晶交织结构（粗面结构）	块状构造，杏仁构造	斜长石 15%~20%，角闪石 1%~2%，黑云母约 3%	斜长石>50%，碱性长石<3%，副矿物 1%~2%	碳酸盐化、绿泥石化、绢云母化、帘石化
粗安岩	紫红-深红色	斑状结构，基质粗面结构	块状构造	斜长石约 5%，碱性长石约 3%，黑云母约 1%	斜长石+碱性长石约 90%，副矿物约 1%	硅化及碳酸盐化较发育
球粒流纹岩	紫红色、灰色、紫灰色	斑状结构，基质球粒结构、微晶结构	流纹构造	斜长石（钠长石）约 7%，钾长石约 3%，黑云母 2%~3%，长英质球粒（呈圆形）40%~45%	钾长石+石英 45%，磷灰石<1%	碳酸盐化
条带状流纹岩	浅灰色	斑状结构，基质霏细结构	条带状构造	石英约 8%，碱性长石约 8%	长英质物质约 82%	高岭土化、绢云母化

2. 火山爆发相

张家口火山阶段火山爆发作用总体较强烈，且第一、第三喷发周期最为强烈。不同区域爆发相的发育特征有差异（表 16-12）。火山爆发岩浆以流纹质、碎屑流堆积相熔结凝灰岩为主，与其他阶段（以中性岩浆、火山崩落堆积相为主）形成鲜明对比。

（1）火山崩落堆积相

形成的岩石主要为集块角砾岩，呈环状分布于火山岩筒周围。岩石呈暗紫色，角砾状结构，块状构造。角砾棱角状—次棱角状及不规则状，大者 150~180cm，一般 2~5cm，无分选，成分为黑色和褐色安山岩，含量可达 80%。胶结物为安山质熔浆。

（2）火山碎屑流堆积相

形成的岩类主要为角砾熔岩和熔结凝灰岩。

角砾熔岩：呈环状分布于火山机构附近，杂色、紫色、紫红色，角砾状结构，块状构造。角砾成分随岩性段不同而有所差异，包括碳酸盐岩、安山岩、粗安岩、凝灰质粉砂岩和流纹岩角砾。角砾棱角状-次棱角状，大小 1~5cm，少数>10cm，含量 25%~65%。根据胶结物类型可进一步划分为安山质、粗安质和流纹质三小类，以流纹质发育最多。

熔结凝灰岩：空间分布广，厚度大，类型多，是张家口期火山岩的主体。根据结构及碎屑特征，可分出流纹质熔结凝灰岩、流纹质熔结角砾凝灰岩、含球粒熔结凝灰岩和英安质熔结凝灰岩。岩石凝灰结构，假流纹构造，物质组成包括塑变岩屑（浆屑）、塑变玻屑、角砾、岩屑、晶屑及火山尘等。

（3）潜火山岩相

是张家口期火山活动阶段最为发育的亚相，多呈脉状、岩枝或岩床产出，岩石类型包括安山玢岩、石英斑岩、闪长玢岩和霏细斑岩、流纹岩等，特征见表 16-14。值得注意的是，潜火山岩多呈侵入状分布在髫髻山组地层中，且岩性同髫髻山组火山岩具有相似性。因此前人通过岩性对比和接触关系，认为潜火山岩多数为髫髻山期火山活动产物。最新潜火山岩同位素年龄揭示其多为早白垩世产物，隶属于张家口期晚期中性火山活动。

3. 火山沉积岩类

发育于张家口火山喷发阶段第二间歇周期和东狼沟火山喷发阶段第一间歇周期等，前者以溢流相、爆发相夹层出现，后者则以主导沉积相出现。发育岩性复杂，包括沉集块岩、沉火山角砾岩、沉凝灰岩、凝灰质砾岩、凝灰质砂岩、凝灰质泥岩等。

砾岩砾石大小不等，多为 0.5~5cm，大者 30~40cm，次棱角状—次圆状，分选性差，成分复杂，以玄武安山岩为主，其次为安山岩、粗安岩、碳酸盐岩、燧石和凝灰岩、石英二长岩等，含量 70%~90%。胶结物为晶屑、碳酸盐、火山灰及黏土矿物，胶结类型为孔隙式和接触式，少数基底式。

凝灰质粉砂岩为粉砂质结构，块状构造，由碳酸盐岩岩屑（约 5%）和石英（约 8%）、长石（约 5%）、钾长石（约 3%）和少量云母等碎屑组成，胶结物主要为黏土质、碳酸盐、铁质和少量细小鳞片状绢云母组成，含量达 80%以上。

表 16-14 张家口期主要潜火山岩岩石特征简表

Table 16-14 Major characteristics of Zhangjiakouan subvolcanic rocks

岩石	颜色	结构	构造	斑晶特征	基质特征
安山玢岩	紫灰色	斑状结构，基质交织结构	块状构造，局部流纹构造	斜长石自形、长板柱状，聚斑状，可见环带，粒径 0.5~4mm；碱性长石具卡钠复合双晶；角闪石黄绿色，长柱状，粒径 0.3~0.5mm，含量 1%~2%	斜长石粒径<0.1mm，含量约 80%；角闪石黄绿色，长柱状，粒径<0.1mm，含量约 2%；磁铁矿占 2%~3%；磷灰石自形柱状，粒径<0.15mm，含量约 1%；次生矿物有绿泥石、绿帘石、碳酸盐矿物
石英斑岩	浅灰色—灰白色	斑状结构	块状—角砾状构造	碱性长石约占 5%~10%；石英边部大小约 0.5mm，含量 5%~15%；中部大小约 1~3mm，含量 1%~2%	基质为隐晶质，含安山岩、碳酸盐等棱角状角砾

岩石	颜色	结构	构造	斑晶特征	基质特征
霏细斑岩	灰白色	稀斑隐晶结构	块状构造	碱性长石，含量3%～5%	以显微长英质矿物为主，含黄铁矿，含量约1%
闪长玢岩	浅灰色—灰绿色	斑状结构，基质显微粒状结构	块状构造	斜长石自形，长板柱状，粒径1～6mm，具卡钠复合双晶，弱绢云母化，含量约40%	细粒斜长石自形长柱状，粒径0.05～0.15mm，含量约50%，具卡钠复合双晶；黑云母褐色，粒径<0.1mm，含量约2%；磷灰石短柱状，微量；磁铁矿他形粒状，含量约3%；碳酸盐矿物+绿泥石含量4%～5%

（据《北京市区域地质志》，1991）

（三）火山韵律与火山旋回

一次火山碎屑流喷发活动所形成的岩石组合构成火山岩的基本层。张家口期火山岩以熔结凝灰岩为主，在垂向剖面上出现熔结强度、岩石外貌、结构构造、碎屑含量及其形态等多种特征有规律的变化（图16-24）。每个基本层的中部为强熔结带，向上、向下熔结强度逐渐减弱。强熔结带中塑变岩屑含量高（*Fi*=14%～20%），且塑变岩屑的扁度值（*FL*=8.5～20）较大，而弱熔结带中*Fi*值低（约9%），*FL*值小（约5）；强熔结岩石致密，相对密度较大（最大为2.76），向上向下逐渐减小；强熔结带中碱性长石和石英晶屑含量低于顶底弱熔结带；各个带主要氧化物的含量总体变化不大，但K_2O、Na_2O在强熔结带稍高，可能与弱熔结带岩石致密度稍差，K、Na易于流失有关；顶部氧化较强，Fe_2O_3略高而FeO略低；岩浆房内发生分异作用，分异指数自下而上逐渐增高（朱勤文，1994）。

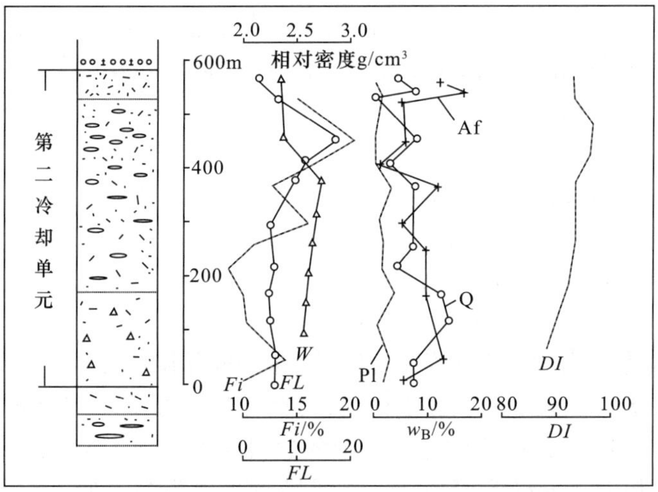

图16-24 岔道张家口组一段第二基本层成分变化图

（据鲍亦冈等，2001）

Fig. 16-24 Schematic map showing composition variation of the second fundamental layer of first member, Zhangjiakou formation

W—岩石相对密度；*DI*—分异指数；Pl—斜长石；Af—碱性长石；Q—石英

Fi—塑变岩屑含量；*FL*—塑变岩屑扁度

（四）火山构造

参考《中国区域地质志·河北志》（2017）火山构造单元划分，本区位于承德-武安火山喷发带（Ⅱ级），内含东岭等多个火山-沉积盆地（Ⅲ级）。

1. Ⅱ级火山构造单元——火山喷发带（鲍亦冈等，1995）

晚燕山期火山喷发带呈北北东向展布，隶属于承德-武安火山喷发带（Ⅰ级），受控于北北东向深断裂构造，带内限于紫荆关、高丽营深断裂带间，宽7～8km。境内起自房山良乡与门头沟杜家庄一带，NNE向延伸至怀柔七道河与延庆千家店一带，长约140km。其向SSW可延伸至河北易县与满城一带，NNE 延伸向河北丰宁、围场一带直达辽西。带中含有东岭、杜家庄-镇厂、怀柔等晚燕山期火山-沉积盆地，发育和生成则受其具体部位特定盖层断裂构造的控制，单个盆地的延长方向各异。盆地叠覆于不同时代构造单元之上，表现为晚燕山期新生性上叠式断陷火山-沉积盆地。

2. Ⅲ级构造单元——火山-沉积盆地（鲍亦冈等，1995）

境内晚燕山期火山喷发带内发育东岭等6个火山-沉积盆地（表16-15），对典型东岭火山-沉积盆地特征描述如下：

东岭盆地位于门头沟区沿河城，北、西侧分别受北东向沿河城断裂和北西向横岭断裂控制，盆地伴随断裂活动同生演化，发育北西向呈掀斜箕状断陷特性，属于上叠式新生性断陷火山盆地，出露面积为 $18km^2$（图16-25）。盆地基底为下侏罗统窑坡组及南大岭组地层，火山岩与下伏地层呈角度不整合接触。

东岭盆地中堆积张家口阶段和东狼沟阶段两期火山岩，组成流纹质熔结凝灰岩、流纹岩、粗安岩为主体的火山-沉积岩系，总厚度大于2130m。火山岩自东向西层位渐新，各阶段火山岩层序清晰，标志明显。张家口阶段第一喷发周期的火山活动以中—偏基性粗安质、玄武安山质熔岩溢流为主，夹有少量火山角砾岩的堆积，底部断续分布厚薄不等的底砾岩，总厚度为197.7m。随后构造活动趋于缓和，喷发活动基本停止，形成张家口阶段第二间歇周期火山-沉积碎屑岩的堆积，岩石以冲洪积相复成分砾岩、砂砾岩、凝灰质砂岩为主，夹薄层粗安岩及流纹质熔结凝灰岩，呈半环状分布于盆地的东部，总厚度496m。第三喷发周期火山活动达到高潮，形成巨厚的流纹质熔结凝灰岩的堆积，厚达927.6m，分布于盆地的中部，在盆地中出露面积最大，约占盆地总面积的50%。东狼沟喷发阶段的初始期，盆地中火山活动出现短暂的间歇，形成厚度不大的凝灰质砂岩。随之发育形成以粗安质熔岩和正常火山碎屑岩为主，夹有透镜状流纹岩、流纹质熔结凝灰岩的火山岩层，厚496.1m，分布于盆地的西北部。

表16-15　北京地区晚燕山期火山-沉积盆地特征简表

Table 16-15　Characteristics of volcanic-sedimentary basins of Late Yanshanian in Beijing

序号	盆地名称	出露面积/km^2	原始分布面积/km^2	盆地内部构造形态	盆地长轴方向	主要岩石类型	地层总厚度/m	盆地下伏最新地层	时期
1	东岭	18	20	向斜	NE50°	流纹质熔结凝灰岩、流纹岩、粗安岩、凝灰质砂岩	>2130	窑坡组	张家口-东狼沟阶段
2	上大水	7		向斜	NW320°	流纹质熔结凝灰岩、流纹岩、粗安质角砾凝灰岩、复成分砾岩	>599	蓟县系	张家口阶段
3	怀柔	65		向斜	NE55°，向斜轴向 NW320°	流纹质熔结凝灰岩、粗面岩、粗安岩、凝灰质砂岩	>4341	九龙山组	张家口-东狼沟阶段
4	岔道	14	110	单斜	NW320°	流纹质熔结凝灰岩、流纹岩、粗面岩、凝灰质砾岩	>1191	髫髻山组	张家口阶段
5	杜家庄-镇厂	200		向斜	NE30°	流纹岩、流纹质熔结凝灰岩、粗安岩、凝灰质砾岩	>2449	髫髻山组	张家口-东狼沟阶段
6	坨里-大灰厂	180	280	单斜	NE55°	粉砂岩、泥质页岩、复成分砾岩、凝灰质砂岩、玄武岩	>1806	石炭系	东狼沟阶段

（据鲍亦冈等，1995）

总体上看，盆地中火山岩岩石成分以酸性流纹质为主，具有从中偏基性—中性—酸性—中偏碱性方向演化的特点。火山活动早期以熔岩喷溢为主，爆发指数仅为3.5%；中期构造活动强烈，爆发指数达77%。晚期又以中性及酸性熔岩溢流为主，爆发指数为6.9%。火山活动呈现喷溢—宁静—强烈爆发—宁静—喷溢为主的变化过程。

3. Ⅳ级构造单元——火山机构

张家口期火山机构发育较多，保存也较好。仅对典型沿河城卫生院古火山口和小围场火山岩管特征描述如下（鲍亦冈等，1995）：

（1）沿河城卫生院古火山口

位于东岭台盆地北缘的沿河城村西南，处于横岭断裂与沿河城-南口断裂的交汇部位。平面上为不规则的椭圆形，类型为中心式火山机构，生成于东狼沟火山喷发阶段晚期，出露面积<$1km^2$。

火山机构的地貌景观特征明显，整个火山机构相对处于负地形，而中心的火山岩塞突起。火山机构发育于张家口喷发阶段第三喷发周期形成的火山灰流相流纹质熔结凝灰岩中，与围岩主要为断层接触，东、西侧断层倾向东，倾角较陡，南侧断层倾向南西，倾角中等。北侧粗安质集块角砾岩直接覆盖于流纹质熔结凝灰岩之上。火山机构由火山颈相粗安玢岩和近火山口相粗安质集块角砾岩组成（图 16-25）。

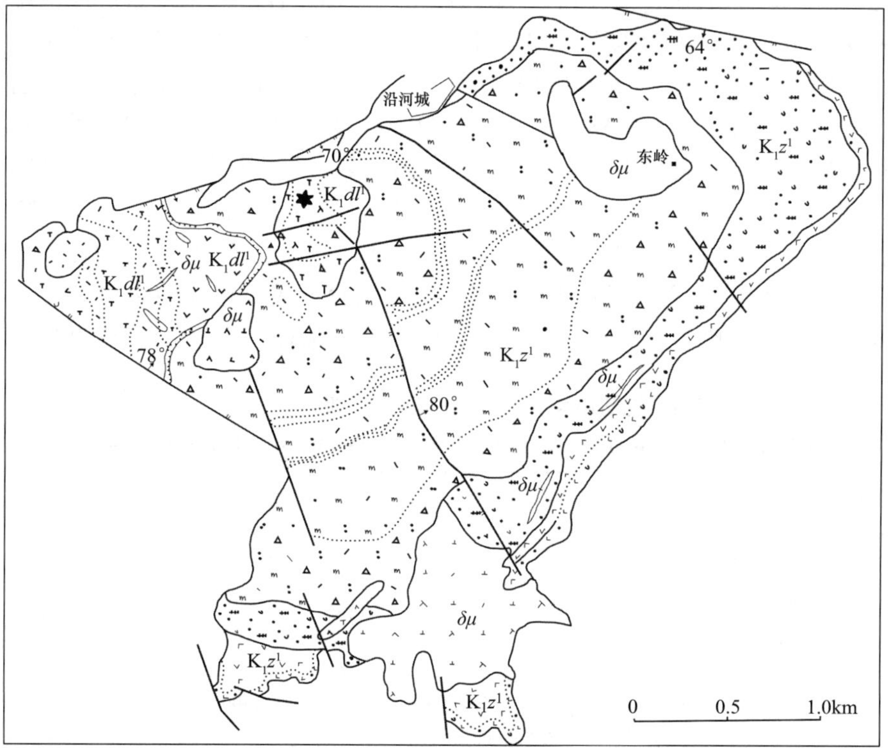

图 16-25　东岭火山-沉积盆地岩性岩相地质图

（据鲍亦冈等，1995）

Fig. 16-25　Lithologic and volcanic lithofacies map of Dongling volcannic-sedimentsry basin

粗安玢岩外倾呈岩塞状、岩枝状侵入于粗安质集块角砾岩中，岩石致密坚硬，性脆，边部有时可见流纹，中部发育原生环形节理，在剖面上呈弧形。岩石构造节理发育，被切割成不规则的尖棱角状碎块。岩性与地球化学特征同该火山机构西部东狼沟阶段第二喷发周期的粗安岩特征相似。

粗安质集块角砾岩呈环状分布于火山颈相粗安玢岩的周围，产状外倾，西、东及南部侵角约 30°，北部倾角约 60°。岩石呈暗紫色，角砾以棱角状-次棱角状粗安岩为主，粒径 2~5cm，大者 180cm，含量可达 80%。岩石结构较疏松，节理不发育。岩石特征与粗安玢岩截然不同。

（2）小围场火山岩管

小围场火山岩管位于怀柔区西部，西栅子-梁根火山-侵入杂岩带的东南端。该杂岩带位于凤驼梁-四海火山-沉积盆地的东南侧，呈北西 330°方向展布，带中见小围场等三个特征基本相似的火山角砾岩管及多个中酸性浅—中深成小型侵入体，受岩管边部环状断裂的控制，各岩管边部均有超浅成-潜火山岩相侵入体出露（图 16-26）。

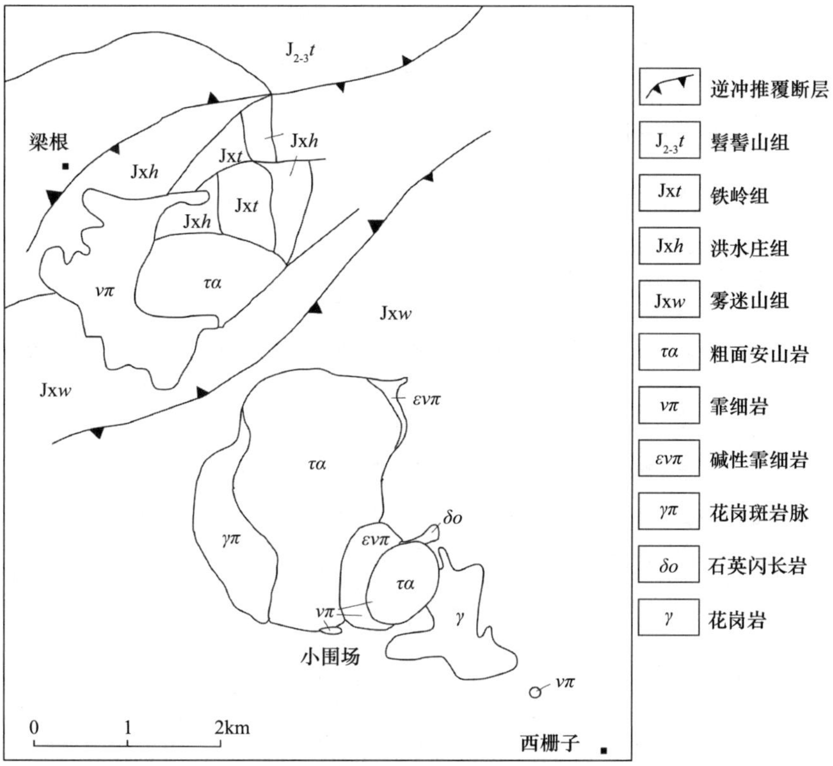

图 16-26　小围场火山岩管地质略图

（据鲍亦冈等，1995）

Fig. 16-26　Structural map of volcanic pipe of Xiaoweichang

岩管东部与蓟县系雾迷山组白云岩、西栅子花岗岩接触，西部与岩管有密切成因联系的浅成侵入体——石英二长岩接触。单个岩管平面上呈椭圆形，出露面积 0.69km^2。岩管边界主内倾，局部外倾，倾角 75°~85°，在剖面上呈垂深较大（>800m）的倒锥状。小围场岩管的北、西和南部，石英二长岩体呈半环状围绕岩管产出，接触带见霏细岩脉沿环状裂隙贯入。

小围场岩管为中心式开放型火山通道，喷发次数较少，剥蚀深度较大。岩管中岩石种类单一，且地表与深部岩性变化不大。岩性蚀变绢云母化、黑云母化、硅化、黄铁矿化，并见有碳酸盐化、绿帘石化，中部及外围岩体中见铜、铅、锌矿化。发育岩石类型主要为粗安质角砾熔岩。角砾成分复杂，以粗安岩、石英岩、花岗岩、石英砂岩、页岩、碳酸盐岩、霏细岩角砾为主，含量 10%~40%，大小 0.2~5cm，大者 30~40cm。不同成分的角砾磨圆度不同，火山岩角砾多为棱角状，碳酸盐岩、石英岩角砾磨圆度较好。角砾成分、含量及粒径的变化在岩管中无明显规律，但岩管南部角砾含量略高，一些石英岩角砾呈卵形，磨圆度极好。晶屑含量约 25%，以碱性长石、斜长石晶屑为主。基质绝大多数已经蚀变成绢云母及黑云母、方解石等。副矿物成分主要为磷灰石、锆石、金红石，并见微量自然铜、自然锡、刚玉、硅灰石、钙铁榴石等。

（五）岩石化学与地球化学特征

张家口期火山岩化学成分及参数见表 16-16。与中国同类岩石均值相比，流纹岩 SiO_2 含量偏高 1%，FeO^T 含量偏低，其余氧化物含量相近；石英粗面岩 SiO_2 含量偏低，碱性含量相当；安山岩类 SiO_2、MgO、Al_2O_3 及 CaO 含量偏低，而 Na_2O 含量偏高 1%。全岩主要氧化物变化趋势图（图 16-27）上，SiO_2、MgO、Al_2O_3 及 CaO 含量与对比系列一及系列二（即平均碱玄岩-响岩系列和粗面玄武岩-粗安岩-安粗岩-碱性流纹岩系列）具有明显的一致性。时间序列上，由张家口阶段至东狼沟阶段，火山岩 CaO 含量逐渐增多，(Na_2O+K_2O) 含量逐渐减少，明显不同于髫髻山期火山岩特性（图 16-28）。

表 16-16　张家口期火山岩岩石化学成分及特征参数一览表

Table 16-16　The chemical composition and main parameters of Zhangjiakou volcanic rocks

序号	火山盆地	火山期次	氧化物含量/%													特征参数					
			SiO_2	TiO_2	Al_2O_3	Fe_2O_3	FeO	MnO	MgO	CaO	Na_2O	K_2O	P_2O_5	LOS	总和	*DI*	*SI*	σ	*A/CNK*	*A/NK*	*m/f*
1	东岭台	张家口阶段	64. 51	0. 56	16. 46	3. 22	0. 47	0. 1	1. 18	1. 67	6. 62	3. 17	0. 21	2. 48	100. 65	85. 54	8. 12	4. 37	0. 95	1. 15	0. 61
2			62. 57	0. 6	16. 26	3. 92	1. 07	0. 13	1. 49	2. 61	4. 74	4. 1	0. 34	2. 16	99. 99	77. 97	9. 81	3. 89	0. 96	1. 33	0. 56
3			63. 95	0. 59	16. 4	2. 89	1. 38	0. 12	1. 31	2. 7	5. 17	3. 38	0. 33	2. 36	100. 58	78. 82	9. 32	3. 43	0. 96	1. 35	0. 57
4			53. 89	1. 28	17. 84	4. 71	2. 12	0. 16	3. 48	7. 17	3. 92	1. 99	0. 52	0. 9	97. 98	51. 12	21. 70	2. 95	0. 82	2. 07	0. 95
5			60. 32	0. 94	17. 72	5. 21	0. 68	0. 12	2. 06	2. 85	5. 35	2. 38	0. 53	1. 74	99. 9	72. 51	13. 34	3. 35	1. 07	1. 56	0. 67
6			54. 39	1. 35	18. 13	8. 45	0. 72	0. 12	2. 13	3. 81	6. 13	2. 21	0. 58	0. 91	98. 93	66. 53	11. 08	5. 74	0. 93	1. 45	0. 45
7			73. 41	0. 4	14. 37	1. 68	0. 05	0. 03	1. 44	0. 37	4. 46	1. 66	0. 08	4. 33	102. 28	87. 74	15. 64	1. 22	1. 47	1. 57	1. 61
8			57. 97	0. 91	17. 56	5. 66	0. 54	0. 14	1. 35	3. 4	5. 3	3. 41	0. 5	1. 5	98. 24	73. 01	8. 44	4. 77	0. 94	1. 42	0. 42
9			49. 85	1. 25	16. 32	4. 45	4. 16	0. 21	3. 93	7. 41	3. 4	2	0. 44	0. 66	94. 08	45. 59	22. 04	3. 21	0. 77	2. 10	0. 84
10	上清水		69. 5	0. 44	15. 26	2. 77	0. 12	0. 09	0. 56	0. 89	4. 38	4. 9	0. 12	3. 35	102. 38	89. 71	4. 44	3. 23	1. 08	1. 22	0. 37
11	岔道		75	0. 17	12. 41	1. 02	0. 82	0. 06	0. 43	0. 59	3. 62	4. 68	0. 04	4. 51	103. 35	93. 38	4. 07	2. 14	1. 03	1. 13	0. 43
12			71	0. 3	14. 75	1. 62	0. 74	0. 1	0. 44	0. 42	4. 58	4. 45	0. 03	3. 74	102. 17	92. 18	3. 73	2. 89	1. 12	1. 19	0. 34
13	汤河口		75. 4	0. 18	11. 49	1. 9	0. 72	0. 04	0. 18	0. 48	3. 2	5. 4	0. 05	4. 57	103. 61	94. 86	1. 58	2. 28	0. 96	1. 03	0. 13
14			72. 97	0. 38	13. 15	1. 63	0. 56	0. 04	0. 3	0. 62	3. 5	5. 6	0. 14	4. 08	102. 97	93. 54	2. 60	2. 75	1. 02	1. 11	0. 26
15	杜家庄		74. 31	0. 2	13. 4	1. 17	0. 71	0. 07	0. 31	0. 65	3. 06	4. 87	0. 02	4. 38	103. 15	91. 68	3. 07	2. 00	1. 17	1. 30	0. 30
16	北宅		74. 74	0. 16	13. 21	1. 11	0. 57	0. 08	0. 65	0. 06	3. 6	4. 59	0. 03	4. 46	103. 26	93. 75	6. 19	2. 10	1. 20	1. 21	0. 70
17			62. 12	0. 64	17. 25	4. 98	0. 98	0. 09	1. 53	1. 7	5. 48	2. 94	0. 37	2. 06	100. 14	78. 71	9. 74	3. 62	1. 13	1. 41	0. 49
18	北宅	东狼沟阶段	60. 38	0. 88	16. 82	4. 19	1. 5	0. 08	2. 5	3. 27	5	3. 11	0. 28	1. 76	99. 77	71. 01	15. 47	3. 67	0. 96	1. 45	0. 83
19	杜家庄		60. 96	0. 78	16. 02	4. 88	2. 02	0. 1	1. 85	2. 87	4. 49	3. 47	0. 24	1. 89	99. 57	72. 58	11. 17	3. 42	0. 98	1. 44	0. 51
20	东岭台		61. 28	0. 7	17. 33	4. 12	1. 88	0. 15	1. 63	2. 24	5. 3	2. 7	0. 31	1. 95	99. 59	75. 50	10. 51	3. 39	1. 10	1. 49	0. 51
21			71. 41	0. 28	14. 36	1. 08	0. 96	0. 1	0. 53	0. 66	3. 76	5. 72	0. 08	3. 74	102. 68	92. 00	4. 40	3. 15	1. 06	1. 16	0. 46
22			66. 68	0. 46	16. 03	2. 62	0. 81	0. 1	1. 24	1. 82	5. 11	2. 52	0. 17	2. 98	100. 54	81. 81	10. 15	2. 41	1. 11	1. 44	0. 68
23	岔道		69. 43	0. 4	15. 65	2. 13	0. 52	0. 12	0. 54	0. 5	4. 84	4. 29	0. 17	3. 39	101. 98	90. 91	4. 41	3. 13	1. 16	1. 24	0. 38
24	大灰厂		49. 59	1. 86	15. 94	4. 63	4. 33	0. 17	5. 37	7. 56	2. 33	3. 72	1. 02	0. 49	97. 01	43. 91	26. 48	4. 67	0. 74	2. 03	1. 10
25			53. 72	1. 54	15. 45	5. 28	2. 32	0. 11	3. 57	7. 19	3. 38	3. 4	0. 93	0. 89	97. 78	55. 06	20. 11	3. 92	0. 69	1. 67	0. 89
26			48. 36	1. 99	15. 79	4. 48	4. 22	0. 19	5. 72	8. 3	2. 38	3. 83	1. 04	0. 4	96. 7	44. 46	27. 85	5. 73	0. 68	1. 96	1. 21
27			54. 55	1. 6	15. 66	6. 1	1. 73	0. 06	2. 91	5. 75	3. 34	4. 7	0. 99	0. 96	98. 35	62. 03	15. 70	5. 21	0. 74	1. 48	0. 71
28			52. 15	2. 07	18. 92	9. 44	0. 64	0. 13	2. 55	6. 65	3. 31	3. 02	1. 23	0. 56	100. 67	51. 25	13. 84	4. 31	0. 91	2. 17	0. 49

注：数据引自鲍亦冈等，1995；李伍平等，2001；李晓勇等，2004；汪洋等，2009；王蕊等，2007。

表 16-17　张家口期火山岩稀土元素成分及特征参数一览表

Table 16-17　REE contents and main parameter of Zhangjiakou volcanic rocks

序号	岩性	火山期次	稀土元素含量/10^{-6}															特征参数				
			La	Ce	Pr	Nd	Sm	Eu	Gd	Tb	Dy	Ho	Er	Tm	Yb	Lu	Y	ΣREE	LREE/HREE	$(La/Yb)_N$	δEu	δCe
1	潜角砾流纹岩	张家口阶段	75.40	132.00	16.05	57.80	9.30	2.35	7.80	1.11	5.60	1.01	2.92	0.45	2.87	0.43	28.46	315.09	12.62	16.30	0.97	1.03
2	流纹岩		62.40	112.00	13.52	50.10	8.36	2.58	7.06	1.01	5.04	0.92	2.69	0.39	2.58	0.38	26.44	269.03	13.20	18.84	0.84	0.93
3	流纹岩		37.49	82.10	9.02	31.37	5.30	0.28	3.90	0.57	3.17	0.67	1.83	0.28	1.69	0.29	17.78	177.96	12.34	16.39	0.57	0.96
4	流纹岩		46.09	90.70	8.85	27.86	4.29	0.57	3.27	0.50	2.95	0.56	1.54	0.25	1.69	0.25	16.74	189.37	12.40	17.35	1.03	0.95
5	英安岩		83.60	143.00	16.76	59.60	9.24	2.93	7.87	1.08	5.23	0.92	2.56	0.38	2.25	0.32	25.60	335.74	15.29	26.65	1.05	0.94
6	粗面岩		51.90	101.68	11.35	40.83	6.57	1.46	5.17	0.73	3.91	0.72	2.03	0.31	1.94	0.29	22.08	228.87	12.77	27.45	0.09	0.61
7	安山岩		38.41	89.60	9.81	36.68	5.63	1.55	4.37	0.60	3.09	0.64	1.75	0.29	1.66	0.31	18.34	194.39	12.85	16.18	0.90	1.05
8	粗安岩		58.22	124.70	14.09	51.03	7.77	1.76	5.52	0.76	3.96	0.79	2.02	0.33	1.75	0.30	21.12	273.00	13.22	16.50	0.86	1.04
9	橄榄玄武岩	东狼沟阶段	67.37	142.10	19.16	69.44	11.49	3.01	7.22	1.06	4.46	0.82	2.01	0.32	1.64	0.22	20.37	330.32	13.20	18.84	0.84	0.93
10	橄榄玄武岩		83.12	178.90	24.09	86.71	14.27	3.68	8.76	1.29	4.89	0.87	2.00	0.33	1.48	0.21	21.38	410.60	12.40	17.35	1.03	0.95

注：数据为均值，引自鲍亦冈等，1995；李伍平等，2001；邵济安等，2003；李晓勇等，2004；汪洋等，2009。

全岩里特曼指数 $\sigma=1.22\sim5.74$，跨碱性岩和亚碱性岩区；固结指数 $SI=1.58\sim26.48$，分异指数 $DI=43.91\sim94.86$，变化范围大，总体分异程度高。$DI<90$ 时，Fe_2O_3 含量较对比岩系高，而 $DI>90$ 时，Fe_2O_3 含量较对比岩系低；$DI>90$ 的酸性岩石，FeO 含量明显较比岩石低；$DI<85$ 或 SiO_2 含量<70%时，Na_2O 及 K_2O 含量变化基本服从对比岩石系列一及系列二的趋势；$DI>85$ 时，即使岩石的 SiO_2 含量（或 DI 值）仅有较小的变化，Na_2O 及 K_2O 含量也会出现较大的起伏。这种特点可能与酸性岩浆上升喷发过程中受到上部地壳岩石的同化混染作用影响有关，也与该组岩石发育爆发相熔结凝灰岩且含有大量上地壳刚性岩屑的特点相一致（鲍亦冈等，2000）。$SiO_2=49.59\%\sim75.4\%$，早期属于硅饱和质岩石，中、晚期属于硅过饱和质岩石。镁铁比值 $m/f=0.13\sim1.61$，属于铁质-富铁质基性岩；$A/CNK=0.68\sim1.47$，$A/NK=1.03\sim2.17$，属于准铝质—过铝质。其中，酸性岩属于过铝质岩。

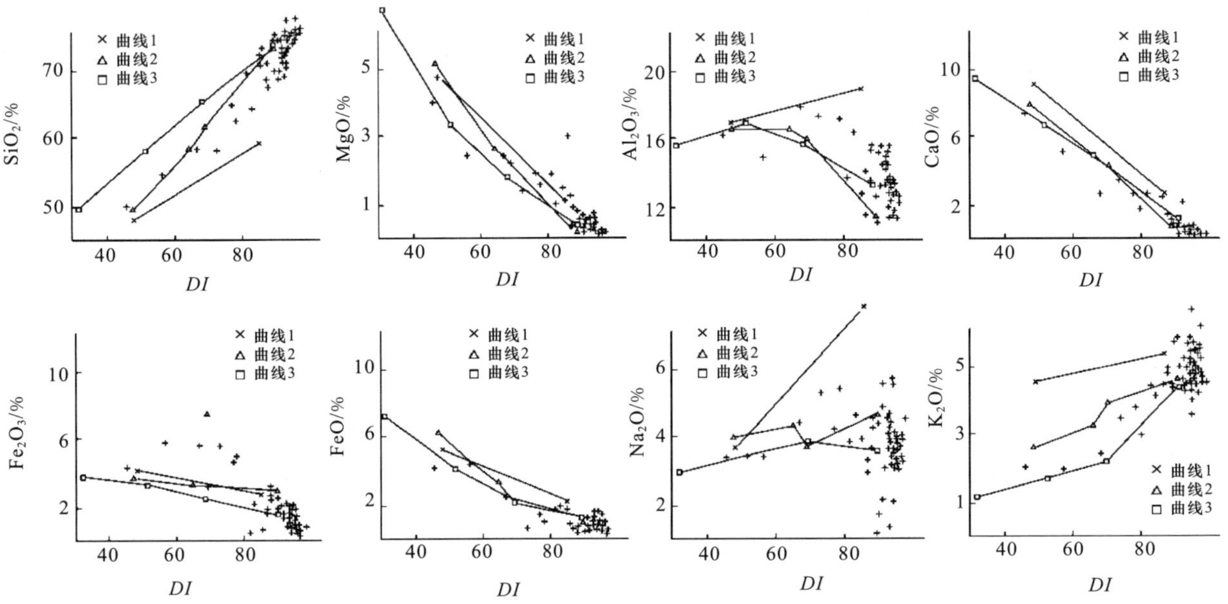

图 16-27　张家口期火山岩主要氧化物-DI 变化趋势图

（据鲍亦冈等，2000）

Fig. 16-27　Oxides differentiation index plots of Zhangjiakouan volcanic rocks

曲线 1 为平均碱玄岩-响岩趋势线（系列一）；曲线 2 为平均粗面玄武岩-粗安岩-安粗岩-碱性流纹岩趋势（系列二）；曲线 3 为平均玄武岩-安山岩-英安岩-流纹岩趋势线（系列三）

稀土元素成分及参数见表 16-17。张家口、东狼沟火山阶段 ΣREE 分别为 $177.96\times10^{-6}\sim335.74\times10^{-6}$ 和 $330.32\times10^{-6}\sim410.60\times10^{-6}$，稀土总量差异较大，东狼沟阶段高于张家口阶段；LREE/HREE 分别为 12.40~116.69 和 17.61~19.83，$(La/Yb)_N$ 分别为 15.91~26.65 和 29.47~40.29，东狼沟阶段分馏程度略高于张家口阶段，但均表现出轻稀土弱富集，重稀土相对亏损，且轻稀土的分馏程度高于重稀土特征；东狼沟阶段 $\delta Eu=1.01$，无明显铕正或负异常；而张家口阶段 δEu 出现分化，$\delta Eu=0.19\sim1.05$，呈现铕的负异常-无异常反映岩浆经历了分离结晶作用。张家口和东郎沟火山阶段配分曲线特征为右倾光滑曲线（图 16-29），但形态略有所差异，反映了可能具有不同的岩浆来源。同为张家口期火山岩，各岩性稀土元素配分曲线基本相似，但稀土元素总量和 δEu 差异较大，反映岩浆混合或分离结晶分异作用。

本期火山岩微量元素特征基本同髫髻山期，大离子亲石元素和过渡族元素含量变化大，含量上东狼沟阶段略高于张家口阶段。同原始地幔相比，Ti 元素含量明显偏高，Cr、Ni 元素含量均偏低。大离子亲石元素 Sr、Ba、Zr 和高场强元素 P、Ti 元素含量有所波动，体现了张家口期火山岩可能经历了长石、磷灰石、锆石和含 Ti-Fe 副矿物的分离结浆演化。在微量元素蛛网图（图 16-30）上，所有岩性均表现出富集 Rb 及 LREE 和亏损 Nb、Ta 和 HREE。

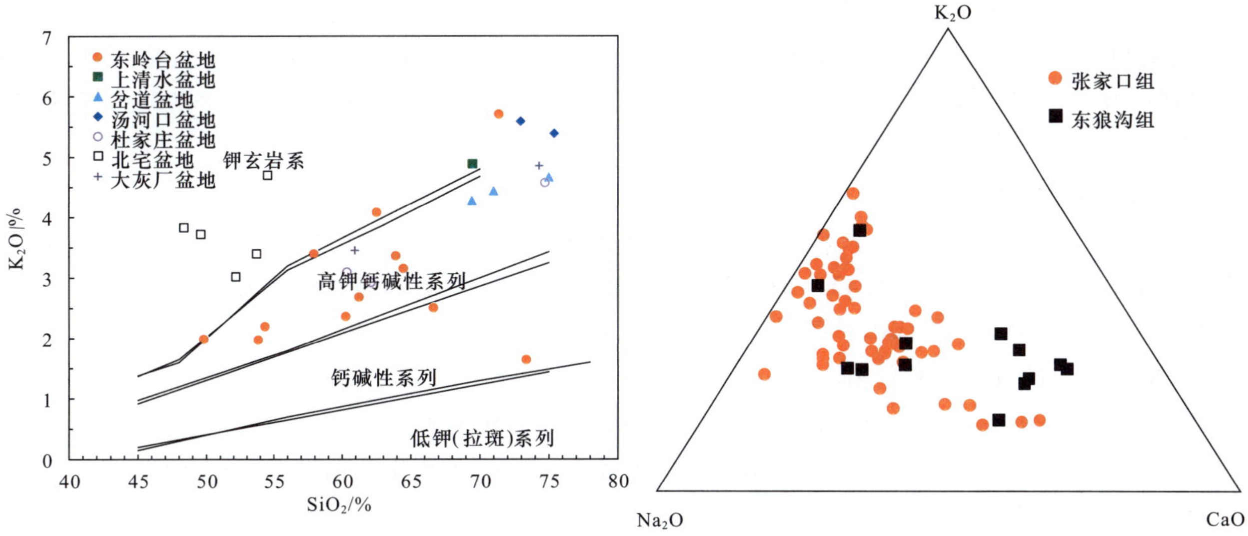

图 16-28　张家口期火山岩 Na_2O-K_2O-CaO 图解

Fig. 16-28　The Na_2O-K_2O-CaO diagram of Zhangjiakou volcanic rocks

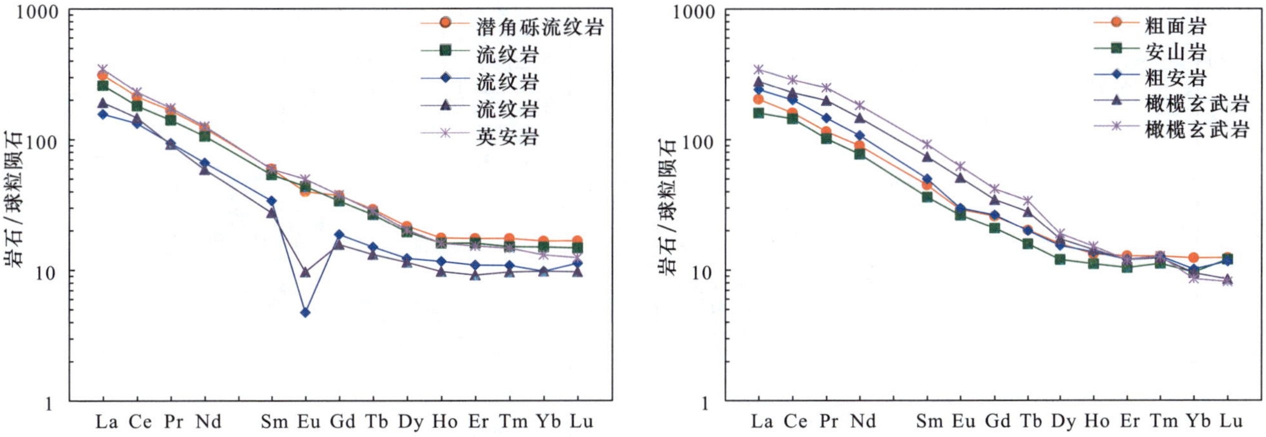

图 16-29　张家口期火山岩稀土元素配分模式图

Fig. 16-29　Chondrite-normalized REE patterns of Zhangjiakouan volcanic rocks

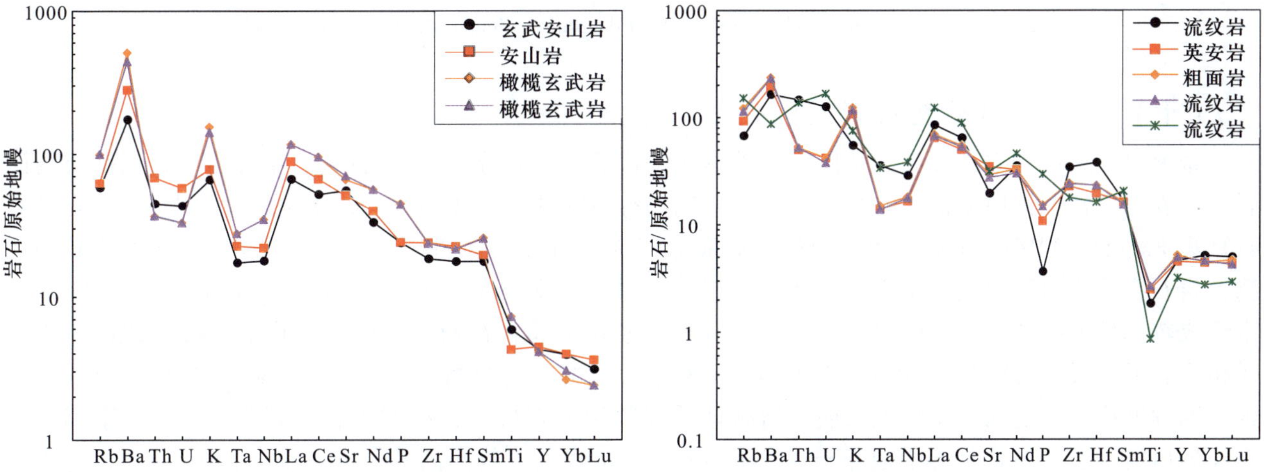

图 16-30　张家口期火山岩微量元素蛛网图

Fig. 16-30　Primitive mantle-normalized spidergrams for trace elements of Zhangjiakouan volcanic rocks

（六）Sr-Nd 同位素特征

张家口组火山岩 $^{143}Nd/^{144}Nd$ = 0.511665 ~ 0.511945，$^{147}Sm/^{144}Nd$ = 0.0939 ~ 0.1152，$^{87}Sr/^{86}Sr$ = 0.706161 ~ 0.718917，ε_{Nd}（t）= -11.99 ~ -17.09；东狼沟组火山岩 $^{87}Sr/^{86}Sr$ = 0.705690 ~ 0.705847，ε_{Nd}（t）= -14.6 ~ -12.3（李晓勇等，2004；王蕊等，2007；汪洋等，2009）。

（七）Pb 同位素特征

张家口组火山岩 Pb 同位素特征同髫髻山组相似，$^{206}Pb/^{204}Pb$ = 16.291 ~ 18.477，$^{207}Pb/^{204}Pb$ = 15.147 ~ 15.731，$^{208}Pb/^{204}Pb$ = 36.430 ~ 38.756，火山岩源区放射成因 Pb 较低，与燕山造山带前寒武纪结晶基底基性麻粒岩和中生代花岗岩 Pb 同位素特征相似（李伍平等，2001）。

（八）Lu-Hf 同位素特征

冀北张家口-宣化地区张家口组流纹岩锆石颗粒 $^{176}Lu/^{177}Hf$ < 0.002，$^{176}Hf/^{177}Hf$ = 0.28217 ~ 0.28227，ε_{Hf}（t）= -18.5 ~ -15.1，锆石形成以后有较少的放射性成因 Hf 的积累（杨进辉等，2006）。

（九）岩石系列划分

该套岩系 σ = 1.22 ~ 5.74，AR = 1.49 ~ 3.94，跨碱性和亚碱性区。在 AR - SiO_2 和 K_2O - SiO_2 图解（图 16-31）上，主要成分点分别位于碱性-钙碱性和高钾钙碱性-钾玄岩系列内。

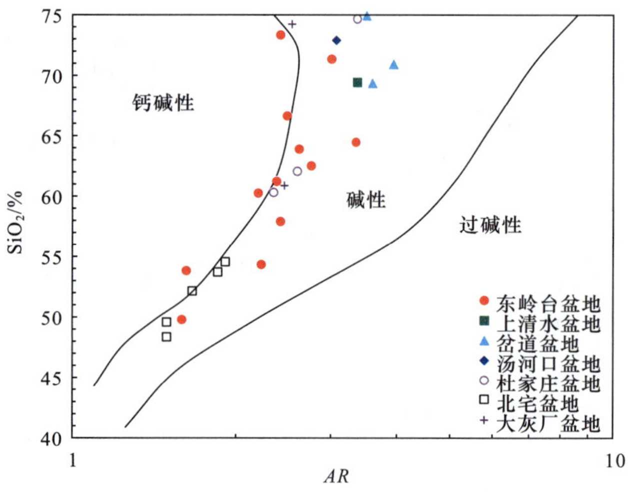

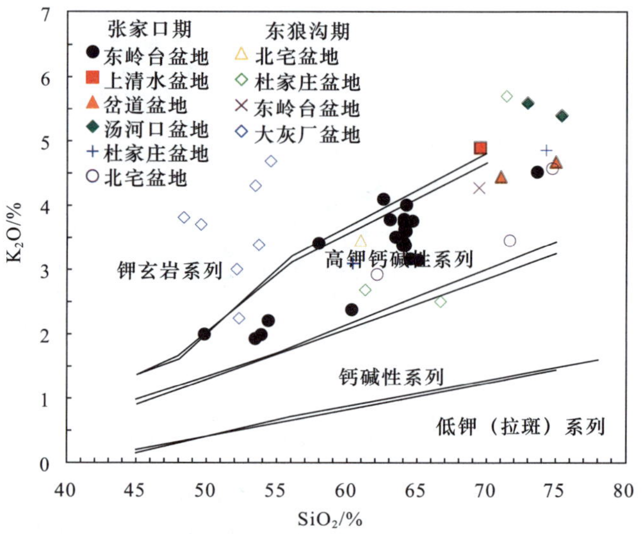

图 16-31　张家口期火山岩 SiO_2-AR 和 SiO_2-K_2O 图解

Fig. 16-31　The SiO_2-AR and SiO_2-K_2O diagram of Zhangjiakouan volcanic rocks

（十）岩石成因类型及岩浆演化

张家口期火山岩岩性复杂，按照岩石地球化学特征，可划分为中-基性岩、中性岩、富铝钠质粗面岩或英安岩、富铝钾质粗面岩和酸性岩类，它们具有不同的岩石成因及岩浆演化特性。中-基性岩是具有富集 Ba、Sr、LREE 和 K 特征的幔源玄武质岩浆在中等压力条件下（1.0 ~ 1.2GPa）分离结晶的产物，岩石在成岩过程中受到了下地壳物质的混染；中性岩富集 Th、U，且 Th 相对 U 略亏损，Nd 同位素组成较基性岩略富集，可能是受到华北克拉通下地壳下部岩石的部分熔融产物的混染所致；富铝钠质粗面岩或英安岩可能是玄武质岩浆固结后又发生含水熔融的产物，高 Sr、相对高 Y 含量与重稀土元素分馏程度中等的地球化学特征表明，熔融体系残留相无斜长石、石榴子石，而含角闪石；富铝钾质粗面岩可能是玄武质岩浆与中地壳岩石发生熔融反应的产物，较高的 Sr、Y 含量，弱 Eu 负异常与重稀土元素分馏程度中等的地球化学特征表明，熔融体系残留相无石榴子石、含角闪石，含少量斜长石；酸性火山岩（尤以流纹岩）出现较大岩石化学差异，体现了不同岩石成因及岩浆演化。早期酸性火山岩形成于高氧逸度条件下的低程度部分熔融，残留相为斜长石+角闪石+辉石+含 Ti-Fe 副矿物+磷灰石，源岩是低放射性成因 Pb 同位素组成的下地壳低钾岩石；晚期酸性火山岩是中性火山岩分离结晶演化而来，分离结晶矿物相是长石+角闪石+磷灰石+锆石+Ti-Fe 副矿物（汪洋等，2009）。

（十一）构造环境分析

杨进辉等（2006）认为张家口组火山岩形成于区域拉张构造背景。然而，区内地质迹象表明：髫髻山组和土城子组形成褶皱构造，且有更老地层推覆到该两组地层之上。而张家口组虽然在部分火山-沉积盆地中发育褶皱构造，东狼沟组为单斜构造，未见更老地层推覆到该两组之上。最新资料研究表明，燕山西段的伸展构造发育于 125Ma 之后，如云蒙山变质核杂岩活动于 125Ma 之后，本区多地见有张家口组火山岩被后期正断层截切，如怀柔区北宅一带。因此，区域性大规模地壳伸展活动发育在张家口组火山岩形成之后。不仅如此，燕山西段广大地区晚侏罗世髫髻山组具有类埃达克岩特征，出露的大量晚侏罗世—早白垩世侵入岩体也具有埃达克岩特征，如云蒙山、碓臼峪花岗岩岩基等。所以，张家口组火山岩并非形成于区域性地壳减薄的伸展构造背景。

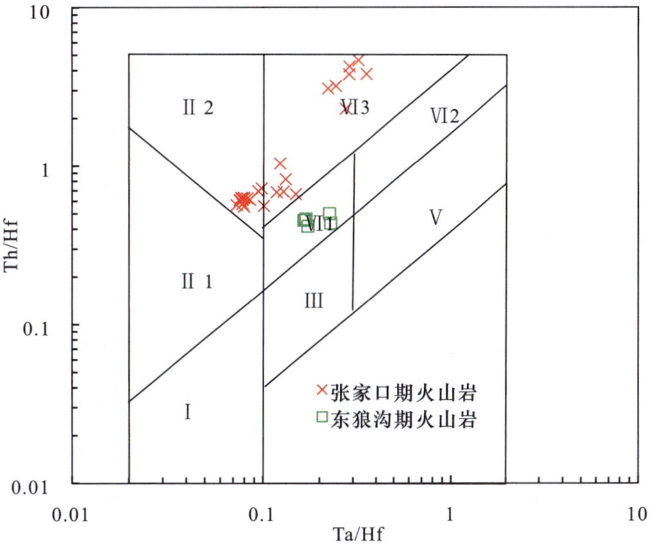

图 16-32　张家口阶段和东狼沟阶段火山岩 Ta/Hf-Th/Hf 构造环境判别图解

Fig. 16-32　The Ta/Hf-Th/Hf diagram of tectonic discrimination of Zhangjiakouan and Donglanggouan volcanic rocks

在 Ta/Hf-Th/Hf 构造环境判断图解（图 16-32）中，本期两组火山岩地层处于完全不同构造背景环境区内。张家口阶段成分处于陆缘岛弧或陆缘火山弧区，而东狼沟阶段成分处于陆内裂谷或陆缘裂谷区。因此，张家口阶段和东狼沟阶段火山岩形成于不同的成因环境。前者形成于燕山造山挤压构造背景环境，而后者则形成于造山晚期伸展构造背景环境，至东狼沟阶段火山岩喷发时期，燕山运动真正的进入造山后阶段。

（十二）火山岩形成时代

张家口期火山岩与沉积岩交替互层产出，碎屑岩中化石和火山岩大量同位素年龄测试（表 16-18）结果表明火山活动时间应属早白垩世，大量同位素年龄介于 137.3~97.74Ma 之间，差异极大，反映了火山-沉积地层可能在区域上具有穿时性。

表 16-18　张家口期火山岩同位素年龄一览表

Table 16-18　Isotopic ages of Zhangjiakouan volcanic rocks

序号	采样地点	岩石名称	测试方法	年龄值/Ma	资料来源
1	门头沟东岭台	流纹岩	K-Ar 全岩	97.74	北京市地质调查所，1995
2	门头沟东岭台	安山岩	K-Ar 全岩	108.40	北京市地质调查所，1995
3	门头沟东岭台	安山岩	K-Ar 全岩	118.40	北京市地质调查所，1995
4	门头沟东岭台	玄武安山岩	K-Ar 全岩	134.00	北京市地质调查所，1995
5	门头沟杜家庄	斑流岩	K-Ar 黑云母	135.00	北京市地质调查所，1995

续表

序号	采样地点	岩石名称	测试方法	年龄值/Ma	资料来源
6	延庆羊头山	流纹质熔结凝灰岩	K-Ar 全岩	119.40	北京地质学院，1964
7	延庆小张家口	流纹质熔结凝灰岩	K-Ar 全岩	122.60	北京地质学院，1964
8	门头沟东岭台	流纹质熔结凝灰岩	Rb-Sr 等时线	135.90	中国地质大学，1995
9	虎什哈头石门	安山岩	K-Ar 全岩	129.12	虎什哈幅区域地质调查报告，2000
10	门头沟杜家庄	多斑流纹岩	锆石 SHRIMP U-Pb	136.2±3	赵兴国，2009
11	门头沟沿河城	粗安岩	锆石 SHRIMP U-Pb	137.3	赵兴国，2009
12	门头沟杜家庄	多斑流纹岩	全岩 K-Ar	135	赵兴国，2009
13	门头沟东岭台	安山岩	全岩 K-Ar	118.4	中南所，1978
14	门头沟东岭台	流纹质熔结凝灰岩	全岩 Rb-Sr	135.9	地科院地质所，1987
15	沿河城	流纹岩	锆石 SHRIMP U-Pb	133.8±4.7	袁洪林等，2005
16	沿河城	流纹岩	锆石 LA-ICP-MS U-Pb	130.1±4.6	袁洪林等，2005
17	沿河城	粗面岩、流纹岩	全岩 Rb-Sr	131.2±12.3	汪洋等，2003
18	张家口-宣化	流纹岩	锆石 U-Pb	126±1	杨进辉等，2006
19	延庆燕羽山	流纹质熔结凝灰岩	全岩 K-Ar	112	北京地质学院，1964
20	延庆羊头山	流纹质熔结凝灰岩、安山质熔结凝灰岩、粗安质熔结凝灰岩	全岩 K-Ar	125、119、117、128、60、107、103、108、109	北京地质学院，1964

第三节　新生代火山岩与火山作用

一、时空分布特征

本期火山岩地表没有出露，只在平原区钻孔中见及，分布于新生代北京断陷、廊固断陷盆地内，称之为前门期火山岩。基于钻孔资料，北京断陷内火山岩分布范围：南起龙潭湖，北到朝阳门，西起中山公园，东到高碑店附近，分布面积约 40km^2。此外，在朝阳区的六里屯和太阳宫，丰台区金家村，海淀区西钓鱼台等地钻孔中也曾见有该期火山岩。基于火山岩分布推测火山喷发方式为裂隙式喷发，可能与南苑-通县、夏垫和礼贤（又称大兴隆起东缘断裂）深断裂密切有关。

前门期火山岩以玄武质熔岩为主，并有玄武质角砾凝灰岩夹层。火山岩层产于古近系前门组地层中，其上下顶底板岩性多为灰黑色、灰绿色砂页岩、砂砾岩等以细碎屑为主的沉积岩层。各钻孔中，玄武岩的层数多寡不一，厚度变化较大，埋藏深度也不相同。多数钻孔中可见 3 层火山岩，少数钻孔（如京热 7 号孔、新 5 孔）有 5 层火山岩，单层玄武岩厚度 10~50m（鲍亦冈等，2001）。区域上，火山岩层厚度呈现北厚南薄变化趋势。

二、火山岩相与岩石学特征

（一）溢流相

为本期火山活动主要岩相，为玄武岩。岩石呈灰绿色、绿色，粗玄结构、拉斑玄武结构，块状、杏仁构造。斑晶矿物以橄榄石、单斜辉石为主，有少量斜长石；基质由斜长石、单斜辉石、橄榄石、磁铁矿、磷灰石、火山玻璃等组成。岩石帘石化、绿泥石化和碳酸盐化蚀变。

橄榄石：呈斑晶和微晶两种形式存在，含量<5%。斑晶多数伊丁石化仅保留假象形态；微晶橄榄石蚀变相对较弱，种属为透铁橄榄石。

单斜辉石：呈斑晶产出，含量约10%，种属均为普通辉石。

斜长石：多呈自形长柱状分布于基质，少与单斜辉石镶嵌形成嵌晶结构。类型为拉长石；有序度0~0.3，属高温无序型。

基性火山玻璃：褐、黑褐、紫褐色，分布于斜长石微晶间隙，脱玻化后呈纤维状雏晶。

（二）火山爆发相

主要为火山爆发空落相，岩性为玄武质晶屑凝灰岩和玻屑晶屑凝灰岩，呈夹层状产出。岩石呈黑绿、灰绿色。岩屑为棱角状的玄武岩，晶屑有破碎的斜长石、辉石、橄榄石等，凝灰结构，胶结物为浅褐、无色火山尘。

三、岩石化学与地球化学特征

在化学成分分类图解上，成分点主要位于玄武岩区，个别分散于玄武岩与碱性玄武岩分界线附近。与中国玄武岩平均成分相比，新生代玄武岩 SiO_2 = 44.3~50.8，均值为48.5，含量相当；而（K_2O+Na_2O）= 2.14~5，平均值偏低，为2.5。与大红峪期玄武岩相比，新生代玄武岩 K_2O 含量偏低，为3.8。与南大岭期玄武岩相比，新生代玄武岩碱质总量也偏低，为1.8。岩石分异指数 DI = 25.7~43.6，均值为32.7，岩浆分异程度较差。

前门期火山岩 ΣREE = 82.97~91.98，稀土元素总量较低；δEu = 1.18~1.20，呈弱正铕异常；δCe = 0.95~0.96，揭示生成于较弱的氧化环境；ΣCe/ΣY = 5.67~5.84，轻稀土弱富集；与稀土元素配分曲线特征一致（图16-33）。

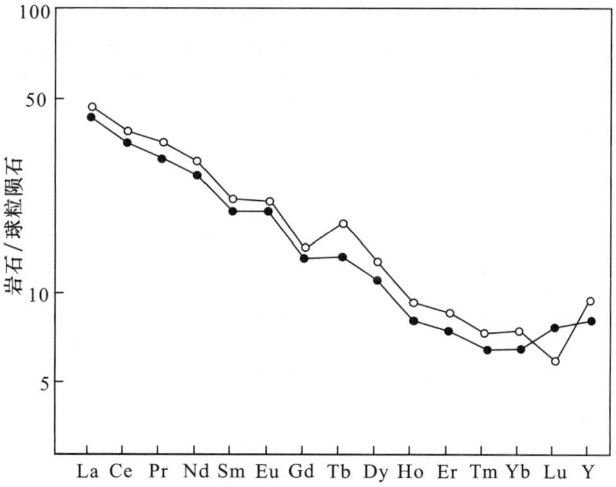

图16-33 前门期火山岩稀土元素配分模式图

Fig. 16-33 Chondrite-normalized REE patterns of Qianmenan volcanic rocks

四、岩石系列

前门期火山岩 σ = 1.4~4，最高值6.98，平均为2.62。在（Na_2O+K_2O）-SiO_2 图解上（图16-34），本期火山岩的样点多数落在亚碱性岩区，少部分落在碱性岩区，显示出由亚碱性岩石向碱性玄武岩过渡的趋势。在 FeO-（FeO/MgO）图解上（图16-34），本期火山岩集中在拉斑玄武岩区。

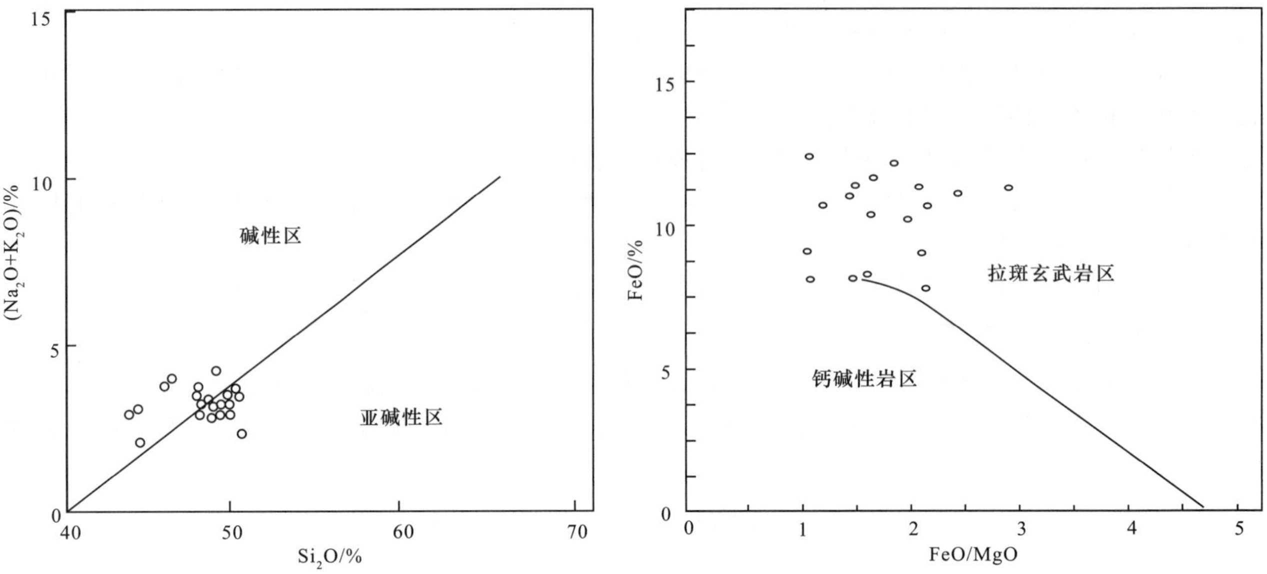

图 16-34　前门期火山岩 Na_2O+K_2O-SiO_2 和 FeO-FeO/MgO 图解

Fig. 16-34　The SiO_2-（K_2O+Na_2O） and FeO-（FeO/MgO） diagram of Qianmenan volcanic rocks

五、火山岩形成时代

该期火山岩夹于沉积岩中，碎屑岩中化石和火山岩大量同位素年龄测试结果（表 16-19）表明火山活动时间应在新生代始新世晚期—渐新世。

表 16-19　新生代玄武岩 K-Ar 法（全岩）年龄数据

Table 16-19　K-Ar isotopic ages of Cenozioc volcanic rocks

样号	取样深度/m	年龄/Ma	资料来源
JL-8	834.53~837.38	71.03	北京地震地质会战，1980
JL-2		64.52	北京地震地质会战，1980
JL-16	740~820	57.13	北京地震地质会战，1980
JL-17	977~993	44.50	北京地震地质会战，1980
JL-7	677.36	28.50	北京地震地质会战，1980
新 5-522	522	53.46±2.26	北京市地质调查研究院，2004
新 5-711	711	58.77±2.60	北京市地质调查研究院，2004

第十七章　火山活动与成矿作用

北京地区与火山活动有关的矿产很少，仅发育少许与中元古代基性、偏碱性、碱性火山活动有关的金属矿产和中生代燕山期火山作用有关的矿产，多表现为潜火山岩或隐爆角砾岩相关的矿产。

第一节　火山活动与金属矿产

一、与中元古代基性、偏碱性、碱性火山活动有关的金属矿产

主要分布于平谷将军关一带，为火山热液型矿床。矿床规模均为小型矿床，矿产类型为铜矿。

二、与燕山期火山岩-潜火山岩有关的金属矿产

主要分布于延庆八达岭、门头沟清水一带，包括热液型铅、锌、银矿和斑岩型和接触交代型铜、钼、铁、锌矿。矿床规模多数为小型矿床，仅发育一个中型大庄科多金属矿床，矿化点较多。

第二节　火山活动与非金属矿产

一、建筑材料

建筑材料仅发育铸石原料，为铸石玄武岩类。主要分布在北京西山一带，产出于南大岭组。先期开发有高井、草帽山等小型矿床，今已封闭。

二、冶金辅料

仅发育普通萤石矿，主要分布于怀柔北部，北京境内未形成矿床，仅见矿化点。

第四篇

侵入岩与深成作用

第十八章　概　　述

地质历史上，北京地区的岩浆侵入活动较为强烈，以太古宙和中生代最为强烈，岩浆活动不仅规模大，而且岩浆成分及演化很有特色，具有代表性。中生代的岩浆侵入活动延续时间超过了100Ma，形成了大小侵入体230余个（鲍亦冈等，2001）。此外，古元古代、中元古代和晚古生代也有相对较弱岩浆活动，形成规模不大的基性或碱性侵入体。本区的侵入岩主要出露在北部的延庆区、昌平区、怀柔区、密云区和西部的门头沟区、房山区，平原区新生代沉积物覆盖下也有一定分布。侵入岩出露面积约2682.37km^2。

最早对北京地区侵入岩开展研究的是叶良辅等（1920）编著的《北京西山地质志》，主要是对阳坊和房山两地的花岗岩做了描述性研究，并通过接触关系和侏罗系砾岩中的砾石成分分析，推断花岗岩的侵入时代为早侏罗世到晚侏罗世之间。之后，李四光（1924）、杨曾威（1928）、何作霖（1937）先后对房山岩体进行了研究；黄汲清等（1928）也对阳坊岩体与髫髻山组火山岩的接触关系进行了专门研究。这些研究工作在岩体地质、侵位时代、结构构造、变质作用等方面提出的见解，迄今仍有参考利用价值。尤其是李四光先生在1924年就利用氧化物等值线法对房山岩体进行了研究，是我国从岩石化学角度对侵入体进行研究的开端。

1958年至1962年，北京市地质局和北京地质学院联合组队开展山区第一轮1：5万区域地质调查，通过对一手资料的总结，从岩浆活动的时空变化以及岩浆演化规律分析入手，创造性地提出了“旋回、阶、期、次、岩体”五级划分方案，将本区内燕山旋回侵入岩划分为4个阶、7个期、十多个次、几十个大小侵入体。该项成果强调了岩浆侵入多期、多阶段、多次性特点，突出了一定范围内同期岩浆活动产物的相似性、可比性。在“阶”和“期”的命名上，采用典型岩体的地理名称，更有独到之处。

20世纪60年代以来开展的山区第二轮1：5万区域地质调查，获取了大量的岩石地球化学及同位素年龄数据，经总结对比，建立了本区岩浆活动序列，划分为前长城纪、中元古代、海西、燕山及喜马拉雅5个侵入岩浆旋回，并对各旋回岩浆作用特点、成岩地质特征及地球化学特征进行了全面总结和研究。1987年以后的1：5万区域地质调查，主要采用单元-超单元研究方法对各个调查图幅内的侵入体进行了划分。

20世纪90年代，许多学者针对区内重要或独特的构造-热事件的侵入活动从不同角度进行了更加深入系统的研究，大大提高了本区侵入岩的研究程度。代表性成果主要有：①郁建华等（1990、1993、1994、1996）、Ramo等（1995）在岩浆岩体系划分中采用了“岩系、岩套、岩组、岩体”4级划分方案，在岩浆成因来源等方面提出了新的认识。其主要成果在《北京地区岩浆深成作用》（1992）和《华北地台北部非造山环斑花岗岩及其有关岩石》（1996）两部专著中得到了充分体现；②白志民等（1991）运用单元-超单元方法对区内侵入岩进行了系统研究，将中生代侵入岩划分为8个超单元、20个单元，并探讨了岩浆来源、性质以及成岩地质环境及其演化规律，出版了《八达岭花岗杂岩》专著；③鲍亦冈等（2001）编著的《北京地质百年研究》，按岩石谱系单位建立花岗岩类等级体制，对本区中生代约计230余个侵入岩体作了全面的分析研究对比，建立了22个侵入单元和6个超单元，分析了岩浆活动的构造环境及岩体侵位机制。另外，张吉顺等（1990）、阎国翰等（1995）、张建新等（1997）、宋鸿林（1996）、郑亚东（1999）分别对房山岩体、云蒙山岩体的的岩石成因学、变形构造、侵位机制进行了深入研究，提出了有意义的见解。

进入21世纪，伴随着同位素测试技术的进步，尤其是高精度锆石测年技术发展，侵入岩研究又一次掀起新一轮热潮。关于北京地区的侵入岩的出版专著有《大庙斜长岩和密云环斑花岗岩的岩石学和地球化学》（谢广轰，2005）和《北京西山地区中生代岩浆作用、变质作用及其对华北东部岩石圈厚度变化的约束》（张树明，2010）。一些科研单位和地质院校开展了国土资源部科技计划专项、中国地质调查局研究项目以及国家自然科学基金项目，如邓晋福等（2004）主持的“中国花岗岩重大地质问题研究项目课题 4——燕山造山带花岗岩与大陆地壳改造和生长”“中国东部中生代大陆岩石圈剧变过程动力学机制与矿产资源响应”等项目，在本区获得了一批高质量的地球化学和锆石测年数据，基本建立了构造-岩浆年代格架。通过上述项目的实施，研究人员发表了大量的研究论文，包括 Davis（2002）、邵济安（2001）、李伍平等（2001）、钱青等（2002）、邓晋福等（2004）、李晓勇等（2004）、袁洪林（2005）、苏尚国等（2006）、马寅生（2007）、和政军等（2011）、汪洋等（2013，2014，2015）、陈印（2014）等。

第一节　构造-侵入岩区带划分

依据区域构造-岩浆活动特点，参考大区域（华北板块北缘）岩浆岩带时空分布的特征、演化规律和构造背景，以及《中国区域地质志·河北志》（2017）构造-岩浆岩带划分方案，对本区新太古代至中生代构造-岩浆岩（侵入岩）区带进行系统划分（表 18-1、图 18-1），共划分为 2 个一级构造单元、5 个二级构造单元和 7 个三级构造单元。

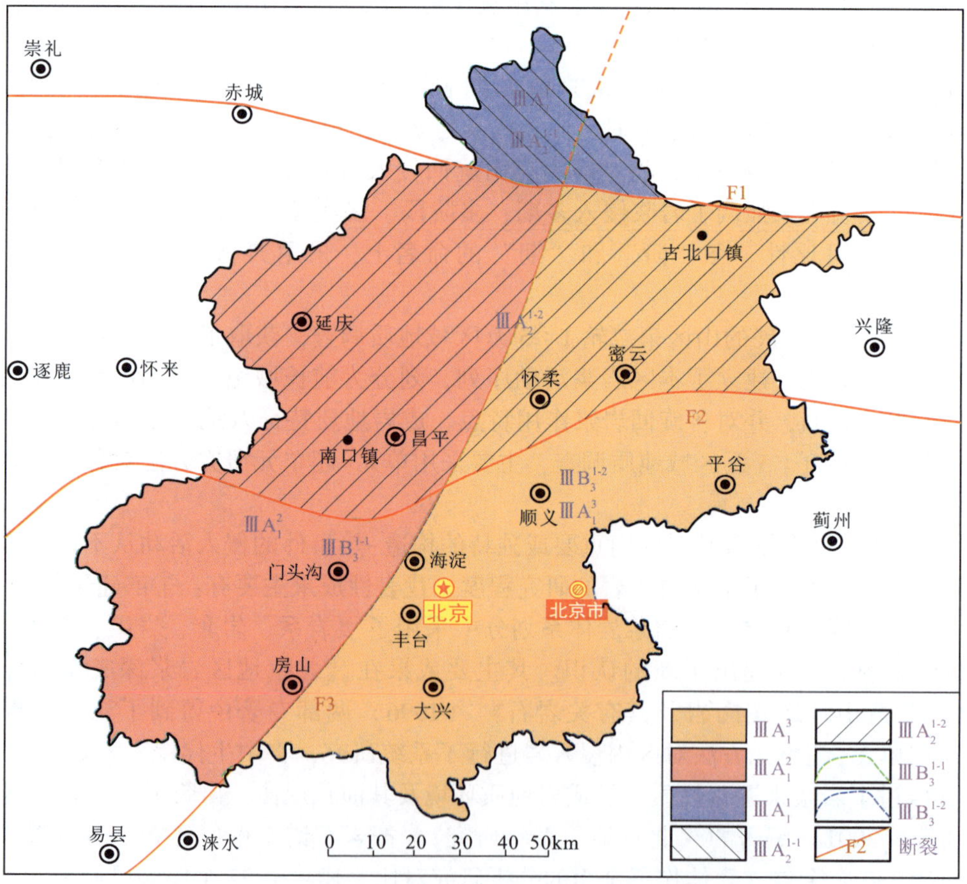

图 18-1　北京市侵入岩构造单元划分图（图例解释见表 18-1）

Fig. 18-1　Division of tectonic units of intrusive rocks in Beijing

F1—赤城-古北口断裂；F2—南口-兴隆断裂；F3—黄庄-高丽营-琉璃庙断裂

表 18-1　侵入岩构造单元划分表

Table 18-1　Division of tectonic units of intrusive rocks

时段	一级	二级	三级
T_3-E_1	晚三叠世—古新世大兴安岭-太行山叠加构造岩浆岩区（IIIB）	冀北-太行山岩浆岩带（$IIIB_3^1$）	房山-延庆岩浆岩亚带（$IIIB_3^{1-2}$）
			大兴-密云岩浆岩亚带（$IIIB_3^{1-3}$）
Pt_2-T_2	新太古代—中三叠世华北陆块构造岩浆岩区（IIIA）	张北-承德岩浆岩带（$IIIA_2^1$）	崇礼-古北口北岩浆岩亚带（$IIIA_{1-2}^1$）
			延庆-密云岩浆岩亚带（$IIIA_{1-2}^2$）
Ar_3-Pt_1		冀中-冀东岩浆岩带（$IIIA_1^3$）	大兴-密云岩浆岩亚带（$IIIA_1^3$）
		冀西冀南岩浆岩带（$IIIA_1^2$）	房山-延庆浆岩亚带（$IIIA_1^2$）
		阴山-冀北岩浆岩带（$IIIA_1^1$）	崇礼-古北口北岩浆岩亚带（$IIIA_1^1$）

新太古代—古元古代侵入岩二级构造单元呈近东西向和北北东向展布，经历了阜平期和吕梁期多期岩浆活动，规模较大。中元古代—中三叠世侵入岩二级构造单元因受西伯利亚板块与柴达木-华北板块俯冲碰撞制约，基本上保持了近东西向展布格局，经历了晋宁期、华力西期和印支期的多期岩浆活动，规模较小。晚三叠世-古新世侵入岩二级构造单元，因受西太平洋板块向欧亚板块俯冲挤压的制约，主要呈北东-北北东向展布，并叠加在早期近东西向构造岩浆岩带上。

第二节　侵入岩期序及年代格架

统计区内侵入岩各类同位素年龄，筛选出高精度可靠同位素年龄（表 18-2），建立侵入岩年代格架。本区侵入岩划分为阜平旋回、吕梁旋回、晋宁旋回、华力西旋回、印支旋回和燕山旋回 6 期岩浆期序（表 18-3、表 18-4）。燕山、晋宁和华力西旋回可进一步分别划分为 4 个、2 个和 2 个亚旋回。

表 18-2　北京市侵入体同位素年龄一览表

Table 18-2　Isotopic ages of intrusions in Beijing

序号	岩体/岩脉名称	测定岩性	年龄/Ma	测定方法	来源
1	高家窑	辉绿岩	95	全岩 K-Ar 法	北京市地质研究所，1993
2	四桥子	正长花岗岩	116±2.3	锆石 U-Pb 法	郁建华等，1994
3	阳坊	花岗岩	118±2	锆石 U-Pb 法	Davis 等，2001
4	大海陀	花岗岩	119±2	锆石 U-Pb 法	Davis 等，2001
5	黑熊山	花岗岩	123.7±1.1	锆石 U-Pb 法	邓晋福等，2004
6	西沟	花岗岩	122.5±1.8	锆石 U-Pb 法	欧阳志侠，2010
7	南口	花岗岩	127.03±0.87	锆石 U-Pb 法	潘颖，2010
8	南口	辉绿岩	126.58±0.77	锆石 U-Pb 法	潘颖，2010
9	白查	碱长花岗岩	127±0.7	锆石 U-Pb 法	邓晋福等，2004
10	千层背	碱长花岗岩	128.8±1.5	锆石 U-Pb 法	Davis 等，2001
11	转山子	正长花岗岩	126.25	黑云母 K-Ar 法	北京市地质调查所，1997
12	卯镇山	正长花岗岩	131.11	锆石 U-Pb 法	北京市地质调查所，1997
13	炭厂	花岗斑岩	123	全岩 K-Ar 法	北京市地质调查所，1993
14	林字台	石英斑岩	129.3±3	锆石 U-Pb 法	核工业部三所，1987
15	大东沟	二长花岗岩	127±2	锆石 U-Pb 法	Davis 等，1996
16	关上	二长花岗岩	129±2.5	锆石 U-Pb 法	郁建华等，1994
17	南窑	花岗岩	128.4±1.5	锆石 U-Pb 法	Davis 等，2001
18	八达岭	斑状花岗岩	138.9	黑云母 K-Ar 法	中国科学院地球化学研究研究所，1962

续表

序号	岩体/岩脉名称	测定岩性	年龄/Ma	测定方法	来源
19	莲花山	二长花岗岩	125.7±1.8	黑云母 K-Ar 法	冶金地质研究所，1981
20	南猴顶	二长花岗岩	127.229	黑云母 K-Ar 法	北京市地质调查所，1997
21	月亮鞍梁	二长花岗岩	126.25	黑云母 K-Ar 法	河北地质三队，1997
22	铁炉子村	花岗岩	137±1.2	锆石 U-Pb 法	焦守涛，2013
23	冯家峪	二长花岗岩	127.4±1.2	锆石 U-Pb 法	欧阳志侠，2010
24	石湖峪	二长花岗岩	128.5±1.2	锆石 U-Pb 法	焦守涛，2013
25	湖门	正长岩	124.2±1.8	锆石 U-Pb 法	邓晋福等，2004
26	雾灵山	石英正长岩	131.7±1.5	锆石 U-Pb 法	Davis 等，2001
27	房山	花岗闪长岩	128.5±1.5	锆石 U-Pb 法	Davis 等，2001
28	西白莲峪	石英二长闪长岩	126.6±1.8	锆石 U-Pb 法	欧阳志侠，2010
29	薛家石梁	闪长岩	127±1.5	锆石 U-Pb 法	邓晋福等，2004
30	薛家石梁	闪长岩	130±1.7	锆石 U-Pb 法	Davis 等，2001
31	黑山寨	石英二长岩	125.1±1.5	锆石 U-Pb 法	邓晋福等，2004
32	南窖	石英二长岩	134±1Ma	锆石 U-Pb 法	张璐等，2008
33	南窖	石英二长岩	128.4±1.5	锆石 TIMS 法	Davis 等，2001
34	铁炉子	石英二长岩	133.2±2.6	锆石 U-Pb 法	郁建华等，1994
35	碾台	石英二长岩	133.6±2.7	锆石 U-Pb 法	郁建华等，1994
36	东老峪	石英二长岩	133±1.3	锆石 U-Pb 法	焦守涛，2013
37	黑峪口	石英二长岩	132.2	黑云母 K-Ar 法	中科院地球化学研究所，1962
38	棋盘岩	辉长岩	132.9±0.7	锆石 U-Pb 法	邓晋福等，2002
39	上庄	辉长岩	128.8±1.7	锆石 U-Pb 法	苏尚国等，2006
40	芹峪	闪长岩	138	全岩 K-Ar 法	北京市地质调查所，1993
41	汉家川	石英二长闪长	147.2	黑云母 K-Ar 法	地质矿产部宜昌地质研究所，1978
42	五道河	花岗闪长岩	141±2	锆石 U-Pb 法	Davis 等，2001
43	大河南	花岗闪长岩	136±2	锆石 U-Pb 法	Davis 等，2001
44	东三岔	二长闪长岩	135.42	全岩 K-Ar 法	河北地质学院，1982
45	上苇甸	花岗闪长岩	138.2±1.5	锆石 U-Pb 法	Davis 等，2001
46	仙人洞	花岗闪长岩	151.2	角闪石 K-Ar 法	北京大学，1975
47	杨麻地	花岗闪长岩	133.65	黑云母 K-Ar 法	地质矿产部宜昌地质研究所，1978
48	黄场	石英二长斑岩	142.6	黑云母 K-Ar 法	地矿部宜昌地质研究所，1978;
49	白羊石虎	石英二长岩	140.6±3.7	锆石 U-Pb 法	郁建华等，1994
50	慈母川	石英二长岩	146.6	黑云母 K-Ar 法	北京地质调查所，1985
51	塘石	辉钼矿	137.6±3.7	Re-Os 法	刘舒波，2012
52	大钟山	石英二长岩	139.9	黑云母 K-Ar 法	地矿部宜昌地质研究所，1978
53	云蒙山	片麻状花岗岩	144.7±2.7	锆石 U-Pb 法	邓晋福等，2004
54	云蒙山	花岗岩	145±5	锆石 U-Pb 法	Shi 等，2009
55	碓臼峪	二长花岗岩	137.7±3.6	锆石 U-Pb 法	邓晋福等，2004
56	房耳峪	二长花岗岩	145	黑云母 K-Ar 法	中国地质科学院地质研究所，1989
57	小豹子港	花岗斑岩	141±2	锆石 U-Pb 法	Davis 等，2001
58	官地南	片麻状花岗岩	143±3	锆石 TIMS 法	Davis 等，2001

续表

序号	岩体/岩脉名称	测定岩性	年龄/Ma	测定方法	来源
59	大石门子	花岗岩	141±2	锆石 TIMS 法	Davis 等，1996
60	四干顶	二长花岗岩	159.3±1.9	锆石 U-Pb 法	邓晋福等，2004
61	沙陀子	二长花岗岩	151±2	锆石 TIMS 法	Davis 等，1996
62	花塔	石英二长岩	152.1	黑云母 K-Ar 法	中国科学院地球化学研究所，1963
63	石城	闪长岩	159±2	锆石 TIMS 法	Davis 等，1996
64	石城	闪长岩	159±5	锆石 U-Pb 法	Shi 等，2004
65	石城	闪长岩	155±1	角闪石 Ar-Ar 法	Wang 等，2012
66	长园	闪长岩	152.7±3.2	锆石 U-Pb 法	邓晋福等，2004
67	上花园	二长闪长岩	154.9	全岩 K-Ar 法	中科院地球化学研究所，1964
68	玉皇山	石英闪长岩	158.2±1.8	黑云母 K-Ar 法	中国地质科学院，1985
69	里长沟	二长闪长玢岩	147	全岩 K-Ar 法	河北地质学院，1979
70	青白口	闪长玢岩	177±1	锆石 U-Pb 法	叶浩，2014
71	黑山坨	花岗岩	201±1	锆石 U-Pb 法	焦永玲，2014
72	大石坡	正长岩	196.5±1.9	锆石 U-Pb 法	邓晋福等，2004
73	大石坡	正长岩	200±1	锆石 U-Pb 法	焦永玲，2014
74	平谷	花岗岩	205.8±2.3	锆石 U-Pb 法	马寅生，2007
75	东窝铺	二长花岗岩	205.3±2.5	锆石 U-Pb 法	马寅生，2007
76	莲花岭	石英二长岩	207.6±3.5	锆石 U-Pb 法	马寅生，2007
77	莲花岭	石英二长岩	203±3	全岩 Rb-Sr 法	王季亮等，1994
78	东窝铺	二长花岗岩	210±4	锆石 U-Pb 法	叶浩，2014
79	石塘峪	花岗岩	213±4	锆石 U-Pb 法	叶浩，2014
80	七道河西	石英二长岩	207	锆石 U-Pb 法	核工业三所，1988
81	水磨湾	二长花岗岩	245.7	锆石 U-Pb 法	北京市地质调查所，1997
82	小梁后	二长花岗岩	247.5±0.62	锆石 U-Pb 法	何付兵等，2018
83	西府营	辉长岩	283.0 土 2	锆石 SHRIMP 法	Zhang 等，2009
84	下窝铺	辉长岩	288±4.8	锆石 U-Pb 法	王惠初等，2007
85	老公营	正长岩	384.0±5.0	锆石 U-Pb 法	何付兵等，2019
86	荞麦峪	正长岩	385.7±4.1	锆石 U-Pb 法	何付兵等，2019
87	提辖庄	花岗岩	1683.8±2.2	锆石 U-Pb 法	郁建华等，1996
88	达峪	似斑状花岗岩	1690.5±2.4	锆石 U-Pb 法	郁建华等，1996
89	碰河寺	花岗斑岩	1673±10	锆石 U-Pb 法	相振群，2014
90	沙厂	环斑花岗岩	1685±15	锆石 U-Pb 法	高维等，2008
91	沙厂	环斑花岗岩	1716±21	角闪石 K-Ar 法	胡世玲等，1990
92	沙厂	环斑花岗岩	1693±110	钾长石 Pb-Pb 法	Ramo 等，1994
93	兰营	石英正长岩	1790.9±24	锆石 U-Pb 法	郁建华等，1990
94	兰营	石英正长岩	1697.3±0.9	锆石 U-Pb 法	Mortensen 等，1996
95	高岭	辉长岩	1731±4	锆石 U-Pb 法	彭彭，2011
96	大庙	石英纹长二长岩	1757±38	锆石 U-Pb 法	谢广轰，2005
97	放马峪	橄榄角闪二辉岩	2010.54±155	Sm-Nd 法	陈森煌等，1991
98	旗杆顶	花岗质片麻岩	2511±15	锆石 U-Pb 法	本志，2015

续表

序号	岩体/岩脉名称	测定岩性	年龄/Ma	测定方法	来源
99	康各庄	奥长花岗质片麻岩	2981	锆石 TIMS 法	金文山等，2000
100	荆子峪	花岗闪长岩	2627±80	锆石 U-Pb 法	金文山等，2000
101	放马峪	辉石岩	2458±22	锆石 U-Pb 法	本志
102	宋营子	英云闪长质片麻岩	2544±35	锆石 U-Pb 法	本志
103	宋营子	英云闪长质片麻岩	2567±8	锆石 SHRIMP 法	Shi et al.，2012
104	苇子峪	花岗质片麻岩	2526±6	锆石 SHRIMP 法	Shi et al.，2012
105	苇子峪	次透辉斜长角闪岩	2486±40	锆石 U-Pb 法	金文山等，2000
106	沙厂	英云闪长质片麻岩	2553±74	锆石 U-Pb 法	金文山等，2000
107	四合堂	石榴角闪斜长辉石岩	2449	锆石 TIMS 法	贺高品等，1993

表 18-3　侵入岩岩浆旋回及构造阶段划分表

Table 18-3　Division of magmatic cycles and tectonic phases of intrusive rocks

时代			旋回	亚旋回	侵入岩建造	典型岩体	构造阶段	构造背景
新生代	第四纪	Q	喜马拉雅 HM					太平洋-欧亚大陆构造
	新近纪	N						
	古近纪	E_3						
		E_2						
		E_1						
中生代	白垩纪	K_2	燕山 YS	YS3	北北东向基性侵入岩	高家窑岩体	非造山	
		K_1^3			南北向基性岩墙	南口岩墙	后造山	
					北北东向碱性侵入岩	李家营岩体		
		K_1^2			北北东向酸性、碱性侵入岩	千层背岩体		
					北东向—北北东向中性、酸性侵入岩	碓臼峪岩体	同造山(板内)	
		K_1^1		YS2	北东向基性、中性、酸性侵入岩	云蒙山岩体		
	侏罗纪	J_3			东西向—北东向基性、中性、酸性侵入岩	石城岩体		
		J_2		YS1	东西向酸性侵入岩	四道穴岩体	前造山	
		J_1			东西向基性、中基性侵入岩	青白口岩脉		
					东西向中性、酸性侵入岩	大石坡岩体	非造山	
	三叠纪	T_3			东西向酸性侵入岩	东窝铺岩体		
		T_2	印支 IS		东西向中性侵入岩	八道河西岩体	碰撞后-晚造山	西伯利亚-华北板块构造
		T_1			东西向中酸性侵入岩	胡营子岩体		
					东西向酸性侵入岩	小梁后岩体		
古生代	二叠纪	P_3					同碰撞	
		P_2	华力西 HR					
		P_1		HR2	东西向基性侵入岩	下窝铺岩体	前碰撞	
	石炭纪	C_2						
		C_1						
	泥盆纪	D_3						
		D_2		HR1	东西向中性侵入岩	老公营岩体	前造山(岛弧)	
		D_1						

续表

时代			旋回	亚旋回	侵入岩建造	典型岩体	构造阶段	构造背景
古生代	志留纪	S_{1-4}	加里东 JR					华北克拉通构造
	奥陶纪	O						
	寒武纪	∈						
			扬子 YZ					
新元古代	震旦纪	Z						
	南华纪	Nh						
	青白口纪	Qb	晋宁 JN					
中元古代	末期	Pt_2^4						
	晚期	Pt_2^3						
	中期	Pt_2^2						
	早期	Pt_2^1		JN2	北东向中基性侵入岩	熊儿寨火山岩颈	非造山	
					东西向中酸性侵入岩	沙厂岩体		
					东西向基性侵入岩	兰营岩体		
				JN1	北东向基性岩墙	高岭岩墙		
古元古代	末期	Pt_1^4	吕梁 LL				后造山	华北克拉通形成期（Columbia 超大陆形成及裂解）
	晚期	Pt_1^3					后碰撞	
	中期	Pt_1^2					同碰撞	
	早期	Pt_1^1			北东向超基性岩脉	红石湾岩体	前碰撞	
新太古代	晚期	Ar_3^3	阜平 FP		东西向、北北东向基性、中性、酸性、碱性侵入岩	茅子峪辉长闪长岩体等	前造山（岛弧）	
	中期	Ar_3^2						
	早期	Ar_3^1						

第三节 侵入岩岩浆组合、成因类型及岩石系列

在区域大地构造背景框架下，根据岩石成因类型、岩石系列、岩石物源及侵入岩岩浆组合，划分本区侵入岩岩浆组合、成因类型及岩石系列（表 18-4）。

表 18-4 侵入岩岩浆组合、成因类型及岩石系列划分表

Table 18-4 Magmatic combinations, genetic types and rock series of intrusive rocks

时代			旋回	岩石名称	代号	岩浆组合	岩石系列	岩石成因类型	构造阶段
中生代	白垩纪	K_2	燕山 YS	辉绿岩	$\beta\mu K_2$	基性-超基性	钙碱性	M 型	非造山
				辉长岩	νK_2				
				玻基辉橄岩	$\psi\sigma\psi K_2$				
		K_1^3		花岗岩	γK_1^3	基性-酸性	碱性、钙碱性	I 型	后造山
				正长岩	ξK_1^3				
				辉绿岩	$\beta\mu K_1^3$				

续表

时代			旋回	岩石名称	代号	岩浆组合	岩石系列	岩石成因类型	构造阶段
中生代	白垩纪	K_1^3	燕山YS	石英斑岩	$\lambda\pi K_1^3$	酸性、碱性	钙碱性	I型、A型	后造山
				细粒花岗岩	γK_1^3				
				花岗斑岩	$\gamma\pi K_1^3$				
				正长花岗岩	$\xi\gamma K_1^3$				
				钾长花岗岩	$\xi\gamma K_1^3$				
				石英正长斑岩	$\xi o\pi K_1^3$				
				石英正长岩	$\xi o K_1^3$				
		K_1^2		霏细岩	ρK_1^2	酸性	钙碱性、碱性	I型、A型	
				碱长花岗岩	$\rho\gamma K_1^2$				
				花岗斑岩	$\gamma\pi K_1^2$				
				正长花岗岩	$\xi\gamma K_1^2$				
				二长花岗岩	$\eta\gamma K_1^2$				
				石英正长斑岩	$\xi o\pi K_1^2$	中酸性	钙碱性、碱性	M型、I型	造山期
				石英正长岩	$\xi o K_1^2$				
				花岗闪长岩	$\gamma\delta K_1^2$				
				二长闪长岩	$\eta\delta K_1^2$				
				正长岩	ξK_1^2				
				闪长岩	δK_1^2				
				石英二长岩	$\eta o K_1^2$				
				二长岩	ηK_1^2				
		K_1^1		二长闪长岩	$\eta\delta K_1^1$	中基性	钙碱性	M型	
				辉长闪长岩	$\nu\delta K_1^1$				
				辉长岩	νK_1^1				
				石英斑岩	$\lambda\pi K_1^1$	酸性	钙碱性	I型	
				花岗斑岩	$\gamma\pi K_1^1$				
				二长花岗岩	$\eta\gamma K_1^1$				
				花岗闪长岩	$\gamma\delta K_1^1$	中酸性	钙碱性	I型、M型	
				石英二长斑岩	$\eta o\pi K_1^1$				
				石英二长岩	$\eta o K_1^1$				
				二长岩	ηK_1^1				
				闪长玢岩	$\delta\mu K_1^1$				
				闪长岩	δK_1^1				
				二长闪长岩	$\eta\delta K_1^1$				
				石英闪长岩	$\delta o K_1^1$				

续表

时代			旋回	岩石名称	代号	岩浆组合	岩石系列	岩石成因类型	构造阶段
中生代	侏罗纪	J_3	燕山YS	花岗斑岩	$\gamma\pi J_3$	中性-酸性	钙碱性	I型	造山期
				二长花岗岩	$\eta\gamma J_3$				
				石英闪长岩	$\delta o J_3$				
				石英二长岩	$\eta o J_3$				
				闪长岩	δJ_3				
				二长闪长岩	$\delta\eta J_3$				
				辉长岩	νJ_3	基性	碱性	I型	
		J_2		二长花岗岩	$\eta\gamma J_2$	酸性	钙碱性	I型	
		J_1		闪长玢岩	$\delta\mu J_1$	基性、中基性	钙碱性、碱性	M型	前造山
				闪长岩	δJ_1				
				辉绿岩	$\beta\mu J_1$				
				辉长闪长岩	$\nu\delta J_1$				
				花岗岩	γJ_1	中性-酸性	碱性、钙碱性	M型、I型	前造山
				石英正长岩	$\xi o J_1$				
				正长岩	ξJ_1				
	三叠纪	T_3		花岗斑岩	$\gamma\pi T_3$	中酸性	钙碱性	I型、A型	非造山
				二长花岗岩	$\eta\gamma T_3$				
				石英二长岩	$\eta o T_3$				
		T_2	印支IS	石英二长岩	$\eta o T_1$	中性	钙碱性	I型	碰撞后-晚造山
		T_1		石英闪长岩	$\delta o T_1$	中性-酸性	钙碱性	I型	
				二长花岗岩	$\eta\gamma T_1$				
				二长闪长岩	$\delta\eta T_1$				
				花岗闪长岩	$\gamma\delta T_1$				
晚古生代	二叠纪	P_1	华力西HR	辉长苏长岩	$\sigma\nu P_1$	中基性	碱性	M型	前碰撞
				辉长岩	νP_1				
				辉长闪长岩	$\nu\delta P_1$				
				石英闪长岩	$\delta o P_1$				
	泥盆纪	D_2		正长岩	ξD_2	中性	碱性	M型、I型	前造山
				正长斑岩	$\xi\pi D_2$				
中元古代	早期	Pt_2^1	晋宁JN	辉绿岩	$\beta\mu Pt_2^1$	中基性	碱性	M型	大陆裂谷
				正长斑岩	$\xi\pi Pt_2^1$				
				斜长岩	$\nu\sigma Pt_2^1$	基性-超基性	钙碱性	M型	稳定克拉通
				辉长岩	νPt_2^1	基性	拉斑玄武岩		
				石英正长岩	$\xi o Pt_2^1$	中酸性-酸性	钙碱性	I型、A型	
				石英二长岩	$\eta o Pt_2^1$				
				环斑花岗岩	$\gamma\theta Pt_2^1$				
				斑状花岗岩	$\gamma p Pt_2^1$				
				花岗岩	γPt_2^1				
				花岗斑岩	$\gamma\pi Pt_2^1$				

续表

<table>
<tr><th colspan="3">时代</th><th>旋回</th><th>岩石名称</th><th>代号</th><th>岩浆组合</th><th>岩石系列</th><th>岩石成因类型</th><th>构造阶段</th></tr>
<tr><td rowspan="2">古元古代</td><td rowspan="2">早期</td><td rowspan="2">Pt_1^1</td><td rowspan="2">吕梁
LL</td><td>变质辉石岩</td><td rowspan="2">$\Psi\iota Pt_1^1$</td><td rowspan="2">基性-超基性</td><td rowspan="2">拉斑玄武岩</td><td rowspan="2">M 型</td><td rowspan="2">俯冲岩浆弧</td></tr>
<tr><td>变质橄榄岩</td></tr>
<tr><td rowspan="9">新太古代</td><td rowspan="9">晚期</td><td rowspan="9">Ar_3^3</td><td rowspan="9">阜平
FP</td><td>花岗质片麻岩</td><td>$Ar_3^3Qgn^{\gamma}$</td><td rowspan="3">基性-酸性</td><td>钙碱性</td><td>I 型</td><td rowspan="7">岛弧</td></tr>
<tr><td>变质辉石岩</td><td rowspan="2">$Ar_3^3Fgn^{\Psi\iota}$</td><td rowspan="2">拉斑玄武岩</td><td rowspan="2">M 型</td></tr>
<tr><td>变质橄榄岩</td></tr>
<tr><td>奥长花岗质片麻岩</td><td>$Ar_3^3Kgn^{\gamma}$</td><td rowspan="3">TTG 组合</td><td rowspan="3"></td><td rowspan="3">I 型</td></tr>
<tr><td>花岗闪长质片麻岩</td><td>$Ar_3^3Jgn^{\gamma\delta}$</td></tr>
<tr><td>英云闪长质片麻岩</td><td>$Ar_3^3Sygn^{\gamma\delta o}$</td></tr>
<tr><td>辉长闪长质片麻岩</td><td>$Ar_3^3Wgn^{\nu\delta}$</td><td>中性</td><td>钙碱性</td><td>M 型</td></tr>
<tr><td>辉长质片麻岩</td><td>$Ar_3^3Hgn^{\nu}$</td><td>基性</td><td>碱性</td><td>M 型</td><td>大陆裂谷</td></tr>
<tr><td>英云闪长质片麻岩</td><td>$Ar_3^3Sgn^{\gamma\delta o}$</td><td>中酸性</td><td>钙碱性</td><td>I 型</td><td>基底形成期</td></tr>
</table>

第十九章　中基性—超基性侵入岩

第一节　概　　述

一、侵入岩期序及岩石单位划分

区内基性—超基性侵入岩规模小，主要呈岩株、岩脉或岩枝状产出，形成时代有新太古代、古元古代、中元古代、古生代和中生代。根据其构造地质背景、时空分布、岩石组合及变质变形特点，参考同位素测年数据和地球化学特征，建立了阜平、吕梁、晋宁、华力西和燕山5个构造岩浆旋回（表18-2、表18-4）。

阜平构造岩浆旋回发育四期次岩浆活动构成两期亚旋回。新太古代黑卧辉长质片麻岩系列和苇子峪辉长闪长质片麻岩系列组合构成第一亚旋回，呈岩基状产出。新太古代放马峪岩墙构成第二亚旋回，呈岩枝或岩脉状产出。吕梁构造旋回为古元古代变质基性—超基性岩组合，呈岩瘤状产出。

晋宁构造岩浆旋回发育三期次岩浆活动构成两期亚旋回。斜长岩-辉长岩组合构成第一亚旋回，呈岩基、岩脉产出。玄武岩-辉绿岩-正长岩组合构成第二亚旋回，呈岩株、岩枝产出。

华力西期构造岩浆旋回发育两期次岩浆活动构成两期亚旋回。中泥盆世辉长岩-正长岩组合构成第一亚旋回，区内仅发育正长岩。早二叠世辉长岩-闪长岩组合构成第二亚旋回。

燕山期发育四期次岩浆活动构成四期亚旋回。早侏罗世辉绿岩-闪长岩组合构成第一亚旋回，呈岩枝、岩脉产出。晚侏罗世辉长岩-闪长岩组合构成第二亚旋回，呈岩枝产出。早白垩世早期辉长岩-二长岩组合构成第三亚旋回，呈岩基产出。晚白垩世辉绿岩-正长岩组合构成第四亚旋回，呈岩脉产出。

二、侵入岩区（带）

（一）新太古代晚期—古元古代早期基性侵入岩

分布于华北陆块构造岩浆岩区（ⅢA）内，分属于冀西-冀南岩浆岩带（$ⅢA_1^2$）和冀中-冀东岩浆岩带（$ⅢA_1^3$），二者以北东向黄庄-高丽营-琉璃庙断裂为界。前者区内出露有限，面积约0.37km^2，分布于房山周口店及门头沟碾台，主要岩性为黑云母角闪斜长片麻岩、斜长角闪岩、黑云母角闪石变粒岩等；后者区内出露面积约81.5km^2，分布于密云黑卧、苇子峪、苍术会、西湾子，怀柔琉璃庙和平谷镇罗营等地，密云、怀柔一带主要岩性为含紫苏黑云辉石片麻岩、黑云辉石角闪片麻岩、角闪黑云二辉片麻岩。西湾子及平谷关山等地区由于受后期角闪岩相退变质作用，岩性为黑云角闪片麻岩。

（二）中元古代早期基性侵入岩

分布于华北陆块构造岩浆岩区（ⅢA）内，属于张北-承德岩浆岩带（$ⅢA_2^1$）。区内分布零散，均位于延庆-密云岩浆岩亚带（$ⅢA_{1-2}^2$）内，为三期岩浆活动产物。一期分布于密云高岭，岩性为辉长岩，呈北东向岩墙状产出；二期分布于怀柔长哨营，沿赤城-古北口区域大断裂东西向岩基产出，发育岩性为斜长岩-石英正长岩组合；三期分布于平谷熊儿寨，顺大红峪期火山通道发育，岩性为玄武岩-辉绿岩-正长岩组合。

（三）古生代晚期基性侵入岩

分布于华北陆块构造岩浆岩区（ⅢA）内，属于张北-承德岩浆岩带（$ⅢA_2^1$）、崇礼-古北口北岩浆岩亚带（$IIIA_{1-2}^2$），分布于怀柔喇叭沟门，赤城-古北口断裂以北，由下窝铺等 4 个辉长岩体组成。

（四）中生代晚期基性侵入岩

分布于华北陆块构造岩浆岩区（IIIA）内，分属于张北-承德岩浆岩带（$IIIA_2^1$）、房山-延庆岩浆岩亚带（$IIIB_3^{1-2}$）。早侏罗世侵入岩仅出露于门头沟青白口和怀柔长哨营一带，呈岩脉状东西向侵位，岩性为辉长岩-闪长岩组合。晚侏罗世侵入岩分布零散，出露于密云朱家湾、延庆太平庄、昌平阳聂村等地，呈北东向岩枝状，岩性为辉长岩-辉长闪长岩组合。早白垩世侵入岩发育规模最大，出露于门头沟棋盘岩、昌平上庄和平谷熊儿寨等地，呈北西向岩株状，岩性为辉长岩-正长岩组合。晚白垩世侵入岩仅出露于延庆青龙桥，呈岩枝或岩脉北北东向侵位，岩性为辉绿岩-辉绿玢岩组合。

第二节　岩石类型及岩石构造组合类型

一、新太古代变质中基性侵入岩

（一）黑云辉长质片麻岩（$Ar_3^3Hgn^{\nu}$）

1. 岩石学特征

发育为密云水库南部的黑卧岩体，出露面积约 1.5km²。岩石呈暗灰色，岩性为黑云二辉斜长片麻岩和黑云辉石角闪斜长片麻岩。岩体内部可见少量斜长角闪岩和角闪岩包体，见大量片麻状二长花岗质脉体顺片麻理侵入。岩体被宋营子英云闪长质片麻岩（穆家峪岩体）和放马峪基性岩体侵入包裹。

（1）黑云二辉斜长片麻岩

仅发育于岩体东侧，呈深灰色，柱粒状变晶结构，主要由暗色矿物紫苏辉石、透辉石、黑云母和浅色矿物斜长石和石英组成。紫苏辉石多呈假象出现，绿泥石化、滑石化或阳起石化，含量约 13%；透辉石短柱状，大多呈集合体状产出，含量 3%~5%；黑云母红褐、黄褐色片状，定向排列，无蚀变；斜长石他形粒状，具绢云母化、钠黝帘石化，反条纹发育，偶见蠕英结构，含量约 70%；石英他形粒状，定向排列，粒度多为 0.5~1.0mm，含量约 4%。副矿物有磁铁矿、磷灰石、锆石等。

（2）黑云辉石角闪斜长片麻岩

是构成岩体的主要岩性，呈暗灰色，粒状变晶结构，主要由暗色矿物角闪石、透辉石、黑云母和浅色矿物斜长石和石英组成，粒度多为 0.2~0.6mm，少量可达 1mm。角闪石半自形柱状-他形粒状，绿色，多色性明显，含量约 15%；透辉石粒状，沿解理有绿帘石化蚀变，并有铁质析出，含量 10%~15%；黑云母棕色，片状，多色性明显，含量约 5%；斜长石粒状，聚片双晶发育，弱绢云母化，含量 55%~60%；石英他形粒状，充填于其他矿物空隙中，含量约 5%。

2. 岩石化学与地球化学特征

岩石化学成分及特征参数见表 19-1。里特曼指数 $\sigma=5.31$，$Na_2O>K_2O$，表明岩石为钠质碱性岩系列；固结指数 $SI=19.71$，分异指数 $DI=39.39$，均反映岩浆分异程度高；镁铁比值 $m/f=0.14$，属于富铁质基性岩；$A/CNK=0.58$，$A/NK=1.79$，属于准铝质。

稀土元素成分及特征参数见表 19-2。$\Sigma REE=177.91\times10^{-6}$，$LREE/HREE=11.09$，$(La/Yb)_N=19.17$，稀土总量较高，轻稀土相对富集，重稀土相对亏损；$\delta Eu=0.94$，呈极弱铕负异常；配分曲线为右倾光滑曲线（图 19-1）。

微量元素成分及特征参数见表 19-3。与中基性侵入岩均值相比，Sr、Ba、V、Y、Sc 含量较高，Cr、Zr 含量较低，Co、Ni 含量基本相近；与上地幔岩石相比，Ba、Sr、Y、Sc、Ti、V 相对富集，Cr、Mn、Co、Ni 相对亏损，Rb/Sr=0.07，接近上地幔均值（0.022），反映成岩物质来源于上地幔且受到地壳物质的混染。微量元素蛛网图（图 19-2）上，强烈富集大离子亲石元素 Rb、K、Sr、Ba，相对亏损高场强元素 Nb、Zr。

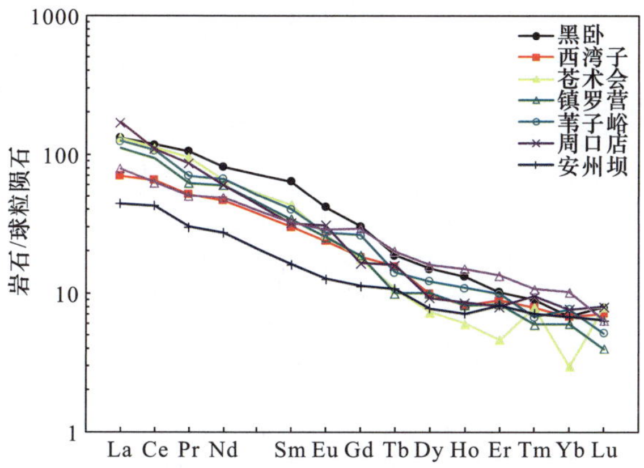

图 19-1　太古宙基性岩稀土元素配分模式图

Fig. 19-1　Chondrite-normalized REE patterns of Archean intrusive rocks

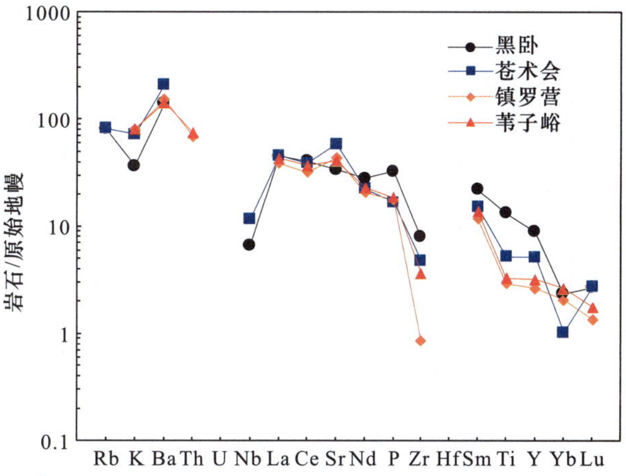

图 19-2　太古宙基性岩微量元素蛛网图

Fig. 19-2　Trace element spider diagram of Archean intrusive rocks

3. 变质变形特征

岩体片麻状、条带状构造高度发育，浅色矿物和暗色矿物相间分布组成约 1mm 条带，镜下可见辉石长轴定向，表明岩体经较强麻粒岩相变形变质。岩体内部还可见条带形成的小型褶皱构造。

4. 原岩恢复

上述地质产状、岩石学和岩石地球化学等特征，特别是残留的原生辉石结构等，揭示了该变质岩原岩为正变质岩。在变质岩原岩恢复 Si-[(al+fm)-(c+alk)]（图 19-3）和 Si-Mg（图 19-4）图解上，成分点也落在火成岩区内。

5. 岩石系列

该套岩系贫硅（SiO_2=47.86）富碱（$Na_2O+K_2O>5\%$）、富钛，且 $Na_2O>K_2O$，矿物成分上出现碱性长石和碱性暗色矿物（富钛辉石），σ=5.31>4，为钠质碱性玄武岩系列。

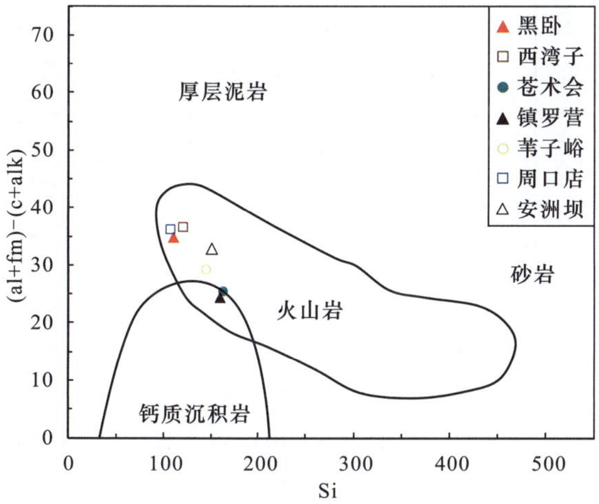

图 19-3　太古宙变质基性岩 Si-[(al+fm)-(c+alk)]原岩恢复图解

Fig. 19-3　Si-[(al+fm)-(c+alk)]diagrams of Archean intrusive rocks

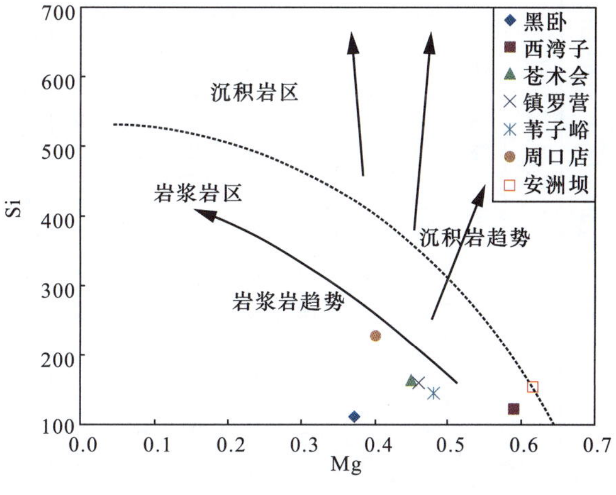

图 19-4　太古宙变质基性岩 Si-Mg 原岩恢复图解

Fig. 19-4　Si-Mg diagrams of Archean intrusive rocks

表 19-1　太古宙变质基性岩岩石化学成分及特征参数一览表

Table 19-1　The chemical composition and main parameters of Archean intrusive rocks

序号	岩体	岩性	氧化物含量/%													特征参数					
			SiO_2	TiO_2	Al_2O_3	Fe_2O_3	FeO	MnO	MgO	CaO	Na_2O	K_2O	P_2O_5	Los	总和	*DI*	*SI*	σ	*m/f*	*A/CNK*	*A/NK*
1	黑卧	黑云辉石角闪片麻岩	47.86	1.78	13.90	6.84	9.43	0.22	5.21	8.93	4.01	1.10	0.71	—	99.99	39.39	19.70	5.31	0.59	0.58	1.79
2	西湾子	黑云角闪片麻岩	51.49	0.64	13.12	6.80	3.92	0.33	8.44	8.38	3.68	1.50	0.38	0.82	99.50	40.67	35.14	2.97	1.45	0.57	1.71
3	苍术会	紫苏角闪片麻岩	56.30	0.56	16.65	4.37	3.75	0.13	3.56	6.08	5.20	2.15	0.36	0.38	99.49	57.98	18.82	3.97	0.81	0.76	1.53
4	镇罗营	含云角闪片麻岩	55.38	0.66	16.44	4.38	3.69	0.16	3.69	5.80	5.65	2.40	0.38	1.18	99.81	62.53	18.72	5.05	0.84	0.73	1.38
5	苇子峪	角闪黑云二辉片麻岩	54.44	0.76	14.64	6.16	4.01	0.19	5.06	6.98	4.50	2.40	0.40	0.42	99.96	52.62	23.10	4.09	0.93	0.65	1.46
6	周口店	角闪黑云斜长片麻岩	53.21	0.68	13.16	4.25	5.03	0.15	8.23	8.96	2.79	1.64	0.22	0.53	98.85	35.41	37.73	1.82	1.63	0.58	2.07
7	安州坝	辉石斜长角闪岩	55.95	0.42	14.41	2.18	4.85	0.24	6.84	7.99	3.86	0.33	0.14	0.4	97.61	41.80	37.87	1.28	1.73	0.68	2.15

注：1—5 引自 1∶5 万密云幅、高岭幅、墙子路幅区域地质调查报告，1994；6 为均值，引自王方正等，1990；7 引自 1∶5 万琉璃庙幅区域地质调查报告，1990。

表 19-2　太古宙变质基性岩稀土元素成分及特征参数一览表

Table 19-2　REE contents and main parameters of Archean intrusive rocks

序号	岩体	岩性	稀土元素含量/10^{-6}										
			La	Ce	Pr	Nd	Sm	Eu	Gd	Tb	Dy	Ho	Er
1	黑卧	黑云辉石角山片麻岩	31.00	72.00	10.00	38.00	9.80	2.40	6.20	0.70	3.80	0.74	1.68
2	西湾子	黑云角闪岩	16.50	39.74	4.86	21.58	4.54	1.36	3.70	0.58	2.48	0.46	1.45
3	苍术会	紫苏角闪片麻岩	31.00	68.00	9.00	30.00	6.60	1.52	3.60	0.40	1.84	0.34	0.76
4	镇罗营	含云角闪片麻岩	26.30	56.50	5.83	27.70	5.15	1.45	3.81	0.37	2.51	0.46	1.36
5	苇子峪	角闪黑云二辉片麻岩	29.30	64.80	6.60	30.70	6.07	1.57	5.31	0.52	3.06	0.61	1.62
6	周口店	角闪黑云斜长片麻岩	27.62	49.05	6.18	20.63	3.65	1.39	2.56	0.44	1.82	0.36	0.99
7	安州坝	辉石斜长角闪岩	10.37	25.78	2.82	12.62	2.45	0.73	2.30	0.40	1.95	0.40	1.34

续表

序号	岩体	岩性	稀土元素含量/10^{-6}							特征参数			
			Tm	Yb	Lu	Y	ΣREE	LREE	HREE	LREE/HREE	$(La/Yb)_N$	δEu	δCe
1	黑卧	黑云辉石角山片麻岩	0.23	1.16	0.20	40.66	177.91	163.20	14.71	11.09	19.17	0.88	1.00
2	西湾子	黑云角闪岩	0.20	1.14	0.18	12.68	98.77	88.58	10.19	8.69	10.38	0.98	1.08
3	苍术会	紫苏角闪片麻岩	0.20	0.50	0.20	8.60	153.96	146.12	7.84	18.64	44.47	0.86	0.99
4	镇罗营	含云角闪片麻岩	0.15	1.01	0.10	11.00	132.70	122.93	9.77	12.58	18.68	0.96	1.07
5	苇子峪	角闪黑云二辉片麻岩	0.17	1.29	0.13	14.40	151.75	139.04	12.71	10.94	16.29	0.83	1.10
6	周口店	角闪黑云斜长片麻岩	0.18	0.98	0.15	10.00	116.00	108.53	7.47	14.53	19.91	1.39	0.89
7	安州坝	辉石斜长角闪岩	0.18	1.13	0.16	11.08	62.63	54.77	7.86	6.97	6.58	0.93	1.15

注：数据来源同表19-1。

表 19-3　太古宙变质基性岩微量元素成分及特征参数一览表

Table 19-3　Trace element contents and main parameters of Archean intrusive rocks

岩体（地区）	岩性	微量元素含量/10^{-6}														特征参数	
		Rb	Ba	Th	Nb	Sr	P	Zr	Ti	Sc	V	Cr	Mn	Co	Ni	Rb/Sr	Sc/Ni
黑卧	黑云辉石角闪片麻岩	50.96	971.6	—	4.732	715	3852	88.74	17210	34.84	291.6	139	221	47.72	130.6	0.07	0.27
苍术会	紫苏角闪片麻岩	50.96	1438	—	8.233	1220	2102	52.84	6697	21.59	166.8	70.11	1331	21.17	26.37	0.04	0.82
镇罗营	含云角闪片麻岩	—	1048	5.82	—	910.1	1337	9.43	3775	12.5	160.7	117.4	595.9	15.34	30.51	—	0.41
苇子峪	角闪黑云二辉片麻岩	—	985.6	6.3	—	854	1538	40.37	4222	21.6	251.4	137	975.5	22.99	21.8	—	0.99

注：引自1∶5万密云幅、高岭幅、墙子路幅区域地质调查报告，1994。

6. 成因类型

黑卧岩体低 SiO_2 高 CaO，Na_2O/K_2O 比值较大（3.65），这与徐克勤（1982）划分的幔源型岩石特征相似。而 $SI=19.71<40$，$DI=39.39$，揭示了幔源原生岩浆经历了分异作用或同化混染作用；稀土配分曲线图（图 19-1）中 LREE 富集，HREE 相对亏损，微量元素蛛网图（图 19-2）中 LILE 富集，HFSE（Nb、Zr）相对亏损，这一特征反映黑卧岩体岩浆来源于上地幔且受到地壳物质的同化混染。

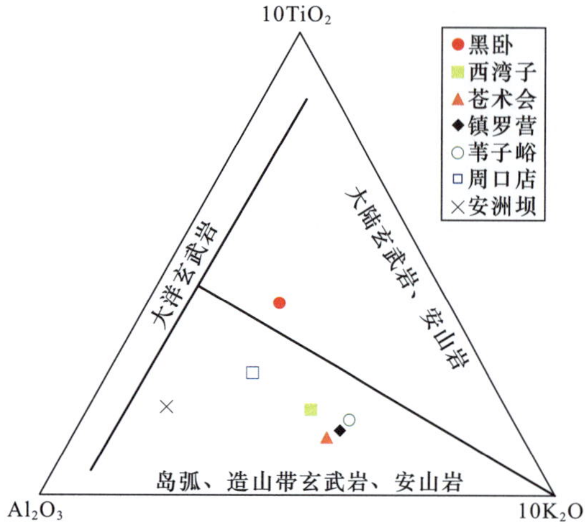

图 19-5 太古宙基性岩 $10TiO_2-Al_2O_3-10K_2O$ 构造环境判别图解

Fig. 19-5 $10TiO_2-Al_2O_3-10K_2O$ diagram of tectonic discrimination of Archean intrusive rocks

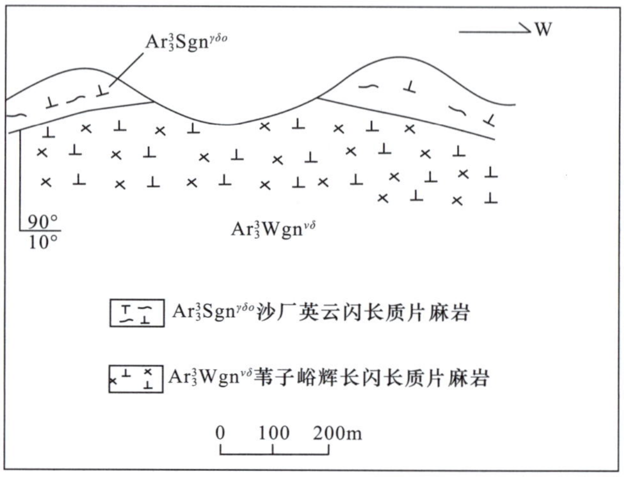

图 19-6 密云苇子峪辉长闪长质片麻岩与沙厂英云闪长质片麻岩侵入接触关系示意图

（据鲍亦冈等，2001）

Fig. 19-6 Geological sketch showing the intrusive contact between Weiziyu gabbro-diorite gneiss and Shachang tonalite gneiss in Weiziyu, Miyun

7. 构造环境分析

黑卧岩体为碱性玄武岩系列，岩浆来源于上地幔且受到来自地壳物质的混染；在 $10TiO_2-Al_2O_3-10K_2O$（图 19-5）构造环境判别图解上，成分位于大陆玄武岩区；表明其侵位于大陆裂谷伸展构造背景环境。

8. 时代归属

黑云岩体遭受高级麻粒岩相变质和复杂褶曲变形；虽然该岩体尚未取得准确同位素测年资料，但被穆家峪英云闪长质岩体和放马峪岩体侵入。地球化学特征揭示其同苇子峪岩体系列可能为同源岩浆演化，时代上可能稍早于苇子峪岩体。根据苇子峪岩体等最新高精度同位素年龄（(2526±6) Ma，Shi et al.，2012），将黑卧岩体时代归属为新太古代晚期。

（二）苇子峪辉长闪长质片麻岩（$Ar_3^3Wgn^{\nu\delta}$）

1. 地质产状与岩石学特征

发育苇子峪、苍术会、镇罗营、西湾子、安州坝和官地（又称周口店）6 个岩体，出露面积约 128km²。岩体内部发育极少量长度<1m 的角闪石岩、斜长角闪石岩包体，普遍发育变质基性岩脉、奥长花岗质片麻岩脉、二长花岗质片麻岩脉和钾长伟晶岩脉等。

（1）苇子峪片麻岩体

分布于密云水库东北苇子峪，被晚侏罗世四干顶岩体破坏呈半环状产出，出露面积约 57km²。岩体向西侵入沙厂英云闪长质片麻岩（图 19-6）和表壳岩密云岩群中，界线清晰呈不规则状。岩体边部常见有规模较小密云岩群和沙厂英云闪长质片麻岩的包体，长度一般不超过 10m。岩体北、东分别被中生界髫髻山组及中元古界长城系不整合覆盖。主要岩石类型为黑云角闪二辉片麻岩、黑云母角闪辉石片麻岩。

黑云角闪二辉片麻岩：灰黑色，粒状镶嵌变晶结构，粒径一般为0.5~1.0mm，条带状-片麻状构造，由暗色矿物紫苏辉石、透辉石、角闪石、黑云母和浅色矿物斜长石、石英组成。紫苏辉石多色性，平行消光或小角度斜消光，含量6%~10%；透辉石淡绿色，柱状-粒状，与紫苏辉石共生，呈条带状分布，含量约10%；黑云母片状，粒度不均，含量2%~10%；角闪石绿色，柱状，含量5%~12%；斜长石他形粒状，聚片双晶发育，多数为奥长石（An=23~25），有些为反条纹长石，见不规则状钾长石客晶分布其中，含量约65%；石英他形，含量2%~5%。副矿物有磁铁矿、磷灰石、锆石等。

黑云角闪辉石片麻岩：灰黑色，柱粒状变晶结构，条带状-片麻状构造，粒径一般为0.5~1.0mm，由暗色矿物普通辉石、角闪石、黑云母和浅色矿物斜长石、石英组成。普通辉石粒状，含量约16%；角闪石黄绿色，长柱状，定向排列，产于辉石颗粒边部，与辉石呈港湾状、锯齿状接触，且有的角闪石中包含辉石颗粒，反映退变质所生特性，含量约10%；黑云母棕红色，定向排列，个别绿泥石化蚀变，含量约5%；斜长石他形粒状，类型为反条纹长石，含量约60%；石英他形粒状，含量约5%。副矿物有磁铁矿、锆石等。

（2）苍术会片麻岩体

分布于密云墙子路至苍术会一带，出露面积约32km^2，周边与其他地质体呈断层接触。主要岩石类型为角闪二辉片麻岩和紫苏角闪片麻岩。

角闪二辉石片麻岩：柱粒状变晶结构，条带状-片麻状构造，由暗色矿物紫苏辉石、透辉石、角闪石、黑云母和浅色矿物斜长石、石英组成。紫苏辉石粒状-短柱状，定向排列，含量0%~10%；透辉石短柱状，斜消光，含量5%~10%；角闪石黄绿色，长柱状，多色性，常沿辉石边缘产出并见辉石包裹体，反映晚期退变质所生特性，含量同辉石呈反比关系；黑云母黑色，片状，含量极少；斜长石他形粒状，聚片双晶发育，含量约75%，类型为奥长石（An=25）；石英他形粒状，常分布在角闪石内和角闪石与辉石接触部位，含量<2%。副矿物主要有磁铁矿、锆石、磷灰石等。

紫苏角闪片麻岩：针、粒状变晶结构，条带状-片麻状构造，由暗色矿物紫苏辉石、角闪石和浅色矿物斜长石、钾长石、石英组成，粒径一般为0.3~1.0mm。紫苏辉石他形粒状-半自形短柱状，浅粉色，平行消光，含量约10%；角闪石绿色，他形粒状-半自形柱状，相对集中呈条带状分布，含量约15%；斜长石他形粒状，含量约75%；钾长石肉红色短柱状，分布不均，含量与后期钾化程度有关；石英他形粒状，充填于其他矿物颗粒之间。副矿物有磷灰石、锆石等。

（3）镇罗营片麻岩体

分布于平谷镇罗营东、北部，出露面积约44km^2，周边被长城系不整合覆盖。岩体岩性变化较大，主要岩石类型为角闪斜长片麻岩及辉石角闪斜长片麻岩。

角闪斜长片麻岩：分布于镇罗营东，针状变晶结构，条带状构造，粒径一般为0.5mm，由暗色矿物角闪石、黑云母和浅色矿物斜长石、石英组成，多发生绿泥石化、绿帘石化蚀变。角闪石绿色，半自形短柱状，内常有榍石、磁铁矿、锆石等包体，并发生纤闪石化、呈不规则状，含量10%~15%；黑云母绿色，鳞片状，含量5%~8%；斜长石他形粒状，聚片双晶发育，具弱绢云母化，含量约75%；石英他形粒状，充填于其他矿物空隙，含量一般<5%。副矿物有磁铁矿、榍石、磷灰石、锆石等。

辉石角闪斜长片麻岩：分布于镇罗营北，针状变晶结构，条带状构造，粒径一般为0.1~0.5mm，由暗色矿物角闪石、辉石和浅色矿物斜长石、石英组成。角闪石绿色，他形粒状-半自形短柱状，多色性，集中呈条带状，部分已纤闪石化和绿泥石化，含量为15%；辉石绿色，他形粒状，裂隙发育，沿裂隙纤闪石化、绿泥石化，含量3%~5%；斜长石他形粒状，含量约75%；石英他形粒状，含量极少。副矿物有磁铁矿、绿帘石、榍石、磷灰石和锆石等。

（4）西湾子片麻岩体

分布于密云四合堂，宏观呈东西向展布，出露面积约4.5km^2。岩体南与长城系呈断层接触，北则与沙厂英云闪长质片麻岩呈侵入或断层接触。侵入界线不明显，露头尺度上岩体与围岩构成“似层状”，片麻理发育程度由围岩向岩体内部明显减弱。

岩体主要岩石类型为黑云角闪片岩或斜长角闪片岩，受后期角闪岩相退变质作用改造明显。斜长石含量36%~56%，石英含量0%~10%，角闪石含量25%~61%，黑云母含量1%~15%，部分岩石可见钾长石，含量0%~6%。副矿物为磁铁矿等。

（5）安州坝片麻岩体

分布于怀柔琉璃庙，不规则状产出，出露面积约2.5km²。岩体宏观侵入特征明显，侵入沙厂英云闪长质片麻岩中，其内部发育长条形、楔状或不规则团块状磁铁石英岩、磁铁片麻岩包体，黄泉峪沟口还见透镜状黑云变粒岩包体；虽然片麻理同围岩一致，但发育程度明显弱。

岩体主要岩石类型为辉石斜长角闪岩，呈黑色-灰黑色，略带绿色，柱粒状变晶结构，致密块状构造。主要由暗色矿物角闪石、透辉石、黑云母和浅色矿物斜长石组成。角闪石绿色，多色性，半自形柱状，粒径0.5~1.2mm，边缘强烈纤闪石化、绿泥石化，含量约78%；透辉石无色，短柱状，干涉色较低，受退变质作用强烈角闪石化呈残余状，且在残余辉石边缘发育一圈铁质氧化物薄膜，解理清晰并沿解理析出铁质物，粒径0.2~0.5mm，含量<6%；斜长石粒状，粒径0.2~0.4mm，聚片双晶，强烈绢云母化，含量约15%；黑云母黄褐色，粒径0.1~0.2mm，晶形不完全，含量极少。副矿物为磷灰石、磁铁矿、石榴子石等。

（6）官地（周口店）片麻岩体

分布于房山周口店，沿白垩纪房山岩体南、北、东三边呈环状产出，出露面积<0.5km²。岩体与其他地质体接触以断层或侵入接触，岩体内广泛发育透镜状或条带状斜长角闪岩脉。岩石类型主要为黑云斜长片麻岩和黑云母角闪石变粒岩，并受到中等强度的混合岩化作用和强烈动力变质作用，形成各种类型的混合岩，普遍发生糜棱岩化。

黑云斜长片麻岩：是岩体主要岩性，呈浅灰色、灰色，镶嵌变晶结构，片麻状构造，主要由暗色矿物黑云母、角闪石和浅色矿物斜长石、微斜长石、条纹长石、石英组成。黑云母黑色鳞片状，粒径一般为0.2~0.6mm，含量10%~15%；角闪石绿色针状，含量2%~3%；斜长石半自形-他形板状，大小不一，大者2~3mm，类型为奥长石，小者1mm，类型为中长石，含量约60%；微斜长石和条纹长石呈他形粒状，含量<5%；石英他形粒状，含量15%~20%。

黑云母角闪石变粒岩：分布于东岭子、南观及官地等，多有程度不等的混合岩化。岩石呈灰绿色、深绿色、墨绿色，细粒变晶结构，片麻状-块状构造，主要由暗色矿物黑云母和浅色矿物斜长石、石英组成。黑云母黑色鳞片状，强烈风化蚀变为绿泥石，含量约15%；长石他形粒状，绢云母及高岭土化，含量约60%；石英他形粒状，含量约25%。部分岩性中可见少量白云母。副矿物为磁铁矿、磷灰石、榍石等。

2. 岩石化学与地球化学特征

岩石化学成分及特征参数见表19-1。里特曼指数σ=1.28~5.05，$Na_2O>K_2O$，表明岩石为钠质钙碱性岩系列；固结指数SI=18.72~37.87，分异指数DI=33.64~62.54，均反映岩浆分异程度高；镁铁比值m/f=0.24~0.53，属于富铁质基性岩；A/CNK=0.57~0.76，A/NK=1.38~2.15，属于准铝质。

稀土元素成分及特征参数见表19-2。ΣREE=62.63×10^{-6}~177.91×10^{-6}，LREE/HREE=6.97~18.64，$(La/Yb)_N$=6.58~44.47，稀土总量较高，轻稀土相对富集，重稀土相对亏损；δEu=0.93~1.27，极弱铕正或负异常；配分曲线为右倾光滑曲线（图19-1）。

微量元素成分及特征参数见表19-3。与中基性侵入岩均值相比，Zr、Ti、Co、Ni含量较低，Sc、V、Y基本相当，Th、Sc、Ba、Cr含量略高；与上地幔岩石相比，Ba、Y、Sc、V相对富集，Cr、Co、Ni相对亏损。微量元素蛛网图（图19-2）上，强烈富集大离子亲石元素K、Ba，相对亏损高场强元素P、Nb、Zr。

3. 变质变形特征

该套岩系普遍发育条带状或强片麻状构造，变形作用较强烈。从岩石矿物特征上看，经历了三次重要变质变形作用。早期发育麻粒岩相变质变形，形成辉石链及辉石条带，残留体或残留辉石形态予以揭示。后期广泛发育退变质角闪岩相变质变形，新生的角闪石及黑云母定向排列，并形成线理，宏观上形成层厚约1~10cm似层状。在变形强带上，辉石等早期麻粒岩相矿物组合全部退变质为角闪岩相矿物组合。晚期发育钾质混合岩化现象，尤以密云苍术会和苇子峪、房山周口店岩体为典型。含量不均的钾质和硅质成分呈不规则状分布，与主岩为过渡关系，含量少时岩石为灰黑色，含量多时为红色。岩石中暗色矿物含量随混合岩化程度不同而变化，局部混合质成分可占60%以上。

4. 原岩恢复

上述地质产状、岩石学和岩石地球化学等特征，特别是残留的原生结构、构造，揭示了该变质岩原岩为正变质岩。在变质岩原岩恢复Si-[(al+fm)-(c+alk)]（图19-3）和Si-Mg（图19-4）图解上，成分点也落在火成岩区内。

5. 岩石系列

该套岩系虽富碱（绝大多数$Na_2O+K_2O>5\%$）且$Na_2O>K_2O$，具有和黑卧岩体相似特征，但岩石贫钛，且碱性长石和碱性暗色矿物较少；$\sigma=1.28\sim5.05$，绝大多数<4，为亚碱性系列；在*A-F-M*图解（图19-7）上，成分点位于钙碱性系列区。岩石*A/CNK*<1，为准铝质。因此，该套岩系为准铝质钙碱性玄武岩系列。

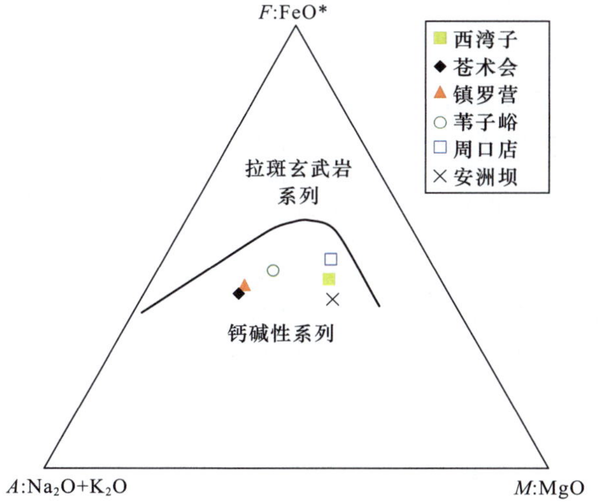

图19-7 火成岩*A-F-M*图解

Fig. 19-7 *A-F-M* diagram of igneous rocks

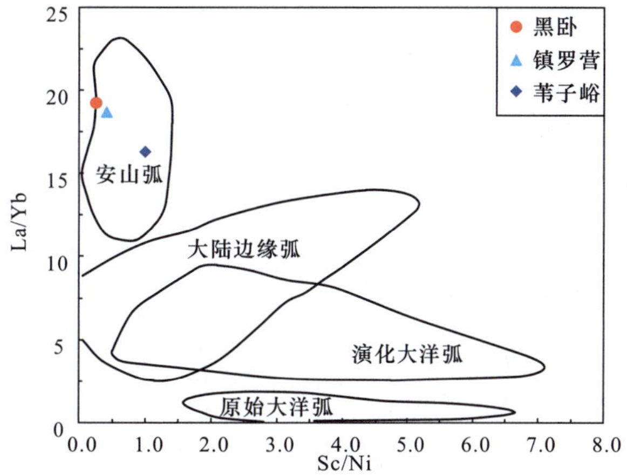

图19-8 Sc/Ni-La/Yb构造环境判别图解

Fig. 19-8 Sc/Ni-La/Yb diagram of tectonic discrimination

6. 成因类型

该套岩系地球化学特征同黑卧岩体极度相似，可能为同源岩浆演化。岩石化学上呈低SiO_2高CaO特征，Na_2O/K_2O比值较大，属于徐克勤（1982）划分的幔源型。*SI*<40，*DI*值高，反映幔源原生岩浆还经历了同化混染作用或岩浆分异作用。岩石LI值低；稀土配分曲线图（图19-1）中LREE富集，HREE相对亏损；微量元素蛛网图（图19-2）中LILE富集，HFSE（Nb、Zr）相对亏损；这些特征又与俯冲环境下受到俯冲板片混染的富集地幔具有类似性。因此，该套岩系岩浆来源于富集型上地幔且受到地壳物质的同化混染。

7. 构造环境分析

该套岩系为钙碱性玄武岩系列，且表现出低钾、钛，高镁特征；富集大离子亲石元素，相对亏损高场强元素和重稀土元素；在$10TiO_2-Al_2O_3-10K_2O$（图19-5）构造环境判别图解上，成分点位于岛弧、造山带玄武岩区内，这些特征反映其侵位于岛弧环境。构造环境图解（图19-8）判别其侵位于洋-陆俯冲型活动大陆边缘安第斯型岛弧环境，而$TiO_2<1.2\%$、低K_2O含量，揭示其可能发育为不成熟的岛弧，与俯冲大洋板片重熔交代地幔楔部分重熔有关。

8. 时代归属

本期岩浆侵位时代厘定为新太古代晚期，主要依据包括：①岩体遭受高级麻粒岩相变质，部分岩体还经后期角闪岩相退变质；②岩浆分异形成的条带经受褶曲复杂变形；③岩体内发育大量变质基性、酸性岩脉和表壳岩密云岩群包体；④多个岩体被中元古代长城系不整合覆盖；⑤苍术会岩体中、基性岩锆石 U-Pb 法和 t_{DM} 法同位素年龄分别为 2486Ma 和 2764Ma（金文山等，1999），锆石 SHRIMP 法同位素年龄为（2526±6）Ma；茅子峪岩体基、中性岩锆石 U-Pb 和 t_{DM} 同位素年龄分别为（2365±37）Ma 和 2745Ma（金文山等，1999）；周口店（官地，亦称关坻）岩体（宋鸿林称为“关坻杂岩”）锆石 SHRIMP 法同位素年龄为（2512±20）Ma（颜丹平等，2005）。

二、古元古代超基性—基性侵入岩

（一）地质产状与岩石学特征

区内古元古代超基性—基性岩体数量多、规模小，单个岩体出露面积<0.1km²，广泛分布于古老变质岩中，以密云放马峪、平顶山、东老窝、四合、潮河主坝、白河涧、蔡家洼和延庆红石湾为典型代表。岩体岩石类型复杂，可划分为三大类（表 19-4）。

表 19-4 古元古代超基性—基性侵入岩岩石类型

Table 19-4 Rock types of Paleoproterozoic intrusive rocks

岩石类型		岩性描述	岩体
橄榄岩类	纯橄榄岩、辉橄岩	灰白色、灰绿色、浅灰黑色等，残余变晶结构，块状构造，由橄榄石和辉石组成，多数叶蛇纹石、纤维蛇纹石化，保留橄榄石、辉石晶形假象，极少量辉石不规则残留。副矿物呈自形半自形粒状，粒径一般 0.3~0.6mm，少数<0.2mm，部分表现为暗色矿物蚀变析出物，常沿橄榄石、辉石残晶边缘及解理裂隙处分布，或被暗色矿物假像包裹，含量 1%~3%	放马峪、平顶山、东老窝
辉石岩类	橄榄斜方辉石岩、二辉石岩、橄榄二辉石岩、角闪斜方辉石岩	黑灰、绿黑或黑褐色，细-中粒等粒状结构、嵌晶结构、残余变晶结构，块状构造。橄榄斜方辉石岩中橄榄石呈星散状分布；辉石主要为半自形的顽火辉石及古铜辉石，均被蛇纹石化后呈半自形残晶，包裹近圆形溶蚀状橄榄石呈包含结构。二辉石岩由斜方辉石（顽火辉石、古铜辉石、紫苏辉石，含量约 45%）及具希列构造的单斜辉石（普通辉石、易变辉石、透辉石，含量约 40%）组成；呈短柱状的辉石构成自形半自形粒状镶嵌结构，含少量橄榄石，含量<3%。角闪斜方辉石岩以普通辉石和角闪石为主，含少量黑云母、石榴子石、磁铁矿等	放马峪、红石湾、四合、潮河主坝、白河涧
角闪石岩类	角闪石岩	黑色，粒状变晶结构，块状构造。主要矿物角闪石，淡黄-草绿色，他形粒状，多色性，不均匀消光，56°解理，正交镜下呈二级黄-二级绿干涉色，不均匀消光，粒径一般 2~4.5mm，个别<0.7mm，弱绿泥石化，含量>95%。次要矿物包括磁铁矿、黑云母和斜长石。磁铁矿黑色，粒状粒径 0.2~0.5mm，含量<1%；黑云母红褐-无色，片状，多色性，粒径 0.2~1.0mm，含量约 1%；斜长石他形粒状，分布在角闪石颗粒的空隙内，粒径<0.1mm，含量约 2%	蔡家洼

1. 放马峪岩体

分布于密云放马峪村东山梁上，由东西两个毗连小岩瘤组成，单个岩体平面呈透镜状，剖面呈楔状，走向 NE 65°，倾向 NW 65°~70°（图 19-9）。岩体侵入新太古代沙厂英云闪长质片麻岩中并以极小角度切割围岩片麻理。西岩体以二辉石岩为主，内岩浆分异辉橄岩等异离体。东岩体以橄榄岩为主，由中心向边缘基性程度连续降低，划分为：①纯橄榄岩相（β_1），已全蛇纹石化；②斜辉橄榄岩相（β_2），蛇纹石化，有辉石残晶；③橄榄辉石岩相（β_3），橄揽石呈浸染状；④辉石岩相（ω），由斜方辉石岩及单斜辉石岩组成（图 19-9、图 19-10）。

岩体中发育岩浆分异富集形成的条带状铬铁矿体（图 19-9），晚期还发育角闪二辉石岩、辉绿玢岩、煌斑岩脉。二辉石岩中包裹大小不等圆形-椭圆形橄榄岩捕虏体，捕虏体边部发育一层厚约 1~5mm 的锈红色烘烤边。岩体底盘尤其集中，呈现橄榄岩和辉石岩互相包裹的椭球状（2~3cm）构造。

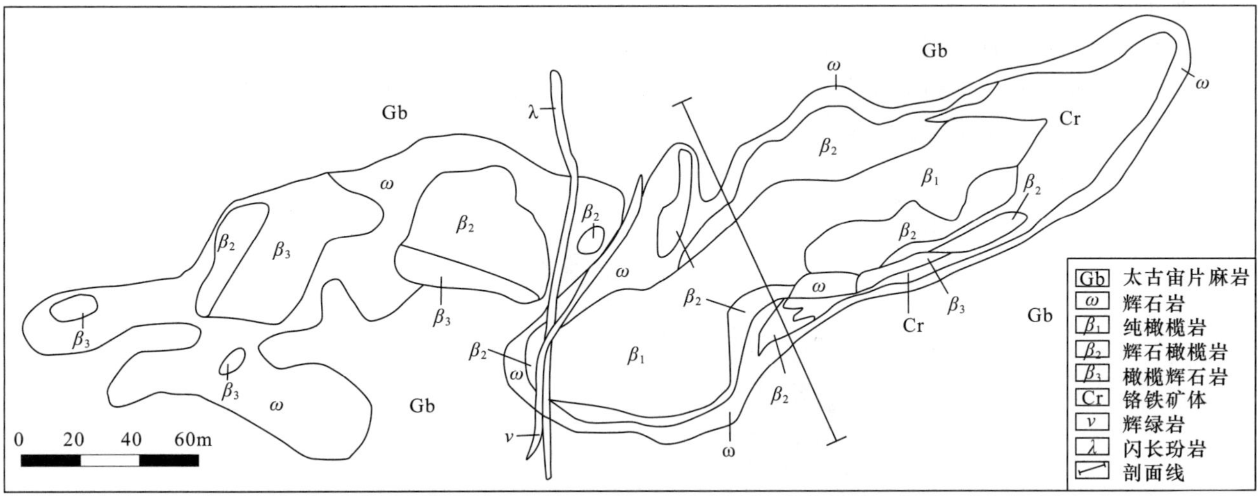

图 19-9 密云放马峪含铬橄榄辉石岩体地质图

（据北京市地质矿产局，1991）

Fig. 19-9 Schematic geological map of Fangmayu intrusion in Miyun

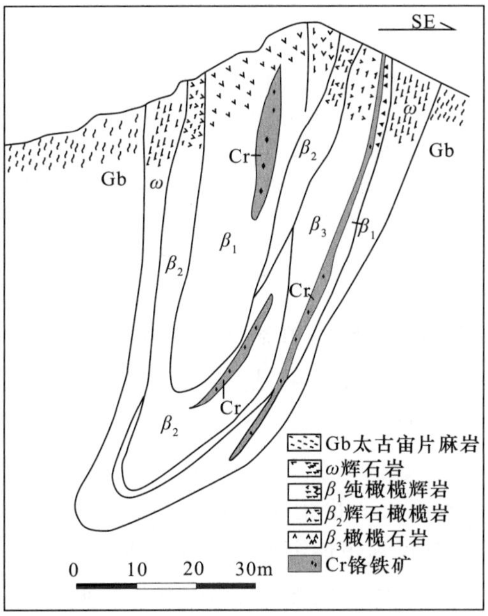

图 19-10 密云放马峪侵入体剖面图

（剖面位置见图 19-9；据北京市地质矿产局，1991）

Fig. 19-10 The geological section of Fangmayu intrusion in Miyun

2. 东老窝岩体

分布于密云东老窝村西，为蝌蚪状岩瘤，出露面积约 0.07km^2。岩体长轴走向约 120°，侵位于太古宙苇子峪辉长闪长质片麻岩中。

岩体可划分为两个相带：西南为 1~3mm 宽辉石岩和橄榄岩相间排列混合岩相带；东北为未分异的纯橄榄岩相。岩体内还发育宽 5~20cm 角闪岩脉，脉体走向与岩体延长方向一致或斜交，倾角平缓。

3. 平顶山岩体

分布于密云水库西南，为长约 335m，宽 105~110m 的岩瘤（图 19-11）。平面呈透镜状，剖面呈扁豆体状。岩体南西被中元古界长城系不整合覆盖，并被更晚期的正长岩脉侵入破坏，其余三面见其顺围岩片麻理侵入表壳岩密云岩群黑云斜长片麻岩。据钻探资料，岩体深度>500m，由地表向深部逐渐变薄直至尖灭。

岩体环形相带发育，划分为中心辉石岩相和边缘橄榄岩相（图 19-11）。橄榄岩相岩性复杂，包括方辉橄榄岩、橄榄岩、单斜橄榄岩、二辉橄榄岩、纯橄岩。辉石岩相由斜方辉石岩及单斜辉石岩组成。岩体中发育岩浆分异富集形成的铬铁矿体（图 19-11）。

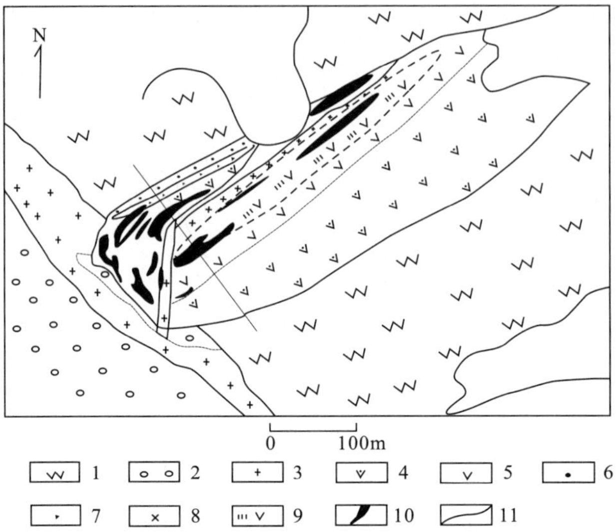

图 19-11　平顶山岩体地质图

（据周永璋等，1988）

Fig. 19-11　The geological sketch map of Pingdingshan intrusion

1—黑云斜长片麻岩；2—石英岩；3—正长岩脉；4—橄榄岩；5—方辉橄榄岩或橄榄岩；6—单斜橄榄岩；7—二辉橄榄岩；8—纯橄岩；9—辉石岩；10—铬铁矿；11—地质界线

4. 潮河主坝岩体

分布于密云水库潮河主坝，为多条岩脉。单条脉宽 1～4m，长 $n\times10$m～$n\times100$m。岩脉主要侵入于宋营子英云闪长质片麻岩中，在潮河主坝北还可见其穿切康各庄奥长花岗质片麻岩脉（图 19-12）。岩石类型为石榴角闪斜长辉石岩和角闪石岩。

5. 四合岩体

分布于密云贺家沟水库南-马家峪，平面上呈脉状，长约 2000m，宽 20～100m。走向北东 50°，倾向北西，倾角 50°～65°，自北东至南西倾角逐渐变陡；剖面呈楔形，最大延深约 210m。岩脉侵入于沙厂英云闪长质片麻岩中，侵入接触关系界线截然，边部具数厘米冷凝边，并见棱角状变质岩捕虏体（图 19-13）。岩脉南西弱分异，为橄榄辉石岩和角闪二辉石岩；北东基本无分异，为角闪二辉石岩，局部为橄榄辉石岩异离体。

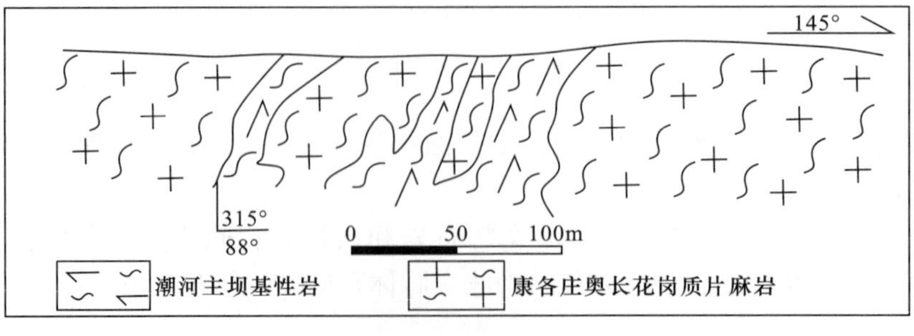

图 19-12　潮河主坝基性岩墙与康各庄奥长花岗质片麻岩侵入接触关系

（据鲍亦冈等，2001）

Fig. 19-12　Geological sketch showing the intrusive contact between Chaohe Dam basic dyke swarm and Kanggezhuang trondhjemitic gneiss intrucsion

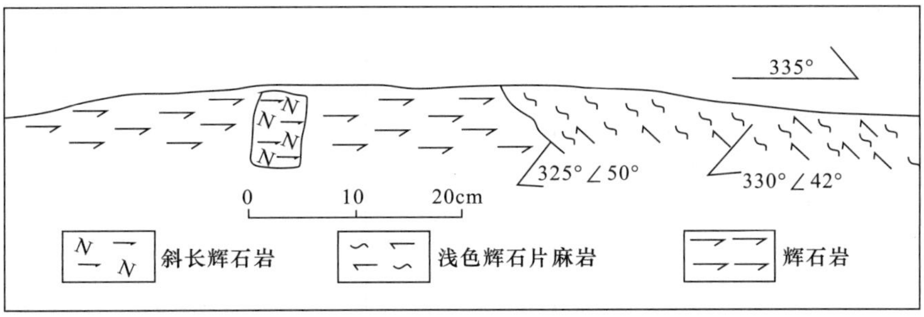

图 19-13　贺家沟北辉石岩岩脉与太古宙变质岩侵入接触关系

Fig. 19-13　The geological sketch showing the intrusive contact between pyroxenite vein and Archean gneiss in Hejiagoubei

6. 红石湾岩体

分布于延庆红石湾村，为长约 550m、宽约 350m 岩枝，平面呈椭圆形，长轴北西向展布。岩体北西被中元古界长城系地层不整合覆盖，钻探揭示风化壳特征清晰（图 19-14），其余三面见其顺围岩片麻理或斜切围岩片麻理侵入于太古宙宋营子英云闪长质片麻岩。

岩体侵位环形相带发育，划分为中心角闪斜方辉石岩相和角闪单斜辉石岩相（图 19-14），中心相中发育呈稀疏或稠密浸染状含铂钯的铜镍金属硫化矿物。

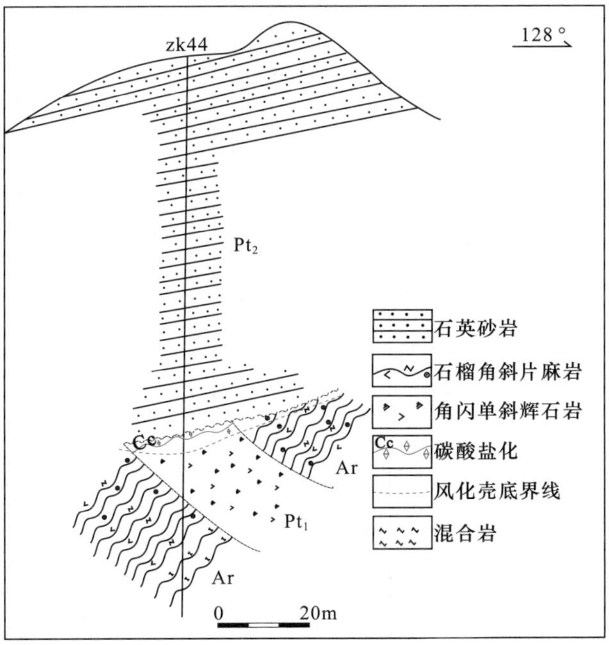

图 19-14　延庆红石湾侵入体地质剖面图

（据北京市地质矿产局，1991）

Fig. 19-14　The geological section of Hongshiwan intrusion in Yanqing

（二）岩石化学与地球化学特征

岩石化学及特征参数见表 19-5。SiO_2 = 31.69～45.96，属于超基性—基性岩类。多数岩体里特曼指数 $\sigma<4$，$Na_2O>K_2O$，表明岩石为钠质亚碱性岩系列；固结指数 SI = 13.87～72.7，分异指数 DI = 0.98～34.58，不同岩体岩浆分异程度不同。同一岩体，岩浆分异形成岩石岩浆演化序列，岩石化学成分由超基性向基性演化，如放马峪岩体由中心向边部 SiO_2 成分逐渐增多而 MgO 成分渐减少，Mg^{2+}/Fe^{2+} = 4.03～4.9，向边部降低；镁铁比值 m/f = 0.08～1.32，属于富铁质超基性或基性岩；A/CNK = 0.47～0.67，A/NK = 1.99～13.36，属于准铝质。

表 19-5　古元古代超基性岩—基性岩岩石化学成分及特征参数一览表

Table 19-5　The chemical composition and main parameters of Paleoproterozoic intrusive rocks

序号	岩体	岩性	氧化物含量/%														特征参数					
			SiO_2	TiO_2	Al_2O_3	Fe_2O_3	FeO	MnO	MgO	CaO	Na_2O	K_2O	P_2O_5	Cr_2O_3	Los	总量	*DI*	*SI*	σ	*m/f*	*A/CNK*	*A/NK*
1	潮河主坝	含石榴角闪斜长辉石岩	45. 96	2. 80	13. 04	11. 87	8. 04	0. 28	3. 79	7. 95	3. 55	0. 67	0. 96			98. 91	34. 58	13. 87	4. 92	0. 36	0. 62	1. 99
2	放马峪	橄榄岩	36. 63	0. 12	1. 41	6. 92	5. 98	0. 11	33. 64	0. 99	0. 15	0. 05	0. 04		13. 08	99. 10	1. 81	72. 7	−0. 28	4. 87	0. 67	4. 74
3	四合	辉石岩	44. 91	0. 21	2. 73	6. 65	5. 90	0. 09	26. 96	1. 99	0. 21	0. 09	0. 09		9. 29	99. 12	2. 58	68. 44	0. 02	4. 01	0. 67	6. 17
4	白河涧	辉石岩	45. 02	0. 60	5. 72	5. 06	6. 27	0. 21	26. 95	5. 60	0. 42	0. 09	0. 04		2. 60	98. 58	4. 27	69. 97	0. 07	4. 35	0. 52	7. 26
5	蔡家洼	角闪石岩	41. 47	1. 12	11. 13	4. 81	8. 71	0. 34	14. 56	10. 91	2. 02	0. 70	0. 34		2. 42	98. 53	15. 96	47. 48	0. 44	1. 94	0. 47	2. 73
6	红石湾	辉石岩	45. 05	0. 72	8. 27	6. 02	7. 62	0. 20	18. 90	7. 22	1. 38	0. 83	0. 17		4. 15	100. 50	17. 28	54. 8	1. 37	2. 55	0. 51	2. 61
7	平顶山	橄榄岩	31. 69	0. 07	1. 93	8. 51	8. 17	0. 16	26. 58	1. 84	0. 04	0. 08	0. 21	8. 29	3. 57	91. 15	0. 98	62. 1	−0. 01	2. 96	0. 55	13. 36

注：2、6、7 为均值；1-5 数据引自 1 : 5 万密云幅、高岭幅、墙子路幅区域地质调查报告，1994；6 数据引自 1 : 5 万永宁幅、四海幅区域地质调查报告，1999；7 数据引自周永璋等，1988。

表 19-6　古元古代超基性岩—基性岩稀土元素成分及特征参数一览表

Table 19-6　REE contents and main parameters of Paleoproterozoic intrusive rocks

序号	岩体	岩性	稀土元素含量/10^{-6}																		特征参数			
			La	Ce	Pr	Nd	Sm	Eu	Gd	Tb	Dy	Ho	Er	Tm	Yb	Lu	Y	ΣREE	LREE	HREE	LREE/HREE	$(La/Yb)_N$	δEu	δCe
1	潮河主坝	含石榴角闪斜长辉石岩	58. 00	120. 00	16. 00	62. 00	15. 00	4. 20	10. 80	1. 60	9. 40	1. 86	4. 80	0. 66	3. 20	0. 38	44. 00	307. 90	275. 20	32. 70	8. 42	13. 00	0. 96	0. 95
2	白河涧	角闪石岩	44. 00	120. 00	21. 00	76. 20	26. 00	2. 40	19. 40	3. 60	18. 40	3. 80	10. 00	1. 46	6. 80	0. 88	100. 00	353. 94	289. 60	64. 34	4. 50	4. 64	0. 31	0. 96
3	红石湾	辉石岩	25. 74	52. 58	6. 70	26. 78	5. 41	1. 63	5. 04	0. 60	3. 32	0. 67	1. 76	0. 24	1. 52	0. 16	17. 44	132. 11	118. 82	13. 29	8. 94	12. 18	0. 94	0. 96
4	放马峪	橄榄岩	1. 43	3. 00	0. 40	1. 61	0. 47	0. 61	0. 40	0. 06	0. 31	0. 06	0. 16	0. 02	0. 12	0. 03	1. 50	8. 68	7. 52	1. 15	6. 53	8. 34	4. 18	0. 96
5	四合	辉石岩	1. 20	2. 50	0. 34	0. 90	0. 21	0. 05	0. 21	0. 12	0. 19	0. 12	0. 12	0. 02	0. 10	0. 03	0. 97	6. 09	5. 20	0. 90	5. 80	9. 06	0. 65	0. 95

注：数据来源同表 19-5。

稀土元素成分及特征参数见表19-6。ΣREE=6.09×10^{-6}~353.94×10^{-6}，橄榄岩-辉石橄榄岩-辉石岩-角闪辉石岩-角闪石岩ΣREE逐渐增大。LREE/HREE=4.50~8.94，$(La/Yb)_N$=4.64~13.00，轻稀土极弱富集，重稀土相对弱亏损；放马峪超基性岩δEu=4.18呈正异常外，其余岩体δEu=0.31~0.96，呈铕负异常，尤以角闪石岩呈明显铕负异常，揭示超基性岩向基性岩演化过程中源区可能发生斜长石分离结晶；稀土元素配分曲线为略右倾光滑曲线（图19-15）。

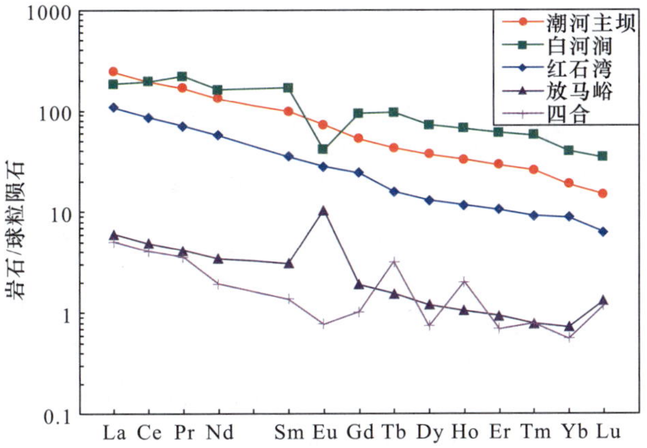

图19-15　古元古代侵入岩稀土元素配分模式图

Fig. 19-15　Chondrite-normalized REE patterns of Paleoproterozoic intrusive rocks

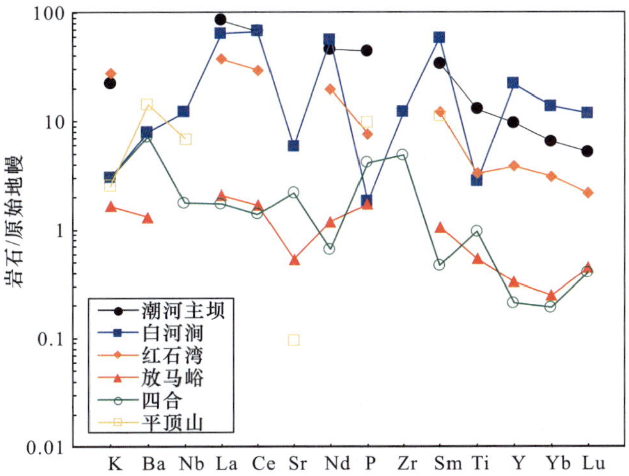

图19-16　古元古代侵入岩微量元素蛛网图

Fig. 19-16　Trace element spider diagram of Paleoproterozoic intrusive rocks

微量元素成分与维氏（1962）超基性岩元素含量相比，普遍Pb、Ta含量较高，橄榄岩或辉石岩Cr、Cu、Co、Ni等元素岩浆分异富集成矿，含量明显偏低，而角闪石岩正好相反。微量元素蛛网图（图19-16）上，各岩体图形型式各有不同，也揭示了岩浆侵位经历了复杂岩浆分异。

（三）原岩恢复

古元古代超基性岩普遍遭受强烈的蛇纹石化，众多岩石薄片中几乎见不到变余橄榄石、斜方辉石残晶，单斜辉石残晶也少见（周永璋等，1988）。岩石地球化学组成和深部发现的含硼水镁石等揭示其是超基性岩类。

除岩浆分异演化形成不同岩石类型外，古元古代超基性岩缺乏蛇绿岩套伴生的其他岩性组合；岩体侵位特征明显，如发育岩浆分异相带，而缺乏典型科马奇岩结构、构造；在Al_2O_3-CaO-MgO图解（图19-18）上，成分位于超镁铁质或镁铁质堆积岩区；重力分异在成矿作用中起到重要作用，且矿石在岩浆上侵过程中或就位后先行晶出（周永璋等，1988）；伴生铬尖晶石富铁，成分同球粒陨石相近，而不同于蛇绿岩套变质橄榄石、布什维尔德层状超镁铁质及碱性玄武岩中橄榄石包体中铬尖晶石。因此，古元古代超基性岩属于超镁铁质或镁铁质堆积岩。

（四）岩石系列

岩石地球化学参数揭示古元古代超基性岩为钠质亚碱性系列，$K_2O<1$，*A-F-M*图解（图19-17）进一步厘定其属于拉斑玄武岩系列。

（五）Sm-Nd同位素特征

放马峪超基性岩全岩Sm=0.303×10^{-6}~1.128×10^{-6}，Nd=0.503×10^{-6}~6.545×10^{-6}，$^{143}Nd/^{144}Nd$=0.511286~0.511760，$^{147}Sm/^{144}Nd$=0.098217~0.134156（陈森煌等，1991）。$\varepsilon_{Nd}(t)$=-0.27~1.37（*t*=2100Ma），I_{Nd}=0.509904~0.509988，与球粒陨石初始值接近。

（六）成因类型与岩浆演化

该套岩系岩浆来源于原始上地幔，可能是原始地幔岩部分熔融析出榴辉岩成分后的地幔难熔残留

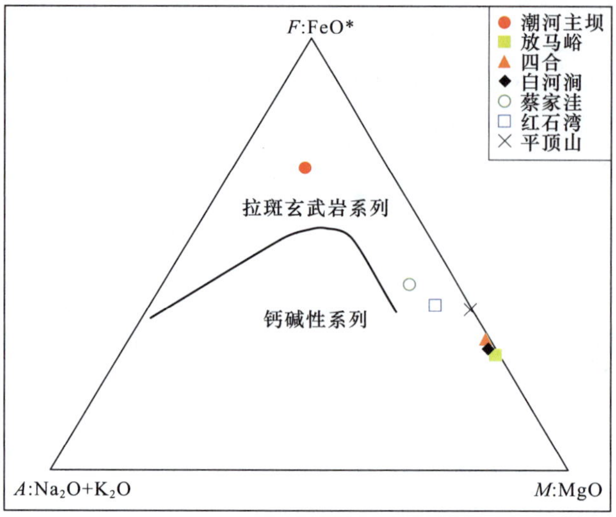

图 19-17　古元古代侵入岩 *A-F-M* 图解

Fig. 19-17　*A-F-M* ternary diagram of Paleoproterozoic intrusive rocks

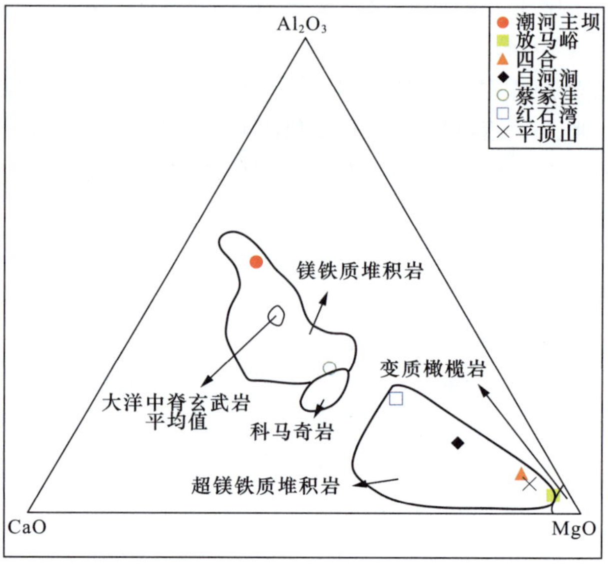

图 19-18　古元古代侵入岩 Al_2O_3-CaO-MgO 图解

Fig. 19-18　Ternary diagram of Al_2O_3-CaO-MgO of Paleoproterozoic intrusive rocks

体进一步熔融的产物（周永璋等，1988）。主要依据包括：①Sm-Nd 同位素与球粒陨石初始值接近；②岩浆演化趋势向贫铁富镁的方向演化；③岩石富铝、铁，贫碱，这与洋中脊深海拉斑玄武岩富镁低铁及低硅富碱的特点是不同的，也不同于岛弧拉斑玄武岩；④REE 配分模式与原始地幔岩特征相似或一致；⑤以岩脉或岩枝侵位，未形成规模较大的岩浆岩体，且顺着围岩片麻理或呈小角度斜切围岩片麻理侵位，有些边部发育冷凝边并发育明显分异条带。岩石具有低 SiO_2、高温、含大量挥发组分和较低黏度，具有强力快速侵位特征。基于单斜辉石化学成分采用麦瑟尔法计算岩体形成温度 $T=1305℃$，估算形成压力 $P=10kPa$（周永璋等，1988）。

古元古代岩体侵位顺序为橄榄岩-辉石岩-角闪石岩，岩石向贫铁，富镁、硅的方向演化。由早至晚岩石中组成暗色矿物的硅原子数比例逐渐增高，岩石类型由硅不饱和向硅过饱和方向发展，而岩石中 Mg/Fe 比值逐渐降低，由早期的富铁质超基性岩演化为富铁质基性岩。由早至晚，岩浆的分异指数也逐渐增大，表明岩浆向分异程度增高的方向演化。在微量元素中，整个古元古代侵入岩普遍富集 Sn、Pb、Ta、Be、Ga 等元素，其中以 Sn、Ta 富集程度最高。Co/Ni 值由早期橄榄岩向晚期的角闪石岩逐渐增高，总体反映了岩浆由超基性向基性演化的特征。

（七）构造背景环境

古元古代超基性岩浆来自于上地幔，岩浆具有低 SiO_2、高温、含大量挥发组分和较低黏度和快速侵位特征；岩体规模较小，多呈上大下小尖灭的无根扁豆体产出；地质产状和地球化学不同于洋中脊或深海大洋盆地的“蛇绿岩建造”。因此，该套岩石形成的构造背景环境不可能是古板块边缘地缝合线处上地幔物质上升（郁建华等，1994）。

区内该套岩体沿张家口-平泉深大断裂带发育，地质历史时期该断裂经历多期构造活化。区域拉张构造背景环境下，伴随着区域大规模构造运动发生和发展，在地幔强大的动力驱使之下，上地幔的超基性岩浆完全可沿切壳深大断裂或已愈合的隐伏深大断裂上升侵位。因此，古元古代超基性岩可能是上地幔的一些碎块呈固体塑性状态沿深大断裂被挤压上来的。

（八）时代归属

本期岩浆侵位时代厘定为古元古代早期，主要依据包括：①均赋存于太古宙中-高级变质岩，潮河主坝岩脉侵入到新太古代康各庄奥长花岗质片麻岩，并可见片麻岩捕虏体；②岩石变质作用表现为绿泥石化、蛇纹石化或是后期热液变质，反映最高遭受绿片岩相变质，其变质程度较太古宙变质要低；③岩体中的岩浆分异条带没有构造变形；④延庆红石湾辉石岩脉等多处见中元古界长城系不整合覆盖其上；⑤放马峪岩体全岩 Sm-Nd 法同位素年龄（2010±155）Ma（陈森煌等，1991），K-Ar 法同位素年龄 2000Ma（鲍亦冈等，2001），四合岩脉角闪石 K-Ar 法同位素年龄 1573.9Ma（北京市地质矿产局，1991），延庆红石湾岩脉辉石 K-Ar 法同位素年龄 1253.6Ma（北京市地质矿产局，1991）。K-Ar 法封闭温度低，易受后期热事件影响，使得同位素年龄偏小，岩体侵位时间应在 2000Ma 或更早一些。

三、中元古代基性侵入岩

区内中元古代基性侵入岩规模小，发育 1 处新地斜长岩体。

（一）地质产状与岩石学特征

分布于怀柔新地村，长 15km，宽 300～500m，出露面积 3.5km^2，呈东西向岩脊状（图 19-19）。岩体南被兰营石英正长岩侵入。其余三面主要以断层同其他地质体接触，北部零星可见其高倾角侵入太古代宋营子英云闪长质片麻岩（图 19-20）。

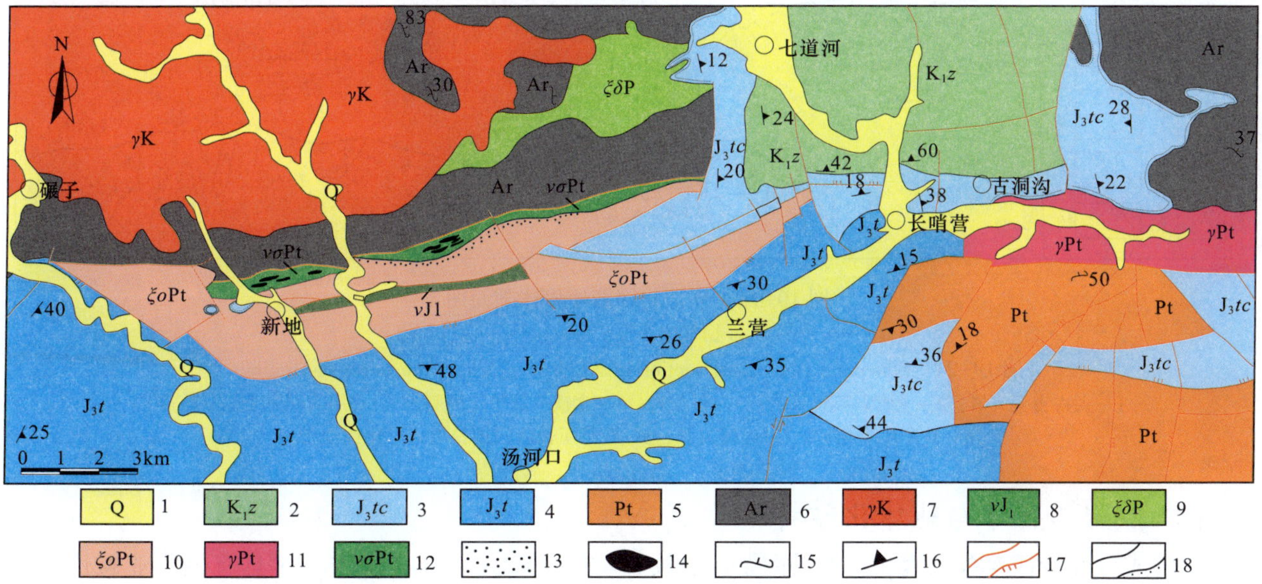

图 19-19 怀柔北部新地岩体地质略图

（据郁建华等，1994）

Fig. 19-19 Schematic geological map of Xindi complex in the north of Huairou

1—第四系；2—下白垩系张家口组；3—上侏罗系土城子组；4—上侏罗系髫髻山组；5—中元古界长城系；6—太古宇；7—早白垩世花岗岩；8—早侏罗世辉长岩；9—早二叠世正长闪长岩；10—中元古代石英正长岩；11—中元古代富钾花岗岩；12—中元古代斜长岩；13—钠长绢云母化蚀变带；14—矿体；15—地质产状；16—片麻理产状；17—断裂；18—地质界线

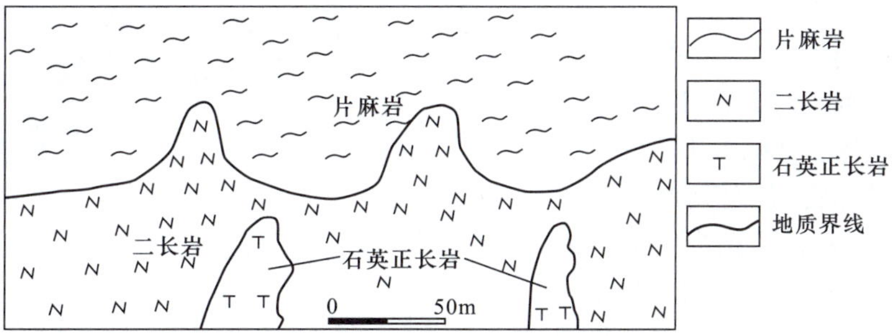

图 19-20 新地村北斜长岩与变质岩接触关系示意图

Fig. 19-20 Geological sketch showing the intrusive contact between Xindi anorthosite and gneiss at the north of Xindi village

岩体虽规模不大，岩浆侵位相带发育，中心发育为巨斑状（3~5cm）堆晶结构斜长岩，边缘则发育为细粒斜长岩。岩石呈白-灰白色，中粗粒自形镶嵌堆积结构。斜长石自形粒状，粒径一般为3~5mm，大者达1cm，具钠长、卡纳及肖钠双晶，强烈钠黝帘石化蚀变，蚀变矿物顺斜长石双晶纹定向排列，类型为中-拉长石（An=30~36），含量80%~85%。暗色矿物分布不均匀，多蚀变被绿泥石、黝帘石及铁质析出物替代，从析出铁质的排列推测可能原矿物为辉石、角闪石，含量5%~15%。岩体内部还发育大量石英正长岩脉、煌斑岩脉和辉长岩脉，脉岩局部钒钛磁铁矿化（图19-21），磁铁矿呈浸染状、稠密浸染状散布于暗色矿物中。此岩体可同河北省大庙斜长岩体相互对比，均沿赤城-长峪营深断裂分布，由于此断裂经历多次活动使岩石强烈破碎。

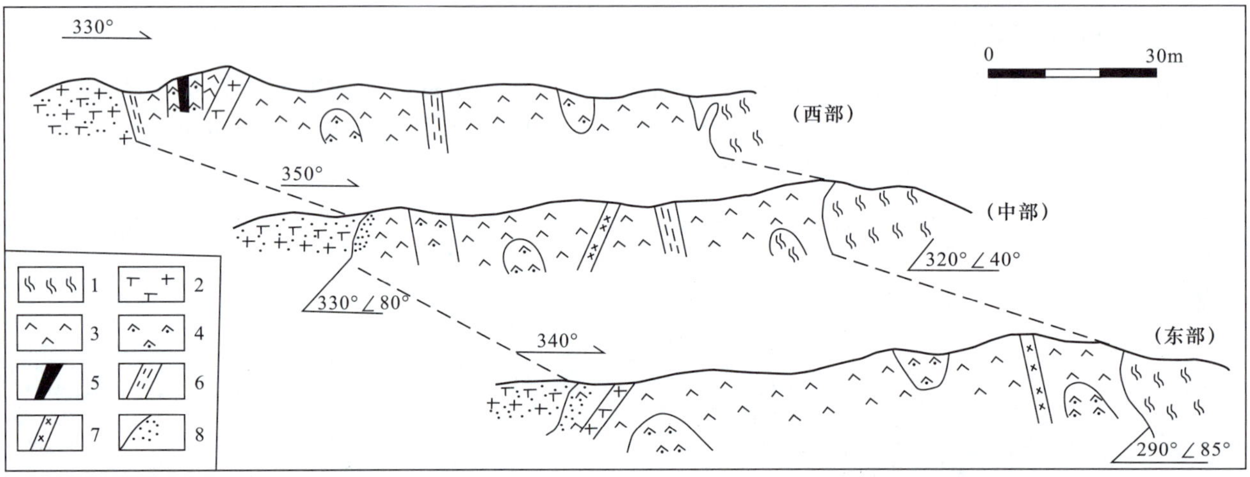

图19-21　怀柔北部新地岩体剖面图
（据郁建华等，1994）
Fig. 19-21　Schenatic Geological section of Xindi complex in the north of Huairou
1—太古宙变质岩；2—兰营石英正长岩；3—斜长岩；4—含铁、钛矿化的斜长岩；5—含钒铁磁铁矿矿体；6—破碎片理化带；7—煌斑岩脉；8—石英正长岩的钠长绢云母化蚀变带

（二）岩石化学与地球化学特征

化学成分及特征参数见表19-7。里特曼指数 $\sigma=3.62\sim3.64$，$Na_2O>K_2O$，具有钠质亚碱性岩特性；固结指数 $SI=10.26\sim10.64$，分异指数 $DI=47.77\sim50.27$，均反映岩浆分异程度高；镁铁比值 $m/f=0.11\sim0.16$，属于富铁质基性岩；$A/CNK=0.49\sim0.58$，$A/NK=0.91\sim0.98$，属于准铝质、过碱质。

稀土元素成分及特征参数见表19-8。$\Sigma REE=100.59\times10^{-6}$，高于大庙斜长岩而低于二长岩，LREE/HREE=9.6，$(La/Yb)_N=13.19$，稀土总量较低，轻稀土相对富集，重稀土相对亏损；$\delta Eu=1.39$，呈弱铕正异常；稀土元素配分曲线为右倾光滑曲线（图19-22）。

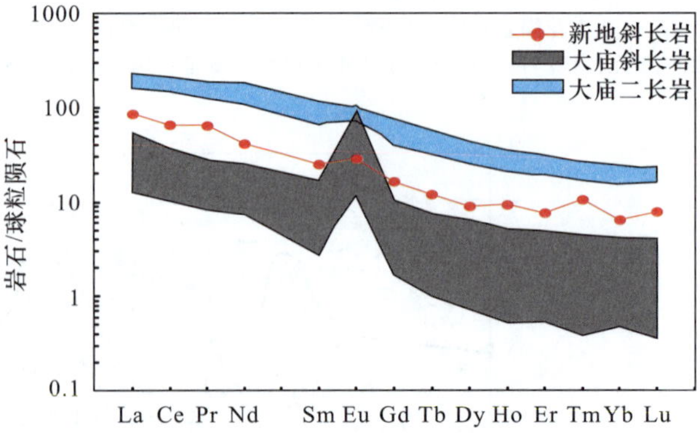

图19-22　中元古代侵入岩稀土元素配分模式图
Fig. 19-22　Chondrite-normalized REE patterns of Mesoproterozoic intrusive rocks

微量元素成分及特征参数见表19-9，同大庙二长岩相似，而略有不同于斜长岩。与维氏基性岩（黎彤等，1962）均值相比，Ba、Sr、Hf、U含量较高，其余元素含量较低；与上地幔岩相比，K含量相当，Rb、Zr含量较低。微量元素蛛网图（图19-23）上，富集大离子亲石元素元素Ba、K，强烈亏损高场强元素Th、U、P和重稀土元素。

表 19-7　中元古代基性岩岩石化学成分及特征参数一览表

Table 19-7　The chemical composition and main parameters of Mesoproterozoic intrusive rocks

序号	岩体	岩性	氧化物含量/%													特征参数				
			SiO_2	TiO_2	Al_2O_3	Fe_2O_3	FeO	MnO_2	MgO	CaO	Na_2O	K_2O	P_2O_5	Los	总和	*DI*	*SI*	σ	*A/CNK*	*A/NK*
1	新地	石英二长岩	52.71	0.8	23.45	2.78	1.88	0.05	1.23	7.98	4.71	1.5	0.25	1.33	98.67	50.26	10.26	3.64	0.49	0.98
2		二长岩	51.49	0.58	23.87	1.81	1.79	0.05	1.13	9.42	4.59	1.35	0.18	2.54	98.80	47.77	10.64	3.62	0.58	0.91
3	大庙	石英二长岩	61.96	0.95	16.17	2.31	4.46	0.11	0.49	3.18	4.10	5.45	0.29	1.84	99.77	76.52	2.91	4.78	0.13	0.88
4		斜长岩	54.19	0.31	26.12	1.77	0.85	0.05	1.35	7.35	5.19	1.18	0.07	1.25	99.68	52.91	13.16	3.47	0.97	1.13
5		二长岩	54.25	1.65	17.19	4.47	6.62	0.19	1.16	4.54	3.84	4.51	0.74	0.96	100.11	62.47	5.63	6.06	0.19	0.88

注：1 数据引自 1∶5 万番字牌幅、汤河口幅区域地质调查报告，1996；2-5 为均值，数据引自谢广轰，2004。

表 19-8　中元古代基性岩稀土元素成分及特征参数一览表

Table 19-8　Trace element contents and main parameters of Mesoproterozoic intrusive rocks

序号	岩体	岩性	稀土元素含量/10^{-6}																		特征参数			
			La	Ce	Pr	Nd	Sm	Eu	Gd	Tb	Dy	Ho	Er	Tm	Yb	Lu	Y	ΣREE	LREE	HREE	LREE/HREE	$(La/Yb)_N$	δEu	δCe
1	新地	二长岩[1]	20.22	40	6.09	19.29	3.83	1.67	3.38	0.45	2.29	0.53	1.27	0.27	1.1	0.2	12.73	100.59	91.10	9.49	9.60	13.19	1.39	0.87
2	大庙	二长岩[2]	41.68	92.39	12.05	54.81	11.16	5.37	9.33	1.32	7.00	1.30	3.58	0.48	3.06	0.46	36.10	243.98	217.46	26.52	8.20	9.78	1.57	1.00
3		斜长岩[2]	4.47	8.23	0.93	3.67	0.57	0.93	0.47	0.06	0.29	0.06	0.14	0.02	0.14	0.02	1.69	19.99	18.80	1.19	15.84	23.45	5.33	0.94

注：1 数据引自 1∶5 万番字牌幅、汤河口幅区域地质调查报告，1996；2-3 为均值，数据引自谢广轰，2004。

（三）岩石系列

新地斜长岩富 Fe、Al，贫 Mg、碱，CIPW 标准矿物计算中出现刚玉，显示钙碱性特征；$\sigma=3.62\sim3.64$，也显示为亚碱性岩系列，这与大庙斜长岩不同。大庙斜长岩在不同岩石系列判别图解上，成分点分散于碱性和亚碱性区内，并长期被认为是碱性岩系列（郁建华，1994）。谢广轰（2004）研究认为大庙斜长岩不同判定方法获得岩石系列得到不同结果，这与 Emslie（1973）统计的北欧和北美岩体型斜长岩相似，究其原因可能与岩体型斜长岩的堆积成因有关。斜长岩中碱质 Fe、Mg 较高，使得相关图解向碱质端元偏移，并不能代表母岩浆特性。岩体斜长岩类较二长岩类相比，铕正异常更加明显，稀土总量也明显偏低，较好显示堆晶作用影响。因此，新地斜长岩同大庙斜长岩属于钙碱性岩系列。在 An-Ab’-Or 图解（图 19-24）中，成分位于普通钙碱性系列。

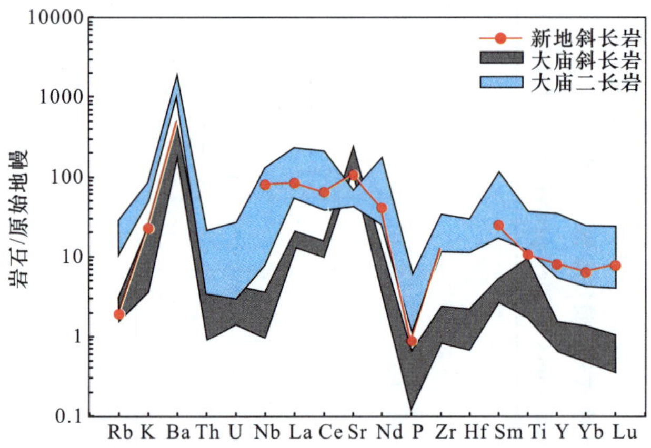

图 19-23 中元古代侵入岩微量元素蛛网图

Fig. 19-23 Trace element spider diagram of Mesoproterozoic intrusive rocks

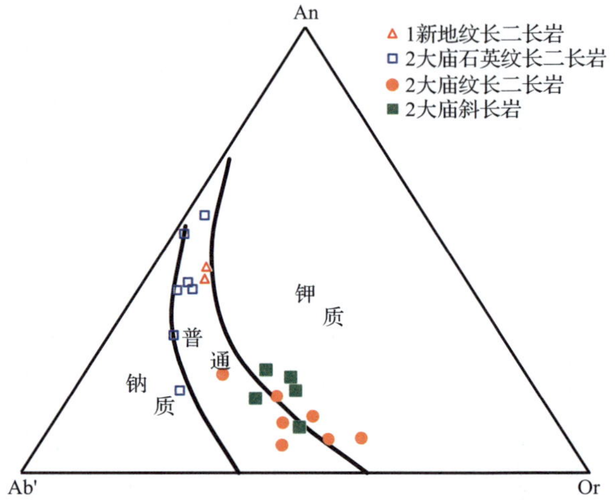

图 19-24 中元古代侵入岩 An-Ab’-Or 图解

Fig. 19-24 An-Ab’-Or ternary diagram of Mesoproterozoic intrusive rocks

表 19-9 中元古代基性岩微量元素成分及参数一览表

Table 19-9 Trace element contents and main parameters of Mesoproterozoic intrusive rocks

序号	岩体	岩性	微量元素含量/10^{-6}									特征参数	
			Rb	Ba	Th	U	Nb	Sr	Zr	Hf	Ta	Rb/Sr	Nb/Ta
1	新地	二长岩	4.50	1165.00	—	—	20.00	780.00	48.00	—	—	0.01	—
2	大庙	二长岩	45.17	3256.51	0.27	0.08	16.85	409.81	76.20	2.00	0.65	0.11	26.04
3		斜长岩	5.29	816.91	0.07	0.02	0.46	1234.66	6.20	0.17	0.03	0.004	17.55

注：1 数据引自 1：5 万番字牌幅、汤河口幅区域地质调查报告，1996；2-3 为均值，数据引自谢广轰，2004。

（四）Sr-Nd 同位素

新地斜长岩同大庙斜长岩特征一致。大庙斜长岩类全岩 $Rb=5.12\times10^{-6}\sim64.30\times10^{-6}$，$Sr=359.40\times10^{-6}\sim1169.25\times10^{-6}$，$Sm=0.53\times10^{-6}\sim11.42\times10^{-6}$，$Nd=3.61\times10^{-6}\sim66.27\times10^{-6}$，$^{87}Rb/^{86}Sr=0.01268\sim0.38520$，$^{143}Nd/^{144}Nd=0.511138\sim0.511414$，$^{147}Sm/^{144}Nd=0.08792\sim0.11182$（谢广轰，2004）。同位素与上地幔接近，而 ε_{Nd}（t）$=-4.76\sim-5.03$（t 取 1757Ma），$I_{Nd}=0.08792\sim0.11182$，$I_{Sr}=0.704327\sim0.713307$，揭示了上地幔源区处于亏损状态或经历了地壳混染。

（五）成因类型与岩浆演化

新地斜长岩岩浆来源于原始上地幔，主要依据包括：①主要造岩矿物为中-拉长石、辉石、角闪石、磁铁矿和钛铁矿，具有明显玄武岩特征；②地球化学成分富 Fe 贫 Mg、碱；③稀土配分曲线和微量元素蛛网图几乎平行于大庙斜长岩，显示其为同源岩浆演化。大庙岩体 Sr-Nd 同位素与上地幔接近，而 $\delta^{34}S=0.9‰\sim2.4‰$，$^{34}S/^{32}S=22.14‰\sim22.1‰$，也和陨石硫非常接近（郁建华等，1994）。

上述大庙岩体 Sr 呈负异常且 Sr 元素存在较大差异，斜长岩类呈现明显正异常而二长岩类则出现弱亏损，反映地壳物质的混染或岩浆发生了斜长石的分离结晶作用。而其上述岩石学特征揭示其具有深源浅侵位特征，且在岩浆房内经结晶分异作用，携有高压矿物巨晶的粥状熔体侵位到上地壳进一步在重力作用下结晶分异（谢广轰，2004）。

（六）构造环境分析

新地及大庙斜长岩出现于华北克拉通的边缘，赋存于变质地层中，成带状分布，受控于区域性深大断裂带，这一特征揭示其形成于古元古代华北板块完成克拉通化并固结稳定后，这和该期构造热事件形成时间至少滞后于变质基底（太古宙变质岩）的年龄 600~800Ma 相吻合。新地及大庙斜长岩还在时空上与形成于拉张环境下的密云花斑花岗岩密切共生，构成双峰式岩浆活动。Emslie（1991）首次将这些在时空上密切共生的岩石组合称为 AMCG 组合，即斜长岩-纹长二长岩-紫苏花岗岩-花岗岩组合；它们共同形成于元古宙，具拉张性质，最后可发育为大洋的裂谷或夭折的大陆裂谷。这种构造环境可能与洋中脊相类似且处于早期或慢扩张的条件下（谢广轰，2004），而在克拉通化之前则不具备产生这种构造-热事件的前提条件。因此，新地和大庙斜长岩形成于非造山构造背景环境，为中元古代燕山裂陷槽早期慢性扩展条件下岩浆活动产物。

（七）时代归属

新地岩体未遭受高级变质和复杂褶曲变形，具有和大庙岩体相似的地质产状、岩石学及地球化学特征，为大庙岩体同期岩浆产物。虽然尚未取得该岩体准确同位素测年资料，但其侵入到新太古代宋营子英云闪长质片麻岩中，又被稍晚兰营石英正长岩侵入。根据兰营石英正长岩锆石 U-Pb 同位素年龄（1790.9±24）Ma（郁建华等，1996）或（1697.3±0.9）Ma（Mortensen，1996），将新地岩体归属为中元古代早期。

四、古生代中基性侵入岩

（一）地质产状与岩石学特征

区内古生代中基性侵入岩均分布于赤城-古北口断裂以北，以怀柔喇叭沟门下窝铺岩体、李家湾岩体为主，还有老米沟和对角沟门 2 个小岩体，出露面积 84km^2。岩体组成岩性及其接触关系复杂，意味着本身可能为短时间多期岩浆侵位的杂岩体。岩石矿物定量统计类型为辉长岩-闪长岩组合（郁建华等，1994），具体岩石特征见表 19-10。

1. 下窝铺岩体

分布于怀柔喇叭沟门下窝铺村，呈 NWW 向、长 13km、宽 2.5~3km 长条形，出露面积约 33km^2。岩体南、北顺围岩片麻理或小角度斜交片麻理侵入太古宙宋营子英云闪长质片麻岩，接触界线模糊，部分发育为条带状脉体呈梳状贯入围岩的渐变接触界线，部分发育为深浅相间的条带过渡型接触界线；西部被燕山期酸性岩侵入接触并在接触带产生热变质的粗晶角闪石化及长英质脉体的贯入交代现象；东部被上侏罗统土城子组及白垩系张家口组不整合覆盖。岩体内普遍发育片麻状构造，边部尤为明显，向岩体中心片麻状构造渐变为条带状、扁豆状暗色矿物集中的析离体，岩石粒度加粗，中心相中粒不等粒结构特征明显。岩体内部还可见大量被揉褶和扭曲的片麻岩捕掳房体、岩浆分异产生的辉石岩、角闪辉石岩等超镁铁质团块，并发育后期花岗岩、花岗斑岩岩脉（图 19-25）。

2. 李家湾岩体

分布于怀柔碾子北西，呈 NW 向长带状，出露面积约 25km^2。岩体北西顺围岩片麻理侵入太古宙宋营子英云闪长质片麻岩，并被白垩系张家口组地层不整合覆盖。侵入界线模糊但不整合地貌特征清晰。其余三面与燕山期岩体呈侵入接触，接触带发育大量岩枝岩脉。岩体内部片麻理发育，特征同下窝铺岩体。该岩体出露破碎，明显受后期断裂构造破坏，内部发育多条糜棱岩带。

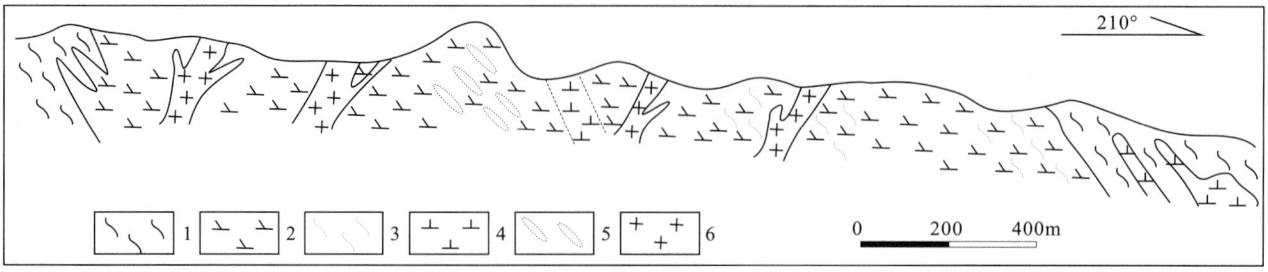

图 19-25　下窝铺岩体地质剖面示意图

（据郁建华等，1994）

Fig. 19-25　Schematic geological section of Xiawopu complex

1—太古宙片麻岩；2—变质辉长闪长岩；3—辉长闪长岩中片麻状构造；4—闪长岩脉；5—暗色矿物集中的析离体；6—燕山期花岗岩

表 19-10　古生代下窝铺岩体岩石类型及特征

Table 19-10　Rock types and its characteristic of Palaeozoic Xiawopu complex

岩石类型	岩性描述	分布
辉长苏长岩	灰黑色，中粒辉长结构。斜长石半自形短柱状，类型为拉、培长石（An=65），粒径 1~2mm，个别 5mm，复合双晶发育，含量 40%~50%；紫苏辉石半自形短柱状-板状，粒径 1~3mm，内含斜长石、单斜辉石包体，部分中心堆积蠕粒状磁铁矿析离体，普遍蚀变呈浑圆状，解理裂隙处出现滑石化，含量 20%~25%；单斜辉石半自形板状，粒径 1~2mm，简单双晶发育，含量 15%~20%；普通角闪石粗粒变晶柱状，粒径大小不一，大者达 5mm，发育变嵌晶结构，内常嵌辉石、斜长岩及磁铁矿等矿物，含量 10%~15%；少量次闪石，绿色，沿辉石边发育，绿泥石化	中心相
辉长岩	灰黑色，辉长结构，块状构造。斜长石短柱状-板状，粒径一般 1~3mm，少量 5mm，发育复杂双晶，类型为拉、培长石，含量 55%~65%；普通辉石半自形短柱状，简单双晶，粒径 1~4mm，普遍退变质为角闪石，常见被角闪石包嵌熔蚀呈残余状及具闪石次变边，含量 5%~30%；普通角闪石黄绿色短柱状，简单双晶，纤闪石化、绿泥石化，退变质重结晶变晶体粒径达 7mm，内包含斜长石、辉石残余体，含量 10%~35%；紫苏辉石含量 0%~3%；副矿物为他形粒状磁铁矿和钛铁矿，熔蚀状磷灰石、锆石、榍石等	边缘相
辉长闪长岩	灰黑色，辉长结构，块状构造。斜长石半自形板状-柱状，粒径 1~2mm，含量 45%~55%；普通辉石无色-淡黄绿色半自形短柱状，粒径 0.5~1mm，含量 40%~45%，普遍退变质为角闪石；普通角闪石黄褐、棕绿色柱状，部分为巨晶筛状变晶体，粒径 8mm，内包嵌大量辉石、斜长石残晶，纤闪石化、绿泥石化、绿帘石化	边缘相
闪长岩	灰黑色，半自形不等粒状结构，块状构造。斜长石半自形柱状，粒径 0.5~4mm，部分晶纹弯曲或断裂具波状消光，发育聚片双晶，类型为中长石（An=45~55），绢云母化，含量 65%~70%；普通角闪石黄绿色柱状，多色性，普遍蚀变分解成细晶柱粒集合体，并与残余辉石、磁铁矿、黑云母、绿泥石等聚集呈团状，含量约 20%；黑云母黄褐色片状，多由角闪石蚀变而来，粒径<0.5mm，含量 0%~5%；单斜辉石残余状发育于角闪石晶内，含量<3%。偶见粒径 0.5mm 熔蚀状柱粒状磷灰石	脉状

（二）岩石化学与地球化学特征

化学成分及参数见表 19-11。岩石呈现 SiO_2 含量变化大（38.63~55.67），低 K_2O、高 Al_2O_3、MgO、CaO、Na_2O 特征，且伴随 SiO_2 含量增多，K_2O、Na_2O 含量减少；里特曼指数 σ=0.46~3.66，Na_2O>K_2O，表明岩石为钠质亚碱性岩系列；固结指数 SI=18.9~30.27，分异指数 DI=10.6~51.13，反映岩浆分异程度一般；镁铁比值 m/f=0.57~1.09，属于铁质基性岩；A/CNK=0.77~0.86，A/NK=1.92~6.13，属于准铝质。

表 19-11 早二叠世中基性岩岩石化学成分及特征参数一览表

Table 19-11 The chemical composition and main parameters of Early Permian

序号	岩性	氧化物含量/%													特征参数					
		SiO_2	TiO_2	Al_2O_3	Fe_2O_3	FeO	MnO	MgO	CaO	Na_2O	K_2O	P_2O_5	Los	总和	*DI*	*SI*	*σ*	*A/CNK*	*A/NK*	*m/f*
1	闪长岩	55. 67	1. 11	16. 31	3. 80	4. 82	0. 14	3. 84	6. 33	3. 50	2. 23	0. 38	1. 61	99. 73	51. 12	21. 21	2. 48	0. 83	2. 00	0. 82
2	闪长岩	52. 63	0. 85	17. 56	2. 53	5. 62	0. 12	4. 91	6. 33	3. 95	2. 46	0. 20	2. 60	99. 75	49. 37	25. 2	3. 9	0. 85	1. 92	1. 09
3	辉长岩	38. 63	1. 88	18. 35	17. 01	8. 95	0. 11	8. 40	11. 15	1. 84	0. 60	0. 03	1. 97	108. 92	10. 61	23. 52	0. 82	0. 77	4. 99	0. 61
4	闪长岩	52. 85	0. 65	18. 56	9. 15	5. 04	0. 14	4. 26	8. 63	3. 69	0. 87	0. 32	0. 84	104. 99	35. 5	18. 9	2. 42	0. 82	2. 65	0. 57
5	辉长闪长岩	45. 84	0. 85	23. 56	6. 10	4. 67	0. 08	5. 69	13. 00	1. 69	0. 98	0. 04	2. 13	104. 63	18. 69	30. 27	3. 65	0. 86	6. 13	0. 99

表 19-12 早二叠世中基性岩稀土元素成分及特征参数一览表

Table 19-12 REE contents and main parameter table of Early Permian

序号	岩性	稀土元素含量/10^{-6}																特征参数			
		La	Ce	Pr	Nd	Sm	Eu	Gd	Tb	Dy	Ho	Er	Tm	Yb	Lu	Y	ΣREE	LREE/HREE	$(La/Yb)_N$	δEu	δCe
1	闪长岩	27. 46	71. 58	9. 35	39. 91	7. 32	2. 06	6. 55	0. 94	5. 33	1. 04	2. 79	0. 50	3. 02	0. 36	26. 57	178. 22	7. 68	6. 52	0. 89	1. 09
2	闪长岩	13. 73	36. 49	5. 04	22. 55	4. 45	1. 43	4. 04	0. 61	3. 52	0. 67	1. 84	0. 33	2. 05	0. 25	17. 20	97. 00	6. 29	4. 80	1. 01	1. 07
3	辉长岩	6. 03	16. 40	2. 73	12. 10	2. 93	0. 64	2. 75	0. 44	2. 56	0. 46	1. 34	0. 20	1. 27	0. 18	20. 01	50. 02	4. 44	3. 41	0. 68	0. 99
4	闪长岩	21. 50	40. 90	5. 27	22. 40	4. 17	1. 35	3. 37	0. 58	2. 98	0. 58	1. 62	0. 27	1. 63	0. 24	15. 40	106. 85	8. 49	9. 46	1. 07	0. 91
5	辉长闪长岩	7. 07	15. 10	2. 33	11. 50	2. 81	0. 97	2. 62	0. 46	2. 54	0. 49	1. 27	0. 21	1. 16	0. 15	12. 80	48. 68	4. 47	4. 37	1. 07	0. 91

表 19-13 早二叠世中基性岩微量元素成分及特征参数一览表

Table 19-13 Trace element contents and main parameters of Early Permian

序号	岩性	微量元素含量/10^{-6}									特征参数	
		Rb	Ba	Th	U	Nb	Sr	Zr	Hf	Ta	Rb/Sr	Nb/Ta
1	闪长岩	60. 07	1702. 80	2. 32	0. 41	8. 98	798. 91	293. 53	17. 46	0. 45	0. 08	19. 95
2	闪长岩	72. 41	1514. 70	2. 10	0. 27	4. 54	1154. 03	108. 03	5. 63	0. 44	0. 06	10. 23
3	辉长岩	5. 26	91. 90	0. 27	0. 12	1. 04	62. 40	27. 30	0. 99	0. 07	0. 08	14. 25
4	闪长岩	12. 40	744. 00	1. 42	0. 43	4. 16	1103. 00	14. 10	0. 62	0. 18	0. 01	23. 77
5	辉长闪长岩	31. 30	383. 00	0. 44	0. 29	2. 62	1106. 00	25. 20	0. 87	0. 104	0. 03	25. 19

稀土元素成分及参数见表 19-12。$\Sigma REE = 48.68 \times 10^{-6} \sim 178.21 \times 10^{-6}$，LREE/HREE = 4.44 ~ 8.48，$(La/Yb)_N = 3.41 \sim 9.46$，稀土总量较低，轻稀土弱富集，重稀土相对弱亏损；$\delta Eu = 0.91 \sim 1.09$，基本无铕正或负异常；配分曲线为右缓倾光滑曲线（图 19-26）。

微量元素成分及参数见表 19-13。岩石呈现高 Sr，低 Y、Yb 和高 Sr/Y、Th/Ta 特征。微量元素蛛网图（图 19-27）上，强烈富集大离子亲石元素 Rb、K、Ba，相对强烈亏损高场强元素 Nb、Th、U、Ta、P，不同岩性 Sr、Zr、Hf 出现分化。

（三）Sr-Nd 同位素

五道营子岩体为邻区河北省天桥镇发育的晚古生代基性侵入体，位于喇叭沟门下窝铺岩体北。闪长岩全岩 $Rb = 1.84 \times 10^{-6} \sim 24.31 \times 10^{-6}$，$Sr = 867.40 \times 10^{-6} \sim 1282.3 \times 10^{-6}$，$Sm = 3.74 \times 10^{-6} \sim 6.11 \times 10^{-6}$，$Nd = 19.88 \times 10^{-6} \sim 31.42 \times 10^{-6}$，$^{87}Rb/^{86}Sr = 0.0042 \sim 0.0811$，$^{143}Nd/^{144}Nd = 0.511794 \sim 0.512016$，$^{147}Sm/^{144}Nd = 0.1046 \sim 0.1138$（Zhang et al.，2009）。同位素与上地壳接近，而 $\varepsilon_{Nd}(t) = -9.3 \sim -13.2$（$t$ 取 280Ma），$I_{Nd} = 0.511602 \sim 0.511804$，$I_{Sr} = 0.70587 \sim 0.70602$，$t_{DM} = 1793 \sim 2060$。

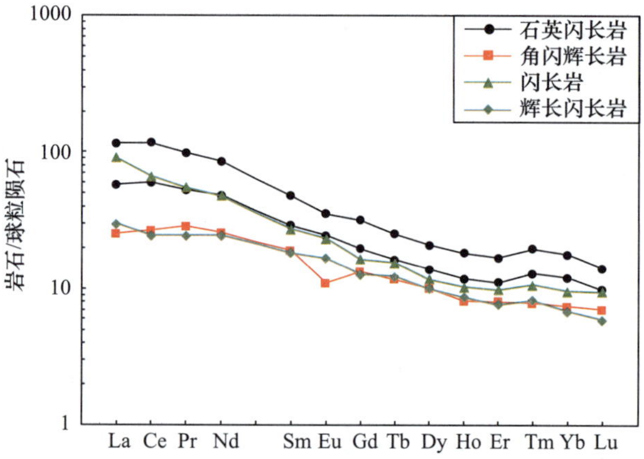

图 19-26　早二叠世侵入岩稀土元素配分模式图

Fig. 19-26　Chondrite-normalized REE patterns of Early Permian intrusive rocks

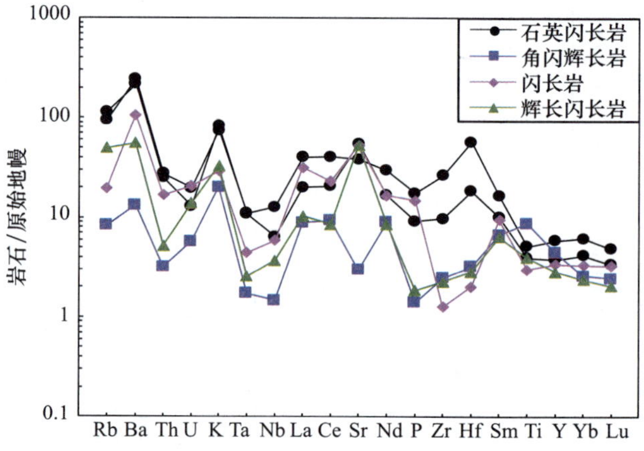

图 19-27　早二叠世侵入岩微量元素蛛网图

Fig. 19-27　Trace element spider diagram of Early Permian intrusive rocks

（四）锆石 Hf 同位素

五道营子闪长岩锆石 $\varepsilon_{Hf}(t) = -15.1 \sim -17.9$，$t_{DM} = 1.51 \sim 1.61$（Zhang et al.，2009），表现出低 $^{176}Hf/^{177}Hf$、负 $\varepsilon_{Hf}(t)$ 值、较老 Hf 同位素模式年龄特征。

（五）岩石系列

该套岩系 $\sigma = 0.46 \sim 3.9$，属于钙碱性类。SiO_2-K_2O 图解（图 19-28）厘定为钙碱性-高钾钙碱性系列。

（六）变质变形特征

早二叠世侵入岩遭受变质变形，使得早期认定为新太古代侵入岩。残留变余辉石和变晶角闪石或黑云母交代现象，揭示岩体变质矿物组合主要为斜长石+角闪石，少数变质矿物组合为斜长石+角闪石+黑云母，岩体发生角闪岩相变质作用。同时，斜长石、角闪石又普遍发生绿帘石化、绿泥石化，说明岩体晚期还遭受绿片岩相变质作用。这一变质矿物演化均反映两期变质形式为退变质作用。

早二叠世岩体多部位发育片麻状构造，尤以岩体边部为最，而岩体内则多呈长条或带状分布，如喇叭沟门片麻理走向 NW-SE，多倾向 NE，局部倾向 SW；杨木栅子片麻理走向 NEE-SWW，倾向 NWW。这些特征揭示了岩体侵位伸展构造背景并出现韧性剪切变形。

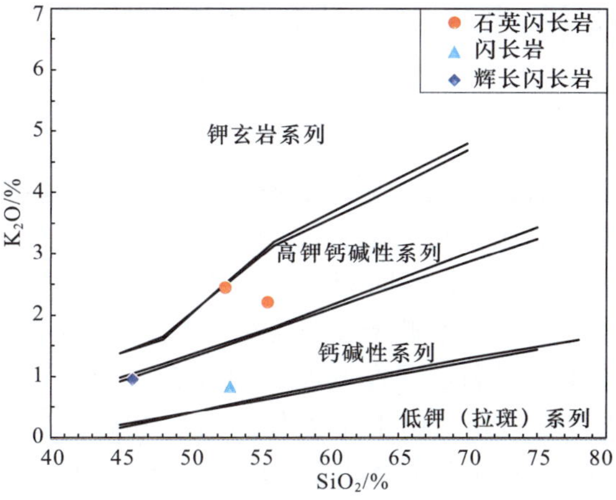

图 19-28 早二叠世侵入岩 SiO_2-K_2O 图解

Fig. 19-28 SiO_2-K_2O diagram of Early Permian intrusive rocks

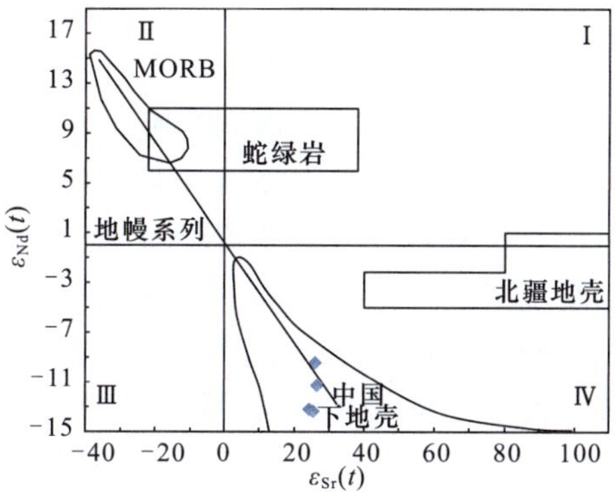

图 19-29 早二叠世侵入岩 ε_{Sr}（t）-ε_{Nd}（t）图解

Fig. 19-29 ε_{Sr}（t）-ε_{Nd}（t）diagram of Early Permian intrusive rocks

（七）成因类型与岩浆演化

早二叠世侵入岩岩浆主要来源于古老下地壳，主要依据包括：①具有 SiO_2 含量变化范围广，高 Al_2O_3、Sr、Ba，低 Y、Yb 等地球化学特征；②轻稀土弱富集且无明显 Eu 正或负异常；③蛛网图中强烈亏损 Nb、Ta 和 Ti；④全岩 Sr-Nd 同位素和锆石 Hf 同位素（图 19-29）与下地壳接近；⑤岩体结晶温度约 750~770℃，侵位压力约 0.38~0.45GPa，计算岩浆侵位深度 12~15km，属于被动式侵位（王惠初等，2007）。同时，该期岩体中发育残留超基性镁铁质团块，表现为多期岩浆活动的杂岩体特征；岩石类型中超基性岩 Sr-Nd 同位素又与富集型地幔接近（Zhang et al.，2009）；富集岩石圈地幔物质可能是早期基性岩浆的重要物质来源。

早二叠世侵入岩岩石化学揭示众多成分随 SiO_2 含量变化呈线性特征演化（如 K_2O、MgO、CaO、Y），揭示了主要岩浆演化过程是分离结晶作用，且角闪石、磁铁矿、钛铁矿可能是主要的分馏结晶矿物，而磷灰石、斜长石、锆石分馏作用并不强（Zhang et al.，2009）。

（八）构造环境分析

区内早二叠世侵入体均位于怀柔喇叭沟门，即赤城-古北口断裂带沿线，呈东、西延伸至河北境内。区域上，该期岩体广泛分布于古元古代末形成的赤城-平泉深断裂及丰宁-隆化深断裂之间。岩体规模较大，虽然受后期构造破坏，岩体原始形态难以恢复，但近东西向带状展布特征依然明显。因此，岩体的形成应受东西向断裂带控制，与古亚洲洋板块俯冲有关。上述矿物岩石、地球化学及同位素等特征与大陆弧侵入岩特征基本一致，在构造环境判别图解（图 19-30）中，成分也位于火山弧玄武岩或岛弧区，揭示了华北克拉通北缘早二叠世活动大陆边缘大地构造背景。研究区为华北克拉通北缘最南端，构造位置相当于北美西部中生代内华达山脉岩基（Zhang et al.，2009）。在古亚洲洋地壳向南俯冲影响下，幔源的

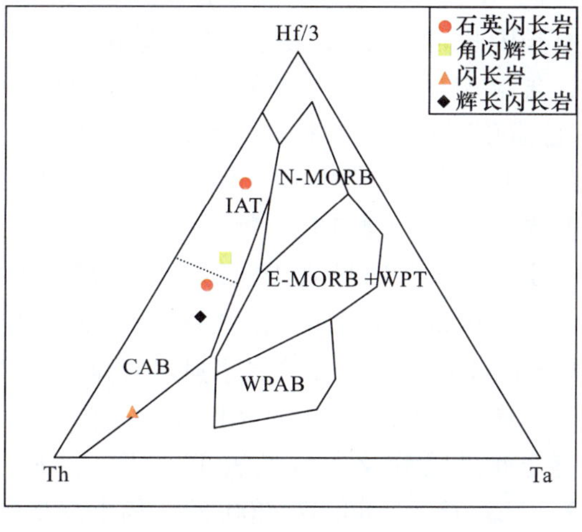

图 19-30 早二叠世侵入岩 Hf/3-Th-Ta 图解

Fig. 19-30 Hf/3-Th-Ta diagram of tectonic discrimination of Early Permian intrusive rocks

WPAB—板内拉斑玄武岩；N-MORB—N 型洋中脊；E-MORB+WPT—E 型洋中脊和板内拉斑玄武岩；IAT—岛弧拉斑玄武岩；CAB—火山弧玄武岩

玄武岩岩浆在古下地壳之下沿着早期深大断裂带上升发展，导致增厚大陆地壳平流加热、重熔形成岩浆源，最终大范围侵位。

（十）时代归属

本期岩浆侵位时代厘定为晚古生代早二叠世，主要依据包括：①该系列岩体虽遭受强烈变质变形，但变质类型为退变质和局部糜棱岩化作用，未发生高级变质变形；②侵入到新太古代宋营子英云闪长质片麻岩，广泛被燕山期侵入岩吞蚀破坏；③邻区河北省同套岩系同中元古界地层接触之处，发现多处大理岩及石英岩捕虏体（1∶5 万喇叭沟门、杨木栅子、三道营子幅区域地质调查报告，1999）；④岩浆岩带近东西向空间展布及化学特性共同揭示同古亚洲洋俯冲有关构造背景；⑤本次获得下窝铺岩体中角闪辉长岩和角闪闪长岩锆石 U-Pb 同位素年龄分别为（284.1±2.6）Ma 和（281.4±1.7）Ma；⑥对角沟门岩体角闪闪长岩锆石 U-Pb 同位素年龄为（288.0±4.8）Ma（王惠初等，2007）；⑦邻区河北省同期五道营子岩体闪长岩锆石 SHRIMP 同位素年龄为（283.0±2）Ma（Zhang et al.，2009）。

五、中生代中基性侵入岩

（一）晚侏罗世中基性侵入岩

1. 地质产状与岩石学特征

分布于密云抗峪、朱家湾、下庄头峪、太平山，延庆太平庄、五座山和昌平阳聂村，由分散零星小岩株组成，多数出露面积<0.5km^2，以抗峪岩体面积最大（1.5km^2）。由于岩体规模较小，岩体内无岩相变化，接触带附近也未发育明显接触变质。平面形态不规则或透镜状，多侵位于太古宙片麻岩（如朱家湾岩体）或中元古界长城系（如太平庄岩体），少侵入侏罗系髫髻山组（如太平山岩体）中。岩石类型有辉长岩和辉长闪长岩。

辉长岩：灰-深灰绿色，辉长结构，块状构造。斜长石自形-半自形板条状，粒径 0.2~1.7mm，钠长双晶、卡钠复合双晶发育，类型为拉长石，强烈绢云母化，含量约 55%；普通辉石淡绿色-无色，短柱状，粒径 0.2~1.5mm，角闪石化、绿泥石化、黑云母化蚀变，并有铁质析出，含量约 32%；斜方辉石无色，短柱状，粒径 0.3~1mm，强烈次闪石化、滑石化、黑云母化呈残余状，含量约 6%；黑云母黄绿色-褐色，粒径 0.2~2mm，含量 1%~12%。

辉长闪长岩：灰-灰褐色，细粒结构，块状构造。斜长石、普通辉石特征同辉长岩，组合构成辉长结构，含量 55%~70%；条纹长石他形粒状，粒径 0.3~1.2mm，与石英交代生成文象结构，含量约 3%；普通角闪石绿色，柱状，粒径 0.1~0.4mm，含量 10%~30%；石英他形粒状，粒径 0.1~0.3mm，含量约 2%；黑云母片状，多色性，粒径 0.2~2mm，含量 5%~12%；副矿物为磁铁矿、磷灰石。

2. 岩石化学与地球化学特征

化学成分及参数见表 19-14。与黎彤（1962）中国辉长岩相比，岩石呈富 SiO_2、MgO、碱，贫 CaO 特征；里特曼指数 σ=4.28~8.78，表明岩石为碱性岩系列；固结指数 SI=27.69~32.94，分异指数 DI=56.84~57.89，反映岩浆强烈分异作用；镁铁比值 m/f=1.31~1.77，属于铁质基性岩；A/CNK=0.66~1.03，A/NK=1.21~1.61，属于偏铝质。

稀土元素成分及参数见表 19-15。ΣREE = 221.61 $\times10^{-6}$ ~ 371.55 $\times10^{-6}$，LREE/HREE = 24.71 ~ 47.92，$(La/Yb)_N$=80.07~200.84，稀土总量高，轻稀土强富集，重稀土相对亏损；δEu=0.74~0.99，极弱铕负异常；配分曲线为右陡倾光滑曲线（图 19-31）。

微量元素成分及参数见表 19-16。与维氏（1962）基性岩相比，Sn、Pb、Ta、Be、Li、Rb、Sr、Zr、Ga 含量偏高，Mo、Cu 含量偏低，其余元素含量基本相当；微量元素蛛网图（图 19-32）上，强烈富集大离子亲石元素 Rb、K、Sr、Ba，相对亏损高场强元素 Nb、Ta。

表 19-14　晚侏罗世基性岩岩石化学成分及特征参数一览表

Table 19-14　The chemical composition and main parameters of Late Jurassic

序号	岩体	岩性	氧化物含量/%													特征参数					
			SiO_2	TiO_2	Al_2O_3	Fe_2O_3	FeO	MnO	MgO	CaO	Na_2O	K_2O	P_2O_5	Los	总和	*DI*	*SI*	σ	*A/NK*	*A/CNK*	*m/f*
1	朱家湾	辉长岩	50. 89	0. 92	14. 28	5. 26	3. 73	0. 18	8. 57	5. 41	4. 4	4. 2	0. 76	0. 58	99. 18	56. 84	32. 94	8. 78	1. 21	0. 66	1. 77
2	太平山	辉长岩	54. 8	1. 3	15. 36	4. 66	3. 99	0. 12	6. 08	2. 96	2. 85	4. 5	0. 72	0. 81	98. 15	57. 89	27. 69	4. 28	1. 61	1. 03	1. 31

注：数据引自 1∶5 万密云幅、高岭幅、墙子路幅区域地质调查报告，1994。

表 19-15　晚侏罗世基性岩稀土元素成分及特征参数一览表

Table 19-15　REE contents and main parameter table of Late Jurassic

序号	岩体	岩性	稀土元素含量/10^{-6}																特征参数			
			La	Ce	Pr	Nd	Sm	Eu	Gd	Tb	Dy	Ho	Er	Tm	Yb	Lu	Y	ΣREE	LREE/HREE	$(La/Yb)_N$	δEu	δCe
1	朱家湾	辉长岩	96. 0	156. 0	22. 8	66. 0	13. 0	3. 3	6. 2	0. 75	4. 0	0. 7	1. 56	0. 18	0. 86	0. 2	20. 6	371. 55	24. 71	80. 07	0. 99	0. 79
2	太平山	辉长岩	56. 0	106. 0	11. 6	38. 0	4. 60	0. 88	2. 2	0. 23	1. 0	0. 2	0. 3	0. 2	0. 2	0. 2	3. 52	221. 61	47. 92	200. 84	0. 74	0. 97

注：数据引自 1∶5 万密云幅、高岭幅、墙子路幅区域地质调查报告，1994。

表 19-16　晚侏罗世基性岩微量元素成分及特征参数一览表

Table 19-16　Trace element contents and main parameters of Late Jurassic

序号	样品号	样品名	微量元素含量/10^{-6}															特征参数	
			Rb	Ba	Th	U	Ta	Nb	Sr	Zr	Hf	V	Cr	Mn	Co	Ni	Cu	Rb/Sr	Nb/Ta
1	朱家湾	辉长岩	—	2501	—	—	2. 61	18. 29	2506	486. 20	—	175. 7	—	—	—	—	44. 55	—	7. 01
2	太平山	辉长岩	70. 08	1480	—	—	2. 14	26. 41	1073	303. 10	—	149. 4	—	—	—	—	39. 21	0. 07	12. 34

注：数据引自 1∶5 万密云幅、高岭幅、墙子路幅区域地质调查报告，1994。

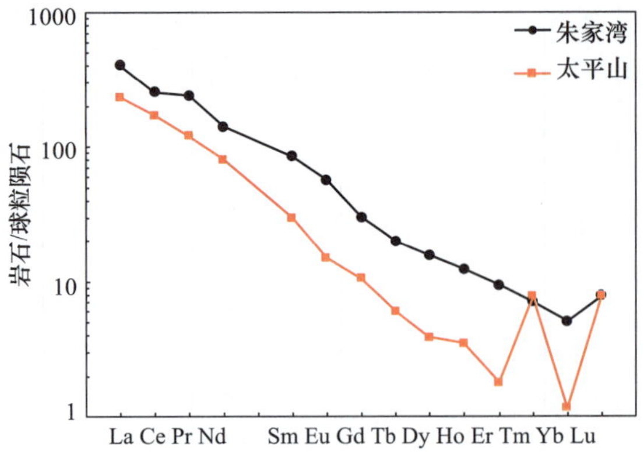

图 19-31 晚侏罗世侵入岩稀土元素配分模式图

Fig. 19-31 Chondrite-normalized REE patterns of Late Jurassic intrusive rocks

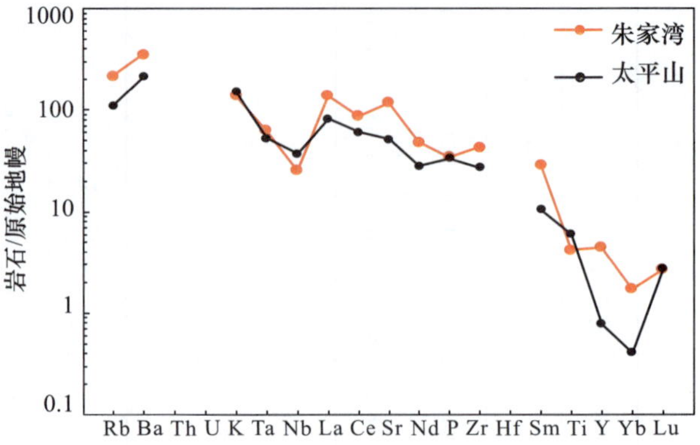

图 19-32 晚侏罗世侵入岩微量元素蛛网图

Fig. 19-32 Trace element spider diagram of Late Jurassic intrusive rocks

3. 岩石系列

该套岩系 $\sigma = 4.28 \sim 8.78$，属于碱性岩系列。P_2O_5-Alk 图解（图 19-33）进一步厘定为强碱性岩系列。

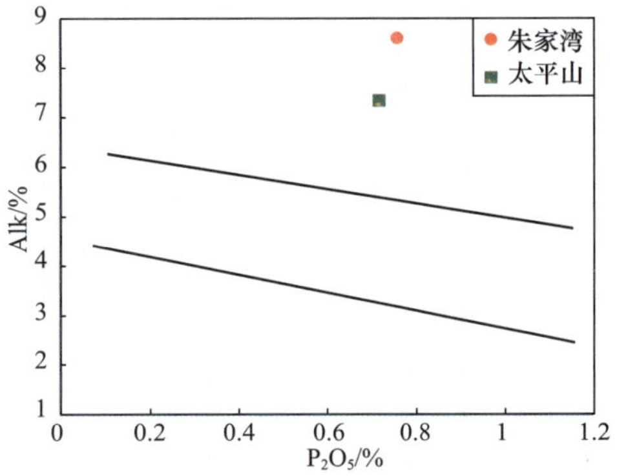

图 19-33 晚侏罗世侵入岩 P_2O_5-Alk 图解

Fig. 19-33 P_2O_5-Alk diagram of Late Jurassic intrusive rocks

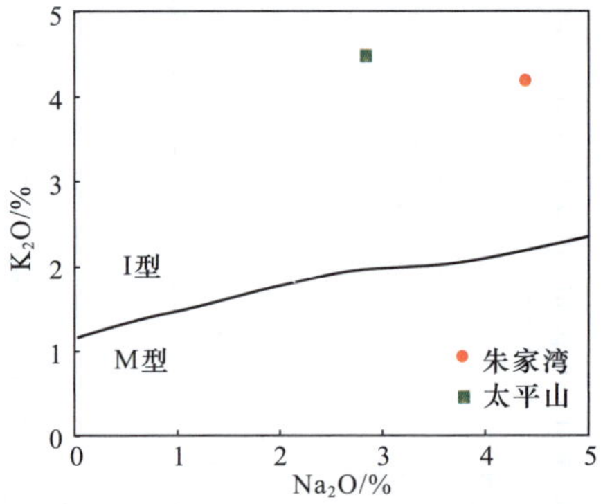

图 19-34 晚侏罗世侵入岩 K_2O-Na_2O 图解

Fig. 19-34 K_2O-Na_2O diagram of Late Jurassic intrusive rocks

4. 成因类型

本套岩系岩浆主要来源于下地壳，主要依据包括：①具有富 SiO_2、MgO、碱，贫 CaO，低 Y、Yb 等地球化学特征；②轻稀土富集且呈极弱 Eu 负异常；③蛛网图显示富集大离子亲石元素，亏损 Nb、Ta；④具有同髫髻山期火山岩高度相似稀土和微量元素特性，可能为同源岩浆产物。髫髻山期火山岩全岩 Sr-Nd 同位素揭示其岩浆起源于下地壳的部分熔融作用，可能有少量上地幔熔体加入（孙善平等，1995；李伍平等，2001）；⑤CIPW 标准矿物中均出现碱性长石，且 Or、Ab、An 含量较高，显示出深源岩浆的特性；⑥K_2O-Na_2O 图解（图 19-34）中，成分点位于 I 型区。

5. 构造环境分析

晚侏罗世基性侵入岩形成于板内伸展造山构造背景环境，具有伸展和造山两大特有的构造属性，主要依据包括：①岩石 Y、Yb 亏损，在富集背景下 Nb、Ta 相对弱亏损等特征同大陆边缘弧或与俯冲作用有关的岩浆弧类似。而强碱性岩石系列又不同于典型的挤压岩浆弧以发育钙碱性或高钾钙碱性系列为主特性；②在 TiO_2-$MnO*10$-P_2O_5*10 图解（图 19-35）上，成分也位于钙碱性玄武岩和洋岛碱性玄武岩之间；③岩体区内分布零散，发育规模较小，形态多不规则状，无明显构造走向，在邻区河北省也具有相似特性。

6. 时代归属

本期岩浆侵位时代厘定为晚侏罗世，主要依据包括：①该系列岩体岩石仅遭受蚀变，未遭受变质变形；②太平山岩体见岩枝侵入侏罗系髫髻山组地层；③广泛被燕山期白垩世侵入岩吞蚀破坏；④抗峪岩体全岩 K-Ar 同位素年龄为160Ma（1∶5万密云、高岭、墙子路幅区域地质调查报告，1990）。

（二）早白垩世中基性侵入岩

1. 地质产状与岩石学特征

早白垩世中基性侵入岩由棋盘岩、上庄、银冶岭、西葫芦峪和三白山5个岩体组成，出露面积10.5km²。

（1）棋盘岩辉长岩体

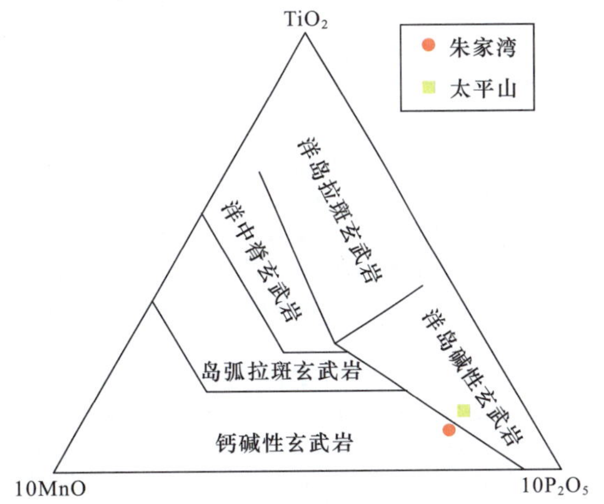

图19-35 晚侏罗世侵入岩 $TiO_2-10MnO-10P_2O_5$ 图解

Fig. 19-35 $TiO_2-10MnO-10P_2O_5$ diagram of Late Jurassic intrusive rocks

分布于门头沟棋盘岩村，呈NW向、长5km、宽1~2km椭圆形，出露面积7km²。岩体北西侧与燕山期闪长玢岩呈侵入接触，其余三面与中元古代待建系下马岭组至古生代寒武系张夏组呈侵入接触（图19-36）。接触带附近发育明显烘烤、冷凝现象，宽度达80m。外带碳酸盐岩围岩大理岩化，碎屑岩地层红柱石角岩化，内带发生强烈混染，并可见大量围岩捕虏体。岩体岩性主体划分为辉长岩和辉长闪长岩两个相带。岩体内还发育大量岩浆分异演化橄榄辉长岩团块及二长闪长岩脉，具体岩石特征见表19-17。

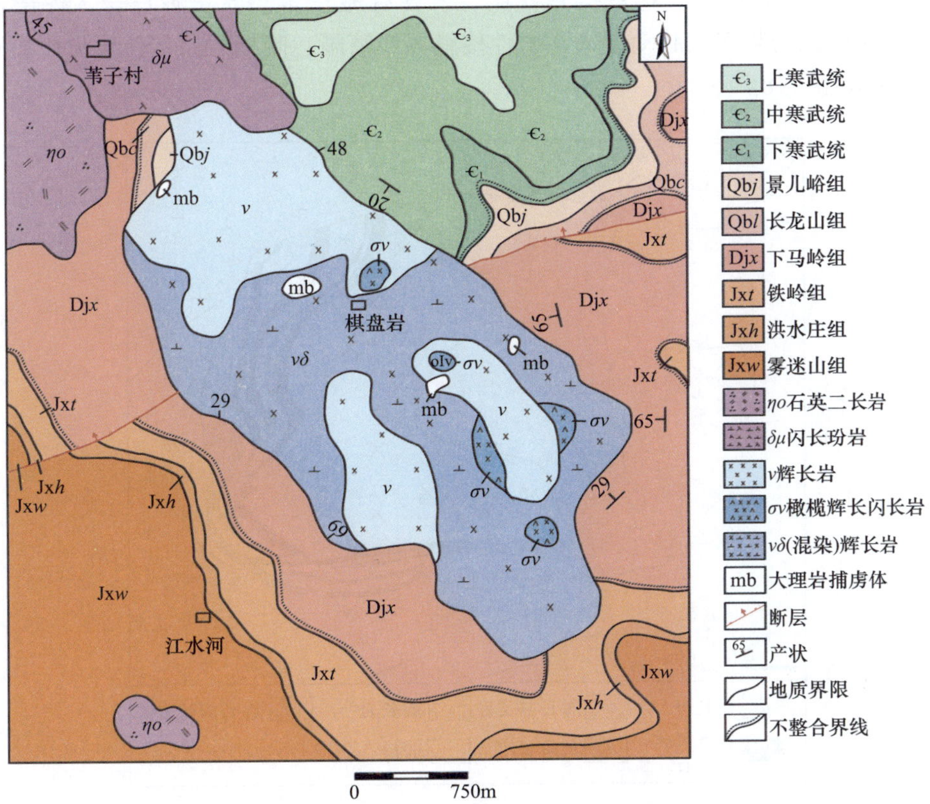

图19-36 棋盘岩辉长岩体地质略图

（据郇建华等，1994，略有修改）

Fig. 19-36 Schematic geological map of Qipanyan gabbro intrusion

表 19-17　棋盘岩辉长岩体岩石类型

Table 19-17　Rock types of Qipanyan gabbro intrusion

岩石类型	岩性描述	分布
橄榄辉长岩	黑-灰黑色，中粗粒辉长结构。橄榄石半自形—自形短柱状，常被辉石及角闪石包嵌，含量 10%~15%；紫苏辉石自形短柱状，内常有橄榄石包体，与普通辉石、磁铁矿共生，含量 10%~15%；普通辉石无色，自形状短柱状，含量约 25%；角闪石半自形针状，多分布在辉石边部，并与黑云母共生，部分包嵌橄榄石、辉石及斜长石，含量约 20%；黑云母半自形片状，含量约 10%；斜长石半自形，粒径 1~1.5mm，充填于暗色矿物间，类型为中长石或倍长石，含量 15%	团块状，尤以岩体东南部更多
辉长岩	黑灰色，中粒辉长结构，局部似斑状构造。普通辉石与紫苏辉石共生，粒径 1~2mm，含量 20%~25%；普通角闪石和黑云母半自形，在辉石边部发育，表现次生特性，含量均为 3%~5%；斜长石半自形柱状，发育聚片、卡钠复合双晶，常镶嵌自形普通辉石，类型为中长石和拉长石，含量 55%~65%；钾长石他形粒状，含量<3%；石英他形粒状，含量<3%	西北部、中部
辉长闪长岩	灰黑色，变斑状结构、条纹状结构。普通辉石残余状，多被角闪石交代，含量 30%~35%；斜长石半自形状，发育环带，晶纹常不平直，部分斜长石见两次晶出现象，含量 55%~65%；钾长石他形粒状，含量 3%~5%；石英他形粒状，含量 3%~5%。岩石普遍碳酸盐化，出现大量次生榍石、绿帘石、褐帘石。在与围岩或近捕虏体附近岩性中常出现混染团粒状透辉石，巨晶筛状变晶角闪石中包裹大量透辉石和斜长石等矿物，呈团状出现，构成直径 1~2cm 近球状巨斑	东南部
二长闪长岩	灰色，细粒结构，块状构造，由斜长石（约 68%）、钾长石（约 15%）、石英（<5%）和黑云母（约 12%）等组成	脉状

（2）上庄辉长岩体

分布于昌平上庄，隶属于薛家石梁杂岩体，位于此杂岩体南部，呈豆荚状展布在黑云母闪长岩岩体中，出露面积 2km^2。岩体内含层状钒钛磁铁矿。岩性有橄长岩、苏长辉长岩、辉长岩和辉长闪长岩（图 19-37），为同源岩浆多次脉动侵入，岩性接触界线清晰，晚期岩浆侵位可见冷凝边。各岩性矿物定量统计结果见表 19-18。

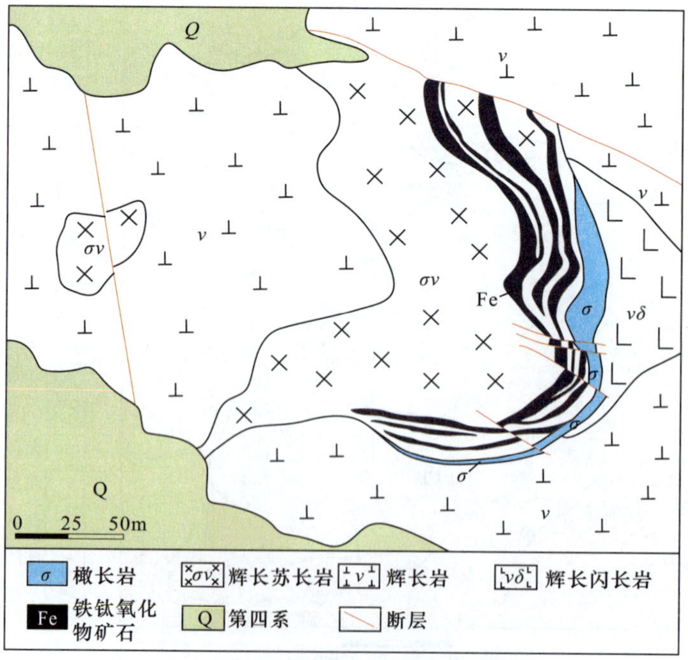

图 19-37　上庄辉长岩体地质略图

（据 Liu et al.，2015）

Fig. 19-37　Schematic geological map of Shangzhuang gabbro intrusion

（3）银冶岭岩体

由银冶岭、西葫芦峪、三白山三个小型岩体组成，分布于密云银冶岭至平谷三白山一带，出露面积2.8km^2。岩体主体侵入于中元古界高于庄组和雾迷山组地层中。接触带附近发育明显烘烤、冷凝现象，外带碳酸盐岩局部大理岩化，内带发生强烈混染，可见大量围岩捕虏体，尤以银冶岭岩体为最。接触面呈锯齿状或港湾状，产状多外倾，倾角中等或陡倾。

岩体主要岩性为辉长闪长岩、闪长岩及石英闪长岩。岩石呈灰绿色或浅灰绿色，全晶质细—中细粒结构、斑状结构，块状构造。斜长石自形板状或柱状，发育聚片、卡钠复合双晶和环带结构，粒径0.5~2mm，含量55%~75%；钾长石粒状，粒径0.1~0.3mm，含量1%~5%；黑云母片状，绿泥石化，内嵌磁铁矿、榍石、磷灰石，粒径0.1~0.5mm，含量8%~15%；角闪石半自形—他形粒状，不均匀绿泥石化，粒径0.1~0.8mm，含量约7%；石英他形粒状，充填于其他矿物间隙，粒径0.1~2mm，含量3%~5%。

表19-18　上庄岩体岩石矿物组成统计表

Table 19-18　Real mineral contents of Shangzhuang intrusion

单位:%

	橄榄石	单斜辉石	斜方辉石	角闪石	云母	斜长石	副矿物
橄长岩	20		10	10	9	40	磁铁矿、钛铁矿、尖晶石
苏长辉长岩		10~18	12	10~20		50~65	磁铁矿、钛铁矿、云母
辉长岩		25		10	5	60	磁铁矿、钛铁矿
辉长闪长岩		10		15	15	55	磁铁矿、钛铁矿

（据腾雪明，2017）

2. 岩石化学与地球化学特征

化学成分及参数见表19-19。岩石呈SiO_2、CaO、MgO含量变化大，高K_2O、FeO^*、TiO_2特征；里特曼指数σ=2.49~2.9，表明岩石为亚碱性岩系列；固结指数SI=13.56~48.38，分异指数DI=20.2~64.02，反映岩浆分异程度一般；镁铁比值m/f=0.59~2.12，属于铁质-富铁质基性岩；A/CNK=0.47~0.94，A/NK=1.34~3.54，属于准铝质。

稀土元素成分及参数见表19-20。ΣREE = 106.66 ×10^{-6} ~ 214.77 ×10^{-6}，LREE/HREE = 10.79 ~ 22.07，$(La/Yb)_N$ = 15.65 ~ 32.06，稀土总量较高，轻稀土强富集，重稀土相对亏损；δEu = 0.86 ~ 1.18，极弱铕正异常；配分曲线为右陡倾光滑曲线（图19-38）。

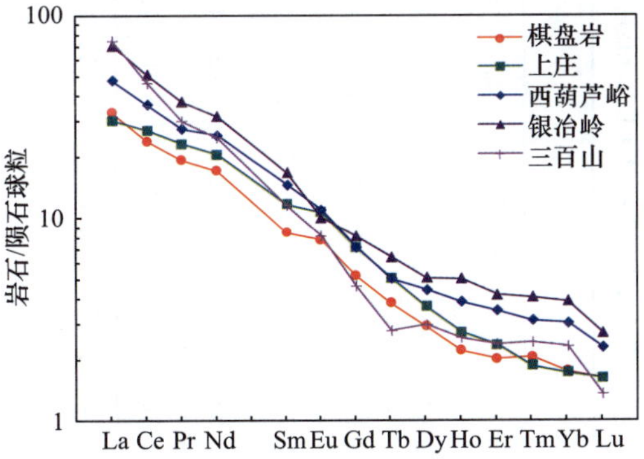

图19-38　早白垩世侵入岩稀土元素配分模式图

Fig. 19-38　Chondrite-normalized REE patterns of Early Cretaceous intrusive rocks

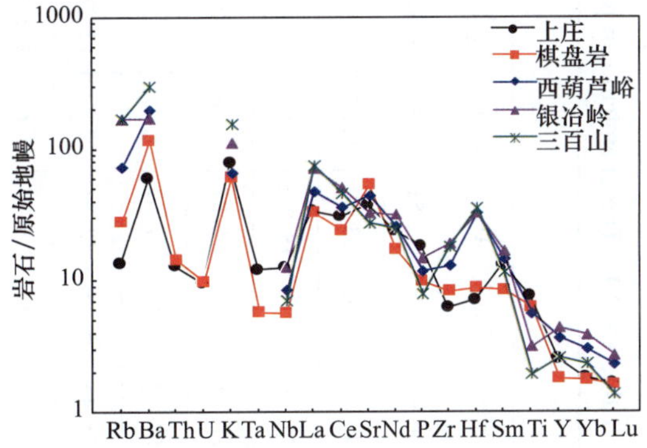

图19-39　早白垩世侵入岩微量元素蛛网图

Fig. 19-39　Trace element spider diagram of Early Cretaceous intrusive rocks

微量元素成分及参数见表19-21。岩石呈现高Sr，低Y、Yb和高Sr/Y特征。微量元素蛛网图（图19-39）上，强烈富集大离子亲石元素Sr、K、Ba，相对弱亏损高场强元素Nb、Th、U、Ta，不同岩性Zr、Hf出现分化。

表 19-19　早白垩世中基性岩岩石化学成分及特征参数一览表

Table 19-19　The chemical composition and main parameters of Early Cretaceous intermediate-basic intrusive rocks

序号	岩体	岩性	氧化物含量/%													特征参数					
			SiO_2	TiO_2	Al_2O_3	Fe_2O_3	FeO	MnO	MgO	CaO	Na_2O	K_2O	P_2O_5	Los	总和	*DI*	*SI*	σ	*A/CNK*	*A/NK*	*m/f*
1	上庄	辉长岩	46.43	1.65	12.51	5.70	7.31	0.15	9.39	12.73	0.71	2.37	0.40	0.44	99.78	20.20	37.15	2.51	0.47	3.35	1.33
2	棋盘岩	辉长岩	45.74	1.36	13.80	6.84	5.89	0.10	14.43	8.06	1.17	1.83	0.21	1.07	100.50	20.90	48.38	2.90	0.74	3.54	2.12
3	西葫芦峪	辉长岩	51.96	1.22	15.925	7.01	6.38	0.185	4.98	7.98	2.73	1.97	0.26	0.79	101.38	37.07	21.84	2.49	0.75	2.41	0.69
4	银冶岭	二长闪长岩	60.71	0.68	16.81	2.16	4.45	0.13	2.15	4.50	3.77	3.32	0.32	0.95	99.95	64.02	13.56	2.80	0.94	1.72	0.59
5	三白山	石英闪长岩	64.09	0.42	17.09	1.95	2.17	0.09	0.93	3.35	4.71	4.64	0.17	0.34	99.95	78.40	6.46	4.13	0.91	1.34	0.41
6	兴隆城	辉绿岩	51.46	2.15	12.34	2.5	12.23	0.22	8.62	5.83	0.82	2.32	0.16	0.73	99.38	27.10	32.54	1.11	0.85	3.20	1.05
7	高家窑	辉绿岩	50.14	1.84	15.39	4.64	5.905	0.13	6.775	5.045	2.05	3.665	0.445	3.22	99.24	42.31	29.53	3.83	0.93	2.10	1.18

注：1 为均值，数据引自苏尚国等（2006）和汪洋等（2014）；2 为均值，数据引自刘翠等（2007）和李芳凝（2006）；3-5 数据引自 1：5 万木林幅、大华山幅、平谷幅区域地质调查报告，2001；6~7 数据引自 1：5 万青龙桥幅、延庆幅区域地质调查报告，1984。

表 19-20　早白垩世中基性岩稀土元素成分及特征参数一览表

Table 19-20　REE contents and main parameter of Early Cretaceous intermediate-basic intrusive rocks

序号	岩体	岩性	稀土元素含量/10^{-6}																特征参数			
			La	Ce	Pr	Nd	Sm	Eu	Gd	Tb	Dy	Ho	Er	Tm	Yb	Lu	Y	ΣREE	LREE/HREE	$(La/Yb)_N$	δEu	δCe
1	上庄	辉长岩	20.62	47.65	6.38	27.74	5.16	1.79	4.23	0.54	2.68	0.45	1.13	0.14	0.85	0.12	11.07	119.48	10.79	17.46	1.17	1.02
2	棋盘岩	辉长岩	22.63	42.42	5.32	23.13	3.75	1.31	3.08	0.41	2.15	0.36	0.97	0.15	0.87	0.12	8.21	106.66	12.15	18.72	1.18	0.95
3	西葫芦峪	辉长岩	32.48	63.86	7.54	34.57	6.45	1.84	4.24	0.54	3.25	0.63	1.67	0.23	1.49	0.17	16.64	158.96	12.01	15.65	1.08	1.00
4	银冶岭	二长闪长岩	48.66	89.59	10.27	42.67	7.42	1.68	4.83	0.69	3.73	0.82	2.00	0.30	1.91	0.20	19.86	214.77	13.83	18.29	0.86	0.98
5	三白山	石英闪长岩	51.36	81.62	8.23	33.53	5.10	1.37	2.73	0.30	2.19	0.42	1.14	0.18	1.15	0.10	11.76	189.42	22.07	32.06	1.12	0.97

注：1 为均值，数据引自苏尚国等（2006）和汪洋等（2014）；2 为均值，数据引自刘翠等（2007）和李芳凝（2006）；3-5 数据引自 1：5 万木林幅、大华山幅、平谷幅区域地质调查报告，2001。

3. Sr-Nd-Pb 同位素特征

棋盘岩辉长岩$^{87}Sr/^{86}Sr$=0.705986~0.706988，$^{87}Rb/^{86}Sr$=0.02045~0.06808，$^{147}Sm/^{144}Nd$=0.1017~0.1471，$^{143}Nd/^{144}Nd$=0.511564~0.511692，$\varepsilon_{Nd}(t)$ （t=132.9Ma） =-15.44~-17.94，$\varepsilon_{Sr}(t)$ （t=132.9Ma） =21.89~39.4，$^{206}Pb/^{204}Pb$=16.4339~17.00804（李芳凝，2006；曹文燕，2010）；上庄辉长岩$^{87}Sr/^{86}Sr$=0.705357~0.70566，$^{143}Nd/^{144}Nd$=0.511845~0.512129，$\varepsilon_{Nd}(t)$ （t=125Ma） =-9.6~-15.47，$\varepsilon_{Sr}(t)$ （t=125Ma） =9.65~16.47，$^{206}Pb/^{204}Pb$=16.173~16.529，$^{207}Pb/^{204}Pb$=15.138~15.241，$^{208}Pb/^{204}Pb$=36.02~36.507（苏尚国等，2006；汪洋等，2013；Liu et al.，2015）。

表 19-21 早白垩世中基性岩微量元素成分及特征参数一览表

Table 19-21 Trace element contents and main parameters of Early Cretaceous intermediate-basic intrusive rocks

序号	岩体	岩性	微量元素含量/10^{-6}								特征参数		
			Rb	Ba	Th	U	Ta	Nb	Sr	Zr	Hf	Rb/Sr	Nb/Ta
1	上庄	辉长岩	8.52	419.67	1.10	0.20	0.49	8.93	793.67	69.60	2.20	0.01	17.78
2	棋盘岩	辉长岩	17.7	814.12	1.22	0.21	0.24	4.02	1123.47	93.0	2.71	0.02	16.15
3	西葫芦峪	辉长岩	40.85	1199.50	1.44	0.49	1.04	13.70	788.90	125.00	6.43	0.05	20.58
4	银冶岭	二长闪长岩	107.00	1181.00	—	—	—	9.00	690.20	211.00	10.00	0.16	—
5	三白山	石英闪长岩	106.00	2073.00	—	—	—	5.00	573.10	202.00	11.00	0.18	—

注：数据来源同表 19-22。

4. 岩石系列

该套岩系 σ=2.49~2.9<4，为亚碱性系列。碱含量较高，长石类型为中-拉长石，不同于低钾拉斑玄武岩系列。SiO_2-K_2O 图解（图 19-40）厘定为高钾钙碱性-钾玄岩系列。

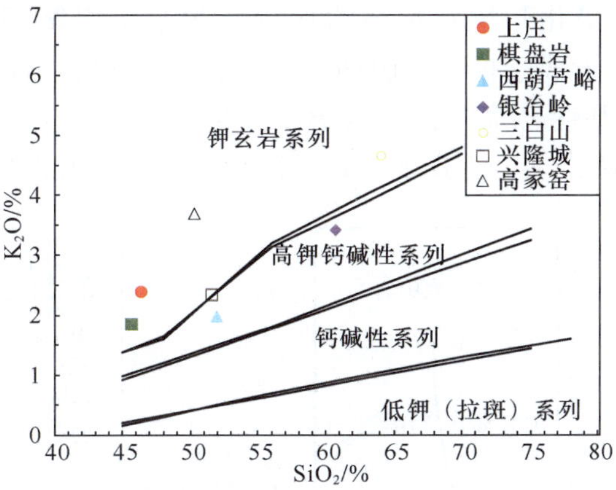

图 19-40 早白垩世基性岩 SiO_2-K_2O 图解

Fig. 19-40 SiO_2-K_2O diagram of Early Cretaceous basic intrusive rocks

5. 成因类型与岩浆演化

该套岩系岩浆主要来源于原始地幔，主要依据有：①角闪石等矿物中内嵌磁铁矿、钛铁矿和斜方辉石矿物存在分别揭示了岩浆源富铁钛、富镁熔体特性；②岩石高 TiO_2、FeO^*，低 Al_2O_3 地球化学特征；③蛛网图中显示富集 Ti，弱亏损 Nb、Ta，不相容元素 LREE、LILE 也与 EMI 源岩浆演化特征相似；④全岩 Sr-Nd 同位素成分（图 19-41）位于 EM I 型地幔区附近，且具有较高负 ε_{Nd} 值；⑤岩体结晶温度约 850~872℃，侵位压力为 3.0~3.3kPa，计算岩浆侵位深度为 13~14km（Liu et al.，2015）。

该套岩系岩浆结晶分异作用并不明显，但岩浆可能受到一定程度的地壳混染。这一特性重要表现在：①存在原生斜方辉石、角闪石、黑云母等富水矿物揭示了熔体岩浆中 H_2O 富集；②具有变化范围

极广的 SiO_2 含量；③哈克图解结晶演化趋势总体并不明显（李芳凝，2006；曹文燕，2010）；④微量元素中 Th、U 元素相对亏损，且富集 LILE、LREE 元素，与地壳源岩浆特征相似；⑤弱 δEu 正异常与高 Sr 元素含量不一致；⑥同位素 Sr-Nd 成分点也未位于典型 EMI 源岩浆区。

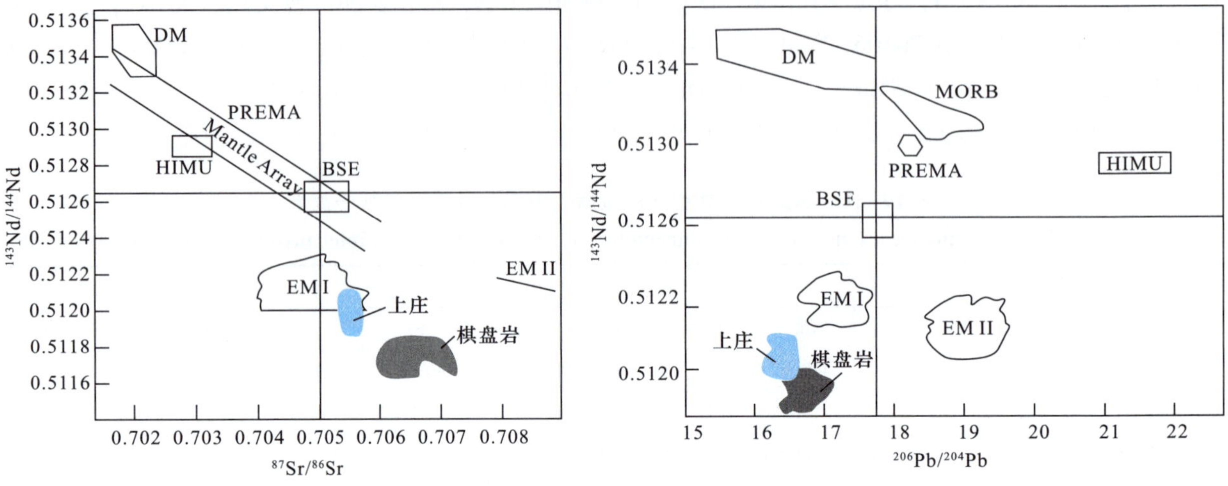

图 19-41　早白垩世基性岩 $^{87}Sr/^{86}Sr-^{143}Nd/^{144}Nd$ 和 $^{206}Pb/^{204}Pb-^{87}Nd/^{86}Nd$ 图解

（数据据苏尚国等，2006；李芳凝，2006；曹文燕，2010；汪洋等，2013；Liu et al.，2015）

Fig. 19-41　$^{87}Sr/^{86}Sr-^{143}Nd/^{144}Nd$ and $^{206}Pb/^{204}Pb-^{87}Nd/^{86}Nd$ diagram of Early Cretaceous basic intrusive rocks

6. 构造环境分析

中生代，尤其是早白垩世，华北克拉通东部经历了广泛的与岩石圈减薄有关的岩浆活动。这一时期也是燕山板内主要造山运动阶段。在 lgδ-lgτ 图（图 19-42）上，成分位于造山带及其派生火山岩区。早白垩世岩体铁钛氧化物矿化与层状火成岩序列还揭示了岩浆侵位伸展背景环境（Liu et al., 2015）。因此，早白垩世基性岩形成于板内造山期伸展构造背景环境。伸展作用导致的岩石圈地幔叠加 H_2O 作用发生减压熔融，产生基性岩浆并在壳内形成高位岩浆房发生混染演化，在短期多幕次侵位。该岩浆活动后续可能持续对壳幔产生巨大影响，以至随后大量发育酸性岩。

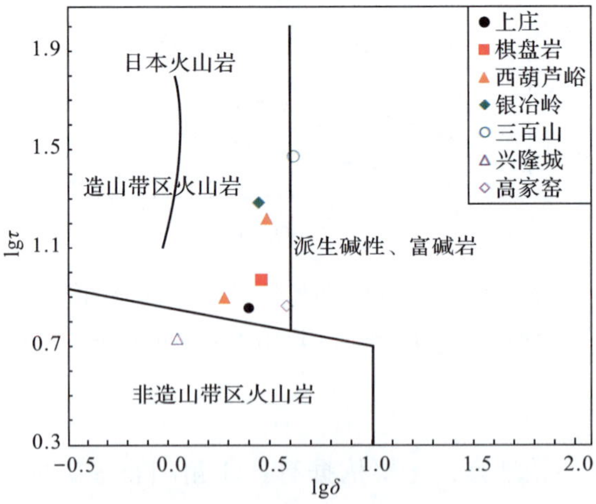

图 19-42　白垩纪基性岩 lgδ-lgτ 图

Fig. 19-42　lgδ-lgτ diagram of Cretaceous basic intrusive rocks

7. 时代归属

本期岩浆侵位时代厘定为早白垩世，主要依据包括：①岩体未发生变质，且蚀变较弱，基本无变形；②虽然侵入围岩地层较老，但棋盘岩岩体切割了晚侏罗世九山向斜，可能受控于九龙山向斜伴生的寺上断裂。银冶岭岩体侵入到黑云母花岗岩，该套黑云母花岗岩发育位置和岩石特征类似晚三叠世盘山黑云母花岗岩；③与伸展造山运动有关的岩石地球化学特征；④上庄辉长岩锆石 SHRIMP U-Pb 年龄为 128.8Ma（苏尚国等，2006）；⑤棋盘岩辉长岩锆石 SHRIMP U-Pb 年龄为（132.9±0.66）Ma（邓晋福等，2002；曹文燕，2010）。

（三）晚白垩世基性侵入岩

1. 地质产状与岩石学特征

区内晚白垩世基性侵入岩均分布于黄庄-高丽营-琉璃庙断裂以西，即房山-延庆岩浆岩亚带中，由高家窑、兴隆城 2 个岩体组成。

高家窑岩体：出露于延庆高家窑村，呈 SN 向蝌蚪状，出露面积 $9km^2$。岩体侵入于中生代髫髻山组中，未发育变质变形，但风化强烈。岩体岩性较为单一，为辉绿岩，内无岩相变化。岩石呈灰黑色，辉绿结构，块状构造，矿物组成为斜长石（约 70%）、角闪石（3%~5%）、辉石（约 20%），副矿物为磷灰石、磁铁矿。

兴隆城岩体：出露于怀柔兴隆城村，呈弓背朝西的弓形，长约 2km，宽约 100m，出露面积 $0.2km^2$。岩体主体顺层侵入于中元古界待建系下马岭组地层，接触界线明显，围岩角岩化；西可见其侵入到九渡河花岗斑岩中，接触面侵向东，倾角约 27°。岩体为辉绿岩。岩石呈灰绿色，辉绿结构，块状构造。斜长石自形-半自形板状，发育双晶，类型属于中长石（An=39.56），含量约 50%；单斜辉石他形粒状，强烈纤闪石化、绿泥石化，含量约 35%。副矿物为磁铁矿、钛铁矿和磷灰石。

2. 岩石化学与地球化学特征

岩石化学成分及参数见表 19-19。岩石呈现高 SiO_2、Al_2O_3，低 TiO_2、MgO、碱特征。里特曼指数 $\sigma=1.11\sim3.83$，$K_2O>Na_2O$，表明岩石为钾质亚碱性岩系列；固结指数 $SI=29.53\sim32.54$，分异指数 $DI=27.1\sim42.31$，反映岩浆发生弱分异作用；镁铁比值 $m/f=1.05\sim1.18$，属于铁质基性岩；$A/CNK=0.85\sim0.93$，$A/NK=2.1\sim3.2$，属于准铝质。

岩石 $\Sigma REE=129.4\times10^{-6}$，LREE/HREE=1.82，$(La/Yb)_N=5.25$，稀土总量较低，轻稀土弱富集，重稀土相对弱亏损；$\delta Eu=0.96$，极弱铕负异常；配分曲线为右缓倾光滑曲线。

与北京境内基性岩相比，Mn、Cu、Pb、V、Sc、Gd、Ta 含量最高，Ba 含量仅次于上庄辉长岩。

3. 岩石系列

该套岩系 $\sigma=1.11\sim3.83$，平均值<4，为亚碱性系列。SiO_2-K_2O 图解（图 19-40）厘定为高钾钙碱性-钾玄岩系列。

4. 成因类型

本套岩系岩浆来源于上地幔，主要依据包括：①晚白垩世华北克拉通东部岩石圈已经减薄；②稀土配分曲线表现为极缓右倾型，同地幔岩浆特征相似；③微量元素含量 V、Sc、Gd 等是区内含量最高；④微量元素蛛网图中 Nb 弱亏损。

5. 构造环境分析

本套岩系形成于燕山板内造山后造山环境，重要依据包括：①岩浆活动规模很小；②侵位时仅发育基性岩，缺乏中酸性岩浆活动；③矿物粒径较小，且发育气孔构造等揭示岩体侵位深度较浅；④岩体呈近南北向展布，同燕山末期构造线一致；⑤在 $\lg\delta-\lg\tau$ 图解（图 19-42）上，成分位于造山带与非造山带过渡地带。这些特征反映了华北克拉通东部岩石圈已经减薄，板内造山进入后造山伸展阶段。本区在拉张主应力作用下，上地幔物质发生局部熔融，形成基性岩浆并超浅成侵位。

6. 时代归属

本期岩浆侵位时代厘定为晚白垩世，主要依据包括：①岩体未发生变质、变形，且蚀变较弱；②岩体吞蚀破坏早白垩世酸性岩，还见其侵入中生代髫髻山组地层；③岩体呈近南北向展布，同燕山末期构造线一致；④岩浆活动规模很小，构造环境研究厘定与后造山伸展背景有关；⑤高家窑辉绿岩全岩 K-Ar 同位素年龄为 95Ma（1∶5 万永宁幅、四海幅区域地质调查报告，1993）。

第三节 脉 岩

区内除上述超基性—中基性岩体外，还发育一系列超基性—中基性岩墙（脉），对研究区构造-岩浆事件研究具有重大意义，本节总结如下。

一、新太古代变质基性岩脉群

以放马峪地区最为典型，区内统一命名为放马峪基性岩脉群（$Ar_3^3Fgn^{\psi\mu}$）。

（一）地质产状与岩石学特征

以密云水库周边、房山周口店和怀柔小东沟最为集中，是新太古代一次规模较大的基性岩浆侵入事件。

1. 放马峪地区

岩脉多数顺围岩片麻理侵入，局部小角度切割片麻理。平面呈透镜状，长百米至千米，宽数米至数百米，发生角闪岩相-绿片岩相变质变形，程度弱于围岩，使得同围岩接触界线清晰。岩脉岩石类型特征见表19-22。

表19-22 放马峪基性岩脉群岩石类型

Table 19-22 Rock types of Fangmayu basic dykes

岩石类型	岩性描述	分布
石榴角闪斜长二辉石岩	灰黑色，粒状变晶结构，“红眼圈”构造。紫苏辉石残余状，被透辉石、石榴子石、角闪石交代，粒径1~2cm，含量40%~60%，解理裂隙还交代皱晶状黑云母，并析出大量不规则状磁铁矿；石榴子石粉红色细粒状，集合体围绕辉石或斜长石分布，类型为铁铝榴石，常含不规则石英包裹体，含量10%~15%；斜长石他形粒状，相互紧密镶嵌，晶面干净，发育聚片双晶，少反条纹，类型为中长石（An=35~38）；石榴子石围绕绿色透辉石、斜长石形成环状“红眼圈”构造，核部辉石和环状石榴子石间发育变质反应生成的细小且干净斜长石和石英，形成圆形、椭圆形或不规则状双圈构造，双圈直径一般为1~1.5cm，最大达3~4cm	密云水库周边
含辉石榴斜长角闪岩	深绿—灰绿色，柱状变晶结构，弱片麻理构造。透辉石淡绿色粒状，晶面干净，含量>10%；角闪石黄绿色柱状，光性、晶形完善，晶体新鲜洁净，具弱阳起石化，显示新生矿物的特征，含量60%~80%；石榴子石粉红色细粒状，含量8%~15%；少量黑云母和析出磁铁矿呈他形粒状分布透辉石、斜长石边部	小东沟、密云水库周边
斜长角闪岩	深绿—灰绿色，粒、柱状变晶结构，弱片麻理构造。角闪石粒、柱状两类，粒径0.4~0.6mm，含量64%~90%。粒状晶色调不均匀，解理不发育，常与黑云母伴生，具强烈阳起石化、绿泥石化，推测为辉石类矿物退变质产物，含量5%~10%。柱状角闪石黄绿色，光性、晶形完善，晶体新鲜洁净，具弱阳起石化，显示新生矿物的特征，含量60%~80%；斜长石他形粒状，粒径0.2~0.4mm，具反条纹结构，绢云母化和钠黝帘石化，含量10%~35%。副矿物为磁铁矿、磷灰石、榍石、锆石和绿帘石。周口店地区岩性还发生糜棱岩化，角闪石由粗大的变形残斑和亚颗粒组成	周口店、小东沟

2. 密云水库周边

分布于密云水库周边，侵入于康各庄奥长花岗质片麻岩及更老变质体中，如东穆家峪顺片麻理侵入宋营子英云闪长质片麻岩，放马峪村和安子沟铁矿区分别可见其斜切片麻理侵入于沙厂英云闪长质片麻岩（图19-43）和密云岩群磁铁石英岩中。规模较大岩脉发育弱相变：中心相矿物颗粒粗，紫苏辉石含量高，且普遍发育“红眼圈”构造；边缘相矿物颗粒细，紫苏辉石含量低。岩石类型以石榴角闪斜长二辉石岩为主，其次为含辉石榴斜长角闪岩，变质矿物组合为Hy+Di+Hb+Gt+Pl。

3. 小东沟地区

分布于怀柔东石门-小东沟、西驼古、牤牛沟-邓厂等，岩脉顺围岩片麻理侵入于宋营子英云闪长质片麻岩中，受晚期变形影响，发育石香肠化、布丁化现象。岩石类型为斜长角闪岩和含辉石榴斜长角闪岩。

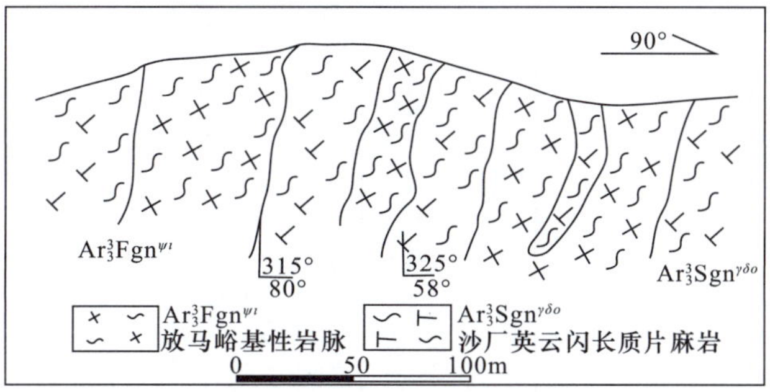

图 19-43 放马峪村西基性岩与沙厂英云闪长质片麻岩侵入接触关系示意图

（据鲍亦冈等，2001）

Fig. 19-43 Geological Sketch showing the intrusive contact between Fangmayun basic rocks and Shachang tonalitic gneiss

4. 周口店地区

分布于周口店官地，岩墙或岩脉顺片麻理侵入于宋营子英云闪长质片麻岩中，岩石类型为角闪斜长岩。脉群均受晚期变形影响，发育石香肠化、布丁化现象，分布于韧性剪切带中脉岩平面呈带拖尾的旋转透镜状。

（二）岩石化学与地球化学特征

化学成分及参数见表 19-23。岩石呈高 SiO_2、CaO、MgO、FeO^*，低碱、FeO^*、TiO_2 特征；里特曼指数 σ=0.31～1.18，表明岩石为亚碱性岩系列；固结指数 SI=19.36～51.69，分异指数 DI=18.76～31.05，反映岩浆分异程度低；镁铁比值 m/f=0.51～2.23，属于铁质-富铁质基性岩；A/CNK=0.42～0.62，A/NK=2.55～3.16，属于准铝质。

稀土元素成分及参数见表 19-24。ΣREE = 97.36 $\times10^{-6}$ ～ 191.95 $\times10^{-6}$，LREE/HREE = 5.36～10.34，$(La/Yb)_N$=7.50～15.19，稀土总量较低，轻稀土弱富集，重稀土相对亏损；δEu=0.74～0.92，弱铕负异常；配分曲线为右缓倾光滑曲线（图 19-44）。

微量元素成分及参数见表 19-25。与基性岩相比，Sr、Y、Ba 含量低，其余元素含量相当。微量元素蛛网图（图 19-45）上，富集大离子亲石元素 Rb、K、Ba，亏损高场强元素 Nb、Zr、U、Ta 和 Ti。

（三）锆石 Hf 同位素特征

周口店斜长角闪岩锆石 $^{176}Hf/^{177}Hf$ = 0.281256～0.281395，$\varepsilon_{Hf}(t)$ = 0.07～4.76（t = 2350～2510Ma），t_{DM}=2550～2720（苑东洋，2016）。

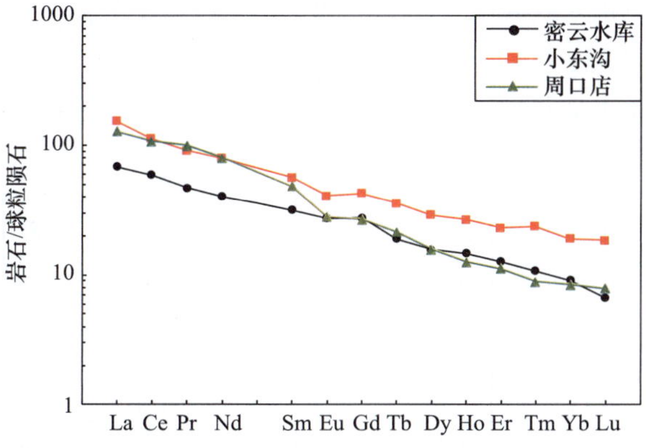

图 19-44 放马峪基性侵入岩稀土元素配分模式图

Fig. 19-44 Chondrite-normalized REE patterns of the Fangmayun basic rocks

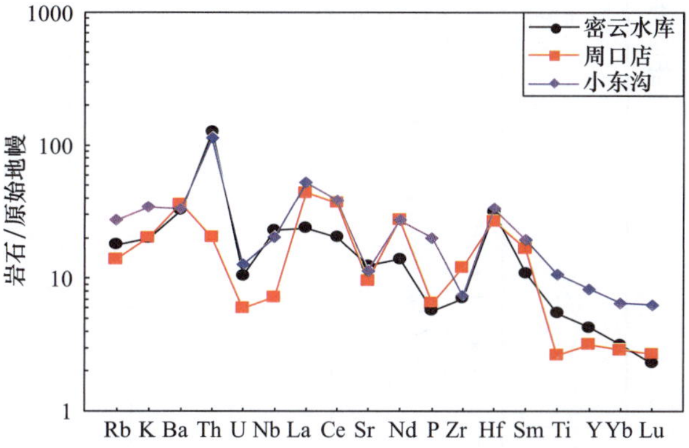

图 19-45 放马峪基性侵入岩微量元素蛛网图

Fig. 19-45 Trace element spider diagram of the Fangmayun basic rocks

表 19-23 放马峪基性岩岩石化学成分及特征参数一览表

Table 19-23 The chemical composition and main parameters of the Fangmayun basic rocks

序号	地点	岩性	氧化物含量/%													特征参数					
			SiO_2	TiO_2	Al_2O_3	Fe_2O_3	FeO	MnO	MgO	CaO	Na_2O	K_2O	P_2O_5	Los	总和	*DI*	*SI*	*σ*	*A/CNK*	*A/NK*	*m/f*
1	密云水库	石榴斜长辉石岩	52.15	1.19	12.13	4.18	7.76	0.22	7.60	8.67	2.49	0.60	0.12	1.18	98.30	31.05	33.68	0.95	0.59	2.55	1.15
2	周口店	含辉石石榴斜长角闪岩	50.76	0.58	7.99	3.67	6.87	0.26	13.03	8.96	1.13	0.61	0.14	1.53	95.55	18.76	51.69	0.31	0.42	3.16	2.23
3	小东沟	斜长角闪岩	49.42	2.37	12.64	6.30	10.39	0.22	4.66	8.98	1.85	1.05	0.44	1.22	99.52	29.44	19.36	1.18	0.62	3.03	0.51

注：1、3 数据引自 1∶5 万琉璃庙幅、密云幅、番字牌幅区域地质调查报告，1990、1993、1996。2 数据引自苑东洋等，2016。

表 19-24 放马峪基性岩稀土元素成分及特征参数一览表

Table 19-24 REE contents andmain parameters of the Fangmayun basic rocks

序号	地点	岩性	稀土元素含量/10^{-6}																特征参数			
			La	Ce	Pr	Nd	Sm	Eu	Gd	Tb	Dy	Ho	Er	Tm	Yb	Lu	Y	ΣREE	LREE/HREE	$(La/Yb)_N$	δEu	δCe
1	密云水库	石榴斜长辉石岩	16.25	36.05	4.47	18.85	4.86	1.58	5.68	0.71	3.99	0.83	2.10	0.28	1.56	0.17	19.40	97.36	5.36	7.50	0.92	1.02
2	小东沟	含辉石石榴斜长角闪岩	36.80	69.97	8.80	37.70	8.82	2.39	8.90	1.36	7.44	1.53	3.88	0.61	3.27	0.48	38.60	191.95	5.99	8.07	0.82	0.92
3	周口店	斜长角闪岩	30.71	66.61	9.61	37.96	7.52	1.64	5.55	0.82	4.05	0.72	1.88	0.23	1.45	0.20	14.75	168.96	10.34	15.19	0.74	0.94

注：1、3 数据引自 1∶5 万琉璃庙幅、密云幅、番字牌幅区域地质调查报告，1990、1993、1996。2 数据引自苑东洋等，2016。

表 19-25 放马峪基性岩微量元素成分及特征参数一览表

Table 19-25 Trace element contents and main parameters of the Fangmayun basic rocks

序号	地点	岩性	微量元素含量/10^{-6}											特征参数	
			Rb	K	Ba	Th	U	Nb	Sr	P	Zr	Hf	Ti	Rb/Sr	Nb/Ta
1	密云水库	石榴斜长辉石岩	11.30	5017.67	226.55	10.78	0.22	16.37	262.75	543.15	79.31	9.60	7120.73	0.04	—
2	周口店	斜长角闪岩	8.97	5091.46	255.33	1.75	0.13	5.22	205.80	625.59	136.70	8.40	3497.08	0.04	—
3	小东沟	含辉斜长角闪岩	17.60	8688.69	235.40	9.80	0.27	14.70	243.80	1934.97	83.70	10.50	14208.15	0.07	—

注：1、3 数据引自 1∶5 万琉璃庙幅、密云幅、番字牌幅区域地质调查报告，1990、1993、1996。2 数据引自苑东洋等，2016。

（四）变质变形特征

该套侵入岩紫苏辉石、普通辉石较为常见，反映侵入后曾遭受麻粒岩相变质作用。同时，紫苏辉石残余状退变成石榴子石、角闪石、磁铁矿等，又受后期退变质作用，且变质形成的角闪石定向排列，反映了角闪岩相变质叠加。周口店地区脉岩还遭受糜棱岩化变质变形。

（五）原岩恢复

上述地质产状、岩石学和岩石地球化学等特征，特别是变质矿物组合和残留的原生结构、构造，揭示了该变质岩原岩为正变质岩。在变质岩原岩恢复 Si−[(al+fm)−(c+alk)]（图 19−46）图解上，成分点也落在火山岩区。

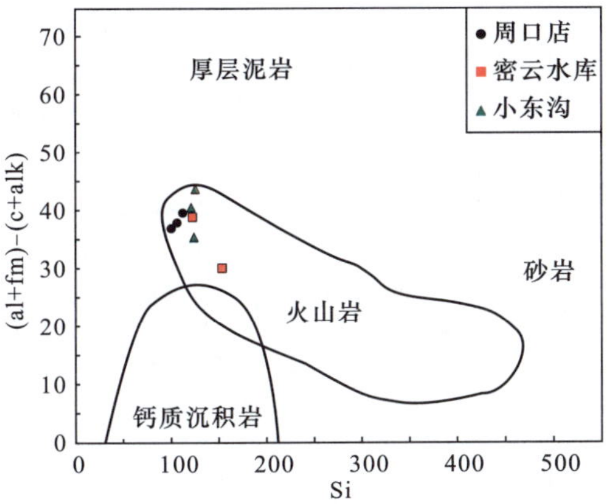

图 19−46　放马峪基性岩 Si−[(al+fm)−(c+alk)]变质岩原岩恢复图解

Fig. 19−46　Si−[(al+fm)−(cl+alk)] analysis diagram of Fangmayu basic rocks

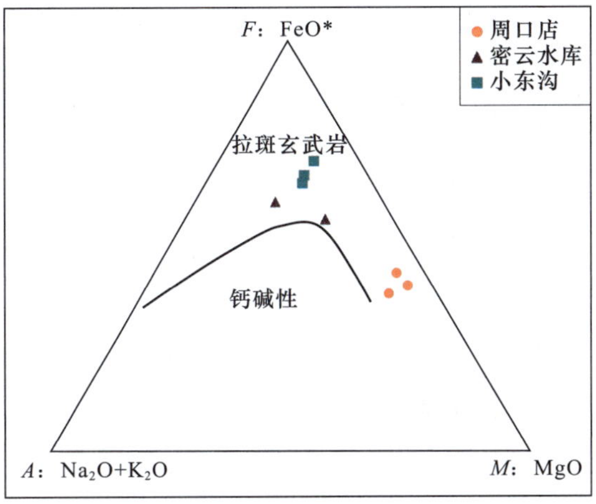

图 19−47　放马峪基性岩 *F−A−M* 图解

Fig. 19−47　The ternary plot *F−A−M* of Fangmayu basic rocks

（六）岩石系列

本套岩系 $\sigma=0.31\sim1.18<4$，为亚碱性系列。岩石由基性斜长石和紫苏辉石、普通辉石和透辉石组成，呈现富硅（SiO_2>49%）、贫碱（Na_2O+K_2O<5%）、贫钛（TiO_2<3%）特征，与拉斑玄武岩相似。在 *F−A−M* 图解（图 19−47）上，成分点位于拉斑玄武岩系列。

（七）成因类型与岩浆演化

本套岩系岩浆可能来源于富集型上地幔玄武质岩部分熔融，主要依据包括：①辉石、角闪石和基性斜长石矿物组成，标型矿物计算也包含透辉石和紫苏辉石两类辉石；②Na_2O/K_2O 比值较大，高 CaO 含量岩石地球化学特征；③稀土元素总量低，且轻重稀土弱分馏与地幔源岩浆特征相似；④La/Ta 值高同俯冲大洋板片流体富集型地幔源区相似；⑤锆石 Hf 同位素成分点主要处于亏损地幔演化线和球粒陨石之间（图 19−48）。同时，周口店岩墙微量元素与稳定元素具有较好正相关揭示了岩石形成经历了分离结晶作用，橄榄石和斜长石是其重要分离结晶矿物，而地壳混染程度不高（苑东洋等，2016）。

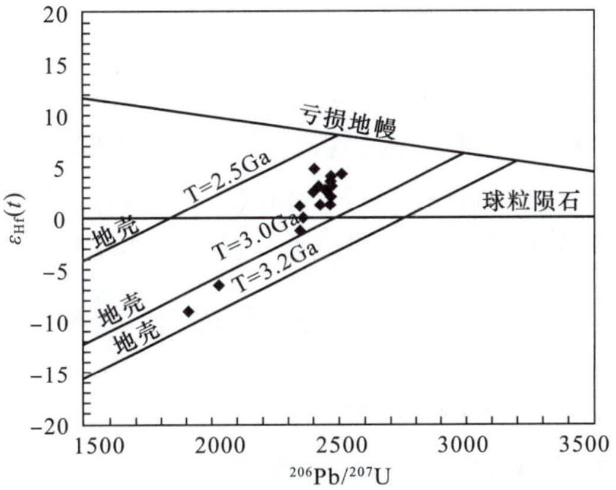

图 19−48　周口店斜长角闪岩锆石 Hf 同位素特征图（据苑东洋等，2016）

Fig. 19−48　Zircon Hf isotopic compositions of amphibolite in Zhoukoudian

（八）构造环境分析

放马峪基性岩墙侵位时代稍晚于新太古代晚期具有岛弧成因的大规模 TTG（宋营子英云闪长质、荆子峪花岗闪长质和康各庄奥长花岗质片麻岩组合）侵位时代；这一岩浆事件在华北板块中部广泛发育，如河北东部曹庄杂岩中橄榄辉长岩岩墙（Li et al., 2010）、遵化杂岩中斜长角闪岩等（Polat et al., 2006）等，研究表明它们多数与弧后盆地伸展作用有关；上述地球化学和 Lu-Hf 同位素特征与岛弧环境下岩石特征也具有一定相似性；在 F1-F2 和 F2-F3 构造环境判别图（图 19-49）中，成分点远离板内玄武岩而接近岛弧玄武岩。因此，放马峪基性岩墙形成于活动大陆边缘岛弧构造背景。

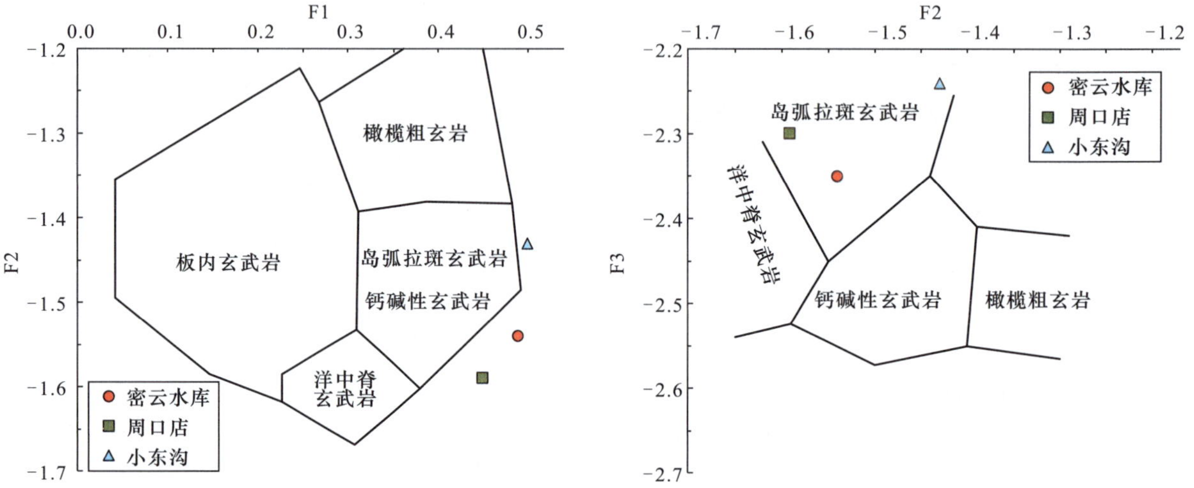

图 19-49　F1-F2 和 F2-F3 构造环境判别图

（据 Pearce，1976）

Fig. 19-49　F1-F2 and F2-F3 diagram of tectonic discrimination

（九）时代归属

本期岩脉侵位时代厘定为新太古代晚期，主要依据包括：①侵入宋营子等新太古代片麻岩岩体或密云岩群表壳岩中；②遭受强烈高级角闪岩相变质及复杂构造变形；③岩石地球化学特征揭示与活动大陆边缘岛弧有关的构造背景环境；④周口店脉岩（斜长角闪岩）锆石 U-Pb 同位素年龄为（2548±23）Ma、（2538±34）Ma 和（2551±37）Ma（刘兵等，2008；苑东洋等，2016）；放马峪脉岩石榴斜长辉石岩锆石 U-Pb 同位素年龄为（2497±15）Ma（方同明等，2016）。

二、中元古代中基性岩墙——高岭岩墙与熊儿寨岩颈

（一）高岭岩墙

1. 地质产状与岩石学特征

以密云高岭最为集中，故命名为高岭岩墙，又称之为密云岩墙（彭澎等，2011）。岩脉多数顺围岩片麻理侵入，走向 25°～45°。平面呈透镜状，长百米至千米，宽数米至数百米，如规模较大高岭栗榛寨-大开岭辉长岩脉全长达 9km，宽 600～800m。岩脉未受区域变质作用，仅弱绢云母化、绿泥石化，变质变形程度明显弱于围岩。一般岩脉少有岩相变化，但大型岩脉仍可依照矿物颗粒大小划分出中粒中心相和细粒边缘相。

岩脉岩性为辉长岩，深灰色，辉长结构，致密块状构造。斜长石自形-半自形长柱状，粒径 1～2mm，类型为中长石-拉长石，含量约 65%；普通辉石半自形柱状，粒径 1.5～2.5mm，弱绿泥石化，内包裹斜长石细晶，含量约 30%；黑云母片状，粒径 0.3～0.5mm，含量约 2%；绿泥石鳞片状，分布于黑云母、辉石边部，揭示其蚀变成因，含量 1%～2%。副矿物为磁铁矿和榍石。

2. 岩石化学与地球化学特征

化学成分及参数见表 19-26。岩石呈高 SiO_2、FeO^*、CaO，低碱、TiO_2、Al_2O_3 特征；里特曼指数 $\sigma=1.12$，表明岩石为亚碱性岩系列；固结指数 $SI=26.84$，分异指数 $DI=26.83$，反映岩浆分异程度低；镁铁比值 $m/f=0.53$，属于铁质基性岩；$A/CNK=0.65$，$A/NK=3.26$，属于准铝质。

稀土元素成分及参数见表 19-27。$\Sigma REE=73.78\times10^{-6}$，LREE/HREE = 4.72，$(La/Yb)_N=4.74$，稀土总量较低，轻稀土弱富集，重稀土相对亏损；$\delta Eu=1.06$，极弱铕正异常；配分曲线为右缓倾光滑曲线（图 19-50）。

微量元素成分及参数见表 19-28。与黎彤（1962）辉长岩相比，各元素含量偏低。微量元素蛛网图（图 19-51）上，富集大离子亲石元素 Ba，极弱亏损高场强元素 Nb、Th 和 U。

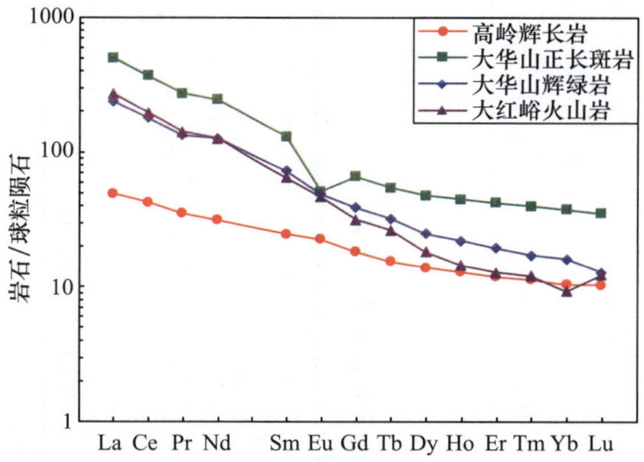

图 19-50 中元古代基性岩墙稀土元素配分模式图

Fig. 19-50 Chondrite-normalized REE patterns of Mesoproterozoic basic dike swarm

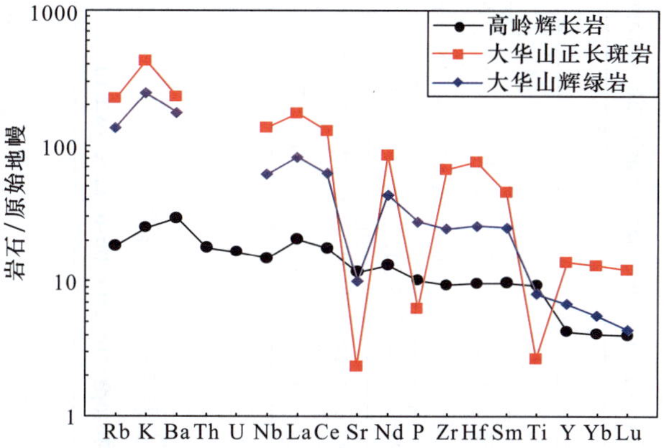

图 19-51 中元古代基性岩墙微量元素蛛网图

Fig. 19-51 Trace element spider diagram of Mesoproterozoic basic dike swarm

3. 岩石系列划分

本套岩系 $\sigma=1.12<4$，为亚碱性系列。岩石地球化学呈现富硅（$SiO_2>49\%$）、贫碱（$Na_2O+K_2O<5\%$）、贫钛（$TiO_2<3\%$）特征，与拉斑玄武岩相似。在 *F-A-M* 图解（图 19-52）上，成分点位于拉斑玄武岩系列。

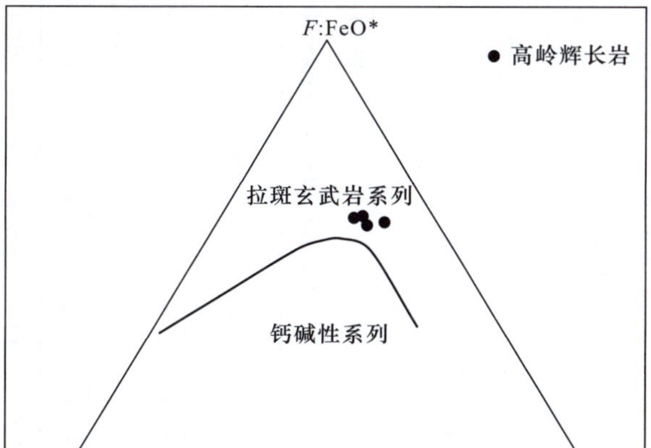

图 19-52 高岭基性岩墙 *F-A-M* 图

Fig. 19-52 The ternary plot *F-A-M* of Gaoling basic dike swarm

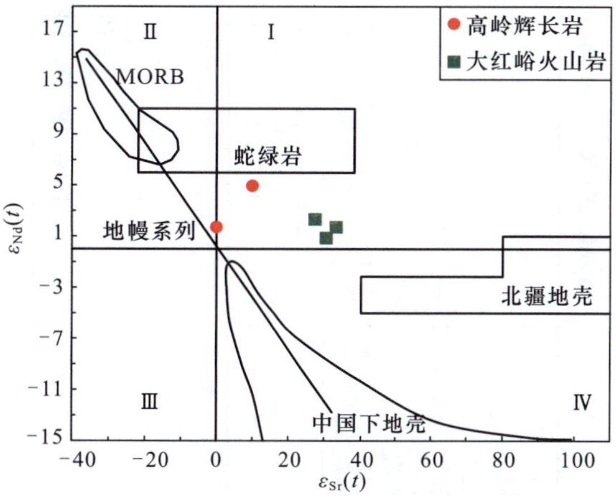

图 19-53 中元古代基性岩墙和火山岩 $\varepsilon_{Sr}(t)-\varepsilon_{Nd}(t)$ 图

Fig. 19-53 $\varepsilon_{Sr}(t)-\varepsilon_{Nd}(t)$ diagram of Mesoproterozoic basic dike swarm and volcanic rock

表 19-26 中元古代基性岩墙岩石化学成分及特征参数一览表

Table 19-26 The chemical composition and main parameters of Mesoproterozoic basic dike swarm

序号	岩体	岩性	氧化物含量/%													特征参数					
			SiO_2	TiO_2	Al_2O_3	Fe_2O_3	FeO	MnO	MgO	CaO	Na_2O	K_2O	P_2O_5	Los	总和	*DI*	*SI*	σ	*A/CNK*	*A/NK*	*m/f*
1	高岭	辉长岩	49. 36	2. 02	13. 77	4. 91	9. 12	0. 19	6. 14	9. 40	2. 08	0. 75	0. 22	2. 04	99. 99	26. 83	26. 84	1. 12	0. 65	3. 26	0. 53
2	熊儿寨	正长斑岩	58. 55	0. 58	16. 08	3. 48	1. 62	0. 12	1. 19	1. 83	0. 37	12. 73	0. 14	1. 39	98. 07	83. 92	6. 17	10. 45	0. 91	1. 12	0. 43
3	熊儿寨	辉绿岩	49. 52	1. 75	15. 93	4. 68	5. 93	0. 13	4. 51	3. 09	1. 55	7. 38	0. 60	5. 31	100. 37	59. 65	18. 76	9. 69	0. 99	1. 51	0. 57
4	大红峪期	玄武岩	47. 70	2. 00	16. 28	7. 29	4. 30	0. 12	5. 02	3. 91	1. 96	5. 61	0. 56	5. 94	100. 69	52. 64	21. 00	8. 57	0. 99	1. 75	0. 75
5	大红峪期	粗面岩	54. 48	1. 12	15. 71	6. 41	1. 75	0. 08	2. 52	2. 27	0. 64	11. 70	0. 36	3. 65	100. 69	77. 00	11. 05	12. 24	0. 88	1. 15	0. 70

注：数据为均值；1 引自彭彭，2011；2~3 引自 1∶5 万大华山幅、木林幅、平谷幅区域地质调查报告，1999。

表 19-27 中元古代基性岩墙稀土元素成分及特征参数一览表

Table 19-27 REE contents andmain parameter of Mesoproterozoic basic dike swarm

序号	岩体	岩性	稀土元素含量/10^{-6}																特征参数			
			La	Ce	Pr	Nd	Sm	Eu	Gd	Tb	Dy	Ho	Er	Tm	Yb	Lu	Y	ΣREE	LREE/HREE	$(La/Yb)_N$	δEu	δCe
1	高岭	辉长岩	11. 70	26. 03	3. 36	14. 68	3. 79	1. 31	3. 76	0. 58	3. 53	0. 73	1. 97	0. 29	1. 77	0. 26	18. 00	73. 78	4. 72	4. 74	1. 06	1. 02
2	熊儿寨	正长斑岩	119. 04	228. 33	26. 02	115. 07	20. 04	2. 97	13. 53	2. 05	12. 18	2. 53	7. 03	1. 02	6. 39	0. 89	0. 00	557. 10	11. 21	13. 35	0. 55	1. 01
3	熊儿寨	辉绿岩	56. 87	110. 55	12. 75	59. 04	11. 11	2. 85	7. 99	1. 21	6. 33	1. 25	3. 22	0. 44	2. 74	0. 33	0. 00	276. 66	10. 78	14. 91	0. 92	1. 01
4	大红峪组	碱玄岩	65. 23	120. 90	13. 44	59. 16	9. 90	2. 70	6. 53	0. 99	4. 66	0. 82	2. 12	0. 31	1. 58	0. 31	0. 00	288. 67	15. 66	29. 55	1. 03	1. 00

注：数据为均值；1 引自彭彭，2011；2~3 引自 1∶5 万大华山幅、木林幅、平谷幅区域地质调查报告，1999。

表 19-28 中元古代基性岩墙微量元素成分及特征参数一览表

Table 19-28 Trace element contents and main parameters of Mesoproterozoic basic dike swarm

序号	岩体	岩性	微量元素含量/10^{-6}											特征参数	
			Rb	K	Ba	Th	U	Nb	Sr	P	Zr	Hf	Ti	Rb/Sr	Nb/Ta
1	高岭	辉长岩	11. 55	6205. 22	203. 30	1. 50	0. 35	10. 49	245. 00	960. 21	104. 65	2. 96	12094. 91	0. 05	—
2	熊儿寨	正长斑岩	142. 33	105703. 22	1606. 33	—	—	96. 67	49. 56	596. 50	747. 33	23. 33	3477. 10	2. 87	—
3	熊儿寨	辉绿岩	86. 00	61222. 09	1225. 40	—	—	44. 00	211. 68	2618. 76	273. 50	7. 93	10491. 25	0. 41	—

注：数据为均值；1 引自彭彭，2011；2~3 引自 1∶5 万大华山幅、木林幅、平谷幅区域地质调查报告，1999。

4. Sr-Nd 同位素特征

辉长岩全岩 Rb=$7.81\times10^{-6}\sim17.6\times10^{-6}$，Sr=$129.4\times10^{-6}\sim394.6\times10^{-6}$，Sm=$2.671\times10^{-6}\sim5.876\times10^{-6}$，Nd=$9.483\times10^{-6}\sim27.54\times10^{-6}$，$^{87}Rb/^{86}Sr=0.1264\sim0.1747$，$^{87}Sr/^{86}Sr=0.706308\sim0.706828$，$^{143}Nd/^{144}Nd=0.512116\sim0.512420$，$^{147}Sm/^{144}Nd=0.1290\sim0.1703$（彭彭等，2011），与常规地幔接近。而 $\varepsilon_{Nd}(t)=1.61\sim4.87$（$t$=1731Ma），$I_{Nd}=0.510481\sim0.510647$，$I_{Sr}=0.70248\sim0.70316$，$t_{DM}$=1859~2560。

5. 成因类型与岩浆演化

本套岩系岩浆可能来源于常规地幔-弱亏损型地幔，主要依据包括：①辉石和中-基性斜长石矿物组成，标型矿物计算中缺乏刚玉而富含透辉石和紫苏辉石矿物；②Na_2O/K_2O 比值较大，高 CaO 含量岩石地球化学特征；③稀土元素总量低，且轻重稀土弱分馏与地幔源岩浆特征相似；④蛛网图中富集 Ba，极弱亏损 Nb、Ta 和 Ti 高场强元素；⑤全岩 Sr-Nd 同位素位于常规地幔区（图 19-53）。同时，高岭岩墙 *SI*、*DI* 指数较小，无明显 Eu 正或负异常，且微量元素蛛网图中高场强元素和大离子亲石元素亏损或富集并不明显，揭示了岩石形成过程中可能未受到地壳混染及分离结晶作用。

6. 构造环境分析

高岭岩墙脉岩宽度较大，类似与地幔柱有关的岩墙；岩墙走向北东，同区内中元古代燕山裂陷槽构造线方向一致；岩石微量元素 Nb、Ta、Ti 极弱亏损，明显不同于岛弧玄武岩；在常量、微量元素构造环境判别图解（图 19-54）中，成分点位于板内玄武岩区；地质产状、地球化学等特性还不同于长城群所在的裂谷有成因联系的斜长岩-环斑花岗岩杂岩体（彭彭等，2011）或大红峪期火山岩。因此，高岭岩墙形成于华北克拉通板内非造山构造环境，可能是燕山裂陷槽最早起始裂解伸展背景下一次岩浆侵位。

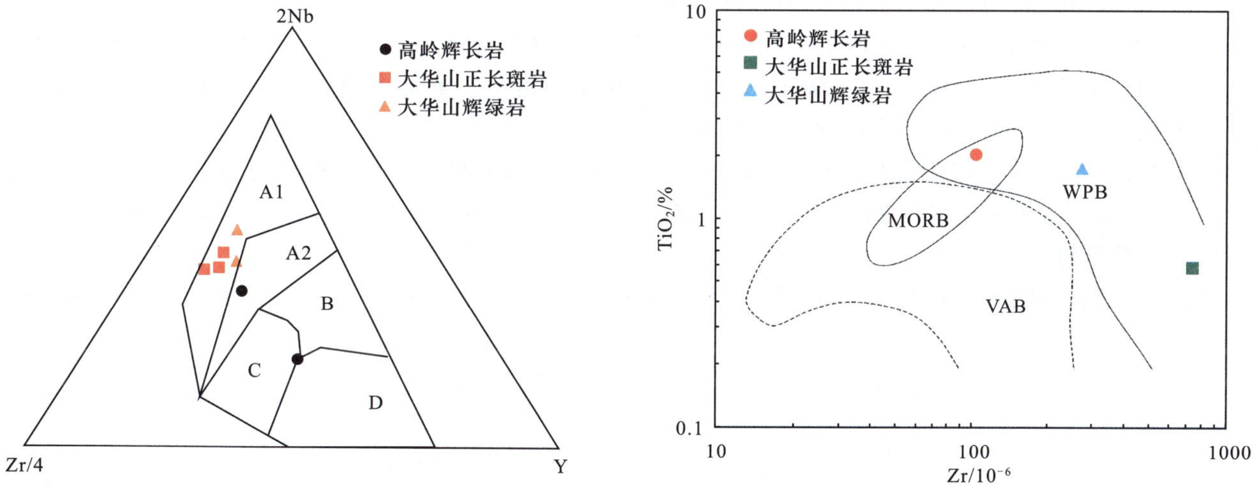

图 19-54 Zr-TiO_2 和 2Nb-Zr/4-Y 构造环境判别图

Fig. 19-54 Zr-TiO_2 and 2Nb-Zr/4-Y diagram of tectonic discrimination

A1+A2—板内碱性玄武岩；A2+C—板内拉斑玄武岩；B—P 型 MORB；D—N 型 MORB；C+D—火山弧玄武岩；WPB—板内玄武岩；MORB—洋中脊玄武岩；VAB—火山弧玄武岩

7. 时代归属

本期岩脉侵位时代厘定为中元古代早期，主要依据包括：①广泛侵入新太古代片麻岩岩体或密云岩群表壳岩中；②未遭受强烈高级变质及复杂构造变形；③高岭桑园村辉长岩脉被中元古界长城系常州沟组底砾岩不整合覆盖（彭彭等，2011）；④高岭桑园村脉岩（辉长岩）锆石 U-Pb 同位素年龄为（1731±4）Ma（彭彭等，2011）。

（二）熊儿寨岩颈

1. 地质产状与岩石学特征

分布于平谷白云寺–将军关一带，邻区河北境内也广泛发育，由数十个直径约几十米至几百米近椭圆形或圆形岩瘤组成，单个出露面积<0.5km^2。以熊儿寨最为集中，故命名为熊儿寨岩颈。岩颈受控于断裂构造，产出于NW、NNW与NE向断裂交切部位，近于直立侵入到中元古界团山子组及更老地层，接触面平整，围岩蚀变与机械破碎均不明显，表现出沿早期火山通道侵位特性。在平谷黄松峪见其被中元古界铁岭组覆盖。岩颈弱岩相变化，中心相发育少量斑晶，发育岩石类型为辉绿岩及其分异的正长斑岩。

辉绿岩：灰绿色–墨绿色，斑状结构，基质辉绿结构，致密块状构造。斑晶为斜长石（主）和辉石，含量5%~20%。其中，斜长石自形板状，发育聚片、卡钠复合双晶，粒径3~5mm，类型为中长石—培长石；基质斜长石细小板条状，类型为中长石，粒径约0.1mm，含量45%~50%；基质绿泥石为辉石等暗色矿物蚀变矿物，少部分保留辉石假象，粒径0.05~0.1mm，含量35%。副矿物为磁铁矿等，含量约15%。

正长斑岩：表现出辉绿岩岩浆演化分异岩性特征，岩石肉红色，碎斑结构，似粗面结构，块状构造。斑晶正长石，碎斑状，发育卡式复合双晶，粒径1~3mm，含量3%~5%。基质正长石细小板条状，杂乱或半定向排列，含量75%~80%；基质绿泥石片状，少部分细小纤维状，绿–淡绿色多色性，由角闪石或黑云母蚀变而成，含量约10%；黑云母片状，浅棕色多色性，部分蚀变绿泥石化，含量约5%；石英他形粒状，极少量。

2. 岩石化学与地球化学特征

化学成分及参数见表19–26，与大红峪期火山岩高度相似，表现出高SiO_2、Al_2O_3、K_2O，低Na_2O、CaO、TiO_2特征；里特曼指数σ=9.69~10.45，为碱性岩系列；固结指数SI=6.17~18.76，分异指数DI=59.65~83.92，反映岩浆高度分异；镁铁比值m/f=0.43~0.57，属于铁质—富铁质基性岩；A/CNK=0.91~0.99，A/NK=1.12~1.51，属于准铝质。

稀土元素成分及参数见表19–27。ΣREE = 276.66 $\times10^{-6}$ ~ 557.10 $\times10^{-6}$，LREE/HREE = 10.78 ~ 11.21，$(La/Yb)_N$ = 13.35~14.91，稀土总量高，轻稀土强富集，重稀土相对亏损；δEu = 0.55~0.92，铕负异常；配分曲线为右陡倾光滑曲线（图19–50）。

微量元素成分及参数见表19–28，与大红峪期火山岩基本相似。微量元素蛛网图（图19–51）上，富集大离子亲石元素K、Rb、Ba，强烈亏损大离子亲石元素Sr和高场强元素Ti、P，弱亏损Nb。

3. 岩石系列

该套岩系σ=9.69~10.45>4，为碱性系列。在SiO_2–Alk图解（图19–55）上，成分点位于碱性岩系列。

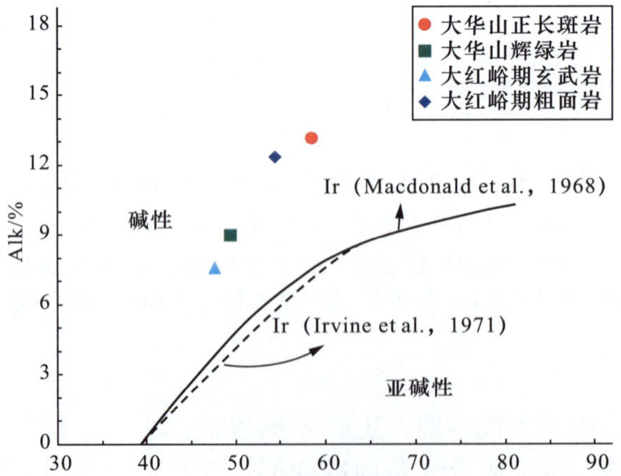

图19–55 熊儿寨基性岩墙SiO_2–Alk图

Fig. 19–55 The ternary plot SiO_2–Alk of the Xiongerzhai basic dike swarm

4. 成因类型

该套岩系岩浆可能来源于富集型地幔，主要依据包括：①沿大红峪期火山通道侵位；②辉石和基性斜长石矿物组成，标型矿物计算包含透辉石和紫苏辉石两类辉石；③ $Al_2O_3/TiO_2<30$ 和 La/Nd 比值等特征与幔源岩浆相似；④稀土元素和微量元素与大红峪期火山岩相似，大红峪期火山岩来源于富集型地幔。与原始地幔相比，该套岩系发育较高 K、Rb、Ti 和较低 Cr、Co、Ni、Sr，表明富集地幔可能经历了低度部分熔融。

该套岩系高 K_2O、ΣREE 岩石地球化学及其相关参数（如 Ce/Nd）比值特性还揭示了岩浆侵位过程中受到地壳物质混染。而辉绿岩至正长斑岩岩性演化、岩石化学参数 *DI* 或 *SI*、哈克图解高线性相关等还揭示了岩浆经历了分离结晶作用。Al_2O_3、TiO_2、P_2O_5 与 SiO_2 相关性揭示了岩浆演化过程中单斜辉石、磷灰石及钛铁矿分离结晶，δEu 负异常则揭示了斜长石的分离结晶。

5. 构造环境分析

该套岩系沿大红峪期火山通道侵位，同大红峪期火山岩一致的岩石地球化学特性，表现出火山活动晚期或末期岩浆侵位特征；岩石微量元素 Nb、Ta 极弱亏损，明显不同于典型岛弧玄武岩；在常量、微量元素构造环境判别图解（图 19-54）中，成分点位于板内玄武岩区。1800~1600Ma 间，在华北陆块发生了张裂为主的地壳运动，形成了一次重要的夭折裂谷或裂陷槽盆地（Zhao et al., 2002）。因此，熊儿寨岩颈侵位形成于华北克拉通板内非造山构造环境，是燕山裂陷槽伸展背景下裂陷夭折末期一次岩浆侵位。

6. 时代归属

本期岩脉侵位时代厘定为中元古代早期，主要依据包括：①侵入到中元古界团山子组及更老地层中；②沿大红峪期火山通道侵位，平谷黄松峪见其被中元古界铁岭组覆盖；③具有和大红峪期火山岩相似岩石学、地球化学演化特征，形成时代稍晚于大红峪期火山岩。

三、中生代中基性岩脉及与其共生岩脉

（一）早侏罗世岩脉

广泛分布于前中生代地层或侵入体中，以怀柔长哨营、门头沟妙峰山最为集中，是中生代早期重要的基性岩浆侵入事件。

1. 地质产状与岩石学特征

（1）银河沟辉长岩

分布于怀柔长哨营西，呈近东西向岩墙或断块产出，岩脉与赤城-古北口断裂带密切有关，周缘主体被此断裂围限，局部见其侵入中元古代石英正长岩，可见围岩顶垂体或捕虏体（图 19-56）。岩脉无岩相变化，岩性为辉长岩，灰色，辉长结构，块状构造。斜长石半自形粒状，含量约 65%；辉石半自形粒状，部分蚀变为角闪石、黑云母，含量约 35%。

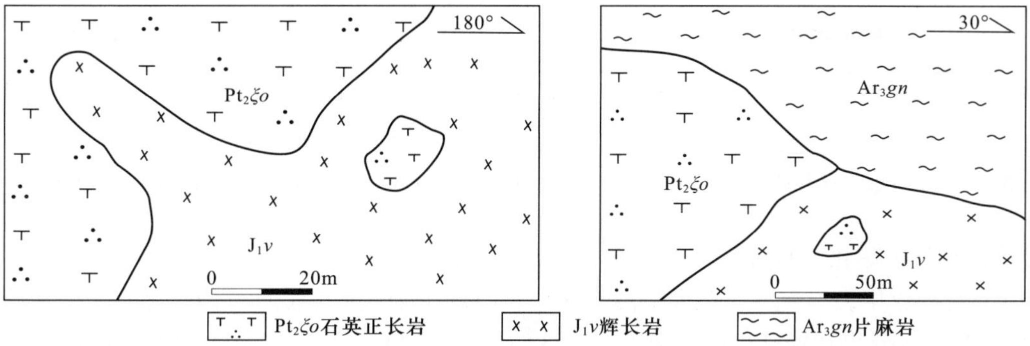

图 19-56 银河沟辉长岩与石英正长岩侵入接触关系示意图（长哨营西）

Fig. 19-56 Geological Sketch showing the intrusive contact between gabbro and quartz syenite in Yinhegou

（2）黄土贵辉长岩

广泛分布于门头沟青白口、大村、下苇甸等，由分散零星系列辉长岩脉组成。单条岩脉宽1~2m，长约200m。岩脉群顺层侵入中元古界下马岭组地层中，走向取决于地层产状。由于脉体规模不大，内无岩相变化，外接触带发育弱烘烤现象。岩石类型为辉长岩，灰黑-黑色、弱蚀变浅灰绿色，辉长结构，致密块状构造。斜长石板条状、柱状，发育两组近直交解理，强钠黝帘石化、绢云母化，粒径0.1~1mm，含量50%~60%；单斜辉石长柱状，发育两组直交解理，简单双晶，部分呈聚合体形式，粒径0.1~0.5mm，含量45%~50%。副矿物为磁铁矿、磷灰石、榍石、黄铁矿。

（3）青白口闪长岩、闪长玢岩

分布于门头沟青白口一带，围绕青白口穹隆环状产出，由规模不等不同走向系列闪长岩脉组成，称之为青白口岩脉群。岩脉群侵入于中元古界铁岭组中，内无岩相变化，外接触带无明显接触变质现象。岩石类型为闪长玢岩，灰绿色，斑状结构，致密块状构造。斑晶斜长石不规则粒状，发育聚片双晶、熔蚀痕迹和环带结构，弱绢云母化，粒径1~2mm，最大达5mm，含量20%~25%；斑晶角闪石不规则粒状，绿—黄绿色多色性，强绿泥石化、绢云母化、碳酸盐化，边部析出铁质，粒度1~1.5mm，个别达8mm，含量约10%。基质由微晶斜长石、角闪石、石英组成，有时见云母矿物。副矿物为磁铁矿、绿泥石、磷灰石、黄铁矿。

（4）桑峪辉长-闪长岩

分布于门头沟桑峪至田庄一带，顺中生代南大岭期盆地边缘产出，由规模不大一系列辉长岩脉和闪长岩脉或组成，称之为青白口岩脉群。平面呈脉状或岩颈状，顺层或火山口充填状侵入于晚侏罗世南大岭组中（图19-57），接触界线清晰，产状陡倾，局部内接触带见围岩捕虏体。

辉长岩：绿黑色，辉长结构。斜长石自形板条状，发育聚片双晶，粒径0.05~1mm，含量约65%；辉石自形-半自形粒状，粒径0.05~0.5mm，含量15%~20%；绿泥石片状，集合体产出，由暗色矿物交代而成，含量10%~15%。

闪长岩：灰绿色，斑状结构，块状构造。斑晶斜长石自形-半自形板状，发育聚片双晶，弱绢云母化、绿泥石化，粒径0.5~1.5mm，含量约20%；斑晶角闪石粒状，绿泥石化，边部析出铁质，含量5%~10%。基质为微晶斜长石（约70%）、角闪石（约5%）、石英（约2%）及少量绿泥石、绿帘石。

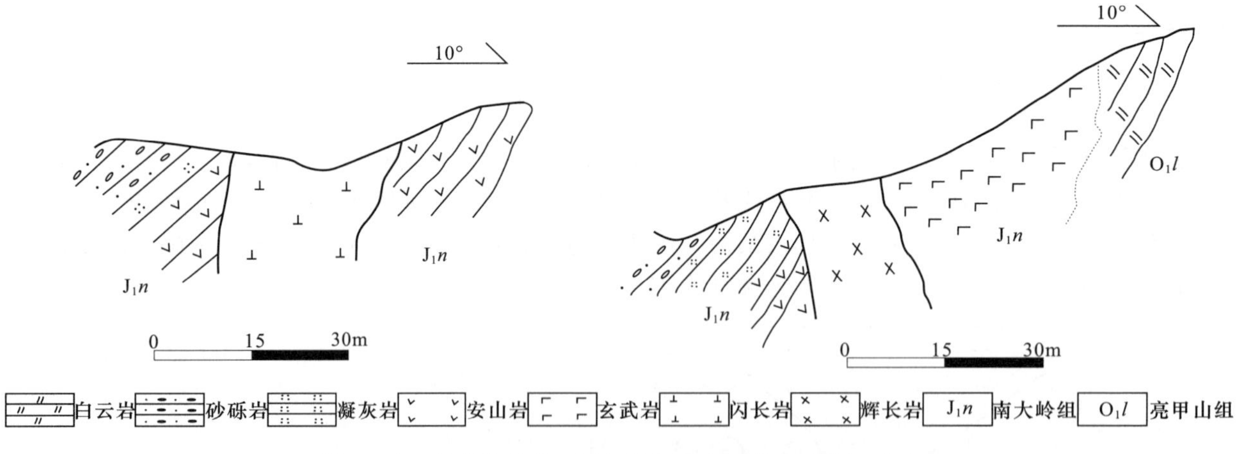

图19-57　桑峪辉长岩、闪长岩与南大岭组地层接触关系

Fig. 19-57　Geological Sketch showing the intrusive contact between gabbro, diorite and Nandaling farmation in Sangyu

2. 岩石化学与地球化学特征

化学成分及参数见表19-29。岩石呈高SiO_2、Na_2O、FeO^*，低K_2O、TiO_2特征；里特曼指数σ=2.51~4.96，跨亚碱性岩和碱性岩区；固结指数SI=10.93~21.94，分异指数DI=42.99~70.18，反映岩浆分异程度较高；镁铁比值m/f=0.38~0.50，属于铁质基性岩；A/CNK=0.63~0.93，A/NK=1.29~2.18，属于准铝质。

稀土元素成分及参数见表 19-30。$\Sigma REE = 172.61 \times 10^{-6} \sim 616.72 \times 10^{-6}$，LREE/HREE = 10.22～17.04，$(La/Yb)_N = 12.29 \sim 41.84$，稀土总量较高，轻稀土强富集，重稀土相对亏损；$\delta Eu = 0.76 \sim 0.89$，弱铕负异常；配分曲线为右陡倾光滑曲线（图 19-58）。

与原始地幔相比，微量元素 Ti、Fe 含量较高。微量元素配分曲线（图 19-59）中，富集 Ti、Mn、Cu 而亏损 Cr、Ni。

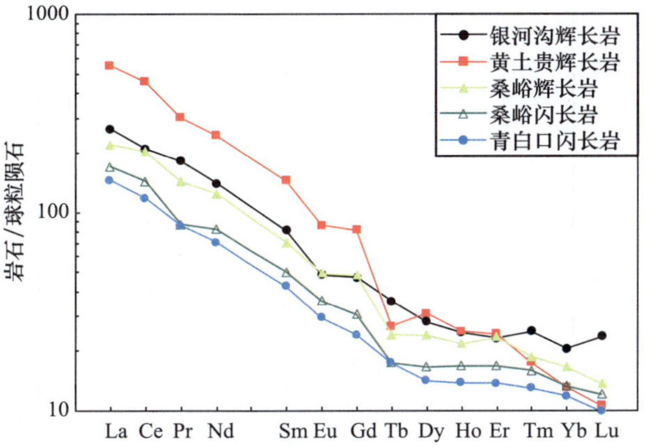

图 19-58 早侏罗世基性岩脉稀土元素配分模式图

Fig. 19-58 Chondrite-normalized REE patterns of Early Jurassic basic dyke vein

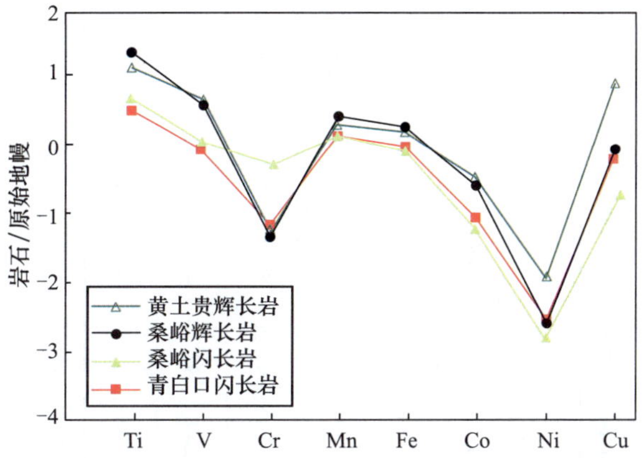

图 19-59 早侏罗世基性岩脉微量元素蛛网图

Fig. 19-59 Trace element spider diagram of Early Jurassic basic dyke vein

3. 岩石系列

该套岩系 $\sigma = 2.51 \sim 4.96$，跨亚碱性岩和碱性岩区。在 SiO_2-Alk 图和 Ol'-Ne'-Q' 图解（图 19-60）上，岩石也跨亚碱性岩和碱性岩系列。因此，早侏罗世基性岩包含碱性岩和亚碱性岩两类岩浆岩系列。

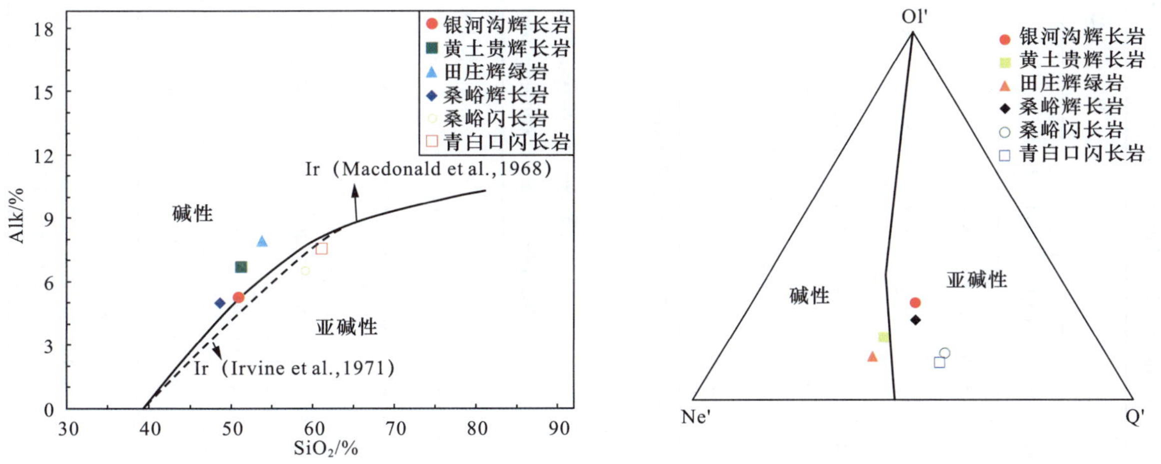

图 19-60 早侏罗世基性脉岩 SiO_2-Alk 和 Ol'-Ne'-Q' 图解

Fig. 19-60 The ternary plot SiO_2-Alk and Ol'-Ne'-Q' of Early Jurassic basic dyke vein

4. 岩脉侵位机制

本期岩脉呈现被动侵位特征：①岩脉多不规则，受构造或地层产状控制，如银河沟岩脉受控于赤城-古北口断裂带，黄土贵、桑峪等岩脉顺层侵位；②围岩无明显挤压变形等现象；③岩性均一，缺乏定向构造；④岩石细粒，斑状或似斑状结构，分异程度低；⑤桑峪岩脉可能同南大岭期火山构造密切有关，形成机制类似于火山塌陷或火山活动后期被动岩浆就位。

表 19-29　早侏罗世基性岩脉岩石化学成分及特征参数一览表

Table 19-29　The chemical composition and main parameters of Early Jurassic basic dyke vein

序号	岩体	岩性	氧化物含量/%													特征参数					
			SiO_2	TiO_2	Al_2O_3	Fe_2O_3	FeO	MnO	MgO	CaO	Na_2O	K_2O	P_2O_5	Los	总和	*DI*	*SI*	σ	*A/CNK*	*A/NK*	*m/f*
1	银河沟	辉长岩	51.08	1.50	13.46	2.06	8.44	0.10	4.42	7.76	3.01	2.22	0.52	2.50	97.07	42.99	21.94	2.78	0.63	1.83	0.44
2	黄土贵	辉长岩	51.49	1.08	17.27	3.23	5.46	0.16	3.35	6.38	4.49	2.19	0.45	1.37	96.91	53.32	17.90	4.49	0.81	1.77	0.48
3	桑峪	辉长岩	53.93	0.78	16.33	2.71	4.82	0.14	3.07	3.54	7.21	0.75	0.40	3.61	97.29	69.47	16.54	4.96	0.85	1.29	0.50
4		闪长岩	48.80	2.70	16.19	6.77	5.61	0.26	3.10	6.73	3.53	1.51	0.86	2.48	98.54	43.95	15.31	3.48	0.82	2.18	0.38
5	青白口	闪长玢岩	59.16	0.54	16.68	2.59	3.88	0.18	1.99	4.80	3.88	2.65	0.42	2.54	99.31	62.99	13.28	2.51	0.93	1.80	0.39

注：数据为均值，1 数据引自 1∶5 万番字牌幅、汤河口幅区域地质调查报告，1995；2~5 引自 1∶5 万雁翅幅、阳坊幅区域地质调查报告，1993。

表 19-30　早侏罗世基性岩脉稀土元素成分及特征参数一览表

Table 19-30　REE contents and main parameter of Early Jurassic basic dyke vein

序号	岩体	岩性	稀土元素含量/10^{-6}																特征参数			
			La	Ce	Pr	Nd	Sm	Eu	Gd	Tb	Dy	Ho	Er	Tm	Yb	Lu	Y	ΣREE	LREE/HREE	$(La/Yb)_N$	δEu	δCe
1	银河沟	辉长岩	62.30	128.00	17.20	65.20	12.40	2.82	9.67	1.33	7.09	1.40	3.84	0.64	3.48	0.60	37.80	315.97	10.26	12.84	0.76	0.94
2	黄土贵	辉长岩	131.25	280.31	28.80	114.91	22.25	5.02	16.84	1.00	7.89	1.43	4.05	0.45	2.25	0.27	37.29	616.72	17.04	41.84	0.76	1.07
3	桑峪	辉长岩	52.75	125.60	13.66	58.47	10.92	2.88	9.98	0.91	6.14	1.24	3.91	0.48	2.85	0.35	34.74	290.14	10.22	13.28	0.83	1.12
4		闪长岩	40.57	88.32	8.35	38.86	7.73	2.10	6.36	0.66	4.26	0.96	2.80	0.41	2.29	0.31	25.13	203.98	10.30	12.71	0.89	1.11
5	青白口	闪长玢岩	34.80	72.96	8.21	33.36	6.56	1.73	5.00	0.66	3.64	0.79	2.29	0.34	2.03	0.26	20.40	172.61	10.51	12.29	0.89	1.02

注：数据为均值，1 数据引自 1∶5 万番字牌幅、汤河口幅区域地质调查报告，1995；2~5 引自 1∶5 万雁翅幅、阳坊幅区域地质调查报告，1993。

5. 成因类型

该套岩系岩浆可能主要来源于富集型地幔，主要依据包括：①深源浅就位侵位特征；②辉石和基性斜长石矿物组成，标型矿物计算包含透辉石和紫苏辉石两类辉石；③辉长岩 $Al_2O_3/TiO_2<30$ 和 La/Nd 比值等特征与幔源岩浆相似；④稀土配分曲线同幔源岩浆特征相似；⑤地球化学特征与南大岭期火山岩相似。南大岭期火山岩由玄武岩向安粗岩岩浆演化，构成较完整的同源岩浆演化序列，早期玄武岩起源于富集型地幔，晚期安粗岩有壳源物质参与（汪洋等，1994）。而桑峪辉长岩和闪长岩为同源岩浆演化，二类岩性侵位揭示了岩石向贫铁、钛，富硅、碱的方向演化特征；且由早至晚，岩浆的分异指数也逐渐增大，表明岩浆向分异程度增高的方向演化。

6. 构造环境分析

该套岩系沿断裂或地层裂隙薄弱带被动式侵位，揭示了岩浆侵位时构造环境相对平静。基于变形与岩浆作用研究表明华北克拉通北缘中生代早期（晚三叠世）已进入构造、岩浆活动相对平静期（蒋小洁，2019）。该套岩系发育碱性岩和钙碱性岩系列，在 $\lg\delta-\lg\tau$ 图（图 19-61）上，成分点位于非造山带火山岩及其派生碱性岩区内。因此，本期岩浆形成于非造山伸展构造环境，为燕山运动前幕伸展构造产物。

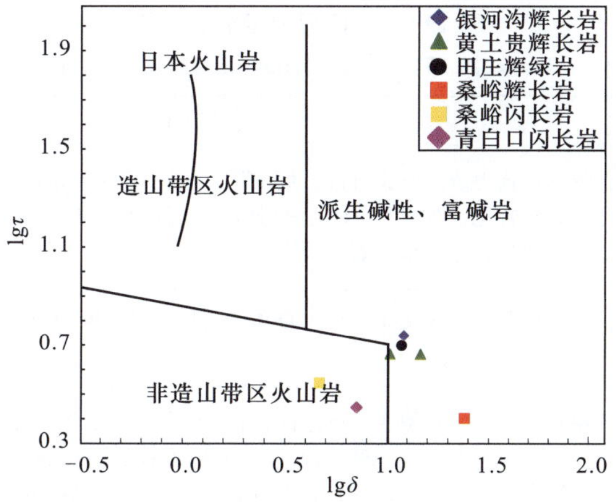

图 19-61　早侏罗世基性岩脉 $\lg\delta-\lg\tau$ 图

Fig. 19-61　$\lg\delta-\lg\tau$ diagram of Early Jurassic basic dyke vein

7. 时代归属

本期岩脉侵位时代厘定为早侏罗世，主要依据包括：①侵入到南大岭组及更老地层中；②桑峪辉长岩、闪长岩地球化学特征同南大岭期火山岩，侵位与南大岭期火山机构密切有关；③黄土贵辉长岩脉顺层侵入下马岭组地层，同围岩一起发生揉皱，说明岩浆侵位时印支期青白口穹隆可能仍在活动；④非造山构造形成背景；⑤青白口闪长岩锆石 U-Pb 同位素年龄为 177~170Ma（叶浩，2014）；⑥田庄辉绿岩和黄土贵辉长岩全岩 K-Ar 法同位素年龄分别为 183Ma 和 229Ma（1：5 万雁翅幅、阳坊幅区域地质调查报告，1993）。

（二）早白垩世双峰式岩墙——南口岩墙

1. 地质产状与岩石学特征

出露于昌平南口花塔、桃洼等，向北延伸至延庆，由一系列 NNW（主）、NE、NNE 向岩脉组成宽 7km，长 40km 的岩墙群，密集集中占有率达 35.65%。岩墙近直立侵入于太古宙片麻岩、中元古界长城系和白垩系东狼沟组，以及燕山期中、酸性侵入岩中，锯齿状接触边缘，内接触带有些发育明显冷凝边，同碳酸盐岩外接触带接触面见黄铁矿颗粒。岩墙局部存在向上翻转展平现象（图 19-62）。单条岩脉一般宽 0.5~3m，最宽达 60m，长 1.5~2km。

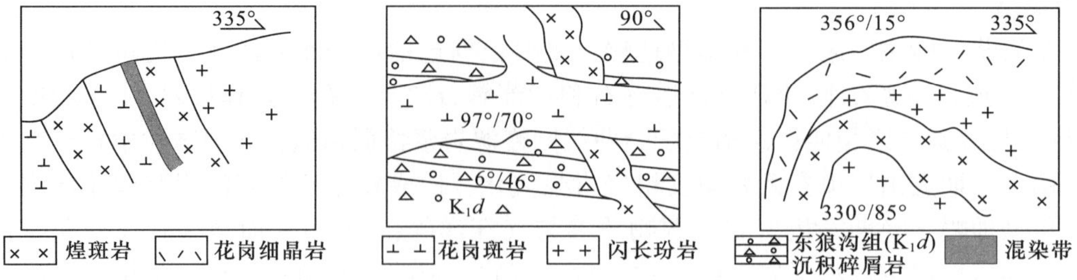

图 19-62　南口岩墙混染带及其与围岩侵入接触关系

（据邵济安等，2001）

Fig. 19-62　Hybridized zone of the Nankou dikes and sketch showing the intrusive contact between dike swarm and surrounding beds

岩墙岩性复杂，未发生明显变质变形，由红黑相间基性岩和酸性岩组成。基性岩为煌斑岩、辉绿岩，酸性岩为花岗岩、花岗斑岩。二者过渡地带渐变混染或相互包裹，形成中性闪长玢岩、花岗闪长岩。具体岩性特征见表 19-31。

表 19-31　南口岩墙岩石类型

Table 19-31　Rock types of Nankou dyke

岩石类型	岩性描述
辉绿岩	灰绿色，辉绿结构，致密块状构造。斜长石自形板条状，粒径 0.1~1.0mm，发育卡纳复合双晶，类型为拉长石或培长石，弱绢云母化、碳酸岩化，含量约 71%；普通辉石短柱状，粒径 0.4~0.8mm，晶面新鲜，个别弱绿泥石化，含量约 24%。副矿物为磁铁矿、磷灰石
煌斑岩	红灰色，斑状结构，块状构造。斑晶主要为角闪石和黑云母，极少钾长石，含量约 30%。其中黑云母六边形自形状，多褐色，少绿色
花岗（斑）岩	灰白色、浅肉红色，斑状结构，块状构造。斑晶钾长石肉红色他形，粒径 2~5mm，边缘多被溶蚀，外围被细晶斜长石包裹形成环带，内包裹角闪石、黑云母，含量约 25%；斑晶石英含量约 10%；斑晶斜长石含量约 5%；基质为斜长石（约 35%）、角闪石（约 10%）和钾长石（约 10%）。副矿物为磁铁矿、磷灰石、锆石、榍石等
闪长玢岩	灰黑色，斑状结构，块状构造。斑晶斜长石半自形，粒径 2~5mm，发育聚片双晶、环带结构，边部弱绢云母化、高岭土化，含量 17%~25%；斑晶角闪石长柱状，粒径 2~4mm，严重蚀变，含量 5%~13%；斑晶钾长石他形椭圆状，粒径一般 5~10mm，最大达 25mm，边缘多被溶蚀，外围被细晶斜长石包裹形成环带，部分内含自形蚀变斜长石、角闪石和黑云母颗粒，含量 2%~3%；斑晶石英他形，粒径 0.5~5mm，含量<1%，部分闪长玢岩还见辉石、黑云母、斜长石斑晶。基质为细粒或微晶斜长石、角闪石。副矿物为黄铁矿（约 1%）、榍石、磷灰石、锆石、磁铁矿、绿泥石等
花岗闪长岩	灰色，似斑状结构，块状构造。斑晶斜长石自形板条状，粒径 2~6mm，蚀变绢云母化、高岭土化，发育聚片双晶、环带结构，内见蚀变绿泥石化角闪石包裹体，含量约 35%；斑晶角闪石自形长柱状，粒径 1~3mm，蚀变绿帘石、绿泥石化，含量约 10%；斑晶石英他形，粒径 2~6mm，含量约 8%；基质细粒斜长石（约 35%）、角闪石（约 10%），粒径 0.5~2mm。副矿物为黄铁矿、磷灰石、锆石等

2. 岩石化学与地球化学特征

岩石化学成分及参数见表 19-32。岩石呈现高 SiO_2、Al_2O_3、碱，低 TiO_2、FeO^*、CaO 特征，且伴随 SiO_2 含量增多，TiO_2、FeO、MgO、CaO 含量减少；中基性、酸性岩里特曼指数 σ 分别为 4.79~7.36 和 2.52~2.81，分属于亚碱性和碱性岩系列；中基性、酸性岩固结指数 *SI* 分别为 21.2~26.69、3.46~8.3，分异指数 *DI* 分别为 48.87~55.48、82.04~92.88，反映岩浆强烈分异作用；中基性岩镁铁比值 m/f=1.21~1.67，属于铁质基性岩；中基性岩 *A/CNK*=0.68~0.90，*A/NK*=1.44~1.85，属于准铝质；酸性岩 *A/CNK*=1.03~1.12，*A/NK*=1.14~1.48，属于偏铝质—过铝质。

稀土元素成分及参数见表 19-33。中基性、酸性岩 ΣREE 分别为 $240.5\times10^{-6}\sim312.77\times10^{-6}$、$172.93\times10^{-6}\sim223.87\times10^{-6}$，LREE/HREE 分别为 14.43～18.82、20.46～21.51，$(La/Yb)_N$ 分别为 22.53～30.12、30.51～31.95，稀土总量较高，轻稀土强富集，重稀土相对亏损；中基性、酸性岩 δEu 分别为 0.84～0.97、0.52～0.87，分别属于铕弱负异常和铕负异常；配分曲线为右陡倾光滑曲线（图 19-63）。

微量元素成分及参数见表 19-34，微量元素蛛网图（图 19-64）上，中基性岩强烈富集大离子亲石元素 Sr、Rb、Ba、Th，相对亏损高场强元素 Nb、Ta。酸性岩则强烈富集 Rb、Th、U，强烈亏损高场强元素 Nb、Ta、Ti、P，弱亏损 Sr。

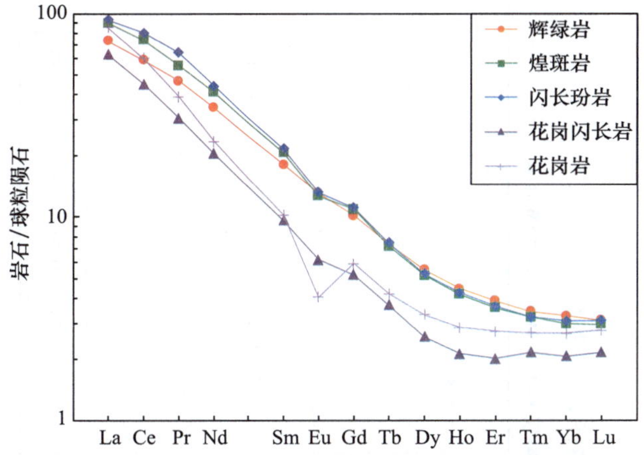

图 19-63　早白垩世岩墙稀土元素配分模式图

Fig. 19-63　Chondrite-normalized REE patterns of Early Cretaceous dike swarm

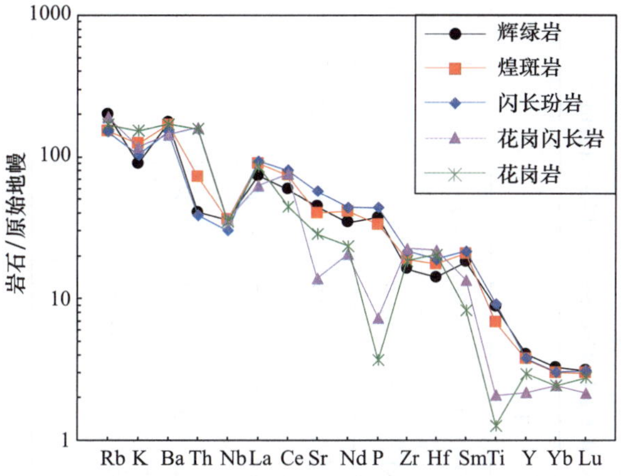

图 19-64　早白垩世岩墙微量元素蛛网图

Fig. 19-64　Trace element spider diagram of Early Cretaceous dike swarm

3. 岩石系列

南口岩墙中基性、酸性岩里特曼指数 σ 分别为 4.79～7.36 和 2.52～2.81，分属于亚碱性和碱性岩系列。在 SiO_2-K_2O 图（图 19-65）上，中基性岩属于钾玄岩系列，酸性岩属于高钾钙碱性系列。

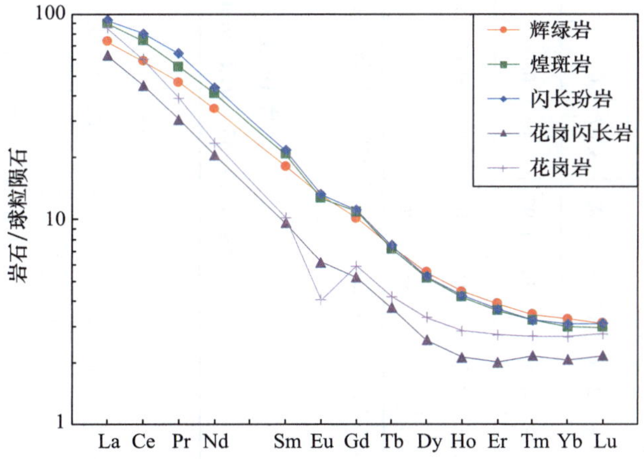

图 19-65　早白垩世岩墙 SiO_2-K_2O 图

Fig. 19-65　SiO_2-K_2O diagram of Early Cretaceous dike swarm

4. Sr-Nd 同位素特征

南口岩墙基性、中性和酸性岩 Sr-Nd 同位素特征高度相似。$^{87}Sr/^{86}Sr=0.706017\sim0.721024$，$^{87}Rb/^{86}Sr=0.1806\sim8.2540$，$^{147}Sm/^{144}Nd=0.0875\sim0.0995$，$^{143}Nd/^{144}Nd=0.511849\sim0.512121$，$\varepsilon_{Nd}(t)=-8.4\sim-13.7$，$\varepsilon_{Sr}(t)=16\sim25.2$，$t_{DM}=1454\sim1702$（邵济安等，2001；潘颖，2010；李德东，2011）。

表 19-32　早白垩世岩墙岩石化学成分及特征参数一览表

Table 19-32　The chemical composition and main parameters of Early Cretaceous dike swarm

序号	地点	岩性	氧化物含量/%													特征参数					
			SiO_2	TiO_2	Al_2O_3	Fe_2O_3	FeO	MnO	MgO	CaO	Na_2O	K_2O	P_2O_5	Los	总和	*DI*	*SI*	*σ*	*A/CNK*	*A/NK*	*m/f*
1	南口	辉绿岩	48.44	1.91	16.40	6.96	3.86	0.27	5.16	5.20	3.60	2.70	0.81	4.25	99.54	48.87	23.49	5.5	0.90	1.85	1.21
2	南口	煌斑岩	53.14	1.47	14.92	5.76	3.20	0.11	5.81	5.64	3.48	3.74	0.72	1.72	99.69	52.69	26.69	4.79	0.75	1.53	1.67
3	南口	闪长玢岩	49.09	2.00	15.26	6.82	3.07	0.12	4.60	6.46	4.39	3.11	0.95	3.30	99.17	55.48	21.2	7.36	0.68	1.44	1.38
4	南口	花岗闪长岩	67.79	0.45	16.36	1.97	1.10	0.06	0.99	1.95	4.40	3.54	0.16	0.96	99.73	82.04	8.3	2.52	1.12	1.48	0.82
5	南口	花岗岩	72.33	0.28	14.15	1.06	0.59	0.02	0.39	0.74	4.51	4.60	0.08	0.74	99.47	92.88	3.46	2.81	1.03	1.14	0.60
6	青龙桥	玻基橄辉岩	48.48	2.80	13.85	3.87	8.47	0.18	6.75	9.37	3.85	2.85	0.66	3.35	104.48	42.28	26.17	8.89	0.52	1.47	0.74

注：1~5 数据为均值，引自潘颖，2010；6 引自 1∶5 万青龙桥幅区域地质调查报告，1984。

表 19-33　早白垩世岩墙稀土元素成分及特征参数一览表

Table 19-33　REE contents and main parameters of Early Cretaceous dike swarm

序号	岩性	稀土元素含量/10^{-6}																特征参数			
		La	Ce	Pr	Nd	Sm	Eu	Gd	Tb	Dy	Ho	Er	Tm	Yb	Lu	Y	ΣREE	LREE/HREE	$(La/Yb)_N$	δEu	δCe
1	辉绿岩	50.55	104.63	12.85	46.65	8.04	2.21	6.04	0.80	4.06	0.73	1.87	0.26	1.61	0.23	18.31	240.50	14.43	22.53	0.97	1.01
2	煌斑岩	61.73	131.61	15.33	55.71	9.23	2.14	6.52	0.78	3.82	0.69	1.73	0.24	1.48	0.22	16.94	291.19	17.83	30.03	0.84	1.05
3	闪长玢岩	64.21	143.28	17.81	59.84	9.62	2.23	6.64	0.81	3.88	0.70	1.75	0.24	1.53	0.23	17.26	312.77	18.82	30.12	0.85	1.04
4	花岗闪长岩	43.36	79.83	8.49	27.87	4.28	1.04	3.10	0.40	1.90	0.35	0.97	0.16	1.02	0.16	9.88	172.93	20.46	30.51	0.87	1.02
5	花岗岩	59.00	107.00	10.79	31.90	4.55	0.68	3.53	0.46	2.45	0.47	1.32	0.20	1.33	0.21	13.40	223.87	21.51	31.95	0.52	1.04

注：1~5 取样地点为南口，数据为均值，引自潘颖，2010。

表 19-34　早白垩世岩墙微量元素成分及特征参数一览表

Table 19-34　Trace element contents and main parameters of Early Cretaceous dike swarm

序号	岩性	微量元素含量/10^{-6}								特征参数
		Rb	Ba	Th	U	Nb	Sr	Zr	Hf	Rb/Sr
1	辉绿岩	125.80	1217.75	3.45	0.10	25.45	935.50	181.25	4.35	0.13
2	煌斑岩	96.40	1173.00	6.15	0.10	25.55	841.75	212.10	5.40	0.11
3	闪长玢岩	96.00	1056.00	3.30	0.10	21.50	1208.00	243.00	5.90	0.08
4	花岗闪长岩	121.70	1001.00	13.70	0.10	25.50	294.00	252.10	6.80	0.41
5	花岗岩	107.15	1187.50	13.40	0.10	24.90	602.75	207.60	6.30	0.18
6	玻基橄辉岩	135.00	768.20	5.38	—	42.84	1037.00	387.70	—	0.13

注：1~5 取样于南口岩脉，数据为均值，引自潘颖，2010；6 取样于青龙桥，数据引自 1∶5 万青龙桥幅区域地质调查报告，1984。

5. 成因类型与岩浆演化

南口岩墙同期发育有基性脉岩和酸性脉岩及过渡性质的中性脉岩，显示出主量和微量元素宽谱系的特征。基性岩和酸性岩彼此接触部位发生一定程度混染作用，但总体是相互独立的（邵济安等，2001；潘颖，2010），中基性岩稀土元素含量高于酸性岩稀土元素含量，显示了二者不具有同源岩浆演化或岩浆混合特性，而具有相互独立的岩浆来源，且侵位过程中经历了快速冷却结晶作用。然而，中基性、酸性岩 Sr-Nd 同位素特征相似，其岩浆应该来源于具有成因联系的同一岩浆房不同岩浆层，Sr-Nd 同位素和酸性岩的偏铝质—过铝质特性揭示可能形成于下地壳低度部分熔融（图 19-66）。微量元素蛛网图上 LILE、LREE 富集，稀土元素强烈分异等是其重要的表现。

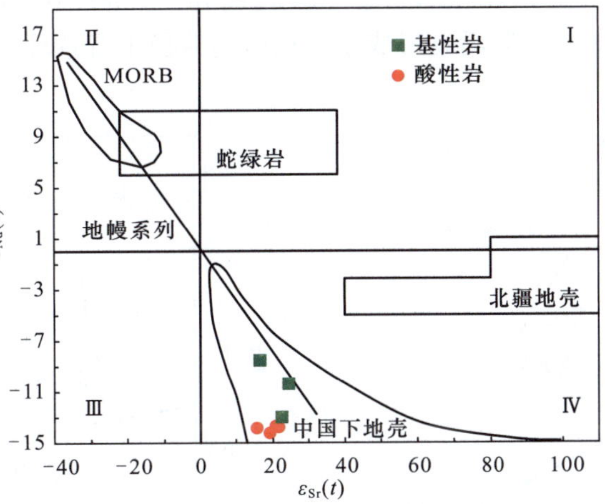

图 19-66　南口岩墙 $\varepsilon_{Sr}(t)$ -$\varepsilon_{Nd}(t)$ 图

Fig. 19-66　$\varepsilon_{Sr}(t)$ -$\varepsilon_{Nd}(t)$ diagram of Nankou dike swarm

6. 构造环境分析

该期脉岩活动后区内处于岩浆活动相对平稳时期，没有大规模的构造、岩浆运动。而脉岩侵位前是区内最强烈岩浆活动期；岩浆双峰式侵位，且成分变化谱系很宽，发育小规模岩浆活动。地球化学成分显示为钾玄岩或高钾钙碱性系列。这些特征揭示了南口岩墙群形成于燕山板内造山运动末期伸展（造山后）构造背景。

7. 时代归属

本期岩墙侵位时代厘定为早白垩世晚期，主要依据包括：①侵入早白垩世早期酸性侵入体及东狼沟组地层中；②晚造山构造背景环境；③岩墙全岩 K-Ar 同位素年龄为 114~120Ma，全岩 Rb-Sr 同位素年龄为 128Ma（邵济安等，2001）；④岩墙煌斑岩、辉绿岩和花岗岩锆石 U-Pb 同位素年龄分别为（127.03±0.87）Ma、（127.91±0.82）Ma 和（127.91±0.82）Ma。

（三）晚白垩世超基性岩脉

1. 地质产状与岩石学特征

以青龙桥岩脉典型，分布于延庆区八达岭长城城门东南，呈 NE 向长 300m、宽 1m 的岩脉状。岩脉近直立侵位于早白垩世早期八达岭似斑状花岗岩体中，界线清晰。岩性为玻基辉橄岩，灰黑色，玻基斑状结构，气孔状、杏仁状构造。斑晶为普通辉石、橄榄石和顽火辉石，基质隐晶质，杏仁体为方解石。普通辉石自形柱状，斑晶形式，含量约 75%；橄榄石自形粒状，粒径 0.15mm，含量约 5%；顽火辉石自形短柱状，粒径 0.15~0.2mm，含量 2%~3%。副矿物为尖晶石、磷灰石等。

2. 岩石化学与地球化学特征

化学及特征参数见表 19-33。岩石呈现低 SiO_2，高碱、TiO_2、FeO^*、CaO、MgO 特征。里特曼指数 $\sigma=8.89$，为碱性岩系列；固结指数 $SI=26.17$，分异指数 $DI=42.28$，反映岩浆弱分异作用；中基性岩镁铁比值 $m/f=0.74$，属于铁质基性岩。与北京境内基性岩相比，微量元素 Ti、Zn、Nb、Zr 含量较高，Cu、Ni 含量相对较低。

3. 岩石系列

岩石里特曼指数 $\sigma=8.89$，为碱性岩类。SiO_2-K_2O 图（图 19-65）揭示为钾玄岩系列。

4. 成因类型与构造环境分析

该套岩系岩浆可能来源于富集型地幔，主要依据包括：①低 SiO_2 和高 TiO_2、FeO^*、CaO、MgO 岩石地球化学特征；②辉石和基性斜长石矿物组成，标型矿物计算包含透辉石和紫苏辉石两类辉石；③深源浅就位侵位特征；④燕山板内造山运动后，华北克拉通东部地幔类型为 EMI 型富集地幔。

该期岩脉活动强度极弱，在 $10TiO_2-Al_2O_3-10K_2O$ 图和 $\lg\delta-\lg\tau$ 图（图 19-67）上，成分点位于大陆玄武岩及非造山带玄武岩区，岩浆活动形成于非造山构造稳定背景环境。

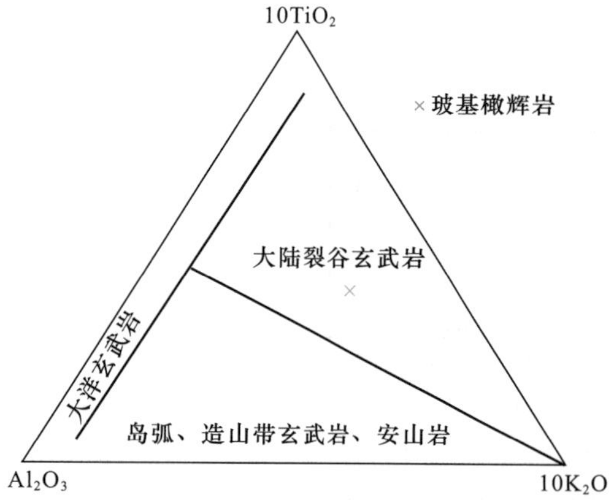

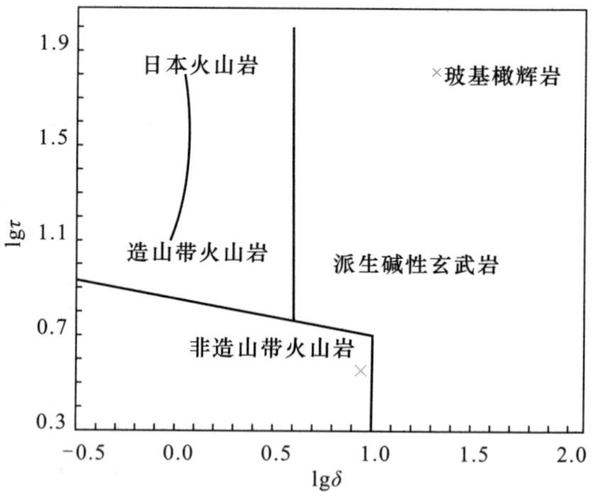

图 19-67　晚白垩世基性岩脉 $10TiO_2-Al_2O_3-10K_2O$ 和 $\lg\delta-\lg\tau$ 图

Fig. 19-67　$10TiO_2-Al_2O_3-10K_2O$ and $\lg\delta-\lg\tau$ diagram of the Late Cretaceous basic dyke vein

5. 时代归属

虽然该期脉岩在青龙桥岩脉中曾获得 12.34Ma 全岩 K-Ar 同位素年龄（鲍亦冈等，2000），但迄今为止，区内未发现露头尺度新近纪岩浆活动。本书基于以下四点认识，将本期岩脉侵位时代厘定为晚白垩世：①侵入早白垩世早期酸性侵入体；②非造山构造背景环境；③岩脉北东向分布特征；④岩性特征及岩石地球化学特征不同于平原区埋藏新生代玄武岩。

第四节 基性—超基性岩浆活动与岩石圈演化及对成矿作用的制约

一、基性—超基性岩浆活动与岩石圈演化

本区位于华北克拉通（板块）内，其岩石圈结构可参考华北克拉通岩石圈结构。华北克拉通是我国最大、也是最古老的一个克拉通，它的发展演化经历了基底形成、盖层稳定发展和活化三个阶段（阎国瑜，2007；朱日祥，2008，2011）。华北克拉通岩石圈物质时空结构可划分为三类，即老物质新结构、老物质老结构和新物质新结构（李廷栋等，2013）。目前，华北克拉通基底的形成过程和华北克拉通破坏过程都还存在不同认识，使得三个阶段划分的时间段存在较大争议。多数学者倾向于华北克拉通在古太古代就已经开始形成陆核，其后大小不等的陆块在不同时代有不同规模的拼接，最终经吕梁运动形成统一的华北克拉通基底（阎国瑜，2007）。而华北克拉通破坏则可能不是一个极为短暂的过程，经历了长达100Ma的漫长过程，最终在早白垩世达到破坏（徐义刚等，2001，2004）。

基性—超基性岩岩浆来源于地幔，记录有丰富地幔岩石圈地质信息可作为地幔软流圈与岩石圈在不同运动机制的重要重用信息载体。根据本区所处的大地构造位置和构造演化历史，基性-超基性岩浆揭示的地幔源区特征，可以把北京地区的岩石圈划分为三大类型，即古岩石圈（中太古代—古元古代）、新岩石圈（中元古代—晚古生代）和现今岩石圈（晚古生代—新生代）。

本区太古宙晚期岩石变质变形强烈，后期改造使得岩石宏观分布特征难以辨认，分布零星的黑卧辉长质片麻岩、苇子峪辉长闪长质片麻岩和放马峪基性岩墙，地球化学特征揭示该阶段古岩石圈呈现富集型特征。这一岩石圈特征一直持续到古元古代，被古元古代超基性镁铁质岩所记录。

古元古代，区内经历一次岛弧俯冲-碰撞类喜马拉雅式造山过程，也是一次重要的地壳生长过程，富集型古岩石圈地幔演化形成全新常规-弱亏损型过渡型岩石圈地幔。区内缺乏可揭示新岩石圈形成演化过程的基性—超基性岩浆岩，但邻区河北省等地古元古代多期基性岩墙记录了华北陆块东、西块体古元古代俯冲碰撞作用模式（王岳军等，2007）和岩石圈改造过程。

中元古代—晚古生代区内长期处于稳定克拉通阶段，由于缺乏强烈构造-岩浆活动，早元古代形成演化的新岩石圈总体得以保持。华北克拉通古生代金伯利岩和其中地幔橄榄岩及金刚石包体研究发现，该阶段岩石圈（新岩石圈）是由大约200km厚、具低热流值及铕亏损的方辉橄榄岩和二辉橄榄岩为主的物质组成（路凤香等，1991；郑建平，1996、1999），即广大的华北板块克拉通化，形成稳定、巨厚的岩石圈。然而，中元古代早期，华北克拉通经历了一次夭折裂谷发育过程，与地幔柱有关的岩墙侵位可能对局部岩石圈结构有所改造。区内发育有高岭辉长质岩墙、新地斜长岩-环斑花岗岩、大红峪期辉长岩-正长岩侵位，记录了多期次与伸展或与古深大断裂有关的岩浆侵位事件，新岩石圈常规-弱亏损型地幔类型通过这些幔源物质得以反映出来。

从晚古生代二叠纪下窝铺等地辉长岩类开始，到侏罗纪、白垩纪基性岩类，其Sr-Nd-Hf同位素特征发生明显变化，岩石圈类型明显不同于前中元古代—早古生代，岩石圈结构再次发生变化，岩石圈壳幔结构可能继承中元古代—早古生代某些特征，但岩石圈地幔已经开始发生转变，由地幔弱亏损转变为不同程度富集。

玄武岩和其中地幔橄榄岩包体及地球物理资料揭示，今华北克拉通岩石圈（今岩石圈）由地幔具有高热流值及大洋型的尖晶石、石榴子石二辉橄榄岩构成，厚度为80~120km（Xu et al.，1996，1999，2000，2001）。因此，新岩石圈和今岩石圈的厚度、物质组成和热状态发生了剧烈的变化，广大华北克拉通东部发生去克拉通化。

二、对成矿作用的制约

由于基性-超基性侵入岩主要形成于地壳伸展机制下拉张构造环境，物质主要来源于上地幔，因而

对成矿作用具有明显的制约性。主要表现在：①与基性—超基性侵入岩有关的矿化，多数为岩浆型矿床，可细分为早期岩浆矿床、晚期岩浆矿床和熔离矿床及部分热液矿床，成岩与成矿具一致性；②岩浆岩的专属性明显，矿床往往在某一特定相带中，矿石成分与含矿岩浆成分具有一定的依从关系；③成矿物质主要来源于上地幔基性-超基性母岩浆；④含矿岩浆和有关矿床的产出，往往与地质构造发展一定阶段相联系，分布于一定的构造单元中；⑤基性—超基性侵入岩的含矿性与岩体的产出状态（层状、非层状）密切有关；⑥基性—超基性侵入岩穿越、切割早期矿化，使得矿石富化。

本区不同时代基性-超基性侵入岩对成矿作用的制约，均具有上述特征。矿床或矿化点成因类型可以进一步划分为岩浆岩型（岩浆结晶分异作用）和岩浆热液型两大类，前者为主。

（一）古元古代橄榄岩型

见于密云平顶山、放马峪和延庆红石湾，形成与超基性岩侵入活动有关的铬铁矿床，钴、钼、镍、硼为伴生的有用成分。矿体赋存于超基性岩体中下部及边部（构造封闭端），与橄榄岩、辉石橄榄岩关系密切，不严格受岩相控制。矿体展布常与相带呈一定角度相交，一般受岩体底盘形态控制，矿体与岩体产状基本一致。岩体原生流动构造（条带状、椭球状构造）与岩体、矿床走向一致，说明在岩浆冷凝及铬尖晶石结晶过程中，由于重力及动力作用影响，显示出定向运移特点。

矿床规模为小型矿床，矿体形态一般呈似层状、透镜状、似脉状等，剖面上呈楔形，与围岩渐变或迅速过渡。矿石中金属矿物主要为铬尖晶石及磁铁矿，非金属矿物有蛇纹石、滑石、碳酸盐岩类、蛭石及绿泥石等呈交代残余状的橄榄石、古铜辉石及透辉石。矿石中铬尖晶石常呈自形或半自形晶粒状结构，易于岩浆结晶早期阶段富集。矿体主要形成于岩浆结晶的早期阶段，为早期岩浆矿床。

（二）中元古代斜长岩型

见于怀柔新地，形成与基性岩侵入活动有关的钛磁铁矿矿床，钒、钛为铁的伴生有用组分。含矿岩石为斜长岩，成矿物质主要来源于上地幔基性—超基性母岩浆，同赤城-长哨营-古北口东西向深大断裂（深达上地幔）在中元古代早期活动密切有关，为岩浆活动分异作用产物。与斜长岩伴生岩石组合为石英正长岩和钾长花岗岩，构成基性—超基性、中酸性双峰式岩浆活动特征。矿床规模为小型矿床，矿体呈似层状、扁豆状、透镜状产出，受控于岩体产出形态，矿床类型属岩浆晚期型矿床。矿石矿物为磁铁矿、钛磁铁矿，脉石为铁闪石、斜长石等。

（三）晚古生代（变质）辉长岩型

见于怀柔区喇叭沟门下窝铺和长哨营一带，形成与基性岩侵入有关的金矿床（矿化点）。古老变质岩是金矿的赋矿围岩，也是可能的矿源层，但橄榄辉长岩侵入岩体周围是最佳金成矿区域。没有形成矿床规模，仅为金矿化点，形成类型为岩浆热液型。大地构造环境为华北板块北缘微小板块拼接后短时间构造伸展环境。

（四）中生代辉长岩型

见于昌平上庄，形成与基性岩有关的钛磁铁矿矿床，钒、钛为铁的伴生有用组分，含矿岩石为橄榄辉长岩，成矿物质来源于地幔基性—超基性母岩浆（苏尚国等，2006）。成矿作用与薛家石梁复式岩体形成时代一致，为早白垩世中期。矿石具粒状、海绵陨铁结构，具条带状、浸染状及致密块状构造，说明磁铁矿、钛铁矿等有用成分的晶出时间晚于硅酸盐类矿物（晚期岩浆矿床）。辉长岩体短期内共发生三次岩浆侵入，对应形成三种类型辉长岩，与辉长岩伴生的岩石组合为正长岩、二长岩，为幔源岩浆底侵作用引起加厚陆壳（下地壳）部分熔融形成。矿床规模为小型矿床，矿体呈簸箕状、扁豆状产出。矿石矿物以磁铁矿为主，次为钛铁矿，极少量磁黄铁矿、黄铁矿、黄铜矿。脉石矿物以辉石、斜长石为主，次为角闪石、橄榄石、黑云母。成矿作用同该阶段伸展构造背景强烈底侵作用有关，而矿体赋存则被后期构造破坏。

第二十章　中酸性侵入岩

第一节　太古宙中酸性侵入岩

太古宙中—酸性侵入岩是在密怀地区第二轮 1：5 万区调工作中从“变质地层”中解体出来，广泛分布于境内北部山区。按照接触关系或侵位时代划分为沙厂英云闪长质片麻岩系列、宋营子英云闪长质片麻岩系列、荆子峪花岗闪长质片麻岩系列、康各庄英云闪长质片麻岩系列和旗杆顶花岗质片麻岩系列。

一、沙厂英云闪长质片麻岩

（一）地质产状与岩石学特征

发育大漕、沙厂、阳坡地、琉璃庙 4 大岩体，是区内太古宙侵入岩主体，分布于密云水库西、北侧和密云四合堂至琉璃庙一带，出露面积约 1100km^2（含密云水库面积）。岩体内发育晚期超镁铁质、镁铁质、长英质系列岩脉，还普遍发育表壳岩包体（3%～10%），尤以大漕岩体最为密集（15%～50%）。包体总体顺片麻理分布，局部带状集中（图 20-1）。包体大小毫米级至百米级，边部甚至内部多被交代形成模糊边界，形成包体重熔残余特征。被交代的包体边部形成含铁石榴子石晕圈，可根据含铁石榴子石集中分布形态还原包体大小。岩体普遍变形变质作用较强，发育宽 0.5～2.0cm 条带状片麻理变形构造，又叠加后期不均匀角闪岩相退变质作用。对 4 个主要岩体地质产状与岩石学特征总结如下：

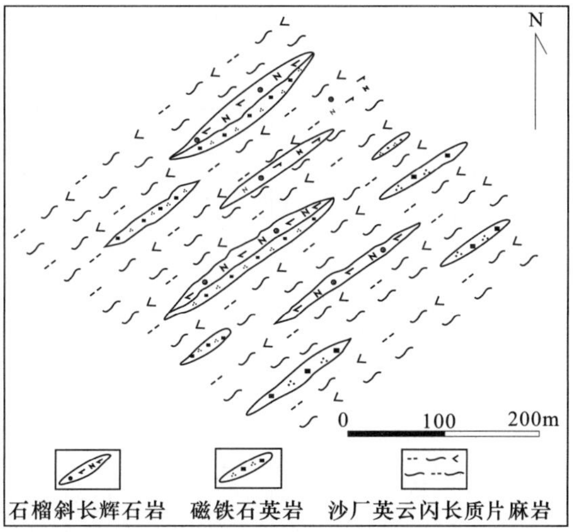

图 20-1　表壳岩包体分布特征示意图

（密云放马峪，据鲍亦冈等，2001）

Fig. 20-1　Geological sketch showing the distribution characteristics of supracrustal rock enclaves

1. 大漕岩体

分布于密云高岭，东与中元古界长城系和中生界髫髻山组不整合或断层接触，南被苇子峪辉长闪长质片麻岩体侵入破坏，北与阳坡地岩体渐变过渡。岩体岩石类型复杂，主要岩性组成及其矿物成分、含量统计见表 20-1。暗色矿物普遍蚀变蛇纹石化、绿泥石化或碳酸盐化，浅色矿物弱蚀变绢云母、高岭土化。

2. 沙厂岩体

分布于密云大城子、巨各庄，南与康各庄奥长花岗质片麻岩岩体被侵入或断层接触，北东被中元古界长城系不整合覆盖，北西被荆子峪花岗闪长质片麻岩体和燕山早期四干顶岩体侵入，其内还被古元古代沙厂环斑花岗岩体侵入破坏。岩体岩石类型复杂，主要岩性组成及其矿物成分、含量统计见表 20-1。

表 20-1　沙厂英云闪长质片麻岩主要岩石类型和矿物定量统计表

Table 20-1　Main rock types and its mineral contents of Shachang tonalite gneiss

岩体	岩性	矿物成分及含量/%										
		斜长石	钾长石	石英	紫苏辉石	透辉石	石榴子石	黑云母	角闪石	磁铁矿	磷灰石	锆石
大漕	黑云石榴紫苏片麻岩	63		20	6		5	5		1.5	0.5	
	含榴云二辉片麻岩	65		15	8	3	4	5		0.5	0.5	
	角闪辉石片麻岩	65		12	2	10			10	0.4	0.5	0.1
	石榴黑云片麻岩	65		15			10	10				
	石榴二辉片麻岩	66		4	6	15	5		3	0.5	0.5	
	二辉角闪片麻岩	49.89		5.61	6.91	13.71	1.78	0.43	20.76	0.89	0.01	
沙厂	含榴角闪二辉片麻岩	74		7	3	6	1		5	3	0.1	
	角闪透辉片麻岩	53		15		15	3	3	10	0.8	0.1	
	黑云片麻岩	85		5				10				
	含云辉角闪片麻岩	73		8		2		3	13	1	0.1	
	含榴角闪辉石片麻岩	63		15		10	1		5	5	0.1	
	含闪榴二辉片麻岩	66		4	22	3	3		1	0.9	0.1	
阳坡地	石榴辉石片麻岩	66		22		6	3	1		2	0.01	0.01
	黑云角闪片麻岩	65		10			3	5	15	1	0.1	
	石榴二辉角闪片麻岩	58		7	1	5	3	5	20	1	0.1	0.01
	黑云母片麻岩	80		10			.	8		1	0.1	
	黑云辉石片麻岩	66		20		8		5		1	0.1	0.01
琉璃庙	石榴角闪斜长片麻岩	40			3		10		45	0.8	0.2	0.5
	黑云斜长片麻岩	50		25				20	5	0.01	0.01	0.01
	角闪斜长片麻岩	60		2				4	33	0.5	0.05	0.01

注：数据资料引自周绍林等（1993）、1∶5 万密云幅、高岭幅、墙子路幅区域地质调查报告（1994）、1∶5 万琉璃庙幅、范各庄幅区域地质调查报告（1990）。

3. 阳坡地岩体

分布于密云团山子、半城子，南与大漕岩体渐变过渡关系，东与中元古界长城系不整合或断层接触，北东被中生界髫髻山组不整合覆盖。岩体岩石类型复杂，主要岩性组成及其矿物成分、含量统计见表 20-1。岩石中单斜辉石多被绿色纤维状角闪石交代呈残晶假象或残留体，含量<5%。其他暗色矿

物多蚀变绿泥石化、绿帘石化、绢云母化，且愈接近半城子韧性剪切带蚀变愈强烈。

4. 琉璃庙岩体

分布于密云四合堂、琉璃庙，东与阳坡地岩体呈渐变过渡关系，北被宋营子英云闪长质片麻岩体侵入切割，南被燕山期云蒙山花岗岩体侵入破坏，局部同中元古界长城系不整合或断层接触。岩体内发育多个燕山期酸性岩体和古元古代超基性岩岩体。岩体中部还被北东向深大崎峰茶断裂切割破坏，并以此断裂为界，两盘变质岩性组合有所差异：西部以石榴角闪斜长片麻岩和黑云斜长片麻岩为主的高级变质麻粒岩相岩性组合；东部以黑云斜长片麻岩和角闪黑云斜长片麻岩为主的中级变质角闪岩相岩性组合。具体主要矿物成分及含量统计见表 20-1。

（二）岩石化学与地球化学特征

岩石化学成分及参数见表 20-2。里特曼指数 $\sigma=1.36\sim3.77$，$Na_2O>K_2O$，表明岩石为钠质亚碱性岩系列；固结指数 $SI=15.71\sim20.65$，分异指数 $DI=56.67\sim71.21$，长英指数 $FL=45.80\sim71.40$，反映岩浆分异程度较高；$A/CNK=0.84\sim0.95$，$A/NK=1.36\sim2.02$，属于偏铝质。

稀土元素成分及参数见表 20-3。$\Sigma REE=48.85\times10^{-6}\sim89.09\times10^{-6}$，$LREE/HREE=6.61\sim13.02$，$(La/Yb)_N=10.27\sim16.33$，稀土总量低，轻稀土强烈富集，重稀土相对亏损；$\delta Eu=0.93\sim1.05$，无明显铕正或负异常；配分曲线为右陡倾光滑曲线（图 20-2）。

微量元素成分及参数见表 20-4。与冀东三屯营英云闪长质片麻岩相似，Ba、Sr、Y 元素含量高于上地幔，Sc、Ti、V 元素含量相当，Zr 和其他元素则含量偏低。微量元素蛛网图（图 20-3）中，富集大离子亲石元素 Sr、Ba 而相对亏损高场强元素 Zr。

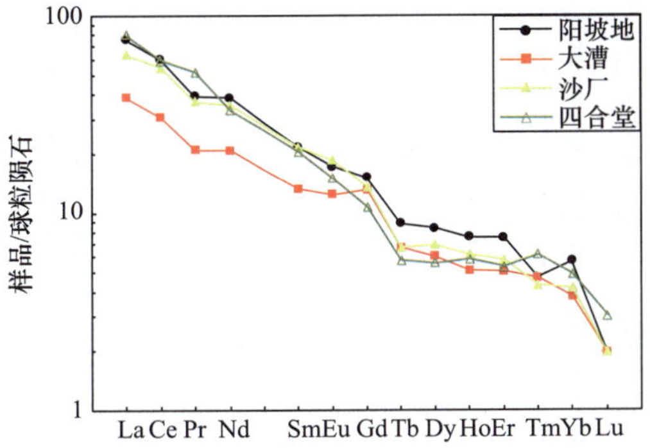

图 20-2　太古宙沙厂英云闪长质片麻岩稀土元素配分模式图

Fig. 20-2　Chondrite-normalized REE patterns of Archean Shachang tonalite gneiss

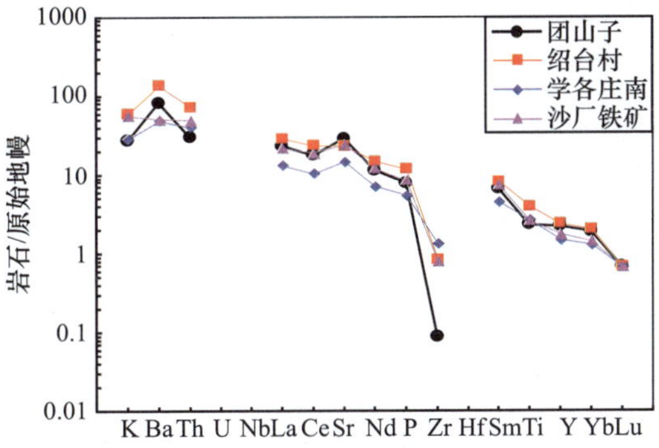

图 20-3　太古宙沙厂英云闪长质片麻岩微量元素蛛网图

Fig. 20-3　Trace element spider diagram of Archean Shachang tonalite gneiss

（三）变质变形特征

该套岩系普遍发育条带状和强片麻状构造，尤以大漕、高岭地区最为强烈。在被后期构造强烈改造的四合堂、琉璃庙地区，已被角闪岩相韧性变形所替代。其他地区亦零星发育角闪岩相和绿片岩相变形，但部分地带仍可见到条带状和强片麻状构造的残余。

（四）原岩恢复

上述地质产状、岩石学和岩石地球化学等特征，特别是残留的矿物原生结构、构造，如包含大量密云岩群包体、岩性中保留的普通辉石或紫苏辉石残留晶体等，揭示了该变质岩原岩为正变质岩。在变质岩原岩恢复 SiO_2-TiO_2 和 Si-[(al+fm)-(c+alk)] 图解（图 20-4）上，成分点也落在火成岩区内。

（五）岩石系列

该套岩系 $\sigma=1.36\sim3.77<4$，为亚碱性系列。岩石富碱、镁而同钙碱性岩相吻合，在 *F*-*A*-*M* 图（图 20-5）上，成分点位于钙碱性系列。

表 20-2 太古宙沙厂英云闪长质片麻岩岩石化学成分及特征参数一览表

Table 20-2 The chemical composition and main parameters of Archean Shachang tonalite gneiss

序号	产地	岩性	氧化物含量/%													特征参数					
			SiO_2	TiO_2	Al_2O_3	Fe_2O_3	FeO	MnO	MgO	CaO	Na_2O	K_2O	P_2O_5	Los	总和	*DI*	*SI*	σ	*FL*	*A/CNK*	*A/NK*
1	阳坡地	黑云片麻岩	65. 19	0. 47	14. 84	2. 18	3. 46	0. 11	2. 21	4. 21	3. 95	2. 25	0. 34	1. 77	100. 96	68. 23	15. 75	1. 71	59. 52	0. 90	1. 66
2	团山子	石榴辉石片麻岩	62. 76	0. 50	15. 83	2. 46	3. 22	0. 12	2. 68	5. 39	4. 92	0. 82	0. 17	1. 98	100. 85	62. 76	19. 05	1. 65	51. 57	0. 84	1. 76
3	邵台村	石榴黑云片麻岩	58. 74	0. 86	15. 67	2. 97	5. 46	0. 17	3. 69	4. 62	3. 95	1. 80	0. 26	2. 88	101. 07	56. 67	20. 65	2. 04	55. 45	0. 93	1. 86
4	高岭	辉石片麻岩	64. 91	0. 45	15. 51	3. 49	1. 76	0. 05	2. 10	4. 15	4. 50	1. 30	0. 14	1. 55	99. 91	67. 80	16. 15	1. 51	58. 32	0. 95	1. 76
5	学各庄南	黑云石榴辉石片麻岩	60. 74	0. 58	15. 67	1. 49	4. 58	0. 15	2. 58	5. 94	4. 15	0. 87	0. 12	4. 50	101. 37	57. 68	18. 87	1. 36	45. 80	0. 84	2. 02
6	沙厂	黑云角闪片麻岩	63. 25	0. 56	15. 33	2. 54	3. 38	0. 10	2. 67	3. 70	4. 56	1. 72	0. 19	2. 98	100. 97	67. 27	17. 98	1. 90	62. 90	0. 95	1. 64
7	琉璃庙	黑云斜长片麻岩	62. 70	0. 59	16. 00	2. 12	3. 66	0. 10	2. 68	3. 45	4. 40	4. 21	0. 16	0. 86	100. 93	71. 21	15. 71	3. 77	71. 40	0. 89	1. 36

注：1、4、6、7 数据为均值，引自 1∶5 万密云幅、高岭幅、墙子路幅区域地质调查报告，1994 和 1∶5 万琉璃庙幅、范各庄幅区域地质调查报告，1990。1、2、3 为阳坡地岩体，4、5 为大槽岩体；6 为沙厂岩体；7 为琉璃庙岩体。

表 20-3 太古宙沙厂英云闪长质片麻岩稀土元素成分及特征参数一览表

Table 20-3 REE contents and main parameters of Archean Shachang tonalite gneiss

序号	岩体	岩性	稀土元素含量/10^{-6}																特征参数			
			La	Ce	Pr	Nd	Sm	Eu	Gd	Tb	Dy	Ho	Er	Tm	Yb	Lu	Y	ΣREE	LREE/HREE	$(La/Yb)_N$	δEu	δCe
1	阳坡地	黑云片麻岩	18. 00	36. 75	3. 72	17. 95	3. 32	1. 00	3. 11	0. 33	2. 12	0. 43	1. 24	0. 12	0. 97	0. 05	10. 45	89. 09	9. 66	13. 38	0. 93	1. 04
2	大漕	黑云石榴辉石片麻岩	9. 16	18. 80	2. 00	9. 72	2. 03	0. 72	2. 70	0. 25	1. 53	0. 29	0. 84	0. 12	0. 64	0. 05	6. 76	48. 85	6. 61	10. 27	0. 94	1. 03
3	沙厂	黑云角闪片麻岩	15. 10	33. 50	3. 49	16. 50	3. 32	1. 08	2. 80	0. 25	1. 75	0. 35	0. 96	0. 11	0. 71	0. 05	8. 05	79. 97	10. 46	15. 26	1. 05	1. 09
4	四合堂	黑云斜长片麻岩	19. 13	36. 51	4. 97	15. 57	3. 16	0. 87	2. 22	0. 22	1. 43	0. 33	0. 89	0. 16	0. 84	0. 08	7. 99	86. 38	13. 02	16. 33	0. 96	0. 90

注：1、4 数据为均值，引自 1∶5 万密云幅、高岭幅、墙子路幅区域地质调查报告，1994 和 1∶5 万琉璃庙幅、范各庄幅区域地质调查报告，1990。

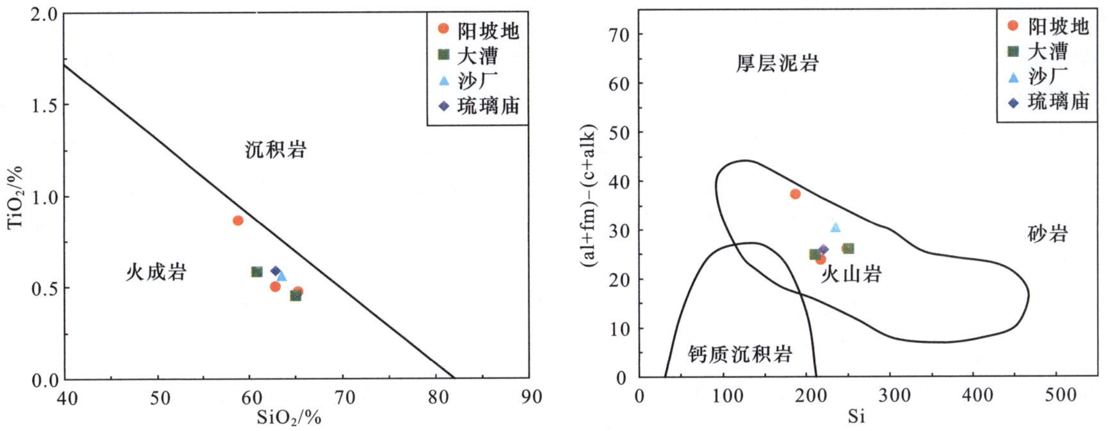

图 20-4 太古宙沙厂英云闪长质片麻岩 SiO_2-TiO_2 和 Si-[(al+fm)-(c+alk)]原岩恢复图解

Fig. 20-4 SiO_2-TiO_2 and Si-[(al+fm)-(c+alk)] diagram of Archean Shachang tonalite gneiss

表 20-4 太古宙沙厂英云闪长质片麻岩微量元素成分及特征参数一览表

Table 20-4 Trace element contents and main parameters of Archean Shachang tonalite gneiss

序号	产地	岩性	元素含量/10^{-6}									
			Sc	V	Cr	Mn	Co	Ni	Ba	Sr	Zr	Th
1	团山子	石榴辉石片麻岩	9.75	61.6	10.83	199.6	<1.00	<4.00	577.70	614.30	<1.00	<3.00
2	邵台村	石榴黑云片麻岩	15.6	184.40	200.00	700.30	19.31	47.41	962.70	492.40	9.40	6.21
3	学各庄南	黑云石榴辉石片麻岩	7.60	127.20	99.61	617.70	15.63	41.24	344.50	314.30	15.01	3.48
4	沙厂铁矿	角闪辉石片麻岩	9.90	122.10	122.80	624.00	15.49	42.01	356.40	518.80	8.87	4.26

注：数据引自 1∶5 万密云幅、高岭幅、墙子路幅区域地质调查报告，1994。1、2 为阳坡地岩体，3 为大槽岩体；4 为沙厂岩体。

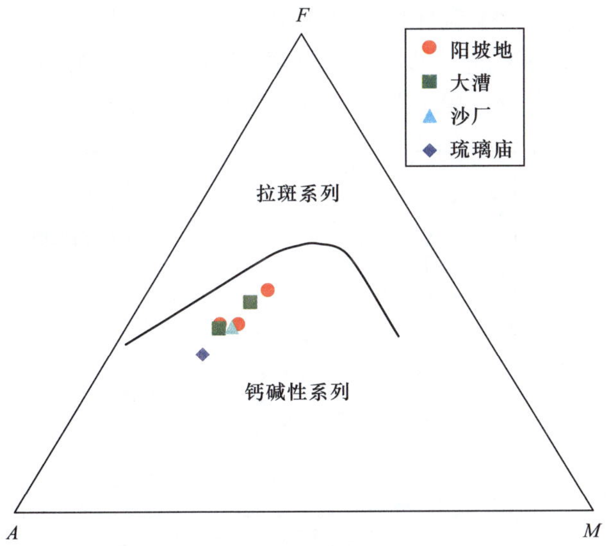

图 20-5 太古宙沙厂英云闪长质片麻岩 *F*-*A*-*M* 图

Fig. 20-5 *F*-*A*-*M* diagram of Archean Shachang tonalite gneiss

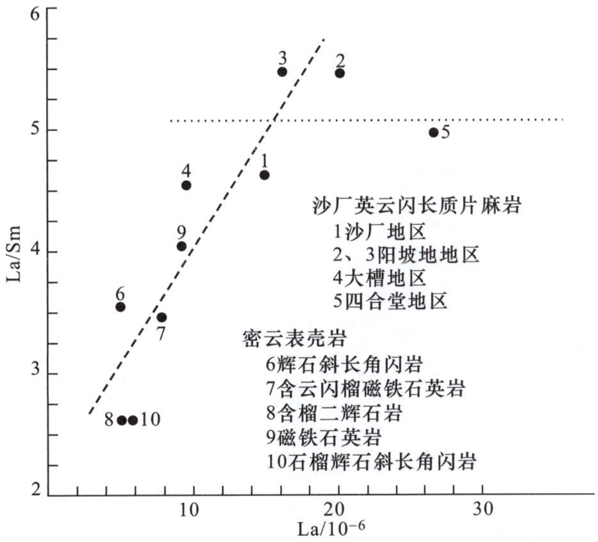

图 20-6 密云地区太古宙变质岩 La/Sm-La 图

Fig. 20-6 La/Sm-La diagram of Archean metamorphic rocks in Miyun area

（六）成因类型

稀土元素配分曲线图（图 20-2）中，该套岩系 LREE 富集，HREE 相对亏损，微量元素蛛网图（图 20-3）中，该套岩系 LILE 富集，HFSE（Nb、Zr）相对亏损，这一特征反映岩浆来源于地壳物质。沙厂英云闪长质片麻岩与表壳岩之间存在重要的成因联系。主要依据有：①空间分布呈面积广大面型展布。虽然岩性复杂，但矿物成分变化的重要因素是多期变质变形改造所致，总体成分仍在英云闪长岩范围内，且各岩性组成内部很难找到进一步细分的证据，以及较明显的接触关系；②普遍含大量具有明显重熔残余构造的表壳岩包体，在经过多次强烈抬升剥蚀之后，仍有如此之多的表壳岩包体散布其中，反映其物质未进行长距离运移。而所含包体重熔残余特征，反映了英云闪长质熔浆在未完全消化表壳岩残块之前即开始进入成岩阶段；③发育交代成因石榴子石晕圈构造；④La/Sm-La 图解（图 20-6）上，沙厂英云闪长质片麻岩投影于密云岩群（表壳岩）的右上方，与密云岩群形成一条斜线。这些特征与程裕淇认为冀东三屯营片麻岩具有原地、半原地生成的特征一致。因此，壳源岩浆侵位并对表壳岩原地—半原地重熔改造机制可能是沙厂英云闪长岩的成因模式。

在重熔程度上，沙厂片麻岩各地表现出一定的不均一性。在大漕至高岭地区（大漕岩体），重熔程度相对较弱，表壳岩残留体相对较多，片麻岩稀土总量也最低（图 20-2）。La/Sm-La 图解（图 20-6）中表壳岩成分点与沙厂片麻岩成分在一条倾斜直线附近，大漕岩体的成分点更接近于表壳岩成分点。沙厂地区（沙厂岩体）重熔层次相对偏下，表壳岩被重熔的程度较大漕地区高，表壳岩包体含量明显减少，稀土总量也略有增高，La/Sm-La 图解（图 20-6）上成分点在大漕片麻岩体之上。阳坡地片麻岩体重熔层次较沙厂片麻岩体偏上，表壳岩被重熔程度较沙厂地区高，除局部有较大规模的表壳岩残留体外，大部分地区仅见零星分布的规模较小的表壳岩包体，稀土总量进一步增高，La/Sm-La 图解（图 20-6）上成分点与沙厂片麻岩体相邻。四合堂地区（琉璃庙岩体）片麻岩可能因为发生韧性剪切变形，成分点偏离拟合曲线，但稀土总量最高，包体含量少的特征仍得以体现。

沙厂英云闪长质片麻岩 La/Sm-La 图（图 20-6）自身还能拟合呈平坦分离结晶曲线，说明岩体侵位过程还经历了分离结晶作用，这和分异指数高等特征参数揭示结果一致。

（七）构造环境分析

该套岩系为钙碱性系列，且表现出低钾、钛，高镁特征；富集大离子亲石元素 Sr、Ba 和高场强元素 P、轻稀土元素，相对亏损高场强元素 Zr 和重稀土元素；在 R1-R2（图 20-7）构造环境判别图解上，成分点（琉璃庙可能受后期韧性剪切变形影响）主要位于破坏性活动大陆边缘区。这些特征反映其侵位于岛弧环境。Sc/Ni-La/Yb 构造环境图解（图 20-8）进一步判别其主要侵位于洋-陆俯冲型活动大陆边缘安第斯型岛弧环境，而 TiO_2<1.2%、低 K_2O 含量，揭示其可能发育为不成熟的岛弧。

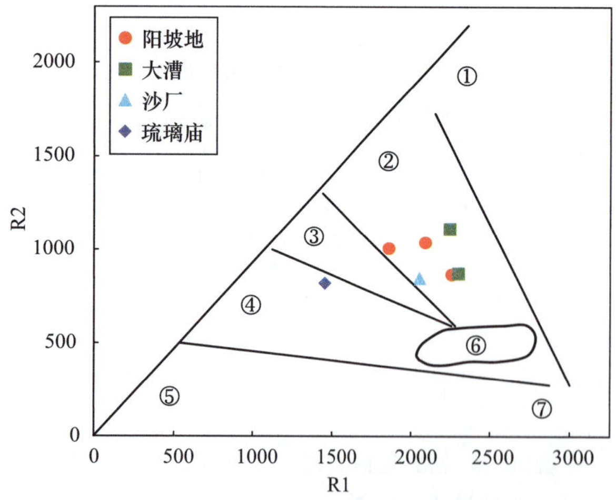

图 20-7 沙厂英云闪长质片麻岩 R1-R2 构造环境判别图

Fig. 20-7 R1-R2 diagram of tectonic discrimination of Shachang tonalite gneiss

①地幔斜长花岗岩；②破坏性活动板块边缘（板块碰撞前）花岗岩；③板块碰撞后隆起期花岗岩；④晚造山期花岗岩；⑤非造山区 A 型花岗岩；⑥同碰撞（S 型）花岗岩；⑦造山期后 A 型花岗岩

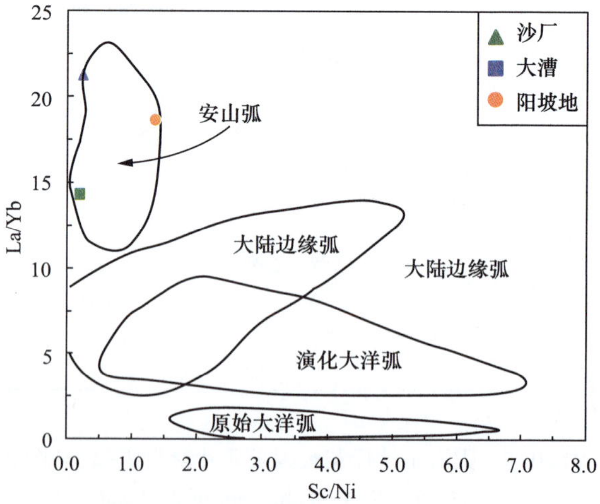

图 20-8 沙厂英云闪长质片麻岩 Sc/Ni-La/Yb 构造环境判别图

Fig. 20-8 Sc/Ni-La/Yb Tectonic diagram of tectonic discrimination of Shachang tonalite gneiss

（八）形成时代

本期岩浆侵位时代厘定为新太古代晚期，主要依据包括：①该系列岩体遭受多期次强烈变质变形；②侵入到新太古代表壳岩密云岩群中，且包含大量密云岩群表壳岩包体；③发育大量变质变形镁铁质或长英质等岩脉；④被新太古代宋营子英云闪长质片麻岩等岩体侵入切割，又被中元古代长城系地层不整合覆盖；⑤获得该套岩系中不同岩性锆石 U-Pb 同位素年龄为 2690~2360Ma（金文山等，1999）。

二、宋营子英云闪长质片麻岩

（一）地质产状与岩石学特征

发育大北湾、穆家峪等 7 大岩体，是区内太古宙仅次于沙厂英云闪长质片麻岩体外另一重要侵入岩单元，分布于密云穆家峪，怀柔崎峰茶、宋营子、喇叭沟门，昌平王家元至下口村西和延庆营门，出露面积 245km^2。侵入体内普遍发育长百米至千米级、宽厘米至米级变质超镁铁质或镁铁质岩脉，以及变质奥长花岗质片麻岩岩脉，见有表壳岩包体。

1. 大北湾岩体

位于怀柔孙胡沟、得田沟，平面形态呈 NNE 向楔形，SW 宽 3.5km，NE 宽 1km，出露面积 13km^2。岩体顺围岩片麻理或小角度切割围岩片麻理侵入新太古代表壳岩和沙厂英云闪长质片麻岩体，界线因后期变质变形（混合岩化）模糊，但边部可见透镜状或不规则状表壳岩包体（图 20-9），以及侵位时斜切割围岩片麻理现象。岩体岩性简单，为黑云角闪斜长片麻岩，各矿物组成及含量见表 20-5。

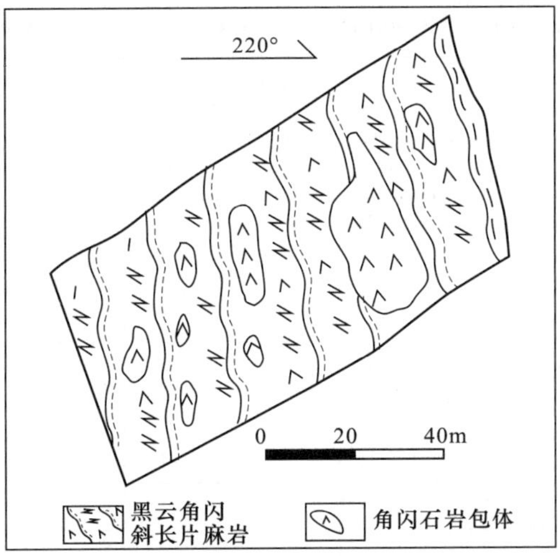

图 20-9 表壳岩包体分布特征示意图（怀柔大北湾）

Fig. 20-9 Geological sketch showing the distribution characteristics of supracrustal rock enclaves

2. 穆家峪岩体

位于密云穆家峪，出露面积 32km^2。岩体北、东分别侵入沙厂英云闪长质片麻岩体和苇子峪辉长闪长片麻岩体中，侵入地貌和界线清晰（图 20-10），边部发育少量新太古代表壳岩、黑卧辉长质片麻岩和苇子峪辉长闪长质片麻岩包体。岩性主要为角闪二辉片麻岩和辉石片麻岩，紫苏辉石蚀变绿泥石化保留假象，并可见角闪石交代及铁质析出现象，各矿物组成及含量见表 20-5。

3. 大甸子岩体

位于怀柔喇叭沟门大甸子、下弯子等地，由 3 个分散小岩体组成，出露面积 22km^2。岩体东被中生界土城子组不整合覆盖，其余周缘被晚古生代早二叠世和中生代白垩世侵入体侵位切割，可见大量晚期不同走向未变质正长岩、花岗岩脉。岩体内发育少量透镜状新太古代表壳岩（斜长角闪岩或黑云角闪变粒岩）包体，则地沟还发现二辉麻粒岩包体。岩体岩性单一，为灰色黑云角闪片麻岩，各矿物组成及含量见表 20-5。

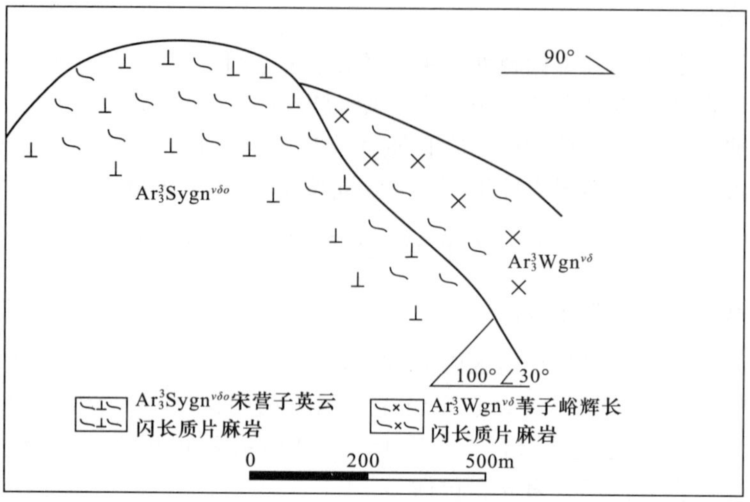

图 20-10　宋营子英云闪长质片麻岩与苇子峪辉长闪长质片麻岩侵入接触关系示意图

（潮河大坝东，据鲍亦冈等，2001）

Fig. 20-10　Geological sketch showing the intrusive contact between Songyingzi tonalitic gneiss and Weiziyu gabbro diorite gneissin

表 20-5　宋营子英云闪长质片麻岩主要岩石类型和矿物定量统计表

Table 20-5　Main rock types and its mineral contents of Songyingzi tonalite gneiss

岩体	岩性	矿物成分及含量/%								副矿物
		斜长石 Pl	钾长石 Or	石英 Q	紫苏辉石 Hy	透辉石 Di	石榴子石 Gt	黑云母 Bi	角闪石 Hb	
大北湾	黑云角闪斜长片麻岩	60~80	0~10	10~25				10~20	5~15	Zr、Ap、Sp、Mt
穆家峪	角闪二辉片麻岩	70		5	3~5	10		2	8	Mt、Ap、Zr
	辉石片麻岩	62		18		12	1	4	4	Mt、Ep、Zr
大甸子	黑云角闪片麻岩	70~80		0~10		1		10	10	Zr、Ap、Sp
麻峪房-王家元	角闪斜长片麻岩	45	5	11~20				6~10	20	Zr、Ap、Se、Mt
	辉石斜长片麻岩	45	2~3	10~15		5~10		6~10	15	Zr、Ap、Se、Mt
许家营	黑云角闪斜长片麻岩	42.5	25					8	25	Ap、Zr、Mt
杨树底下	含辉石石英角闪斜长片麻岩	50	12.5	5		3			20	Mt、Ap、Zr、Ch、Ep
东石门	透辉角闪黑云斜长片麻岩	65		8	5		15	7		Zr、Mt、Ap
	黑云角闪斜长片麻岩	65		10			10	15		Ap、Zr、Mt

注：数据资料引自 1∶5 万密云幅、高岭幅、墙子路幅区域地质调查报告，1994；1∶5 万琉璃庙幅、范各庄幅区域地质调查报告，1990；1∶5 万昌平幅、小汤山幅区域地质调查报告，1986；1∶5 万番字牌幅、汤河口幅区域地质调查报告，1996；1∶5 万杨木栅子幅、喇叭沟门、三道营子幅区域地质调查报告，1999。Zr—锆石，Ap—磷灰石，Sp—榍石，Mt—磁铁矿，Ep—绿帘石，Se—绢云母，Ch—绿泥石。

4. 麻峪房-王家元岩体

位于昌平麻峪房村、德胜口、臭泥坑、响潭和王家元水库一带，呈 NE 向长条状，由 3 个分散岩体组成，出露面积 29km^2。岩体东被中元古界长城系不整合覆盖，西与中元古界蓟县系呈断层接触，其余被中生代燕山期中酸性岩侵入破坏。内部发育大量透镜状新太古界表壳岩包体。岩体岩性主要为角闪斜长片麻岩，其次为辉石斜长片麻岩，各矿物组成及含量见表 20-5。

5. 杨树底下岩体

位于怀柔北干沟至邓厂一带，延伸至河北，出露面积 31km^2（含河北省）。岩体西与中生界土城子组呈断层或不整合接触；南与中元古代东辛店钾长花岗岩呈断层或被侵入接触；北与沙厂英云闪长质片麻岩断层接触。岩性为含辉石石英角闪斜长片麻岩，各矿物组成及含量见表 20-5。

6. 许家营岩体

位于怀柔许家营至三道卧铺一带，呈近 EW 向长条状，出露面积 23km^2。岩体北被古生代和中生代侵入体切割破坏；南与中元古代斜长岩呈断层或被侵入接触。岩体内普遍不均匀混合岩化，含透镜状新太古界表壳岩包体。岩体岩性为黑云角闪斜长片麻岩，各矿物组成及含量见表 20-5。

7. 东石门岩体

位于怀柔东石门、小东沟、龙潭沿子和西坨古等地，由 3 个分散小岩体组成，出露面积 49km^2。岩体南西侵入于沙厂英云闪长质片麻岩体及表壳岩密云岩群中，界线模糊；南东被中生代燕山期侵入体切割破坏；北与中元古界长城系呈断层接触。内部发育透镜状新太古代表壳岩包体和超基性岩脉。岩性为黑云角闪斜长片麻岩、透辉角闪黑云斜长片麻岩，各矿物组成及含量见表 20-5。

（二）岩石化学与地球化学特征

岩石化学成分及参数见表 20-6。岩石比沙厂英云闪长质片麻岩低镁、铁，高钾。里特曼指数 σ = 0.94～3.27，$Na_2O>K_2O$，为钠质亚碱性岩系列；固结指数 SI = 13.44～24.55，分异指数 DI = 38.52～71.268，长英指数 FL = 35.18～67.80，岩浆分异程度较高；A/CNK = 0.72～1.13，A/NK = 1.30～2.33，主要属偏铝质。

稀土元素成分及参数见表 20-7。ΣREE = 87.46 ×10^{-6} ～ 227.46 ×10^{-6}，LREE/HREE = 10.02～26.36，$(La/Yb)_N$ = 12.16～77.11，稀土总量较高，轻稀土强烈富集，重稀土相对亏损；δEu = 0.56～1.19，除许家营岩体弱铕正异常外，其余极弱铕负异常；配分曲线为右陡倾光滑曲线（图 20-11）。

微量元素成分含量与冀东秋花峪英云闪长质片麻岩相似，K、Ba、Sr、Th 元素含量高于上地幔，Sc、Cr、Co、Ni 元素含量偏低，其余元素含量相当。微量元素蛛网图（图 20-12）中，富集大离子亲石元素 K、Ba、Sr 和高场强元素 Th，而相对亏损高场强元素 Zr 和重稀土元素。

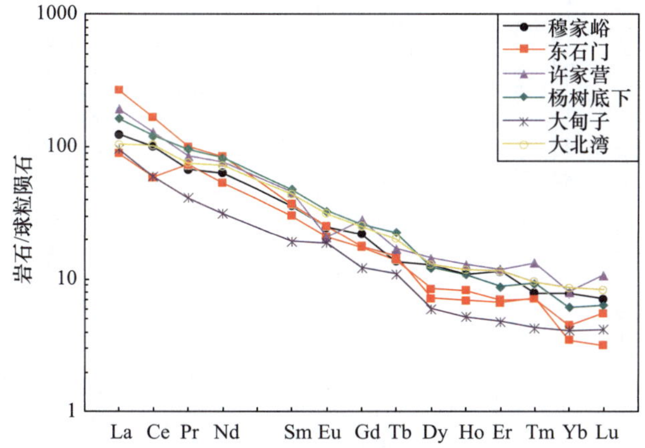

图 20-11 太古宙宋营子英云闪长质片麻岩稀土元素配分模式图

Fig. 20-11 Chondrite-normalized REE patterns of Archean Songyingzi tonalite gneiss

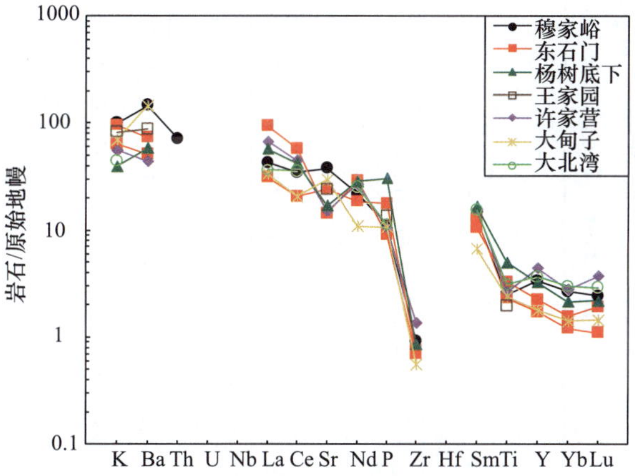

图 20-12 太古宙宋营子英云闪长质片麻岩微量元素蛛网图

Fig. 20-12 Trace element spider diagram of Archean Songyingzi tonalite gneiss

表 20-6　太古宙宋营子英云闪长质片麻岩岩石化学成分及特征参数一览表

Table 20-6　The chemical composition and main parameters of Archean Songyingzi tonalite gneiss

序号	岩体	岩性	氧化物含量/%													特征参数					
			SiO_2	TiO_2	Al_2O_3	Fe_2O_3	FeO	MnO	MgO	CaO	Na_2O	K_2O	P_2O_5	Los	总和	*DI*	*SI*	σ	*FL*	*A/CNK*	*A/NK*
1	穆家峪	含云角闪片麻岩	62. 34	0. 54	14. 90	3. 22	3. 50	0. 12	2. 28	3. 80	5. 00	3. 00	0. 24	1. 10	100. 04	71. 68	13. 44	3. 27	67. 80	0. 81	1. 30
2	东石门	黑云角闪斜长片麻岩	64. 94	0. 49	16. 08	2. 20	2. 71	0. 06	2. 40	3. 26	3. 82	1. 87	0. 19	1. 60	99. 61	68. 79	18. 51	1. 45	63. 57	1. 13	1. 94
3		透辉角闪黑云斜长片麻岩	59. 94	0. 70	16. 48	3. 00	4. 02	0. 11	2. 77	4. 66	3. 66	2. 82	0. 38	0. 90	99. 44	61. 36	17. 05	2. 42	58. 17	0. 94	1. 82
4	许家营	黑云角闪斜长片麻岩	62. 38	0. 56	14. 55	2. 44	3. 98	0. 08	3. 14	5. 19	2. 72	1. 64	0. 23	1. 90	98. 78	57. 61	22. 60	0. 94	45. 63	0. 93	2. 33
5	杨树底下	含辉石石英角闪斜长片麻岩	50. 66	1. 06	16. 55	4. 95	5. 37	0. 14	4. 85	8. 73	3. 57	1. 17	0. 65	2. 24	99. 92	38. 52	24. 55	2. 64	35. 18	0. 72	2. 32
6	大甸子	黑云角闪斜长片麻岩	59. 95	0. 51	15. 86	2. 51	4. 29	0. 07	3. 17	5. 66	4. 42	2. 06	0. 23	2. 13	100. 85	60. 17	19. 25	2. 43	53. 38	0. 80	1. 67
7	王家元	角闪斜长片麻岩	62. 05	0. 43	15. 39	2. 30	3. 72	0. 10	3. 04	4. 51	4. 36	2. 38	0. 27	0. 84	99. 39	65. 14	19. 27	2. 34	59. 91	0. 86	1. 58
8		辉石斜长片麻岩	60. 51	0. 40	15. 19	2. 34	4. 28	0. 11	3. 80	3. 80	4. 80	2. 45	0. 30	0. 94	98. 92	65. 08	21. 51	2. 92	65. 61	0. 87	1. 44
9	大北湾	黑云角闪斜长片麻岩	63. 57	0. 66	15. 63	2. 94	2. 91	0. 11	2. 50	4. 34	4. 70	1. 30	0. 24	0. 71	99. 61	65. 87	17. 51	1. 73	58. 03	0. 92	1. 71

注：数据为均值，引自1∶5万密云幅、高岭幅、墙子路幅区域地质调查报告，1994；1∶5万琉璃庙幅、范各庄幅区域地质调查报告，1990；1∶5万昌平幅、小汤山幅区域地质调查报告，1986；1∶5万番字牌幅、汤河口幅区域地质调查报告，1996；1∶5万杨木栅子幅、喇叭沟门、三道营子幅区域地质调查报告，1999。

表 20-7　太古宙宋营子英云闪长质片麻岩稀土元素成分及特征参数一览表

Table 20-7　REE contents and main parameters of Archean Songyingzi tonalite gneiss

序号	岩体	岩性	稀土元素含量/10^{-6}																特征参数			
			La	Ce	Pr	Nd	Sm	Eu	Gd	Tb	Dy	Ho	Er	Tm	Yb	Lu	Y	ΣREE	LREE/HREE	$(La/Yb)_N$	δEu	δCe
1	穆家峪	含云角闪片麻岩	29. 40	61. 40	6. 45	29. 60	5. 50	1. 44	4. 51	0. 51	3. 24	0. 62	1. 89	0. 20	1. 32	0. 18	15. 20	146. 26	10. 73	15. 98	0. 86	1. 05
2	东石门	黑云角闪斜长片麻岩	63. 07	100. 60	9. 50	38. 90	5. 62	1. 45	3. 62	0. 56	1. 81	0. 39	1. 10	0. 18	0. 59	0. 08	7. 65	227. 46	26. 36	77. 11	0. 92	0. 90
3		透辉角闪黑云斜长片麻岩	21. 00	35. 90	6. 94	24. 80	4. 60	1. 21	3. 54	0. 52	2. 13	0. 46	1. 15	0. 18	0. 76	0. 14	10. 00	103. 33	10. 64	19. 82	0. 88	0. 73
4	许家营	黑云角闪斜长片麻岩	45. 41	78. 13	7. 99	35. 77	6. 90	1. 20	5. 73	0. 63	3. 63	0. 73	1. 95	0. 34	1. 35	0. 27	19. 60	189. 99	12. 00	24. 13	0. 56	0. 93
5	杨树底下	含辉石石英角闪斜长片麻岩	38. 45	73. 35	9. 04	38. 10	7. 26	1. 89	5. 26	0. 84	3. 08	0. 61	1. 45	0. 24	1. 04	0. 16	14. 50	180. 74	13. 28	26. 65	0. 89	0. 93
6	大甸子	黑云角闪斜长片麻岩	22. 55	36. 09	3. 90	14. 52	2. 93	1. 08	2. 49	0. 41	1. 51	0. 29	0. 79	0. 11	0. 69	0. 11	7. 96	87. 46	12. 70	23. 39	1. 19	0. 87
7	大北湾	黑云角闪斜长片麻岩	24. 58	62. 46	7. 11	33. 48	6. 69	1. 81	5. 13	0. 75	3. 25	0. 66	1. 89	0. 24	1. 45	0. 21	16. 48	149. 71	10. 02	12. 16	0. 91	1. 14

注：数据为均值，来源同表 20-6。

（三）变质变形特征

岩体普遍发育较强变形变质作用，并形成紧闭-同斜褶皱变形。紫苏辉石或透辉石矿物的存在揭示岩石经历高级麻粒岩相变质作用。在昌平王家元岩体、密云穆家峪岩体保存较好，其余岩体辉石则多被角闪石或黑云母交代、蚀变以残晶形式存在，如许家营岩体，揭示了后期叠加多期次不均匀角闪岩相、绿片岩相退变质作用。

（四）原岩恢复

宋营子英云闪长质片麻岩为正变质岩，证据包括：①岩性虽有所变化，但岩石保留粒状变晶结构，矿物组成简单，以斜长石、角闪石、黑云母为主，其次为石英、钾长石矿物组合，缺乏或少有泥质原岩变质矿物存在；②副矿物组合为磷灰石、磁铁矿、锆石、榍石组合，具有典型岩浆岩副矿物组合特征，且多数自形，少有磨圆现象；③岩石组成虽有片麻理，但野外产状成层性并不明显，且众多地质体边界具有侵入体边界特点；④在变质岩原岩恢复 SiO_2-TiO_2 和 Si-[(al+fm)-(c+alk)]图解（图 20-13）上，成分点落在火成岩区内。

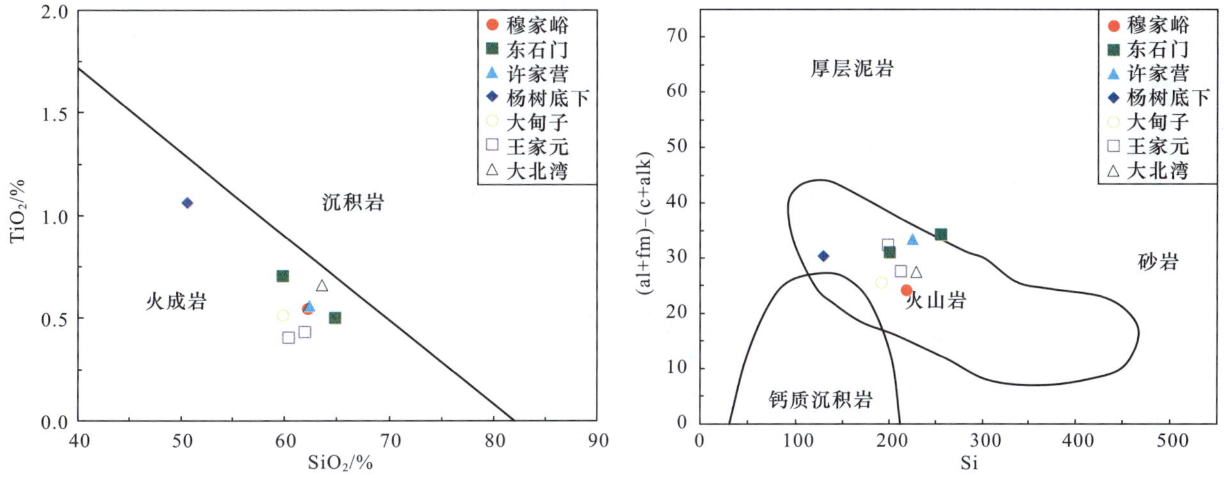

图 20-13 太古宙宋营子英云闪长质片麻岩 SiO_2-TiO_2 和 Si-[(al+fm)-(c+alk)]原岩恢复图解

Fig. 20-13 SiO_2-TiO_2 and Si-[(al+fm)-(c+alk)] diagram of Archean Songyingzi tonalite gneiss

（五）岩石系列

该套岩系 $\sigma=0.94\sim3.27<4$，为亚碱性系列。岩石贫硅，富碱、镁，高钠低钾，同钙碱性岩相吻合，在 *F-A-M* 图（图 20-14）上，成分点位于钙碱性系列。

（六）成因类型

该套岩系低铁、镁和高钙、铝，富集大离子亲石元素，轻重稀土元素高分异等特征揭示了其岩浆来源于地壳，而非深源地幔。地质产状和地球化学特征不同于沙厂英云闪长质片麻岩，各岩体 La-La/Sm 也不能拟合成直线，同沙厂英云闪长质片麻岩及密云表壳岩均无线性关系，又不是浅源表壳岩（围岩）部分熔融产物。在 K_2O-Na_2O 图（图 20-15）上，成分点位于 I 型花岗岩区。因此宋营子英云闪长质片麻岩可能来源于较深下地壳物质的部分熔融。

（七）构造环境分析

该套岩系表现出低钛、高镁特征，$SiO_2=50.66\sim64.94$，多数 $FeO^*/MgO<2$、$K_2O/Na_2O<0.6$；富集大离子亲石元素 K、Sr、Ba 和高场强元素 P、轻稀土元素，相对亏损高场强元素 Zr 和重稀土元素，同岛弧环境火成岩特征相似。在 TiO_2*10-Al_2O_3-K_2O*10 和 FeO^*-MgO-Al_2O_3 构造环境判别图

解（图 20-16）上，成分点位于岛弧、造山带玄武岩（安山岩）区。岩石 TiO_2<1.2%，属于钙碱性系列（图 20-17），且 SiO_2-FeO^*/MgO>50%，较高 K_2O 含量，揭示其可能形成于成熟的岛弧环境，是在早期沙厂英云闪长质片麻岩揭示不成熟洋-陆俯冲型活动大陆边缘安第斯型岛弧成熟演化。

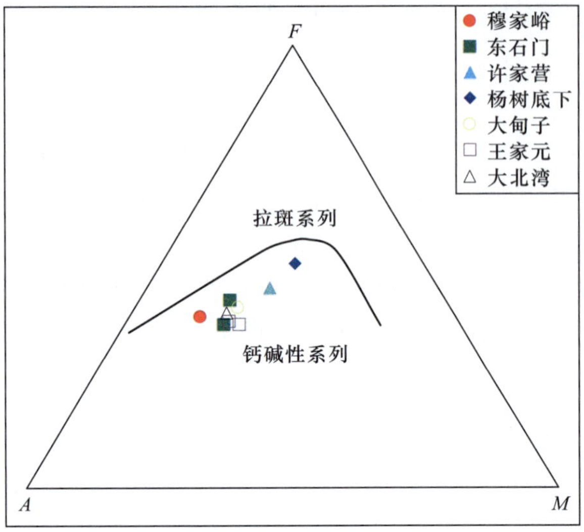

图 20-14 太古宙营子英云闪长质片麻岩 *F-A-M* 图

Fig. 20-14 *F-A-M* diagram of Archean Songyingzi tonalite gneiss

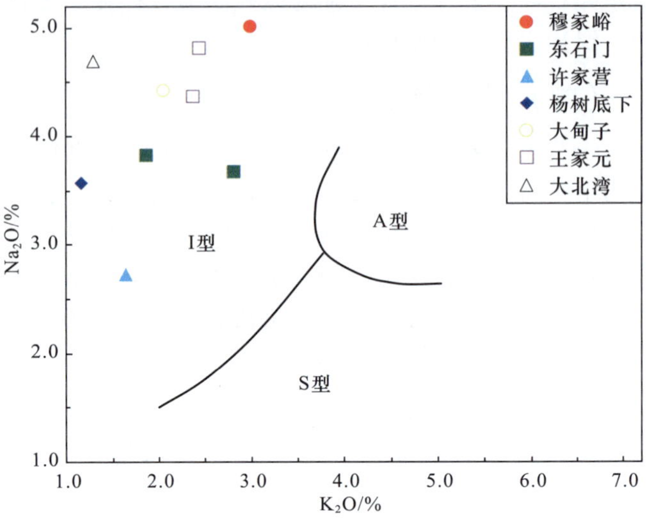

图 20-15 太古宙宋营子英云闪长质片麻岩 K_2O-Na_2O 图

Fig. 20-15 K_2O-Na_2O diagram of Archean Songyingzi tonalite gneiss

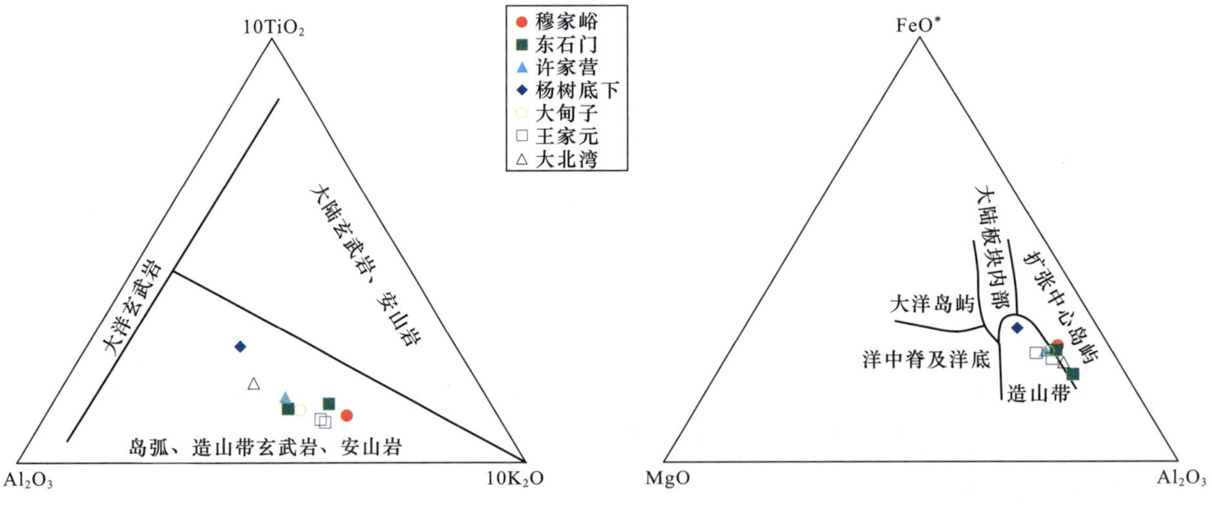

图 20-16 宋营子片麻岩 $10TiO_2$-Al_2O_3-$10K_2O$ 和 FeO^*-MgO-Al_2O_3 图解

Fig. 20-16 $10TiO_2$-Al_2O_3-$10K_2O$ and FeO^*-MgO- Al_2O_3 diagram of Songyingzi gneiss

（八）形成时代

本期岩浆侵位时代厘定为新太古代晚期，主要依据包括：①该系列岩体遭受多期次强烈变质变形；②侵入苇子峪辉长闪长质片麻岩岩体和新太古界密云岩群表壳岩中；③内部包含密云岩群表壳岩包体，并发育大量镁铁质或长英质等岩脉；④被中元古界长城系地层不整合覆盖；⑤该套岩系中不同岩性锆石 U-Pb 同位素年龄为 2405~2539Ma（金文山等，1999；任荣等，2011）；⑥穆家峪岩体含石榴角闪英云闪长质片麻岩锆石 U-Pb 同位素年龄为（2544±35）Ma。

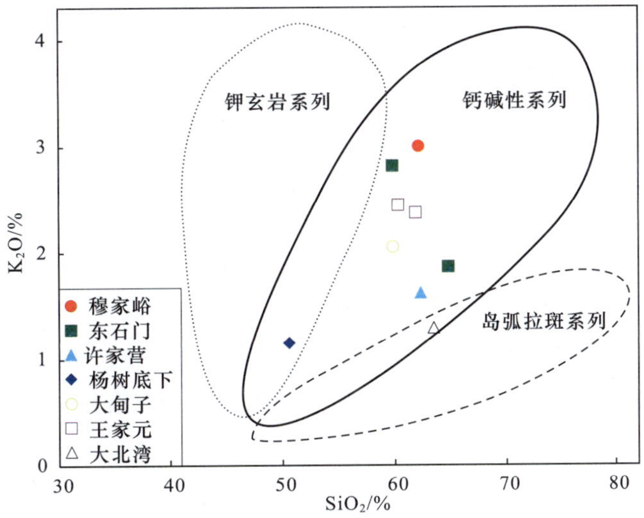

图 20-17 宋营子英云闪长质片麻岩 SiO_2-K_2O 图

Fig. 20-17 SiO_2-K_2O diagram of Archean Songyingzi tonalite gneiss

三、荆子峪花岗闪长质片麻岩

(一) 地质产状与岩石学特征

发育荆子峪和王家窝铺两大岩体，分布于密云水树峪、上庄头峪和怀柔王家窝铺，出露面积 39km²。岩体仅在边部见有规模较小、数量极少包体。

1. 荆子峪岩体

位于密云水树峪、上庄头峪，呈不规则状，出露面积 17km²。岩体抗风化能力较强，比围岩地貌高 100~200m，宏观上外倾产状，侵入特征明显。南侧侵入沙厂英云闪长质片麻岩，接触界线清晰，内接触带混合岩化强烈，并可见围岩包体，外接触带发育大量岩枝（图 20-18）；北侧侵入苇子峪辉长闪长质片麻岩，又被中生代燕山期四干顶花岗岩体侵入破坏。按照矿物成分组成划分中心相和边缘相。中心相岩石类型为黑云母片麻岩，边缘相岩石类型为含云闪透辉片麻岩，由中心向边缘暗色矿物含量增加，石英含量减少。

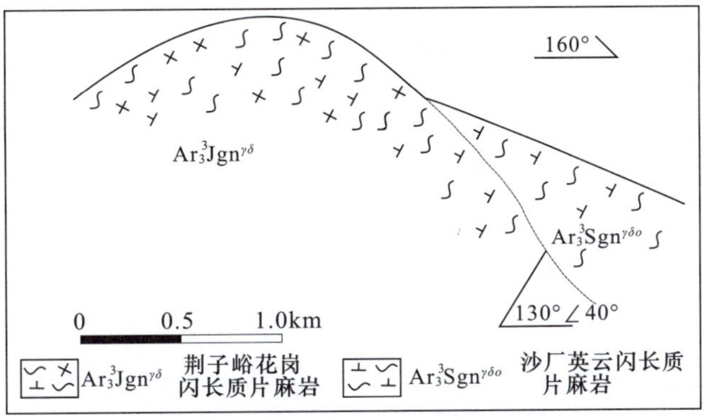

图 20-18 荆子峪花岗闪长质片麻岩与沙厂英云闪长质片麻岩侵入接触关系示意图

（据鲍亦冈等，2001）

Fig. 20-18 Geological sketch showing the intrusive contact relation between Shachang tonalitic gneiss and Jingziyu granodioritic gneiss

（1）黑云母片麻岩

为岩体中心相主要岩石类型，浅灰色，鳞片粒状变晶结构，片麻状构造。斜长石他形—半自形粒状，粒径0.5~1.0mm，发育聚片双晶和钠质交代，钾长石多补片状，部分呈反条纹长石，蚀变绢云母，含量70%~75%；石英他形粒状，波状消光，有的发育变形条带，相对集中呈条带状构造，含量10%~15%；暗色矿物主要为黑云母，粒径0.5~1.0mm，强烈蚀变绿泥石化、绿帘石化；副矿物为绿帘石、榍石、磷灰石、锆石等。其中，锆石发育生长环带。

（2）含云闪透辉片麻岩

为岩体边缘相主要岩石类型，浅灰色，鳞片粒状变晶结构，片麻状构造。透辉石淡绿色他形粒状、短柱状，强烈绿泥石化、次闪石化，部分已呈假象，含量5%~12%；普通角闪石绿色，多分布于辉石边缘，为辉石退变质成因，含量3%~7%；黑云母红褐色—黄褐色片状，定向排列，边缘绿泥石化；斜长石他形粒状，发育聚片双晶、反条纹结构，绢云母化、钠黝帘石化；石英可识别出0.1~0.2mm和0.5~1.0mm两个粒级。

2. 王家窝铺岩体

位于怀柔王家窝铺，出露面积22km²。周缘同中元古界长城系和宋营子英云闪长质片麻岩断层接触，局部侵入到宋营子英云闪长质片麻岩中。岩体岩性为含辉角闪斜长片麻岩，呈浅灰色，矿物组成：钾长石含量约40%，斜长石含量约25%，石英含量约15%，角闪石含量约12%，透辉石含量约5%。副矿物为磁铁矿、磷灰石、锆石，及蚀变绿泥石和绿帘石。

（二）岩石化学与地球化学特征

岩石化学成分及参数见表20-8。岩石呈现高硅、铝、碱，低钛、铁、镁特征。里特曼指数 σ = 2.37~2.80，$Na_2O>K_2O$，为钠质亚碱性岩系列；固结指数 SI = 12.05~12.01，分异指数 DI = 71.80~76.53，长英指数 FL = 66.98~70.22，岩浆分异程度较低；A/CNK = 0.86~1.07，A/NK = 1.31~1.77，属于偏铝质。

稀土元素成分及参数见表20-9。$\Sigma REE = 92.74\times10^{-6}\sim336.91\times10^{-6}$，LREE/HREE = 14.24~33.46，$(La/Yb)_N$ = 28.85~122.23，稀土总量差异较大，轻稀土强烈富集，重稀土相对亏损；δEu = 1.06~1.25，铕弱正异常；配分曲线为右陡倾光滑曲线（图20-19）。

与原始地幔相比，微量元素成分K、Ba、Sr高，其余元素含量低，尤其Co、Ni元素。微量元素蛛网图（图20-20）中，富集大离子亲石元素K、Ba和轻稀土元素，相对亏损高场强元素Th、Zr和重稀土元素。

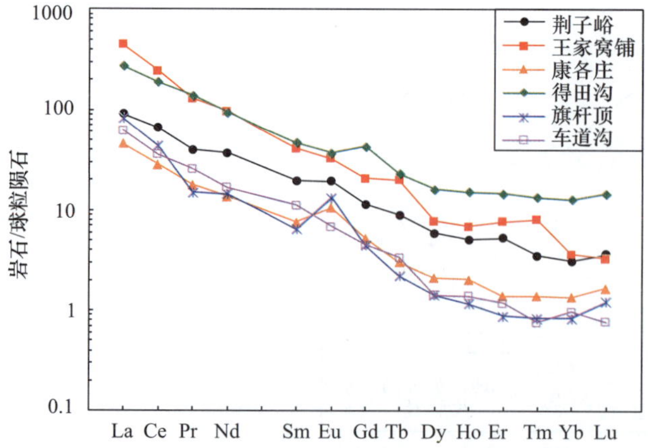

图20-19 太古宙荆子峪、康各庄和旗杆顶片麻岩稀土元素配分模式图

Fig. 20-19 Chondrite-normalized REE patterns of Archean Jingziyu, Kanggezhuang and Qiganding gneiss

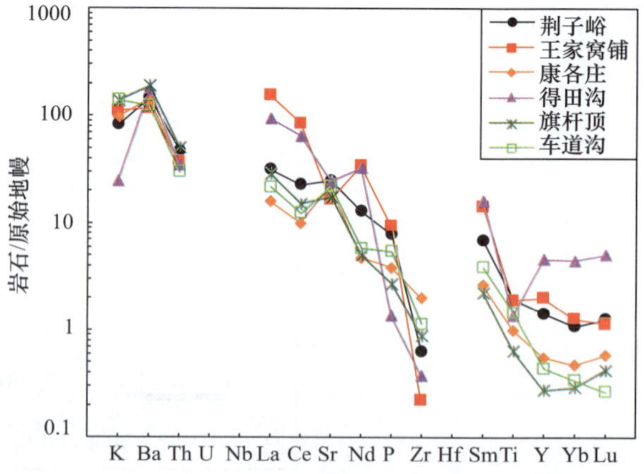

图20-20 太古宙荆子峪、康各庄和旗杆顶片麻岩微量元素蛛网图

Fig. 20-20 Trace element spider diagram of Archean Jingziyu, Kanggezhuang and Qiganding gneiss

表 20-8　太古宙荆子峪、康各庄和旗杆顶片麻岩岩石化学成分及特征参数一览表

Table 20-8　The chemical composition and main parameters of Archean Jingziyu, Kanggezhuang and Qiganding gneiss

序号	岩体	岩性	氧化物含量/%													特征参数					
			SiO_2	TiO_2	Al_2O_3	Fe_2O_3	FeO	MnO	MgO	CaO	Na_2O	K_2O	P_2O_5	Los	总和	*DI*	*SI*	*σ*	*FL*	*A/CNK*	*A/NK*
1	荆子峪	黑云母片麻岩	65.56	0.42	15.41	2.56	2.08	0.08	1.72	3.38	5.50	2.47	0.17	0.98	100.33	76.53	12.05	2.80	70.22	0.86	1.31
2	王家窝铺	含辉角闪斜长片麻岩	63.90	0.42	17.37	1.66	2.23	0.06	1.63	3.51	3.74	3.37	0.21	1.30	99.38	71.80	12.91	2.37	66.98	1.07	1.77
3	康各庄	奥长花岗质片麻岩	69.42	0.22	14.18	1.42	1.52	0.06	1.57	3.01	4.55	2.95	0.09	0.85	99.81	80.15	13.06	2.12	71.39	0.88	1.33
4	得田沟	奥长花岗质片麻岩	76.39	0.30	11.00	2.21	2.09	0.12	0.75	2.08	3.58	0.75	0.03	0.33	99.63	80.99	8.06	0.56	67.55	1.05	1.64
5	旗杆顶	二长花岗质片麻岩	72.45	0.14	13.87	1.08	1.05	0.02	0.51	0.96	4.27	4.26	0.06	0.80	99.47	90.70	4.57	2.46	89.88	1.04	1.19
6	车道沟	含辉榴黑云斜长片麻岩	67.88	0.32	15.23	1.19	2.19	0.01	1.37	2.84	3.63	4.28	0.12	0.35	99.41	78.34	10.82	2.50	73.58	0.97	1.44

注：2、3 数据为均值，引自 1∶5 万密云幅、高岭幅、墙子路幅区域地质调查报告，1994；1∶5 万番字牌幅、汤河口幅区域地质调查报告，1996。

表 20-9　太古宙荆子峪、康各庄和旗杆顶片麻岩稀土元素成分及特征参数一览表

Table 20-9　REE contents and main parameters of Archean Jingziyu, Kanggezhuang and Qiganding gneiss

序号	岩体	岩性	稀土元素含量/10^{-6}																特征参数			
			La	Ce	Pr	Nd	Sm	Eu	Gd	Tb	Dy	Ho	Er	Tm	Yb	Lu	Y	ΣREE	LREE/HREE	$(La/Yb)_N$	δEu	δCe
1	荆子峪	黑云母片麻岩	21.40	40.00	3.80	17.30	3.02	1.13	2.36	0.34	1.51	0.29	0.87	0.09	0.53	0.09	6.49	92.74	14.24	28.85	1.25	1.00
2	王家窝铺	含辉角闪斜长片麻岩	107.35	152.75	12.50	46.20	6.41	1.93	4.36	0.77	2.03	0.40	1.31	0.21	0.63	0.09	9.17	336.91	33.46	122.23	1.06	0.85
3	康各庄	奥长花岗质片麻岩	11.05	17.85	1.74	6.52	1.19	0.63	1.11	0.12	0.55	0.12	0.23	0.04	0.24	0.04	2.56	41.40	15.97	33.59	1.64	0.90
4	得田沟	奥长花岗质片麻岩	65.86	117.29	13.44	44.65	7.34	2.21	9。66	0.88	4.22	0.88	2.47	0.35	2.23	0.38	21.32	262.20	21.98	21.18	0.83	0.91
5	旗杆顶	二长花岗质片麻岩	19.80	27.10	1.46	6.96	1.01	0.79	0.94	0.08	0.37	0.07	0.15	0.02	0.15	0.03	1.28	58.93	31.47	96.62	2.43	0.90
6	车道沟	含辉榴黑云斜长片麻岩	15.07	22.26	2.50	8.04	1.75	0.41	0.95	0.13	0.37	0.08	0.20	0.02	0.17	0.02	2.04	51.97	25.79	63.59	0.88	0.81

注：2、3 数据为均值，来源同表 20-6。

（三）变形与变质特征

岩石中可见少量紫苏辉石，反映早期曾经过麻粒岩相变质作用改造，又由于岩体保留了较弱的片麻理构造，说明麻粒岩相变质变形作用较弱。受后期角闪岩相变形作用叠加，岩石发生角闪岩相退变，并形成板状片麻岩和角闪石线理，岩体中形成近南北向角闪岩相变形带。同宋营子英云闪长质片麻岩一致，王家窝铺岩体内部也发育宽度100~400m不等的韧性剪切变形，斜长石和石英均被拉长并定向排列。

（四）原岩恢复

该套岩系地貌、地质产状和岩性等具有典型侵入体特征，特别是荆子峪岩体还发育侵入体岩相变化，与围岩接触侵入迹象清晰。在变质岩原岩恢复 SiO_2-TiO_2 和 Si-[(al+fm)-(c+alk)]图解（图20-21）上，成分点落在火成岩区内。因此，该套变质岩为正变质岩。

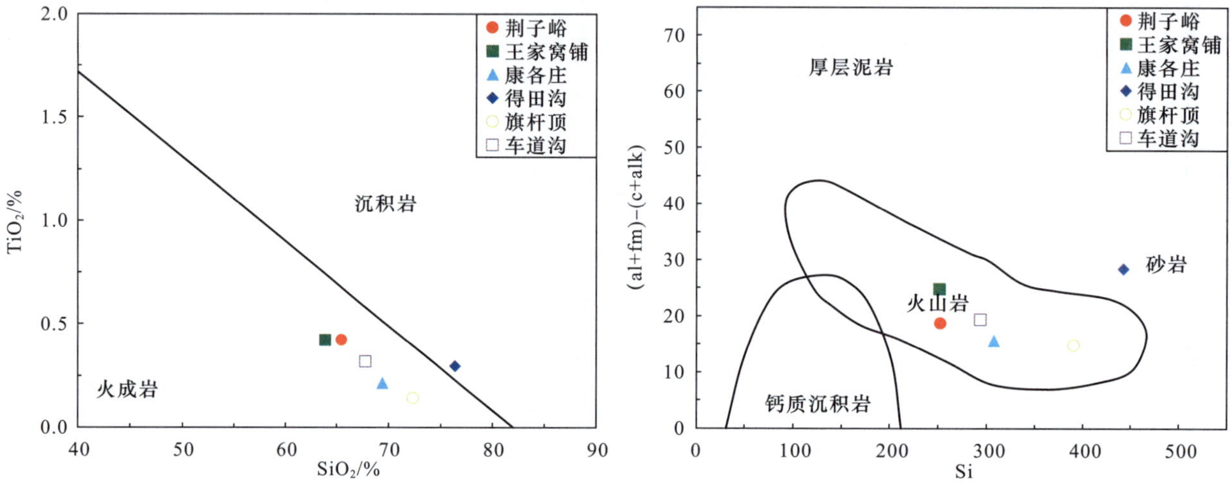

图20-21 太古宙荆子峪、康各庄和旗杆顶片麻岩 SiO_2-TiO_2 和 Si-[(al+fm)-(c+alk)]原岩恢复图解

Fig. 20-21 SiO_2-TiO_2 and Si-[(al+fm)-(c+alk)] diagram of Archean Jingziyu, Kanggezhuang and Qiganding gneiss

（五）岩石系列

该套岩系 $\sigma=2.37\sim2.80<4$，为亚碱性系列。在 SiO_2-K_2O 图（图20-22）上，成分点位于高钾钙碱性系列。

（六）成因类型

荆子峪花岗闪长质片麻岩矿物组成和化学成分与花岗闪长岩相同，*DI*、*SI* 等特征参数、稀土和微量元素成分含量等揭示其经历较弱岩浆分异演化。岩石低铁、镁、钛和高硅、铝，富集大离子亲石元素，轻重稀土元素强分异，不同岩体岩石化学特征差异较大等特征揭示了其岩浆来源于地壳。在 Na_2O-K_2O 和 *C/MF-A/MF* 图（图20-23）上，成分点分别位于I型花岗岩区和基性岩与变质砂岩部分熔融过渡区，也说明了其成岩物质来源于地壳。

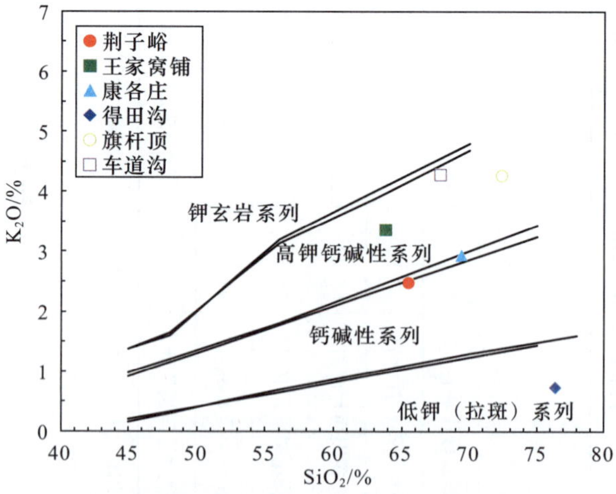

图20-22 太古宙荆子峪、康各庄和旗杆顶片麻岩 SiO_2-K_2O 图

Fig. 20-22 SiO_2-K_2O diagram of Archean Jingziyu, Kanggezhuang and Qiganding gneiss

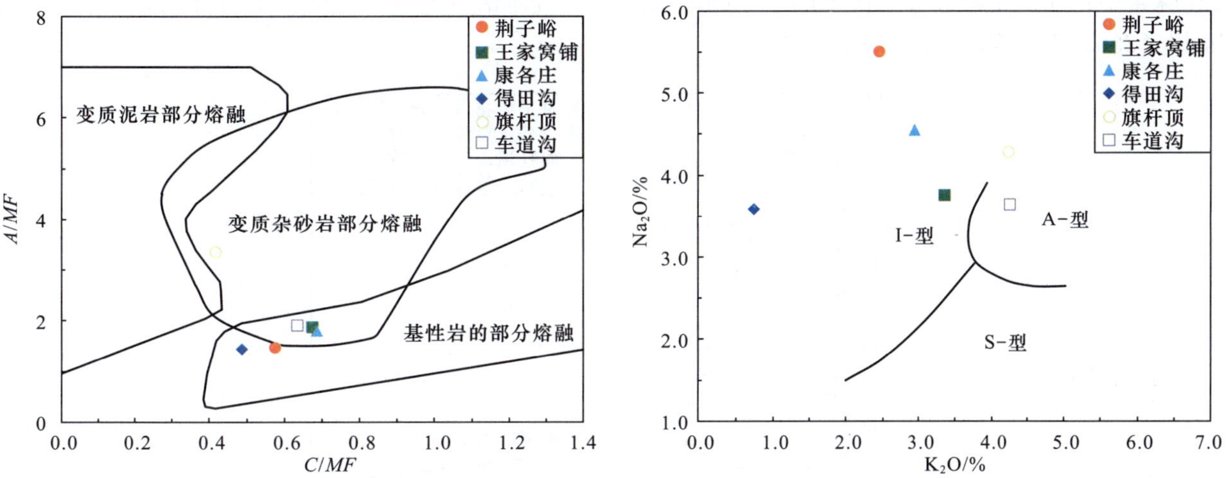

图 20-23 太古宙荆子峪、康各庄和旗杆顶片麻岩 SiO_2-K_2O 图

Fig. 20-23 SiO_2-K_2O diagram of Archean Jingziyu, Kanggezhuang and Qiganding gneiss

（七）构造环境分析

该套岩系富集大离子亲石元素 K、Sr、Ba 和轻稀土元素，相对亏损高场强元素 Zr 和重稀土元素，同活动大陆边缘岛弧环境火成岩特征相似。在 $MgO-Al_2O_3$ 和（FeO^*+MgO）－ CaO 构造环境判别图解（图 20-24）上，成分点位于岛弧花岗岩+大陆弧花岗岩+大陆碰撞花岗岩区。R1＝1715～2065，不同于同碰撞花岗岩（R1>2000）。也不同于宋营子英云闪长质片麻岩。该套岩系属于高钾钙碱性系列，虽 $SiO_2-FeO^*/MgO>50\%$，但 $SiO_2=63.90\sim65.56$，$FeO^*/MgO>2$、$K_2O/Na_2O>0.6$，呈现弱 Eu 正异常，表现出硅、钾含量增加，与成熟活动大陆边缘弧或与俯冲作用有关的岩浆弧特征相似，意味着地壳成熟度与厚度可能进一步增加。在 Sr/Y-Y 图（图 20-25），成分点也逐渐由中酸性宋营子英云闪长质片麻岩经典岛弧特征岩石向更加酸性荆子峪、康各庄和旗杆顶岩石具埃达克岩特征过渡。

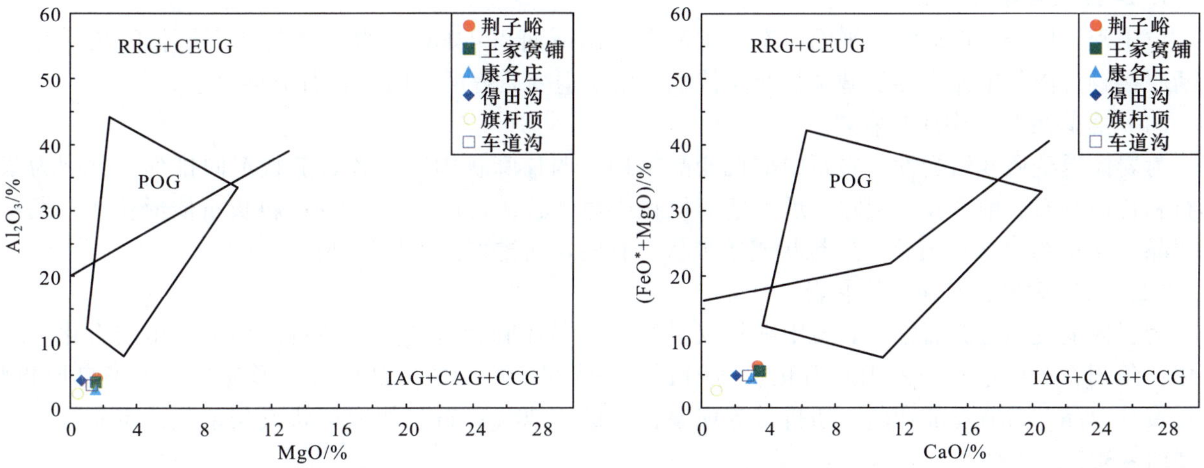

图 20-24 荆子峪、康各庄和旗杆顶片麻岩 $MgO-Al_2O_3$ 和（FeO^*+MgO）－ CaO 图解

Fig. 20-24 $MgO-Al_2O_3$ and（FeO^*+MgO）－CaO diagram of Archean Jingziyu, Kanggezhuang and Qiganding gneiss

（八）形成时代

本期岩浆侵位时代厘定为新太古代晚期，主要依据包括：①岩体遭受弱麻粒岩相变质，还经后期角闪岩相退变质；②岩浆分异形成的条带经受褶曲复杂变形；③侵入到新太古代晚期宋营子英云闪长

质片麻岩中；④荆子峪和王家窝铺两个花岗闪长质片麻岩体锆石 U-Pb 同位素年龄分别为（2627±80）Ma 和（2662±24）Ma（金文山等，1999）。

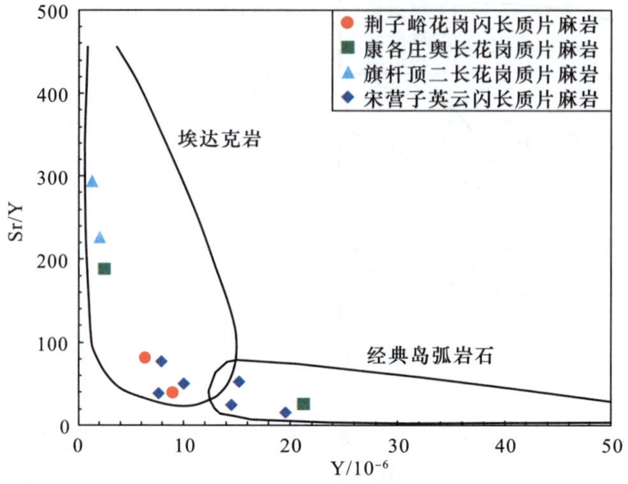

图 20-25　宋营子、荆子峪、康各庄和旗杆顶片麻岩 Sr/Y-Y 图

Fig. 20-25　Sr/Y-Y diagram of Archean Songyingzi, Jingziyu, Kanggezhuang and Qiganding gneiss

四、康各庄奥长花岗质片麻岩

（一）地质产状与岩石学特征

发育康各庄和得田沟两大岩体，分布于密云返迴峪和怀柔得田沟，出露面积 $48km^2$。此外，在除旗杆顶花岗质片麻岩以外的上述各太古宙片麻岩体中广泛发育有奥长花岗质片麻岩岩脉，可能也与这期岩浆侵位事件有关。这些岩脉宽数十厘米至数米，长数米至上百米，脉状侵入接触关系清晰。该期岩体内极少见有表壳岩包体和辉石角闪岩包体。

1. 康各庄岩体

位于密云墙子路返迴峪，平面形态呈 EW 向，出露面积 $44km^2$。岩体北与沙厂辉长闪长质片麻岩呈断层接触，南与中元古界长城系呈断层接触。由于退变质程度不同，岩石类型复杂。

（1）含辉奥长花岗质片麻岩

为岩体退变质残留岩性。岩石中细粒变晶结构，弱片麻状构造。辉石链状定向排列，类型为紫苏辉石和普通辉石，前者发育裂隙，后者他形粒状或短柱状，含量 5%~10%；斜长石他形粒状，发育聚片双晶，含量约 70%；石英他形粒状或条带状，有些定向排列，含量约 20%。

（2）含闪辉奥长花岗质片麻岩

为岩体退变质过渡岩性。岩石中细粒变晶结构，弱片麻状构造。透辉石浅绿色，角闪石化、绿泥石化；角闪石黄绿色，强烈阳起石化、透闪石化，含量 1%~4%；斜长石他形粒状，少半自形板状，发育聚片双晶，常有双晶弯曲、扭折带等现象，含量约 70%；石英他形粒状或长条状，定向排列，含量 20%~30%。

（3）含云奥长花岗质片麻岩

为岩体退变质岩性。岩石鳞片状变晶结构，弱片麻状构造。黑云母半自形片状，多色性，多绿泥石化保留黑云母假象，并沿节理析出铁质，含量 15%~25%；斜长石他形-半自形粒状，发育聚片双晶，强烈绢云母化，含量 55%~65%；石英他形粒状，有些拉长定向排列，含量 20%~30%。

2. 得田沟岩体

位于怀柔得田沟东，平面形态呈 NNE 向不规则楔状，出露面积 $4km^2$。岩体北西与沙厂英云闪长质片麻岩呈断层接触，南东与大北湾岩体呈渐变过渡关系，界线模糊。岩体主体岩性为长英质片麻岩，呈灰绿色，粒状变晶结构，弱片麻状构造。斜长石他形粒状，粒径0.4～1mm，强烈绢云母化、钠黝帘石化致使边缘不规则，发育钠长石净边结构和微斜长石反条纹结构，含量59%～63%；石英他形粒状，粒径0.1～1.5mm，条纹状集合体定向分布，含量20%～40%；黑云母黄绿色、绿褐色自形板状，粒径0.1～0.3mm，含量<5%；正长石他形粒状，粒径<0.2mm，含量<2%。部分岩石含针状角闪石，含量<5%。副矿物为磷灰石、榍石、磁铁矿等。

（二）岩石化学与地球化学特征

岩石化学成分及参数见表20-8。岩石呈现高硅、铝，低钛、铁、镁特征。里特曼指数 $\sigma=0.56\sim2.12$，$Na_2O>K_2O$，为钠质亚碱性岩系列；固结指数 $SI=8.06\sim13.06$，分异指数 $DI=80.15\sim80.99$，长英指数 $FL=67.55\sim71.39$，岩浆分异程度较低；$A/CNK=0.88\sim1.05$，$A/NK=1.33\sim1.64$，属于偏铝质。

稀土元素成分及参数见表20-9。$\Sigma REE=41.41\times10^{-6}\sim262.20\times10^{-6}$，$LREE/HREE=15.97\sim21.98$，$(La/Yb)_N=21.18\sim33.59$，稀土总量差异较大，轻稀土强烈富集，重稀土相对亏损；康各庄岩体 $\delta Eu=1.64$，铕正异常；得田沟岩体 $\delta Eu=0.83$，铕弱负异常；配分曲线为右陡倾光滑曲线（图20-19）。

微量元素特征同荆子峪花岗闪长质片麻岩和旗杆顶二长花岗质片麻岩。与宋营子英云闪长质片麻岩相比，Sr、Ba元素含量偏高，其余元素含量明显偏低；与密云表壳岩相比，Ba、Sr、Y等元素含量偏高，过渡元素含量低。与上地幔岩石相比，Ba、Sr、K富集，其余元素均亏损，尤以Co、Cr、Ni元素。微量元素蛛网图（图20-20）中，富集大离子亲石元素K、Ba和轻稀土元素，相对亏损高场强元素Th、Zr和重稀土元素。

（三）变质变形特征

岩石早期经过麻粒岩相变质作用，后期遭受角闪岩相退变质作用，基本特征同荆子峪花岗闪长质片麻岩。

（四）原岩恢复

在变质岩原岩恢复 SiO_2-TiO_2 和 Si-[(al+fm)-(c+alk)]图解（图20-21）上，得田沟岩体成分点位于副变质岩区，康各庄岩体落在火成岩区内。然而，综合得田沟岩体①强烈混合岩化，且缺乏表壳岩特征变质矿物；②岩性组成为暗色矿物和长石、石英，副矿物为锆石、榍石、磁铁矿组合，具有典型岩浆岩特性；③野外地质产状上同大北湾岩体渐变过渡，无明显地层成层性，认为该套变质岩为正变质岩。

（五）岩石系列

该套岩系 $\sigma=0.56\sim2.12<4$，为亚碱性系列。在 SiO_2-K_2O 图（图20-22）上，康各庄岩体成分点位于高钾钙碱性系列，得田沟岩体成分点位于低钾拉斑系列。

（六）成因类型

康各庄奥长花岗质片麻岩矿物组成和化学成分与花岗岩相同，*DI*、*SI* 等特征参数、稀土和微量元素成分含量等揭示其经历较弱岩浆分异演化。与荆子峪花岗闪长质片麻岩相同，岩体内无暗色包体，岩石表现出低铁、镁、钛和高硅、铝，富集大离子亲石元素，轻重稀土元素高分异特征，且不同岩体岩石化学特征差异较大，揭示了其岩浆来源于地壳。在 Na_2O-K_2O 和 *C/MF-A/MF* 图（图20-23）上，成分点分别位于I型花岗岩区和基性岩与变质砂岩部分熔融过渡区，也说明了其成岩物质来源于地壳。陈能松等（1993）研究认为，其可能来源于地壳变质英安岩-钠质流纹岩或火山-沉积岩的部分熔融。

（七）构造环境

康各庄奥长花岗质片麻岩体表现出同荆子峪花岗闪长质片麻岩体几乎一致的地质产状、岩石特征与地球化学特征。还同宋营子英云闪长质岩体直接渐变过渡接触，岩石特征和化学特征也相似。该三者组成太古宙典型 TTG 组合，是岛弧构造背景环境重要岩浆岩组合类型（图 20-24）。

（八）形成时代

本期岩浆侵位时代厘定为新太古代晚期，主要依据包括：①岩体遭受弱麻粒岩相变质，还经后期角闪岩相退变质；②新太古代晚期系列片麻岩体（如宋营子英云闪长质片麻岩）中广泛发育与此岩性岩石化学特征相似的岩脉；③被中元古界长城系不整合覆盖；④同宋营子英云闪长质片麻岩、荆子峪花岗闪长质片麻岩组成太古宙典型 TTG 组合，形成于成熟大陆边缘岩浆弧或与俯冲作用有关的岩浆弧构造背景环境；⑤宋营子英云闪长质片麻岩中奥长花岗质岩脉锆石 t_{DM} 同位素年龄为 2981Ma（金文山等，1999）。

五、旗杆顶花岗质片麻岩

（一）地质产状与岩石学特征

发育旗杆顶、和尚峪、土粉沟、车道沟 4 个岩体，分布于密云独角峪、和尚峪、放马峪和怀柔红石湾，总出露面积 9.4km^2。太古宙片麻岩岩体中广泛发育有二长花岗质、钾长花岗伟晶质片麻岩脉，可能也与这期岩浆侵位事件有关。

1. 旗杆顶岩体

位于密云独角峪至渠家沟，呈椭圆状，出露面积 4km^2。岩体地貌特征明显，高出周围围岩度程 150m 以上。岩体侵入于宋营子英云闪长质片麻岩中，围岩遭受明显同化作用，形成较窄过渡带（图 20-26）。岩体可根据岩性划分中心相和边缘相，中心相为含云二长花岗质片麻岩，边缘相为透辉角闪片麻岩，由边缘相向中心相暗色矿物逐渐减少，石英含量逐渐增多。岩体岩性矿物组成及含量见表 20-10。

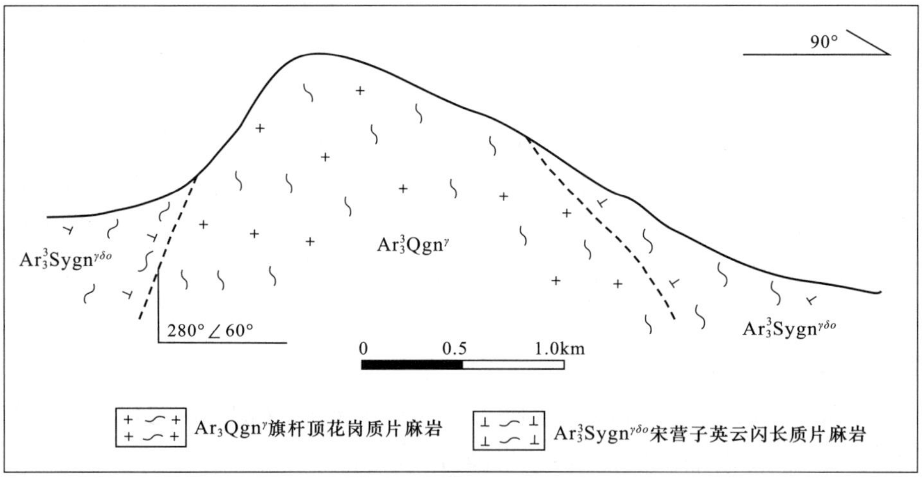

图 20-26　旗杆顶花岗质片麻岩与宋营子英云闪长质片麻岩侵入接触关系示意图
（据鲍亦冈等，2001）
Fig. 20-26　Geological sketch showing the intrusive contact relation between Songyingzi tonalitic gneiss and Qiganding granitic gneiss

表 20-10　旗杆顶花岗质片麻岩单元主要岩石类型和矿物定量统计表

Table 20-10　Main rock types and its real mineral contents of Qiganding granitic gneiss

岩体	岩性	矿物成分及含量/%							副矿物
		Pl	Or	Q	Di	Gt	Bi	Hb	
旗杆顶	含云二长花岗质片麻岩	35~40	35~40	20~30			3~7		Ch、Mt、Ap、Zr
	透辉角闪片麻岩	30~40	20~25	10~15	3~7		1~3	3~10	Ch、Mt、Ap、Ep、Zr

续表

岩体	岩性	矿物成分及含量/%							副矿物
		Pl	Or	Q	Di	Gt	Bi	Hb	
和尚峪	含透辉角闪黑云片麻岩	25	35	20	3		10	5	Ch、Mt、Ap、Ep、Zr
土粉沟	角闪黑云片麻岩	35	35	20			5	5	Ch、Mt、Ap、Ep、Zr
车道沟	含辉榴黑云片麻岩	85	0.5	10	0.5	1	3		Zr、Mt、Ap
	石榴子石黑云斜长片麻岩	75		15	1	4	5		Ap、Zr、Mt

注：数据资料引自1：5万密云幅、高岭幅、墙子路幅区域地质调查报告，1994；1：5万永宁幅、四海幅区域地质调查报告，1999。Ch—绿泥石，Mt—磁铁矿，Ap—磷灰石，Zr—锆石，Ep—绿帘石，Pl-斜长石，Or—钾长石，Q—石英，Di—透辉石，Gt—石榴子石，Bi—黑云母，Hb—角闪石。

2. 和尚峪岩体

位于密云和尚峪，呈长条状，出露面积4km^2，比围岩地貌高100~150m，宏观产状外倾，倾角60°，侵入特征明显。北部侵入苇子峪辉长闪长质片麻岩体中；西南被中生代燕山期四干顶花岗岩体切割破坏；东南与中元古界长城系呈断层接触。岩体内岩性特征同旗杆顶岩体，可按照岩性划分中心相和边缘相。岩体岩性矿物组成及含量见表20-10。

3. 土粉沟岩体

位于密云放马峪水库东北，呈小岩瘤状，出露面积0.2km^2。岩体侵入沙厂英云闪长质片麻岩，侵入特征明显。岩石类型为角闪黑云片麻岩，主要矿物组成及含量见表20-10。

4. 车道沟岩体

位于怀柔车道沟村附近，呈不规则状，出露面积1.2km^2。岩体周缘被中元古界长城系常州沟组不整合覆盖，内被古元古代红石湾橄榄辉长岩侵入破坏，并发育环状角闪石榴斜长辉石岩脉、斜长角闪岩脉及片麻状钾长花岗岩脉，局部还可见表壳岩（磁铁石英岩）包体。岩体岩性为浅色含辉榴黑云片麻岩和条带状石榴子石黑云斜长片麻岩，主要矿物组成及含量见表20-10。

（二）岩石化学与地球化学特征

岩石化学成分及参数见表20-8，特征同康各庄花岗闪长质片麻岩相似。里特曼指数σ=2.46~2.50，Na_2O、K_2O含量相当，为普通质亚碱性岩系列；固结指数SI=4.57~10.82，分异指数DI=78.34~90.70，长英指数FL=73.58~89.88，岩浆分异程度较低；A/CNK=0.97~1.04，A/NK=1.19~1.44，属于偏铝质。

稀土元素成分及参数见表20-9。$\Sigma REE = 51.97\times10^{-6} \sim 58.93\times10^{-6}$，LREE/HREE = 25.79~31.47，$(La/Yb)_N$=63.59~96.62，稀土元素总量低，轻稀土强烈富集，重稀土相对亏损；旗杆顶岩体δEu=2.43，铕正异常；车道沟岩体δEu=0.88，铕弱负异常；配分曲线为右陡倾光滑曲线（图20-19）。

微量元素特征同荆子峪花岗闪长质片麻岩和康各庄奥长花岗质片麻岩。微量元素蛛网图（图20-20）中，富集大离子亲石元素K、Ba、Sr和轻稀土元素，相对亏损高场强元素Th、Zr和重稀土元素。

（三）原岩恢复

该变质岩为正变质岩，主要依据包括：①缺乏表壳岩特征变质矿物；②岩性组成为暗色矿物和长石、石英，副矿物为锆石、榍石、磁铁矿组合，具有典型岩浆岩特性；③野外地质产状无明显地层成层性；④宏观地貌侵入特征明显；⑤在变质岩原岩恢复SiO_2-TiO_2和Si-[(al+fm)-(c+alk)]图解（图20-21）上，成分点位于正变质岩区。

（四）岩石系列

该套岩系σ=2.46~2.50<4，为亚碱性系列。在SiO_2-K_2O图（图20-22）上，成分点位于高钾钙碱性系列。

（五）成因类型与构造环境

成因类型及构造环境与荆子峪花岗闪长质片麻岩一致，共同组合构成太古宙典型 TTG 组合。

（六）形成时代

本期岩浆侵位时代厘定为新太古代晚期，主要依据包括：①本次获得旗杆顶花岗质片麻岩锆石 U-Pb 同位素年龄为（2511±15）Ma；②岩体侵入到宋营子、沙厂英云闪长质片麻岩岩体中；③岩体被燕山期早期侵入体切割破坏，被中元古界长城系地层不整合覆盖；④岩体普遍发育弱片麻状构造，遭受区域高级变质变形作用；⑤同宋营子英云闪长质片麻岩、荆子峪花岗闪长质片麻岩组成太古宙 TTG 组合，形成于成熟大陆边缘岩浆弧或与俯冲作用有关的岩浆弧构造背景环境。

第二节　中元古代中酸性侵入岩

一、地质产状与岩石学特征

发育涝洼、古洞沟、东辛店、兰营、沙厂 5 个岩体，出露面积 103km^2。受赤城-古北口和密云-墙子路-兴隆深大断裂控制呈近 EW 向脊状分布，是区内中元古代侵入体主体。岩体侵入新太古代变质岩中，呈岩株状产出。此外，在岩体附近太古宙片麻岩体中发育有大量未变质的细粒斑状花岗岩或花岗斑岩脉，尤以沙厂岩体周边地区最为发育，可能也与这期岩浆侵位事件有关。

（一）涝洼岩体

位于密云古北口镇，向东延伸至河北涝洼乡，沿赤城-古北口深大断裂分布，平面形态呈 WE 向长 30km，宽 2~6km 的长条形。北侧侵入新太古代宋营子英云闪长质片麻岩中，并被侏罗系不整合覆盖；南侧与晚侏罗统土城子组和中元古界长城系呈断层接触；东侧被早白垩世张家口组地层不整合覆盖（图 20-27）。岩体岩性复杂，宏观上由西向东依次为石英正长岩、石英二长岩、二长花岗岩。岩石呈肉红色，组成矿物弱碳酸盐化、绢云母化、白云母化或高岭土化蚀变，部分被构造碎裂岩化或发育裂隙。钾长石类型为条纹长石，发育钠质补片或暗色矿物包体；斜长石类型为更长石，发育聚片双晶、交代净边结构；石英具波状消光。具体矿物成分及含量见表 20-11。

（二）古洞沟岩体

位于怀柔古洞沟至朝岭湾，沿赤城-古北口深大断裂分布，是河北省涝洼岩体向西的延伸，平面形态呈 WE 向长条形，出露面积 5km^2。岩体南、西与中元古界长城系和中生界侏罗系土城子组呈断层接触；北、东被稍晚期东辛店岩体侵入。岩体岩性为含石英二长岩，具体矿物成分及含量见表 20-11。

表 20-11　中元古代侵入岩主要岩石类型和矿物定量统计表

Table 20-11　Main rock types and its real mineral contents of Middle Proterozoic intermediate acidity rock

岩体	岩性	粒径	主要矿物含量/%						副矿物含量/%
			钾长石 Qr	斜长石 Pl	石英 Q	角闪石 Hb	黑云母 Bi	辉石 Py	
涝洼	石英二长岩	中粗粒	57~65	13~33	5~20	4	4		Ap、Mt、Zr、Sp
	二长花岗岩	粗粒	50	28	20		2		Ap、Mt、Zr、Sp
	石英正长岩	中粒	55	35	4	5		<1	Ap、Mt、Zr、Sp
兰营	石英正长岩	中粒	55~65	15~20	10~15	5~10		<1	Ap、Zr、Sp、Rt
古洞沟	含石英二长岩	中粗粒	55	35	4	5			Ap、Mt、Zr、Sp
东辛店	钾长花岗岩	中粗粒	65	8~10	25		3		Mt、Zr、Ap

续表

岩体	岩性	粒径	主要矿物含量/%						副矿物含量/%
			钾长石 Qr	斜长石 Pl	石英 Q	角闪石 Hb	黑云母 Bi	辉石 Py	
沙厂	环斑花岗岩	粗粒	42	25	24	3.5±	3.5		表 20-12
	似斑状黑云母花岗岩	中粗粒	45	19	30		4.5		表 20-12
	中细粒黑云母花岗岩	中细粒	49	10	35		5		表 20-12

注：数据资料引自 1∶5 万密云幅、高岭幅、墙子路幅区域地质调查报告，1994；1∶5 万番字牌幅、汤河口幅区域地质调查报告，1996。Qr—钾长石，Pl—斜长石，Q—石英，Hb—角闪石，Bi—黑云母，Py—辉石，Ap—磷灰石，Mt—磁铁矿，Zr—锆石，Sp—榍石，Rt—金红石。

（三）东辛店岩体

位于怀柔古洞沟、朝岭湾，沿赤城-古北口深大断裂分布，与涝洼、兰营岩体构成近东西向延伸，平面形态呈 EW 向长条形，出露面积 $28km^2$。南、西侵入稍早期古洞沟岩体，接触带可见含石英二长岩捕虏体；其余周缘与围岩多为断层接触。岩体岩性为中粗粒状钾长花岗岩，具体矿物成分及含量见表 20-11。

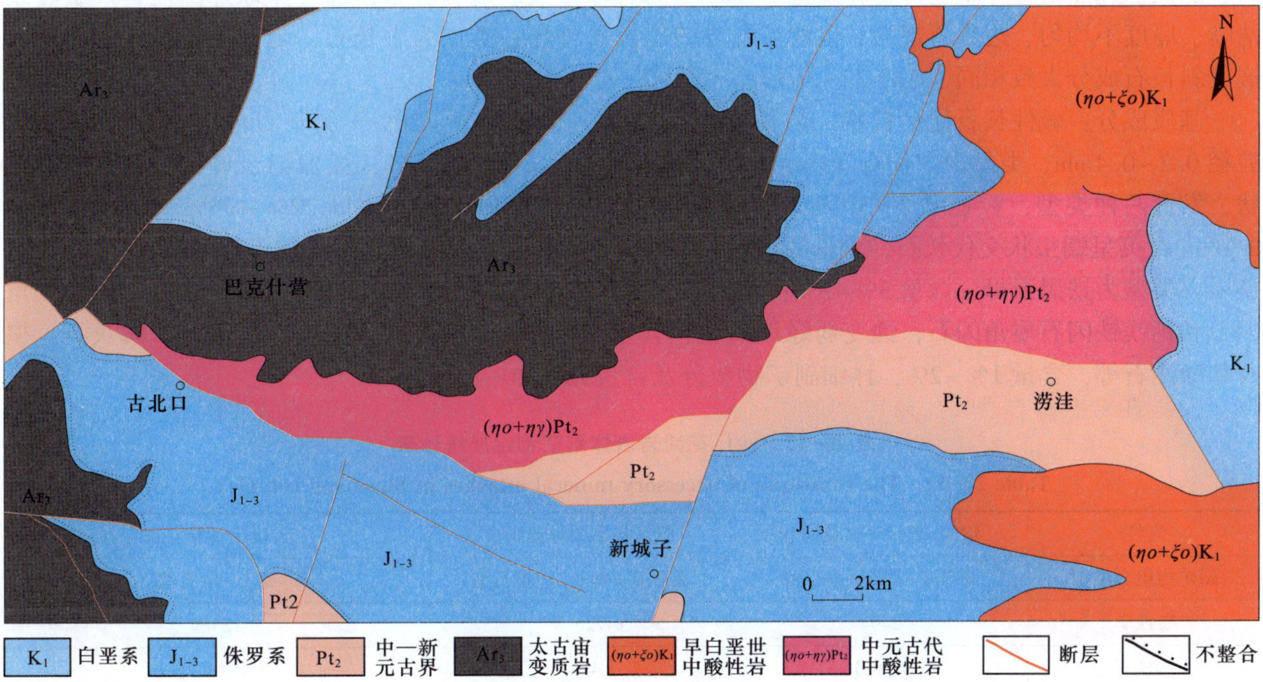

图 20-27　涝洼岩体地质简图

（据 1∶25 万承德幅区域地质图，略有修改）

Fig. 20-27　The geological sketch map of Laowa complex

（四）兰营岩体

位于怀柔下湾子、银河沟、兰营村、大石门，沿赤城-古北口深大断裂分布，与古洞沟、东辛店岩体构成近东西向延伸，平面形态呈 EW 向长条形，出露面积 $15km^2$。岩体南、北同中生界土城子组、太古宙宋营子英云闪长质片麻岩体或中元古代新地斜长岩体断层接触，局部可见其侵入于太古宙宋营子英云闪长质片麻岩体中；西被上侏罗统土城子组不整合覆盖，接触面发育厚约 5m 的古风化壳，并在底砾岩中见大小不一岩体角砾。岩体具岩相变化，边缘相粒度细，色率增高，内还发育大量围岩捕虏体。捕虏体可见强烈同化混染痕迹及毫米级细脉。岩体岩性为肉红色石英正长岩，斜长石除更长石外，还发育钠长石，且边部被石英交代发育为缝合线状边界，局部呈似文象结构，可见少许希列结构辉石和铁质橄榄石。具体矿物成分及含量见表20-11。

（五）沙厂岩体

位于密云邓家湾、大城子，沿密云-墙子路-兴隆深大断裂分布，平面形态呈EW向长12km，宽2～3km长条形，出露面积25km^2。岩体周缘与围岩多数为断层接触，明显受后期断层再活动影响，但局部仍可见其斜切片麻理侵入新太古代片麻岩中。外接触带由于后期断层活动影响，热变质或同化混染现象较弱，挤压变形现象也不明显，但局部仍可见烘烤现象，且发育大量与岩体相连的小岩枝（图20-28）；内接触带发育大量平行于接触面的长条状围岩片麻岩捕虏体。

岩体沿断裂构造短期发生3次侵位。第一次侵位形成环斑花岗岩，主要分布于岩体西或呈岩块被后来侵入岩包裹；第二次侵位形成似斑状黑云母花岗岩；第三次侵位形成中细粒黑云母花岗岩、二云母花岗岩和细粒花岗岩、花岗斑岩，呈岩枝状分布于岩体边部（图20-28）。

1. 环斑花岗岩

又称球斑花岗岩，灰白色、暗灰色，粗粒似斑状结构，块状构造。斑晶呈卵球状，粒径一般5～7cm，最大为10～12cm，小者2～4cm，含量25%～30%；卵球成分由单个或几个晶体组成，类型为碱性长石，化学成分计算为钠-正长石（Or=60.6，Ab=35.4，An=4.0），内常包裹一些细小半自形状角闪石、磁铁矿、黑云母等暗色矿物，暗色矿物常呈同心环带或放射状分布；部分卵球外围发育一层界线清晰、厚度不均匀、边界不规则、多颗粒组成的斜长石外环，类型为中长石，与基质斜长石及卵球内嵌晶斜长石成分大致相同（An=32～36）。

基质成分：碱性长石他形粒状，粒径0.7～2.5mm，含量15%～20%；斜长石他形—半自形柱状，粒径0.2～0.4mm，类型为中长石（Or=0.92，Ab=56.34，An=42.74），强烈绢云母化；石英他形粒状，发育两种类型，粒径0.2～3.5mm的石英晶面干净，发育裂纹，含量22%～25%。粒径0.05～0.1mm石英呈蠕虫状交代长石，含量约3%；黑云母半自形片状，粒径0.5～1.5mm，强烈蚀变，析出铁质，类型为铁黑云母，含量3%～5%；角闪石半自形柱状，粒径0.3～1.5mm，多色性，类型为铁角闪石和钙镁铁闪石质角闪石，蚀变弱绿泥石化，含量4%～5%。副矿物为磁铁矿、锆石、磷灰石、榍石和绿泥石等，含量1%～2%，详细副矿物成分及含量见表20-12。

表20-12　沙厂花岗岩岩体副矿物组成统计表

Table 20-12　The constitute of accessory mineral attaches in Shachang complex

单位：g/t

岩性	角闪-黑云环斑花岗岩	似斑状黑云母花岗岩	二云母中细粒花岗岩
副矿物组合类型	磁铁矿-榍石-萤石型	磁铁矿-磷灰石-萤石型	锆石-萤石型
磁铁矿	2494	725.8	125.9
钛铁矿	651.8	42.8	12
铁橄榄石或紫苏辉石	2.5	87.9	+
普通辉石	1.1	54	+
角闪石	8630.3	253	+
榍石	2078.7	4.8	+
锆石	728.8	124.8	148
磷灰石	361.2	937.2	0.1
萤石	317.1	510.4	1332.5
石榴子石	+	57	+
绿帘石	5665.7	134.3	+
褐帘石	23.3	+	30.4
独居石	+	69.1	+

续表

岩性	角闪-黑云环斑花岗岩	似斑状黑云母花岗岩	二云母中细粒花岗岩
黄铁矿	3	+	3
总量	21035.1	3031.8	1653.4
其他矿物	自然铅、刚玉、辉钼矿、赤铁矿、磁黄铁矿	黝帘石、铅丹、褐铁矿、辉钼矿、自然锡	赤铁矿、软锰矿、铅丹、自然锡、磁赤铁矿、钍石、方铅矿、辉钼矿、斑铜矿

注：+表示极微量。 （据郁建华等，1994）

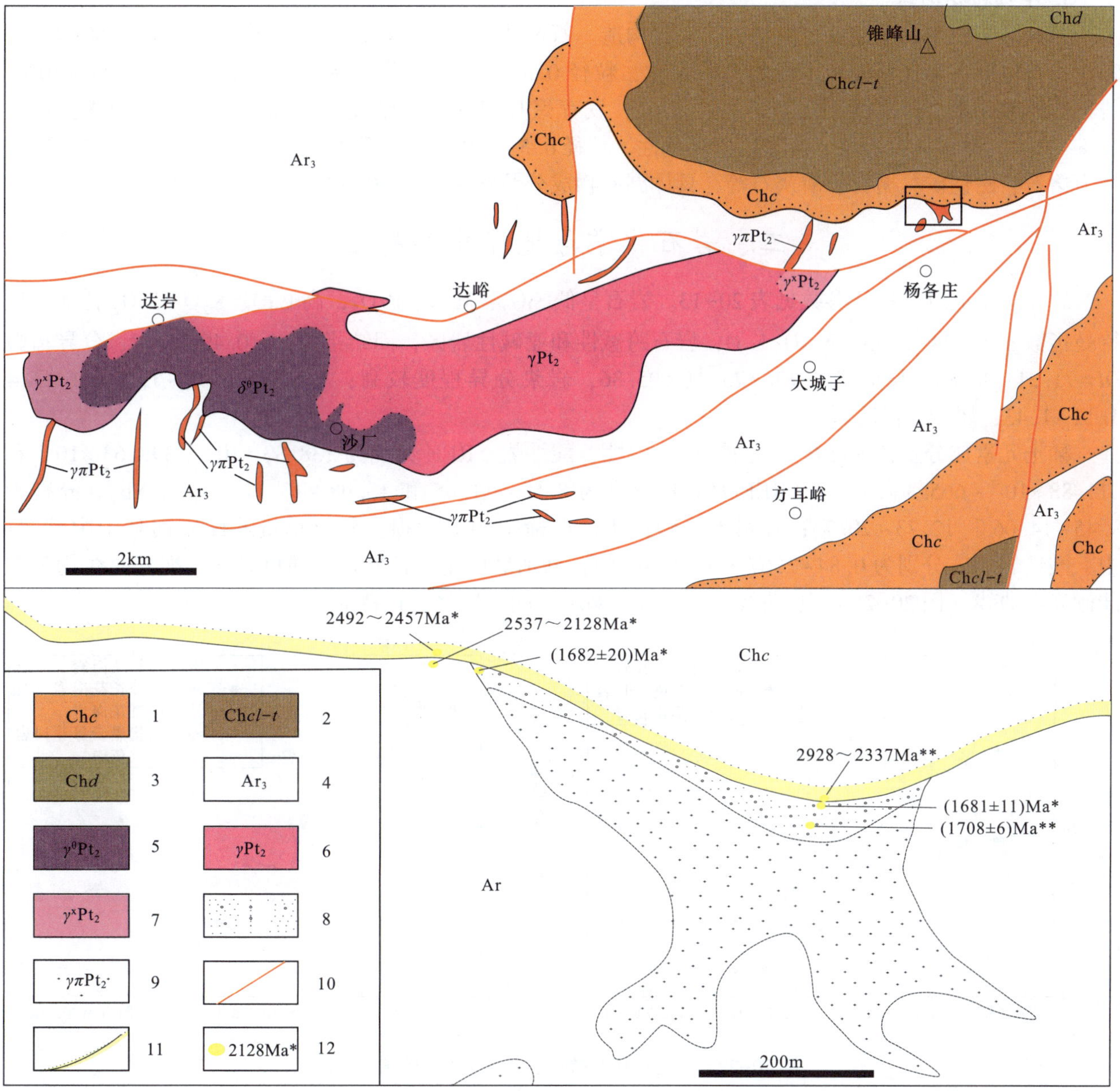

图 20-28 密云沙厂一带地质图

（据和政军等，2011）

Fig. 20-28 Geologial map of Shachang in Miyun

1—常州沟组；2—串岭沟组和团山子组；3—大红峪组；4—新太古代变质岩；5—环斑花岗岩；6—似斑状黑云母花岗岩；7—中细粒黑云母花岗岩；8—古风化壳；9—花岗斑岩；10—断层；11—不整合界线；12—SHRIMP 年龄（*）或 LA-ICP-MS 年龄（**）

2. 似斑状黑云母花岗岩

灰白色，似斑状结构，块状构造。斑晶他形粒状，粒径 1~2cm，类型为碱性条纹长石和微斜条纹长石（Or=96，Ab=4），发育条纹构造、格子双晶，含量 20%~30%。斑晶条纹长石弱绿泥石化，包裹嵌晶斜长石、石英。基质条纹长石他形粒状，粒径 0.7~2.5mm，含量约 20%；基质斜长石他形-半自形柱状，粒径 0.3~2.5mm，个别>5mm，类型为更长石（Or=2，Ab=77，An=21），发育聚片双晶，强烈绢云母化，含量约 25%；基质石英他形粒状，粒径 0.2~2.0mm，不均匀消光，含量 25%~30%；基质黑云母半自形片状，强烈蚀变，类型为铁黑云母，含量 4%~5%。副矿物为锆石、磁铁矿、磷灰石等，具体矿物成分及含量见表 20-12。

3. 中细粒花岗岩

白色—灰白色，中细粒花岗结构，块状构造。石英他形粒状，粒径 0.2~2.0mm，含量 35%~40%；碱性长石他形—半自形板状，强烈高岭土化，粒径 0.5~2.5mm，类型为条纹长石，含量 28%~30%；斜长石他形—半自形柱状，粒径 0.3~2.5mm，发育聚片双晶、卡钠复合双晶，强烈绢云母化，含量 27%~35%；黑云母黄褐色、淡黄色，多色性，粒径 0.5~1.0mm，强烈绿泥石化，含量 1%~2%。副矿物为锆石、磁铁矿、榍石、磷灰石等，具体副矿物成分及含量见表 20-12。

二、岩石化学与地球化学特征

岩石化学成分及特征参数见表 20-13。岩石呈低 SiO_2、MgO、P_2O_5，高 CaO、K_2O、TiO_2 特征。里特曼指数 $\sigma=2.24\sim5.53$，$Na_2O<K_2O$，岩石跨碱性和亚碱性岩区；固结指数 $SI=0.49\sim4.57$，分异指数 $DI=71.21\sim92.86$，长英指数 $FL=73.21\sim95.86$，岩浆分异程度较高；$A/CNK=0.85\sim1.07$，$A/NK=1.12\sim1.43$，属于偏铝质。

稀土元素成分及特征参数见表 20-14。中性、酸性岩 ΣREE 分别为 $138.83\times10^{-6}\sim398.63\times10^{-6}$ 和 $611.88\times10^{-6}\sim665.36\times10^{-6}$；LREE/HREE 分别为 8.50~11.65 和 13.98~18.17；$(La/Yb)_N$ 分别为 8.45~14.06 和 12.73~20.75；轻稀土强烈富集，重稀土相对亏损；稀土总量酸性岩远高于中性岩；中、酸性岩 δEu 分别为 0.84~2.67 和 0.24~0.62，分别呈现铕弱正异常和强负异常；配分曲线均为右陡倾光滑曲线（图 20-29），且曲线基本平行，揭示了同源岩浆演化特征。

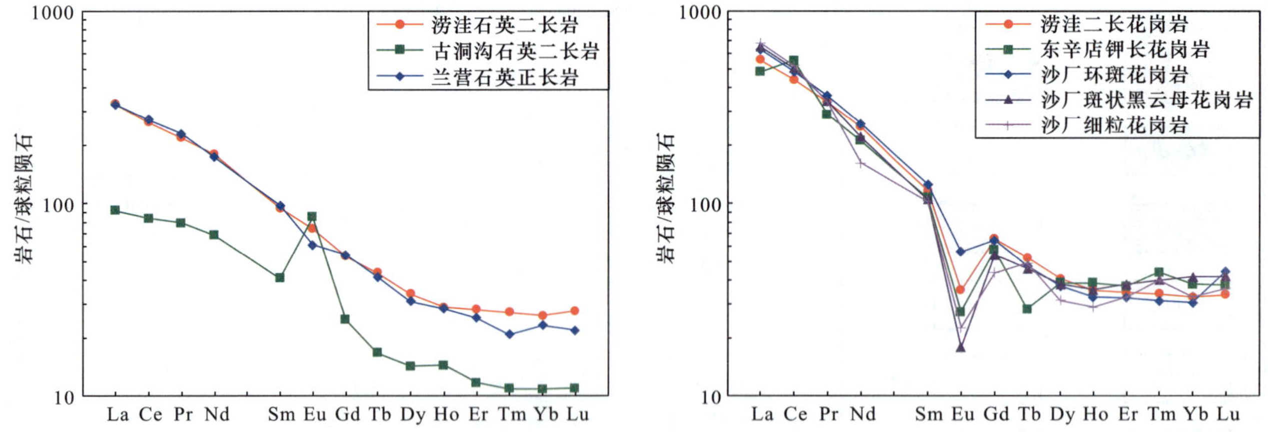

图 20-29　中元古代中酸性岩稀土元素配分模式图

Fig. 20-29　Chondrite-normalized REE patterns of Mesoproterozoic intermediate-acid rocks

微量元素成分及特征参数见表 20-15。除 Sr、Ba、Zr、Hf 元素外，其余各微量元素含量中性岩均低于酸性岩。在酸性岩的环斑花岗岩-似斑状黑云母花岗岩-中细粒花岗岩演化序列中，Sr、Ba、Zr、Hf 元素含量降低，而其余元素含量增高。这一特征揭示伴随该期岩浆演化岩石向酸性富硅方向演化，且微量元素也向着 Sr、Ba、Zr、Hf、Ga、Zn 和 Sc 元素含量逐渐降低，而 Rb、Cs、Li、Nb、Ta、Y、

表 20-13　中元古代中酸性岩岩石化学成分及特征参数一览表

Table 20-13　The chemical composition and main parameters of Mesoproterozoic intermediate-acid rocks

序号	岩体	岩性	氧化物含量/%													特征参数					
			SiO_2	TiO_2	Al_2O_3	Fe_2O_3	FeO	MnO	MgO	CaO	Na_2O	K_2O	P_2O_5	Los	总和	*DI*	*SI*	*σ*	*FL*	*A/CNK*	*A/NK*
1	涝洼	石英二长岩	60. 26	0. 73	16. 05	1. 77	3. 49	0. 09	0. 60	2. 53	4. 21	5. 84	0. 19	1. 96	97. 72	80. 19	3. 77	5. 53	79. 89	0. 90	1. 21
2		二长花岗岩	58. 35	0. 94	16. 92	4. 06	4. 94	0. 18	0. 76	3. 18	4. 28	4. 41	0. 46	1. 38	99. 86	71. 21	4. 12	4. 79	73. 21	0. 96	1. 43
3	古洞沟	石英二长岩	61. 02	1. 02	16. 36	4. 81	2. 46	0. 09	0. 79	1. 74	3. 48	5. 89	0. 27	1. 87	99. 80	78. 89	4. 57	4. 73	84. 34	1. 07	1. 35
4	兰营	石英正长岩	72. 47	0. 54	12. 88	2. 82	0. 64	0. 02	0. 06	0. 38	3. 45	5. 34	0. 06	2. 02	100. 68	92. 86	0. 49	2. 61	95. 86	1. 06	1. 12
5	东辛店	钾长花岗岩	70. 17	0. 35	12. 83	1. 84	1. 09	0. 08	0. 28	2. 00	3. 26	5. 61	0. 05	3. 70	101. 24	88. 81	2. 33	2. 86	81. 60	0. 85	1. 12
6	沙厂	环斑花岗岩	68. 17	0. 59	14. 67	1. 44	2. 27	0. 05	0. 48	1. 76	3. 43	5. 44	0. 12	3. 15	101. 56	84. 80	3. 68	3. 09	83. 44	1. 00	1. 27
7		似斑状黑云母花岗岩	73. 22	0. 23	13. 04	1. 31	1. 03	0. 03	0. 18	1. 04	3. 20	5. 56	0. 03	2. 13	101. 00	91. 48	1. 60	2. 53	89. 39	0. 99	1. 16
8		细粒花岗岩	74. 81	0. 13	13. 09	0. 98	0. 65	0. 03	0. 21	0. 84	3. 25	5. 21	0. 03	2. 41	101. 63	92. 61	2. 04	2. 24	90. 97	1. 05	1. 19

注：1、2、6、7、8 数据为均值；1、2 据相振群（2014）；3、4、5 数据据 1∶5 万番字牌幅、汤河口幅区调报告，1996；6、7、8 数据据 1∶5 万密云幅区调报告（1994）和谢广轰（2004）。

表 20-14　中元古代中酸性岩稀土元素成分及特征参数一览表

Table 20-14　REE contents and main parameters of Mesoproterozoic intermediate-acid rocks

序号	岩体	岩性	稀土元素含量/10^{-6}																特征参数			
			La	Ce	Pr	Nd	Sm	Eu	Gd	Tb	Dy	Ho	Er	Tm	Yb	Lu	Y	ΣREE	LREE/HREE	$(La/Yb)_N$	δEu	δCe
1	涝洼	石英二长岩	78. 00	162. 00	20. 91	84. 11	14. 52	4. 31	10. 95	1. 64	8. 66	1. 65	4. 67	0. 70	4. 47	0. 71	43. 61	397. 27	10. 88	12. 52	1. 04	0. 98
2		二长花岗岩	132. 40	270. 00	32. 02	116. 70	17. 87	2. 06	13. 48	1. 96	10. 37	2. 01	5. 73	0. 87	5. 57	0. 85	52. 41	611. 88	13. 98	17. 04	0. 41	1. 02
3	古洞沟	石英二长岩	21. 90	51. 50	7. 58	32. 00	6. 28	4. 96	5. 14	0. 63	3. 65	0. 82	1. 95	0. 28	1. 86	0. 28	19. 40	138. 83	8. 50	8. 45	2. 67	0. 98
4	兰营	石英正长岩	77. 80	167. 30	21. 88	81. 59	15. 02	3. 54	11. 15	1. 56	7. 91	1. 62	4. 22	0. 54	3. 97	0. 56	36. 74	398. 63	11. 65	14. 06	0. 84	0. 99
5	东辛店	钾长花岗岩	115. 00	339. 00	27. 60	99. 30	16. 50	1. 59	11. 80	1. 06	9. 87	2. 19	6. 16	1. 12	6. 48	0. 96	51. 90	638. 63	15. 11	12. 73	0. 35	1. 48
6	沙厂	环斑花岗岩	150. 23	297. 57	34. 57	121. 48	19. 31	3. 27	13. 30	1. 77	9. 48	1. 86	5. 38	0. 80	5. 22	1. 13	52. 64	665. 36	16. 09	20. 66	0. 62	1. 01
7		似斑状黑云母花岗岩	155. 11	308. 77	32. 31	103. 62	16. 12	1. 05	11. 17	1. 72	9. 68	2. 02	6. 32	1. 02	7. 09	1. 06	68. 68	657. 04	15. 40	15. 69	0. 24	1. 07
8		细粒花岗岩	162. 00	320. 00	32. 00	76. 00	15. 80	1. 32	9. 00	1. 84	8. 00	1. 64	5. 40	1. 02	5. 60	0. 92	56. 00	640. 54	18. 17	20. 75	0. 34	1. 09

注：1、2、6、7、8 数据为均值；1、2 据相振群（2014）；3、4、5 数据据 1∶5 万番字牌幅、汤河口幅区调报告，1996；6、7、8 数据据 1∶5 万密云幅区调报告（1994）和谢广轰（2004）。

Th、U、Pb、Cd、Sn、Cl 元素含量逐渐增高演化（谢广轰，2004），与芬兰东南部 Suomenniemi 环斑花岗岩复式岩体特征一致。微量元素蛛网图（图 20-30）中，中性、酸性岩曲线基本相似，富集大离子亲石元素 Sr、Ba、K，相对强烈亏损元素 Sr、Ti、P，弱亏损 Nb、Ta。

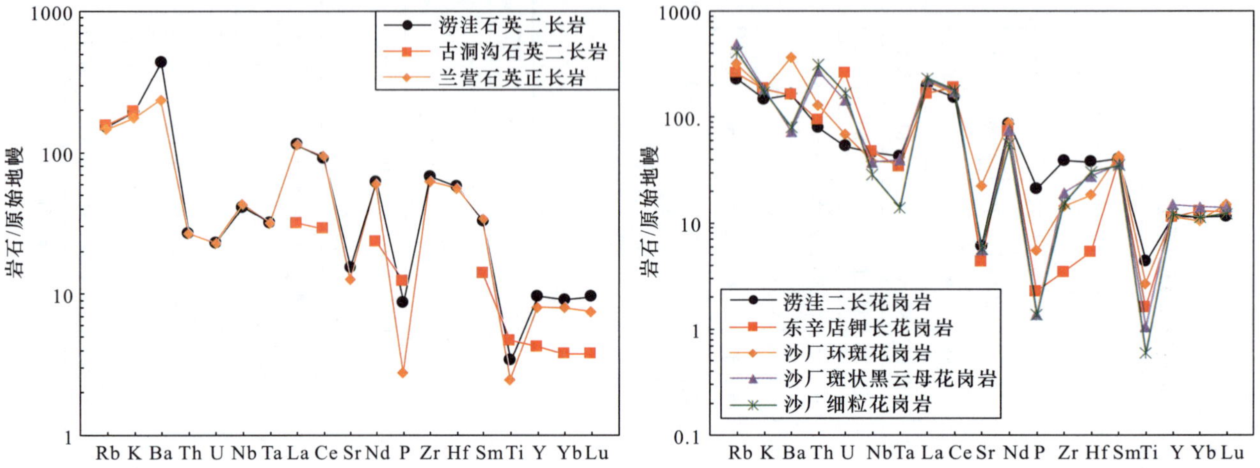

图 20-30　中元古代中酸性岩微量元素蛛网图

Fig. 20-30　Trace element spider diagram of Mesoproterozoic intermediate-acid rocks

表 20-15　中元古代中酸性岩微量元素成分及特征参数一览表

Table 20-15　Trace element contents and main parameters of Mesoproterozoic intermediate-acid rocks

序号	岩体	岩性	微量元素含量/10^{-6}									特征参数	
			Rb	Ba	Th	U	Nb	Ta	Sr	Zr	Hf	Rb/Sr	Nb/Ta
1	涝洼	石英二长岩	96.95	3053.00	2.27	0.48	29.34	1.30	323.25	754.75	17.72	0.30	22.52
2		二长花岗岩	144.40	1139.17	6.72	1.11	32.80	1.75	126.97	435.35	11.63	1.14	18.79
3	兰营	石英正长岩	93.00	1653.00	2.27	0.48	31.00	1.30	269.00	703.00	17.30	0.35	23.85
4	东辛店	钾长花岗岩	166.00	1133.00	7.94	5.51	34.00	1.40	90.90	38.90	1.66	1.83	24.29
5	沙厂	环斑花岗岩	204.14	2566.58	11.08	1.46	27.74	1.55	475.92	163.67	5.79	0.43	17.87
6		似斑状黑云母花岗岩	309.99	514.75	23.13	3.05	27.07	1.64	119.27	219.08	8.70	2.60	16.53
7		中细粒花岗岩	263.00	566.00	26.80	3.56	20.64	0.58	122.85	174.55	9.50	2.14	35.59

注：数据来源同表 20-13。

三、岩石系列

该套岩系 σ=2.24~5.53，跨越碱性岩和亚碱性岩系列。在不同岩石系列判别图解上，成分点也分散分布于碱性和亚碱性区内（谢广轰等，2004）。由于该套岩石和大庙或新地斜长岩特征相似，侵位前深部明显发生结晶分异作用（如长石堆晶作用），使得熔浆中元素丰度和元素比值发生改变，全岩化学分析不能代表原始熔体的组成，岩石系列判别的图解不具有统计意义。该套岩石富碱高钾、低镁等特征势必使得其相关岩石系列判别向碱性岩区偏离（谢广轰等，2004）。该套岩石富 Si、K，贫 Mg，酸性岩石 CIPW 标准矿物计算中出现刚玉，具有钙碱性特征。在 SiO_2-K_2O 图（图 20-31）上，成分点位于钾玄岩系列。

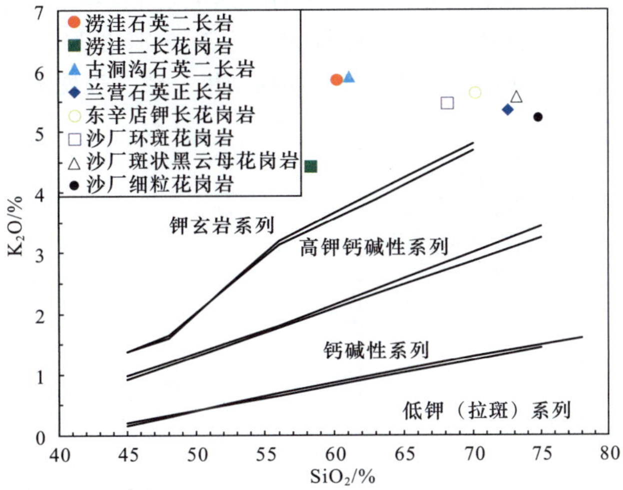

图 20-31　中元古代中酸性岩 SiO_2-K_2O 图

Fig. 20-31　SiO_2-K_2O diagram of Mesoproterozoic intermediate-acid rocks

四、Sr-Nd-Pb 同位素特征

兰营石英正长岩全岩 Rb = 119.6，Sr = 167.3，$^{87}Sr/^{86}Sr$ = 0.753358，$^{87}Rb/^{86}Sr$ = 2.064；Sm = 18.66，Nd = 121.4，$^{143}Nd/^{144}Nd$ = 0.51118，$^{147}Sm/^{144}Nd$ = 0.093；t 取 1697Ma 时，ε_{Nd}（t） = −5.89，ε_{Sr}（t） = 7.7（任康绪等，2006）。沙厂复合花岗岩体全岩 Rb = 152.992 ~ 327.394，Sr = 54.017 ~ 550.183，$^{87}Sr/^{86}Sr$ = 0.7259 ~ 1.11248，$^{87}Rb/^{86}Sr$ = 0.9261 ~ 18.238；Sm = 13.05 ~ 29.96，Nd = 76.73 ~ 200.1，$^{143}Nd/^{144}Nd$ = 0.51098 ~ 0.51139，$^{147}Sm/^{144}Nd$ = 0.0758 ~ 0.1093；t 取 1673Ma 时，ε_{Nd}（t） = −5.11 ~ −6.43，ε_{Sr}（t） = −24.5 ~ 40.5（黄少雄等，1990；Ramo et al.，1995；郁建华等，1996）。

沙厂复合花岗岩体全岩和钾长石 Pb 同位素比值均较低，尤其钾长石 Pb 同位素比值更低。全岩 $^{206}Pb/^{204}Pb$ = 15.180 ~ 18.752，$^{207}Pb/^{204}Pb$ = 15.040 ~ 15.427，$^{208}Pb/^{204}Pb$ = 35.11 ~ 39.69；钾长石 U = 0.0295 ~ 0.0618，U/Pb = 0.0006 ~ 0.0017，表明放射成因 Pb 很少；钾长石 $^{206}Pb/^{204}Pb$ = 14.717 ~ 15.254，$^{207}Pb/^{204}Pb$ = 14.970 ~ 15.083，$^{208}Pb/^{204}Pb$ = 34.45 ~ 35.06。

五、O 同位素特征

沙厂环斑花岗岩 $\delta^{18}O$ = 7.62‰，似斑状黑云母花岗岩 $\delta^{18}O$ = 8.28‰，均大于上地幔派生的岩浆氧同位素值 5.5‰ ~ 7‰（黄少雄等，1988）。

六、Lu-Hf 同位素特征

沙厂环斑花岗岩 $^{176}Yb/^{177}Hf$ = 0.011605 ~ 0.152269，$^{176}Lu/^{177}Hf$ = 0.000378 ~ 0.001368，$^{176}Hf/^{177}Hf$ = 0.281512 ~ 0.281845。t = 1680Ma 时，ε_{Hf}（t） = −8.0 ~ −2.1，t_{DM} = 2240 ~ 2400Ma，与围岩片麻岩单阶段模式年龄相同（杨进辉等，2005；相振群，2014）；涝洼二长花岗岩 $^{176}Yb/^{177}Hf$ = 0.011632 ~ 0.074372，$^{176}Lu/^{177}Hf$ = 0.000309 ~ 0.001837，$^{176}Hf/^{177}Hf$ = 0.281461 ~ 0.281539。t = 1695Ma 时，ε_{Hf}（t） = −5.8 ~ −9.1，t_{DM} = 2350 ~ 2.48Ma（相振群，2014）；涝洼石英二长岩 $^{176}Yb/^{177}Hf$ = 0.011873 ~ 0.097331，$^{176}Lu/^{177}Hf$ = 0.000281 ~ 0.001049，$^{176}Hf/^{177}Hf$ = 0.281474 ~ 0.281607。t = 1696Ma 时，ε_{Hf}（t） = −3.9 ~ −9.1，t_{DM} = 2270 ~ 2520Ma（相振群，2014）。

七、成因类型与岩浆演化

该套岩系矿物主要由碱性长石、富铁质暗色矿物和石英组成，发育环斑构造；地球化学上呈现富硅、碱，贫钙、镁，富 Rb、Th，贫 Sr，酸性岩还具有显著的负 Eu 异常，这些特征与典型 A 型花岗岩一致。在（Zr+Nb+Ce+Y）-[（Na_2O+K_2O）/CaO]和 Nb-Y-Ce 判别图解（图 20-32）上，成分点还位于 A1 和 A2 型花岗岩过渡地带，与该套岩性碱性含量偏高结果一致，表现出 A1-A2 的延续性。

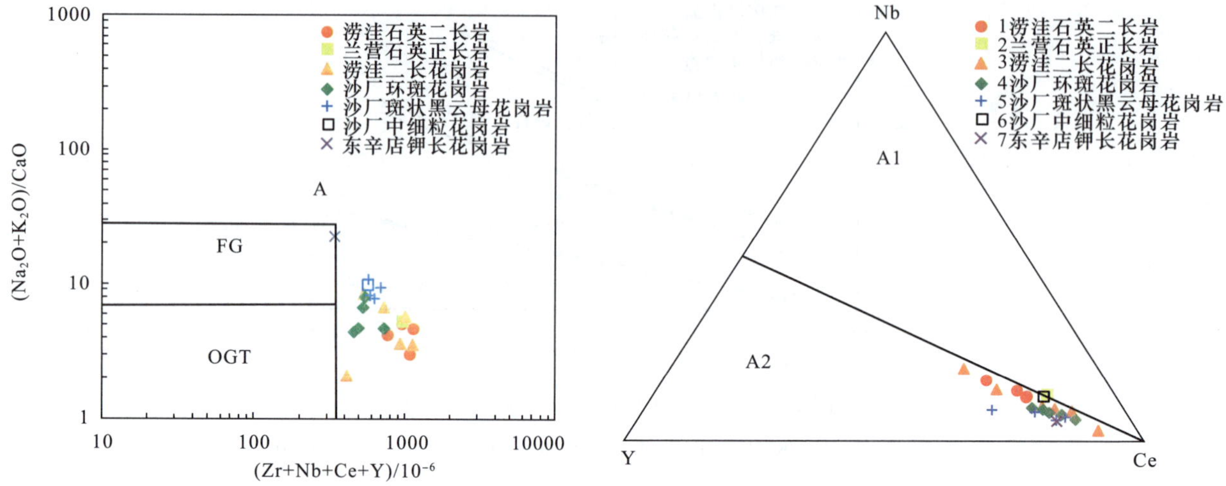

图 20-32　中元古代中酸性岩（Zr+Nb+Ce+Y）-[（Na_2O+K_2O）/CaO]和 Nb-Y-Ce 图解

Fig. 20-32　(Zr+Nb+Ce+Y)-[(Na_2O+K_2O)/CaO] and Nb-Y-Ce diagram of Mesoproterozoic intermediate-acid rocks

该套中酸性岩伴随 SiO_2 含量增加，*DI* 值逐渐递增，共同构成相互衔接的演化系列，表现出一致的岩石化学特征（谢广轰，1980），也揭示了岩性同源岩浆演化过程。该套岩石暗色矿物均表现出富铁特征，岩浆富铁本质是中元古代中酸性岩类区别于其他岩类的标志之一（郁建华等，1994），揭示岩浆相对深成侵位还原条件下的分异演化作用，也与环斑花岗岩中赋存大量钾长石巨晶和高温石英相吻合。在 ε_{Sr}（t）-ε_{Nd}（t）图解（图 20-33）上，成分点位于下地壳区。Sr-Nd-Pb 同位素与围岩晚太古界变质岩具有明显继承性，表明这些岩体的原始岩浆可能来源于华北克拉通太古宙期间生长下地壳物质部分熔融（郁建华等，1996；谢广轰，2004；杨进辉等，2006）。

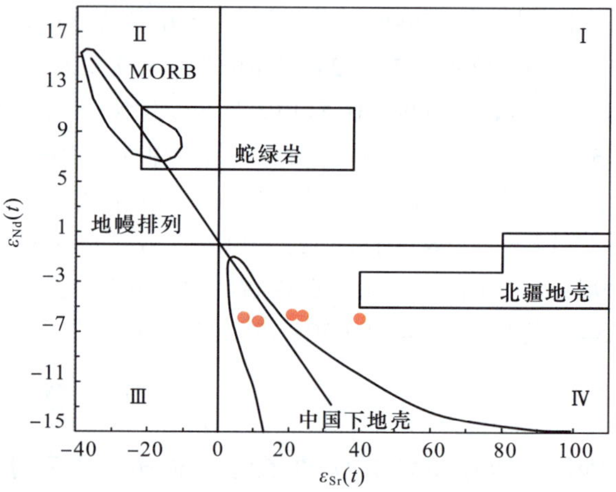

图 20-33　中元古代中酸性岩 ε_{Sr}（t）-ε_{Nd}（t）图

Fig. 20-33　ε_{Sr}（t）-ε_{Nd}（t）diagram of Mesoproterozoic intermediate-acid rocks

该套岩石发育的巨大斑晶和 *DI*、*SI* 特征参数等共同揭示了岩浆侵位过程中经历了强烈结晶分异作用。稀土铕正、负异常演化反映了斜长石可能是最重要的分离结晶矿物，而 FeO^*、MgO 组分同 SiO_2 相关性还反映了镁铁质矿物分离结晶作用。

谢广轰（2004）系统总结了密云沙厂环斑花岗岩和大庙、新地斜长岩的有关地质特征，提出了 AMCG 组合成因模式的 5 个重要过程：①大陆地壳克拉通化（刚性化）后，地幔上涌、岩石圈减薄和破裂，导致上地幔减压熔融产生玄武质岩浆；②玄武质岩浆汇聚在壳-幔边界附近，形成巨大的镁铁质岩浆房并发生底侵作用，导致下地壳深熔作用，产生花岗质熔体并汇聚在下地壳底部形成长英质岩浆房，并叠置在镁铁质岩浆房的上方；③镁铁质岩浆主要结晶出橄榄石、富铝辉石和斜长石（三元长石），长英质岩浆中主要是透长石结晶，由于结晶时的物理化学条件相对稳定，使矿物得以长大为巨大单体；④镁铁质岩浆房中，由于矿物比重的差异，在高压条件下斜长石巨晶将上浮，在岩浆房的上部形成巨大的斜长石漂浮堆积体（粥状熔体），而镁铁硅酸盐矿物因相对密度较大，下沉到岩浆房底部，并脱离岩浆体系，重新进入地幔；长英质岩浆房中的透长石巨晶也将相对集中于岩浆房上部，形成粥状岩浆；⑤在深部岩浆房中，由于富长石的漂浮堆积体同地壳的密度差，导致重力的不稳定性；更由于在伸展体制下产生的刚性破裂，使粥状熔体（包括残余岩浆）以底劈形式沿深大断裂侵位到上地壳浅部并固结。

八、构造环境分析

该套岩系在时空上总是同岩体型斜长岩密切共生，有时甚至共存于一个杂岩体内（如新地-古洞沟-兰营杂岩体），暗示它们之间必然有内在联系。中酸性岩和斜长岩密切共生而又不同源，只能视二者为相同地质作用下产物（郁建华等，1996；谢广轰，2004），具有共同的大地构造背景环境，基性岩和酸性岩共生构成双峰式岩浆侵位，形成于非造山构造背景环境，为燕山裂陷槽裂陷早期慢性扩展条件下岩浆活动产物。

九、时代归属

本期岩浆侵位时代厘定为中元古代早期，主要依据包括：①岩体未遭受高级变质和复杂褶曲变形；②广泛侵位到新太古代变质岩中；③同期花岗斑岩脉被中元古界长城系不整合覆盖；④岩体沿深大断裂发育，被深大断裂后期复活所破坏；⑤同斜长岩构成 AMCG 组合，形成于非造山构造背景环境；⑥兰营石英正长岩锆石 U-Pb 同位素年龄为（1790.9±24）Ma（郁建华等，1990）；⑦沙厂杂岩体不同岩性锆石 U-Pb 同位素年龄为（1658±5）Ma 至（1706±15）Ma（黄少雄，1990；宋彪，1992；Ramo et al.，1994；郁建华等，1996；相振群，2014），全岩或矿物 Rb-Sr 等时线年龄（1670±240）Ma（黄少雄，1990；郁建华，1996），钾长石卵球中角闪石包裹体 $^{40}Ar-^{39}Ar$ 坪年龄（1716±21）Ma（胡世玲等，1990）。

第三节　晚古生代中性侵入岩

一、地质产状与岩石学特征

发育梨树沟、荞麦峪、栗树洼、老公营 4 个岩体，沿赤城-古北口深大断裂带呈近 EW 向分布，出露面积 22.51km²。沿此深大断裂邻区河北省还发育水泉沟、石踩口岩体。此外，在 4 个岩体附近新太古代片麻岩和中元古代地层中发育有大量未变质的细粒正长岩、正长斑岩脉，可能也与这期岩浆侵位事件有关。

岩性主要为正长岩，呈砖红色、肉红色，中细粒或斑状结构，致密块状构造，多受应力作用发育碎裂结构。斑晶钾长石为主，呈自形粒状，粒径 5~10mm，个别 10~30mm，部分具黑色或灰黑色内心，类型为正长石-钠正长石，含量 5%~30%；少量斑晶为斜长石和角闪石；基质钾长石半自形—他形，发育格子、卡式双晶，部分边缘发育钠长石环边，形成“净边结构”，类型为微斜长石、正长石；基质斜长石半自形长条状，发育聚片双晶，多绢云母化，类型为钠长石-更长石-中长石；基质普通角闪石自形长柱状，深绿-浅黄多色性，类型为钙闪石中镁-绿钙闪角闪石；基质黑云母半自形片状，绿泥石化，类型为镁黑云母类。副矿物为磁铁矿、磷灰石、锆石、榍石、黄铁矿。

（一）梨树沟岩体

分布于密云水库西，由连续 5 个小岩珠沿石城断裂呈近 SN 向分布，出露面积 6km²。岩体东、南、北侵入新太古代变质岩中，侵入迹象明显。外接触带普遍发育混杂、混染现象，局部可见明显烘烤现象；内接触带见大量厘米至百米级不规则状变质岩捕虏体。岩体西与中生代燕山期石城闪长岩体断层接触，破碎带中见此岩体角砾。岩体岩性没有发生变质变形，内部无岩相变化，但可见大量石英脉、辉绿岩脉和煌斑岩脉。

岩体短期发生3次侵位，组成岩性复杂杂岩体，分别发育为正长岩、正长斑岩和伟晶正长岩。正长斑岩和正长岩构成岩性主体，二者呈岩枝状穿插，局部过渡关系，并偶见正长岩以捕虏体赋存于正长斑岩中，二者可能为同源同期两次岩浆脉动产物。伟晶正长岩以岩脉状产出，侵入正长斑岩及正长岩中。

（二）荞麦峪岩体

分布于密云水库白河主坝，同梨树沟岩体隔密云水库，可能为同一岩体，出露面积3km^2。岩体西淹没入密云水库下，南被第四系覆盖，南西和北东分别侵入中元古界长城系和新太古代片麻岩中。岩体岩性组成及特征同梨树沟岩体。

（三）栗树洼岩体

分布于密云栗树洼村，由连续3个小岩瘤（石门、栗树洼和大坡南峪）组合呈近EW向分布，出露面积1.51km^2。岩体南东侵入到中元古界蓟县系高于庄组中，接触面内倾，产状320°~340°∠40°~50°；南西侵入到晚太古界变质岩中；岩体北东沙沟峪附近被中生界侏罗系髫髻山组地层不整合覆盖，接触面产状0°∠38°；南被中生代燕山期高石虎-张家断层切断破坏，并被中生界侏罗系髫髻山组不整合覆盖。岩体岩性为正长岩和正长斑岩，二者无明显界线。

（四）老公营岩体

位于密云琉璃庙西沟门、前安岭，岩体平面形态呈NE向长条形，附近皮条沟、槟榔沟还发育多个小规模小岩瘤，出露面积12km^2。岩体北、南侵入于新太古代变质岩中，接触界线清晰，接触面内陡倾，外接触带热接触变质现象不明显，但可见大量正长斑岩脉贯入，内接触带发育10~30cm宽冷凝边和岩浆混染现象。岩体内发育大量毫米级至百米级棱角状或次棱角状的新太古代变质岩（角闪斜长片麻岩）捕虏体。岩体南西侵入中元古界长城系中，围岩中发育大量正长斑岩脉。岩体东被崎峰茶断裂错断。

岩体岩性为正长岩和石英正长斑岩，二者呈不规则团块状穿插，逐渐过渡无明显界限。岩体边部普通还发育灰绿色、暗红色混杂特征混染岩，岩性为正长闪长岩或闪辉正煌岩，是正长岩浆与外来岩基（太古宙变质岩）组分以无固定比例混合而成。

二、岩石化学与地球化学特征

岩石化学成分及特征参数见表20-16。与河北省石垛口、水泉沟岩体相似，岩石呈现富Al_2O_3、碱，贫SiO_2、MgO、FeO^*特征。里特曼指数$\sigma=6.74\sim7.81$，为碱性岩系列；固结指数$SI=1.93\sim5.29$，分异指数$DI=84.32\sim88.65$，长英指数$FL=82.02\sim90.03$，岩浆分异程度较高；$A/CNK=0.86\sim1.02$，$A/NK=1.11\sim1.19$，属于偏铝质。

稀土元素成分及特征参数见表20-17。$\Sigma REE=152.23\times10^{-6}\sim198.54\times10^{-6}$；LREE/HREE = 11.26~35.16；$(La/Yb)_N=12.79\sim66.65$；轻稀土强烈富集，重稀土相对亏损；$\delta Eu=0.81\sim0.99$，呈现铕弱负异常；配分曲线为右陡倾光滑曲线（图20-34）。

微量元素含量及特征参数见表20-18。与中国中性岩（黎彤，1962）相比，K、Rb、Sr、Ba含量偏高，Nb、Ta、Hf元素含量偏低，其余元素含量相当。在微量元素配分曲线图（图20-35）上，表现富集大离子亲石元素Ba、K、Rb、Sr，强烈亏损高场强元素P、Ti、Ta、Nb。

三、岩石系列

该套岩系$Na_2O+K_2O=11.41\sim12.24$，$\sigma=6.74\sim7.81$，为强碱性岩系列。

四、Sr-Nd同位素特征

老公营和梨树沟正长岩全岩$^{87}Sr/^{86}Sr=0.710477\sim0.708203$，$^{87}Rb/^{86}Sr=0.8897\sim0.4874$，$^{143}Nd/^{144}Nd=$

表 20-16　晚古生代中性岩岩石化学成分及特征参数一览表

Table 20-16　The chemical composition and main parameters of Late Paleozoic intermediate rocks

序号	岩体	岩性	氧化物含量/%													特征参数					
			SiO_2	TiO_2	Al_2O_3	Fe_2O_3	FeO	MnO	MgO	CaO	Na_2O	K_2O	P_2O_5	Los	总和	*DI*	*SI*	σ	*FL*	*A/CNK*	*A/NK*
1	石垛口	正长岩	62.35	0.25	18.32	1.37	1.37	0.10	0.56	2.42	6.53	4.84	0.08	1.21	99.40	86.04	3.82	6.54	82.45	0.90	1.15
2	荞麦峪	正长岩	61.67	0.37	16.44	2.81	1.10	0.12	0.85	2.50	3.78	7.63	0.14	2.02	99.43	84.32	5.29	6.74	82.02	0.86	1.14
3	梨树沟	正长岩	62.01	0.37	18.20	2.10	1.58	0.10	0.85	1.64	5.51	6.73	0.15	0.67	99.92	87.04	5.07	7.81	88.18	0.94	1.11
4	老公营	正长岩	63.60	0.23	17.35	1.41	1.27	0.09	0.29	1.34	2.74	9.32	0.02	1.25	98.88	88.65	1.93	6.88	90.03	1.02	1.19

注：数据为均值；1 引自《中国区域地质志·河北志》，2017；2、3、4 为本次实测。

表 20-17　晚古生代中性岩稀土元素成分及特征参数一览表

Table 20-17　REE contents and main parameter of Late Paleozoic intermediate rocks

序号	岩体	岩性	稀土元素含量/10^{-6}																特征参数			
			La	Ce	Pr	Nd	Sm	Eu	Gd	Tb	Dy	Ho	Er	Tm	Yb	Lu	Y	ΣREE	LREE/HREE	$(La/Yb)_N$	δEu	δCe
1	石垛口	正长岩	63.60	126.10	13.15	57.60	12.00	3.80	8.00	1.10	6.00	1.30	2.80	0.40	2.40	0.35	20.40	298.6	12.36	19.01	1.12	1.01
2	荞麦峪	正长岩	30.66	67.61	8.22	32.11	5.39	1.60	4.17	0.59	3.13	0.56	1.62	0.27	1.72	0.28	15.70	157.93	11.80	12.79	0.99	1.02
3	梨树沟	正长岩	31.72	63.70	7.48	29.47	5.80	1.64	4.39	0.61	3.19	0.59	1.63	0.24	1.54	0.23	15.99	152.23	11.26	14.77	0.95	0.98
4	老公营	正长岩	49.25	95.61	9.83	33.35	4.08	0.93	2.65	0.27	1.10	0.18	0.51	0.08	0.53	0.17	3.95	198.54	35.16	66.65	0.81	1.00

注：数据为均值；1《中国区域地质志·河北志》，2017；2、3、4 为本次实测。

表 20-18　晚古生代中性岩微量元素成分及参数一览表

Table 20-18　Trace element contents and main parameters of Late Paleozoic intermediate rocks

序号	岩体	岩性	微量元素含量/10^{-6}										特征参数		
			Rb	Ba	Th	U	Ta	Nb	Sr	Ti	Zr	Hf	Rb/Sr	Nb/Ta	Ti/Zr
1	荞麦峪	正长岩	153.32	3843.85	25.79	3.27	0.45	11.94	1553.18	2238.13	209.13	5.31	0.10	26.83	10.70
2	梨树沟	正长岩	186.00	3323.00	6.22	1.99	0.38	7.39	1699.50	2238.13	115.40	2.86	0.11	19.71	19.39
3	老公营	正长岩	213.01	4714.03	13.99	2.83	0.62	12.27	1321.03	1348.88	144.04	3.60	0.16	19.89	9.36

注：数据为均值，为本次实测。

0.511706 ~ 0.511678，$^{147}Sm/^{144}Nd$ = 0.0968 ~ 0.0652；取 t = 385.1Ma 时，$\varepsilon_{Sr}(t)$ = 21.08 ~ 22.04，$\varepsilon_{Nd}(t)$ = -18.18 ~ -18.73，t_{DM} = 1880 ~ 1510Ma。

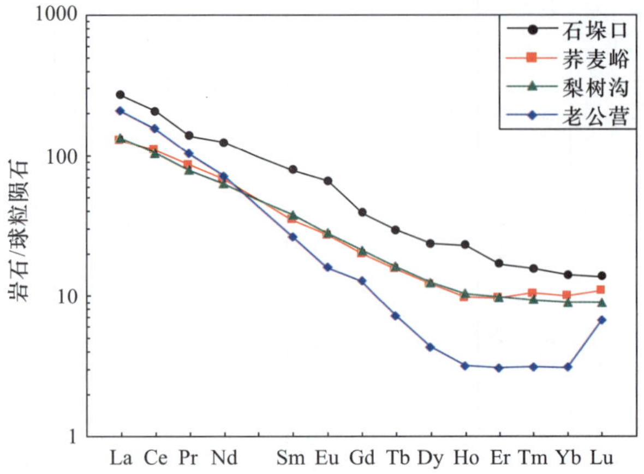

图 20-34　中泥盆世中性岩稀土元素配分模式图
Fig. 20-34　Chondrite-normalized REE patterns of Middle Devonian intermediate rock

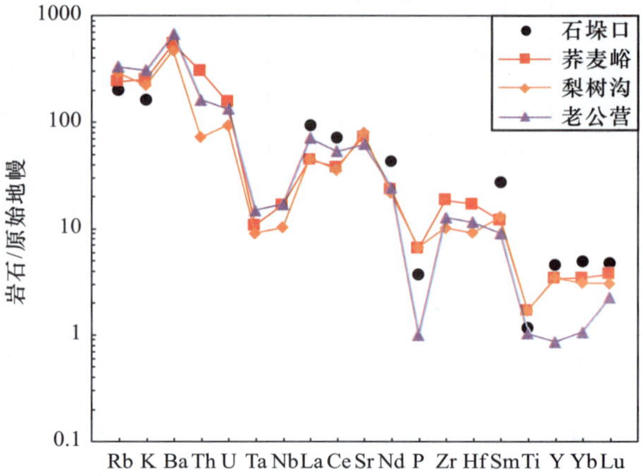

图 20-35　中泥盆世中性岩微量元素蛛网图
Fig. 20-35　Trace element spider diagram of Middle Devonian intermediate rock

五、Lu-Hf 同位素特征

老公营正长岩锆石$^{176}Yb/^{177}Hf$ = 0.000576 ~ 0.030086，$^{176}Lu/^{177}Hf$ = 0.000021 ~ 0.001099，表明锆石在形成以后仅有较少放射成因 Hf 的积累。$^{176}Hf/^{177}Hf$ = 0.281977 ~ 0.282029，$f_{Lu/Hf}$ = -0.97 ~ -1，小于镁铁质地壳（-0.34）和硅铝质地壳（-0.72）值（Vervoorrt et al.，1996）；取 t = 385.1Ma 时，$\varepsilon_{Hf}(t)$ = -19.92 ~ -17.96，t_{DM} = 1795 ~ 1691Ma。

六、成因类型与岩浆演化

该套岩系富 Al_2O_3、Sr，低 FeO^*、TiO_2、Cr、Ni；$\varepsilon_{Nd}(t)$ = -12.26 ~ -13.28，Nb/U = 2.34 ~ 10.48，$^{176}Hf/^{177}Hf$ = 0.281977 ~ 0.282029；在 $\varepsilon_{Sr}(t) - \varepsilon_{Nd}(t)$ 和锆石 $\varepsilon_{Hf}(t)$ -年龄图（图 20-36）上，成分点位于下地壳和华北克拉通北部 2.5Ga 平均地壳演化线与古元古代地壳演化线之间，这些特征反映其岩浆可能来源于下地壳。又该套岩系岩体内部发育大量新太古代变质岩捕虏体，并有正长岩与新太古代变质岩的混染岩存在，且正长岩稀土元素分布模式与围岩新太古代中酸性变质岩类极度相似（He et al.，2019）；正长岩锆石模式年龄和全岩 Sr-Nd 模式年龄 t_{DM} 分别为 1691 ~ 1795Ma 和 1510 ~ 1880Ma，两个阶段模式年龄略小于围岩晚太古界基底岩石锆石 U-Pb 年龄；钙质角闪石计算结晶温压条件分别为 948 ~ 960℃（Green et al.，1967）和 0.863 ~ 0.868GPa（Schmidt et al.，1992），锆石饱和温度计算 667 ~ 914℃（Watson et al.，1983），揭示具有深源岩浆及侵位特征。因此，下地壳太古宙变质基底可能广泛的参与了正长岩的成岩过程。

同时，正长岩还具有低 SiO_2 和富 K_2O 特征；与冀北水泉沟正长岩相似，全岩 Nd 同位素两阶段模式年龄变化范围大，且模式年龄小于围岩晚太古界变质岩的年龄，说明可能有年轻地幔物质的加入。梨树沟岩体中出现暗色微粒包体也暗示其形成过程中存在岩浆混合作用。因此，该套岩系源区物质主要是壳源的，同时又混合有幔源碱性玄武质岩浆，而其成因模型可能为 MASH 模型（李长民等，2014）。

七、构造环境分析

该套岩系岩性组合具有富碱，贫铁、镁的特征，属于碱性岩系列，并具有明显火山弧火成岩特性（图 20-37），与区域上中-晚泥盆世古蒙古洋板块向南俯冲、使华北克拉通北缘表现为沟-弧-盆安第斯型活动大陆边缘（孟祥化，2001；Cope 等，2005；Jiang，2005；Zhang et al.，2010）的构造背景有关。在西伯利亚板块向南俯冲后短期伸展构造背景环境下，位于弧后的尚义-赤城-承德-北票超壳深大断裂复活，使得壳幔混合正长质岩浆沿此断裂快速侵位，形成具有火山弧特性正长岩（李长民等，2014）。

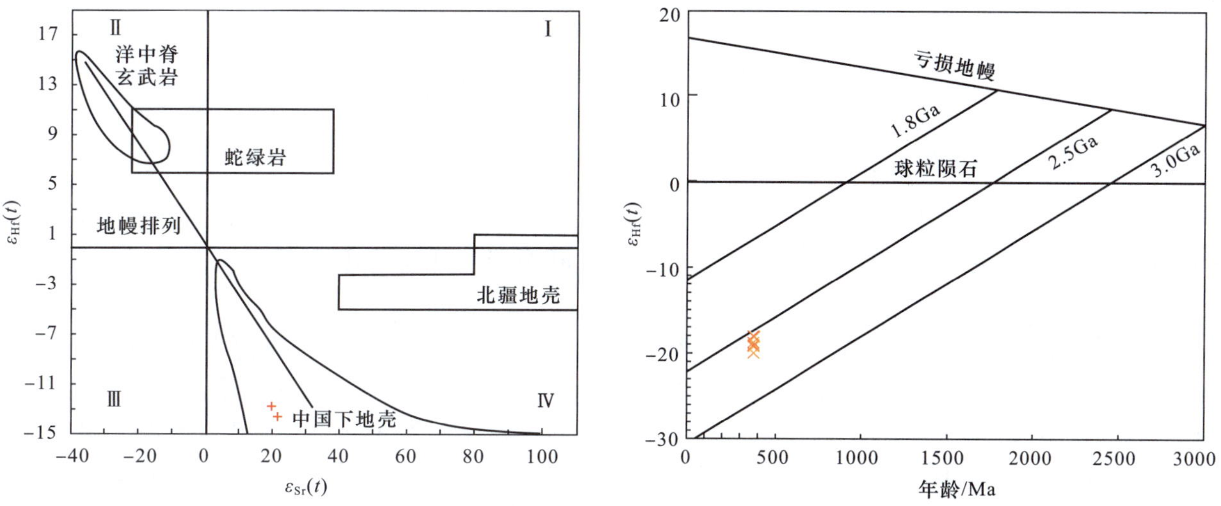

图 20-36　晚古生代中性岩全岩 ε_{Sr}（t）$-\varepsilon_{Nd}$（t）和锆石 ε_{Hf}（t）-年龄图

Fig. 20-36　ε_{Sr}（t）$-\varepsilon_{Nd}$（t）and zircons ε_{Hf}（t）-age diagram of Late Paleozoic intermediate rock

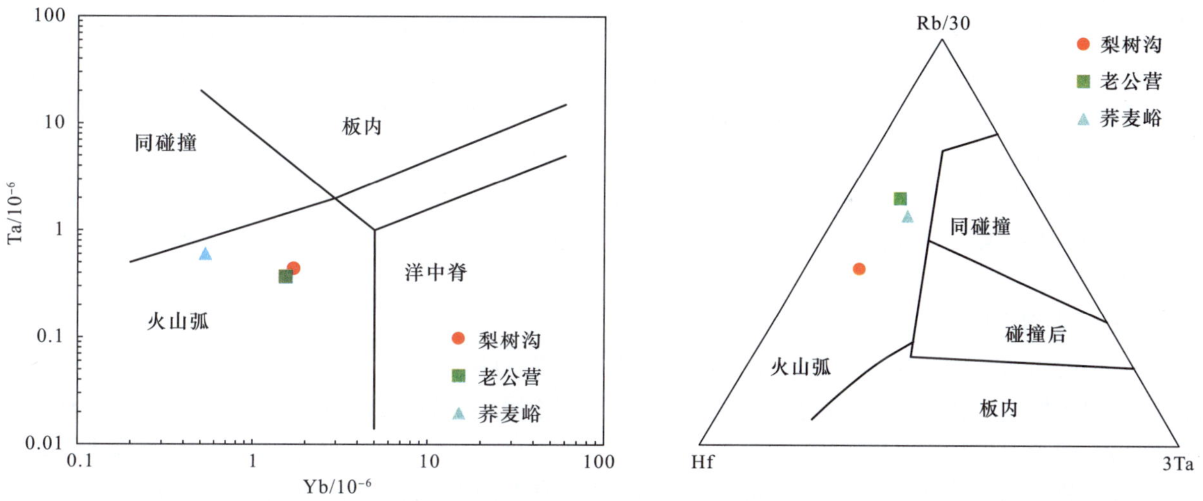

图 20-37　中泥盆世 Yb-Ta 和 Rb/30-Hf-3Ta 图解

Fig. 20-37　Tectonic discrimination diagram of Yb-Ta and Rb/30-Hf-3Ta of Middle Devonian intermediate rock

八、时代归属

本期岩浆侵位时代厘定为中泥盆世，主要依据包括：①岩体未遭受高级变质和复杂褶曲变形；②广泛侵位到晚太古界变质岩和中元古代地层中；③岩体沿大断裂发育，这些断裂中生代强烈活动，岩体广泛被此断裂所破坏；④特征同河北省中泥盆世水泉沟、石踩口岩体；⑤梨树沟正长岩锆石 U-Pb 年龄为（384.2±4.0）Ma 和（384.0±5.0）Ma；⑥老公营正长岩锆石 U-Pb 年龄为（385.1±4.7）Ma。

第四节　中生代中酸性侵入岩

中生代中-酸性侵入岩按形成时代可划分为早三叠世、中三叠世、晚三叠世、早侏罗世、中侏罗世、晚侏罗世、早白垩世早期、早白垩世中期、早白垩世晚期，分属印支构造-岩浆旋回和燕山构造-岩浆旋回。

一、早三叠世中酸性侵入岩

（一）地质产状与岩石学特征

发育水头、胡营子、田营北沟、小梁后、下窝铺、碾子、上马铺、鹿角沟和四道甸 9 个岩体，沿赤城-古北口深大断裂带近 EW 向分布，可能受控于此断裂带，出露面积 78.5km²。

1. 水头岩体

位于怀柔水头村西，呈长轴走向近 SN 椭圆形，长 1km，宽 0.65km，出露面积 0.65km²。岩体顺层侵入中元古界蓟县系高于庄组地层中，外接触带局部大理岩化，内接触带无明显蚀变。岩性为二长闪长岩，细粒结构，块状构造。斜长石自形-半自形板状，粒径 0.2~1mm，发育聚片双晶，偶见格子、卡纳复合双晶，个别具环带结构，含量约 84%；辉石他形粒状，粒径 0.2~1mm，含量约 12%；次透辉石他形粒状，粒径 0.2~1mm，含量约 2%；黑云母半自形片状，含量约 1%；副矿物为磷灰石、磁铁矿等。

2. 胡营子岩体

位于怀柔胡营子村，由不规则的 4 个小岩瘤组成，出露面积 10.5km²。岩体以顶垂体形态残留出露于中生代燕山期酸性侵入体中，接触界线清晰。与围岩接触边部发育后期岩浆侵位流面构造或冷凝边，局部可见梳状后期岩浆侵位岩枝，而围岩边部可见胡营子岩体捕虏体。岩体岩性为石英闪长岩，灰白色，花岗结构，块状构造。斜长石他形-半自形板状，粒径 1~3mm，发育聚片双晶，绢云母化，内嵌榍石，含量 53%~58%；钾长石他形粒状，内嵌副长石、黑云母矿物，含量 25%~30%；石英他形粒状，粒径 1~3mm，含量 5%~10%；角闪石他形，粒径 1~3mm，绿泥石化析出铁质，含量约 5%；黑云母他形片状，粒径 1~2mm，绿泥石化，含量约 7%。副矿物为磁铁矿、磷灰石、榍石等。

3. 田营北沟岩体

位于怀柔田营北沟，呈不规则状小岩瘤状产出，出露面积 1km²。岩体以顶垂体形态残留出露于中生代燕山期酸性侵入体中，因同围岩颜色、粒度差异较大，接触界线清晰。岩体岩性为花岗闪长岩，灰白色，中粒含斑结构，块状结构。斜长石自形板状，发育聚片双晶、环带结构，大者构成斑晶，含量约 50%；钾长石半自形短柱状，边部交代斜长石形成蠕虫结构，大者构成斑晶，含量 15%~20%；石英他形粒状，波状消光，含量 15%~20%；黑云母半自形片状，绿泥石化、碳酸盐岩化，含量约 10%；角闪石半自形，含量约 5%。副矿物为磷灰石、榍石、磁铁矿等。

4. 小梁后岩体

位于怀柔小梁后村，呈近圆形岩珠产出，出露面积 35km²。岩体以顶垂体形态残留出露于中生代燕山期酸性侵入体中，北、西接触带与围岩颜色差异明显，东、南与围岩粒径差异明显。岩体岩性为似斑状中粗粒二长花岗岩，灰白色-浅肉红色，斑状结构，块状构造。斑晶碱性长石半自形板状，发育卡钠复合双晶、条纹结构，类型为条纹长石和正长石，粒径 8~40mm，含量 10%~30%；斑晶斜长石半自形板状，粒径 7~30mm，发育聚片、卡钠复合双晶，含量约 5%；斑晶石英浑圆粒状，粒径 5~7mm，含量 10%~20%。基质石英他形粒状，粒径 2~7mm，含量约 40%；碱性长石半自形-他形板状，特征同斑晶，内嵌石英矿物，粒径 2~8mm，高岭土化，含量 20%~30%；基质斜长石半自形-他形板状，粒径2~7mm，发育聚片、卡钠复合双晶，绢云母化，含量25%~40%；黑云母他形片状，绿泥石化，含量3%~5%。副矿物为榍石、磁铁矿，常与黑云母伴生。

5. 下窝铺岩体

位于怀柔下窝铺村，长条带状产出，出露面积$2km^2$。岩体北被断层切割，其余周缘侵入于晚古生代早二叠世辉长闪长岩体中，接触面平直外倾，倾角约70°，内接触带见宽度不等细粒冷凝边，并发育与接触面平行的弱片麻状构造。岩体岩性为二长花岗岩，特征同小梁后岩体。

6. 碾子岩体

位于怀柔碾子村，呈不规则状，出露面积$12km^2$。岩体北侧被中生代燕山期早白垩世苗营细粒黑云母花岗岩侵入，接触界线清晰。与围岩接触边部发育后期岩浆侵位形成宽约1m冷凝边。西南侧侵入于新太古代变质岩中，并被中生代燕山期早白垩世抱榆树沟细粒花岗岩呈枝杈脉状侵入穿插。

岩体可根据岩性特征划分中心相和边缘相，中心相粒径大，发育斑晶，且暗色矿物含量高，岩性特征同小梁后岩体。边缘相为中细粒二长花岗岩，浅肉红色，中细粒花岗结构，块状构造。斜长石半自形板状，粒径2~3.3mm，发育聚片双晶，黏土化，零星绢云母化、绿帘石化，含量35%~40%；钾长石半自形板状，粒度2~5mm，类型为微斜长石（主）和微斜条纹长石，高岭土化，零星绢云母化，边部交代斜长石（粒径0.4~2mm）形成蠕虫结构，含量约35%~40%；石英他形粒状，多集合体状产出，部分波状消光，晶内见被碳酸盐等填充愈合裂隙，粒径0.2~2mm，含量约20%；黑云母片状，粒径0.2~1mm，被白云母、绿泥石、碳酸盐、绿帘石等交代呈假象，含量<5%。副矿物为磷灰石、磁铁矿等。

7. 上马铺岩体

位于怀柔上马铺村，呈近EW向长条带状产出，出露面积$7km^2$。岩体侵入晚古生代早二叠世辉长闪长岩体中，岩体及发育岩性特征同下窝铺岩体。

8. 鹿角沟岩体

位于怀柔鹿角沟村，呈不规则状产出，出露面积$8km^2$。岩体以顶垂体形态残留出露于中生代燕山期酸性侵入体中，同围岩岩性相似，接触带界线模糊。

9. 四道甸岩体

位于怀柔四道甸村村呈近EW向长条状产出，出露面积$3km^2$。岩体东被中生代燕山期早白垩世正长花岗岩体侵入破坏，其余周缘侵入晚古生代早二叠世辉长闪长岩体中，内接触带发育弱片麻理构造。岩体内无岩性岩相变化，发育岩性为似斑状中粗粒二长花岗岩，特征同小梁后岩体。

（二）岩石化学与地球化学特征

岩石化学成分及特征参数见表20-19。岩石呈高SiO_2、Al_2O_3及碱度，低MgO、P_2O_5、MnO特征。里特曼指数$\sigma=2.33\sim3.87$，为亚碱性系列；固结指数$SI=2.16\sim15.18$，分异指数$DI=53.46\sim90.84$，长英指数$FL=49.38\sim88.06$，岩浆分异程度较强；中、酸性岩A/CNK分别为0.77~0.96和0.97~1.11，A/NK分别为1.44~1.74和1.22~1.32，分属于准铝质和过铝质。

稀土元素成分及特征参数见表20-20。中、酸性岩ΣREE分别为$164.16\times10^{-6}\sim198.57\times10^{-6}$和$62.75\times10^{-6}\sim150.03\times10^{-6}$；LREE/HREE分别为19.49~26.28和14.04~24.45；$(La/Yb)_N$分别为37.55~49.30和14.87~41.54；轻稀土强烈富集，重稀土相对亏损；稀土总量酸性岩低于中性岩；中、酸性岩δEu分别为1.01~1.58和0.57~1.03，分别呈现铕弱正异常和弱负异常；配分曲线均为右陡倾光滑曲线（图20-38），且曲线基本平行，揭示了同源岩浆演化特征。

微量元素含量及特征参数见表20-21。与稀土元素特征相反，微量元素含量中性岩均低于酸性岩。微量元素蛛网图（图20-39）中，中性、酸性岩曲线基本相似，富集大离子亲石元素Rb、Sr、Ba、K，相对强烈亏损元素Tu、U、Nb、Ta、P、Zr。

表 20-19　早—中三叠世中酸性岩岩石化学成分及特征参数一览表

Table 20-19　The chemical composition and main parameters of Early to Middle Trassic intermediate-acid rocks

序号	岩体	岩性	氧化物含量/%													特征参数					
			SiO_2	TiO_2	Al_2O_3	Fe_2O_3	FeO	MnO	MgO	CaO	Na_2O	K_2O	P_2O_5	Los	总和	*DI*	*SI*	*σ*	*FL*	*A/CNK*	*A/NK*
1	水头	二长闪长岩	54.44	1.10	17.10	6.11	3.56	0.14	2.89	6.89	4.53	2.19	0.56	0.13	99.64	53.46	15.18	3.87	49.38	0.77	1.74
2	胡营子	石英闪长岩	61.51	0.78	16.07	3.26	3.18	0.06	1.97	4.55	4.80	2.71	0.29	0.22	99.39	68.43	12.39	3.01	62.25	0.84	1.48
3	田营北沟	花岗闪长岩	66.52	0.64	15.63	2.36	2.12	0.07	1.07	3.02	4.82	2.68	0.21	0.34	99.48	77.21	8.23	2.37	71.29	0.96	1.44
4	四道甸	二长花岗岩	70.53	0.34	14.39	1.54	1.54	0.06	0.56	1.89	3.90	4.13	0.24	0.25	99.38	85.31	4.84	2.33	80.94	1.00	1.32
5	小梁后	二长花岗岩	71.50	0.19	15.29	0.69	0.62	0.03	0.23	1.30	4.65	4.50	0.04	0.77	99.81	90.84	2.16	2.92	87.60	1.03	1.22
6	碾子	二长花岗岩	72.10	0.18	14.06	1.01	1.73	0.02	0.34	1.07	3.88	3.99	1.08	0.38	99.84	89.70	3.13	2.12	88.06	1.11	1.31
7	鹿角沟	二长花岗岩	72.76	0.12	14.30	0.42	1.28	0.02	0.29	1.84	3.77	4.86	0.03	0.17	99.86	88.25	2.73	2.50	82.43	0.97	1.25
8	七道河西	石英二长岩	66.27	0.46	16.23	1.27	1.85	0.03	0.72	2.45	5.58	3.20	0.17	0.97	99.20	82.84	5.71	3.27	78.18	0.95	1.28
9	八道河西	石英二长岩	67.24	0.34	15.47	1.33	1.74	0.03	0.68	1.76	4.42	4.96	0.12	0.88	98.97	85.78	5.18	3.58	84.20	0.98	1.22

注：数据为均值；1、2、3、4、7、8、9 引自 1∶5 万杨木栅子幅、喇叭沟门幅、三道营子幅区域地质调查报告，1999；5、6 为本次实测。

表 20-20　早—中三叠世中酸性岩稀土元素成分及特征参数一览表

Table 20-20　REE contents and main parameter of Early to Middle Trassic intermediate-acid rocks

序号	岩体	岩性	稀土元素含量/10^{-6}																特征参数			
			La	Ce	Pr	Nd	Sm	Eu	Gd	Tb	Dy	Ho	Er	Tm	Yb	Lu	Y	ΣREE	LREE/HREE	$(La/Yb)_N$	δEu	δCe
1	水头	二长闪长岩	52.35	85.30	9.12	34.80	7.78	1.94	1.80	0.30	2.26	0.44	1.22	0.16	1.00	0.10	11.84	198.57	26.28	37.55	1.58	0.96
2	胡营子	石英闪长岩	42.95	70.53	8.29	27.63	5.37	1.39	3.28	0.39	1.98	0.38	0.87	0.18	0.63	0.31	7.35	164.16	19.49	49.30	1.01	0.92
3	田营北沟	花岗闪长岩	49.70	76.99	9.89	29.50	6.78	1.62	2.53	0.35	1.97	0.40	0.84	0.12	0.74	0.22	0.10	181.65	24.33	48.18	1.20	0.85
4	四道甸	二长花岗岩	41.11	67.67	6.89	23.76	3.87	0.84	1.95	0.35	1.44	0.33	0.73	0.13	0.84	0.12	7.92	150.03	24.45	35.19	0.93	0.99
5	小梁后	二长花岗岩	14.42	36.51	4.24	15.70	2.46	0.65	2.00	0.26	1.27	0.23	0.61	0.11	0.70	0.10	6.38	79.24	14.04	14.87	0.90	1.15
6	碾子	二长花岗岩	24.90	38.49	4.13	12.76	2.40	0.59	1.29	0.30	0.76	0.15	0.39	0.09	0.43	0.08	7.04	86.76	23.84	41.54	1.03	0.93
7	鹿角沟	二长花岗岩	17.21	27.51	2.74	9.14	2.02	0.34	1.67	0.30	0.63	0.14	0.39	0.08	0.49	0.09	3.34	62.75	15.56	25.19	0.57	0.98
8	七道河西	石英二长岩	67.26	108.70	10.80	41.02	7.52	1.60	2.34	0.47	0.90	0.20	0.38	0.07	0.52	0.07	5.00	241.85	47.86	92.78	1.17	0.99
9	八道河西	石英二长岩	41.68	64.23	7.23	22.86	3.92	0.67	1.55	0.30	0.37	0.10	0.21	0.04	0.41	0.05	1.55	143.62	46.40	72.92	0.83	0.91

注：数据为均值；1、2、3、4、7、8、9 引自 1∶5 万杨木栅子幅、喇叭沟门幅、三道营子幅区域地质调查报告，1999；5、6 为本次实测。

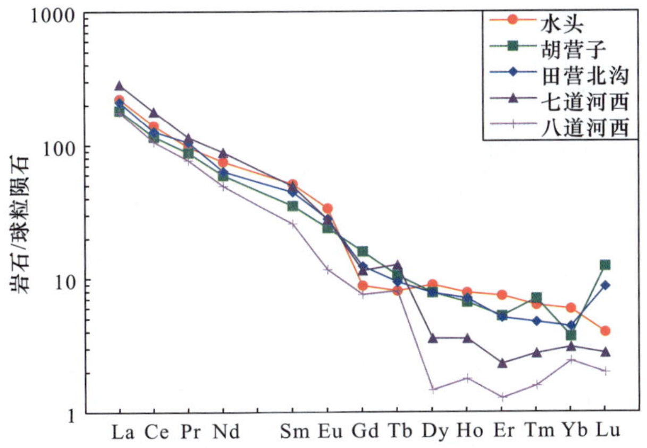

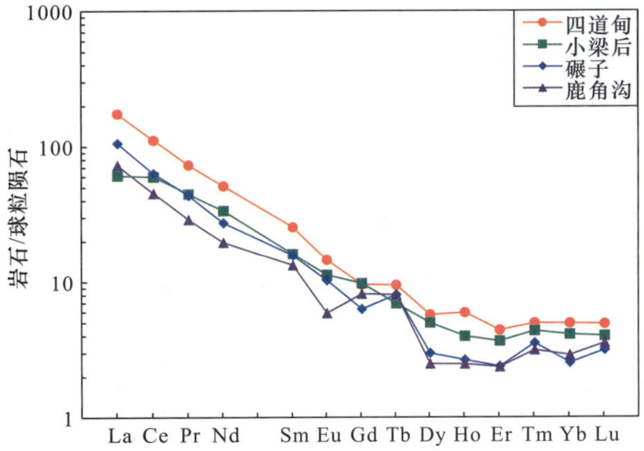

图 20-38　早—中三叠世中酸性岩稀土元素配分模式图

Fig. 20-38　Chondrite-normalized REE patterns of Early to Middle Trassic intermediate-acid rocks

表 20-21　早—中三叠世中酸性岩微量元素成分及特征参数一览表

Table 20-21　Trace element contents and main parameters of Early to Middle Trassic intermediate acidity rock

序号	岩体	岩性	微量元素含量/10^{-6}									特征参数	
			Rb	Ba	Th	U	Nb	Ta	Sr	Zr	Hf	Rb/Sr	Nb/Ta
1	水头	二长闪长岩		973.90	5.45				900.10	22.95			
2	胡营子	石英闪长岩		1306.75	7.88				867.55	15.02			
3	田营北沟	花岗闪长岩		1424.00	9.96				684.00	26.88			
4	四道甸	二长花岗岩	85.74	1145.00	2.98	0.13	0.47	4.12	550.90	18.12	4.17	0.16	0.11
5	小梁后	二长花岗岩	99.95	1537.80	3.09	0.12	0.56	4.86	486.45	11.79	3.94	0.21	0.12
6	碾子	二长花岗岩	103.20	1373.00	3.86	0.12	0.63	4.78	610.70	16.62	3.69	0.17	0.13
7	鹿角沟	二长花岗岩		1030.00	4.30				462.00	12.00			
8	七道河西	石英二长岩		1850.00	3.40				1050.00	9.40			
9	八道河西	石英二长岩		810.00	0.40				574.00	3.80			

注：数据来源同表 20-19。

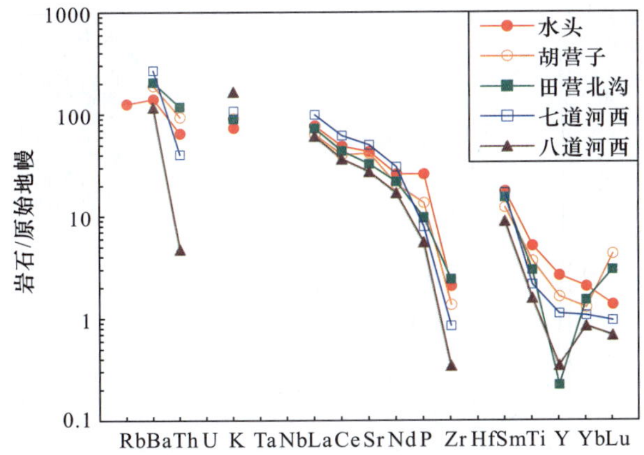

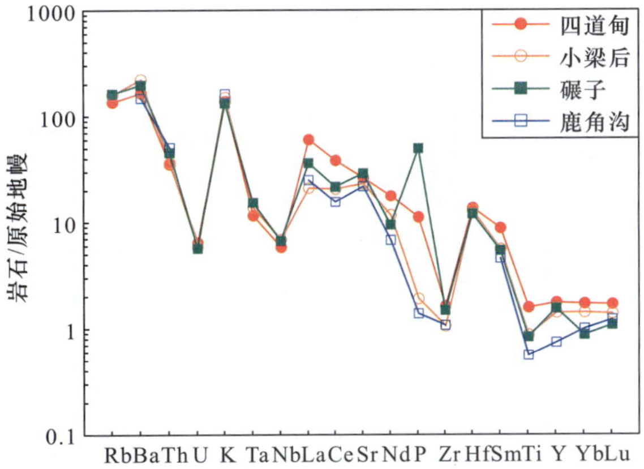

图 20-39　早—中三叠世中酸性岩微量元素蛛网图

Fig. 20-39　Trace element spider diagram of Early to Middle Trassic intermediate-acid rocks

(三) 岩石系列

该套岩系 σ=2.12~3.87<4，属于亚碱性岩系列。$FeO^*/MgO-SiO_2$ 和 SiO_2-K_2O 图解（图 20-40）

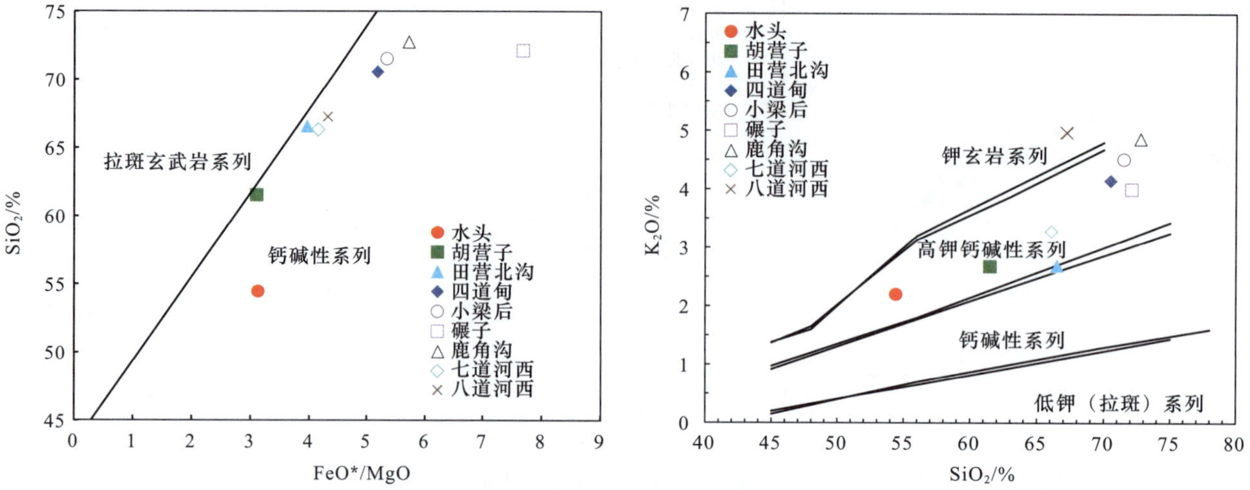

图 20-40 早—中三叠世中酸性岩 $FeO^*/MgO-SiO_2$ 和 SiO_2-K_2O 图解

Fig. 20-40 $FeO^*/MgO-SiO_2$ and SiO_2-K_2O diagram of Early to Middle Trassic intermediate-acid rocks

中位于高钾钙碱性系列。

（四）Sr-Nd 同位素特征

小梁后斑状二长花岗岩全岩 $^{87}Sr/^{86}Sr$ = 0.70823，$^{87}Rb/^{86}Sr$ = 0.7063，$^{143}Nd/^{144}Nd$ = 0.511563，$^{147}Sm/^{144}Nd$ = 0.0876；取 t = 247.5Ma 时，$\varepsilon_{Nd}(t)$ = −20.97，t_{DM} = 1915Ma。

（五）成因类型与岩浆演化

该套岩系属于 I 型花岗岩，岩浆来源于新元古代增生加厚基性镁铁质下地壳的部分熔融，主要依据有：①二长闪长岩、石英闪长岩、花岗闪长岩和二长花岗岩岩石组合；②造岩矿物石英、斜长石、碱性长石和暗色矿物黑云母、角闪石组合；③$A/CNK<1.1$，富碱且 $Na_2O>3.2\%$，为高钾钙碱性系列；④岩石轻稀土富集且 Eu 亏损不明显，Rb/Sr = 0.16～0.21；⑤富集大离子亲石元素元素 Rb、Ba、K 而亏损 Ti、P、Nb、Ta；⑥副矿物为富磁铁矿、磷灰石和榍石，缺乏白云母、石榴子石等 S 型花岗岩矿物；⑦SiO_2-Ce 图解（图 20-41）中，成分点位于 I 型花岗岩区；⑧埃达克岩判别图解（图 20-42）

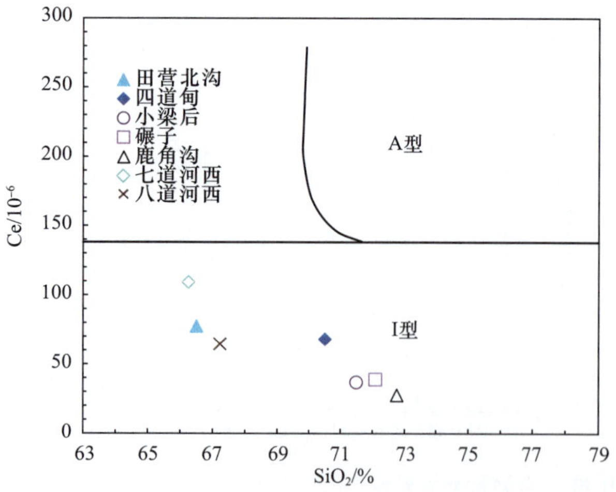

图 20-41 早—中三叠世中酸性岩 SiO_2-Ce 图解

Fig. 20-41 SiO_2-Ce diagram of Early to Middle Trassic intermediate-acid rocks

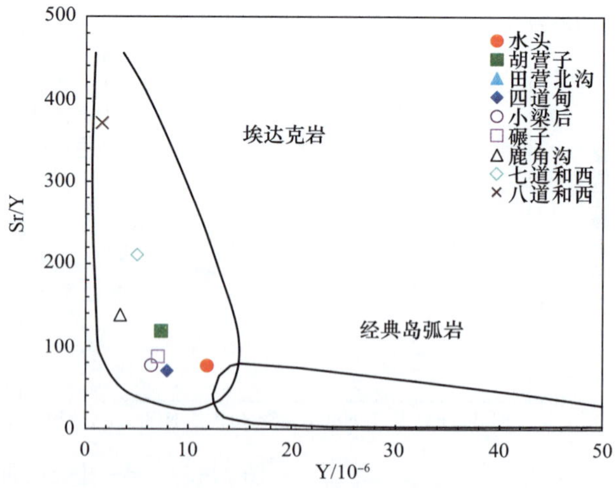

图 20-42 早—中三叠世中酸性岩 Sr/Y-Y 图解

Fig. 20-42 Sr/Y-Y diagram of Early to Middle Trassic intermediate-acid rocks

上，成分点位于埃达克岩区，但 K_2O 含量和 Rb/Sr 比值高，$Mg^{\#}$值和 Cr、Ni 含量较低，且在区域内缺乏同期火山岩组合，又不同于典型俯冲洋壳部分熔融形成的埃达克岩（Defant et al.，1990，1993；肖庆辉，2002）和拆沉下地壳物质熔融形成的埃达克岩（Matin，1999；肖庆辉，2002）；⑨上述 Sr-Nd 同位素特征。

该套岩系的高 Sr 和无负 Eu 异常特征可能暗示这种基性岩浆源岩中的斜长石已进入熔体，而并不是以残留相形式存在，而低 Y 及 HREE 表明其残留相中存在石榴子石或者角闪石（肖庆辉，2002）。二长花岗岩 Y/Yb 比值接近 10，稀土配分曲线表现为陡倾型，轻重稀土强烈分馏，进一步表明其岩浆源区残留相的主体以石榴子石和角闪石为主，且其岩浆源区部分熔融的残余相更可能是石榴子石角闪岩而不是榴辉岩。

（六）构造环境分析

区域上该套岩系沿赤城-古北口深大断裂带呈近 EW 向分布，明显与蒙古增生褶皱带碰撞造山（中亚造山）运动引起的华北板块北缘地壳增厚有关。本区早、中三叠世花岗岩体系列虽然处于华北克拉通内部，但毗邻华北克拉通北缘，说明了中亚造山运动还对华北板块内部产生一定影响，可能造成了强烈的基底岩石再造。

（七）时代归属

本期岩浆侵位时代厘定为早三叠世，主要依据有：①岩体未遭受高级变质和复杂褶曲变形；②广泛侵位到晚古生代早二叠世侵入体中，并被中生代燕山期中酸性侵入体切割破坏；③岩体沿大断裂发育，这些断裂中生代强烈活动，岩体广泛被此断裂所破坏；④小梁后二长花岗岩锆石 U-Pb 同位素年龄为（247.5±0.62）Ma（何付兵等，2018）。

二、中三叠世中性侵入岩

（一）地质产状与岩石学特征

发育七道河西、八道河西 2 个岩体，沿赤城-古北口深大断裂带近 EW 向分布，可能受控于此断裂带，出露面积 2.5km²。

1. 七道河西岩体

位于怀柔七道河村西，呈长轴近 EW 向椭圆形产出，出露面积约 1.5km²。岩体东被晚侏罗系土城子组不整合覆盖；其余周缘侵入晚古生代早二叠世辉长闪长岩体和中生代早三叠世二长花岗岩中。岩体岩性为石英二长岩，普遍发育碎裂构造和片麻状构造，并可见晚太古界片麻岩（黑云母片麻岩及斜长角闪岩）捕虏体。

石英二长岩，灰白—浅肉红色，中细粒结构，块状构造。各组成矿物普遍碎裂化，见两组相交被碳酸盐矿物充填碎裂纹。石英他形粒状，波状消光，粒径 0.1~2mm，含量约 15%；碱性长石半自形—他形，粒径 0.1~0.2mm，个别 3mm，类型为微斜长石，含量约 25%；斜长石半自形—他形，发育聚片双晶，双晶纹构造破碎错开不连续，局部被钾长石交代，粒径 0.1~2mm，含量约 50%；角闪石他形，蚀变为黑云母，粒径 0.1~2mm，含量约 8%；副矿物为榍石、磁铁矿等。

2. 八道河西岩体

位于怀柔八道河村西，呈近 EW 向延长不规则状产出，出露面积 1km²。岩体南、西分别侵入晚古生代早二叠世辉长闪长岩体中和中生代早三叠世二长花岗岩中，北被第四系覆盖。岩体岩性特征同七道河西岩体。

（二）岩石化学与地球化学特征

岩石化学成分及特征参数见表 20-19，基本同早三叠世中性岩。里特曼指数 $\sigma=3.27\sim3.58$，为亚碱性系列；固结指数 $SI=5.78\sim5.18$，分异指数 $DI=82.84\sim85.78$，长英指数 $FL=78.18\sim84.20$，岩浆分异作用较强；$A/CNK=0.95\sim0.98$，$A/NK=1.22\sim1.28$，属于偏铝质。

稀土元素成分及特征参数见表20-20。$\Sigma REE=143.62\times10^{-6}\sim241.85\times10^{-6}$；LREE/HREE=46.40~47.86；$(La/Yb)_N=72.92\sim92.78$；轻稀土强烈富集，重稀土相对亏损；$\delta Eu=0.91\sim0.99$，呈现铕弱负异常；配分曲线为右陡倾光滑曲线（图20-38）。

微量元素含量及特征参数见表20-21，基本同早三叠世中性岩。微量元素蛛网图（图20-39）中，与早三叠世中酸性岩曲线基本相似，富集大离子亲石元素Rb、Sr、Ba、K，强烈亏损元素Tu、U、Nb、Ta、P、Zr。

（三）岩石系列

该套岩系$\sigma=3.27\sim3.58<4$，属于亚碱性岩系列。$FeO^*/MgO-SiO_2$和SiO_2-K_2O图解（图20-40）中，位于高钾钙碱性-钾玄岩系列。

（四）成因类型与构造环境

该套岩系和早三叠世中性岩特征相似，表现为相同成因类型，且形成于相同的大地构造背景环境，即形成于中三叠世中亚造山后碰撞构造背景环境下，由新元古代增生加厚基性镁铁质下地壳发生部分熔融形成。

（五）时代归属

本期岩浆侵位时代厘定为中三叠世，主要依据有：①岩体未遭受高级变质和复杂褶曲变形；②侵位到晚古生代早二叠世侵入体和中生代早三叠世侵入体中；③岩体沿大断裂发育，这些断裂中生代强烈活动，岩体广泛被此断裂所破坏，发育碎裂构造和片麻状构造；④七道河西岩体全岩K-Ar同位素年龄为227.03Ma（北京市地质调查所，1999），锆石U-Pb同位素年龄为207Ma（核工业三所，1988）。

三、晚三叠世中—酸性侵入岩

（一）地质产状与岩石学特征

发育盘山、玻璃台、东白山、珠窝4个岩体，境内出露面积53km²。此外，在各岩体附近地质体中发育有大量花岗岩、花岗斑岩脉，可能也与这期岩浆侵位事件有关。

1. 盘山岩体

位于平谷南山村，向南延伸至天津盘山、官庄（图20-43），发育为南、北两个长轴近SN向椭圆形复式杂岩体，出露面积60km²，北京境内30km²。根据重力和钻探资料，南、北岩体在深部相连。南岩体半隐伏产出，南东侧被第四系覆盖；北岩体产状复杂，界面较陡，倾角50°~85°。两个岩体顺层侵入中元古界长城系和蓟县系中，接触界线清晰，外接触带发育大量梳状岩脉，内接触带发育冷凝边。岩体岩性岩相复杂，表现为短期多次岩浆侵位杂岩体。依据岩石类型及其相互间穿切关系等，可将岩体划分为四个侵入期次岩性单元（图20-43）：狼家峪细粒花岗岩单元、官庄含斑石英二长岩单元、东罗庄中粗粒花岗岩单元和小盘山中细粒黑云母花岗岩单元。其中，官庄单元中含边界平滑，弱交代混染，形态呈圆状、椭圆状（主）和透镜状、不规则状（少）暗色细粒闪长岩同源包体。各单元岩性矿物组成及含量特征见表20-22。

表20-22 盘山岩体主要岩石类型及其特征

Table 20-22 Main rock types and its characteristics of Panshan complex

岩体	期次单元	岩性	特征描述
盘山	狼家峪	中细粒花岗岩	中细粒黑云母二长花岗岩为主，中细粒结构，块状构造。主要矿物组成为钾长石（30%~40%）、斜长石（25%~35%）、石英（25%~30%）、黑云母（5%~7%）及少量绿色柱状角闪石，粒径0.25~2mm。副矿物为磁铁矿、钛铁矿、榍石、锆石、磷灰石、褐帘石，可见黄铁矿、辉钼矿等
			黑云正长花岗岩为次，中细粒结构，块状构造。主要矿物组成为钾长石（55%~60%）、斜长石（15%）、石英（20%~25%）和黑云母。副矿物为磁铁矿、锆石、榍石、磷灰石、褐帘石等

续表

岩体	期次单元	岩性	特征描述
盘山	官庄	含斑石英二长岩	浅灰—灰白色，含斑结构，块状构造。斑晶斜长石半自形板状，粒径1~3cm，含量<5%。基质成分：钾长石他形粒状，发育格子、卡式双晶，类型为条纹长石，粒径1~5mm，弱高岭土化，含量33%~37%；斜长石半自形板状，发育聚片双晶、环带结构，粒径1~3mm，含量33~38%；石英他形粒状，波状消光，粒径1~3mm，含量15%~20%；黑云母半自形片状，多色性，粒径1~2mm，含量约5%；角闪石半自形—他形粒状，发育解理，多色性，粒径0.1~0.6mm，含量1%~2%；副矿物为磁铁矿、锆石、磷灰石、榍石、钛铁矿、褐帘石，少量赤褐铁矿、黄铁矿等
	东罗庄	中粗粒花岗岩	粗中粒花岗结构、似斑状结构，块状构造。主要矿物组成为钾长石（40%~45%）、斜长石（20%~25%）、石英（30%~35%）及黑云母（<5%）；副矿物为磁铁矿、赤褐铁矿、白钛石、锆石、磷灰石、钍石、黄铁矿、钛铁矿等
	小盘山	黑云母花岗岩	细粒结构、似斑状结构，块状构造。斑晶矿物为斜长石（10%~15%）和钾长石（<5%），粒径约2mm。基质主要矿物组成为钾长石（35%~40%）、斜长石（20%~25%）和石英（25%），粒径0.2~1mm。副矿物磁铁矿、赤褐铁矿、锆石为主，少量磷灰石、黄铁矿、钍石、榍石、白钛石、钛铁矿等

2. 玻璃台岩体

位于平谷玻璃台，呈近EW向长条形产出，出露面积1.0km^2。岩体西侵入新太古界片麻岩中，接触面产状207°∠20°，接触界线模糊；其余周缘高角度外倾侵入中元古界长城系中，外接触带局部大理岩化、矽卡岩化。岩体岩性为灰黄色花岗斑岩，斑状结构，块状构造。斑晶石英他形粒状，粒径0.2~0.3mm，含量2%~3%；斑晶碱性长石他形粒状，粒径0.4~1.5mm，弱高岭土化，含量5%~6%；斑晶斜长石长柱状，粒径0.3~1.5mm，强烈绢云母化，含量1%~2%。基质霏细结构，由石英、碱性长石和斜长石组成，并有少量（含量<1%）黑云母和磁铁矿，可见石英交代碱性长石所形成的蠕虫状构造，矿物具不均匀高岭土化、绢云母化。

3. 东白山岩体

位于平谷河南寨村南，呈SN向不规则状产出。岩体南被北东向断裂切割破坏，向北侵入中元古界蓟县系雾迷山组中；东、西被中生代燕山期早白垩世潜火山岩或超浅成侵入岩侵入破坏。岩体岩性为斑状黑云母花岗岩和花岗斑岩，灰白色-浅粉红色，中细粒—细粒似斑状结构，块状构造。钾长石自形板状，粒径0.2~5mm，大者15mm形成斑晶，类型为条纹长石，含量45%~55%；斜长石短板状—粒状，发育聚片双晶，粒径0.2~4mm，大者10mm形成斑晶，类型为更-中长石，含量15%~50%；石英他形粒状，粒径0.1~5mm，边缘参差不齐，呈紧密镶嵌状，含量25%~30%；黑云母片状，多色性，粒径1mm，含量2%~6%。副矿物为磁铁矿、钛铁矿、磷灰石、榍石、锆石等。

4. 珠窝岩体

位于门头沟雁翅村，出露于印支期青白口穹隆核部，由一系列放射状岩脉组成，深部相通，组合发育为规模较大隐伏岩体。岩体与NE向张性断层关系密切，均侵入到中—新元古代及古生代地层，边部发育有宽约30m的破碎带，内发育闪长岩包体，并同中—新元古代地层一起发生后期变形。

岩体岩性为石英二长质花岗斑岩，呈灰黄-灰绿色，有时呈肉红色，致密块状构造，斑状结构。斑晶有斜长石（粒径1~1.5mm，含量10%~15%）、钾长石（粒径0.5~1.5mm，含量5%~10%）、石英（粒径0.8~1.2mm，含量10%~15%）。基质微晶结构，矿物主要为钾长石、石英，有时可见少量角闪石和黑云母。副矿物为磁铁矿、黄铁矿、磷灰石和锆石等。

（二）岩石化学与地球化学特征

岩石化学成分及特征参数见表20-23。岩石呈高SiO_2、碱，贫Al_2O_3、FeO^*、MnO特征。里特曼指数σ=1.68~3.17，为亚碱性系列；固结指数SI=1.80~11.21，分异指数DI=81.94~94.61，长英指数FL=76.09~93.33，岩浆经历强烈分异作用；A/CNK=0.09~1.22，A/NK=1.09~1.91，属于准铝质

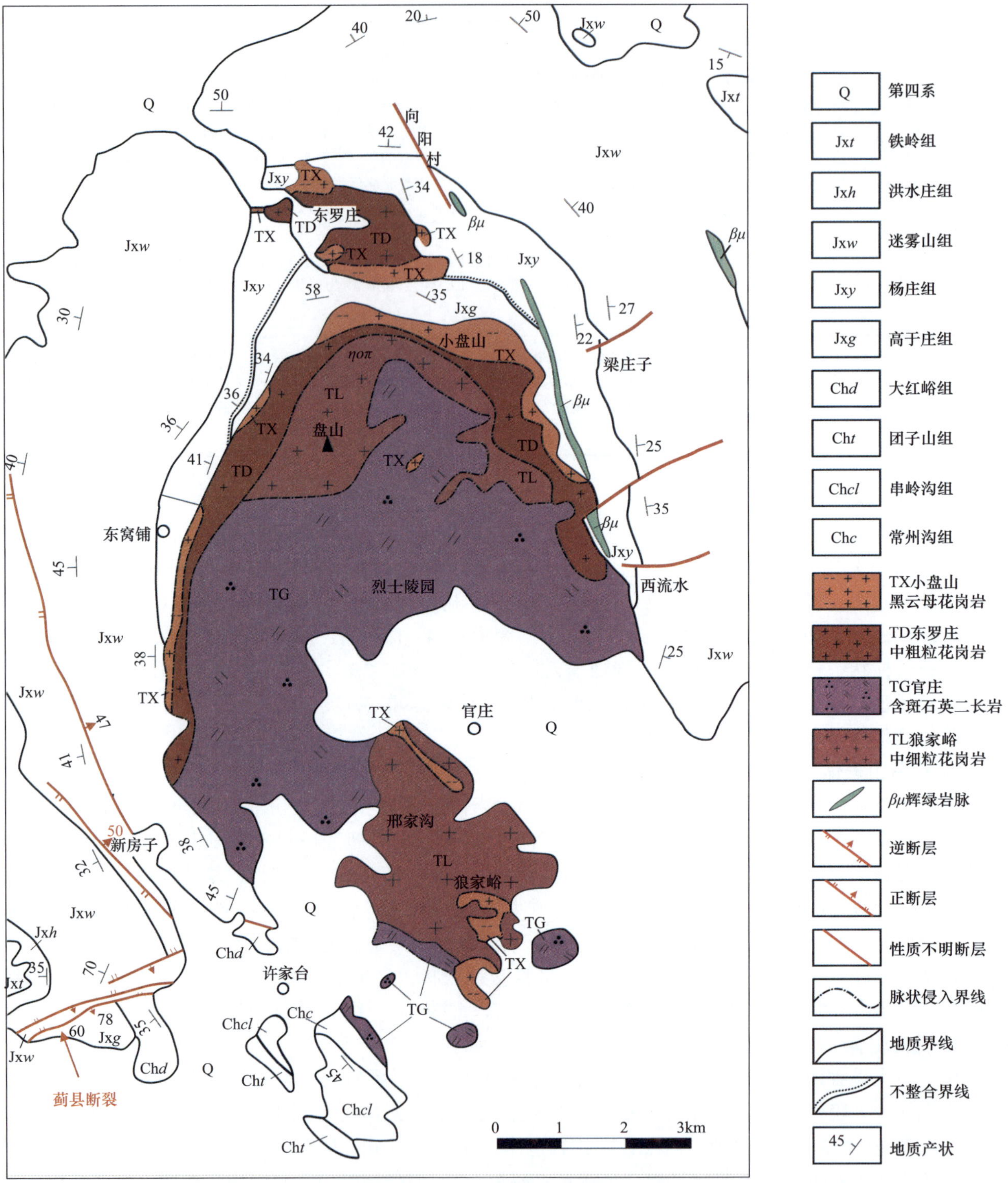

图 20-43　盘山岩体地质简图

（据 1∶5 万蓟县幅地质图，略有修改）

Fig. 20-43　The geological sketch map of the Panshan complex

向过铝质过渡。

稀土元素成分及特征参数见表 20-24。$\Sigma REE = 72.89 \times 10^{-6} \sim 166.76 \times 10^{-6}$；LREE/HREE = 5.31～19.74；$(La/Yb)_N = 3.24 \sim 29.34$；轻稀土弱富集，重稀土相对亏损；稀土总量酸性岩低于中性岩；$\delta Eu = 0.29 \sim 0.94$，呈现铕负异常；配分曲线为右缓倾光滑曲线（图 20-44）。

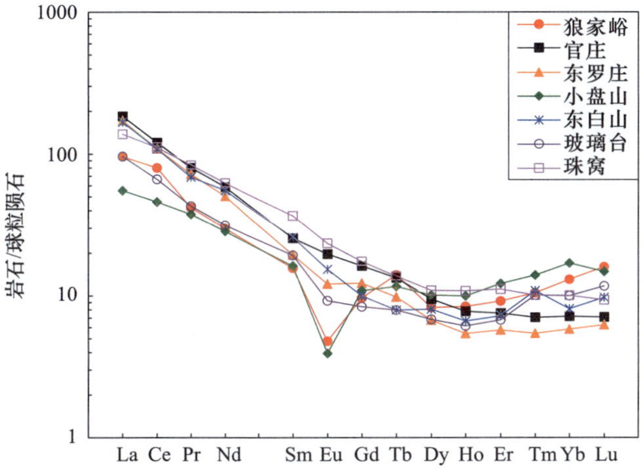

图 20-44　晚三叠世中酸性岩稀土配分模式图

Fig. 20-44　Chondrite-normalized REE patterns of Late Trassic intermediate-acid rocks

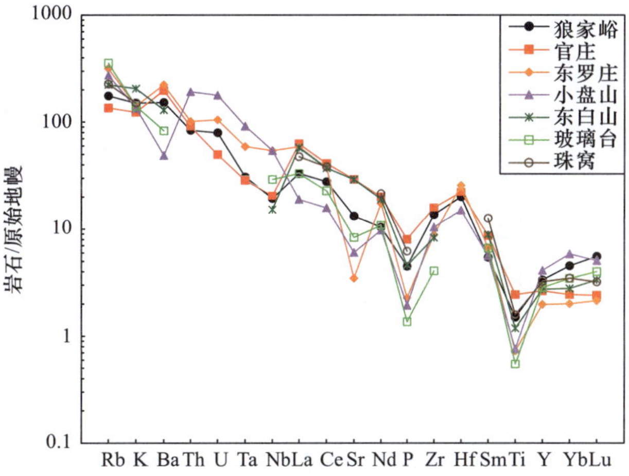

图 20-45　晚三叠世中酸性岩微量元素蛛网图

Fig. 20-45　Trace element spider diagram of Late Trassic intermediate-acid rocks

微量元素成分及特征参数见表 20-25。各岩体微量元素差异较大，揭示了岩浆经历了强结晶分异作用。微量元素蛛网图（图 20-45）中，富集大离子亲石元素 K、Rb，高场强元素 Zr、Hf、Th、U，相对亏损 Ti、P 和重稀土元素。元素 Ba、Nb、Ta 和 Sr 出现分化。盘山岩体中性岩和粗粒花岗岩出现极弱 Nb、Ta 亏损，细粒花岗岩则出现 Ba、Sr 相对弱亏损。

（三）岩石系列

该套岩系 $\sigma=1.68\sim3.17<4$，属于亚碱性岩系列。岩石富碱贫 FeO^*、MgO，属于钙碱性系列。SiO_2-K_2O 图解（图 20-46）中属于高钾钙碱性-钾玄岩系列。

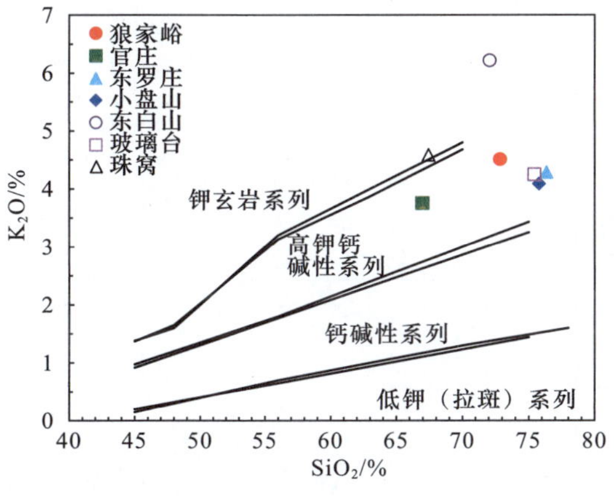

图 20-46　晚三叠世中酸性岩 SiO_2-K_2O 图解

Fig. 20-46　SiO_2-K_2O diagram of Late Trassic intermediate-acid rocks

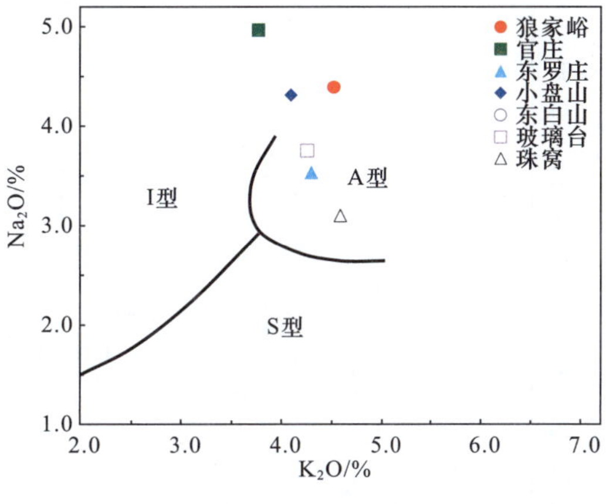

图 20-47　晚三叠世中酸性岩微量元素蛛网图

Fig. 20-47　Na_2O-K_2O diagram of Late Trassic intermediate-acid rocks

（四）Sr-Nd 同位素特征

盘山杂岩体不同岩性 Sr-Nd 同位素特征略有差异。$^{87}Sr/^{86}Sr=0.70538\sim0.783191$，$^{87}Rb/^{86}Sr=0.2971\sim28.92$，$^{87}Sr/^{86}Sr=0.70538\sim0.783191$，$^{147}Sm/^{144}Nd=0.08533\sim0.1186$，$^{143}Nd/^{144}Nd=0.511571\sim0.511722$；取 $t=205Ma$ 时，$\varepsilon_{Nd}(t)=-12.73\sim-15.67$，$t_{DM}=2070\sim1581Ma$（王玉富等，1988；杨富全等，2007）。

表 20-23 晚三叠世中酸性岩岩石化学成分及特征参数一览表

Table 20-23 The chemical composition and main parameters of Late Trassic intermediate-acid rocks

序号	岩体	岩性	氧化物含量/%													特征参数					
			SiO_2	TiO_2	Al_2O_3	Fe_2O_3	FeO	MnO	MgO	CaO	Na_2O	K_2O	P_2O_5	Los	总和	*DI*	*SI*	σ	*FL*	*A/CNK*	*A/NK*
1	狼家峪	斑状二长花岗岩	72. 87	0. 33	13. 30	0. 88	1. 01	0. 04	0. 74	1. 27	4. 40	4. 52	0. 10	0. 87	100. 33	91. 10	6. 41	2. 66	87. 53	0. 92	1. 10
2	官庄	斑状石英二长岩	66. 99	0. 54	15. 37	1. 64	1. 63	0. 06	1. 51	2. 75	4. 97	3. 77	0. 18	0. 84	100. 24	81. 17	11. 21	3. 17	76. 09	0. 89	1. 25
3	东罗庄	粗粒花岗岩	76. 36	0. 16	12. 50	0. 55	0. 89	0. 03	0. 17	0. 56	3. 54	4. 30	0. 05	0. 33	99. 44	93. 75	1. 80	1. 84	93. 33	1. 09	1. 19
4	小盘山	细粒花岗岩	75. 80	0. 17	12. 58	0. 65	0. 44	0. 04	0. 28	0. 73	4. 32	4. 10	0. 04	0. 76	99. 92	94. 61	2. 86	2. 15	92. 02	0. 98	1. 09
5	东白山	中细粒斑状黑云母花岗岩	72. 08	0. 26	15. 38	1. 37	0. 27	0. 07	0. 20	0. 62	0. 80	6. 23	0. 10	1. 32	98. 70	87. 82	2. 27	1. 68	91. 90	1. 67	1. 91
6	玻璃台	中粗粒斑状花岗岩	75. 44	0. 12	13. 48	0. 43	1. 06	0. 11	0. 41	0. 80	3. 76	4. 26	0. 03	0. 24	100. 14	91. 38	4. 13	1. 98	90. 93	1. 10	1. 25
7	珠窝	花岗斑岩	67. 42	0. 35	16. 17	2. 24	0. 99	0. 13	0. 73	1. 75	3. 10	4. 58	0. 14	1. 68	99. 28	81. 94	6. 35	2. 37	81. 48	1. 22	1. 61

注：1~4 数据为均值，引自杨富全等，2007；5 数据为本次实测；6 数据引自 1∶5 万木林幅区域地质调查报告，1996；7 为均值，数据引自 1∶5 万雁翅幅区域地质调查报告，1996。

表 20-24 晚三叠世中酸性岩稀土元素成分及特征参数一览表

Table 20-24 REE contents and main parameters of Late Trassic intermediate-acid rocks

序号	岩体	岩性	稀土元素含量/10^{-6}																特征参数			
			La	Ce	Pr	Nd	Sm	Eu	Gd	Tb	Dy	Ho	Er	Tm	Yb	Lu	Y	ΣREE	LREE/HREE	$(La/Yb)_N$	δEu	δCe
1	狼家峪	斑状二长花岗岩	22. 84	49. 20	4. 01	14. 06	2. 42	0. 28	2. 00	0. 53	2. 12	0. 48	1. 53	0. 27	2. 24	0. 41	15. 26	102. 39	9. 69	7. 31	0. 38	1. 16
2	官庄	斑状石英二长岩	43. 60	73. 60	7. 60	27. 40	3. 90	1. 14	3. 34	0. 50	2. 41	0. 44	1. 25	0. 18	1. 22	0. 18	12. 20	166. 76	16. 52	25. 63	0. 94	0. 91
3	东罗庄	粗粒花岗岩	40. 90	67. 20	6. 90	23. 80	3. 00	0. 71	2. 55	0. 37	1. 73	0. 31	0. 96	0. 14	1. 00	0. 16	9. 10	149. 73	19. 74	29. 34	0. 76	0. 90
4	小盘山	细粒花岗岩	13. 20	28. 40	3. 60	13. 40	2. 50	0. 23	2. 26	0. 44	2. 59	0. 57	2. 04	0. 36	2. 92	0. 38	18. 90	72. 89	5. 31	3. 24	0. 29	0. 99
5	东白山	中细粒斑状黑云母花岗岩	40. 04	66. 70	6. 57	26. 11	3. 99	0. 90	2. 07	0. 30	2. 06	0. 38	1. 21	0. 28	1. 39	0. 25	12. 60	152. 26	18. 17	20. 65	0. 86	0. 92
6	玻璃台	中粗粒斑状花岗岩	23. 00	40. 99	4. 12	14. 85	2. 98	0. 54	1. 74	0. 30	1. 75	0. 35	1. 13	0. 26	1. 72	0. 30	13. 12	94. 04	11. 45	9. 60	0. 67	0. 95
7	珠窝	花岗斑岩	32. 90	68. 50	7. 99	29. 40	5. 67	1. 37	3. 62	0. 52	2. 80	0. 62	1. 86	0. 26	1. 73	0. 24	14. 80	157. 48	12. 52	13. 64	0. 86	1. 00

注：数据来源同表 20-23。

表 20-25 晚三叠世中酸性岩微量元素成分及特征参数一览表

Table 20-25 Trace element contents and main parameters of Late Trassic intermediate-acid rocks

序号	岩体	岩性	微量元素含量/10^{-6}									特征参数	
			Rb	Ba	Th	U	Nb	Ta	Sr	Zr	Hf	Rb/Sr	Nb/Ta
1	狼家峪	斑状二长花岗岩	111.33	1063.00	7.13	1.67	1.26	13.80	279.33	152.67	6.21	0.40	0.09
2	官庄	斑状石英二长岩	87.00	1392.00	7.80	1.06	1.19	14.70	622.00	179.00	6.94	0.14	0.08
3	东罗庄	粗粒花岗岩	204.00	1573.00	8.70	2.23	2.45	39.00	74.00	100.00	8.00	2.76	0.06
4	小盘山	细粒花岗岩	175.33	346.00	16.50	3.77	3.79	39.00	129.00	118.00	4.71	1.36	0.10
5	东白山	中细粒斑状黑云母花岗岩	143.00	912.00				11.00	620.70	94.00		0.23	
6	玻璃台	中粗粒斑状花岗岩	230.00	586.50				21.00	177.60	46.00		1.30	

注：数据来源同表 20-23。

（五）成因类型与岩浆演化

该套岩系全岩 SiO_2 含量分布范围广，矿物成分也有所不同；在 Na_2O-K_2O 图（图20-47）上，该套岩系分散分布于 I 型花岗岩和 A 型花岗岩区；$(^{87}Sr/^{86}Sr)_i-\varepsilon_{Nd}(t)$ 图（图20-48）上，成分点也分散位于原始地幔和 EM Ⅰ间，更接近 EM Ⅰ区；盘山杂岩体露头尺度上还可见不同岩性间明显侵入接触关系。这些特征揭示了该套岩系侵入年代与成因类型的复杂性。

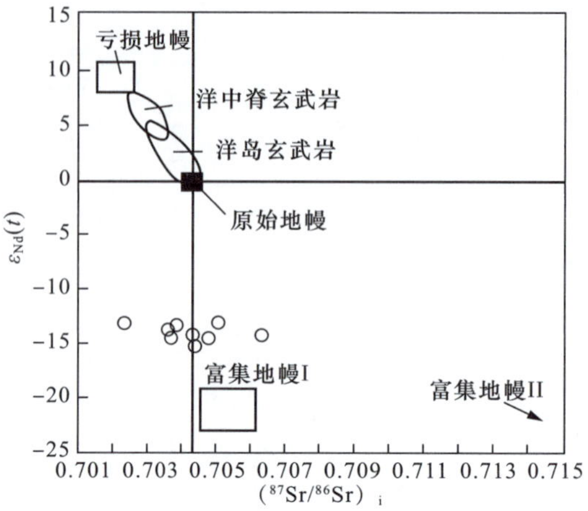

图20-48 晚三叠世中酸性岩 $(^{87}Sr/^{86}Sr)_i-\varepsilon_{Nd}(t)$ 图（据杨富全等，2007）

Fig. 20-48 $(^{87}Sr/^{86}Sr)_i-\varepsilon_{Nd}(t)$ diagram of Late Trassic intermediate-acid rocks

按照 SiO_2 含量可将该套岩系划分为中酸性和酸性岩，二者分属于 I 型花岗岩和 A 型花岗岩，形成于不同的两个岩浆房（陈一笠等，1993；杨富全等，2007）。I 型中酸性岩发育黑云母暗色矿物，显示 K、Rb、Sr、Ba 正异常，亏损 Nb、Ta，且在盘山、珠窝岩体中均发育暗色矿物包体，暗示其壳幔混合源成因，即岩浆起源与富集岩石圈地幔有关，同时也有古老下地壳源物质的贡献（杨富全等，2007）。A 型酸性岩少见暗色矿物，Sr、Ba、P、Ti 元素相对亏损，稀土元素配分曲线还显示岩浆形成时斜长石、角闪石等矿物的强分离结晶，在源区残留了大量的角闪石，造成中稀土元素亏损和 Lu 元素含量增高（杨富全等，2007），这些特征揭示了其岩浆源于特性不同的壳源。

（六）构造环境分析

该套岩系富 SiO_2、碱，贫 Fe、Mg；在 R1-R2 构造环境判别图解（图20-49）中，I 型中酸性岩位于同碰撞或晚造山区，A 型酸性岩位于造山后 A 型花岗岩区内。在（Y+Nb）-Rb 构造环境判别图解（图20-49）上，成分均位于碰撞后花岗岩区。因此，该套岩系记录了构造背景转换过程，即早期 I 型花岗岩记录了晚造山碰撞挤压构造环境，晚期的 A 型花岗岩则形成于造山后伸展的构造环境（杨富全等，2007）。

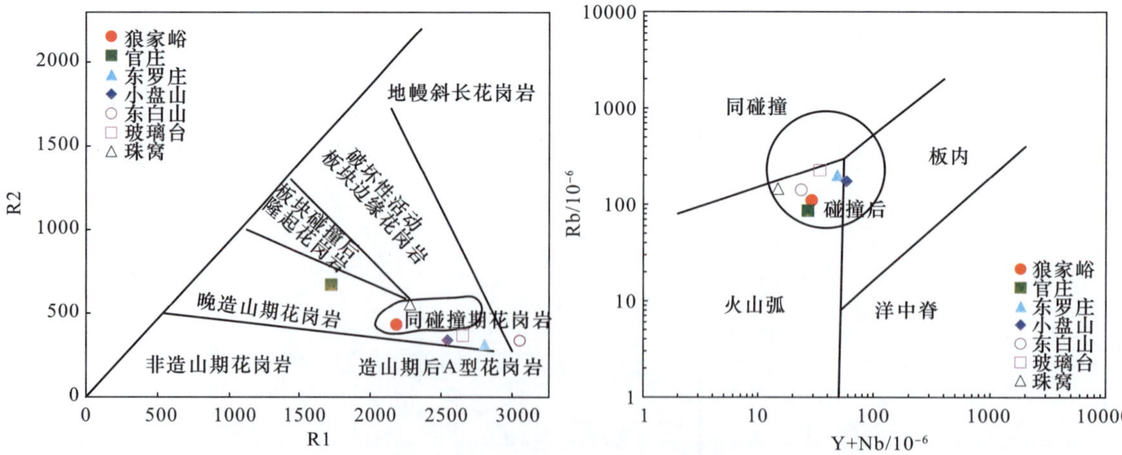

图20-49 晚三叠世中酸性岩 R1-R2 和（Y+Nb）-Rb 图解

Fig. 20-49 Tectonic discrimination diagram of R1-R2 and (Y+Nb) -Rb for Late Trassic intermediate-acid rocks

（七）时代归属

本期岩浆侵位时代厘定为晚三叠世，主要依据有：①岩体未遭受后期变质变形仅岩性发生弱蚀变；②具有造山后A型花岗岩特性；③侵入晚古生代地层中；④珠窝岩体伴生印支期青白口穹隆构造变形同步侵位；⑤盘山杂岩体石英二长岩锆石U-Pb同位素年龄（208±4）Ma至（217.1±2）Ma，花岗岩锆石U-Pb同位素年龄（203±5）Ma至（222.7±1.5）Ma（陈一笠等，1991；马寅生等，2007；叶浩，2014）；盘山岩体附近花岗岩脉锆石U-Pb同位素年龄为（213±4）Ma；全岩或矿物K-Ar、Ar-Ar同位素年龄和Rb-Sr同位素年龄与此相当（叶伯丹等，1986；王玉富等，1989；吴珍汉等，1999）。

四、早侏罗世中酸性侵入岩

（一）地质产状与岩石学特征

发育大石坡-黑山坨、老君堂2大岩体，出露面积27km²。此外，在门头沟青白口、桑峪等地发育一系列中性、基性岩脉群（称之为青白口岩脉群），也是早侏罗世重要一期岩浆侵位活动（详细见第十九章第三节）。

1. 大石坡-黑山坨岩体

位于昌平大石坡，八达岭侵入杂岩带西南缘，呈长轴NNE向椭圆状产出，出露面积18km²（图20-50）。岩体中部被中生代燕山期断裂切割，周缘侵入到太古宙变质岩和中元古界蓟县系高于庄组中，接触面外侵，倾角约60°。

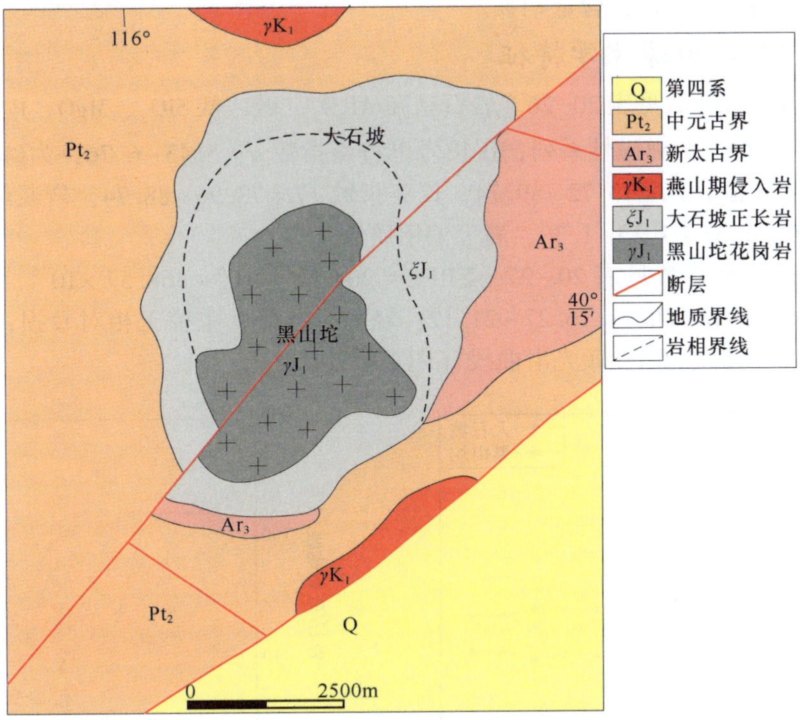

图20-50　大石坡-黑山坨岩体地质简图

（据汪洋等，2015，略有修改）

Fig. 20-50　The Geological sketch map of Dashipo-Heishantuo complex

岩体由大石坡正长岩和黑山坨花岗岩两部分组成（图20-50）。前者构成岩体外环，由外向内依次发育为角闪石正长岩和黑云母正长岩；后者构成岩体中心，发育为花岗岩。大石坡正长岩和黑山坨花岗岩呈侵入接触，接触界线清晰，内接触带花岗岩中发育冷凝边和正长岩捕虏体，外接触带正长岩中发育梳状花岗岩枝。

大石坡正长岩内还发育暗色和浅色两类椭圆或不规则状包体，大部分直径2~5cm，大者可达20~30cm。正长岩灰白色，中粗粒不等粒结构，块状构造。碱性长石自形-他形板柱状，发育环带，部分包裹黑云母、不透明矿物微晶，含量63%~64%；角闪石他形-半自形柱粒状，粒径较大呈斑晶态产出，含量2.5%~3%；黑云母半自形-他形片状，部分绿泥石化，含量0%~11%，向岩体中心含量增加；斜长石他形粒状，含量约18%；石英他形粒状，充填于长石间隙，含量约2%；其他不透明矿物多以包裹体形态分布于碱性长石内，含量约2.5%。副矿物为榍石、磁铁矿、磷灰石、锆石等，多包裹于暗色矿物中。

黑山坨花岗岩为灰白色，中-细粒等粒结构，块状构造。石英他形粒状，粒径1~3mm，含量30%~35%；钾长石自形板柱状，粒径1~3mm，个别7mm，发育条纹结构，含量30%~55%；斜长石半自形粒状，粒径1~2mm，发育聚片双晶，含量15%~35%；镁质黑云母半自形片状，含量<3%。副矿物为榍石、锆石、磷灰石、磁铁矿等。

2. 老君堂岩体

位于昌平定陵北、黄泉寺、老君堂一带，呈半隐伏产出，出露面积3km^2。岩体南侧被第四纪覆盖，其余周缘侵入于中元古界雾迷山组中，接触面产状变化较大，宏观向SE侧伏。岩体侵位可能受控于NE向庆陵-龙虎台断裂和NW向上寺-上天桥断裂构造，分布于二组断裂交切部位。岩体岩性为石英正长岩，浅紫灰色—灰褐色，中粒—中细粒结构。钾长石他形粒状，粒径0.2~4mm，类型为条纹长石，黏土化、绢云母化，含量70%~90%；斜长石半自形柱状，粒径0.2~3mm，含量10%~15%；石英他形粒状，粒径0.5~2mm，分布不均匀，含量5%~10%；暗色矿物角闪石和黑云母含量1%~3%；副矿物为榍石、磷灰石、磁铁矿，含量<1%。

（二）岩石化学与地球化学特征

岩石化学成分及特征参数见表20-26。岩石呈高Al_2O_3、碱，低SiO_2、MgO、P_2O_5、MnO特征。花岗岩里特曼指数$\sigma=2.66$，为亚碱性系列，正长岩里特曼指数$\sigma=5.43\sim6.76$，为碱性系列，固结指数$SI=5.17\sim13.18$，分异指数$DI=73.72\sim89.74$，长英指数$FL=73.96\sim88.74$，岩浆经历较强分异作用；$A/CNK=0.87\sim1.06$，$A/NK=1.13\sim1.30$，属于偏铝质。

稀土元素成分及特征参数见表20-27。$\Sigma REE=204.38\times10^{-6}\sim368.57\times10^{-6}$，总量较高；LREE/HREE=12.80~16.66；$(La/Yb)_N=16.27\sim21.12$；轻稀土富集，重稀土相对亏损；$\delta Eu=0.74\sim0.86$，呈现铕弱负异常；配分曲线为右陡倾光滑曲线（图20-51）。

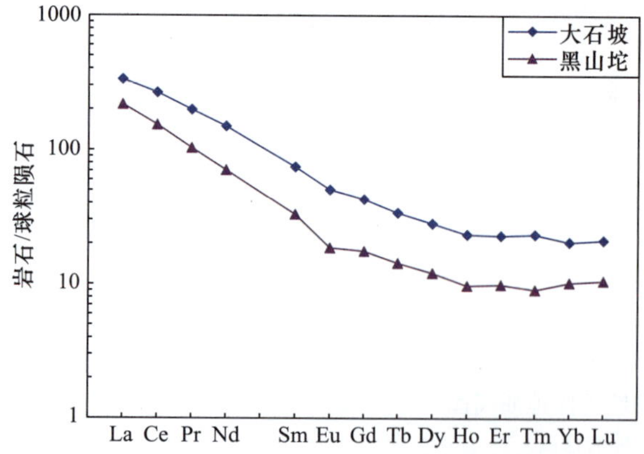

图20-51 早侏罗世中酸性岩稀土配分模式图

Fig. 20-51 Chondrite-normalized REE patterns of Early Jurassic intermediate-acid rocks

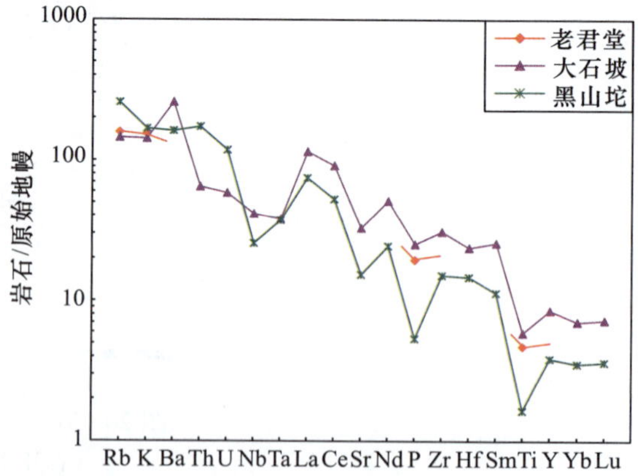

图20-52 早侏罗世中酸性岩微量元素蛛网图

Fig. 20-52 Trace element spider diagram of Early Jurassic intermediate-acid rocks

微量元素含量及特征参数见表20-28。各岩体微量元素含量相当，特征相似。大离子亲石元素中性岩低于酸性岩，其余元素正好相反。微量元素蛛网图（图20-52）中，曲线形态基本相似，富集大离子亲

表 20-26　早—中侏罗世中酸性岩岩石化学成分及特征参数一览表

Table 20-26　The chemical composition and main parameters of Early to Middle Jurassic intermediate-acid rocks

序号	岩体	岩性	氧化物含量/%													特征参数					
			SiO_2	TiO_2	Al_2O_3	Fe_2O_3	FeO	MnO	MgO	CaO	Na_2O	K_2O	P_2O_5	Los	总和	*DI*	*SI*	*σ*	*FL*	*A/CNK*	*A/NK*
1	老君堂	石英正长岩	60.35	1.02	17.33	3.06	1.92	0.15	1.22	2.56	6.32	4.60	0.42	1.13	100.08	82.00	7.15	6.76	81.01	0.87	1.13
2	大石坡	正长岩	58.53	1.27	16.86	3.65	1.83	0.12	2.24	3.30	5.11	4.26	0.54	1.60	99.32	73.72	13.18	5.43	73.96	0.89	1.30
3	黑山坨	花岗岩	71.12	0.35	14.30	1.28	0.64	0.04	0.58	1.10	3.67	5.01	0.12	1.40	99.61	89.74	5.17	2.66	88.74	1.06	1.25
4	抱榆沟	二长花岗岩	75.33	0.03	12.97	0.34	1.25	0.01	0.19	0.84	4.32	4.28	0.02	0.11	99.67	93.29	1.83	2.28	91.10	0.98	1.11
5	四道穴	二长花岗岩	73.40	0.08	13.86	0.73	1.51	0.01	0.18	1.14	4.80	3.70	0.02	0.15	99.58	90.77	1.65	2.37	88.17	0.99	1.16

注：数据为均值；1 引自 1∶5 万雁翅幅区域地质调查报告，1996；2、3 引自焦永玲，2014；4、5 引自 1∶5 万杨木栅子幅、三道营子幅、喇叭沟门幅区域地质调查报告，1999。

表 20-27　早—中侏罗世中酸性岩稀土元素成分及特征参数一览表

Table 20-27　REE contents and main parameters of Early to Middle Jurassic intermediate-acid rocks

序号	岩体	岩性	稀土元素含量/10^{-6}																特征参数			
			La	Ce	Pr	Nd	Sm	Eu	Gd	Tb	Dy	Ho	Er	Tm	Yb	Lu	Y	ΣREE	LREE/HREE	$(La/Yb)_N$	δEu	δCe
1	大石坡	正长岩	78.50	161.29	18.70	69.14	11.32	2.90	8.72	1.25	7.09	1.31	3.75	0.59	3.46	0.53	38.44	368.57	12.80	16.27	0.86	1.00
2	黑山坨	花岗岩	51.06	93.20	9.74	32.74	5.00	1.07	3.58	0.53	3.05	0.55	1.63	0.23	1.73	0.27	17.50	204.38	16.66	21.12	0.74	0.96
3	抱榆沟	二长花岗岩	4.93	8.90	1.31	5.48	1.98	0.33	2.27	0.35	2.15	0.43	1.18	0.17	1.09	0.14	12.51	30.71	2.95	3.24	0.47	0.84
4	四道穴	二长花岗岩	9.34	27.13	3.31	15.01	5.56	0.23	6.72	1.26	8.68	2.18	6.36	1.02	7.35	1.22	63.28	95.37	1.74	0.91	0.11	1.19

注：数据为均值，来源同表 20-26。

表 20-28　早—中侏罗世中酸性岩微量元素成分及特征参数一览表

Table 20-28　Trace element contents and main parameters of Early to Middle Jurassic intermediate-acid rocks

序号	岩体	岩性	微量元素含量/10^{-6}									特征参数	
			Rb	Ba	Th	U	Nb	Ta	Sr	Zr	Hf	Rb/Sr	Nb/Ta
1	大石坡	正长岩	91.29	1790.57	5.47	1.22	29.34	1.55	687.00	343.29	7.29	0.13	18.91
2	黑山坨	花岗岩	161.66	1120.80	14.61	2.46	18.08	1.52	320.00	168.00	4.51	0.51	11.93
3	抱榆沟	二长花岗岩		90.93	6.55				113.90	23.44			
4	四道穴	二长花岗岩		757.10	4.60				320.10	14.89			

注：数据为均值，来源同表 20-26。

石元素 Rb、Ba、K，相对亏损元素 Sr、P、Ti，弱亏损 Nb、Ta。正长岩还表现出相对亏损 Th、U 元素。

（三）岩石系列

该套岩系中，花岗岩 $\sigma=2.66<4$，为亚碱性岩系列，正长岩 $\sigma=5.43\sim6.76>4$，为碱性岩系列，与判别图解（图 20-53）揭示结果一致。在 K_2O-SiO_2 图解（图 20-54）中，花岗岩为高钾钙碱性系列，正长岩、石英正长岩为钾玄岩系列。

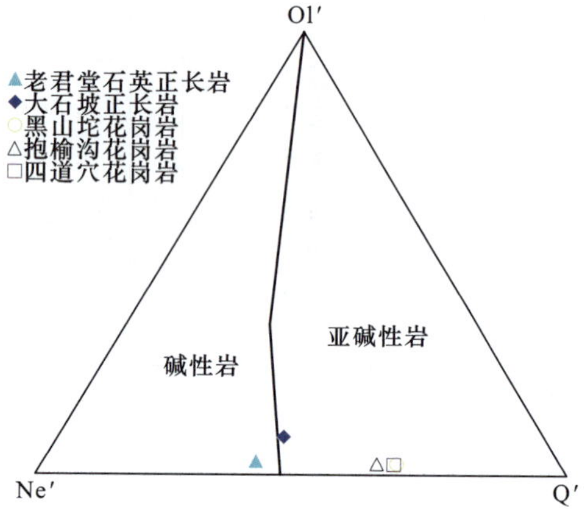

图 20-53　早—中侏罗世中酸性岩 Ol'-Ne'-Q' 图解

Fig. 20-53　Ol'-Ne'-Q' diagram of Early to Middle Jurassic intermediate-acid rocks

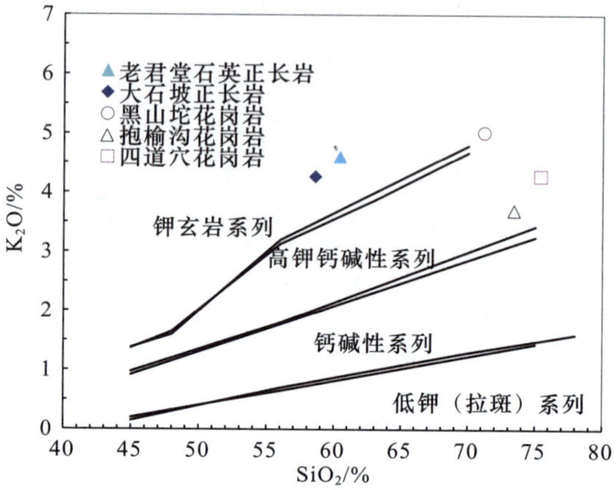

图 20-54　早—中侏罗世中酸性岩 SiO_2-K_2O 图解

Fig. 20-54　SiO_2-K_2O diagram of Early to Middle Jurassic intermediate-acid rocks

（四）Sr-Nd 同位素特征

大石坡正长岩和黑山坨花岗岩 Sr-Nd 同位素组成相近。正长岩 $Sm=11.38\times10^{-6}\sim11.69\times10^{-6}$，$Nd=66.40\sim70.30$，$^{147}Sm/^{144}Nd=0.100579\sim0.103689$，$^{143}Nd/^{144}Nd=0.511892\sim0.511901$；$Sr=661.4\sim714.8$，$Rb=87.2\sim108.9$，$^{87}Rb/^{86}Sr=0.381344\sim0.441031$，$^{87}Sr/^{86}Sr=0.70586\sim0.70614$；取 $t=200Ma$ 时，$\varepsilon_{Nd}(t)=-12.1\sim-12.2$，$t_{DM}=1730\sim1690Ma$；花岗岩 $Sm=4.03\times10^{-6}\sim5.71$，$Nd=26.74\times10^{-6}\sim37.92\times10^{-6}$，$^{147}Sm/^{144}Nd=0.091158\sim0.091258$，$^{143}Nd/^{144}Nd=0.511581\sim0.511711$；$Sr=308.1\times10^{-6}\sim345.4\times10^{-6}$，$Rb=147.2\times10^{-6}\sim152.1\times10^{-6}$，$^{87}Rb/^{86}Sr=1.233830\sim1.429339$，$^{87}Sr/^{86}Sr=0.708603\sim0.709918$；取 $t=201Ma$ 时，$\varepsilon_{Nd}(t)=-15.39\sim-17.91$，$t_{DM}=1950\sim1790Ma$（焦永玲，2014）。

（五）Pb 同位素特征

大石坡正长岩和黑山坨花岗岩 Pb 同位素组成也相近，都表现出放射性 Pb 显著亏损的特征。正长岩 $^{206}Pb/^{204}Pb=16.23\sim16.27$，$^{207}Pb/^{204}Pb=15.20\sim15.22$，$^{208}Pb/^{204}Pb=36.24\sim36.46$；花岗岩 $^{206}Pb/^{204}Pb=16.22\sim16.37$，$^{207}Pb/^{204}Pb=15.20\sim15.22$，$^{208}Pb/^{204}Pb=36.30\sim36.45$（汪洋等，2015）。

（六）成因类型与岩浆演化

在 $\varepsilon_{Nd}(t)-(^{87}Sr/^{86}Sr)_i$ 图（图 20-55）中，大石坡正长岩位于 EMI 型富集地幔源区，黑山坨花岗岩与之接近，但更靠近华北下地壳；$(^{207}Pb/^{204}Pb)_t-(^{206}Pb/^{204}Pb)_t$ 和 $(^{208}Pb/^{204}Pb)_t-(^{206}Pb/^{204}Pb)_t$ 图（图 20-56）中，正长岩与花岗岩均位于华北克拉通岩石圈地幔和下地壳间，更接近下地壳区。Sr-Nd-Pb 同位素揭示了大石坡-黑山坨杂岩体岩浆来源于富集地幔镁质富钾基性岩浆。相比于大石坡正长岩，黑山坨花岗岩具有更低 Nb/Ta 值和 $(^{87}Sr/^{86}Sr)_i$ 值，高 Th、U 含量和 Th/U 值，还揭示黑山坨花岗岩不是正长岩浆分离结晶的产物（汪洋等，2015），而形成于幔源富钾镁质基性岩浆与壳源岩石的反应熔融过程，表现为地壳源岩的成分特征体。

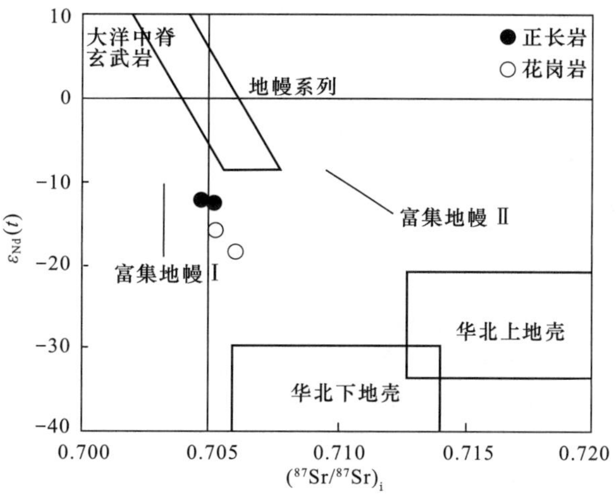

图 20-55 大石坡-黑山坨杂岩体 ε_{Nd} (t) ($^{87}Sr/^{87}Sr$)$_i$ 图

（据汪洋等，2015）

Fig. 20-55 ε_{Nd} (t) - ($^{87}Sr/^{86}Sr$)$_i$ diagram of Dashipo-Heishantuo complex

大石坡-黑山坨杂岩体富集 LILE、亏损 HFSE，且随着 SiO_2 含量的增加，MgO、FeO*、CaO、TiO_2、MnO、P_2O_5 含量和 CaO/Al_2O_3 比值逐渐降低，V、Ni、Cr 等过渡族元素含量及 Mg#降低，K_2O、Ba 含量及 K_2O/Na_2O、La/Sm、Eu/Eu* 比值增加，也较好揭示了其岩浆来源于富集地幔的镁质富钾基性岩浆，还揭示了岩浆演化过程中经历了橄榄石、辉石、磷灰石、钛铁矿等矿物分离结晶作用。大石坡正长岩中出现微粒包体等揭示岩浆还发生混合现象。

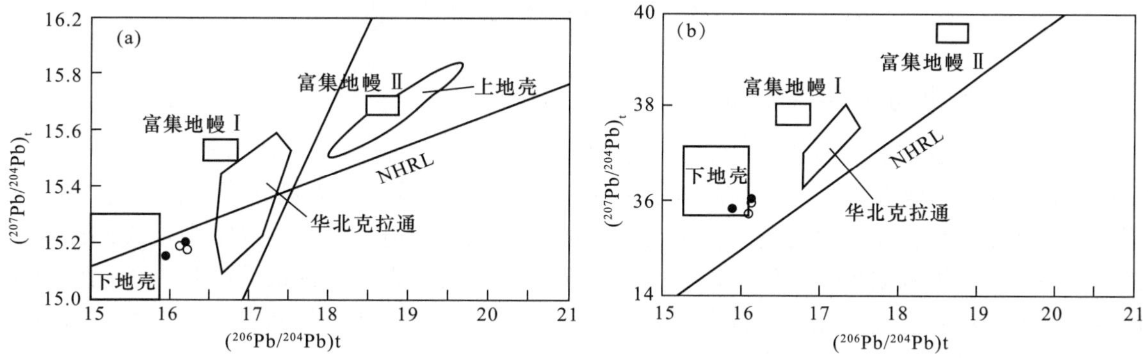

图 20-56 大石坡-黑山坨杂岩体 ($^{207}Pb/^{204}Pb$)$_t$- ($^{206}Pb/^{204}Pb$)$_t$ 和 ($^{208}Pb/^{204}Pb$)$_t$- ($^{206}Pb/^{204}Pb$)$_t$ 图

（据焦永玲，2014）

Fig. 20-56 ($^{207}Pb/^{204}Pb$)$_t$- ($^{206}Pb/^{204}Pb$)$_t$ and ($^{208}Pb/^{204}Pb$)$_t$- ($^{206}Pb/^{204}Pb$)$_t$ diagram of Dashipo-Heishantuo complex

（七）构造环境分析

基于角闪石温压计算，岩体侵位深度约 18km，时处于较厚上地壳状态（焦永玲，2014）。构造地质学研究表明，由于华北板块与西伯利亚板块碰撞（古亚洲洋闭合），三叠纪时期燕辽地区发生过一次强烈的上盘向南的推覆运动，导致地壳增厚（Davis et al.，2001；Wang，2013）。变形与岩浆作用研究表明华北克拉通北缘中生代早期（晚三叠世）已进入造山后构造、岩浆活动相对平静期（蒋小洁，2019）。该套岩系发育碱性岩和钙碱性岩系列，在 Rb-（Y+Nb）和 R1-R2 构造环境判别图解（图 20-57）上，成分点位于同碰撞隆起或碰撞后隆起花岗岩区，揭示了本期岩浆形成于华北板块和西伯利亚板块碰撞后相对稳定构造背景环境，处于非造山背景下伸展构造环境。

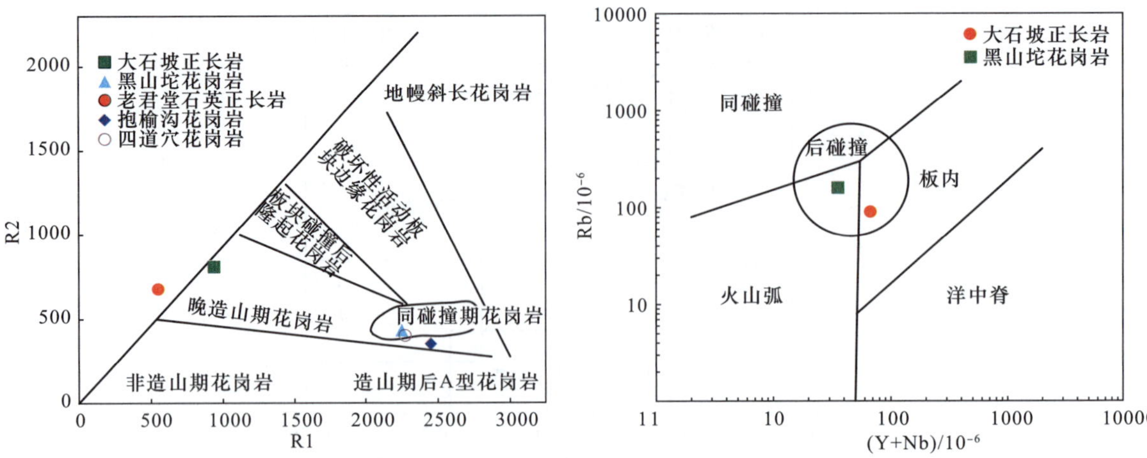

图 20-57　早—中侏罗世中酸性岩 R1-R2 和（Y+Nb）-Rb 图解

Fig. 20-57　Tectonic discrimination diagram of R1-R2 and（Y+Nb）-Rb of Early to Middle Jurassic intermediate-acid rocks

（八）时代归属

本期岩浆侵位时代厘定为早侏罗世，主要依据有：①岩体未遭受变质和变形；②侵位到中元古界地层中；③岩体被中生代燕山期北东向断裂切割破坏；④大石坡正长岩、花岗岩锆石 U-Pb 同位素年龄分别为（200±1）Ma、（196.5±1.9）Ma 和（201±1）Ma（Deng et al., 2004；焦永玲，2014）；⑤侵位于非造山稳定构造背景环境。

五、中侏罗世酸性侵入岩

（一）地质产状与岩石学特征

发育抱榆沟、四道穴和红石山 3 个岩体，位于怀柔北安营、石洞子、转山子，呈不规则状产出，境内出露面积 23km²。抱榆沟、红石山岩体分别向河北延伸，统称之为大栅子岩体，境内小范围出露，其南侵入中生代早三叠世田营北沟花岗闪长岩中，边部发育冷凝边，其余周缘被中生代燕山期早白垩世岩体侵入破坏。四道穴岩体南东侵入古生代早二叠世辉长闪长岩中，其余周缘被中生代燕山期早白垩世侵入岩侵入破坏。岩体岩性为二长花岗岩，肉红色，中细粒结构，块状构造。石英他形粒状，粒径 0.1~3.5mm，含量 22%~25%；碱性长石他形粒状，类型为微斜长石（主）和正长石（次），粒径 0.5~4.5mm，含量 35%~45%；斜长石半自形-他形粒状，发育聚片双晶，含量 30%~40%；黑云母棕色片状，粒径 0.2~1mm，含量约 1%；白云母半自形片状，粒径 0.2~1mm，含量约 0.2%。副矿物为磁铁矿和磷灰石等。

（二）岩石化学与地球化学特征

岩石化学成分及特征参数见表 20-26。岩石呈高 SiO_2、碱，低 MgO、Fe_2O_3、MnO 特征。里特曼指数 σ=2.28~2.37，为亚碱性系列；固结指数 *SI*=1.65~1.83，分异指数 *DI*=90.77~93.29，长英指数 *FL*=88.17~91.10，岩浆经历强烈分异作用；*A/CNK*=0.98~0.99，*A/NK*=1.11~1.16，属于偏铝质。

稀土元素成分及特征参数见表 20-27。ΣREE=30.71×10^{-6}~95.37×10^{-6}，总量低；LREE/HREE=1.74~2.75；$(La/Yb)_N$=0.91~3.24；轻重稀土极弱分馏作用；δEu=0.11~0.47，呈现铕强负异常；配分曲线为“海鸥式”近平坦曲线（图 20-58）。

微量元素含量及特征参数见表 20-28。各微量元素含量低于同类岩类。微量元素蛛网图（图 20-59）中，富集大离子亲石元素 Rb、Ba、K，相对亏损高场强元素 P、Ti。

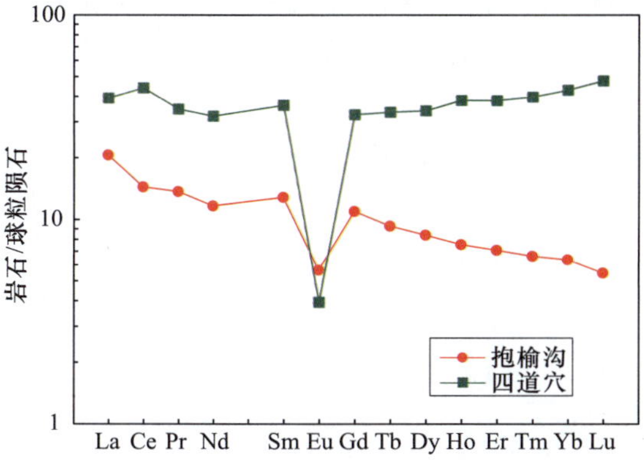

图 20-58　中侏罗世酸性岩稀土配分模式图

Fig. 20-58　Chondrite-normalized REE patterns of Middle Jurassic acid rocks

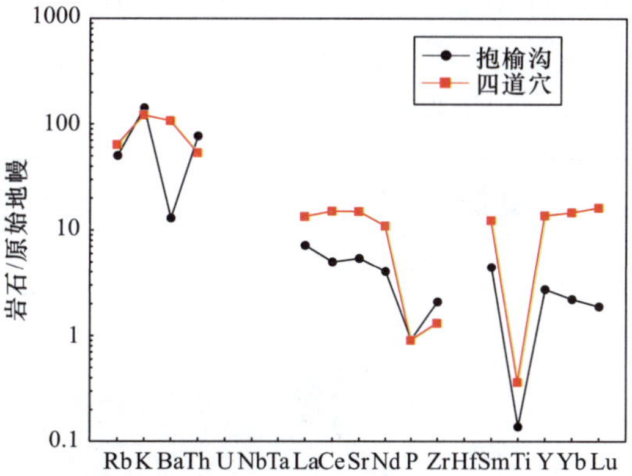

图 20-59　中侏罗世酸性岩微量元素蛛网图

Fig. 20-59　Trace element spider diagram of Middle Jurassic acid rocks

（三）岩石系列

该套岩系 σ=2.28~2.37，为亚碱性系列，与判别图解（图 20-53）揭示结果一致。在 K_2O-SiO_2 图解（图 20-54）中，为高钾钙碱性系列。

（四）成因类型与构造环境

本套岩系为富 SiO_2、碱，贫 CaO、MgO、TiO_2 的高钾钙碱性系列，具有显著的负 Eu 异常和典型的“海鸥式”稀土元素配分曲线，同典型 A 型酸性火山岩特征相似。轻重稀土弱分馏揭示了原始岩浆中缺乏橄榄石、斜方辉石和单斜辉石，又矿物组成中含富水白云母、黑云母，说明其岩浆更可能来源于地壳玄武安山质或碎屑沉积岩物质熔融。

在 R1-R2 构造环境判别图解（图 20-57）中，成分点位于造山后 A 型花岗岩区，揭示了岩体侵位时，华北板块和西伯利亚板块碰撞造山结束，区内处于相对稳定构造背景环境，表现为非造山背景下伸展构造背景环境。

（五）时代归属

本期岩浆侵位时代厘定为中侏罗世，主要依据有：①岩体未遭受变质和变形及明显蚀变现象；②侵位到古生代早二叠世侵入体中，又被中生代燕山期系列侵入体切割破坏；④与邻区河北省大栅子岩体相邻，具有完全一致地质与地球化学特性。大栅子二长花岗岩锆石 U-Pb 同位素年龄为（168±3）Ma 和（172±2）Ma（《中国区域地质志・河北志》，2017）；⑤岩体规模较小，岩浆活动强度较弱，侵位于非造山构造背景环境。

六、晚侏罗世中酸性侵入岩

晚侏罗世中酸性侵入岩区内广泛分布，发育岩体数量众多，单个岩体规模也较大，宏观呈 NE 向产出，同早白垩世侵入岩一起构成境内最重要的中生代八达岭侵入杂岩带。

（一）地质产状与岩石学特征

发育长园-官山-沙峪、石城-云蒙山、四干顶-曹庄子、平义分、沙岭、玉皇山、果庄、南峪、花塔等系列岩体。下面对典型岩体地质产状与岩石学特征总结如下：

1. 长园-官山-沙峪杂岩体

位于昌平区与怀柔区交界处，侵入于黑坨山背斜南翼和三渡河倒转向斜轴部，切割破坏了下李家村-神堂峪村 NE 向推覆构造体系。岩体内被中生代早白垩世西后峪细粒花岗岩体侵入穿插，平面形态呈长轴走向 NEE 不规则环形产出。其中，内环为沙峪斑状花岗岩，出露面积 $23km^2$；中环发育为官山石英二长岩，出露面积 $18km^2$；外环为不完整半环状，仅在南部出露，为长园石英二长闪长岩，出露面积 $56km^2$（图 20-60）。

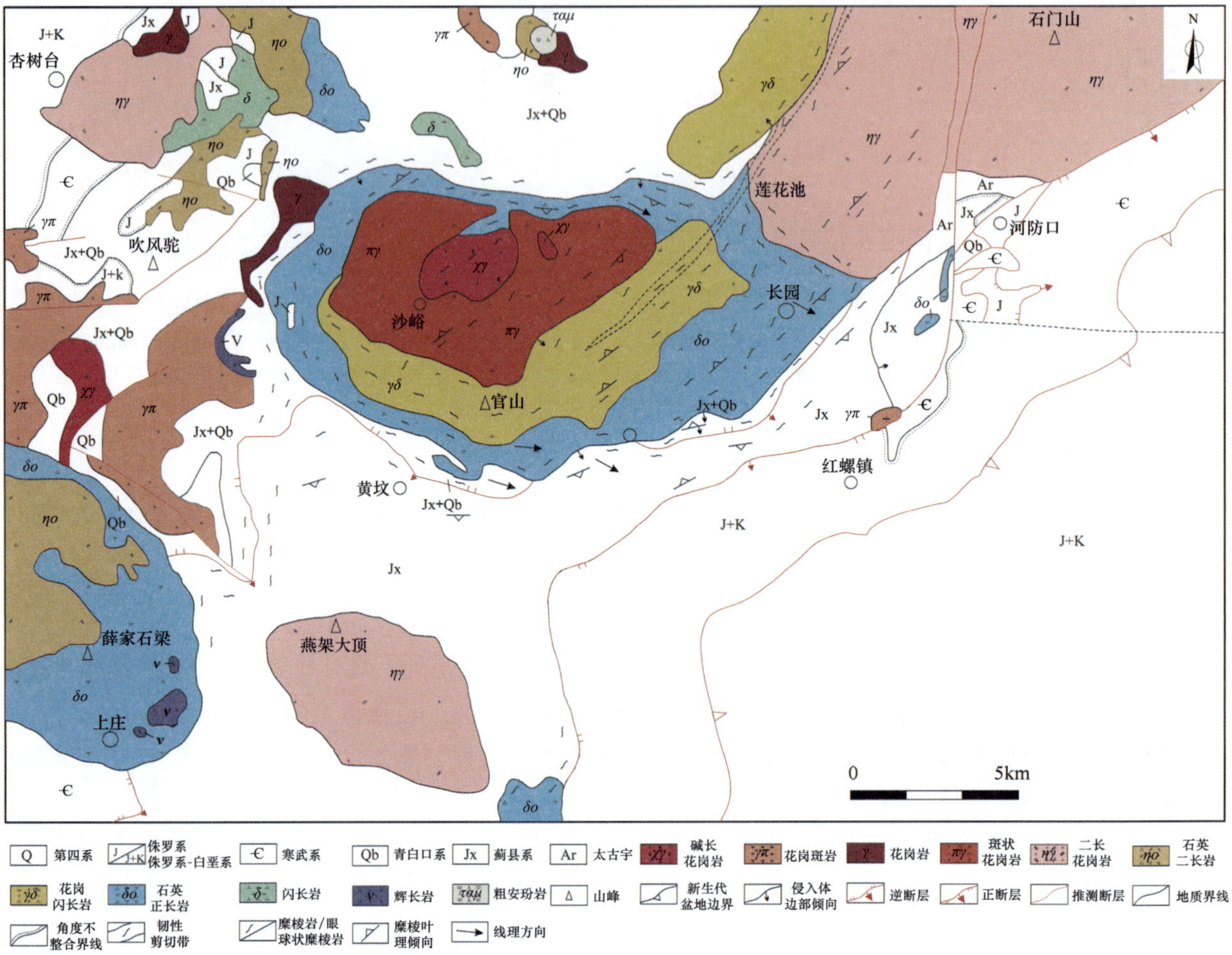

图 20-60　长园-官山-沙峪杂岩体地质简图

（据汪洋等，2004，略有修改）

Fig. 20-60　The geological sketch map of the Changyuan-Guanshan-Shayu complex

长园石英闪长岩东、中部强烈变形，发育片麻理构造；南、北内接触带发育糜棱岩带。岩体东北部受云蒙山杂岩体侵位和后期莲花池-苇店韧性剪切影响，片麻理还发生偏转或被切割。岩体边部普遍发育镁铁质暗色微粒包体，包体同寄主岩一起发生压扁拉长。岩体岩性为石英二长闪长岩，灰色-深灰色，中细粒结构，块状构造。斜长石板柱状，粒径 1~2mm，少数达 4mm，发育聚片双晶、波状消光，含量 60%~65%；钾长石半自形粒状，粒径 0.5~2mm，发育卡式双晶，含量约 10%；石英他形粒状，粒径 0.2~1mm，波状消光，含量约 10%；黑云母半自形粒状，粒径 0.2~1mm，局部可见晶棱弯曲、膝折现象，含量约 10%；角闪石，粒径 0.2~0.5mm，普遍被黑云母交代保留角闪石晶形，发育解理，含量约 5%。

官山石英二长岩呈向南凸出不对称环包围沙峪斑状花岗岩体。岩体岩性同长园岩体岩性呈渐变过渡接触。岩性为石英二长岩，灰色，中细粒结构，块状构造。斜长石发育聚片双晶，粒径 0.5~2mm，含量约 35%；条纹长石他形粒状，粒径 0.5~2mm，含量 30%~35%；石英他形粒状，粒径 0.2~1mm，波状消光，含量 10%~15%；黑云母半自形片状，粒径 0.5~1mm，含量 15%~20%。

沙峪斑状花岗岩位于杂岩体内环，与长园石英二长闪长岩、官山石英二长岩界线清晰侵入接触，内接触带发育 5~10cm 冷凝边，并可见长园石英二长闪长岩捕虏体。岩性为斑状花岗岩，粉红色，中细粒似斑状结构，局部糜棱结构，块状构造，局部片麻状构造。钾长石半自形粒状，粒径 0.1~4mm，波状消光，类型为条纹长石，内嵌斜长石、磁铁矿、石榴子石、石英矿物；斜长石半自形板状，粒径 0.1~4mm，波状消光，部分晶纹弯曲；石英半自形—自形，粒径 0.5~1mm，强波状消光，集合体或定向排列，内嵌尘点状微粒；黑云母半自形片状，粒径 1~2mm，被绿泥石交代呈残余状；石榴子石自形粒状，粒径 0.1~0.2mm，被磁铁矿包裹成“黑眼圈”状。

2. 石城-云蒙山杂岩体

位于怀柔区和密云区北交界处，由 15 个大小不等岩体组合而成杂岩体，平面呈长轴 NE 向卵圆形产出，构成 NNE 向云蒙山背形（又称云蒙山短轴背斜）的核部，出露面积 300km^2。岩体普遍不同程度发育片麻状构造，内发育未变质变形网、脉状伟晶岩、花岗岩、闪长玢岩脉和少量辉绿岩脉，局部发育强韧性变形带。岩体周缘侵入新太古界变质岩和中元古界地层中，内接触带局部可见围岩捕虏体。岩体岩性主要为闪长岩类和花岗岩类（表 20-29）。

表 20-29　石城-云蒙山杂岩体主要岩石类型及其特征

Table 20-29　Main rock types and its characteristics of Shicheng-Yunmengshan complex

序号	岩体期次单元	岩性	结构	构造	主要矿物及含量/%					
					斜长石	钾长石	石英	黑云母	角闪石	辉石
1	三神庙	微细粒二长花岗岩	微细粒变晶结构、糜棱结构	微片麻状构造	20~25	30~39	20~35	2~6	<1	
2	五道河	中细粒斑状花岗闪长岩	中细粒变晶结构、糜棱结构	微片麻状构造	41.86	21.45	31.66	3.35	1~2	
3	大石门子	中细粒含斑花岗闪长岩	中细粒变晶结构、糜棱结构	眼球状构造、片麻状构造	44.77	21.82	29.45	4.5	1~2	
4	云蒙山	中细粒少斑二长花岗岩	中细粒变晶结构、糜棱结构	眼球状构造、片麻状构造	20~30	20~30	25~35	3~8	1~2	
5	官地	中细粒斑状二长花岗岩	中细粒变晶结构、糜棱结构	眼球状构造、片麻状构造	32	35	30	3~8		
6	大水峪	中细粒多斑二长花岗岩	中细粒变晶结构、糜棱结构	眼球状构造、片麻状构造	30~40	36	30~40	3~8		
7	沙坨子	中细粒少斑二长花岗岩	中细粒变晶结构、糜棱结构	眼球状构造、片麻状构造	35	32	25	5~8	1~2	
8	石城	辉石闪长岩	中细粒变晶结构、糜棱结构	眼球状构造、片麻状构造、窗棂构造	47~65		2~6	14~31	15~30	5

（据姬广义等，2004）

3. 四干顶杂岩体

位于密云韩庄、白龙潭、四干顶、龙潭沟，发育为 NNE 向椭圆形复式杂岩体，出露面积 43km^2。岩体中部被中生代四干顶断裂切割破坏。依据岩石类型及其相互间穿切关系等，可将岩体划分为四个岩性侵位单元：白龙潭石英（二长）闪长岩、曹庄子石英二长岩、四干顶二长花岗岩和小豹子将花岗斑岩，具体岩性特征见表 20-30。

表 20-30　四干顶杂岩体主要岩石类型及其特征

Table 20-30　Main rock types and its characteristics of Siganding complex

序号	岩性期次单元	岩性	结构构造	斜长石	碱性长石	石英	黑云母	角闪石	副矿物
1	白龙潭	石英闪长岩	半自形柱粒状结构，块状构造。	半自形板状、宽板状、板条状，粒径0.8~1.8mm，中长石，发育环带结构，绢云母化，含量约73%	他形粒状，类型正长石和条纹长石，粒径0.4~0.8mm，交代斜长石，含量约6%	他形粒状，粒径0.3~0.6mm，熔蚀现象，含量约7%	棕褐色，半自形片状或条形，粒径0.5~1.0mm，含量约5%	绿色柱状，粒径0.5~1.0mm，含量约7%	磁铁矿，粒径0.1~0.3mm，含量约2%；榍石、磷灰石、绿帘石等
2	白龙潭	石英二长闪长岩	中粒花岗结构，块状构造。	半自形板状、粒状，粒径1.0~3.2mm，类型为中长石，强烈绢云母化，发育钠长、卡钠复合双晶、环带和反条纹结构，含量50%~60%	板状，类型为正长石、条纹长石，粒径1.0~2.0mm，交代、吞蚀包裹斜长石，黏土化，含量20%~25%	他形粒状，粒径0.4~1.6mm，被溶蚀现象，含量8%~12%	红褐—淡黄色，半自形条状，粒径0.5~1.8mm，绿泥石化，含量4%~5%	绿色柱状，粒径0.5~2.0mm，黑云母化、绿泥石化、次闪石化，含量3%~6%	磁铁矿，粒径0.1~0.3mm，含量约2%；榍石、磷灰石、帘石及透辉石等
3	曹庄子	石英二长岩	中细粒花岗结构，块状构造。	自形板状、板条形，粒径0.5~2.0mm，类型为更长石，发育钠长、复合双晶，绢云母化、钠黑黝帘石化，含量35%~40%	他形粒状或半自形板状，类型为正长石、条纹长石，粒径0.5~3.0mm，黏土化，发育格子双晶，交代斜长石，含量38%~43%	他形粒状，粒径0.4~2.0mm，被熔蚀现象，含量15%~18%	黄褐色，自形板条形，粒径0.4~0.7mm，含量约2%	浅绿色自形柱状，粒径0.2~0.5mm，绿泥石化、黑云母化，1%~3%	磁铁矿，粒径0.1~0.3mm，含量约1%；榍石、磷灰石、锆石、白云母、绿帘石等
4	四干顶	二长花岗岩	中粒半自形粒状结构，块状构造。	半自形板状、宽板状，粒径1.2~3.2mm，类型为更-中长石，绢云母化、钾化，发育钠长双晶，含量约35%	他形粒状，类型为正长石、条纹长石，粒径1.0~2.5mm，高岭土化，交代斜长石，含量约33%	他形粒状，粒径0.5~2.0mm，波状消光，含量约22%	红褐条形，粒径1.0~1.8mm，含量约6%	绿色柱状，粒径1.0~2.0mm，含量约3%	榍石、磁铁矿、磷灰石等
5	小豹子将	花岗斑岩	斑状结构，块状构造	自形板状，粒径0.5~2.5mm，发育双晶，轻度绢云母化，含量约5%	他形粒状或半自形板状，类型为条纹长石，粒径0.5~2.5mm，强黏土化，含量约6%	自形—半自形粒状，粒径1.0~2.5mm，被熔蚀现象，含量约9%	基质微晶结构，粒径<0.04mm，部分交代呈细小叶片状白云母和尘状铁质物，含量约80%，副矿物为磁铁矿等		

白龙潭石英（二长）闪长岩单元呈半环状分布于四干顶杂岩体南西边部，出露面积11km²。单元北东被曹庄子石英二长岩单元脉动侵入，形成宽10~20m渐变过渡带，其余周缘侵入新太古代变质岩和中元古界长城系中，与中元古界接触的外接触带内发育梳状岩脉，并形成宽700m的石英岩化、角岩化或大理岩化热变质带，同太古宙变质岩接触的外接触带发育宽30~50m的混染岩化带。

曹庄子石英二长岩单元呈宽1~2.5km椭圆环形产出，出露面积25km²。单元中部被四干顶二长花岗岩单元侵入，渐变过渡接触关系；北侧侵入新太古代片麻岩中，接触界线清晰，内接触带还发育宽约 50m 混染岩化带；南侧与白龙潭石英（二长）闪长岩单元渐变过渡接触。

四干顶二长花岗岩单元组成四干顶杂岩体中心，呈长轴 NE 向不规则状，出露面积 $8km^2$。单元侵入曹庄子石英二长岩单元中部，呈渐变过渡关系。雷家沟、小木魁峪一带发育石英（二长）闪长岩残留体，岩性同白龙潭单元。

小豹子将花岗斑岩单元由 2 个长轴近 SN 向椭圆形和蝌蚪形小岩株组成，出露面积 $0.3km^2$。单元侵入于曹庄子石英二长岩单元，地貌、颜色等界线清晰，内接触带可见石英二长岩捕虏体，外接触带发育梳状岩枝。

4. 沙岭岩体

位于密云前沙岭、后沙岭，呈长轴 NE 向椭圆形，出露面积 $2.4km^2$。岩体南侧侵入中元古界长城系地层中，接触带未见明显接触变质现象；其余周缘侵入于新太古代变质岩中，接触界线清晰，发育宽 10～50m 混染岩化带。岩性为闪长岩和石英闪长岩，岩石灰色，中细粒结构，块状构造。斜长石半自形板状，发育净边、蠕英、反条纹结构，绢云母化，含量约 70%；钾长石他形粒状，多交代斜长石成条纹长石，还见正长石，含量约 5%；石英他形粒状，含量约 6%；角闪石半自形柱状，不均匀分布，含量约 15%；黑云母半自形片状，与角闪石伴生产出，含量约 4%。副矿物为钛铁矿、磷灰石、锆石、榍石、黄铜矿、黄铁矿等。

（二）岩石化学与地球化学特征

岩石化学成分及特征参数见表 20-31。岩石 SiO_2 含量变化范围大，呈高 Al_2O_3、碱，低 MgO、P_2O_5、FeO^*、MnO 特征。里特曼指数 $\sigma=2.00\sim4.15$，跨碱性与亚碱性岩区；固结指数 $SI=2.27\sim21.86$，分异指数 $DI=59.21\sim95.87$，长英指数 $FL=57.52\sim97.53$，岩浆分异程度较强；$A/CNK=0.82\sim1.26$，$A/NK=1.07\sim1.71$，跨准铝质和过铝质岩区。同一杂岩体，伴随岩浆侵位演化，SiO_2、K_2O、Na_2O+K_2O 含量逐渐增高，Al_2O_3 及其他氧化物含量均逐渐降低；且随 SiO_2 含量的增高，莱特碱度率 *AR*、分异指数 *DI*、长英指数 *FL* 及 K_2O/Na_2O 比值均随之增高，反映岩浆向富硅、富碱、贫铝方向演化，且由富钠贫钾向富钾贫钠方向演化。

稀土元素成分及特征参数见表 20-32。中—酸性岩 ΣREE 分别为 $128.97\times10^{-6}\sim240.96\times10^{-6}$ 和 $48.75\times10^{-6}\sim178.85\times10^{-6}$；LREE/HREE 分别为 14.60～35.28 和 14.93～40.63；$(La/Yb)_N$ 分别为 20.60～64.16 和 18.91～122.36；轻稀土强烈富集，重稀土相对亏损；稀土总量酸性岩低于中性岩；中、酸性岩 δEu 分别为 0.85～1.08 和 0.66～1.01，呈铕弱正异常至负异常，揭示了岩浆演化经历了斜长石的分离结晶作用；配分曲线均为右陡倾光滑曲线（图 20-61）。同一杂岩体，伴随岩浆侵位演化，δEu、La/Sm 增大，LREE/HREE、Ce/Yb、La/Yb、Eu/Sm 值逐渐减小，揭示岩浆向重稀土相对富集、轻重稀土分馏程度减弱的方向演化。

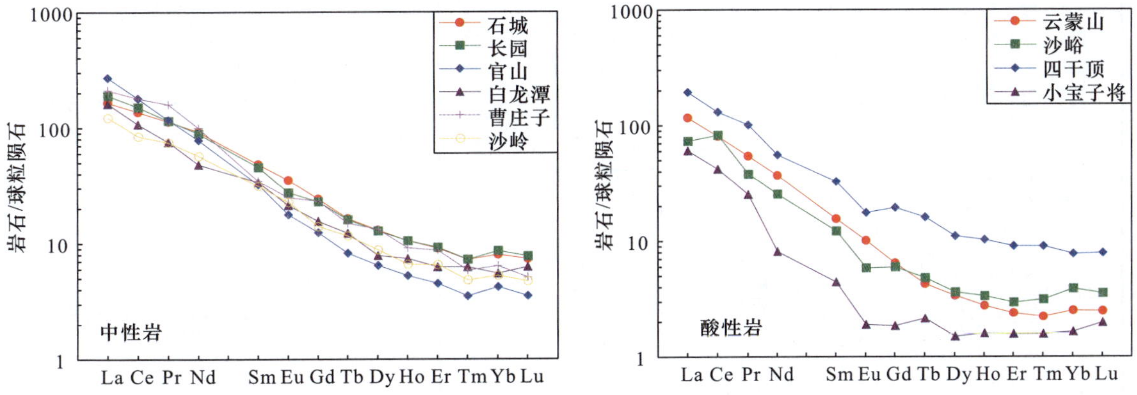

图 20-61　晚侏罗世中酸性岩稀土元素配分模式图

Fig. 20-61　Chondrite-normalized REE patterns of Late Jurassic intermediate-acid rocks

表 20-31　晚侏罗世中酸性岩岩石化学成分及特征参数一览表

Table 20-31　The chemical composition and main parameters of Late Jurassic intermediate-acid rocks

序号	岩体	岩性	氧化物含量/%													特征参数					
			SiO_2	TiO_2	Al_2O_3	Fe_2O_3	FeO	MnO	MgO	CaO	Na_2O	K_2O	P_2O_5	Los	总和	*DI*	*SI*	σ	*FL*	*A/CNK*	*A/NK*
1	石城	闪长岩	55.50	1.02	18.41	4.68	2.69	0.10	2.82	5.35	5.04	2.31	0.54	1.23	99.68	59.21	16.23	4.15	57.86	0.90	1.71
2	云蒙山	花岗岩	70.97	0.16	15.64	1.36	0.13	0.03	0.27	1.46	4.86	3.93	0.07	0.59	99.45	89.46	2.55	2.74	85.75	1.05	1.28
3	长园	石英闪长岩	58.38	1.03	16.21	3.31	3.02	0.10	3.81	5.14	4.47	2.89	0.44	1.05	99.81	61.67	21.86	3.44	58.86	0.82	1.55
4	官山	石英二长岩	63.11	0.74	16.28	2.25	2.38	0.07	1.88	3.52	4.49	3.78	0.23	0.65	99.38	73.86	12.74	3.35	70.14	0.91	1.42
5	沙峪	花岗岩	70.87	0.28	14.69	1.16	0.78	0.06	0.51	1.59	4.52	3.92	0.08	0.78	99.22	88.08	4.70	2.53	84.18	1.01	1.26
6	白龙潭	石英闪长岩	66.04	0.54	15.00	2.73	1.83	0.12	1.41	3.37	4.42	3.32	0.23	0.58	99.59	76.55	10.34	2.58	69.67	0.88	1.38
7	曹庄子	石英二长岩	67.26	0.50	15.22	1.60	2.05	0.06	1.33	2.62	4.17	4.19	0.21	0.22	99.43	80.12	9.97	2.86	76.14	0.94	1.34
8	四干顶	二长花岗岩	68.08	0.49	14.66	1.30	2.00	0.16	0.92	1.84	3.98	4.87	0.12	0.83	99.25	84.28	7.04	3.09	82.79	0.97	1.24
9	小宝子将	花岗斑岩	73.02	0.12	12.32	1.31	0.65	0.01	0.30	2.01	2.40	5.60	0.02	1.33	99.09	88.65	2.94	2.11	79.92	0.90	1.23
10	沙岭	石英闪长岩	60.59	0.78	16.61	2.91	3.11	0.13	2.27	4.32	4.84	3.21	0.33	0.18	99.27	68.74	13.93	3.63	65.08	0.86	1.45
11	果庄	花岗岩	69.08	0.39	15.09	1.39	1.65	0.06	1.07	2.25	4.32	4.17	0.15	0.11	99.73	83.05	8.49	2.76	79.05	0.96	1.30
12	南峪	花岗斑岩	75.64	0.12	12.78	0.27	0.24	0.01	0.20	0.64	0.40	7.70	0.02	1.05	99.07	92.90	2.27	2.00	92.68	1.26	1.42
13	花塔	花岗岩	75.31	0.17	12.14	1.08	0.46	0.06	0.27	0.21	4.33	3.95	0.05	1.33	99.36	95.87	2.69	2.11	97.53	1.03	1.07

注：数据均为均值；31 自 1∶5 万琉璃庙、范各庄幅区域地质调查报告，1990；1∶5 万密云、高岭、墙子路幅区域地质调查报告，1994。

表 20-32　晚侏罗世中酸性岩稀土元素成分及特征参数一览表

Table 20-32　REE contents and main parameters of Late Jurassic intermediate-acid rocks

序号	岩体	岩性	稀土元素含量/10^{-6}																特征参数			
			La	Ce	Pr	Nd	Sm	Eu	Gd	Tb	Dy	Ho	Er	Tm	Yb	Lu	Y	ΣREE	LREE/HREE	$(La/Yb)_N$	δEu	δCe
1	石城	闪长岩	39.20	83.98	10.79	43.20	7.44	2.05	5.00	0.62	3.32	0.60	1.51	0.19	1.37	0.19	15.98	199.43	14.60	20.60	1.03	1.00
2	云蒙山	花岗岩	27.75	49.50	5.18	17.33	2.39	0.59	1.33	0.16	0.86	0.16	0.40	0.06	0.43	0.06	5.01	106.18	29.81	46.47	1.01	1.01
3	长园	石英闪长岩	45.18	92.52	10.97	41.68	7.00	1.60	4.72	0.60	3.26	0.60	1.54	0.19	1.47	0.20	16.14	211.51	15.84	22.05	0.85	1.02
4	官山	石英二长岩	64.40	110.00	11.15	36.70	4.95	1.03	2.56	0.31	1.65	0.30	0.75	0.09	0.72	0.09	8.00	234.70	35.28	64.16	0.88	1.01
5	沙峪	花岗岩	17.40	50.60	3.62	11.90	1.87	0.34	1.23	0.18	0.92	0.19	0.49	0.08	0.66	0.09	5.70	89.57	22.33	18.91	0.69	1.56
6	白龙潭	石英闪长岩	38.40	66.00	7.20	22.60	5.20	1.24	3.20	0.46	2.00	0.42	1.04	0.16	0.94	0.16	10.60	149.02	16.78	29.30	0.93	0.97
7	曹庄子	石英二长岩	50.09	110.10	15.19	46.74	5.33	1.45	4.79	0.56	3.37	0.52	1.45	0.15	1.09	0.13	14.56	240.96	18.98	32.96	0.88	0.98
8	四干顶	二长花岗岩	46.00	80.00	9.60	26.00	5.00	1.02	4.00	0.60	2.80	0.58	1.50	0.23	1.32	0.20	16.20	178.85	14.93	25.00	0.70	0.93
9	小宝子将	花岗斑岩	14.40	25.80	2.40	3.80	0.68	0.11	0.38	0.08	0.38	0.09	0.26	0.04	0.28	0.05	3.60	48.75	30.25	36.89	0.66	1.08
10	沙岭	石英闪长岩	29.00	51.83	7.12	26.72	4.81	1.32	2.89	0.44	2.24	0.37	1.09	0.12	0.90	0.12	12.77	128.97	14.79	23.09	1.08	0.88
11	南峪	花岗斑岩	29.00	57.00	5.20	16.00	2.60	0.30	1.05	0.13	0.72	0.13	0.26	0.20	0.17	0.05	2.94	112.81	40.63	122.36	0.56	1.14

注：数据来源于同表 20-31。

微量元素含量及特征参数见表 20-33。微量元素 Rb 含量中性岩均低于酸性岩，Sr 元素正好相反，其余元素含量富集或分散特征不强。微量元素蛛网图（图 20-62）中，中、酸性岩曲线基本相似，富集大离子亲石元素 Rb、Sr、Ba、K，相对强烈亏损元素 Nb、重稀土元素，弱亏损元素 Tu、U、Ta。酸性岩还表现出相对亏损元素 P、Ti。同一杂岩体，伴随岩浆侵位演化，V、Ga、Cr、Sr、Nb、Ta 含量减少，Rb/Sr、Nb/Ta 比值增大。

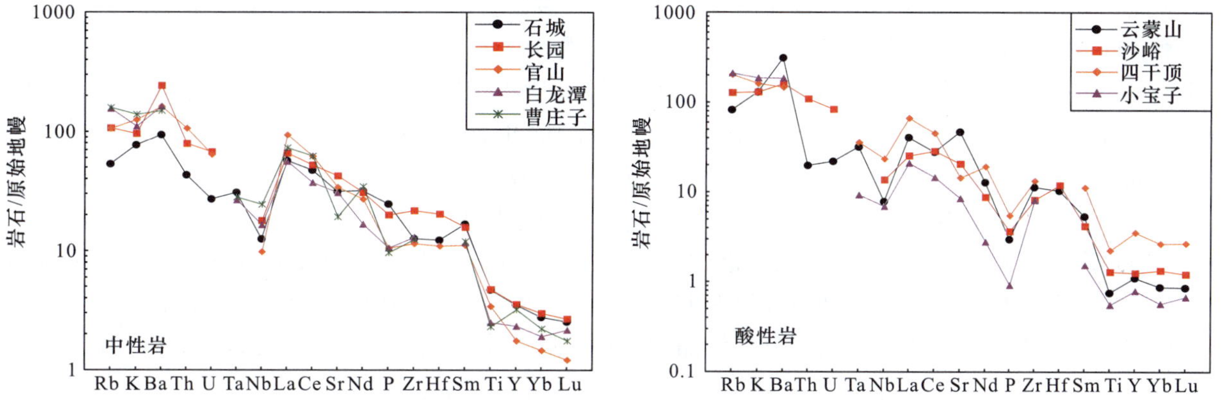

图 20-62　晚侏罗世中酸性岩微量元素蛛网图

Fig. 20-62　Trace element spider diagram of Late Jurassic intermediate-acid rocks

表 20-33　晚侏罗世中酸性岩微量元素成分及特征参数一览表

Table 20-33　Trace element contents and main parameters of Late Jurassic intermediate acidity rock

序号	岩体	岩性	微量元素含量/10^{-6}									特征参数	
			Rb	Ba	Th	U	Nb	Ta	Sr	Zr	Hf	Rb/Sr	Nb/Ta
1	石城	闪长岩	33.81	657.14	3.67	0.57	8.97	1.26	662.93	141.86	3.81	0.05	7.10
2	云蒙山	花岗岩	52.40	2180.00	1.68	0.46	5.60	1.30	983.00	127.00	3.20	0.05	4.31
3	长园	石英闪长岩	67.88	1691.60	6.70	1.41	12.74		895.60	243.20	6.29	0.08	
4	官山	石英二长岩	67.30	1640.00	9.03	1.34	7.00		718.00	129.00	3.40	0.09	
5	沙峪	花岗岩	81.33	1125.33	9.28	1.76	9.77		434.33	92.00	3.70	0.19	
6	白龙潭	石英闪长岩	98.76	1132.00			11.74	1.09	646.60	145.49		0.15	10.77
7	曹庄子	石英二长岩	100.70	1043.00			17.40	1.15	405.15	139.44		0.25	15.13
8	四干顶	二长花岗岩	129.00	1026.00			16.70	1.46	305.00	149.56		0.42	11.44
9	小宝子将	花岗斑岩	133.70	1295.70			4.94	0.38	179.10	90.99		0.75	13.00
10	沙岭	石英闪长岩	54.30	1491.00			16.66	1.56		68.24			10.68
11	南峪	花岗斑岩	172.00	807.30			0.56	0.13	76.76	128.60		2.24	4.31

注：数据来源同表 20-31。

（三）岩石系列

该套岩系 σ=2.00~4.15，仅石城闪长岩 σ=4.15>4 外，其余岩体 σ<4。石城-云蒙山杂岩体经过后期变形作用，特别是石城闪长岩，岩浆侵位还经历强烈岩浆分异作用，岩浆分异可能使得熔浆中元素丰度和元素比值发生改变，全岩化学分析不能代表原始熔体的组成。石城闪长岩富碱高钾、低镁等特征会使得其岩石系列判别向碱性岩区偏离，岩石仍表现出钙碱性特征。在 Ol’-Ne’-Q’ 判别图解（图 20-63）上，成分点也位于钙碱性区。因此，该套岩系为钙碱性系列。在 SiO_2-K_2O 图解（图 20-63）上，成分点位于高钾钙碱性-钾玄岩系列。

（四）Sr-Nd 同位素特征

石城闪长岩和云蒙山二长花岗岩$^{87}Sr/^{86}Sr=0.70493\sim0.70601$，$^{143}Nd/^{144}Nd=0.511681\sim0.511653$；取 $t=159$Ma 时，ε_{Nd}（t）$=-16.89\sim-17.16$，$t_{DM}=1742\sim2123$Ma（赵美涛等，2014）。

（五）成因类型与岩浆演化

该套岩系富硅，属于偏铝质—过铝质的高钾钙碱性系列，酸性岩普遍含角闪石矿物，而无原生富铝的矿物，如白云母、堇青石等，也无碱性暗色矿物，多数岩体 $Na_2O>3.2\%$，$A/CNK<1.1$，亏损高场强元素，同高钾钙碱性 I 型花岗岩特征一致，岩浆来源于未经受变质的火成岩再次熔融（康月蓝等，2018）。该系列岩体还多表现为脉动式多期次岩浆侵位，形成闪长岩-石英二长岩-花岗岩的中酸性岩组合。中性岩形成时代稍早，岩石 SiO_2 含量低，较高 $Mg^{\#}$值，$A/CNK<1$，Rb/Sr 值接近于地幔值；酸性岩形成时代稍晚，SiO_2 含量高，具有较低 $Mg^{\#}$值，$A/CNK>1$，Rb/Sr 值接近于下地壳值；在（$^{87}Sr/^{86}Sr$）$_i$-ε_{Nd}（t）图（图 20-64）中，成分点位于富集 I 型地幔和下地壳之间，酸性岩接近下地壳；在 SiO_2-MgO 图（图 20-65）中，成分点分散位于俯冲洋壳熔融形成的埃达克质岩和增厚下地壳熔融形成的埃达克质岩区。因此，中性岩和酸性岩两期岩浆作用反映了岩浆物质源区由地幔/下地壳物质向地壳物质转化，前者来源于壳幔基性岩浆分离结晶的产物，后者是华北克拉通下地壳部分熔融的产物。

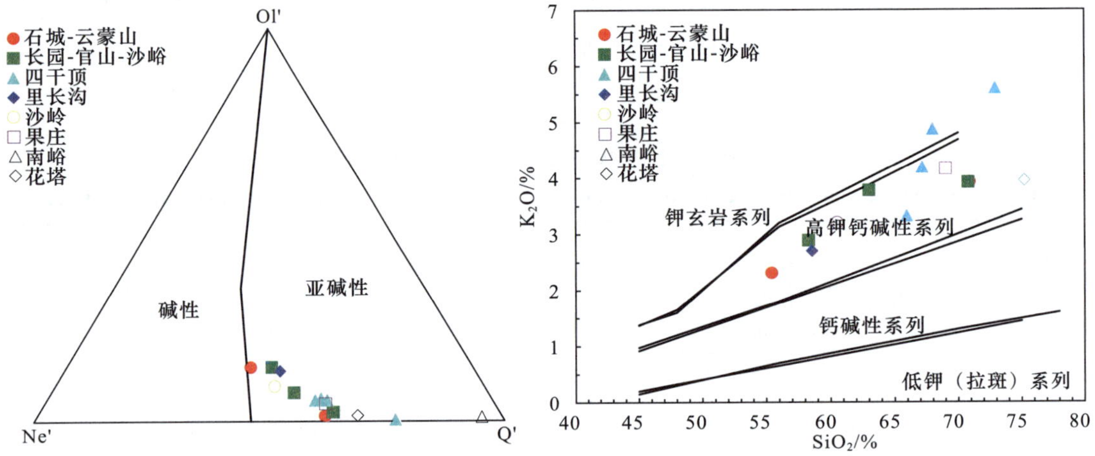

图 20-63　晚侏罗世中酸性岩 Ol'-Ne'-Q' 和 SiO_2-K_2O 图解

Fig. 20-63　Ol'-Ne'-Q' and SiO_2-K_2O diagram of Late Jurassic intermediate-acid rocks

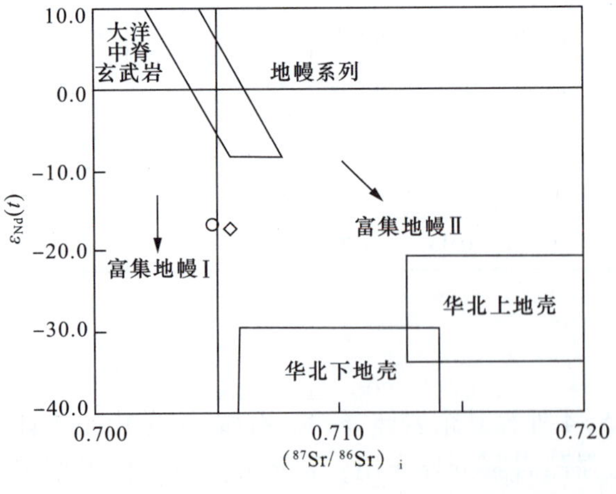

图 20-64　石城-云蒙山杂岩体（$^{87}Sr/^{86}Sr$）$_i$-ε_{Nd}（t）图

（据赵美涛等，2014）

Fig. 20-64　（$^{87}Sr/^{86}Sr$）$_i$-ε_{Nd}（t）diagram of Shicheng-Yunmengshan complex

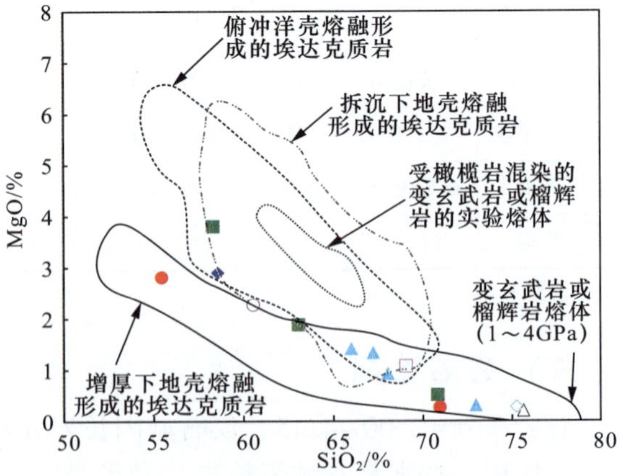

图 20-65　晚侏罗世中酸性岩 SiO_2-MgO 图

（图例同图 20-63）

Fig. 20-65　SiO_2-MgO diagram of Late Jurassic intermediate-acid rocks

（六）构造环境分析

该套岩系富集大离子亲石元素，亏损 Ta、Nb、P 和 Ti 等高场强元素，具有高 Sr、低 Y 含量，显示埃达克岩特性。在 Sr/Y-Y 和 Yb_N-（La/Yb）$_N$ 埃达克岩判别图解（图 20-66）上，多数成分位于埃达克岩区，揭示了该期岩浆侵位活动期间，存在或正在发育增厚下地壳。在 CaO-FeO*+MgO 和（Y+Nb）-Rb 构造环境判别图解（图 20-67）中，成分点分别位于岛弧+大陆弧+碰撞花岗岩类和火山弧花岗岩类。依据岩浆岩构造组合识别标志（邓晋福等，2007），该套岩系为 G_1G_2 组合，属于大陆边缘弧环境岩浆岩组合，形成于燕山造山挤压构造背景环境。

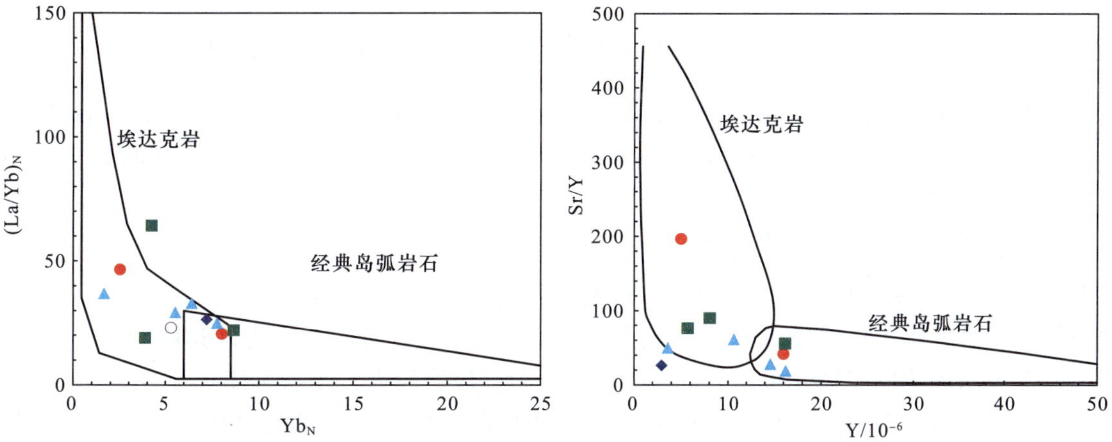

图 20-66　晚侏罗世中酸性岩（La/Yb）$_N$-Yb_N 和 Sr/Y-Y 图解

（图例同图 20-63）

Fig. 20-66　（La/Yb）$_N$-Yb_N and Sr/Y-Y diagram of Late Jurassic intermediate-acid rocks

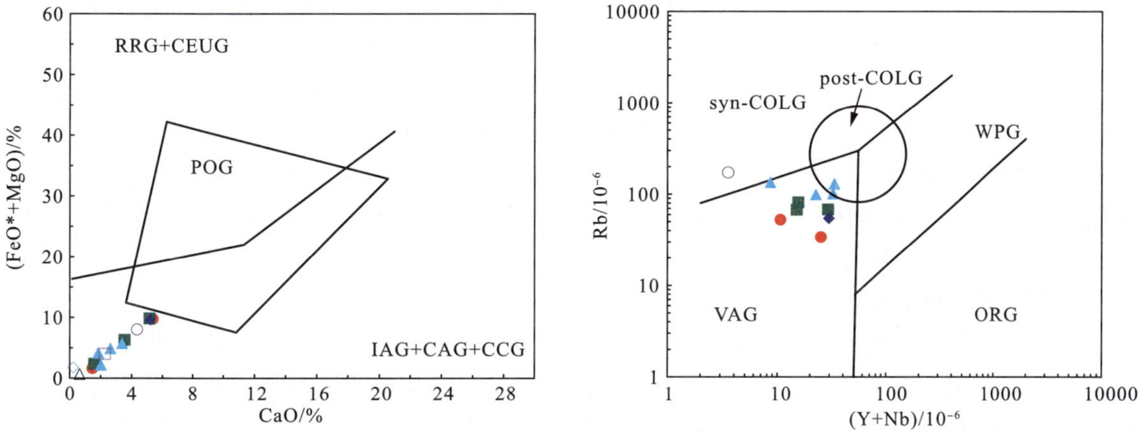

图 20-67　晚侏罗世中酸性岩 CaO-FeO*+MgO 和（Y+Nb）-Rb 图解

（图例同图 20-63）

Fig. 20-67　Tectonic discrimination diagram of CaO-FeO*+MgO and（Y+Nb）-Rb for Late Jurassic intermediate-acid rocks

IAG—岛弧花岗岩类；CAG—大陆弧花岗岩类；CCG—碰撞花岗岩类；POG—造山后花岗岩类；RRG—裂谷花岗岩类；CEUG—陆内造陆花岗岩类；VAR—火山弧花岗岩；ORG—大洋脊花岗岩；WPG—板内花岗岩；syn-COLG—同碰撞花岗岩；post-COLG—碰撞后花岗岩

（七）时代归属

本期岩浆中性岩侵位时代厘定为晚侏罗世，酸性岩侵位时代厘定为晚侏罗世晚期至早白垩世早期，主要依据有：①岩体组合呈 NEE—NE 向中酸性岩浆岩带，同燕山期早期构造形迹特征一致；②虽发生一定程度变形，但岩体变形不均一，变质作用较低；③形成于造山增厚地壳（高 Sr 低 Y）构造背景环境；④侵入到中元古界等较老地质体，又遭受同造山作用有关变形改造；⑤石城闪长岩和云蒙山花岗岩锆石 U-Pb 同位素年龄分别为（151±5）Ma~（162±5）Ma 和（141±2）Ma~（152±4）Ma（Davis et al.，1996；邓晋福等，2001、2004；Shi 等，2004；Wang et al.，2012；康月蓝等，2018）；长园石英二长闪长岩和官山石英二长岩锆石 U-Pb 同位素年龄为（152. 7±3. 2）Ma~（155±3）Ma 和（151±2）Ma~（153±2）Ma（邓晋福等，2004）；花塔花岗岩黑云母 K-Ar 同位素年龄为 152. 1Ma（鲍亦冈等，2001）；上花园二长闪长岩全岩 K-Ar 同位素年龄为 148. 4~154. 9Ma（鲍亦冈等，2001）；四干顶二长花岗岩锆石 U-Pb 同位素年龄为（159. 3±1. 9）Ma。

七、早白垩世早期中—酸性侵入岩（145~132. 9Ma）

早白垩世早期中—酸性侵入岩区内分布最为广泛，发育岩体数量最多，单个岩体规模也较大，是区内岩浆活动最强烈阶段。岩体宏观呈 NE—NS 向产出，与晚侏罗世、早白垩世中期侵入岩一起构成境内最重要的中生代八达岭侵入杂岩带。

（一）地质产状与岩石学特征

发育汉家川-东三岔、房耳峪、分水岭-石湖峪、磨石口-白羊石虎、上苇甸、大庄科、碾台、芹峪、铁炉子等众多岩体，集中分布于延庆八达岭至怀柔黑坨山一带，称之为八达岭杂岩带。门头沟斋堂、清水、东灵山等地也零星分布。单个岩体形态复杂，规模较大，多数岩体长轴 NE30°~45°和近 SN 向展布，组合岩浆岩带 NE35°展布。岩体多数被后期早白垩世中、晚期岩体破坏，一般完整性较差。岩体有中性、中酸性、酸性各类岩性，且同一岩体一般也有岩性、岩相变化。

1. 汉家川-东三岔岩体

位于延庆龙庙沟、汉家川、东三岔、香屯，呈长轴近 SN 向不规则状。岩体北侧侵入中生代侏罗系髫髻山组中，内接触带同化混染，外接触带蚀变；其余周缘被早白垩世中、晚期岩体侵入破坏。岩体内石英斑岩、正长斑岩、花岗岩、闪长玢岩、煌斑岩等脉岩广泛分布，原生流线、流面构造和次生节理也较为发育。岩体岩性岩相复杂，边缘相为细粒闪长岩，中心相为中-细粒含石英闪长岩，龙泉峪还发育有石英二长岩。岩石深灰色-灰绿色，强烈蚀变绿泥石化、钠长石化，似斑状结构，块状构造。斜长石半自形板状，发育聚片、卡纳双晶和环带构造，类型为更-中长石，含量 70%~80%；钾长石他形粒状，发育条纹结造和钠长石交代边缘，类型为条纹长石，含量 3%~10%；石英他形粒状，集合体产出，含量 3%~12%；普通角闪石，发育净边结构，含量 15%~20%；黑云母半自形片状，含量 5%~10%；副矿物为榍石、磷灰石、锆石等。

2. 分水岭-石湖峪杂岩体

位于怀柔黄花城、石湖峪，呈长轴近 SN 向不规则状，出露面积 23. 5km^2。岩体北东侵入中生代侏罗系髫髻山组中；东、西两侧侵入于中元古界蓟县系中；南、南东被早白垩世中期岩体侵入破坏；北西以断层与西水峪花岗岩体接触。岩体内发育花岗斑岩、花岗细晶岩、正长斑岩、闪长玢岩、石英斑岩、煌斑岩等脉岩。岩体岩性岩相复杂，中央相为花岗岩，边缘相为中、细粒似斑状石英二长岩，二者渐变过渡，仅石英含量有所差异。岩石似斑状、花岗结构，块状构造。钾长石半自形粒状，发育格子双晶和交代条纹结构，类型为微斜条纹长石，含量 50%~70%；斜长石半自形板状，类型为更长石，钠长石化、弱绢云母化，含量 5%~30%；石英他形粒状，普遍强波状消光和发育碎裂重结晶现象，含量 17%~27%；角闪石含量 1%~4%；黑云母半自形片状，弱绿泥石化，含量 2%~6%。副矿物为锆石、榍石、磷灰石和少量钛铁矿、白钛石、金红石、白钨矿、辉钼矿、孔雀石等。

3. 上苇甸岩体

位于门头沟下苇甸，沿下苇甸穹隆核部发育。岩体西被黄土台-妙峰山断层切割与中生界侏罗系土城子组接触；其余周缘侵入中元古界待建系、青白口系和古生界寒武系地层中，接触界线清晰，外接触带红柱石角岩化、石榴子石矽卡岩化、透辉石大理岩化、磁铁矿化，内接触带常见围岩捕虏体和顶垂体。岩体岩性为二长闪长岩和花岗闪长岩，前者分布于岩体西、南边部，内发育辉长岩包体。后者是岩体主体岩性，与前者接触部位可见二长闪长岩捕虏体，揭示为稍晚期脉动侵位（图20-68）。

图20-68　上苇甸岩体地质略图

（据1∶5万雁翅幅区域地质图）

Fig. 20-68　The geological sketch map of Shangweidian complex

二长闪长岩：灰白—灰色，不等粒结构，致密块状构造。斜长石自形—半自形板柱状，发育聚片双晶和环带结构，黝帘石化、绢云母化，粒径1~1.5mm，含量10%~15%；石英他形粒状，斑状消光，集合体产出，充填于其他矿物颗粒间，粒径0.2~1.5mm，含量5%~10%；角闪石半自形柱状，多色性，集合体产出，边部蚀变铁质析出，粒径0.4~1.1mm，含量5%~10%；辉石他形粒状，墨绿色，微弱多色性，与角闪石密切伴生集合体产出，粒径0.5~1.5mm，含量2%~15%。有时可见到云母类矿物。副矿物为磁铁矿、磷灰石，榍石等。

花岗闪长岩：灰黄—灰白色，不等粒状镶嵌结构，致密块状构造。斜长石自形板条状，发育港湾状结构和聚片双晶，轻绢云母化，边缘常钾长石化，粒径0.6~1.3mm，大者达2.5mm，含量45%~

50%；条纹长石半自形-他形短柱状、粒状，发育条纹结构、文象结构，解理不发育，轻钠黝帘石化，粒径0.4~1.5mm，含量30%~35%；石英他形粒状，集合体状填充于其他矿物间，波状消光，部分发育六边形截面，粒度0.4~1.1mm，大者达2mm，含量约20%；角闪石长柱状，多色性，边部铁质析出，内嵌石英矿物，粒径0.4~1.5mm，含量5%~10%；黑云母半自形片状，褐黑色，强绿泥石化，边部铁质析出，内嵌磷灰石矿物，含量<5%。副矿物为磁铁矿、磷灰石、榍石等。

4. 磨石口-白羊石虎岩体

位于延庆磨石口、岔石口，呈长轴NNW向长条形。岩体南西与房耳峪岩体过渡渐变接触，接触界线模糊。其余周缘侵入中生界侏罗系髫髻山组和中元古界蓟县系中，接触界线清晰；外接触带大理岩化、绿泥石化、角岩化、矽卡岩化。岩体内部石英斑岩、正长斑岩、花岗岩、闪长玢岩、煌斑岩等脉岩广泛分布，原生流线、流面构造和次生节理也较为发育。岩体岩性岩相复杂，南东边缘为二长闪长岩，中部、北西部为石英二长岩。岩石肉红色-灰白色，不等粒结构，块状构造。石英他形粒状，粒径0.1~0.8mm，少数1~1.5mm，含量10%~22%；斜长石半自形板状，少数发育环带结构，粒径0.5~2mm，少数3~4mm，含量35%~55%；角闪石半自形柱状，多色性，沿解理缝被黑云母、石英、榍石、绿泥石等交代，粒径0.1~0.4mm，少数达0.8mm，含量2%~5%；黑云母半自形片状，沿解理被绿泥石交代，内嵌磷灰石包裹体，粒径0.15~0.6mm，少数0.8~1.2mm，含量2%~10%。副矿物为磷灰石、榍石、磁铁矿等。

5. 房耳峪杂岩体

位于延庆房耳峪、水泉沟，呈长轴近SN向不规则形，出露面积13.5km^2。岩体基本特征同相邻磨石口-白羊石虎岩体。岩体多处被晚期碱性、酸性岩体岩侵入破坏，侵入痕迹清晰。岩体岩性岩相复杂，中央相为花岗岩，过渡相为石英二长岩，边缘相为二长闪长岩。过渡相和边缘相岩性特征同磨石口-白羊石虎岩体。中央相花岗岩浅肉红色，不等粒结构，块状构造。钾长石半自形短柱状，发育格子双晶、条纹结构，类型为微斜条纹长石，含量30%~60%；斜长石半自形板状，发育蠕虫结构，类型为更长石，边缘钠长石化，中心绢云母化，含量25%~40%；石英弱波状消光，集合体产出，含量25%~40%；黑云母半自形片状，含量<2%。副矿物为榍石、锆石、磷灰石、钛铁矿、白钨矿、金红石、电气石、孔雀石等。

6. 芹峪-苇子水岩体

位于门头沟芹峪、田庄、苇子水，由呈近SN向不规则状三个小岩珠组成，分布于青白口穹隆东南翼与中生代髫髻山盆地西。芹峪岩珠顺层侵入中元古界待建系和蓟县系地层，呈枝杈不规则状产出；田庄和苇子水岩珠顺层侵入中生界侏罗系九龙山组，呈条带不规则状产出；岩体侵入接触界线清晰，边部发育小破碎带和泥化带，产状变化较大，底部直立陡倾，向上产状变为同围岩产状一致。

岩体主要发育两种岩性。其一为二长闪长玢岩，灰绿色-黄绿色，斑状结构，致密块状构造。斑晶斜长石自形板条状，发育聚片双晶、环带结构，强钠长石化、绢云母化，粒径0.3~3.5mm，含量15%~20%；斑晶角闪石半自形柱状，发育解理，绿泥石化、黑云母化，常见铁质析出物，粒径0.25~1.75mm，含量5%~10%。基质由斜长石（55%~60%）、角闪石（15%~20%）、石英（5%~10%）、黑云母（约3%）组成。副矿物为磁铁矿、磷灰石、榍石。其二为花岗斑岩，灰白、浅肉红色，斑状结构，致密块状构造。斑晶斜长石自形板条状，发育聚片、卡钠复合双晶，局部有环带结构，弱绢云母化，粒径0.3~1.7mm，含量10%~15%；黑云母黄褐色，边部常见铁质析出，含量约3%；角闪石他形粒状，多色性，多蚀变为黑云母，含量<5%。基质为细晶—隐晶质结构，由长石、石英和隐晶质矿物组成，含量>80%。副矿物为磁铁矿、磷灰石、锆石等。

7. 大庄科杂岩体

位于延庆大庄科村，由里长沟、大沙岭、桃树湾等多个不规则小岩枝、岩株组成，沿近EW、NW和NE向三组断裂交汇古火山机构侵位，出露面积20km^2。岩体北、南分别侵入中生界侏罗系髫髻山组和中元古界长城系地层中，表现为在陷落破火山口内发育岩浆杂岩体（图20-69）。岩体岩性岩相复杂，生成发育过程见示意图（图 20-70），各岩体岩石特征见表 20-34。

表 20-34　大庄科杂岩体基本特征统计表

Table 20-34　Main characteristics of Dazhuangke complex

序号	岩体	岩相	岩性	结构构造	粒径	蚀变	矿物成分及含量	其他
1	里长沟	中央相Ⅰ	闪长岩	花岗结构，块状构造	2~6mm	无	斜长石 40%~60%，黑云母和角闪石 20%~30%，少量钾长石和石英	
		边缘相Ⅱ	闪长玢岩	斑状结构，块状构造	斑晶 1 ~ 3mm；基质<0. 5mm	无	斑晶斜长石 20%~30%，斑晶角闪石和辉石约 10%；基质斜长石、角闪石和黑云母（40%~60%）	
2	大钟山		石英二长岩	花岗结构，块状构造	<2mm	弱绿泥石化、青磐岩化	斜长石约 40%，钾长石约 30%，石英 15% ~ 20%，黑云母约 5%；角闪石 3%~6%	
3	大沙岭	中央相Ⅰ	粗粒花岗岩	不等粒结构，块状构造	2~13mm	无	钾长石 35%~40%，斜长石 20%~30%，石英 25% ~ 35%，黑云母约 5%	
		过渡相Ⅱ	斑状花岗岩	似斑状结构，块状构造	斑晶 2 ~ 5mm，基质<1mm	无	钾长石 35%~40%，斜长石 25%~35%，石英 20%~30%，黑云母和角闪石 5%~10%	
		边缘相Ⅲ	细粒花岗岩	斑状结构，块状构造	斑晶 2 ~ 5mm，基质<1mm	无	斑晶石英为主，含量 5% ~ 10%；基质钾长石 30% ~ 40%，斜长石 5% ~ 20%，石英 40% ~ 50%，黑云母<5%	混杂角砾边部发育
		边缘相Ⅱ	花岗斑岩			弱绢云母化、硅化、钾化		
4	桃树湾		微晶闪长玢岩	斑状结构，块状构造	<1mm	绿泥石化、青磐岩化	斜长石、钾长石、黑云母、石英	流动构造

（据杨开辉，1990）

杂岩体具清楚的环状构造，其边界是一条环状陷落断层，东半部已被晚期含角砾石英二长斑岩岩墙充填（图 20-69）。该环状岩墙长约 9. 5km，宽数十米至数百米，产状向中心内倾，倾角 50°~70°。西半环主要由流纹斑岩岩墙充填，断续分布，长约 5km，亦向环内倾斜。在环形构造东南外缘有一条弧形分布的似斑状二长花岗岩岩枝侵入，同为环形构造侵位。

杂岩体侵位表现出破火山口塌陷充填侵位特性，火山岩浆喷发之后，岩浆房空虚，位于其上方的地壳块体在重力作用下发生沉陷，形成环状裂陷。新的岩浆再次活动时，将乘虚侵入充填环状裂陷中，形成近圆形分布、且有环状岩墙伴生的复式杂岩体（鲍亦冈等，2001）。这种定位方式可以由火山喷发通道扩展而成，也可以是在构造交汇处，早期岩浆侵入到地壳高位后冷凝下沉，促使较晚活动的岩浆沿环状裂隙上侵定位而成。这种多期次复杂岩浆活动形成的杂岩体中，既有中浅带定位的岩体，又有次火山相岩石，还经常见有含较多挥发分的岩浆在热蒸气流的带动下冲碎已冷凝岩石和顶盖围岩，再将它们胶结在一起形成的岩浆角砾岩体（图 20-70）。

杂岩体各岩性组合和切割关系揭示了该杂岩体经历了两期岩浆脉动侵位，前期为闪长岩、石英二长岩、闪长玢岩和花岗斑岩组合，后期为细粒花岗岩、石英二长斑岩和花岗斑岩组合，后期侵位岩性普遍发生矿化蚀变，并最终成矿，即大庄科辉钼矿。

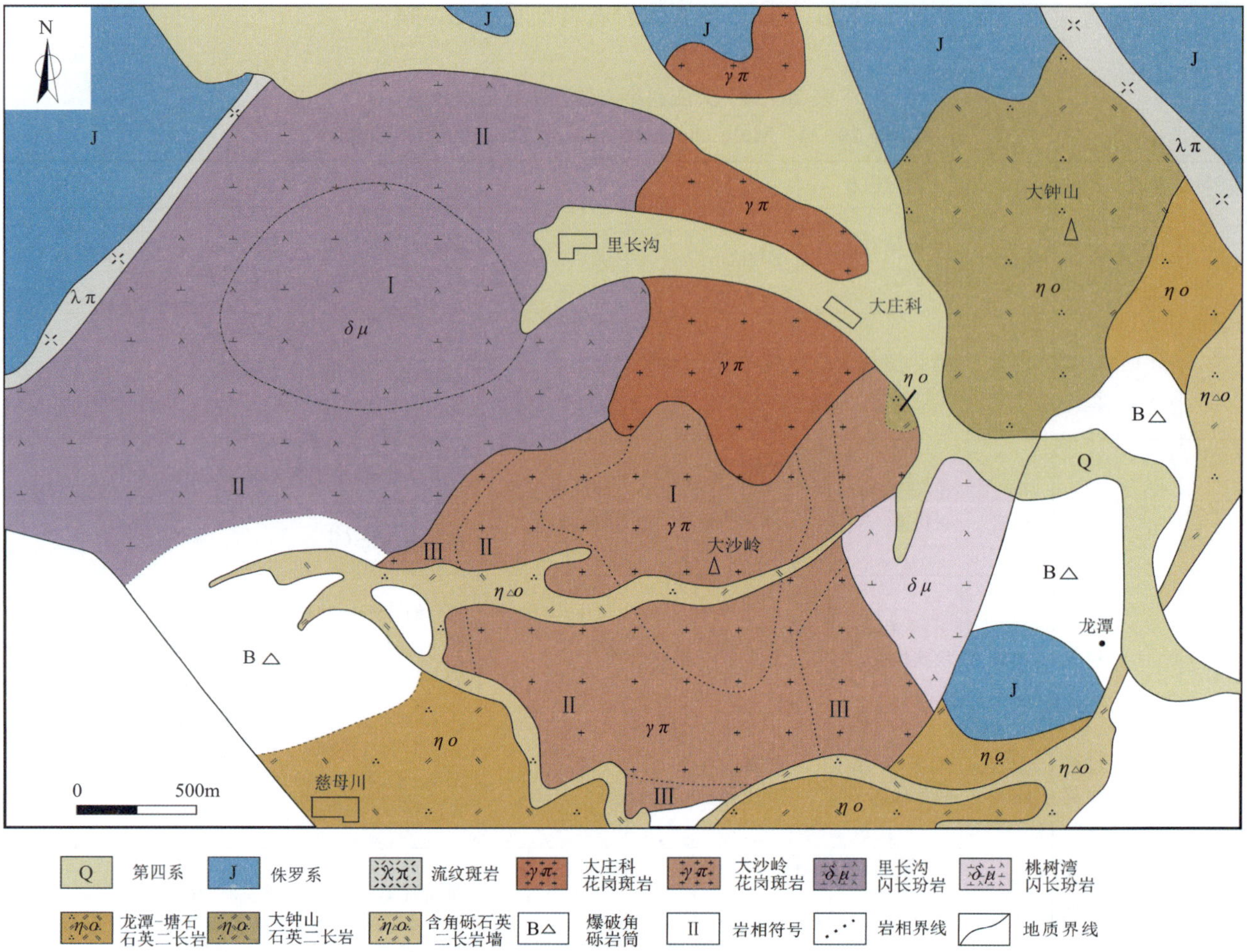

图 20-69　大庄科杂岩体地质构造略图

（据杨开辉，1990，略有修改）

Fig. 20-69　The geological sketch map of Dazhuangke complex

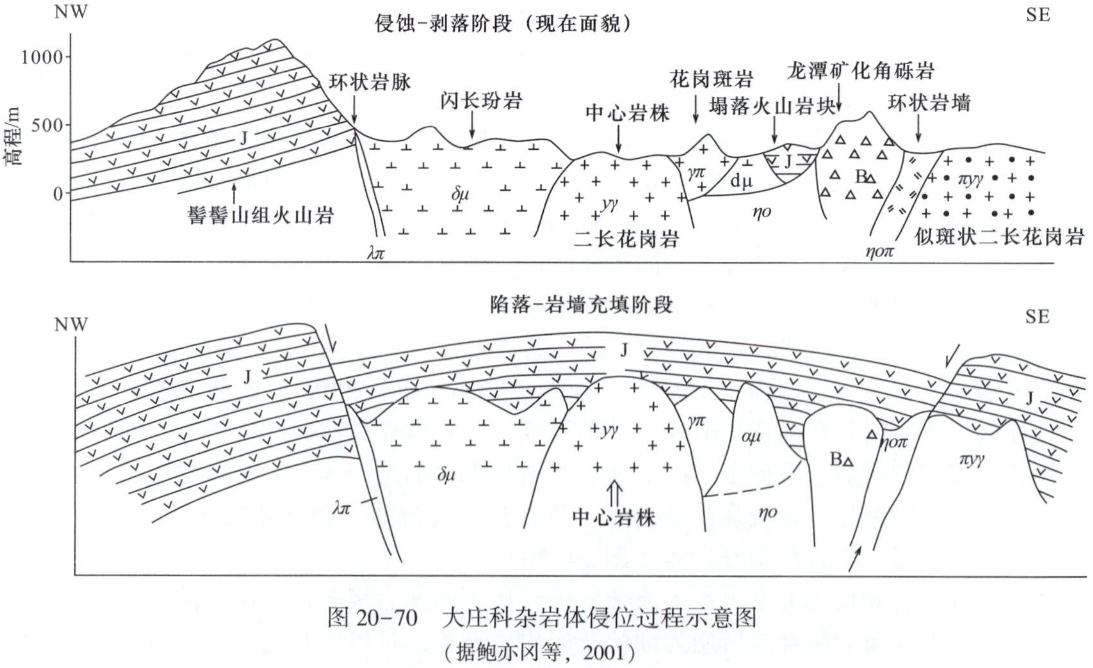

图 20-70　大庄科杂岩体侵位过程示意图

（据鲍亦冈等，2001）

Fig. 20-70　Invased processing sketch of Dazhuangke complex

图例同 20-69

8. 黄场-仙人洞杂岩体

位于昌平西黄场、仙人洞、王家元，由仙人洞、十二道梁、碾台等不同岩体组合呈不规则状复式杂岩体，出露面积 13km^2。岩体西、北西顺层侵入中生界白垩系张家口组火山岩中，周缘侵入中生界侏罗系髫髻山组或中元古界长城系地层中，接触界线清晰，外接触带发育梳状岩脉。岩性岩相复杂，发育有闪长岩、石英二长斑岩和花岗斑岩等。

杂岩体受控于 NNW 向断裂构造，呈现岩墙式侵入扩张。以中部碾台石英二长岩最为典型，在两条平行的 NNW 向断裂间石英二长斑岩岩体呈墙状侵位，形成长 3.5km，宽 0.7~0.8km 的斑岩墙，将早期的石英闪长岩推挤到两侧，并形成很宽的混染岩带。东侧的混染岩石（粗粒斑杂状石英二长闪长岩）沿断裂线分布，西侧的石英闪长岩也大部分被混染成为石英二长闪长岩。石英二长斑岩体的后期热液活动还在岩体内形成细脉浸染型铜钼矿化及广泛的蚀变带，且矿化带及钾质蚀变带的空间形态也呈向南东延伸。

（二）岩石化学与地球化学特征

岩石化学成分及特征参数见表 20-35。岩石 SiO_2、FeO^*、Al_2O_3 含量变化范围大；与同类岩石均值相比，呈高 SiO_2、碱，低 CaO、MgO、MnO 特征。里特曼指数 σ=1.41~4.67，跨碱性与亚碱性岩系列；固结指数 SI=0.60~17.90，分异指数 DI=60.72~94.13，长英指数 FL=54.01~94.58，岩浆经历较强分异作用；A/CNK=0.82~1.13，A/NK=1.05~2.40，跨准铝质和过铝质岩区。同一杂岩体，岩石成分演化趋势明显，特征基本同晚侏罗世杂岩体，如 SiO_2、K_2O、Na_2O+K_2O 含量逐渐增高等，揭示了不同岩类之间可能具有成因关系，而且岩浆演化分离结晶起着重要作用。

稀土元素成分及特征参数见表 20-36。中、酸性岩 ΣREE 分别为 136.73×10^{-6}~268.67×10^{-6} 和 50.48×10^{-6}~132.01×10^{-6}；LREE/HREE 分别为 13.48~23.20 和 8.43~18.78；$(La/Yb)_N$ 分别为 17.46~40.13 和 9.62~36.38；轻稀土中等—较强富集，重稀土相对亏损；稀土总量酸性岩显著低于中性岩；中、酸性岩 δEu 分别为 0.55~1.17 和 0.33~1.00，显示弱铕负异常—弱铕正异常和铕强负异常—无铕正或负异常特征，揭示杂岩体岩浆演化经历了斜长石的分离结晶作用；配分曲线为右陡倾-缓倾光滑曲线（图 20-71）。同一杂岩体，与晚侏罗世杂岩体特征基本相似，伴随岩浆侵位分异演化特征明显，如 La/Sm 增大，LREE/HREE、δEu、Ce/Yb、La/Yb、Eu/Sm 值逐渐减小，揭示岩浆向重稀土相对富集，轻重稀土分馏程度减弱的方向演化。

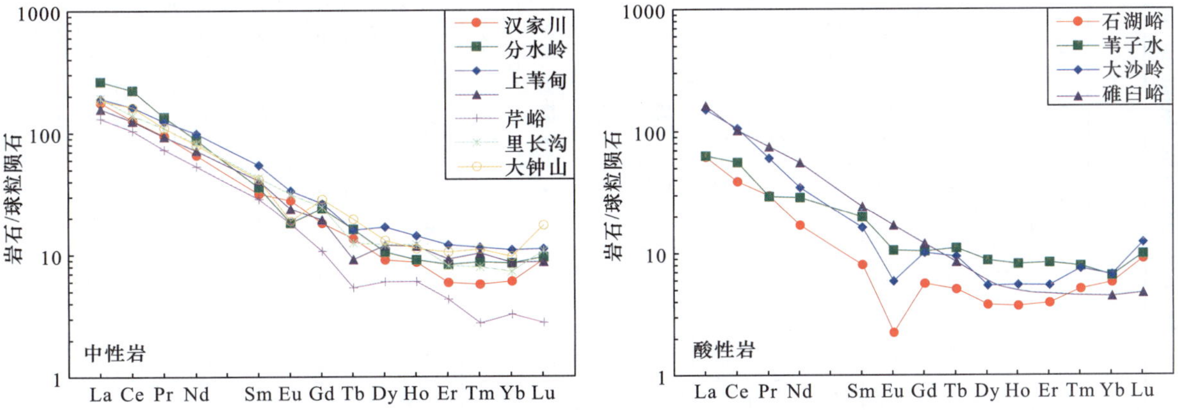

图 20-71　早白垩世早期中酸性岩稀土元素配分模式图

Fig. 20-71　Chondrite-normalized REE patterns of Early Cretaceous intermediate-acid rocks

微量元素含量及特征参数见表 20-37。相比于中国同类岩性均值，微量元素含量偏高，且各杂岩体差异很大，揭示了岩浆来源及其演化复杂性。根据微量元素含量特征，可划分为高 Ba-Sr 和低 Ba-Sr 岩类。前者杂岩体中发育相对稍早，以中性、中酸性岩为主，具有高 LREE，低 Y 和 HREE，高 Sr/Y 和 La/Yb 比值，低 Rb/Sr 比值；在微量元素蛛网图（图 20-72）中，具有明显的 Nb、Ta 和 Ti 负异

表 20-35　早白垩世早期中酸性岩岩石化学成分及特征参数一览表

Table 20-35　The chemical composition and main parameters of Early Cretaceous intermediate-acid rocks

序号	岩体	岩性	氧化物含量/%													特征参数					
			SiO_2	TiO_2	Al_2O_3	Fe_2O_3	FeO	MnO	MgO	CaO	Na_2O	K_2O	P_2O_5	Los	总和	*DI*	*SI*	σ	*FL*	*A/CNK*	*A/NK*
1	汉家川	石英闪长岩	61.69	0.84	16.97	3.46	3.08	0.09	2.21	4.41	2.63	2.55	1.00	0.25	99.18	62.75	16.00	1.41	54.01	1.12	2.40
2	分水岭	石英二长岩	73.04	0.06	14.405	0.85	0.265	0.045	0.19	0.79	4.71	4.84	0.04	0.44	99.67	94.21	1.76	3.03	92.36	1.00	1.11
3	石湖峪	二长花岗岩	75.26	0.05	13.56	0.27	0.19	0.03	0.06	0.54	4.59	4.84	0.01	0.27	99.67	96.87	0.60	2.75	94.58	0.98	1.06
4	上苇甸	二长闪长岩	55.89	0.84	18.30	2.43	4.77	0.14	2.32	5.23	5.02	2.66	0.48	1.03	99.10	60.72	13.46	4.38	59.49	0.89	1.64
5		花岗闪长岩	59.46	0.70	18.35	2.62	3.16	0.11	1.73	5.20	5.02	1.93	0.30	0.30	98.90	63.65	12.01	2.87	57.19	0.93	1.77
6	磨石口	石英闪长岩	56.52		16.62	3.47	5.08	0.12	3.61	5.02	4.49	3.52	0.75	0.75	99.95	60.27	17.90	4.67	61.47	0.82	1.48
7	白羊石虎	石英二长岩	67.75	0.39	15.07	1.75	1.76	0.06	1.12	2.90	4.18	4.05	0.20	0.37	99.59	80.37	8.69	2.72	73.92	0.91	1.34
8	房耳峪	石英闪长岩	59.83	0.90	17.20	3.23	3.51	0.08	2.36	4.31	4.15	2.96	0.28	0.86	99.67	64.12	14.60	2.95	62.26	0.96	1.72
9		二长花岗岩	69.13	0.36	15.26	1.57	1.58	0.07	0.88	2.30	4.23	3.72	0.12	0.24	99.46	82.10	7.36	2.41	77.56	1.01	1.39
10	九仙庙	石英二长岩	60.32	0.86	17.57	4.15	1.66	0.08	1.62	3.71	4.75	3.81	0.27	0.35	99.15	71.39	10.23	4.15	69.76	0.94	1.47
11	苇子水	花岗斑岩	68.47	0.23	16.09	0.90	0.95	0.12	0.78	1.07	5.57	2.95	0.11	1.67	98.88	88.21	6.96	2.80	88.89	1.13	1.30
12	芹峪	闪长岩	65.22	0.30	15.02	1.65	2.81	0.14	1.08	2.47	4.03	3.48	0.17	1.53	97.90	77.91	8.28	2.46	75.25	1.01	1.45
13	里长沟	闪长玢岩	58.64	0.98	16.86	3.53	3.50	0.10	2.90	5.20	4.34	2.70	0.38	0.66	99.79	61.19	17.16	3.12	57.52	0.86	1.68
14	大钟山	石英二长岩	66.06	0.60	14.48	2.52	2.51	0.10	1.43	2.68	3.99	5.04	0.20	0.13	99.74	80.30	9.24	3.52	77.11	0.86	1.20
15	大沙岭	花岗岩	74.48	0.185	13.08	0.95	0.51	0.045	0.41	0.925	4.665	4.4	0.055	0.11	99.82	94.13	3.76	2.61	90.74	0.93	1.05
16	桃树湾	闪长岩	65.26	0.52	16.31	2	2.03	0.05	1.27	2.34	5.12	3.98	0.25	0.49	99.62	80.93	8.83	3.69	79.55	0.96	1.28
17	碓臼峪	二长花岗岩	71.21	0.39	14.11	1.70	0.77	0.05	0.98	2.10	4.24	3.55	0.12	0.51	99.73	84.57	8.75	2.14	78.74	0.96	1.31

注：数据为均值，引自 1 : 5 万沙峪幅区域地质调查报告，1966；1 : 5 万永宁幅、四海幅区域地质调查报告，1993；1 : 5 万雁翅幅、阳坊幅区域地质调查报告，1996；1 : 5 万青龙桥幅、延庆幅区域地质调查报告，1984；杨开辉，1990；焦守涛等，2013；王焰等，2001；钱青等，2002；汪洋等，2009。

表 20-36　早白垩世早期中酸性岩稀土元素成分及特征参数一览表

Table 20-36　REE contents and main parameters of Early Cretaceous intermediate-acid rocks

序号	岩体	岩性	稀土元素含量/10^{-6}																特征参数			
			La	Ce	Pr	Nd	Sm	Eu	Gd	Tb	Dy	Ho	Er	Tm	Yb	Lu	Y	ΣREE	LREE/HREE	$(La/Yb)_N$	δEu	δCe
1	汉家川	石英闪长岩	41. 63	77. 51	8. 96	30. 64	4. 82	1. 62	3. 72	0. 51	2. 29	0. 49	0. 98	0. 15	1. 03	0. 23	11. 90	174. 54	17. 62	29. 13	1. 17	0. 98
2	分水岭	石英二长岩	61. 71	135. 12	12. 71	40. 75	5. 43	1. 05	4. 89	0. 6	2. 65	0. 51	1. 36	0. 22	1. 43	0. 24	12. 6	268. 67	21. 58	30. 95	0. 62	1. 18
3	石湖峪	二长花岗岩	14. 63	23. 75	2. 82	7. 92	1. 23	0. 13	1. 16	0. 19	0. 96	0. 21	0. 65	0. 13	0. 98	0. 23	6. 41	54. 99	11. 19	10. 71	0. 33	0. 91
4	上苇甸	二长闪长岩	44. 80	98. 39	11. 64	45. 74	8. 26	1. 93	5. 39	0. 59	4. 25	0. 80	1. 97	0. 29	1. 84	0. 28	19. 96	226. 17	13. 68	17. 46	0. 88	1. 06
5		花岗闪长岩	36. 90	75. 83	8. 79	33. 19	6. 18	1. 37	3. 96	0. 34	3. 01	0. 66	1. 51	0. 26	1. 45	0. 22	15. 40	173. 67	14. 22	18. 25	0. 85	1. 03
6	苇子水	花岗斑岩	14. 88	34. 02	2. 77	13. 34	3. 02	0. 61	2. 12	0. 41	2. 21	0. 46	1. 38	0. 20	1. 11	0. 25	13. 94	76. 78	8. 43	9. 62	0. 74	1. 30
7	芹峪	闪长岩	30. 77	63. 32	6. 90	24. 68	4. 37	1. 04	2. 19	0. 20	1. 52	0. 34	0. 71	0. 07	0. 55	0. 07	5. 85	136. 73	23. 20	40. 13	1. 03	1. 07
8	里长沟	二长闪长玢岩	45. 18	84. 42	10. 31	37. 99	6. 51	1. 81	5. 01	0. 46	3. 05	0. 68	1. 33	0. 20	1. 23	0. 27	15. 17	198. 44	15. 23	26. 42	0. 97	0. 96
9	大钟山	石英二长岩	43. 41	98. 53	10. 39	37. 06	6. 34	1. 09	5. 85	0. 73	3. 31	0. 63	1. 73	0. 28	1. 63	0. 44	16. 87	211. 42	13. 48	19. 10	0. 55	1. 14
10	大沙岭	花岗岩	35. 47	64. 22	5. 69	16. 11	2. 47	0. 34	2. 06	0. 35	1. 37	0. 31	0. 90	0. 19	1. 13	0. 31	8. 52	130. 92	18. 78	22. 52	0. 46	1. 11
11	碓臼峪	二长花岗岩	38. 25	62. 13	—	25. 78	3. 69	0. 98	—	0. 32	—	—	—	—	0. 75	0. 12	6. 50	132. 007	—	36. 38	1. 00	0. 92

注：数据为均值，数据来源同表 20-35。

常，轻重稀土强烈分离，缺乏明显的 Sr 和 Eu 负异常；后者在杂岩体中发育稍晚，主要为酸性岩类，具有较低 LREE，高 Y 和 HREE，低 Sr/Y 和 La/Yb 比值，高 Rb/Sr 比值；在微量元素蛛网图（图 20-72）中，既亏损 Nb、Ta 和 Ti，还亏损 P、Ba、Sr、Eu。

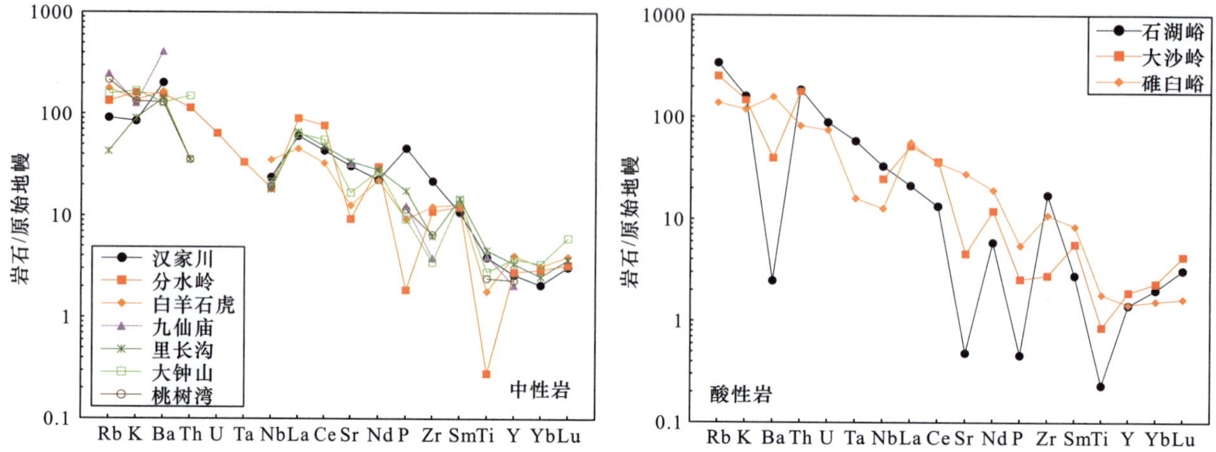

图 20-72 早白垩世早期中酸性岩微量元素蛛网图

Fig. 20-72 Trace element spider diagram of Early Cretaceous intermediate-acid rocks

表 20-37 早白垩世早期中酸性岩微量元素成分及特征参数一览表

Table 20-37 Trace element contents and main parameters of Early Cretaceous intermediate-acid rocks

序号	岩体	岩性	微量元素含量/10^{-6}									特征参数	
			Rb	Ba	Th	U	Nb	Ta	Sr	Zr	Hf	Rb/Sr	Nb/Ta
1	汉家川	石英闪长岩	58.00	1425.00				17.00	650.00	246.00		0.09	
2	分水岭	石英二长岩	84.00	1079.00	9.61	1.34	1.36	13.20	195.00	122.60		0.43	0.10
3	石湖峪	二长花岗岩	218.00	17.30	15.89	1.86	2.39	23.50	10.20	193.00		21.37	0.10
4	白羊石虎	石英二长岩	113.00	1145.00				25.00	265.00	137.00		0.43	
5	九仙庙	石英二长岩	157.00	2841.00				14.76	686.40	42.53		0.23	
6	里长沟	二长闪长玢岩	27.00	979.30	3.00			12.81	710.20	69.45		0.04	
7	大钟山	石英二长岩	100.00	890.50	12.63			15.23	353.20	38.52		0.28	
8	大沙岭	花岗岩	160.00	274.70	15.06			17.48	95.14	30.78		1.68	
9	桃树湾	闪长岩	137.00	900.30	3.00			13.69		72.01			
10	碓臼峪	二长花岗岩	86.75	1093.25	6.90	1.55	0.65	9.00	578.25	120.75		0.15	0.07

注：数据来源同表 20-35。

（三）岩石系列

该套岩系 $\sigma=1.41\sim4.67$，上苇甸二长闪长岩、磨石口石英闪长岩、九仙庙石英二长岩 $\sigma>4$ 外，其余岩体 $\sigma<4$。考虑该套杂岩系经历了较强岩浆分异作用，可能使得熔浆中元素丰度和元素比值发生改变。在 Ol’-Ne’-Q’ 判别图解（图 20-73）上，成分点位于钙碱性区。因此，该套岩系为钙碱性系列。在 K_2O-SiO_2 图（图 20-73）上，成分点位于高钾钙碱性-钾玄岩系列。

（四）Sr-Nd 同位素特征

碓臼峪花岗岩 $^{87}Sr/^{86}Sr=0.70675\sim0.70811$，$^{87}Rb/^{86}Sr=0.2511\sim0.7463$，$^{143}Nd/^{144}Nd=0.511512\sim0.511808$，$^{147}Sm/^{144}Nd=0.0820\sim0.0934$；取 $t=150$Ma 时，$\varepsilon_{Nd}(t)=-14.5\sim-20.2$，$t_{DM}=1620\sim2040$Ma（钱青等，2002）。

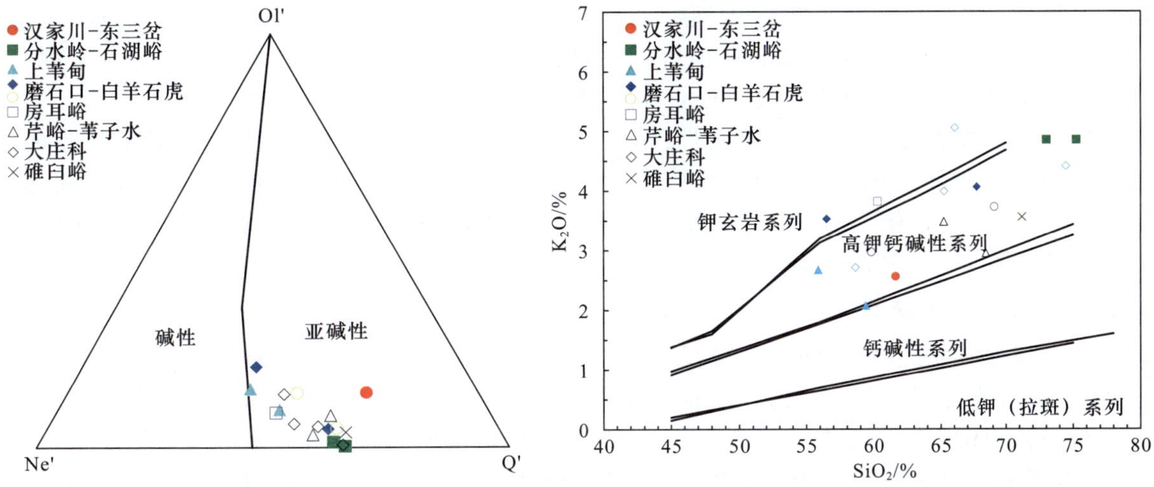

图 20-73　早白垩世早期中酸性岩 Ol’-Ne’-Q’ 和 K_2O-SiO_2 图

Fig. 20-73　Ol’-Ne’-Q’ and SiO_2-K_2O diagram of Early Cretaceous intermediate-acid rocks

（五）S 同位素特征

大庄科杂岩体铜钼矿床中黄铁矿和辉钼矿硫同位素变化范围窄，$^{32}S/^{34}S=22.1\sim22.19$，均值 22.14，$\delta^{34}S=1.4‰\sim5.4‰$，均值 3.5‰，明显不同于陨石硫同位素（$^{32}S/^{34}S=22$，$\delta^{34}S=0‰\sim0.6‰$）组成，而与我国长江中下游中生代 I 型闪长岩-花岗闪长岩类岩体及有关铜铁矿床硫同位素（$\delta^{34}S=0.74‰\sim4.25‰$）类似（郁建华等，1992）。

（六）成因类型与岩浆演化

该套岩系属于高钾钙碱性-钾玄岩系列，酸性岩普遍含角闪石矿物，而无原生富铝的矿物，如白云母、堇青石，也无碱性暗色矿物，SiO_2 含量变化大，$Na_2O>3.2\%$，*A/CNK* 和 Na/K 比值等同 I 型花岗岩相似。在 K_2O-Na_2O 图解（图 20-74）上，成分点位于 I 型花岗岩区，揭示岩浆来源于未经受变质的火成岩再次熔融（康月蓝等，2018）。与晚侏罗世侵入岩相似，该系列岩体也多表现为脉动式多期次岩浆侵位，发育岩性为闪长岩-石英二长岩-花岗岩中—酸性岩组合。岩石普遍亏损 Nb、Ta、Ti 等高场强元素，揭示其与富集大陆岩石圈地幔岩浆或俯冲板块岩浆岩有关。在 $(^{87}Sr/^{86}Sr)_i$-$\varepsilon_{Nd}(t)$ 图（图 20-75）中，成分点位于富集型 I 地幔区（EMI）附近，富集 EMI型地幔是该套岩系岩浆重要来源。

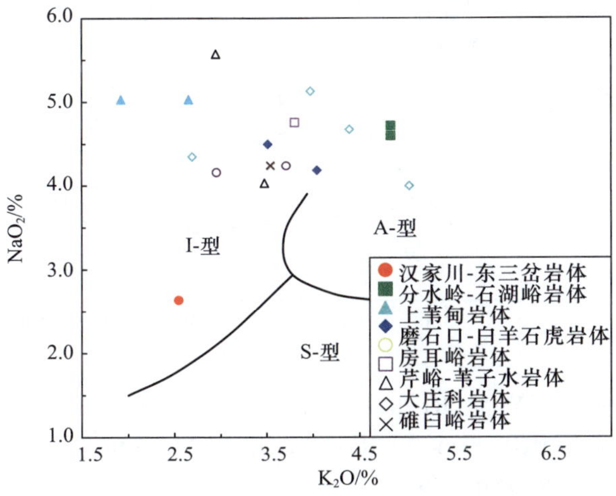

图 20-74　早白垩世早期中酸性岩 K_2O-Na_2O 图

Fig. 20-74　K_2O-Na_2O diagram of Early Cretaceous intermediate-acid rocks

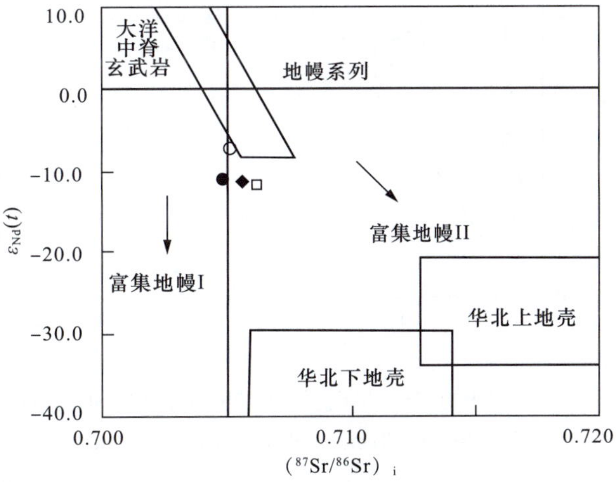

图 20-75　碓白峪岩体 $(^{87}Sr/^{86}Sr)_i$-$\varepsilon_{Nd}(t)$ 图

（据钱青等，2002）

Fig. 20-75　$(^{87}Sr/^{86}Sr)_i$-$\varepsilon_{Nd}(t)$ diagram of Duijiuyu complex

同时，该套岩系特征又与典型 EMI 型岩浆岩不完全相同，如低 TiO_2，高 Sr、成分点不完全位于 EM Ⅰ区、酸性岩类接近于下地壳 Rb/Sr 值等，说明了下地壳物质也是该套岩系重要岩浆来源，钱青等（2002）研究认为地壳物质加入量可能是 20%～50%。

该套岩系主、微量元素普遍具有相关性，揭示了分离结晶作用在岩浆演化过程中起到了重要作用。早期暗色矿物（如辉石、角闪石）的分离结晶可能导致了 SiO_2 与 MgO、FeO^* 和 Co、Ni 之间的相关，P_2O_5、TiO_2 同 SiO_2 负相关，而同 REE 正相关，表明了含 P 和 Ti 的副矿物相的分离对岩浆演化中 REE 含量的逐渐降低起了重要作用（钱青等，2002）。

（七）构造环境分析

该套岩系部分中-酸性岩（如汉家川石英闪长岩、铁炉子二长花岗岩）富集大离子亲石元素，亏损 Ta、Nb 等高场强元素，高 Sr、低 Y，显示埃达克岩特性。在 Sr/Y-Y 埃达克岩判别图解（图 20-76）上，成分点也位于埃达克岩区，揭示了这些岩浆侵位活动期间，存在或正在发育增厚下地壳。在 MgO-FeO^* 构造环境判别图解（图 20-76）中，成分点位于岛弧+大陆弧+碰撞花岗岩类区；在（Y+Nb）-Rb 构造环境判别图解（图 20-76）中，成分点位于火山弧和同碰撞交汇区；在 R1-R2 构造环境

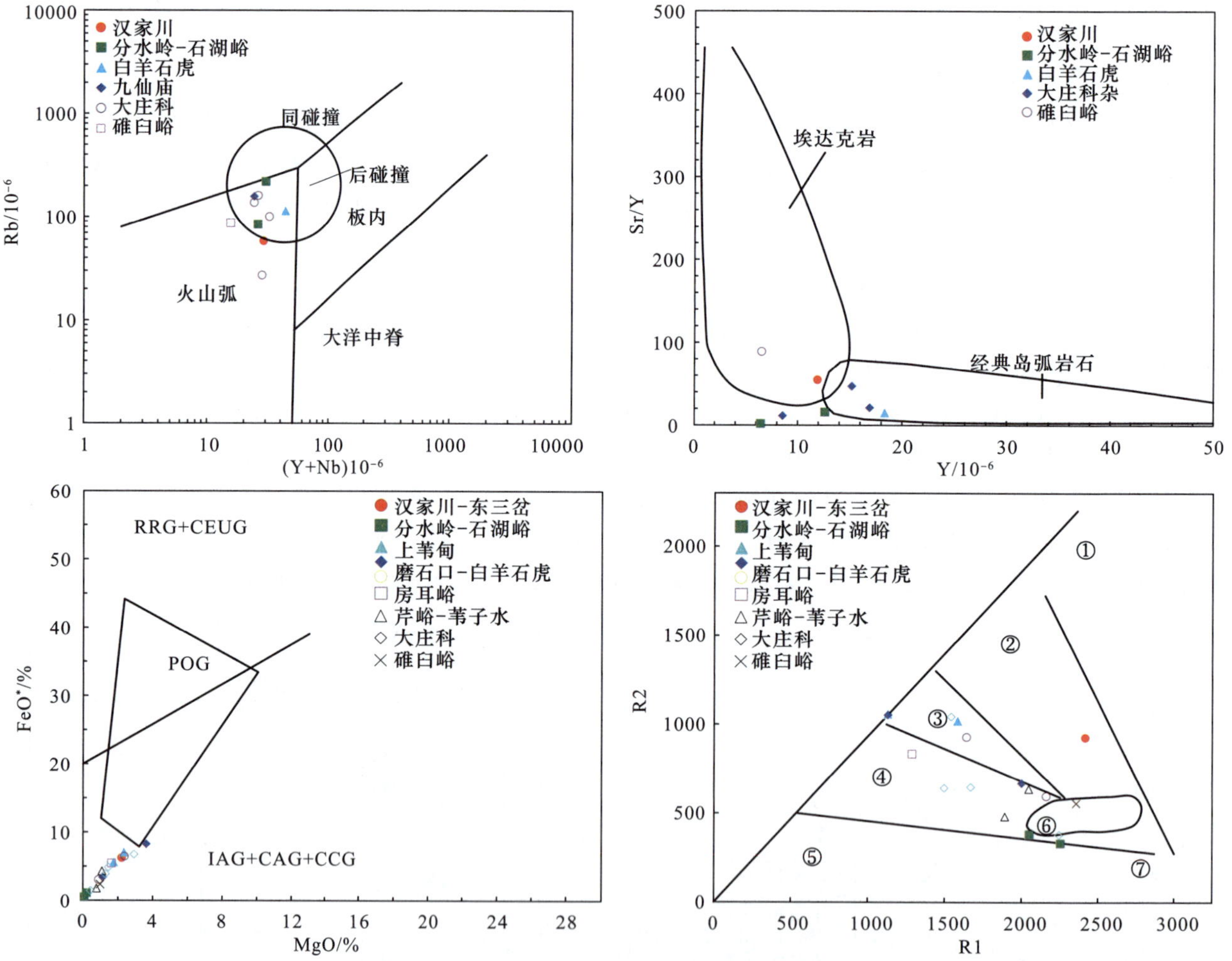

图 20-76　早白垩世早期中酸性岩 CaO-FeO^*+MgO、Sr/Y-Y、R1-R2 和（Y+Nb）-Rb 图解

Fig. 20-76　Tectonic discrimination diagram of CaO-FeO^*+MgO, Sr/Y-Y, R1-R2 and (Y+Nb) -Rb of Early Cretaceous intermediate-acid rocks

IAG—岛弧花岗岩类；CAG—大陆弧花岗岩类；CCG—碰撞花岗岩类；POG—造山后花岗岩类；RRG—裂谷花岗岩类；CEUG—陆内造陆花岗岩类；VAR—火山弧花岗岩；ORG—大洋脊花岗岩；WPG—板内花岗岩；syn-COLG—同碰撞花岗岩；post-COLG—碰撞后花岗岩；①地幔斜长花岗岩；②板块碰撞前消减地区花岗岩；③板块碰撞后隆起期花岗岩；④晚造山期花岗岩；⑤⑦造山期后 A 型花岗岩；⑥同碰撞花岗岩

判别图解（图 20-76）中，成分点分散位于同碰撞 S 型花岗岩+碰撞后隆起花岗岩+晚造山期花岗岩区。因此，该套岩系多期次岩浆侵位，形成于燕山造山重要构造转换期，经历了燕山造山运动主过程向后造山转换，区域构造应力场可能是先挤压后伸展。

（八）时代归属

该套岩系侵位时代厘定为早白垩世早期，主要依据有：①单个岩体主要呈呈近 SN 向分布，岩体组合成 NE 向中酸性岩浆岩带，同燕山期中期构造形迹特征一致；②侵位于中生界侏罗系及更老地层或侏罗纪及更老侵入体中；③岩体内普遍发育与非造山有关的碱性、基性岩脉；④芹峪闪长玢岩全岩 K-Ar 同位素年龄为 138Ma（1∶5 万雁翅幅、阳坊幅区域区域地质调查报告，1996）；汉家川-东三岔石英闪长岩锆石 U-Pb 同位素年龄为（137.6±1.2）Ma，全岩和黑云母 K-Ar 同位素年龄分别为 135.42Ma、147.2Ma（鲍亦冈等，2000；邓晋福等，2004）；大河南花岗闪长岩锆石 U-Pb 同位素年龄为（136±2）Ma（邓晋福等，2004）；上苇甸花岗闪长岩锆石 U-Pb 同位素年龄为（138.2±1.5）Ma（Davis 等，2001）；大庄科杂岩体中，里长沟二长闪长玢岩和塘石石英二长岩全岩 K-Ar 同位素年龄分别为 143.9~168Ma 和 132.22Ma（鲍亦冈等，2000），塘石辉钼矿 Re-Os 同位素年龄为（137.6±3.7）Ma（刘舒波，2012），而碌碡沟石英二长岩锆石 U-Pb 同位素年龄为（140.6±3.7）Ma（郁建华等，1994）；九仙庙、黄场和慈母川石英二长岩黑云母 K-Ar 同位素年龄分别为 139.9Ma、142.6Ma 和 146.6Ma（鲍亦冈等，2000）；碓臼峪二长花岗岩锆石 U-Pb 同位素年龄为（137.7±3.6）Ma，黑云母和全岩 K-Ar 同位素年龄分别为（129.4±1.7）Ma 至（135.5±1.7）Ma 和 144.7Ma，全岩 Rb-Sr 同位素年龄为（128.7±3.8）Ma（鲍亦冈等，2000；邓晋福等，2004、焦守涛等，2013）；铁炉子石英二长岩和花岗岩锆石 U-Pb 同位素年龄分别为（133.2±2.6）Ma 和（137±1.2）Ma（郁建华等，1994；焦守涛，2013）；八达岭斑状花岗岩黑云母 K-Ar 同位素年龄为 138.9Ma（鲍亦冈等，2000），石湖峪二长花岗岩锆石 U-Pb 同位素年龄为（128.5±1.2）Ma（焦守涛，2013）；碾台石英二长岩锆石 U-Pb 同位素年龄为（133.6±2.7）Ma。

八、早白垩世中期中—酸性侵入岩（132.9~125Ma）

早白垩世中期中—酸性侵入岩区内同样广泛发育，岩体数量较多，单个岩体规模较大，形态复杂，受控于早期构造及围岩地质体。岩体宏观呈 NE—NNE 向产出，与晚侏罗世、早白垩世早期侵入岩一起构成境内最重要的中生代八达岭侵入杂岩带。

（一）地质产状与岩石学特征

发育薛家石梁、房山、大羊山、冯家峪、闷虎峪、千层背、双石头、炭厂、河南台等众多岩体，集中分布于密云新城子、冯家峪、西水峪，昌平黑山寨、阳坊、德陵，延庆玉皇山、八达岭、石槽，房山南窖、周口店，怀柔怀北、卯镇山、月亮鞍梁，门头沟斋堂、张家山等，单个岩体规模较大，形态复杂，长轴多呈 NW 或近 SN 向，主要受控于早期构造及围岩地质体。岩体岩性包含中性、中酸性、酸性，以酸性岩为主，岩相多发育边缘相、过渡相和中央相，岩性岩相复杂。下面对典型岩体地质产状与岩石学特征总结如下：

1. 薛家石梁-黑山寨杂岩体

位于昌平黑熊山、九渡河、薛家石梁、下庄，隶属于中生代燕山褶皱-逆冲推覆带西段，呈长轴 NW 向卵圆状，出露面积 17km^2。杂岩体北西内倾约 45°，侵入中生代早白垩世早期分水岭中酸性岩体；其余周缘外倾 45°~65°，侵入古生界寒武系或中元古界蓟县系雾迷山组地层，接触界线清晰，外接触带发育宽约 200m 的大理岩化、蛇纹石化带。

杂岩体岩性岩相复杂，由上庄辉长岩、薛家石梁二长辉长岩、黑山寨二长岩、湖门正长岩和黑熊山花岗岩组成（图 20-77），具体各岩性单元特征见表 20-38。杂岩体岩性呈不规则同心圆状或环状分布，二长辉长岩、二长岩和正长岩呈渐变过渡（涌动）接触关系，发育宽数十厘米至数十米过渡带。过渡带发生 NW 向宽缓“热褶皱”现象，内发育平行于过渡带定向分布的形态多样、大小差异基性包体和流面构造（图 20-78），揭示短期强力就位特性。花岗岩与其他岩性侵入接触，接触界面截然，揭示形成时代可能稍晚。

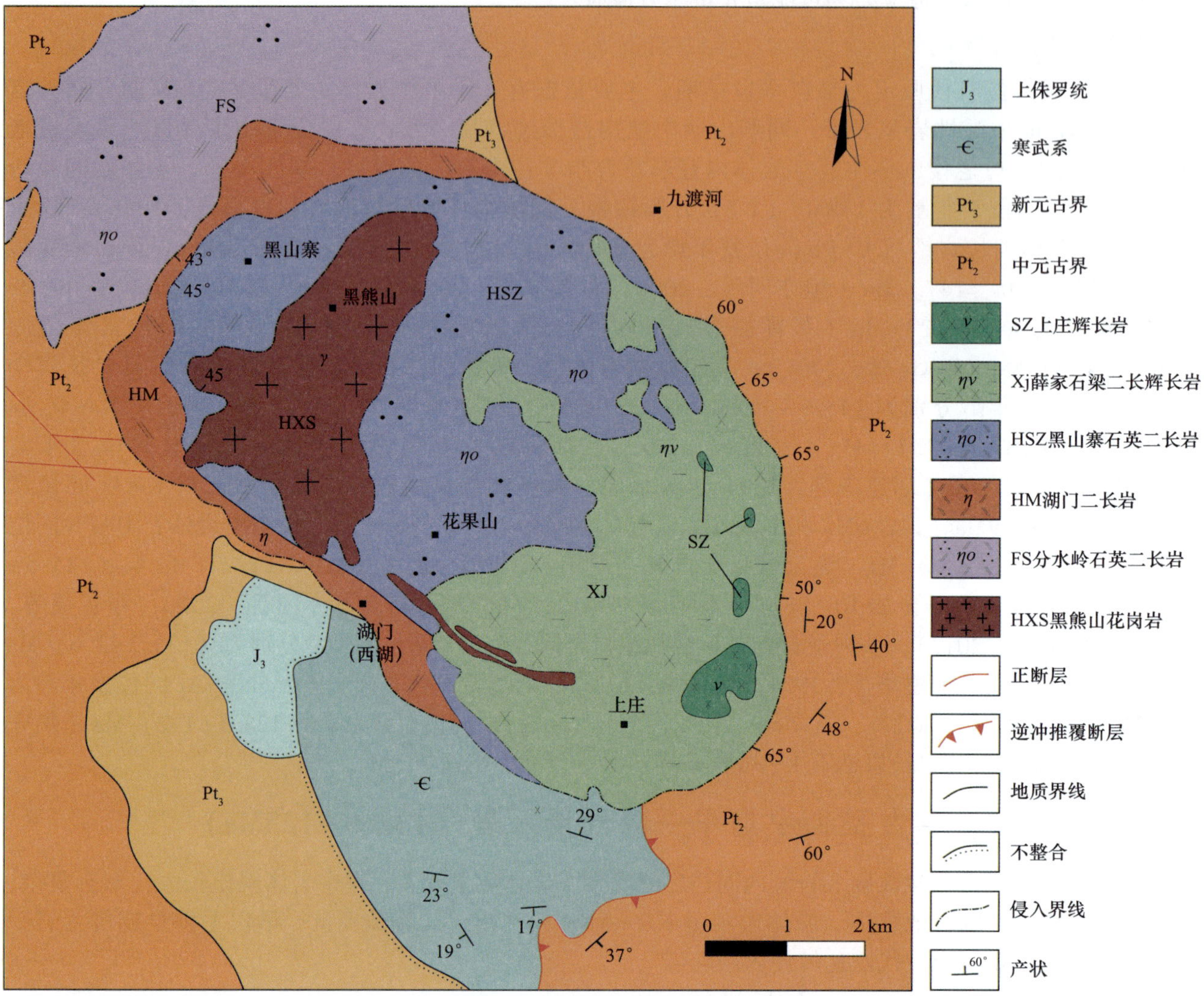

图 20-77　薛家石梁杂岩体地质图

（据 Wang et al.，2013）

Fig. 20-77　The geological sketch map of Xuejiashiliang complex

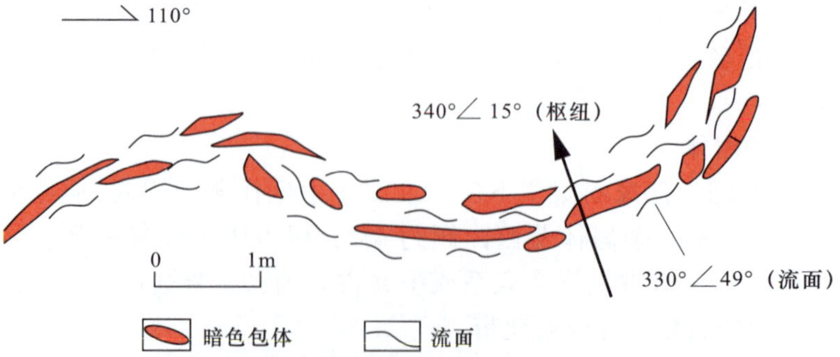

图 20-78　薛家石梁杂岩体内部发育的“热褶皱”平面示意图

（据陈亦寒，2006）

Fig. 20-78　Plane sketch of “thermal fold” in Xuejiashiliang complex

表 20-38 薛家石梁杂岩体基本特征

Table 20-38 Main characteristics of Xuejiashiliang complex

序号	岩体	岩性	结构构造	粒径	矿物成分及含量	岩体位置
1	黑熊山	花岗岩	花岗结构、块状构造	细粒—中粒	钾长石类型为微斜长石和条纹长石，含量约 35%；斜长石半自形板柱状，含量约 30%；石英他形粒状，含量约 30%；黑云母、角闪石含量约 5%	中心
2	湖门	正长岩	等粒结构，片麻状构造	中粒	钾长石半自形-他形柱状，发育卡式双晶，含量约 75%；斜长石他形粒状，分布于钾长石边部，含量约 10%；辉石蚀变为角闪石残留态，含量约 5%；黑云母半自形鳞片状，含量约 6%；石英他形粒状，含量约 4%	北西
3	黑山寨	二长岩	斑状结构、弱片麻状构造、块状构造	粗粒不等粒	斜长石自形-半自形柱状，发育聚片、卡钠复合双晶和环带结构，类型为更长石，含量约 45%；钾长石半自形-他形板状，发育卡式双晶，内嵌斜长石微晶，含量约 35%；黑云母片状，多色性，内嵌榍石、磁铁矿微晶，含量约 9%；角闪石短柱状、粒状，多色性，类型为镁角闪石，含量约 7%；石英他形粒状，填隙状分布，含量约 4%；副矿物为榍石、磁铁矿、磷灰石等	主体
4	薛家石梁	二长辉长岩	中细粒结构，块状构造	中细粒	斜长石半自形板状，发育聚片双晶，类型为更长石，含量约 55%；单斜辉石半自形柱状，含量约 20%；角闪石不规则状，多色性，沿辉石边部分布，含量约 10%；黑云母半自形鳞片状，多色性，类型为镁黑云母，含量约 10%；条纹长石不规则状，含量约 5%；少量石英和磁铁矿	南部
5	上庄	辉长岩	辉长结构，块状构造	细粒	单斜辉石半自形柱状，含量 30%；角闪石他形-半自形柱状，多色性，类型为镁角闪石，含量 10%；黑云母鳞片状，多色性，类型为镁黑云母，含量 5%；斜长石半自形板柱状，发育聚片双晶，类型为中长石，含量 50%；少量橄榄石。副矿物为钛磁铁矿等	东南部

2. 房山杂岩体

位于房山周口店，组成房山穹隆核部。平面形态呈长轴 NW 向半隐伏近圆状，出露面积 47km^2。钻孔和航磁资料显示剖面形态呈微倾斜的蘑菇状。岩体南东被第四系覆盖；其余周缘外倾近顺层侵入太古宙变质岩或古生界寒武系—三叠系地层，接触界面截然，产状与围岩产状近一致，外接触带发育宽约 500m 的矽线石片麻岩或片理化角岩、片理化大理岩热接触变质带，并发育有与岩体形态协调的环状断裂和放射状断裂，揭示岩体侵位时热动力对围岩构造的强迫一致影响（图 20-79）。

岩体中发育平行于接触面的面状组构及压扁的捕虏体，岩体边部，向中央逐渐减弱并消失。这些面理组构在西北部倾角较陡（80°~90°），东南侧较缓（40°~60°）。岩体中的捕虏体以暗色闪长岩最发育（约占 97%），它们也不同程度地发生压扁变形，且压扁面与岩体接触面近于平行。在岩体中央部位，捕虏体基本没有变形，而越靠近岩体边部压扁程度越高，且在西部边缘为最强。

岩体西北缘还发育一条呈新月形的强变形带（韧性剪切带），长约 6km，最宽处 0.7km。该变形带略呈对称分布，以花岗闪长质糜棱岩带为中心，向两侧依次为糜棱岩化花岗闪长岩和花岗闪长岩微变形带。带内还发育同构造透入性片理和片麻理。在靠近岩体的围岩及岩体内接触带上，也发育一系列与岩体侵位密切相关的变形构造及变质岩石，如靠近岩体的围岩中普遍发育一组与岩体边界平行的呈环状分布的挤压片理；近岩体围岩普遍变薄，砂砾岩中的砾石明显压扁拉长；岩浆侵位时使先期褶皱紧闭，产生轴面劈理；受热变质和动力变质共同作用的影响，形成热动力变质带等。

岩体岩性岩相复杂，由中酸性岩浆四次脉动侵位形成，呈同心环带状分布。最外侧为暗色细粒石英闪长岩，断续分布在复式岩体边部，但在西北部缺失。向内依次为中粒花岗闪长岩、似斑状花岗闪长岩，中心部位为粗粒巨斑状花岗闪长岩，揭示了气球膨胀式主动强力侵位机制。即岩浆侵位犹如吹

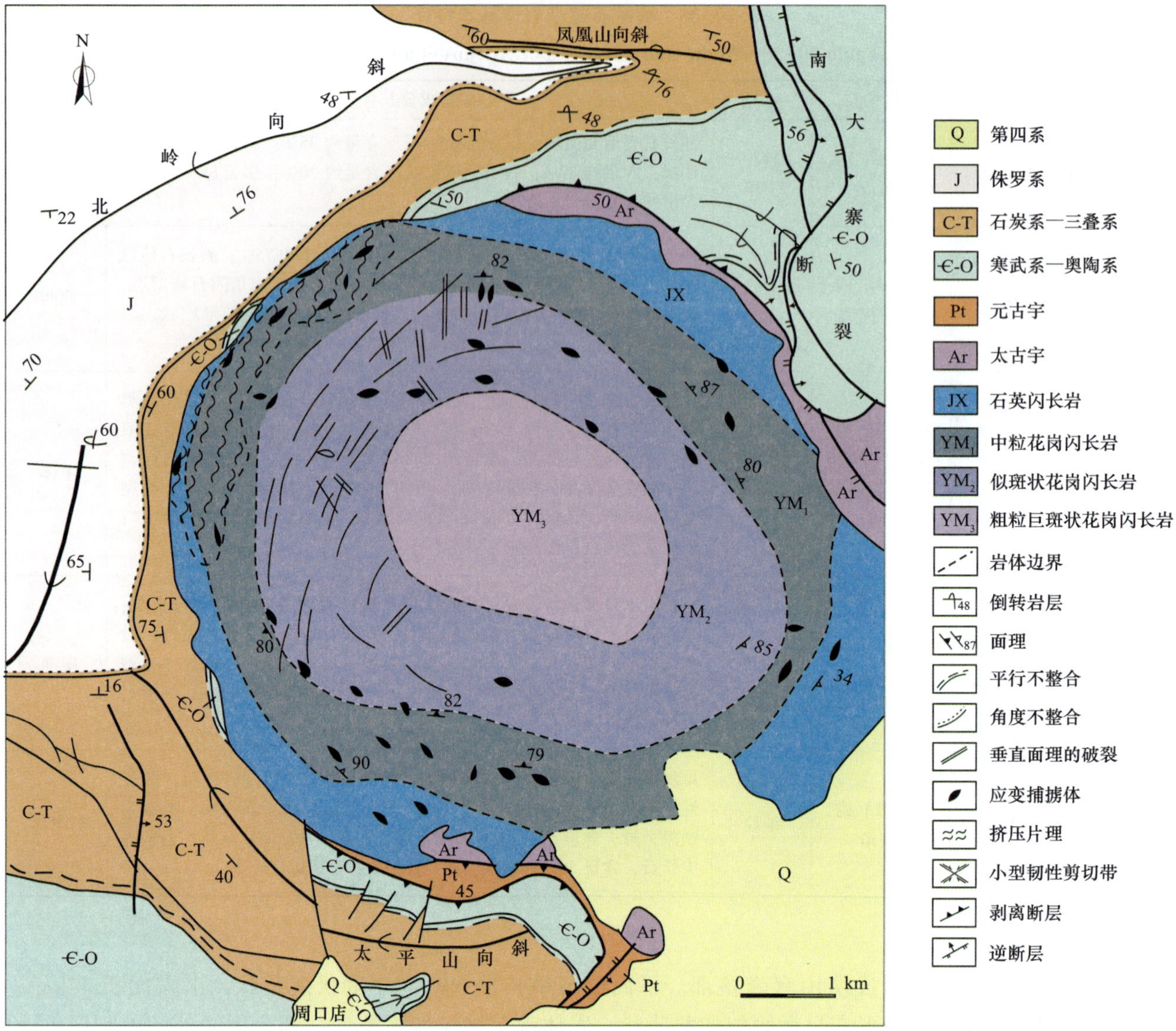

图 20-79　房山杂岩体地质构造略图

（据王人镜等，1990，略有修改）

Fig. 20-79　Geological and structural schematic map of Fangshan complex intrusions

1—倒转岩层；2—面理；3—平行不整合；4—角度不整合；5—垂直面理的破裂；6—应变捕虏体；7—挤压片理；8—小型韧性剪切带；9—剥离断层；10—逆断层；JX_7—房山石英闪长岩；YM_{9-1}—中粒花岗闪长岩；YM_{9-2}—似斑状花岗闪长岩；YM_{9-3}—粗粒巨斑状花岗闪长岩

气球一样，岩浆不断上侵膨胀，由中心向四周推挤围岩或较早侵位的岩体，主动占据空间而侵位（鲍亦冈等，2001）。

3. 大羊山岩体

位于昌平上铺、连石山村，组成大羊山短轴背斜核部，呈长轴 NW 向近椭圆形，出露面积 $20km^2$。岩体侵入中元古界蓟县系雾迷山组中，接触界线清晰，外接触带大理岩化，可见少量白云岩捕虏体，并发育一系列平行于接触带的弧形褶皱（图 20-80）。围岩近接触带处褶皱强烈紧密，围岩地层直立或倒转；由接触带向外，褶皱构造逐渐变得平缓，甚至完全消失，揭示了岩浆侵位强力挤压作用。

岩体岩性岩相复杂，为成分相近的 3 次岩浆脉动侵位形成，由西向东，由外向内，依次为边缘相细粒花岗闪长岩、过渡相中粒黑云母花岗闪长岩、中心相似斑状黑云母花岗闪长岩，各岩相岩性特征见表 20-39。根据岩体结构、产状反映岩体岩浆侵位由东南向西北。后期不均匀剥蚀作用使得平面上岩相带分布不对称。

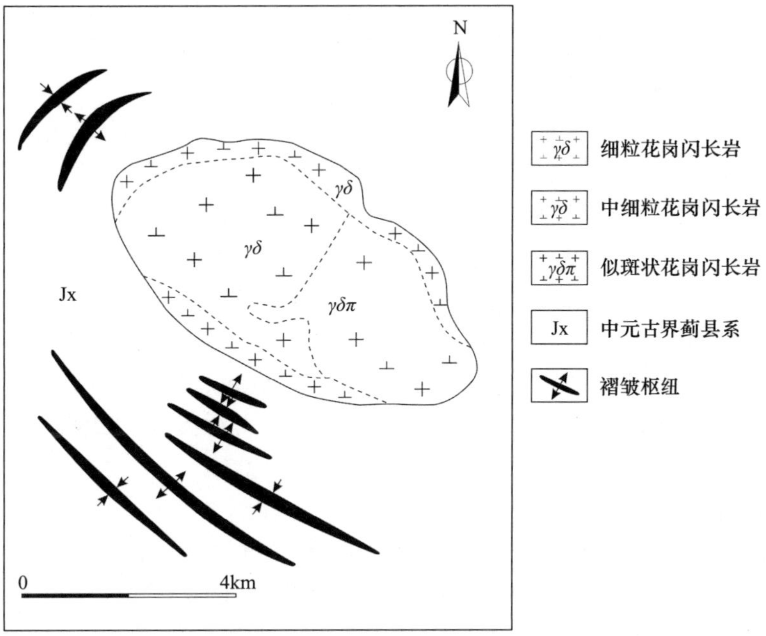

图 20-80　大羊山复式岩体及其围岩地质构造略图
（据鲍亦冈等，2000）
Fig. 20-80　Geological and structural sketch map of Dayangshan complex and its surrounding bed

表 20-39　大羊山杂岩体基本特征

Table 20-39　Main characteristics of Dayangshan complex

序号	岩性	岩相	结构构造	粒径	矿物成分及含量	其他
1	花岗闪长岩	边缘相	花岗结构，块状构造	细粒	钾长石类型为正长石和微斜长石，含量 50%～55%，其中正长石高岭土化，边部被钠长石交代；斜长石发育聚片双晶，类型为更长石，含量 15%～20%；石英他形粒状，含量 20%～25%；黑云母含量<5%；副矿物为榍石、磷灰石、磁铁矿和绿帘石等	流动构造
2	黑云母花岗闪长岩	过渡相	花岗结构、块状构造	中细粒	钾长石类型为正长石、微斜长石和条纹长石，含量约 50%，常交代斜长石；斜长石含量约 20%；石英含量 20%～25%；黑云母含量 5%～15%。副矿物为榍石、磁铁矿、磷灰石、绿帘石、白云母等	
3	似斑状黑云母花岗闪长岩	中心相	似斑状结构，块状构造	中粒—中细粒	斑晶钾长石，大小 0.5～2.0cm，类型为条纹长石，内嵌黑云母、斜长石和榍石微晶，黏土化，含量 5%～10%；基质中细粒，组成及含量基本同过渡相	暗色包体

4. 冯家峪杂岩体

位于密云冯家峪、不老屯，由 10 个小岩珠、体组合而成，呈长轴 NW 向近椭圆形产出，出露面积约 80km^2。岩体南东为密云水库，其余周边侵入新太古宙沙厂英云闪长质片麻岩中。内接触带发育围岩残留体、顶垂体，揭示岩体剥蚀深度有限。岩体岩性岩相复杂，可划分出有冯家峪闪长岩、燕落石英二长闪长岩、南桶子二长花岗岩和柏岔山花岗岩等。

冯家峪闪长岩位于杂岩体中，多呈不规则状，内广泛发育石英二长闪长岩、花岗闪长岩、二长花岗岩、花岗岩、闪长玢岩和煌斑岩等岩脉，以富集角闪石、黑云母和榍石为特征。岩石呈灰黑色，中细粒结构，主要由斜长石（约 60%）、角闪石（约 20%）、黑云母（约 8%）、条纹微斜长石（约 8%）、石英（约 3%）组成，偶见普通辉石残留于角闪石中，副矿物为榍石（约 2%）、磷灰石和磁铁矿等。

燕落石英二长闪长岩位于杂岩体西南至东南边缘，呈半封闭外环，出露面积约 30km^2。内富含塑性变形、定向分布的微粒黑云母闪长质包体。岩石呈灰白色，中粒结构，块状构造，主要由斜长石（约54%）、钾长石（条纹微斜长石，约24%）、黑云母（约7%）、角闪石（约5%）和石英（约9%）组成，副矿物为榍石、磷灰石等。

南桶子二长花岗岩环绕柏岔山（柏岔山、云峰山）花岗岩呈连续环状分布，出露面积20km²。岩体侵入燕落石英二长闪长岩中，外接触带发育岩枝，内接触带可见燕落石英二长闪长岩和冯家峪闪长岩捕虏体，偶见微粒闪长质包体，以边部最为发育。岩石呈浅肉红色，粗粒似斑状结构，斑晶自形条纹长石，粒径约15mm，含量7%~32.7%；基质由斜长石（约38%）、条纹长石（约33%）、石英（约22%）、黑云母（约5%）组成，副矿物为榍石、磷灰石和锆石等。

柏岔山花岗岩由柏岔山、云峰山等小岩株产出于杂岩体内，出露面积共约20km²，地貌上显示为陡峻山峰。岩体侵入上述各岩性中，界线截然。岩石呈浅肉红色，中细粒结构，块状构造，主要由斜长石（约28%）、微斜长石（约21%）、条纹长石（约21%）、石英（约26%）、黑云母（约5%）组成，副矿物为磁铁矿等。

5. 炭厂-河南台岩体

位于门头沟张家山，沿中生代地层间不整合面侵位于中生代髫髻山盆地边部，平面形态呈长轴NE向不规则状，由炭厂、河南台等3个小岩瘤组成，各岩瘤长轴连线平行，揭示了岩浆来源一致性。岩体顺不整合界面侵入中生界侏罗系九龙山组、髫髻山组和土城子组地层中，侵位明显受构造控制。岩体产状变化较大，底部直立陡倾，边部具有小的破碎带和泥化带，向上部逐渐呈顺层状分布，与围岩中生代地层基本平行。

岩体岩性为花岗斑岩，灰白-浅肉红色，斑状结构，致密块状构造。斑晶斜长石自形板条状、长柱状，发育聚片、卡钠复合双晶，局部见环带结构，弱绢云母化，粒径0.3~1.7mm，含量10%~15%；斑晶黑云母黄褐色，微褪色性，边部常析出铁质，含量约3%；角闪石他形粒状，多色性，常蚀变成黑云母，含量<5%。基质细晶—隐晶质，由长石、石英和隐晶质矿物组成，含量>80%。副矿物为磁铁矿、磷灰石、锆石等。

6. 雾灵山岩体

位于密云新城子，向东延伸至河北境内，呈长轴NNE向不规则椭圆形，出露面积230km²。岩体北西、北侵入中生界侏罗系土城子组中，其余周缘侵入中元古代、古生代地层中，接触界线清晰，外接触带发育热接触变质现象，内接触带发育围岩捕虏体和顶垂体。

岩体岩性岩相复杂，划分为中心相似斑状碱长正长岩、边缘相碱性石英正长岩和呈岩墙状碱长正长斑岩（图20-81）。三类岩桐岩石特征见表20-40。似斑状碱长正长岩内还发育少量呈浑圆状包体，具有等粒结构和斑状结构两种类型。前者主要分布在边部，后者多出现在中心。

表20-40 雾灵山杂岩体基本特征

Table 20-40 Main characteristics of Wulingshan complex

序号	岩性	岩相	结构构造	粒径	矿物成分及含量	其他
1	似斑状碱长正长岩	中心相	斑状结构，块状构造	中粒—中粗粒	斜长石（<5%）；普通辉石（约2%）；碱性长石（约80%）；角闪石（3%~5%）；黑云母（3%~6%）；石英（3%~7%）	
2	碱性石英正长斑岩	岩脉	斑状结构，块状构造	中—中细粒	斑晶碱性长石（20%~30%）、普通辉石（约2%）、铁橄榄石微量；基质碱性长石（>50%）、霓辉石（约3%）、亚铁钠闪石（约5%）、石英（5%~10%）	
3	碱性石英正长岩	边缘相	等粒结构，块状构造	中粗粒	反条纹长石（>80%）、石英（8%~15%）、亚铁钠闪石（约8%）、霓辉石	萤石、锆石、磁铁矿
4	碱长（石英）正长斑岩	岩墙状	斑状结构，块状构造	微晶	斑晶碱性长石（30%~50%）；基质碱性长石（>85%）、普通辉石（<2%）、角闪石（<2%）、黑云母（<2%）、石英（<5%）和斜长石（<1%）	

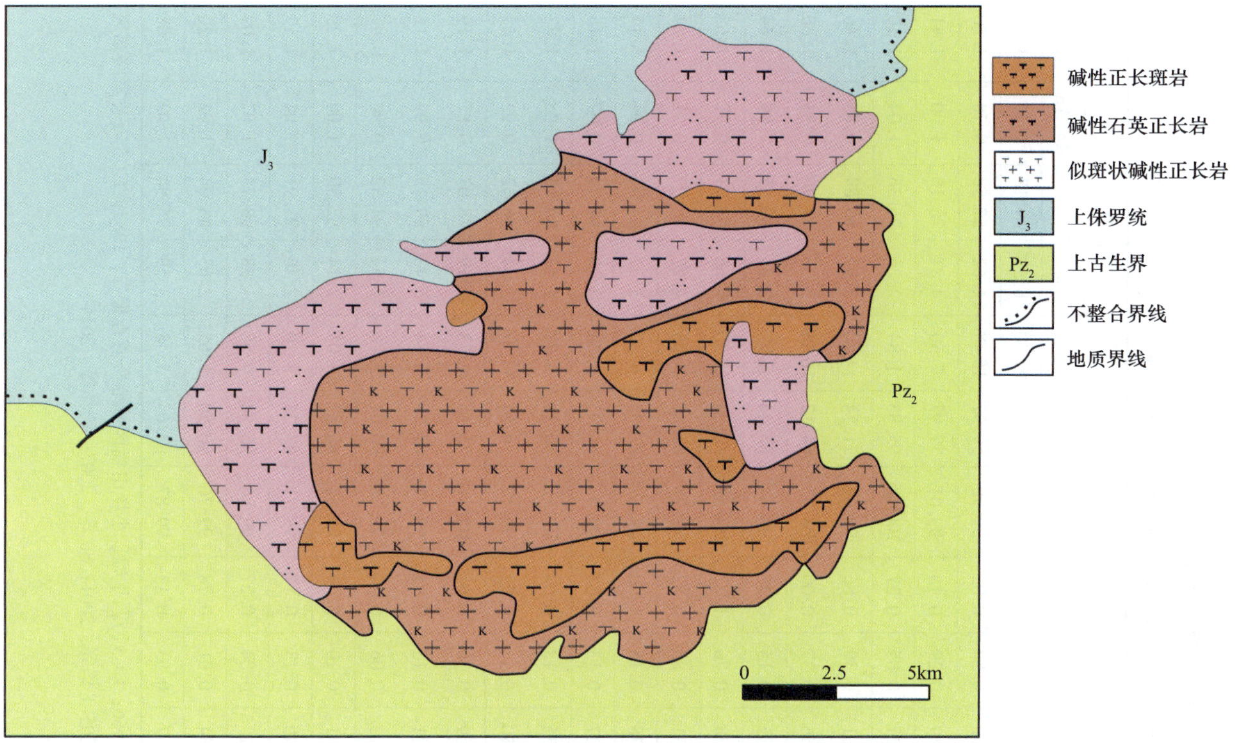

图 20-81　雾灵山杂岩体地质略图

（据许保良等，1995）

Fig. 20-81　Geological Sketch map of Wulingshan complex

（二）岩石化学与地球化学特征

岩石化学成分及特征参数见表 20-41。岩石 SiO_2、FeO^*、Al_2O_3 含量变化范围大；与同类岩石均值相比，呈高 SiO_2、碱，低 CaO、MgO、MnO 特征。里特曼指数 σ=1.88~6.27，跨碱性与亚碱性岩区；固结指数 SI=0.58~23.81，分异指数 DI=47.71~95.83，长英指数 FL=48.91~96.36，岩浆经历较强分异作用；A/CNK=0.78~1.09，A/NK=0.97~1.84，跨准铝质和过铝质岩区，主要属于偏铝质。不同杂岩体，岩浆分异程度不同；同一杂岩体，岩石成分演化趋势明显，SiO_2、K_2O、Na_2O+K_2O 含量和 DI 分异指数逐渐增高，TiO_2、FeO^*、MnO、MgO、P_2O_5 含量和 SI 指数逐渐降低，岩石向酸性、富 K、Al 质演化，也揭示了不同岩类之间可能具有成因关系，而且分离结晶可能起着重要作用。

稀土元素成分及特征参数见表 20-42。中、酸性岩 ΣREE 分别为 200.91 $\times10^{-6}$~386.39 $\times10^{-6}$ 和 97.00 $\times10^{-6}$~352.96 $\times10^{-6}$；LREE/HREE 分别为 11.38~27.19 和 5.85~29.45；$(La/Yb)_N$ 分别为 14.89~78.62 和 4.98~73.37；轻稀土中等-较弱富集，重稀土相对亏损；中、酸性岩 δEu 分别为 0.29~2.22 和 0.33~1.10，显示岩性发育铕负异常—铕正异常。碱性岩（σ>4）δEu 值呈现铕正异常或无铕异常，亚碱性（σ<4）δEu 呈现铕负异常或无铕异常。不同杂岩体稀土总量差异较大，体现了岩浆来源复杂性；同一杂岩体，配分曲线基本相似：表现为右倾-缓倾光滑曲线（图 20-82）；酸性岩稀土总量明显低于中性岩；δEu 值经历了显著变化，这些特征均揭示了岩浆经历了强结晶分异作用，且斜长石可能是重要的结晶分异矿物。不仅如此，岩浆分期侵位在稀土元素特征上也有所差异：同期侵位岩浆稀土元素特征相似，且分异演化特征明显，多数杂岩体（雾灵山岩体除外）伴随岩浆演化 La/Sm、δEu、LREE/HREE 增大，Ce/Yb、La/Yb、Eu/Sm 值逐渐减小，揭示岩浆向重稀土相对富集、轻重稀土分馏程度减弱的方向演化。而杂岩体稍晚期岩浆侵位，如薛家石梁黑熊山、冯家峪柏岔山花岗岩侵位，稀土元素特征同先期明显不同，揭示了独立的岩浆来源。

表 20-41　早白垩世中期中酸性岩岩石化学成分及特征参数一览表

Table 20-41　The chemical composition and main parameters of Early Cretaceous intermediate-acid rocks

序号	岩体	岩性	氧化物含量/%													特征参数					
			SiO_2	TiO_2	Al_2O_3	Fe_2O_3	FeO	MnO	MgO	CaO	Na_2O	K_2O	P_2O_5	Los	总和	*DI*	*SI*	*σ*	*FL*	*A/CNK*	*A/NK*
1	薛家石梁	二长辉长岩	48.77	2.31	16.51	5.76	6.41	0.13	4.96	6.50	3.85	2.44	1.02	0.90	99.54	47.71	21.31	6.27	49.18	0.79	1.84
2	黑山寨	二长岩	55.65	1.66	16.48	4.83	4.13	0.15	3.16	5.65	4.75	2.92	0.50	0.02	99.90	59.32	16.06	4.62	57.58	0.78	1.50
3	湖门	正长岩	61.67	0.80	18.87	2.05	1.51	0.06	1.06	2.99	5.36	4.19	0.29	0.16	99.01	77.97	7.50	4.81	76.16	1.00	1.41
4	黑熊山	花岗岩	76.17	0.17	12.99	0.80	0.30	0.12	0.18	0.33	4.39	4.27	0.05	0.27	100.05	95.83	1.84	2.26	96.36	1.04	1.10
5	房山	花岗闪长岩	63.29	0.69	15.76	1.96	2.89	0.08	2.13	3.87	4.20	3.61	0.30	0.86	99.64	71.74	14.42	2.97	66.89	0.88	1.46
6		石英闪长岩	59.76	0.72	16.73	2.29	3.70	0.05	2.60	5.18	4.02	3.56	0.37	0.00	98.95	64.12	16.09	3.36	59.40	0.84	1.60
7	大羊山	花岗闪长岩	69.12	0.46	16.07	1.04	1.63	0.03	0.79	1.92	4.60	3.35	0.15	0.48	99.64	83.44	6.92	2.41	80.55	1.09	1.44
8	冯家峪	闪长岩	50.40	1.48	16.09	3.84	6.68	0.15	5.38	7.00	4.12	2.58	0.76	1.52	100.00	50.28	23.81	5.66	48.91	0.72	1.68
9	燕落	石英二长闪长岩	63.80	0.66	16.28	2.45	1.56	0.07	1.37	2.66	5.16	4.31	0.82	0.82	99.96	81.17	9.26	4.27	78.07	0.91	1.24
10	南桶子	二长花岗岩	69.42	0.36	14.87	1.41	1.11	0.05	0.80	1.67	4.61	4.78	0.15	1.00	100.23	87.85	6.30	3.32	84.90	0.94	1.17
11	柏岔山	花岗岩	71.10	0.28	14.40	1.01	1.16	0.05	0.51	1.30	4.08	5.27	0.13	0.64	99.93	89.95	4.24	3.10	87.79	0.97	1.16
12	大海陀	花岗岩	72.55	0.58	14.55	1.00	1.23	0.08	0.42	1.06	4.34	4.35	0.18	0.00	100.32	90.11	3.71	2.56	89.13	1.06	1.23
13		二长闪长岩	61.96	0.08	17.36	2.55	2.50	0.10	1.82	4.11	4.63	3.50	0.34	0.22	99.17	70.57	12.16	3.44	66.42	0.92	1.52
14	雾灵山	似斑状碱长正长岩	61.75	1.11	16.50	1.36	3.81	0.87	1.62	2.77	5.78	4.09	0.36	0.65	100.64	76.55	9.73	5.19	78.07	0.87	1.19
15		碱长正长斑岩	64.55	0.51	17.21	1.77	1.90	0.13	0.58	1.33	5.66	5.21	0.13	0.95	99.91	87.29	3.84	5.43	89.13	0.99	1.15
16		碱性石英正长岩	68.71	0.40	14.51	1.52	2.13	0.15	0.41	0.63	5.88	4.93	0.06	0.44	99.75	90.61	2.76	4.53	94.52	0.90	0.97
17	关上	二长花岗岩	72.27	0.18	13.89	1.10	1.25	0.08	0.66	1.68	3.92	4.01	0.10	0.44	99.58	86.82	6.01	2.14	82.50	1.00	1.29
18	卯镇山	正长花岗岩	74.02	0.11	14.15	0.54	0.72	0.03	0.32	0.77	4.75	4.00	0.13	0.17	99.71	93.24	3.10	2.46	91.91	1.04	1.17
19	白查	碱长花岗岩	77.52	0.12	11.92	1.50	0.26	0.03	0.06	0.10	3.91	4.15	0.04	0.27	99.86	96.34	0.58	1.88	98.73	1.07	1.09
20	转山子	正长花岗岩	74.04	0.08	13.43	0.60	1.59	0.03	0.17	1.08	4.61	3.34	0.01	0.78	99.76	90.50	1.65	2.03	88.04	1.02	1.20
21	麻峪房	钾长花岗岩	71.74	0.26	14.61	1.24	0.53	0.04	0.69	0.43	4.50	5.67	0.03	0.13	99.87	93.65	5.48	3.59	95.94	1.02	1.08

注：数据为均值，引自 1∶5 万雁翅幅、阳坊幅区域地质调查报告，1996；1∶5 万沙峪幅区域地质调查报告，1966；1∶5 万昌平幅、小场山幅区域地质调查报告，1987；1∶5 万曹家路幅区域地质调查报告，1997；1∶5 万密云幅、高岭幅、墙子路幅区域地质调查报告，1994；Yang 等，2008；蔡剑辉等，2005 王焰等，2001；Su 等，2007；汪洋等，2014。

表 20-42　早白垩世中期中酸性岩稀土元素成分及特征参数一览表

Table 20-42　REE contents and main parameters of Early Cretaceous intermediate-acid rocks

序号	岩体	岩性	稀土元素含量/10^{-6}															特征参数				
			La	Ce	Pr	Nd	Sm	Eu	Gd	Tb	Dy	Ho	Er	Tm	Yb	Lu	Y	ΣREE	LREE/HREE	$(La/Yb)_N$	δEu	δCe
1	薛家石梁	二长辉长岩	83.10	150.00	20.00	84.60	15.30	3.94	13.00	1.50	7.21	1.34	3.20	0.42	2.42	0.36	16.08	386.39	12.12	24.63	0.85	0.90
2	黑山寨	二长岩	57.91	115.80	13.76	52.40	8.61	2.81	6.63	0.93	4.98	0.81	2.24	0.30	1.86	0.18	21.20	269.22	14.02	22.33	1.14	1.01
3	湖门	正长岩	46.72	87.09	10.22	38.68	5.84	3.85	4.80	0.59	2.95	0.49	1.34	0.18	1.16	0.18	13.20	204.09	16.46	28.89	2.22	0.98
4	黑熊山	花岗岩	29.39	43.08	4.42	11.86	1.55	0.16	1.39	0.23	1.36	0.32	1.10	0.22	1.67	0.27	10.95	97.00	13.79	12.64	0.33	0.93
5	房山	花岗闪长岩	49.82	90.36	11.40	38.88	6.00	1.68	3.75	0.47	2.18	0.41	1.09	0.16	0.93	0.13	9.77	207.26	21.73	38.43	1.08	0.93
6		石英闪长岩	49.32	92.28	12.10	42.67	6.44	1.67	3.69	0.53	1.73	0.30	0.64	0.12	0.45	0.06	7.06	212.00	27.19	78.62	1.05	0.93
7	冯家峪	闪长岩	57.60	114.00	12.20	43.40	6.68	1.81	5.01	0.59	3.26	0.57	1.52	0.20	1.32	0.20	15.40	248.36	18.60	31.30	0.96	1.05
8	燕落	石英二长闪长岩	47.13	90.26	9.99	35.88	5.85	1.66	4.47	0.60	2.75	0.53	1.33	0.21	0.99	0.16	12.64	201.79	17.31	34.32	0.99	1.02
9	南桶子	二长花岗岩	36.20	64.00	7.77	30.60	5.42	1.53	3.92	0.65	2.24	0.49	1.14	0.21	0.65	0.11	9.88	154.93	15.46	39.95	1.01	0.94
10	柏岔山	花岗岩	35.00	71.10	6.49	20.60	3.07	0.62	2.34	0.32	1.79	0.30	0.93	0.13	0.78	0.13	8.96	143.60	20.37	32.19	0.71	1.16
11	大海陀	花岗岩	53.60	96.80	11.20	40.70	6.27	1.86	4.29	0.60	3.08	0.70	2.11	0.27	1.49	0.34	7.77	223.31	16.34	25.80	1.10	0.97
12		二长闪长岩	52.50	91.30	9.73	31.60	4.59	0.90	2.94	0.42	2.70	0.55	1.76	0.24	1.41	0.27	6.63	200.91	18.52	26.71	0.75	0.99
13	雾灵山	似斑状碱长正长岩	51.90	109.00	11.60	51.50	9.40	3.20	7.50	1.10	5.50	1.00	2.60	0.30	2.50	0.30	23.00	257.40	11.38	14.89	1.17	1.09
14		碱长正长斑岩	50.90	93.80	9.60	38.90	6.90	1.90	5.70	0.90	4.10	0.80	2.10	0.30	2.70	0.30	24.00	218.90	11.95	13.52	0.93	1.04
15		碱性石英正长岩	78.30	142.00	13.90	53.20	9.40	0.80	7.50	1.20	6.70	1.40	3.70	0.50	3.20	0.50	33.00	322.30	12.05	17.55	0.29	1.06
16	卯镇山	正长花岗岩	36.15	85.57	9.53	37.48	8.71	0.42	6.93	1.60	8.17	1.72	4.88	0.75	5.21	1.12	46.08	208.24	5.85	4.98	0.17	1.13
17	白查	碱长花岗岩	75.70	162.55	18.35	58.95	9.57	0.09	6.42	1.29	6.88	1.37	4.21	0.86	5.93	0.81	32.45	352.96	11.72	9.16	0.04	1.07
18	转山子	正长花岗岩	38.87	57.76	5.90	19.99	4.27	1.03	2.07	0.30	0.75	0.17	0.49	0.09	0.38	0.09	3.91	132.16	29.45	73.37	1.06	0.94

注：数据为均值，数据来源自同表 20-41。

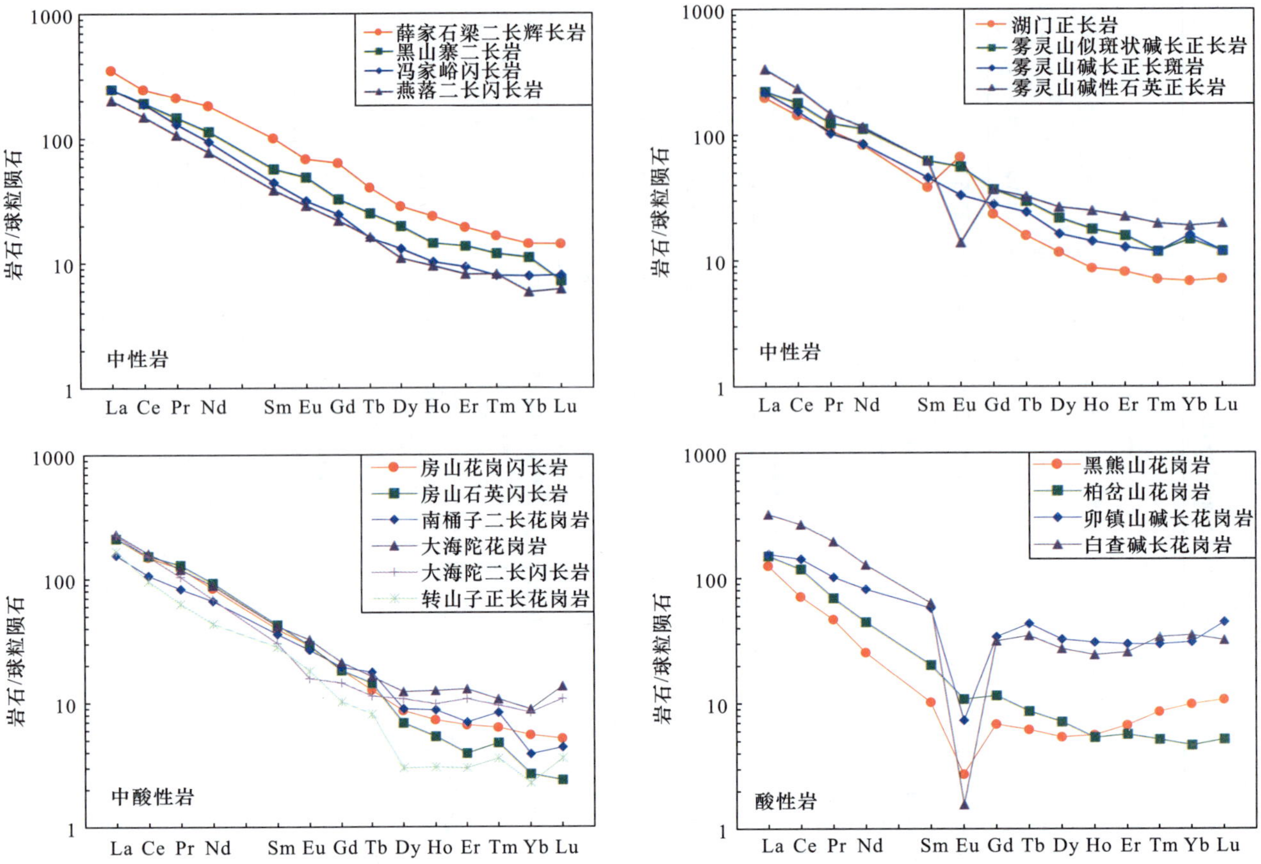

图 20-82　早白垩世中期中酸性岩稀土元素配分模式图

Fig. 20-82　Chondrite-normalized REE patterns of Early Cretaceous intermediate-acid rocks

微量元素含量及特征参数见表 20-43。相比于中国同类岩性均值，微量元素 Rb、Nb 含量偏低，Ba、Hf、Ta 含量偏高，且各杂岩体差异很大，揭示了岩浆来源及其演化复杂性。同一杂岩体，往往伴随 SiO_2 含量增加，各微量元素含量减小，也揭示了岩浆强烈分离结晶演化作用。在微量元素蛛网图（图 20-83）中，碱性岩类（$\sigma>4$）具有相似特征曲线，表现为富集大离子亲石元素 Ba、Sr、K、Rb，尤其是 K 元素，而相对亏损高场强元素 Th、U、Nb、Ta 和重稀土元素，晚期演化后岩性还相对亏损 P、Ti，揭示了磷灰石、钛铁矿矿物可能是重要的分异矿物。亚碱性岩类（$\sigma<4$）则表现出明显不同的微量元素特征（图 20-83），导致这一现象的根本原因是岩浆侵位形成于不同时代且具有不同岩浆来源。黑熊山、柏岔山和白查酸性岩类形成时代较晚，富集大离子亲石元素 Rb、K，亏损 Ba、Sr 和高场强元素 P 和 Ti；而其余中酸性或酸性岩具有和碱性岩相似的微量元素特征，形成时代稍早，富集大离子亲石元素 Ba、Sr、K、Rb，而相对亏损高场强元素 Nb、Ta 和重稀土元素，不同在于 Th、U 亏损不明显。

表 20-43　早白垩世中期中酸性岩微量元素成分及特征参数一览表

Table 20-43　Trace element contents and main parameters of Early Cretaceous intermediate-acid rocks

序号	岩体	岩性	微量元素含量/10^{-6}									特征参数	
			Rb	Ba	Th	U	Nb	Ta	Sr	Zr	Hf	Rb/Sr	Nb/Ta
1	薛家石梁	二长辉长岩	40.65	2128.50	3.20	0.78	21.70	1.13	1492.50	222.00	5.41	0.03	19.20
2	黑山寨	二长岩	40.60	1363.00	3.07	0.75	19.90	1.09	711.00	166.00	3.95	0.06	18.26
3	湖门	正长岩	53.50	4362.00	4.04	0.94	17.30	1.08	1093.00	155.00	3.96	0.05	16.02

续表

序号	岩体	岩性	微量元素含量/10^{-6}									特征参数	
			Rb	Ba	Th	U	Nb	Ta	Sr	Zr	Hf	Rb/Sr	Nb/Ta
4	黑熊山	花岗岩	185.00	19.60	24.39	4.39	42.10	2.72	19.00	125.00	6.19	9.74	15.48
5	房山	花岗闪长岩	70.72	2056.20	10.65	1.18	4.39	0.73	1178.00	187.00	1.10	0.06	6.01
6		石英闪长岩	52.05	1278.50	10.65	0.80	10.09	0.71	1376.50	170.00	1.93	0.04	14.30
7	冯家峪	闪长岩	48.30	1446.00	6.20	0.87	15.70	0.92	1006.00	399.70	7.74	0.05	17.07
8	燕落	石英二长闪长岩	58.00	1728.00	4.30	0.79	17.40	1.04	840.00	319.00	7.35	0.07	16.73
9	南桶子	二长花岗岩	117.00	1341.00	14.90	1.78	20.50	2.04	506.00	239.00	6.36	0.23	10.05
10	柏岔山	花岗岩	147.00	941.00	12.10	2.71	20.40	1.60	389.00	172.00	4.62	0.38	12.75
11	大海陀	花岗岩	97.50	1061.00	10.60	1.85	14.00	0.96	399.50	174.50	5.15	0.24	14.66
12		二长闪长岩	87.00	1084.00	9.10	1.70	26.40	1.78	889.00	302.00	6.40	0.10	14.83
13	雾灵山	似斑状碱长正长岩	92.30				18.00	1.36	0.00	350.00			13.24
14		碱长正长斑岩	83.60				20.50	1.11	0.00	310.00			18.47
15		碱性石英正长岩	101.00				30.70	1.92	0.00	380.00			15.99
16	白查	碱长花岗岩	239.00	28.50	31.97	6.43	99.68	6.27	8.00	466.50	17.44	29.88	15.90

注：数据来源同表20-41。

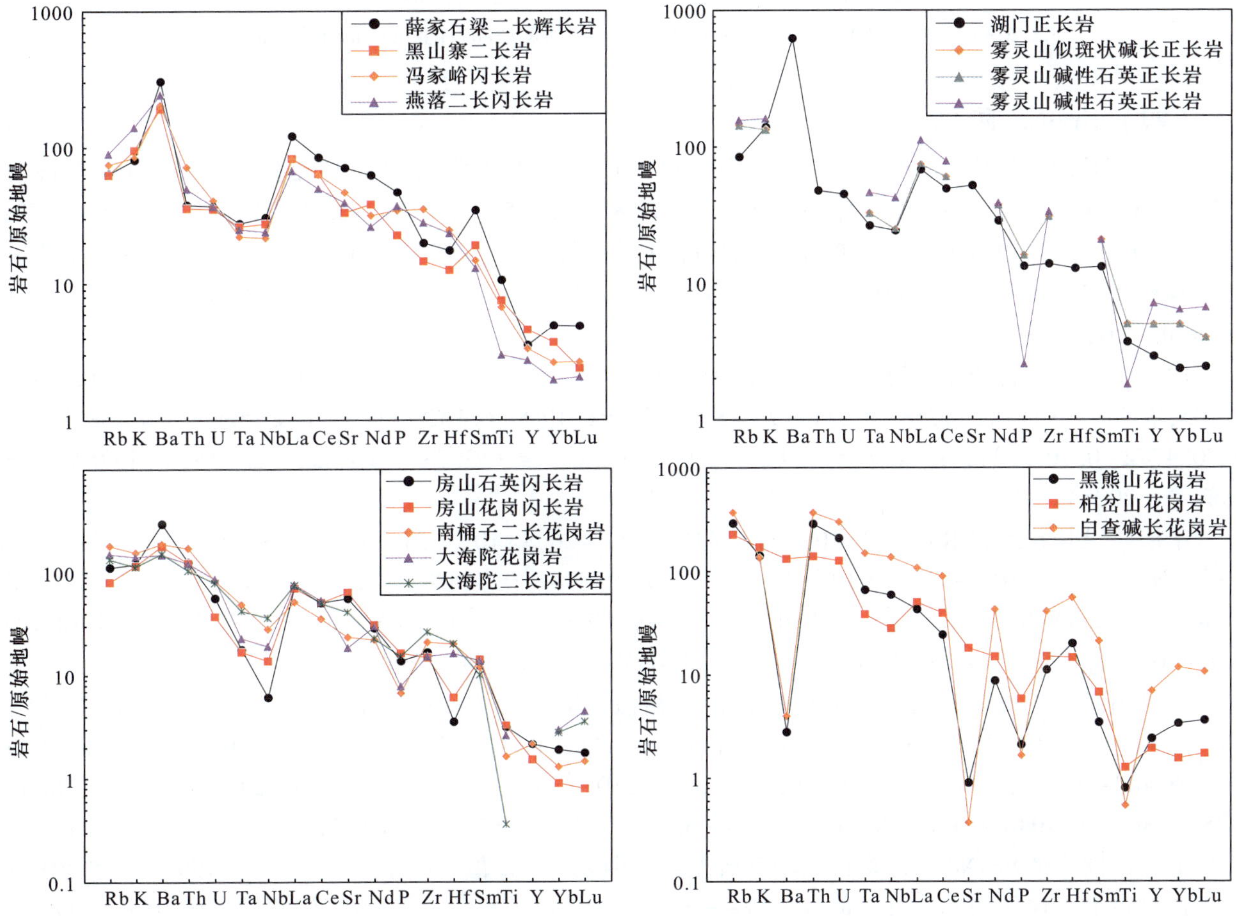

图 20-83　早白垩世中期中酸性岩微量元素蛛网图

Fig. 20-83　Trace element spider diagram of Early Cretaceous intermediate-acid rocks

（三）岩石系列

该套岩系 σ=1.88~6.27，跨碱性与亚碱性岩区，与 SiO_2-Alk 碱图揭示特征一致。根据岩石 SiO_2 含量和主要矿物组成，可划分为中-中酸性类和酸性（偏酸碱性岩）类。前者包括闪长岩、花岗闪长岩、正长岩、石英闪长岩、二长岩和二长花岗岩等，后者则包括碱长、钾长花岗岩和石英碱性正长岩。在 SiO_2-K_2O 图（图 20-84）上，成分点位于高钾钙碱性-钾玄岩系列。

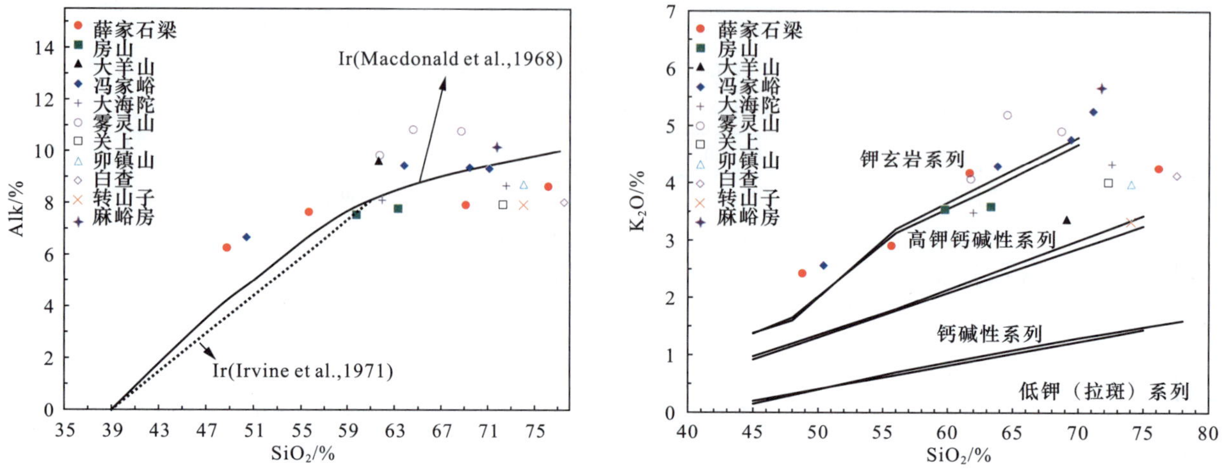

图 20-84　早白垩世中期中酸性岩 SiO_2-Alk 和 SiO_2-K_2O 图

Fig. 20-84　SiO_2-Alk and SiO_2-K_2O diagram of Early Cretaceous intermediate-acid rocks

（四）Sr-Nd 同位素特征

薛家石梁杂岩体中二长辉长岩、二长岩和正长岩同上庄辉长岩具有相似 Sr-Nd 同位素特征：$^{143}Nd/^{144}Nd$=0.511845~0.512129，$^{87}Sr/^{86}Sr$=0.705180~0.705677；取 t=125Ma 时，$\varepsilon_{Nd}(t)$ =-6.79~-12.33，t_{DM}=730~1524Ma；而黑熊山花岗岩特征略有所不同：$^{143}Nd/^{144}Nd$=0.511876~0.511868，$^{87}Sr/^{86}Sr$=0.750406~0.75048；取 t=121Ma 时，$\varepsilon_{Nd}(t)$ =-11.83~-13.10，t_{DM}=910~1483Ma（王焰等，2001；Su et al.，2007；汪洋等，2014）。

房山石英闪长岩和花岗闪长岩 Sr-Nd 特征相似：$^{87}Sr/^{86}Sr$=0.705858~0.705931，$^{87}Rb/^{86}Sr$=0.1496~0.205，$^{143}Nd/^{144}Nd$=0.511858~0.511819，$^{147}Sm/^{144}Nd$=0.1088~0.1025；取 t=130.7Ma 时，$(^{87}Sr/^{86}Sr)_i$=0.7054772~0.7053931，$\varepsilon_{Sr}(t)$ = 14.86~16.06，$\varepsilon_{Nd}(t)$ =-13.75~-14.42，t_{DM}=1822~1875Ma（蔡剑辉等，2005）。

千层背碱长花岗岩 Rb=114×10^{-6}~172×10^{-6}，Sr=26.8×10^{-6}~31.8×10^{-6}，Sm=12.0×10^{-6}~15.4×10^{-6}，Nd=81.8×10^{-6}~97.2×10^{-6}；$^{87}Sr/^{86}Sr$ = 0.724844~0.740888，$^{87}Rb/^{86}Sr$ = 10.35~18.55，$^{143}Nd/^{144}Nd$=0.512069~0.512093，$^{147}Sm/^{144}Nd$=0.0889~0.0959；取 t=128.8Ma 时，$(^{87}Sr/^{86}Sr)_i$=0.70572~0.70836，$\varepsilon_{Nd}(t)$ =-9.0~-9.3，t_{DM}=1318~1365Ma（Yang et al.，2008）。

雾灵山碱长正长岩和碱性石英正长岩特征相似：Rb=49.2×10^{-6}~155×10^{-6}，Sr=14×10^{-6}~353×10^{-6}，Sm=5.98×10^{-6}~15.3×10^{-6}，Nd=34.3×10^{-6}~82.1×10^{-6}；$^{87}Sr/^{86}Sr$=0.706292~0.748840，$^{87}Rb/^{86}Sr$=0.402~20.54，$^{143}Nd/^{144}Nd$=0.512182~0.512250，$^{147}Sm/^{144}Nd$=0.1017~0.1243；取 t=131Ma 时，$(^{87}Sr/^{86}Sr)_i$=0.70467~0.71089，$\varepsilon_{Nd}(t)$ =-6.1~-7.5，t_{DM}=1265~1560Ma（Yang et al.，2008）。

上述不同岩体 Sr-Nd 同位素特征略有所差异，但 $\varepsilon_{Nd}(t)$ 值远高于 Jahn et al.（1999）归纳出的华北克拉通古老下地壳的 $\varepsilon_{Nd}(t)$ 值（-44~-32），而与张国辉等（1998）汉诺坝二辉麻粒岩包体的 $\varepsilon_{Nd}(t)$ 值（-18~-8）比较接近。

（五）Pb 同位素特征

薛家石梁杂岩体各岩性 Pb 同位素特征相似：$^{206}Pb/^{204}Pb$=16.125~16.529、$^{207}Pb/^{204}Pb$=15.152~

15.241、$^{208}Pb/^{204}Pb$=36.02~36.3934（汪洋等，2014）。白查碱性花岗岩 Pb 同位素特征与薛家石梁岩体相似：$^{206}Pb/^{204}Pb$=16.131，$^{207}Pb/^{204}Pb$=15.188，$^{208}Pb/^{204}Pb$=36.146（汪洋等，2009）；而房山岩体 Pb 同位素略有所不同：$^{206}Pb/^{204}Pb$ = 16.5207 ~ 16.6792、$^{207}Pb/^{204}Pb$ = 15.0672 ~ 15.2664、$^{208}Pb/^{204}Pb$ = 35.5131~36.6057（蔡剑辉等，2005）。

上述岩体间 Pb 同位素虽略有所差异，但它们都具有放射性 Pb 同位素强烈亏损特征，其值均低于洋岛玄武岩和洋中脊玄武岩，而高于下地壳麻粒岩 Pb 同位素比值范围的下限。

（六）C-H-O 同位素特征

房山岩体石英闪长岩和花岗闪长岩碳、氢、氧同位素特征相似：$\delta^{13}C$=-4‰~-8‰，与火成碳酸岩和金伯利岩 $\delta^{13}C$ 相吻合，属深成岩；含水矿物氢同位素 δD=-30‰~-70‰，与 Sheppard et al.（1970）确定的地幔中原生水 δD=-48‰比较接近；$\delta^{18}O$=8.5‰~10‰，位于 Taylor（1978）正常花岗岩变化范围内（6‰~10‰）（郑斯成等，1987）。

（七）Lu-Hf 同位素特征

雾灵山岩体不同岩性 Lu-Hf 同位素特征相似：$^{176}Yb/^{177}Hf$ = 0.064789 ~ 0.236016；$^{176}Lu/^{177}Hf$ = 0.001312~0.003888，$^{176}Hf/^{177}Hf$=0.282475~0.282740，$\varepsilon_{Hf}(t)$ = -7.8 ~ -0.4，T_{DM2}=1130~1750Ma，$f_{Lu/Hf}$=-0.97~0.89，揭示岩浆来源于富集地幔与地壳熔融岩浆的混合源区（Yang et al.，2008；陈春良等，2014）。

（八）成因类型与岩浆演化

该套岩系表现为多期脉动侵入接触关系，中性岩、中酸性岩间表现为渐变过渡，具有同源岩浆演化特征。而中性岩（中-酸性岩）与酸性岩、碱性岩间具有痕迹清晰的侵入接触关系，如黑熊山、柏岔子花岗岩分别同薛家石梁、冯家峪闪长岩侵入接触。因此，中性岩、中酸性岩和酸性岩、碱性岩之间可能具有相对独立的岩浆来源。

中性岩和中酸性岩具有典型 I 型花岗岩特征，如 SiO_2 含量变化大，Na_2O>3.2%，*Al/CNK* 和 Na/K 比值等。在 K_2O-Na_2O 和 FeO^*/MgO-（Zr+Nb+Ce+Y）图解（图 20-85）上，成分点均位于 I 型花岗岩区。在（$^{87}Sr/^{86}Sr$）$_i$-$\varepsilon_{Nd}(t)$、$^{206}Pb/^{204}Pb(t)$ -$^{207}Pb/^{204}Pb(t)$ 和 t-$\varepsilon_{Hf}(t)$ 同位素图解（图 20-86）中，成分点均位于华北新生下地壳（t_{DM}<2000Ma）和富集 I 型地幔间，更接近华北下地壳，揭示了其岩浆主要来源于下地壳未经受变质的火成岩再次熔融。同时，中性岩或中酸性岩发育岩性为闪长岩-石英二长岩-花岗岩中酸性岩组合，普遍含有暗色包体，Pb 同位素、$\varepsilon_{Nd}(t)$ 又不同于典型下地壳比值，岩石化学普遍亏损 Nb、Ta 等高场强元素，揭示富集大陆岩石圈地幔也提供了重要的岩浆来源。

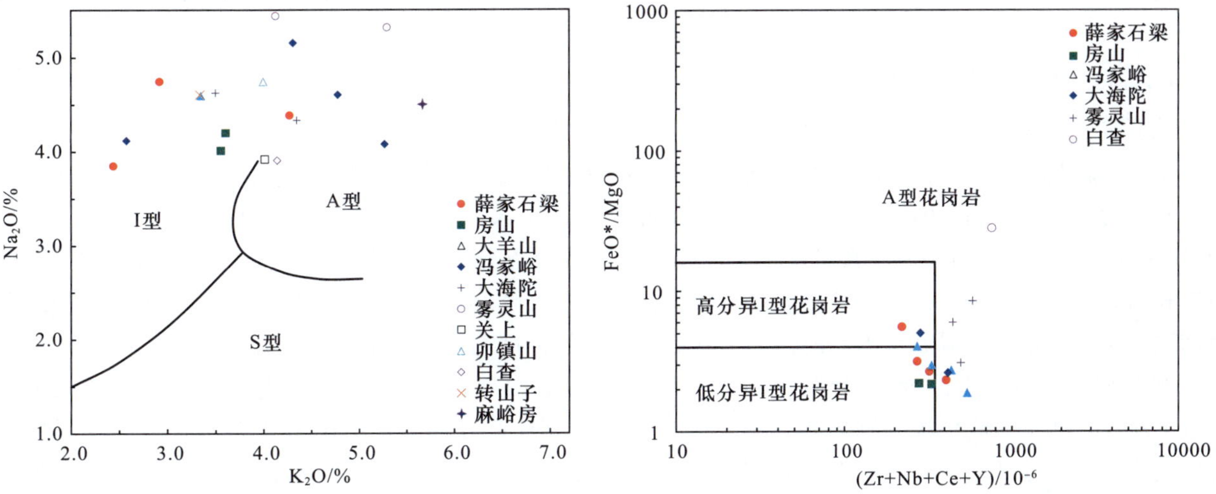

图 20-85 早白垩世中期中酸性岩 K_2O-Na_2O 图解和 FeO^*/MgO-（Zr+Nb+Ce+Y）图解

Fig. 20-85 K_2O-Na_2O and FeO^*/MgO-（Zr+Nb+Ce+Y）diagram of Early Cretaceous intermediate-acid rocks

酸性岩和碱性岩具有典型 A 型花岗岩特征，如发育他形霓辉石、钠闪石等矿物，两种类型条纹长石，高碱含量和 FeO^*/MgO 比值等，且具有还原型 A 型花岗岩特征（汪洋，2009）。在 K_2O-Na_2O 和 $FeO^*/MgO-(Zr+Nb+Ce+Y)$ 图解（图 20-85）上，成分点均位于 A 型花岗岩区。在 $(^{87}Sr/^{86}Sr)_i-\varepsilon_{Nd}(t)$ 和 $t-\varepsilon_{Hf}(t)$ 同位素图解（图 20-86）中，成分点均更接近华北新生下地壳（t_{DM}<2000Ma）区，揭示了其岩浆来源于下地壳未经受变质的火成岩再次熔融。

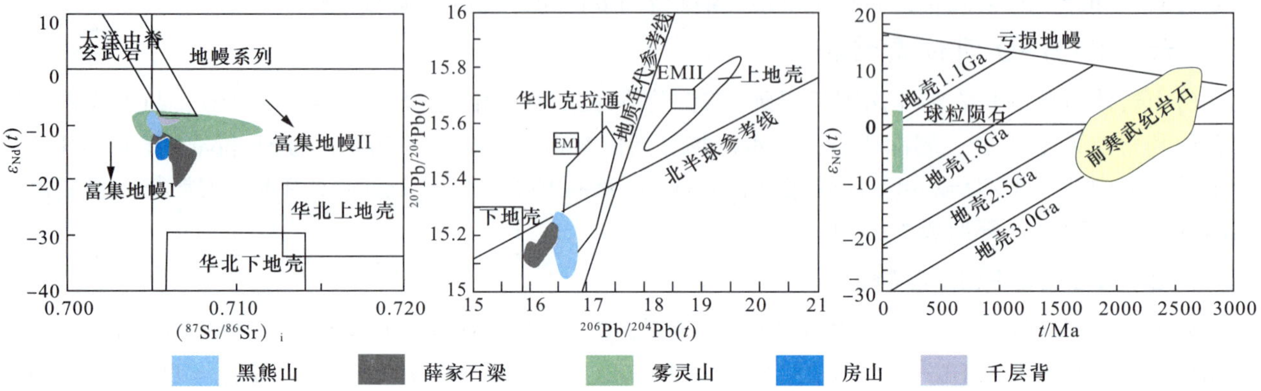

图 20-86 早白垩世中期中—酸性岩 $(^{87}Sr/^{86}Sr)_i-\varepsilon_{Nd}(t)$，$^{206}Pb/^{204}Pb(t)-(^{207}Pb/^{204}Pb(t))$ 和 $t-\varepsilon_{Hf}(t)$ 图解

（数据引自王焰等，2001；蔡剑辉等，2005；Su et al.，2007；Yang et al.，2008；汪洋等，2014；陈春良等，2014）

Fig. 20-86 $(^{87}Sr/^{86}Sr)_i-\varepsilon_{Nd}(t)$，$^{206}Pb/^{204}Pb(t)-^{207}Pb/^{204}Pb(t)$ and $t-\varepsilon_{Hf}(t)$ diagrams of mid-term of Early Cretaceous intermediate-acid rocks

与早白垩世早期杂岩体相似，该套岩系主、微量元素普遍具有相关性，揭示了分离结晶作用在岩浆演化过程中起到了重要作用。如薛家石梁杂岩体早期基性、中性和中酸性岩间呈现经典 AFC 演化过程（汪洋等，2014），即原始幔源岩浆在岩浆房内分异，其早期分离结晶的辉石、斜长石和磁铁矿等下沉形成上庄堆晶含钒钛磁铁矿辉长岩，残余岩浆继续演化形成低 Ti 辉长岩、低 Ti 闪长岩（SiO_2>50%）、二长岩和正长岩；其中湖门正长岩包含了大量岩浆结晶过程中上浮堆积的长石，显示了明显 Eu 的正异常。其后，高 Ti 闪长岩浆注入岩浆房，包裹堆晶含钒钛磁铁矿辉长岩和低 Ti 辉长岩；同时与岩浆房内未固结的中性岩浆发生同位素的交换，导致 Nd 同位素组成的相对均一化（汪洋等，2014）。同时，Sr-Nd 同位素特性还揭示了成岩过程中明显受到上地壳物质的（不均匀）混染和地幔物质的混合。如房山岩体不同岩性斜长石结晶环境研究表明了富钙基性岩浆的混合作用（覃锋等，2006）。

（九）构造环境分析

该套岩系中-中酸性岩和酸性、碱性岩形成于不同构造背景环境。

中-中酸性岩与同期东狼沟组火山岩同期伴生发育，形成钾玄岩和高钾钙碱性火山岩与侵入岩组合，广泛发育为无 Eu 异常的 REE 分布模式，岩浆来源于下地壳部分熔融，并有地幔物质的参与。在 R1-R2 和（Y+Nb）-Rb 构造环境判别图解（图 20-87）中，成分点分别位于晚造山花岗岩+板块碰撞后隆起花岗岩和后碰撞环境区内。因此，该套岩系形成于板块碰撞后隆起-晚造山伸展构造背景环境，是造山运动晚期地壳增厚、地温增高和重力势能作用下开始发生强烈伸展垮塌致使下地壳部分熔融的产物。

酸性和碱性岩属于还原型 A 型花岗岩，形成于后造山（PA）伸展构造背景环境：①在 R1-R2 和（Y+Nb）-Rb 构造环境判别图解（图 20-87）中，成分点分别位于造山后花岗岩和板内花岗岩+后碰撞花岗岩区；②Zr、Nb、Ce 和 Y 含量较低，不同于典型非造山（AA）花岗岩；③短期持续发育，岩浆侵入新生的大陆地壳中；④野外地质产状不同于非造山 AA 型花岗岩，长条状、狭窄的岩石圈裂谷；⑤缺乏与之伴生的超基性、基性岩浆活动，如斜长岩、正长岩；⑥缺乏非造山 A 型花岗岩发育构造背景环境，即岩浆侵位后燕山造山运动还发育一期重要的构造挤压活动（汪洋，2009）。

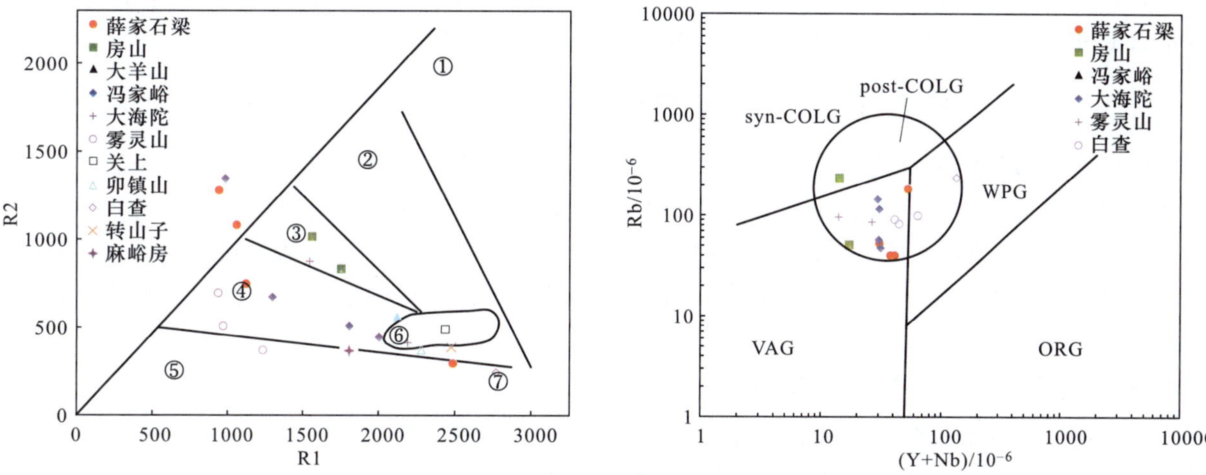

图 20-87　早白垩世中期中酸性岩 R1-R2 和（Y+Nb）-Rb 图解

Fig. 20-87　R1-R2 and（Y+Nb）-Rb diagrams of mid-term of Early Cretaceous intermediate-acid rocks

syn-COLG—同碰撞花岗岩；post-COLG—碰撞后花岗岩；①地幔斜长花岗岩；②板块碰撞前消减地区花岗岩；③板块碰撞后隆起期花岗岩；④晚造山期花岗岩；⑤⑦造山期后 A 型花岗岩；⑥同碰撞花岗岩

（十）时代归属

该套岩系侵位时代厘定为早白垩世中期，部分岩体延续至早白垩世晚期。主要依据包括：①岩体规模巨大，是后造山-晚造山岩浆强烈活动期；②单个侵入杂岩体多呈北西向或南北向，有别于燕山期早期岩体。而岩体组合构成北东向带状中酸性岩浆岩带，同燕山期中期构造形迹特征一致；③岩体普遍未发生变质变形作用，且后期蚀变都很少；④形成于后造山-晚造山地壳逐渐减薄伸展构造背景环境；⑤侵位于中生代燕山早期岩体或晚古生代及其以老地层中；⑥岩体内发育与非造山有关的碱性、基性或超基性岩脉；⑦房山石英二长闪长岩和花岗闪长岩锆石 SHRIMP U-Pb 同位素年龄分别为（131±0.1）Ma 和（130.7±1.4）Ma，花岗闪长岩黑云母 K-Ar 同位素年龄为 132Ma，黑云母$^{40}Ar-^{39}Ar$ 同位素年龄为（132.1±0.4）Ma，花岗闪长岩锆石 U-Pb 同位素年龄为（128.5±1.5）Ma（朱杰辰等，1987；Davis et al.，2001；蔡剑辉等，2005；陶继东等，2006、2009）；上庄辉长岩锆石 SHRIMP U-Pb 同位素年龄为 120.4~129.7Ma，薛家石梁二长辉长岩锆石 SHRIMP 同位素年龄为（130±1.7）Ma，黑山寨二长岩锆石 SHRIMP U-Pb 同位素年龄为（125.1±1.5）Ma，黑熊山花岗岩锆石 SHRIMP U-Pb 同位素年龄为（123.7±1.1）Ma，湖门正长岩锆石 SHRIMP U-Pb 同位素年龄为（124.2±1.8）Ma（Davis et al.，2001；邓晋福等，2004；Su et al.，2007）；双窑沟二长花岗岩锆石 U-Pb 同位素年龄为（129±2）Ma 和（127±1.5）Ma（Davis et al.，1996；孙会一等，2016）；闷虎峪花岗岩锆石 SHRIMP U-Pb 同位素年龄为（130±1）Ma（孙会一等，2016）；冯家峪闪长岩和似斑状花岗岩锆石 SHRIMP U-Pb 同位素年龄均为（127±1）Ma（孙会一等，2016），二长花岗岩锆石 U-Pb 同位素年龄为（127.4±1.2）Ma（欧阳志侠，2010）；白查碱长花岗岩锆石 U-Pb 同位素年龄为（127±0.7）Ma（邓晋福等，2004）；千层背碱长花岗岩锆石 U-Pb 同位素年龄为（128.8±1.5）Ma（Davis et al.，2001）；转山子正长花岗岩黑云母 K-Ar 同位素年龄为 126.25Ma（1∶5 万杨木栅子幅区域地质调查报告，1997）；卯镇山正长花岗岩锆石 U-Pb 同位素年龄为 131.11Ma（1∶5 万杨木栅子幅区域地质调查报告，1997）；林字台石英斑岩锆石 U-Pb 同位素年龄为（129.3±3）Ma（鲍亦冈等，2000）；大东沟二长花岗岩锆石 U-Pb 同位素年龄为（127±2）Ma（Davis et al.，1996）；关上二长花岗岩锆石 U-Pb 同位素年龄为（129±2.5）Ma（郁建华等，1994）；南窑石英二长岩和花岗岩锆石 U-Pb 同位素年龄分别为（128.4±1.5）Ma 至（134±1）Ma 和（128.4±1.5）Ma（Davis et al.，2001；张璐等，2008）；莲花山、南猴顶、月亮鞍梁二长花岗岩黑云母 K-Ar 同位素年龄分别为（125.7±1.8）Ma、（127.23±1.5）Ma

和（126.25±2.3）Ma（鲍亦冈等，2000）；雾灵山石英正长岩和碱长正长岩锆石 U-Pb 同位素年龄分别为（131.7±1.5）Ma 和（130±1）Ma（Davis et al.，2001；Yang et al.，2008）；西白莲峪石英二长闪长岩锆石 U-Pb 同位素年龄为（126.6±1.8）Ma（欧阳志侠，2010）。

九、早白垩世晚期中酸性侵入岩（125～100.5Ma）

区内发育较多早白垩世晚期中酸性侵入岩，岩体数量仍较多，但分布零散，且单个岩体规模一般较小，多以岩脉、岩枝产出，少数岩珠状产出，岩体形态更为复杂，受控于早期构造及围岩地质体。岩体宏观多呈 NNE 产出；侵位深度不大，多为中浅成-浅成相；岩石结构较细、致密坚硬，抗风化能力强，地貌上多形成小的高凸山峰或山脊；岩体一般岩性较为单一，且普遍不发育脉岩及暗色包体。

（一）地质产状与岩石学特征

发育阳坊、四桥子、李家营、上帽山、九渡河、半沟、红螺寺等众多岩体，分散分布于区内山区。岩体岩性酸性岩为主，岩性岩相简单。对主要岩体分述如下：

1. 阳坊岩体

位于昌平南安河、车耳营、阳坊，发育近 SN 向新月形复式杂岩体（图 20-88），出露面积 10km^2。根据航磁和钻探资料，岩体东第四系堆积物之下隐伏产出，组成完整长轴 NNE 向椭圆形杂岩体。其余周缘侵入中生界侏罗系土城子组、髫髻山组和古生界奥陶系地层中，接触界线清晰，内接触带发育冷凝边和边缘相细粒岩带，外接触带角岩化、透辉石大理岩化。

岩体由大觉寺闪长岩、车耳营黑云母花岗岩和白虎涧白岗岩组成。其中，大觉寺闪长岩中发育片麻状构造；车耳营黑云母花岗岩分布范围最广，又可划分出边缘相中细粒黑云花岗岩和中心相中粗粒似斑状黑云母花岗岩，两者之间无明显的界线，呈现涌动接触关系。各岩性矿物组成及含量特征见表 20-44。

表 20-44　阳坊杂岩体主要岩石类型及其特征

Table 20-44　Main rock types and its characteristics of Yangfang complex

岩体	期次	岩性	特征描述
阳坊	大觉寺	闪长岩	深灰—灰黑色，不等粒镶嵌结构，致密块状构造。斜长岩自形长柱状、板条状，发育聚片双晶，弱绢云母化，边部多被钾长石交代，粒径 0.5～1.5mm，含量 65%～70%；角闪石自形柱状，多色性，解理发育，内嵌榍石包裹体，纤闪石化，常与黑云母呈团块状分布，0.5～1.5mm，含量 25%～30%；黑云母自形片状，具褪色性，微具定向，边部铁质析出，含量<5%。副矿物为磁铁矿、磷灰石、榍石等
	车耳营	黑云母花岗岩	灰白—白色，不等粒镶嵌结构，致密块状构造。钾长石半自形板粒状，发育条纹结构，常见钠化边，粒径普遍 3～5mm，小者仅 0.5mm，含量 50%～55%；斜长石半自形板条状，发育聚片双晶、环带结构，微弱绢云母化，粒径 0.1～0.8mm，大者达 4mm，含量 10%～15%；石英他形粒状，粒径 0.8～3mm，局部填隙结构，有时可见波状消光，多呈集合体产出，含量 30%～35%；黑云母半自形片状，褐色或金黄色，边部常有铁质析出，内嵌磷灰石包裹体，含量<5%。副矿物为磁铁矿、磷灰石、榍石、锆石等
	白虎涧	白岗岩	白色—红色，粒状镶嵌结构，致密块状构造。钾长石半自形板状、短柱状，发育条纹、文象结构，粒径 0.5～1.0mm，轻微绢云母化，含量 35%～40%；斜长石自形板条状、长柱状，发育聚片双晶，集合体产出，粒径 0.5～1.0mm，轻微绢云母化，含量 25%～30%；石英他形粒状，波状消光，多呈聚斑出现，发育云棱锥状截面，粒径 0.5～1.0mm，含量 20%～25%。副矿物为磁铁矿、磷灰石、榍石、锆石、黄铁矿、黄铜矿、方铅矿、闪锌矿等

2. 四桥子岩体

位于延庆居庸关四桥子，呈近 SN 向长条形，出露面积 6km^2。岩体南侧侵入中元古界蓟县系高于庄组中，其余周缘侵入早白垩世早期八达岭斑状花岗岩、九仙庙石英二长岩等岩体和下白垩统期张家口组中，岩体还切断了NNW向南口岩脉群（图20-89），接触界线清晰，外接触带岩墙群发育明显烘烤边，内接触带发育冷凝边，并见不规则花岗质细脉侵入岩脉群中。

岩体岩性为浅灰白色—粉红色细粒花岗岩，中细粒花岗结构，块状构造。钾长石他形粒状，粒径1~3mm，类型多为条纹长石，含量约55%；斜长石他形粒状，多被条纹长石交代成残余状，类型为更长石，粒径1~2mm，含量<10%；石英他形粒状，粒径0.5~2mm，含量25%~35%；黑云母半自形片状，粒径<0.1mm，含量1%~3%。副矿物为榍石、磁铁矿等。

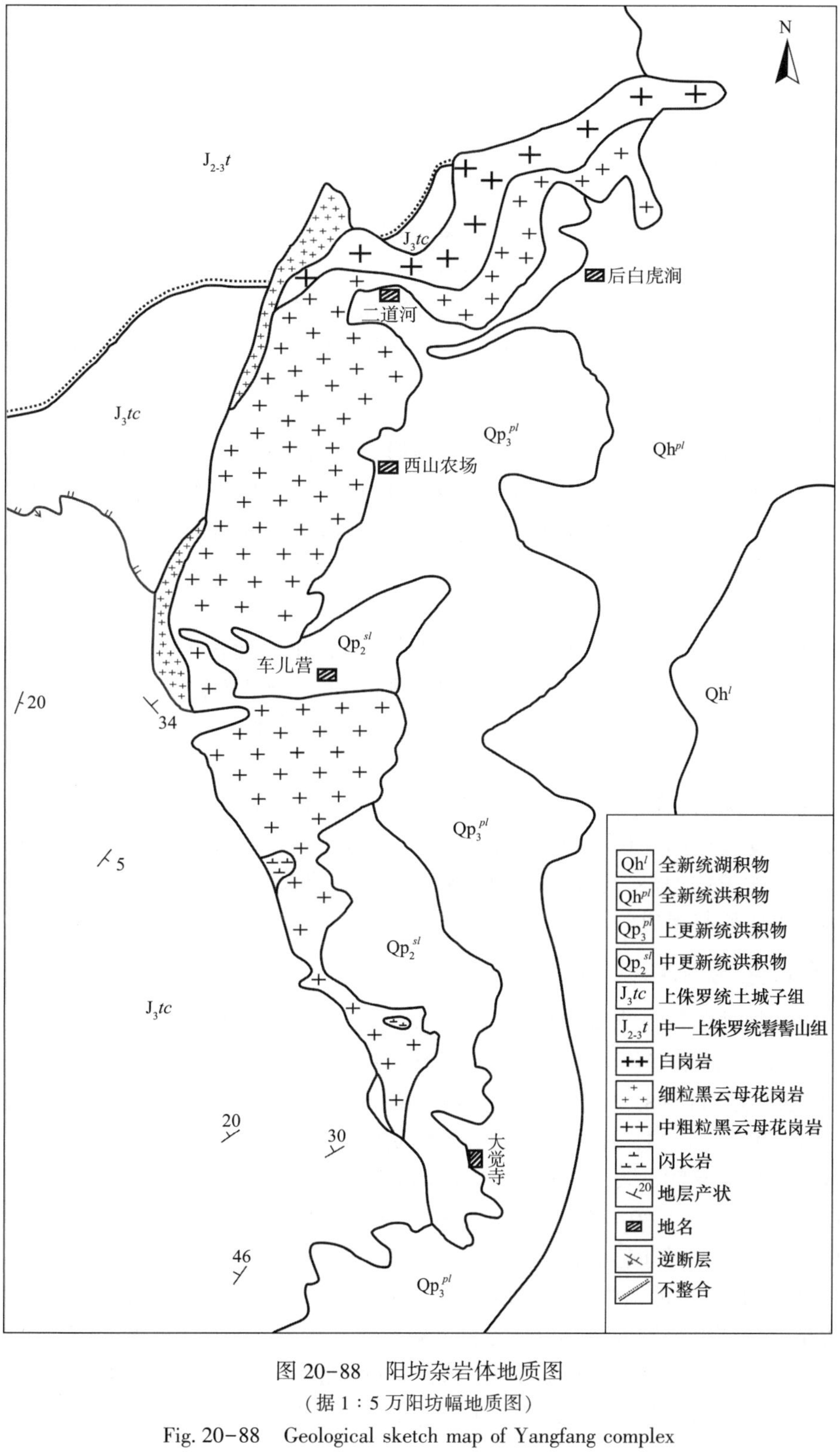

图20-88 阳坊杂岩体地质图

（据1：5万阳坊幅地质图）

Fig. 20-88 Geological sketch map of Yangfang complex

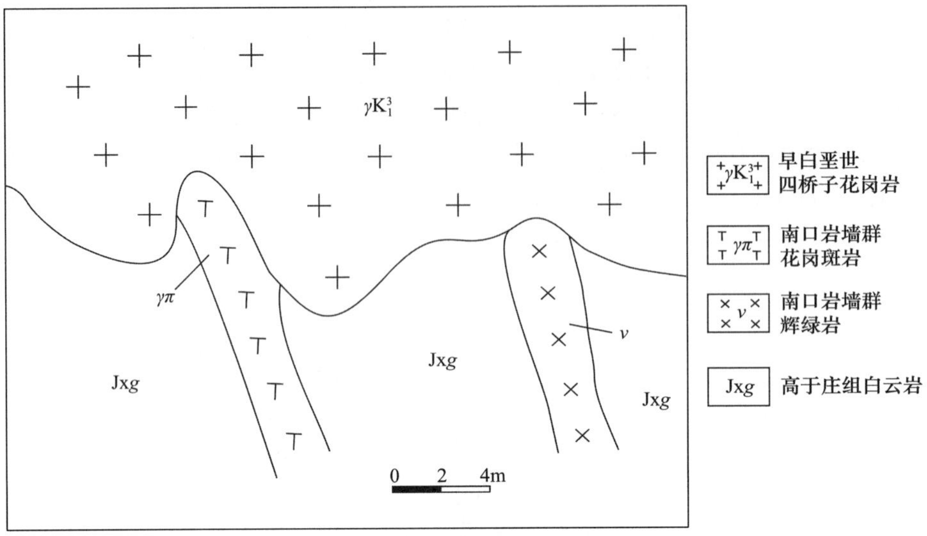

图 20-89 四桥子岩体与南口岩墙群关系

Fig. 20-89 Geological Sketch showing the intrusive contact between Nankou dyke swarms and Siqiaozi intrusion

3. 西石虎岩体

位于昌平大水泉、西石虎村，呈长轴 NE 向椭圆形，出露面积 6km^2。岩体侵位受控于九渡河-十三陵断裂，以倾向 SE、倾角约 50°分别侵入中元古界青白口系、古生界寒武系地层（NW）和中生界侏罗系九龙山组（SE）中。岩体岩性为霏细岩，呈浅灰色，微晶霏细结构，致密块状构造，主要由长英质矿物组成，粒径 0.1～0.5mm。

4. 九渡河岩体

位于怀柔九渡河、梁根，呈 NNE 向不规则状岩株产出，出露面积 12.5km^2。岩体南东侧侵入晚侏罗世阳聂村西辉长岩中，使得辉长岩体支离破碎不规则；北东侧被晚白垩世兴隆城辉绿岩脉侵入；其余周缘侵入中元古界青白口系和古生界寒武系中，侵入界线清晰，接触面多顺层，外接触带局部角岩化。

岩体由中心相花岗斑岩和边缘相石英斑岩相带组成，局部还发育正长斑岩。岩石斑状结构，块状构造，弱绢云母化蚀变。斑晶钾长石半自形板状，类型为条纹长石和正长石，含量 3%～5%；斑晶斜长石类型为更长石，含量约 1%；斑晶石英他形粒状，发育高温熔蚀现象，含量 4%～10%。基质隐晶质结构，主要由长英质矿物组成。

（二）岩石化学与地球化学特征

岩石化学成分及特征参数见表 20-45。岩石呈高 SiO_2、碱，低 CaO、MgO、MnO 特征。里特曼指数 $\sigma=2.27\sim4.21$，除三道梁岩体 $\sigma>4$ 外，其余岩体 $\sigma<4$，主体为亚碱性系列；固结指数 $SI=0.38\sim17.47$，分异指数 $DI=68.06\sim98.39$，长英指数 $FL=66.59\sim98.59$，岩浆经历较强分异作用；$A/CNK=0.91\sim1.17$，$A/NK=0.99\sim1.53$，属于偏铝质—过铝质岩。

稀土元素成分及特征参数见表 20-46。$\Sigma REE=83.44\times10^{-6}\sim251.72\times10^{-6}$，酸性岩稀土总量明显低于中酸性岩；LREE/HREE=9.70～40.10；$(La/Yb)_N=17.48\sim59.76$，轻重稀土中等—强分馏；$\delta Eu=0.21\sim1.45$，发育铕负异常—铕正异常，揭示了岩浆经历了分异结晶作用；配分曲线表现为右倾—缓倾光滑曲线（图 20-90），轻稀土中等—较强富集，重稀土相对亏损。

微量元素含量及特征参数见表 20-47，各岩体微量元素含量差异不大。相比于中国同类岩性均值，微量元素 Ba 含量偏高，其余元素偏低或相当。根据微量元素特征可分为两种类型，SiO_2 含量偏低中酸性岩富集 Rb、K、LREE，强烈富集元素 Ba，相对亏损 Ta、Nb、HREE，闪长岩还富集元素 Zr、Hf；

表 20-45　早白垩世晚期中酸性岩岩石化学成分及特征参数一览表

Table 20-45　The chemical composition and main parameters of Early Cretaceous intermediate-acid rocks

序号	岩体	岩性	氧化物含量/%													特征参数					
			SiO_2	TiO_2	Al_2O_3	Fe_2O_3	FeO	MnO	MgO	CaO	Na_2O	K_2O	P_2O_5	Los	总和	*DI*	*SI*	σ	*FL*	*A/CNK*	*A/NK*
1	大觉寺	闪长岩	60.53	1.09	16.55	3.77	1.69	0.07	2.80	3.98	3.96	3.97	0.48	0.56	99.45	68.06	17.47	3.52	66.59	0.92	1.53
2	车耳营	黑云母花岗岩	74.20	0.22	13.69	1.34	0.28	0.03	0.80	0.41	4.52	4.24	0.06	0.43	100.22	93.09	7.19	2.46	95.55	1.07	1.14
3	白虎涧	白岗岩	77.55	0.08	12.08	0.50	0.27	0.02	0.04	0.13	4.15	4.70	0.02	0.18	99.72	98.39	0.38	2.27	98.59	0.99	1.01
4	三道梁	石英正长斑岩	65.55	0.60	16.17	2.15	1.70	0.08	0.67	1.94	6.45	3.35	0.16	0.38	99.20	86.10	4.69	4.21	83.48	0.91	1.14
5	上帽山	石英正长斑岩	63.44	0.80	16.59	3.40	2.23	0.10	1.20	3.78	5.06	2.49	0.31	0.12	99.52	72.58	8.41	2.77	66.64	0.93	1.51
6	九渡河	花岗岩	75.79	0.08	12.93	0.88	0.48	0.06	0.19	0.34	4.57	4.40	0.04	0.38	100.14	96.07	1.81	2.45	96.35	1.00	1.05
7	四桥子	花岗岩	72.42	0.15	13.76	1.34	0.98	0.07	0.38	0.50	5.04	4.71	0.05	0.20	99.60	94.42	3.06	3.22	95.12	0.96	1.03
8	闷虎峪	钾长花岗岩	71.78	0.25	14.89	0.89	0.94	0.03	0.08	1.31	4.14	5.24	0.10	0.29	99.90	91.05	0.71	3.05	87.78	1.00	1.19
9	歪脖沟	花岗岩	74.35	0.12	12.66	0.86	1.72	0.11	0.26	0.53	5.15	4.00	0.03	0.11	99.90	93.83	2.17	2.67	94.52	0.92	0.99
10	西石虎	霏细岩	73.13	0.18	13.52	1.74	0.50	0.14	0.29	0.21	5.35	4.20	0.07	0.27	99.60	95.57	2.41	3.02	97.85	0.98	1.01
11	红螺寺	花岗岩	73.71	0.22	14.15	1.25	0.68	0.88	0.37	0.13	4.14	4.64	0.03	0.91	101.11	92.35	3.35	2.51	98.54	1.17	1.20

注：数据为均值，来源于1∶5万雁翅幅、阳坊幅区域地质调查报告，1996；1∶5万杨木栅子幅、三道营子、区域地质调查报告，1999；1∶5万昌平幅、小汤山幅区域地质调查报告，1986；1∶5万沙峪幅区域地质调查报告，1966；程素华等，2010；孙会一等，2016。

表 20-46　早白垩世晚期中酸性岩稀土元素成分及特征参数一览表

Table 20-46　REE contents and main parameters of Early Cretaceous intermediate-acid rocks

序号	岩体	岩性	稀土元素含量/10^{-6}															特征参数				
			La	Ce	Pr	Nd	Sm	Eu	Gd	Tb	Dy	Ho	Er	Tm	Yb	Lu	Y	ΣREE	LREE/HREE	$(La/Yb)_N$	δEu	δCe
1	大觉寺	闪长岩	59.10	132.00	16.19	61.58	8.74	2.29	6.01	0.79	3.49	0.58	1.65	0.22	1.31	0.20	15.77	19.63	32.36	0.97	1.05	19.63
2	车耳营	黑云母花岗岩	55.78	88.00	8.90	28.19	3.55	0.73	2.62	0.35	1.54	0.27	0.81	0.12	0.77	0.11	8.51	28.05	52.19	0.73	0.97	28.05
3	白虎涧	白岗岩	32.60	41.00	2.69	4.70	0.39	0.03	0.51	0.07	0.29	0.07	0.30	0.07	0.60	0.12	4.03	40.10	38.97	0.21	1.07	40.10
4	三道梁	石英正长斑岩	49.45	80.07	9.32	32.54	6.04	2.57	4.85	0.57	2.48	0.48	1.30	0.13	0.88	0.10	11.58	16.68	40.31	1.45	0.91	16.68
5	上帽山	石英正长斑岩	56.15	90.92	11.05	36.91	8.11	2.25	4.16	0.48	2.34	0.47	1.23	0.16	0.94	0.15	11.74	20.68	42.85	1.18	0.89	20.68
6	九渡河	花岗岩	39.00	36.31	3.19	12.04	3.18	0.26	2.57	0.50	2.63	0.35	1.45	0.20	1.60	0.39	5.60	9.70	17.48	0.28	0.80	9.70
7	闷虎峪	钾长花岗岩	66.65	124.05	11.40	35.70	4.78	0.90	3.40	0.41	2.03	0.35	1.02	0.13	0.80	0.13	9.48	29.49	59.76	0.68	1.10	29.49
8	歪脖沟	花岗岩	49.70	76.99	9.89	29.50	6.78	1.62	4.53	0.35	1.97	0.40	1.22	0.13	0.64	0.12	0.10	18.64	55.70	0.89	0.85	18.64

注：数据为均值，数据来源同表 20-45。

而 SiO_2 含量偏高酸性岩富集 Rb、K、LREE，相对亏损 Sr、P、HREE，弱亏损 Ta、Nb，白岗岩还强烈富集元素 Th、U 而亏损元素 Ba（图 20-91）。

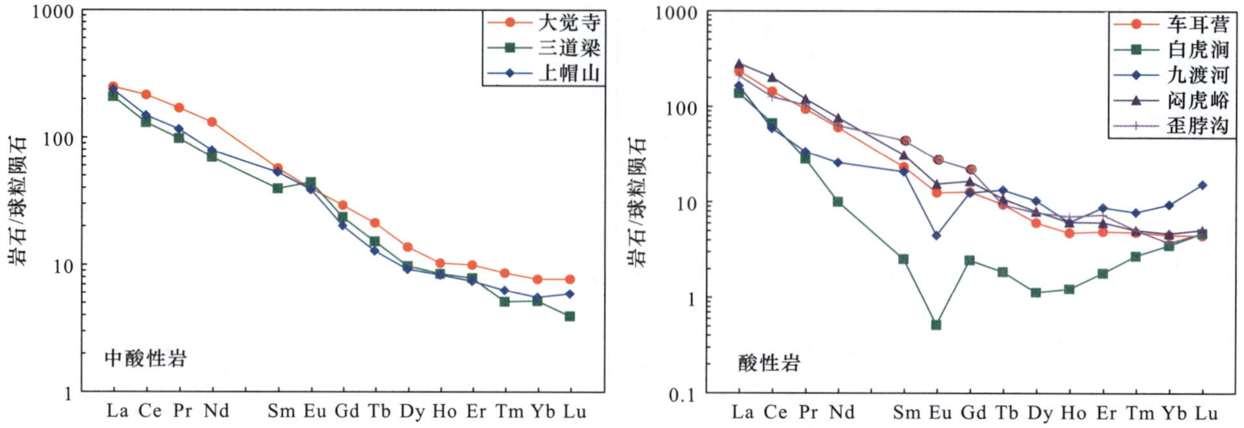

图 20-90　早白垩世晚期中酸性岩稀土元素配分模式图

Fig. 20-90　Chondrite-normalized REE patterns of Early Cretaceous intermediate-acid rocks

表 20-47　早白垩世晚期中酸性岩微量元素成分及特征参数一览表

Table 20-47　Trace element contents and main parameters of Early Cretaceous intermediate-acid rocks

序号	岩体	岩性	微量元素含量/10^{-6}									特征参数	
			Rb	Ba	Th	U	Nb	Ta	Sr	Zr	Hf	Rb/Sr	Nb/Ta
1	大觉寺	闪长岩	69.33	2029.33	5.10	0.53	21.90	1.31	1018.00	276.00	14.50	0.07	16.72
2	车耳营	黑云母花岗岩	108.00	686.67	8.17	1.28	17.20	1.30	260.67	202.33	10.30	0.4	13.26
3	白虎涧	白岗岩	179.00	10.00	18.19	5.12	39.30	2.19	14.00	108.00	6.10	12.79	17.95
4	三道梁	石英正长斑岩	97.40	3736.00	3.75				439.90	74.45		0.22	
7	上帽山	石英正长斑岩	89.90	836.60	16.35				263.90	48.24		0.34	
8	九渡河	花岗岩	114.00										
9	闷虎峪	钾长花岗岩	123.50	1657.00	14.75	1.16	21.75	2.01	437.00	257.00	6.32	0.28	10.82
10	歪脖沟	花岗岩	108.00										

注：数据来源同表 20-45。

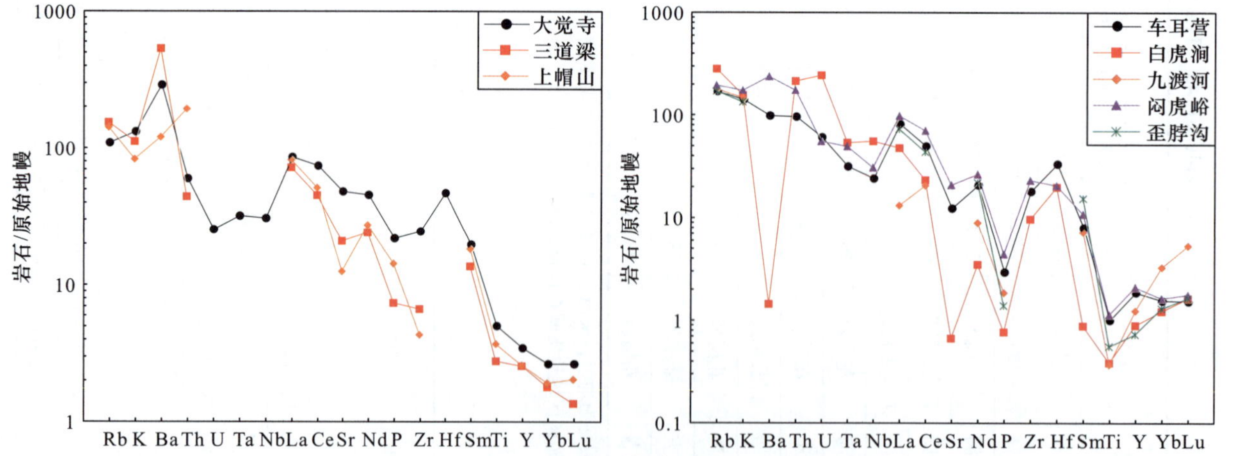

图 20-91　早白垩世晚期中酸性岩微量元素蛛网图

Fig. 20-91　Trace element spider diagram of Early Cretaceous intermediate-acid rocks

（三）岩石系列

该套岩系 $\sigma=2.27\sim4.21$，除三道梁岩体 $\sigma>4$ 外，其余岩体 $\sigma<4$，主体为亚碱性系列；在 Ol'-Ne'-Q'图（20-92）中，成分点均位于亚碱性系列。岩石 $K_2O>3.2\%$，属于钙碱性系列；在 SiO_2-K_2O 图（图 20-92）上，成分点位于高钾钙碱性-钾玄岩系列。

（四）成因类型与岩浆演化

该套岩系具有典型 A 型花岗岩特征，如发育条纹长石，高碱含量和 FeO^*/MgO 比值等；岩体普遍缺乏包体；SiO_2 含量变化小；在 K_2O-Na_2O 和 $FeO^*/MgO-(Zr+Nb+Ce+Y)$ 图解（图 20-93）上，成分点均位于 A 型花岗岩区。岩石地球化学特性，如高 Th/U 值、极弱 Eu 负异常、低 Sr 含量等，揭示了其岩浆可能来源于下地壳基性岩在相对低压下发生失水熔融，且经历了长石和磷灰石的分离结晶作用（程素华等，2010）。

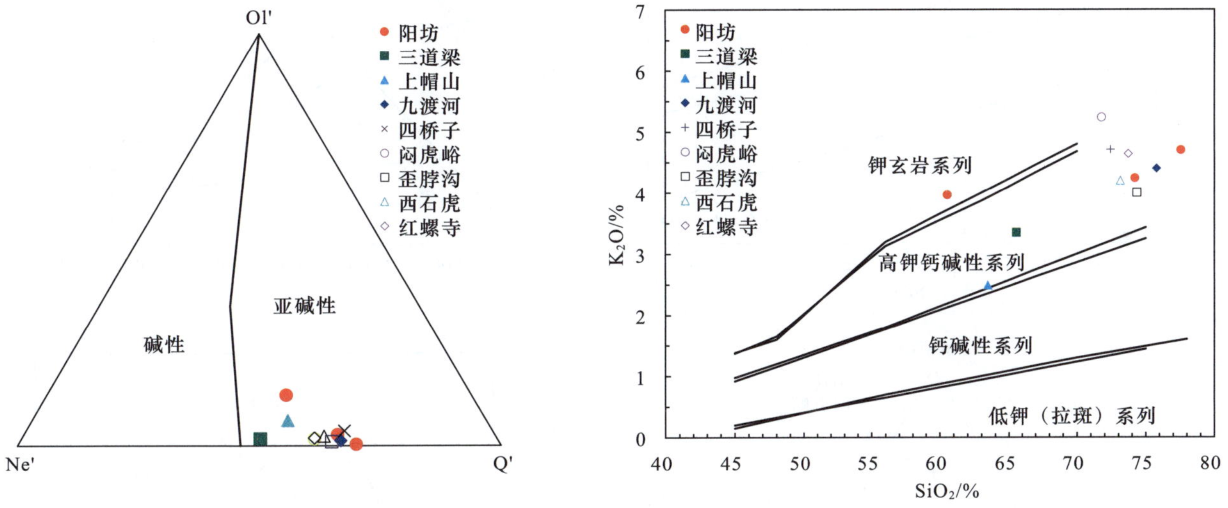

图 20-92 早白垩世中期中酸性岩 SiO_2-Alk 和 SiO_2-K_2O 图

Fig. 20-92 SiO_2-Alk and SiO_2-K_2O diagram of Early Cretaceous intermediate-acid rocks

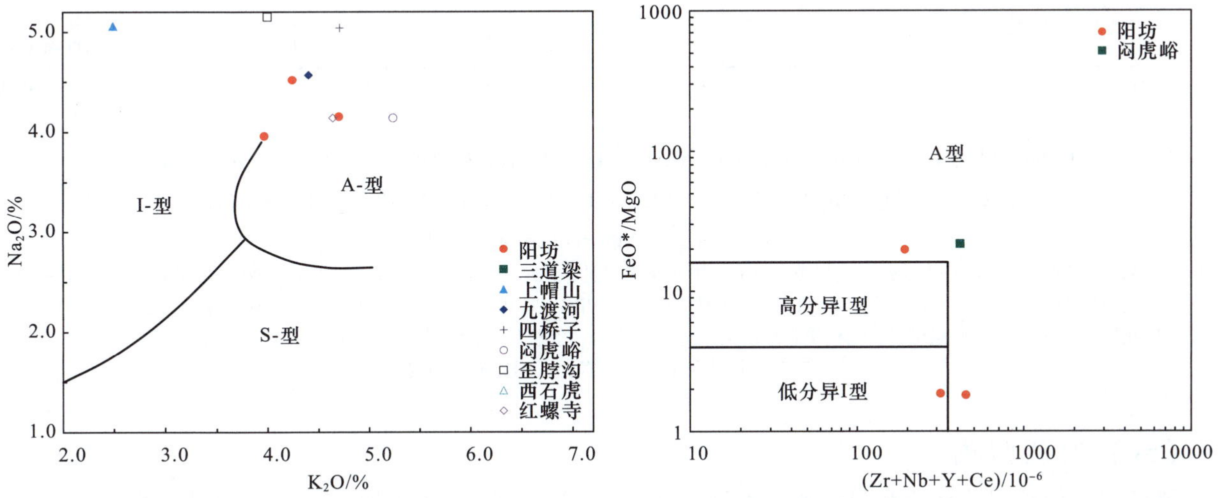

图 20-93 早白垩世晚期中酸性岩 K_2O-Na_2O 图解和 $FeO^*/MgO-(Zr+Nb+Ce+Y)$ 图解

Fig. 20-93 K_2O-Na_2O and $FeO^*/MgO-(Zr+Nb+Ce+Y)$ diagram of Early Cretaceous intermediate-acid rocks

（五）构造环境分析

该套岩系发育为中酸性和酸性岩组合，主要岩性为酸性岩类，为高钾钙碱性-钾玄岩系列，且属于后造山 A 型花岗岩（PA）。在 SiO_2-Al_2O_3、R1-R2、Rb/30-Hf-Ta×3 和（Y+Nb）-Rb 构造环境判别图解（图 20-94）中，酸性花岗岩位于造山后花岗岩或后碰撞花岗岩区，中酸性岩位于后碰撞花岗岩+晚造山花岗岩区。因此，该套岩系主要形成于造山后伸展构造背景环境，表现为燕山造山运动后地壳已经伸展垮塌阶段的岩浆侵位。

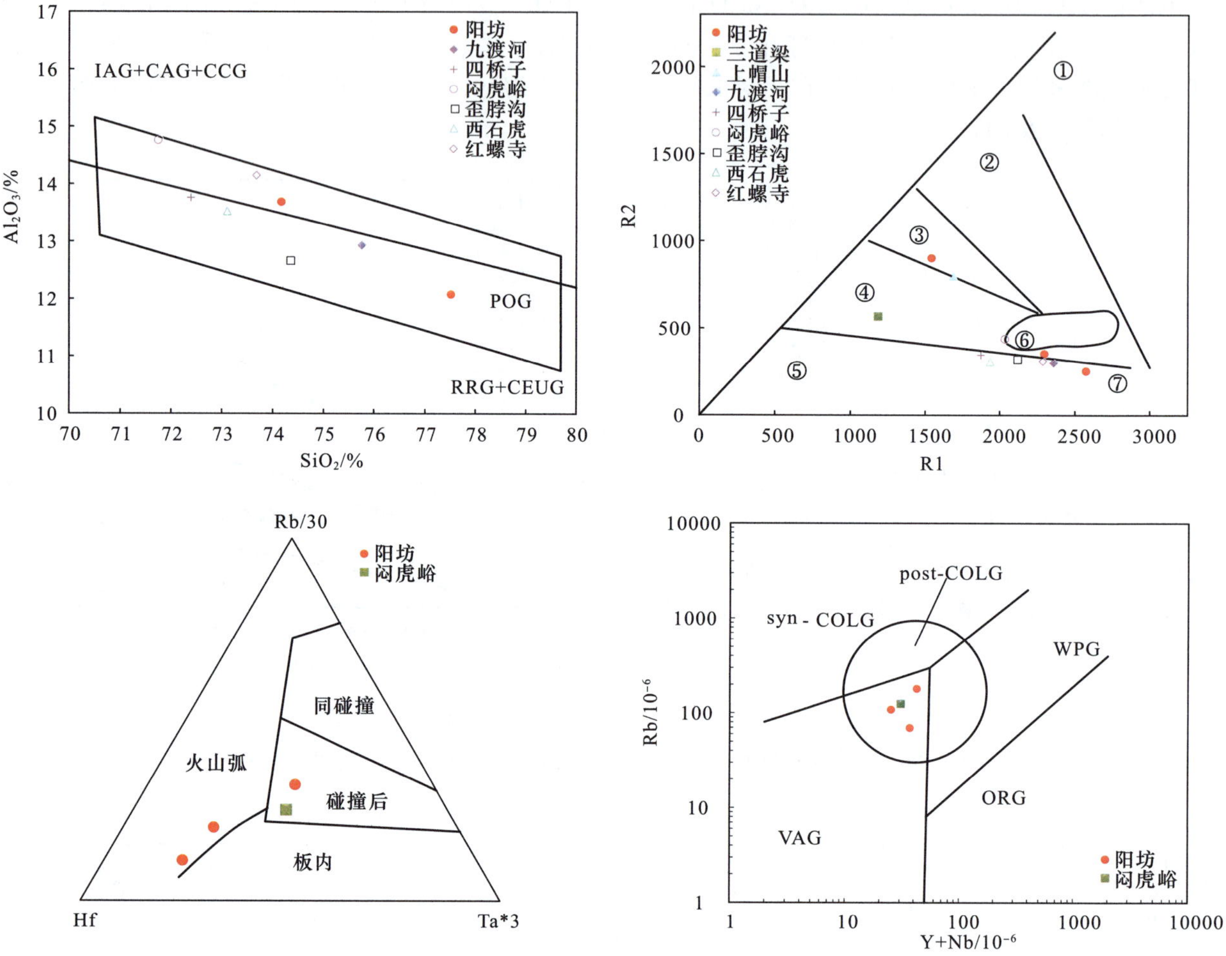

图 20-94　早白垩世晚期中酸性岩 SiO_2-Al_2O_3、R1-R2、Rb/30-Hf-Ta＊3 和（Y+Nb）-Rb 图解

Fig. 20-94　SiO_2-Al_2O_3, R1-R2, Rb/30-Hf-Ta＊3 and（Y+Nb）-Rb diagrams of Early Cretaceous intermediate-acid rocks

IAG—岛弧花岗岩类；CAG—大陆弧花岗岩类；CCG—碰撞花岗岩类；POG—造山后花岗岩类；RRG—裂谷花岗岩类；CEUG—陆内造陆花岗岩类；VAR—火山弧花岗岩；ORG—大洋脊花岗岩；WPG—板内花岗岩；syn-COLG—同碰撞花岗岩；post-COLG—碰撞后花岗岩；①地幔斜长花岗岩；②板块碰撞前消减地区花岗岩；③板块碰撞后隆起期花岗岩；④晚造山期花岗岩；⑤⑦造山期后 A 型花岗岩；⑥同碰撞花岗岩

（六）时代归属

该套岩系侵位时代厘定为早白垩世晚期，主要依据有：①岩体规模较小，呈岩脉、岩瘤或岩枝产出，与造山后弱岩浆活动特性一致；②单个岩体呈 NNE 向，有别于燕山期早期、中期岩体，岩体组合仍成 NE 向中酸性岩浆岩带，同燕山期晚期构造形迹特征一致；③岩体普遍未发生变质变形作用，后期蚀变都很少；④形成于造山后减薄地壳构造背景环境；⑤侵位于中生代燕山早期、中期岩体或中生界白垩系张家口组及更老地层中；⑥岩体内少有岩脉发育，岩性较为单一，酸性岩具有造山后 A 型花

岗岩特性；⑦西沟花岗岩锆石 U-Pb 同位素年龄为（122.5±1.8）Ma（欧阳志侠，2010）；炭厂花岗斑岩全岩 K-Ar 同位素年龄为 123Ma（鲍亦冈等，2000）；四桥子正长花岗岩锆石 U-Pb 同位素年龄为（116±2.3）Ma（郁建华等，1994）；阳坊花岗岩锆石 U-Pb 同位素年龄为（118±2）Ma（Davis et al.，2001）；大海陀花岗岩锆石 U-Pb 同位素年龄为（119±2）Ma（Davis et al.，2001）。

第五节　脉　　岩

一、中元古代与晚古生代脉岩

中元古代、晚古生代岩浆活动范围局限于区内北部山区，规模较小，主要以深成侵入岩为主，少有浅成岩浆活动，脉岩不甚发育。

（一）中元古代脉岩

发育有花岗斑岩、细粒花岗岩，呈细小岩脉状侵入太古宙地质体或同期稍早环斑花岗岩及斑状花岗岩中。岩脉主要分布于同期大型岩体附近，如密云环斑花岗岩体周缘。脉岩走向近 SN，宽 1~3m，长 0.5~1.5km。岩脉未发育蚀变，脉岩岩石学和地球化学特征基本同同期环斑花岗岩、斑状黑云母花岗岩。

（二）晚古生代脉岩

仅发育正长岩、正长斑岩脉，主要分布于同期岩体周缘，如梨树沟正长岩体，在远离大型岩体同一构造线方向也零星产出，受控于先期构造裂隙，呈走向近 SN 或 NNE 向岩枝、岩墙状。岩脉宽数米至数十米，长数十米至百余米。岩脉未见明显蚀变，岩性及岩石地球化学特征同中泥盆世正长岩。

二、中生代脉岩

本区中生代岩浆活动强烈，持续时间长，伴随着岩浆结晶分异、构造与燕山造山和壳幔岩石圈的演化，发育众多中-酸性岩脉。

（一）正长斑岩

为本区广泛分布的一类脉岩，多数呈 NE 向产出。岩脉宽数厘米至数百米，长数十米至数千米。在密云地区可见正长斑岩脉切断了中元古代高岭辉长岩墙，又被后期中生代未变质的辉绿岩脉所切断。岩石呈灰黄色，斑状结构，致密块状构造。斑晶正长石半自形板状，粒径一般 0.4~1.6mm，大者>2mm，发育卡式双晶，强烈高岭土化，弱绢云母化，内嵌细晶正长石矿物，含量约 18%；斑晶斜长石自形板状、板条形，粒径 0.4~1.5mm，发育聚片双晶，弱绢云母化、高岭土化，含量约 2%；基质细晶结构，粒径<0.12mm，正长石含量约 75%；斜长石强烈绢云母化，含量约 1%；石英他形粒状，次生成因，含量约 4%。副矿物为磁铁矿、磷灰石、锆石等。

（二）安山玢岩、闪长玢岩

也是本区广泛分布一类脉岩，在各类地质体中均有出露，尤以中-上侏罗统髫髻山组和下白垩统张家口组分布尤为广泛。岩石呈灰黑色，斑状结构，块状构造。斑晶总含量约 30%，粒径 2~5mm，成分以斜长石为主，发育聚片双晶、环带结造，沿边缘环带常发生弱绢云母化和高岭土化；还发育少量角闪石、石英及钾长石斑晶。基质细晶—微晶结构，含量约 70%，主要组成为斜长石和角闪石。副矿物为磁铁矿、锆石等。

（三）闪长岩

在各类地质体中均有分布，尤以早白垩世东狼沟组和张家口组中最为常见。岩石呈灰绿色—灰黑色，半自形粒状结构，块状构造。斜长石半自形板状、柱状，粒径 0.15mm，常绢云母化、高岭土化，少量绿帘石化，含量约 81%；角闪石自形柱状，普遍绿泥石化或被绿帘石、黑云母交代，粒径<

0.25mm，含量约10%；黑云母鳞片状，绿泥石化，粒径0.5mm，含量约4%；石英他形粒状，粒径<0.1mm，多充填于斜长石空隙中，含量约2%。副矿物为磁铁矿、锆石等。

（四）花岗细晶岩、花岗斑岩

在各类地质体中均有分布，尤其在中生代酸性侵入岩体内最为集中。尤以南口岩墙群最为典型，同基性岩共生，构成双峰式岩浆侵位，称之为南口岩墙群，具体岩脉及岩性特征见第十九章脉岩部分。

（五）石英霏细斑岩、流纹斑岩

分布于八达岭一带，多数在大型霏细岩或花岗斑岩体附近产出。脉岩规模不大，一般走向NNW向，脉宽50~150m，长数十米至数千米。岩石呈浅褐红色，斑状结构，基质具霏细结构，致密块状构造。斑晶石英他形—半自形粒状，粒径0.5~1.4mm，被基质矿物熔蚀，边缘呈圆滑状或港湾状，含量约6%；斑晶斜长石半自形—自形板状，粒径0.5~1.5mm，发育聚片双晶，强烈绢云母化，有时聚斑状产出，含量约3%；斑晶正长石他形—半自形板粒状，粒径0.5~1.0mm，高岭土化，含量约2%；斑晶黑云母半自形片状，粒径0.4~1.0mm，内部铁质析出，含量约1%。基质霏细结构，由他形粒状石英、斜长石、正长石细微晶粒组成，粒径<0.06mm，彼此镶嵌聚集，局部可见少量叶片状云母类矿物。副矿物为磷灰石、绿泥石、锆石等。

第六节　中酸性侵入岩浆活动与成矿作用

区内中酸性侵入岩浆活动发育于新太古代、中元古代、晚古生代、中生代，尤以中生代三叠纪—白垩纪极其强烈，已发现与岩浆侵入活动有成因联系的矿产多处，包括内生金属矿产（包括矿床和矿点）、非金属矿产。在各个地质时期中，由于矿床成因不同，形成了岩浆矿床、热液矿床、接触交代（矽卡岩）矿床、斑岩型矿床和非金属矿床五种具有工业价值的工业矿床。

一、热液型矿产

中元古代早期有与伸展环境下环斑花岗岩-花岗岩侵入活动有关的热液型钨矿，如密云沙厂、墙子路、吴家峪热液型钨矿点；早侏罗世有与前造山作用的浅成花岗斑岩-闪长玢岩侵入活动有关的热液型金矿，如门头沟区青口乡珠窝金矿点；早白垩世有与分水岭-石湖峪等系列花岗岩、花岗斑岩侵入活动有关的热液型金矿、钼矿、锰（银）等复合型矿产，如怀柔区黄花城乡石湖峪小型热液型钼矿、分水岭峪儿崖式小型岩浆热液改造型金矿、得田沟大冰沟小型热液型金矿、杨树底下中型热液型金矿、万庄小型热液型金矿、上口含金石英脉、西湖中低温热液型锰（银）矿点、怀柔区黄花城乡杏树台区热液型金矿点、怀柔对石小型热液型金矿、密云区冯家峪热液型金矿点等。

二、矽卡岩型矿产

晚三叠世有与前造山作用的盘山钙碱性花岗岩侵入活动有关的石塘峪小型接触交代型钼矿；晚侏罗世有与太平庄辉长岩-闪长岩侵入活动有关的五座山磁铁矿矿化点；早白垩世有与分水岭-房耳峪-杨麻地-八达岭等系列花岗岩侵入活动有关的系列多金属矿产，如西栅子小型接触交代型铁锌矿、斋堂镇碾台接触交代型铜钼矿点、门头沟榆木沟小型铁多金属矿、门头沟上苇甸铁矿点、延庆石槽小型接触交代型铜矿、怀柔蓬华池中型接触交代型铁锌矿、怀柔河防口接触交代型铜矿点、平谷翟各庄小型接触交代型铜钼矿、延庆大东三岔中型接触交代型钼矿、延庆东三岔矿区香屯中型接触交代型铁锌矿等；早白垩世还有与薛家石梁-湖门晚造山过碱性-碱性正长岩-闪长岩侵入活动有关的东西红山小型接触交代型铁锌矿。

三、斑岩型矿产

与具有斑状结构的中酸性浅成和超浅成小岩体有关，矿产的形成均产于斑岩中及其内外接触带附近。早白垩世有与石窑等花岗斑岩、石英斑岩有关的多金属铜钼矿，如昌平黄场小型斑岩型铜钼矿；早白垩世有与隐爆角砾岩有关的钼矿，如延庆董家沟-大庄科中型斑岩型钼矿，钼矿与浅成相-超浅成相-隐蔽爆破相的各种斑岩和隐爆角砾岩密切相关，其中石英二长岩是主要的含钼岩体，也是含矿角砾岩的母体。

四、非金属矿产

中元古代早期有与新地-古北口大陆伸展双峰式石英正长岩-斜长岩-似斑状花岗岩侵入活动有关的兰营中型萤石矿。早侏罗世有与老君堂正长岩侵入活动有关的老君堂玉石矿点。晚侏罗世-早白垩世有与长园-官山、黑熊山等系列石英二长岩-闪长岩-花岗岩中酸性侵入活动有关的建筑石材，如官山石英二长岩侵位到二长闪长岩中，具有似斑状结构并普遍遭受绿帘石化，致使绿色斜长石与红色钾长石斑晶交相辉映组成美丽图案，形成建筑石材；再如岩体本身岩性组成的石材，如黑熊山石材、莲花山硅石石材等。早白垩世还有与石窑花岗岩侵入活动有关西湖玉石矿，细晶岩与围岩接触带向岩体内部倾斜的矽卡岩带内由蔷薇辉石矽卡岩、透辉石-石榴子石矽卡岩等组成，当蔷薇辉石富集时形成玉石矿。

第二十一章　岩浆活动总体特征及演化规律

第一节　岩浆活动与岩石圈演化

岩浆活动与岩石圈演化总的变化规律是随着岩浆由基性至酸性向偏碱性，即由低钾向高钾钙碱性演化，地壳的成熟度也随之增高。

一、岩浆活动与大陆壳的演化

关于地壳的成因问题，有两种不同的观点。第一种认为地壳在地球历史上很早就已经形成，其成因与引起地核与地幔分凝期的主要分异作用有关，因而推测地球早期就存在一个硅铝层，并不断地受晚期的改造。第二种则认为地球早期为硅镁质，以后硅铝质岩石不断地从地幔分异出来，并在太古宙晚期形成高潮，现今上地壳的2/3的分布面积是在新太古代（28~25Ga）花岗岩侵位期或经地壳增生分异超级事件形成的。

本区乃至邻近的河北地区尚未发现有确凿证据的中太古代及其以前的侵入岩。多种古老岩石的同位素年龄显示了两组明显的峰值，即2.6~2.4Ga（新太古代）和2.0~1.6Ga（古元古代晚期—中元古代中期）。很显然这是华北地块岩石圈形成演化过程中非常重要的两个时期。

新太古代中期—晚期，本区发育二幕三期八类基性—酸性侵入岩，包括 T_1（英云闪长质）、T_2（奥长花岗质）、G_1（花岗闪长质）和 G_2（花岗质）中酸性侵入岩，主要由 $T_1T_2G_1$ 组成，代表初始刚性陆壳形成。同时，这一阶段，并未发现大量的钾质花岗岩类（G_1 和 G_2），表明该阶段陆壳并未发育成熟。但通过对二幕三期八类基性—酸性侵入岩研究认识，新太古代是本区大陆根（地幔）和陆壳基底的形成时期，也成为本区古老岩石圈生长的最主要时期。二幕三期八类侵入岩记录了本区古老地壳由不成熟的洋内岛弧（沙厂英云闪长质片麻岩侵位）-半成熟的大陆岛弧（黑卧辉长质片麻岩和苇子峪辉长闪长质片麻岩侵位，T_1T_2）-成熟的大陆边缘岛弧（宋营子英云闪长质片麻岩、荆子峪花岗闪长质片麻岩、康各庄奥长花岗质片麻岩侵位，$T_1T_2G_1$ 或 $T_1T_2G_2$）-大陆碰撞造山带岩浆弧（旗杆顶花岗质片麻岩侵位，G_1G_2），反映了初始玄武质地壳转化为半成熟的大陆地壳的整个过程。

古元古代本区仅发现存在于表壳岩及英云闪长质片麻岩中的为数不多的超镁铁质岩墙，是上地幔的一些碎块在伸展构造背景下呈固体塑性状态沿深大断裂被挤压上来的。岩墙呈无根包体形式出现，地质特征及岩石化学、地球化学等特征揭示其是在地壳较薄、热流值高并呈塑性的环境下发生的，稀土元素特征和球粒陨石基本一致，可能代表了本区早期常规原始型岩石圈地幔组分。同时，Sr-Nd同位素揭示其还具有一定程度的地壳物质混染。因此，虽然经历新太古代地壳高速生长，在广大华北地区可能已经存在发育成熟的大陆地壳，但本区古元古代仍然是大陆地壳生长的重要阶段，仍然经历了一定程度的壳幔演化，陆壳产生一定分异作用，岩石圈地幔则由早期富集型向原始或弱亏损型转变，而期间缺乏与之相应的大规模岩浆活动及沉积地层，本区可能表现为各微陆块（成熟陆核）构造拼合（造山带）交界地带。

中元古代本区发育早期基性岩墙（高岭岩墙）、早期斜长岩-环斑花岗岩和中期大红峪组钾玄岩系列火山岩石，且斜长岩-环斑花岗岩受到不同程度的壳源物质混染。因此，经新太古代末期—古元古代漫长的地质历史演化，岩石圈地幔类型发生明显变化，在中元古代最终已形成成熟岩石圈壳幔结构。

通过对中元古代早期—中期本区二幕三期五类基性-酸性侵入岩研究，认为该阶段岩石圈演化表现为成熟古老大陆地壳和岩石圈一次伸展裂解，但其并未发育演化为大洋-岛弧等后期板块演化阶段，而在经历短期伸展裂解事件后就夭折了，形成稳定克拉通化壳幔结构，即发育一次坳拉谷系。仅发育早期大规模基性岩墙的侵位和稍晚非造山型岩浆活动（斜长岩类、环斑花岗岩类及伟晶岩脉等）及玄武质岩浆侵位，以及裂陷早期形成的变质基底隆升和退变质构造事件等（李江海等，2000）。

中元古代晚期—晚古生代，本区进入克拉通盖层演化阶段，处在相对稳定构造背景环境，没有形成区域性岩浆活动事件。区内除因地壳垂直升降运动于上地壳近地表接受沉积或隆起引起短期剥蚀外，下地壳及地幔可能并没有较大变化。

晚古生代—新生代这一漫长地质历史时期，本区经历了中泥盆世、早二叠世、早中晚三叠世、早中晚侏罗世、早晚白垩世及新生代多期次构造-岩浆热事件。这些构造-岩浆热事件对本区克拉通化岩石圈进行了长达100Ma改造（华北克拉通活化）。自此，本区乃至更大的华北地区具有的稳定性质便不复存在，有人称之为地幔置换（郑建平，1999；Gao et al.，2002；Zheng et al.，2007），也有学者将其称之为岩石圈转型（周新华，2006）。即太古宙形成并逐渐成熟的克拉通化岩石圈遭受破坏，表现为软流圈上涌及可能发生的壳幔物质部分熔融，诱发地壳范围内的一系列岩浆活动、变质作用（吴福元等，2008），岩石圈厚度发生了巨大减薄。从改造的角度看，软流圈的上涌对应地表的隆升和地壳表层的伸展，从而导致大规模岩浆活动、地壳深部岩石向浅部的剥露和地表沉积盆地的广泛发育（吴福元等，2008）。这里，关于华北克拉通活化时间上限、下限，主要破坏时限及破坏的垂向幅度，形成机制争议很大。但有一点毋庸置疑，本区地壳形成于新太古代—古元古代，并经过中元古代—早古生代演化和晚古生代—中生代的两次复杂演化过程是肯定存在的，并最终形成现今的复杂壳幔结构，即古岩石圈、新岩石圈和现今岩石圈发育过程。根据动力来源，本区晚古生代—新生代壳幔演化进程又可划分两幕板块造山旋回，即中泥盆世—中三叠世中亚造山构造旋回和晚三叠世—早白垩世燕山造山构造旋回。不同规模、特性侵入岩组合形成特征鲜明岩浆岩带，记录了造山运动和壳幔演化的重要构造形迹，记录了岛弧—碰撞—碰撞后伸展垮塌壳幔作用过程。总之，大陆壳的形成、生长和演化主要受岩浆作用的控制，是地球内部向外释放热能的过程，它导致地球阶段性的快速冷却（邓晋福等，2009）。

二、岩浆活动与造山运动

本区新太古代—古元古代具有构造拼合造山特性。但由于本区区域过小，且区内太古代基底出露面积有限，加上地质年代久远，岩浆活动与造山运动关系尚不明确。但通过区内出露有限的变质岩体的综合研究，新太古代不成熟岛弧—成熟岛弧—古元古代碰撞造山运动进程仍得以体现。

中元古代以来，虽然全区大地构造位置处于华北克拉通内部，但伴随华北克拉通活化，区内北山区（燕山）和西山区（太行山区）均具有造山带性质，具有造山带的典型演化过程，即经历拉张-挤压-抬升复杂过程。北山地区岩浆活动与中亚造山密切有关，而西山区岩浆活动与燕山板内造山密切有关。

（一）中亚造山运动与岩浆活动

中亚造山运动的3个阶段及其岩浆活动特征如下：

1. 早期拉张阶段

中泥盆世，受中亚造山早期影响，区内中—新元古代形成的近东西向深大断裂（包括赤城-古北口断裂带、密云-墙子路断裂带）复活，壳幔混合源岩浆沿着这些断裂快速上涌，侵位于不同深度的上地壳。由于本区距离中亚造山带较远，仅仅表现为造山带远程效应影响。因此，岩浆活动强度有限，且岩浆岩性质并不具有典型造山带内前期岩浆岩特性，也不存在陆壳加厚、山体和山根形成。但这一伸展事件对岩石圈地幔具有一定的改造作用，并为岩石圈最终破坏起到诱发作用。

2. 中期挤压造山阶段

晚泥盆世—晚石炭世，表现为中亚造山主峰期，构造应力场转换为挤压。同样，由于本区距离中亚造山带较远，挤压造山影响力明显减弱，使得区内并不能产生因强烈挤压而形成的双层岩浆房及壳幔混合型的大规模岩浆侵位。反而挤压构造背景封闭了早期拉张阶段复活的深大断裂，即岩浆侵位通道，使得该期缺乏与之相对应的典型挤压造山岩浆活动。

3. 晚期抬升阶段

早二叠世—晚三叠世，中亚造山进入晚期或后造山阶段，构造应力场进入晚造山—后造山的伸展垮塌。区内北山再次演化为类似于早先拉张阶段的特征。即中—新元古代形成的近东西向深大断裂（包括赤城-古北口断裂带、密云-墙子路断裂带）再次复活，先是早二叠世基性—超基性幔源岩浆小规模侵位，地壳增厚，山体整体隆升（区内海相沉积逐渐演化为陆相沉积）。而后幔源岩浆侵位使得地幔热量向地壳输入，发生地壳局部熔融，形成早三叠世壳幔混合源岩浆活动。至晚三叠世，热影响深度达到上地壳，以盘山岩体为代表的壳源A型花岗岩也开始形成。同样本区距离中亚造山带较远，仍表现为造山带远程效应影响，岩浆活动强度有限，且形成区内特有的I、A混合特征型的后造山花岗岩（盘山岩体）。

（二）燕山板内造山运动与岩浆活动

通过对中生代典型岩体侵位机制（详见本章第三节）剖析，燕山期岩浆由源区上升、运移直至最后定位的全部动力学过程中，以岩浆自然浮力、区域构造应力为驱动力。区内岩体多为复合型（包括岩性、岩相及侵位期次），揭示燕山期来自深部的大量花岗岩岩浆是以多源、多阶段脉动方式到达上地壳，又以深源高位为特征的，它们多定位于4~10km。在漫长的多期定位过程中，复式岩基多沿壳断裂带、超壳断裂带定位，构成本区与构造应力场演化特征相一致、关系极为密切的构造-岩浆岩带。燕山期应力场的NEE-NE-NNE三幕次演变，形成本区3个方位的构造-岩浆岩带。侵位方式不尽相同，且与造山作用过程，包括前造山期、同造山期、造山后期相伴，其力学机制和动力学过程各异。按照岩浆成因演化特征，燕山板内造山大致经历了早侏罗世及造山后—中侏罗世（160Ma）前造山伸展性质地壳增厚、晚侏罗世—早白垩世早期（140Ma）同造山挤压地壳增厚、早白垩世中期—早白垩世晚期（125Ma）晚造山及造山后伸展拉张地壳变薄3个过程。

1. 前造山期的侵位（挤压-拉张过渡）

三叠纪末期至早侏罗世早期区内北山以及华北克拉通北缘经历了一次短期构造挤压事件，在华北克拉通北缘广泛发育向南长距离的逆掩推覆，导致晚古生代钙碱性侵入岩以及三叠纪正长岩类与围岩一起成为异地体（赵越，1990；郑亚东等，2000；Davis et al.，2001）。本次构造挤压使得陆壳加厚，伴生小规模挤压性质的岩浆侵位。区内发育大石坡-黑山坨复式杂岩体沿太古宇和长城系不整合面侵位，拉开了燕山板内造山序幕。汪洋等（2015）对大石坡-黑山坨复式岩体角闪石全铝压力计推算大石坡正长岩侵位深度约18km，早侏罗世时期太古宇与长城系不整合面以上地层的累积厚度不超过7km，表明该阶段区内上部地壳已经增厚了约11km。随后，构造应力反转，区域构造应力场由挤压转换为伸展，才真正开始了燕山运动前造山幕（燕山构造I幕）。区内形成一系列近东西向和北东东向、与构造应力场一致的、具有岩石圈或超岩石圈性质的断裂，幔源岩浆沿断裂快速上涌或喷出地表，或在地壳的不同深度定位，从而使陆壳加厚，初步形成山体和山根。该阶段岩浆活动形式表现为大规模的南大岭期基性火山喷发活动。

2. 同造山侵位（挤压型）

由于长期断裂活动和岩浆活动造成的应力场的转换，强烈的侧向挤压封闭了早期拉张形成的岩浆通道，幔源岩浆由于密度较大而快速上升运动的能量减弱，只能定位或封存于下部地壳的有利空间。在挥发性组分的帮助下，炽热的岩浆把大量热能通过传导和对流的方式传递到下地壳岩石中，引起下地壳岩石的塑性化和部分熔融，形成双层岩浆房，即壳源岩浆之下覆盖着大量幔源岩浆，或者中下地壳岩浆覆盖着下地壳岩浆。两种岩浆按照不同比例混合形成的混合岩浆在地壳韵律性收缩过程中不断排出，形成大规模分布的高钾钙碱性侵入岩和火山岩，并使陆壳急剧增厚，燕山山体的发育渐趋成熟。随着山体外侧的岩石圈物质不断向热轴运移，受热的部分加入到山体之中，未受热的部分向下俯冲，加入山根或岩石圈根。这一复杂过程即燕山造山运动主过程。

燕山主造山期岩浆侵位与褶皱作用同步，构成燕山构造Ⅱ幕。形成的岩体长轴与褶皱轴同侵入期构造带平行；伴有接触变质作用，往往形成接触动力变质片岩带（如石城-云蒙山岩体）；岩体边部有强烈的叶理和线理、片麻状构造或保存有内应力作用的构造形迹；成分以中性和中偏碱性为主，边部多发育早期闪长岩、石英闪长岩侵位，晚期为酸性侵入岩侵位；岩体内具暗色闪长质包体，其流面产状定向排列，显示同心环状构造。总之，从岩体内部到围岩，构造效应比热变效应改造的范围更大些，其岩浆侵位条件显然与高温和强大的定向压力有关。

3. 晚造山—造山后期侵位（伸展型）

地壳不能无限制加厚，当浮力大到足以抵消侧向压力之后，山体便在山根作用下发生应力的均衡调整，使得山体整体隆升。同样，由山体外侧插入的下地壳岩块由于重力的原因可能造成岩石圈根的崩塌，并使其得以与软流圈进行层圈置换，较重的岩块下沉进入软流圈，较轻的软流圈物质迅速上升，加速了山体的抬升。下沉的山根或岩石圈物质接触炽热的物质圈之后，发生部分熔融，形成以 A 型为主的碱性正长岩-花岗岩类。

晚造山—造山后岩浆侵位构成于燕山构造Ⅲ幕，其间伴随断裂活动，导致形成大规模构造-岩浆岩带。岩浆侵位表现为拉张型充填式侵位，顶蚀、塌陷、破火山口式充填侵位最为常见，岩体形态与围岩的边界不规则，接触带构造、变质效应较弱，同多表现为岩性演化复杂的复合型岩体，如薛家石梁复式岩体、冯家峪复式岩体等。至造山后，岩浆活动转弱直至停歇。

总之，岩浆活动（如定位机制、岩性组合、岩浆系列等）与陆内造山作用息息相关。中生代岩浆活动不仅是构造运动的表现，而且也是陆内（燕山板内）造山带动力来源之一。

第二节　壳幔作用与岩浆活动

在现有资料和研究成果的基础上，结合 Picher（1983、1987）依据板块构造理论在 Chappell 和 White 的成因分类基础上将岩浆岩划分五种类型的相关认识，对本区不同时代的基性、超基性和花岗岩类型及相关标志，即时空分布、岩石类型、主要造岩矿物、岩石化学、地球化学以及矿化特征等方面进行总结和概括，划分 3 种岩浆物质来源（壳源、壳幔混合源和幔源）和 4 种成因类型（同熔型、陆壳改造型、幔源型和壳幔混合型），以反映本区岩浆活动的壳幔作用的特点。

一、太古宙—古元古代壳幔作用与岩浆活动

考虑到太古宙岩浆活动受到强烈的区域变质变形作用遮盖和复杂化，及太古宙特殊的壳幔结构等因素，使其难以用现代板块来完全与之对比。按照侵入岩体接触关系和同位素年龄数据，可以划分为 9 期岩浆活动事件（详见第十九章第四节）。其构成岩石类型多样，包括辉长闪长质片麻岩、英云闪长质片麻岩、花岗闪长质片麻岩、奥长花岗质片麻岩、花岗质片麻岩、超镁铁质岩和变质基性岩墙。

根据同位素、岩石组合及地球化学的多方面特征判断，成因类型包括壳源、壳幔混合源及幔源，形成于岛弧-大陆弧-前碰撞-同碰撞-大陆裂解复杂构造背景环境。

二、中元古代壳幔作用与岩浆活动

中元古代岩浆活动按照时间序列可以划分为三期，分别为早期辉长岩岩墙、第一期斜长岩-石英二长岩-花岗岩组合和第二期辉绿岩-粗安岩-正长岩组合，早期（高岭）辉长岩墙为拉斑玄武岩系列，成因类型为幔源型。随后的斜长岩-石英二长岩-花岗岩组合中中性岩为钙碱性岩系列，成因类型为壳幔混合源型，经历了复杂岩浆演化过程。而第二期（熊儿寨）辉绿岩-粗安岩-正长岩组合又为碱性岩系列，成因类型也为壳幔混合源，岩浆同样经历了复杂的演化过程。

中元古代岩浆活动形成构造背景环境均为非造山背景下裂陷槽演化早期伸展拉张构造环境。三次

岩浆活动是全球性发生在1.8～1.4Ga的中元古代时期的非造山环境下岩浆活动的一个组成部分。太古宙-古元古代强烈变质变形作用之后，基底固结，华北板块克拉通化，在非造山时期克拉通边缘或内部的巨大断裂带所派生的拉张性构造环境下的裂解作用与岩浆活动有直接关系。

三、晚古生代壳幔作用与岩浆活动

晚古生代岩浆活动按照时间序列可划分为两期，分别为中泥盆世正长岩侵位和早二叠世辉长闪长岩侵位。一期岩浆活动中，中泥盆世正长岩为碱性岩系列，成因类型为壳幔混合源型。二期岩浆活动中早二叠世辉长岩-闪长岩为钙碱性系列，其成因类型为幔源型。二期岩浆活动经历了一定的岩浆演化过程，且经历了区域变质变形作用。

晚古生代二期岩浆活动形成于伸展拉张构造环境，是发生在晚古生代中亚造山背景下远程造山效应下的岩浆活动。伴随着中亚造山活动，对华北克拉通边缘壳幔产生不同程度破坏。华北克拉通的活化在华北板块内部由北向南、由东向西发展，影响到本区北部的燕山地区。本阶段岩浆活动对本区壳幔结构影响主要表现为幔源岩浆侵位使得地壳厚度不同程度减薄，地壳地温梯度增高。

四、早中生代壳幔作用与岩浆活动

早中生代岩浆活动时间序列上发育三叠纪一期，由早三叠世二长闪长岩、花岗闪长岩、石英闪长岩、二长花岗岩，中三叠世石英二长岩和晚三叠世石英二长岩、花岗岩侵位构成。岩石均为高钾钙碱性系列，主要成因类型为壳幔混合源型，仅中二叠世石英二长岩和晚三叠世花岗斑岩、细粒花岗岩为壳源型，而岩浆成因机制均为部分熔融作用。

早中生代三叠纪岩浆活动处于挤压构造环境向伸展构造环境演化过渡背景。期间，岩浆活动总体规模较小，岩体为早中生代中亚造山大环境下造山末期—后造山稳定—非造山构造背景下侵位产物。该期岩浆活动是晚古生代中亚造山的继承和延续。伴随中亚造山运动结束，区内最终进入非造山构造背景环境。该期岩浆活动对区内壳幔结构进一步改造，华北克拉通被进一步活化。通过Sr-Nd同位素揭示，该阶段地幔性质同晚古生代及中上元古代基本相似，未发生本质性变化。但地壳厚度变化可能巨大，张旗（2001）等认为加厚地壳东部高原的存在，致使岩石Sr、Y、Yb等元素含量较古生代发生较大变化。

五、中—晚中生代壳幔作用与岩浆活动

中—晚中生代岩浆活动按照时间序列可划分为两期，第一期与非造山有关的早侏罗世正长岩-花岗岩和第二期与燕山板内造山有关的三幕期次岩浆亚旋回。两期岩浆（包括亚旋回）均由小规模的碱性系列幔源中基性岩起始，后开始发育大规模壳幔混合源的高钾钙碱性系列中性和中酸性岩，最后又发育小规模的壳源碱性岩系列中酸性岩和酸性岩，说明壳幔作用过程从深部开始，并最终影响到浅部上地壳。本期火成岩岩石类型多样，岩浆演化机制主要同分离结晶密切有关，少有部分熔融作用。

中—晚中生代岩浆活动是在活化的华北克拉通燕山造山背景下发育的。区内该阶段岩浆岩特征包括：偏基性岩石所占比例较小；同位素组成变化幅度稍大；洋壳物质特点不明显；绝大多数岩石为典型的钙碱系列岩石组合；岩浆活动具有连续性、多期多幕性；完整的侵入及火山岩系有连续的基性-中性-中酸性-酸性岩的旋回性演化及共生组合，显示了火山岩和侵入岩同源特征及岩浆岩I型钙碱性岩浆的分异演化规律；富镁而贫铁的暗色矿物组合；以I型岩石类型为主，典型的中性岩石副矿物类型及标型特征都显示I型钙碱性岩系的共性；上地幔及下地壳岩浆物质来源在岩浆侵位过程中，由于地壳上部物质的不断混入而具有向S型地壳重熔岩浆转变的趋势，同时由于花岗岩浆的分异作用又有向A型碱性花岗岩演化的趋势。因此，该期岩浆活动对区内壳幔结构造成巨大影响，伴随燕山运动多幕次构造演化、大规模岩浆活动，华北克拉通遭到严重破坏，区内地幔类型也可能发生根本性改变，由EMI型地幔演化为EMII型地幔，地壳厚度则经历了巨大的减薄演化过程，并形成燕山地区壳幔结构不均一性。

第三节　岩浆岩成因探讨

本区岩浆岩十分发育，不但岩石类型多样，而且成岩时代众多。对不同时代、不同类型的岩浆成因，本节拟定按照时间序列从岩石构造组合、岩石化学、地球化学和稳定同位素等多方面做综合探讨和初步总结。

一、太古宙—古元古代岩浆岩的成因及其演化规律

本区太古宙发生过不同规模的多期次岩浆活动。不同时期岩浆起源和演化、岩浆活动方式，岩浆成分及稀土、微量元素特点均存在一定变化规律。

（一）岩浆活动期次及成分的演化、岩浆活动方式

新太古代晚期有2期岩浆活动，第一期岩浆为英云闪长质成分（沙厂）；第二期总体成分为中性偏基性，并形成一个从基性到中酸性的岩浆分离结晶系列（黑卧-苇子峪-宋营子）。

新太古代末期发生3期岩浆活动：第一期小规模基性岩墙（放马峪）侵位；第二期为花岗闪长质岩浆（荆子峪）侵入；第三期为奥长花岗质-二长花岗质（康各庄-旗杆顶）岩浆侵入。大规模岩浆活动后，还发育一次钾长花岗质岩浆脉状侵入活动。二期岩浆侵入活动，岩性形成较完整的基性-中酸性-酸性演化序列，化学成分上全碱含量逐渐增高，CaO 含量则逐渐减少（图 21-1），形成由钙碱性向碱性的岩浆演化序列。

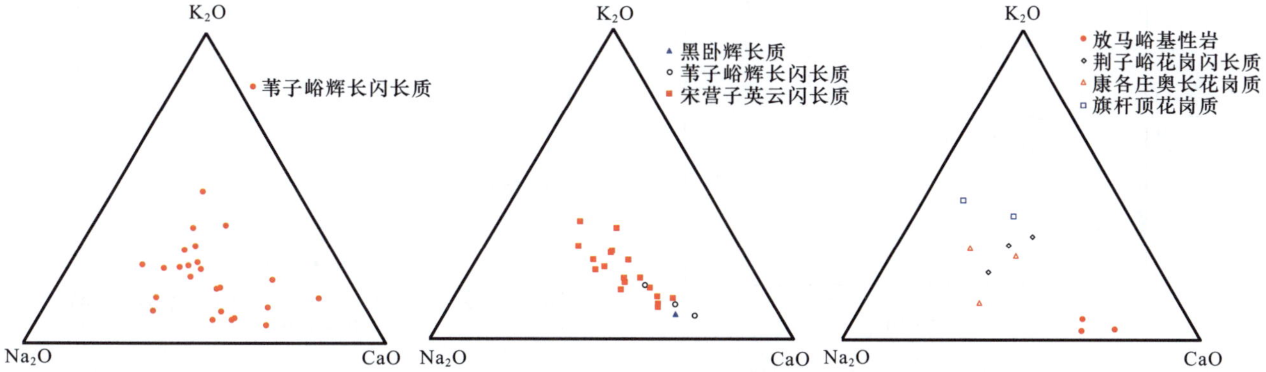

图 21-1　新太古代片麻岩岩石成分 K_2O-Na_2O-CaO 图解

Fig. 21-1　Rock composition K_2O-Na_2O-CaO diagrams of Neoarchean gneiss

古元古代发育一期小规模基性岩浆活动，活动方式为岩浆小规模伸展型岩墙。岩浆成分有三种：超镁铁质、镁铁质、硅铁质。

（二）太古宙岩浆作用的稀土演化特征

新太古代晚期第一期英云闪长质（沙厂）岩浆中稀土总量较高，与幔源基性岩原岩部分熔融理论吻合；新太古代末期第二期中基性岩浆的稀土总量更高，说明岩浆可能来源于深部地壳或地幔的稀土富集区。新太古代末期基性岩墙稀土总量较低，具有幔源岩浆特征，而酸性岩浆的稀土总量较低，并呈现越到晚期越低的明显变化趋势，这种现象与通常所见的岩浆稀土演化规律不同。古元古代基性岩浆中稀土总量最低，与地幔超基性岩特征一致。

由稀土总量与磷含量∑REE-P 图解（图 21-2）可见，本区新古代片麻岩的稀土总量与磷的含量之间表现出极好的线性相关性，新太古代晚期二者呈现小角度正交关系。而新太古代末期二者呈现高角度正交关系，这表明磷灰石是稀土元素的主要载体矿物，在后期岩浆重熔作用中，磷灰石进入熔体

的比例较少，所以，末期岩浆中稀土总量较低。黑卧-苇子峪-宋营子片麻岩系列和荆子峪-康各庄-旗杆顶片麻岩系列各自形成独立的∑REE-P 演化曲线，说明是两期独立的岩浆演化序列。

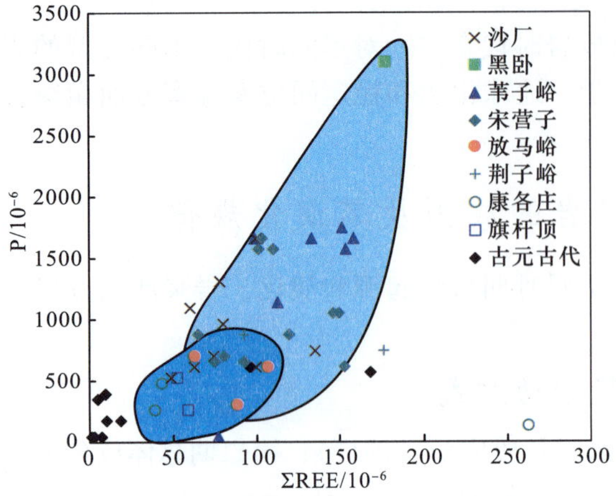

图 21-2　太古宙片麻岩∑REE-P 图解

Fig. 21-2　Provenance distribution Plot REE-P of the gneiss in Archaen

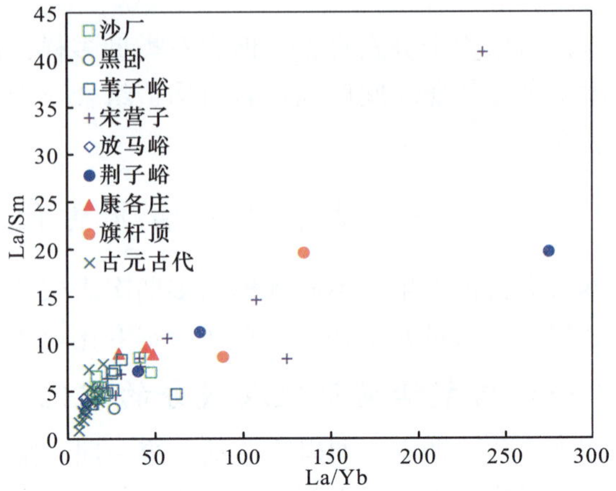

图 21-3　太古宙片麻岩 La/Sm-La/Yb 图解

Fig. 21-3　La/Yb-La/Sm analysis plot of the gneiss in Archaen

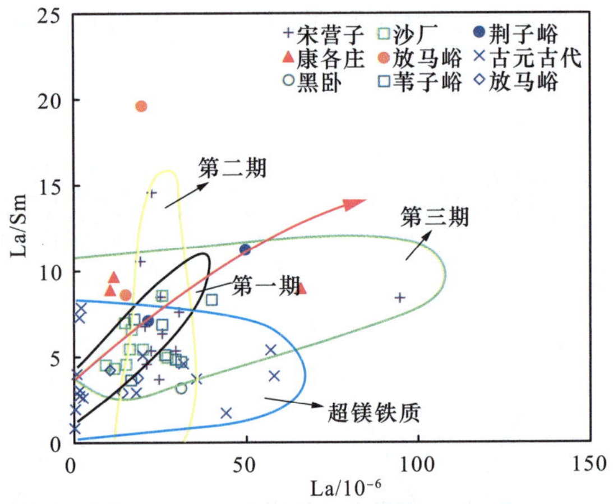

图 21-4　太古宙片麻岩 La/Sm-La 图解

Fig. 21-4　La vs. La/Sm analysis plot of the gneiss in Archaen

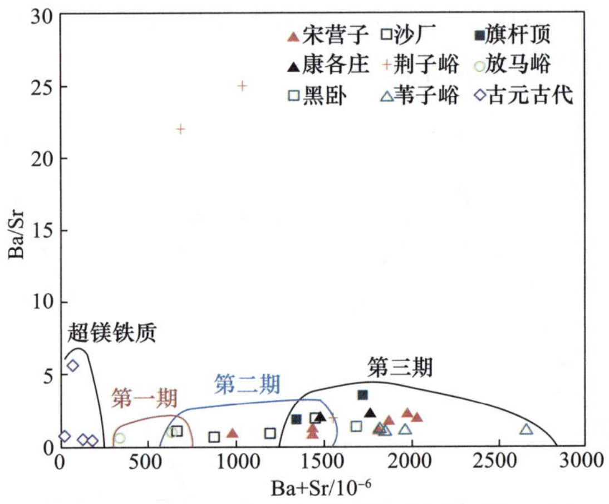

图 21-5　太古宙片麻岩（Ba+Sr）-Ba/Sr 图解

Fig. 21-5　（Ba+Sr）-Ba/Sr analysis plots of the gneiss in Archaen

La/Sm、La/Yb 的演化呈现出很好的规律性（图 21-3、图 21-4）。新太古代晚期基性岩浆的 La/Sm<4，La/Yb<10，这与轻重稀土分馏程度低相吻合，也与岩浆地幔来源相吻合。新太古代晚期、末期片麻岩 La/Sm、La/Yb 比值则比新太古代基性岩浆明显高。岩性上，酸性岩 La/Sm、La/Yb 比值高于中酸性，中酸性又高于中基性（图 21-3、图 21-4）。即新太古代晚期沙厂英云闪长质岩浆 La/Sm、La/Yb 比值反而高于苇子峪辉长闪长质和放马峪基性岩脉群，因此，沙厂片麻岩岩浆具有其自身独立的稀土演化规律。不仅如此，新太古代三期岩浆的轻重稀土分馏普遍明显。且 La/Sm、La/Yb 比值在每一期次岩浆演化过程中分别沿着斜率约为 45°线性方向有序演化。在 La/Sm-La/Yb 图解（图 21-3、图 21-4）中，沙厂片麻岩的成分点与超镁铁质及表壳岩系成分点大致在一条斜线附近，表明其岩浆是由新太古代基性岩部分熔融而来。而新太古代晚期花岗闪长质-奥长花岗质岩浆、二长花岗质岩浆中的 La/Sm、La/Yb 比值向高数值方向演化，呈现很有规律的单调增加，同时新太古代各期岩浆中轻重稀土分馏程度越来越高的规律十分明显，这些事实与新太古代岩浆的地壳起源相吻合。在 La-La/Sm 图解（图 21-3、图 21-4）上，沙厂英云闪长质片麻岩构成较好部分熔融曲线，也显示其部分熔融成因

类型，黑卧-苇子峪-宋营子片麻岩系列则不能构成分离结晶、部分熔融的成因曲线，显示其岩浆成因复杂性，而放马峪-荆子峪-康各庄-旗杆顶片麻岩系列既形成一条分离结晶曲线，又发育一条 La 值无变化、而 La/Sm 值变化骤增的稀土分异演化很强的曲线，说明了岩浆演化的特殊性（图 21-4）。

稀土配分曲线中，新太古代基性岩浆中除超镁铁质岩浆出现较明显的负铕异常外，其他类型岩浆中铕异常不明显。新太古代中酸性岩浆中则出现不同程度正异常，且伴随着岩浆活动由早期到晚期，其正铕异常越来越明显。

（三）太古宙岩浆作用的微量元素演化

从整体上看，与太古宙岩浆成分由基性向酸性成分演化的规律一致，太古宙岩浆中微量过渡元素（V、Cr、Co、Ni 等）也呈早期含量高、晚期含量低的趋势。而大离子亲石元素则相反，如 Sr、Ba 等，早期含量低，晚期含量高。在 Ba/Sr-（Ba+Sr）图解（图 21-5）上，（Ba+Sr）含量由新太古代晚期到新太古代末期，其含量明显增加。而 Ba/Sr 比值则呈现小幅度增加趋势，尤其是太古代末期。在 Y-Sr/Y 图解（图 21-6）上，超铁镁质和基性岩 Y 含量及 Sr/Y 比值均较低。中—酸性岩 Y 含量从新太古代晚期—末期开始，其含量小幅度逐渐减小，而 Sr/Y 比值则有大幅度增高，特别是新太古代末期，康各庄、旗杆顶花岗质片麻岩具有类埃达克岩特征。这充分说明太古代晚期伴随着构造演化，地壳不仅完全刚性化，而且出现了地壳增厚的现象。

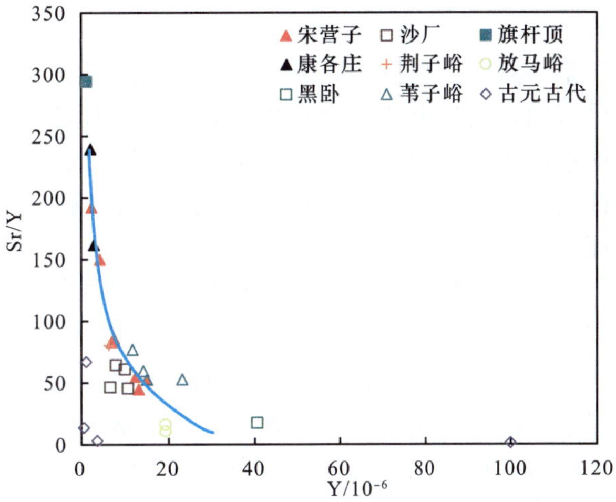

图 21-6 太古宙片麻岩 Y-Sr/Y 图解

Fig. 21-6 Y-Sr/Y analysis plots of the gneiss in Archaen

（四）太古宙岩浆起源及变化

根据上述分析，新太古代超基性岩浆（镁铁质）来源于原始超基性岩浆的分异。新太古代晚饭期英云闪长质（沙厂）岩浆来源于地壳基性岩的部分熔融。新太古代中基性岩浆（放马峪）则来源于深部地壳或地幔中的稀土富集区。新太古代晚期中酸性岩浆来源于不同深度地壳的重熔作用，新太古代末期酸性岩浆来源于浅部地壳重熔作用。由此可见，太古宙岩浆活动的岩浆来源区伴随着地壳演化，大致有由深部向浅部发展的趋势。

（五）新太古代岩浆活动规模

从目前岩体出露情况看，新太古代超镁铁质岩仅仅以残片状产于新太古代沙厂英云闪长片麻岩岩石中，主要受后期侵入岩体破坏影响，但通过本区及大区域上新太古代表壳岩系的分布分析，新太古代基性岩浆活动规模巨大、可能为太古宙时期规模最大的岩浆活动。新太古代晚期第一期英云闪长质（沙厂）片麻岩分布广泛，构成复杂的杂岩系，且岩浆侵位浅等特征，说明该期岩浆活动的规模较大。而新太古代中-酸性岩体分布有限，岩浆形成深度有限，因此，其岩浆活动规模与新太古代晚期第一期相比明显变小，并有逐渐减小的趋势。

二、中—新元古代岩浆岩的成因及其演化规律

本区中—新元古代发生过不同规模的多期次岩浆活动。不同时期岩浆起源和演化、岩浆活动方式，岩浆成分及稀土、微量元素特点均存在一定变化规律。

（一）岩浆活动期次及成分的演化、岩浆活动方式

区内中—新元古代共发育高岭基性岩墙、新地斜长岩、古北口石英二长岩、沙厂环斑花岗岩和大

红峪组火山岩、熊儿寨岩颈6类岩浆岩。岩性组成包括基性-中性-酸性全部岩石单元。

郁建华（1992）、谢广轰等（2004）综合研究认为，中元古代岩浆活动具有双峰式岩浆活动特征，为伸展型大陆裂解（燕山裂陷槽）背景下产物，并建立了中元古代3个阶段的岩浆演化和壳幔结构变化过程。但最新的研究成果，高岭岩墙、环斑花岗岩均被长城系不整合覆盖，其形成时代明显早于大红峪组火山岩-辉绿岩。因此，这两期岩浆活动为不同构造背景下产物。但是，高岭岩墙和大庙斜长岩、沙厂环斑花岗岩在形成时代上具有一致性。因此中元古代3个阶段岩浆活动中，大红峪组火山岩-辉绿岩-正长岩则构成另一个独立的岩浆岩演化序列单元。而高岭岩墙和斜长岩-环斑花岗岩的岩浆活动可能在成因上存在关联性，即高岭岩墙-大庙斜长岩-沙厂环斑花岗岩可组成其岩浆演化的3个阶段模式。

第一阶段在克拉通固结后，深切上地幔的深大断裂发育形成拉张性裂谷环境。深大断裂控制的裂陷槽区内正常型上地幔物质产生部分熔融形成基性玄武质岩浆，并沿着裂谷区内（高岭一带）构造薄弱地带侵入到上地壳，形成大规模的基性岩墙。彭彭等（2011）认为该期克拉通的张性裂解动力与地幔柱密切有关。正是由于这种热地幔柱影响，使得该期形成的玄武质岩浆表现出岛弧型拉斑玄武岩特征。

第二阶段裂谷作用在张性构造环境进一步发育，地幔演化为富集型地幔，而地壳持续伸展，上地幔物质产生部分熔融形成基性的玄武质岩浆。沿张性构造上升的基性岩浆活动带来巨大热流促使固结程度还不高的克拉通基底下部地壳物质部分熔融，在这种部分熔融比例不高的情况下出现的相对缺水的熔浆中富含钾、铁大离子亲石元素及不相容元素，形成富钾花岗质干熔浆，并与基性岩浆形成混合岩浆，上升至岩浆房内。聚集在岩浆房中的混合岩浆由于张性构造发育的压力释放作用、岩浆间的不混熔性等而产生分离结晶作用。在分离结晶作用比较彻底的深成宁静构造环境下，基性斜长石的分离和不断晶出堆集而形成斜长岩或形成含暗色矿物较多的苏长岩、辉长岩等岩体。受张性断裂诱发而快速侵位时，分离结晶作用持续影响，可形成较接近混合岩浆组成的石英二长岩类。

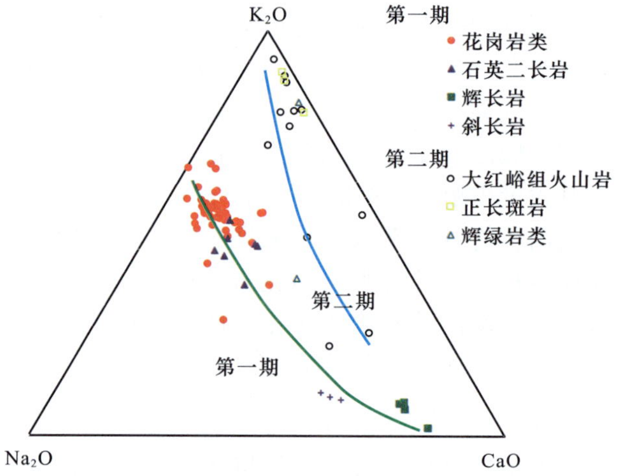

图 21-7　中—新元古代侵入岩岩石成分 K_2O-Na_2O-CaO 图解

Fig. 21-7　Rock classification plot K_2O-Na_2O-CaO of the intrusive rocks in Meso-Neoproterozoic

第三阶段是残留在岩浆房中的主要为下地壳来源的拉帕克维花岗岩浆，这是一种酸度不高而粘度很大、缺水相对干的并富含碱质和氟的岩浆。这种花岗岩浆在张性断裂多次活动的诱发下形成脉动式侵位，在漫长的侵位过程中不断混入大量地壳物质，并进行大规模结晶分异作用。晚阶段花岗岩浆已逐渐转变为酸度很高、含水矿物增多、饱含挥发分的碱性花岗岩浆，总体上形成一系列A型花岗岩侵入体。球斑结构是该阶段岩石特征结构，实质上它是这种富含钾质及挥发分氟的高粘度花岗岩浆的产物，并且还要求有一定的较快冷凝结晶的构造环境。

因此，按照岩浆活动期次，共划分为两期：第一期辉长岩-斜长岩-石英二长岩-环斑花岗岩组合，由基性-中性-酸性构成一期次完整岩浆构造旋回，岩浆活动方式为深部侵位方式，原始岩浆来源于下地幔，后期有壳源物质加入。第二期为玄武岩-辉绿岩-正长斑岩组合，主要为基性岩岩浆分异演化而来，岩浆活动方式为浅成—超浅成侵位或喷发，原始岩浆也来源于下地幔。化学成分上（图21-7），两期岩浆均向CaO含量减少，碱质含量方向增大的方向演化，即由钙碱性向碱性方向演化。同时，二期岩浆活动分别构成自身岩浆演化曲线，第一期向碱性方向演化，第二期则更向钾质方向演化。

（二）中—新元古代岩浆作用的稀土演化特征

中—新元古代各岩类中稀土总量由小到大分别为高岭辉长岩、新地斜长岩、大红峪组玄武岩、涝洼石英二长岩、大红峪组安粗岩和辉绿岩、正长斑岩、环斑花岗岩。基性岩稀土总量低于酸性岩，同它们形成的岩浆来源关系密切。

在∑REE-La/Sm 图解（图 21-8）和 La/Sm-La/Yb 图解上（图 21-9），两期岩浆均朝着稀土元素总量增加方向演化，但又构成自身岩浆演化特征，也说明它们形成两期独立的岩浆演化序列。第一期还朝着 La/Sm 比值及 La/Yb 比值增大的方向演化，说明随着岩浆演化，稀土分馏程度增高。而第二期伴随岩浆演化，岩性变化较大，但稀土元素分馏程度基本未发生变化，说明第三期岩浆活动结晶分异作用要弱。

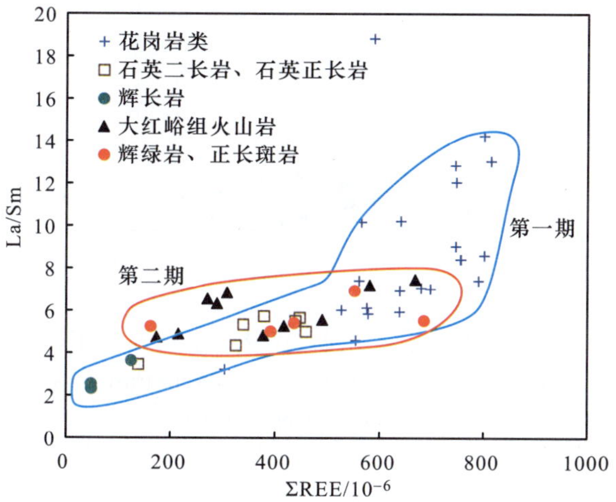

图 21-8　中—新元古代侵入岩∑REE-La/Sm 图解
Fig. 21-8　∑REE-La/Sm analysis diagram of the intrusive rocks in Meso-Neoproterozoic

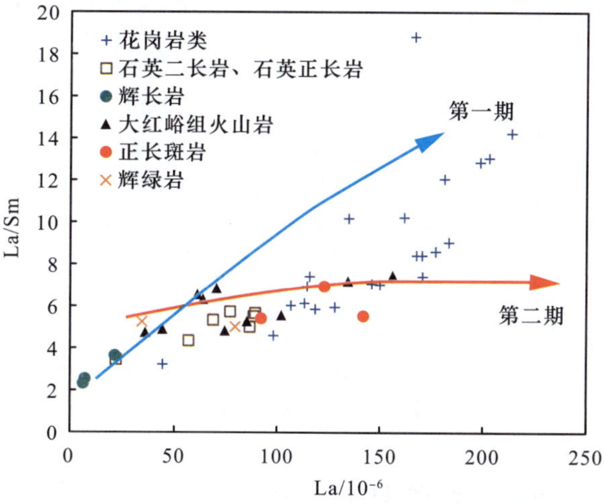

图 21-9　中—新元古代侵入岩 La/Yb-La/Sm 图解
Fig. 21-9　La/Yb-La/Sm analysis diagram of the intrusive rocks in Meso-Neoproterozoic

稀土配分曲线中，基性岩表现为铕无异常—弱正异常，而中性岩—酸性岩呈现不同程度的铕负异常，而 Ce 异常在各岩类中变化不大（图 21-10）。因此，本时间段两期岩浆均朝着铕元素负异常方向演化，且负异常程度越来越强，这与岩浆分异作用有关。

（三）中—新元古代岩浆作用的微量元素演化

整体上与中—新元古代岩浆成分由基性向中性或酸性演化的规律一致，该阶段岩浆中微量过渡元素（V、Mn、Fe、Cu 等）也呈早期含量高、晚期含量低的趋势。而大离子亲石元素则相反，如 Sr、Ba 等，早期含量低，晚期含量高。

在 Ba/Sr-（Ba+Sr）图解（图 21-11）中，各类岩性（Ba+Sr）含量基本相似，但两期岩浆活动具有自身演化特征。第一期岩浆朝着（Ba+Sr）含量增大、Ba/Sr 比值略增高的方向演化，而第二期（Ba+Sr）含量变化不大，但 Ba/Sr 比值朝着含量变大的趋势骤变演化，二者形成各自的演化变化规律。

在 Y-Sr/Y 图解（图 21-12）上，两期岩浆活动均朝着 Y 元素含量减小、Sr/Y 比值减小的方向演化，明显不同于一般造山带岩浆演化规律，说

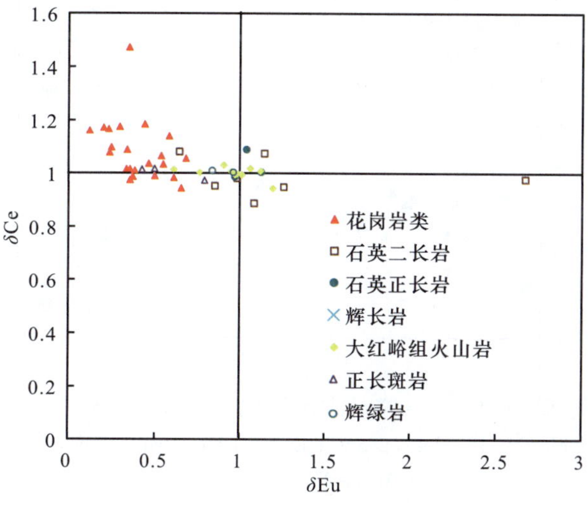

图 21-10　中—新元古代侵入岩 Ce-Eu 图解
Fig. 21-10　Ce-Eu analysis diagram of the intrusive rocks in Meso-Neoproterozoic

明该期伸展构造背景下地壳逐渐减薄的特征。同时，两期岩浆演化各自组成自身岩浆演化曲线，体现在演化曲线斜率上。第一期岩浆演化 Sr、Y 变化快，第二期则相对变化较慢，推测可能和岩浆活动方式有关和岩浆结晶分异程度有关。

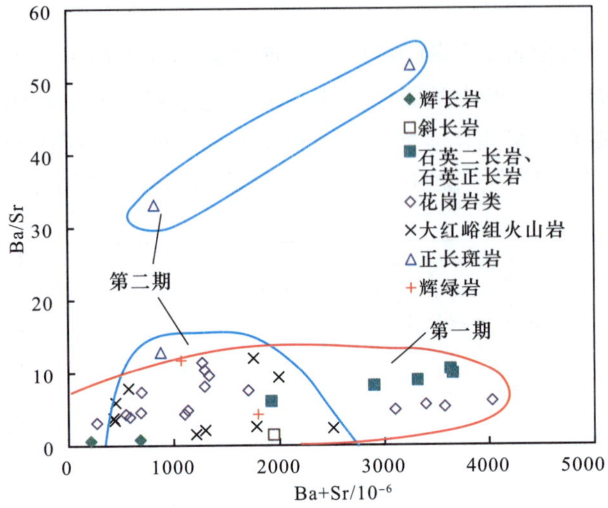

图 21-11　中—新元古代侵入岩 Ba/Sr-（Ba+Sr）图解

Fig. 21-11　Ba/Sr-（Ba+Sr）diagram of the instrusive rocks in Meso-Neoproterozoic

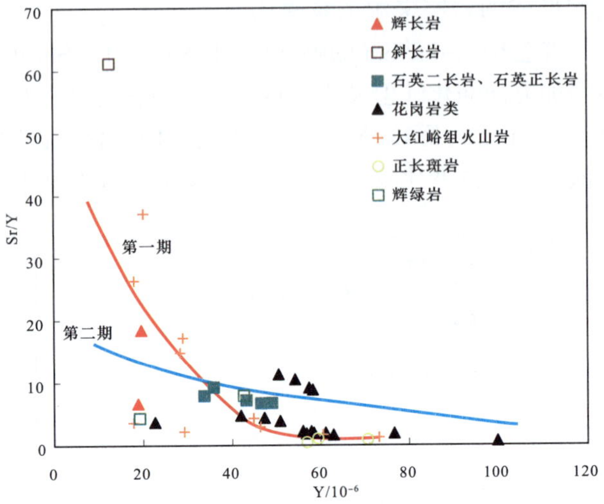

图 21-12　中—新元古代侵入岩 Y-Sr/Y 图解

Fig. 21-12　Y-Sr/Y diagram of the instrusive rocks in Meso-Neoproterozoic

（四）岩浆起源及形成机制

根据上述分析，两期岩浆均来源于上地幔，并伴随着岩浆演化，有地壳物质混合。因此，中—新元古代岩浆来源区伴随着地壳演化，大致也具有由深部向浅部发展的趋势。

在 La-La/Sm 图解（图 21-13）和 Ce-Y 图解（图 21-14）上，第一期岩浆由分离结晶作用控制，而第二期则具有部分熔融作用机制。两期岩浆具有不同的形成机制，可能是导致该阶段岩浆活动形式、岩体特征差异的原因之一。

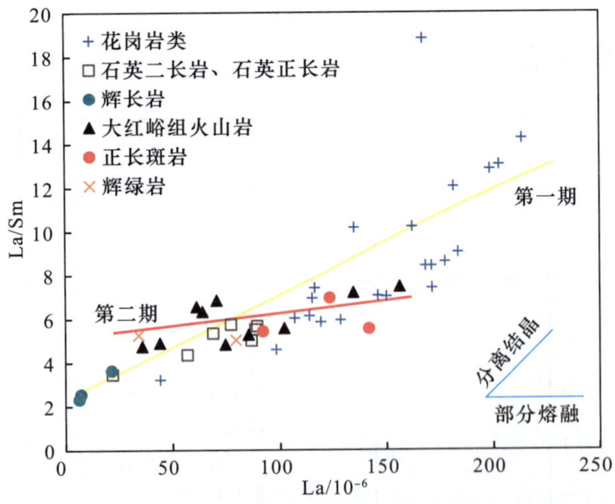

图 21-13　中—新元古代侵入岩 La-La/Sm 图解

Fig. 21-13　La-La/Sm diagram of the intrusive rocks in Meso-Neoproterozoic

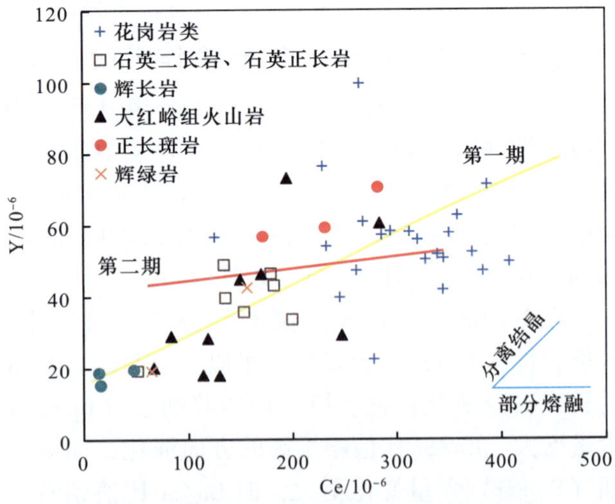

图 21-14　中—新元古代侵入岩 Ce-Y 图解

Fig. 21-14　Ce-Y diagram of the intrusive rocks in Meso-Neoproterozoic

（五）中—新元古代岩浆活动规模

中—新元古代两期岩浆活动中，第一期岩性为基性—中性—酸性，说明岩浆演化完整，岩浆活动强度大。从目前岩浆岩出露情况看，环斑花岗岩、新地斜长岩分布广泛，密云沙厂、古北口、河北均有大量分布，且岩体多呈超大复合型岩体出露，同期还有众多具有关联的花岗斑岩脉出露于太古宙片麻岩中，说明下部可能有大型隐伏岩体。因此，推测中元古代第一期岩浆活动强度大，岩浆规模大。而第二期岩浆活动仅发育大红峪组火成岩及其后发育的辉绿岩、正长岩两类岩性，分布区域也仅局限于平谷熊儿寨—天津蓟县附近，说明该期岩浆活动的规模相对较小。因此，中—新元古代岩浆活动规模也有由大变小的变化规律。

三、古生代岩浆岩的成因及其演化规律

（一）岩浆活动期次及成分的演化、岩浆活动方式

本区古生代经历中泥盆世正长岩、早二叠世辉长闪长岩岩浆侵位，分别形成于不同的构造背景环境。岩浆活动具有共同的特征：①沿深大断裂带侵位，受控于复活的中元古代断裂带；②对围岩有不同程度的同化混染作用和穿插现象，岩体中发育大量围岩捕虏体。

中泥盆世岩石组合为碱性岩系列，化学成分同我国正长岩平均成分相比，SiO_2、CaO 含量接近，TiO_2、MgO、K_2O、P_2O_5、Na_2O 偏低，而 Al_2O_3、FeO^* 偏高。早二叠世岩石组合属钙碱性系列，化学成分与我国辉长岩类平均成分相似，各种特征指数均为正常太平洋型钙碱系列，辉长岩中角闪石的成分也与造山带中钙碱系列岩石相似；正长岩属高温高压型正长岩。因此，其同为深源深成岩体。其岩浆来源于壳幔混合物质，为上地幔和下地壳物质部分熔融产物侵位形成，可能受控于赤城-古北口深大断裂的大规模伸展诱导的地幔物质上涌有关。早二叠世辉长闪长岩为钠质钙碱性系列，岩体侵位深度大（12~15km），具有深源岩浆演化特征。岩浆来源为岩石圈地幔和下地壳的部分熔融，为壳幔混合源。岩体的形成受东西向断裂带控制，属被动式侵位。化学成分上（图 21-15），两期岩浆分别位于 CaO 端元和碱性端元，显示其具有独立的岩浆演化特征。

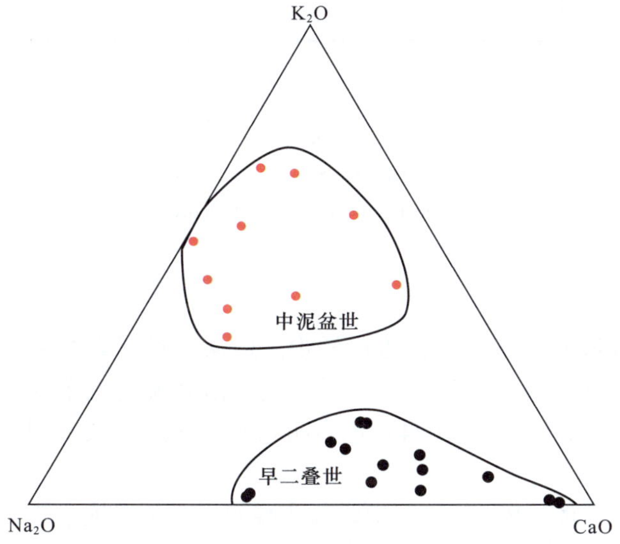

图 21-15　古生代侵入岩 Na_2O-K_2O-CaO 图解

Fig. 21-15　The ternary plots Na_2O-K_2O-CaO of the intrusive rocks in Paleozoic

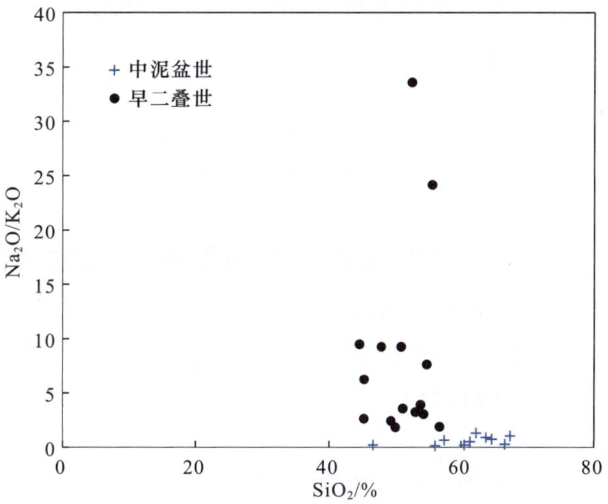

图 21-16　古生代侵入岩 SiO_2-Na_2O+K_2O 图解

Fig. 21-16　SiO_2 vs. Na_2O+K_2O diagram of the intrusive rocks in Paleozoic

中泥盆世侵入岩单元岩性单一，正长岩岩石化学不能拟合成较好曲线，说明岩浆侵位过程中自身侵位演化较弱，也间接体现其侵位-冷凝时间较短。而早二叠世辉长闪长岩单元岩性组合由角闪辉长岩-橄榄辉长岩-辉长岩-二辉辉长岩-正长闪长岩演化，岩石成分表现出规律变化。自早至晚，SiO_2 含量略有增加（图 21-16），但 FeO、P_2O_5、Na_2O、K_2O 含量逐渐增高，而 MgO、CaO、Al_2O_3 含量则逐

渐降低，Na_2O/K_2O 比值逐渐增大。说明岩石向富硅、富碱性方向演化。

（二）古生代岩浆作用的稀土演化特征

两期岩浆稀土元素配分曲线证实，有的代表了未分异亏损的地幔玄武岩组成，有的是亏损的分异玄武岩组成。早二叠世辉长岩分异形成的闪长岩还具有大陆拉斑玄武岩的配分特点。古生代两期岩浆活动具有相似的稀土元素总量、配分曲线和 La/Sm 比值、La/Yb 比值，考虑两期岩浆均为壳幔混合深成型，均受控于早期断裂带，具有较多相似性。虽然其形成时代相差较远，但区内构造背景环境主体仍然为较为稳定克拉通环境，壳幔岩石圈结构基本未发生变化。因此推测该两期岩浆源区可能具有某种构造成因上关联性。

在∑REE-La/Sm 图解（图 21-17）上，两期岩浆特征点相互穿插，也很好的揭示了上述特征。在 La/Sm-La/Yb 图解上（图 21-18），两期岩浆均朝着稀土元素 La/Sm 和 La/Yb 比值增加方向演化，岩浆演化拟合曲线基本重合，说明稀土分馏程度相似且伴随岩浆演化逐渐增强。

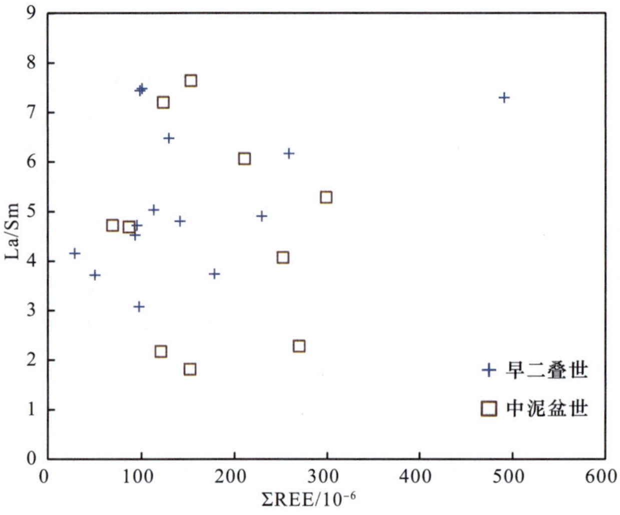

图 21-17　古生代侵入岩∑REE-La/Sm 图解
Fig. 21-17　∑REE-La/Sm diagram of the intrusive rock in Paleozoic

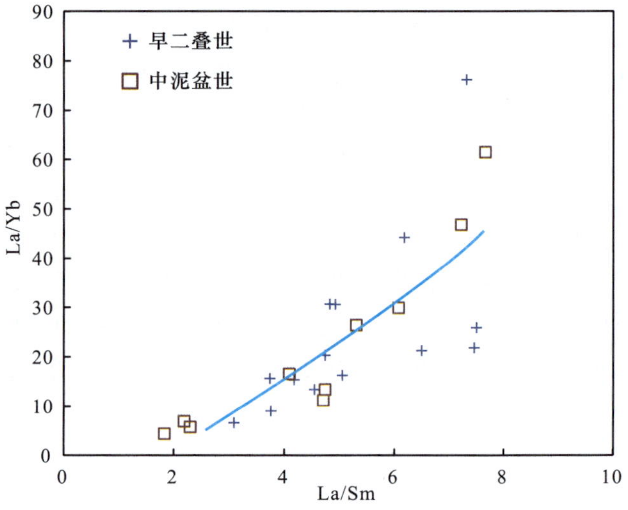

图 21-18　古生代侵入岩 La/Yb-La/Sm 图解
Fig. 21-18　La/Sm-La/ Yb diagram of the intrusive rock in Paleozoic

（三）古生代岩浆作用的微量元素演化

由于微量元素样品数量有限，古生代岩浆作用微量元素演化特征难以辨识。但从微量元素蛛网图上明显看出，中泥盆世正长岩具有地壳微量元素分布型式特征，而早二叠世接近火山岛弧钙碱系列型式。早二叠世辉长岩中个别磷灰石的锶同位素组成（$^{87}Sr/^{86}Sr=0.7074$）也表明它比较接近钙碱系列岩石的组成。

在 Ba/Sr-（Ba+Sr）图解（图 21-19）上，两期岩浆（Ba+Sr）含量分属两端元，正长岩远小于辉长闪长岩，而 Ba/Sr 比值基本相当，均值上正长岩略高于辉长闪长岩单元。在 Y-Sr/Y 图解（图 21-20）上，古生代两期岩浆 Y、Sr 元素含量也不在一个数量级上，正长岩高于辉长闪长岩。但单期岩浆演化特征完全相似，均朝着 Y 逐渐升高、Sr/Y 比值减小的方向演化。且单期岩浆演化特征也明显不同于一般造山带岩浆演化规律，说明该期本区伸展构造背景下地壳逐渐减薄。

（四）古生代岩浆起源及形成机制

根据上述分析，古生代两期岩浆均来源于壳幔混合型岩浆，岩浆侵位受构造控制。在 La-La/Sm 图解（图 21-21）和 Ce-Y 图解（图 21-22）上，两期岩浆均显示具有分离结晶岩浆形成机制，还揭示其受地壳物质混合影响。

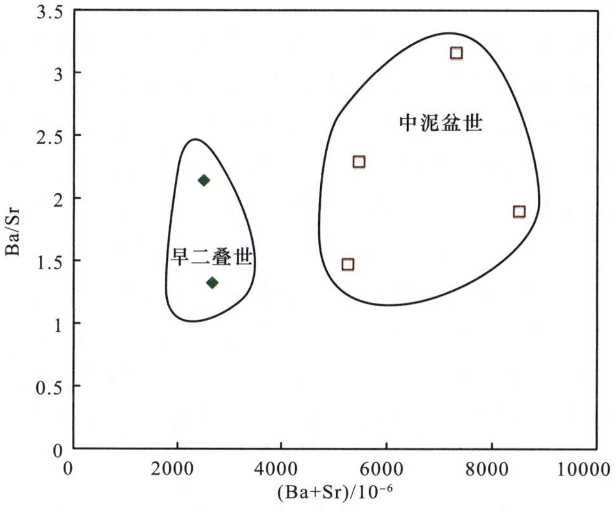

图 21-19　古生代侵入岩 Ba/Sr-（Ba+Sr）图解

Fig. 21-19　Ba/Sr-（Ba+Sr）diagram of the intrusive rocks in Paleozoic

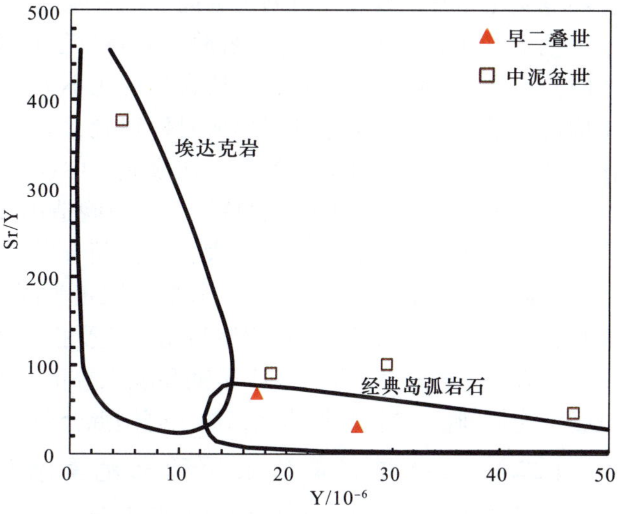

图 21-20　古生代侵入岩 Y-Sr/Y 图解

Fig. 21-20　Y-Sr/Y diagram of the intrusive rocks in Paleozoic

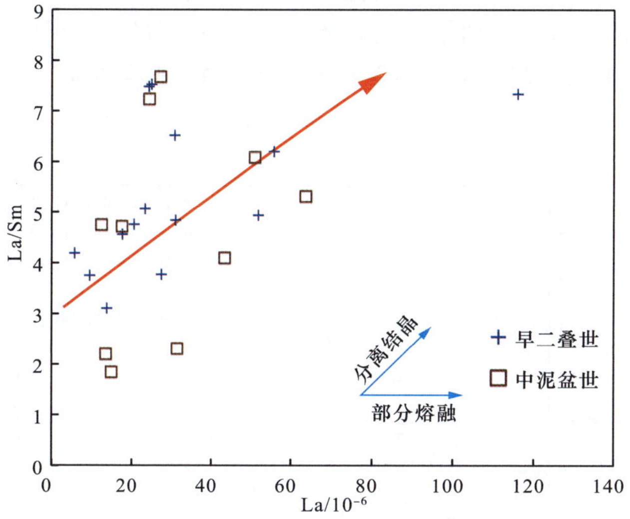

图 21-21　古生代侵入岩 La-La/Sm 图解

Fig. 21-21　La-La/Sm diagram of the intrusive rocks in Paleozoic

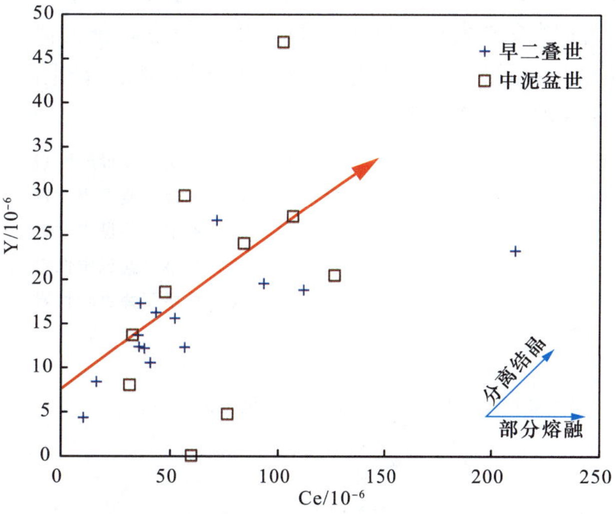

图 21-22　古生代侵入岩 Ce-Y 图解

Fig. 21-22　Ce-Y diagram of the intrusive rocks in Paleozoic

（五）岩浆活动规模

由于古生代两期岩浆活动时间较短，总体岩性变化小，岩浆演化也不完整，岩浆活动强度也不大，仅沿着赤城-古北口断裂带和密云-墙子路断裂带零星侵位。因此，该阶段岩浆活动规模小，尤其是中泥盆世，也使得现今出露地表岩体数量有限，分布面积有限。

四、三叠纪岩浆岩的成因及其演化规律

（一）岩浆起源、活动期次与成分的演化、岩浆活动方式

本区三叠纪经历早三叠世石英闪长岩、花岗闪长岩、二长花岗岩，中三叠世石英二长岩和晚三叠世花岗岩、石英二长岩岩浆侵位，其形成于中亚同造山-晚造山-造山后的复杂构造背景环境。由于构

造环境演化具有连续性，可划定为同一期岩浆活动期次。

化学成分上（图 21-23），各阶段岩石位于自身相对独立的空间分布区。早三叠世酸性岩属碱度高亚碱性系列，伴随岩浆演化，钙质含量变化不大，而由钠质向钾质方向演化，形成于造山作用有关挤压构造背景。早三叠世中性岩形成时代稍晚，和中三叠世、晚三叠世中性岩位于中间区，伴随岩浆演化，共同由钙质向碱性方向演化，构成完整中性—酸性岩浆演化序列。晚三叠世部分酸性岩（黑云母二长花岗岩），具有 I 型花岗岩特征，对应晚造山伸展构造背景。而至晚三叠世末期花岗斑岩、细粒花岗岩，化学成分又再次碱含量增大，并伴随着岩浆演化，再次向钾质方向演化，CaO 含量变化不大。

因此，三叠纪岩浆活动虽然岩浆期次限定为同一期，但岩浆演化特征可以划分为三个阶段，每个阶段具有自身岩浆演化特征。而岩浆活动方式受控于大地构造背景环境，岩浆活动引起的壳幔结构变化使得侵位方式差异较大。早三叠世早期和晚三叠世末期岩浆来源于单一壳源，岩体表现为浅成侵位，而早三叠世晚期-晚三叠世晚期岩浆来源于壳幔混合源，地壳厚度也因造山运动增厚，岩体表现为深成侵位。

（二）三叠纪岩浆作用的稀土元素演化特征

三叠纪岩浆岩总体表现出相似的稀土元素总量和 La/Sm 比值，中性岩类稀土总量稍高，而 La/Yb 比值稍低。但 La/Yb 比值差异较大，中性岩类明显高于酸性岩类。这些特征同一般造山带中酸性岩对比特征完全一致。因此，稀土元素较好的揭示了其类似造山带岩浆演化过程。

在 La/Sm-La 图解（图 21-24）、∑REE-La/Sm 图解（图 21-25）、La/Sm-La/Yb 图解（图 21-26）上，按照时间序列对稀土元素进行梳理，可以总结出两个阶段的演化规律。第一阶段为早三叠世酸性岩浆岩表现出随着岩浆演化，稀土总量增高、且 La/Sm 比值增加的趋势，说明稀土分馏程度逐渐增高。而第二阶段为早三叠世中性岩-晚三叠世酸性岩岩浆演化，伴随着岩浆演化，稀土总量增高、但 La/Sm 比值变化不大。

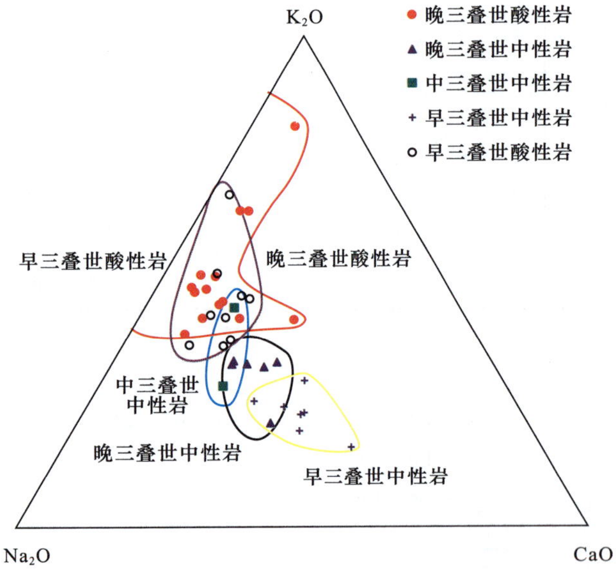

图 21-23　三叠纪侵入岩 Na_2O-K_2O-CaO 图解

Fig. 21-23　The ternary plots Na_2O-K_2O-CaO of the instrusive rocks in Triassic

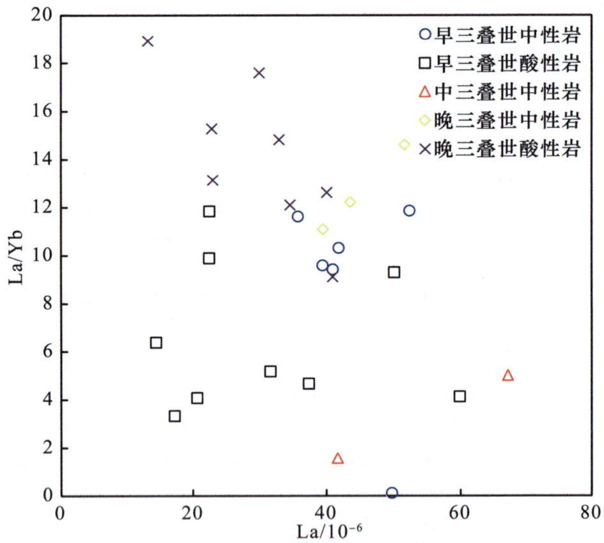

图 21-24　三叠纪侵入岩 La-La/Yb 图解

Fig. 21-24　La-La/Yb diagram of the instrusive rocks in Triassic

（三）三叠世岩浆作用的微量元素演化特征

在 Y-δEu 图解（图 21-27）上，仅晚三叠世酸性岩表现出铕负异常，其余无铕异常。这说明早、晚三叠世酸性岩岩浆演化存在差异，晚三叠世岩浆源区可能存在斜长石残留。时间序列上，早三叠世酸性岩 Sr、Y 元素含量分布零散，独成特征。而早三叠世中性岩—中三叠世中性岩—晚三叠世中性岩、酸性岩，随着岩浆演化，Y、Yb 元素含量先减小后增高（图 21-28），可能同造山地壳厚度变化密切有关。在 Ba/Sr-（Ba+Sr）图解（图 21-29）上，各阶段不同岩性 Ba/Sr 相当，而 Ba+Sr 含量晚三叠世酸性岩明显偏低，体现了不一致的岩浆构造背景环境。在 Y-Sr/Y 图解（图 21-30）上，按照时间序列，形成两个时间段两条演化曲线，第一阶段为 Y 元素含量变化不大，但 Sr/Y 比值急剧增大，岩性均表现出埃达克岩特征。第二阶段为 Y 元素含量急剧降低，且 Sr/Y 比值也逐渐降低的阶段，这一岩浆演化规律同早三叠世-晚三叠世中亚造山构造岩浆背景完全一致。

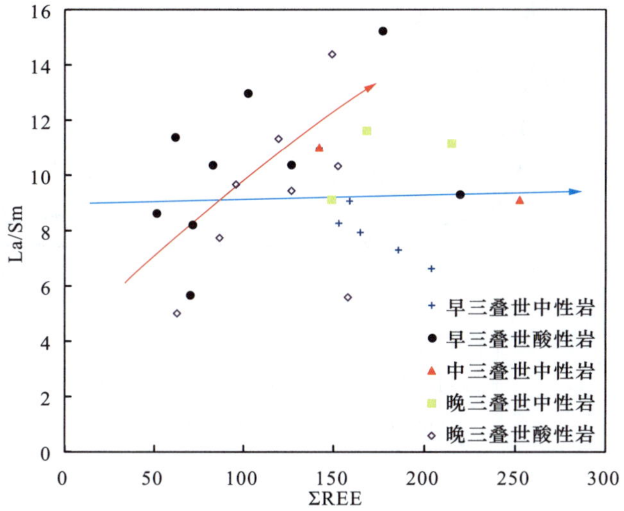

图 21-25　三叠纪侵入岩 ΣREE-La/Sm 图解

Fig. 21-25　La-La/Yb analysis diagram of the instrusive rocks in Triassic

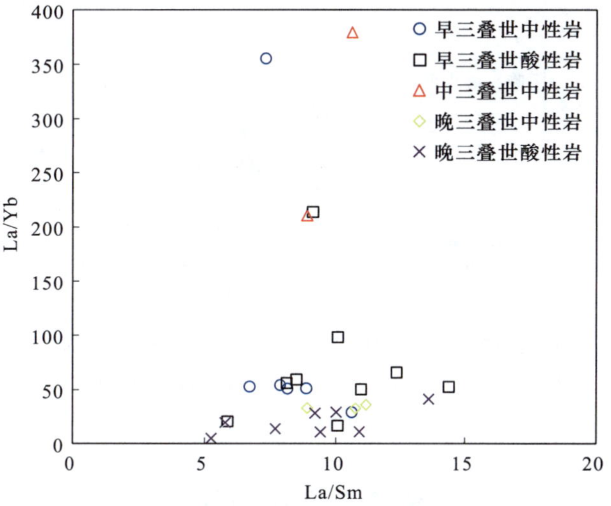

图 21-26　三叠纪侵入岩 La/Sm-La/Yb 图解

Fig. 21-26　La/Sm-La/Yb analysis diagram of the instrusive rocks in Triassic

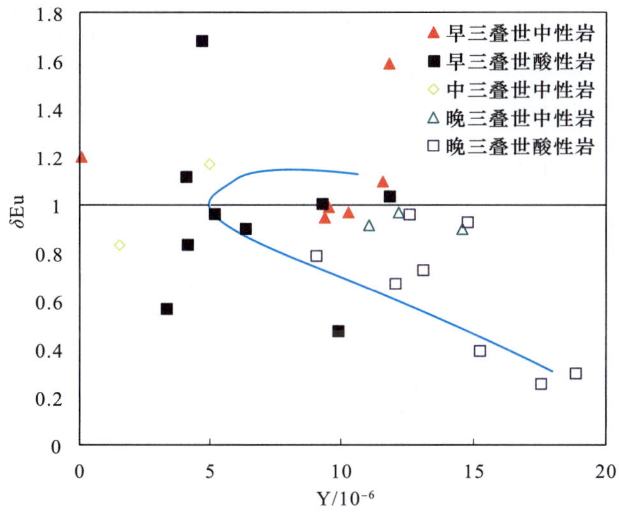

图 21-27　三叠纪侵入岩 Y-δEu 图解

Fig. 21-27　Y-δEu analysis diagram of the instrusive rocks in Triassic

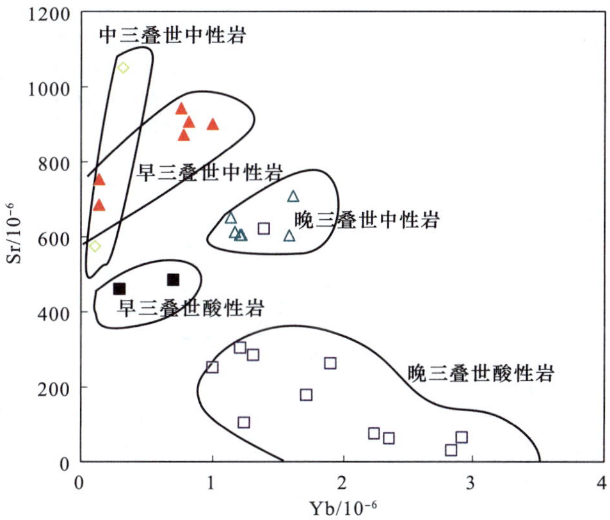

图 21-28　三叠纪侵入岩 Sr-Yb 图解

Fig. 21-28　Yb-Sr analysis diagram of the instrusive rocks in Triassic

（四）三叠纪岩浆起源与形成机制

早三叠世岩石均为高钾钙碱性系列，但中性岩和酸性岩岩浆来源不具有同源岩浆演化特征，中性岩为壳幔混合源成因，酸性岩为壳源成因。中三叠世岩石为碱性岩系列，为壳源成因。晚三叠世岩石为高钾钙碱性-钾玄岩系列。在 La-La/Sm（图 21-31）和 Ce-Y 图解（图 21-32）上，各类岩性各阶段岩浆均显示出部分熔融的岩浆形成机制。

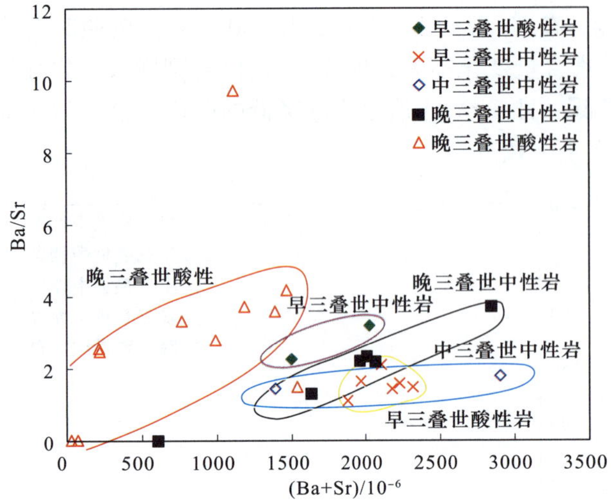

图 21-29 三叠纪侵入岩 Ba/Sr-（Ba+Sr）图解

Fig. 21-29 （Ba+Sr） -Ba/Sr analysis diagram of the instrusive rocks in Triassic

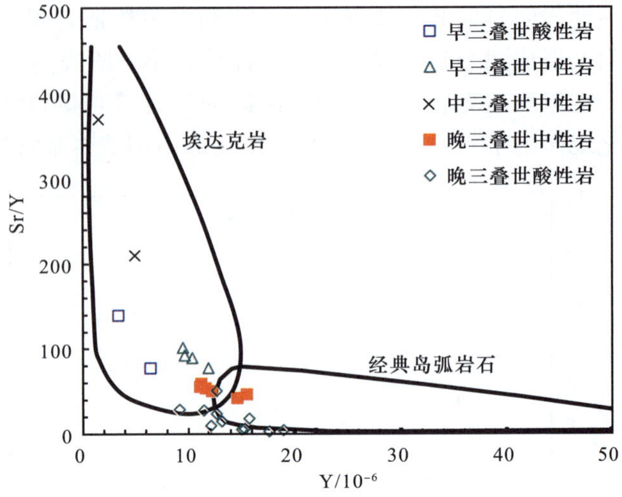

图 21-30 三叠纪侵入岩 Y-Sr/Y 图解

Fig. 21-30 Y-Y-Sr analysis diagram of the instrusive rocks in Triassic

（五）三叠纪岩浆活动规模

三叠世岩浆虽然具有造山带演化过程岩浆特征，由于并非位于造山带内部，只是中亚造山对其远程影响。因此，整个三叠纪岩浆活动规模并不大。

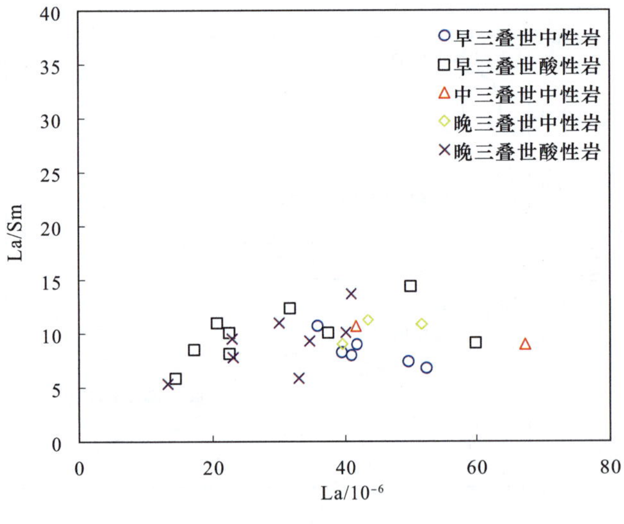

图 21-31 三叠纪侵入岩 La-La/Sm 图解

Fig. 21-31 La-La/Sm analysis diagram of the instrusive rocks in Triassic

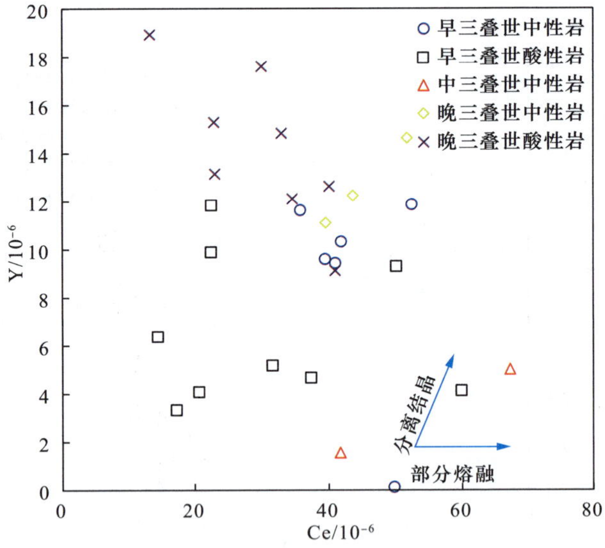

图 21-32 三叠纪侵入岩 Ce-Y 图解

Fig. 21-32 Ce-Yanalysis diagram of the instrusive rocks in Triassic

五、侏罗纪岩浆岩的成因及其演化规律

（一）岩浆起源、活动期次与成分的演化

本区侏罗纪经历早侏罗世早期辉长岩-闪长岩、早侏罗世晚期正长岩-花岗岩，中侏罗世期花岗岩，晚侏罗世辉长岩-闪长岩-石英闪长岩-石英二长岩-花岗岩岩浆侵位。按照构造-岩浆演化，可划

分为两期构造-岩浆事件。第一期为早侏罗世正长岩-花岗岩系列，为中亚造山晚期构造岩浆事件。第二期为早侏罗世晚期—晚侏罗世（延续到早白垩世早期）辉长岩-闪长岩-石英二长岩-花岗岩复杂岩性组合系列，形成于燕山板内造山的复杂构造背景环境，同早白垩世系列岩体一起构成中生代燕山运动完整构造-岩浆岩带。由于构造环境演化具有连续性，先后经历前造山-同造山-造山晚期-晚造山期等系列岩浆演化，因此虽然该阶段岩体众多、岩性复杂，但仍和早白垩世岩浆事件可划定为同一期，即燕山板内造山岩浆构造事件。

化学成分图上（图 21-33、图 21-34），两期岩浆伴随岩性演化，CaO 逐渐减少，而碱质成分尤其是 K_2O 逐渐增高。因此，岩石向钙碱性方向、钾质方向演化。岩石成分演化同构造背景环境密切有关，碱性基性岩形成于伸展构造背景，而钙碱性中性岩和酸性岩是构造反转挤压背景下岩浆侵位的产物。侏罗纪两期不同岩石组合的碱性—钙碱性岩浆演化，构成燕山板内造山 I 幕和 II 幕运动的重要岩浆演化特征。燕山运动两幕次先伸展后挤压，最终造成该阶段岩浆侵位、岩性差异和岩体规模差异等。

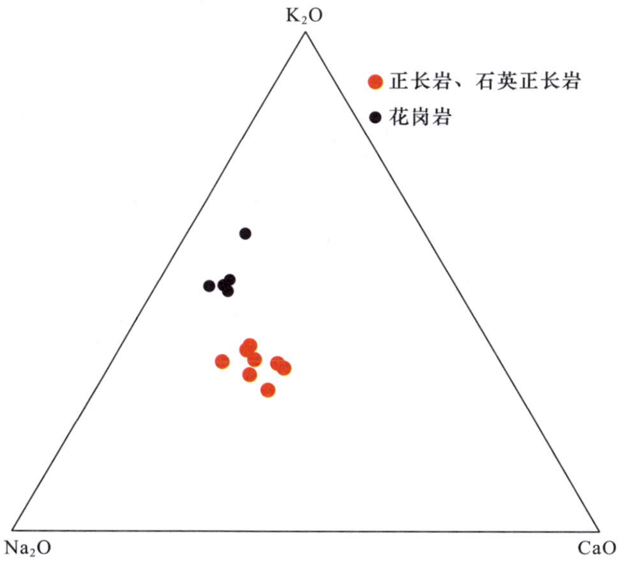

图 21-33 早侏罗世侵入岩 Na_2O-K_2O-CaO 图解

Fig. 21-33 The ternary Plots Na_2O-K_2O-CaO of the instrusive rocks in Early Jurassic

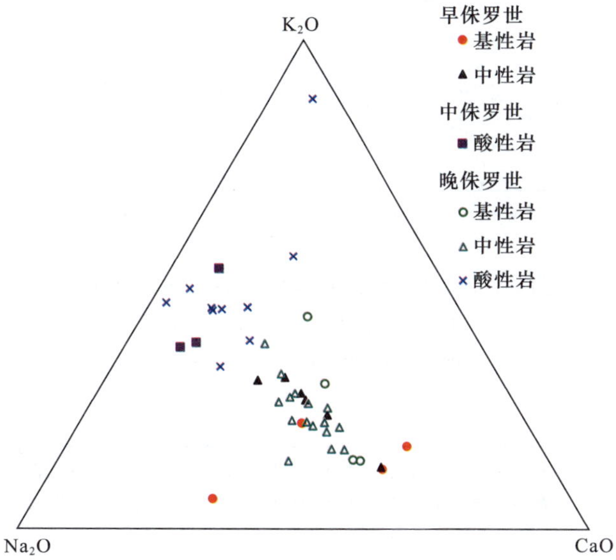

图 21-34 侏罗纪侵入岩 Na_2O-K_2O-CaO 图解

Fig. 21-34 The ternary Plots Na_2O-K_2O-CaO of the instrusive rocks in Jurassic

（二）侏罗纪岩浆作用的稀土元素演化规律

在 $\sum$REE-La/Sm 图（图 21-35）上，第一期正长岩-花岗岩 La/Sm 比值高于第二期各类岩性，尤其是花岗岩。由正长岩到花岗岩，La/Sm 比值增大，说明稀土分馏程度逐渐递增，而稀土总量则明显减小。而第二期岩浆事件可分为两幕，两幕岩浆岩稀土总量相当，但 La/Sm 比值第 I 幕明显低于第 II 幕。两幕岩浆岩从基性岩—中性岩—酸性岩，稀土总量逐渐降低，特征明显不同于第一期，但 La/Sm 比值变化不大，略降低。在 La/Sm-La/Yb 图（图 21-36）上，第一期由正长岩至花岗岩 La/Sm 比值增幅大，但 La/Yb 比值变化不大，说明轻稀土产生分馏作用，但轻重稀土分馏程度基本不变。而第二期两幕则表现明显不同，从基性岩—中性岩—酸性岩，轻稀土及轻重稀土分馏程度都逐渐降低，且变化幅度也较大。

（三）侏罗纪岩浆作用的微量元素演化特征

在 Ba/Sr-（Ba+Sr）图解（图 21-37）上，基性岩 Ba+Sr 值高于中性岩和酸性岩。岩浆演化时间序列上，两期岩浆 Ba+Sr 值差不多，但第一期 Ba/Sr 比值明显高于第二期，第二期两幕 Ba/Sr 比值则相当。在 Y-Sr/Y 图解和 Sr-Yb 图解（图 21-38）上，两期岩浆及不同岩性 Sr、Y、Yb 含量或比值几乎各自位于相对独立的空间区域，少有重叠，揭示该阶段复杂的壳幔演化过程。第一期岩石具有低 Sr 和高 Y、Yb 类南岭型特征，且随着岩浆演化，Sr、Y 具有同步降低趋势，而 Sr/Y 比值变化不大。而第二期岩石总体则具有高 Sr 和低 Y、Yb 类埃达克岩或喜马拉雅型花岗岩特征（图 21-39），总体 Y、Yb 含量变化不大，Sr 含量和 Sr/Y 比值随着岩浆演化逐渐降低，原因可能与燕山期板内造山引起的地壳厚度变化有关。

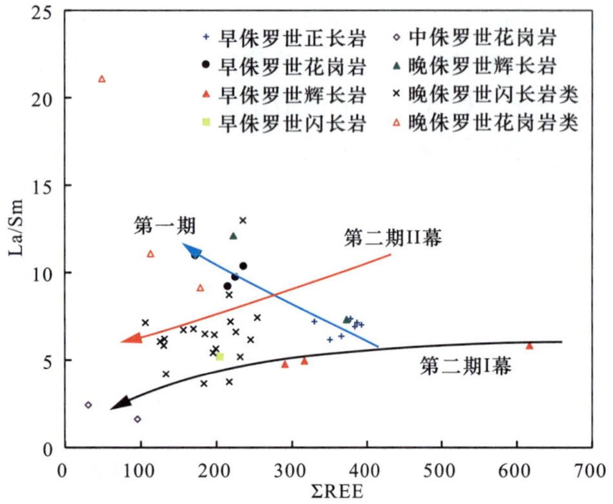

图 21-35　侏罗纪侵入岩∑REE-La/Sm 图解

Fig. 21-35　∑REE-La/Sm diagram of the instrusive rocks in Jurassic

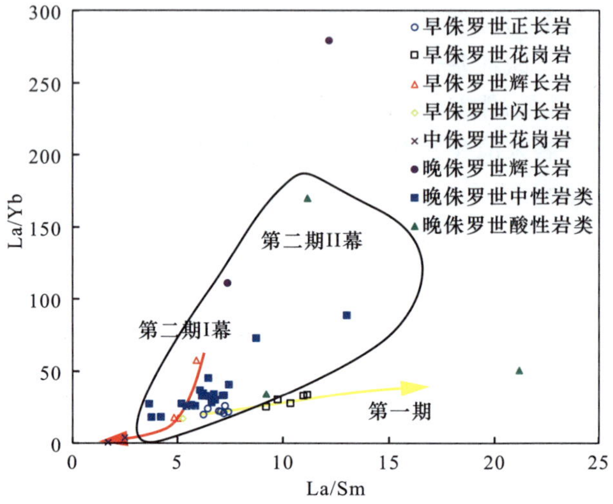

图 21-36　侏罗纪侵入岩 La/Sm-La/Yb 图解

Fig. 21-36　La/Sm-La/Yb diagram of the instrusive rocks in Jurassic

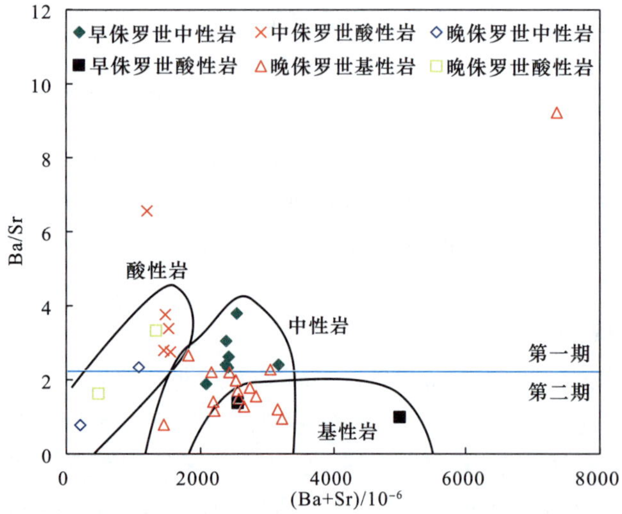

图 21-37　侏罗纪侵入岩 Ba/Sr-（Ba+Sr）图解

Fig. 21-37　Ba/Sr-Ba/Sr diagram of the instrusive rocks in Jurassic

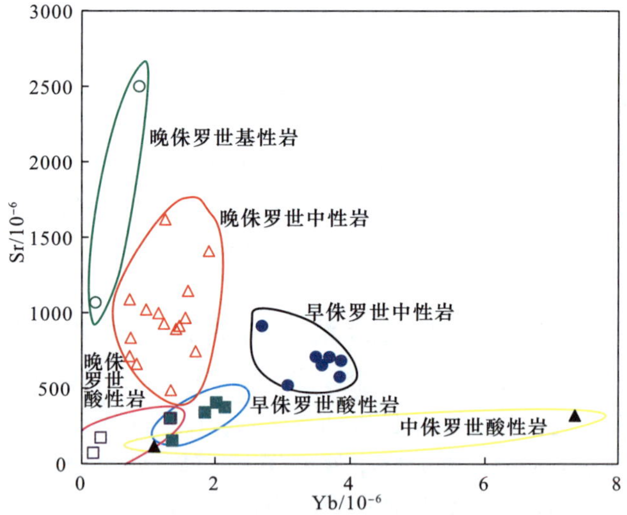

图 21-38　侏罗纪侵入岩 Sr-Yb 图解

Fig. 21-38　Yb-Sr diagram of the instrusive rocks in Jurassic

（四）岩浆起源与形成机制

第一期正长岩为碱性岩系列，花岗岩为钙碱性系列。正长岩起源于富集地幔，是幔源钾质基性岩浆在高压下分离结晶的产物。而幔源岩浆底侵加热并交代下地壳，促使其部分熔融产生花岗质岩浆，形成后期酸性岩。该期岩浆为早侏罗世独立一期岩浆事件，岩浆顺太古宙和中—新元古代地层接触带侵入，侵位深度超过 11km，为深成岩浆侵位。

第二期岩石组成复杂，按照时间先后分别形成两幕：即由早侏罗世基性岩—早侏罗世中性岩—早侏罗世酸性岩—中侏罗世酸性岩和晚侏罗世基性岩—晚侏罗世中性岩—晚侏罗世酸性岩。早期辉长岩或辉长闪长岩分布范围有限，为碱性岩系列。其后的闪长岩、石英闪长岩、石英二长岩及花岗岩等岩石均为钙碱性系列，且多数样品为高钾钙碱性系列。岩浆成因起源上，除中、晚侏罗世酸性岩外，该阶段基性岩、中性岩岩浆均来源于 I 富集型下地幔，并伴随岩浆演化，后期有壳源物质加入。

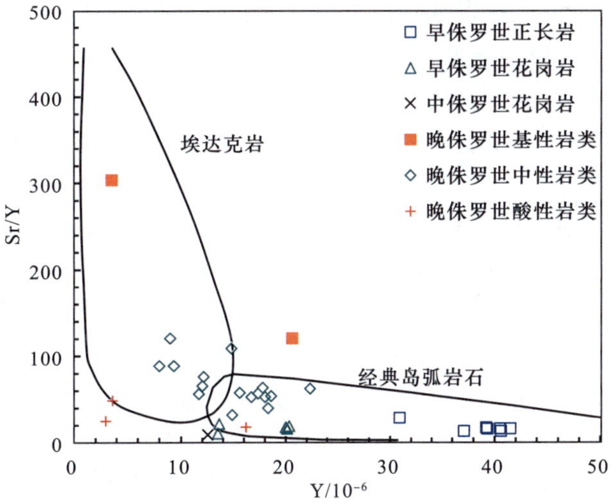

图 21-39　侏罗纪侵入岩 Sr-Sr/Y 图解

Fig. 21-39　Yb-Sr/Y diagram of the instrusive rocks in Jurassic

在 La-La/Sm（图 21-40）和 Ce-Y 图解（图 21-41）上，第一期岩浆和第二期 I 幕表现出分离结晶岩浆形成机制，而第二期 II 幕则表现出部分熔融岩浆形成机制。

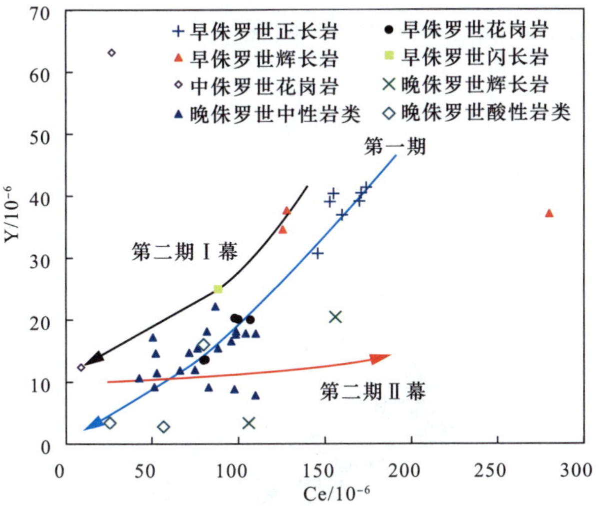

图 21-40　侏罗纪侵入岩 Ce-Y 图解

Fig. 21-40　Ce-Y analysis diagram of the instrusive rocks in Jurassic

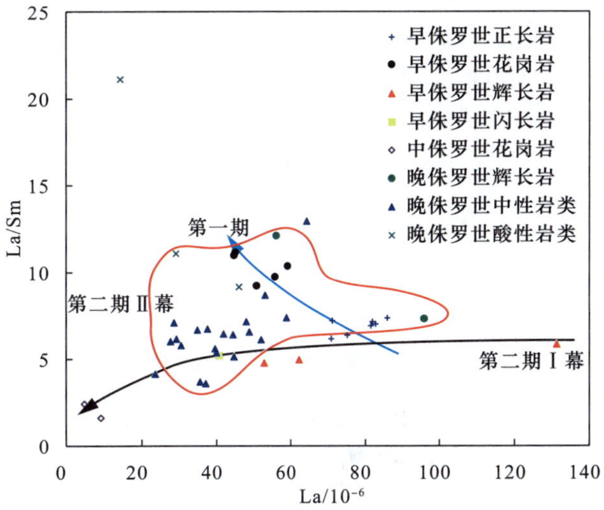

图 21-41　罗纪侵入岩 La-La/Sm 图解

Fig. 21-41　La-La/Sm analysis diagram of the instrusive rocks in Jurassic

（五）岩浆活动方式和活动规模

侏罗纪岩浆具有造山带演化过程中岩浆典型特征。燕山运动 I 幕和 II 幕，尤其是 II 幕主幕次运动，使得本区岩浆活动规模巨大。燕山运动单幕先伸展后挤压，构造应力场的不同，使岩浆活动规模差异较大。伸展背景下，岩浆活动规模相对较小，形成岩枝或岩脉状小型岩体（脉），岩浆侵位深度一般较浅，表现为浅成岩浆作用。而挤压背景下，岩浆活动规模大，形成区内大型岩体（如长园岩体、石城岩体等），岩体侵位深度大，表现为深成岩浆作用。

六、白垩纪岩浆岩的成因及其演化规律

（一）岩浆起源、活动期次与成分的演化

本区早白垩世是岩浆活动最发育时期，岩性几乎包括基性岩-中性岩-酸性岩的所有岩类。按照时间演化序列，可划分为早白垩世早期花岗岩-石英二长岩-闪长（玢）岩序列，早白垩世中期辉长岩-闪长岩-石英二长岩-花岗岩和早白垩世晚期—晚白垩世煌斑岩（辉绿岩）-二长花岗岩（白岗岩）-正长（碱长）花岗岩，和晚侏罗世岩浆事件一起构成中生代区内燕山运动3个幕的完整构造-岩浆岩带。其中，早侏罗世晚期-中侏罗世构成燕山运动Ⅰ幕，为燕山板内造山前造山阶段；晚侏罗世—早白垩世早期构成燕山运动Ⅱ幕，为燕山板内造山（同碰撞）期；早白垩世中期—晚期构成构成燕山运动Ⅲ幕，为燕山板内造山晚期—后造山期。本区燕山旋回的碱性—钙碱系列岩浆活动在中国东部具有典型意义。它可以和环太平洋的安第斯山为代表的Ⅰ型（科迪勒拉）岩浆活运相对比，并且包括了整个燕山旋回各阶段演化的系列产物，形成深成侵入与火山喷发相对应的完整岩石系列。

早白垩世早期花岗岩-石英二长岩-闪长（玢）岩侵位和晚侏罗世辉长岩-闪长岩-石英二长岩-花岗岩一起，为燕山板内造山同碰撞挤压构造背景下岩浆活动产物，为一个相对独立的构造-岩浆演化亚旋回。早白垩世中期辉长岩-闪长岩-石英二长岩-花岗岩则是燕山板内造山晚期伸展构造背景下岩浆活动产物，为另一个相对独立的构造-岩浆演化亚旋回。而早白垩世晚期煌斑岩（辉绿岩）-二长花岗岩（白岗岩）-正长（碱长）花岗岩则是燕山板内造山后造山相对稳定阶段伸展构造背景下岩浆活动产物，又为一个相对独立的构造-岩浆演化亚旋回。总之，早白垩世岩浆活动可划分为一个岩浆活动旋回（期次）的三个亚旋回。化学成分图（图21-42、图21-43）上，单次亚旋回的岩性由基性—中性—酸性演化，CaO成分逐渐减少，而碱质（Na_2O+K_2O）成分逐渐增高。因此，早白垩世岩石总体向钙碱性方向、钾质方向演化。

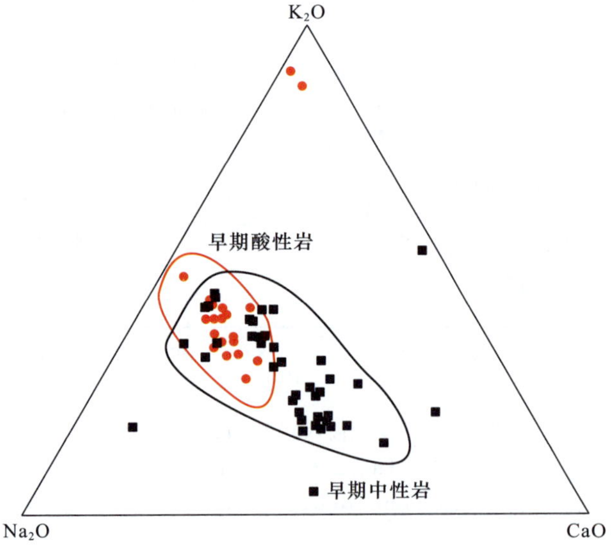

图21-42 早白垩世侵入岩 Na_2O-K_2O-CaO 图解

Fig. 21-42 The Na_2O-K_2O-CaO ternary Plot of the instrusive rocks in Early Cretaceous

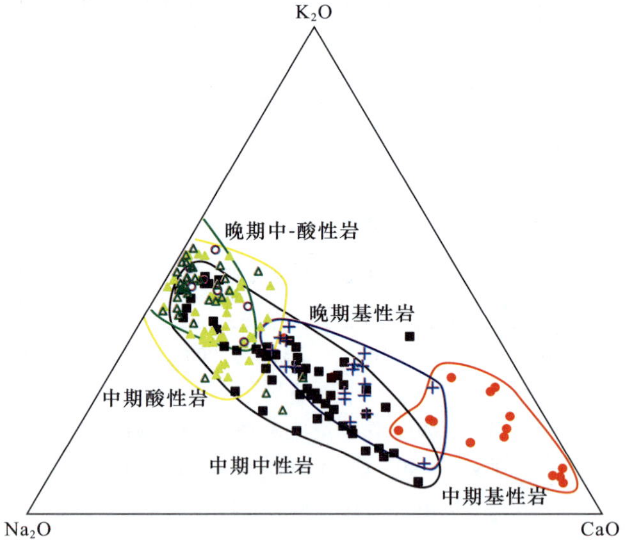

图21-43 早白垩世侵入岩 Na_2O-K_2O-CaO 图解

Fig. 21-43 The Na_2O-K_2O-CaO ternary Plot of the instrusive rocks in Early Cretaceous

（二）白垩纪岩浆作用的稀土元素演化特征

在∑REE-La/Sm图（图21-44）上，早白垩世早期（左）花岗岩和花岗闪长岩La/Sm比值明显高于同期闪长岩、石英二长岩，尤其是花岗岩，说明稀土分馏程度逐渐递增，而稀土总量则明显减小。早白垩世中期（中）岩浆活动基性岩和中性岩能构成较好岩浆演化序列，而酸性岩独立形成

单独岩浆演化，具有不同的岩浆来源。早白垩世晚期和晚白垩世岩石（右）具有和中期类似特征，酸性-碱性花岗岩和基性岩独立发育为具有不同岩浆来源的双峰式岩浆活动特征，揭示燕山期造山运动已经结束。

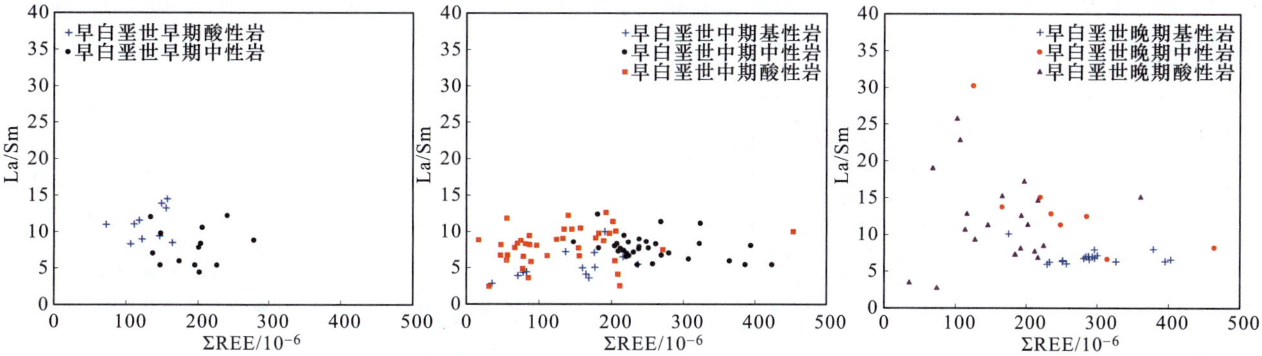

图 21-44 白垩纪侵入岩∑REE-La/Sm 图解

Fig. 21-44 ∑REE-La/Sm analysis diagram of the instrusive rocks in Early Cretaceous

（三）白垩纪岩浆作用的微量元素演化特征

在 Ba/Sr-（Ba+Sr）图解（图 21-46）上，Ba+Sr 值、Ba/Sr 比值变化幅度较大。早白垩世早期和中期特征相似，Ba+Sr 值基本位于同一中—高值区域，Ba/Sr 比值则略低，而早白垩世晚期及晚白垩世 Ba+Sr 值相对较低，Ba/Sr 比值略高。早白垩世早期和中期岩石 Ba+Sr 值高，为高 Sr、Ba 型岩石。钱青（2002）把该阶段酸性岩的这一特征统称为八达岭高 Ba、Sr 花岗岩类，并对其成因进行了探讨，认为同富集型幔源岩浆和壳源物质混合、岩石分离结晶作用有关。该阶段基性岩 Ba+Sr 值、Ba/Sr 比值略低于中-酸性岩，这一特征同侏罗纪及晚三叠世等早期岩石特征完全不同。早白垩世晚期及晚白垩世岩石 Ba+Sr 值低，同其壳源岩浆成因及减薄的岩石圈厚度有关。

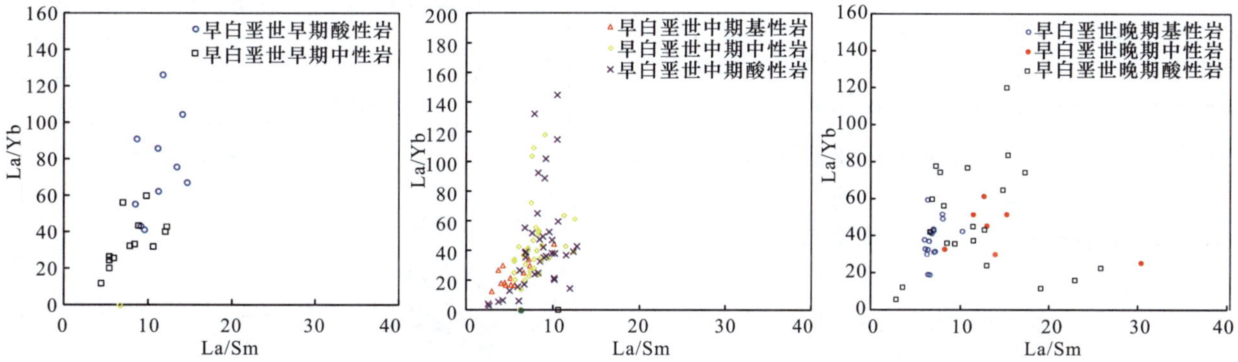

图 21-45 白垩纪侵入岩 La/Yb-La/Sm 图解

Fig. 21-45 La/Sm-La/Yb analysis diagram of the instrusive rocks in Early Cretaceous

在 Y-Sr/Y 图解（图 21-48）和 Sr-Yb 图解（图 21-47）上，早白垩世三幕岩浆亚旋回岩石具有大致相似的 Y、Yb 元素含量，但 Sr 元素含量和 Sr/Y 比值却有所不同。早白垩世早期和中期具有大致相似的 Sr 元素含量和 Sr/Y 比值，高于早白垩世晚期岩石，具有典型埃达克岩特征。而 K_1^3 期 Sr 元素含量及 Sr/Y 比值稍低，绝大多数具有喜马拉雅型花岗岩特征，超酸性石英斑岩等还具有南岭型花岗岩特征。

（四）岩浆起源与形成机制

经历燕山板内造山期—造山晚期—造山后期的复杂岩浆演化过程，伴随燕山板内造山，壳幔结构发生重要变化。地壳厚度、地温梯度等变化是最典型表现形式之一，正是这些壳幔结构变化，使得岩浆起源深度在早白垩世变化较大。第一个亚旋回早白垩世早期花岗岩、石英二长岩、闪长玢岩等全部岩类均为高钾钙碱性系列，岩浆起源于壳幔混合源。第二个亚旋回早白垩世中期基性岩为碱性岩系列，岩浆起源于亏损地幔，而中-酸性岩类为高钾钙碱性-钾玄岩系列，岩浆起源于壳幔混合源。第三个亚旋回早白垩世晚期-晚白垩世期基性岩为碱性岩系列，中酸性岩为高钾钙碱性系列，岩浆起源于不同深度地壳。

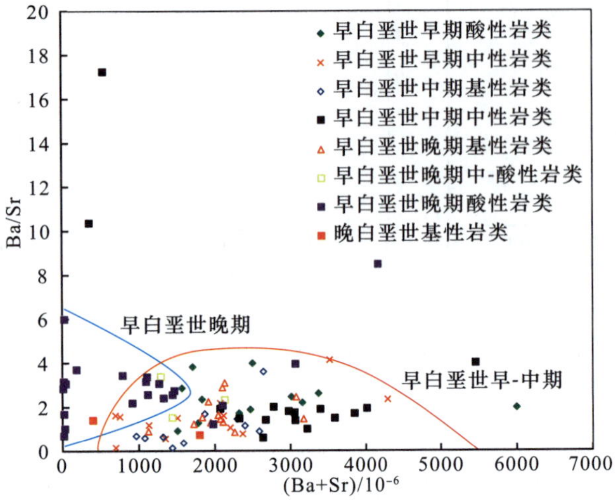

图 21-46　白垩纪侵入岩 Ba/Sr-（Ba+Sr）图解

Fig. 21-46　（Ba+Sr）-Ba/Sr analysis diagram of the instrusive rocks in Cretaceous

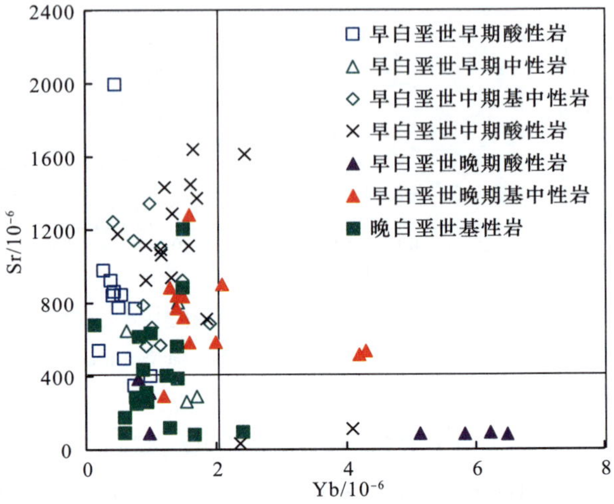

图 21-47　白垩纪侵入岩 Sr-Yb 图解

Fig. 21-47　Yb-Sr analysis diagram of the instrusive rocks in Cretaceous

不仅如此，壳幔结构的变化还使得岩浆侵位深度也发生较大变化。第一、二亚旋回为早白垩世早期和中期因造山强烈活动构造背景（挤压和伸展垮塌）环境，在地壳厚度较大情况下，以深成岩浆作用为主。同时，在每个亚旋回的末期也有浅成岩浆作用发育。而早白垩世晚期—晚白垩世早期第三亚旋回期因后造山相对较稳定的构造背景，且地壳也因伸展垮塌而减薄情况下，以浅成岩浆作用为主，少有深成岩浆侵位。

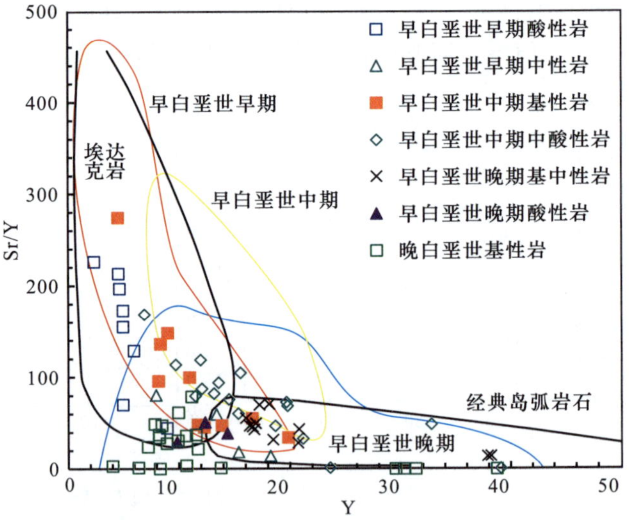

图 21-48　早白垩世侵入岩 Sr/Y-Y 图解

Fig. 21-48　Sr/Y-Y analysis diagram of the instrusive rocks in Cretaceous

在 La-La/Sm（图 20-49）和 Ce-Y 图解（图 21-50）上，三个亚旋回岩浆均表现出不同程度分离结晶岩浆形成机制。因此，早白垩世岩浆形成机制主要表现为分离结晶作用，其同三叠世及早侏罗世岩体表现出不一致的形成机制。早白垩世三个岩浆亚旋回分离结晶程度又有所不同，表现为 La-La/Sm 和 Ce-Y 图解上拟合曲线斜率不同，早白垩世早期和中期特征相似，而晚期明显不同，这同上述特征基本一致。

（五）岩浆活动方式和活动规模

早白垩世岩浆具有造山带演化过程中岩浆典型特征。燕山运动旋回的第三幕运动，主要集中于早白垩世，使得本区岩浆活动规模巨大。正如前述，燕山运动单幕先伸展后挤压，其构造应力场的不同，使得岩浆活动规模差异较大，具有和中—晚侏罗世一致的岩浆活动特征。即岩浆活动规模伴随燕山运动 III 幕经历多幕弱-强交替演化。至早白垩世晚期—晚白垩世期燕山板内后造山阶段，岩浆活动规模逐渐减小，并最终趋于稳定，岩浆活动规模又再次降低，且多以浅层岩浆活动为主。因此，早白垩世岩浆活动规模总体表现为先强后弱，岩浆活动方式由深成向浅成方式转变的特征。

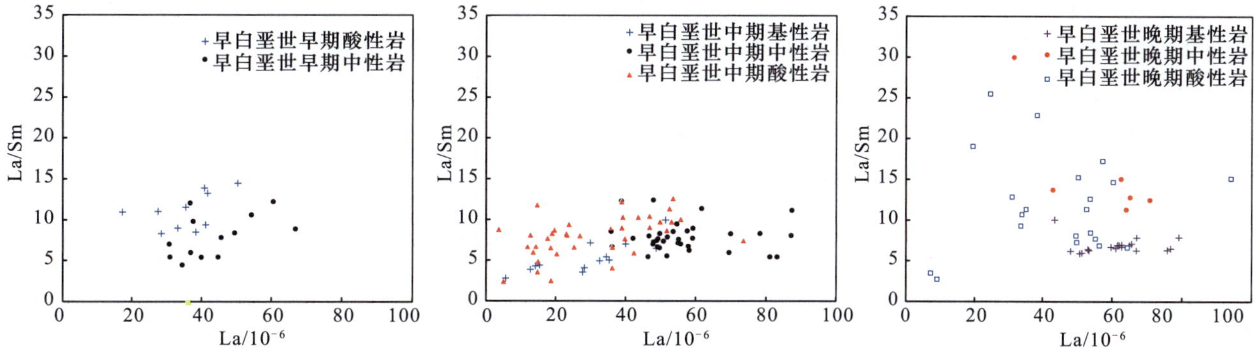

图 21-49　早白垩世侵入岩 La-La/Sm 图解

Fig. 21-49　La-La/Sm analysis diagram of the intrusive rocks in Cretaceous

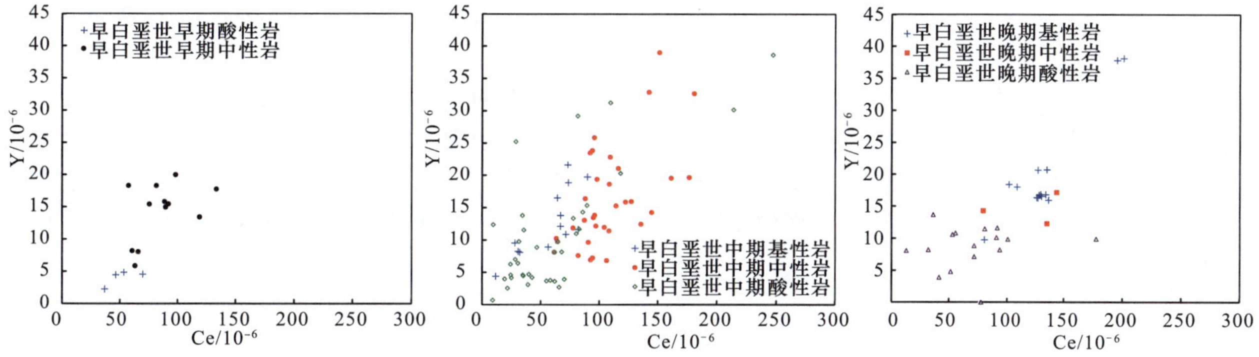

图 21-50　早白垩世侵入岩 Ce-Y 图解

Fig. 21-50　La-La/Sm analysis diagram of the intrusive rocks in Cretaceous

第四节　构造-岩浆演化特点和规律

岩浆岩的形成、发展与构造的形成、发展和演化具有内在的成因联系。

一、太古宙—早元古代古陆拼合造山

新太古代早期岩浆活动（超镁铁质）是在地壳较薄、热流值高，并呈塑性的环境下发生的，其形成构造背景尚不明确。至新太古代中期本区陆壳已初步具有刚性特征，伴随大地构造演化，本区逐渐向大陆环境发展，以沙厂英云闪长质片麻岩为主的复合型杂岩体侵位时期构成一个复杂的碰撞前-晚造山复杂的构造背景演化过程。而黑卧-苇子峪-宋营子系列侵位时期形成的大陆可能再次裂解，构成大陆裂解-岛弧构造演化过程。至新太古代晚期，放马峪-荆子峪-康各庄-旗杆顶系列侵位时期岛弧逐渐演化为大陆弧，并开始新一轮碰撞造山，经历前碰撞-同碰撞地壳增厚演化过程，最终形成具有类似 A 型花岗岩特征的旗杆顶系列岩浆侵位。

二、中元古代非造山裂陷槽

区内中元古代岩浆活动可划分为两期，第一期高岭基性岩墙-大庙斜长岩-沙厂花岗岩，根据同位素、岩石组合及地球化学等各方面特征，证实该期富钾碱性岩系-富钾钙碱性岩系的岩浆作用既有上地幔派生的基性岩浆，又有下地壳部分熔融的富钾碱性的花岗岩浆的参与。岩浆演化经历三个阶段，这三个阶段构成中元古代早期由基性-中性-酸性完整构造岩浆演化序列，是中元古代一次重要的非造山型岩浆活动。

第二期大红峪期火山岩-辉绿岩-正长斑岩，则是燕山裂陷槽在第一期岩浆活动基础上继续发展的又一期岩浆活动。发育于裂谷作用之后，这一点可以从宏观大红峪期火山岩不整合覆盖于因裂谷作用堆积的滨、浅海相沉积岩之上得到证实。区内大红峪期火成岩具有被动裂谷式裂谷-地幔上隆-火山作用的活动模式。也正是如此，区内该期火山喷发活动具喷发物体积小、地壳延伸速率低、火山作用不连续、并伴随沉积作用，且火山作用持续时间短等特征，这同第一期岩浆活动有着较大的差异。

三、晚古生代造山远程效应

本次通过整合各类资料，并补充大量实物工作，在区内发现中泥盆世、早二叠世岩体。对区内稳定克拉通破坏的起始时间和大地构造背景环境有了不同的认识。晚古生代区内具体构造背景环境有待进一步研究。通过岩浆活动及构造-岩浆岩初步分析，认为区内该阶段发育的小规模岩浆活动构造环境同中亚造山（多幕碰撞拼合）后短期伸展构造应力场有关。在伸展构造应力场背景下，华北克拉通北部早期深大断裂复活形成岩浆通道，使得本期岩浆侵位，并对壳幔结构产生破坏影响。

四、中生代板内造山

对燕山地区中生代（燕山运动）岩浆活动的构造背景，目前较为一致的认识是其属于大陆板块内部的陆内造山背景。燕山中—新生代陆内造山既不同于陆缘俯冲造山，又不同于陆间碰撞造山，也不同于地槽封闭造山。该区的构造应力场分布、深部热动力状态、深部滑脱及重力作用过程、岩浆上升侵位、岩浆演化等多方面有其自身的规律和特点，并与燕山中—新生代陆内（板内）造山作用存在紧密的内在成因联系。

第五篇

变质地质

第二十二章　区域变质岩系

北京地区的变质岩主要是区域变质岩。岩石类型包括麻粒岩、片麻岩、变粒岩、斜长角闪岩、斜长辉石岩、板岩、千枚岩、片岩及磁铁石英岩等。在一些侵入体与围岩接触带部位产出接触变质岩或热液蚀变岩；动力变质岩则多见于重要构造带上，但二者分布范围远小于区域变质岩。

区内变质岩以新太古代变质岩为主，集中分布于密云—怀柔隆起地区，以结晶基底形式出现，遭受多期变形变质作用，在平谷、昌平、延庆、门头沟、房山等地区零星出露。新太古代变质作用是本区已知最早的区域热事件，主要发育有密云岩群麻粒岩相变质作用和四合堂岩群角闪岩相变质作用，二者发生时间难以严格区分，均为2500~2800Ma，后者可能代表了区内晚一期的重要变质作用。

变质作用使该时期岩石发生变形和变质，形成一套麻粒岩相—角闪岩相的区域变质岩石，在空间上呈面形、条带状分布。其中密云岩群变质岩以麻粒岩相为主，区域热流空间分布与变质强度关系密切，该期变质岩后期发生退化变质作用及叠加变质。岩石中矿物组合常呈现多期次，并出现多种反应和交代现象。退化变质原因为同期退化或多期区域变质，局部退变质现象可能与混合岩化作用有关。变质作用类型属区域中高温变质作用。

本区变质岩大致可划分为变质表壳岩（包括密云岩群和四合堂岩群）、变质深成侵入体、动力变质岩、接触热变质岩、接触交代变质岩等几种类型。其中前2种属于区域性变质岩，可以形成大面积分布；而后3种属于局部性变质岩，往往仅局限于局部地质条件上的有利地段。

密云岩群属麻粒岩相变质岩，主要岩性包括磁铁石英岩类、斜长角闪岩、斜长辉石岩、石榴二辉石岩、角闪二辉石岩、角闪石岩、二辉角闪石岩、变粒岩等，以包体形式多种岩性集中成片分布。根据化学和矿物成分及原岩特征可划分为超镁铁质岩类、镁铁质变质岩类及长英质变质类三大系列岩石；重熔混合岩化作用普遍，混合程度不均匀，不能进行分带，大部分属于混合质变质岩及条带状、条痕状混合岩，级别较高的混合岩及花岗质岩类少见。

四合堂岩群属角闪岩相变质岩，岩石类型包括斜长片麻岩、变粒岩、斜长角闪岩等。与密云岩群相比较，岩石以细粒变晶结构为主。原岩建造中，基性火山岩比例减少，碎屑岩类增多。变质程度低于密云岩群。混合岩化作用相对较弱，以钠质交代为主。

变质深成侵入体与密云岩群、四合堂岩群岩性相近，并共同经历了区域变质作用，可进一步划为英云闪长质—奥长花岗质—花岗闪长质片麻岩和石英二长质—奥长花岗质—花岗质片麻岩两种岩性组合。前者包括沙厂英云闪长质片麻岩、黑卧辉长质片麻岩、苇子峪辉长闪长片麻岩、宋营子英云闪长质片麻岩、放马峪基性岩墙群（鲍亦冈等，2000；陈森煌，1994），后者包括荆子峪花岗闪长质片麻岩、康各庄奥长花岗质片麻岩、旗杆顶花岗质片麻岩（鲍亦冈等，2000）。

北京西山地区变质岩主要发育在中元古界上部至中生界下侏罗统中，主要为一套变质程度较浅的片岩、板岩、千枚岩及变质砂岩、砾岩，应变矿物硬绿泥石分布普遍，属中低压绿片岩相变质或低角闪岩相变质，总体呈NNW向展布，受构造环境差异热流活动强弱影响，该类型变质岩在时空上表现为不同程度变质带分布。

第一节 新太古代变质岩系

一、变质岩特征

（一）片岩类

1. 分布

主要分布于密云四合堂和怀柔西湾子等地区。为中低级区域变质作用产物，片状矿物含量>30%，变晶粒度>0.1mm，具定向特征。属新太古代高级深变质岩后期退变质作用产物，受后期韧性剪切带控制，呈条带状展布。

2. 岩石特征

区内片岩类型主要有黑云斜长片岩和角闪斜长片岩。具鳞片粒状变晶结构、纤状细粒变晶结构，片状构造及变余微层理构造，主要矿物成分为斜长石（40%~55%）、角闪石（25%~40%）、黑云母（约10%）和石英（<5%），有时含钾长石、白云母等，副矿物为磁铁矿、榍石和磷灰石，斜长石含量大于石英。其原岩主要为安山质、安山玄武质硬砂岩、凝灰岩或层凝灰岩。

（二）片麻岩类

1. 分布

片麻岩类是本区变质岩中最常见的岩石类型之一，在密云、怀柔、平谷、昌平、房山等地均有出露。平原区兴热1井1610m、兴热2井2238m见角闪斜长片麻岩、花岗质片麻岩、辉石角闪斜长片麻岩、二辉斜长片麻岩、含辉石黑云母角闪斜长片麻岩，证实在大兴地区深部存在片麻岩类（吕金波等，2008）。片麻岩类具片麻状构造，岩性单调稳定，区域上厚度变化大，片麻岩常与变粒岩共生，呈互层状或渐变过渡，不易区分，在成因上有一定共性。

2. 岩石特征

片麻岩呈灰绿色、灰黑色，暗色矿物与浅色矿物常形成不连续的条痕状、条带状及片麻状构造。粒状变晶、鳞片粒状变晶、柱状及不均匀粒状变晶结构。矿物粒度一般>0.3mm，略粗于变粒岩。密云沙厂地区以中细粒为主，粒度0.3~1mm；四合堂地区以细粒变晶为主，粒度0.3~0.5mm，主要有粒状矿物（斜长石、石英等）和暗色片、柱状矿物（辉石、角闪石及黑云母等）组成。根据暗色矿物的不同划分为辉石斜长片麻岩、角闪斜长片麻岩和黑云斜长片麻岩三大类。

据主要变质矿物共生组合可划分为10个主要类型（表22-1）。

表22-1 片麻岩主要岩石名称与矿物共生组合

Table 22-1 The main rock names and paragenetic mineral associations of gneiss

岩石名称	矿物共生组合
透辉斜长片麻岩	Di+Pl+Qtz+Hbl+Grt
二辉斜长片麻岩	Di+Hy+Pl+Qtz±Hbl+Bt
石榴透辉斜长片麻岩	Di+Pl+Qtz+Grt±Bt
角闪透辉斜长片麻岩	Di+Pl+Qtz+Hbl±Grt
角闪斜长片麻岩	Hbl+Pl+Qtz
透辉角闪斜长片麻岩	Hbl+Di+Pl+Qtz±Grt
黑云角闪斜长片麻岩	Hbl+Bt+Pl+Qtz
黑云斜长片麻岩	Bt+Pl+Qtz

续表

岩石名称	矿物共生组合
透辉黑云斜长片麻岩	Bt+Di+Pl+Qtz
石榴黑云斜长片麻岩	Bt+Pl+Qtz+Grt

注：Di—透辉石；Pi—斜长石；Qtz—石英；Hbl—角闪石；Grt—石榴子石；Hy—紫苏辉石；Bt—黑云母。

斜长石（45%~60%）是斜长片麻岩类最主要的矿物，呈他形、半自形板柱状、粒状结构。矿物为奥长石和中长石（An23~50），一般为中长石（An30~50），常具反条纹，部分斜长石具次圆状碎屑外形，有些岩石含少量微斜条纹长石。

石英（5%~15%）多呈他形不规则粒状，含量相比变粒岩较少，有时构成条痕状、透镜状或板条状集合体。

黑云母呈褐色-棕褐色片状，大小0.3~1.0mm，吸收性为棕褐色-浅黄色，从角闪岩相—麻粒岩相MgO的含量变化为12.67%~24%，TiO_2为2.59%~3.27%至4%~5%。呈细鳞片状、雏晶状或集合体状者为最晚期产物，明显交代早期辉石、角闪石、黑云母及石榴子石等。

角闪石呈柱粒状变晶，粒度0.4~1.2mm，因变质程度差异，颜色表现为棕绿色-绿色-蓝绿色多色性。有时呈粗大筛状变晶，含细微粒状石英、长石等包体或被黑云母、绿泥石、绿帘石等交代。

紫苏辉石仅见于二辉斜长片麻岩中。呈半自形粒状、短柱状，粒径0.3~1mm，平行消光或斜消光，消光角10°~15°。

透辉石常被次闪石、绿泥石、黑云母及碳酸盐等交代，呈半自形柱粒状，粒径0.3~1.0mm。

铁铝榴石一般多呈粉红色、粒状或自形晶单体、粒状集合体及变斑晶出现。粒径一般0.2~0.6mm，具筛状变晶结构者含石英、斜长石、辉石及角闪石等微粒包体，说明形成较晚，常为黑云母、绿泥石交代或水化产生铁锈物质。此矿物在密云沙厂-大漕地区分布较普遍。

（三）变粒岩及浅粒岩类

1. 分布

分布于密云和怀柔地区变质岩系中，为本区主要变质岩石。根据暗色矿物种类和含量可分为黑云变粒岩、角闪变粒岩、辉石变粒岩和浅粒岩四类（表22-2）。

表22-2 变粒岩及浅粒岩类主要岩石名称与矿物共生组合

Table 22-2 The main rock names and paragenetic mineral associations of granulite and leptite

岩石名称	矿物共生组合
黑云变粒岩	Pl+Qtz+Bt±Grt±Mgt
角闪变粒岩	Pl+Qtz+Hbl±Bt
透辉变粒岩	Pl+Qtz+Di±Bt
角闪透辉变粒岩	Pl+Qtz+Di+Hbl±Bt
石榴辉石变粒岩	Pl+Qtz+Di+Grt
石榴黑云透辉变粒岩	Pl+Qtz+Di+Bt±Grt
石榴（透辉）角闪变粒岩	Pl+Qtz+Hbl+Di+Grt
磁铁石榴辉石变粒岩	Pl+Qtz+Bt+Pl+Q
斜长浅粒岩	Pl+Qtz
二长浅粒岩	Pl+Kfs+Qtz
含透辉（磁铁）浅粒岩	Pl+Qtz+Di±Mgt
石榴浅粒岩	Pl+Qtz+Grt

注：Mgt—磁铁矿；Kfs—钾长石；Di—透辉石；Pi—斜长石；Qtz—石英；Hbl—角闪石；Grt—石榴子石；Hy—紫苏辉石；Bt—黑云母。

2. 岩石特征

（1）变粒岩岩石特征

变粒岩沿走向呈薄层–巨厚层状稳定产出，厚度自数十厘米至数十米不等，具明显层状特征。

岩石呈灰黄色、灰绿色细–中粗粒变晶结构，块状、变余层理、弱定向构造，矿物粒度 0.1～1.2mm，一般 0.3～0.8mm。斜长石（50%～60%）和石英（5%～20%）为主要矿物，部分含微斜条纹长石、次量矿物为片柱状黑云母、角闪石、透辉石及石榴子石等，暗色矿物含量<30%。

斜长石主要为奥长石–中长石（An20～30），粒状、板柱状构造，表现为变形拉长定向，优势粒度不均匀或具反条纹结构，含量一般>25%。不含石英时，浅色矿物由斜长石组成，为斜长变粒岩。

黑云母呈棕褐色，片状、板状，含量 10%～20%，主要由含铁较高的黑云母变质形成，其中细鳞片状黑云母由黑云母化交代暗色矿物、石榴子石、磁铁矿变质生成。

角闪石为半自形粒状、柱状，属钙质角闪石，颜色因变质程度差异而略有不同。

透辉石呈淡绿色，被次闪石、绿泥石及碳酸盐等交代，呈多色性。

个别岩石中含微斜条纹长石，表现为次圆状碎屑特征。副矿物为锆石、磷灰石，呈浑圆粒状，磁铁矿呈星点状。

（2）浅粒岩岩石特征

一般呈层状、薄层状分布于变质岩中，层位较稳定。常见有斜长浅粒岩（斜长石 50%～60%）和二长浅粒岩（钾长石 15%～20%）。

岩石呈浅灰色、灰白色、浅红色，粒度一般 0.1～0.5mm，细粒状变晶结构、镶嵌粒状变晶结构，块状构造、弱定向构造。含石英较多的浅粒岩（20%～30%）常显条纹状构造等。石英变形拉长呈条纹状。部分浅粒岩含少量透辉石、紫苏辉石、石榴子石及磁铁矿等，当含量到一定程度时变为变粒岩或片麻岩等。局部地段可见石榴子石变粒岩与浅粒岩韵律层理。

（四）斜长角闪岩和斜长辉石岩

1. 分布

该类岩石在密云沙–大漕地区和四合堂地区分布广泛。以斜长辉石岩、石榴辉石斜长角闪岩及斜长角闪岩等角闪质岩石为主（表 22–3）。

一般呈薄层—厚层状，呈夹层或透镜体产出，宽数米至十余米，亦有数厘米至数十厘米厚脉状似脉状侵入体形态产于角闪质斜长片麻岩中。岩性稳定，与围岩成分上呈现过渡关系，界线清楚。如斜方辉石和透辉斜长角闪岩在共存，反映原岩类型比较复杂。

表 22–3 斜长角闪岩和斜长辉石岩主要岩石名称与矿物共生组合

Table 22–3 The main rock names and paragenetic mineral associations of amphibolite and plagioclase pyroxenite

岩石名称	矿物共生组合
斜长角闪岩	Hbl+Pl±Qtz
含石榴辉石斜长角闪岩	Hbl+Pl+Grt±Qtz
透辉斜长角闪岩	Hbl+Pl+Di+Bt±Qtz
石榴斜长角闪岩	Hbl+Pl+Grt±Qtz
黑云斜长角闪岩	Hbl+Pl+Bt±Qtz
斜长辉石岩	Di+Pl+Qtz
石榴斜长辉石岩	Di+Pl+Grt+Hbl
石榴角闪斜长辉石岩	Di+Hbl+Pl+Grt

注：Di—透辉石；Pi—斜长石；Qtz—石英；Hbl—角闪石；Grt—石榴子石；Hy—紫苏辉石；Bt—黑云母。

2. 岩石特征

岩石中暗色矿物含量30%~80%，多数含量超过50%，属较暗色岩石。呈灰绿色、灰黑色，细—中粗粒状变晶、变余状变晶结构，块状弱定向—定向构造。

矿物成分以斜长石（一般30%~40%）、角闪石和透辉石为主，后两者含量变化不定，次为石榴子石、黑云母及紫苏辉石等，石英有或无。角闪石属普通角闪石，呈蓝绿色、绿色（角闪岩相），或棕绿色（麻粒岩相）与辉石平衡共生或交代，晚期角闪石晶形常较粗大，并含斜长石、石英等包体，退变生成次闪石、绿泥石、绿帘石、黑云母及碳酸盐等，次闪石在角闪石边缘形成次变边。

斜长石半自形板状或粒状变晶结构，粒径0.5~1mm，An约为40，斜长石与暗色矿物之间无关系，有时具反条纹结构。

石榴子石呈淡粉色或无色，粒状变晶，自形结构，含量最高达20%。属含钙的镁铝-铁铝榴石。

单斜辉石淡绿色或无色，属透辉石-次透辉石，粒度0.5~1mm。呈变斑时，显筛状变晶结构，斜方辉石多为紫苏辉石，分布有限，部分被角闪石、黑云母等交代。四合堂岩群仅在阳坡地地区斜长角闪岩中出现单斜辉石。

斜长辉石岩中可见棕绿色角闪石和石榴子石次要矿物。密云大漕地区斜长辉石岩中磁铁矿含量达3%~5%以上，并与磁铁石英岩呈渐变过渡，反映原岩铁质含量高。而四合堂地区多为细粒斜长角闪岩，属变质玄武岩、变质辉绿岩，顺层产出，呈现正变质特征。怀柔西湾子地区斜长角闪岩中黑云母含量较高，具片麻理构造及变余微层理。

（五）麻粒岩类

1. 分布

麻粒岩是在麻粒岩相温压条件下形成的区域变质岩，以出现紫苏辉石为标准，含不同数量黑云母、角闪石等与紫苏辉石共生。

主要分布于密云沙厂—大漕一带变质地层中，其中沙厂地区数量相对较多且厚，大漕地区变质岩中辉石质岩石总量仅占20%~30%，其中紫苏麻粒岩数量很少。

表22-4 麻粒岩类岩石名称与矿物共生组合

Table 22-4 Rock names and paragenetic mineral associations of granulites

岩石名称	矿物共生组合
二辉麻粒岩	Hy+Di+Pl±Qtz+Alm
黑云角闪二辉麻粒岩	Hy+Di+Hbl+Pl+Bt
角闪石榴二辉麻粒岩	Hy+Di+Pl+Alm+Hbl
石榴黑云紫苏麻粒岩	Hy+Pl+Bt+Alm+Mgt
黑云紫苏麻粒岩	Hy+Pl+Bt+Mgt
石榴紫苏麻粒岩	Hy+Pl+Alm

注：Alm—铁铝榴石；Di—透辉石；Pi—斜长石；Qtz—石英；Hbl—角闪石；Grt—石榴子石；Hy—紫苏辉石；Bt—黑云母。

2. 岩石特征

根据暗色矿物（主要为辉石）与浅色矿物（包括斜长石和石英）的量比划分为暗色麻粒岩（暗色矿物含量85%~50%），麻粒岩（暗色矿物含量50%~25%）和浅色麻粒岩（暗色矿物含量<25%）。本区麻粒岩中暗色矿物含量多在25%~30%之间，浅色麻粒岩较少。

岩石呈似层状、透镜状、脉状产出，延伸不稳定，沿走向或倾向变化大，多过渡为变粒岩等其他类岩石，层状者产出时常与上下岩层渐变过渡或互层，脉状者呈侵入形态产出穿切围岩，界线清楚。

岩石呈灰黑色、绿黑色，粒状变晶、镶嵌粒状变晶结构，块状、片麻状、条带状构造，矿物粒度一般0.3~1.0mm。矿物组成主要为Hy+Pl±Di±Hbl±Bt±Grt。

紫苏辉石是麻粒岩代表性矿物，多色性明显，含量10%~15%。粒度0.4~0.6mm，N_p=粉红色、

N_m=浅粉色、N_g=浅绿—近无色，平行消光，部分斜消光、最大 $N_g \wedge C=20°$，（-）$2V=62°\sim80°$。前期常被角闪石、黑云母或石榴子石等交代，沿晶粒边部解理和裂纹常析出微细粒状磁铁矿，晚期被细鳞片状黑云母交代。

透辉石在麻粒岩中广泛分布，呈淡绿色柱状、粒状变晶，与紫苏辉石平衡共生，属次透辉石，常被次闪石、黑云母等交代，$N_g \wedge C=42°\sim44°$，（+）$2V=80°$±。

普通角闪石呈绿色柱状、棕绿色，有时与辉石平衡共生，属高温角闪石。部分角闪石交代辉石，属退变质产物。

黑云母有两种，早期原生者 N_g=红褐色；N_p=淡黄色，晶形完整，与紫苏辉石、角闪石共生。晚期黑云母交代早期矿物，多呈不规则鳞片状集合体。

石榴子石以铁铝榴石成分为主，他形或自形，粉红色，少数近无色，粒径一般0.2~0.4mm，常含石英、斜长石及透辉石等不规则包体，在麻粒岩中较均匀分布或集中成团粒、环状或条带。石榴子石含量一般较少。

斜长石呈板状或粒状晶形，双晶反条纹结构常见，反条纹可能系交代而成。由于混合岩化作用影响，岩石中 An 值变化较大，一般为中长石（An31~43）。含石榴子石的麻粒岩中，斜长石常表现为不等粒状，在石榴子石分布处斜长石粒度普遍细小。

石英和微斜长石含量较少，部分和混合岩化作用关系密切或为后期变质反应产物。麻粒岩中的副矿物以磷灰石、锆石为主，锆石常为浑圆粒状，粒度约0.05mm，以浅—深玫瑰色及橘红色最常见，部分无色。磷灰石的粒度及含量均大于锆石。

（六）磁铁石英岩

1. 分布

磁铁石英岩是一种变质岩石，同时也是区内重要的铁矿石类型。集中分布在密云沙厂、大漕、小漕、山神庙以及怀柔马圈子等地区表壳岩系中。呈透镜状产出，产状和围岩一致，厚度变化不一，一般在向斜转折端磁铁石英岩明显增厚，铁含量增高。

2. 岩石特征

岩（矿）石呈粒状变晶、齿状变晶及镶嵌粒状变晶结构，条纹状、条带状、片麻状及块状构造。浅色条带为石英，暗色条带由磁铁矿和含铁硅酸盐类矿物组成，片麻状构造主要为变质重结晶矿物（表22-5）。

表22-5 磁铁石英岩类岩石名称与矿物共生组合

Table 22-5 Rock names and paragenetic mineral associations of magnetite quartzite

岩石名称	矿物共生组合	
辉石磁铁石英岩	Mgt+Qtz+Cpx+Ocx+Hbl+Grt	
角闪磁铁石英岩	Mgt+Qtz+Hbl±Py+Bt+Grt	
磁铁石英岩	Mgt+Qtz	
含磁铁石英岩	Mgt+Qtz+Pl	
磁铁片麻岩	Mgt+Qtz+Pl±Py+Hbl±Bt	

注：Mgt—磁铁矿；Cpx—单斜辉石；Opx—斜方辉石；Py—辉石；Qtz—石英；Hbl—角闪石；Grt—石榴子石；Hy—紫苏辉石；Bt—黑云母。

金属矿物以磁铁矿（25%~35%）为主，脉石矿物以石英为主，含铁硅酸盐矿物有辉石、角闪石、黑云母和石榴子石等。有时含少量斜长石晶屑，其他矿物有磷灰石、锆石等，有时见极少量赤铁矿、磁赤铁矿、钛铁矿及硫化物。叠加蚀变矿物有绿泥石、绿帘石、碳酸盐及绢云母等。硅酸盐类矿物因变质程度不同而异。密云地区沙厂磁铁石英岩中含辉石普遍，四合堂地区以含普通角闪石为主，有时含镁铁闪石（见于富含石榴子石的磁铁石英岩中），局部含钠铁闪石及含霓辉石组成的透辉石。

磁铁矿呈半自形—他形晶粒状或不规则状集合体，呈条纹状、条带状或分布于其他矿物粒间，粒径一般0.1~0.6mm。部分呈完好八面体自形晶包于石英晶粒中，粒度细微，粒径0.05~0.1mm。因变质及混合岩化作用影响，使磁铁矿重结晶粒度变粗，大者可达1~2mm以上。据统计，大漕地区磁铁石英岩中磁铁矿粒度>0.1mm者占93%。

沙厂地区磁铁矿的化学成分为 $Fe_2O_3$66.71%~69.97%；FeO 26.63%~30.06%；MgO 0.15%~0.38%；CaO 0.1%~0.36%；TiO_2 0.07%~0.69%；微量元素甚微，其中Ni含量高于Co。

石英呈粒状变晶、镶嵌粒状变晶，粒度0.1~0.5mm，组成浅色条带，常具波形消光、拉长及变形纹等。

副矿物以锆石、氟磷灰石为主。锆石多呈碎块状、角砾状、裂纹发育，少量呈浑圆状、椭圆状。其中多数锆石呈玫瑰紫色，分布普遍且相对含量较高，为原生碎屑物质，部分锆石呈玫瑰紫色-粉红色者常具麻坑及擦痕，不含包体，次生加大普遍，可能为经过短距离搬运的火山产物。锆石的粒度一般为0.09~0.13mm。

氟磷灰石中含氟量为1.82%~2.64%，平均2.38%；MnO、CO_2 亦较高，平均含量分别为0.04%和0.72%。

磁铁石英岩来源于海底火山喷发的产物，当火山活动减弱或间歇时期，硅铁质及火山碎屑物混入发生沉积，经变质重结晶形成，火山碎屑物含量高时则形成含磁铁岩石。

二、原岩恢复和大地构造环境分析

（一）密云岩群（$Ar_3My.$）

1. 主要岩石类型及变质建造

密云岩群主要有辉石斜长片麻岩+二辉麻粒岩、榴辉黑云变粒岩+石榴斜长辉石岩建造2类。

（1）辉石斜长片麻岩+二辉麻粒岩

该建造主要分布在沙厂一带的密云岩群表壳岩中。变质岩石以辉石斜长片麻岩和二辉麻粒岩为主，夹斜长辉石岩、辉石变粒岩和条带状、片麻状辉石磁铁石英岩。辉石斜长片麻岩含一定量的角闪石、黑云母或紫苏辉石。麻粒岩以二辉麻粒岩为主，黑云紫苏麻粒岩次之。该建造原岩为中基性火山沉积岩+含铁硅质岩建造。

建造的变质程度达到了麻粒岩相，普遍发育中等程度混合岩化。早期顺层钠质混合岩化形成了条痕-条带状顺层混合岩；晚期钾质混合岩化，但混合岩化作用较弱，肉红色混合岩化脉体可平行或斜交片麻理围岩。在该建造长英质混合岩化脉体中见有角闪石、透辉石或石榴子石粗大变斑晶，大部分未发现紫苏辉石。

（2）榴辉黑云变粒岩+石榴斜长辉石岩建造

主要分布在密云大漕-小漕-高岭地区。岩石组合主要是厚层榴辉黑云变粒岩与薄层石榴斜长辉石岩（角闪）岩互层，夹辉石斜长片麻岩、石榴二辉麻粒岩、浅粒岩及数层辉石磁铁石英岩。石榴二辉麻粒岩内多见有后成合晶结构，其中紫苏辉石呈被交代置换的残粒状态。变余层状构造和韵律层发育，岩石中普遍含有较多黑云母和石榴子石，反映了原岩沉积特征，并且岩石中混入了泥砂质成分。形成于火山活动相对宁静期，近火山沉积环境。

变质程度中下部为麻粒岩相，上部为高级角闪岩相的过渡性特征。混合岩化程度较弱，以钠质条带状混合岩为主，极少见到晚期钾质混合岩化现象。

原岩建造和变质建造特征表明，该建造是本区新太古代晚期形成的变质表壳岩。

2. 岩石化学特征与原岩恢复

区内变质表壳岩岩石化学成分见（表22-6），综合前人成果，SiO_2 含量变化范围：变质超镁铁质岩在44%~48%之间，变质镁铁质岩（基性岩）在47%~51%之间，变粒岩类为53%~55%，磁铁石英岩类为47.28%~85%，最高可达90%以上。除磁铁石英岩较特殊外，其余均与岩浆岩中的超基性岩、

表 22-6　密云表壳岩化学分析结果表

Table 22-6　Chemical analysis results of Miyun supracrustal rock

序号	样品编号	SiO_2	TiO_2	Al_2O_3	Fe_2O_3	FeO	MnO	MgO	CaO	Na_2O	K_2O	P_2O_5	H_2O^+	烧失量	总和	地点
1	（石榴）角闪斜长二辉麻粒岩	50. 82	0. 59	14. 50	3. 87	7. 10	0. 22	7. 67	8. 87	3. 14	1. 14	0. 24	1. 39	1. 38	99. 54	瑶亭
2	（石榴）角闪斜长二辉麻粒岩	50. 29	0. 98	18. 43	2. 93	9. 14	0. 15	4. 78	7. 51	4. 06	1. 32	0. 15	0. 79		100. 53	不详
3	（石榴）角闪斜长二辉麻粒岩	52. 85	1. 49	14. 21	4. 43	8. 21	0. 20	4. 88	8. 53	3. 18	0. 60	0. 12	1. 16		99. 86	放马峪
4	（石榴、角闪）斜长次透辉岩	50. 94	0. 97	16. 68	4. 69	6. 97	0. 26	5. 29	8. 57	3. 94	1. 22	0. 10	1. 10	0. 97	100. 60	小开岭
5	（石榴、角闪）斜长次透辉岩	48. 02	0. 74	14. 88	2. 74	9. 67	0. 20	8. 73	10. 22	2. 73	0. 93	0. 06	0. 92		99. 84	霍各庄
6	（石榴、角闪）斜长次透辉岩	50. 87	1. 74	13. 01	3. 44	12. 57	0. 26	5. 38	9. 22	1. 90	0. 76	0. 23	0. 76		100. 14	霍各庄
7	（石榴、角闪）斜长次透辉岩	48. 84	1. 87	13. 11	8. 82	7. 87	0. 30	5. 11	9. 96	3. 02	0. 60	0. 18	0. 67		100. 35	高岭
8	（石榴、角闪）斜长次透辉岩	49. 44	1. 16	14. 30	6. 71	7. 45	0. 22	5. 60	12. 02	3. 09	0. 64	0. 14	0. 16		100. 93	大漕
9	（石榴、角闪）斜长次透辉岩	47. 91	1. 27	14. 72	5. 61	7. 38	0. 25	5. 96	12. 27	3. 08	0. 71	0. 11	0. 47		99. 74	小漕
10	石榴次透辉斜长角闪岩	48. 12	0. 99	14. 72	4. 74	9. 27	0. 25	6. 91	9. 77	2. 36	0. 80	0. 10	1. 50		99. 94	黑古沿
11	石榴次透辉斜长角闪岩	48. 37	1. 13	14. 58	4. 02	9. 83	0. 22	7. 85	10. 85	2. 53	0. 51	0. 11	0. 72		100. 72	高岭
12	石榴次透辉斜长角闪岩	48. 91	0. 79	14. 91	3. 41	9. 25	0. 23	7. 18	10. 32	3. 06	0. 90	0. 12	1. 06		100. 14	大漕
13	石榴次透辉斜长角闪岩	48. 82	0. 94	14. 26	3. 86	9. 46	0. 19	7. 15	10. 32	3. 45	0. 94	0. 10	0. 30		99. 75	大漕
14	角闪斜长辉石岩	48. 26	0. 74	16. 34	3. 28	7. 37	0. 18	6. 88	10. 63	2. 74	1. 34	0. 10	1. 74	0. 12	99. 30	密云
15	含石榴次透辉黑云斜长变粒岩	64. 70	0. 39	16. 49	0. 64	3. 38	0. 05	2. 20	4. 12	4. 52	1. 30	0. 17	1. 52	1. 57	99. 53	小开岭
16	含黑云石榴次透辉斜长变粒岩	59. 03	1. 08	11. 32	7. 41	7. 02	0. 17	3. 45	5. 22	3. 23	0. 75	0. 17	1. 06		99. 91	小漕
17	含石榴黑云二辉斜长变粒岩	61. 02	0. 66	14. 69	3. 06	5. 76	0. 13	3. 67	3. 81	4. 27	1. 78	0. 22	0. 53	0. 60	99. 67	小漕

注：氧化物含量单位为%。　　（据金文山，2000）

基性岩相当，少量与中性岩相当。总体看，变质表壳岩的岩石化学成分变化较大，其中 TiO_2-0.36%～1.90%，Al_2O_3 2.68%～16.75% ，Fe_2O_3+FeO 10.83%～43.89%，MgO 1.16%～18.10%，CaO 1.45%～11.12%，Na_2O 0.10%～3.30% ，K_2O 0.09%～1.35% 。岩性复杂性反映了原岩成分变化较大。

表壳岩的原岩恢复主要依据：在区内许多地点可以见到表壳岩的原生层状特征，特别是石榴斜长辉石岩类与磁铁石英岩的紧密伴生关系，说明表壳岩具有原始沉积特征；表壳岩具有相对较细的结构，且结构特征相对均匀，反映了原岩结构较细的火山喷出岩结构特征；岩石化学成分，是原岩岩石类型恢复的主要依据；各种原岩恢复图解。

根据上述几点，总体上看区内变质基性岩恢复为玄武岩类，并依据 AFM 图解将其原岩判定为拉斑玄武岩系列，根据其轻稀土相对富集的特点，其原岩与 K. Condie 划分的 TH_2 型拉斑玄武岩相当。

超镁铁质岩类化学成分、结构特点与变质镁铁质基性岩相似，稀土元素含量也大体相似。一般呈规模不大的透镜体、夹层或不规则团块状产出。按其产状特征分为两类，一类是超基性顺层小侵入体和岩脉，另一类是成层的火山喷发成因的超基性熔岩。根据其负铕异常的特点分析，可能为玄武质岩浆经过斜长石结晶分离的残余岩浆的产物，其原岩应为超镁铁质火山岩。

镁铁质岩类：主要为麻粒岩类、斜长辉石岩及斜长角闪岩。

长英质变质岩类：主要包括（角闪）辉石斜长片麻岩和变粒岩及黑云变粒岩等。

根据以上特征，将变质基性岩恢复为玄武岩类，并依据 AFM 图解将其原岩判定为拉斑玄武岩系列（图 22-1），根据其轻稀土相对富集的特点，其原岩与 K. Condie 划分的 TH_2 型拉斑玄武岩相当。

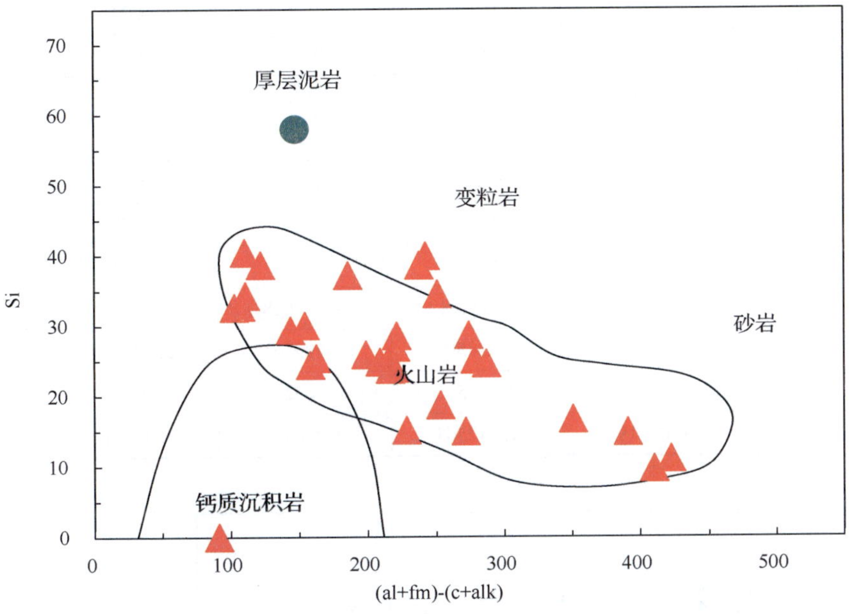

图 22-1　密云表壳岩（al+fm）－（c+alk）-Si 图解

Fig. 22-1　（al+fm）－（c+alk）-Si analysis diagram of Miyun supracrustal rock

变粒岩：具有岩石结构细、平缓的稀土配分曲线外，在尼格里四面体图解中位于火成岩区，在西蒙南的（al+fm）－（c+alk）-Si 图解中位于泥质岩附近（图 22-2）。根据其稀土总量较变质基性岩高，但其图谱与基性岩相似性的特点，变粒岩可能是由变质基性岩的原岩-拉斑玄武岩风化堆积的产物。

3. 地球化学特征

ΣREE 除样品 XT35（变粒岩）稀土总量为 93.12×10^{-6} 外，其他样品稀土总量均在 25.88×10^{-6}～53.61×10^{-6}（图 22-2）之间，具有超基性岩—基性岩类的特征，变粒岩 LREE/HREE 为 1.9，La_N/Yb_N 为 1.4，初始分异程度差，后期可能经历混合岩化作用改造。其余样品 LREE/HREE 在 3～6 之间，La_N/Yb_N 在 2.4～5.6 之间，轻重稀土分异程度一般。样品 XT16 出现明显的 δEu 异常，反应出壳幔混合的特征。其余样品均表现轻微铕正异常。图谱整体表现出向右微倾的分布形式，具有活动大陆边缘-岛弧火山的特征（表 22-7）。

表 22-7 密云表壳岩岩石稀土元素分析结果

Table 22-7 REE element analysis results of Miyun supracrustal rock

样品号 采样地点		XT24 半截峪	XT3 贾峪北西	XT35 土粉沟	XT36 土粉沟	XT2 放马峪北	XT16 陡岭子
稀土元素含量 $/10^{-6}$	La	7.2	8.99	9.2	4.6	5.87	5.75
	Ce	14.6	20.9	25.4	8.6	14.6	13.9
	Pr	1.53	2.42	3.4	1.1	1.71	1.56
	Nd	8.24	10.8	16	4.2	8.88	8.77
	Sm	2.13	2.27	5.8	1.34	2.31	2.25
	Eu	0.8	0.86	2	0.56	0.86	0.49
	Gd	2.18	2.11	8	1.6	2.91	3.45
	Tb	0.33	0.3	1.26	0.24	0.47	0.3
	Dy	2.3	2.06	8.4	1.42	3.01	1.79
	Ho	0.57	0.46	2	0.32	0.68	0.39
	Er	1.64	1.32	5.6	0.84	1.95	1.06
	Tm	0.23	0.17	0.82	0.2	0.24	0.14
	Yb	1.66	1.15	4.6	0.66	1.7	0.91
	Lu	0.1	0.1	0.64	0.2	0.23	0.1
	Y	14.5	10.7	42	8	16.4	9.17
	ΣREE	43.51	53.61	93.12	25.88	45.42	40.56
	LREE	34.5	46.24	61.8	20.4	34.23	32.72
	HREE	9.01	7.37	31.32	5.48	11.19	7.84
特征参数	LREE/HREE	3.829079	6.274084	1.97318	3.722628	3.058981	4.173469
	La_N/Yb_N	3.111179	5.607411	1.434599	4.999361	2.476793	4.532387
	δEu	1.124598	1.181281	0.897545	1.167207	1.013553	0.536497
	δCe	1.026404	1.077192	1.112568	0.906944	1.115612	1.116565

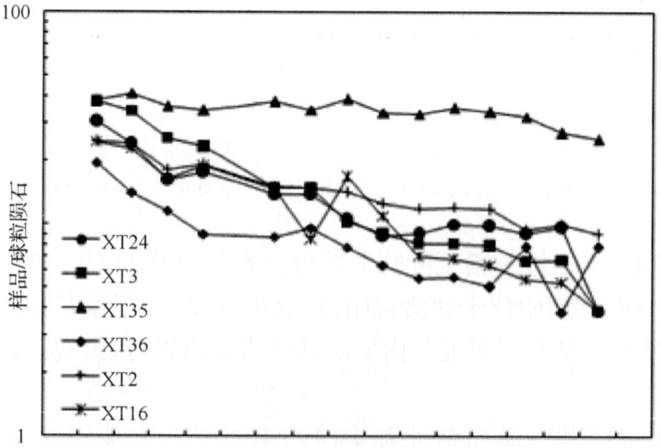

图 22-2 密云表壳岩稀土元素球粒陨石标准化分布型式图

Fig. 22-2 REE elements chondrite-normalized distributed diagram of Miyun supracrustal rock

4. 原岩建造及其大地构造环境分析

密云岩群原岩建造为超基性、基性-中酸性火山岩夹含泥质碎屑沉积岩和铁硅质岩的沉积-火山岩建造。其中磁铁石英岩与变质基性岩成因被认为早期岩浆侵入时形成大量硅铁质物质，经后期混入火山碎屑叠加改造重熔结晶而成。

由图 22-2 判断，Eu 出现一个富集和一个亏损现象。变质岩存在两期，早期以基性火山岩为主的火山活动频繁，晚期中酸性火山活动加强，在火山活动宁静时期，形成了含铁硅质岩、凝灰质半黏土岩和砂质黏土岩，但始终缺乏碳酸盐岩。整个岩群构成了一个完整的火山-沉积旋回。据金文山（1999a，1999b，2000）研究认为，沙厂-大漕地区变质表壳岩原岩为杂砂岩+英安质、安山质火山岩及其碎屑岩+基性火山岩及其碎屑岩+硅铁质沉积岩两大类，形成于活动大陆边缘海相盆地。

根据图 22-3，1 件样品落在钙质沉积岩区，3 件落在砂岩区，说明其原岩应为杂砂岩、岩屑砂岩，而岩石地球化学样品总体落在火山岩区，说明密云岩群（石榴）角闪斜长二辉麻粒岩原岩应是火山产物，其中石榴次透辉斜长角闪岩具有玄武岩类和泥质岩类变质特征。

综合分析认为，密云岩群表壳岩形成于活动大陆边缘海相盆地环境，具有裂谷型拉斑玄武岩特点。

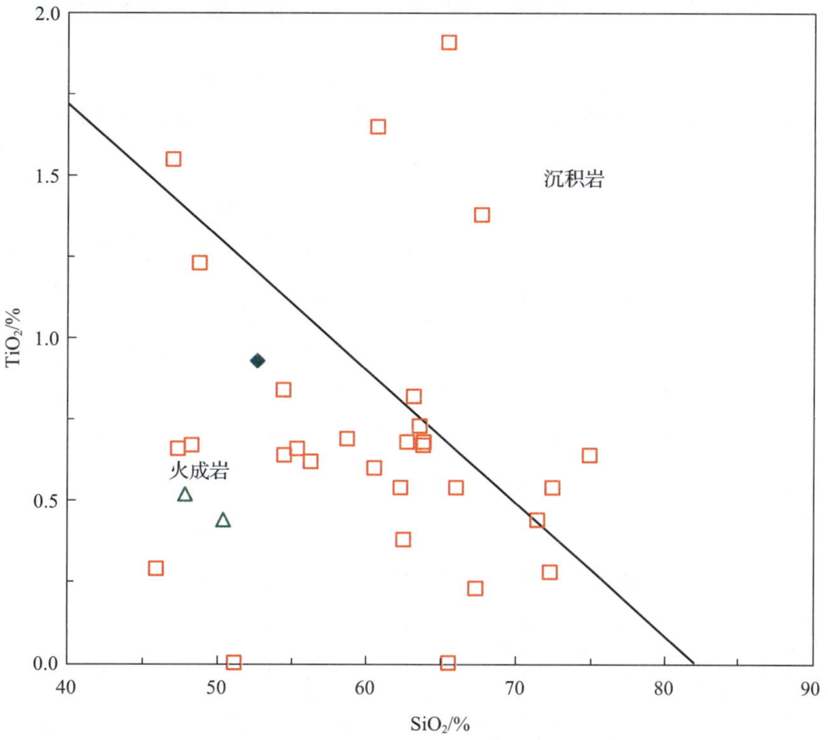

图 22-3 密云表壳 TiO_2-SiO_2 图解

（据姬广义，1990；塔尼，1976）

Fig. 22-3 TiO_2-SiO_2 diagram of Miyun supracrustal rock

（二）四合堂岩群（Ar_3*Sh.*）

1. 主要岩石类型及变质建造

（1）黑云角闪斜长片麻岩+角闪黑云变粒岩+黑云浅粒岩建造

集中于半城子断裂和石城断裂之间的阳坡地地区。该建造下部是黑云角闪片麻岩与斜长角闪岩互层，中部是黑云角闪斜长片麻岩夹角闪黑云变粒岩，上部是黑云变粒岩与黑云浅粒岩互层。

变质程度达角闪岩相。混合岩化现象普遍，强度较弱。混合岩化岩石以条带状、片麻状黑云角闪质混合岩为主。在长英质混合岩化脉体中，黑云母、角闪石或阳起石变斑晶较多，未见到辉石或石榴子石。

（2）黑云角闪斜长片麻岩+磁铁石英岩建造

该建造主要分布山神庙地区。在区内出露不全，变质岩石仅见到斜长角闪岩与黑云角闪斜长片麻岩互层，夹多层条带、片麻状角闪磁铁石英岩。原岩为基-中性火山碎屑岩夹含铁硅质岩。变质程度达角闪岩相。混合岩化作用微弱，形成条带状混合质的变质岩石。

2. 岩石化学特征与原岩恢复

区内四合堂岩群区域变质岩中角闪石岩、斜长角闪岩、斜长角闪片岩的 SiO_2 = 46.80% ~ 53.22%，Al_2O_3 = 6.28% ~ 16.33%（表 22-8）。尼格里参数中 *Si* = 98.3 ~ 204.5，*fm* = 31.6 ~ 62.48，*c* = 14.1 ~ 28.89，*al* = 7.25 ~ 31.1，*alk* = 2.5 ~ 23.2。在 TiO_2-SiO_2 图解中，部分样点因 TiO_2 较高落于沉积区，大部分样点落于火成岩区。在（al+fm）-（c+alk）-Si 图解中，样点主要投于火山岩区和接近火山岩的厚层泥岩区，少数样点落于火山岩与钙质沉积岩重叠区（图 22-4）。图解反映出本区斜长角闪（片）岩、角闪石岩等的变质原岩可能为基性火山岩和接近火山岩成分的火山~沉积岩。在（Fe_2O_3+FeO+TiO_2）-Al_2O_3-MgO 图解中，样点主要落于科马提岩（苦橄拉斑玄武岩）区内（图 22-5）。按布鲁克和哈特所提供的科马提岩化学成分范围（SiO_2 = 46% ~ 53%、CaO/Al_2O_3 > 1、MgO > 9%、TiO_2 < 0.9%、K_2O < 0.9%）来衡量，有 1 件样品（1 号样品辉石角闪岩）的化学成分完全与之相当。

表 22-8　四合堂表壳岩岩石化学分析结果

Table 22-8　Petrochemistry analysis results of Sihetang supracrustal rock

序号	岩石名称	主要氧化物含量/%							
		SiO_2	TiO_2	Al_2O_3	Fe_2O_3	FeO	MnO	MgO	CaO
6	斜长角闪岩	51.93	0.67	15.04	4.18	6.28	0.14	6.8	7.23
8	斜长角闪岩	46.8	1.91	16.33	3.66	10.24	0.1	5.86	9.04
11	斜长角闪岩	51.2	0.69	15.84	3.46	6.91	0.17	6.47	9
13	斜长角闪岩	53.06	0.67	14.87	3.65	6.26	0.15	6.64	7.73
14	斜长角闪岩	53.22	0.82	16.07	3.69	5.79	0.15	4.52	7.16
15	斜长角闪岩	47.95	1.65	12.99	6.03	10.8	0.44	6.34	9.37
16	黑云角闪变粒岩	59.7	0.54	16.17	3.01	3.89	0.08	3.56	6.5
17	黑云变粒岩	55.54	0.68	18.02	1.95	4.36	0.09	3.42	4.53
18	黑云变粒岩	66.54	0.38	14.93	1.71	2.08	0.06	1.96	4.9
19	黑云角闪变粒岩	58.01	0.6	16.43	3.67	3.99	0.1	3.69	6.05
20	黑云变粒岩	61.53	0.54	17.21	1.42	3.94	0.05	2.6	3.7
21	黑云变粒岩	56.48	0.62	16.69	2.76	2.64	0.06	2.38	3.38
22	黑云角闪变粒岩	55.97	0.66	14.64	3.35	5.83	0.09	3.37	4.06
23	黑云角闪变粒岩	61.21	0.84	15.2	2.48	5.26	0.12	2.54	3.14
24	黑云变粒岩	63.73	0.64	15.53	2.3	3.19	0.08	3.1	2.58
25	黑云变粒岩	56.28	0.66	16.82	1.95	4.41	0.17	3.28	4.21
26	黑云二长变粒岩	66.06	0.28	16.87	1.03	1.68	0.08	1.06	3.02
27	黑云变粒岩	62.79	0.64	16.23	1.77	4.29	0.12	3.15	2.01
28	黑云角闪斜长片麻岩	59.7	0.54	16.17	3.01	3.89	0.08	3.56	6.5
29	黑云浅粒岩	70.18	0.29	14.98	0.97	1.61	0.03	0.85	2.6
30	黑云二长浅粒岩	72.6	0.23	12.99	1.62	1.05	0.05	0.89	1.13
31	黑云二长浅粒岩	67.38	0.44	14.9	2.4	1.49	0.05	1.96	4.05

续表

序号	岩石名称	主要氧化物含量/%							
		Na_2O	K_2O	P_2O_5	H_2O^+	H_2O^-	CO_2	烧失量	
6	斜长角闪岩	2.88	3.41	0.36	0.65	0.02			
8	斜长角闪岩	3.04	0.63	0.37	0.98	0.09	0.45		
11	斜长角闪岩	3.94	0.93	0.39	0.28	0.09	0.55		
12	斜长角闪岩	3.14	0.74	0.18	0.75	0.03	0.26		
13	斜长角闪岩	3.92	1.63	0.38	0.51	0.07			
14	斜长角闪岩	3.6	3.14	0.69	0.42	0.11	0.47		
15	斜长角闪岩	2.18	0.68	0.24	0.6	0.21			
16	黑云角闪变粒岩	4.49	6.69	0.25	0.57	0.14			
17	黑云变粒岩	5.27	4.44	0.12	0.94	0.07			
18	黑云变粒岩	1.03	5.06	0.19	0.38	0.11	0.55		
19	黑云角闪变粒岩	4.45	1.39	0.35	1.29	0.12			
20	黑云变粒岩	5.19	2.83	0.77	0.12	0.43			
21	黑云变粒岩	4.67	6.63	0.17	1.19	0.07		3.86	
22	黑云角闪变粒岩	6.22	4	0.26	1.03	0.06		1.17	
23	黑云角闪变粒岩	3.85	4.11	0.16	1.21	0.11			
24	黑云变粒岩	2.71	5.04	0.14	0.74	0.15			
25	黑云变粒岩	4.51	4.69	0.34	0.42	0.1			
26	黑云二长变粒岩	5.55	2.72	0.11	0.39	0.11			
27	黑云变粒岩	5.4	2.5	0.14	0.75	0.15			
28	黑云角闪斜长片麻岩	4.49	6.69	0.25	0.57	0.14			
29	黑云浅粒岩	4.82	2.43	0.13	0.28	0.11	0.5		
30	黑云二长浅粒岩	3.3	5.55	0.09	0.48	0.05			
31	黑云二长浅粒岩	4.38	0.17	1.05	0.06				

（据姬广义，1990）

区内变粒岩、片麻岩类的 SiO_2 为 55.54～66.54%、Al_2O_3 = 14.93%～18.02%、TiO_2 = 0.38%～0.84%。尼格里值中，Si = 110.1～334.9，fm = 15.5～53.7，c = 6.0～23，al = 17.5～42.9，alk = 5.8～34.1。据图 22-3 样点绝大多数落于火成岩区（仅 1 件样品落于沉积岩区），图 22-4 中样点除少数落于厚层泥岩与火山岩与钙质沉积岩重叠区外，大多数分布在火山岩区内，在图 22-5 中，样点部分落于拉斑玄武岩系列区内，部分落于钙碱性玄武岩系列区内，均反映区内变粒岩、片麻岩类岩石的原岩建造可能为中性火山岩和相应的火山—沉积岩，其化学成分相当于拉斑玄武岩系列向钙碱性玄武岩系列的过度类型。

区内浅粒岩类岩石化学成分中，SiO_2 = 67.38%～72.60%、Al_2O_3 = 12.99%～14.98%、TiO_2 = 0.23%～0.44%。尼格里值中，Si=297.1～381.4，fm=15.6～22.3，c=6.3～13.1，al=38.7～41.8，alk=29.5～35.5。在 TiO_2-SiO_2 图解（图 22-3）中，样点落于火成岩区。在（al+fm）－（c+alk）－Si 图解中，样点全部落于火山岩区。

在（Fe_2O_3+FeO+TiO_2）－Al_2O_3－MgO 图解（图 22-5）中，样点全部投于拉斑玄武岩系列区内。结合人工重砂样品鉴定成果分析，区内浅粒岩原岩可能为火山-沉积岩。其化学成分相当于钙碱性玄武岩系列的酸性火山岩。

3. 地球化学特征

收集分析了区内 5 件稀土样品，从上述表格中所列数据可知，本区变粒岩、浅粒岩、黑云片岩的稀土元素具有如下特征：

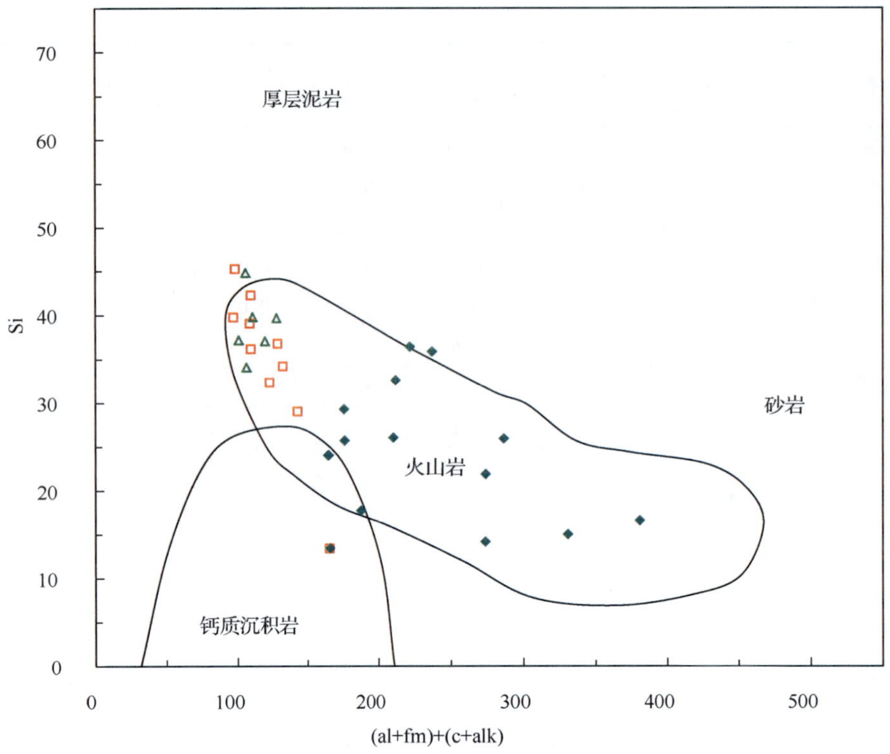

图 22-4 （al+fm）-（c+alk）-Si 图解

（据姬广义，1990）

Fig. 22-4 （al+fm）-（c+alk）-Si analysis diagram

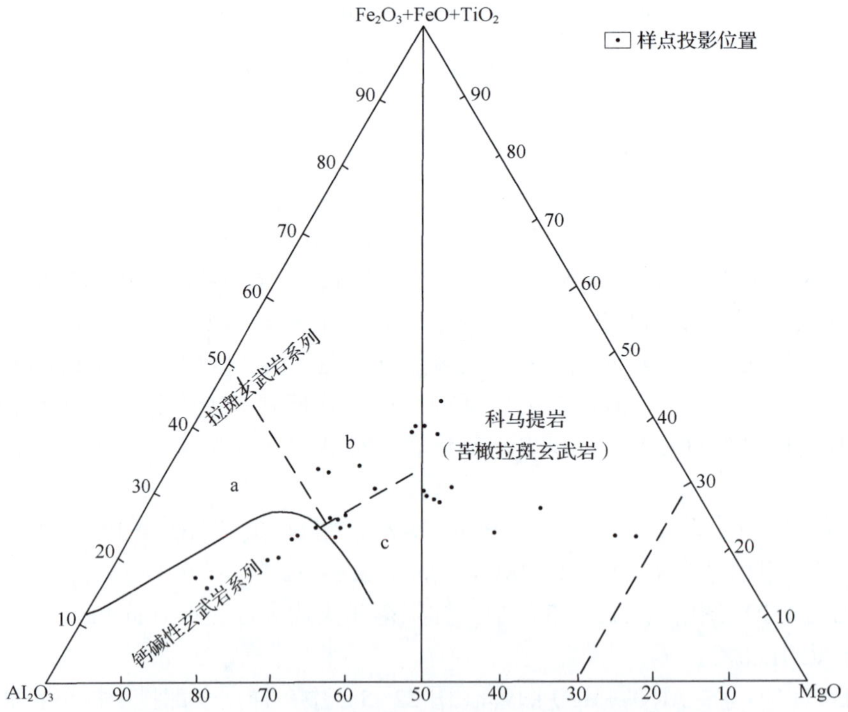

图 22-5 （Fe_2O_3+FeO+TiO_2）-Al_2O_3-MgO 阳离子三角图

（据姬广义，1990）

Fig. 22-5 The ternary plots of（Fe_2O_3+FeO+TiO_2）-Al_2O_3-MgO

1）LREE/HREE 为 4.02~9.17，均为轻稀土富集型，重稀土亏损较明显。

2）Eu/Sm 为 0.236~0.338，其比值与沉积岩、玄武岩和粗玄岩（沉积岩 0.20；玄武岩 0.32；粗玄岩 0.36）的稀土元素比值大体相当。

3）δEu 值为 0.87~1.33，样式具弱的正铕或负铕异常，与平坦型较为接近。

区内变质岩、浅粒岩、黑云片岩和稀土分配型式（图 22-6，图 22-7），曲线均向右中等倾斜，1~5 号样曲线均呈镥（Lu）负异常；3~5 号样曲线呈铽（Tb）负异常。

稀土元素的以上特征显示出四合堂岩群变粒岩、浅粒岩、构造片岩的原岩可能是大陆拉斑玄武岩系列的火山岩和沉积岩以及它们之间的过渡类型（表 22-9）。

表 22-9 四合堂表壳岩岩石稀土元素分析结果

Table 22-9 REE elements analysis results of Sihetang supracrustal rock

编号	岩石名称	含量/10^{-6}								
		La	Ce	Pr	Nd	Sm	Eu	Gd	Tb	Dy
1	变粒岩	11.19	23.43	3.43	12.25	2.71	0.64	2.06	0.35	1.58
2	浅粒岩	14.77	27.16	3.86	10.99	2.12	0.63	1.39	0.13	0.78
3	黑云变粒岩	16.69	31.58	4.16	12.69	2.54	0.78	1.83	0.05	1.56
4	黑云片岩	26.94	55.96	7.94	25.48	5.45	1.31	4.03	0.42	3.03
5	黑云变粒岩	25.54	44.4	5.46	16.46	2.99	1.01	1.79	0.13	0.19

编号	岩石名称	含量/10^{-6}					
		Ho	Er	Tm	Yb	Lu	Y
1	变粒岩	0.32	0.77	0.11	0.64	0.04	7.49
2	浅粒岩	0.17	0.36	0.13	0.31	0	3.72
3	黑云变粒岩	0.31	0.98	0.17	1.01	0.09	8.37
4	黑云片岩	0.61	1.62	0.27	1.62	0.2	14.81
5	黑云变粒岩	0.25	0.73	0.11	0.62	0.05	5.58

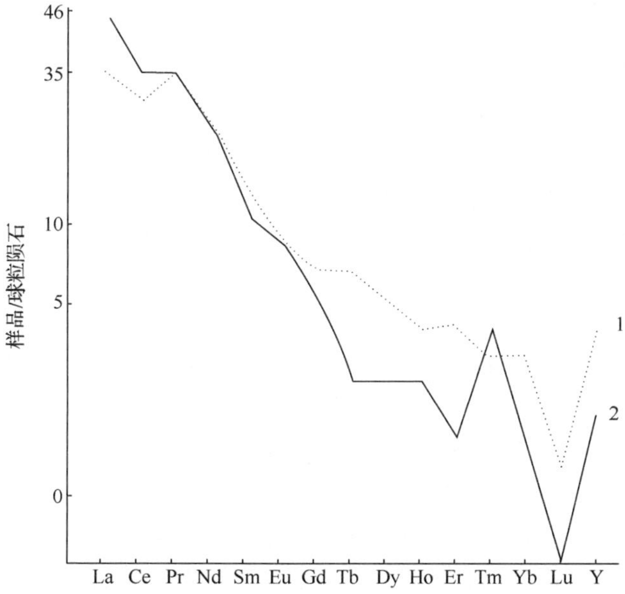

图 22-6 四合堂岩群变粒岩、浅粒岩稀土配分曲线

Fig. 22-6 REE elements distributed diagram of leuco-granulitite granulite and leptite of Sihetang rock group

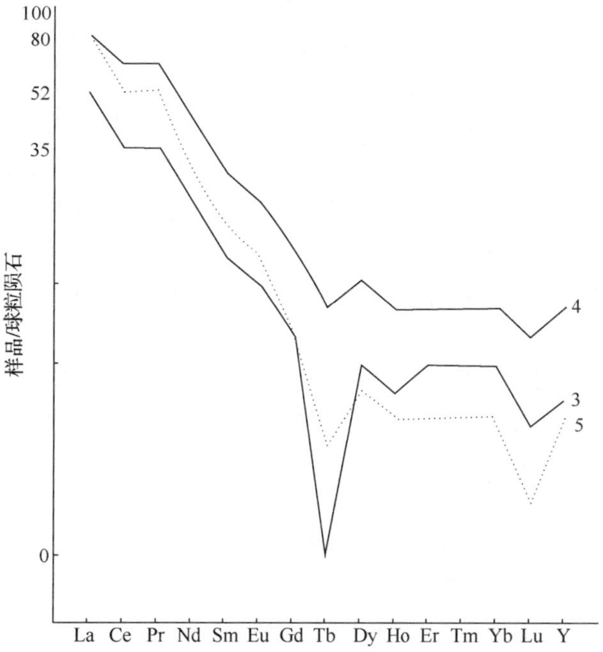

图 22-7 四合堂岩群黑云变粒岩、黑云片岩稀土配分曲线

Fig. 22-7 REE elements distributed diagram of biotite granulite and biotite schist of Sihetang rock group

4. 原岩建造及其大地构造环境分析

综上所述，四合堂岩群建造主要原岩为基性-中基性-中酸性熔岩及其火山碎屑岩，超基性岩甚少。时间上，早期区内海底火山作用强烈，形成了钙碱系列的玄武岩、安山岩等，火山作用间歇期间凝灰质砂岩、粉砂岩、硅质岩沉积，后期基性-中性-中酸性凝灰质砂岩交替沉积。四合堂岩群基本上构成了一个完整的以中性-中酸性火山碎屑岩为主的火山-沉积旋回。火山岩系列主要为钙碱玄武岩系列，Na_2O/K_2O 较高，普遍比密云岩群 Fe、Mg 含量高，反映其形成环境可能为岛弧或岩浆弧环境，但更近于大陆边缘。分析认为四合堂岩群原岩应形成于挤压造山环境下的岛弧或岩浆弧及相关盆地环境。

第二节　晚古生代—早中生代变质岩系

一、变质岩特征

区内晚古生代—早中生代变质岩系主要分布于北京西部的房山和门头沟等地。按原岩性质，分为泥质岩类、长英质碎屑岩类、碳酸盐岩和基性火山岩类。

（一）泥质岩类变质岩

该类岩石在区内分布最广，按结构构造主要为角页岩类、片岩类、千枚岩类。

1. 角页岩类

（1）分布

常靠近岩体产出，一般距离不超过 350m。按组成矿物，有石榴红柱角页岩、石榴硅线红柱角页岩、碳质红柱角页岩、黑云红柱角页岩、含石榴子石十字石铝直闪红柱角页岩、二云红柱角页岩、黑云石榴红柱角页岩、白云母红柱角页岩、十字红柱硬绿泥角页岩和红柱硬绿泥角页岩等。

（2）岩石特征

岩石呈暗灰色、灰黑色、浅黄色、黄褐色、暗棕黑色，一般较致密，块状，坚硬或异常坚硬。在细粒的基质中通常见到灰色、暗灰色的柱状红柱石或空晶石，铁褐色的铁铝榴石，黄绿色的硬绿泥石及浅黄色或黄褐色的放射状铝直闪石。它们一般呈板状变晶，无定向排列。

显微镜下岩石具花岗变晶结构、角页岩结构、细鳞片花岗变晶基质的斑状变晶结构、鳞片花岗变晶结构和放射花岗变晶结构。

红柱石是岩石的主要组成矿物，一般在28%~35%之间，少数达70%~80%，表现为四种形态：晶体仅边部干净，内部充满碳质包裹体；晶体中心出现碳质小方块，并通过碳质连接四角；晶体内的碳质包裹物组成十字状对角线；晶体中心出现碳质空心方块，并与碳质线与四个顶角连接。

有的红柱石十分干净，中心往往有一粉红色或玫瑰色的方块，外围则为灰色。两种色调的红柱石具有某些不同的光学和化学性质，红色部分的折光率、相对密度和 TiO_2、Fe_2O_3 的含量都略大于无色部分，但 $2V$ 相反。

在不同的角页岩中，尚有矽线石、铁铝榴石、十字石、铝直闪石、硬绿泥石、云母和石英等矿物。

矽线石，无色，呈细柱状、纤状，往往取代黑云母和红柱石，有时出现分带，由内向外依次为红柱石-白云母-矽线石，白云母边宽约 0.55mm，矽线石边宽 0.03mm，含量 7%~15%。

铁铝榴石，无色，往往呈较好的板状变晶，粒径一般 0.3~1.1mm，一些颗粒内部出现放射状、十字状及线状的铁质和石英包粒，含量 2%~7%。

铝直闪石，呈放射状、蒿束状和柱状。新鲜时呈浅蓝绿色，多色选比较明显，N_g'浅蓝绿色，$N_m'=N_p'$浅黄色。纵切面平行消光，正延性，二轴正光性，属铝斜方闪石。容易氧化，变为黄褐色，含量 10%~21%。

十字石、细柱状，黄色，结晶好而干净时具多色性，N_g'橙黄色，N_p'浅黄色或无色。往往与铝直闪石共存，但后者结晶稍早，故出现十字石依附于铝直闪石发育的现象，由于析出多余的石英，十字

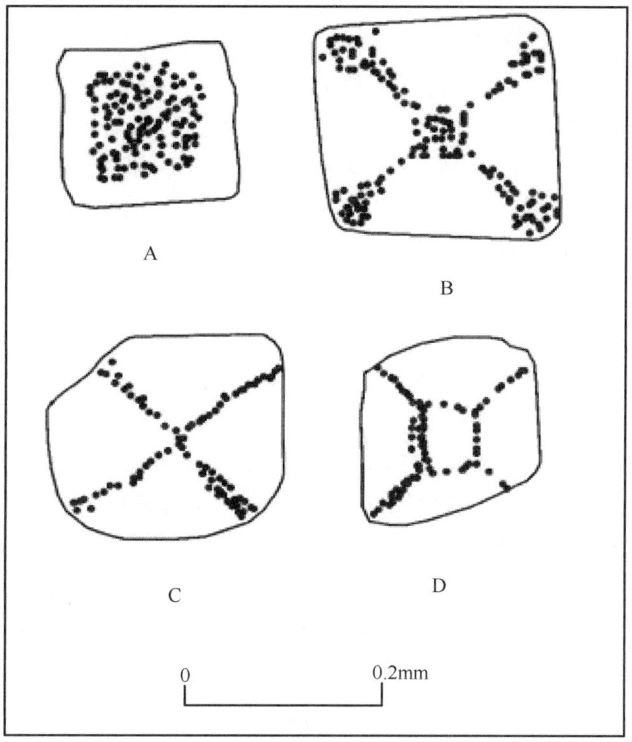

图 22-8　红柱石内碳质包裹体

（据刘国惠，1987）

Fig. 22-8　Carbonaceous inclusion in the andalusite

石内部呈现多种形态的残留石英蠕虫体，含量20%左右。

云母类有黑云母和白云母，均为较大鳞片。黑云母具较浓的色调、多色性明显，N_g'棕褐、红褐色、$N_m'=N_p'$黄色、浅黄色，一般定向不明显，含量变化大，通常2%~10%，有时达20%；白云母有两种，一种为单独的鳞片，另一种由交代红柱石或黑云母而成。以前者为主，含量5%~15%。

硬绿泥石，在角页岩中为不稳定的矿物，呈鳞片状、放射状。浅黄绿色，常具聚片双晶。容易褐铁矿化，一般存在于基质之中。含量3%~17%。

矿物组成主要是细粒石英和碳质，充填于结晶矿物之间，石英含量5%~25%，碳质含量5%~15%。

2. 片岩类

（1）分布

普遍产于西山地区岩石中，数量上多于角页岩，多产于离岩体300m范围内。

（2）岩石特征

该类型片理发育，呈暗灰色、深灰色、黄褐色、浅黄灰色和灰紫色。特征矿物是红柱石、十字石、矽线石和石榴子石。野外露头上红柱石常呈长柱状变斑晶，一般长达3~8mm，在富含碳质的岩石中，最长可达15mm，断面直径一般0.8~1.5mm，长轴往往定向排列，新鲜的颗粒，中心呈浅红色，受矽线石化时呈灰白色。十字石在岩石中呈暗红或黑红色小柱状体，少数具十字双晶，长度一般1mm左右。矽线石主要发育于紧靠岩体的泥质岩石中，有的呈白色菊花状，称为菊花石。石榴子石呈暗褐色，粒径1mm左右，在一些地方正常层序的露头中见到石榴子石呈粒级递变的现象。石榴子石由下至上，粒径成倍增大，反映了在一定温压条件下，某些变质矿物的粒级变化主要与原岩中该矿物化学成分含量的多少和富集程度。

显微镜下，片岩类岩石具斑状变晶的鳞片变晶结构、鳞片花岗变晶结构和纤状变晶结构，平行构造和片状构造。

云母类以黑云母为主，白云母次之，靠近岩体时白云母增多。黑云母常呈较大的变斑状鳞片，多色性明显，N_g'暗红色、暗黄色，N_p'黄色、浅黄色，含量5%~18%。白云母常为细鳞片，一般无色，有时具淡绿色，含量2%~13%。

红柱石往往呈方形或长方形的自形晶，常具一组解理，平行消光。新鲜而干净的结晶体，微具多色性，N_p'粉红色，N_g'淡绿或无色。有的结晶具筛孔结构。有的晶体内含有黑云母小鳞片，有的黑云母定向明显与岩石片理垂直，表明存在二期构造。有的红柱石边部被绢云母交代，形成交代环。红柱石含量5%~17%。

矽线石常与石榴子石及黑云母共生，呈细柱状及放射状。有两种产出方式一种是由红柱石转变或被交代而成的矽线石，有时伴生白云母；另一种是矽线石附着黑云母发育。岩石中矽线石分布不广，含量多时达40%，少时1%~3%。

十字石呈浅黄色、橙黄色，呈短柱状或六边形，粒径长0.3~0.8mm，少数达1mm，宽0.2~0.6mm，解理不明显，长轴具平行片理特征，也有斜交或垂直片理的。晶体内部常含有较多暗色尘点，表面模糊，有的十字石内含有褪了色的巨沙钟构造的硬绿泥石，含量2%~5%。

石榴子石呈变斑晶，粉色，粒径0.2~1mm，有的颗粒内部含有较多的黑云母鳞片和石英包粒，呈放射状排列。石榴子石的组成是铁铝榴石分子82.1%，钙铁榴石分子8.5%，镁铝榴石分子7%，锰铝榴石分子2.4%，片岩中石榴子石含量1%~5%。

石英，细粒，等轴状、菱角状和拉长状，有时构成小条带或豆荚体，与片理一致。粒径一般0.1mm左右，少数达3mm，含量18%~35%。

一些片岩中，碳质甚高，可达30%。此时往往构成平行条带，与其他矿物组成的条带相间。少数片岩中，偶见到少量中长石。

3. 板岩–千枚岩类

（1）分布

本类岩石主要产于上古生界太原组、山西组及侏罗系窑坡组地层中。多产于离岩体200~300m以外的地段。

（2）岩石特征

岩石呈薄板状，具丝绢光泽。含红柱石多的时候，呈黑色、暗灰色和灰黄色；含硬绿泥石含量增多时，表现为浅绿色、黄绿色和紫灰色。

特征矿物是红柱石、硬绿泥石、黑云母和绢云母，因含量不同而组成不同的千枚岩和板岩。显微镜下岩石主要呈显微鳞片状、斑状变晶结构、放射球颗粒斑状变晶结构，片状构造和条带构造。

红柱石大多为充满碳质尘点的空晶石，或中心为碳质方块，四角有碳质线连接。结晶程度低时，呈雏晶状。有的红柱石含有较多量的石英碎屑，甚至包含细长鳞片的硬绿泥石，并保存沙钟构造。红柱石含量变化较大，10%~40%。

硬绿泥石，鳞片状或球颗状，具多色性，N_g'浅黄色或无色，N_p'浅蓝色或黄绿色，负延性，具聚片双晶或标准的沙钟状，象脚鼓状构造。含量8%~25%。

绿泥石，呈黄绿色鳞片，N_g'绿黄色，N_p'淡黄色，多色性明显较角页岩和片岩中的弱。含量3%~7%。

黑云母，呈较大鳞片，N_g'绿黄色，N_p'淡黄色，多色性明显较角页岩和片岩中弱。含量3%~7%。

石英是基质的主要组分，细粒，碎屑状，往往集结成透镜体、扁豆状或长条状，与片理一致，含量10%~30%。

此外尚有或多或少的碳质，分散状或条带状，含量5%~25%。

（二）长英质变质岩

该类岩石多保留沉积碎屑岩的特征，由长石和石英组成。变质作用发生时，胶结物首先生成变质矿物，如云母类，帘石类和角闪石等。变质强时，石英重结晶。除一部分为石英岩和长石石英岩外，基本上分为两类：

1. 变粒岩类

（1）分布

此类岩石主要产于三叠系双泉组，一般产于离岩体边界450m之外。常为薄层状，由于暗色矿物含量不同或粒度的差异，形成层状或纹层构造，有时见很好的交错层。在流体的影响下，暗色矿物及长英矿物呈现分异结集的现象。

（2）岩石特征

重要变质矿物主要为黑云母和普通角闪石，两者或共存，或分别组成不同的变粒岩，两者含量均为10%~30%，有时含较多帘石类，达5%~20%。长石和石英总量为50%~70%，一般长石多于石英。岩石具细粒花岗变晶结构，不等粒花岗变晶结构和显微鳞片不等粒变晶结构，部分为变余碎屑结构。

2. 变质砂岩类

（1）分布

此类岩石产于侏罗系南大岭组和窑坡组，岩石为中层至厚层状。碎屑结构明显，主要是长石和石英碎屑，往往粗细不均，分选甚差，细粒者比较坚硬，中至粗粒者比较疏松。

（2）岩石特征

重要矿物由胶结物变成黑云母、绿泥石、绿帘石、硬绿泥石和绢云母组成，总含量达70%~90%，碎屑物为长石和石英。岩石主要为变余砂状结构，原岩岩性为砂岩或凝灰质砂岩。

（三）碳酸盐岩类变质岩

（1）分布

本类岩石主要产于中—新元古界、奥陶-寒武系碳酸盐岩地层中，多离岩体较近。

（2）岩石特征

岩石为全晶质、中厚层状。由于氧化钙含量和杂质的不同，按特征变质矿物可分为橄榄大理岩、透辉大理岩和透闪大理岩。

岩石具粒状变晶结构，部分岩石具拉长定向构造。

常见的重要矿物，除方解石外，有镁橄榄石、透辉石和透闪石，其次是钙铁榴石、符山石、白云母、帘石类和长石。矿物多呈分散状，有时透辉石与符山石和帘石等构成条带状或团块状，矿物粒径0.1~0.8mm，各矿物含量3%~15%。

该类变质岩中部分岩石岩性比较纯，含硅和杂砂质少；另一些则相反，含有较高的硅、铝和铁（表22-10）。

（四）变质基性火山岩

（1）分布

主要分布于二叠系—侏罗系变火山岩系中，以侏罗系南大岭组变玄武岩为典型代表。

（2）岩石特征

呈薄层至中厚层状，岩石为绿色、灰绿色至暗绿色，常见较多暗色杏仁体，呈椭圆形或压扁形，粒径大小不等，大者5mm×8mm，小者1mm×3mm，杏仁体约占岩石体积的10%~30%。

岩石具变余辉绿结构或变余含长结构。斑晶矿物以完全暗化或蚀变的橄榄石、辉石为主，还有斜长石。斑晶含量不等，最高15%，斜长石斑晶的含量<8%，为自形板状晶体，长约2mm，弱绢云母化，探针结果计算的端员组成为Or 0.10、Ab 99.17、An 0.73，属钠长石。橄榄石已全部变为铁叶绿泥石及磁铁矿，而保留原矿物的外形及解理特点。

原先的斜长石和角闪石斑晶，大多发生不同程度的变质。斜长石变成绢云母和碳酸盐或黝帘石的假晶，普通角闪石变为绿泥石或阳起石。杏仁体由绿泥石与方解石组成。基质中有绿帘石、黝帘石和绿泥石，有时出现少量硬绿泥石。基质中斜长石微晶的原生外形大都得以保留，但已被钠长石、绿帘

表 22-10 晚古生代—早中生代变质岩化学成分

Table 22-10 Metamorphic rock chemical composition of Late Paleozoic to Early Mesozoic

序号	岩石	样号	含量/%																		尼格里值						
			SiO_2	TiO_2	Al_2O_3	Fe_2O_3	FeO	MnO	MgO	CaO	Na_2O	K_2O	H_2O^+	H_2O^-	C	CO_2	SO_3	Cr_2O_3	P_2O_5	总和	al	fm	c	alk	sl	mg	k
1	红柱硅线片岩	62149C	55.66	1.41	36.56	0.57	0.94	0.02	0.05	0.46	0.30	0.36	1.39	0.20	1.41	0.09	0.01	—	0.04	99.47	91	5	2	2	235	0.05	0.42
2	含石榴红柱硅线片岩	62149A	60.18	1.56	29.30	2.27	3.74	0.08	0.25	0.54	0.08	0.23	1.06	—	0.28	0.34	0.01	—	0.08	100.00	74	22	3	1	258	0.07	0.65
3	碳质红柱角页岩	62150E	51.97	1.22	29.46	1.15	0.40	0.02	0.05	0.70	2.20	1.38	3.97	0.31	6.72	0.09	0.01	—	0.07	99.72	78	5	3	14	233	0.05	0.29
4	含石榴硅线红柱角页岩	62150D	59.48	1.34	27.43	3.52	0.70	0.04	0.05	0.48	0.26	0.75	3.89	0.40	0.69	0.09	0.01	—	0.14	99.27	78	16	2	4	287	0.02	0.65
5	含硅线红柱角页岩	63235A	64.37	1.25	24.92	1.67	0.89	0.03	0.18	0.36	0.33	1.93	2.32	0.68	0.56	0.16	—	0.02	—	99.67	78	12	2	8	342	0.13	0.79
6	含石榴黑云红柱角页岩	63237C	55.89	1.38	28.56	6.40	1.73	0.12	0.12	0.48	0.22	1.40	2.17	0.74	0.47	0.25	—	0.03	—	99.96	68	26	2	4	224	0.02	0.8
7	红柱云母石英片岩	63008B	69.26	0.53	17.42	4.03	0.71	0.08	0.87	0.22	0.26	2.60	2.20	0.68	1.22	—	—	—	0.36	100.44	59	29	1	11	400	0.25	0.87
8	云母红柱碳质片岩	63089C	63.76	1.04	20.88	5.54	1.10	0.03	0.84	0.13	0.55	2.58	2.70	0.18	0.71	0.08	—	—	—	100.12	58	30	1	21	306	0.19	0.75
9	白云红柱碳质片岩	63089D	65.76	0.70	19.58	4.31	0.76	0.03	0.79	0.16	1.66	3.00	2.98	0.66	0.83	—	—	—	0.12	101.34	57	25	1	17	324	0.23	0.54
10	含红柱十字云母片岩	63010	71.01	0.60	16.17	4.05	1.46	—	0.47	0.46	0.35	3.60	1.46	0.16	—	—	—	—	—	99.79	54	28	3	15	404	0.14	0.86
11	石榴红柱石英云母片岩	63274C	65.94	0.80	17.35	5.76	1.38	0.03	1.58	0.18	0.28	3.53	2.42	0.42	—	—	—	—	0.04	99.76	49	38	1	12	318	0.30	0.89
12	含石榴红柱十字石英角页岩	63237D	63.73	1.50	19.72	6.83	4.95	0.20	0.37	0.48	0.10	0.35	1.58	0.50	—	0.21	—	—	0.08	100.55	52	45	2	1	284	0.05	0.7

续表

序号	岩石	样号	含量/%																		尼格里值						
			SiO_2	TiO_2	Al_2O_3	Fe_2O_3	FeO	MnO	MgO	CaO	Na_2O	K_2O	H_2O^+	H_2O^-	C	CO_2	SO_3	Cr_2O_3	P_2O_5	总和	al	fm	c	alk	sl	mg	k
13	十字铝直闪红柱角页岩	63217I	52.20	3.60	29.28	3.38	3.82	0.07	0.38	0.40	0.15	0.96	1.93	0.43	2.76	0.66	—	0.06	—	100.08	70	25	2	3	211	0.09	0.8
14	矽线石榴红柱角页岩	62150B	46.65	1.29	22.62	15.37	1.84	0.06	1.00	0.32	0.45	1.78	7.48	1.35	0.59	0.09	0.01	—	0.13	101.03	45	49	1	5	156	0.10	0.72
15	石榴红柱角页岩	63275C	56.88	0.68	13.28	11.38	4.51	0.11	3.39	0.79	0.10	2.70	5.16	1.55	—	0.10	—	—	0.04	100.67	28	62	3	7	204	0.28	0.95
16	矽线石榴子石英角页岩	63228A	42.77	2.95	29.10	3.45	19.25	0.13	0.39	0.55	0.15	0.40	0.74	0.28	—	0.25	—	0.03	—	100.44	46	52	1	1	114	0.03	0.64
17	石榴铝直闪石岩	63222D	44.51	1.25	15.98	3.65	29.66	0.33	0.64	0.88	0.38	0.40	1.22	0.48	0.14	0.45	—	0.03	—	100.00	24	72	2	2	112	0.03	0.39
18	碳质红柱角页岩	63171A	40.04	0.35	41.83	0.38	1.63	—	0.30	0.30	2.26	2.20	3.84	0.84	5.53	—	—	—	0.04	99.54	80	7	1	12	131	0.2	0.39
19	红柱角页岩	63171B	37.78	0.08	55.03	1.75	2.95	0.02	—	0.89	0.32	0.70	1.20	—	—	0.07	—	—	0.01	100.82	84	10	2	4	98	0	0.33
20	红柱角页岩	63305A	55.50	0.90	38.61	0.81	0.45	—	0.30	0.05	0.96	0.50	1.48	0.26	0.22	—	—	—	0.04	100.08	89.4	5.4	0.2	5	81	0.3	0.24
21	红柱角页岩	62027B	51.56	3.18	40.21	0.53	0.69	0.02	0.19	0.67	0.87	1.10	1.56	0.04	0.40	0.13	0.01	—	0.04	101.20	87	4	3	6	189	0.22	0.44
22	含硬绿红柱角页岩	63076B	46.82	2.60	41.42	0.85	2.76	0.01	0.44	0.31	0.52	0.34	3.28	0.15	0.08	—	—	0.03	0.04	99.65	84	12	1	3	161	0.17	0.29
23	红柱千枚岩	62159D	54.52	2.30	33.43	1.88	1.25	0.03	0.30	0.05	1.00	3.40	2.10	0.16	0.45	—	—	—	0.08	100.95	67	15	1	17	269	0.14	0.7
24	红柱硬绿泥石岩	63225C	52.63	3.00	27.43	2.51	10.37	0.08	0.30	0.36	0.10	0.22	2.66	0.16	—	0.68	—	0.04	—	100.54	58	40	1	1	81	0.03	0.6
25	碳质红柱千枚岩	63050D	60.98	0.80	22.62	3.20	1.21	—	0.78	0.09	1.04	4.30	3.16	0.70	1.90	—	—	—	0.08	100.86	61	21	1	17	281	0.24	0.73
26	红柱硬绿泥石岩	63226C	42.62	2.95	30.25	2.73	14.07	0.06	0.60	0.16	0.32	0.90	4.16	0.16	—	0.82	—	0.04	—	99.84	53	44	1	2	127	0.06	0.6
27	含红柱硬绿泥石岩	63075	43.30	3.50	29.48	2.21	14.29	0.07	0.88	1.10	0.57	0.63	3.98	0.03	—	—	—	0.10	0.04	100.18	51	43	3	3	126	0.08	0.4

续表

序号	岩石	样号	含量/%																	尼格里值						
			SiO_2	TiO_2	Al_2O_3	Fe_2O_3	FeO	MnO	MgO	CaO	Na_2O	K_2O	H_2O^+	H_2O^-	C	CO_2	SO_3	Cr_2O_3	P_2O_5	总和	al	fm	c	alk	sl	mg	k
28	红柱硬绿泥石岩	63226C	49.44	1.00	29.95	3.91	6.42	0.07	1.17	0.26	1.25	1.93	4.18	0.25	0.29	0.07	—	0.01	0.05	100.85	58	33	1	8	163	0.17	0.5
29	绢云硬绿泥石岩	63107A	45.20	1.80	29.54	1.85	12.39	0.25	0.82	0.81	1.22	0.65	4.96	0.28	0.13	—	—	0.03	0.07	100.00	53	40	2	5	137	0.09	0.26
30	硬绿泥千枚岩	63203B	39.74	3.26	33.12	6.00	4.43	—	0.44	0.84	0.95	5.78	5.32	0.28	—	—	—	0.04	—	100.20	58	26	3	13	112	0.07	0.8
31	硬绿泥千枚岩	63105B	61.22	1.18	21.69	3.92	4.87	0.06	0.76	1.15	1.00	1.48	3.12	0.08	0.32	0.11	—	—	—	100.96	53	34	5	8	254	0.13	0.49
32	硬绿泥千枚岩	63063B	58.44	0.92	20.40	4.65	5.62	0.15	1.34	1.10	1.08	2.32	3.42	0.34	—	0.07	—	—	—	99.85	46	40	4	10	225	0.19	0.58
33	硬绿泥千枚岩	62074C	63.04	0.80	19.51	3.38	4.52	0.12	1.15	1.16	0.99	2.32	2.73	0.30	0.52	0.09	0.01	—	0.12	100.76	49	35	5	11	272	0.21	0.60
34	含橄榄石大理岩	62014A	2.16		0.45	3.98	2.32		7.75	43.80	0.03	0.03								60.49	0.42	25.7	73.7	0.07	3.3	0.7	0.37
35	含透辉石大理岩	62063	8.92		1.15	2.13	1.22		1.23	47.90	0.19	0.23								62.97	1.21	7.86	90.36	0.56	15.6	0.4	0.43
36	透辉钙铁榴石大理岩	62071A	24.24		1.37	16.72	4.35		2.05	35.84	0.14	0.37								85.08	1.4	33	65	0.6	41.1	0.15	0.58
37	透闪石大理岩	62015A	1.62		0.44	0.25	0.28		20.62	27.30	0.05	0.07								50.63	0.43	5.12	48.2	0.14	2.7	0.98	0.46
38	含白云母大理岩	62017A	0.56		0.57	0.15	0.26		0.62	54.39	0.02	0.07								56.64	0.57	2.1	97.2	0.1	0.9	0.74	0.7
39	条纹状大理岩	62207	8.66		0.71	0.53	0.20		3.28	43.67	0.04	0.10								58.14	1.8	10	88	0.2	16.2	0.89	0.62
40	含透闪石大理岩	62208	2.32		0.39	0.55	0.22		7.17	45.17	0.02	0.09								56.21	0.38	18.69	8.08	0.11	3.8	0.94	0.75
41	含符山透辉大理岩	62235	38.60		6.89	2.76	1.69		14.03	24.51	0.41	0.25								89.14	7.4	44.1	47.75	0.90	0.90	0.85	0.28

（据刘国惠，1987）

石和绢云母所交代，有些微晶已全部转变为钠长石，对斜长石微晶的成分测定结果表明，均为 An<6 的钠长石，其成分特点明显与岩石的化学成分特点不协调。基质中辉石微晶多已绿帘石化，偶见未蚀变者，经成分测定为 Ca-Mg-Fe 辉石组的透辉石。

王人镜等（1990）称之为细碧岩，认为它是由玄武岩等基性火山岩经低级区域变质作用形成的。原生的玄武岩在绿片岩相区域变质作用中，在有 CO_2 和 Na_2O 成分的水溶液作用下，斜长石和辉石分解，斜长石去钙长石化，辉石变成绿泥石、绿帘石、阳起石。鲍亦冈（2000）采用“变玄武岩”这一名称，认为变玄武岩中除发育变生矿物外，无明显交代结构，上覆、下伏地层及火山沉积夹层中也未见明显交代现象。此外变玄武岩中硬绿泥石的晶体形态（车轮状、放射状集合体）、化学成分与区内下伏受印支期作用影响而形成的绿片岩相区域变质的浅变质岩系中的硬绿泥石在晶体形态（束状）和化学成分等方面有着明显差异，说明它们不是同一变质机制作用下的产物。此外，整个区域内变玄武岩的平均化学成分与未变质岩石以及世界范围内同类岩石的平均成分并无特别明显的差异，说明变玄武岩自身是一个相对封闭的体系，在自变质置换作用过程中并无明显的物质带入带出现象。因此在尚未取得深入的研究成果之前，本志仍以“变玄武岩”称之。

二、原岩恢复和大地构造环境分析

北京地区晚古生代—早古生代变质岩均为浅变质岩，岩石类型主要有角页岩、片岩、板岩-千枚岩、变质粉砂岩、碳酸盐质变质岩以及基性火山岩类。原岩多数保留原始沉积特征，主要为一套陆源-滨浅海相沉积岩，后期受区域变质和接触变质作用，形成浅变质岩。

（一）泥质岩类变质岩

1. 角页岩类

此类岩石主要赋存于石炭系，其次是二叠系和侏罗系。据 12 个岩石化学分析（表 22-10），SiO_2 多数为 40%～60%；TiO_2 多数>1%，少数>3%；Al_2O_3 25%～40%，个别达 55%；K_2O 1%～2%，Na_2O<1%，全铁量<10%。均属于泥质岩类，投影后大多落入高铝泥质岩（或黏土岩）区（Ⅰ区），少数属于含铁泥质岩（Ⅱ区），个别属富铁泥质岩（铁泥岩）（Ⅲ区）。

2. 片岩类

本类片岩主要产于中—新元古界、石炭系、二叠系和侏罗系。据 10 个岩石化学样分析（表 22-10），它们主要化学特征是 SiO_2 55%～71%，Al_2O_3 多数为 15%～30%，TiO_2 0.5%～1.4%，全铁 2%～6%，K_2O 多数为 2%～4%，Na_2O<0.4%，也即 K_2O 比 Na_2O 大几倍至几十倍。其他金属氧化物含量不超过 1%。这些都是泥质岩石的特征，在图 22-9 上多数落入含铁铝硅酸盐岩石（II）区，少数落入富铝铝硅酸盐岩石区（I），它们的原岩大致属黏土质岩石、含铁页岩、粉砂页岩和碳质页岩。

3. 板岩-千枚岩类

据岩石化学分析，一类是富铝的页岩，其化学特征是 SiO_2 50%～60%，Al_2O_3 22%～33%，TFe < 5%，K_2O 3%～4%，Na_2O 1%。另一类是铁质页岩或铁泥岩（表 22-10），其化学性质是 SiO_2 40%～63%，Al_2O_3 一般<22%，TFe 8%～10%，K_2O 2%～6%，Na_2O 1%。在图 22-9 中，大多数落入含铁铝硅酸盐岩石（II）区。

（二）长英质变质岩

1. 变粒岩类

根据产出特征和结构构造，原岩属中酸性凝灰质砂岩。

2. 变质砂岩类

原岩属砂岩或凝灰质砂岩。

（三）碳酸盐质变质岩

原岩主要为一套沉积碳酸盐岩，为陆表海沉积环境，从变质矿物组合分析，与接触变质作用关系密切。

（四）变质基性火山岩

原岩属玄武岩或安山玄武岩。产于侏罗系底部，与硬绿泥石岩或硬绿千枚岩互层。

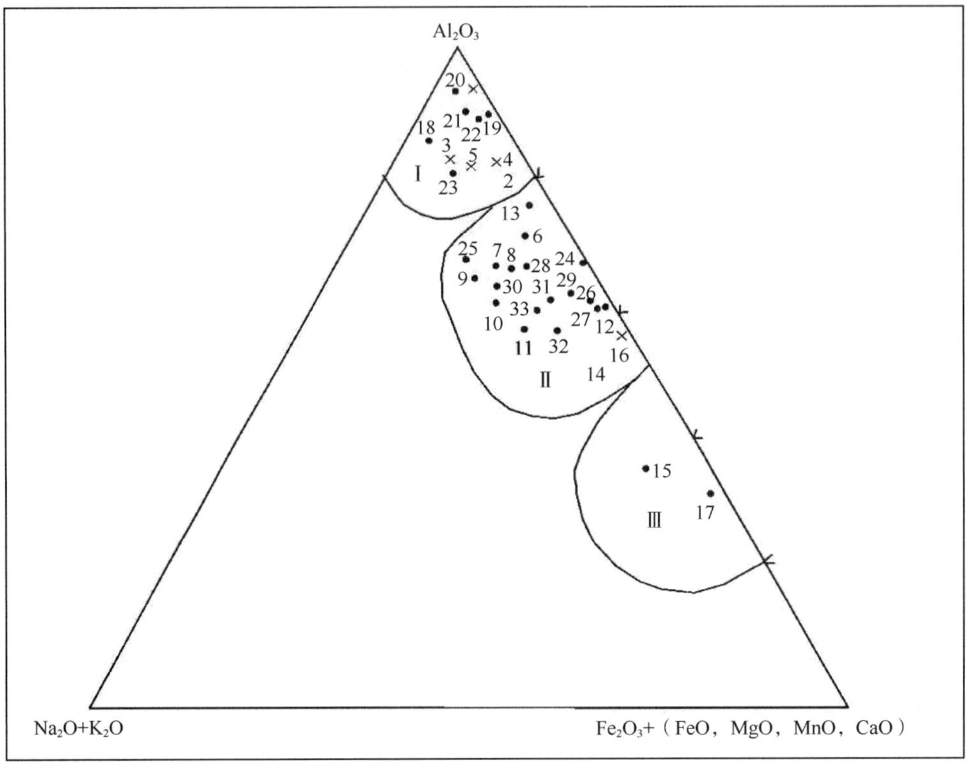

图 22-9　泥质岩石的划分

（据刘国惠，1987）

Fig. 22-9　Division of argillaceous rocks

Ⅰ—富铝泥质岩或富铝硅碳酸盐岩；Ⅱ—含铁泥质岩或含铁铝硅酸盐岩；Ⅲ—富铁泥质岩或铁铝硅酸盐岩

×高角闪岩相岩石；Δ 低角闪岩相岩石；• 高绿片岩相岩石

第二十三章　区域变质作用

第一节　变质作用分期

根据不同时期变质岩系中变质岩典型变质矿物共生组合、矿物时代和相互交代关系，主期变质与后期叠加变质矿物特征、变质相温压条件等资料，结合变质地质单元划分等相关资料综合研究，将本区区域变质作用发生期划分为阜平期（P $€_{B2}$，新太古代）、五台期（P $€_{B3}$，新太古代）、吕梁期（P $€_{C2}$）、华力西期（V）、印支期（I）、燕山期（Y）（表 23-1，表 23-2）。从新太古代区域变质作用开始，后期区域变质作用依次叠加于早期区域变质作用之上，形成相应共生矿物组合。

表 23-1　区域变质作用分期表

Table 23-1　Stages of regional metamorphism

地质时代	变质作用期次	名称	代号	主要变质相	期序
早侏罗世	燕山期	燕山期区域变质作用	Y	绿片岩相	6
中三叠世	印支期	印支期区域变质作用	I	绿片岩相	5
晚石炭世	华力西期	华力西期区域变质作用	V	绿帘角闪岩相-绿片岩相	4
古元古代	吕梁期	吕梁期区域变质作用	P $€_{C2}$	角闪岩相—绿帘角闪岩相	3
新太古代晚期	五台期	五台期区域变质作用	P $€_{B3}$	（麻粒岩相）角闪岩相—绿帘角闪岩相	2
新太古代晚期	阜平期	阜平期区域变质作用	P $€_{B2}$	麻粒岩相—角闪岩相	1

表 23-2　区域变质作用 *P-T* 分析表

Table 23-2　*P-T* analysis results of regional metamorphism

温度/℃	压力/GPa	主期变质地质体	期序
350~450	0.3~0.8	侏罗纪地质体	6
450~575	0.25~0.7	印支期地质体	5
575~700	0.5~0.7	华力西期地质体	4
760~810	0.96~1.04	古元古代侵入岩	3
540~585	0.4~1	新太古代四合堂岩群、新太古代 TTG 杂岩	2
782~849	0.8~1	新太古代密云岩群	1

第二节　变质作用分析

北京地区位于华北陆块相系，晋冀古陆核变质大相（陆松年，2015）。其中晋冀古陆核是我国最古老的陆块，初始陆块早在 3000Ma 以前即以出现，结晶基底于 1800Ma 前固结形成。根据北京地区地质历史发展特征差异，对北京地区新太古代—中生代变质相进行划分。北京地区太古宙地层可分为基底杂岩相和古弧盆系相。

其中基底杂岩相主要由新太古代古变质表壳岩组成，其产出形式是在相对年轻的变质地质单元中以残留体形态出现，岩石出露范围很小，成分复杂，以含 BIF 型硅铁建造为代表，难以恢复形成时的构造环境，因其形成时期处在地壳垂向增生阶段，多形成了系列古老穹隆构造，区内包括了阜平期和五台期变质岩石地层单位。

古弧盆系相进一步划分为英云闪长质-奥长花岗质-花岗闪长质片麻岩组合（TTG）和石英二长质-奥长花岗质-花岗质片麻岩两个岩石组合，广泛分布于太古宙古陆核区，使太古宙古陆核普遍遭受区域变质变质重熔作用，与基底杂岩相伴而生。

一、阜平期区域变质作用（$P\!\!\!\!\!-\!\!\in_{B2}$）

本次根据变质作用形成时间，将新太古代密云岩群（2900~2600Ma）划为阜平期区域变质作用，该期变质岩经历了多期区域叠加改造作用，形成了特定的典型变质矿物组合。

1. 变质矿物共生组合

（1）变超镁铁质岩类的矿物共生组合

紫苏辉石（或古铜辉石）+次透辉石（或透辉石）——二辉石岩；

古铜辉石+透辉石+普通角闪石——角闪二辉石岩；

紫苏辉石+次透辉石+普通角闪石——角闪二辉石、二辉角闪岩；

次透辉石+普通角闪石——次透辉角闪石岩、角闪次透辉石岩。

（2）变质基性岩类的矿物共生组合

紫苏辉石+次透辉石+斜长石——斜长二辉麻粒岩；

紫苏辉石+次透辉石+普通角闪石+斜长石——斜长角闪二辉麻粒岩、二辉斜长角闪岩；

次透辉石+斜长石-斜长次透辉岩；

次透辉石+普通角闪石+斜长石——角闪斜长次透辉岩、次透辉斜长角闪岩；

普通角闪石+斜长石——斜长角闪岩。

（3）SiO_2 含量相当于中性—中酸性的变质岩类（变杂砂岩、变安山岩——英安质岩类）的矿物共生组合

黑云母+紫苏辉石+斜长石+石英—黑云紫苏斜长变粒岩；黑云母+紫苏辉石+次透辉石+斜长石+石英——黑云二辉斜长变粒岩；黑云母+次透辉石+斜长石+石英——黑云次透辉斜长变粒岩；（石榴子石）+黑云母+斜长石+石英——（含石榴）黑云斜长变粒岩。

（4）变铁硅质岩类的矿物共生组合

紫苏辉石（或亚铁紫苏辉石）+次透辉石（或亚铁次透辉石）+磁铁矿+石英——二辉磁铁石英岩；

石榴子石+次透辉石（或亚铁次透辉石）+磁铁矿+石英——含石榴单斜辉石磁铁石英岩；

石榴子石+磁铁矿+斜长石+石英——含石榴磁铁斜长石英岩；

磁铁矿+石英——磁铁石英岩。

2. 变质温压条件分析

据金文山（2000）采集二辉石岩样品采用四种方法进行温压测试，并与周鸿勋（1983）和卢良兆等（1984）的研究数据进行对比分析，本次对金文山（2000）获得的二辉石岩样品电子探针数据采用 Lindsley（1985）基于实验测定的辉石温度计进行温度计算，并运用吴春明（2013）对辉石温度计极限值的估算方法对计算结果进行校正，测得区内变质温度在 772~995℃之间。

卢良兆（1984）据变质岩的石榴子石—斜方辉石地压计及单斜辉石+石榴子石+斜长石矿物组合，推测原密云岩群第一期变质压力为 1000MPa 左右，变化范围为 900~1350MPa。1000MPa 相应的地壳深度约 35km，地热梯度为 22~25℃/km，属于中压相系。由于本区基性深变质岩类的石榴子石成因复杂，大量属于后生成因，上述估计压力的可靠性有待进一步研究。周绍林等（1993）据 Yagi 等人的研究认为，辉石中钛辉石分子含量随压力增大而降低，当压力大于 1000MPa 时，钛辉石不再出现。密云地区单斜辉石普遍含低量的（1.02%~1.11%）钛辉石分子，推测其形成压力接近于 1000MPa。金文山（2000）总结前人工作认为，本区未出现超高压相的榴辉岩，以及单斜辉石的成分指示形成压力较

高特点，即 n（$NaAlSiO_2O_6$+$CaAl_2SiO_2$）/n（$CaTiAlO_6$+$CaFe_2^3$+SiO_6）>1（卢良兆，1984），推测该表壳岩系麻粒岩相变质时的压力为 0. 8~1GPa，其进行了 4 个角闪石样品再 Si-Al^{IV} 图解中 3 个样品位于麻粒岩相区，1 个位于高级角闪岩相区，认为其属于高温区域变质作用产物。

根据冀东麻粒岩相变质作用的研究结果表明（贺高品等，1991；贺高品，1992），该区第一期变质作用经历了一个长期的演化过程，变质作用从大约 3000Ma 开始，直到约 2500Ma 才最后结束，期间经历了多期构造和热事件的影响，因此区内同位素年龄数据集中在 2500Ma 左右。密云地区与冀东地区的情况基本一致，2500Ma 应为密云地区第一期变质作用的结束时间，第二期变质作用主要表现为基性岩墙的变质以及部分表壳岩和早期侵入岩中的叠加变质现象。因为基性岩墙切穿了遭受第一期变质作用强烈改造的表壳岩和早期侵入岩，因此它们的形成时间应在 2500Ma 之后。贺高品曾对密云地区的变质基性岩墙（石榴二辉麻粒岩）进行了同位素年龄取样，并获得了（1801±57）Ma 的 Sm-Nd 变质矿物和岩石等时线年龄（贺高品等 1993），因此区内第二期变质作用的时间应为 1800Ma 左右，其研究结果与赵国春（2001，2002，2009）的研究成果在华北克拉通基底拼合时间上基本吻合。

赵国春（2001，2002，2009）认为华北克拉通基底东部陆块冀东-密云地区，在 1850Ma 前后与东部陆块沿中部带发生碰撞拼合，冀东-密云基底岩石主要以高级—中低级花岗绿岩地体形式出露，时代上以新太古代变质岩石组合为主。新太古代时期，大量 TTG 片麻岩、镁铁质—超镁铁质层状侵入。在北京密云-怀柔地区太古宙基底岩石以 TTG 片麻岩穹隆构造为主要特征、间夹小规模表壳岩线性带，如密云穹隆。

在变质作用演化上，新太古代基底岩石表现为近等压冷却（IBC）型逆时针 $P-T$ 演化特征（图 23-1）。在北京密云、怀柔等地区广泛发育铁镁质麻粒岩、斜长角闪岩和泥质片麻岩保存早期进变质、峰期和峰期后退变质阶段矿物组合特征。早期进变质矿物组合主要以矿物包裹体形式存在于峰期

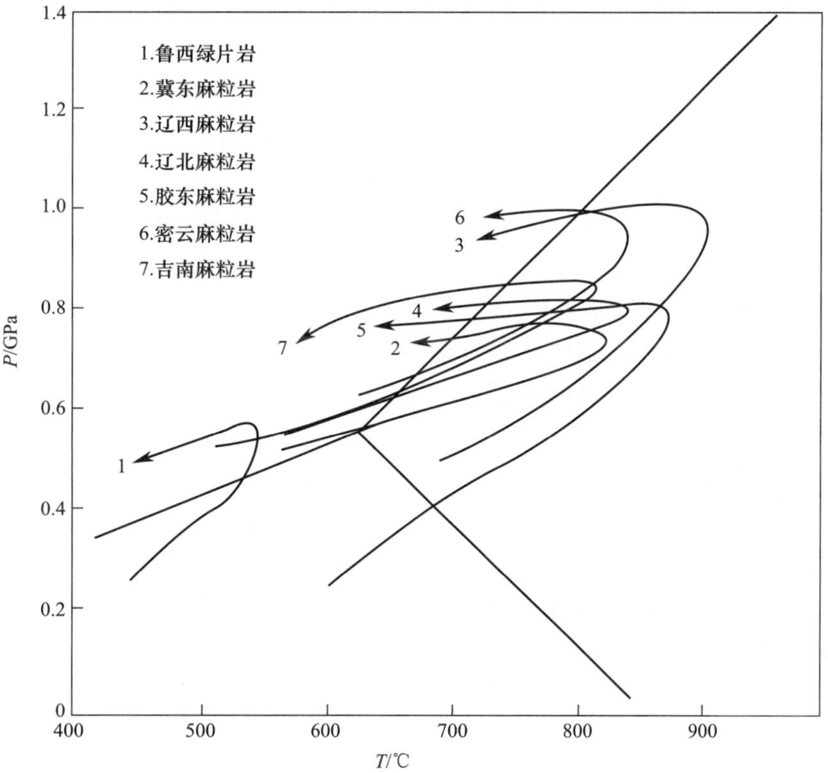

图 23-1　华北克拉通东部陆块基底岩石变质作用 $P-T$ 轨迹

（据赵国春，2001）

Fig. 23-1　$P-T$ paths of metamorphic rocks in the Eastern Block of the North China Craton

阶段矿物中，如在镁铁质麻粒岩中早期角闪石+斜长石+石英+黑云包裹斜方辉石和石榴子石中，使岩石在峰期阶段发生了麻粒岩相的进变质作用，形成了紫苏辉石、透辉石和石榴子石等高温变质矿物，而普通角闪石、黑云母和斜长石等仍为稳定矿物，同时在变质峰期阶段发生了广泛的混合岩化作用和长英质组分的加入，早期形成的普通角闪石、黑云母和斜长石，多数转变为稳定条件下的新矿物，少数在紫苏辉石、透辉石和石榴子石中呈微粒包体被保留下来。

在斜长角闪岩中，早期组合绿泥石+阳起石+绿帘石+钠长石+石英包裹于斜长石、角闪石和石榴子石中；在泥质片麻岩中，早期组合黑云母+斜长石+石英±红柱石包裹于石榴子石和硅线石中，变质过程中有大量地壳下部重熔的英云闪长质-花岗闪长质岩浆以及地幔分熔的基性和超级性岩浆侵位，并由于重力均衡作用使增厚的地壳发生快速隆起和侵蚀作用，使温度和压力急速下降到600~700℃和0.4~0.5GPa，在变质峰期后在侵入岩及其围岩附近发生高角闪岩相的变质作用，在侵入岩中形成了不含紫苏辉石的角闪斜长片麻岩、斜长角闪岩和角闪石岩。

结合变质温压条件计算结果，这些矿物组合转变限定了具有近等压冷却（IBC）特征的逆时针 *P*-*T* 演化轨迹，反映了变质作用与大量地幔岩浆底侵作用有关。综上所述，研究表明密云地区新太古代变质岩石发生近等压冷却型（IBC）逆时针 *P*-*T* 演化的变质作用，其与岛弧、大陆裂谷或地幔柱（热点）环境中的大量地幔岩浆底侵作用有关。

二、五台期区域变质作用（$P\text{-}Є_{B3}$）

本次将区内新太古代晚期四合堂岩群（2600~2500Ma）变质岩划为五台期区域变质作用，代表区域角闪岩相变质特征。

1. 变质矿物共生组合

（1）变中基性—基性变质岩类矿物共生组合

次透辉石+普通角闪石+斜长石——次透辉斜长角闪岩；

普通角闪石+斜长石——斜长角闪岩。

（2）变粒岩类的矿物共生组合

普通角闪石+黑云母+斜长石+石英——含角闪黑云斜长变粒岩；

黑云母+斜长石+石英——黑云斜长变粒岩；

石榴子石+普通角闪石+黑云母+斜长石+石英——含石榴黑云斜长变粒岩，含石榴角闪黑云斜长变粒岩。

（3）变铁硅质岩类矿物共生组合

阳起石+磁铁矿+石英——阳起磁铁石英岩；

黑云母+磁铁矿+石英——黑云磁铁石英岩。

2. 变质温压条件分析

据金文山（2000）采集的6个片状（黑云母化）斜长角闪岩的角闪石样品，经分析均为角闪岩相变质岩，属角闪岩相区。斜长角闪岩的绿帘石化和黑云母化很普遍，未见石榴子石。晚期的变质基性岩墙中局部可见强烈的石榴子石化现象。黑云斜长变粒岩类偶尔见到石榴子石。个别含石榴黑云斜长变粒岩样品中的石榴子石的镁铝榴石组分含量为14%，铁铝榴石为62%，锰铝榴石为11%，钙铝榴石为13%，属于钙锰质铁铝榴石，其含铁度为82%，位于蓝晶石带。该样品中的石榴子石-黑云母对其所得变质温度为540~585℃，据姬广义（1994）在四合堂地区开展变质岩研究认为变质压力不低于0.8GPa。

三、吕梁期区域变质作用（$P\text{-}Є_{C2}$）

新太古代—古元古代（2500~1800Ma）深成侵入体经历了多期区域变质作用的叠加改造，形成了多个变质矿物世代和典型变质矿物，该时期区域变质作用是区内最重要的一期变质作用。

1. 变质矿物共生组合

（1）变超镁铁质岩类的矿物共生组合

（磁铁矿）+单斜辉石——（含磁铁）次透辉石岩；

（次透辉石）+普通角闪石——次透辉角闪石岩、角闪石岩。

（2）变质基性岩类的矿物共生组合

紫苏辉石+次透辉石+角闪石+斜长石——角闪斜长二辉麻粒岩；

次透辉石+斜长——斜长次透辉岩；

次透辉石+普通角闪石+斜长石——角闪斜长次透辉岩；

普通角闪石+斜长石——斜长角闪岩。

（3）SiO_2 含量相当于中性—中酸性变质岩类的矿物共生组合

黑云母+次透辉石+斜长石+石英——黑云次透辉斜长变粒岩；

黑云母+斜长石+石英——黑云斜长变粒岩。

（4）富含磁铁矿和石英岩类的矿物共生组合

磁铁矿+石英+单斜辉石——含磁铁石英单斜辉石岩；

磁铁矿+石英+石榴子石+次透辉石+石英——磁铁次透辉石榴斜长变粒岩；

角闪石（阳起石）+磁铁矿+石英——角闪磁铁石英岩、含磁铁阳起石英岩。

2. 变质温压条件分析

据贺高品（1992）对放马峪岩墙的研究，其变质温度已达 760~810℃，压力达到 0.96~1.04GPa。

四、华力西期区域变质作用（V）

华力西期区域变质作用使区内中元古代—晚石炭世地质体发生了区域变质作用，形成了相应的典型变质矿物组合。该期区域变质作用在北京西山地区表现明显，由于变质程度低不易辨别。

1. 变质矿物共生组合

（1）变质砂岩泥岩

绢云母+绿泥石+石英；

白云母+黑云母+石英+石榴子石；

白云母+黑云母+石英+绿泥石。

（2）变质泥碳质砂岩

石英+石墨+白云母+黑云母+斜长石+石英；

绢云母+石英+石墨。

（3）变质钙硅酸盐岩类

方解石+透闪石+石英。

（4）变质碳酸盐岩类

方解石+石英+白云母；

方解石+透闪石+石英+黑云母。

2. 变质温压条件分析

石榴子石呈半自形—他形粒状，粒径 0.1~0.5mm；透闪石呈半自形-他形柱状，粒径 0.1~0.2mm；黑云母、白云母呈鳞片状，片径 0.05~0.5mm；绢云母呈白色，鳞片状，结晶细小，多在 0.1mm 以下，个体边界不清晰；绿泥石黄色，鳞片状，结晶细小，多在 0.01mm 以下；石墨呈鳞片状，片径 0.01~0.2mm；斜长石他形粒状，以钠长石为主，部分为更长石，粒径 0.05~0.15mm；石英他形粒状，部分呈圆粒状，粒径 0.05~3mm，波状消光；方解石无色，半自形-他形粒状，粒径 0.1~0.25mm，个别达 0.3mm，具干涉色，有时见聚片双晶。

根据变质矿物的研究，其区域变质作用的温压条件为 $T=350\sim570$℃，$P=0.2\sim0.8$GPa，变质作用以绿帘角闪岩相-绿片岩相为主，属中低压变质相（图 23-2，表 23-3，表 23-4）。

表 23-3 红柱石环带内外电子探针分析结果

Table 23-3 Electron microprobe analysis results of the inner and outer girdle of andalusite

点号	矿物	Na_2O	Al_2O_3	SiO	TiO	MnO	FeO	总量
①	An	0. 35	61. 69	37. 38	0. 13	0. 21	0. 47	100. 27
②	An	0. 24	61. 37	37. 13	0	0	1. 14	99. 89
③	An	0. 47	62. 04	36. 5	0	0	0. 56	99. 67
④	An	0. 46	61. 89	37. 29	0. 28	0. 05	0. 26	100. 23

注：氧化物含量单位为%。（据万永平，2007）

表 23-4 包裹体内磁黄铁矿和钛铁矿电子探针分析结果

Table 23-4 Electron microprobe analysis results of pyrrhotite and ilmenite in inclusion

含量/%	S	Fe	Co	Ni	Cu	Se	总量
	38. 93	60. 72	0. 12	0. 51	0. 01	0. 16	100. 45
含量/%	NaO	MgO	SiO	TiO_2	MnO	FeO	总量
	0. 33	0. 44	0. 34	54. 17	1. 52	43. 64	100. 43

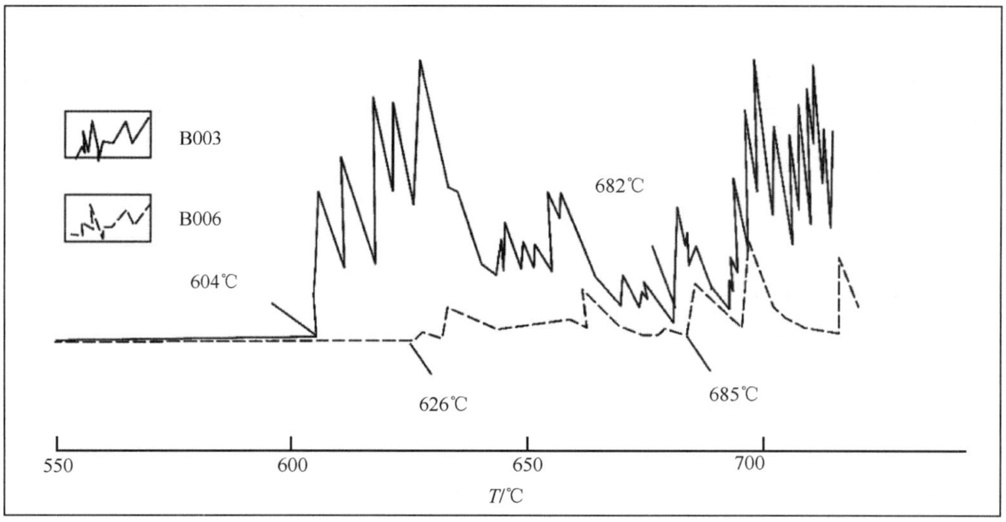

图 23-2 包裹体内磁黄铁矿和钛铁矿电子探针示意图

（据万永平，2007）

Fig. 23-2 Electron microprobe schematic diagram of pyrrhotite and ilmenite in inclusion

五、印支期区域变质作用（I）

该时期变质作用主要分布于北京北山新太古代变质岩带南部和西山周口店一带。绿片岩相是西山地区最发育的变质相，主要岩石类型有片岩、板岩、千枚岩及变质砂岩、砾岩等。

1. 变质矿物共生组合

按硬绿泥石、黑云母等变质矿物在变质泥质岩中首次出现，进一步划分为硬绿泥石带和黑云母带。

硬绿泥石带以硬绿泥石首次出现为标志，并且十字石、蓝晶石、黑云母普遍存在，某些岩石中出现石榴子石与硬绿泥石共生，此时 Al_2SiO_5 同质多象变体为红柱石。

矿物组合有：①叶蜡石+绢云母+石英+石墨；②绢云母+石英；③绢云母+绿泥石+绿帘石+石英；④硬绿泥石+绢云母+绿泥石+石英；⑤硬绿泥石+黑云母+白云母（绢云母）+钠长石+石英；⑥黑云母+绢云母+绿泥石+石英；⑦红柱石+硬绿泥石+绢云母+绿泥石+石英；⑧石榴子石+黑云母+白云母+石英；⑨石榴子石+黑云母+绿泥石+石英。

2. 变质温压条件分析

地壳运动可划分为早期运动（褶皱运动、水平运动）和造陆运动（垂直运动、升降运动）两类，其中造陆运动包括大规模的褶皱运动和断裂运动，地壳表层造成了定向和成列出现的褶皱与断裂，但造陆运动中造山运动常形成孤立而平缓开阔的褶皱、穹隆和断裂，但不具备造山运动的方向性和系统性。各种运动发生时间和运动性质均需要依靠地层中的不整合（角度不整合、构造不整合）和假整合（平行不整合）来鉴定。

通过对北京地区沉积建造和构造形态分析，北京地区自吕梁运动以后，直到侏罗纪沉积之前，区内只发生过一次呈南北向挤压的褶皱运动，其发生时间在双泉组形成之后。因此，认为区内侏罗系杏石口组和双泉组之间属角度不整合的关系。

北京地区印支期主要为一套由砂岩和页岩组成的陆源碎屑岩建造，相互间均为平行不整合接触。

印支运动及其前后地壳运动的关系，目前未有定论。北京地区印支运动和燕山运动无论就其所产生的构造形式，还是其运动的力学方式来说，都各具特色，截然不同。北京地区双泉组与其下伏二叠系石盒子组之间为连续沉积，就更大的区域而言，山西省河北省的二叠—三叠系也都是连续沉积关系。可以说北京地区，甚至整个华北地区自长城纪起，至三叠纪双泉期，区内只有升降运动的海陆变迁，没有强烈的构造变动。从岩层、岩体的形变角度看，印支运动所产生的构造形变及其加力方式与燕山运动有本质上的不同，与吕梁运动有可能是一致的，它的存在标志着区内陆块稳定发展阶段的结束。

受印支构造运动的影响，石炭系和二叠系的岩石遭受了不同程度的区域变质作用。变质岩类型有变质砂岩、变质粉砂岩、变质石英砂岩、板岩、千枚岩、片岩等，新生变质矿物有绢云母、硬绿泥石、叶蜡石、红柱石。变质作用造成岩石颜色的改变，如硬绿泥石含量高时岩石绿色比例加重（柯柏林，2008）。

区域变质作用的温压条件为325~570℃，压力为0.2~0.7GPa。结合该区域变质特点，综合分析认为，本区印支期区域变质类型，属于造山带区域变质作用——区域中低温—低温中压型绿片岩相区域变质作用。主体具有单相变质，在区域上呈狭长带状分布的特征。

六、燕山期区域变质作用（Y）

1. 变质矿物共生组合

主要位于北京北山新太古代变质岩带南部和西山周口店一带。按特征矿物组合和特征矿物首次出现，划分为低角闪岩相的十字石带。

此带以泥质原岩中出现十字石为特征，硬绿泥石开始消失，十字石、蓝晶石、黑云母及石榴子石共生，但不出现红柱石与十字石共生。一般矿物共生组合为：

1）十字石±蓝晶石+黑云母+白云母±石榴子石+石英；

2）蓝晶石+黑云母+白云母+石英。

2. 变质温压条件分析

本区燕山期区域变质作用主要为燕山运动前和初始造山幕（J_1），形成谷积山背斜等重要构造，并使区内火山岩岩系呈线状、新月形分布，反映挤压和伸展诱发的岩石圈尺度的破裂和玄武质岩浆沿破裂底侵于壳底，受燕山运动影响在区域上沉积形成一套薄层变玄武岩。伸展运动快速结束后，紧接着发生平静的煤系沉积期，后期燕山运动开始发生区域性收缩挤压，形成类磨拉石沉积和低温、中—高温变质作用，后期地壳隆升剥蚀，形成与后期建造之间的角度不整合。

随着变形的增强和古地温梯度的增加（赵越，1990），在京西出现低绿片岩相的变质。主要为一套纯陆相含煤沉积建造，厚约200~600m。根据研究，北京地区早侏罗世含煤沉积建造，总厚度等值线和相间的富贫煤带排列的总体方向是一致的，它们均呈东西向或近东西向延伸，反映了该时期聚煤盆地的形成和发展是受东西向构造体系控制的。早侏罗世以后，发生中侏罗世和晚侏罗世之间的燕山运动第一幕，使区内早、中侏罗世一起发生了形变，形成了一系列轴向作北东东展布的褶皱断裂群。

该套变质岩系区内研究的较少，温压条件为350~700℃，0.25~0.8GPa。

第三节　变质相带

一、划分依据

变质带指在特定时期形成具有某种共性的变质岩系，在空间上往往呈有规律的分布。本次划分的依据主要包括：①大地构造单元划分；②重要断裂和褶皱构造；③变质作用、变质相；④特征变质矿物在区域上呈达到一定规模；⑤具有特殊意义的变质带；⑥重要变质带的命名以特征矿物和具有代表性的地名命名。

二、划分方案

（一）密云岩群重要变质相带

北京地区变质岩石类型（不同原岩类型）及矿物组合的变化，以石榴角闪辉石质变质岩的等化学系列原岩的矿物组合的空间变化，由北向南基本可划分出两个变质带（单斜辉石-石榴子石带和二辉石带），反映了在一定变质强度基础上由深部热流值变化而出现的变质强度和*P-T*条件的空间变化。

1. 单斜辉石-石榴子石带

主要分布于半城子—大漕地段，总体上呈北东向带状展布，大致位于大漕—小漕—高岭一带与原大漕组岩层基本一致，在崎峰茶断裂以西的变质岩地区也属于该带。代表性岩石类型为石榴斜长辉石岩，共生变质岩有石榴辉石斜长角闪岩、黑云辉石角闪斜长片麻岩、石榴黑云变粒岩及斜长片麻岩、石榴辉石变粒岩及浅粒岩。含钙的铁铝榴石-透辉石共生体的开始出现时为此带的标志。

代表性平衡共生的矿物组合：ⓐ Pl+Grt+Cpx±Qtz；ⓑ Pl+Grt+Cpx+Hbl±Qtz±Scp；②Pl+Grt±Qtz；③Pl+Grt+Bt±Qtz；④Pl+Grt+Cpx+Bt±Qtz；⑤Pl+Cpx+Bt±Qtz；⑥Opx+Cpx±Hbl。

矿物共生ACF图解（图23-5）表明各组合均处于平衡状态，组合①、②、③是此带代表组合，Pl+Grt+Cpx±Qtz是标志性组合，出现角闪石和辉石平衡共生是和岩层中H_2O的含量较高有关。当含一定量K_2O时，出现含黑云母的组合④、⑤。组合⑥不含石英或长石，属于SiO_2不饱和组合，Al_2O_3很低，Opx代替Grt提前出现，为比较特殊的组合。

2. 二辉石带

主要分布于沙厂—霍各庄一带，大量紫苏辉石出现时进入
此带的标志。典型岩石类型为黑云紫苏麻粒岩、二辉麻粒岩、石榴紫苏麻粒岩和角闪二辉麻粒岩，它们成夹层或透镜体分布于石榴黑云辉石斜长片麻岩、石榴黑云斜长片麻岩及黑云变粒岩为主的岩层中。

代表性平衡共生矿物组合：①Pl+Hy+Di+Bt；②Pl+Hy+Grt+Di±Bt±Mgt；③Pl+Di±Bt±Hbl；④Pl+Hy±Di±Hbl；⑤Hy+Di±Hbl；⑥Pl±Grt+Bt±Qtz。

上述矿物组合的变化主要是受原岩成分的控制。组合中紫苏辉石具明显粉红色—淡绿色多色性，Fs=30~36，其他矿物光性特点及斜长石成分和前一带无明显差别。

矿物共生ACF图解（图23-3）表，组合①、②是标志性组合，组③、⑥是和单斜辉石-石榴子石带贯通组合，(4) 的原岩和 (1) 相似，但不含K_2O，因此，角闪石代替黑云母出现。结构分析表明，

和辉石共生的黑云母和角闪石为稳定矿物，其出现与否取决于原岩中 K_2O 及 Na_2O、Al_2O_3 的含量。二辉石带是在单斜辉石-石榴子石带矿物组合基础上发展起来的，紫苏辉石出现的变质反应是：Cpx+Grt+SiO_2⇌Opx+An，由于远未达到新的完全平衡，所以上一带有关组合仍以残存状态存在（图 23-4）。

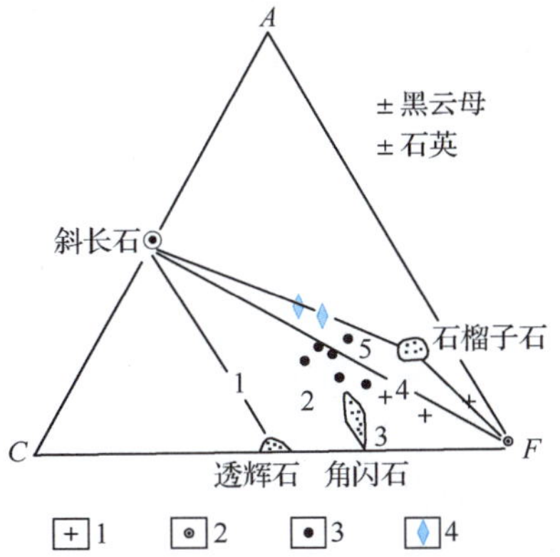

图 23-3　密云岩群麻粒岩相区 *ACF* 图解

Fig. 23-3　*ACF* analysis diagram of granulite facies of Miyun rock group

1—二辉石岩；2—石榴紫苏麻粒岩；3—二辉麻粒岩；4—黑云紫苏斜长片麻岩；

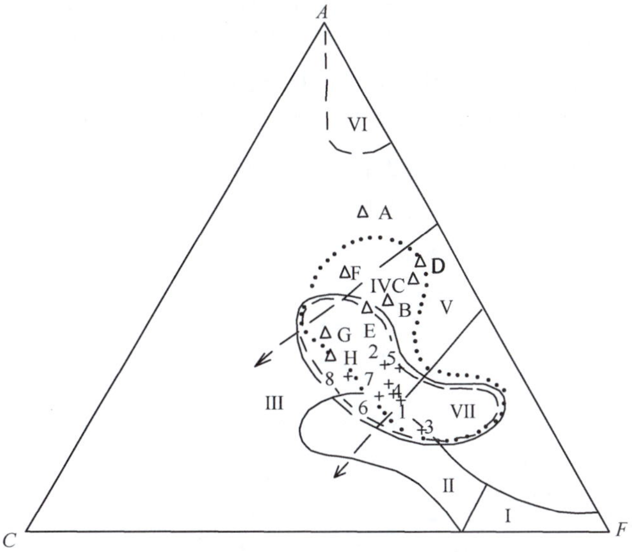

图 23-4　（角闪）黑云斜长变粒岩类的 *ACF* 图解

（据金文山，2000）

Fig. 23-4　*ACF* analysis diagram of hornblende biotite plagioclase granulite（amphibole）

（二）四合堂岩群重要变质相带

四合堂岩群主要属于角闪岩相变质地区，划分为低角闪岩相（包括山神庙、西湾子、宋营子及阳坡地等地区）变质岩系和高级角闪岩相（单斜辉石-角闪石带），后者相当于角闪岩相的高温阶段。

1. 低角闪岩带

主要岩石类型有黑云斜长片麻岩（变粒岩）、（黑云）角闪斜长片麻岩、斜长角闪岩、角闪石岩、角闪磁铁石英岩及浅粒岩等。

代表性矿物组合：①Hbl+Pl±Qtz；②Hbl+Pl+Bt+Qtz；③Pl+Bt+Qtz±Mgt；④Mgt+Qtz+Hbl；⑤Pl+Kfs+Qtz。

蓝绿色角闪石是本区低温角闪岩相的产物，岩石结构以细粒变晶为主，中粒结构仅见于某些岩石。由于缺少变质泥质岩，可以斜长角闪岩中斜长石成分（一般 An=30）作为划分标准。本区该类岩石中斜长石成分为 An=35~37，岩石中基本不出现含辉石类的矿物组合，共生矿物为黑云母、角闪石及石英的组合，结合其他矿物共生关系，说明变质程度属于低角闪岩相，温度大致为 500~600℃（图 23-5）。

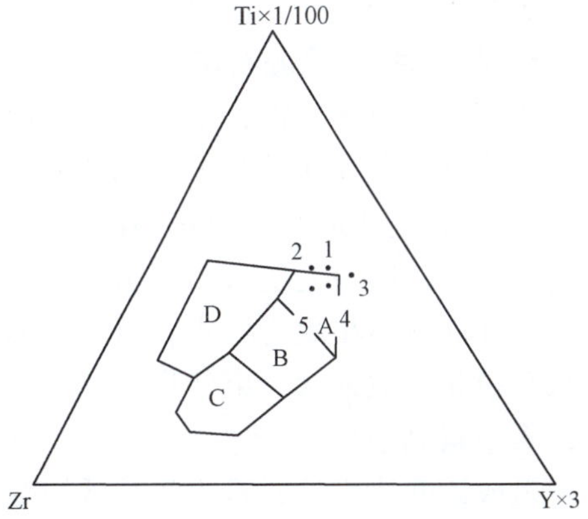

图 23-5　斜长角闪岩类的 Ti×1/100-Zr-Y×3 图解

（据金文山，2000）

Fig. 23-5　The ternary plot Ti×1/100-Zr-Y×3 of Plagioclase amphibolite class

2. 高级角闪岩带（单斜辉石-普通角闪石带）

分布于车道岭西北-阳坡地地区，呈北东、北东东狭长带状展布，平谷关上一带有零星分布。

主要岩石类型有细粒斜长角闪岩、黑云角闪斜长片麻岩、角闪变粒岩、黑云变粒岩、浅粒岩及磁铁石英岩、长石石英岩等组成，有些含透辉石。

代表性矿物共生组合：①Cpx+Hbl+Pl；②Hbl+Pl±Qtz；③Bt+Hbl+Pl+Qtz；④Grt+Bt+Pl+Qtz；⑤Bt+Pl+Qtz；⑥Cpx+Hbl。

上述矿物共生关系属于高角闪岩相。此带角闪质岩石中，主要表现为普通角闪石和辉石共生，一般不含石榴子石。组合①、②、③属于 SiO_2 不饱和组合，一般不含石英或有些含石英。普通角闪石和黑云母共生，可能与原岩中 K_2O/Al_2O_3 比值较高有关。组合①中单斜辉石与斜长石等矿物平衡共生，可能反映原岩钾钠组分较低，铝、铁、镁组分较高。

角闪石具棕绿色多色性，与单斜辉石、石榴子石共生时，接触界线平直，处于平衡状态。石英、斜长石和黑云母是区内不同变质相贯通矿物组合。推测高角闪岩相变质温度大致为600~700℃，属中压相系（图23-6，图23-7）。

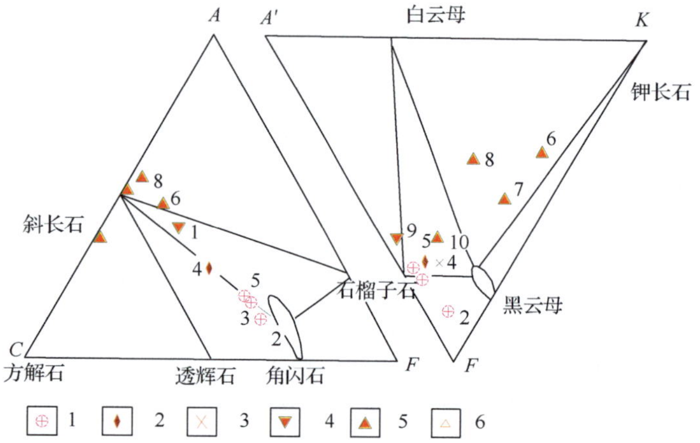

图23-6　四合堂岩群高级角闪岩相区 *ACF* 和 *A'KF* 相图

Fig. 23-6　*ACF* and *AKF* facies diagram of higher amphibolite facies of Sihetang rock group

1—斜长角闪岩；2—黑云角闪斜长片麻岩；3—角闪黑云斜长片麻岩；4—角闪变粒岩；5—黑云二长变粒岩；6—黑云斜长变粒岩

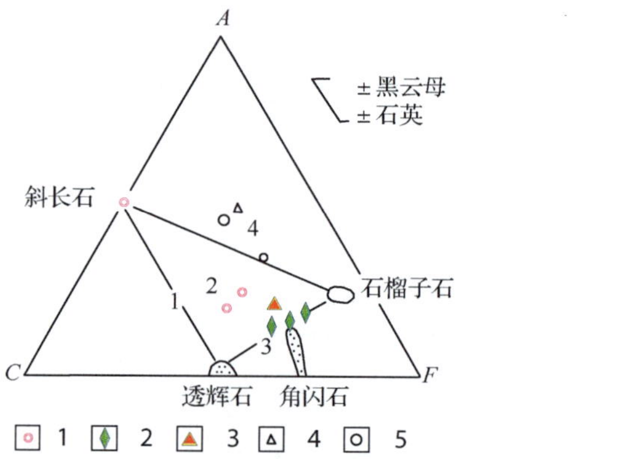

图23-7　四合堂岩群高级角闪岩相区 *ACF* 相图

Fig. 23-7　*ACF* facies diagram of higher amphibolite facies of Miyun rock group

1—石榴斜长辉石岩；2—榴辉斜长片麻岩；3—榴辉变粒岩；4—榴辉黑云变粒岩；5—榴辉浅粒岩

（三）晚古生代—早中生代重要变质相带

根据变质带的空间分布与区域构造及矿物之间的关系，晚古生代—早中生代区内先后经历了区域浅变质与接触变质两种性质的变质作用，以区域浅变质为主，在房山岩体周边发生接触变质。具体划分了5个变质带。

1. 硬绿泥石-绿泥石带

该变质带具区域性，主要出露在西山大安山地区及周口店西部中上元古界至侏罗系的板岩、千枚岩、变质砂岩及变质辉绿岩等岩石中，呈面状分布特征，变质带自房山沿北东东向至香山、黑龙潭，长约50km，宽10km。它与区域上侵入体的分布没有直接的依存关系，在区内多产于离岩体500m以外的岩石中。推断该变质带与燕山运动形成的西山褶皱带一致，是区域动力低温热流变质作用用的产物。

以硬绿泥石出现为标志，且不出现黑云母、石榴子石等变质矿物。硬绿泥石一般呈黑色，柱状、板状及粒状，放射状、束状集合体，大小一般1~2mm，多色性明显，无色—浅黄绿色，聚片双晶发育，属单斜晶系。

组成该变质带的岩石主要是低级变质的泥灰岩、长英质碎屑岩和变质基性火山岩。具有意义的变质矿物共生组合有两类，一类泥质岩石中绢云母+硬绿泥石、绢云母+绿泥石+硬绿泥石、绢云母+绿泥石（+绿帘石）；另一类在基性火山岩中硬绿泥石+绿泥石+绢云母（+钠长石）、绿泥石+绿帘石（+钠长石）、阳起石+绿泥石+绿帘石（+钠长石）。根据变质矿物组合应属低绿片岩相。

根据哈克尔试验：在封闭系统、还原条件和压应力下，高岭土+氧化亚铁→硬绿泥石+石英+水，设定温度 350~450℃，当温度升高时，会导致硬绿泥石消失；然而硬绿泥石形成的压力适应范围较大，最大达 1GPa 以上，说明硬绿泥石在低温低压下是稳定的，若压力降至 0.5GPa，则温度必须低于 500℃，大于 500℃同时压力较高时，硬绿泥石变得不稳定并开始分解；另外硬绿泥石带出现蓝晶石、刚玉等深度变质矿物，说明硬绿泥石封闭系统已进入半开放状态，部分硬绿泥石被氧化后形成的新矿物。根据上述 3 点，推测本区硬绿泥石带温压条件为 350~450℃，0.25~0.8GPa。

2. 黑云母带

主要分布于西山变质砂岩、板岩和千枚岩中。起初黑云母鳞片细小，黄绿色调，与岩体关系密切，随着靠近岩体，岩石变为片岩和角页岩时，黑云母色调变浓，鳞片增大，表明温度增高，黑云母含有更多的钛、铁元素。黑云母带是岩体影响的早期产物，该带宽多为 100~400m。

岩石主要为泥砂质变质岩，矿物共生组合为黑云母+绢云母、黑云母+绿泥石+绢云母、黑云母+绿帘石+绿泥石、黑云母+绢云母+绿帘石，其中黑云母+绿泥石的矿物组合一在石炭系太原组地层中现象较明显。该带与硬绿泥石带的不同在于普遍出现了黑云母，在某些岩石中亦出现石榴子石或蓝晶石，但它们与硬绿泥石共生，说明该带中硬绿泥石仍然作为一个稳定的矿物而存在。

据分析，硬绿泥石带相当于中压绿片岩相的绿泥石带，而黑云母带则包括两个不同类型的矿物组合，一是含蓝晶石的黑云母带，在石炭系太原组中现象较明显，二是与红柱石发生共生的黑云母带，多发育在石炭系山西组中。前者属中压相系，而后者为低压相系。

由于变质矿物组合是较低温的产物，在一些富铁的岩石中，常见到少量硬绿泥石存在。此带属绿片亚相，属接触热变质作用的第一阶段产物，并由外向岩体内依次产生渐进变质带。黑云母带 $T=450 \sim 575$℃，$P>2.5$GPa。

3. 红柱石带

由于热流温度而升高，泥质岩石受到进一步改造，成为千枚岩、片岩和角页岩，并伴随出现了典型的接触热流变质矿物红柱石，在显微镜下，包括了早期的硬绿泥石或黑云母，表明红柱石生成晚于黑云母，但硬绿泥石仍有发育。红柱石带的宽度因地而异，窄的约 30m，宽的可达 1000m。与此带相当，在一些大理岩或结晶灰岩中则构成白云母带。

该带的主要变质矿物共生组合，在泥砂质变质岩中表现为红柱石+绢云母、红柱石+绢云母+硬绿泥石、绢云母+硬绿泥石、红柱石+硬绿泥石岩石组合；碳酸盐变质岩中矿物共生组合是白云母+斜长石+方解石、白云母+石英+方解石、白云母+方解石岩石组合。

按泥砂质岩石的矿物组合，该带属高绿片岩相。

4. 铁铝榴石-十字石带

由红柱石带向内，千枚岩消失，大量出现片岩和角页岩及变粒岩，以产出铁铝榴石，十字石及角闪石为特征。

显微镜下可以见到铁铝榴石包含无色细粒红柱石及十字石包含硬绿泥石的现象。铝直闪石常与十字石共生，两者紧相伴随，从两者之间的关系看，十字石稍晚于铝直闪石，但它们仍属同一变质阶段的产物。

片岩和角页岩中的主要矿物共生组合表现为红柱石+白云母、十字石+红柱石+白云母、十字石+黑云母、十字石+白云母、十字石+白云母+黑云母、十字石+铁铝榴石+黑云母、十字石+铁铝榴石（+红柱石）、铝直闪石+铁铝榴石、铝直闪石+黑云母（+红柱石）岩石组合。后四种组合主要产于富铁铝的泥质岩石中。

此变质带中由凝灰质粉砂岩变质而成的变粒岩，主要共生矿物组合由绿帘石+黑云母+斜长石、角闪石+绿帘石+黑云母+斜长石、角闪石+绿帘石+斜长石、角闪石+黑云母+斜长石。

5. 矽线石-白云母带

紧靠岩体的片岩和角页岩，矽线石和白云母共生，构成宽约 30~100m 的变质带。矽线石一般是由红柱石转变而成，有的呈放射状，有时伴随有白云母。也有黑云母被矽线石和白云母替带的现象。

变质泥砂页岩中的主要矿物共生组合有矽线石+白云母、矽线石+铁铝榴石、矽线石+白云母+铁铝榴石、矽线石+白云母+黑云母、矽线石+黑云母+铁铝榴石。

变质相的变化反映了温压条件和原岩性质的差异。按划分出的原岩类型，在不同变质相条件下，同类岩石中的 Al_2O_3、Fe_2O_3 和 FeO 的含量呈现不同的变化特征。如富含泥质岩类和含铁泥质岩类，由高绿片岩相到角闪岩相，随着变质程度增高，Fe_2O_3 含量增大，而 Al_2O_3 和 FeO 含量减少。大致证明，由于温度急剧增高，岩石中的铝质可能出现一定程度的活化并发生部分迁移，致使铝含量降低，而铁则由于氧逸度增高，一部分低价铁变为高价铁，因而 Fe_2O_3 含量增大。

第四节　变质岩相分析

一、麻粒岩相区域变质作用

麻粒岩相区变质岩系主要组成矿物是紫苏辉石、透辉石、石榴子石、斜长石、普遍角闪石，金云母含量较低，石英含量普遍偏低。不同层位变质矿物组合尚有差异：沙厂、苇子峪和大漕地区中下部变质岩石组合具麻粒岩相特征，划归麻粒岩相区；大漕地区上部变质岩石组合缺少紫苏辉石，相应划归高级角闪岩相区。

1. 麻粒岩相区

以紫苏辉石普遍存在为标志的麻粒岩，集中分布在沙厂、苇子峪和大漕等地区，其面积约占密云岩群的87%。

麻粒岩相区经常见到的代表性岩石有二辉麻粒岩、黑云（角闪）二辉麻粒岩、黑云石榴子苏麻粒岩、辉石磁铁石英岩和角闪二辉石岩等。其矿物共生组合如下：①斜长石+紫苏辉石+透辉石±角闪石；②紫苏辉石+透辉石±斜长石；③斜长石+石英+紫苏辉石+黑云母；④斜长石+石英+紫苏辉石+石榴子石+黑云母。

在黑云石榴紫苏麻粒岩和榴辉黑云变粒岩中，可见到黑云母被石榴子石置换呈残片状形态的现象。石榴子石呈串珠状、树叉状集合体继承了黑云母的某些晶形特征。由于上述矿物共生组合紫苏辉石和水同时出现，紫苏辉石多已绿泥石化。其中钾长石不是以半自形或他形的变晶颗粒形式出现，而多成为反条纹长石中的钾长石条纹。

本区形成麻粒岩矿物组合的所有可能的化学反应都有角闪石或黑云母的参加。温克勒（1973）认为包含着普通角闪石和（或）黑云母的所有反应都是“连续”的反应，换言之，是滑动平衡反应，即在给定总压和水压下，反应在一定温度下开始，而所有的平衡相都会存在于某一温度范围内。镁铁矿物的 Fe/Mg 比值将随温度而改变。虽然斜方辉石由角闪石或黑云母分解产生，因其数量不断减少达到平衡，反应不需要进行到完全。密云岩群麻粒岩常含有一定量含水矿物普通角闪石或黑云母，紫苏辉石+透辉石+普通角闪石+黑云母+斜长石的五相共存的岩石多见。这表明本区的麻粒岩相变质过程中 P_{H_2O} 偏高，温度还没有达到足以使含水矿物最后消失的程度。

采用 Lindsley（1985）基于实验测定的辉石温度计进行温度计算，并运用吴春明（2013）对辉石温度计极限值的估算方法对计算结果进行校正（表 23-5）。

表 23-5　辉石电子探针测试数据及其计算一览表

Table 23-5　The electron probe test datum and its calculation results

样品号	DQ13		DQ02		px60		px89		q1		q5		q8		056	
岩石名称	角闪二辉岩		角闪二辉岩		榴辉麻粒岩		角闪麻粒岩		麻粒岩		麻粒岩		麻粒岩		榴辉磁铁石英岩	
矿物	Opx	Cpx	Opx	Cpx	Opx	Cpx	Opx	Cpx	Opx	Cpx	Opx	Cpx	Opx	Cpx	Opx	Cpx
SiO_2	54.45	50.40	52.62	49.90	49.80	51.20	50.60	49.80	52.10	49.80	51.15	52.05	50.90	50.70	50.80	52.10

续表

样品号	DQ13		DQ02		px60		px89		q1		q5		q8		056	
TiO_2	0.08	0.30	0.06	0.14	0.05	0.39	0.07	0.11	0.06	0.11	0.09	0.29	0.06	0.27	0.02	0.06
Al_2O_3	3.74	3.64	3.01	4.32	3.24	4.16	2.73	3.54	2.37	3.54	2.57	4.06	2.29	3.89	2.27	3.66
Cr_2O_3			0.06	0.17					0.05		0.79	0.20	0.04	0.09	0.06	0.01
FeO	6.97	2.42	16.68	6.44	16.93	5.47	17.97	6.36	17.83	6.36	18.05	6.03	19.94	6.22	22.02	7.45
MnO	0.19	0.11	0.46	0.40	0.28	0.11	0.45	0.22	0.37	0.22	0.35	0.15	0.42	0.17	0.36	0.14
NiO	0.00	0.00			0.00	0.00	0.00	0.00	0.00	0.00	0.00	0.00	0.00	0.00	0.00	0.00
MgO	32.50	17.40	23.38	14.33	23.80	14.25	20.10	13.15	19.75	13.15	19.50	12.25	20.75	12.25	19.75	12.00
CaO	0.64	21.50	0.38	20.95	1.27	18.90	1.25	19.75	1.10	19.75	1.10	19.75	1.60	20.25	0.80	18.75
Na_2O	0.10	0.59	0.08	0.61	0.07	1.06	0.54	1.40	0.28	1.40	0.23	1.18	0.20	0.98	0.35	1.08
K_2O	0.02	0.00	0.01	0.05	0.00	0.00	0.03	0.00	0.08	0.00	0.23	0.09	0.05	0.04	0.04	0.04
总量	100.29	99.78	100.09	99.01	99.93	99.56	99.57	99.86	99.77	99.86	100.35	100.60	100.46	99.45	100.10	99.86
以6个氧原子为基础计算离子数																
Si	4.01	4.03	4.01	4.03	1.90	1.96	1.99	1.93	2.05	1.93	2.02	2.00	1.96	1.97	1.97	2.02
Ti	0.00	0.00	0.00	0.00	0.00	0.01	0.00	0.00	0.00	0.00	0.00	0.01	0.00	0.01	0.00	0.00
Al	1.91	1.88	1.91	1.88	0.15	0.19	0.13	0.16	0.11	0.16	0.12	0.18	0.10	0.18	0.10	0.17
Cr	0.00	0.01	0.00	0.01	0.00	0.00	0.00	0.00	0.00	0.00	0.02	0.01	0.00	0.00	0.00	0.00
Fe	0.15	0.16	0.15	0.16	0.06	0.00	0.00	0.07	0.00	0.07	0.00	0.00	0.00	0.00	0.00	0.00
Mn	0.00	0.00	0.00	0.00	0.48	0.17	0.59	0.14	0.58	0.14	0.59	0.19	0.64	0.20	0.71	0.24
Ni	0.04	0.08	0.04	0.08	0.01	0.00	0.01	0.01	0.01	0.01	0.01	0.00	0.01	0.01	0.01	0.00
Mg	0.17	0.00	0.17	0.00	0.00	0.00	0.00	0.00	0.00	0.00	0.00	0.00	0.00	0.00	0.00	0.00
Ca	0.01	0.00	0.01	0.00	1.35	0.81	1.18	0.76	1.16	0.76	1.15	0.70	1.19	0.71	1.14	0.69
Na	0.00	0.00	0.00	0.00	0.05	0.77	0.05	0.82	0.05	0.82	0.05	0.81	0.07	0.84	0.03	0.78
K	1.70	0.97	1.70	0.97	0.01	0.08	0.04	0.11	0.02	0.11	0.02	0.09	0.01	0.07	0.03	0.08
总量	7.98	7.13	7.98	7.13	4.00	4.00	4.00	4.00	3.99	4.00	3.98	3.99	4.00	4.00	4.00	4.00
X_{Fe}^{opx}					0.26	0.18	0.33	0.15	0.33	0.15	0.34	0.21	0.35	0.22	0.38	0.26
Na^{M2}	10.00	0.00	20.00	0.00	0.01	0.08	0.04	0.11	0.02	0.11	0.02	0.09	0.01	0.07	0.03	0.08
Ca^{M2}	0.09	0.00	0.09	0.00	0.05	0.77	0.05	0.82	0.05	0.82	0.05	0.81	0.07	0.84	0.03	0.78
Ca^*	0.01	0.04	0.01	0.04	0.05	0.84	0.05	0.92	0.05	0.92	0.05	0.89	0.07	0.91	0.03	0.85
K_D^*	0.02	0.86	0.02	0.86	0.17	0.00	0.09	0.00	0.09	0.00	0.11	0.00	0.09	0.00	0.16	0.00
温度计算结果																
TBKN/K	1223	1229	1128	1138	1232	1244	1037	1047	1036	1045	1100	1112	1052	1064	1138	1151
TBKN/℃	986	956	891	865	995	971	800	774	799	772	863	838	815	791	901	878

注：数据采用Lindsley（1985）基于实验测定的辉石温度计进行温度计算，并运用吴春明（2013）对辉石温度计极限值的估算方法对计算结果进行校正；氧化物含量单位为%。

2. 高级角闪岩相区（$P\text{-}€_{B3}$）

该岩相区以棕绿色普通角闪石、石榴子石和透辉石的平衡共生为标志，在区域上分布于密云西北部的妈妈头山、不老屯、车道岭、田庄、上甸子一带和密云东南部的墙子路及平谷关上一线以东的地带，总面积相当于密云岩群的12%。

高级角闪岩相区常见有石榴斜长辉石岩、榴辉斜长角闪岩、角闪榴辉斜长片麻岩、榴辉黑云变粒岩和角闪辉石岩等。代表性矿物共生组合为：①斜长石+透辉石+棕绿色普通角闪石；②斜长石+透辉石+石榴子石+棕绿色普通角闪石；③透辉石+棕绿色普通角闪石±斜长石；④斜长石+石英+石榴子石+黑云母。

在高级角闪岩相区的榴辉斜长角闪岩中，石榴子石与透辉石处于共生状态，棕绿色普通角闪石和透辉石、石榴子石之间虽然多是平直或圆滑的共生边界，但也常见到石榴子石或透辉石晶体包裹角闪石或斜长石残片的现象。并且，透辉石+石榴子石总量与角闪石之间在数量上存在着反消长关系，即角闪石含量越高，石榴子石+透辉石总量就越低，反之亦然。

这一反应具有滑动平衡的性质，因此造成相应地区的变质基性岩种类繁多，如斜长角闪岩、榴辉斜长角闪岩、石榴角闪斜长辉石岩、石榴斜长辉石岩等（表23-6）。

表23-6　石榴子石-黑云母对温度压力计计算结果

Table 23-6　Calculation results of thermo-pressure meter of garnet and biotite

样品 98717	含量/%							
	SiO_2	TiO_2	Al_2O_3	FeO	MnO	MgO	CaO	Na_2O	K_2O
石榴子石	36.2	0.3	20.14	29.38	5.32	3.72	4.69		
黑云母	36.94	2.3	16.45	18.83	0.16	13.96	0.7	0.32	6.96
温度	572℃								
压力	445MPa								

（据金文山，2000）

其中运用了吴春明等（2004）采用经验标定方法的黑云母石榴子石-黑云母压力计，计算压力为445MPa；运用Holdaway（2000）的黑云母石榴子石地质温度计计算温度为572℃。

由于温度、压力计算采用的黑云母—石榴子石温度计和黑云母—石榴子石压力计中的主要矿物黑云母是岩石退变质过程中出现的，其所标定的温压范围应为退变质过程中的温度和压力条件，由此也可以推断峰期变质作用的温压条件应略高于地质温压计的计算结果（表23-7）。

表23-7　角闪石的化学成分

Table 23-7　The chemical components of amphibole

序号		Q-10	DQ-11	97175	97170	$PZII-D^1_{56}$	$PZVD^1_{28}$	$PZVD^1_{18}$	$PZII-D^1_{258}$	97252	7233
SiO_2	含量/%	41.1	41.7	42.09	41	46.9	44.6	45.1	44.1	40.35	41.67
TiO_2		1.88	2.35	1.63	2.29	0.56	0.62	0.4	0.73	0.64	0.71
Al_2O_3		12.27	11.87	11.06	11.47	8.9	12.87	13.33	10.58	11.92	13.83
Fe_2O_3		3.85	3.7			6.39	5.96	4.81	6.94		
FeO		10.48	12.51	15.48	16.95	11.35	11.73	10.34	11.94	22.62	18.71
MnO		0.12	0.1	0.08	0.08	0.32	0.28	0.4	0.32	0.39	0.26
MgO		11.33	10.83	11.89	10.59	11.75	9.9	11.6	10	7.98	8.61
CaO		11	10.5	11.69	11.48	10.6	11	10	10.5	12.06	12.03
Na_2O		2.05	1.85	1.64	1.47	1.084	1.3	1.64	1.452	1.29	1.3
K_2O		1.5	1.3	1.58	1.72	0.9	0.81	0.74	1.08	0.68	0.83
H_2O^+		1.43	1.12			1.59	1.37	1.36	1.64		
F		0.288	0.238								
O~F		0.12	0.1								
总和		97.43	97.93	97.14	97.05	100.34	100.44	99.72	99.28	96.92	97.95
按24个（O、OH、F）原子计算离子数											

续表

Si	6.21	6.38	6.6666	6.5703	6.9	6.63	6.58	6.63	6.5623	6.4525	
Al^{IV}	1.79	1.62	1.3334	1.4297	1.1	1.37	1.42	2.37	2.4377	1.5475	
Al^{VI}	0.36	0.51	0.7237	0.7282	0.44	0.94	0.82	0.5	0.8508	0.9857	
Ti	0.26	0.27	0.1904	0.2697	0.06	0.04	0.07	0.08	0.0782	0.0967	
Fe^{3+}	0.43	0.42	—	—	0.71	0.53	0.66	0.78	—	—	
Fe^{2+}	1.31	1.6	2.0476	2.2736	1.4	1.27	1.45	1.5	2.9437	2.4208	
Mn	0.02	0.01	0.0095	0.0105	0.04	0.05	0.03	0.04	0.0528	0.0335	
Mg	2.54	2.49	2.7523	2.4277	2.58	2.54	2.18	2.24	1.9166	1.9832	
Ca	1.77	1.73	1.9809	1.9749	1.67	1.57	1.74	2.69	2.0929	2.0391	
Na	0.39	0.55	0.4952	0.4624	0.31	0.47	0.37	0.42	0.4107	0.391	
K	0.29	0.26	0.3238	0.3468	0.17	0.14	0.15	0.22	0.1369	0.1675	
Si	6.21	6.38	6.6666	6.5703	6.9	6.63	6.58	6.63	6.5623	6.4525	
Al^{IV}	1.79	1.62	1.3334	1.4297	1.1	1.37	1.42	2.37	2.4377	1.5475	
Al^{VI}	0.36	0.51	0.7237	0.7282	0.44	0.94	0.82	0.5	0.8508	0.9857	
Ti	0.26	0.27	0.1904	0.2697	0.06	0.04	0.07	0.08	0.0782	0.0967	
Fe^{3+}	0.43	0.42	—	—	0.71	0.53	0.66	0.78	—	—	
Fe^{2+}	1.31	1.6	2.0476	2.2736	1.4	1.27	1.45	1.5	2.9437	2.4208	
Mn	0.02	0.01	0.0095	0.0105	0.04	0.05	0.03	0.04	0.0528	0.0335	
Mg	2.54	2.49	2.7523	2.4277	2.58	2.54	2.18	2.24	1.9166	1.9832	
Ca	1.77	1.73	1.9809	1.9749	1.67	1.57	1.74	2.69	2.0929	2.0391	
Na	0.39	0.55	0.4952	0.4624	0.31	0.47	0.37	0.42	0.4107	0.391	
K	0.29	0.26	0.3238	0.3468	0.17	0.14	0.15	0.22	0.1369	0.1675	

注：样品1~4属大漕-沙厂表壳岩系；样品5~10属四合堂表壳岩系。（据金文山，2000）

二、角闪岩相区变质作用

被限定在半城子断裂和崎峰茶断裂之间的四合堂-山神庙地区，以黑云角闪质岩石为主。岩石组合主要有角闪石岩、斜长角闪岩、黑云角闪斜长片麻岩、角闪黑云斜长片岩、角闪黑云变粒岩和角闪磁铁石英岩等，在西湾子地区还见有黑云二长变粒岩、浅粒岩（图23-8）。代表性的矿物共生组合为：①斜长石+绿色普通角闪石+透辉石；②斜长石+绿色普通角闪石+黑云母；③斜长石+绿色普通角闪石+石榴子石；④斜长石+石英+绿色普通角闪石；⑤斜长石+石英+黑云母；⑥斜长石+石英+绿色普通角闪石+黑云母。其中组①和③较少见。

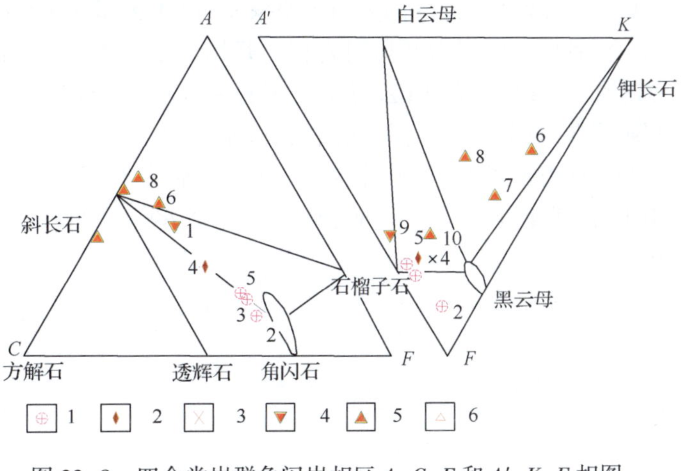

图23-8 四合堂岩群角闪岩相区 $A-C-F$ 和 $A'-K-F$ 相图

Fig. 23-8 $A-C-F$ and $A'-K-F$ facies diagram of amphibolite facies of Sihetang rock group

1—斜长角闪岩；2—黑云角闪斜长片麻岩；3—角闪黑云斜长片麻岩；4—角闪变粒岩；5—黑云二长变粒岩；6—黑云斜长变粒岩

在变质基性岩中，绿帘石、阳起石消失，普通角闪石开始出现的温度超过500℃。四合堂岩群变质岩基性岩主要由斜长石和绿色普遍角闪石组成，含有少量温度更高的透辉石，反映变质温度已远超500℃，可能在600℃左右。

正常情况下较高压力有利于变质基性岩中铁铝榴石的形成，反之石榴子石在低压相系的变质基性岩中非常少见的，或者不存在。四合堂地区变质岩中很少见到石榴子石的出现，说明其变质压力偏低。总之，四合堂地区变质岩发生了中等偏低压力温度偏高的角闪岩相区域变质作用。

伴随着角闪岩相的变质作用，四合堂地区变质岩变形强烈，形成了大型复式向斜构造。在变质变形过程中，变质火山岩达到了塑性状态，但温度和压力条件却没有使其发生塑性流动。因此，岩层厚度在褶皱过程中变化不大。如果把条带状硅铁建造作为标志层，那么在一个较小的范围内，可看出铁矿层在走向和倾向上延伸稳定，剖面上具有比较明显的板状、层状、似层状特征。褶皱过程中岩层厚度变化小且次级褶皱宽度大的特点使四合堂地区变质岩具有单斜构造的假象。

三、晚古生代—早中生代变质作用

变质岩石学研究认为变形作用与变质作用相互制约，二者对应关系十分明显，变形越严重，变质相越高，二者处于统一热应力条件下相伴进行。

随着印支旋回构造演化，岩石韧性变形程度由增到减，相应岩石伴随发生着进变质到退变质过程。印支初期岩石中出现前构造期蓝晶石和十字石变斑晶，并具强塑性变形，硬绿泥石、黑云母和白云母的优选方位均平行于褶叠层的顺层面理，硬绿泥石具定向排列。变形带强度不同控制了不同变质强度带的空间展布。伴随变质作用，岩石动力变质现象十分明显，如深部（下古生界及更早）地层遭受强固态流变及相当部分岩层的糜棱岩化。印支晚期变质作用与面理褶皱的陡倾轴面褶劈理有关，变质矿物以云母类为主，间或有硬绿泥石，规模小、变质程度低、具退变质性质。

受燕山期变质作用变质泥岩中亦可划分为硬绿泥石-绿泥石带、黑云母带、红柱石带，空间分布受北岭向斜控制，蓝晶石、硬绿泥石、红柱石等特征变质矿物均具定向性。硬绿泥石多呈放射状、蒿状集合体。岩石塑性变形不明显。

该期变质作用的多期性与构造旋回演化对应关系十分明显，特征变质矿物同变形期变晶矿物特征，主变质期之后随构造作用岩石发生退变质作用，其与主变质期叠加变质的关系尚不清楚。

北京地区晚古生代—早中生代各类浅变质岩的形成与深部热液活动有关。构造环境的变化导致构造变形特点的不同，也控制着热液活动的强弱和分布，表现在不同变质带内变质程度不一样。

本区绿片岩相硬绿泥石带变质作用的温压条件，以叶蜡石、硬绿泥石首次出现为代表。变质反应为：①高岭石+石英=叶蜡石+水；②绿泥石+高岭石=硬绿泥石+石英。

黑云母带以绿泥石+绢云母=硬绿泥石+黑云母+石英+水的反应表示。

以十字石出现作为低角闪岩相的开始，十字石与蓝晶石的共生，硬绿泥石消失，十字石形成的变质反应为：①绿泥石+白云母=十字石+黑云母+石英+水；②硬绿泥石+红柱石=十字石+石英+水；③硬绿泥石+石英=十字石+铁铝榴石+水。

各变质相带的变质条件为：①硬绿泥石带 $T=350\sim450$℃；②黑云母带 $T=450\sim575$℃，$P>0.25$GPa；③十字石带 $T=592$℃，深度2.69～3.11km，变质相可达角闪岩相（据万永平，2007），$P>0.5$Gpa。

第二十四章　其他变质作用

区内涉及的其他变质作用主要有接触变质作用、低温动力变质和气液变质作用。

第一节　接触变质作用

接触变质作用包括热接触变质作用、低温交代变质作用及气-液变质或蚀变作用三类，均与岩浆活动相关，与内生矿产的成矿作用关系密切，同时也是重要的找矿标志。

北京地区岩浆活动具多期次和多旋回活动的特点，除在太古宙变质深成侵入岩及其围岩中有多期中高级区域变质作用的叠加改造不易被鉴别外，在元古宙及其之后的各期次岩浆活动中均有不同程度的热力变质岩形成，呈规则和不规则分带特点。其中以晚三叠世—早白垩世热力变质作用表现最为强烈，同时该时期也是区域上最重要的成矿期。

一、热接触变质作用

热接触变质作用是叠加在区域变质岩之上的，大部分接热变质岩都发育有明显的片理定向组构，形成的各种角（页）岩、片岩、大理岩与区域变质岩很难区分，只能从产状及矿物组合上加以判别。其特征描述如下：

1. 角（页）岩类

（1）红柱石角（页）岩类

暗灰色—灰黑色，块状构造，肉眼可见红柱石变斑晶，大小在（2～3）mm×（7～8）mm之间，排列无定向，有的红柱石变斑晶，呈放射状集合体，形如菊花，故有人称之为菊花石。岩石可见变基质，其为斑状变晶结构，可见显微鳞片花岗变晶结构或放射状花岗变晶结构。

（2）硬绿泥石化角岩

岩石新鲜面为灰绿色到暗绿色，风化呈褐红色。硬绿泥石为变斑晶，镜下为蒿束状集合体，排列无定向性，经常风化呈赤铁矿，变基质以石墨尘点及细粒石英为主。

2. 接触片岩

（1）红柱石云母片岩

其与区域变质的红柱石片岩的区别主要是：红柱石粒度大，具环带结构，核心是具有粉红色多色性的红柱石，外环是不带多色性的无色红柱石。可与矽线石共生，而不与蓝晶石共生。另外变基质中黑云母鳞片也较大。红柱石普遍发育叶蜡石化。

（2）矽线石云母片岩

该类岩石可逐渐变为石榴子石红柱石片岩。有一系列过渡的岩石种属；含矽线石的石榴子石云母片岩、石榴子石红柱石矽线石片岩等。岩石呈暗灰色，灰黑色，具片状构造至片麻状构造。中细粒纤维-鳞片变晶结构或斑状变晶结构。

3. 大理岩类

区内离岩体较远处，出现透闪石大理岩、透辉石大理岩和含橄榄石大理岩等。

（1）透闪石大理岩

岩石为灰色-暗灰色，透闪石灰白色纤维状，常为0.5～1cm的条带或直径为2～10cm大小不等团块。中、细粒纤维状变晶结构，主要矿物是白云石，方解石次之，透闪石含量在10%左右，分布不均，少者不足5%。偶见白云母及斜长石，含量不超过2%。其原岩为硅质白云岩。

（2）透辉石大理岩

岩石呈浅绿、浅灰、浅黄褐，条带状或块状、斑杂状构造。中细粒变晶结构。矿物成分以方解石为主，有时有白云石，碳酸盐矿物一般80%以上。特征矿物是透辉石，为柱状晶体，常相对集中成条带或形状各异的团块，透辉石柱粒度在0.3~0.8mm，含量在10%左右。另含有透闪石、绿帘石、黑云母、钙铁榴石、符山石，斜长石等，含量不超过15%。

（3）含橄榄石大理岩

岩石呈暗灰色、浅黄褐色。块状构造可见变余层理。花岗变晶结构，主要矿物是方解石，占比约60%，其次是白云石，约20%。典型的变质矿物是镁橄榄石，无色圆形及他形粒状，粒径0.1~0.3mm，含量变化较大，在8%~15%之间。常沿裂隙发育蛇纹石化，形成蛇纹石大理岩。

二、接触交代变质作用

主要由中酸性岩浆侵入围岩温度升高引起热变质，当条件有利时，结晶晚期析出的挥发组分，聚集能量，携带物质成分进入围岩中，并与围岩发生交代作用，改变了围岩的化学成分，变质生成新矿物。常见于侵入岩接触带及其附近，除围岩外（外接触带），有时侵入岩体本身的边缘部分（内接触带）也发生变质作用。

接触带的形态及裂隙构造特征是聚集产生接触交代作用的必要条件之一，但更主要的是侵入体和围岩的成分及性质。一般花岗闪长岩，花岗岩等中酸性侵入体和碳酸盐类接触时，最易发生交代作用。中生代中酸性侵入岩体与碳酸盐岩围岩，在侵入过程中发生交代变质作用，形成矽卡岩带。该带中蚀变矿物组合、蚀变带形态，规模，并常叠加有后期岩浆热液蚀变作用，伴有Fe、Mo、Cu和Pb、Zn矿化。如大榆木沟铁铜矿、延庆石槽铜钼矿、平谷石塘峪铜钼矿、延庆东三岔锌铁矿。

矽卡岩是由中酸性侵入岩与碳酸盐类岩石（石灰岩或白云质石灰岩）接触时，发生接触交代作用形成的岩石，可进一步划分为钙质矽卡岩和镁质矽卡岩。

1. 钙质矽卡岩

主要由钙铝石榴子石—钙铁石榴子石系列岩石组成，是在浅处及中等深度条件下，由石灰岩经交代作用后形成。

矽卡岩表征变化很大，颜色取决于矿物成分及粒度，常见为暗色、暗绿色或暗棕色等，少数含硅灰石等浅色矿物较多者为浅灰色。结构变化复杂，为不均匀粒状变晶结构、纤状变晶结构。矽卡岩多由1~3种矿物组成，晶形完整，颗粒粗大，相对密度较大，有时呈疏松多孔，但有时呈细粒致密状。其构造一般为块状或斑杂状，少数情况下，由于顺层交代形成条带状、角砾状，如昌平西湖锰矿发现角砾状矽卡岩。

矽卡岩常出现于侵入体的内接触带某一地段，或其附近200~400m左右范围内，与岩体形态空间形态关系密切，主要发育在接触带内凹部位或围岩体下垂处地段（如石槽铜矿）。与矽卡岩形成有关的侵入体，最常见的为花岗闪长岩、花岗岩，其次为石英闪长岩，正长岩等，多数为中浅成的小侵入体或大侵入体的分支凸出部位。

主要矿物有：透辉石、钙铝榴石、符山石、绿帘石、绿泥石等。以石榴子石矽卡岩为代表，常呈粉红色，或因绿帘石化常呈绿色或暗绿色，致密块状，粒状变晶结构或交代嵌晶结构。主要由钙铝石榴子石及其蚀变矿物绿帘石组成，含量70%~80%，其次有透辉石、磷灰石、榍石、金红石、钛铁矿以及长石交代残骸。钙铝石榴子石为无色至淡黄色，粒径0.1~1mm不等，常被绿帘石、金云母交代。石榴子石矽卡岩中经常含有透辉石，含量5%~10%不等，其含量大于50%时构成石榴子石透辉石矽卡岩。这里的透辉石交代斜长石和钾长石，呈粒状或他形粒状，本身又被石榴子石交代。此外尚有钙矽卡岩化石英二长岩或称蚀变石英二长岩，其特征钾长石具固溶分解条纹，黏土化，斜长石双晶基本消失，边部为钠长石成分，中心黏土化或绢云母化，两种长石外形均模糊，暗色矿物褪色，并常出现矽卡岩矿物：透辉石、石榴子石、符山石等。

该类矽卡岩是钼矿化的最好蚀变围岩，铁、铜矿化选择透辉石交代，未发现锌矿化。

2. 镁质矽卡岩

主要分布在矽卡岩带外接触带，以镁橄榄石矽卡岩为代表，绿黑色、暗绿色，蚀变轻者呈褐色、灰白色。块状构造，粒状变晶结构。主要有橄榄石、蛇纹石组成，含量70%～80%，其次为绿泥石、磁铁矿、方解石。镁橄榄石在镜下呈无色透明，方形圆粒状，具裂理，粒径0.01～0.8mm不等，交代、包裹方解石，其自身又被纤维状蛇纹石交代，但保留其假象，交代成分较多时形成蛇纹岩。

此外，尚有透辉石矽卡岩，镁橄榄石透辉石矽卡岩以及镁矽卡岩化大理岩。本类岩石中除镁橄榄石、透辉石外，尚有镁铁尖晶石、水镁石、硼镁铁矿等。镁矽卡岩是铁、铜锌矿化的最好蚀变围岩。

在形成时间上镁质矽卡岩要先于钙质矽卡岩；矽卡岩后期，由于水热气化作用分别产生变矽卡岩，形成一系列含水矽卡岩矿物。在空间上，镁矽卡岩主要分布于外带，钙矽卡岩主要分布于内带，尤其在围岩蚀变发育地段，水平分带比较明显（表24-1），由于岩枝沿接触带穿插，两类矽卡岩局部相互交替产出，其余多数地段矽卡岩发育不完全。

表24-1　大榆木沟地区矽卡岩水平分带表

Table 24-1　The skarn horizontal zoning of Dayumugou region

<table>
<tr><td colspan="2">矽卡岩分带</td><td colspan="3">镁矽卡岩（外接触带）</td><td colspan="3">钙矽卡岩（内接触带）</td></tr>
<tr><td colspan="2">分带编号</td><td>6</td><td>5</td><td>4</td><td>3</td><td>2</td><td>1</td></tr>
<tr><td colspan="2">矽卡岩分带</td><td>镁橄榄石矽卡岩化大理岩</td><td>磁铁闪锌矿镁橄榄石矽卡岩化大理岩</td><td>磁铁黄铜矿化带</td><td>透辉石石榴子石矽卡岩</td><td>石榴子石透辉石矽卡岩</td><td>绿泥石化蚀变石英二长岩</td></tr>
<tr><td rowspan="4">含矿性</td><td>Fe</td><td rowspan="4"></td><td style="background:#ccc"></td><td style="background:#ccc"></td><td style="background:#ccc"></td><td rowspan="4"></td><td rowspan="4"></td></tr>
<tr><td>Cu</td><td></td><td style="background:#ccc"></td><td style="background:#ccc"></td></tr>
<tr><td>Zn</td><td style="background:#ccc"></td><td style="background:#ccc"></td><td style="background:#ccc"></td></tr>
<tr><td>Mo</td><td></td><td style="background:#ccc"></td><td style="background:#ccc"></td></tr>
</table>

第二节　低温动力变质作用

北京地区低温动力变质作用主要发生在较大规模的断裂带，诸如半城子断裂、石城断裂和崎峰茶断裂都是形成时代较早并且多期活动的古断裂。晚期断裂活动使岩石发生脆性变形，形成以粒径减小为特征的碎裂岩、碎斑岩和糜棱岩。早期断裂产生于曾经的地下深部，那些部位因机械破碎作用所生成的热量难以发散，结果又使破碎岩石发生了一定程度的重结晶，形成糜棱片岩和各种变晶糜棱岩。断裂带及其两侧的热液活动强烈，使变质岩系一般发生了相当于绿片岩相的叠加变质。

密云地区半城子断裂、石城断裂及崎峰茶断裂带内岩石中产生交代作用，最强部位常见单一矿物的多晶条带，其中的矿物常具动态重结晶特征，这反映了在退变质作用过程中岩石曾经历过韧性变形阶段。而区内大量的糜棱岩产出，显示曾遭受过强烈的韧性剪切作用的改造。

低温动力变质岩的变质矿物组合，以云蒙山北部的四合堂、青石岭、四道沟、柏查子一带的糜棱岩中发育最典型。在这一地区，斜长角闪岩被改造成各种构造片岩（黑云片岩、角闪片岩等）；变粒岩被改造成黑云斜长片麻岩；磁铁石英成为磁铁石英片岩；浅粒岩被改造成长英质片麻岩。岩石中所有组成矿物都发生了重大变化。糜棱岩矿物组合与变形强度关系密切。在弱应变域，主要矿物组合为：角闪石+斜长石+黑云母+石英，其中部分角闪石、斜长石晶体出现破碎，石英定向拉长，长宽比例为1/3～1/5。岩石具片麻状构造。在强应力域，角闪石明显向黑云母转变，长石产生核幔构造，石英定向拉长，长宽比例最大可达1/20。最为强烈部位，黑云母逐渐向绿泥石转化，斜长石向绢云母和石英转化的现象，这时的矿物组合为：角闪石+黑云母+绿泥石+绢云母和斜长石+石英。即暗色矿物与浅色矿物发生了明显的分集作用。弱应变带岩石中硅质交代的蠕英结构、穿孔结构多见，石英多晶条带较粗，晶体粒度 0.8mm 左右，直线型接触，具静态重结晶特点；微斜长石交代斜长石强烈；斜长石为更长石（An=15 左右）；普通角闪石呈黄绿色—蓝绿色，部分具钠质交代的蓝色反应边。各种矿物均显定向，部分斜长石边缘圆化，产生细化基质。强应变带，石英、长石核幔构造发育；长石大量出现机械双晶；石英、长石发生重结晶，晶体内亚颗粒结构发育；角闪石产生机械双晶；黑云母产生膝折。硅质交代的蠕英结构、穿孔结构消失，代之以细而长的石英、长石的多晶条带。晶体粒度小于 0.03mm，具动态重结晶的多边形弯曲的外形特征。

韧性剪切带中矿物的变质特征表明，随着韧性变形的加强，硅质、钾质、钠质交代作用均增强，以致发生局部重熔作用。反映出韧性剪切条件下各种交代作用明显增强。

在云蒙山岩体北缘，石城闪长岩和云蒙山花岗岩的相继侵入，使中元古界长城系强烈变形，形成了紧闭的倒转向斜构造。褶皱同时，沉积岩系受接触热变质的影响产生了接触变质带，由内向外依次是橄榄石带、透辉石带和透闪石带。

接触变质作用涉及西湾子地区变质岩系，使黑云角闪岩又发生了一定程度的重结晶——在片岩片理面上形成由粗大黑云母叶片和蓝绿色普通角闪石构成的长柱状变斑晶。变斑状角闪石大多平行片理方向，也有斜交片理的现象。

本区晚期岩体很多。一般晚期岩体的侵入多会引起早期地层的隆起、褶皱或断裂，并往往形成宽窄不同的角岩化带及热液蚀变带。

1）接触变质岩石和岩体的侵入围岩产生不同程度的变质，而且距离岩体愈近变质程度愈深，反之愈浅。

2）主要变质矿物为长石、石英、透闪石、透辉石，其次是黑云母、白云母、绿帘石、阳起石，偶见方柱石和堇青石。在内接触带的大理岩捕虏体中见斜硅镁石和镁橄榄石等接触变质的表性矿物。

3）从外接触带到内接触带透闪石、透辉石、橄榄石依次出现，反映了越接近岩体温度越高，形成由低温到高温矿物的递增变质特征。变质作用由岩体向外具有接触变质晕分带性。由于本区岩石对变质作用中温度、压力条件改变而反映的灵敏度由硅质白云岩→泥质粉砂岩→长石石英砂岩→石英砂岩依次降低。根据反映最灵敏的硅质白云岩变质过程中依次出现的矿物共生组合，说明岩体内接触带到外接触带可分为 3 个等温变质带：即橄榄石带、透辉石带、透闪石带。

①$5CaMg(CO_3)_2+8SiO_2+H_2O=Ca_2Mg_5(Si_4O_{11})_2(OH)_2+3CaCO_3+7CO_2$

白云石　　石英　　透闪石　　方解石

②$Ca_2Mg_5(Si_4O_{11})_2(OH)_2+3CaCO_3=4CaMgSi_2O_6+CaMg(CO_3)_2+H_2O+CO_2$

透闪石　　方解石　　透辉石　　白云石

③$CaMgSi_2O_6+CaMg(CO_3)_2=2Mg_2SiO_4+4CaCO_3+2CO_2$

透辉石　　白云石　　镁橄榄石　　方解石

本区橄榄石带中仍然存在着透辉石带和透闪石带的矿物组合，可能是因岩体内基础的 CO_2 分压较高而又无法大量释放所致。外接触带透辉石带和透闪石带中的矿物组合基本达到平衡，可能是在相对开放体系中 CO_2 不断散失的结果。

4）由于碳酸盐岩变质作用的体系是双变平衡体系，在总压力固定不变时，体系的平衡取决于温度和 CO_2 分压。在一定的流体压力下，硅质白云岩随着温度的升高能依次生成滑石、透闪石、透辉石和镁橄榄石等钙硅酸盐矿物。以上橄榄石带、透辉石带、透闪石带表现出由岩体外向温压逐渐降低的趋势。

5）变质温度梯度较大。据文献资料，生成透闪石、透辉石和橄榄石所需要的温度、压力实验资料，可粗略地估计本区温压条件，估计橄榄石带的温度为 700℃左右，透闪石带的形成温度为 600℃左右。在接触热变质岩系的最大出露宽度不超过 3km 的情况下，变质温度相差 100℃，说明如此大的温度梯度是区域变质作用所不能达到的。

接触变质岩具有两期构造变形作用。第一期是在近南北向挤压应力作用下产生的压缩性变形，产生轴向近东西、轴面向北倾的倒转向斜构造，使岩层强烈透镜体化，底砾岩中的砾石被压扁，大量的石香肠和砾石的长轴平行于褶皱轴面。向斜本身被挤压后呈尖棱状柔皱特征。一般背斜较宽缓，向斜紧闭，类似隔槽式构造。第二期变形形成横跨叠加第一期褶皱之上的枢纽为北北东向的小型紧闭褶皱，它们围绕云蒙山岩体北侧呈发散扇状分布。由于伴随有岩体向北上侵的顶托作用，岩层和早期向斜被进一步压扁，轴面也发生了弯曲。同时产生的东西向层间滑动，使岩层中的一些石香肠、砾石或其他坚硬的薄岩层、岩块发生碾滚，形成大量的窗棱构造、石英杆、砾石杆等（表 24-2）。

表 24-2　接触变质岩主要特征一览表

Table 24-2　The main characteristics of contact metamorphic rock

<table>
<tr><th colspan="2">建造类型</th><th>主要变质岩石组合</th><th>矿物共生组合</th><th>结构构造</th><th>变质带</th><th colspan="3">推测温压条件</th></tr>
<tr><td colspan="2">变质粗碎屑岩</td><td>变质底砾岩、长石石英岩、石英岩、长石石英片岩</td><td>长石+石英</td><td>花岗变晶结构</td><td></td><td colspan="3"></td></tr>
<tr><td colspan="2">变质泥质岩-泥质粉砂岩</td><td>变质粉砂岩、千枚状黑云片岩、黑云片岩</td><td>长石+石英+黑云母+（白云母+绿帘石+阳起石）±透闪石+透辉石±方柱石±堇青石</td><td>微-细鳞片状变晶结构、变余微层状构造普遍</td><td></td><td colspan="3"></td></tr>
<tr><td rowspan="4">变质碳酸盐岩</td><td rowspan="2">透闪石大理岩</td><td rowspan="2">金云母大理岩、透闪石大理岩、金云母透闪石大理岩</td><td rowspan="2">白云石+方解石+透闪石±金云母</td><td rowspan="4">粒状变晶结构、蒿束状变晶结构、块状、条带状构造、团块状构造、豆状构造、角砾状构造</td><td rowspan="2">透闪石带</td><td>1×10^8 Pa</td><td>5×10^8 Pa</td><td>6×10^8 Pa</td></tr>
<tr><td>580℃</td><td>605℃</td><td>615℃</td></tr>
<tr><td>透闪石辉石大理岩</td><td>白云大理岩、金云母大理岩、透闪石大理岩、透辉石大理岩、透闪石大理岩、透闪透辉石岩、透辉石岩</td><td>白云石+方解石+透辉石±石英±金云母</td><td>透闪石-透辉石带</td><td>620℃</td><td>635℃</td><td>660℃</td></tr>
<tr><td>透辉石橄榄石大理岩</td><td>白云石大理岩、透辉大理岩、透闪透辉大理岩、橄榄透辉大理岩、金云母大理岩、透闪透辉石岩</td><td>白云石+方解石+透闪石+斜硅镁石+镁橄榄石±金云母</td><td>透辉石-橄榄石带</td><td>660℃</td><td>685℃</td><td>715℃</td></tr>
</table>

第三节　气-液变质作用

热的气体及溶液作用于已形成的岩石，使已有岩石发生矿物成分、化学成分及结构构造的变化，多沿构造破碎带及矿脉边缘发育，作为一种找矿标志，又称近矿蚀变（围岩蚀变），简称蚀变岩。岩浆期后热液蚀变作用，主要发育于中酸性侵入岩体内及边部（或发育于爆破岩浆角砾岩体中），主要蚀变岩石具有硅化、绢云母化、钾化、绢-云英岩等特征，常形成浅成脉状体，细-网脉状，少见面型蚀变，伴有 Fe、Mo、Pb、Zn 等硫化物矿化作用。

1. 硅化

岩浆活动使围岩中石英或隐晶质二氧化硅含量增加，形成硅化的二氧化硅一般是由热液带入，部分是由原岩中二氧化硅残留、相对富集而成的。也可由长石或其他矿物经蚀变后形成。硅化几乎在任何岩石中都可发育，从高温到低温条件下均可产生。由于硅化可以在广泛的环境中由热液作用形成，因此硅化蚀变常与其他蚀变，如绢云母化、绿泥石化、碳酸盐化、云英岩化、泥化、长石化等相伴产出。与硅化有关的矿床很多，其中主要是铜、铅、锌、钼、金等。

如在延庆里长沟、怀柔石湖峪等地。区域内主要发育有高温、中温两种硅化类型。高温硅化现象发育于怀柔石湖峪—北庄一带，受西台、麻子将断裂的次级断裂控制，蚀变带宽1~2m。此蚀变带与云英岩化共生，并与云英岩化呈逐渐过渡，属高温类型，是寻找辉钼矿的重要标志；中温硅化现象在分水岭地区，发育于含金石英脉两侧，蚀变带宽仅几厘米至30cm。

2. 绢云母化

广泛发育的中—低温热液蚀变，在中性和酸性火成岩及板岩等富铝岩石中最常见。单矿物的绢云母岩，一般少见。绢云母化常伴随有石英和黄铁矿的产生，因而可称为绢英岩化，若黄铁矿含量超过5%时，则称为黄铁绢英岩化。绢英岩化与云英岩化过程在本质上相同，只是前者形成温度较低，它们之间可存在着过渡关系，即云英-绢英岩化。在金、铜、铅、锌、钼、铋以及萤石、红柱石、刚玉等矿床中都能见绢云母化现象。特别是斑岩型铜、钼矿床、黄铁矿型铜矿床和金多金属矿床。

3. 钾化

钾长石化的结果是蚀变岩石中形成微斜长石、透长石、正长石、冰长石等矿物。原岩以酸性火成岩为主，其次是中性火成岩及较富长英质的沉积岩、沉积变质岩。形成条件多为高温（冰长石为低温）。有关的矿床有金、银、铜、铅、锌等。

为钾质交代的产物，包括微斜长石化、正长石化、透长石化和冰长石化。由于它们不易区别，且成分几乎完全相同故统称钾长石化。在与花岗岩有关的钨、锡、铍、铌、钽以及斑岩铜、钼矿床等的下部，经常发生有大规模的钾长石化带。低温热液的钾长石化，以冰长石化为主，多发生在中性、弱酸性火山岩中，也可在基性或酸性岩中发生，有时与青盘岩化有关。与其有关的矿产主要为火山岩系中的一些金属矿床，如昌平黄场铜钼矿、大庄科钼矿。

4. 绢-云英岩化

该类岩石是酸性侵入岩或其他成因的类似的长英质岩石，在高温气体及热液交代作用的产物。云英岩主要是花岗质岩石遭受气液变质的产物。

岩石外貌特征为浅色（灰色、浅灰绿色或浅粉红色），块状构造。中粗粒粒状变晶结构、鳞片花岗变晶结构。

矿物成分以石英、云母（白云母、锂云母）、萤石、黄玉及电气石为主，有时可见金红石、微斜长石、钠长石、绿柱石、石榴子石等。常见金属矿物有铁矿、辉钼矿及黄铁矿等。此外，尚有云英岩化晚期叠加的蚀变矿物，如绿泥石、绢云母、水云母及高岭土等矿物。该类型蚀变岩分布明显受断裂的次级断裂控制，为热液沿断裂裂隙中更次级小裂隙活动交代的产物。云英岩一般规模较小，与围岩呈渐变过渡。组成云英岩的白云母，一般片径仅0.2mm，故区内的云英岩化非属典型，或称绢英岩化。

第二十五章　混合岩化作用

按照混合岩化作用的成因和形成方式，北京地区混合岩化主要位于密云-怀柔地区，可划分为岩体边部由注入式混合岩化形成的边缘混合岩和深熔作用形成的各种区域性原地混合岩。区内混合岩化以重熔作用为主，也有沿层理、片麻理、片理贯入，交代比较普遍，属于区域混合岩化。由于后期岩浆活动影响，局部也出现岩体边缘混合岩化（如沙厂）。混合岩化发育程度中等，分布范围较广但不均衡。在混合岩化作用发育的地段，混合脉体的贯入及硅铁质的重熔、迁移和重新分配，可形成较粗大的磁铁矿颗粒形成富铁矿体。在铁矿体中常因混合物的渗入而使铁矿品位降低。

第一节　混合岩化作用基本特征

一、注入式边缘混合岩化作用

在四干顶、柏岔山、云蒙山的边部，多有花岗质、花岗闪长质或伟晶质岩脉沿着围岩片麻理顺层贯入到变质岩地层中的现象，向里具有不同发育程度的边缘混合岩。

云蒙山花岗岩体边部的注入式混合岩化比较典型。在岩体与石城片麻岩的接触部位，与岩体的矿物成分和结构几乎相同的花岗闪长质脉体沿着片麻理贯入到片麻岩中，形成了黑白相间的注入混合岩。黑云角闪斜长片麻岩与花岗闪长质脉体间界线清楚，二者间没有物质成分上的交换。这种注入的边缘混合岩仅分布于岩体的外接触带中，由此向外，混合岩化程度迅速减弱直至最后消失。

二、区域性混合岩化作用

区域性混合岩化，广泛出露在新太古代地层中，一方面和区域变质岩呈过渡关系，另一方面又与某些花岗质岩石互相过渡。岩石中各种交代结构及重熔现象很普遍，并且产生了酸性斜长石-钾长石-石英三元低熔共结物。沙厂地区混合岩化程度大于四合堂岩群，其次为苇子峪地区，大漕地区相对较弱，未见紫苏花岗岩或较大面积的混合花岗岩出露。

1. 重结晶作用

混合岩化对变质岩的重要影响之一，就是广泛的重结晶作用使矿物粒度增大，这在长英质岩石中表现的尤为明显。变粒岩和浅粒岩本来具典型的细粒粒状镶嵌变晶结构，受混合岩化影响变成细—中粒花岗变晶结构。在长英质混合岩化条带中，长石和石英重结晶成中—粗粒晶体，其中还往往可见到暗色矿物的粗大变斑晶。

2. 石英含量增加及石英的重熔条纹

混合岩化造成的变质岩石的碱性长石作用与硅化作用是同步的。在低熔点的斜长石、钾长石和石英当中，受混合岩化的影响，石英是最先活动的组分。变质岩中共生的石英一般呈他形粒状略为压扁的形态，混合岩化的选择性重熔作用则使石英转变成条纹、条痕或条带状集合体。集合体内石英呈压扁拉长状，彼此之间边界相接触，波状消光现象普遍而强烈。对于石英含量较高的变粒岩和浅粒岩，重熔的石英条痕或条纹平行排列，导致岩石中条纹状构造的发育。条纹状浅粒岩、变粒岩，虽然以其明显的条纹状构造区别于不受混合岩化影响的相应变质岩石，但就其内部矿物间的相互关系而言，还没有形成广泛的交代结构，仅是长英质矿物重结晶而已。

3. 钾质交代

混合质岩和混合岩中，钾长石是形成温度较低的微斜长石，具格子状双晶，并往往交代斜长石。

钾质交代作用在区内有几种表现。

（1）钾长石含量增加

变质岩经混合岩化作用后，其钾长石含量显著增加。无论在野外露头上，还是在岩石薄片中，钾长石分布不均，多聚集成团块状或条带状，有的尚可呈眼球状变斑晶存在。

（2）斜长石的“暗化”

所谓斜长石的“暗化”，是指重折率较高的斜长石被重折率较低的钾长石置换而发生的干涉色降低的现象。由于钾交代的程度不同，斜长石常表现出不同的“暗化”中间产物：交代开始，钾长石呈孤立分散的小斑点稀疏地分布斜长石之上；继之，钾长石暗色斑点面积逐渐扩大并慢慢连成一片，致使斜长石反成为散布于钾长石之中的绢云母化的残片；最后，整个斜长石被钾长石交代完毕，钾长石继承了斜长石晶形，有的尚发育了完好的格子状双晶。

（3）蠕英石结构

钾长石交代斜长石游离出来的 SiO_2，在斜长石一侧形成蠕虫状石英嵌晶，构成美丽的蠕英石结构。虽然花岗岩中常出现蠕英石结构，但它仍不失为混合岩的重要结构之一。

（4）缝合线结构

钾长石交代斜长石，使本来比较平滑的斜长石边缘呈现锯齿状或曲线状复杂形态，曲线的弧形凹面指向被交代的斜长石。

4. 钠质交代结构

混合岩化过程中的钠质交代作用除造成斜长石牌号降低外，还形成了一系列可以观察到的现象。

（1）净边结构

富含钠质的粒间溶液的活动，使斜长石边缘首先发生钠长石化，形成洁净透明的钠长石净边。混合岩中有些斜长石边缘透明，而中心部位则高岭土化、绢云母化强烈，矿物变得模糊不清。

（2）渗透状交代结构

在混合岩化变质岩里，常出现处于生长过程中的更长石变斑晶，变斑晶包含着大量乳滴状石英嵌晶以及光性上尚未完全趋于一致的斜长石细粒。当两个或多个变斑状斜长石聚集于一起时，彼此的接触界线弯曲不平，呈缝合线状。更有甚者，变斑状斜长石之间出现相互渗透的现象。变斑晶主体与渗透到相邻斜长石边缘中去的细粒保持着光性方位上的一致性。更长石变斑晶可撑开由黑云母或角闪石组成的微弱片麻理形成黑云角闪质片麻状混合岩。

全岩 An/Ab 比值较高的块状斜长角闪岩或斜长辉石岩混合岩化的程度通常较低。此类岩石受强烈混合岩化作用时往往伴随有岩石的破裂，形成角砾状混合岩。角砾是基体成分，“腹结”角砾者为更长石或更钠长石细粒。混合岩化作用稍弱时，角砾边缘界线清晰，相邻角砾尚可相互拼结；混合岩化强烈时，角砾边缘变得模糊，数量减少，并可发生一定程度的旋转。角砾状构造向着网脉状构造演化过程中，暗色组分减少，更长石增加，An/Ab 比值降低。这种特殊的混合岩化是一种钠质交代过程。

5. 铁镁质硅酸盐岩矿物的蚀变

在混合岩化过程中，除了广泛的钾质和钠质交代作用以及大量矿物的重结晶外，许多铁镁硅酸盐矿物被含水矿物交代和置换，也是混合岩化的重要表现之一。紫苏辉石、石榴子石被黑云母和绿泥石交代，透辉石和普通角闪石被阳起石、绿泥石交代等现象，说明混合岩化过程中水是相当活跃的组分。混合岩化过程中新生的钾长石、斜长石和石英组成的长英质、花岗闪长质或花岗质脉体的数量比例，反映出混合岩作用的强度。脉体物质的聚集状态是混合岩分类的重要依据。

三、密怀变质岩区混合岩化特征

1. 混合岩分布特征

在北京密云-怀柔地区混合岩化现象是相当普遍的，通过对密云岩群和四合堂岩群变质岩与混合岩主要特征对比（表 25-1）认为，区内变质岩和混合岩化强度有所不同，其中密云岩群混合岩化强度高于四合堂岩群。从全区范围看，不存在一个混合岩化中心，只有混合程度不同的层位。

在密云岩群中，沙厂地区变质岩系混合岩化作用较强，苇子峪和大漕地区相对较弱。沙厂地区混合岩以早期钠质交代的条带状混合岩为主，也有片麻状混合岩和少量均质混合岩，未见紫苏花岗岩或较大面积的混合花岗岩出露。局部地区的晚期钾质交代作用较强，例如霍各庄—达岩一带，部分角闪辉石质混合岩具有角闪辉石二长片麻岩的外貌。混合岩的基体物质成分一般并不难确定，其较细粒的暗色条带部分往往能在一定程度上保留着变质岩石的结构、构造和矿物成分特征。苇子峪和大漕地区混合程度稍弱，以条带状混合质变粒岩和片麻岩为主，条带状混合岩次之。大漕地区变质岩石的沉积构造发育，顺层的混合岩化现象明显，长英质混合岩化条带延续性好，有的貌似沉积作用形成的长英质薄层。

在阳坡地和宋营子地区四合堂岩群中的混合岩化强度高于西湾子和山神庙地区。山神庙地区只有些条带状混合岩化现象，西湾子地区混合岩化作用微弱。可能是由于处于构造活动带上的缘故，宋营子地区混合岩化程度高于阳坡地地区。阳坡地地段所见混合岩化为条带状混合岩、部分片麻状混合岩，大量的是各种条带状混合质变质岩。在宋营子地区内，黑云角闪片麻状混合岩普遍，条带状混合岩相对较少，有些层位甚至出现阴影状混合岩。

表 25-1　变质岩矿物与混合岩主要矿物特征对比

Table 25-1　Comparison of main mineral characteristics metamorphic rocks and migmatite

矿物	密云岩群麻粒岩相变质岩系		四合堂岩群角闪岩相变质岩系	
	变质岩	混合岩	变质岩	混合岩
斜长石	细粒—中粗粒—中更长石，中长石具反条纹结构	中—粗粒更长石，具蠕英石结构、净边结构、缝合线结构	细粒更-中长石、中长石	中—粗粒，变斑状更长石具穿孔结构、渗透状交代结构
钾长石	细—中粒，他形，不交代其他矿物	中—粗粒，交代斜长石，具微文结构、条纹结构和格子状双晶	细粒他形粒状，不交代斜长石，为共生矿物，可作为微层状结构	中—粗粒、眼球状变斑晶，交代斜长石或使其“暗化”
石英	细—中粗粒，他形，扁长状	中—粗粒条带状集合体或团块状集合体	细粒变晶，偶见浑圆状变余砂状	中—粗粒条纹、条痕或条带状集合体
紫苏辉石	细—中粒半自形柱粒状	多倍镁铁闪长石、滑石、黑云母和绿泥石强烈交代		
次透辉石	浅绿色细—中粒半自形柱粒状	少数呈变斑晶，常为阳起石和绿泥石交代	浅绿色细粒半自形柱粒状，较为少见	
石榴子石	粉红色细粒等轴状晶体	少呈变斑晶，常被黑云母和绿泥石交代	粉红色自形细粒，较为少见	
普通角闪石	棕绿色细粒等轴状晶体	少呈变斑晶，多为阳起石、黑云母、绿泥石交代	绿色细—中粒长柱状	蓝绿色粗粒变斑晶，或组成条带状暗色体，阳起石化普遍
黑云母	深褐色叶片状	交代其他暗色矿物，也可被绿泥石交代	深棕绿色鳞片状	重结晶的黑云母常构成豆荚状暗色体
绿帘石		交代辉石、角闪石和斜长石等矿物	变余残粒状，少有环带状结构	交代斜长石和角闪石等
榍石			微—细粒状集合体，可组成钛磁铁矿的镶边	重结晶的条带状集合体与角闪石、黑云母暗色体紧密伴生

2. 混合岩化期次

在密云岩群及四合堂岩群的条带状混合质变质岩或条带状混合岩中，条带状物质总与基体的层状构造平行，并与基体物质一起褶皱，这说明混合岩化作用发生于变质之后、主要褶皱构造运动之前。前已叙述，密云岩群和四合堂岩群的变质作用发生于新太古代，与变质作用时限相差不远的混合岩化作用也属于新太古代，但应稍晚于前两者。两期混合岩脉体物质成分有所不同：新太古代早期密云岩群混合岩化脉体为花岗闪长质成分，其中多有透辉石、石榴子石及角闪石变斑晶；新太古代晚期四合堂岩群混合岩化脉体为长英质成分，其中有绿色闪长石或黑云母变斑晶。

3. 影响混合岩化强度的控制因素

（1）变质强度

混合岩化是一种超变质作用，其强度与变质作用强度密切相关：变质的温压条件越高，混合岩化作用也就越强烈。本区密云岩群麻粒岩相变质作用之后发生的区域性混合岩化在区域上形成了大量各种形态的混合岩。混合岩中交代结构发育，此外，还有净边结构、蠕英石结构、缝合线结构等。四合堂岩群角闪岩相变质作用之后发生的混合岩化现象，没有上次的普遍，混合岩中交代结构不发育，混合作用以注入式为主，并且有的地段内的混合岩化作用相当微弱。

（2）层位高低

在同一个变质岩群的地层内，往往下部层位比上部层位的混合程度深。密云岩群下部沙厂英云闪长质片麻岩比上部苇子峪奥长花岗质片麻岩、大漕地区变质岩系的混合作用强烈，四合堂岩群阳坡地和宋营子地段混合作用强度也明显高于西湾子地区和山神庙地区。

（3）变质岩的基性程度

由温度上升而引起的深熔作用的初熔温度与变质岩的 An/Ab 比值有着重要联系：An/Ab 比值高的，初熔温度也高；An/Ab 比值低的，初熔温度则偏低。在某一特定温度下，基性岩的深熔作用刚刚开始，中酸性岩石则早已形成一定数量的熔体。因此，在混合岩化过程中，基性程度低的片麻岩、变粒岩和浅粒岩等岩石的混合岩化作用强烈，而基性程度高的斜长角闪岩、斜长辉石岩的混合岩化作用则较弱。辉石岩、角闪石岩以及磁铁石英岩更难发生混合岩化。

（4）构造部位

在混合岩化作用发生的同时，如果伴随有断裂活动，则断裂附近的围限压力迅速降低，从而降低了变质岩石的初熔温度，结构使得深熔作用开始于比别处温度偏低的条件下。断裂构造本身又是熔体物质上涌的通道，这样，较高层位的岩石就容易发生较强的混合岩化作用。宋营子地区变质基性岩脉很发育，说明这一部分的构造活动曾有过较活跃的阶段。宋营子地区混合岩化强烈的现象可能与其处于构造活动部位有关。宋营子地区与山神庙地区呈断层接触。在这个接触带上，难于混合岩化的斜长角闪岩和角闪石岩也发生了很强烈的混合岩化，形成延伸稳定的角砾状混合岩化带。

4. 混合岩化成因探讨

据任留东（2012）混合岩成因解释，目前尚未取得一致认识，主要存在两种观点，而且两种认识在自然界均存在，且相互之间存在一定联系。以 Sederholm（1907）为代表认为这种岩石是由较老的岩石（原岩）和外来的花岗质岩浆混合形成，并以脉状体分布于岩石中，其强调源自于地壳深处的深熔作用形成的外来岩浆同原岩混合、相互作用，并发生注入变质（injection metamorphism）；Holmquist（1916）认为混合岩中的脉状体不是外来物质的加入，而是原岩通过变质分异作用产生，后来有人认为原岩经熔（溶）融作用生成。

在诸多关于混合岩化成因的假设中，任留东（2012）认为从成因上形成机制有 4 种，分别为外来岩浆注入、原地深熔作用、交代作用和变质分异作用，实际上以前两种成因占主导。而在我国普遍认为“深熔作用”具有充分的说服力和坚实的物理化学基础。按照深熔作用理论，在超变质的条件下，片麻岩发生选择性重熔，形成长石花岗质、花岗闪长质成分的熔体；熔体物质又分凝成透镜状、条痕状或条带状聚集体，导致片麻岩中出现不同的混合岩化现象。在深熔性质的混合岩化过程中，片麻岩部分熔融产生的花岗质成分的熔体和结晶残余体总的化学成分基本相当于原来的片麻岩，说明深熔作用混合岩化是一种原地混合岩化。

密怀地区变质岩属于变质中、基性火山岩，原岩建造中缺少钾长石。那么深熔作用过程中形成的花岗质熔体内的钾长石来源于何处？本区缺少碱性长石的变质中、基性火山岩系，在深熔作用过程中完全可以提供混合岩化所必要的物质来源，其中包括钾长石。

混合岩化岩石内基体与脉体的数量比例可反映出混合岩化的强度。脉体物质中钾长石和斜长石之间的数量关系也是混合岩化强弱的重要标志。本区密云岩群比四合堂岩群混合岩化程度高，前者混合岩化脉体相当于花岗闪长质成分，而后者则主要由更长石和石英组成，极少见到钾长石。根据混合岩化脉体的矿物成分把混合岩化分为两期：早期为钠质交代作用，混合岩化脉体平行于区域片麻理，矿物成分是斜长石、石英和少量钾长石，混合岩在野外露头上呈白色；晚期为钾质交代作用，混合岩化脉体平行或斜交片麻理，其矿物成分是斜长石、石英和含量较高的交代作用，基岩露头呈粉红色色调。早期钠质交代强烈而普遍，晚期钾质交代相对微弱，分布也不普遍。

由深熔作用引起的混合岩化，经历了发生、发展到终结的各个阶段。在初始熔融阶段，地壳褶皱运动刚刚开始，尚未大规模回返和隆起，故地壳应力小，温度也不太高。部分熔融阶段首先进入熔体的是低熔点的组分斜长石和石英，分凝出的脉体是长英质的。当温度进一步升高，一部分黑云母将被分解和消耗，从而生成钾长石。此时钾长石与斜长石+石英相比，数量还是少的，熔体成分相当于花岗闪长质成分。在强烈褶皱阶段，应力和温度急剧增高，混合岩化进入旺盛阶段，更多的黑云母参与反应，生成了更多的钾长石，深熔熔体具有花岗质成分。褶皱主期之后的花岗质混合岩化脉体常常斜交区域片麻理，并可沿一些断裂形成花岗质伟晶岩。

按照分析，混合岩化脉体物质中钾长石的有无，以及它与斜长石+石英的数量比例关系，可以反映出混合岩化过程所处的不同阶段。钾长石含量越高，表明混合岩化作用的温度越高，混合岩化作用的强度相应也就越强烈。四合堂岩群混合岩化岩石中，大量出现的长英质（更长石+石英）脉体表现混合岩温度只略微超过片麻岩初始熔融温度；密云岩群混合岩的脉体物质主要是花岗闪长质成分，含有一定量的钾长石，表明混合岩化温度比片麻岩初始熔融温度高得多（图 25-1）。

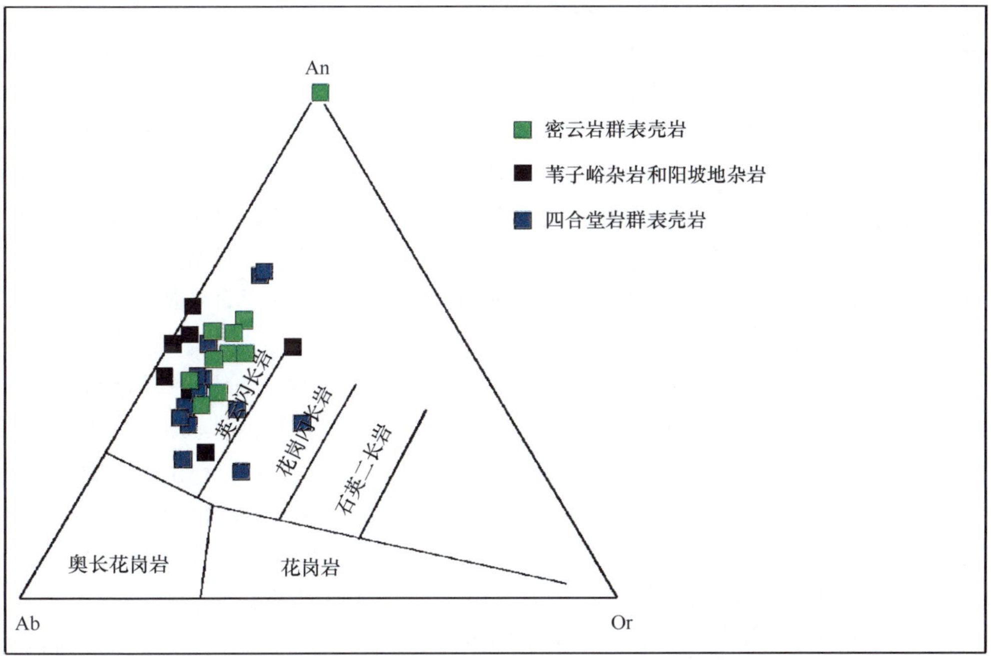

图 25-1　TTG 岩系中混合岩化作用的 An-Ab-Or 图解

Fig. 25-1　The ternary plot An-Ab-Or analysised diagram of migmatization of the TTG rock series

在变质火山岩系经历了麻粒岩相变质作用，但还没有或尚未发生强烈褶皱的情况下，深熔作用导致片麻岩部分熔融，生成了花岗闪长质熔体，分凝出的花岗闪长质脉体平行于片麻岩片理；密云岩群强烈褶皱时，应力和温度增加，深熔体由黑云母分解而提供的钾长石含量增加，熔体转变成花岗质成分，分凝出的花岗质脉体平行于片麻理或与之斜交。早期钠质交代、晚期钾质交代只不过是同一期混合岩化不同发展阶段的表现形式。

第二节　混合岩—花岗质岩石演化特征

混合岩化变质岩是岩石在固态或塑性状态下，由于外来岩浆或稀薄岩浆及相伴生的富碱流体发生渗透、注入、重熔以及交代作用下，岩石部分或全部重结晶，形成具有混合成分的、粒度常粗于变质岩的岩石。由变质岩向混合岩转化的过程称为混合岩化作用，该作用的高级阶段使混合岩趋向均匀，成分和结构转变为类似岩浆型花岗岩的过程称为花岗岩化。

古老的花岗质岩类广泛分布于早前寒武纪结晶基底中，尤其在高级变质区，它的总体含量超过区域变质岩类，一般占70%以上。尽管其中可见到一些小型晚期花岗质岩体穿插早期花岗质岩的现象，但由于大规模的早期花岗质岩一般与围岩（变质岩）没有明显的界线，含有许多大小不等的围岩变质岩的残留体，一部分还显示花岗岩与变质岩呈互层状（条带状）的岩相特征，而有别于典型的岩浆型花岗岩体。在20世纪70年代中期之前，我国多数地质学家在混合岩化作用和花岗岩化作用的理论指导下研究分布于变质岩区中的花岗质岩成因，以及混合花岗岩与岩浆花岗岩的区别标志。主要区别标志包括野外标志（空间关系、残余特征、岩相特点、矿产类型）、矿物学标志、地球化学标志和岩相学标志等。该阶段对混合岩类研究成果基本反映了早前寒武纪结晶基底中多数古老花岗质岩的成分及其特征。然而，过去对古老花岗岩研究忽略了太古宙变形（变质）或未变形的岩浆型花岗岩存在的事实。

20世纪70年代以来，我国许多地质学家受到“北大西洋岩浆派”的影响，由“混合派”转向“岩浆派”，将许多相当于国外常用的“长英质片麻岩”或“灰色片麻岩”的混合岩类称之为奥长花岗岩-英云闪长岩-花岗闪长质岩系（TTG），认为它们属于古深成花岗质侵入岩定位后经变形、变质后的产物。该阶段在我国早前寒武纪结晶基底中经同位素地质测年研究确定了一批太古宙岩浆型花岗质岩侵入体。贺同兴等（1992）、林强（1992）分别对冀东和华北地台东部的太古宙花岗岩进行了详细研究，根据岩浆来源、构造变形及深熔作用等特点，将太古宙花岗岩划分为：幔源岩浆分异系列（幔源分异型花岗岩）、幔-壳源混合系列（硅质渗透型花岗岩）、壳源岩浆分异系列（壳源分异型花岗岩）、深熔岩浆分凝系列（深熔型花岗岩）。这四个成因系列是太古宙上地幔-下地壳-变质岩系。但应该指出，实际工作中TTG质岩石研究存在“扩大化”和“简单化”现象，不论以往所认为的较均一花岗质岩体是变质变形的深成“TTG”岩体还是具有相应成分的均质混合岩-混合花岗岩体，将大量条带状混合岩认作“TTG”及其剪切成因的构造岩无疑是一种误解（任留东等，2012；刘兵等，2008）。

就密云-怀柔地区而言，古老的花岗质岩的研究也经历了上述两个阶段，目前也存在TTG质岩石成因研究的扩大化和简单化现象，这主要表现为三个方面：①将条带状混合岩认作变质的深成“TTG”；②忽略了呈片状、条痕状和斑杂状的非层状（非条带状），成分相当于中性（SiO_2含量为53%~63%）的混合岩的普遍存在；③忽略了TTG质岩石中相当一部分包含变质岩的剩余物。

王世发等（1994）将原密云岩群沙厂组和大漕组的主体部分称为沙厂片麻岩系，并认为它是遭受变质变形的经原地-半原地重熔形成的英云闪长质岩体，其中含有基性变质岩和磁铁石英岩。实际上原沙厂和大漕两地区域变质岩主体是SiO_2含量多为59%~65%的（含石榴、辉石）黑云斜长变粒岩，其次是基性变质岩和磁铁石英岩。条带状混合岩是原大漕组和沙厂组的主体。虽然其总体成分相当于TTG质岩，但真正的TTG质岩呈脉状产出，即构不成独立的规模较大的岩体。金文山（2000）主张将原大漕组和沙厂组的主体称为混合岩化表壳岩系，局部发育有一定规模的非层状TTG地质体。

王恩林（1987）认为原张家坟群不具有地层含义，是一个受到强烈剪切作用的夹表壳岩构造残留体的巨大岩浆杂岩体（英云闪长岩、花岗闪长岩、角闪石岩和闪长岩），并称其为三棱山杂岩体。麦延庆（1987）称原四合堂群为三岔口英云闪长岩–花岗闪长岩杂岩体或三岔口片麻岩。实际上原四合堂群除阳坡地组外，其余三个组区域变质表壳岩相当发育，其主要是 SiO_2 含量为 55%～65%的（含石榴、角闪）黑云斜长变粒岩和基性的斜长角闪岩类，其次为磁铁石英岩。变粒岩类与斜长角闪岩大多也与 TTG 质岩呈互层状产出，构成条带状混合岩；只是有一部分在强烈韧性剪切带遭受动力变质叠加作用变成糜棱岩类。条带状混合岩虽然总体成分也相当于 TTG 质岩，鉴于与沙厂岩系同样，金文山（2000）将原四合堂群的宋营子组、西湾子组和山神庙组称为混合岩化的表壳岩系，在其分布区发育一些闪长岩和闪长玢岩，但分布零星。

在各个岩石、地层单元中均见有呈斑杂岩或条痕状或片麻状的，SiO_2 含量相当于中性的非似层状混合岩类，常与 TTG 质岩混杂产出。这类混合岩在苇子峪地区变质杂岩中最为发育，确定其为混合岩的重要依据是，无论从露头上还是岩石薄片中，均含有相当比例（一般为 5%～30%）的基性变质岩类的残留体。在薄片中这些残留体呈规则斑点，或呈片麻理一致的条痕。其属于基性变质岩类遭受壳源深熔岩浆及有关的碱性流体的混合岩作用，向混合花岗岩演化的过渡产物，其成分相当于二长闪长（辉长质、石英二长闪长（辉长）质及少量的闪长（辉长）质、石英奥长质和石英二长质岩。

本区 TTG 质岩中相当部分含有基性变质岩类的斑点和条痕，或者残留或已重结晶的基性变质岩中的暗色硅酸盐矿物（辉石、角闪石、石榴子石）和磁铁矿等副矿物，但其含量比上述混合岩类明显减少。这些暗色硅酸盐矿物有些被次闪石化和绿泥石化，反映了部分 TTG 质岩石产生高度混合岩化（花岗岩化）作用，应属于混合花岗岩类。近年来，一些对太古宙 TTG 质岩的研究忽略了花岗岩化作用在太古宙花岗质岩形成过程中的重要作用。

第二十六章　变质地质构造分区

北京地区变质岩主要包括密云-怀柔地区新太古代变质岩和西山晚古生代—早中生代变质岩。根据地壳演化、构造旋回及地质作用（如深成侵入作用、变质作用、构造作用形式和变形等）差异，将北京市变质岩系划分为北山和西山两个构造分区，建立了全区变质岩基本构造轮廓。

本次在划分变质构造分区基础上进一步对密云-怀柔地区变质岩系进行了变质小区、变质构造带的划分。对于早前寒武纪变质地体构造区的划分，主要是采用地质历史分析和对比方法进行的。根据地壳演化、构造旋回及其主要地质作用（如深成侵入作用、变质作用、构造作用形式和变形等）的特征和差异，将本区地质构造划分为变质分区、变质小区以及更小的构造单元，从而建立全区基本构造轮廓。

第一节　构造区划分的原则及其边界

北京基底变质岩区是华北陆块区冀东-密云陆核组成部分。按照不同构造旋回（包括阜平旋回和五台旋回）及其形成的构造层，依其构造环境、沉积作用、岩石组合类型、变质作用及变质相、构造运动形式及形态构造特征、同位素年龄等时空一致的综合标志，建立构造、变质作用及地质年龄统一的构造区。

在构造区内，按不同构造层次的变质建造和原岩建造、构造型式和主导构造线方向的差别可划分亚区或小区。它主要反映不同变形相和褶皱群（带）的分布规律及局部地质环境和物化条件的某些差异。

张贻侠（1981）曾提出四大主要构造区分界标志：①线型活动带；②元古宙线型海相裂陷槽；③年轻的盆地；④多期活动的大型断裂。前三者发育于太古宙克拉通、地块之间或者其周围边缘地带，往往是太古宙构造分区的良好标志。有的线型活动带也可以发育于太古宙构造区内部，成为构造亚区的分界。后者多适宜于次级构造单元的划分。

北京地区太古宙基底构造区的接合边界大致有两类：

1）多期活动的古断裂带。这种断裂带往往是在更古老的大型韧性剪切带（或古滑断构造）基础上发育起来的，它的断距较大，多期活动特点明显。在它们之中有大量典型的古糜棱岩、古角砾岩或是被混合岩化作用胶结、熔合的角砾状混合岩、古变质脉岩群等的线型密集带。沿断裂还发育有线型浅变质带。在密怀地区，多条这样相互平行的大型古断裂带将麻粒岩相地块和角闪岩相地块拼接起来，形成两种地块彼此相间排列的构造格局。半城子古断裂带和崎峰茶古断裂带为构造区的分界。石城断裂、黑古沿断裂和火郎峪断裂则是亚区或次一级构造单元的分界。

2）年轻的线型海相裂谷和盆地。在本区太古宙基底周围分布有中—新元古代和显生宙的年轻线型海相裂谷和盆地。基底东侧、南侧、西侧的海相裂谷和沉积盆地都较开阔，构造简单；而基底北侧的海相裂谷较狭长，构造复杂，产状陡。这说明南、北两部分基底活动性的差异较大。深部构造研究也表明，基底周围年轻裂谷和盆地中盖层之下与基底边缘的接触地带，亦是受深断裂控制的边界，如东、西两侧为兴隆-蓟县断裂和汤河口-通县断裂控制，南、北两侧为墙子路-八达岭断裂和赤城-古北口断裂控制。

不同构造区的差别主要表现在以下几个方面。

1）岩石、地层及其组合差异。不同构造区独特的岩石组合及地层组合表明它们的原岩建造、成岩环境、成岩时限是不一致的。

2）变质建造差异。麻粒岩相和角闪岩相变质建造可明显地分为两大区，代表不同的变质温压环境。

3）构造变形差异。麻粒岩相带发育了阜平运动期紧贴式褶皱及华北运动期多级别褶皱和同期透入性变形及面状构造置换。角闪岩相带仅发育了华北运动期的多级别褶皱和同期透入性变形，有一期主要的面状构造置换。两套变质岩系在构造走向、褶皱样式等方面都明显不同。

4）混合岩化差异。麻粒岩相区混合岩化以深熔作用为主，伴有钠质、钾质交代作用。角闪岩相区混合岩化较弱，以斜长石的更-钠长石化及石英的选择性压熔为主，交代作用次之。

5）岩浆岩差异。麻粒岩相区有大量的古老超镁铁质、镁铁质岩浆岩出现并多已变形变质，另有重熔型云英闪长岩、奥长花岗岩侵位。角闪岩相区有较多的基性、中性古岩浆岩侵位，如变形变质辉长岩、闪长岩、中性斜长岩等层状岩石组合，局部有变质碳酸盐岩脉穿入。

6）其他差异。两构造区的原岩建造、含矿性、地球物理和地球化学特征也明显不同。不同变形历史阶段形成的断裂构造的发育程度及性质也颇不一样。在地貌上，麻粒岩相区的低缓丘陵也与角闪岩相区的陡峻单面山形成鲜明的对比。

第二节　构造区的划分

北京地区变质大区、变质区、变质分区划分总体上沿用陆松年（2015）划分方案，根据岩石构造组合、变质相、变质时代、岩石地球化学及原岩建造进行分析研究，区分地质演化阶段中的变质旋回进行划分。北京地区变质岩分别范围有限，本次重点划分了本地区的变质小区和变质构造带。

这一划分方案主要表现了基底岩系中不同时代岩石地层单位所组成的构造层或构造阶层的变形相和变质相的统一。方案的实质就是构造-岩群的划分。它基本与地层单元的划分是一致的。

（一）II-3-2-1SC 沙厂-大漕变质小区

该区位于冀东变质地带-陆核（II-3-2）区内构造单元-密云隆起的东段，主要分布在密云水库周围地区，北以 NE 向崎峰茶、石城大型古断裂带与 II 区接合，东、南、西三面均与元古宙以来的年轻盆地或坳陷槽毗邻，以沉积盖层下的不整合面或构造软弱带为界。

该构造区是由密云岩群麻粒岩相、高级角闪岩相岩石所组成的地块。地块内部构造为一个穹状（短轴）复式背斜和一个长条状复式向斜。褶皱构造轴面在南部倾向西，向北逐渐转为倾向北西，片麻理走向呈北东向且弧形弯曲。南部褶皱枢纽系统地向南倾伏。北部褶皱枢纽向 NE 和 SW 两个方向倾伏。该区整体上是中间上隆而四周下坳，向北东收敛、仰起，向南西撒开、倾伏的片麻岩褶皱隆起。岩层经受三次区域性构造变动的改造：早期为近东西向塑性折叠层式褶皱，中期为北东向至北北东向大型紧闭同斜褶皱，晚期为近南北向开阔褶皱。本区在太古宙末就基本形成了地块陆核，同时差异性运动使南部出现断块并使其上升。沙厂一带受后期东西向应力的影响，断块产生左旋扭动，使先期褶皱向北偏转，从而使其与区域构造线方向极不协调（表 26-1）。

表 26-1　北京市变质构造分区表

Table 26-1　The metamorphic structure partition of Beijing

一级	二级	三级	四级（变质小区）	五级（变质构造带）
Ⅱ华北变质域（陆块区）	Ⅱ-3 晋冀变质区（陆块区）	Ⅱ-3-1 遵化-建平变质地带-古岩浆岩弧 Ar_B（Ar_3）	Ⅱ-3-1-1 西山变质岩区	
		Ⅱ-3-2 冀东变质地带-陆核 Ar_A（Ar_{1-2}）	Ⅱ-3-2-1 沙厂-大漕变质小区	Ⅱ-3-2-1SC 沙厂褶断带
				Ⅱ-3-2-1DC 大漕线状紧密同斜褶皱带
				Ⅱ-3-2-1MJ 马圈子-南口复杂紧闭褶皱带
			Ⅱ-3-2-2 四合堂变质小区	Ⅱ-3-2-2YP 阳坡地尖愣褶皱带
				Ⅱ-3-2-2SH 四合堂半环状紧密同斜褶皱带
				Ⅱ-3-2-2GS 关上倒转褶皱带 Ⅲ A_1^{3-2c}

（二）Ⅱ-3-2-2 四合堂变质小区

该区主要分布在沙厂-大漕变质小区的西北侧，是由四合堂岩群角闪岩相岩石组成的地块。地块内部的构造为一个大型的形态复杂的复式倒转向斜。它的北部受东西向古断裂带控制，南部被后期岩体侵位而受强烈改造（杨振升，1986）；东、西两侧以石城、崎峰茶北东向大型古断裂带与Ⅰ区结合。在太古宙末，该区主体部分已基本固结。五台期、吕梁期构造运动影响到本区深部岩浆底劈上拱，使原来的复式向斜构造进一步压紧和穹起，致使区域构造线围绕云蒙山花岗岩体呈半环状分布。

总之，在整体上，密怀变质地区的构造格局是沙厂-大漕变质小区和四合堂变质小区沿北东向大型断裂带相拼接的格局。下构造层与上构造层顺北西-南东方向彼此相间出现。两构造层均呈北东向延伸。这种近于指状穿插的拼接关系，表明在太古宙末期至早元古宙晚期，具麻粒岩相的下部构造层相对具角闪岩相的上部构造层而言发生了一系列逆冲和长距离的迁移变化，最终形成了太古宙不同的构造层之间的不协调关系。

（三）Ⅱ-3-1-1 西山变质岩区西山变质小区

北京地区晚古生代—早中生代变质岩系主要分布于周口店—香山—黑龙潭一线，其范围长约50km，宽约10km，总体呈NNE向展布。带内褶皱、断裂复杂，发育中低压类型的低绿片岩相、低角闪岩相变质作用，属于不均一多相变质地带，该变质小区总体上表现为中低压变质相作用。变质作用时期主要为印支期和燕山期，基本反映了本区盖层中变质作用的主要特点。

受这一时期变质作用影响，除中生界外，古生界和元古宇均受到不同程度的影响。变质作用与构造变形、变形相与变质相关系十分密切，应力作用强，应力矿物硬绿泥石分布普遍是其突出特点。

伴随构造变形演化，至少有两次区域变质作用发生，一般进变质期与早期变形一致，形成多种类型的变质岩石，主要岩石类型为一套变质程度较浅的片岩、板岩、千枚岩及变质砂岩、砾岩等。变质原岩的矿物成分、结构构造常得以保留，大量煤层经变质形成无烟煤。区内混合岩化作用不发育，而同构造期的基性、中酸性岩浆活动又造成了一些地区热事件的复杂性。

印支期变质作用涉及主要为中下二叠统。受燕山期区域变质影响的地层，最上部地层为下侏罗统窑坡组，下限有待进一步研究。

第六篇

岩石圈结构构造

岩石圈结构构造研究主要依靠物探资料，本区的主要物探工作如下：

北京地区重力工作程度山区低、平原高，山区主要以1：100万和1：20万为主，平原区则在不同年代开展过1：10万和1：5万的重力测量工作。1954年，地质部华北石油普查大队物探队在华北平原（包括北京地区）进行了1：100万重力普查；1959年，北京市地质局物探大队在北京平原区进行了1：10万重力测量工作，测量面积为4960km^2；1959年，石油部科学研究队和北京石油学院在北京地区进行了1：20万重力测量；1960年，地质部中原物探大队完成北纬39°以北地区1：10万重力测量，包括北京平原区；1960年，地质部综合大队开展了燕山地区1：100万重力测量。1966—1969年，石油部物探局646厂在北京平原区及相邻部分地区完成了1：5万和1：10万重力测量，工作面积8210km^2，资料较系统完整；1982年，北京市地质局物探四分队对北京平原区进行了1：5万磁法、重力区域调查工作，面积包括北京幅和通州幅；1986年，北京市地矿局物探大队以1969年石油部物探局646厂1：10万重力测量资料为基础，补充了1961年地质部中原物探队1：10万重力测量部分资料，加以系统整理，统一归算至波斯坦系统，控制面积12000km^2，本次工作提高了北京地区重力资料的研究程度和资料的系统完整性；1988年，地质部第二综合物探大队系统整理了1：100万区域重力测量张家口幅，包括北京地区，布格重力异常精度为$\pm2.0\times10^{-5}m/s^2$；1990年，河北地矿局物探大队完成了密云幅、北京幅1：20万区域重力测量；1992年，地矿部第二综合物探大队完成全国1：20万区域重力测量，该成果为北京地区深部地质构造、区域地质构造研究及矿产预测提供了基础资料。2006年，北京市地质调查研究院完成了1：1万重力剖面测线116.4km；2009年，北京市地质勘察技术院在平谷地区完成重力面积工作360km^2，测网500m×250m；2010—2011年，北京市地质调查研究院在“北京平原区活动断裂监测专项地质调查项目”中完成1：1万重力剖面测线553km，点距50m；1：5000重力面积测量44km^2，布格重力异常总精度为79μGal，测点平面位置中误差为11.4cm，测点高程中误差为2.1cm。

从1956年开始，北京地区共进行了六次系统的航空物探测量。1956年，二机部三局182队在北京山前地区完成了1：2.5万航空磁力测量（简称航磁）和航空放射性测量（简称航放），航磁测量精度为±100nT，这次测量没有编写文字报告；1959年，地质部航测大队901队在北京山区完成了1：10万航磁和航放测量，航磁测量精度为±25nT；1972年，地质部航测大队901队在西山地区完成1：5万航磁测量工作，测量精度为±10nT；1974年，航空物探大队904队在北山地区完成了1：5万航磁测量6860km^2，测量精度为±10nT；唐山地震后，地质部航测大队于1978年在京津唐地区进行了1：20万航空磁法和航空遥感测量；1983年，由航空物探总队在北京完成了1：10万航空遥感综合调查和高精度航磁测量。上述六次航空物探测量覆盖了北京全区，为北京地区基础地质研究、成矿规律研究、城市规划管理和环境保护工作提供了基础资料。特别是在大地构造划分、深部磁性断块构造研究和普查找矿方面起了重要作用。

北京地区地磁工作开展较早，20世纪50年代起各个矿区陆续开展铁矿磁法勘查工作，北部山区不同比例尺的资料比较完备、勘查结果较详细，特别是密云和怀柔大部分铁矿区都开展过航磁异常查证和矿区磁测详查工作；平原区也做过比较详细的地面磁法工作，如1981—1984年北京市物化探队进行了北京平原区1：5万地面磁法区域测量工作，实测面积5404km^2，剖面测量36.5km，涉及昌平、小汤山、怀柔、木林、阳坊、沙河、顺义、杨镇、马驹桥、香河、琉璃河、庞各庄、安次等21个1：5万图幅，该成果报告中圈定了平原区隐伏侏罗系火山岩分布，半定量估算了火山岩的厚度，圈定了隐伏或半隐伏的房山、阳坊、辛相庄、固安、礼贤、草厂等6个岩体边界，推断和修改了部分隐伏断裂构造。

国土资源部于2000年制定并实施的“中国岩石圈三维结构”专项研究计划的成果《中国华北地区岩石圈三维结构及演化》于2007年出版，为全面了解和认识北京地区岩石圈三维结构特征，提供了重要资料。

本篇主要搜集整理已覆盖北京市全区的1：20万区域重力调查、1：5万航空磁测、1：20万化探扫面以及全区岩石物性、地球化学调查等资料；搜集整理穿过或位于北京境内的深地震测深、大地电磁测深剖面等资料，包括国家地震地质研究所在华北地区开展的大地电磁测深、转换波测深和大地热流的探测研究资料；国家地震局地球物理研究所等多家单位分别用工程爆破、天然地震观测、重力和磁测等不同方法对华北地区地壳-上地幔结构进行探测研究的成果资料等，系统研究本区深部地质构造特征。

第二十七章　区域地球物理场特征

第一节　岩石圈速度结构

近年来，许多学者在首都圈利用人工地震测深、地学大断面探测、地震体波层析成像等方法开展岩石圈亦即地壳和上地幔速度结构研究，取得了诸多重大成果。研究发现首都圈速度分布表现为强烈的横向不均匀性，且浅部的速度分布同地表地质结构分布有很好的对应关系。

一、深地震剖面地壳-上地幔速度结构

近些年来，国家地震局地球物理勘探大队先后在北京及周边地区布置的深地震测深剖面，其中经过北京地区的剖面有 16 条（图 27-1，表 27-1），为研究北京及周边地区壳幔结构特征提供了基础资料。

表 27-1　北京地区深地震测深剖面统计表

Table 27-1　The table of deep seismic reflection profiles in Beijing

剖面原始编号	剖面名称	测线长/km	施工时间	参考文献
Ⅱ	北京-张家口-化德	310	1984 年	赵金仁等，2005
H-01	柏各庄-丰宁	410	1979 年，1981 年	嘉世旭，2005
H-02	塘沽-密云	220	1979 年	徐锡伟等，2002
H-03	南堡-康保	约 437	1981 年	嘉世旭，2005
H-04	盐山-大兴-延庆	476	1977 年	任清芳等，1997
H-05	乐亭-张家口	440.95	1981 年	嘉世旭，2005
H-06	宁河-北京-涿鹿	348	1978 年	赵金仁等，1999
H-11	石家庄-喀喇沁旗	746	1982 年	王帅军等，2005
H-20	北京-怀来-丰镇	342	1993 年	祝治平等，1997；宋松岩等，1998
H-22	文安-察右中旗	420	1994 年	张成科等，1997；刘宝峰等，2000
	武清-北京-赤城	320	2005 年	王帅军等，2007
YH-Ⅰ	延庆-怀来	50	1997 年	刘保金等，1999
YH-Ⅱ	延庆-怀来	40	1997 年	
SP-Ⅰ	香河-牛栏山	80	1998 年	张先康，2002
SP-Ⅱ	凤河营-平谷	60	1998 年	张先康，2002
	昌平-宝坻	100	2009 年	刘保金，2009

注：剖面均由中国地震局地球物理勘探中心完成。

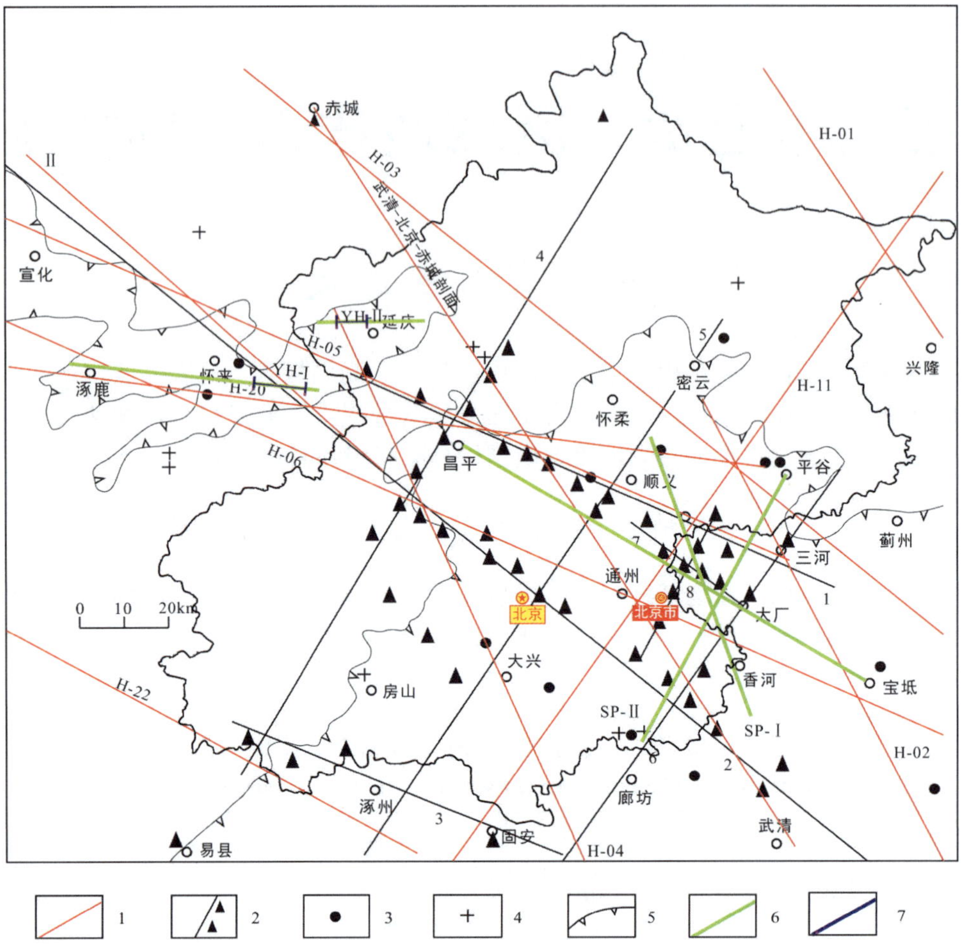

图 27-1 通过北京地区地球物理探测剖面位置图

Fig. 27-1 The location of the geophysical profiles in Beijing

1—人工地震测深测线；2—天然地震转换波测线及测点；3—大地电磁测深测点；4—地热测点；5—山区和平原界限；6—深地震反射剖面测线；7—高分辨率浅层地震测线

（一）宁河-北京-涿鹿剖面（H-06）

1. 剖面概况

剖面由北西向南东测量，依次通过怀安县、涿鹿县、北京市和宁河县，终点与柏各庄-丰南-丰宁-正蓝旗剖面起点重合，全长348km。剖面位于燕山造山带桑干-承德构造带与冀西-冀南微陆块交汇部位，东位于燕山山区与华北盆地过渡带，地质构造复杂，沿剖面地壳横向不均匀性明显。

剖面共布设10个炮点，观测点距为2~5km。西部山区的3个炮点，由于激发条件差，仅在炮点SP152.64获得了较好的地震记录。另外，北京颐和园至通州之间，由于强烈的工业和人为干扰，没有布设观测点。

2. 地壳-上地幔速度结构

沿剖面具有明显的层状结构，地壳结构分为上、中、下三层（图27-2）。

1）上部地壳。基于回折波 P_g 所穿透的地层及 P_g 波之后出现的一组反射波 P_2 所反映的界面作为上部地壳的下界，速度结构大体上由两个速度间断面及正、负梯度层所组成（赵金仁等，1999）。该层顶面地表速度在东部华北平原区即香河以东段为2.00~2.40km/s，且从地表向下0~3.0km深速度为2.0~4.4km/s的低速层基本反映了新生界沉积盖层的地层速度。北京平原区的地表速度为2.50~3.50km/s，西部山区（桩号265）地表速度为3.00~3.50km/s。上地壳0~5.0km深度速度梯度较强，速度变化于 2.00~6.10km/s 之间，而 5.0km 深度至上地壳底则发育为较弱速度梯度，速度变化于 6.10~6.25km/s 之间。上地壳速度结构特征揭示地壳厚度沿剖面具有明显的变化，总体上东薄西厚，香河以东为 14.0~15.0km，香河以西为 15.0~17.5km；境内西山下方最厚，达 17.5km，平原区最薄约为 14km，由东向西变化幅度达 4.0km。

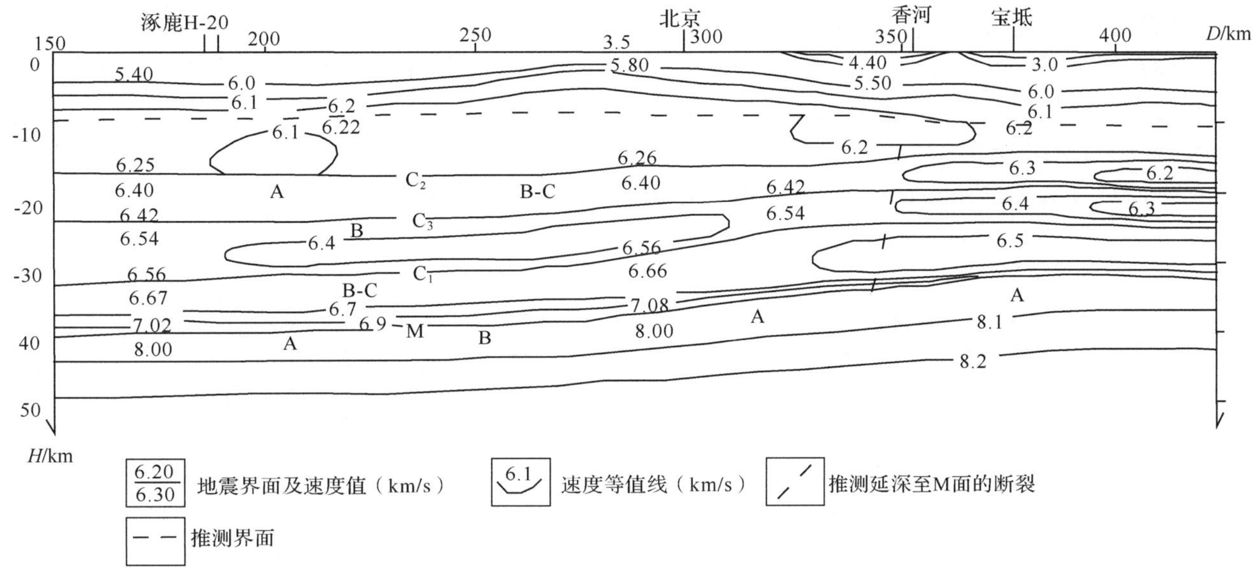

图 27-2　宁河-北京-涿鹿剖面壳幔速度结构图

（据赵金仁等，1999）

Fig. 27-2　Velocity structure of the crust and upple mantle along the Ninghe-Beijing-Zhuolu profile

C_1、C_2、C_3—壳内界面；M—莫霍界面；A—强反射段；B—弱反射段；C—模糊不清反射段；H-20—地震剖面

2）中部地壳。基于壳内反射波 P_2 所反映的界面作为中部地壳的底界，由一个速度间断面和正、负速度梯度层组成。中部地壳的层间速度为 6.2~6.45km/s，剖面西端至香河（桩号 150~350）间由一个较弱的正梯度层组成，速度结构较为简单，没有明显的横向变化，而香河以东段存在一个 6.20~6.30km/s 的低速异常体，比周围正常速度低约 2.3%。中地壳厚度为 4.0~7.0km，在华北平原与燕山褶断带的过渡带附近（桩号 300~360）最薄，厚约 4.0km。中部地壳底界埋深在北京西山至香河（桩号 265~350）间变化较大，由 20km 迅速加深到 24km，揭示了该区段复杂的壳幔构造特征。而北京西山以西和香河以东段横向变化不大，分别处于 20km 和 24km。

3）下部地壳。下部地壳由壳内反射波 P_4 和反射波 P_m 共同反映的地层结构，层内存在着正、负梯度层和两个速度间断面。第一速度间断面速度为 6.56km/s，层内速度 6.30~6.56km/s，在桩号 190~310 和 340~420 的范围内存在着 6.30~6.40km/s 的低速异常体，比周围正常速度低 2.3%~3.8%，其他区段为较弱的速度梯度带。第二速度间断面速度为 7.9km/s，层内速度具有和中部地壳相似的结构特征，即香河以西表现为正梯度点，而香河以东存在一个 6.50km/s 的低速异常体。下部地壳埋深具有和中部地壳完全相似的特征，香河以东及西山以西起伏变化不大，深度分别在 32~34km 和 39~40km 之间，而在香河至西山段则迅速发生变化，在 70~80km 范围内变化达 6km，揭示北京平原区和西山地区所表现的复杂的浅部地质构造在深部地壳结构上也存在较大的差异。

4）壳幔分界面。基于 P_n 波所反映的界面。上地幔表现为一组正速度梯度层，顶部的速度为 7.9~8.0km/s，向下至 50km 深度范围为 8.2 km/s。作为壳幔分界面莫霍面，其埋藏深度在桩号 450 附近为 33km，桩号 360 附近为 33.5km，到通州附近（桩号 310）减少为 31km，这是本剖面中地壳最薄的地方。在北京以西，莫霍面的埋藏深度急剧增加，至白家疃附近深度达 37~38km，再往西似略有抬升，至桩号 200 的涿鹿县小矾山，深度为 37km。

此外，此剖面在通州至香河间（桩号 320）分别存在着切穿莫霍面的深大断裂。

（二）盐山-大兴-延庆剖面（H-04）

1. 剖面概况

剖面起于山东省滨县里则乡大井河（桩号 179. 70），经过盐山、黄骅、静海、大兴至赤城县瓦房沟（桩号 296. 30），全长 476m，方位 333°。剖面的南段位于平原区，北段进入燕山山区，穿过一系列的基底断裂，基岩起伏较大，表现出强烈的横向非均匀特征。

剖面在 8 个爆炸点进行了 13 次吨级爆炸，选用本剖面中 4 炮资料，综合与本剖面相交的 5 条剖面交点附近单炮资料，采用垂向非均匀模式进行地质解释。

2. 地壳-上地幔结构

沿剖面具有明显的层状结构，地壳结构分为表层、上、中、下四层（图 27-3）。

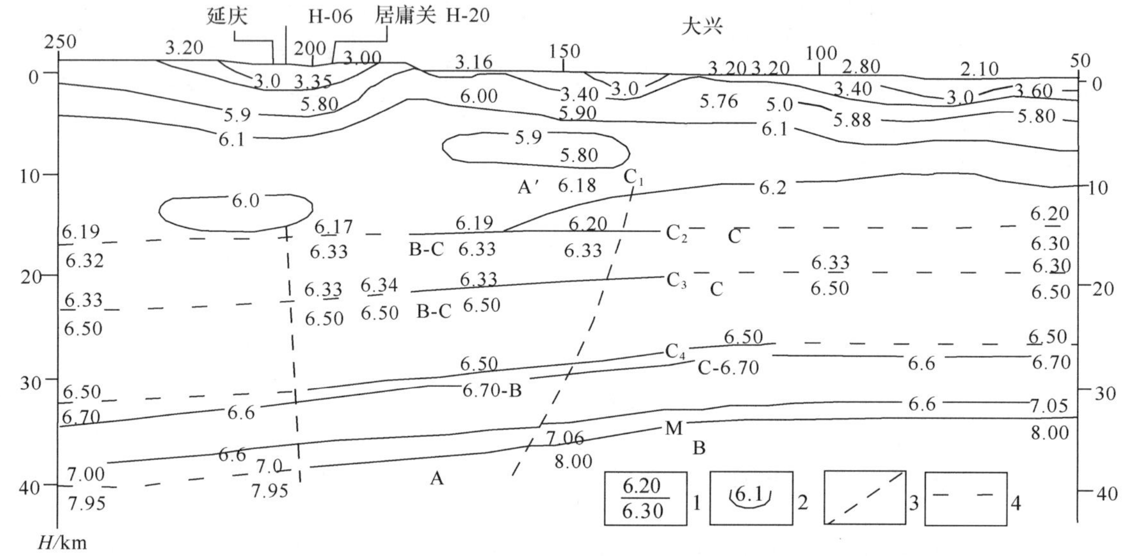

图 27-3 盐山-大兴-延庆剖面壳幔速度结构图

（据任清芳等，1997）

Fig. 27-3 Velocity structure of the crust and upple mantle along Yanshan-Daxing-Yanqing profile

1—地震界面及速度值（km/s）；2—速度等直线（km/s）；3—推测延伸至 M 面的断裂；

4—推断界面；A—强反射段；B—弱反射段；C—模糊不清反射段；

H-06，H-20—地震剖面；C1，C2，C3，C4—壳内界面

1）地壳表层。地壳表层速度值介于 2～4. 8km/s，埋藏深度和高电阻标准层相当。地壳表层厚度变化比较大，北京西北地区表层很薄甚至缺失，往东南方向厚度逐渐增加。在大兴凸起、牛坨镇凸起上厚 0. 5～1. 5km，在廊坊凹陷、武清-霸州凹陷、火山-沉积盆地内，厚度达 3～4. 5km。

2）上部地壳。上部地壳为一高速层，由一个速度间断面和一组正速度梯度层组成，层速度为 4. 8～6. 2km/s，层厚度 9. 0～12km，平原区相对较薄，山区较厚，底界埋深为 12. 5～14. 5km。

3）中部地壳。中部地壳是一个低速层，由一个速度间断面和一组匀速层和弱的梯度层组成，层厚度 11～12. 5km，最低速度为 5. 6～5. 81km/s。在剖面的桩号 100 至桩号 72 附近，低速层厚度显著地减薄为 7. 5km，这里低速层下部可能存在着一个不很明显的高速层。在部面的南段，炮点 SP26. 38 的记录显示在中部地壳与上部地壳之间也存在一个高速夹层，其速度为 6. 8km/s，厚约 3km。

4）下部地壳。下部地壳为一正速度低梯度层，速度梯度 0. 11～0. 16km/s，其底部速度为 7. 6km/s，层厚度为 9～13km。

5）壳幔分界面。图 27-4 是盐山-大兴-延庆剖面的莫霍面深度图，计算得到上地幔顶部速度为 8. 0～8. 3km/s。

地壳厚度变化的总趋势是由东南部向西北山区厚度逐渐增加，盐山县附近为 31km，京西高崖口附

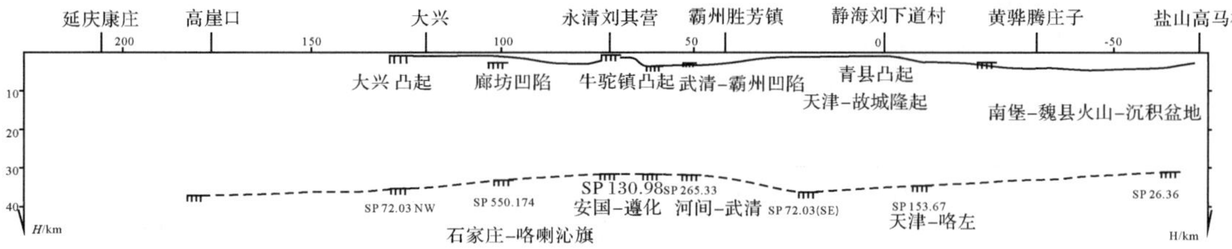

图 27-4　盐山-大兴-延庆剖面地壳厚度图

Fig. 27-4　Crustal thickness of Yanshan-Daxing-Yanqing profile

近达到 37km。剖面中段莫霍面呈波浪形起伏，大兴至霸县为莫霍面隆起区，霸县-静海为下凹区，最深点位于静海县梁头乡（桩号 22）附近，深度为 36.3km。

大兴-霸县莫霍面隆起区的位置，大体上与廊坊-衡水火山沉积盆地的位置相当，而霸县-静海莫霍面的相对下凹则与天津-故城隆起相对应，静海以东莫霍面的抬升，又呈现与南堡-魏县火山-沉积盆地反向对应的关系。

（三）塘沽-密云剖面（H-02）

1. 剖面概况

剖面全长 220km，方位 338°。剖面东南段为华北盆地，西北段位于燕山山区。塘沽-密云剖面通过天津和北京的东部地区，同时穿越了宁河震区、三河-平谷历史大震区和宝坻小震活动区，并且在三河、宝坻附近分别与乐亭-张家口剖面和宁河-北京-涿鹿剖面相交。

2. 地壳上地幔结构

利用垂向非均匀介质的地壳模式，通过在 TQ-16 机上对炮点 SP3.56、SP69.38、SP101.5 的实测走时进行理论走时曲线拟合，计算得到了剖面地壳速度结构图（图 27-5）。

1）地壳表层及上部地壳。大致可分为 3 层，第一层的速度为 2.20～4.60km/s，厚度约 0.2～4km；第二层的速度为 3.20～6.00km/s，厚度 2～3km，第三层的速度为 5.9～6.48km/s，厚度为 6.4～7km。上部地壳总厚度为 10～13km，南部较厚，北部逐渐变薄。

2）中、下部地壳。总厚度为 20～25km，由南向北厚度逐渐增大。该层的上部为一低速层，厚约 7～10km，最低速度值 5.9km/s。其下部位一速度梯度层，速度梯度为 0.148～0.158km/s/km。梯度层底部的速度为 6.75～7.80km/s。根据炮点 SP3.56 的资料计算，中下部地壳中间存在着一个高速夹层，该高速层的埋藏深度为 20km，厚约 6～7km；剖面的北部，该高速层不明显（图 27-5）。

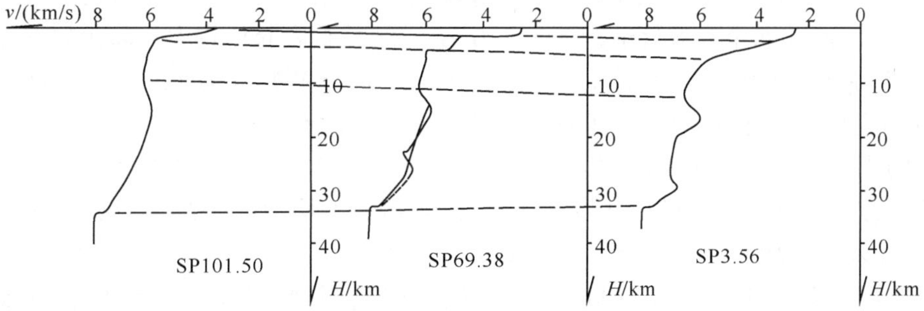

图 27-5　塘沽-密云剖面速度结构图

Fig. 27-5　Velocity struction of Tanggu-Miyun profile

3）壳幔分界面。由炮点 SP3.56 获得的上地幔顶部的速度为 8.10km/s。利用 t^2-x^2 法求取的 P_m 波的平均速度在剖面的南部为 6.10km/s，向北逐渐增大到 6.20km/s，炮点 SP101.50 计算出的 P_m 波波速达到 6.30km/s。分别用 6.10km/s 和 6.20km/s 的速度值计算了剖面上莫霍面的深度值，其结果与理论走时曲线拟合的结果基本一致（图 27-6）。

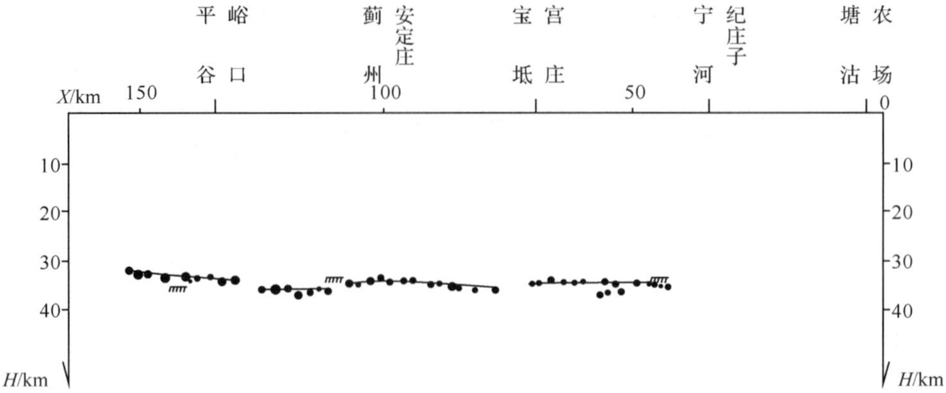

图 27-6　塘沽-密云剖面地壳厚度图

Fig. 27-6　Crustal thickness of Tanggu-Miyun profile

4）从图 27-7 可以看到，从宁河纪庄子至宝坻宫庄，莫霍面的深度为 33～34km。在宝坻区石桥林附近（桩号 76）莫霍面有一局部变化，由桩号 70 的 33km 突变为桩号 77 的 35.5km，该突变带与山前断裂的位置相当，因此，山前断裂可能是一个断至莫霍面的区域性断裂。从山前断裂向北，莫霍面逐渐抬升，在天津蓟州安定庄附近为 33km，北京平谷峪口附近也是 33km 左右。在它们之间的地区，莫霍面可能存在着局部的下凹，最大深度为 35km。1679 年三河-平谷大地震正是位于莫霍面埋深发生较大变化的地方。

（四）武清-北京-赤城剖面

武清-北京-赤城高分辨深地震测深探测剖面由武清穿过香河、北京、顺义、延庆，止于河北省赤城县，全长 220 余千米。共有观测点 205 个，观测点距为 0.5～2.50 km。

1）该宽角反射/折射剖面揭示了冀中坳陷至太行山隆起区为地壳厚度变异带，其厚度变异带展布为东浅西深。二维速度结构与深部构造（图 27-7）特征表明，沿剖面具有明显的层状结构，并且在不同的地质构造单元内各层界面表现出复杂多变的起伏状态，总体上自南东向北西呈逐渐下倾趋势，莫霍面表现尤为明显。莫霍面自北京向北西至 330 km 桩号左右，由 32.0 km 迅速加深至 40.0 km，莫霍面起伏变化的幅度达 8.0km 左右，表明了华北平原、燕山褶断带和太行山隆起交汇区及其周围复杂的浅部地质构造现象与其深部复杂的构造环境有密切的关系。

2）结晶基底及上地壳的厚度变化较大，结晶基底由华北平原区向燕山山区逐渐减薄，武清下方基底的厚度为 7.3 km，向北西至赤城逐渐减薄为 1.0 km。结晶基底起伏较大，起伏幅度达 6.3km。上地壳的厚度由华北平原区向燕山山区逐渐加厚，武清下方上地壳的厚度为 13.0 km，向北西至延庆逐渐加厚为 16.5 km，变化幅度达 3.5 km。其平均速度为 5.80～6.05km/s。

3）中地壳的厚度为 5.0～8.0 km，其底部深度在 18.0～23.5 km 之间变化，由东部平原向西部山区逐渐加厚，该层为一弱梯度层或匀速层，横向存在强烈的不均匀性，该层把脆性的上地壳和相对塑性的下地壳区分开来。平均速度为 5.85～6.19 km/s。

4）下地壳厚度一般为 13.0～17.0 km，南东薄北西厚，其底界面（莫霍面）的深度变化范围为 31.5～40.0 km，可见下地壳的厚度和深度的变化幅度均大于中地壳。在冀中坳陷区下方该层的厚度和莫霍面的埋深明显小于燕山隆起区，表明华北新生代坳陷的形成与地幔隆起有直接的关系。其平均速度为 6.10～6.42 km/s。

5）速度结构在纵向与横向上具有明显的非均匀性，呈现出南东低北西高的基本结构特征，各区段之间在横向上存在的差异特征进一步表明华北平原区与燕山褶断带过渡区的壳幔构造的复杂性。这些异常的壳幔速度结构和构造的形成与地球内部的地质演化过程有关，即上地幔隆起引起上地幔物质侧向流动并沿地壳深断裂或不连续界面的地段入侵到中下地壳，致使壳内介质温度升高，甚至部分块体出现熔融现象，可能导致壳内速度异常体和复杂的构造格局的形成。

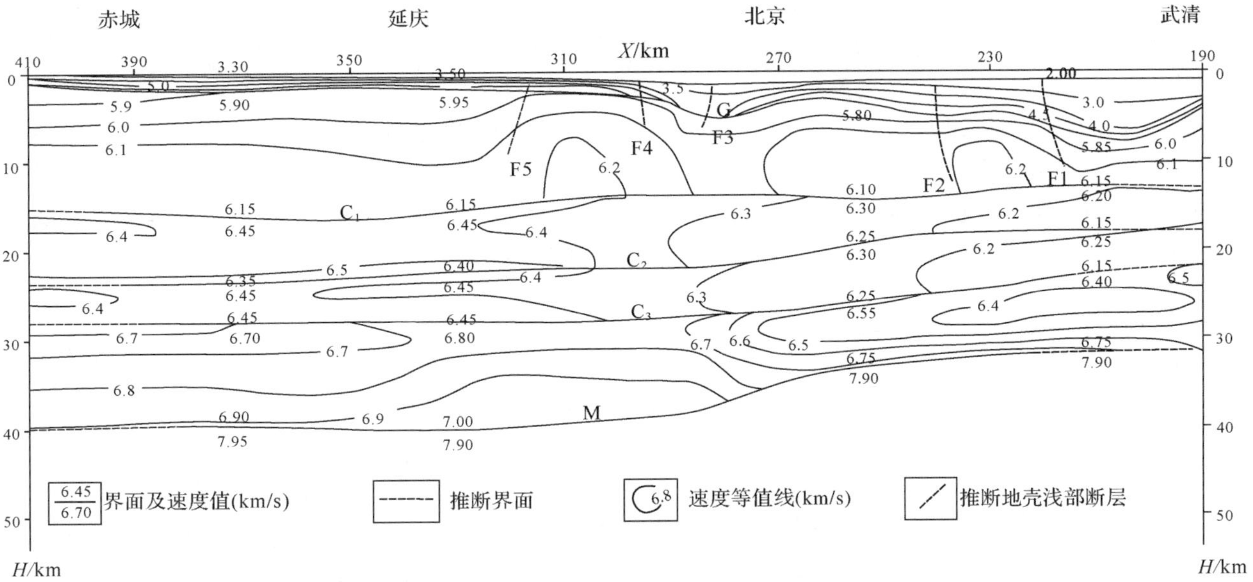

图 27-7　武清-北京-赤城剖面二维速度结构与构造图

（据王帅军，2007）

Fig. 27-8　2-D crustal velocity structure and tectonic along Wuqing-Beijing-Chicheng profile

G—结晶基底；C1，C2，C3—壳内反射界面；M—莫霍面；F1—宝坻断裂；F2—夏垫-凤河营断裂；F3—顺义断裂；F4—黄庄-高丽营断裂；F5—紫荆关断裂

（五）北京-怀来-丰镇剖面（H-20）

北京-怀来-丰镇（H-20）剖面，东起北京市平谷区峪口（东经 116°59′48″，北纬 40°11′57″），向西经顺义、昌平、延庆、怀来、天镇至内蒙古自治区丰镇县（东经 113°06′00″，北纬 40°29′11″），全长 342km。在剖面中段的延怀盆地，该剖面与反射剖面 YH-Ⅰ基本重合。剖面自东向西经过了燕山与华北平原的过渡带，八达岭、延怀盆地及深井、怀安、天镇-阳高等山间盆地。该剖面位于华北裂谷、燕山褶皱带与山西断隆的接合部分。穿过了太行山重力梯度带、张家口-渤海地震带和一系列北北东、北东向为主的断裂。图 27-8 为速度结构图。

上地壳上层厚度自东向西逐渐加大，由顺义、昌平的 8.0 km 增加到天镇附近的 10.5km。该层地表部分为新生代和部分中生代的沉积盖层，厚度变化较大。山区很薄或缺失，盆地则较厚，可达 0.2 km。S 波的速度变化较大，在盆地内一般由 1.20 km/s 增至 1.70km/s 左右，在山区则由 1.65 km/s 增至 2.00 km/s。在中、新生代沉积层下，为一速度梯度层。延庆以东地区，速度由 3.25 km/s 增至底部的 3.52 km/s；在剖面中部的怀来—涿鹿一带，速度由 3.15km/s 增至 3.45 km/s，比两侧偏低；剖面西部，速度由 3.38 km/s 增至 3.51 km/s。上地壳下层，厚约 6.5～8.5 km，为一匀速层，速度为 3.45～3.52km/s。在天镇以西地区，C1 界面下存在一低速体，最低速度为 3.40 km/s。C2 界面从整体上看，基本上是东浅西深。在昌平附近深度为 15.5 km，向西缓慢增加，至天镇东，深达 17.5 km，而后又抬升至 17.0 km 左右。

中地壳内，昌平至涿鹿西间存在一低速体。其核心部位位于延庆—涿鹿间，速度为 3.50km/s。与周边相比，降低了 0.20 km/s。剖面的东、西两段基本上为匀速层，东部速度为 3.70 km/s，西部速度为 3.83 km/s。C3 界面的起伏与 C2 界面基本一致，东浅西深。东部深约 22.5 km，天镇以东深为 25.0km，以西界面上抬至 24.0 km。

下地壳厚度由东端平谷附近的 12.0 km 左右增至西端丰镇的 18.0 km 左右。下地壳顶部速度较为稳定，为 3.65 km/s。只在怀来下方，速度略有减小，为 3.60 km/s；而在下地壳底部速度由东向西从 3.85 km/s 逐渐增到 3.96 km/s。

本剖面的莫霍面形态与壳内各界面形态基本一致，东浅西深。昌平附近深为 37.5 km，向西缓慢加深，到剖面西部天镇附近，深约 42.5 km。上地幔顶部的 S 波速度由东向西逐渐升高，在东部的昌平—延庆一带为 4.27 km/s，至西部的天镇附近达 4.45 km/s。

该剖面所揭示的地壳厚度由东端的 34.0 km 向西逐渐增大到 44.0 km，怀来地区地壳厚度比以往推测的 35.0~36.0 km 要大，为 39.0~40.0 km。地壳中存在 3 处低速体，它们是上地壳下层，天镇附近最低速度为 3.0 km/s 的低速体；中地壳下层，延庆—怀来—涿鹿附近的最低速度为 3.50 km/s 的低速体和延庆以东地带壳幔过渡带内，速度为 3.70 km/s 的低速体。

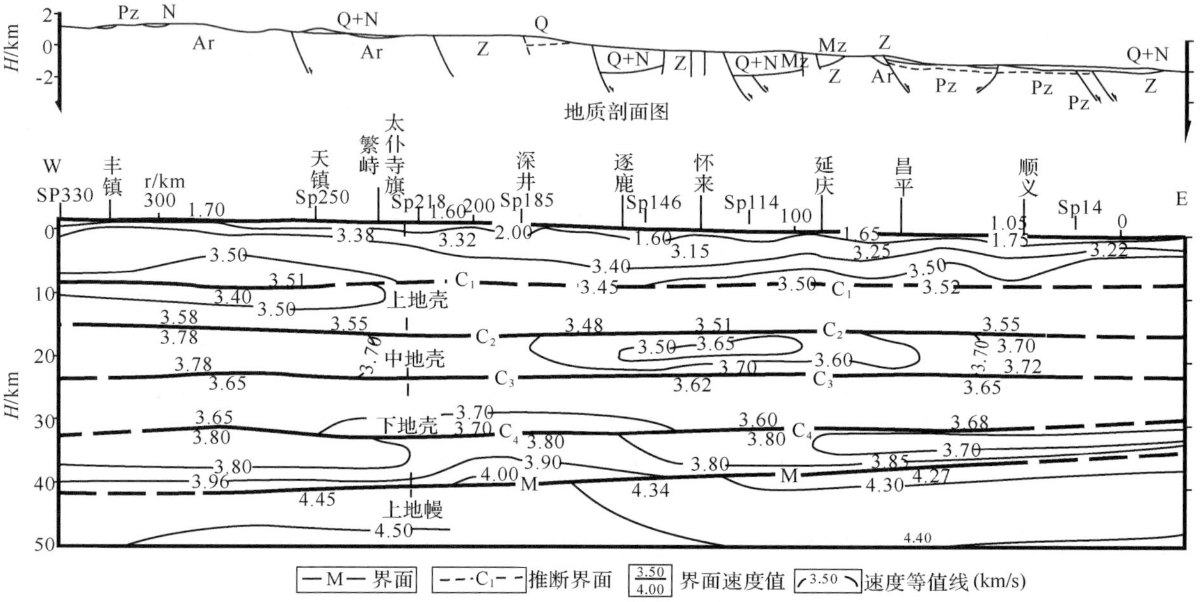

图 27-8　北京-怀来-丰镇剖面地壳—上地幔 S 波二维速度结构图

（据宋松岩，1998）

Fig. 27-8　Three-dimensional S-wave velocity structure of the crust and upper mantle along Beijing-Huailai Fengzhen profile

（六）昌平-宝坻深地震反射剖面

该剖面的东南端点位于天津市宝坻区北约 2 km 的西河务村（117°18′07″E，39°46 ′19.4″N），终点位于北京昌平区何营村附近（116°17′35.5″E，40°13′01.2″N），长度为 100 km。深地震反射剖面的数据采集采用道间距 20 m、炮间距 100 m、600 道接收、60 次覆盖的观测系统，采用了在排列内部激发、双边不对称零偏移距接收的工作方法。剖面自西向东穿过的断裂主要有：小汤山-东北旺断裂、黄庄-高丽营断裂、顺义-良乡断裂、通县-南苑断裂、夏垫断裂和香河断裂等。

该剖面获得了浅至第四纪覆盖层底界、深达莫霍面的地壳精细结构（图 27-9）。地壳结构在纵向上具有明显的分带性，在横向上显示出以断裂为界的块状结构特点。

1. 上地壳特征

深度约 18~21km 以上为上地壳。上地壳以 TWT2.5~4s 的反射波 TG 为界还可进一步分为上、下两个部分，即上地壳上部和上地壳下部。在上地壳上部，剖面反射界面形态和反射波组构，以桩号 40km 为界具有明显不同的特征，桩号 40km 以西，剖面揭示了 2~3 组反射能量较强、横向连续性较好的反射波；桩号 40km 以东，出现在剖面上的是一套产状自东向西倾伏的密集反射层，这套反射具有典型的沉积盆地特征，沉积盆地最深处约为 8~9km. 从上地壳上部出现的一系列层组分明、横向分段连续性较好的地层反射波场特征来看，这部分地壳物质可能为新生代、中生代和古生代的沉积岩系.

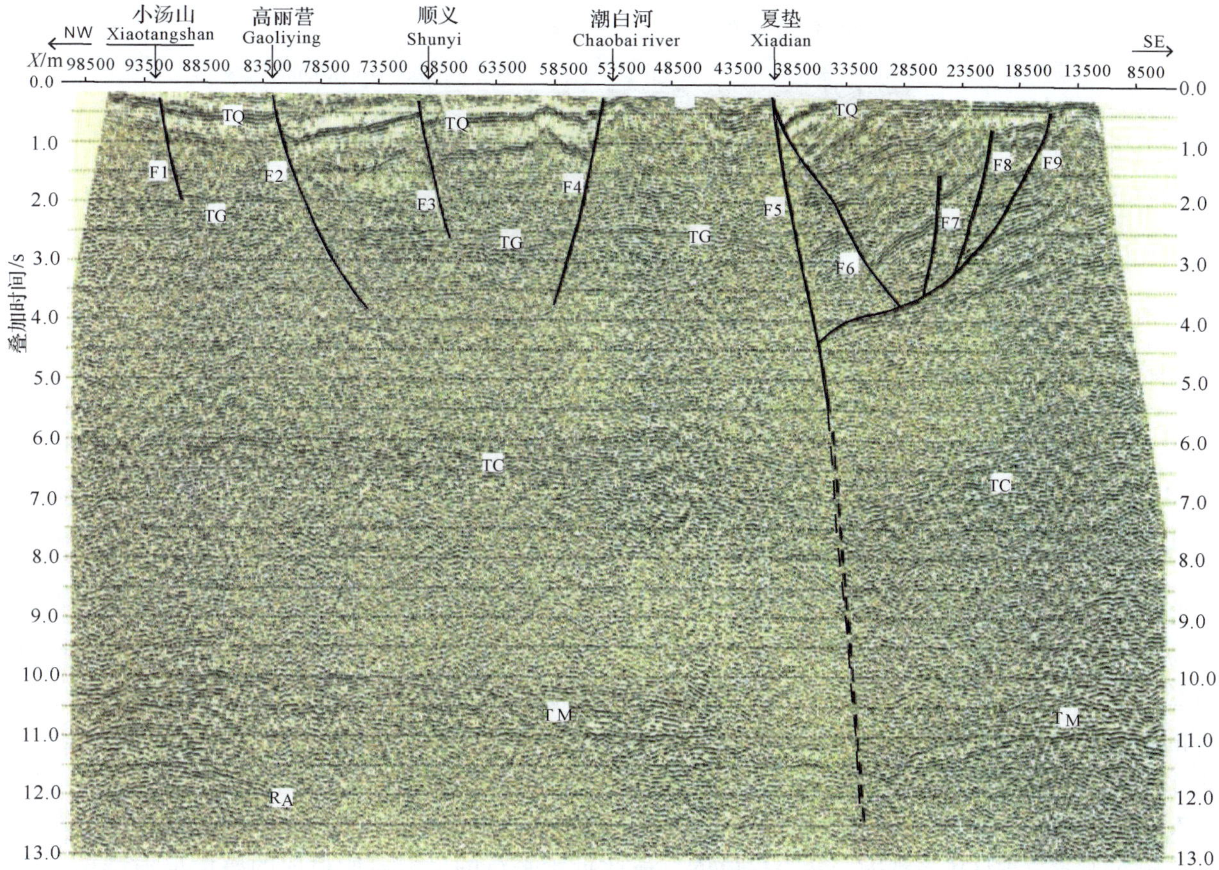

图 27-9 昌平-宝坻深地震反射叠加时间剖面和解释结果图

（据刘保金，2009）

Fig. 27-9 Stacked time section and its interpretation along Changping-Baodi deep seismic reflection profile

在 TWT4~6s 之间的上地壳下部，剖面揭示了一些延续长度较短、反射能量较弱、且无规律可寻的凌乱反射事件，表明这部分地壳结构具有明显的“反射透明”性，推测该部分地壳物质可能为元古代及其以前的变质岩系。

深地震反射剖面自西向东依次穿过顺义盆地、通州隆起、大厂盆地和香河凸起。剖面浅部的一系列反射层也清楚地显示出了这种盆地和隆起相间的构造格局。在盆地和隆起的交接部位，剖面反射层还出现有反射波同相轴的明显错断以及反射波能量的突然变化，表明这些隐伏盆地均受到断裂的影响与控制。根据顺义和夏垫附钻孔地质资料，把剖面浅部的反射波 TQ 解释为第四纪覆盖层的底界，其厚度沿剖面具有较大的起伏变化，在高丽营、顺义及夏垫附近，第四系厚度约为 600~700m，在香河凸起（桩号 29 以东）、通州隆起（桩号约 40~54）和京西隆起（桩号 93 以西）等地段上，其厚度小于 300~350m。

反射波 TG 被解释为来自结晶基底的一组反射，该反射波在剖面上起伏变化形态清楚，横向上可被连续追踪，且具有明显的横向分段性，局部地段上，基底面被断裂切割，地层显得比较破碎。与深地震反射剖面近于重合的高分辨地震折射剖面结果（图 27-10）表明，结晶基底（P 波速度 5.8~6.0km/s 的等值线）沿剖面具有较大的起伏变化，在夏垫以东的大厂隐伏盆地内，基底埋深约为 7~8km，顺义盆地内深度约为 5.5~6.0km，在通州隆起上（桩号约 40~54），其深度约为 4km，在高丽营以西的剖面段上，基底面自西向东倾伏，其深度约在 2.5~3km 之间变化。对比图 27-9 和图 27-10 可以看出，二者所反映的基底面形态和埋深具有较好的一致性。

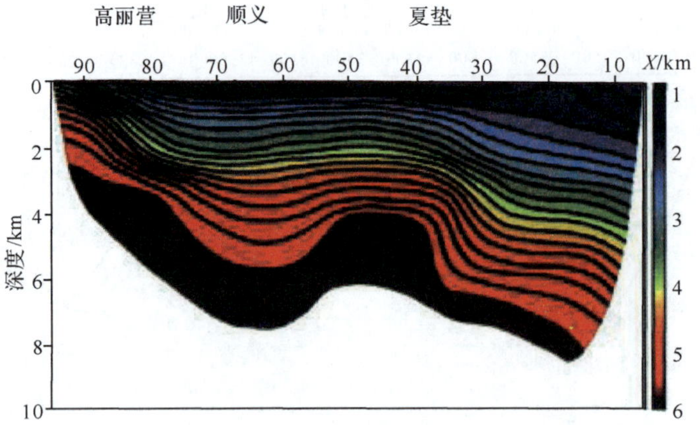

图 27-10　昌平-宝坻剖面高分辨地震折射 P 波速度结构图
（据刘保金，2009）
Fig. 27-10　The P-wave velocity structure of high resolution seismic

2. 下地壳特征

位于反射叠层 TC 和 TM 之间的部分为下地壳。与结构复杂的上地壳相比，下地壳结构相对简单。在夏垫（三河-平谷地震区）以西，下地壳总体表现为弱的反射性质；夏垫以东，下地壳主要以倾向北西的强反射为主，从 TWT6.5~11.5s，该区段下地壳内均有不同强度的倾斜反射存在，此外，还夹有近水平的层状反射和少量南倾的反射震相，构成了该段剖面上最明显的反射特征。

3. 壳幔过渡带

壳幔过渡带反射 TM 表现为一套小尺度的、彼此近于平行的反射段所组成的叠层反射结构，剖面上大约在 TWT 10s 左右开始出现，对应顶界面深度约为 30~31km；壳幔过渡带在纵向上的持续时间约为 1~1.2s，对应壳幔过渡带厚度约为 3~3.6km，其底界对应于莫霍面的位置，其深度约为 34~35km。壳幔过渡带反射波在剖面上比较清晰，横向上可被连续追踪，在夏垫（三河-平谷地震区）附近，壳幔过渡带反射能量明显变弱或出现漫散射。在壳幔过渡带的内部，反射震相的多相位具有明显的横向变化，且内部反射同相轴之间多呈近水平关系，并伴有相交、错断、弱振幅或无反射。表明剖面经过地区的壳幔过渡带存在着强烈的横向不均匀性。PmP 震相大约从折合走时 2.8s 一直延续到 4s 左右，其持续时间为 1.2s，若按平均速度 6km/s 计算，则壳幔过渡带厚度为 3.6km。

二、三维速度结构特征

（一）研究主要进展

刘昌铨等（1982）在京津及其外围地区用点测深方法观测宽角度反射波，莫霍界面反射波 P_m、壳内中间层 P_c 反射波组，以及结晶基底首波 P_g 与莫霍面折射波 P_n，选用水平均匀分层介质地壳结构模式，应用阻尼最小二乘法反演速度结构，结果表明该区速度随深度有较明显的成层分布规律，同时也说明该区的速度横向不均匀性是存在的。

李松林等（2001）对多条二维剖面资料进行统一分析和解释，通过建立统一的偏导矩阵，将各剖面资料联立求解，利用该方法对中国地震局地球物理勘探中心在首都圈内 6 条人工地震宽角反射/折射剖面资料进行了重新处理和解释，得到首都圈的地壳三维速度结构和构造特征并结合其他地球物理研究成果，确定了壳内存在的 4 处低速层（体）的空间范围。

嘉世旭等（2005）通过对首都圈地区不同时期的 18 条深地震测深测线资料的重新统一处理，形成采样网格密度 0.25°×0.25°×（2~5）km 的速度网格化数字地壳，通过对三维数据的可视化，得到了首都圈地壳不同走向、不同圈层的截面图像。

谭毅培（2009）综合基础地理信息数据、数字化地质图、沉积层的地脉动探测结果、人工地震测深和宽角折射等多种资料，在对数据进行分析和整理的基础上，利用建模工具软件 GeoCAD，建立了北京地区（东经 115.5°~117.5°，北纬 39°~41°）地壳及上地幔顶部 P 波速度结构和地质界面模型。

另外，还有很多专家学者对华北地区进行过地震测深剖面的综合分析等工作，得到了多个华北地区地壳及上地幔的速度结构，为充分总结北京地区的地壳上地幔的速度结构提供了资料支持，并且可以对结果进行相互验证。张成科等（2002）利用渤海湾及其邻区的 10 多条（通过北京 3 条）地震测深剖面段观测资料，对部分剖面进行二维射线追踪、走进拟合及合成地震图计算，获得了渤海湾及邻区地壳上地幔速度结构。段永红等（2002）采用正方形网格对华北地区速度模型进行参数化，通过对华北地区 30 条深地震测深剖面的 P_g 波三维有限差分速度层析成像反演，获得了该地区上部地壳的三维速度结构。魏文博等（2007）在发表的深地震测深剖面资料的基础上，选择了 14 条测线的资料（其中包括一条穿过北京的北京-怀来-丰镇深地震测深剖面），利用地理信息系统（ARCINFO）的"矢量化"功能，以及克里格数据网格化技术构建华北区域性的地壳三维速度结构模型。张亚明等（2008）综合京津唐及邻区的 8 条宽角反射地震剖面地壳速度结构，计算了该区域不同构造域上、中及下地壳平均速度和厚度。

（二）主要成果

1. 层圈速度结构特征

李松林等（2001）对多条二维剖面资料进行统一分析和解释，通过建立统一的偏导矩阵，将各剖面资料联立求解，利用该方法对中国地震局地球物理勘探中心在首都圈内 6 条人工地震宽角反射/折射剖面资料进行了重新处理和解释，得到首都圈的地壳三维速度结构和构造特征并结合其他地球物理研究成果，确定了壳内存在的 4 处低速层（体）的空间范围。

地壳大致为两层结构。由地表至 P_c 波的反射界面（康氏界面）之间为上地壳，康氏界面以下到莫霍界面之间为下地壳。地壳基本速度结构和构造特征以及低速体特征介绍如下：

（1）上地壳

上地壳包括沉积盖层和其下的结晶基岩，它由 3 个不同的速度梯度层组成：第一层为新生代低速盖层，速度一般为 2.00~3.50km/s；第二层为一强速度梯度层，速度从盖层以下的 3.00~4.20km/s 急增至结晶基底顶部的 5.70~5.90km/s；第三层为下部的弱速度梯度层，其速度由 6.10km/s 左右增至底部的 6.40km/s。结晶基底起伏变化较大，最深处为 7.00km，最浅处接近地表。区域内上地壳厚度变化不大，总的趋势是东薄西厚。最浅部位于东部的三河、宝坻一带，为 20~21km，北京附近为 22km，至张家口北西加厚到 23~24km。在北京-涿鹿间上地壳较厚，为 24km。由蔚县向涞源方向上地壳迅速变厚，至研究区边缘约为 26km。其他地区上地壳厚度大致在 22~ 24km 之间变化，且存在着局部起伏。

（2）下地壳

下地壳基本上由两层组成，上层为一弱速度梯度层，速度为 6.5~6.6km/s。下层为一强速度梯度层，在靠近莫霍界面 3.0~4.0km 左右，速度值由 6.7km/s 迅速增加到 7.1km/s，反映了壳幔过渡的特征。研究区下地壳厚度自东向西逐渐加厚，东部下地壳厚约 10km，西部最厚为 19km。涞水、北京、密云的东南下地壳等厚线走向为北东向，由天津附近的 10km 向北西逐渐变厚，涞水、北京、密云一线为 14km。西部以昌平、怀来、宣化为界分为南北两部分：北部地区下地壳等厚线走向为北东东方向，其东部的昌平北部下地壳厚约 15km，向西逐渐变厚；其下地壳等厚线的走向呈南东方向，下地壳厚度由昌平、怀来附近的 16km，向南西逐渐变薄。北京西至涿鹿间有一下地壳较薄区，其度约 15km。

（3）莫霍界面

区域内地壳由南东向北西方向逐渐变厚。东南部华北坳陷区地壳厚 31~34km，由天津向北西逐渐变厚，天津地壳厚 31~32km。由大兴、保定向北西地壳急剧变厚，由 33~34km 很快变为 37km，说明平原区临近山区的边缘地带为平原区地壳与山区地壳的过渡带。北京、易县、涿鹿之间莫霍面较平缓，地壳厚度约为39km。由该带向北西地壳变厚，张家口一带厚42km。蔚县、涞源向南西地壳厚度变化不大，一般为40km。

（4）壳内异常体分布

1）延庆怀来地区。其范围与延怀盆地基本一致，顶面深度为11.0~13.0km，底面深度为15.0~17.0km。速度为6.00~6.10km/s，与围岩的相对速度差为2%~3%（图27-11）。

2）太行山燕山山前地带。该低速区位于紫荆关断裂带和太行山前断裂带之间，呈北东向条带状。上地壳内有两个低速层：第一个低速层的顶面深度为10.0~11.5km，厚度为2.0~4.5km，速度为6.00km/s，主要分布在昌平、易县、容城一带；第二个低速层的顶面深度为16.2~17.5km，厚度为3.0~3.8km，速度为6.20~6.30km/s，主要分布在易县、涿州、容城、定兴、雄县一带。上地壳低速层的速度与围岩的相对速度差为2%（图27-11）。

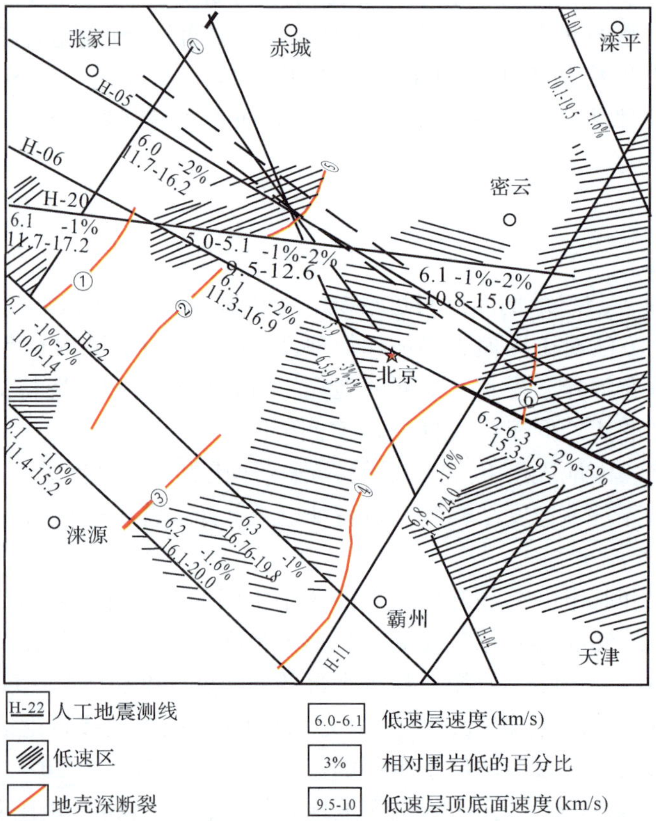

图27-11　首都圈低速层（体）分布

（据李松林，2001）

Fig. 27-11　Distribution of low velocity layers（bodies）in Capital area

①涿鹿-阳原深断裂带；②怀来-灵丘深断裂带；③紫荆关深断裂带；④太行山前深断裂带；⑤延庆深断裂带；⑥夏垫深断裂带；⑦张家口-渤海深断裂带

2. 不同走向不同深度截面的速度结构特征

嘉世旭等（2005）通过对首都圈地区不同时期的18条深地震测深测线资料的重新统一处理，形成采样网格密度0.25°×0.25°×（2~5）km的速度网格化数字地壳，通过对三维数据的可视化，得到了首都圈地壳不同走向、不同圈层的截面图像。成果如下：

（1）结晶基底特征

北京西部山区、北部燕山隆起区结晶基底埋深一般约1~2km，延庆-怀来盆地约3km；东南部裂陷盆地结晶基底埋深3~7km，差异较大，盆地内裂陷区（冀中坳陷）6~7km，位于冀中坳陷北部的大兴凸起埋深约3km；裂陷盆地西缘沿太行山前（怀柔-涿县）断裂东西两侧，盆地北缘沿燕山山前（宝坻-桐柏、宁河-昌黎）断裂南北两侧等值线密集，基底由隆起区的约1km迅速加深到坳陷区的约4km。结晶基底埋深在隆起的造山带较为一致，主要变化集中在裂陷盆地，体现了新生代以来太行山、燕山隆起抬升和裂陷盆地内部不同块体的沉降差异（图27-12a）。

（2）莫霍界面深度构造

首都圈区域地壳厚30~42km，由东南向西北逐渐增厚。东南部盆地区域地壳厚30~34km，冀中、黄骅坳陷地壳较薄，为30~31km，沧县隆起带地壳厚度约32~33km，唐山西北莫霍面局部隆起、厚约31km，太行山前地带迅速增厚至36km，太行隆起区为38~40km，燕山地壳显示了由东（渤海边）约32km向西逐渐加深至（张家口东侧）约41km，西北部地壳厚度最大约41~42km。首都圈地壳厚度等值线主要显示为北东向构造，在张家口-渤海地震带西段（北京—怀来）等值线向西局部凸出，北京以东等值线由北东向变为北东东走向，东南部盆地内的等值线圈闭及走向与不同地质构造单元相对应（图27-12b）。

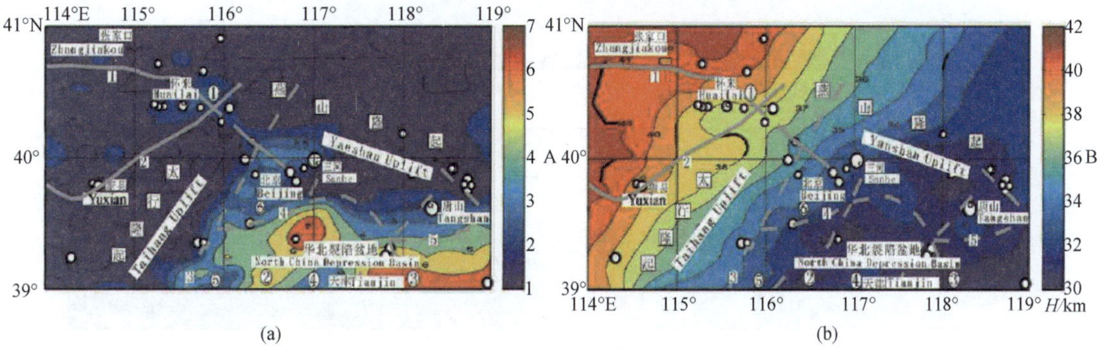

图27-12 首都圈结晶基底（a）和莫霍界面深度结构（b）

（据嘉世旭，2005）

Fig. 27-12 Crystalline basement (a) and the Moho discontinuity depth (b) beneath the Capital area

①延怀盆地；②冀中拗陷；③黄骅拗陷；④沧县隆起；⑤容县凸起；⑥大兴凸起

1—怀来-宣化断裂；2—怀来-蔚县断裂；3—怀柔-涿州断裂；4—宝坻-桐柏断裂；5—宁河-昌黎断裂

（3）不同走向P波速度剖面

剖面图（图27-13）主要显示了结晶基底、莫霍界面埋深和壳内速度构造。一般来说，沉积岩层介质速度小于5.4km/s，壳内介质速度为6.0~6.8km/s，壳幔过渡带速度小于7.4km/s。

剖面A：由西南沿北东走向经太行隆起穿越延怀盆地、燕山隆起，长约320km。剖面中段延怀盆地上地壳存在约7~15km深向西南倾斜的低速构造；太行山西侧、燕山地壳仅在上地壳约15km深处有轻微速度逆转，深度约17~23km处出现密集等值线，速度由6.2km/s增加至6.4km/s，显示了上、下地壳的分界，上、下地壳内部速度分别为6.0~6.2km/s和6.4~6.6km/s，为相对稳定的地壳构造特征。

剖面B：沿太行山前，由北东穿越三河-平谷强震区、燕山隆起东部，长约350km。南部裂陷盆地西侧边缘沉积厚度明显变薄，约2~5km，北部燕山隆起小于2km；地壳厚度沿剖面变化不大，约34~35km，与太行山重力梯度带等值线穿过燕山向北东方向平稳延伸、向北西方向迅速减小对应。在三河-平谷强震区上、中部地壳约8~22km深处显示了向南倾斜的低速6.0~6.1km/s块体构造。

剖面C：沿燕山南缘、张家口-渤海断裂带，穿越唐山、三河-平谷强震区和延怀盆地，长约500km。剖面显示出明显的分段构造特征：大致可分为延怀盆地以西，延怀盆地至三河及三河至唐山等三段，显示了由西北向东南上地壳低速体逐渐加深、加厚的构造特征。北京主要位于中段延怀盆地至三河段，上地壳低速层下移（6.1km/s、深约8~15km）、厚度增加（约5km），地壳厚度由延怀盆地（38km）向西北（张家口41km）迅速加大。

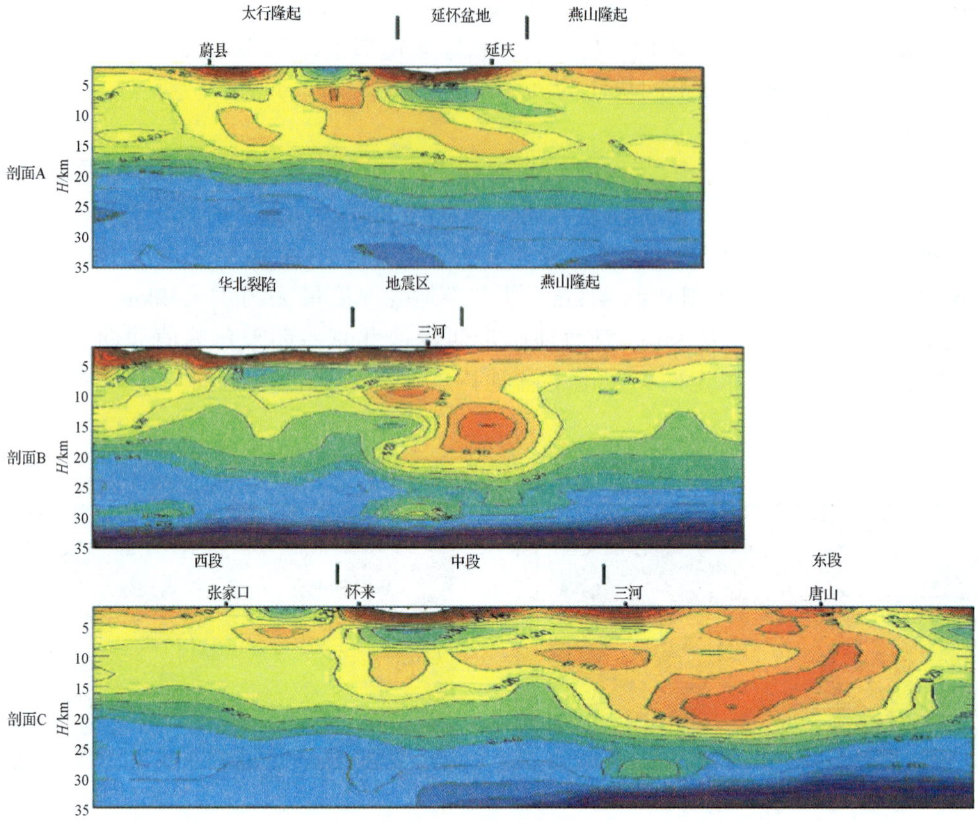

图 27-13　首都圈地壳不同走向的垂直剖面图

（据嘉世旭，2005）

Fig. 27-13　Vertical sections of the crust along different strikes in the Capital area

（4）不同深度横向截面图

图 27-14a~f 分别是 2、10、15、25、30km 和 35km 深度的横向截面图，图中的白圆圈代表 1985 年以前发生的 $M \geqslant 5.0$ 的地震。

图 27-14a~c 显示地壳上部构造。在 2km 深度上，在西部太行、北部燕山隆起区，除西北延怀及蔚县盆地等局部地区，整个隆起区为结晶基岩覆盖，速度约 6.0km/s。东南部裂陷盆地，冀中坳陷东侧、黄骅坳陷速度约 3.0~3.8km/s，属于新生代沉积构造。在盆地内的隆起区大兴凸起速度约 4.0~5.0km/s，显示为古老岩层构造。约 5.3km/s 速度等值线清晰勾划出盆地边缘，边缘带等值线密集，速度横向变化约 4.5~6.0km/s，较好地反映了盆山交界边缘断裂及次级构造边缘断层的形态。10km 和 15km 处的速度结构反映了北部燕山隆起、西南部太行隆起和东南部裂陷盆地等三大构造单元以及接触过渡带张家口-渤海断陷带在上部地壳的延伸构造特征。10km 截面图突出显示了张家口-渤海断陷带的低速（6.0~6.1km/s）特征，三河、唐山西北存在明显的低速（约 6.0km/s）圈闭，这种低速圈闭在 15km 深处北移并扩展为唐山-密云北东向低速条带，而断陷带西段（北京-怀来）在 15km 深处低速构造则不明显。西部延怀盆地与蔚县盆地连通形成北东向相对低速区（约 6.1~6.2km/s）。

图 27-14d~f 显示了地壳下部构造。25km 截面图突出了太行隆起区的高速（6.5~6.6km/s）结构，周缘为深大断裂所围，与燕山约 6.4km/s 速度结构形成明显差异。燕山东部局部仍延续了地壳上部的低速结构。东南部裂陷区的构造单元显示为高低速相间的北东向构造特征。30km 截面图显示：燕山隆起东侧速度为 6.9~7.2km/s，壳幔过渡带构造；东南部裂陷区冀中、黄骅坳陷为高速区（7.2~7.8km/s），体现了坳陷区地幔隆升、接近地壳底部的速度值；燕山中、西部及太行隆起速度为 6.6~6.8km/s，为地壳下部速度。35km 截面图清晰勾划出首都圈地壳以北东走向为界的东西分块特征，东侧已进入上地幔顶部，西侧为地壳下部及壳幔过渡带。

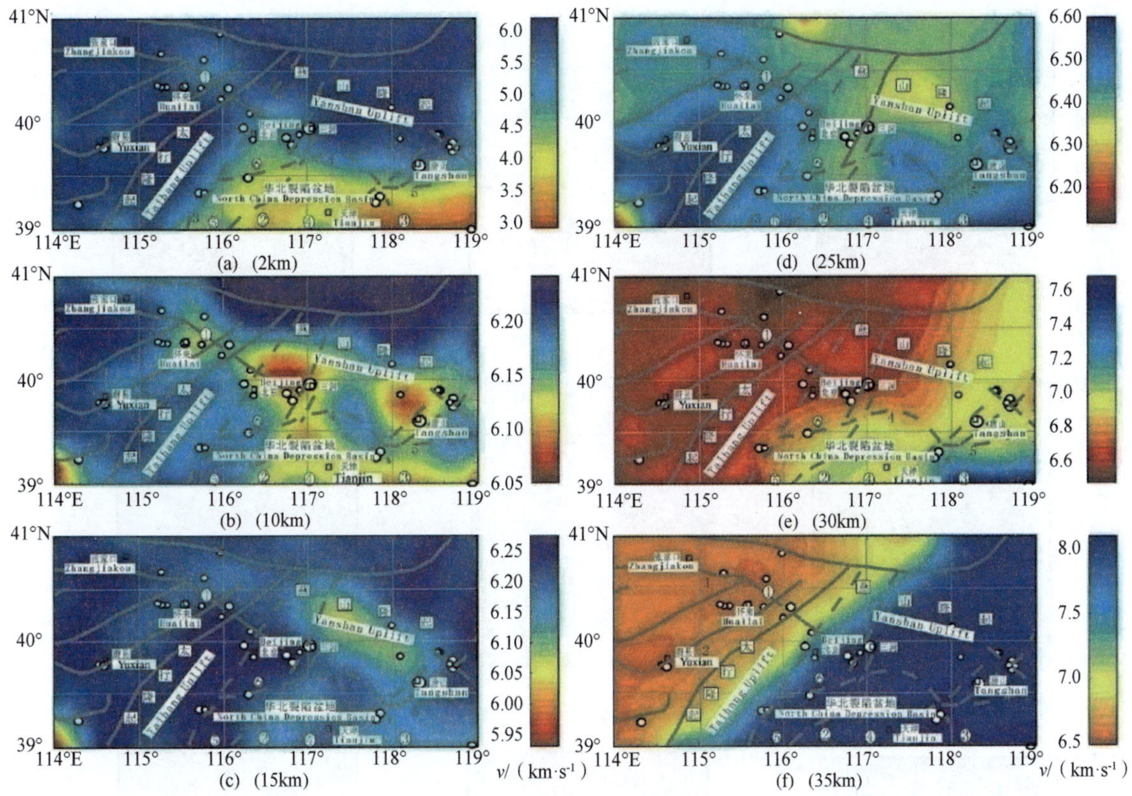

图 27-14　首都圈不同深度截面图

（据嘉世旭，2005）

Fig. 27-14　Sectional views of the crust at different depths in the Capital area

①延怀盆地；②冀中坳陷；③黄骅坳陷；④沧县隆起；⑤容县凸起；⑥大兴凸起

1—怀来-宣化断裂；2—怀来-蔚县断裂；3—怀柔-涿州断裂；4—宝坻-桐柏断裂；5—宁河-昌黎断裂

3. 地壳及上地幔顶部 P 波速度结构和地质界面模型

谭毅培（2009）综合基础地理信息数据、数字化地质图、沉积层的地脉动探测结果、人工地震测深和宽角折射等多种资料，在对数据进行分析和整理的基础上，利用建模工具软件 GeoCAD，建立了北京地区（东经 115.5°～117.5°，北纬 39°～41°）地壳及上地幔顶部 P 波速度结构和地质界面模型（图 27-15）。

模型包括整个北京地区的高程面和莫霍面、东南部盆地的第四纪中全新世、晚更新世、中更新世及早更新世的底界面，前新生界基岩基底界面，以及 0～45km 埋深范围的 P 波速度值。北京地区北部和西部以结晶基岩构成的低山区为特征，东部和南部由向南东方向呈缓慢倾变的平原区构成。地壳厚度由东南部向西北方向逐渐加大，表现为莫霍 面由东南向西北深度增加。在东南部的盆地地区，莫霍面深度普遍在 30～34km 之间，北部的燕山山脉和西部的太行山脉地区的莫霍面深度达到 36～40km，变化趋势与表层地形成镜像关系。地壳 P 波速度总体比较稳定，但还是可以比较清楚地观察到地壳的分层，以及中下地壳内速度梯度倒转区域的存在。这些速度梯度倒转区域近似层状分布，在中上地壳层面基本水平，而在下地壳出现东南高西北低的趋势，与莫霍面的分布状态接近。

三、地震层析成像岩石圈速度结构特征

（一）地震层析成像主要研究进展

金安蜀等（1980）用分布在北京地区各方位上，震中距在 20°～100°范围内的 120 次远震在本区 15 个地震台上的 P 波到时，研究了北京地区地壳和上地幔的三维速度结构。结果表明北京地区地壳和上

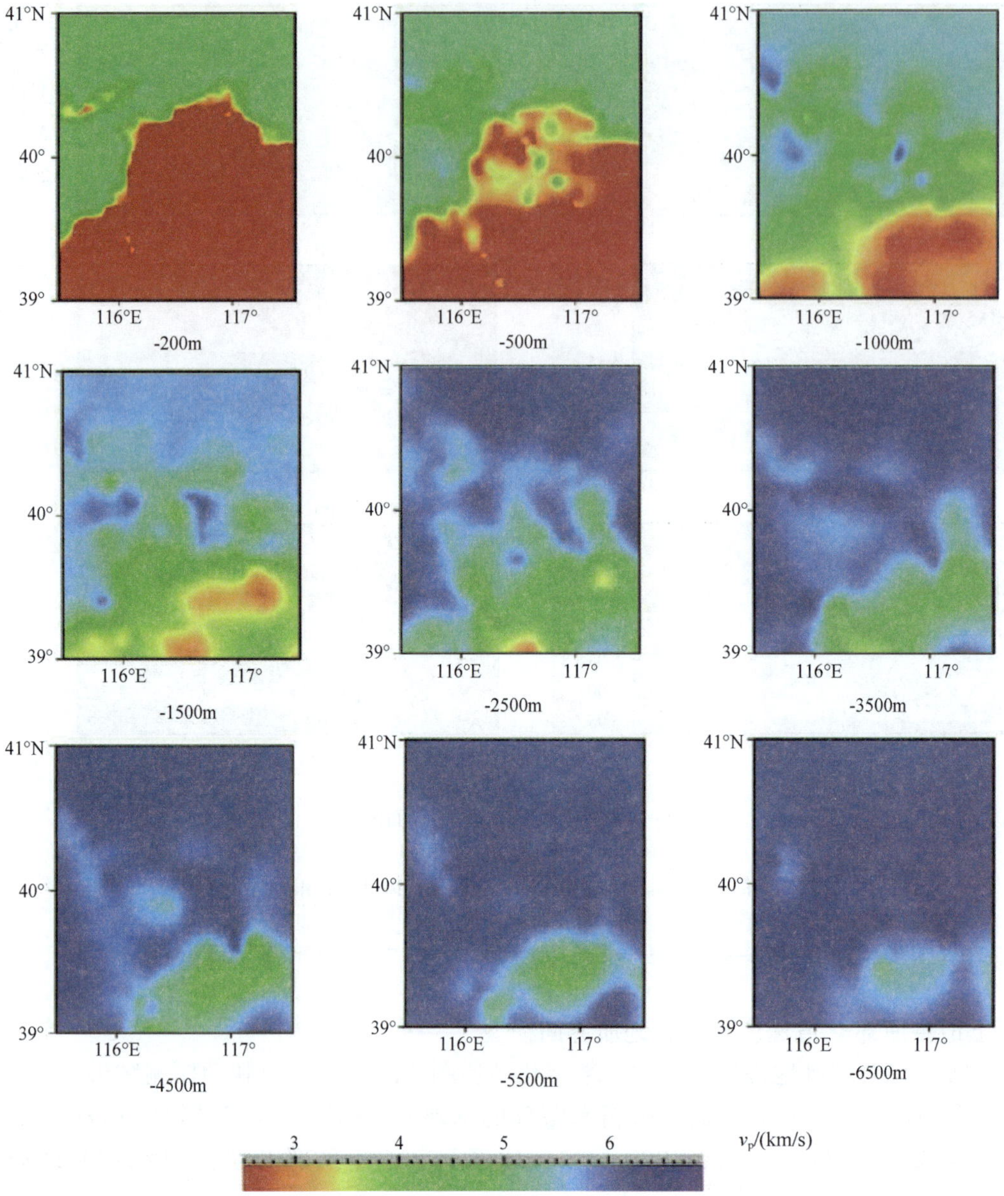

图 27-15 P 波速度模型在浅部的水平速度分布图，展示了沉积盆地形状随深度的变化

Fig. 27-15 Horizontal distribution of P-wave velocity in shallow area with color bar shown at the bottom, which displays the variation of the shape of the basin with depth

图下部的色标表示了 P 波速度值；在埋深为 200m 截面图中能够很直接地辨认出北部的燕山山脉、西部的太行山脉、东南部的平原地区和西北部的延庆-怀来盆地；埋深为 1000m、1500m 和 2500m 截面图清晰地显示出了华北盆地中冀中凹陷、仓县隆起、黄骅凹陷等Ⅱ级构造单元

地幔的 P 波度存在明显的横向差异，东南部为相对低速区（地壳中波速度低 10%~14%，在上地幔中低 8%~9%），西北部为相对高速区（在地壳中波速度高 9%左右，进入上地幔后差异逐渐消失），中部是正常区。在地表，速度差异带的分布与覆盖层的分布大体一致，而在深部，这种差异带的分布与地震活动性分布相当符合，而且本区的几次强震如三河-平谷地震（1679 年，$M=8$）年、沙城地震（1730 年，$M=6.75$）和唐山地震（1976 年，$M=7.8$）的震源位于速度差异带的边界附近。

朱露培等（1990）采用频谱参数化方法，对北京电信传输台网记录的 2447 个远震波到时残差数据进行了反演，结果表明京津唐张地区深部 100 多公里内存在明显的速度横向不均匀性；50km 以上部分的结构与地表的地质构造有较好的对应关系；在燕山隆起区下部表现为高速异常区，而凹陷区下部则呈现低速异常区；深度100km以下，地表构造的痕迹逐步消失，异常幅度也相应降低。研究区域范围内存在三个明显的低速异常块体，其中以天津和唐山附近的水平尺度为100km左右的低速异常区幅度最大，向下一直延伸到约100km的深度，而且在以下存在一低速异常构造与之对应。研究区域内发生的强震大部分都位于低速异常区边缘向高速区过渡的梯度带上。

孙若昧和刘福田（1995）用1984—1991年发生在京津唐地区（37°~41°N，112.5°~120°E）的2505个地震的22234条P波到时进行的层析成像，主要揭示了速度图像与地震活动性间的联系：①公元1000年以来该区强震震中在上地壳的投影大多分布在高速块体内或高速块体与低速块体相交地带，偏高速体的一侧；②介质速度变化剧烈的部位易发生地震，如唐山地震和三河-平谷地震。

于湘伟等（2003）据华北遥测台网和延庆-怀来数字地震台网记录的京津唐地区的P波首波资料，利用近震层析成像方法反演了地壳三维P波速度结构。研究该地区P波速度结构横向不均匀性较明显，绝大部分地震发生在P波低速异常块体与高速异常块体的交界地带。

黄金莉（2005）等利用首都圈密集数字地震台网123个台站记录的2973个区域地震48750个精确P波到时数据采用地震层析成像方法得到了研究区三维P波地壳速度结构。结果指出华北平原、太行山和燕山隆起区速度结构存在明显差异上地壳速度图像与地表地质、地形和岩性密切相关，隆凹相间的华北盆地呈现出地震速度快慢交替的北东-南西向异常带，太行山和燕山隆起表现为高速异常，而山间盆地显示为小范围的低速异常。多数大地震（$M \geqslant 6.0$）都发生在高速块体边侧，而在这些强震下面存在明显低速和高导岩层。

齐诚等（2006）选用华北地震遥测台网和首都圈数字化地震遥测台网1993—2005年纪录的2866个地震事件中的33487条P波和31822条S波的到时资料，计算得到了水平分辨率25km到50km之间的首都圈地区（38.5°N~41°N，114°E~120°E）地壳三维P波和S波速度结构，并进一步获得泊松比分布。首都圈地区P波和S波速度分布表现出强烈的横向不均匀性，浅层速度分布同地表地质结构分布相一致，分析得出震区强震多发生在低速体与高速体之间、低泊松比地区，且震源下部存在低速、高泊松比异常体。

Lei et al.（2008）利用高质量的1500条P波和38500条S波走时资料研究了首都圈地区地壳三维P波速度结构。主要成果：张家口-渤海断裂带从上地壳至下地壳都表现为低速异常，不过至下地壳低速异常幅度变小，该断裂带从东南至西北低速异常特征不同，反映了太行隆起、燕山隆起和华北盆地动力学演化的差异。

（二）三维速度模型结构特征

王夫运（2005）利用三维不分块地震层析成像技术，分析了北京东北部地区的三维覆工地震莫霍界面反射波PmP和SmS走时数据，重建了北京地区上地壳三维速度结构图像。在此基础上，并结合前人的结构模型来分析三维速度模型结构特征。

P波慢度扰动切片图，在3km、7km深度切片上可知：北北东向或近南北向的高、低速带（具有一定走向、长宽比例大）呈相间分布，从西往东分别有：顺义-密云低速带、三河-平谷西侧高速带、大厂-三河东侧串珠状低速带、平谷东侧高速带。在11km深度切片上，P波慢度扰动分布的形态发生变化：以高丽营为中心的北东向高速带范围扩大；顺义-良乡断裂东侧低速带南部依然存在，但北部为高速体代替；夏垫断裂西侧高速带走向从北北东向改变为近南北向，且向北扩展；北东向连续的大厂-三河低速带演化为近南北向的三河-平谷串珠状低速带；平谷东侧的高速体范围缩小。

S波慢度扰动图像，3km深度切片显示出台阵中心部位的密云、通州、三河、平谷之间较大范围的低速扰动块体，为顺义-良乡断裂和夏垫断裂所围限。夏垫断裂东侧被高速带所环绕，这种格局几乎一直延续到7km。随着深度的增加，位于台阵中心部位的低速异常体被高速扰动体侵入。在11km深度切片上逐渐被高速体（长宽比例接近于1）分割成3个低速体。环绕低速体的高速带也逐渐演变为蓟县附近的局部高速体。

不同深度v_p/v_s扰动图像特征：在3km深度的切片上，高、低值扰动条带相间排列并呈北东方向展布，位于台阵中心部位的高值扰动条带，可能与断层构造引起的层位错动有一定的关系。随着深度的增加，高、低值扰动条带相间排列的格局发生了变化，在7km深度切片上，形状不规则的高、低值扰动体零乱排列。在11km深度切片上，又呈现高、低值扰动条带相间的特征。

四、本区地壳和上地幔速度结构综合分析

由于北京地区地质构造、地壳结构及其复杂性，要从整体上提高认识该地区地壳结构的可靠性，需要使用更多的分析方法和更丰富的人工地震记录资料从不同方面对该区的地壳结构进行分析研究。

基于上述剖面分析以及取得建模等多种方法，对本区的地壳和上地幔速度结构进行综合分析，可知北京地区地壳和上地幔的P波和S波速度分布表现出强烈的横向不均匀性，浅层速度分布同地表地质结构分布相一致。50km以上部分的结构与地表的地质构造有较好的对应关系，在燕山隆起区下部表现为高速异常区，而平原凹陷区下部则呈现低速异常区，特别是在王夫运（2005）所做的上地壳速度结构模型可知地表构造与P波扰动图像存在明显的相关性，高、低速扰动带展布方向和活动断裂走向一致。深度100km以下，地表构造的痕迹逐步消失，异常幅度也相应降低。地壳下部出现大规模的低速异常与高导层相对应。大多数研究认为区域内发生的强震大部分都位于低速异常区边缘向高速区过渡的梯度带上。

目前，首都圈速度结构模型结果已应用于震源的精确定位（刘福田等，1986）、地壳形成和演化机制（张先康等，1996）、强地震动模拟（高孟潭等，2002；潘波等，2006，2009；刘博研等，2007）等研究中，特别是为地震的成因、孕育和发生等方面的研究提供更加可靠的证据。由于人工勘探的高成本及探测区域有限，而天然地震层析成像反演得到的速度结构分辨率低且明显依赖于初始模型，以及地质调查、地球物理勘探和反演方法的不同与实施目的和时间的差异，这些地下结构模型的建立，存在比较突出的问题是结构模型比较简单，对于各种类型数据的融合程度不够。随着观测条件的改善，特别是空间分辨率逐步提高，大量的三维地壳结构的研究可以更加清楚地了解上地幔速度结构与构造特征。

第二节　区域重磁特征

一、区域重力场特征

（一）岩（矿）石密度特征

北京地区多年来积累了不同时期、不同工作单位测定的丰富的岩（矿）石物性资料。各类岩石密度变化范围较大，为1.75~3.3g/cm^3（表27-2，表27-3）。其中沉积岩属于低密度岩类，密度为1.75~2.8g/cm^3，岩石密度按黏土岩—碎屑岩—灰岩依次增大，也随埋深和成岩固结程度增加而增大；岩浆岩密度为2.52~3.3g/cm^3，其中侵入岩密度略大于喷出岩。岩浆岩的密度值具有从酸性岩—中性岩—基性岩—超基性岩依次增大的规律。基性岩、超基性岩属于高密度体，酸性侵入岩属于低密度体。中酸性侵入岩密度与碳酸盐岩类沉积岩密度相当或略低。变质岩类密度一般较大，为2.8~3.0g/cm^3。

表27-2　北京地区侵入岩、喷出岩岩石密度一览表

Table 27-2　The density of intrusive and volcanic rocks in Beijing area　　单位：g/cm^3

侵入岩类		喷出岩类	
岩性	密度	岩性	密度
酸性、偏碱性：似斑状花岗岩、中细粒花岗岩、二长花岗岩	2.58~2.60	酸性、中酸性：流纹岩、霏细岩、粗面岩	2.52

续表

侵入岩类		喷出岩类	
中酸性：花岗闪长岩	2.65	中性：安山岩、安山玢岩、角闪石安山岩等	2.53
中性：闪长岩类	2.70~2.74		
基性—中基性岩：辉石闪长岩、灰长岩、辉石岩	2.90~3.33	基性：玄武岩	2.59

表 27-3　北京地区地层岩石密度值统计表

Table 27-3　The density of sedimentary rocks in Beijing area

单位：g/cm^3

界		系	代号	主要岩性	密度范围	平均值	分层值
新生界		第四系	Q	砂、砂砾石层、黏土	1.75~2.01	2.00	2.12
		新近系	N	泥岩、粉砂岩、砂砾岩	2.10~2.45	2.15	
		古近系	E	泥岩、砂岩、页岩	2.15~2.50	2.20	
中生界		下白垩统	K_1	砂岩、粗安岩	2.37~2.50	2.50	2.54
		侏罗系	J	安山岩类、凝灰岩、火山角砾岩	2.41~2.72	2.58	
古生界	上古生界	二叠系	P	砂岩、砾岩	2.61~2.74	2.62	2.61
		石炭系	C	页岩、粉砂岩	2.60~2.62	2.61	
	下古生界	奥陶系	O	灰岩、豹皮灰岩、白云质灰岩	2.66~2.73	2.71	
		寒武系	€	泥质灰岩、灰岩、页岩	2.56~2.72	2.68	
中—新元古界		青白口系	Qb	泥质灰岩、页岩、粉砂岩	2.62~2.75	2.64	2.77
		蓟县系	Jx	白云质灰岩、灰岩	2.69~2.80	2.80	
		长城系	Ch	砾岩、石英砂岩、灰岩	2.60~2.77	2.70	
新太古界		四合堂岩群	Ar*Sh*.	角闪岩相变质岩系	2.75~2.88	2.87	
		密云岩群	Ar*My*.	麻粒岩相变质岩系		2.99	

由表 27-2 和表 27-3 可见，各地质年代地层密度特征主要取决于岩石密度及其分布组合状况。时代由老到新，地层密度由大到小。太古宇密度最大，新生界密度最小。老地层密度较稳定，新地层密度变化大。根据地层变化特征，北京地区可划分 4 个密度层和 3 个密度界面，即新生界密度层，密度变化范围为 1.75~2.53g/cm^3，平均密度为 2.12g/cm^3；中生界密度层，密度变化范围为 2.37~2.72g/cm^3，平均密度为 2.54g/cm^3；上古生界密度层，密度变化范围为 2.60~2.74g/cm^3，平均密度为 2.61g/cm^3；下古生界密度层，密度变化范围为 2.56~2.99g/cm^3，平均密度为 2.77g/cm^3。与密度层相对应的密度界面为新生界与中生界之间的密度界面，中生界与古生界之间的密度界面，下古生界奥陶系及其以下地层与其上沉积地层之间的密度界面。

由于同一时代地层的不同岩性或不同岩性组合可能存在着密度差异，因而增加了密度界面的复杂性。当侏罗系以安山岩为主时，其密度值与下伏古生界密度差异不明显，而与上覆白垩系存在 0.2g/cm^3 的密度差，此密度界面则不在中生界与古生界之间，而存在于中生界内部侏罗系与白垩系之间。新生界与其下伏任何老地层均存在明显密度差，新生界断陷盆地能产生明显重力低异常。中—新元古界蓟县系与其上覆地层存在明显密度差，隐伏蓟县系隆起可产生局部重力高异常。

（二）区域重力场特征

华北北部重力异常东高西低，渤海海域异常值最高，向西逐渐降低，重力场总体走向为北东向，北京地区位于北北东向大兴安岭-太行山-武陵山重力梯度带上，东南部平原区在正异常背景基础上出现正负异常相间排列现象：负异常区对应内部隐伏断陷盆地，正异常对应隐伏隆起；重力梯度带则反映次级块体间或不同新构造单元间具有一定规模的边界断裂带；西北部山区为负异常背景，在太行山地区水平梯度变化最大，达 1×10^{-5}m/（s^2·km）。这种重力异常趋势变化反映出北京地区地壳厚度由东向西逐渐变厚。北部受燕山构造带的影响，出现东西走向的构造，其更多地表现为镶边特征，进入华北陆块影响逐渐减弱。受华北陆块北缘燕山构造带重力场影响，在重力场转折处，北西向断裂发育，并出现走向北西的小型张性断陷盆地，形成了北东、东西、北西等多向构造交汇格架。其中以北东向构造为主体，形成隆凹相间的盆地分布。

（三）区域重力异常分区

北京市布格重力异常等值线在北京地区呈密集与舒缓相间分布，重力异常值由南东向北西逐渐降低，即由 $5\times10^{-5}m/s^2$，逐渐降为 $-120\times10^{-5}m/s^2$。该异常梯级带在向北东方向延展时，在昌平南口以东发生局部分枝变化，一部分继续向北东方向延展，形成张家口–承德重力梯级带；另一部分向东弯曲，形成近东西向的昌平–遵化–秦皇岛重力梯级带。根据重力场的异常形态和特征，北京地区的重力场可划分为 3 个区，即西山重力梯级带异常区、平原重力高异常区和北山重力低异常区。

（1）西山重力梯级带异常区

西山重力梯级带异常区位于大兴安岭–太行山–武陵山重力异常梯级带上，南起张坊经门头沟、昌平十三陵到延寿镇，异常呈北东向展布，西北低、东南高，布格重力异常变化于 $-122\times10^{-5}\sim-22\times10^{-5}m/s^2$ 之间，梯度变化较大。异常区内布格重力异常极小值出现在西北部，异常表现为北东走向的椭圆状异常，由延庆中生代盆地引起。

（2）平原重力高异常区

平原重力高异常区表现为西北重力低、东南重力高的特点，异常呈北东走向，布格重力异常变化范围在 $-46\times10^{-5}\sim10\times10^{-5}m/s^2$ 之间。西北为丰台–顺义重力低异常，东南为大兴重力高异常区，两个异常区分界线表现为受礼贤断裂控制的密集梯度陡变带。

（3）北山重力低异常区

北山重力低异常特征表现为西低东高的特点，异常区由多个椭圆状和似圆状的异常组成，布格重力异常变化范围在 $-121\times10^{-5}\sim-28\times10^{-5}m/s^2$ 之间，梯度变化较大。其中两个较大的异常中心分别位于云蒙山岩体和怀柔北部的花岗岩体。

二、区域重力资料地质解释

利用重力资料，结合航磁、地震、电测深及地质成果资料系统地对北京市盆地、断裂和侵入岩体进行综合推断解释，结果如图 27–16 所示。

（一）重力推断盆地

1. 丰台古近纪盆地（P–7）

丰台古近纪盆地在剩余重力异常图上表现为椭圆状的负异常，异常中心幅值为 $-12\times10^{-5}m/s^2$，根据重力异常方向导数、电测深剖面和地震资料综合圈定丰台盆地的边界，盆地面积约为 $433km^2$。由多个钻孔证实盆地基底为中生界及古生界，盆地内沉积古近纪、新近纪和第四纪地层，钻孔揭露盆地新生代最大沉积厚度可达 1300m。

2. 采育古近纪盆地（P–9）

采育古近纪盆地位于通州区东南，与河北接壤地段。剩余重力异常图上表现为椭圆状的负异常，异常中心幅值为 $-4\times10^{-5}m/s^2$，根据重力异常方向导数、电测深剖面和地震资料综合圈定采育盆地的边界，面积约为 $217km^2$。盆地内沉积有古近纪、新近纪和第四纪沉积物，基底为古生界和中—新元古界，钻孔揭露盆地新生代最大沉积厚度为 3200m。

3. 大厂古近纪盆地（P–8）

大厂古近纪盆地剩余重力异常图表现为条带状的负异常，异常中心幅值为 $-6\times10^{-5}m/s^2$，盆地面积约为 $67km^2$，基底为中—新元古界，以及古生界—中生界，新生代沉积地层为古近纪、新近纪和第四纪地层，根据重力异常方向导数、电测深剖面、地震剖面和钻孔资料证实大厂古近纪盆新生代沉积厚

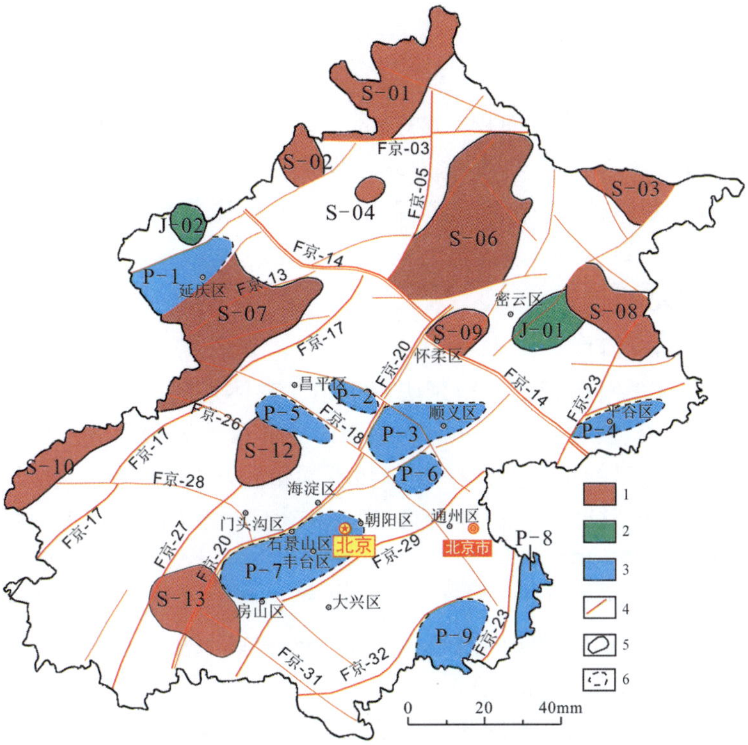

图 27-16　北京市重力推断地质构造图

（据北京市矿产资源潜力评价成果报告）

Fig. 27-16　The geological structure map inferred through gravity in Beijing

1—重力推断酸性—中酸性岩体及编号；2—重力推断基性岩体及编号；3—重力推断盆地及编号；
4—重力推断断层及编号；5—岩体边界；6—盆地边界

度为 5000~7000m。

4. 东坝-天竺新近纪盆地（P-6）

东坝-天竺新近纪盆地南起朝阳区东坝北至顺义天竺，总面积约为 80km^2，剩余重力异常图上表现为椭圆状的负异常，异常中心幅值为-10×10^{-5}m/s^2，根据重力异常方向导数、电测深剖面和地震资料综合圈定东坝-天竺盆地的边界，盆地内沉积为新近纪和第四纪沉积物，基底为中生界及元古宇，钻孔揭露盆地新生代最大沉积厚度约为 700m。

5. 后沙峪-俸伯第四纪盆地（P-3）

后沙峪-俸伯第四纪盆地在剩余重力异常图表现为椭圆状的负异常，异常中心幅值为-11×10^{-5}m/s^2，盆地面积约为 257km^2，基底为古生界和中生界，盆地内堆积第四纪沉积物，根据重力异常方向导数、电测深剖面、地震剖面和钻孔资料证实大厂古近纪盆沉积厚度为 600~1300m。

6. 兴寿第四纪盆地（P-2）

兴寿第四纪盆地位于昌平区小汤山以北，兴寿—崔村一带，剩余重力异常为负异常，异常延伸方向为北西向，根据重力异常方向导数和电测深曲线的梯级带综合圈定盆地的边界范围，计算盆地面积约为 58km^2，钻孔揭示基底为中生界及元古宇，盆地内沉积第四纪地层，沉积厚度为 300~600m。

7. 马池口第四纪盆地（P-5）

马池口第四纪盆地剩余异常表现为等轴状的负异常，异常有两个中心组成，异常中心幅值分别为-9×10^{-5}m/s^2 和-11×10^{-5}m/s^2，异常走向北西向。马池口盆地基底为中—新元古界和中生界，盆地内沉积物为第四纪地层，盆地面积约为 116km^2，钻孔揭露最大沉积厚度可达 800m。

8. 平谷第四纪盆地（P-4）

平谷第四纪盆地位于平谷区小汤山夏各庄、韩庄一带，剩余重力异常为椭圆状负异常，异常延伸方向为北东向，异常中心幅值为$-4\times10^{-5}m/s^2$。平谷盆地基底为中—新元古界，面积约 $110km^2$，沉积物为第四纪沉积地层，最大沉积厚度为 600m。

9. 延庆第四纪盆地（P-1）

延庆第四纪盆地位于延庆县境内，面积约 $237km^2$，基底为中生界及中—新元古界，盆地内沉积第四纪地层，厚度达 600~800m。剩余重力异常为北东向椭圆状异常，异常中心幅值为$-17\times10^{-5}m/s^2$，由于盆地北西和南东两侧受断裂控制，异常边缘表现为明显的密集梯级带，根据重力垂向一阶导数确定了其边界范围。

（二）重力推断断裂构造

1. 永宁-怀柔断裂（F 京-14）

永宁-怀柔断裂南东起怀柔马昌营，北至延庆永宁镇，整体走向为北西向，全长约 100km，断裂在怀柔雁栖镇附近被黄庄-高丽营断裂错断，分为南北两段，北段为地表出露，南段为隐伏断裂。该断裂在布格重力异常图上表现为明显的重力异常梯级带和不同重力场的分界线，重力梯级带在怀柔雁栖镇产生扭曲，根据断裂的重力场特征和断裂延伸的长度，将永宁-怀柔断裂定为一级断裂。

2. 黄庄-高丽营断裂（F 京-20）

黄庄-高丽营断裂南西起自涿县西城坊，经坨里、黄庄、洼里、高丽营至怀柔一线，全长约 110km，走向为北东 20°~50°，断裂面倾向南东，倾角 55°~75°，表现为陡倾的断裂面，性质为高角度正断裂，是北京平原区主要的隐伏活动断裂。

黄庄-高丽营断裂在布格异常图上表现为明显的重力异常梯级带，且在洼里以北和高丽营附近异常发生扭曲，后经推断证实被南口-孙河断裂和十三陵-李桥断裂错断，另根据钻孔、地震、地球化学等勘查方法证实，该断裂为全新世以来表现为活动断裂。

3. 夏垫-马坊断裂（京 F-23）

夏垫-马坊断裂是指大厂凹陷西边界的控制断裂，其分布从平谷区的夏垫、西集，在大杜社—牛堡屯一带向西北偏移后经长子营、礼贤至固安，呈北北东向延伸，全长 90km 左右，其显著特征是控制了大厂凹陷和廊坊-固安凹陷巨厚的第三系堆积，在重力异常图上表现为明显的梯级带，为隐伏的活动断裂。由于该断裂所处的夏垫一带在公元 1679 年曾发生过八级地震，根据向宏发等（1988）在夏垫一带所做的人工地震剖面解释，在古近纪的夏垫-马坊断裂基础上形成了第四纪的夏垫新断裂。

4. 礼贤断裂（F 京-32）

礼贤断裂南西起礼贤北东至漷县，全长约为 56km，整体走向为北东向，在重力场特征为正负异常的分界线，为平原区隐伏的活动断裂。

5. 南苑-通县断裂（F 京-29）

南苑-通县断裂西南起自涿州，经刁窝、陈家房、南苑、通州，向东北方向延伸。断裂总的走向北东 50°~30°，断裂面倾向北西，倾角约 50°左右，北西盘下降，南东盘上升，性质为正断层。向东可能延伸到夏垫或沿燕郊向东北延展。该断裂主要是地面电测深、重力、人工地震等物探手段结合钻孔资料确定的。地表水系明显受到断裂影响，如温榆河及潮白河经过断裂时迅速汇合，箭杆河在断裂附近明显转折，说明该断裂第四纪有活动。

6. 佛子庄-北安河断裂（F 京-27）

佛子庄-北安河断裂南起张坊北到上庄以北，全长约为 84. 8km，延伸方向为北北东向，布格重力异常图表现为明显的梯级带，张坊-北安河段为地表出露，北安河以北为隐伏断裂。该断裂控制房山花岗岩体，并穿过阳坊花岗岩体。

7. 斋堂-高崖口断裂（F 京-17）

斋堂-高崖口断裂全长约为 83km，南西从黄塔起，经斋堂、杨村、高崖口至黄花城以北一线，整体走向为北东向，倾向南东，断裂由北向南分别被永定河断裂、刘村—上庄断裂和南口-孙河断裂错断，共分为四段。布格重力异常表现为明显的梯级带，并在错断的地方异常等值线发生扭曲。

8. 崎峰茶–汤河口断裂（F 京–5）

崎峰茶–汤河口断裂南起大榛峪，北到喇叭沟门以南，总长约 49km，整体南北方向延伸，倾向西，断裂控制云蒙山岩体，在布格异常图上表现为不同重力场的分界线，东部布格异常相对较低，西部偏高。

9. 花盆–长哨营断裂（F 京–3）

花盆–长哨营断裂东自花盆西至长哨营以东，全长约为 48km，走向近东西向，在汤河口以北被崎峰茶–汤河口断裂错断，分为东西两段，布格重力异常图上表现为明显的梯级带。

根据重力异常特征还推断出永定河断裂（F 京–28）、南口–断裂孙河（F 京–18）、窦店–榆垡断裂（F 京–31）、刘村–上庄断裂（F 京–26）、八达岭–四海断裂（F 京–13）等 23 条断裂，其中隐伏断裂 4 条，半隐伏断裂 2 条。这些断裂的重力异常特征主要表现为重力异常梯度带或不同重力场的分界线，规模较小。

（三）重力推断侵入岩体

1. 房山花岗岩体（S–13）

房山花岗岩体位于房山区东南，是一个半隐伏岩体，岩性以花岗岩为主，剩余重力异常图上表现为在正异常背景下出现一个孤立的圆状负异常，异常中心位于燕化小区，幅值为 $-1\times10^{-5}m/s^2$，根据重力垂向一阶导数等值线图，结合航磁 ΔT 垂向一阶导数等值线图综合圈定出岩体的在地表的投影面积约 $320km^2$。

2. 洪水口石英二长岩体（S–10）

洪水口石英二长岩体位于门头沟区西北，为一出露岩体。岩体中部为中性岩体，岩性为闪长岩，周围为斑岩体，岩性为石英二长斑岩。剩余重力异常图上表现为正负伴生的异常，中间为椭圆状正异常，两边为条带状负异常，异常范围在 -16×10^{-5} ~ $10\times10^{-5}m/s^2$ 之间。

3. 对臼峪花岗岩体（S–07）

对臼峪花岗岩体分布南自昌平区南口以北，北至延庆县沈家营。整体走向为北东，岩性为花岗岩，剩余重力异常图表现为负异常，由四个椭圆型异常组成，异常极值为 $-15\times10^{-5}m/s^2$，为一出露岩体，岩体面积约 $654km^2$。

4. 大海坨山基性岩体（S–05）

大海坨山基性岩体位于延庆靳家堡北，剩余重力异常表现为椭圆状的正异常，异常中心幅值为 $8\times10^{-5}m/s^2$，根据航磁 ΔT 垂向一阶导数等值线图推测岩体在地表的投影面积约为 $58km^2$，该岩体地表被元古宙地层覆盖，推断岩性为辉长岩，对于寻找钒钛磁铁矿具有一定的意义。

5. 宰相庄花岗岩体（S–09）

宰相庄花岗岩体位于怀柔区内，剩余重力异常特征表现为圆状的负异常，异常极值为 $-10\times10^{-5}m/s^2$。该岩体地表被第四系覆盖，受断裂构造控制走向为北东，为隐伏岩体。钻孔资料显示岩体埋深在 50~350m，自西南向东北埋深逐渐变浅，岩性为花岗岩。

6. 阳坊花岗岩体（S–12）

阳坊岩体位于海淀与昌平交界处，剩余重力异常图表现为负异常，岩体仅在西部有小部分出露，大部分被第四系覆盖，埋深 200~300m，根据钻孔资料显示，岩体的岩性为花岗岩。

7. 穆家峪基性岩体（J–01）

穆家峪岩体位于密云水库北，剩余重力异常图上表现为正异常，异常极值为 $8\times10^{-5}m/s^2$，其与密云水库铁矿航磁异常距离较近，异常形态不明显，利用航磁 ΔT 垂向一阶导数圈定出岩体边界范围，该岩体侵入新太古代的变质岩中，有超镁铁质脉岩出露，走向北东，岩体中发现有钒钛磁铁矿，是寻找钒钛磁铁矿是一个指示标志。

8. 大城子花岗岩体（S–08）

大城子花岗岩体位于密云东南与平谷交界处，在剩余异常图上表现为两个相距不远的椭圆状负异常，异常极值为$-14\times10^{-5}m/s^2$。该岩体侵入中—新元古界及新太古代变质岩中，为隐伏岩体，走向北西，推断岩性为花岗岩。

9. 云蒙山花岗岩体（S-06）

云蒙山花岗岩体位于密云西部与怀柔交界，重力剩余异常表现为由三个北北东向排列分布负异常组成，整体形态不规则，异常极值为$-14\times10^{-5}m/s^2$，为出露岩体，岩体分布范围约825km^2，岩性为花岗岩。

10. 古北口花岗岩体（S-03）

古北口花岗岩体位于密云高岭以北，剩余异常特征表现为北西走向的负异常，异常极值为$-5\times10^{-5}m/s^2$，为隐伏岩体，推断岩性为花岗岩。

11. 小川花岗岩体（S-04）

小川花岗岩体位于延庆小川西北，剩余重力异常在正异常背景下出现孤立的椭圆状负异常，异常中心幅值$-6\times10^{-5}m/s^2$，为隐伏岩体，综合重力剩余异常、航磁异常和周围地质情况，推断岩体为花岗岩。

12. 红旗田花岗岩体（S-02）

红旗田花岗岩体位于延庆千家店以西，剩余重力异常在正异常背景下出现孤立的椭圆状负异常，异常中心幅值$-3\times10^{-5}m/s^2$，岩体零星出露，岩性为花岗岩，利用重力垂向一阶导数和航磁ΔT垂向一阶导数圈定岩体的面积约为95km^2。

13. 苗营二长花岗岩体（S-01）

苗营二长花岗岩体位于怀柔北部，剩余重力异常由两个椭圆状的负异常组成，整体走向北东，异常极值为$-13\times10^{-5}m/s^2$，为出露岩体，岩性为花岗岩，利用重力垂向一阶导数和航磁ΔT垂向一阶导数圈定岩体的面积约为出露面积418km^2。

三、区域磁场特征

（一）岩（矿）石磁性特征

北京地区各时代地层除普遍缺失的震旦系、上奥陶统、志留系、泥盆系、下石炭统、上白垩统（部分）及古近系古新统外，从新太古界的古老变质岩系直至第四系都有沉积，侵入岩和火山岩出露亦很广泛。各类岩（矿）石的分布发育和磁性特征如下（表27-4）。

表27-4　北京市岩（矿）石参数统计表

Table 27-4　The rock（ore）stone parameter statistics in Beijing

顺序号	岩矿石名称	标本数量/个	$K/10^{-5}$SI	$Jr/$（10^{-3}A/m）	采集地区
			平均值	平均值	
1	沙厂式变质铁矿（包括部分磁铁石英岩、磁铁片麻岩）	200	121265	133300	冯家峪达峪达岩西峪
2	硅卡岩型铁锌矿	20	224938	129000	汉家川、石窑、三岔、磨石口
3	硅卡岩型含铜磁铁矿	36	34935	10500	石槽、石展台
4	橄榄岩	123	9173	5400	石城、沙峪里、养鱼池
5	辉石岩	31	9048	8800	沙峪里、石城、大牛房
6	角闪岩	168	8922	3600	沙峪里，西庄子、石城、三岔、磨石口、大牛房
7	片麻岩	237	3644	1300	西峪、石城、养鱼池、水库南、西庄子、冯家峪等
8	混合岩	166	2513	800	西峪、沙峪里、石城、西庄子

续表

顺序号	岩矿石名称	标本数量/个	$K/10^{-5}$SI	$Jr/$ (10^{-3}A/m)	采集地区
			平均值	平均值	
9	辉长岩	92	19227	4100	密云水库南、石城、上庄
10	变质闪长岩	123	10179	11500	长哨营—下窝铺一带
11	闪长岩	633	5529	1200	大牛房、上庄、辛开沟、汉家川、房西峪、长园、庄户三岔
12	正长岩	227	1885	1100	密云水库南、延庆太平庄三岔、磨石口、西水峪、分水峪
13	花岗岩	229	754	100	莲花山、九渡河、房耳峪、三岔、磨石口、黑熊山
14	石英斑岩	44	63	0	九渡河
15	玄武岩（磁铁矿）	153	9676	1500	年丰、高家两河
16	安山岩（包括粗安岩、英安岩和辉石安山岩）	174	1634	5300	塔河、百花山、公涧铺、西水峪
17	辉绿岩	33	251	100	密云水库南、石城、九渡河
18	安山质火山角砾岩	76	1885	12600	百花山、磨石口
		231	1759	600	罐尔塘一带

1. 沉积岩磁性特征

沉积岩表现为无磁性或弱磁性，在航磁图上均表现为平稳的负磁场或极不明显的正磁场。

2. 变质岩磁性特征

这里主要指新太古代表壳岩，各类斜长片麻岩及混合岩一般为中磁性岩石，K 值平均变化为 $1683\times10^{-5}\sim4898\times10^{-5}$SI，$Jr$ 值平均变化为 $780\times10^{-3}\sim1960\times10^{-3}$A/m，片麻岩、石英岩及大理岩等一般无磁性或为磁性极微弱的岩石。前者一般在航磁异常上常表现为400~500nT波状起伏的正磁场，如分布于密云水库四周、怀北大甸—下窝铺、平谷大城子—墙子路等地的背景磁场，即为此类片麻岩系引起；而片岩、石英岩及大理岩等则表现为负磁场，如西驼古、阳坡地、半城子的东北部负磁场地区，即与之有关。

3. 火山岩磁性特征

火山岩类的磁性变化很大，K 值从 0 到（$n\times10^2\sim n\times10^3$）$\times10^{-5}$SI，$Jr$ 值从 $600\times10^{-3}\sim12600\times10^{-3}$A/m，$Jr\geqslant Ji$，岩石随基性成分的增加其磁性亦有逐渐增强的趋势。在磁场上多表现为低缓的负磁场背景上有较频繁的弱小跳起，部分地区如百花山、髫髻山等航磁异常有时还有200~300nT的正异常带出现，年丰一带的蚀变玄武岩则形成了500~800nT的高磁异常，而大红峪组安山玄武岩则在区域正磁场背景上显示出100~300nT的跳跃。

4. 侵入岩磁性特征

据磁参数测定，侵入岩绝大多数均具一定的磁性。按照超基性岩—基性岩—中性岩—酸性岩成分变化，磁性表现为由强到弱，可以分为强、中、弱三类。

（1）基性—超基性岩属强磁性岩类

主要分布于密云水库一带，延庆、昌平、平谷仅有零星出露。主要岩石为橄榄岩、辉石橄榄岩、辉石岩、辉长岩及角闪岩等。其 κ 值平均变化范围为 $3516\times10^{-5}\sim22106\times10^{-5}$SI，$Jr$ 为 $730\times10^{-3}\sim15600\times10^{-3}$A/m，$Ji\geqslant Jr$。磁场上由于它们多侵入于太古代的片麻岩中，与沙厂式变质铁矿经常紧密伴生，本身的规模较小，因此其异常特征常被干扰，表现不甚清楚。仅个别规模稍大，干扰较小的水漳辉石岩体、穆家峪辉石岩体、上庄辉长岩体等在磁场上有较明显的显示。其异常常呈现规模较大、但强度稍大的椭圆状、圆状异常，极大值可达500~2000nT。曲线较为圆滑。

（2）中性岩至酸性岩属中至弱磁性岩类

包括辉石闪长岩、闪长岩、正长闪长岩、花岗闪长岩、二长岩、花岗岩、石英斑岩等岩石类型。空间上多以杂岩体形式产出，分布于延庆、昌平和怀柔地区，其次在西山、密云东北、密云南部和平谷南部也有分布。其中闪长岩类的岩性变化较大，磁性亦存在相应的变化。从表 27-4 中可看出：以上庄黑云母闪长岩磁性最强，κ 值可达 10000×10^{-5}SI，Jr 达 5450×10^{-3}A/m；而长园—沙峪的闪长岩、石英闪长岩、正长闪长岩、黑云母闪长岩磁性最弱，κ 为 376×10^{-5}SI，Jr 为 270×10^{-3}A/m；闪长岩的总平均磁化率（κ）为 5526×10^{-5}SI，总平均剩余磁化强度（Jr）为 1200×10^{-3}A/m。

（3）花岗岩、石英斑岩及石英正长岩等岩石磁化率（κ）变化范围为：$113\times10^{-5}\sim2198\times10^{-5}$SI，$Jr$ 变化范围为 $30\times10^{-3}\sim420\times10^{-3}$ A/m。

5. 磁铁矿的磁性特征

北京市磁铁矿按成因类型可分为两类，即变质型铁矿和侵入岩体型铁矿，它们均具较强的磁性。

1）变质型铁矿（磁铁石英岩和磁铁片麻岩）的 κ 值变化范围为 $21854\times10^{-5}\sim226080\times10^{-5}$SI，平均值为 121204×10^{-5}SI。Jr 变化范围为 $1800\times10^{-3}\sim240000\times10^{-3}$A/m，平均值为 13300×10^{-3}A/m。磁场上表现为狭长带状，幅度和梯度变化极大，北侧常伴生有明显负值，呈尖锐的高磁异常。航磁异常强度可达 1000~3000nT 以上，如东庄禾-太师屯-学各庄异常，半城子-殿臣峪-董各庄异常，冯家峪-四堂合异常，后安岭-琉璃庙异常，以及沙厂、霍各庄、达岩-水树峪异常等。

2）接触交代型铁矿 κ 值变化范围为 $34917\sim224824\times10^{-5}$SI，$Jr$ 变化范围为 $10500\times10^{-3}\sim129000\times10^{-3}$A/m。由于矿体规模较小，磁场上表现为在强度不大的圆状、椭圆状、不规则状岩体异常背景上呈数纳特左右的叠加小异常。

3）岩浆型铁矿 κ 值变化范围为 $50240\times10^{-5}\sim125600\times10^{-5}$SI，$Jr$ 变化范围为 $10000\times10^{-3}\sim20000\times10^{-3}$A/m，磁场上常叠加于强磁性基性超基性岩体异常或新太古代片麻岩系异常之上，铁矿异常与背景异常很难区分。

（二）区域磁场特征

北京地区航磁异常属冀北地区高强度正负磁异常复杂交变区，异常分布呈北东向、北西向和东西向三组不同走向的高强度正异常带的互相交叉叠加，反映出区内多期次多组合断块式地质构造特征。区域磁场从密怀平的高异常区向四周低异常或平稳正负磁场区逐渐降低，正磁场强度一般在 100~400nT 左右，负磁场强度一般在-100~200nT 左右。区域异常轴向除怀北地区主要为东西向外，其余地区有从北京西部到东北部由北东向至北东东至东西向变化的趋势。在此背景上，不同地区尚有北北东向、近南北向和北西—北西西向的异常与之交会重叠。

根据航磁异常类型、强度、走向和范围等特征可以将区域场大致分为 4 个磁场区：怀北磁场区、密怀磁场区、延昌西山磁场区和平原磁场区（图 27-17）。

1. 怀北磁场区

怀北磁场区位于怀柔北部，其南界大致在万家寺—长哨营—古北口—新城子一带。以磁场突变带与南部两个磁场区相邻。不同时期侵入岩浆岩形成的局部高强度的正异常呈椭圆状、扁圆状或宽带状，以近东西向延伸、略有起伏和跳跃的正负磁场为背景，呈东西向叠加于背景场之上。

2. 密怀磁场区

密怀磁场区包括密云和怀柔区南部，该异常区北部与怀北异常区以异常梯级带相连，西部与延昌西山异常区相邻，东部延伸至河北省。异常区除东北边出现负磁场外，均以 200~400nT 的高磁场为背景，新太古代磁铁石英岩引起的局部异常以强度大幅度大梯度陡的线状、条带状异常为主，呈北东、北北东、北西和东西向叠加于背景场之上。其次也有各种不规则圆状、椭圆状、强度低梯度缓的异常较有规律地分布其中。

3. 延昌西山磁场区

延昌西山磁场区包括延庆部分地区、昌平部分地区和全部西山地区。区域以北东向规律展布的宽

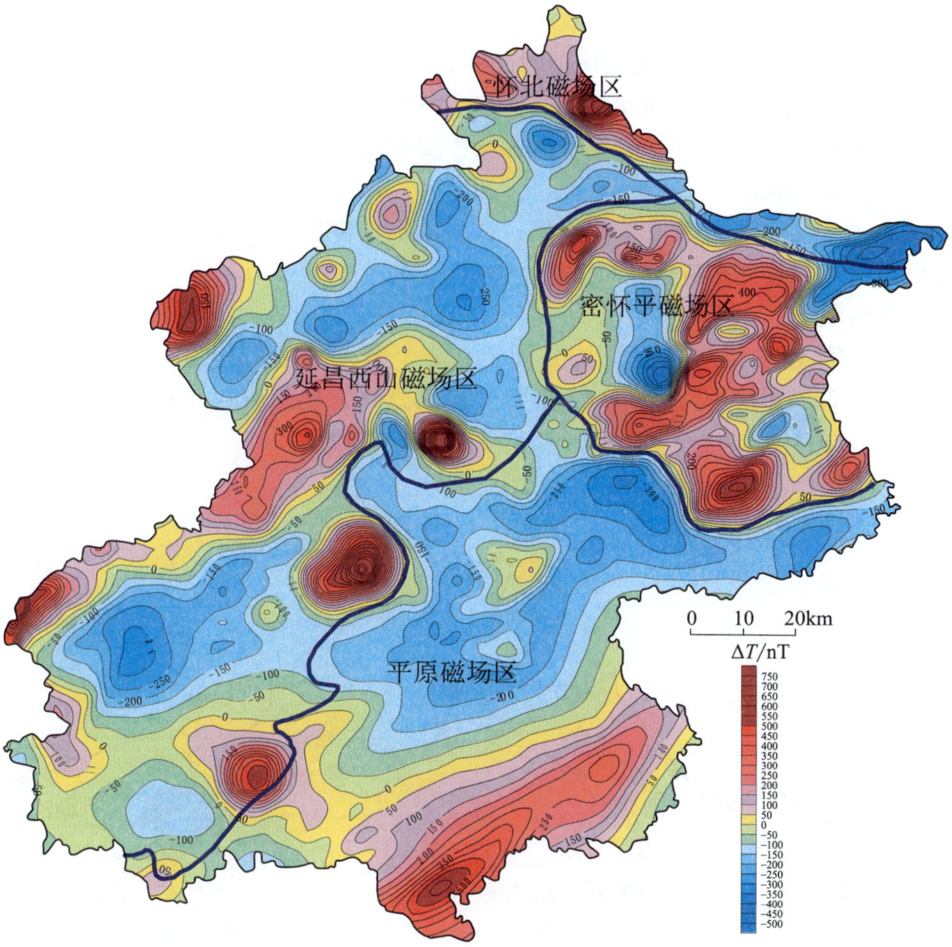

图 27-17 北京地区磁异常分区图

Fig. 27-17 The Magnetic anomaly zoning map in Beijing area

大平稳的正负磁场为背景，自西南向东北数个不同大小规模的圆状、似圆状或椭圆状高强度异常像“珍珠”一样呈北东向、北北东向，北西向和北西西向叠加于背景场之上，局部异常强度可达500nT以上，异常的整体走向呈北东向，反映区域岩浆侵入特征。

4. 平原磁场区

平原磁场区异常较为平缓，表现为-100nT左右的负磁场背景，以礼贤—安定—漷县—西集为界，可进一步划分两个亚区。西北部表现北东走向、宽缓的椭圆状负异常，分布范围与北京凹陷基本一致，东南部表现为北东走向、狭长的椭圆状正异常，异常中心极值达600nT，由太古宙变质基底隆起引起，东北和西南两端延伸至河北省。

（三）磁异常分类及其分布特征

1. 磁异常分类原则

根据1975年北京市地质研究所物化探室提供的《北京市航磁异常综合整理报告》为基础，编号由北至南，由西至东，依次由小到大顺序递增，按 ΔT 异常所处的地质环境、找矿意义和以往工作程度，对磁异常进行分类，划分为甲、乙、丙、丁四大类，其中甲类异常、乙1类和乙2类异常为矿致异常。划分原则如下：

1）甲类异常：为矿致异常，可分两个亚类。

甲1类异常：已知矿引起、推断还有找矿潜力的异常。

甲2类异常：已知矿引起、推断进一步找矿潜力不大的异常。

2）乙类异常：推断具有找矿意义的异常。分三个亚类：

乙 1 类异常：推断矿体引起的异常。

乙 2 类异常：推断含矿地质体或地质构造引起的异常。

乙 3 类异常：推断具有找矿意义的地质体或构造引起的异常。

3）丙类异常：找矿前景不明异常。按目前工作程度和认识水平，无法判明是否由具找矿意义的地质体或地质构造等引起的异常。

4）丁类异常：按目前工作程度和认识水平，认为由不具备找矿意义的岩性体引起的异常。

2. 磁异常分类结果及其分布特征

（1）异常分类结果

根据磁异常的分类原则，北京市范围内共圈定 239 个航磁异常，其中经过查证的异常有 114 个，包括 28 个钻探查证、22 个面积型查证和 64 个野外踏勘查证，确定与已知矿或矿点有关的矿异常 22 处（图 27-18）。

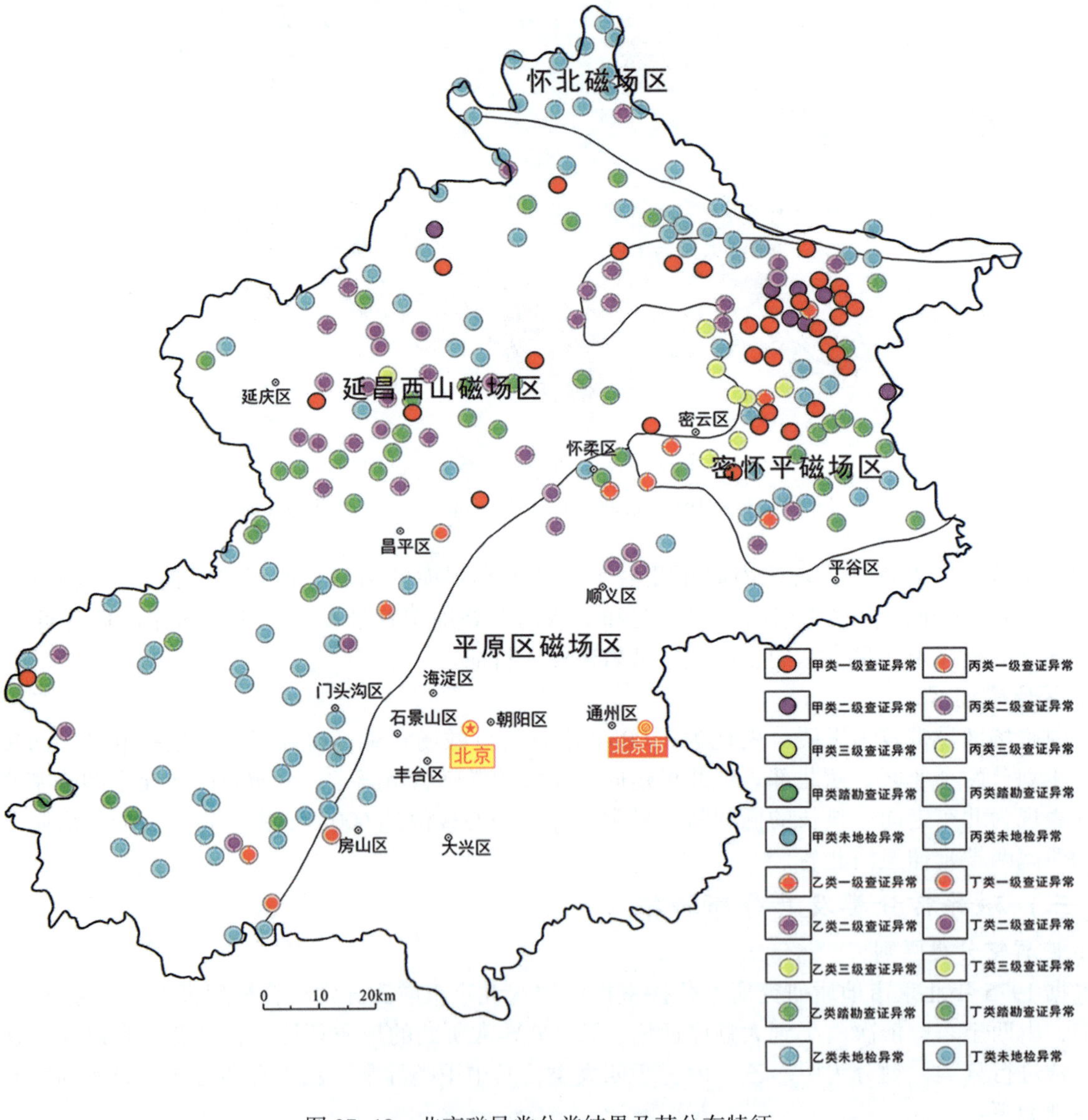

图 27-18 北京磁异常分类结果及其分布特征

（据北京市矿产资源潜力评价成果报告）

Fig. 27-18 The classification aeromagnetic and distribution of Beijing

（2）异常分布特征

北京市航磁异常分布主要集中在四大磁场区，即怀北磁场区、密怀磁场区和延昌西山磁场区。

a. 甲类异常分布特征

甲类异常主要分布在密云、怀柔、平谷等地的太古代变质岩地层中，该地区也是沉积变质型磁铁矿分布区。其次，延庆和昌平地区也有一些甲类异常零星分布，主要由一些接触交代型蚀变的铁矿和多金属矿引起。

b. 乙类异常分布特征

乙类异常集中分布在密云、怀柔、平谷等地的太古宙变质岩地层中，异常对应于磁铁矿石英岩和老地层中的多金属矿。

c. 丙类异常分布特征

丙类异常分布范围较大，在怀北磁场区、延昌西磁场区、密怀平磁场区均有分布，这些异常主要对应中酸性侵入岩体内部的多金属矿化。

d. 丁类异常分布特征

丁类异常数量较少，零星分布在怀北磁场区、延昌西磁场区和密怀平磁场区，目前认识水平认为是由侵入岩体引起，可能与多金属矿有关。

四、区域磁异常地质解译

结合重力、地震、电测深及地质成果资料，对本区重要断裂、侵入岩体和变质岩地层进行了推断解释。

（一）磁法推断断裂构造

本区内断裂构造类别较多，性质复杂，各种断裂之间常表现出一定成生联系，分布上几乎都有一定间距的规律性，一些区域性的断裂带具有长期活动特点，且山区和平原的断裂是连续的。按展布方向分为东西向，北东向、北北东向、北西向、北北西向、北西西向和近南北向等七组，其中以前三组较发育（图 27-19）。对本区金属矿产和区域构造格架重要的断裂构造有：

1. 赤城-长哨营-古北口深断裂带（京 F0001）

赤城-长哨营-古北口断裂是本区重要金矿成矿区带，控制了本成矿区带内金矿床的分布和密怀成矿区带沉积变质型铁矿的分布。该断裂经过汤河口北—赤城—古北口一线，东西均延出北京市区，整体走向为近东西向，全长约 140km，密云-怀柔地区长约 60km。该断裂在磁异常图上表现为明显的正负异常分界线和梯度带异常，断裂为密怀穹隆北边界。沿断裂带有太古宙（及华力西期）基性、超基性岩体分布，中元古代有钾质花岗岩、石英正长岩与斜长岩等侵入，自晚古生代至燕山期有大规模、多期次岩浆活动。

2. 沙厂-墙子路大断裂（京 F0015）

位于赤城-长哨营-古北口深断裂之南约 40km，走向东西，往东延入河北省青龙一带，全长大于 200km。断层面总体南倾，倾角 40°~50°。中元古代密云沙厂斜长环球斑花岗岩沿断裂侵入，其切割面在康氏面附近，为四级构造单元分界线。断裂南北两侧（尤其南侧）次级断裂构造控制了一些金、银矿床、矿点的分布，是金、银重要成矿区之一。

3. 沿河城-南口-琉璃庙大断裂（京 F0016）

斜贯本区中部，走向北东 50°~60°，延伸约 140km 以上。自西南向东北由沿河城断裂、南口山前断裂和奇峰茶-琉璃庙断裂三段相接而成。该断裂通过部位在地壳深处为扭转的北东向坡级带，并深切莫霍面为岩浆活动提供了有利通道。进入中生代燕山运动期间，由于应力场发生重大变化，尤其燕山中、晚期，在八达岭一带于北东向、北西向、东西向及北北东向等构造交汇部位发生频繁火山活动和岩浆岩侵位，形成北东向八达岭构造岩浆岩带，它是本区最重要内生金属矿产（Cu、Pb、Zn、Mo、Au 等）成矿带，伴随火山喷发形成的浅成—超浅成相岩体常与金矿化关系密切。

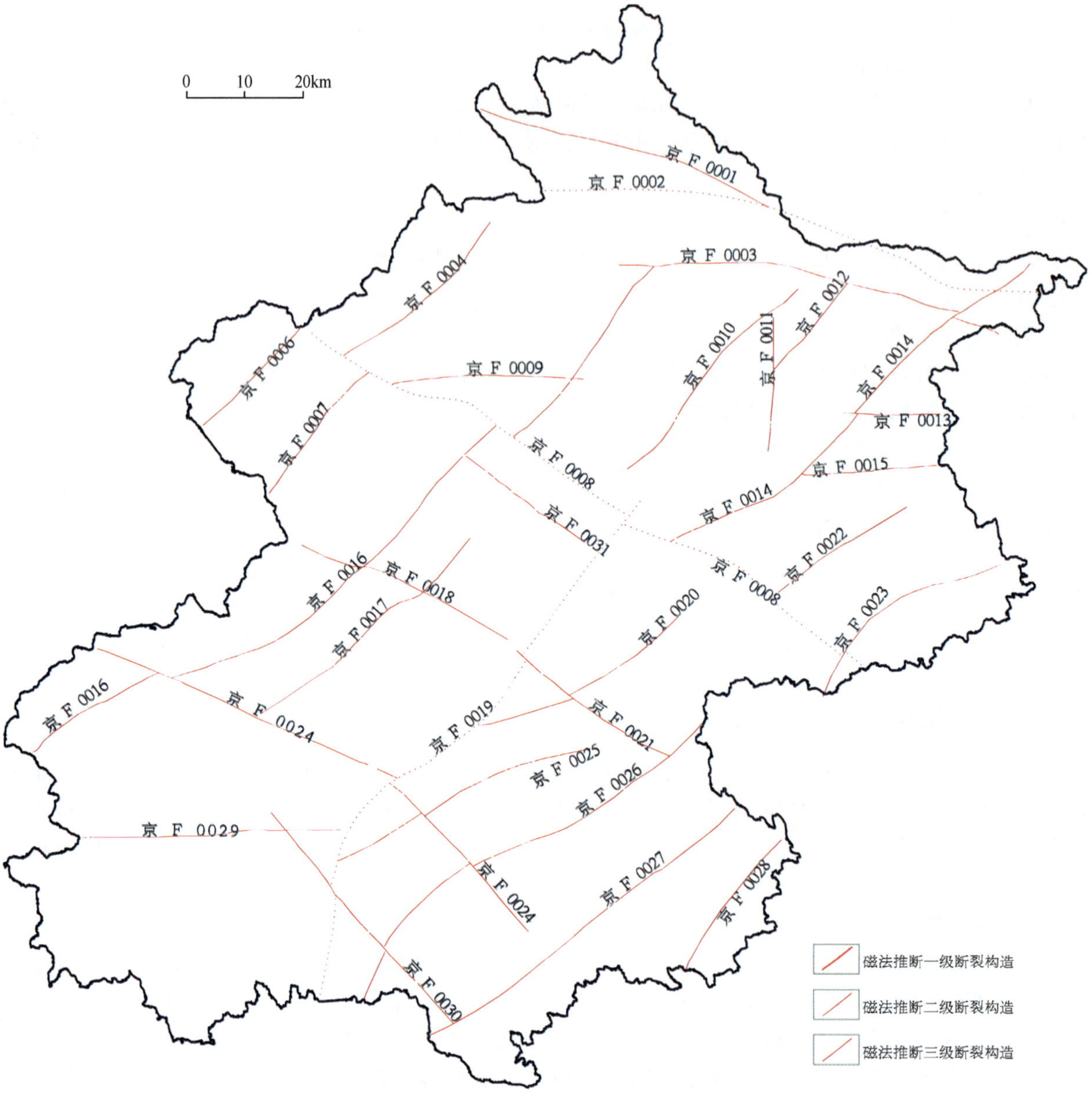

图 27-19　北京市磁法推断断裂构造分布图

Fig. 27-19　The fracture structure distribution deduced with magnetic method in Beijing

该大断裂的东北段为奇峰茶-琉璃庙断裂，断裂走向由北东 50°左右逐渐偏转至北东 30°左右，断面向西北倾斜，倾角 60°左右，并主要分布于太古宙变质岩系中，岩浆活动造成矿源层中金元素进一步活化、运移和聚集，或在其外围脆-韧性剪切带中形成矿体，该断裂及其两侧构造显示了良好的导矿和容矿作用。

4. 北北东向断裂（京 F0006）

紫荆关-大海坨深断裂出露于本区西部边界附近，沿紫荆关、双塘沟、大海坨呈北东 25°~30°左右延伸，长度大于 800km，主体部位在河北省境内。该断裂为不同磁异常区的分界及重力异常的梯度带，切割深度已达上地幔，经历了张、张剪—压剪—张剪性过程，具多期活动特点。该断裂导控岩体，岩体加大断裂，构成北北东向构造岩浆带，本区大海坨、大河南花岗岩受其控制，但与其有关的岩浆热液型金矿化现象较弱。此外，北北东向涿县-丰台-怀柔-白马关深断裂带，大部分为隐伏状态，在东北段地表为北北东向石城-冯家峪断裂破碎带，在变质岩分布区一些金矿化点分布与其有关。

次一级北北东向断裂区内也很发育。走向一般为北东 20°~30°，倾角 30°~60°不等，长度多为 20~30km，具压剪性特征，常截切东西向、北东向、北西向断裂，形成时期大致为燕山中晚期。控制了一些金矿（化）点的分布，如关上、冯家峪及密云北部地区等。

5. 黄庄-高丽营断裂（京 F0019）

黄庄-高丽营断裂是一条隐伏的深大断活动断裂，也是北京凹陷和西山隆起的分界构造。南起涿县西城坊，经坨里、黄庄、八里庄、高丽营至怀柔一线，总长 110km，走向北东，断裂面倾向南东，倾角约 50°~80°。性质为正断层，最大断距在 1000m 以上。断裂于燕山运动末期形成，切割了侏罗系及其以前的地层，控制了白垩系和古近系、新近系及第四系沉积。是Ⅱ级构造单元的分界线。据钻孔资料，断裂两盘新生界厚度相差近千米。在高丽营一带沿断裂发育地裂缝，说明该断裂最近时期还在活动。

该断裂在布格重力异常图和航磁 ΔT 化极水平方向（135°）一阶导数等值线图也有所反映。在航磁 ΔT 化极水平方向（135°）一阶导数等值线图上，断裂表现为串珠状的高值正异常；在布格重力异常图上表现为明显的重力异常梯级带，且在洼里以北和高丽营附近异常发生扭曲，推断被南口-孙河断裂和十三陵-李桥断裂错断。

6. 顺义断裂（京 F0020）

该断裂延展约 40km，其西南起自健翔桥附近，经天竺、首都机场附近，经顺义北、北小营到焦庄户一带，走向由南西段的近东西向转为北东向，倾向北西，倾角 60°左右，为正断层。地表表现为线状、沟状地，沿断裂在顺义一带发育地裂缝，表明这一断裂的展布和活动性。该断裂控制了顺义凹陷的发育。

顺义断裂在物探成果上反映较明显，布格重力异常图上表现为明显的梯级带，航磁 ΔT 化极水平方向（135°）一阶导数等值线图上表现为串珠状的正异常，另据前人所做的光电地震资料显示，断裂位置的反射层明显发生错动，证明了断裂的存在。

7. 夏垫-马坊断裂（京 F0027）

夏垫-马坊断裂是北京平原区一条重要的活动断裂，北起平谷马坊、经夏垫、西集、凤河营到廊坊以西一带。走向 NE30°，倾向 SE，倾角 65°~70°，为正断层。在重力异常图上表现为明显的梯级带，航磁 ΔT 化极等值线平面图上表现为正负磁场的分界线，南东为椭圆状的正异常，北西为大面积的负异常。它是大兴隆起与大厂凹陷的分界构造。据钻孔证实断层两盘第四系厚度相差达 300m。1679 年平谷马坊 8 级大地震就发生在该断裂带上，是一条规模较大的活动断裂。

8. 南口-孙河断裂（京 F0018）

南口-孙河断裂是北京地区一条重要的活动断裂。其总体呈 SE130°~140°方向展布。其北西端起自昌平南口镇，向南东方向经百泉、孙河，长约 50km。

航磁 ΔT 化极水平方向（45°）一阶导数等值线图表现为梯级带和正负磁场的分界线，断裂北东为串珠状的正异常，南西为串珠状的正异常。剩余重力异常图上表现为明显梯级带。断裂北西段（南口至北七家段）倾向南西，上盘下落，形成一个 NW 向的陷落带，控制北西向马池口第四纪凹陷，凹陷内第四系厚达 600m 以上。而在断层的下盘第四系极薄，并有基岩残丘裸露地表。断裂南东段（孙河至通县段）倾向北东，控制顺义第四纪凹陷南部的东坝沉积中心。上盘下降，如后沙峪、古城一带堆积了厚达 800 余米的第四系，下盘相对抬升，如来广营一带第四系厚度仅 200~300m。该断裂形成的时间应在新近纪末—第四纪初。

（二）磁法推断岩浆岩

北京地区岩浆侵入和火山喷发活动较为强烈，自太古宙到新生代均有不同规模的岩浆活动，形成了大小侵入岩体 230 余个，各期火山岩的出露分布总面积约为 2000km^2，本次工作利用磁测资料综合推断侵入岩体 16 处，其中中酸性岩体 14 处，基性岩体 2 处；推断火山岩地层 5 处（图 27-20）。

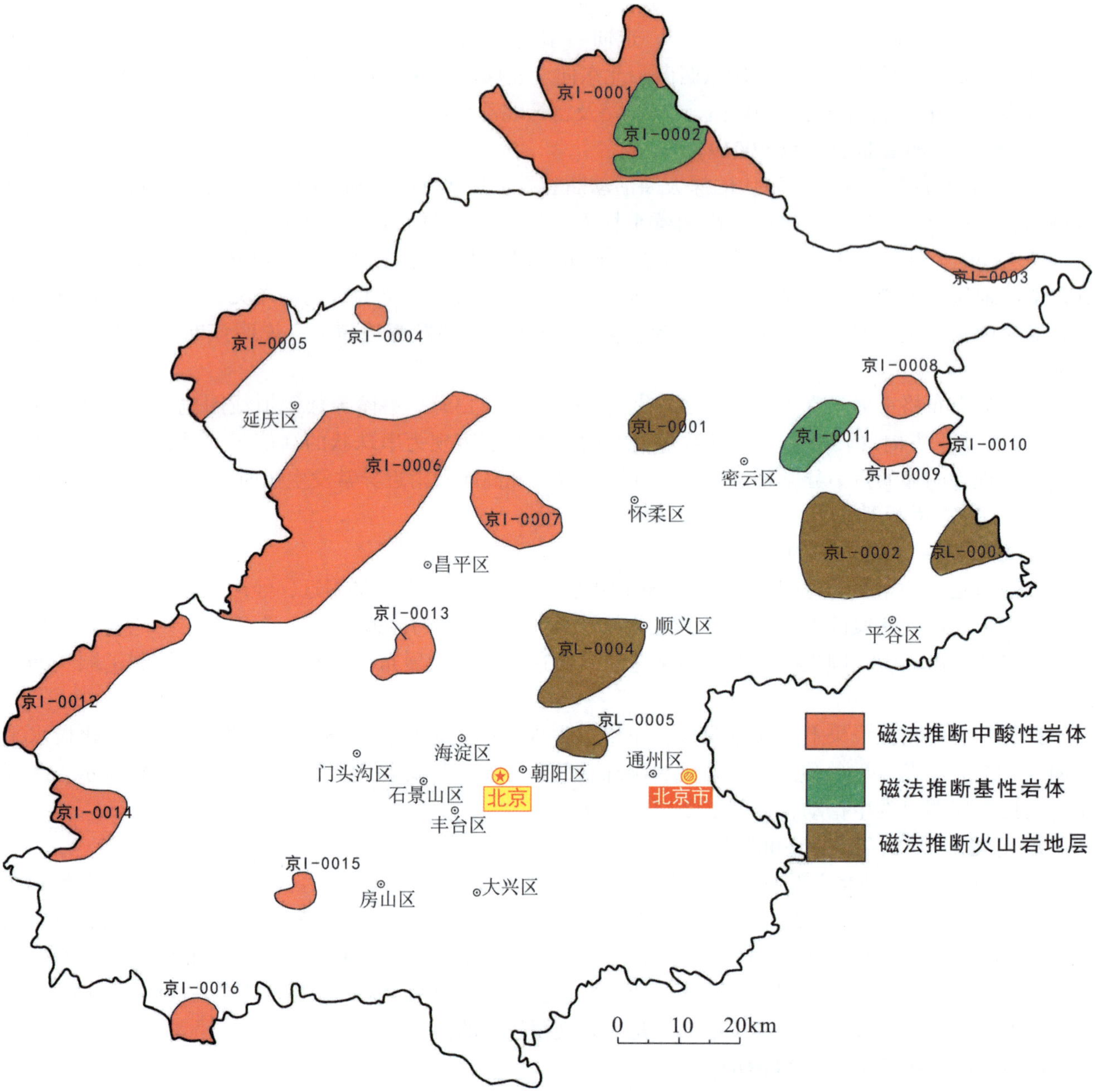

图 27-20 北京市磁法推断岩浆岩分布图

（据北京市矿产资源潜力评价成果报告）

Fig. 27-20 The magmatic rocks distribution deduced with magnetic method in Beijing

1. 房山花岗岩体

房山花岗岩体位于房山区东南，是一个半隐伏岩体，岩性以花岗岩为主，剩余重力异常图上表现为在正异常背景下出现一个孤立的圆状负异常，异常中心位于燕化小区，幅值为$-1\times10^{-5}m/s^2$，航磁ΔT化极垂向一阶导数等值线平面图上，在平缓的背景场上，出现一个极值为350nT的圆状正异常，重力场低、磁场高，利用重磁资料综合圈定岩体的在地表的投影面积约$320km^2$。

2. 阳坊花岗岩体

阳坊岩体位于海淀区与昌平区交界处，剩余重力异常图上编号为L京-0031，异常极值为$-11\times10^{-5}m/s^2$，走向近东西。航磁ΔT化极垂向一阶导数等值线平面图上异常整体呈北东走向，由两个椭圆状的异常中心组成，极值达500nT，呈北西缓、南东陡的特点。岩体仅在西部有小部分出露，大部分被第四系覆盖，埋深200~300m，根据钻孔资料显示，岩体的岩性为花岗岩。

（三）磁法推断变质岩地层

北京市发育的变质岩主要是太古宙的片麻岩和磁铁石英岩，磁性较强，航磁异常一般在500nT以上，集中分布在密云和怀柔地区。利用其磁异常特征，以化极后的垂向二阶导数零值线，并结合化极后不同高度的延拓平面等值线图，对北京市的变质岩地层进行圈定，共圈定变质岩地层三处（图27-21），分别位于密云、怀柔、延庆和平原区南部，其中结合前人的钻孔资料首次圈定了平原区南部隐伏的变质岩地层的分布范围。下面以密怀变质岩地层为例举例说明磁法推断变质岩地层的分布特征。

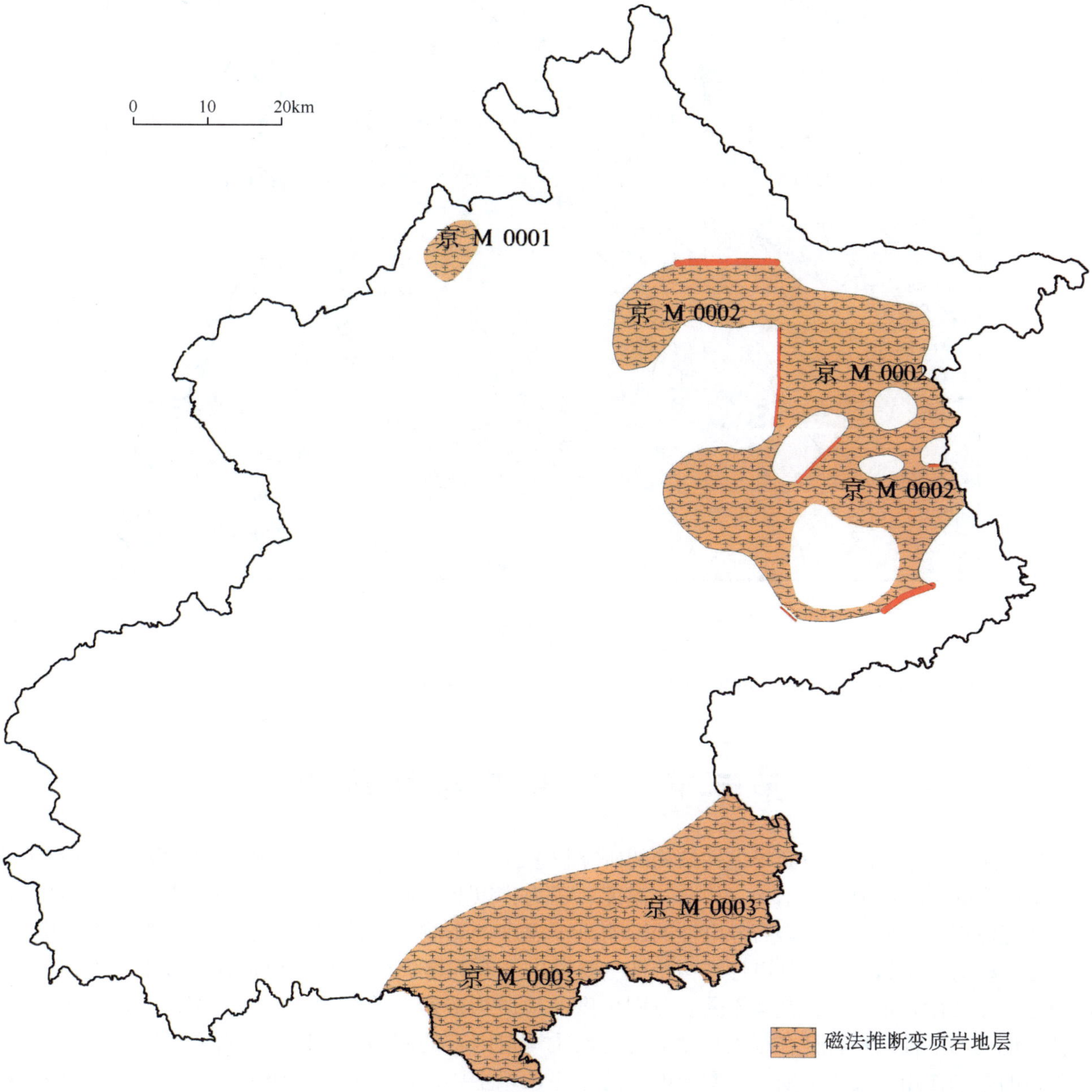

图27-21　北京市磁法推断变质岩地层分布图

Fig. 27-21　Distribution map of metamorphic stratigraphic area inferred with magnetic method in Beijing

密怀变质岩分布于密云和怀柔两地（京M 0002），变质岩岩石主要为太古代沙厂云英闪长质片麻岩、黑卧辉长质片麻岩、苇子峪辉长闪长片麻岩、苇子峪辉长闪长片麻岩，以及磁铁石英岩、磁铁片麻岩和沙厂式铁矿等，这些岩（矿）石的 k 值平均变化为 $1683\times10^{-5}\sim4898\times10^{-5}$SI，$Jr$ 值平均变化为 $780\times10^{-3}\sim1960\times10^{-3}$A/m，在 1：200 000 航磁 ΔT 异常表现为 400～500nT 波状起伏的正磁场（图 27-22）。

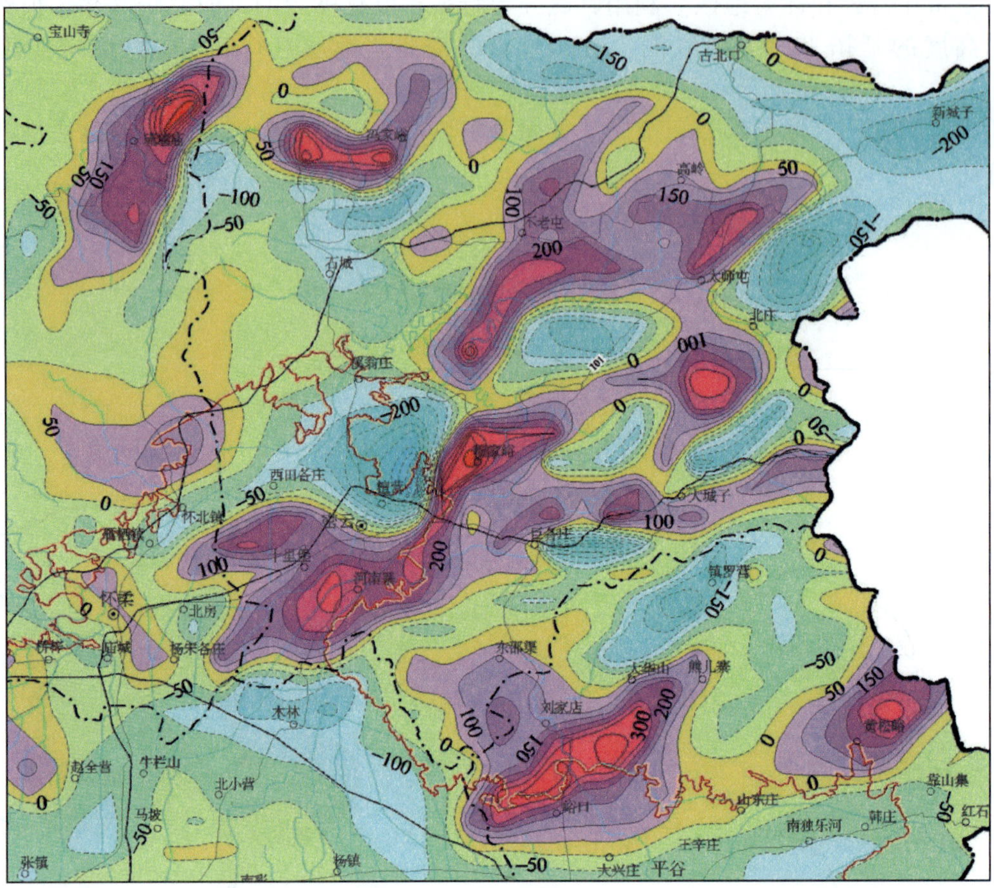

图 27-22　密云沉积变质岩地层航磁 ΔT 化极垂向一阶导数图

（据北京市矿产资源潜力评价成果报告）

Fig. 27-22　Sedimentary metamorphic strata in Miyun aeromagnetic ΔT vertical first derivative graph

第三节　岩石圈电性结构

近年来，关于大陆岩石圈导电性结构的研究较多，大陆岩石圈的导电性除了与其物质成分有关外，还取决于大陆岩石圈内部的结构、构造、温度、压力，以及物质成分的物理状态等因素。其中压力对大陆地壳深部和上地幔顶部岩石导电性的影响小于温度的影响，因此在正常状况下，岩石圈的电阻率与温度有极为密切的关系，其电性结构特征主要取决于地下热状态。在 200 km 深度以下，上地幔范围内地温分布规律趋于一致，基本不受岩石圈尺度的地质构造运动的影响。而软流圈、上地幔盖层，以及整个地壳的电性结构特征在很大程度上受到地质构造因素的制约，特比是岩石圈尺度的构造运动有更明显的控制作用。关于大陆岩石圈导电性的这些认识主要来源于两方面研究数据，一方面是岩石导电性的高温、高压实验室测量数据；另一方面则是岩石圈电性结构的野外实际探测结果，即宽频带大地电磁测深的结果。本部分内容主要收集了中国地震局等单位在本区的大地电磁测深剖面资料，以论述北京地区岩石圈电性结构特征。

1. 北京–柏各庄剖面

剖面沿北京南郊狼堡至渤海北岸柏各庄，穿越冀中坳陷、沧县隆起和黄骅坳陷，共计 7 个大地电磁测深点（图 27-23）。地壳上地幔电性分层如下：

1）地壳表层高导层：主要发育在平原地区，其电阻率值为几到十几欧姆·米，此层厚度变化大，在山区和山区边缘缺失。

2）壳上部高阻层：此层电阻率在 $10^2 \sim 10^3 \Omega \cdot m$ 范围内变化，越接近山区电阻率值越大，此层与高速的上地壳相当。

3）壳内高导层：各视电阻率曲线中部出现的明显的极小值，可认为是壳内高导层反映，反演结果表明该高导层的电阻率值为几到十几欧姆·米，埋深在十几到二十几千米，厚度 3~4km 或更大些，有时与壳内低速层位置相当。

4）下部高阻层：该层电阻率值一般为 $10^2 \sim 10^5 \Omega \cdot m$，山区大，平原小，厚度变化很大，在平原区为 30km，山区厚度增大，可达 60~70km。

5）上地幔高导层：此层电阻率值为一到几欧姆·米，但埋深变化很大，在平原区最浅处 50km 左右，山区埋深可达 100km 以上。一般认为上地幔高导层大致和上地幔低速层相当。

从剖面分层电性特征分析，上地幔低阻区有明显的横向起伏，波动幅度由西向东变大，波峰与北京凹陷、武清凹陷和黄骅坳陷分别对应；波谷与大兴隆起和沧县隆起对应；在沧县隆起的下地壳、上地幔内有较大范围的高阻圈闭。另一特点是地壳内出现低阻带，电阻率小于 25Ω·m 的电磁的范围呈“香肠状”结构，坳陷区厚，隆起区薄。地壳表层的低阻分布范围与区内的新生代沉积发育状况基本一致（赵国泽等，1986）。

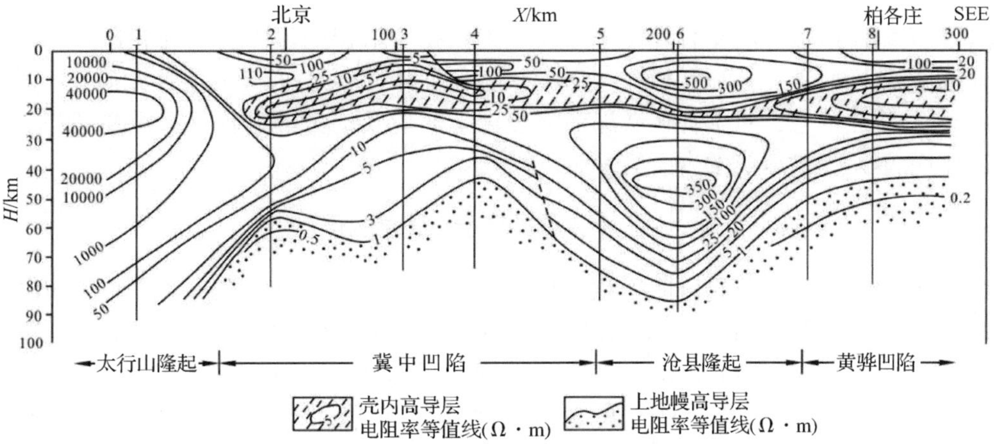

图 27-23　北京-柏各庄地壳上地幔电阻率结构剖面

（据赵国泽，1986）

Fig. 27-23　Electrical structure of the crust and upper mantle along Beijing-Baigezhuang profile

2. 怀来-马兰峪剖面

西起新生代的怀来-延庆盆地，穿过廊坊-衡水火山-沉积盆地北缘。并通过三河-平谷大震区，全长 140km。沿该剖面普遍存在壳内高导层，厚 3~5km，电阻率为几至十几欧姆·米，埋藏深度在怀来测点为 28km，其他测点为 20km 左右，上地幔高导层在平原北部约为 60~70km，向两侧显著加深，西部的怀来测点和东部的马兰峪测点达 90~95km。1679 年的三河-平谷 8 级大地震大致位于上地幔高导层埋深显著变化的王辛庄和马兰峪测点之间（图 27-24）。

大地电磁探测结果证明，太行山前断裂为一组向东倾斜，岩石圈尺度的深断裂。以太行山前断裂为界划分为东、西两区。东区为低阻区，其岩石圈电导率由浅到深逐渐减小，与大陆区岩石圈的导电性结构相反；西区为高阻区，地壳表层岩石的电导率很低，但随着深度增大，岩石圈电导率逐渐增大，表现出正常大陆区岩石圈导电性结构的特点。

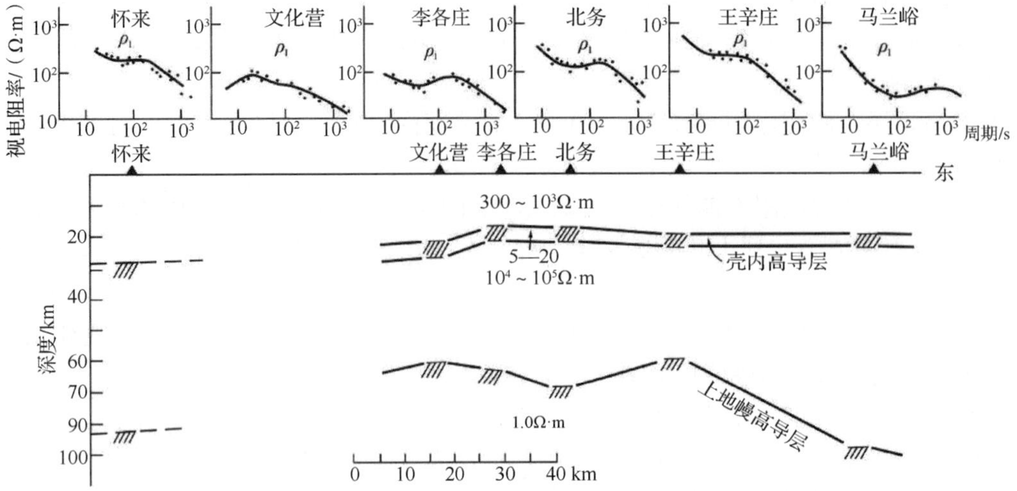

图 27-24　怀来-马兰峪大地电磁测深剖面图

Fig. 27-24　Magnetotelluric sounding profile of Huailai-Malanyu

第四节　岩石圈热结构

地球内部的热能对地球的演化起着决定性的作用，是构造运动、火山活动和地震等地质作用的基本能源。岩石圈热结构又称壳-幔热结构，主要研究岩石圈的温度分布及其特性。研究区域地壳及上地幔地热场的分布及其热状态，对了解区域深部物理化学环境及动力学过程具有重要意义，将为划分该区岩石圈结构构造单元提供地热学方面的参照。

一、大地热流值特征

华北地区自燕山期以来大陆岩石圈发生了巨大的减薄作用，可能丢失了约 100~150 km 的岩石圈根带（邓晋福，2013）；新生代以来华北东部地区陆壳厚度也明显减薄。地球内部的热量通过介质传导、对流等形式不断的传输到地壳浅部。大地热流值的变化与上地幔的隆起、坳陷相间排列的构造格局密切相关（邓晋福，2013）。北京地区西北部山区热流值较低，一般低于 60mW/m^2，在背斜的轴部和活动断裂带热流值增高。东南部平原区热流值较高，一般高于 60mW/m^2（图 27-25）。

二、地温场特征

（一）山区地温场

本区山区和山间盆地地热点，多分布在区域断裂和伴生的次级构造以及岩体附近，在断裂的交汇部位和岩体与围岩的接触带，多以单泉和泉群的形式集中放热，形成小型热田。燕山山区的温泉温度一般 50~60℃，大地热流值一般为（0.61~1.84）×41.686mW/m^2。

（二）平原区地温场

北京平原区为上地幔隆起区，软流圈大面积上涌，浅部可至到 60~70km，地壳厚度 35~37km，活动构造发育，大地热流值为（1.7~2.0）×41.686mW/m^2，形成了大面积的地热异常区。

1. 地温场平面变化

平原区地温场的分布特征与构造密切相关，不同地区不同构造部位，地温场不同，主要是受凹凸相间的地质构造格局控制，地温明显呈高低相间带状展布，盖层地温梯度等值线与区域北东、北北东、北西西及近东西向的构造走向及基底凹凸构造轮廓基本一致，基岩隆起区为相对高温区，凹陷区为相应地温区。并且在平原区特定的构造格局控制下，新生界盖层地温梯度大小与基岩顶面埋深密切相关，一般地温梯度随基岩埋深变浅而增大。大兴隆起由于上覆新生界盖层厚度一般为100~300m，部分地段厚度仅数十米，实质上与基岩裸露区相当，具有低地温的特点，1000m埋深地温小于30℃。

2. 地温场垂向变化

地温场在垂向上的变化规律，不仅与其所在的构造部位有关，而且与岩石性质相关。一般情况下，在不富水的泥岩、页岩段，地温梯度略有递增的趋势，地温梯度为1.5~2.5℃的变化规律；而在白云岩、砂岩等富水地段由于水动力循环的影响，地热流体在三维空间内的局部对流作用加强，地温梯度明显变低（≈1℃），白云岩甚至更低。

图27-25 北京地区大地热流值分布情况图

（据邓晋福，2013）

Fig. 27-25 The distribution of terrestrial heat flow values in Beijing

三、岩石圈温度结构

（一）岩石圈剖面热结构

利用热流资料，结合地质、地球物理资料和岩石的放射性元素资料计算深度-温度曲线，可以了解岩石圈的热状态。根据张家口-塘沽剖面（高文学，1993）大地热流和地壳各界面及莫霍面温度（图27-26），本区岩石圈温度结构特征如下：

1）西部、北部山区（燕山-太行山）隆起区地壳各界面和莫霍面的热流和温度低于平原区。

2）在平原区内，地壳各界面和莫霍面的热流和温度显示相对隆起区高于凹陷区，即大兴隆起的地壳各界面和莫霍面的热流值和温度高于北京凹陷。

3）山间断陷盆地（延庆盆地）基底面和地壳界面的温度高于平原区，莫霍面温度高于平原区的凹陷区，低于平原区内的隆起区。

另外，据邓晋福（2007）关于华北地区温度结构的研究成果，可知北京地区岩石圈热状态整体上相对较“热”，其中以东部的华北盆地最热。北部和西部山区相对较“冷”（表27-5）。

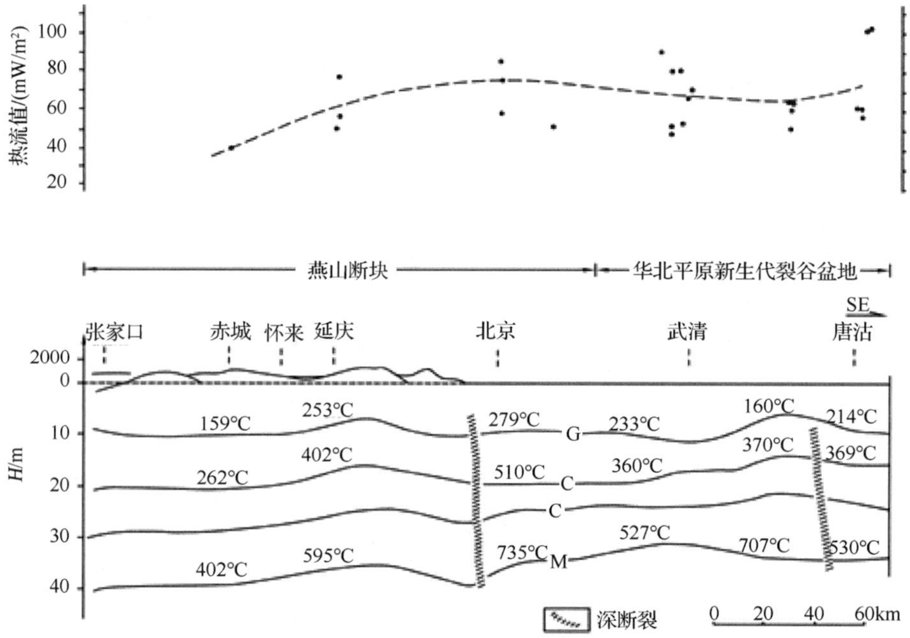

图27-26　张家口-塘沽剖面大地热流和地壳各界面及莫霍面温度

Fig. 27-26　Terrestrial heat flow and temperature at different interfaces in the crust and at the Moho discontinuity along Zhangjiakou-Tanggu profile

表27-5　北京主要构造单元平均热流值和深部地热特征参数

Table 27-5　Main tectonic unit average heat flow value and characteristic parameters of deep geotherm in Beijing

构造单元	Q	Q_c/Q_m	T_{40km}/km	H/km	H的误差/km
华北盆地	67（96）	0.54	857	71	+4/-3
燕山-阴山造山带	56（26）	0.92	583	116	+12/-10

注：Q为平均热流值（mW/m²），括号内的数字为构造单元内热流观测值的数目。Q_c/Q_m为壳幔热流比值；T_{40km}为40km深度温度；H为热岩石圈厚度。

（据邓晋福，2007）

第五节　岩石圈应力场特征

一、应力场与地壳形变

根据北京华北地区20多年来的地壳垂直形变（图27-27）可见：①整个华北坳陷相对于周围山区继续下沉，与新构造运动的趋势是一致的；②形变等值线与形变趋势图等值线的延伸方向为北北东向，沉降幅度自西向东有逐渐增大的趋势。反映了北北东向断裂的活动对现今地壳运动仍起主要的控制作用。另外燕山南麓北西向的现代沉降区的分布与第四纪凹陷区也具有一致性，反映了北西向构造现今运动的继承性特征。此外典型的沙河—通州一带沉降区的长轴方向为北西向，反映北京地区现今主压应力方向为N70°～90°E，主张应力方向为0°～20°W。

二、区域应力场与震源机制

华北及北京地区部分强震的宏观考察结果反映，极震区等震线的长轴方向为N15°～30°E或N60°～70°W，地裂缝分布方向为北北东，多为顺扭（表27-4）。根据P波初动解资料得出的两组节面的走向是北北东向和北西向，倾角较陡，其中北北东向一组与地震宏观考察结果和余震分布一致性较好。在地震发生时，北北东向断裂多表现为顺扭兼压性，如邢台、唐山、和林格尔地震；而北西向断裂多为反扭并明显张性，如海城地震。根据震源机制解得出的本区现今主压应力为北东东-南西西向，仰角不大，应力以水平作用为主。这种应力作用方式与地震的宏观考察结果是吻合的（表27-6）。

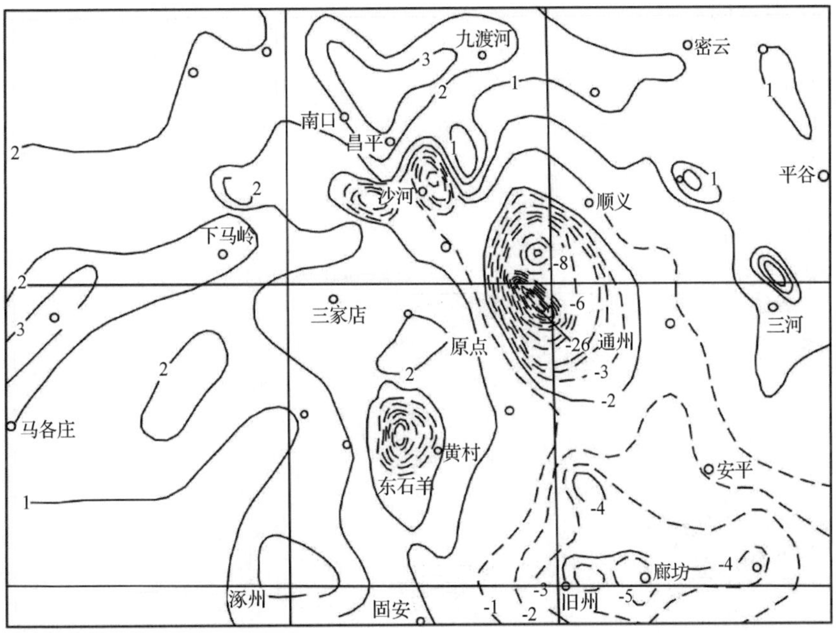

图27-27　北京地区地壳垂直形变速率（1950—1994年）

Fig. 27-27　The vertical crustal deformation rate in Beijing area

图中实线为上升速率，虚线为下降速率，单位：mm/a

表27-6　北京外围地区近期几次主要强震宏观考察表

Table 27-6　Macroscopic inspection of several recent major earthquake in Beijing

时间	地点	震级	极震区长轴方向	地裂缝分布方向	余震分布方向	发震构造
1966年3月8日	邢台	6.8	N28°E	NNE，顺扭	N30°～40°E	NNE

续表

时间	地点	震级	极震区长轴方向	地裂缝分布方向	余震分布方向	发震构造
1966 年 3 月 22 日	邢台	7.2	N28°E	NNE，顺扭	N30°~40°E	NNE
1967 年 3 月 27 日	河间	6.3	N60°E~65°W	N40°~50°W	NWW	NWW
1975 年 2 月 4 日	海城	7.3	N60°E~70°W	N70°W 反扭	N67°W	NWW
1976 年 4 月 6 日	和林格尔	6.3	N25°E	NNE 顺扭兼压	N30°E±	NNE
1976 年 7 月 28 日	唐山	7.8	N25°E~30°W	N30°E 顺扭	N25°E	NNE
1976 年 7 月 28 日	滦县	7.1	N15°E	N15°~20°E 反扭	N30°E	NNE

（据谭成轩，2014）

三、实测地应力状态

北京及其外围附近地区水压致裂和应力解除地应力绝对值测量结果反映本区多数测点近期的主压应力方向为北西—北西西方向，少数测点为北东—北东东向（表 27-7，图 27-28）。主压应力值在不同测点变化很大。北京地区总体上是山区的应力值较平原区高。需要说明的是，对于现今区域应力场，这些实测地应力的测点仍嫌稀少，测量深度大部分较浅，并受局部构造影响，分布也不均匀，它们主要是代表了测点附近的局部地应力状态。北京及其外围附近地区深部钻孔崩落法确定最大水平主压应力方向为北东东向，与研究区震源机制、地壳形变场等反映的区域最大水平主压应力方向较为一致，可以代表区域应力作用方向。

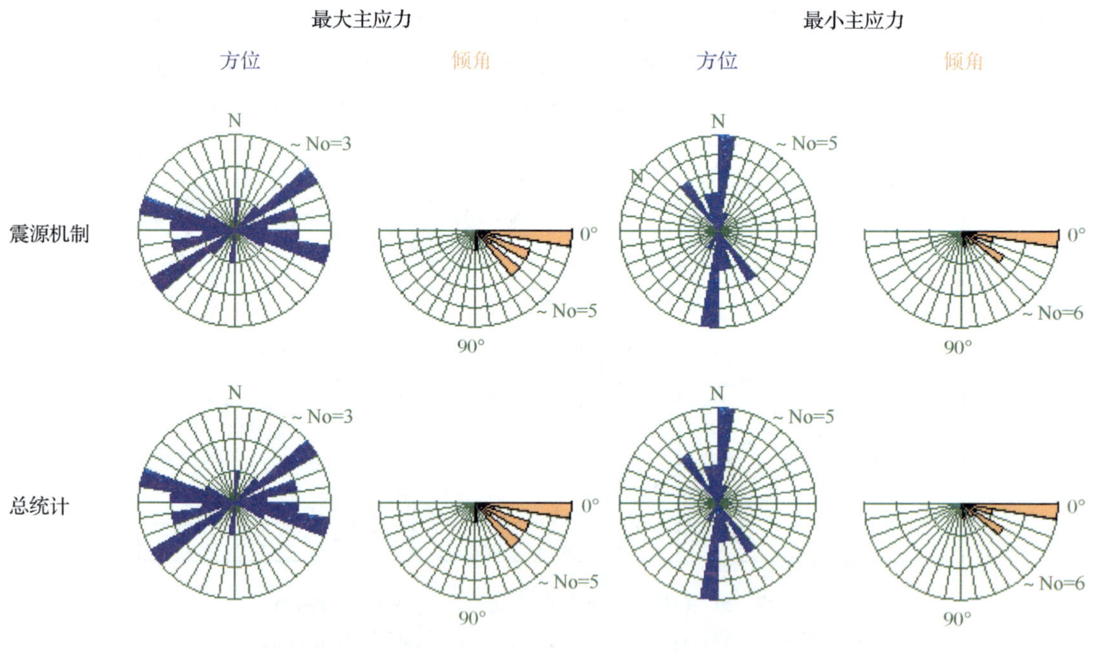

图 27-28　北京及其外围附近地区主应力方向统计玫瑰花图

Fig. 27-28　The principal stress direction statistical figure rose in Beijing and its peripheral region near

表 27-7　北京及其外围附近地区地应力状态测量数据

Table 27-7　The table of in-situ stress measurement data in Beijing and its peripheral region

序号	纬度/°	经度/°	测量深度/m	最大水平主应力方位/°	最大水平主应力大小/MPa	最小水平主应力大小/MPa	地点	测量方法
1	40.393	116.956	46	346.2			北京邓家湾	水压致裂
2	39.717	115.948	127	48			房山塔湾	
3	39.293	115.498	85	80			河北易县	
4	40.26	116.28	344	315			十三陵 ZK36 钻孔	
5	40.26	116.279	344	300			十三陵 ZK38 钻孔	
6	40.055	116.16	20	293	5.3	3.04	北京温泉 No. 1	应力解除
7	39.786	115.973	7	341	5.7	2.35	房山北车营	
8	39.786	115.973	16	66	6.3	0.49	房山东庄子	
9	39.3	115.492	32	310	4.1	3.6	河北易县 No. 2	
10	39.786	115.973	25	39	2.25	0.78	房山三福村	
11	39.786	115.973	12	74	0.84	0.55	房山煤矿	
12	39.721	115.927	19	50	8.5	4.4	房山歇息岗 No. 1	
13	40.14	116.934	17	287	2.25	0.2	顺义吴雄寺 No. 1	
14	39.721	115.927	17	15	5.3	2.55	燕山动力厂	
15	40.14	116.934	18	292	2.75	0.49	顺义吴雄寺 No. 2	
16	39.721	115.927	20	59	5.78	4.2	房山歇息岗 No. 2	
17	40.14	116.934	19	282	2.65	0.69	顺义吴雄寺 No. 3	
18	39.721	115.927	21	55	7.25	4.3	房山歇息岗 No. 3	
19	40.14	116.934	19	287	2.55	0.39	顺义吴雄寺 No. 4	
20	40.14	116.934	12	285	3.04	1.76	顺义吴雄寺 No. 5	
21	39.3	115.492	16	272	4	2.7	河北易县 No. 1	
22	40.055	116.16	21	295	3.53	2.16	北京温泉 No. 2	
23	39.3	115.492	47	286	4.5	3.9	河北易县 No. 3	
24	40.233	116.252	15	285	1.18	0.78	昌平	
25	39.3	115.492	65	316	5.4	4.3	河北易县 No. 4	
26	39.887	116.171	12	325	2.06	0.88	北京大灰厂	
27	40.377	115.213	13	287	2.55	1.37	河北涿鹿尤家园	
28	40.14	116.934	20	277	3.53	1.67	顺义吴雄寺 No. 6	
29	39.964	116.895	13	291	2.06	0.49	河北三河孤山	
30	40.35	116.626	17	277	4.02	1.08	怀柔坟头村	
31	40.14	116.934	15	285	2.65	2.06	顺义吴雄寺 No. 7	
32	40.141	116.133	10	60	0.92	0.65	昌平阳坊	
33	39.981	116.029	800	75			门头沟煤矿	钻孔崩落
34	39.065	116.197	2587	81			任丘油田岔 49 井	
35	39.049	116.623	4870	95			任丘油田苏 401 井	
36	39.041	116.253	2350	83			任丘油田岔 12-9 井	
37	39.086	117.586	2417	111			大港油田塘 20 井	

第二十八章　大陆岩石圈地球化学特征

大陆地壳岩石组成具有复杂性和不均匀性，因此很难了解大陆地壳不同结构层的物质组成与性质。北京地区属于大面积盖层区，只能根据不同研究者的推断或据多解性的地球物理资料来解释深部岩石的组成。在估计大陆地壳成分时，有三方面基本限制条件：一是地壳的生长历史；二是地球物理对地壳结构及物质组成的限制；三是出露的大陆地壳剖面和地壳包体，超深钻及实验岩石学限制。北京地区大陆岩石圈的地球化学特征主要依据大区域尺度的大陆超深钻、大陆地壳剖面等研究成果来进行总结。

第一节　岩石圈结构层元素丰度

区域地壳岩石元素丰度的研究，是指在系统的岩石样品采集和分析之后，进行区域地壳结构模型、区域地壳岩石和地层学模型及元素丰度值的检验。

一、区域地壳结构模型

北京地区及华北地区多种方法的探测与研究成果表明本区地壳结构具有以下特征：大陆地壳划分为上部地壳、中部地壳和下部地壳 3 部分，地壳为非均匀多层叠置结构。上部地壳包括地壳表层和下面的高速层，总厚度 9~15km；中部地壳主要是花岗闪长质层，为一组低速层或高低速相间层，层厚度 10~20km；下部地壳主要是玄武质层，为速度随深度显著增加的梯度层，层厚度 5~11km；地壳横向上具显著不均匀性，莫霍面具波浪起伏的状态；地壳总体平均厚度为 34km，东南部平原区 33km，越往西部和北部，地壳越厚，燕山山区可达 40km。

吴宗絮等（1994）基于出露的早前寒武纪变质岩系和中新生代岩浆岩的岩石学与地质压力计研究，结合地球物理测深资料与高温高压下岩石中地震波传播速度的实验结果，认为冀东和阜平-龙泉关-五台山-繁峙地区的古老变质岩系是各自构造单元出露的大陆地壳断面，并提出上述地区地壳-上地幔岩石学和地层学结构（表 28-1）。

表 28-1　华北地区地壳—上地幔岩石学结构模型表

Table 28-1　The table of crust and upper mantle petrological structure model in North China

分层	太行山-五台地区			燕山地区		
	岩石	v_p/（km · s^{-1}）	v_s/（km · s^{-1}）	岩石	v_p/（km · s^{-1}）	v_s/（km · s^{-1}）
上地壳	沉积盖层 绿片岩相岩、花岗闪长质变粒岩 花岗岩-花岗闪长岩岩体	5.95~6.1	3.42~3.42	沉积盖层 绿片岩相片岩、花岗闪长质变粒岩-花岗闪长岩岩体	6.1	3.45
中地壳	花岗岩（低速体）	5.8	3.70	角闪岩相片岩、花岗闪长质片麻岩、花岗闪长岩体	6.4	3.75
	角闪岩相片岩、片麻岩	6.5	3.61	含水低速体	6.1	3.62

续表

分层	太行山-五台地区			燕山地区		
	岩石	v_p/（km·s^{-1}）	v_s/（km·s^{-1}）	岩石	v_p/（km·s^{-1}）	v_s/（km·s^{-1}）
下地壳	TTG（低速体）	6.4	3.80	基性麻粒岩	7.0	4.45
	中酸性麻粒岩	6.6	4.40			
	基性麻粒岩、石榴子石-歪长石-辉石堆晶岩	7.0	4.40	石榴子石-歪长石-辉石堆晶岩		
上地幔	尖晶石相二辉橄榄岩，少量透辉石橄榄岩、纯橄榄岩	8.0~8.14	4.40	尖晶石相二辉橄榄岩，少量方辉石橄榄岩、纯橄榄岩	7.9~8.1	4.45

二、地壳化学成分垂向变化

地壳化学组成垂向变化是地壳时空演化的共时态结构。境内燕辽裂陷带内，从下地壳、中地壳到上地壳和沉积盖层，亲酸性热液元素含量呈逐步增高趋势，而亲基性超基性元素含量则呈逐步降低的趋势，上地幔成分与上述变化趋势协调一致。这种规律性的地球化学行为表明了地壳内部分异和均衡调整程度较高。

三、剖面岩石学结构模型

张家口-塘沽剖面岩石学结构模型（高文学等，1993；图28-1）表明：

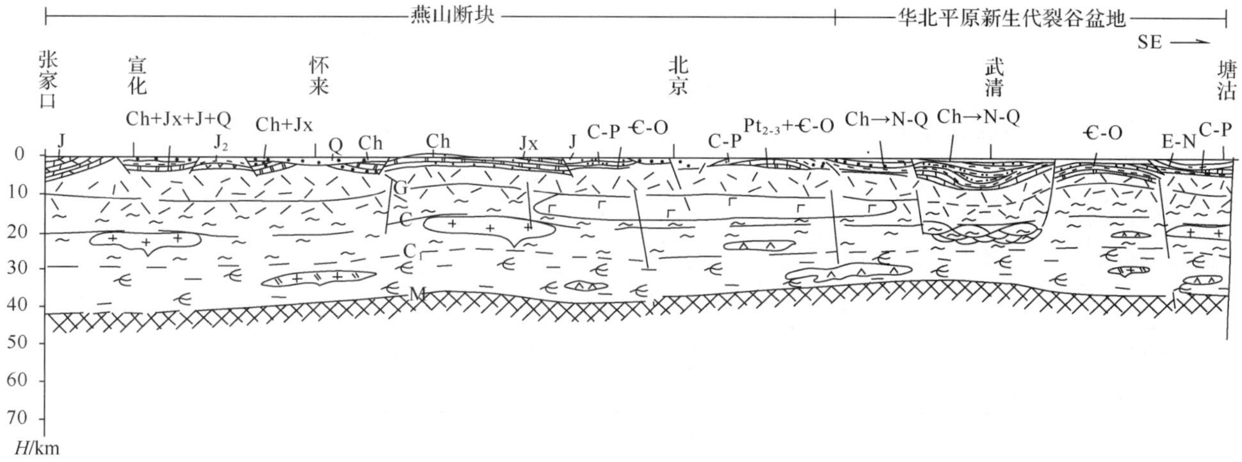

图28-1 张家口-塘沽剖面陆壳的岩石学结构模型

（据高文学等，1993）

Fig. 28-1 A petrologically structural model for the crust along Zhangjiakou-Tanggu profile

壳幔之间表现高阻高速高密度的特征，可能是古近纪已固结的玄武岩岩浆房。

下地壳主要由麻粒岩相岩石组成。下部以基性麻粒岩为主，并夹有许多透镜状、似层状超镁铁岩岩体。上部在首都圈东、西部略有所差异，东部由中性成分的麻粒岩和紫苏花岗岩等组成，西部则以中酸性成分的片麻岩、浅粒岩为主。

中地壳以角闪岩相岩石组成，主要由中酸性成分的混合岩化的各种片麻岩、浅粒岩、变粒岩和少量角闪岩、英云闪长岩、花岗闪长岩组成。首都圈西部还可能含少量大理岩和斜长角闪岩。

上地壳以绿片岩相岩石组成，部分为低角闪岩相，主要由中酸性成分的变粒岩、绿泥片岩类和绢云片岩、板岩、千枚岩、大理岩、石英岩以及花岗岩等组成，顶部为沉积盖层。

地壳内部局部发育有层状低速体（v_p = 6.1 ~ 6.0km/s），与花岗岩高压实验测定波速（v_p = 5.99 ~ 6.14km/s）基本一致，这些低速体可能为盖层发育花岗岩体的深部岩浆房。

第二节　地幔地球化学成分推断

地球壳层是地球在漫长的地质发展历史中由地幔物质演化而成的。因此了解区域地幔特征，特别是了解区域上地幔的地球化学特征，可以进一步深入认识区域地壳的组成和演化、区域构造活动和区域矿产分布等特征。

区域上地幔地球化学成分的估计除了利用玄武岩中的地幔岩包体外，就是通过阿尔卑斯超镁铁质岩的成分来推算，即林伍德地幔模型法。该模型的关键是确定地幔部分熔融上涌后形成的熔融体（玄武岩）和残余固相（超基性岩）的代表及相对比例。

表 28-2　华北陆块北部上地幔岩石主量元素含量表

Table 28-2　Major element contents of the upper mantle in northern NCC

单位:%

岩性及地区	SiO_2	TiO_2	Al_2O_3	Fe_2O_3	FeO	MnO	MgO	CaO	Na_2O	K_2O	P_2O_5
高寺台橄榄岩	39.02	0.02	0.25	11.05	2.53	0.17	45.50	1.07	0.04	0.03	0.01
蕨菜沟角闪石岩	41.90	1.25	8.52	3.84	6.30	0.13	13.92	10.77	1.39	0.42	0.03
燕山地区	40.46	0.64	4.39	7.47	4.42	0.15	29.71	5.92	0.72	0.23	0.02
尖晶石二辉橄榄岩	44.41	0.16	2.97	1.90	6.16	0.23	40.50	3.70	0.32	0.07	0.07
华北陆块南缘①	43.74	0.84	7.18	5.94	6.60	0.17	26.06	5.21	0.94	0.34	0.15
地幔岩②	43.95	0.57	3.88	0.75	7.50	0.13	39.00	2.60	0.60	0.22	

注：①据韩吟文等，1990；②据 Ringwood 等，1966。

表 28-3　华北陆块北部上地幔岩石微量元素丰度表

Table 28-3　Trace element content of the upper mantle in northern NCC

岩性及地区	Sr	Ba	Sc	V	Cr	Co	Ni	Cu	Zn	Pb	Mo	W	Au
高寺台橄榄岩	35	60	3.2	13	3276	252	1593	13	52	5.4	0.19	0.18	1.02
蕨菜沟角闪石岩	339	179	77.8	394	535	51	144	35	50	7.2	0.16	0.10	0.90
燕山地区	187	120	41	204	1906	152	869	24	51	6.3	0.18	0.14	0.96
华北陆块南缘①	140	184	23	148	1979	59	846	16	97	13	1.7	1.13	1.76
世界上地幔②	120	76	18	80	1600	160	1500	40	60	2.1	0.6	0.30	5.00

注：①据韩吟文等，1990；②据黎彤，1981；微量元素含量单位 Au 为 10^{-9}，其余为 10^{-6}。

根据区内基性—超基性岩浆岩元素含量统计可知（表 28-2，表 28-3），本区地幔由尖晶石/石榴子石二辉橄榄岩、各类辉石岩、二辉岩等组成。REE 及微量元素弱富集。这类地幔的岩浆以拉斑玄武岩为主，少数地区可以出现碱性的中酸性火成岩。

第三节　铅同位素组成特征

目前岩石圈铅同位素的研究主要是分析花岗岩中长石铅及玄武岩中的全岩铅。因为，花岗岩是陆壳形成演化过程中地质构造运动的产物，可视为从地壳深部或基底中采集到的样品。而玄武岩是来自于地幔的岩浆岩，可视为从上地幔中采集到的样品（张理刚，1995）。

岩石中铅同位素的含量，取决于其最终来源物质的 U-Th-Pb 体系及其生成时代等诸多因素，而其 U-Th-Pb 体系又受最初来源物质的 U-Th-Pb 体系及后期地质构造热动力作用等制约。总体来说，上地壳极强的富集 U、Th，因此上地壳物质中富集放射性成因 Pb；下地壳通常由片麻岩和 U、Th 剧烈亏损的高变质麻粒岩组成，缺乏放射性成因铅；上地幔的 U/Pb 和 Th/Pb 正好介于上述两者之间。

收集了本区中生代岩体中花岗岩中铅同位素数据（表 28-4）。可以看出，^{204}Pb 丰度较高，变化范围为 1.4%～1.481%，$^{206}Pb/^{204}Pb$ 较低，值域为 15.701%～17.406%，平均值为 16.488%。$^{208}Pb/^{204}Pb$ 值较低，值域为 30.606%～37.593%，平均值为 36.755%。通过 $^{206}Pb/^{204}Pb$ 与 $^{207}Pb/^{204}Pb$ 对比可知，本区岩浆岩大部分数据值都位于下地壳和地幔之间，说明了这些样品的源区物质更多来自下地壳及地幔。

表 28-4　北京地区岩体中铅同位素组成

Table 28-4　The lead isotope composition of plutons in Beijing area

编号	产地岩性	$^{204}Pb/\%$	$^{206}Pb/^{204}Pb$	$^{207}Pb/^{204}Pb$	$^{208}Pb/^{204}Pb$	资料来源
1	怀柔兰营环斑花岗岩	1.481	15.770	15.127	35.606	张理刚，1995
2	怀柔兰营石英正长岩	1.463	15.701	15.135	36.537	张理刚，1995
3	密云四干顶二长花岗岩	1.417	16.990	15.406	37.190	张理刚，1995
4	昌平对臼峪花岗岩	1.434	16.625	15.343	36.783	张理刚，1995
5	密云铅锌矿花岗斑岩	1.464	16.240	15.109	35.972	张理刚，1995
6	门头沟白羊石虎二长岩	1.400	17.406	15.434	37.593	余昌涛，1989
7	延庆大海坨斑状花岗岩	1.430	16.680	15.310	36.940	罗殿文，1991
8	延庆大海坨斑状花岗岩	1.420	16.490	15.470	37.420	罗殿文，1991

第四节　层圈相互作用的事实和特点

地球形成初期，物质分异作用形成了层圈结构。而后由于各层圈在物质组成、能量和状态方面的差异及同一圈层内成分和能量均一性，必然导致圈层的相互作用——物质和能量交换及动量传递，以及层圈内的物质和能量对流，推动着地球及其圈层的发展演化。全球构造运动表现为层圈间的动量传递与板块的离散与汇聚，即开-合规律，地球化学运动表现为层圈间的物质交换和循环及层圈内的物质分异和演化；地球物理运动表现为层圈间和层圈内的热传导和热对流，生物学运动表现为生物圈与水圈、大气圈和岩石圈间的物质交换与循环机生物圈自身的发展演化。地球层圈相互作用构成了一个完整的地球动力学系统。路凤香（2013）等在研究岩石圈三维结构时，发现在壳-幔之间及岩石圈-软流圈之间存在相互作用带（过渡带），它们不仅是重要的岩浆源区，与不少固体矿产和能源资源有着极为密切的联系，而且通过它们能进一步认识岩石圈动力学过程。在层圈相互作用中，热和物质的交换及其动力学过程是引起中、新生代岩石圈内部厚度调整、岩石圈不均匀减薄以及区域构造-岩浆-成矿作用的重要机理。

一、早前寒武纪层圈相互作用与大规模硅铁建造

早前寒武纪岩石圈以结晶分异、克拉通形成为主要特征。以超基性（少量）、基性和中酸性钙碱性火山岩最发育，从早到晚伴有越来越多的沉积作用，并有多次 TTG 型岩浆侵入，导致地壳逐渐增厚，是基底和岩石圈形成阶段。由火山作用带来丰富的 Fe、Au 等元素，形成了原始矿源层，经区域变质作用形成了硅铁建造（BIF）和沉积变质铁矿，是早前寒武纪最突出的特点之一。

二、中元古代—中三叠世层圈相互作用与成矿特点

中元古代—中三叠世为稳定盖层形成和早前寒武纪岩石圈相对稳定发展阶段。该阶段早期主要是在地幔软流圈反向对流条件下开始了近南北向伸展拉张运动，晚期阶段在地幔软流圈相向对流条件下变为近南北向的挤压造山运动。

三、中、新生代层圈相互作用事实与特点

（一）壳-幔相互作用及特征

北京地区在中、新生代处于陆内构造活化期，构造环境处于挤压—弱拉伸状态。路凤香等（2003）认为广泛发育的钙碱性火成岩和壳-幔作用带是壳-幔相互作用发生的证据。壳-幔作用带主要组成为：下地壳基性（酸性）麻粒岩、底侵的基性岩、镁铁质—超镁铁质堆晶岩（如辉石岩）和岩石圈地幔±软流圈地幔。

深部参与壳-幔作用的3个端元主要是：下地壳、岩石圈地幔和软流圈地幔（匡少平，1999；黄方，李曙光等，2003；路凤香等，2005）。软流圈除了起到提供热源的作用外，在少数地区，也有软流圈来源的岩浆出现，但源区几乎都加入了地壳组分，这是壳-幔相互作用最基本的标志。导源的岩浆以高钾钙碱性火成岩系列为主，地球化学特征表现为：适度富集LREE，亏损HFSE（特别是Nb），ε_{Nd}为负值，其中基性岩该值为0~-10，中基性—中性岩低于-10。壳-幔相互作用主要为地幔来源的底侵熔体与下地壳的作用，过程是幔源岩浆侵位于地壳底部，为下地壳提供了热源，诱发下地壳及少量岩石圈地幔组分熔融。导源的岩浆为中性安山岩、粗安岩等，因处于地壳挤压的环境，地壳缩短，加之岩浆底侵作用，造成了加厚的下地壳。北京西山中上侏罗统髫髻山组粗安岩属于这种类型（李伍平等，2001）。火山岩总体分布呈近EW向，具体的火山盆地呈近NE向，火山岩最大厚度为4900m，形成高山，地壳厚度经估算约45km（方法据Kay，1991）。髫髻山组岩石组合以粗安岩及同成分的火山碎屑岩为主体，粗安岩岩性变化比较小，$(La/Yb)_N$变化范围2~42，$(La/Nb)_N$地区平均变化范围4.44~12，Sr/Y变化范围167.67~36.42，HFSE具有明显的负异常，$\varepsilon_{Nd}(t)$变化范围-14.8~-16.7。源区成分主要为含角闪石石榴子石麻粒岩+底侵的基性岩+岩石圈地幔组分（较少）的壳-幔过渡带。髫髻山组火山岩最高的$Mg^{\#}$为50，峰值为40，推测其源区含地幔组分<10%（据Rapp 1997、1999年实验资料）。岩浆作用过程可以用MMSH（熔融、混合、储存、均一化）过程进行描述，即上述源区成分发生熔融后，在岩浆房中还经历过均一化的过程。所需的热量由底侵岩浆和构造运动的剪切摩擦热所提供。

（二）岩石圈-软流层相互作用及特征

岩石圈-软流圈相互作用带位于二者的过渡位置。地球物理资料表明，该界面形态复杂，出现水平互层与垂直并置接触的格局。据路凤香等（2013）的研究，在中国东部（100Ma以后），该带深度大约在60~120km之间，带内两圈层发生相互作用，并成为这一时期拉斑—碱性玄武岩岩浆的主要源区。岩石圈-软流圈的相互作用带的组成为岩石圈地幔+软流圈地幔，后者为主要组成部分。但是这两者之间的相互作用目前还难以辨认。

第二十九章　岩石圈结构构造

第一节　莫霍面结构特征

一、莫霍面结构特征

根据《中国区域地质志·河北志》中采用重力数据计算的全区莫霍面深度等深图，本区莫霍界面的等深线走向明显地呈北东—北北东向及近东西向（图 29-1）。其埋深由东南向西北方向渐渐加大。东南部平原区埋深相对较浅，一般在 34～37km 之间变化，东南牛堡屯一带埋深约 32km，北部密云—张山营一带埋深分别为 38km 及 42km，平原区变化于 32～36km 之间，山区变化于 37～42km 之间。主要的隆坳和结晶基底顶界面成大致相反的镜像对应关系。平谷与墙子路间等深线为近东西向、局部扭曲呈北西走向之埋深变异带，其梯度达 7m/km 左右。汤河口至昌平以西的北东—北北东向埋深变异带与康拉德界面近似，即在一定地段其走向表现出扭曲，变异带的宽度基本未变，但其梯度却明显加大，约 10m/km。这说明前述的本区几条北北东向和东西向深断裂已切穿“玄武岩”层，深达莫霍界面。另外在平谷一带近东西向等深线也发生较明显的弯曲，反映此南北一线可能有深断裂存在。在西部与北部太行山-燕山地区，莫霍面等深线以线性梯级带为主要特征，莫霍面的这一陡变特征，反映出上地

图 29-1　北京地区莫霍面等深线图

（据《中国区域地质志·河北志》，2017）

Fig. 29-1　Moho depth contour map of Beijing

幔为陡坡状或陡坎状。根据莫霍面等深线图，大致以张家口—昌平一线为界，可分为东西两部分，东部比西部莫霍面埋深稍浅，相差 1~2km。而且东西两区的中东部和中西部各出现一处莫霍面等深线相对宽缓的区域，反映出在幔坡带中有局部的幔凸存在。

二、深部构造分区

根据本区莫霍面埋深的变化特征，以及早前寒武纪结晶基底埋深、布格重力异常、深地震剖面及航磁等资料，可将本区划分为两个深部地质构造区。（图 29-2）

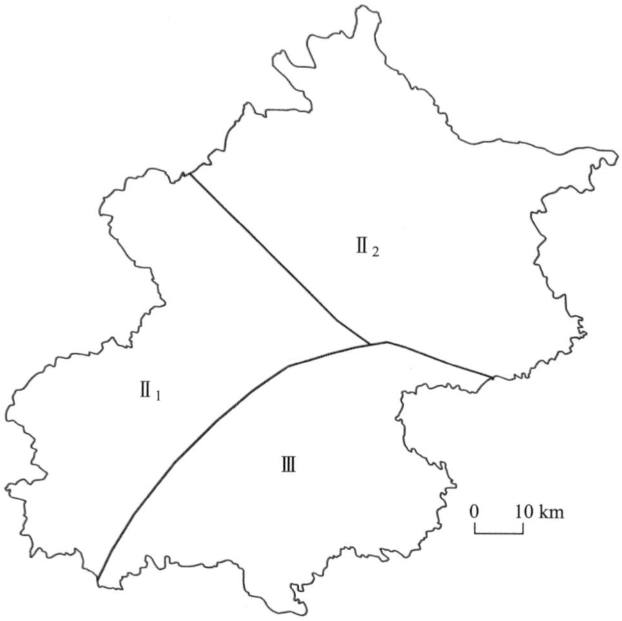

图 29-2 北京地区深部构造分区图
（据《中国区域地质志·河北志》，2017）
Fig. 29-2 Deep tectonic zoning map of Beijing

（一）太行山-燕山幔坡带（Ⅱ）

浅层构造特征基本上是以燕山期为主的强烈的构造岩浆活动。该区布格重力异常值为-4×10^{-5}~$-138\times10^{-5}m/s^2$。航磁异常总的看来是一个正负异常相排列的磁场区。重磁局部异常走向多变，反应本区是活动剧烈的构造单元。莫霍面埋深 36~42km，变异落差 6km。根据莫霍面等深线形态可划分为两个亚区：

1. 西山幔坡（$Ⅱ_1$）

位于昌平以西，莫霍面埋深 36.5~42.5km，梯度变化较大，整体反应为幔坡特征。但莫霍面等深线形态及走向在北部发生明显改变，由北东向梯度带转变为近东西向宽缓状。北京位于怀来-阜平幔阶小区，位于怀来-蔚县以南，莫霍面埋深 36~40.5km，等深线呈北东向带状展布，梯度变化相对较大，反映为幔阶特征。

2. 北山幔坡（$Ⅱ_2$）

位于昌平、崇礼以北，莫霍面埋深 36~42.5km，梯度变化较大，整体反应为幔坡特征。在幔坡的中东部莫霍面等深线形态发生明显改变，也出现局部宽缓的区域，北京位于冀东幔阶，莫霍面等深线呈带状分布，梯度变化较大，反映为幔阶特征。

（二）平原幔隆区（Ⅲ）

北京山前断裂之东南平原区，该区主体为新生代坳陷区。早前寒武纪结晶基底埋深一般为 1.5~6km，最大约 10km，布格重力异常值在-46×10^{-5}~$10\times10^{-5}m/s^2$之间变化，平原磁场区异常较为平缓，平均为100nT，平原磁场区的负磁场背景。重磁异常方向均已北东向、北北东向为主。莫霍面埋深34~37km，该区整体为地幔隆起区。

第二节　地壳结构及厚度变化

北京地区所在的华北地震构造区北部是新生代裂谷作用十分强烈的地区，华北平原区经历了古近纪差异性裂陷形成盆岭构造和新近纪以来区域性整体沉降坳陷形成平原等两期新构造演化阶段，而西部山区（太行-五台山断块隆起）在新生代持续抬升的基础上，上新世—第四纪时期经历裂谷作用，形成山西地堑系。总结前文中提到的大量人工地震测深获得的地壳上地幔结构速度分析可知（滕吉文等，1974，1975；曾融生等，1988；孙武城等，1988；林真明等，1990，1992；张先康等，1996；张成科等，1997，1998；祝治平等，1995，1997；徐锡伟等，2002），北京地区地壳上地幔深部构造具有纵向分层和横向分块的基本特征，且许多地方存在低速层（体）。

北京城区地壳厚度约为35.5km，向西地壳厚度逐步加大至怀来盆地地壳厚度达40~41km，等深线走向基本与太行山脉平行，为北东向。北京至玉田一带莫霍界面深度为35km，向北逐渐加深，平谷—遵化一带为35.5~36km，燕山地区为37~38km，(近东西向)，在昌平至八达岭一带。莫霍界面等深线由北东走向转为东西走向。大兴—固安一带属小型上地幔隆起区，最薄处为34.5km。

1. 西山幔坡带

（1）重力

重力异常呈北东向展布，西北低、东南高，梯度变化较大。异常区内布格重力异常极小值出现在西北部，异常表现为北东走向的椭圆状异常，由延庆断陷盆地引起。

（2）磁异常

区域以北东向规律展布的宽大平稳的正负磁场为背景，自西南向东北数个不同大小规模的圆状、似圆状或椭圆状高强度异常像“珍珠”一样呈北东向、北北东向，北西向和北西西向叠加于背景场之上，局部异常强度可达500nT以上，异常的整体走向呈北东向，反映区域岩浆侵入特征。

（3）速度结构

基底埋深1~2km，速度3.0~5.2km/s；上地壳厚度约24~27km，速度6.0~6.4km/s；下地壳厚约12~15 km，速度6.4~6.8km/s；下地壳下部（14~17km）以及下地壳壳幔过渡带（33~36km）分别存在约6.1km/s和6.4km/s轻微的速度逆转层。地壳厚度38~42km，平均速度6.2~6.3km/s。

（4）电性特征

为高阻区，地壳表层岩石的电导率很低，随着深度增大，岩石圈电导率逐渐增大。

（5）温度结构

西北部山区热流值较低，一般低于60mW/m^2，在背斜的轴部和活动断裂带热流值增高。地壳各界面和莫霍面的热流和温度低于平原区。

（6）岩石圈结构

属于造山型岩石圈。太古代陆壳受到强烈改造，形成以花岗质为主的新的陆壳。它们是从地幔中分出的新生物质（辉长岩和闪长质）与太古宙—古元古代陆壳再熔融产生的再生花岗质岩浆混合的产物。岩石圈地幔由方辉橄榄岩和二辉橄榄岩构成。

2. 北山幔坡带

（1）重力

北山重力低异常特征表现为西低东高的特点，异常区有多个椭圆状和似圆状的异常组成，梯度变化较大。其中两个较大的异常中心分别由为云蒙山岩体和怀柔北部的花岗岩体引起。

（2）磁异常

不同时期侵入岩浆岩形成的局部高强度的正异常呈椭圆状、扁圆状或宽带状，以近东西向延伸、略有起伏和跳跃的正负磁场为背景，呈东西向叠加于背景场之上。太古宙磁铁石英岩引起的局部异常以强度大幅度大梯度陡的线状、条带状异常为主，呈北东、北北东、北西和东西向叠加于背景场之上。其次也有各种不规则圆状、椭圆状、强度低梯度缓的异常较有规律的分布于其中。

（3）速度结构

燕山块体地壳结构：基底埋深1~2km，速度3.2~5.4km/s；上地壳厚度约22~26km，速度6.0~6.4km/s；下地壳厚约12~15km，速度6.4~6.8km/s；下地壳下部（16~20km），下地壳壳幔过渡带（32~40km）分别存在约4km、3km，速度约6.1km/s和6.4km/s的低速度层。地壳厚度约34~40km，平均速度6.2~6.3km/s。

（4）电性特征

为高阻区，地壳表层岩石的电导率很低，随着深度增大，岩石圈电导率逐渐增大。

（5）温度结构

西北部山区热流值较低，一般低于60mW/m^2，在背斜的轴部和活动断裂带热流值增高。地壳各界面和莫霍面的热流和温度低于平原区。

（6）岩石圈结构

属于造山型岩石圈。太古代陆壳受到强烈改造，形成以花岗质为主的新的陆壳。它们是从地幔中分出的新生物质（辉长岩和闪长质）与太古宙—古元古代陆壳再熔融产生的再生花岗质岩浆混合的产物。岩石圈地幔由方辉橄榄岩和二辉橄榄岩构成。

3. 平原幔隆区

（1）重力

平原重力高异常区表现为西北重力低、东南重力高的特点，异常呈北东走向，西北为丰台-顺义重力低异常，东南为大兴重力高异常区，两个异常区分界线表现为受礼贤断裂控制的密集梯度陡变带。

（2）磁场

平原磁场区异常较为平缓，以-100nT异常较为平的负磁场背景，以礼贤—安定—漷县—西集为界，可进一步划分两个亚区。西北部表现北东走向、宽缓的椭圆状负异常，分布范围与北京凹陷基本一致，东南部表现为北东走向、狭长的椭圆状正异常，异常中心极值达600nT，由太古宙变质基底隆起引起。

（3）速度结构

基底埋深2~6km，其中大兴凸起基底埋深约3km。速度1.6~5.0km/s，结晶基底由速度5.6~6.2km/s的强速度梯度层构成；上地幔厚度约20~22km，由两层低速（5.8km/s，6.2km/s）占主导的层位构成；下地壳厚约6~8km，速度6.0~6.6km/s。壳幔过渡带由一低速层和一厚约5km的高低速互薄层构成。地壳厚度约28~30km，平均速度甚低，约5.5~5.8km/s；显示了裂陷盆地坳陷区的地壳上隆减薄，地表巨幅张裂下陷，壳内低速占主导地位的破碎松散构造特征。

（4）电性特征

为低阻区，其岩石圈电导率由浅到深逐渐减小。上地幔低阻区有明显的横向起伏，波动幅度由西向东变大，波峰与北京坳陷对应；波谷与大兴隆起对应；地壳内出现低阻带，呈“香肠状”结构，坳陷区厚，隆起区薄，与壳内低速带相当。地壳表层的低阻分布范围与区内的新生代沉积发育状况基本一致。

（5）温度结构

热流值较高，一般高于60mW/m^2。相对隆起区高于凹陷区，即大兴隆起的地壳各界面和莫霍面的热流和温度高于北京凹陷。

（6）岩石圈结构

陆壳组成以花岗闪长质为主。岩石圈地幔由弱亏损的二辉橄榄岩构成，其亏损程度很弱，地壳和岩石圈地幔也是最薄的，是整个岩石圈厚度尺度上伸展减薄的结果。

第三节　岩石圈类型及岩石学结构

20 世纪 80 年代以来，研究岩石圈结构的一个新趋势是岩石学模型与地球物理模型的结合，两者互为补充。据邓晋福等（2007）的研究，采用的方法是收集已有的各类岩石纵波速度测量数据，建立岩石成分与速度 v_p 的拟合方程，结合出露地表的深部陆壳岩石（主要是前寒武纪变质岩系）及火成岩（火山岩与侵入岩）包含的岩浆源区信息，建立了华北地区的陆壳岩石学结构和化学结构。在此研究成果上可以看出北京地区现今岩石圈三维结构可识别和划分为 2 个类型，分别为造山型岩石圈和裂谷盆地型岩石圈，其岩石圈综合结构见图 29-3，壳幔物质时空结构见表 29-1。

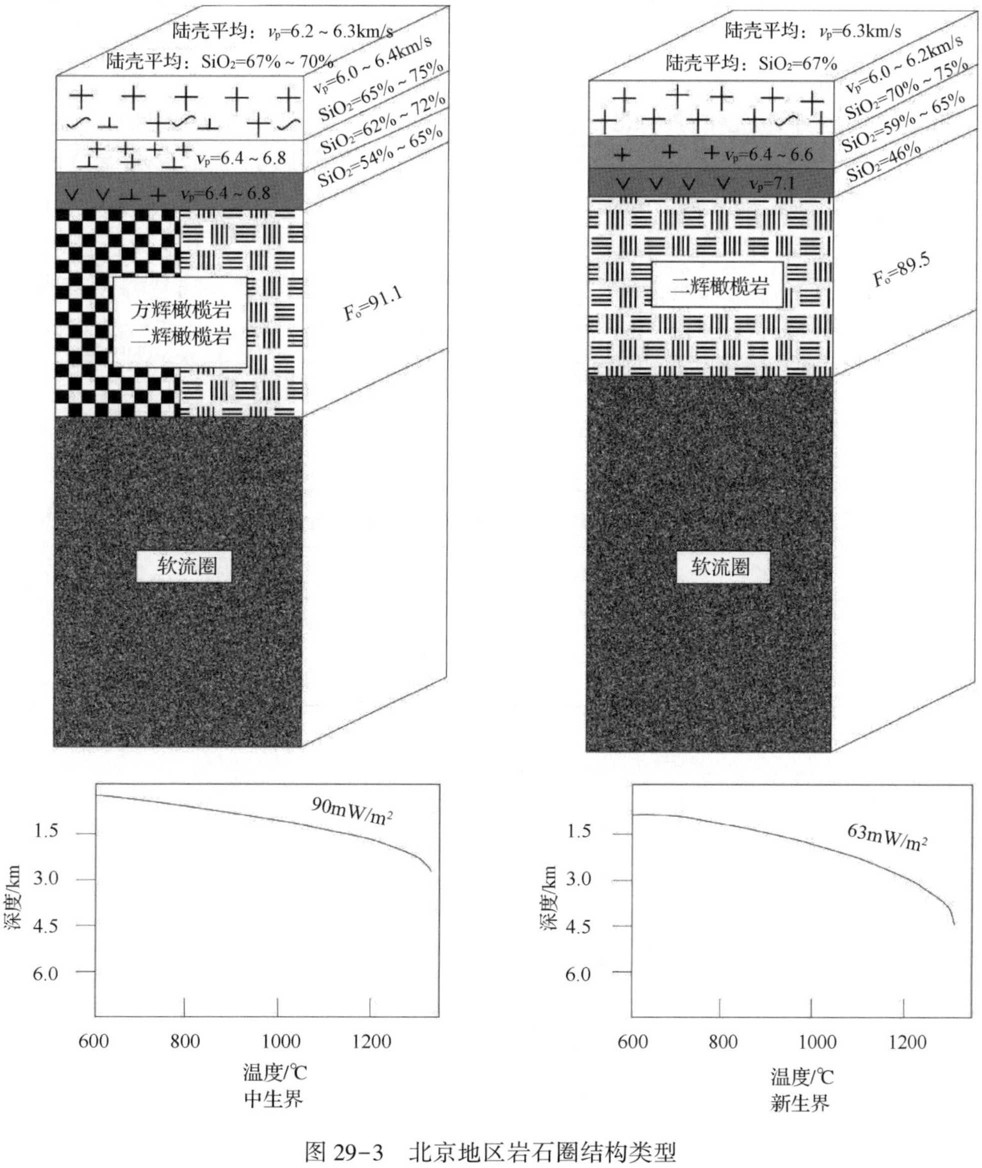

图 29-3　北京地区岩石圈结构类型

（据李廷栋等，2013）

Fig. 29-3　The lithosphere structure types in Beijing

1. 造山型岩石圈

北京山区由燕山和太行山山脉的组成部分，是燕山运动的产物，属于造山型岩石圈。在燕山运动过程中华北克拉通“活化”，大量对流地幔的物质与热注入，使地壳原来的 TTG 陆壳组分熔融上侵，强烈的以花岗质为主的岩浆侵入活动，使太古陆壳受到强烈改造，形成以花岗质为主的新的陆壳（SiO_2=67%~70%），它们是从地幔中分出的新生物质（辉长岩和闪长质）与太古宙—古元古代陆壳再熔融产生的再生花岗质岩浆混合的产物，也就是说，J-K 形成的新陆壳既有从地幔中分出的新生陆壳的贡献，也有原有古老陆壳再熔融产生的再生陆壳的贡献。岩石圈地幔由方辉橄榄岩和二辉橄榄岩构成，其亏损程度弱于克拉通型岩石圈的大陆根。

就地壳厚度和岩石圈地幔的厚度来说，造山带型岩石圈因为经历了新生代的伸展减薄，所以不能代表燕山期造山时的地壳厚度和岩石圈地幔的厚度，但由于总体上喜马拉雅期时没有玄武岩浆活动，即没有地幔物质的注入，所以，岩石圈地幔和陆壳的物质及其结构仍然是燕山运动期间造山时形成的。

华北平原是喜马拉雅期岩浆-构造事件的产物，它们原来可能已经经过燕山期的一次改造，但是，喜马拉雅期事件又对燕山期的结果有重大改造，所以，岩石圈的地幔和陆壳的物质和结构是喜马拉雅期形成的。

2. 裂谷盆地型岩石圈

北京的平原区部分位于华北平原东北部，属于裂谷盆地型岩石圈，陆壳组成以花岗闪长质为主。燕山运动是一次广泛的造山运动，所以，可能包括华北东部平原在内的华北东部克拉通都已被改造和“活化”，形成类似于燕山-太行造山带的陆壳，其陆壳在燕山期时的平均组成应该与燕山-太行山造山带型一样为花岗质，现在地球物理表现出较高的陆壳平均 v_p 值（相对于燕山-太行造山带）和较低的 SiO_2 含量（平均组成变为花岗闪长质，SiO_2=67%），是由于喜马拉雅期大陆裂谷发育时又有对流地幔新注入大陆的结果（邓晋福等，1996），亦即陆壳在喜马拉雅期又被“基性化”。岩石圈地幔由弱亏损的二辉橄榄岩构成，其亏损程度很弱，地壳和岩石圈地幔也是最薄的，是整个岩石圈厚度尺度上伸展减薄的结果。

表 29-1　北京大陆岩石圈壳-幔物质时空结构

Table 29-1　The time and space structure of continental lithosphere crust-mantle material in Beijing area

岩石圈类型	代表性地区	壳幔结构形成时代	壳幔物质形成时代	现今岩石圈根状态
造山带型	北京西部山区、北部山区	中生代	中生代	已去根
裂谷岩石圈	北京平原区	新生代	新生代	

第三十章　深部地质作用及其对资源环境的制约

地质历史时期中的多次地质构造变动形成了现今的区域地质构造格局和大地构造体系，而实测的区域重力、磁场等地球物理场就是这些区域地质构造深部的反映，同区域成矿等资源环境关系密切。

邓晋福等（2008）分析重力异常场、航磁异常场与矿产资源的分布关系表明：①矿产资源的分布与重力异常梯度带有密切的关系；②矿产资源的分布和重力异常相关的地壳厚度分布有着相应的关系；③矿产资源的分布与高磁异常带的分布也有相关性。在重力异常梯度带、航磁高异常带及其附近地带，都有系列的矿产赋存与分布，从而形成矿产资源呈条带状的分布状态，即矿产资源成矿聚集带，这是金属矿产、非金属矿产的重要成矿机理。

第一节　深部地质作用与地质资源分布的关系

在大陆地壳的演化过程中，重力及重力的均衡调整在构造运动中起着重要的作用。由于地壳中质量分布的不均匀与内外营力作用使这种不均匀性继续扩大，必然导致上地幔所受到的静压力的不平衡，而引起上地幔的下沉和上隆的重力均衡过程，使地壳中温度压力发生变化、岩石的再熔、花岗质岩浆的形成和物质的重新分布，以及上地幔玄武质岩浆的上涌；这些构造变动和岩浆活动对于区域成矿作用起着一定的控制作用。研究深部地质作用与资源矿产生成和空间分布的相互关系，是研究区域成矿规律、开展成矿预测工作的一个新课题，自20世纪60年代中期以来在国内外日益受到重视。目前的基本做法是，运用区域重力测量、地震测深、大地电磁测深、天然地震转换波、航空磁测和超深探等手段，查明地壳各层的厚度以及康氏面、莫霍面的变化与构造特征，然后研究已知成矿带、成矿区、矿田、矿床与上述深部构造的关系。

一、深部构造与区域构造的关系

空间上，深部华北平原幔隆区与浅部华北盆地基本一致。深部太行山-燕山幔坡带则比较复杂。太行山区浅部构造与深部构造走向一致，一系列北东—北北东向区域断裂带发育；而燕山地区深部构造变异带斜切了东西向燕辽裂陷带、华北北缘隆起带，使深部构造变异带明显加大，并形成了一些次级变异带和上地幔凹陷与凸起，它们与浅部构造也能很好的对应。因此，现阶段北京市深部构造表现为中生代以来地壳变动与演化的结果，其对新生代和中生代沉积有着明显的控制作用，并斜切中生代以前的古构造单元，动力来源可能与太平洋板块向欧亚板块俯冲有关。

二、深部构造与岩浆活动的关系

深部构造主要是中生代以来地壳变动与演化的结果，所以对中生代以来的岩浆活动有明显的控制作用，对中生代以前的岩浆活动的影响仅限于破坏和改造。燕山期侵入岩是区内最主要的成矿母岩，大量分布在太行山-燕山幔坡带深部构造变异带中。北京地区的八达岭岩浆岩带就包含其中。平原区的岩浆活动以喜马拉雅期为主。

岩浆岩的岩性也与深部构造有明显的关系。太行山-燕山幔坡带（深部构造变异带）基性—酸性岩均很发育，侵入岩以闪长岩、石英闪长岩、花岗闪长岩、花岗岩为主，火山岩从基性到酸性均有大面积分布。华北平原幔隆区的岩浆活动以基性岩为主。

新生代玄武岩活动延续了很长时间，从始新世一直到更新世都有喷溢活动，始新世—渐新世是岩浆活动的高峰时期。本区新近纪玄武岩以裂隙式喷溢为主，明显地受深断裂的控制，第四纪玄武岩以中心式喷发为主，但也受深断裂控制。从岩石化学性质分析，古近纪玄武岩多属于富碱性的过渡型拉斑玄武岩。原始岩浆来自35~50km深处；新近纪和第四纪玄武岩多属碱性玄武岩，原始岩浆来自50~60km深处。

三、深部构造对区域成矿的制约

对流地幔物质和热输入是大陆成矿作用的直接驱动力（邓晋福，2003，2004）。地壳的深部构造在地壳演化中并不是一成不变的，而是处在不断的运动和发展中，从总体上来看是太平洋板块向欧亚板块俯冲在大陆地壳上的反映，它控制了本区中生代以来的构造运动、岩浆活动和成矿作用。

第二节　深部地质作用与环境地质的关系

地质灾害是环境地质的重要构成内容，是指地质作用为主要原因引起的自然灾害，因为地质物质运动或自然地理条件变化造成的环境破坏过程，主要包括火山、地震、滑坡、泥石流、地面塌陷、地面沉降、地裂缝等。

一、深部地质作用与地震的关系

（一）历史地震与现代地震

北京地区有历史记录的古地震自公元274年起。受观测条件的限制，早期的记录遗漏较多，其中最大的一次强震是1679年发生的夏垫8.0级地震。对现代小震震源深度信息进行统计，北京地区的震源深度分布规律大体上服从正态分布：震源以6km深度的居多，8~9km的次之。对地震事件震级信息进行统计，近10年来，北京地区的震级分布也具有正态分布的规律：震级以1级以下者居多，大于3级的地震只占很少的比例。

（二）深部地质与地震形成条件

1. 重磁场特征

强震带在航磁异常上表现为线性梯度带，陡变带及不同走向的磁场的接触带，在航磁的化极上延图上为一负异常带，航磁弱磁性反应构造应力的集中地段。在重力异常上表现为不同走向的线性异常接触带，是密度强烈反差的接触带。

2. 岩石圈深部结构特征及强震带内多震层特征

据万天丰（2011）研究认为中国大陆地震主要是壳内地震，尤其集中在尚未愈合的中地壳低速高导层以及地壳底面（莫霍面）。本区90%以上的地震发生在5~20km深度范围内。这一地震集中的深度大致相当于基底界面与地壳界面之间，即“花岗岩”层。地震与该层地壳的关系十分密切。本区“花岗岩”层顶面埋深4~12km，底界埋深14~22km，平均厚度8~12km。在横向上厚度变化不大，但埋深起伏大。基底界面大致以昌平—宁河一线为界，分为两大部分。东北部基底界面为北西西向，埋深自南向北逐渐加大。在昌黎、唐山、宝坻、玉田、密云一带为北西西向的基底界面隆起带，埋深4~6km。西南部基底界面为北北东向，且自西北向东南呈凹隆相间分布。这两部分之间是基底界面构造的转变带和埋深变化的梯度带。断裂构造发育，是地质构造复杂的地区。地壳界面的构造形态与基底界面的相似。而地震与二者所围限的中层即“花岗岩”层的关系十分密切。

转换波法提供的剖面也清楚地反映了地震震中在垂向上大部分分布“花岗岩”层内，尤其是在其顶、底面附近。在此层范围内外地震频数锐减，随着“花岗岩”层埋深在横向上的起伏变化，震源深度相应地也有深浅变化。

震中在基底界面构造平面图上的位置表明，本区6级以上强震主要分布在北西西向的基底界面埋深梯度带和北北东向的基底界面隆起与凹陷区的转变带上。

地震主要与“花岗岩”层地壳的构造特征和岩石的物理性质有关。根据大地电磁测深结果，本区地壳在10~20km深度处有一电性高导层，电阻率数欧姆·米至十余欧姆·米，厚度约10km，据人工地震及转换波测深资料，这一深度大致相当于康氏面深度，相当于地震测深的壳内低速层，与航磁资料计算的居里面深度（16~26km）大体一致。按大地热流计算的500℃等温面深度也大致在该层地壳的下部。据研究，在温度500~600℃，压力（4~6）$\times 10^8$Pa条件下，如果有自由水存在，花岗岩类可以出现蠕变和部分熔融。另外，地震测深资料还表明低速层以下的地壳波速和密度均随深度加大，介质的流变性也有显著增强。这样本区地壳在物理状态上大体可分成两部分。低速层及其以下的地壳为韧性岩层，具有蠕变性；低速层以上的地壳为脆性岩层。在地应力作用下，上部脆性地壳遭受挤压，并在低速层上侧向运动，当运动受到阻碍时应力会在该层地壳厚度或埋深变化的梯度带积累，这些部位断裂构造也很发育，岩石脆性强，因而易于发生破裂形成地震。下部低速层和下地壳层由于具有韧性，在应力作用下可以通过缓慢的蠕变来实现调整，达到新的平衡和稳定，因而大震不出现在低速层内。本区地震在20km深度以下已大为减少，30km以下未见强震发生。地震集中的这个深度范围的下界可能主要与壳内低速层的存在有关。

二、深部构造与地质灾害关系

岩石圈结构主要通过构造运动对地质环境产生影响，主要表现为对壳表层新构造运动、活动断裂、活动盆地起直接控制作用，控制地震带的空间展布，制约环境地质分区格局，对滑坡、崩塌、泥石流、地面沉降、地裂缝等地质灾害易发区的分布格局具有空间对应关系，对地质灾害的发生具有基础性作用。深部构造的研究成果对深入研究地质环境特征具有重要意义。

1. 崩塌、滑坡、泥石流易发区地质构造背景

据孟晖（2004）编制的中国崩塌、滑坡易发程度分布图可以看出，崩塌、滑坡和泥石流易发区的空间分布格局基本相同。北京地区存在一定范围的崩滑流易发区，地处活动构造带以及强震带，与深部燕山、太行山中生代造山带型岩石圈和华北平原裂谷型岩石圈结构不连续对应。除个别地区由浅部构造为主导因素外，崩塌、滑坡、泥石流易发区大体上都与深部岩石圈结构相对应。

2. 地面沉降易发区的地质构造因素

北京地处华北平原西北部，而华北平原是地面沉降易发区，主要为巨厚层第四纪沉积发育区，目前把形成地面沉降的主导因素归于地下水超采和地面负荷加载。但从宏观格局分析，地面沉降高发区，与中国东部新生代地堑盆地直接相关，深部与华北平原裂谷型岩石圈对应，说明地面沉降易发区的分布受到岩石圈结构影响。地裂缝在北京地区也较发育，主要原因是由于地下水超量开采以及活动构造，一般与地堑构造密切相关（孟晖等，2004）。

第七篇

地质构造

按照板块学说观点，北京地区大地构造位置处于华北陆块中北部燕山中新生代陆内造山带西段。在漫长的地质历史发展过程中，区内经历了多次构造运动和多阶段多旋回的地质构造演化。

百年以来，众多中外学者在区内地质构造方面开展了重要研究。李希霍芬（1869，1871）曾提出西山存在南口、恒山两个山带的认识。叶良辅等（1920）通过系统填图首次提出了北京西山地质构造总体格局。翁文灏（1926，1927，1929）根据北京及中国东部有关调查资料，率先提出了中生代侏罗纪—白垩纪时存在大规模的造山运动，命名为燕山运动。陈恺等（1935）通过 29 条剖面的研究对西山的地质构造作了全面分析，尤其在逆掩断层分布和特征方面作了系统总结。谢家荣（1936，1937）发表了《北京西山地质构造概论》，对造山运动时代加以分析，将燕山运动划分为五幕。

1949 年至 20 世纪末，北京地区的构造地质研究有 3 次较大的飞跃（鲍亦冈等，2001）：①20 世纪 50 年代末至 60 年代初，在山区开展的第一轮 1∶5 万区调工作，较系统地总结了本区的地质构造特征，对地质构造单元作了划分。同期，北京地质学院区域地质教研室在马杏垣教授领导下，对大灰厂、磁家务、周口店一带各式小型构造进行了研究。②70 年代中后期，由地勘、科研及院校众多单位共同参加的北京地区地震地质会战，投入了大量实物工作，使地质构造的研究出现了空前活跃的局面。③80 年代至 90 年代，新一轮 1∶5 万区调和地勘、科研、教学单位构造专题研究的开展，大大加深了对本区构造地质特征的认识。

21 世纪以来，北京地区的地质工作主要集中在平原地区，地质构造研究集中于隐伏断裂调查领域。尤以 2012—2014 年投入大量物探和钻探，对黄庄-高丽营、南口-孙河、顺义和夏垫 4 条主要活动断裂进行了精确定位和活动性研究，提高了平原区活动构造研究程度。

本志以全球构造观点和板块构造学说为基础，以大陆动力学的最新研究进展为理论指导，对区域地质构造的形成演化进行全面系统总结。

第三十一章　构造单元及其主要特征

第一节　构造单元划分原则

本志中构造单元划分主要采用了全球构造观点和板块学说。在地质历史中，板块的状态和格局是在不断变化的，不同地质历史时期的板块构造格局或面貌不同，甚至可能有很大的变化（刘训，2015）。本次工作按照地质志编制指南，在全国构造单元划分的基础上，参考河北省的划分方案进行划分。按照全国二级单元划分出了的古陆块、地块等，确定北京地区在地块的位置，相对来说境内都是一些相对稳定的地区。但在不同历史时期，它们的不同地段又可以表现出不同活动性，从而有一些稳定区与活动区的区别。因此在古陆块内部，作为三级构造单元，可以区分出一些不同时期的隆起区和坳陷区；也有一些长期稳定的陆核和陆块；还有一些在不同地质时期出现的裂谷带或裂陷槽。同时一些大型中新生代的内陆盆地作为独立的一个构造单元。因此三级构造单元按照不同时代以及表现出的活动性为准则来划分，即构造单元的划分以详细的基础资料作为支撑，构造演化是构造单元划分的重要依据，不同时代的构造单元划分是不同的。本次构造单元划分一二级构造单元是以古生代时（具体应为南华纪至中三叠世）的状态为基础的，主要考虑北京地区在古生代的全国构造图中所处的位置，三级单元是在二级单元内的进一步划分，主要考虑他们在二级单元中的位置和作用，同时以时代限定。同时还要考虑一些“古板块”，有一定的时间范畴。即确定那些曾经存在过的不同板块，它们的边界在何处以及如何发展变化。

构造单元级别划分的原则：地质年代老的构造单元级别高于地质年代新的构造单元，但上叠构造单元中规模大的构造单元级别高。构造单元合并的原则：构造单元规模较小或单元数量较多，合并为较高级别构造单元时，按照以基底构造单元为主，或者以优势构造单元为主的原则。

第二节　构造单元划分及构造演化

北京地区位于一级构造单元柴达木-华北板块、二级构造单元华北陆块（克拉通）内部，基本属于一个较稳定的构造单元。由于本区后期构造单元多叠加于早期不同构造单元之上，各构造演化隶属关系与叠置关系清楚、层次分明、阶段性与旋回性明显，在构造演化基础上进行更详细的构造单元划分。

一、太古宙—古元古代构造特征

北京地区该时代地质体主要位于密云以北地区，大致呈北西西走向的长圆状。岩石主要为新太古代岩石组合，主要包括TTG片麻岩、镁铁质层状侵入岩、脉岩以及密云岩群和四合堂群表壳岩。其中表壳岩在约2.5Ga经历了绿片岩相至麻粒岩相区域变质作用和透入性变形作用，同时形成同构造花岗岩和紫苏花岗岩。该区域的构造样式特点是自晚古生代以来，区域隆起的强度和幅度较大，致使顶部各构造层、亚构造层几乎剥蚀殆尽，造成太古界构造层的直接暴露，仅在边缘有长城系及上侏罗统保留，中部及西南部有厚约200~400m的第四系沉积。由于该区长期的不断上隆及后期构造的不断叠加，因此在基底的线型褶皱上复合了各种形态的晚期褶皱和断裂，使其发生重组和明显错位，造成构造特征的复杂化，同时，不同时期不同构造层次的变形构造也得以充分揭露，形成了密怀穹隆构造。在变质作用演化上，新太古代基底岩石无论在高级区还是在花岗-绿岩地体中都表现为近等压冷却（IBC）型逆时针 $P-T$ 演化特征。铁镁质麻粒岩、斜长角闪岩和泥质片麻岩保存早期进变质、峰期和峰期后退变质阶段矿物组合。早期进变质矿物组合主要以矿物包裹体形式存在于峰期阶段矿物中。如在镁铁质麻粒岩中，早期组合角闪石+斜长石+石英+黑云包裹斜方辉石和石榴子石中；在斜长角闪岩中，早期组合绿泥石+阳起石+绿帘石+钠长石+石英包裹于斜长石、角闪石和石榴子石中；峰值组合主要以变斑晶和基质形式存在，峰期后退变质组合主要以后成合晶或退变边形式存在，如在镁铁质麻粒岩中可见石榴子石+石英呈后成合晶围绕斜长石和紫苏辉石生长，在泥质片麻岩中可见蓝晶石交代硅线石现象。结合变质温压条件计算结果，这些矿物组合转变限定了具有近等压冷却（IBC）特征的逆时针 $P-T$ 演化轨迹，反映了变质作用与大量地幔岩浆底侵作用有关。因此该区构造环境可能与岛弧、大陆裂谷或地幔柱（热点）环境中的大量地幔岩浆底侵作用有关。据第五篇中变质岩部分所述，北京地区该时代地层主要集中在新太古代，尚未发现古元古代地层。另外在北京平原区底部钻孔研究中发现有片麻岩。据目前研究成果，赵国春（2002）认为华北克拉通基底是由东部陆块、西部陆块和中部带 3 个主要构造单元所组成，而东、西陆块在古元古宙（~1.85Ga）沿中部带拼合导致华北克拉通基底的最终形成。北京地区位于东部陆块范围内。因此北京地区全区范围内在该时段一级构造单元为柴达木-华北板块，二级构造单元为华北陆块，三级构造单元属于冀中-冀东微陆块，按照河北省划分方案北京地区位于四级构造单元沧州-迁西断块（$ⅢA_1^{5-1}$）内。五级构造单元称为密云变质基底杂岩（旧称密云凸起或密云隆起）。

二、中元古代—中三叠世构造特征

北京地区乃至整个华北地区自中元古代起到三叠纪中—晚期都发育一套稳定型沉积盖层，与整个华北地区有着共同的海陆变迁史，中元古代到晚奥陶世为海相，缺失晚奥陶世—晚石炭世的中古生代沉积。

在空间位置上属于三级构造单元中元古代—中三叠世燕山-辽西裂陷带（燕辽裂陷槽）的一部分。约在 1800Ma，克拉通基底在构造应力场的作用下，发生断裂，形成北东向的裂谷，发展为燕辽陆内裂谷，形成燕辽裂陷海。北界为尚义-赤城-隆化断裂，东界大致在抚宁—滦县—塘沽—衡水—永年一线，坳陷最强烈地段位于蓟县—宽城一带，南部以岔头-新乐区域断裂为界与晋中南-邢台沉降（坳陷）区相邻，西部、东部延入邻区。其底界不整合在太古宙及古元古代变质岩之上，顶界被寒武纪地层覆盖。由一套未变质的板内海相、潟湖相富镁碳酸岩和少量碎屑岩、黏土岩组成，含微体古生物和叠层石，出露良好，分布广泛，发育齐全，蓟县—宽城一带厚度近万米。整体呈西部窄、东部宽的东西向带状展布，区内南北宽 265~362km，东西长约 500km。在中元古代—中奥陶世为间歇性裂陷海槽盆地，晚石炭世—中三叠世叠加了海陆交互相-陆相盆地沉积。据《中国区域地质志·河北志》（2017）中，将燕山-辽西裂陷带划分为蓟县-唐山裂谷盆地、宣化-易县盆地、承德北盆地和秦皇岛盆地 4 个次级盆地构造单元。北京位于蓟县-唐山裂谷盆地、宣化-易县盆地和承德北盆地 3 个盆地（Ⅳ级）范围内。

1. 宣化-易县盆地

宣化-易县盆地位于河北省中西部，属于燕山-辽西裂陷带中西段的边缘盆地，以断层和水下基底隆起与周围其他构造单元相隔，北部与华北北缘隆起带及承德北盆地相邻，东部与蓟县-唐山裂谷盆地相邻，南部与晋中南-邢台沉降（坳陷）区相邻。区内东西宽 100~215km，南北长约 310km。由中元古代-中三叠世地质体组成，受后期改造露头不太连续。地层厚度 1788~7286m。发育褶皱与断裂构造，受后期改造，构造线呈北西向、北东向及近东西向展布。该盆地目前显示位于重力负异常区，南东部边缘地区位于重力梯度变化带上；航磁（ΔT）以正异常为主，局部间夹负异常；地壳（莫氏面）深度在 35.5~42km 之间，具由南东向北西及由东向西变深的趋势。

北京在该构造单元范围主要地层为中—新元古代—古生代沉积地层以及燕山期叠加改造形成的火山盆地和侵入岩。结合后面晚三叠世—早白垩世的构造特征，将该构造单元划分为 5 个五级构造单元。

（1）房山碳酸盐岩陆表海（PZ）

地质时代从寒武纪至早三叠世。早古生代，华北克拉通进入了陆表海碳酸盐台地演化阶段，早期（ϵ_1）为陆缘碎屑岩-碳酸盐岩组合，中期（ϵ_1—O_1）为缓坡-斜坡碳酸盐岩组合；晚期（O_2）潮坪-局限台地碳酸盐岩组合。晚古生代为海陆交互陆表海，沉积一套陆表海沼泽含煤碎屑岩组合（C_2—P_{1-2}），本溪期初始沉积铝土质、铁质岩系，随后沉积砂岩夹紫红色页岩和煤层，太原期早期，海侵范围扩大，形成一套海陆交互相含煤碎屑岩-碳酸盐岩建造。二者是以太原组和山西组的平行不整合界线划分的。沉积环境也由陆表海向海陆交替的过渡环境发展。

（2）八达岭陆内造山岩浆杂岩（J—K）、门头沟断陷盆地（J—K）和四海断陷盆地（J—K）

这3个构造单元主要是燕山期叠加改造形成的侵入岩和火山盆地。晚三叠世华北板块的构造强烈活动期，侵入岩和火山岩广泛分布，在北京地区形成了盘山、珠窝前造山侵入岩组合。北东东向或近东西向较大型掀斜或单断、地堑型内陆盆地（洼地）发育，基本继承了早中三叠世构造格局。早侏罗世门头沟群形成时期是一个火山作用活动较弱时期，在南大岭晚期才有中基性火山岩喷发。地层断续分布于百花山-妙峰山沉积洼地的边缘，侵入活动较弱。中侏罗世在各中生代洼地中喷发的酸性物质呈夹层或夹正常沉积层，构成九龙山组。之后，中性岩浆的喷发，形成了髫髻山组。这次火山喷发在本区火山构造中多可见到，堆积了喷溢相、爆发相为主的大量火山岩及潜火山岩。岩浆侵入活动较强，长园、云蒙山、石窑等同造山侵入岩组合，晚侏罗世在断陷或坳陷盆地中堆积了巨厚的类磨拉石建造。早白垩世强烈的造山运动，产生了以北东、北北东向为主的一系列不同方向断裂构造，同时部分的古断裂也开始复活。受这些断裂控制，形成了规模不等的断陷盆地。这些盆地叠覆于侏罗纪盆地之上，多为北东、北北东向。早白垩世张家口期火山构造洼地，多分布于侏罗纪火山-沉积洼地的西侧或中部，而且规模小、分布零星。不同程度地改造了侏罗纪形成的洼地展布方向。这个时期侵入活动相当强烈，形成了房山、大海坨、对臼峪等一系列岩体，为晚造山和后造山岩石组合。根据侵入岩的分布和火山盆地的分布范围划分了八达岭陆内造山岩浆杂岩（J—K）、门头沟断陷盆地（J—K）以及四海断陷盆地（J—K）2个五级构造单元。

（3）延庆凹陷盆地

位于本区西北延庆—康庄一带，地貌上为北东-南西向延伸盆地。早古生代末期本区逐渐隆起，至中生代初期仍继续隆升。中生代中晚期有上侏罗统陆相火山喷发沉积外，核部地带仍保持隆起状态，直至新生代中晚期形成断陷，接受沉积。因此划分为一个独立的五级构造单元。

2. 承德北盆地

承德北盆地位于承德北一带，属于燕山-辽西裂陷带北东边缘盆地，以断层和水下基底隆起与周围其他构造单元相隔，北部与华北北缘隆起带相邻，南部与宣化-易县盆地及蓟县-唐山裂谷盆地相邻，东部及东北部延入邻区。由中元古代-中三叠世地质体组成，受后期改造露头不连续。地层厚度1384~2787m。发育褶皱与断裂构造，主构造线呈近东西向，受后期改造，局部呈北东向展布。在二叠纪时期，叠加的云雾山火山盆地，具有弧后火山盆地的性质。该盆地目前显示位于重力负异常区，中北部位于重力梯度变化带上；航磁（ΔT）以正异常为特征；地壳（莫氏面）深度在37~41km之间，具由东向西变深的趋势。

北京在该构造单位内主要是位于北部三块石、汤河口、长哨营以北地区，在赤城-长哨营-古北口东西向深断裂带以北，是华北北缘隆起带的一部分。本区自元古宙后，逐渐上升隆起为陆，直至中生代，在其断陷盆地中才有侏罗-白垩系的陆源碎屑岩、火山岩及火山碎屑岩堆积。因此，其基本形态于元古宙时期已经形成，中生代时期仅使其构造复杂化。中生代构造层主要由上侏罗统髫髻山组、后城组的安山岩、安山质火山碎屑岩、火山碎屑沉积岩及下白垩统张家口组流纹岩等组成。区内褶皱构造不发育。断裂构造较发育。南部边界有长期活动的东西向正断层和逆断层，如银河沟-下孟背断层，庄户沟-南沟断层等，规模较大，延伸长度可达数10km以上。它们都是赤城-长哨营-古北口东西向深断裂的浅部构造，组合形态呈地垒式、阶梯式及对冲式等，其存在对本区地层和岩浆活动的分布和发育有明显的控制作用。横切东西向断裂带有一些规模较小的水平剪切断层，使东西向断裂发生错位。该区东南及北部有近南北向断裂及其相伴的北东、北西向次级断裂发育，如帽山-喇叭沟门断裂等。它们均展布于近南北向的丰宁-怀柔西-采育深断裂带中，常以破裂带形式出现，规模通常小于近东西向的断裂。沿断裂带常有中生代时期各种岩脉侵入。区内岩浆活动强烈。中元古代时期有兰营石英正长岩、古洞沟花岗岩，华力西期—印支期有下窝铺-南台子、大西沟、上甸子等变质辉长闪长岩，七道河西沟、东湾子、对角沟门及喇叭沟门花岗岩，燕山中、晚期有下帽山石英闪长岩，孙家栅子花岗闪长岩、石英二长岩、转山子花岗岩、石洞子石英正长岩等中性—中酸性岩类及与之相对应的火山喷发活动，形成了结构较为复杂的侵入岩和喷发岩分布区。划分为喇叭沟门大陆边缘岩浆杂岩五级构造单元。

3. 蓟县-唐山裂谷盆地

蓟县-唐山裂谷盆地属于燕山-辽西裂陷带的中心盆地，以断层和水下基底隆起与周围其他构造单元相隔，北部与承德北盆地相邻，西部与宣化-易县盆地相邻，南西部与晋中南-邢台沉降（坳陷）区相邻，东部与秦皇岛盆地相邻，北东与南东延入邻区。区内北西-南东向宽 80~215km，南西-北东向长约 440km。由中元古代—中三叠世地质体组成，受后期改造露头不太连续。地层厚度 5117~13464m，并含有裂谷型火山岩地层。发育褶皱与断裂构造，受后期改造，构造线呈北西向、北东向及近东西向展布。该盆地目前显示重力异常以正负异常相间为特征，顺义-遵化一线位于重力梯度变化带上；航磁（ΔT）以正负异常相间为特征；地壳（莫氏面）深度在 34~37km 之间，具由南东向北西变深的趋势。

北京在该构造单元内可划分为密云变质基底杂岩和平谷夭折裂谷 2 个五级构造单元。其中密云变质基底杂岩特征前文已叙述。

平谷夭折裂谷北侧以密云-墙子路断裂和新城断裂为界，西、南部以荆园断裂、二十里长山断裂为界。本区太古界仅在局部地区出露，中上元古界遍布全区，是燕辽裂陷槽沉积中心，其最大沉积厚度近一万米，并伴有碱性玄武岩和粗面岩的火山喷发。古生代除有近千米的浅海相碳酸盐岩沉积外，尚有 500 余米厚的海陆交互相含煤碎屑岩沉积。中生代强烈隆起，致使古生界及部分中—新元古界遭受剥蚀。新生代晚期，平谷一带发生断陷，沉积了 350~500 余米的第四系。区内褶皱和断裂构造较发育。除结晶基底中褶皱较紧密呈等斜线形外，盖层中一般多为平缓开阔之波状褶皱，其中以北东向者规模较大，延续性较好，葫芦峪、孔城峪一带局部有规模较小的次级褶皱发育。区城断裂构造既有东西向、北东向、北北东向及近南北向者，也有北西向者，但以前三组较发育，规模较大，延伸亦较稳定。岩浆活动除中元古代大红峪期有钾质玄武岩、粗面岩的喷溢，将军关、钵尔塘、黄松峪一带有同期的中基性或超基性火山角砾岩体沿断裂带侵位外，还有沿东西向断裂侵入的独具特点的沙厂球斑花岗岩岩体、燕山期有中浅成含超浅成的中性、中酸性和酸性岩体，它们主要分布于东北、西部及南部等处的多组断裂构造复合部位。

三、晚三叠世—早白垩世构造特征

晚三叠世—晚白垩世早期是燕山构造旋回阶段，该时期的构造形成演化构成了区域上华北板内造山演化旋回，形成大兴安岭-太行山板内造山带。该造山带大致以黄旗-灵山区域断裂为界，划分出 2 个三级构造单元，即火山喷发带，主要构造线呈北北东向—北东向。本区处于承德-武安火山喷发带三级构造单元中，又分 3 个阶段形成了 16 个构造-火山盆地（Ⅳ级构造单元）。

晚三叠世形成了盘山、北京地区珠窝前造山侵入岩组合，北东东向或近东西向较大型掀斜或单断、地堑型内陆盆地（洼地）发育，基本继承了早—中三叠世构造格局。

早侏罗世初期是一个火山作用活动较弱时期，在南大岭晚期才有中基性火山岩喷发。地层断续分布于百花山-妙峰山沉积洼地的边缘，形成西山火山盆地，侵入活动较弱。中侏罗世在各中生代洼地中喷发的酸性物质呈夹层或夹正常沉积层，构成龙门-九龙山组。之后，中性岩浆的喷发，形成了髫髻山组。这次火山喷发在本区火山构造中多可见到，堆积了喷溢相、爆发相为主的大量火山岩及潜火山岩，形成了髫髻山盆地岩浆侵入活动较强，长园、云蒙山、石窑等同造山侵入岩组合，晚侏罗世在断陷或坳陷盆地中堆积了巨厚的类磨拉石建造——土城子组。西南起自百花山，往东北方向经髫髻山、妙峰山、十三陵、怀柔山前延至密云新城子一带，宽约20km，发育有百花山、髫髻山、新城子3个火山-沉积盆地。北部处于居庸关深断裂的西北侧，形成了燕羽山、凤驼梁-四海、千家店、后城等火山盆地。

早白垩世强烈的造山运动，产生了以北东、北北东向为主的一系列不同方向断裂构造，同时部分古断裂也开始复活。受这些断裂控制，形成了规模不等的断陷盆地。主要有坨里-大灰厂、东岭、七道河等火山-沉积盆地。这些盆地叠覆于侏罗纪盆地之上，多为北东、北北东向，总的特点是：①早白垩世张家口期火山构造洼地，多分布于侏罗纪火山-沉积洼地的西侧或中部，而且规模小、分布零星。②后者叠加在前者之上，以25°~40°相交。③不同程度地改造了侏罗纪形成的洼地的展布方向。这个时期侵入活动相当强烈，形成了房山、大海坨、堆臼峪等一系列岩体，为晚造山和后造山岩石组合。

四、晚白垩世—全新世构造特征

隶属于华北板块（II级）新生代弧后渤海湾裂谷盆地（III级，廊坊-衡水火山-沉积盆地），平原区位于盆内冀中坳陷（IV级）西北边部，其东连沧县凸起，南抵邢衡隆起，西接太行山隆起，北邻燕山褶皱带。区域重力场为正负重力高区，布格重力异常值均高于周围山区，磁场上除市区东南边界为高正磁异常外，其余均为稳定的负磁场区。

新生代古近纪始新世至渐新世时期，在强烈伸展作用背景下，区内北东向断裂相应发生了不同规模的引张正断层活动，形成“两隆、两凹”盆地与山岭相间排列的盆岭构造（马杏垣等，1987；北京市地质矿产局，1991；徐锡伟等，2002；崔盛芹等，2002；Qi et al.，2010）。宏观上，它们呈北北东方向展布，可进一步划分为京西断隆（又称西山凸起，太行山凸起）、大兴断隆（V-2-1-2）、北京断陷（V-2-1-1）、大厂断陷（V-2-1-3）和固安-武清断陷（V-2-1-4，又称廊固凹陷）等四级构造单元，它们的分界断裂由西向东分别为黄庄-高丽营断裂、南苑-通县断裂、夏垫断裂和大兴断裂（又称礼贤断裂、涿县-凤河营-宝坻断裂）。

新近纪晚期至第四纪，北京平原区构造运动则表现为持续下降接受沉积，使得古近纪盆地-山岭和广泛发育的断裂构造均埋藏于厚层的新近纪和第四纪沉积层之下（徐杰等，2001，2003；Xu et al.，2004；徐锡伟等，2002）。同时，由于新生北西向活动断层或北西向老断裂复活，北京平原区中生代以来形成的“两隆、两凹”的构造特征和堆积环境彻底解体，沉积中心的长轴方向发生改变，从古近纪的北北东向转为北西向，形成了马池口凹陷、后沙峪凹陷等新的沉积中心。

（1）北京断陷（V-2-1-1）

该构造单元走向北东向，下伏地层主要为中生代、古生代及元古宙地层。始新世时，受北北东向的黄庄-高丽营断裂、顺义-良乡断裂、南苑-通县断裂的控制，地层沉积主要在黄庄-高丽营断裂和南苑-通县断裂之间，成北北东向展布，沉积中心大致在丰台一带。渐新世时由于南苑-通县断裂活动加强，与始新世相比沉降中心向东南移动，以北京市区为沉积中心（厚度达300m），成北北东向展布，展布范围比始新世更广，北端已越过来广营、平房，但未到天竺一带。到新近纪时地层分布范围基本在黄庄-高丽营断裂和南苑-通县断裂之间，与始新统—渐新统相比，西边界市区段和琉璃河段向西北移动；以北京市区为沉积中心（厚度达1100m），但沉积厚度远超过始新统—渐新统，呈北北东向展布，北端已越过天竺到相各庄、牛栏山一带。另外在古近纪时岩浆活动强烈，始新统—渐新世约有5次玄武岩喷发，多在凹陷的中心沿断裂带溢出。到新近纪，仍伴随有陆相玄武岩的喷发，但其规模及厚度都比古近纪要小。

（2）大兴断隆（V-2-1-2）

位于北京通州南至大兴一线。西北为北京断陷，以南苑-通县断裂为界，东南以涿县-凤河营断裂、夏垫断裂为界，与固安-武清断陷和大厂断陷相邻。其基底以中、上元古界及下古生界为主。是中、新生代相对隆起幅度较大的区域。晚三叠世末期开始稳定上升遭受剥蚀，新生界沉积厚度一般50~160m 不等。

（3）大厂断陷（Ⅴ-2-1-3）

位于北京市东南边界一带，呈北东向延伸，西邻大兴断隆，以夏垫断裂为界，东北与东南延入河北省，基底岩系南、北段可能为古生界、中部为中—新元古界，新生界地层发育较全，厚度达5000余米以上。此断陷地处北北东向断裂活动带南段上盘，为断裂带转折较大部位，常有不同级别地震发生。

（4）固安-武清断陷（Ⅴ-2-1-4）

位于北京市区南部边界附近，大部分在河北省范围内，是渤海湾盆地冀中坳陷北部的一个北东走向的古近纪箕状断陷。北与大厂断陷相接，西靠大兴断隆，南为牛驼镇凸起，东邻武清凹陷，南北长约90km，东西宽约20~40km，分布面积2600km^2（金凤鸣等，2006；王志宏等，2008）。其基底由前中元古代结晶基底和中元古代—中生代沉积基底组成。现今构造格局主要是燕山期和喜马拉雅期构造运动综合作用的结果，中生代末的燕山运动以挤压、褶皱变形和差异隆升剥蚀为主，发育一系列北东、北北东向走滑断裂及其配套的北西、北西西向断层（高慧君，2004；杨军侠，2004）。其古近纪的基本沉积构造格局表现为：以凹陷西缘具拆离滑脱性质的走向北北东倾向南东东的大兴断裂带（礼贤断裂）为主导，并与断面倾向北西的正断层联合构成“反Y字形”伸展断裂系统，呈现为靠近大兴断裂一侧强烈沉降、向东逐渐掀斜抬升的箕状断陷，发育西深东浅、西厚东薄的孔店组—沙四段、沙三段、沙二段—沙一段和东营组四套伸展断陷型内陆河湖相沉积层系。东营组沉积末发生区域抬升剥蚀，新近纪以来进入区域坳陷阶段，发育河流漫滩相—沼泽相沉积建造组合（高慧君，2004；杨军侠，2004）。

第三节　构造单元综合划分方案

本综合划分方案是在上文的基础上对北京地区现存地质构造进行的划分。一、二级构造单元按照全国尺度进行划分，三、四级构造单元划分参照河北省构造单元划分方案。划分为2个三级构造单元和4个四级构造单元。在四级构造单元划分基础上进一步划分了12个5级构造单元（表31-1，图31-1）。

表31-1　大地构造单元划分表

Table 31-1　The division of tectonic units

一级	二级	三级	四级	五级
柴达木-华北板块Ⅰ	中太古代—第四纪华北陆块Ⅱ	中元古代—中三叠世燕山-辽西裂陷带Ⅲ-1	宣化-易县盆地Ⅳ-1-1	Ⅴ-1-1-1 房山碳酸盐岩陆表海（PZ）
				Ⅴ-1-1-2 八达岭陆内造山岩浆杂岩（J-K）
				Ⅴ-1-1-3 门头沟断陷盆地（J-K）
				Ⅴ-1-1-4 延庆凹陷盆地（E-Q）
				Ⅴ-1-1-5 四海断陷盆地（J-K）
			承德北盆地Ⅳ-1-2	Ⅴ-1-2-1 喇叭沟门大陆边缘岩浆杂岩（P）
			蓟县-唐山裂谷盆地Ⅳ-1-3	Ⅴ-1-3-1 密云变质基底杂岩（Ar）
				Ⅴ-1-3-2 平谷夭折裂谷（Pt）
		晚白垩世—全新世华北盆地Ⅲ-2	廊坊-衡水火山-沉积盆地Ⅳ-2-1	Ⅴ-2-1-1 北京断陷（E-Q）
				Ⅴ-2-1-2 大兴断隆（E-Q）
				Ⅴ-2-1-3 大厂断陷（E-Q）
				Ⅴ-2-1-4 固安-武清断陷（E-Q）

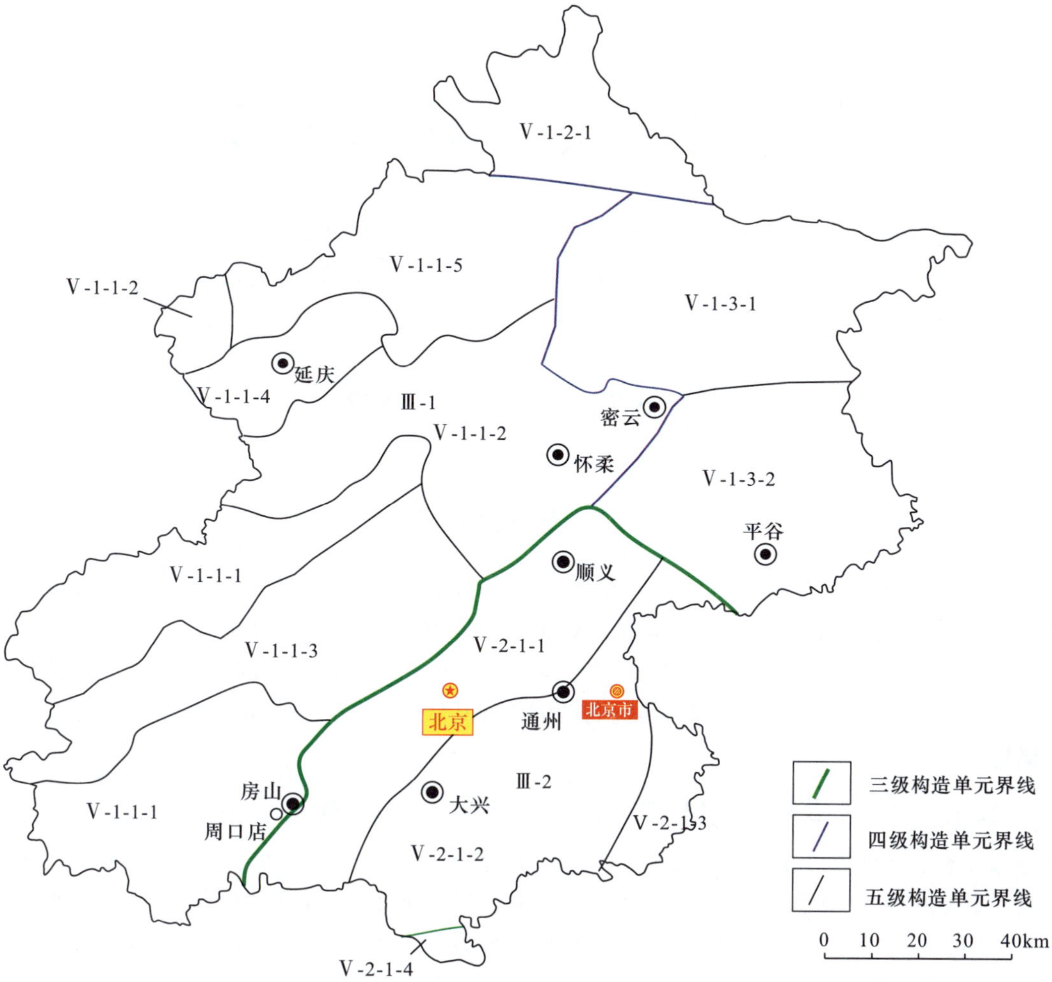

图 31-1　北京市构造单元划分综合图

Fig. 31-1　The division map of integrated tectonic units in Beijing

本次构造单元划分是主要以地质演化历史为主要线索，结合建造与构造进行综合分析，但是受资料所限，特别是目前在该区针对地质体变形、变位的运动学和动力学方面的研究是比较欠缺的，同时也没有足够的构造年代学数据的支撑，因此不能从根本上解决建造与后期改造的问题，留待后续工作来解决。

第三十二章　构造形变

构造形变是地壳发展演变的重要记录之一，不同阶段和发展时期的构造形变特征，必然反映相应阶段和时期构造演化特点及其彼此间的差异和关系。多旋回多期次的构造运动在本区地壳岩层中塑造了各种规模级别不等、形态复杂多样、但有一定规律可寻的构造形迹和形变样式。

第一节　褶皱构造

一、基底褶皱

本区克拉通结晶基底最终形成于古元古代末期，主要由深变质的密云岩群（表壳岩）、四合堂岩群（表壳岩）和变质深成岩体构成。在经历了多期次的韧性剪切、变形变质作用后，形成为各类片麻岩。结晶基底在密怀地区成片出露，其他地区仅零星分布。结晶基底的构造形变以韧性变形为特征，变质和变形同步进行。在1∶20万怀柔幅区调报告中对基底的褶皱变形分了两期，并对每期的构造特征进行了描述。

（一）发育于密云岩群中的褶皱构造——第一期褶皱（f1）

主要分布于密云岩群中，四合堂岩群局部地段也可见。褶皱样式为紧闭的同斜褶皱和转折端加厚的相似褶皱，其规模多为中小尺度，波长短、波幅大，轴面平卧或斜卧，枢纽近水平，并且两翼岩层常被拉长后拉断，逐步发展为片内无根褶皱。轴面片理发育，具透入性，强烈置换先期面理。本期褶皱原始轴迹为近东西向（周绍林，1984）。该期构造是目前能够识别出的北京地区地壳演化过程中最早的构造变形，是岩石处于塑性—半塑性状态下产生的固态流变和深层次韧性剪切作用的产物。

伴随本期变形，密云岩群麻粒岩相岩石发生角闪岩相退化变质作用，表现在麻粒岩相变质矿物被改造或交代，紫苏辉石、石榴子石等矿物出现碎裂结构，解理面弯曲或被交代，斜长石普遍发育反条纹长石出溶结构，并有单斜辉石、石榴子石、角闪石、石英和少量黑云母生成。

（二）复式褶皱构造——第二期褶皱（f2）

该期褶皱变形广泛分布于太古宙变质岩中（图32-1），以半城子断裂带以东地区最为发育。受后期构造破坏和岩体侵入的影响不大，其形迹保留较为完整。该地区内由北向南依次发育有前宝岭复向形、栗榛寨-西恒河复背形、高岭复向形、阁老峪背形、穆家峪背形与向形、达岩西向形、水峪复向形和郝家庄复背形等，构成了结晶基底的基本格架。第二期褶皱具有如下特征：

1）主体褶皱（复式）以大、中型尺度为主，一般宽10~15km，长20km以上。次级褶皱宽数十米至上百米，延长数千米不等。

2）褶皱轴向总体呈北东-南西向展布，相邻背向形轴迹基本平行。因受后期构造活动等因素影响，各地的轴向由北西向南东逐渐向北偏转，即由前宝岭、西恒河到穆家峪、郝家庄，轴向由NE60°依次变化为NE55°、NE30°、NE15°。

3）褶皱样式为多级次的同斜或倒转复式背形、复式向形的组合。轴面倾向北西、北西西或近于直立。倾角在40°~80°之间，一般背形紧闭，向形宽阔，属对称或不对称顶厚翼薄褶皱（图32-2a）。其枢纽常呈舒缓波状，如沙厂向形（图32-2b），是后期东西向褶皱构造叠加在南北向的向形之上形成的横跨褶皱。

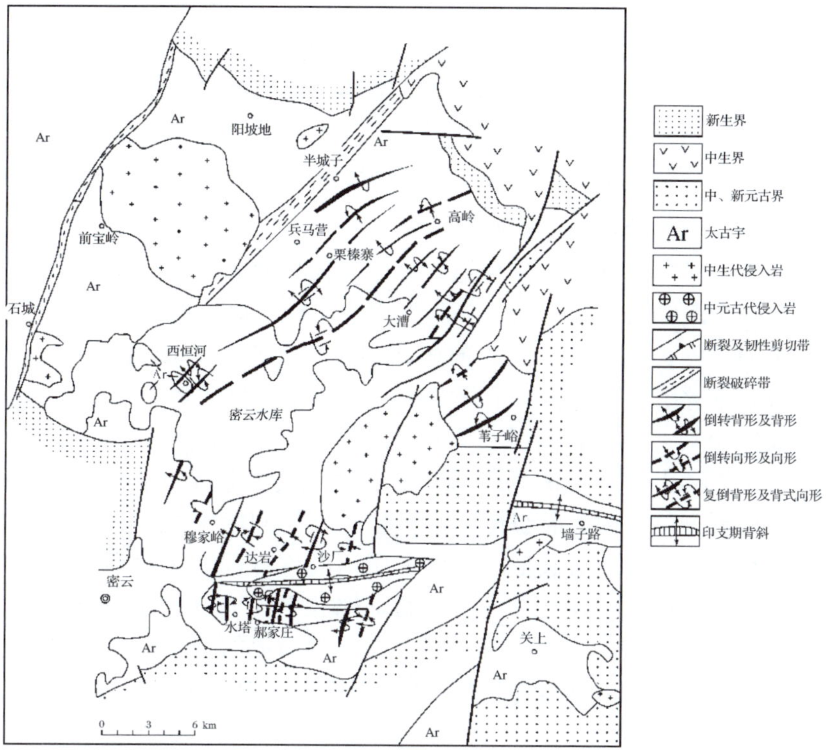

图 32-1　密云地区太古宙变质岩中的复式褶皱

（据周绍林，1987）

Fig. 32-1　Composite folds within Archean metamorphic rocks in Miyun area

4）轴面片理发育程度较弱，不具透入性。本期褶皱是在第一期褶皱构造基础上发生的一次强烈的面理褶皱，前期形成的深层次（麻粒岩相）韧性剪切带构造面理和第一期褶皱面理均构成了本次褶皱的变形面理。其变形时代应为新太古代末期。

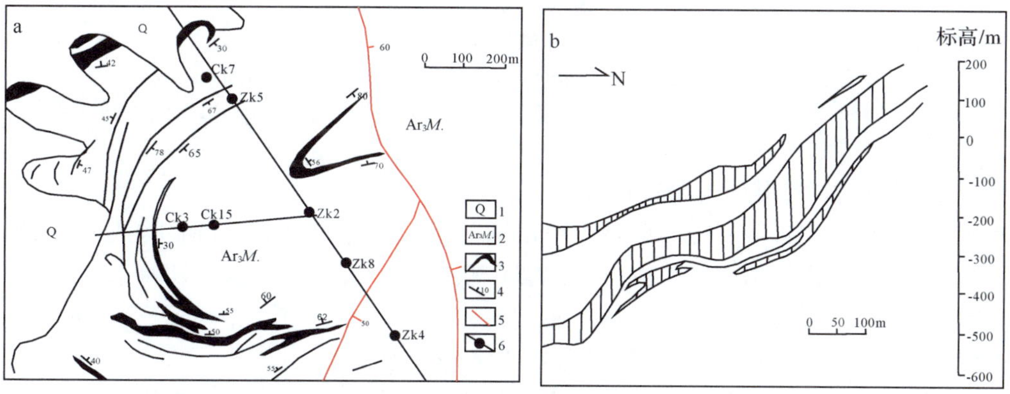

图 32-2　第二期褶皱构造的顶厚褶皱特征和沙厂向形构造特征

（据周绍林，1985，修改）

Fig32-2　The second phase of the fold structure and structure characteristics of Shachang synform

a—兵马营铁矿区向斜转折端处矿体平面略图；b—沙厂向形矿体纵剖面图

1—第四系；2—密云岩群；3—铁矿体；4—产状；5—断层；6—勘探线及钻孔

二、盖层褶皱

本区古元古代末的吕梁运动使结晶基底最终固结，吕梁期以后，自中元古代至早二叠世，经历了长期的盖层平稳发育阶段，表现为升降运动和轻微的隆坳，褶皱变形不明显。晚二叠世以后由于经历了陆内（板内）造山复杂构造变形阶段，因而本区盖层具有强烈的构造变形，且具有多期次、多类型、多层次和多体制的变形特点，形成了较为复杂盖层构造组合。

（一）印支期褶皱——第三期褶皱（f3）

本期褶皱主要发育于中元古代早期（长城纪）—中生代早期（中三叠世）地层之中，形成于晚二叠世—中三叠世双泉组的杂色碎屑岩沉积以后，印支构造运动（晚二叠世—中三叠世）近南北向挤压构造环境。该期褶皱走向东西，横剖面上背、向斜构造呈较典型箱状或宽缓的波状，枢纽近于水平，轴面近于直立，褶皱被上三叠统杏石口组、下侏罗统南大岭组角度不整合覆盖（图 32-3）。平面上明显被燕山期褶皱或断裂斜交裁切，二者间明显不协调。该期褶皱构造一般规模不大，稳定延伸约 6～7km。空间上自北至南，大致以 10～15km 间距沿一定纬度呈大致平行的带状断续分布。受后期构造干扰破坏，部分地段褶皱延展方位、构造形态特征常发生程度不同的改变。但是所不同的是有的褶皱是以原生层理变形弯曲而形成，而有的褶皱是以置换了层理的流劈理为次生面理变形弯曲而形成，这是印支期褶皱构造的一大特点。这一差异反映了当时北京地区不同地段地壳热条件存在差异，且该期构造变形可分为两个世代的构造变形活动，早期为横弯隆起和局部的固态流变；晚期在近南北向挤压应力场的作用下在盖层中广泛发育近东西向褶皱构造。下面选取规模较大且具代表的褶皱构造简述其特征。

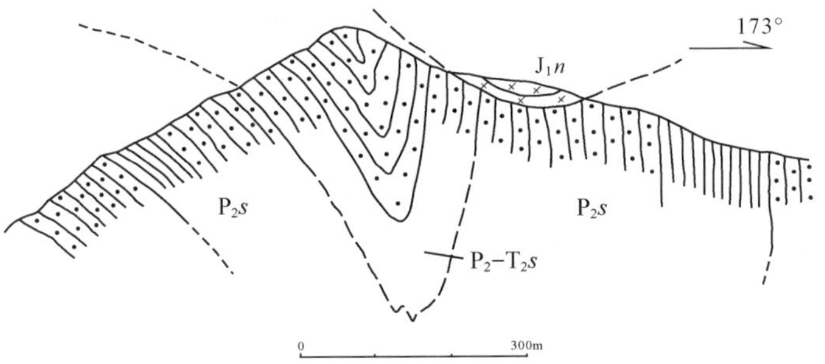

图 32-3　房山凤凰山 640 高地剖面

（据宋鸿林，1984）

Fig. 32-3　The section of 640 highland in Fenghuangshan，Fangshan Area

J_1n—南大龄组辉绿岩；P_3-T_2s—双泉组变质细砂岩及板岩夹砾岩；P_2s—石盒子组变质砂岩夹板岩

（1）大白石尖-谷积山箱状背形带（$f3_1$）

位于房山西部长操、北直河一带，褶皱主要由蓟县系、待建系及青白口系构成，核部岩层产状一般较平缓，由蓟县系雾迷山组第四段组成。南北翼产状较陡，由洪水庄组、铁岭组、下马岭组及青白口系组成，翼部发育次级褶皱。走向东西向，枢纽近水平，轴面近直立，两翼产状 40°～70°，东南部被霞云岭推覆构造切割，北邻百花山向斜，南部地层产状渐变舒缓。

谷积山背斜位于门头沟中部鲁家滩以南。褶皱主要由青白口系及下古生界构成，褶皱面为深度置换原生层理的流劈理。发育扇形轴面劈理，走向东西向，轴面直立，枢纽近于水平，两翼产状中等。其东南部被八宝山推覆构造切割，北翼与九龙山向斜相邻。谷积山背斜实际为一箱形复背斜，由两个次级背斜和一个次级向斜组成，背斜核部为青白口系，两翼为寒武系、奥陶系，向斜由寒武系组成。谷积山背斜往东延伸的倾伏端又称为马鞍山背斜。据宋鸿林等（1966）研究，此带褶皱构造主体走向近东西，往东至马鞍山背斜走向受后期构造改造干扰略向北偏转，呈北东东向。褶皱构造内部固态流变构造十分发育，岩层强烈拉伸变薄，不同尺度的“顺层”掩卧褶皱、“顺层”韧性剪切带，窗棂构造及黏滞型石香肠、楔入褶皱、楔入岩脉、透入性“顺层”破劈理、拉伸线理等普遍可见。这不仅反映本区东西向箱形褶皱形成之前或同时，已经历了一系列强烈变形，而且还指示变形时变形-变质构造环境处于地壳较深层次。

（2）石槽复合背斜（$f3_2$）

主要分布在延庆石槽一带，长36km，宽12km。褶皱主要由雾迷山组构成。由东部的大滩向斜和西部的桃蹄沟向斜以及石槽背斜组成。中间被中生代火山沉积岩系隔断。褶皱的总体形态较为开阔，略呈箱形，走向东西向，石槽背斜北东东向，枢纽水平，轴面近于直立，向斜两翼对称，倾角24°~45°，背斜两翼不对称，倾角45°~79°。褶皱的中段及东段都被北北东向沙梁子-上花楼断裂破坏，东部在褶皱的核部发育燕山期中酸性侵入体，并伴随多金属矿化，主要成矿元素铜、铁、铅、锌、金银等，是石槽铜矿区的控岩控矿构造。

（3）尚庄子-仓术会向斜（$f3_3$）

分布在密云水库以东一带，褶皱规模不大，核部由蓟县系雾迷山组第三段组成，南翼由雾迷山组第一、二段与杨庄组和高于庄组组成，岩层产状向北倾斜，倾角30°~40°。北翼由于断裂破坏，产状陡且不稳定。走向北东东，岩层向南东东倾斜，倾角20°。向斜的翼部及东端均被后期断裂程各庄断裂切割并发生牵引，尤其受东端的断裂活动造成向斜枢纽弯曲，走向偏转呈弧形。

（4）宝金山背斜（$f3_4$）

分布在房山周口店以西，为穹状背斜，主要由雾迷山组、洪水庄组及铁岭组构成。总体轴向接近东西向，岩层产状较平缓，标志褶皱形态的面理为置换了原生层理的早期流劈理，其形态较为舒缓，顶部开阔而略呈穹状褶皱。从宏观上看，背斜是非常简单的构造变形，但是从内部结构解析上，有着十分复杂的多期变形的叠加关系，如顺层韧性剪切带及固态流变构造（后文详述），并且夹在霞云岭推覆构造和黄山店推覆构造之间，包容了两个推覆构造的堆叠岩席和冲断层，使褶皱成为多重变形面的共轴叠加变形。

（5）太平山向斜（$f3_5$）

分布在房山周口店一带房山穹隆南缘，向斜核部为石炭系和二叠系，翼部为奥陶系，在北翼可明显见到马家沟组以下各组地层厚度变薄。向斜轴向近东西向，枢纽波状起伏，总体上向东扬起，轴面直立，北翼产状较陡，为50°~80°，南翼倾角较缓，为30°~50°，并发育顺层断裂；核部地层局部地段产状陡倾，甚至直立，向斜总体呈漏斗形。向斜的北翼被房山岩体侵位，南翼及东端被第四系覆盖。

（6）沙厂-墙子路背斜（$f3_6$）

位于密云东部，背斜核部为新太古界，翼部为长城系及蓟县系，规模较大，是中元古代同侵入褶皱再叠加所致，定形和加强期属印支期。其核部为新太古代变质岩，内有中元古代沙厂环斑花岗岩侵入。在后沙岭一带被花岗闪长岩体（同位素年龄值为195Ma）侵入。背斜西端被后期断裂切割，并隐伏于新生代地层之下。背斜遭受北北东向及近东西向断裂的破坏，枢纽沿北北东向断裂左行错开约3km。

（二）燕山期褶皱

燕山期褶皱是本区盖层强烈褶皱变形时期的产物。它以褶皱的规模大、延伸远、褶皱较强烈与断裂构造密切相伴，空间上呈北东向斜列排布等特征与印支期褶皱相区别。按其发育顺序、构造特征和方位等差异，可划分为早、中、晚3个时期的褶皱。

1. 燕山早期褶皱——第四期褶皱（f4）（J_2）

燕山早期褶皱卷进的最上部地层为中侏罗统九龙山组，被上侏罗统髫髻山组不整合覆盖。据前文限定的髫髻山组年龄160Ma，可基本推出该挤压变形发生在160Ma（张岳桥，2007），主要受北北西-南南东向挤压，形成北东东-南西西向褶皱构造。据统计，北京地区该期褶皱主要为走向北东60°的斜列式褶皱，规模较大多在10km以上，剖面上呈连续波状，两翼产状为缓至中等倾角，相伴的逆冲断层发育。主要展布于北京的西部和北部。西部山区自西北至东南主要有燕家台复式背斜、门头沟复式背斜、北岭向斜。北部山区自西北至东南有五里坡背料、红旗甸背斜、柏木井背斜、南口背斜、八达岭箱式背斜、大地背斜。京东有坡子山向斜、大华山背斜。另外，京东北西走向的松各庄向斜、西智向斜，京西北西走向的棋盘山向斜、了思台南背斜及洪水口—寺台子间的北西向褶皱，也可能为燕山早期产物。因为它们均被燕山中晚期甚至燕山早期的北东向构造切割改造，致使其构造形态、轴向常发生改变。

（1）燕家台复式背斜（$f4_1$）

出露于门头沟区最西部。其轴部是燕家台背斜，西北翼发育烟筒山向斜，东南翼发育梁家山向斜、传子岭-川底下背斜、梁家庄南向斜。各背斜核部均由中—新元古界组成，向斜核部由寒武系、奥陶系组成。梁家山向斜核部尚有石炭系，为髫髻山组不整合覆盖。背、向斜两翼产状一般 30°~40°，部分地段可达 55°~60°。轴面近于直立，枢纽水平，南翼被髫髻山组不整合覆盖，北西翼则被北东向断裂破坏。晚期受同方位沿河城断裂等切割破坏，局部褶皱构造形态较复杂。

（2）门头沟复式背斜（$f4_2$）

出露于门头沟区东部。自西北至东南有王平村背斜、九龙山-香峪大梁向斜、红庙岭背斜、石景山向斜、石门营背斜。背斜由早古生代地层组成，向斜核部为九龙山组，翼部由三叠系和中下侏罗统组成。褶皱舒缓波状。北翼和核部产状较平缓，南翼往往伴有断层产状陡立，甚至倒转。

（3）北岭叠加向斜（$f4_3$）

位于北京西南房山区（图 32-4）。向斜的核心部位在连三顶—猫耳山—上寺岭一带。凤凰山及太平山分别为此“弯月”形大向斜的北、南两个扬起端部位，南窖—银水一带则为其向西凸出的一个分支。向斜由九龙山组、龙门组、窑坡组及南大岭组构成，向斜总体呈北东向展布，只在东北端转向近东西向。由于中段略向北西突出，平面上为月牙形。受房山岩体的拱托，向斜的南东翼产状较陡，倾角约 60°~70°，个别甚至倒转，北西翼较缓，倾角 20°~40°，轴面总体向南东倾斜，倾角约 60°。向斜具有宽大的槽部，东北端形态较简单，东南端构造形态较复杂，发育次一级或后期叠加形成的褶皱。核部由九龙山组组成，翼部出露下侏罗统至早古生代地层。剖面上呈西北翼平缓东南翼陡峻之“箕状”。向斜不整合叠覆盖在由古生界构成的印支期向斜构造之上，在东南端二者呈共轴不共面状态。其中比较著名的太平山向斜即位于此叠加向斜东南扬起端。太平山向斜轴向近东西，与山脊线大致吻合，枢纽呈波状起伏，总体表现为向东扬起。太平山向斜北翼较陡，倾角为 50°~80°，南翼较缓，倾角 30°~50°。局部地段向斜核心岩层近于直立，向斜两翼受到其他构造影响的地段，出现构造叠加和强化现象，形成小型的次级褶皱，如向斜北翼 194 高地南侧发育一系列轴面直立的小褶皱（图 32-5A）。向斜南翼刀背梁一带，出现一系列轴面北倾的斜歪甚至倒转褶皱（图 32-5B）。总体来说，太平山向斜无论是核部还是翼部，岩层产状都是由上向下迅速变陡，北翼甚至出现倒转的趋势，次级褶皱陡立、紧闭，岩层遭受挤压破碎较强，从横剖面上看具有类似于漏斗的形态。

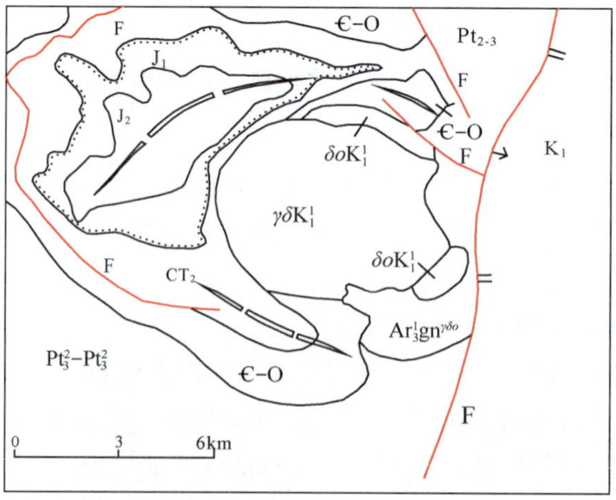

图 32-4　北岭叠加向斜区域地质简图

Fig. 32-4　The brief geological map of Beiling synform

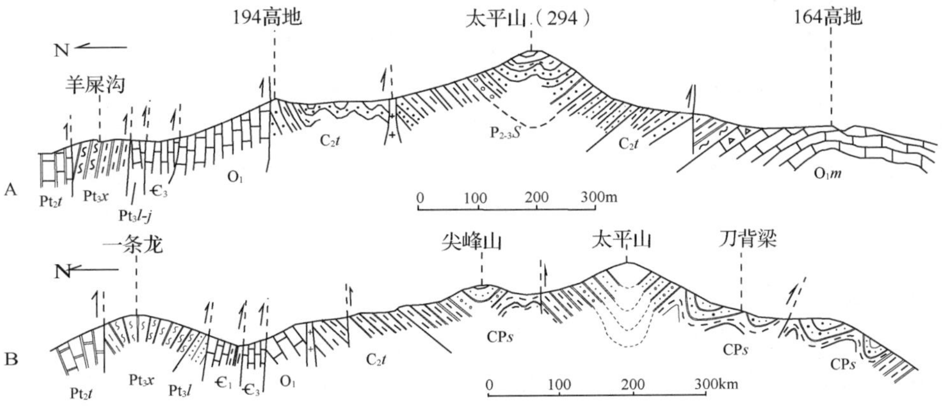

图 32-5 太平山一带信手地质剖面图

（据谭应佳，1986，修改）

Fig. 32-5 The hand-painting geological profile of Taipingshan area

A—194 高地-294 高地-164 高地信手地质剖面图；B——条龙-尖峰山-刀背梁信手地质剖面图

$P_{2-3}s$—二叠纪石盒子组砂岩夹板岩、含砾砂岩；CPs—山西组砂岩、板岩、含砾砂岩；

C_2t—太原组板岩角岩及砂岩；O_1m—马家沟组灰岩；ϵ_1、ϵ_3—寒武系灰岩；

Pt_3l+j—龙山组及景儿峪组大理岩、千枚岩；Pt_3l—龙山组石英砂岩；

Pt_3x—下马岭组片岩、千枚岩；Pt_2t—铁岭组大理岩

（4）坡子山向斜、大华山背料（$f4_4$）

出露于关上一镇罗营地区，与新太古代发育于结晶基底内的复式向斜带大致重叠。坡子山向斜在张家台一庙峪间，轴向北东 50°~60°，向南西倾伏，核部为高于庄组，翼部为常州沟组，东北段被程各庄断裂截切后向北错位，并使褶皱轴向略向东偏转，由常州沟组及串岭沟组组成。大华山背斜核部出露四合堂岩群变质岩，两翼由长城系组成，褶皱的形态较为宽缓，枢纽近于水平但已发生弯曲，轴面近于直立，并呈波状弯曲，轴向 50°~60°，与坡子山向斜彼此平行延伸，背斜发育区内断裂构造极为发育，尤其是北北东向程各庄断裂将背斜切割成三段，枢纽的左行位移总量可达 5km，至关上一带轴向转为东西，再往东则呈北西走向，形成略向北突出的弧形。

（5）红旗甸背斜（$f4_5$）

分布在延庆红旗甸一带，褶皱由长城系和蓟县系构成，核部出露太古宙变质岩。枢纽呈北东东向，轴面近于直立，翼部产状较缓，伴生构造不发育。背斜的东端被北北东向断裂切割，同时由于断裂的左行扭动使原本近东西向的枢纽偏转成北北东向。

（6）八达岭箱式背斜（$f4_6$）

八达岭箱式背斜从老峪沟往东，向北经王家元、磨盘山延至松树沟村北，全长约 35km，宽 8~20km。背斜核部由新太古代变质岩构成，两翼为长城系和蓟县系，东南翼有青白口系、下古生界和中上侏罗统分布（部分被第四系覆盖），北西翼倾向北西，倾角 20°~35°，南东翼倾向南东，倾角 40°~45°。翼部地层无论沿走向还是倾向，均呈平直的单斜构造，核部宽缓，长城系发育舒缓波状褶曲。背斜西南倾伏端位于王家元村南一带，倾向南西，倾角 10°。转折端圆滑开阔并发育有北东向次级褶皱。该期褶皱有长期发展史，初始形成于燕山早期，基本定形于燕山中期。燕山早期受沿河城-南口断裂的影响，断裂西北侧发生隆起褶皱变形，形成宽缓的背斜；燕山中期在北西-南东向挤压应力作用下，沿河城-南口断裂再次逆冲，背斜又一次隆褶，两侧火山岩系发生变形。到新生代，南口山前断裂表现为正断活动性质，南东盘下降，北西盘抬升，形成八达岭背斜持续隆起，形成现今的构造与地貌特征。

（7）其他

五里坡背斜、柏木井背斜、南口背斜及大地背斜实际上是复式背斜带上的次级背斜。背向斜均由中—新元古界和部分下古生界组成，无上古生界及早中生代地层。其岩层强度大，褶皱和缓，波长和

幅度较大，与西山地区前述同一尺度的复式向斜带中的褶皱在构造特征上形成鲜明的对比，后期构造破坏干扰严重，被新地层覆盖，需仔细鉴别。上述背斜两翼大致对称，倾角中等。

松各庄向斜、西智向斜及北京东部分布于古生界、中—新元古界及太古界中的北西向背向斜构造，褶皱形态宽缓，翼部受断裂影响岩层产状变陡，二十里长山一带可见地层倒转。由于缺乏直接的上覆地层证据，其时代尚难准确确定。目前仅根据构造特征、规模，以及为晚期构造叠加、切割关系等大致判定，特别是与西山地区的北西向褶皱构造对比，初步判断其可能为燕山早期或更早的时期形成。

2. 燕山中期褶皱——第五期褶皱（f5）（J_3 之后）

该期褶皱卷入的最新地层为上侏罗统土城子组，形成于北西-南东向挤压构造环境，空间上呈斜列形式斜跨于早期北东向褶皱之上，剖面上向斜呈斜歪宽大的槽状，背斜呈狭窄的脊状，轴面略向北西倒转，倾向东南，规模较大，延伸 10~15km，倒转翼逆冲断层发育。主要有燕羽山向斜、百花山向斜、髫髻山向斜、九龙山向斜、四海复向斜、千家店向斜、云蒙山背斜等。

（1）燕羽山向斜（$f5_1$）

位于延庆中部玉皇山—燕羽山一带。褶皱发育于髫髻山组中，为开阔向斜，核部产状近水平，轴向北东向，枢纽近水平，轴面近于直立，两翼近对称，倾角 20°~35°。向斜的两翼均被北东向断裂切割，两端则被北西向断裂破坏。南东翼被后期岩体侵入，北西翼多被第四系覆盖。

（2）百花山向斜（$f5_2$）

位于西山百花山一带。为一略向南西缓倾伏，轴面向南东陡倾的斜歪箱状向斜。北东端于斋堂西南略翘起，西南倾伏端为燕山晚期镇厂向斜截切掩覆。东南翼伴有马兰倒转脊状背斜及逆冲断层，产状稍陡，西北翼及向斜核部岩层产状较缓，核部由髫髻山组组成，翼部为九龙山组、龙门组、窑坡组及南大岭组。褶皱为侏罗山式，其轴向为北东向，近水平，两翼产状较缓。翼部的次级褶皱和纵向伴生断裂较为发育。该向斜受控于北东向的基底断裂活动，实际上是上叠式复合褶皱，髫髻山组不整合覆盖在早—中侏罗世地层之上。

（3）髫髻山向斜（$f5_3$）

位于北京西山髫髻山—妙峰山—阳坊一带，形态与百花山向斜类似，呈北东走向，轴面向南东陡倾，为一斜歪箱状向斜（图 32-6）。西北翼产状较缓，核部开阔平坦，东南翼伴有斜歪并倒转的脊状背斜及逆冲断层。隔此脊状背斜及逆冲断层，东侧有盆状的涧沟向斜相伴。主要由髫髻山组、土城子组及下伏的九龙山组、龙门组、窑坡组、南大岭组构成。向斜核部由髫髻山组组成，西北及东南翼与庙安岭向斜重叠部分有中下侏罗统出露。东北段髫髻山组则直接不整合于中侏罗统及古生界之上。其西南端具侏罗山式褶皱特征，永定河以东则为单斜构造，西南段的翼部伴生褶皱和顺层滑覆构造较为发育。西南与百花山向斜相衔接，两翼与青白口半穹隆和下苇甸穹隆相接，并被北东向黄土贵推覆断裂切割。受早期构造限制和晚期构造改造，东北和西南端褶皱轴方位略有偏转，总体上呈现出“S”形特征，具有同沉积褶皱特征（赵越，2002）。

图 32-6 髫髻山向斜构造剖面示意图

（据宋时雨，2014）

Fig. 32-6 The tectonic profile of Tiaojishan syncline

（4）九龙山向斜（$f5_4$）

位于门头沟区九龙山—香峪一带。褶皱由九龙山组、龙门组、窑坡组及南大岭组构成。其枢纽水平，轴面直立，轴向北东 65°，枢纽近水平，略向北东倾伏，两翼产状倾角约 30°，为侏罗山式褶皱。向斜不整合叠覆在下苇甸穹隆和谷积山背斜的翼部上，中部被北西向永定河隐伏断裂左行切割，造成两侧岩层的褶皱程度和形态略有不同。西南侧的向斜构造相对宽缓，东北侧相对狭窄，在核部发育次级背斜构造形式复式褶皱，具有同沉积褶皱特征（赵越，2002）。

（5）千家店向斜（$f5_5$）

出露于北京北部延庆石槽—千家店—花盆一带，自北至南可进一步划分为黄土梁、河口及上下虎叫 3 个次级向斜。核部由上侏罗统土城子组组成，走向北北东，枢纽为北北东向，近于水平，轴面倒向南东；南东翼产状较缓，北西翼陡，褶皱形态较为简单。南东翼土城子组直接不整合于中—新元古界之上，北西翼受断层控制，土城子组与中—新元古界间呈断层接触。近断层在东北茨顶及西南刘斌堡等地均有更次级背斜、向斜相伴。向斜东侧为红果寺背斜，主要由中—新元古界组成，北段斜跨于燕山早期大荒地背斜及柏木井背斜之上，核部受稍晚形成的断层切割，南段受晚期近南北向构造干扰破坏，构造形态保存不好，为一南东翼稍陡、北西冀倾角中等之不对称背斜。

（6）四海复向斜（$f5_6$）

位于北京北部延庆和怀柔的四海—汤河口一带。向斜主要由髫髻山组及土城子组构成，为内部构造较复杂、受多次构造重叠的复合构造。褶皱枢纽为北东向，略呈“S”形扭曲，且在汤河口—宝山寺一带呈明显的复式褶皱。由西南至东北，褶皱走向由北东东—北东—北北东—近南北，再转至北北东—北东—北东东。南北段及中段均有更次一级的背斜、向斜构造。该向斜的东西两侧均被北北东向断裂切割，北端被东西向断裂破坏，南端被岩体侵位，汤河口以南一带组成向斜的火山-沉积岩系被自东而来的元古宙飞来峰压盖。

（7）云蒙山背斜（$f5_7$）

出露于怀柔北部，呈一规模较大略向北倾伏的短轴背斜。核部由燕山晚期云蒙山花岗岩体所占据，翼部为中元古界及太古宙变质岩系。南端翘起并受北东向断裂限制截切。受稍晚期北北东—近南北向断裂穿切，具有同侵入褶皱性质。目前主要根据同褶皱期侵入的云蒙山花岗岩体的同位素年龄（144Ma）及其构造特征等，确定为燕山中期。

另外在西灰岭、凤驼梁、十三陵水库、怀柔—河防口，东庄禾—新城子等地为中上侏罗统分布区，其原始形成时期可能也为向斜或复式向斜构造，但由于后期构造干扰破坏使得现今构造形态保存不完整或呈单斜构造形态。

3. 燕山晚期褶皱——第六期褶皱（f6）

该期褶皱形成于早白垩世中期末或中晚期的北西西-南东东向或北西-南东向挤压构造环境。此期褶皱构造轴向北北东—近南北向，轴向近直立，以形态宽大、舒缓的鼻状（半弯窿状）构造盆地为特征，有时也在早期褶皱构造上镶嵌呈近南北向的花边，或在断裂构造旁侧呈牵引褶皱。卷入的最上部地层为下白垩统夏庄组，自西南至东北有镇厂-杜家庄向斜、下苇甸穹窿、坨里背斜、夏庄向斜、云蒙山倾伏短轴背斜等。

（1）镇厂-杜家庄向斜（$f6_1$）

出露于北京西山最西部，市区所见部分仅为向斜北段翘起端，由下白垩统张家口组构成。东冀产状倾向北西西，倾角 30°～45°，局部可达 70°～80°，并有石英斑岩小岩体侵入，核部岩层产状平缓，西至西北翼倾角较小，但被相继形成的断裂切割、破坏，为一轴面略向东陡倾的不对称向斜。

（2）坨里向斜（$f6_2$）

位于北京西南房山区东北部，主要由白垩纪早期的碎屑岩组成。其枢纽呈近南北向，近水平略向西南扬起，轴面直立略向东陡倾，两翼产状舒缓，倾角一般<20°。褶皱的主体被第四系和古近系、新近系覆盖，西翼和北端被黄庄-高丽营断裂切割。向斜西北部的断裂附近，由于受黄庄—高丽营断裂的持续活动的影响，形成了反牵引次级背斜构造，平行断裂的长轴约 5km，枢纽与断裂同步弯曲，即坨里反牵引向斜构造。

(3) 夏庄向斜 ($f6_3$)

据目前研究，向斜南端略翘起封闭，东北为古近系长辛店组不整合掩覆，轴面近南北向，呈大致对称直立开阔的盆状向斜。核部由夏庄组构成，西南翼由石盒子组（原卢尚坟组）组成，并向坨里背斜缓慢过渡。

另外，在马兰—斋堂、马跑泉—老峪沟、王家园等地有一些规模较大的南北向褶皱。在金鸡台、潭柘寺等地，于早期北东向庙安岭向斜、九龙山向斜南翼有一些花边的南北向小褶皱。它们显然不是局部的次级构造，而有一定的区域性，是燕山晚期褶皱构造在此一带叠加的结果。另外，密云东北程各庄断裂北段旁侧，伴有一系列南北向的小褶皱，它们可能是与晚燕山期近南北向断裂扭动有关。

（三）喜马拉雅期断陷盆地

喜马拉雅期本区主要表现为改变早期断裂构造的性质，垂直北东向构造发生伸展断陷，形成断陷盆地。

1. 北京东南巨型断陷盆地

东南巨型断陷盆地属华北盆地的一部分，地处华北平原西北隅。自古近纪始新世中晚期开始裂陷伸展，逐渐形成北东向的两“隆”一“凹”格局（图 32-7），东南为大兴断隆，西北侧为京西北隆起，其间称北京断陷。始新世至渐新世时期沉积局限在北北东向地堑式断陷盆地内，这一断陷盆地沿走向方向逐渐发生掀斜，因此始新世长辛店组的分布偏向盆地的西南端，而其上的前门组则逐渐向北东方向超覆古近纪断陷堆积时受同生断裂控制较为明显，盆地西北侧断裂活动性强、沉降幅度大，空间上略成箕状。新近纪时构造性质转化为整体坳陷，沉积范围超越地堑的控制，向四周扩展，在空间上显示出顺义、丰台、琉璃河-涿州 3 个斜列的沉降中心，其中丰台凹陷沉降幅度较大，可达 1000 余米。第四纪时盆地发展渐趋稳定，在南口—顺义、怀柔—东田阳、宰相庄、燕落—石匣、平谷等地略形成南北成串、东西成排的凹陷中心。

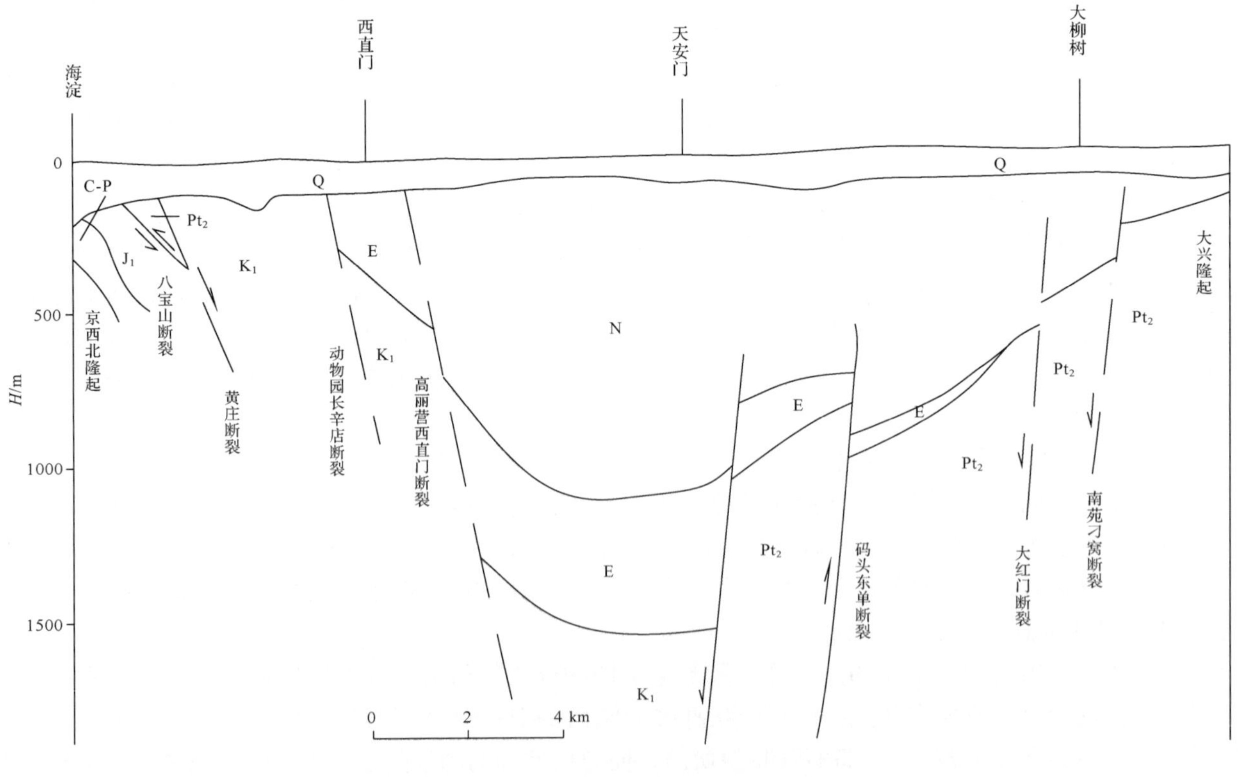

图 32-7 北京断陷海淀-大柳树构造剖面

Fig. 32-7 The Haidian-Daliugou tectonic profile of Beijing

2. 延庆断陷盆地

属喜马拉雅旋回晚期山间断陷盆地，其西南与河北省阳原盆地断续相连。晚古近纪开始断陷伸展，现今仍在继续发展，呈北东-南西向之“槽”状，内部为断裂构造进一步复杂化，形成规模更小、沉降幅度各异、长轴为北北东向的菱形凹凸。沉积中心或深凹部位一般靠近西北侧，南侧逐渐翘起。第四纪以来，整个盆地沉降幅度和速度逐渐协调一致，进入统一坳陷发展时期。

第二节 断裂构造

本区断裂构造较发育，不同方向、不同期次、不同类型的断裂交错切割，形成较为复杂的断裂构造格局，将本区划分为一系列大小不等、形状各异的断块体。按照断裂构造的规模、性质、切割深度、产状及其在区域地质构造发展中的重要作用，可将本区断裂构造划分为重要区域断裂和一般断裂两级。

一、重要区域断裂

重要区域断裂主要是指对区域沉积作用、沉积建造、各种岩浆活动等均有明显的控制作用，且规模大、切割深的断裂带。沿深断裂带有较强烈的动力变质现象，形成动力变质带，在区域性地球物理场特征上常有较显著的反映。这些断裂均经历了长期的发展演化过程，具有多期活动和性质多变的特点，由单一或成束、成带的断裂形式表露，或以岩浆岩的连续分布间接显示，具有重要的导岩、导矿和控岩、控矿作用，是各类构造单元的分界线，控制断裂两侧地体的发展演化等。根据上述条件和原则，初步确定北京地区的重要区域断裂主要有9条。其分布见（图32-8）。

（一）赤城-长哨营-古北口-承德深断裂（F6）

呈近东西向展布于本区北部，主体位于河北省境内，长度>800km，本区内断续出露长度约30km。地表由多条断裂、挤压动力变质带和线状褶皱组成。区内展布宽度7~8km。对该深断裂的研究较早，不同的研究者从不同的实际资料出发已先后做过很多工作。此断裂自元古宙以来就已经存在，特别是中元古代断裂的南侧燕山裂陷槽开始接受盖层沉积以来，堆积了近万米的中—新元古代地层。沿深断裂带前长城纪时期有大量的基性—超基性岩体侵入。中元古代时期则有钾质花岗岩、斜长岩侵入。晚古生代至早中生代的华力西—印支期有中酸性—酸性侵入岩，局部可能有基性、超基性岩的侵入活动。中生代燕山期以来，活动性加剧，沿断裂带除有下中侏罗统地层分布外，燕山期的花岗岩、花岗闪长岩和碱性岩亦广泛发育。河北省境内沿断裂有喜马拉雅期的玄武岩喷发。

（1）构造特征

发育于本区的主要为该深断裂的东段，即三块石以东地段。由于中生代火山-沉积盆地和穹隆核部上覆中元古代地层由南向北的滑覆型地块（飞来峰）的压覆，很多早期东西向深断裂形迹被掩盖，仅能从残露的构造形迹分析其构造特征。

1）深断裂带东段总体呈东西向展布，由中元古代侵入岩体和伴生的多期次断裂组成的断裂岩浆岩带，东西向延伸长度42km，向东延出本区。

2）以新地斜长岩、兰营石英正长岩、偏桥含石英二长岩和古洞沟钾长花岗岩为代表的中元古代侵入岩体呈东西向狭长带状岩脊，侵入于太古宙变质岩中。在新地斜长岩体北侧、古洞沟钾长花岗岩体南侧变质岩中，发育有近东西向绿片岩相韧性剪切带，应属于深断裂带早期活动形成的韧性剪切带构造岩。

3）中元古代侵入岩体内部和边部，发育一系列近东西向平行排列的脆性断层，大多显示为挤压逆冲性质，为岩体侵入后深断裂再次活动的表现，并且有东西向展布的早侏罗世辉长岩体，侵入到中元古代岩体中。

4）密怀穹隆北翼，汤河口—番字牌一带，发育有近东西向断层束，由4~8条断层组成宽3~5km的断裂带。它们彼此近于平行，倾向北或北北东，其倾角自南向北由50°到80°逐渐变陡，向北延深可能归并到深断裂的主断裂带上。空间上呈叠瓦状，其内可见到变质岩长条状岩片逆冲于盖层之上。沿断裂带发育挤压片理、挤压透镜体，并伴有绿片岩相变质作用，断裂性质主要为逆冲性质，部分断裂显示先正后逆的活动过程，形成于印支期，对密怀穹隆的形成有控制作用。

（2）航磁异常特征

在航磁 ΔT 等值线平面图上，断裂带东段表现为近东西向正磁异常带，其南侧总体为低缓正、负磁场区。在航磁 ΔT 化极上延 1000m 等值线平面图上，断裂带显示更加明显。北侧为正磁异常带，南侧为负磁异常带，且正、负磁场变异带向北偏移 1km，也反映了深断裂带总体向北倾斜的特点。

（3）断裂形成、发展历史过程分析

在西段河北境内中元古代侵入杂岩体内部岩体接触部位存在糜棱岩化变质岩。东段中元古代侵入岩边侧和断裂带上，也都发现断续分布的发育于太古宙变质岩中的绿片岩相韧性剪切带（局部有角闪岩相剪切带）。这些遗存的剪切带形迹是深断裂早期活动的产物，形成于中元古代岩浆侵入之前，时间限定于中元古代早期或更早的新太古代—古元古代。推断太古宙末至古元古代为断裂的发生期（《中国区域地质志·河北志》，2017）。

中元古代时期为断裂的强烈活动期。主要表现在两个方面：第一，对中元古代地层沉积有明显的控制作用。该深断裂与北侧崇礼-赤城-丰宁深断裂和南侧密云-墙子路断裂，是控制燕山裂陷槽发展的北部深断裂。受南北向拉张应力场控制，由北而南发生阶梯状下落。崇礼-古北口断裂从中元古代开始，南盘大幅度坳陷，北盘相对隆起，区内在海侵初期虽遭超覆，但也仅具“陆棚”性质。沉积厚度与南侧相比相差悬殊。东侧常州沟组至大红峪组累计厚度，在断裂以北的隆化地区仅 86m，断裂南侧的承德地区，厚度突变为 750m。陈晋镳（1980）认为断裂活动具有裂谷特征。第二，断裂控制了中元古代岩浆侵入活动。以小张家口辉石岩和新地斜长岩为代表的幔源深源超基性—基性岩浆，沿深断裂侵位，于中段和东段形成岩脊状岩体。以二长岩、正长岩和富钾花岗岩为代表的硅铝质重熔岩浆，沿深断裂上升侵位，在与崇礼-赤城-丰宁深断裂小角度交汇部位，形成近东西向大型复式岩基，在断裂东段也形成岩基状岩体。本时期断裂活动持续到晚古生代。

古生代，断裂依然呈现南拗北隆的差异活动，阻隔了南侧陆表海的北进，至今在断裂北侧未发现古生代地层沉积。断裂带以北到崇礼-赤城-丰宁深断裂间，晚古生代变质辉长岩体和二长花岗岩体发育构成近东西向岩浆岩带，但岩体分布向南并未越过崇礼-古北口深断裂带，显示深断裂对其有控制作用。

中生代以来断裂新生性、继承性活动剧烈而频繁，主要表现在：

1）印支期断裂强烈活动，在南北向挤压收缩应力场控制下，北盘向南强烈逆冲，形成大规模的压性逆断层带并伴有绿片岩相动力变质作用。而南侧的密云-墙子路断裂，则表现为由南向北的逆冲活动。两断裂之间的地块发生褶皱变形，形成近东西向展布的背斜和向斜。在白河堡水库以东，总体出现正向隆起（如密怀穹隆），白河堡水库以西总体发生负向坳陷。

2）燕山运动早期，在印支运动构造背景下，曾发生过近南北向拉张作用，受其影响，沿深断裂带出现了北升南降的裂陷活动，在断裂中段表现明显。首先，南翼的青白口系、寒武系地层呈断块或断片滑落于裂陷槽内。之后，发生了南大岭组基性玄武岩喷溢和窑坡组含煤陆相碎屑岩的堆积，形成了沿断裂带南侧分布的早期后城盆地。

3）燕山运动中期，南大岭组和窑坡组堆积之后，在北北西-南南东向挤压应力作用下，控制坳陷盆地北界的东西向断裂再次活动，北盘向南东斜冲挤压，形成了强烈片理化带、构造岩片和韧、脆性兼有的动力变质岩带。

4）燕山运动中晚期，早阶段在北西-南东向伸展体制控制下，发生了北东—北北东向张性裂陷，导致了土城子组、张家口组火山-沉积盆地的形成。晚阶段在北西-南东向挤压收缩机制下，断裂再次活动，显示张扭性正断特征。断裂带南盘地块在北西-南东向挤压力作用下，以深断裂带为边界，发生北西-南东向挤压变形，形成北东—北北东向褶皱和逆断层，其构造形迹一般不越过近东西向深断裂带，或进入东西向深断裂带即减弱消失。这种构造活动的结果，造成深断裂带南北两盘形成左行或右行的水平扭动效应。

（二）南口-平谷-山海关深断裂

呈东西向展布于燕山南麓。地貌上作为现代山区和平原分界。沿蔚县—南口—密云—兴隆—喜峰口—青龙一带呈近东西向蛇曲状展布，横贯北京，两端均延入河北。从露头特征分析，为一较复杂的强变形带，具有长期多期活动性，倾向与性质多变的特点，发育多期构造岩类。该断裂形成于吕梁晚期（古元古代晚期）较早阶段近南北向挤压构造环境，以中浅层次脆韧性变形及挤压断裂活动为特征。该期形成的构造形迹与构造岩在后期多期构造叠加改造中大部分地段被破坏或被覆盖，仅在河北省青龙县八道河一带有保存，构造线呈北东东向展布，构造面理以向北北西倾斜为主，局部向南南东倾斜，倾角80°；发育于太古宙变质地质体之中。中元古代—中泥盆世（晋宁期—华力西早期）在密云以东区段有活动显示，以伸展拉张韧性变形与正断层活动为特征，为近南北向拉张构造环境，构造线呈近东西向展布，构造面理向北及向南倾斜，倾角70°~90°。在密云一带断裂带及附近有中元古代非造山型岩体、岩脉分布，以南地区有长城纪大红峪期非造山型偏碱性粗面岩分布，显示了该断裂带对岩浆活动有重要控制作用。同时对中元古代发育的裂谷及长城系沉积有一定的控制作用。中生代时期，断裂北侧也有较明显的岩浆活动。上新世—全新世早期，在南口一带有正断层活动显示，切割了上更新统马兰组，被全新世沉积物覆盖，构造线呈北东向展布，构造面理向南东倾斜，倾角60°，沿断裂带与其他断裂交汇复合部位，时有大小不等的地震发生。

重力图上为东西向带状阶梯异常，磁场上为不同异常区的分界。切割深度达康拉德界面和莫霍界面，反映具岩石圈断裂性质。

（三）紫荆关-大海坨深断裂

此深断裂出露于本区西部边界附近，主体部分在河北省境内，亦称之为太行山深断裂西带，是大兴安岭-太行山-武陵山构造带中段太行山地区的主体构造，总体走向20°~30°，断裂带在卫星遥感影像上表现为鲜明的北北东向线性构造带。断裂带内构造岩主要是碎裂化岩石（包括构造透镜体和构造角砾岩）、初碎裂岩和碎裂岩，显微变形机制主要有破裂作用、摩擦滑动和压溶作用。《中国区域地质志·河北志》（2011）称之为上黄旗-灵山区域断裂。该断裂分为南北两支，其南支称之为紫荆关-灵山区域断裂，总长约有580km。带内发育多期构造岩，为一较复杂的变形带，具有长期多期活动性、倾向与性质多变的特点。其中紫荆关-大海坨断裂便是其中的南支的一部分。这里所指的紫荆关-大海沱深断裂是燕山旋回中晚期强烈活动并基本定型的断裂，延伸自紫荆关—双塘涧—大海坨，呈北北东向25°~30°，总体向东南倾斜，倾角50°~70°。对此深断裂带的存在、特征、延伸方向及在地质构造发展演化中的作用，不同研究者尚有不同认识。

鞠紫云（1982）认为紫荆关断裂在中元古代时期已开始活动，对中—新元古界的沉积分布起控制作用。断裂主要表现为断裂岩浆活动带，沿断裂带有极发育的中酸性岩体侵入，大海坨花岗岩体与大河南正长岩体侵入明显受其控制。

陈桂华等（2003）在断裂变形带内发育两期典型的共轭剪节理，但区域分布各有特点，反映出断裂变形的空间演化。由两期共轭剪节理恢复了两期古构造应力场：中侏罗世—早白垩世的北西-南东向近水平挤压，晚白垩世—始新世的北东-南西向近水平挤压。而一组与断裂近平行的非共轭性密集剪节理指示始新世后北西-南东向拉张。在岩体内，岩石矿物由位错滑移而产生一定程度塑性变形，出现波状消光、机械双晶等显微构造现象。构造岩中胶结物和微裂隙充填脉反映了岩石碎裂变形的3次叠加。结合区域资料分析，紫荆关断裂带自古生代末开始活化以来有三次主要构造活动：第一次是中侏罗世—早白垩世，在北西-南东向近水平挤压作用下，断裂左行压扭性活动，控制太行山地区岩浆岩带及与其相关的接触交代型矿床的形成。其后的继承性活动在断裂带内形成碎裂程度最强达到碎裂岩的碎裂岩系列构造岩，变形带两侧发育两组典型的共轭剪节理；第二次在晚白垩世—始新世，断裂在北东-南西向近水平挤压作用下右行张扭性活动，断裂带两主断裂在斜列重叠的端部由于局部应力集中进一步扩展形成节理带，而断裂在岩体区也发育两组典型的共轭剪节理，断裂内形成初碎裂岩为主的构造岩；第三次是喜马拉雅期以西侧抬升、东侧下降的强烈升降为主的活动，在基岩中，主断层两侧围岩出现微破裂。

地球物理场上也为不同磁异常区的分界及重力异常的梯度带，是本区北北东向地壳厚度变异带的最西边界，在康拉德界面和莫霍界面上均有反映，切割深度已明显达到上地幔。

新生代以来，沿断裂带不同地段有较明显的活动性，对新生界沉积、地震活动都有一定的控制作用。

（四）涿县-丰台-怀柔-白马关深断裂

断裂带除东北段地表有北北东向石城-冯家峪断裂破碎带出露外，其余地段均被新生代沉积物所覆盖。石城-冯家峪断裂带地表地质特征不少专家先后做过较多的研究（王嘉荫等，1973；杨主恩等，1981；史兰斌等，1984）。怀柔-丰台段隐伏于平原之下者称高丽营断裂。南段经丰台、良乡、涿县出本区，断裂带继续沿太行山东侧山前延伸，为太行山与平原之分界。区内断裂长度>400km，与前述紫荆关-大海坨断裂带平行延伸，走向北东 20°~30°，总体向东南倾斜，倾角较陡。对本区燕山中晚期火山喷发及岩浆侵入活动有一定的控制作用，特别是张家口期的酸性火山喷发基本被限制于断裂带之西侧，而新近纪的顺义凹陷、丰台凹陷和琉璃河-涿县凹陷，则主要展布于断裂带东侧。北端石城-冯家峪断裂带为压剪性逆掩断层，并具多期活动特点，西侧有云蒙山杂岩体侵入，现今显示正断层性质，并有地震事件发生的反映。往南在怀柔宰相庄一带有花岗岩体侵入。

该深断裂在重力场、磁场及人工地震等地球物理资料上均有反映，构成各种地球物理场的明显界面。康拉德界面和莫霍界面均呈明显的梯度变异带，是本区北北东向地壳厚度变异带的最东部边界。

深断裂活动的强盛时期为燕山中晚期，表现为压剪性逆时针扭动，东南盘向北西逆冲，新生代以来，断裂表现为张剪性，顺时针扭动正断。中、南段平原下与其他断裂交会处，常伴有玄武岩的喷发。

（五）黑峪口-居庸关-良乡西深断裂

断裂带呈近南北向，展布于本区延庆—良乡西一线。延伸>100km，宽度约 6~8km。北部山区可见南北向断裂成束、成带分布，如营门-吴家沟断裂、延庆-样田断裂、西二道沟-黑峪口断裂等，沿断裂带燕山晚期酸性熔岩、火山碎屑沉积岩发育，厚度较大。往南与八达岭岩浆岩带交会部位有各类酸性、偏碱性岩体及各类超浅成（或潜火山）岩体的侵入。

门头沟坳褶区，有近南北向的褶皱、短轴背斜或规模相对较小的穹隆，空间上相继呈带状集中分布于一定范围内。房山—良乡间，西侧有南观断裂，东侧良乡附近也有近南北向的断裂反映，它们共同控制着燕山晚期及新生代构造盆地的发展。

区域磁场通过此深断裂带，其两侧磁异常区的宽度、强度均发生显著变化。重力场沿此深断裂方向北北东向的梯度带发生较显著的南北向扭曲。康拉德界面、莫霍界面等值线也发生类似的扭曲现象。人工地震剖面，包括天然地震转换波资料均反映出断裂的存在，切割深度已达上地幔（部学忠等，1979，1980；腾吉文等，1979）。断裂活动历史较复杂，中—新元古代时期可能已开始活动，沉积受其一定程度控制，断裂力学性质可能为张剪性正断层，东盘阶梯状下掉。燕山晚期至喜马拉雅晚期断裂力学性质由压剪性演变为张性，东盘由递冲转为正断层。

（六）丰宁-怀柔-采育深断裂

此深断裂的存在及其性质，近年来逐渐为人们所认识（张之一，1979），往北已进入河北省境内，长度达 100km 以上，宽度 7~8km。地表以构造岩浆活动带、断裂破碎带、结晶基底埋深突变带、新生代断陷等呈线状分布为特征。怀柔北部为转山子花岗岩岩体侵入，燕山晚期沿此方向在岩体中仍有断裂发生。四海向斜与云蒙山倾伏背斜间，结晶基底落差截然，东侧“X”形交叉断裂显著发育。云蒙山花岗杂岩体与长园花岗闪长岩体长轴方位以此为界表现出重大差异。怀柔一带中生代火山沉积盆地东、西边界呈南北向伸展。

往南新生代第四纪的怀柔凹陷、顺义凹陷、通县-牛堡屯凹陷，在南北方向上成串分布，沉积厚度为100~600m。反映了第四纪时期该深断裂也有一定的活动性。

地球物理场上，该深断裂以各类异常带的明显扭曲为特征，指示断裂带性质以剪切为主。人工地震资料及天然地震转换波资料均反映此深断裂的存在，切割深度达上地幔。

该断裂活动时期较早，具多期活动的特点。中—新元古代时期该断裂可能为张剪性，对沉积有一定影响；中生代为压剪性，对火山喷发及岩浆侵入活动有控制作用，新生代表现为剪性或张剪性，断裂在穿切其他构造时常发生走向上的错位和强度的减弱。

（七）黄庄-高丽营断裂带

黄庄-高丽营大型断裂带呈波状弯曲展布于北京西山山麓和山前平原地带（徐杰等，1992）。南起涞水，向北经石楼、辛开口、晓幼营、大灰厂、辛庄西、芦井、过永定河，再经黄庄、洼里、北七家，继续北延经高丽营、怀柔庙城、大屯至密云西田各庄一带，全长130余千米。总体走向北北东，倾向南东，是控制新生代古近纪—新近纪北京凹陷西界主要断裂。该断裂在第八篇第四纪地质中有更为详细的介绍。

（八）沙厂-墙子路大断裂（F36）

1. 基本特征

位于赤城-长哨营-古北口深断裂之南约40km的沙厂—墙子路一线，两者平行排列，走向东西，往东延入河北省青龙一带，区域上称为“密云-喜峰口大断裂”。全长>300km，宽约6km。往西在密云、河防口一带被新生界和侏罗系掩覆，青石岭断裂以西长园岩体轴向呈东西向，沿河城-南口-琉璃庙大断裂在黄花城一带受该东西向断裂的影响发生扭曲和偏转，西二道河-康庄东西向断裂再度出现等，均反映沙厂-墙子路大断裂向西延伸的踪迹。1：20万怀柔区幅区调和怀来-延庆地区1：5万区调工作认为该断裂带向西已延至张山营—麻峪口一线，再向西被中生代北北东向梓田火山断陷盆地（上黄旗-乌龙沟深断裂带）、新生代北东向延怀断陷盆地和北东向八达岭构造岩浆岩带截切或掩盖，早期构造形迹已难寻觅。东段有中元古代沙厂环斑花岗岩体的侵入，其切割面在康氏面附近。中段为燕山期长园岩体的侵入，西二道河一带东西向断裂对侏罗系的分布有限制作用，康庄东西向断裂对延庆新生代盆地边界起控制作用。以上特征指示了断裂的存在和活动具多期次的特点。断裂在区域重力场、磁场上有一定的显示，但在康拉德界面和莫霍面上的反映不及南北侧深断裂显著，总体向南倾斜，倾角40°~50°，为四级构造单元分界线。在化极上延20km的航磁图上，断裂南侧为正磁场区，以北为负磁场区，界线截然，推测其切割深度在康氏面附近。

印支期断裂活动强烈，在断裂西段和东段，沿深断裂带发育多条近东西向具规模的逆断层，断面倾向南或北，但多数倾向南，倾角50°~80°，基本反映了断裂带总体向南高角度倾斜的产状特征。断裂南北两侧（尤其南侧）次级断裂构造控制了一些金、银矿床、矿点的分布。

2. 活动特征及发展历史分析

（1）中元古代时期

该断裂带初始活动可能发生于吕梁运动期末，即古元古代末期。中元古代初期断裂开始强烈活动，受区域南北向伸展机制控制，与兴蒙造山带南缘东西向断裂同步，由北向南阶梯状正断活动，成为燕山裂陷槽沉降中心北缘的一条深断裂。它控制了长城纪大红峪期钾质基性岩浆和碱质岩浆的喷发和沙厂超单元岩体的侵位，对中元古代沉积地层有明显的控制作用。至蓟县纪末期伴随芹峪上升燕山裂陷槽的封闭而暂停活动。

（2）印支运动时期

早期活动受南北向拉张应力场控制，断裂带依然显示正断活动特征，与崇礼-古北口深断裂一起控制了所夹断块差异性升降活动，出现了局部隆升（密云-怀柔地区）和坳陷（宣龙地区），沿盖层和基底间发生南北向滑脱，形成剥离断层。

晚期活动在南北向收缩挤压应力作用下，断裂带发生强烈挤压逆冲活动，南盘老地层（以变质岩为主）冲覆于北盘新地层（以中—新元古界为主）之上，形成多条近东西向排布的、长度达10~20km的压性逆断层。同时，受本次活动影响，地层褶皱变形，沿断带及两侧，形成一系列近东西向褶皱。在西段发育有水口山向斜、营门背斜；中东段有沙峪向斜、大黄山向斜和沙厂背斜、墙子路背斜等。

值得指出的是，这次构造活动对赤城-古北口深断裂以南，本断裂带以北的宽约42km的断块构造变形影响极大。在早期差异性升降活动基础上进一步褶皱变形，形成一系列近东西向褶皱，密云-怀柔地区进一步隆起成为密怀穹隆，宣龙地区进一步坳陷形成宣龙复向斜。在以后的地质发展时期，坳陷地带控制了中生代侏罗纪地层堆积，隆起区遭受剥蚀。

（3）燕山运动时期

断裂带活动依然明显，但强度不如印支期，主要表现在以下几个方面：

1）断裂带中西段再次活动，于中生代火山岩地区和侵入岩体中发育小规模近东西向压扭性逆断层，以及与岩体侵入同生的逆冲式韧性断层（龙潭韧性断层）。

2）对燕山期岩浆侵入活动控制明显。晚侏罗世沙峪复式岩体、分水岭岩体和早白垩世对臼峪岩体等沿断裂带分布，其长轴呈近东西向伸展；早白垩世大海陀复式岩基侵位在该断裂带与上黄旗-乌龙沟深断裂交汇部位；早白垩世云蒙山复式岩体向南的延伸终止于断裂带上。

3）限制了燕山早、中期北东向褶皱向北东方向的延伸（如八达岭背斜、密云茶棚—葫芦峪—三百山一带的褶皱），并使其轴向向东偏转及背斜发生倾伏，同时也显示了断裂带右行扭动的活动特征。

（4）喜马拉雅期

断裂活动目前还不很清楚，延庆新生代断陷盆地斜跨在本断裂带上，但其多被北东向断裂控制明显；密云以西至河防口一线山区与平原分界线是否标志断裂带新生代再次活动，值得探讨。

（九）沿河城-南口-琉璃庙断裂带（F1）

位于门头沟区双塘涧—沿河城—溜石港—南口—琉璃庙一线，在构造上位于北东向太行山构造带与东西向燕山构造带的交接部位，地质构造环境较为复杂。该断裂带在本区内延伸>140km，自西南向东北由沿河城断裂、南口山前断裂、崎峰茶-琉璃庙断裂三段相接而成。破碎带较宽且组成较为复杂，包括断层角砾岩、碎裂岩、断层泥等。两盘发育平行的劈理带和小断裂。局部发育拖曳褶皱。断裂落差巨大，断裂切割了梁家庄向斜、东岭台向斜、青白口穹隆、居庸关背斜的南东翼及云蒙山隆起的北西（翼）部。

沿河城断裂分布于沿河城—白羊城一带，构造上位于北东向燕家台复式背斜与髫髻山-百花山复式向斜的过渡带，其走向总体呈北东向，个别地段分枝呈近东西向，断面倾向南东，倾角60°，总体显示为南东盘下降的正断性质，但是雁翅阳坊幅区调报告认为该断裂带初期活动为逆冲性质，而这种矛盾是由于断裂后期活动效应引起的。在溜石港一带最为明显，其岩石破碎后隐约保留有原生沉积构造，但已被多组节理切割成1~5cm菱形碎块，破碎带边部，岩石趋于完整，中部则有多条小断层、碎裂岩及断层泥组成。断层上盘主要为中元古界高于庄组白云岩，下盘可见上更新统黄色砂土夹碎石块及全新统残坡积物，断层破碎带宽2m，可见清晰的断层滑动面，滑动面向上未切穿上更新统地层，由此推断断裂最晚活动时期为晚更新世晚期（刘广平，2001）。

南口山前断裂断层走向北东，倾向南东，倾角45°~60°，断带宽度20~100m，形成破碎带最宽可达300m（图32-9）。带内岩石强烈破碎，形成碎裂岩和断层角砾岩带，局部见糜棱岩和断层泥。断裂下盘较为完整，上盘较破碎（图32-10）。东北部上口一带由于受岩浆活动和其他构造干扰，以及早期东西向构造的限制，断裂发生了较大的偏转和扭曲，地表出露时隐时现。断层泥中测得TL年龄为（3.598±0.306）$\times10^4$a和（1.341±0.114）$\times10^4$a（刘广平等，2001），最晚活动年代为晚更新世晚期至全新世早期。

崎峰茶-琉璃庙断裂位于密怀隆起西部，南起永安堡，向北经崎峰茶，至琉璃庙以北的干沟门，全长约37km，为发育于太古宙变质岩内的古断裂带。总体走向北东，并略向东南突出。得田沟以南走向北东58°，以北转向北东20°~25°，倾向北西，倾角>45°。在地表以破碎带形式出现。断带宽度

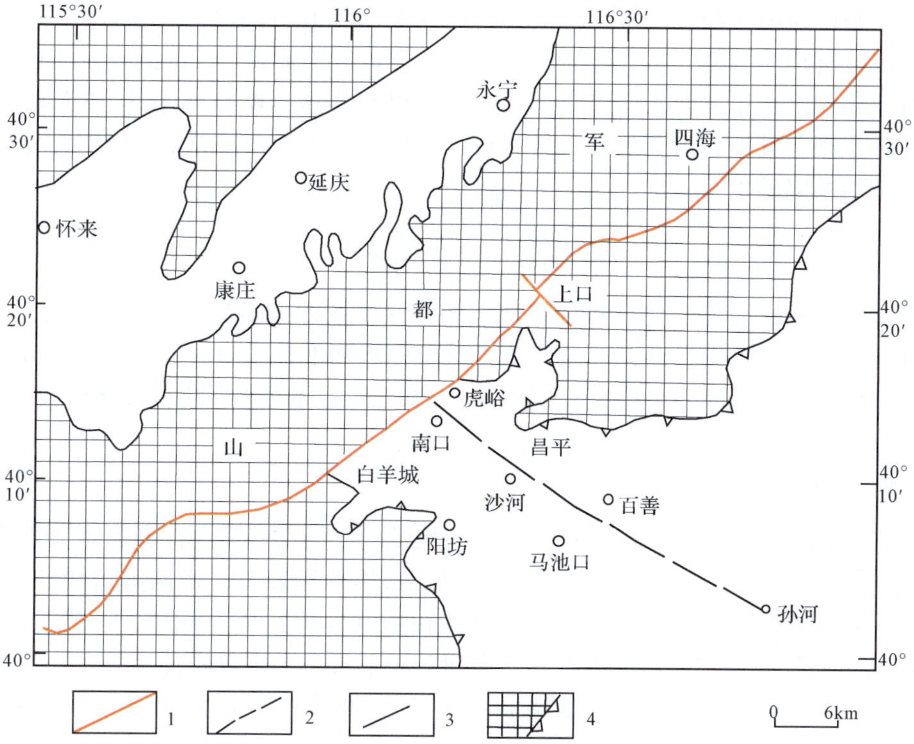

图 32-8 北京南口山前断层的几何展布

（据高文学等，1993；刘广平，2001）

Fig. 32-8 Geometric distribution of the Nankou Mount frontal fault in Beijing

1—正断层；2—隐伏断层；3—断层分段边界；4—盆地与基岩山区边界

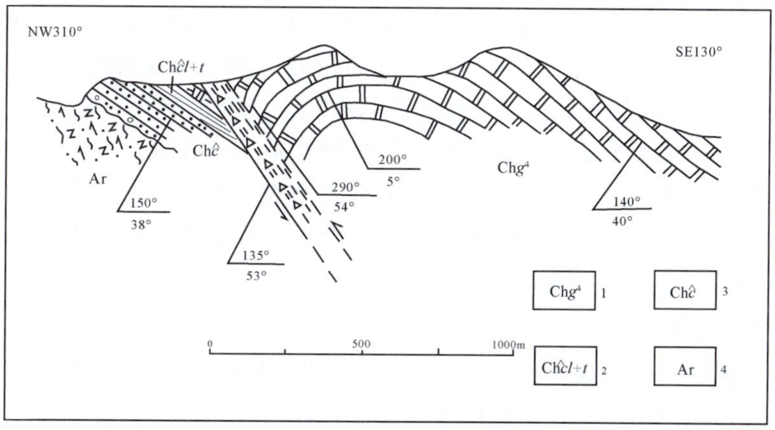

图 32-9 昌平虎峪村东北 1. 5km 处南口山前断裂构造剖面图

（据昌平、小汤山幅 1：5 万区调报告）

Fig. 32-9 Profile of the Nankou piedmont fault structure at 1. 5km northeast of Huyu Village in Changping area

1—高于庄组四段；2—串岭沟组、团山子组；3—常州沟组；4—太古宇

箭头指向为断裂早期活动两盘运动方向，上盘新地层与下盘老地层接触显示后期正断活动特征

150~450m。

该断裂通过部位在地壳深处为扭转的北东向幔坡带，并深切莫霍面，有较复杂的相对运动和变形过程。该断裂经历了多期次的活动，最早形成于燕山早期，在北北西-南南东方向挤压应力作用下，为逆断裂；燕山中期在北西-南东向挤压应力作用下，断裂东盘向西盘逆冲；燕山晚期在区域性北西西-南东东向挤压应力作用下，断裂表现为伸展性活动，东南盘正断下降；新生代以来，断裂的部分地段

仍表现出活动性，呈明显的正断层性质。整个大断裂在西山隆升、四海向斜盆地的形成与发展，以及燕山期的火山喷发—沉积等方面起了重要的控制作用，且该断裂限制或切割了北西向断裂及印支期近东西向断裂，其北东端终止于东西向构造带，其南西端延出本区，断裂两端分别控制了燕山中—晚期的火山活动；而中段则与其他北东向和北西向断裂一起控制了八达岭杂岩体的产出，新生代仍是活动断裂，控制了马池口断陷的发育。

二、一般断裂

一般断裂指规模较小，延伸多在100km以下，切割深度较浅的盖层断裂。按其方位可以分为东西向、北东向、北北东向、北西向、北北西向、北西西向和近南北向等7组，以前3组较为发育，其次为第4组，最后3组不同地段发育的疏密程度不尽一致。下面简述各组断裂分布情况，选部分断裂进行详细描述（表32-1）。

（一）东西向断裂

本区除前述东西向深大断裂外，与之平行延伸、规模较小的次级东西向断裂十分发育。其特征是：走向上呈舒缓波状，形成于挤压环境，沿断裂糜棱岩、挤压片理发育，走向剪切滑动、变质现象常可见及。由于形成时期相对较早，后期在伸展环境下往往被其他构造成分截切、干扰、掩盖，甚至卷入其他构造。致使其走向发生偏转或错位不连续，空间分布上常具一定等距性。

区内出露较好、具一定规模的东西向断裂主要有：旺泉峪断裂（F35）、浇花峪断裂（F34）、芦子水断裂（F33）等。

（二）北东向断裂

本区北东向断裂基本定形于燕山早期，常被形成时代稍晚的北北东向断裂迁就利用，使其延展方位常发生向北北东向偏转，正常情况下走向多为北东50°~60°，向东南或西北倾斜，倾角20°~45°不等。沿走向舒缓波状弯曲，糜棱岩、挤压片理化及破碎带常发育，常伴有明显的动力变质现象，具较明显的挤压构造特征，经常以逆冲甚至逆掩推覆构造形式出现。部分地区由于后期构造的不断隆起抬升，在断裂的中深构造层次常可见向韧性剪切断裂逐渐过渡的现象。该类断裂对燕山早期的岩浆侵入活动、燕山中期的火山喷发-沉积活动有较明显的控制作用。断裂的规模一般较大，长度可达10km以上，延伸较稳定，空间上常成束成带，并与北东向的褶皱构造密切伴生。

山区出露较好的北东向断裂主要有：磨盘山-王家元断裂（F7）、梁家庄断裂（F2）、梨园岭-煤窑断裂（F3）、斋堂-军响断裂（F4）、新城子断裂（F21）、东庄禾-太师屯断裂（F10）、霞云岭断裂（F37）、八宝山断裂（F42）、大华山断裂（F11）等。

（三）北北东向断裂

除前述北北东向深断裂外，区内与之平行延伸的次级断裂也很发育。走向一般为北东20~30°，倾向南东或北西，倾角30°~60°不等。延伸较稳定，规模多在20~30km之间。形态舒缓波状，常呈雁行斜列，具明显的压剪性特征。其形成时期大致为燕山中晚期，一般均截切东西向断裂，现今仍有较大的活动性，特别是在平原区与其他构造交接部位，常有不同级别的地震发生。新生代以来，断裂性质由挤压性向伸展性正断层转化。空间上呈带状分布。

北北东向断裂在山区出露较好的断裂主要有：白河堡断裂（F14）、红石湾-营盘断裂（F15）、沙梁子-上花楼断裂（F17）、崎峰茶断裂（F9）、四干顶-葡萄园断裂（F22）、杏树台断裂（F16）、江米涧-西牛角峪断裂（F23）、珠窝断裂（F13）、马栏断裂（F12）等。

（四）北西向断裂

区内比较发育。走向北西315°~320°，倾向东北或西南，倾角50°~70°。断面一般较为平直，略显舒缓波状，延伸较大，长度约20km。其发育时间有的可能较早，燕山早期以挤压特征出现，燕山中期断裂面转化为伸展作用，燕山晚期以来复又转为以挤压特征为主。

规模较大、延伸亦较稳定的北西向断裂主要集中分布于本区燕山南麓，主要有：施庄-镇边城断裂（F27）、禾子涧断裂（F28）、南口-孙河断裂、上口-下庄断裂（F30）（林傅勇，1980）、二十里长山断裂（F31）等。

（五）北北西向断裂

该方向断裂近年来不断有所发现，并为人们所重视，其活动时期可能较北西向断裂更早，中生代时期对本区地质构造的发展起过一定的作用，新生代以来断裂仍有一定的活动性。断裂走向北西325°~335°，倾向东北或西南，倾角一般较陡70°~80°。断裂面平直，穿切能力强，具明显的剪切性质。延伸稳定，长度一般可达10~30km。截切东西向和北东向断裂，被北西向、北北东向和近南北向断裂所切割。沿断裂可见偏碱性基性小岩体和煌斑岩脉贯入。

山区出露较好的北北西向断裂主要有：六渡-张坊剪切劈理带、牛战村断裂、大安山-红煤厂断裂、黄楼院-仙人洞断裂、小川东断裂、茶棚山前断裂、牛角峪-白云寺断裂等。

（六）北西西向断裂

该方位断裂构造近年在陆地卫星象片的研究中有所发现，目前认为在昌平—顺义—三河一线存在一隐伏活动断裂带。走向290°~300°。卫片上表现为色调深、形迹十分清晰的地下水富集带。地面主要特征是河流沿此断裂带发生同步弯曲，对水系的发育、地下水活动有控制作用。潮白河流至顺义一带突然发生引人注目的拐弯，往东南入河北省后，河流的这种曲折或顺其方位伸展的现象尤为明显。从水系的扭曲特征可判断断裂具左旋平移特点。

断裂在重力场、磁场、地壳厚度变化上也有一定的反映，此外，近年来发现沿此断裂带三级以上的地震有密集成带的分布特点。

（七）近南北向断裂

区内近南北向断裂略显示沿一定经度成带分布，沿走向可出现较大的曲折，往往迁就、追踪北北东向和北北西向两组断裂延伸，局部甚至呈菱形结环。断裂经历了较为复杂的发展过程，基本定形时期约在较晚的燕山晚期，新生代第四纪的凹陷盆地常沿此构造方向成串或成带分布，反映断裂在较近时期以来仍具明显的活动性。断裂早期可能为张性至张剪性，然后经历了由剪性至晚期以压剪性特征为主的演化。此方向的断裂对燕山期岩浆活动、第四系分布、近代的地震活动都有一定的控制和影响作用。

出露较好的近南北向断裂主要有：靳家堡断裂（F25）、张山营断裂（F26）、青石岭断裂（F18）、山神庙断裂（F19）、程各庄断裂（F24）等。

由于该方位断裂与现今地震的活动性有较密切的关系，今后应加强对其展布的确切位置、分布规律及其与其他构造复合关系的研究。

表 32-1 北京地区主要断裂构造特征一览表

Table 32-1 Characteristics of main fault structures in Beijing

断裂名称	分布位置	断裂产状	性质	主要伴生构造	与其他构造关系	与岩浆活动关系	其他
沿河城-南口-琉璃庙断裂带（F1）	门头沟区双塘涧—沿河城—溜石港—南口—琉璃庙一线，本区内延伸140km以上	走向50°±，倾向南东，倾角60°~70°	正断层	断裂两盘岩石组成复杂。破碎带较宽且组成较为复杂，包括断层角砾岩、碎裂岩、断层泥等。两盘发育平行的劈理带和小断裂。局部发育拖曳褶皱。断裂落差巨大	断裂切割了梁家庄向斜、东岭台向斜、青白口穹隆、居庸关背斜的南东翼及云蒙山隆起的北西（翼）部；限制或切割了北西向断裂及印支期近东西向断裂，其北东端终止于东西向构造带，其南西端延出本区	断裂两端分别控制了燕山中—晚期的火山活动；而中段则与其他北东向和北西向断裂一起控制了八达岭杂岩体的产出	该断裂具明显的多期活动特征，新生代仍是活动断裂。控制了马池口断陷的发育过程

续表

断裂名称	分布位置	断裂产状	性质	主要伴生构造	与其他构造关系	与岩浆活动关系	其他
梁家庄断裂（F2）	门头沟区西部的齐家庄—梁家庄一带，长约7km	走向大约60°，倾向南东，倾角约为70°	逆断层	断裂两盘岩石均为古生界，破碎带较窄，由断层角砾岩和断层泥组成。断裂垂直落差约500m	断裂发育在梁家庄背斜和百花山向斜的共同翼上	断裂两端分布有燕山期的潜火山岩体	
梨园岭－煤窑断裂（F3）	门头沟区西部的梨园岭一带，长约7km	走向大约60°，倾向南东，倾角>70°	正断层	断裂上盘的南西段为髫髻山组，北东段为古生界；下盘的南西段为古生界，北东段为中—新元古界。破碎带较窄，发育张性构造角砾岩。断裂垂直落差达600m	断裂发育在梁家山向斜的北西翼，西北紧邻沿河城断裂带，并与北西向小断层互为切割	断裂对梁家庄盆地内的火山喷发及相应的潜火山岩的产出具控制作用	
斋堂－军响断裂（F4）	门头沟区西部的清水—斋堂—青白口一线	走向大约65°，倾向南东，倾角40°。长约十几千米	逆断层	断裂的南西段发育在髫髻山组分布区内，北东段的上盘为窑坡组，下盘为九龙山组。沿断裂发育断层角砾岩。断裂的垂直落差约100m	该断裂切割了北北东向的马栏断裂	该断裂控制了髫髻山期早期的裂隙式喷发，是火山熔浆的通道	
黄柏寺断裂（F5）	延庆区北部张山营—黄柏寺一线	走向60°，倾向南东，倾角中等长约20km	正断层	断裂上盘主要是第四系，下盘是中生代火山岩及元古宇。沿断裂发育断层角砾岩，但厚度不大	是延庆盆地的控制性边缘断裂，受到近南北向断裂的切割		新生代活动断裂
磨盘山－王家园断裂（F7）	昌平区磨盘山—王家园一带	走向大约50°，倾向南东，倾角约50°。断裂长约10km	逆断层	断裂两盘均为太古宇、元古宇以及中生代的侵入体。沿断裂发育宽度不大的碎裂岩和断层角砾岩，断裂的垂直落差达几百米	断裂发育在居庸关背斜的核部	该断裂的中部及北东端均发育中生代中酸性侵入体	
崎峰茶断裂（F9）	位于密怀隆起西部，南起永安堡，向北经崎峰茶，至琉璃庙以北的干沟门	全长约37km。宽150~400m，总体走向北东，并略向东南突出。得田沟以南走向NE58°，以北转向NE20°~25°，倾向北西，倾角>45°	燕山期活动显示正断层性质。而在吕梁期以前的古老断裂活动则表现逆断层性质	为发育于太古宙变质岩内的古断裂带，长约30km。沿断裂出现较宽的构造破碎带，以发育碎裂岩、角砾岩、挤压片理带为特征。构造岩有强烈的绿泥石化、绢云母化、碳酸盐化。断带中的角砾、岩块未见有韧性形变特征，总体显示脆性破裂特点	断裂带两侧的长城系各组地层，其岩性与厚度均无明显差异，它们沉积覆盖于断裂两侧麻粒岩相和角闪岩相变质岩之上	沿断裂带有燕山晚期霏细岩和正长岩等岩体侵入并被破碎	现有资料表明，该断裂具有长期活动性
东庄禾－太师屯断裂（F10）	密云区的东北部	走向45°，长约12km		断裂的南西段两盘均为太古宙变质岩系，北东段两盘均为元古宇	断裂受到东西向断裂的限制以及北北东向断裂的切割	断裂控制并破坏了髫髻山期火山沉积盆地	
大华山断裂（F11）	平谷区东部	走向45°，倾向南东，倾角大约80°。断裂长约10km	逆断层	断裂上盘为太古宙片麻岩，下盘为常州沟组。破碎带宽5~10m，发育断层角砾岩、碎裂岩、断层泥等。其垂直落差可达600m			

续表

断裂名称	分布位置	断裂产状	性质	主要伴生构造	与其他构造关系	与岩浆活动关系	其他
马栏断裂带（F12）	百花山北麓	走向30°，倾向既有北西，也有南东，倾角35°~70°不等。整个断裂带长约15km	逆断层为主，个别为正断层	卷入断裂带的地层主要是中生界窑坡组、龙门组、九龙山组和髫髻山组。窑坡组常逆冲于九龙山组之上。靠近断裂的地层常直立或倒转形成强烈挤压带。断裂下盘的劈理构造很发育，但向两端断裂消失于中生界中，因而表现为表层滑动的应力集中带。地层的最大断距为250m	断裂带斜穿百花山向斜	断裂带对中生代的火山活动具控制作用	
珠窝断裂（F13）	门头沟珠窝一带	由三条雁行式斜列的断裂组成，走向约25°，倾向南东，倾角陡立，总体长约10km	正断层	断裂主要发育在元古宇分布区，南西段横切古生界。沿断裂发育不宽的断层角砾岩。断裂的断距很小	断裂是青白口穹隆形成时伴生的顶部横张裂隙		
白河堡断裂（F14）	延庆区北部白河堡一带	走向大约30°，倾向北西，倾角85°。长约20km	逆断层	断裂基本发育在土城子组中，局部切割铁岭组。沿断裂发育构造角砾岩	断裂切割近东西向和北西向断裂		
红石湾-营盘断裂（F15）	延庆区北部的红石湾一带	走向25°，倾向北西，倾角85°。长约15km	逆断层	断裂上盘为元古宇，下盘主要为土城子组。发育约30m宽的破碎带，主要为构造透镜体、构造角砾岩、碎裂岩、断层泥。垂直落差达几百米	断裂受到东西向断裂的限制，切割了近东西向的红旗甸背斜以及北北东向的花盆向斜		
杏树台断裂（F16）	怀柔区西部的杏树台一带	走向20°，倾向北西，倾角80°。长约10km	逆断层	断裂发育在髫髻山组和杏树台霏细斑岩体中，沿断裂发育挤压透镜体和构造角砾岩。断裂的断距不大		断裂控制了杏树台霏细斑岩的侵入	
沙梁子-上花楼断裂（F17）	延庆区东北部的柏木井—沙梁子一带	走向25°，倾角近于直立。长约20km	逆断层	断裂上盘为元古宇，下盘为雾迷山组及中生界。破碎带宽50m，由构造角砾岩和碎裂岩组成，断距>30m	断裂切割了东西向断裂，并破坏了四海向斜的北西翼		燕山中期活动
青石岭断裂（F18）	怀柔区中部	走向8°，倾向北西，倾角60°~75°，局部直立。长约42km	平移断层	断层两侧地质体具明显位错。断裂在地表以破碎带形式产出，且具见有分支复合现象。断距达500~1000m。断带内广泛发育碎裂化岩石、碎裂岩、断层角砾岩和断层泥，表现为脆性构造岩特征	断裂切割了除第四系外的所有地层和早期断裂	断裂中有花岗岩、辉长辉绿岩脉，并均有不同程度的碎裂	燕山晚期活动

续表

断裂名称	分布位置	断裂产状	性质	主要伴生构造	与其他构造关系	与岩浆活动关系	其他
山神庙断裂（F19）	密云区西北部	走向 15°，倾向北西，倾角 80°	正断层	断裂南部切穿云蒙山岩体，中部发育在太古宇中，北部两盘则主要为元古宇和中生界。断裂带中发育构造角砾岩及大理岩	断裂切穿了云蒙山隆起、近东西向的褶皱和断裂		
石城断裂（F20）	密云水库西侧	走向 25°，倾向南东，倾角 30° ~ 70°。长约 24km	逆断层	断裂两盘基本上由太古宙变质岩、元古宇及中生代的侵入岩组成。断裂带较宽，可达上百米，由碎裂岩、构造透镜体、断层泥、构造角砾岩等组成	断裂截断了河防口推覆断裂以及北西向和近东西向断裂构造，破坏了云蒙山隆起的东翼	断裂控制了燕山晚期正长岩的侵入	断裂具多期活动性
新城子断裂（F21）	密云新城子—董家坟一带	总体倾向北西，局部倾向北东，倾角中等—直立。长度＞11km，宽 2~50m	左行平移断层	断裂两盘为中元古界和中生界，断层两侧地质界线发生左行位错。碎裂岩、断层泥及构造透镜体发育	断裂切割了古北口断裂和司马台断裂		新生代仍在活动
四干顶－葡萄园断裂（F22）	密云区东部	走向 30°，倾向南东，倾角中等。断裂长约 14km	左行平移断裂	断裂两盘均由太古宙变质岩和四干顶岩体组成。沿断裂发育糜棱岩、碎裂岩及断层泥。平移距离几百米	断裂切割了近东西向的断裂	断裂是四干顶岩体侵位的控制构造	
江米洞－西牛角峪断裂(F23)	平谷区西北部	走向 30°，倾向北西，倾角 70°。长约 13km	左行平移断裂	断裂南段主要发育在中元古界分布区内，北段两盘为太古宙变质岩。破碎带宽 3 ~ 10m，发育碎粉岩、断层泥等。水平断距 1500m	断裂切割了早期的东西向断裂和北东向的城墙顶向斜		
程各庄断裂（F24）	平谷区北部	走向近南北，断面陡立。长约 20km。向南被第四系掩盖	左行平移断裂	断裂两盘主要为中元古界和太古宇。破碎带宽仅几米，发育挤压片理、断层角砾岩及断层泥。水平断距达 1500m	断裂切割了东西向断裂和北东向的镇罗营背斜等早期构造		多期活动断裂，1679 年的马坊 8 级地震即在此断裂的南延部位发生
靳家堡断裂（F25）	延庆区西北部	走向近南北，断面倾向西，倾角陡立，长约 10km，向南被第四系掩盖	正断层	断裂在基岩中的两盘地层主要为土城子组和元古宇。破碎带由碎裂岩和构造角砾岩组成	断裂切割了早期的北东向断裂，是延庆盆地的控制构造		是新生代活动断裂
张山营断裂（F26）	延庆区西北部	走向近南北，倾向不定，倾角近于直立。长约 10km	正断层	断裂东盘主要为土城子组，西盘为太古宇和元古宇。破碎带包括碎裂岩和构造角砾岩	断裂切割了北东向构造和近东西向断裂		是新生代活动断裂

续表

断裂名称	分布位置	断裂产状	性质	主要伴生构造	与其他构造关系	与岩浆活动关系	其他
施庄－镇边城断裂（F27）	昌平区西北部	走向北西，倾向北东，倾角上陡下缓，上部倾角50°。断裂在本区长约7km	逆冲断裂	断裂上盘为高于庄组、杨庄组和雾迷山组，下盘为雾迷山组、青白口系和古生界。沿断裂发育宽度不大的碎裂岩和断层泥，上盘发育拖曳背斜。断裂最大垂直落差达3000m	断裂破坏了青白口穹隆的北翼；在市界附近被北东向断裂切割，其东端被中生代火山-沉积岩系掩盖	沿断裂发育基性岩脉	
禾子涧断裂（F28）	昌平区西北部	走向大约120°，倾向北东，倾角约60°。本区长仅4km	逆断层	断裂上下盘均为中—新元古界，破碎带包括断层泥、断层角砾岩等。断距不详	被北东向和北北东向断裂切割		
南庄－寨口断层（F29）	门头沟区南庄—寨口一线	东西向延伸约7km，断层面总体倾向170°～190°，倾角60°～70°	逆断层	断层发生于古生界及中生界三叠系双泉组内，断距最大处（南庄）上盘为中寒武统毛庄组，下盘为下奥陶统冶里组，垂直地层断距近400m	断层西段被后城组掩盖，东段至南安河一带被中生代杏石口组—南大峪组掩盖		印支期断裂
上口－下庄断裂（F30）	昌平区北部	走向大约120°，倾向北东，倾角50°～80°。长约10m	正断层	西段横切蓟县系，东段沿薛家石梁岩体与蓟县系的接触带发育。断裂带宽数米至200m，包括断层泥、构造角砾岩以及碎裂岩。视断距达1km	该断裂被北东向断裂切割	断裂控制了薛家石梁岩体的侵位	具多期活动性
二十里长山断裂（F31）	顺义区东南部	走向大约130°，倾向北东，倾角70°以上。全长约40km	高角度逆断层	断裂的大部分被第四系掩盖，出露部分两盘为古生界。沿断裂发育以角砾岩为主的破碎带。断裂附近岩石直立甚至倒转。断距不详	断裂被北东向高丽营断裂截断	燕山中期控制了侏罗系陆相碎屑岩、火山岩的堆积	燕山期活动强烈；新生代断裂南部活动，为山区和平原的分界线
荆栗园－圣水头断裂（F32）	位于平谷区荆栗园一带	倾向北西或南东，倾角80°断裂长10km，宽3～5m具右滑特征	逆断层	断裂为北西盘为太古宙片麻岩，南东盘为高于庄组。北西盘片麻理产状90°∠45°			
芦子水断裂（F33）	房山区西部	走向近东西，倾向北，倾角约30°，本区长约8km	逆断层	断裂上盘主要为髫髻山组，下盘为古生界。破碎带狭窄，包括断层泥和构造角砾岩。断距不详	断裂发育在百花山向斜的南翼		
浇花峪断裂（F34）	昌平区西部	走向近东西，倾向北，倾角约为70°。本区长约10km	正断层	断裂两盘主要是高于庄组，西段发育晚侏罗世火山岩。断距约几百米	断裂被北东向断裂切割	沿断裂充填有石英二长岩脉	

续表

断裂名称	分布位置	断裂产状	性质	主要伴生构造	与其他构造关系	与岩浆活动关系	其他
旺泉峪断裂（F35）	怀柔区西部	走向近东西，倾向南，倾角 > 60°。长约7km	逆断层	断裂上盘为雾迷山组和铁岭组；下盘为青白口系。发育不宽的碎裂岩带。断距上百米	断裂两端均受到北东向断裂的切割		
沙厂—墙子路断裂（F36）	怀柔—密云东部	走向近东西，断面波状，倾角近于直立，长约40km	压性断裂	除部分地段切割古生界和中生界外，断裂两盘基本由太古宇组成，东段切割元古宇	断裂受到北北东向断裂的切割	断裂控制了沙厂岩体的侵位	为多期活动断裂

（据鲍亦冈，2001，修改）

第三节 其他构造类型

本区其他构造类型主要有逆冲推覆构造、剥离断层、变质核杂岩、韧性剪切带、固态流变构造群落、穹隆和半穹隆构造等6种。下面分别作简要叙述。

一、逆冲推覆构造

北京地区对逆冲推覆构造的研究早自20世纪30年代即已开始，陈恺等（1935）首次对北京西山地区的推覆构造进行了系统的描述。通过20世纪50年代末至60年代初，以及70年代中期至90年代两轮1∶5万区调资料的积累，特别是在80年代后资料进一步丰富，认识逐步深化，目前认为北京地区中浅层次的逆冲推覆构造共发生过两次：一次是在晚侏罗世土城子组火山-沉积物堆积之后，形成规模较大的北东向推覆构造带，斜跨整个北京地区，由霞云岭、八宝山、十三陵等一系列逆冲推覆断层所组成；另一次发生在晚白垩世，形成东西向推覆构造带，发育在长哨营—白马关—古北口—新城子一带，表现为由北向南推覆。此外在北京西部还发育零星的印支期东西向推覆构造，如门头沟区北部的杨村断裂。下面分别对两个推覆构造带的典型代表进行描述。

（一）北东向推覆构造带

北京地区的北东向推覆构造带分布自西南部房山区张坊北部，向北东经房山北部、门头沟西部、昌平东北部、怀柔西北部到汤河口一带，总体走向约NE40°，长约160km，宽约32km。其中包含一系列逆冲断层，例如北京西山的南大寨-八宝山断层、黄山店逆冲断层、长操-霞云岭逆冲断层、曹家堡—黄土梁逆冲断层、北京北部昌平十三陵逆冲断层、上庄-红螺寺逆冲断层及怀柔汤河口逆冲断层等。单条断层多呈现为向北西凸出的弧形形态，但整个构造带具有比较清晰的北东向带状展布特征。张长厚（2006）认为在北京西山乃至更大范围的燕山西段，发育有分布于3个北东向带状区域的北东向逆冲推覆构造，3个带状区域的间隔约为60km，延伸长度自东向西依次减小，呈现出明显的逆冲构造发育的三角形区域。逆冲构造具有基底卷入的厚皮构造与盖层内部的薄皮构造共存的构造属性，上盘运动方向总体指向北西，逆冲构造变形主要发生在140~130Ma。逆冲后伸展构造变形以发育在主要逆冲构造后侧为主，并利用先存构造薄弱带。北京地区的北东向推覆构造带属于该构造带的一部分。

1. 十三陵逆冲推覆构造

（1）十三陵逆冲推覆断裂的发育位置

该断裂发育在北京十三陵景区的东侧，十三陵水库大坝附近。自昌平城北侧起，经水库大坝向北东延伸，被薛家石梁岩体侵位而表现不清，长约10km，过薛家石梁岩体再往北东追索，相应地发育黄坎推覆断层。而从昌平城往西南延伸，被大片第四系掩盖，与阳坊以西的黄土台推覆断层遥相呼应。

（2）断裂构造特征

推覆断裂的上盘主要为蓟县系雾迷山组（图 32-10），其岩石组合及基本层序特征与著名的蓟县剖面基本一致。而断裂的下盘则主要为以元古宇和古生界为基底的髫髻山组火山岩系。下盘所见的雾迷山组与上盘相同层位的岩石组合和基本层序特征有较大的差异。

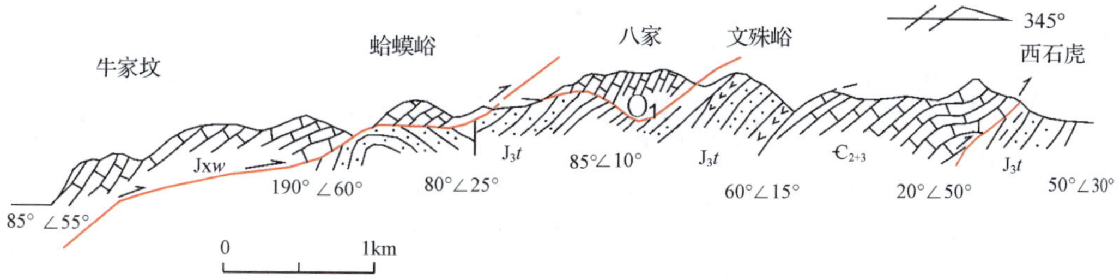

图 32-10　昌平十三陵逆冲推覆构造剖面图

Fig. 32-10　The structure section of nappe foult in Ming Tombs in Changping area

Jxw—蓟县系雾迷山组；J_3t—上侏罗统髫髻山组；ϵ_{2+3}—中—上寒武统；O_1—下奥陶统

断裂总体走向北东 50°，倾向南东，倾角 30°~50°，前锋断面最大倾角 60°（图 32-11），最小倾角 15°，断面呈舒缓波状（图 32-12）。

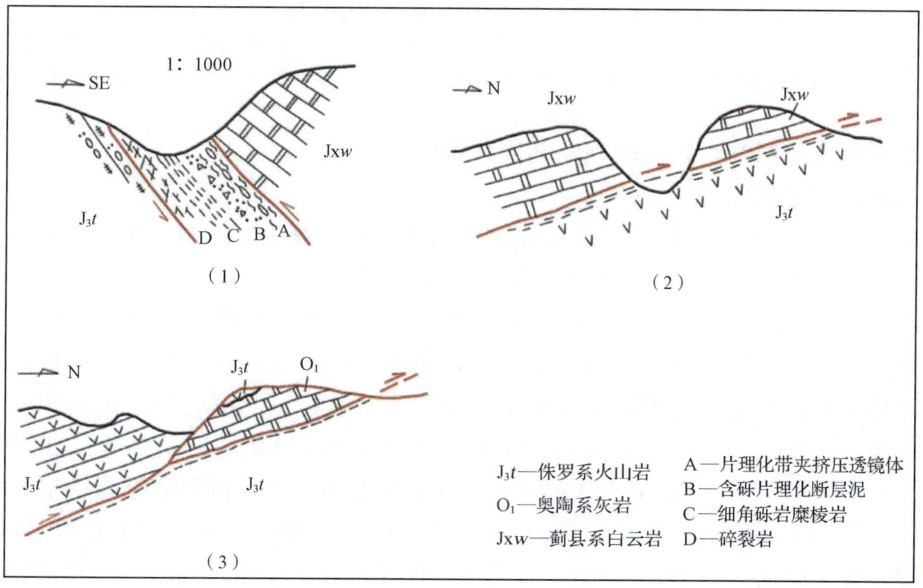

图 32-11　十三陵逆冲推覆构造剖面图

（据小汤山幅 1：5 万区调报告）

Fig. 32-11　Sketch showing Ming Tombs nappe fault

断裂破碎带较为狭窄，最宽处仅几米，一般不到 1m，主要为碎裂岩，前锋发育角砾岩，碎裂物主要是下盘的古生代和中生代岩石。上盘岩系在前锋带发育倒转的拖曳褶皱。断裂上盘岩块中发育北西向的横张裂隙，并充填有闪长岩脉。下盘岩系中还发育与主断裂平行的规模较小的推覆断裂。除此之外，断裂两盘的变形作用并不明显。

（3）断裂的推覆距离问题

根据断裂的几何特征推断，该断裂的推覆距离应在 10km 以上。而从断裂上下盘相同层位（主要是雾迷山组）的差异性推断，十三陵推覆构造的根带似应在孙河一带。据此推断，则其推覆距离可达

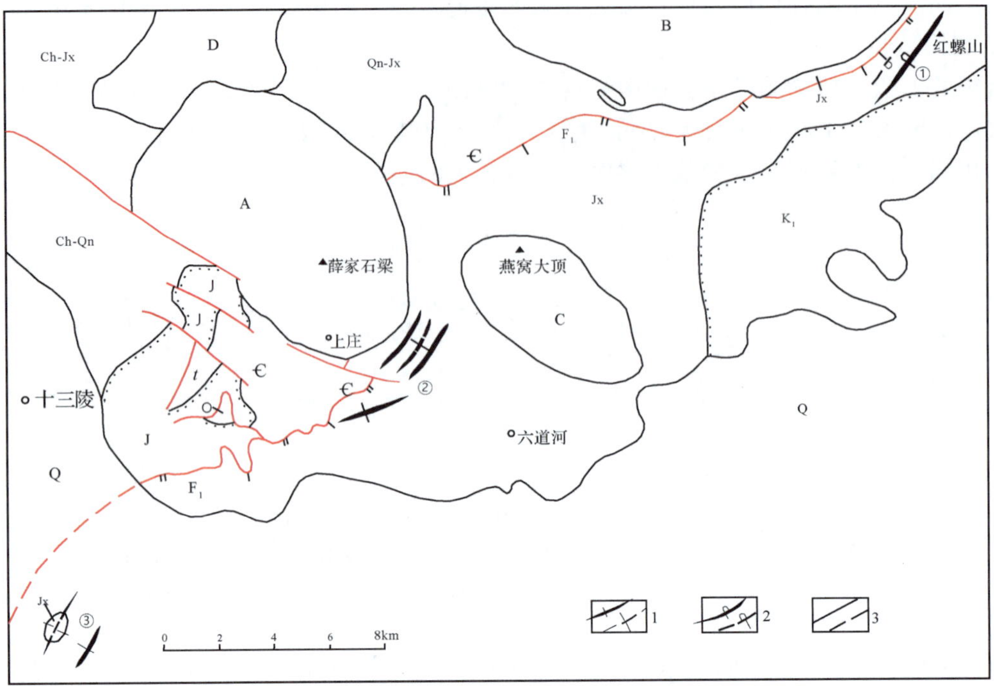

图 32-12 十三陵逆冲推覆构造上盘前锋褶皱示意图

Fig. 32-12 The Sketch for the frontal fold of nappe fault in Ming Tombs

1—背斜与向斜；2—倒转背斜与向斜；3—断层与推测断层（①红螺山背、向斜；②牛蹄岭背、向斜；③化庄背、向斜）；F1—十三陵逆冲推覆构造；A—薛家石梁岩体；B—沙峪岩体；C—大羊山岩体；D—分水岭岩体

60km。但就目前掌握的资料尚难以准确判断其推覆距离。

（4）推覆构造的形成时代讨论

小汤山幅 1∶5 万区调报告（1986）认为该推覆构造形成于燕山运动早期偏晚阶段，具体在髫髻山火山岩喷发之后，断裂活动性质为逆冲性质。吴新国等（1992）认为该期推覆构造发生在燕山运动中期，是一次强烈水平挤压运动的产物，推覆距离一般为几十千米。推覆断裂切割的最新地层为髫髻山组，沿断裂带有燕山中晚期岩浆侵入活动，薛家石梁岩体、长园岩体等晚侏罗世—早白垩世侵入杂岩体或以断裂为边界侵入，或侵位破坏断裂的连续延伸，其轴向均与推覆构造的走向相交。薛家石梁岩体的 SHRIMP 年龄（127±1.5）Ma（苏尚国，2006）属早白垩世早期。袁洪林等（2005）对北京西山髫髻山组的火山岩锆石 U-Pb SHRIMP 和 LA-ICP-MS 定年及地球化学的研究结果表明，髫髻山组上部的火山时代为（137.1±4.5）Ma。长园岩体的 SHRIMP 年龄 152Ma（Davis，2002），岩体边部闪长岩碎裂严重，劈理发育，劈理产状 230°∠75°，受到非常明显的强烈挤压作用。因此可以推断长园岩体侵位在推覆构造之前，并且岩体的存在导致逆冲推覆构造的走向由北东向变为近北西西向，即岩体阻隔了推覆断层，断裂前锋带部位发育倒转褶皱（见图版 7-1）。而在薛家石梁岩体边部，未发现有挤压的现象。这些事实表明断裂形成于晚侏罗世土城子组沉积以及长园岩体侵入之后，薛家石梁岩体侵入之前。在沿该推覆断层的一个地点（凤山采石场，Cui et al.，1996，点位 2）所采集的断层泥的全岩 $^{40}Ar-^{39}Ar$ 坪年龄为（140±6）Ma（Hodges，1999），因此基本上可以断定该年龄值是比较可靠的，因此该期推覆构造的形成时代应该是早白垩世早期，活动时限为 140～130Ma。

2. 长操-霞云岭逆冲推覆构造（F37～F38）

该推覆构造冲断面北部起始于房山红煤厂北侧，向西延伸，在长操发生转折，进而向南延伸至南窑。全长>30 km，冲断面总体分布形态呈北西向略突出的弧形。冲断面的产状总体向南东倾斜，冲断层倾角上部陡，向下变缓，呈铲状。逆冲推覆构造推覆体由中元古界和待建系下马岭组构成，推覆构造下盘沿断层线出露的是下马岭组。该冲断层总体几何模式为断坪-断坡组成的平滑曲线状几何形态。

长操-霞云岭逆冲断层与教军场-大安山逆冲断层、马兰-胡林逆冲断层共同构成了具有典型、连续的断坪-断坡结构的逆冲构造系统，形成于髫髻山期之后，南窑闪长岩（128Ma）侵入之前（葛孟春等，1990；张长厚，2006）。

3. 黄山店逆冲推覆构造（F41）

该推覆构造位于北岭向斜西南的黄山店、上方山一带（图 32-13），分布范围约 40km^2。卷入地层主要为中—新元古代地层，总体特征表现为两个相互叠置的照明天平卧背斜和黑头山平卧向斜，铲式冲断层发育在平卧背斜倒转翼。背斜下翼地层大规模倒转。冲断面或韧性剪切带与相对软弱的雾迷山组大理岩与钙质千枚岩互层段、洪水庄组黑色千枚岩、下马岭组千枚岩及砂质千枚岩互层段等处发育宽大的断坪，切过其间的强硬岩层时，则形成断坡。该推覆构造冲断面呈北东向展布。褶皱推覆体在稍晚时期发生了再褶皱，卷入了近东西向的迎风峪向斜及其南北侧的宝金山向斜和三岔向斜。在推覆断层的西界上方山一带，倒转的雾迷山组硅质条带大理岩和洪水庄组千枚岩呈飞来峰叠置于正常的铁岭组之上。在照明天平卧背斜转折端可见叠层石平卧及下翼叠层石倒置现象，并发生了剪切变形。冲断层的主断面发育于雾迷山组厚层白云岩之内。平卧褶皱的枢纽方位为北东东 75°~80°，轴面倾向南南东，稍显斜卧状态，褶皱内部的小褶皱和断层伴生的小褶皱的枢纽平均方位为北东向，逆冲推覆方向总体为北北西向，据研究其位移量达 3~5km（任建业，1987；葛孟春等，1990）。

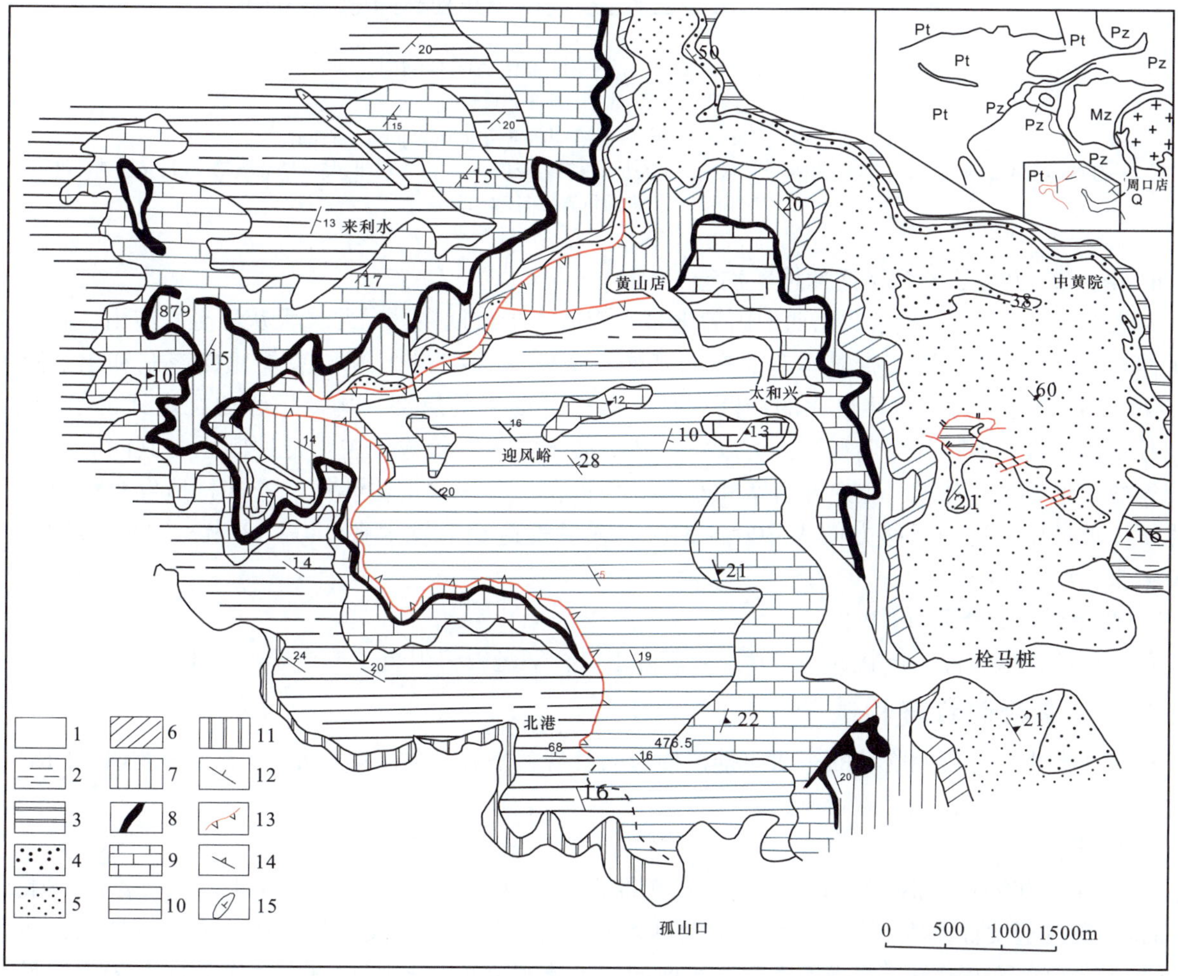

图 32-13 北京西山黄山店逆冲推覆构造地质构造图

Fig. 32-13 Huangshandian nappe tectonic map in the West Hills area of Beijing

1—第四系；2—寒武系；3~5—景儿峪组、长龙山组、下马岭组；6~11—铁岭组第二段、铁岭组第一段、洪水庄组、雾迷山组第四亚段、雾迷山组第三亚段雾迷山组第二亚段；12，14—产状；13—逆断层；15—闪长岩体据周口店幅区调中对黄山店地区的照明天横剖面（图 32-14，A-A’）进行了长度平衡复位，求得褶皱-冲断构成的地壳缩短量 e=-33%。对其他未经复原的剖面（图 32-15，B-B’，C-C’，D-D’，E-E’），根据上、下盘中同一标志层（洪水庄组）的相对错断距离估算其最大位移距离，可以看出冲断层的位移距离自东而西逐渐增大，冲断层向东逐渐消失。

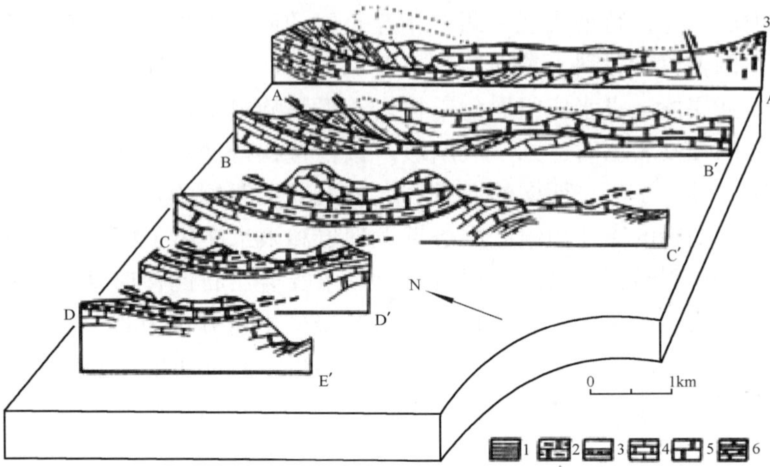

图 32-14 黄山店逆冲推覆构造联合剖面

（据 1∶5 万周口店幅区调报告）

Fig. 32-14 Huangshandian fold-thrust structure joint section

1—雾迷山组第二亚段；2—雾迷山组第三亚段；3—雾迷山组第四亚段；4—洪水庄组；5—铁岭组第一段；6—铁岭组第二段。剖面位置见图 32-15

4. 南大寨-八宝山断裂及逆冲推覆构造（F42）

南大寨-八宝山断裂位于北京西山山麓地带，南起八十亩地南，向北经大灰厂至八宝山，有可能继续延伸至太平庄附近。断裂带基本呈北东向延伸，全长 50 多千米。断层总体倾向南东和东，倾角在 20°~40°之间变化。其中位移最大的冲断层位于整个冲断系的后部，构成后缘叠瓦扇式的组合型式。根据其结构特点和断片的变形特征，可以确定其扩展方式为背状式。主干逆冲断系与其间所夹的断块共同构成背形叠置双冲式；主逆冲岩席和断夹块叠置在一起构成推覆体堆。逆冲构造发育在地壳上部的脆性变形域中，逆冲运移方向为 300°~310°，逆冲构造变形发生在 140~130Ma 之间（赵占元等，1994；张长厚，2006）。

5. 黄土台推覆断裂（F39）

位于门头沟区东部，产状 SE∠30°，上盘古生界，下盘髫髻山组，形成于燕山中期。

（二）东西向推覆构造带

形成于燕山构造旋回晚期的东西向推覆构造的发现，是北京地区第二轮 1∶5 万区调取得的主要成果之一。

1. 发育位置

北京地区的东西向推覆构造主要发育在东北部的怀柔、密云一带，在古北口至曹家路一带长约 20km，推覆构造发育最清晰。从古北口向西延入河北省境，到白马关一带再次进入北京地区，在元古宇和太古宇发育区内形成一些规模不大的东西向逆冲推覆断层。长哨营以西的东西向断裂带仍具有逆冲推覆性质，但其构造表现不如古北口—曹家路一带清晰。

2. 构造特征

在古北口—曹家路一带的推覆构造由两条推覆断裂组成，北支为古北口推覆断裂、南支为司马台推覆断裂。

（1）古北口推覆断裂（F43）

该断裂的西段，上盘为古北口富钾花岗岩体（目前认为形成于元古宙），下盘为中生代土城子组火山-沉积岩；中段和东段上盘为中—新元古界，盖在下盘土城子组之上。由于上下盘岩性的差异较大，沿断裂形成陡峭的山脊，尤其在司马台一带表现尤为明显。

断裂破碎带非常宽，在司马台一带破碎带宽度超过 1km，主要由碎裂化的白云岩组成，且胶结较好，白云岩碎裂后形成粒径为几厘米的碎块，但岩系的总体层序并没有搅乱。碎块之间往往被铁质和方解石的细脉充填，在岩石表面形成网状花纹。破碎带中部发育十几米宽的断层泥和揉皱带，其中的小破裂面均为低角度北倾。断裂面倾向北，断裂面的前锋产状较陡，一般可达 60°，而根部则趋缓，剖面上呈犁形。断裂的前锋发育紧闭且倒转的拖曳褶皱，其轴面倒向南，倾角与断面一致。断裂西端截止于半城子断裂。

古北口断裂是一条古老断裂，最早的活动期为元古宙，中生代时又有较强的活动，具有明显的多期活动性。

（2）司马台推覆断裂（F44）

该断裂东起曹家路一带，向西在新城子与古北口断裂一起被北北东向断裂切割，再向西在古北口以东与古北口断裂汇为一支。

断裂的上下盘均为土城子组二、三段紫红色砂砾岩。上盘岩系明显较下盘的相同层位薄，二者相差十余倍，反映出虽是同一盆地的相同层位，但在盆地中的沉积位置不同，引起相同层位厚度上的差异。由于上下盘之间不存在岩性差异，沿断裂没有明显的地貌特征。断裂的产状很缓，其倾角一般<30°。断裂面较为平直。断裂破碎带很窄，一般仅几十厘米，局部可达 10m。主要由碎裂岩、碎粒岩及断层泥组成，胶结不好。伴生构造主要表现为断裂面两侧岩石中发育一些小断层和破裂面。断裂上盘发育拖曳背斜构造，宽度不大，延续性较好，褶皱较为紧闭，断裂曾经受过强烈的挤压作用（图 32-15）。

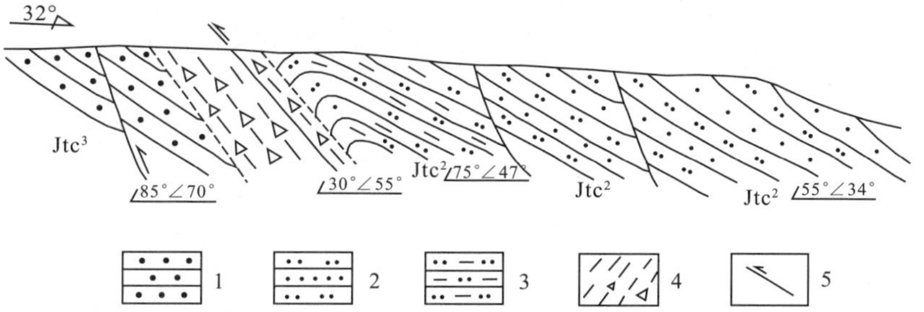

图 32-15　司马台断裂剖面图

（据曹家路幅 1∶5 万区调报告）

Fig. 32-15　The profile of Simatai fault

1—土城子组三段砾岩；2—土城子组二段粉砂岩；3—土城子组二段泥质粉砂岩；4—断裂碎裂岩；5—断层

司马台断裂是新产生的断裂，不具多期活动性，影响的最新地层是晚侏罗世土城子组。据 1∶5 万曹家路幅区调资料，该断裂切割了雾灵山岩体（同位素年龄为 100~70Ma），因此该断裂的形成时间应为燕山晚期，这与区域上东西向断裂的最后一次逆冲推覆活动的时间基本一致。而且雾灵山岩体为碱性花岗岩，代表区域的伸展环境，且受到了断裂的控制，因此司马台断裂早期形成于伸展环境，晚期经历了挤压收缩环境，它的形成表明区域上的东西向构造在燕山晚期有活动，且具有右行斜冲特征。根据断裂两盘土城子组二、三段的厚度与邻区的对比结果，估算其垂直断距>850m、水平断距则可达 5km 以上。

二、剥离断层

对北京地区剥离断层的识别是在 20 世纪 80 年代新一轮 1∶5 万区调中取得的重要成果之一。宋鸿林等（1987，1990）也对此作过较多研究。根据前人的研究成果，本区的剥离断层主要发育于房山岩体南北两侧、北岭向斜西侧的南窖一带，此外密云、怀柔地区沿太古宙基底变质岩与盖层之间的不整合面也被初步确定有多条剥离断层。剥离断层是伸展构造在脆-韧性过渡域中的主要表现，是一种平缓铲状大型正断层，使地壳浅层次年轻地层直接覆盖在深层次老地层之上。

（一）房山岩体南北两侧剥离断层（F46）

发育在房山岩体南北两侧，呈近东西向延伸。原始主断面表现为基底与盖层之间的剥离，出露于房山岩体周围太古宙变质岩的外缘，构成一个小型变质核杂岩体的顶面。受后期岩浆活动房山岩体底劈式侵入的上拱作用及构造活动的改造和剥蚀，使原来统一的低角度断面分解成南北两支与岩体边界平行的近东西向断层。

断层上盘的沉积盖层与下盘的太古宙变质岩系直接接触，存在地层的非正常缺失，如北支断层上盘的岩层自西向东南岩层越来越老，依次为奥陶系、寒武系各组直到待建系下马岭组（图32-16）；与太古宇直接接触的最新地层为下奥陶统冶里组及亮甲山组。断层造成的地层缺失包括全部寒武系及中—新元古界，本区寒武系及蓟县系雾迷山组以上中—新元古界的总厚度仅为1500m。而北京典型区域上仅中—新元古界总厚度就达4561m，假定剥离断层原始倾角为10°或更小，则可估算出断层上盘向南东东滑动距离达28km。断层下盘的太古宙岩石主要为透入性的糜棱岩化，岩性为角闪斜长质变余糜棱岩、长英质变余糜棱岩及角闪构造片岩等，具流状构造和拉伸线理，为典型的S-L构造岩，偶可见糜棱面理的片内无根褶皱。显微镜下石英呈丝带构造，可能受到后期的静态重结晶，形成不同粒度的多晶条带，属变余糜棱岩。从长石的变形和石英的静态重结晶判断，它反映了角闪岩相条件下的糜棱岩化。

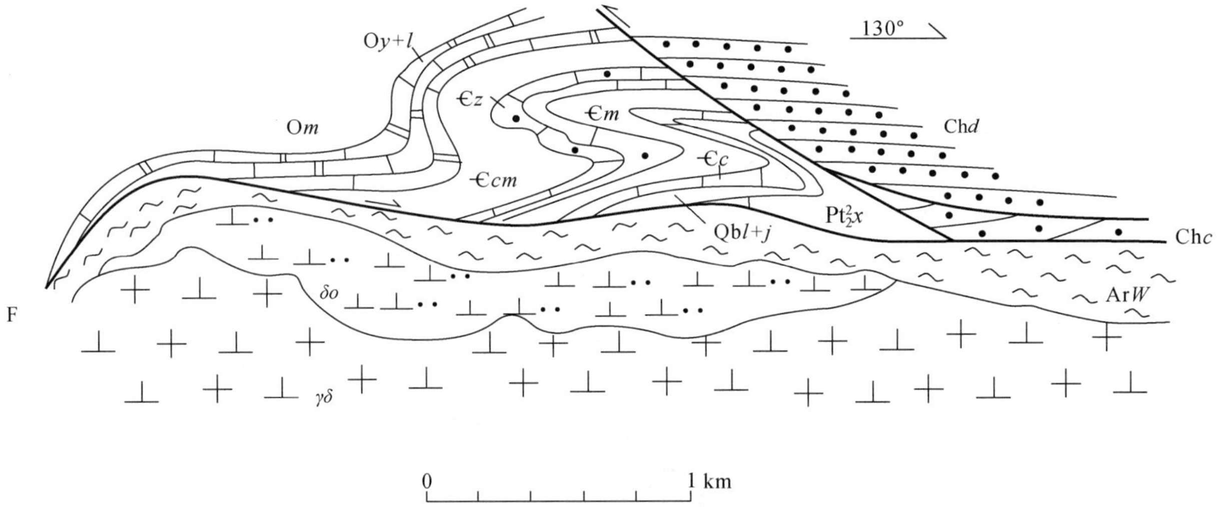

图32-16　房山中窖剥离断层构造综合剖面图

（据宋鸿林，1996）

Fig. 32-16　Comprehensive structure profile of detachment fault in Zhongjiao，Fangshan

γδ—花岗闪长岩；δo—石英闪长岩；ArW—苇子峪辉长闪长质片麻岩；Chc—常州沟组；Chd—大红峪组；Pt_2^2x—下马岭组；Qnl+j—龙山组及景儿峪组；€c—昌平组；€m—馒头组；€z—张夏组；€cm—炒米店组；Oy+l—冶里组及亮甲山组；Om—马家沟组

断层上盘的岩石也受到了强烈的韧性变形。但其变质作用明显低于下盘的岩石。以龙山组石英砂岩和砾岩为例，它们也受到了普遍的糜棱岩化，依各地的变形强度不同而形成初糜棱岩、糜棱岩或超糜棱岩。其中砂岩的显微构造以丝带及核幔构造为特征，反映大量的亚晶粒和动态重结晶的新晶粒，新生晶粒粒度在1~2μm。岩石只在局部地区或后期岩体附近经动态重结晶成变余糜棱岩，这反映其主要是绿片岩相条件下的糜棱岩化。与基底岩系不同的是，盖层岩石具初始的成层性，在近水平的韧性剪切下形成了复杂的多级顺层掩卧褶皱，其中的各类指向构造的指向都指示上盘向南东东正向滑移，这与依据地层展布所推测的剥离断层的运动方向非常吻合。

断层上盘岩系内发育有次一级的剥离断层。其中较明显的是南观以南发育在下马岭组顶部的剥离断层，已被卷入印支期东西向褶皱中。其上下盘之间缺失地层的总厚度在150m以上，且上盘地层与断层面的交角在5°左右，据此估计上盘沿断层面向南东东滑移的距离在1.7km以上。

（二）南窖剥离地层（F47）

南窖剥离断层由两条断层组成：水峪-中窖断层及奥陶系顶板剥离断层。前者上盘为产状简单的奥陶系白云岩及灰岩，下盘为强烈变形的寒武系及青白口系，断层面与上下盘岩层一起构成一个近东西向的向斜（印支期北岭向斜的西端）。后者断层面作为上下古生界的分界面围绕北岭向斜分布，该断层亦以地层缺失为特点：在中窖北山，其下盘奥陶系灰岩（已糜棱岩化）直接与上盘的二叠系接触；向东断层断距逐渐变小，直至基本上不缺失地层为止。断层面的产状与奥陶系灰岩的糜棱面理及上古生界以上的顺层板劈理近于平行，在西部，地层倾角为30°~40°，到东部房山岩体的边缘，倾角可近于直立甚至倒转，应该是受岩体的热动力构造的改造所致。根据断层上下盘的地层缺失情况推算，该断层倾向南东，倾角仅2°，滑距>4km，为上盘南东向或南东东向运动的低角度正断层。

关于该断层的形成时代，宋鸿林（1990）等认为应属印支晚期，是再一次发生的近南北向伸展作用的标志。本志认为这些剥离断层应该是与同一地区发育的固态流变构造形成于同一构造环境，即为印支期早期的产物。

（三）大黄木厂-孙胡沟-霸王山剥离断层（F45）

分布在北京北部云蒙山岩体的两侧，呈北西向展布。断层面以低角度向南西倾斜，断层上盘由中—新元古界组成，下盘则为太古宙变质岩系。断层出露部分被北东向构造切割成三段，即云蒙山北西侧的大黄木厂-崎峰茶段、孙胡沟段及云蒙山南东侧的梨树沟-霸王山段，向南东断裂没有延伸。断面总体倾向南西，受后期构造活动影响倾角较陡，一般>50°，局部达80°。沿断裂常州沟组底部的部分岩性顺层缺失，孙胡沟一带甚至高于庄组与变质岩接触；并发育断层泥、构造角砾岩及片理化带；下盘发生糜棱岩化，其面理与断面平行。断裂两盘发育平行的滑脱面。由于后期构造的强烈影响，断层的成因和形成时代均难以判断。

（四）密怀隆起周围的剥离断层

主要见于太古宙基底变质岩与盖层之间的不整合面及其附近，分布于密怀隆起的四周，呈不规则的环状展布。由于中生代侏罗系火山-沉积地层超覆掩盖，燕山期侵入岩体的破坏，后期推覆-滑覆地块的压覆以及后期断裂的破坏，在隆起的西北部、西南部和东北部，剥离断层已难以见到，目前能鉴别出剥离性质的断层主要见于隆起的西部、南部、东南部和北部。

隆起西部的崎峰茶地段，神洞沟-杏树地断层走向北北西，倾向225°~280°，长约12km，宽2~16m，断层上盘为长城系，发育糜棱岩化，下盘为英云闪长质构造片岩（太古宙变质岩），断裂带中发育构造透镜体、构造片岩和碎裂糜棱岩化石英岩，断层具脆韧性变形特征。

隆起北部的白马关断裂发育在古北口近东西向断裂带的南侧，并与断裂带平行产出，呈近东西向至南东东向，倾向北至北北东，倾角50°~70°，出露长度30km，宽2~5m。断裂带早期活动上盘向北正断滑脱，之后上盘向南逆冲，最后为右行平移、斜冲性质。

隆起东部、东南部，受北东、北北东及东西向断层的切割，剥离断层在下甸子—贾峪、下窝铺—程各庄、大梯子峪和东庄禾等地被分割呈若干片断。断面总体倾向北东、东或北，倾角20°~40°，受后期断裂的改造，局部倾角可达70°~80°。在隆起内部的顶盖部位，剥离断层比较平缓，而隆起边部，断层面则较陡。断层上下盘岩石都发生了碎裂岩化—超碎裂岩化，局部地段出现断层泥，微角砾岩或很窄的片理化。受断层影响，常缺失常州沟组一段地层，局部缺失二、三段。

隆起南部以密云-墙子路东西向断裂为界，以北有南东东向梨树沟-南石副坝断层，以南有五亩地-牛角峪-吴家峪断层，断层断续发育，局部保留了原角度不整合界面。梨树沟-南石副坝断层由于隆起后期变形强烈，地层产状南倾陡立或局部倒转，剥离断层也随之变为陡倾。五亩地-牛角峪-吴家峪断层走向呈北东东向，断续长度>20km，倾向南东，倾角 45°。断层构造岩特征与隆起东部、东南部，断层构造特征相同。

此外，在隆起内部云蒙山岩体北侧的贾峪—鲇鱼沟以南，沿贾峪-黄峪口倒转向斜南翼中元古代盖层与太古宙变质岩的不整合面，发育有韧性剪切断层。该断层原始走向应为近东西向，与倒转向斜轴近于平行。受云蒙山岩体破坏，向东延伸被石城断裂截断。现保留部分呈北西走向，长约 1.5km。断面倾向北—北东，倾角 20°~35°。断层下盘太古宙变质岩具糜棱岩化和片理化，上盘长城系常州沟组下部砾石层厚度大大减小，砾石被挤压后呈三轴不等长的椭圆饼状和细长杆状，拉长方向与断面倾向一致。该断层具有多次活动特征，早期显示正断滑脱性质，具剥离断层特征，上盘向北位移；晚期为逆冲推覆性质，并有热动力变形变质作用的叠加改造。

发育在密怀隆起盖层与结晶基底角度不整合面处的剥离断层具有脆-韧性变形特征，其形成于中—浅构造层次；从隆起内（靠近核部）保留的盖层底界不整合界面产状显示出断层原始产状比较平缓。为低角度正断滑脱性质。而隆起周边断层产状变陡，并随翼部盖层产状作同步改变，这是由于隆起形成过程中对剥离断层改造的结果。

（五）剥离断层的成因分析

分布于密云、怀柔地区，发育于盖层与结晶基底之间角度不整合界面的剥离断层，与密怀隆起有密切的成生联系。已有翔实的资料证实，密怀隆起的主体形成于印支期。关于印支运动，近年来的调查研究普遍认为有两次主要活动期，早期活动是在南北向伸展体制构造环境下，发生引张应力作用造成；晚期活动则受南北向收缩体制控制，在南北向挤压应力作用下，造成盖层褶皱形变，并发生近东西向以压性逆断为主的断裂活动。

就密怀隆起而言，其北界为赤城-古北口断裂，南界为密云-墙子路断裂，这两条断裂均为近东西向展布，是区内长期活动的古深断裂。印支运动早期，在区域南北向引张应力作用下，两条断裂中间的基底断块缓慢抬升，相邻南北两侧的基底断块下降。抬升地块上覆巨厚的盖层，在重力作用下沿着结晶基底顶部不整合面向两侧低角度滑脱，这就是剥离断层的形成机制。一般情况下，在上升断块与下降断块边界（古断裂带）地带，由于两侧地块升降幅度大，断层倾角较陡，而上升地块中间，断层倾角较缓，这与本区剥离断层的产状变化是相符的。

三、变质核杂岩

变质核杂岩（metamorphic core complex）是被构造上拆离及伸展的未变质沉积盖层所覆盖的、呈孤立的平缓弯形或拱形强烈变形的变质岩和侵入岩构成的隆起。变质核杂岩包含 3 个必不可少的构造要素：①主拆离断层为一大规模位移（几十千米）的准区域至区域性的低角度（典型情况下倾角<40°）正断层；②下盘为与断层相关的糜棱状片岩、片麻岩和可能出露的更深层次的非糜棱状结晶岩；③上盘为含多世代正断层的上地壳基底岩和（或）表壳岩层（Davis et al.，2002）。本区发现的变质核杂岩主要有房山变质核杂岩和云蒙山变质核杂岩。

（一）房山变质核杂岩

1. 空间展布特点

房山变质核杂岩出露于房山镇西北，其核部由太古宙变质岩及侵入其中的房山花岗闪长岩体组成，其空间形态为呈近等轴的穹状背斜。从房山镇向西北到东方红炼油厂的西界直径约 9km，大部分为房山岩体，只在岩体的北部和西部边缘出露少许太古宙变质岩。变质核杂岩体与周围的盖层以基底剥离断层接触，由西向东，与杂岩直接接触的地层越来越老，从奥陶系、寒武系直到新元古界青白口系，越过南大寨断层后，杂岩直接与中元古界长城系呈断层接触。杂岩体东侧被第四系覆盖。

2. 核部岩石的岩石特点及变形特点

变质核杂岩的核心主体是太古宙变质岩，主要由片麻岩、斜长角闪岩、黑云母角闪变粒岩等组成。岩石普遍受到程度中等到强的混合岩化作用；斜长角闪岩属苦橄岩—科马提质玄武岩浆系列，具太古宇绿岩带的岩石特征。各种片麻岩均为明显的 Eu 正异常，与世界太古宙花岗质岩石相似，而与下马岭组泥质岩及寒武系含泥质碳酸盐岩的 Eu 负异常明显不同，副矿物中锆石含量较多，且锆石经过了多种变质作用、重结晶作用及熔融作用。上述资料说明这些太古宙变质岩不是下马岭组片岩或寒武系岩石经房山花岗闪长岩的混染而成的，其生成早于长城纪，应属太古宙的变质杂岩。

太古宙变质岩普遍受到强烈的动力变质作用，形成一套以变余糜棱岩为主的，包括糜棱岩及碎裂糜棱岩的动力变质岩。岩石中发育透入性的糜棱面理及拉伸线理，形成典型的 S-L 构造岩，且发育 S-C 组构、片内小褶皱及小型韧性剪切带，所有这些都是在角闪岩相条件下的动力变质作用所形成的变余糜棱岩的岩石变形特点。受后期房山岩体侵入的影响，加强了其重结晶作用，但在露头上糜棱岩的流状构造仍能保存，可辨认出其早期糜棱岩化的痕迹。

在流状构造保存比较完整的露头上，可以看到糜棱面理的产状从杂岩体中心向外陡倾，与房山岩体的接触面平行，且与基底剥离断层的产状一致，显示出穹状分布的几何特点；线理产状大致平行于面理倾向，向四周倾斜，可能是房山岩体底劈式改造的结果。

3. 顶部变质剪切滑脱带特征

变质剪切滑脱带发育于杂岩与盖层之间。断层面与杂岩的糜棱面理近于平行，而与上盘岩层的层理及顺层面理以小角度斜交，因此在图上可见杂岩与盖层不同时代的地层相接触（图 32-11）。根据杂岩体四周盖层岩石的分布，可以判断杂岩体顶面的基底剥离断层原始产状为向南东东缓倾的正断层。平衡后的基底剥离断层平均倾角为 6°，向南东东正向滑动距离至少 28km。

由于剥离断层上下盘岩石的面理近于平行，因而在露头上很难发现其踪迹。通常下盘近断层处的糜棱岩退变质成绿片岩相，经碎裂化形成绿泥石化的碎裂糜棱岩，在断层面处还形成薄层断层泥。另外，还有席状岩体沿断层侵入。

盖层岩石也受到了强烈的韧性变形，形成各种形式的褶叠层。其最大特点是普遍发育的早期面理既是顺层掩卧褶皱的轴面面理，又与岩性层的界面平行。而广泛发育的拉伸线理及各种运动标志所显示的运动方向与基底剥离断层的运动方向一致，即上层向南东东正向滑动。与美国盆岭区变质核杂岩的古近系—新近系盖层的脆性剪切变形不同，本区盖层仍反映了较深层次的韧性变形。

房山岩体的侵入是造成房山变质核杂岩穹状隆起的最终因素，它使房山岩体内部的糜棱面理与房山岩体的接触面，以及变质岩的糜棱面理与基底剥离断层的产状都近于一致。其总体形态呈穹状外貌，这是底劈式侵入岩体的特点之一。但房山岩体又不完全位于杂岩核部，它向西刺穿基底剥离断层而侵入到上古生界的地层之中，并在边部形成了强烈的边缘动力变质带。

4. 形成的地质时代和有关问题的讨论

由于变质核杂岩构造与周口店固态流变构造具有密切的成因联系，因而其生成时代在印支期。单文琅等（1990）认为，由于房山穹隆周围石炭系太原组及以后各组地层的岩相具有以穹隆为轴的对称发育的现象，因而穹隆的初始隆起应发生在晚石炭世之前；同时基底剥离断层的形成与穹隆的形成有成因关系。因而基底剥离断层的发生和盖层的伸展很可能早在印支运动前就开始了。鲍亦冈（2001）认为固态流变以及变质核杂岩的形成应该发生在稳定盖层沉积发育完成之后。认为地壳经历着穹隆的构造改造，地表又同时接受稳定的盖层沉积，这是不可能发生的，且卷入剥离断层的脉岩年龄值约为 200Ma，因此将固态流变构造及变质核杂岩确定为印支运动的产物。

（二）云蒙山变质核杂岩（或称云蒙山地区挤压伸展体系构造）

云蒙山变质核杂岩位于北京北部密云与怀柔交界处，构造上表现为北东向云蒙山大背形叠置在东西向密云隆起上，其东南边界为一倾向东南（倾角 15°~25°）断距很大的正断层（河防口低角度正正断层）；北缘和东缘分别出现了四合堂剪切带和大水峪剪切带。王玉芳（1981）在该区发现水峪低角韧性正断层，研究了下盘的糜棱岩石英组构特征。郑亚东（1985）对云蒙山北西侧的推覆构造进行了研究，根据云蒙山区构造格局和运动方式的对称性，提出“热隆-滑覆-推覆”构造模式（郑亚东等，1986，1988），此后与 Davis 等合作研究（1996，1998，2002），认为云蒙山变质核杂岩在组成上与北美科迪勒拉变质核杂岩类似，提出云蒙山地区存在变质核杂岩构造，由核部岩体、河防口拆离断层带及未变质的上盘岩系组成（图 32-17）。河防口拆离断层带将核部杂岩体与东南部的丘陵、平原区隔开。本志初步认为云蒙山变质核杂岩构造存在，但仍需要进一步甄别各构造形迹的相互关系。

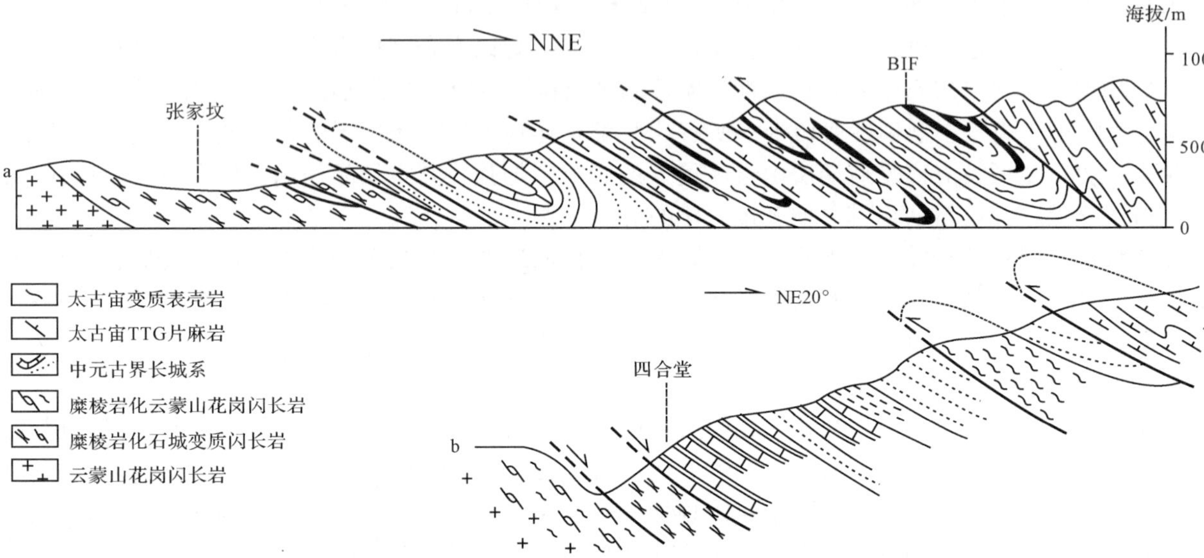

图 32-17　云蒙山北部张家坟（a）和四合堂（b）附近 NE 向岩石-构造剖面图

Fig. 32-17　Lithological and NE structural cross sections near Zhangjiafen（a）and Sihetang（b）in northern Yunmengshan

1. 地层和岩性

云蒙山地区的地层主要包括：新太古界，岩性为高级片麻岩，云母斜长片岩，变粒岩夹角闪岩；中—新元古界，包括长城系、蓟县系和青白口系，岩石类型主要有大理岩、浅变质灰岩、浅变质白云岩及浅变质砂岩等；寒武系，底部为昌平组豹皮灰岩，其上馒头组主要为紫红色泥岩，中部张夏组以鲕粒灰岩为主，上寒武系主要为炒米店组泥质条带灰岩夹竹叶状灰岩；未见奥陶系。该地区的中生代陆相盆地主要有西北部四海盆地、东南部密云-怀柔盆地和东北部古北口附近盆地，西北部四海盆地和东北部古北口附近盆地出露侏罗纪髫髻山组和后城组火山岩及火山角砾岩，属粗安岩-粗面岩-流纹岩组合，多以断层与元古宙盖层相接触；东南部密云-怀柔盆地内出露中侏罗统九龙山组砂、砾岩及下白垩统张家口组火山岩，多被第四系覆盖。

云蒙山岩浆杂岩体宏观上呈卵圆形，由大小不等的 15 个岩浆岩侵入体组成，出露总面积约 300km^2。这些岩体普遍的特征是不同程度地发育片麻状构造，并普遍遭受了强烈韧性变形改造，构成了北北东（近南北）向展布的短轴背形核部。与新太古代杂岩和变形变质的中—新元古代沉积岩共同卷入了云蒙山背形的构造系统中，各侵入体在形成时代上存在明显的先后关系，与围岩的变形是统一而协调的。

许继峰（1994）对云蒙山岩浆杂岩体 6 个全岩 Rb-Sr 的同位素分析表明成岩年龄为（126±21.2）Ma。Davis et al.（1996）在该区进行了详细的野外工作和锆石 U-Pb 年龄测定，提出云蒙山岩体的侵位年龄（143~141Ma）已逐渐被人们所接受（张建新等，1997；朱大岗等，2002）。刘翠（2004）应用锆石 SHRIMP 定年方法对云蒙山片麻状花岗岩进行年代学研究，认为（144±4）Ma 代表了云蒙山岩浆杂岩体的侵位时间。

2. 主要构造要素

（1）四合堂韧性剪切带

四合堂韧性剪切带是云蒙山岩基北部发育的一条逆冲型韧性剪切带，该剪切带沿梧桐豪—四合堂—捧河岩呈弧状展布，剪切带上盘包括太古代变质基底与中—新元古代盖层，而下盘为云蒙山花岗岩及石城闪长岩岩体。早期研究称为三棱山韧性剪切带（王恩林等，1987），现称四合堂韧性剪切带（Davis et al.，1994，1996；鲍亦冈等，2001）。该韧性剪切带由糜棱岩、千糜岩与片岩组成。剪切带内岩体和基底变质岩一同被强烈面理化，广泛发育糜棱面理，并出现透入性的矿物拉伸线理。糜棱岩面理缓倾（倾角为10°~30°），倾向从西部的NW300°渐变到东部的NE20°。线理倾伏方向变化于345°~25°之间，倾伏角一般为10°~30°。这些强烈韧性变形构造带形成围绕云蒙山岩基向外缓倾的同心环状构造，后期被数条北北东走向的高角度脆性断裂切割破坏。卷入韧性剪切变形的新太古代变质岩发生了退变质改造，而中元古代沉积岩则发生绿片岩相动力变质作用。

关于剪切带的剪切指向和形成时代，认识则很不统一。Davis et al.（1996）根据四合堂韧性剪切带位于四合堂倒转背斜的下翼，剪切指向应与褶皱的倒向一致，认为是一向南逆冲的韧性剪切带，沿韧性剪切带同构造侵入的变闪长岩和花岗闪长岩年龄（159~141Ma）代表剪切带的活动时代。Davis等（1996）判断的早期向南逆冲剪切指向有一定依据，但似乎忽略了晚期花岗岩就位、云蒙山快速隆起所导致的反向剪切。另一观点注意到上部向北剪切的证据（王恩林等，1987；鲍亦冈等，2001）。张家声等（2007）根据沙坨子剖面特征，认为该剪切带原为一向南缓倾、上部向南剪切的剥离断层，后因云蒙山岩体就位上隆，将剪切带倾向北倾，导致逆冲运动的假象。这种解释可说明早后期剪切方式的不一致和伸展导致的地层缺失，然而自北向南的伸展性滑脱不能解释剪切带内地层的多次重复，即所指的“褶皱层”。陈印（2014）沿剪切带采集了一系列定向样品，其显微照片显示，大量的S-C组构、长石或角闪石残斑拖尾等表明四合堂剪切带上盘向南南西运动，对长英质变形定向样品进行石英C轴组构测量结果也指示上盘向南南西运动。陈印（2014）通过同构造岩脉的锆石U-Pb定年，限定了四合堂剪切带的活动时间为143~138Ma。

（2）云蒙山背形

云蒙山背形北东向倾伏，倾伏角25°。现在只能确定其相对年代，即四合堂背斜之后，河防口断层形成之前或与之同期。核部主要为花岗闪长岩，锆石U-Pb年龄为151~127Ma，代表岩浆形成时代。岩体冷却史表明在118Ma冷却速率急剧加快，暗示岩体在中白垩世的快速隆升（Davis，1996）。吴珍汉等（2000）分析云蒙山岩体等热历史结果表明，106~103.93Ma期间速度快速隆升。

（3）大水峪韧性剪切带

大水峪韧性剪切带沿河防口—石城—冯家峪一线展布，分别穿过长园、云蒙山与石城岩体东南缘及北部变质基底与中—新元古代盖层之间。该剪切带宽约1km，所切割的岩体面理化显著，矿物拉伸线理发育，常发育初糜棱岩、糜棱岩甚至超糜棱岩。大水峪韧性剪切带由南向北，走向上呈北东—北北东向变化，指示了受到后期不均匀隆升的影响。剪切带一致向南东倾，具有北缓南陡的变化规律，优势倾角由北向南30°向60°变化。矿物拉伸线理一致向南东倾，以倾向或近倾向线理为主，倾伏角随面理产状变化。剪切带内部一系列不对称长石碎斑、S-C组构、C面理/残斑拖尾、不对称布丁和其他运动学标志，均指示上部向下和向南东的剪切方式（陈印，2014）。根据糜棱岩和矿物组合、亚晶粒和重结晶晶粒的平均粒度（史兰斌等，1984）与氧同位素的研究（上官志冠，1985），估计该剪切带的形成深度为10~15km。陈印（2014）结合前人资料和实验定年结果限定了四合堂剪切带的活动时代为143~138Ma，略晚于云蒙山岩基的侵位。野外观测表明，大水峪剪切带切割云蒙山岩基和四合堂剪切带，指示其活动时间晚于138Ma。而结合锆石U-Pb定年与角闪石、黑云母$^{40}Ar-^{39}Ar$定年表明，126Ma时大水峪剪切带仍在活动，而120~116Ma期间该剪切带快速抬升至300℃等温面。

（4）河防口低角度正断层

河防口低角度正断层位于云蒙山的南东翼，沿河防口—大水峪—黑龙潭—石城一线展布，南段（河防口段）走向北东，北段（石城段）走向北北东，倾向南东，倾角为25°~70°。断层面出露于河防口村北和溪翁庄北（梨树沟）至黑龙潭（大关桥）公路的路堑两侧。断层面上一般见发育黄色或灰黑色富伊利石断层泥，厚5~6m，其内部具有剪切组构（由条带层和择优取向的透镜状岩片组成）。剥去断层泥，下盘为坚硬的微角砾岩的断层面上，显示倾向擦痕，指示断层上盘下滑的正断运动。已知该断层错断的最新岩层为侏罗系。在河防口采集的富伊利石断层泥的K-Ar年龄为晚白垩世（（89±4）Ma~（72±2）Ma）（王玉芳等，1989）。推断该断层运动的末期或重新活动的时期为晚白垩世。王巧云（2006）从黑云母的变形特征和形成环境分析入手，对云蒙山东南侧剥离断层带形成时的稳压条件进行研究，认为河防口低角度正断层是在地壳伸展背景下形成的，大水峪韧性剪切带及云蒙山变质核杂岩是在花岗岩冷却和隆升过程中不断向浅层次过渡的产物。

3. 几种成因认识

关于云蒙山隆起的成因，郑亚东等（1986）最早提出热隆-滑覆-推覆构造模式，其注意到云蒙山区构造格局与运动学的对称性，将其内侧低角正断层构造和外侧逆冲断层的形成归因于云蒙山核部花岗岩基的热力隆起，在重力驱动下的内侧伸展，外侧缩短。

Davis（1996）根据野外系统观测与填图，运用构造分析与年代学研究相结合的方法，查明了云蒙山地区主要构造或地质体的产状，运动学方式与形成时代，如：晚侏罗世四合堂向南倒转背斜、下翼韧性剪切带、变闪长岩、云蒙山核部早白垩世花岗闪长深成岩体的形成、花岗闪长岩体的快速隆升并伴随河防口断层及其下盘韧性剪切带的形成等，认为云蒙山地区是一种叠加于南北缩短的北西-南东向科迪勒拉型变质核杂岩模式。

Passchie 等（2005，2007）深入系统地研究了云蒙山花岗闪长岩体周边侵入在变闪长岩中同构造花岗岩脉体几何形态、产状、岩石化学与变形运动特征，认为云蒙山隆起属于一种叠加于南倾伸展拆离的热隆模式。云蒙山地区的变形构造表现出早期的中高温韧性变形和晚期的中低温韧-脆性变形，反映了挤压-伸展体系具有复杂的构造演化历史：①早期（晚侏罗世—早白垩世）的地壳缩短，即四合堂及稍后的大水峪韧性剪切带、云蒙山背形的形成；②晚期（早白垩世—晚白垩世）的地壳伸展，即花岗岩体隆升产生的崎峰茶-琉璃庙滑脱带及更晚些的低角度正断层。环云蒙山韧性剪切带从下至上其变形温度逐渐降低，表明变质核杂岩—花岗岩穹隆的构造演化过程中，花岗岩冷却的热历史及其对韧性剪切带的作用和影响。环云蒙山剪切带为变质核杂岩—花岗岩穹隆燕山期隆升时变质岩基底（Ar）、沉积盖层（Pt_2）与花岗岩之间滑脱作用的产物。根据X射线岩石组构类型导出的S面推断，剪切运动主应力方向总体从早期的北西-南东向正剪切运动，到晚期的上层向下、向外的环形韧-脆性滑脱，且在隆升过程中，叠加了相同运动方向的浅层次脆性构造及与韧性剪切带产状近于一致的河防口低角度正断层。四合堂韧性剪切带形成阶段的区域伸展的拉张环境，使得地壳岩石发生部分熔融（图32-18A）。随着花岗闪长岩向上侵位所造成的浮力增加（岩浆底辟），上覆岩石受重力驱动开始向四周拆离，云蒙山剪切带形成，逐渐冷却的岩基顶部也参与了变形。岩浆源继续分异的物质得以沿张裂隙以岩脉的形式侵入到活动的剪切带中，并随即发生了变形。与此同时，上覆的脆性地壳以发育高角度正断层的方式不断垮塌和剥蚀，使下部地壳实现重力卸载。原来处在下地壳层次的太古宙变质结晶基底得以在下部岩浆底辟和上部卸载的联合作用下，逐渐抬升出露地表（图32-18B），这一模型提供了变质核杂岩抬升的动力学证据。

郑亚东（2010）基于前人提出的模式最新的研究成果，对原热隆-滑覆-推覆构造模式做了一些修正，认为云蒙山地区在晚侏罗世（159Ma）南北缩短作用下，形成南翼倒转的密云大背斜，倒转翼应变软化形成四合堂韧性剪切带，并推测其浅层前缘可能为十三陵水库一带逆冲断层；早白垩世早期（142~144Ma），因前述缩短事件地壳增厚，下地壳熔融形成云蒙山花岗闪长岩；早白垩世中晚期（118~104Ma），花岗岩体快速隆起。在隆升力和重力的作用下，地壳侧向伸展，云蒙山北西翼和南东翼分别形成琉璃庙伸展型飞来峰和河防口低角度正断层。

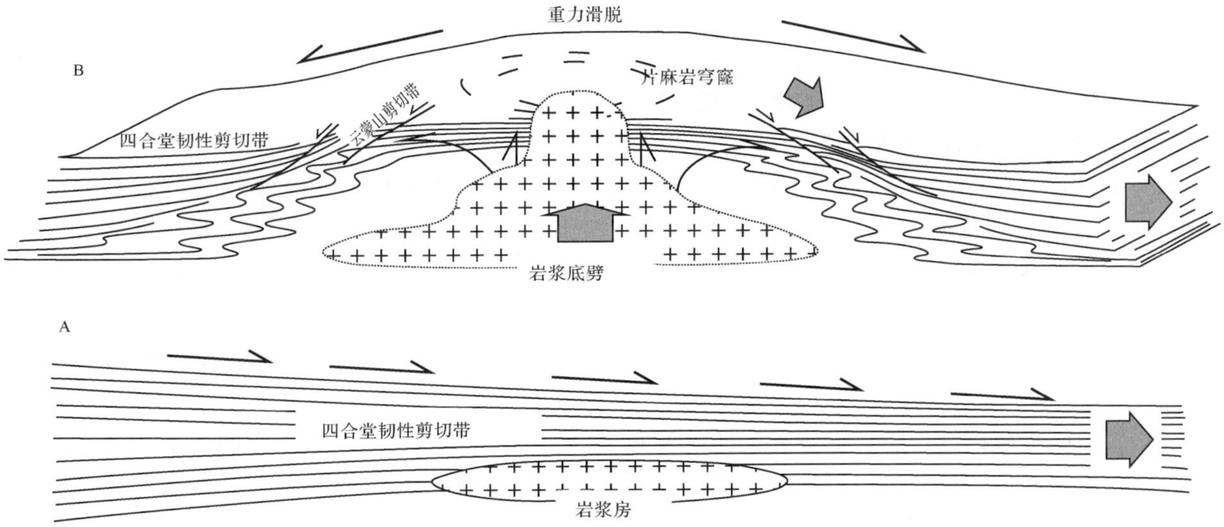

图 32-18　云蒙山核杂岩抬升的动力学模型
（据张家声，2007）
Fig. 32-18　Geodynamic model for Yunmengshan metamorphic core complex uplifting

陈印（2014）认为在变质核杂岩形成之前，于 148～141Ma 期间（早白垩世初）发生了大型云蒙山岩基侵位，后因区域性挤压在岩体北缘形成了原始近东西向的逆冲型四合堂韧性剪切带（143～138Ma），并断定变质核杂岩核部大型的云蒙山岩基属于变质核杂岩发育之前侵位的岩体；在 135～125Ma 期间，云蒙山变质核杂岩发生了较强的岩浆活动，该岩浆活动诱发了变质核杂岩的均匀隆升，大水峪韧性剪切带强烈活动；在云蒙山变质核杂岩演化后期（125Ma 以后），该变质核杂岩在 125～114Ma 期间发生了快速隆升，并使石城、云蒙山及北部的一系列岩体逐渐剥离地表，早期的四合堂剪切带在此期间也与大水峪剪切带一同被抬升剥露；114Ma 之后的早白垩世可能为该核杂岩缓慢抬升与弱伸展活动阶段，此后则停止了活动。

四、韧性剪切带

本区韧性剪切带的研究工作始于 20 世纪 80 年代，王玉芳（1981）厘定了怀柔区大水峪低角度韧性正断层，揭开了本区对地壳中—深层次构造的研究的序幕，随后有很多学者对本区深入研究，不断有成果发表。在 1∶20 万怀柔幅区调中，按照基底和盖层不同时期发育的韧性剪切带进行了描述。鲍亦冈（2001）中针对基底的韧性断层划分不同的变形层次。本次工作将按照韧性剪切带研究历史、空间分布、基本地质构造特征、形成机制和形成时代进行叙述。

（一）基底发育的韧性剪切带

结晶基底岩石的韧性剪切构造比较发育，由韧性剪切带形式表现出来。依据剪切带内构造岩变形强度、变质程度，可以划分出深层次麻粒岩相韧性剪切带、中层次角闪岩相韧性剪切带和浅层次绿片岩相韧性剪切带 3 种类型。

1. 深层次麻粒岩相韧性剪切带

广泛分布于太古宙岩石中，可进一步划分为强变形带和弱变形带两个级别。

1）强变形带主要发育在中太古代中基性侵入岩体和密云岩群残片中。侵入岩体经韧性剪切变形变质作用的有沙厂英云闪长质片麻岩和苇子峪片麻岩套。其中以沙厂英云闪长质片麻岩变形最为强烈。构造岩以条带状和强片麻状构造为特征，变质矿物发生了强烈的变形改造，发育辉石链、石榴子石条带链及磁铁矿拉伸线理。长石定向拉长形成线理且不具波状消光，石英形成单晶或多晶条带。剪切褶皱和塑性流变褶皱发育，常见 A 型褶皱。表壳岩被拉长拉断，形成布丁石香肠构造，总体上呈断续条带状分布。

2）弱变形带主要反映在新太古代形成的岩石单位中，构造岩以弱片麻状构造为特征，除局部见到透辉石链状结构外，其他麻粒岩相矿物变形特征不明显。

2. 中层次角闪岩相韧性剪切带

此类型韧性剪切带出现在大中型复式褶皱翼部。已发现规模较大的韧性剪切带分布在关上、聂家峪、沙厂、穆家峪—潮河主坝、上庄头峪等地。带宽数十米至数千米，长数百米至数千米。规模最大的关上韧性剪切带宽度可达5km，可见长度达10km以上。构造岩具典型的板状构造和条带状构造特征。变形分异条带的暗色矿物主要由角闪石和黑云母组成。变形带中已见不到早期麻粒岩相变质矿物。新生角闪石、黑云母沿条带强烈定向分布，并发育生长线理，局部见有A型褶皱和箭鞘褶皱。

中层次角闪岩相韧性剪切带呈带状叠加于麻粒岩相韧性剪切带上，剪切带走向一般为北东东—北北东向，由剪切面理产状显示出剪切带以高角度倾向北西。面理上发育的角闪石、黑云母线理倾伏方向以北东向为主，倾伏角10°~25°。

本期韧性剪切带可能形成于新太古代末期的阜平运动。本次变形过程中，早期麻粒岩相岩石普遍发生了不同程度的角闪岩相退变质作用，且以韧性剪切带内岩石退变质最为彻底。除此之外，在崎峰茶断裂与石城断裂之间的地堑断块中，太古宙变质岩普遍为角闪岩相，并被后期绿片岩相韧性剪切带叠加改造。此处角闪岩相变质岩应为早期深层次麻粒岩相韧性剪切带上部层次韧性剪切变形的产物。

3. 浅层次绿片岩相韧性剪切带

这类构造呈带状分布于杨家台、沙厂、半城子等地。形成于不同时期、不同的构造环境。

（1）杨家台、沙厂韧性剪切带

呈北北东向展布，宽达百米、长数百米。构造带以含硬绿泥石、阳起石为特征，并形成线理。叠加于角闪岩相韧性剪切带上。这次绿片岩相变形变质可能是古元古代拉伸环境下造成地幔上隆、热流上升，在区域性绿片岩相退变质作用下，局部韧性变形的结果。

（2）半城子韧性剪切带

分布于沙厂英云闪长质片麻岩中，总体呈北东向楔形展布，向南尖灭于半城子水库以南，北端复合于近东西向古北口-赤城断裂带上。该带北宽南窄，出露长度7km。带内发育糜棱岩、初糜棱岩、千糜岩、构造片岩，并出现了硬绿泥石化现象。

由南西向北东，剪切带倾向320°~360°，倾角35°~75°，拉伸线理指向20°~90°。运动方向为右行斜冲。该剪切带与古北口-赤城断裂带平面上构成“入”字形，可能为东西向断裂右行斜冲运动造成的分支韧性断层。剪切带向南延伸被燕山期柏岔山岩体（早白垩世）所截，推测剪切带产生时代为晚侏罗世—早白垩世，是后期构造叠加在古老基底上的韧性断层。

（二）盖层发育的韧性剪切带

主要分布于沙峪-云蒙山燕山期侵入岩体的边侧、密怀隆起北侧东西向构造带内。主要断层有四合堂韧性剪切带、龙潭韧性剪切带、上李家韧性剪切带、莲花池韧性断层、水堡子韧性剪切带、水峪韧性剪切带、云蒙山花岗岩体周边韧性剪切带、水泉沟-石洞沟韧性剪切带等。

1. 四合堂韧性剪切带

该韧性剪切带位于密怀隆起的西部，崎峰茶断裂与石城断裂之间，西自兰旗棚子，向北经琉璃庙、四合堂转向东至石炮沟，沿云蒙山复式岩体北部周缘分布，呈向北突出的弓形展布。全长约30km，宽3~8km。韧性剪切带发育于太古宙变质岩与中元古界长城系中，由规模较大的6~8条彼此近于平行的强应变带夹弱应变域组成，组合形态平面上呈平行束状，剖面上呈叠瓦状。王恩林（1987）提出在该区域发育的变质岩是韧性变形的TTG，而不是前人认为的表壳岩，并称之为三棱山剪切带。在1：20万怀柔区调报告中，该描述了该剪切带在图幅范围内的发育特征，称之为琉璃庙-石炮沟韧性逆断层。

在鲍亦冈（2001）将该韧性剪切带称为四合堂韧性剪切带。陈印（2013）重新厘定了四合堂剪切带的时空演化规律。他认为四合堂剪切带受后期均衡隆升影响而呈弧状展布于云蒙山岩基的西缘、北缘及东北缘。在云蒙山岩基东北缘，该剪切带被后期大水峪剪切带所叠加。在该剪切带内部发育北北东-南南西向矿物拉伸线理，面理倾角较小（20°）。露头构造、显微构造及石英 C 轴组构均指示该剪切带上盘向南南西运动。剪切带上盘太古宙变质基底及下盘云蒙山岩基均卷入韧性变形。剪切带内部发育大量的同构造岩脉，受剪切带变形影响发生不同程度的置换。认为云蒙山岩基为四合堂剪切带同构造侵位岩体。

本韧性剪切带具有如下特点：

1）该韧性剪切带可分为三段：西段（梧桐豪附近）走向北北东向，倾向北西西向；北段（四合堂附近）走向近东西，倾向北；东段（沙陀子附近）走向近南北向，倾向北东。剪切带分别切过太古宙基底片麻岩、中元古界长城系和中生代岩体。根据陈印等（2013）等野外实地观察，北段与东段剪切带沿长城系盖层顶、底延伸，主体出现在变质基底和中生代岩体内。在四合堂村附近，该剪切带为 2~3 条糜棱岩带组成的复合韧性剪切带。四合堂韧性剪切带主要由初糜棱岩和糜棱岩组成，局部出现超糜棱岩，其中没有发育断层角砾岩、断层泥等脆性构造系列。西段剪切带主体位于基底片麻岩内，向南消失于元古宙盖层之下。梧桐豪以北的西段剪切带，云蒙山岩体受剪切带影响而发生塑性变形，但在梧桐豪以南，云蒙山岩体基本上未卷入剪切带变形。卷入剪切带的变质岩和长城系各组地层，其层序倒置且重复出现或缺失，显示上盘上升的构造特征（图 32-19）。

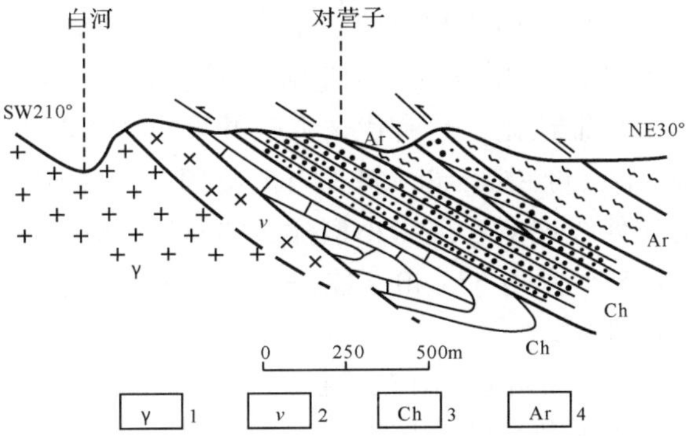

图 32-19　四合堂韧性剪切带剖面示意图

Fig. 32-19　Sketch showing the profile of Sihetang ductile shear zone

1—云蒙山花岗岩；2—石城辉长岩；3—长城系石英砂岩、粉砂岩、白云岩（糜棱岩化、片理化）；4—太古宙变质岩（糜棱岩）

2）该剪切带内岩体和基底变质岩被强烈的面理化，广泛发育糜棱面理，并出现了透入性的矿物拉伸线理，其中在云蒙山花岗岩内线理主要为拉长的石英和定向排列的长石。在西端剪切带内，糜棱面理倾向 280°~350°，倾角 20°~50°。矿物拉伸线理倾伏向 40°~320°，倾伏角 25°~45°。北段剪切带的糜棱面理和线理较为稳定，线理倾伏向北北东，在赤平投影上呈点密集。其中糜棱面理的倾向 340°~65°，倾角 12°~36°。线理倾伏向 2°~37°，倾伏角 10°~35°。东段剪切带产状变化较大，糜棱面理极点呈大圆环带分布。由于叠加了晚期大水峪韧性剪切带变形，线理的赤平投影总体也呈大圆环带分布，但是两期线理之间具有明显的差异，其中东段四合堂韧性剪切带的面理倾向和线理倾伏向均为北东，而晚期大水峪韧性剪切带叠加的矿物拉伸线理倾伏向为南东。总之，弯曲的四合堂剪切带虽然糜棱面理产状多变，但其中具有较为稳定北北东—南南东向矿物拉伸线理。该剪切带沿云蒙山复式岩体北部边缘延伸，指示其弯曲是该岩体上拱所致。

3）由磁铁石英岩形成的石香肠构造和长石碎斑构成的旋转构造、显微尺度上的 S-C 组构、石英 C 轴组构、旋斑拖尾、标志体（主要为岩脉）的褶皱及布丁化等现象一致揭示四合堂韧性剪切带的运动方向为上盘向南南西运动。四合堂剪切带的形成过程中，垂直于面理方向具有很强的正压力，其旋度值较小，以纯剪变形为主。

4）四合堂剪切带内岩体、盖层和基底变质岩转变成了初糜棱岩或糜棱岩。基底内剪切带糜棱岩残斑主要为角闪石和长石，基质中多为动态重结晶的角闪石和长石及角闪石退变的黑云母，长石呈SR（亚颗粒旋转）型动态重结晶。剪切带内的糜棱岩化花岗岩脉残斑主要为长石及石英集合体，基质主要为动态重结晶的长石与石英，前者呈SR与BLG（嵌入式）型重结晶共存，而后者为GBM（颗粒边界迁移）型重结晶。卷入剪切带的盖层同样也发育了糜棱岩，其中的长石与石英广泛出现了动态重结晶。

从以上资料分析来看，四合堂韧性剪切带原始走向为近东西向，是在由北向南逆冲推覆作用下，发生于中浅层次的低角度韧性剪切的产物。关于其形成时代有太古宙、印支期和燕山中晚期等不同观点。王继明（2001）认为其形成于印支期的可能性较大。陈印等（2013）通过对该剪切带的构造和年代学研究，认为该韧性剪切带为逆冲型韧性剪切带，上盘向南南西运动。并测定了剪切带内3阶段同构造岩脉锆石的LA-ICP-MS定年结果，可靠地限定了四合堂韧性剪切带的活动时间为138～143Ma。

2. 龙潭韧性断层（F55）和上李家韧性断层（F56）

这两条韧性断层分别出露于长元二长闪长岩岩体的北、南东侧内接触带附近。龙潭韧性断层长12km，宽数十米至400余米，呈东西向展布，倾向165°～190°，倾角40°。上李家韧性断层（即上李家-环坨峪韧性断层，简称上李家韧性断层）断续出露长度11km，宽约200m，总体走向北东东，倾向120°～170°，倾角45°～60°。剪切带内二长闪长岩岩石强烈糜棱岩化，以形成纹层状闪长质糜棱岩为特征。岩石具糜棱结构，条带状、眼球状构造和剪切褶皱，发育S-C组构及拉伸线理，塑性变形矿物为石英、黑云母和部分斜长石。石英呈多晶条带、构造亚颗粒，具波状消光。斜长石可见变形纹弯曲，边缘轻度粒化，组成碎斑；黑云母细粒化，具有优选方位并聚集成带与石英构成黑白相间的条带构造。新生矿物有绢云母，形成石英+绢云母+黑云母+斜长石矿物组合，变质条件相当于绿片岩相，推测形成于中—浅层次。龙潭断层的拉伸线理多表现为小型揉皱的枢纽，倾伏向105°～110°，倾角10°；下李家断层的拉伸线理倾伏向80°～95°，倾角5°～40°根据S-C组构剪切褶皱指向构造指示剪切方向由东往西，断层性质属顺扭逆冲和斜冲性质。

断层形成与长元岩体（锆石U-Pb年龄（152.7±3.2）Ma）侵位密切相关，该岩体位于密云-墙子路近东西向断裂带和十三陵逆冲推覆断裂的交接部位，剪切带位于岩体内接触带附近，走向与接触面平行。在燕山运动的中期末，受北西-南东向挤压应力场控制，近东西向断裂发生右行平移和斜冲运动。同期侵位的长元岩体受到同构造活动的影响，在受应力作用最强的北侧和东南侧发生顺扭斜冲，形成近东西向的龙潭韧性剪切带和北东东向上李家韧性剪切带，由于二者所处的构造部位不同，龙潭断层的规模和变形强度大于上李家断层。

3. 莲花池韧性断层（F57）

位于怀柔慕田峪、莲花池、交界河一线，走向北东30°，倾向北西，倾角60°～70°，长6km，推测原始长度约20km。断层发育在长元二长闪长岩和官山二长岩体中，向北被五道河、大石门子、沟脑岩体侵位破坏。但沿云蒙山岩体与五道河、大石门子等岩体的接触面两侧，有太古宙变质岩和长城系岩块呈狭长楔形捕虏体（或顶蚀体）带状分布，岩块发育挤压破碎带，产状与莲花池韧性断层相一致，表征韧性断层原始长度大于现存长度，可达20km。控制了大石门子、五道河等岩体的侵位。断带中发育糜棱岩，其特点基本同龙潭韧性断层的糜棱岩，不同的是同构造新晶出现绿帘石，具优选方位，倾伏方向同剪切面理倾向。莲花池韧性剪切带向南延伸切割了龙潭韧性剪切带，向北控制了燕山晚期大石门子、五道河岩体的侵位边界。其形成时代应为燕山运动的中、晚期。是受区域北西-南东向挤压应力场控制，在中浅层次热动力作用下形成的线型逆冲性韧性剪切带。

4. 水堡子韧性剪切带（F53）

发育于石城辉长岩岩体中，分布在云蒙山花岗岩体的东、北东侧，南起北白岩村北经梨树沟、水堡子、北石城、二平台至对营子村西，全长约20km，宽约1km，呈向北东突出的弯月形。辉长岩普遍糜棱岩化，形成眼球状初糜岩和条带、条纹状糜棱岩，在二平台—对营子一带可见黑云母构造片岩。岩石片理、片麻理发育，岩石变形强度各地不一。剪切带宏观构造中普遍发育S-C组构、眼球构造、杆状构造、拉伸线理和长英质脉体拉断的石香肠构造，微观构造有石英单晶、多晶条带、变形纹、机械双晶、波状消光等。塑性变形矿物主要是石英，发育石英拔丝、多晶条带亚颗粒。斜长石既有塑性变形，又有脆性变形，颗粒边缘轻度粒化形成碎斑。角闪石部分退变成黑云母，黑云母强烈定向，具有透入性。剪切面理在大观桥以南倾向60°~110°，倾角14°~30°，角闪石拉伸线理南东135°~170°倾伏，倾角0°~20°。在大观桥以北倾向北东10°~20°，倾角10°~25°，由花岗岩脉构成的杆状构造向北东10°~15°倾伏，倾角26°。指向构造有剪切褶皱、S-C组构、伟晶质脉石香肠等（图32-20，图32-21），指示上盘下滑，具正断层性质。拉伸线理显示韧性剪切带具有北段向北北东、南段向南东的斜向滑脱特征。

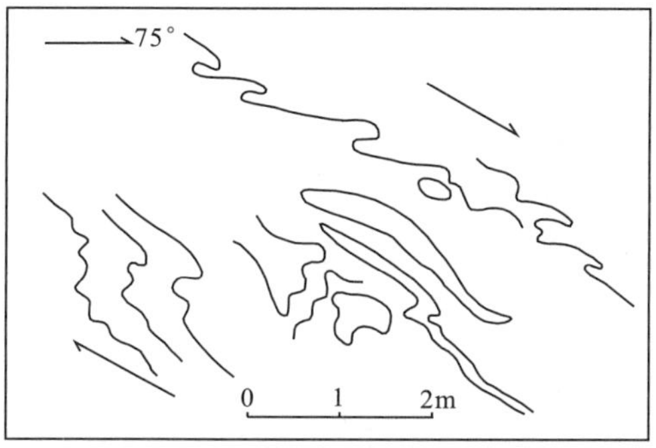

图32-20　石城辉长岩体中不同成分条带表现出的滑覆型弯褶素描图
（据鲍亦冈，2001）
Fig. 32-20　Sketch showing flexural-slip folds of Shicheng gabbro indicated by mineral strips
（素描点位于水堡子西600m处）

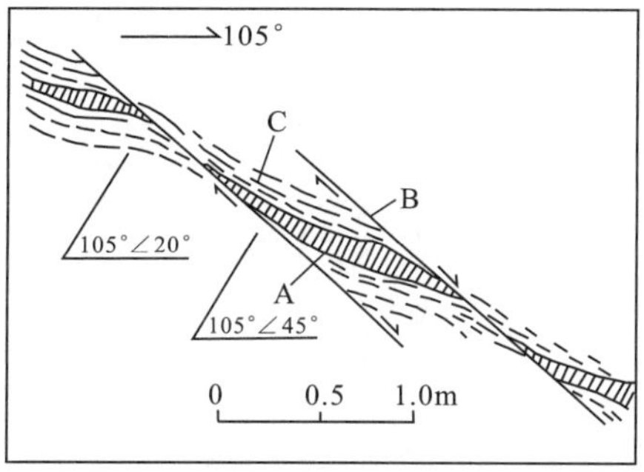

图32-21　石城辉长岩体内伟晶岩脉与韧性剪切面理关系素描图
（据鲍亦冈，2001）
Fig. 32-21　Sketch showing relation between pegmatite dyke and ductile shear foliation within Shicheng gabbro
（素描点位于水堡子西700m处）
A—伟晶岩脉；B—剪切面理；C—岩体片理

5. 大水峪韧性剪切带（F51）

发育于怀柔区北河防口至密云县黑山寺一线，大水峪中粗粒似斑状花岗岩体的南侧，长度>10km，宽1~2km，倾向南东，倾角35°~45°。带内岩石变形强烈，表现为花岗质糜棱岩、糜棱岩化花岗岩（图32-22）。岩体外侧围岩为太古宙变质岩、蓟县系铁岭组白云岩和小闪长岩体，也见有糜棱岩化和鞘褶皱等韧性变形构造。韧性剪切带内岩石S-C组构极为发育，指向构造表明该断层具上盘下降的正断层性质。据王玉芳、史兰斌等研究，大水峪韧性剪切带是在围压40MPa，深度10~12km，温度320~450℃，缺水环境下形成的，并估算其总位移量达5.3km。带内岩石发生强烈的塑性变形，长英岩岩脉纵向拉长形成韧性石香肠，岩石强烈揉皱并形成小型的韧性剪切带（图32-23）。垂直剪切带自北至南，即自边界至近中心部位，岩石和矿物的塑性变形、矿物化学成分、单矿物内部微化学组成的变化，均随剪切应变的递增而逐渐增强，原来的片麻状花岗岩等依次渐变为糜棱岩化花岗岩、长英质初糜棱岩和糜棱岩。王玉芳等（1989）从大水峪韧性剪切带糜棱岩中获得过113~106 Ma的黑云母K-Ar年龄。Davis等（1996b）报道了大水峪韧性剪切带内钾长石$^{40}Ar-^{39}Ar$年龄为118Ma。Wang等（2012）从大水峪剪切带糜棱岩中获得的角闪石$^{40}Ar-^{39}Ar$年龄为125 Ma，黑云母年龄为120Ma，钾长石年龄为114 Ma；而旁侧未变形岩体的锆石U-Pb年龄为127 Ma，黑云母$^{40}Ar-^{39}Ar$年龄为118 Ma和122 Ma。

6. 云蒙山花岗岩体周边韧性剪切带（F54）

分布于云蒙山花岗岩体的边部，空间展布呈宽1~2km北东向略长的环带状，出露长度>50km。剪切面理向四周倾斜，北部—南西部倾角10°~20°，北西侧及南东侧倾角25°~45°。剪切带内岩石具强弱不一的糜棱岩化，变形强度由内向外增大，形成糜棱岩化花岗岩、花岗质初糜岩、糜棱岩，局部发育超糜棱岩。糜棱岩以发育长石碎斑眼球体、石英拔丝条带以及黑云母呈细条纹定向分布为特征。糜棱岩叶理和剪切面理发育，从S-C组构指向构造判断，剪切带具有正断滑覆性质，属韧性正断层（图32-23）。形成机制是云蒙山花岗岩体的侵位上升和膨胀作用，形成时代为燕山期晚期（云蒙山岩体年龄：锆石TIMS年龄为（143±2）Ma、（142±2）Ma；锆石U-Pb年龄为（144.7±2.7）Ma）。

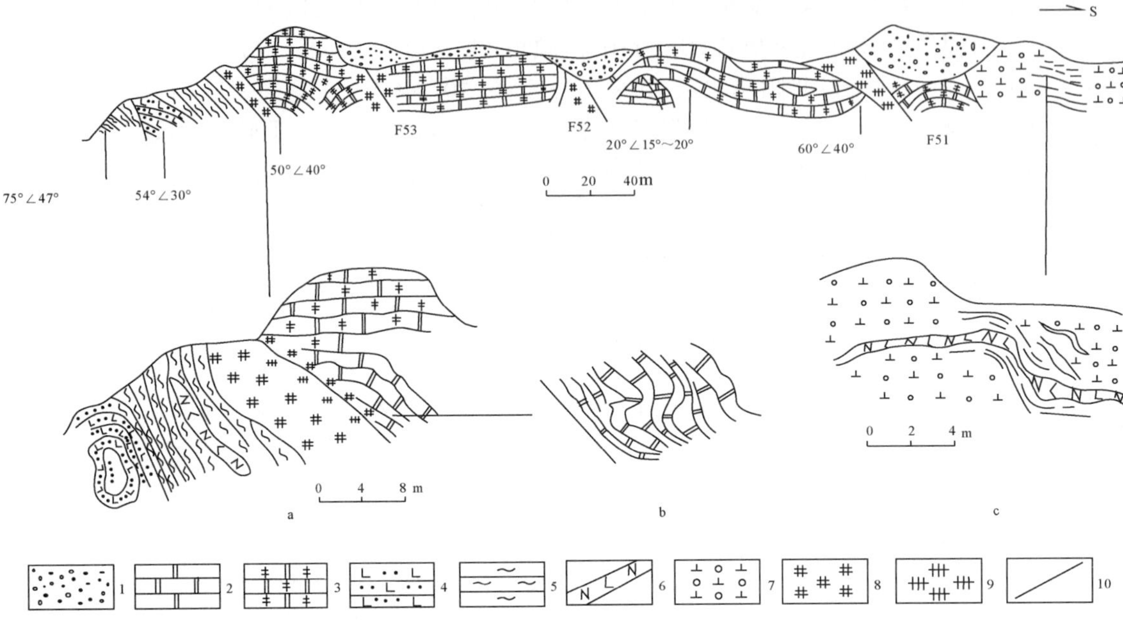

图 32-22 怀柔大水峪断层河防口剖面

（据史兰斌，1984）

Fig. 32-22 Hefangkou profile of Dashuiyu fault in Huairou

a—断层接触关系；b—面附件大理岩层的褶皱；c—石英闪长岩中长英岩岩脉的韧性剪切变形；
1—第四系堆积物；2—大理岩；3—碎裂大理岩；4—钙质石英岩；5—片麻岩；6—长石石英脉；7—石英闪长岩；
8—碎裂岩；9—超碎裂岩；10—断层

7. 水泉沟韧性剪切带

位于赤城-长哨营-古北口-承德深断裂上，近东西向展布，长度>14km，剪切带总体倾向北，倾角70°。其西段位于北京市内，出露长度约7km，东段进入河北省地界。

1∶5万番字牌区调报告中对该剪切带进行了调查，认为该韧性剪切带内主要岩石类型由两部分构成。其一为太古宙变质岩，经韧性剪切变形形成糜棱岩，构造岩由碎斑和基质组成。碎斑一般以石英和斜长石为主，常被压扁拉长，基质由微细粒状斜长石、石英、黑云母等组成。其二为长城系硅质泥质岩石和碳酸盐岩，经变形变质形成黑云石英片岩和绢云母石英片岩，岩石主要矿物为石英，少量斜长石；新生矿物为黑云母和绢云母，含量可达25%，片状定向分布。碳酸盐岩呈粒状集合体定向条带状。上述构造岩特征，表明其形成于中浅层次绿片岩相韧性剪切变形环境。韧性剪切带北侧与长城系和中生代火山沉积岩呈断层接触，南侧进入太古宙变质岩中。剪切带以南，发育近东西向高角度向北倾斜、彼此平行或微角度斜交的6~7条逆断层，共同组成叠瓦式逆冲断层系。剪切带局部可见被燕山期推覆-滑覆的长城系飞来峰地块压覆。报告中认为水泉沟韧性剪切带形成时代应为印支期。是在区域南北向挤压应力（主压应力由北向南）作用下形成的逆冲性韧性剪切带，其位于叠瓦式逆冲断层系的北侧根部，构造环境已达中浅层次。

东段在河北省的部分被命名为南沟门-四方地韧性变形带，长度约17km。构造岩为糜棱岩系列，绿片岩相，构造岩内发育S-C组构、不对称残斑、旋转石香肠构造及剪切构造，这些指向标志标明该剪切带为左行逆冲型（据1∶5万古北口幅区调报告）。报告中根据带内脉体产状及交切关系，糜棱岩

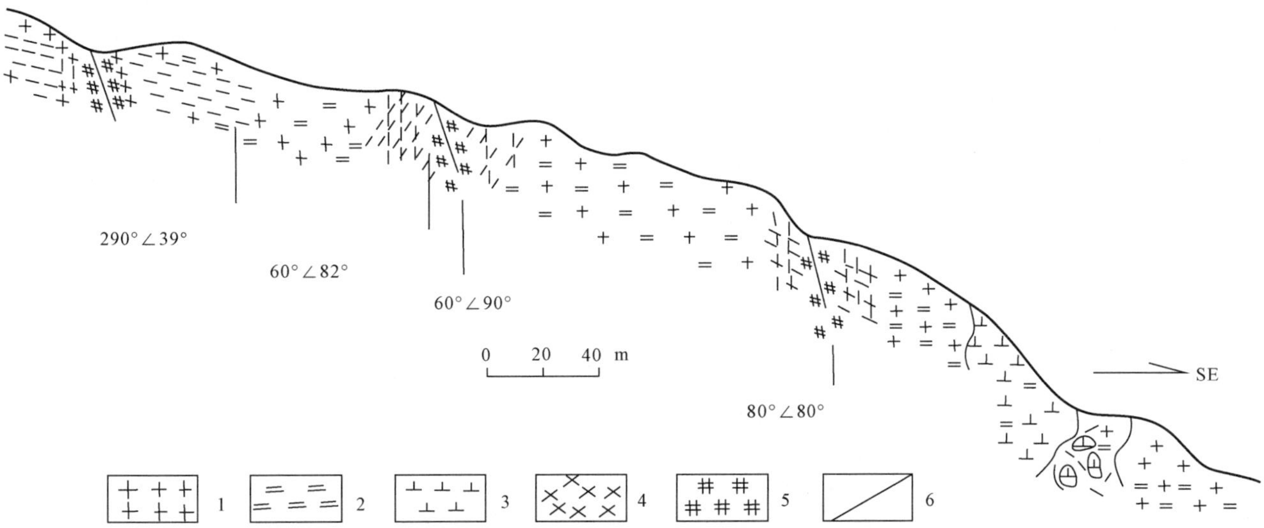

图 32-23 怀柔大水峪断层带大水峪剖面

（据史兰斌，1984）

Fig. 32-23 Dashuiyu tectonic profile of Dashuiyu fault in Huairou

1—花岗岩；2—糜棱岩化；3—闪长岩；4—劈理带；5—碎裂岩；6—推测断层

带的限制和切割特征，揭示该剪切带至少经历两次运动方向基本一致的剪切运动。早期高温剪切变形发生于新太古代末期，晚期中低温剪切变形发生于古元古代晚期，以后又遭受印支及燕山期脆性断裂活动的叠加改造。

综上，该剪切带的特征基本都认为是中浅层次的逆冲韧性剪切带，但是具体的形成时代还有待进一步的工作来进行深入研究。

五、固态流变构造群落

“褶叠层”的概念是单文琅等（1984，1987，1991）在研究周口店地区浅变质盖层构造时提出的，是指在伸展构造体制和一定温压条件的构造环境下，岩层发生水平分层固态流变剪切作用，形成了以顺层韧性剪切带和顺层掩卧褶皱为代表的固态流变构造群落。该群落的其他构造要素还包括透入性顺层流劈理、石香肠构造、窗棂构造以及各类拉伸线理等。虽然构造现象丰富，且伴生有绿片岩相的变质作用，但是盖层的基本层序并没有发生变化。这一奇特的构造-地层的变形-变质组合被称之为固态流变构造群落。

（一）分布

周口店固态流变构造群落大体上围绕房山岩体分布，在岩体旁侧的盖层中变质变形表现最为明显和强烈；向四周变质变形作用随着构造层次和涉及地层的改变而逐渐减弱或终止。北侧截止于九龙山向斜南翼；西侧限于霞云岭推覆构造断层面；南侧于黄山店推覆构造的下盘迅速变形减弱；东侧受到八宝山断裂的破坏，其断裂面东侧上盘的盖层岩石中未见类似构造发育。从整体上看流变构造发育的区域为呈北东向展布的矩形区。

对于周口店固态流变构造群落的分布范围有不同看法，宋鸿林等认为其北侧限于谷积山背斜的南翼；而西侧霞云岭推覆构造的上下盘其变质变形是渐变过渡的。在鲍亦冈（2001）对其分布进行了探讨：

1）根据龙门台幅区调资料，霞云岭推覆构造上下盘岩石的变形强度具有较明显的差异。上盘雾迷山组岩层中顺层掩卧褶皱和顺层流劈理仍较发育，而下盘铁岭组和下马岭组中没有类似构造，仅在下马岭组岩层中发育板理构造。所以霞云岭推覆构造的发育位置基本上是周口店固态流变构造群落的西部边界。

2）霞云岭断裂上下盘在断面附近的变质作用都不很明显，其中所含的变质矿物主要是绿泥石。

3）谷积山背斜北翼的上古生界岩层主要是由碎屑岩类组成，在同样的伸展体制构造作用下，由于与中心区的变形变质条件相比温压条件有一定的减弱，加之碎屑岩的能干性较灰岩类强，因此其变质变形强度明显不如中心区的青白口系和下古生界；岩层的减薄现象也不如中心区明显。但是其中仍然广泛发育顺层掩卧褶皱、顺层流劈理、各类拉伸线理等构造形迹，且其指向方向也与整体群落的活动方式相配套。因而谷积山背斜的北翼同样属于固态流变构造群落的分布范围。

（二）基本构造要素特征

固态流变构造群落由多种构造要素组成（图32-24）。它们各具特点，但又互为补充，是一个统一的伸展构造体制的产物。

图32-24　周口店固态流变构造群落主要构造要素

（据鲍亦冈，2001）

Fig. 32-24　Main structural elements of Zhoukoudian solid flow domain

1—顺层韧性剪切带；2—半协调顺层掩卧褶皱；3—不协调顺层掩卧褶皱（窗棂构造）；4—协调顺层掩卧褶皱；5—透入性流劈理（完全置换层理）

1. 顺层韧性剪切带

此类构造产生于相对较厚的能干层夹薄层非能干层的岩性组合中，在较深构造层次环境的水平剪切应力作用下，能干层因为较厚而不易屈服，使得薄弱层发生强烈应变形成韧性剪切带，伴随的变质作用也发生在薄弱层中。顺层韧性剪切带是固态流变构造群落中具有标志意义的构造，其中S-C组构具有明显的指向意义。如黄山店铁岭组内发育的韧性剪切带中S-C组构指示其上盘相对向110°方向滑动。

2. 顺层掩卧褶皱

顺层掩卧褶皱在组成周口店固态流变构造群落的各个层位都有不同程度的发育。在层系厚度较薄且均匀、岩石能干性较差的碳酸盐岩建造和细薄层为主的碎屑岩系中发育程度最高。前者如景儿峪组、雾迷山组部分层位及下古生界；后者如下马岭组以及上古生界。同时顺层掩卧褶皱的形成又受到主体构造群落形成时各层位所处的局部环境影响。由于内外因素共同作用的结果，形成大小不同、形态多样且分布广泛的顺层掩卧褶皱，成为识别构造群落发育与否及发育程度的显著标志。

由于顺层掩卧褶皱属韧性变形构造，形成于较高的温压环境，因而普遍伴生绿片岩相的变质作用，构造群落的中心地带甚至发育角闪岩相的动力变质作用，主要的标志矿物包括绿泥石和硬绿泥石等，局部见蓝晶石；另一个显著特点是褶皱的轴面流劈理非常发育，并且随褶皱程度的增高，流劈理置换或完全置换层理构造，形成新生面理构造。变质矿物也顺新生面理定向发育，矿物生长线理与褶皱的倒向一起构成固态流变构造的指向构造。大量的统计结果同样表明组成周口店固态流变构造群落的岩层或构造岩层其上层面向南东东方向滑动，与其他指向构造的统计结果一致。

3. 透入性流劈理

在周口店固态流变构造发育区域内，流劈理是最为常见的构造要素。在劈理面上常常发育明显的矿物生长线理，具有指向意义。此外，作为置换了层理的新生面理构造，流劈理成为后期构造变形的标志性构造，形成固态流变构造发育区内独具特色的地质景观。其发育程度可以作为同物异相的标志。

4. 石香肠构造及窗棂构造

在岩层中发育石香肠构造或窗棂构造，此类构造具有一定的指向意义。

（三）固态流变构造群落与其他地质作用的关系

周口店固态流变构造群落影响的地层包括雾迷山组以上直至早中生代的双泉组，受影响地层累积厚度达5000m以上，影响范围大约640km^2，在北京西山的构造格架中占有重要地位，与其周边其他构造亦有联系。

1）从流变构造的分布形态及与其他印支期构造的相互关系看，流变构造的形成可能受到北东向和东西向基底构造的控制。区域上基底断裂为北东向；而印支期本区，受到南北向构造应力场的控制，形成近东西向的构造。两者共同作用使得固态流变构造群落呈北东向展布，但其各项指向构造指示的流动方向却为近东西向的格局。

2）固态流变构造群落的形成又与后期构造的活动形式和构造形迹有着密切的关系。主要表现是后期的褶皱作用在周口店地区是以置换了原生层理的流劈理（S1）作为变形面，形成面理褶皱，如谷积山背斜。其核部为老地层，南北两翼依次出露较新层位，由S1面理的产状标识出的构造形态是明显的箱形背斜构造；但其岩层中发育的小构造的构造指向大多不能反映背斜形成的应力场，尤其是由原生层理标识出的小褶皱的倒向在背斜的两翼是相同的，而不是像通常背斜两翼的伴生褶皱那样相向而生。因此，这些小构造归属于固态流变构造群落，而不是背斜的伴生构造。

此外，固态流变构造群落形成于一个特殊的构造薄弱地带，其形成又使得这一地区的地壳进一步弱化，成为后期构造-岩浆活动的活跃地带。

（四）形成机制探讨

从组成周口店固态流变构造群落的构造要素所代表的构造环境以及伴生的变质相看，其形成于中深构造层次，相当于Mattauer（1980）界定的流劈理上限界面以下深度范围的p-t条件，其所处的环境温度为300~500℃，压力为2×10^8~3×10^8Pa。在明显的热异常的作用下，局部地区盖层岩石受到异常温度和静压力的双重影响而发生一定程度的塑化，进而产生较深构造层次的伸展式地壳圈层水平剪切拆离，在基本不影响盖层的整体层序的情况下形成丰富多彩的构造形迹。

六、穹隆和半穹隆构造

北京地区尚见有一些穹隆和半穹隆构造，它们同变质核杂岩构造一样，是伸展构造体制背景下形成的，所不同的是穹隆的形成没有伴生较深构造层次的构造改造，卷入的岩石以脆性变形为主。其成因主要是受伸展向引张应力的作用，不同基底块体之间发生差异性升降，在一些特定地段，盖层受到上升块体的顶托作用（有些地段还伴有岩浆上升侵位的顶托），形成穹隆和半穹隆构造。北京地区已知的穹隆和半穹隆构造包括下苇甸穹隆、青白口半穹隆等，主要分布在西山地区。此外，有些地区的地层微呈隆起状，但并未构成明显的穹隆构造。

（一）密怀穹隆

1. 位置、规模及形态

区域上位于赤城-长哨营-古北口-承德深断裂和密云-墙子路断裂两条东西向深大断裂带之间。穹隆长轴呈北西西-南东东向展布，长约60km，宽约40km，平面形态呈卵圆形（图32-25）。

穹隆核部主要由新太古界密云岩群变质岩组成，内有云蒙山、柏岔山等燕山期侵入杂岩体侵位。变质岩片麻理、片理走向北东、北北东或近东西向。翼部主要由长城系和蓟县系等组成，在穹隆东北部和西部有中生代上侏罗统髫髻山组角度不整合覆于其上。盖层与结晶基底的接触关系主要有以下4种表现形式：①正常的角度不整合接触关系，如杏树地、五亩地、高岭东北等地。②脆性断裂接触关系，如大黄木厂、宁村、溪翁庄等地。③沿不整合面有正长斑岩侵入接触关系，如溪翁庄、长岭沟门

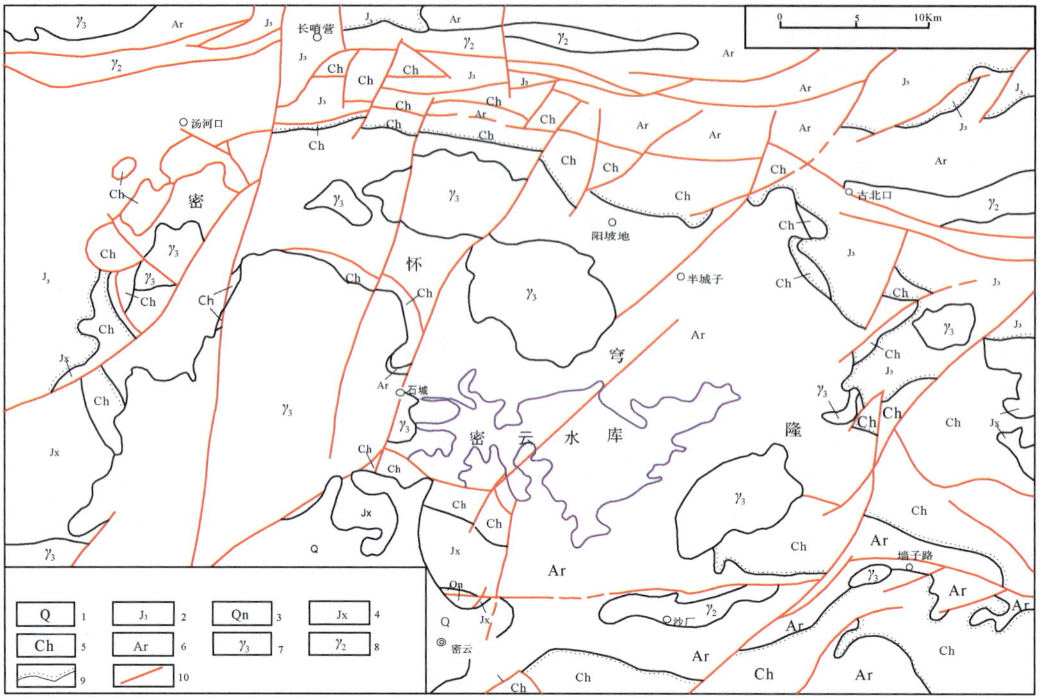

图 32-25　密怀穹隆地质略图

（据 1∶20 万怀柔幅区调报告修改）

Fig. 32-25　The brief geological map of Mihuai dome structure

1—第四系；2—上侏罗统；3—青白口系；4—蓟县系；5—长城系；6—太古宇；7—燕山期侵入体；
8—长城期侵入岩体；9—角度不整合界线；10—断裂

西等地。④以韧性断层相接触，如年鱼沟、沙坨子北一带。其中以前三种为主，更以第一种多见。

2. 产状特征

沉积盖层除局部倒转外，均向外倾斜。倾角西，南翼陡，北、东翼缓。北翼一般倾向北东 20°～30°，倾角 50°；南翼一般倾向南西 190°～210°，倾角 70°～80°，并常见直立或倒转。穹隆西部受云蒙山杂岩体强力侵位影响及大黄木厂滑覆断层的改造和石窑-大地背斜、四海向斜的叠加，致使地层高角度向西倾斜或近直立产出并局部翻转，推测原始倾伏角约在 50°。穹隆的长轴展布方向约为北西 295°，推测原始走向近东西向，轴向偏转可能是由于燕山晚期云蒙山杂岩体在穹隆西南部自南往北强力侵位后改造的结果。

3. 叠加褶皱

主要有贾峪-黄峪口倒转向斜、石窑-大地背斜和云蒙山背形。其中贾峪-黄峪口倒转向斜可能是穹隆内部的同期次级褶皱，位于云蒙山杂岩体的北侧，分布在梧桐豪、东峪、四合堂到黄峪口一线，受燕山晚期云蒙山杂岩体侵位和韧性断层改造，呈向北凸出的弧形。原始轴向近东西向，长度>10km，宽约 2km。向斜出露不完整，东部被石城断裂截切，中西部被云蒙山杂岩体侵入蚕蚀，仅残存其倒转的北翼或其中的一部分，仅在贾峪到黄峪口一带保存完整。向斜核部由长城系常州沟组、串岭沟组、团山子组组成，二平台村附近出露有大红峪组和高于庄组；翼部由密云岩群组成。构成向斜的新老地层普遍遭受燕山期韧性断层的剪切作用，发生不同程度的变形变质，轴面片理及糜棱面理极为发育，并强烈置换了先期面理和层理。向斜为同斜紧闭倒转类型，两翼产状均倾向北东或北北东，倾角 20°～35°。构成轴面向北北东斜卧的倒转向斜。

（二）下苇甸穹隆

1. 分布位置及组成地层

该穹隆发育在门头沟下苇甸地区。近核心部位发育中生代侵入体；顶部为下马岭组，向外依次为龙山组、景儿峪组、寒武系、奥陶系、石炭系和二叠系；其北、东、南三面均被中生界不整合覆盖，色树坟—龙泉务一带可以看到南大岭组直接不整合超覆在石炭系之上；西面古生界作为推覆断层的上盘压覆在髫髻山组之上。穹隆平面上略呈东西向展布的卵形，东西长约12km，南北宽约8km（图32-26）。

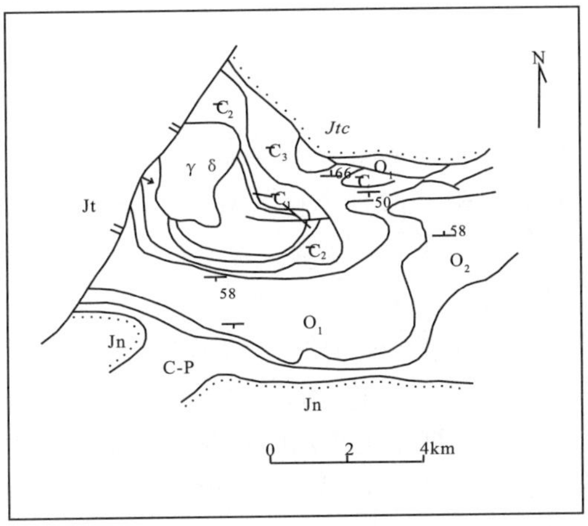

图32-26　下苇甸穹隆构造特征平面示意图
（据鲍亦冈，2001）
Fig. 32-26　Plane schematic map showing the structural characteristics of Xiaweidian dome
Qn—青白口系；Є_1—下寒武统；Є_2—中寒武统；Є_3—上寒武统；O_1—下奥陶统；
O_2—中奥陶统；C-P—石炭系—二叠系；Jn—南大岭组；Jt—髫髻山组；Jtc—土城子组；
γδ—上苇甸花岗闪长岩

2. 主要构造特征

青白口系作为穹隆的核部，其产状较为平缓，局部地区部分块体受岩体的顶托与其他部分分开，产状陡立，实际上已成为岩体的顶棚悬挂体。外缘的上古生界构造形态简单，为围绕穹隆的低角度单斜岩层。

构成穹隆主体的下古生界构造形态较为丰富。从局部看，寒武系和奥陶系的岩层组成一系列东西向展布的褶皱构造，主要分为两类：一类分布于东侧军庄、灰峪一带，东西向褶皱其轴面直立，两翼产状相近，属于受近南北向挤压而成的纵弯褶皱，它们不是下苇甸穹隆的伴生构造，而是穹隆形成后经受挤压作用形成的后期构造；另一类分布于丁家滩—下苇甸一带，其褶皱轴面一律倒向南，背斜的北翼长且产状平缓，南翼则短而陡，褶皱的包络面向南倾，层间节理的排列方式表明岩层的上层面是向南（下）滑动的。以上特征表明褶皱为横弯成因，属于穹隆形成时的伴生构造。此外在妙峰山北麓中生代地层中残存有一块古生代地层，其岩层产状表明是一个轴面倒向北的不对称背斜。

总体来看，下苇甸穹隆的其他的伴生构造较为简单，主要以节理和裂隙为主，岩石的原生构造保存完整，没有受到明显的变质变形作用的影响。侵位于下苇甸穹隆核部的中生代侵入体，有辉长岩和酸性岩类。其中辉长岩的同位素年龄为229Ma，应为三叠纪的侵入体。

3. 形成时代以及形成机制探讨

由于穹隆核部最早的侵入岩属印支期，其东侧的后期东西向褶皱经区域对比应是印支晚期的构造，加之不整合覆于穹隆的组成地层之上的最老层位是早侏罗世南大岭组，因而下苇甸穹隆的形成时代应是印支早期。

从区域上看，北京西山于印支早期其地壳处于伸展活动期，不同基底块体之间发生差异性升降。有些地段伴有热隆起，形成固态流变构造，如周口店地区；而下苇甸地区则表现为“冷隆起”，形成了相对简单的穹隆构造。

4. 有关问题讨论

关于下苇甸穹隆的成因及其构造形态存在不同的认识。有人将前述两种不同成因的东西向褶皱作为一个整体来考虑，认为其总体构造形态为一复式背斜，是燕山期形成的纵弯褶皱。对此本志认为应该充分考虑不同褶皱类型所反映的不同成因方式，它们显然应该是不同阶段、不同成因的产物。

（三）青白口半穹隆

1. 分布及组成

该构造展布在北京西山的青白口、沿河城、大村、雁翅一带。其形状为北西向的半椭圆形，分布面积约 70km^2。半穹隆的核部为雾迷山组，向外依次出露洪水庄组、铁岭组、下马岭组、青白口系及下古生界。核部岩层中广泛发育顺层的中酸性岩脉。

2. 构造特征

组成半穹隆核部的雾迷山组岩层产状平缓，倾角一般<20°，核部占据了整个半穹隆的绝大部分面积；翼部出露自洪水庄组层位向上的岩层，产状变陡，其中西南部产状近于直立，东南部倾角最大可达 44°，东北部则没有明显的陡立带；北西部受北东向断裂的切割，使穹隆的翼部缺失。东南西 3 面均被中生界不整合超覆，地表没有上古生界出露；从平面上看是一个没有完全封闭、不完整的穹隆构造，故称之为半穹隆，或称之为青白口背斜（叶浩，2014）。

边部下马岭组岩系中发育复杂、丰富的小型构造变形。核部发育北东向和北西向的裂隙构造北东和南西地层中发育次级褶皱，具有不对称性，反映出半穹隆具有北东-南西向横弯成因。半穹隆的核部和翼部发育大量的脉岩，且这些脉岩具有从翼部向核部由基性向酸性过渡、出露位置越来越高但侵入层位越来越低的变化特点。虽然都是脉岩，但构成了较完整的岩浆演化系列，表明半穹隆下面可能有隐伏侵入体。

3. 与其他构造的关系

前已述及，青白口半穹隆的西南部、南部及东部都被中生代的火山-沉积岩系所不整合覆盖；北西侧被具多期活动性的沿河城断裂切割；北侧被印支期的东西向推覆构造切割。从平面上可以清楚地看出，髫髻山火山-沉积盆地夹持在青白口半穹隆、下苇甸穹隆等印支期构造高地之间，而东岭台火山盆地则发育在半穹隆与沿河城断裂交接部位。

4. 形成时代及有关问题讨论

核部侵入的闪长玢岩脉与围岩一起发生褶皱，并有层间滑动，说明闪长玢岩脉侵位于褶皱变形之前，闪长玢岩锆石 SHRIMP 年龄约为 177Ma（叶浩，2014），基本限定了青白口穹隆起始变形的时代下限。

另外，沿河城断裂截切了青白口半穹隆，该断裂现今总体表现为倾向南东的正断层。在断层的北侧应找到穹隆的另一部分，但在断裂的西北盘并未发现半穹隆其余部分的痕迹，说明沿河城断裂具有多期活动性，当青白口半穹隆的基底地块抬升活动时，地块的西北侧受到沿河城断裂的限定，表现出自北向南斜向抬升的特点，导致盖层变形呈半穹隆状态。

第三十三章　构造旋回、构造层

地壳和岩石圈构造的发展演变表现出相对平静、和缓、漫长的量变期与相对集中、短暂、强烈活动的质变期的交替。这种交替发展过程，即为构造发展阶段或旋回。一定的构造发展阶段或旋回，必有与之相对应的构造层形成。通过对一定构造单元构造层的沉积建造、岩浆建造及构造形变特征，变质作用和成矿作用等的综合分析，可以认识、恢复或再造特定地质时期该构造单元的大地构造性质、类型及其发展演化过程。其在时间上代表了地壳发展历史的一定构造阶段，空间分布上代表着相应地质时期构造运动所影响的范围。因此，构造旋回和构造层是地壳构造发展的不可分割的时、空范畴。

本志划分构造旋回时主要考虑以下原则：

1）按照构造单元划分构造旋回的原则；

2）以区域性角度不整合，平行不整合的存在及其强度为标志的原则，注重不同时期的主要构造事件，特别是地壳重大的开裂与拼合、重组；

3）沉积建造、沉积旋回、沉积岩相类型的变化原则；

4）构造变形特征的构造格局、变形类型、变形强度原则；

5）岩浆建造的岩石类型、活动强度与方式原则；

6）变质作用类型与变质程度原则；

7）适当考虑成矿作用特点、同位素年龄资料及古生物特征的变化等。

按照上述原则，本区地质构造发展可划分为4个大阶段、9个旋回及相应的构造层（表33-1），整个地质构造演化经历了多阶段的发展过程。

第一节　新太古代—古元古代构造阶段

目前多数观点认为新太古代—古元古代是华北克拉通基底最终形成时期（王鸿祯等，1985；Zhao et al.，2001，2002，2003，2005；Lu et al.，2002，2008）。新太古代晚期，TTG岩系和变质表壳岩广泛形成（2.50~2.60Ga，刘富等，2009），构成了华北克拉通的大部分结晶基底。古元古代末期，东、西两大陆块发生碰撞拼合（翟明国，2007），导致克拉通统一基底的最终形成。

本志将北京地区太古宙表壳岩划分为密云岩群和四合堂岩群（约2500Ma），认为成岩时代为新太古代早、晚两期，其形成时代应早于其他各类变质地质体。由于岩石地层变质变形强烈，统称为阜平旋回（新太古代构造层）。古元古代时期，本区均为遭受剥蚀的陆地，缺失沉积地层记录。但根据区内密云、怀柔地区发现的古老岩浆活动，划分为吕梁旋回。

一、阜平旋回（新太古代构造层）

地壳形成初始阶段的状况目前还知之甚少，近年来的调查研究表明密云岩群和四合堂岩群（表壳岩）是北京地区所见到的最古老的大陆地壳。它是一套变质的火山-沉积岩系，以基性火山喷发堆积为主，夹硅铁质岩。原岩为一套基性—中基性—中酸性的熔岩及其火山碎屑岩、凝灰质沉积岩夹多层薄—厚层铁质岩建造，形成时代为新太古代。

二、吕梁旋回

古元古代时期，本区为遭受剥蚀的陆地，缺失沉积地层记录，仅发育较弱岩浆活动和断裂活动，区域变质作用同样较弱。岩浆活动以橄榄岩、辉石岩、角闪岩、辉长岩及辉绿岩为代表的超基性—基

表 33-1 北京市构造旋回划分与综合特征表

Table 33-1 Tectonic cycle division and integrated characteristics in Beijing

地质时代			构造旋回			构造层			沉积建造	火山岩建造	侵入岩建造	区域变质作用	构造环境	构造性质
			阶段	旋回	期	类型	构造层	亚构造层						
新生代	第四纪	Q	晚白垩世—第四纪阶段	喜马拉雅旋回(HM)	晚期	上叠构造层	喜马拉雅构造层	晚喜马拉雅亚构造层					陆内裂陷、断陷盆地，拉张环境。	非造山
	新近纪	N			早期			早中喜马拉雅亚构造层	河湖相碎屑岩建造	基性、碱性				
	古近纪	E_3												
		E_2												
		E_1	晚三叠世—早白垩世阶段	燕山旋回（YS）	晚期								大陆整体隆起过渡环境	后造山
中生代	白垩纪	K_2					燕山构造层	上燕山亚构造层	内陆河湖相碎屑岩建造		偏碱性、碱性		过渡环境	
		K_1^2								中性、酸性、偏碱性	基性、中酸性、偏碱性		陆内挤压继承性和上叠性盆地，陆内或板内造山环境	
		K_1^1									基性、中性、酸性			
	侏罗纪	J_3			中期			中燕山亚构造层	内陆河湖相碎屑岩含煤建造与红色碎屑岩建造	基性、中性、酸性	基性、中性、酸性			造山期
		J_2									酸性			
		J_1			早期			下燕山亚构造层	山麓冲洪积扇相、湖相碎屑岩建造	基性、中性、酸性	基性、中性、酸性	绿片岩相		
	三叠纪	T_3									偏碱性、碱性		过渡环境	前造山
		T_2^2	南华纪至中三叠世阶段	印支旋回（IS）		盖层构造层	印支构造层					绿片岩相		后碰撞
		T_2^1							海陆交互相过渡为陆相含煤碎屑岩及不含煤的杂色碎屑岩建造		基性、中性、酸性		挤压型坳陷盆地	
		T_1									中性、中酸性			
古生代	二叠纪	P_3												同碰撞
		P_2		华力西旋回	晚期		华力西构造层	上华力西亚构造层	滨浅海碎屑岩碳酸盐岩建造				板内挤压型凹陷盆地	前碰撞
		P_1									基性	角闪岩相—绿片岩相		
	石炭纪	C_2											克拉通盆地	
		C_1			中								大陆整体隆起	

续表

地质时代			构造旋回			构造层			沉积建造	火山岩建造	侵入岩建造	区域变质作用	构造环境	构造性质
			阶段	旋回	期	类型	构造层	亚构造层						
	泥盆纪	D_3			期									
		D_2			早期						中性			前造山
		D_1		加里东旋回	晚期		加里东构造层						大陆整体隆起拉张环境	非造山
	志留纪	S_{2-4}												
		S_1												
	奥陶纪	O_3												
		O_{1-2}			早期			下加里东亚构造层	潟湖—浅海陆棚碎屑岩、碳酸盐岩建造				继承克拉通盆地拉张环境	
	寒武纪	$€_{2-4}$												
		$€_1$		扬子旋回									大陆整体隆起拉张环境	
新元古代	震旦纪	Z												
	南华纪	Nh												
	青白口纪	$Pt_3^1-Pt_3^3$	中元古代—新元古代阶段	晋宁旋回	晚期		晋宁构造层	上晋宁亚构造层	潟湖—浅海陆棚碎屑岩、碳酸盐岩建造				大陆—陆缘裂谷裂陷盆地，拉张环	非造山（或超大陆裂解）
中元古代	待建纪	Pt_2^2												
	蓟县纪				中期			中晋宁亚构造层						
	长城纪	$Pt_2^1-Pt_2^2$			早期			下晋宁亚构造层		双峰式火山岩	基性、中基性、中酸性			
古元古代	晚期	Pt_1	新太古代—古元古代阶段	吕梁旋回			吕梁构造层				基性、超基性	绿片岩相变质作用	微陆块拼合，基底形成	造山期（参考河北）
	中期													
	早期													
新太古代	晚期	Ar_3^3		阜平旋回		基底构造层	阜平构造层				TTG 组合	（麻粒岩相）角闪岩相		非造山
	中期	Ar_3^2												
	早期	Ar_3^1								中基性火山-沉积岩				

性岩体和岩脉沿北东向张性断裂构造带侵位，形成小型岩体和岩墙群。同时发生浅层次韧性剪切和绿片岩相区域变质作用。需要指出的是，这些岩浆岩的年龄值不同学者测试结果差异较大，是古元古代侵位，还是太古宙侵位尚难下确切结论。古元古代出现明显的脆性破裂，受北西-南东向挤压作用，形成了北东向崎峰茶断裂和石城断裂的早期逆冲性质断裂活动，使不同构造层次的麻粒岩相变质岩和角闪岩相变质岩直接接触。古元古代末期的构造-热事件（1900~1800Ma）使结晶基底全部固结，形成了本区的基底构造层，导致了包括燕山地区在内的统一华北陆块的形成。李四光将这期重要的地壳运动称为“吕梁运动”。

第二节　中元古代—新元古代构造阶段

吕梁运动以后，至新元古代，本区地壳进入稳定的克拉通（准地台）发展阶段，普遍接受了地台型沉积。该期是整个陆块由不成熟到成熟，地壳由活动转化为稳定的过渡时期。其沉积厚度在蓟州为9200m，昌平为4561m，冀西怀来一带为3949m，由东到西变薄。地壳只有升降运动而引起的海陆变迁，各系、组之间均为整合或平行不整合接触，火山活动不强烈。从中元古代常州沟期开始，海水由北东方向侵入，在经历几次海进、海退过程后，到新元古代景儿峪期，沉积地层累计厚度约万米。按其发展特点和差异，划分为一个旋回——晋宁旋回，根据中、上元古界各系之间存在的因区域性地壳抬升而造成的平行不整合面和各系沉积特征的差异，又将其划分为3个亚构造层。

一、晋宁旋回早期（Pt_2^1）

进入中元古代，华北陆块已固结形成统一的陆壳，但因地壳厚度薄，刚性强度弱和固结程度低，地壳受引张破裂拉开，断块陷落，形成中元古代初裂谷堆积。本时期北京地区受北部近东西向的丰宁-隆化断裂、尚义-平泉断裂、密云-喜峰口断裂和北东向太行山断裂、青龙-滦县断裂等古断裂带控制，在地壳南北向拉张作用下，出现了北升南降的断块阶梯状陷落，裂陷方向大致呈东西向，以尚义-平泉断裂带为北界，南侧的北京地区形成了燕山裂谷。该期地壳受拉张因而裂陷的性质表现得较为明显，形成了滨海—浅海相以碎屑岩为主夹碳酸盐岩沉积建造，受古同生断裂的控制，各地岩性岩相和厚度有很大差异。该旋回内的岩浆作用可以分为2期：①火山岩分布在长城系的团山子组和大红峪组，锆石U-Pb年龄为1680~1620Ma，表现为碱性玄武质和富钾粗面质熔浆的喷溢；②非造山侵入岩（斜长岩—奥长环斑花岗岩—斑状花岗岩）的同位素年龄为1700~1670Ma，主要分布在密云、怀柔北部，沿赤城-古北口东西向断裂带。

二、晋宁旋回中期（Pt_2^2）

中元古代高于庄期开始，华北陆块稳定克拉通化，地壳活动由南北向的拉张裂陷转化为区域性的整体升降，缺乏相应岩浆活动和变质作用，沉积物由以陆源碎屑为主转化为以内源碳酸盐岩为主。蓟县期发育了一套巨厚的具有各种浅水动荡标志和多种形态叠层石的滨海—浅海碳酸盐相沉积，为中元古代以来最为广泛的一次海侵。

三、晋宁旋回晚期（Pt_2^3—Pt_3^1）

新元古代，华北陆块继续保持稳定状态。境内总体处于稳定克拉通型沉积盆地环境。待建纪至青白口纪时海域收缩，沉降和沉积中心转移到北京西山一带，发育了下马岭期的滨岸含铁砂泥相—潮间、潟湖相—潮下含碳含硅泥岩相沉积，龙山期的滨岸三角洲分支河道粗碎屑沉积和潮间—潮下泥岩相沉积，以及景儿峪期浅海陆棚碳酸盐岩相沉积。然而这一期间，深部岩石圈可能变围期造山构造远程影响，发生一定改造，伴随形成极弱基性岩席侵位，获得1380~1320Ma的锆石和斜锆石U-Pb同位素年龄同时，远程火山灰物质进入下高岭组地层演化为极薄层致五脱岩。

第三节 南华纪—中三叠世构造阶段

一、扬子旋回（Nh—ϵ_1）

青白口纪末，本区整体上升为陆。抬升时限大致相当于800~570Ma，境内经历了长达230Ma的风化剥蚀，形成了新元古界青白口系景儿峪组与下古生界寒武系昌平组之间的平行不整合接触。

二、加里东旋回（ϵ_2—D_1）

1. 加里东旋回早期（ϵ_2—O_2）

自早寒武世昌平期至中奥陶世马家沟期，本区再次下降，发生大面积海侵，其时限大致为570~450Ma，经历了约120Ma。期间，境内仍处于稳定克拉通背景，在陆表浅海环境下形成了一套以碳酸盐岩为主夹部分细碎屑岩的沉积建造——下寒武统昌平组至中奥陶统上马家沟组，共2个系5个组，地层总厚853~1774m。当时地势比较平坦，没有明显的地形分异，区内各组地层岩相变化不大，组间接触关系为整合过渡或平行不整合。反映这一时期地壳升降缓慢，沉积环境稳定。单独划分为一个亚构造层——下加里东构造层。

2. 加里东旋回晚期（O_3—D_1）

中奥陶世至晚奥陶世，全区又上升为陆，造成晚奥陶世至早石炭世近150Ma的沉积缺失。

三、华力西旋回（D_2—P_2）和华力西构造层

1. 华力西旋回早期（D_2）

中泥盆世为华力西早期亚旋回（D_2），本区整体仍处于隆起状态，缺失沉积记录。

2. 华力西旋回中期（D_3—C_1）

本区总体仍处于隆起状态，没有任何沉积建造与火山岩建造记录。然而，受北部西伯利亚板块向华北板块俯冲影响，在北京北部及河北地区发育有较多的侵入岩，呈近东西向展布，具有陆缘岩浆弧特征。

3. 华力西旋回晚期（C_2—P_2）

晚石炭世—中二叠世即华力西晚期亚旋回，其形成的各种地质建造组成晚华力西亚构造层，该亚旋回西伯利亚板块向华北陆块之下俯冲加剧。本区转变为在挤压环境，形成继承性盆地沉积了晚石炭世太原组至石盒子组下部地层，为一套海陆交互转河湖相碎屑岩含煤建造。该亚旋回在北京北部地区始发育有少量侵入岩，充分显示北京北部地区已受到西伯利亚板块向华北陆块俯冲作用的强烈影响。

四、印支旋回（P_3—T_2）

晚二叠世—中三叠世划分为印支旋回，本区仍处华北稳定克拉通环境，区域沉积环境由滨海平原向内陆河湖盆地发展，沉积建造为双泉组的岩屑砂岩、粉砂岩夹砾岩的岩石组合，并发育少量的中酸性侵入体，以双泉组与杏石口组之间存在的角度不整合或平行不整合为标志的印支运动最终结束了该构造层，本区从此进入陆内适山复杂构造变形阶段。

第四节 晚三叠世—早白垩世构造阶段

晚三叠世—早白垩世构造阶段是“燕山运动”重要发展阶段。“燕山运动”这个术语的提出，是世界地质科学史的一件大事，它打破了华北克拉通数亿年的稳定格局，引起了一系列北东向深大断裂形成和剧烈的岩浆活动，揭露了中国东部大部分地区于中生代中后期发生了强烈的属于造山型的地壳

运动的事实，推动了大地构造学及有关学科大大向前发展。它的提出不止是阐明了一个地壳运动期，而且从学术理论及找矿实践上说，还有更高层次的意义。

一、研究历史回顾

（一）燕山运动提出

北京西山地质研究始于1916年，在章鸿钊、翁文灏和丁文江的组织和领导下，由叶良辅、王竹泉、朱庭祜等13个学员分段对北京西山地质情况进行勘测，最终由叶良辅1919年执笔汇编而成《北京西山地质志》，再由丁文江、翁文灏、章鸿钊及安特生等人校正（朱庭祜，1951），正式发布于1920年（叶良辅，1920），这是中国第一部地质调查报告。在分析研究北京西山地质成果的基础上，1926年翁文灏在东京泛太平洋科学大会上发表了题为《中国东部的地壳运动》一文，首次提出了“燕山运动”。

次年，他对文章作少量修改后，发表了《中国东部中生代以来的地壳运动及岩浆活动》（Wong，1927），提出“北京西山九龙山和髫髻山岩系之间倾角的差别也许并不很显著，但在我看来，不会有多大疑问，在髫髻山岩系沉积之前，曾发生过重大的变形和剥蚀”，进一步将“髫髻山岩系沉积之前”细化为“燕山运动年代应限于北京西山的九龙山系与髫髻山系之间”，并注意到西山髫髻山组火山岩覆盖在不同时代的地层之上，这表明在髫髻山火山岩沉积之前，区域上曾发生过重大的变形和剥蚀，继而论述了这一事件在中国东部的巨大影响。1929年翁文灏根据北票地区所见加以修改，将燕山运动分为A、B两幕：A幕指上侏罗统后或下白垩统初期的宽广褶皱；B幕指白垩纪间的急激褶皱及逆掩断层。在此两幕之间有强烈的火山喷发和深成侵入活动。

（二）燕山运动主要研究阶段

丁文江（1929）在翁文灏（1929）研究基础上，将燕山运动划分为3个不同时期褶皱幕，即翁文灏（1929）提出的A、B幕之间还存在一期中间幕。

谢家荣（1933）对中生代不整合提出与叶良辅类似的质疑，对不整合进行重新划分，并对燕山运动进行重新分期研究（1936，1937），分别以前门头沟组、前九龙山组、前东岭台组、前坨里组、前长辛店组的不整合或假整合为代表，将燕山运动划分为五期（表33-2）。

黄汲清（1945）认为谢家荣的第一幕应属印支运动（旋回），并将北京西山地区分为九龙山组与髫髻山组、坨里组、长辛店组3幕。随后，黄汲清（1960）总结了过去地质工作的成果，主张燕山运动是独立的和统一的构造旋回，活动时限为早三叠世—晚白垩世。并将燕山运动修订为五期，分别以5个不整合面为界（表33-2），认为其中的1期（即九龙山组下不整合）及3期（即南天门组下不整合）最为重要。同时确定1期（即翁氏燕山运动或其A期）应发生在中侏罗世初，指出四川运动或5期的确切时期尚待解决。

赵宗溥（1963）将燕山运动划分为初幕、次幕、晚幕、末幕四幕，分别以4个不整合面为界划分（表33-2），分别以髫髻山组、张家口组、坨里组和长辛店组不整合或假整合为代表，活动时限为早三叠世—晚白垩世。

崔盛芹等（1981）、张之一（1981）继承黄汲清观念，将燕山运动划分为1～5幕次，但在幕次划分的不整合层位上有争议。崔盛芹等（2002）结合板内造山认识，对燕山运动进行修正，认为燕山构造带既是一条中、新生代的陆内或板内造山带，也是一条重要的构造-岩浆-成矿带与现今活动构造带。燕山及邻区处于远离当时陆洋边界的陆内环境，形成典型的古克拉通基础上的陆内造山带，陆内造山作用过程可划分为：①始造山期，包括自三叠纪中、晚期至早侏罗世前的印支期，经过本阶段的构造变形序列，构成“古燕山”陆内造山带与构造-岩浆带的初步轮廓，例如北京地区八达岭复背斜雏型的出现；②主造山期，包括侏罗纪—白垩纪燕山期更为强烈的构造变动和广泛的岩浆活动和成矿作用在内，通过这一阶段多期次、多方式的构造变形序列，塑造成“中燕山”陆内造山带与构造-岩浆-成矿带的基本格局，这一时期也是八达岭复背斜、髫髻山复向斜及其有关构造-岩浆带的主要成型

表 33-2 燕山运动划分沿革表

Table 33-2 Division history of the Yanshan Movement

地质时代：古近-新近纪（第三纪）；白垩纪（晚、早）；侏罗纪（晚、中、早）；三叠纪（晚、中、早）

研究者	年代	
翁文灏	1927	南岭运动 髫髻山组 下不整合 ～ 燕山运动
翁文灏	1929	王氏统 下不整合 ～B期 北票上火山岩系 下不整合 ～A期
黄汲清	1960	喜马拉雅运动；燕山运动 官庄组 下不整合 ～5期 王氏组 下不整合 ～4期 南天门组 下不整合 ～3期 张家口组 下不整合 ～2期 髫髻山组（九龙山组）下不整合 ～1期
赵宗溥	1963	长辛店组 下不整合 ～末幕 燕山运动 坨里组 下不整合 ～晚幕 张家口组 下不整合 ～次幕 髫髻山组（九龙山组）下不整合 ～初幕
崔盛芹等	1981	喜马拉雅运动；燕山运动（晚期、中期、早期）；晚期、早期（三叠纪） 前长辛店组幕 ～5幕 前王氏组幕 ～4幕 前坨里组幕 ～3幕 前东岭台组幕 ～2幕 前九龙山组幕 ～1幕 前郭家店组幕 ～晚幕 前老虎沟组幕 ～早幕
张之一	1981	长辛店组 下不整合 ～5幕 燕山运动 坨里组 下不整合 ～4幕 东岭台组 下不整合 ～3幕 ～2幕 ～1幕
鲍亦刚等	1983	燕山运动 前长辛店组构造幕 ～3幕 前东岭台组构造幕 ～2幕 前髫髻山组构造幕 ～1幕
徐正聪等	1983	喜马拉雅运动；燕山运动；印支运动 汉诺坝组 下不整合 ～4幕 土井子 下不整合 ～3幕 白旗组 下不整合 ～2幕 九龙山组 下不整合 ～1幕
聂宗笙	1985	燕山运动（晚期、中期、早期） 下第三系 下不整合 ～5幕 王氏组 下不整合 ～4幕 莱阳组 下不整合 ～3幕 白旗组 下不整合 ～2幕 九龙山组 下不整合 ～1幕
赵越	1990	喜马拉雅运动；燕山运动（晚期、中期、早期）；印支运动（晚期、早期） 早白垩世晚期幕（坨里期为主） 晚侏罗世期间幕（土城子晚期为主） 中侏罗世期间幕（龙门期为主） 晚三叠世末幕（邓杖子期） 中三叠世期间幕（胡杖子晚期）
任纪舜	1999	阿尔卑斯旋回 喜马拉雅旋回 燕山旋回 印支旋回
Davis	2001	左旋走滑 伸展 挤压 向北逆冲 伸展 向南逆冲
崔盛芹等	2002	燕山运动；印支运动 ～4幕 坨里组 ～3幕 辛庄组 张家口组 ～2幕 后城子组 九龙山组 ～1幕 龙门组 ～2幕 ～1幕
邓晋福等	2005	K_2 K_1^2 后造山 K_1^1 晚造山 J_3 峰期造山幕 J_2 早期造山幕 T_3-J_1 前造山
黄树文等	2007	燕山运动：弱挤压变形期 主伸展-岩石圈减薄期 强挤压陆内造山期 印支运动

期；③重造山期，在中生代末至新生代初，始造山期—主造山期既成的山岳地貌基本遭到剥蚀，形成五台期夷平面之后，又经过古近纪—新近纪—第四纪喜马拉雅期以伸展方式为主的重造山作用，构成“新燕山”陆内造山带不同级别盆-岭构造与现今地貌景观。

鲍亦冈（1983）认为北京地区早侏罗世早期—早白垩世末为燕山构造旋回阶段。其下界为晚三叠世杏石口组与下伏二叠纪—三叠纪双泉组间之平行或微角度不整合构造面，其上界为下白垩统夏庄组与上覆古近系始新统长辛店组间所存在的角度不整合构造面。将燕山构造旋回划分为3个褶皱幕：第一幕，中侏罗世末—晚侏罗世前，构造方向为北东东向；第二幕晚侏罗世末—早白垩世前，构造方向为北东向；第三幕为早白垩世前—晚白垩世末，构造方向为北北东向。

徐正聪等（1983）将燕山运动划分为4幕次，分别以九龙山组、白旗组、土井子组和汉诺坝组不整合为代表。

聂宗笙等（1985）将燕山运动划分为5幕次，时限包括整个侏罗纪和白垩纪。

任纪舜（1990）指出：“在中国东部以来的历次地质事件中，燕山造山旋回是一次最重要的，压倒一切的构造运动。”中国的主要大地构造单元雏形基本形成，并将燕山运动分为早、中、晚3个旋回，时限为整个侏罗纪和白垩纪，认为同太平洋板块低角度俯冲有关。

赵越等（1990，2004）认为作为燕山运动主幕构造转换时代为中侏罗世。燕山板内变形带的主要构造事件发生在中侏罗世，时代应在（160±5）Ma前。这期间，东亚构造体制开始发生重要转变，其主要标志是代表活动陆缘的增生杂岩带和火山弧的出现。东西向燕山板内变形带为代表的南北向地壳缩短持续到（135±1）Ma前，而后开始逐渐被北北东向的表壳构造岩浆作用取代，说明燕山运动记录了东亚构造体制的转变。

邓晋福等（1996，2005）、吴福元等（1999）认为燕山运动的运动机制是岩石圈伸展减薄，类似裂谷动力学机制。

张宏仁（1998，2000）认为燕山运动是中国东部遭受来自外部的快速打击的后果，以脆性变形为主的构造运动，并将该快速打击事件定义为燕山运动的“绪动”或“燕山事件”，认为燕山运动是一次强烈构造事件的结果。

Davis（2001）将侏罗纪—白垩纪变形划分为：①>180Ma的向南逆冲推覆作用；②中侏罗世—晚侏罗世早期的伸展断层作用；③晚侏罗世—早白垩世早期的向北逆冲作用；④早白垩世晚期的挤压作用；⑤中白垩世伸展变形；⑥晚白垩世左旋走滑作用。提出燕山运动是蒙古-鄂霍次克海关闭形成的南北向挤压与太平洋板块向东亚大陆俯冲形成的向西北方向挤压的交替作用的结果。

邓晋福等（2004）采用造山带结构要素组合的概念，对华北燕山造山带进行了研究。燕山造山带各演化阶段的结构要素组合特征如下：①前造山和初始造山幕（J_1）。其中，早侏罗世早期为前造山伸展构造，结构要素组合有三叉式裂谷带、板内型玄武岩、含煤建造；早侏罗世晚期为初始造山收缩构造，结构要素组合有向北倾伏的褶曲与逆冲、九龙山组类磨拉石建造，硬绿泥石-十字石-蓝晶石为标志的低温、中—高压变质带。②早期造山幕（J_2）。其中，中侏罗世早期为同造山伸展构造，结构要素组合有岩石圈上隆伸展有关的火山盆地及可能的同期侵入岩，火山岩线型分布；中侏罗世晚期为收缩构造，有关的结构要素组合为逆冲推覆和褶曲变形、磨拉石建造、同构造侵入体和角闪岩相变质岩。③峰期造山幕（J_3）。其中，晚侏罗世早期同构造伸展构造，结构要素组合有岩石圈上隆伸展有关的火山盆地与同期侵入岩，火山岩面型分布，火成岩组合中出现高压粗面岩类，较大量的流纹岩；晚侏罗世晚期收缩构造，有关结构要素组合为逆冲推覆和褶曲变形、磨拉石建造、同构造侵入体和角闪岩相变质岩，侵入岩中出现高压正长岩类。④晚造山幕（K_1^1）发生于早白垩世早期，有关的结构要素组合为收缩变形分布较局限，湖相沉积建造替代磨拉石建造，侵入岩组合中出现过碱性石英正长岩，大晶洞构造的花岗岩及科马提质辉长岩等。⑤后造山幕（K_1^2）为早白垩世晚期伸展构造，有关的结构要素组合为正断层、变质核杂岩、双峰式岩墙辟、典型的过碱性花岗岩和含煤建造。

翟明国等（2004）强调华北中生代构造转换不是陆内造山结果，而是与周边块体夹击引起华北地幔大规模上隆有关。

董树文等（2007）认为燕山运动主幕主要发生在晚侏罗世（（165±5）Ma~136Ma），以燕山地区髫髻山组火山岩和九龙山砾岩下的角度不整合为界，在缺失髫髻山组和九龙山组的地区，以后城子组砾岩层与下伏地层的不整合为界。即燕山运动起始于（165±5）Ma之前，结束于83Ma。将燕山运动分为3个时期：燕山运动主幕强挤压陆内造山期（（165±5）Ma~136Ma），主伸展垮塌与岩石圈减薄期（135~100Ma），燕山运动晚幕弱挤压变形期（100~83Ma），并且认为中—晚侏罗世中国东部甚至东亚大陆不仅受到太平洋板块俯冲边界动力系统的控制，同时受控于来自北部的西伯利亚板块与古亚洲构造带沿蒙古-鄂霍次克洋汇聚碰撞产生的远程构造作用，还有来自拉萨地块与羌塘地块的碰撞产生的来自西南方向的挤压，构成中—晚侏罗世多向挤压汇聚的板块动力学格局（多向挤压汇聚的动力学）。

（三）燕山运动关键问题

综上研究历史回顾，燕山运动关键问题包括燕山运动起始时间、分期与主幕次时间、燕山运动构造机制（挤压与伸展作用）和动力学问题。

二、燕山（构造）运动阶段时间约束

构造运动是由地球内部能量引起的、导致地壳或岩石圈物质发生变形和变位的机械运动，具有阶段性和周期性。“（角度）不整合”代表一次大的地壳变动，是划分构造运动阶段的重要依据。（角度）不整合之上最底部地层年代是新一轮构造运动阶段的起始。因此，确定各组地层之间接触关系和确定各不整合之上的地层年代是梳理燕山运动构造阶段时间约束的关键。

基于区域地层清理，北京境内区域性不整合自下而上依次位于：①杏石口组和双泉组之间；②髫髻山组或龙门-九龙山组底部（存在争议）；③土城子组与张家口组之间；④夏庄组与长辛店组之间。区域上，第1个和第4个不整合分别缺失中—晚三叠世和晚白垩世沉积，不整合面的时限间隔较长，可作为约束燕山构造运动阶段时间约束。因此，杏石口组沉积时代可作为燕山构造运动阶段的起始，而长辛店组沉积时代可作为喜马拉雅构造运动阶段的起始，即燕山构造运动阶段的终止。

杏石口组地层沉积时代归属争议较大。用于确定地层时代的古植物化石既有晚三叠世分子，也有侏罗纪分子；发现的*Euestheria*? sp. 叶肢介化石（米家榕，1984）在2016年最新国际地层表中为早侏罗世化石代表之一，且地层中发育鱼化石（刘宪亭，1988）；而西山盆地和下板城盆地碎屑锆石测年结果中最年轻的锆石加权平均年龄为（197±6）Ma和（206±1.5）Ma，具有晚三叠世末期火成岩物源，表现为北东东-南西西向的开阔谷地曲流河和冲积扇沉积环境，盆地因伸展拉张而处于欠补偿沉积状态。

不仅如此，晚三叠世盘山、珠窝岩体和早侏罗世大石坡-黑山坨岩体记录了晚三叠世末期—早侏罗世早期挤压构造向伸展拉伸构造背景环境转变过程，处于非造山构造背景，是本区中亚构造域向燕山板内构造域转化前过程阶段。在盘山岩体东罗庄单元及大石坡-黑山坨单元伸展背景下侵位时，区域上方才开始沉积杏石口组地层。自此，也基本结束中三叠世以来的印支运动引起的地壳增厚、区域抬升剥蚀（地层缺失）。大石坡-黑山坨岩体角闪石全铝压力计推算地壳增厚11km（汪洋等，2015），同中三叠世出现小规模高Sr低Y类埃达克岩和自北向南逆冲推覆构造、近东西向褶皱完全吻合。综合古生物化石、同位素数据、沉积特征和构造-岩浆演化厘定杏石口组为晚三叠世-早侏罗世，时间约为盘山岩体乐罗压单元或大石坡-黑山坨岩体侵位时间，即203~195.5Ma。

自Anderson（1923）于长辛店地区建立长辛店组地层以来，该组地层时代经历始新世、古近纪、早更新世等不同争议（详见第一篇）。大量哺乳动物化石、淡水腹足类化石的发现，确定其形成年代为晚始新世。

综上，燕山构造运动阶段时间约束为晚三叠世或早侏罗世—晚始新世，时代约为203~56Ma。

三、构造阶段划分

燕山运动是整个侏罗纪、白垩纪期间广泛发生于中国全境的重要构造运动。其在不同构造部位的强度和表现形式有所差别，如中国东部和东亚濒太平洋地区，其构造变形和岩浆活动具有自西向东愈加强烈的演变规律，地壳运动与构造变动具有长期性与多幕性相统一、渐进与激化相交替的特点，岩浆喷发和侵入活动具有多期次性。

（一）燕山运动构造阶段划分依据

梳理前人工作基础上，借助板块构造学说观念，综合北京境内燕山运动阶段构造、沉积、岩浆活动、变质作用特征，对燕山运动构造阶段进行划分，并对燕山运动特征进行简单总结。

（二）构造阶段划分

1. 前造山阶段（晚三叠世—早侏罗世）

（1）地层年代确定与各地层接触关系

该阶段由老到新发育杏石口组、南大岭组、窑坡组、龙门组和九龙山组火山-沉积地层，化石和同位素年龄揭示地层沉积时代为晚三叠世—早侏罗世。小范围内可见杏石口组和南大岭组呈小角度不整合接触，但杏石口组为山间陆相河流沉积，且南大岭组底部为溢流相火山岩沉积，存在局部喷发不整合可能。而区域上，该三套地层结伴出现，且后期构造变形一致。因此，其地层之间接触关系总体为整合或假整合接触。

（2）构造组合及形成时代

该套地层由陆相基性火山岩-含煤岩系-粗碎屑岩系构成，含有重要的生物化石“门头沟植物群”，构造变形以北东东向（NE60°~70°）箱形向斜为主。

（3）岩浆岩带分布与岩浆演化

该阶段岩浆活动表现为南大岭期火山活动和以盘山岩体、东白山岩体、珠窝岩体、大石坡-黑山坨岩体等组合而成的晚三叠世—早侏罗世侵入岩带。时间序列上，基于同位素年龄统计，二者形成于不同时间阶段，且具有不同的大地构造背景环境。侵入岩带早于火山岩形成时代。

侵入岩带由系列岩体组成，单个岩体呈南北向或东西向展布，岩体组合近东西向展布，形成岩浆岩带。岩性为花岗岩、石英二长岩、正长岩组合的高钾钙碱性—碱性岩系列，岩体记录了印支运动挤压构造背景向伸展构造背景环境转变复杂岩浆侵位和演化过程，具有中亚晚造山—后造山构造属性，也是燕山运动构造阶段前造山（非造山）阶段。

南大岭期火山活动早期（一段）为裂隙式喷发，火山活动具有明显的脉动式特点，数十次基性玄武质熔浆溢流，形成了宽厚的火山熔岩被。而晚期（二段）为点式喷发，在构造（断裂）交接部位形成了小面积以中性粗安质岩浆爆发作用为主的第二喷发周期火山活动，喷发中心在法城—桑峪附近。与南大岭期火山活动晚期同时间段还有小规模顺层闪长玢岩岩脉（位置相当）侵位（青白口岩脉群）。

综上，前造山阶段，岩浆活动以伸展型火山活动为主，末期有小规模的脉状岩脉侵位，总体岩浆活动强度较弱。

（4）火山-沉积盆地对比

通过地层分布恢复该阶段火山-沉积盆地，北京境内发育西山盆地和昌怀盆地。该盆地性质为伸展型继承性（断陷）盆地，形成于燕山板内造山前北东东向伸展构造背景条件下，盆地可类比于河北下花园盆地。由于受后期构造强烈破坏，原始沉积盆地面貌丧失殆尽，难以恢复，且控制盆地的断裂也不能确定。鲍亦冈等（2000）认为盆地可能受密云-墙子路断裂活化控制，但南大岭组裂隙式火山喷发特征在地层中较为明确。而可类比的河北下花园盆地表现为由上黄旗-乌龙沟断裂控制新生的断陷盆地（王立峰等，1998）。该期盆地沉积除裂陷作用影响外，盆地的发育和展布明显受印支期构造影响和限定，印支运动形成的构造格架近东西向，限定了南大岭组（溢流相沉积）印模式展布亦呈近东西向。

2. 造山阶段（中侏罗世—早白垩世早期）

（1）地层年代确定与各地层接触关系

该阶段由老到新发育髫髻山组、土城子组火山-沉积地层，化石和同位素年龄揭示地层沉积时代为中侏罗世—晚侏罗世（详见中生代多重地层划分对比表）。该套地层接触关系极为混乱，髫髻山组之下、九龙山组之下、龙门组之下均有不整合面依据。陆相地层的“同时异相”和不整合面的穿时性是地质构造最为重要的特征，由于变形递进、盆地迁移等因素，构造运动可形成一个或多个分布有限的不整合面（李海龙等，2014）。但这些不整合面在空间上最终可收敛于一个主不整合面。

前人（翁文灏，1928；黄汲清，1945；朱庭祜，1951；赵宗溥，1963；鲍亦冈等，1995）研究认为髫髻山组与其下地层的不整合面在区域上最为明显，它覆盖于不同地层之上。根据火山喷发带划分，北京境内可划分为南、北两个亚带。南亚带在早侏罗世盆地内继承演化，北亚带具有新生性。因此，髫髻山组底部不整合具有区域上可对比性。

值得注意的是，在北京新生的火山喷发带北亚带中，所有火山-沉积盆地一段以火山岩为主，但在火山岩底部局部也发育不同厚度、不同粒径碎屑岩沉积，由于其发育厚度差异较大，且不像龙门组巨大砾岩特征鲜明，未有引起重视。但在 1∶5 万区域地质填图时，多图幅圈定有小面积的早侏罗世地层，如永宁幅、后城幅、三道营子幅等。陆相盆地沉积相对独立，使得该套碎屑岩也未开展区域性对比，或建立相应地层单元，仅一并归入髫髻山组或土城子组，或直接定为窑坡组。梳理北亚带中碎屑岩和火山岩接触关系，具有完全类似于南带龙门-九龙山组同髫髻山组地层接触关系（图 33-1）。不仅如此，髫髻山组地层在区域上被划分为四段，二—四段区域上可对比，火山-沉积厚度基本相当，但一段沉积厚度在北亚带巨大，致使整个北亚带地层总厚度均大于南亚带。鲍亦冈等（1995）也发现此规律，认为北亚带火山活动强度更大。这里，是否可以参考南亚带，把此

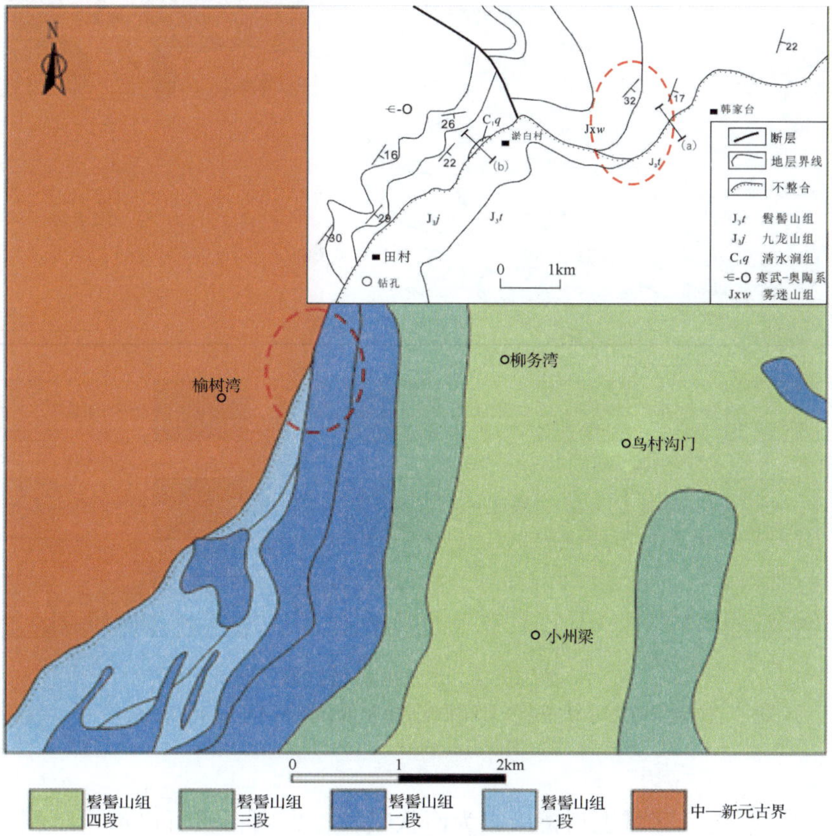

图 33-1 髫髻山组不同段接触关系图

（据 1∶5 万四海幅地质图；角图据李海龙等，2014）

Fig. 33-1 The contact relation of members of Tiaojishan formation

套碎屑岩单独划分出来，建立可同南亚带以及更大范围内（河北、山西、辽宁等）可对比的一套地层。如此，龙门-九龙山组碎屑岩沉积在北京境内的火山喷发带北亚带中也是可能存在的北亚带中火山岩地层之下碎屑岩建议单独划分出来同南亚带龙门组和九龙山组进行对比，以进一步明确燕山造山阶段的起始重大地质问题。

九龙山组和龙门组为髫髻山组火山岩系下部层位（赵宗溥，1963；翁文灏，1928）。实际上九龙山组底部和顶部各发育一套砾岩，叶良辅（1920）将底部砾岩划分为九龙山组，下部砾岩划分为窑坡组，但未给出划分这两套砾岩的理由。王竹泉（1933）谨慎起见，将下部砾岩从窑坡组上部划分开，命名为龙门组砾岩。同时，他还指出两套砾岩并未看出差别，建立龙门组仅仅是谨慎的考虑。1∶5 万石景山幅区域地质填图时，基于岩性对比把该套上部砾岩也划归龙门组，使得图面上出现产状冲突，九龙山向斜核部地层又出现老地层尴尬解释（图 33-2）。张宏仁等（2013）、李海龙等（2014）野外详细考证并认为龙门组与九龙山组本就是一套地层，本次工作认可该认识。但考虑岩性有所区分，地层研究意义较大，仍保留龙门组地层。

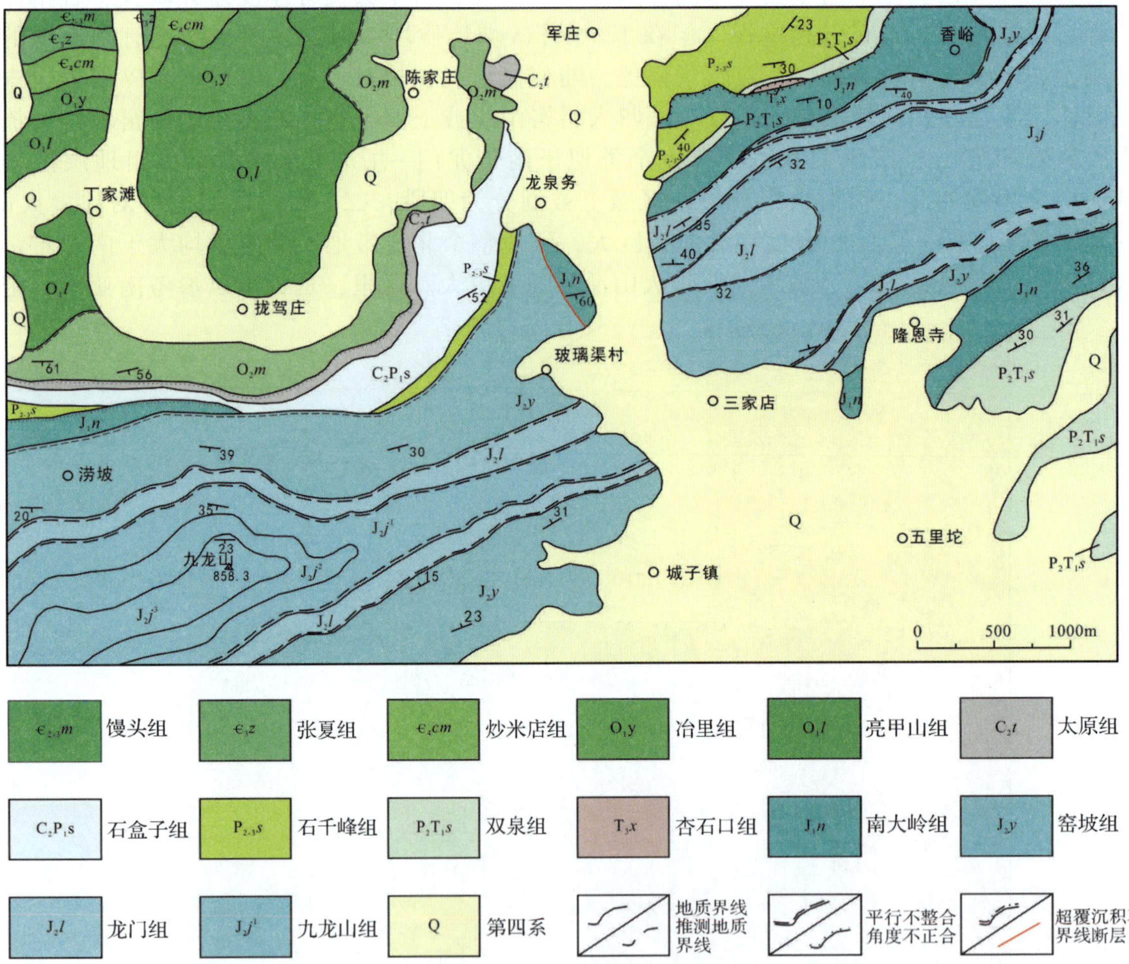

图 33-2　九龙山向斜构造

（据 1∶5 万石景山幅地质图）

Fig. 33-2　The Jiulongshan syncline structure

李海龙等（2014）认为，区域上髫髻山组为面状喷发，火山物质可以最早沉积于盆地低洼处并完好保存，然而在盆地边缘的剥蚀区火山物质未能保留或仅部分保留。所以，九龙山组自下而上火山物质是逐渐增多的，而髫髻山组底部的沉积物质是逐渐减少的。基于此认识，其还建立了龙门组、九龙山组和髫髻山组时空分布模式图，并总结出九龙山组与龙门组之下的不整合面实际上也是同一界面（黄汲清，1945；赵宗溥，1963；朱庭祜，1951；张宏仁等，2013；李海龙等，2014），龙门组、九龙山组、髫髻山组之下的三个不整合面最终均收敛于髫髻山组下的不整合面。因此，燕山运动的不整合面首先是一个穿时面，应从龙门-九龙山组开始沉积到髫髻山组喷发结束为止。

但是，李海龙等（2014）认为的穿时不整合面与沉积盆地和岩浆活动仍存在冲突。龙门组—土城子组沉积阶段经历至少两次构造反转，构造应力场的多幕次变化才会引起该阶段既有大规模中—基性岩浆活动，又出现岩浆活动的薄弱期的可能，以及沉积盆地中心迁移等特征。

综上，北京境内龙门组、九龙山组并非仅仅发育于髫髻山火山活动期火山喷发带南亚带，北亚带中也有发育。龙门组和九龙山组为一套陆相河流碎屑岩沉积组合，相变较大。在小范围内可见龙门组—九龙山组和髫髻山组不同地层小角度不整合接触，同样考虑髫髻山组地层火山岩沉积，存在局部喷发不整合可能，区域性的大不整合必须寻找地层宏观展布和组合特征。北京境内髫髻山期火山岩带北亚带存在，具有新生性，说明区域性的大型不整合是存在的。把北区带中髫髻山组地层底部碎屑岩单独区分出来，作为龙门-九龙山组地层，也基本符合岩石地层清理规律。如此，龙门-九龙山组和髫髻山组地层区域上结伴出现，龙门-九龙山组底部不整合运动“A幕/绪动”的真正不整合面，特征上指的是上火山岩系和下部砾岩等碎屑岩组合与下部窑破组煤系地层之间的角度不整合面。而髫髻山组和下部砾岩等碎屑岩从这个角度而言，仅仅就是指这一局部不整合事件。“A幕/绪动”的角度不整合面在北京西山应位于窑坡组煤系地层之上，在北京火山喷发带北亚带中则为髫髻山底部一套碎屑岩组合，在山西位于大同煤系地层之上，在冀东位于下花园组之上，在辽西位于北票组之上。煤系地层之上为一套区域上可供对比的砾岩，山西为云岗组，北京西山为龙门组，辽西为海房沟组（赵宗溥，1963；张宏仁等，2013；李海龙等，2014）。本志关于此不整合面未有开展新研究，地层划分时仍保留龙门组和九龙山组，也未从北亚带中髫髻山组中将底部碎屑岩分离出来对比龙门组和九龙山组，考虑髫髻山组底部不整合面具有更广区域性，是新生火山岩喷发带、盆地新生特征等，故将此作为造山运动起始。

（2）构造组合及形成时代

该套地层由中性火山岩-杂色粗碎屑岩系构成，变形以区域性逆断层发育为主，并发育北东向箱形向斜。

A. 区域性逆断层——北东向逆冲推覆构造

由北京西山的南大寨-八宝山断层、黄山店逆冲断层、长操-霞云岭逆冲断层，曹家堡-黄土梁逆冲断层，北京北部昌平十三陵逆冲断层、上庄-红螺寺逆冲断层以及怀柔汤河口逆冲断层等组合而成。自西南部房山区张坊北部，向北东经房山区北部，门头沟区西部，昌平区东北部，怀柔区西北部到汤河口一带，总体走向约北东40°，长约160km，宽约32km。单条断层多呈现为向北西凸出的弧形形态，但整个构造带具有比较清晰的北东向带状展布特征。

上述逆断层，除八宝山逆断层因其推覆到石炭-二叠纪地层之上，推覆形成时间是燕山运动A幕（张宏仁等，2013）还是B幕（Hodges，1999）存在争议。其余逆断层，如十三陵逆断层均切割的最新地层为中生代髫髻山组，表现为早白垩世早期燕山运动B幕（主幕）产物。

该套逆冲推覆系统具有近距离推覆特征（见第三十二章第三节）。张宏仁等（2013）认为该期推覆构造断距极大，理由是基于龙门组砾岩砾石统计（图33-3）表明，砾石组成中见有大量元古宙地层的砾石，以说明从那时起，超大规模的逆断层已经发生，使得属于雾迷山组的硅质白云岩和燧石被推到地表，并受到剥蚀。从被剥蚀掉的地层厚度可以推断，逆断层的垂直断距可达几千米，而龙门组砾岩和洼里砾岩同属于同构造砾岩。

最新研究发现，在八宝山断裂北东方向延伸北京平原地带，地热井勘探也揭示小汤山一带存在寒武纪地层推覆到中生代地层之上的现象（张柏林等，2005），其和北部十三陵逆断层（部分隐伏）可构成区域性北东向逆冲推覆系统。因此，八宝山逆断层发育于早白垩世早期，即燕山运动B幕（主幕）产物。

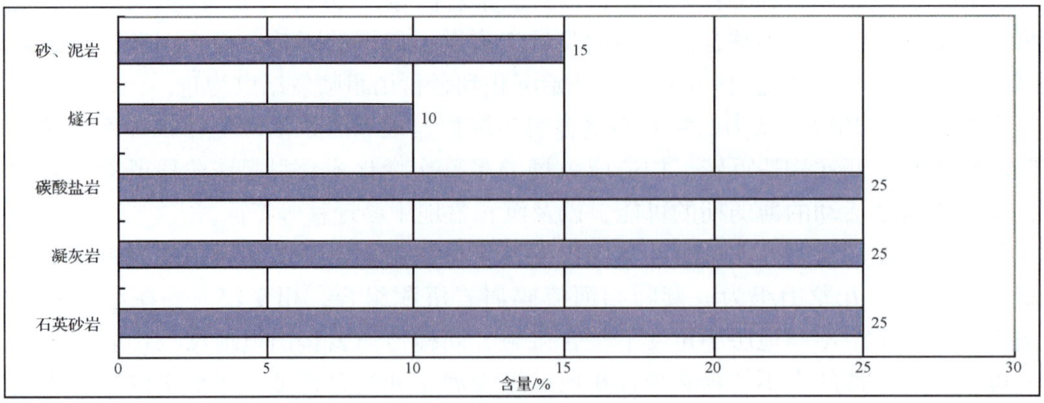

图 33-3　龙门组砾岩组成统计图

Fig. 33-3　Statistic of Longmen formation conglomerate

B. 龙门、洼里砾岩是否为同构造砾岩？

龙门组厚 10~30m，以连续—微角度不整合关系覆盖于窑坡组之上，与上覆九龙山组在岩性上呈渐变过渡关系。沉积特点是颜色以紫红色、灰绿色为主，含较多砾岩，砂岩以粗、中粒砂岩居多，细砂岩次之。空间区域上，西部大安山、北部法城一带，以粗碎屑岩为主，砾岩丰度为 39%，粗砂岩 15%，细砂岩 15%，粉砂岩 31%；往东、南沉积粒度变细，妙峰山一带砾岩丰度为 25%，粗砂岩 13%，细砂岩 16%，粉砂岩、泥岩丰度为 46%，岳家坡一带砾岩丰度为 19%，粗砂岩、细砂岩为 31%，粉砂岩、泥岩 50%。时间序列上，早期发育为砾岩、粗砂岩组合，粗砂岩中发育斜层理，为强水流河流相沉积；晚期发育为粉砂岩、泥岩组合，且可见水平层理（纹理）及菱铁质结核，为湖泊相沉积。

洼里砾岩主要特征有：①砾石为滚圆—半滚圆状，岩性较为单一，绝大多数角砾为雾迷山组燧石条带白云岩、燧石团块、含砂泥质白云岩、隐晶和微晶白云岩，以及少量棱角状的石英砾石成分。此外，尚可见奥陶系灰岩和石炭系石英砂岩。②所见砾石直径不一，大者可达 40~50cm，小者 2mm。③胶结物由泥质、硅质、碳酸盐质（白云质和灰质）和中、小砾石组成，胶结类型属孔隙式。④凡属硅质的砾石都为棱角状，而碳酸盐岩的砾石均有一定磨圆度。⑤镜下或手标本观察发现，碳酸盐岩的砾石边缘常有一圈铁质氧化圈，显示出成岩前，砾石遭受了风化作用，经历过一定距离的流水搬运，砾石分选性差，不同粒径的砾石堆积在一起，呈颗粒支撑或颗粒-杂基支撑，充填物为小砾石、细砾和粗砂，总之大砾石由中砾石支撑着，而中砾石又由小砾石支撑着，这是洼里砾岩的一大特点。这种颗粒-杂基支撑的砾岩反映着一种特殊的沉积环境——泥石流沉积。

综上，龙门砾岩和洼里砾岩可能均为髫髻山组底部一套山地河流相沉积地层，底部具有角度不整合特性，均表现为残存盆地边缘（军响一带或洼里一带）与盆缘褶皱、断裂相关的河流冲洪积扇状砾岩体沉积。在这些粗碎屑冲积扇向盆地内快速充填后，盆地中心依旧保持较深水湖相环境。因此，该两套地层沉积上的确具有挤压构造背景下抬升同造山砾岩、前陆盆地砾岩，或者磨拉石特征。张宏仁等（2013）强调该阶段地壳变动以断裂为主，褶皱不发育，逆断层上盘上升过程中重力坍塌在下盘上，并基于砾石组成、地层厚度推断逆断层的垂直断距可达 4km。这一推理必须有一个前提条件，即龙门组沉积前区域上完全稳定克拉通化，地层未有任何褶皱变形。近些年来，北京境内及更大区域内印支运动相关构造（逆冲、褶皱）的新证据越来越多，与之相对应的岩浆活动也多有发现，印支运动引起的地壳抬升可达 11km 以上。在龙门期沉积前，中元古界完全有可能已经局部被剥蚀到地表。这和西山龙门组沉积同样具有西部物源，砾石组成跨度大，有新生火山岩组成，且沉积物由西向东逐渐变细结论一致，而并非张宏仁等（2013）认定的沉积物来自逆断层的上盘即北京市的平原区。

（3）岩浆岩带分布与岩浆演化

该阶段岩浆活动表现为髫髻山期火山活动和以石城-云蒙山、长圆杂岩体等组成的侵入岩带。二者共性特征为面状大范围分布，说明该期岩浆活动强烈，而其展布方向经历由北东东向转变为北东向，经历一个重要转变期。

时间序列上，基于同位素年龄统计，二者也形成于不同时间阶段，且具有不同的大地构造背景。髫髻山期火山活动稍早于侵入活动，表现为以伸展构造背景环境为主大规模基中性岩浆活动，晚期土城子期具有构造转换（伸展—挤压）的特性，而侵入岩带则稍晚于火山岩形成时代，主体表现为以挤压构造背景环境为主的中—酸性岩浆活动。这一特征表明该阶段岩浆活动先期以火山活动为主，后期以侵入活动为主。

髫髻山期火山活动表现为中心式爆发与溢流方式同时存在。髫髻山火山活动期内有两次连续而强烈的火山喷发活动，构成两个厚大的火山岩层。在这两次大规模火山喷发活动之后，均出现较长时间的火山作用间歇期，形成两段火山沉积碎屑岩堆积。因此，在剖面上髫髻山组表现为两个火山岩段与两个沉积碎屑岩段相间出现的地层结构。

作为髫髻山期火山活动尾声的产物，土城子组中仍见有少量火山岩夹层。其中，第一岩性段中普遍发育有厚度不等的流纹质角砾凝灰岩、流纹质凝灰岩及粗安质凝灰岩等，与凝灰质砂岩、砾岩呈互层或夹层状产出。第二岩性段的中、上部，于白河堡和千家店地区发育数层熔岩和角砾凝灰岩，自下而上岩性由粗面质变为粗安质。而白河堡地区最上部的火山岩夹层为玄武岩，反映出土城子组时期火山活动的岩浆化学成分有从酸性向中性、基性方向变化的趋势。

侵入岩岩浆岩带岩性组成复杂，早期侵入岩类以石城、太平庄等岩体为典型代表，岩性则为辉长岩、辉长闪长岩、石英闪长岩、石英二长岩基性—中性岩类，侵位时代约153Ma，相当于土城子组地层沉积时期，单个岩体主要呈近南北向或近东西向展布，岩体组合宏观呈北东东向展布，仍为伸展构造背景下岩浆侵位产物。稍晚期侵入岩类以云蒙山、八达岭等岩体为典型代表，岩性则为闪长岩、石英二长岩、花岗闪长岩、花岗岩中性—酸性岩类，侵位时代约145Ma，单个岩体有北东东向（次）和北东向（主），岩体组合宏观呈北东向展布，为强烈挤压构造背景下岩浆侵位产物。两期岩浆侵入活动Sr-Y元素特征揭示，岩石经历了低Sr低Y到高Sr低Y的转变，可能揭示地壳厚度变化，为燕山运动造山产物。

岩浆岩带由北东东向转变为北东向后，该期岩浆活动进入短暂的间歇期，间歇时间约8~10Ma，直至135Ma后张家口期火山活动和小规模的侵入活动。

（4）火山-沉积盆地对比

通过地层分布恢复该阶段火山-沉积盆地，该期盆地具有明显转换性质。

首先，龙门-九龙山组碎屑岩盆地具有挤压构造性质盆地。龙门期沉积盆地南亚带分布在残存西山盆地边缘（军响一带为代表），北亚带分布在山间新生盆地底部，盆地空间分布上总体仍具有北东东向展布特征，但又具有和盆缘褶皱、断裂相关的河流冲洪积扇状砾岩体沉积特征。这一特征说明该阶段具有构造转换演化特征，一方面使得盆地局部具有新生性特征，另一方面使得沉积地层接触关系复杂，总体地层分布零散。宏观上，沉积物厚度由盆地沉积中心向盆地边部急剧减薄，明显有别于窑坡组盆内沉积厚度变化不大的沉积特征。而在这些粗碎屑冲积扇向盆地内快速充填后，盆地中心依旧保持较深水湖相环境。而九龙山期沉积盆地在沉积物宏观分布上，基本特征同龙门期，即由盆地沉积中心向盆地边缘厚度急剧变薄，盆地呈现挤压同坳陷充填特征。同时，九龙山期盆地区域沉积过程中，还具有逐渐向北西掀斜特征，如使得西山一带沉积中心由门头沟区岳家坡向北西方向迁移至大台—板桥一带；盆地西北边缘青白口—田庄一带，仅出露九龙山组三段地层，且超覆于寒武纪或中—新元古界地层之上。

龙门期和九龙山期沉积盆地因强烈挤压抬升而处于超补偿填充状态，并最终因燕山运动A幕强烈挤压而接近消亡。构造上燕山运动A幕区域构造应力场持续挤压收缩，在使得（杏石口期—南大岭期—窑坡期）燕山运动第一世代伸展盆地（南大岭期—窑坡期盆地）消亡同时，并使盆地中的堆积物发生纵弯褶皱和剪切滑脱作用（构造应力场由北东东向北东向转化引起），形成了髫髻山、九龙山-香峪大梁和北岭等侏罗山式向斜构造，轴向北东 65°~70°。

再次，髫髻山火山-碎屑盆地具有伸展构造性质盆地。在经历龙门-九龙山期短暂挤压后，构造体制转换，构造背景由挤压再次转变为伸展环境。受北东方向居庸关深断裂、南口-沿河城深断裂及颐和园深断裂（推测）影响，形成该期断裂控制的张性裂陷盆地。裂陷伸展方向北东向，使得盆地宏观展布方向北东 50°~60°。根据北京境内盆地的展布情况，可分为两个带。南带西南起自百花山，往东北方向经髫髻山、妙峰山、十三陵、怀柔山前延至密云新城子一带，分布于南口-沿河城断裂和颐和园断裂之间宽约 20km 的地区，依次发育有百花山、髫髻山、新城子 3 个火山-沉积盆地。东北端的新城子盆地因受到东西向赤城-古北口深断裂的制约，盆地向东西方向偏转延伸。北带处于居庸关深断裂的西北侧，发育有燕羽山、凤驼梁-四海盆地 2 个火山-沉积盆地，总延展方向与南带平行。该期盆地内堆积髫髻山组火山岩-碎屑岩堆积物。盆地在龙门-九龙山期盆地基础上演化：①同上个世代的盆地展布方向明显不同，岩石地层亦有不同于上个世代的延伸展布格局；②盆地分布范围及数量明显增大、增多。这一特征说明该期盆地已经完成了构造体制转换。

最后，土城子组火山-碎屑岩盆地具有挤压构造性质盆地。区内构造背景环境再次反转，再次由伸展转变为挤压构造环境。盆地表现为燕山运动造山 B 幕（主幕）板内造山挤压型（继承）坳陷型盆地。该期燕山运动时间短但强度大，形成区域上的最大角度不整合，为燕山期主幕。北京境内发育有髫髻山盆地、新城子盆地、后城盆地、千家店盆地和凤驼梁-四海盆地 5 个火山-碎屑岩盆地。髫髻山盆地和新城子盆地为髫髻山期南带继承型盆地，凤驼梁-四海盆地为髫髻山期北带继承型盆地，而后城盆地、千家店盆地为髫髻山期北带新生型盆地。该期盆地继承性特征表现在：①髫髻山组和土城子组地层接触关系上，当盆地内部两组地层均有发育时，二者表现为产状一致的整合接触；②土城子期盆地群展布方向同髫髻山期一致；③土城子组地层展布也同髫髻山组基本一致。而新生性特征则表现在：①至土城子期，在中—新元古代地层基底上，新生千家店盆地和后城盆地；②髫髻山组和土城子组同时发育盆地，髫髻山组四段与其上覆的土城子组，在分界线附近岩石类型相近，主要为复成分砾岩、凝灰质砂岩等，但其成因特征差异明显，前者主要是火山泥石流相或冲洪积相，砾石成分多为不同类型的火山物质，外来碎屑物质相对较少，砾石大小混杂，分选极差，多数为次棱角状—次圆状，无层理，反映出堆积物搬运距离短、堆积速度快的成因特征，而后者的碎屑岩，碎屑成分复杂，以外来碎屑为主，火山物质较少，分选性及磨圆度均较好，岩层层理发育，韵律结构清晰，显示碎屑搬运距离长、物质来源广的河流相沉积特征，沉积时期构造环境相对稳定。因此，二者之间虽为整合接触，但具短时间的沉积间断，构造环境发生新的变化；③随后燕山运动 B 幕区域构造应力场持续挤压收缩，因强烈挤压抬升盆地处于超补偿填充状态，最终使得燕山运动伸展型第二世代盆地（髫髻山期盆地）消亡，同时造成盆地内地层发生纵弯褶皱形成了千家店等向斜、沙梁子-上花楼逆冲断层等挤压构造，轴向北东 55°~60°；④上覆的地层同土城子组角度不整合接触；⑤土城子期，火山活动强度逐渐减弱，且具有由北部向南部减弱趋势，而髫髻山期则表现为由西南向东北喷发强度逐渐转弱特征，与土城子组沉积时代相当的侵入岩岩浆岩带逐渐形成，岩浆侵入活动越来越强，最终形成区内最大的北东向八达岭杂岩带。

综上，燕山造山期盆地具有伸展和挤压型两类，伸展型盆地时间跨度长，挤压型盆地相对时间跨度小。燕山造山运动具有短时快速造山特征。

3. 造山后阶段（早白垩世早期—早白垩世晚期）

（1）地层年代确定与各地层接触关系

该阶段由老到新发育张家口组、东狼沟组、九佛堂组、坨里组和夏庄组火山-沉积地层，化石和同位素年龄揭示地层沉积时代为早白垩世早期—早白垩世晚期。区域上可见张家口组不整合覆盖在土城子组及更老地层之上，其余各组地层之间为整合接触。

（2）构造组合及形成时代

该套地层由酸性火山岩与湖沼相和河流相的碎屑岩系组成，褶皱变形为北北东向箱形向斜（北北东 25°~30°）叠加于早期形成的褶皱构造之上，含有著名的“热河生物群”化石。

（3）岩浆岩带分布与岩浆演化

该阶段岩浆活动表现为张家口期火山活动和以周口店、阳坊、南猴顶岩体等组成的侵入岩带。具有和造山期相似的特征，即面状大范围分布，说明该期岩浆活动依然强烈，而其展布方向经历由北东向转变为北东东向。

时间序列上，基于同位素年龄统计，二者同样形成于不同时间阶段，且具有不同的大地构造背景。张家口期火山活动稍早于侵入活动，表现为以挤压构造背景环境为主大规模中酸性岩浆活动，晚期东狼沟组具有构造转换（挤压—伸展）的特性，形成酸性—基性（底部—顶部）岩浆喷发。而侵入岩带则稍晚于火山岩形成时代，表现为以伸展构造背景环境为主的中—酸性岩浆活动。因此，造山后阶段岩浆活动也表现出同造山阶段一致的特征，即先期以火山活动为主，后期以侵入活动为主。

张家口期火山活动亦以中心式喷发为主，显示了层火山的特征。活动强度十分强烈，爆发作用强，爆发指数高，平均爆发指数为 44%。火山岩中流纹质岩石平均占 64.7%以上，特别是火山碎屑流相流纹质熔结凝灰岩的大量出现是本阶段火山活动最主要的特征。火山活动从早到晚有一定规律：第一火山喷发周期的早期为熔浆喷溢，稍后喷溢与爆发交替发生，喷发强度逐渐加大，且岩性以玄武安山岩、粗安岩、粗安质角砾凝灰岩等中性火山喷出岩为主。在第一火山喷发周期以后，出现火山活动间歇期，形成一套以复成分砾岩为主的火山-沉积碎屑岩，并有少量凝灰砂岩和薄层喷出岩夹层。第二火山喷发周期是张家口期火山活动最强盛的时期，以酸性火山喷出岩占绝对优势，形成以流纹质熔结凝灰岩、流纹质角砾凝灰岩及流纹岩为主体的巨厚层堆积。这种以早、晚两个火山喷发周期，中间夹一个火山活动间歇期的火山活动特点，构成了该期火山岩层剖面上极为清晰的火山-沉积地层三分结构。

与髫髻山期火山活动相类似，在张家口期整个火山活动过程中也有一个“尾声”阶段，表现在张家口组的上覆地层——东狼沟组第二岩性段中夹有玄武粗安岩（或粗安岩）和火山碎屑沉积岩。岩浆活动具有明显基性活动特征，火山活动进入深源伸展侵位特征。

同造山期相似，较早期侵入岩岩浆岩带岩性组成复杂，以薛家石梁、棋盘岩、周口店、大羊山等岩体为典型代表，岩性包括辉长岩、辉长闪长岩、石英闪长岩、石英二长岩、正长岩、石英正长岩、花岗闪长岩基性—中性岩类，侵位时代约 130Ma，相当于东狼沟组沉积时期，单个岩体主要呈北西向展布，岩体组合宏观仍呈北东向展布，成因为幔源岩浆的底侵作用（以辉长岩岩浆为代表）引起加厚陆壳（下地壳）的部分熔融。而稍晚期侵入岩类以阳坊等岩体为典型代表，岩性则为花岗岩酸性岩类，单个岩体呈北东东向（主）或近南北向（次），岩体组合宏观呈北东东向展布，成因为地壳物质的部分熔融。因此，两幕次岩浆侵位均表现为强烈伸展构造背景下侵位特征，岩浆活动强度逐渐减弱，为造山后伸展垮塌岩浆活动。且岩浆岩带由北东向转变为北北东向后，并逐渐进入平静期，在晚白垩世出现小规模的基性辉绿岩侵位。

（4）火山-沉积盆地对比

通过地层分布恢复该阶段火山-沉积盆地，该期盆地也具有明显转换性质。

首先，张家口期发育北北东走向受大断裂控制的火山喷发带及喷发带内火山-碎屑岩盆地。由于主要受控于挤压构造背景环境中心式火山喷发活动，致使该期火山-碎屑岩盆地表现为小型火山盆地，发育有东岭盆地、杜家庄-镇厂盆地以及上大水、岔道、怀柔、七道河等断陷盆地。昌平棉山、密云五指山也有小面积地层出露，可能也发育有小型盆地。这些盆地断续分布，组成燕山晚期北北东向火山喷发带。可划分为东、西两个喷发带，西部喷发带受控于紫荆关深断裂，东部喷发带受控于黄庄-高丽营深断裂，整体看受北北东向深断裂控制，盆地产出的方向性极强，呈北东 30°方向延展，这一喷发带在北京境界内延长约 140km，带宽 70~80 km，两端向外均延出本市。但带内每一个火山盆地的发育和生成则受其具体部位特定盖层断裂构造的控制，盆地的延长方向各异。各火山盆地内张家口组的厚度变化较大（600~4300m），且起始时间略有差异，大致由南向北发展，也说明该期盆地具有转向特征。

其次，东狼沟期和九佛堂期火山-沉积盆地，造山后构造伸展背景下，火山活动逐渐转弱，岩浆以大规模侵位活动为主，盆地数量及规模减少。北京境内残留有坨里盆地、东岭盆地、怀柔盆地、杜家庄-镇厂盆地。盆地宏观走向转向为北北东向，盆内沉积主要为含集块角砾凝灰岩、凝灰岩及沉凝灰岩，夹有玄武岩或粗安岩。沉积碎屑岩沉积厚度不大，碎屑物的粒度由下往上逐渐变细，并夹有间歇流水环境中形成的沉凝灰岩，岩层中产有鱼的鳞片及骨骼化石，并见较多腹足类、双壳类和介形虫化石，为火山间歇期河湖相沉积。总体盆地内部处于超补偿沉积，致使盆地逐渐萎缩。

最后，坨里组、夏庄组沉积时期，区域构造再次反转，由伸展转变为挤压构造环境，使得燕山运动伸展阶段（张家口组晚期—东狼沟组—九佛堂组沉积）形成的第三世代盆地逐渐萎缩，仅在山间局地形成挤压型同构造沉积（坨里组—夏庄组），其具有类似于龙门-九龙山组的特征。该期区域上盆地数量少，规模均较小，零星分布，盆地发育于燕山期向喜马拉雅期过渡过程中相对平静的构造背景下，不像龙门-九龙山组地层广泛发育。

四、动力学探讨

关于燕山运动构造动力学争议极大，主要的观点有：①张宏仁等认为属天体事件；②任纪舜等主流观点认为同太平洋板块低角度俯冲有关；③张岳桥等（2007）归纳为多向挤压汇聚的动力学；④邓晋福等（1996，2005）、吴福元等（1999）认为其属于岩石圈伸展减薄，类似裂谷动力学机制；⑤徐嘉伟等（1985）认为剪切大陆边缘学说。

由于北京辖区范围有限，在小范围内讨论动力学依据尚且不足。但值得注意的是，基于燕山运动构造-沉积-岩浆综合分析和燕山运动的特征总结，燕山运动三幕次均具有伸展—挤压构造反转现象。伸展期时间跨度大，而挤压期时间跨度小。这与万天丰（2004）通过东西方向剖面，概略地计算燕山期褶皱变形所需时间约为 3.66Ma 特征吻合。

燕山运动伸展期大规模正断层发育、沉积-火山盆地发育，而每一幕挤压构造形成不同方向箱形褶皱，主幕还发生区域性逆冲推覆构造，形成大规模具有方向性特征的岩浆岩带，并对先期盆地进行改造。总体上，三幕挤压构造应力场由北东东（近东西向）—北东向—北东东向呈逆时针方向旋转，表明燕山期构造应力场的近水平的最大主应力方向也是逐渐转变的。这似乎类似于李四光（1973）所述的从古华夏、中华夏到新华夏构造的变化。

值得注意的是，侏罗纪时期，华北陆块发生了古磁北方位从北东 30°显著地转成与现代的磁北比较接近的状态，即古地磁方位发生了逆时针转动 20°~30°的情况（万天丰，2004）。古生物分布与古气候带的特征也揭示了燕山期中国大陆地块的逆时针转动（顾知微，1982；刘本培等，1995；庞其清等，1995）。广泛的中国大陆在侏罗纪时期发生显著的逆时针转动，可能与北美板块西部鄂霍次克部分朝西和伊佐奈岐板块朝西北方向运移、俯冲或碰撞有关（Maruyuama et al.，1986，1997；Moore，1988）。因此，古太平洋板块低角度俯冲对燕山运动力可能起到至关重要作用，而岩石圈伸展减薄则本身就是燕山造山运动过程的表现。

五、燕山运动与华北克拉通活化讨论

克拉通是指地壳形成之后（至少自显生宙以来）保持稳定状态、极少经受强烈构造变形的构造单元。华北克拉通具有太古宙—古元古代结晶基底和中元古代—古生代盖层，总体上长期保持构造变形与岩浆活动等极其微弱的构造稳定状态，是中国境内保存较好的三个克拉通之一。

华北克拉通活化的科学命题，主要是基于岩石学和地球化学的研究，通过对古生代时期和新生代以来华北东部岩石圈厚度、热状态、岩石圈地幔组成与时代等特征的比较研究逐渐认识到的一个重大大陆构造动力学问题（张长厚等，2009）。

华北克拉通古生代时期曾具有典型古老克拉通的全部特征。与世界上典型克拉通不同的是，华北克拉通东部中生代以来遭到了强烈破坏。多数学者认为华北克拉通活化范围的西部边界位于近南北向重力梯度带。

北京境内，燕山运动前出现小规模的岩浆活动，包括中泥盆世正长岩，早二叠世变质辉长岩、石英闪长岩，早三叠世花岗岩，中三叠世石英二长岩和晚三叠世石英二长岩、花岗岩侵位。空间分布上，中三叠世之前侵入岩仅分布在燕山褶皱带，位于华北克拉通北缘和克拉通分界线附近（槽台学说内蒙地槽和华北地台分界线附近）。不仅如此，岩浆侵位还具有由北向南发展特征，即越北岩浆侵位越早，直到晚三叠世在燕山褶皱带靠近平原一带才开始出现岩浆活动（盘山岩体）。

北京境内，中泥盆世正长岩具有典型克拉通型岩浆活动特征，如深源富碳酸盐的碱性岩石，侵位深度深、富集型地幔等。Tappe 等（2007）认为克拉通活化最早期的现象是，克拉通岩石圈厚度至少减薄了 60km。早二叠世—晚三叠世岩浆表现为钙碱性系列，成因类型为富集型地幔或新元古代增生的加厚基性镁铁质下地壳发生部分熔融，不具有典型克拉通岩浆活动特征，但富集型地幔和巨厚的岩石圈仍然存在，岩浆活动规模小、分布有限等指示华北克拉通仍然较为稳定。因此，华北克拉通活化可能在中泥盆世就已经开始了，但主峰期肯定在侏罗纪—白垩纪，同燕山运动发生的时间具有一致性。这里就提出燕山运动与华北克拉通活化之间的两个问题：

（一）燕山运动与华北克拉通破坏关系

通过北京境内岩浆岩研究，境内乃至更大区域范围内深部中泥盆世岩体侵入揭示存在软流圈主动上涌，这一过程持续到晚三叠世，在长达 150Ma 的软流圈持续上涌对岩石圈加热过程中，在遭受加热区域的上方岩石圈变薄并隆起，即造陆运动。造陆运动持续性可通过①早三叠世—中三叠世类埃达克岩侵位，②早白垩世大石坡正长岩侵位抬升，以及③广大华北克拉通北缘缺失泥盆世—早石炭世地层，④这一时间段区域内构造格架变化剧烈，由稳定克拉通海相沉积演变为半稳定的陆相沉积得以体现。燕山运动起始于广大华北克拉通北缘的软流圈主动上涌和造陆运动背景下的，而这一过程本身就是去克拉通化。

（二）克拉通活化过程

华北克拉通的活化是在古老克拉通基础上的直接活化，还是在古老克拉通基础上经历了造山作用过程之后的造山带伸展垮塌活化这是涉及华北克拉通破坏动力学模型建立的根本问题之一（张长厚等，2009），也即燕山运动研究的关键问题——燕山运动是否为造山运动。

在长期软流圈上涌、岩石圈隆起和造陆运动构造背景下，华北克拉通活化的峰期阶段，也就是燕山构造运动阶段。燕山运动首先表现为南大岭期火山作用，而后是窑坡组地层伸展裂陷沉积和地块隆起与沉降作用。由于在此之前受印支运动影响，发育伸展前隆起导致已有地质体遭受剥蚀。因此，南大岭组火山岩直接覆盖在伸展前隆起剥蚀面上，而且其分布不受裂陷盆地（谷地）的限制。上述这些特征均表现出前造山（主动裂谷）演化特征。

在中侏罗世—早白垩世早期境内经历了两幕（窑坡组和龙门-九龙山组之间及土城子组—张家口组之间，以后者为主）区域上北东东向和北东向收缩变形和逆冲构造，尤其是北东向构造，两幕构造运动（燕山运动 A、B 幕）形成髫髻山组火山岩、云蒙山岩体等高 Sr 低 Y 埃达克岩火成岩，具有区域上地壳增厚的指示意义（张旗等认为的东部高原），并发生下地壳榴辉岩化（Deng et al.，2003），即板内造山作用引起的地壳加厚。

早白垩世中期—晚白垩世，华北克拉通首先遭受快速、脉冲式的酸性和中性岩浆侵入作用，形成区内早白垩世早期—晚期北东向—北北东向的八达岭岩浆岩带，而后在很短的时间里，区域上经历幅度极大的均匀快速隆升，随后又再次发生伸展作用，挤压—伸展多次构造反转，最终使得早期的中、酸性岩浆活动消失，代之以中等碱性到过碱性的岩浆作用直至无岩浆活动。这些特征与造山后期—后造山期，造山带内岩石圈拆沉作用及其地表响应一致。即由于岩石圈内部密度差异造成的重力不稳定性，加上伊佐奈岐板块碰撞触发下，岩石圈发生大规模拆沉作用，即相对较冷、密度较高的岩石圈底部与上覆岩石圈发生拆离并沉入软流圈地幔当中，软流圈占据拆沉岩石圈的相应位置，使上覆岩石圈的浮力增大，进而造成剩余的岩石圈部分隆起，并在势能增加下而发生类似伸展（主动裂谷）作用（Sengör et al.，1978）。

综上，燕山板内造山是存在的，这与广泛学者（Gao et al.，2002，2004；Deng et al.，2003，2006，2007；Wu et al.，2003；Zhai et al.，2005，2007）关于燕山运动板内造山认识一致。因此，古老克拉通活化属于在经历了造山作用过程之后的造山带伸展垮塌作用。

六、燕山运动特征

综上，对燕山运动特征进行简单总结：

（一）旋转性

燕山运动岩浆岩带、火山-沉积盆地空间展布揭示燕山运动表现出三次逆时针旋转特征。

（二）少区域变质作用

整个燕山运动阶段沉积地层，仅南大岭组和窑坡组发生低绿片岩相的变质，变质温压条件为350～700℃，0.25～0.8GPa。而更老石炭-二叠纪地层及底部地层也仅仅发育绿片岩相变质。这一特性说明燕山运动具有少区域变质作用特征。

（三）断裂为主构造运动

燕山运动褶皱不甚发育，致使“岩系之间倾角的差别也许并不很显著”。燕山运动三幕次运动形成三个构造方向箱形褶皱。相比较褶皱构造，燕山运动断裂发育，尤其是北东向断裂，且断裂性质复杂，具有断裂发育为主的构造运动特征。

（四）伸展—挤压构造多次反转

燕山运动阶段，区域构造应力场经历三次伸展—挤压的构造反转。伸展期形成区域裂陷型盆地，岩浆活动以火山活动为主伴随有少许岩浆侵位，挤压期形成凹陷型盆地，岩浆活动以侵入活动为主伴随有少许火山活动。

（五）深部地质决定作用

燕山运动动力来源于岩石圈深部，具有由深部向浅部发展的特征。

第五节　晚白垩世—第四纪构造阶段

燕山运动之后，晚白垩世期间，华北克拉通总体为岩浆与构造活动的平静期，晚白垩世-古新世，本区整体上呈隆起状态，长期风化剥蚀，属于“华北准平原”的一部分，即北台期夷平面（吴忱，1999）。古近纪期间，华北克拉通处于伸展构造背景。已存的北东向断裂沿河城断裂，黄庄-高丽营断裂等发生强烈的正断活动，华北准平原解体，形成一系列断陷盆地。本区形成了延庆-怀柔盆地、北京断陷盆地，盆地中堆积了巨厚的河湖相碎屑沉积物。新近纪时期，华北克拉通处于弱拉张环境，本区主要发育继承性的拗陷型盆地叠加在早期盆地之上，北京冲积平原初步形成。第四纪以来，北东向—北北东向断层和北西向断展共同活动，破坏了本区两隆-凹的构造格局，形成了以马池口凹陷。顺义凹陷等新的沉积中心，主要为河湖相碎屑沉积物。

据此将喜马拉雅旋回划分为2个亚构造层：下部亚构造层包括上始新统长辛店组及始—渐新统前门组，沉积时限从58～25Ma，地层厚度320～3000余米，在下部亚构造层前门期堆积时有偏碱性玄武岩浆的喷溢；上部亚构造层包括新近系的天坛组、天竺组，以及第四系。自新近纪至今，经历了25Ma，地层总厚度852～1900余米。

第三十四章　构造活动与成矿作用

北京地区矿产资源丰富，种类繁多，截至2014年底，共发现各类矿产127种（含亚矿种），其中固体矿产121种。已经查明资源储量并编入《北京市矿产资源储量表》的固体矿产有67种354处矿产地，其中能源矿产1种29个矿产地，金属矿产19种116个矿产地，冶金辅助原料非金属矿产9种44处矿产地，化工原料非金属矿产6种43处矿产地，建筑材料及其他非金属矿产32种122处矿产地。煤炭、铁、钼、石灰岩、白云岩、建筑用砂及汉白玉等矿产是北京地区的优势矿产资源。成矿作用主要包括沉积成矿作用和内生成矿作用（岩浆成矿作用和内生成矿作用），各种成矿作用与相应的构造环境和构造活动密切相关。

第一节　构造阶段与成矿作用

北京地区大地构造演化经历了新太古代—古元古代结晶基底与岩石圈形成阶段、中元古代—中三叠世盖层形成与岩石圈稳定阶段、晚三叠世—早白垩世板内造山和岩石圈减薄阶段和晚白垩世—第四纪造山后伸展拉张与岩石圈再造、现代地貌形成阶段，形成了明显与构造阶段有关的成矿作用及矿产类型。

一、新太古代—古元古代

该阶段为初始陆核形成期，根据变质作用程度不同，可将本期沉积建造划分为两个岩石构造组合，是富含铁矿的重要的岩石构造组合。密云岩群变质表壳岩-片麻岩组合，是新太古代变质沉积型铁矿的重要含矿层位，位于密云的沙厂、大槽、马圈子等地。主要岩石组合为辉石变粒岩、二辉麻粒岩、黑云辉石斜长片麻岩、含辉石磁铁石英岩。变质相主要为麻粒岩相。主要代表性铁矿床有沙厂铁矿、大槽铁矿以及马圈子铁矿。四合堂岩群斜长角闪岩-磁铁石英岩组合，是新太古代变质火山-沉积型铁矿的重要金矿层位，位于四合堂地区，介于半城子断裂带和崎峰茶断裂带之间，沿云蒙山岩体北侧呈半环状展布，原始地层近东西向延伸。岩石组合为一套角闪岩相变质的斜长角闪岩、角闪斜长片麻岩、黑云变粒岩和斜长片麻岩、浅粒岩及磁铁石英岩，斜长角闪岩类与厚层磁铁石英岩互层，夹角闪黑云变粒岩。原岩主要为基性、中基性、中酸性熔岩及火山碎屑岩，超基性原岩很少，部分为偏碱性的火山沉积岩和少量的陆缘碎屑沉积岩。主要代表性铁矿床有冯家峪铁矿和上峪铁矿。

二、中元古代—中三叠世

盖层形成期，为稳定沉积阶段，构造和岩浆活动很少，主要形成与沉积作用有关的矿床，为产于滨海中的砂岩矿、板岩、石英砂岩矿床和产于浅海中的石灰岩矿床等。

1. 铁、钾、硅石等矿产

铁矿，一般形成于海盆下降，海侵旋回的初期（长城纪、青白口纪早世形成赤铁矿点或矿化点），含矿岩石组合为滨浅海的砂岩粉砂岩泥岩组合以及滨浅海的碳酸盐岩组合（例如蓟县系铁岭组）；在长城纪和蓟县纪晚期形成铁锰矿（矿化）或含锰赤铁矿-菱锰矿化。钾矿（含钾岩石）的成矿时代短暂，成型产地集中分布在长城纪早期，主要含矿岩石组合为相当于串岭沟组的滨浅海砂岩粉砂岩泥岩组合，其他时代不含矿。硅石（石英砂岩、石英岩）比较有规律的分布于长城纪、蓟县纪和青白口纪早世，成矿作用随时间序列的演化具有由强到弱的特点。

2. 碳酸盐岩矿产

北京市与沉积作用有关的碳酸盐岩矿床包括白云岩、石灰岩、天然水泥石和大理石矿产。白云岩大多产于中元古代，含矿层位有规律地在长城纪晚期、蓟县纪中晚期和青白口纪晚期，即海侵沉积旋回中晚期的富镁碳酸盐岩地层，相对应的岩石构造组合为滨浅海碳酸盐岩岩石组合，一般矿床规模大、矿石质量好。到早古生代寒武纪和奥陶纪也有出现，但并不呈单矿种矿床出现，绝大多数与石灰岩成复合矿种的共（伴）生矿床出现，至晚古生代以后，白云岩矿化层位消失。大理石矿床是本市优势非金属沉积矿产，其产出层位及成矿时代与白云岩、石灰岩一样，主要在中—新元古代及下古生代。此外，驰名中外的汉白玉等大理岩（变质层控），控矿层位比较狭窄，主要分布在遭受变质作用的雾迷山组的滨浅海碳酸盐岩组合中。

3. 石灰岩（制碱用灰岩、熔剂用灰岩、水泥用灰岩）矿产

作为工业用石灰岩矿层主要赋存于碳酸盐陆表海亚相中。其中昌平组是本市富含电石、熔剂及水泥用石灰岩的主要层位之一。馒头组陆源碎屑-白云岩组合是本市熔剂用、水泥用石灰岩的重要含矿岩系，含矿性较好，与昌平组灰岩组合相比，电石用石灰岩的含矿性较差。该类岩石组合以鲕状灰岩大量出现为主要特征。张夏组陆表海陆源碎屑-灰岩组合主要呈带状分布在西山地区，赋存熔剂或水泥用灰岩，含矿层岩性主要为鲕状灰岩。冶里陆表海灰岩组合是本市赋存石灰岩矿产的重要含矿岩系之一，主要为电石、制碱用灰岩为主。岩性以泥质条带灰岩、镁泥质斑状豹皮灰岩和碳酸钙含量较高的岩石为主要岩石类型。马家沟陆表海灰岩组合是本市电石、制碱用灰岩、水泥用灰岩的主要含矿岩系。岩石组合主要由豹皮灰岩、石灰岩、含燧石团块灰岩、白云质灰岩等海相碳酸盐岩组成，陆缘碎屑岩类极少。

4. 建筑石料

在北京碳酸盐岩台地上的各种岩石，如各种石灰岩、白云岩、藻灰岩、竹叶状灰岩等均可作建筑石料，也是目前主要的建筑用石料。

5. 与海陆交互陆表海沉积环境有关的矿床

海陆交互陆表海亚相（C_2—P_1）包含太原陆表海砂泥岩组合（C_2）、山西陆表海沼泽含煤碎屑岩组合（C_2—P_1）、石盒子陆表海海陆交互砂泥岩夹砾岩组合（C_2—P_1）和双泉湖泊砂岩—粉砂岩组合（P_1）等4个岩石构造组合。在沼泽含煤泥砂岩组合和沼泽含煤黏土岩粉砂岩组合中含有煤矿及少量的油页岩矿。

三、晚三叠世—早白垩世

中生代，本区形成了一系列受构造控制的陆相盆地、火山喷发和岩浆侵入活动，构成盆岭相间的构造格局，特别是早白垩世强烈的伸展变形和裂谷作用，形成了广泛发育的断陷盆地。构造体制发生的重大转折是导致矿床形成的内因，不同的构造环境，即岩石构造组合和建造制约了各自不同矿床的形成，而不同的岩石建造是矿床形成的具体载体，因而形成了金、铁、铜、铅、锌、钼等多金属矿、煤等丰富的矿产。

1. 与陆内断陷盆地有关的矿产

陆内断陷盆地中窑坡组河湖相含煤碎屑岩组合，是本区侏罗纪煤矿重要的含矿建造。为一套以灰绿色或灰黑色砂岩、粉砂岩及泥岩等细碎屑岩为主的陆相含煤地层，总厚度440~500m。总体上粒度较细，粉砂岩占地45%~48.5%。沉积旋回结构清晰，每一旋回结构的下部多为深灰色—灰黑色中粒或细粒砂岩，上部为灰黑色—黑色粉砂岩、泥岩，并产有煤层。主要按照含煤情况的差异分为两个岩性段：下段含煤5~12层，多数可采，为重要含煤层段；上段含煤性差，多不可采，岩层中普遍发育菱铁矿结核。

2. 与造山早—中期岩浆岩有关的金属矿产

燕山板内前造山期（早侏罗世—早白垩世），伴随地幔物质上涌，下部地壳初步加热，在局部地段发生熔融作用，这一阶段矿化较弱，未形成矿点。至晚侏罗世碰撞造山期大量玄武质岩浆的又一次底侵，在陆壳温度已经比较高的背景下，岩浆熔融区向浅部发展，陆壳升温达峰期使上地壳岩石大规模熔融（邓晋福等，2007）。碰撞造山早—中期由不同成分的岩浆多次脉动或涌动侵位形成的复式岩体，岩浆成分连续演化，形成一套近乎完整的中酸性岩石组合，伴生铅、锌、钼、金、铜矿。

3. 与造山晚期岩浆岩有关的多金属矿产

燕山板内造山晚期（早白垩世）形成的浅成—超浅成侵入岩组合，虽出露面积不大，然而却形成了可观的铜、钼、锰、金和非金属矿产。当中酸性花岗岩类围岩为碳酸盐岩时，则伴生矿化或接触交代型矿产，而浅成斑岩或玢岩的岩体内部或接触带形成斑岩型矿化或矿产。

4. 与造山后伸展晚期岩浆岩有关的矿产

燕山板内造山后期（早白垩世晚期），“过热”的地壳开始冷却，冷却是从地壳上部开始的，底侵的玄武岩和玄武质—安山质岩浆又大量喷发，并形成不少辉长岩侵入体，伴生铁矿，如上庄矿床；侏罗纪形成的新的玄武质下地壳熔融形成二长岩、石英二长岩，并伴生钼、铁锌矿。后造山期，陆壳逐渐冷却，岩浆活动减弱，使“过热”陆壳变为温热环境（邓晋福等，2007），同时伴生金矿。

四、晚白垩世—第四纪

晚白垩世—全新世，北京地区主要为断陷盆地发育阶段，形成于与断陷盆地有关的矿产。断陷区内河湖发育，沉积相以河湖沼泽相发育为主，形成的矿产主要包括砂金、铸型用砂矿、建筑砂砾石、砖瓦（水泥）黏土矿产。这类矿产主要产于新生代第四纪的洪、冲积相的松散堆积物之中。泥炭主要分布于河湖沼泽相和山前平原的前缘洼地之中。

第二节　区域断裂构造带的控矿作用

区内断裂构造类别多，性质复杂，各断裂之间常表现出一定成生联系，分布上几乎都有一定规律性，一些区域性的断裂带具长期活动特点。区域上断裂按展布方向分为东西向、北东向、北北东向、北西向、北北西向、北西西向和近南北向等 7 组，其中以前三组较发育，控制了本区基本地质构造格架，也是重要的控岩、控矿构造。

本区控制金属矿产的主要断裂有：

1. 赤城-长哨营-古北口-承德深断裂

呈近东西向展布于本区北部，为区域性尚义-平泉深断裂带（长度>800km）中段向东南弯曲段。深断裂带对中元古代和晚古生代华力西期岩浆侵入活动控制明显。中元古代斜长岩、石英二长岩、石英正长岩和钾长花岗岩体沿断裂带侵位，构成近东西向长条岩脊状侵入岩带；断带北侧控制了华力西期变质辉长岩类和印支期二长花岗岩的侵位。大量找矿成果证实断裂带两侧，尤其是南侧 30km 范围内，是本区重要金矿成矿区带，控制了本区一些重要金、钒钛磁铁矿、铬铁矿、铂、钯及萤石等矿床的分布。

2. 沙厂-墙子路（密云-墙子路）大断裂

位于赤城-长哨营-古北口-承德深断裂南侧约 40km，两者平行排列，走向东西，往东延入河北省青龙一带，全长>200km。中元古代密云沙厂斜长环球斑花岗岩沿断裂侵入，其切割面在康氏面附近，为四级构造单元分界线。断裂南北两侧（尤其南侧）次级断裂构造控制了一些金、银矿床、矿点的分布，是金、银重要成矿区之一。

3. 沿河城-南口-琉璃庙大断裂

斜贯本区中部，走向北东 50°~60°，延伸>140km。该断裂带具有多期次活动历史，早期主要显示逆冲性质，后期主要表现为正断性质。在断裂两侧形成次级北东向、北北东向和北西向断裂，并控制附近中生代向斜构造盆地和背斜隆起带的形成和分布。在上述多方向构造交汇复合部位，于燕山运动中晚期发生频繁火山活动和岩浆岩侵位，形成北东向八达岭构造岩浆岩带，它是本区最重要内生金属矿产（Cu、Pb、Zn、Mo、Au等）成矿带，伴随火山作用形成的浅成-超浅成相岩体常与金矿化关系密切。

该大断裂的东北段为崎峰茶-琉璃庙断裂，断裂走向由北东50°逐渐偏转至北东30°，断面向西北倾斜，倾角60°，并主要分布于太古宙变质岩系中，岩浆活动造成矿源层中金元素进一步活化、运移和聚集，或在其外围脆-韧性剪切带中形成矿体，该断裂及其两侧构造显示了良好的导矿和容矿作用。

4. 北西向断裂构造

该组构造在本区比较发育，规模较大者十余条，其长度>20km，走向北西315°~320°，倾向225°~320°，倾角50°~70°，具多期构造的形迹。本区不少重要金、银矿床（杨树底下、大桥-筒子、杨家洼、银冶岭）及多金属矿等与北西向构造关系密切。北西向构造与北东向（北北东向）及东西向断裂构造交汇处，对燕山期岩浆活动及与其有关内生成矿作用有明显控制作用。如分水岭一带、银冶岭地区、大榆木沟、碾台、黄场、石槽、三百山、河防口、石窑等内生金属矿床、矿（化）点均处于上述构造交汇部位。

利用遥感图象解译成果，结合地质、化探、矿产分布等分析，认为本区存在一条北西向石槽-石窑-银冶岭构造岩浆岩带。该构造岩浆岩带走向310°~315°，有12个环形影像，其中6个有已知金属矿点和蚀变岩体，多发育于几组构造交汇部位与侵入岩体关系密切。

5. 紫荆关-大海坨深断裂

沿紫荆关、双涧沟、大海坨呈北东25°~30°延伸，长度>800km，主体构造在河北省境内。该断裂为不同磁异常区的分界及重力异常的梯度带，切割深度已达上地幔，经历了伸展—挤压—伸展的发展过程，具多期活动特点。该断裂导控岩体，构成北北东向构造岩浆带，本区大海坨、大河南花岗岩受其控制，但与其有关的岩浆热液型金矿化现象较弱。

此外，北北东向涿县-丰台-怀柔-白马关深断裂带，大部分为隐伏状态，在东北段地表为北北东向石城-冯家峪断裂破碎带，在变质岩分布区一些金矿化点分布与其有关。区域内北北东向小规模断裂也很发育，较重要者达25条。走向为北东20°~30°，倾向30°~60°，长度多在20~30km之间，属挤压构造环境，常截切东西向、北东向、北西向断裂，形成时期大致为燕山中晚期，也控制了一些金矿（化）点的分布，如关上、冯家峪及密云北部地区等。

第八篇

第四纪地质与地貌

第三十五章　概　　述

1949年以前，先后有维理士（B. Willis，1904）、安特生（J. G. Andersson，1919）、叶良辅等（1920）、王竹泉（1926）、巴尔博（G. B. Barbour，1929、1935）等多位学者阐述了对地文期的认识。安特生（1918）、师丹斯基（O. Zdansky，1923）、李捷、步林、杨钟健、裴文中（1927，1928，1929）等对北京地区古人类学及其化石进行了调查研究，1929年裴文中发现了轰动世界的第一个北京猿人头盖骨轰动世界。叶良辅（1920）、李四光（1922）、安特生（1939）等学者分别对北京地区黄土、古冰川开始了相关研究。

1949年至20世纪70年代，不同学者（李四光、孙殿卿等）和单位（地质力学研究所、北京市水文地质工程地质第一大队等）先后在西山地区开展了冰川遗迹调查与研究工作，曾以冰川观点对第四纪地层进行划分；安芷生（1979）、李鼎容等（1979）、陈方吉（1979）、张子斌（1981）等学者对北京平原地区第四系进行了划分。

20世纪80年代至20世纪末，利用利用延庆盆地和北京平原区1：5万青龙桥幅、怀柔幅和通县幅区调工作中施工的钻孔开展第四纪地层、气候环境演化、古地磁等综合性研究；在山区1：5万沿河城幅、周口店幅区调工作中更加注重了对第四纪河谷阶地和洞穴的调查研究；在上述工作的基础上北京地区第四纪综合研究有了新进展，于1985、1991、1995年分别出版了《北京猿人遗址综合研究》《北京市区域地质志》《北京地区第四纪古地理研究》等专著，对第四纪以来地层划分、植被演化、古地理环境、新构造运动等方面的研究做出了重要贡献。

21世纪以来，本区第四纪地质工作进一步发展。一是平原区1：5万区调工作，对第四纪地层采用了岩性（年代）+微地貌单元进行划分，划分结果更趋合理。二是城市地质调查工作，在2005—2008年期间开展了大规模的城市地质调查工作，北京市作为全国城市地质调查试点地区之一，进行了多参数城市综合地质调查工作，其中施工了29个第四纪钻孔，结合以往资料，建立了北京地区第四纪地层三维地质结构模型，对北京平原冲洪积扇发育区的地质结构有了新的认识。2012—2014年，开展了北京平原区活动断裂监测专项地质调查，对主要活动断裂（黄庄-高丽营断裂、南口-孙河断裂、顺义断裂和夏垫断裂）进行系统调查和活动时限研究，施工钻孔24个，并进行了11个钻孔的古地磁测试，获取了丰富的第四纪地质研究成果。三是学术研究方面工作新进展，如2007年郭旭东编著出版了《北京第四纪地质导论》，较系统论述了北京地区第四纪冰川与气候问题、冰缘作用与环境、黄土结构、新构造与地震预报等，探讨了第四纪地质基本特征和演变规律。

第四纪在地质历史时期中虽短暂，但这一时期地貌演化显著，导致沉积地层同期异相现象普遍存在，如山区的沟谷和洞穴堆积物，与同一时期平原地区的堆积物特征存在较大差异。近年来调查发现在一个较小的地域内也可能同时存在冲积的砂砾石堆积和湖沼相泥质沉积物。可见，在山区某些露头和洞穴建立的第四纪岩石地层单位，已不适用于平原区第四纪地层划分与对比，而平原区基于某个钻孔建立的岩石地层单位的可对比性也存在局限性。

关于北京地区第四纪岩石地层的划分，前人也做了大量工作（表35-1），而岩石地层单位的建立是一项很规范、严肃的工作，在缺乏系统的划分与对比情况下，为了避免造成混乱，本志不建立新的岩石地层单位，在山区及洞穴第四纪地层中保留使用普遍认可的、研究较系统的岩石地层单位，平原区采用年代地层（下更新统、中更新统、上更新统和全新统）序列进行梳理总结。

表 35-1　北京地区第四纪地层划分沿革表

Table 35-1　History of stratigraphic classification of Quaternary in Beijing area

层位（系）	层位（统）	德日进（1941）地文期 堆积期侵蚀期	裴文中（1957）时期 文化期	裴文中（1957）古人类 人类化石	华北地层表 北京市分册（1977）	李鼎容等（1979）	李鼎容等（1979）底界年龄/10^4a	北京市区域地质志（1991）地层	北京市区域地质志（1991）地层划分 系	北京市区域地质志（1991）地层划分 统	怀柔幅（1990）	怀柔幅（1990）	怀柔幅（1990）底界年龄/10^4a	通县幅（1990）	通县幅（1990）	通县幅（1990）底界年龄/10^4a	蔡向民（2016）年龄/Ma	蔡向民（2016）地层划分 系	蔡向民（2016）地层划分 统	蔡向民（2016）岩石地层单位	蔡向民（2016）极性世	本志 年龄/Ma	本志 山区	本志 洞穴	本志 平原	本志 极性世
第四系	全新统	次生黄土	新石器时代 中石器时代	现代人	刘斌屯组	全新统	1~1.2	刘斌屯组	第四系	全新统	上	现代河床沉积	0.2	上	刘斌屯组	0.25		第四系	全新统		布容		全新统	全新统（洞穴石笋沉积）	全新统	布容
		板桥			尹各庄组			尹各庄组			中	Ⅰ阶地堆积	0.6	中	尹各庄组	0.75										
					肖家河组			肖家河组			下	Ⅱ阶地堆积	1	下	肖家河组	1										
	上更新统	马兰 辽河	晚期 山顶洞文化	山顶洞人 资阳人 长阳人 马坝人 河套人	百花山冰期堆积	军营组	39±5.8	马兰组		上更新统	上	上张喜庄组	5	上	通县组	7	0.011		上更新统	马兰组		0.011	上更新统（马兰组）	上更新统（羊耳峪组）	上更新统	
			中期 河套文化		马兰组						中		12	中		12										
		马兰 清水			碧云寺冰期堆积						下		15	下		15										
	中更新统	周口店第1地点	早期 中国猿人文化	中国猿人	周口店组	翟里组	70	周口店组		中更新统	上	中张喜庄组	73	上	翟里组	73	0.126		中更新统	周口店组		0.126	中更新统（周口店组）	中更新统（周口店组）	中更新统	
		周口店第15地点			龙骨山冰期堆积						下			下												
	下更新统	泥河湾		“巨猿”？	泥河湾组	夏垫组	240	泥河湾组		下更新统	上	下张喜庄组	248	上	夏垫组	95	0.781		下更新统	泥河湾组	松山	0.781	下更新统（泥河湾组）	下更新统（太平山组）	下更新统	松山
		汾河			朝阳冰期堆积						中			中		170										
											下			下		248										
新近系	上新统	榆社 未定名；保德 唐县			天竺组	顺义组		天坛组	新近系	上新统	怀柔组			天坛组和天竺组			2.588	新近系	上新统		高新	2.588	鱼岭组	鱼岭组	天坛组	高新

第三十六章　山区及洞穴第四系

第一节　山区第四纪地层

一、下更新统（泥河湾组（Qp^1n））

1924年巴尔博（G. B. Harbour）将河北省阳原县泥河湾村的一套湖相地层称为“泥河湾层”。1964年编制的《1∶20万北京市地质图》和1977年出版的《华北地区区域地层表·北京市分册》均使用了“泥河湾组”这一名称。1991年出版的《北京市区域地质志》将泥河湾组定义为：“含早更新世泥河湾动物群（相当于欧洲维拉方动物群）化石的不同类型地层或可与之对比的地层，古地磁测量位于马莫斯事件底部以上以及布容极性世之下（包括松山反极性世及高斯正极性世上部）的沉积层。”其含义实质上属年代地层单位。鲍亦冈（2001）将泥河湾组定义为：“以淡黄绿色-灰白色及少量淡黄褐色粉砂、粉砂质黏土为主，含少量的粗砂、砂砾石（岩）夹泥灰岩等，普遍含钙质结核，底部为灰白色-黄绿色砂、砾岩互层。”

（一）代表剖面

以延庆杨户庄剖面1~7层为代表，描述如下（图36-1）。

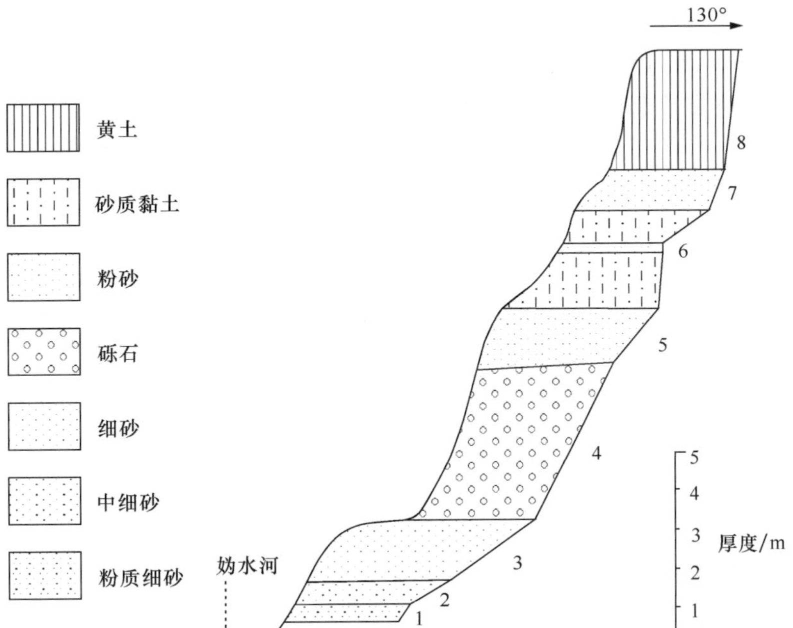

图36-1　杨户庄更新世地层沉积剖面

（据萧宗正，1996，修改）

Fig. 36-1　Measured section of Pleistocene strata in Yanghuzhuang

8. 棕黄色黄土，含钙质柱状结核，具柱状节理，含大量孢粉　3.1m

7. 土黄色细砂，含钙质板状结核，产丰富的有孔虫 *Evolutononin shanxiense* 及大量介形虫　0.8m

6. 上部浅灰黄色砂质黏土，层理薄，中部浅灰绿色粉砂，中层状；底部为浅灰黄色薄层砂质黏土含丰富的有孔虫 *Evolutononin shanxiense*、介形虫、腹足类、轮藻及孢粉　2.5m
5. 土黄色粉砂，含大量介形虫及孢粉　1.5m
4. 杂色砾石层　4.0m
3. 土黄色细砂，含钙、泥质结核，含大量介形虫、个别有孔虫、腹足类、双壳类及孢粉　1.5m
2. 土黄色中—细砂，含海相介形虫 *Sinocytheridea impressa* 及陆相介形虫、双壳类、孢粉　0.5m
1. 淡黄色粉质细砂，含大量介形虫及孢粉　0.4m

（以下掩盖，未出露）

（二）地质特征及区域变化

泥河湾组主要分布于山区河谷的Ⅳ级阶地上。在延庆盆地其主要分布于东南侧边缘沿军都山、八达岭山麓一带，以杨户庄、阜高营地区出露相对较好，岩性以砂、亚砂、亚黏土为主，夹砂砾石层，颜色灰绿、灰白、灰褐及土黄、黄褐色等（鲍亦冈，2001），其中在阜高营地区发育明显钙质结核。西山地区主要分布于长沟峪升平山、破窑山东坡一带、娄子水栓马庄一带、施庄以及雁翅的Ⅳ级阶地上，主要岩性为红棕色、棕黄色砂质黏土、黏质砂土与夹砾石层，厚4~7m，砾石风化程度较高，具定向排列特征。在昌平虎峪、龙虎台一带也有出露（魏兰英等，1982），岩性主要为下部棕红色黏土含砾石，中部棕红色黏土含钙质、锰核、虫孔等，上部棕红色（杂色）黏土夹砾石层，风化程度较高，砾石成分复杂（花岗岩、灰岩、砂岩等）。北山地区在密云河谷Ⅳ阶地上岩性主要为砂砾石，砾石磨圆度较好，局部见紫褐色黏土及泥砾层，在前焦家坞一带可见有黄褐色、褐色牛轭湖积沉积（图36-2）。

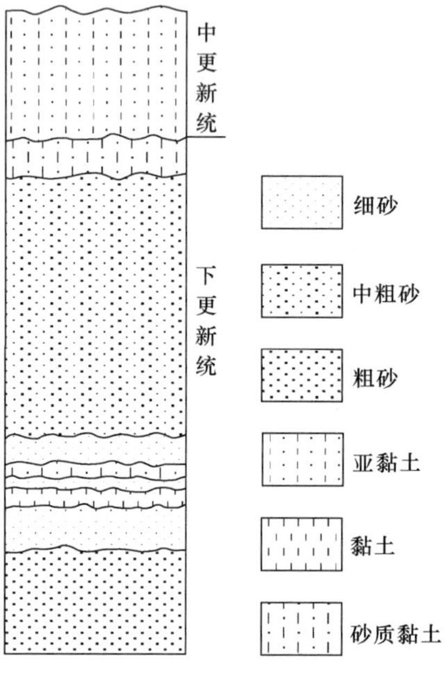

图36-2　焦家坞村柱状图
（据1∶5万密云幅，1994）
Fig. 36-2　Column in Jiaojiawu Village

上覆地层：中更新统褐红色亚黏土

---------------- 平行不整合 ----------------

8. 黄白色含黏质砂土　0.40m
7. 浅黄色中、粗砂，发育大型交错层理，并含碳化植物碎片　3.00m
6. 黄白色细砂　0.30m
5. 褐色黏土，中间夹少量黏质砂土　0.15m
4. 黄白色细砂，发育水平层理　0.10m
3. 褐色黏土　0.20m
2. 黄白色细砂层，下部发育小型斜层理，上部发育波状层理和水平层理　0.50m
1. 浅黄色粗砂，夹有中、细砂　1.00m

未见底

（三）生物组合特征及时代归属

山区（除洞穴）出露的下更新统中生物资料较少，孢粉以松、桦栎属及草本蒿属菊科为主。其中在杨户庄剖面上含介形类化石比较丰富，主要有：*Ilyocypris salebrosa stepanaitys*，*I. bradyi* Sars，*I. gibba*（Romdohr），*Cyprideis littoralis*（Brady），*Cyprinotus chiuhsienenensis* Lee，*Eucypris inflate*（Sars），*Physocypria fadevi* Dub.，*Limnocythere binoda* Huang，*L. dubiosa* Daday，*Darwinula stevensoni*（Brady et

Robertson), *Candoniella albicans*（Brady), *Candona sangganheensis* Huang 等。第 7 层、第 8 层含丰富的山西外旋九字虫（*Evolutononin shanxiense* Wang），1~3、5~6 层含有大量介形虫，其中 3 层有中华美花介（*Sinocytheridea impressa*（Brady））, 1~8 层含有孢粉组合显示大多属喜冷类型。

二、中更新统（周口店组（Qp^2z））

1959 年杨钟健、裴文中、贾兰坡等正式创建周口店阶，《华北地区区域地层表·北京市分册》（1977）称周口店组，《北京市区域地质志》（1991）沿用这一概念。鲍亦冈（2001）将周口店组定义为："发育于河流Ⅲ级阶地上的堆积物，岩性以红色或浅棕黄色黏质砂土、砂质黏土为主，常夹数层棕褐色埋藏土壤或风化层，底部可见风化砾石或褐红色泥砾层，厚 10~20m。"

（一）代表剖面

以房山区龙骨山西北老牛沟口周口店河Ⅲ级阶地剖面作为正层型，为 1988 年开展 1：5 万周口店幅区调工作时测制。

周口店组（Qp^2z）	14. 1~>16. 4m
3. 红褐色含角砾亚黏土层。亚黏土与棱角—次棱角状原地斜坡岩屑构成的角砾层，呈韵律性堆积	2. 9~5. 2m
2. 红黄色亚砂土层。粉砂含量 30%~42%，黏土含量 6%~30%，垂直节理发育	4. 2m
1. 红色砂砾层。砾石岩性为砂岩、千枚岩、碳质板岩、石英砂岩及角岩，其主要来源于北部石炭纪岩层。砾石 ab 面呈叠瓦状排列，多倾向北或北西。其砾间充填物和砂质透镜体的正态概率曲线呈三段式，反映其沉积的水动力条件不是冰川沉积，而是流水成因	>7. 0m

（二）岩性岩相变化

周口店组主要分布于山区河谷的Ⅲ级阶地上，以坡积、残坡积和洪积为主，岩性以褐红色黏质砂土，砂质黏土为主。前人所称的"下砾石层"与"离石黄土"相当于本组。"下砾石层"主要出露于房山老牛沟沟口，断面出露约 1km，厚度不一，最厚达 13. 4m，岩性由下至上为风化砾石层、黄红色含碎石黄土状土、红土-碎石堆积层。"离石黄土"（又称老黄土或红色土 C 带）主要分布于山区河谷两侧，如周口店、南口、永定河的Ⅲ级阶地上更高地貌部位，或冠于丘顶，或存于山鞍，或被覆于斜坡，其产状受原始地形起伏控制。以雁翅车站铁路西北侧雁翅剖面（图 36-3）（赵希涛，1981）为代表，岩性以红黄色或浅棕黄色黏质砂土或砂质黏土为主，常夹数条棕-褐色埋藏土壤或风化层，其下部常分布有钙质结核，质地较坚硬，略显层理，垂直节理不发育，厚度一般约 10m。

延庆盆地边缘刘家堡、玉皇山、小张家口等地，沟谷中也见有棕红、浅黄灰绿色黏质砂土、砂质黏土层（徐志华等，1985）。在八达岭一带主要为棕红、褐红色泥砾层，其中夹有棕红色较致密的黏质砂土、砂质黏土，中粗砂及砂砾石层，厚 10~20m。

在密云镇罗营一带为褐红色亚砂土与角砾混杂堆积，潮河下游楼子峪、杨家堡清水河Ⅲ级阶地见浅褐色、褐红色砂质黏土，见有铁锰质淋滤物。

（三）生物特征及时代归属

斋堂黄土剖面底部的"离石黄土"，古地磁测得离石黄土底界，即 B/M（布容期/松山期）界线年龄为 0. 78Ma。1：5 万雁翅幅区调工作对离石黄土的孢粉结果显示以灌木及草本花粉占优势，如榆（*Ulmus*）、椴（*Tilia*）、桦（*Betula*）和松（*Pinus*）等，说明当时气候较湿润，水热条件好，利于森林生长。1：5 万雁翅幅区调中对"离石黄土"进行了热释光测年，年龄分别为（282. 05±33. 84）×10^3a（杨家庄）、（196. 58±39. 31）×10^3a（王峪）、（315. 78±63. 15）×10^3a（溜石港）、（126. 02±25. 02）×10^3a（前桑峪），其形成时代属于中更新世。

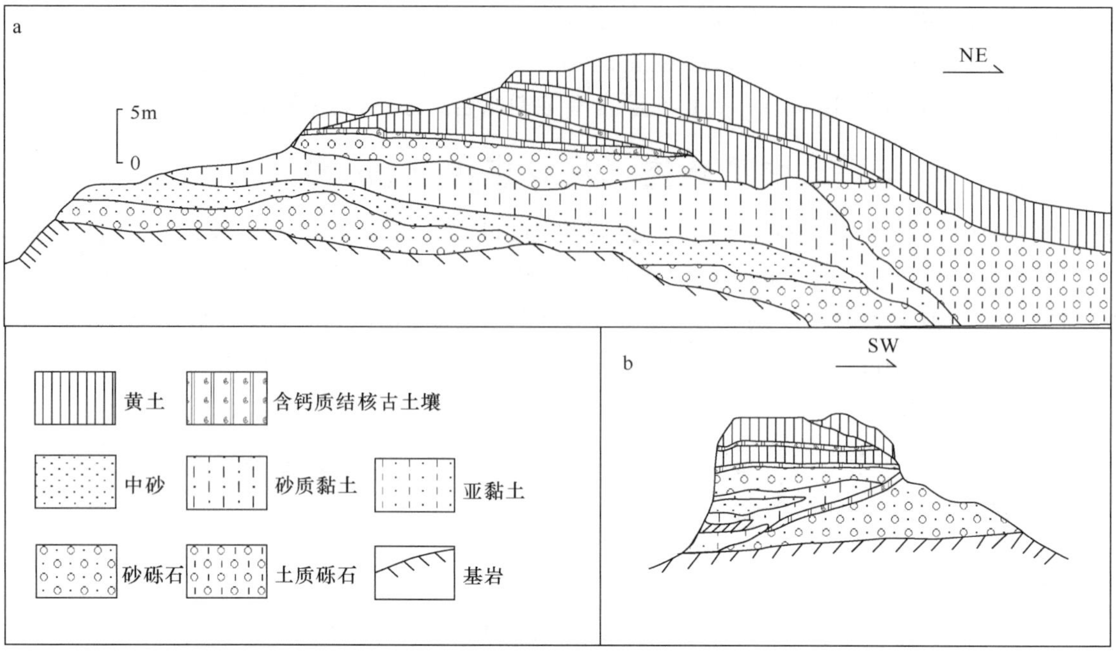

图 36-3 雁翅车站铁路西北侧（a）和东南侧（b）剖面
（据赵希涛等，1981）
Fig. 36-3 Section of northwest（a）and southeast（b）Yanchi Station

三、上更新统（马兰组（Qp^3m））

《北京西山地质志》中将发育在马兰阶地的黄土，按时代划分为晚更新世“古黄土”和晚更新世“新黄土”。德日进等（1933）将华北广泛分布的晚更新世黄土归为马兰期堆积，自此以后马兰黄土被视为华北晚更新世的标准层位之一。此后各种地质报告和文献广泛使用马兰组这一地层名称。鲍亦冈（2001）将马兰组定义为：“岩性以灰黄或棕黄色粉砂质黏土为主，粒度度均匀，纯净，结构疏松，无节理，多大孔隙，常含钙质结核，有的为白色菌丝体或白色斑点，在一些剖面中发育 1~3 层埋藏土壤层，垂直节理和陡直的沟堑较发育，形成了黄土发育区的独特地貌景观。”

（一）代表剖面

北京西山东斋堂剖面（安芷生等，1984），见图 36-4。

上覆地层：全新统残坡积土壤和耕作土

--------------- 平行不整合 ---------------

马兰组		8.5~9.5m
4. 上部黄土（L_{1-3}）	黄色、褐黄色粉砂质亚黏土，岩性均一，疏松、多孔、大孔和垂直节理发育。距今 1 万~3 万年	3.0~3.5m
3. 中部古土壤（L_{1-2}）	浅灰色或灰棕色，棱状结构，白色假菌丝发育，底部有钙结核，碳酸盐 5.48‰。距今 3 万~4 万年	1.0m
2. 下部黄土（L_{1-1}）	浅灰黄色，粗粉砂质亚黏土，质地均匀、疏松、多孔，大孔发育，偶见细小碎石。距今 4.5 万~8.5 万年	3.5~4.0m
1. 底部古土壤（S_1）	具棱块状或柱状结构，底部发育钙质结核。距今 8.5 万~12.5 万年	约 1.0m

--------------- 平行不整合 ---------------

下伏地层：中更新统周口店组

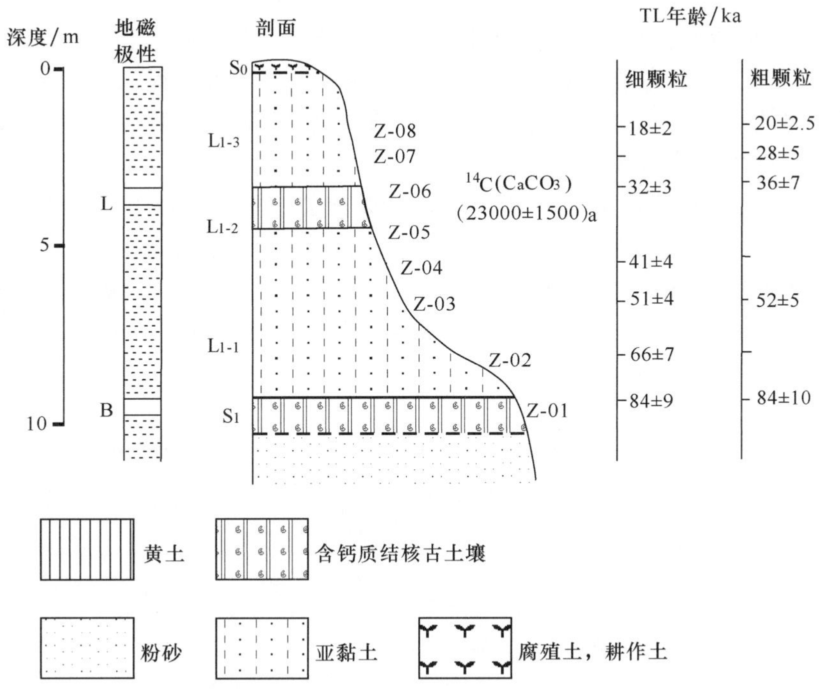

图 36-4 北京西山东斋堂马兰黄土剖面及磁性地层和年代
（据安芷生，1984；卢演俦，1986）
Fig. 36-4 Malan loess section of magnetic stratigraphy and age in the Dongzhaitang of Xishan Hills, Beijing area

（二）地质特征及区域变化

马兰组主要分布于各河流Ⅱ级阶地和谷坡上，岩性以黄土状土（次生黄土）、黄色黏质砂土、砂质黏土夹数层透镜状砾石层为主，一般厚数米至十余米，最厚可达 30~40m。成因类型有洪积、冲积、坡积、残坡积等。

北京西山清水河流域的斋堂盆地是马兰黄土首次确定之处。桑峪、燕翅一带是马兰黄土最为发育的地区之一。山区永定河右岸堆积阶地上（图 36-5），岩性为砂砾岩，具韵律，砾石为次圆状，粒径以 2~6cm 为主，成分为白云岩、粉砂岩、闪长玢岩、流纹岩、安山岩等。房山张坊-石亭村一带以及车厂一带出露厚度 3.5~20.0m 不等，岩性为夹有砾石的次生黄土。昌平西峰山一带岩性为灰黄色粉砂夹砾石，砂砾石呈透镜状展布，粉砂土为马兰期次生黄土，较硬，有虫孔，发育垂直节理，含小砾石。门头沟军庄-寨口一带，岩性为马兰黄土夹砾石，黄土为灰黄色粉砂，虫孔发育，结构松散。平谷罗营村出露岩性为砂、角砾及黏土混合物。密云黑山寺一带发育大面积洪积扇，堆

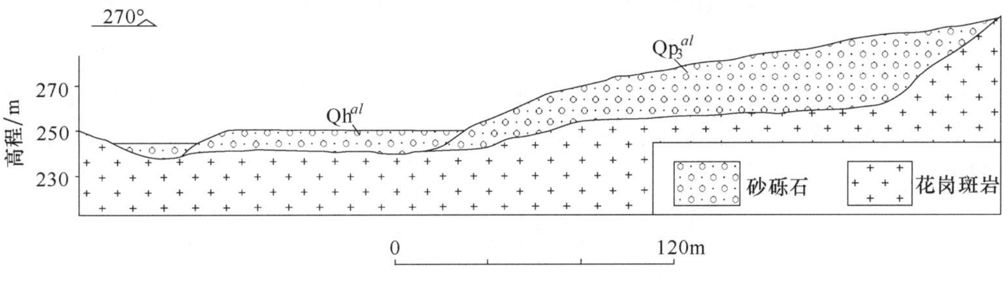

图 36-5 河南台永定河阶地剖面
（据 1∶5 万雁翅幅，1993）
Fig. 36-5 Terrace section of Yongding river in Henantai

积物主要由砂砾石组成，大小混杂、分选差，显示急剧堆积；新城子、搭梁地一带出露的最大厚度为8m，岩性为褐黄色亚黏土、亚砂土夹砾石层，局部可见2~3层灰黑色古土壤层；高岭田庄村南东出露岩性为砂砾石层、浅黄色含黏砂土等。延庆米家堡-中羊坊-程家窑一带的Ⅱ级阶地上砂砾石、砂、黏砂组成的韵律层，二元结构普遍可见；在妫水河流域两侧为黄色—黄白色砂质黏土沉积，常见钙质结核，直径0.5~1.0cm。

清水河南岸马兰峪沟口马兰黄土之上可见“马兰砾石层”（图36-6），该词系安特生（1923）首先提出的。“马兰砾石层”呈黄褐色，结构松散，成分较均匀，该砾石层由砾石，粗砂与黄土等组成，小的不连续夹层层理较清晰。砾石成分主要为火山岩、凝灰岩、辉绿岩、火山角砾岩及少量砂岩、花岗岩和页岩。砾石分选性较差，一般砾径为5~10cm，较大为50~80cm，最大达100cm以上。砾石磨圆度和扁平度都较高，具定向排列。

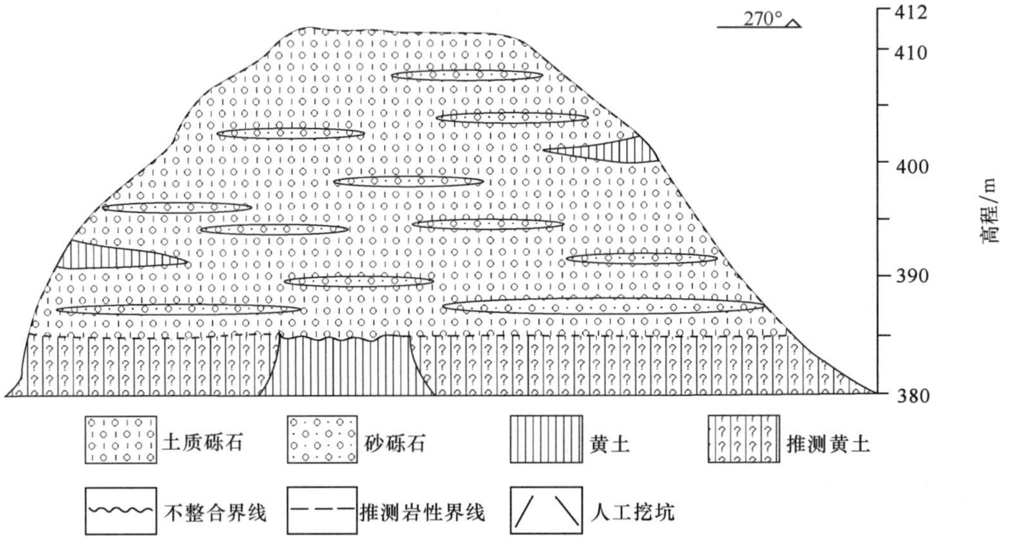

图36-6　斋堂镇马兰峪沟口马兰砾石层地质剖面

（据郭旭东，2007）

Fig. 36-6　Geological section of Malan gravel bed in Malanyu valley mouth, Zhaitang County

（三）生物特征及时代归属

东斋堂马兰黄土剖面孢粉以草本植物花粉为主，其中尤以耐干旱的蒿属（84.6%~91.9%）为主，藜科和禾本科为次，木本植物花粉约占草本植物花粉的1/3，在古土壤层中孢粉较丰富（严富华等1986）。此外发现脊椎动物化石 *Cervus*（*P.*）*hortulorum*，在房山张坊村北的含砾石层黄土中见有脊椎动物化石碎块（脊骨和股骨）。密云苇子峪清水河Ⅱ级阶地中产出的原始牛腿骨、青羊、鹿；密云城关产出的纳玛古象下颌骨；李各庄、达峪沟、蔡家峪产出的原始牛角等化石。马兰黄土（东斋堂）中灰棕色古土壤（L_{1-2}）记录了一次相当于拉尚事件的暂短磁极性倒转事件，其年代约距今3万年前，可与山顶洞形成时代相当，底部褐土型古土壤（S_1）中记录了相当于布莱克事件的磁极性倒转事件，其年代为（120~80）$\times 10^3$a。在延庆刘家堡、玉皇山、姜家台、小张家口、王木营、营城等（1：5万青龙桥幅，1984）地表出露剖面上进行了热释光测年，结果在（83±4.1）$\times 10^3$a~（710±3.5）$\times 10^3$a，均属于晚更新世。

四、全新统（Qh）

山区全新统分布相对较广，主要在永定河、潮白河、大石河、清水河、错河及山区间歇性河床、漫滩、Ⅰ级阶地以及山麓地带，沉积物以冲洪积、冲积、坡积为主，岩性以砂砾石、砂及黏土类沉积为主，颜色为褐黄色、灰色、灰褐色等，厚度差异较大，一般在2~10m之间。

第二节　洞穴第四纪堆积

洞穴堆积是第四纪一种特殊的沉积类型。其生成环境有一定的特殊性，而且堆积物中不仅见有人类化石及遗迹（尤其是石器），而且含有丰富而连续演替的哺乳动物化石。对它们进行系统地发掘与研究，为古人类学、古脊椎动物学、第四纪古环境及其生态变化等研究，均具重大意义。北京山麓地带洞穴多且堆积类型复杂，从 20 世纪 20 年代就开始进行了调查研究，对第四纪洞穴堆积地层的划分，主要是以生物地层和气候地层为划分依据，属年代地层划分范畴。杨子赓等（1985）、1∶5 万周口店幅区调报告（1988）、曹伯勋等（1991）对部分洞穴堆积建立了岩石地层单位，萧宗正（1995）认为本区第四纪洞穴地层基本上是连续的，并建立了一套完整的洞穴岩石地层层序（表 36–1）。

表 36–1　北京地区第四纪洞穴地层划分沿革表

Table 36–1　History of stratigraphic classification of Quaternary caves in Beijing area

<table>
<tr><th colspan="3">作者
地层</th><th>杨钟健
（1948）</th><th>袁复礼
（1955）</th><th>杨子赓等
（1985）</th><th>1∶5 万周口店幅区
调报告（1988）</th><th>曹伯勋等
（1991）</th><th>萧宗正
（1995）</th><th>本志</th></tr>
<tr><td rowspan="7">第四纪</td><td colspan="2">全新统
Qh</td><td></td><td></td><td></td><td></td><td></td><td></td><td>全新统
（洞穴沉积）</td></tr>
<tr><td rowspan="6">更新统</td><td rowspan="2">上更新统
Qp^3</td><td rowspan="2">山顶洞
堆积</td><td rowspan="2">山顶洞
堆积</td><td rowspan="2">山顶洞</td><td rowspan="2">东岭子洞穴地层
新洞堆积
山顶洞堆积</td><td>山顶洞</td><td>山顶洞组
（山顶洞）</td><td rowspan="2">上更新统
（羊耳峪组）</td></tr>
<tr><td>向阳洞</td><td>羊耳峪组
（新洞）</td></tr>
<tr><td rowspan="2">中更新统
Qp^2</td><td rowspan="2">第 3 地点
第 1 地点
第 13 地点</td><td rowspan="2">第 15 地点
第 1 地点
第 13 地点</td><td rowspan="2">周口店组
（第 1 地点
1～13 层）
龙骨山组
（第 1 地点
14～15 层）</td><td rowspan="2">第 1 地点 1～13 层
太平山北坡 1～3 层
周口店组</td><td rowspan="2">新洞
第 1 地点
第 13 地点
太平山北坡
西洞</td><td>新洞组
（新洞）</td><td rowspan="2">中更新统
（周口店组）</td></tr>
<tr><td>周口店组
（第一地点）</td></tr>
<tr><td rowspan="2">下更新统
Qp^1</td><td rowspan="2"></td><td rowspan="2">第 18 地点</td><td rowspan="2">第 1 地点
16～17 层
第 12 地点</td><td rowspan="2">太平山北坡 4～13 层
第 12 地点
第 1 地点 14～17 层
太平山组</td><td rowspan="2">第 12 地点
太平山组</td><td>太平山组
（太平山北坡东洞）</td><td rowspan="2">下更新统
（太平山组）</td></tr>
<tr><td>灰峪洞组
（第 18 地点）</td></tr>
</table>

一、下更新统（太平山组（$Qp^1 t$））

1985 年在 1∶5 万周口店幅填图工作中，在周口店太平山北坡大砾岩山下发现的一套洞穴堆积物剖面上建立了“太平山组”。

（一）代表剖面

以周口店太平山北坡大砾岩山下太平山组为代表剖面（图 36–7）。

上覆地层：　中更新统周口店组红色含钙质结核亚黏土

———— 侵　蚀　面 ————

下更新统太平山组（Qp^1t）　9. 47m

9. 红色亚黏土，含少量钙质结核，顶部有古冰楔　0. 70m

8. 红色砂土夹 3～4 层薄层灰色、黄色黏土，发育古冰楔　0. 50m

7. 棕红色与黑灰色薄层黏土互层，发育古冰楔，含小型哺乳动物化石 *Allocricetus ehiki* *Schaub*（艾克氏变异仓鼠），*Apodemus* sp.（姬鼠）及蛇化石；孢粉　0. 65m

6. 红色含砾砂层与黏土组成韵律层，含次生钙质结核及少量钟乳石块，化石丰富主要有：*Alloricetus ehiki* Schaub，*Cricetulus* cf. *varians* Zdansky，*Lepus wongi* Young（翁氏野兔），*Microtus* sp.（田鼠），*crocidura wongi* Pei（麝鼠）等，顶部被冰楔切入　0. 85m

——————侵　蚀　面——————

5. 黄土状亚黏土，较松散，含钙质结核，呈透镜状分布于剖面东部　0. 32m
4. 巧克力色薄层黏土，底部有一层后 20cm 圆形相连的钙质结核层，含少量 *Allocrecetus* sp.　0. 40m
3. 棕黄色薄层亚砂土，含不连续钙质结核 *Vemes* sp.　0. 30m
2. 灰黑色薄层有机质黏土与粉砂互层，具水平层理，含次生钙质结核与石灰岩巨块。含 *Artemisia*（70%），*Pinus*，*Picea*，*Abies*，Cupressaceae（5% ~ 20%），*Quercus*，*Celtis*，*Betula*，*Juglans*（5%~40%），*Pteris*（40%~80%）　4. 50m
1. 红棕色黏土，层理不明显，含次生钙质结核及灰岩角砾，未见底　>1. 25m

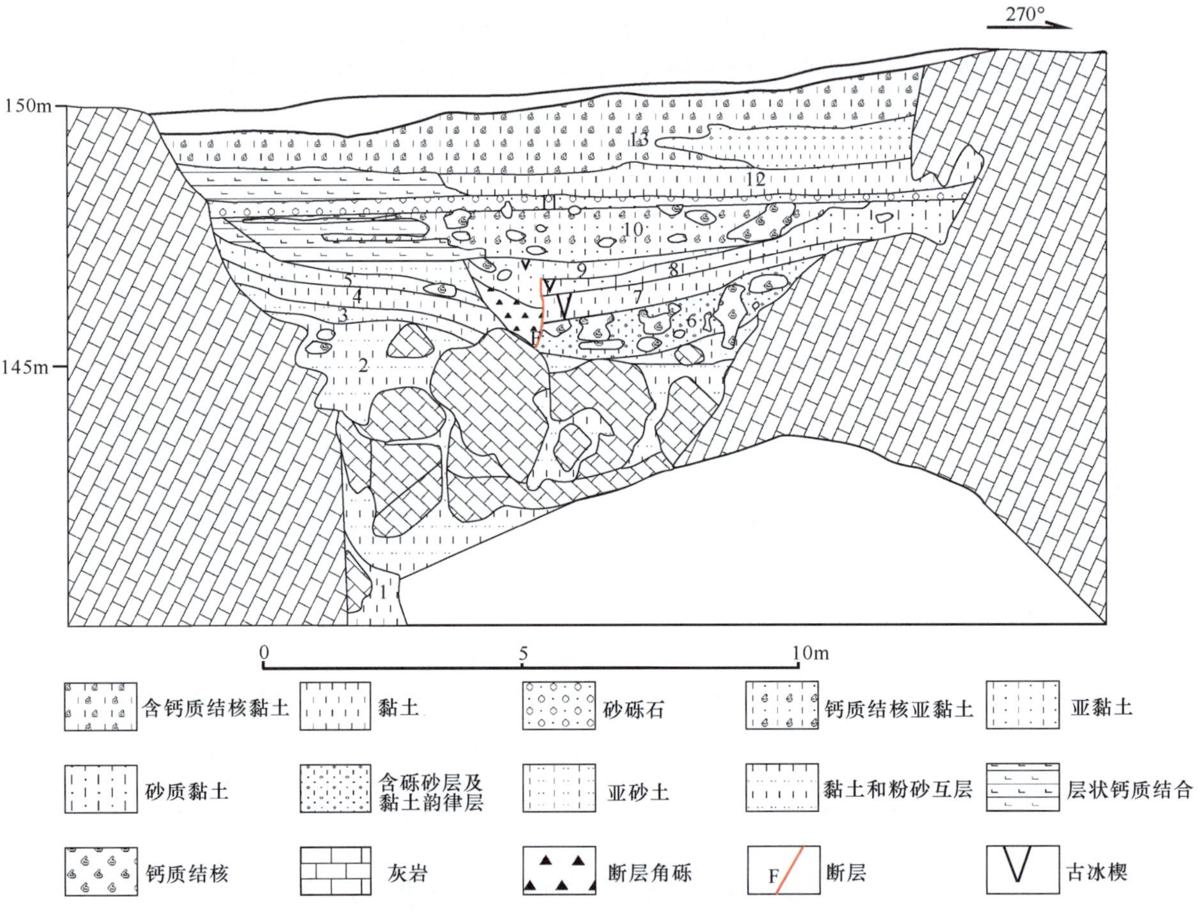

图 36-7　太平山北坡东洞下更新统太平山组剖面图
（据 1∶5 万周口店幅，1988）
Fig. 36-7　Profile of lower Pleistocene in the east hole at the northern slope of the Taiping hill

（二）岩性岩相变化

太平山组主要分布于太平山北坡大砾岩山下、周口店第 1 地点猿人洞探井中第 14~17 层、军庄灰峪第 18 地点以及怀柔龙牙洞等地。周口店猿人洞探井 14~17 层的主要岩性为棕黄色含砾粉砂质粗砂、灰色薄层黏土质粉砂及红色黏土砾石，未发现动物化石，厚约 10m。军庄灰峪 18 地点也有分布，共厚约 8m，划分 3 层，下部为红黏土，含大小不等的灰岩碎块，厚约 4m；中部含碎石红色黏土，含大量啮齿动物化石，厚约 0. 8m；上部黄色黏土含化石甚少，厚约 3. 2m（贾兰坡等，1982）。怀柔龙牙洞堆积下部为浅黄色砂质黏土含角砾与薄层方解石互层，厚 0. 9~1. 0m；上部为黄色砂质黏土，钙质胶结，厚 1. 45~1. 95m。

（三）生物特征及时代归属

太平山北坡洞穴堆积中发现的主要是啮齿类小型哺乳动物化石，其中以 *Allocricetus ehiki* Schaub 最为重要，最早见于陕西洛川午城黄土底部。该化石层古地磁年龄在松山反极性期，距今（73~92）× 10^4a，属早更新世晚期动物群。怀柔龙牙洞哺乳动物化石见于剖面中的第 4、6 层的黄色砂质黏土及钙质胶结角砾层中，动物化石有燕山尖齿鼠、红土异费鼠、杨氏原鼢鼠、拉氏豪猪、鬣狗、熊、双角犀、三门马、羚羊、鹿和牛科未定种。周口店第 18 地点，第 2 层中产小哺乳动物化石，第 3 层产大化石，其中包括白猬、杨氏原鼢鼠、艾克氏异仓鼠、早成岩松属、林姬鼠、肖氏东方鼠、硕旱獭、复齿拟鼠兔、西藏鼠兔、师氏次兔、联合翼兔、埃楚斯坎熊、直隶犬、疑豺、鸡骨山狐、贾氏獾、貂、香鼬、鬣狗、山西猞猁、三门马、香麝、布氏麂、鹿。

孢粉资料显示，阔叶树中以栎、榆、豆科、柳、桑等为主，针叶树主要是松及柏科。草本植物中草原类型的藜、蒿、禾木科含量较高。从孢粉上看，未发现上新世的亚热带植物孑遗，推测为早更新世堆积。灰峪第 18 地点和怀柔龙牙洞发现哺乳动物化石其时代相当于早更新世。

二、中更新统（周口店组（Qp^2z））

由杨钟健、裴文中、贾兰坡等 1959 年作为年代地层周口店阶确定下来，后在 1977 年《华北地区区域地层表·北京市分册》称周口店组，指周口店猿人洞第一地点从上至下 1~13 层的洞穴堆积，自上而下厚约 40m，是著名北京人（*Homo erectus pekinensis*）产地。本次工作继续沿用该岩石地层名称（图 36-8）。

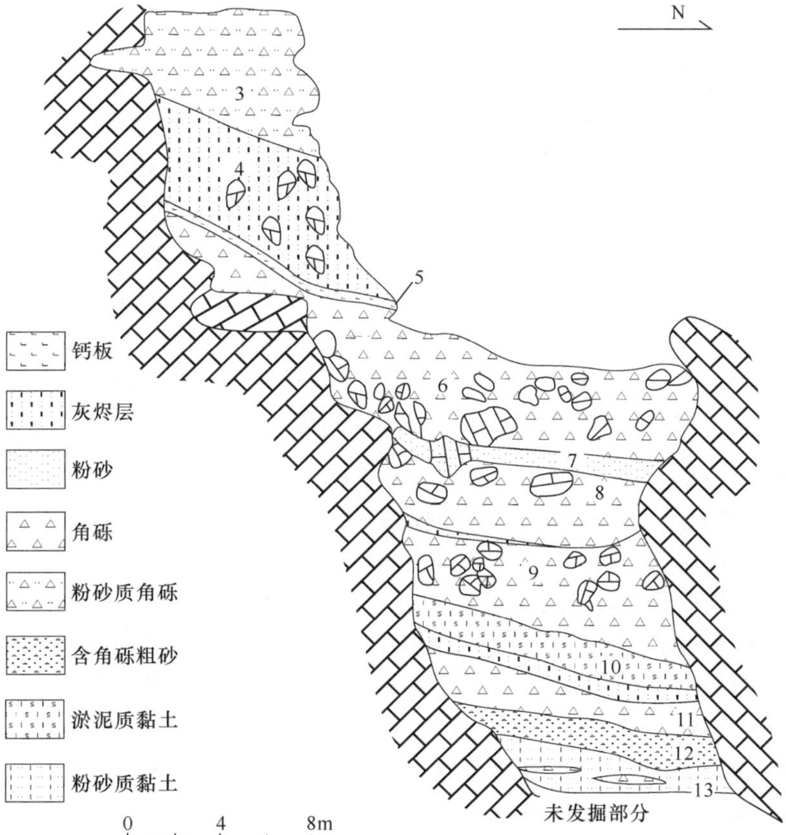

图 36-8　北京中国猿人化石产地南北剖面

（据贾兰坡，1959）

Fig. 36-8　Section of the Zhoukoudian Sinanthropus site in Beijing

（一）代表剖面

中更新统洞穴堆积以周口店猿人洞之堆积物为代表。

上覆地层：上更新统山顶洞堆积

——————— 侵　蚀　面 ———————

周口店组		38.25m
1-2.	粉砂角砾岩夹钙板层	1.70m
3.	巨大角砾岩及钟乳石，1966年在此层发现北京猿人头盖骨	3.50m
4.	灰烬层（上文化层）	6.90m
5.	黑青色硬钙板，向西渐变为钟乳石层	0.65m
6.	角砾石	8.70m
7.	灰色含云母碎屑细粉砂层	1.80m
8-9.	角砾岩夹灰烬层（下文化层）、1929年在此层发现北京猿人第一头盖骨	6.90m
10.	红色泥土及灰烬层，1959年再次发现北京猿人下颌骨	0.80m
11.	角砾岩层	0.90m
12.	棕褐色含角砾粗砂层	1.50m
13.	红色粉砂质黏土夹角砾层	4.90m

——————— 侵　蚀　面 ———————

下伏地层：下更新统第14层红色黏土质粉砂砾石层

（二）岩性岩相变化

中更新世洞穴堆积除第1地点周口店猿人洞外，在第4地点新洞也有堆积，堆积物为黄色砂质土，含大量石灰岩块，总厚度2m以上，上部为松散堆积，厚约1m；其下为钙质胶结，比较坚硬。堆积物上、下均含动物化石，并有一厚层灰烬层。第2地点龙骨山旧猿人陈列馆西约500m，堆积物为红土及角砾岩；第3、5地点堆积的角砾岩；第7地点、第8地点（东洞）、第9地点、第13地点、第15地点、第19地点、第20地点、第21地点、第22地点、第23地点、第24地点，上述位置均在第1地点附近，堆积物为红色黏土及角砾岩。其他区域亦有同期的洞穴堆积物，如房山燕山石化工厂北上店村的上店洞，堆积物为亚砂土，亚黏土及角砾；海淀四季青镇采石场的四季青洞，填充裂隙状堆积物；密云溪翁庄的溪翁庄洞，堆积物为棕红色黏土、砂土和角砾。

（三）生物特征及时代归属

周口店组时期生物繁盛，第1地点除发现 *Homo erectus pekinensis* 和大批石器外，哺乳动物化石丰富，据统计约有100种哺乳类动物化石，其中包括54种大型哺乳类（食肉目、长鼻目、奇蹄目和偶蹄目）。化石名单如表36-2所示。周口店组所含如此丰富的哺乳类化石组合，被命名为周口店动物群，是我国北方中更新世动物群的主要代表，此外堆积物中含大量的刮削器等石器。周口店第1地点中更新世早期的鸟类动物群化石其数量之大、种类之多也在世界上极为少见，计有9目19科48属62种，以鸡形类和雀形类最多（侯连海，1985），但无真正的水鸟类化石发现。

周口店组中蕨类植物、灌木、草本植物与乔木的孢粉发育，见图36-9及表36-3，它大体反映了距今（70~23）$\times 10^4$a期间周口店地区的植物类型及其演化状况。根据骨化石^{230}Th-^{234}U年龄测试结果，北京猿人年龄为50×10^4a左右，并应用古地磁方法测得第1地点洞穴堆积第13层以上，属布容正极性时，年代小于70×10^4a。

表 36-2 周口店第一地点含哺乳动物化石洞穴堆积层、年龄及哺乳动物群

Table 36-2 Mammal fossil cave deposit, age and mammalian fauna in Locality 1 at Zhoukoudian Site

层位	岩性	哺乳动物群	年龄/10⁴a
第 1~3 层	角砾岩夹砂土钟乳石混合层	Homo erectus pekinensis, Macaca robustus, Neomys sinensis, Neomys bohlini, Crocidura sp., Erinaceus olgae, Ia io, Miniopterrus schreberii, Cricetinus varians, Cricetulus cf. grieus, Crietulus cf. obsucrus, Micromys cf. minutus, Canis lupus, Vulpes cf. corsac, Ursus thibetanus kokeni, Ursus arctos, Meles cf. leucurus, Crocuta ultima, Panthera cf. tigris, Dicerorhinus choikotienensis, Equus sanmeniensis, Cervus (Pseudaxis) grayi, Megaloceros pachosteus, Cervus sp.	23~25.6（铀系法）
第 4 层	灰烬层	Homo erectus pekinensis, Macaca robustus, Scaptochirus primitivus, Neomys sinensis, Neomys bohlini, Crocidura sp. , Erinaceus olgae, Rhinoiophus pleistocaenicus, Miniopterus schreberii, Ia io, Cricetinus varians, Cricertulus cf. griseus, Cricetulus cf. obscurus, Micromys cf. minutus, Rattus rattus, Gerbillus roborowskii, Microtus brandtioides, Microtus epiratticeps, Ochotona koslowi, Canis lupus variabilis, Ursus thibetanus kokeni, Ursus arctos, Palaeoloxodon cf. namadicus, Dicerorhinus thoukoutienensis, Equus sanmeniensis, Moschus moschiferus pekinensis, Megaloceros pachyosteus, Ovis sp.	
第 5 层	黑青色硬钙板	Homo erectus pekinensis, Macaca robustus, Scaptochirusprimitivus, Erinaceus olgae, Trogontherium cuvieri, Cricetinus varians, Cricetulus obscurus, Micromys cf. minutus, Rattus rattus, Microtus brandtioides, Microtus epirattice, Canis lupus, Canis lupus vriabilis, Vulpes cf. corsac, Ursus thibetanus kokeni, Ursus arctos, Ailuropoda? sp., Meles cf. leucurus, Mustela cf. sibirica, Hyaena brevirostris sinensis, Machairodus inexpectatus, Panthera cf. tigris, Panthera cf. pardus, Felis teihardi, Felis sp. A, Felis sp. B, Felis cf. microtis, Cynailurus sp., Moschus moschiferus pekinensis, Cervus (Pseudaxis) grayi, Megaloceros pachyosteus, Gazella sp., Osvis cf. ammon	
第 6 层	角砾岩层	Homo erectus pekinensis, Macaca robustus, Erinaceus olgai? Castor sp., Trogontherium curvieri, Cricetinus varians, Cricetulus cf. obscurus, Micromys cf. minutus, Canis lupus variabilis, Nyctereutes sinensis, Vulpes cf. corsac, Ursus thibetanus kokeni, Ursus arctos, Meles cf. leucurus, Hyaena brevirostris sinensis, Panthera cf. tigris, Dicerorhinus choukoutienensis, Equus cf. sanmeniensis, Moschus moschiferus pekinensis, Cervus (Pseudaxis) grayi, Megaloceros pachyosteus, Gazella sp., Ovis cf. ammons	
第 7 层	灰色细砂层	Macoca robustus, Marmota complicidens, Trogontherium cuvieri, Cricetulus cf. griseus, Cricetulus cf. obscurus, Micromys cf. minutus, Canis lupus, Nyctereutes sinensis, Lutra melina, Hyaena brevirostris, Panthera cf. tigris, Felis sp. B, Dicerorhinus choukoutienensis, Equus sanmeniensis, Suslydekkeri, Moschus moschiferus pekinensis, Cervus (Pseudaxis) grayi, Megaloceros pachyosteus, Ovis cf. ammon, Bubalus teilhardi	37~40（古地磁法）
第 8 层	角砾岩夹灰烬层	Homo erectus pekinensis, Macaca robustus, Marmota sp., Trogontherium cuvieri, Hystrix cf. subcristata, Canis lupus, Canis lupus variabilis, Canis cyonids, Nyctereutes sinensis, Cuon antiquus, Vulpes cf. corsac, Ursus thibetanus kokeni, Ursus arctos, Ursus cf. spelaeus, Meles cf. lecurus, Gulo? sp. , Hyaena brevirostris sinenis, Machairodus inexpectatus, Panthera cf. tigris, Felis teillardi, Felis sp. A. , Palaeoloxodon cf. namadicus, Dicerorhinus choukoutienensis, Coelodonta antiquitatis yenshensis, Cervus (Pseudaxis) gravi, Megaloceros pachyosteus, Spirocerus peii, Ovis cf. ammon, Bubalus teilhardi	
第 9 层			
第 10 层	红色土及灰烬层	Homo erectus pekinensis, Scaptochirus primitives, Neomys sinnensis, Neomys bohlini, Cricetulus cf. obscurus, Mus sylvaticus, Mus musculus, Gerbillus roborowskii, Microtus epiratticops, Ochotona koslowi, Canis lupus variabilis, Nyctereutes sinensis, Vulpes cf. vulpes, Ursus arctos, Ursus cf. spelaeus, Nyaena brevirostris sinensis, Machairodus inexpectatus, Panthera, cf. pardus, Felis teilhardi, Dicerorhinus choukoutienensis, Equus sanmeniensis, Sus lydekkeri, Cervus (Pseudaxis) gravi, Megaloceros pachyosteus	46.2（裂变径迹法）
第 11 层	角砾岩		
第 12 层	棕褐色含角砂粗砂岩	Dicerohinus choukoutiesensis	

续表

层位	岩性	哺乳动物群	年龄/10^4a
第 13 层	红色粉砂质黏土角砾岩	*Ngospalax epitingi*, *Hyaena brevirostris sinentsis*, *Equus sanmeniensis*, *Megaloceros flabellatus*	

注：根据周口店综合研究资料（胡长康，补充 12、13 层资料）。（《北京市区域地质志》（1991））

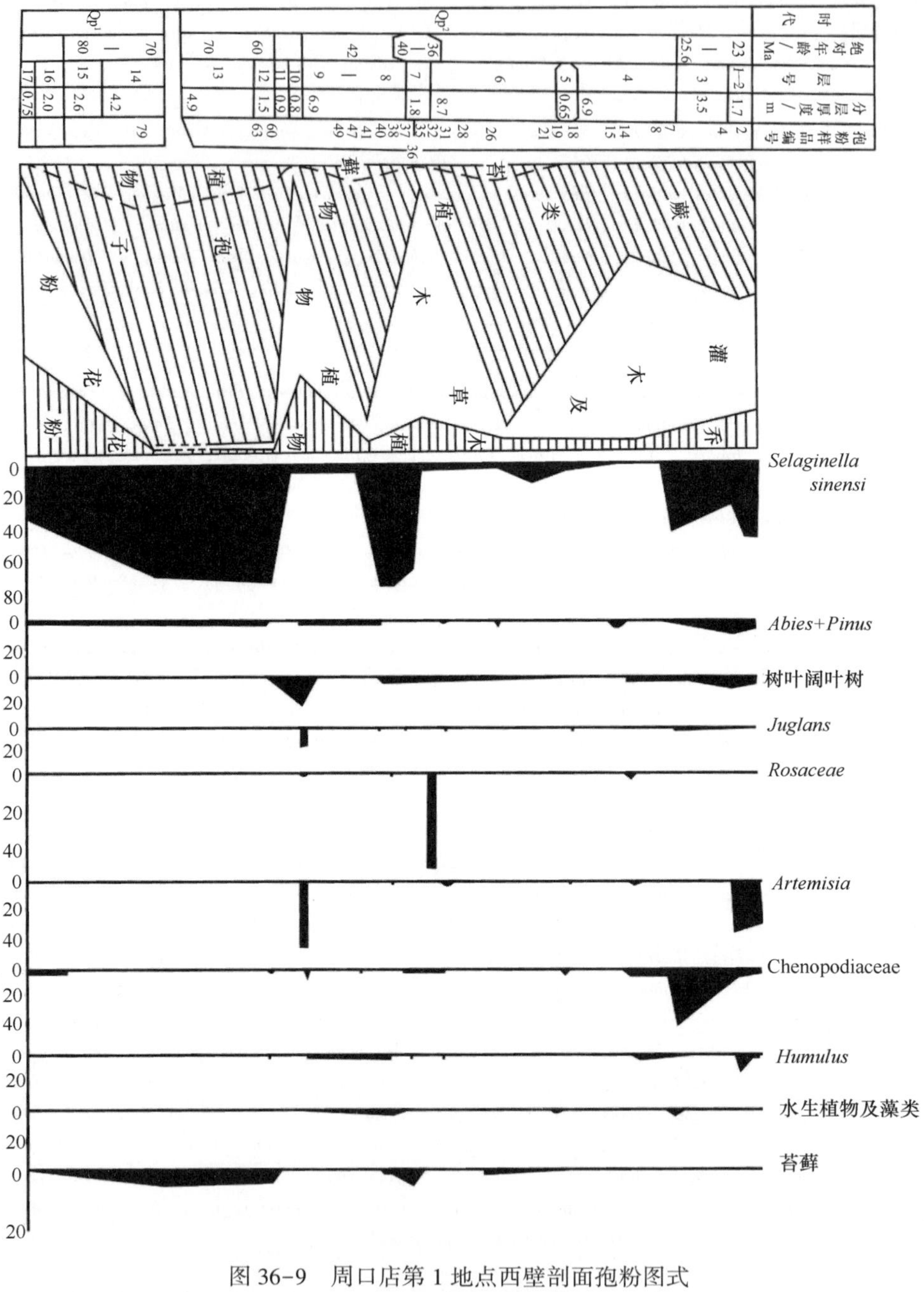

图 36-9 周口店第 1 地点西壁剖面孢粉图式
（据孔昭宸等，1985）
Fig. 36-9 Pollen diagram of west wall section in Locality 1 at Zhoukoudian Site

三、上更新统（羊耳峪组（Qp^3y））

晚更新世洞穴堆积以东岭子洞穴堆积物为代表，堆积厚度较大（约 30m），化石较丰富，距今（1.9~11.4）×10^4a（图 36-10）。

表 36-3　周口店中更新统地层对比表

Table 36-3　Correlation of Middle Pleistocene strata in Zhoukoudian

地层		年龄/10^4a	第一地点分层		岩性	动植物类	重矿物组合	黏土矿物	气候特点	下砾石层分层	
中更新统	周口店组	$23^{+3}_{-2.3}$ $25.6^{+6.1}_{-4}$	第1～2层	第一段	粉砂角砾岩夹钙板层	针阔叶混交林或落叶阔叶林夹草原，草原植物较多。多干旱草原动物为主	褐铁矿-绿帘石-绿泥石为主，普通角闪石次之。风化系数 K=3.88	伊利石为主，蒙脱石、高岭石次之，下部有蛭石和绿泥石	半干旱温带气候	上段	红土坡积层
			第3层		巨大角砾岩及钟乳石层						
		29～31	第4层	第二段	灰烬层	针阔叶混交林，有静水池沼分布，草本植物增多。草啮齿类动物增多	普通角闪石-褐铁矿-黑云母-绿泥石。风化系数 K=0.20～0.90	蒙脱石为主，伊利石、高岭石次之	向干旱方向转变的温带气候		
			第5层	第三段	“钟乳石层”	温带蕨类植物较多。林栖动物比例较高有喜水及近水种类	褐铁矿-黑色金属矿物-黑云母-绿泥石。风化系数 K=3.35～3.55	伊利石为主，蒙脱石、高岭石、蛭石和绿泥石次之	温带半湿润气候，偏暖	中段	红色土层
			第6层		角砾岩						
		37～40	第7层	第四段	含云母碎屑细粉砂岩	温带落叶阔叶林夹草原，8～9层温带蕨类植物大量出现，有高山藓类，下部偏凉。林栖和草原类型的动物，下部草原类型动物稍多，第7层喜水、近水动物较突出	普通角闪石-褐铁矿-绿帘石。风化系数 K=0.36～1.64	伊利石为主，高岭石、蒙脱石、蛭石和绿泥石次之，上部蒙脱石略有增加	趋向干冷的半干旱温带气候，中部有短时间温湿阶段		
		$42^{+>18}_{-10}$	第8～9层		角砾岩						
		46.2±4.5	第10层		红色泥土及灰烬层						
		70	第11层		角砾岩						
			第12层	第五段	棕褐色含角砾粗砂层	温带针阔叶混交林有静水分布，喜湿的及温带蕨类较多。温带草原湖沼类型动物群	褐铁矿-黑色金属矿物-绿帘石-云母。风化系数 K=2.03～3.70	伊利石为主，高岭石、蛭石和绿泥石次之	半湿润温带气候，偏暖	下段	砾石层
			第13层		红色粉砂质黏土角砾层						
下更新统	龙骨山组	70	第14层		红色黏土质粉砂砾石层	暖温带落叶阔叶及针阔混交林，有大量温带蕨类植物	榍石-褐铁矿-锆石	伊利石为主高岭石较多，蛭石和绿泥石次之	半温润暖温带气候	下段	砾石层
			第15层		棕红色粗砂砾石层、含砾细砂层		榍石-褐铁矿-锆石	伊利石为主高岭石次之有蒙脱石	偏干的温带气候		

（据杨子赓，1985）

（一）代表剖面

该剖面为 1986 年 1：5 万周口店幅区调填图工作组成员曹树钊发现，位于羊耳峪村北东岭子果园西北，在海拔约 220m 山坡上的奥陶系灰岩之中，其洞穴堆积剖面出露厚度约 7.20m，自上而下为：

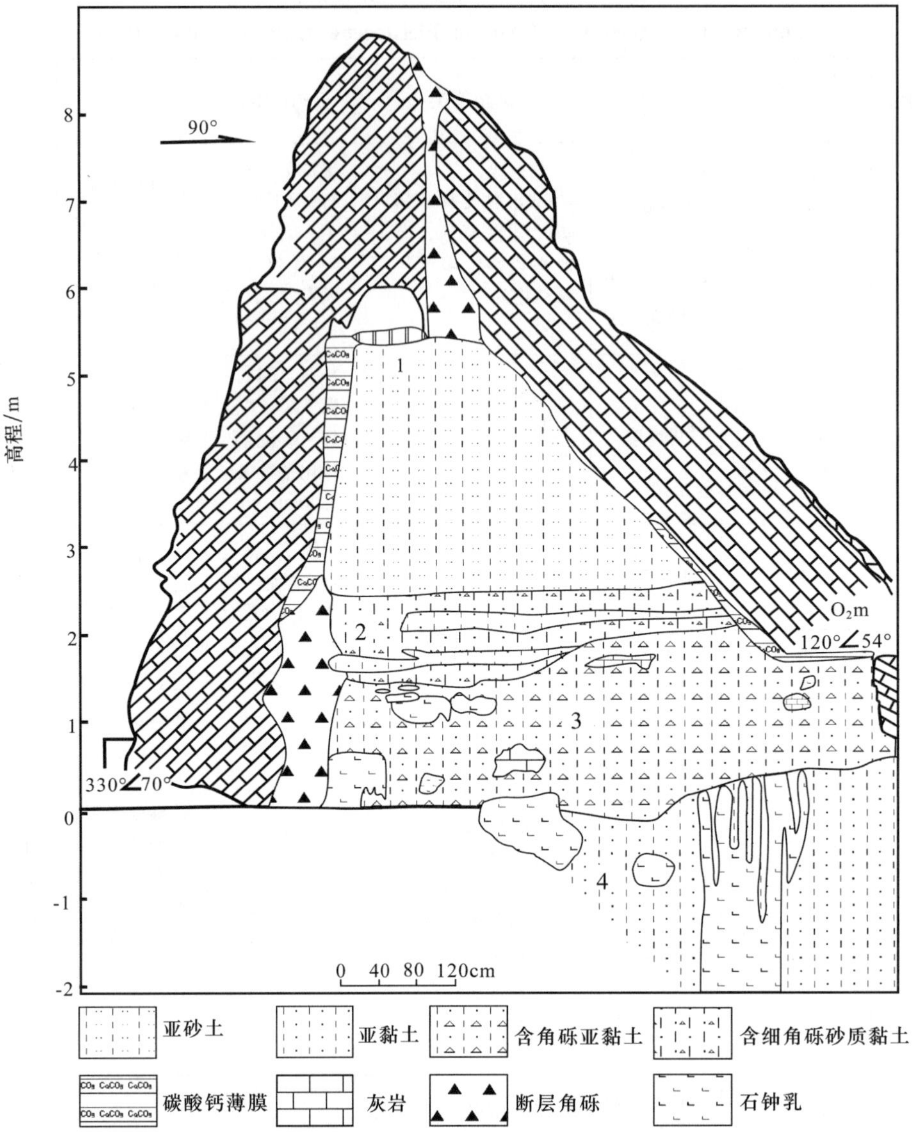

图 36-10　晚更新世周口店东岭子洞穴堆积剖面图

（据 1∶5 万周口店幅，1988）

Fig. 36-10　Profile of cave deposits in Late Pleistocene in Donglingzi, Zhoukoudian

1. 灰黄色亚砂土，含 *Hypolagus* sp.，*Cervus*（*Pseudaxis*）sp.，顶部热释光年龄（19.4±1.4）×10^4a　2.8m
2. 灰黄色钙质胶结的砂土与含细灰岩角砾亚砂土互层　0.8m
3. 黄色角砾亚黏土，灰岩角砾一般 5~6cm，松散。含哺乳动物化石碎片及石钟乳碎块，钟乳石碎块大者达 0.6m，灰岩砾大者达 40~50cm　2.0m
4. 棕红色亚黏土层，含石钟乳碎块，含哺乳动物化石 *Microtus* sp.，*Allocricetus teilhardi* Zheng，*Cricetulus* sp.，*Cervus*（*Pseudaxis*）sp. 及鹿的二支角、上颌骨等，石钟乳铀系年龄上限为（114.7±4.9）×10^3a，年龄下限>350×10^3a　>1.6m

（二）岩性岩相变化

晚更新世洞穴堆积以除羊耳峪东岭子洞穴外，还有周口店第 26 地点的山顶洞，堆积物为灰色砂质土夹大量未胶结的石灰岩碎块，最下部稍微胶结；昌平龙骨洞，堆积物为黄棕色角砾及砂质黏土，厚度约 40m（黄万波，1981）；房山云水洞，堆积物为粉砂夹角砾，在宝儿水村峭壁上洞穴，堆积褐黄色的黏土和砾石；门头沟珠窝洞，堆积物为砂质黏土、粗砂层及砾石层。还有海淀田家疃洞及昌平高崖口镇韩家台洞等洞穴堆积角砾、砂质黏土沉积。

（三）生物特征及时代归属

晚更新世是早期智人向高级阶段发展的时期，晚更新世中期堆积含有“新洞人”的牙齿及半生哺乳动物化石（赵树森等，1985），顶部钟乳石经铀系法测定其年龄为（9.8±0.5）$\times 10^4$a。晚更新世晚期堆积含“山顶洞人”化石及其文化遗迹和伴生哺乳动物化石，其下窨 TL 年龄为（32.5$\times 10^4$~5$\times 10^4$a），上室^{14}C 年龄为（1.8340±0.041）$\times 10^4$a，同时期洞穴外在斋堂马兰阶地还发现原始牛、赤鹿等脊椎动物化石。

东岭子洞为一巨大漏斗状充填洞穴，产有大量啮齿类动物化石、少量哺乳动物化石及少量其他动物化石（表 36-4）。在 29 种哺乳动物化石中，绝灭种只有翁氏貉、翁氏野兔、异含鼠、变异含鼠、拟布氏田鼠、日进鼠等，占总数的 21.4%。该动物群大多数为现生种属。如黑线包鼠、花鼠、岩松鼠、黑鼠、似浣熊貉、沙狐、北京斑鹿等，均延续到现代。其中，个别在周口店动物群和山顶洞动物群中有不同分布。如：变异仓鼠在两群中都存在；拟布氏田鼠只在前者动物群中；北京斑鹿等只在后者动物群中。根据古地磁、热释光法和铀系等年代资料综合分析，羊耳峪组洞穴沉积物形成时间距今（1.9~11.4）$\times 10^4$a。

表 36-4　东岭子动物群及其相似种属在华北地区的时代分布

Table 36-4　The age distribution of Donglingzi animal groups and their similar species in North China

层位	哺乳动物化石	Qp^1			Qp^2			Qp^3			Qh
		Qp^{1-1}	Qp^{1-2}	Qp^{1-3}	Qp^{2-1}	Qp^{2-2}	Qp^{2-3}	Qp^{3-1}	Qp^{3-2}	Qp^{3-3}	
(3)	*Cervus* (*Pseudaxis*) *hortulorum* *Lepuswongi* Young, 1927										
(5)	*Microtus* sp. *Sciurotamias* sp. *Micromys* sp.										
(6)	*Lepus wongi* Young, 1927 *Allocricetus* sp. *Cricetulus varians* Zdansky *Cricetulus barabensis griseus* (Mile-Edwards), 1876 *Microtus brandtioides* Young *Chardinomys* sp. *Cervus* (*Pseudaxis*) *hortu rorum* *Gazella* sp. *Nyctereutes procyonoides* Gray *Cervua* (*Elaphus*) sp.										
(7)	*Rattus rattus* Linnaeus *Lepus wongi* Yong *Sciurotamias praecox* Teihard *Eutamias sibiricus* *Chardinomys* sp. *Micromys minutus* Pallas *Vulpes corsac* Linnaeus *Crocidura wongi* Pei *Hipposideros* sp.										

续表

层位	哺乳动物化石	Qp^1			Qp^2			Qp^3			Qh
		Qp^{1-1}	Qp^{1-2}	Qp^{1-3}	Qp^{2-1}	Qp^{2-2}	Qp^{2-3}	Qp^{3-1}	Qp^{3-2}	Qp^{3-3}	
(8)	*Sciurotamis praecox* Teihard										
	Allocricetus sp.										
	Cricetulus varians Zdansky										
	Cricetulus varians griscus										
	Myospalax sp.										
	Hyperacrius sp.										
	Apoderacrius sp.										
	Chardinomys sp.										
	Crocidura wongi Pei										
	Hipposideros sp.										

注：————华北地区；════本地区；--------推断。　　（据曹伯勋，1991）

洞穴沉积物含脊椎动物化石，有山顶洞人（*Homo sapiens*），洞熊（*Ursus spelaeus*），最后鬣狗（*Hyaena ultina*），犀牛（*Rhinoceros* sp.），野驴（*Equus hemionus*）等48种哺乳类化石。山顶洞动物群中的绝灭种有鸵鸟、鬣狗、洞熊等，占12.1%，现生种类占87.9%。山顶洞动物群中有若干代表南方气候的动物，如香猫、猎豹等。

四、全新统（洞穴石笋沉积）

目前，第四系剖面的建立主要类型有黄土沉积剖面、湖泊沉积剖面、大洋沉积剖面、冰芯剖面和滩涂剖面等。对于洞穴碳酸钙沉积研究，在1952年美国人Moore提出speleothem（cave calcite）术语后，开始了钟乳石沉积研究。北京地区洞穴碳酸钙沉积较发育，从20世纪90年代开始对石花洞从地质学、气象学和环境学方面进行了较系统的研究。通过20多年的研究（刘东生等，1997；谭明等，1997，2002，2006，2014；秦小光等，1998；吕金波等，1999），石花洞洞层多达8层，钟乳石种类丰富，发育有石盾、石珊瑚、石瀑布、钙板、石笋、石钟乳、石柱等，在洞内首次发现月奶石、碳酸钙微层理以及发现膜足硬肢马陆新种穴居动物。

经调查人员的研究（赵树森等，1981，1985，1990；邢如连等，1989；吕金波等，2007），根据石花洞碳酸钙沉积物测得丰富的年代学数据（表36-5），结合其沉积形态，认为在气候湿热、滴水较多时石花洞中形成的粗犷石笋为中更新世沉积；在气候干冷、滴水较少时洞中形成的杆状石笋为晚更新世沉积。全新世为短柱状石笋沉积，顶部有未饱和水，尤其以石笋剖面中发育典型微层理为代表特征。

表36-5　石花洞钟乳石^{210}Pb、^{14}C、$^{230}Th/^{234}U$、ESR年代测定结果

Table 36-5　^{210}Pb, ^{14}C, $^{230}Th/^{234}U$ and ESR dating of speleothems in the Shihua Cave

测试方法		测试时代/ka			备注
		中更新世	晚更新世	全新世	
^{210}Pb				<0.10	李红春，1995
^{14}C	常规方法		11.72、13.78、31.05	0.63、1.21、2.51、3.31、5.01、7.43、9.66	尹金辉等，1995
	AMS法			0.13、0.80	鲁向阳等，1995
$^{230}Th-^{234}U$		169.0、178.0、183.0、235.0、240.0、241.0、>350.0	14.9、42.9、47.5、50.4、51.8、54.7、64.6、65.5、69.5、78.5、81.5、98.8、	5.7、8.9、8.1	赵树森，1981，1995

续表

测试方法		测试时代/ka			备注
		中更新世	晚更新世	全新世	
ESR		130、179、235、245、267、316、472、518	14		邢如连等，1989

（据吕金波等，2007）

全新世洞穴石笋沉积，以发育典型微生长层（微层理）为特征（图 36-11）。一般微层厚度为 $15\times10^{-6}\sim150\times10^{-6}$m，在透射显微镜下表现为亮带与暗带相间构成的亮暗变化序列，1 个亮带与 1 个暗带组成 1 个“微层理对”，每个“微层理对”不仅厚度不同，亮度也存在差异（秦小光，1998），其中发育大量纹层的石笋，对重建气候历史具重要的价值。

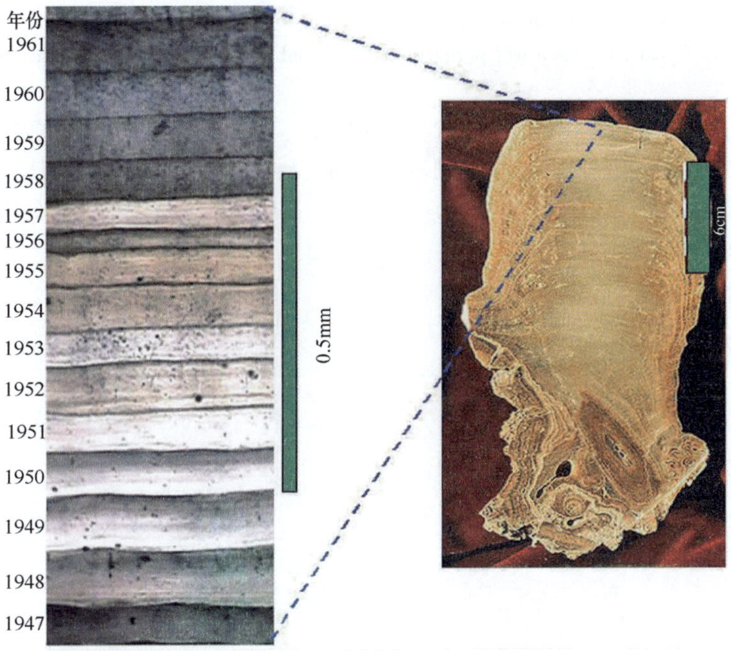

图 36-11　石花洞全新世石笋微层理

（据吕金波等，2007）

Fig. 36-11　Holocene stalagmite in the Shihua Cave

北京洞穴石笋微层的厚度与降水和温度有着密切的关系，其中受降水影响较大（刘东生等，1997）。秦小光等（1998）利用显微结构特征区分年层，提取了石笋样品 1131 层的微层厚度曲线（图 36-12），并确定了其顶部年龄为 1980a。微层厚度曲线显示出在公元前 1000a 左右和 1900a 以后微层厚度大，而 1500a 左右的微层厚度小，整体演化趋势是一种近千年的起伏变化。其中又有约 130~150a 的明显次级周期波动，共可划分出 7 个相对湿润高峰期和 7 个相对干旱低谷期，表明近 1ka 以来有至少有 7 次较大的气候波动。在 1130a 时段中，有明显的约 136a 较长周期和 50a、16~18a、11a、5. 8a 的较短周期。可见，微层厚度的变化基本反映了北京地区 1ka 以来的气候波动变化。不同阶段有不同的主导周期，在微层厚度大的时期（W 阶段），反映了夏季风加强时，降水增多、温度升高，促使微层厚度加大，夏季风对碳酸钙沉淀产生较大影响，留下较强的波动变化印迹；在厚度小的时期（D 阶段），夏季风减弱后，降水减少、温度下降，石笋微层厚度减小，夏季风的影响相应降低，其波动变化留下的印迹也较弱。因此，石笋微层厚度可能是一种中国北方地区较好的气候替代性指标。

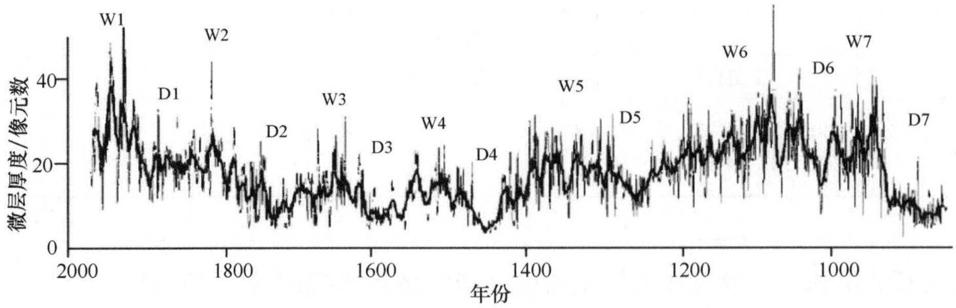

图 36-12　石花洞石笋微层（年层）厚度曲线（W1 和 D1 等分别表示层厚的峰谷区编号）和滑动功率谱图象（滑动窗口宽 80a）

（据秦小光，1999）

Fig. 36-12　The thickness series of laminae in stalagmite from Shihua Cave and the result picture of its moving spectral analysis (the bin width = 80 years)

W1, number of peak, means a wet period and D1, number of valley, a dry period, respectively

除石花洞外，北京地区还发育大量的洞穴石笋沉积，其中多数集中在北京西山大石河流域，从上游到下游依次发育有鸡毛洞、银狐洞、石花洞和孔水洞，由一条地下暗河将其连为一个洞系，为多层溶洞。随着北京西山间歇性地上升，在不同地壳相对稳定时期形成上下不同海拔高度的溶洞，代表了与之对应的新构造隆升期次。

第三十七章　平原区第四系

平原区第四系研究所依赖的重要资料即为钻孔数据。近几十年来北京平原区施工了大量的钻孔，其中有20世纪80年代地震地质会战、90年代通县幅和怀柔幅1∶5万区调、2005~2008年的多参数立体地质调查、2012—2014年的北京平原区活动断裂专项调查2014—2016年北京1∶5万琉璃河等三幅区调等项目，钻孔第四纪地层编录详细，其中大部分钻孔尝试开展了第四纪层序地层划分，采集了年龄、古地磁、孢粉及微体古生物、黏土矿物、粒度、氧化物和微量元素等样品，取得了丰富的实验测试数据。本志着重对上述资料进行总结。

本志将平原区第四系划分为更新统和全新统，并依据全国地层委员会《中国地层表》（2013），将古地磁松山反极性世（M）和高斯正极性世（G）界面作为第四系的下限，地质年龄确定在2.588Ma，多个钻孔M/G界面上下的第四纪沉积物岩性、古气候和微古生物群落有较明显变化：界面之下沉积物的胶结程度较好，普遍呈棕红色黏土类沉积；之上堆积物则较松散，普遍为灰、黄褐色砂类沉积。孢粉资料表明界面以上气候变冷，$CaCO_3$含量的差异在界面上下有较大变化。古地磁松山反极性世（M）与布容正极性世（B）的界线作为中更新统的底界，地质年龄为0.781Ma，上更新统和全新统底界年龄分别为0.126Ma和0.0117Ma。

北京平原区第四系结构较复杂，北部山区有延庆第四纪沉积凹陷，平原地区北部主要有马池口、后沙峪-俸伯、平谷等第四纪凹陷。由于同一时期不同地区所处地理、构造位置，物源、成因等因素有差异，造成岩性组合特征上存在较大差异，给地层横向对比带来较大困难。本志将北京地区划分为几个典型第四纪沉积凹陷（图37-1），在每个凹陷内选择一个数据详细的钻孔作为代表剖面，进行凹陷内地层划分与对比，为今后的区域地层对比奠定基础。

第一节　更新统

一、岩性地层

（一）马池口沉积凹陷（Ⅰ）

位于平原西北部南口-马池口-沙河一带，由北东向的南口山前断裂及北西向的南口-孙河断裂所控制，凹陷呈现向西南翘起的形态。选取ZK3孔为代表性钻孔，根据磁性地层学结果建立ZK3孔的磁极倒转序列，结合岩性特征与测年资料综合划分，第四系底界于孔深493.8m处，中更新统底界于157.4m处，上更新统底界置于87.7m处，全新统地层底界在孔深7.2m处。岩性描述如下。

1. 代表剖面（ZK3孔）

钻孔ZK3位于北京昌平百善镇东沙屯村西农田，孔深616.20m，描述如下：

上覆地层：全新统灰绿色中粗砂

———— 整　合 ————

上更新统　　总厚80.5m

5. 灰绿色粉砂质黏土，黏粒组分70%，粉粒组分30%，以长石、石英为主，岩芯中在5.35m处夹3cm厚钙质结核层。局部夹有杂色砾石层，砾石以花岗岩、石英岩、闪长岩等为主，分选性好，磨圆度高，大小5~7cm　　4.0m

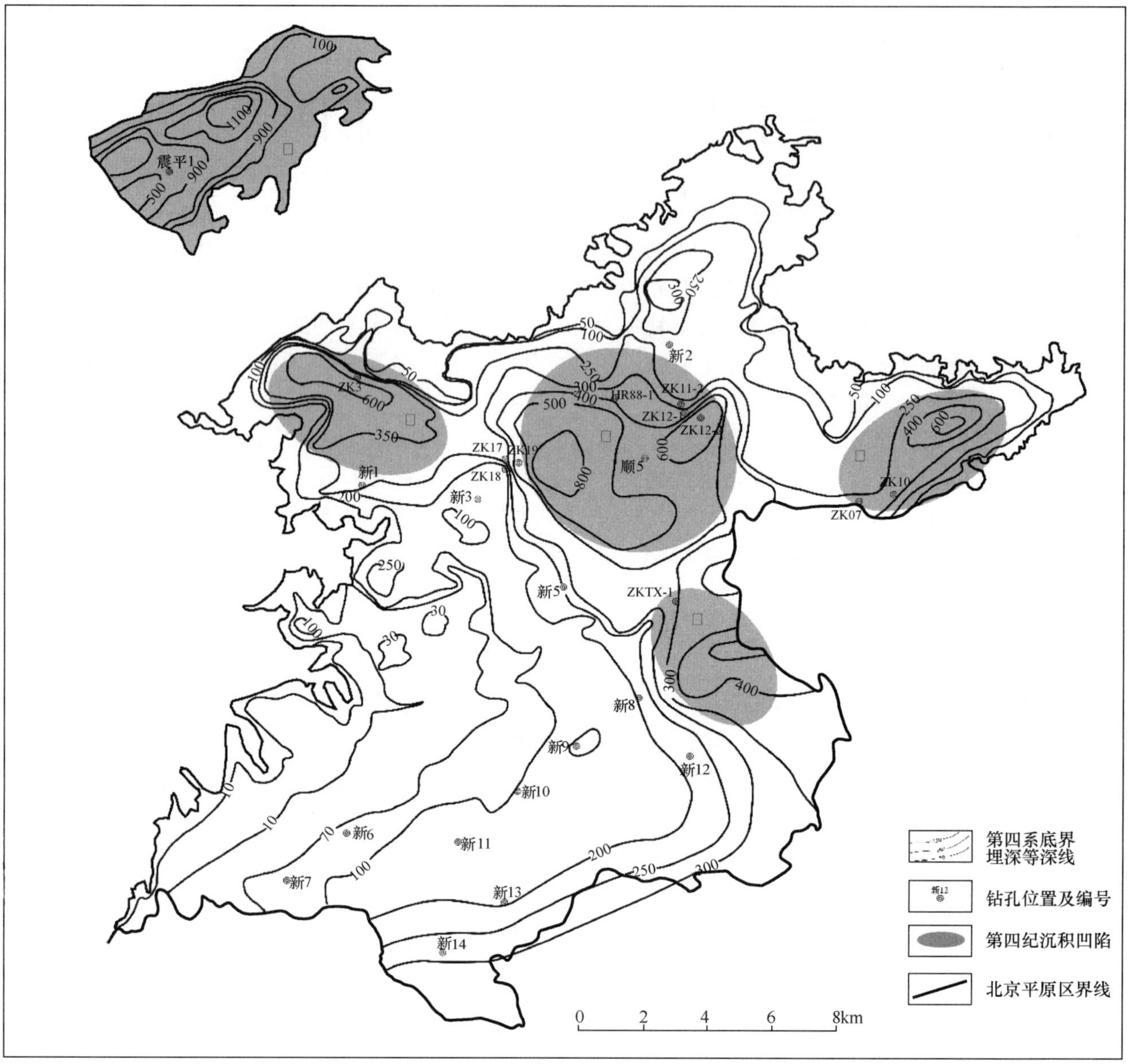

图 37-1　北京地区第四纪主要沉积凹陷分布简图

Fig. 37-1　Distribution of Quaternary major sedimentary depressions in Beijing region

Ⅰ—马池口凹陷；Ⅱ—后沙峪-俸伯凹陷；Ⅲ—甘棠凹陷；Ⅳ—平谷凹陷；Ⅴ—延庆凹陷

6. 上部为棕黄色粉砂质黏土，黏粒组分 60%，粉粒组分 40%，以长石、石英为主，岩芯见大量灰白色斑点，大小 2mm 左右，厚 0.6m；中部为黄绿色黏质细砂，黏粒组分 45%，砂粒组分 55%，以长石、石英为主，大小 0.10~0.25mm，偶见锈斑，厚 2.9m；底部为灰色粉砂质黏土，黏粒组分 75%，粉粒组分 25%，以长石、石英为主，偶见中砂颗粒，厚 0.7m　　4.8m
7. 上部为黄色、灰绿色、黄绿色黏质粉砂，黏粒组分 40%，粉粒组分 60%，以长石、石英为主，长石 30%~40%，石英 50%~60%，云母次之，含量 10%，片状，大小 0.005~0.100mm；中部为黑灰色粉砂质黏土，黏粒组分 75%，粉粒组分 25%，岩芯中见少量钙质结核，大小 2mm 左右；下部为棕黄色黏质细砂，黏粒组分 40%，砂粒组分 60%，以长石、石英为主，长石 40%~50%，石英 50%~60%，局部粒度变化为粗砂，偶见锈斑　　5.0m
8. 棕黄色黏质粗砂，黏粒组分 40%，砂粒组分 60%，以长石、石英为主。长石 40%~50%，石英 50%~60%，大小 0.5~2.0mm，见少量锈斑　　2.2m

9. 杂色砂砾石，结构松散，砂以粗砂为主，含量70%，以长石、石英为主，长石40%~50%，石英50%~60%，大小0.5~2.0mm，分选性好，磨圆一般，砾石成分复杂，以花岗岩、正长岩、石英岩等为主，大小不一，0.5~7.0cm，分选性差，磨圆度高 5.7m

10. 上部为棕黄色粉砂质黏土夹粉砂层及团块，粉砂质黏土。黏粒组分75%，粉粒组分25%，以长石、石英为主，粉砂层厚1~3cm，团块多成“月牙状”，局部夹有薄层棕红色黏土条带，多具纹层状；下部为棕灰色砂质黏土，黏粒组分70%，砂砾组分30%，以长石、石英为主，大小0.10~0.25mm 1.7m

11. 棕黄色黏质中砂。黏粒组分30%，砂粒组分70%，以长石、石英为主，长石30%~40%，石英50%~60%，云母片状，含量10%，大小0.2~0.5mm，见少量锈斑 1.9m

12. 杂色砂砾石，结构松散，砂以粗砂为主，含量70%，以长石、石英为主，长石40%~50%，石英50%~60%，大小0.5~2.0mm，分选性好，磨圆一般，砾石成分复杂，以花岗岩、正长岩、石英岩等为主，大小不一，一般0.5~7.0cm，分选性差，磨圆度高。底部见约1m厚的棕红色黏土 4.1m

13. 棕黄色粉砂质黏土，黏粒组分70%，粉粒组分30%，以长石、石英为主，见少量锈斑 2.8m

14. 黑灰色粉砂质黏土，黏粒组分70%，粉粒组分30%，岩芯中见大量小螺残片，少量炭斑 3.1m

15. 上部为灰绿色黏质粉砂。黏粒组分45%，粉粒组分55%，以长石石英为主，岩芯中见少量锈斑；下部为灰绿色粉砂质黏土。黏粒组分70%，粉粒组分30%，以长石、石英为主，见大量灰白色斑点，大小2mm左右 3.5m

16. 棕黄色中粗砂，结构松散，颗粒以长石、石英为主，长石40%~50%，石英50%~60%，大小0.5~2.0mm，分选性好，磨圆度高，偶见小砾石 2.1m

17. 棕黄色粉砂质黏土，黏粒组分70%，粉粒组分30%，以长石、石英为主，偶见锈斑 2.5m

18. 杂色砂砾石，结构松散，砂以粗砂为主，含量90%以上，长石20%~30%，石英40%~50%，岩屑30%左右，大小0.5~2.0mm，分选性好，磨圆度低，砾石含量10%，以花岗岩、石英岩等为主，大小不一，一般2~5cm，分选性差，磨圆度高 3.6m

19. 上部为棕黄色、灰色砂质黏土，黏粒组分65%，砂粒组分35%，以长石、石英为主，大小0.2~0.5mm，在54.7m处见一15°角倾斜的锈色条带；中部为深灰色粉砂质黏土，黏粒组分75%，粉粒组分25%以下，见少量灰白色斑点，大小2mm左右。下部为棕黄色砂质黏土黏粒组分70%，砂粒组分30%，已长石、石英为主，见少量锈斑 2.7m

20. 棕黄色中粗砂，局部粒度有变化。结构松散，颗粒以长石、石英为主，长石30%~40%，石英50%~60%，云母20%，大小0.5~2.0mm，底部见粉砂质黏土团块，团块边缘见少量锈斑 0.5m

21. 棕黄色、黄绿色粉砂，结构松散，颗粒以长石、石英为主，长石30%~40%，石英50%~60%，云母次之，含量20%，大小0.005~0.100mm，夹有黏土薄层，见大量锈斑。中部夹有约1m的粉砂质黏土 3.2m

22. 黑灰色砂质黏土，黏粒组分70%，砂粒组分30%，以长石、石英为主，大小0.1~0.3mm，碳质成分较高。局部夹有灰白色粉细砂层 10.7m

23. 灰绿色黏质粉砂与粉砂质黏土互层。黏质粉砂，黏粒组分45%，粉粒组分55%，以长石、石英为主，长石30%~40%，石英50%~60%，云母10%，大小0.005~0.100mm，偶见炭斑。粉砂质黏土，黏粒组分75%，粉粒组分25%，见大量炭斑 2.4m

24. 棕黄色中粗砂，结构松散，强风化，颗粒以长石、石英为主，岩屑次之，长石30%~40%，石英40%~50%，岩屑20%，大小0.5~2.0mm，含少量小砾石，大小1~2cm，以花岗岩、石英岩、正长岩等为主，见大量锈斑，底部接触面见大量炭质成分。局部见有约10cm左右的黏质粉砂 2.05m

25. 上部为棕黄色、灰绿色粉砂质黏土，手搓有轻微涩感，捻面不光滑，湿时不易成条，黏粒组分70%，粉粒组分30%，以长石、石英为主，见少量锈斑。在79.70~79.75m处见钙质结核层；下部为黑灰色粉砂质黏土，黏粒组分70%，粉粒组分30%，见大量炭斑，少量小螺残片 11.95m

——————— 整　合 ———————

中更新统 总厚 69.7m

26. 深灰色细砂，结构松散，颗粒以长石、石英为主，长石 30%~40%，石英 50%~60%，云母 10%，大小 0.10~0.25mm，岩芯底部多纹层理，与下伏地层呈不整合接触。88.1~88.7m 处见灰黑色黏土层 2.0m
27. 黑色黏土，手搓有涩感，捻面光滑，湿时可搓成条，不易断，黏粒组分 95%以上，见大量炭斑，偶含钙质结核，底部为厚约 30cm 的杂色砾石层 1.0m
28. 灰绿色黏质粉砂，黏粒组分 40%，粉粒组分 60%，以长石、石英为主，长石 30%~40%，石英 50%~60%，云母 10%，大小 0.005~0.075mm，见少量锈斑 3.0m
29. 灰绿色中粗砂，结构较松散，颗粒以长石、石英为主，长石 30%~40%，石英 50%~60%，云母 10%，大小 0.10~0.25mm，分选性好磨圆度高，局部夹有粉砂质黏土条带，见大量炭斑 3.5m
30. 灰黑、浅灰色粉砂质黏土，手搓有轻微涩感，捻面不光滑，湿时不易成条，黏粒组分 70%，粉粒组分 30%，以长石、石英为主，见少量小螺残片 4.8m
31. 灰绿色粉砂质黏土，黏粒组分 70%，粉粒组分 30%，见少量小螺残片，偶见锈斑。下部见有 50cm 灰绿色细砂 6.3m
32. 灰黑色、棕黄色粉砂质黏土，黏粒组分 70%，粉粒组分 30%，以长石、石英为主，见少量锈斑 4.1m
33. 棕黄色粗砂，结构松散，颗粒以长石、石英为主，长石 30%~40%，石英 50%~60%，岩屑 20%，大小 0.5~2.0mm，分选性好，磨圆度低，含 10%左右小砾石，以花岗岩、石英岩、安山岩等为主，大小 0.5~1.0cm，底部见大量锈斑 1.1m
34. 上部为棕黄色、灰绿色粉砂，结构松散，颗粒以长石、石英为主，长石 30%~40%，石英 50%~60%，云母 10%，大小 0.005~0.075mm，见少量锈斑；下部为黑灰色粉砂质黏土，黏粒组分 70%，粉粒组分 30%，偶见炭斑 1.8m
35. 黄绿色、灰白色黏质细砂，黏粒组分 30%，砂粒组分 70%，以长石、石英为主，长石 30%~40%，石英 50%~60%，云母 10%，大小 0.10~0.25mm，见大量锈斑。底部为 80cm 的黑色粉砂质黏土 1.2m
36. 杂色砂砾石，结构松散，砂以粗砂为主，含量 80%，颗粒以长石、石英为主，长石 30%~40%，石英 50%~60%，岩屑 20%，大小 0.5~2.0mm，分选性好，磨圆度低，砾石成分复杂，以花岗岩、石英岩、安山岩等为主，大小不一，一般为 0.5~4.0cm，分选性差，磨圆度高 2.1m
37. 黑灰色砂质黏土，黏粒组分 70%，砂粒组分 30%，以长石、石英为主，底部见大量锈色条带，多具斜层理构造。上部见有 30cm 的灰白色中砂 3.4m
38. 棕黄色中细砂，结构松散，颗粒以长石、石英为主，长石 30%~40%，石英 50%~60%，岩屑 10%，大小 0.25~0.50mm，分选性好，磨圆度较高，见少量锈斑，自 124m 向下，多夹有粉砂质黏土薄层，厚 2~5cm，薄层边缘为锈色条纹 2.5m
39. 棕黄色粉砂质黏土与黏质粉砂互层。粉砂质黏土，黏粒组分 70%，粉粒组分 30%，见大量锈斑及条带。黏质粉砂，黏粒组分 30%，粉粒组分 70%，颗粒以长石、石英为主，长石 30%~40%，石英 50%~60%，岩屑 10%，大小 0.005~0.075mm，局部见大量锈斑 6.9m
40. 上部为灰白色中粗砂，结构松散，颗粒以长石、石英为主，长石 30%~40%，石英 50%~60%，岩屑 20%左右，大小 0.3~2.0mm，分选性好，磨圆度高；中部为棕黄色黏质粉砂，黏粒组分 40%，粉粒组分 60%，以长石、石英为主，长石 30%~40%，石英 50%~60%，云母 10%，大小 0.005~0.100mm，见大量锈斑；下部为灰绿色粉砂质黏土，黏粒组分 60%，粉粒组分 40%，以长石、石英为主，偶见锈质结核 7.8m
41. 灰绿、黑灰、棕黄色粉砂质黏土，手搓有轻微涩感，捻面不光滑，湿时不易成条，黏粒组分 70%，粉粒组分 30%，以长石、石英为主，偶见锈斑 3.5m
42. 杂色砂砾石，结构松散，砂含量 80%，以长石、石英及岩屑为主，长石 20%~30%，石英 40%~50%，岩屑 20%~30%，大小 0.5~2.0mm，分选性好，磨圆度高，砾石含量 20%，成分复杂，以花岗岩、石英岩、正长岩等为主，大小均一，在 3~5cm 之间，磨圆度高 3.7m

43. 棕黄色粉砂质黏土，手搓有轻微涩感，捻面不光滑，湿时不易成条，黏粒组分 70%，粉粒组分 30%，以长石、石英为主，见大量锈斑及钙质结核 2.2m

44. 上部为棕黄色黏质细砂，黏粒组分 40%，砂粒组分 60%，颗粒以长石、石英为主，长石 30%~40%，石英 40%~50%，岩屑 10%，云母 10%，大小 0.10~0.25mm，分选性好，磨圆度高，见少量锈斑；中部为灰白色粗砂，结构松散，颗粒以长石、石英及岩屑为主，长石 20%~30%，石英 40%~50%，岩屑 20%，大小 0.5~2.0mm，分选、磨圆差；下部为棕灰色粉砂质黏土黏粒组分 70%，粉粒组分 30%，见少量锈斑 2.7m

45. 上部为棕黄色细砂，结构松散，颗粒物以长石、石英为主，长石 20%~30%，石英 40%~50%，岩屑 20%，石英 10%，大小 0.10~0.25mm，分选性好，磨圆程度高，底部接触带见大量锈色条带；下部为灰绿色粉砂质黏土，手搓有轻微涩感，捻面不光滑，湿时不易成条，黏粒组分 70%，粉粒组分 30%，以长石、石英为主，偶见炭斑，底部见有 40cm 细砂层 3.7m

46. 灰绿色粉砂质黏土黏粒组分 70%，粉粒组分 30%，以长石、石英为主，偶见炭斑，在 160.55m 处夹 1cm 厚细砂，具斜层理构造，在 160.70~160.75m 夹粗砂层 2.4m

——————— 整　合 ———————

下更新统 总厚 336.4m

47. 灰绿色粉砂质黏土黏粒组分 70%，粉粒组分 30%，以长石、石英为主，偶见炭斑，在 160.55m 处夹 1cm 厚细砂，具斜层理构造，在 160.70~160.75m 夹粗砂层 4.5m

48. 上部为灰绿色粗砂，结构松散，颗粒以长石、石英及岩屑为主，长石 20%~30%，石英 40%~50%，岩屑 30%~40%，大小 0.5~2.0mm，见少量砾石，以花岗岩为主，大小 0.5cm 左右；下部为灰色粉砂质黏土，黏粒组分 70%，粉粒组分 30%，以长石、石英为主，偶含小螺残片 2.9m

49. 灰色厚层状粉砂质黏土夹棕黄色薄层状粉砂质黏土。粉砂质黏土，黏粒组分 70%，粉粒组分 30%，以长石、石英为主，偶含小螺残片 9.0m

50. 杂色砂砾石，结构松散，砂含量 80%以上，以长石、石英、岩屑为主，长石 20%~30%，石英 40%~50%，岩屑 30%~40%，大小 0.5~2.0mm，分选性好，磨圆较差，砾石含量 20%以下，成分复杂，以花岗岩、闪长岩、正长岩等组成，大小不一，一般 1~5cm，分选性差，磨圆度高 3.6m

51. 灰绿色粉砂质黏土，黏粒组分 75%，粉粒组分 25%，以长石、石英为主。局部见有灰绿色中细砂层 5.6m

52. 灰白色粗砂，结构松散，颗粒物以长石、石英、岩屑组成，长石 20%~30%，石英 40%~50%，岩屑 30%~40%，大小 0.5~2.0mm，分选性好，磨圆度低，偶见小砾石 1.8m

53. 灰绿色粉砂质黏土，手搓有轻微涩感，捻面不光滑，湿时可搓成条，易断，黏粒组分 75%，粉粒组分 25%，以长石、石英为主，局部见小螺残片，局部含薄层中细砂 13.7m

54. 棕黄色粉砂质黏土，黏粒组分 70%，粉粒组分 30%，以长石、石英为主，偶见锈斑 5.2m

55. 黑灰色粉砂质黏土，手搓有轻微涩感，捻面不光滑，湿时不易成条，黏粒组分 70%，粉粒组分 30%，以长石、石英为主，偶见钙质结核。局部夹有 10~20cm 的棕黄色粉砂层 6.5m

56. 上部为棕黄色粉砂质黏土，黏粒组分 70%，粉粒组分 30%，见少量锈斑；下部为棕黄色黏质细砂，黏粒组分 45%，砂粒组分 55%，以长石、石英为主，长石 30%~40%，石英 40%~50%，云母 10%，岩屑 10%，大小 0.10~0.25mm，见少量锈斑 2.1m

57. 杂色砂砾石，结构松散，砂含量 80%左右，颗粒以长石、石英及岩屑为主，长石 20%~30%，石英 40%~50%，岩屑 20%~30%，大小 0.5~2.0mm，分选性好，磨圆较差，砾石含量 20%左右，成分复杂，以花岗岩、辉长岩、闪长岩、石英岩等为主，大小不一，一般 2~4cm，分选性差，磨圆较好，呈次圆状 3.8m

58. 棕黄色粉砂质黏土，手搓有轻微涩感，捻面不光滑，湿时不易成条，黏粒组分 65%，粉粒组分 35%，以长石、石英为主，见大量锈斑，偶见钙质结核，大小 1~2cm 6.1m

59. 上部为灰绿色黏质粉砂，黏粒组分 40%，粉粒组分 60%，以长石、石英为主，长石 30%~40%，石英 40%~50%，云母 10%，岩屑 10%，大小 0.005~0.100mm，见少量锈斑，局部夹有细砂薄层；下部为灰绿色中细砂，结构松散，颗粒物以长石、石英为主，长石 30%~40%，石英 40%~50%，岩屑 10%，大小 0.1~0.5mm，分选性好，磨圆度高 11.45m

60. 灰色粉砂质黏土，手搓有轻微涩感，捻面不光滑，湿时不易成条，黏粒组分 70%，粉粒组分 30%，见少量小螺残片 8.15m

61. 灰白、灰绿色中细砂，结构松散，颗粒物以长石、石英为主，长石 30%~40%，石英 40%~50%，云母 10%，岩屑 10%，大小 0.1~0.5mm，分选性好，磨圆度高，见少量砾石，大小 0.5cm 左右，顶部夹 3cm 厚粉砂质黏土 2.7m

62. 灰色粉砂质黏土，手搓有轻微涩感，捻面不光滑，湿时不易成条，黏粒组分 70%，粉粒组分 30%，见少量小螺残片 3.9m

63. 上部为棕黄色粉砂质黏土，黏粒组分 70%，粉粒组分 30%，偶见炭斑及钙质结核；下部为棕黄色黏质粉砂，黏粒组分 40%，粉粒组分 60%，以长石、石英为主，长石 30%~40%，石英 50%~60%，云母 10%，大小 0.005~0.075mm，见大量锈斑 1.2m

64. 灰白色中砂，结构较松散，颗粒物以长石、石英为主，长石 30%~40%，石英 40%~50%，岩屑 20%，大小 0.3~0.5mm，分选性好，磨圆一般，底部见少量锈斑 0.9m

65. 上部为棕色黏土与黏质粉砂互层，黏土黏粒组分 95%以上，见少量炭斑，层厚 2~3cm，黏质粉砂呈薄层状，厚 0.5cm 左右，与黏土层韵律互层，且具斜层理沉积构造，黏质粉砂层边缘见大量锈斑；中部为灰色粉砂质黏土，黏粒组分 70%，粉粒组分 30%，偶见炭斑；下部为灰色黏质粉砂，黏粒组分 30%，粉粒组分 70%，以长石、石英为主，长石 30%~40%，石英 40%~50%，岩屑 10%，云母 10% 12.5m

66. 棕黄色粉砂质黏土，黏粒组分 60%，粉粒组分 40%，以长石、石英为主，见大量锈斑 0.6m

67. 灰色粉砂质黏土，黏粒组分 65%，粉粒组分 35%，以长石、石英为主，偶见锈斑 1.3m

68. 青灰色粉砂质黏土，黏粒组分 65%，粉粒组分 35%，以长石、石英为主 7.1m

69. 棕黄色砂质黏土，黏粒组分 60%，粉粒组分 40%，以长石、石英为主，大小 0.3~0.5mm，见大量锈斑 1.0m

70. 棕灰色粗砂，结构松散，颗粒物以长石、石英为主，长石 30%~40%，石英 40%~50%，岩屑 20%左右，大小 0.5~2.0mm，分选性好，磨圆一般，风化较强，见大量锈斑 1.0m

71. 上部为棕黄色粉砂质黏土，黏粒组分 70%，粉粒组分 30%，以长石、石英为主，见少量锈斑；中部为灰绿色粉砂质黏土，黏粒组分 65%，粉粒组分 35%，以长石、石英为主，见少量锈斑；下部为灰绿色黏质粉砂，黏粒组分 40%，粉粒组分 60%，以长石、石英为主，长石 30%~40%，石英 40%~50%，岩屑 10%，云母 10%，大小 0.005~0.075mm 7.8m

72. 棕黄色黏质粉砂，黏粒组分 40%，粉粒组分 60%，以长石、石英为主，长石 20%~30%，石英 40%~50%，岩屑 10%~20%，云母 5%~10%，大小 0.005~0.075mm，见大量锈斑 2.1m

73. 灰色粉砂质黏土，黏粒组分 70%，粉粒组分 30%，见少量钙质结核，偶见锈斑 6.2m

74. 棕黄色粉砂质黏土，黏粒组分 70%，粉粒组分 30%，见大量锈斑。局部见有棕黄色黏质粉砂 6.7m

75. 青灰色粉砂质黏土，黏粒组分 70%，粉粒组分 30%，以长石、石英为主，见大量锈斑 5.6m

76. 黑灰色粉砂质黏土，黏粒组分 70%，粉粒组分 30%，见少量炭斑。局部见有 10~30cm 的灰黑色粉砂层 4.0m

77. 杂色砂砾石，结构松散，砂含量 80%以上，颗粒以长石、石英为主，长石 20%~30%，石英 30%~40%，岩屑 30%~40%，大小 0.5~2.0mm，分选性好，磨圆一般，砾石含量小于 20%，成分复杂，以花岗岩、正长岩、闪长岩等为主，大小不一，一般 2~4mm，磨圆度较好。局部见有细砂层 4.0m

78. 灰色粉砂质黏土，手搓有轻微涩感，捻面不光滑，湿时不易成条，黏粒组分 70%，粉粒组分 30%，见大量炭斑，少量锈斑 1.9m

79. 杂色砂砾石，半固结状，砂以中粗砂为主，含量 60%，以长石、石英为主，长石 30%~40%，石英 40%~50%，岩屑 20%，大小 0.5~2.0mm，长石风化较强，砾石含量 40%，成分复杂，以花岗岩、辉长岩、闪长岩、石英岩等为主，大小不一，一般 1~9mm，磨圆度较好 6.4m

80. 上部为棕黄色砂质黏土，黏粒组分 70%，砂粒组分 30%，以长石、石英为主，大小 0.10～0.25mm，见少量锈斑；中部为棕色粗砂，结构松散，颗粒物以长石、石英为主，长石 30%～40%，石英 40%～50%，岩屑 10%～20%，大小 0.5～2.0mm，分选性好，磨圆一般，长石风化较强，偶含小砾石；下部为棕黄色粉砂质黏土，黏粒组分 70%，砂粒组分 30%，见少量锈斑 7.6m

81. 杂色砂砾石，结构松散，砂含量 80%左右，以长石、石英为主，长石 20%～30%，石英 40%～50%，岩屑 20%～30%，大小 0.5～2.0mm，分选性差，磨圆一般，砾石含量 20%，成分复杂，以花岗岩、安山岩、正长岩、石英岩等为主，大小不一，一般 1～8mm，磨圆度较好 5.5m

82. 棕黄色粉砂质黏土，手搓有轻微涩感，捻面不光滑，湿时可搓成条，易断，黏粒组分 75%，粉粒组分 25%，见少量锈斑 2.6m

83. 青灰色粉砂质黏土，手搓有轻微涩感，捻面不光滑，湿时不易成条，黏粒组分 70%，粉粒组分 30%，偶见小螺残片 3.6m

84. 青灰色粉砂质黏土，手搓有轻微涩感，捻面不光滑，湿时不易成条，黏粒组分 70%，粉粒组分 30%，偶见小螺残片，顶部见有 20cm 细砂 4.5m

85. 棕黄色粉砂质黏土，黏粒组分 70%，粉粒组分 30%，偶见钙质结核 0.4m

86. 灰色粉砂质黏土，黏粒组分 70%，粉粒组分 30%，偶见小螺残片。中部夹有 40cm 青灰色黏质中砂 4.8m

87. 灰绿色黏质细砂，黏粒组分 30%，砂粒组分 70%，以长石、石英为主，长石 30%～40%，石英 40%～50%，岩屑 10%，云母 10%，大小 0.10～0.25mm 2.0m

88. 上部为灰绿色黏质粉砂，黏粒组分 40%，粉粒组分 60%，以长石、石英为主，长石 30%～40%，石英 40%～50%，岩屑 10%，云母 10%，大小 0.005～0.075mm，厚度为 1.3m；下部为灰色粉砂质黏土，黏粒组分 70%，粉粒组分 30%，偶见钙质结核 4.4m

89. 上部为棕黄色粉砂质黏土，黏粒组分 70%，粉粒组分 30%，以长石、石英为主，见大量锈斑，顶部 10cm 内裹有中粗砂颗粒；中部为棕黄色砂质黏土，黏粒组分 60%，砂粒组分 40%，以长石、石英为主，大小 0.10～0.25mm，见大量钙质结核，少量炭斑；下部为棕黄色粗砂，半固结状，砂粒以长石、石英、岩屑为主，长石 30%～40%，石英 40%～50%，岩屑 20%，大小 0.5～2mm，分选性差，偶含小砾石，长石风化较强 8.7m

90. 杂色砂砾石，结构松散，砂含量 80%，以长石、石英、岩屑为主，长石 20%～30%，石英 40%～50%，岩屑 20%～30%，大小 0.5～2.0mm，分选性差，磨圆一般，砾石含量 20%，成分复杂，以花岗岩、安山岩、闪长岩、石英岩等为主，大小不一，一般 1～8cm，磨圆较好。364.5～366.7m 处见有灰黄色粉质黏土，黏粒组分 60%，粉粒组分 40%，在 365.25～365.30m，365.55～365.60m 及 365.95～366.00m 见钙质结核层 12.3m

91. 上部为灰黄色粉质黏土，粒组分 60%，粉粒组分 40%；中部为薄层的中细砂；下部为灰黑色粉砂质黏土，黏粒组分 60%，粉粒组分 40%，以长石、石英为主，见大量锈斑 6.0m

92. 棕黄色砂层，半固结状，砂粒以长石、石英、岩屑为主，长石 30%～40%，石英 40%～50%，岩屑 20%，大小 0.5～2.0mm，分选性差，偶含小砾石，长石风化较强，该层位局部粒度有变化 2.0m

93. 棕灰色粉砂质黏土。黏粒组分 70%，粉粒组分 30%，以长石、石英为主，上部见大量锈斑，底部见少量炭斑。局部夹有薄层中砂 6.3m

94. 杂色粗砂，结构松散。颗粒物以长石、石英为主，长石 20%～30%，石英 40%～50%，岩屑 20%～30%，大小 0.5～2.0mm，分选性差，磨圆一般，偶含小砾石，大小 1cm 左右。局部夹有灰绿色博层状粉砂质黏土 1.3m

95. 灰绿色粗砂，结构松散，局部呈固结状，似砂岩，颗粒以长石、石英为主，长石 20%～30%，石英 40%～50%，岩屑 20%～30%，大小 0.5～2.0mm，分选性差，磨圆一般，在 395.5～395.6m 处夹棕色粉砂质黏土薄层 9.1m

96. 棕黄色粉砂质黏土黏粒组分 70%，粉粒组分 30%，见大量锈斑，偶见炭斑 5.3m

97. 上部为杂色砾石，结构松散，砾石成分复杂，以花岗岩、安山岩、石英岩、闪长岩、砂岩等为主，大小不一，一般 2~9cm，分选性差，磨圆较好。下部为灰黄色粉砂质黏土，黏粒组分 70%，粉粒组分 30% 6.3m
98. 灰色粉砂质黏土，黏粒组分 70%，粉粒组分 30%，见少量植物根系及小螺残片 4.1m
99. 灰色粉砂质黏土，黏粒组分 70%，粉粒组分 30%，见少量植物根系及小螺残片，在 419.2~419.3m 及 419.5~419.6m 处夹灰黄色粉砂层 8.3m
100. 灰色粉砂质黏土，黏粒组分 70%，粉粒组分 30%，见大量小螺残片 10.9m
101. 灰黄色粉砂质黏土，黏粒组分 70%，粉粒组分 30%，见少量锈斑 4.3m
102. 灰色粉砂质黏土，黏粒组分 70%，粉粒组分 30%，见大量小螺残片 6.2m
103. 杂色砾石，结构松散，砾石成分复杂，以安山岩、花岗岩、闪长岩、石英岩等为主，大小不一，一般 1~4cm，分选性差，磨圆较好。中部见有约 0.9m 厚的灰黄色粉砂质黏土，黏粒组分 70%，粉粒组分 30%，见少量锈斑 3.1m
104. 灰黄色、灰色粉砂质黏土，黏粒组分 75%以上，粉粒组分 25%以下，见少量小螺残片 9.1m
105. 棕黄色粉砂质黏土，黏粒组分 70%，粉粒组分 30%，见大量锈斑 2.9m
106. 灰色粉砂质黏土黏粒组分 70%，粉粒组分 30%，以长石、石英为主，偶见锈斑。 12.7m
107. 杂色砾石，结构松散，砾石成分复杂，以安山岩、花岗岩、闪长岩、石英岩等为主，大小不一，一般 1~4cm，分选性差，磨圆较好 1.5m
108. 黑灰色粉砂质黏土，手搓有轻微涩感，捻面不光滑，湿时不易成条，黏粒组分 70%，粉粒组分 30%，见少量锈斑 1.7m
109. 灰色粉砂质黏土，黏粒组分 70%，粉粒组分 30%，见大量小螺残片 12.8m
110. 棕黄色、黄绿色黏质粉砂，黏粒组分 45%，粉粒组分 55%，以长石、石英为主，长石 30%~40%，石英 40%~50%，岩屑 10%~20%，大小 0.005~0.075mm，见少量锈斑及钙质结核 4.8m

--------------- 平行不整合 ---------------

下伏地层：上新统棕黄色、黄绿色黏质粉砂，杂色砾石，结构松散，砾石成分复杂，以安山岩、花岗岩、石英岩、闪长岩等为主，大小不一，一般 1~6cm，分选性差，磨圆较好。

2. 岩性岩相变化

马池口凹陷东部以 ZK18 孔、东南部以新 3 孔、西部以新 1 孔、北部以 ZK3 孔地层为代表，凹陷内岩性岩相变化特征如下（图 37-2）。

早更新世，东部总沉积厚度 127.61m 分为上、下两段。下段岩性为粉砂、粉砂质黏土、中砂、细砂、粗砂、中粗砂、砂质黏土、粉细砂、黏土互层沉积，形成多个沉积旋回。沉积物颜色以褐黄色、灰绿色、灰褐色、棕黄色为主，以滨湖相、浅湖相、深湖相沉积为主。上段发育砂砾石、粉砂、粉砂质黏土、细砂、粉砂、黏土、粉砂质黏土与粉砂互层、砂质黏土，沉积物颜色以褐黄色、灰绿色、黄绿色为主，以滨湖相、浅湖相、水下河道相沉积为主。东南部总沉积厚度为 94m，分为上、下两段。下段为粉砂、粉砂质亚黏土、粉质黏土互层沉积，沉积物颜色以黄橙色、黄棕色、淡黄色、淡灰色、橄榄灰色为主，以河漫滩相沉积为主；上段为粗砂、卵石、中砂、细砂、粉砂质亚黏土、粉砂、粉质黏土沉积，混泥砂砾卵石、粉砂质亚黏土，沉积物颜色以暗棕色、黄棕色、浊棕色、黑棕色为主，河道边滩相、河漫滩相沉积，短期为洪积扇扇中相、河床相沉积。西部地区总沉积厚度为 71.42m，分为上、下两段。下段岩性为砂卵石、粉砂质黏土、砂卵石、粉砂质黏土、粉质黏土、粉砂质黏土沉积，沉积旋回明显，沉积物颜色以杂色、暗棕色、黄棕色、棕色为主，以河漫滩相、河床相沉积为主；上段发育泥包卵石、含砾砂卵石、砂卵石、砂砾、粉砂沉积，沉积物颜色以暗棕色、黄棕色、浊棕色为主，以河道边滩相、河床相、冲洪积扇中相沉积为主（图 37-2）。

中更新世，东部总沉积厚度 56.05m，岩性自下而上发育有粗砂、砂砾石、细砂、粉砂质黏土、黏土沉积，颜色以灰褐色、黄绿色、褐黄色为主，以河床相、河道边滩相、河漫滩相、牛轭湖相沉积为主。东南部总沉积厚度 53.53m，下部岩性为粉砂质亚黏土、粉质黏土、细砂、粉砂质亚黏土与粉质黏

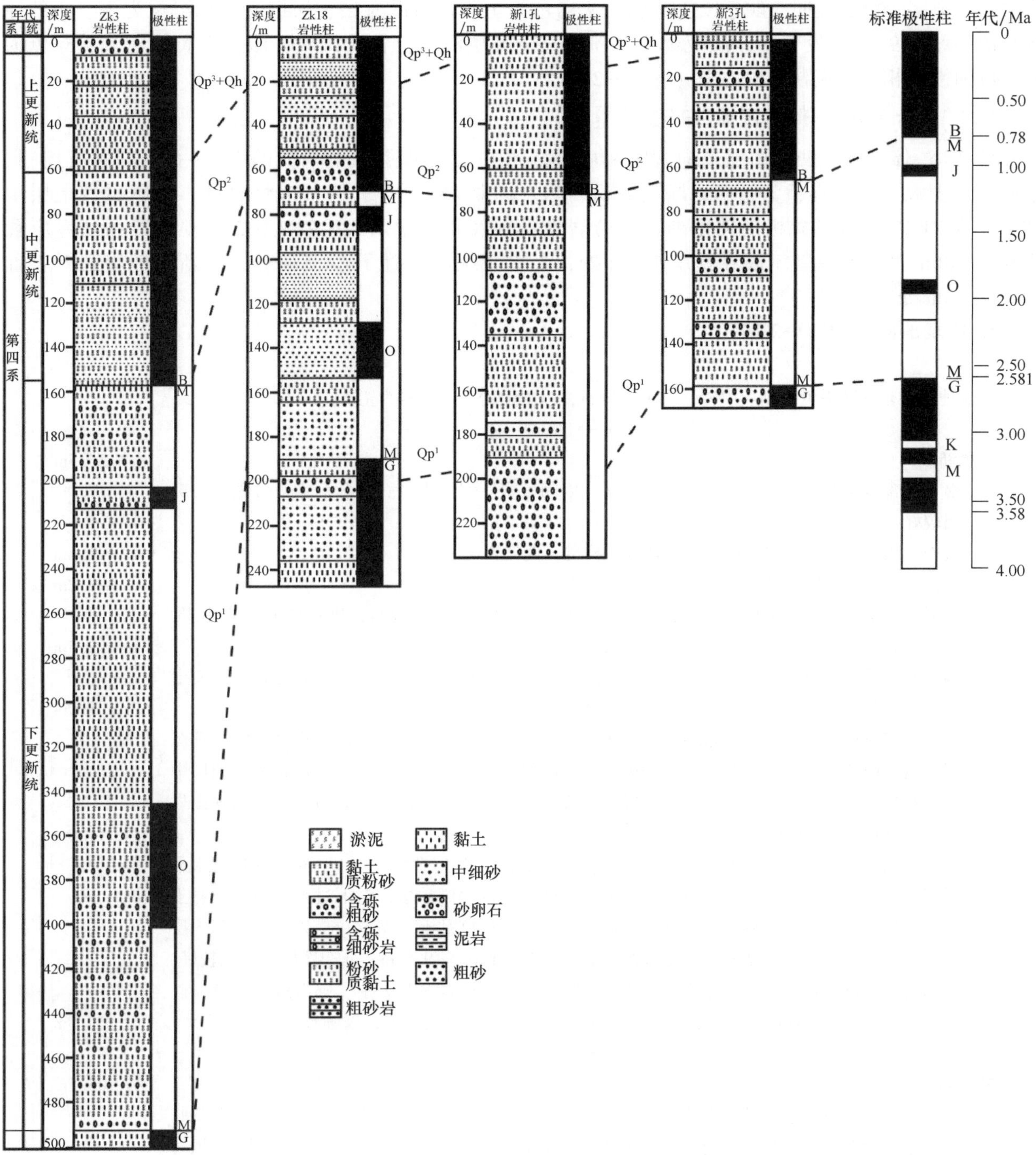

图 37-2　马池口沉积凹陷典型钻孔柱状对比图

Fig. 37-2　Columnar correlation of the typical holes in Machikou sedimentary depression

土互层沉积，上部岩性为砂砾、中砂、细砂、粉砂质亚黏土、粉质黏土、粉砂沉积。颜色以黄棕色、暗棕色、浊棕色为主，以河道边滩相、河漫滩相沉积为主。西部总沉积厚度 60.94m，岩性为粉砂、粉砂质亚黏土、粉砂质黏土、砂砾互层沉积，沉积物颜色以黄棕色、浊棕色、灰橄榄色、黑橄榄色为主，总体以浅湖相沉积为主。北部地区总沉积厚度为 96.7m，岩性为粗砂、粉砂质黏土、细砂、粉砂、砂砾石、砂质黏土、中粗砂、中砂、中细砂及黏土互层沉积，形成多个沉积旋回。颜色以棕黄色、灰绿色、黑灰色、灰黑色、灰色、杂色为主。以河道边滩相、河漫滩相沉积、河床相沉积为主（图 37-2）。

晚更新世，地层横向变化相对较小。东部地区总沉积厚度 9.15m，岩性以粉砂、粉砂质黏土互层为主，形成多个沉积旋回。颜色以褐黄色、灰褐色为主。以河漫滩相、河道边滩相沉积为主。西部地区总沉积厚度9.2m，发育粉砂质亚黏土、粉质黏土互层沉积，含有机质，颜色以黄棕色、灰棕色为主，以河漫滩相、牛轭湖沉积为主（图37-2）。

（二）后沙峪-倴伯沉积凹陷（Ⅱ）

后沙峪-倴伯沉积凹陷由北东向黄庄-高丽营断裂、顺义断裂及北西向的南口-孙河断裂所控制，基底为侏罗系及古生界，为第四纪断陷沉积凹陷。选取ZK12-2孔为代表性钻孔，根据磁性地层学结果建立ZK12-2孔的磁极倒转序列，结合岩性特征及年代学研究，ZK12-2钻孔第四系底界应置于孔深706m处，中更新统底界置于170m处，上更新统底界置于66.5m处，全新统地层底界应置于孔深10m处（图37-3）。

1. 代表剖面（ZK12-2钻孔）

上覆地层：全新统褐黄色黏质中砂，含少量砾石，以安山岩、闪长岩为主。

———— 整　合 ————

上更新统	总厚56.5m
4. 褐灰色含砂砾石砂质黏土，砾石含量40%，以闪长岩、安山岩、石英脉为主，分选差，磨圆较好	7.0m
5. 灰褐色黏质粉砂，地层中多具斜层理沉积构造，偶见炭斑	2.8m
6. 褐色淤泥层，有较强臭味，湿时呈“果冻”状	5.2m
7. 黑褐色黏土，黏粒95%以上，地层中含少量灰色粉砂团块及小砾石，砾石成分以安山岩、石英脉岩居多，大小0.5~1.0cm，磨圆较好	2.5m
8. 褐色砂砾石层，结构松散，砂以中砂为主，主要矿物为石英长石。砾石成为复杂，以安山岩、石英脉岩、花岗岩为主，大小不一，分选性差，磨圆较好	8.9m
9. 上部为褐黄色粉砂夹薄层褐色黏土，粉砂结构松散。局部见有锈色条纹，条纹具斜层理沉积构造；中部为黑褐色黏土，黏粒95%以上，局部见灰白色砂质黏土团块及条带，地层中见少量炭斑；下部为褐色砂砾石层，结构松散，砂以中砂为主，砾石成为复杂，以安山岩为主，分选性差，磨圆较好	11.0m
10. 上部为褐黄色粉砂夹薄层褐色粉砂质黏土，粉砂结构松散，层厚0.5~1.0cm，多具近水平层理，地层上部见斜层理沉积构造，在49.7m处夹2cm厚细砂层，向下过度为粉砂质黏土局部可见“月牙状”粉砂团块及薄层粉砂；中部为黑灰色黏土，黏粒95%以上，局部地层中含小砾石，砾石成分复杂，以安山岩为主，大小一般0.3~0.5cm，分选性好，磨圆较好；下部为褐黄色砂砾石层。砂为细砂，结构松散，砂粒以长石石英为主。砾石成分复杂，以安山岩、角闪岩、花岗岩、石英脉岩等为主，大小不一，分选性差，磨圆较好	19.1m

———— 整　合 ————

中更新统	总厚103.5m
11. 褐黄色黏土与中细砂互层，地层含少量砾石，砾石成分复杂，以安山岩、闪长岩、石英脉岩为主，分选性差，磨圆较好。褐黄色黏土，黏粒95%以上，局部见“月牙状”沉积构造。褐色中细砂，结构松散，主要矿物为长石石英，含少量的云母	8.1m
12. 褐色黏土，黏粒95%以上，地层底部含大量钙质结核，大小0.5~5.0mm	3.4m
13. 灰绿色粉砂质黏土与黏质粉砂互层，地层中见少量斜层理沉积构造。偶见腹足类残片。底部见有薄层的褐黄色粉细砂层	14.3m
14. 褐色砂砾石层，结构松散，砂以中砂为主。砾石成分复杂，以安山岩、闪长岩、石英脉岩等为主，大小不一，分选性差，磨圆较好	15.1m
15. 褐黄色黏质粉砂，局部见大量锈斑	3.4m
16. 褐黄色中细砂，结构松散	4.2m
17. 褐色粉砂质黏土，地层中包裹黏土，粉砂质黏土中见大量锈斑	9.0m
18. 褐色粗砂，结构松散。地层中夹杂少量砾石，砾石成分以安山岩、石英脉岩居多，磨圆较好	7.5m

19. 灰褐色粉砂质黏土，地层底部夹粉细砂团。夹中砂团，中砂中见大量腹足类残片。局部见少量炭斑，含少量小砾石 12.5m

20. 褐色中细砂，结构松散 7.0m

21. 褐色砂砾石层，结构松散，砾石成分复杂，以安山岩、角闪岩、石英脉岩等为主，大小不一，分选性差，磨圆度好 2.3m

22. 褐色细砂，结构松散 4.0m

23. 黑褐色黏土，地层中见少量腹足类残片，偶见炭斑 7.7m

24. 黑褐色细砂，结构松散，底部见斜层理沉积构造 3.2m

25. 灰绿色粉砂质黏土 1.8m

———— 整 合 ————

下更新统 总厚 536.0m

26. 灰白色粗砂，结构松散。底部粒度变细，发育斜层理沉积构造 8.0m

27. 褐色粉砂质黏土与黏质粉砂互层，地层中见少量锈斑。地层中见大量黏土条纹，多具斜层理沉积构造 21.5m

28. 褐色粉砂，结构松散，局部见锈斑 3.4m

29. 褐色中粗砂，结构松散 8.3m

30. 褐色粉砂质黏土与黏质粉砂互层，夹薄层粉砂质黏土条带，粉砂质黏土条带厚 3~5mm，具斜层理沉积构造 8.6m

31. 灰白色中粗砂，结构松散。底层 1m 为黏质粗砂，风化较强 8.4m

32. 灰白色细砂，结构松散。局部夹有黑褐色黏土和灰绿色黏土 13.3m

33. 褐色粉砂质黏土与黏质粉砂韵律互层，地层多具斜层理沉积构造，各层厚 3mm 左右。地层中见大量腹足类动物残片 9.3m

34. 灰褐色细砂，结构松散，偶见钙质结核 2.4m

35. 褐色黏质粉砂，地层具斜层理沉积构造，局部包裹粉细砂团块，地层中见少量钙质结核，大小 0.5~5mm 6.6m

36. 浅灰白色细中砂，结构松散，局部见少量黑褐色磁铁矿富集 2.0m

37. 灰绿色粉砂质黏土，局部见少量炭斑 3.1m

38. 褐色砂砾石层。砂为中粗砂，结构松散，砾石成分复杂，以安山岩、正长斑岩等为主，分选较好，磨圆较差，呈次棱角状 3.7m

39. 浅灰色粉砂与粉砂质黏土互层，偶见锈斑，新鲜面上见少量的腹足类残片 25.6m

40. 灰白色含砾粗砂，砾石主要成分为花岗岩，斑岩，磨圆较好 7.8m

41. 灰黑色粉砂质黏土，含少量腹足类残片 2.0m

42. 灰白色砂砾石层，砂以粗砂为主，砾石成分为花岗岩、斑岩等，磨圆较好 12.6m

43. 浅黄色中粗砂，砾石成分为花岗岩，磨圆较好，偶见云母 4.4m

44. 浅黄灰白色含砾中粗砂，砂为中粗砂，局部粒度较细，主要成分为长石、石英。砾成分为花岗岩、石英脉、斑岩等，磨圆较好，呈次棱角状—次圆状 33.0m

45. 浅灰色粉砂与浅灰色黏质粉砂互层，发育斜层理，见有大量锈斑 8.4m

46. 灰白色砂砾石层，砂以粗砂为主，砾石成分为花岗岩、花岗斑岩、安山岩、泥岩等，磨圆较好 19.6m

47. 黄褐色粉砂与浅灰色黏质粉砂互层，呈微层理构造，含大量锈斑。粉砂中发育清晰的交错层理 14.0m

48. 上部为灰白色含砾粗砂，砾石成分以花岗岩为主，分选差，磨圆一般；下部为灰褐色粉砂质黏土，偶见炭斑 11.0m

49. 上部为黄褐色粉砂质黏土，偶见炭斑；下部为灰白色粗砂，松散，主要成分为长石石英，含少量的云母 20.0m

50. 灰绿色黏土与褐黄色黏土互层，含大量锈斑及腹足类残片 4.0m

51. 灰白色含砾粗砂，砾以花岗岩、砂岩、斑岩为主，磨圆较好。底部为棕黄色黏土，含大量炭斑 6.0m

52.	黑灰色粉砂质黏土，含腹足类残片	10.0m
53.	灰白色含砾粗砂，层间夹两层黏土条带	4.0m
54.	黑灰色粉砂质黏土与黏质粉砂互层，地层中偶见炭斑，含腹足类残片及钙质结核	5.9m
55.	棕黄色粉砂与浅黄色粉砂质黏土，含大量锈斑；中间夹有薄层细砂，发育斜层理	5.1m
56.	灰白色砂砾石层，砂为粗砂，砾石成分复杂，为花岗岩、斑岩、安山岩、石英砂岩等，磨圆较好	11.0m
57.	浅黄色粉砂与黏质粉砂互层，含大量锈斑，夹细砂层	19.0m
58.	灰白色含砾粗砂，砾石成分复杂，花岗岩、安山岩、石英岩、斑岩等，分选一般，本层502.7~502.8m处见宽约3cm左右的黑色条带，砂及砾石均被碳化，染手，局部见有斜层理	13.2m
59.	黄褐色含砾粗砂，剖面上部含1.2m厚的黄褐色粉砂质黏土。夹有薄层棕黄色粉砂质黏土，底部见大量炭斑。含大量铁质结核，大小1~2mm，结核外部为碳质包裹	35.9m
60.	褐色砂砾石层，结构松散，砂为粗砂为主，砾石成分复杂，以安山岩、石英脉、花岗岩等为主。584.6~585.6m砂砾石外部多见碳化	51.5m
61.	棕黄色砂质黏土，偶见砂粒。新鲜面上见植物根系	5.3m
62.	棕红色砂砾石层，半固结，砾石成分以花岗岩、安山岩、页岩等为主。次棱角—次圆状。局部较破碎，新鲜面上见高岭石化现象。中部夹有厚约4m的棕红色黏土，含大量锈斑	24.1m
63.	灰褐色砂砾石层，本层局部见半固结得地层，钙质胶结，风化严重，高岭石化、绿泥石化明显	7.0m
64.	棕红色砂质黏土，中间夹薄层黏土条带，偶见炭斑	4.3m
65.	灰褐、棕红色砂砾石层（风化壳），局部见强烈的高岭石化现象	72.7m

--------------- 平行不整合 ---------------

下伏地层：上新统棕红色半固结状泥岩。

2. 岩性岩相变化

后沙峪–俸伯凹陷第四系岩性特征西部以ZK19孔、中部以顺5孔、东部以ZK12–2孔、北部以新2孔、HR88–1孔为代表。

早更新世，西部和东部有2个沉积中心。东部沉积中心厚度约536m，分为上、下两段。下段为厚层砂砾石及砾石与薄层黏土、砂质黏土、细砂、粗砂、中砂、粉砂质黏土互层沉积。颜色以棕红色、褐色、褐黄色、棕黄色、灰白色为主。为河床相沉积，短期为为河道边滩相、河漫滩相、牛轭湖相沉积。上段为薄层砂质黏土、黏土、粉砂质黏土、细砂、粗砂、含砾粗砂、中砂、粉砂、黏质粉砂、含砾细砂、中粗砂、砂砾石互层沉积。颜色以褐黄色、黄褐色、灰白色、浅灰色、灰黑色、浅黄色、棕黄色为主。为滨湖相、浅湖相沉积，水下河道相沉积。西部沉积中心厚约302m，总体上为一套河流–湖泊相的砂砾石、中粗砂、黏质粉砂、粉砂质黏土、黏土、淤泥质黏土等沉积。沉积物总体颗粒较细，粉砂成分含量较高，局部可见二元结构，颜色以灰绿色、黄绿色、灰黑色、褐黄色为主，沉积物中偶见钙质结核。下段以河床相、河道边滩相、河漫滩相沉积为主，上段以滨湖、浅湖相、深湖相、水下河道相沉积为主。此外，北部沉积厚度64m，主体为河道边滩相沉积为主（图37–3图例同37–2，以下同）。

中更新世，沉积中心延续早更新世沉积面貌，位于凹陷东部，总沉积厚度145m，分为上、下两段。下段为粉砂质黏土、粉砂、砂质黏土、细砂、黏土、细砂、中砂、砂砾石、粗砂互层沉积，形成多个沉积旋回。颜色以褐色、褐黄色、灰褐色、灰绿色为主。为滨湖相、浅湖相、深湖相、水下河道相沉积。上段自下而上发育砂砾石与细砂互层、粉砂与砂质黏土、黏土、粉砂质黏土互层沉积，形成四个沉积旋回，颜色以褐黄色、褐色、灰绿色、灰黑色、黑褐色为主。为河床相、河道边滩相、河漫滩相、牛轭湖相沉积。西部地区沉积物颜色主要为灰绿色、浅灰黄色、褐灰色，为一套河湖相沉积，沉积物的颗粒总体呈现粗细变化的韵律特征，并见有钙质结核及螺类化石碎片，属水动力较弱河流相及湖沼边缘相（图37–3）。

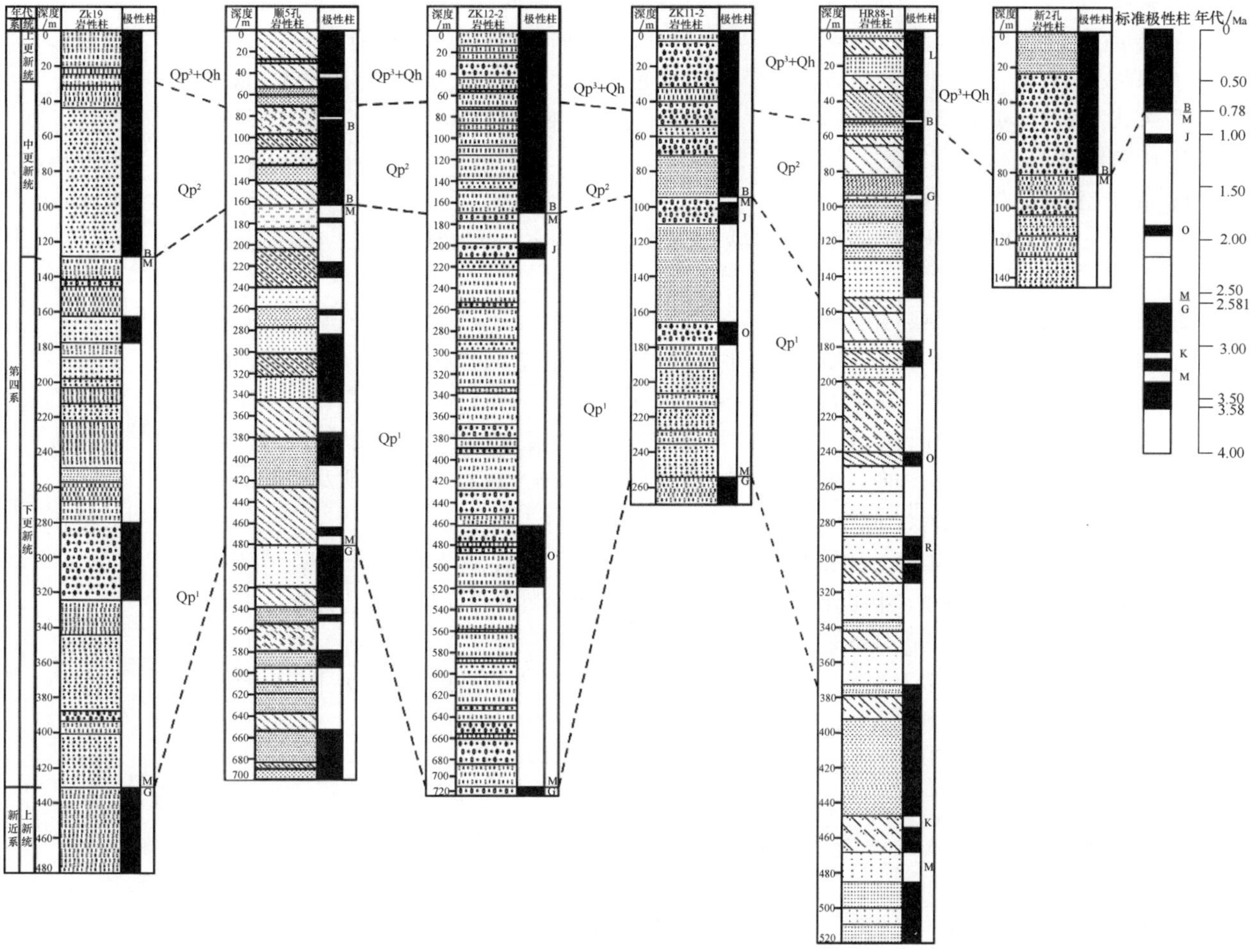

图 37-3　后沙峪-俸伯沉积凹陷典型钻孔柱状对比图

Fig. 37-3　Columnar correlation of the typical holes in Houshayu-Fengbo sedimentary depression

晚更新世，沉积中心位于东北部地区，总沉积厚度 23.7m，岩性自下而上发育淤泥、黏土与粉砂、粉砂质黏土互层沉积，砾石与黏土互层、中砂与含砂砾石砂质黏土、细砂互层、粉砂质黏土、砂质黏土、黏土、砂质黏土、粉砂沉积形成两个沉积旋回，颜色以灰褐色、黑褐色、褐色、褐黄色为主。为河床相、河道边滩相、河漫滩相、牛轭湖相沉积。中部总沉积厚度 14.85m，岩性总体上以黄褐色、褐黄色粉砂质黏土、黏质粉砂、粉砂为主，组成几个向上变细的韵律层，表明当时水动力环境是有规律的变化，总体表现为河流相与浅湖沼相交替出现（图 37-3）。

（三）甘棠沉积凹陷（Ⅲ）

1. 代表剖面（ZKTX-1 钻孔）

以通县双埠头 ZKTX-1 钻孔（据 1∶50000 通县幅，1991）为代表，描述如下。

上覆地层：全新统灰白色粗中砂。

———— 整　合 ————

上更新统　　总厚 58.17m

5. 灰黑色淤泥，具水平纹层，发育虫孔和炭屑　　1.47m
6. 灰褐色、黄灰色砂质黏土夹黄灰色粉砂，有机质含量高，底部见钙质结核，粉砂层中具铁锈斑　　4.69m
7. 黄褐、浅黄绿、褐灰色粉砂、中砂、含砾粗砂夹砾石层，具水平层理，见铁锈染，砂以石英、长石为主，分选磨圆不好，砾石成分为泥质团块，火山岩等，砾径约 1cm。底部见螺类化石：*Bithynia fuchsiana*（Moellendorf）、*Gyraulus chihliensis*（Ping）　　7.52m

8. 褐灰、浅绿灰、暗绿灰色淤泥、粉砂层，淤泥层有机质含量高，具团粒结构，砂层中云母含量高。本层含螺类化石：*Galba pervia*（Martens）、*Gyraulus chihliensis*（Ping） 5. 32m
9. 绿灰、灰色中、粗砂，砂以长石、石英为主，暗色矿物含量较高，分选、磨圆中等。见螺类化石：*Gyraulus chihliensis*（Ping）、*Hippeutis* spp.。本层热释光测年结果（9. 59 ± 0. 73）$\times 10^4$a 1. 83m
10. 暗灰色淤泥，有机质含量高 3. 46m
11. 黄灰、灰色含砾粗砂夹灰色砂质黏土，砂以石英、长石为主，分选磨圆不好，砾石成分以火山岩为主，砾径 2~20mm 1. 80m
12. 暗灰色、浅褐灰色淤泥，含有机质，见钙质沉淀和锰结核 3. 69m
13. 暗灰、灰色砂质黏土，含有机质，具水平纹层。含螺类化石：*Galba pervia*（Martens）、*Gyraulus chihliensis*（Ping）、*Gyraulus* spp.、*Succinea hopeiensis*（Ping） 7. 86m
14. 浅绿灰、浅灰、灰色中、粗砂及含砾粗砂，砂以石英、长石为主，分选磨圆不好，砾石以火山岩为主，粒径 2~10mm。本层夹暗灰色砂质黏土，具水平层纹，含有机质及植物碎片 4. 30m
15. 暗褐灰色淤泥，有机质含量较高 2. 85m
16. 灰色中砂、浅绿灰色粉砂，具水平层理，砂以石英、长石为主，含岩屑，分选磨圆不好 2. 57m
17. 深灰、暗灰色黏质砂土、中粗砂。具水平纹层，见碳化的植物碎片和软体动物化石碎片。砂以石英、长石为主，分选不好 2. 50m
18. 灰色砂质黏土、黏质砂土及绿灰色、灰白色细、中、粗砂。砂以石英、长石为主，含岩屑，分选中等，具水平层理，云母含量高。见化石碎片 8. 31m

——————— 整　　合 ———————

中更新统 总厚 104. 07m

19. 绿灰色黏土、浅黄灰色中砂，顶部具团粒结构，见虫孔及钙质结核和螺类化石：*Gyraulus* spp. 3. 56m
20. 顶部绿灰色砂质黏土，中部黄灰、绿灰色中细砂，下部黄灰色含砾粗砂，砂质成分以石英、长石为主，分选磨圆不好，粒径 2~5mm，成分以火山岩为主 4. 65m
21. 黄灰、灰、浅绿灰色砂质黏土、黏质砂土，具水平层纹，下部富集螺类化石：*Radix lagotis*（Schrank）、*Radix plicatula*（Benson）、*Galba pervia*（Martens）、*Gyraulus chihliensis*（Ping）、*Gyraulus compressus*（Hutton）、*Succinea* spp. 3. 44m
22. 黄灰色含砾粗砂及中砂。砂以石英、长石为主，暗色矿物较多 5. 96m
23. 黄灰、黄褐、灰褐、褐色砂质黏土，见铁染现象、钙质结核，含有机质及螺类化石：*Galba pervia*（Martens）、*Gyraulus chihliensis*（Ping）、*Gyraulus compressus*（Hutton）、*Vallonia tenera*（Reinhardt），本层热释光测年为（28. 12±2. 01）$\times 10^4$a 9. 78m
24. 暗灰色淤泥，具团粒结构，有机质含量高，见钙质结核。本层含螺类化石：*Gyraulus chihliensis*（Ping）、*Vallonia Costata*（Müller） 4. 47m
25. 顶部为浅黄灰色黏质砂土，中部为浅绿灰黄灰色中细砂，见水平层理。下部为粗砂细砾层，砾径约 1cm，见螺类化石：*Parafossarulus striatulus*（Benson）、*Semisulcospira cancellata*（Benson）、*Radix swinhoei*（H. Adams）、*Gyraulus chihliensis*（Ping）、*Hippeutis* spp.；双壳类化石：*Unio* spp.、*Lamprotula* spp. 9. 53m
26. 上部灰色黏土，具团粒结构，含钙质结核及虫孔，见软体动物化石碎片。中部为黄褐色、暗灰色砂质黏土，具团粒结构，见钙质结核及虫孔，具锈斑。下部为黄灰色含砾粗砂，砂质成分以石英、长石为主，含岩屑，砾石粒径约 1cm 大小 8. 16m
27. 灰色砂质黏土，粗、中、细砂，砂以石英、长石为主，含砾石及钙质结核 4. 89m
28. 上部为暗灰色砂质黏土，有机质含量较高，含小钙质结核，富含螺类化石：*Radix plicatula*（Benson）、*Gyraulus* spp.，中部为灰色粉砂，云母含量较高，为本区第二海相层，含海相介形虫：*Sanyuania psaronius* Huang，下部为黄灰色中粗砂、砾石层及暗灰色淤泥，见螺类化石及螺类口盖化石：*Bithynia* spp.，*Gyraulus* spp.，*Operculum* 5. 32m

29. 上部为黄褐色砂质黏土与棕黄色黏土互层，下部为暗灰、褐灰、灰色砂质黏土，含钙质结核及螺类化石：*Bithynia fuchsiana*（Moellendorff）、*Galba elegans*（Ping）、*G. pertens*（Martens）、*Gyraulus chiliensis*（Ping）、*Gastrocopta*（*Sinalbimula*）*Subfusiformis* Guo。本层上部热释光测年结果为（60.12±4.33）$\times 10^4$a　　14.66m

30. 灰色中砂、细砂，以石英、长石为主，分选磨圆不好，含岩屑和钙质结核，产丰富的螺类化石：*Bithynia fuchsiana*（Meollendorff）、*B. misella*（Gredler）、*Parafossarulus striatulus*（Benson）、*Radix ovata*（Drapareaud）、*R. swinhoei*（H. Adams）、*Gyraulus chihliensis*（Ping）、*H. minutus*（Ping）、*Cathaica* spp.、*Vallonia costata*（Müller）　　1.65m

31. 灰色黏质砂土夹黄褐色黏质砂土，具水平层理，含铁锈染条带及钙质结核　　8.58m

32. 灰、黄灰色黏质砂土夹灰黄色细砂，见斑状及条带状铁锈染色，发育铁质结核，产软体动物化石：*Bithynia* spp.、*Stenothyra glabra*（A. Adams）、*Gyraulus chihliensis*（Ping）、*Operculum*、*Unio* spp.　　1.68m

33. 黄灰色、浅灰色中、细砂，砂质成分以石英、长石为主，分选磨圆中等　　3.52m

34. 黄灰色、灰褐色黏质砂土夹灰褐色粉砂，锈斑发育，见虫孔及钙质结核　　1.68m

35. 褐色黏土，褐灰、灰、暗灰色砂质黏土、黏质砂土，具团粒结构和水平层理，见虫孔，含钙质结核、锈斑，产螺类化石：*Gyraulus chihliensis*（Ping）　　5.16m

36. 灰色粉砂，云母含量高，具虫孔呈锈色，见化石碎片　　1.55m

37. 灰色黏质砂土、细砂、中粗砂，含钙质结核及炭屑，见水平层纹及化石碎片　　5.83m

———— 整　合 ————

下更新统　　总厚 125.59m

38. 上部暗灰、灰绿色砂质黏土，含有机质、铁质结核、钙质结核：下部黄灰、灰、褐色黏质砂土夹浅黄灰色中砂，具水平层纹及铁染　　7.95m

39. 顶部褐灰色黏土，虫孔发育具锈斑及钙质沉淀。中下部为暗褐灰、灰色砂质黏土，结构致密，含钙质结核。本组含软体化石：*Bithynia fuchsiana*（Moellendorff）、*Parafossarulus striatulus*（Benson）、*Radix* spp.、*Gyraulus chihliensis*（Ping）、*Operculum*　　4.75m

40. 暗灰、褐灰色黏质砂土、砂质黏土夹粉砂层，云母含量多，见水平层理　　2.88m

41. 灰、黄灰色中细砂、粗砂，具水平层理，以石英、长石为主，分选磨圆不好　　2.85m

42. 上部暗灰色砂质黏土，见铁锰结核，中部暗灰色细砂．具水平层理，具铁染；下部灰、黄灰色中、粗砂夹暗灰色砂质黏土，成分以石英、长石为主，分选磨圆不好　　3.73m

43. 上部暗灰色、褐灰色砂质黏土，富含动物化石，顶部见姜石；下部黄灰色细砂，分选磨圆不好，砂以石英、长石为主，本层软体化石为：*Parafossarulus striatulus*（Benson）、*Gyraulus chihliensis*（Ping）、*Lampratula teai*（Gray）、*Lamprotula* spp.　　1.97m

44. 褐灰、灰褐、灰色砂质黏土、黏质砂土，具锈斑，中部夹绿黄色中砂，底部为浅灰色含砾粗砂　　2.70m

45. 锈黄、灰色中砂、中细砂与暗灰色、褐灰色黏土、砂质黏土互层。黏土层中发育铁锈斑及水平层理；砂层中具水平层理，见螺类化石：*Gyraulus chihliemis*（Ping）　　3.88m

46. 上部灰色黏质砂土，具锈斑，下部绿灰色中砂。以石英、长石为主，含岩屑，分选磨圆中等　　2.68m

47. 上部暗灰色砂质黏土，下部浅灰色含砾中砂，砂以石英、长石为主，分选磨圆较好，含炭屑。本层见蚌类化石及其碎片：*Unio* spp.　　2.10m

48. 浅灰色中砂、中粗砂，以石英、长石为主，分选磨圆不好，见蚌类碎片。本层中夹暗灰色砂质黏土，具锈斑　　3.42m

49. 浅灰色细砂、含砾粗砂，砂以石英、长石为主，含炭屑较多，分选磨圆不好，砾石成分主要为泥质灰岩，砾径约 5mm 大小，见螺类化石：*Gyraulus chihliensis*（Ping）　　3.40m

50. 灰褐、棕褐色砂质黏土，夹黄褐色粉砂，含钙质结核。见螺类化石：*Bellamya* spp.、*Parafossarulus striatulus*（Benson）、*Stenothyra glabra*（A. Adams）、*Galba pertens*、*Gyraulus chihliensis*（Ping）、*Sucine* spp.　　2.90m

51. 绿灰色黏土，含钙质结核与软体动物化石碎片　　3.02m

52. 灰褐、黄绿色、土黄色黏质砂土，具锈斑，铁锰结核呈豆状富集 4.26m

53. 圆棕红、黄褐色黏土、砂质黏土，顶部团粒结构，具虫孔并富含钙质结核 1.87m

54. 棕黄色粉砂，分选好 1.08m

55. 棕褐、棕、黄棕色砂质黏土，团粒结构，含钙质结核，顶部见水平纹层 5.37m

56. 上部棕褐、灰黄色黏质砂土，下部土黄色粉砂，具水平层理，铁染发育，含钙质结核 3.78m

57. 棕灰色、暗灰色黏质砂土、砂质黏土，具水平层理，含钙质结核及软体化石 *Bellamya* spp.、*Ggraulus chihliensis* (Ping)、*Andonta* spp. 6.90m

58. 灰、褐灰色砂质黏土夹灰色细砂，团粒结构，含钙质结核 3.07m

59. 灰黄、浅灰、灰绿色黏质砂土，见铁染，富含螺类化石：*Parafossarulus striatulus* (Benson)、*Radix lagotis* (Schrank)、*Radix plicatula* (Benson)、*Radix swinhoei* (H. Adams)、*Gyraulus chihliensis* (Ping) 4.16m

60. 灰黄色细砂、粉砂，具水平层理 3.29m

61. 灰黄、灰、黄灰色中砂，水平层理发育，具锈斑 5.67m

62. 暗灰、深灰、褐灰色黏质砂土，含钙质结核，见软体动物化石及碎片：*Bellamya* spp.、*Parafossarulus striatulus* (Benson)、*Stenothyra glabra* (A. Adams)、*Semisulcospira elegans* Lee、*Radix swinhoei* (H. Adams)、*Gyraulus chihliensis* (Ping)，本层夹海相夹层，含半咸水介形虫类：*Callistocythere trokis* Huang 4.52m

63. 灰黑色黏质砂土，暗灰色中砂、细砂，分选磨圆不好 2.05m

64. 褐灰、黄绿色黏砂，具团粒结构，见钙质结核、铁染及螺类化石：*Radix* spp.、*Gyraulus chihliensis* (Ping)、*Vertigo* (*V.*) spp.。本层有机质含量高 3.18m

65. 褐灰色砂质黏土及棕黄色粉砂，含钙质结核，具铁染 3.91m

66. 上部浅灰绿、黄色相间黏质砂土，含锰结核、钙质结核，铁染发育，具团粒结构，下部浅灰、黄灰、灰黄色含砾粗砂、细砂、粉砂，具水平层理 5.48m

67. 深灰、浅褐灰色黏质砂土，铁染发育，见炭屑 1.91m

68. 灰白、灰、灰绿色粗、中、细砂，具水平层理，见锈斑 4.66m

69. 暗绿灰色砂质黏土夹黏质砂土，具锈斑、钙质结核 3.88m

70. 暗绿、锈黄、灰白色粗砂、含砾粗砂，具铁染，底部为粗砂砾石，砾石直径 3~5mm，砾石成分为火山岩、燧石、石英、灰岩等 8.32m

--------------- 平行不整合 ---------------

下伏地层：上新统棕红、棕黄、黄棕、锈黄色为主的中粗砂、含砾粗砂、砾卵石层夹黏质砂土、砂质黏土，半固结。

2. 岩性岩相变化

该沉积凹陷下更新统总体上为一套河湖相砾、砂、砂质黏土、黏土沉积，其中在 ZKTX-1 孔中于孔深 265.85~267.26m 处，发现一海侵层位。沉积物粒度相对较细，砂及砂质成分含量较高，其下段由 3 个沉积韵律旋回明显，颜色以浅灰、灰、绿灰色为主，见钙质结核，上部典型特点为出现较富集的铁锰结核。上段为中粗砂与砂质黏土、黏质砂土互层，颜色以浅灰、暗灰、绿灰为主（图 37-4）。

中更新统沉积物粒级较细，砂以中、细砂为主，铁染发育，软体动物化石在许多层位富集，沉积旋回明显。下段以黄灰色为主，夹灰、暗灰、黄褐色、褐灰色砂质黏土、黏质砂土，中、细砂为主，见多层钙质淀积层，发育钙质结核及软体化石。上段以黄灰色为主，见绿灰、黄褐色砂质黏土、黏质砂土及砂，沉积旋回较明显，有机质含量较高，动物化石较多（图 37-4）。

上更新统以砂、砂质黏土、黏质砂土及淤泥为主，沉积韵律明显，含植物化石碎片。下段主要为砂、砂质黏土、黏质砂土及黏土，由几个沉积韵律层组成，颜色主要为黄色、黄棕色和灰、绿灰色为主，为河漫滩相沉积；中段颜色以暗灰、绿灰、褐灰、黄灰色为主，沉积以淤泥及中粗砂为主，有机质含量较高，底部见锰结核，主体为湖沼相沉积；上段颜色为黄褐、黄灰、棕黄和棕红色为主，沉积以砂质黏土、黏质砂土、砂、含砾粗砂和淤泥为主，组成几个不同的韵律层，以河道夹小型湖沼沉积为主（图 37-4）。

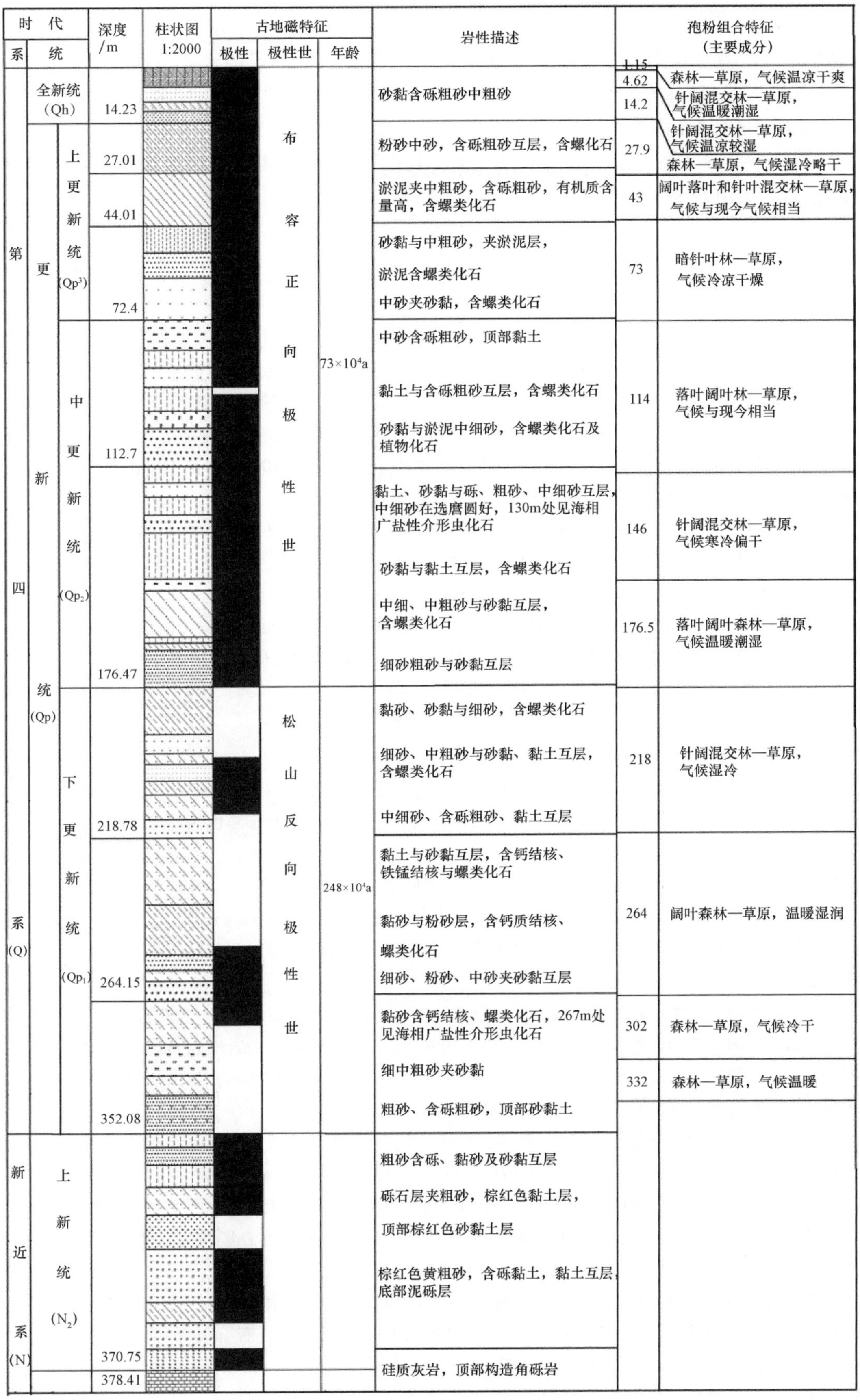

图 37-4　双埠头 ZKTX-1 孔综合柱状图

Fig. 37-4　Composite columnar section of ZKTX-1 in Shuangbutou

（四）平谷沉积凹陷（Ⅳ）

平谷沉积凹陷地处燕山南麓与华北平原北端的相交地带，其东、南、北三面环山。平谷盆地东西长约 32km，南北宽约 15km。本次选取凹陷内 ZK10 钻孔作为标准剖面进行研究，描述如下。

1. 代表剖面（ZK10 钻孔）

ZK10 钻孔位于平谷区马坊镇经济技术开发区，孔深 255.2m，揭露第四纪沉积物厚度 253.8m，下伏地层为新近系，下更新统剖面自上而下岩性描述如下：

上覆地层：全新统黄褐色黏质粉砂

———— 整　合 ————

上更新统 总厚 25.6m

3. 黄褐色黏质粉砂，粉砂质组分约占 70%～80%，黏土质组分约占 5%～10%，砂粒由长石和石英颗粒构成，磨圆度、分选性和成熟度均较好，可见灰白色团块状钙质结核，含量约 5%～10%，大小不一，大者 2cm，一般 0.5～1.0cm，见少量炭斑 2.9m
4. 灰褐色粉砂质黏土，黏土质组分占 80%～90%，粉砂质组分占 5%～10%，砂粒主要由石英颗粒构成，磨圆度、分选性和成熟度均较好，含少量炭斑 9.0m
5. 浅灰、褐黄杂色粉砂质黏土，粉砂质组分占 5%～10%，黏土质组分占 80%～85%，砂粒由长石和石英颗粒构成，磨圆度、分选性和成熟度均较好，可见灰白色钙质结核团块，含量约在 5%左右，大小不一，大者 1cm，小者 0.1cm，一般 0.5～0.8cm，并见少量炭斑 3.8m
6. 灰黑色粉砂质黏土，黏土质组分约占 80%～90%，粉砂质组分约占 5%～10%，砂粒由长石和石英颗粒构成，磨圆度、分选性和成熟度均较好，局部含极少的砾石，含量小于 5%，0.5～0.8cm 大小，成分多为石英岩，呈次圆状 9.9m

———— 整　合 ————

中更新统 总厚 58.8m

7. 浅灰色中砂，砂粒由长石和石英颗粒为主构成，少量岩屑及云母碎屑颗粒，中砂为主，少量细、粉砂，磨圆度、分选性和成熟度均一般，见有锈黄色细砂条带与黑色炭质条带状夹层，斜层理发育 3.8m
8. 褐黄色粉砂质黏土，粉砂质组分占 5%～10%，黏土质组分占 80%～85%，砂粒由长石和石英颗粒构成，磨圆度、分选性和成熟度均较好，可见少量炭斑，含量约 5%左右 4.1m
9. 褐黄色细砂，砂粒由长石和石英、云母和岩屑颗粒构成，磨圆度较好，分选性较好，成熟度一般 3.4m
10. 浅灰色含砾粗砂，磨圆度较好，分选性和成熟度一般，砂组分占 80%～85%，砂粒以长石、石英和岩屑碎屑为主，云母少量，砾石组分占 10%～15%，成分复杂，岩性以石英岩、脉石英、砂岩和泥岩、火山岩等为主，大小不一，大者 2cm，一般 0.5～1cm，呈次棱角状—次圆状，砾石边部往往有锈黄色氧化圈、风化晕 8.0m
11. 灰褐色黏质粉砂，粉砂质组分占 80%～90%，黏土质组分占 5%～10%，砂粒由长石和石英颗粒构成，磨圆度、分选性和成熟度均较好，含腹足类化石残片 11.6m
12. 褐黄色中砂，中砂组分占 70%～80%，黏土组分占 10%～20%，砂粒由长石和石英、云母和岩屑颗粒构成，磨圆度较好，分选性和成熟度均一般，局部黏土质组分集中构成团块状，并黏土质组分占中偶见锈斑 6.7m
13. 灰白色含砾粗砂，砂组分占 70%～80%，砂粒由长石和石英、岩屑和云母碎屑颗粒构成，磨圆度，分选性和成熟度均一般，砾石组分占 10%～20%，成分复杂，有石英岩、花岗岩、花岗斑岩、砂岩等为主，大者 8cm，一般 2～3cm，次圆状 0.5m
14. 褐黄色黏土层，黏粒组分 95%，含锈斑 0.3m
15. 褐黄、灰白色杂色细砂，砂粒由长石和石英、云母和岩屑颗粒构成，磨圆度较好，分选性较好，成熟度一般，水平层理发育 1.6m

16. 灰白色砂砾，磨圆度、分选性和成熟度均较差，砂组分占 50%~60%，砾石组分占 30%~40%，砂以粗砂为主，由长石、石英、岩屑和云母类碎屑颗粒构成，砾石成分复杂，石英岩、石英砂岩居多，花岗岩、斑岩次之，火山角砾少量，大者 10cm 以上，次圆状—次棱角状 16.3m

17. 褐黄色粉砂，粉砂质组分占 90%以上，砂粒由长石和石英颗粒构成，磨圆度、分选性和成熟度均较好，新鲜面上见有不规则锈斑，含量约 5%~8% 2.5m

———— 整 合 ————

下更新统 总厚 167.8m

18. 褐黄色粉砂，粉砂质组分占 90%以上，砂粒由长石和石英颗粒构成，磨圆度、分选性和成熟度均较好，新鲜面上见有不规则锈斑，含量 5%~8% 2.6m

19. 灰白色砂砾，磨圆度、分选性和成熟度均较差，砂组分占 50%~60%，砾石组分占 30%~40%，砂以粗砂为主，由长石、石英、岩屑和云母类碎屑颗粒构成，砾石成分复杂，石英岩、石英砂岩居多，花岗岩、花岗斑岩次之，火山角砾少量，大者 10cm 以上，次圆状—次棱角状 17.7m

20. 灰白色砂砾，磨圆度、分选性和成熟度均较差，砂组分占 25%~30%，砾石组分占 60%~70%，砂以中砂为主，由长石、石英、岩屑和云母类碎屑颗粒构成，砾石成分复杂，石英岩、花岗岩和石英砂岩居多，少量斑岩及火山角砾，大者 8cm 以上，一般 1~3cm，呈次棱角状—次圆状 122.3m

21. 棕红色含砾砂质黏土，磨圆度、分选性和成熟度均较差，黏土质组分占 70%~80%，砂组分占 10%~15%，砾石占 5%，黏土中见有灰白色条带，水平层理发育，砂粒以粗砂为主，颗粒多为石英；砾石成分复杂，有火山岩、石英岩和泥岩等，大小 0.5~1.0cm，呈棱角状 25.2m

~~~~~~ 角度不整合 ~~~~~~

下伏地层：灰白色白云质灰岩，风化面多浅灰色，新鲜面为灰白色，发育条带状方解石。

2. 岩性岩相变化

平谷沉积凹陷内地势平坦，堆积了巨厚的新生代松散沉积物，直接不整合覆盖于基岩地层之上。整个盆地内沉积物岩性较为简单，相变不大，但沉积厚度变化较大。沉积物岩性以砾石、砂质黏土为主。凹陷东部发育为以砾石层为主，以河流相沉积为主（图 37-5）。

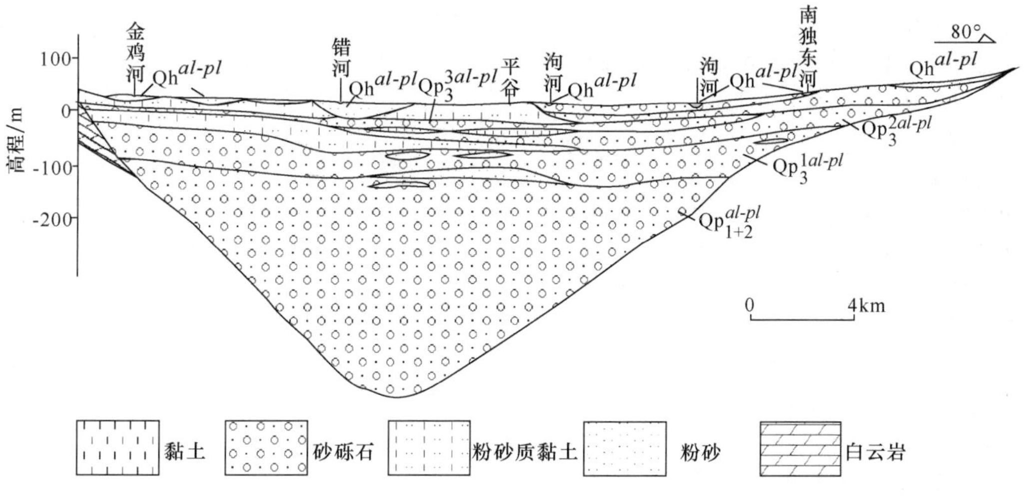

图 37-5 平谷沉积凹陷地质剖面图

（据吕晓俭，2000）

Fig. 37-5 Geological section of Pinggu sedimentary depression
~~~~~~

中更新统，岩性主要为粉砂、砂砾、砾石、粗砂、细砂、中砂、粉砂与黏土互层沉积，颜色以褐黄色、灰白色、灰褐色、灰黑色为主，以河床相、河漫滩相、牛轭湖相沉积为主。

上更新统，岩性主要为砂质黏土、粉砂、黏质粉砂沉积，颜色以黄褐色、灰褐色为主，灰黑色黏土中含腹足类化石残片，以河漫滩相、牛轭湖相沉积为主。

（五）延庆沉积凹陷（V）

延庆沉积凹陷属于山间断陷盆地，其西南与河北省阳原盆地断续相连。新近纪开始断陷伸展，现今仍在继续发展，呈北东-南西向展布。沉积中心或者深凹部位靠近西北侧，南侧呈逐渐翘起趋势。本次选取盆地内震平1孔作为标准剖面，其岩性描述如下。

1. 代表剖面（震平1孔）

震平1孔位于延庆县官厅水库东岸延庆农场西南角。钻孔标高为50.0m，终孔深度为1002.2m（图37-6）。描述如下：

上更新统	总厚 42.20m
1. 浅褐色黏土质砂，多植物根系和虫洞	0.50m
2. 灰、灰黄色砂质黏土	13.50m
3. 褐色砂质黏土，成层现象明显	25.60m
4. 灰色黏土质砂，成层现象明显	1.40m
5. 深灰色泥灰，含螺壳	1.20m
———— 整　合 ————	
中更新统	总厚 81.70m
6. 灰绿、黄白色砂质黏土，含钙质结核	25.46m
7. 灰绿色砂质黏土	9.64m
8. 绿灰色砂质黏土	14.30m
9. 灰色、灰绿色砂质黏土	32.30m
———— 整　合 ————	
下更新统	总厚 741.1m
10. 灰绿色砂质黏土夹薄层粉砂	36.10m
11. 灰绿、黄白色砂质黏土	40.00m
12. 灰绿色砂质黏土，黏性较大，具水平层理，下部夹有细砂层	30.11m
13. 灰绿色砂质黏土，夹灰色粉细砂和少量粉砂	66.50m
14. 灰绿色砂质黏土，具水平层理	46.04m
15. 灰色粉砂	4.80m
16. 灰色砂质黏土	26.30m
17. 灰色中细砂，主要成分为石英、长石、云母	4.94m
18. 灰色粉砂、黏土质砂夹少量砂质黏土	15.61m
19. 灰色粉砂、中砂	5.32m
20. 灰色砂质黏土、粉砂	6.66m
21. 灰色淤泥质、砂质黏土夹薄层粉砂	36.52m
22. 灰色粉砂夹淤泥质、黏土质砂	15.53m
23. 灰色黏土	17.01m
24. 灰色砂质黏土夹薄层粉砂	44.92m
25. 灰色粉砂	1.00m
26. 灰绿色砂质黏土，在535.22m处夹10cm厚的泥炭	57.23m
27. 灰绿色砂质黏土，含有孔虫化石：山西外旋九字虫 *Evolutononion shanxiensis* Wang	3.71m
28. 灰绿色砂质黏土，偶夹褐色薄层砂质黏土	85.39m
29. 棕褐色砂质黏土夹薄层粉砂	5.55m

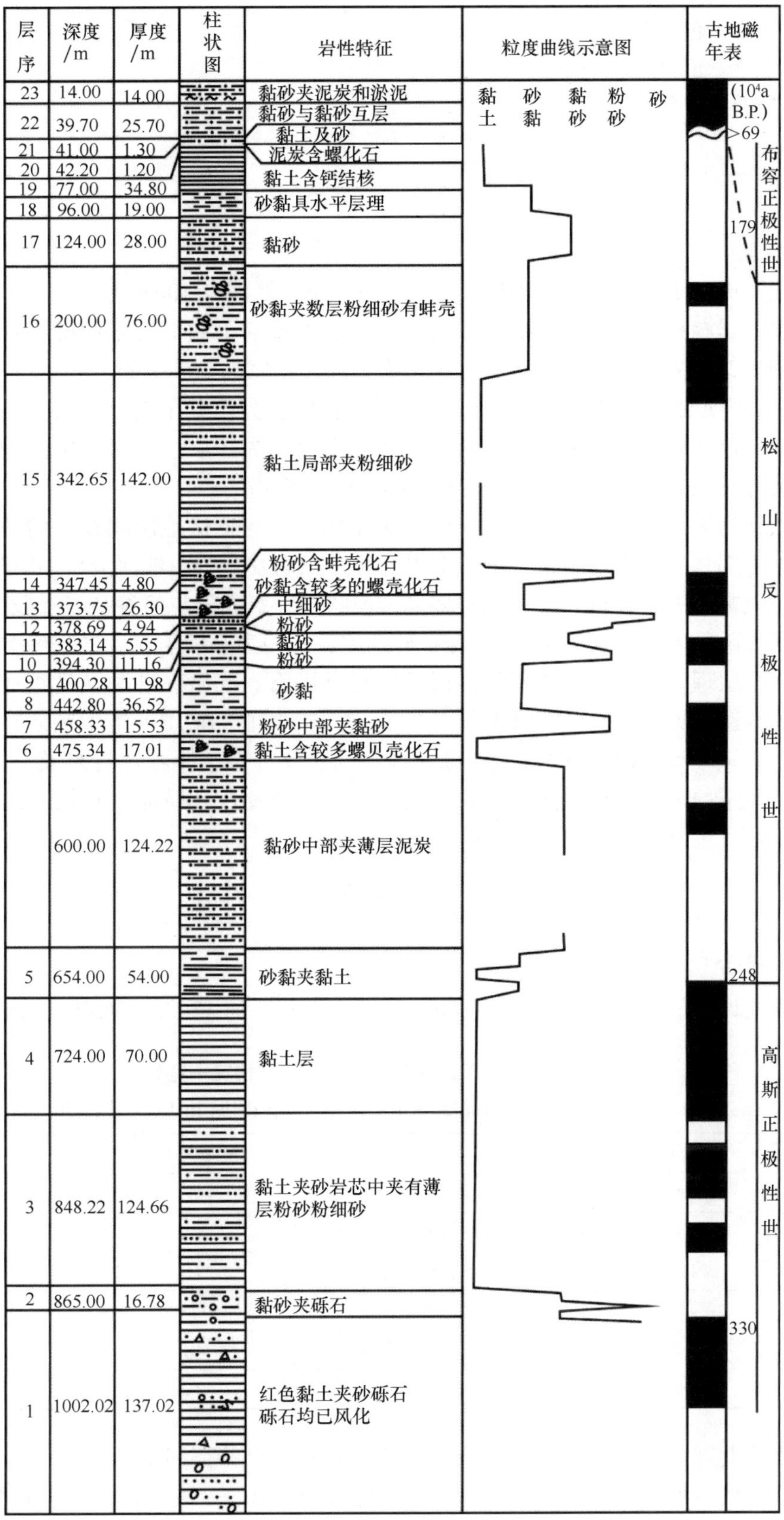

图 37-6 延庆沉积凹陷震平 1 孔综合柱状图

（据庞其清，1984）

Fig. 37-6 Integrated columnar section of Zhenping 1 hole in Yanqing Depression

30. 灰绿色砂质黏土夹薄层粉砂 50.85m

31. 棕褐色局部呈灰绿色粉砂质黏土 89.23m

32. 灰绿色局部为棕灰色砂质黏土夹薄层粉细砂 35.00m

33. 浅黄色粉砂夹砾石，砾石成分主要为火成岩、沉积岩 16.78m

--------------- 平行不整合 ---------------

下伏地层：上新统红色黏土夹砾石和砂

2. 岩性岩相变化

延庆沉积凹陷下更新统下部为灰绿色砂质黏土，在535m处夹有10cm泥炭层，上部为灰色、灰绿色黏土夹粉砂互层，属湖相层。中更新统沉积物按其成因类型的不同具有不同的特征，地表露头显示为冲积型和洪积型，钻孔揭露盆地中央埋藏有湖积物。冲积物底部为砂砾石，含砂砾石层；中部为粗-中-细砂层，水平层理或交错层理发育；上部为棕黄色-棕红色黏质砂土夹有3层古土壤，钙质结核较发育。洪积物以棕红色含砾、夹透镜状砾石的黏质砂土或砂质黏土为主。凹陷湖相沉积物主要为灰-灰绿色砂质黏土，孔深100m左右测古地磁年龄为（80~90）$\times10^4$a。

上更新统以洪积物为主，岩性为砂砾层和砂质黏土沉积，与下伏地层间存在清楚的侵蚀面，分布于Ⅲ级阶地或台地上部，局部出现湖相黏土。上更新统洪积物发育于山地河流出山口处，岩性为黄-黄白色黏质砂土、砂质黏土夹砂砾石，砾石夹层多呈透镜状。砾石成分与河谷基岩一致，次棱角状-次圆状。扇根部砾石多、砾径大，最大约1.5m。砾石分选差，砾间为黏砂填充。向扇缘分选渐好，沉积物粒度变细且发育水平层理。上更新统冲积物由砾石、粗砂砾石、黏砂、砂黏及砂组成韵律层。水平层理发育，二元结构普遍可见。砂层中有时见板状、槽状交错层理，黏质砂土中钙质结核发育。热释光测年为3.17$\times10^4$a（东桑园东RS1159）。上更新统风积物主要分布于小呼坨村东及羊儿岭村东山凹处，面积约2km^2，呈现为高10余米的黄色黏质砂土丘，垂直节理发育，局部见有钙质结核，覆盖于中更新统或基岩之上。

二、生物地层

（一）孢粉组合

第四纪孢粉研究详细的钻孔有震平1孔、ZK3孔、新5孔、通13孔、ZKTX-1孔等（表37-1）。

表37-1　北京平原区主要钻孔孢粉测试情况一览表

Table 37-1　Pollen test results of the main deep drills in Beijing plain

钻孔编号	孔深/m	构造单元	孢粉样品	测试时间（年）
震平1孔	1002.2	延庆凹陷	—	1988
ZKTX-1孔	376.44	甘棠凹陷	246	1989
通13孔	405.75	后沙峪-俸伯凹陷	91	1978
顺1孔	857.5	后沙峪-俸伯凹陷	425	1988
新5孔	862.66	后沙峪-俸伯凹陷	401	2006
ZK3孔	616	马池口凹陷	262	2014

本志以北京平原北部马池口凹陷ZK3钻孔为代表，孢粉组合特征如下：

ZK3钻孔位于北京昌平百善镇东沙屯村西农田，孔深616.20m。该钻孔岩芯中有97个样品中发现较为丰富的孢粉颗粒，共统计29057粒孢子或花粉，平均300粒/样品。其余100余个样品，每个样品含十几到几十粒不等，未达到统计数量。

孢粉百分比计算以所有孢粉总和为基数，总体上以草本植物花粉为主，蕨类孢子和藻类在某些层位含量较高，木本花粉含量较低。草本类型以蒿属（*Artemisia*），藜科（Chenopodiaceae），禾本科（Poaceae），菊科（Asteraceae）如紫菀属（*Aster*），蒲公英型（*Taraxacum*），风毛菊型（*Saussurea*），蓼属（*Polygonum*，包括三孔沟和散孔类型），唐松草属（*Thalictrum*），石竹科（Caryophyllaceae），大戟科（Euphorbiaceae），十字花科（Cruciferae），唇形科（Labiatae），葎草属（*Humulus*），伞形科（Umbelliferae），地榆属（*Sanguisorba*）等比较丰富。水生或沼泽类型如莎草科（Cyperaceae），香蒲属（*Typha*），黑三棱属（*Sparganium*），狐尾藻属（*Myriophyllum*），眼子菜科（Potamogetonaceae）等较常见。灌木类型有麻黄属（*Ephedra*），白刺属（*Nitraria*），忍冬科（Caprifoliaceae），榛属（*Corylus*），柳属（*Salix*）等。蕨类孢子以水龙骨科（Polypodiaceae）和卷柏属（*Selaginella*）为主。藻类主要有环纹藻（*Concentrisystes*），水绵（*Spirogyra*），盘星藻（*Pediastrum*），双星藻（*Zygnema*）等。木本花粉类型经常出现的有松属（*Pinus*），云杉属（*Picea*），冷杉属（*Abies*），柏科（Cupressaceae），桦木属（*Betula*），栎属（*Quercus*），栗属（*Castanea*），榆属（*Ulmus*），椴属（*Tilia*）等。另外，在组合中还出现了少量不能鉴定的孢粉，含量一般比较低。

早更新世划分为 4 个孢粉组合带。

组合 1：（540.0～483.6m，2.9～2.5Ma，B356-B329），该组合中，草本与灌木花粉含量为 92.1%～98.8%，平均 96.7%。木本花粉含量为 0.0%～4.8%，平均 2.1%。蕨类孢子和藻类的含量为 0.0%～3.2%，平均 1.2%。反映的植被类型是以蒿属为主的草原，气候冷干。

组合 2：（483.6～311.3m，2.5～1.6Ma，B325-B223），该组合中，草本与灌木花粉含量为 9.5%～98.7%，平均 79.0%。木本花粉含量为 0.3%～31.4%，平均 3.7%。蕨类孢子和藻类的含量为 0.3%～90.1%，平均 17.3%。反映的植被类型仍然是以蒿属为主的草原，在某些时段出现水生植物含量高值、蕨类孢子与藻类占优势，反映了植被的波动和湖泊的多期次扩张。

组合 3：（311.3～204.2m，1.6～1.0Ma，B222-B149），该组合中，草本与灌木花粉含量为 49.1%～98.9%，平均 76.8%。木本花粉含量为 0.0%～26.0%，平均 12.5%。蕨类孢子和藻类的含量为 0.9%～27.7%，平均 10.7%。反映的植被类型为以蒿属为主的草原，气候较上一个阶段稍暖，在某些层位出现水生植物花粉含量高值，反映了气候与植被的波动和湖泊的 2 次扩张。

组合 4：（204.2～157.5m，1.00～0.78Ma，B147-B124）该组合中，草本与灌木花粉含量为 23.4%～94.9%，平均 77.3%。木本花粉含量为 2.8%～75.5%，平均 18.0%。蕨类孢子和藻类的含量为 0.0%～15.0%，平均 4.7%。本带孢粉浓度较高，以出现柏科花粉峰值为特征，喜湿类型的百分含量较高，说明植被在某些时段一度由草原演化为以柏科为建群种的疏林草原，气候变得较为冷湿，这与中更新世气候转型期全球气候变冷、变幅增大相呼应。

中更新世孢粉组合带：（157.5～86.6m，0.780～0.124Ma，B122-B74），该组合中，草本与灌木花粉含量为 20.1%～100.0%，平均 85.0%。木本花粉含量为 0.0%～27.6%，平均 5.9%。蕨类孢子和藻类的含量为 0.0%～78.4%，平均 9.2%。本带木本植物花粉含量下降，草本花粉含量升高，植被为以蒿属为主的草原，气候冷干。在某些层位出现水生植物含量高值、蕨类孢子与藻类占优势，反映了植被的波动和湖泊的两次扩张。

晚更新世孢粉组合带：（86.6～0.7m，0.124Ma 至今，B72-B1），该组合中，草本与灌木花粉含量为 39.5%～100.0%，平均 91.9%。木本花粉含量为 0.0%～22.1%，平均 2.6%。蕨类孢子和藻类的含量为 0.0%～38.4%，平均 5.5%。本带反映的植物类型为以菊科蒿属、藜科为主的草原，气候冷干。在某些层位出现水生植物含量高值，反映了植被的波动和湖泊的多次扩张。

（二）哺乳动物化石

平原区下更新统、中更新统中哺乳动物化石研究资料相对甚少。

平原区哺乳动物化石主要分布于晚更新世地层和全新世地层中，平原区地面以下 10～20m 的砂层或砂砾石层中可以找到我国北方晚更新世的典型化石：原始牛、普氏羚羊以及披毛犀、德永象、赤鹿等。在全新世地层中，发现有螺类、碎骨和用梅花鹿角做的工具，此外还发现有家禽、牲畜的杂骨、陶片等，反映出人类活动的影响。

（三）软体动物

现仅就化石组合及其反映的环境概述如下（1：25 万延庆幅，2002）：

1. 早更新世

早期（Qp^{1-1}）：以腹足类肺螺亚纲陆生种为主，其中有直隶小旋螺、旋转螺为主，分布于干旱灌木丛、草丛及石块下生活，为耐干寒种类，反映气候较为寒冷。

中期（Qp^{1-2}）：以前鳃亚纲种类为主，其中有环棱螺、纹沼螺以及双壳类无齿蚌，多生活于湖沼、池塘、河流水域中，为喜温型软体动物，反映气候温暖湿润。

晚期（Qp^{1-3}）：前期以小旋螺为主，种属单调，反映环境又趋恶化；后期以前鳃亚纲为主，其中有纹沼螺、小旋螺，还有双壳类丽蚌，反映为温暖湿润的湖沼环境。

2. 中更新世

早期（Qp^{2-1}）：以前鳃亚纲为主，其中有豆螺、狭口螺、小旋螺等，多生活在水草茂盛，淡水河湖环境之中，气候比较温暖。

中期（Qp^{2-2}）：以陆生肺螺亚纲为特征，其中有华蜗牛、小旋螺和土螺等，多为耐干寒的种属，反映气候向干冷演化。

晚期（Qp^{2-3}）：以前鳃亚纲水生种为主，其中有萝卜螺、纹沼螺等，并有双壳类丽蚌和少量肺螺亚纲水生种属如瓦娄蜗牛，反映当时为温和多水环境，但气候有偏干趋势。

3. 晚更新世

早期（Qp^{3-1}）：见腹足类化石，属种单调，生物个体小，数量也较少，主要有土蜗、琥珀螺、小旋螺，多生活于河流之中，反映当时环境较前期有所恶化。

中期（Qp^{3-2}）：属种单调，个体小，主要有小旋螺、豆螺、圆扁旋螺及土蜗，此外还有喜温湿的圆扁螺及豆螺，反映当时为温湿环境，但已向干凉转化。

晚期（Qp^{3-3}）：多为耐干寒的陆生肺螺亚纲，如条华蜗牛、粉华蜗牛，这是黄土地区常见的软体动物化石，表明该时期为干冷的草原环境，而在局部低洼地区可见少量喜湿种类。

（四）微体古生物

北京平原地区实施了众多的钻孔，对若干钻孔进行了微体古生物的系统分析，特别是在下更新统中发现两层海相层，为平原区第四纪古地理环境研究提供丰富资料。现就顺 1 孔、顺 5 孔、通州双埠头 ZKTX-1 孔的微体古生物组合进行分析。

早更新世地层中，在后沙峪-俸伯沉积凹陷钻孔底部的浅灰色中砂、细砂层中，发现海相有孔虫、介形虫与钙质超微化石。此外，在凹陷的边缘地区有时可发现广盐性海相介形虫化石，如双埠头钻孔等。在顺 5 孔埋深 428.6m 处，为灰色、灰绿色砂层夹粉砂、黏土层中发现丰富的有孔虫化石群，为一海相层，古地磁测年为 226×10^4a。经鉴定，该化石群中浮游类约占个体总数的 35%，底栖类约占个体总数的 65%。其中包括波罗的透明虫、泡抱球虫、胖抱球虫、变异抱球虫比较种、两侧平坦虫、面包虫、光辉面包虫、圆辐虫未定种等。根据有孔虫生态分析，波罗的透明虫是冷水水域中常见种，而浮游有孔虫中以抱球虫为主，它是正常盐度喜暖种，而底栖有孔虫中面包虫占绝对优势，它属喜暖种，由此可见，该动物群既有喜暖分子，也有喜冷分子，但喜暖分子占优势，显示气候较为温暖。早更新世中期，在海相层之上有介形类化石出现，其中包括有柯氏土星介、开封土星介、土星介未定种、纯净小玻璃介、玻璃介、金景介亚种，大多生活于淡水环境，气候温暖湿润。早更新世晚期，又发育一海相层，其中含有孔虫化石，如面包虫（夏 2 孔）及报球虫（顺 3 孔），但数量较少，种属也较单调，反映当时气候较温暖。

在通州双埠头孔更新世地层中，不仅发现了介形类化石，在孔深 267m 及 130m 处发现半咸水广盐性海相介形虫化石，可与顺 5 孔海相层位相对比。在孔深 267.26~265.85m 处发现广盐性海相介形虫化石隆脊花介与陆相介形虫纯净小玻璃介、土星介未定种共生，属海陆过渡相半咸水水域的介形类化石层。孔深 274~255m 之间，介形类有普蓝白花介、桑千河玻璃介、瘤湖花介、土星介、玻璃介、纯净小玻璃介、金星介和真星介等，此外还有少量上新世晚期属种，显示为淡水到半咸水环境，气候呈现温凉半湿润-半干旱的波动性特征。孔深255~219m之间，介形类以窑子头玻璃介、扁湖花介为主，还有玻璃介、纯净小玻璃介、土星介、金星介、真星介等，反映当时水域由淡水向半咸水环境演化，气候由温暖半湿润向温和半干旱方向变化。

中更新世，北京平原地区堆积物以冲洪积物为主，具有明显的韵律变化，在冲积扇前缘地带常发育小型湖沼，沉积了河湖相地层。

在通州双埠头钻孔，可划分出早、中、晚3期。早期（孔深165~157m）：介形类以双瘤湖花介、近侯氏玻璃介为主，还有玻璃介、小玻璃介、真星介、土星介等属，该组合中属种单调，分异度低，属微咸-半咸水环境；中期（孔深146~127m）：介形类有斑纹三原介、压缩真星介，并与纯净小玻璃介、奇妙小玻璃介、哈萨克斯坦小玻璃介共生，属海陆过渡相向半咸水相环境过渡的类型，代表一次海侵过程，根据三原介生态分析，它是生活在温暖水体中广盐性海相介形虫，反映当时气候较温暖；晚期（孔深117~83m）：介形类以玻璃介、小玻璃介、土星介为主，其中包括丰县玻璃介、粗糙上星介、纯净小玻璃介、双折土星介、隆起土星介、布氏土星介等，该组合中因含有喜暖属种，属温暖的淡水环境。

在顺1孔、顺5孔中，一些层段含有丰富的介形类化石，其中有开封土星介、近东山土星介、柯氏土星介、具瘤土星介、近美丽土星介、布氏土星介、隆起土星介、土星介未定种、玻璃介未定种、纯净小玻璃介、球星介未定种和真星介未定种等，反映中更新世时是以淡水环境为主，但有时为半咸水的湖沼环境。

晚更新世，北京地区气候环境发生明显变化，突出表现在气候由温湿向冷干演化，平原地区积水洼地逐渐缩小，并使水质发生变化。在通州区双埠头孔孔深64~30m之间，含介形类化石，其下部以双瘤湖花介为主，还有小玻璃介属、真星介属，反映当时水温在10℃以下，呈微咸水环境；上部以窄骊山介为主，并伴有纯净小玻璃介、柯氏土星介、双折上星介、东山土星介等，反映为淡水或微咸水环境，气候为温凉半湿润。晚更新世晚期气候冷干，并有黄土堆积，未见介形类化石。

三、化学地层

对沉积层进行微量元素、碳酸钙、矿物、氧化物分析，可以作为地层划分和对比的依据，并可以从中提取古气候、古环境变化信息（李华章，1991）。

（一）黏土矿物分析

对不同地质时期的沉积地层中的黏土矿物进行X射线衍射分析，可以得出黏土矿物组合类型及所含高岭石、绿泥石、伊利石及蒙脱石相对的百分含量，可作为反映古气候变迁的标志之一。不同的黏土矿物组合反映不同的气候类型，在亚热带地区的土壤或风化壳中黏土矿物组合是以高岭石为主，在中纬度干旱、半干旱地区的黑钙土、栗钙土中黏土矿物组合是以伊利石和蒙脱石为主，因此剖面中黏土矿物组合类型的变化，反映出气候环境的变化。

1. 北京平原区

在怀柔张喜庄孔（HR88-1）、通县双埠头孔进行了系统黏土矿物分析，表明了不同地质时期黏土矿物组合的垂向变化差异，依此推断了古气候的演变。

李龙吟等（1993）对张喜庄钻孔不同黏土进行了X射线衍射测定和分析，对其相对百分含量作了分析，从变化曲线显示了，不同沉积时期古气候的演变（图37-7）。

（1）下更新统（370~144m）

下更新统下部（370~333m）高岭石、伊利石、绿泥石含量相比上新统呈下降趋势，只有蒙脱石含量上升到70%，说明当时处于较为干寒的气候环境；下更新统中-上部（333~144m）高岭石、绿泥石含量均小于10%，上下变化不明显，但伊利石、蒙脱石含量变化显著，前者为8%~30%，后者为50%~80%，两者互为消长，出现3个明显波峰曲线，反映出3次冷暖、干湿的波动。

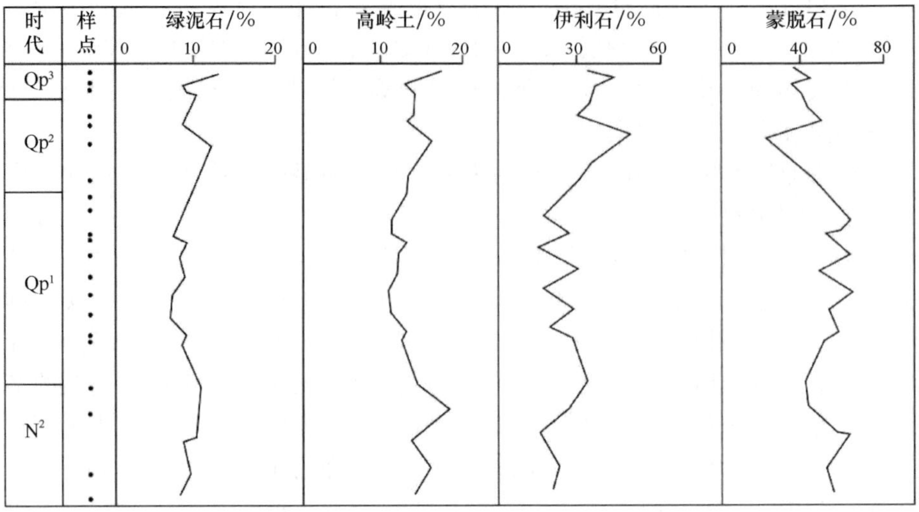

图 37-7　张喜庄钻孔黏土矿物百分含量变化曲线

（据李龙吟等，1993）

Fig. 37-7　Percent curve of clay minerals in borehole of Zhangxizhuang Village

（2）中更新统（144~46m）

中更新统下部，绿泥石、高岭石、伊利石含量均呈逐步增加的趋势，其中绿泥石由10%增至15%，高岭石由10%增至16%，伊利石由30%增至60%，但蒙脱石含量由60%减少到8%，说明中更新统早期气候为暖干环境；中更新统上部，绿泥石、高岭石、伊利石含量明显减少，但蒙脱石含量由8%增加到50%，说明中更新世晚期气候向温带干旱草原气候环境变化。

（3）上更新统（46~30m）

在孔深46~30m之间伊利石和高岭石含量相对较高，而蒙脱石含量较低，出现明显波峰曲线，说明当时存在着较大的冷、暖气候波动，随后伊利石、蒙脱石有所减少，而高岭石、绿泥石有所增加，反映气候又向温湿演化，时代相当于晚更新世中期。

2. 延庆盆地

官厅南侧出露了早更新世晚期的湖相沉积，以棕黄色、棕色为主的亚砂土和亚黏土层，层中含有丽蚌碎片。如黏土矿物X射线衍射曲线图所示（图37-8），A线中伊利石峰底部较为扩散，说明其层间的K^+淋失更多，反映较为温暖湿润的环境，从4种主要黏土矿物的质量相对百分含量看，蒙脱石的含量比伊利石还要多，也说明了同样的气候条件；B线中伊利石峰底扩散比A要弱，且伊利石含量增至60.4%，蒙脱石下降至18.4%，表明气温偏低，湿润程度明显下降，C线中黏土矿物的重量相对百分含量伊利石由60.4%下降为54.7%，蒙脱石由18.4%增至29.9%，表明气候变暖变湿，蒙脱石含量比A线要小，说明气温虽有回升，但幅度明显减小，由此说明早更新世晚期气候总的来说是向凉的方向演变。

中更新世早、中、晚期代表层的黏土矿物X射线衍射曲线（图37-9）表明，从A线至C线，伊利石峰的底部扩散越来越小，且晶面间距也变小，说明伊利石越来越多的保留了云母的特性，可见化学风化越来越弱，气候向干凉方向发展。从4种主要黏土矿物的质量相对百分含量变化来看，蒙脱石含量由中更新世早期的25.3%，下降至中期的11.56%和晚期的2.9%。而绿泥石含量由早期的9.77%增至晚期的27.8%，说明当时风化强度不大，气候有逐渐变为干凉的趋势。

（二）沉积物化学元素分析

根据前人的研究，将地壳中含量大于1%的元素称为常量元素，小于1%的称为微量元素。地层中主要的造岩元素Si、Al、Fe、Mn、Ca、Mg、K、Na等为常量元素，它们多以氧化物形式存在，其他元素为微量元素。对北京地区若干剖面进行系统分析，对不同化学元素的迁移与聚集的规律研究，可以

	蒙脱石	伊利石	高岭石	绿泥石
曲线C	29.9%	54.7%	7.0%	8.4%
曲线B	18.4%	60.4%	9.6%	11.6%
曲线A	44.5%	44.2%	5.8%	5.5%

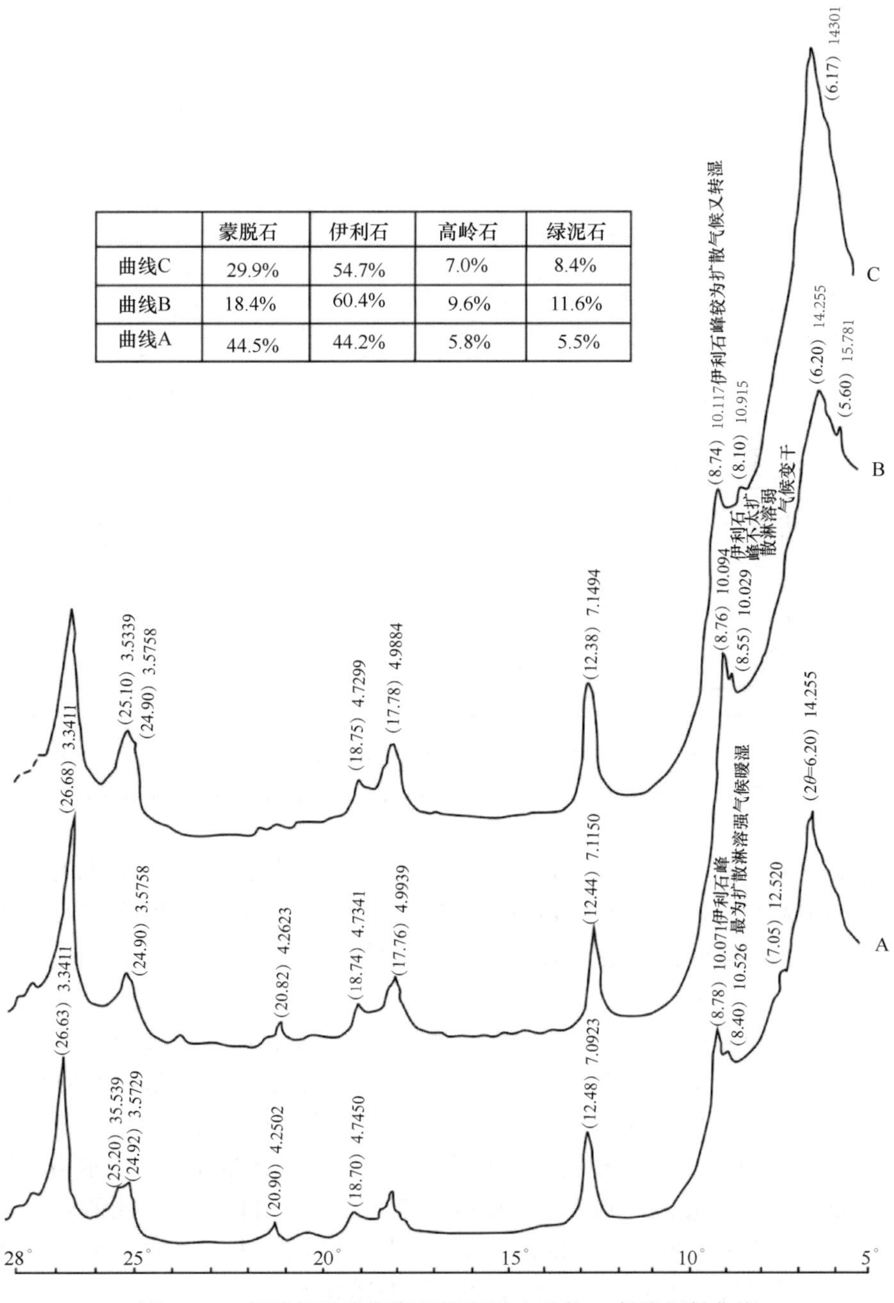

图 37-8　早更新世晚期湖相沉积黏土矿物 X 射线衍射曲线

Fig. 37-8　X diffraction curve of clay minerals in lacustrine deposits in late Early Pleistocene

提供不同地质时期的环境演变特征。

沉积层中化学元素的迁移与聚集是由元素本身的物理化学性质所决定，也与所处地貌部位及气候状况也有着密切关系。铝元素的化学性质最稳定，它多以 Al_2O_3 的形式保留在地层之中，在温暖潮湿的气候条件下，水介质为酸性，沉积物中易溶元素溶解迁移，而 Al_2O_3 相对富集，在干旱气候条件下相对含量则降低。铁是变价元素，多以 Fe_2O_3 与 FeO 形式出现，FeO 含量变化是随 Fe_2O_3 的含量增高而降低，二者呈负相关关系，可作为判断氧化或还原环境。钙、镁是活动性中等或较强的元素，在地层中多以 CaO 与 MgO 形式出现，在半湿润、半干旱气候条件下，对钙、镁的溶解、迁移和富聚是有利的。钾、钠是化学性质最活跃的元素，在干旱气候条件下对钾、钠富聚最为有利。硅在地层中含量最多，并以 SiO_2 形式保留在沉积层中，其变化周期与 Al_2O_3、Fe_2O_3 的含量变化周期呈反相关关系，铝、铁在潮湿气候条件下相对富集，而硅在干旱气候条件下相对富集。因此，可以通过氧化物比值计算，

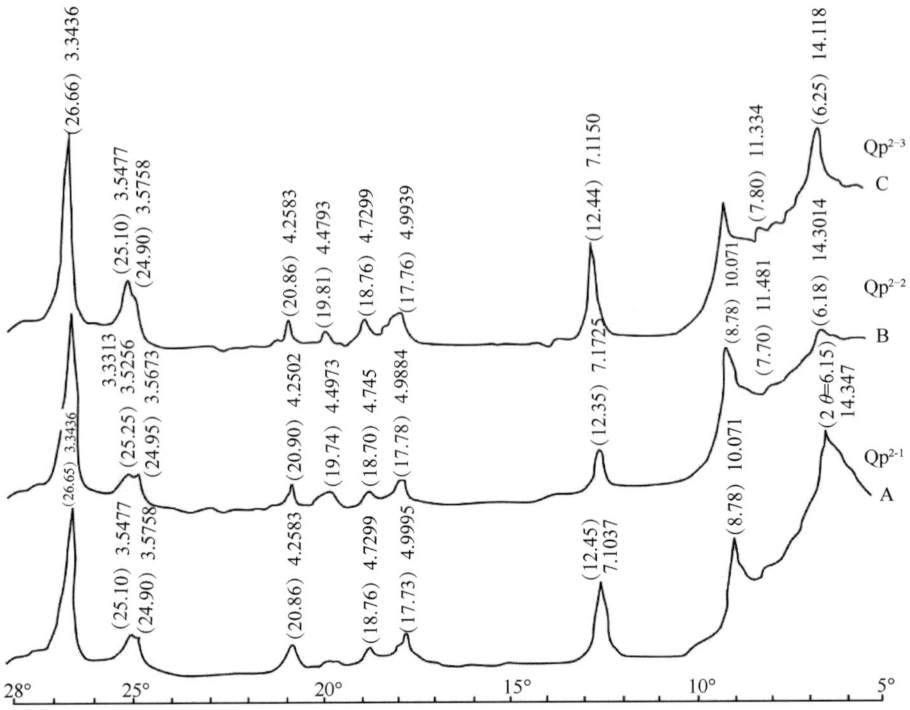

图 37-9　中更新统地层黏土矿物 X 射线衍射曲线

Fig. 37-9　X diffraction curve of clay minerals in Middle Pleistocene strata

获得 T 值（$T=Fe_2O_3/FeO$）和 G 值（$G=SiO_2/Al_2O_3+Fe_2O_3$）说明沉积层堆积时期的氧化条件和古气候状况。也就是说，T 值越大，说明高价铁含量高，处于强烈的氧化环境，反之则处于还原环境。而 G 值含量越大，说明化学作用微弱，气候较为干冷；含量数值越小，化学作用越强烈，气候较为温暖湿润。

沉积层中的微量元素种类及含量变化，亦可作为判断环境变化的指标之一。微量元素的变化与沉积物粒度、生物活动有关，凡粒度变细，生物活动强度加大的，有利于微量元素的富集。通过光谱半定量测定，Be、Ba、Cr、Cu、Mn、Ni、Pb、Sr、Ti、V、Zn 等元素变化明显，而其他元素变化较小。有些微量元素如 Cu、Zn、Mn、V 等与植物生长有关，如上述元素含量高时，表示当时植物繁茂的温暖湿润的气候环境，而 Sr、Ba 含量越高则表示气候越干冷，因此，可用 Sr/Ba 比值表示气候的干湿。沉积层中可溶性盐的总含量亦可作为气候干湿的指标，含盐量越高，表示气候越干燥，反之表示气候湿润。现就北京平原地区及延庆盆地的沉积物化学特征，概述如下：

1. 北京平原区

现将通州双埠头钻孔（ZKTX-1）、张喜庄钻孔（HR88-1）概述如下：

通州双埠头钻孔，孔深 376.44m，包括新近系、下更新统、中更新统、上更新统及全新统，是北京平原区中研究较为详细的钻孔之一。北京地质测试中心对该孔进行了氧化物及微量元素分析，对不同地质时期氧化物含量分析及其比值计算，可以获得不同时期的环境信息（图 37-10）。

下更新统，从氧化物含量变化曲线看，具有明显波动，反映环境的变化。SiO_2 含量为 41.76%~59.69%，Fe_2O_3 为 4.03%~5.88%，FeO 为 0.25%~0.93%，K_2O 为 2.9%~2%，TiO_2 为 0.8%~1%。G 值为 2~3.4，T 值为 9.49~20。下更新统下部，T 值只有 10 左右，而 G 值为 2.2 左右，属弱氧化环境，表明早更新世早期气候较为干冷；下更新统中部，SiO_2、Fe_2O_3、Al_2O_3、TiO_2 百分含量明显增加，而 FeO、K_2O、CaO、MgO 的含量明显减少，属较强的氧化环境，表明早更新世中期气候温暖湿润。下更新统上部，T 值为 6~11，G 值为 3.2~2.1，SiO_2 含量减少，而 CaO、MgO、K_2O 等含量有所增加，表明早更新世晚期气候为温干环境。

中更新统，SiO_2 含量为42%~50%，Fe_2O_3 为4.5%~6.16%，FeO为0.7%~1.1%，K_2O 为2%~2.2%，TiO_2 变化不大，在0.6%左右。*T*值在下部为4~10，上部为10~35，反映为弱氧化环境，*G*值在下部为3.8~2.2，上部逐渐加大到3.6，反映化学作用由稍强→弱→稍强的变化过程，气候由温湿到干冷到湿润的变化，从各氧化物含量变化也表现出这种演化趋势。

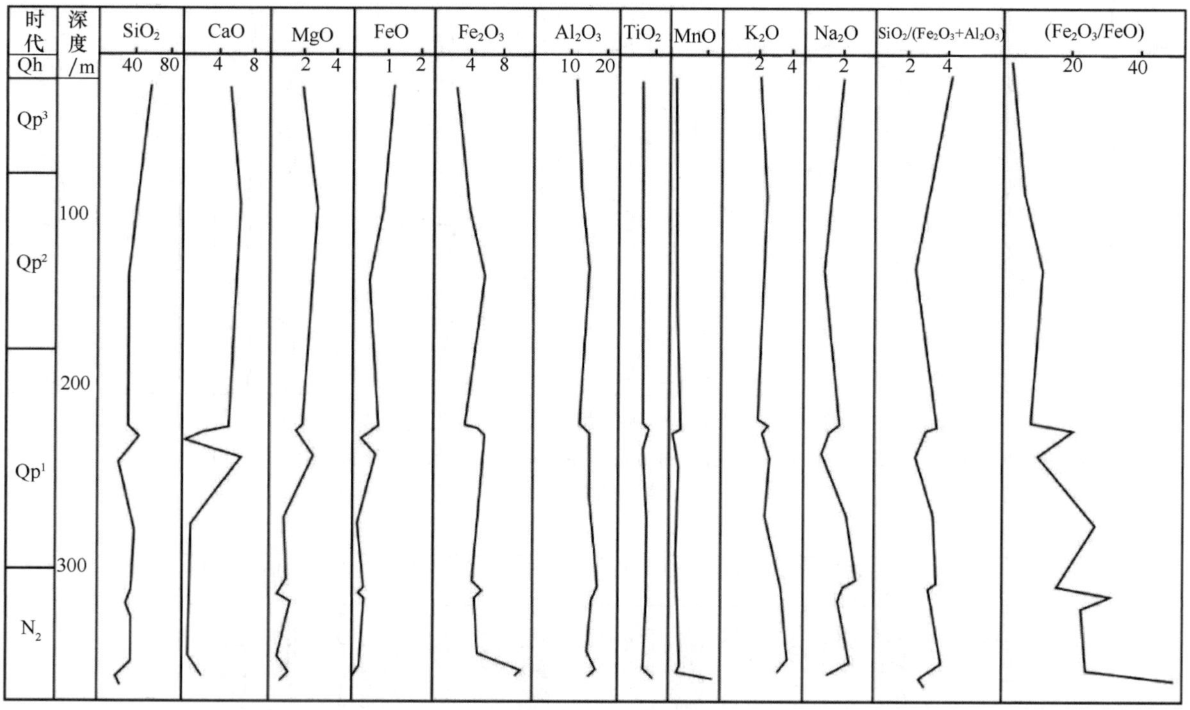

图37-10 双埠头钻孔氧化物百分含量变化曲线

Fig. 37-10 Percent change curve of oxide in borehole of Shuangfutou Village

氧化物含量单位为%

上更新统，SiO_2 含量为46.8%~62%，Al_2O_3 为12%左右，Fe_2O_3 为2.57%~5.67%，FeO为0.79%~1.32%，而CaO、K_2O、MgO含量均有逐渐减少的趋势，属弱氧化环境，但不同阶段气候有所不同。

对不同时期沉积物所作的微量元素含量测定，可以得出在垂直方向上各元素的富集与散失的基本特征，从而获得不同时期环境演变的过程（图37-11）。

下更新统，根据各元素富集与散失的变化规律，可将下更新统划分为下、中、上3部分：下部，大部分元素处于富集状态，而Mn、Ni处于散失状态，Sr/Ba比可达0.7~0.9，反映早更新世早期气候冷干；中部，多数元素含量变化较大，除Mn、Ni相对富集外，其他元素多处于散失状态，Sr/Ba比也在0.4~0.8之间波动，反映早更新世中期气候以温湿为主，但有冷干波动；上部，多数元素处于富集状态，其中Mn、Cu、Ni、Sr等更为明显，反映早更新世早期晚期气候潮湿寒冷。

中更新统，多数元素处于散失状态，反映出中更新世以温暖气候为主，但其中有2~3次波动，Sr/Ba比多在0.2~0.8之间。

上更新统，Cu、Cr、Sr、Ti、Zn等元素均处于富集状态，而Mn元素则为散失状态，反映晚更新世气候以干寒为主，但其中也有相对温湿的时期。

怀柔区张喜庄钻孔，李龙吟等（1993）对该孔作了氧化物测定（图37-12）。

下更新统，SiO_2 含量为54%~68%，Al_2O_3 为12%~15.5%，Fe_2O_3 为2.5%~4.5%，*G*值为2.88~3.77，*T*值为1.04~3.6，MgO含量为1.8%~3.2%，而CaO含量下部较低，多在1.2%~1.4%之间，而上部在2.7%~5%之间，反映早更新世气候以温暖湿润为主，而上部向冷干方向演化，但其间也有小的气候波动。

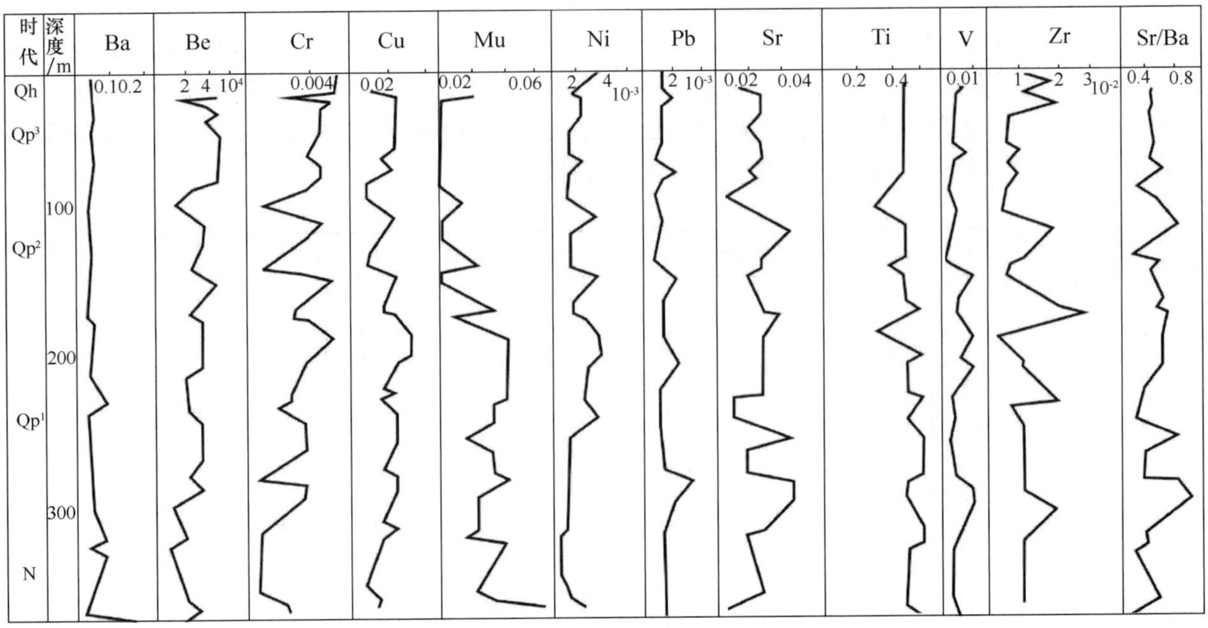

图 37-11　双埠头钻孔微量元素含量变化曲线

Fig. 37-11　Content change curve of trace element in borehole of Shuangfutou Village

元素含量单位为%

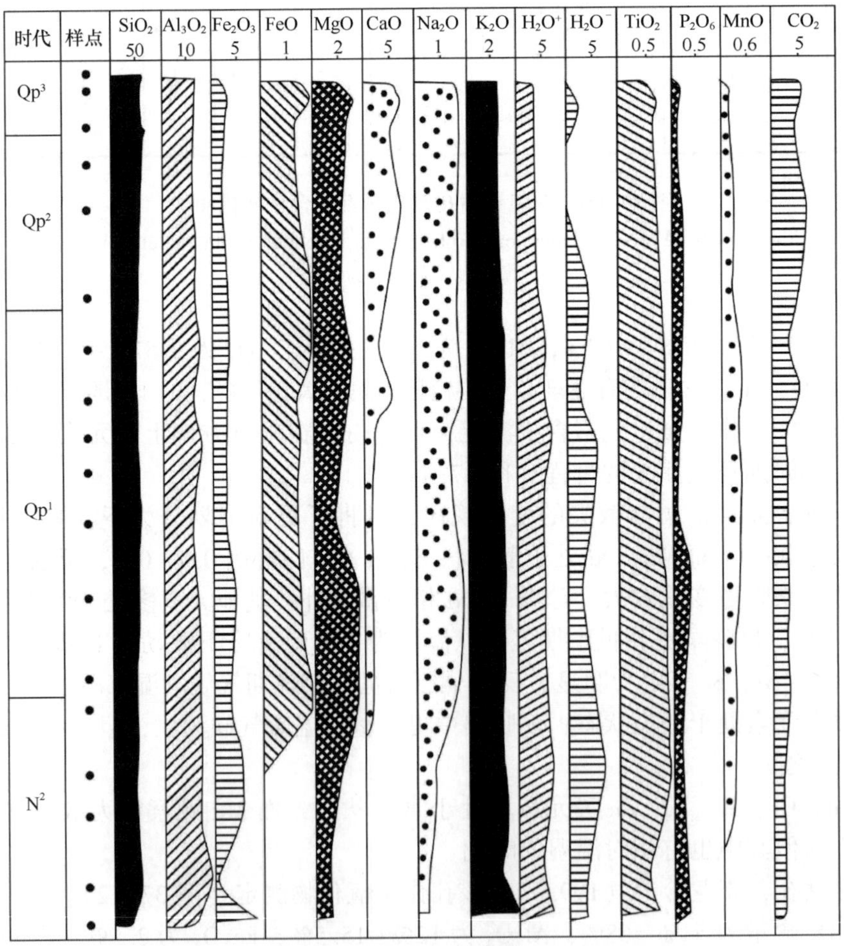

图 37-12　张喜庄钻孔氧化物含量变化曲线

（据李龙吟等，1993）

Fig. 37-12　Percent change curve of oxide content in borehole of Zhangxizhuang Village

氧化物含量单位为%

中更新统，SiO_2 含量为 60%~68%，Al_2O_3 为 12%~13%，Fe_2O_3 为 2%~2.4%，*G* 值为 4~4.7，*T* 值为 0.9~1.84，反映中更新世气候温润，但从 CaO 含量变化可以看出在中部曾出现过冷干的时期。

上更新统，SiO_2 含量为 60%~63%，Al_2O_3 为 11%~12%，Fe_2O_3 为 2.5%~4.5%，*G* 值为 4.28~4.84，*T* 值为 1.19~1.66，CaO 为 4.8%~6.2%，属弱氧化环境，气候由温干向冷干方向演化。

从上述钻孔岩石化学分析，由于取样密度不足，利用地球化学元素数据对第四纪环境演变还难于准确而详尽复原，但在不同地质时期的环境变化趋势则基本相同，因而为北京地区环境演变研究提供了有价值的分析资料。总体来看，化学元素分析结果与孢粉、生物特征所反映的气候变化较吻合。

2. 延庆盆地

早更新世晚期至中更新世中期的湖相沉积在官厅水库南侧出露较好，本志采用该剖面样品数据对延庆盆地化学地层进行描述。在该剖面的湖相层中采集了 25 块样品，分别做了黏土颗粒含量测定（图 37-13）、可溶性盐含量及含盐量测定（表 37-2，图 37-14）。

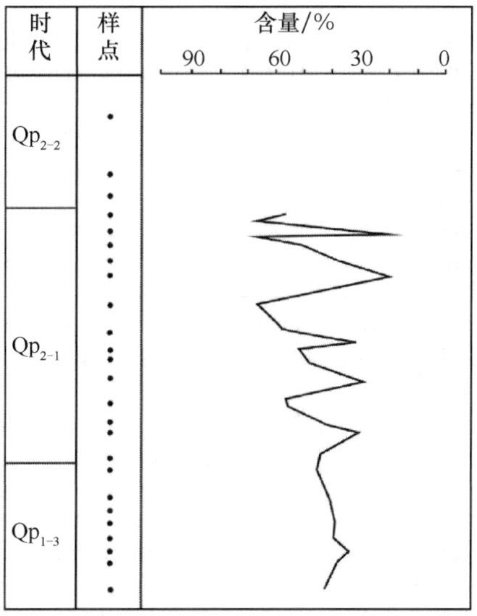

图 37-13　官厅水库南侧剖面黏土颗粒含量分布曲线

Fig. 37-13　Percent content distribution curve of clay particles of southern Guanting Reservoir section

从上述图表资料显示，早更新世晚期可划分 3 个阶段：第 1 阶段（Qp^{1-3} Ⅰ）古湖水含盐量为 1.077~3.864g/L，为淡水环境，显示当时气候温暖偏湿；第 2 阶段（Qp^{1-3} Ⅱ）古湖水平均含盐量最高，可达 5.275g/L，湖水属微咸环境，与生活湖中的网状白花介、粗糙土星介等微体化石生态习性相吻合，气候为温和偏干；第 3 阶段（Qp^{1-3} Ⅲ）古湖水的含盐量有所下降，为 2.630g/L，显示出降水增多，湖水变淡，气候为温暖偏湿的环境。表中八大离子含量中以 SO_4^{2-} 最高，这可能与张家口下花园一带的煤系地层有关，因为煤系地层中含有 S 元素，在延庆古湖发育时期，这些煤系地层正好处在入湖河流附近。CO_3^{2-} 含量最低，几乎近于零，根据化学原理分析，CO_3^{2-} 仅在溶液的 pH 值大于或等于 10 时，才会在溶液中单独存在，当溶液的 pH 值小于 10 且大于 8 时，CO_3^{2-} 与 HCO_3^- 并存，当 pH 值小于或等于 8 时，溶液中的 CO_3^{2-} 完全转变为 HCO_3^-。湖水的 pH 值往往与气候条件紧密相关，鉴于官厅水库南侧剖面的 CO_3^{2-} 含量极微，推断气候不是十分干旱。当时延庆古湖的化学性质以硫酸盐和氯化物型为主。

表 37-2　官厅水库南侧剖面化学成分分析表

Table 37-2　Table of chemical composition analysis of southern Guanting Reservoir section

样品号	离子质量及浓度 样品间距/m	CO_3^{2-}	HCO_3^-	SO_4^{2-}	Cl^-	Ca^{2+}	Mg^{2+}	K^+，Na^+	含盐量/$(g \cdot L^{-1})$	时代
1025		0.000 0.000	0.075 1.229	0.971 20.228	0.023 0.650	0.006 0.300	0.002 0.140	0.673 21.670	1.077	
1027		0.000 0.000	0.089 1.457	0.942 19.618	0.021 0.572	0.011 0.550	0.001 0.060	0.653 21.040	1.706	
1031	1.6	0.000 0.000	0.060 0.982	1.210 25.210	0.964 27.200	0.039 1.941	0.001 0.050	1.596 51.4	3.864	
1034		0.000 0.000	0.054 0.889	1.107 23.059	1.244 35.100	0.029 1.471	0.005 0.415	1.775 57.160	4.215	
1039	2.0	0.000 0.000	0.054 0.878	1.063 22.149	1.110 31.300	0.045 2.261	0.005 0.430	1.603 51.640	3.880	
1041	0.8	0.000 0.000	0.038 0.630	1.520 31.663	1.489 42.000	0.113 5.652	0.010 0.860	2.104 67.780	5.275	Qp^{1-3}
1043	1.1	0.000 0.000	0.086 1.413	0.906 18.868	0.116 3.270	0.013 0.670	0.001 0.105	0.707 22.780	1.830	
1046	0.9	0.000 0.000	0.091 1.498	0.896 18.627	0.090 2.550	0.008 0.410	0.003 0.265	0.682 21.980	1.771	
1048	1.0	0.000 0.000	0.050 0.951	1.035 21.559	1.007 28.440	0.058 2.831	0.010 0.835	1.467 47.24	1.771	
1050	1.0	0.000 0.000	0.056 0.920	0.913 19.016	0.574 16.200	0.025 1.245	0.027 2.209	1.023 32.950	2.630	
1052	1.0	0.000 0.000	0.064 1.044	0.783 16.315	0.280 7.910	0.037 1.861	0.004 0.290	0.716 23.060	1.884	
1054	1.0	0.000 0.000	0.076 1.240	0.852 17.751	0.893 25.200	0.045 2.239	0.066 5.432	1.196 38.530	3.123	
1058	1.6	0.000 0.000	0.062 1.023	0.833 17.360	0.514 14.500	0.024 1.205	0.025 2.018	0.921 29.660	2.379	
1059	0.5	0.000 0.000	0.078 1.281	0.723 15.070	0.347 9.800	0.022 1.084	0.019 1.596	0.729 23.470	1.919	Qp^{2-1}
1060	0.5	0.000 0.000	0.102 1.674	0.783 16.315	0.667 18.800	0.026 1.305	0.032 2.651	1.081 34.830	2.692	
1061	0.5	0.000 0.000	0.048 0.796	0.730 15.201	1.074 30.300	0.029 1.446	0.025 2.038	1.329 42.810	3.241	
1062	0.5	0.000 0.000	0.095 1.550	0.691 14.396	0.542 15.300	0.024 1.220	0.005 0.405	0.920 29.620	2.277	
1066	2.0	0.000 0.000	0.100 1.632	0.822 17.118	0.168 4.730	0.040 2.008	0.013 1.084	0.667 21.810	1.820	

续表

样品号	离子质量及浓度 样品间距/m	CO_3^{2-}	HCO_3^-	SO_4^{2-}	Cl^-	Ca^{2+}	Mg^{2+}	K^+，Na^+	含盐量 /$(g\cdot L^{-1})$	时代
1069	1.5	0.000 0.000	0.066 1.075	0.828 17.259	0.433 12.200	0.021 1.064	0.094 7.871	0.671 21.600	2.112	Qp^{2-1}
1071	1.0	0.000 0.000	0.075 1.229	0.927 17.219	0.329 10.700	0.073 2.120	0.079 3.996	0.715 23.930	2.087	
1072	0.5	0.012 0.413	0.074 1.219	0.899 10.755	4.155 32.890	0.023 1.306	0.043 3.504	1.468 47.060	3.651	
1078	2.5	0.000 0.000	0.100 1.632	0.302 16.707	0.372 10.500	0.060 3.992	0.057 4.579	0.660 21.270	2.050	
1082	2.0	0.000 0.000	0.063 1.033	0.956 19.919	0.415 11.700	0.012 0.592	0.070 1.677	0.991 31.414	2.410	
1083	0.6	0.000 0.000	0.062 1.043	0.812 16.902	0.348 9.600	0.070 3.514	0.034 2.800	0.661 21.300	1.879	
1084	0.7	0.004 0.145	0.115 1.800	0.852 17.751	0.033 0.940	0.006 0.301	0.001 0.018	0.617 19.930	1.629	Qp^{2-2}
1087	1.5	0.000 0.000	0.109 1.707	0.536 11.917	0.055 1.550	0.022 1.894	0.021 1.757	0.386 12.420	1.127	
1088	0.4	0.007 0.217	0.107 1.787	0.876 10.253	0.039 1.840	0.008 0.382	0.008 0.683	0.621 20.020	1.666	
1090	1.3	0.007 0.217	0.076 1.250	0.671 13.776	0.038 1.870	0.054 1.677	0.032 2.661	0.371 11.840	1.229	
1093	0.7	0.000 0.000	0.144 2.356	0.649 13.524	0.037 1.876	0.017 0.764	0.013 1.054	0.464 14.932	1.325	

注：格内上、下两行数据分别为离子质量（g/L）和浓度（mg/L）。

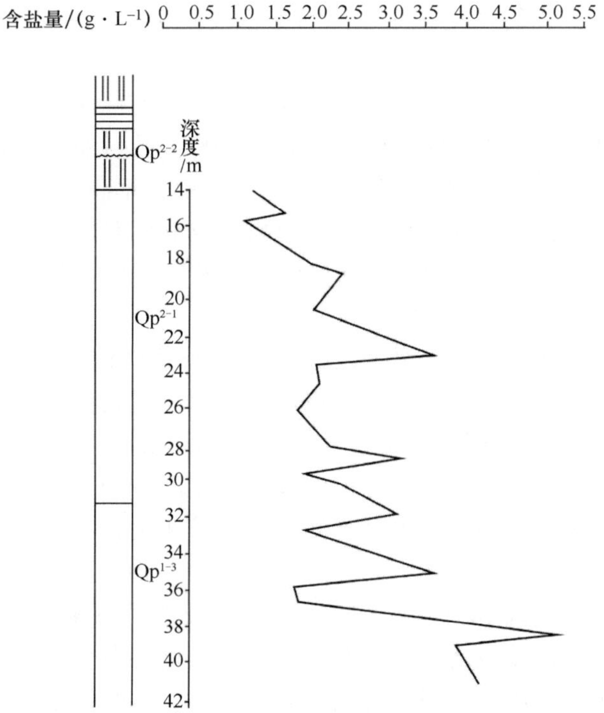

图 37-14　官厅水库南侧剖面古湖水矿化度曲线

Fig. 37-14　Mineralization degree curve of ancient lake water of southern Guanting Reservoir section

四、磁性地层

环境磁学是介于地学、磁学和环境科学之间的新兴边缘学科，它是通过对不同空间和时间范围内的沉积物中磁性参数的测定，提取古气候和环境变化的信息。各类沉积物中磁性矿物在迁移、沉积过程中，由于环境的变化，其磁性矿物的赋存状态（如成分、粒度、含量等）随之变化，从而记录了气候和环境变化的信息。因此，通过对某些沉积物如风成黄土、湖泊沉积、海洋沉积以及大气中的漂浮物质等的各种磁学特征的系统测试和研究，寻找其环境意义，从中获取规律性的认识，获取环境变化研究新的手段，开拓多学科间相互配合。由于磁记录的稳定性以及磁性参数测量的快速、灵活性和无破坏性，使得环境磁学很快成为认识和研究古环境演化规律的重要手段之一。

样品沉积物磁化率是衡量沉积物在外磁场作用下被磁化难易程度的物理量。沉积物是在特定的沉积环境中形成的，记载了环境条件的变化，其所携带的磁性矿物则因其对环境的灵敏反映和记录的稳定性而成为较好的环境指示物质。磁化率作为表征物质磁学特征的物理量，能够判断样品记载的环境变化信息、分析古气候变化规律及其细节、推断样品形成过程的环境条件、为古环境研究提供可靠的磁学证据。

在湖相沉积物中，磁化率反映了沉积物所含磁性矿物类型、含量和磁颗粒的总体水平。胡守云等（1998）对呼伦湖沉积物的环境磁学研究揭示，磁化率的高（低）相应指示湿润（干旱）的气候及较高（低）的湖面，在高湖面、高有机碳含量的情况下，沉积环境相对还原，自生的亚铁磁性铁硫化物成为导致沉积物磁化率增强的主要磁性矿物。湖泊沉积物磁化率的变化特征，常常同相应的孢粉学、地球化学特征相一致，说明磁化率可作为一个环境代用指标。研究认为，在相对暖湿时期具较强氧化条件的环境中，形成磁化率较高、粒度较细的沉积物，指示较高的沉积水位；反之，在相对干冷具较强还原条件的环境中，形成磁化率较低、粒度较粗的沉积物，指示较低的沉积水位。

从新5孔来看，在早更新世早期，磁化率出现1个高值，反映了冲积成因的粗粒沉积，在早更新世中期，出现1个磁化率高值，同样反映冲积成因的粗粒沉积，到了晚期，出现3个磁化率高值，为以粗粒为主的冲洪积物；中更新世出现8个高值，为以粗粒为主的冲洪积沉积。仔细比对磁化率高值出现的深度和所对应的岩性，我们发现磁化率高与颗粒较粗相对应，这可能与物源区岩性特征有关，总体反映暖湿气候。

根据研究区内14个钻孔岩芯沉积物磁化率-深度曲线（图37-15）特征，可见磁化率值总体偏低，波动相对较小，其中新2孔、新3孔、新4孔、新5孔、新10孔、新11孔、新12孔、新14孔中可见2~3个磁化率高峰值，反映当时气候波动性相对不大，3个磁化率高峰值可能代表3次暖湿期-干冷期的变化旋回；而中更新世以来，磁化率值总体相对偏高，波动相对较大，除钻孔2、6、7外，其余钻孔都可见5个以上的磁化率高峰值，反映中更新世以来气候变化波动较大，至少出现了5次暖湿期-干冷期的变化旋回。

选择代表性的钻孔4和钻孔5中磁化率-深度曲线与深海氧同位素进行对比分析，2个钻孔中更新世以来的磁化率变化曲线有规律波动，其曲线变化规律与深海氧同位素（$\delta^{18}O$）波动曲线变化规律一致，其磁化率峰值分别对应深海氧同位素1~19阶，反映了北京市平原地区中更新世（0.78Ma）以来出现了9次暖湿期-干冷期的变化旋回，磁化率高峰值对应温度曲线的暖峰位置。以上磁化率曲线分析表明北京平原地区中新世以来的气候变化主要受到全球气候变化影响，而磁化率的次一级变化则反映了区域气候变化的影响。

综上所述，通过北京平原地区14个钻孔岩芯磁化率变化曲线与深海氧同位素（$\delta^{18}O$）波动曲线的分析对比，可以得出气候变化的相似结论，从而表明气候变化具有全球性特点，只是由于区域性的气候差异，在程度上及时间上有微小的不同。由此可见全球性气候变化是引起北京市平原区环境变化的主导因素，根据众多钻孔岩芯沉积物中的孢粉组合、生物演化、地球化学特征分析，并辅以磁性年代测定，可以得出北京平原地区第四纪以来的气候变化规律和冷暖干湿变化周期。

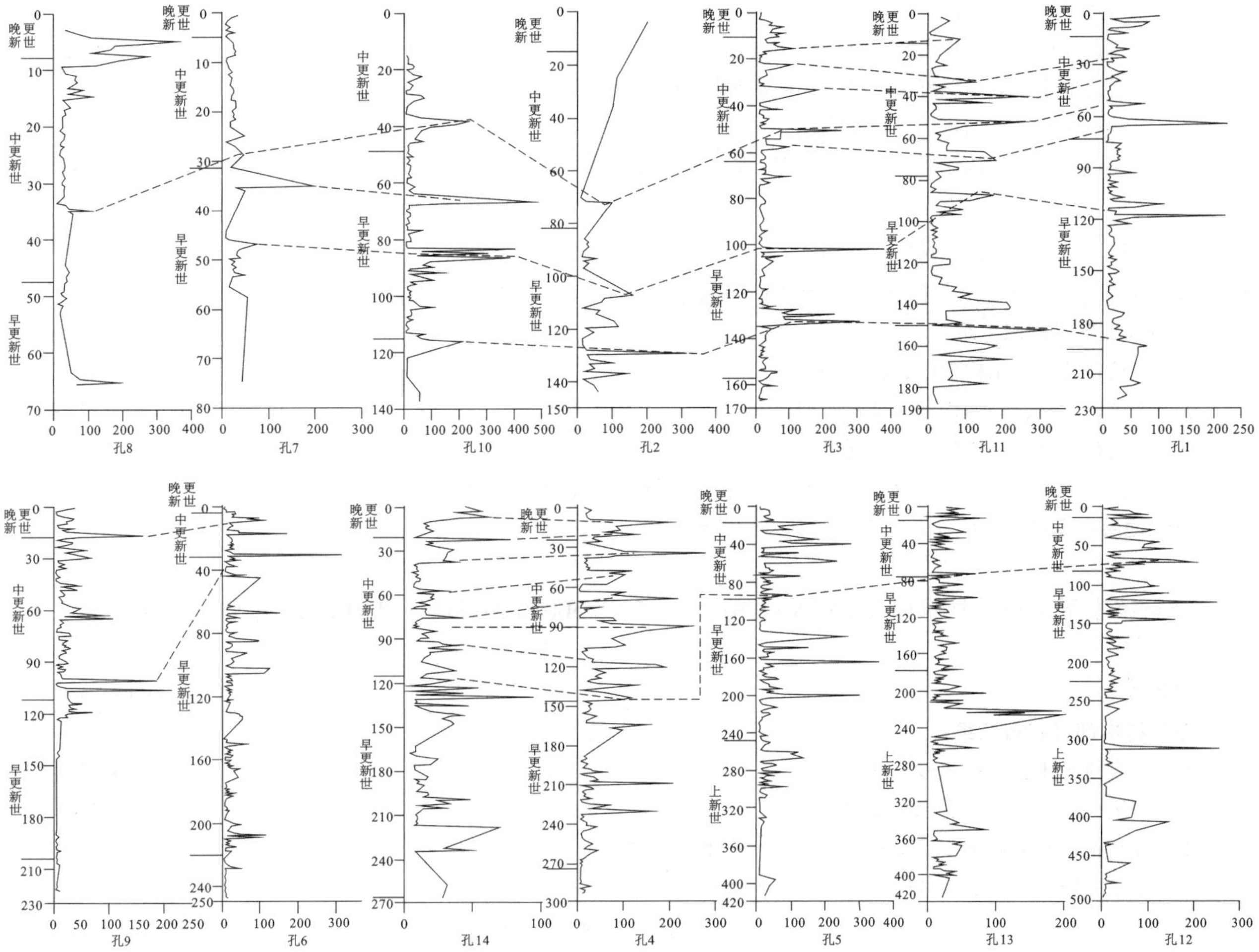

图 37-15　钻孔岩芯沉积物磁化率-深度曲线

Fig. 37-15　Curves of magnetic susceptibility versus depth for sediments from drilling cores

深度单位为 m，磁化率单位为 κ（$10^{-3}m^3 \cdot kg^{-1}$）

钻孔岩芯磁化率曲线变化规律与深海氧同位素（$\delta^{18}O$）波动曲线变化规律一致，反映了北京市平原地区中更新世（0.78Ma）以来出现了 9 次暖湿期-干冷期的变化旋回，早更新世可能出现了 3 次暖湿期-干冷期的变化旋回，这和孢粉分析的结果吻合性比较好。以上磁化率曲线分析表明北京平原地区中新世以来的气候变化主要受到全球气候变化影响，而磁化率的次一级变化则反映了区域气候变化的影响。

第二节　全新统

一、岩性地层

（一）代表剖面

全新统上部按照堆积类型分为两个代表剖面，冲积型以黎辛庄段河漫滩刘斌屯组剖面为代表；风积型以丁各庄北风成砂丘刘斌屯组剖面为代表。描述如下：

北运河冲积型，黎辛庄段河漫滩剖面：

全新统上部（刘斌屯组） 总厚 2. 00m

1. 黄褐色耕植土，松散 0. 20m
2. 土黄色粉砂，结构松散，见楔状交错层理 0. 50m
3. 灰褐色黏土，固结较好，见锈斑、及水平层理 0. 10m
4. 土黄色黏砂，结构松散，见斜层理 0. 40m
5. 灰黄色细砂，结构松散，见水平层理，石英含量高 0. 80m

风积型，丁各庄北风成砂丘剖面：

全新统上部（刘斌屯组） 总厚 11. 2m

1. 土黄色粉砂，结构松散，粒度均一 5. 0m
2. 土黄色粉砂，结构松散，见斜层理，含瓦块 1. 2m
3. 土黄色粉砂，结构松散，含瓦块 5. 0m

--------------- 平行不整合 ---------------

下伏地层：全新统中部（尹各庄组）黄褐色砂质黏土

全新统中下部以通县双埠头 ZKTX-1 钻孔（据 1：50000 通县幅，1991）为代表，描述如下。

全新统 总厚 14. 23m

全新统中部（尹各庄组）

1. 地表耕植土，灰褐色砂质黏土 0. 92m
2. 灰褐色砂质黏土，见水平纹层 0. 53m
3. 浅灰褐色黏质砂土，结构松散，普遍发育水平纹层 0. 83m
4. 褐色砂质黏土，底部铁染发育，见水平纹层 2. 34m

———— 整　合 ————

全新统下部（肖家河组）

5. 浅黄灰色粉砂，分选中等，磨圆不好，石英、长石为主，见云母和暗色矿物，水平层理发育 0. 67m
6. 褐灰色黏质砂土，有机质含量高，见大量植物根系 0. 21m
7. 灰色含粗砂细砂，砂质成分以石英、长石为主，次为岩屑和暗色矿物，分选磨圆不好，有机质含量高 1. 87m
8. 褐灰色砂质黏土，含植物根系 0. 20m
9. 灰色、浅灰色中、粗砂，分选磨圆不好，以石英、长石为主，暗色矿物含量 5%～7% 0. 35m
10. 暗灰色砂质黏土，砂为中砂，含量 18%左右，有机质含量高，见炭屑和贝壳碎片 0. 15m
11. 灰色中、细砂 0. 55m
12. 浅灰色中砂，松散，磨圆不好，分选中等，见泥质团块 1. 00m
13. 灰褐色含砾粗砂，砾石含量 3%～4%，砾径 3～10mm，磨圆较好。砂质成分以石英、长石为主，次为暗色矿物，结构松散，分选磨圆不好 0. 73m
14. 灰色中砂，结构松散，石英、长石为主，次为暗色矿物，分选磨圆不好 0. 67m
15. 灰白、浅灰色含砾粗砂，结构松散，砾石含量约 3%，砾径 2mm×3mm～10mm×20mm，分选差，磨圆较好 2. 02m
16. 浅灰色中-粗砂，结构较致密，略固结，分选磨圆不好，以石英、长石为主，次为暗色矿物，含有机质较多 1. 19m

--------------- 平行不整合 ---------------

下伏地层：下更新统通县组上段下部：灰黑色淤泥

（二）岩性岩相变化

全新世堆积物主要分布在河流、河漫滩、低洼地沉积，横向变化较大，据勘探资料显示，其沉积厚度的变化是受古地形和构造的控制，各处相差悬殊，总的规律是由西北部向东南部逐渐增厚的趋势，一般从3～20m不等，在一些古河道的牛轭湖或低洼地有淤泥埋藏或泥炭沉积。地表因人类长期活动，堆积厚度不等的人工填土。

全新世地层由明显的三段结构组成。下段厚度在0.5～12.0m之间，主要由灰黄、灰褐色粉细砂、黏质粉砂、灰色黏土类等组成，以河流相为主，局地出现湖沼相，时代属全新世早期；中段厚度在1.5～9.0m之间，以灰色、灰褐色、灰黑色黏质粉砂、粉质黏土、黏土、淤泥质等为主，粒度较细，以湖沼相为主，河流相次之，时代属全新世中期；上段厚度为3～7m，沉积物颜色较中段发生明显的变化，逐渐演变为黄褐色色调为主，属于泛滥平原沉积环境，河流沉积次之，局部地段出现决口扇堆积，时代属全新世晚期。

（1）全新统下部（肖家河组）分布及岩性特征

全新统下部（肖家河组）广泛分布在平原地区，除局部出露地表外，皆被全新世中、晚期沉积物掩埋于地下，多埋藏较深，一般为2～10m或更深。其厚度一般2～6m，最厚达二十余米。为一套以冲积为主的堆积物，岩性一般为砂砾石层和粗、中、细砂及黏质砂土、砂质黏土。沉积物色调以灰色为主。局地出现泛滥平原亚相（王府井东方广场、大兴小红门），岩性为亚黏土、亚砂土，有时可见到棕褐色黏土层，亚黏土、亚砂土层中可见到黄色锈斑，有大量根孔发育。

（2）全新统中部（尹各庄组）分布及岩性特征

全新统中部（尹各庄组）大面积出露地表，发育于全新世河谷地区，构成Ⅰ级阶地。其主要分布于河流域的Ⅰ级阶地上。厚度一般为2～6m。

全新统中部（尹各庄组）主要有三种沉积类型：冲积型河床沉积、冲积型河漫滩沉积和古湖沼型沉积，局地可见决口扇相（王府井东方广场、大兴小红门）。河床沉积主要分布于Ⅰ级阶地古河床中，在野外剖面上见岩性多为粉、细砂沉积，结构松散，并发育水平及旋卷层理，局部见交错层理。河漫滩沉积范围较广，主要分布在Ⅰ级阶地之上，发育典型二元结构，于下部砂层中见水平及交错层理。古湖沼型沉积主要分布于清华园火车站—政法大学、王四营—小郊亭—南花园—白鹿司—南四营及昆明湖一带，延庆局部地区可见。古湖沼型沉积沉积了一套以褐色黏土、灰黑色淤泥为主，局部夹泥炭（海淀区北坞、延庆东门营村附近）的古湖沼堆积。决口扇相仅王府井东方广场剖面有所报道，显示了至少三次洪水快速堆积过程（袁宝印等，2002）。

（3）全新统上部（刘斌屯组）分布及岩性特征

全新统上部（刘斌屯组）仅分布在部分河槽及河漫滩范围内，还见于一些已消亡的古河道中和一些发育于Ⅰ级阶地上的小型洼地中。沉积物成因类型以冲积为主，局部为湖沼沉积，岩性为土黄色、灰黄色、灰褐色黏质砂土、黏土及粉、细砂，水平层理、斜层理和交错层理发育，厚2～3m；部分地区见河流相沉积（延庆张山营村、军营村附近），岩性为砂砾石夹细砂，灰褐色黏砂沉积物；还有个别地区见风积物（王府井东方广场、大兴小红门、祁家豁子），岩性以亚砂土-黄土层为特征，显示为风成黄土堆积。

二、生物地层

（一）孢粉组合

1. 延庆沉积盆地

以延庆西五里营钻孔剖面为代表：

延庆西五里营钻孔孔深7.40m。该孔可划分出3个沉积韵律层，反映出洪积扇的发育过程。根据孢粉图示分析（图37-16），自7.00～5.35m之间是以松、蒿花粉组合带，木本植物以松属为主，在5.78m处松粉占孢粉总量的48%，草本植物中以蒿属、菊科及水龙骨为主，从蒿属含量变化看由

77.5%降至41%（5.78m处），反映当时环境的变化，山地生长着松林，而盆地以干旱草原为主，但在此期间气候是以温凉偏干为主，亦有干湿的波动。在5.35~3.50m之间，松属花粉含量下降到10%，而蒿属花粉含量出现3次多达90%以上的时段，反映当时是以草原为主，而气候曾有过2~3次波动，表现为气温下降的冷干环境，推测该带的底界为距今5000a前后。3.5~1.0m之间为松属花粉占优势，1.7~1.8m之间松粉占孢粉量95.1%，出现时间为距今4000~3600a。而草本植物中蒿属和水龙骨在该层段的底部和顶部形成峰值，而在中部为低值，但从孢粉组合中水生、湿生花粉含量少，而旱生的麻黄连续出现，反映当时气候温凉偏干。在1.0~0.3m处（^{14}C年龄测定为1895±100a B.P.）孢粉组合是以松粉减少，蒿粉增加为特征，0.30~0.35m之间的样品中几乎全部为蒿粉，反映当时为草原植被，气候已向干旱转化。

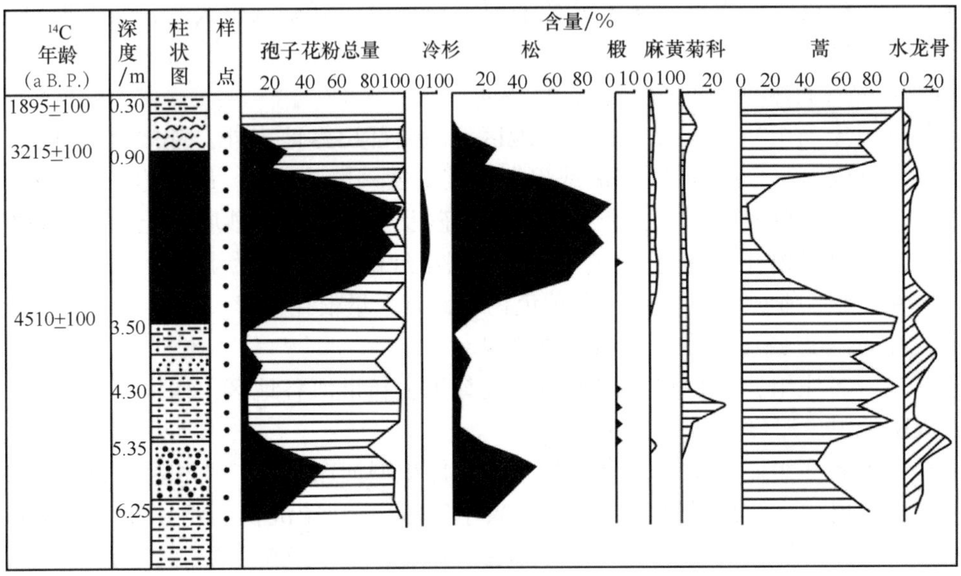

图37-16　延庆盆地西五里营钻孔孢粉图示

（据孔昭宸等，1982）

Fig. 37-16　Pollen diagram of Xiwuliying hole in Yanqing Basin

2. 北京平原

以坟庄剖面孢粉组合（图37-17）为代表：

早全新世（距今12000~8500a）：距今12000a前后，孢粉组合中木本植物花粉占孢粉总量的79.4%，而草本植物花粉含量大量减少。距今（11850±200）a时，云杉、冷杉、松的花粉占孢粉总量的33%，阔叶树有椴、榆、栎等花粉，草本植物中藜科和唐松草比例下降为1%~3%，蒿属与香蒲也呈下降趋势。由于阔叶树的出现和数量的增多，反映气温的回升，使古地理环境发生了明显变化，从而进入全新世时期，其时限可定为距今12000a前后。

早全新世，依据孢粉组合中木本植物、草本植物和蕨类植物中属种及数量的变化，反映当时植被演替和气候波动的状况。距今11819~10860a之间（孔深6.81~5.51m），可划分出前后两期，早期虽仍有不少针叶树种的存在，但阔叶树种的数量有所增加，当时植被是以针阔叶混交林为特征，沼泽发育，林下灌丛和草被不很发育，反映气候温暖湿润；晚期时，是以椴、松为主的针阔混交林，沼生植物数量减少，反映当时沼泽面积缩小，气候为温暖偏干。距今11000a前后，以松、椴、栎组成的温带针阔混交林又有所减少，而蒿属、藜科花粉数量有所增加，显示气候向干凉转化。距今10000a前后，松属花粉有所增加，而蒿属、藜科等继续存在，沼泽洼地缩小，气候冷干。距今10000~8500a时期，以草本植物蒿属、菊科、藜科等占优势，山地区有温带针叶林。当时气候干凉，随后冷杉、松粉数量增加，并有水龙骨孢子，反映为冷湿气候。

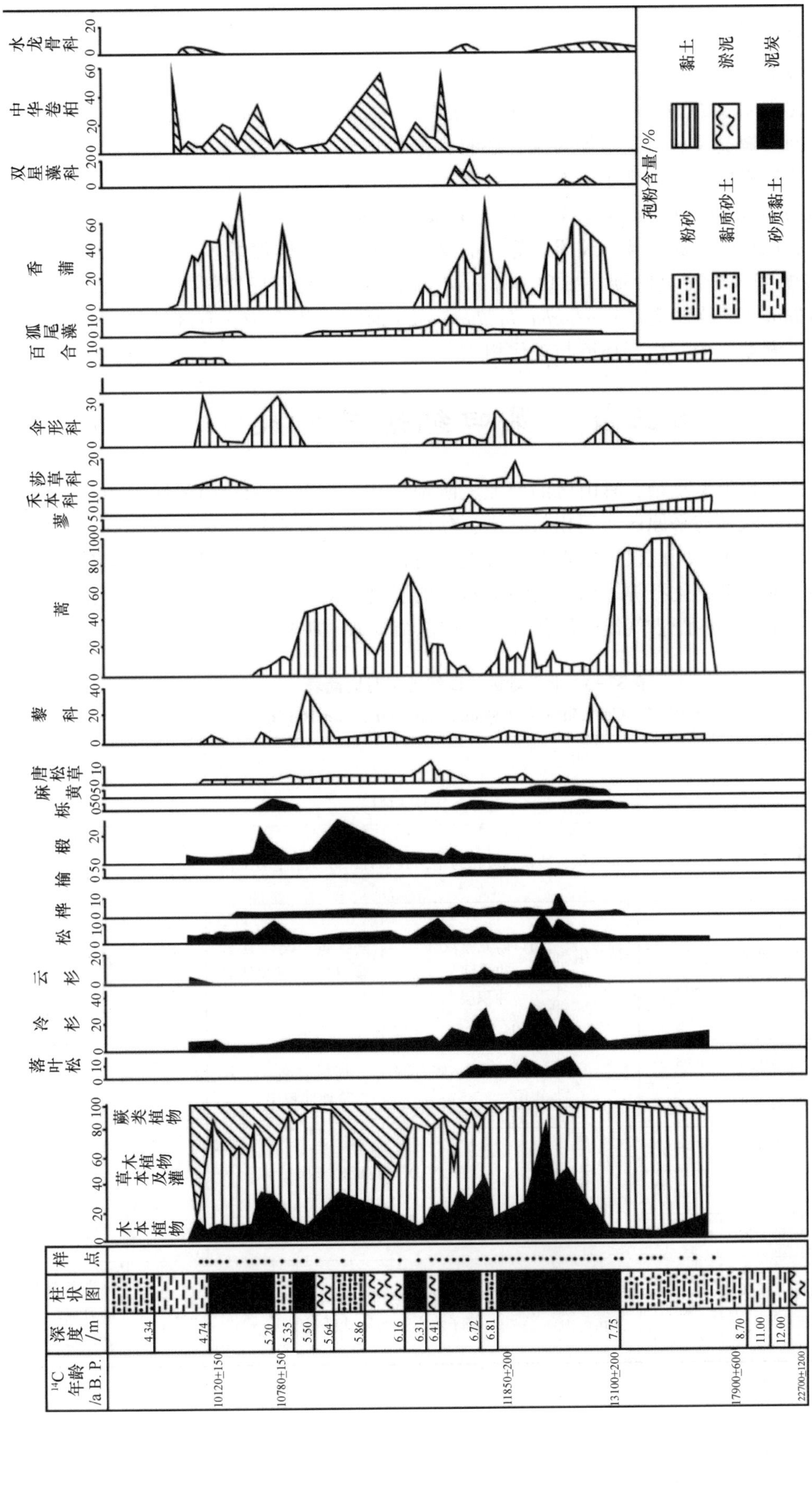

图 37-17 坟庄剖面孢粉图示

（据孔昭宸等，1982，修改）

Fig. 37-17 Pollen diagram of Fenzhuang section

（二）哺乳动物化石

平原区在全新世地层中发现的哺乳动物甚少，仅发现碎骨和用梅花鹿角做的工具，此外还发现有家禽、牲畜的杂骨等。

（三）软体动物

全新世早期，出现 *Metodontia bersotoski*（狭长间齿螺）-*Galba truncatula*（截口土蜗）组合带，以肺螺亚纲的陆生狭长间齿螺类及截口土蜗等为主，还有少量粉华蜗牛和萝卜螺等，化石属种单调，多生活于温凉环境中。全新世中期，出现 *Tricula*（拟钉螺）-*Bithynia fuchsiana*（赤豆螺）-*Segmentina*（隔扁螺）组合带、*Parafossarulus striatulus*（纹沼螺）-*Bithynia fuchsiana*（赤豆螺）组合带，以水生前鳃亚纲类为主，其中有拟钉螺、赤豆螺、纹沼螺等，不仅属种丰富，个体数量多，并且多为喜暖湿属种，反映当时气候温暖湿润。全新世晚期，出现 *Bellamya purificata*（梨形环棱螺）-*Radix*（萝卜螺）-*Semisulcospira cancellata*（方格短沟蜷）-*Corbicula fluminea*（河蚬）组合带，有环棱螺、蓝蚬等，现多分布于河流、水渠、湖塘之中，生活环境与现代相似。

第三节　第四纪地层对比

北京地区第四系沉积类型复杂，有山区的河流、阶地堆积、洞穴堆积，山麓地带冲洪积相的砂砾石及黏性土沉积。山间盆地与平原区河湖相粉砂质、黏性土、细砂（粗砂）夹砾石层沉积，也有湖沼相有机质黏土及泥炭沉积等，因此仅根据岩性进行地层对比存在一定困难，根据孢粉组合特征、^{14}C 年龄测定、磁性地层、哺乳动物化石等综合因素将第四纪地层进行对比（表 37-3），同时，综合本区生物地层特征，对北京地区第四纪古气候的演化进行梳理总结（表 37-4）。

表 37-3　北京地区第四纪地层对比简表

Table 37-3　Correlation of Quaternary strata in Beijing area

<table>
<tr><td rowspan="2">时代</td><td colspan="7">山区</td><td rowspan="2" colspan="2">平原地区代表地层</td></tr>
<tr><td colspan="4">洞穴代表地层</td><td colspan="3">山区代表性地层</td></tr>
<tr><td>全新世</td><td colspan="4">石笋微层理沉积</td><td colspan="3">冲洪积、冲积、坡积为主，砂砾石、砂及粘土类沉积</td><td>全新统
Qh</td><td>砂质粘土、
细砂夹泥炭层</td></tr>
<tr><td>晚更新世</td><td>羊耳峪组
Qp_3y</td><td colspan="3">羊耳峪村东岭子
洞穴堆积物为代表</td><td>马兰组
Qp_3m</td><td>周口店龙家坟
牧牛河
冲积层</td><td>马兰黄土与二层棕壤型古土壤雁翅红黄土</td><td>上更新统
Qp_3</td><td>洪积砂卵砾石层、河湖相黄土质粘性土夹砾石，局地含钙质结核。</td></tr>
<tr><td rowspan="5">中更新世</td><td rowspan="5">周口店组
Qp_2z</td><td rowspan="5">第 1
地点</td><td>第 1 段</td><td>1~3 层</td><td rowspan="5">周口店组
Qp_2z</td><td rowspan="2">上段：红土-碎石坡积层</td><td rowspan="2">三福村红色土、第二阶地冲积层</td><td rowspan="5">中更新统
Qp_2</td><td rowspan="5">棕红色河湖相砂质粘土、粘质砂土夹砾石</td></tr>
<tr><td>第 2 段</td><td>4 层</td></tr>
<tr><td>第 3 段</td><td>5~6 层</td><td colspan="2" rowspan="2">中段：黄红色亚砂土层</td></tr>
<tr><td>第 4 段</td><td>7~11 层</td></tr>
<tr><td>第 5 段</td><td>12~13 层</td><td colspan="2">下段：红色砾石层</td></tr>
<tr><td>早更新世</td><td>太平山组
Qp_1t</td><td colspan="2">太平山北
坡东洞</td><td>第 1 地点
探井
14~15 层</td><td>泥河湾组
Qp_1n</td><td colspan="2">官厅水库，阜高营，杨庄户，
篓子水，长沟峪砂砾石层</td><td>下更新统
Qp_1</td><td>河湖相粘质砂土、砂质粘土互层含砂砾石</td></tr>
</table>

表 37-4　北京地区第四纪以来古气候的演化对比简表

Table 37-4　A brief comparison of paleoclimate evolution since Quaternary in Beijing area

时代		年龄	孢粉组合	植被类型	气候状况	沉积相
全新世	晚	3000a 至今	松、栎、蒿、菊	针阔叶林、草原发展	温干	泥炭停止发育
	中	8000~3000a	松、栎、榆、桑、中华卷柏	落叶阔叶林夹针叶林	温暖，距今 7000 和 5000 年短期降温	泥炭湖沼地区域缩小
	早	12~8ka	松、冷杉、桦	针阔叶混交林	开始转暖，较今稍凉湿	泥炭湖沼开始发育
晚更新世	晚	30~12ka	松、云杉、冷杉、栎、桦、蒿、菊、藜	针阔叶混交林—草原	干冷，曾比现今低 7~8℃	有融冻褶曲现象发现披毛犀化石
	中—早	150~30ka	栎、榆、松、桦、蔷薇、藜、菊、蒿	阔叶阔叶林草原	温暖，曾比今高 2℃	现代水文网成型，永定河扇地发育
中更新世	晚	230~150ka	蒿、藜为主、少量桦、栎	疏林草原	干凉	湖沼消失，坡洪积物广泛发育
	中	250~230ka	藜、蒿、松、云杉、桦、栎	含着阔叶树的温带草原	温带半干旱气候干湿冷暖波动明显	湖沼广泛发育，剥蚀丘陵区缩小，沉积平原区扩大
		310~250ka	藜、桦、松、冷杉、栎、榆	针阔叶混交林—草原	趋向干旱的温带气候	
		370~310ka	栎、榆、胡桃、桦等	阔叶阔叶林草原	温带半湿润，偏暖	
		600~370ka	松、栎、榆、胡桃、桦、蒿等	落叶阔叶林夹针叶树草原	趋向干冷的温带半干旱气候，距今 40 万年温暖	
		700~600ka	松、冷杉、桦、栎、胡桃、榆、朴	针阔叶混交林	温带半湿润，偏暖	
	早	800~700ka	桦、栎、榆、松	阔叶阔叶及针阔混交林	由偏干的温带气候向半湿润暖温带气候发展	冲积物发育，八达岭古河消失
早更新世	晚	1950~800ka	榆、栎、椴、桦、松	暖温带落叶阔叶林，含针叶树	潮湿，向上变干凉	湖沼广泛发育
	早	2480~1950ka	藜、菊、蒿、桦、榆、栎、朴，底部为孢粉贫乏带	温带针阔混交林草原	干冷，向上转为温润	坡洪积、冲洪积发育，在顺义凹陷带中发生海侵

第三十八章　沉积物与沉积作用

第一节　第四纪沉积作用及沉积类型

一、河流沉积作用及沉积类型

北京地区河流作用是第四纪沉积物搬运和沉积的主要地质营力。河流沉积作用主要有以下两种沉积类型。

（一）曲流河沉积

在山区和平原均较发育，山区主要沉积物为砂砾石层，且粒度越往上游越粗，递变为砾石层，近山麓地带则以中细砂、砂砾石沉积为主，发育边滩（点砂坝）沉积，下部发育板状交错层理，上部发育槽状交错层理，局部见有水平层理。局部地带可出现牛轭湖沉积的砂夹砂质黏土、黏质粉砂，如晚更新世和全新世沉积物。平原区沉积物多由砂砾石层、中粗砂、粉细砂组成，多数分布在永定河与潮白河流域，温榆河流域亦有少量分布，平谷盆地内钻孔揭露的砾石层厚度可达几十米，为泃河和错河沉积物。北京平原多为冲洪积扇形成的泛滥平原，其上发育有多期曲流河。

（二）辫状河沉积

辫状河沉积主要发育在平原地区，具有河床宽浅、多河道、侧向迁移等特点，发育河床和泛滥平原沉积亚相，与曲流河相比，不发育堤岸和牛轭湖沉积亚相。心滩沉积物下部为早期砂坝残留，上部为泛滥平原，由黏土和粉砂组成。钻孔中揭露沉积构造较少，集中在后沙峪-俸伯凹陷内，属于北京凹陷北缘，大兴隆起上亦有分布，属潮白河水系和永定河水系。

二、湖泊沉积作用及沉积类型

北京平原主要的第四纪沉积盆地中均有湖泊沉积发育，如马池口凹陷、后沙峪-俸伯凹陷及平谷凹陷等。湖泊沉积物以灰绿色、灰色、棕灰色或棕红色等为主，黏土呈块状结构，含钙、铁、锰质结核，具有完整旋回韵律，富含生物化石或者炭化植物根系。深湖相以厚层黏土-砂质黏土为主，夹有薄层砂；滨湖相以厚层砂层为主，夹薄层黏土，常见二者呈交替沉积。滨湖-深湖沉积自下而上为厚层砂-泥质砂-黏土；深湖-滨湖沉积自下而上为厚层黏土-砂质黏土-黏质粉砂-砂层。沉积物组合主要见于马池口凹陷、平谷凹陷及延庆凹陷内，厚度不一，多为几米至几十米。砂层之上多见淤泥层，属河漫湖沉积。

三、沼泽沉积作用及沉积类型

平原区该类型主要沉积物为泥炭层或腐殖质黏土、砂质黏土（局部含泥质细条带），厚度较薄。下部具有水平层理和低角度的斜层理，上部层理不发育、多植物根系、虫孔、干裂。该沉积类型多见于全新世地层。北京地区该沉积类型主要分布在西南部（如房山坟庄地区），主要位于河流水系的一级阶地附近，为一套扇缘洼地沉积的褐色淤泥和泥炭，局部为灰黑色砂质黏土。

四、风的沉积作用及沉积类型

在地貌上形成风积地貌和风蚀地貌，风积地貌的代表类型有砂丘、砂垄和砂链，风蚀地貌是由于风力挖蚀作用形成的凹形地貌。风积物岩性以灰白、黄灰色中细砂为主，砂粒的分选及磨圆良好，具有明显的风成交错层理。风积物中轻矿物占90%~95%，以石英、钾长石为主，含少量斜长石及贝壳碎屑；重矿物含量大于2%，以角闪石、绿帘石、磁铁矿和褐铁矿为主，其次为石榴子石、锆石及云母，含少量电气石、金红石和钛铁矿。

五、冲洪积扇沉积作用及沉积类型

冲洪积扇主要分布于河流的出山口地带，以永定河和潮白河冲积扇最发育，其中永定河冲积扇包括南北两个扇体，南部扇体形成时代较早，北部扇体形成时代较晚。上述两个冲积扇是北京平原的主要组成部分。此外，西山和北山山前，均发育有规模较小的冲洪积扇，岩性以棕色、棕褐色、褐色的中粗砂为主，含砾石及不定量的黏土、粉砂质黏土，具有多阶小旋回的特征。砂层底部常见砂和黏土的混合物，黏土中含钙质结核以及砂粒，具有砾石-泥质砂、砂质黏土的二元结构，黏土底部有时可见小砾石。

六、冲积平原沉积作用及沉积类型

北京平原即永定河和潮白河两大水系形成的冲积平原。沉积物较细，由灰黄色、黄色含黏土细粉砂与黏土不等厚互层组成，局部夹有中细砂，靠近叉道的部分粉细砂增厚，黏土变薄；远离河道黏土层增多加厚，砂泥之间呈突变关系，植物碎屑丰富，常构成纹层。沉积物中常夹有铁锈斑、铁染条带等，反映沉积环境为氧化环境。沉积构造发育，当泥质含量升高时，主要为砂泥互层水平层理；细粉砂层内层理类型较多，有水平层理、交错层理、爬升层理、透镜状层理等，层理厚度一般为3~5cm。常含大量的陆相淡水腹足类、双壳类生物化石，整体构成曲流河和辫状河发育的冲积平原类型。

七、复合沉积作用及沉积类型

复合沉积作用见于北京平原区低洼地带，如马池口凹陷、后沙峪-俸伯凹陷等，为随季节和气候变化而湖泊与沼泽交替变化形成的沉积类型，由浅灰、灰、深灰、灰绿色黏土和粉砂质黏土和砂质黏土或黏质粉砂组成，局部夹有泥炭层，发育水平层理。以富含属种单调的陆相淡水腹足类、双壳类生物化石及有机质为特征，地表发育喜水耐碱的草本植物。

八、洞穴沉积作用及沉积类型

洞穴堆积是第四纪一种特殊的地层类型。它伴随着第四纪某一地史阶段地壳的稳定下降，气候的冷暖变化，人类与脊椎动物的活动以及岩溶、流水或崩塌等地质作用，逐渐形成了有特定生物的角砾状碎屑与黏土堆积。多以中粗砂、角砾及棕红色黏土为主，并见人类化石及其遗迹（尤其是石器和灰烬层等），而且含有丰富而连续演替的哺乳动物化石。另一类特殊的洞穴沉积类型为碳酸钙沉积，如北京石花洞内的钟乳石沉积，从沉积形态来看，包括粗犷石笋、杆状石笋及微层理石笋等。

第二节　第四纪主要沉积物岩石类型

本区第四纪沉积物主体为陆相沉积，包括河流、湖泊、沼泽、风积、冲洪积等形成的沉积建造。不同时期、不同地区的主要沉积组合类型见表38-1。

表 38-1　北京地区不同时期主要沉积类型表

Table 38-1　Main sedimentary types in Beijing area

地质时期			太行山-燕山地区	北京山前平原地区
第四纪	全新世	晚期	河流、湖泊相砂质、砂泥质沉积建造，间夹砂砾石沉积建造为主，局部见风积砂土建造	湖相，沼泽相砂质黏土，灰黑色黏土，淤泥夹泥炭沉积建造
		中期		以湖沼相砂泥质沉积建造为主，局部地区出现河流、湖沼相砂质、砂泥质、泥质沉积建造，间夹河流相砂质沉积建造
		早期		以河流、湖沼相砂质、砂泥质沉积建造为主，部分地段夹有河流相砂质沉积建造
	更新世	晚更新世	河流、湖泊、湖沼相及残坡积砂石、砂、砂泥及泥质沉积建造，部分地质发育风积砂土建造	以河流相砂质、砂泥质、泥质沉积建造为主，间夹湖相砂泥质沉积建造，部分地质有河流相砂砾质沉积建造
			冰碛相岩块、角砾、砾石与砂泥混合堆积建造及冰水相含砾砂泥质沉积建造	冲洪积相含细砂、砾、砂泥质（泥包砾）、泥质沉积建造
		中更新世	河流、湖泊、湖沼相及残坡积砂砾石、砂、砂泥及泥质沉积建造	以河流相砾石、含砾砂质、砂泥质、泥质沉积建造为主，间夹湖相砂泥质、泥质沉积建造，部分地段有泥炭沉积建造
			冰碛相岩块、角砾、砾石与砂泥的混合堆积建造及冰水相含砾砂泥质建造	以河流相含细砾砂泥质、泥质沉积建造为主，山前地区有少量冰碛相砾石混合堆积
		早更新世	河流、湖泊、湖沼相砂砾石、砂、砂泥及泥质沉积建造	以河湖相砂质、砂泥质、泥质沉积建造为主，部分地段夹有河流相砂砾质沉积建造
			冰碛相岩块、角砾、砾石与砂泥的混合堆积建造及冰水相含砾砂泥质、泥质沉积建造	以冲洪积相砾石、含砾砂泥质、泥质沉积建造为主，部分地区有少量湖沼相泥质的扇缘洼地沉积建造
			以河流和湖泊相砂砾石、砂、砂泥及泥质沉积建造为主，个别盆地中夹有海相砂泥质、泥质沉积建造	以河湖相砂质、砂泥质、泥质沉积建造为主，间夹冲洪积相砾石、含砾砂质、泛滥平原相砂泥质、泥质沉积建造，部分地段夹河流相砂砾质沉积建造及湖相泥质建造

第三十九章　第四纪沉积演化

本志对北京地区延怀盆地和北京平原进行了古地理环境演化梳理和总结。

第一节　延怀盆地沉积演化

一、早更新世

早更新世初期，延怀盆地古地理环境发生较大的变化。由于构造运动使北东向断裂活动进一步加强，本区地貌分异更为显著，东部延庆湖和官厅湖成为一体，西部涿鹿湖未与东部古湖贯通，只以河道相连。由于老君山的隆起，使得原来的上新世晚期的古河道改道而行。延庆-官厅古湖继续沿着刘斌堡-黑海岭-四海流入古白河水系，另一支经八达岭流经南口河进入北京平原（图 39-1）。

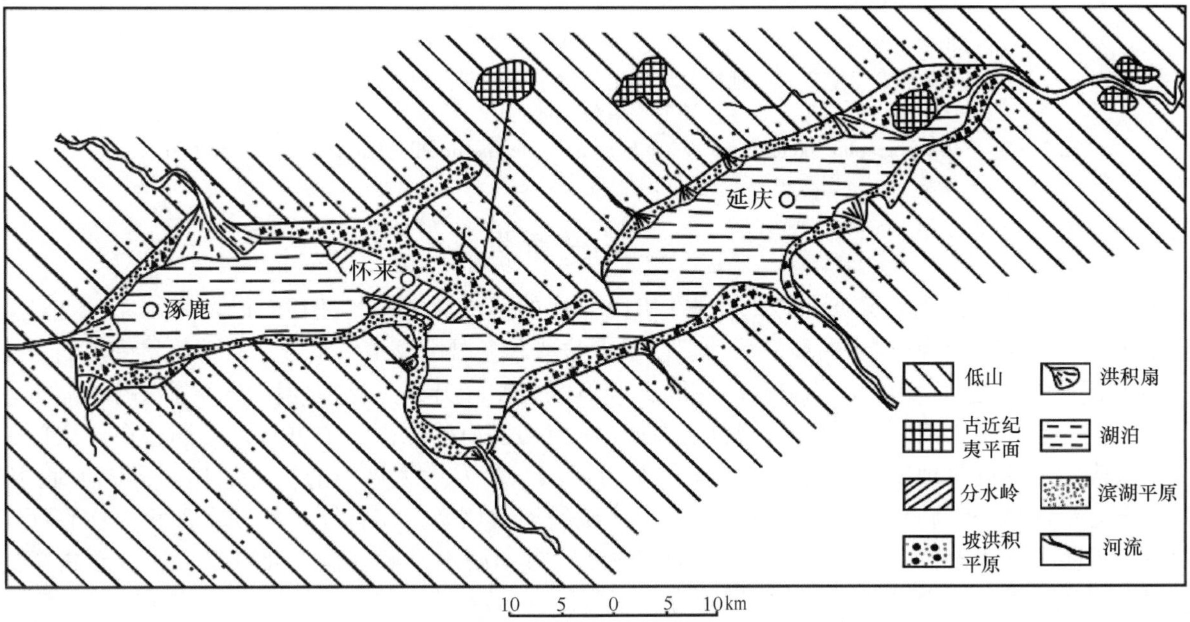

图 39-1　延怀盆地早更新世早期古地理略图

（据李华章，1995）

Fig. 39-1　Paleogeographic sketch map of Yanhuai Basin in the early stage of Early Pleistocene

由图 39-1 可见，在湖泊所及范围内的湖相沉积，表现为粗颗粒的滨湖相砂砾石向湖心相细颗粒黏土类递变的规律。在湖泊范围以外，表现为向山前坡、洪积相堆积过渡。在入湖河流的河口处，常为水下冲、洪积相沉积。

早更新世中—晚期，古湖面达到最大。据钻孔分析，延怀盆地的两个古湖已连成一体，延庆盆地西北缘湖底下陷。延庆-官厅古湖的南缘此期间湖相地层出露广泛，在颜家沟、北寨、施庄、鹿叫坡南东一带均有出露。在矾山盆地东端，钻孔揭示了从何家寨至桑园堡一线以东埋藏着含丽蚌化石的湖相沉积。在老君山紧邻延怀盆地一侧的山坡，南起万家窑西至温泉屯一带均可见到此期的湖相层分布（图 39-2）。

早更新世晚期湖相沉积在官厅水库南侧剖面出露较好，岩性为棕黄色、棕色黏砂及砂黏层，其中含丽蚌化石碎片。从早更新世晚期孢粉组合看，虽有松、桦、漆属等木本植物，但草本植物含量高，并有卷柏出现，为疏林草原景观，气候总体上向凉干方向转变，但其中有小的气候波动。

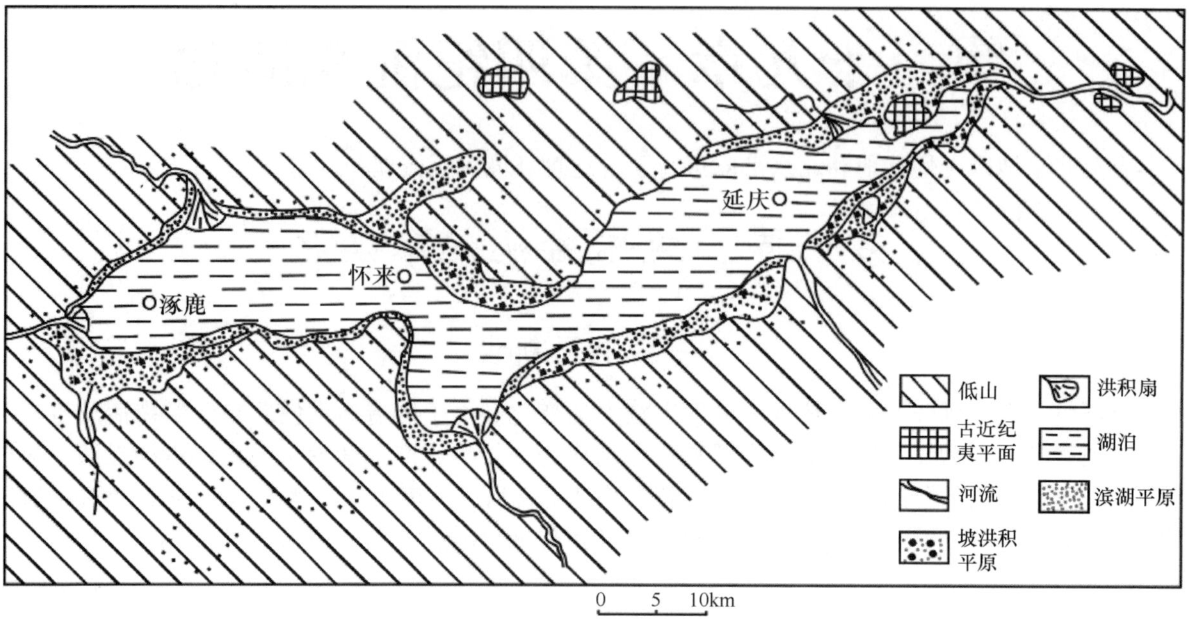

图 39-2 延怀盆地早更新世中—晚期古地理略图

（据李华章，1995）

Fig. 39-2 Paleogeographic sketch map of Yanhuai Basin in the middle-late stage of Early Pleistocene

二、中更新世

中更新世，延怀盆地沉积演化主要表现在矾山盆地的形成和古湖的迁移，以及古水系的变迁。中更新世早期，由于黑山寺-矾山断裂的活动，使矾山盆地和老君山及歪头山分离，下沉接受离石黄土堆积。中更新世中、晚期，由于新构造运动使黑汉岭地区急剧隆升，黑汉岭地区成为分水岭，靠近盆地一侧的河流一反原来的流向，西流注入延庆盆地，成为妫水河的源头。中更新世中期，在广安门一带发生地块隆起，使高阶地明显抬升，在广安门至沿河城一带发生大规模的重力崩塌，形成了堰塞湖，大坝以下至沿河城一带湖相层广泛发育。由于沿河城的河道堵塞，使官厅湖的湖水加积，在官厅至后郝窑一带发生一次大湖侵。中更新世晚期，延庆-官厅古湖湖水经峡谷泄入北京平原，导致古湖的消亡。可见，该古湖仅在中更新世早、中期时存在于官厅一带（图 39-3）。

中更新世植被特征显示，早期为针阔混交林及草原景观；中、晚期喜寒树种及草本植物含量增多，反映了气候由温带气候向寒温带气候演化。晚期有疑湖花介、奇异白花介等，表明气候有明显变凉的趋势，湖水的含盐度也有所增加，呈稍咸水或微咸水环境。

三、晚更新世

晚更新世，延怀盆地主要体现为统一水系的形成，盆地边缘的洪积台地和洪积扇发育，马兰黄土广泛堆积。晚更新世早期，盆地边缘发育了规模不等的洪积扇群，盆地南缘受军都山隆起山前洪积扇抬升，形成洪积台地。在盆地北缘除怀来至狼山一带形成洪积台地外，其他地区以下沉为主，形成一系列完整的洪积扇群。晚更新世中期，是洪积扇发育最盛的时期，矾山盆地因构造相对稳定，洪积扇相对不太发育。晚更新世晚期，石匣山地被切穿，桑干河水系开始形成，绕过老君山北麓再折向南流，然后与妫水河合流，进入官厅峡谷，形成延怀盆地统一水系（图 39-4）。

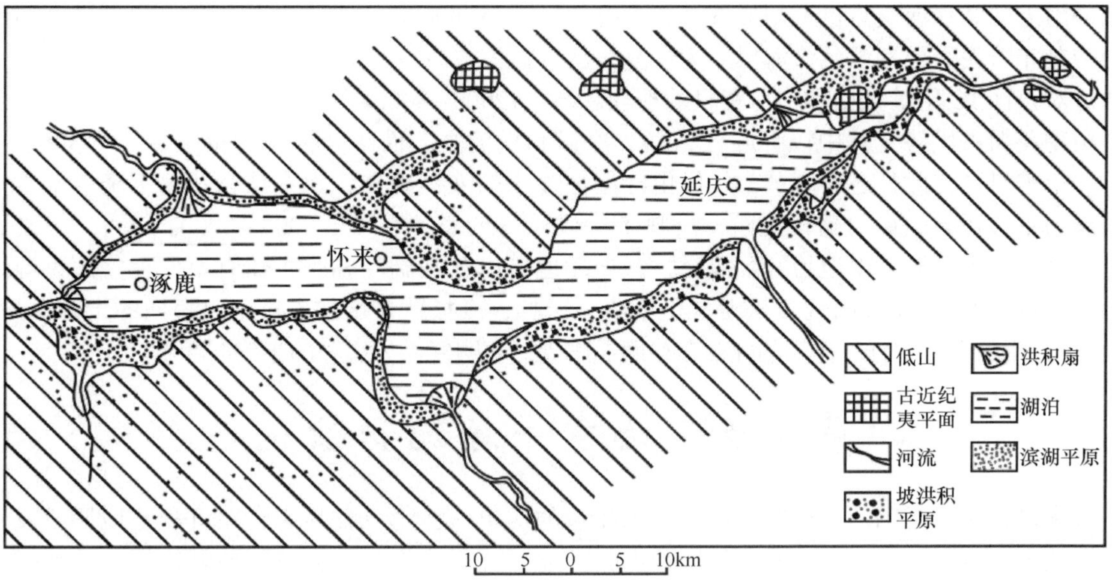

图 39-3　延怀盆地中更新世中期古地理略图
（据李华章，1995）
Fig. 39-3　Paleogeographic sketch map of Yanhuai Basin in the middle stage of Middle Pleistocene

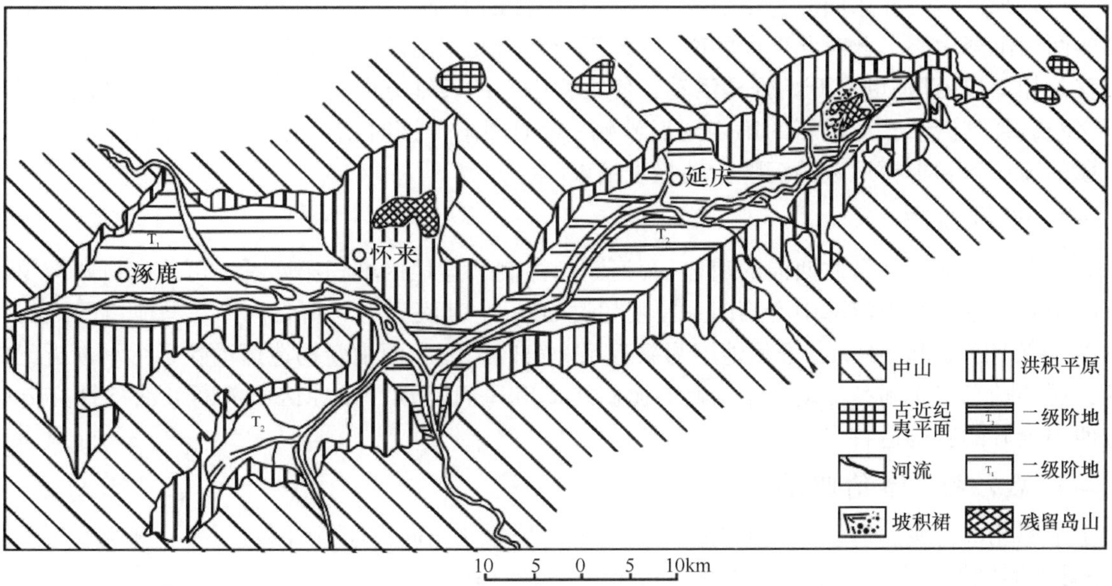

图 39-4　延怀盆地晚更新世晚期古地理略图
（据李华章，1995）
Fig. 39-4　Paleogeographic sketch map of Yanhuai Basin in the late stage of Late Pleistocene

根据官厅水库南侧剖面晚更新世地层孢粉分析，在晚更新世早、中期山地有少量松、桦等木本植物，而草本植物含量高，显示为疏林草原景观；晚期气候寒冷而干旱，盆地中为草原或干草原景观，发掘的动物化石有鸵鸟蛋、原始牛等，反映了干旱-半干旱的气候。

四、全新世

延庆盆地全新世古地理环境主要表现为：一是老君山山前断裂及施庄断裂的活动，使官厅一带的早中更新世湖相地层遭受剥蚀；二是永定河官厅峡谷段继续下切并形成全新世阶地，在延庆盆地的北缘方家冲附近的妫水河支流也下切形成全新世阶地；三是在延庆盆地的古河道及山前洪积扇前缘地下水溢出带分布着泥炭层和黑色淤泥层，为恢复全新世古植被演替、气候变迁提供了地层条件。

通过对该区^{14}C 年龄数据和孢粉资料的整理总结，12000~8000a 期间，气候有开始转暖的趋势，早期木本植物花粉占优势，其中以针叶树云杉、冷杉和松为主，而阔叶树种含量很少，草本植物有蒿、唐松草、禾本科等，此外还有麻黄，气候由冷湿向冷干演变；晚期，木本植物花粉下降，草本植物花粉以蒿属为主，还有少量的菊科花粉，反映了气候干冷的草原环境。8000~3000a 期间，气候温暖湿润，进入全新世大暖期，山地为针阔混交林，丘陵地带以落叶阔叶林为主兼有草原，盆地为湿地草原景观，但 5600~5000a 间曾出现过松粉含量下降，云杉、冷杉数量增加，显示出短期的干冷草原环境。3000a 至今，孢粉组合中栎、榆等阔叶树花粉明显减少，表明气候为温和的半湿润环境。

第二节　北京平原沉积演化

一般来讲，构造运动的过程同时也是沉积建造的历史。作为控盆断裂，在沉积盆地形成发育的过程中，盆地不断沉降接受沉积，盆地外侧则不断隆起，沉积速率的差异特征能够很好地反映控盆断裂的活动状况和规律。北京平原区由于受新构造运动（如断裂活动）的影响，盆地不断演化接受沉积，进入第四纪，形成了新的马池口凹陷、后沙峪-俸伯凹陷及平谷凹陷等。

一、盆地沉积演化

第四纪以来，本区北东向应力场引发了北西向张性断裂活动，形成以北西向为主，北北东向仍继续活动的两组沉降中心。此时北西向的南口-孙河断裂与北东向的顺义断裂活动显著，沉积中心也随之发生转移。如南口-孙河断裂开始强烈活动，使沉积中心的长轴方向转为北西向，形成马池口凹陷；北东向顺义断裂强烈活动，形成新的第四纪后沙峪-俸伯凹陷。该时期，北京凹陷南部沉积中心逐渐北移，南段的琉璃河和丰台地区的沉降活动趋于结束。

（一）沙河凹陷

由北东向的山前断裂及北西向的南口-孙河断裂控制，盆地内堆积第四系，可分南口西南及马池口两个沉降中心，南口西南沉降中心第四系厚度超过 600m，基底为中—新元古界。马池口凹陷在百善地区 ZK3 揭露第四系厚度 616m，下伏地层为新近系。第四纪北京西山新构造运动除了继承新近纪的上升运动外，还伴随有多次断裂活动、不均衡的掀斜运动和垂直运动，平原区山前断裂活动性亦能够客观的反应西山隆升的新构造特点。

（二）后沙峪-俸伯凹陷

位于平原北部古城—后沙峪—俸伯—南彩一带，是由北东向黄庄-高丽营断裂、顺义断裂及北西向的南口-孙河断裂所控制，基底为侏罗系及古生界及中—新元古界，西深东浅，靠近黄庄-高丽营断裂和顺义断裂附近，沉积厚度大，盆地呈箕状形态，盆地内堆积第四系厚度 600~1300m。据 JY-138 孔揭露第四系厚度 1283m，下伏地层为侏罗系；JY-488 孔揭露第四系厚度 1355m，下伏地层为侏罗系；JY-518 孔揭露第四纪沉积物厚度 608m，下伏地层为蓟县系；JY-180 孔埋深 677m 未穿透第四系；ZK19 孔揭露第四系厚度 500.9m，下伏地层为新近系；ZK12-2 揭露第四系厚度为 706m，下伏地层为新近系。

凹陷内钻孔 ZK12-2 和 ZK19 磁性地层年代学框架研究表明，受活动断裂的影响，凹陷内第四纪不同时期沉积物厚度和沉积速率具有显著的变化。通过对钻孔 ZK12-2 和 ZK19 剖面对比可知，顺义断裂在第四纪早更新世的强烈活动，最终造成了北京凹陷北部盆地的解体，发生沉降并接受第四纪沉积。晚更新世，顺义断裂再次强烈活动。全新世，顺义断裂强烈活动并造成了地表地裂缝的发育和建筑物墙体等的破坏。

（三）平谷凹陷

位于东部平谷一带，由夏垫断裂、东高村断裂等所控制，沉积中心在平谷城区。据平热井揭露第四系厚度690m，JY-510孔揭露第四系厚度557m，下伏地层均为蓟县系。

平谷地区区域重力场呈西南高东北低、北西西向密集梯度带展布特征，布格异常等值线在平谷城区附近发生同行扭曲揭示平谷盆地重力低异常特征。平谷盆地周缘及内部发育的断裂构造的相关研究最早可追溯到20世纪70—80年代北京市地震地质会战时期，经研究认为该盆地内发育有马昌营断裂、夏垫断裂、二十里长山断裂带（包括灵山断裂）及四道岭-上营断裂平原隐伏部分，并认为4条断裂中夏垫断裂和二十里长山断裂为第四纪活动断裂，其余2条为非第四纪活动断裂，断裂性质均为正断层。北京市多参数立体地质调查（2007）搜集并新增部分实物工作量，认为盆地内主要发育有夏垫-马坊断裂、马昌营断裂、二十里长山断裂带及东高村断裂，并且强调夏垫-马坊断裂和东高村断裂均为第四纪活动断裂，控制该区第四纪地层沉积，是平谷盆地的控盆断裂。

二、岩相古地理

（一）早更新世

在继承新近纪晚期古水系的基础上，发育了八达岭、高崖口、古白河、大石河及拒马河等古河道。北山与西山山前发育了以粗砾为主的冲洪积扇，形成了扇形堆积平原。山前平原区受断裂活动控制进一步形成第四纪马池口凹陷、后沙峪-俸伯凹陷等构造单元，各个凹陷及隆起区内沉积作用受到气候、构造运动的强烈控制，横向上岩性、岩相变化较大，沉积环境多变的古地理格局。

（二）中更新世

古地理格局与早更新世基本一致。广安门至八达岭一带发生隆起，使得八达岭古河道消失。由于河流冲积扇的不断扩展，早更新世的一些残丘被覆盖，地表微隆起面积缩小。二十里长山地块向西掀斜，在其边缘地带有坡洪积物堆积。在广大的山前地带发育有洪积扇裙和扇缘洼地，平原低洼地区有湖沼发育，继续沉降接受砂质黏土、黏质砂土沉积，形成洪冲积平原。

（三）晚更新世

晚更新世早、中期是北京地区地壳强烈活动的时期，也是现代地貌的形成时期。早期，气候温湿，为马兰黄土与古土壤的堆积期。永定河上下贯通，自三家店出山后形成以石景山为顶的冲积扇，其前缘可达清河、通州一带；古潮河沿北东向断裂在焦家务附近袭夺了石峨古河道的上游段，形成了潮河冲积扇；平谷盆地原来的石峨古河道的中、下游形成了错河、泃河水系。至此，平谷地区的完整水系形成，流出盆地后形成了泃河冲积扇；温榆河基本形成，其流向与现今相似，是永定河的支流之一。

晚更新世晚期，永定河、潮河、白河、泃河等冲积扇的形成使大兴隆起地块面积大大缩小。大兴隆起地块活动微弱，受前期堆积作用，永定河向西南方向迁移，形成巨大的冲积扇。古白河和古潮河出山后，将晚更新世早、中期形成的冲积扇切割成宽浅的槽谷。古白河向西流经西智到神山，折向西南经北房村于龙王头入小中河，南流至通州古城村；古潮河向西南流至河南寨折向西南，经木林和孤努山向东流经小店村于荆坨村入鲍邱河；古温榆河沿永定河冲积扇边缘流至通州注入古白河。

（四）全新世

全新世早期（12000~7500a），永定河由石景山沿现在的清河注入温榆河，即永定河清河段古河道；古白河向东摆动，脱离小中河河道，古潮河向西摆动，脱离鲍丘河河道，并在现今潮白河宽广谷地中摆动，而小中河、鲍河形成独立的水系；温榆河切割了晚更新世的黄土台面，形成浅槽形河床，河床中堆积了砂、淤泥和泥炭；泃河切割了晚更新世的冲洪积平原，形成稳定的河道、阶地、河漫滩和废弃的牛轭湖等地貌形态。

全新世中期（7500~2500a），永定河由石景山直泻南下，形成以衙门口为顶点的新冲积扇，南苑一带处于老扇与新扇之间，形成一系列扇间洼地，发育了大小湖沼；潮河、白河汇合于通州东，然后沿北运河向东南流入渤海。

全新世晚期（2500a 以来），根据文献记载，潮河、白河汇流于通州东部，此后逐渐向北移动，到了公元 900—1000 年，汇流点移至牛栏山北大胡营村附近，沿白河古道南下。公元 1555 年，人工开挖白河新河，使得潮河、白河汇流于河槽村东。永定河当时沿㶟水故道经大兴机场、采育向东南流去。但是永定河由于多次泛滥，使其成为“无定河”。由于永定河的多次溃决，在大兴区境内留下多个决口扇，受风的吹扬形成多种形态的砂丘和砂地。

第四十章　第四纪地貌

第一节　概　述

北京位于中国二级台阶和三级台阶分界处，地貌形态差异明显，由华北中低山区（西部山地、北部山地）和华北平原区（东南部平原）两个地貌单元构成。西部统称西山，属于太行山山脉，由一系列北东向褶皱山系组成。北部统称北山（军都山），属于燕山山脉，其中的延庆盆地属断陷盆地。

最早对本区进行地貌研究的学者是美国的维理士，他1903年来中国进行地质调查时，便提出华北地区的地文期可分为北台期、唐县期、临州期和汾河期4期，并根据本区山地与平原的地貌特征，首次提出“北京湾”的概念。1919年安特生赴斋堂盆地进行地质调查时首先发现了位于清水河南岸大墩台梁上的砾石层，并与唐县砾石层的时代相比，定名为北京西山的上新世“唐县砾石层”。

1920年，叶良辅等编著的《北京西山地质志》对北京西山地区的山脉、平原、河流等地貌进行了概要描述，其中地文期采用了安特生的划分（唐县期之地面、汾河期自上而下之侵蚀、马兰期之梯地、板桥期自上而下之侵蚀）。

20世纪50年代末至60年代初，北京地区开展的第一轮1：5万区调，编制了部分图幅第四纪地貌图。80年代初，航空物探遥感中心开展了航空遥感地貌调查，编制了北京市1：10万航空遥感地貌图。

1986年出版的《北京市农业资源与区划图集》采用形态和成因相结合的原则，划分了2个一级地貌单元和9个二级地貌单元。

20世纪80年代以来，吴忱对华北地区的地貌进行了系统研究，认为华北山地和盆地抬升与坳陷、侵蚀与堆积、暴露与埋藏的过程，代表了地貌演化的全过程，是地貌形成演化理论研究的理想平台，出版了《华北平原古河道研究》《华北山地地貌演化与岩穴的形成》《华北山地地形面、地文期与地貌发育史》《华北地貌环境及其形成演化》等专著。

第二节　地貌总体特征

本区地貌由西北山地和东南平原两大地貌单元组成。山区剥蚀切割强烈、发育多期夷平面和河流阶地，平原堆积厚度大，地貌类型复杂多样。

一、地势呈现明显的西北高、东南低的特征

本市总的地势是西北高，东南低。百花山、白草畔、东灵山、海坨山山峰高度接近或超过2000m，东灵山2303m，为全市的最高点。平原面积海拔绝大部分为30~50m，最低点位于东南部通州区的柴厂屯附近，其海拔仅8m。平原区地势由西北向东南倾斜。拒马河、大石河、永定河、温榆河、潮白河、蓟运河后，形成众多冲洪积扇，其中以永定河冲洪积扇和潮白河冲洪积扇最大，约占整个平原面积的70%以上。

二、山区剥蚀切割强烈、平原堆积厚度大

受内、外地质作用的共同影响，山区剥蚀作用强烈，峰谷高差大，沟谷众多，河谷结构复杂。新生代以来，在新构造运动作用下，山区隆升和平原区沉降的趋势一直在持续，山区的剥蚀物在流水作

用下，堆积于平原地区，形成了巨厚的新生代沉积物。

三、北部军都山（属燕山山脉）和西部西山（属太行山脉）地貌形态有所差异

由于中生代褶皱构造和新生代差异隆升作用的影响，北京西部和北部山区地貌有所不同。北部军都山多呈断块状，山脉走向以不明显的近东西向或北东东-南西西向为主。断陷特征明显，形成延怀盆地（海拔475~510m）。由于各断块在抬升量方面略有差异，各地山地的剥蚀面也有起伏变化，河流往往有沿两组断层发育的现象，如妫河、潮河、潮白河干流都取北东-南西流向，而怀河、汤河则是北西-南东流向，这些河流都显著受构造线的控制。

西山受燕山期褶皱构造的影响，形成了一系列北东-南西向山脉，主要有笔架山-东灵山、老龙窝-百花山-白草畔、上寺岭-猫耳山等几组山系。

四、岩性不同导致地貌形态各异

不同的岩性造就不同类型的地貌。西山地区广泛分布着新元古界和下古生界灰岩，中生界的砂页岩也占有很大面积，花岗岩的出露很零星。北山地区出露在地表的主要是太古宙的变质岩以及中-新元古界坚硬的石英砂岩和含硅质条带白云岩，花岗岩出露面积大。

质地坚硬的中元古界长城系石英砂岩，抗风化能力强，常突出于周围地面之上。如平谷黄松峪一带出露的中元古界长城系长石石英砂岩，岩层产状平缓，岩石硬度大，抗风化力强，节理发育的岩层在外动力地质作用的影响下，形成了峰林、峰丛、孤峰及障谷等地貌景观。

花岗岩在地貌上产生宽谷与缓坡。花岗岩地区的山峰，除云蒙山因隆升速度快而呈现山高陡峻外，一般常发育为比较浑圆的形态，如房山区北部有小规模的花岗闪长岩体，经风化剥蚀，呈现球状风化地貌景观，形成了较典型的花岗岩丘陵地貌。

本区火山岩形成的山岳地貌景观，多是海拔千米以上的高山地带，1000m以上的山峰多由中生代火山岩组成，其火山岩堆积厚度达4000多米。如西山区东灵山，海拔2303m，北山延庆盆地东南缘的燕羽山，海拔1278m，其火山岩堆积厚度达4971m。

虽然北京地区的石灰岩溶蚀作用不如南方湿润地区普遍，但是在西山碳酸盐岩分布地区，发育典型的北方岩溶地貌，如房山拒马河流域的似峰丛状岩溶地貌、大石河流域以石花洞为代表的岩溶洞穴群等。

五、山区地貌面呈明显的“层状”结构

地貌面“层状”结构是不同时期夷平面的表现形式。根据现有资料的分析，本区分布广泛的夷平面主要有两期，即北台期和唐县期。郭旭东（2007）在这两期之间划分出百花期、东灵期和狼虎期夷平面。形成的“层状”面最上层海拔2000m左右，如东灵山、白草畔、百花山平台；第二层是1400~1600m的山顶台面，主要分布在黄草梁-南山鞍一线和老龙窝-庙安岭-髫髻山一线；第三层为1000~1200m，地貌面较平坦，分布于柏峪西、梁家山、妙峰山、张家山、九龙山、猫耳山、上寺岭、大洼尖等地，分布范围宽广，是构成西山山地地貌面的主体部分；第四层为唐县期夷平面，海拔较低，主要分布在河谷两侧的低山地带。

受北山断陷盆地和断块差异升降作用的影响，夷平面抬升的幅度在各地有所差异，如青龙桥所普遍存在的1070m夷平面，在关沟两侧则以920m，甚至830m出现。参考西山夷平面划分的依据，密云水库一带的北台期夷平面海拔800~1000m，唐县期夷平面海拔400~600m。

六、其地貌类型多样复杂

由于北京地区地质构造和岩性复杂，导致了多种多样的地貌类型。按地貌成因分类，可划分为侵

蚀构造山地、剥蚀构造台地、丘陵和堆积构造平原等类型。按地貌形态分类，可划分为中山、低山、丘陵、台地和平原等类型。

第三节　地貌单元划分及其特征

一、地貌单元划分

本区地貌单元划分原则：一级和二级地貌单元划分遵照《中国区域地质志工作指南》的划分方案，三级和四级地貌单元划分参考河北省地貌单元划分成果，结合本区的特点，依据成因和形态分类相结合的原则进行划分（表40-1）。

表40-1　北京地区第四纪地貌单元划分表

Table 40-1　Division of Quaternary geomorphic units in Beijing area

一级	二级	三级	四级
西北山盆和中部高原区（Ⅱ）	黄土高原和华北中、低山区（II_6）	华北中、低山和丘陵区（II_6^1）	燕山西段中、低山和丘陵区（II_6^{1-1}）
			太行山北段中低山和丘陵区（II_6^{1-2}）
			延庆山间盆地（II_6^{1-3}）
东部平原和中、低山、丘陵区（Ⅲ）	松辽-华北平原区（III_1）	华北平原区（III_1^1）	大清河流域（III_1^{1-1}）
			永定河流域（III_1^{1-2}）
			北运河流域（III_1^{1-3}）
			潮白河流域（III_1^{1-4}）
			蓟运河流域（III_1^{1-5}）

二、各级地貌单元主要特征

（一）华北中、低山和丘陵区（II_6^1）

1. 燕山西段中、低山和丘陵区（II_6^{1-1}）

该单元主要是北京北部的军都山。受古北口-长哨营断裂带、沙厂-墙子路断裂带和密云大背斜等东西向构造的控制，局部山系呈东西向。伴随燕山运动产生的一系列走向北东或北东东的复式褶皱（如红旗甸向斜、千家店背斜、四海向斜、南口复式背斜）和断裂带（如半城子断裂、琉璃庙断裂、延矾断裂）也控制着北东向山系演化。受南北向断裂带（如二道河断裂、青石岭断裂、娘子水断裂、黄崖关断裂）和北西向断裂带（如南口-孙河断裂）影响，形成断块状山系。

中山区主要分布于延庆西北及怀柔北部，主要山峰有海坨山（2241m）、黑坨山（1534m）、凤坨梁（1529m）、云蒙山（1414m）、燕羽山（1278m）、佛爷顶（1252m）、桃山（1180m）、四座楼山（1060m）、八达岭（1015m）。低山区呈环带状，展布于中山带之间或近山前地带。丘陵主要分布于昌平区南口—九里山山前、怀柔区庙城至密云区西智山前和密云水库周围地带，顺义区二十里长山分布有岛状残丘。丘陵一般表现为馒头状，丘顶浑圆，丘谷交错，无明显脉络，坡度7°~15°，少数大于25°，主要由花岗岩、片麻岩、火山碎屑岩组成，土层厚度较大。

该单元中的主要河流有白河和潮河。白河发源于河北省沽源县，在延怀盆地的北部自西向东流动，至白河堡水库附近入京，发育穿洞；至蔡木沟附近，北面的黑河汇入白河；至帽湾东，东北部的汤河汇入，河流总体方向改为自西北向东南流动。由汤河口—四合堂，流经变质岩区和云蒙山岩体，河曲发育，特别是在四合堂一带，深切河曲地貌非常壮观。

潮河发源于河北省丰宁满族自治县草碾子沟南山下，至古北口进入北京，至太师屯汇入密云水库。潮河主要流经变质岩地区，河谷宽阔（图 40-1），一般发育二级阶地。

白河与潮河在密云区城西南的河漕村汇合后，进入北京平原，下游称为潮白河。

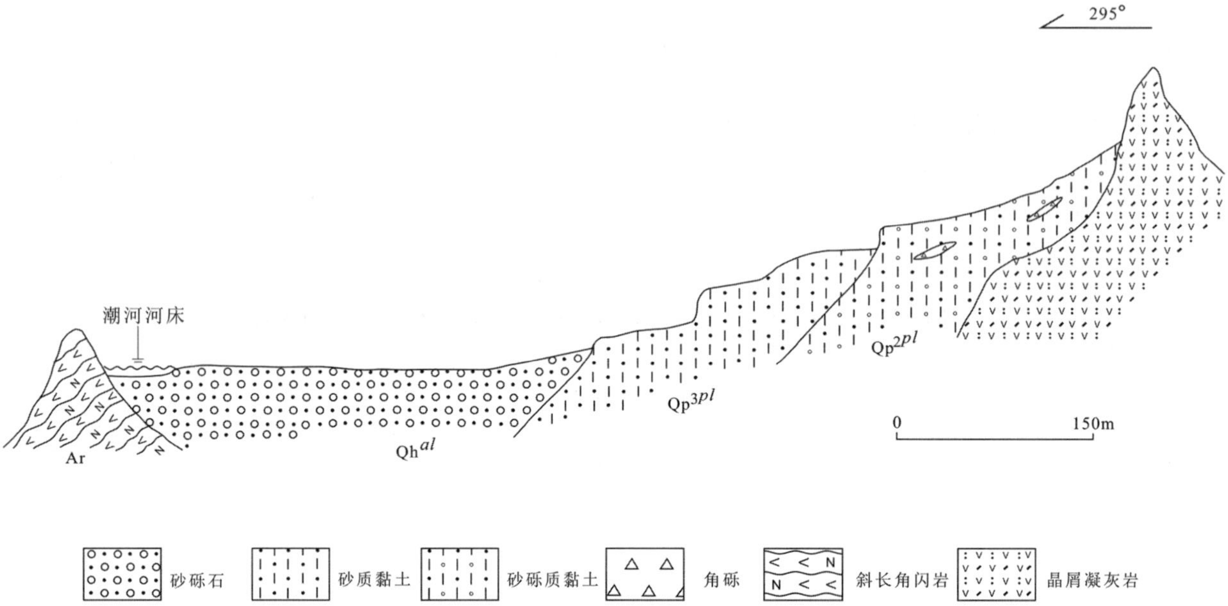

图 40-1　密云前南台潮河河谷剖面图

Fig. 40-1　The Qiannantai Chaohe River Valley section in Miyun

本单元燕山期侵入岩出露面积近 2000km²，岩石以花岗岩类为主，出露面积最大的有八达岭花岗杂岩体和云蒙山杂岩。岩体受内、外地质作用的影响，形成了众多的花岗岩地貌景观（表 40-2）。

表 40-2　北京地区主要花岗岩类地质地貌特征表

Table 40-2　Geological and geomorphological characteristics of main granitoids in Beijing area

编号	岩类	侵入体名称	形成时代/Ma	地貌遗迹景区	岩石地貌特征
I	八达岭花岗杂岩	黑熊山花岗岩体	120.9	黑熊山花岗岩地貌	为中细粒碱长花岗岩，节理发育，常形成典型的花岗岩“石蛋”地貌。有学者认为该地貌特征是冰川活动的证据，有待研究探讨
		大杨山花岗闪长岩体	中生代	大杨山花岗岩地貌	隆起地貌特征显著，形成陡峻的花岗岩石壁和孤峰，具有典型的花岗岩地貌特征，山间有一座“万人楼”，为高达 200 多米的花岗岩孤峰。有“京北世外桃源”之称
		莲花山花岗岩体	131.0~125.7	莲花山花岗岩地貌	为早白垩世侵入体。沿垂直节理在山体上形成条条沟隙，远看似含苞欲放的莲花，故称“莲花山”
		碓臼峪花岗岩体	135.5~128.0	碓臼峪花岗岩地貌	花岗岩体节理发育，沿纵节理和横节理风化，常形成孤峰、悬石等地貌形态，具有山岳型地貌特征
		石洞子钾长花岗岩体	128.5	云峰山花岗岩地貌	为早白垩世侵入体，花岗岩呈孤峰挺立、峰奇峻险、怪石林立，具有高凸的地貌形态特征，顶部的岩体常见球状风化形成的各种怪异形态

续表

编号	岩类	侵入体名称	形成时代/Ma	地貌遗迹景区	岩石地貌特征
II	云蒙山变质核花岗杂岩	大水峪中粗粒似斑状花岗岩体	113.9~106.2	青龙峡花岗岩地貌	为似斑状粗粒花岗岩体，常形成断壁和陡峭的山岳
		云蒙山花岗岩体（有北京小黄山之美称）	119.3~114.0	云蒙山花岗岩地貌	山岳型花岗岩地貌
				紫云山花岗岩地貌	山岳型花岗岩地貌
				幽谷神潭花岗岩地貌	球状风化地貌、孤峰、陡崖
				神堂峪花岗岩地貌	山岳型花岗岩地貌、象形石、孤峰
				三峪花岗岩地貌	山岳型花岗岩地貌、陡崖
				清凉谷花岗岩地貌	山岳型花岗岩地貌、陡崖
				精灵谷花岗岩地貌	山岳型花岗岩地貌、陡崖、孤峰
				桃园仙谷花岗岩地貌	似片麻状层理、陡崖、象形石
				云蒙峡花岗岩地貌	山岳型花岗岩地貌、陡崖
III	延庆西北独立岩体	大海坨花岗岩体	133.7~123.9	松山花岗岩地貌	呈北北东向展布的纺锤状粗粒花岗岩体，表层风化剥蚀严重，球形风化明显
				古崖居花岗岩地貌	为粗粒花岗岩，表层岩石易风化剥蚀，形成半球状风化山体地貌。古人开凿于花岗岩壁上的洞穴群是我国迄今为止已发现的最大古崖居遗址
	平谷南山岩体	盘山花岗岩体	222.7~214.0	盘山花岗岩地貌	晚三叠世花岗岩体在北京地区发育良好，具有典型花岗岩球状风化的地貌形态特征，主要岩性为二长花岗岩

2. 太行山北段中低山和丘陵区（II_6^{1-2}）

该单元中山地貌展布比较有序，多呈北东向延伸，由北向南平行相间排列，地势逐渐降低。主要包括下述4列：东灵山-黄草梁、金树塔-梁家山、白草畔-百花山-清水尖、上寺岭-猫耳山。这些中山主要由中生代火山岩组成，山体形态表现为山顶宽平舒缓，山坡坡度较大，沟谷深切且密度大（表40-3）。山体层状地貌明显，状似多层台阶，反映了地壳运动强烈的间歇性隆升。很多中山山顶被认为是北台期夷平面。

表 40-3　西山主要中山地貌特征指标

Table 40-3　Geomorphological characteristics of Xishan hills

序号	山峰名称	海拔高度/m	相对高度/m	山坡坡度/（°）	沟谷密度/（条·km^{-1}）	主要岩性
1	东灵山	2303	1200	26	2.4	J_2t 火山岩
2	白草畔	2046	1350	32	1.9	J_2t 火山岩
3	百花山	1991	1400	35	2.6	J_2t 火山岩
4	黄草梁	1733	830	22	1.6	Qbj 白云岩
5	大背尖	1718	620	27	3.0	J_2t 火山岩
6	灰金坨	1685	890	39	2.4	Jxw 白云岩
7	老龙窝	1649	1250	32	1.8	J_2l 砂岩
8	金树塔	1611	870	31	3.4	J_2t 火山岩
9	清水尖	1528	1080	35	3.0	J_2t 火山岩
10	髽髻山	1524	1060	33	2.1	J_2t 火山岩

续表

序号	山峰名称	海拔高度/m	相对高度/m	山坡坡度/（°）	沟谷密度/（条·km^{-1}）	主要岩性
11	梁家山	1406	750	31	3.8	J_2t 火山岩
12	水峪东尖	1361	910	28	3.6	ϵ_2 石灰岩
13	上寺岭	1307	1150	26	2.5	J_2j 砂岩

低山主要分布于中山的外围，以百花山-清水尖中山带外围的低山带面积最大，连绵不断。除了中山外围之外，九龙山也是本区重要的非岩溶低山地貌系统。低山地貌除了地势上低于中山地貌之外，山体坡度也相对较缓，一般15°~20°，但河网切割密度相对较大，可达4条/km以上。其纵剖面的阶梯特征更加明显，不但有平缓的山顶和山梁，而且河谷两侧还发育有宽谷和多级阶地，反映了地貌发育的多期性和比较强的流水切割侵蚀作用。

丘陵分布于房山和海淀的山前地带，一般呈台地状展布于河谷两侧。展布于辛庄—黄元井—罗家峪—孤山口一线的丘陵，海拔在100m以下，岩性组成主要为碳酸盐岩和片岩等，受风化剥蚀及溶蚀作用的共同影响，多形成浑圆的小型山丘，并保留一定厚度的残坡积层，因此与平原地貌的分界线不很明显。

该单元地貌除受构造控制明显外，由于岩性的不同而导致地貌差异化特征亦很明显：①拒马河流域和大石河流域岩溶分布区。拒马河流域地层为雾迷山组硅质条带白云岩，呈岩溶蛇曲，似峰丛地貌，岩溶石柱为其特色。北岸发育仙栖洞和云水洞等唐县期夷平面洞穴。孤山寨内可以见到8级错落峰丛。大石河流域地层主要为古生代石灰岩，上游北岸发育圣莲山岩溶地貌，下游南岸发育鸡毛洞、银狐洞、石花洞和孔水洞，形成石花洞系，发育8个洞层。大石河出山口南部发育房山花岗岩地貌。此外，在鲁家滩一带寒武纪—奥陶纪灰岩出露区，发育小型岩溶盆地。②永定河、清水河流域火山岩地貌区。北京地区中生代时期和我国东部一样，是大规模构造岩浆激烈活动的时期，其主要喷溢活动是在侏罗纪—白垩纪，形成的火山岩厚达4000m以上，在内外地质营力的共同塑造下，形成了现今西山地区的火山岩地貌，最显著的特点是保留了一系列北东向的山系和各期夷平面，为研究中生代以来造山运动和山区平原形成机理提供了重要依据，具有很高的科学考察价值。

该单元内主要河流有永定河、清水河、大石河及拒马河。永定河山峡（官厅大坝—石景山）全长约110km，是沟通延怀盆地与北京平原的河道。地质构造限定了永定河山峡的位置和形状。由于幽州背斜核部的张裂隙抗冲力较弱，使得怀来盆地中的桑干河、洋河和妫水河3条水系汇于官厅水库，沿着幽州背斜核部进入西山，至沿河城改为顺沿河城断裂流动，至向阳口改为顺着青白口穹隆的张裂隙流动，至雁翅横跨髫髻山向斜，至清水涧沿着下苇甸穹隆与九龙山向斜之间呈深切河曲流动，最后沿着永定河断裂流入北京平原。

表 40-4　永定河山峡阶地发育状况

Table 40-4　Development status of gorge terraces of Yongding River

阶地级数	官厅大坝		幽州背斜		沿河城断裂		青白口穹隆		下苇甸穹隆		丁家滩蛇曲		阶地时代
	海拔高度/m	结构	海拔高度/m	结构	海拔高度/m	结构	海拔高度/m	结构	海拔高度/m	结构	海拔高度/m	结构	
Ⅷ							440	基					N_1
Ⅶ			535	嵌					290	侵	248	侵	N_2
Ⅵ			505	基	450	嵌	359	基	256	基			Qp^1
Ⅴ	479	基	475	基	423	嵌	329	基			217	基	Qp^1
Ⅳ	464	基			405	基	305	基	222	堆			Qp^2
Ⅲ							298	基			175	基	Qp^2
Ⅱ					377	堆			198	堆			Qp^3
Ⅰ							271	堆	185	堆	156	堆	Qh

注：基—基座阶地；堆—堆积阶地；侵—侵蚀阶地；嵌—嵌入阶地。

吕金波（2015）认为永定河发育8级阶地（表40-4）：Ⅶ-Ⅷ级阶地为宽谷地貌，说明上新世的永定河为老年型河流，河曲发育；Ⅴ-Ⅵ级阶地的阶坡较高，为基座阶地，说明进入第四纪，山体隆升，河流下切的速度加快，将永定河限定在今天的峡谷中；Ⅲ-Ⅳ级阶地的阶面较宽，有的阶面砾石之上有离石黄土出现，说明中更新世西山隆升速度减慢，较为稳定；Ⅱ-Ⅲ级阶地，主要为韵律明显的马兰砾石层堆积，说明天气干凉，永定河堆积的速度大于山体抬升的速度，形成加积型阶地。

3. 延庆山间盆地（$Ⅱ_6^{1-3}$）

延庆盆地与怀来盆地、涿鹿盆地、矾山盆地4个相互连通的盆地共同组成延怀盆地，是冀北中山内部北东向断裂控制下的新生代断陷盆地。盆地内地势平坦，海拔高度在480~600m之间，两侧山地海拔1000~2241m，呈四周高、中间低的形态，自盆地边缘至盆地中心，依次发育中山、低山、丘陵、黄土台地、洪积扇、妫水河阶地。盆地内主要河流为妫水河。地表主要为第四纪堆积物所覆盖，盆地内发育岛山，较大的有永宁镇以西的团山，高度为791m，由中元古代硅质灰岩组成，其他岛山分布在东山和南山的山麓带，高度不等，相对高度从十几米到一百余米，属丘陵范围。盆地边缘以断层与山地相接，界线很清晰。盆地内的地貌类型，山前为洪积扇所构成，以古城洪积扇为最大，其次是康庄洪积扇、西拨子洪积扇、大榆树洪积扇、张山营和井庄洪积扇等。盆地南部西拨子、高庙屯、井庄、永宁等乡镇山前有较厚的次生坡洪积黄土堆积，由于受流水的切割，黄土沟谷十分发育，沟谷深度一般都在10~30m之间。

受黄柏寺断裂和南部康庄-八里店断裂的控制，整个盆地呈北东向延伸，东北高、西南低，形若袋状。盆地自新生代以来强烈下陷，北部下陷深，南部下陷浅，有自北而南逐渐翘起之势。盆地内，新生代沉积厚度达2300m左右，其中古近系和新近系厚度达1300m，第四系厚达800m左右。沉积中心位于盆地北缘，即延庆东北部的单家营及西南部的太平庄一带。在盆地南缘，沉积较薄，新生代沉积厚度400~500m，其中第四系只有100~200m左右。其岩相变化自南而北由粗变细，自南部的砂、砾石及黄土状沉积向北过渡为细砂、粉砂、淤泥质沉积，表示盆地北陷南掀，南仰北俯的不对称特点。从地球物理场来看，盆地内重力场平均分布，电测深曲线类型一致，地震低速层位连续平稳，这些特征一致表明，延庆盆地新生代沉积比较完整，各层变化稳定，新构造运动的性质以持续下沉为主。

该单元内主要河流是妫水河，其上游为盘肠河，源于大海坨山，河道在山地内迂回曲折，下切强烈，形成嶂谷，反映山地强烈抬升。河道的弯曲可能受岩石裂隙控制。盘肠河于古城出山后称为妫水河，笔直地流向东南，于西龙湾拐向西南。延庆盆地向西南方向降低，然而妫水河出山后流向东南显示出强烈地受断层控制。延庆盆地发育两级阶地，一级阶地高3~5m，宽500m，出现于妫水河两岸狭窄的范围内。二级阶地高10~15m，宽10km，构成延庆盆地的平原。阶地剖面底部已见湖相层黏土出露，向上逐渐过渡为河流相黄棕色黏土质粉砂，说明湖泊延续时间较长。

（二）华北平原区（$Ⅲ_1^1$）

根据北京市平原区流域划分为5个四级地貌单元，大清河流域（$Ⅲ_1^{1-1}$）、永定河流域（$Ⅲ_1^{1-2}$）、北运河流域（$Ⅲ_1^{1-3}$）、潮白河流域（$Ⅲ_1^{1-4}$）和蓟运河流域（$Ⅲ_1^{1-5}$）。

1. 大清河流域（$Ⅲ_1^{1-1}$）

大部分位于房山区和丰台区，在北京市界内有大小支流10余条，主要支流有拒马河、大石河和小清河，在山前地带形成冲积扇。大清河流域塑造的冲积平原面积较大，平坦广阔，略有起伏，有切割，冲沟发育，地势由西北向东南倾斜，物质组成具有一定的分选性，在河流出山口物质较粗，堆积物除了上部的黄土状冲洪积物外，还有砂、卵石层，距山麓越远，堆积的物质粒度越细。

拒马河是塑造北京平原的主要河流之一，发源于河北省涞源县，张坊镇以上流域面积为4810km^2。拒马河在张坊镇出山之后，在镇江营一带分为两条分支河流：北支沿镇江营—高村向南流出北京市，在镇江营一带形成“U”形槽谷，受当时冲积扇的地形地貌影响，向下游方向辫状河流的特征很明显，形成多处心滩。由于流水作用的逐渐减弱，使心滩得以保留，并形成高于河床的微地貌单元；南支沿镇江营—南河—演武庄分布，常年无水，成为一条干河床。区域上拒马河所形成的冲积扇面积很大，形成的冲积泛滥平原，地势起伏不大，地面总体向东南方向倾斜。拒马河冲积扇可划分为4个次级地貌单元：河漫滩、故道洼地、砂垄、现代河道。向阳村和三合庄村北，受大石河、小清河、拒马河三河汇流及拒马河河流改向影响，地势相对较低，呈现洼地地貌，而沙窝村周围地势高且地表分布粉细砂，为砂垄地貌；北拒马河以南地区地势平坦为河漫滩地貌。

大石河发源于北京房山区百花山南麓，干流河道长120km，于坨里以南进入平原，经窦店、琉璃河流出本市。甘池—岳各庄—娄子水一线以东、长沟镇—尚庄—下胡良一线以北的河流如周口店河、胡良河等均为大石河的支流，属大石河的流域范围。大石河冲洪积扇总体上呈一倾斜平原，依据地形、微地貌等差异大石河冲洪积扇可划分为6个三级地貌单元：微斜坡台地、低平地、高平地、故道洼地、泛滥平原、古河道及现代河道地貌形态类型。大石河冲积平原地势较拒马河冲积扇平原高，由高台地与河谷地貌组成，河流侵蚀下切作用较明显，高台地边缘陡坎地貌特征较突出，形成河流阶地。

2. 永定河流域（Ⅲ_1^{1-2}）

历史上永定河流域面积很广，形成的冲积扇平原是北京平原的主要组成部分。永定河于三家店出山后，在漫长的地质历史时期频繁迁移，形成的冲积平原地势西北高、东南低，地表和缓，平均坡度1‰左右，由三级阶地、二级阶地、一级阶地、河漫滩等地貌基本形态及一系列微地貌组成。

孙秀萍（1982）经调查研究认为永定河形成的古河道主要有古清河、古金沟河、古高粱河、古蓟河、古㶟水河及古未名河，并将永定河水系变迁的历史概括划分为如下4个阶段：

晚更新世晚期，永定河冲积扇发育时期：在晚更新世晚期玉木冰期最盛期，气候寒冷，寒冻风化强烈，岩屑物质大量产生，河流的搬运能力不足以将其全部带走，故为河流沉积物的加积时期。因此，在北京山区形成了由深厚砾石层组成的马兰阶地。在山前，则形成了宽广的冲积扇。

晚更新世末至全新世早期，永定河水系形成时期：随着全球性气候转暖冰川消退，岩屑物质来源大大减少，河流水量增加提高了搬运能力，同时玉木冰期最盛期海平面大幅度降低，所造成的河流侵蚀基准面降低的影响已上溯到本区，因此，永定河冲积扇停止发育，开始了河流深切的时期。上述各古河道就是在此时期形成的。

全新世中期，永定河水系的繁盛时期：全新世中期气候温暖，水量丰沛，永定河冲积扇上水网密布，植被茂盛，泥炭发育。由于当时海面已回升到现今平均海面附近或稍高，致使侵蚀基准面相应抬高，有利于河床的堆积。

全新世晚期，永定河古河道消亡时期：由于人类生产活动等因素对山区植被的破坏，水土流失现象日益严重，故永定河一度称为浑河。据历史记载，永定河古称㶟水，在汉代活动于北京城南，称清泉河，到北魏时从城西南隅流过至辽代河道，明初称浑河，南下袭夺白沟河。17世纪初永定河改走今道，1698年筑堤并由清康熙皇帝赐名。在短短的几百年后，永定河现已成为地上河。

3. 北运河流域（Ⅲ_1^{1-3}）

北运河上游温榆河是西山及山前溢出带发源的多条河流汇聚而成，形成向东南部呈缓倾斜的冲洪积扇平原，是北京平原的重要组成部分。

温榆河水系呈树枝状。上游沟谷较发育，除在较大沟谷中有溪流外，不少地段呈潜流状，特别是出山口多形成规模较大的洪积扇，沉积物以砂砾石为主，枯水期时多断流，只有在扇缘溢出带有水流出时才形成固定河道。上游主要有东沙河、关沟、北沙河和南沙河，于沙河镇东1km处汇聚，自此以下称温榆河。自沙河镇以下至曹碾村，温榆河河道呈曲流状，各沟河从北部沟谷注入，在古城附近又有清河由西南流入，形成较大的汇流点。古城以下沿顺义、朝阳区界向东南流经通州区注入北运河。

北运河流域山前洪积扇极为发育，如南口洪积扇、白羊城洪积扇、西峰山洪积扇、北安河洪积扇及温泉冷泉洪积扇等。这些洪积扇除白羊城洪积扇是由砂砾石（包括巨砾）组成外，其余都是以细粒物质为主，形成山前马兰黄土台地。由于洪积扇堆积物孔隙率较高，赋存地下水，是温榆河的主要水源。

根据北安河附近的高里掌下层泥炭的^{14}C年龄（9930±150）a推断，全新世早期，温榆河古河道切割了晚更新世黄土台地的台面，并形成了浅槽型河床，切割深度一般4~7m，沉积物以砂为主，局部地区含有泥炭或淤泥，在吕各庄一带温榆河河谷宽约1.5km，河曲发育。河流的侧向侵蚀作用比较明显，凸岸被侵蚀，凹岸再堆积。河流可能是在全新世开始形成，并切割晚更新世的高台地。在吕各庄村南形成类似二级阶地的阶梯状地貌，阶梯高约1.5~2.0m。此地貌的原始形态由于被人类活动所改造，现只在吕各庄村保留有一个缓坡。全新世中期，随着河流侧向侵蚀作用的加强，温渝河河谷逐渐变窄，在吕各庄南畜牧场处形成一级阶地，阶地面平坦，阶地坡陡立，高度为2.5m左右。堆积物韵律性沉积很明显，具有二元结构特征。全新统晚期河流进一步萎缩，侧向侵蚀作用增强，在凸岸形成宽阔的河漫滩。堆积物的颗粒较细，下部为细砂，上部为粉砂，具有平原区现代河漫滩的二元结构特征。现代温渝河河床的宽度约为50m。

4. 潮白河流域（$Ⅲ_1^{1-4}$）

潮白河是北京平原上第二大河，出山后在山前形成广阔的冲积扇平原，其沉积厚度可达200~300m。现在的潮白河河道是长期演变的结果。在全新世早期，潮河与白河单独入海，白河流入现在的小中河，沿北运河入海；潮河流入鲍邱河再入海。全新世中期，白河东移，潮河西迁，二河合流后在现在潮白河谷地流动。全新世晚期，潮白河汇合点不断北移，最终在二者的不断迁移沉积中形成了当今的潮白河冲积平原。平原地势由北而南倾斜，其组成物质在该河两侧的高位冲积平原上，分布着黄褐色或黄棕色为主，结构坚硬、致密，有明显的垂直节理，以黏土质砂或砂质黏土为主，并夹有粉砂和黏土薄层的沉积层，局部地区有砾石层，表层发育褐色土。

在通州一带，潮白河冲积扇与永定河老冲洪积扇相连。潮白河冲洪积扇受新构造运动影响，河槽下切，新期古河道大多在槽中变动，这是与永定河冲洪积扇不同的。潮白河冲洪积扇的形成时代与永定河老冲洪积扇一样，为晚更新世。自燕郊以南地势低平，并有泛滥而成的砂地。

5. 蓟运河流域（$Ⅲ_1^{1-5}$）

位于平谷境内，在北京市境内有泃河、错河和金鸡河三大支流，3条河流在平谷汇合后出境。该区北、东、南三面环山，地势东高西低，呈半封闭状山间盆地，属山间冲积洪积倾斜平原。沿山前地带为洪积扇相连。北山山麓的峪口、峨嵋山、黄松峪等洪积扇面积较大，而南山山麓洪积扇都较小。洪积扇具有新老不同特征，老期洪积扇由地壳抬升而被切割，地表覆盖着黄土。新期的洪积扇由于发育时间短，地表主要是砂砾石。泃河横贯平原的东部和中部，在西沥津以下，河道曲折，形成冲积平原，地势平坦。在靠山集、土门和祖务一带因地势抬升和外营力的相互作用，出现了一些基岩台地。

第四十一章　活动断裂与新构造运动

第一节　平原区活动断裂

本志中活动断裂指全新世以来发生过活动的断裂。北京平原区主要活动断裂有黄庄-高丽营断裂、南口-孙河断裂、顺义断裂和夏垫断裂等（图 41-1）。

图 41-1　北京平原区主要活动断裂分布图

Fig. 41-1　Distribution map of major active faults in Beijing Plain

F_1—黄庄-高丽营断裂；F_2—顺义断裂；F_3—夏垫断裂；F_4—南口-孙河断裂

一、黄庄-高丽营断裂

黄庄-高丽营断裂是 1966 年石油部 646 厂进行人工物探时发现的，为北京凹陷和西山隆起的分界构造。在北京地区该断裂南起房山长沟一带，经坨里、黄庄、八里庄、高丽营沿至怀柔一线，总长近 110km。断裂走向为北北东-北东，倾向南东，倾角 50°~80°，正断层，最大断距在 1000m 以上，断裂于燕山运动末期形成，切割了侏罗系及更老地层，主要控制了白垩系、古近系、新近系及第四系。

（一）分段特征

参照活动断层分段原则与方法（丁国瑜等，1982，1984，1990，1992，1993；邓起东等，1995；张培震等，1998），根据黄庄-高丽营断裂对新生代地层的控制作用，结合其与北西向断裂的交切关系，将该断裂自西南向东北划分为 3 段，即南段（房山段）、中段（黄庄段）和北段（高丽营-怀柔段）。其中南段北起永定河一带，向西南经芦井、辛庄西、大灰厂、晓幼营、辛开口至长沟一带，调查发现，在芦井、大灰厂至晓幼营一带，局部地区可见有断裂的地表露头；中段北起洼里一带，经西南经黄庄鲁谷、到衙门口一带，控制了丰台凹陷的沉积；北段北起怀柔，向西南经庙城、高丽营至太平庄，控制了怀柔凹陷的沉积。

1. 南段

黄庄-高丽营断裂南段的南、北端走向为北东向，中部房山岩体以东走向为近南北向。据北京市地质调查研究院（2012）岳各庄联合剖面显示，断裂两端的次级断裂很发育，有 2~4 条，断裂带宽度较宽，可达 2000m 左右。主断裂上断点埋深为 5m 左右，中部地段出露地表。

2. 中段

黄庄-高丽营断裂中段据北京市地质调查研究院（2012）八宝山西-八宝山东联合剖面显示，其次级断裂不发育，仅确定一条，相应断裂带宽度较小，为 500m 左右。

3. 北段

黄庄-高丽营断裂带北段据北京市地质调查研究院（2012）庙城联合剖面显示，北部宽度较大，可达 2500~3000m，在高丽营一带宽度仅为几百米，次级断裂 2~3 条。

（二）结构特征

现有资料表明，黄庄-高丽营断裂由一系列铲状、雁列式断层组成。据邵学钟等（1979，1980），断裂切割深度可达 50km 左右；据刘保金等（2009），该断裂向下可延伸到 8~9km 的深度上，可见，在深部延伸上还有待进一步勘查和证实。北京市地质调查研究院（2011）调查资料显示，黄庄-高丽营断裂带在中浅深度结构复杂，显示出不同的构造样式或构造组合形式（图 41-2）。

（三）活动性分析

黄庄-高丽营断裂活动性总体表现为北段较强，中段、南段活动性相对较弱特征。

1. 南段

通过资料收集和整理分析，在晓幼营、芦井、大灰厂、辛庄北一带剖面反映了黄庄-高丽营断裂的活动特征（徐杰，1992；陈国星，1994），其最新活动在晚更新世末期，活动频度不大，在中更新世断裂活动较强烈，在郭家坟—房山一带上新世至早更新世有明显活动，中更新世以来无活动迹象。

2. 中段

据地震会战资料分析，断裂下盘在首都体育场钻孔仅 89m 处就见到花岗岩，而上盘位于西四自然资源部院内的京热-1 井揭露古近系和新近系厚度达 1000m 以上。上述两孔说明该断裂新近纪还在强烈活动。据北京市地震局（2001—2003）、北京市地质矿产勘查开发局（2007）在奥林匹克公园区域开展断裂探测表明，在该区域断裂最新活动时间为早更新世。

3. 北段

断裂活动性在探槽中揭露十分直观。贾三满等（2007）在高丽营镇西王路村西南地裂缝带上开挖探槽，其东壁断裂在深 6m 左右存在约 1.56m 的断距，断面直达地表（图 41-3），现今仍在蠕动，在地面形成地裂缝，并对居民住房、墙体等造成破坏，出现不同程度开裂现象，研究发现其为全新世断层，其活动性延续至今。

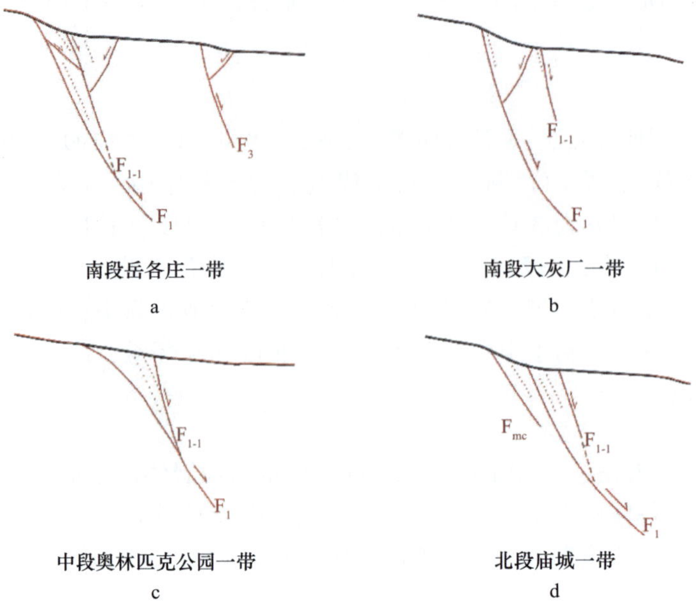

图 41-2　黄庄-高丽营断裂带结构样式示意图

Fig. 41-2　Sketch map of structural patterns of Huangzhuang-Gaoliying fracture zone

a、b—南段断裂结构；c—中段断裂结构；d—北段断裂结构；

F_1、F_2—黄庄-高丽营及其次级断裂；F_3—石楼断裂；F_{mc}—庙城断裂；

……　推测（隐伏）断裂　……　断裂带

二、南口-孙河断裂

南口-孙河断裂北西端起自昌平南口镇，向南东方向经七间房、百泉庄、东三旗、孙河至通州，总体走向 315°（见图 41-1），是北京平原区一条重要的北西向活动断裂。

（一）分段特征

南口-孙河断裂大致以北七家为界，分为两段。北西段：南口至北七家，断裂面倾向南西，两盘基岩面落差 200~400m，断裂的北东侧（下盘）局部基岩呈孤岛状残山裸露。断裂南西侧（上盘）明显下降，形成马池口凹陷。南东段：北七家至通州，断裂面倾向北东，倾角 60°~70°，两盘基岩面落差 50~100m，上盘下降形成后沙峪凹陷。

1. 北西段

南口-孙河断裂基岩上错断了青白口系，断裂带宽度约 1000m，该段可解译出 1 条次级断裂，主断裂倾向南西，倾角约为 76°，上断点埋深约 60m，根据现有地质资料显示，该断裂错断了晚更新世晚期地层。

2. 南东段

南口-孙河断裂在基岩上错断了蓟县系，该段断裂带宽度约 300m，主断裂倾向北东，倾角约为 65°，上断点埋深约 23m，根据现有区域地质资料显示，该断裂错断了晚更新世地层。

（二）结构特征

南口-孙河断裂是由主断层和一系列次级断层组成的断裂带，断裂带宽度为 500~2000m。北西段次级断裂发育，畜仓屯一带发育 2 条次级断裂，断裂带宽度约 1000m，百善镇一带发育 4 条次级断裂，断裂带宽度约 2400m。南东段次级断裂在北七家一带共发育有 3 条次级断裂，断裂带宽度约 1600m，在孙河至金盏一带发育 1 条次级断裂，断裂带宽度约 500m。

（三）活动性分析

近年来地形变观测结果表明，该断裂带现今仍具有缓慢形变的特性，且地表地裂缝、地面塌陷等

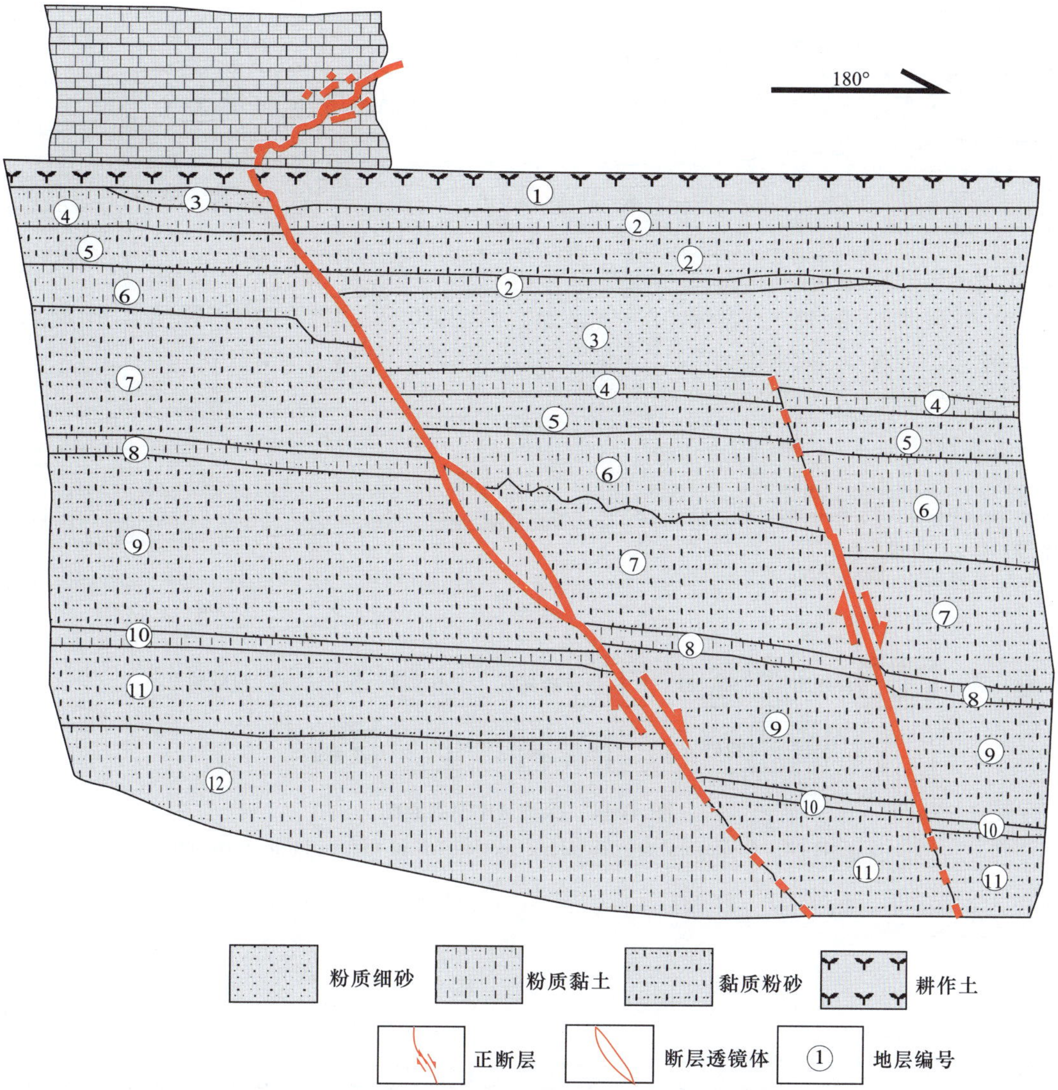

图 41-3　黄庄-高丽营断裂北段西王路探槽东壁简图
（据贾三满等，2007）
Fig. 41-3　Sketch map showing east wall of Xiwanglu exploratoty trench, northern segment of Huangzhuang-Gaoliying fracture

灾害现象较为发育。

1. 北西段

（1）联合钻孔剖面分析

在南口-孙河断裂两盘实施的 ZK109 与 ZK108，间距为 90m，基岩埋深落差达 312.50m，通过钻孔岩性组合对比，结合磁性地层及年代学数据，建立了钻孔年代地层格架，结果表明，南口-孙河断裂中更新世、晚更新世和全新世以来的垂直活动速率分别为 0.12mm/a、0.58mm/a、0.31mm/a，活动规律总体表现为弱—强—弱的特点。

（2）人工探槽揭露

在上述钻孔剖面上开挖了长约 150m、宽约 3m、深约 4m 的人工探槽，总体走向为北东 40°。探槽可分为 3 段，其中 AB 段剖面（图 41-4）上见有明显的地层错断现象，如剖面层 4 中褐黄色粉砂层中存在一明显的小型断裂 F_2，上下盘厚度差异约 13cm，应为断裂造成的坎前堆积。AB 段剖面揭示断裂最新的一次造成地表破裂的古地震事件时间为（9230±45）a B. P.，所影响到的地表最浅层位深度约 1.2m。BC 段剖面（图 41-5）上未见明显地层错断现象。CD 段剖面（图 41-6）上则发育较多的小型

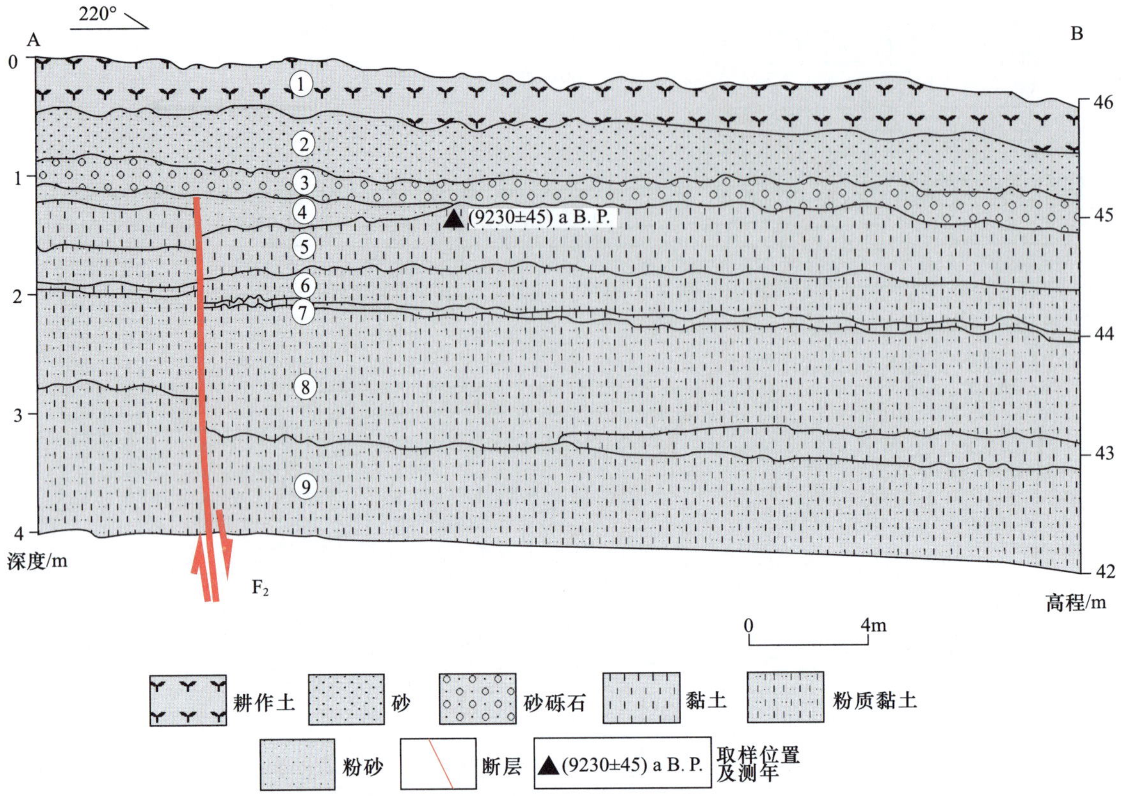

图 41-4　探槽 AB 段剖面图

Fig. 41-4　Geological section along section AB of the exploratoty trench

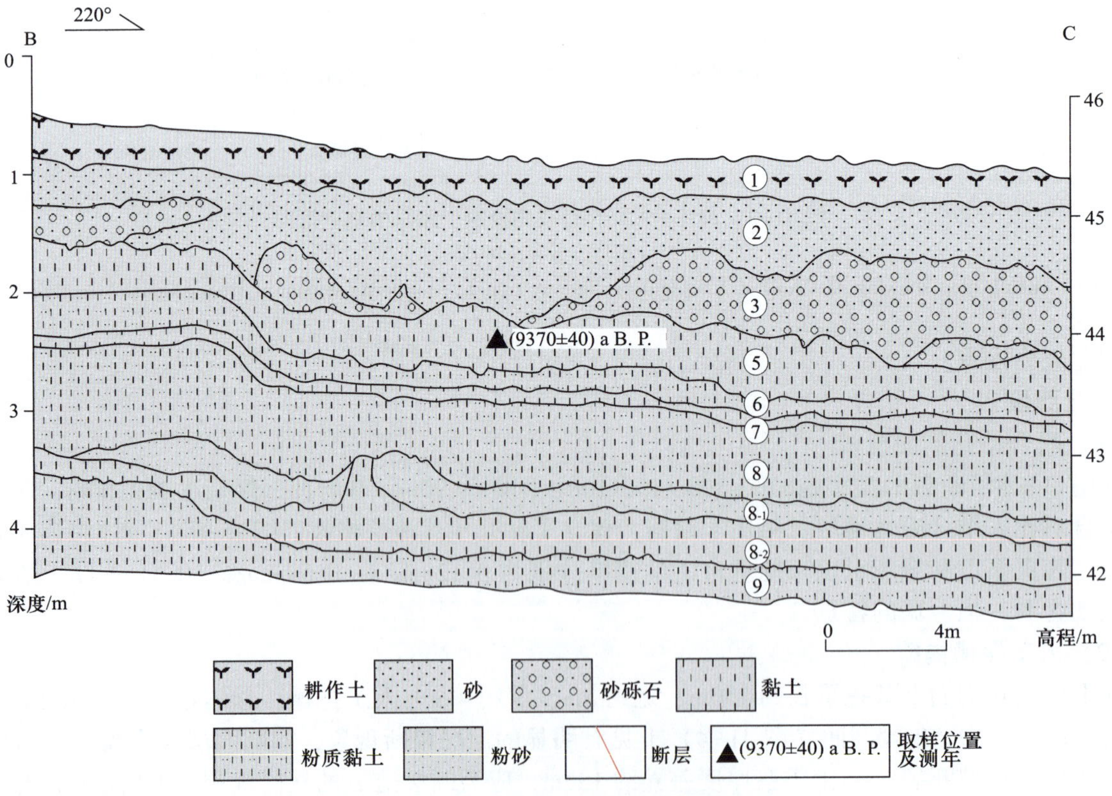

图 41-5　探槽 BC 段剖面图

Fig. 41-5　Geological section along section BC of the exploratoty trench

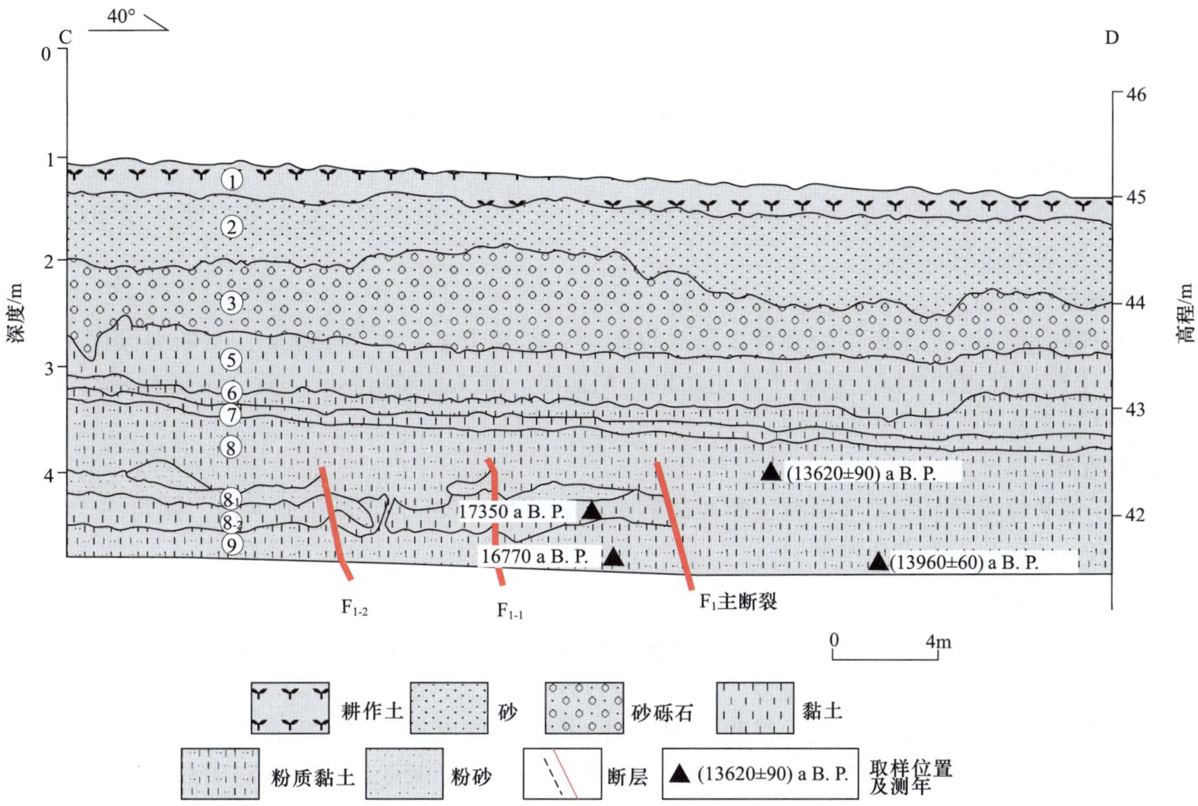

图 41-6 探槽 CD 段剖面图

Fig. 41-6 Geological section along section CD of the exploratoty trench

构造现象，F_1 为探槽揭露的主断裂，其最新活动时间为（13620±60）a B. P.。

综合分析，百善探槽中所揭露的地质构造现象表明南口-孙河断裂北段至少可识别出两次古地震事件，时间分别为（9230±45）a B. P. ~（9370±40）a B. P. 和（13620±90）a B. P. ~（13960±60）a B. P.，错断地层的最浅深度为 1.2m。

2. 南东段

南东段于燕丹村一带在断裂两盘施工钻孔 ZK17 和 ZK18，两孔口标高落差为 1.0m。通过对钻孔 ZK17 和 ZK18 联合剖面地层落差计算，南口-孙河断裂早、中、晚更新世和全新世以来的垂直活动速率分别为 0.065 mm/a、0.046 mm/a、0.030 mm/a 和 0.110mm/a。可见，南东段早更新世—晚更新世活动较弱，全新世活动强烈（表 41-1）。

表 41-1 南口-孙河断裂南段第四纪活动速率统计表

Table 41-1 Statistical table of Quaternary activity rate in the south segment of Nankou-Sunhe fracture

断裂分段	时代	垂直位移量/m	平均活动速率/（$mm \cdot a^{-1}$）
南口-孙河断裂南东段	全新世	1.2	0.110
	晚更新世	4.1	0.030
	中更新世	30.0	0.046
	早更新世	116.5	0.065

三、顺义断裂

顺义断裂西南起健翔桥附近，向东北经天竺、首都机场附近、顺义、北小营到焦庄户一带，走向40°~60°，倾向南东，倾角60°左右，延伸约40km，为正断层（见图41-1）。

（一）分段特征

顺义断裂以南口-孙河断裂为界可分为西南段、北东段，两段倾向一致，基岩面落差200~400m，其中北东段次级断裂较发育。

1. 北东段

据南彩联合剖面，顺义断裂带发育3条次级断裂，断裂带宽度约为1500m，主断裂倾向南东，倾角约49°，上断点埋深约61m，错断了晚更新世晚期地层。顺义县城南到北小营一带沿断裂发育地裂缝表明断裂已断至地表，现今仍在活动。

2. 南西段

据来广营联合剖面显示顺义断裂主断裂在第四系中断距约23m，该断层倾向南，倾角约75°，上断点深度约13m，错断了晚更新世地层。

（二）结构特征

顺义断裂是由主断裂和一系列次级断裂所组成的断裂带，整个断裂带宽度400~2500m，次级断裂发育，在蒋各庄一带发育1条次级断裂，南彩一带发育4条。石油工业部646厂二大队于1967年横跨顺义断裂实施了6条人工地震剖面，探测深度达到1000m，剖面显示该断裂在军营以南分为两支，二者相距1000m左右，错断了新近纪和第四纪下部地层，在200m深度以下地层有明显的错断。该断裂经来广营折向南西西，与黄庄-高丽营断裂相交，最大落差达400m，呈阶梯状断块。

（三）活动性分析

该断裂第四纪以来长期活动，在顺义苇沟、橡胶二厂、军营村、北小营等地造成明显的地面开裂、地面塌陷等现象。

1. 联合钻孔剖面分析

在断裂下盘、上盘分别实施的钻孔ZK12-1、ZK12-2，位于顺义前郝家疃村西，孔深分别为298.9m、716.9m，经钻孔地层（据白凌燕等，2014）对比分析得出，顺义断裂早、中、晚更新世和全新世以来的垂直活动速率分别为0.280mm/a、0.090mm/a、0.120mm/a和0.140mm/a（表41-2）。可见，顺义断裂早更新世活动较强烈，中更新世断裂活动性减弱，到晚更新世开始活化，从晚更新世到全新世活动性有逐渐增强趋势，这与断裂带沿线所见建筑墙体开裂和地裂缝发育等地表形变相吻合。

表41-2 顺义断裂第四纪活动速率统计表

Table 41-2 Statistical table of Quaternary movement rate of the Shunyi fracture

断裂分段	时代	垂直位移量/m	平均活动速率/（$mm \cdot a^{-1}$）
顺义断裂北段	全新世	1.4	0.140
	晚更新世	13.1	0.120
	中更新世	57.6	0.090
	早更新世	289.9	0.280

2. 钻孔旋回地层分析

利用钻孔沉积物磁化率数据进行旋回地层分析，所建立的精度为0.1Ma的连续天文年代标尺，为顺义断裂第四纪以来活动性精细刻画提供了年代学证据（据张磊等，2015）。由此推算出顺义断裂两盘在不同地质历史时期地层的落差（表41-3），得出顺义断裂第四纪以来活动性变化规律。

表 41-3 顺义断裂每 100ka 活动速率统计表

Table 41-3 The movement rate per 100 ka of the Shunyi fault zone

ZK12-1				ZK12-2				活动速率/(mm·a^{-1})
年龄/ka	深度/m	埋深/m	沉积速率/(mm·a^{-1})	年龄/ka	深度/m	埋深/m	沉积速率/(mm·a^{-1})	
100	7.75	7.75	0.12	100	8.41	8.41	0.22	0.10
200	19.97	12.22	0.14	200	30.69	22.28	0.28	0.14
300	33.86	13.89	0.19	300	58.54	27.85	0.25	0.06
400	53.19	19.33	0.14	400	83.41	24.87	0.25	0.11
500	67.10	13.91	0.13	500	108.45	25.04	0.20	0.07
600	80.20	13.10	0.18	600	128.39	19.94	0.20	0.02
700	97.81	17.61	0.19	700	148.3	19.91	0.21	0.02
800	116.74	18.93	0.11	800	169.54	21.24	0.15	0.04
900	127.93	11.19	0.10	900	185.02	15.48	0.12	0.02
1000	138.13	10.20	0.04	1000	197.49	12.47	0.24	0.20
1100	142.44	4.31	0.05	1100	221.86	24.37	0.43	0.37
1200	147.72	5.28	0.04	1200	264.49	42.63	0.44	0.40
1300	151.72	4.00	0.03	1300	308.17	43.68	0.18	0.15
1400	154.73	3.01	0.04	1400	326.63	18.46	0.30	0.26
1500	158.89	4.16	0.07	1500	356.61	29.98	0.40	0.33
1600	165.63	6.74	0.07	1600	396.34	39.73	0.39	0.32
1700	172.20	6.57	0.04	1700	435.37	39.03	0.38	0.34
1800	176.63	4.43	0.05	1800	473.64	38.27	0.29	0.24
1900	181.64	5.01	0.10	1900	502.16	28.52	0.42	0.32
2000	191.46	9.82	0.07	2000	544.24	42.08	0.29	0.23
2100	198.07	6.61	0.05	2100	573.51	29.27	0.22	0.17
2200	203.48	5.41	0.07	2200	595.69	22.18	0.40	0.34
2300	210.31	6.83	0.08	2300	636.11	40.42	0.23	0.15
2400	218.25	7.94	0.08	2400	658.79	22.68	0.40	0.32
2500	226.47	8.22	—	2500	698.56	39.77	—	—

由上述分析可见，早更新世期间，曲线所反映的活动速率背景值总体较高，波动明显，并出现5次明显的波峰（谷）变化，表明该时期断裂活动强度较大，分别为2.4~2.3Ma、2.2~2.1Ma、1.9~1.8Ma、1.6~1.3Ma和1.2~0.9Ma。断裂最大活动速率约0.4mm/a，最小活动速率约0.1mm/a，平均活动速率约为0.3mm/a，整个凹陷盆地受顺义断裂影响，两盘地层差异性升降明显，反应出顺义断裂在早更新世时期强烈活动的特点；早更新世晚期到中更新世早期（0.9~0.6Ma），断裂的活动速率曲线无明显的波动，总体表现较平缓，平均活动速率0.03mm/a，表明在该段时期内，顺义断裂无大强度的活动迹象，活动性相对较弱；中更新世晚期到晚更新世初，活动速率曲线再次

出现波动，0.5~0.1Ma 之间出现两个波峰，两个活动强烈期分为对应 0.2~0.1Ma 和 0.4~0.3Ma 之间，最大活动速率 0.14mm/a，最小活动速率 0.06mm/a，平均活动速率 0.1mm/a，由此可见顺义断裂在该段时期内再次活化，断裂活动性强弱交替变化，且具有逐渐增强的趋势，但活动速率总体较早更新世弱。

3. 人工探槽

在顺义五里仓大桥南中石化油库开挖的探槽，揭露了顺义断裂，该探槽东西向长 15m、南北向宽 6m、深 7m，由深到浅共揭露 7 套岩性地层单元（图 41-7）。据探槽研究显示，晚更新世以来活动强度较大；全新世（约 7.8ka 之后）发生过一次黏滑事件，体现在顺义油库发生的 5cm 左右的相对位移量。

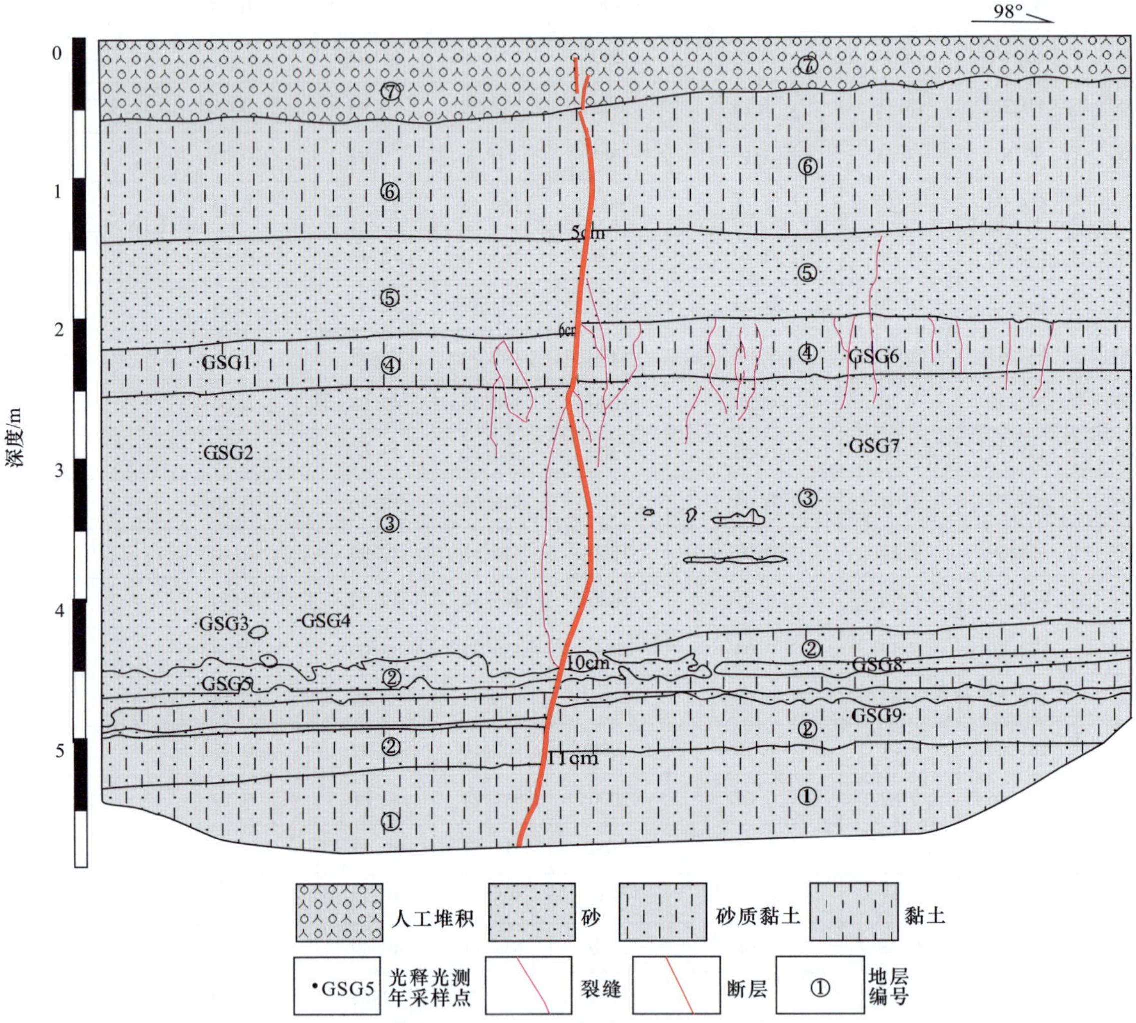

图 41-7 顺义油库探槽南剖面图

Fig. 41-7 Geological section of south wall of Youku exploratory trench in Shunyi

四、夏垫断裂

夏垫断裂南起凤河营，向北东经通州永乐店、西集、河北三河至平谷马坊，北东向延伸约 60km，走向 40°~45°，倾角 30°~80°。北京市区域内断裂长度约为 35km（见图 41-1）。据史料记载，公元 1679 年 9 月 2 日在河北三河一带发生了一次 8 级地震，该地震对北京地区破坏较大，因此该断裂被受到高度重视。

（一）分段特征

夏垫断裂大致以西集–溯县一带为界分为南北两段。北段：在构造形式上其主次断裂同西集断裂带共同构成夏垫断裂带，控制大厂凹陷（北部），整个断裂带断层产状均较为平缓，断裂带断距较大，最大可达4000~5000m；南段：在构造形式上其主次断裂同永乐店断裂带共同构成夏垫断裂带，控制大厂凹陷（中部），整个断裂带断层产状均较为陡立，最大断距可达3000m。夏垫断裂南段宽度大于北段，新生代地层中分支断裂也较多。

1. 北段

综合地质、地震、高密电及重力方法解译夏垫断裂切入到第四系，为第四纪以来活动断裂。

2. 南段

综合地质、地震及重力方法共解译出夏垫断裂及其次级，断裂为第四纪以来的活动断裂，二者共同构成夏垫断裂带。

（二）结构特征

夏垫断裂带可分解为不同深度的断裂，即下地壳–地幔深部断裂带、基岩断裂带和新生代断裂带3类。

1. 下地壳–地幔深部断裂带

张先康等对通过夏垫断裂带的多条深部探测剖面，表明该断裂带附近上地壳深度为21~23km，下地壳深度为36~37km；上地壳发育为相对脆性特征，仅在底部深度15~21km处发育低速异常体。

2. 基岩断裂带

夏垫断裂带在深部基岩中表现为一个较为宽广的断层构造破碎带，构成了大厂凹陷与大兴隆起的分界线。整个基岩断裂带呈宽带状产出，北北东走向，宽2~4km。

3. 新生代断裂带

夏垫断裂带第四系浅部由一条主干断层和几条北北东向，具不同产状的次级断层组成，大致由北向南宏观上被切割成3段。Ⅰ段：大致从平谷马坊经河北燕郊至通州西集一带，由一条主断层和众多分支断层构成，宽度约500m；Ⅱ段：从通州西集至觅子店一带，单条断层产出；Ⅲ段：由通州觅子店至风河营一带，由一系列小角度交叉的北北东向断层组成，断层分布密集，宽度约2000m。

基于夏垫断裂深部至浅部的构造特征，夏垫断裂带地壳浅层次以各种正断层及低角度拆离带组合为主，中深层次以塑性变形为主，表现为塑性伸展流动，地壳深部除塑性伸展流动变形外，尚发育大量基性岩墙群，形成该区域的低速带区域。

（三）活动性分析

夏垫断裂带是一条集黏滑和蠕滑为一体的活动性断裂，北段活动强度大，以黏滑为主，而南段活动强度相对较弱，以蠕滑为主。

1. 北段

据钻孔ZK07、ZK08、ZK09和ZK10联合剖面显示，夏垫断裂主断裂位于钻孔ZK08和ZK09之间（图41–8），结合地层岩性及年代学可知，0.0117Ma至今夏垫断裂带主断裂垂直相对运动速率为0.236mm/a；0.126~0.0117Ma，夏垫断裂带主断裂垂直相对运动速率为0.032mm/a；0.78~0.126Ma，上下两盘垂直相对运动速率为0.031mm/a；2.58~0.78Ma，上下两盘垂直相对运动速率为0.006mm/a。第四纪以来，夏垫断裂带活动特征在全新世活动强度相对最大（表41–4），与历史记载的多次强震活动相一致。

中国地震局分别于夏垫断裂东柳河屯、潘各庄、齐心庄等地开挖人工探槽及钻孔，揭露14ka以来的破坏性古地震事件有5次。全新世有4次古地震事件，其发生的时间分别为10.85~9.71ka、7.39~6.68ka、5.416~2.233ka以及1679年，4次事件的平均间隔为（3.336±0.396）ka。

表 41-4 夏垫断裂北段第四纪活动速率统计表

Table 41-4 Statistical table of Quaternary movement rate in the northern segment of the Xiadian fault

断裂分段	时代	累积位移量/m	平均活动速率/（mm·a^{-1}）
夏垫断裂北段	全新世	2.6	0.236
	晚更新世	3.7	0.032
	中更新世	20.3	0.031
	早更新世	11.2	0.006

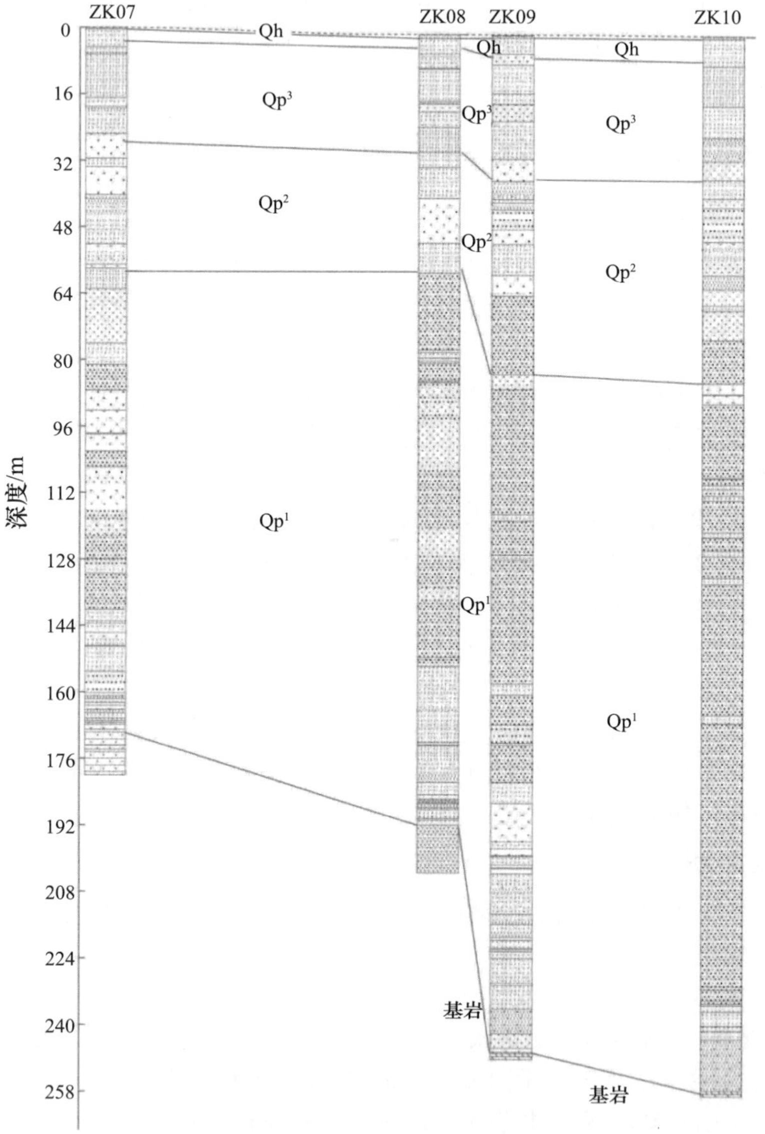

图 41-8 ZK08 和 ZK09 对比图

Fig. 41-8 Columnar correlation of ZK08 and ZK09 drill holes

2. 南段

该段进行过地表调查研究，即安次第一砖厂采土坑，采坑四壁保存较好，采壁陡直，深约 9m。为了确定该地区断裂活动时间特征，经对采坑内相关现象进行详细编录并取样分析（图 41-9），结果表明错断发生在距今 6700~2600a 之间。

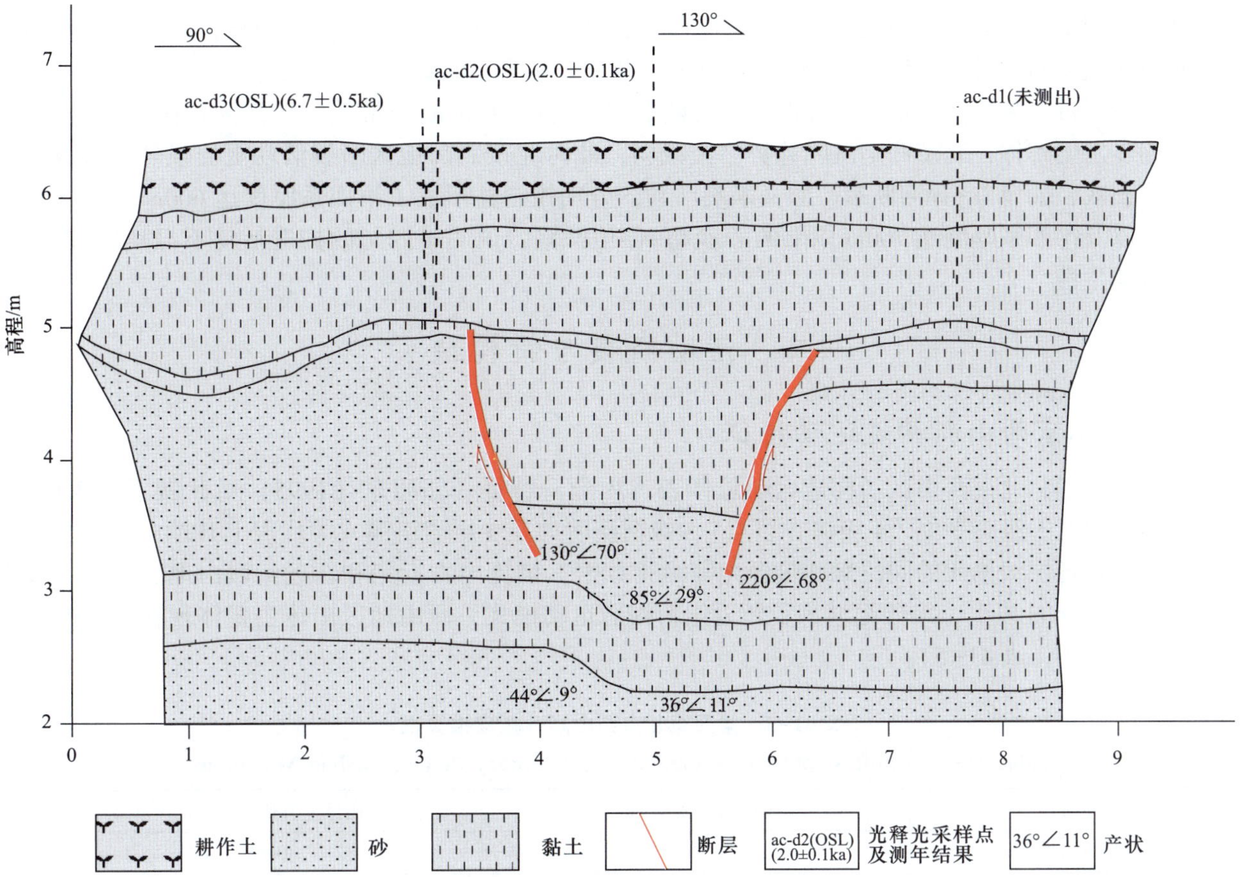

图 41-9 安次砖厂采坑北壁素描图

Fig. 41-9 Sketch map of the northern wall in Anci brick factoty pit

第二节 北京地区新构造运动

一、新构造运动特征

北京地区进入新生代以来，新构造活动总体特征表现为以伸展作用为主，持续时间较长，并以活动断裂为表现形式。受断裂活动的影响，断裂两侧发生显著的差异沉降。各条活动断裂在不同历史时期活动强弱不一，同一断裂不同部位在相同时期的活动性也存在差异。由此控制了平原区不同区域新生代沉积物的分布及其厚度差异，也显示出了新构造运动的特征及其演化。

古近纪—新近纪，区域构造主应力方向为北东-南西向，先存燕山运动时期所形成的一系列北北东向断裂的结构面再次发生活化，沿先前断裂发生明显的拉张与断陷作用，受黄庄-高丽营、南苑-通县断裂活动的控制，北京平原区出现了“两隆一凹”的构造格局，即京西隆起、北京凹陷、大兴隆起。断裂构造控制了平原区古近纪、新近纪沉积物的分布及其厚度。在古近纪，北京凹陷、大厂凹陷堆积了巨厚沉积物，其中北京凹陷形成了沉积中心长轴为北北东向的构造沉积面貌。新近纪先存的北京凹陷进一步发育，在其内沉积了较厚的新近系，该时期北京凹陷以外沉积物厚度不一、分布不均，总体上厚度不大。此时断裂活动表现出两个特征：一是北京凹陷强烈沉陷，该时期的沉积地层（天坛组和天竺组）厚度达 700~1000m；二是断裂活动强度存在差异，如控制北京凹陷的边界断裂，其活动性从西南向东北方向表现为增强趋势，北京凹陷内的沉积物向东北方向发生超覆，如始新世的长辛店组只分布在北太平庄附近，渐新世的前门组向东北拓展到来广营，而到中—上新世的天坛组和天竺组其沉积范围更向东北延展到顺义以东的程各庄一带。

进入第四纪，主应力方向变为北东东-南西西，新的近北西向的断裂作用明显，北京地区从西向东，北东向主要断裂段带有紫荆关-怀来断裂带、黄庄-高丽营断裂带、顺义断裂带、南苑-通县断裂带、礼贤断裂带、夏垫断裂带等，北西向主要断裂带有南口-孙河断裂带等。平原区先存“两隆一凹”的格局逐步发生解体，北京凹陷沉降趋势已经减弱，此时北京北部地区断裂活动更加强烈，形成新的第四纪沉积区，如南口-孙河断裂、顺义断裂活动，分别控制并形成了第四系厚度达600m以上的马池口凹陷，以及第四系厚度达700m以上的后沙峪-俸伯凹陷，说明第四纪的构造活动不但具有继承性，并且具有新生性。

二、活动断裂与新构造

（一）北京西山新构造运动特征

新生代时期，北京西山区新构造运动总体表现为隆升特征。郭旭东（1995）对此做了深入研究，通过对西山地区5级夷平面、4级河谷裂点、3级河流阶地及5层岩溶洞穴等代表上升标志的地貌单元的调查，结合各层级单元的同位素和热释光等测年数据，推算出古近纪和新近纪西山地区山体主脊带总上升量达1700m以上。第四纪继承了新近纪新构造运动模式，西山地区山体主要体现在山体的加速上升（图41-10，表41-5）。

表41-5　第四纪北京西山隆升幅度推算表

Table 41-5　Uplift amplitude estimation of Quaternary Beijing-Xishan Mountains

地质时代	山体上升幅度/m	山体上升速度/（$mm \cdot a^{-1}$）
早更新世	120	0.05~0.07
中更新世	60	0.05~0.1
晚更新世	72	0.82
全新世	3~5，最大达8	0.66

（据郭旭东，2007）

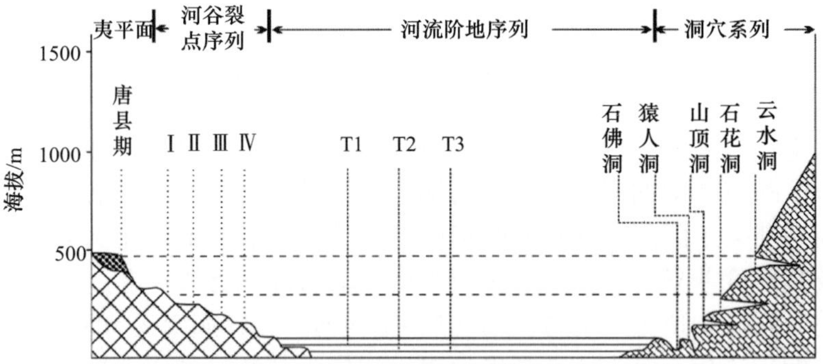

图41-10　北京西山新构造上升标志示意图

（据郭旭东，1995）

Fig. 41-10　Scheme showing the indicatiors of netectonic uplifting in Xishan Hills, Beijing

由表41-5分析，西山地区早、中更新世山脉上升幅度在第四纪占主要地位，但是上升速度相对较小，到了晚更新世，北京西山上升速度加快，达到新生代以来最大值，全新世仍呈上升趋势。

（二）平原区第四纪新构造运动

第四纪时期，平原区“两隆一凹”的地貌格局解体，受北东向和北西向活动断裂的控制，形成了马池口、后沙峪-俸伯等第四纪凹陷。由于断裂活动的差异性，致使不同凹陷中堆积物的厚度存在差异，同期异相现象显著。通过平原区主要活动断裂上下两盘的钻孔资料对比分析，得出主要活动断裂的活动速率（表 41-6），总体上表征了平原地区第四纪各个时期的新构造活动特征。

表 41-6　主要活动断裂第四纪以来活动速率表

Table 41-6　Active rate table of major active faults since Quaternary

单位：mm/a

地质时代	黄庄-高丽营断裂	南口-孙河断裂	顺义断裂	夏垫断裂
早更新世	0. 03	—	0. 28	0. 01
中更新世	0. 05	0. 12	0. 09	0. 03
晚更新世	0. 01	0. 58	0. 12	0. 03
全新世	0. 12	0. 31	0. 14	0. 24

从整体来看，各地质历史时期断裂活动规律既具有相对一致性，又存在特殊性，反映出在统一的地壳应力场下各活动地块的形变差异，由于垂向结构和水平边界条件的差异，使得各地块应变反应时间、活动强度、活动方式及运动方向有所不同。

西山隆升幅度变化特征与平原区第四纪沉积中心沉降速率总体上表现出相对吻合的趋势（表 41-7），西山在持续上升，呈现出上升速度越来越快的趋势，平原区发生沉降，主要活动断裂在全新世以来其相对垂向速率差明显较大，反应地壳运动也越来越强烈。体现了北京平原活动断裂与盆山耦合作用、沉积建造存在紧密的联系。

表 41-7　平原区第四纪沉降（中心）速率和北京西山隆升速率对比表

Table 41-7　Comparison of Quaternary subsiding（center）rate and Beijing Xishan Hills uplifting rate in plain area

地质时代		平原区第四纪主要沉积盆地中心沉积速率/（mm·a^{-1}）				西山隆升速率/（mm·a^{-1}）（郭旭东，2007）
		ZK19（后沙峪-俸伯凹陷）	ZK12-2（后沙峪-俸伯凹陷）	ZK3（马池口凹陷）	ZK10（平谷凹陷）	
早更新世	2. 58～0. 78Ma	0. 16	0. 32	0. 19	0. 06	0. 05～0. 2
中更新世	0. 78～0. 126Ma	0. 14	0. 17	0. 11	0. 12	0. 10
晚更新世	0. 126～0. 01Ma	0. 32	0. 40	0. 70	0. 25	0. 82
全新世	0. 01Ma 至今	0. 53	0. 97	0. 63	0. 51	0. 66

三、地震活动特征

北京市行政区域内有记录的历史地震自 274 年起（表 41-8），其中最大的一次强震是 1679 年发生的夏垫 8 级地震。

（一）地震活动时间分布特征

本区地震活动在时间上有活跃期和平静期之分。地震的震级和 $M_s \geq 4$ 级地震年频数随着时间的推移有强弱和高低的变化，呈现波浪状起伏。在活跃期内地震强度大，频度也高。根据这种变化特征，本区的地震活动自 12 世纪末期至今大致可以分出 3 个活跃期和 2 个平静期。第一个活跃期自公元 1284—1356 年，历时 72 年。第二个活跃期自 1476—1730 年，历时 254 年，1679 年夏垫 8 级地震即发生在这个活跃期里。自 1867 年至今为第 3 个活跃期。

在地震活跃期里，震级变化的总趋势呈现一强一弱的变化过程。如在第 2 个活跃期里，1484 年在北京居庸关发生 6¾级地震，其后 1679 年夏垫 8 级地震达到最高峰。以后震级很快降低，于 1720 年和 1730 年发生沙城 6¾级和北京西郊 6½级地震后进入平静期。反映了应力经过几次大的应力释放后，区域地壳变形活动进入相对稳定的阶段，总的变化过程是由若干个（小的）地震活动幕呈波浪式发展完成的。

在地震平静期里，震级和地震年频数较之活跃期显著降低。在两个平静期里都没有发生 $M_s \geq 6$ 级的地震，5 级地震也不多。$M_s \geq 4$ 级地震的最高年频数没有超过 5 次。

地震活动在时间上活跃期与平静期的交替，反映了地震的周期性活动特征。若干个以幕次划分的短周期组成以期次划分的长周期，地震活动的周期性反映了区域地应力积累-释放交替进行的过程和构造活动的阶段性变化特征。地应力积累过程对应地震活动平静期和新构造活动减弱阶段，地应力释放过程对应地震活动活跃期和新构造活动增强阶段。在更长的时间尺度上，还有以若干个活动期与平静期组成的更长地震周期的趋势，这可能与全球的应力变化和构造运动规律有关。

（二）地震活动的空间分布特征

北京平原区的地震分布具有明显的规律性。地震主要分布在第四纪以来有明显活动的断裂附近，尤其是第四纪以来断裂两盘差异运动幅度较大的活动断裂附近（表 41-8）。这些断裂两盘第四系落差多在 100m 以上。断裂切割深度也很大，如夏垫断裂是切割莫霍面的深断裂。此外，地震一般发生在几组断裂交汇部位附近、断裂拐弯处或端部附近。如 1057 年固安 6¾级地震、1536 年郎府 6 级地震、1665 年通县 6½级地震、1679 年夏垫 8 级地震及 1730 年东北旺 6½级地震等均发生在这些特殊部位。从等震线形态上看，上述几个强震与北北东向和北西向断裂关系密切。从地震震中与第四系沉积分布的关系看，大于或等于 4 级的地震大部分发生在以活动断裂为边界的第四系凹陷处。

表 41-8　北京地区若干历史地震发震部位简表

Table 41-8　The places of historical earthquakes in Beijing area

地震发生时间/年	震级	参考地名	与地震关系密切的断裂	走向	第四系落差/m	地震发生构造部位
1057	6¾	西胡村	燕郊、 涿县-宝坻	NNE-EW	100~200 150~200	与北东向断裂交汇部位
1076	5	北京前门	南苑-通县	EW	50~100	—
1532	5½	邵府	夏垫-马坊	NNE	>300	断裂附近夏垫凹陷南缘
1536	6	郎府	夏垫-马坊	NNE	>300	—
1586	5	莲花池	黄庄-高丽营 南苑-通县	NNE EW	400 50~100	断裂拐弯、交汇部位
1658	6	涞水	房山-涞水 涿县-宝坻	NW-NNE EW	50~800 150~200	与北东向断裂交汇部位
1665	6½	通县	南苑-通县	EW	50~100	与北西向断裂交汇部位
1679	8	夏垫	夏垫-马坊 三河-邦均	NNE EW	>300 50~100	断裂交汇部位
1746	5	昌平	南口-孙河	NW	400~500	—

（三）强震活动区的构造背景

北京地区 90%以上地震发生在 5~20km 深度范围内，大致相对于 G 界面与 C 界面之间，即“花岗岩”层，与地壳关系十分密切。本区“花岗岩”层顶面埋深 4~12km，底界埋深 14~22km，平均厚度 8~12km。埋深及起伏较大，震源深度相应也有深浅变化，该“花岗岩”层底界深度大致相当于康拉德面深度，为壳内低速层，低速层以上的地壳为脆性岩层，断裂构造发育，岩石脆性强，因而易发生破裂。下部低速层和下地壳具有韧性，发生缓慢蠕变，逐步达到新的平衡。由表 41-9 可见，北京及其周边地区在 33km 以下未见强震发生，可能与壳内低速层的存在有关。

表 41-9 北京及其周边地区 $M_S \geq 4$ 级地震

Table 41-9 Earthquakes ($M_S \geq 4$) in Beijing and its surrounding areas

发震时刻（国际时）	东经/（°）	北纬/（°）	震源深度/km	震级
1970-5-24 16：33	118.12	39.55	15	4.2
1972-1-30 3：56	118.9	39.1	20	4.3
1972-3-25 14：22	116.6	40.4	30	4.2
1972-4-30 3：56	118.9	39.1	20	4.3
1973-9-21 4：22	116	38.6	15	4.1
1973-10-21 4：22	116	38.6	15	4.1
1974-12-15 9：13	117.7	39.5	27	4.1
1976-4-22 1：54	117.1	38.7	33	4.1
1976-7-27 19：42	118	39.4	22	7.8
1976-7-27 21：04	117.7	39.1	33	4.7
1976-7-27 21：11	118.2	39.4	33	5.3
1976-7-27 22：31	118	39.3	33	4.1
1976-7-27 22：39	118.4	39.8	33	5.3
1976-7-27 23：03	117.8	39.5	33	5
1976-7-27 23：17	117.8	39.2	19	6.2
1976-7-27 23：36	117.9	39.6	33	4.3
1976-7-27 23：40	118.1	39.7	33	4.8
1976-7-27 23：56	117.9	39.4	33	4.3
1976-7-28 0：04	118	39.4	33	4.1
1976-7-28 0：56	118	39.4	33	4.1
1976-7-28 0：58	117.7	39.2	33	5.2
1976-7-28 1：13	118.4	39.7	33	4.5
1976-7-28 1：21	118.3	39.5	33	5
1976-7-28 1：45	117.9	39	21	5.5
1976-7-28 2：15	118.3	39.7	33	4.1
1976-7-28 2：44	118	39.4	33	4.4
1976-7-28 2：45	118.2	39.6	33	4.3
1976-7-28 3：19	118.4	39.7	33	4.1
1976-7-28 3：29	118.4	39.7	33	4.2
1976-7-28 3：33	118.1	39.3	33	4.1
1976-7-28 4：06	118.2	39.5	33	4.3
1976-7-28 5：08	118.1	39.5	33	4.7
1976-7-28 6：06	118.3	39.6	33	4.4
1976-7-28 6：59	118.2	39.7	33	4.8
1976-7-28 7：14	118.2	39.6	33	4.8
1976-7-28 9：28	118.2	39.4	33	4.6
1976-7-28 9：42	117.9	39.6	33	4.8
1976-7-28 9：48	117.9	39.6	33	5
1976-7-28 9：56	117.9	39.3	33	4.1

续表

发震时刻（国际时）	东经/（°）	北纬/（°）	震源深度/km	震级
1976-7-28 10：23	118.3	39.6	33	4.1
1976-7-28 10：45	118.5	39.7	22	7.1
1976-7-28 10：58	118.6	39.7	33	5.5
1976-7-28 11：58	118.3	39.7	33	4.2
1976-7-28 12：20	118.4	39.7	33	4.7
1976-7-28 12：40	118.5	39.7	33	4.1
1976-7-28 13：49	118.3	39.5	33	4.2
1976-7-28 14：15	118.6	39.8	33	4
1976-7-28 15：35	118.8	39.7	24	5.5
1976-7-28 16：08	118.2	39.6	33	4
1976-7-28 16：23	117.8	39.2	33	4.5
1976-7-2816：27	118.5	39.9	33	4.1
1976-7-28 18：49	118.3	39.3	33	4.4
1976-7-28 18：57	118.6	39.8	33	4.4
1976-7-28 23：08	118.1	39.5	33	4.2
1976-7-28 23：24	118.3	39.7	33	4.7
1976-7-29 1：01	118.9	39.9	33	4.9
1976-7-29 4：40	118.5	39.8	33	4.4
1976-7-29 4：55	118	39.3	33	4.2
1976-7-29 5：02	118.4	39.8	33	4.5
1976-7-29 5：30	118.2	39.8	33	4.7
1976-7-29 5：43	118.2	39.5	33	4.7
1976-7-29 7：49	118.1	39.4	33	4.1
1976-7-29 11：05	117.9	39.3	33	4.6
1976-7-29 11：09	118.6	39.9	33	4.3
1976-7-29 12：29	117.8	39.5	33	4.5
1976-7-29 12：47	118.4	39.6	33	4.5
1976-7-29 14：16	118.7	39.8	33	4.2
1976-7-29 18：26	117.9	39.3	33	4.3
1976-7-29 20：15	118.3	39.6	33	4.1
1976-7-29 23：53	118.7	39.8	33	4.3
1976-7-30 0：45	118.1	39.5	33	4.1
1976-7-30 6：12	118.5	39.8	33	4.2
1976-7-30 9：22	118.9	39.8	33	4.1
1976-7-30 15：41	118.1	39.4	33	4
1976-7-30 17：02	118.4	39.7	33	4
1976-7-30 20：06	118.7	39.7	33	4.3
1976-7-30 20：17	118.4	39.7	33	4
1976-7-30 21：23	118.4	39.7	33	5
1976-7-30 22：26	118.5	39.7	33	4.2

续表

发震时刻（国际时）	东经/（°）	北纬/（°）	震源深度/km	震级
1976-7-30 23：52	117.9	39.4	33	4.1
1976-7-31 7：23	118.7	39.7	33	4.2
1976-7-31 7：42	118.5	39.7	33	4
1976-7-31 15：20	118.6	39.9	33	4.1
1976-7-31 17：51	118.7	39.9	33	4
1976-7-31 18：01	118.3	39.5	33	4
1976-7-31 22：26	118.2	39.4	33	4
1976-8-1 20：53	118.3	39.5	33	4.6
1976-8-1 23：58	118	39.4	33	4
1976-8-2 8：38	118.7	39.9	33	4.1
1976-8-2 9：15	118.2	39.4	33	4.3
1976-8-2 10：26	118.9	39.9	33	4.1
1976-8-2 10：49	118.5	39.5	33	4.1
1976-8-3 1：43	118.6	39.8	33	4.2
1976-8-3 5：55	118.5	39.7	33	4.3
1976-8-3 6：54	118.4	39.9	33	4.2
1976-8-3 7：40	118.8	40	33	4
1976-8-3 9：03	118.5	39.6	33	4
1976-8-3 10：16	118.1	39.6	33	4.3
1976-8-3 12：21	118.2	39.4	33	4.1
1976-8-3 12：40	118.8	39.8	33	4.8
1976-8-4 1：24	118.7	39.8	33	4.2
1976-8-4 5：04	117.9	39.2	33	4.2
1976-8-4 13：44	118	39.4	33	4.2
1976-8-5 15：31	118.4	39.6	33	4.3
1976-8-5 19：12	118.5	39.8	33	4.1
1976-8-6 11：49	118.3	39.5	33	4.1
1976-8-6 13：17	118.3	39.7	33	4
1976-8-6 17：34	117.9	39.3	33	4
1976-8-6 20：40	117.9	39.4	33	4
1976-8-7 6：45	118.5	39.8	33	4
1976-8-8 11：09	118.7	39.6	33	4.8
1976-8-8 22：41	118.8	39.9	33	5.3
1976-8-8 23：36	118.9	39.9	33	4.3
1976-8-9 1：28	118.5	39.8	33	4.4
1976-8-9 2：10	118.3	39.4	33	4.5
1976-8-9 5：47	117.9	39.5	33	4.4
1976-8-9 7：58	118.4	39.7	33	4.2
1976-8-9 19：53	118.3	39.8	33	4.3
1976-8-10 5：28	118.2	39.4	33	4.4

续表

发震时刻（国际时）	东经/（°）	北纬/（°）	震源深度/km	震级
1976-8-10 14：23	118.9	39.9	33	4.3
1976-8-11 1：36	118.5	39.8	33	4.5
1976-8-12 23：56	118.9	39.9	33	4
1976-8-14 5：40	118.8	39.9	33	4.2
1976-8-14 7：10	118.8	39.8	33	4.1
1976-8-14 16：02	118.7	39.9	33	4.5
1976-8-15 10：21	118.3	39.7	33	4.6
1976-8-15 10：28	118.4	39.8	33	4.4
1976-8-15 10：53	118.4	39.7	33	4.2
1976-8-15 14：33	118.1	39.4	33	4
1976-8-15 14：53	118.5	39.7	33	4.2
1976-8-15 21：27	118.9	39.9	33	4.2
1976-8-15 22：39	118	39.5	33	4.4
1976-8-16 9：58	118.4	39.8	33	4.3
1976-8-16 16：42	118.1	39.4	33	4
1976-8-16 21：20	117.9	39.4	33	4.2
1976-8-18 15：37	118.3	39.6	33	4
1976-8-21 20：02	118.9	39.9	33	4
1976-8-23 17：15	118.3	39.7	33	4.2
1976-8-24 10：52	117.5	39.6	33	4.1
1976-8-25 17：31	118.2	39.7	33	4.1
1976-8-28 6：34	118.9	39.9	33	4.5
1976-8-31 3：25	118.9	39.7	33	5.6
1976-8-31 3：27	118.4	39.9	33	5.3
1976-8-31 3：56	118.9	39.8	33	4.4
1976-8-31 13：04	118.9	39.9	33	4.1
1976-9-2 8：49	118.9	39.9	33	4.3
1976-9-6 17：01	119.1	39.9	33	4.5
1976-9-6 20：02	119.1	39.9	33	4
1976-9-12 18：39	119	39.9	33	4.2
1976-9-15 6：08	118.9	39.7	33	4.3
1976-9-18 19：15	118.7	39.6	33	4.9
1976-9-25 11：50	118.5	39.6	33	5
1976-9-28 23：12	118.9	39.9	33	4.6
1976-9-30 20：05	118.6	39.6	33	4.4
1976-10-7 11：33	118.4	39.7	33	4.5
1976-10-9 17：13	118.4	39.8	33	4
1976-10-23 11：06	118	39.4	33	4.2
1976-11-12 13：13	118.8	39.9	33	5
1976-11-15 13：53	117.7	39.4	17	6.9

续表

发震时刻（国际时）	东经/（°）	北纬/（°）	震源深度/km	震级
1976-12-2 0：42	117.5	39.6	33	4.7
1976-12-13 3：48	118.5	39.9	33	4.5
1976-12-27 9：01	118.6	39.6	33	4.2
1976-12-31 11：17	118.2	39.6	33	4.1
1977-1-3 5：36	118.6	39.6	15	4.2
1977-1-14 22：46	118.8	39.7	15	4.5
1977-1-30 4：11	118.2	39.5	25	4.7
1977-3-7 0：28	119	40	18	5.7
1977-3-23 23：14	118.6	39.8	15	4.4
1977-4-10 21：10	118.1	39.6	15	4.2
1977-4-29 6：00	118.2	39.5	15	4.6
1977-5-12 11：17	117.7	39.2	19	6.2
1977-5-17 17：23	118.8	39.7	15	4.6
1977-5-26 3：22	118.7	39.8	15	4.3
1977-6-10 0：40	118.2	39.4	15	4.7
1977-6-13 4：25	118.1	39.1	15	4
1977-6-17 17：55	118.7	39.3	15	4
1977-6-19 0：19	118.1	39.4	15	4
1977-7-9 20：27	119	40	19	4.7
1977-7-14 20：32	118.2	39.5	15	4.5
1977-7-14 20：36	118.4	39.7	15	4.5
1977-11-26 22：46	118	39.4	16	5.1
1977-12-20 16：23	118.8	39.6	15	4.3
1977-12-20 19：43	118.7	39.6	15	4.1
1977-12-21 22：27	118.7	39.6	33	4.2
1978-5-9 2：28	117.8	39.3	15	4
1978-8-7 12：03	118.1	39.3	15	4.4
1978-8-7 13：27	118.1	39.3	15	4.1
1978-12-6 7：30	118.4	39.9	10	4.1
1980-2-7 5：31	117.87	39.45	10	4.6
1983-10-4 16：25	119.06	39.79	10	4
1984-1-7 11：18	118.76	39.76	15	4.5
1985-4-22 3：31	118.9	39.8	33	4.3
1985-10-5 4：01	118.6	39.74	10	4.3
1988-8-3 9：44	118.9	39.65	10	4.3
1989-10-18 21：02	113.95	39.97	9	4.6
1989-12-13 1：10	113.97	39.89	8	4
1990-7-3 19：39	118.65	39.74	7	4.1
1990-7-21 0：41	115.87	40.65	12	4.5
1990-9-22 3：02	116.55	40.09	12	4.2

续表

发震时刻（国际时）	东经/（°）	北纬/（°）	震源深度/km	震级
1991-5-29 11：02	118.41	39.68	12	4.5
1991-5-29 23：06	118.38	39.66	24	4.9
1991-7-27 9：54	118.85	39.72	6	4.1
1992-7-21 21：43	117.95	39.29	32	4.3
1993-2-3 16：55	118.61	39.81	12	4.6
1995-10-5 22：26	118.55	39.72	20	4.8
1996-12-15 21：36	116.7	40.23	10	4.2
1998-1-10 3：50	114.51	41.12	15	6.3
1998-4-14 2：47	118.51	39.65	23	4.6
1998-6-2 1：32	114.46	41.11	10	4.3
1998-7-27 1：05	114.52	41.25	15	4.2
1998-7-27 1：17	114.56	41.21	33	4.5
1998-8-13 10：21	114.48	41.19	15	4.5
1999-3-11 13：18	114.57	41.22	17	5.6
1999-3-12 15：16	114.45	41.34	18	4.1
1999-5-12 0：55	118.97	40.3	15	4
1999-11-1 13：20	113.99	39.87	15	4.1
1999-11-1 13：25	113.98	39.91	10	5.6
2000-4-9 16：17	114.52	41.29	32	4.2
2001-3-2 4：29	115.02	40.41	19	4.1
2003-11-14 21：43	119.03	39.74	12	4.1
2003-12-11 14：48	117.68	39.32	19	4
2003-12-11 14：49	118.03	39.58	12	4
2005-8-31 15：48	114.76	39.99	14	4.1
2006-7-4 3：56	116.35	38.97	14	5
2008-3-10 20：24	118.97	39.99	15	4
2010-3-6 3：00	118.5	39.7	10	4.3
2012-5-28 10：22	118.5	39.7	8	4.8
2012-6-18 3：05	117.6	39.6	9	4
2014-9-6 18：37	115.4	40.3	10	4.3
2015-9-14 18：10	118.8	39.72	14	4.2
2016-6-23 8：37	114.2	40.95	14	4
2016-9-10 18：09	118.32	39.69	10	4

注：数据引自中国地震台网速报地震目录，1970—2017年。

第九篇

平原区三维地质结构

以往平原地区地质调查都是二维平面（或剖面）的，在水文地质调查工作中，有的以栅格图的方式来表达含水层和隔水层的三维结构，可以说是平原地区三维地质结构的雏形。北京市平原区地质结构复杂，新生界之下的基岩被北东向和北西向的断裂切割成众多块体，块体之间差异沉降明显（图Ⅸ-1），使不同地区新生代堆积物在厚度和岩性上都存在较大差异。2004—2008年原北京市地质矿产勘查开发局组织开展的“北京市多参数立体地质调查”项目，设立了“城市三维地质结构调查”专题，该专题在全面系统收集研究分析钻孔和物探等资料的基础上，补充了钻探和物探工作，综合运用现代勘查技术，开展了平原区基岩层、新生界和工程层3个层面的立体地质调查，重点调查了基岩面的三维特征，建立了新生界和工程层三维立体地质结构，并建立三维可视化城市地质数据管理和服务系统，为城市规划、城市建设与管理以及社会公众信息需求提供服务。以此项目为基础，本志仅对基岩地层、新生界和平原区浅表层（工程地质层）三维地质结构进行总结。

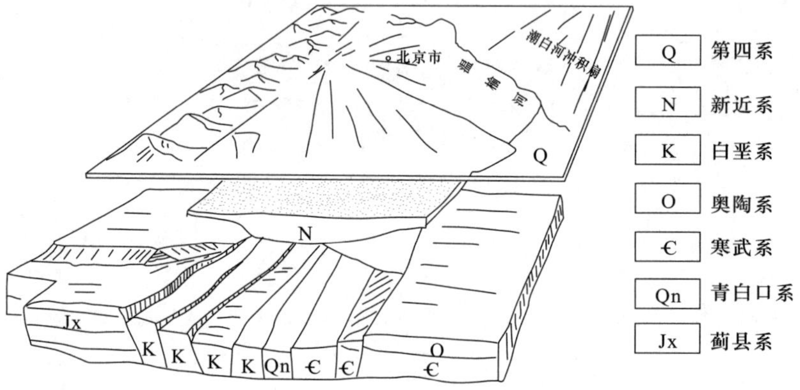

图Ⅸ-1　北京平原区三维地质结构简图
（据蔡向民，2009）
Fig. Ⅸ-1　The sketch map of bedrock geological 3D model of Beijng Plain

第四十二章　基岩三维地质结构

第一节　平原区基岩地层三维地质结构

一、太古宙地层

太古宙变质岩在山区主要分布于密云，昌平、房山有零星分布。在平原区太古宙变质岩主要分布在密云城区以东的平原和沟谷中，基岩大致埋深在 0～100m 之间，岩性与密云山区变质岩一致。

另外，位于北京平原东南部采育一带的桐热 51 钻孔在揭穿新生界后见变质岩，大兴榆垡一带的兴热 1 孔和安定一带的兴热 2 孔在揭穿中—新元古界后钻遇变质岩，埋深分别是 1610m 和 2050m，岩性从上到下依次为角闪斜长片麻岩、花岗质片麻岩、辉石角闪斜长片麻岩、二辉斜长片麻岩、含辉石黑云母的角闪斜长片麻岩、角闪石的黑云母斜长辉石岩等。取自安定镇兴热 2 井孔深 2238m 处岩粉（二辉斜长片麻岩）的地球化学特征，与北京北山密云地区麻粒岩相变质岩有相似之处。该片麻岩成分显示英云闪长岩和奥长花岗岩的过渡类型（吕金波等，2008）。

前人利用榆垡—安定一带的航磁异常图推测过北京城南地区片麻岩埋深的相对起伏情况。航磁异常图显示在大兴区固安、榆垡、安定地区及河北省廊坊一带为片麻岩的隆起区，在大兴城区至通州城区一带和廊坊城区东南一带为两个沉积盖层较厚、片麻岩结晶基底埋藏较深的地区。

二、中—新元古代地层

北京地区中—新元古代地层出露广泛，约占山区面积的 1/3，平原区钻孔中也多次揭露，分布面积大，几乎覆盖全平原区，为一套滨海-浅海相的碳酸盐岩、碎屑岩和黏土岩，夹少量碱玄武岩系。

（一）长城系（Pt_2^1）

长城系包括常州沟组、串岭沟组、团山子组、大红峪组，与下伏太古宙地层呈不整合接触。平原区长城系主要在平谷和延庆的山前少量出露。另据大兴区安定、榆垡两地的钻孔揭露，其顶界埋深分别是 640m 和 880m，残留厚度分别是 1335m 和 1060m。

（二）蓟县系（Pt_2^2）

1. 空间分布特征

蓟县纪地层分布范围最大，占平原区面积的 95%。埋藏深度变化大，除北山山前、良乡附近呈残山出露外，顺义板桥（埋深 300m）、顺义米各庄（埋深 600m）、顺义李遂（埋深 400m）、朝阳呼家楼（埋深 900m）、通州何家坟（埋深 500m）、大兴庞各庄—魏善庄—马驹桥一带（埋深 300～400m）、二十里长山以东（埋深 100m 左右）埋深浅。埋藏较深的地区主要有：东北旺—回龙观—来广营—古城—西马坡一带，从西向东埋深增大，一般埋深 2800～5000m；云岗—丰台—地坛一带，埋深 3500～4000m，而蓟县系顶部缺失铁岭组和洪水庄组；采育—凤河营一带埋深大于 4000m，地层较齐全。

蓟县纪地层，空间形态呈现轻微舒缓的褶皱，在向斜核部存在青白口系，埋藏深度较大，厚度正常，一般厚 2000～2500m；分布在背斜顶部时，往往出现削顶现象，地层不全，缺少青白口系，多出现在断陷内，厚度小于 2000m；最浅处厚度小于 1000m，如南彩—李遂和庞各庄地区；山前地带局部地区地层不全，厚度变薄。

蓟县纪地层中白云岩岩溶裂隙发育，是北京地区的热水和凉水储层，尤其是雾迷山组是最好的地热储层。

蓟县纪地层在平原区的分布见图 42-1，中心城区的空间分布模型见图 42-2。

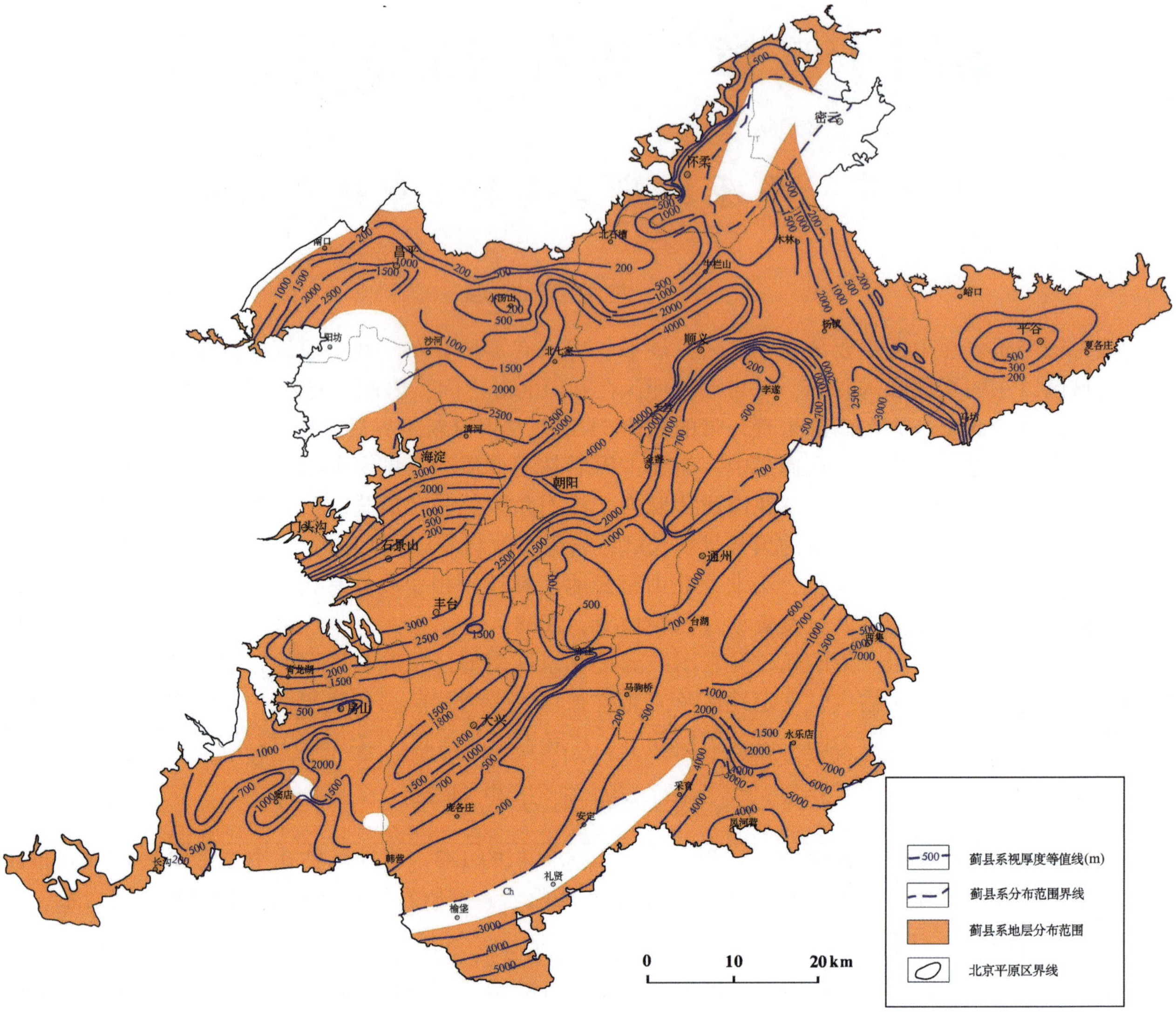

图 42-1　平原区蓟县纪地层分布范围及视厚度图

Fig. 42-1　The distribution and apparent thickness of Jixian system in plain area of Beijing

2. 地层岩性特征

1）高于庄组（Pt_2^2g）：曾称“高于庄灰岩”，是以碳酸盐岩占绝对优势，为厚层至薄层白云岩和白云质灰岩，叠层石较为发育，与下伏大红峪组呈整合接触或平行不整合接触，与上覆杨庄组呈整合接触或平行不整合接触。高于庄组在大兴区的安定、榆垡两地顶界埋深分别是 640m 和 880m，残留厚度分别是 1335m 和 1060m。区内揭露的大多是高于庄组白云岩及白云质灰岩等。

2）杨庄组（Pt_2^2y）：主要由紫红色—灰白色含砾石英砂岩、白云质石英砂岩、砂质白云岩和泥晶、粉晶白云岩组成，厚度一般小于 100m，与高于庄组呈整合或平行不整合接触。杨庄组钻孔在昌平镇有揭露，视厚度约 110m。

3）雾迷山组（Pt_2^2w）：为一套富镁碳酸盐岩，夹少量碎屑岩和黏土岩。钻孔揭露的该组局部地段厚度大于 1000m，不易分组和详细分层，除顺义区李遂观测井见到小叠层石（北京市顺义区南彩—李遂地热田地热详查地质报告，1990）可以视为雾迷山组外，通常用电测井方法帮助识别：雾迷山组白云岩视电阻率曲线呈现了相当高的幅值（约 1000Ω · m），放射性自然伽马曲线幅值低，且稳定平直，

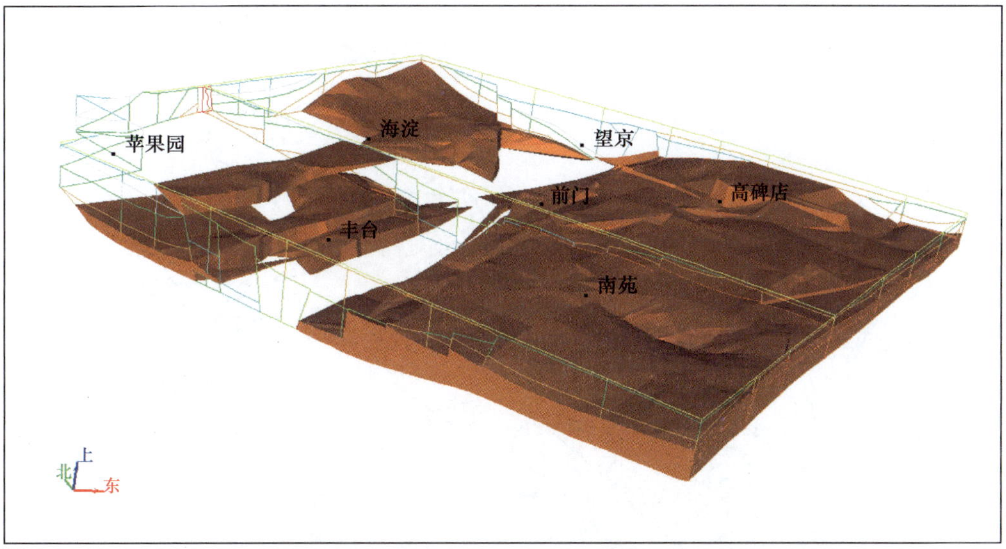

图 42-2　中心城区 2000m 以浅蓟县纪地层空间分布模型示意图

Fig. 42-2　The spatial distribution model sketch of Jixian system within 2000m from the surface in central urban area of Beijing

没有高低起伏，局部有略高些的抖动，整体几乎呈一条直线，这种曲线组合特征在北京地区甚至在华北地区区别于下界。本组在平原区分布广泛，隐伏在不同地层之下，是北京最著名的凉水、热水储层。

4）洪水庄组（Pt_2^2h）：以黑色和深灰色页岩、粉砂质页岩为主，夹薄层白云岩和少量砂岩，发育黄铁矿、菱镁矿及锰质结核，厚约 100m。平原区揭露厚度 50～100m，一般厚 90m。视电阻率曲线呈现低阻形态，放射性自然伽马曲线高幅值特征尤为突出：曲线中间幅值相当高，形成一个上陡下缓的高峰，向上、向下逐渐变低，呈渐变过渡关系，直至与上下白云岩对应的曲线形态相一致，有一定的对称性，这一曲线特征与平原区的钻孔十分相似，还有页岩岩性，通常被视为蓟县系中的最重要标志层。JY-482 井埋深 925 m 处的黑色页岩中发现有丰富的藻类碎片 *Laminarites* sp.（北京市平原区地热普查地质报告，1990）；JY-468 井埋深 1185m 的页岩中还发现含有 *Trachysphaeridium* sp. 粗面球形藻；JY-467 井的黑色页岩中含有原始光球藻、六面球形藻、古巢球面藻等，展现了洪水庄组古植物群的特征。

5）铁岭组（Pt_2^2t）：分为两段。平原区揭露厚度一般为 340m，以白云岩为主，中间夹有 20m 左右的蓝绿色、绿灰色页岩可作为分层标志。如果没有见到此页岩，可借助于放射性曲线形态特征来识别：上部为一较平直低幅值段，中—下部在较平直的、逐渐抬升的幅值背景中有一段较高幅值段出现，完整的铁岭组中部都具有这一放射性曲线特征，可以作为辅助标志层。

（三）待建系（Pt_2^3）和青白口系（Pt_3^1）

1. 空间分布特征

主要分布于北山山前昌平南口—怀柔怀北以南的平原地区、房山石楼—大兴—通州宋庄一带、顺义木林—杨镇—大孙各庄及通州西集—永乐店地区，面积约占平原区的 50%。北山山前除百善残山出露外，孟祖村、板桥地区埋深 200～300m，厚度 500～1000m。山前向平原区整体趋势由西往东、由北向南埋深增大，一般埋深 3000～4000m，西部厚度 300～500m；东部古城—西马坡一带没有钻孔揭露，厚度不清。在房山石楼—大兴—通州宋庄一带，受大兴隆起的影响，在向斜核部黄村、东高地、双桥地区埋深 1000m 左右，厚度 500～600m；向斜两翼宋庄、台湖、瀛海地区埋深相对浅，一般为 200～400m，而且地层不齐全，厚度小于 500m。木林西—杨镇—大孙各庄一带沿北西方向分布，在二十里长山向斜核部埋深 1000～1500m，两翼较浅，埋深 400～500m。西集-永乐店地区埋藏比较深，推测大于 5000m，目前尚没有钻孔揭露。

待建纪地层和青白口纪地层在平原区的分布见图 42-3，中心城区空间分布模型见图 42-4。

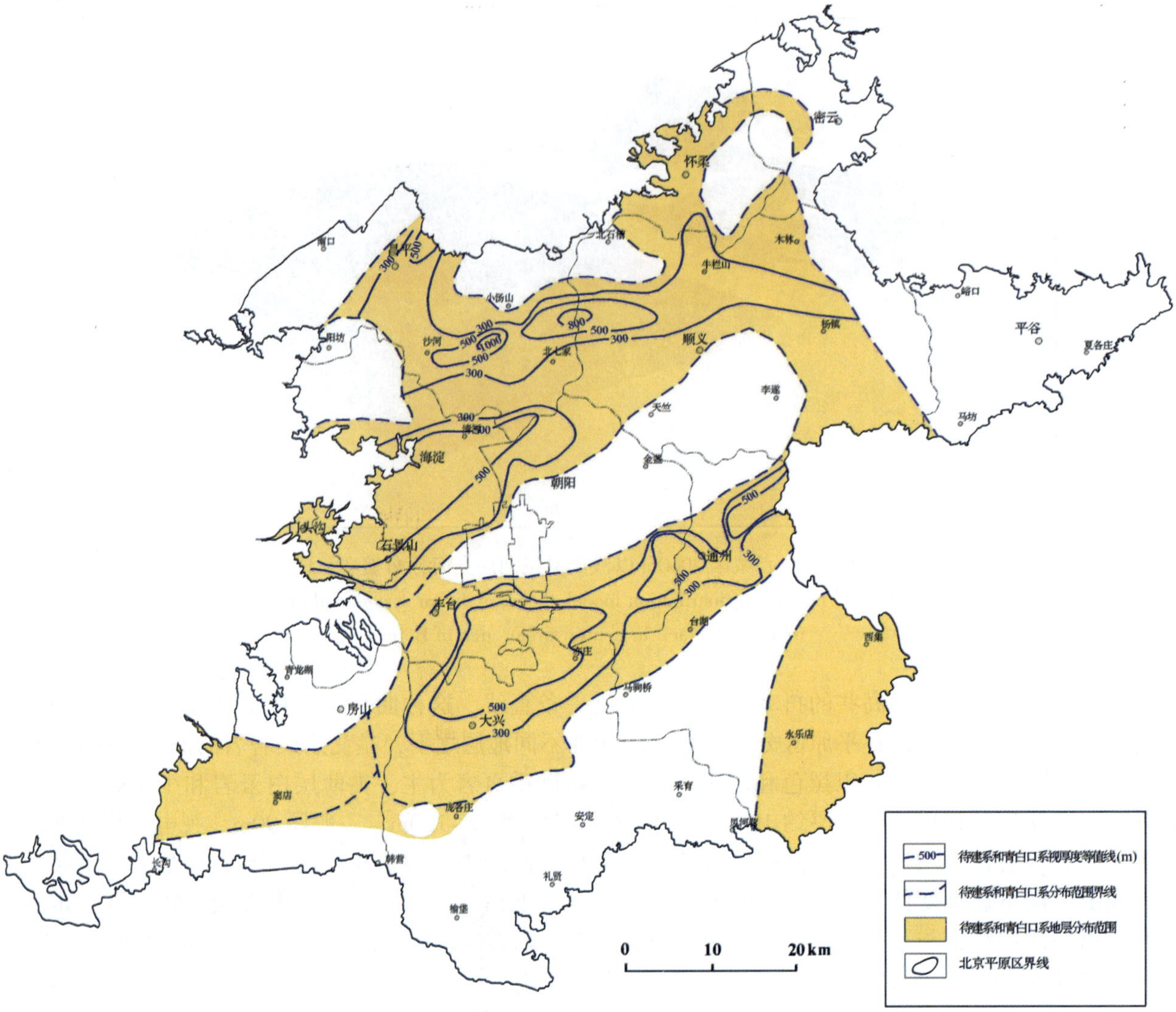

图 42-3　平原区待建纪和青白口纪地层分布范围及视厚度图

Fig. 42-3　The distribution and apparent thickness of to be established system and Qingbaikou system in plain area of Beijing

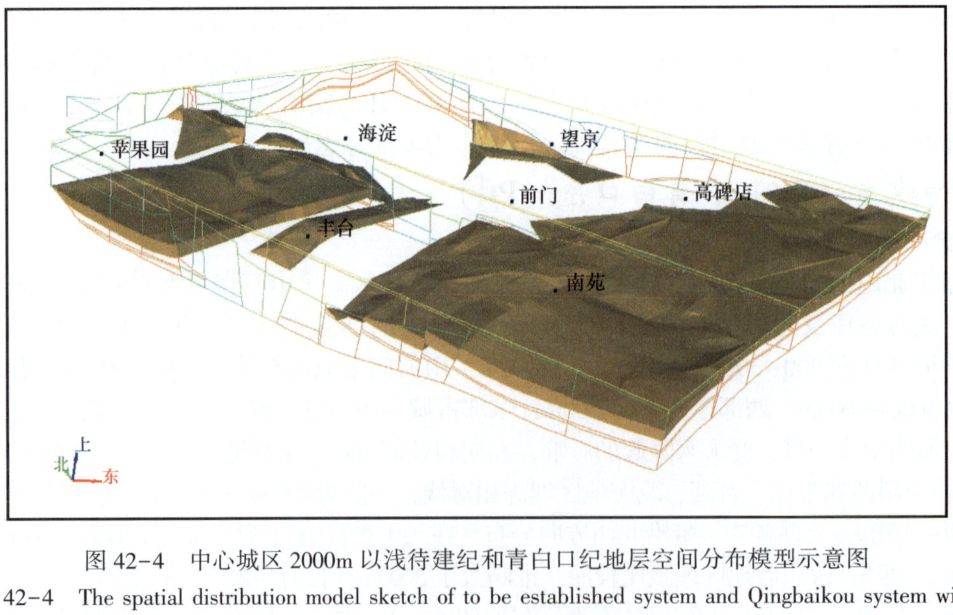

图 42-4　中心城区 2000m 以浅待建纪和青白口纪地层空间分布模型示意图

Fig. 42-4　The spatial distribution model sketch of to be established system and Qingbaikou system within 2000m from the surface in central urban area of Beijing

2. 地层岩性特征

1）下马岭组（Pt_2^3x）：由灰色、灰绿色、灰黑色页岩和粉砂岩组成，一般分为3个岩性段。平原区广泛分布，钻孔揭露视厚度150~296m，一般为280m左右，岩性以深灰、灰、灰绿色页岩为主，中上部夹砂岩，含古植物化石（北京市东南城区地下热矿水水文地质勘察总结报告，1978）：京热2号井埋深1012m处页岩中含有古片藻 *Laminarites antiquissimus*，坚壁原始光球藻 *Protoleiosphaeridium solidum*，粗面球形藻 *Trachysphaeridium tim*。该段测井曲线特征明显，容易识别，以低幅值（150Ω·m）直线形的视电阻率曲线形态和高幅值大段放射性曲线形态为特征，城区、小汤山地区的天然放射性曲线底部有一段幅值更高。

2）龙山组（Pt_3^1l）：在平原区钻孔中常见，视厚度100m，岩性为石英砂岩和页岩互层，部分地区见海绿石，视电阻率曲线和天然放射性曲线呈现高低相反形态对应，顶部和底部天然放射性曲线出现两个高峰，且顶面比底面高。

3）景儿峪组（Pt_3^1j）：在平原区该组钻孔揭露视厚度130m，相当稳定，岩性为蛋清色、黄绿色、暗紫红色灰岩互层，整体呈现为杂色，是北京地区重要的标志层之一。泥质含量由下向上逐渐增多，相应的视电阻率曲线幅值由下向上降低，天然放射性曲线幅值则由下向上逐渐升高。这一特征明显区别于昌平组或龙山组。

三、古生代地层

北京地区早古生代地层出露完好，主要分布在西山，少量分布在北山及顺义区二十里长山。晚古生代地层主要分布在北京西部的门头沟区、房山区、石景山区及海淀区。另外，在北京平原区中生代、新生代地层之下也隐伏有古生代地层。

（一）寒武系（∈）

1. 空间分布特征

分布与青白口系大致相同，但分布范围小于青白口系，约占青白系的1/4，面积约占平原区的40%。除顺义区北小营附近的呼奴山、怀柔李史山残山出露外，其他地区都被中生界或新生界掩盖，向斜核部埋深大，两翼浅，埋深在200~400m之间。来广营、南法信、西马坡等地埋深大于4000m，其余地区如小汤山南部1000~1400m，香山植物园附近1800m，东北旺—立水桥（含北京大学、中国农业大学、北沙滩）1600m，龙庭侯2000m，古城约2600m。

平原区寒武纪地层厚度变化比较大，大部分地区顶面遭受剥蚀，地层不全，一般小于500m。沿槐树岭—衙门口—八里庄—北京大学一带和大东流—张喜庄—衙门村一带视厚度800~1200m。南部的黄村、东部通州镇残留厚度550~750m，海淀区东北旺厚690m，门头沟区永定镇厚530m。岩屑录井通常难以识别寒武系各组地层，目前常用方法是用测井曲线组合特征来划分层位，取得良好效果。

平原区寒武纪地层分布见图42-5，由于中心城区有很多钻孔寒武系与奥陶系难以区分，建立三维地质模型时将寒武系与奥陶系合为一层，中心城区寒武纪—奥陶纪地层空间分布模型见图42-6。

2. 地层岩性特征

1）昌平组（$\in_1c$）：主要岩性为灰白色—灰色豹斑状灰岩和白云岩、细晶灰岩。平原区揭露昌平组的视厚度一般为45~75m，测井曲线呈现较高的视电阻率幅值和较低的平直天然放射性幅值形态，与上、下地层的测井曲线形态容易区分。

2）馒头组（$\in_{1-2}m$）：平原区揭露的馒头组以紫红色页岩为主，个别地区为黑灰色页岩（有一种观点认为是构造运动引起的褪色作用形成的，是否属馒头组，尚有待研究）。该组泥质含量普遍高，局部较低，但钙质含量较高，相应的视电阻率曲线幅值一般比较高，不稳定，高中夹低；天然放射性曲线形态呈大段的高幅值，其中夹有多个（有时多达7~8个）幅值相对较低的低谷。馒头组平原区厚度不太稳定，一般视厚度200m，但通州地区则逾300m。

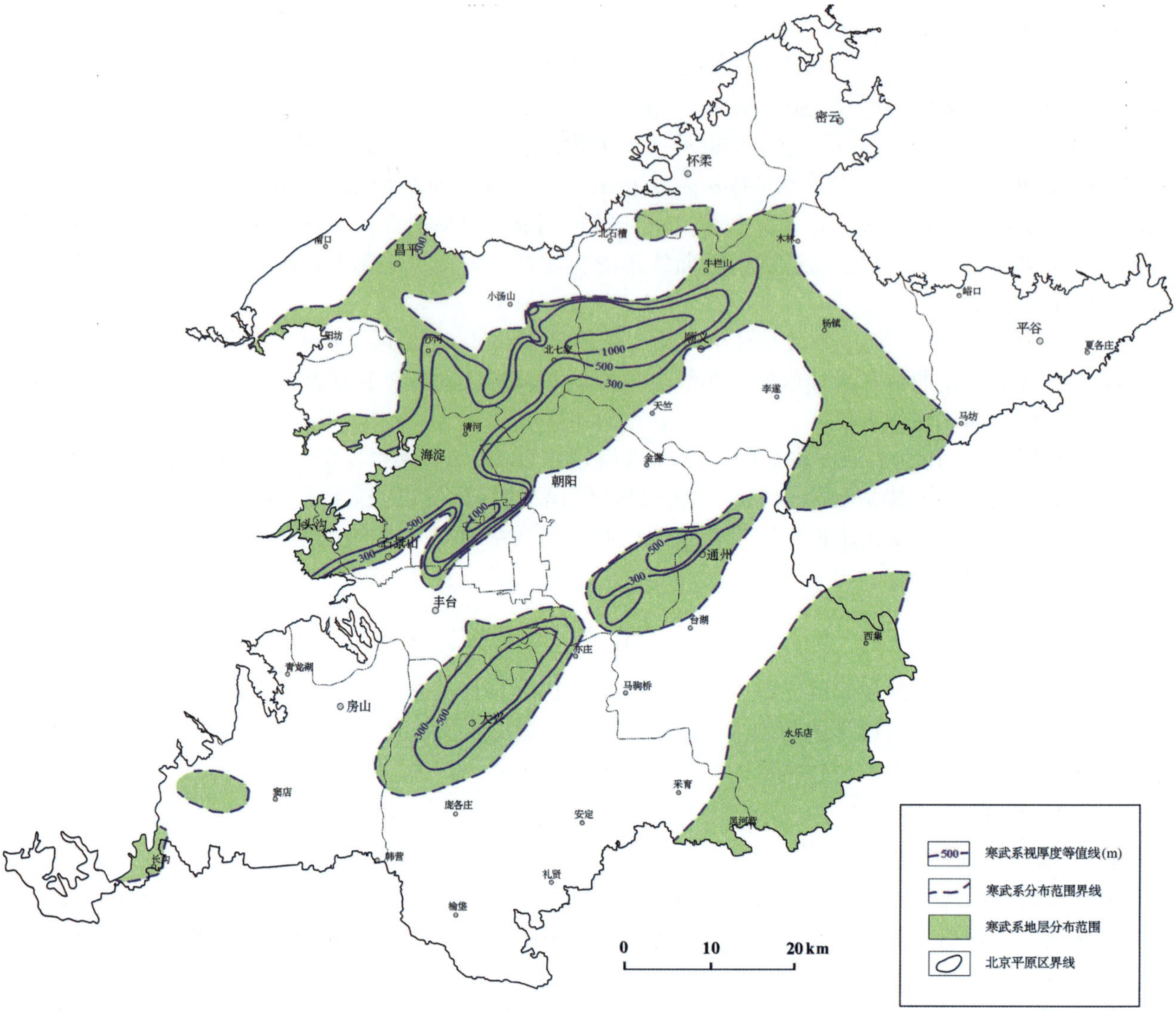

图 42-5　平原区寒武纪地层分布范围及视厚度图

Fig. 42-5　The distribution and apparent thickness of Cambrian system in plain area of Beijing

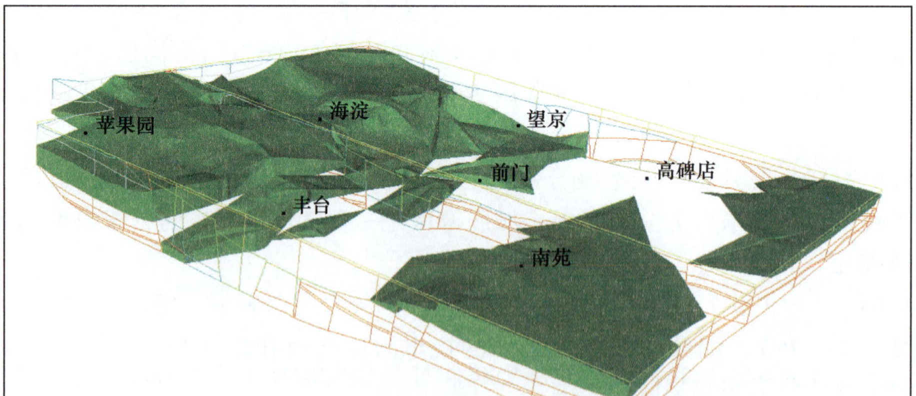

图 42-6　中心城区 2000m 以浅寒武纪—奥陶纪地层空间分布示意图

Fig. 42-6　The spatial distribution model sketch of Cambrian to Ordovician within 2000m from the surface in central urban area of Beijing

3）张夏组（ϵ_2z）：山区以发育厚层鲕粒灰岩为特征。平原区该组钻孔揭露视厚度较薄，北七家附近的京东-1井为80m，其他地区在75m左右，岩性以灰岩为主，夹泥质岩，钻井岩屑中很难见到“鲕”，且假“鲕”为多。天然放射性曲线幅值上部低，中下部较高。

4）炒米店组（ϵ_3cm）：以中、薄层状含泥云质条带灰岩、竹叶状灰岩为主，泥质含量不均匀，岩屑中识别不出“竹叶”。JY-462井该段中部天然放射性曲线为高幅值段，向下逐渐降低，向上则呈现两峰夹一谷的形态，谷的宽度各井不一，总的趋势中间高两头低，高幅值段厚度不足百米，在该段内岩性出现红色调。见*Anomocarid*、*Cyclolorenjeua* sp.，属中晚寒武世。

（二）奥陶系（O）

1. 空间分布特征

主要分布在海淀山前—昌平回龙观—顺义马坡一带平原地区，在大兴隆起的局部及大孙各庄向斜中部和凤河营向斜一带也少量分布，面积约占平原区面积的20%。在海淀山前昆明湖南侧的背斜轴部埋深约200~300m；阳坊岩体南部埋深100~200m；顺义牛栏山、二十里长山残山出露；在大兴旧宫—念坛一带埋深60~70m。东北旺—古城一带，是九龙山向斜东延的轴部，隐伏于侏罗系、三叠系及石炭系—二叠系之下，埋深从肖家河、东北旺的1000m左右，向东逐渐加深到2400m左右；来广营—南法信向斜轴部最深3500~4000m，两翼未有揭露；大孙各庄向斜中部埋深1000~1200m，其上覆盖石炭系—二叠系，厚度约600m，北东侧上翘，产状直立，出露地表；凤河营向斜中部由石炭系—二叠系覆盖，埋深最深处大于3000m，厚度最大600m，两翼埋深约2500m，因剥蚀地层不全，厚度逐渐变薄。

平原区奥陶系灰岩岩溶、裂隙发育，是北京地区岩溶、裂隙水的良好储层，水质好，适合生活饮用；在局部地区也是比较好的地热水储层，如立水桥、后沙峪等地区。平原区奥陶纪地层的分布见图42-7。

2. 地层岩性特征

1）冶里组（O_1y）：由灰色厚层、中厚层状泥晶灰岩组成，中上部夹有钙质页岩和竹叶状灰岩。含有三叶虫、牙形石、棘皮类化石。山区主要分布在西山，北山出露不全，主要岩性是豹斑状灰岩。与下伏炒米店组呈整合接触。

2）亮甲山组（O_1l）：下部以深灰—灰黑色泥晶、粉晶灰岩为主；中、上部以细—粉晶白云岩为主，以含有大量燧石条带和团块为特征，含有正常海相生物化石。山区主要分布在西山地区，京东二十里长山有少量出露。

平原区大兴区黄村钻孔397.4~401.3m深段发现有圆货贝*Obolus*。与下伏冶里组呈整合接触。

3）马家沟组（O_2m）：厚度大，主要为灰、深灰及灰黑色厚层—巨厚层泥晶灰岩。平原区主要分布在北部，中部、南部零星分布，大兴区黄村钻孔在埋深98~313m段发现有箭牙形石、三角牙形石、唐山牙形石和分离斜牙形石，属早寒武世；顺义区后沙峪钻孔在2400~2580m段发现针锐牙形石（未定种）*Acodus* sp.，冠齿牙形虫（未定种）*Cristodus* sp.，尖牙形石（未定种）*Scolopodus* sp.，属早奥陶世。

岩屑录井不容易见到含泥质灰岩，而且钻孔揭露奥陶纪地层不连续，难以划分各组，借助于电测井曲线特征不仅可以划分到组，还可以找到与寒武系的界线。天然放射性曲线在对应的各组分界处都有一段高幅值段，具有区域性规律，特别是在一组从上至下相继升高的低值背景下，出现3个小尖峰和4个由上至下逐渐降低的小尖峰形态，其高尖峰底处分别对应亮甲山组与冶里组的界线和冶里组与炒米店组的界线。

（三）石炭系—二叠系

1. 空间分布特征

平原区分布局限，西山褶皱带东部隐伏于三叠系之下，面积约700km²。大孙各庄向斜石炭系—二叠系面积约100km²，向斜核部埋深约400m，四周埋深200~300m，厚度100~600m。凤

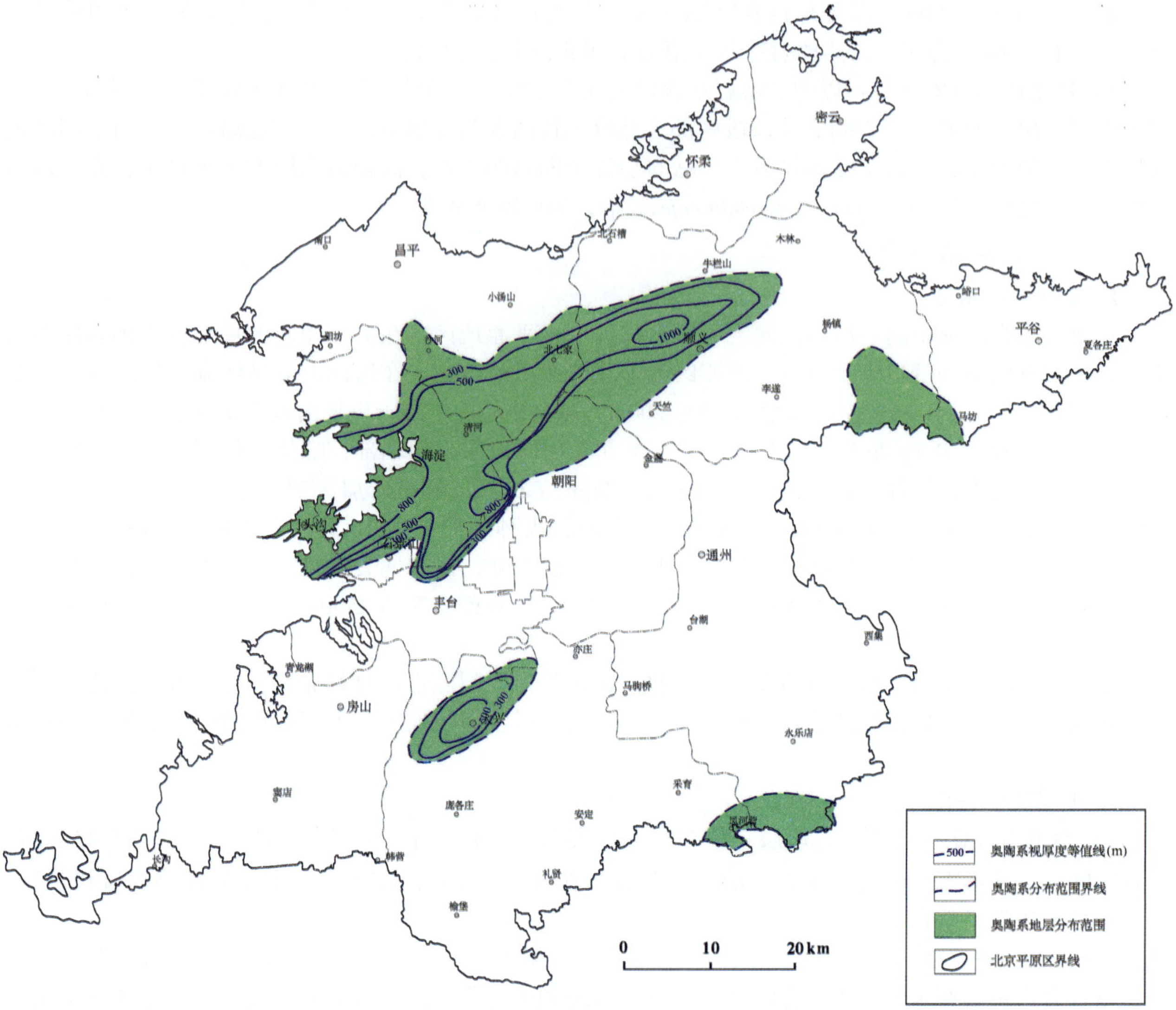

图 42-7　平原区奥陶纪地层分布范围及视厚度图

Fig. 42-7　The distribution and apparent thickness of Ordovician system in plain area of Beijing

河营地区面积约 60km^2，向斜核部埋深 2000m，厚度 500~600m，两翼遭受剥蚀地层不齐全，厚度逐渐变薄。

平原区石炭系—二叠纪地层分布见图 42-8，中心城区空间分布模型见图 42-9。

2. 地层岩性特征

1）太原组（C_2t）：是一套海陆交互相沉积。主要由灰色—灰黑色粉砂岩、细砂岩、泥质岩、泥灰岩组成。平原区主要分布在京东的顺义区、通州区，严格受大孙各庄向斜等构造控制。

2）山西组（CP_1s）：是一套陆相含煤地层。以深灰、灰黑色砂岩、页岩、黏土岩及砾岩为主。

3）石盒子组（$P_{2-3}sh$）：岩性以含砾砂岩、粗砂岩、粉砂岩、粉砂质黏土岩为主，中粒级碎屑岩较少，下部夹煤线。

晚古生代地层的中下部为黑色泥岩夹煤与砂岩互层组合，在岩屑录井中很容易发现；上部石盒子组以长石石英砂岩和砖红色、肉红色为特征，也容易识别；电测井方面由于电性差异明显，界面清楚，其中在含煤层段视电阻率曲线呈现低幅值到几乎为零的直线，到砂岩层对应视电阻率曲线呈刺刀状的尖峰，处在泥岩、页岩相间。

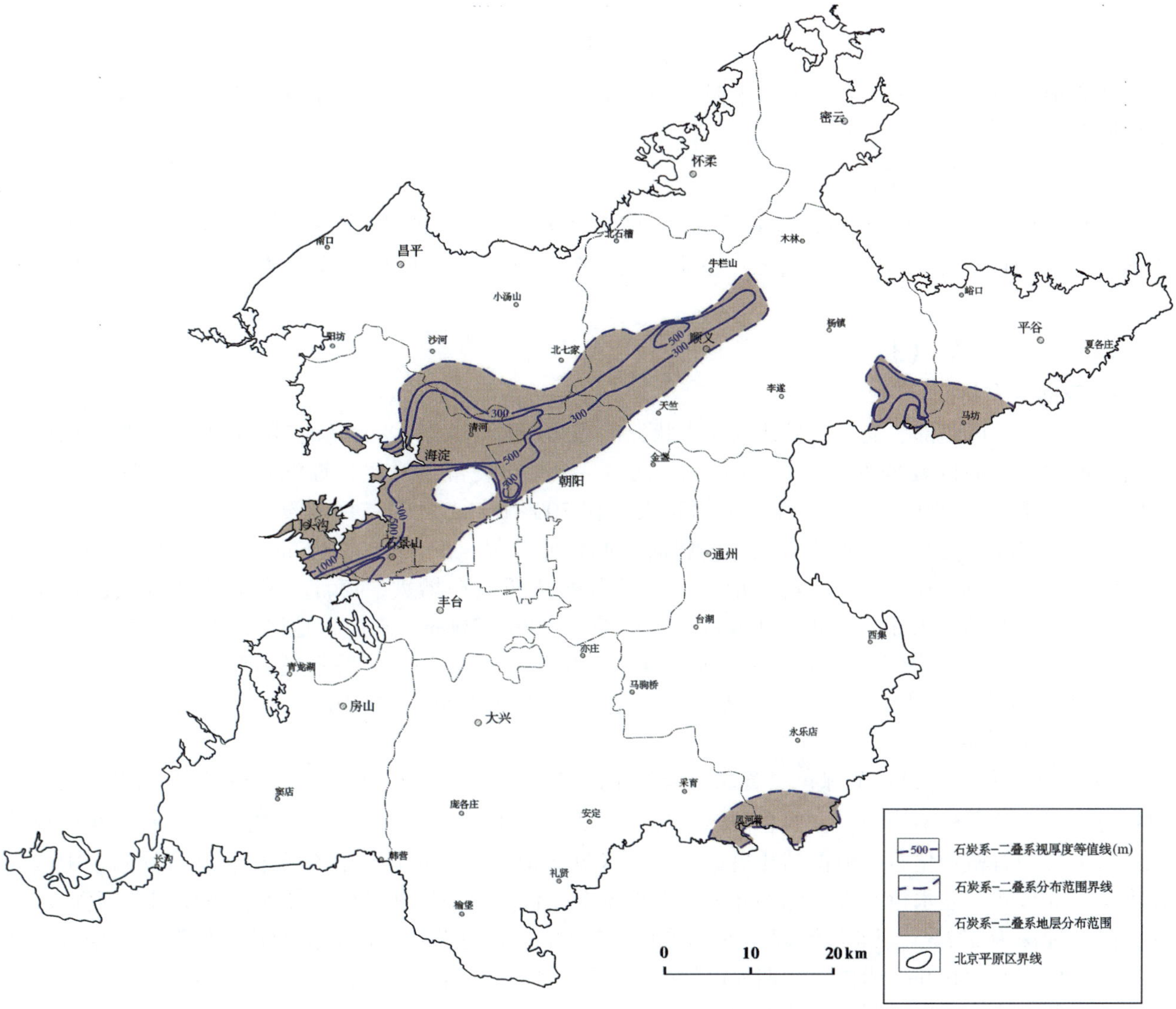

图 42-8　平原区石炭纪—二叠纪地层分布范围及视厚度图

Fig. 42-8　The distribution and apparent thickness of Carboniferous system and Permian in plain area of Beijing

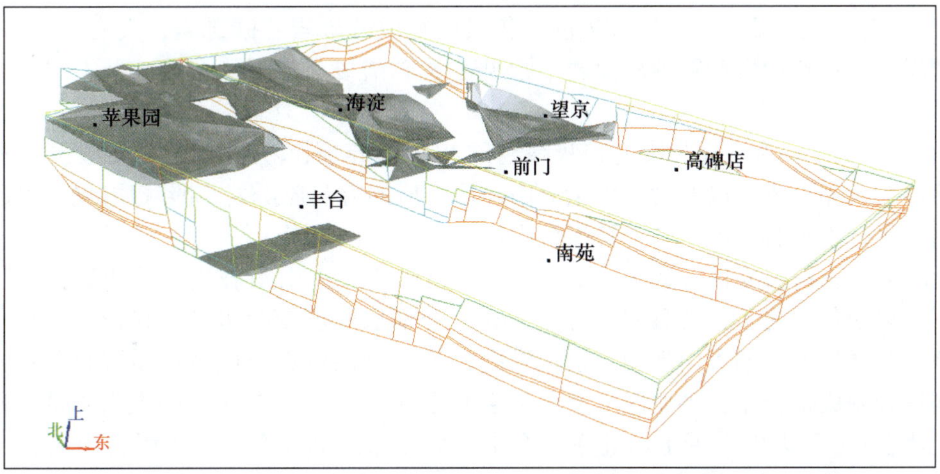

图 42-9　中心城区 2000m 以浅石炭纪—二叠纪地层空间模型分布示意图

Fig. 42-9　The spatial distribution model sketch of Carboniferous and Permian within 2000m from the surface in central urban area of Beijing

四、中生代地层

北京地区早中生代地层主要分布于西部的门头沟区、房山区、石景山区和海淀区。中—晚中生代地层则分布很广。

（一）三叠系（T）

双泉组（P_3Ts）和杏石口组（T_3x）在北京地区分布较局限，仅出露于海淀区、门头沟区、房山区、石景山区的北岭向斜、九龙山-香峪大梁向斜的两翼。平原区的资料有限，岩性特征与山区一致。

（二）侏罗系（J）

1. 空间分布特征

分布在平原区西部、北部，呈北东向带状，面积占平原区的30%。马池口—昌平西一带，属西山髫髻山向斜向平原区延伸部分，宽2~4km，南口-孙河断裂带上升盘，埋深较浅，一般埋深不超过100m，局部有零星残山出露；下降盘埋深稍大，在300~600m之间，目前没有钻孔揭穿厚度。东北旺-回龙观-北七家地区，是西山九龙山向斜向平原延伸部分，埋深150~400m，厚度300~1200m。海淀-来广营-后沙峪-南法信-西马坡地区，埋藏深度从西向东增大，海淀、来广营地较浅，埋深为200m；后沙峪地区埋藏最深，可达1200m，一般厚度500~1500m，来广营地区最厚，可达3000m。苹果园—田村一带，埋深较浅，一般50~100m，厚度300~500m。与下伏三叠系呈不整合接触。以京热-14井和永热-1井为代表性钻孔。侏罗纪地层分布范围及厚度分布见图42-10。

2. 地层岩性特征

1）南大岭组（J_1n）：灰绿色、黄绿、绿色气孔与杏仁状玄武岩、灰绿色、灰紫色玄武岩。南大岭组在北京西山分布极不稳定，玄武岩在南大岭、模式口一带最厚。该段地层一般不含化石，在底部黄色页岩夹层中有植物化石。根据钻孔揭露，平原区见有南大岭期火山岩，分布在北京植物园、中国农业大学、北沙滩、立水桥一带，此外怀柔年丰也见有发育。火山岩埋深变化较大，植物园北埋深54m，农大附中埋深为232m。至北沙滩埋深突然增大至2300m，可见此带火山岩由西往东埋深逐渐增大。怀柔年丰火山岩埋深在几百米内。本期火山岩的厚度变化也大，最厚为植物园北钻孔，厚度为400m，最薄处钻孔厚度为75m。控制面积约150km^2。

2）窑坡组（J_1y）：为一套以灰绿色、灰色砂岩、粉砂岩、泥岩等细碎屑岩为主的陆相含煤地层，总厚度440~500m。北京西山分布面积约1800km^2。京东平原区在顺义、牛栏山以东与二十里长山西侧之间的龙庭候地区新生化沉积层之下都见有窑坡组地层分布，面积约45km^2。

3）龙门组（J_2l）：龙门组为一套深灰色至灰黑色复成分砾岩、砂砾岩，夹有岩屑砂岩、粉砂岩、粉砂质泥岩和煤线。在粉砂岩和粉砂质泥岩中常见菱铁矿结核。与下伏窑坡组及上覆九龙山组均为平行不整合接触。

4）九龙山组（J_2j）：为一套灰紫色和灰绿色陆相火山碎屑沉积岩。岩石类型包括不同粒级的砾岩、砂岩、粉砂岩及黏土岩，以及含有大量火山碎屑物质的凝灰质砾岩、凝灰质砂岩和凝灰质粉砂岩。与下伏龙门组呈平行不整合接触，与上覆髫髻山组呈不整合接触。

5）髫髻山组（$J_{2-3}t$）：岩性分为3段。上段凝灰质砾岩、火山角砾岩夹凝灰岩，中段安山岩与凝灰岩互层，下段以凝灰岩为主，夹凝灰质砾岩。这一套地层厚度变化极其不稳定，缺少化石资料。髫髻山期是平原区最强烈的一期火山活动。根据钻孔揭露、物探资料及区域地质资料分析，受北东向构造和东西向构造控制成带分布，平原区主要有两条北东向及一条近东西向的火山岩带，其他地区零星分布。其中北小营-昌平火山岩带走向北东，宽1~2km，长15km左右，为髫髻山向斜进入平原区部分，被北西向南口断裂切割，断裂东北侧埋深100m左右，断裂西南侧最大埋深超过600m，没有钻孔揭露该带火山岩厚度。来广营-顺义火山岩带沿安贞、来广营、孙河、顺义、马辛庄一带分布，走向北东，宽度3~6km，长约37km，该带火山岩埋深变化较大，安贞埋深1500m，来广营埋深250m左右，

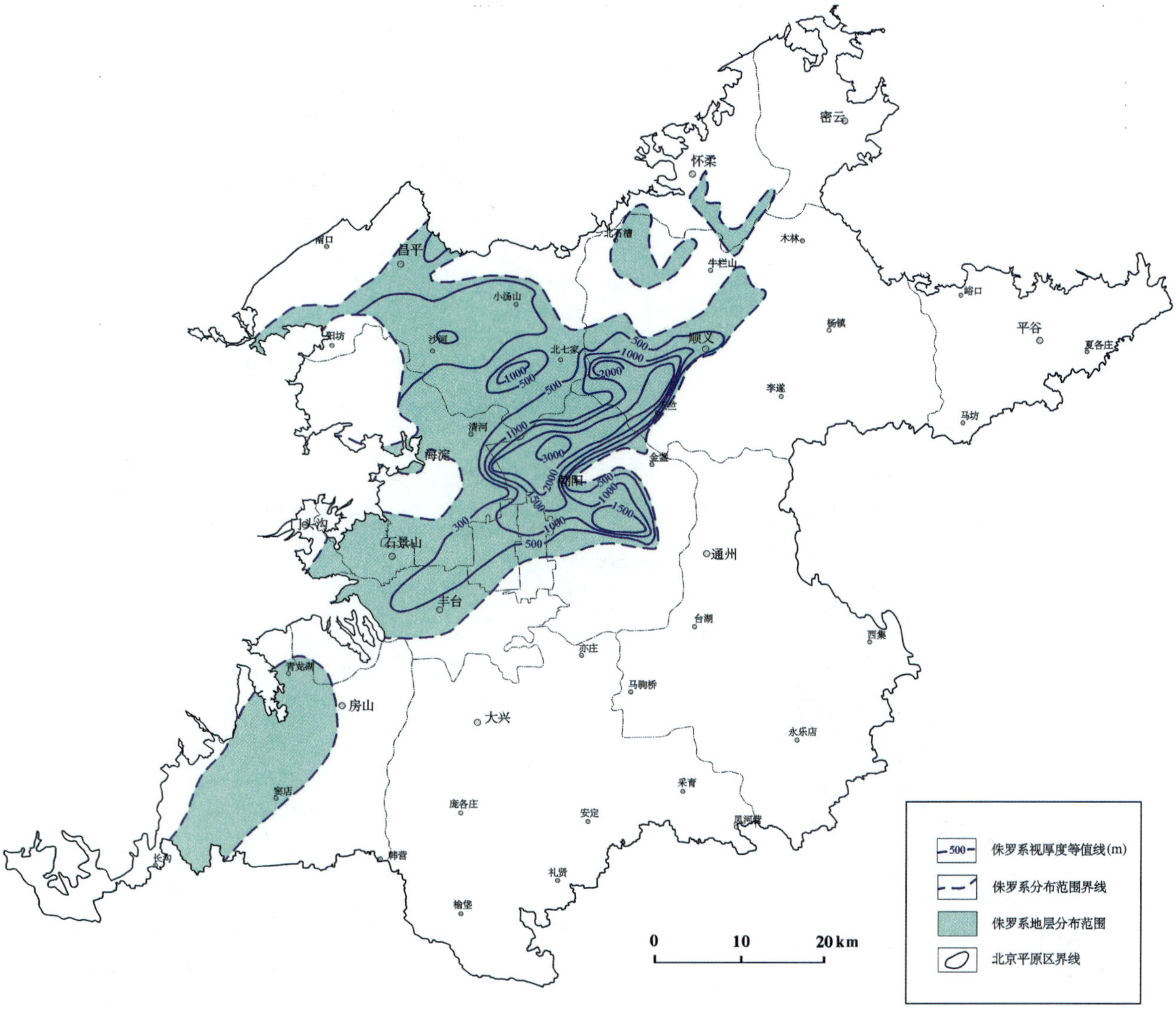

图 42-10　平原区侏罗纪地层分布范围及视厚度图

Fig. 42-10　The distribution and apparent thickness of Jurassic in plain area of Beijing

孙河埋深 800m，顺义埋深 500m，来广营来-2 井钻孔揭露火山岩厚度达 3000m。古城-后沙峪火山岩带分布在沙子营、古城、后沙峪，走向近东西，约 6km，长约 12km，埋深 1000m 左右，后沙峪附近的 JY-181（顺后热-6 井）孔揭露火山岩厚度近 1400m。

6）土城子组（J_3tc）：原称“土城子砾岩层”，土城子组主要是一套紫红—灰绿色河湖相沉积碎屑岩，以凝灰质砂岩、复成分砾岩为主，其次为紫红—灰绿色砂岩、粉砂岩及页岩，夹有多层基性—酸性火山岩层。

（三）白垩系（K）

1. 空间分布特征

平原区白垩系张家口组、东狼沟组、九佛堂组地层分布在北山山前、南邵、北石槽、棉山、怀柔水库及北京断陷的丰台、朝阳区酒仙桥一带，面积占平原区的 10%左右。北山山前有零星残山出露，其余埋深 100~400m，厚度变化很大，一般为几十至上千米。在北京断陷中呈带状分布，走向北东，埋深变化大。呼家楼地区埋深 500m 左右，向西至金融街埋深 2800m，厚度超过 800m；丰台云岗一带埋深近 3000m，厚度约 700m 左右；东郊丽都公园埋深 2600m，厚度 230m；平房地区最厚，可能达

900m。平原区白垩纪地层分布见图 42-11，中心城区白垩系张家口组、东狼沟组、九佛堂组空间分布见图 42-12。

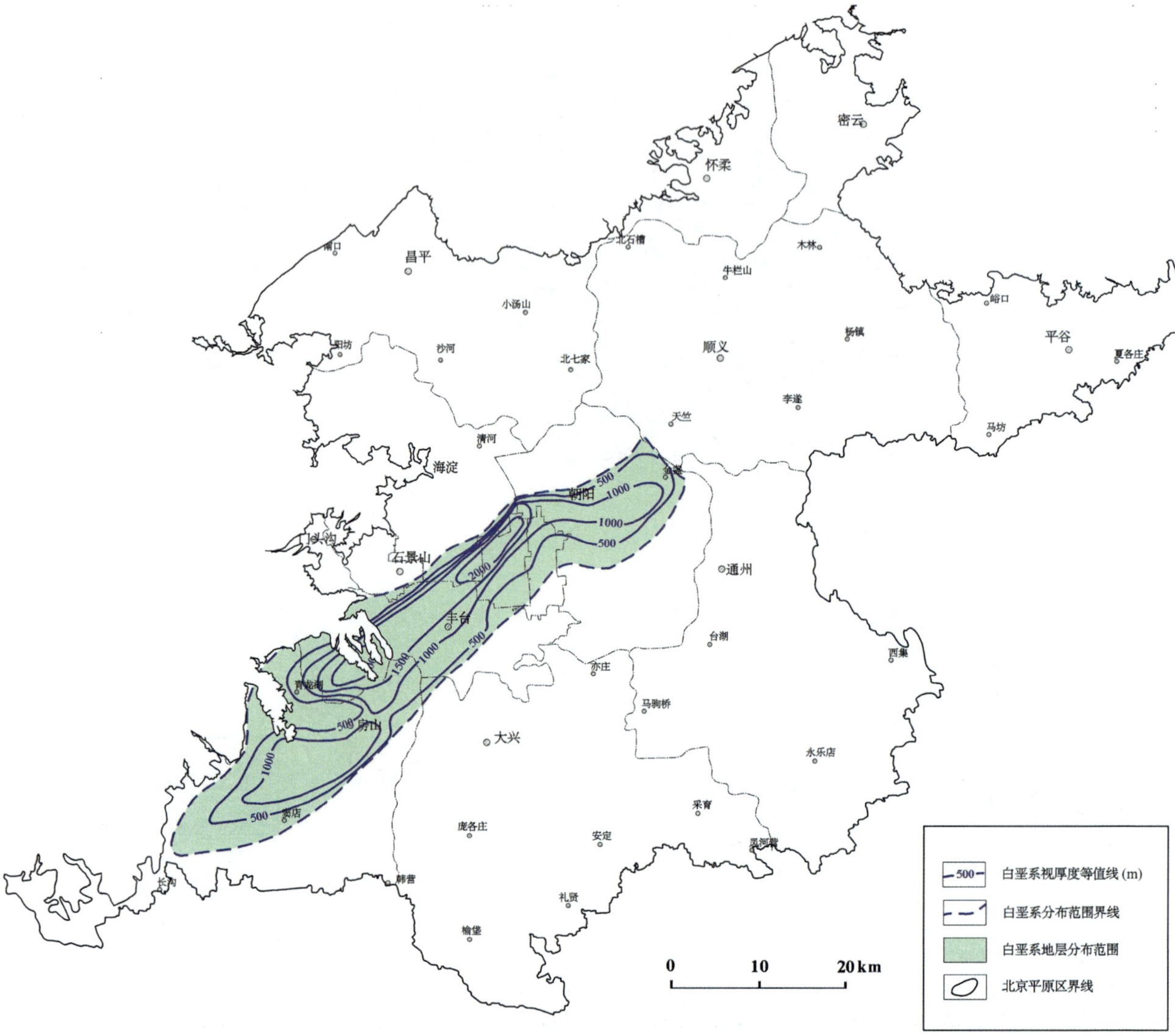

图 42-11　平原区白垩纪地层分布范围及视厚度图

Fig. 42-11　The distribution and apparent thickness of Cretaceous in plain area of Beijing

平原区白垩系坨里组、夏庄组主要分布于北京断陷内的石碣、坨里、王佐、丰台、朝阳、酒仙桥、南皋地区，呈北东向带状，面积约占平原区的 10%。在坨里出露，向西南和东北方向埋藏深度增大，郭公庄、右安门、西单一带最深，一般为 1600~1800m；厚度坨里-云岗地区最厚，可达 2000~3000m，往东北方向变薄，金融街厚度 1000m，蟹岛厚度 800m，阎村—窦店厚度 500~1000m。京热-14 井与 JR-73 井为代表性钻孔。中心城区白垩系坨里组、夏庄组空间分布模型见图 42-13。

2. 地层岩性特征

1）张家口组（K_1z）：根据山区该组岩性组合结合平原区的钻探资料，本期火山岩以中心式喷发为主，见有大量流纹质熔结凝灰岩，是本期火山岩最主要的岩性特征。其岩石类型：火山熔岩类有玄武安山岩、安山岩、粒安岩和流纹岩；火山碎屑岩类有角砾熔岩、熔结凝灰岩、凝灰岩等。主要分布在平原区北部山前、兴寿、石槽、桥梓、泉河、杨宋一带及平房、团结湖、西单、卢沟桥、丰台一带，面积控制在 600km^2 左右。山前一带火山岩埋深变化较大，由数十米至数百米，其厚度未见钻孔揭露，

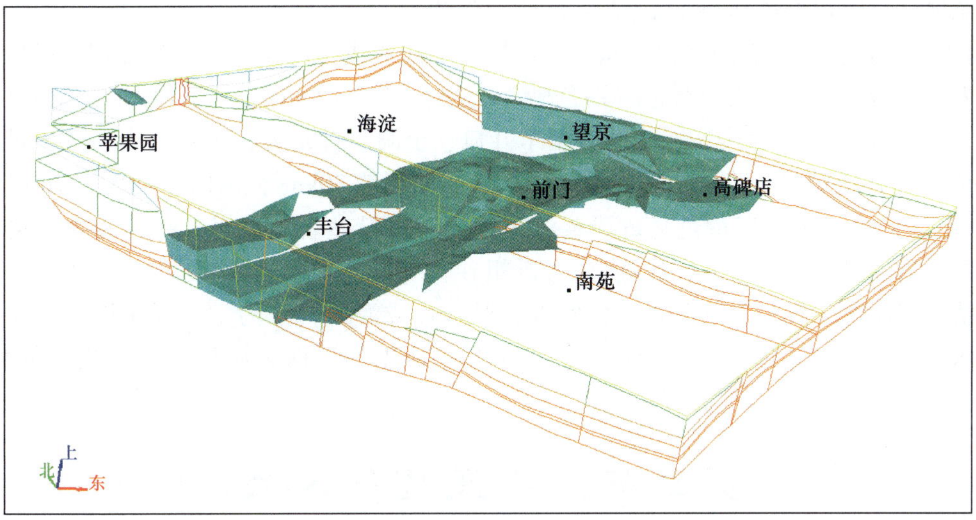

图 42-12　中心城区 2000m 以浅白垩系张家口组、东狼沟组、九佛堂组空间分布模型示意图
Fig. 42-12　The spatial distribution model sketch of Cretaceous Zhangjiakou formation, Donglanggou formation, Jiufotang formation within 2000m from the surface in central urban area of Beijing

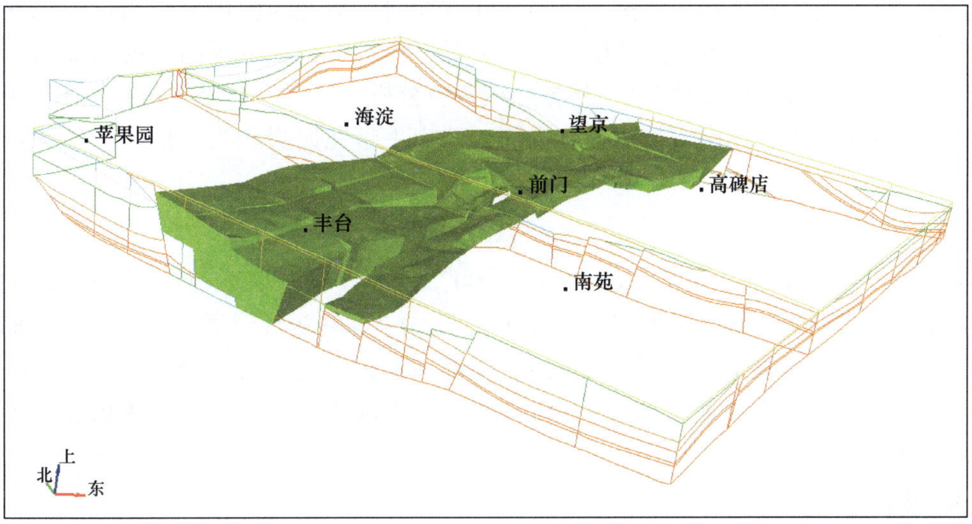

图 42-13　中心城区 2000m 以浅白垩纪地层坨里组、夏庄组空间分布模型示意图
Fig. 42-13　The spatial distribution model sketch of Cretaceous Tuoli formation, Xiazhuang formation within 2000m from the surface in central urban area of Beijing

推测 300~400m。平房—丰台一带火山岩走向北东，宽约 5km，长约 30km，埋深 500~1000m，钻孔揭露火山岩厚度超过 1000m。

2）东狼沟组（K_1d）：为一套火山沉积岩系，主要岩性为含集块角砾凝灰岩、角砾凝灰岩、沉凝灰岩、凝灰质砂岩、复成分砾岩夹有数层玄武岩或粗安岩。本组出露总面积约为 10km^2。主要产于坨里火山沉积盆地北部的大灰厂附近。坨里盆地岩层厚 420.3m。东岭盆地岩层厚 507.5m，怀柔盆地岩层厚 134.87m，杜家庄-镇厂盆地岩层厚 175.22m。

3）九佛堂组（K_1j）：岩性比较简单，是一套以浅湖相为主的细碎屑岩沉积岩。主要为黑色、灰色、黄绿色页岩、粉砂岩、泥岩夹岩屑砂岩、细砂岩。与上下岩层均呈整合接触。九佛堂组仅分布于丰台区坨里盆地，总面积不足0.5km²，厚度不大，约70m。

4）坨里组（K_1t）：为一套同沉积断陷盆地内快速堆积形成的河流相至冲洪积扇相的巨厚层粗碎屑岩系，可分为3个岩性段。一段为黄色—褐色砂砾岩层，厚约120m；二段分布于坨里—公主坟一带，厚约300m；三段分布广泛，在辛庄、晓幼营至辛开口、土洞、青龙头、岗上、小马庄一带，厚度大于300m。坨里组分布在琉璃河断陷盆地内坨里一带，出露面积约30km²，剖面厚度大于760m。

5）夏庄组（K_1x）：为一套黄色、黄绿色沉积岩层，以粉砂岩、页岩、砂岩为主，夹有砾岩及泥灰岩，分为两个岩性段。一段由复成分砾岩-含砾粗砂岩-砂岩-粉砂岩-泥岩和含砾砂岩-粉砂岩-泥岩组成旋回性基本层序；二段主要以砂岩、粉砂岩、泥岩（或泥灰岩）为主。夏庄组分布在琉璃河断陷盆地内的常乐寺—青龙头水库—大苑村以东，沙锅村—西王佐以西近圆形分布，出露面积约10km²，剖面厚度575m。

第二节　平原区侵入岩

平原区隐伏于新生界之下的侵入体主要有宰相庄岩体、阳坊岩体、交道岩体和窑上岩体等。

（一）宰相庄隐伏花岗闪长岩、花岗岩体

据钻孔揭露和物探资料推测，宰相庄岩体在杨宋各庄、北房一带隐伏于第四系之下，岩体埋深约100~300m，面积约140km²。钻孔揭露在大周各庄见花岗闪长岩，在王化村、南房村见花岗岩。在航磁异常图上表现为正异常，东南侧异常极大值为800~900nT，向北西方向异常值逐渐降低。推测该岩体边部为花岗闪长岩、中部为花岗岩。经航磁化极上延，异常中心随上延高度的增加向杨宋各庄一带汇集，推测该岩体向南东倾，倾角约70°。根据区域资料分析，花岗闪长岩与长元岩体、平义分岩体相当，推断其形成于燕山中期，花岗岩与范各庄幅的云蒙山一带花岗岩相当。推断该隐伏岩体形成于燕山中晚期。

（二）阳坊花岗岩体

该岩体大部分掩埋于第四系之下，埋深可达100~300m，根据航磁ΔT异常圈定岩体的面积约150km²。出露部分为其西部边缘，沿温泉—北安河—阳坊一线呈新月形展布，面积约10km²。岩体北部为白岗岩，岩石为全晶质似斑结构，块状构造；中部为黑云母花岗岩，分为细粒黑云母花岗岩相和粗粒斑状黑云母花岗岩相（平原区未划分）；岩体南部及东部为闪长岩，岩石为黑色—灰黑色，全晶质等粒结构，块状构造。

（三）交道闪长岩体

为新生界掩盖的隐伏岩体，在航磁ΔT异常图上显示近等轴状异常，ΔT最大值仅140nT。根据钻孔揭露和物探资料推测，岩体形状近圆形，窦店附近钻孔揭露为闪长岩，最浅埋深60m左右，面积约9km²。依据航磁及区域地质资料分析，岩体与房山早期侵入体相当，推测为燕山早期产物。

（四）窑上闪长岩

窑上闪长岩全被新生界掩盖，根据钻孔揭露埋深在100m左右。航磁ΔT异常为等轴状异常，ΔT最大值为180nT。依据航磁资料推断面积约20km²。根据区域地质资料分析，窑上闪长岩与房山早期侵入体相当，推测为燕山早期产物。

第三节　平原区基岩地质构造

结合钻孔及物探资料，重点总结平原区褶皱构造和断裂构造的空间展布特征。

一、褶皱构造

在漫长构造发展过程中，本区经历了多期的构造运动。根据目前的资料，平原地区隐伏的褶皱构造主要有印支期和燕山期两类。

（一）印支期褶皱构造

北京地区印支期褶皱构造早期发育为横弯隆起和局部的固态流变，晚期发育为近东西向褶皱构造。其特点是形态较为宽缓，枢纽近于水平，轴面近于直立，卷入的地层最高层位为早—中三叠世的双泉组。但有所不同的是，有的褶皱是以原生层理变形弯曲而形成，而有的褶皱是以置换了层理的流劈理为次生面理变形弯曲而形成，这是印支期褶皱构造的一大特点。平原区褶皱构造受后期构造影响，构造形态不完整。

1. 小汤山背斜

分布于分布在小汤山、高丽营一带，大部分被第四系覆盖，背斜主要由中—新元古界及古生界构成。轴向大致呈东西向，轴部分布蓟县系，有多个钻孔揭露蓟县系埋深在100~350m之间。两翼分布青白口系，北翼有白垩系覆盖，青白口系是否连续尚未查清；南翼青白口系和古生界齐全，有多个钻孔揭露证实。背斜东段被北西向的小汤山断裂切割破坏。

2. 板桥背斜

分布于赵全营、板桥一带，全被第四纪沉积地层覆盖，主要由蓟县系、青白口系及寒武系构成。背斜轴向大致呈东西向，轴部分布蓟县系，北石槽附近钻孔揭露埋深249m遇见蓟县系；北翼为青白口系及寒武系钻孔在埋深220m揭露青白口系。桥梓附近钻孔在埋深77m揭露寒武系，南翼钻孔在埋深445m见寒武系。背斜被北东向的黄庄-高丽营断裂带切割。

3. 昆明湖背斜

属北京西山斜河涧-颐和园褶皱构造带（西段有南庄-大牛道背斜、灰峪向斜、下苇店-军庄背斜、丁家滩-野溪背斜、斜河涧向斜等；东段由老府村向斜、香山背斜、陈家沟-团城向斜、八大处背斜等）向东延伸进入平原区部分，褶皱主要由奥陶系及石炭系—二叠系组成，背斜轴向近东西，核部为奥陶系，多眼钻孔揭露奥陶系埋深在200~300m之间；两翼为石炭系—二叠系，多眼钻孔揭露石炭系—二叠系埋深一般在100~200m之间。背斜东段被北东向的八宝山断裂切断。

4. 凤河营向斜

属华北凹陷盆地的一部分，是大厂凹陷的边缘，位于平原区的东南部，全被新生界覆盖。向斜主要由石炭系—二叠系、奥陶系及寒武系构成。向斜轴向在测区内为东西向，伸出区外转为北东向。核部钻孔埋深1851m见石炭系—二叠系。两翼为奥陶系及寒武系，北翼地层宽缓，南翼分布地层狭陡，与桐柏背斜衔接。南翼钻孔揭露埋深2035m见奥陶系。向斜西端被北北东向的东庄断裂接截，南翼被东西向的桐柏断裂切割破坏。

（二）燕山期褶皱构造

燕山期褶皱构造从山区延展至平原区，具有山区的构造特点。燕山构造旋回经历了明显挤压收缩期，形成了主要有侏罗纪—白垩纪火山-沉积物建造组成的醒目的褶皱构造。构造线方向从早期的北东东向到中期的北东向（在平原区出现短轴近东西向），晚期则转为北北东向。褶皱的形态由于各期的应力强度和边界条件的不同有很大的差异。燕山期褶皱的最大特点是每一期褶皱作用的对象都是在其之前发生的山间盆地中接受的火山-沉积建造。

1. 平原区中南部大兴隆起区

大兴隆起区位于平原区的中南部，沿北东方向分布，西南及东北部伸出区外，北西界线为南苑-通县断裂，东南界线为礼贤断裂及夏垫断裂，宽12~20km，长度90km。布格重力异常显示为区域的正异常；航磁ΔT化极异常表现为与布格重力异常形状相似，位置向南偏移的区域正异常，最大值可达560nT。西南段包括黄村向斜、榆垡-礼贤背斜；东北段包括通州向斜、张家湾背斜。

（1）黄村向斜

位于平原区中南部，沿罗奇营、黄村镇、金星乡一线分布，全被第四系覆盖。主要由青白口系及下古生界构成，向斜以北东方向展布。核部分布奥陶系，多眼钻孔揭露埋深在80m左右。目前还没有收集到钻孔揭露奥陶系的厚度，依据物探及区域地质推测奥陶系厚度600m左右。两翼分布寒武系及青白口系，西北翼被南苑-通县断裂带破坏，地层及其埋深出现突变，南苑-通县断裂带的北侧，黄村附近的兴6井孔埋深812m见寒武系。东南翼与榆垡-礼贤背斜相连，但两者之间受北东向的瀛海断裂的影响地层连续性不好。

虽然组成该褶皱的最新地层为古生界，但与山区印支期褶皱轴向比较不相符，而在大兴隆起的北东端（石佛寺）见有侏罗纪地层残留，说明这里的褶皱卷入了侏罗纪地层。由于大兴隆起历了长期严重剥蚀阶段，中生界几乎被剥蚀殆尽，致使这一带现今揭露的为古生代地层，所以把这里的褶皱划为燕山期褶皱（后述的榆垡-礼贤背斜、通州向斜及张家湾背斜都有共同的特点）。

（2）榆垡-礼贤背斜

位于平原区南部，沿榆垡、礼贤、安定、长子营一带分布，全被新生界覆盖。背斜主要由长城系、蓟县系、青白口系构成。轴向北东，轴部分布长城系，榆垡附近的JY-530孔埋深880m见长城系。安定附近钻孔埋深640m见长城系（其分布范围主要依据物探资料推测）。两翼分布蓟县系及青白口系。西北翼与黄村向斜相接，潘马房附近的JY-528钻孔埋深624m见蓟县系；庞各庄附近的兴热4井孔埋深358m见蓟县系。背斜的东南翼被北东向礼贤断裂带（至少两支断裂组成的断裂带，后述）沿背斜轴向切割，榆垡附近的兴热7井孔埋深1742m才见长城系。南各庄附近钻孔埋深1940m见到蓟县系。南各庄东钻孔埋深3597m还未揭穿新生界。

（3）通州向斜

位于平原区中东部，沿黑庄户、通州镇、疃里一线分布，全被新生界覆盖。由寒武系及中—新元古界构成。向斜轴向北东，核部分布寒武系，北东窄，南西宽，定辛庄附近钻孔埋深298m见寒武系，半壁店孔埋深274m见寒武系，疃里孔埋深305m见寒武系。两翼分布青白口系及蓟县系，有多眼钻孔控制。西北翼受南苑-通县断裂带影响，地层连续性差，管庄附近钻孔埋深700m见青白口系，朝阳农场孔埋深700m见蓟县系。东南翼与张家湾背斜衔接。潞城北钻孔埋深276m见青白口系，宋庄附近钻孔埋深191m见青白口系。

（4）张家湾背斜

位于平原区东南部，沿陆辛庄—上店一带分布，东北部伸出北京市辖区，全被第四系覆盖。背斜轴向北东，轴部为蓟县系，张家湾附近的JY-306孔埋深504m见蓟县系，柴府附近钻孔埋深586m未见蓟县系，说明蓟县系埋深不稳定。两翼分布青白口系及寒武系，西北翼与通州向斜衔接，潞城附近钻孔埋深276m见青白口系，梨园附近钻孔埋深273m揭露寒武系；东南翼为青白口系，揭露钻孔不多，仅在牛堡屯钻孔埋深695m见青白口系，张家湾东南孔埋深204m揭露寒武系。

2. 平原区西北部

在平原区西部可划分为北小营-昌平向斜、回龙观-平西府向斜及来广营-南法信向斜。

（1）北小营-昌平向斜

位于平原区西北部，属百花山-髫髻山复式向斜向平原延伸部分。向斜轴向北东，核部分布髫髻山组。昌平城南钻孔揭露见有髫髻山组。两翼没有钻孔揭露，东南翼在影碧山推覆构造作用下，蓟县纪地层盖在侏罗纪地层之上，向斜形态不清。

（2）回龙观-平西府向斜

位于平原区西北部，被第四系覆盖。属门头沟复式背斜（自西北至东南有王平村背斜、九龙山-香峪大梁向斜、红庙岭背斜、石景山向斜、石门营向斜）之九龙山-香峪大梁向斜向平原区延伸部分。航磁图上除出现局部的正异常外，表现为宽阔的负异常带，磁场强度-80～-160nT。向斜沿东北旺、霍营、平西府、东三旗一带分布；轴向北东，核部分布中侏罗统九龙山组，东北旺钻孔在埋深185m处见中侏罗统九龙山组。回龙观东北SR-13井孔揭露埋深258m见侏罗九龙山组。西北翼为石炭系—二叠系、奥陶系及寒武系，多眼钻孔揭露。西南翼在青龙桥—清华园出现石炭系—二叠系。根据航、地磁资料，结合钻孔资料分析，在中关村、奥运村、燕丹一带分布有髫髻山组，向斜形态不清。

（3）来广营-南法信向斜

在来广营、孙河、南法信、马坡一带出现醒目的航磁 ΔT 正异常带。异常带中间宽、两端窄，宽度达4~12km，异常值在-80~-140nT范围内变化。多眼钻孔揭露，该磁异常带为髫髻山期安山岩及髫髻山组（磁性地层）引起。后沙峪附近的顺参1井孔埋深720m见髫髻山组，马坡附近钻孔埋深372m见髫髻山组。因此，把上述有磁性的地层划分为髫髻山组，而将回龙观—平西府一带无磁性地层划为九龙山组，作为中侏罗统与晚侏罗统的区分标志，依据这种特征圈定中侏罗统与晚侏罗统的分布范围。该带向斜的东北翼，钻孔揭露有石炭系—二叠系、奥陶系、寒武系，出现正常的向斜地层序列；向斜轴部的来广营—南法信一带侏罗系厚度大，靠近北部逐渐变浅，而且埋深浅，西南翼由于受顺义断裂的影响，向斜形态不清。

3. 平原区东北部

主要是大孙各庄向斜，位于平原区东部大孙各庄—石佛寺一带。向斜主要由古生界组成，轴向北西。核部主要分布石炭系—二叠系，埋深100~200m，多眼钻孔证实。两翼为奥陶系、寒武系。东北翼在二十里长山出露奥陶系和寒武系，西南翼多眼钻孔揭露分布奥陶系、寒武系，埋深200~300m。向斜北西窄，南东宽，被北西向的二十里长山断裂带、北东向的夏垫断裂及东西向的皮各庄断裂切割破坏。因南聂庄—刘辛庄一带见有侏罗系，故将本向斜划归为燕山期。

二、断裂构造

根据断裂构造的规模、控制深度及其在区域地质构造发展中的重要作用，将北京平原区断裂划分为控制区域性地质构造单元的断裂和一般断裂两个级别。确定断裂的依据主要是物探资料和钻孔资料。

（一）控制区域性地质构造单元的断裂

控制区域性地质构造单元的断裂对区域性的沉积作用、沉积建造、岩浆活动有控制作用。平原区控制区域性地质构造单元断裂主要有东西向、北东向及北西向3组断裂构造，主要的东西向断裂有桐柏断裂、皮各庄断裂；主要的北东向断裂有八宝山断裂、影碧山断裂、黄庄-高丽营断裂带、南苑-通县断裂带、礼贤断裂带、夏垫断裂带；主要的北西向断裂有南口-孙河断裂带。其中黄庄-高丽营断裂带、顺义断裂带、夏垫断裂带、南口-孙河断裂带作为活动断裂在第四十一章中已详细描述，其他断裂带分别叙述如下。

1. 东西向断裂

（1）桐柏断裂

西起通州马房，经肖家务、沙窝店、小甸屯伸出区外，然后，沿利尚、羊房方向延伸。大部分在市区外，区内长度达26km；总的走向近东西转向北东。断裂显示明显的重力梯级带。桐柏断裂属规模大、切割深的大断裂，是大厂凹陷与固安-武清断陷的分界断裂。根据地震资料及钻孔揭露，武清盆地沉积中心，新生界厚度可达10000m，大厂断陷盆地新生界厚度2000~5000m，两盆地新生界落差数千米。可见桐柏断裂断距较大，推测为印支期形成的挤压断裂，喜马拉雅期又开始活动，形成正断裂。

（2）皮各庄断裂

位于平原区东部，大部分分布在市辖区外，西起平家疃，经南各庄，伸出北京市区，经皮各庄、程官营、李蔡街，过夏垫断裂继续往东延伸，走向近东西，区内长约20km。显示明显的重力梯级带。皮各庄断裂属燕山裂陷槽与华北凹陷盆地的分界线，规模大、切割深，推测为印支期形成的断裂，后期继续活动，被北北东向夏垫断裂错断，断裂的西段断层面向北倾，东段断层面向南倾。

2. 北东向断裂

（1）八宝山断裂

八宝山断裂南起涞水县城西北，沿北东方向，经南尚乐、长沟、牛口峪，至磁家务方向骤变，再经晓幼营、大灰厂、八宝山，止于海淀镇附近。该断裂呈波状弯曲展布于北京山麓和山前平原地带，总体走向北北东至北东，倾向南东，倾角35°~45°。断裂总长约90km，八宝山断裂自中侏罗世开始发育，至早白垩世，主要表现为张裂运动，基本上控制了中生代盆地的西部边界。早白垩世末的燕山运动，使之转变为挤压逆冲断层。至中生代末被后来形成的黄庄-高丽营断裂所切割，故新生代以来其对北京凹陷的控制作用退让给陡倾的黄庄-高丽营断裂。根据其构造特点，以永定河为界，可将其分为南北两段，南段大部分出露于地表，北段则隐伏于平原之下。

（2）影碧山断裂

位于昌平城区东部，是十三陵水库东-下庄逆冲推覆断裂向平原区延伸部分。断层走向北东转北北东，断层面倾向东南，平原区分布长度15km。该断裂物探资料反映不明显，主要根据钻孔资料确定断裂。平原区多个钻孔揭露，南东盘为蓟县系，北西盘为侏罗系。在中段被南口-孙河断裂带切断。

（3）南苑-通县断裂

南苑-通县断裂为平原区的隐伏断裂，是北京凹陷与大兴隆起的分界线。断裂南起码头，经窑上、公义庄、葫芦垡、西芦城、西红门、高碑店、通州、富豪至南庄头附近。断裂在通县以东地表有显示，总体走向北东50°~30°，断裂面倾向北西，倾角约50°，北西盘下降，南东盘上升，为一正断层。现已经基本查明，该断裂由两支间距1~2km、大致平行的断裂组成，南支断裂东南侧第四系之下为古生界，而其西北侧第四系之下为新近系。

该断裂是根据地面电测深、重力、人工地震等物探手段结合钻孔资料确定的。根据钻孔剖面的分析，在通过断裂两侧的基底高度变化不大，例如西红门基底高度为56m，断裂另一侧的大兴县委礼堂基底高度为55m。在另一断面上，左安门基底高度为83m，而在断裂的南侧花园子村基底高度为86m。由此说明该断裂形成于第四纪以前，第四纪以来断裂两侧没有大的差异性运动。

南苑-通县断裂带，在公义被北西向的公义断裂错断，在西芦城被永定河断裂错断，在南苑被南北向断裂错开，在小南庄被东风断裂错断，除了高碑店至通州段为东西向展布，整体走向北东40°~50°，根据物探资料，断层倾向北西，倾角大于60°，北支断裂比南支断裂断距大，燕山晚期受北西-南东方向挤压，形成逆断层，喜马拉雅期应力场由北西-南东向挤压转变为北西-南东向张引，上盘沿老断裂下滑，转变为正断层，控制了古近系及新近系的分布。

（4）礼贤断裂

在平原区沿榆垡、礼贤、安定南、长子营方向展布，西南伸出区外，往宫村镇方向延伸，由两条断裂组成的断裂带（位于断裂带南部的断裂称为礼贤断裂，北部的断裂称为安定断裂），总的走向为北东东转为北东向，区内长达45km。

礼贤断裂是固安-武清凹陷与大兴隆起的分界线，其控制了固安凹陷古近纪—新近纪以来的地层沉积。断裂向东南倾斜，倾角较缓，一般倾角在40°左右，新生界覆盖层厚度很大，断裂在平面上的投影是一条带，而不是一条线。断裂北西盘钻孔揭露基岩为中—新元古界，埋深仅有数百米至近千米，上覆第四系；断裂的南东盘沉积了巨厚的古近系和新近系，断裂对断陷盆地古近纪—新近纪的沉积地层有明显的控制作用。据人工地震资料，断陷盆地内古近系—新近系的沉积最大厚度大于10000m，下伏基岩为中—新元古界和古生界。推断断裂形成时代为古近纪。

（二）一般断裂

一般断裂相对于控制区域性地质构造单元的断裂，规模小，切割深度较浅。在北京平原区这类断裂相当发育，此类断裂的分布特征及其产状要素见表42-1。

表 42-1 平原区主要断裂构造特征一览表

Table 42-1 The primary structure characteristics of the fault in plain area

断裂名称	分布位置	产状	性质	与其他构造的关系	与岩浆活动的关系	钻孔揭露情况	备注
大柳树断裂	昌平区东部小汤山镇东	走向为北西 35°，断层面倾向北东，倾角大于 60°，长度 10km，被第四系覆盖	正断层	属北山褶皱带的断裂，断层两盘分为中元古界、古生界、中生界，切割小汤山背斜，控制大东流小盆地西界		JY-513 孔与 JY-488 孔地层有明显落差	
牛栏山断裂	怀柔区东南部	走向为北东，倾向北西，长达 14km，全被第四系掩盖	正断层	属北山褶皱带的断裂，断层西北侧为寒武系和侏罗系，东南侧为奥陶系、侏罗系	控制了宰相庄岩体的南界分布	JY-43 孔与 JY-29 孔第四系落差 210m	
张喜庄断裂	怀柔区南部	走向为北东，倾向南东，长达 16km，全被第四系掩盖	正断层	属北山褶皱带的断裂，断层西北侧为中生界，东南侧为古生界及中生界，南段被西官庄断裂切断		JY-177 孔与 JY-511 孔石炭系—二叠系落差超过 1000m	
昆明湖断裂	海淀区北部	走向近南北，长度达 13km	平移断层	属西山褶皱带的断裂，断裂切割海淀背斜			
白家疃断裂	海淀区北部	山区向平原区延伸的断裂，近东西向分布，断层面倾向南，倾角 60°左右，进入平原区长达 11km	逆断层	属西山褶皱带的断裂，断层两侧均分布古生界、中生界，切割海淀背斜，中段被南北向的冷泉村断裂切断	有酸性岩侵入		
西沙屯断裂	昌平区东南部	北段山区出露，走向为北东，断层面倾向东南，长达 22km	逆断层	属西山褶皱带的断裂，断层西段切割阳坊岩体，中间被北西向的上庄断裂及北西向的南口断裂切断			
冷泉断裂	海淀区北部	山区出露，近似南北走向，长度 17km	平移断层	属西山褶皱带的断裂，断层两盘为阳坊花岗闪长岩，中生界及古生界，东盘分布中生界及古生界。断层切割九龙山向斜及海淀背斜			
沙河断裂	昌平区南部	全被第四系掩盖，断裂走向为北东，断层面倾向北西，长达 16km	正断层	属西山褶皱带的断裂，断层西段切割阳坊岩体，中间被北西向的上庄断裂切断，东段被北西向的郑各庄断裂切断		JY-116 孔与 JY-132 孔奥陶系落差近 100 m	
兴寿断裂	昌平区兴寿	走向北西西，长约 19km，倾向南东东	正断层	属北山褶皱带的断裂，断层全被第四系覆盖，东段南东东盘为蓟县系，北西西盘为白垩系			

续表

断裂名称	分布位置	产状	性质	与其他构造的关系	与岩浆活动的关系	钻孔揭露情况	备注
郑各庄断裂	昌平区东南部	全被第四系掩盖，走向为北西，断层面倾向北东，长达13km	正断层	属北山褶皱带中的断裂，断层中、西段两盘均为古生界，东段为侏罗系，切割沙河断裂		JY-484孔与JY-486孔侏罗系落差70m	
莲花池断裂	中心城区附近	大致以车公庄断裂平行，间距2km，走向为北东60°，断层面倾向东南，倾角约70°，长达22km	正断层	属北京断陷盆地内断裂			
车公庄断裂	中心城区附近	走向为北东60°，断层面倾向东南，倾角70°左右，长度达25km	正断层	属北京断陷盆地内断裂		JY-455孔与JY-259孔白垩系落差305m；JY-535孔与JY-265孔白垩系落差220m	
良乡-前门断裂带	西起南上岗，经长阳、郭公庄、宣武山、大栅栏、左家庄	由良乡北断裂及前门断裂组成，断裂走向由近东西转为北东，长约40km，倾向北西，倾角较陡，大于70°	正断层	属北京断陷内规模较大的断裂，断距大、切割深，推断为燕山晚期形成，喜马拉雅期有活动；断裂的两侧地层差异大，古近系—新近系厚度相差500m。良乡-前门断裂被北西向的永定河断裂及南北向的东铁匠营断裂切割		JY-459与JY-266为分布断裂两侧的一对孔，揭露地层变化大。JY-459孔见侏罗系，JY-266孔未见侏罗系，蓟县系落差1315m	断裂多期活动
崇文门断裂	沿良乡南、狼垡村北东、白盆窑、右安门、崇文门、六里屯一线分布	走向北东，长约39km，倾向西北，倾角小于70°	正断层	属北京断陷内断裂，形成于燕山晚期，控制白垩系地层分布；被北西向的永定河断裂及南北向铁匠营断裂切割		JY-266孔与JY-268孔相距1km左右，位于断裂东南侧JY-268孔古近系下伏为青白口系，断裂西北侧JY-266孔古近系下伏白垩系，厚度185m。JY-349及JY-456为断裂两侧的一对孔，JY-456孔未见白垩系，古近系—新近系落差450m	断裂多期活动
楼梓庄断裂	沿东风、东坝、楼梓庄、李桥、王家坟一线分布	走向北东，长约28km，倾向北西，倾角大于70°	正断层	属北京断陷内规模较大的断裂，孙河断裂以西控制白垩系的分布。推断形成燕山晚期，喜马拉雅期断裂继续活动。断裂被北东向的东坝断裂、孙河断裂、李桥断裂切割错动		JY-463与JY-464为断裂两侧的一对孔。JY-464孔见白垩系，厚度773m。JY-463未见白垩系，两孔蓟县系落差1050m	原为九仙桥-李家桥断裂
良乡南断裂	西起黄庄断裂经过紫草坞、良乡南	走向近东西，长19km，倾向南南东，倾角70°左右	正断层	北京断陷内的断裂，被北西向的公义断裂切割错动		JY-452及JY-453为断裂两侧一对孔，蓟县系落差465m	

续表

断裂名称	分布位置	产状	性质	与其他构造的关系	与岩浆活动的关系	钻孔揭露情况	备注
永定河断裂	东南起南苑-通县断裂，经大宁水库、长辛店、首钢进入山区	走向北西，长约30km，倾向东北，倾角70°左右	正断层	跨越北京断陷的北西向断裂，西北进入山区，为丰台断陷盆与琉璃河断陷盆地的分界线，推断断裂属喜马拉雅期形成的断裂		断裂的西南侧长辛店一带古近系出露，JY-323孔揭露厚度300m左右，断裂东北侧JY-263孔、JY-265孔及JY-245孔揭露古近系底界为750~1035m，断裂两侧古近系落差450~735m	
公义庄断裂	北起黄庄断裂，经青龙湖、良乡、公义	走向北西，长26km，良乡以北段倾向东北；良乡以南倾向西南，倾角小于60°	平推断层	属北京断陷内的隐伏断裂，错断了南苑-通县断裂及良乡南断裂，推断断裂为第四纪形成	断裂东南端与南苑-通县断裂相接处，揭露有闪长岩	赵营钻孔及JY-320孔为断裂两侧的钻孔，JY-320孔埋深102m见蓟县系，而赵营钻孔埋深620m才见蓟县系，第四系落差518m	
琉璃河断裂	沿闫村、窦店西、琉璃河一线分布，西南伸出区外	走向北北东，区内长17km，倾向东北，倾角大于70°	正断层	属北京断陷内规模较大的隐伏断裂，推断为第四纪形成的断裂		JY-405和JY-505为断裂两侧的钻孔，JY-405钻孔埋深43m见白垩系，琉-1孔埋深904m还未见白垩系，白垩系落差近1000m	
望泉寺断裂	沿花梨坎、冯家营、望泉寺、两树行一线分布	走向北东，长约22km，倾向南东，倾角50°左右	逆断层	该断裂和顺义断裂作为北京断陷与西山褶皱带东段分界线。两者相距1.0~1.5km，相伴而行。推测花梨坎断裂与八宝山断裂同时期形成，有可能为同一条断裂		JY-189孔、JY-517孔揭露元古宇盖在髫髻山组上	
南彩断裂	经过吴家营、平各庄、南彩、李各庄、龙湾屯附近	走向为北东转北北东，长24km	正断层	与顺义断裂共同控制了第四纪俸伯盆地的形成			
李桥断裂	北起顺义断裂，经李桥镇、止于南苑-通县断裂	走向北西。长约16km，倾向西南，倾角大于	正断层	李桥断裂是东坝-天竺断陷盆的东界断裂，属喜马拉雅期形成的断裂，错断了楼梓庄断裂		JY-182与JY-171为断裂两侧的钻孔，第四系落差441m	
东风断裂	北起望京南，经酒仙桥西、东风、八里庄	走向北北西，长约11km，倾向南东，倾角大于70°	正断层	属北京断陷内的北西向隐伏断裂，是丰台断陷盆地与东坝-天竺断陷盆地分界线，断裂错断南苑-通县断裂，属喜马拉雅期形成的断裂		JY-273与JY-256为断裂两侧的钻孔，新近系—古近系落差大于200m	
燕郊断裂	西南起田府，经张家湾北、东堡北伸出区外	走向北东，长13km，倾向南东	正断层	大兴隆起内规模较大的隐伏断裂，属喜马拉雅期形成的断裂，被张家湾断裂错断		JY-134与JY-286为断裂两侧的钻孔，第四系落差280m	

续表

断裂名称	分布位置	产状	性质	与其他构造的关系	与岩浆活动的关系	钻孔揭露情况	备注
姚辛庄断裂	南起牛堡屯、经中辛庄、姚辛庄、卜落垡伸出本区	走向北东，长度18km，倾向西南	正断层	为大兴隆起内隐伏断裂，东南侧为寒武系，西北为青白口系，第四系落差200~500m，推断为第四纪形成的断裂，被北东向张家湾断裂错断			
瀛海断裂	南起皮各庄，经北藏村、孙村至瀛海北	走向北东，长度25km，倾向西北	正断层	属大兴隆起内隐伏断裂	其与公义断裂交接部位有闪长岩侵入		
杨镇断裂	南起太平庄，经田各庄、谢辛庄、杨镇东、李各庄西止于顺义断裂	走向北西，长达23km。东北盘为活动盘	逆断层	属北山褶皱带的隐伏断裂。二十里长山的断裂带由多条断裂组成，杨镇断裂是其中一条			
马昌营断裂	东南起夏垫断裂，经马昌营、张镇、龙湾屯西，唐指山小库西、两河、止于牛栏山断裂	走向北偏西40°，长约31km	逆断层	属北山褶皱带内隐伏断裂，在唐指山水库附近可见长城系掩盖蓟县系			

第四节　平原区基岩三维地质结构模型

2008年完成的“北京市多参数立体调查”项目，综合平原区地层、地质构造成果，基于GIS平台，利用ACCESS、MapGIS、三维地质建模等软件，首次建立了平原区基岩地质构造模型（反映基岩表层信息）及中心城区基岩三维地质结构模型，可以实现三维地质信息输入、数据库管理和三维地质数据建模与可视化分析等多种功能。

一、基岩三维地质结构模型

以中心城区基岩三维地质结构建模过程为例，首先在中心城区布设8条北西向剖面（剖面间距3km左右）和9条北东方向剖面。应用MapGIS软件，开展二维地质剖面的系统处理，建立剖面图层组（图42-14，图42-15）并设置好相关参数。其次，应用二维剖面系统建立地质结构三维栅格图（图42-16），建立3个层次的数据库，分别为基础地质数据库、成果数据库及模型资料数据库。最后，根据北京市平原区基岩地质构造图的信息，如地质界线、断裂线和基岩等深线的信息等，利用三维软件处理，建立了北京市平原区基岩地质构造模型（图42-17）和中心城区基岩三维地质结构模型（图42-18）。

北京市平原区基岩地质构造模型，反映了全平原区基岩表层地质构造特征，面积约6400km^2，其地层划分：中—新元古界分为长城系、蓟县系、待建系和青白口系；古生界分为寒武系、奥陶系、石炭系—二叠系；中生界分侏罗系及白垩系。

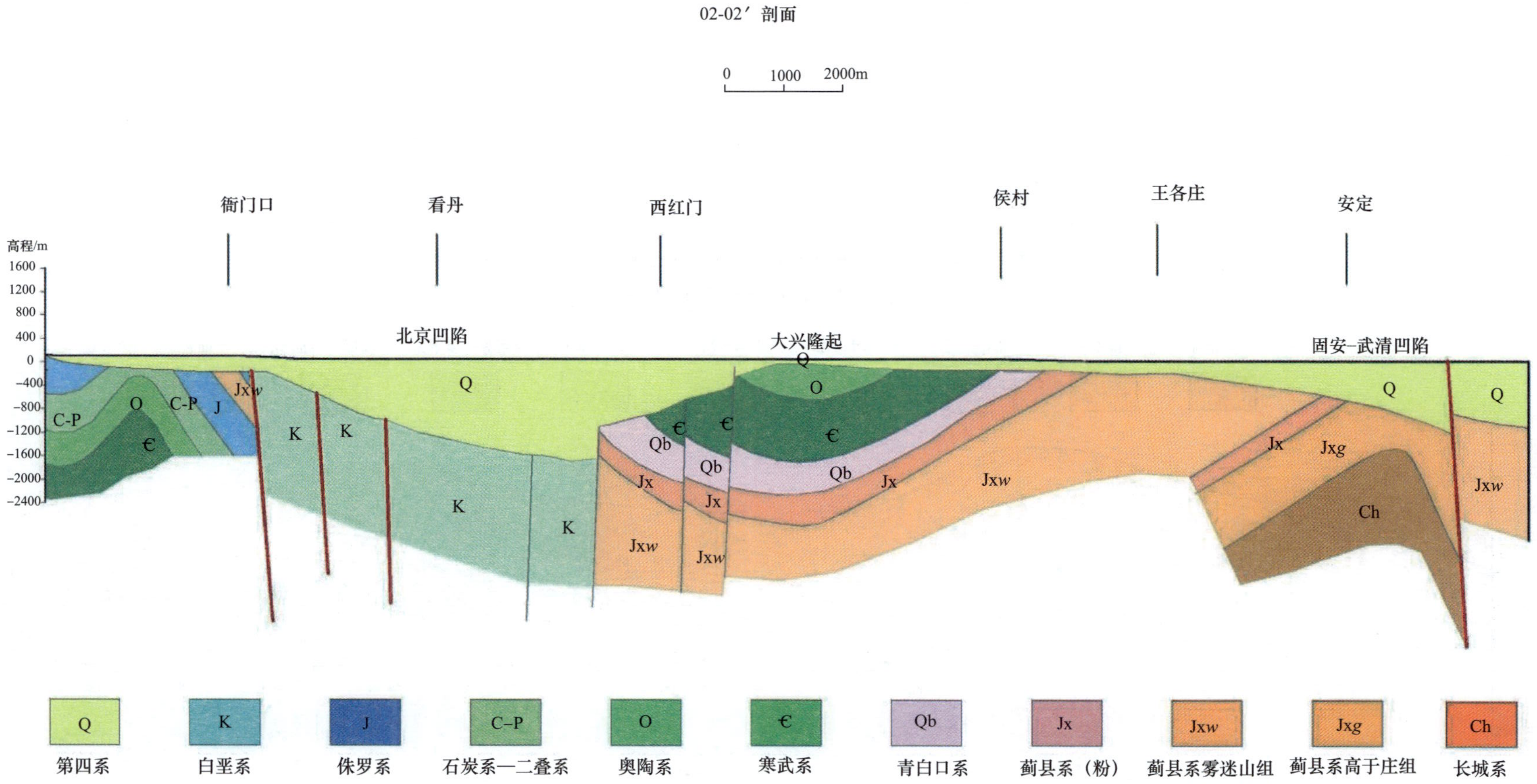

图 42-14　02 剖面的二维地质剖面示意图

Fig. 42-14　The two-dimensional geological profile sketch of "02" profile

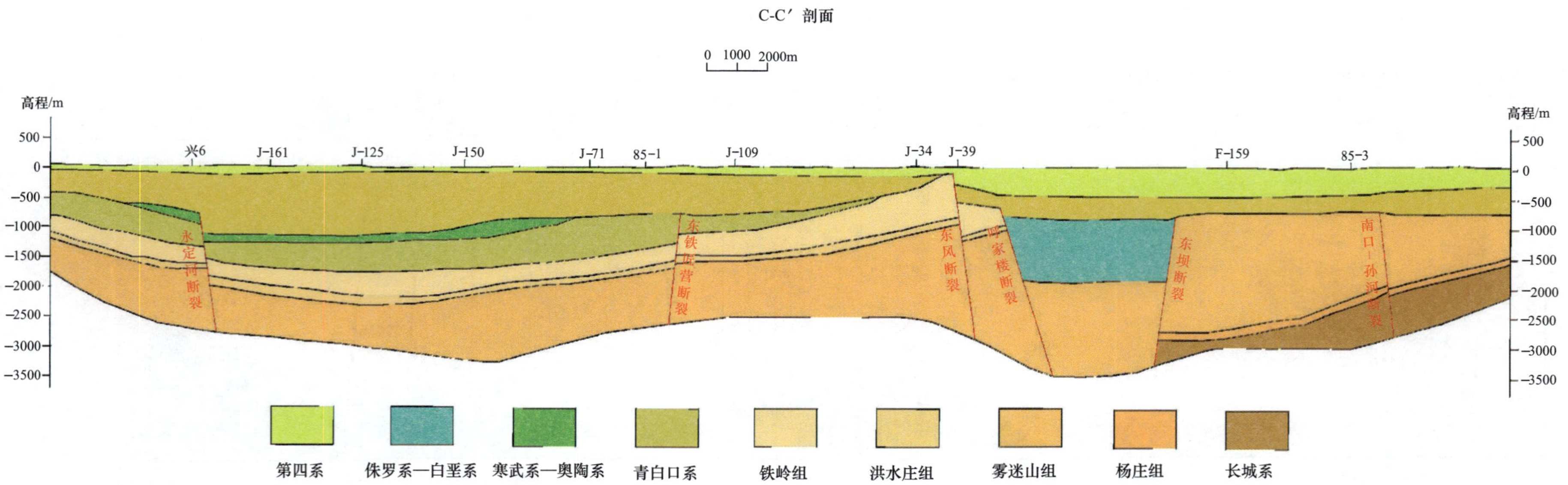

图 42-15　C 剖面的二维地质剖面示意图

Fig. 42-15　The two-dimensional geological profile sketch of “C” profile

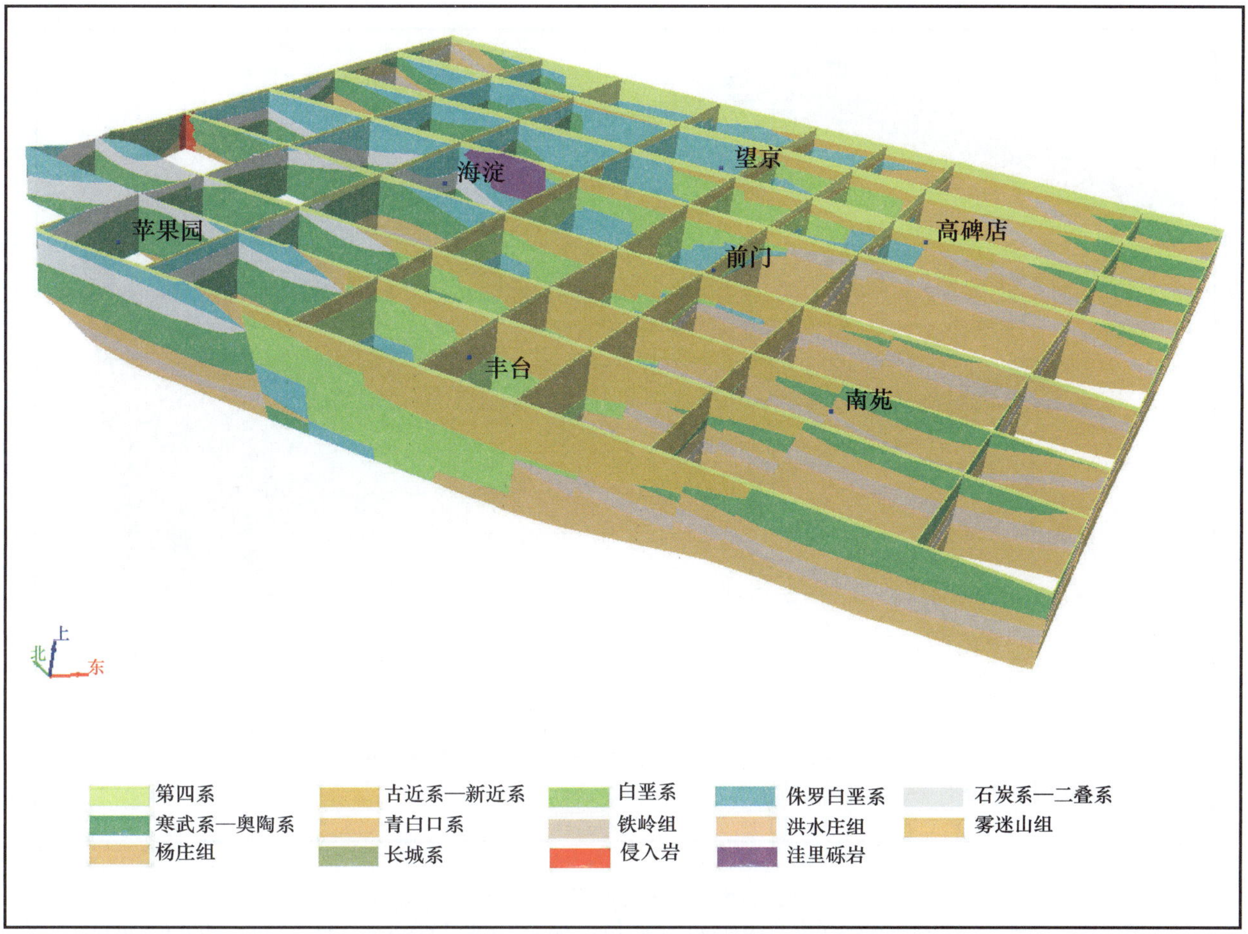

图 42-16 中心城区基岩三维地质结构模型栅格图

Fig. 42-16 Grid graphics for three-dimensional geologic structure model of bedrock in central urban area

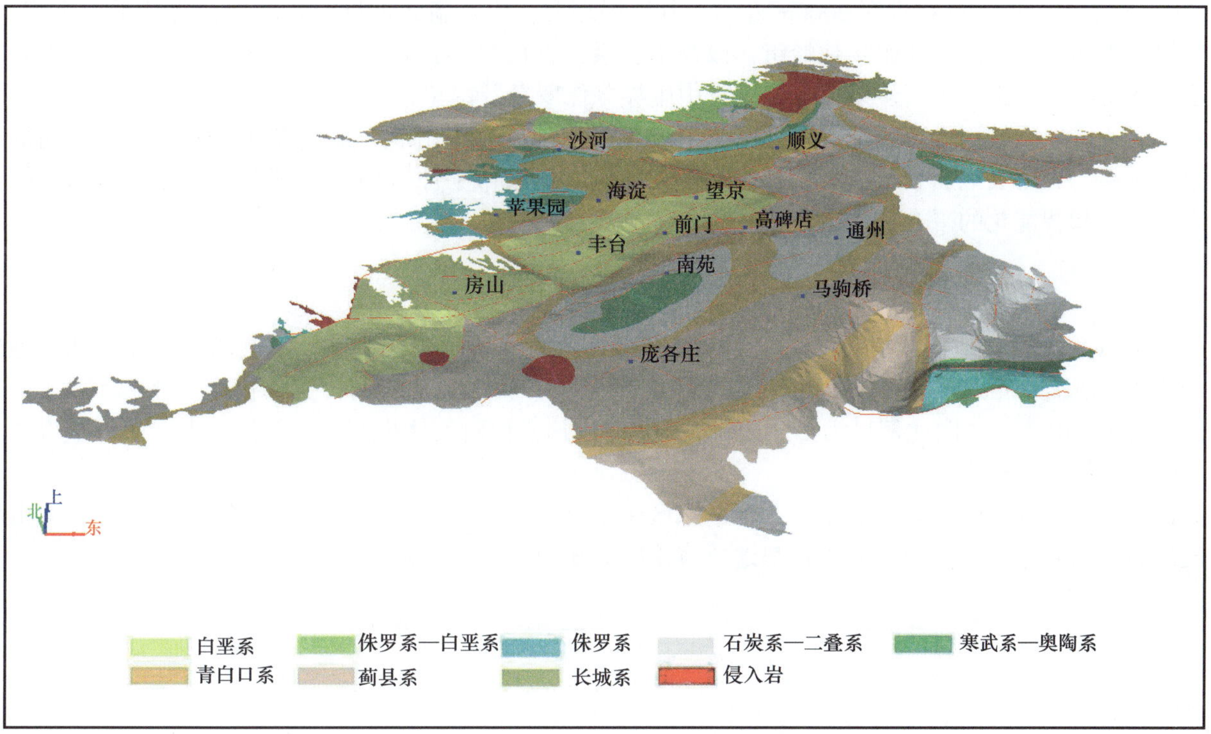

图 42-17 平原区前新生界地质构造模型

Fig. 42-17 Geologic structure model of Pre-Cenozoic in plain area

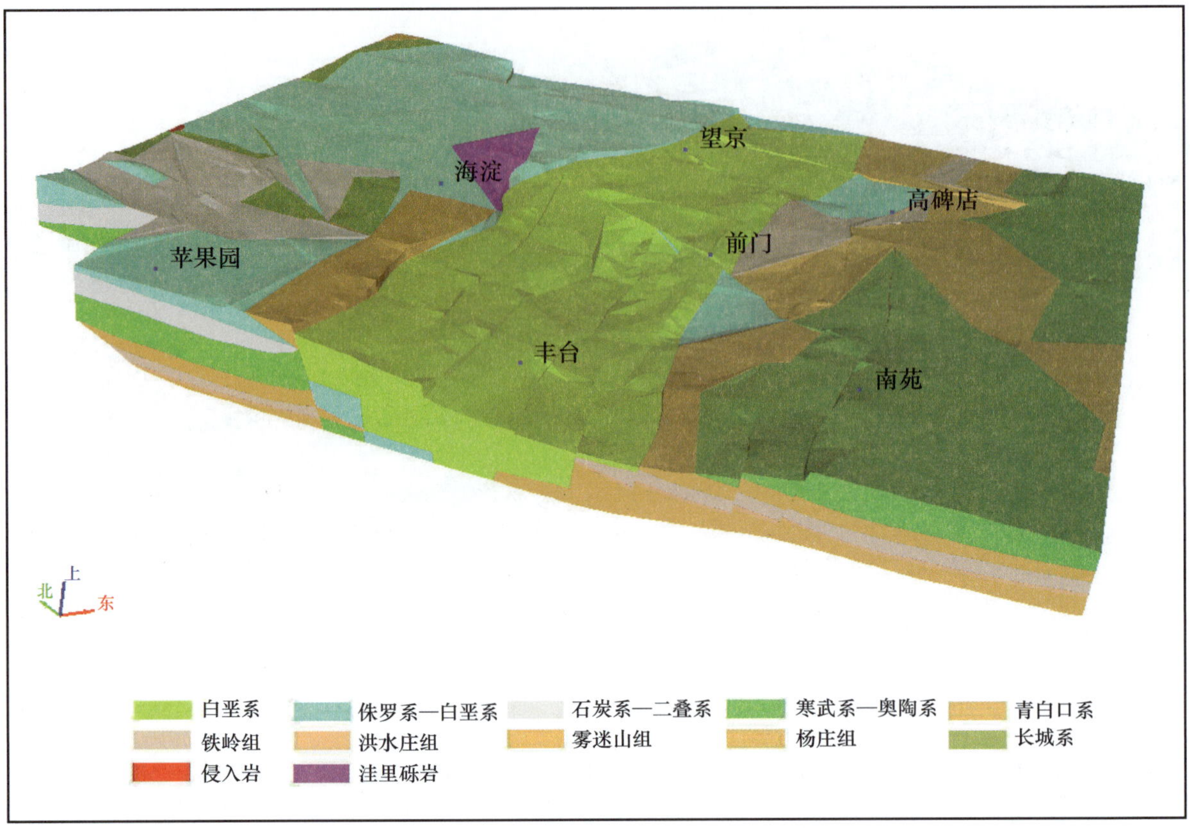

图 42-18 剥去新生界的中心城区基岩三维地质结构模型

Fig. 42-18 Three-dimensional geological structure model of bedrock with Cenozoic removed in central urban area

中心城区基岩内部三维地质构造模型，反映了中心城区基岩内部三维地质结构特征，面积约 $1000km^2$，底界深度控制在埋深 2000m 左右。其地层划分：中—新元古界分长城系，蓟县系高于庄组、杨庄组、雾迷山组、洪水庄组、铁岭组，以及待建系、青白口系；古生界上、下两分，即下古生界寒武系—奥陶系、上古生界石炭系—二叠系；中生界分侏罗系及白垩系。

基岩三维地质结构模型用三维图形图像的方式，生动地表现出地下地质界面和地质体的空间形态和组合关系，为广大地质技术人员和管理决策人员研究北京市平原区地质问题和实际工程应用提供了一个综合化和智能化的平台。

二、基岩实体模型

2014 年，北京市地质调查研究院以“北京市多参数立体地质调查项目”编制的 1∶10 万基岩地质图为基础，综合考虑基岩埋深、构造演化等因素，制作了平原区基岩结构实体模型。该模型 X、Y 向比例尺为 1∶10 万，Z 向比例尺为 1∶12500。利用 3D 扫描仪对实体模型进行扫描，生成了可全方位自由旋转的三维数字模型。

该三维数字模型较直观地展示了北京平原区主要断裂的展布、基岩地貌起伏等特征，对北京市城市规划具有指导意义，同时有利于首都地质演化历史知识的宣传和普及。

第四十三章　新生界三维地质结构

城市地区新生代地质与人类生产活动的关系最为密切，建立其三维地质模型，对于展示城市地质概况、分析地质环境、指导城市规划和工程部署、合理利用地下水资源、防治城市地质灾害等有着重大意义。

第一节　古近纪地层三维地质结构

北京地区经历了强烈的燕山运动之后进入了古近纪，尔后经历了一个明显的剥蚀过程。随着京西北隆起的继续抬升，本区出现了山区和平原的分化，平原区的轮廓开始展现出端倪。古近纪沉积在这一基础上发展演化，沉积范围向北东扩展，从下白垩统在平原区的分布来看，北京凹陷原先大致局限在八宝山断裂与良乡-前门-顺义断裂之间，古近纪时向东迁移至南苑-通县断裂附近。北京凹陷内，古近系的沉积中心大致位于宛平、卢沟桥、丰台一带，最厚处沉积厚度超过400m。岩性在八里庄—丰台小井一带为紫红色半胶结砂砾岩，上部岩石呈砖红色，胶结程度较差，下部呈紫色，胶结程度略好。

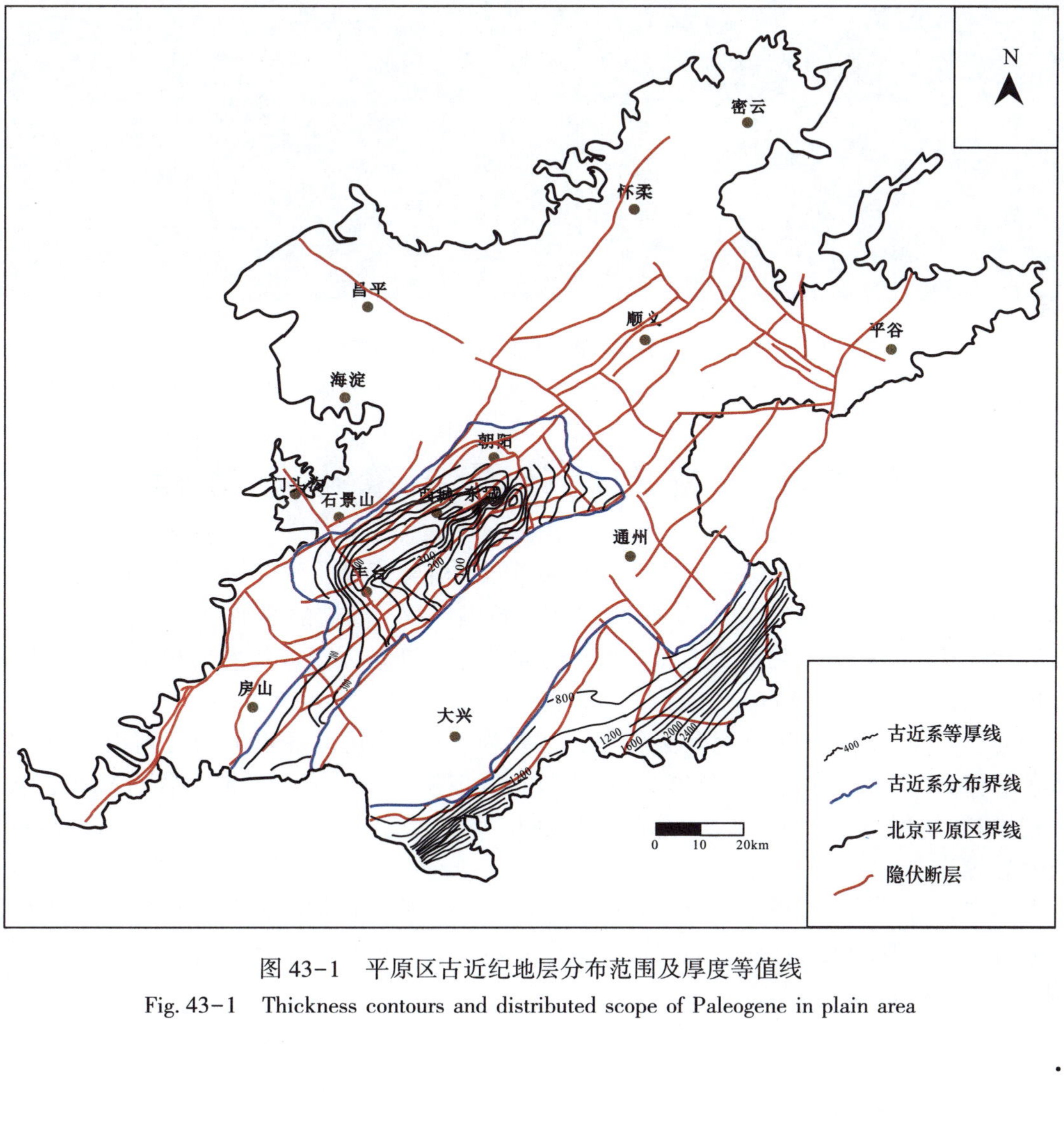

图 43-1　平原区古近纪地层分布范围及厚度等值线

Fig. 43-1　Thickness contours and distributed scope of Paleogene in plain area

在大厂凹陷内则沉积了巨厚的古近系，主要为沙河街组1～3段和东营组，为一套湖相和过渡相（海泛和海水影响地层为过渡相）暗色沙泥岩沉积，化石丰富，为淡水、半咸水共生，与下伏的孔店组纯陆相淡水沉积呈平行不整合接触。大厂凹陷最大沉积厚度达3000m。主要岩性：上部为深灰、灰黑色泥岩夹灰白色粉砂岩，局部地区夹黑色玄武岩，厚度400～1000m，中部为绿灰、青灰、深灰色泥岩与灰白色粉、细砂岩不等厚互层，局部地区夹数层玄武岩另外，在古近纪的岩浆活动十分引人注目，始新世—渐新世约有5次玄武岩喷发，多在凹陷的中心沿断裂带溢出。就古近纪沉积物的岩性、岩相而言，自下而上为陆相沉积紫红色砂砾岩、泥岩，向上过渡为灰褐色泥岩，在大厂凹陷中局部还有油页岩及煤线，说明有机界的渐趋繁茂，古近系分布范围及厚度见图43-1。

利用三维建模技术，根据厚度等值线和分布范围对古近系进行了三维构建，并将地表数据进行了叠加，建立的三维结构图如图43-2所示。古近系厚度在北京凹陷和大厂凹陷内差别巨大，北京凹陷沉积中心在凹陷的东北部，而大厂凹陷内则越往东南越增厚。

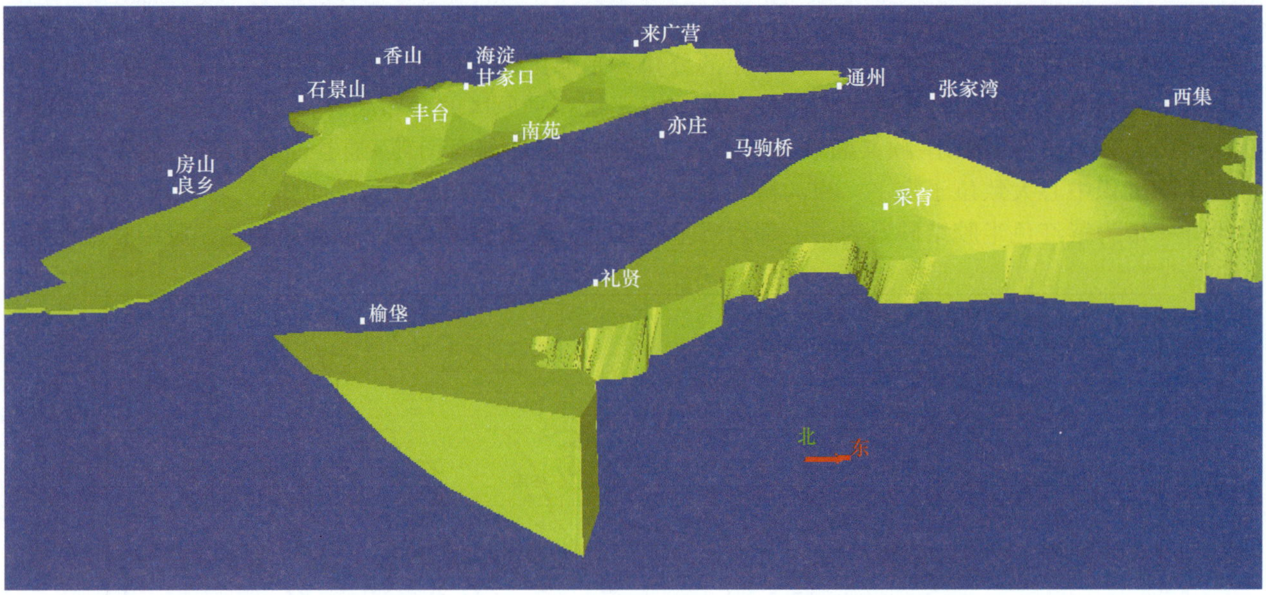

图43-2　平原区古近纪地层三维地质结构图

Fig. 43-2　Three-dimensional geological structure of Paleogene in plain area

第二节　新近纪地层三维地质结构

新近纪，北京平原地区湖盆进一步扩大，北东向断裂继续活动，沿断裂下盘沉降尤为明显，填充了巨厚的泥砂堆积物。北京凹陷在原有的基础上沉积范围展宽，隆起的范围则相对缩小变窄。受良乡-顺义断裂继承性活动的影响，北京凹陷沉降中心向北东方向迁移，新近系的沉积中心位于北天堂—葆台—四合庄一线，厚度超过1700m。

新近纪北京凹陷内发育了天坛组与天竺组，主要为还原-弱还原环境下的河湖相沉积，岩性为灰色、灰黑色及棕黄色半胶结泥岩、粉砂岩及砾岩。最大厚度超过千米，并同时伴有陆相玄武岩喷发，但其喷发规模和厚度都较古近纪时小。岩性仍以陆相沉积盆地的紫红色、灰白色、灰绿色砂质泥岩、粉砂岩及砾岩等为主。

大厂凹陷内，新近系的沉积中心位于安次、武清一带，其沉积物以紫色、红棕色泥岩为主，其次为砂岩及砂砾岩，反映了氧化条件下的河流及浅湖相沉积，有的地区夹劣质煤线，最厚处超过400m。沉积范围如图43-3所示。

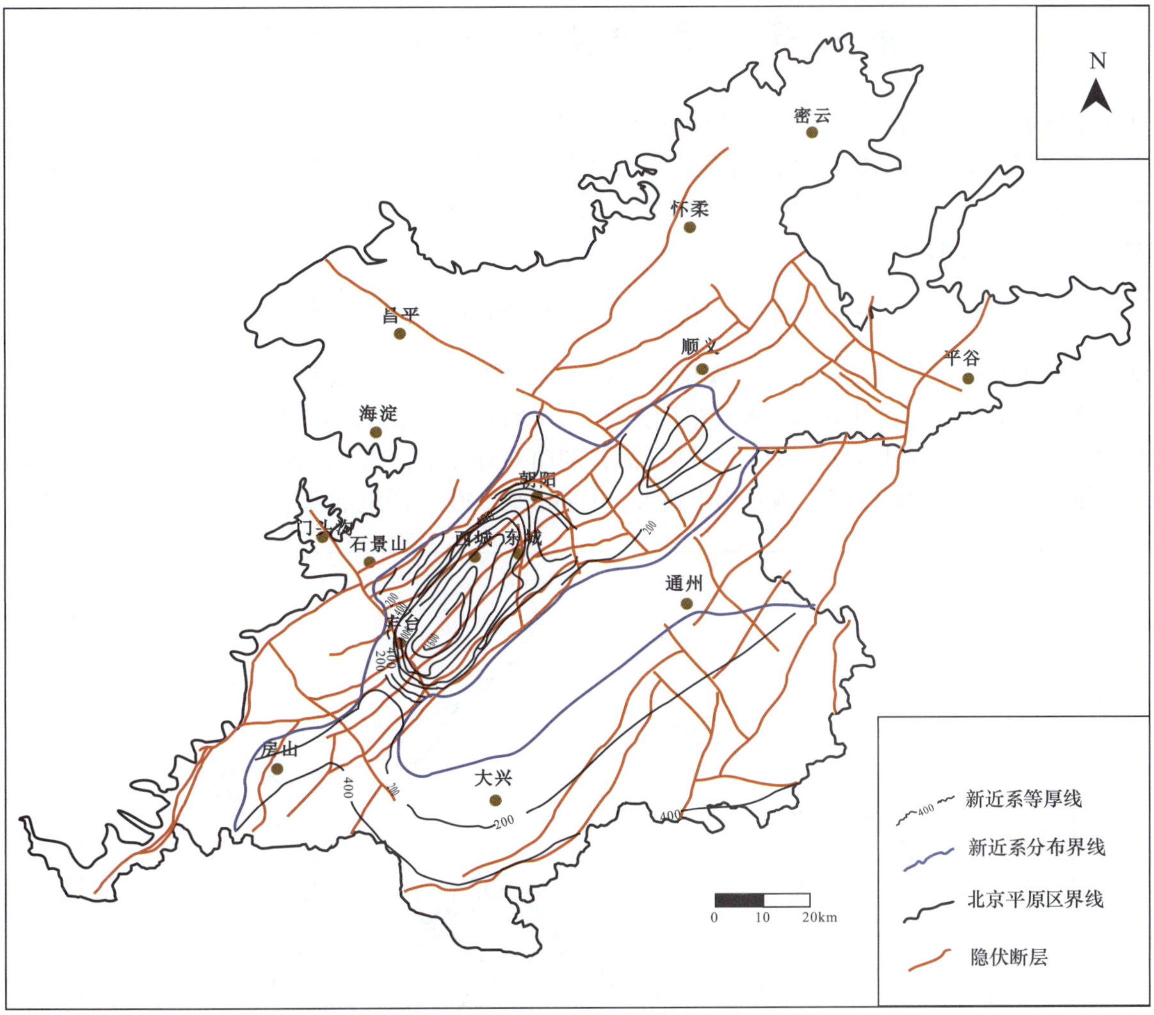

图 43-3 平原区新近纪地层分布范围及厚度等值线图

Fig. 43-3 Thickness contours and distribution of Neogene in plain area

利用空间计算机技术，将沉积范围和厚度等值线和转化为立体结构模型。见图 43-4。

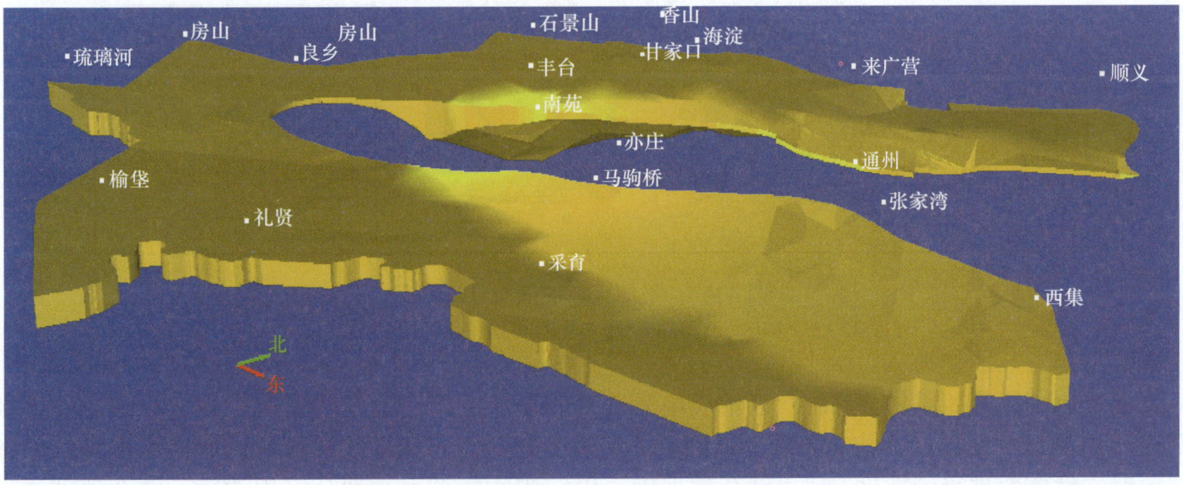

图 43-4 平原区新近纪地层三维地质结构图

Fig. 43-4 Three-dimensional geological structure of Neogene in plain area

第三节　第四系三维地质结构

北京地区具有典型的山前冲洪积平原的特点，多条河流的摆动、交叠使得地层在横向上存在强烈的相变，使得仅依靠丰富的钻孔资料刻画整个北京平原区第四系立体结构难度很大。根据国内外目前的技术手段和研究现状，结合北京地区的实际，用剖面和钻孔相结合的方法建立北京平原区第四系立体模型是一种比较好的方法。

一、第四纪地层三维地质结构模型

新生界基准格架的搭建是三维地质模型的基础，而格架主要是通过利用所选的基准孔资料绘制的基准剖面来体现，由于基准孔基本只揭穿第四纪地层，所以该格架主要反映第四纪的地层结构。在平原区共布设了 17 条由基准孔控制的第四系基准剖面，剖面共有 2 组，北东向剖面共 7 条，编号由 A 到 G，走向北东 40°；北西向剖面 10 条，编号由 I 到 X，走向北西 320°，两组剖面呈正交关系（图 43-5）。

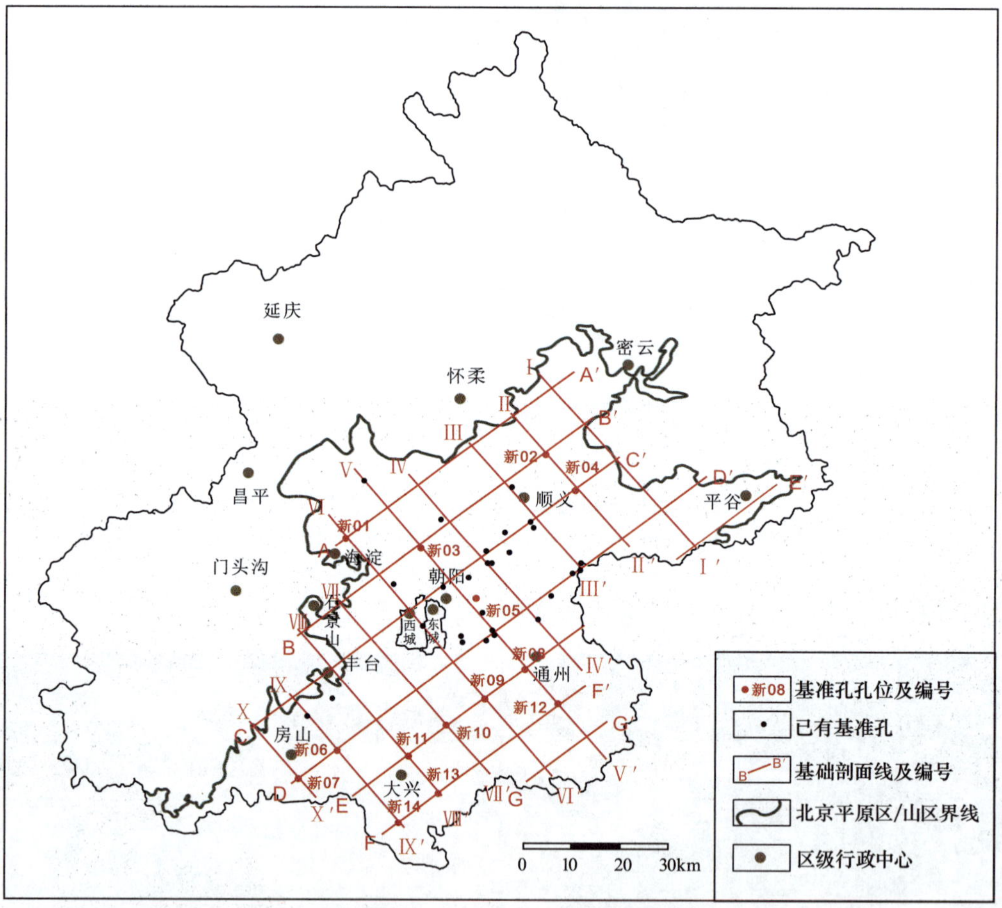

图 43-5　平原区标准剖面分布图

Fig. 43-5　distribution map of the standard profile in plain area

剖面绘制中，第四系划分到统；以地质单元（冲积扇、沉积凹陷）为宏观把握，以遥感、区域物探资料（电测深、地震）为控制条件，达到表征河道变迁和沉积相界线的目的；严格依据真实的钻孔资料，对岩性资料进行标准化，建立钻孔数据库；以沉积物颗粒大小划分沉积韵律，结合地球物理测井曲线划分大的岩性分界线；最后考虑相变，以反映古河道变迁、沉积物相变为原则连接岩性分界线。

剖面绘制完成后，经过扫描矢量化、坐标校正、格式转化、系统导入等数据处理加工，开展地层岩性赋属性、三维地质结构成体等过程，建立三维立体模型图（图43-6）。

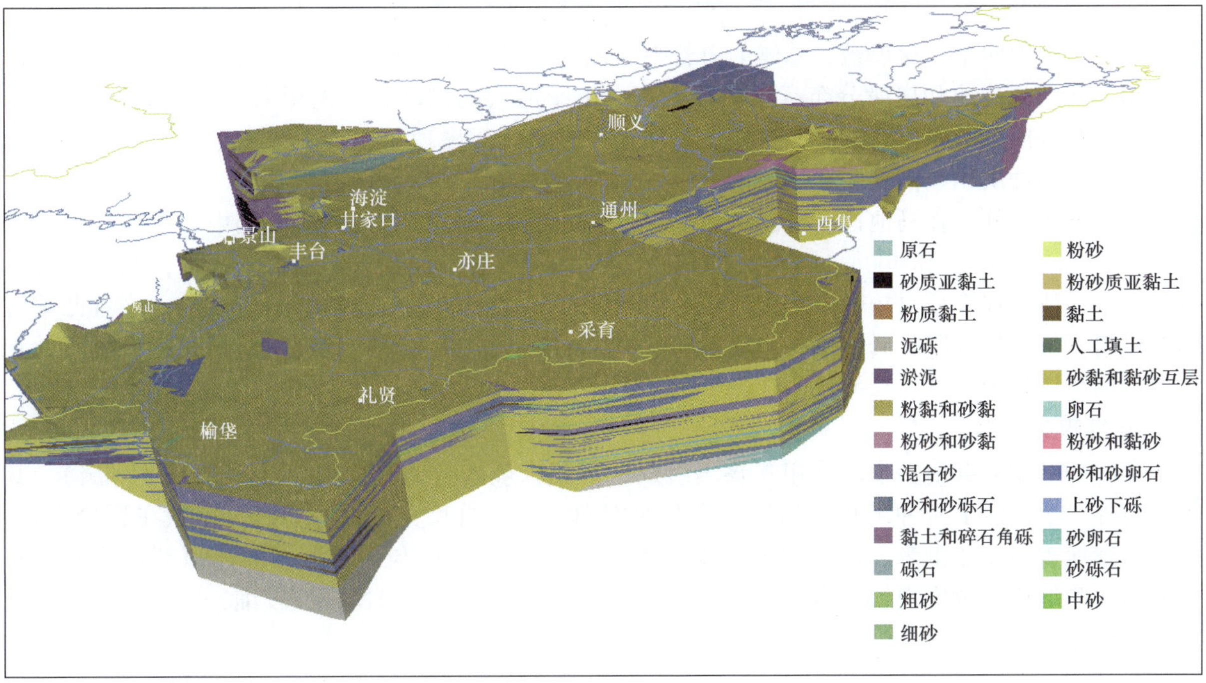

图43-6　北京市平原区第四纪地层立体模型

Fig. 43-6　Stereomodel of Quaternary system in Beijing Plain

该模型具有任意切割剖面、切割不同的体块、人工随意输入切面位置和法向、开挖功能。并可以设置任意的网格数目和网格方向（图43-7）。

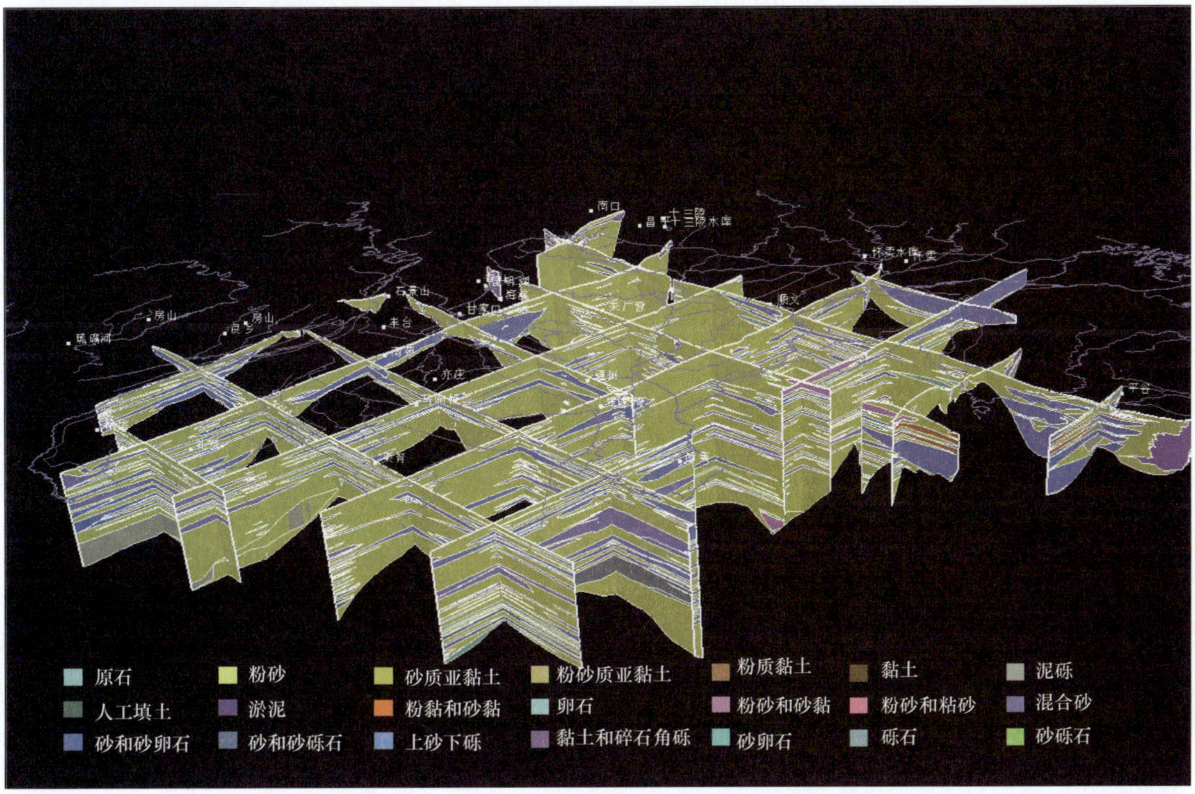

图43-7　北京市平原区新生代地层立体模型剖切图

Fig. 43-7　Section cutting plane of Cenozoic Stereomodel in Beijing Plain

二、第四纪地层三维地质结构特征

北京平原被第四系广泛覆盖，总体上由若干个巨大的冲洪积扇和沉积凹陷构成。在冲洪积扇内发育河流相，在扇间洼地和扇缘分布有湖泊沼泽相。由于河流的摆动，冲洪积扇内岩性横向变化迅速，对比困难。在垂向上，砂砾石、砂、粉砂、亚黏土、黏土交互出现。在扇中、扇缘部位可见由粗到细的韵律，可作为冲洪积扇内岩层连接的依据之一。

除了冲洪积扇外，在马池口、后沙峪和平谷存在3个第四纪沉积凹陷，凹陷中心第四纪沉积物厚度分别为597m、1000m和400m。此外在通州东南有大厂凹陷，其主体延至市辖区外，凹陷中心第四系厚数百米，研究程度相对较低。沉积凹陷的岩性特征明显，靠近山前的马池口凹陷、平谷凹陷沉积物颗粒较粗，沉积物较厚。后沙峪凹陷沉积物颗粒较细。这些冲洪积扇和沉积凹陷共同构成了北京平原第四系的总体轮廓。

1. 冲洪积扇

永定河冲洪积扇群由多个扇体相互叠压组成。在北京境内有两个扇体（Ⅰ号扇、Ⅱ号扇），其余分布在境外。永定河Ⅰ号冲洪积扇的轴线为东西方向，大致沿长安街由西向东。扇顶在石景山地区，砂砾石出露地表厚度数十米。冲洪积扇的中部为中砂-细砂-粉砂与黏土互层。河道部位为砂砾石和砂。河道两侧沉积物粒度变细。通州地区是该冲洪积扇的扇缘，沉积物颗粒较细，以黏土、亚黏土为主，见湖沼相沉积（图43-8，图43-9）。

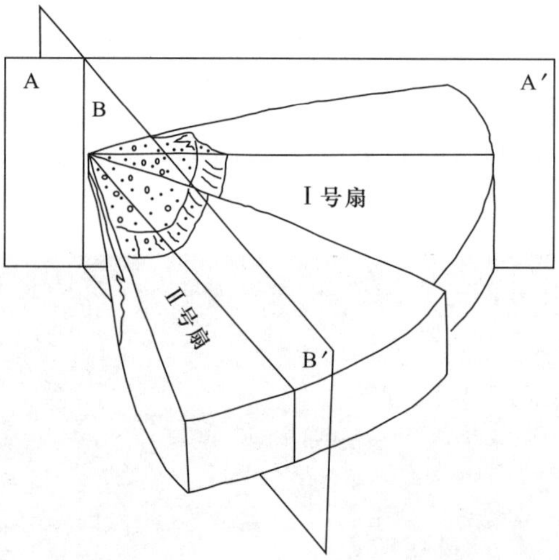

图43-8　永定河冲洪积扇剖面位置示意图

Fig. 43-8　The section position sketch of alluvial-proluvial fan in Yongding River

通州向东至宋庄地区是永定河冲洪积扇与潮白河冲洪积扇的结合部位。两个扇的沉积物相互叠压比较复杂。永定河冲洪积扇的形成时间较早，经新5孔揭露，永定河冲洪积扇的底界埋深422m，底部见厚度90余米的泥砂砾岩，分选性差，砾石占20%～30%，卵石占10%～50%，最大砾径大于10cm。经古地磁测试，其形成时代为上新世中期，距今3.58～3.33Ma。冲洪积扇之下为上新世湖相沉积。在大兴隆起部位冲洪积扇覆盖在寒武系、奥陶系之上。永定河Ⅱ号冲洪积扇形成时间较晚，其轴线为南东方向，与古漯水、古浑河、古永定河走向一致。Ⅱ号扇的规模较大，扇缘分布在与河北省交界附近。扇顶在石景山附近，砂砾石层沿古河道走向向东南方向凸出。扇体的北东侧叠置在Ⅰ号扇之上。剖面上有侵蚀Ⅰ号扇的现象。在通州区永乐店小甸屯一带见有泥炭堆积，为扇缘湖沼相沉积（图43-10）。

潮白河冲洪积扇的规模比永定河冲洪积扇小。由于潮白河河道相对稳定，所以其扇体无大的迁移，其结构也相对简单。扇体形状呈狭长的窄扇形。在北京境内潮白河河道位于扇的轴部，呈南北向。沉积物粒度沿轴部由扇顶向南由砂砾石渐变为中细砂、粉砂，南部在通州与永定河冲洪积扇相互叠置交

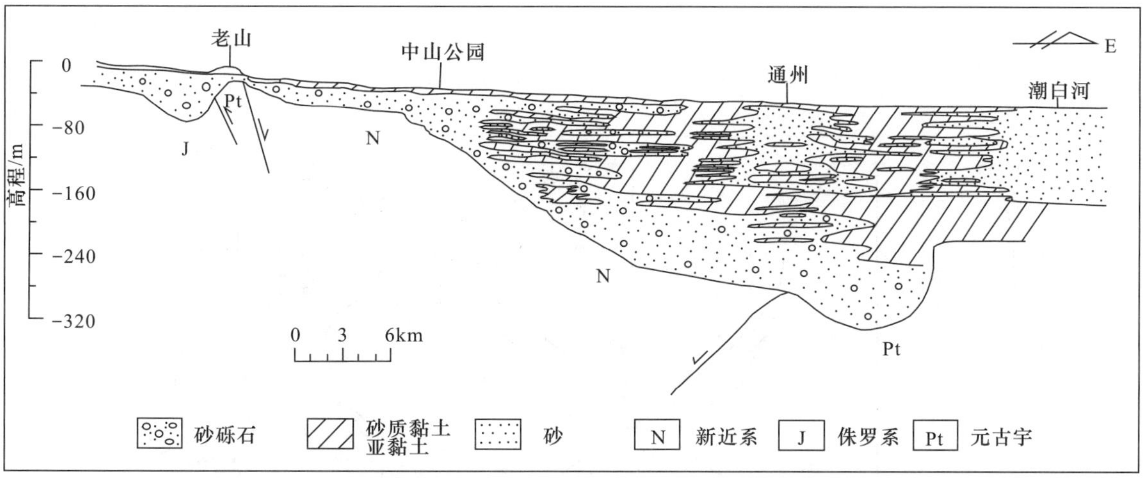

图 43-9 冲积扇中 A-A′地质剖面示意图

Fig. 43-9 The sketch of A-A' geological section of alluvial fan

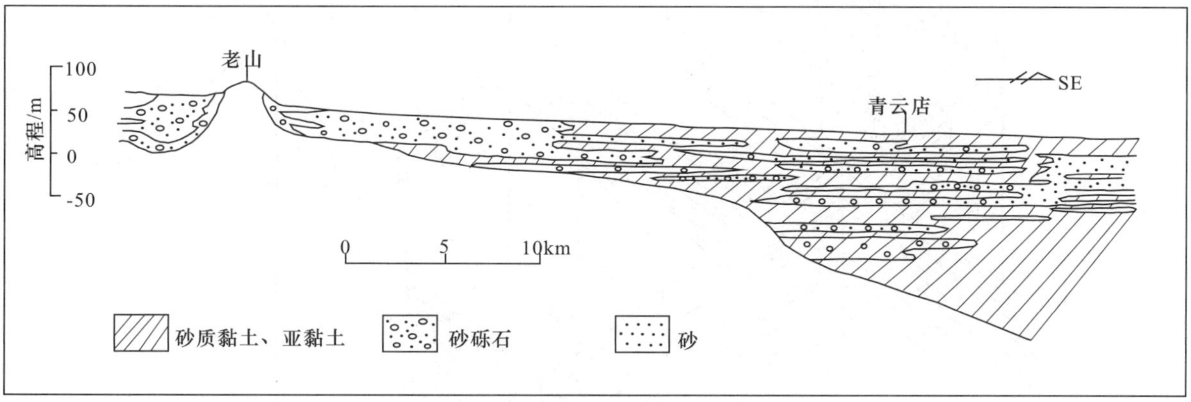

图 43-10 冲积扇中 B-B′地质剖面示意图

Fig. 43-10 The sketch of B-B' geological section of alluvial fan

错。潮白河河道砂石蕴藏丰富，可作为重要的建筑用砂产地。河道两侧地下水丰富，是北京重要的地下水源地。潮白河冲洪积扇与永定河冲洪积扇以温榆河为界。由于北京地区的供水很大程度上来源于第四系中松散孔隙水，潮白河第四系三维结构模型的构建，克服了传统的“平面二维”方式的认知局限，可为决策者开展地表水、地下水联合调蓄等活动提供科学依据，见图 43-11。

2. 沉积凹陷

马池口凹陷位于南口-孙河断裂西北端南侧，因断裂上盘下降形成。震平 2 孔揭露的第四系厚 597.6m。由于地处山麓，沉积物颗粒较粗，主要沉积物为砂质黏土、中细砂、砂砾石互层。见 10 余层泥砾岩，每层厚 6~38m，颜色以黄色，杂色为主，砾石大小不一，见磨圆较好的砾石，砾石多被泥砂包裹。砾石成分简单，下部以花岗岩为主，上部主要为花岗岩和火山岩。附近山区为大面积花岗岩和火山岩出露区，是砾石的主要物源区。越靠近山区，沉积物粒度越大。凹陷基底为紫色中生代火山角砾岩，有风化现象。如图 43-12 所示，剖面左起马池口凹陷的山前地带，横切该凹陷的南部，剖切了沙河水库、温榆河河道和潮白河河道。凹陷两侧为坡积物，主要由砂质黏土、黏质砂土夹碎石和角砾组成，形成犬牙状堆积形态，凹陷中部由黏质砂土和砂质黏土夹厚层的砂卵石。剖面中部横穿怀柔水库，右侧横切了潮白河冲洪积扇的顶端，冲洪积扇主要由砂砾石、砂卵石组成，局部夹薄层的黏性透镜体，砂卵石厚度近 100m，是潮白河冲洪积扇良好的补给区和储水层。

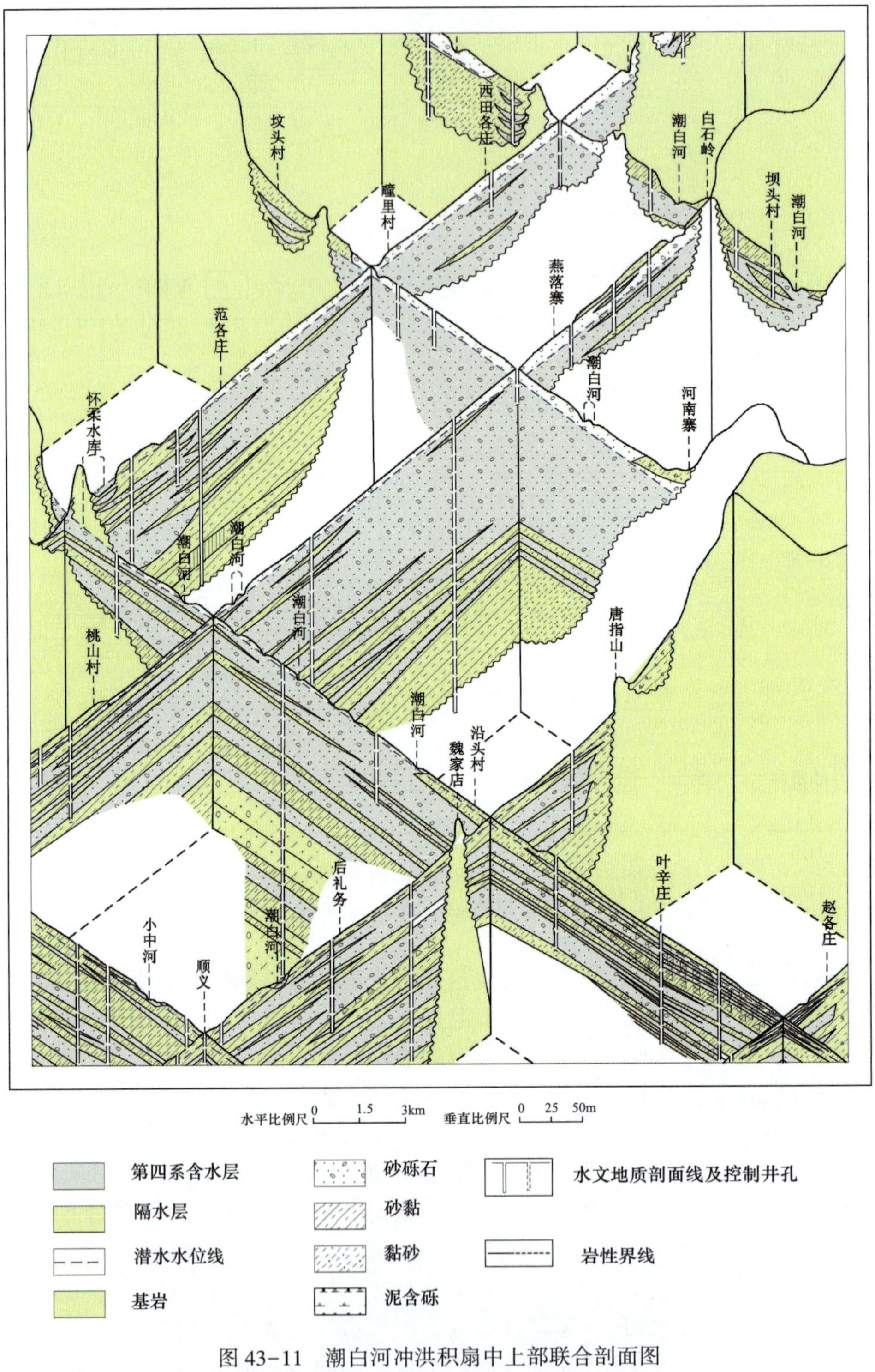

图 43-11　潮白河冲洪积扇中上部联合剖面图

（据多参数新生界立体地质调查专题报告）

Fig. 43-11　The joint profile of the middle and upper part of the alluvial fan of Chaobai River

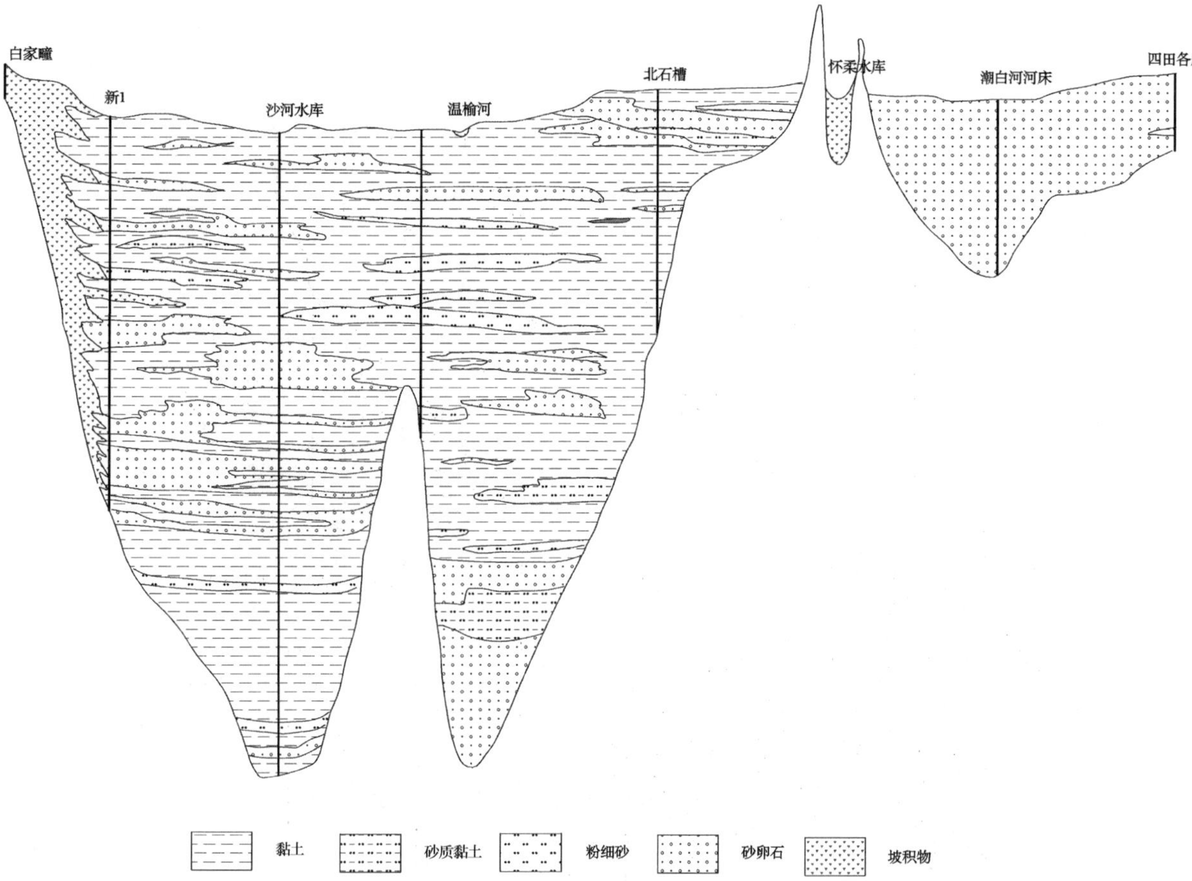

图 43-12　马池口凹陷北东向剖面

Fig. 43-12　The northeast strike profile of Machikou Depression

后沙峪凹陷位于南口-孙河断裂北侧顺义附近，因断裂上盘下降形成。凹陷中心新生界厚约1000m。总体上沉积物粒度为上细下粗。上部为灰色、杂色砂砾石、砂、砂质黏土。灰黄色粉砂层（天竺剖面），含黏质砂土，灰褐色与深灰色砂质黏土互层。灰白色细砂层，灰褐色、褐色深灰色黏土夹砂质黏土，黏土断面具油脂光泽。下部为灰黄色、杂色砂砾石。底部浅棕红色泥岩，色匀质纯性软，成岩性差，易造浆。底部见巨厚的泥砾岩。砾石色杂，以灰、红、黑色为主，成分以石英质为主，其次为中基性火成岩。根据其他钻孔资料，泥砾岩应为上新世产物。后沙峪凹陷基底为中生代凝灰岩、凝灰质角砾岩（图 43-13）。

平谷凹陷位于平谷盆地，凹陷中心基岩埋深 600 余米，主要为中—新元古界长城系、蓟县系白云岩。基岩之上为巨厚第四系单一结构的砂卵砾石层，主要由洵河冲洪积物组成。砂砾石层砾径由东向西减小，东部以漂砾为主，西南出山部位以砂为主，砾石较少。砾石成分主要为砂岩、片麻岩，少量白云岩、火山岩。砾石磨圆中等。砾石层中夹有厚度不等的透镜状、层状砂层。图 43-14 剖面北起北台上水库山前，横切了潮白河密云扇，横过潮白河上游的两条主河道，向南穿过龙湾屯、张镇和平谷马坊。北部山前有薄层坡积物，在北房附近有一局部的第四系凹陷，最大沉积厚度达 300m。有上下两层厚层的砂卵砾石，下层厚 80 余米；上层砂卵石厚达 60m 左右，向东南部逐渐过渡为多层，多夹砂质黏土、黏质砂土透镜体。由张坊的西北往东南，第四系厚度由小增大，厚又逐渐减小。

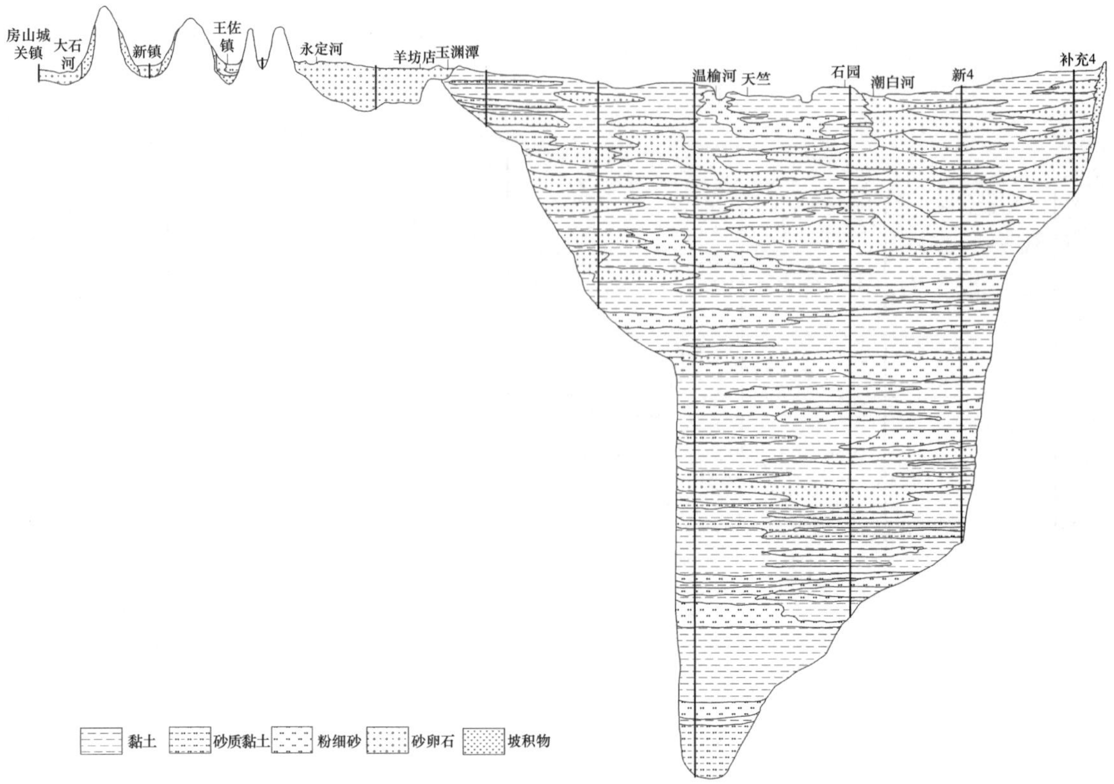

图 43-13　后沙峪凹陷北东向剖面

Fig. 43-13　The northeast strike profile of Houshayu Depression

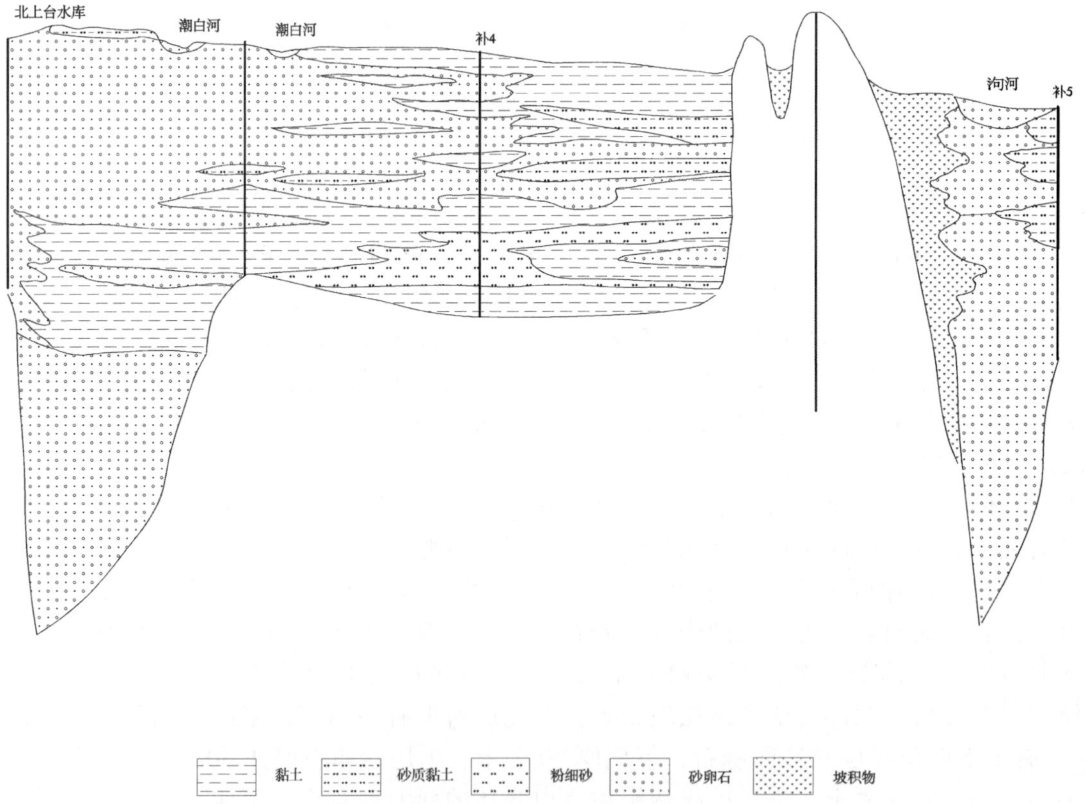

图 43-14　平谷凹陷北西向剖面

Fig. 43-14　The northwest strike profile of Pinggu Depression

第四十四章　平原区浅表层（工程地质层）三维地质结构

本志所称的浅表层主要是指与北京平原区城市工程建设相关的地表30m以浅的工程建设层。一方面，城市工程建设以浅表层的岩土体为基础或建筑环境，可以作为建筑物地基、场地和天然建材；另一方面，城市大规模的开发建设，又干扰了浅表层的地质环境，导致诸多岩土体工程地质问题的发生，并随着城市建设规模的增大而造成的地质问题会更加突出。为科学地进行城市规划和建设，合理利用天然地基，最大限度地利用岩土体工程能力和节约土地资源，调查研究区域上浅表层（工程地质层）三维地质结构、了解岩土体工程地质特征是很重要的。根据北京地区城市规划建设特点，按照建设重点划分为“五环内、顺义区、通州区、昌平区、大兴区、房山区、平谷区”7个分区进行描述，本志从平原区宏观总体三维结构到分区具体三维结构，对平原区整体的工程地质进行总结评价，对不利于规划建设的软弱地层进行了重点阐述，以求更加精细有效地为规划建设服务。

第一节　平原区浅表层三维地质结构

一、平原区浅表层的三维地质结构

北京平原第四系主要由永定河、潮白河等河流的冲洪积扇和沉积凹陷构成，沉积物成因类型主要有冲积相、洪积相、湖沼相，有些钻孔发现有海相层。冲洪积扇相互交错，叠压关系复杂，在扇间洼地和扇边缘存在湖泊和沼泽沉积，如昆明湖、玉渊潭、莲花池等。地层主要是黏性土及粉、细砂层，主要为交互出现的多层结构。在河流故道及泛滥区内分布有疏松粉细砂层，地下水位一般埋藏较浅，整体地形较为平坦，坡度小于1‰。

平原区浅表层（30m以浅）自西部山前至东侧通州，主要的地层结构分布有一定的规律，山前至西三环附近，地层结构比较简单，为均一结构，厚层的卵石填充砂砾，规划建设时，需要针对地下水开展防护措施。西三环至东三环中部地区，为砂土、黏土、卵砾石2~3种地层交互结构，规划建设时，需要注意地面沉降梯度带和尽量避免破坏隔水层（黏土）。东三环至通州东部地区，为多种地层复杂交互结构，卵砾石层逐渐消失，以细颗粒沉积地层为主（图44-1）。

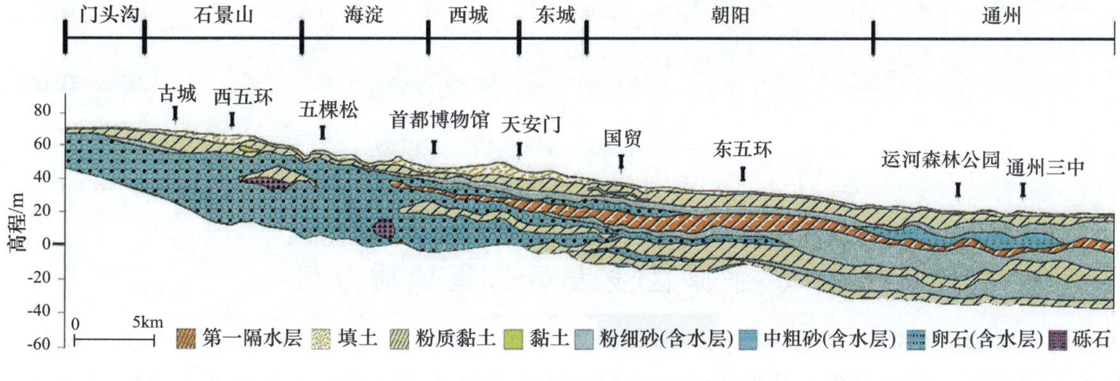

图44-1　平原区浅表层地层剖面（以长安街沿线剖面为例）

Fig. 44-1　The shallow layer stratigraphic section in plain area（take Chang'an street extension section as example）

第四纪早期在平原北部的马池口、后沙峪和平谷形成3个沉积凹陷，沉积物巨厚，最深的后沙峪凹陷厚度在1000m以上。随着河流的摆动，冲洪积扇内岩性横向不断变化，城市建设不断发展，人工堆积情况也随之变化，平原区地质三维结构组成相当复杂。

比较有利的条件是北京平原区分布着大量的钻孔，依托钻孔数据的支撑，可以利用计算机和三维建模技术，把已掌握的不均匀、不连续、散乱的地质信息，通过数学曲面和拟合，形成可视的、连续的、形象直观的三维地质模型和图形图像，处理岩层界面和结构面的组合关系，尽量真实地反映地下地质结构全貌，而且随着地下钻孔或其他地质信息越来越丰富，三维地质模型也会越来越准确，可以给城市规划、建设管理等诸多应用部门提供一个直观明了的决策依据和技术支持。

平原区浅表层的三维地质结构建模，总体思路是以钻孔数据为基础，其他勘探手段数据为辅助。工作思路是“取点-连线-建面-成体”。第一，保证工作区内钻孔数量满足精度要求。第二，按照横、纵、斜交叉体系建立二维剖面网。第三，在地质变化规律指导下，用数学方法修正、插值、拟合，形成三维立体结构。

以平原区整体为例，平原区共利用钻孔161个，一般孔深30~40m，局部山前地带钻孔深度至基岩层，孔深15~30m；最深钻孔达50m。建立不同方向的二维地质剖面38个，钻孔和剖面覆盖六环内和六环外，六环外精度为1∶10万，六环内精度为1∶5万。

按上述方法，最终构建了基本的平原区地质结构三维模型（图44-2），从模型可见平原区浅表层（30m以浅）地层结构具有明显的分区差异。

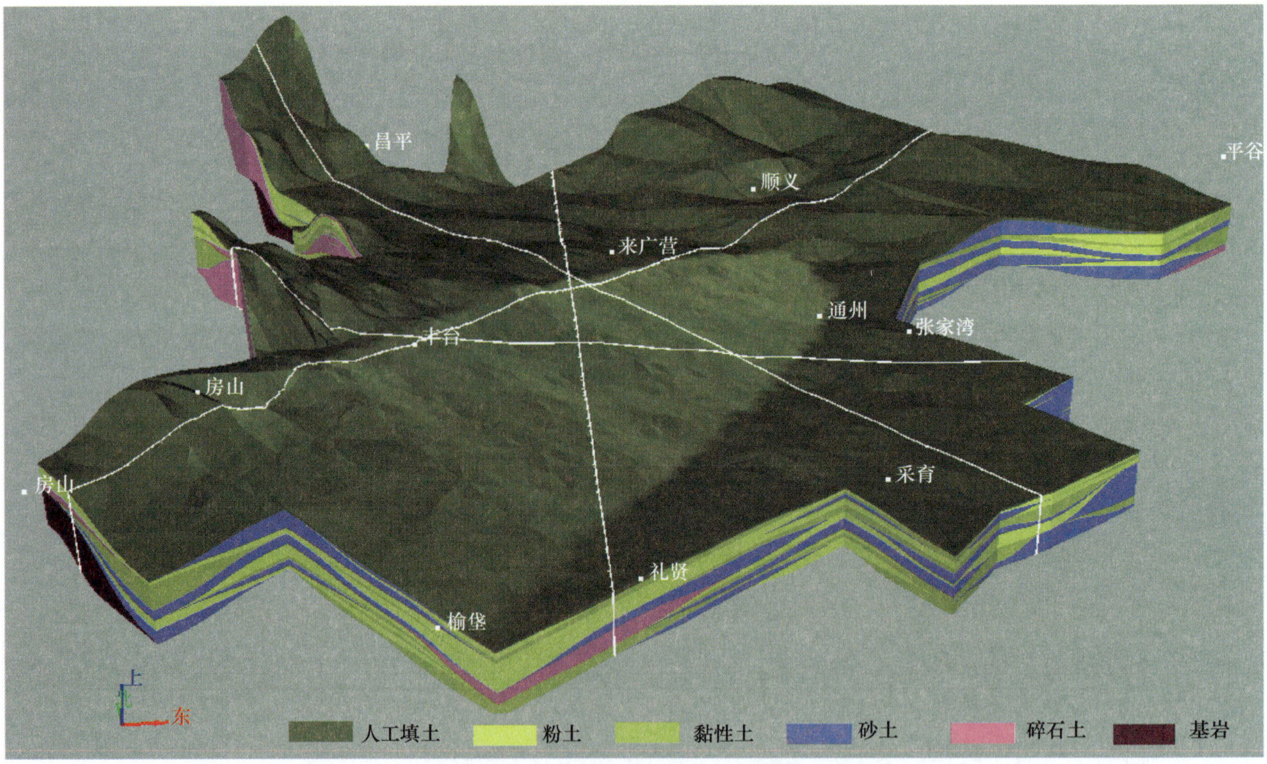

图44-2　平原区地质结构三维模型

Fig. 44-2　The 3D model of geological structure in plain area

二、平原区浅表层工程地质分区

通过对区域地壳稳定性、地貌特征、浅表层的承载力特征、浅表层的压缩性特征、地下水条件、不良地质作用等要素综合分析，从地质学的角度开展了平原区适宜建设的工程地质分区（表44-1，图44-3）。

表 44-1 北京市平原区工程地质分区简表

Table 44-1 Brief table of engineering geological zoning in plain area of Beijing

区	亚区	段
大清河水系堆积工程地质区［Ⅰ］	房山较稳定工程地质亚区［Ⅰ$_{1}$］	山前工程地质条件良好段［Ⅰ$_{1}^{1}$］
		沿山工程地质条件较好段［Ⅰ$_{1}^{2}$］
		南部工程地质条件较差段［Ⅰ$_{1}^{3}$］
永定河水系堆积工程地质区［Ⅱ］	永定河冲积扇后缘较稳定—较不稳定工程地质亚区［Ⅱ$_{1}$］	厚层非黏性土工程地质条件良好段［Ⅱ$_{1}^{1}$］
		沿山工程地质条件良好段［Ⅱ$_{1}^{2}$］
	永定河阶地较稳定—较不稳定工程地质亚区［Ⅱ$_{2}$］	城郊二级阶地工程地质条件较好段［Ⅱ$_{2}^{1}$］
		城区填土工程地质条件较差段［Ⅱ$_{2}^{2}$］
		海淀洼地工程地质条件差段［Ⅱ$_{2}^{3}$］
	永定河冲积平原较稳定工程地质亚区［Ⅱ$_{3}$］	冲积平原中上部工程地质条件较好段［Ⅱ$_{3}^{1}$］
		冲积平原中下部工程地质条件较差段［Ⅱ$_{3}^{2}$］
		低位平原工程地质条件差段［Ⅱ$_{3}^{3}$］
北运河水系堆积工程地质区［Ⅲ］	温榆河冲积扇后缘较不稳定工程地质亚区［Ⅲ$_{1}$］	沿山工程地质条件良好段［Ⅲ$_{1}^{1}$］
	温榆河冲积平原较稳定—不稳定工程地质亚区［Ⅲ$_{2}$］	河间地块工程地质条件较好段［Ⅲ$_{2}^{1}$］
		山前二级阶地工程地质条件差段［Ⅲ$_{2}^{2}$］
		现代河谷边缘工程地质条件差段［Ⅲ$_{2}^{3}$］
潮白河水系堆积工程地质区［Ⅳ］	潮白河冲积扇后缘较稳定工程地质亚区［Ⅳ$_{1}$］	沿山工程地质条件良好段［Ⅳ$_{1}^{1}$］
		冲积扇后缘工程地质条件较好段［Ⅳ$_{1}^{2}$］
	潮白河冲积平原较稳定—较不稳定工程地质亚区［Ⅳ$_{2}$］	山前二级阶地工程地质条件较好段［Ⅳ$_{2}^{1}$］
		河间地块工程地质条件较好段［Ⅳ$_{2}^{2}$］
		山前低洼地带工程地质条件较差段［Ⅳ$_{2}^{3}$］
		冲积平原工程地质条件较差段［Ⅳ$_{2}^{4}$］
		下游低位平原工程地质条件差段［Ⅳ$_{2}^{5}$］
泃河水系堆积工程地质区［Ⅴ］	泃河冲积平原较稳定工程地质亚区［Ⅴ$_{1}$］	沿山工程地质条件良好段［Ⅴ$_{1}^{1}$］
		山前二级阶地工程地质条件较好段［Ⅴ$_{1}^{2}$］
	泃河冲积平原较不稳定工程地质亚区［Ⅴ$_{2}$］	泃河两岸洼地工程地质条件较差段［Ⅴ$_{2}^{1}$］
剥蚀构造工程地质区［Ⅵ］	孤山、残丘	

（一）大清河水系堆积工程地质区（Ⅰ）

该区位于北京平原区的西南部。东部基本以现永定河冲积扇西部边缘为界，西部则以山区平原区界线为界。本区仅含房山较稳定工程地质亚区（Ⅰ$_{1}$），分为 3 个工程地质段。

（1）山前工程地质条件良好段（Ⅰ$_{1}^{1}$）

主要分布在房山区东部平原区的中部，曹章、交道、窦店、大紫草坞、赵各庄一带，位于大石河Ⅱ级阶地上。地形坡度 1‰~5‰，地面起伏变化较大。

地基土体上部为黏性土，其下为含砾石、碎石的黏性土。本段土体承载力较高，为工程地质条件良好段，但由于地处山前，地基土多具有典型的黄土湿陷性状。地下水埋深变化较大，多在 5~15m 之间。

（2）沿山工程地质条件较好段（Ⅰ$_{1}^{2}$）

主要位于平原区西南部山前地带，北起丰台区长辛店，南至房山区张坊。地形坡度 1‰~3‰，地形起伏变化较大。

丰台基岩台地区位于崇各庄—长辛店一带，基底基岩为白垩系和古近-新近系，以残丘台地形式出

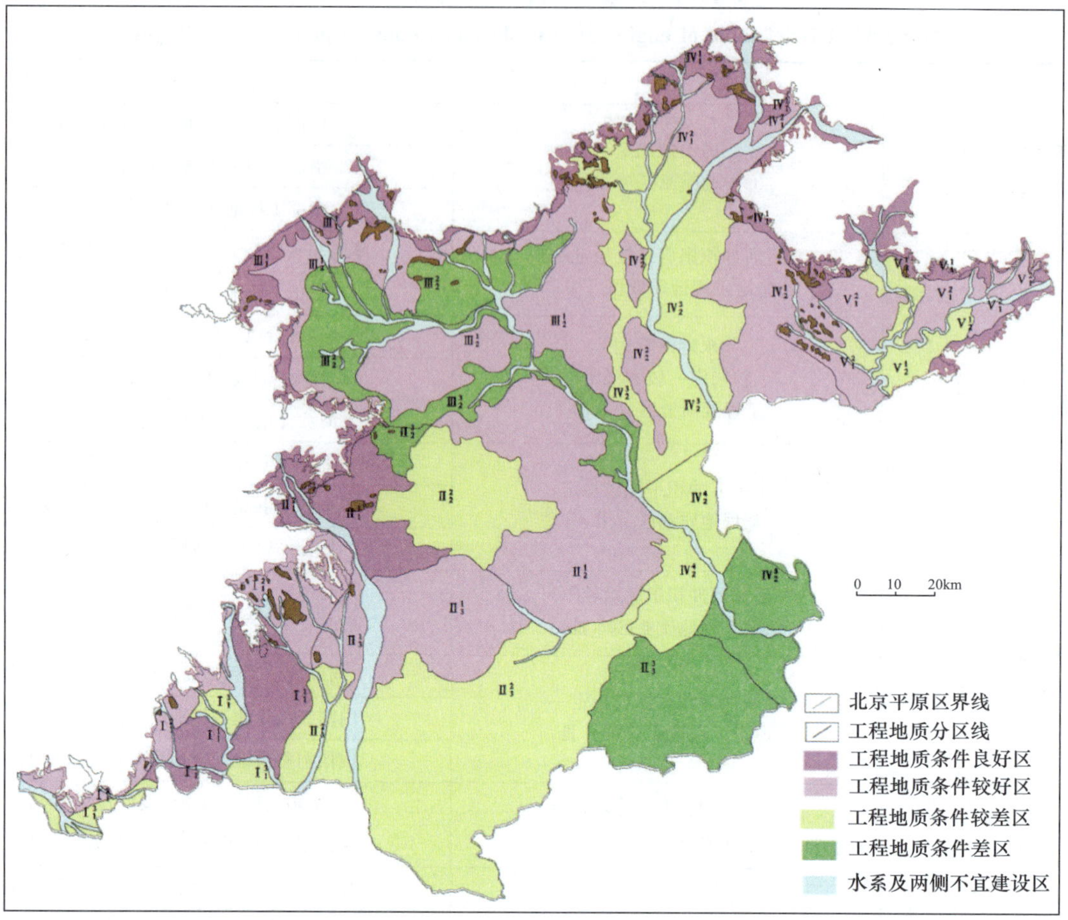

图 44-3　平原区工程地质区划示意图

Fig. 44-3　The schematic diagram of engineering geological zoning in plain area

现，基岩与第四系松散层间隔出露，海拔 50～100m。地基土体多呈双层结构，上部多为黏性土，层薄，厚度小，多改造成平缓梯田；下部多为坡残积碎石土，土层不均一，密度较差，局部具软湿陷性，工程地质性质变化较大。地下水埋深无规律，变化很大。

张坊、西甘池、房山南部等地多为山前斜坡和黄土类土台地。上部土层为黄土类黏性土夹砂卵石透镜体，下部为红色黏性土和基岩，工程地质条件较好，其主要工程地质问题是起伏变化较大的地形坡度和山前红土的胀缩性。

（3）南部工程地质条件较差段（I_1^3）

主要分布在拒马河口、原房山县城南和琉璃河等 3 处。地面标高变化大（30～70m）。表层为新近沉积黏性土，局部含碎石或泥炭；下部土体为砂砾石层，厚度变化较大，地下水位埋深较浅（5～10m），本段地质条件较差。

（二）永定河水系堆积工程地质区（Ⅱ）

位于北京平原区的中部，西部基本以大石河为界，东部则以南口-孙河断裂为界。本区分为 3 个工程地质亚区，并在此基础上分为 8 个工程地质段。

1. 永定河冲积扇后缘较稳定—较不稳定工程地质亚区（II_1）

（1）厚层非黏性土工程地质条件良好段（II_1^1）

主要分布在石景山区约古城、丰台区的黄土岗等地。地形由西北向东南降低，坡度 1‰～5‰，地面标高 80～45m。

地基土体以砂卵石为主，排水条件良好，压缩性小，地下水埋藏较大，是工程地质条件良好段。但北辛安一带由于采石作用形成巨大的砂石坑，是工程建设应该注意并采取相应措施处理的地段。

（2）沿山工程地质条件良好段（II_1^2）

主要分布在永定河出口处，地面高差较大，山前扇地、台地边缘呈锯齿状，宽度较窄。地基土体上部为粉质黏土，下部为含砂卵石层，地下水埋深较大，是工程地质条件较好段。

2. 永定河阶地较稳定—较不稳定工程地质亚区（II_2）

（1）城郊二级阶地工程地质条件较好段［II_2^1］

主要分布在朝阳区来广营、东坝、管庄和通县麦庄、双桥一带。地势平坦，坡度为1‰，地面标高45~20m，为永定河冲积扇中下部。局部池塘分布较多。地基土体以黏性土与非黏性土互层为主，为多层结构土体。地下水位5~15.0m，局部小于2m。本段部分地区为地面沉降中心地区且东南部存在液化砂土和软土。

（2）城区填土工程地质条件较差段［II_2^2］

大致以三环划界，形态不规则，人工改造已使旧地形面目全非，不少沟、河、坑、渠已被淹埋，古河道变化也较大。地形坡度大致为1‰~3‰，地面标高55~25m，市中心区为45m。地基土体以多层结构为主，土体上部为人工填土层，下部为一般第四纪沉积的黏性土和砂卵石层，局部有淤泥层。城区的人工填土虽然堆积时间相对较长，但仍存在不均匀沉陷问题，因此不宜直接作为一般建筑的天然地基持力层。

（3）海淀洼地工程地质条件差段［II_2^3］

主要分布在万泉庄、北坞村、挂甲屯、清河一带，为永定河故道之一，是著名的山前洼地和风景区。地基土体为多层结构。地基土体上部为黏性土及淤泥质粉质黏土，富含淤泥质和泥炭，下部为河床相砂卵石，埋深不等。本段工程地质条件差，主要工程地质问题是软土地基。

3. 永定河冲积平原较稳定工程地质亚区［II_3］

（1）冲积平原中上部工程地质条件较好段［II_3^1］

主要分布在房山区长阳、丰台区南范、大兴区黄村一带，为永定河冲洪积扇中部。地形平缓，高程45~35m，坡度1‰~2‰。地基土层以多层结构为主。本段工程地质条件较好，但局部存在液化砂土。

（2）冲积平原中下部工程地质条件较差段［II_3^2］

主要分布在房山区小清河两岸和永定河泛滥平原区，地面标高40~20m，坡降1‰左右，地面平坦，偶有砂丘起伏。地基土体以多层结构为主，黏性土中普遍夹有砂层。本段工程地质条件较差，液化土层几乎覆盖全区。地面砂化虽经治理，尚存在面积不小的砂地，著名的赤鲁砂丘就在该段。

（3）低位平原工程地质条件差段［II_3^3］

主要位于大兴区采育和通县永乐店一带。地面标高8.8~20m，坡降小于1‰。地表有砂地及砂丘分布，柴厂屯—敖硝营是著名的低洼地带。地基土体以多层结构为主，多为新近沉积土层，局部有淤泥质土和泥炭分布。本段工程地质条件差，大部分地区砂土液化，地基土体压缩性较高，地震时亦易于产生震陷。

（三）北运河水系堆积工程地质区［Ⅲ］

本区西部和北部以山区平原界限为界、南部边界基本沿永定河清河古道北岸为界，东部则大致沿黄庄-高丽营断裂东北部为界。本区分为2个工程地质亚区，并在此基础上分为4个工程地质段。

1. 温榆河冲积扇后缘较不稳定工程地质亚区［III_1］

本亚区工程地质段为沿山工程地质条件良好段［III_1^1］

主要分布在海淀区和昌平区山前地带，呈条带状，由各山区沟谷口处的洪冲积扇和台地组成，宽度1~6km，地形坡度较大，一般为1‰~5‰，地面标高55~260m，最高处位于白羊城一带。地形由山前向平原降低，地面高差变化大，冲沟发育，切割较深，一般深达3m。在河流及冲沟两则形成陡壁，可高达20m。

近山土体以风化残积碎石、黏性土为主，在台地及洪积扇上部，表层出露次生黄土，局部地区为红土，其下部为卵砾石及黏性土互层。

本段工程地质条件良好，但地形较复杂、沟壑纵横，土具湿陷性和膨胀性。

2. 温榆河冲积平原较稳定—不稳定工程地质亚区［III_2］

（1）河间地块工程地质条件较好段［III_2^1］

南邵、南口农场、白虎涧、北安河一带位于洪积扇前缘地带。坡度1‰～3‰，地面标高60～100m。沿河流出现陡岸及沟壑。表层岩性多为新近沉积的黏性土，土层薄，其下为卵砾石层，厚度较大。地下水埋深变化较大，含水层埋深10～40m不等。

六里屯、回龙观、白庙村、张喜庄一带位于Ⅱ级阶地上，地形较平坦，地面标高30～50m，坡度1‰～3‰。地基土体多为黏性土，地下水位埋深5～10m，在东三旗附近存在液化砂土区，在六里屯、东北旺等西部及西南部有软土分布。

（2）山前二级阶地工程地质条件差段［III_2^2］

主要位于温榆河中上游及Ⅱ级阶地，地势相对低洼，坡度1‰～3‰。地基土体变化较大，黄土类土及淤泥质土广泛分布，地基承载力较低，地下水位较浅，工程地质条件较差，存在软土地基、砂土液化、地裂缝等地质灾害。

（3）现代河谷边缘工程地质条件差段［III_2^3］

分布在清河—温榆河—小中河两侧的超河漫滩地带。地面标高20～30m，坡度1‰～3‰。地基土体为多层结构，上部土体为新近沉积的细砂及粉质黏土，多处含有泥炭层，下部为砂卵石夹黏性土。工程地质条件较差，主要地质灾害为砂土液化和地裂缝，此外还应注意防洪。

（四）潮白河水系堆积工程地质区［Ⅳ］

该区位于北京平原区的东部，基本呈条状南北向展布，由北往南逐渐变窄。西部大致以南口–孙河断裂为界，东部则以二十里长山断裂断裂为界，南部则基本以潮白河为界。本区分为2个工程地质亚区，并在此基础上分为7个工程地质段。

1. 潮白河冲积扇后缘较稳定工程地质亚区［IV_1］

（1）沿山工程地质条件良好段［IV_1^1］

主要分布在山前地带，地形变化较大，黄土台地常形成陡坎，高度达10m左右。扇地及台地部沟发育。地下水位埋深较大，供水条件差。地基土体常呈双层结构，上层为黄土类黏性土，下层多为碎卵石，工程地质条件良好，但地形变化大，黄土类土具有湿陷性。

（2）冲积扇后缘工程地质条件较好段［IV_1^2］

主要分布在密云大辛庄、韩各庄、西田各庄、双井、统军庄和怀柔的胜利、范各庄一带。地貌上多为潮白河Ⅰ级阶地。地面标高变化较大，为50～70m，坡度5‰左右。地基土体多为黏性土与砂砾石层互层。本段渗透条件较好，是地下水补给区之一。

2. 潮白河冲积平原较稳定—较不稳定工程地质亚区［IV_2］

（1）山前二级阶地工程地质条件较好段［IV_2^1］

主要分布在顺义李各庄、杨各庄、孙各庄、北务一带，自东北向西南微倾，地面标高20～40m，坡度1‰～3‰，局部坡降稍大（主要为山前及二十里长山附近）。地面有砂地分布。土体呈多层结构，上部为黄土类土，下部土体有软土出现，但分布面积很小。本段工程地质条件较好，但存在软基和液化砂层。

（2）河间地块工程地质条件较好段［IV_2^2］

主要分布在顺义县的牛栏山、平各庄、李家桥一带，为潮白河、小中河之间的Ⅱ级阶地，呈南北向长条状，地形在阶地面上较平整，坡度1‰～3‰，地面标高45～20m。土体为多层结构，10m深度左右有细砂层，为可液化区。

（3）山前低洼地带工程地质条件差段［$Ⅳ_2^3$］

主要分布在北房、木林、南彩、沿河一带，潮白河、小中河、箭杆河的Ⅰ级阶地，地势低平，沟谷纵横交错，地面标高45~20m，坡降1‰~3‰，局部地区有砂丘。地基土体以多层结构为主，表层为新近沉积的黏性土，下部为细砂层及卵石层和粉质黏土，含少量有机质。工程地质条件较差，存在软基和液化砂土。

（4）冲积平原工程地质条件差段［$Ⅳ_2^4$］

主要分布在宋庄、胡各庄、张家湾、牛堡屯一带，为潮白河与北运河间Ⅰ级阶地。地形平坦，偶有砂地及砂丘，地面标高20~30m，坡度1‰。此段与［$Ⅳ_2^3$］工程地质段特征相似，不同处在于震害稍轻，可液化砂层分布面积较大。

（5）下游低位平原工程地质条件差段［$Ⅳ_2^5$］

主要分布在通州侉子店、济县、西集、马头、觅子店一带，地面积水洼地、砂地、砂丘零星分布。地面标高10~20m，坡度1‰。土体属多层结构，工程地质条件差，其性质与［$Ⅱ_3^3$］工程地质段相似。

（五）泃河水系堆积工程地质区［Ⅴ］

该区西部以二十里长山断裂为界，北部与东部则以山区边界为界。分为2个工程地质亚区，并在此基础上分为3个工程地质段。

1. 泃河冲积平原较稳定工程地质亚区［$Ⅴ_1$］

（1）沿山工程地质条件良好段［$Ⅴ_1^1$］

主要分布在峪口、黄松峪、豹峪等山前地带。宽0.5~5km不等，坡度1‰~3‰。地面标高35~100m。由山区向平原中心斜倾，局部地区有3~5m高陡坎。

地基土体表层为黄土类土夹碎石层，下部土层为粉质黏土夹砂及碎石，工程地质条件良好，但地形变化大，黄土类土具湿陷性。

（2）山前二级阶地工程地质条件较好段［$Ⅴ_1^2$］

主要分布在王辛庄、大兴庄一带，为泃河的Ⅱ级阶地和河间地块。地形由四周向中心缓倾斜，坡度1‰~3‰，局部达1‰，地面标高35~60m。地基土体以单层黏性土结构为主，夹砂层，工程地质条件较好，但局部地区存在软基和砂土液化问题。

2. 泃河冲积平原较不稳定工程地质亚区［$Ⅴ_2$］

本亚区工程地质段为泃河两岸洼地工程地质条件差段［$Ⅴ_2^1$］

主要分布在平谷东高地、英城一带，为泃河Ⅰ级阶地。坡度1‰，地面标高30~40m。地基土体以粉质黏土为主，含粉砂及有机质。淤泥质土和泥炭分布点较多，土体具中等压缩性。工程地质条件较差，存在砂土液化、地裂缝及软土震陷等灾害。

（六）剥蚀构造工程地质区［Ⅵ］

主要分布在山区与平原区交接附近，呈不连续明显转折的系列状较圆滑的孤山、残丘，主要有：老山、玉泉山、龙山，卧虎山、九里山、大汤山、牛栏山、西智山、唐指山、庞里山等。上述地区工程地质条件良好，但建筑物布局时宜考虑地形或坡度的限制。

第二节　平原区浅表层三维地质结构分区域研究

一、五环内浅表层三维地质结构

北京市五环内地区，西距太行山脉约50km，北距燕山山脉约50km，东南距渤海湾约100km，地势西北略高，东南稍低，地面标高30~70m，平均坡度约为0.6‰。位于永定河古河道二级阶地，浅表层 30m 范围内地层主要为第四纪松散土体，其地层按沉积年代可分为人工堆积层、新近沉积层和一般第四纪沉积层（图 44-4，图 44-5）。

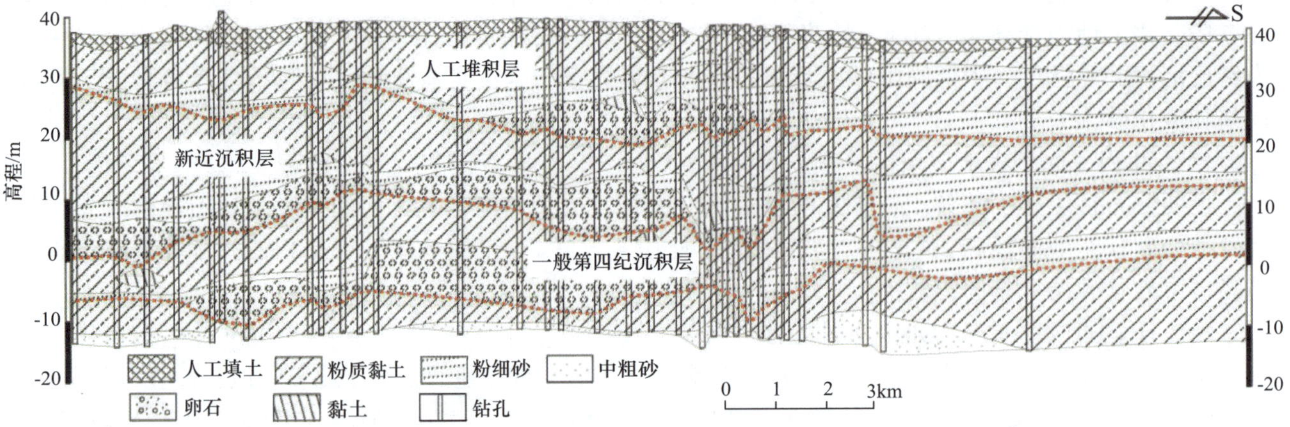

图 44-4　五环内浅表层地层剖面

Fig. 44-4　The shallow strata profiles inside the Fifth Ring Road of Beijing

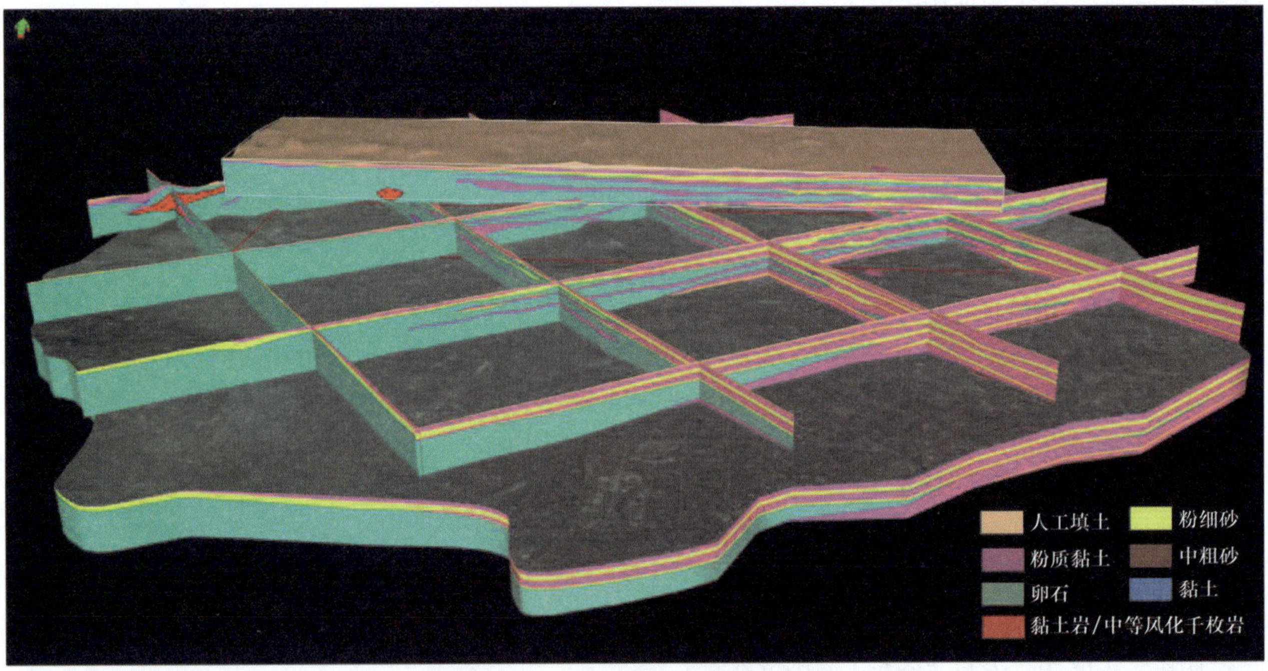

图 44-5　五环内地质结构三维模型

Fig. 44-5　The 3D model of geological structure inside the Fifth Ring Road of Beijing

1. 人工堆积层

由人类活动堆积而成，土质均匀性差、强度低、压缩性高，常具湿陷性。主要分布在人口密集区，在原人工采坑、垃圾填埋场和湖、塘及低洼地区域也有零星成片分布。一般厚 1.0~5.0m，局部 3.0~7.0m，最大约 13m（三环至四环局部采坑）。表层为杂填土，下伏素填土。

2. 新近沉积层

主要地层为粉质黏土层、粉土层、黏土层、粉细砂层、中粗砂层、圆砾-卵石层、角砾-碎石层、漂石层、块石层。

3. 一般第四纪沉积层

主要地层为粉土层、粉质黏土层、黏土层、粉细砂层、中粗砂层、圆砾-卵石层、角砾-碎石层、漂石层、块石层。

二、顺义区浅表层三维地质结构

顺义区位于北京平原的东北部，地处燕山南麓，华北平原北端，属潮白河冲积扇中下段，主要由发源于北部山区的潮白河、温榆河等河流联合冲积而形成的冲洪积平原为主，辅以中低山及零星分布的基岩残丘，其中平原面积占95.7%。地势总体上北高南低，坡度为1‰～3‰。区内平原海拔高度25～45m，平均海拔约35m。

区内30m以内浅表层为多层土体结构，以黏性土为主，全区除潮白河河道外，其他区域上部均为黏性土，中下部有少量砂土，底部均以碎石土为主。

三、通州区浅表层三维地质结构

通州区地处永定河、潮白河及温榆河冲洪积平原，地势西北略高、东南稍低，地面标高16～29m，坡度为0.6‰左右，地势相对平坦。主要发育有永定河Ⅱ级阶地、潮白河冲积平原、潮白河Ⅰ级阶地和温榆河Ⅰ级阶地。受古河道变迁的切割，北部有起伏不平的坡岗地块；东部北运河与潮白河之间地区，由于近代河流泛滥堆积作用，其地势表现为近河床高，远河床低，形成顺河延伸的条形洼地；南部及西南部为永定河作用地区，地势在西北高东南低的基础上呈现由东北至西南向上的波状起伏之势。

浅表层地层以粉土、黏性土和砂类土为主，局部夹碎石类土，具有典型的多层土体结构；地表下30m深度范围内存在两层较厚的砂类土，中间夹黏性土层，且从西到东砂类土增多变厚，埋深相对变浅（图44-6）。

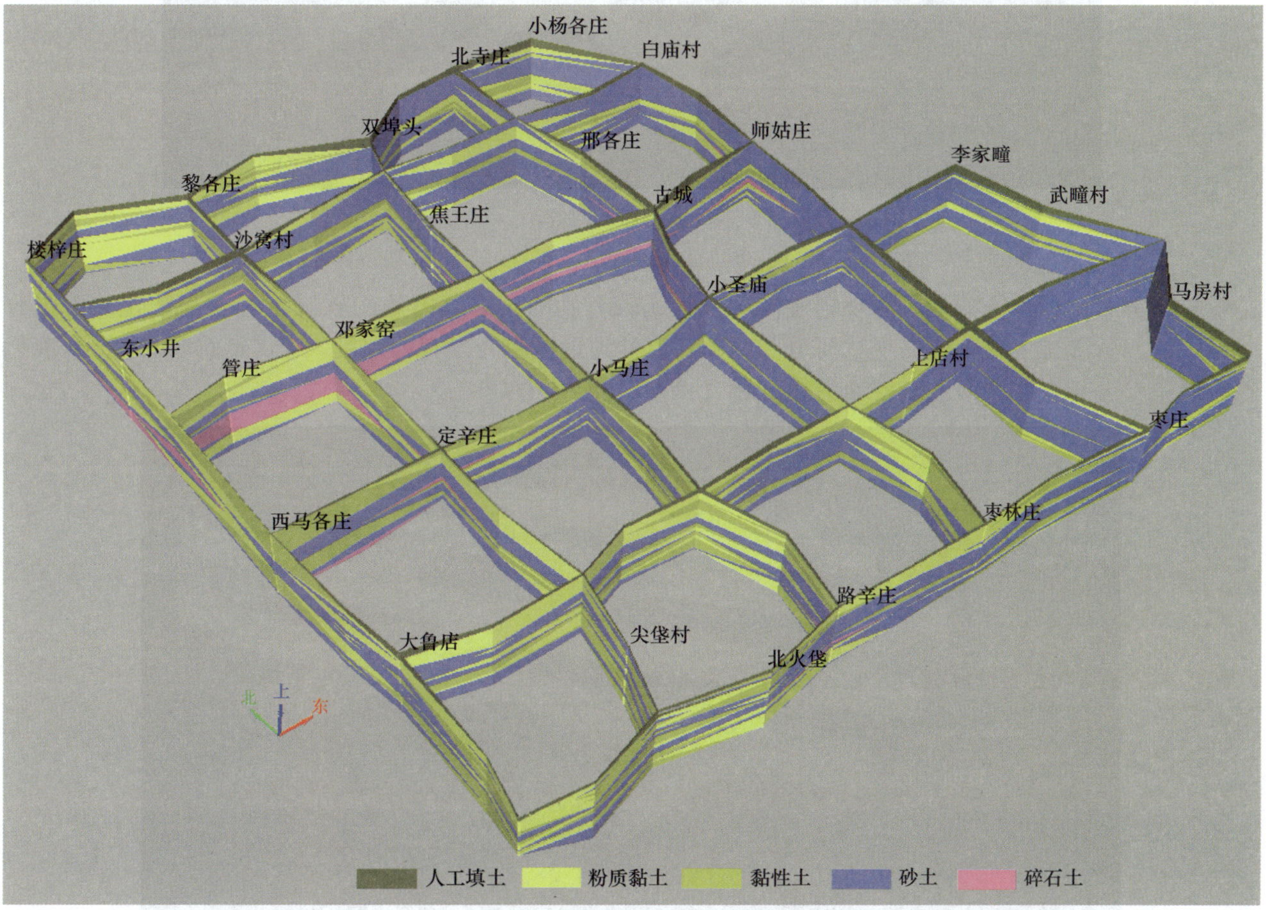

图44-6 通州区浅表层地层结构三维栅格图

Fig. 44-6 The 3D grid map of shallow stratigraphic structure in Tongzhou District

通州区广泛存在厚度不等的软黏土地层，持力层较低，属于规划建设的不良地质体，尤其在三间房西北、大辛庄以东小部分地区，软黏土厚度超过 7m。

四、昌平区浅表层三维地质结构

昌平区整体地势上由北北西、向南南东地形呈缓倾斜状，高程从约 300m 降至约 35m，地面坡度约 2‰。地处南口山前冲洪积扇、龙山山前洪积扇、温榆河冲积扇结合地带，具有典型的多层岩土体结构。

浅表层 30m 深度范围内地层分布来看，除人工填土层外，主要存在两层较厚的粗粒土和两层较厚的细粒土，呈典型的互层状态。除古河道和砂丘、砂地分布区为粉细砂外，各大层均主要以黏性土、粉土为主，不同层都有不同厚度的卵砾石层。粗粒土大层中夹着细粒土薄层和透镜体，细粒土大层中亦混夹着粗粒土薄层和透镜体，充分反映了洪冲积平原的沉积韵律（图 44-7）。

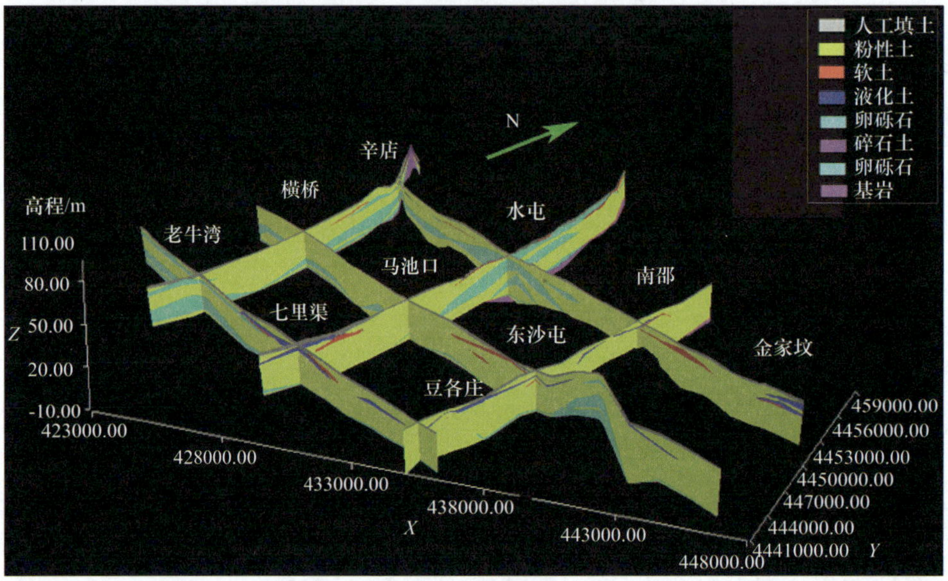

图 44-7 昌平区浅表层地层结构三维栅格图

Fig. 44-7 The 3D grid map of shallow stratigraphic structure in Changping District

总体上看，昌平区的地层对规划建设来说是比较有利的，承载力地的软黏土和容易液化的砂土分布不广泛，软黏土主要分布在现状河道及两侧，以沙河地区较为发育，厚度大于 5m（图 44-8）。液化砂土层主要分布在北沙河、东沙河沿线区域（图 44-9）。

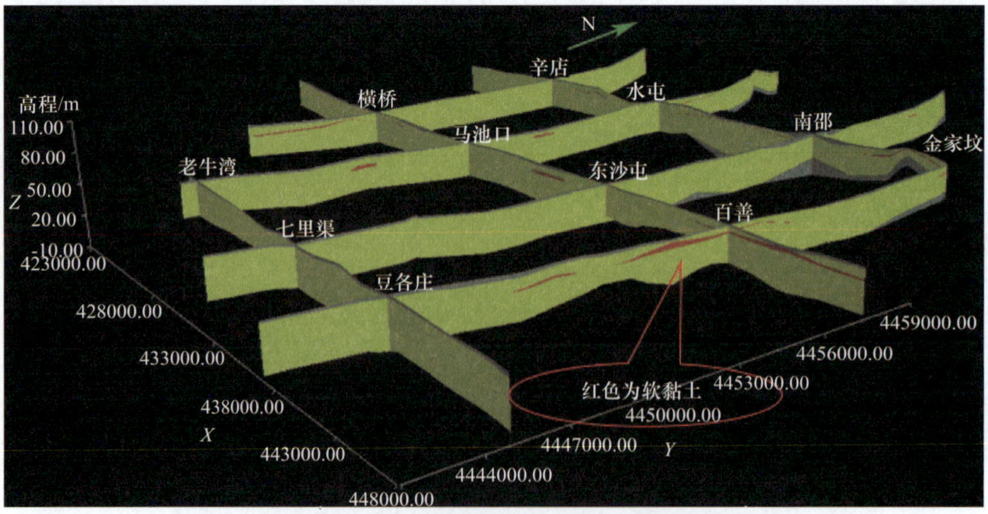

图 44-8 昌平区浅表层软黏土分布三维栅格图

Fig. 44-8 The 3D model of shallow clay layer in Changping District

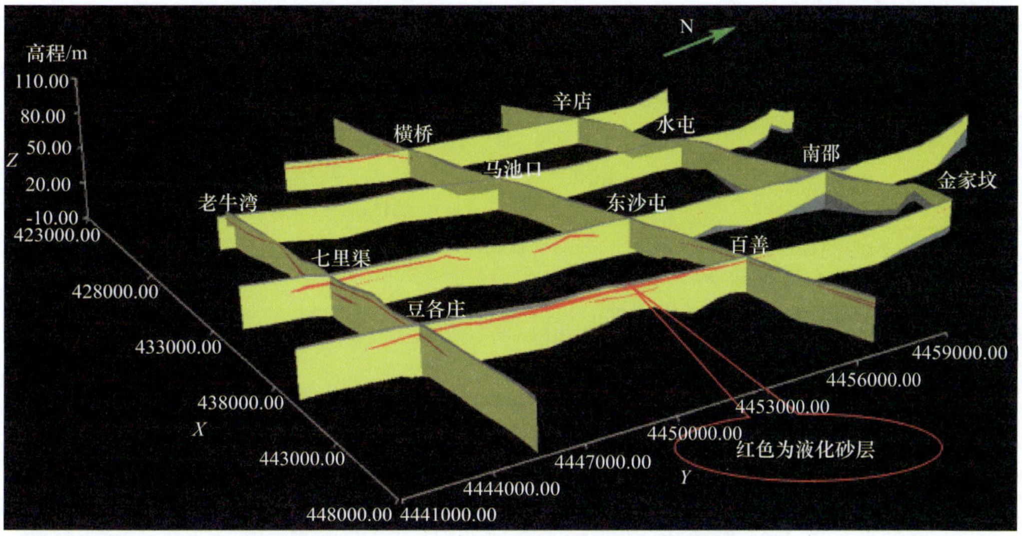

图 44-9　昌平区浅表层液化砂土分布三维栅格图
Fig. 44-9　The 3D model of shallow liquefied sand in Changping District

五、大兴区浅表层三维地质结构

大兴区地处北京平原区永定河洪冲积平原的中下部，地势总体较平坦，西北略高，东南稍低，坡度在1‰~1.5‰之间，地貌主要受永定河古河道变迁的切割作用、风力作用、洪泛作用及人类工程活动等多种作用影响，主要为河流（侵蚀）堆积地貌，其次为风成地貌和人工地貌。主要有一级阶地平原、洼地、河流决口扇地、砂丘与砂地、河床及漫滩、现代河道与古河道、水坑和鱼塘、采土和采砂坑、人工土堆等微地貌单元。

30m 内浅表层地层岩性主要为卵石、砂、黏性土及粉土。由西北向东南，厚度从 40~50m 逐渐增加到 250m 左右，地层颗粒由粗变细，结构由单一卵石层渐变成多层砂砾石与黏性土互层（图 44-10）。

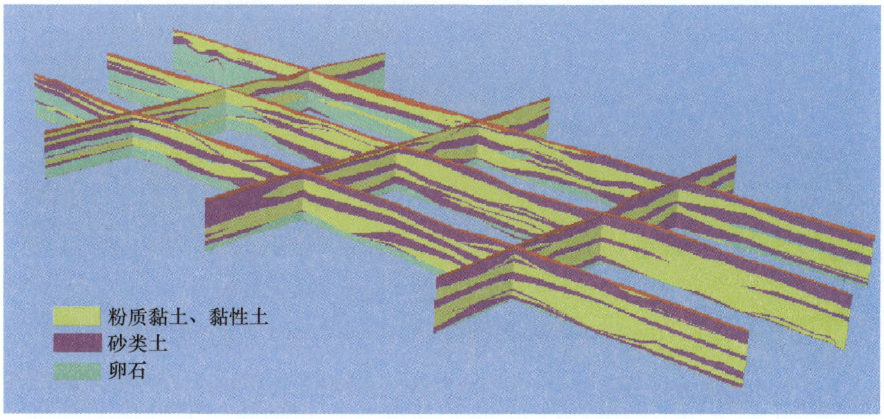

图 44-10　大兴区浅表层地层结构三维栅格图
Fig. 44-10　The 3D grid map of shallow stratigraphic structure in Daxing District

大兴区大面积分布有软黏土和砂土液化层，软黏土在大兴区东部和东北部区域厚度超过 5m，砂土液化层在全区均广泛分布，厚度为 6~8m，对城镇规划和工程应用来说属于不良地层（图 44-11，图 44-12）。

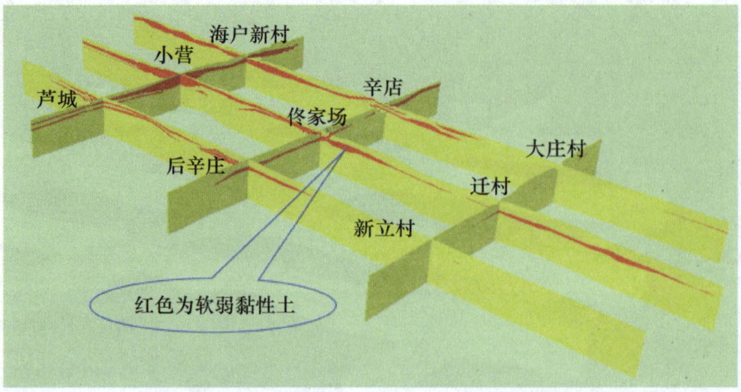

图 44-11 大兴区软黏土分布三维栅格图

Fig. 44-11 The 3D model grid map of clay layer in Daxing District

六、房山区浅表层三维地质结构

房山区地形总体上西北高东南低，海拔 26～1242.7m，猫耳山为本区最高峰，最低处为琉璃河镇立教村。存在山区地貌和平原区地貌两种类型，重点介绍平原区浅表层相关三维地质结构内容。房山平原区由永定河、大石河等河流冲积形成的，平原地势平坦和缓，总体上由西北向东南缓倾斜，平均坡度 1‰。

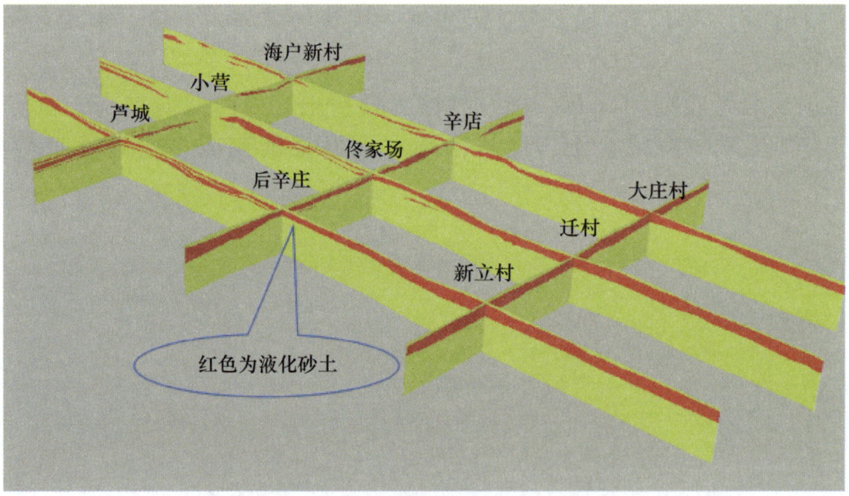

图 44-12 大兴区液化砂土分布三维栅格图

Fig. 44-12 The 3D model grid map of liquefied sand in Daxing District

30m 内浅表层地层主要为粉土、粉质黏土-重粉质黏土、粉细砂、中粗砂、圆砾、卵石层、黏土-重粉质黏土和角砾、碎石。岩土体为多层结构（图 44-13，图 44-14）。

房山区除部分地区存在一定的软黏土和液化砂土外，区域内广泛分布有卵石层，对城镇规划建设和工程应用来说是良好的地质条件。30m 地层范围内有两层卵石，主要分布在大石河沿岸和小清河沿岸，部分地区厚度达 20m（图 44-15）。

七、平谷区浅表层三维地质结构

平谷区位于燕山南麓与华北平原北端的相交地带的平谷盆地，盆地北、东、南三面环山，仅西南

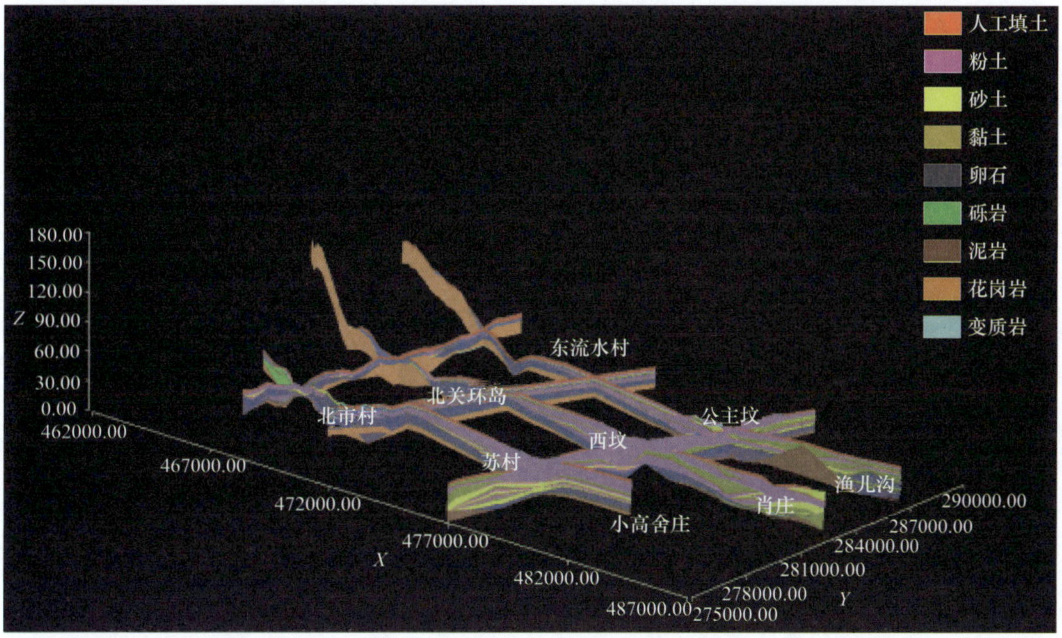

图 44-13　房山区浅表层地层结构三维栅格图

Fig. 44-13　The 3D grid map of shallow stratigraphie structure in Fangshan District

开口与北京平原相连接。工作区地形东北高，西南低，地面标高 22~85m，地形坡度 1.8‰~2.3‰。地势平坦，并有沟河、泇河、小辛寨石河等河流环绕。

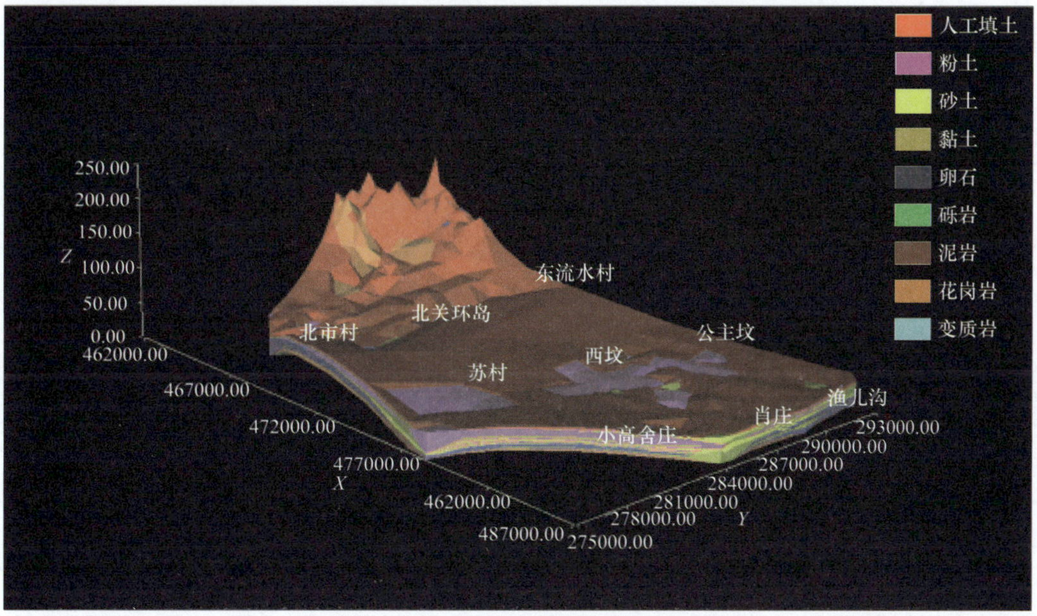

图 44-14　房山区浅表层地质结构三维图

Fig. 44-14　The 3D model grid map of shallow geological structure of Fangshan District

工作区地貌类型可分为低山、残丘、坡洪积台地、洪积扇、冲洪积平原、河床、河漫滩及古河道等地貌单元，另外还有人类活动形成的采砂坑、采石场、垃圾场、人工水塘、河渠等人工地貌。

平谷区北部一部分位于山前洪坡积台地和冲洪积扇上，其余大部分都位于洵河、泇河形成的冲洪积平原，地层具有多层岩土体结构（图 44-16），细粒土（粉土、黏性土）与粗粒土（砂类土、碎石

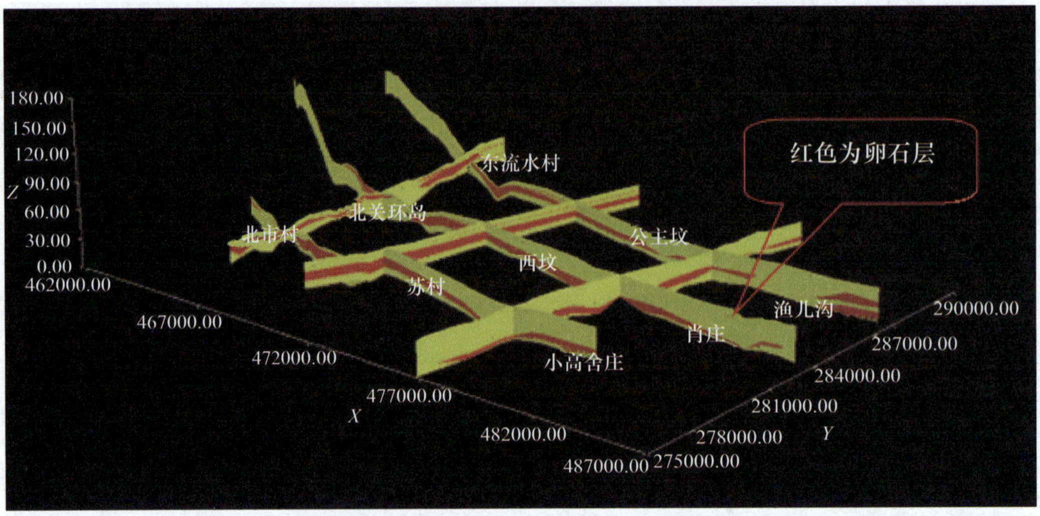

图 44-15　房山区浅表层卵石层分布三维栅格图

Fig. 44-15　The 3D model of grid map of shallow cobble layer in Fangshan District

土）互层。地表下 30m 深度范围内地层从北向南，碎石、卵石的埋深逐渐变大，且颗粒直径逐渐变小，由碎石、卵石逐渐变为细砂；从北向南，黏性土地层厚度加大。卵石层厚度在区南部较薄，北部及东南部靠近山前一带相对较厚（图 44-17）。

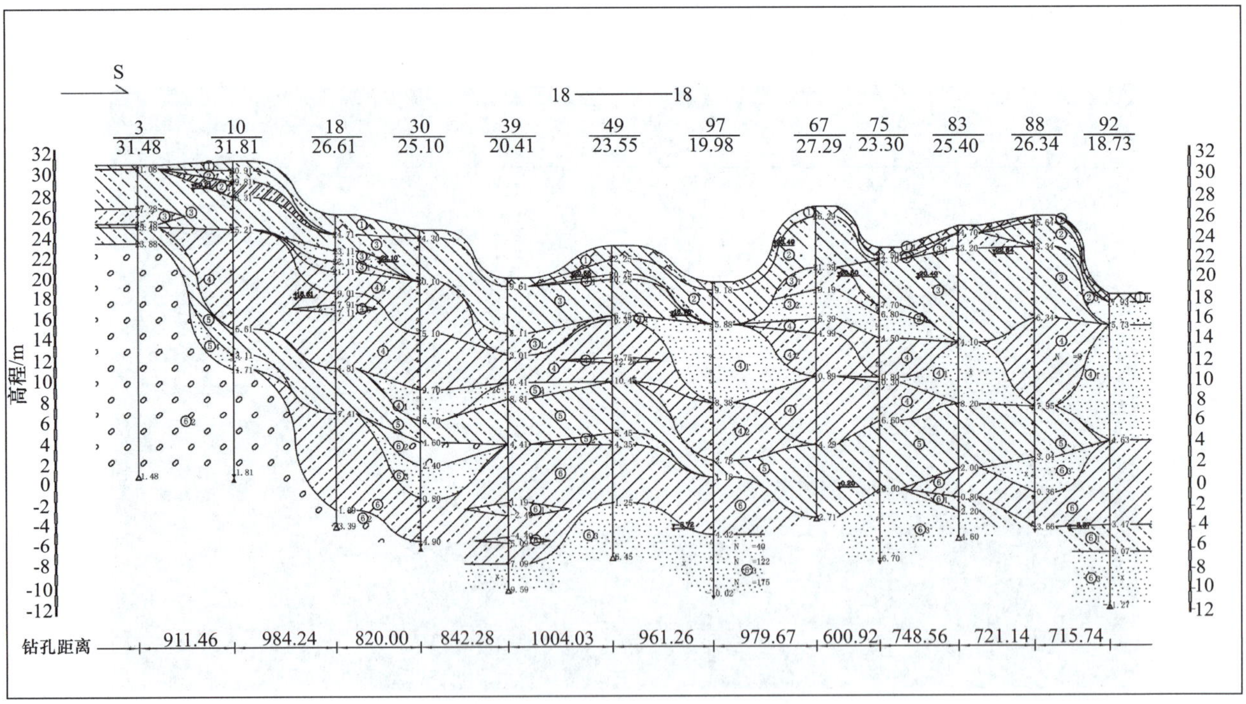

图 44-16　平谷区浅表层地层剖面图

Fig. 44-16　The profiles of shallow strata of Pinggu district

平谷区浅表层新近系沉积的黏土层性质比较软弱，对规划建设属不利地层，厚度 1~5m，在全区普遍分布（图 44-18）。

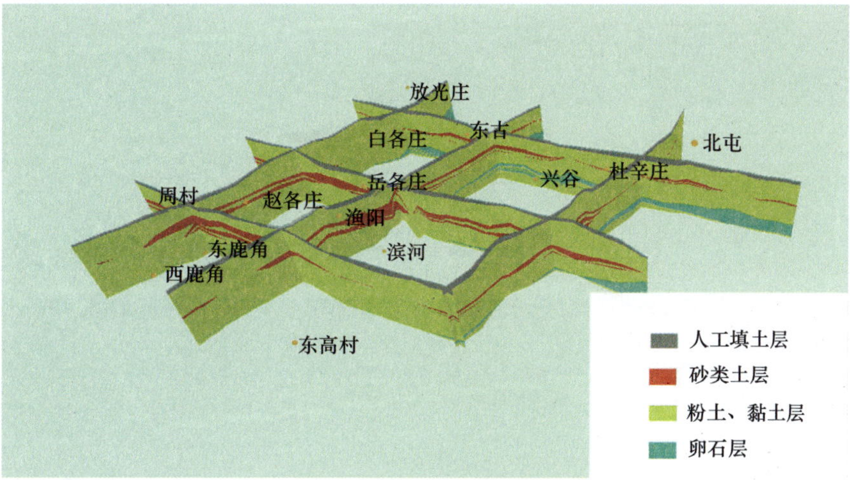

图 44-17　平谷区浅表层地层结构三维栅格图

Fig. 44-17　The 3D grid map of shallow stratigraphic structure in Pinggu District

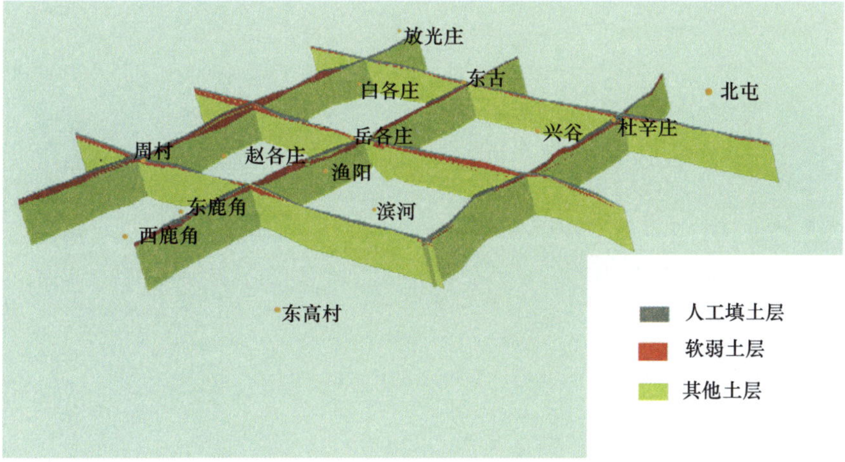

图 44-18　平谷区软黏土分布三维栅格图

Fig. 44-18　The 3D model grid map of soft clay layer in Pinggu District

第十篇

城市地质资源与环境

地质资源与环境，是制约城市发展的重要因素。就北京而言，主要的地质资源包括地下空间资源、地下水资源、固体矿产资源、地热和浅层地热能资源及地质遗迹资源等。城市规划、建设主要受到土壤地质与环境、地质灾害等环境地质条件的影响。城市发展须充分考虑到资源与环境容量，只有在地质环境系统承载能力范围内进行合理规划，才能实现人与自然的和谐可持续发展。

20世纪60年代起，城市地质工作逐渐成为北京地质工作的中心，进入21世纪后城市地质工作得到了快速发展。近年来，北京地区先后开展了多参数立体地质调查、通州等11个规划新城前期区域工程地质勘查、城市活动断裂及地壳稳定性调查、典型区小城镇水工环地质综合调查、城市地质信息管理与服务系统建设、首都地质资源环境承载力监测预警平台建设等工作，综合运用现代勘查技术，开展了城市三维地质结构调查、活动断裂及地壳稳定性调查、水土地球化学环境、工程地质和地质灾害调查，建立了三维可视化城市地质数据管理和服务系统，为城市规划、建设与管理及社会公众信息需求提供服务。开展了城市地下空间资源开发利用战略及可行性研究、地下空间资源调查评价及关键技术研究、地下空间资源安全监测等工作，进一步研究分析了北京地区地下空间资源开发影响因素及变化规律，为城市规划建设、保障城市地质安全提供了参考依据。先后完成了不同比例尺城市供水、农田供水、厂矿供水等水文地质勘查工作，长期开展地下水动态监测、水质监测等工作，为保护地下水资源、防治地下水污染、保障饮水安全提供了有力支撑。开展了北京市矿产资源潜力评价、浅层地温能地热以及再生水热能综合利用规划研究、重要地质遗迹调查、城市地下资源环境三维模型建设等工作，对区域地质资源分布、数量、矿产储量等情况进行了调查。完成了区县地质灾害区划和防治规划，开展了突发地质灾害防治技术服务等工作，为保障城市地质安全提供了基础数据。

在上述城市地质工作的基础上，本志对北京市城市地质资源与环境条件进行了梳理和总结。

第四十五章　地下空间资源

城市化进程使得城市建设用地的供需矛盾愈发尖锐。地下空间资源的合理开发利用则可减轻地表土地供应压力，提高城市交通疏导能力，修复地表自然生态环境，保障城市公共防灾能力，是城市可持续规划与发展的重要环节。

北京地下空间资源开发已进入快速发展阶段，建成面积已达到 $3000\times10^4m^2$。充分利用地下空间集约化城市开发建设，节省了土地资源，缓解了城市发展对土地的占用；维护了地面环境，保护了历史风貌；发展地下轨道交通、停车场，成为了解决北京交通问题重要手段；部分城市功能转入地下，改善了地面环境（石晓冬，2006）。

第一节　地下空间资源特征

一、地下空间资源

地下空间资源是迄今尚未被充分利用的一种自然资源，具有很大的开发潜力。地下空间资源包括3个方面的含义：一是天然存在的资源蕴藏总量；二是在一定技术条件下可供合理开发的资源总量；三是在一定历史时期内可供有效利用的地下空间总量（齐伟，2013）。本志所指城市地下空间资源是指在城市地表以下一定深度范围内，在现有经济条件、地质条件约束下，能开发利用的土层或岩层中天然形成或经人工开发而成的空间（郑桂森等，2017）。

二、地下空间资源特征

（一）地下空间资源有限性

地下空间资源开发利用是随着社会经济的发展，在城市建成区或城市规划区开展，其资源属性的有限性显而易见：其一数量有限，地下空间开发平面受城市布局限制，纵向受经济技术水平限制；其二质量有限，既受地表以下岩土体的性能限制，又受到周边地质环境的影响。

（二）地下空间资源关联性

地下空间资源开发利用在地质体中进行，必然与地质体及赋存其中的地温场、地下水流场、应力场、地下生态系统相关联。一方面，地质条件对地下空间施加着持续的作用，有些甚至极具破坏性；另一方面，其开发利用尤其是大型的线性建（构）筑物也对地质环境可能存在持续影响。

（三）开发利用不可逆性

地下空间由于六面受限，且往往作为地表建筑的承载体，需要更高的坚固性，加之往往单体规模较大，难以进行二次规划和拆除重建，开发利用具有不可逆性。有些不合理的地下建筑不得不废弃，造成有限资源的浪费。

第二节　影响地下空间资源开发利用的地质条件

地下空间资源具备资源属性，开发利用需要查清地下空间三维地质结构，并评估其数量、质量。从地质角度而言，地下空间资源质量是指开发利用深度范围内各类地质要素对城市地下空间资源开发利用的影响程度，以影响程度大小刻画地下空间资源质量的高低。北京平原区冲洪积扇主要由砂砾石、砂、黏土、亚黏土构成，其分布在垂向上总体呈互层状；横向变化较大，以永定河冲积扇为例，从扇顶到扇缘通州，粒度由砂砾石逐渐变为黏土或亚黏土，其厚度从几十米增加至几百米。现有经济技术条件能大规模开发的城市地下空间资源主要分布于此。冲洪积扇的结构特点、物质组成以及其物理力学特性，决定了平原区松散层地下水赋存以及运营，同时对软土分布、砂土液化以及地面沉降等地质灾害起着控制作用，直接影响着地下空间资源的开发利用（蔡向民等，2010），是资源开发基础条件，也是资源具体载体，决定了资源利用整体格局。

一、基础性地质条件

岩土体的结构及属性特征是基础条件，决定了地下空间资源开发利用的总体布局（表 45-1，表 45-2）。冲洪积扇是北京城市地下空间资源开发利用的主要载体，它的组成、结构、构造及物理力学参数性质共同构成了地下空间资源开发利用的基础条件。由于各类型土体形成时代、作用和环境的不同，各自具有不同的结构特性和物理力学参数，对地下空间资源开发利用相关的物理力学参数主要有地基承载力、压缩模量、黏聚力和内摩擦角等（吴立新等，2004）。

表 45-1　不同地层结构特点对地下空间开发利用影响一览表

Table 45-1　The influence of different stratum structure characteristics on the development and utilization of underground space

结构类型	示例	特点	对地下空间资源开发利用影响
单层	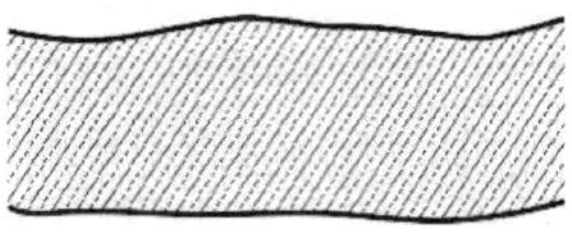	地层均一	施工条件单一，控制难度小，不可预见地质问题少
双层	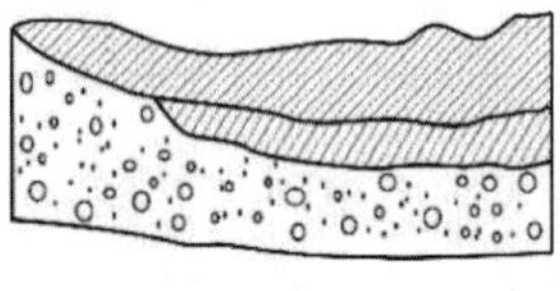	两种地层交互出现，地质均一性较差，可能存在软弱下卧层，施工时可能同时遭遇含水层和隔水层	施工条件较复杂，施工不确定因素较多，可能施工遭遇含水层和隔水层而引发突涌，可能连同多层含水层，引起地下水污染，如果遭遇上软下硬这类复合地层结构，因软弱层排土过多引发地层下沉，造成盾构在线路上偏离
多层	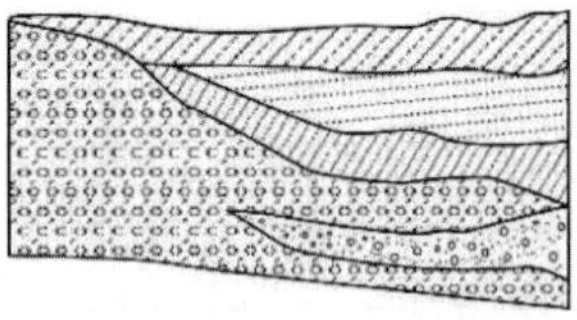	多种地层交互出现，地质体均匀性差，可能存在软弱下卧层，施工时可能同时遭遇多个含水层和隔水层。侧壁受力不均匀，基床系数变异大	施工条件复杂，施工不确定因素多，可能施工遭遇含水层和隔水层而引发突涌，可能连同多层含水层，引起地下水污染，如果遭遇上软下硬这类复合地层结构，因软弱层排土过多引发地层下沉，造成盾构在线路上偏离

表 45-2　不同土体类型对地下空间开发利用影响一览表

Table 45-2　The influence of different soil types on the development and utilization of underground space

土层类别	结构特征及物理参数特点	对地下空间开发的影响
填土	土质不均匀、湿陷性、自重压密性及低强度、高压缩性	可能会引起较大的地层位移和累计地面沉降差异
砂土	透水性强、压缩速度快、内摩擦角大、抗剪强度较高等	在地下水水位高的地区进行地下空间开发时候可能产生涌水或渗漏，塌方、塌陷等地质灾害
粉土	介于砂类土和黏性土之间	在地下水条件不利时可能发生坍塌

续表

土层类别	结构特征及物理参数特点	对地下空间开发的影响
可塑、软塑状黏性土	黏性土中黏粒含量较大，压缩量大，抗剪强度主要取决于黏聚力，内摩擦角较小	变形量大，对地下空间开发影响较大
坚硬、硬塑状黏性土	压缩量小，抗剪强度较大，自稳能力好，内摩擦角较大	变形量较小
卵石	具有空隙大、透水性强、压缩性低、抗剪强度大。但由于透水性强，常使基坑涌水或渗漏	对开挖的成本和运营后的防水要求高。如果存在漂石，则对盾构施工影响很大

资料来源：《岩土工程勘察设计规范》（GB 50021—2001）。

二、约束性条件

地下空间资源开发利用在地质体中进行，地质性质决定地下空间资源的开发利用，影响地下空间资源的地质约束条件主要有活动断裂及地裂缝、水文地质条件、地面沉降、地面塌陷砂土液化、隐伏岩溶塌陷等（吴立新等，2002；表 45-3）。活动断裂及地裂缝、地面沉降等地质因素引起地下设施受损的事件时有发生，且有加剧趋势，如频繁发生的城市道路路面塌陷、管线开裂以及房屋开裂等事件，多数情况下与地质条件有直接关系。

表 45-3　约束条件对地下空间开发利用影响一览表

Table 45-3　The influence of constraints on the development and utilization of underground space

约束条件	影响机理	影响范围
地面沉降	沉降跨梯度带发生不均匀沉降，交汇处土体不稳定；沉降严重地区压缩层厚度变化大	不均匀沉降带分布地区的条带状影响；沉降严重地区的面状影响
活动断裂及地裂缝	活动断裂及地裂缝在地表错动，导致土体形变；活动断裂在地下深部错动，导致岩体形变；活动断裂活动诱发地震	断裂带和地裂缝分布地区的条带状影响，以及地震发生时面状影响
地面塌陷	土体受不同影响因素制约，产生的破坏现象，主要发生在城市建设地区	发生地面塌陷的位置，呈点状分布
砂土液化	可液化土层丧失抗剪强度，成为液体状态，破坏土体结构	可液化层的分布不同，常导致地面开裂、喷水冒砂、地基不均匀沉降等，一般呈点状或面状分布
水文地质条件	地下水的埋深，影响开发深度；地下水类型，影响开发技术的选择；地下水流场方向影响分布规划；地下水的化学性质，对土体和建筑本体，均产生影响	根据不同分布情况产生不同的影响，一般为条带状或面状
隐伏岩溶塌陷	岩溶塌陷易发区土体结构不稳定，在外力作用下破坏土体结构	多为漏斗状至块状，影响范围呈点状或面状分布

（一）地面沉降

地面沉降对地下空间的影响主要体现在跨沉降梯度带进行地下空间建设时，由于不均匀沉降在梯度带的交汇处造成土体不稳定。沉降量的不断累积，会对地下空间设施，特别是线性工程造成影响甚至破坏。近年来，地面沉降发生面积和沉降速率都呈上升趋势，构成了地铁运行和地下管网等诸多设施的安全隐患。

（二）活动断裂及地裂缝

北京平原隐伏断裂比较发育，断裂两盘的位移引发的地面错动和岩土体形变，都会对跨断裂修建或建于其附近地下空间设施造成拉裂、变形甚至破坏。根据北京市多参数立体地质调查报告，黄庄-高丽营断裂已造成近地表地层严重错断。

（三）地面塌陷

北京地区已发生过多起地面塌陷事故。人工填土、流砂的分布，地下管线破损产生的渗流，以及已存在的地下空洞均可造成地面塌陷。塌陷具有隐发性、突发性和群发性的特点，对地下空间设施构成较大威胁。

（四）砂土液化

砂土液化对地下空间资源开发利用产生不利影响。根据研究，地震液化后可能产生大变形，饱水砂土地基强度极度降低，在建筑物荷载或土体自重作用下，地表出现大的垂向或侧向变形的现象，它会对液化区的各种地下结构产生巨大的破坏（刘汉龙等，2002）。

（五）水文地质条件

水文地质条件是影响地下空间开发利用的主要制约条件。地下水水位、水压力及水质等都从不同的方面对其开发利用和维护使用过程产生影响。随着北京近年来地铁的大规模建设，开发深度越来越深，在建设时则可能遭遇承压水的影响，尤其北京地铁主要在土层中进行，围岩稳定性不如岩体，如果在承压水埋深浅、水头压力大的地区，且没有彻底查清承压水位和水压而盲目设计施工，则可能遭遇突涌、施工中断甚至人员伤亡等严重后果，也可能导致整个地铁线路的重新规划和设计。

另外地下水的腐蚀性对地下空间的影响主要通过其对地下建筑的腐蚀作用表现，具有较强腐蚀性的地下水（例如含有较多 HSO_4^-、HCO_3^- 的地下水），对钢筋混凝土有较强的腐蚀性，降低建筑构件的强度，影响安全性和耐久性。

（六）隐伏岩溶塌陷

北京平原部分地区松散层下存在石灰岩、白云岩等容易形成岩溶的岩体。埋藏较浅的隐伏岩溶塌陷对地下空间开发产生影响，主要体现在地下空间开发、建设过程中，对溶洞上覆岩层进行开挖，破坏了溶洞的受力均衡状态，导致溶洞的覆盖层坍塌，并在溶洞上方形成漏斗状土坑，连锁破坏周边建筑物，破坏性极大。北京地区曾经发生过隐伏岩溶塌陷事故，在平原区北部的洼里地区、东部通州等地均有隐伏岩溶存在。

第三节　地下空间资源量

北京城区（主要指北京城市五环以内约 750km² 范围）是地下空间资源开发利用重要区域，本节主要阐述该区域地下空间三维地质结构、数量及质量。

按对地下空间资源开发利用程度的不同，地下空间资源划分为四大类：已开发利用资源、不可开发利用资源、局部可开发利用资源和可充分开发利用资源。

1）已开发利用资源：已有建筑物地基基础影响区域（包括基础、桩基的影响区域和地下室）、地铁及大型单体地下空间建筑（图 45-1）。

$$V_1 = V_1^1 + V_1^2 + V_1^3 \tag{45-1}$$

式中：V_1^1——已开发地铁地下线路及其直接影响区域体积；

V_1^2——根据建筑规范推出的建筑有地下空间体积；

V_1^3——根据人防数据等获得的已开发的城市地下空间资源量的体积。

2）不可开发利用资源：包括水源保护地、文物保护区及严重地质灾害区。

$$V_2 = V_2^1 + V_2^2 + V_2^3 + V_2^4 + V_2^5 + V_2^6 \tag{45-2}$$

式中：V_2^1——文物保护区下限制利用的地下空间资源量的体积；

V_2^2——河流下利用的地下空间资源量的体积；

V_2^3——水源保护地周边一定范围内利用的地下空间资源量的体积；

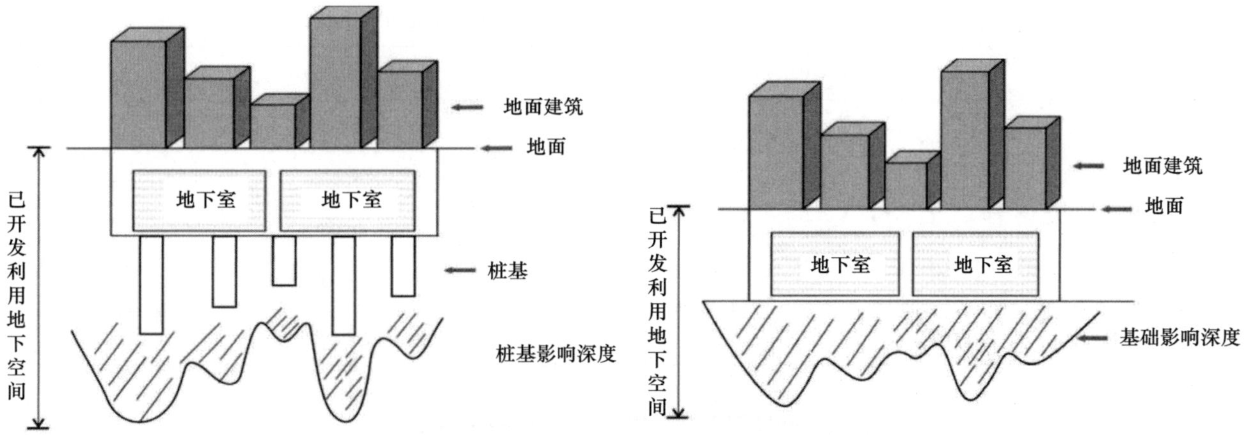

图 45-1 已开发利用空间示意图

Fig. 45-1 Developed and utilized space diagram

左图为桩基及 CFG 桩（水泥-粉煤灰-碎石桩）；右图为天然地基及换填地基

V_2^4——垃圾填埋区周边一定范围内利用的地下空间资源量的体积；

V_2^5——城市绿地周边一定范围内利用的地下空间资源量的体积；

V_2^6——城市道路周边利用的地下空间资源量的体积。

3）局部可开发利用资源：建筑基础底面影响深度以下的深层空间、生态地及景区绿地等。

$$V_3 = V_3^1 + V_3^2 \quad (45-3)$$

式中：V_3^1——建筑基础底面影响深度以下的空间；

V_3^2——生态地、景区绿地范围地下空间资源量体积。

4）可充分开发利用资源：可供人类充分开发利用的地下空间资源。从功能利用角度来讲包括计划改造拆除重建地区、规划新增用地、未开发利用的空地、广场和普通绿地等。

$$V_4 = V_{总} - (V_1 + V_2 + V_3) \quad (45-4)$$

式中：$V_{总}$——地下空间资源量总体积。

依据 1∶1 万地形图，1∶5000 遥感影像图以及 5000 余钻孔作为地下空间资源量开发利用现状利用依据，辅助实地调查验证，对已开发利用地下空间资源量进行统计，总面积约 750km^2，其中地块单元总面积为 523.26km^2，道路面积约 226.74km^2。在计算的地块单元内，地下空间资源量分布现状如图 45-2、表 45-4 所示，假定地下空间建设地下建筑，层高为 5m，则可折算建筑面积当量值如表 45-5 所示。

表 45-4 地下空间资源量

Table 45-4 Resources of underground space

单位：$10^4 m^3$

深度类型	0~10m	10~30m	30~50m	合计
已利用	255950	204390	49720	510060
可充分利用	161680	628150	782030	1571860
局部可开发利用	99120	200960	201750	501830
不可利用	6510	13020	13020	32550
总计	523260	1046520	1046520	2616300

表 45-5 地下空间资源量估算

Table 45-5 Estimation of resources of underground space

单位：$10^4 m^2$

深度类型	0~10m	10~30m	30~50m	合计
已利用	51190	40878	9944	102012
可充分利用	32336	125630	156406	314372
局部可开发利用	19824	40192	40350	100366
不可利用	1302	2604	2604	6510
总计	104652	209304	209304	523260

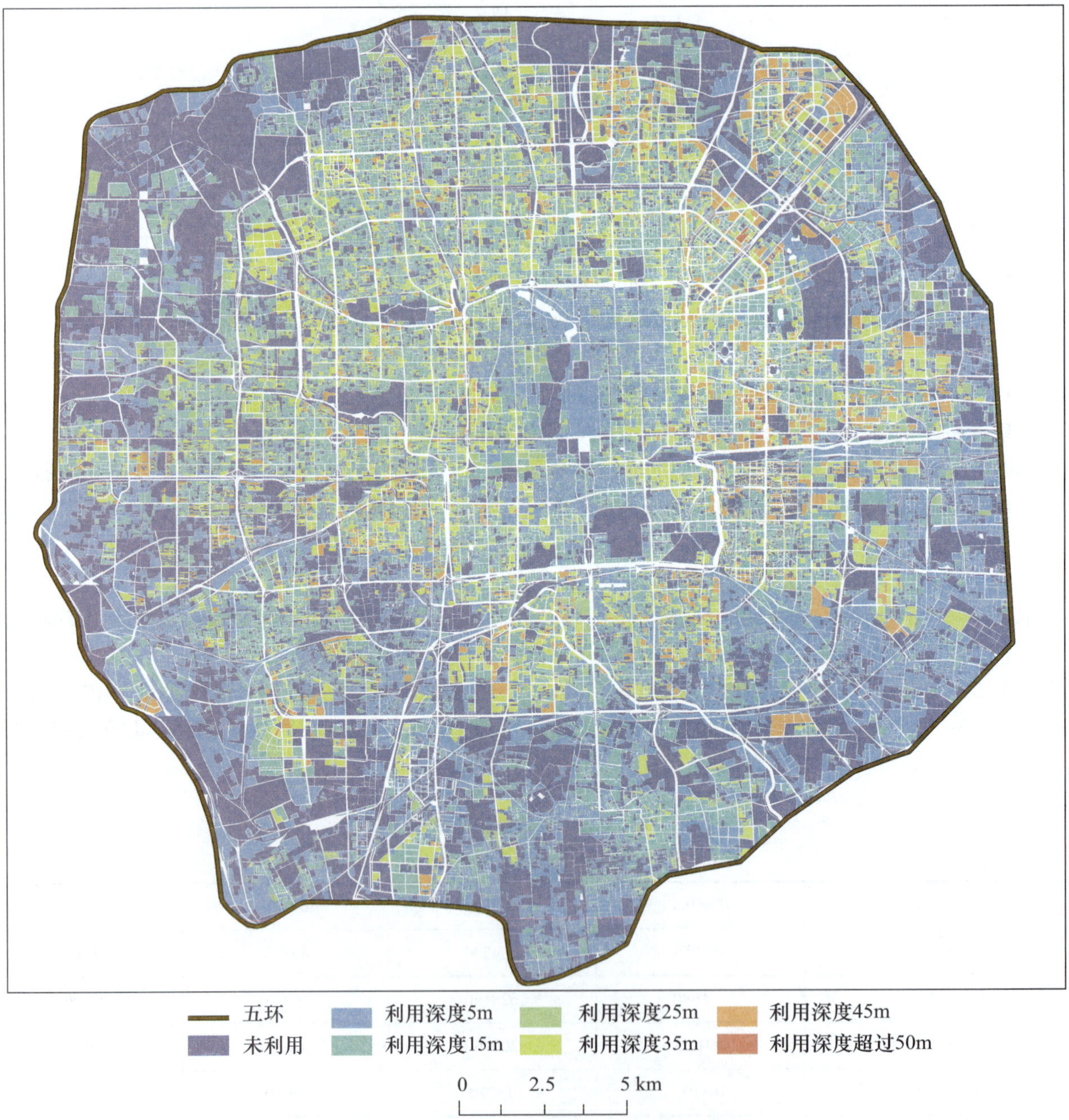

图 45-2 北京五环内已开发利用地下空间资源深度

Fig. 45-2 Depth of underground space resources developed and utilized within the Fifth Ring urban area of Beijing

由于地下管线和地铁主要分布于道路以下，可以将地下空间管线已占用的地下空间容量及分布，在道路下地下空间资源量中排除。现状市政基础设施中各种管线都基本敷设城市道路红线下 0~10m 以内。因此假定道路浅层 0~10m 的空间 80%已被市政管线所利用，则道路下 0~10m 已利用的资源量为 $181392\times10^4m^3$。

第四节　地下空间资源质量

本志所指地下空间资源质量是指在现代化经济技术条件下，安全利用地质资源环境是否产生负作用，或产生负作用是否可控制。质量评价包括影响地下空间资源开发利用的基础条件和约束条件在内的多项评价单因子：前者包括土体类型结构及物理力学参数，后者包括活动断裂及地裂缝、地面沉降、砂土液化、隐伏岩溶塌陷及水文地质条件。依据各类地质条件进行宏观评价，可将北京平原区地下空间质量等级划分为一级、二级和三级（图 45-3~图 45-5，表 45-6）。总体而言，北京平原区地下 0~50m 范围内 70%以上区域，地下空间资源开发利用等级都为适建区，适宜进行地下空间资源开发利用；五环内地下 0~50m 范围的地下空间资源质量等级 80%以上都集中在一级和二级，地下空间资源较为适宜开发利用，局部地区需要注意地面沉降、活动断裂对地下空间（工程）的持续性不利影响。

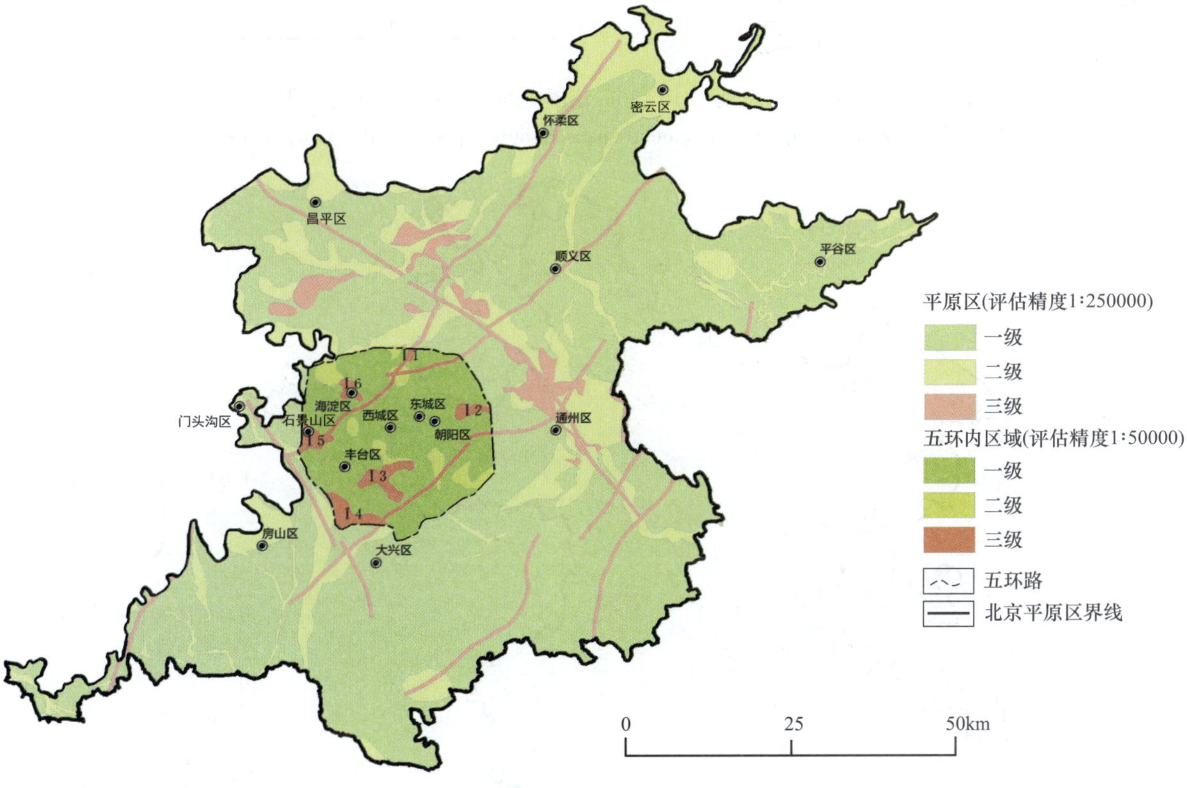

图 45-3　北京市地下 0~10m 范围空间资源质量分区图

Fig. 45-3　Zoning map of underground space resource quality（0-10m）of Beijing

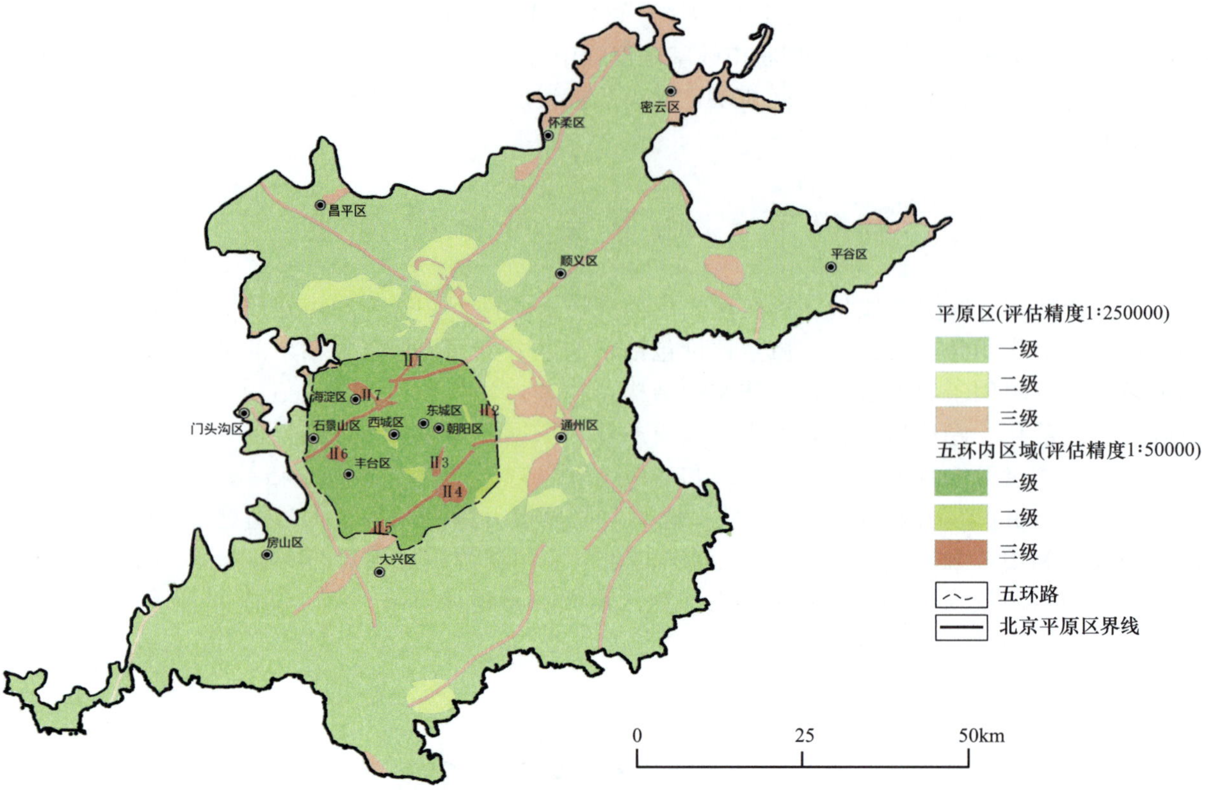

图 45-4 北京市地下 10~30m 范围空间资源开发利用分区图

Fig. 45-4 Zoning map of underground space resource quality (10-30m) of Beijing

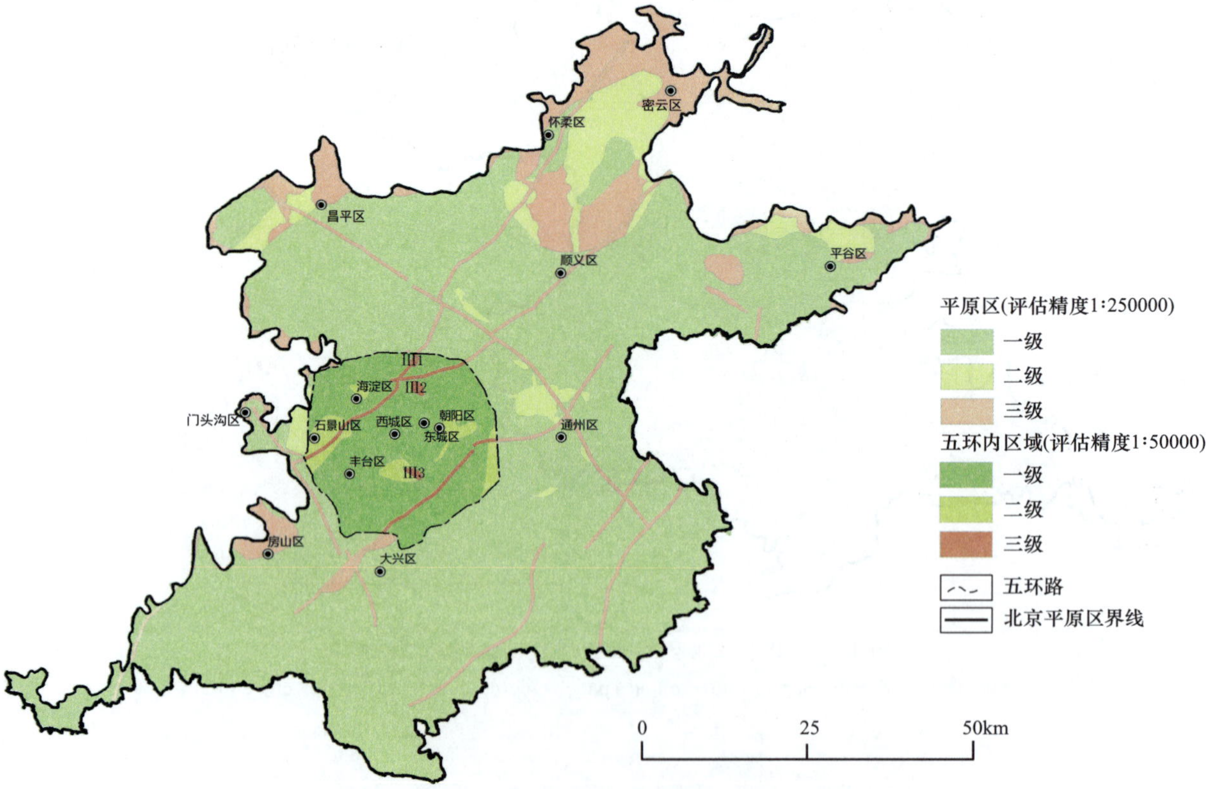

图 45-5 北京市地下 30~50m 范围空间资源开发利用分区图

Fig. 45-5 Zoning map of underground space resource quality (30-50m) of Beijing

表 45-6　地下空间资源质量等级统计表

Table 45-6　Quality grade of underground space resources

单位：km^2

序号	质量分级	等级说明	平原区（除五环城区外）（地表面积 $5588km^2$）			五环内城区（地表面积 $750km^2$）			地质影响因素
			地下 0~10m	地下 10~30m	地下 30~50m	地下 0~10m	地下 10~30m	地下 30~50m	
1	一级	区内整体上基本无影响地下空间资源开发利用的地质因素，或存在影响地下空间资源开发利用地质因素，但影响程度较低（适宜区）	4119	4690	4462	601.53	684.06	649.9	地质因素影响较小
2	二级	区内存在影响地下空间开发利用的不利地质条件，通过现有技术方法可以处理，或通过现在技术手段可以规避不利地质条件的继发性危险（一般区）	1159	401	364	60.39	19.45	91.28	存在一定程度地质灾害，工程地质条件、水文地质条件复杂。应尽量控制地下空间开发利用规模，在开发利用前进行详细工程勘察并做好防护措施
3	三级	区内存在多种对地下空间开发利用影响程度较大的不利地质条件，通过现有技术方法很难处理且不利地质条件有继发性危险（敏感区）	310	497	762	88.08	46.49	8.82	地面沉降、砂土液化、隐伏岩溶塌陷等多种地质灾害易发区，且工程地质条件、水文地质条件复杂。应控制地下空间开发利用规模，在开发利用前进行详细工程勘察并进行专家论证。活动断裂分布位置，在活动断裂 300m 以内范围避免进行地下空间开发利用

一、平原区地下空间资源质量

对平原区分布的活动断裂及地裂缝、地面沉降、砂土液化、隐伏岩溶塌陷、水文地质条件、工程地质条件进行综合分析，依据各类地质条件做出宏观评价，评价深度50m以浅，分为0~10m、10~30m及30~50m 3个层次进行。

（一）地下0~10m范围空间资源质量

1. 三级区

主要包括通州区西北部及昌平区、密云区、平谷区、房山区部分地区。

其中，昌平、密云、平谷以及房山地区属于工程地质条件较差地区，山前平原地下水埋深较浅。因此在进行地下空间开发利用时，需要针对工程地质条件进行更为详细的勘察，同时做好地下水设防，防止在开挖施工或盾构施工时突水、涌水发生。

在通州区西北部附近地区由于工程地质条件较差，同时存在明显的地面沉降现象以及砂土液化易发区，在进行地下空间开发利用时，需要针对工程地质条件进行更为详细的勘察，在进行线性地下空间工程时避免垂直沉降梯度带方向建设，在进行地下空间施工建设时做好砂土液化防护措施。

2. 二级区

主要位于山前地区以及国家奥林匹克体育中心、西红门等地区，存在隐伏岩溶塌陷发生的可能性。进行地下空间开发利用时，需开展隐伏岩溶塌陷的调查工作，进行详细的工程地质勘察工作。

（二）地下10~30m范围空间资源质量

1. 三级区

西红门镇、国家奥林匹克体育中心以及昌平区、密云区、平谷区山前地区存在隐伏岩溶塌陷发生的可能性，在上述地区进行地下空间开发利用时，需要针对隐伏岩溶塌陷开展详细调查。

海淀上庄、苏家坨、昌平北七家、东小口地区，工程地质条件较差，且地下水埋深为10~30m，在进行地下空间开发利用前应开展详细的工程地质勘察工作，同时做好地下水设防，防止在开挖施工或盾构施工时突水、涌水发生。

大兴区榆垡、礼贤地区，工程地质条件较差，潜水埋深为20~30m，且隔水层埋深为10~30m，在进行地下空间开发利用前应开展详细的工程地质勘察工作，同时做好地下水设防，防止在开挖施工或盾构施工时突水、涌水发生。在隔水层内尽量避免较大规模的地下空间开发利用，防止由破环隔水层导致的水体污染及地下水压力失衡。

通州西北部附近地区由于工程地质条件较差，同时存在明显的地面沉降现象，并存在砂土液化易发区，在进行地下空间开发利用时，需要针对工程地质条件进行更为详细的勘察。在进行线性地下空间工程时避免垂直沉降梯度带建设。在进行地下空间施工建设时注意因砂土液化引起喷水冒砂，做好砂土液化防护措施。

2. 二级区

北京市平原区南部地区工程地质条件较差、潜水埋深为20~30m，在进行地下空间开发利用前应开展详细的工程地质勘察工作，同时做好地下水设防，防止在开挖施工或盾构施工时突水、涌水发生。

通州西北部—海淀上庄镇一带，大部分地区工程地质条件较差，建议在进行地下空间开发利用前进行详细的工程地质勘察工作。

（三）地下30~50m范围空间资源质量

1. 三级区

西红门镇、国家奥林匹克体育中心、昌平区、密云区、平谷区、房山区山前地区存在隐伏岩溶塌陷发生的可能性，山区大部分地区潜水埋深为30~50m，加剧了隐伏岩溶塌陷发生的可能，因此在上述地区进行地下空间开发利用时，需要针对隐伏岩溶塌陷开展详细调查，同时做好地下水设防，防止在开挖施工或盾构施工时突水、涌水发生。上述部分地质条件较差地区还存在隔水层埋深的问题，密云区山前地区、房山区山前地区及平谷部分地区隔水层埋深在 30m 以上，在隔水层内尽量避免规模较大的地下空间开发利用，防止由破环隔水层导致的水体污染及地下水压力失衡。

2. 二级区

通州西北部附近地区存在明显的地面沉降现象，并存在砂土液化易发区。因为在 30～50m 深度内砂土液化影响不大，因此只需考虑地面沉降的问题，在进行线性地下空间工程时应避免垂直沉降梯度带建设。

石景山地区潜水埋深为 35～40m，同时隔水层埋深在 50m 左右，在地下空间开发利用时应做好地下水设防，防止在开挖施工或盾构施工时突水、涌水发生。在隔水层内尽量避免规模较大的地下空间开发利用，防止由破环隔水层导致的水体污染及地下水压力失衡。

密云山前、平谷山前、昌平山前地区隔水层埋深为 40～50m，在隔水层内尽量避免规模较大的地下空间开发利用，防止由破环隔水层导致的水体污染及地下水压力失衡。

除以上问题外，在活动断裂带两侧 300m 范围内属于活动断裂影响带，该区域内应避免地下空间开发利用，同时线性地下空间建设也应避让活动断裂影响带。

二、城区地下空间资源质量

通过梳理五环内地区（地表面积 750km^2）原始钻孔数据、地质报告、遥感数据、二维地质图件、物探数据等，最终将 50m 以浅的岩土体类型概化为 6 类，即：人工填土、粉质黏土、黏土、粉细砂、中粗砂及卵砾石，局部地区偶见基岩，如黏土岩、风化千枚岩，以薄夹层或透镜体状产出，分布范围普遍较小。采用 Creatar Xmodeling 平台搭建了五环内地区浅部（50m）松散层的三维地质模型（图 45-6，图 45-7），直观展示了区内岩土体的空间分布规律：①横向上，西部地层略单一，发育厚层卵石，东部地层复杂多变，发育厚层粉质黏土；②垂向上，东部地区发育 3 个韵律层，各韵律按粉质黏土→粉细砂→（中粗砂）→卵砾石的规律变化，大层内零星夹杂黏土等透镜体，局部偶见薄夹层。此外，三维属性模型集成了诸如岩土体含水量、液塑限、压缩模量等重要的地质数据，为后续地质研究，尤其是预测与评价，提供了关键的基础数据与结构框架，如地质监测点的布设、地下空间资源的适宜性评价等。

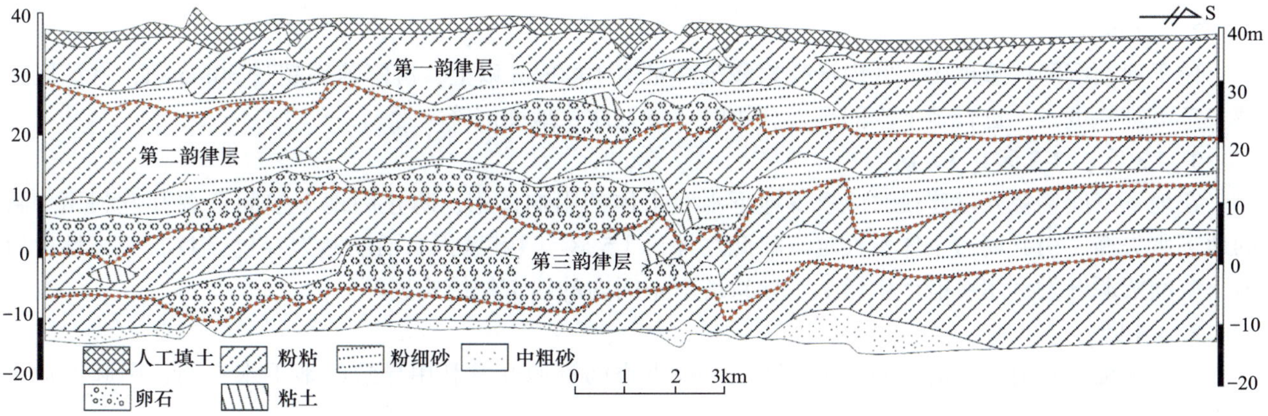

图 45-6 北京五环内区域地下 0～50m 浅部地层剖面

Fig. 45-6 Sub-bottom profile within the Fifth Ring urban area of Beijing（0～50m）

北京五环城区地下空间资源质量评估包含基础条件和约束条件在内的多项评价单因子：前者包括土体类型结构及物理力学参数，后者包括活动断裂及地裂逢、地面沉降、砂土液化、隐伏岩溶塌陷及水文地质条件等。经评价，将该区地下空间质量等级划分为一级、二级和三级。

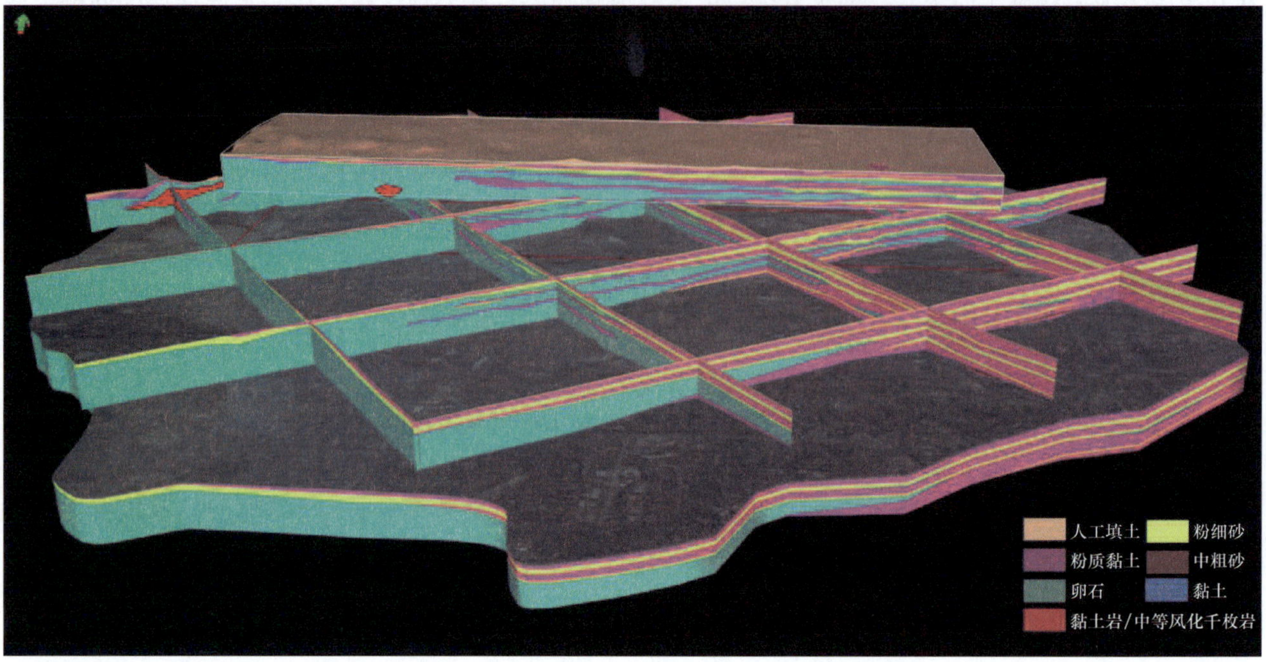

图 45-7 北京五环内区域浅部松散层三维地质结构模型

Fig. 45-7 3D geological structure model of shallow loose layer within the Fifth Ringurban area of Beijing

(一) 0~10m 地下空间资源质量

该层内地下空间资源质量分级主要为一级区和二级区。

1. Ⅰ1 区

该区内分布有“洼里砾岩”，特别是在洼里龙王堂附近“洼里砾岩”埋深较浅（埋深 30~50m），应避免地下空间资源的开发和建设，在施工时应防范隐伏岩溶塌陷的发生。另外，该区内地下水埋深较浅（地下 8m 以内），在地下空间资源开发建设时应重视地下水设防，防止在开挖施工或盾构施工时突水、涌水发生。

2. Ⅰ2 区

位于地面沉降严重地区，区内最高累积沉降量达 1000mm 以上，沉降速率超过 30mm/a，且该区穿越多条沉降梯度带，进行地下空间开发利用时要特别注意地面沉降问题。

3. Ⅰ3 区，Ⅰ5 区，Ⅰ6 区

在地下空间资源开发建设时应重视地下水对土体的影响作用，防止因地下空间资源开发利用导致的地下水失衡，引起周边土体不稳定。除此之外，Ⅰ5 区、Ⅰ6 区内有活动断裂穿过，具有破坏性，对地下空间资源开发利用可能产生影响，导致地下空间内的建筑物结构破坏。

4. Ⅰ4 区

应重视地下水对土体的影响作用，防止因地下空间资源开发利用导致的地下水失衡，引起周边土体不稳定。

(二) 10~30m 地下空间资源质量

该层内地下空间资源质量分级主要为一级区和二级区，另有三级区 7 个（Ⅱ1、Ⅱ2、Ⅱ3、Ⅱ4、Ⅱ5、Ⅱ6、Ⅱ7）。

1. Ⅱ1 区

位于岩溶塌陷易发区，该区内分布有“洼里砾岩”，特别是在洼里龙王堂附近“洼里砾岩”埋深较浅（埋深 30~50m），应避免地下空间资源的开发和建设，在施工时应防范隐伏岩溶塌陷的发生。另外，该区内地下水埋深较浅（地下 8m 以内），在地下空间资源开发建设时应重视地下水设防，防止在开挖施工或盾构施工时突水、涌水发生。

2. Ⅱ2 区

位于地面沉降严重地区，区内最高累积沉降量达 1000mm 以上，沉降速率超过 30mm/a，且该区穿越多条沉降梯度带，在进行地下空间开发利用时要特别注意地面沉降问题。区内地下水埋深 22m 左右，在地下空间资源开发建设时应重视地下水设防，防止在开挖施工或盾构施工时突水、涌水发生。

3. Ⅱ3 区，Ⅱ4 区

区内地下水埋深较浅（地下 22m 以内），地下空间资源开发建设时应重视地下水设防，防止在开挖施工或盾构施工时突水、涌水发生。

4. Ⅱ5 区，Ⅱ6 区，Ⅱ7 区

区内基础地质条件一般，应注意地下空间资源开发利用时容易发生的工程施工问题。地下水埋深 20~30m，在地下空间资源开发建设时应重视地下水设防，防止在开挖施工或盾构施工时突水、涌水发生。重视地下水对土体的影响作用，防止因地下空间资源开发利用导致的地下水失衡，引起周边土体不稳定。除此之外，Ⅱ6 区、Ⅱ7 区内有活动断裂穿过，具有破坏性，对地下空间资源开发利用可能产生影响，导致地下空间内的建筑物本体结构破坏。Ⅱ5 区砂土液化严重，进行地下空间资源开发利用时，要特别注意砂土液化问题，进一步开展工程地质勘察工作。

（三）30~50m 地下空间资源质量

该层内地下空间资源质量分级主要为一级区和二级区，另有三级区 3 个（Ⅲ1 区、Ⅲ2 区、Ⅲ3 区）。

1. Ⅲ1 区

位于岩溶塌陷易发区，该区内分布有“洼里砾岩”，特别是在洼里龙王堂附近“洼里砾岩”埋深较浅（地下 30~50m），应避免地下空间资源的开发和建设，施工时应防范隐伏岩溶塌陷的发生。

2. Ⅲ2 区

地下空间资源开发建设时应重视地下水设防，防止在开挖施工或盾构施工时突水、涌水发生。另外还要重视地下水对土体的影响作用，防止因地下空间资源开发利用导致的地下水失衡，引起周边土体不稳定。

3. Ⅲ3 区

在地下空间资源开发建设时应重视地下水对土体的影响作用，防止因地下空间资源开发利用导致的地下水失衡，引起周边土体不稳定。

除以上问题外，在活动断裂带两侧 300m 范围内属于活动断裂影响带，该区域内应避免地下空间开发利用，同时线性地下空间建设也应避让活动断裂影响带。

第五节　可持续开发利用建议

北京平原区地下 50m 以上的空间是地下空间资源利用主体，地下 30m 以上是大规模开发利用的重点深度。从地下空间资源天然蕴藏量来看，资源量是丰富的，但是从长远发展而言，资源量又是十分有限的，且受到地质环境条件制约，必须要精心探测、科学规划利用这种宝贵资源，避免资源浪费和过度开发，保持城市空间可持续发展。

一、地质条件不利地区的地下空间资源利用建议

由于活动断裂上下两盘的相对运动造成地层水平及垂直位移，会对跨断裂建设的地下空间造成破坏。如果按照抗震标准，建议重大工程在断裂 200m 范围采取避让措施。从活动断裂引发的地质灾害而言，大型线性工程应尽量减少穿过断裂带，如必须穿过，建议采取工程措施以减少断裂上下两盘的

运动而造成水平及垂直位移带来的影响。如穿过其他未确定性质的断裂，建议应对其进行进一步调查和评价。

地面沉降产生的地层形变会对穿越沉降梯度带的线性工程造成影响。跨越地面沉降区建设的轨道交通线、对沉降有严格要求的重要建筑、管线等工程应该开展地面沉降对工程影响的勘察与研究，要采取相应工程措施防止地面沉降长期发展对地下线性工程的影响。如已建成，要注意地面沉降容易对地下建筑结构产生破坏，需要对其定期进行检测，必要时开展实时监测工作。

在砂土液化中等及严重地区地下20m以内（易液化层主要为粉细砂层，埋深在15~20m之间）进行地下空间开发利用（尤其是地铁隧道工程）时，需要进行详细的工程勘察，同时合理选择施工手段，避免因震动造成砂土液化现象。

二、不同深度的地下空间资源开发利用建议

北京平原冲洪积平原横向岩性变化较大。以北京长安街沿线地质结构剖面为例，由东至西大致分为三类结构，分别为均一地层结构、2~3层交互地层结构以及3层以上多层地层结构（图45-8）。在均一地层结构地区深度35m以内，需要针对地下水开展防护措施，深度超过35m时要考虑大厚度含水层涌水问题；中部2~3层地层交互地区，深度30m以内需避免穿越活动断裂以及地面沉降梯度带，在隔水层内尽量避免大规模利用，防止破坏隔水层导致的水体污染及地下水压力失衡。多种地层复杂交互结构地区，深度25~30m以内需避免穿越活动断裂及地面沉降梯度带，在隔水层内亦需注意水体污染及压力失衡，对含水层需注意地下水防护措施及大厚度含水层涌水问题。

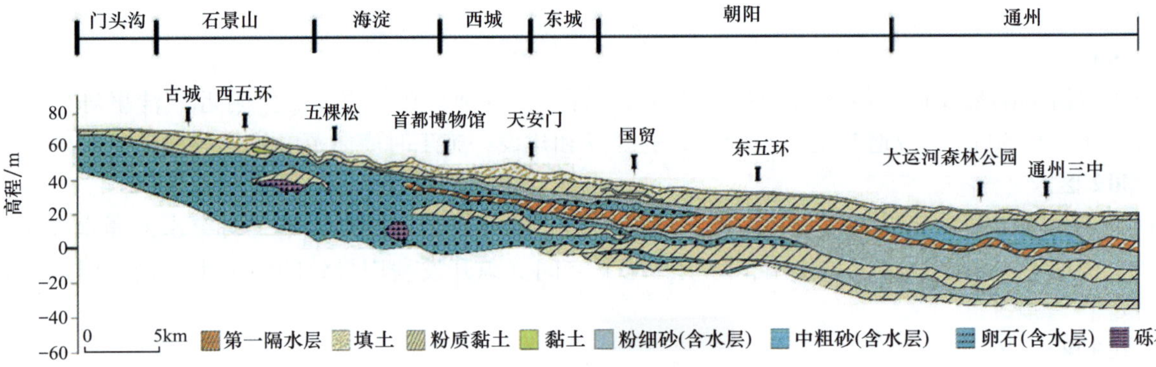

图45-8 长安街沿线地质结构剖面示意图

Fig. 45-8 Geological structure profile along the Chang'an Street

三、地下水流场保护建议

地下水流场受阻会导致阻碍地下水流场的设施上方地下水位升高，下方地下水位降低，可能对地下工程产生破坏，甚至导致地基失衡。建议地下线性工程建设方向避免大范围切割地下水流场。

永定河、潮白河等河流的冲积扇地区的砂砾石层具有较好的透水性，是城市地下水的重要补给区（“海绵体”）。该地区开发利用深度不宜大面积超过30m，且平面范围不宜超过地面建筑投影范围，以保障地下水源的有效补给及地下水质安全。

第四十六章　地下水资源

北京是以地下水作为主要供水水源的超大型城市，在目前北京的多水源供水格局中，地下水约占全市供水量的60%，地下水在支撑北京的供水中，发挥着不可替代的作用。

地下水系统是一个错综复杂的，包括各种天然因素、人为因素所控制的，具有不同等级的，互相联系又互相影响，在时空分布上具有四维性质和各自特征、不断运动演化的若干独立单元的统一体。

一、地下水系统划分

（一）区域水流系统划分

北京处于海河流域地下水系统之中，依据水系的分布特征，按河流的水动力能力、搬运能力、区域地下水流动状况和地下水补排关系，划分为潮白－蓟运－温榆河、永定河和大石河－拒马河3个相对独立的区域地下水系统（图46-1）。

潮白－蓟运－温榆河地下水系统包含了潮白河、蓟运河和温榆河3条河流流域范围内的地下水；永定河地下水系统包含了永定河山区、山间盆地和冲洪积扇、冲洪积平原地区的地下水；大石河和拒马河两条河流的山区地下水与平原松散孔隙水之间的水力联系密切，故划分为大石河－拒马河地下水系统。

边界上，在山区，温榆河水系和永定河水系的地表分水岭为永定河地下水系统和潮白－蓟运－温榆河地下水系统的边界，百花山向斜、髽髻山向斜、北岭向斜北翼、谷积山背斜、九龙山向斜、香峪大梁向斜南翼及八宝山断裂中段和东段控制的岩溶水边界为永定河地下水系统和大石河－拒马河地下水系统的边界；在平原区，永定河地下水系统与潮白－蓟运－温榆河地下水系统的边界为永定河冲洪积扇和潮白河、温榆河冲洪积扇的交界部位；永定河冲洪积扇和大石河冲洪积扇的交界部位为永定河地下水系统与大石河－拒马河地下水系统的边界。

（二）子系统划分

区域水流系统中的次一级子系统划分考虑到含水介质类型、含水介质空间分布以及地下水流动特征，划分出3个地下水子系统，即基岩裂隙水子系统、岩溶裂隙水子系统和第四系（含山间盆地第四系）松散孔隙水子系统。

基岩裂隙水主要赋存于沉积岩、岩浆岩和变质岩构造裂隙及风化裂隙中，大气降水和流经山区的地表水系是地下水的主要补给来源，其排泄以自然排泄为主，一部分以泉的形式或泄流的形式向河排泄成为地表水，一部分向平原区径流，并有一定的人工开采量。该子系统地下水具有硬度低、低溶解性总固体的特点，水质比较好，但含水量较小，供水意义一般不大，常作为山区缺水村镇的供水来源。

岩溶裂隙水子系统依据岩溶分布情况、水动力条件划分出山区岩溶地下水子系统、山前岩溶裂隙水子系统以及隐伏岩溶水子系统，具体包括北山和西山山区岩溶地下水子系统，昌平高崖口－南口、昌平十三陵－桃峪口、顺义二十里长山－平谷、西山鲁家滩－玉泉山、延庆旧县－石槽、房山长沟－周口店山前岩溶裂隙水子系统，以及大兴隆起隐伏岩溶水子系统。按地层岩性可以划分为奥陶系、寒武系和青白口系、蓟县系，以及长城系岩溶裂隙水含水岩组，其中长城系、蓟县系碳酸盐岩以裂隙含水为主，寒武系、奥陶系碳酸盐岩以溶洞含水为主，岩溶水赋存条件有所差异。该子系统地下水主要接受大气补给，径流方向受地形控制，山区岩溶水地下水子系统以泉和泄流形式向河流排泄，山前岩溶水地下水子系统以泉、向平原排泄及人工开发利用为主。山区、山前岩溶水裂隙水子系统地下水化学类型单一，是低硬度、低溶解性总固体的优质饮用水源，北京地区许多供水厂以此作为主要供水水源，例如

图 46-1　北京地区地下水系统分区图

（据北京市多参数立体地质调查报告）

Fig. 46-1　Groundwater system zoning map in Beijing area

水源三厂、张坊应急供水水源地、石景山杨庄水厂等。

第四系松散孔隙水主要赋存于河流冲洪积作用形成的砂及砂卵石含水层中，不同地区富水特征各有不同。北京平原冲洪积扇顶部地区，含水层为单层砂卵砾石，富水条件好；冲洪积扇中上部地区，含水层为 2~3 层砂卵石，富水条件较好；冲洪积扇中下部至冲洪积平原区，含水层结构由冲洪积扇中上部单、双层结构向下逐渐过渡到多层结构，颗粒由砂卵石逐渐过渡到粗砂、中砂、细砂、粉细砂为主，富水条件较前两者差。总的来说，北京市平原区由西北向东南，由潜水区逐渐过渡为潜水-承压水区，透水性及富水性由强变弱（图 46-2~图 46-4），单井出水能力在 500~5000m^3/d 之间，其中永定河、潮白河等冲洪积扇单井出水量在 3000~5000m^3/d 以上（北京市地质勘察技术院，2012）。该系统地下水补给来源主要是降水入渗、侧向径流、农田灌溉及人工回灌等，其径流方向与地形地貌变化一致，即由山前向平原、由北西向南东流动。冲洪积扇中上部地区分布着数十处水源地，例如市第一水厂、石景山水厂、丰台水厂、密云水厂、水源八厂、怀柔应急水源地等。

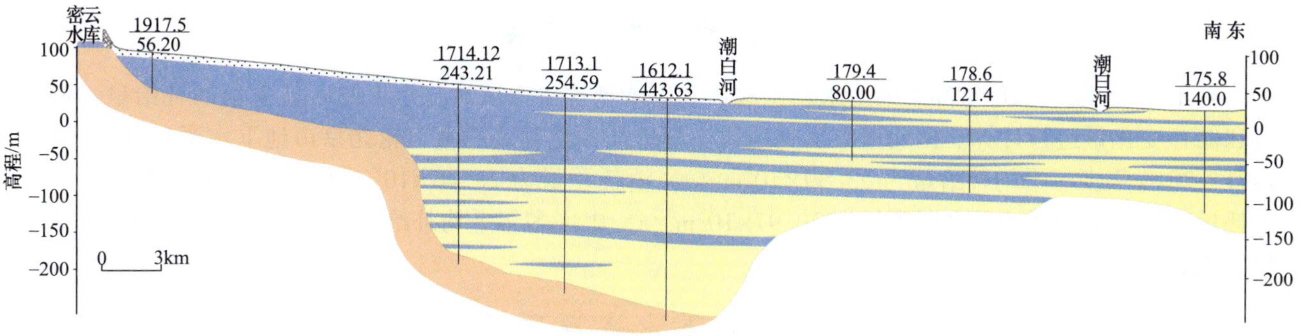

图 46-2　自密云水库南东向水文地质剖面

（据北京市地质勘察技术院，2012）

Fig. 46-2　Southeast hydrogeological profile from the Miyun Reservoir

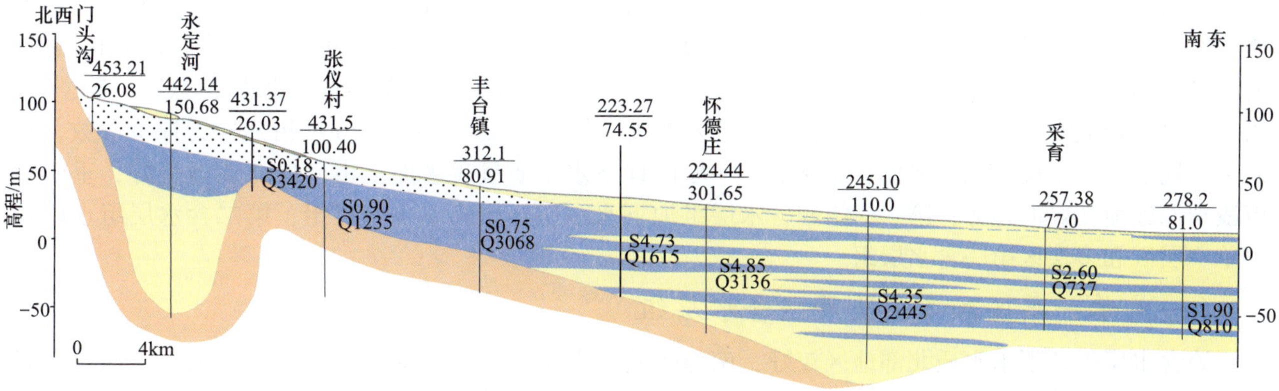

图 46-3　自门头沟南东向水文地质剖面

（据北京市地质勘察技术院，2012）

Fig. 46-3　Southeast hydrogeological profile from Mentougou

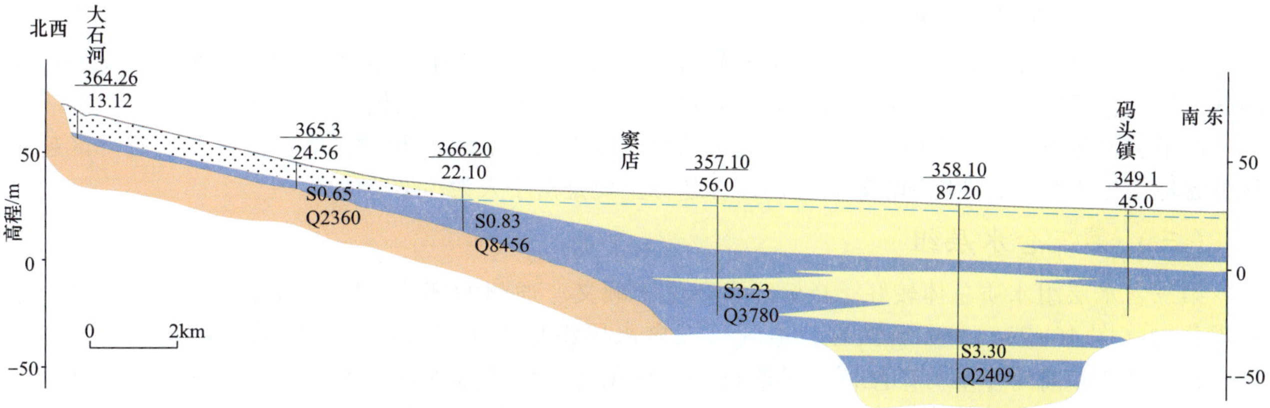

图 46-4　自大石河南东向水文地质剖面

（据北京市地质勘察技术院，2012）

Fig. 46-4　Southeast hydrogeological profile from the Dashi River

二、地下水资源量

自1980年以来，每年都会采用水均衡法、动态均衡法等方法对北京平原区地下水的补给量、消耗量和储存变化量进行计算。据北京市水文地质工程地质大队开展的第三轮北京市地下水资源评价工作，1991—2014年多年平均补给量为$23.59\times10^8 m^3/a$，消耗总量为$28.46\times10^8 m^3/a$，扣除难以袭夺的自然消耗量，北京平原区地下水资源量为$20.97\times10^8 m^3/a$。山区多年平均补给量为$11.89\times10^8 m^3/a$，消耗总量为$11.43\times10^8 m^3/a$。

三、地下水资源质量

北京市地下水环境监测网的区域环境监测网主要监控4个含水层组。第一含水层组主要为潜水及埋深小于50m的浅层或具有微承压性质的含水层组；第二含水层组分布于平原承压水区，含水层组底界深度100m；第三含水层组广泛分布于冲洪积扇中下部地区，含水层组底界埋深150~180m；第四含水层组第四承压含水层组底界埋深为300m。

第四系地下水是北京市主要的供水水源，原北京市地质矿产勘查开发局、北京市水文地质工程地质大队联合编著的《北京地下水》以《地下水质量标准》（GB/T 14848—1993）为标准，以pH、总硬度、氯化物、硫酸盐、硝酸盐氮、亚硝酸盐氮、氨氮、溶解性总固体、高锰酸盐指数、总铁、锰、镉、铅、阴离子合成洗涤剂、砷、汞、六价铬、挥发酚类、氰化物及氟化物共20项指标作为评价参数，对北京市地下水质量进行综合评价。总体上，平原区地下水水质呈现平面上由南向北逐渐变好，垂向上由浅到深逐渐变好的趋势；第三、第四含水层组地下水水质好于第二含水层组，第二含水层组水质好于单一层区和多层区第一含水层组。

（一）单一层区和多层区第一含水层组

该含水层组地下水水质北部地区较好，南部较差（图46-5）。平原区南部房山、大兴、通州及城近郊区水质较差；北部的昌平、怀柔、密云、平谷、顺义部分地区、延庆部分地区水质符合国家饮用水标准。

从利用角度来讲，单一层区地下水总体上水质较好，均应作为地下水饮用水源功能区。多层区第一含水层地下水超标指标较多，污染较为严重，在使用功能上，可作为城市绿化用水等以及农村生产用水，不宜作为生活饮用水水源。

（二）第二含水层组

该含水层组同样是北部地区地下水水质较好，南部较差。平原区北部的平谷、昌平、顺义的部分地区水质符合国家饮用水标准；南部大兴、通州、房山及城近郊区水质较差（图46-6）。

昌平、通州东部、顺义南部超标区的超标指标主要为锰、铁和氟化物。大兴西部及城近郊区超标区的超标指标主要为总硬度和溶解性总固体，超标主要是由于人为污染造成的。

（三）第三含水层组

第三含水层组水质总体较好，超标区主要位于顺义、通州的部分地区，超标指标主要为铁、锰和氟化物（图46-7）。平原区大部分地区第三含水层组地下水可直接作为生活饮用水水源；对于铁、锰天然劣质地下水超标的地区，可通过曝气氧化等处理措施进行处理后使用；对于氟化物超标的地下水，作为饮用水源处理成本相对较高，可作为对水质要求不高的工业用水，以及农业灌溉用水。

（四）第四含水层组

第四含水层组地下水总体水质较好，未出现超标区，最主要的功能是作为城乡居民生活饮用水水源（图46-8）。

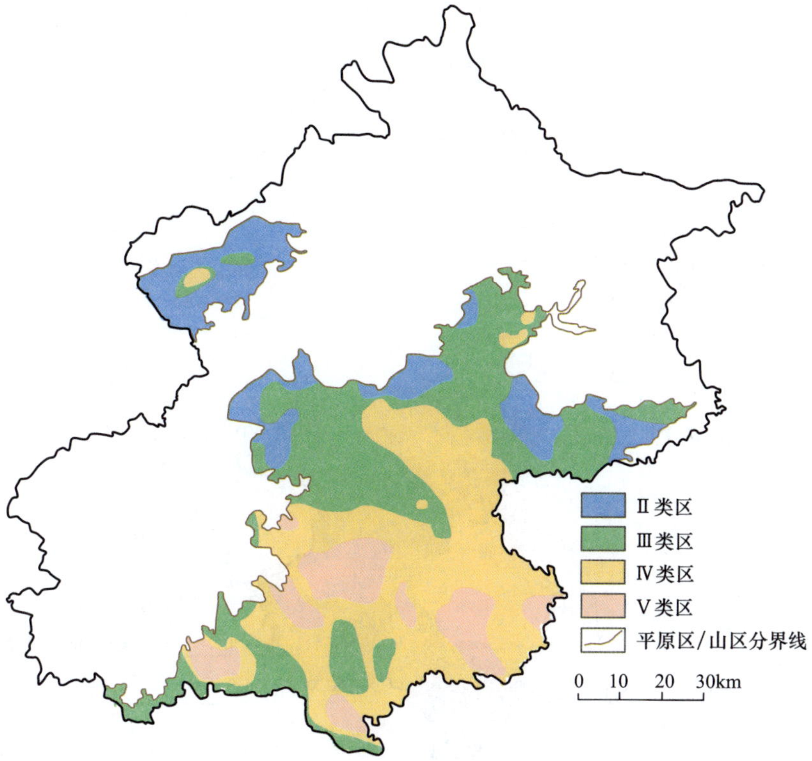

图 46-5 平原区地下水水质分区图（第一含水层组）
（据北京市总体规划修编地质基础支撑体系研究）
Fig. 46-5 Water quality distribution map of underground water in plain area (the first layer)

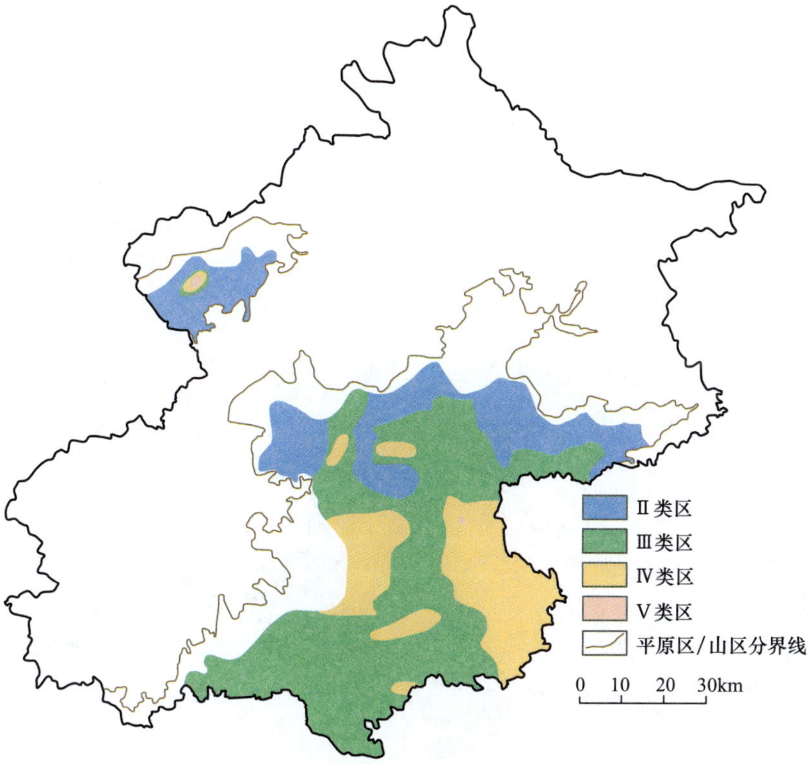

图 46-6 平原区地下水水质分区图（第二含水层组）
（据北京市总体规划修编地质基础支撑体系研究）
Fig. 46-6 Water quality distribution map of underground water in plain area (the second layer)

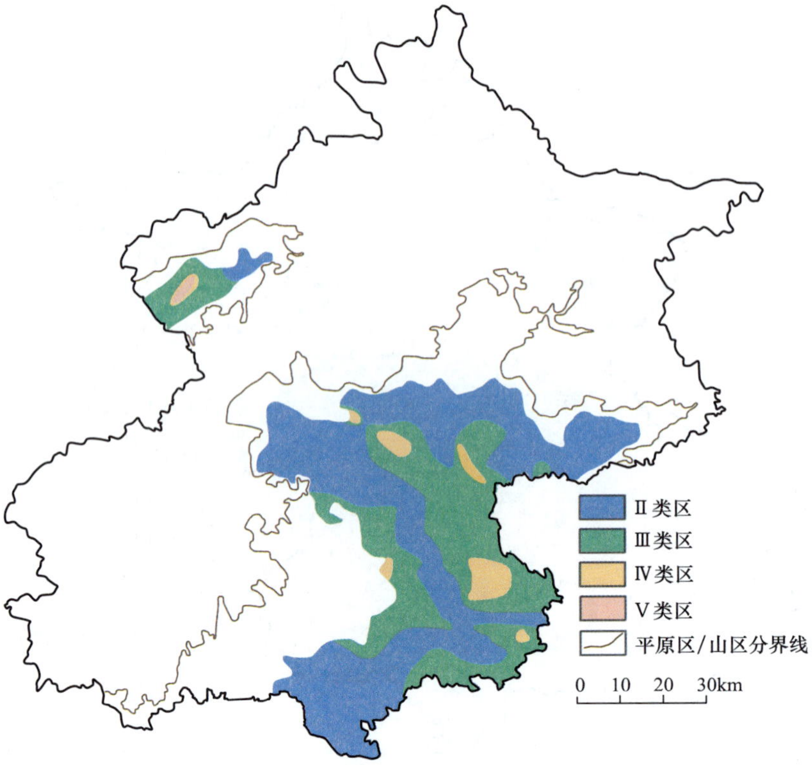

图 46-7 平原区地下水水质分区图（第三含水层组）
（引自北京市总体规划修编地质基础支撑体系研究）
Fig. 46-7 Water quality distribution map of underground water in plain area（the third layer）

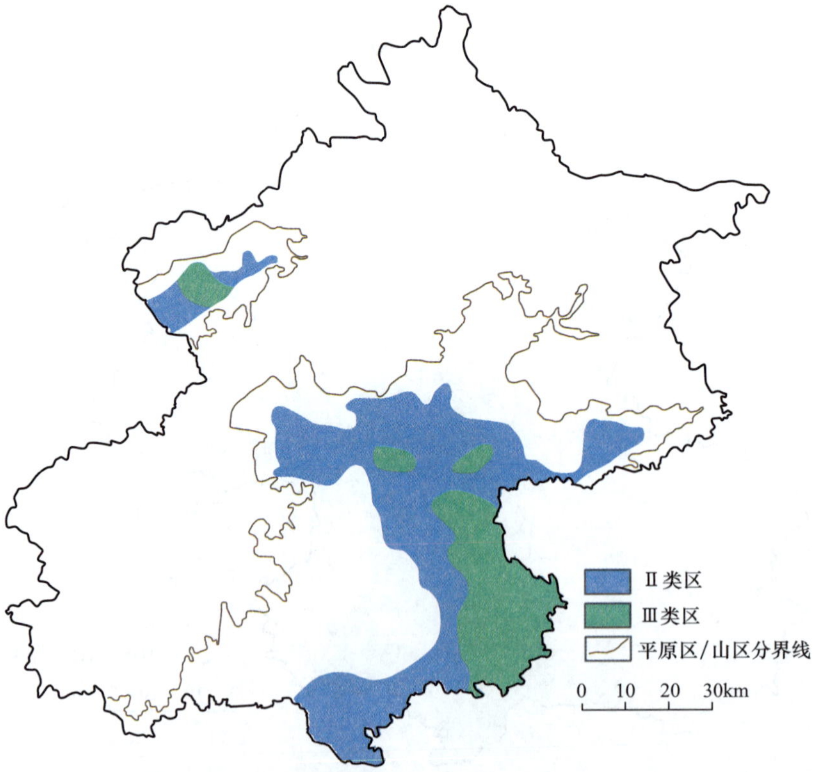

图 46-8 北京市平原区地下水水质分区图（第四含水层组）
（据北京市总体规划修编地质基础支撑体系研究）
Fig. 46-8 Water quality distribution map of underground water in plain area（the fourth layer）

四、可持续利用对策

（一）地下水利用

为保证地下水资源的可持续利用，可采取以下一些对策，例如调整地下水开采布局，减少严重超采区、超采区地下水开采，调整工业自备井、农业自备井的开采，分质供水、优质优用，合理利用地下水；保护地下水环境，完善地下水环境监控体系，控制地下水污染；研究地下水污染修复技术，加强地下水污染治理；科学利用地下储水空间，完善地下水战略储备体系；建立地下水资源科学利用的新模式，以实现地下水资源的可持续利用。

南水北调江水进京，给北京地下水涵养保护提供了契机。加强地下水资源保护和利用工作，改善地下水环境，促进地下水资源可持续利用，确保城市供水安全，事关首都经济社会发展和生态环境建设的全局。

（二）地下水调蓄

地下水调蓄是一种主动性的、有目的地储存、调节和利用地下水资源的工程措施，尤其是可以利用多余的洪水和外调水，改善干旱地区的缺水问题，提高水资源的利用率，还可以从多方面来改善生态环境。从水文地质条件分析，北京市五大水系冲洪积扇顶部地区具有较好的调蓄条件（图 46-9）。

其中，潮白河调蓄区面积 355km^2，调蓄空间 11.7×10^{8}m^3；永定河调蓄区面积 333km^2，调蓄空间 13.5×10^{8}m^3；泃河调蓄区面积 170km^2，调蓄空间 5.7×10^{8}m^3；温榆河调蓄区面积 173km^2，调蓄空间 7.3×10^{8}m^3；大石河调蓄区面积 106km^2，调蓄空间 1.2×10^{8}m^3，5 个调蓄区总调蓄空间为 39.4×10^{8}m^3。

充分利用南水北调通水契机，多调水，还欠账，优先回补水源地和应急水源地，加速集中水源地的地下水涵养。采取停采、压采等措施，限制超采区的地下水开采，遏制地下水位继续下降趋势。按照地下水“采补平衡”的原则，“有序压采”和“人工回补”相结合，结合南水北调水源，开展地下水资源养蓄及战略储备工程规划方案编制工作，勘察划定回补区，优化开采布局。加快实施密云、怀柔、顺义、西郊等有利地段地下水源地回补工程，建立地下水资源战略储备机制，增加地下水资源战略储备，保障非常时期的用水需要。采取“地下水、地表水联动”方式促进地下水资源涵养，采取多种回灌措施实现地下水位止跌回升，逐步填平区域地下水漏斗。建立与完善地质环境影响评价及监测体系，实现水资源高效利用，提升首都水资源科学化管理的水平，实现地下水资源的可持续利用。

（三）地下水保护

根据《饮用水水源保护区划分技术规范》（HJ 338—2018）相关要求，对中小型水源地采用公式法，大型水源地采用数值模拟法进行保护区划分，包括一级保护区、二级保护区和准保护区。综合考虑各水源地的水文地质条件、供水规模及供水地位，对第一、第二、第五水厂只设定了一级保护区，未设定二级保护区和准保护区，岩溶水源地划定了二级保护区，将碳酸盐岩裸露山区划为岩溶水补给区。最终划定一级保护区面积 11.17km^2，二级保护区面积 317km^2，准保护区面积 1786.3km^2，岩溶水补给区面积 2822km^2（图 46-10）。

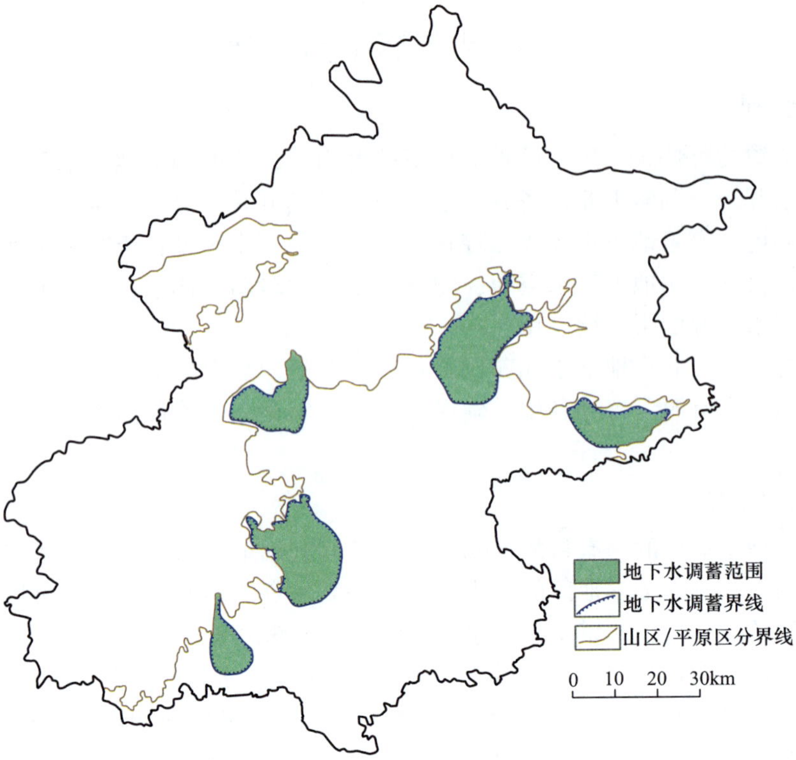

图 46-9 平原区地下水调蓄范围分布图
（据北京市总体规划修编地质基础支撑体系研究）
Fig. 46-9 Distribution map of groundwater regulation scopes in plain area

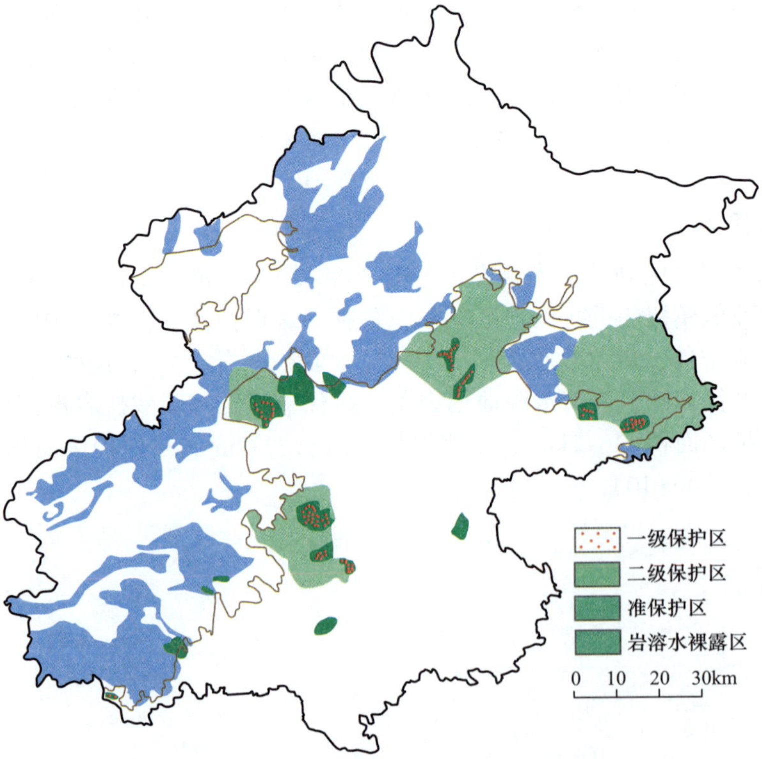

图 46-10 平原区地下水保护范围分布图
（据北京市总体规划修编地质基础支撑体系研究）
Fig. 46-10 Distribution map of groundwater protection scopes in plain area

第四十七章　固体矿产资源

第一节　资源概况

一、资源分布概况

北京地区矿产种类繁多，截至2014年底，共发现各类矿产127种（含亚矿种），其中固体矿产121种，查明资源储量并已编入《北京市矿产资源储量表》的固体矿产有67种，354处矿产地。矿床规模以中、小型为主，其中大型矿产地42处，中型矿产地101处，小型矿产地196处，矿点15处。铁的矿产地最多，共46个；然后为煤炭和泥炭，矿产地分别为29个和28个；其次为冶金用白云岩13个，熔剂用灰岩12个，制灰用灰岩11个，其余矿种矿产地小于10个。

北京市固体矿产资源分布不均衡，具有分布广泛、矿种相对集中，以远郊区为主的特点。煤炭90%以上的查明资源储量分布于京西门头沟区和房山区；铁矿93%以上的查明资源储量分布于密云区；有色金属矿产主要集中分布于密云区、延庆区及怀柔区；化工、冶金及建筑用各类石灰岩、白云岩等矿产主要分布于山区与平原交界的西部与北部山区地带（图47-1）。

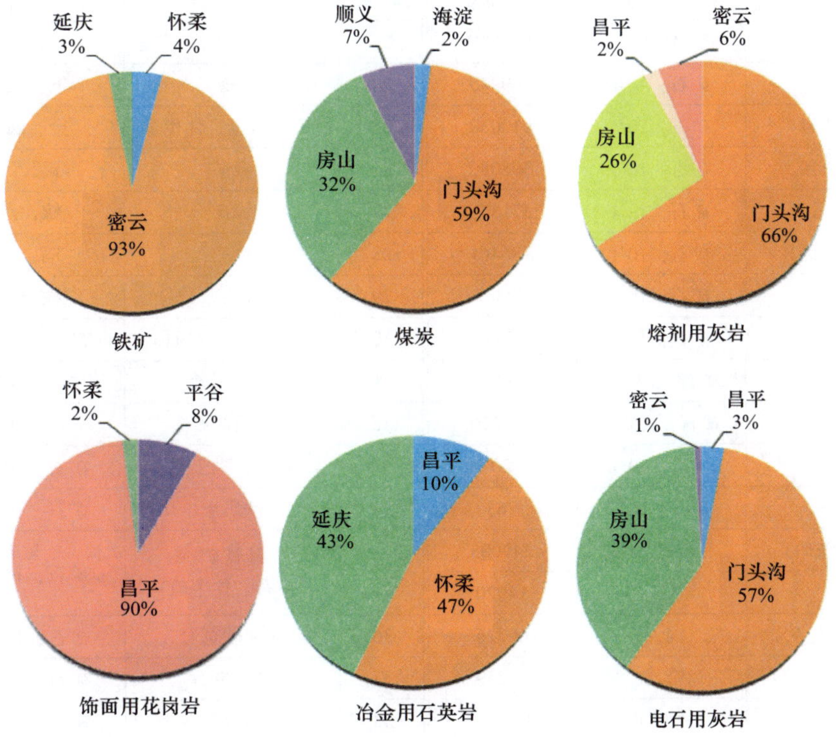

图47-1　北京市重要矿产资源在各区县的分布情况

Fig. 47-1　Important mineral resource distribution information of dističts and counties in Beijing

二、资源量

北京市矿产资源勘查程度总体偏低，基础储量所占比例不高（图 47-2），67 种固体矿产中仅有煤炭、铁、水泥用灰岩、冶金用白云岩、建材非金属等矿种总量比较丰富，并且基础储量所占比例较高，其他矿种则总量较少，基础储量也偏低，更有冶金用石英岩、制碱用灰岩等 32 个矿种基础储量为零。截至 2015 年底，北京市矿产资源各矿种保有资源储量见表 47-1。

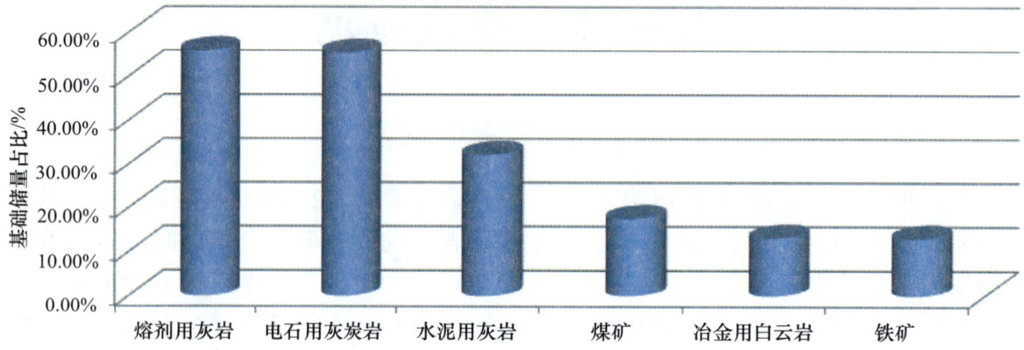

图 47-2 北京市主要矿产的基础储量在其资源储量中所占比例示意图

Fig. 47-2 Proportion graph for basic reserve occupied in resource reserve of main mineral

表 47-1 北京市矿产资源保有资源储量一览表（截至 2015 年底）

Table 47-1 List of mineral resources reserves by the end of 2015 in Beijing

序号	矿种	计量单位	资源储量	序号	矿种	计量单位	资源储量
1	煤炭	10^3t	2090391	23	铂矿	铂，kg	1018
2	铁矿	矿石，10^3t	969945	24	钯矿	钯，kg	975
3	水泥用灰岩	矿石，10^3t	910030	25	金矿（岩金、伴生金）	金，kg	6361
4	熔剂用灰岩	矿石，10^3t	308030	26	银矿	银，t	419
5	冶金用白云岩	矿石，10^3t	371871	27	镓矿	镓，t	41
6	冶金用石英岩	矿石，10^3t	186461	28	镉矿	镉，t	140
7	制碱用灰岩	矿石，10^3t	72767	29	红柱石	红柱石，t	256642
8	电石用灰岩	矿石，10^3t	96351	30	普通萤石（矿石）	矿石，10^3t	311
9	饰面用大理岩	矿石，10^3m^3	34829	31	铸型用砂	矿石，10^3t	3092
10	饰面用花岗岩	矿石，10^3m^3	222110	32	冶金用脉石英	矿石，10^3t	1159
11	锰矿	矿石，10^3t	20	33	耐火黏土	矿石，10^3t	23883
12	铬矿	矿石，10^3t	768	34	铁矾土	矿石，10^3t	412
13	钛矿（钛铁矿）	TiO_2，t	249005	35	硫铁矿（矿石、伴生硫）	矿石，10^3t	178
14	钒矿	V_2O_5，t	14920				
15	铜矿（非伴生矿）	铜，t	66326	36	含钾砂页岩	矿石，10^3t	208260
16	铅矿	铅，t	34259	37	含钾岩石	矿石，10^3t	34354
17	锌矿	锌，t	148716	38	泥炭	矿石，10^3t	9809
18	铝土矿	矿石，10^3t	420	39	石墨（隐晶质石墨）	隐晶质石墨，10^3t	102
19	镁矿（炼镁白云岩）	矿石，10^3t	18039				
20	钨矿（原生矿）	WO_3，t	1583	40	滑石	矿石，10^3t	9
21	铋矿	铋，t	488	41	石棉	石棉，10^3t	40
22	钼矿	钼，t	73164	42	长石	矿石，10^3t	1600

续表

序号	矿种	计量单位	资源储量	序号	矿种	计量单位	资源储量
43	叶蜡石	矿石，10^3t	3698	56	陶粒页岩	矿石，10^3t	28340
44	透辉石	矿石，10^3t	3073	57	砖瓦用页岩	矿石，10^3m^3	44780
45	玉石	矿石，t	63826	58	水泥配料用页岩	矿石，10^3t	43760
46	建筑石料用灰岩	矿石，10^3m^3	27792	59	陶瓷土	矿石，10^3t	914
47	制灰用石灰岩	矿石，10^3t	328458	60	砖瓦用黏土	矿石，10^3m^3	4010
48	泥灰岩	矿石，10^3t	5030	61	水泥配料用黏土	矿石，10^3t	13140
49	玻璃用石英岩	矿石，10^3t	7450	62	饰面用角闪岩	矿石，10^3m^3	2000
50	玻璃用砂岩	矿石，10^3t	8495.41	63	铸石用辉绿岩	矿石，10^3t	3090
51	水泥配料用砂岩	矿石，10^3t	22240	64	饰面用辉长岩	矿石，10^3m^3	2470
52	建筑用砂	矿石，10^3m^3	129500	65	饰面用闪长岩	矿石，10^3m^3	6370
53	砖瓦用砂	矿石，10^3m^3	14990	66	建筑用花岗岩	矿石，10^3m^3	7149
54	水泥配料用脉石英	矿石，10^3t	213	67	饰面用板岩	矿石，10^3m^3	10773
55	天然油石	矿石，10^3t	538				

注：数据来源为北京市规划和自然资源委员会。

第二节　区域成矿规律

一、成矿区带划分

成矿区带（又称成矿单元）是具有较丰富矿产资源及潜力的成矿地质单元。在某一成矿区带内往往具有主导的成矿地质环境、地质演化历史及与之相应的区域成矿作用，其内各类矿床组合往往有规律地集中分布，反映了矿产资源的区域性宏观分布特征及受控因素。北京市矿产资源潜力评价项目成果对北京市进行了五级成矿区带划分（图 47-3，表 47-2）。

表 47-2　北京市成矿区带初步划分表

Table 47-2　Preliminary division of metallogenic belts in Beijing

<table>
<tr><th>Ⅰ级</th><th>Ⅱ级</th><th>Ⅲ级</th><th>Ⅳ级</th><th colspan="2">Ⅴ级</th></tr>
<tr><td rowspan="9">太平洋成矿域</td><td rowspan="9">华北(陆块)板块成矿省(Ⅱ-15)</td><td rowspan="8">华北陆块北缘东段 Fe-Cu-Pb-Zn-Au-U-硼-菱镁矿-滑石-石墨-金刚石成矿带（Ⅲ-57）</td><td rowspan="8">燕辽（次级坳陷、拉张）Cu-Mo-Zn-Ag-Au-Fe-Mo 煤成矿亚带（Ar_2、Pt_2）（Ⅲ-57-2）</td><td colspan="2">怀柔北 Fe、Au、萤石成矿带（Ⅲ-57-2-①）</td></tr>
<tr><td rowspan="4">八达岭 Fe、Au、Cu、Mo、Pb、Zn、煤炭、石灰岩成矿带Ⅲ-57-2-②</td><td>石槽-石窑-银冶岭 Cu、Ag、Pb、Zn 成矿亚带（北西向）（Ⅲ-57-2-②a）</td></tr>
<tr><td>薛家石梁 Au、Mo、Cu、Pb、Zn 成矿亚带（北西向）（Ⅲ-57-2-②b）</td></tr>
<tr><td>大榆木沟-黄场-大庄科 Mo、Cu、Pb、Zn 成矿亚带（北东向）(Ⅲ-57-2-②c)</td></tr>
<tr><td>延庆怀来新生代断陷盆地地热成矿亚带（北东向）（Ⅲ-57-2-②d）</td></tr>
<tr><td colspan="2">京西煤、石灰岩、Ag、Pb、Zn 成矿带（Ⅲ-57-2-③）</td></tr>
<tr><td colspan="2">密怀地区 Fe、Cu、Au、Cr 成矿带（Ⅲ-57-2-④）</td></tr>
<tr><td colspan="2">密云-平谷 Ag、Au 成矿带（Ⅲ-57-2-⑤）</td></tr>
<tr><td>华北盆地（断坳）石油天然气成矿区（Kz）（Ⅲ-62）</td><td></td><td colspan="2"></td></tr>
</table>

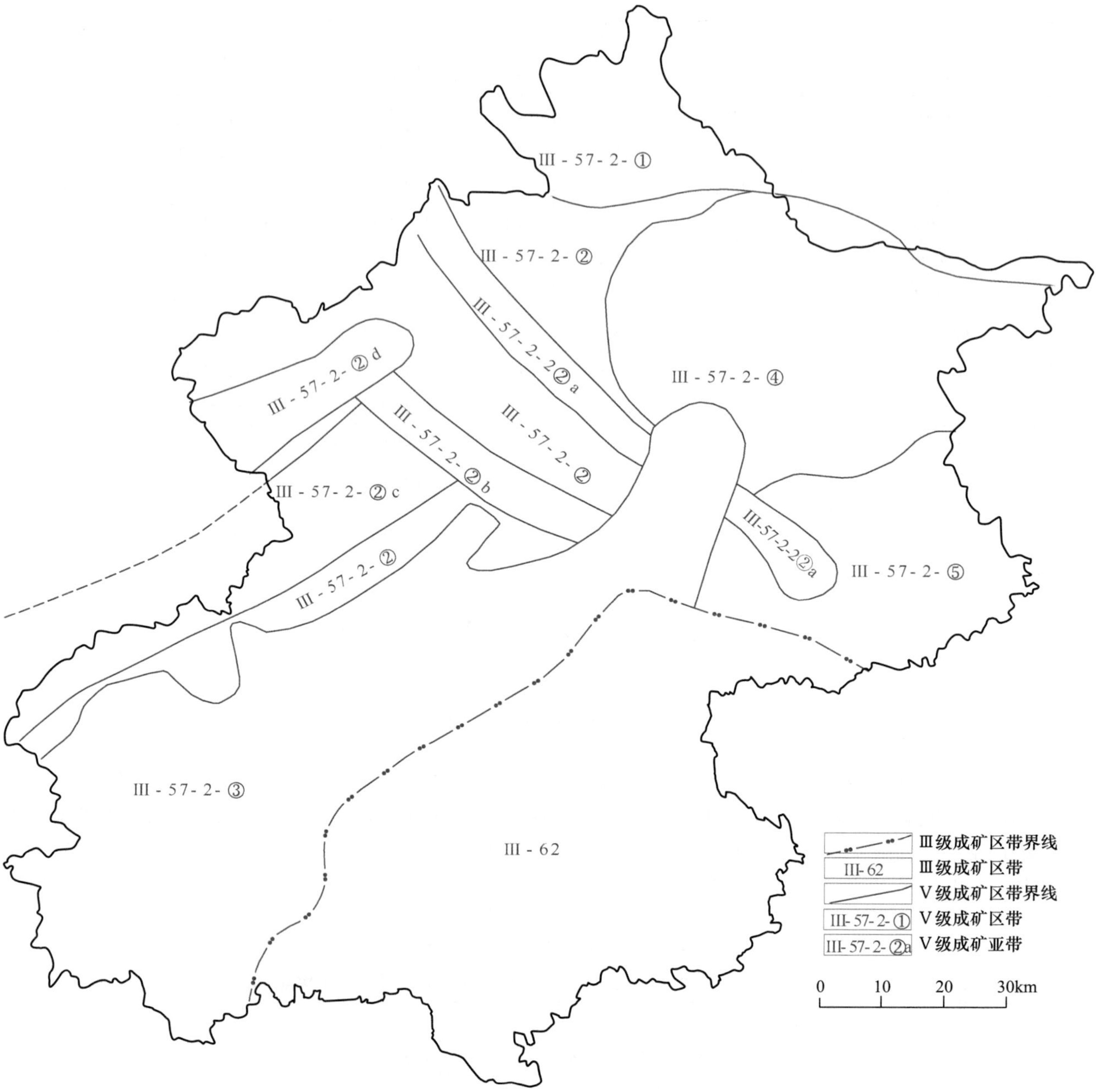

图 47-3　北京市成矿区带划分图

Fig. 47-3　Division map of metallogenic belts in Beijing

1. 怀柔北 Fe、Au、萤石成矿带（Ⅲ-57-2-①）

南界以赤城-长哨营-古北口近东西向断裂带为界，因受深断裂控制，中元古代岩浆侵位形成近东西向断裂岩浆岩带，故南界位置南移到岩浆岩带边界，其界线向东应有延伸。属承德后造山岩浆杂岩亚相。南界近东西向深断裂带有中元古代斜长岩、石英正长岩、斜长花岗岩体侵入，在深断裂带北侧有晚古生代变质辉长岩类为代表的基性岩体和二长花岗岩为代表的酸性岩体呈近东西向分布，并叠加有晚三叠世、燕山期酸性岩体的侵入活动。

区内已发现矿种主要为金矿，在大甸子北沟、东岔、八道河等地，均发现沿北东向构造破碎带充填热液型、破碎蚀变带、石英脉型金矿化点、矿点。兰营地区发现低温热液型萤石矿。

2. 八达岭 Fe、Au、Cu、Mo、Pb、Zn、煤炭、石灰岩成矿带（Ⅲ-57-2-②）

该成矿带西南自大榆木沟向北东经黄场、浇花峪、大庄科、崎峰茶至古北口、白马关一线，跨越门头沟、昌平、延庆、怀柔、密云等区，呈北东向带状展布。长 100km 以上，宽 30km 左右，面积约 $3000km^2$。其范围大体相当西山-八达岭燕山期火山-侵入杂岩带的主体部位，是北京地区内生金属矿产主要产出地区。

构造位置跨冀东古陆核、遵化岩浆弧和太行山岩浆弧 3 个Ⅲ级构造相。

出露地层主要有太古宙变质岩系，中—新元古代和古生代海相碳酸盐岩、碎屑岩及黏土岩，中生代火山-沉积岩系。

区内断裂构造发育，主要有东西向、北东向、北北东向及北西向。东西向构造多形成于燕山期之前，后期有继承性活动；沿河城-南口-琉璃庙大断裂呈北东 30°~50°斜贯成矿带，该断裂经历了多期次活动，变形过程复杂。上述各组不同方向构造交汇部位控制了区内不同时期岩浆活动。此外，在古老变质岩系中发育不同规模、不同方向的韧性剪切带，是矿带内金成矿作用重要控矿和赋矿构造。

本区是燕山期陆块活化强烈地区之一，不仅燕山期断裂构造发育，而且岩浆活动强烈频繁，侏罗纪—白垩纪发生了 3 期相对应的火山喷溢和岩浆侵入活动，中部为八达岭侵入杂岩带，两侧为断续相连的凹陷火山盆地。燕山期大规模的构造-岩浆活动，不同期次成矿元素的叠加进一步增加了成矿可能性，提供了良好的成矿条件，形成了铜、铅、锌、钼、铁、金等一系列内生金属矿产。燕山期也是北京地区最重要的内生金属矿产成矿期。燕山期构造-岩浆活动也为古老变质岩系（矿源层）中金的活化、迁移、聚集成矿提供了大量热能，也可能带来部分成矿物质。

在八达岭成矿带内，依据内生金属的矿化强度，控岩控矿构造条件，又可划分出 4 个重要成矿亚带。

1）石槽-石窑-银冶岭 Cu、Ag、Pb、Zn 成矿亚带（北西向）（Ⅲ-57-2-②a），该带为北西向隐伏构造岩浆岩带，控制燕山中期（晚侏罗世）一系列中浅成—超浅成中酸性—酸性小岩体侵位，与其成因密切的内生多金属有铜、钼、金、银、铅、锌。

2）薛家石梁 Au、Mo、Cu、Pb、Zn 成矿亚带（北西向）（Ⅲ-57-2-②b）：此带分布于大庄科—东三岔—分水岭—上庄—大羊山一线，位于北西向薛家石梁构造岩浆岩带内，岩带内广泛出露燕山早、中、晚期侵入岩体，与其关系密切的内生多金属有钼、铅、锌、金、锰、铁等。

3）大榆木沟-黄场-大庄科 Mo、Cu、Pb、Zn 成矿亚带（北东向）（Ⅲ-57-2-②c）：此带分布于大榆木沟—黄场—大庄科一线，带内主要发育中浅成—超浅成中酸性—酸性岩体侵位，发育有钼、铜、铅、锌等矿种。

4）延庆怀来新生代断陷盆地地热成矿亚带（北东向）（Ⅲ-57-2-②d）：位于河北省怀来县—延庆新生代北东向断陷盆地内，主要蕴含地热资源。

3. 京西煤、石灰岩、Ag、Pb、Zn 成矿带（Ⅲ-57-2-③）

根据区内地质作用、古地理古构造、岩性岩相、赋矿岩系层位及岩性条件、经构造岩浆岩活动改造、成矿特征的一致性划分为一个大的成矿带。该成矿带主要位于北京西山房山、门头沟区域内，属西山断陷盆地、房山碳酸盐岩台地和十渡-房山陆表海盆地Ⅳ级构造亚相。

区内褶皱发育，印支期褶皱呈近东西向—北西西向，为宽缓的向斜构造，控制了古生界、中生界的分布，主要构造行迹有百花山-庙安岭向斜、九龙山向斜、北岭向斜等。

成矿带内的石灰岩矿层赋存于寒武系和奥陶系中，主要分布于向斜的两翼。煤层主要赋存于石炭系—二叠系和下侏罗统门窑坡组中，主要分布于向斜近核部两翼和核部。因此成矿带内的北东向向斜是控制了石灰岩和煤矿的主要构造。印支期宽缓褶皱隆起带内，常见有晚三叠世（印支期）中酸性侵入岩体（小岩体、岩床），其深部可能有隐伏大岩体。在隆起内核部雾迷山组、铁岭组中，多见有顺层产出的中低温—低温充填型金矿（珠窝金矿）、重晶石、铅锌矿、银（北直河、大西沟）矿化、矿点和小型矿床。

4. 密怀地区 Fe、Cu、Au、Cr 成矿带（Ⅲ-57-2-④）

位于赤城-古北口和密云-墙子路两条断裂之间的密怀穹隆核部，是区内中生代以来隆升幅度最大的地区，控制了太古宇密云岩群表壳岩含铁建造和片麻岩古老侵入体的集中成片分布，属密怀高级变质基底杂岩构造亚相。

成矿带位于东西向深大断裂和北东向崎峰茶大断裂的交汇部位，带内近东西向、北东向、北北东向次级断裂发育。带内岩浆活动频繁，从太古宙、中元古代到中生代各时期都有岩浆侵入活动。在构造、岩浆岩多期、多次叠加活动的热动力作用下，重要的金矿源层（太古宙变质结晶基底）中金元素得以活化，并伴随岩浆热液迁移富集成矿。

5. 密云–平谷 Ag、Au 成矿带（Ⅲ–57–2–⑤）

位于平谷杨家洼、关上、孔城峪和密云银冶岭、冶仙山一带，处于河北省冀东金成矿带的西端。

大地构造位置处于平谷大陆边缘裂谷构造亚相，密云–墙子路东西向大断裂南北两侧（以南侧为主）。

区内大面积出露中元古界长城系、蓟县系，北部分布有新太古代变质岩，西部零星出露上侏罗统。褶皱、断裂构造较发育。褶皱构造以北东向褶皱规模较大，其轴部位于关上—大华山一线，有古老变质岩出露，两翼为中元古代地层；断裂构造除密云–墙子路大断裂外，南口–平谷–山海关隐伏深断裂穿过本区，对区内构造–岩浆活动有一定的控制作用。此外，还发育次级北东向、北北东向、北西向构造。燕山裂陷槽强烈下陷的中心部位即位于区内，中元古代有富钾玄武质、粗面质熔浆喷溢，以及火山角砾岩管、辉长岩等侵位，火山活动中心在平谷熊儿寨一带。北部沿东西向大断裂有中元古代沙厂球斑花岗岩侵入。燕山期中性—酸性中浅成侵入体产于断裂构造或构造复合部位，多呈规模较小的岩株、岩墙或岩脉，如银冶岭闪长岩，三百山石英闪长岩，盘山花岗岩，关上、玻璃台一带花岗斑岩等，常与金、银矿化有较密切关系。倪家洼、将军关一带分布前长城期基性火山角砾岩、超铁镁质岩等。

二、主要矿产资源产出的时间规律

北京地区对已发现固体矿产成矿年代学研究相比邻区较弱。近 10 年来，万渝生、金文山、王慧初、邓晋福、汪洋、王登红等对区内地层、岩体及矿床采取样品，获得了可观的测年数据，并进行了成因方面探讨研究，取得了丰硕的成果。对煤炭、铁、金、铜、铅、锌、钨、锰、钼、铬、银、萤石等 12 个矿种进行统计分析，北京地区成矿总体上呈现一老一新的特点：太古宙是本市沉积变质型铁矿、岩浆分异型铬铁矿主要成矿期，矿床数达 41 个；燕山期是本市内生金、铜、铅、锌、多金属矿产最重要成矿期，矿床数达 53 个（图 47–4）。

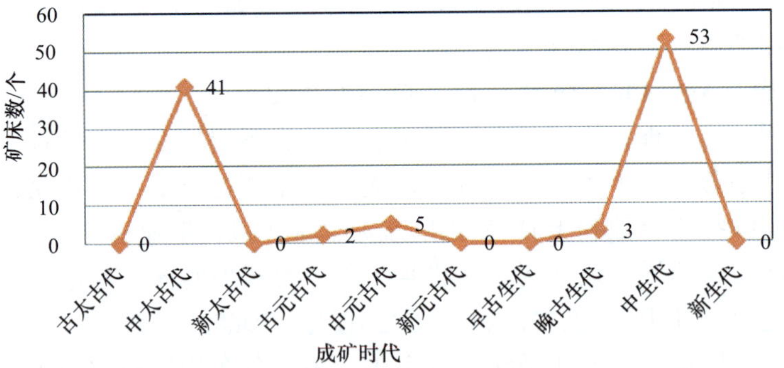

图 47–4　北京地区煤炭、铁等 12 矿种成矿时代统计表

Fig. 47–4　Statistical table of coal, iron and the other twelve mineral metallogenic epoches in Beijing area

三、主要矿产资源空间分布规律

1. 黑色金属

铁矿是北京市优势矿产资源之一，分布较为广泛、类型多，成矿时间表现为一老一新的特点。其

中以沉积变质型铁矿最为重要，具有典型的 BIF 铁矿特征，分布相对集中，资源潜力大，主要位于密怀地区成矿带；岩浆型铁矿多为钒钛磁铁矿，位于昌平上庄和怀柔新地两地；接触交代型铁矿主要由中酸性岩体侵入到蓟县系形成矽卡岩接触带铁矿床，主要位于八达岭成矿带内。

锰矿：位于八达岭成矿带内的昌平西湖地区，为接触交代型锰银矿，为小型矿床。

铬铁矿：分布于密怀地区成矿带的密云平顶山、放马峪等地，铬铁矿主要赋存于岩体的方辉橄榄岩相（第一次）和纯橄榄岩相（第二次）中，每个主要矿体形成一个聚合群，中心部位为致密块状铬铁矿床，两侧分别是稠密、中等、稀疏浸染状矿石。

2. 有色金属

以铜、钼、铅、锌矿等最重要，常形成中—小型矿床，主要分布于八达岭成矿带黄场和石槽地区。斑岩型铜钼矿分布于北东向沿河城-南口大断裂的旁侧，与该地段斑岩体侵入有关；接触交代型铜钼矿在延庆石槽地区，为小型矿床。

钼矿：北京地区目前已知钼矿主要富集于延庆区董家沟—大庄科—石窑一带。成因类型有：岩浆角砾岩型，如延庆区董家沟、大庄科；热液型，如延庆区东三岔、怀柔区石湖峪；接触交代型，如平谷区石塘峪、延庆石槽。上述成因类型中最具有工业意义的为岩浆角砾岩型，达到中型规模，其余均为小型或矿化点。

钨矿：主要位于密云成矿带内，与中元古代沙厂-墙子路断裂带和沙厂环斑花岗岩侵位有关。矿床类型为岩浆热液型，已发现小型矿床。

铅锌矿：主要为接触交代型，大部分位于延庆东三岔、门头沟大榆木沟、密云银冶岭等地区，其余地区零星分布。其中，延庆东三岔地区铅锌矿床达到中型规模，其余均为小型或矿化点。

3. 贵金属

金矿：金矿分布广泛，5 个 V 级成矿区带内均有分布。在各种构造环境中，几乎都可见到金矿床（点）。

成因类型有变质构造蚀变岩型，常形成于密云太古宙陆核边缘，已发现矿床规模多为中型，可实现突破达到大型规模。如得田沟金矿、怀柔东岔金矿。

岩浆型金矿：位于八达岭成矿带的昌平分水岭和延庆石窑地区，主要与燕山期石英二长岩、二长花岗岩关系密切，已发现矿产地有分水岭金矿、怀柔杨树底下金矿、延庆石窑金矿等。

银矿：本市银矿多为银铅锌共伴生矿床，成因类型为接触交代型，主要位于密云地区银冶岭和房山北直河等地，与蓟县系碳酸盐地层关系密切，已发现有银冶岭银铅锌矿、北直河银铅锌矿等。

4. 非金属（萤石）

北京地区萤石矿规模较小，仅在怀柔兰营地区出现，为低温热液型矿床。

5. 煤炭

煤炭为北京市优势矿产资源之一，分布相对集中，主要位于京西门头沟区和房山区，顺义龙庭侯地区有少量分布，主要与石炭-二叠系海陆交互相地层及侏罗系窑坡组地层关系密切。

第三节　开发利用状况

北京山区的矿山开采已经有数百年历史。由于改革开放初期过分追求经济效益，忽略了环境保护与治理，造成了矿山环境的持续恶化。西部房山和门头沟地区、北部怀柔和密云等地，是矿山环境问题严重区；西山山前及永定河流域、昌平山前及怀柔等地区为矿山环境问题较严重区，以中小型矿山为主，开采矿种主要为煤炭、金矿、非金属建材矿等；昌平、延庆、房山等区县以开采石灰岩、花岗岩等非金属矿为主的矿山环境问题一般。常见的矿山环境问题主要包括废渣废水粉尘等污染环境、矿区地下水位下降、地形地貌景观破坏、矿山土壤环境破坏等，还可能诱发崩塌、泥石流等次生地质灾害。

根据《北京市总体规划（2004—2020 年）》，北京定位于建设空气清新、环境优美、生态良好的宜居城市。2008 年以来，随着产业结构调整和城市功能定位的改变，北京一直实施关矿、限采政策，2009 年以来，在生产固体矿山企业数量大幅度减少，自 2004 年至 2012 年底，全市累计关闭固体矿山 808 家，剩余在生产矿山 34 家（图 47-5）。2013 年，通过对 1443 处（含在生产矿山 34 处）矿山调查，其中：407 处矿山已经得到治理；289 处矿山改建成道路、公园、房屋、池塘等已经综合利用；121 处可以自然恢复不用治理；尚有 538 处未治理且存在不同程度的矿山地质环境问题；88 处局部已经治理，仍需进一步治理。2014 年度在生产的固体矿产地共 21 处，其中昌平 4 处、密云 5 处、顺义 2 处、怀柔 1 处、门头沟 4 处、房山 5 处，开采量较大的主要为煤矿、铁矿等。截至 2014 年，市辖区内已无有色金属、贵金属及稀有元素矿产、化工原料矿山企业（图 47-6）。

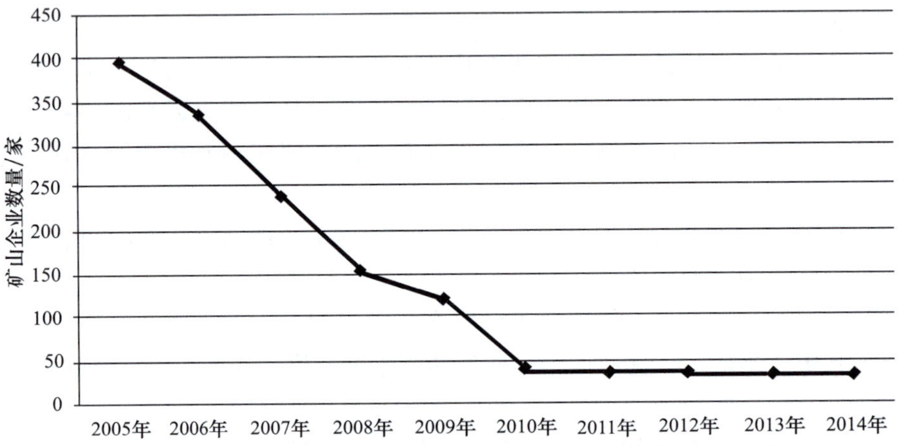

图 47-5　固体矿山企业数量变化情况

Fig. 47-5　Changing situation of solid mine enterprises number

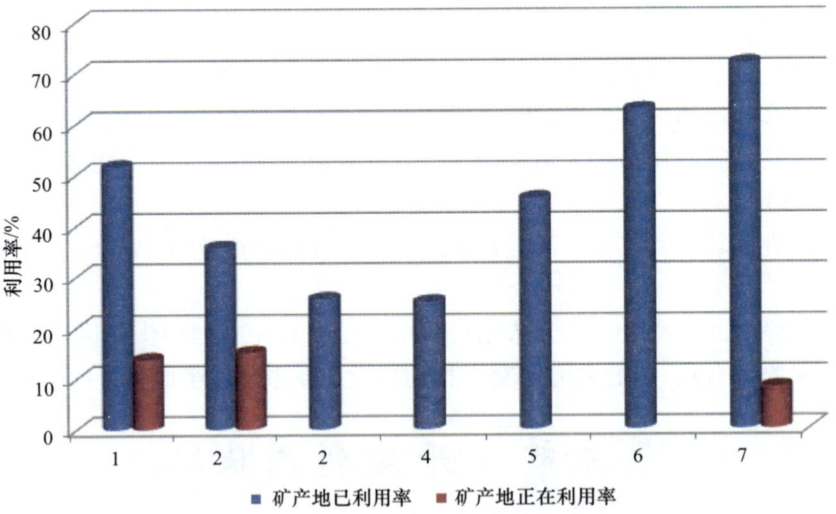

图 47-6　截至 2014 年底矿产利用情况统计表

Fig. 47-6　Statistical table of mine utilization situation by the end of 2014

1—能源矿产；2—黑色金属矿产；3—有色金属矿产；4—贵金属及稀有元素矿产；5—冶金辅助原料矿产；6—化工原料非金属矿产；7—建筑非金属矿产

近几年，北京市对山区矿山环境恢复治理工作十分重视，加大了山区矿山环境恢复治理的力度，并且启动了一系列的矿山环境恢复治理项目，取得了初步治理效果，例如房山霞云岭乡四马台煤矿、房山史家营乡大村涧煤矿治理工程等。建立了北京市矿山地质环境监测系统，对矿山地质环境现状进行动态分析评价，监测治理成果，预测发展趋势，研究治理恢复措施。随着矿山的逐步关闭，矿山环境破坏程度将逐渐减轻，再加上矿山环境治理工作的落实，北京市矿山地质环境将逐渐改善。

第四十八章　地热与浅层地热能资源

北京地区的可再生能源主要包括地热与浅层地热能（亦称浅层地温能）资源。根据地热资源勘查规范的定义，地热资源是指埋藏于地表以下、含有一定温度的自然资源（温度大于25℃），包含蒸汽型、干热岩型、热水型等多种形态。北京市的地热资源以中低温亚经济的热水型为主，热水型地热资源是指赋存于天然地下水及其蒸汽中的地热资源，是一种可以替代传统化石燃料的清洁能源，北京市的热水型地热资源温度范围为25~117℃，存在方式有两种，一种是自然出露的温泉，另一种是人工钻凿的地热井。浅层地热能是指地球浅表层一定深度范围内（一般为200m）的岩土体、地下水和地表水中所蕴藏的一种低温热能，其能量主要来源于太阳辐射和地球梯度增温，在当前经济技术条件下具备开发利用价值的低温地热资源，其温度一般低于25℃，是一种可再生、环保、储量巨大的新型能源。地热与浅层地热能资源的开发利用，对缓解首都能源供应压力、保障能源供应安全、促进地区经济发展具有重要意义。

第一节　地　　热

北京市的地热开发利用历史悠久，据《水经注》记载，早在北魏时期就有了延庆佛峪口塘子庙温泉可沐浴治病的记载。而最有名的当数小汤山温泉，从宋元时代就已经开发利用。20世纪70年代初，随着常规能源短缺，地热悄然兴起。1969—1970年，开始物探测温工作。1970年4月—1971年4月，在地质部大院开凿京热1井。1971年4月29日，地质部长李四光先生因病逝世，在他遗物中发现一张纸条，上面写着："在我们这样一个伟大的社会主义国家里，我们中国人有志气、有力量克服一切科学技术上的困难，去打开这个无比庞大的热库，让它为人民所利用。如果我们不这样做，而还是走资本主义陈腐的老路，把地球交给我们珍贵的遗产——煤炭之类内容极其丰富的财富，不管青红皂白一概当作燃料烧掉，不到一千年，我们的后代，对我们这种愚蠢和无所作为的行径，是不会宽恕的。"在李四光先生的倡导下，北京兴起了地热勘查开发的热潮。

一、地热田分布

《北京市地热资源2006—2020年可持续利用规划》将北京平原区深度3500m内，划分出相对独立又有一定水力联系的10个地热田，总面积约2760km²，其中，开发研究程度较高的为东南城区、小汤山、良乡及李遂地热田。北京市地热研究院对地热田的综合研究成果显示，最高温度地热井为位于凤河营地热田的兴热-12号井，出水温度为117℃（表48-1，图48-1）。

延庆地热田主要是依据其东南侧北东走向的康庄-旧县断裂、西北侧北东走向的古城-狼山断裂及东侧南北走向的断裂划分的，位于延庆凹陷盆地内，地热田西南地区较东北地区富水性差。

小汤山地热田在构造上位于门头沟断陷盆地和八达岭陆内造山岩浆杂岩接触地带，处于北西向南口-孙河断裂与北东向黄庄-高丽营断裂交汇的三角地带，内部被多条断裂切割，北西向断裂具有明显的导水导热作用，为热田内温度较高的地带。

后沙峪地热田在构造上处于北京断陷北部，西侧以南口-孙河断裂为界，北侧以黄庄-高丽营断裂为界，具有深部热储层和较高的热储层温度，具备开发深部地热资源的条件。

京西北地热田位于门头沟断陷盆地，东北以南口-孙河断裂为界与小汤山地热田相连，热田内深大断裂较多，主要有八宝山断裂、黄庄-高丽营断裂等，具有一定的开发前景。

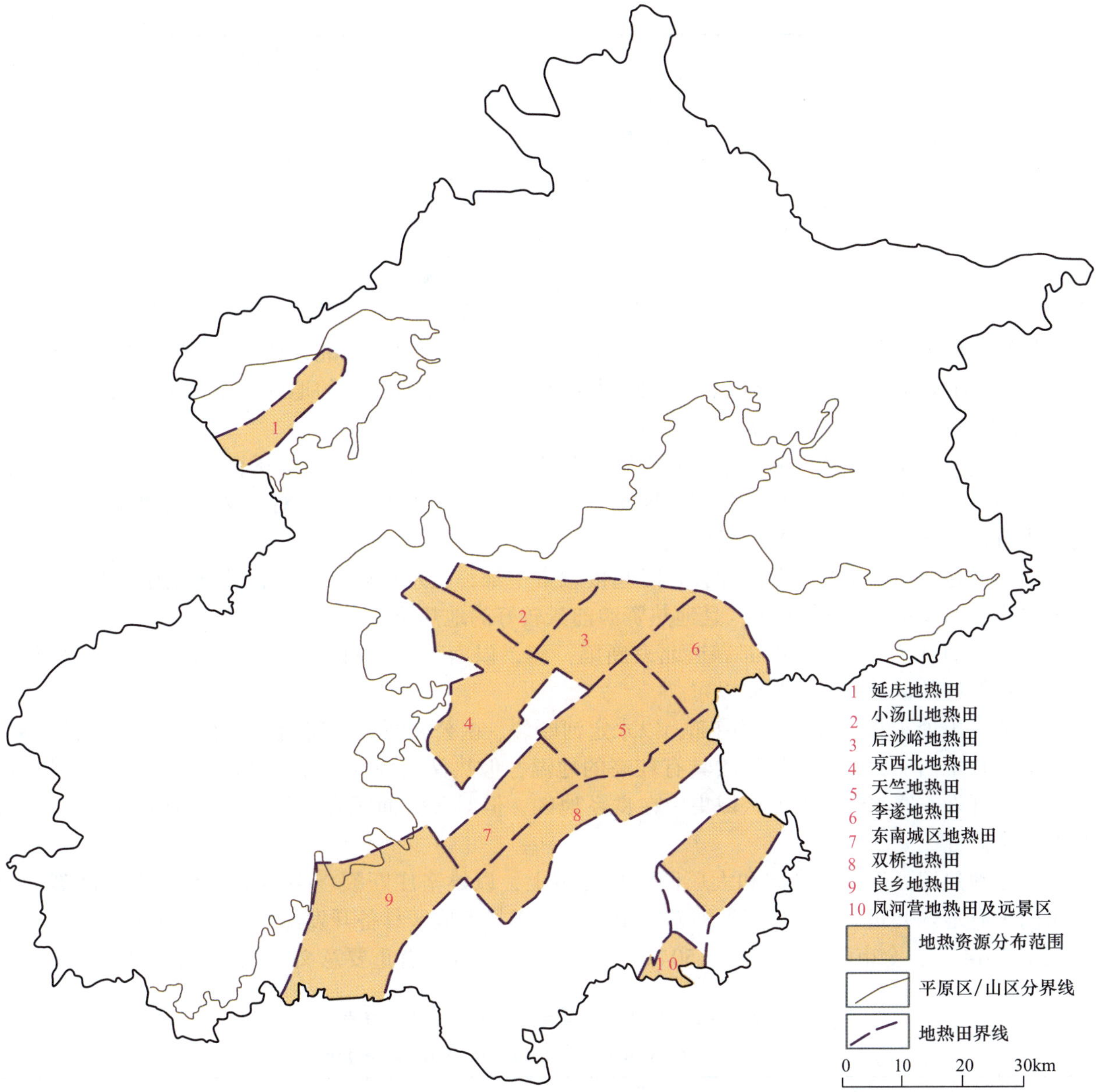

图 48-1 北京市地热田分布图

Fig. 48-1 Distribution map of geothermal field in Beijing

表 48-1 平原区地热田基本情况表

Table 48-1 Basic information of geothermal fields in plain area

序号	地热田	面积/km²	热储主要特征	
			热储埋深/m	最高温度/℃
1	延庆地热田	121.88	2000	80
2	小汤山地热田	186.42	0~3000	70
3	后沙峪地热田	239.85	2000~4000	78
4	京西北（沙河）地热田	363.21	2000~4000	78
5	天竺地热田	290.75	1000~2600	89
6	李遂地热田	273.04	400	72

续表

序号	地热田	面积/km²	热储主要特征	
			热储埋深/m	最高温度/℃
7	东南城区地热田	207.44	100~1000	88
8	双桥地热田	339.00	1500	57
9	良乡地热田	475.77	0~2000	69
10	凤河营地热田	262.51	2000~3500	117

天竺地热田处于北京断陷中北部，其西南以太阳宫断裂为界，西北以南皋断裂和顺义断裂为界，东南以南苑-通县断裂为界，东北以李家桥断裂为界，热田内富水性好的地区主要集中在太阳宫断裂与南苑-通县断裂交汇附近的高碑店地区。

李遂地热田处于北京断陷和大兴隆起接触带北部，以顺义断裂、李家桥断裂、南苑-通县断裂及皮各庄断裂为界，热田内大部分地区热储层埋藏较浅，热井出水温度较低，对于热储层埋藏较深的地区目前尚未有地热井，有待进一步开发研究。

东南城区地热田处于北京断陷和大兴隆起接触带中部，富水性较好的地区主要集中在热田东北部的东八里庄、六里屯-东铁营地区，是地热资源前景较好的地热田之一。

双桥地热田处于大兴隆起中部靠近北京断陷一侧，以南苑-通县断裂、永定河断裂、瀛海断裂为界，富水性较好的区域为通州梨园以东。

良乡地热田处于北京断陷的西南部，以永定河断裂、南公义断裂、琉璃河断裂、南苑-通县断裂为界。热田内长辛店—大富庄一带虽然具有较高的地温，但热储层埋深大于3500m，在当前经济技术条件下，不宜开发；富水性较好的区域集中在良乡地区，但对于大面积开发该地区的地热资源，还需进一步的技术研究。

凤河营地热田处于大兴隆起和大厂凹陷接触带上，以姚辛庄断裂、牛堡屯断裂和夏垫断裂为界，地热田内现有地热井相对较少，但凤河营、永乐店、漷县等地区具备开发深部地热资源的前景。

截至2014年，全市已钻探地热井504眼，其中多孔钻井具有重要意义（表48-2，图48-2）。

表48-2　北京地区具有重要意义钻井一览表

Table 48-2　List of significant drilling in Beijing

地区	钻井编号	时间	产能	意义
小汤山	汤5-2号	1974年	出水温度42℃ 出水量968m³/d	突破前人圈定的低温范围
	汤热7井	1981年	出水温度59.4℃ 出水量1380m³/d	证实小汤山南部具有良好的地热条件
城区	JR_3	1971年3月	出水温度39.2℃ 出水量1144m³/d	北京城区第一眼成功的地热井
	JR_5	1971年8月	出水温度53℃ 出水量1144m³/d	引起北京市民的轰动
	JR_{117}	2002年	最高热恢复 水位41.8m	北京地区热恢复水位最高的地热井
城区	JR_{120}	2003年	出水温度89℃ 出水量1019m³/d	北京地区出水温度最高的地热井

续表

地区	钻井编号	时间	产能	意义
延庆	庆 2	1977 年	出水温度 32.5℃ 出水量 810m³/d	预示延庆盆地的地热前景
	新延热 1	1993 年	出水温度 52.5℃ 出水量 1850m³/d	打开延庆地区地热开发的局面
良乡	碧 1	1995 年	出水温度 54.6℃ 出水量 2163m³/d	开拓了良乡地区地热开发新局面
京西北	香热 1	1999 年	出水温度 42℃ 出水量 1500m³/d	“门头沟叠陷褶”内第一眼地热井
	沙热 5	2000 年	出水温度 69℃ 出水量 887m³/d	开拓沙河地区的地热开发新局面
	JR_{167}	2006 年	出水温度 64℃ 出水量 2324m³/d	北京地区最深的成功地热井，井深 4088.88m
顺义	顺热 1	1983 年	出水温度 48℃ 出水量 2900m³/d	天竺地区第一眼地热井
	208-4	1985 年	出水温度 43℃ 出水量 1143m³/d	李遂地区第一眼地热井
	顺后热 2	2002 年	出水温度 75℃ 出水量 1965m³/d	北京地区第一眼侏罗系地热井
通州-南苑	双深 1	1989 年	出水温度 43℃ 出水量 1800m³/d	为通州-南苑地区地热开发奠定基础
北部山前	昌热 1	1997 年	出水温度 40℃ 出水量 1200m³/d	昌平西南部地区第一眼地热井
	怀热 3	2006 年	出水温度 40℃ 出水量 603m³/d	怀柔地区第一眼地热井
	平热 1	1998 年	出水温度 41.2℃ 出水量 2301m³/d	平谷地区第一眼地热井

（据《北京地热》，2010）

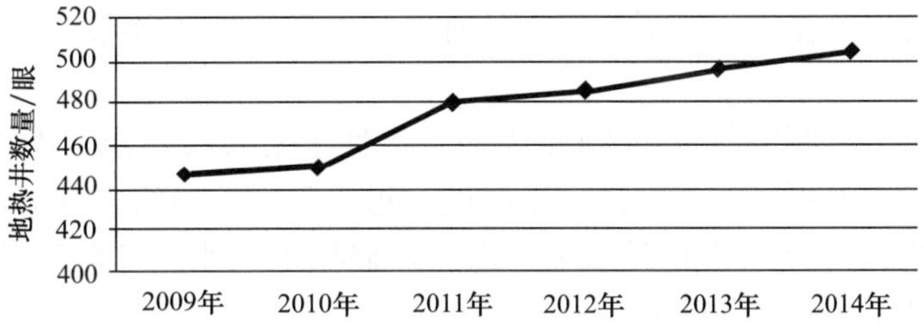

图 48-2　北京市近年来钻探地热井数量

Fig. 48-2　Number of geothermal wells drilled in Beijing in recent years

二、热储层

热储层是指地热流体相对富集，具有一定渗透性并含载热流体的岩层或岩体破碎带。

北京平原区基岩广泛分布着碳酸盐岩地层，它们经历了多次构造变动和溶蚀，裂隙和岩溶发育，是地下水良好的储存场所。在这套地层之上普遍沉积了数百米至数千米的新生代—中生代地层，形成了相对隔水、隔热的保温盖层。从平原区多眼地热井的统计资料来看，岩性以泥岩、砂岩、砾岩、安山岩、凝灰岩及玄武岩为主的有古近系、新近系、白垩系和侏罗系，岩性以页岩、粉砂岩及细砂岩为主的有石炭系、二叠系，岩性以页岩为主的有待建系下马岭组和蓟县系洪水庄组等。渗透性一般来说很差，可以限制地下水的对流，构成了保温盖层。在上述盖层中，各地层的地热增温率也有一定的差别，其中以页岩和泥岩保温效果最佳。

北京地区主要热储层为蓟县系，其中以雾迷山组最好，铁岭组次之；寒武纪和奥陶纪灰岩在盖层条件具备的情况下，也是较好的热储层。另外，在个别地区高于庄组白云岩和裂隙发育的侏罗纪安山岩热水储集也较为丰富（表48-3）。

表48-3 各热田主要热储层一览表

Table 48-3 List of thermal reservoirs in each geothermal field

序号	地热田	热储层
1	延庆地热田	雾迷山组
2	小汤山地热田	铁岭组、雾迷山组
3	后沙峪地热田	铁岭组、雾迷山组、奥陶系
4	京西北（沙河）地热田	奥陶系、铁岭组、雾迷山组
5	天竺地热田	雾迷山组
6	李遂地热田	铁岭组、雾迷山组
7	东南城区地热田	铁岭组、雾迷山组
8	双桥地热田	铁岭组、雾迷山组
9	良乡地热田	雾迷山组
10	凤河营地热田	奥陶系、雾迷山组

（一）蓟县系热储层

蓟县系热储层主要为铁岭组和雾迷山组，两者之间被一套厚80m左右的洪水庄组隔热保温层分开。蓟县系储量占据了北京市地热资源总储量的80%以上，是主要的热水储层。

雾迷山组是北京地区分布面积最广的热储层，在平原区约占90%以上，岩性主要为硅质白云岩、燧石条带白云岩、纹层状泥晶白云岩等，厚度一般大于2000m。由于受多次地质构造运动的影响，雾迷山组裂隙较为发育，为岩溶作用创造了条件，具有较好的储存空间和渗透能力，形成北京地区主要的热储层。

铁岭组在洪水庄组之上，厚度一般为350m左右，岩性主要为白云岩、灰质白云岩与泥晶白云岩。在同一地点雾迷山组储层水温更高，而且铁岭组厚度远远小于雾迷山组，前者的厚度不足后者的1/5，加之地质构造的原因，绝大部分地热井开采雾迷山组热储层地热水，部分地热井采用两层合并开采，仅有很少的地热井开采铁岭组热储层热水。

（二）寒武系与奥陶系热储层

寒武系热储层主要分布在南苑-通县断裂以北地区，分布几乎和断裂平行，呈带状展布，最大埋深不超过1500m。寒武系昌平组灰岩储层厚度较薄，一般不超过80m，而且相对储层的温度也较低，因此不作为主要热储层。炒米店组及张夏组也具有一定的赋水性。

奥陶系热储层在黄庄-高丽营断裂和顺义断裂之间分布最广，在来广营及其附近地区储层埋藏最深。在黄庄-高丽营断裂的北侧与八宝山断裂之间的地域内也分布有奥陶系热储层，埋深基本表现是由北东向南西渐深。奥陶系热储层的水温和水量都较理想。在永定河断裂以东黄庄-高丽营断裂的南侧分布有奥陶系热储层，延断裂呈北东向带状展布。奥陶系在良好盖层具备的条件下是除蓟县纪热储层之外的主要热储层，储层全厚度达750m左右。

（三）其他热储层

高于庄组热储层埋藏较深，主要分布在密云变质基底杂岩南部的平原区和大兴隆起南部构造抬升较高的地区。目前钻探能够钻到并成功开凿的昌热1、昌热3、兴热4及平热1井均取自高于庄组。

在地热井的钻探施工过程中曾多次在侏罗纪安山岩的裂隙中发现较好的热水储集井段，这些井一般靠近断裂构造。如顺后热2井在钻探过程中发现侏罗纪安山岩有很好的地热显示，于2919m深度终孔，成井于侏罗系，获得了日出水量近2000m^3、出水温度达75℃的热矿水，而且自流，水头高出地面17m。另一眼取侏罗系构造水的是金融街京热91井，该井3545m未见蓟县系储层，后在侏罗系中成井，日出水量接近500m^3。可见盖层在储存空间和渗透性的条件下，可以构成热储层。古近系和新近系热储层主要分布在北京平原区的东部及南部边缘、觅子店凹陷及固安新凹陷内，在华北平原地区此类孔隙性热储层显示良好的潜力，在北京的部分只是边缘地带，地热地质条件相对较差。

三、地热水化学特征

北京地区的地热水多为矿化度在500～700mg/L之间的重碳酸、硫酸钠型水，氟、偏硅酸含量较高，多数为氟、偏硅酸医疗热矿水，另外地热水中还含有一定量的其他微量元素。总体说来，北京地区地热水水质良好。

1）绝大多数地热井水中氟及偏硅酸含量达到医疗价值标准，有的达到命名浓度。

2）部分井地热水中偏硼酸含量达到有医疗价值，个别达到命名浓度。

3）部分地热水中有硫化氢味，少部分达到矿水浓度。

4）个别井水中氡、镭、碘达到了命名标准。

5）大部分地热井水中的矿化度小于1g/L，属于淡温泉水，少部分大于1g/L，最高者达到6.7g/L。

6）北京地区的地热水一般属微硬或软水。

北京地区各地区地热水水化学类型见表48-4。

表48-4 北京市各区域水化学类型表

Table 48-4 Hydrochemical types of each region in Beijing

序号	地热田（地区）名称	热储层	水化学类型
1	良乡	蓟县系雾迷山组	HCO_3-SO_4-Ca-Mg-Na HCO_3-SO_4-Ca-Na HCO_3-SO_4-Na-Ca
2	东南城区	蓟县系雾迷山组	HCO_3-SO_4-Na-Ca HCO_3-SO_4-Na HCO_3-Na HCO_3-SO_4-Cl-Na HCO_3-Cl-Na Cl-HCO_3-SO_4-Na SO_4-Na
3	双桥	蓟县系雾迷山组	HCO_3-SO_4-Na HCO_3-Na HCO_3-SO_4-Na-Ca HCO_3-Cl-Na

续表

序号	地热田（地区）名称	热储层	水化学类型
4	天竺	蓟县系雾迷山组	HCO_3-Cl-Na Cl-HCO_3-Na SO_4-Cl-Na SO_4-Na
5	李遂	蓟县系雾迷山组	HCO_3-Na
6	小汤山	蓟县系雾迷山组	HCO_3-Na-Ca HCO_3-Ca-Mg
7	京西北（沙河）	蓟县系雾迷山组	HCO_3-Na-Ca HCO_3-Ca-Mg
8	延庆	蓟县系雾迷山组	HCO_3-Na-Ca

（据北京市地热资源潜力评价成果报告）

四、地热资源量

2015 年，北京市水文地质工程地质大队对北京市地热资源现状进行了调查评价与区划，据其评价结果，北京市地热资源总量为 9.94×10^{19}J，折合标准煤 3.39×10^{9}t（1t 标准煤大约为 2.93×10^{10}J，燃煤锅炉的热效率按 0.6 计算）；地热资源可开采量为 1.49×10^{19}J，折合标准煤 5.09×10^{8}t。

第二节　浅层地热能

1997 年，中国科技部和美国能源部正式签署《中美能源效率及可再生能源合作协定书》是我国浅层地热能勘查开发起步的标志性事件，双方从国家政府层面对地源热泵进行扶持和引导，建立了中美两国政府地源热泵技术合作项目北方示范工程即北京朝阳区的嘉和丽园地下水热泵工程。北京地区众多的科研机构、地质勘查单位、勘察设计单位、民营企业纷纷看好浅层地热能资源开发利用前景，大力引进、消化与吸收、研究浅层地热能资源开发利用核心技术。经过多年科研积累和工程实践，北京地区浅层地热能资源开发利用与研究已走在了全国的前列。

一、浅层地热能开发利用现状及资源潜力

地源热泵系统是北京浅层地热能资源开发利用的主要形式，主要包括地下水地源热泵、地埋管地源热泵和地表水地源热泵，其中地下水地源热泵项目主要分布在海淀、丰台、朝阳、顺义 4 个区，地埋管式主要分布在顺义、昌平南部、海淀北部、大兴等地区，地表水地源热泵项目较少。结合《北京市城市总体规划（2005—2020 年）》《北京市土地利用总体规划（2006—2020 年）》和北京市各区国民经济和社会发展第 12 个五年规划纲要，北京市浅层地热能资源开发利用重点区域包括中心城区、规划新城、重点镇和新农村。

浅层地热能开发利用工程使用范围较广，建筑类型既包括办公楼、商业建筑、教学楼、医疗、宾馆酒店等，其中办公楼和商业建筑所占比例较大，超过了 50%。其服务功能以冬季供暖和夏季制冷为主，兼顾提供生活热水，少量项目有加热游泳池、加热景观水池和温室大棚等功能。但北京地区浅层地热能开发利用过程中也存在一些问题，例如在冲洪积扇下部或水文地质条件较差等不适宜的地区建设了地下水地源热泵系统，部分项目前期未进行详细的浅层地热能资源勘查造成无水可抽或回灌困难。

截至 2012 年底，北京市建成 1042 个地源热泵项目（图 48-3），其中地下水地源热泵项目 876 个，实现供暖面积约 $2506\times10^{4}m^{2}$；地埋管地源热泵项目 157 个，供暖面积约 $702\times10^{4}m^{2}$；地表水地源热泵项目 9 个，供暖面积约 $68\times10^{4}m^{2}$。卢惠华（2013）统计表明，北京有国内运行最大的地源热泵系统，

如中石化管理干部学院地源热泵系统项目和国家北方林木基地地源热泵系统，也有全国最大的燃煤锅炉改为水源热泵的供暖和制冷项目。

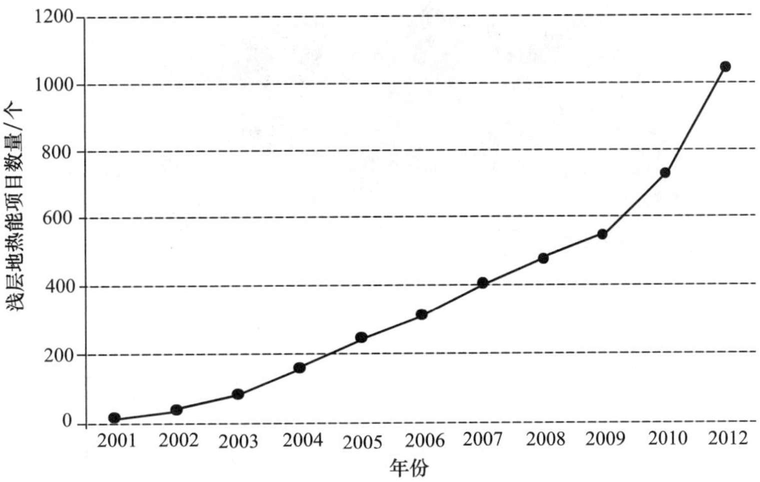

图 48-3　北京市近年来浅层地热能项目数量

Fig. 48-3　Number of shallow geothermal energy project in Beijing in recent years

2013 年，北京市地质勘察技术院开展了浅层地热能资源调查评价工作，计算出平原区浅层地热能资源量折合标准煤 857×10^4t，可为 7.21×10^9m² 建筑供暖。

二、影响浅层地热能赋存与开发的因素

浅层地热能资源主要赋存在地表以下 200m 内，受地层结构、岩性、构造、水文地质条件等因素影响显著。

（一）地层结构

第四系岩土体的热导率、天然含水率、天然密度、相对密度、饱和度、孔隙比等热物性特征，是影响浅层地热能资源赋存和开发利用适宜性的重要因素。2012 年，北京市地质勘察技术院在平谷、顺义、大兴、通州、房山、延庆、昌平和密云等地区开展了取芯钻探工作，取样 301 件，对粉质黏土、黏质粉土、黏土、重粉质黏土、砂质粉土及砂质黏土进行了热物性分析。根据其研究成果，黏土、粉质黏土、重粉质黏土的孔隙比、天然含水率和相对密度要略高于黏质粉土和砂质粉土，黏土、粉质黏土、重粉质黏土的热导率要略大于黏质粉土和砂质粉土，热导率值随着土体天然含水率和天然密度的增加而增加。

（二）水文地质条件

水量、水温、水质和供水稳定性是影响地下水地源热泵系统运行效果的重要因素，水量充足、水温适度、水质适宜、供水稳定、含砂量 $<5\times10^{-6}$、浑浊度 <20mg/L、固体颗粒物粒径 <0.5mm、pH 6.5~8.5、CaO 含量 <200mg/L、矿化度 <3g/L 的环境有利于地下水地源热泵系统的运行。

地下水渗流对地埋管换热能力有着非常重要的影响，由于地下水的热容量大，吸收或散发热量也大，在有地下水渗流的情况下，热量或冷量容易被地下水带走，形成另外一条热流通道，大大降低传热热阻。

（三）构造条件和地温场条件

北京平原区 50m 深地温主要集中在 13~18℃之间，异常区主要集中在城北地区，城南区异常区较少，呈点状无规则分布，异常区的地温梯度多在 2.5℃/100m。地温最高约 25℃，出现在小汤山地区（图 48-4）。

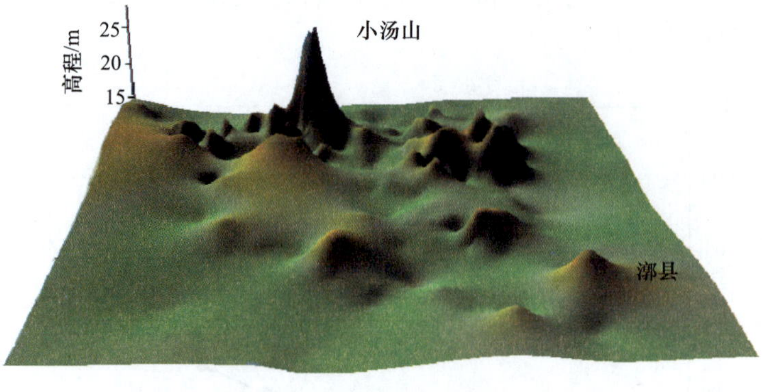

图 48-4　平原区地下 50m 深地温 3D 表面图
（据北京市平原区浅层地热能资源潜力评价）
Fig. 48-4　Geothermal three dimentional surface map of plain areas below 50m depth

70m 深地温异常区主要分布在平原区以北地区，并呈东西带状分布。阳坊-上庄-洼里异常区呈南北向条状分布，地温值在 16℃以上；小汤山异常区呈北东向片状分布，局部地温值超过 24℃；李桥异常区位于天竺的东部，呈放射性分布，地温值多集中在 19℃以上；亦庄、张家湾、永乐店异常区均位于城南地区，地温值集中在 17~19℃之间。

地温场分布与地质条件关系密切，地壳深处的热量以深大断裂为主要通道向上传导，对流过程中被具有隔热的盖层阻挡，聚集在具有赋水条件的地层中和周围的岩土体内。松散层内岩土体的热导率低于基底岩石热导率，因此浅部地温梯度高于深部基岩的地温梯度。一般情况下，离断裂构造部位越近，地温场出现异常的可能性越大，浅部产生地温异常的概率也就越大（图 48-5）。受岩石圈较薄和隐伏活动断裂的影响，北京平原区大地热流值（67 mW/m^2）略高于全球平均值（62.79mW/m^2）和我国大陆地区平均值（66 mW/m^2）。

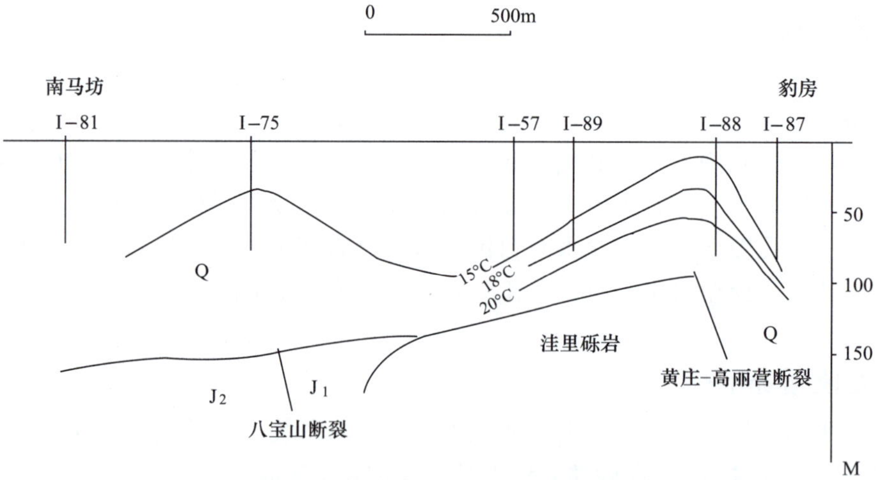

图 48-5　地温与断裂关系示意剖面图
（引自《北京市平原区浅层地热能资源潜力评价》）
Fig. 48-5　Profile sketch for geothermal and fracture relations

1. 小汤山地温异常区

小汤山地区寒武系和中—新元古界的碳酸盐岩埋深小，一般在 100~400m 之间，基本位于灰岩突起地带；小汤山断裂具有良好的导热导水性，热水井距该断裂越近，热储层的顶板温度越高，出水温度越高，出水量越大。另外，地下水的溢流导致土壤中氟元素富集，根据土壤地球化学背景分析结果，该地区土壤氟元素含量较高。

2. 洼里地温异常区

洼里地温异常区浅部地温高于临区，50m 深度温度一般为 17～19℃；70m 处温度为 18～21℃；200m 深度温度达到 22.5℃。异常区内地温梯度多在 5℃/100m 左右，远高于周围地区的 2.5 ℃/100m。异常区位于黄庄–高丽营断裂带西侧，平面上等温线与梯度线的闭合形态呈北东向条带状，基本上沿断裂延伸方向展布；剖面上断裂上方的地温在同一深度上达到最高值，向断裂两侧温度迅速下降。这些特征表明，黄庄–高丽营断裂对洼里地区的地温异常起控制作用。

3. 凤河营地区

该地区位于大兴隆起南部，区域内断裂构造发育，浅层地热能资源供暖潜力较好。

三、平原区浅层地热能开发适宜性区划

北京市地热研究院、北京市地质勘察技术院在浅层地热能赋存条件研究、影响因素分析、资源潜力评价、开发利用效益分析以及热能开采区地质环境监测等方面，开展了大量工作，并对平原区浅层地热能开发适宜性进行了区划。

（一）地下水地源热泵系统适宜区划

地下水地源热泵系统适宜区主要分布在平原区各冲洪积扇上部，面积约为 1523km^2。区内含水层富水性好，含水层颗粒等级在中砂以上，单井出水量大于 3000m^3/d（水位降深在 5m 以内），深度 100m 以内有效含水层累积厚度大于 10m，回灌能力强，回灌量与出水量之比大于 0.5。

地下水地源热泵系统较适宜区总面积约 4663km^2，主要位于平原区各冲洪积扇中下部以及山前河流冲洪积扇上部地区。其中，平原区各冲洪积扇中下部地区含水层富水性一般，含水层颗粒等级在中砂以下，单井出水量大于 500～3000m^3/d（水位降深在 5m 以内），深度 100m 以内有效含水层累积厚度小于 10m，回灌能力一般，回灌量与出水量之比小于 0.5；山前河流冲洪积扇上部地区含水层颗粒较粗，回灌能力强，回灌量与出水量之比大于 0.5，但由于长期超采地下水使地下水位下降明显，导致有效含水层累积厚度小于 10m，若将来地下水位恢复，则该区域可作为地下水地源热泵系统适宜区进行开发。

地下水地源热泵系统不适宜区总面积约 397km^2，主要分布在山前及残山附近。含水层富水性差，单井出水量小于 500m^3/d（水位降深在 5m 以内）。

（二）地埋管地源热泵系统适宜区划

地埋管地源热泵地质条件适宜区面积约为 3196km^2，主要位于平原区各冲洪积扇中下部。区内第四系厚度大于 100m，岩性主要为黏性土、粉细砂或砂层夹少数砾石层，卵砾石层厚度小于 5m，地埋管孔成井成本低。

地埋管地源热泵地质条件较适宜区面积约为 3387km^2，主要位于平原区各冲洪积扇上部。区内第四系厚度不等，岩性主要为砂、砂砾石层夹少数卵砾石，卵砾石层厚度大于 5m。

第四十九章　地质遗迹资源

第一节　研究简史

地质遗迹是指在地球漫长的地质演化历史中，由内、外动力地质作用形成、发展并遗留下来的、不可再生的各种地质体的总和，是人类认识地质现象、推测地质环境和演变条件的重要依据。北京地区的地球演化过程是华北地壳运动和演化历史的缩影，有着漫长悠久的地质发展演化历史和强烈的构造岩浆活动，造就了丰富的地质遗迹资源和多种多样的自然景观。

北京市专门针对地质遗迹的调查评价工作始于1988年，北京市地质矿产局水文地质工程地质公司根据地质矿产部和城乡建设环境保护部要求，对北京地区比较珍贵的、有较重大地质研究意义和保护价值的自然地质景观以及旅游地质资源进行科学考察，并进行地质自然保护区区划和旅游地质资源开发规划，于1990年完成了北京市旅游地质资源考察报告及相关图件，首次全面、系统地考察了北京山区地质自然资源情况，并在此基础上进行了地质自然保护区区划和旅游地质资源分区，提出国家级保护区（点）2处、市级保护区区划和自然区（点）14处。1996年，由北京市地质矿产局和北京市计划委员会共同开展“保护北京地区若干重要地质遗迹的调查”项目，对北京地区17处重要地质遗迹资源进行了专门调查，提交了保护北京地区若干重要地质遗迹的调查报告，对北京地区17处重要地质遗迹资源的特征进行了论述，并报道了这些地质遗迹资源遭受破坏的情况。

2000年，北京市地质矿产勘查开发局对北京市9类67处地质遗迹进行调查，编写了《北京市地质遗迹保护工作概况》。2007年以来，由中央及地方财政支持开展的北京市地质遗迹资源及其保护现状综合调查、北京市门头沟地质遗迹调查、北京市地质遗迹保护规划、华北地区重要地质遗迹调查（北京）等项目，针对北京市地质遗迹资源进行了全面的调查工作，基本摸清了北京市地质遗迹的家底，编制的调查报告还对地质遗迹分布情况、保护现状及资源价值等进行了评述，提出了地质遗迹资源保护及合理利用建议，为全面深入开展北京市地质遗迹资源的保护工作奠定了扎实的基础。2016年，本市统计重要地质遗迹50处，其中世界级2处、国家级7处、省级41处。

第二节　重要地质遗迹

一、遗迹类型及重要遗迹点

2014年，北京市地质研究所开展了华北地区重要地质遗迹调查（北京）项目，成果表明，北京地区分布有重要地质遗迹点50处，可分为3大类、12类、24亚类（表49-1）。

表49-1　北京地区重要地质遗迹一览表

Table 49-1　List of important geological relics in Beijing area

序号	大类	类	亚类	遗迹点	备注
1	基础地质类	地层剖面	层型（典型剖面）	马兰组层型	马兰组层型剖面
2				周口店组层型	富含古人类化石
3				密云岩群	北京最老变质岩
4			地质事件剖面	下苇店寒武纪地质事件剖面	受风暴事件影响

续表

序号	大类	类	亚类	遗迹点	备注
5	基础地质类	岩石剖面	沉积岩剖面	干沟沉积岩剖面	顶面痕
6				六渡沉积岩剖面	含硅质结核
7		构造剖面	不整合面	下苇店古生界与新元古界不整合面	
8				黄松峪中元古界与太古宇不整合面	吕梁运动期形成
9			褶皱与变形	七渡背斜	两翼对称的背斜
10				排字岭单斜	燕山期地层变形
11				孤山口固态流变构造	顺层掩卧褶皱
12			断裂	霞云岭逆冲推覆构造	北东向
13		重要化石产地	古人类化石产地	周口店古人类化石产地	人类化石
14			古生物群化石产地	大灰厂白垩纪热河生物群	
15				灰峪石炭纪—二叠纪古生物化石	
16			古植物化石产地	千家店侏罗纪木化石群	硅化木
17				岳家坡侏罗纪门头沟植物群	窑坡组蕨类植物
18			古动物化石产地	周口店动物群	中更新世动物群
19			古生物遗迹化石产地	千家店侏罗纪恐龙足迹化石产地	恐龙足迹
20		重要岩矿石产地	典型矿物岩石命名地	密云沙厂斜长环球斑花岗岩体	华北地区中元古代沉积的起点（不整合）
21				大石窝汉白玉产地	
22			矿业遗址	沙厂铁矿产地	
23				兰营萤石矿产地	
24				塔洼金矿产地	
25				杨树底下金矿产地	
26		冰川	古冰川遗址	模式口-八大处古冰川遗迹	擦痕、漂砾
27	地貌景观类地质遗迹	岩土体地貌	碳酸盐岩地貌	十渡岩溶地貌	峰丛、岩溶陡壁
28				旧县镇龙庆峡岩溶地貌	峡谷岩溶地貌
29				东关上-上方山洞穴群岩溶地貌	仙栖洞
30				佛子庄洞穴群岩溶地貌	石花洞、银狐洞
31				圣莲山岩溶地貌	圣米
32				黄松峪京东大溶洞岩溶地貌	规模最大、赋存地层最老（北京地区）
33			侵入岩地貌	房山花岗岩地貌	岩浆呈气球膨胀式侵位
34				云蒙山花岗岩地貌	发育最好、保存最完整（北京地区）
35				莲花山花岗岩地貌	八达岭杂岩的组成部分
36			碎屑岩地貌	黄松峪石英砂岩地貌	中国北方砂岩地貌的典型代表
37				六道河石英砂岩地貌	小昆仑山砂岩地貌
38		水体地貌	潭	大庄科潭	
39			泉	珍珠泉乡珍珠泉	
40				潭柘寺镇潭柘寺泉	
41				河北镇河北泉	

续表

序号	大类	类	亚类	遗迹点	备注
42	地貌景观类地质遗迹	火山地貌	火山机构	黄松峪火山机构	三羊古火山口、梨树沟火山机构
			火山岩地貌	灵山火山岩地貌	顶部为北京最高的一级夷平面
43		构造地貌	峡谷	鱼子山京东大峡谷地貌	在河流长期的溶蚀、侵蚀-冲蚀、冲刷切割作用下形成
44				沙梁子乌龙峡谷地貌	火山熔岩峡谷地貌
				白河峡谷地貌	百里山水画廊
45				永定河峡谷地貌	我国屈指可数的大峡谷之一
46				龙门洞峡谷地貌	
47	地质灾害类	地震遗迹	地裂缝	高丽营镇西王路村地裂缝	新构造运动所致
48		地质灾害	泥石流	番字牌西沟泥石流	

（据北京市地质研究所）

二、遗迹分级及遗迹点价值评述

北京地区有世界级地质遗迹2处，分别为周口店古人类化石产地和千家店侏罗纪恐龙足迹化石产地；国家级地质遗迹7处，分别为下苇店寒武纪地质事件剖面、千家店侏罗纪木化石群、周口店古动物群、十渡岩溶地貌、东关上-上方山洞穴群岩溶地貌、佛子庄洞穴群岩溶地貌和云蒙山花岗岩地貌；省级地质遗迹41处。

（一）世界级地质遗迹

周口店古人类化石产地位于房山区周口店镇，以龙骨山为中心的4km^{2}范围内，分布有北京猿人遗址（距今约77万年）、新洞人遗址（距今约10万年）、田园洞人遗址（距今4.2万~3.85万年）和山顶洞人遗址（距今约3万年），出土了“北京人”头盖骨（图49-1，图49-2）、头骨碎片、面骨、下颌骨、股骨断片、胫骨、肱骨、锁骨、月骨、牙齿等古人类化石和砸击石核、石片、刮削器、尖状器、石锥、雕刻器、砍砸器等石器，同时还出土了灰烬等用火遗迹。周口店北京猿人遗址在世界范围内也是一处有关人类起源与演化研究的圣地，是世界上更新世古人类遗址中内涵最丰富、材料最齐全和最有科研价值的、唯一保存了纵贯70多万年的史前人类活动遗迹的遗址，确立了猿人阶段在人类发展史中的地位，并把人类控制用火的记录推到了40万~50万年以前，是我国古人类学、旧石器时代考古学和第四纪地质学的科研基地，北京猿人地点（第一地点）的沉积成为华北中更新统的标准剖面。

图49-1　周口店遗址出土的“北京人”头盖骨

（据《中国房山世界地质公园导游手册》）

Fig. 49-1　The “Peking man” skull-cap cranium unearthed from the Zhoukoudian site

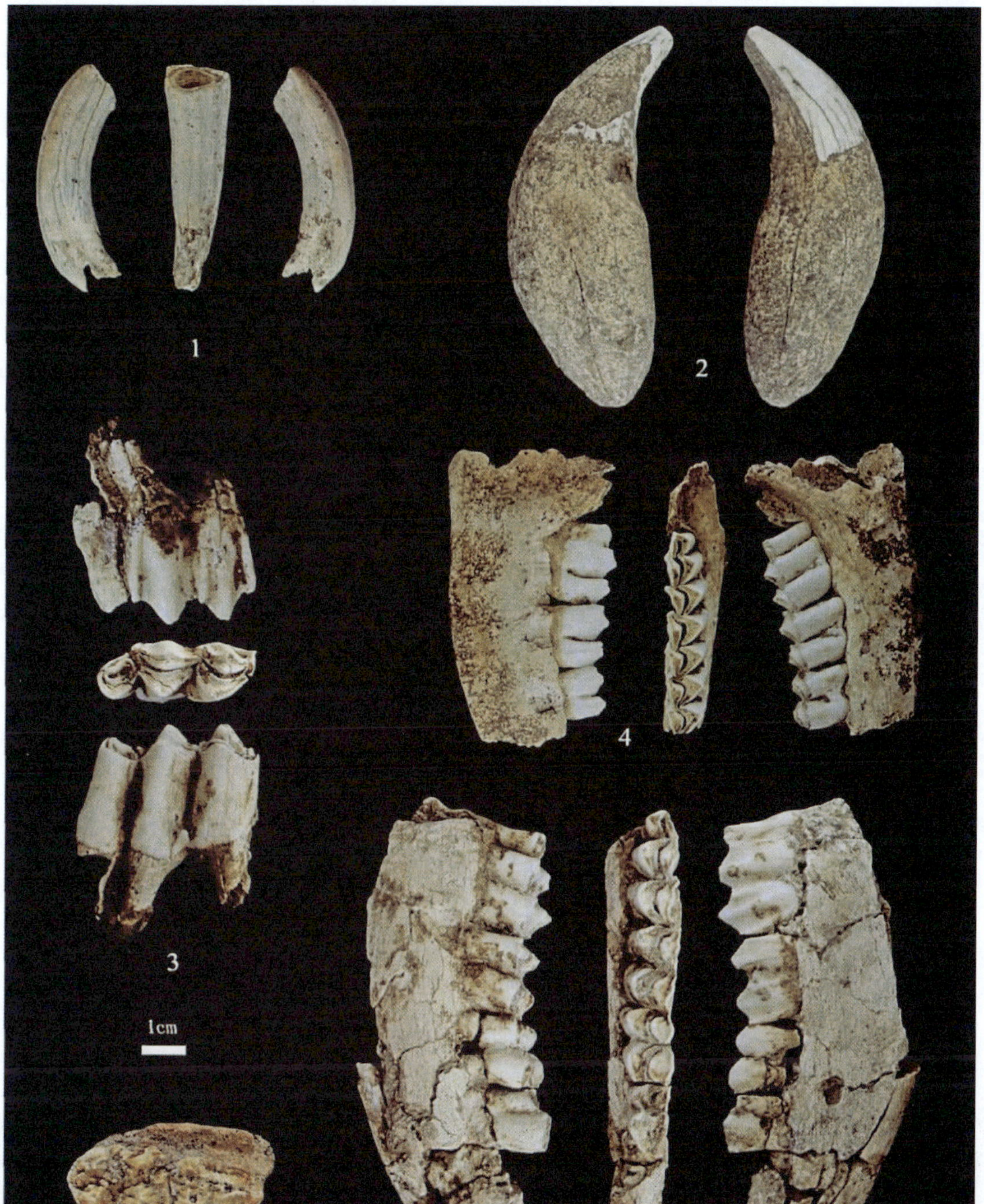

图 49-2　周口店第一地点西剖面第三层 2009—2010 年发掘出土典型动物种属骨骼材料

（据张双权等，2016）

Fig. 49-2　Macro-mammal species from Layer 3 during 2009-2010 field seasons at Zhoukoudian

1—*Equus sanmeniensis*，左侧 I2；2—*Ursus* sp. 右侧 c1；3—*Cervus elaphus* 右侧 m3；4—*Cervus* sp. 左侧下颌；5—*Sus lydekkeri*，左侧 DP3-DP4；6—*Cervus*（*Sika*）*grayi* 右侧下颌

千家店侏罗纪恐龙足迹化石产地共计足迹化石6处，为侏罗纪时期恐龙留下来的分离的单个脚印，形成于未完全固结的土城子组沉积物表面，基本上均为足迹凹坑印痕（图49-3）。足迹化石长度14~67cm、宽9.8~47cm、步间距12.5~36cm，其轮廓和形状分别归属于覆遁甲龙类、蜥脚类、鸟脚类和兽脚类，其中，覆遁甲龙 *Deltapodus* 类型恐龙足迹为中国首次发现。千家店恐龙足迹化石不仅是北京地区有恐龙存活的证据，同时也丰富了土城子组时期恐龙的种类，对分析晚侏罗世—早白垩世华北地区恐龙动物群的构成和习性具有重大的科学价值。

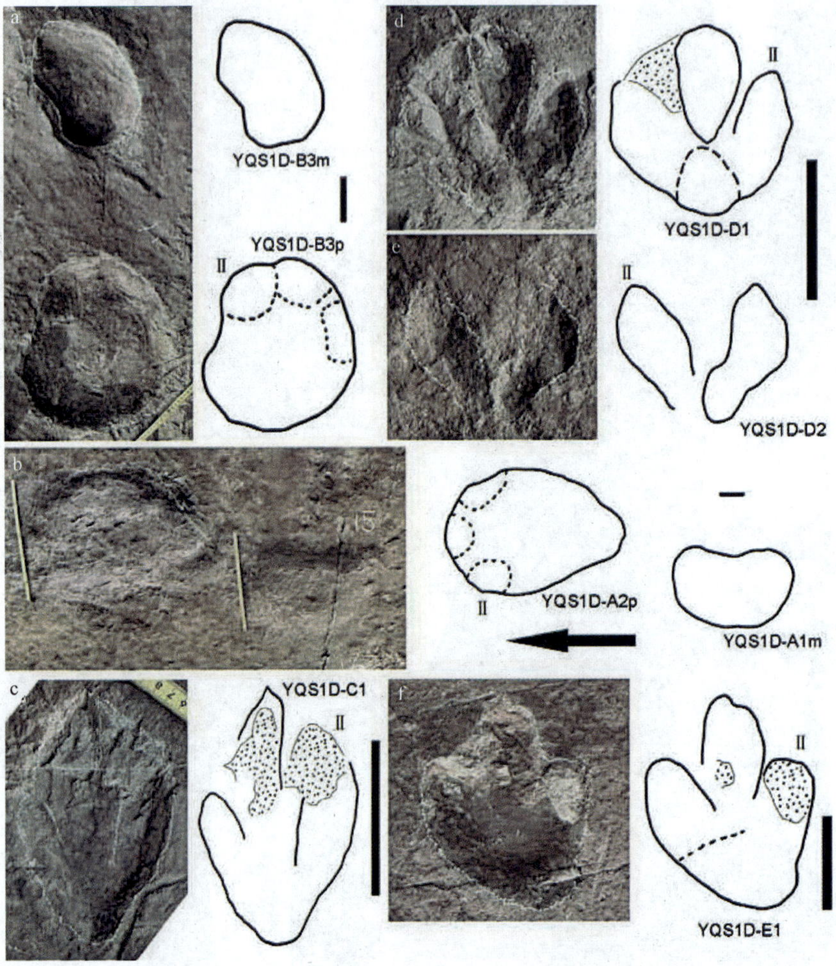

图49-3　化石点保存较好恐龙足迹的照片及轮廓图

（据张建平等，2012）

Fig. 49-3　Photograph and skeleton map of well-preserved dinosaur footprints

a—覆遁甲龙类足迹 YQS1D-B3m（前足迹）和 YQS1D-B3p（后足迹）；b—覆遁甲龙类足迹 YQS1D-A1m（前足迹）和 YQS1D-A2p（后足迹），箭头表示前进方向；c—兽脚类足迹 YQS1D-C1；d 与 e—鸟脚类足迹 YQS1D-D1 和 YQS1D-D2；f—鸟脚类足迹 YQS1D-E1；“Ⅱ”表示第二趾的位置

（二）国家级地质遗迹

下苇店寒武纪地质事件剖面位于门头沟区妙峰山镇下苇店村东，剖面全长3~4km，沉积了昌平组、张夏祖、炒米店组等，具有典型的风暴沉积序列。受蓟县运动影响，北京地区缺失了南华纪和震旦纪地层，从昌平期开始海水入侵，沉积了一套底砾岩和潮上云坪相含露源碎屑含泥晶白云岩平行不整合覆盖在新元古界青白口系景儿峪组上，从而形成了平行不整合面。该剖面是华北地台寒武系沉积现象较为丰富的地层剖面之一，为北京地区蓟县运动发生沉积间断的重要证据。

千家店侏罗纪木化石群位于延庆千家店镇下德龙湾村，裸露地表的硅化木有57株，其中出露完整的约30株，是目前华北地区规模较大、保存完整的硅化木群。硅化木产于侏罗系土城子组第一、二段的浅绿色页岩及粉砂页岩中，大多是裸子植物中的松柏类，具有集中分布、数量多、形态各异的特点，出露的硅化木最长达15m，直径最大达2.5m，表面多呈浅褐黄色或灰白色，质地坚硬。由于原地保存，石化后树木的形态、纹理均保存完好，对研究华北地区中生代地质演化历史和本地区古地理环境、古气候条件等具有重要的参考价值。

周口店古动物群是华北地区更新世中期的一个多种生态类型的哺乳动物群，出土了古脊椎动物化石和鸟类化石，其中，古脊椎动物化石包括灵长类、食肉类、食草类、食虫类、翼手类、啮齿类等；鸟类化石1000多件，以中—小型鸟类为主，包括9目、18科、48属、62种。最富有代表性的动物为肿骨鹿、葛氏斑鹿、洞熊、最后斑鬣狗、最后剑齿虎、杨氏虎、德氏水牛等。

十渡岩溶地貌、东关上-上方山洞穴群岩溶地貌、佛子庄洞穴群岩溶地貌是典型的碳酸盐岩地貌。其中，十渡岩溶地貌位于房山十渡地区拒马河北岸，主要为中元古界蓟县系雾迷山组燧石条带白云岩，属地上岩溶景观，以“峰丛、孤峰、嶂谷以及现代河流冲蚀地貌”为主，例如蝙蝠山、笔架山等；东关上-上方山洞穴群岩溶地貌发育区主要为中元古界铁岭组灰质白云岩和雾迷山组燧石条带白云岩，以石花洞为代表（图49-4），石花洞与桂林芦笛岩、福建玉华洞、杭州瑶琳洞并称我国四大岩溶洞穴，主要由鸡毛洞、银狐洞、石花洞、清风洞和孔水洞5个岩溶洞穴组成，几乎汇聚了自然界岩溶洞穴的各种沉积类型，银狐洞中的石毛（倒挂银狐）是“中华国宝”和世界溶洞奇观；佛子庄洞穴群岩溶地貌以仙栖洞、云水洞为代表，主要为中奥陶统马家沟组灰岩鸡毛洞、银狐洞、石花洞、清风洞和孔水洞，仙栖洞是北京地区继石花洞后发现的又一大型水旱相连的岩溶洞穴。

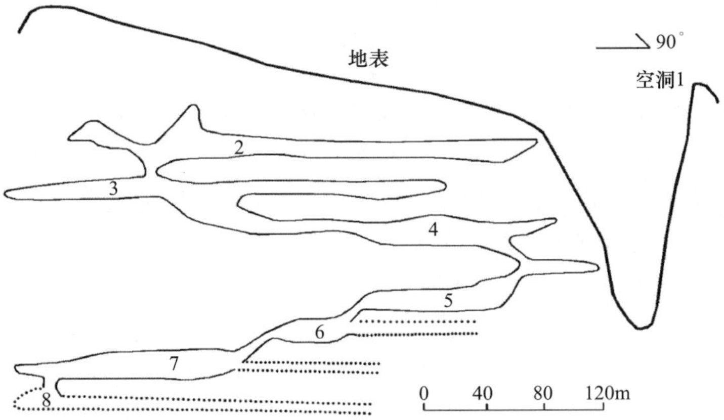

图49-4　石花洞洞层结构图

（据吕金波，2012）

Fig. 49-4　Hole layer structure in Shihua Cave

北京地区燕山期岩浆侵入活动显著，经过多期地壳抬升、下降运动，历经亿万年的风化剥蚀后，形成了形态多样、壮观美丽的花岗岩地质地貌，出露面积约占山区总面积的20%。云蒙山花岗岩地貌是燕山晚期岩浆侵入的产物，出露于密云区石城镇白河以南的云蒙山花岗岩体中，发育有悬崖绝壁、瀑布、壶穴等景观，以4座楼山西壁、天门山西壁规模最大。云蒙山变质核杂岩是中国境内首批确认的3处变质核杂岩之一，由美国地质学家首次在国际上报道而成名，是大陆伸展构造的重要表现形式，是太平洋板块向欧亚板块俯冲在板块内部形成的陆内造山作用的重要表现，是在我国召开的第30届国际地质大会的典型参观路线之一。

（三）省级地质遗迹

密云岩群为北京地区最古老的变质岩地层，同时是华北地区重要的太古宙变质岩单元，在全国范围内具有对比意义，同时，密云岩群赋存着国内重要铁矿类型、北京地区规模最大的变质型铁矿——沙厂铁矿。

干沟沉积岩剖面为典型的顶面痕沉积构造，是判断地层顶底面、是否倒转的重要标志，也是研究沉积环境的重要标志，该遗迹点具有出露面积大、保存完整的特点。

孤山口固态流变构造遗迹点是研究北京地区燕山期陆内造山作用的重要地质载体，小构造群落与复杂的掩卧褶皱相伴生的各种小构造丰富多彩，素有“半部小构造教科书”之称。

密云沙厂斜长环球斑花岗岩体（图 49-5）是燕疗裂谷和哥伦比亚超大陆裂解的重要标志，是华北中元古代沉积的起点，也是北京地区唯一能与全国乃至全球新元古代的斜长岩-奥长环斑花岗岩进行对比的独体杂岩系。因此，该地质遗迹对研究燕辽裂谷、哥伦比亚超大陆裂解、华北中元古代沉积作用有重要的参考价值。

图 49-5　密云沙厂斜长环斑花岗岩

Fig. 49-5　Plagioclase rapakivi at Shachang Town，Miyun

大石窝汉白玉产地是华北地区唯一出产汉白玉大理岩的地方，开采、加工历史悠久，石材品质优良，被称为“中国一号”，并名列“中国名特石材品种”之首，毛主席纪念堂、人民英雄纪念碑基座浮雕、天安门广场国旗基座石栏板、云南世博会华表、澳门回归莲花底座等标志性建筑采用的汉白玉石材均取自大石窝。另外，大石窝祖辈相传的石匠工艺，敲、打、滑、拉、安 5 项基本技能及其文化价值已于 2006 年成功申报为国家非物质文化遗产。

圣莲山地质公园屹立在风景秀美的北京西山腹地百花山东麓的群山丛中，以其奇特多样、分布广泛的地质遗迹享誉京城内外，不仅发育有独特的第四纪（2. 6Ma）洞穴沉积遗迹——“圣米（石英颗粒）”、赋存于寒武纪—奥陶纪（0. 54~0. 45Ga）地层中的海相沉积构造，还有深厚的人文积淀。

永定河峡谷是我国屈指可数的大峡谷之一，景色迤逦，永定河文化与北京的形成、发展息息相关，具有古人类演化、古都城历史变迁等丰富内涵，被称为地质大观园。

第三节　地质遗迹保护

根据联合国教科文组织和自然资源部提出的地质遗迹保护要求和管理框架，北京市已经把地质遗迹保护工作纳入各级自然资源管理部门的职责范围。为了保护不可再生的珍贵地质遗迹资源，北京市积极推进地质遗迹保护工作，从 20 世纪末开始地质公园、地质遗迹自然保护区和矿山公园的申报和建设工作。地质遗迹保护区（地质公园）是地质遗迹保护和利用的有效方式，北京地区 50 处重要地质遗迹点，多数已通过建立地质遗迹自然保护区、公园或景区的形式得到了保护；或者通过已有景区进行保护，例如模式口-八大处古冰川遗迹通过中国第四纪冰川遗迹陈列馆、八大处公园进行保护，鱼子山京东大峡谷地貌通过京东大峡谷景区进行保护，龙门涧峡谷地貌通过龙门涧景区进行保护；截至目前，仅“下苇店寒武纪地质事件剖面”1 处国家级遗迹点及部分省级遗迹点暂时未得到充分利用和保护。

截至 2017 年底，经联合国教科文组织、国土资源部和市国土资源局批准，共建立世界级地质公园 2 处、国家级地质公园 5 处、市级地质公园 1 处，以及国家矿山公园 4 处（表 49-2~表 49-4）。

表 49-2　北京市地质公园一览表

Table 49-2　List of geoparks in Beijing

序号	级别	公园名称	所在区	主要地质遗迹点	面积/km^2	批建时间	揭碑时间
1	世界级	中国房山世界地质公园	房山，河北涞水、涞源	周口店组层型地质剖面、六渡沉积岩剖面、七渡背斜、孤山口固态流变构造、周口店古人类化石产地、周口店动物群、十渡岩溶地貌、东关上-上方山洞穴群岩溶地貌、佛子庄洞穴群岩溶地貌、圣莲山岩溶地貌、房山花岗岩地貌	1045	2006. 9. 17	2010. 5
2		中国延庆世界地质公园	延庆	干沟沉积岩剖面、排字岭单斜、千家店侏罗纪木化石群、千家店侏罗纪恐龙足迹化石产地、旧县镇龙庆峡岩溶地貌、莲花山花岗岩地貌、六道河石英砂岩地貌、大庄科潭、四海珍珠泉、沙梁子乌龙峡谷地貌、白河峡谷地貌	620. 38	2013. 9. 9	2014. 9
3	国家级	北京石花洞国家地质公园	房山	石花洞系洞穴群岩溶地貌	33. 5	2001. 12. 10	2002. 6
4		北京延庆硅化木国家地质公园	延庆	干沟沉积岩剖面、排字岭单斜、千家店侏罗纪木化石群、千家店侏罗纪恐龙足迹化石产地、六道河石英砂岩地貌、沙梁子乌龙峡谷地貌、白河峡谷地貌	141. 4	2001. 12. 10	2002. 9
5		北京十渡国家地质公园	房山	六渡沉积岩剖面、七渡背斜、十渡岩溶地貌、东关上-上方山洞穴群岩溶地貌	290. 37	2004. 1. 19	2005. 6
6		北京平谷黄松峪国家地质公园	平谷	黄松峪中元古界与太古宇不整合界面、黄松峪京东大溶洞岩溶地貌、黄松峪石英砂岩地貌、黄松峪火山机构	36. 4	2009. 8. 19	2013. 9
7		北京密云云蒙山国家地质公园	密云	云蒙山花岗岩地貌、白河峡谷地貌	238. 2	2009. 8. 19	2014. 10
8	市级	北京房山区圣莲山地质公园	房山	圣莲山岩溶地貌	28	2004. 6. 10	

表 49-3　北京市地质遗迹自然保护区一览表

Table 49-3　List of geological relic nature reserves in Beijing

序号	级别	名称	所在区	主要遗迹类型	面积/km^2	批建时间
1	市级	房山石花洞地质遗迹自然保护区	房山	岩溶洞穴	36. 5	2000. 12
2		延庆朝阳寺木化石自然保护区	延庆	古生物化石	20. 5	2001. 12
3	区级	平谷大溶洞地质遗迹自然保护区	平谷	岩溶洞穴	0. 37	1997

表 49-4　北京市矿山公园一览表

Table 49-4　List of mine parks in Beijing

序号	名称	所在区	地质遗迹点	面积/km^2	批建时间	揭碑时间
1	北京平谷黄松峪国家矿山公园	平谷	塔洼金矿产地	1. 86	2005. 9	2012. 9
2	北京首云国家矿山公园	密云	密云沙厂斜长环球斑花岗岩体、沙厂铁矿产地（图 49-6）	3. 58	2010. 6	2012. 9

续表

序号	名称	所在区	地质遗迹点	面积/km^2	批建时间	揭碑时间
3	北京怀柔圆金梦国家矿山公园	怀柔	杨树底下金矿产地	5.56	2010.6	
4	北京史家营国家矿山公园	房山	古煤矿矿硐、现代煤矿矿井、储煤仓	58.5	2013.1	

图 49-6　首云国家矿山公园沙厂铁矿露天采坑

Fig. 49-6　Open-pit in Shouyun national mine park

北京平谷黄松峪国家矿山公园位于平谷区黄松峪乡境内，地处燕山南麓地带，是我国首批国家矿山公园之一，拥有清朝及中华人民共和国建立前后诸多采矿遗址，遗存有少量民采采矿工具和土法选冶设备等珍贵的矿业活动遗迹。北京首云国家矿山公园位于密云区巨各庄镇，拥有垂直深度达 300 多米、开采时长近半个世纪的露天大矿坑（图 49-6），拥有北京市唯一一家铁矿博物馆，区内的斜长环斑花岗岩地质遗迹罕见。北京怀柔圆金梦国家矿山公园位于怀柔区琉璃庙乡杨树下村南，是我国唯一的以看金、淘金、了解黄金文化、掌握黄金知识为特色的旅游景区，拥有一条高差 400 多米，采运工序复杂的采、运矿生产系统，还修建了黄金科普馆，能够现场演示炼金技术，展现黄金生产的全过程。

第五十章　土壤地质与环境

土壤作为陆地生态系统重要组成部分，是人类和动植物生存不可替代的环境因子，也是食品安全与人体健康的基本保障，在保护环境和维持生态平衡中具有重要作用。近年来，由于世界人口的不断增长、工业化进程的快速发展，以及对环境资源的不合理利用，导致土壤地质环境问题日益突出。我国土壤地质环境总体状况也不容乐观，因此，国务院于 2016 年 5 月 28 日印发了《土壤污染防治行动计划》。

北京作为首都，为相应国家号召，在土壤地质环境调查方面开展了大量工作。2003 年以来，先后开展了“北京市平原区土壤环境地球化学调查”“北京城市地质土壤调查与评价”“北京市土地资源质量综合地质评价”等多个项目，调查了土壤重金属元素、有机污染物的环境质量，对工厂及搬迁区、垃圾填埋场、矿山开采土壤污染隐患区、元素富集致病隐患区和新农村建设适宜产业区的土壤地质环境进行了评价。这些工作为优化城乡用地结构布局、协调土地利用规划与生态建设、保护与合理利用农用地、规划集约利用建设用地及土地开发整理等方面提供了地质科学依据。

第一节　北京平原区土壤地球化学特征

一、北京平原区土壤背景值与基准值

对土壤地球化学特征的研究，是了解土壤地球化学作用的前提，是土壤环境调查研究工作的基础。北京市平原区土壤地球化学基准值与背景值，是研究土壤中元素的区域分布特征及集中分散规律的基础，土壤 54 个元素及指标背景值与基准值对比结果见表 50-1。

表 50-1　北京市平原区土壤元素及指标背景值与基准值表

Table 50-1　Background and baseline values of soil elements in Beijing Plain

项目	背景值	基准值	背景值/基准值	项目	背景值	基准值	背景值/基准值
pH	8.2	8.5	0.96	Ba	609	590	1.03
Sc	9.77	10.07	0.97	Ce	60.8	58.7	1.04
Ni	25.2	25.9	0.97	B	33.82	32.61	1.04
V	70.7	72.2	0.98	As	8.34	8.03	1.04
CaO	3.35	3.42	0.98	La	30.9	29.7	1.04
Al_2O_3	12.38	12.63	0.98	Cl	71	68	1.04
Ga	15.9	16.2	0.98	Sb	0.75	0.71	1.06
SiO_2	65.3	66.53	0.98	Zr	273	255	1.07
Mn	547	557	0.98	Mo	0.61	0.56	1.09
Sr	246.5	250	0.99	Cu	22.3	20.4	1.09
Rb	91.6	92.7	0.99	Au	1.27	1.16	1.09
F	517	522	0.99	I	1.5	1.36	1.10
Fe_2O_3	3.98	4.01	0.99	Zn	68.2	61.7	1.11
Na_2O	1.95	1.96	0.99	Sn	2.72	2.35	1.16

续表

项目	背景值	基准值	背景值/基准值	项目	背景值	基准值	背景值/基准值
Co	10.6	10.64	1.00	Bi	0.27	0.23	1.17
Ti	3594	3583	1.00	Pb	23.5	19.9	1.18
MgO	1.82	1.81	1.01	Ag	0.071	0.06	1.18
Nb	13.5	13.4	1.01	U	2.04	1.59	1.28
K_2O	2.46	2.44	1.01	Cd	0.136	0.093	1.46
Cr	58.5	58	1.01	Br	2.7	1.7	1.59
Y	22.4	22.2	1.01	P	845	532	1.59
Be	1.95	1.93	1.01	C	1.31	0.66	1.98
Th	9.7	9.6	1.01	S	207	102	2.03
Li	27.62	27.31	1.01	Se	0.17	0.075	2.27
Ge	1.31	1.29	1.02	N	0.088	0.025	3.52
Tl	0.53	0.52	1.02	Hg	0.057	0.016	3.56
W	1.45	1.41	1.03	C_{org}	0.82	0.2	4.10

注：氧化物、C_{org}、C、N 单位为%，pH 无量纲，Au 单位为 ng/g，其余单位为 μg/g。

对土壤元素背景值、基准值进行比较发现：绝大多数元素背景值与基准值的比值在 0.9~1.1 之间，表明土壤的成土过程中，表层土壤内元素的带入、带出量较少，总体含量变化不大，其元素的丰缺主要受成土母质的控制，表层土壤与深层土壤具有一定的继承性；C、P、Br、Cd、U、Ag 等背景值与基准值之比在 1.1~1.2 之间，元素在表层土壤中略有偏高，表生富集作用不很明显；C_{org}、Hg、N、Se、S、Pb、Bi、Se、Zn 等在表层土壤显著偏高，表生富集作用强烈。C_{org}、Hg、N、Se、S 背景值分别是深层样的 4.1、3.56、3.52、2.27、2.03 倍。反映了表层土壤在成土过程中受人类活动改造的影响，其元素地球化学特征发生了很大程度的变化。

总的来说，在成土过程中，虽然表层土壤受到了一定程度的后期改造，元素有一些带入、带出，但无论是从表层及深层土壤中元素的分布特征，还是从这两个层位元素间的相关系数及背景值与基准值来看，北京平原区表层及深层土壤环境呈现出的一致性还是十分明显的。

二、不同沉积相土壤地球化学特征

北京平原区地表第四系主要为全新世（Qh）沉积物及晚更新世（Qp^3）沉积物。这两个层位土壤元素背景值见表 50-2。

表 50-2　不同沉积相土壤元素及指标背景值统计表

Table 50-2　Background values of soil elements and indexes in different sedimentary facies

项目	晚更新世沉积物		全新世沉积物		项目	晚更新世沉积物		全新世沉积物	
	背景值	样品数	背景值	样品数		背景值	样品数	背景值	样品数
Ag	0.072	255	0.068	1000	Mn	569	266	521	1122
Al_2O_3	13.09	252	12.14	1281	Mo	0.62	269	0.58	1147
As	9.29	261	7.83	1184	N	0.097	259	0.085	1130
Au	1.35	222	1.27	891	Na_2O	1.81	272	1.97	1053
B	37.48	271	31.30	1043	Nb	14.2	239	13.0	1196
Ba	609	257	615	1124	Ni	26.8	245	24.6	1160
Be	2.06	255	1.89	1222	P	780	245	890	1085

续表

项目	晚更新世沉积物		全新世沉积物		项目	晚更新世沉积物		全新世沉积物	
	背景值	样品数	背景值	样品数		背景值	样品数	背景值	样品数
Bi	0.30	236	0.26	1093	Pb	23.9	218	22.9	1037
Br	2.8	239	2.9	1151	Rb	97.4	235	89.5	1323
C	1.18	252	1.38	1121	S	209	240	218	1033
CaO	2.03	249	3.71	1369	Sb	0.85	248	0.73	1138
Cd	0.147	260	0.147	1097	Sc	10.43	238	9.48	1091
Ce	63.6	258	58.8	1157	Se	0.19	248	0.18	1170
Cl	64	182	79	825	SiO_2	67.46	273	64.67	1089
Co	11.39	218	10.30	1094	Sn	2.92	258	2.65	1037
Cr	60.5	224	57.9	1054	Sr	200.2	251	273.8	994
Cu	23.8	225	22.6	1103	Th	10.5	282	9.3	1207
F	539	242	504	1082	Ti	3869	235	3480	1155
Fe_2O_3	4.31	227	3.84	1170	Tl	0.57	279	0.52	1288
Ga	16.7	265	15.8	1244	U	2.09	251	2.01	1214
Ge	1.41	269	1.29	1159	V	76.3	233	68.3	1111
Hg	0.068	222	0.049	747	W	1.56	234	1.38	1173
I	1.62	255	1.46	1135	Y	24.7	234	21.5	1191
K_2O	2.49	268	2.34	761	Zn	70.6	244	68.8	1103
La	32.0	263	29.5	1201	Zr	285	276	265	1217
Li	30.37	246	26.37	1055	C_{org}	0.92	255	0.81	1007
MgO	1.60	207	1.86	1138	pH	8.22	212	8.27	1095

注：氧化物、C_{org}、C、N 单位为%，pH 无量纲，Au 单位为 ng/g，其余单位为 μg/g。

全新世沉积物分布于北京平原的广大地区，以各大河流的冲积物为主。土壤中除 Hg 以外，其余元素与指标的背景值大体接近于北京平原区表层背景值。

晚更新世沉积物在北京平原出露的面积较小，主要分布于山区与平原过渡地带及北京平原北部微高地及阶地之上，以冲洪积物及残坡积物为主。晚更新世沉积物土壤中大多数元素的背景值高于全新世沉积物，Hg、B、As、Sb、Bi、Y、Li、Th、C_{org}、N、W、Fe_2O_3、V、Ti、I、Co 表现为明显偏高和略偏高；明显低于全新世沉积物背景值的仅有 CaO 和 Sr，分别为全新统土壤的 0.55 和 0.73 倍，Cl、C、MgO、P 略有偏低，在晚更新世沉积物土壤分布区呈现低背景。

三、不同成因类型土壤地球化学特征

北京平原区土壤成因类型主要有 5 种，元素与指标的背景值见表 50-3。

表 50-3　不同成因类型土壤元素及指标背景值统计表

Table 50-3　Background values of soil elements and indexes in different genetic types

项目	冲积平原		冲积湖积平原		冲积洪积平原		洪积平原		剥蚀堆积台地		平原区背景值
	背景值	样品数	背景值	样品数	背景值	样品数	背景值	样品数	背景值	样品数	
Ag	0.067	739	0.090	53	0.077	340	0.071	139	0.072	74	0.069
Al_2O_3	12.14	676	12.14	44	12.23	285	12.78	140	12.64	67	12.34
As	7.68	712	7.80	42	8.42	259	9.03	135	8.38	62	8.07

续表

项目	冲积平原		冲积湖积平原		冲积洪积平原		洪积平原		剥蚀堆积台地		平原区背景值
	背景值	样品数	背景值	样品数	背景值	样品数	背景值	样品数	背景值	样品数	
Au	1.26	684	2.63	53	1.56	314	1.32	163	1.24	64	1.28
B	30.79	724	31.88	46	32.91	298	38.56	163	33.88	65	32.10
Ba	615	566	622	39	607	255	592	134	634	35	612
Be	1.86	664	1.90	49	1.91	286	2.03	147	1.96	62	1.93
Bi	0.25	707	0.29	50	0.27	279	0.31	147	0.28	61	0.27
Br	2.9	805	2.9	52	3.0	347	2.8	166	2.9	67	2.9
C	1.32	713	1.98	51	1.49	328	1.18	161	1.99	66	1.34
CaO	3.64	722	4.41	52	3.68	339	1.65	140	3.93	68	3.45
Cd	0.140	707	0.164	47	0.145	301	0.149	126	0.161	72	0.146
Ce	58.9	761	59.1	49	58.4	292	63.2	136	59.1	74	59.7
Cl	77	740	87	45	87	298	69	147	64	70	76
Co	10.41	688	10.11	46	10.28	304	11.15	142	10.45	61	10.46
Cr	58.2	704	57.0	55	58.1	282	59.5	130	56.7	55	58.3
Cu	22.1	831	26.9	36	24.6	362	23.8	165	25.3	76	22.9
F	504	645	500	54	509	295	530	141	524	64	511
Fe_2O_3	3.86	588	3.82	46	3.88	276	4.25	101	4.04	73	3.93
Ga	15.9	714	15.7	47	15.9	343	16.4	157	16.0	78	16.0
Ge	1.29	677	1.29	37	1.30	313	1.45	115	1.37	76	1.30
Hg	0.048	695	0.213	50	0.076	312	0.065	151	0.078	75	0.055
I	1.46	736	1.40	52	1.53	330	1.53	167	1.39	78	1.49
K_2O	2.46	679	2.38	53	2.36	274	2.47	135	2.38	65	2.35
La	29.4	713	29.1	46	29.3	340	31.7	156	31.1	75	29.9
Li	25.74	741	27.38	52	27.33	258	29.97	173	28.24	79	27.06
MgO	1.80	819	1.87	54	1.96	336	1.57	166	1.82	78	1.81
Mn	524	769	519	48	518	317	568	143	547	74	531
Mo	0.57	758	0.65	55	0.63	241	0.62	124	0.71	72	0.59
N	0.083	837	0.086	58	0.088	373	0.093	114	0.095	78	0.087
Na_2O	2.04	752	1.86	43	1.90	291	1.78	145	1.90	33	1.95
Nb	13.2	809	12.6	43	12.8	241	14.3	163	13.4	50	13.3
Ni	24.4	700	24.6	49	25.6	292	26.6	152	24.6	62	25.0
P	900	696	840	50	854	304	807	135	866	61	876
Pb	22.2	733	28.3	49	24.0	238	23.6	153	27.1	64	23.3
Rb	89.9	712	85.4	52	90.9	290	95.5	130	87.9	60	90.8
S	213	703	299	56	249	331	201	161	276	65	213
Sb	0.70	673	0.92	50	0.77	329	0.84	173	0.91	71	0.75
Sc	9.49	821	9.40	52	9.60	321	10.20	165	9.63	75	9.61
Se	0.17	702	0.21	48	0.19	296	0.19	155	0.23	70	0.18
SiO_2	65.34	681	63.03	52	63.67	293	67.99	145	63.78	72	65.09
Sn	2.58	691	3.11	49	2.88	297	2.94	154	2.99	72	2.73

续表

项目	冲积平原		冲积湖积平原		冲积洪积平原		洪积平原		剥蚀堆积台地		平原区背景值
	背景值	样品数	背景值	样品数	背景值	样品数	背景值	样品数	背景值	样品数	
Sr	271.5	635	290.1	54	260.8	258	179.0	163	260.1	70	262.2
Th	9.3	707	9.1	46	9.4	236	10.3	153	9.9	46	9.5
Ti	3500	670	3237	44	3422	243	3848	147	3538	26	3544
Tl	0.51	658	0.53	48	0.52	301	0.57	159	0.52	69	0.53
U	1.92	708	2.08	53	2.04	328	2.09	152	2.30	75	2.02
V	68.3	529	67.1	52	69.0	225	75.3	138	71.3	36	69.7
W	1.35	680	1.55	43	1.51	300	1.56	125	1.50	64	1.43
Y	21.7	737	20.8	48	21.6	301	24.1	146	22.1	73	22.0
Zn	67.4	688	80.9	51	72.1	336	70.7	143	80.5	54	69.5
Zr	268	796	236	54	258	335	295	171	262	73	269
pH	8.32	715	8.30	56	8.21	322	8.18	168	8.20	76	8.26

注：氧化物、C_{org}、C、N 单位为%，pH 无量纲，Au 单位为 ng/g，其余单位为 μg/g。

冲积平原由河流迁徙和泛滥冲积而形成。北京平原区土壤主要为潮白河、永定河、温榆河及泃河冲积物发育而成的土壤，是平原区分布面积最大的土壤成因类型。相对于北京平原土壤元素背景值而言，土壤中 Hg 略有偏低，其他 53 种元素与指标的背景值与北京平原基本接近。由流水和湖沼共同作用堆积而成的冲积湖积平原，主要分布于北京城区北部的清河两岸及东北部来广营—酒仙桥—管庄一带。土壤中 Hg、Au、C、S、Ag、CaO、Sb、Pb 的背景值明显高于北京平原，Se、Cu、Zn、Cl、Sn、Cd、Sr 为略偏高。可以看出，冲积湖积土壤中重金属元素及 C、S、C_{org}、Cl 等趋于富集。由常年流水和季节流水共同堆积而成的冲积洪积平原，分布于北京平原西北部及东南部一带。与北京平原元素背景值相比，土壤中 Hg、Au 明显偏高，S、Cl、Ag、C 为略偏高，其他元素基本相近。洪积平原是由多条来自山地的季节性河流的沉积物所堆积形成的若干个洪积扇联合而组成的山前倾斜平原带，主要分布于北京平原的山麓地带。土壤中 CaO、Sr 明显低于平原区背景值，MgO、C 略低；B、Hg、As、Sb、Bi、Ge、C_{org} 略偏高。剥蚀堆积台地具有坡度较缓的台面和较陡的台坎，相对高度一般在 30～50m 之间。分布于海淀区东部八里庄—东升及朝阳区西部朝外—亚运村一带。相对于北京平原元素背景值来说，土壤中 C、Hg、C_{org}、S、Se、Sb 表现为明显富集；仅 Cl 在台地区略偏低。

四、不同水系流域土壤地球化学特征

水系发源地不同，流域内沉积物成分存在差异，可导致不同流域土壤中元素组合及富集、贫化程度不同。北京平原区土壤元素及指标的含量分布呈现与水系流域相关的特点（表 50-4）。

表 50-4 不同水系流域土壤元素及指标地球化学特征参数统计表

Table 50-4 Statistical table of geochemical characteristic parameters of elements in different river reaches

项目	蓟运河流域		潮白河流域		北运河流域		永定河流域		大清河流域		平原区平均值
	均值	变异系数	均值	变异系数	均值	变异系数	均值	变异系数	均值	变异系数	
Ag	0.090	76	0.069	36	0.121	104	0.062	60	0.074	32	0.097
Al_2O_3	12.67	9	12.86	5	12.38	7	11.3	5	12.53	7	12.38
As	9.81	52	7.07	27	8.42	20	7.29	22	8.58	29	8.24
Au	3.23	227	1.72	126	3.43	266	1.34	66	1.45	41	2.65
B	45.30	25	29.34	23	32.93	19	29	24	37.80	28	33.78
Ba	603	6	712	11	647	25	590	9	604	13	641

续表

项目	蓟运河流域		潮白河流域		北运河流域		永定河流域		大清河流域		平原区平均值
	均值	变异系数	均值	变异系数	均值	变异系数	均值	变异系数	均值	变异系数	
Be	2.03	10	2	8	1.92	10	1.65	9	1.91	9	1.91
Bi	0.36	50	0.28	54	0.33	48	0.23	48	0.35	71	0.32
Br	2.8	18	2.5	24	3.4	29	3.24	35	3.3	30	3.1
C	1.03	22	0.98	37	1.7	36	1.53	35	1.87	39	1.52
CaO	1.59	28	2.07	28	3.97	36	4.91	13	3.88	40	3.54
Cd	0.174	56	0.144	66	0.178	45	0.14	25	0.167	28	0.167
Ce	58.9	12	60.6	14	61	14	55.9	13	65.6	15	60.7
Cl	79	58	87	82	158	124	93	37	91	59	123
Co	11.23	17	10.22	18	10.62	12	10.09	12	11.17	10	10.63
Cr	63.4	15	53.3	16	58.1	12	59.1	8	59.5	9	58.2
Cu	25.7	26	21.3	30	27.8	43	19.6	24	23.7	20	25.1
F	537	14	490	19	529	15	476	12	570	16	523
Fe_2O_3	4.13	15	3.88	15	4	11	3.66	13	4.28	11	3.99
Ga	16.4	5	16.9	6	16.1	7	14.8	6	16.1	7	16.1
Ge	1.44	11	1.31	8	1.33	11	1.24	9	1.45	13	1.34
Hg	0.063	48	0.063	73	0.204	133	0.068	199	0.113	93	0.141
I	1.83	21	1.32	36	1.64	29	2.53	68	1.5	28	1.70
K_2O	2.68	7	2.75	6	2.43	7	2.28	4	2.37	8	2.48
La	30.4	11	31.3	13	30.6	15	27.3	13	32.8	15	30.6
Li	31.20	18	25.62	18	27.53	15	23.23	17	31.33	20	27.57
MgO	1.56	17	1.47	22	1.95	16	1.84	13	1.93	18	1.82
Mn	658	44	555	19	550	14	496	13	566	11	557
Mo	0.59	25	0.59	22	0.67	31	0.62	215	0.64	28	0.64
N	0.098	21	0.082	29	0.090	27	0.064	33	0.094	27	0.087
Na_2O	1.73	12	2.34	15	2.01	13	2.12	12	1.78	21	2.02
Nb	14.3	10	13.9	7	13.4	13	12.6	6	13.8	7	13.6
Ni	26.8	16	22.5	20	25.3	13	24	13	26.1	11	25.0
P	868	29	868	26	987	29	921	21	945	23	944
Pb	25.4	40	23.6	10	27.5	30	24.8	270	24	20	26.0
Rb	100.7	3	97.5	5	89.4	8	80.9	6	89	9	90.7
S	202	22	191	35	303	46	211	44	252	35	258
Sb	0.87	46	0.63	25	0.86	43	0.69	58	0.83	28	0.80
Sc	10.3	13	9.36	16	9.7	11	9.21	10	10.16	9	9.71
Se	0.19	16	0.16	25	0.22	86	0.16	88	0.21	38	0.20
SiO_2	69.85	3	68.55	4	64.30	4	64.76	4	64.41	5	65.58
Sn	2.84	19	2.69	21	3.46	50	2.45	45	2.87	25	3.09
Sr	181.8	12	286.5	20	274.1	17	283.4	13	230.3	34	262.4
Th	10.1	11	9.7	11	9.5	13	8.9	11	10.4	11	9.6
Ti	3767	11	3486	11	3577	11	3464	7	3833	7	3601

续表

项目	蓟运河流域		潮白河流域		北运河流域		永定河流域		大清河流域		平原区平均值
	均值	变异系数	均值	变异系数	均值	变异系数	均值	变异系数	均值	变异系数	
Tl	0.56	9	0.55	11	0.53	13	0.49	10	0.53	11	0.53
U	2.03	13	1.96	24	2.16	25	1.91	23	2.27	20	2.10
V	73.7	14	67.7	14	70.5	10	66.6	9	76.2	9	70.6
W	1.69	20	1.31	27	1.63	111	1.21	17	1.5	19	1.52
Y	23.4	9	21.5	12	21.8	11	21.1	9	23.7	9	22.1
Zn	67.4	15	67.1	19	80.1	25	63.0	20	72.8	16	73.9
Zr	294	7	291	13	263	13	296	16	*282	13	277
C_{org}	0.84	20	0.76	34	0.99	35	0.69	52	1.04	37	0.91
pH	0.71	8	7.8	8	8.21	3	8.58	2	8.37	3	8.16

注：氧化物、C_{org}、C、N单位为%，pH无量纲，Au单位为ng/g，其余单位为μg/g，变异系数单位为%。

潮白河流域：大多数元素在这一流域内的含量平均值低于北京平原均值，仅造岩元素氧化物 Na_2O、K_2O 和微量元素Ba偏高，表现为相对富集，在该流域呈现出高背景分布；Hg、CaO、C、Au、Cl、Ag、S、Br、I、Sb明显低于平原区均值，Se、Br、MgO、C_{org}、Cu、W、As、B、Cd、Sn、Bi略偏低。其中Au、Cl、Hg、Cd在该流域中的分布不均匀，表现为很大起伏到较大起伏，Bi、C、Ag、I、S、C_{org}、Cu等为中等起伏。

北运河流域：这一流域面积较大，是北京平原区人居较为集中的区域。区内绝大部分元素与指标的均值与北京平原平均值接近，仅Hg、Au、Cl、Ag、S、Sn、CaO、C、Cu等高于和略高于平原区，Hg、Au、Cl、Ag的富集特征显著。这一流域含量分布不均匀的元素最多，Au、Hg、Cl、W、Ag等元素离散程度大，分异强烈，Se、Sn、Bi、S、Cd、Cu、Sb、C、CaO、C_{org}、Mo、Pb等元素含量变化范围也较大。这些多为化学活动性较强、受人类生产生活影响较多的元素，在北运河流域内元素表现为不均匀分布，有异常存在。

永定河流域：区内有44种元素和指标低于北京平原均值，Hg、Au、Ag、Bi、N、Cl、C_{org}、Cu、Sn等元素贫乏，为低背景分布区，W、Se、S、Cd、Li、Zn、B、Be、Sb、As、Mn、Rb、La略偏低；I、CaO在这一流域表现为明显富集，其含量均值是北京平原的1.49和1.39倍。Pb、Mo、Hg、Se、I、Au的含量分布起伏较大。

蓟运河流域：与北京平原元素平均值相比，B、Au、As、Mn、Li、Bi、N、W、Rb为明显偏高和略偏高，表现为相对富集，以B、Au富集程度为高。造岩元素氧化物CaO、Na_2O、MgO及微量元素Hg、Cl、Sr、S为明显偏低和略偏低，表现为相对贫乏。这一流域中，Au、Ag的分异性较强，尤其是Au，变异系数高达227%。Cl、Cd、As、Bi、Hg、Sb、Mn、Pb含量分布为中等起伏，有一定程度的分异。

大清河流域：多数元素表现为背景分布，且含量起伏不大。C、C_{org}、Li、B在这一流域含量平均值略有偏高；Au、Cl、Ag则表现为明显偏低，Hg、Sr、Na_2O、I为略偏低。该流域中Hg、Bi的变异系数较大，含量分布不均一。

总的来看，北京平原区五大水系流域表层土壤中元素及指标平均值有一定的差异，在永定河和潮白河流域表现得更加明显，主要是由于这两个流域土壤偏砂性，一些在黏土中富集的元素含量偏低。

第二节　北京土壤地球化学与第四纪地质

北京平原微向渤海倾斜，其上发育的永定河、温榆河、潮白河、拒马河、大石河和泃河等大小河流贯穿全区。北京平原区是流水侵蚀、流水沉积作用最突出、最典型的地区，永定河进入北京平原后，形成巨大的永定河冲积扇，扇体几乎覆盖了整个北京平原。永定河在北京平原区自北东、东西、南东、南西不断改道迁移，其逐渐演变的过程，也是不断地对新、老扇体进行侵蚀、分割、破坏的过程，导致先成扇残缺不全，后成扇相互叠加、掩埋先成扇。永定河上游流经黄土高原，有十分广阔的汇水面积，为永定河搬运提供了丰富的物质来源。北京平原地表物质主要是晚更新世时期形成的马兰期黄土状物质和全新世时期的堆积物。全新世时期的堆积物，包括永定河泛滥平原、河流谷地堆积物、山前近期洪冲积扇。潮白河进入北京平原后，河道变宽，形成曲流、分流，甚至出现游荡河段，但其并未在平原区形成冲积扇，在沿途也未见留下更多冲积物，只是一扫而过。相反它对永定河冲积扇却起到了侵蚀、破坏作用。

一、土壤元素的组合特征

北京平原区土壤元素的地球化学特征，除了受控于沉积物的物质来源（成土母岩及母质）外，地貌、地表水径流以及人类活动等后期改造对其也有很大程度的影响。为了探讨具有类似分布规律的元素组合特征，将北京平原表层、深层土壤样品54项元素及指标进行因子分析，其因子相关元素组合见表50-5。

表50-5　主要元素因子相关组合表

Table 50-5　Related factors of main elements

因子	表层土壤	深层土壤
F1	Fe_2O_3-Ti-Ce-V-La-Ga-Al_2O_3-Nb-Co-Sc-Be-Th-Y-F-Li-Ni-Mn	Na_2O-Sr-(Ni)-(Cr)-(Sc)-(V)-(Co)-(Y)-(Fe_2O_3)-(B)-(Sb)-(Li)-(Ti)-(Mn)-(Th)-(Sn)-(Bi)-(Al_2O_3)-(W)-(Be)-(Cu)
F2	Cd-Cu-Zn	K_2O-SiO_2-Rb-Ba-(CaO)-(MgO)-(pH)-(C)
F3	K_2O-SiO_2-Rb-(CaO)-(MgO)-(pH)-(C)	Cd-Ag-Pb-Cu-Zn-Mo

表层土壤F1因子表现为以铁族元素、微量及稀土等元素的组合为特征。这一因子组合中大多数元素在山区与平原过渡地带，土壤与水系沉积物两种介质中的含量差异是不明显的，只是不同的元素组合，它们的浓集中心有些差异。如稀有、稀土元素Be、Nb、La、Ce等的浓集中心多分布在岩体附近，在阳坊岩体一带表现得尤为明显；而铁族元素Fe_2O_3、Ti、V等在平原北部山前更为浓集。山前土壤多为周边岩石物理、化学风化的产物，表明这些元素的富集与周边物源密不可分。

深层土壤F1因子为两组在含量区域分布上互为反相关的元素组合。一组为Na_2O、Sr，在F1因子中为较高的正因子载荷；另一组为Ni、Cr、Sc、V、Co、Y、Fe_2O_3、B、Sb、Li、Ti、Mn、Th、Sn、Bi、As、W、Cu、F、Al_2O_3、Be，为较高的负因子载荷，主要为以亲铁元素和稀有分散元素、亲硫元素的组合为特征，比表层土壤F1因子包括的元素种类要多。Na_2O、Sr的高背景区同它们在表层土壤中的分布表现一致，主要位于潮白河、永定河两个冲积平原区，在潮白河平原区富集特征更加显著。而在平谷倾斜平原区及西部的山前洪积扇地带，Ni、Cr、Sc、V、Co、Y、Fe_2O_3、B、Sb、Li、Ti、Mn、Th、Sn、Bi、As、W、Cu、F、Al_2O_3、Be大都表现为高背景分布。

表层土壤F2因子元素组合Cd-Cu-Zn，与深层土壤F3因子元素组合Cd-Ag-Pb-Cu-Zn-Mo都代表重金属元素组合。这些元素的明显分布特征是在昌平区南口—流村一带有着很高的含量值。对比这一带水系沉积物中相应元素的分布，Pb、Cu、Zn、Mo等含量明显高于周边水系沉积物中相应元素的含量，呈现富集状态。而在永定河、潮白河冲积平原区，上述多数元素都表现为低背景分布。与表层土壤不同的是，Cd、Ag、Pb、Cu、Zn等元素在城区深层土壤中表现为背景或低于背景分布，说明在城区范围内，人为因素造成的污染中，Cd、Ag、Pb、Cu、Zn还未影响到深层土壤。

表层土壤F3因子与深层土壤F2因子元素组合均为K_2O-SiO_2-Rb-(CaO)-(MgO)-(pH)-(C)，以两组含量分布截然不同的造岩元素组合为主，可以反映土壤母岩的基本信息。K、Si、Ba等的高背景区主要在潮白河冲积平原区，北部靠近山区地带富集趋势更加明显。潮白河冲积平原的物源区以中酸性岩浆岩及变质岩为主，岩石中富含长石、石英等矿物，导致了土壤中K、Si、Ba等呈现高背景分布。Ca、Mg、C、pH 等多种元素及指标在这一区域均为低含量分布，其原因一方面可能与成土母岩物质成分有关，另一方面可能是由于后期表生物理化学作用，土壤中的一些元素或化合物随流水迁移所致。这些区域深层土壤中元素的含量分布与表层土壤总的趋势是对应的，表明了土壤母质成分对不同深度土壤中元素分布具有最重要的控制作用。

二、土壤的母质类型

北京平原区土壤主要来自永定河流水沉积作用。永定河汇水面积十分广阔，上游大大小小的支流经过黄土高原上若干大小盆地，盆地内不但有广布的黄土层，而且还有巨厚的冰碛泥砾沉积层和湖相沉积层，为河流搬运提供了丰富的物质来源。土壤经过反复流水搬运沉积作用已充分均匀化，失去了原有母岩的特性，与周边山区出露的基岩没有亲缘关系。因此说，地壳平均岩石组分——酸性岩石母质类型的土壤，一般不会有自然成因的土壤环境质量问题，也就是说这类母质类型的土壤，具有均匀的元素含量。维诺格拉多夫提出利用岩石、土壤化学元素含量比值，判定其原岩类别的地球化学方法，他列举了 8 个元素对的 7 类原岩（石陨石、超基性岩、基性岩、酸性岩、沉积岩和地壳平均组分）的标准比值。通过对北京平原区表层土壤（0～20cm）的母质类型进行判别，表明 Al/Ga、Rb/Tl、Ca/Sr、Th/U、Ni/Co 比值接近地壳岩石平均组分标准比值，Zn/Cd、K/Rb、Mo/W 比值接近酸性岩石标准比值（图 50-1）。

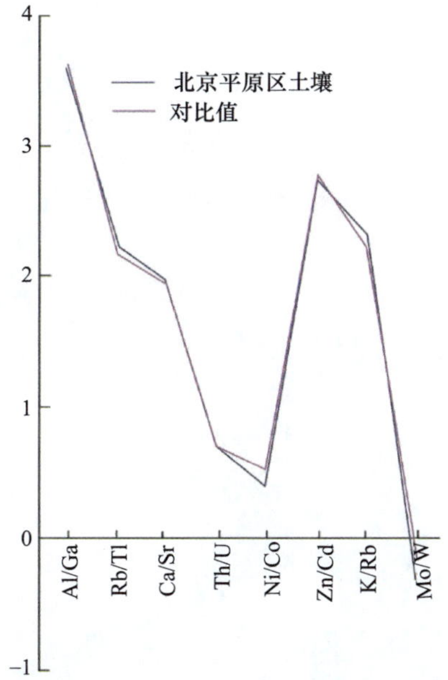

图 50-1　北京平原区土壤母质类型对比图

Fig. 50-1　Comparison of soil parent material types in Beijing Plain

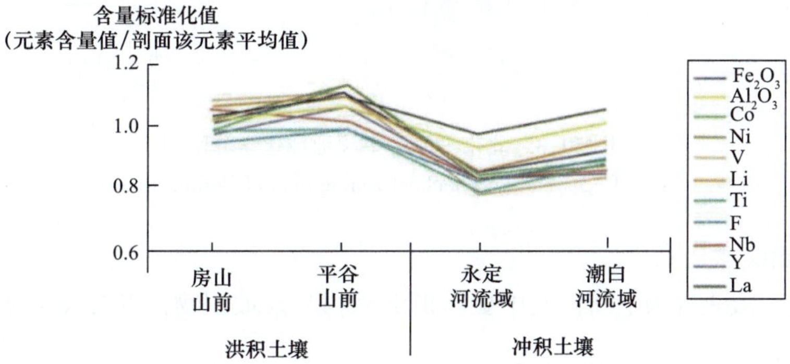

图 50-2　北京平原区土壤元素地球化学演化图

Fig. 50-2　Geochemical evolution of soil elements in Beijing Plain

三、洪积物-冲积物土壤元素演化

北京平原区土壤地质因子元素（氧化物）组合有 Fe_2O_3、Al_2O_3、Ti、Ce、V、La、Co、Y、F、Ni、Li、Mn、Ga、Nb、Sc、Be、Th，在表层土壤中这些元素的含量变化有一个共同特点，即在洪积→冲洪积→冲积的地质演化过程中元素含量发生由高到低的地球化学变化（图 50-2）。土壤地质因子元素含量的这种地球化学变化规律，一方面证实平原区冲积物土壤是后期河流的切割、侵蚀、改造新老冲洪积堆积物的基础上逐渐演变形成的；另一方面也表明平原区冲积物土壤形成过程基本没有其他新的物质参与。

四、土壤地球化学地质分区

根据土壤元素含量因子分析，结合土壤母质类型及土壤元素演化过程，将平原区土壤分为Ⅰ区、Ⅱ区、Ⅲ区 3 个土壤地球化学区（图 50-3）。综合研究发现，Ⅰ区主要是山地、丘陵向平原区的过渡的洪积平原，以黄土、棕色黏性土和不同地质历史时期的坡、残积物土壤为主；Ⅱ区为地势平坦的平原区，以冲积物为主的黄土状物质土壤；Ⅲ区为现代冲积平原，以砂类、砂砾石土壤为主。Ⅰ区被Ⅱ区分割，呈残留体分布。Ⅱ区又被Ⅲ区分割。这形象地反映出北京地区土壤的地质演化过程，也反映了北京平原区流水侵蚀、流水沉积作用的过程。

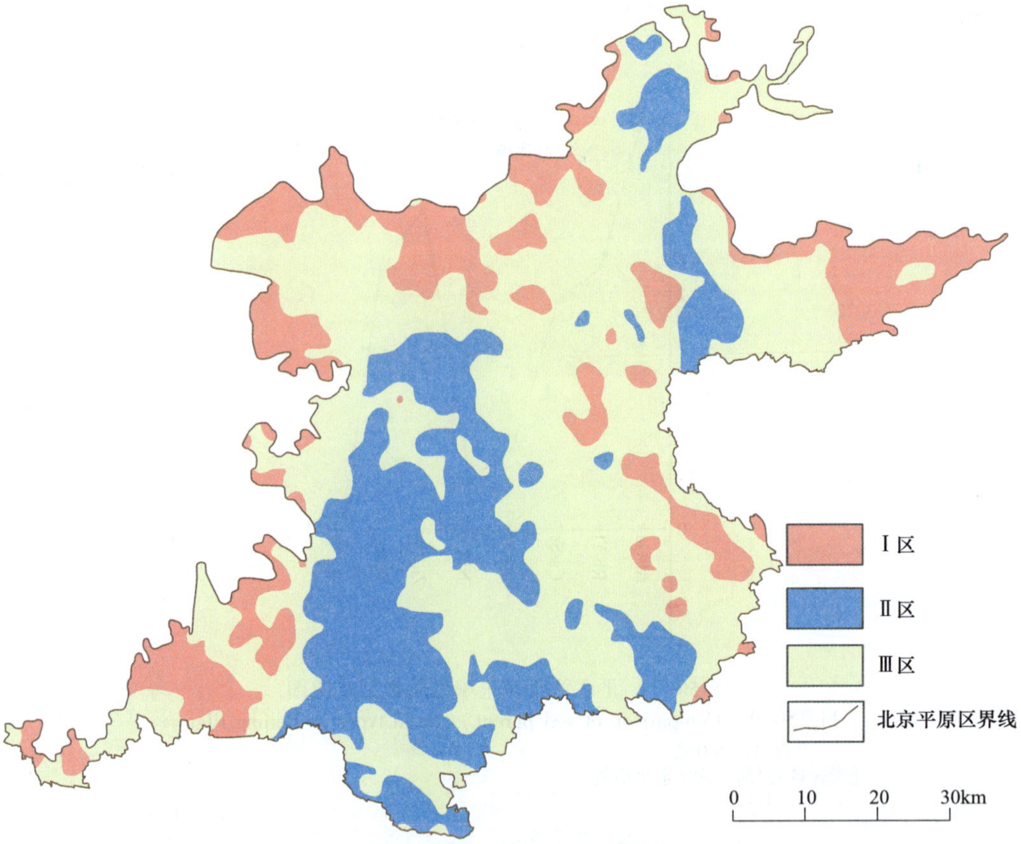

图 50-3　北京平原区土壤地球化学分区图

Fig 50-3　Soil geochemical zoning map of Beijing Plain

1. 土壤地球化学Ⅰ区

本区域在地貌上一般表现为上新世及中更新世洪积物与残坡积物，以及晚更新世的洪积台地与冲积台地。主要特征如下：

土壤中 Fe_2O_3、Ti、Ce、V、La、Ga、Al_2O_3、Nb、Co、Sc、Be、Th、Y、F、Li、Ni、Mn 等含量较高，从这些元素（氧化物）的地球化学图上看，它们的区域含量分布特征大体上相似，主要位于北京平原东部平谷区的大兴庄—韩庄一带、西山及北山的山前、昌平小汤山地区、顺义南彩及李遂一带、通州的牛牧屯—西集一带。土壤中 Al_2O_3、Fe_2O_3 化学成分在山前洪积土壤中含量浓度较高主要是因为这些地区的土壤分选差，各粒级物质混杂，大部分微粒黏土矿物被保留在冲积、洪积土壤中；稀土元素 Be、Nb、La、Ce 等的浓集中心多分布在就近物理、化学分化的岩体附近；铁族元素 Fe_2O_3、Ti、V 等在山前更为富集。

这些地带地貌上表现为坡积裙和山麓冲洪积台地，土壤主要是坡积物、洪冲积物。其中，基岩山地及平原区内残丘向平原过渡的坡积裙，主要由黄棕色沙壤土组成，底部见紫红色黏土，局部地区有残留的风化壳。晚更新世时期形成的洪积台地，几乎分布于所有山麓小河出口处。这些台地在近山部分大多由粗大的砂砾石夹少量黄土状物质组成，砾石分选、磨圆较差，离山越远砾石变稀，分选、磨圆变好，且砾石层厚度变薄，砂、土物质层厚度增大。

2. 土壤地球化学Ⅱ区

本区域地貌单元为全新世洪积扇及前缘洼地、现代河谷的一、二级阶地。由山麓冲洪积台地受后期河流的切割、侵蚀、改造，已逐渐演变成冲积、洪积台地，其中还包括冲洪积扇缘洼地和近代洪积扇缘洼地。主要特征如下：

相较于土壤地球化学Ⅰ区，土壤地球化学Ⅱ区土壤中更富集 Ag、As、B、Br、C、Cd、Cl、Hg、I、Mo、N、Pb、S、Sb、Se、Sn、U、CaO、C_{org}、FeO 等，既有与有机物质密切相关的 CaO、C、C_{org}、N 、Cl 等元素与指标，也有与环境质量相关的 As、Cd、Hg、Pb 等元素。而 Fe_2O_3、Ti、Ce、V、La、Ga、Al_2O_3、Nb、Co、Sc、Be、Th、Y、F、Li、Ni、Mn 等元素（氧化物）含量与北京市土壤基准值基本相当。分布于北京平原的广大地区。永定河、潮白河、温榆河、拒马河、大石河、泃河等较大河流是本区域沉积物的主要输送者。它们大多从西北流向东南，少部分从北向南流或从东北向西南流淌，各自形成了一系列大小不等冲洪积扇体，与山麓冲洪积台地渐变过渡相互连接，形成了本区内的冲积平原及扇缘洼地，沉积物一般以黏性土为主。

3. 土壤地球化学Ⅲ区

土壤地球化学Ⅲ区主要由近代永定河—温榆河流域和潮白河流域两部分组成，之间沿被沿温榆河一带的土壤地球化学Ⅱ区隔开。本区主要在北京城区以南及东南广大的冲积平原上。永定河、温榆河、潮白河以及它们的古河道，流经上游冲积、洪积台地之后，将流水作用所携带的沉积物堆积在台地外围，联合形成了宽阔的近代冲积平原。

相较于其他两个地球化学区，本区土壤中更富集 SiO_2、Zr、K_2O、C_{org}、S、Hg、Pb、Cd、Se、Zn、N、P、MgO、TC、CaO 等，也更贫乏 Fe_2O_3、Ti、Ce、V、La、Ga、Al_2O_3、Nb、Co、Sc、Be、Th、Y、F、Li、Ni、Mn 等元素和指标。本区深受人为活动影响，有益元素 N、P、K、Se 及环境元素 Hg、Pb、Cd 等含量显著高于其他两个地球化学区。本区西北顶点在石景山附近永定河出山口处，东北部与宽广的潮白河谷地连为一体，向东南方向展开，前缘已伸出北京市界。由于地势低平，河道曲折，易于河道的变迁。洪水泛滥后，形成若干沙丘与洼地。沙丘呈长条形分布。

第三节　土壤环境质量

土壤环境质量状况受来自大气、水体、人为活动等多种因素的影响，土壤环境的优劣直接决定了生长于土壤表层的植物体内的污染物含量，决定了人类生活的质量。土壤污染物在土壤圈表层易受后期人类活动干扰而发生积累，当积累到一定程度后就可能对生态环境乃至食物链造成危害。

一、有益元素

（一）富硒土

以大于 0.3mg/kg 圈定富硒土壤，富硒土壤主要分布于昌平区的南口地区和流村镇，石景山区，房山区的周口店地区、石楼镇和琉璃河地区，在故宫西侧、大兴区黄村、通州区台湖镇、朝阳区等地零星分布（图50-4）。

房山区地层岩石中含硒高的煤层和碳质页岩的风化是房山平原区土壤中硒的主要来源之一，由大石河携带风化物从山区流向平原，经过石楼、韩村河、琉璃河流出北京，该流域内的土壤中硒含量均较高。

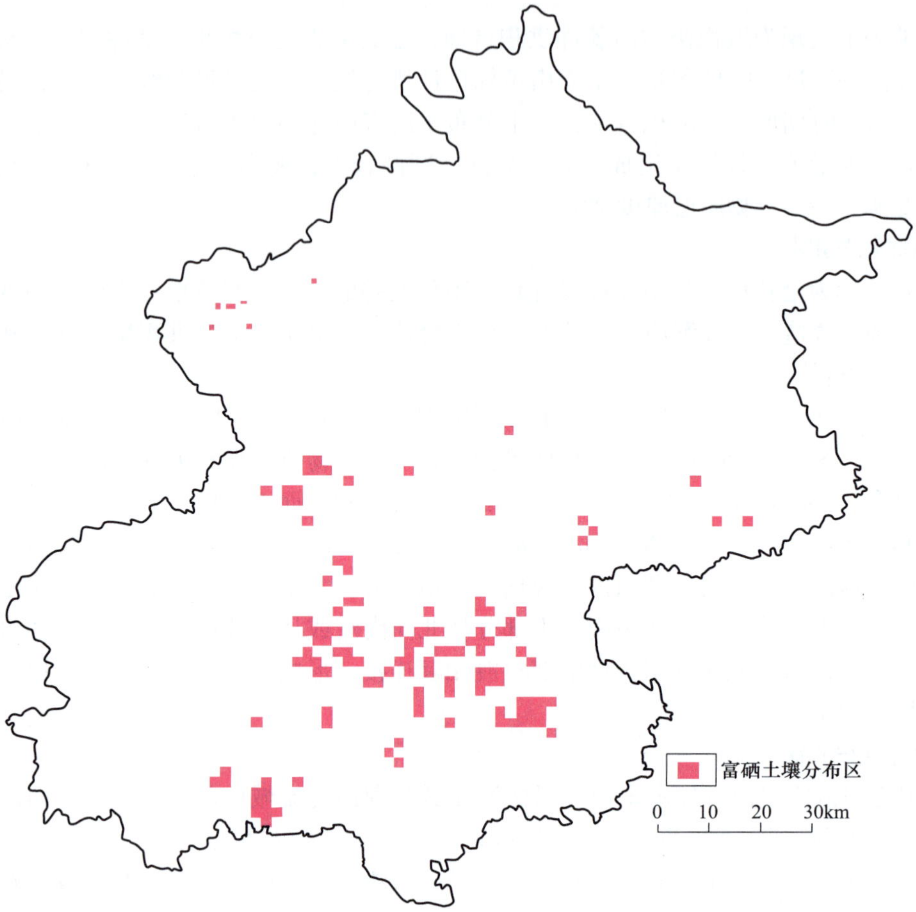

图50-4 北京平原区富硒土壤分布图

Fig. 50-4 Distribution of selenium-rich soil in Beijing Plain

（二）富钾土

土壤全钾能很好地反映土壤母质特征，代表土地长期供给钾素的能力，且受施肥影响较小具有稳定性，并与速效钾有较好的相关性，因此选用土壤全钾作为天然富钾土地资源的评价指标。全国第二次土壤普查把土壤全钾含量大于2.5%确定为一级丰富区，因此本志将表层土壤全钾含量大于2.5%定为富钾土壤资源区标准。

富钾土地资源区分布在潮白河、蓟运河流域，怀柔、密云和平谷山前平原区都为富钾土壤分布区（图50-5）。山区富钾黏土质粉砂岩、页岩以及云蒙山花岗岩，是潮白河和蓟运河流域土壤钾的主要来源。

二、土壤肥力

（一）主量元素

以P、K和TOC作为评估指标，平原区53.06%的表层土壤肥力适中，主要分布在永定河流域的西侧、潮白河流域的东侧；土壤肥力丰富的土壤为16.43%，沿着平原区的边缘分布，通州地区土壤的大量元素肥力状况良好；肥力相对缺乏的土壤面积为20.51%，主要分布在永定河流域的东侧，延庆地区

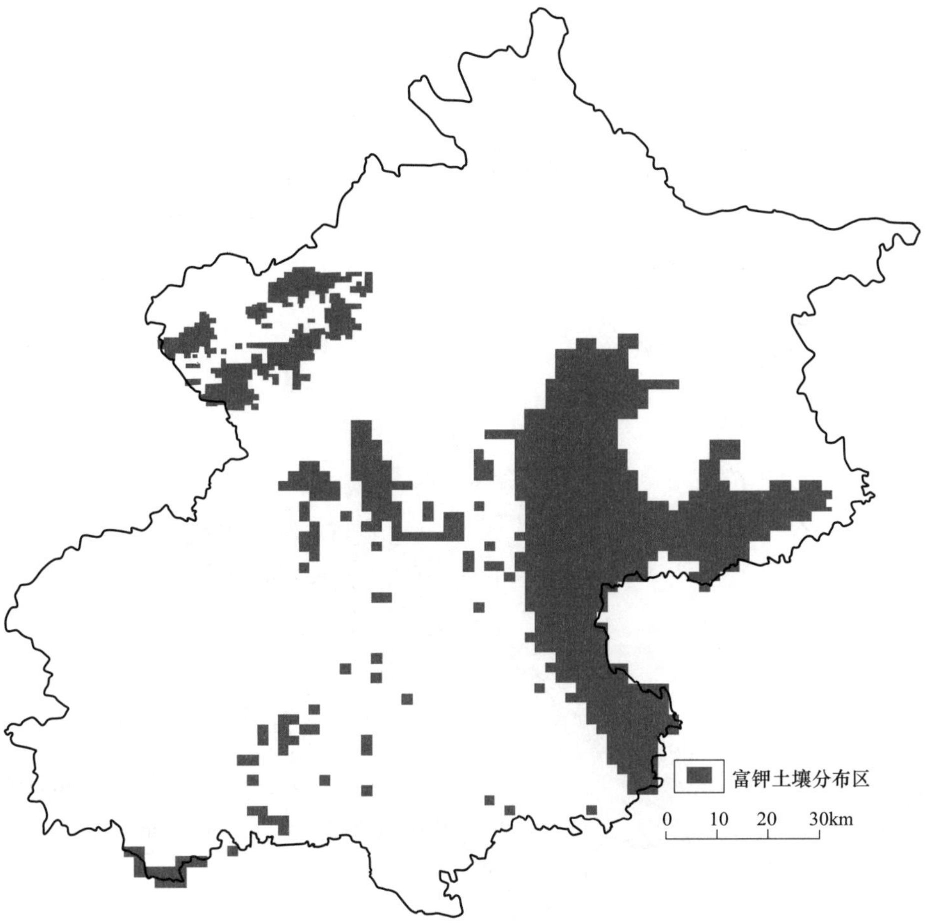

图 50-5 北京平原区富钾土壤分布图

Fig. 50-5 Distribution of potassium-rich soil in Beijing Plain

和密云西北隅。

(二) 微量元素

以有效硼（B）、有效铜（Cu）、有效锌（Zn）、硒（Se）作为评估指标，综合来看平原区土壤微量元素肥力状况良好，二等以上土壤面积占全区的72.52%。平原区的西部边缘（房山、门头沟、昌平）地区微量营养元素丰富；平谷地区和凉水河流域土壤的微量元素肥力状况良好；永定河和潮白河地区微量元素肥力相对匮乏；延庆盆地微量元素肥力适中或匮乏。

硒对人体健康具有重要的意义，北京平原区32%的表层土壤缺硒。

三、土壤环境质量综合评价

8种重金属元素中，平原区表层土壤Ni、Cr元素的含量几乎全部低于国家一级标准，即处于自然背景状态，可以认为Ni、Cr元素不作为有害元素。Cu、Zn元素在平原区郊区土壤中基本上都处于自然背景状态，仅在城区含量略高，但基本上都低于国家二级标准，同时，Cu、Zn是作物生长的必需营养微量元素，故该两元素主要作为营养元素而不作为有害元素来参与评价。因此，影响土壤环境质量的有害元素主要为Hg、Cd、As、Pb。

北京平原区（含延庆盆地）约75.9%的地区为土壤一级质量区，城区周边及郊区部分零散地区为土壤二级质量区，约占总面积的23.5%，而城区中心、石景山北、丰台北、小汤山等少数地区为土壤三级质量区，约占总面积的0.6%。也就是说北京平原区土壤环境总体质量是较好的，约99%的地区环境综合质量为优良，仅在人类活动强烈地区的土壤环境质量较差（图50-6）。

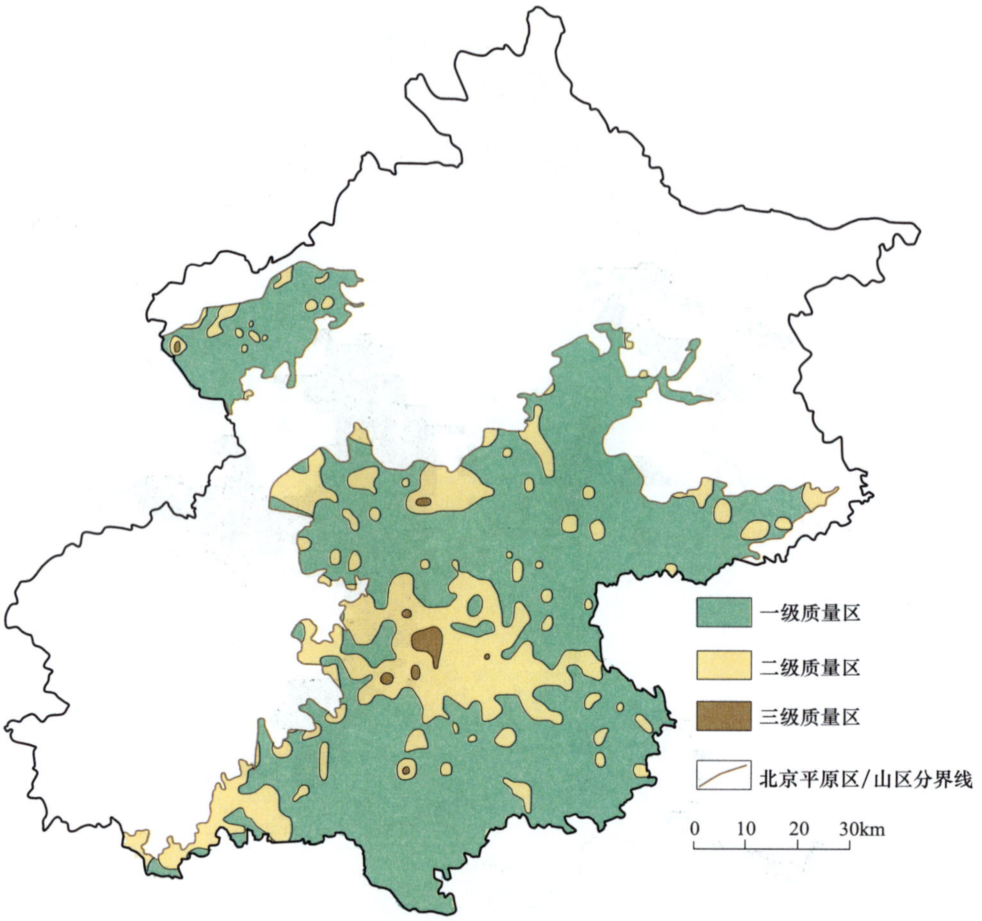

图 50-6 北京市土壤环境质量综合等级图

Fig. 50-6 Comprehensive evolution of soil environmental quality in Beijing

第四节 土地环境质量

一、生态效应

（一）典型地区小麦、玉米安全性评价

小麦和玉米是北京地区的主要农作物，在饮食结构中占很大比例。对昌平、朝阳、大兴、房山、丰台、海淀、怀柔、密云、平谷、顺义、通州和延庆等12个典型地区的小麦、玉米籽实和根系土取样研究后发现，除F、Zn外，小麦籽实中的有害元素平均含量均高于玉米籽实（图50-7）。

除F外，所有有害元素在玉米籽实中含量均低于我国现行的各类食品卫生标准；有两件小麦籽实样品As和Pb含量超过了食品污染物限量标准，其余元素在小麦籽实中含量均低于食品污染物限量标准。玉米籽实和小麦籽实中的F含量均超过了食品污染物限量标准（1mg/kg），其中玉米籽实F含量平均值为6.04mg/kg，小麦籽实中F含量平均值为3.5mg/kg。如此高的超标程度和如此大的超标范围，其确切原因有必要进一步研究。

综上所述，北京市大宗农作物除普遍存在籽实中F含量超标外，As、Cd等有害元素含量均低于现行的我国各类食品卫生标准，安全程度较高。

（二）北京市蔬菜安全性评价

辣椒、小白菜、葱、韭菜、黄瓜、西红柿等不同农作物具有不同的重金属污染耐受能力，除了自

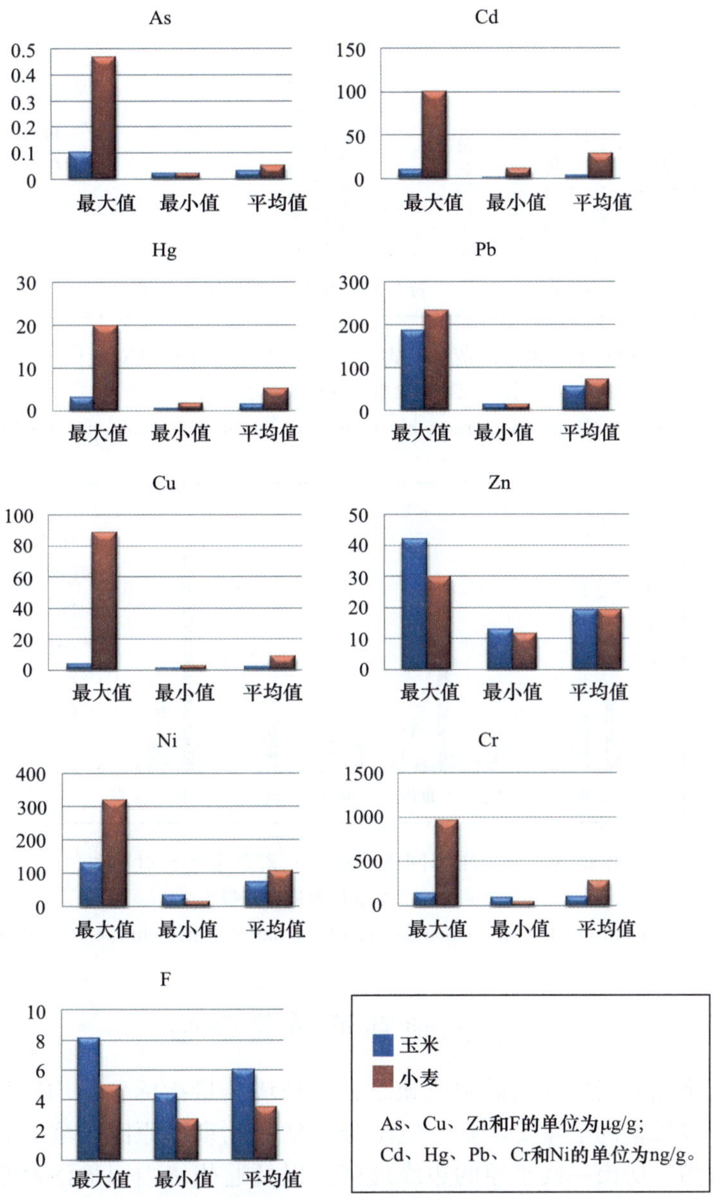

图 50-7　典型地区小麦和玉米籽实中有害元素平均含量特征对比

（据北京市多参数立体地质调查报告）

Fig. 50-7　The average content features comparison for harmful elements of corn seeds and wheat seed in typical areas

身的抗逆能力外，同一种农作物在不同地区所受的污染程度也不一致。按照食品中各元素的限量卫生标准进行判别表明，北京市蔬菜有害元素超标现象比农作物严重，普遍存在 As、F 超标，个别蔬菜样本中 Pb、Cr 和 Ni 超标（图 50-8）。不同种类蔬菜对不同元素的富集系数差异很大，在假定蔬菜中有害元素均来自土壤的前提下，一般情况下，叶菜类蔬菜如小白菜、韭菜等有害元素的富集程度远高于茄果类如西红柿、辣椒等的富集程度（图 50-9）。

从图 50-9 中可以看出，相同的蔬菜种类，不同元素富集程度也有很大差异，一般情况下，Cd 的富集系数最大，且 Cd 的水溶态和离子交换态含量比例较高，如果土壤中 Cd 含量增加较快，潜在的生态危害会很大。因此，在 Cd、Hg、Pb、As 等有害元素异常富集的地方，建议减少种植叶菜类蔬菜，多种植茄果类蔬菜，以进一步提高食品的安全程度，保证人民身体健康，增加蔬菜出口的国际竞争能力。

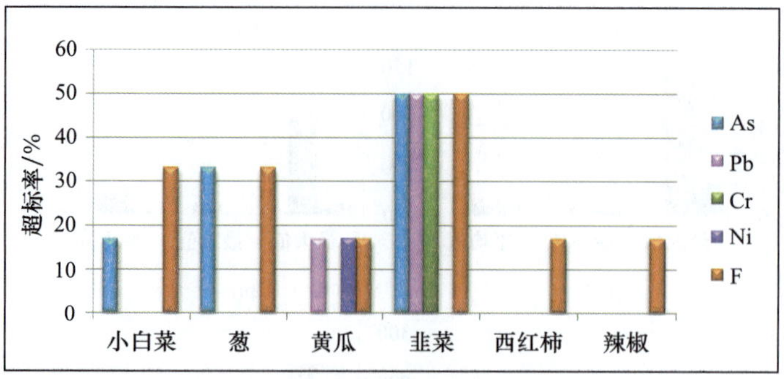

图 50-8　北京市蔬菜有害元素含量的超标率特征对比图
（据北京市多参数立体地质调查报告）
Fig. 50-8　Overstandard rate comparison histograms for harmful elements of vegetable in Beijing

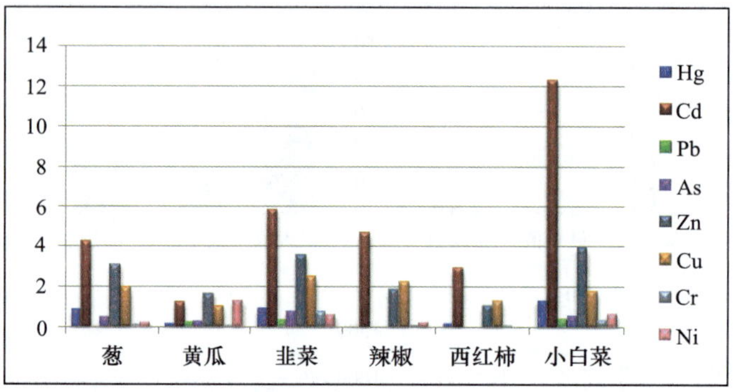

图 50-9　北京市不同蔬菜有害元素富集系数对比图
（据北京市多参数立体地质调查报告）
Fig. 50-9　Enrichment coefficient comparison chart for harmful elements of vegetables in Beijing

二、土地环境质量评估

由于北京平原区土壤肥力指标含量适中或缺乏，而环境质量指标含量均很低，土壤安全性高，故土壤肥力是制约土地环境质量的最重要因素。经综合评估，北京平原区土壤综合质量最好的地区位于平谷地区、昌平—门头沟—房山一线平原的边缘地区，以及通州境内温榆河水系两侧和凉水河、凤港减河流域，这些地区 As 等有害元素含量低，且有益元素丰富，适合于种植无机食品和名优特农产品，进一步挖掘土地经济潜力。永定河水系两侧土壤和延庆地区土壤的综合质量相对较差，这些地区土壤养分相对缺乏，进行农业种植时应适当增施农家肥和钾肥。

三、土地资源可持续利用建议

对城市与农村土地利用的空间布局进行适度调整。平原区及台地丘陵区以土壤一级质量区与土壤二级质量区为主，总体状况优良，适宜城市建设和农用地的开发利用。浅部山区土壤元素分布具有明显的地域性特征，建议农业种植要因地制宜，科学规划，提高农产品附加值。城市部分功能区迁至郊区，有助于减少城市人口的承载量与污染程度，降低城市脆弱度，但要在迁入区注重生态环境的保护，防止出现新的污染。土壤三级质量区对城乡居住、娱乐基本不具有危害，但不适宜农业种植，建议将此类地区规划为植树造林带、绿化景观带等。另外，完善土壤地球化学环境监测系统，实时掌握土壤有益有害元素时空变化规律，为科学管理提供基础支撑。

第五十一章 地质灾害

第一节 概 述

一、地质灾害概况

地质灾害是指在地球内、外动力或人为活动的作用下，地球发生异常能量释放、物质运动、岩土体变形位移及环境异常变化等，危害人类生命财产、生活与经济活动或破坏人类赖以生存与发展的资源、环境的现象或过程。

北京地区发育的地质灾害主要有地面沉降、地裂缝、砂土液化、采空塌陷、崩塌、滑坡和泥石流等，是地质灾害发育较多、较严重的特大城市之一，具有灾种多、活动频繁、群发性强的特点。特别是近几十年，经济建设的高速发展和人口的过快增长，自然环境破坏的加剧以及不合理的人类工程活动等使地质环境不断恶化，地质灾害经常发生。地质灾害给北京社会经济发展、人口安全带来巨大挑战，成为限制北京城市发展的制约因素之一。本章基于前面章节对北京地区地质成果的综合研究，对北京地区内地质灾害发育现状进行研究，对诱发地质灾害的地质背景条件进行分析，为城市发展建设提供基础地质资料。

二、北京地区研究动态

在北京市政府的高度重视和大力支持下，地质、水利、林业等部门以及自然资源部遥感中心、中国科学院成都山地研究所、北京大学、北京师范大学、北京工业大学、北京市政设计院等多所高校和科研部门先后开展了地质灾害调查与重点防治工作。北京市地质矿产勘查开发局是承担北京市地质灾害调查、监测、预警和防治任务的主要业务部门，30 多年来完成了一系列的地质灾害调查、监测、应急和预警工作：1980—1990 年，相继完成了北京市东郊地面沉降观测和全市主要地质灾害调查，编制了北京市第一份 1∶10 万地质灾害分布图；1991—2000 年陆续完成了北部山区和西山门头沟清水河流域泥石流灾害勘查，西山地区和门头沟区门城镇采空塌陷灾害勘查，通过一系列灾害勘查，为灾害治理工作奠定了坚实基础；与此同时，完成了密云区番字牌西沟泥石流小流域综合治理；2001 年以来，不仅完成了 11 个区的地质灾害调查与区划工作，并与北京市气象台共同合作，开展了地质灾害气象预报技术工作，参与编制地质灾害应急预案，实施地质灾害应急调查，在平原区建立了地面沉降监测站，开展了山区突发性地质灾害监测工作。地质灾害调查和监测预警工作为政府制订减灾防灾、国土开发整治、经济建设和社会发展规划及地质环境监督管理提供宏观决策依据。

第二节 平原区地质灾害

一、地面沉降

（一）形成与发展

北京市地面沉降的发现可追溯到 20 世纪 30 年代，此后经历了形成、发展和扩展、快速发展等不同阶段。地面沉降的发育，对城市线性设施安全存在影响，可导致水准点失效，造成高程损失，同时影响建筑物的抗震能力。

1949 年后，随着城市建设和工业发展，地下水开采快速增加，地面沉降区逐步形成。20 世纪 50—60 年代，由于北京东郊八里庄及酒仙桥一带分别建立了纺织工业区和电子工业区，大量开采地下水，使这一带的地下水位逐年下降，形成水位降落漏斗，逐渐形成了东郊地面沉降区。

20 世纪 70—80 年代初，随着地下水的继续开采，地下水位降落漏斗范围扩大，东郊地面沉降也随之发展，累计沉降量大于 50mm 的沉降区面积超过 600km^2，平均沉降速率达到 31mm/a。20 世纪 80—90 年代，地面沉降面积扩展较快，到 1999 年东郊地面沉降漏斗区向北扩展到昌平区南部及顺义区南部；向东扩展到通州区，与通州区地面沉降漏斗区相接，大于 50mm 的沉降区面积以平均每年增加约 138km^2 的速度迅速扩大到了 2815km^2，平均沉降速率达到 36. 1mm/a。

1999 年以后由于连续干旱，加上城市建设快速发展，地下水开采量和开采深度持续加大，导致地面沉降进入了一个快速发展时期，沉降速率较大，累计沉降量持续增加，沉降区面积继续扩大。

（二）发育现状

1. 空间分布

经北京市水文地质工程地质大队长期监测，北京地区地面沉降自 20 世纪 50—60 年代形成发展至 2013 年，沉降区面积已达 6390km^2。其中，1955—2013 年累计沉降量 50mm 以上的区域面积为 4343km^2；累计沉降量 500mm 以上的区域面积为 1349km^2，主要分布在昌平区、顺义区、海淀区、朝阳区、通州区和大兴区（图 51-1，表 51-1）。

表 51-1　北京市 1955—2013 年地面沉降累计沉降量分布特征

Table 51-1　Cumulative subsiding distribution of surface subsidence in Beijing from 1955 to 2013

累计沉降量/mm	面积/km^2	分布区域	区域特征
0~50	2047	主要分布在平原西部、北部山前地带以及平原区西部地区，大兴局部地区	区域内人口、建筑物密度相对较低
50~500	2994	面积约为整个平原区的一半，涉及昌平、顺义、平谷、朝阳、海淀、通州、大兴等大部分地区	区内有多条由中心城区往外延伸的地铁、轻轨等线性工程；涉及多条跨省高速公路或铁路
500~1000	1152	包括北部三大沉降区沉降中心附近地区以及南部大兴北臧村—半壁店—安定一带以南地区	区域内大部分地区属于经济发达、人口密集区
>1000	197	主要分布在昌平八仙庄，朝阳来广营、金盏、三间房—通州管庄—黑庄户，大兴赵村、小马坊、礼贤地区	区域内大部分地区属于经济发达、人口密集区，金盏附近地区建有首都机场及空港经济区

地面沉降已成为北京平原区主要地质灾害之一。自形成至 2013 年，北京地面沉降逐渐发展成为以瀛海庄—长子营—于家务—永乐店一带为界的“北区”和“南区”两个大区。“北区”主要分布于通州区、台湖镇、永顺镇，朝阳区三间房地区、管庄地区、东坝地区、金盏地区、来广营地区、崔各庄地区、孙河地区，顺义区后沙峪镇、高丽营镇，昌平区小汤山镇、北七家镇地区，为累计沉降量（1955—2013 年）大于 650mm 的区域，主要包括昌平沙河-八仙庄沉降区、朝阳来广营-金盏沉降区、朝阳东八里庄-通州台湖沉降区；“南区”分布于平原区南部大兴区庞各庄、榆垡和礼贤地区，包括大兴榆垡-礼贤沉降区，为累计沉降量（1955—2013 年）大于 800mm 的区域。

另外，顺义北务、杨镇地区和平谷王辛庄形成的相对独立的小沉降区，累计沉降量均大于 250mm。

2. 时间演化

（1）沉降中心转移

沉降中心会随时间的推移发生变动，形成于 20 世纪 50—60 年代的东部沉降区逐渐东移，朝阳来广营沉降中心移至金盏地区，东八里庄-大郊亭沉降中心东移至朝阳三间房、黑庄户和通州城区；形成

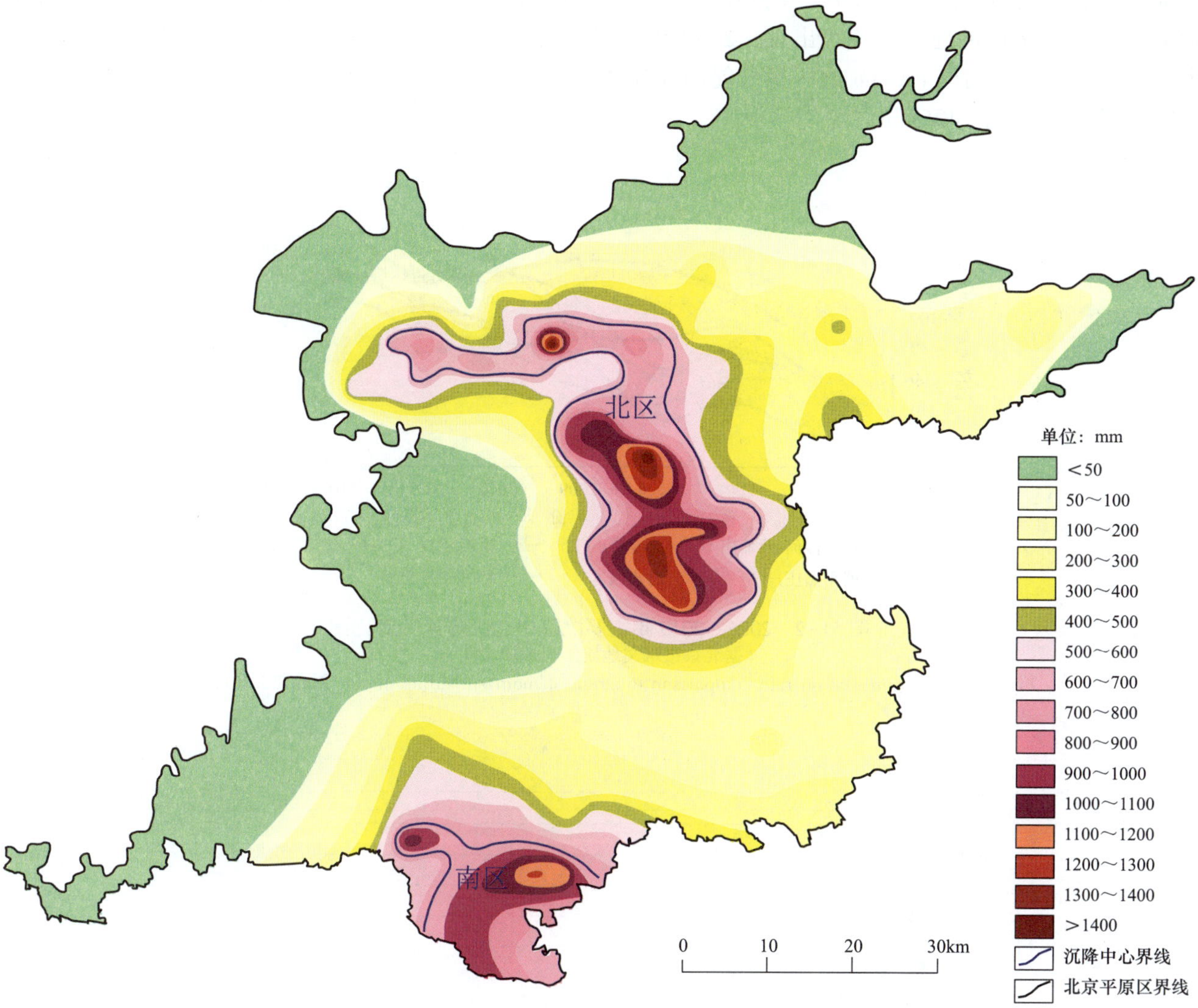

图 51-1　北京市平原区 1955—2013 年累计沉降量图

（据北京市水文地质工程地质大队）

Fig. 51-1　Cumulative subsiding graph of Beijing from 1955 to 2013

于 20 世纪 70—80 年代的北部昌平沙河-八仙庄沉降中心继续发展，新的海淀西小营沉降中心逐渐形成，北部由单个沉降中心的格局向两个甚至多个中心转变；20 世纪 90 年代以后才形成的大兴榆垡-礼贤沉降中心仍然位于礼贤小马坊，赵村和榆垡以南地区沉降发展较快；顺义平各庄地区近年来地面沉降不突出，自 2013 年起不再视为沉降中心。

目前北京市地面沉降有 7 个沉降中心，其中昌平区沙河-八仙庄，海淀西小营，朝阳金盏、三间房、黑庄户，通州城区等 6 个沉降中心位于沉降“北区”；大兴礼贤小马坊沉降中心位于沉降“南区”。

（2）沉降速率

多年沉降速率对比表明，北京地面沉降各沉降中心速率总体呈上升趋势（图 51-2），沉降区目前仍然在持续发展。除大兴榆垡-礼贤沉降中心最大速率出现在 2010 年以外，其余各沉降中心速率最大值均出现在 2012 年，朝阳黑庄户沉降中心沉降速率最大，为 159.6mm/a。

（3）累计沉降量

自形成以来，北京市地面沉降一直持续发展，沉降区累计沉降量呈上升趋势，相同累计沉降量区段的分布面积逐年扩大（图 51-3，表 51-2）。2005 年平原北部和东部累计沉降量大于 1000mm

的地区呈散点分布状态，自2010年开始连成一片，到2013年已接近200km²；到2011年，累计沉降量大于1100mm的地区已成片出现；2012年，累计沉降量大于1300mm的地区已成片出现；2013年，累计沉降量大于1400mm的区域已能明显观测记录，1955—2013年累计沉降量最大值出现在昌平八仙庄地区，为1495mm。

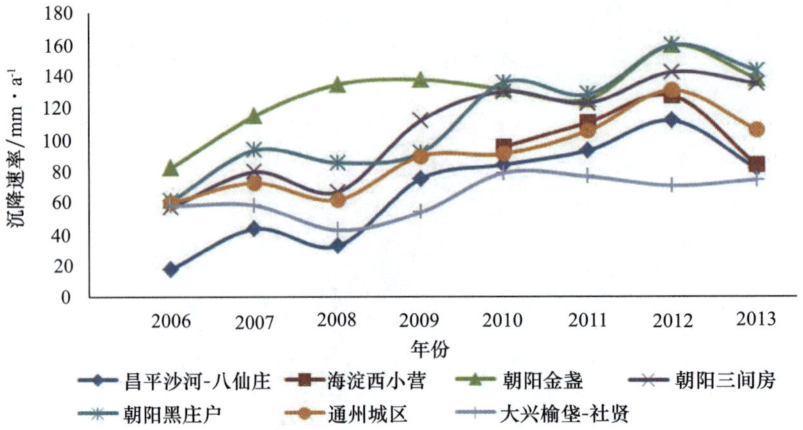

图51-2　2006—2013年主要沉降中心沉降速率对比图
（据北京市水文地质工程地质大队）
Fig. 51-2　Sedimentation rate comparison diagram of main subsiding center from 2006 to 2013

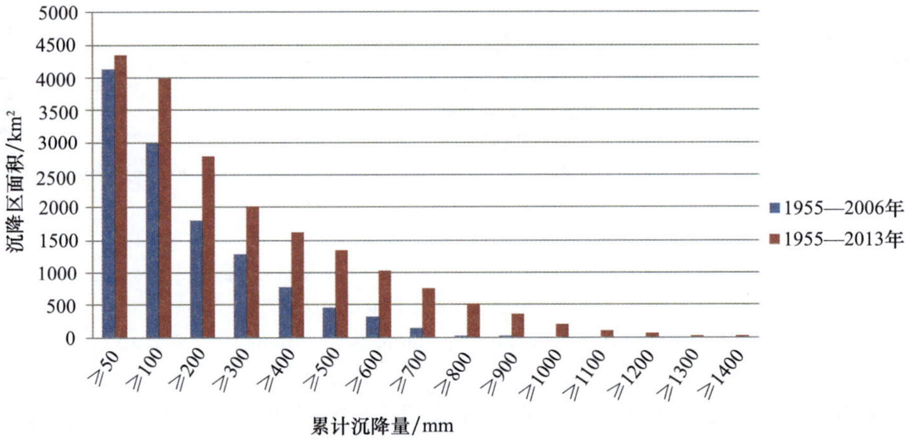

图51-3　2006年与2013年沉降区面积对比直方图
（据北京市水文地质工程地质大队）
Fig. 51-3 Comparision histograms of subsiding area in 2006 and 2013

表51-2　各沉降区1955—2013年最大累计沉降量统计表

Table 51-2　Statistical table of maximum cumulative subsidence in different subsiding regions from 1955 to 2013

沉降区	昌平沙河-八仙庄		朝阳来广营-金盏	朝阳东八里庄-通州台湖	大兴榆垡-礼贤	
累计最大点	八仙庄	沙河	金盏	三间房	小马坊	赵村
沉降量/mm	1495	836	1482	1360	1201	1055

（据北京市水文地质工程地质大队）

（三）主要影响因素

影响北京平原地区地面沉降发展的因素包括自然因素和人为因素，自然因素包括地质构造运动、地层岩性及结构特征和土层次固结沉降等，是产生地面沉降的重要地质背景；人为因素包括超量抽汲地下水、工程建设等。

超量抽汲地下水是地面沉降产生的直接诱因。长期超采地下水，造成地下水位的大幅下降，导致含水层上覆土层孔隙水压力降低，使土层固结失水，土层压缩，同时含水层颗粒排列更趋紧密，形成了地面沉降。分析多年地面沉降和地下水动态监测结果可知，长时间序列下，地下水位变化与地面沉降变化趋势基本一致，地下水位基本处于持续下降状态，大部分地区地面沉降以较快的速度发展，累计沉降量不断增加，地面沉降发育区域与地下水超采严重区域基本吻合，但沉降中心与地下水漏斗中心并非完全吻合。

北京地区区域地质构造运动，表现为山区相对上升，而平原区缓慢下降，沉积了较厚的第四纪地层。基底起伏较大，存在较多断裂。地面沉降区与平原区主要活动断裂的分布具有一定的相关性（图 51-4）。

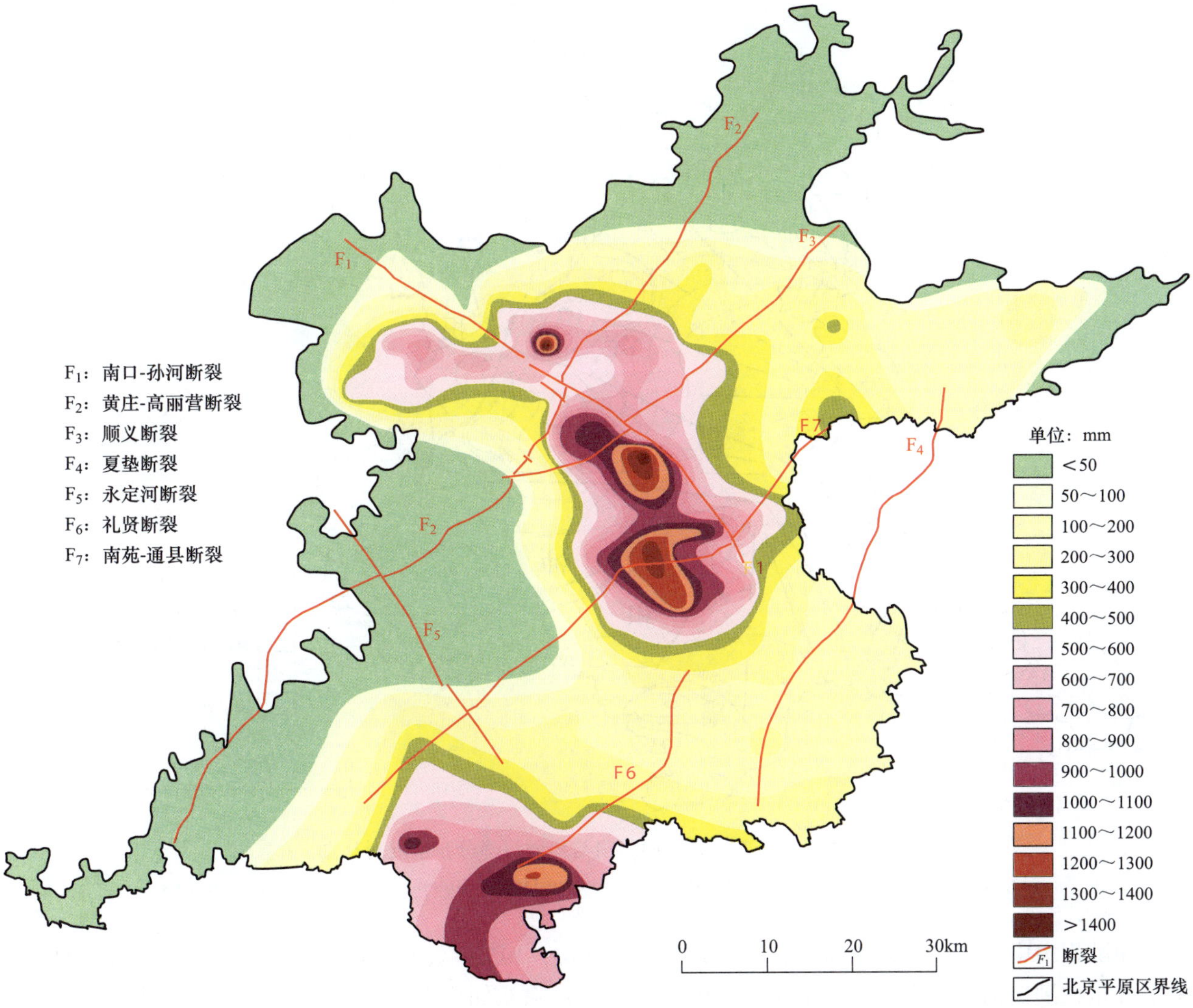

图 51-4　平原区地面沉降区与隐伏断裂分布特征示意图

Fig. 51-4　Distribution schematic of ground subsidence area and buried fault in plain area

地面沉降北区位于南口-孙河断裂与黄庄-高丽营断裂、顺义断裂、南苑-通县断裂交汇的区域；礼贤断裂带经过了地面沉降南区，断裂两侧沉降量较大。西北部地面沉降区主要受南口-孙河断裂和黄庄-高丽营断裂控制，沉降发展趋势与断层方向一致。历年累计沉降量最大的八仙庄沉降中心所在区域断裂活动复杂，南口-孙河断层在此区域发生扭转，断面倾向由南西转变成北东，上盘下降，沙河和八仙庄沉降中心分别位于断裂两侧；黄庄-高丽营断裂在此区域被南口-孙河断裂截断，区域内次级断裂发育。构造运动对地面沉降的影响主要是由于所处的构造单元在区域应力场的作用下整体呈缓慢下降趋势；同时，断层上、下盘的相对升降运动也会对断裂带两侧沉降速率的差异产生一定的影响。

北京平原区由于受新构造运动的影响，古河道变迁十分频繁，构成一个相当典型的泛滥平原，沉积环境复杂，由多个相互叠压、交错的冲洪积扇组成，大致以温榆河为界，东部为潮白河冲洪积扇，西部为永定河冲洪积扇。从西北向东南，从山前向平原，沉积物层次增多，沉积物颗粒由粗变细。地面沉降的分布与冲洪积扇的发育特征有关（图 51-5）。

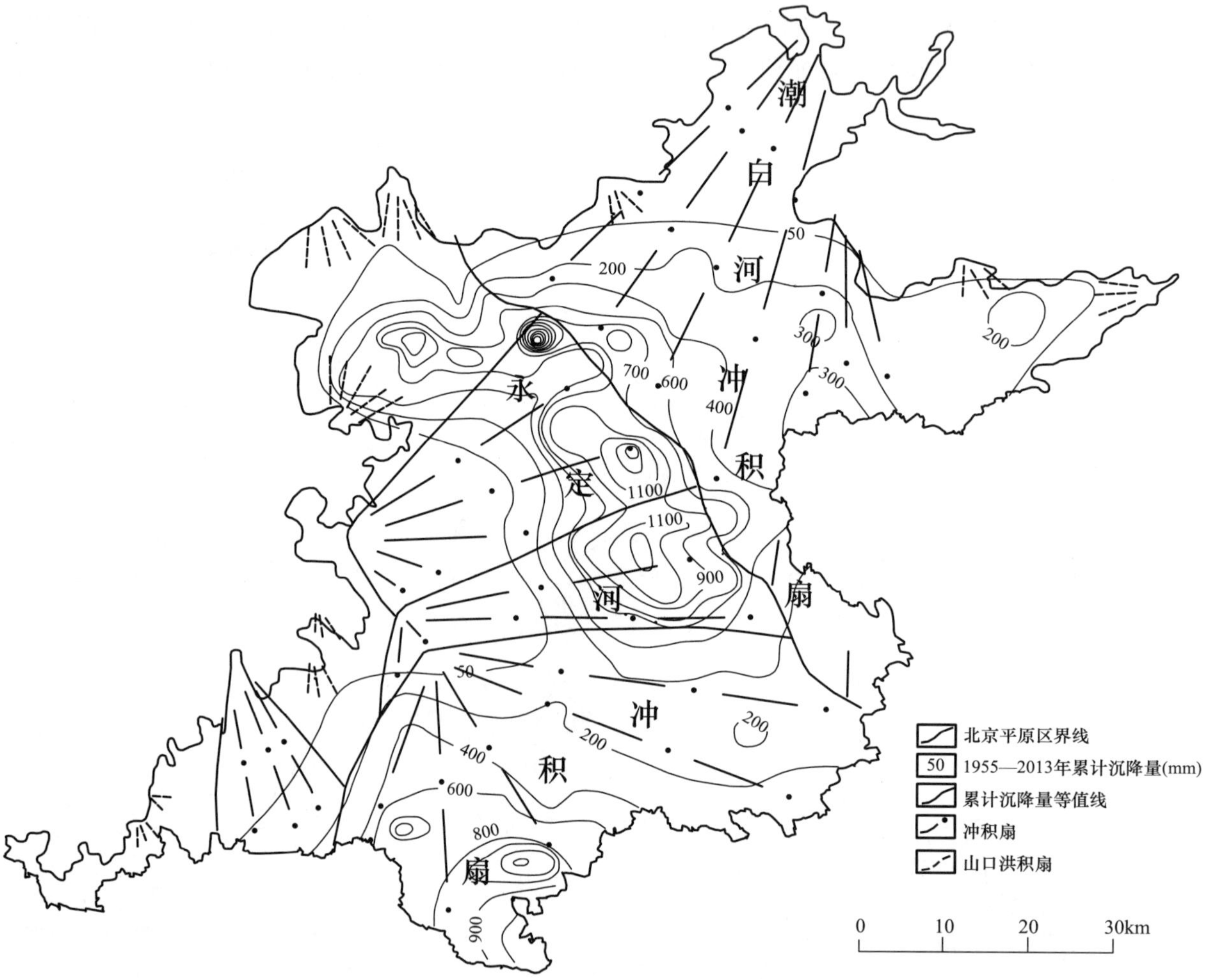

图 51-5 北京平原区冲积扇与地面沉降分布情况对比图

Fig. 51-5 Contrasting pattern of alluvial fan and surface subsidence distribution in plain area

冲洪积扇中下部的累计地面沉降量远大于扇顶部位，1955—2013 年累计沉降量 500mm 以上的区域主要分布在潮白河冲洪积扇和永定河冲洪积扇中下部；除八仙庄沉降中心位于温榆河冲洪积扇外，其余 6 个地面沉降中心全都位于永定河冲洪积扇扇缘部位。中更新世后期以后，永定河逐渐贯穿西山流入平原，受断裂构造的影响，经过了多次改道，逐渐由北向南迁移，形成了复杂的由多个扇体相互叠加组合而成的永定河冲洪积扇；潮白河冲洪积扇的规模比永定河冲洪积扇小，由于潮白河河道相对稳定，所以其扇体无大的迁移，结构相对简单。永定河冲洪积扇复杂的沉积环境更有利于地面沉降的发展扩展，因此，永定河冲洪积扇上的地面沉降较潮白河冲洪积扇更为显著。在西部、北部的山前地带和河流冲洪积扇的顶部及中上部，第四系厚度一般为20~40m，为单一的砂、卵砾石层或砂、卵砾石顶部覆盖薄层黏性土，含水层颗粒粗，富水性好，厚度大，抗压强度高，可压缩性低。因此，虽然北京众多水源地都建在冲洪积扇顶部及中上部，但这些地区地面沉降并不显著。例如位于潮白河冲积扇上部的怀柔应急水源地，2003年供水以来，地下水位累计下降了20~30m，形成了一个大型的地下水降落漏斗，但是监测显示，该地区未发生明显的地面沉降。在冲洪积平原中下部，土体类型为第四系多层砂、多层砾石与多层黏性土，以黏性土为主。黏性土层颗粒较细，富水性差，压缩性高，释水后地面沉降明显，在区域地下水长期超采的情况下，累计地面沉降量远大于扇顶部位。例如昌平沙河一带，百米以内砂层厚度所占比例小于0.2，黏性土层巨厚，虽然地下水未被超采，但土体释水-压缩-变形明显，累计沉降量居北京市前列。另外，在扇缘、扇间洼地、古河道、河漫滩等部位，岩性以冲积相为主夹湖沼沉积，湖沼沉积以富含有机质的黑色淤泥和泥炭为主，这类土体孔隙度较高，渗透性强，压缩性高，利于地面沉降的形成与发展。

二、地裂缝

（一）发育现状

北京地区地裂缝古已有之，历史上曾留下许多有关地裂缝的文字记载，例如1965年邢台地震后在平原地区出现了与主构造方向一致的地裂缝，1976年唐山大地震前后在昌平南新村、顺义俸伯镇等地出现了大量地裂缝。目前，平原区地裂缝主要分布在平谷的马坊、英城、大兴庄、东高村、夏各庄，通州的张家湾、西集、郎府、觅子店、双埠头、辛安屯，顺义的木林、北采、牛山、马坡、塔河、李桥、徐辛庄、焦庄户，昌平的十三陵、南新村、大东流，房山的纸坊村、小次洛村，以及延庆的大柏老等地。目前发现的对城市建设有较大影响的地裂缝主要有顺义地裂缝、高丽营地裂缝、北小营地裂缝、羊房地裂缝、庙卷地裂缝、北彩地裂缝、北甸地裂缝和孙河地裂缝等（图51-6）。

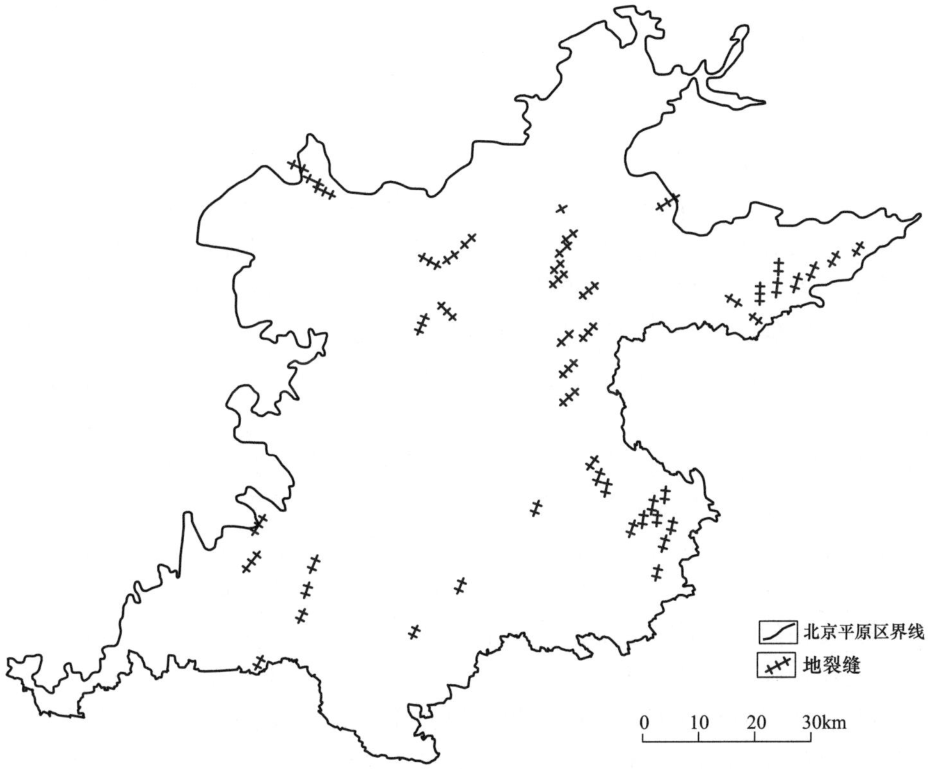

图51-6　北京平原区地裂缝分布

Fig. 51-6　Distribution map of ground fissures in plain area

其中，顺义地裂缝和高丽营地裂缝仍在继续发展。顺义地裂缝由仁和镇塔河村一带向北东延伸至木林地区，影响总长度约25km，对原北京市橡胶二厂造成了严重破坏，工厂外围2004年新建的围墙已出现多处裂缝，附近新建道路也相应发生裂缝，2007年顺义规划新城工程地质勘察可见地裂缝带宽度达200m左右。高丽营地裂缝主要分布在西王路村、唐自头村、土沟一带，延伸2500m左右，地裂缝带宽度达300m左右，对原土沟村、华都肉鸡厂、八仙别墅区破坏严重。

（二）主要影响因素

谢广林（1988）认为地裂缝的成因与自然条件和内外营力作用密切相关，刘传正（1995）按成因将地裂缝分为内动力成因、外动力成因和人类工程参与诱发三大类，王景明（2000）认为地裂缝分为构造成因和非构造成因。实践表明，许多地裂缝并不是单一成因的，而是以一种成因为主，同时又受其他因素影响的综合作用的结果。其中，地层是地裂缝形成的介质因素，地质结构是其构造因素，现代构造应力和现代地壳变动是其动力因素，块体演化是其时间因素。受黄庄-高丽营、顺义、夏垫和南口-孙河等多条隐伏活动断裂的活动影响，北京地区以构造断裂控制的地裂缝比较典型，超量开采地下水对地裂缝活动具有诱发促进作用。

北京平原区有黄庄-高丽营断裂、顺义断裂、南口-孙河断裂等多条隐伏活动断裂发育，地壳稳定性较差，沿这些断裂构成地壳体中的力学薄弱环节，同时也是力学作用的控制性部位。在深部地壳变形过程中，区域构造应力场在活断层部位，不仅使得应力场方向和大小发生变化，还将应力场的作用转化为主要沿活断层传播的表面应力作用，使得断裂带附近的剪切应力发生集中和增强。尤其在活断层的端点、拐点、交汇点、分叉点、弧形转弯、倾向反转等部位应力和应变集中，有利于地裂缝的形成。

北京平原地区沉积层厚度较大，在第四系中大量分布着厚度不均的砾石层、中-粗砂层、细砂层或砂层透镜体等透水性较好、含水率较大的地层，同时分布着大量压缩性较高的黏性土层，具有一定的韵律、旋回特征，这些特征对地裂缝的形成起着重要作用。表层第四纪松散堆积物在不同的含水条件下，对应的物理力学特征尤其是强度具有明显的差异性，在饱水状态下土体易散化与结构软化，致使强度降低，导致了地裂缝具有反复开启活动的基本条件。

三、砂土液化

砂土液化是一种由地震引起的典型突发性地质灾害，是地震引起的最显著的震害形式之一。北京地区砂土液化灾害主要分布在地势低洼、沉积物新、结构松散的潮白河、温榆河、小中河、泃河等河流的中下游沿岸地区。历史上最为详细的灾害记录为1976年唐山大地震时平原区出现的大面积的砂土液化，主要发生在潮白河、温榆河和北运河沿岸一级阶地区域及通州区的西南部地区。通州区西集-觅子店地区液化面积为50~60km^2，平谷马各庄-南宅地区液化面积为10km^2，顺义、通州区的潮白河两岸液化面积为800km^2，大兴区、昌平区和房山区有零星分布，尤以通州区西集-觅子店地区液化程度较重（图51-7）。

砂土液化是多种因素共同作用的结果，其产生、发展和消散受多种因素制约，包括动荷条件、土的物理性质、埋藏条件、受力条件和边界条件等。在北京平原区复杂的沉积环境中，冲洪积扇的顶部的卵石、砾石及粗颗粒砂土等因为透水性好，排水快，抗压缩性强，地下水埋藏较深，故很难发生砂土液化现象；过渡地带及冲积平原的细颗粒砂土由于抗压缩性较弱，地层结构复杂，地下水位较浅且多呈饱和承压状态，受强烈地震时容易发生砂土液化；尤其是现代河床两侧的低洼地带，由于河道摆移，沉积年代新，土质松软，特别是在河流的中下游一带，地下水埋藏较浅，沉积物多为松散粉细砂，地震时更易发生液化，出现喷砂冒水现象；全新世中、晚期沉积的粉质黏土一般颜色较深，中下部密度高，湿润饱和，易发生液化。近年来，平原区地下水超采现象显著，水位的变化对砂土液化起到决定性作用，地下水位的下降不仅使砂土由饱和变为非饱和，而且水位下降使砂土重度增大，砂土经历压密作用，砂土变密实，即使水位回升，砂土也比原来密实，使砂土液化的可能性降低。

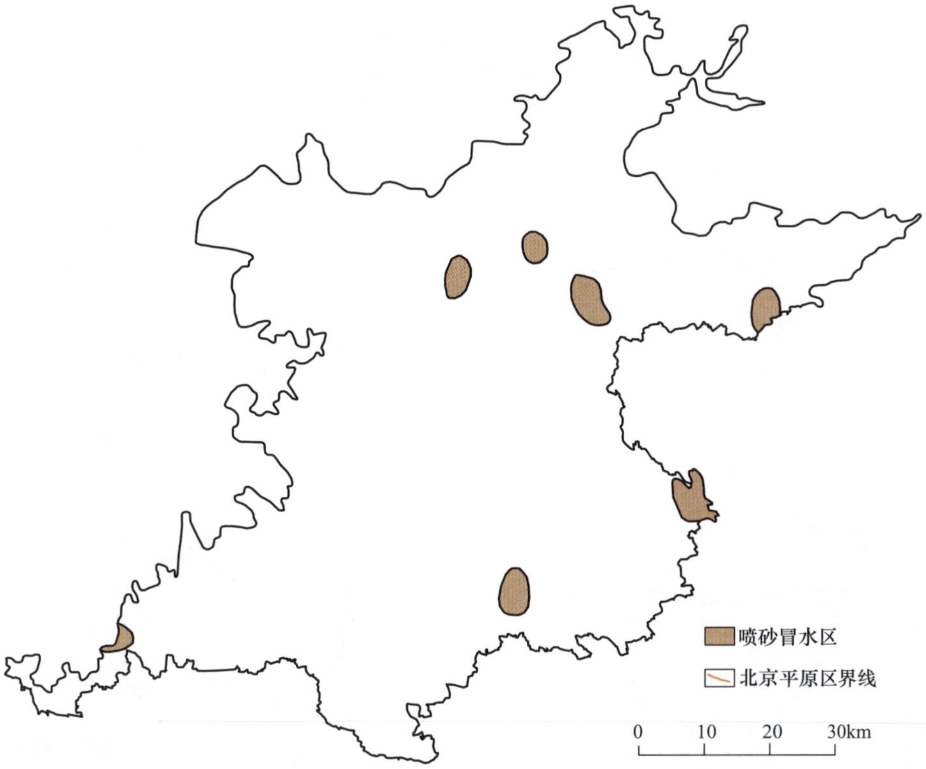

图 51-7　北京平原区 1976 年唐山地震时喷砂冒水区分布示意图
（据《北京地质灾害》，2008）
Fig. 51-7　Distribution map of Tangshan Earthquake sand-blasting water-emitting in plain area，1976

第三节　山区地质灾害

一、采空塌陷

矿山采空塌陷是地面塌陷地质灾害的一种。在北京地区，相对于岩溶塌陷和市政建设地下工程引发的塌陷而言，采空塌陷的分布范围和影响均较大；另外，煤是北京地区的主要矿产，远自辽金时代，京西的煤资源就已开发利用，据统计，西山地区历代开采的老小煤窑达3000多座，由采煤形成的地下采空时间跨度很大，由于历史积累以及采空的不可见性、发生塌陷的不可预见性和查清采空的高难度性，造成目前北京地区采矿塌陷危险性和危害性极大。野外调查和室内综合分析表明，北京西山各主要国营煤矿都已进入开采中—晚期阶段，塌陷坑、地裂缝、山体滑塌、地面不均匀沉降及破坏性矿震等已构成灾害链。本节主要对京西地区采煤引起地面塌陷的地质背景条件进行讨论。

（一）分布特征

北京矿山采空塌陷现象主要分布在西部的房山区和门头沟区，呈北东向展布，在海淀区和丰台区有零星分布（图 51-8），跨越了山前平原、低山丘陵和中高山多个地貌单元。

这些地区广布侏罗纪、石炭纪—二叠纪含煤地层，采空塌陷在地表主要表现为塌陷坑、地裂缝、不均匀沉降和滑坡等。与石炭纪—二叠纪煤系相关的地质灾害以奥灰水透入、煤泥流、近地表井泉疏干、地面塌陷坑、地面不均匀沉降、山体滑坡、地裂缝为主；与侏罗纪煤系相关的地质灾害以地表水直接贯入、古煤窑采空积水泄漏、近地表井泉疏干、顶板崩塌、严重冲击地压、地面塌陷坑、地面不均匀沉降、地裂缝、强烈矿震为主。古塌坑规模较小，数量有限，后又经历上百年的塌实和人工填埋，大部分已难觅踪迹，在门头沟区门城镇地区有200余处古塌陷坑分布。现在的采空

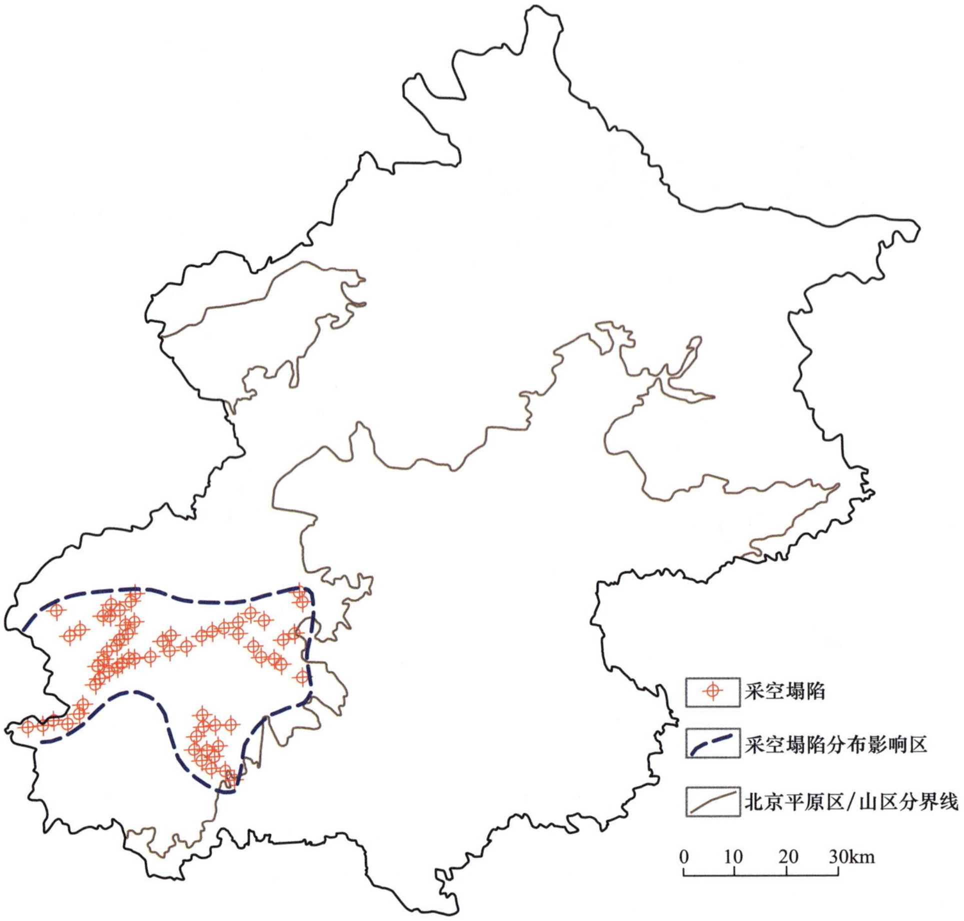

图 51-8 北京市西山地区采煤塌陷分布图

（据《北京地质灾害》，2008）

Fig. 51-8 Distribution sketch of mining subsidence in Xishan Hills, Beijing

塌陷大部分是 20 世纪 70 年代中后期以来形成的，基本上与近几十年的煤层开采区相吻合，国营矿山采空塌陷以木城涧和大台矿区数量最多，乡镇矿山采空塌陷以房山史家营、蒲洼、霞云岭和门头沟斋堂与清水镇数量最多。资料调查统计显示，北京西山已发现塌坑 1232 个，地裂缝 577 条，不均匀地面沉降点 47 处，山体滑塌 84 处，随着现代采煤活动的持续进行，塌坑的分布仍在动态变化中，数量要高于上述统计数字。

采空塌陷地质灾害时有发生。例如受采空影响，2000 年春，房山区史家营村有 30 余户房屋开裂、变形，裂缝最宽达到 100mm，公路也出现严重变形；108 国道房山区蒲洼乡宝水路段受采空影响，1992 年、2002 年发生两次严重塌陷事件，路面全部重修；2004 年 4 月 6 日早 6 点左右，一辆运煤东风重载汽车行至房山区大安山乡西沟公路段时，道路突然出现塌陷，汽车坠入塌坑；2004 年以来，门头沟区戒台寺及周边秋坡村大量建筑物墙体开裂，寺内路面不均匀沉降，大悲殿、牡丹园、大雄宝殿和辽宁塔均遭到了不同程度的破坏，108 国道路面出现裂缝，多处被剪断并下陷，最大错距达 70cm，灾情严重。

（二）主要影响因素

采空塌陷发生发展情况千变万化，影响因素较多，包括地质条件、地下采空区大小、煤层厚度及开采深度、重复采动和地形地势等，自然重力或其他外界触发因素是其发展的动因。

煤系地层岩性条件是塌陷发生的基础地质背景，北京地区主要成煤时代为石炭纪—二叠纪和侏罗纪。其中，石炭纪—二叠纪煤系煤层顶底板均以变质碳质页岩-粉砂岩为主，岩石硬度小，强度较低，顶板一般随开采的进行而塌落，开采过程中难以形成大面积悬顶；受区域变质作用影响，侏罗纪煤系岩石、煤层和顶底板的硬度、强度均有大幅度提高，在煤炭开采时顶板一般不随开采规律正常冒落，从而形成空顶稳定和安全的假象，令人们放松警惕，当煤层及顶底板的应力平衡状态遭到破坏时，多以脆性变形方式释放应力，易伴随矿震产生剧烈冒落。

人为的地下采矿活动造成的采空区是形成采空塌陷的先决条件，其分布特征决定和控制了采空塌陷灾害发生的区域、类型和规模。开采条件和开采手段决定了采空的垂向分布和空间规模。据《北京地质灾害》统计，北京历史老窑及近代个体小窑的开采规模较小，采空距地表深度大多数小于100m，但采空区面积大，在已知的采空区中，垂深小于100m的约占63.6%，垂深大于200m的仅占7.91%；国营矿山开采规模大，采空区面积小，但形成的采空体体积明显大于小窑，采空区距地表深度一般大于100m，其中一半以上的国营矿山垂深大于200m。浅部采空区面积越大，发生塌陷的可能性就越大。

水文条件的变化是采空塌陷的一个重要促发因素。当隐伏采空区上方破碎岩石达到或接近极限平衡状态时，由于水的浸润作用，使其润滑、软化和加荷，导致失稳，发生塌陷。形成塌陷的水文条件影响因素主要有：降水、管道及地表渗漏、地下采矿降排水。

重复采动时，岩层及地表移动过程比初次采动剧烈，表现为地表下沉值增大，地表移动速度加大，移动总时间缩短，地表移动范围扩大。在同样地质条件、开采深度和开采强度条件下，有些采空区发生塌陷，有的则未发生。经调查发现，发生塌陷的采空区上方，常有古采空区分布，而未塌的则没有。另外，许多成灾塌陷都是浅部古空或新近开采的小窑采空与附近或深部国营大矿采空相互叠加，重复采动作用引起的。例如，门头沟门城镇地区和安家滩地区，浅部有古窑分布，深部有国营矿山开采，形成了一定的重复采动问题。

二、崩　　塌

（一）发育现状

崩塌是北京市主要地质灾害类型之一，主要分布在山区，尤以山区道路沿线分布最多，以滑移式、倾倒式和错落式崩塌最为常见，具有发生突然、竖向位移量大、历史短暂、撞击力强的特点，汛期为崩塌地质灾害的高发期。

北京市规划和自然资源委员会公布数据显示，截至2016年，北京共有崩塌地质灾害隐患2423处（图51-9，表51-3），在10个山区均有分布，包括小型2277处、中型140处、大型6处，主要威胁道路和居民点安全。北京市几乎每年都发生崩塌地质灾害事件，规模一般不大，所造成的灾害也是局部的或单一的，造成的直接灾害程度总体相对较小，但由于历史上未对其进行过系统记载，相关灾情记录较少。从21世纪开始，政府有关部门对地质灾害进行了系统调查研究，并在汛期开展应急工作，调查灾害事件的成因，并提出防治建议。

表51-3　北京市崩塌隐患分布特征

Table 51-3　The hidden trouble of collapse in Beijing

地区	崩塌隐患总数/处	各规模隐患点数量/处		
		大型	中型	小型
门头沟	476	0	10	466
怀柔	422	0	3	419
延庆	366	0	2	364
房山	359	0	3	356
密云	344	5	102	237
昌平	235	0	1	234
平谷	176	1	15	160

续表

地区	崩塌隐患总数/处	各规模隐患点数量/处		
		大型	中型	小型
海淀	19	0	4	15
丰台	17	0	0	17
石景山	9	0	0	9
合计	2423	6	140	2277

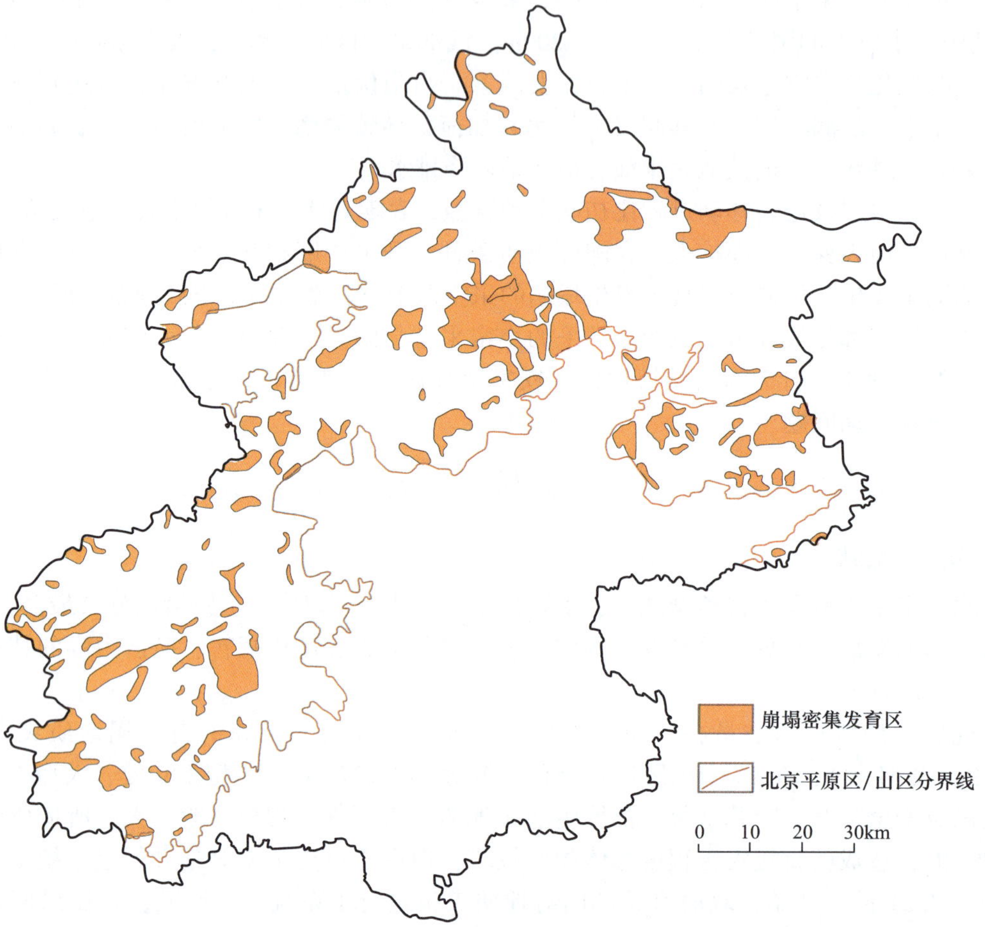

图 51-9 北京市崩塌密集发育区分布图
（据《北京地质灾害》，2008）
Fig. 51-9 Distribution map of collapse intensive development regions in Beijing

（二）主要影响因素

崩塌的形成与岩性、构造、地形、气候等条件，以及人类活动有关。其中，降水是崩塌最直接的触发因素，崩塌易发生于降水过程之中或稍微滞后，北京地区降水主要集中在6—9月份，汛期为崩塌地质灾害的高发期，特大暴雨、大暴雨和较长时间的连续降水天气是崩塌灾害最常见的诱发因素之一，雨水可使边坡岩土体饱和，增强润滑作用，顺岩体裂隙流入软弱面带走充填物中的细粒物质，降低抗剪强度引起崩塌。不合理的人类工程活动是地质灾害形成和发生的主要原因之一，人工削坡形成的不稳定边坡是孕育崩塌隐患的温床，这些边坡在各种因素的影响下一直处于不断的发展和变化之中，受风化、冻胀、植物根系、降雨冲刷、开挖坡脚、爆破振动、地震等因素的影响，可导致崩塌的发生。另外，人类活动范围的不断扩大占据或缩小了崩塌地质过程的活动空间，使原本自然的崩塌地质现象转变成为危害人们生命财产安全的地质灾害。

三、泥石流

（一）发育状况

泥石流是北京地区造成人员伤亡最大的地质灾害，近几十年来，泥石流灾害已造成约 500 人死亡，经济损失数亿元，例如 1972 年怀柔、延庆地区发生泥石流，死亡 55 人，首钢勘探队 3 名成员也在泥石流丧生。北京市规划和自然资源委员会公布数据显示，截至 2016 年，北京共有泥石流地质灾害隐患 879 处，包括小型 735 处、中型 134 处、大型 10 处，主要分布在密云北部、怀柔中部、延庆东部及房山、门头沟西部地区（表 51-4，图 51-10）。以密云冯家峪、石城、不老屯乡，怀柔崎峰茶、八道河、琉璃庙，延庆四海镇，门头沟斋堂、清水镇，房山十渡、霞云岭、蒲洼乡泥石流最为发育。

表 51-4　北京市泥石流隐患分布特征

Table 51-4　The hidden trouble of debris flow in Beijing

地区	泥石流隐患总数/处	各规模隐患点数量/处		
		大型	中型	小型
密云	271	1	19	251
怀柔	223	1	14	208
房山	141	6	56	79
平谷	84	1	7	76
延庆	65	0	4	61
门头沟	42	1	20	21
昌平	42	0	12	30
海淀	7	0	1	6
丰台	3	0	1	2
石景山	1	0	0	1
合计	879	10	134	735

注：数据截至 2016 年。

北京地区区域性泥石流活动相对活跃，但仍属低频；单沟泥石流规模较小，但易成群暴发；群发型泥石流暴发区随暴雨中心移动，重现位置不固定，但历次都集中分布在降水强度超过 200mm 的区域；活动范围广泛，灾害面积大，总体损失严重；泥石流类型以沟谷型为主，常见高容重低黏性泥石流。

（二）致灾因素

泥石流的致灾因素很多，很多灾害的发生不仅取决于自然界，往往与人类不恰当的活动密切相关。

1. 大范围、持续性强降水是导致严重泥石流灾害的重要原因

北京山区泥石流属暴雨型泥石流。泥石流集中分布于降水高值区，受北京午后型和夜雨型影响，大部分泥石流发生于下午 2 时—次日凌晨，或长或短地滞后于高雨强，好发于高雨强稍后或 1~2h。一般暴雨强度越大，笼罩空间范围越广，泥石流的群发性越大，总强度越大，灾害程度也就越严重。尤其是接踵于连阴雨之后的强降水，更有利于泥石流的发生。因此，当泥石流易发区其他条件没有发生改变时，泥石流灾害的严重程度常取决于降水的性质、范围与强度，大范围持续性强降水是引发严重泥石流灾害的重要原因。例如，2012 年，北京“7·21”特大暴雨发生时，房山区平均降水量为 460mm，居全市降水量首位，随后房山区内河北镇西区沟（含晒台沟、口儿沟、老窑沟、杏园沟、朱家沟 5 条支沟）爆发泥石流；2016 年 7 月 20 日，房山区霞云岭乡降水明显，堂上村日降水量达

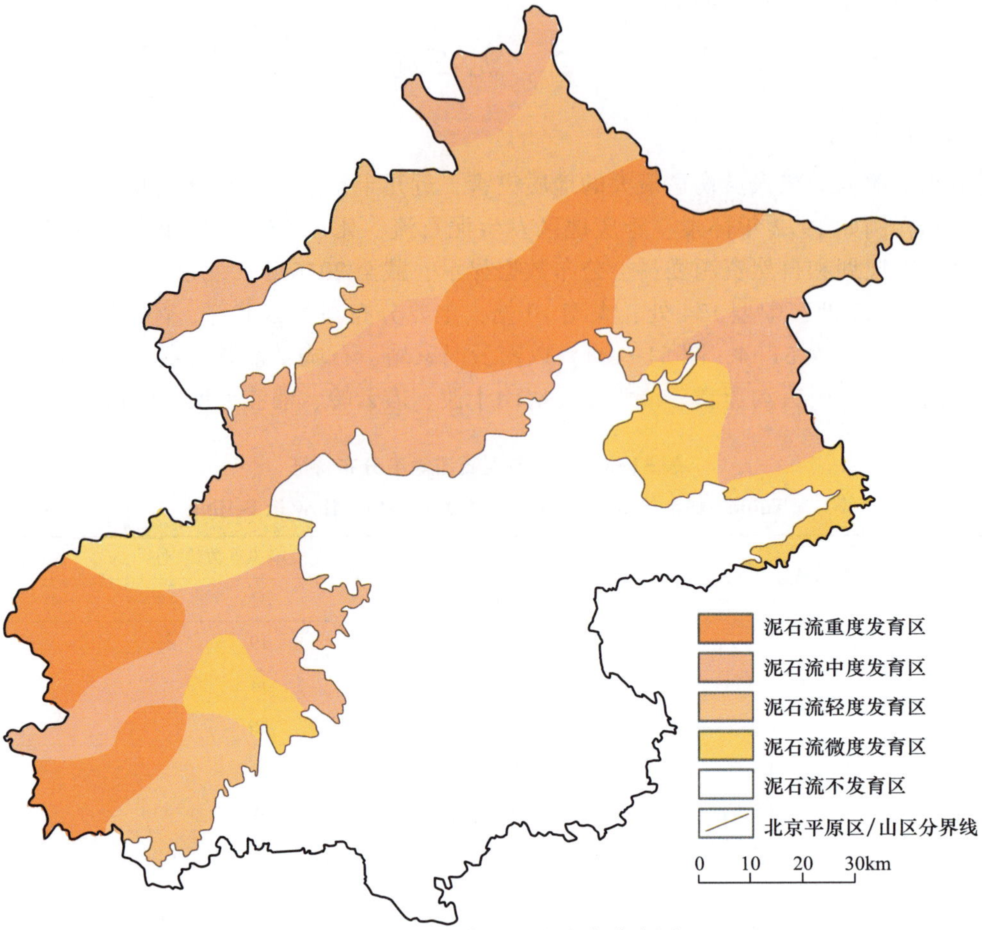

图 51-10 北京市泥石流发育分布图
（据《北京地质灾害》，2008，修改）
Fig. 51-10 Development distributions of debris flow in Beijing

140mm，随后房山区内江心台沟发生泥石流。

2. 人居工程设置不当，小规模泥石流也可导致重大人员伤亡

泥石流本是自然界中的一种正常地质现象，但由于人类人口数量上快速增长，社会经济活动日益活跃，进行各类工程建设，以及呈掠夺式的资源开发，从而使得泥石流这一自然地质现象给人类造成巨大危害。泥石流灾害的发生是各种因素综合作用的结果，其灾害程度既决定于泥石流的危险度、发生时间、规模、性质，也取决于受灾体的性质、规模、所处的地貌部位及预防程度。有时一次大规模泥石流由于远离人类活动区域或因人们采取了有效的预防措施，造成的灾害较低；而有些小规模泥石流却能酿成重灾。北京历史上不乏小规模泥石流导致重大人员伤亡事件的发生：如 1972 年 7 月 30 日夜发生在怀柔碾盘洼村的泥石流，其物质冲出量不过几千立方米，却导致 22 人死亡的巨大灾难。分析灾害原因主要有以下 3 方面：一是泥石流流程极短，不过一、二百米长，属高容重黏性的山坡型泥石流；二是受灾体碾盘洼村建在坡脚下，恰处在山坡泥石流的堆积处，位置极为不利；三是泥石流发生在夜间，居民全部在屋中熟睡，没有任何防备，无法采取任何逃生措施。由此可见，自然条件及一些人类活动导致了泥石流的形成和发生，但并不意味着一定会发生灾害或重大的灾害，灾害的发生及成灾程度在很大程度上取决于我们人类本身，即受灾体相关情况。

3. 山区旅游开发存在巨大灾害隐患

大自然塑造出北京山区峰峦叠障、沟深谷长的景观，近几年山区进行了大规模的景区旅游开发，但由于不了解历史灾害，又无防灾意识，山区旅游开发普遍存在着盲目建设的现象。不少沟谷在其中下游特别是沟口建有人工游泳池、钓鱼池及其他游乐设施，这些地方恰是泥石流物质大量冲淤的场所，是灾害最易出现的地方。沿沟谷的中游陆续建有一些休息亭、瞭望塔等建筑物及一些水、电设施，由于建设位置不当或缺少挡护等措施，当泥石流或崩塌等灾害发生时，这些建筑物就会被毁，易造成较大损失。这些旅游地区也是未来城市人群不断涌入的地方，因此存在巨大的生命安全隐患。

4. 各类不当的工矿与工程建设会诱发或加重灾害的发生

北京山区曾进行或正在进行大量的采矿与筑路工程，今后仍将有大量工程建设活动。有些工程会改变原有的地质地貌及沟坡松散物质条件，加大原泥石流的规模，加重灾害，甚至导致新的泥石流灾害的发生。

四、滑　　坡

滑坡是边坡变形破坏的一种形式，在北京地区不甚发育。由于北京山区广泛分布着灰岩、白云岩、石英砂岩、花岗岩等坚硬岩类，缺少利于滑动的结构面，因此北京地区具有一定规模的天然滑坡较少，一般多为古滑坡或由人类工程活动引起的边坡滑移。可分为土层滑坡和岩层滑坡。

北京市规划和自然资源委员会公布数据显示，截至 2016 年，北京共有滑坡地质灾害隐患 36 处，包括小型 35 处、大型 1 处。主要分布在平谷区、房山区、怀柔区、门头沟区、密云区、延庆区等地，其中平谷区发育有 13 处，房山区发育有 8 处，怀柔区发育有 7 处，门头沟区发育有 3 处，密云区发育有 4 处，延庆区发育有 1 处（图 51-11）。

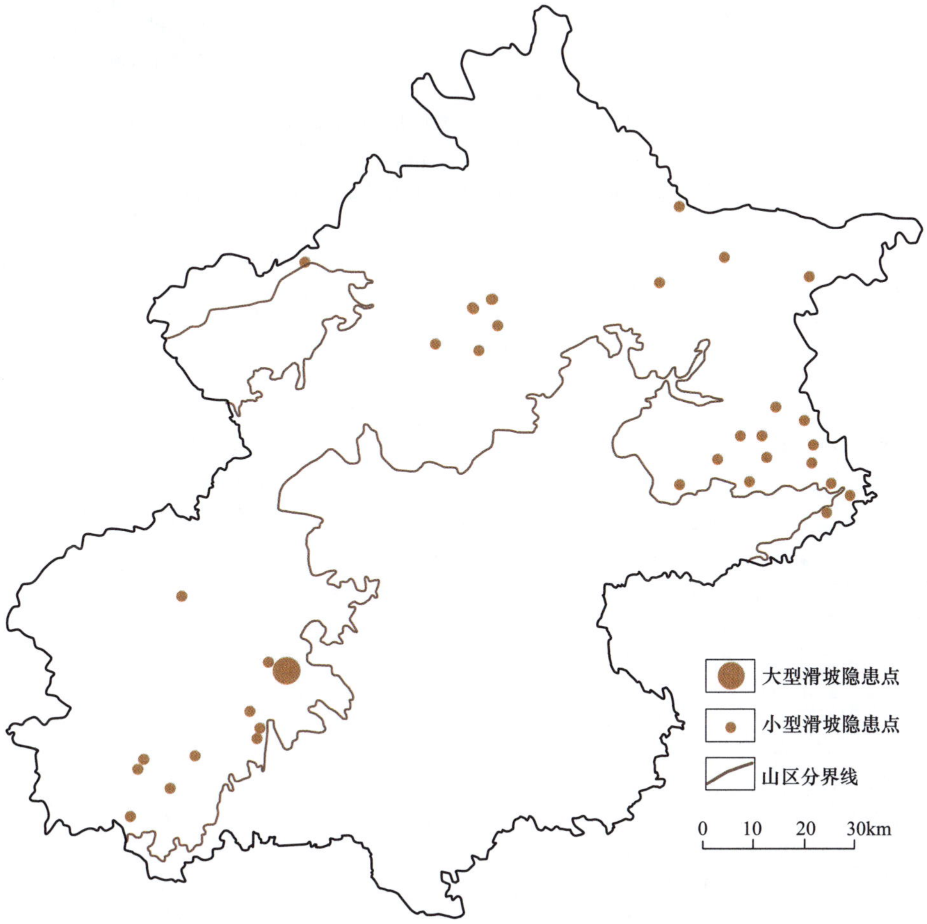

图 51-11　北京市滑坡隐患点分布图

Fig. 51-11　Development distribution of landslides in Beijing

位于门头沟永定地区的戒台寺滑坡为一处大型破碎岩石滑坡群，滑坡前后缘高差约230m，可分为4个主要滑块（图51-12），具有多条、多级和多层滑带的特征，除受构造运动影响外，主要由采空塌陷、戒台寺旅游景点生活用水下渗、爆破震动等人为因素造成，对著名的国家级文物重点保护单位戒台寺构成了严重威胁。

戒台寺院内地坪及部分殿堂存在裂缝，2004年复建千佛阁时发现一条长大裂缝从西围墙进入，分别穿过大悲殿—真武殿—牡丹院—千佛阁遗址—大雄宝殿—迦蓝殿—鼓楼—山门殿后，延伸入东侧自然沟，裂缝所经之处建筑物均出现局部下沉或拉裂；山梁西侧的秋坡村存在地面沉陷、房屋开裂的现象；108国道裂缝密集，存在多处剪断下陷。经治理，戒台寺寺院内变形得到了有效控制，滑坡基本处于稳定，但滑坡前缘仍存在隐患，画家院子内地面仍有拉张变形，秋坡村、石门沟等地区也仍有隐患。

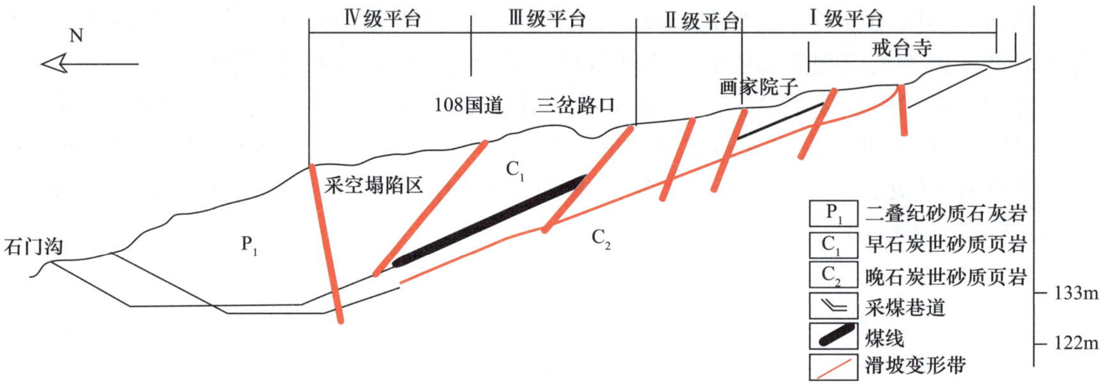

图51-12　戒台寺滑坡地质剖面图

（据《北京地质灾害》，2008）

Fig. 51-12　Geological section of landslide in Jeitai Temple

第五十二章　城市地下资源环境三维模型

为更加精准服务国土资源管理和城市规划建设，在平原区三维地质结构的基础上，2017 年北京市地质调查研究院以通州区为试点开展了城市地下资源环境三维模型建设工作。项目利用数据分析、三维模型建设等现代化信息技术方法，通过对通州区已有地下资源环境大数据进行整合，并进行相应专业的应用分析评价，搭建了通州区城市地下资源环境三维模型，包含：工程建设层、砂土液化、地下空间资源、地下水资源、地下水质量、地面沉降、浅表层（地表以下 20cm 范围）土壤地球化学及浅层地热能资源 8 个专业三维模型。通州区城市地下资源环境三维模型由三维结构模型和属性模型构成，三维结构模型能够立体、清晰地展示出工程建设层、砂土液化、地下空间资源、地下水资源、地下水质量、地面沉降、土壤地球化学等相关信息的空间分布；属性模型可以对空间中任意位置的地下资源环境属性信息进行查询。

一、工程建设层三维模型

通过收集整理通州区重要工程钻孔 1515 个，绘制出 77 条横向、纵向标准地质剖面，按照钻孔建模方法，在地层界面基础上，根据地层沉积韵律，通过不断细化、分析和调整，搭建了通州全区与核心区 50m 以浅工程层三维地质模型。标准化后的地层整体可分 3 个韵律（图 52-1），每一韵律由 3~4 个主要岩性层组成，自上而下依次为（数字为岩性层代号）：黏质粉土层（22/32）、黏质黏土层（23/33/43/53）、粉细砂层（24/34/44）、中粗砂-卵砾石层（25/35HX）。其中，中粗砂-卵砾石层仅发育在区内局部地区，核心地区由于钻孔密集，故可识别出该层；其他区域缺乏相应控制钻孔，底部为粉细砂层，并未划分出中粗砂-卵砾石层。各大层内夹不等量、不同岩性透镜体或薄层。

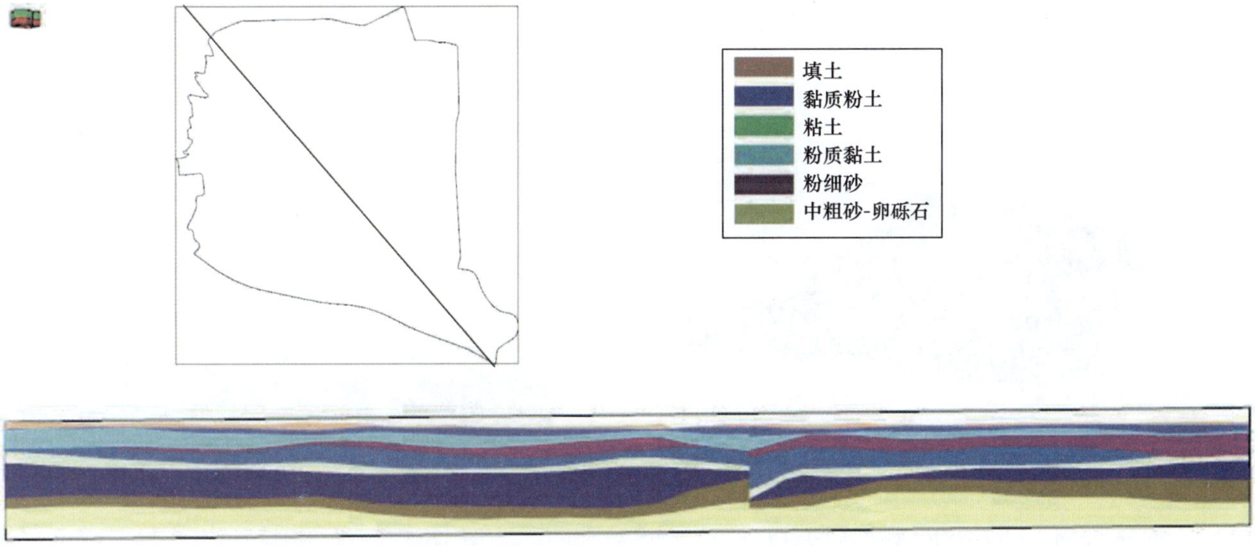

图 52-1　核心区三维地质模型北西-南东向剖面示意图（垂向×30）

Fig. 52-1　NW-SE strike profile of 3D model of core area（Vertical×30）

（一）通州核心区工程建设层三维地质体结构模型

通州核心区 50m 以浅细分为 11 大层（第二层黏质粉土层仅局部出现），自上而下依次为（数字为岩性层代号）：填土层（10）、黏质粉土层（22）、粉质黏土层（23）、粉细砂层（24）、中粗砂-卵砾石层（25）；黏质粉土层（32）、粉质黏土层（33）、粉细砂层（34）、中粗砂-卵砾石层（35）；粉质黏土层（43）、粉细砂层（44）。三维地质结构模型见图 52-2，剖面见图 52-3。

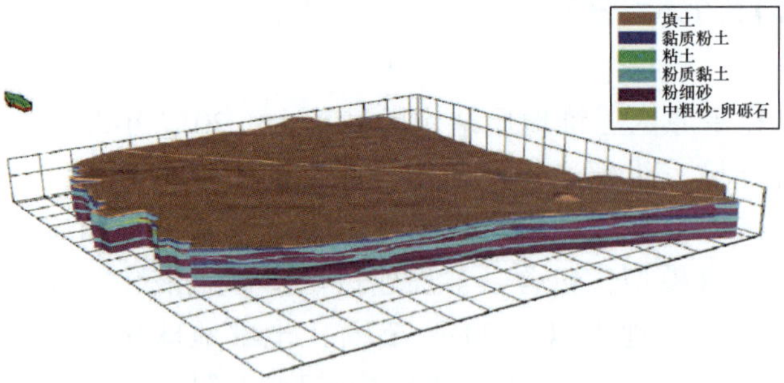

图 52-2　通州核心区工程建设层三维地质结构模型示意图

Fig. 52-2　Sketch map of 3D geological structure model of Tongzhou core area

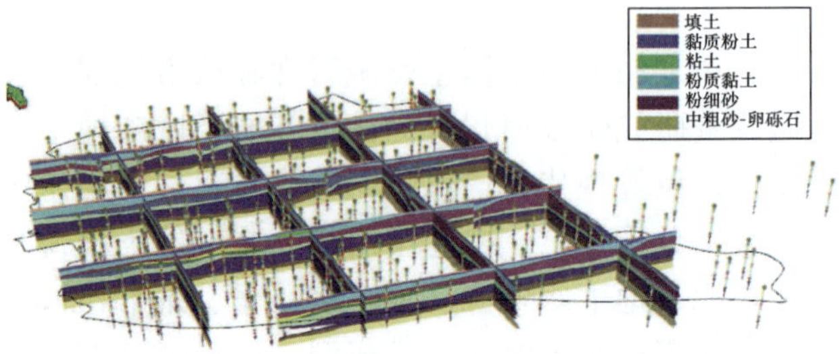

图 52-3　通州核心区地层展布剖面示意图

Fig. 52-3　Sketch map of profile of Tongzhou core area

各层因其在横向上的展布范围与其垂向厚度差异，其内发育不等量不同大小的透镜体，其中粉质黏土层（23）内发育 2 处黏土与 1 处粉细砂透镜体，呈连续状展布；粉细砂层（24）因其层厚较大，其内发育数个小规模透镜体，既有粉质黏土、黏土，亦有粘质粉土和中粗砂；中粗砂-卵砾石层（25）内发育 1 处粉细砂透镜体；粉质黏土层（33）内分布 6 个透镜体；粉细砂层（34）内发育 4 处透镜体；粉质黏土层（43）内发育 1 处粉土透镜体。其中：粉细砂层（24）、中粗砂-卵砾石层（25）具体情况如下。

1. 粉细砂层（24）

第一层粉细砂。层厚在不同地区差异较大，多集中在 4~16m 之间，平均厚约 8.9m。灰-黄灰色，密实-中密，湿度湿-饱和，强度中-较硬，主要成分为长石、石英、云母、氧化铁及有机质等，混黏土团块，偶含呈透镜体状产出的卵石或中粗砂-圆砾，局部夹少量粉土或粉质黏土薄层（图 52-4、图 52-5）。

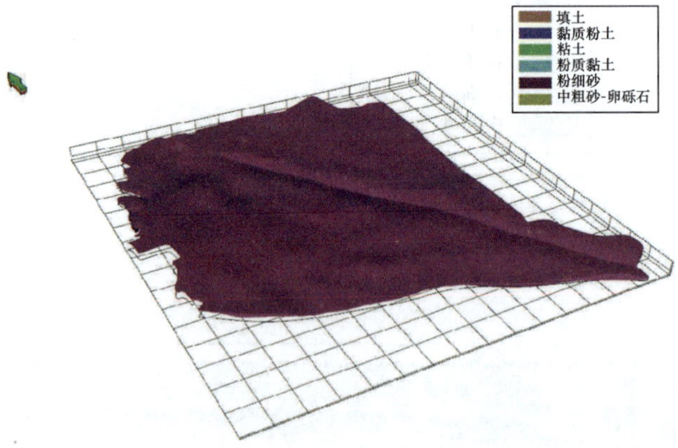

图 52-4　三维结构模型中第一层粉细砂分布示意图

Fig. 52-4　Sketch map of the first layer fine sand distribution in 3D structure model

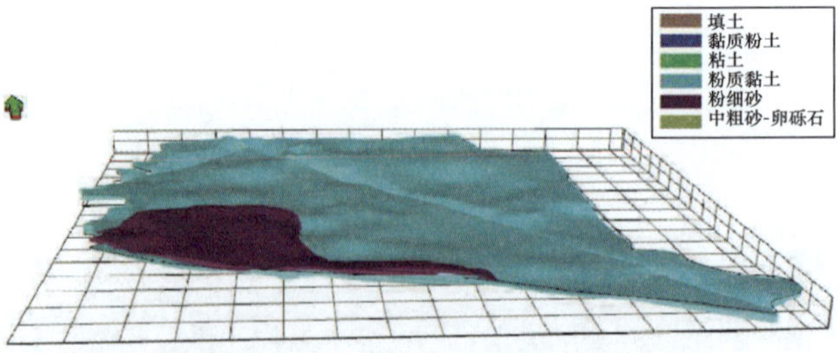

图 52-5　三维结构模型中第一层粉细砂及透镜体分布示意图

Fig. 52-5　Sketch map of the first layer fine sand and lenses distribution

2. 中粗砂-卵砾石层（25）

该层分布于核心区的中部地区，主要为中粗砂、砾砂或夹卵砾石。层厚多介于1～6m之间，均厚约4m。褐灰色-褐黄色，密度中上-密实，湿度湿-饱和，强度较硬-硬，主要成分为石英、长石，含云母、少量氧化铁或有机质。部分地区发育少量细砂、粉砂或黏质粉土，常呈薄条状夹杂在中粗砂层内。局部以圆砾为主，中粗砂充填，偶见砂质或粉土薄层，见图52-6。

至该深度（约17m）以上可视为核心区松散层的第一套旋回（图52-7），自上而下由黏质粉土层、粉质黏土层（二者顺序在局部地区可能交替出现）、粉细砂层过渡为中粗砂-卵砾石层，由于复杂的沉积环境，各层间界线有时并不明显或偶有穿插。大层间较为明显的夹层如粉细砂层内的粉质黏土薄层，在模型创建中，将后者以透镜体形式展示。该旋回除中粗砂-卵砾石层集中于中部地区外，其余大层在区内相对均匀展布，但厚度略有差异。

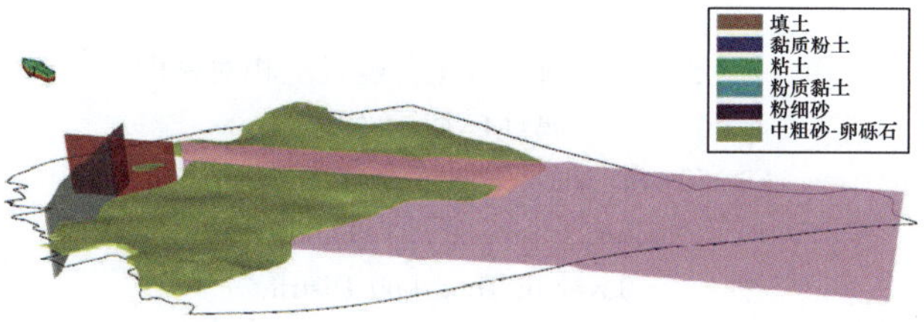

图52-6　三维结构模型中第一层中粗砂-卵砾石分布示意图

Fig. 52-6　Sketch map of the first layer medium coarse sand-Gravel distribution in 3D structure model

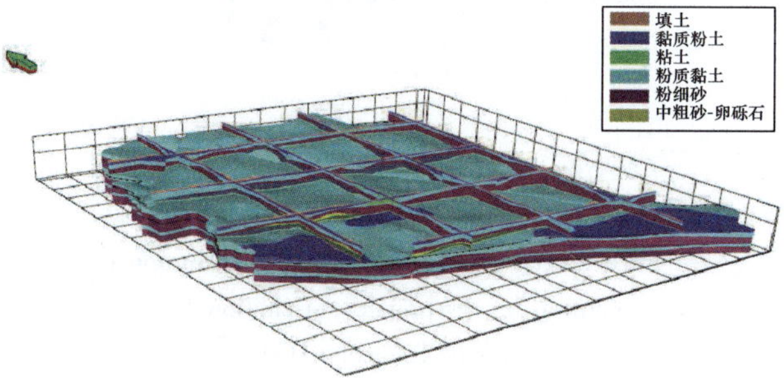

图52-7　三维结构模型中第一旋回各大层剖面示意图

Fig. 52-7　Sketch map of profile of each layer in the first cycle in 3D structure model

（二）通州全区工程建设层三维地质结构模型

通州全区50m以浅细分为10大层（第二层黏质粉土层仅局部出现），自上而下依次为（数字为岩性层代号）：填土层（10）、黏质粉土层（22）、粉质黏土层（23）、粉细砂层（24）、中粗砂-卵砾石层（25）；黏质粉土层（32）、粉质黏土层（33）、粉细砂层（34）；粉质黏土层（43）、粉细砂层（44）。局部地区发育粉质黏土层（53）。三维结构模型见图52-8，剖面见图52-9。

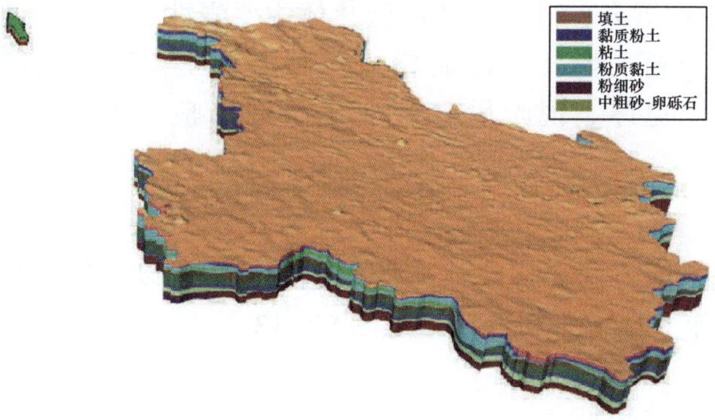

图52-8　通州区工程建设层三维地质结构模型示意图

Fig. 52-8　Sketch map of 3D geological structure model in Tongzhou District

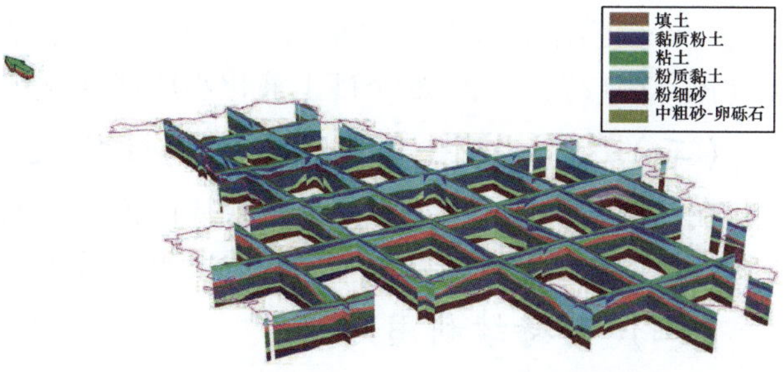

图52-9　通州区三维地质结构模型剖面示意图

Fig. 52-9　Profile sketch map of 3D geological structure model in Tongzhou District

（三）通州区工程建设层三维地质体属性模型

在三维结构模型的基础上，创建各地质体对应的块体模型，采用普通克里金法插值方法完成块体属性赋值后，形成属性模型，其中属性共有 14 大类，包括：含水量、天然密度、干密度、相对密度、孔隙比、饱和度、液限、塑限、塑性指数、液性指数、压缩系数、压缩模量、黏聚力、内摩擦角。天然密度、相对密度三维属性模型见图 52-10、图 52-11。

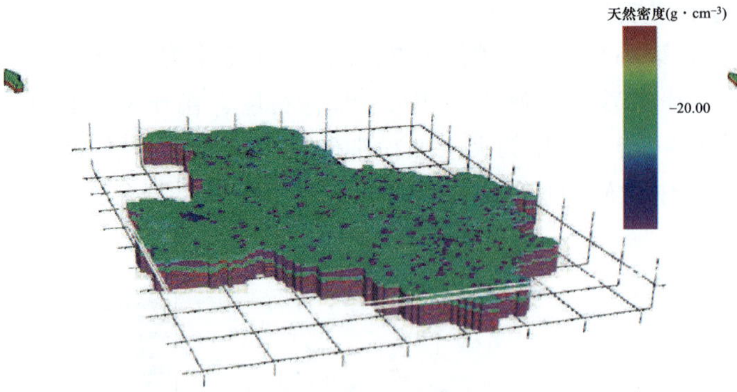

图 52-10　通州区工程建设层天然密度属性模型示意图

Fig. 52-10　Sketch map of the distribution of natural density property of shallow loose layer in Tongzhou District

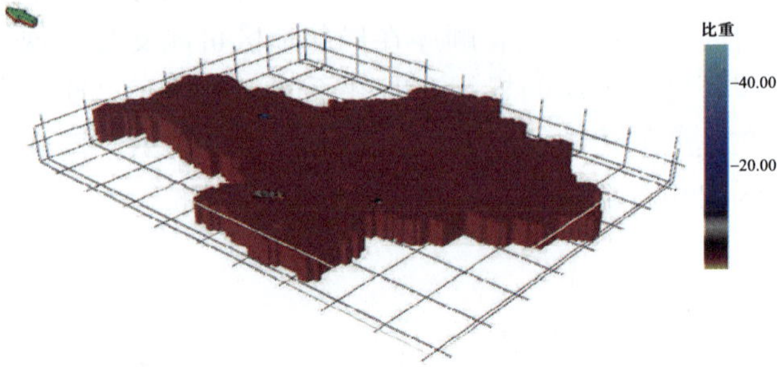

图 52-11　通州区工程建设层相对密度属性模型示意图

Fig. 52-11　Sketch map of the distribution of proportion property of shallow loose layer in Tongzhou District

（四）可持续利用建议

从评价结果看，0~10m 深度范围内，绝大部分地区土体压缩模量小于 7.5MPa，属于中高压缩性，适宜修建多层建筑；地下 10~20m 深度范围内，土体以中—中低压缩性为主，适宜修建中高—高层建筑。总体而言，通州地区若修建超高层建筑，一般均需要进行地基处理或采用桩基础。

二、砂土液化三维模型

通过对 1145 个工程钻孔液化数据进行判别分析，按照钻孔建模方法，在地层界面基础上，根据地层沉积韵律，通过不断细化、分析和调整，搭建出砂土液化三维地质体模型。

（一）砂土液化三维地质体结构模型

通州地区不同等级砂土液化区整体上呈北西-南东向展布，砂土液化层液化体主要岩性为砂和粉质黏土，砂土液化层主体为粉砂、细砂、中砂，其次为粉质黏土、砂质粉土、黏质粉土，砂液化层（体）分布范围比粉土液化层（体）广，液化厚度也更厚。现阶段砂土液化层地层整体可分为 6 层（图 52-12），包含 36 个砂土液化小层或液化体。

（二）砂土液化三维地质体属性模型

在三维结构模型的基础上，创建块体模型，通过插值方法完成块体属性赋值后，形成以液化指数进行表达的砂土液化属性模型。现阶段水位砂土液化第二层三维属性模型见图 52-13。

（三）可持续利用建议

液化砂层一般不宜作为天然地基持力层，工程地基尽量避免使用此层位，控制深度在 17.9m（主体液化层底深）以下。使用液化砂层作为承载地基时需处理避免砂土液化影响，可依据建筑抗震设防类别的不同采用不同的处理措施及程度，例如振动、振动加密、挤密碎石桩、砂石桩法、水泥粉煤灰碎石桩法等。北运河沿岸为砂土液化潜发区兼断裂影响区域，不利于工程建设，建议工程建设前对其砂土液化等级加强精度验证并做好工程防护措施。建设轨道交通等重大工程时建议提高抗震设防烈度，并长期监测地下水水位变动情况。

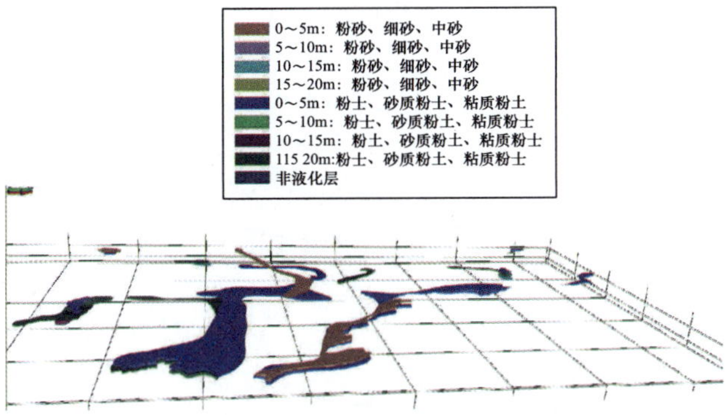

图 52-12　现状水位砂土液化三维结构模型

Fig. 52-12　3D structure model of liquefaction at present water level

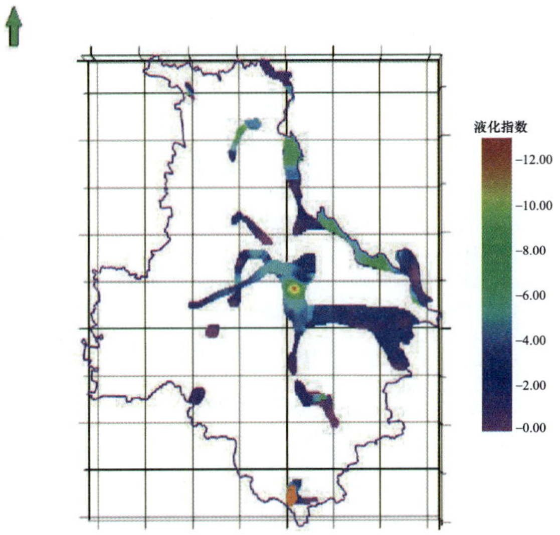

图 52-13　现状水位砂土液化第二层（X2）三维属性模型

Fig. 52-13　3D attribute model of the second liquefaction at present water level

三、地下空间资源三维模型

按照地下空间资源开发利用层次及地质结构分层特点，依次对通州地区地下 0～10m、10～17m、17～25m、25～30m、30～40m 和 40～50m 6 个层次进行综合评价，建立了地下空间三维可视化模型（图 52-14～图 52-17）。经评价，通州地区 80%以上区域，地下空间资源质量等级都为一级，适宜开展地下空间资源开发利用。但北至宋庄美术馆，南至疃里大街，西至疃里桥，东至宋庄当代艺术园一带可能发育岩溶，进行深部开发时宜进一步勘探以明确影响范围和深度；地下 0～10m 深度范围进行地下空间资源开发利用需要注意新近沉积土的影响；在南口-孙河断裂、南苑-通县断裂、张家湾断裂、姚辛庄断裂及夏垫断裂影响带范围进行重大工程建设时，应采取避让或工程措施以减少断裂活动带来的影响；当工程穿过其他未确定性质的断裂时，应对断裂性质进行进一步调查和评价。

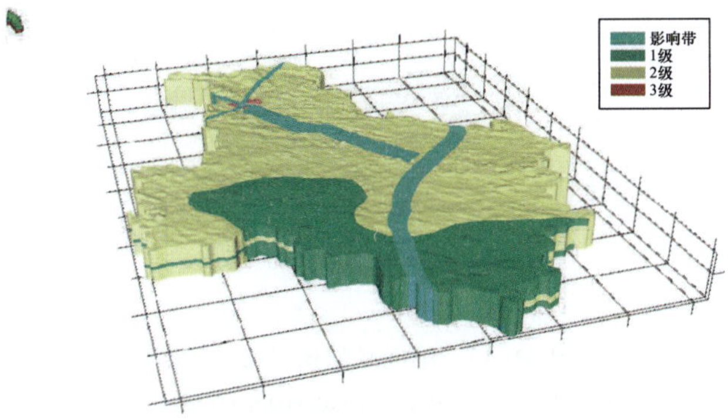

图 52-14　地下空间综合评价三维结构模型

Fig. 52-14　3D structure model for comprehensive evaluation of underground space

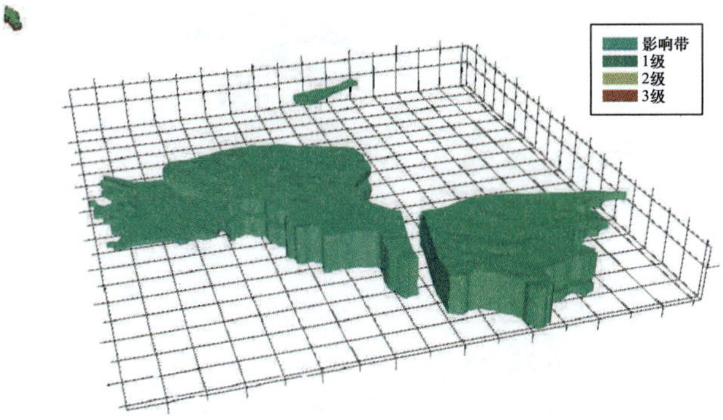

图 52-15　0～50m 地下空间地质适宜性一级分区

Fig. 52-15　First level division of geological suitability of underground space（0-50m）

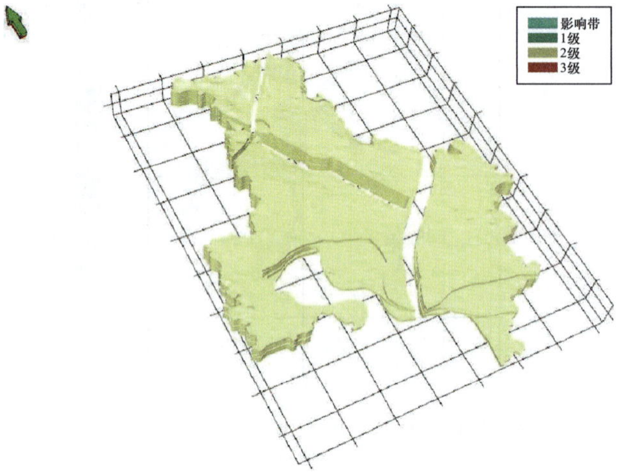

图 52-16　0~50m 地下空间地质适宜性二级分区
Fig. 52-16　Second level division of geological suitability of underground space（0-50m）

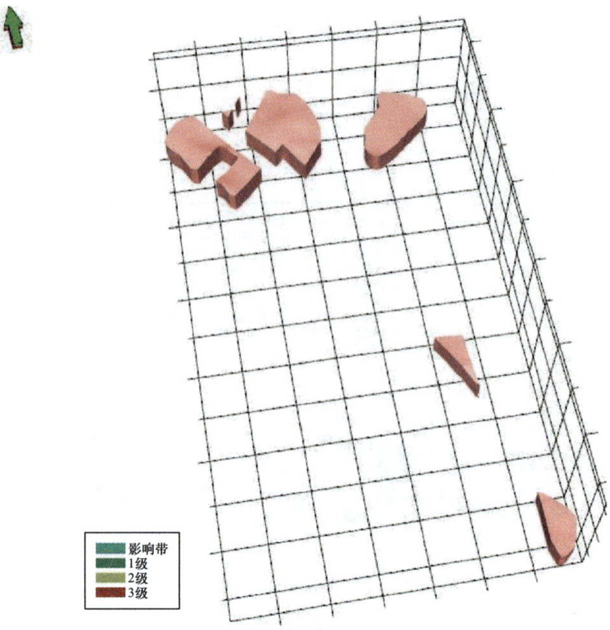

图 52-17　0~50m 地下空间地质适宜性三级分区
Fig. 52-17　Third level division of geological suitability of underground space（0-50m）

四、地下水资源三维模型

依据水文地质条件特征，基于通州区 530 个钻孔（其中水文孔 235 个）资料进行分析整理、岩性概化，划分出 4 层含水层组。利用数值分析和三维建模软件，分别对第四系地下水和岩溶地下水进行数值模拟分析，建立地下水资源三维模型。

（一）第四系地下水三维地质结构模型

钻孔显示，地下水结构共 31 层，结合分层地面沉降形变特征，综合考虑地下水开发利用层位的特点，对通州区内第四系中的含水层、弱透水层进行组合，将地下水系统在垂向上概化为 4 层，搭建第四系地下水资源三维地质结构模型，见图 52-18 至图 52-20。

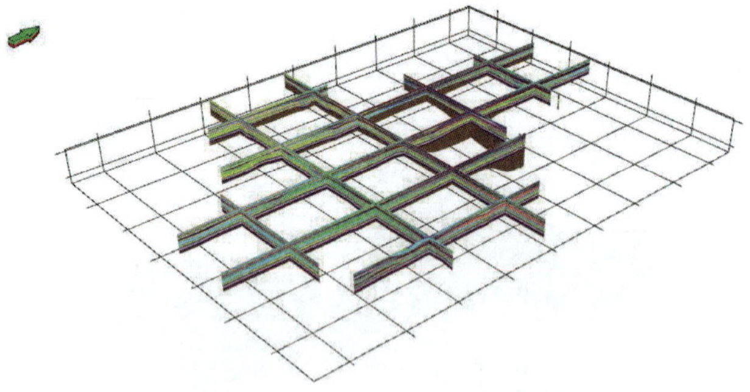

图 52-18　第四系联合剖面示意图
Fig. 52-18　Sketch map of combined profiles of Quaternary

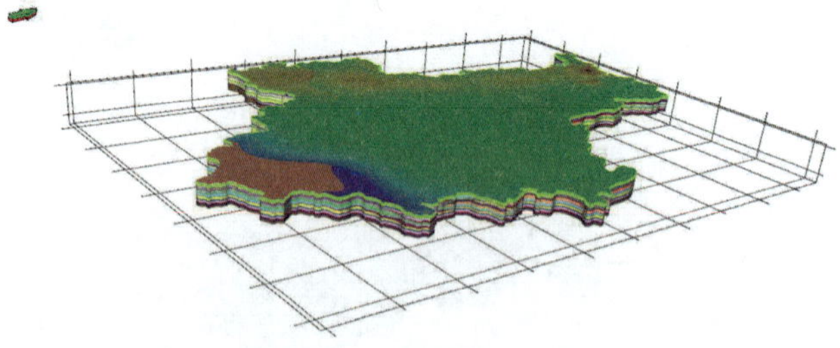

图 52-19　第四系结构模型示意图
Fig. 52-19　Model of Quaternary stratigraphic structure of Quaternary

第一含水岩组（底板埋深 40~50m）和第二含水层组（底板埋深 80~100m）含水层以细砂为主，其次为中砂、粗砂，局部地区含砾石。含水层一般有 5~8 层，累计厚度 30~50m。

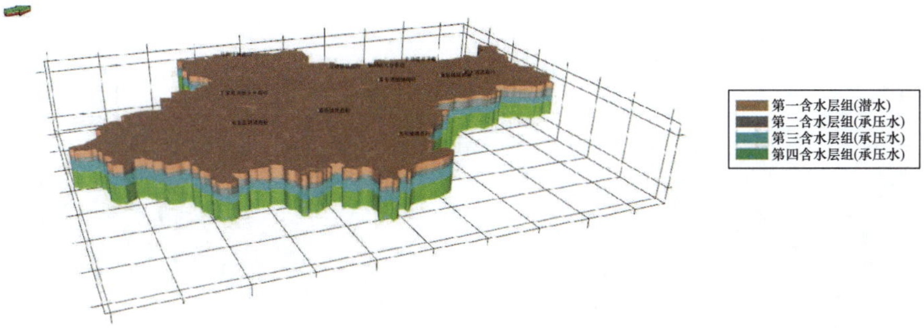

图 52-20　第四系地下水含水层结构模型

Fig. 52-20　Structural model of Quaternary groundwater aquifer of Quaternary

第三含水层组底板埋深 150~180m，第四含水层组底板埋深为 300m，局部地区以基岩为界。通州区深层地下水含水层主要由细砂、中砂和粉砂组成，中粗砂较少，含水层累计厚度在 50m 左右。

（二）第四系地下水等值线模型

将钻孔信息中地下水位数据作为控制数据，模拟生成面数据，形成第四系地下水等值面三维可视化模型，见图 52-21。

（三）第四系地下水资源量可视化模型

将地下水结构模型按照通州行政区界进行区域划分，形成以行政区为单位的模型区块，把地下水资源量分别赋予各模型区块，实现行政区块的地下水资源量可视化，见图 52-22。

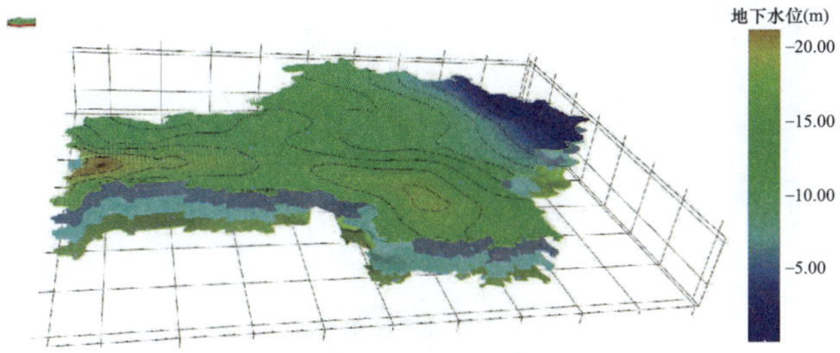

图 52-21　第四系地下水等值面三维模型

Fig. 52-21　3D model of Quaternary groundwater isolines

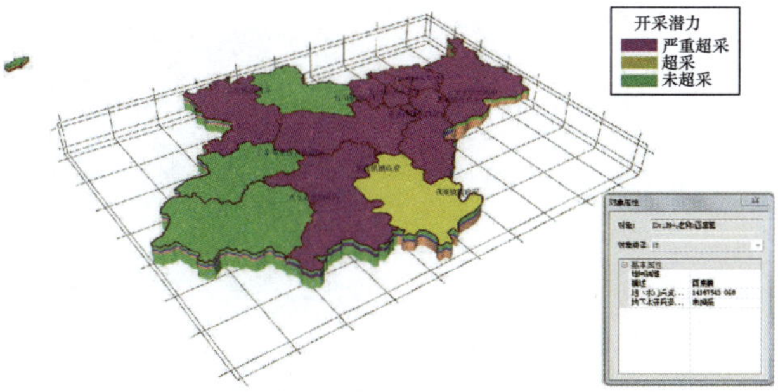

图 52-22　通州区第四系地下水资源量可视化模型

Fig. 52-22　Visual model of Quaternary groundwater resources in Tongzhou District

（四）岩溶水模型

从饮用水开发角度，不考虑蓟县系低温岩溶热水，将地层分为 4 层，分别为第四系松散孔隙水含水层，古近系—新近系、寒武系馒头组组成的相对隔水层，昌平组和景儿峪组组成的主要岩溶裂隙含水层，龙山组和下马岭组组成的隔水层底界。岩溶水三维结构模型见图 52-23 和图 52-24。

（五）可持续利用建议

第四系地下水超采严重，已经形成大范围的降落漏斗，并引发了较为严重的地面沉降问题，建议限制开采第四系地下水，在地下水漏斗中心和地面沉降中心附近通过拦蓄降水、改善河道增加地表水对地下水的补给，或通过人工回灌地下水增加补给量。同时，通州区岩溶地下水有较为丰富的水资源储量，可以进行合理规划开采。

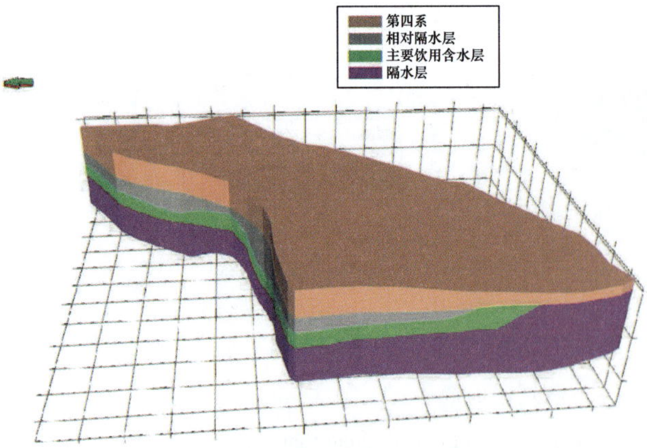

图 52-23 岩溶水三维结构模型（一）
Fig. 52-23 3D model of karst groundwater aquifer, No. 1

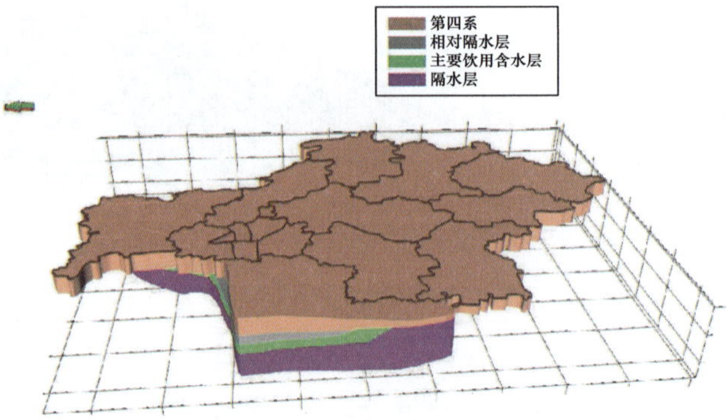

图 52-24 岩溶水三维结构模型（二）
3D model of karst groundwater aquifer, No. 2

五、地下水质量三维模型

依据“北京市平原区地下水环境监测网运行项目”2017 年度 8—9 月份的监测数据，采用《地下水质量标准》（GB/T 14848—2017）对 4 个含水层组的水质进行评价，结合地下水流场等水文地质条件绘制水质分区，搭建了地下水水质量三维可视化模型（图 52-25）。

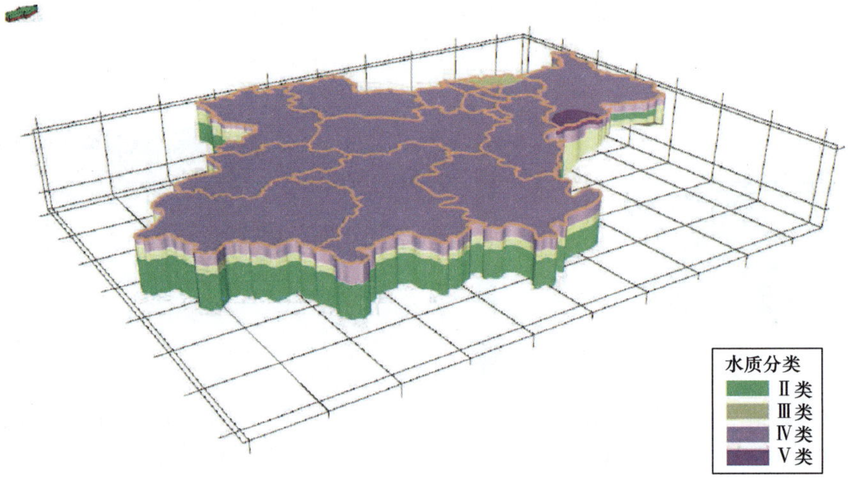

图 52-25 通州区地下水质量三维可视化模型
Fig. 52-25 3D model of groundwater quality in Tongzhou District

通州大部分地区第一、二含水层地下水以Ⅳ类水为主，可用于农业灌溉以及部分工业用水；宋庄镇东南角的摇不动村、师姑庄村附近为Ⅴ类水，该地区浅层地下水不适于农业灌溉，可用于城市绿化；通州大部分地区第三、第四含水层地下水以Ⅲ类水和Ⅱ类为主，适用于生活饮用。由于浅层地下水污染较重，应禁止混层取水，以避免混层开采形成天窗越流情况，减少浅层地下水对深层地下水的污染。

六、地面沉降三维模型

以地下水模型为基础，充分考虑通州地区地层特性、水文地质条件和沉降特点，利用 2010—2017 年地面沉降资料，建立了一个概化为 4 层结构的非均质各向异性、三维非稳定流地下水系统和地面沉降数值模型。通过三维建模软件实现地面沉降三维可视化，能够查看地面沉降累计沉降量、年度沉降速率、监测站地面沉降监测曲线、地面沉降危险性评价结果等，见图 52-26 和图 52-27。

图 52-26　地面沉降三维模型查询累计沉降量

Fig. 52-26　Accumulative land subsidence in Tongzhou District

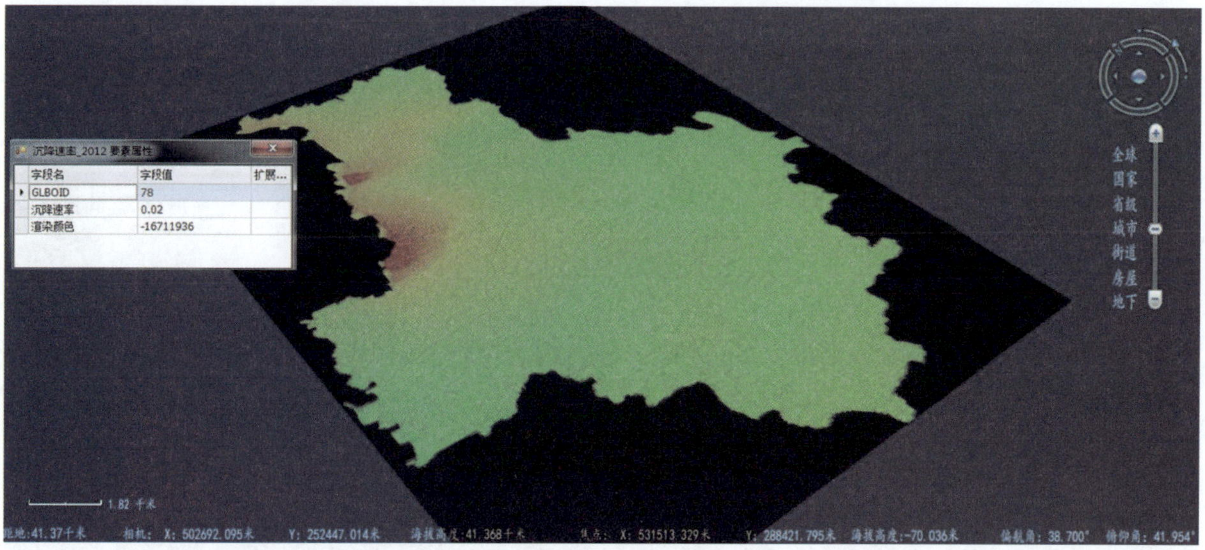

图 52-27　通州区地面沉降速率

Fig. 52-27　The rate of land subsidence in Tongzhou District

地面沉降中心主要位于通州城区，产生的地层形变会对穿越沉降梯度带的线型工程造成影响。因此，跨地面沉降区建设的轨道交通线、对沉降有严格要求的重要建筑、管线等工程施工前应开展地面沉降对工程影响的勘察与研究，采取相应工程措施防止地面沉降长期发展对地下线性工程的影响。

七、土壤地球化学三维模型

针对通州区水田、水浇地、旱地、园地等农用地，以及城市、建制镇、村庄、采矿用地、风景名胜和特殊用地等建设用地土壤环境质量进行等级评价，将评价结果数据赋值到地理图斑要素，建立表层土壤地球化学三维可视化模型（图 52-28，图 52-29）。

通州区土壤环境综合质量总体为一级，在北京化学工业集团有限责任公司东方化工厂南部等地存在土壤环境综合质量为三级和四级污染的情况。针对张家湾镇上马头和张辛庄村土壤环境质量受到东方化工厂东侧小企业生产的影响，建议防止企业生产造成的“三废”排放到农用地；对土壤污染地块

图 52-28　水田土壤环境质量三维可视化模型

Fig. 52-28　3D effect display of paddy soil environmental quality

图 52-29　土壤地球化学评价结果查询

Fig. 52-29　Query of soil geochemical evaluation results

进行土壤环境监测和农产品协同监测，对食用农产品不符合质量安全标准的采取农艺调控、替代种植等安全利用措施。

八、浅层地热能资源三维模型

利用水文地质资料数据和浅层地热能动态监测数据，对通州区浅层地热能进行静态储量评价、可开采资源量评价，并建立了通州区浅层地热能三维可视化模型，见图 52-30 至图 52-32。通州区大部分地区是地埋管地源热泵条件适宜区，各乡镇浅层地热能静态储量、可利用资源量较丰富，通州区静态资源量为 542.15×10^{12}kJ，一个采暖季可提取 49.20×10^{12}kJ 的热量，一个制冷季可排放 81.36×10^{12}kJ 的热量。张家湾监测站的换热孔和监测孔温度变化曲线表明，距换热孔矩阵越远，温度变化幅度和波动越小。地层温度场展现了很好的“弹性”特征，系统停止运行后原始地温场可以迅速恢复，地下没有产生不可逆的热累积，可进一步加大通州区各乡镇浅层地热能开发利用程度。

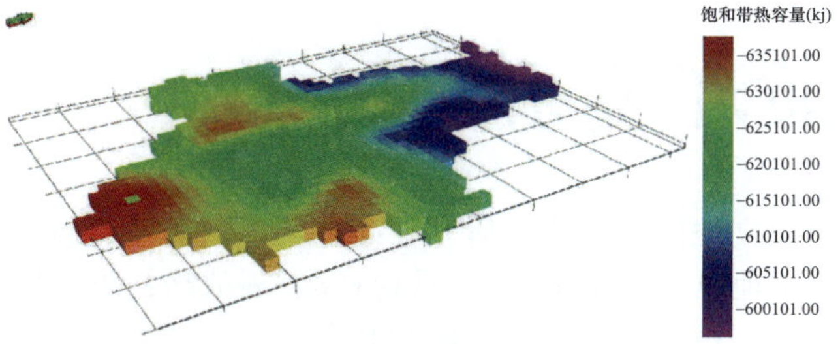

图 52-30　三维阵列点的方式展示饱水带中的静态储量

Fig. 52-30　3D array points to show the static reserves in the saturated zone

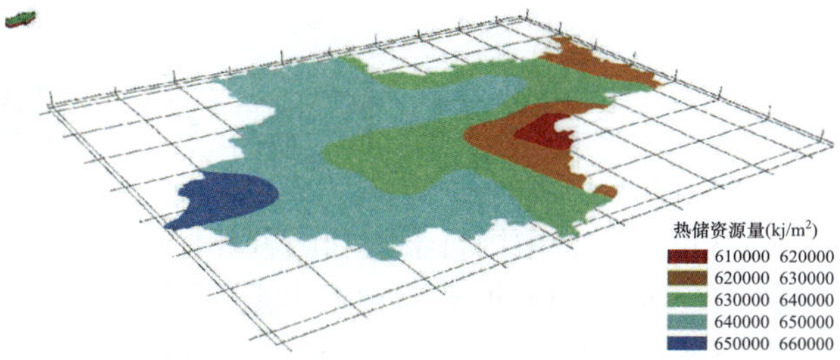

图 52-31　静态资源储量评价三维可视化模型

Fig. 52-31　3D visualization model of static resource reserve evaluation

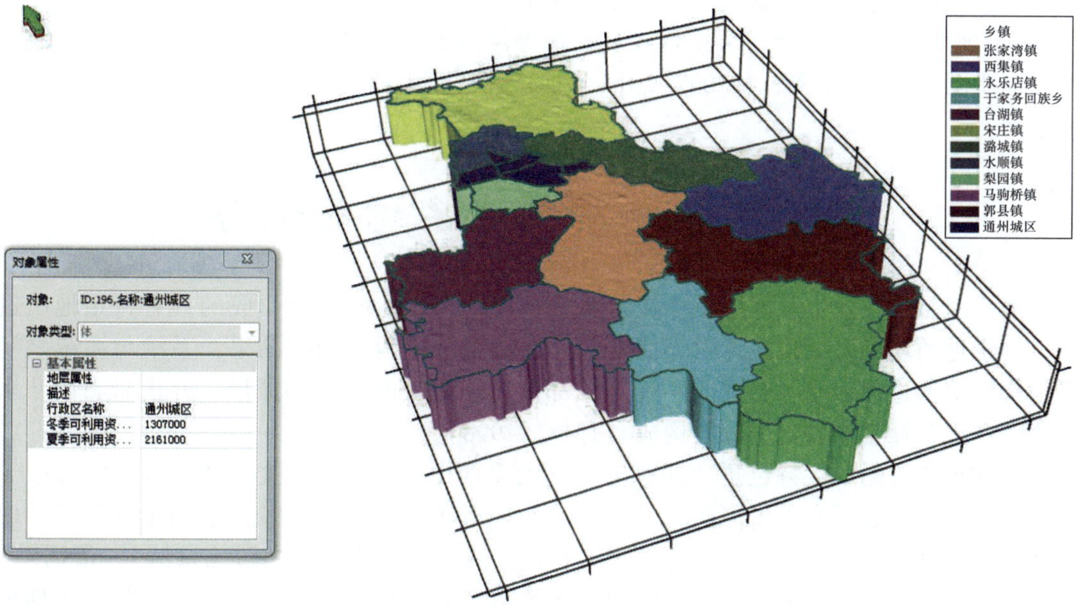

图 52-32　可采资源量分区三维可视化模型示意图

Fig. 52-32　Sketch map of 3D visualization model of recoverable resources partition

第五十三章 地质环境系统划分

20 世纪末，中国科学院和国家环保总局联合提出建立地质环境系统研究中心，刘东生提出加强跨学科综合研究，解决现实环境问题，孙枢（1998）提出地质环境系统研究与可持续发展战略，王思敬提出论人类工程活动与地质环境的相互作用。2008 年 8 月 6—14 日在挪威奥斯陆举行的第 33 届国际地质大会主题为地球科学系统是可持续发展的基础。2016 年 7 月 4—6 日在上海召开了第四届地球系统科学大会。我们在系统科学思想的感召下，结合北京的实际，思考北京的地质环境系统划分。

人类活动是在人地界面进行的，虽然会受到自然地质环境的制约，同时，地质环境也会受到人类活动反馈制约，研究人类活动与地质环境作用的综合环境效应，以期取得人与自然的和谐发展（王思敬等，1998）。

北京是具有山地（包括延庆盆地）和平原的城市，是带动整个国家地质工作发展的地质学摇篮城市。《北京西山地质志》记载："一九一六年夏，辅等任事于地质调查所，所长丁文江首令完成五万分之一之西山地质图"（叶良辅等，1920）。2016 年恰逢北京西山地质调查百年，划分北京的地质环境系统有重要意义。

第一节 地质环境系统的概念

地质环境系统是指在发生和发展上存在联系的各种营力在地球表层形成的具有特定地质过程和边界的区域，一个地质环境系统是地球内外地质营力在特定范围内作用的结果，系统与系统之间存在不断的能量与物质交换过程，每一个地质环境系统都占有地球表面一定的区域，并由岩石圈、水圈、大气圈、生物圈和人类社会的特定部分所构成（孙枢，1998）。

北京在地质地貌上的最大特色是"北京湾"。《北京西山地质志》地文一章记载："维理士有言曰，中国东部自纬度四十度起，有大平原向北入丛山，形如海湾，其东南口径宽达四十五英里，山脉因之被横断。从平原以视其四维之山岭，犹海湾之于石壁。其湾澳之部分名之曰北京湾"。依据"北京湾"山地与平原的交界线，山地与平原所呈现的地质环境系统差别很大，将北京地区划分为山地地质环境系统和平原地质环境系统。

第二节 地质环境系统的三级单元划分

地球本身是最大的地质环境系统，7 大洲 4 大洋分别是相对独立的地质环境系统，我国可划分出在亚洲东部季风作用条件下的三大台阶的地质环境系统，北京恰好处于中国第二台阶和第三台阶过渡带，具有山地和平原两大地质环境系统。

北京山地与平原界线清晰。山地为黄土高原及华北中、低山。平原为松辽-华北平原。地质环境系统详细划分为三级（表 53-1，图 53-1）。

表 53-1　北京地区地质环境系统划分

Table 53-1　The division of systems of geological environment in Beijing

山地地质环境系统（一级）

一级	二级	三级	主要特征
黄土高原及华北中山、低山区	太行山中山、低山区	南部碳酸盐岩区	拒马河与大石河流域。 拒马河流域地层为雾迷山组硅质条带白云岩，地层倾角较小，节理垂直，容易产生岩溶崩塌。河流地貌呈岩溶蛇曲，以似峰丛地貌、岩溶石柱为其特色。北岸发育三清洞、仙栖洞和云水洞等唐县期夷平面洞穴。 大石河流域地层主要为古生代石灰岩，上游北岸发育圣莲山岩溶地貌，下游南岸发育鸡毛洞、银狐洞、石花洞和孔水洞，形成石花洞系地下暗河，有水质较好的地下岩溶水
		中部火山岩区	属永定河流域，清水河支流火山岩抗风化能力强，形成东灵山顶北台期夷平面、百花山唐县期夷平面。 火山岩位于向斜成山是核部，岩性复杂，节理裂隙发育，风化碎石较多，容易产生崩塌、滑坡和泥石流地质灾害。 清水河谷主要为火山岩分布区，在斋堂谷地中的火山岩之上发育马兰黄土。 永定河谷横切庙安岭火山岩向斜，河道平直
		北部碳酸盐岩区	九山向斜为岩溶盆地，使龙门涧常年有水。幽州背斜核部形成永定河岩溶峡谷，青白口穹隆张裂隙构成岩溶峡谷（南石洋大峡谷最典型），高崖口东西向断裂带构成晚更新世马兰台地
	燕山中山、低山区	西部松散沉积物区	盆地南北边缘受北东向断裂控制，妫水河流过，官厅水库坐落其中，从边缘到中心依次发育坡积裙、台地、洪积扇和阶地
		中部岩浆岩区	自西向东依次为八达岭、怀北、大羊山、云蒙山等花岗岩地貌，怀柔中部火山岩
		东部变质岩区	密怀变质岩分布于密云水库周边，抗风化能力弱，山脊线较为圆滑。由于变质岩岩性均匀，故山地变质岩区的白河和潮河呈现曲流河
		东南部碳酸盐岩区	主要为中—新元古代碳酸盐岩分布区，属于岩溶地貌，发育京东大峡谷和京东大溶洞

平原地质环境系统（一级）

一级	二级	三级	主要特征
松辽、华北平原区	山前岛状残丘区	太行山前	石景山、老山、八宝山、田村山
		燕山山前	九里山、棉山、大汤山、小汤山、牛栏山、二十里长山
	大清河冲洪积扇区	拒马河冲洪积扇区	分布北京西南部，主体已出北京市，主要为白云岩砾石，冲洪积扇底部隔水底板为古近纪半胶结砂砾石
		大石河冲洪积扇区	分布于山前地区，砂砾石岩性复杂，化探富 Sr 图很好地显示大体形状为北窄南宽展布
		牤牛河冲洪积扇区	分布于大石河与永定河之间，由多条山谷流出的小型冲洪积扇群构成山前平原
	永定河冲洪积扇区		分为 4 个冲洪积扇，依次为老冲洪积扇、新冲洪积扇、现代冲洪积扇、永清冲洪积扇，第一个分布于北京市，中间 2 个顶部带分布于北京市。 老冲洪积扇为北京中心城区坐落的位置，顶部带位于石景山区，溢出带到达东郊，为马兰堆积期。 新冲洪积扇沿着永定河断裂向南展布。 现代冲洪积扇，位于老、新两个冲洪积扇之间。 永清冲洪积扇已出北京
	北运河山前洪积扇群区	山前洪积扇群区	海淀山后与南口山前地区形成山前洪积扇群，从西南向北东依次有北安河洪积扇、西峰山洪积扇、南口洪积扇、十三陵洪积扇等，形成洪积扇群
		洪积扇群溢出带区	洪积扇群的众多溢出带，在高里掌—辛力屯一带形成扇前洼地，至今地下埋藏大量全新世泥炭
	潮白河冲洪积扇区		广阔的冲洪积扇平原，到达通州附近与永定河老冲洪积扇相连。潮白河冲洪积扇受新构造运动影响，河槽下切，新期古河道大多在槽中变动，这是与永定河冲洪积扇不同的。潮白河冲洪积扇的形成时代与永定河老冲洪积扇一样，为晚更新世。 近代泛滥平原已经出了北京市，自燕郊以下地势低平，并有泛滥而成的砂地
	蓟运河冲洪积扇区		分布于平谷平原区

（据吕金波等，2017）

图 53-1　北京地质环境系统分区
（据吕金波等，2017）
Fig. 53-1　Partitions of systems of geological environment in Beijing

北京地区的大地构造位于柴达木-华北板块华北陆块（克拉通）华北北缘隆起带、燕山中—新元古代断裂带和华北盆地，地质构造的复杂造成地质环境系统也复杂。北京市总面积 16410km^2，山地面积 10072km^2，占 61.4%，海拔一般为 1000~1500m，最高峰为门头沟东灵山，海拔 2303m。平原面积 6338km^2，占 38.6%，海拔一般为 10~60m，最低处在通州柴厂屯，海拔 8m，著名的天安门广场海拔 44.4m。

山地一级地质环境系统为黄土高原及华北中山、低山区。山地二级地质环境系统分为太行山中山、低山区和燕山中山、低山区。山地三级地质环境系统分为西山南部碳酸盐岩区，中部火山岩区，北部碳酸盐岩区；北山西部松散沉积物区，中部岩浆岩区，东部变质岩区，东南部碳酸盐岩区。

平原一级地质环境系统为松辽-华北平原区。平原二级地质环境系统分为山前岛状残丘区、大清河冲洪积扇区、永定河冲洪积扇区、北运河冲洪积扇群区、潮白河冲洪积扇区、蓟运河冲洪积扇区。平原三级地质环境系统分为太行山山前岛状残丘区和燕山山前岛状残丘区，拒马河冲洪积扇区、大石河冲洪积扇区、牤牛河冲洪积扇区、山前洪积扇群区和洪积扇群溢出带区。

第十一篇
区域地质发展史

迄今为止，北京地区尚未发现确切的古—中太古代独立地质体，中太古代前地质演化历史参考邻区资料推演。冀东地区曹庄组是近些年发现的华北地区最早的地质体。该组地层中铬云母石英岩中碎屑锆石类型为岩浆锆石，来源于酸性花岗质至中酸性长英质岩浆岩，锆石 U-Pb SHRIMP 年龄为 3890~3540Ma（刘敦一等，1994；Nutman et al.，2011），表明曹庄岩组形成之前冀东地区就有酸性岩类发育（伍家善等，1991；阎月华等，1991；刘敦一等，1994）。这一发现揭示本区可能在约 4600Ma 原始地壳形成以后，原始地壳至少在 4000Ma 就开始了火山或侵位岩浆分异活动并持续发展，最终在古太古代晚期演化为硅镁质壳（主）与硅铝质壳（次）并存的古地壳格局。

北京地区已知地质记录反映地史发展由第四纪上溯到太古宙，历时约 3000Ma。在这漫长而复杂的地质历史时期中，地壳发生过多次剧烈的变动，在区内形成了一系列不同类型的沉积与变质建造、火山活动与岩浆侵入、构造行迹与构造带等。

《北京市区域地质志》（1991）总结了北京地区在地壳演化过程中经历的 3 个大地构造发展阶段（地台基底形成阶段、地台盖层发育阶段和大陆边缘活动带阶段）和 5 次重要的地质事件（阜平运动、吕梁运动、印支运动、燕山运动和喜马拉雅运动），突出了北京地区多旋回地史发展的特点。鲍亦冈等（2001）也将本区地质构造发展划分为 3 个构造发展阶段（地台结晶基底形成发展阶段、拗拉槽与稳定克拉通盖层发育阶段、陆内（板内）造山复杂构造阶段）。本志在前人工作基础上，按照板块构造学说及大陆动力学的观点，以不同时段的建造与改造为依据，结合区域地质构造与区域地球物理、地球化学资料等综合研究成果，着重于从板块活动及板块拼贴、裂解，以及岩石圈的形成演化等角度来论述区域地质发展变化的过程。最终，本篇将本区分为 4 个各具特色的发展演化阶段：新太古代—古元古代构造阶段（结晶基底与岩石圈形成阶段）；中元古代—中三叠世构造阶段（盖层形成与岩石圈稳定阶段）；晚三叠世—早白垩世构造阶段（板内造山与岩石圈减薄阶段）；晚白垩世—第四纪构造阶段（造山后伸展拉张与岩石圈再造、现代地貌形成阶段）。

第五十四章　新太古代—古元古代构造阶段

北京地区处于华北克拉通中北部，新太古代—古元古代是基底形成时期。新太古代晚期，华北克拉通广泛发育 TTG 岩系和变质表壳岩，构成了其大部分结晶基底，并在古元古代末期，东、西两大陆块发生碰撞拼合（赵国春，2001；翟明国等，2007），最终形成统一华北克拉通（陆块）。古元古代末期碰撞-拼合复杂构造-热事件还使得结晶基底全部固结，形成华北古陆块地幔岩石圈。据目前研究成果，整个岩石圈或大陆根平均厚度达 200km（张文佑等，1983；邓晋福等，1996，2007）。地壳平均厚度 42km，上地壳为花岗闪长质片麻岩与花岗质片麻岩，下地壳为花岗闪长质麻粒岩、TTG、酸性榴辉岩夹基性麻粒岩，整体可称为 TTG 型陆壳；上地幔岩石圈平均厚度为 158km，为方辉橄榄岩型（《中国区域地质志·河北志》，2017）。根据境内出露的地质体，结合华北克拉通基底形成构造岩浆演化，将北京地区该构造阶段划分为两个构造期，即新太古代阜平期和古元古代吕梁期，并简要叙述其演化特点。

一、新太古代地质构造发展期

新太古代地质构造发展期为阜平构造期，形成时代相当于 3200~2500Ma。

约 2800Ma，北京地区及其外围可能处于极不稳定的活动大陆边缘岛弧盆地环境，在薄陆壳和洋壳交界处，海底火山强烈而频繁的间歇性喷发活动。这些火山喷发的熔岩流和火山碎屑与早期侵入于上部地壳的花岗岩类物质混合在一起，组成新太古代晚期密云岩群表壳岩、四合堂岩群表壳岩及磁铁石英岩，构成冀东古陆核的一部分。这一过程持续到约 2700Ma，北京境内进入早前寒武纪变质基底陆壳（新生陆壳）孕育期，壳源岩浆在古陆核中侵位并对表壳岩原地-半原地重熔改造，广泛发育面型展布特征的沙厂英云闪长质片麻岩系列，同残留古陆核共同构成主要基底陆壳。

2700~2500Ma，本区早前寒武纪变质基底陆壳（新生陆壳）开始拉张裂解状态并向盆地发展，大洋形成演变。初期（2700~2600Ma），新生古陆壳在其下部的地幔软流圈对流和上涌作用下裂解，逐步形成了数个相互隔离的初始大洋-陆缘裂谷构造盆地，主要分布在邻区河北省内。境内则发育为古陆地蚀源区，伴随发育少许具有碱性特征的基性岩浆活动（黑卧辉长质片麻岩）。中—后期（2600~2500Ma），伴随地壳持续拉张裂解，大洋形成并演化，最终区域不同微陆块或块体之间俯冲运动开始，再次进入极不稳定的活动大陆边缘岩浆弧环境。境内位于岩浆弧-弧后构造位置，发育强烈深部岩浆侵入和挤压变形变质现象，主体缺乏同期复杂与岩浆弧有关的火山-沉积地层。境外邻区则发育大规模的地表与海底火山喷发，即阜平岩群等由麻粒岩、斜长角闪岩、变粒岩、片麻岩及磁铁石英岩、大理岩或不纯大理岩等组成，其原岩以海相碎屑岩夹硅铁质岩、碳酸盐岩或钙硅酸盐岩与多旋回喷发的基性、中性、酸性火山岩相间产出为特征（《中国区域地质志·河北志》，2017）。复杂的构造变形、变质作用和岩浆活动构成了境内最重要的阜平期构造-热事件，形成现今所观测到的新太古代主要变形变质现象。广泛出露的新太古代 TTG 类侵入体（宋营子英云闪长质片麻岩、荆子峪花岗闪长质片麻岩和康各庄奥长花岗质片麻岩）是主期岩浆活动产物。主岩浆侵位活跃期前后还发育小规模的基性和酸性岩浆活动，早期发育为中基性苇子峪辉长闪长质片麻岩，晚期为放马峪变质超基性-基性岩，末期发育为旗杆顶花岗质片麻岩，并伴随强烈混合岩化现象。同弧后伸展有关的基性-超基性岩脉群也广泛发育，如周口店岩脉群等。这些复杂岩浆演化序列，由基性、中性、酸性及碱性深成侵入岩组合为 $T_1T_2G_1G_2$ 类型，记录了区内中元古代新生陆壳向半成熟、成熟地壳演化过程。构造变形则表现为密云岩群、四合堂岩群褶皱上隆及强韧性剪切变形，形成东西向构造线，近水平的面理褶皱成为大-中型尺度的、多级次同斜或倒转复式背形和复式向形。构造变形伴有强烈区域变质，密云岩群、四合堂岩群和先期 $T_1T_2G_1G_2$ 侵入岩均发生了强烈紫苏麻粒岩相变质作用。

二、古元古代地质构造发展期

古元古代地质构造发展期为吕梁构造期，形成时代相当于2500～1800Ma，是华北克拉通东、西两大陆块俯冲-碰撞-拼合演化阶段。本区发育为遭受剥蚀的陆地，境内缺失沉积地层记录，仅在密怀地区发现活动较弱岩浆活动，并同早期地质体一起记录了期间变质作用和构造形迹，即吕梁期构造-热事件。

古元古代岩浆活动总体较弱，发育以橄榄岩、辉石岩、角闪岩、辉长岩及辉绿岩为代表的超基性-基性岩组合，呈小型岩体和岩墙群产出，形成于碰撞拼合后伸展构造背景下沿张家口-长哨营-古北口-平泉深大断裂带侵位。它们遭受绿片岩相变质作用，可类比更早地质体绿片岩相变质作用，反映古元古代经历了普遍的绿片岩相区域变质作用，有别于早期新大古代麻粒岩相变质作用。此构造发展阶段，境内构造形迹较少。然而，在空间上，不同构造层次的麻粒岩相变质岩和角闪岩相变质岩以北东向崎峰茶断裂和石城断裂为界明显有所区分，揭示了期间这两条断裂可能受北西-南东向挤压作用，经历了一次重要的脆性逆冲性质活动。

第五十五章　中元古代—中三叠世构造阶段

这一阶段相当于吕梁运动以后，晚三叠世杏石口组沉积以前的整个中—新元古代、古生代及早—中三叠世时期，发育为稳定盖层形成和早前寒武纪岩石圈相对稳定发展阶段，可分为中元古代早期—新元古代早期、新元古代中期—早寒武世、中寒武世—中奥陶世、晚奥陶世—早石炭世和中石炭世—中三叠世5个构造发展期。这一阶段从沉积建造而言，区内普遍发育的是属于陆块沉积的碳酸盐岩建造和碎屑岩建造基本类型，总体岩浆作用、火山作用既不强烈，也不普遍。各系、组地层之间均为整合及平行不整合接触。地壳只有升降运动的海陆变迁，没有强烈的构造变形，基本不发育变质作用。

一、中元古代早期—新元古代早期

中元古代早期—新元古代早期为晋宁期，时代约为1800~780Ma。

（一）长城纪（1800~1600Ma）

中元古代初，华北克拉通地幔软流圈经过吕梁晚期较晚阶段的稳定调整过渡后，在1800Ma左右开始了以位于华北北缘隆起带为中心（或中轴）的近南北向反向对流，并在随后的1Ma间经历了一次大规模的上隆，致使多幕镁铁质岩墙群沿着古元古代华北克拉通中部拼合带（软弱带）侵位，分布于华北板块多个地区。北京地区发育著名的高岭岩墙群，主要由辉绿岩和辉长岩组成，锆石U-Pb同位素年龄为1731Ma左右，岩墙直立或近直立。初期幔源岩浆侵位后还引起华北克拉通太古宙期间形成的下地壳物质部分熔融，演化出的中酸性岩和基性岩密切共生而又不同源，为燕山裂陷槽裂陷早期慢性扩展条件下独特的具有非造山特性的岩浆侵位活动，即沙厂环斑花岗岩、兰营石英正长岩和新地斜长岩侵位。

与深部地质作用相对应，华北克拉通地壳长期处于伸展-裂陷的构造环境，基底断裂复活作张性差异活动，尤以尚义-隆化、康保-围场（北京境内称之为长哨营-古北口断裂）两区域断裂活动幅度最大，促使华北北缘隆起带整体上升形成分水岭，其南部与北部地区相对下降，形成两个陆缘裂谷裂陷海盆地（《中国区域地质志·河北志》，2017）。本区位于南部联合盆地内。该盆地总体呈近东西向展布，北以长哨营-古北口深断裂为界与华北北缘隆起带相接，南缘大致位于蓟县—房山一线，邻近五台-太行山古陆西延至冀西北的宣化、尚义一带，而东延出本区与广海相通。盆地内沉积长城系常州沟组至大红峪组。常州沟初期，在密云、平谷地区发育水流方向自西向东的网状河相砾岩和含砾长石石英砂岩沉积，开启了裂陷槽盆地演化进程。伴随着燕山裂陷槽演化发展，海水由东向西逐渐入侵本区，最终在常州沟期早期（二段）开始了中元古代早期海相沉积过程，燕辽海盆逐渐形成，在境内主体发育为一套潟湖-浅海陆棚碎屑岩、碳酸盐岩沉积建造，上部夹有基性、偏碱性火山岩，即大红峪期火山岩。大红峪期火山活动标志着该阶段裂陷槽进入裂陷主发育期，也是最重要的发育阶段。长城纪，地球外圈的含氧性大气圈基本形成，大气圈和水圈中已开始拥有一定数量的游离氧，藻类生物开始繁盛，最终被串岭沟期—大红峪期地层所记录。

中元古代初长城纪变质作用较弱，发育弱蚀变及岩浆侵位时的接触变质（早期），而构造活动主要表现为近东西向的拉张运动，致使东西向的基底断裂活动明显。深大基底断裂不仅控制了早期中酸性和基性岩浆侵位，如密云沙厂斜长球斑花岗岩体沿密云-墙子路断裂侵入，兰营-古北口石英正长岩体及斜长岩体等沿长哨营-古北口断裂的侵入，还控制了后期盆地发展的边界，进而控制了沉积地层分布。伴随裂陷槽发展，晚期在裂陷槽盆地内还可能新生了一系列近东西向的断裂构造，沿此断裂密云、平谷地区大量富钾粗面岩和玄武岩的发生多次喷溢活动。

（二）蓟县纪（1600~1400Ma）

中元古代中期蓟县纪，燕山裂陷槽逐渐消亡，伴随地幔失热和活化能量衰减、区域地壳应力的释放，岩石圈上部的构造活动不再表现裂陷扩展活动，而呈现为挠曲下翘、不均匀沉降。燕山裂陷槽盆地演化为向克拉通盆地过渡的坳陷槽盆地，活动机制由断陷沉降演化为以均衡载荷沉降为主，阵发性断裂活动和局部的沉陷为辅。整个蓟县纪深部能量衰减也导致该阶段岩浆活动较弱，没有经历显著的构造变形和变质作用。境内未发现该时期的侵入岩体，但伴随裂陷槽型北东向狭窄的裂陷活动转变为较为宽缓的坳陷作用，地壳呈现脉动式下降与上升频繁震荡运动，引起沉积范围扩展、收缩。蓟县纪本区经历了 2 次较大规模的海进和 3 次抬升，分别为高于庄期、雾迷山期海进和滦县、铁岭、芹峪上升，形成高于庄组与杨庄组或雾迷山组、杨庄组与铁岭组、铁岭组与下马岭组之间的平行不整合界面。

蓟县纪演化形成的坳陷槽盆地总体继续保持了长城纪时期南北高、中间凹的总体地貌特征。海域的北界为长哨营-古北口深断裂，沉积中心仍然在原燕山裂陷槽的位置上。所不同的是，由于长时期的夷平作用，此时海盆起伏和缓，从中心向西侧坡降不大。期间，盆内沉积发育主体为潟湖-浅海陆棚碎屑岩、碳酸盐岩沉积建造，蓟县纪各个沉积时期，岩性岩相组合略有所差异。

蓟县纪高于庄期，北京地区与整个华北地区一起发生了大幅度的整体沉降，海水迅速由原燕山裂陷槽向南侵漫，区内出现了自常州沟期以来中元古代的第一次广泛海侵超覆。军都山水下高地与太行五台古陆连接，其西侧为宣龙海盆，东侧为东西向马兰峪水下高地，分隔的兴隆-宽城海槽及蓟州-迁西海盆构成了“厂”字形古地理面貌。高于庄期还是整个华北地区沉积作用的一个明显的转折时期。由于华北地区已基本上准平原化，低坡降致使沉积时缺乏陆源碎屑的供给，境内沉积从此由以陆源碎屑沉积为主转化为以内源碳酸盐岩沉积为主。当时气候温暖，海水不深，菌藻类生物大量繁殖，在滨海-浅海相的高于庄组碳酸盐岩建造中，藻叠层石发育。

滦县上升结束了高于庄期广泛海侵，至蓟县纪杨庄期，海水北流，回归“燕山裂陷槽”。北京地区的南部、河北太行山北段及山西等地缺失杨庄组的沉积。当时气候干燥，日照充足，海水被大量蒸发、浓缩，在氧化作用增强的碱性环境下形了一套紫红色非补偿的杨庄组潟湖蒸发相碳酸盐岩建造。坳陷槽海盆的边缘即房山西部地区、古北口一带及军都山水下高地区则发育为石英砂岩及含砂白云岩的沉积。

蓟县纪雾迷山期是中—新元古代最大的一次海侵期，是华北地区中—新元古代第二次大幅度的地壳整体沉降期，海水又一次由“燕山裂陷槽”迅速向南侵漫，形成了一个比高于庄期更加广阔的陆表海。在河北、山西等地雾迷山组直接超覆在高于庄组及更老的地层之上，形成了雾迷山组与高于庄组之间大面积的超覆不整合。这次海侵结束了区内杨庄期咸化潟湖环境的发展，沉积了雾迷山组以滨海相叠层石白云岩为主的藻礁碳酸盐岩建造。当时，区内气候温暖、潮湿，地球外圈的含氧性大气圈已有了进一步的发展，大气圈和水圈中游离氧的成分比以前有了明显的增加。海水的深度不大，具备良好的光合作用条件，为生物的活动提供了良好的繁衍环境，藻类繁茂，藻席、藻礁发育，是中—新元古代一个典型的生物造礁时期。雾迷山组中还发育板刺状角砾构造，说明在中元古代由燕辽坳陷槽轴部的断裂可能仍有触发的古地震活动。

雾迷山期末地壳再次抬升（铁岭上升），至蓟县纪洪水庄期华北地区又一次发生了大面积的海退，海水再次向北回流，回流注入“燕山裂陷槽”。当时，水体相对加深，盐度增高，在接近于还原状态的环境中沉积了以黑色页岩为主的洪水庄组潟湖泥岩相碎屑岩建造。

铁岭期坳陷槽海盆演化为稳定克拉通（陆表海）盆地，并持续至晚古生代。铁岭期海域面积较洪水庄期稍有扩大，沉积中心向西扩展到延庆四海一带。当时的气候温暖、海水相对稳定，是区内的藻类生物大量繁衍的又一个高峰时期，区内普遍发育了一套以叠层石白云质灰岩为主的滨海-浅海相藻礁碳酸盐岩建造。同时在海盆的中心延庆四海等地，浅海相的铁锰碳酸盐岩中形成了“四海式”含锰赤铁矿矿床。铁岭期沉积作用还出现了一次重要的转折，以大量白云质灰岩的出现为特征。从此以后，各地质时代的沉积岩中，叠层石白云岩或白云岩的组分显著减少，石灰岩开始占据主导地位。

铁岭期末地壳上升（芹峪上升），境内演化为陆地，海水从区内全部撤出，陆表遭受剥蚀。当时气候湿热，雨水充足，先期沉积的碳酸盐岩遭到了明显的风化、淋滤，在区内形成了较普遍的风化壳。

（三）待建纪（1400~1000Ma）

待建纪下马岭期，地壳再度构造沉降，海水又一次沿燕山海槽由东向西进入本区。早期区内普遍发育为一段氧化滨岸含铁粉砂泥岩相沉积，中—后期随着水体的不断加深，发育为水体滞流较闭塞的潟湖相环境，形成了下马岭组二、三段黑色页岩和黄绿色伊利石页岩等碎屑岩的沉积。沉积中心则早期由东向西从密云—兴隆一带迁移到青白口—怀来一带，中晚期可粗略划分出青白口海盆和密兴海槽两个古地理单元。围绕青白口海盆、密兴海槽的边缘向周缘，依次发育为滨岸含铁粉砂泥岩沉积和潮间潟湖含灰-泥岩相沉积；在青白口海盆、密兴海槽中心则发育潮下潟湖含硅-泥岩相沉积；怀来—延庆西部发育潮下潟湖藻灰岩相沉积；平谷—蓟州上升为陆，缺失下马岭组二、四段的沉积。该期，古气候环境仍处于温带气候环境，气候温湿。

待建纪晚期发生蔚县上升，海水逐渐从区内退出。“军都山隆起”以东地区缺失下马岭组四段的沉积，军都山隆起以西地区，由于“军都山隆起”的阻隔作用，海水没有能够全部撤出，在河北怀来赵家山及涿鹿黄土港等地形成了下马岭组四段潟湖相含黄铁矿和石膏的页岩和泥灰岩的沉积。泥灰岩中富含各种类型叠层石，是燕山地区最晚的一个叠层石层位。伴随蔚县上升持续发展，本区及广大的华北地区整体抬升为陆遭受剥蚀，形成下马岭组与龙山组之间的平行不整合界面。

待建纪本区处于非造山的拉张构造环境，早、晚期各经历了一次重要的构造抬升运动（芹峪和蔚县上升）。虽然该期未有显著变质变形，但仍发育极弱的岩浆活动，表现为在下马岭组碎屑岩中含火山活动的斑脱岩夹层，在兴隆—平泉一带还可见具有非造山型基性偏碱性岩床、岩脉侵入下马岭组—景儿峪组中，锆石 U-Pb 同位素年龄为 1320Ma。

（四）青白口纪（1000~780Ma）

青白口纪龙山期，地壳再次构造沉降，海水入侵，本区再次发育为盆地接受沉积。同下马岭期相比，龙山期盆地范围有较大的扩展，盆地北部边界位于赤城北—隆化南一线，西部边界位于张家口—蔚县—阜平一线，南部边界位于石家庄—赞皇—衡水—故城之间，东部延拓至渤海湾。沉积中心位于怀来—延庆一带。盆内古地理格局大致分为两个单元，即青白口海盆和怀来-密云海槽。古海域处于海绿石圈定的边界范围之内，边界以外为冲积搬运沉积作用为主的物源供给区。其中易县—涞水一线以北的京西地区，正处于多条古水流汇合入海的喇叭状河口湾环境。其中发育分支河道、分流间湾、河口坝等沉积类型。在离河口稍远的京西、怀来、昌平一带，形成横向展布的三角洲前缘沙坝，它阻隔削减来自东面的潮流-波浪作用能量，使向岸侧较宽阔区域（房山、大兴、通州、城区）成为较闭塞滞流的类似潮下潟湖环境的分流间湾沉积。沙坝北部密云、怀柔北部、延庆北部则发育为前三角洲潮下泥岩相沉积。

青白口纪龙山期本区所处纬度再次增高，处于为亚寒带、寒带气候环境。该期是中国晚前寒武纪生物演化中的一个重要的时期。红藻开始出现，褐藻大量繁殖，是藻类生物由低级向高级演化的重要发展时期。龙山组中以 *Longfeng shania* sp.，*Chuaria* 为代表的宏观藻类化石的出现，是地质史中重大生物事件。这一化石组合中的化石大都成单体存在，具黑色碳质薄膜，直径 3mm 以上，形态多样，特别是其中的 *Lonfengshania* sp.，已有明显茎的分化，有的具假根（固着器）。拟茎也可能是类浮游生物在水中附着的原始器官。杜汝霖等（1990）建议将这一生物群命名为“怀来生物群”。怀来生物群反映了我国晚前寒武纪一个独特生物发育阶段中藻类生物群的组合面貌。

青白口纪景儿峪期，华北地壳再次转化为整体抬升。受此影响，龙山期发展海盆分布范围缩小，但盆地水体深度反而增大。盆地西部边界退缩到蔚县东—阜平东一带，南部与东南部边界退缩到新乐—天津北一带，东部边界处于秦皇岛与滦县之间地区。区内沉积中心位于北京海淀—朝阳一带，表现为长方形海盆状。景儿峪期，由于华北板块的完全克拉通化，海盆沉积环境极度稳定，古地理格局较为单一。北京境内，沉积岩性、岩相变化不大，沉积一套以化学沉积作用为主的富钙贫镁碳酸盐岩序列，均处于浅海潮下低能沉积环境，为潮下盐岩泥沉积。该期本区所处纬度再次增高，古纬度达76°，为寒带近极地气候环境。

景儿峪期末发生蓟县抬升，本区整体为陆，从此结束了中—新元古代时期海相地层的发育史。

二、新元古代中期—早寒武世

新元古代中期—早寒武世为扬子期，形成时代为780~521Ma。

由于青白口纪末期以来长时间的蓟县上升运动，中元古代以来发育演化燕辽盆地在整个新元古代中期南华纪—早古生代早寒武世期间一直处于隆起状态，区域上缺失沉积、岩浆及变质各类建造记录，即缺失扬子期构造旋回记录。然而，新近晚泥盆世、二叠纪等古生代侵入体的新发现，也记录有这一时代的岩浆型锆石，可能指示本区北部岩石圈深部有极弱岩浆活动（陈印，2014；He et al., 2019），且可能是引起蓟县上升重要动力来源。青白口纪末期以来的蓟县抬升一直持续到早寒武世，使得本区和华北地区一起隆起为陆，接受剥蚀、夷平，最终形成新元古界早期（青白口界）景儿峪组与中寒武统昌平组之间的平行不整合界面，即长达260Ma的沉积间断。

三、中寒武世—中奥陶世

中寒武世—中奥陶世为加里东期早期，形成时代为521~460.9Ma。

中寒武世昌平期开始，蓟县上升运动结束，海水开始由东向西进入本区，从此拉开了本区早古生代海进沉积旋回的序幕。早古生代是本区稳定克拉通盖层发育的重要时期之一，也是整个地史发展过程中最稳定的一个阶段，因而缺乏同期岩浆活动与变质变形。该阶段，华北陆表海与兴蒙古海槽可能已相互沟通，地壳升降速度缓慢，沉积厚度不大，岩性、岩相分布均比较稳定，沉积受古地貌影响较大。伴随蓟县抬升，古陆的不断夷平、缩小，区内地势比较平缓，仅在北部保存了一串近东西方向排列的链状岛屿群，致使沉积建造带（即按沉积走向所展示的带状分布）的区域走向以东西向为主（叶连俊，1983）。

中寒武世—中奥陶世本区经历了2次沉降、1次短期局部抬升和1次普遍抬升。第一次沉降发生在中寒武纪昌平期末，第二沉降发生在中寒武世馒头期至中奥陶世马家沟期，缺失中奥陶世晚期和晚奥陶世沉积。在这两次沉降期间则经历了一个极短期的抬升，形成了昌平组和馒头组间的平行不整合。而中奥陶世晚期后，全区经历了一次普遍的抬升，这次抬升自中奥陶世马家沟期末开始，到石炭纪太原期前，本区和华北地区一起整体上升为陆，长期遭受剥蚀和夷平。

中寒武世—中奥陶世阶段沉积演化过程与特征为：

中寒武世昌平期，燕山地区再次沉降发生海侵，海水由东向西侵入，逐渐演化出克拉通型新燕山海盆。燕山海盆北、西、南3面为古陆，一面通海，形如簸箕状，表现为半封闭的海湾状。本区位于燕山海盆西北边部，大致以新城子—四海—鸡鸣驿一线为界，其北为内蒙古陆，其南为海盆。内蒙古陆受长达260Ma的隆起抬升影响，地势平缓，以化学风化侵蚀作用为主，极少能为海盆提供陆源物源，致使盆内沉积以内源碎屑岩为主。宏观上，沉积盆内作用受新燕山海盆盆底地貌控制。东低西高、北低南高的盆底古地貌总趋势致使沉积物东厚西薄、北厚南薄，由西向东海水逐渐加深。沿着海盆边部太师屯—溪翁庄—东花园—沿河城—清水—九龙一线以北、以西，沉积为灰色、灰黄色中厚层-厚层状白云岩、泥质白云岩、灰岩、泥灰岩、粉砂岩、泥岩组合，为混合云坪沉积；紧邻混合云坪相沉积的盆地边部沉积为豹斑状灰岩、白云岩、灰质（白云质）角砾岩组合，为云坪相沉积；庄户台—青白口—大庄科—沙厂一线以东沉积为豹斑状灰岩为主，发育粉（细）晶灰岩、核形石灰岩组合，为灰坪相沉积。

昌平期末，燕山地区地壳出现短期抬升，海水由西向东退却，结束了早古生代第一次海侵，本区上升成陆，并遭受风化剥蚀，古地形趋于夷平。

中寒武世馒头期，燕山地区又再次发生沉降，开始了早古生代第二次海侵，也是显生宙最广泛、经历时间最长、规模最大的一次海侵，本次海侵一直持续到早奥陶世中期才结束。馒头期，海水自北东向南西侵入，海域范围极广，略有别于第一次海侵。北京境内全区被海水覆盖，海水处于半封闭状态，大多数时期水体安静，气候干燥炎热，蒸发作用强烈，海水咸化。古地理环境分割性很大，岩相时间、空间上均有变化，形成复杂沉积相类型。早—中期沉积仍受昌平期古地貌面影响，宏观上大致以门头沟南—昌平东—怀柔—密云西一线为界，该线以西、以东分别发育潮上萨布哈相和潮坪潟湖相沉积；晚期海侵继续扩大，海水逐渐加深，发育浅海陆棚潮坪-潟湖相沉积，可细分为潮坪潟湖相、潮坪云坪相、潮坪泥坪相和潮坪混合坪相。馒头期，沉积地层厚度揭示境内发育西山和平谷-兴隆两个沉积中心，呈北东东向分布。

中寒武世张夏期，为馒头期海侵的继续，区内水体逐渐加深，海底地形起伏日趋夷平，水动力增强，碳酸盐岩大量出现，藻类、生物化石丰富。新燕山海盆海水由北东向南西方向不断扩展，海域范围不断扩大。至张夏期中晚期，北京境内已演化为温暖潮湿的典型开阔碳酸盐台地沉积，沉降速度与沉积物堆积速度已达平衡。延庆—四海—墙子路一线以北发育台地滩前浅缓坡边滩沉积相（滩前斜坡），水体相对较深，沉积内源颗粒，少有生物发育。而该线以南，房山—大兴南一线以北发育为浅滩-滩间海相沉积，可细分暴露鲕滩岛、含砂鲕浅滩及高能鲕滩沉积微相，沉积多种多样的鲕粒以及其他各种内源颗粒，水浅而动荡或宁静而适度的循环，氧气、阳光充足，适宜生物繁盛，三叶虫动物群属种数目发展到高峰阶段，计有50余种属，藻类也很发育；房山—大兴南一线以南，可能沉积厚度相对较薄，遭受剥蚀，无沉积地层残留，但从邻区河北易县一带沉积薄-中厚层泥质条带灰岩、薄层致密灰岩、含鲕粒白云黏土质灰岩、鲕粒灰岩加薄层致密灰岩和少量黄绿色页岩组合推测，可能发育滩间海相沉积。

晚寒武世炒米店期，为早古生代第2次海侵继续演化时期。新燕山海盆古地形受张夏期沉积影响，由早期的北高南低、西高东低转变为北低南高，西高东低，致使海水水深由东向西、由北向南递增，但海侵方向保持为北东向。期间，地壳升降频繁，海水动荡不定。同张夏期相比，总体水体明显加深，沉积环境以台地斜坡相沉积（海泛期）为主，时有抬升作用下水体变浅的潮间、潮下沉积（海退期）。海泛期，沉积为薄层泥质灰岩、风暴砾屑灰岩、含砂细晶灰岩、含砾屑细晶灰岩、钙质页岩、钙质粉砂岩等组合的沉积序列，由南向北依次发育浅缓坡边滩沉积相、中陡坡盆地沉积相和浅缓坡盆地沉积相，最北部可能还发育深水盆地相沉积，以海绿石凝缩段沉积特征；海退期，水体从南向北收缩，沉积为泥质（晶）灰岩、竹叶砾屑灰岩（具氧化圈）、砂屑灰岩、生物屑灰岩等岩性组合序列，局部甚至经过一定时间的沉积暴露，风化作用强烈，形成风化土壤层，表现为局限台地相或局限海相沉积。炒米店期，沉积厚度宏观上南薄北厚，区域差异较大，累计沉积厚度90~360m，是早古生代沉积厚度最大的时期。炒米店期新燕山海盆由于处于低纬度地区，在赤道飓风带控制影响下，气候温暖，风暴作用频率高。风暴沉积在昌平组底部、炒米店组、冶里组均有发现，以各种独特的颗粒灰岩、粉屑灰岩、泥晶灰岩和一些与生物丘共生的岩石组合为特征，其中以炒米店组风暴沉积最发育。同时，较深水沉积环境不利于生物繁衍生息，总体生物不甚发育。但在该期初、末和地壳抬升海退时期，即水体较浅情况下发育比较繁盛的三叶虫动物群，初期出现的三叶虫新种属球接子目，末期则出现笔石。

早奥陶世冶里期，第2次海侵继续演化，本区进入局限海的发育阶段。沉积环境为以低能为主、高能和低能交替的潮下-潮间-潮上环境，沉积了冶里组泥晶灰岩夹竹叶状灰岩，泥质条带泥晶灰岩、细角砾碎屑灰岩等碳酸盐岩沉积。由于气候干燥，海水被大量蒸发，尤其潮上带中高镁卤水的回流作用，不利于海相底栖或底栖浮游生物的繁殖，冶里组除在顶部的竹叶状灰岩透镜体中有化石外，基本上不含化石。

早奥陶世亮甲山期，本区继续处于局限海沉积环境，但由于来自北面的海水不断地贯入，局限海中高镁卤水的形成和回流作用已不甚发育，海相底栖和底栖浮游生物大量繁衍，亮甲山组碳酸盐岩中含有丰富的头足类、腹足类和牙形石等生物化石。亮甲山期还是区内奥陶纪期间白云岩化作用的重要时期。亮甲山早期，本区主体处于潮下低能环境，沉积了一套巨厚层-中层含燧石团块泥晶灰岩，伴随风暴沉积有少量水下浅滩相竹叶状灰岩。亮甲山晚期，地壳开始上升，区内呈现海退迹象，在潮上蒸发环境中形成了亮甲山组上部以白云岩为代表的滨海蒸发相碳酸盐岩建造。伴随海退演化，至亮甲山末期，本区一度露出海面，在高度蒸发条件下形成了亮甲山组顶部角砾状白云岩。

中奥陶世马家沟期，本区扭转亮甲山期末海退趋势，再次经历短期海进。但此次海进规模有限，水体总体较浅，主体仍发育为潮上-潮间交替沉积环境。马家沟期，本区共经历了两次由潮上-潮间的海进型沉积韵律，即划分的下、上马家沟组。海进早期表现为潮上蒸发潮坪相沉积，发育为角砾状灰岩与泥晶白云岩；晚期表现为潮间-潮下沉积，发育为泥质条纹灰岩、泥晶灰岩与细晶白云岩、灰质或泥质白云岩及白云质灰岩互层。两期海进后，至中奥陶世马家沟期末，地壳开始抬升，伴随海水引退，本区发育为潮上蒸发环境，沉积马家沟组顶部的角砾状白云岩和角砾状灰岩。该期，本区气候干燥，利于生物生长，地层中含丰富头足类化石，特别是在海进沉积韵律的中上部，潮间强水动力条件下发育了厚度较大含生物碎屑灰岩，同生破碎及团块构造也有发育。

综上，中寒武世—中奥陶世主要发育为一套滨海-浅海相的碳酸盐岩建造。该建造是区内重要的水泥用、电石用和熔剂用石灰岩的含矿建造，纵向上构成了一个比较完整的海进沉积旋回。这个海进沉积旋回是本区继中—新元古代后的又一个大型海进沉积旋回，也是区内地史发展中最后一次有规模的海进沉积旋回。

四、晚奥陶世—早石炭世

晚奥陶世—早石炭世时代为460.9~318.1Ma，跨越了大约140Ma之久，经历了加里东期晚期—华力西期早期两大构造旋回。本区自中奥陶世马家沟期末整体上升以来，在整个晚奥陶世、志留纪和泥盆纪期间，一直处于陆表剥蚀、夷平阶段，没有沉积记录。马家沟组与太原组之间的平行不整合界面的形成，正是加里东晚期至华力西早期本区整体上升（晚奥陶世—早石炭世上升）的结果。因此，该阶段本区总体仍处于克拉通稳定大地构造背景环境之下。然而，自中泥盆世开始，伴随区域上古亚洲洋板块向华北克拉通俯冲-碰撞，稳定的华北克拉通可能受到远程效应影响，北缘深大断裂构造活化，深部壳-幔物质沿此通道侵位，发育了具有非造山向造山转化的过渡型碱性岩浆岩带，即尚义-大庙岩浆岩带。境内密云琉璃庙—溪翁庄—太师屯一带，也宏观发育呈近东西向展布的碱性正长岩、正长斑岩体，岩石多具浅变质特征，可能与长哨营-古北口大断裂的大规模伸展诱导的地幔物质上涌有关。

五、晚石炭世—中三叠世

晚石炭世—中三叠世可划分为两个小阶段：晚石炭世—中二叠世为华力西晚期，时代为318.1~260.4Ma。晚二叠世—中三叠世为印支期，时代为260.4~228.7Ma。

晚石炭世太原期，本区结束了晚奥陶世以来的抬升运动，再次沉降接受沉积。由于自奥陶纪以来长期的夷平作用，区内地势平坦，自西北向东南方向略有倾斜，同华北地区一起构成了一个广阔的剥蚀平原。这一时期下沉速度缓慢，具有明显的旋回性震荡作用。因此，太原期没有形成有规模的海进沉积旋回。期间，气候逐渐温暖、潮湿，泄水条件良好，区内在频繁的震荡旋回的作用下，近海平原上河流发育、湖沼密布，并不时有海水的侵漫，形成了一套海陆交互相的含煤碎屑岩建造。初期，长期的风化剥蚀作用在太原组底部普遍发育一层底砾岩，与下伏的奥陶系呈平行不整合接触。砾岩多为复成分砾岩，其发育情况及砾石的成分因地而异，变化较大。区域上，长期风化剥蚀还使大量的铁铝质矿物在古风化的低洼部分富集，形成了太原组含煤碎屑岩建造底部的陆相-滨海相沉积型的山西式铁矿和“G层”铝土矿。本区由于距蚀源区相对较近，在相当“G层”铝土矿的层位上只形成了以风化分解不充分的水云母为主要黏土矿物的硬质耐火黏土，未能形成具工业规模的铝土矿床。

晚石炭世—早二叠世山西期，基本保持了太原期沉积特征，延续沉积了一套含煤碎屑岩建造，由多个滨海湖沼相的沉积旋回构成，每一个沉积旋回即相当于一个独立的地壳震荡旋回。单个沉积旋回的基本相素成分和形成的顺序在剖面上由下往上大致为冲积砂（砾）岩、湖沼砂质泥岩、沼泽泥岩、滨海泥灰岩、钙质页岩、湖沼砂质泥岩、细砂岩。起初，区内由于沼泽化尚不是很强烈，成煤条件不佳，未形成工业价值的煤层。之后伴随地壳上升，海岸线南移，区内沼泽发育，造煤植物十分繁盛，形成晚古生代的一个重要的成煤时期。整个山西组含煤建造由砂岩、粉砂岩及泥岩等组成。除在西南部色树坟等地夹有1~2层海相泥灰岩及钙质泥质岩外，主要为一套河、湖沼泽相的碎屑岩沉积，是一个以陆相为主夹海相沉积的含煤碎屑岩建造。该建造的构造及其中单个沉积韵律的基本要素成分和形成顺序与太原组含煤沉积建造大体一致。但由于北部古陆的扩大和沉降幅度能够及时得到沉积物的补偿，区内山西期沉积物颗粒显著加大，粗砂岩的成分明显增多，滨海泥灰岩发育不全，泥炭沼泽甚为发育，具有比太原期良好的成煤地质环境，赋存有工业价值的煤层。

早二叠世山西期末，地壳继续抬升，本区脱离了海洋环境，从此结束了区内海相地层的发育史。该期，大体相当于现今的清水河道—昌平一线以西的军都山地区已明显隆起，同华北北缘高地一起接受剥蚀，构成了一个统一的蚀源区。华北北缘高地以南、军都山隆起东南地区接受沉积，主要为一套泛滥平原上陆相碎屑岩建造。此建造仍含煤，但成因有别于石炭纪，见于河床砂砾岩-河漫滩砂泥岩-湖沼砂泥岩组成的完整河湖、沼泽相沉积旋回的顶部。

中—晚二叠世石盒子期，地壳继续抬升，区内气候变得炎热、干燥，在氧化环境下沉积了一套红色的石英砂岩建造。建造的成熟度较低，主要由河流相砂砾岩，即粗粒石英砂岩、含砾石英砂岩及石英细砂岩组成。砂岩颗粒中，除见长石及岩屑颗粒外，还出现有云母碎屑，分选性和磨圆度均不好。早—晚二叠世，区内气候由温暖潮湿渐趋干热演化，地壳以上升为主，断裂差异性活动明显加强，致使区内河流发育，地形相对复杂。而且沿着老深大断裂壳幔源大规模岩浆侵位，侵入岩分布面积扩大了3.5倍。区内发育了早二叠世—中二叠世具有造山特征的基性、中性、酸性侵入岩，同长哨营-古北口断裂、密云-墙子路断裂有关的一系列岩体呈近东西向展布，如下窝铺岩体等，表明该期岩浆活动不仅仅局限于华北克拉通北缘，已经越过邻区尚义-隆化区域断裂向南推进了105km，深入到了燕山-辽西裂陷带的中北部地区，充分显示了华力西晚期古亚洲洋板块向华北克拉通之下俯冲作用之强烈、波及范围之广阔（《中国区域地质志·河北志》，2017）。

晚二叠世晚期—早三叠世双泉期，区域内基本延续了石盒子期演化，沉积范围与石盒子期大体相当，二者为连续沉积。期间，区内气候更趋炎热、干燥，在河流沉积背景下沉积了双泉组紫红色、灰绿色等杂色碎屑岩建造和火山-沉积建造。除底部发育有二叠纪、三叠纪的混生植物，并出现三叠纪主要分子外，整个中上部地层迄今没有化石的发现。双泉期，伴随地壳继续上升，断裂差异性活动更加强烈，地形差异更大，双泉寺一带还发育有火山活动，形成英安质-安山质火山集块岩和凝灰质砾石等。这种杂色碎屑岩建造实际上是在地壳运动剧烈的时期、火山活动频繁地区沉积物快速堆积，来不及充分分异，使许多岩屑和不稳定矿物与黏土质一起堆积而形成的陆地喷发火山-沉积建造。双泉期是本区自中元古代大红峪期以后的又一个火山活动期。双泉期火山活动反映华北克拉通至此阶段可能不再稳定（活化），早三叠世中酸性、以酸性为主的陆地火山喷发，是又一次剧烈的构造运动（印支运动）即将来临的一个孕育期。除火山活动外，同期还发育来源于新元古代增生加厚基性镁铁质下地壳部分熔融具I型岩浆岩特征的中性、中酸性、酸性侵入岩。它们未遭受高级变质和复杂褶曲变形，沿长哨营-古北口断裂呈近东西向展布，明显仍与华北板块北缘碰撞增生后的伸展环境有关。

总之，中元古代—中三叠世构造运动阶段，区域稳定盖层构造层全面形成，使地壳进一步增厚。地幔岩石圈处于相对稳定状态，仍为方辉橄榄岩型（邓晋福等，2007）。

中三叠世后，本区由中亚造山晚期（碰撞后）转入了后造山稳定过渡阶段，应力场背景由伸展转化为强烈挤压，运动的方向自南向北，致使地壳强烈抬升，全区以整体隆起为特征。这次强烈挤压（印支运动）结束了印支构造旋回，正式进入燕山构造旋回。其不仅使长城系到双泉组（也包括了长城纪前的变质岩系）发生了显著的褶皱，包括区内东西向展布的京西谷积山背斜及灰峪向斜、京东墙子路-遵化复背斜等在内，形成了晚三叠世杏石口组与下伏地质体之间的角度不整合界面；还使得长哨营-古北口断裂、密云-墙子路断裂等东西向基底断裂重新强烈活动。与之相对应，印支运动还可能伴随着一次低绿片岩相的区域变质作用，发育有少量过渡型偏碱性、碱性侵入岩。岩浆活动强度总体有限，表现出特有的I、A混合型岩浆特征的造山后过渡特性。

第五十六章　晚三叠世—早白垩世构造阶段

晚三叠世—早白垩世为燕山期（燕山构造旋回），是本区地壳强烈活动、克拉通分化、岩石圈强烈改造及板内造山阶段，分为板内前造山期、板内造山期、板内造山后期和板内后造山期4个亚构造旋回。

燕山早期（晚三叠世—中侏罗世）为燕山板内前造山阶段，本区主要仍受控于来自北部的西伯利亚板块与古亚洲构造带沿蒙古-鄂霍次克洋汇聚碰撞产生的远程构造作用；晚三叠世—早侏罗世本区在地幔软流圈反向对流条件下开始了近南北向（北北西-南南东向）伸展拉张运动，多数断裂复活张性活动明显；中侏罗世则发生了燕山构造旋回第一次构造反转，区域构造应力场演化为近南北向挤压。

燕山中期—晚期（中侏罗世—早白垩世）依次经历了燕山板内造山、板内造山后阶段，本区乃至广大中国东部甚至东亚大陆不仅处于太平洋板块俯冲边界动力系统的控制下，同时受控于来自北部的西伯利亚板块与古亚洲构造带沿蒙古-鄂霍次克洋汇聚碰撞产生的远程构造作用，还有来自拉萨地块与羌塘地块碰撞产生的西南方向的挤压，构成中侏罗世—早白垩世多向挤压汇聚的板块动力学格局。该阶段既有伸展型火山-碎屑盆地发育，也有岩浆的多期喷发和侵入，还发育北东向、北北东向多组显著褶皱和断裂构造，构造变形较为复杂，局部还有较强的变质作用。同时，受克拉通分化、岩石圈强烈改造与再造、壳幔作用、板内造山强烈活动的制约，成矿作用以内生金属矿产为主，是本区最重要的内生金属矿产成矿期，可谓各类金属矿产的爆发成矿期。

一、晚三叠世—中侏罗世

晚三叠世—中侏罗世为燕山板内前造山期，本区总体处于相对宁静状态。

晚三叠世末期，在经历印支运动后，构造应力场再次转化为伸展，在拉张背景下来源于富集地幔镁质富钾基性岩浆顺着构造薄弱带侵位形成大石坡正长岩，幔源岩浆侵位使得热量向地壳输入，引起下地壳部分熔融，形成几乎同期的壳幔混合源的黑山坨花岗岩。伴随深部岩浆活动，地壳分异演化，区域伸展构造背景环境下的断裂拉张运动，孕育了燕山造山幕早期伸展型盆地，此时本区可能仍受控于来自北部的西伯利亚板块与古亚洲构造带沿蒙古-鄂霍次克洋汇聚碰撞产生的远程构造作用。区内演化出西山和昌怀两个火山-沉积盆地，二者呈东西向和近东西向延伸，沉积了上三叠统杏石口组—下侏罗统窑坡组。

晚三叠世杏石口期，西山盆地初始形成，在印支期形成的褶皱挠曲古地貌低洼区，主要是军都山隆起东南侧的京西和京东地区，发育了开阔谷地曲流河沉积背景下的一套灰色、灰黑色砾岩、砂岩、粉砂岩及黏土岩组合。此时，古气候主体温暖潮湿。

早侏罗世南大岭期，伴随伸展构造持续演化，沉积盆地向东、西方向扩展，西山、昌怀盆地进入成熟演化阶段。盆地内除发育近源碎屑岩河流相沉积外，还堆积有多期次玄武岩和玄武安山岩的溢流、喷发物，构成南大岭期火山-碎屑岩地层。早期，火山活动呈带状喷出，晚期呈点式喷发，溢流玄武岩、玄武安山岩的分布明显受古地形、地貌控制，在喷发中心大台—军庄一带形成盆内火山山地新的古地貌格局。

早侏罗世窑坡期，造山幕早期伸展型盆地得以大发展，沉积了一套暗色含砾粗砂岩、砂岩、粉砂岩、泥岩及煤层组合，沉积厚度大，总体上沉积粒度较细，并含大量植物化石及淡水双壳类化石。该阶段早期发育为陆相湖盆三角洲-沼泽相沉积；晚期，盆地逐渐封闭，演化为平缓内陆河湖相沉积，且盆地逐渐由南向北掀斜，沉积中心则向北迁移，致使斋堂、牛角峪、法城口等地晚期沉积物直接超覆在南大岭组之上。窑坡期古气候温暖潮湿，亦为重要的成煤时期。含煤地层的岩相带、地层厚度等值线和相间的富、贫煤带的总体展布方向是一致的，呈东西向和近东西向延伸。该期，岩浆活动强度转弱，由强烈的火山喷发（南大岭火山岩）演化为小规模的岩墙（青白口岩脉）侵位，顺地层或构造薄弱带类似于火山塌陷被动侵位，发育岩性为中基性辉长岩-闪长岩，具有和南大岭期火山岩同源岩浆演化特征。

中侏罗世，长期孕育的构造变动开始爆发，区域构造应力场由伸展转化为挤压，发生燕山构造旋回首次构造反转，但构造变动强度不大。期间，岩浆活动极弱，境内仅北部邻华北克拉通北缘（怀柔北）发育一套来源于地壳玄武安山质或碎屑沉积岩物质熔融具有非造山特征的 A 型花岗岩，揭示了中亚造山对华北克拉通远程影响已经彻底结束，境内转而主要受古太平洋构造域影响。初期，构造变形、变质作用并不强烈，但沉积盆地性质、沉积环境发生变化，由晚三叠世—早侏罗世伸展型盆地向挤压挠曲型盆地演变，但先期形成的盆地得以保存并持续演化，在盆地内沉积了龙门组和九龙山组。

中侏罗世龙门期，西山盆地萎缩，沉积中心随盆地萎缩向东迁移，由门头沟髫髻山、杨家峪一带迁移至门头沟区岳家坡一带。盆地内沉积了一套杂色复成分砾岩、岩屑砂岩、粉砂岩及粉砂质泥岩组合，早期为强水流河流相沉积，晚期为湖泊相沉积。中侏罗世九龙山期则延续了龙门期构造演化。由于更加强烈的挤压作用，西山盆地逐渐向北西掀斜，一方面使得沉积中心由门头沟区岳家坡向北西方向迁移至大台—板桥一带，另一方面使得盆地北西边缘青白口—田庄一带仅出露九龙山组三段，且超覆于寒武系或中—新元古界之上。盆地内沉积环境早期以湖相沉积为主，晚期演化为冲积扇、辫状河沉积物的快速堆积，北部、东北部多以近端扇顶相沉积为主，中部为扇顶—扇中—扇尾混合相区。向南中段扇中相、远端扇尾相沉积逐渐增多。九龙山组形成之际，大量凝灰质砂岩出现，说明至中侏罗世九龙山期再次开始出现火山活动，开启了燕山板内造山序幕。

二、中侏罗世晚期—早白垩世早期

中侏罗世晚期—早白垩世早期为燕山板内造山期。

中侏罗世晚期—早白垩世早期，由于来自东侧太平洋板块、北侧蒙古-鄂霍次克海、西南侧拉萨地块和特提斯洋向东亚大陆的多向汇聚，形成以华北地块为中心的汇聚构造运动样式与多向挤压变形统一的汇聚构造体系。特别是太平洋洋脊扩张带的扩张幅度与洋脊扩张带以西的太平洋板块俯冲幅度的进一步加大加剧，本区进入了板内造山激化期—鼎盛期。

中侏罗世晚期—晚侏罗世初期，在经历小规模幔源岩浆（太平山、抗峪岩体）侵位后，迎来了较为强烈的多期次脉动式岩浆就位，在境内形成闪长岩-石英二长岩-花岗岩的中-酸性岩组合，以石城-云蒙山复式岩体侵位为典型代表（约 160Ma）。深部岩浆过程使得地壳增厚，褶皱运动加强，断裂活动强烈，产生轴向北东 60°~70°、形态呈箱形的八达岭箱式背斜和北岭叠加向斜等一系列褶皱构造，以及霞云岭推覆断裂、八宝山逆冲断层等。与之相对应，先期龙门期—九龙山期挤压继承性盆地逐渐萎缩，致使区内相当长时间内缺乏沉积，形成九龙山组和髫髻山组火山岩间的角度不整合面。

中侏罗世晚期—晚侏罗世中期，可能是太平洋板块向华北板块俯冲方向转变，板内挤压造山短期调整，燕山构造旋回再次经历一次构造反转，区域构造应力场由挤压转化为伸展。居庸关、南口-沿河城及颐和园深断裂强烈正断活动，形成呈北东 50°~60°向南、北两个火山喷发带，带内发育为伸展型火山-沉积盆地，包括南带的百花山、髫髻山、新城子盆地，北带的燕羽山、凤驼梁-四海盆地。盆内发育有两次连续、强烈、大规模的火山喷发活动，形成厚度较大，以粗安质熔岩、正常火山碎屑岩为主的喷出岩层的堆积。这两次火山喷发活动之后均出现较长时间的火山作用间歇期，发育形成巨厚的火山-沉积碎屑岩层，即髫髻山组。该阶段，与大规模的火山喷发正好相反，岩浆侵入作用极弱，岩浆活动方式发生明显变化。

晚侏罗世晚期（约 145Ma），燕山运动最强构造变动爆发，区域构造应力场经短期伸展调整后再次转化为挤压。与中侏罗世晚期相似，在构造转换初期，构造变形、变质作用及岩浆活动并不强烈，但沉积盆地性质、沉积环境发生变化，由伸展型盆地向挤压挠曲型盆地演变，在先期盆地内沉积了土城子组。

早白垩世早期（约 130Ma），进入燕山板内造山鼎盛期，延续了晚侏罗世晚期强大的南东向挤压应力作用，先期土城子期挤压继承性盆地则逐渐萎缩，致使区内相当长时间内缺乏沉积，形成土城子组或髫髻山组和张家口组火山岩间的角度不整合面。同时，先期沉积的髫髻山组等地层褶皱变形，形成轴向北东 40°的一系列向斜，如燕羽山向斜、百花山向斜、髫髻山向斜、九龙山向斜、四海复向斜、千家店向斜、云蒙山背斜等，斜置在早期北东东向褶皱之上；早期北东东向褶皱也进一步变形强化。在本次褶皱形变过程中，同时发生了一系列大型北东向逆冲-逆掩-推覆断裂带和北西向张性断裂。处于地壳中-浅层次的岩石，在特定的热动力和构造条件下也发生塑性变形，局部还形成韧性断层带。期间，岩浆活动也极为强烈，大规模的壳幔混合源高钾钙碱性-钾玄岩系列的中酸性岩浆多次强烈侵入，形成区内多期次复式杂岩体，如分水岭-石湖峪、磨石口-白羊石虎杂岩体等。

三、早白垩世中期—早白垩世晚期

早白垩世中—晚期为燕山板内造山后期。

早白垩世中—晚期（125~80Ma），在经历燕山板内强烈造山运动后，本区进入晚造山-造山后伸展垮塌构造阶段。期间，中元古代以来长期稳定、巨厚的克拉通型岩石圈大大减薄（华北克拉通破坏），地幔流体活跃上隆导致伸展变形，形成底辟热隆和沿隆起带分布的变质核杂岩构造体系，如云蒙山变质核杂岩。区域上，北西西-南东东向伸展环境控制下还发生正断层活动，如紫荆关断裂、黄庄-高丽营断裂等，并新生上叠式火山-沉积盆地，盆地内连续堆积了厚达 3000m 张家口组和东狼沟组的火山-碎屑岩地层，角度不整合覆盖于下伏地层之上。除张家口期—东狼沟期酸性偏碱性和中酸性偏碱性火山喷发岩浆活动外，沿着大型断裂构造，大量中酸性富碱性岩浆沿北北东 25°~30°方向侵位，形成了杨麻地-八达岭钙碱性-碱性花岗岩组合、薛家石梁-湖门碱性-过碱性正长岩组合、西石虎浅成花岗斑岩-正长斑岩-霏细岩组合、铁炉子-四桥子钙碱性-碱性花岗岩组合等系列具有后造山伸展特征复式侵入体。

四、早白垩世末期

早白垩世末期为燕山板内后造山期。

早白垩世末期，区域进入燕山板内后造山稳定过渡阶段，构造应力场再次发生重大转变，形成北西-南东向挤压和南北向反扭旋转联合应力作用。由于该阶段强烈压扭，区域上少有沉积地层发育，仅在早期（早白垩世末期）断陷盆地残留的山间局地发育一个坨里小盆地。盆地内先沉积了坨里组一套河流相及洪积相砾岩和砂岩为主的粗碎屑岩建造，后沉积了夏庄组一套湖相细碎屑岩夹泥岩、泥灰岩组合沉积。古气候则由温热气候逐渐转变为相对潮湿温暖气候类型，繁殖了大量生物，标志着房山生物群进入繁盛阶段。

综上，燕山板内造山阶段既有伸展体制下形成的各种构造类型，又有挤压体制下形成的各种构造类型。构造形迹轴向先后发育有北东东、北东和北北东向，这些构造形迹组合在一起，形成了本区中生代陆内造山带复杂的构造图像。同时，燕山运动中、晚期还是本区内生金属矿产的主要成矿期，直接控制了内生金属矿产的形成与赋存。

第五十七章　晚白垩世—第四纪构造阶段

晚白垩世以来，本区进入喜马拉雅构造阶段。受乌兰巴托-鄂霍次克海裂谷带（也称之为华北-东海新生代裂陷带）扩张与太平洋板块俯冲制约，以及印度板块与青藏高原俯冲碰撞影响，地幔软流圈以乌兰巴托-鄂霍次克海裂谷带之下为轴作反向对流、以太平洋洋脊为轴作相向对流（《中国区域地质志·河北志》，2017）。本区地壳经历了由南北方向上的逆时针扭动向顺时针扭动演化过程，可划分为晚白垩世—古新世稳定过渡期、始新世—渐新世裂陷期和中新世—第四纪坳陷期。

一、晚白垩世—古新世

晚白垩世初（约80Ma）至古近纪古新世，区内承袭了燕山板内后造山稳定过渡特征。地壳处于强烈隆升剥蚀状态，由于长期剥蚀、均夷作用，地貌呈准平原化，形成北台期夷平面。该阶段在区内乃至整个中国东部还形成了一套独立的呈北北东向展布，具有反扭特征的褶皱、断裂构造体系，奠定了现今构造的基本格架。境内西部、西北部张家口组—东狼沟组及其更老地质体记录了该期走向北东30°的叠瓦式逆冲断层及同方向伴生褶皱，伴随发育小规模具有非造山特性基性岩浆侵位，如白河堡辉绿岩侵位；而东部由于被新生代断坳陷覆盖，构造形迹较难恢复，但钻孔揭露北西走向的反扭断层可能也有发育（柯柏林等，2009）。这次强烈挤压（燕山运动Ⅱ幕）结束了燕山构造旋回，正式进入喜马拉雅构造旋回，造成了古近系始新统长辛店组与下伏地质体之间的角度不整合界面。

二、始新世—渐新世

始新世开始，境内地壳在新生代裂谷作用下，准平原沿老断裂带不断开裂、解体。随着京西北隆起的继续抬升，本区出现了山区和平原的分化，北京平原区的轮廓开始展现出端倪。在南口-孙河断裂以南，以北东向黄庄-高丽营断裂、南苑-通县断裂和夏垫-礼贤断裂为界，形成了“两隆两凹”的基本构造格局。“两隆”分别为京西隆起和大兴隆起，该期继续遭受剥蚀，为盆地提供沉积物源。“两凹”分别为北京凹陷和大厂凹陷，接受沉积，累计沉积厚度达千米以上。

始新世，北京凹陷发育由砾岩、黏土质页岩、粉砂岩透镜体组成的山麓冲洪积相长辛店组，地层中产犀类、鱼鳖类、古钝脚类化石，沉积中心位于丰台一带；大厂凹陷内发育一套相对北京凹陷粒度更细的河流-滨湖相沙河街组沉积。

渐新世，北京凹陷发育灰绿、灰褐色粉砂质泥岩，粉砂岩为主夹黑色碳质页岩和绿灰、灰白色细粒硬砂岩及粉砂岩组合而成的河漫滩相-浅湖相前门组，地层中产腹足类、介形类及脊椎动物的鱼鳖类化石，沉积中心位于前门一带；大厂凹陷内发育为一套浅湖相东营组沉积。该地层沉积期间，沿南苑-通县断裂带和礼贤断裂带，多次发生裂隙式溢流、喷发式火山活动，盆地内堆积了多层玄武岩及火山碎屑岩夹层。

渐新世末，由于裂谷作用减弱或消失，构造应力由拉张转变为挤压环境，断陷盆地进入收缩期，面积不断缩小，区内地壳普遍上升产生沉积间断，形成中新统和下伏渐新统之间的区域不整合。

三、中新世—第四纪

中新世—上新世，构造活动趋于平缓，北东东向断裂继续活动，但活动强度明显减弱，且由断裂控制的地块出现不均衡沉降。京西隆起演化为山地，与北京平原进一步分化。平原内的沉积环境与古近纪类似，伴随北东东向断裂继续活动区域整体下沉，断裂上盘沉降尤为明显，堆积了巨厚的泥砂沉积物。断裂下盘也开始接受沉积，致使断陷湖盆进一步扩大，断陷范围在原有的基础上展宽，断隆的范围则相对变窄。大厂凹陷盆地主要表现为横向扩展，沉积范围向西几乎淹没了大兴凸起，先后沉积了馆陶组—明化镇组，馆陶组沉积物以陆相河湖相的红色砂砾岩、泥岩为主，中下部夹少量灰白、灰绿色砂岩、粉砂岩及砾岩；明化镇组为山前洪积扇-河流相砂砾质、泥砂质岩石沉积组合。北京凹陷盆地则主要表现为纵向扩展，往北扩展越过南口-孙河断裂遍及顺义地区，先后沉积了天坛组—天竺组，天坛组为河湖相红棕色砂质泥岩、泥质粉砂岩、细砂岩夹多层灰白色、灰绿色含砾硬砂岩及细砾岩，并同时伴有陆相玄武岩喷发，但其喷发规模和厚度都较古近纪小；天竺组为河湖相灰色、灰黑色、棕黄色半胶结泥岩、粉砂岩及砾岩沉积组合。

第四纪以来，本区构造活动强度进一步减弱，在古老断裂体系的影响和控制下，继承和新生了一系列第四纪活动断裂，并控制了局部的沉积作用。尤以北西向南口-孙河断裂活动明显，北京市区从中新世—上新世的沉积区转变为早更新世的剥蚀区，而北京平原区北部进入第四纪后构造活动强烈期，堆积了一定厚度的第四系。

早更新世，北京平原的构造格架已基本形成，周边山地不断抬升，遭受强烈剥蚀。北山与西山山前相对下降，发育了以粗砾为主的冲洪积扇，形成了似扇形堆积平原。山前平原区则受活动断裂控制分化形成了沙河凹陷、后沙峪凹陷、北京凹陷、大兴隆起、大厂凹陷、廊固凹陷和平谷凹陷。由于各个凹陷及隆起区内沉积作用受到气候、构造运动的强烈控制，并且沉积区和剥蚀区存在多次交叠和变迁，致使平原区厚度较大的新生代沉积物的相变十分突出。即使在很小的沉积区域内其岩性自下而上均有显著的差异。造成了在同一时期不同地点，由于古地理环境的不同形成了不同的第四纪沉积物。

中更新世，北京地区新构造运动又有了新的变化，区域进入短期的构造活跃期：①山区间歇性缓慢抬升转化为强烈抬升，幽州至八达岭一带发生隆起，使得八达岭古河消失。②平原区呈现整体不均衡下沉，沉积中心由早更新世向北迁移转化为不断向南迁移；由于河流冲积扇的不断扩展，早更新世的一些残丘被覆盖，并使微隆起地块在地表出露面积缩小。③北东向断裂构造出现活动性分段特征，南苑-通县、黄庄-高丽营、顺义、夏垫等断裂以南口-孙河断裂为界，其南段不再活动。④由于新生北西向断裂活动，导致北京平原区中生代以来形成的“两隆两凹”的构造格局和沉积环境彻底解体，沉积中心的长轴方向也发生改变，从古近纪的北北东向转变为北西向，形成马池口凹陷、顺义凹陷等第四纪新凹陷中心。⑤该期沉积范围更广，沉积环境更加复杂。二十里长山地块向西掀斜，在其边缘地点有坡洪积物堆积，在广大的山前地带发育有洪积扇裙和扇缘洼地，平原低洼地区发育有湖沼，平原地区继续下降接受砂质黏土、黏质砂土沉积，形成洪冲积平原。

晚更新世早、中期延续了之前北京地区地壳强烈活动，为现代地貌的形成奠定基础。早期，气候温湿，为马兰黄土与古土壤的堆积期。永定河上下贯通，自三家店出山后形成以石景山为顶点的冲洪积扇，其前缘可达清河、通州一带；古潮河沿北东向断裂在焦家务附近袭夺了石峨古河道的上游段，形成了潮河，潮河出山后流入后沙峪凹陷带，形成了潮河冲洪积扇；平谷盆地原来的石峨古河道的中、下游演化形成了错河、泃河水系，至此，平谷地区的水系基本形成。此时，温榆河基本成形，其流向与现今相似，是永定河的支流之一。

晚更新世晚期，北京平原气候比较寒冷，温度偏低，水量充足，植被覆盖率高，故泥砂供给量较少，而水流的冲刷与搬运能力较强，山前台地上堆积了厚约4m的洪积砂砾石层。机械风化作用与冻融作用十分强烈，在山区产生了大量碎石堆积，在山间盆地形成了一系列的山麓洪冲扇。永定河、潮河、白河、泃河等形成的冲积扇使大兴隆起面积大大缩小，仅限于八宝山以南和南苑以西地区。大兴隆起地块活动微弱，永定河向西南方向迁移，使冲积扇的前缘可达清河、通州、南苑一带，形成巨大的冲积扇。古白河和古潮河出山后，将晚更新世早、中期形成的冲积扇切割成宽浅的槽谷，古白河向西流经西智到神山，折向西南经北房村于龙王头入小中河，然后南流至通州古城村，折向东南入渤海；古潮河向西南流至河南寨折向西南，经木林和孤努山东流经小店村于荆坨村入鲍邱河。古温榆河沿永定河冲积扇边缘流至通州注入古白河。

全新世，由于新构造运动的影响，永定河不断向南移动，使新期冲积扇也随之南移；潮河、白河的汇流点不断北移，由通州附近迁移至牛栏山一带，留下多条古河道；全新世气候转暖，针叶林向高山退缩，而阔叶树逐渐增加，中全新世时，气候温暖湿润，阔叶树较繁茂，晚全新世时，气候以凉干为主，为森林草原环境。

总之，在新生代断裂活动的影响下，从始新世开始，北台期的夷平面出现地形地貌分异，山区抬升，平原区下降，在新近纪阶段基本上形成了现今的地貌格局，京北为雄伟的军都山脉，京西为太行山余脉的末端。进入第四纪，尤其是中更新世以来，山区表现为强烈上升，平原下沉继续接受沉积，推测今后山区仍具有继续上升的趋势，这将从根本上决定和影响着北京地域的生态环境地质问题。

结　语

北京地区开展过两轮系统的 1 : 5 万区域地质调查，1991 年编写出版的《北京市区域地质志》所使用的资料截至 1984 年，基本上是第一轮 1 : 5 万区调资料。1997 年，北京市地质调查所鲍亦冈先生主持的“北京市 1 : 5 万区调片区总结”项目，以第二轮 1 : 5 万区调资料及 20 世纪内几代地质工作者研究成果为基础，对本区地质工作进行了高水平的总结，并于 2001 年编写出版了《北京地质百年研究》。该书对北京地区基础地质研究作了系统的历史回顾与评价，并以现代地学理论为指导，对本区的基础地质特征作了系统探讨和阐述，弥补了 1991 年版《北京市区域地质志》的短板和不足，是编写本志的主要参考成果。21 世纪以来，本区地质工作转向平原地区，北京市地质矿产勘查开发局先后组织实施了“北京多参数立体地质调查”“北京平原区活动断裂监测专项地质调查”等项目，本志涉及的第四纪地质、城市地质等内容，主要参考了此调查研究成果。本志对以往资料和成果系统整理与综合研究，取得了以下十个方面的新认识：

1. 区域地层

基于变质铁矿是赋存于沉积地层之中的基本认识，以及原岩恢复新成果，将太古宙地质体划分为表壳岩和变质侵入体。根据岩石组合、变质程度和变质相、混合岩化强度以及变形等特征，将太古宙表壳岩系按非正式岩石地层单位划分为密云岩群和四合堂岩群。综合新的同位素测年数据以及地层区域对比，将上述 2 个岩群置于新太古代，并初步厘定密云岩群在下、四合堂岩群在上的叠置关系。

采用《北京市岩石地层》（1996）成果，将中元古代至白垩纪地层划分为 35 个岩石地层单位，各组层型描述亦引用该书。根据全国地层委员会颁布的《中国地层表》（2013），将原属长城系的高于庄组划归蓟县系，将原属青白口系的下马岭组划归临时建立的中元古界待建系；综合区域地层对比和最新同位素年龄等将髫髻山组地层形成时代调整为中—晚侏罗世。

沿用《北京市区域地质志》（1991）地层划分方案，分山区和平原区分别对古近系和新近系地层进行总结，建立相应岩石地层单位。基于《北京地质百年研究》（2001）认识，将原上新统鱼岭组（下部）和东岭子组（下部）合并为鱼岭组。

基于区域地层同位素年龄、不整合、特征岩类等最新研究成果，以岩石地层为主线，按照岩石地层序列属性，进行了中—新元古界和古生界年代地层、生物地层、层序地层、事件地层等多重地层划分，并进行区域地层对比研究。

2. 沉积岩与沉积作用

以不同地区的实测剖面为基础，系统总结中元古代—白垩世岩石地层岩性类型、岩性组合和结构构造等，开展沉积相亚相沉积环境分析，按照岩石地层单位修编或编制了其沉积阶段主要岩相古地理图。

依据区域构造-沉积演化特征，系统总结了中—新元古代、早古生代、晚古生代、中生代和新生代北京地区沉积盆地及沉积建造，对本区盆地类型、性质有一定新认识：中—新元古代先后发育裂陷槽、坳陷槽和陆表海盆地；晚古生代先后发育陆表海、板内挤压型挠曲盆地；重新厘定中生代杏石口组-夏庄组沉积阶段火山-沉积盆地属性，认为龙门组—九龙山组等地层沉积阶段为挤压坳陷型盆地。

3. 火山岩与火山作用

采用《北京燕山期火山地质及火山岩》（1995）成果，从岩相、相序列、相组合 3 个方面对火山岩相做了研究和分析，总结了新太古代、中元古代、晚古生代、中生代和新生代火山-沉积序列。

基于同位素年龄、区域地层对比等最新进展，重新厘定中生代火山-沉积盆地中火山岩形成时代，总结火山-沉积盆地演化过程，结合岩石化学与地球化学、区域构造对本区火山岩形成的大地构造背景环境有一定新认识：髫髫山火山喷发阶段处于区域伸展构造背景环境，而土城子期—张家口期火山喷发阶段处于区域挤压构造背景环境，东狼沟期火山喷发阶段又再次转化为伸展构造背景环境。

4. 侵入岩与深成作用

基于北京山区第二轮 1∶5 万区域地质调查成果，从密云岩群和四合堂岩群中划分出新太古代系列侵入岩体，并进行详细时代+岩性划分，厘定了新太古代 4 幕 9 期次岩浆活动。第一幕为发育于新太古代早期表壳岩内基性岩浆喷发或喷溢活动，岩浆成分为超镁铁质、镁铁质和硅铁质；第二幕为晚太古代晚期侵位于表壳岩的沙厂片麻岩，侵位深度大，岩浆成分为英云闪长质；第三幕为新太古代晚期侵位于表壳岩的黑卧-苇子峪-宋营子片麻岩组合，3 期岩浆成分分别为辉长质、辉长闪长质和英云闪长质，构成一次基性到中酸性岩浆演化序列；第四幕为新太古代晚期浅侵位于表壳岩及早期岩体的放马峪-荆子峪-康各庄-旗杆顶片麻岩组合，4 期岩浆成分分别为基性、花岗闪长质、奥长花岗质和二长花岗质，构成典型 TTG 岩浆完整基性—中酸性—酸性演化序列。

基于密云地区花岗斑岩和辉长岩脉与常州沟组接触关系、同位素年龄等最新成果，建立中元古代高岭岩墙群，厘定中—新元古代 2 幕 6 期次岩浆活动。第一幕为发育于中元古代初期侵位于新太古代地质体的高岭辉长岩-新地斜长岩-古北口石英二长岩-沙厂环斑花岗岩组合，构成一次基性到酸性岩浆演化序列；第二幕为发育于中元古代早期大红峪组火山岩-大华山辉绿岩、正长岩组合，构成一次基性—中基性岩浆演化序列。两幕岩浆活动形成于燕山裂陷槽起始和初期阶段，因所处阶段不同，故其岩浆活动方式、强度等均有所不同。

基于岩体地质、变质地质和获取最新同位素年龄、岩石化学等最新成果，将《北京市区域地质志》（1991）前长城期系列侵入岩进行系统梳理，新厘定出中泥盆世正长岩（正长斑岩）和早二叠世辉长岩-闪长岩组合；从八达岭花岗杂岩带中新厘定出早三叠世花岗闪长岩-二长花岗岩组合和中三叠世石英二长岩；对北京地区显生宙晚古生代—早中生代区域处于稳定克拉通大地构造背景（无岩浆事件）提出了新认识。

基于岩体地质和最新同位素年龄成果，修订上庄、棋盘岩、大石坡、四干顶、薛家石梁等一系列岩体侵位时代，建立系统的燕山运动旋回（早侏罗世-晚白垩世）岩浆演化序列，对北京地区中生代燕山运动岩浆岩带、岩浆活动幕次划分上取得了新的认识。按照燕山板内造山运动理论，将八达岭花岗杂岩带解体为继承性近东西向前造山、北东向造山期-造山晚期和北北东向后造山 3 个构造-岩浆岩带。与之相对应构成 3 幕次构造-岩浆演化旋回，单次岩浆演化表现为基性辉长岩-中性闪长岩或石英二长岩-酸性花岗岩演化序列。

基于岩体地质、岩石化学与地球化学等研究成果，对北京地区岩浆侵位方式、岩浆侵位演化和构造背景环境进行了系统梳理，并对北京地区壳幔结构形成与演化进行了简单总结。

5. 变质地质

基于不同时期和不同变质构造单元变质岩系中代表性变质岩的典型变质矿物共生组合、矿物时代和交代关系，主期变质与后期叠加变质矿物特征、变质相温压条件等，将本区区域变质作用发生期划分为阜平期（新太古代早期）、五台期（新太古代晚期）、吕梁期、华力西期、印支期、燕山期，依次形成麻粒岩相-角闪岩相、角闪岩相-绿帘角闪岩相、角闪岩相-绿帘角闪岩相、绿帘角闪岩相-绿片岩相、绿片岩相和绿片岩相变质作用。

基于区域构造、变质作用和变质相特征等，将本区变质相带划分为 3 个重要变质相带、9 个岩性矿物组合变质带：密云岩群重要变质相带（包括单斜辉石-石榴子石带、二辉石带）；四合堂岩群重要变质相带（包括低角闪岩相、高级角闪岩相）；晚古生代-早中生代重要变质相带（包括硬绿泥石-绿泥石带，黑云母带，红柱石带，铁铝榴石-十字石带，硅线石-白云母带）。

基于北京地区变质岩系与变质作用新研究成果，系统总结了北京地区接触变质作用、低温动力变质和气液变质作用等其他变质作用特征和分布规律，梳理了混合岩化作用分布范围、特征及其成因，探讨了混合岩-花岗质岩石演化特征。

6. 岩石圈结构构造

收集了通过北京地区的深地震剖面、电磁测深剖面及解译资料，分析总结了北京地区深部三维结构的相关研究资料，初步总结了北京地区的岩石圈速度结构特征、电性结构、温度结构等特征。

编制了莫霍面等深图和深部构造分区图，区分出幔坡带和幔隆区。总结了本区地壳结构特征，根据本区地壳多层结构且有横向结构变化的特征，分析探讨了地壳厚度变化规律。总结了大陆岩石圈地球化学特征，探讨了层圈相互作用，讨论了深部地质作用及其对资源环境的制约。

7. 地质构造

基于板块构造学说和大陆动力学观点，对构造单元、构造层、构造演化进行系统梳理，划分了构造阶段、构造旋回及相应的构造层。

基于区域构造形迹特征和同位素年龄等资料，系统总结逆冲推覆构造、剥离断层、变质核杂岩、韧性剪切带、固态流变构造群落、穹隆和半穹隆构造等构造组合，依据近年来取得的年龄数据对重要构造体系形成时代、构造背景进行了探讨。

基于燕山运动研究最新进展，按照陆内造山观点将燕山运动划分前造山、造山期、后造山、造山后 4 个阶段，系统总结燕山运动构造旋回沉积演化、变质变形及岩浆活动，认为其起始于广大华北克拉通北缘的软流圈主动上涌和板内造陆运动背景，而这一过程就是华北克拉通东部去克拉通化。

8. 第四纪地质与地貌

基于平原区第四系沉积物源及岩性、隐伏构造、沉积厚度、沉积环境等资料，划分了延庆盆地、马池口凹陷、后沙峪-俸伯凹陷、平谷凹陷 4 个第四系沉积单元，总结了每个单元的岩性地层、磁性地层、孢粉及生物分带、沉积物粘土矿物及化学元素特征，并进行了区域对比研究，分早、中、晚更新世和全新世总结了各时期沉积古地理演化。

基于第二轮 1∶5 万区域地质调查成果和 DEM、遥感影像等，系统总结北京市第四纪地貌特征，进行区域地貌单元划分，编制了第四纪地质与地貌图。

基于平原区隐伏断裂调查研究等成果，详细总结了平原区黄庄-高丽营、顺义、夏垫及南口-孙河等主要全新世活动断裂的几何结构、活动时代、活动强度等特征，以及活动断裂沿线灾害表现及其对地质灾害空间分布的制约关系。

9. 平原区三维地质结构

采用《北京市多参数立体地质调查报告》等成果，基于区域重力、电测深、航磁和地热井等资料，系统总结平原区和延庆盆地基岩面地层、侵入岩及地质构造，修编了基岩地质图；重新厘定古近纪和新近纪地层分布范围和沉积厚度，在新生代沉积与盆地演化方面取得一定新认识。

基于钻孔资料等建立 2 个方向地质剖面，分层次构建平原区基岩、新生界和工程层三维地质结构数字模型，三维展示了各地质体时空分布特征。特别是工程层三维地质结构模型，划分了单层、双层和多层 3 种土体结构类型，将平原区划分为 6 个工程地质区、10 个工程地质亚区、24 个工程地质段，并进行了工程地质条件评价，为城市规划提供了地球科学方面的支撑。

10. 城市地质资源与环境

对北京城市土壤、地下空间、地下水、矿产、能源、地质遗迹资源和崩塌、滑坡、泥石流、地面沉降、地裂缝、采空塌陷等地质灾害等进行系统总结，进行城市地质环境系统划分，并以通州区为例，展示了城市地质资源环境多维数据模型构建，支撑城市规划和管理，服务于首都经济发展和生态文明建设等。

本志认为下列问题还需今后进一步深入研究：

首先，20 世纪 90 年代采用新方法开展的第二轮 1∶5 万区域地质调查，将四合堂岩群和密云岩群合并为密云岩群。本志总结两套地层变质变形特征，认为二者还是有所区别，仍保留二套地层划分和处置方案，并基于新同位素数据，将四合堂岩群置于密云岩群之上。本次厘定方案未有发现显著密云岩群和四合堂岩群接触关系依据，且该地层现有的同位素年龄差别并不大。同时，在原变质地层（表壳岩）中识别出一系列变质深成侵入体，使变质地层的分布范围大大缩小。所识别出的某些变质深成侵入体包含大量表壳岩包体（捕虏体或残留体），甚至包含具有一定规模的磁铁石英岩和BIF铁矿，将其划归为侵入体，与现今公认的“BIF铁矿是火山-沉积作用形成的”这一观点相矛盾。当前本区划分的变质深成侵入体与表壳岩的界线部分很难区分，其中包含大量表壳岩包体和一定规模的BIF铁矿的侵入体。变质地层及深成侵入体划分还需进一步深入研究。

其次，在地层分区、地貌单元划分、构造单元划分等方面，由于划分原则和应用场景的差异，各类单元的划分结果也各异，如本区地貌的单元划分便存在多个版本，原因在于采用了不同的划分原则。本志主张小区域单元划分应与大的区域保持一致，一级和二级划分借鉴《中国区域地质志工作指南》（2012），三级和四级参考《中国区域地质志·河北志》（2017），五级根据本区地质背景进行划分。按照此原则进行划分时，其分区单元界线局部与本区同类地质界线（或地质内容）存在不一致之处，且有些分区与以往划分结果不尽相同。因此，各类地质单元的划分需进一步开展。

再次，由于受现有资料和研究水平制约，平原区古近纪和新近纪地层的代表剖面描述和岩石地层单位划分等沿用《北京市区域地质》（1991）成果。而平原区第四纪地层虽划分为4个区域进行岩性、年代等总结，但尚未进行岩石地层单位划分。新生代基础地质研究有待后续进一步深入开展。

最后，总结城市地质工作成果是本志主要内容之一，鉴于城市地质工作内容广泛性，且其成果偏重于应用，本志将城市地质的内容分置于“第四纪地质与地貌”、“平原区三维地质结构”、“城市地质资源与环境”三个篇章中。随着本区城市地质工作持续开展，新成果和认识不断涌现，城市地质方面研究还需持续深入并进行总结。

主要参考文献

安海波 . 2011. 华北平原地裂缝成因机理研究 . 西安：长安大学 .

安特生，著，袁复礼，节译，1923：中国北部新生代 . 地质专报 . 甲种 3 号 .

安芷生，卢演俦 . 1984. 华北晚更新世与马兰期气候地层划分 . 科学通报，29（4）：228~228.

安芷生，魏兰英，卢演俦，等 . 1979. 顺 5 孔的磁性地层学和早松山世的北京海侵 . 地球化学，4：343~345.

巴尔博 . 1929. 张家口附近地质志 . 地质专报，甲种 6 号 .

巴尔博 . 1934. 中国中部之地文期 . 中国地质学会会志，13（3）.

白凌燕，张磊，蔡向民，等 . 2014. 磁性地层年代对北京平原顺义断裂第四纪活动性的约束 . 现代地质，2014（6）：1234~1242.

白志民，许淑贞，葛世伟，1991，八达岭花岗杂岩 . 北京：地质出版社 .

白志民，等 . 1991. 云蒙山杂岩体的特征及成因再认识 . 北京地质 . 第 3 期：1~10.

鲍亦冈，白志民，等 . 1995. 北京燕山期火山地质及火山岩 . 北京：地质出版社 .

白志民，等 . 1999. 燕山造山带中生代火山喷发及岩浆演化 . 地质论评 . 45（s1）：534~540.

包志伟，等 . 1996. 东秦岭方城新元古代碱性正长岩形成时代及其动力学意义 . 科学通报 . 53（6）：684~694.

包志伟，等 . 1996. 冀西北水泉沟正长岩杂岩体的成因 . 岩石学报 . 55（4）：562~572.

鲍亦冈，刘振锋，王世发，等 . 2001. 北京地质百年研究——北京地区基础地质研究的历史与最新成果 . 北京：地质出版社 .

鲍亦冈，谢德源，陈正邦 . 1983. 论北京地区燕山运动 . 地质学报 . 57（2）：195-204.

鲍亦冈 . 北京市岩石地层 . 武汉：中国地质大学出版社，1996.

北京大学地质地理系古生物地层专业 . 古生物学常用拉丁文 . 北京大学地质地理系，1976.

北京地矿局物化探队综合室 . 1987. 北京平原区 1：5 万电测深区域调查工作报告 .

北京地质局水文地质工程地质大队 . 1979. 北京平原区基岩地质构造图说明书 .

北京地质局水文一大队 . 1965. 北京市密怀顺平原区农田供水水文地质勘察报告 .

北京地质局水文一大队 . 1978. 北京市水文地质图、立体透视图说明书 .

北京地质调查研究院 . 1995. 1：5 万大华山幅、木林幅、杨镇幅区域地质调查报告 .

北京地质调查研究院 . 2002. 1：25 万延庆幅区域地质调查报告 .

北京地质调查研究院 . 2004. 1：5 万沙河镇幅区域地质调查成果报告 .

北京地质调查研究院 . 2010. 昌平新城前期区域工程地质勘查报告 .

北京地质学院 . 1961. 北京的地质 . 北京：北京出版社 .

北京市地方志编纂委员会 . 2001. 北京志之地质矿产志 . 北京：北京出版社 .

北京市地震地质会战办公室 . 1979. 北京地区主要断裂（带）调查报告资料汇编（北京市地震地质会战研究成果汇编（4），2~1）.

北京市地震地质会战办公室 . 1982. 北京平原区全新世构造活动调查研究 .

北京市地震地质会战办公室 . 1983. 北京地区地震地质的估价及新构造运动 .

北京市地震地质会战办公室 . 1987. 北京平原区钻孔柱状图集 .

北京市地质局 . 1979. 北京地区构造体系图说明书（1：10 万）.

北京市地质勘察技术院 . 2012. 北京市华清地热开发有限责任公司 . 北京市浅层地热能调查评价报告 .

北京市地质勘察技术院 . 2014. 北京市浅层地温能、地热以及再生水热能综合利用规划研究报告 .

北京市地质矿产局 . 1991. 北京市区域地质志 . 北京：地质出版社 .

北京市地质矿产局 . 1984. 1：5 万青龙桥幅区域地质调查报告 .

北京市地质矿产局 . 1984. 1：5 万清水幅区域地质调查报告 .

北京市地质矿产局 . 1985. 1：5 万沿河幅区域地质调查报告 .

北京市地质矿产局 . 1988. 1：5 万昌平幅、小汤山幅区域地质调查报告 .
北京市地质矿产局 . 1990. 1：20 万怀柔幅区域地质调查报告 .
北京市地质矿产局 . 1990. 1：5 万琉璃庙幅、范各庄幅区域地质调查报告 .
北京市地质矿产局北京地质调查大队 . 1988. 1：5 万周口店幅区域地质调查报告 .
北京市地质矿产勘查开发局 . 2014. 首都地质资源环境承载能力监测预警平台工程分布图 .
北京市地质矿产勘查开发局 . 2015. 2000—2014 年北京城市地质工作成果通报 .
北京市地质矿产勘查开发局，北京市地热研究院 . 2010. 北京地热 . 北京：中国大地出版社 .
北京市地质矿产勘查开发局，北京市地质勘察技术院 . 2008. 北京浅层地温能资源 . 北京：中国大地出版社 .
北京市地质矿产勘查开发局，北京市地质调查研究院 . 2008. 北京城市地质 . 北京：中国大地出版社 .
北京市地质矿产勘查开发局，北京市地质研究所 . 2008. 北京地质灾害 . 北京：中国大地出版社 .
北京市地质矿产勘查开发局 . 1991. 北京市幅、通县幅 1：5 万区域地质调查报告. 北京：北京市地质矿产勘查开发局 .
北京市地质矿产勘查开发局 . 1991. 怀柔县 1：5 万区域地质调查报告 .
北京市地质矿产勘查开发局 . 2008. 北京市平原区多参数立体地质调查报告 .
北京市地质矿产勘查开发局 . 2011. 北京规划新城前期区域工程地质勘查评价报告 .
北京市地质调查所 . 1994. 1：5 万密云幅、高岭幅、墙子路幅区域地质调查报告 .
北京市地质调查所 . 1995. 1：5 万龙门幅、张坊幅区域地质调查（联测）报告 .
北京市地质调查所 . 1999. 1：5 万三道营子、喇叭沟门、杨木栅子幅区域地质调查报告 .
北京市地质调查所 . 1999. 北京市地质图（1：50000）. 北京：地质出版社 .
北京市地质调查研究院 . 1994. 1：5 万雁翅幅、阳坊幅地质图、地质图说明书 .
北京市地质调查研究院 . 2013. 北京市平原区活动断裂检测专项地质调查 .
北京市地质调查研究院 . 2017. 北京城市地下空间资源调查评价及关键技术研究成果报告 .
北京市地质研究所 . 1992. 1：5 万北京市幅、通县幅地质图、基岩地质图、区域地质调查报告 .
北京市区域地层表编写组 . 1977. 华北区域地层表北京市分册 . 北京：地质出版社 .
北京市水文地质工程地质大队 . 2015. 北京市地面沉降监测年度报告 . 北京：北京市水文地质工程地质大队 .
步达生，德日进，等 . 1933. 中国猿人史要 . 地质专报，甲种 11 号 .
蔡剑辉，等 . 2005. 北京房山岩体锆石 U-Pb 年龄和 Sr、Nd、Pb 同位素与微量元素特征及成因探讨 . 岩石学报 . 21（03）：776~788.
蔡剑辉，等 . 2006. 太行山—大兴安岭东麓晚中生代碱性侵入岩岩石地球化学特征及其意义 . 地球学报，27（5）：447~459.
蔡向民，郭高轩，栾英波，等 . 2010. 永定河形成时代研究 . 第四纪研究，30（1）：174~179.
蔡向民，何静，白凌燕，等 . 北京市地下空间资源开发利用规划的地质问题 . 地下空间与工程学报，2010，12：1105~1111.
蔡向民，黄淇，张磊，等 . 2014. 北京温榆河的成因 . 中国地质，41（5），1515~1521.
蔡向民，栾英波，郭高轩，等 . 2007. 城市地质调查中第四纪松散沉积物的分类研究 . 城市地质，2（3）：21~25.
蔡向民，栾英波，郭高轩，等 . 2009. 北京平原地区地质系统 . 城市地质，4（3）：6~12.
蔡向民，栾英波，郭高轩，等 . 2009. 北京平原第四系的三维结构 . 中国地质，36（5）：1021~1029.
蔡向民，栾英波，梁亚南等 . 2010. 北京市平原区下伏泥砂砾石层特征及成因探讨 . 地质调查与研究，12（4）：309~415.
曹伯勋，李长安，田明中 . 1989. 北京周口店地区早更新世古冰楔遗迹的发现及其研究意义 . 地球科学，14（3）：239~244.
曹伯勋，田明中，袁铃声 . 1994. 北京周口店新发现的洞穴堆积物研究 . 北京：地质出版社 .
曹伯勋 . 1995. 地貌学及第四纪地质学 . 北京：中国地质大学出版社 .
曹芳 . 1992. 北京平谷中元古界高于庄组藻类微化石 . 地质论评，38（4）：382~387.
曹花花 . 2013. 华北板块北缘东段古生代—早中生代火成岩的年代学与地球化学研究 . 长春：吉林大学博士学位论文，1~158.
曹瑞骥 . 1984. 北京十三陵地区前寒武纪雾迷山组叠层石和藻席中的藻类群落 . 微体古生物学报，1（2）：93~100+121~122.
曹文燕，等 . 2010. 北京市门头沟区棋盘岩岩体地质地球化学特征及其成因. 中国地质大学（北京）.

曾令森，张建新，许志琴，邱小平．2006. 北京云蒙山浅色花岗岩脉及韧性剪切变形的地球化学特征．岩石矿物学杂志，25（3）：203～209.
常廷改．2006. 断层活动性评价中关于测年资料的引用．工程地质学报，4（2）：190～195.
常英，张清波．1994. 北京西山坨里组孢粉组合及其地层意义．北京地质，（3）：1～9.
陈春良等．2014. 河北雾灵山杂岩体锆石 Hf 同位素特征及其区域对比研究．现代地质：28（4）：663～673.
陈方吉．1979. 北京地区全新世地层及自然环境的变化．中国科学，8（9）：900～907.
陈芬，窦亚伟，黄其胜．1984. 北京西山侏罗纪植物群. 北京：地质出版社．
陈芬，杨关秀．1982. 河北平泉、北京西山早白垩世植物化石．Journal of Integrative Plant Biology，（6）：85～90+125～126.
陈海燕，张运强，张计东，等．冀北承德盆地侏罗系九龙山组凝灰岩 LA-ICP-MS 锆石 U-Pb 年龄与地球化学特征．地质通报，2014（7）：966～973.
陈红凯，鲜学福，唐红梅．2010. 石质山区崩塌灾害形成机制．四川大学学报（工程科学版），42（3）：1～6.
陈辉，何淑军．2008. 北京西部永定河第四纪冲积扇基坑边坡稳定性分析．地质通报，27（11）：1888～1893.
陈建强，史晓颖．华北地台中寒武统张夏组上部高频层序研究．地层学杂志，1998（2）：109～115.
陈建强，等．2004. 沉积学及古地理学教程．北京：地质出版社．
陈建强，等．2004. 扬子区中、晚奥陶世和兰多维利世四射珊瑚起源、扩散及生物古地理关系（英文）. 古生物学报，43（2）：179～191.
陈晋镳，等．1980. 蓟县震旦亚界的研究. 中国震旦亚界. 天津：天津科技出版社，56～276.
陈晋镳，等．1980. 略论蓟县及邻区震旦亚界的几个问题. 中国地质科学院天津地质矿产研究所文集（1）.
陈能松，王人镜，单文琅，等．1994. 密云杂岩西段等压冷却 P-T-t 轨迹确定及地球动力学成因．地质科学，29（4）：355～365.
陈能松，王人镜，胡鱼华，等．1993. 北京密云群西段灰色片麻岩的地球化学特征和成因．地球科学，18（4）：477～488+527.
陈能松，等．2006. 北京西山周口店官坻杂岩蒸发法锆石 Pb-Pb 年龄：太古宙成因和克拉通化事件证据．地质科技情报，25（3）：41～44.
陈森煌，刘道荣，包志伟，等．1991a. 华北地台北缘几个超基性岩带的侵位年代及其演化．地球化学，6（2）：128～133.
陈森煌，张祯堂，刘道荣，等．1991b. 河北省岩石超基性岩带的岩体类型及其成矿作用．地质找矿论丛，6（3）：46～58.
陈绍海，等．1998. 汉诺坝玄武岩中地幔岩捕掳体 REE 和 Sr，Nd 同位素地球化学．岩石学报，14（3）：366～380.
陈衍景．1998. 影响碰撞造山成岩成矿模式的因素及其机制．地学前缘，5（增刊）：109～118.
陈一笠，等．1991. 试论天津蓟县盘山岩体的侵入时代．天津地质学会志，9（1）：22～26.
陈一笠，等．1993. 津北印支期盘山岩体侵入位时的大地构造环境．天津地质学会志，11（2）：20～25.
陈印．2014. 云蒙山变质核杂岩的演化及其与华北克拉通破坏的关系．合肥：合肥工业大学．
陈印，等．2014. 云蒙山变质核杂岩的变形规律与发育机制．科学通报，83（2）：310～320.
陈玉山．1990. 北京西山晚古生代植物群组合及双泉组时代探讨．中国煤炭地质，（2）：17～19.
陈正邦．1994. 北京地区的构造地质研究进展．北京地质，34（2）：29～32.
程捷，田明中，曹伯勋，等．1996. 周口店地区第四纪哺乳动物群的演化与环境变迁探讨．现代地质，10（2）：202～212.
程捷，王海芝，张西娟，等．2002. 周口店第四纪地质与环境变迁研究进展与展望．地质力学学报，12（4）：347～353.
程绍平，杨桂枝．2008. 国外新构造研究进展述评．地震地质，30（1）：31～43.
程素华，等．2010. 北京阳坊岩体元素地球化学特征、成因及构造背景．56（2）：205～213.
程裕淇主编，1994. 中国区域地质概论. 北京：地质出版社．
池际尚主编．1988. 中国东部新生代玄武岩及上地幔研究．北京：地质出版社．
崔盛芹，吴珍汉，马寅生．2003. 北京地区中、新生代陆内造山过程——以十三陵-龙庆峡-八达岭路线观察为基础．地质力学学报，9（3）：201～219.
崔文元，魏春景，舒桂明．1990. 北京密云变质杂岩矿物学和结晶的 P-T 条件．矿物学报，10（3）：208～220.
代军治，等．2006. 华北地台北缘燕辽钼（铜）成矿带矿床地质特征及动力学背景．矿床地质．矿床地质，25（5）：

598~612.
单青山．1996. 北京西山地面塌陷的形成因素浅析．北京地质，2：1~10.
单青生，潘云唐．1994. 北京早白垩世东狼沟组非海相腹足类化石．北京地质，(2)：8~14.
单青生，杨鸿连，刘连刚．1994. 北京-通县平原地区第四纪环境演变．北京地质，4：1~7.
单青生．2002. 北京地区第四纪软体动物群与环境演变. 中国地质大学．
德日进．1941. 东亚地质及人类原始. 北京：北京地质学生物学研究所出版品 7 号．
邓晋福，莫宣学，罗照华，等．1999. 火成岩构造组合与壳-幔成矿系统．地学前缘，6 (2)：259~270.
邓晋福，苏尚国，刘翠．2006．关于华北克拉通燕山期岩石圈减薄的机制与过程的讨论是拆沉，还是热侵蚀和化学交代?，地学前缘，13 (2)：105-119.
邓晋福，苏尚国，赵国春，等．2004，华北燕山造山带结构要素组合．高校地质学报，10 (3)：315-323.
邓晋福，赵国春，苏尚国，等．2006. 地球史中的构造活动论和发展阶段论，地学前缘，13 (6)：14-21.
邓晋福，赵国春，苏尚国，等．燕山造山带燕山期构造叠加及其大地构造背景．大地构造与成矿学，2005，29 (2)：157-165.
邓晋福，赵海玲，莫宣学，吴宗絮，罗照华．1996. 中国大陆根-柱构造——大陆动力学的钥匙．北京：地质出版社，110.
邓晋福，等．1990. 中国东部新生代上部软流圈性质及变迁历史．中国矿物岩石地球化学学会地幔矿物岩石地球化学专业委员会编：中国上地幔特征与动力学论文集．北京：地质出版社，8~13.
邓晋福，等．1996. 燕辽地区燕山期火成岩与造山模型．现代地质，1996 (2)：137~148.
邓晋福，等．1998. 岩石圈-软流圈系统的形成与演化. 全国矿物岩石地球化学学术交流会．
邓晋福，等．1999. 中国东部燕山期岩石圈-软流圈系统大灾变与成矿环境．矿床地质，18 (4)：309~315.
邓晋福，等．2000. 中国东部燕山期火成岩构造组合与造山-深部过程．地质论评，46 (1)：41~48.
邓晋福，等．2002. 东昆仑印支晚期幔源岩浆活动．地质通报，21 (6)：292~297.
邓晋福，等．2004. 北京云蒙山片麻状花岗岩锆石 SHRIMP 定年及其地质意义．岩石矿物学杂志，23 (2)：141~146.
邓晋福，等．2004. 华北地区三类岩石圈的壳幔岩石学结构与化学结构及其大陆动力学意义．岩石矿物学杂志，23 (2)：127~140.
邓晋福，等．2004. 岩石成因、构造环境与成矿作用．北京：地质出版社，1~381.
邓晋福，等．2007. 华北太行-燕山-辽西地区燕山期 (J-K) 造山过程与成矿作用．现代地质，21 (2)：232~240.
邓晋福，等．2007. 火成岩组合与构造环境讨论．高校地质学报，13 (3)：392~402.
邓晋福，等．2009. 太行山-燕辽地区燕山期造山过程、岩浆源区成矿作用．中国地质，36 (3)：623~633.
邓起东，徐锡伟，张先康，等．2003. 城市活动断裂探测的方法和技术．地学前缘，10 (1)：93~103.
邓起东，张培震，冉康勇，等．2002. 中国活动构造基本特征．中国科学 (D 辑)，32 (12)：1020~1031.
邓起东，等．2002. 中国活动构造研究的进展与展望．地质论评，18 (2)：68~177.
地球科学大辞典编委会．2006. 地球科学大辞典 (基础学科卷)，北京：地质出版社．883.
地质部水文地质工程地质局 901 队．1957. 1956 年北京市附近地下水动态观测年终报告.
丁嘉贤，张子斌，王丁，等．1979. 北京地区泥炭普查、勘探总结报告. 北京：北京地质局 102 队．
董春艳，王世进，刘敦一，等．2010. 华北克拉通古元古代晚期地壳演化和荆山群形成时代制约-胶东地区变质中-基性侵入岩锆石 SHRIMP U-Pb 定年．岩石学报，027 (06)：1699~06.
董德茂，韦京莲，刘汉英．2008. 北京地区花岗岩类地质地貌遗迹资源及开发其潜在价值建议．中国地质学会旅游地学与地质公园研究分会第 23 届年会暨二连恐龙地质公园建设与旅游发展战略研讨会论文集．
董德茂，韦京莲，刘汉英．2010. 北京地区砂岩峰林地貌特征．中国地质学会旅游地学与地质公园研究分会第 25 届年会暨张家界世界地质公园建设与旅游发展战略研讨会论文集．
董梅，胡辉．2018. 城市环境下多尺度三维地质建模方法及应用实例分析．浙江：浙江大学出版社，1-139.
杜建军，马寅生，等．2008. 京津地区区域地壳稳定性评价．地球学报，29 (4)：502~509.
杜汝霖，等．1979. 燕山西段震旦亚界．石家庄经济学院学报．(4)：3~19+89.
杜汝霖，等．1980. 冀西北怀来震旦亚界青白口系长龙山组 CHUARIA 等多种藻类化石的发现及其意义．石家庄经济学院学报．1980 (1)：21~29+84~87.
杜钊锋．2013. 北京典型区域不均匀地面沉降成因机制分析．北京：首都师范大学．
杜子图，李东旭，高德臻．1995. 北京西山水峪复合型旋扭构造研究．地质力学学报，1 (3)：65~71.

段永红，张先康，方盛明．2002. 华北地区上部地壳结构的三维有限差分层析成像．地球物理学报，45（3）：362~369.
范蔚茗，郭锋，2005. 华北地区晚中生代镁铁质岩浆作用及其地球动力学背景．大地构造与成矿学，29（1）：44~55.
方迎尧，范正国，乔春贵，等．2005. 北京南部地区地球物理特征与区域地质构造．物探与化探，29（1）：5~9.
冯小铭，郭坤一，等．2003. 城市地质工作的初步探讨．地质通报，22（8）：571~579.
冯益民，曹宣铎，张二朋，等．2002. 西秦岭造山带结构造山过程及动力学．西安：西安地图出版社，264.
傅昭仁，单文琅，葛蒙春．2013. 变质岩层构造的形变相分析．地球科学-武汉地质学院学报，22（3）：66~69.
高凡，高励．1982. 燕山地区太古代麻粒岩相岩石退变质作用的研究．矿物岩石地球化学通讯，1（1）：10~13.
高健，宋天锐．1985. 北京十三陵上前寒武系高于庄组碳酸盐岩中的压扁-透镜状层理．地质论评，31（4）：316~321.
高景华，徐明才，荣立新，等．2007. 探测城市断裂活动性的浅层地震方法技术．物探与化探，10（31）：4~8.
高林志，丁孝忠，刘燕学，等．2013. 中国中-新元古代地层年表的新标定. 全国地层会议．
高林志，丁孝忠，庞维华，等，2011，中国中-新元古代年表的修正．地层学杂志．35（1）1~7.
高林志，张传恒，刘鹏举，等．2009. 华北—江南地区中、新元古代地层格架的再认识．地球学报，30（4）：433~446.
高林志，张传恒，史晓颖，等．2007. 华北青白口系下马岭组凝灰岩锆石 SHRIMP U-Pb 定年．地质通报，26（3）：249~255.
高林志，张传恒，尹崇玉，等．2008. 华北古陆中、新元古代年代地层框架 SHRIMP 锆石年龄新依据．地球学报，29（3）：366~376.
高孟潭，俞言祥，张晓梅，等．2002. 北京地区地震动的三维有限差分模拟．中国地震，18（4）：356~364.
高明修．1979. 华北块断构造区的现代引张应力场．地震地质，1（2）：1~12.
高文学，马瑾，1994. 首都圈地震地质环境与地震灾害（附图册）. 北京：地震出版社．
高亚峰，高亚伟．2007. 我国城市地质调查研究现状及发展方向．城市地质，2（2）：1~8.
高战武，陈棋福，黄金莉，等．2010. 北京地区主要活动断裂深部速度结构特征及强震构造分析．震灾防御技术，5（3）：271~280.
高振寰．1979. 唐山地震引起北京地区烈度异常区的原因分析．地震地质，1（2）：74~82.
高志勇，郑荣才，罗平．2007. 陆相高分辨率层序地层中洪泛面特征研究．成都理工大学学报，34（1）：47~56.
葛肖虹，王敏沛．2010. 西去泥河湾——解读古人类与燕山隆升的历史．自然杂志，32（5）：294~299.
耿良玉．1984. 北京西山夏庄夏庄组介形类记述．古生物学报，23（6）：39~42+137.
耿元生，陆松年．2014. 中国前寒武纪地层年代学研究的进展和相关问题．地学前缘，21（2）：102~118.
古生物学名词审定委员会．2009. 古生物学名词．北京：科学出版社．
关绍曾，萧宗正，张清波，等．1995. 北京延庆盆地第四纪早期介形类动物群、古生态及对该区古环境的认识．化工矿产地质，17（4）：217~228.
关绍曾，萧宗正，张清波，等．1997. 北京地区第四纪早期海相介形类．微体古生物学报，14（2）：191~202.
郭高轩，栾英波，叶超，等．2008. 北京平原第四纪下限问题研究．城市地质，3（2）：13~16.
郭佳，等．2011. 四山林场金矿床的地球化学特征与成矿时代．矿物学报．2011（s1）：580~580.
郭沫若，杨钟健，裴文中，等，1955，中国人类化石的发现与研究，中国猿人第一个头盖骨发现二十五周年纪念会报告专集. 北京：科学出版社．
郭沫若．1955. 中国猿人第 1 头盖骨发现 25 周年纪念会报告专集．北京：科学出版社．
郭培国，戴志强，聂道忠，等．2014. 城市地质的研究内容和发展现状．资源环境与工程，28（3）：74~77.
郭旭东，王思敬．1995. 北京城市未来发展中地壳环境的潜在危机．现代地质，9（4）：475~487.
郭旭东，严富华，黄秀铭．1995. 北京西山新构造运动的分期．现代地质，9（1）：51~59.
郭旭东，严富华．1993. 北京西山上升及其对自然环境的反馈作用. 纪念袁复礼教授诞辰 100 周年学术讨论会组织委员会．纪念袁复礼教授诞辰 100 周年学术讨论会论文集．北京：地震出版社．
郭旭东，严富华．1993. 北京西山潭柘寺地区第四纪冰川与环境问题研究．冰川冻土，15（4）：574~581.
国家地震局《深部物探成果》编写组．1986. 中国地壳上地幔地球物理探测成果．北京：地震出版社．
韩露，丁毅，张贵宾．2003. 中国岩石圈三维结构数值模拟数据库．地球物理学进展，18（2）：312~316.
何斌，徐义刚，王雅玫，韦必则．2005. 北京西山房山岩体岩浆底辟构造及其地质意义．地球科学，30（3）：298~308.
何付兵，白凌燕，王继明，等．2013. 夏垫断裂带深部构造特征与第四纪活动性讨论．地震地质，35（3）：490~505.
何静，周圆心，郑桂森，等，北京城市地下空间资源开发利用地质适宜性评价体系研究及应用．地下空间与工程学报．
何情，张建平，邢立达，等，2015，北京延庆千家店地区土城子组恐龙足迹点沉积环境．地质通报．34（9）．1726~1734.

何作霖，等 . 1937. 房山花岗闪长岩之特殊斑状结构 . 中国地质学会志 . 16：407~421.
和政军，宋天锐，丁孝忠，等 . 2000. 燕山中元古代裂谷早期同沉积断裂活动及其对事件沉积的影响 . 古地理学报，2（3）：83~91.
和政军，宋天锐，丁孝忠 . 2000. 北京及邻区长城纪火山事件的沉积记录 . 沉积学报，18（4）：500~520.
和政军 . 1990. 燕山地区长城纪沉积盆地分析及沉积建造学研究. 中国地质科学院 .
和政军，等 . 1994. 燕山地区长城纪沉积演化及构造背景 . 沉积学报 . 1994（2）：10~19.
和政军，等 . 1990. 砂岩碎屑组分与板块构造位置关系的研究现状 . 地质科技情报 . 1990（4）：7~12.
和政军，等 . 2011. 北京密云元古宙常州沟组之下环斑花岗岩古风华壳掩饰的发现及其碎屑锆石年龄 . 地质通报，30（5）：798~802.
河北地质矿产局 . 1989. 河北省、北京市和天津市区域地质志 . 北京：地质出版社 .
河北省区域地质矿产调查研究所，2017，河北省北京市天津市区域地质志 .
贺聪，等 . 2013. 京西潭柘寺地区石炭-二叠纪沉积相及沉积环境演化 . 北京工业职业技术学院学报，12（1）：10~14.
贺高品，叶慧文，夏胜利 . 1993. 北京密云地区变质基性岩墙的 Sm-Nd 同位素年龄及其地质意义 . 岩石学报，9（3）：312~317.
贺高品，叶慧文，夏胜利 . 1994. 北京密云麻粒岩相区变质作用演化及 PTt 轨迹的研究 . 岩石学报，10（1）：14~26.
贺同兴，等 . 1987. 混合岩化作用的地质成因分类 . 中国地质科学院院报 . 第 16 号 .
贺同兴，等 . 1992. 冀东太古宙花岗岩成因 . 长春：吉林科学技术出版社 .
赫云兰，刘波，秦善，等 . 2012. 北京西山下苇甸中寒武统张夏组生物丘发育特征及其地质意义 . 地质科技情报，31（1）：9~15.
洪友崇，程政武，王思恩，等 . 1980. 京西晚期中生代地层和古生物的研究 . 地质学报，（2）：98-110+188-190.
洪友崇，萧宗正 . 1997. 北京延庆后城组蜚蠊目、鞘翅目长翅目化石（昆虫纲）. 北京地质，（3）：1~10.
洪友崇，萧宗正 . 1997. 北京延庆县后城组蜚蠊目化石（昆虫纲）. 北京地质，（2）：1~9.
洪友崇 . 1982. 京西早白垩世卢尚坟昆虫群 . 中国地质科学院天津地质矿产研究所文集 .
侯惠菲，高亚峰，王栓庄，等 . 2004. 城市地质调查内容及其发展 . 北京地质，16（3）：24~29.
胡波，翟明国，彭澎，等 . 2013. 华北克拉通古元古代末-新元古代地质时间——来自北京西山地区寒武系和侏罗系碎屑锆石 LA-ICP-MS U-Pb 年代学的证据 . 岩石学报，029（07）：2508~2536.
胡厚田，赵志明 . 2009. 岩浆岩和变质岩边坡岩体结构类型的划分 . 防灾减灾工程学报，29（3）：277~281.
胡厚田 . 2005. 崩塌落石研究 . 铁道工程学报，12 月增刊 378~391.
胡俊良，赵太平，陈伟，等 . 2007. 华北克拉通 1.75Ga 基性岩墙特征及其研究进展 . 大地构造与成矿学，31（04）：457~470.
胡世玲，等 . 1990. 大庙斜长岩同位素地质年龄、稀土地球化学及其地质意义 . 地质科学 . 第 4 期 . 1990（4）：332~343.
胡受奚，赵懿英，徐金方，等 . 1997. 华北地台金矿地质. 北京：科学出版社，220.
黄迪颖，林启彬 . 2001. 北京早白垩世昼蜓稚虫化石 . 科学通报，46（3）：235~240.
黄福生，等 . 1983. 个旧锡矿花岗岩的氧氢碳稳定同位素研究 . 岩石矿物及测试 . 2（4）：241~247.
黄汲清，姜春发 . 1962，从多旋回构造运动观点初步探讨地壳发展规律 . 地质学报，42（2）：105~152.
黄汲清，任纪舜，姜春发，等 . 1977，中国大地构造基本轮廓，地质学报，1977，（2）：117~135.
黄汲清，等 . 1928. 羊坊花岗岩与髫髻山层之接触 . 中国地质学会志 . 7（3~4）.
黄金莉，赵大鹏 . 2005. 首都圈地区地壳三维 P 波速度细结构与强震孕育的深部构造环境 . 科学通报，50（4）：348~355.
黄少雄，等 . 1990. 北京密云环斑花岗杂岩的岩石学及地球化学研究（摘要）. 地质地球化学 . 第 4 期 .
黄万波，关健 . 1983. 京郊燕山早更新世洞穴堆积与哺乳类化石 . 古脊椎动物与古人类，21（1）：71~78.
黄万波 . 1960. 中国猿人洞穴堆积 . 古脊椎动物学报，2（1）：85~97.
黄万波 . 1960. 周口店各洞穴堆积简述 . 地质科学，3（2）：85~90.
黄兴根，赵希涛 . 1987. 北京地区第四纪海侵及新构造运动初析 . 地震地质，9（2）：62~64.
黄秀铭，汪良谋，徐杰，等 . 1991. 北京地区新构造运动特征 . 地震地质，13（1）：43~51.
黄学光，等 . 2006. 燕山中、新元古代沉积盆地构造演化 . 地质调查与研究，29（4）：263~270.
黄勇，杨忠芳，等 . 2014. 北京城区 24 年来土壤 Hg 元素时空变异特征 . 现代地质，28（5）：935~941.

黄宗理，张良弼主编 . 2006. 地球科学大辞典（基础科学卷），北京：地质出版社 .
姬广义，汪洋，夏希凡 . 2005. 燕山中段南麓蓟县北部常州沟杨庄一带地质构造基本特征 . 北京地 质，17（4）：1~33.
姬广义，汪洋，夏希凡 . 2005. 华北燕山期区域变质与内生成矿作用关系雏议 . 北京地质，17（3）：29~33.
姬广义，汪洋 . 2004. 北京东部丰台顶一带地质构造特征 . 北京地质，16（1）：1~15.
姬广义，汪洋，夏希凡 . 2004. 北京怀柔长园花岗杂岩体岩石学和构造变形特征及其地质意义，北京地质，16（4）：1~13.
姬广义，汪洋，夏希凡 . 2005. 论基础地质与城市地质之关系及存在问题 . 北京地质，17（1）：1~11.
姬广义，等 . 2002. 北京市凤鸵梁-四海地区中生代地层基本层序 . 北京地质 . 2002（1）：1~12.
姬广义，等 . 2004. 北京云蒙山岩浆杂岩体的岩石学和构造变形特征 . 北京地质 16（3）：2~11.
纪玉杰 . 2002. 北京西山侏罗纪煤田采空区地质特征与地质灾害 . 北京地质，14（1）：28~36.
纪玉杰 . 2002. 北京西山侏罗纪煤田矿震与地质灾害的关系 . 北京地质，14（3）：11~21.
纪玉杰 . 2003. 北京西山侏罗纪煤系变形特征与地质灾害的关系 . 北京地质，015（02）：1~15.
纪玉杰 . 2004. 北京西山石炭-二叠纪煤系变形变质特征与地质灾害 . 北京地质，16（2）：1~17.
纪玉杰 . 1996. 北京城郊的地面沉降成因浅析 . 北京地质，3：15~19.
嘉世旭，刘昌铨 . 1995. 华北地区人工地震测深震相与地壳结构研究 . 地震地质，17（2）：97~105.
嘉世旭，齐诚，王夫运，等 . 2005. 首都圈地壳网格化三维结构 . 地球物理学报，48（6）：1316~1324.
贾炳文，等 . 1964. 铁鳞石-蛇纹石族矿物新变种 . 地质学报 . 44（1）.
贾建称，王根厚，魏铭等 . 2006. 北京周口店太平山滑坡形成机理与预防措施 . 青海大学学报，24（1）：4~9.
贾兰坡，1954，中国猿人 . 龙门联合书局 .
贾兰坡，1960，关于中国猿人的骨器问题 . 考古学报 . 1（3）. 1~3.
贾兰坡，卫奇 . 1976. 阳高许家窑旧石器石代文化遗址 . 考古学报，2：97~114.
贾兰坡，卫奇 . 1982. 建议用古人类学和考古学的成果建立我国第四系的标准剖面 . 地质学院，3：71~79.
贾兰坡 . 1951. 山顶洞人. 上海：龙门联合书局 .
贾兰坡 . 1964. 中国猿人及其文化. 北京：中华书局 .
贾三满，郭萌 . 2007. 从高丽营探槽分析黄庄~高丽营断裂与地裂缝的关系 . 城市地质，2（7）：24~28.
贾三满，刘明坤，田芳，等 . 2011. 北京地区地裂缝分类及防治措施 . 城市地质，6（2）：4~24.
贾三满，王海刚，赵守生，等 . 2007. 北京地面沉降机理研究初探 . 城市地质，2（1）：20~26.
贾振远，等 . 1986. 华北地台东部奥陶系地层南北向沉积环境特征 . 沉积学报 . 1986（4）：44~54.
江娃利 . 1992. 北京平谷地区地表陡坎的成因识别 . 地震地质，21（4）：309~315.
蒋忠诚，谢运球，章程，等，1996. 北京西山岩溶. 广西：广西师范大学出版社 .
焦青，邱泽华 . 2006. 北京平原地区主要活动断裂带研究进展 . 地壳构造与地壳应力文集，18：72~84.
焦永玲 . 2014. 北京大石坡-黑山坨复式岩体岩石地球化学特征、成因及锆石 U-Pb 年代学研究 . 中国地质大学（北京）.
焦永玲，等 . 2014. 北京大石坡-黑山坨复式岩体岩石地球化学特征及成因 . 矿床地质 . 33（增刊）：209~210.
金安蜀，刘福田，孙永智，1980. 北京地区地壳和上地幔的三维波速度结构地球物理学报 . 23（2）：172~182.
金江军，潘懋 . 2007. 近 10 年来城市地质学研究和城市地质工作进展评述 . 地质通报，26（3）：366~371.
金文山，管爱莲，1999b. 北京地区太古宙岩石同位素年龄及其意义 . 前寒武纪研究进展，22（3）：1~13.
金文山，管爱莲 . 1999a. 北京地区太古宙岩石、地层单元划分及其特征 . 前寒武纪研究进展，22（2）：1~10.
金文山，李双保，管爱莲 . 2000. 北京地区早前寒武纪结晶基底 . 北京：地质出版社 .
金文山 . 1993. 变质岩分类和命名 . 国外前寒武纪地质，（3）：18~33.
金文山，等 . 1999. 北京地区太古宙岩石、地层单元划分及其特征 . 前寒武纪研究进展 . 22（2）：1~10.
金文山，等 . 1999. 北京地区太古宙岩石同位素年林及其意义 . 前寒武纪研究进展 . 22（3）：1~13.
金之钧，李京昌，刘国臣 . 1997. 米兰科维奇旋回识别问题 . 地学前缘，4（3）：4~22.
邝生爱，李晋明 . 1987. 应用遥感多时相处理及多种信息源综合研究北京平原区构造 . 地质学报，3：267~274.
赖星蓉 . 1984. 北京大兴井下奥陶纪牙形石及其意义 . 石油勘探与开发，（5）：26~29.
雷坤超，陈蓓蓓，贾三满，等 . 2014. 基于 PS-InSAR 技术的北京地面沉降特征及成因初探 . 光谱学与光谱分析，34（8）：2186-2189.
黎彤，等 . 1962. 大洋岩石圈和大陆岩石圈的元素丰度 . 地球化学 . 40（1）：1~5.
黎彤，等 . 1962. 地球和地壳的化学元素丰度 . 地质出版社 .

李春昱 . 2004. 板块构造论文选集 . 北京：地质出版社 .
李鼎荣，彭一民，刘清泗，等 . 1979. 北京平原区上新统-更新统的划分 . 地质科学，14（4）：342~351.
李鼎容，谢振钊，王安德，等 . 1982. 北京平原区第四系划分及其下限问题 . 石油与天然气地质，3（4）：379~387.
李鼎容 . 1982. 北京平原第四系划分及其下限问题 . 石油与天然气地质，3（4）：379~387.
李东旭，邓一岗，等 . 1994. 北京西山地质构造系统分析 . 北京，地质出版社 .
李芳凝，等 . 2006. 北京燕山地区中生代棋盘岩岩体特征及成因. 中国地质大学（北京）.
李海龙，张宏仁，渠洪杰，等 . 2014. 燕山运动“绪动/A 幕”的本意及其锆石 U-Pb 年代学制约 . 地质论评，60（5）：1026~1042.
李红丽 . 1997. 北京西山显生宙区域变质岩及接触变质岩造岩矿物矿物化学特征（硕士论文）. 北京：中国地质大学 .
李华章 . 1989. 北京地区新构造运动特征与地震的关系 . 北京师范大学学报（自然科学版），4：78~84.
李华章 . 1991. 北京岩相古地理. 北京：地质出版社 .
李华章 . 1995. 北京地区第四纪古地理研究北京：地质出版社 .
李怀坤，李惠民，陆松年 . 1995. 长城系团山子组火山岩颗粒锆石 U-Pb 年龄及其地质意义 . 地球化学，(1)：43~48.
李怀坤，朱士兴，相振群，等 . 2010. 北京延庆高于庄组凝灰岩的锆石 U-Pb 定年研究及其对华北北部中元古界划分新方案的进一步约束 . 岩石学报，26（7）：2131~2140.
李江海，A Kroner，钱祥麟，等 . 2000. 华北克拉通恒山-承德新太古高压麻粒岩带构造演化 . 安徽地质，10（03）：218~220.
李江海，钱祥麟 . 1998. 华北北部早寒武纪麻粒岩相带研究新认识 . 地学前缘，5（4）：191~192.
李江海，等 . 2000. 华北克拉通中部中元古代岩墙群的古地磁学研究及其地质意义 . 中国科学（D 辑）. 30（6）：602~608.
李俊建，罗镇宽，燕长海，等 . 2010. 华北陆块的构造格局及其演化 . 地质找矿论丛，2010，25（2）：89~100.
李烈荣，王秉忱，郑桂森 . 2012. 我国城市地质工作主要进展与未来发展 . 城市地质，7（3）：3~13.
李龙吟，陈华慧 . 1994. 北京怀柔 HR88~1 钻孔剖面磁性地层学研究 . 地层学杂志，18（1）：39~44.
李龙吟，陈华慧 . 1992. 北京怀柔地区第四纪下限的综合标志 . 地球科学：中国地质大学学报，7（6）：631~637.
李容全，孙秀萍 . 1985. 北京房山县山前平原古水系及其特征初步研究 . 北京师范大学学报，4：85~92.
李容全，徐振溥，等 . 1983. 北京地区泥石流分布特点 . 山地研究，3（1）：42~48.
李容全，郑良美，刘敬中 . 1992. 北京平原山地交界带新构造活动 . 北京师范大学学报（自然科学版），28（4），558~562.
李声之，许洪才 . 1995. 对河北省太古宙地层及主要构造事件的划分 . 北京地质，第四期：8~26.
李思田，1988，断陷盆地分析与煤聚集规律：中国东北部晚中生代断陷盆地沉积、构造演化和能源预测研究的方法与成果，北京：地质出版社，367.
李思田，王华，路凤香等主编 . 1999. 盆地动力学：基本思路与若干研究方法 . 武汉：中国地质大学出版社，200.
李思田，等 . 1998. Late Mesozoic rifting in northeast China and northeast Asia fault basin system . 中国科学：化学 .（2）：246~256.
李四光 . 1937. 冰期之庐山 . 中央研究院地质研究所专刊，乙种 ，2 卷 .
李四光 . 1975. 中国第四纪冰川 . 北京：科学出版社 .
李松林，张先康，宋占隆，等 . 2001. 多条人工地震测深剖面资料联合反演首都圈三维地壳结构 . 地球物理学报，44（3），360~368.
李腾飞 . 2014. 北京昌平东闸村钻孔中更新世气候转型期的孢粉组合特征及其古气候意义. 首都师范大学 .
李廷栋，等 . 2013. 新一代《中国区域地质志》编制与研究的新进展 . 中国地质学会 2013 年学术年会论文摘要汇编 .
李廷栋，袁学诚，肖庆辉，等 . 2013. 中国岩石圈三维结构 . 北京：地质出版社 .
李伍平，章大港，路凤香，等 . 2001. 北京西山髫髻山组火山岩的地球化学特征与岩浆起源 . 岩石矿物学杂志，20（2）：123~133.
李伍平，等 . 2000. 北京西山东岭台组（J_3d）火山岩的成因及其构造环境探讨 . 岩石学报，16（3）：345~352.
李伍平，等 . 2001. 北京西山髫髻山组火山岩的地球化学特征与岩浆起源 . 岩石矿物学杂志 . 20（2）：123~133.
李伍平，等 . 2002. 燕山造山带中生代火山岩地球化学特征及其地球动力学背景 . 博士后研究工作报告 . 中国科学院广州地球化学研究所 .
李现根，等 . 2009. 延庆—怀来地区中元古代沉积地层及微亮晶碳酸盐岩研究. 中国地质大学（北京）.

李小龙，方同明，等 . 2013. 北京四道沟地区变质型铁矿地质特征及找矿远景分析 . 中国矿业，22（3）：66～69.
李晓勇，等 . 2003. 燕山地区中生代火山作用成因及其对深部过程的制约. 中国科学院研究生院（广州地球化学研究所）.
李晓勇，等 . 2004. 北京西山东岭太组粗面质火山岩地球化学：下地壳熔融成因 . 大地构造与成矿学. 28（2）：155～164.
李晓勇，等 . 2004. 古亚洲洋对华北陆缘岩石圈的改造作用：来自于西山南大岭组中基性火山岩的地球化学证据 . 岩石学报 . 20（3）：557～566.
李延河，侯可军，万德芳，等 . 2012. Algoma 型和 Superior 型硅铁建造地球化学对比研究 . 岩石学报，028（11）：3513～3519.
李永化，张小咏，崔之久，等 . 2002. 第四纪泥石流活动期与气候期的阶段性耦合过程 . 第四纪研究，22（4）：340～348.
李友枝，庄育勋，等 . 2003. 城市地质-国家地质工作的新领域 . 地质通报，22（8）：589～596.
李增学，魏久传，韩美莲 . 2000. 华北晚古生代陆表海盆地东南缘高频海侵事件对聚煤作用的影响 . 地学前缘，7（3）：202.
李增学，余继峰，郭建斌，等 . 2003. 陆表海盆地海侵事件成煤作用机制分析，沉积学报，21（2）：288～287.
李长安，何科昭，温长顺 . 1992. 北京怀柔平原区隐伏断层及构造活动性初探 . 地球科学，17（5）：546～554.
李长安，张玉芬 . 1990. 北京平原北部第四纪沉积环境的马尔柯夫链综合分析 . 地质科技情报，4：86～91.
李长安 . 1992. 北京怀柔地区第四纪地层划分与对比 . 中国区域地质，1，118～125.
李长安 . 1993. 北京平原区第四纪古气候变化 . 中国区域地质，4：336～343.
李长安 . 1994. 北京平原东北部晚新生代地层划分及地层层序初建 . 地层学杂志，18（2）：138～145.
李忠，等 . 2003. 中国盆地动力学研究的现状与思考. 中国矿物岩石地球化学学会学术年会 .
李忠，等 . 2007. 华北克拉通东部南北缘中生代火山-沉积格局及其构造转折过程 . 古地理学报，9（3）：227～242.
李祖湘 . 1979. 铜合金蒸汽保护无氧化加热退火 . 金属热处理 .（8）：43.
郦道元，2001. 古典名著聚珍文库：水经注. 陈桥驿，译 . 杭州：浙江古籍出版社 .
梁玉左，曹瑞骥，张录易，等 . 1983. 中国晚前寒武纪 Pseudogymnosolenidae. 中国地质科学院文集 .
林景星 . 1977. 华北平原第四纪海进海退现象的初步认识 . 地质学报，2：109～116.
林景星 . 1981. 北京顺义早更新世有孔虫动物群的古生态及古环境 . 地质科学，1：55～60.
林强，等 . 1992. 华北地台东部太古宙花岗岩 . 北京：科学出版社 .
刘宝峰，张先康，张成科，等 . 2000. 华北文安一蔚县一察哈尔右翼中旗剖面 s 波资料的地质解释 . 地震地质，22（1）：81～88.
刘保金，胡平，孟勇奇，等 . 2009. 北京地区地壳精细结构的深地震反射剖面探测研究 地球物理学报，52（9）：2264～2272.
刘保金，孙振国，赵成斌，等 . 1999. 延庆—怀来地区高分辨率浅地震反射剖面 . 地震地质，21（4）：425～430.
刘本培，蒋子堃，邓胜徽，等 . 2013. 硅化木向光性与华北板块旋转探讨 . 科学通报，（s1）：230～230.
刘本培，于菁珊，杨守仁 . 1982. 北京西山门头沟组双壳类动物群的发现兼论中国北部早期中生代成煤期时代 . 地质学报，（4）：74～82+99.
刘兵，巴金，张璐，等 . 2008. 北京周口店官地杂岩变质-深熔作用的锆石 . 地质科技情报，27（6）：1491～1502.
刘兵，等 . 2008. 北京周口店官地杂岩变质-深熔作用的锆石 LA-ICP-MS U-Pb 定年 . 地质科技情报，27（6）：37～42.
刘昌铨，杨健，1982. 京津及其外围地区地壳速度结构初步研究 . 地震学报，4（3）：217～226.
刘翠，邓晋福，苏尚国，等 . 2006. 华北地区燕山期岩浆活动热供给的初步数值模拟 . 地学前缘，13（2）：158～164.
刘翠，邓晋福，苏尚国，等 . 2007. 北京门头沟地区棋盘岩铁辉长岩的发现及其地质意义 . 现代地质，21（2）：327～331.
刘翠，等 . 2004. 北京云蒙山片麻状花岗岩锆石 SHRIMP 定年及其地质意义 . 岩石矿物学杂志，23（2）：141～146.
刘东生 . 1998. 加强跨学科综合研究，解决现实环境问题//孙枢 . 中国科学院、国家环保总局地质环境系统研究中心年报：地质环境系统研究 . 北京：海洋出版社：10-12.
刘东生，丁梦林 . 1984. 中国早期人类化石层位与黄土-深海沉积气候旋回的对比 . 人类学报，3：93～101.
刘东生，谭明，秦小光，等 . 1997. 洞穴碳酸钙微层理在中国的首次发现及其对全球变化研究的意义 . 第四纪研究，（1）：41～51.

刘福田，曲克信，吴华，等 . 1986. 华北地区的地震层面成像 . 地球物理学报，28（5）：442~449.
刘国惠，伍家善 . 1987. 北京房山地区变质带 . 中国地质科学院报，16（2）：113~137.
刘汉龙，周云东，高玉峰 . 砂土地震液化后大变形特性试验研究 . 岩土工程学报，2002（02）：142-146.
刘和甫 . 1995. 前陆盆地类型及褶皱-冲断层样式 . 地学前缘，2（3-4）：59~68.
刘欢，朱士兴 . 2014. 北京延庆地区雾迷山组燧石相微体古生物研究 . 微体古生物学报，(1)：37~53.
刘健，赵越，柳小明 . 2006. 冀北承德盆地髫髻山组火山岩的时代 . 岩石学报，22（11）：2617~2630.
刘健，赵越，柳小明，等 . 2007. 燕山褶断带下板城盆地杏石口组沉积特征及其构造意义 . 岩石学报，23（3）：639~654.
刘俊来，纪沫，关会梅，等 . 2008. 地壳的拆离作用与华北克拉通破坏：晚中生代伸展构造约束 . 地学前缘，15（3）：72~81.
刘清晓，等 . 2007. 北京市平原区基岩立体地质调查成果报告 .
刘绍龙，等 . 1986. 华北地区大型三叠纪原始沉积盆地的存在 . 地质学报 . 1986（2）：16~26.
刘宪亭 . 1988. 北京西山杏石口组的鱼化石 . 古脊椎动物学报 . 26（4）：40~48+83.
刘小丽，等 . 2015. 花岗质岩浆的变压结晶作用及其动力学意义：以北京房山岩体为例 . 现代地质 . 29（3）：514~528.
刘予，叶超，贾三满 . 2007. 北京市平原地面沉降区含水岩组和可压缩层划分 . 分析研究，2（1）：10~15.
刘玉洲 . 2009. 崩塌灾害链式演化跟踪研究 . 重庆：重庆交通大学 .
刘允良，张德云 . 1982. 密云水库北部变质岩构造遥感图像特征的研究 . 长春地质学院学报，1（1）：41~46.
刘长礼 . 2007. 城市地质环境风险经济学评价 . 北京：中国地质科学院 .
刘自亮，等 . 2009. 燕山中东部高于庄组—雾迷山组层序岩相古地理及臼齿碳酸盐岩研究. 中国地质大学（北京）.
柳永清 . 1991. 北京十三陵前寒武系常州沟组沉积微环境和 T、DI 指数研究 . 中国地质科学院 562 综合大队集刊 .
卢冰，等 . 1995. 山东半岛的地体构造及金矿成矿的区域地质背景 . 地质论评 . 41（1）：7~14.
卢良兆，靳是琴，叶慧文，等 . 1984. 北京密云岩群变质作用的温度、压力研究 . 地球化学，1：50~60.
卢演俦，魏兰英，尹金辉，等 . 2003. 北京西山古山洪堆积-马兰砾石形成环境及年代 . 第四纪研究，23（6）：611~620.
卢演俦，张景昭，赵华 . 1991. 第四纪沉积物热释光测年中古剂量的测定方法 . 核技术，14（2）：109~133.
陆松年，于海峰，李怀坤，郭坤一，等 . 2005. 中国前寒武纪重大地质问题研究——中国西部前寒武纪重大地质事件群及其全球构造意义，北京：地质出版社 . 206.
陆松年，李怀坤，相振群 . 2010. 中国中元古代同位素地质年代学研究进展述评 . 中国地质，37（4）：1002~1013.
陆松年，李惠民 . 1991. 蓟县长城系大红峪组火山岩的单颗粒锆石 U-Pb 法准确定年 . 中国地质科学院院报，2：137-145.
陆松年，杨春亮，李怀坤，等 . 1996. 建议的中国太古宙划分方案 . 华北地质矿产杂志，(1)：37~42.
陆松年，等 . 2015. 中国变质岩大地构造图及说明书（1：2500000）. 北京：地质出版社 .
路凤香，郑建平，邵济安，等 . 2006. 华北东部中生代晚期—新生代软流圈上涌与岩石圈减薄 . 地学前缘，13（2）：86~92.
路凤香，等 . 1991. 金伯利岩中金刚石多源性标志 . 地质科技情报 10：63~142.
路凤香，等 . 1997. 华北克拉通、扬子克拉通与秦岭造山带古地幔组成及状态的对比 . 地球科学 . 第 3 期 .
栾英波，蔡向民，郭高轩，等 . 2011. 北京平原区新 5 孔岩心特征及地层初步划分 . 中国地质，38（2）：495~503.
栾英波，郭高轩，黄骁 . 2008. 北京山前冲洪积缓倾斜平原晚新生代沉积物磁性地层学研究 . 城市地质，3（2）：17~20.
栾英波，郭莉，郭高轩，等 . 2008. 北京地区新 5 孔第四系松散沉积物的微量元素地球化学及古气候、古环境研究 . 分析研究，3（3）：18~23.
罗翠，朱茂炎 . 2010. 华北地台前寒武纪“后生动物化石”评述 . 古生物学报，49（3）：288~307.
罗明辉，张世民，任俊杰，等 . 2008. 北京断陷的新生代沉积与构造演化 . 地壳构造与地壳应力文集 26：62~74.
罗勇，贾三满，赵波，等 . 2011. 北京南部地区地面沉降发育特征及成因分析 . 城市地质，6（3）：1~21.
吕金波，郑桂森，李良景，等，2016，北京平原区地质剖面的初步建立——太古宙结晶基底和元古宙地层 . 中国地质 . 43（3）. 879~889.
吕金波，2014，北京地区基础地质研究史 . 城市地质 . 9（3）8~13.
吕金波，车用太，王继明，等，2006，京北地区热水水文地球化学特征与地热系统的成因模式 . 地震地质 . 28（3）.

419~429.
吕金波，丁孝忠 . 2017. 北京区域地质调查论著沿革 . 城市地质，第 4 期：96~102.
吕金波，李良景，丁长顺，等 . 2008. 北京城南平原区钻孔中发现太古宙片麻岩 . 城市地质 . 3（2）502~509.
吕金波，刘振锋，李良景，等 . 2012. 北京市幅（J50C001002）1：25 万区域地质调查 . 中国科技成果，第 19 期：40~42.
吕金波，卢耀如，郑桂森，等 . 2010. 北京西山岩溶洞系的形成及其与新构造运动的关系 . 地质通报，29（4）：502~509.
吕金波，郑桂森，李安宁，等，2016，北京百年地质调查的传承与发展——《北京市区域地质志》修编 . 地质通报. 35（1）. 24~35.
吕金波，郑桂森，王继明，等 . 2005. 北京市房山区石门花岗岩体特征. 见：陈安泽，等，中国地质学会旅游地学与国家地质公园研究分会成立大会暨第 20 届旅游地学与地质公园学术年会论文集. 北京：林业出版社. 312~315.
吕金波，郑桂森 . 1993. 北京西山哺乳动物化石点的新发现 . 中国区域地质 .（4）. 381~382.
吕金波 . 2007. 北京房山世界地质公园岩溶景观特色 . 城市地质，2（3）：26~30.
吕金波 . 2011. 北京西山圣莲山地质公园的岩溶地貌演化 . 城市地质，6（3）：22~39.
吕金波 . 2015.《水经注》中有关北京地理的科学思想 . 城市地质，10（1）：8~14.
吕金波 . 1994. 北京西山新生代古地理环境的演变 . 河北地质学院学报，17（3）：228~235.
吕作勇，吴建平 . 2010. 华北地区地壳上地幔三维 P 波速度结构 . 地震学报，32（1）：1~11.
马昌前，等 . 1988. 北京周口店岩株侵位和成分分带的岩浆动力学机理 . 地质学报 . 62（4）：329~341.
马昌前，等 . 1992. 花岗质岩浆起源和多次岩浆混合的标志：包体——以北京周口店岩体为例 . 地质论评 . 38（2）：109~119.
马芳，等 . 2003. 北京房山侵入体：微量元素、稀土元素及 Sr 同位素组成及其成因的制约 . 地球学报，24（2）：105~106.
马丽芳主编 . 2001. 中国地质图集 . 北京：地质出版社，113~118.
马胜云，马越，马进 . 李四光和他的时代 . 2012. 北京 . 科学出版社 .
马杏垣，等 . 1951. 平谷县罗峪沟震旦纪的火山颈（节要）. 地质论评，1（1951）：54~55.
马寅生，等 . 2002. 华北北部中新生代构造体制的转换过程 . 地质力学学报 . 8（1）：15~25.
马寅生，等 . 2007. 燕山中段盘山花岗岩体锆石 SHRIMP U-Pb 年龄测定及其构造意义 . 岩石学报 . 23（03）：547~556.
马永生 . 2004. 碳酸盐岩储层沉积学. 北京：地质出版社 . 1999.
马宗晋，林品仁，洪汉净编著 . 2003. 地球构造与动力学 . 广州：科技出版社 .
麦延庆 . 1987. 北京密云太古宙高级区岩石组合及变形-变质作用 . 世界地质（构造地质专辑）1：59~68.
毛景文，张作衡，余立金，等 . 2003. 华北及邻区中生代大规模成矿的地球动力学背景：从金属矿床年龄精测得到启示 . 中国科学（D），33：289~299.
梅冥相 . 2007. 中元古代叠层石—非叠层石碳酸盐岩层序地层序列及其沉积特征——以北京延庆千沟剖面高于庄组为例 . 现代地质，21（2）：387~396.
梅冥相，等 . 2005. 雾迷山旋回层的模式及其垂向叠加类型：前寒武纪与米兰柯维奇旋回相关的旋回性沉积作用 .
梅冥相，等 . 2011. 北京西郊下苇甸剖面寒武系崮山组叠层石生物丘的沉积组构 . 岩石学报，27（8）：2473~2486.
梅仕龙，梅冥相 . 1996. 华北地台中，上寒武统露头层序地层学研究的新认识 . 地层学杂志，（2）：146~152.
孟祥化，葛铭，Tuck M E. 1996. 中国晚寒武世长山期最大海泛事件及其全球对比意义 . 地质学报，（2）：108~120.
孟祥化，葛铭，等 . 2006. 中朝板块新元古代微亮晶（臼齿构造）碳酸盐事件、层序地层和建系研究 . 地层学杂志，30（3）：211~222.
孟祥化，乔秀夫，葛铭 . 1986. 华北古浅海碳酸盐风暴沉积和丁家滩相序模式 . 沉积学报，（02）：4~21+133~136+145.
孟祥化，等 . 1979. 沉积建造及其共生矿床分析. 北京：地质出版社 .
孟祥化，等 . 1979. 燕辽地区蓟县系隐藻叠层石碳酸盐岩沉积环境及沉积模式 . 吉林大学学报（地）. 1982（4）：15~24.
孟祥化，等 . 1989. 中国早震旦世沉积建造及盆地格局 . 现代地质 .（1）：1~16.
孟祥化，等 . 1990.《沉积建造学》讲座 第六讲 稳定陆壳克拉通盆地的沉积建造（二）. 岩相古地理 . 第 5 期 .
孟祥化，等 . 1992. 火山沉积建造—最重要的含锰建造 . 现代地质 .（4）：418~425.
孟祥化，等 . 1993. 内源盆地沉积研究. 北京：石油工业出版社 .

孟祥化，等 . 2001. 中国华北地台二叠纪前陆盆地的发现及其证据 . 地质科技情报 . 20（1）：8~14.
孟祥化，等 . 2004. 中朝板块层序事件演化：天文周期的沉积响应和意义 . 2004（3）：102~102.
孟祥化等 1991.《沉积建造学》讲座 . 沉积与特提斯地质 . 1991（5）：42~51.
米家榕，张川波，孙春林，等 . 1984. 北京西山杏石口组发育特征及其时代 . 地质学报，(4)：4~14+88.
明镜，潘懋，屈红刚，等 . 2009. 北京市新生界三维地质结构模型构建 . 北京大学学报（自然科学版），45（1）：111~119.
莫宣学，赵志丹，邓晋福，等，2003，印度-亚洲大陆主碰撞过程的火山作用响应 . 地学前缘，10（3）：135~148.
倪志云，杨桂芳，黄俊华，等 . 2011. 北京平原区晚更新世以来有机碳同位素特征及其古环境意义 . 地球学报，32（2）：171~177.
牛宝贵，和政军，宋彪，任纪舜 . 2003. 张家口组火山岩 SHRIMP 定年及其重大意义 . 地质通报，22（2）：140~141.
农商部地质调查所 . 1920. 北京西山地质志地质专报，甲种第 1 号，1~92.
欧阳志侠，等 . 2010. 华北克拉通主要变质核杂岩晚中生代花岗岩时代成因类型对比及意义 . 中国地质科学院 .
欧阳自远，张福勤，林祝文，等 . 1995. 华北克拉通基底构造格架及形成模式，科学通报，040（8）：734~736.
潘桂棠，等 . 2015. 中国大地构造图及说明书（1：2500000）. 北京：地质出版社 .
潘桂棠、肖庆辉、陆松年，等 . 2008. 大地构造相的定义、划分、特征及鉴别标志，地质通报，27（10）：33~57.
潘懋，金江军 . 2006. 城市地质学发展前景展望 . 城市地质，1（1）：9~13.
潘颖，等 . 2010. 北京南口脉岩岩石学及锆石 U-Pb 年代学特征 . 中国地质大学（北京）.
潘云唐，单青生 . 1997. 北京平原区第四纪软体动物化石组合 . 大自然探索，16（62）：63~67.
裴文中，1934，周口店洞穴采掘记 . 地质专报乙种第 7 号 .
裴文中，1960，关于中国猿人骨器问题的说明和意见 . 考古学报 . 1（2）.
裴文中，1960，中国原始人类的生活环境 . 古脊椎动物与古人类 . 2（1）.
裴文中 . 1957. 中国第四纪哺乳动物群的地理分布 . 古脊椎动物与古人类，1（1）.
彭澎，刘富，翟明国，等 . 2011. 密云岩墙群的时代及其对长城系底界年龄的制约 . 科学通报，56（35）：2975~2980.
彭澎，翟明国，张华锋，等 . 2004. 华北克拉通 1. 8Ga 镁铁质岩墙群的地球化学特征及其地质意义：以晋冀蒙交界地区为例 . 岩石学报，020（03）：0439~0456.
彭阳，季强 . 1998. 北京西山及邻区寒武系顶部微晶丘特征及层序地层学意义 . 地质论评，44（1）：35~43.
彭阳，章雨旭 . 2000. 北京西山及邻区下古生界白云石化与层序地层 . 地球学报，21（1）：78~85.
彭一民 . 1987. 北京平原地区新生代古地理概貌 . 地球科学：武汉地质学报，12（1）：49~56.
齐诚，赵大鹏，陈额 . 2006. 等首都圈地区地壳波和波三维速度结构及其与大地震的关系地球物理学报，49（3）：805~815.
齐诚，赵大鹏，陈颙，等 . 2006. 首都圈地区地壳 P 波和 S 波三维速度结构及其与大地震/的关系 . 地球物理学报，49（3）：805~815.
齐伟，《隧道与地下工程概论 第 2 版》. 北京：地质出版社，2013：1~205.
齐永安，吴贤涛，张国成 . 1999. 北京昌平青白口系痕迹化石 . 古生物学报，38（4）：517~521.
前寒武纪研究进展 . 1998，21（1）：35~44.
吴昌华，梅华林 . 1998. 晋冀蒙高级地体石榴石基性麻粒岩地质粒子及其构造意义 .
钱迈平 . 2008. 华北古陆东南缘新元古代生物群. 地质出版社 .
钱青，等 . 2002. 八达岭基性岩和高 Ba-Sr 花岗岩的地球化学特征及成因探讨-华北和大别-苏鲁造山带中生代岩浆岩的对比 . 中国科学院地质与地球物理研究所 2002 学术论文摘要汇编 .
钱逸 . 1999. 中国小壳化石分类学与生物地层学. 北京：科学出版社 .
乔秀夫，宋天锐，高林志，等 . 1994. 碳酸盐岩振动液化地震序列 . 地质学报，(1)：16~34.
乔秀夫，王彦斌 . 2014. 华北克拉通中元古界底界年龄与盆地性质讨论 . 地质学报，88（9）：1623~1637.
乔秀夫 . 1976. 青白口群地层学研究 . 地质科学，11（3）：246~265.
乔秀夫，等 . 2002. 中朝板块元古宙板内地震带与盆地格局 . 地学前缘 . 9（3）：141~149.
邱家骧 . 1964. 与房山侵入体有关的几个地质学、岩石学问题的探讨 . 北京地质学院科学研究论文集 3（岩石专辑）.
邱家骧，等 . 1963. 北京密云超基性岩岩石学研究 . 第一届矿物、岩石、地球化学专业学术会议论文摘要汇编（岩石）.
邱家骧，等 . 1998. 北京地区中元古代与中生代火山岩的酸度、系列、构造环境及岩浆成因 . 岩石矿物学杂志，17（2）：104~117.

邱瑞照，邓晋福，周肃．2004. 华北地区三类岩石圈的壳幔岩石学结构与化学结构及其地球动力学意义．岩石矿物学杂志，23（2）：127~140.
全露霞，等．2012. 北京西山地区髫髻山组火山岩典型岩相结构及其成因分析．首都师范大学．
冉勇康，陈立春，徐锡伟．2001. 北京西北活动构造定量资料与未来强震地点的讨论．地震学报，23（5）：502~513.
冉勇康，邓东启，等．1997. 1679 年三河一平谷 8 级地震发震断层的古地震及其重复间隔．地震地质，19（3）：193~200.
任东，等．1995. 北京与邻区侏罗—白垩纪动物群及其地层．北京：地震出版社．
任国玉．1994. 国外末次冰期极盛期以来环境演变研究的主要特点．地球科学进展，9（2）：65~69.
任凯珍．2013. 北京地区地质灾害调查与区划综合研究．北京：中国地质大学．
任康绪，阎国翰，蔡剑辉，等．2006. 华北克拉通北部古-中元古代富碱侵入岩的 Nd、Sr、Pb 同位素地球化学：岩石圈地幔富集的证据．岩石学报，022（12）：2933~2944.
任康绪，等．2006. 华北克拉通北部地区古-中元古代富碱侵入岩年代学及意义．岩石学报，22（2）：377~386.
任留东，耿元生，杜利林，等．2012. 试论阜平杂岩的深熔作用．地质学报，86（9）：1491~1502.
任美锷，刘泽纯，金瑾乐，等．1981. 北京周口店洞穴发育及其与古人类生活的关系．中国科学，3：330~336.
任清芳，张成科，张先康，等．1997. 八达岭及周缘地区壳幔结构与潜在震源探讨．西北地震学报，19（3）：56~62.
任荣，韩宝福，张志诚，等．2011. 北京昌平地区基底片麻岩和中-新元古代盖层锆石 U~Pb 年龄和 Hf 同位素研究及其地质意义．岩石学报，027（06）：1721~45.
邵济安，张履桥，魏春景，等．2001. 北京南口中生代双峰式岩墙群的组成及其特征．地质学报，75（2）：205~212.
邵济安．1993. 伸展构造与造山过程．见肖庆辉主编．当代地质科学前沿．武汉：中国地质大学出版社．54~160.
邵济安，等．2001. 北京南口中生代双峰式岩墙群的组成及其特征．地质学报．75（2）：205~212.
邵济安，等．2001. 华北克拉通元古代的三次伸展事件．岩石学报，18（2）：152~160.
邵济安，等．2001. 南口-古崖居中生代双峰式岩墙群形成机制的地球化学制约．地球化学，30（6）：517~524.
邵颖．苏州地铁 4 号线盾构施工引起地面沉降变形研究．中国矿业大学，2016.
慎乃齐，张杰坤，连建发，等．2000. 京郊门城镇地面塌陷危险性研究．14（4）：479~483.
石敏，冯庆来，朱士兴．华北中元古代燕山盆地生物群演化及其与地质事件的耦合关系．中国科学：地球科学，2014（6）：1124~1141.
石晓冬，北京城市地下空间开发利用的历程与未来．地下空间与工程学报，2006，2（7）：1088-1092.
史洪起．2000. 论北京地区新近沉积土．路基工程，1（4）：28~31.
史晓颖，陈建强，梅仕龙．1999，中朝地台奥陶系层序地层序列及其对比．地球科学-中国地质大学学报，24（6）：573~580.
史晓颖，陈建强．华北地台东部寒武系层序地层年代格架．地学前缘，1997（z2）：161~173.
史晓颖．2002.《中国层序地层研究》评介．中国图书评论，（10）：63~63.
宋彪．1992. 密云环斑花岗岩的同位素地质年代学、稀土地球化学特征及成因讨论．中国地质科学院地质研究所所刊，25：137~157.
宋彪，等．1992. 辽宁弓长岭含铁建造年代学研究．岩石矿物学杂志．11（4）：1~7.
宋鸿林，葛梦春．1984. 从构造特征论北京西山的印支运动．地质论评，30（1）：77~80.
宋鸿林．1999. 燕山式板内造山带基本特征与动力学探讨．地学前缘，6（4）：309~316.
宋鸿林，等．1996. 北京房山变质核杂岩的基本特征及其成因探讨．现代地质．10（2）：149~158.
宋松岩，周雪松，张先康，等．1998. 北京-丰镇 DSS 剖面 S 波资料的解释和推断，地震学报，20（1）：60~67.
宋天锐，高健．1985. 北京十三陵地区的上前寒武系沉积岩中发现 16 亿年的藻类丝状体．科学通报，30（10）：769~769.
宋天锐，高健．1985. 最古老的后生动物痕迹化石？——对北京十三陵前寒武系常州沟组充填管状构造的探讨．沉积学报，3（2）：85~96.
宋天锐，和政军，万渝生，等．2003. 前寒武纪沉积岩中自生独居石的发现及其意义．沉积学报，21（1）：118~124.
宋天锐，和政军．2000. 北京十三陵元古宙大红峪组含事件信息的沉积岩研究．岩石矿物学杂志，19（4）：323~332.
宋天锐，和政军．2000. 北京十三陵中元古代串岭沟期地质事的探索．地质论评，46（4）：400~406.
宋天锐，刘燕学．2009. 古地震记录与岩相古地理．沉积学报，27（5）：872~879.
宋天锐．1988. 北京十三陵前寒武纪碳酸盐岩地层中的一套可能的地震-海啸序列．科学通报，33（8）：609~609.

宋天锐．2007. 北京十三陵地区中元古界长城系沉积相标志及沉积环境模式．古地理学报，9（5）：461~472.
宋武．1984. 北京西山首次在雾迷山组地层中发现微古化石．地质通报，(3)：98.
苏德辰，李庆谋，罗光文，等．1995. Fischer 图解及其在旋回层序研究中的应用——以北京西山张夏组为例．现代地质，(3)：279~283.
苏德辰，孙爱萍，郑桂森，等．2013. 北京西山寒武系滑塌构造的初步研究．地质学报，87（8）：1067~1075.
苏德辰，孙爱萍．2011. 北京永定河谷中元古界雾迷山组软沉积物变形与古地震发生频率．古地理学报，13（6）：591~614.
苏尚国，邓晋福，赵国春，等．2006. 北京燕山地区薛家石梁杂岩体特征、成因、源区性质及岩石圈减薄方式．地学前缘，13（2）：148~157.
苏尚国，等．2009. 太行山-燕辽地区燕山期造山过程、岩浆源区与成矿作用．国防科学技术大学．
苏文博，李怀坤，HUFF W D，等．2010. 铁岭组钾质斑脱岩锆石 SHRIMP U-Pb 年代学研究及其地质意义．科学通报，(22)：2197~2206.
孙大中．1990. 前寒武纪地质年代学问题的探讨．地质通报，(4)：289~300.
孙殿卿，徐煜坚．1944. 广西纪冰川遗迹的初步观察．地质论评，9（3~4）：34~62.
孙殿卿，徐煜坚．1950. 中国南部第四纪冰川分布概况及冰碛物分类．地质论评．15（1~3）.
孙殿卿，杨怀仁．1961. 大冰期时期中国的冰川遗迹．地质学报，41（3~4）：3~14.
孙殿卿．1957. 中国第四纪冰川遗迹纪要．北京：科学出版社．
孙会一，等．2016. 北京密云地区早白垩世花岗岩 SHRIMP 锆石 U-Pb 年龄及其地质意义．地球科学与环境学报．38（1）：43~54.
孙善平，等．1961. 宣化斑脱岩成分及成因问题．北京地质学院学报．1961（00）：74~91. 1964.
孙枢，1998. 中国科学院、国家环保总局地质环境系统研究中心年报：地质环境系统研究．北京：海洋出版社．
孙淑芬，朱士兴，黄学光．2006. 天津蓟县中元古界高于庄组宏观化石的发现及其地质意义．古生物学报，45（2）：207~220.
孙淑芬，朱士兴，黄学光，等．2004. 燕山长城系串岭沟组 Parachuaria 化石的发现及其意义．地质学报，78（6）：721~725.
孙淑芬，朱士兴．2000. 华北地区古元古代真核生物化石．地质学报，74（2）：116~122.
孙淑芬．1989. 河北宽城长城系下统微古植物群．地质科学，(3)：235~243.
孙淑芬．2006. 中国蓟县中、新元古界微古植物. 北京：地质出版社．
孙小华，郝春燕．2007. 北京城市地质工作服务的新方向．城市地质：2（2）：9~13.
孙秀萍，赵希涛．1982. 北京平原永定河古河道．科学通报，27（16）：46~49.
孙颖，刘久荣，韩征，等．2009. 北京市地热资源开发利用状况．安徽农业科学，37（16）：7564~7566.
孙云铸．1957，寒武纪下界问题．地质学报，37（3）：231-239.
孙志明，汪洋，邓晋福，等．2007. 北京地区中生代侵入岩的时间序列及其地质意义．矿物岩石地球化学通报，26（3）：258~262.
覃峰，等．2006. 北京房山岩体形成过程中岩浆混合作用证据．岩石学报，22（12）：957~970.
谭明．2004. 中国高分辨率气候记录与全球变化．第四纪研究，4（4）：455~462.
谭毅培，王伟君，齐诚等，2009. 北京地区地壳三维结构的综合建模．震灾防御技术，4（1）：24~39.
谭成轩，等．2014. 北京地区主要活动断裂研究与地壳稳定性评价．北京：地质出版社．
唐烽，高林志．1999. 北京及邻区青白口纪化石生物相研究．地质论评，45（1）：50~58.
唐烽．北京地区青白口纪长龙山期宏观藻类的新发现. 地层古生物论文集，1995.
陶继东，马昌前，张金阳，等．2009. 北京房山花岗闪长岩体中包体的演化及闪长质微粒包体的成因．地质科技情报，28（2）：33~41.
陶继东，等．2006. 房山侵入体高 Sr 低 Y 地球化学特征及其成因．地质科技情报，25（2）：11~17.
陶继东，等．2007. 房山岩体的岩浆来源、形成机制及动力学意义．信阳师范学院学报：自然科学版，20（2）：252~256.
陶继东，等．2009. 北京房山花岗闪长岩体中包体的演化及闪长质微粒包体的成因．地质科技情报，28（3）：33~40.
田芳，郭萌，罗勇，等．2012. 北京地面沉降区土体变形特征．中国地质，39（1）：236~242.
田树刚，章雨旭．1997. 华北地台北部奥陶纪露头层序地层．地球学报：中国地质科学院院报，(1)：87~97.

同号文，刘金毅，张双权 . 2004. 周口店田园洞大中型哺乳动物记述 . 人类学学报，23（3）：213～223.
同号文，张双权，李青，等 . 2008. 北京房山十渡西太平洞晚更新世哺乳动物化石 . 古脊椎动物学报，46（1）：51～70.
万天丰 . 2011. 中国大地构造学，北京：地质出版社，1～232.
万天丰，等 . 1999. 中国中、新生代板内变形速度研究 . 现代地质 . 13（1）：83～92.
万永平，王根厚，郭春影，等 . 2007. 京西房山岩体南侧青白口系下马岭组构造片岩中红柱石变形特征及其意义 . 地质通报，26（2）：200～205.
万永平，王根厚，郭春影，等 . 2007. 京西房山岩体南侧青白口系下马岭组构造片岩中红柱石的变形特征及其地质意义 . 地质通报，26（2）：200～205.
万渝生，董春艳，等 . 2012. 华北克拉通早前寒武纪条带状铁建造新城时代-SHRIMP 锆石 U-Pb 定年 . 地质学报，86（9）：1448～1478.
万渝生，张巧大，宋天锐 . 2003. 北京十三陵长城系长沟组碎屑锆石 SHRIMP 年龄：华北克拉通盖层物源区及最大沉积年龄的限定 . 科学通报，48（18）：1970～75.
汪洋，姬广义，夏希凡 . 2005. 燕山造山带复杂构造变形型式的新证据及其地质意义 . 北京地质，17（2）：1～13.
汪良谋，徐杰 . 1990. 北京拗陷构造活动性分析 . 中国地震，6（2）：25～36.
汪洋，姬广义，夏希凡 . 2005. 北京西山大灰厂东狼沟组钾质火山岩形成的构造背景探讨 . 北京地质，17（3）：1～10.
汪洋，姬广义 . 2003. 北京四海地区髫髻山组-后城组火山岩岩石化学特征 . 北京地质，15（1）：1～11.
汪洋，姬广义 . 2003. 北京西山东岭台组和髫髻山组火山岩同位素年龄探讨 . 北京地质，15（4）：6～12.
汪洋，姬广义 . 2004. 北京延庆北部地区后城组火山岩岩石化学特征与成因探讨 . 北京地质，16（1）：16～25.
汪洋，李家振，孙善平，等 . 2001. 北京西山髫髻山组火山岩 Sm-Nd 等时线年龄初步研究 . 北京地质，13（3）：18～20.
汪洋，孙金生 . 1994. 北京西山早侏罗世南大岭组玄武岩微量元素特征及成因 . 北京地质，4：16～24.
汪洋 . 2002. 中国东部燕山期强过铝质花岗岩的地球动力学意义 . 北京地质，14（3）：1～10.
汪洋 . 2005. 北京西山南大岭组玄武岩地球化学演化及其地质意义. 火山作用与地球层圈演化—全国第四次火山学术研讨会论文摘要集.
汪洋，等 . 2002. 冀晋辽地区燕山期强过铝质侵入岩岩石化学特征及其动力学意义 . 地质通报 . 21（4）：270～276.
汪洋，等 . 2003. 广东吴川—四会断裂带燕山期强过铝质花岗岩岩石化学特征的地球动力学意义 . 大地构造与成矿学. 27（1）：56～63.
汪洋，等 . 2005. 华北燕山期区域变质与内生成矿作用关系雏议 . 北京地质 .（3）：29～33.
汪洋，等 . 2007. 三论南岭燕山早期花岗岩的成因类型. 全国岩石学与地球动力学暨化学地球动力学研讨会 .
汪洋，等 . 2009. 北京白查 A 型花岗岩的地球化学特征及其成因与构造指示意义 . 岩石学报 . 25（1）：13～24.
汪洋，等 . 2014. 钙碱性火成岩的角闪石全铝压力计——回顾、评价和应用实例 . 大地构造与成矿学 . 39（1）：139～152.
汪洋，等 . 2014. 造山崩塌过程的岩浆作用响应：以北京薛家石梁-黑山寨岩浆杂岩体为例 . 大地构造与成矿学 . 38（3）：656～669.
汪洋，等 . 2015. 北京大石坡-黑山坨复式岩体地球化学特征、岩石成因及其地质意义地质论评 . 60（4）：839～850.
王强 . 2001. 北京地区中生代构造演化特征与控煤作用的探讨 . 北京地质，13（2）：28～31.
王恩林，杨占兴，等 . 1987. 北京密怀隆起张家坟群的解体与三棱山韧性剪切带的确定 . 世界地质（构造地质专辑）1：69～77.
王恩营 . 2006. 北京周口店复杂地质作用关联分析 . 河南理工大学学报 . 25（4）：292～294.
王方正，肖龙，肖文松，等 . 1990. 北京周口店关坻杂岩太古宙成因的岩石学及地球化学证据 . 地球科学-中国地质大学学报，22（3）：529～538.
王芳，等 . 2009. 华北陆块北缘崇礼-赤城地区晚古生代花岗岩类的锆石年龄和 Sr-Nd-Hf 同位素组成 . 岩石学报，25（11）：3057～3074.
王飞燕 . 1983. 周口店洞穴及岩溶发育 . 中国岩溶，1：41～48.
王夫运，张先康，陈棋福，等 . 2005. 北京地区上地壳三维细结构层析成像 . 地球物理学报，48（2）：359～366.
王夫运，张先康，陈颙，等 . 2004. 张渤构造带中东段二维 P 波速度结构-安新-香河-宽城剖面 . 地震学报，26（增刊）：3l～41.
王海芝 . 2008. 北京山区泥石流分布区的地形地貌特征 . 城市地质，3（4）：19～21.

王恒升 1928：北京西山妙峰山髻髻山一带之火成岩，《农矿部地质调查所地质汇报》. 第 11 号 .
王红强 . 2007. 泥河湾盆地东谷坨剖面磁性特征及环境意义 . 第四纪研究，27（6）：1081～1091.
王鸿桢，杨森楠，刘本培，等 . 1990. 中国及邻区构造古地理和生物古地理. 武汉：中国地质大学出版社：35～88.
王惠初，初航，相振群，等 . 2012. 华北克拉通北缘崇礼～赤城地区的红旗营子（岩）群：一套晚古生代的变质杂岩 . 地学前缘，19（5）：100～113.
王惠初，陆松年，赵凤清 . 2005. 华北克拉通古元古代地质记录及其构造意义 . 地质调查与研究 . 28（3）130～143.
王惠初，于海峰，苗培森，等 . 2011. 前寒武纪地质学研究进展与前景 . 地质调查与研究，34（04）：241～252.
王惠初，等 . 2007. 冀北闪长质岩石的锆石 SHRIMP U-Pb 年龄：晚古生代岩浆弧的地质记录 . 岩石学报 . 23（3）. 597～604.
王季亮，等 . 1994. 河北省中酸性岩体地质特征及其与成矿的关系 . 北京：地质出版社 . 34～40.
王嘉荫 . 1951. 北京西山的硬绿泥石带 . 中国地质学会志，Z1 期：200～205.
王景明，王春梅，刘科 . 2001. 地裂缝及其灾害研究的新进展 . 地球科学进展 . 16（3）：303～313.
王可法，陈锦石 . 1993. 燕山地区铁岭组稳定同位素组成及其地质意义 . 地球化学，1：10～17.
王立峰，等 . 1998. 河北省怀来县晚三叠世冲积扇及其构造环境 . 华北地质矿产杂志，(3)：230～235.
王乃文，何希贤 . 1983. 华北地区海相第四系的生物地层系统 . 海洋地质与第四纪地质，3（2）：103～108.
王乃文，何希贤 . 1985. 北京地区早第四纪有孔虫研究 . 地层古生物论文集，47～73.
王巧云，颜丹平，苏本勋，等 . 2006. 京北云蒙山变质核杂岩南东侧剥离断层带的构造热演化 . 地质通报，25（4）：448～453.
王人镜，等 . 1990. 北京周口店岩体热动力构造及定位机制研究 . 中国区域地质 . 1990（第 1 期）：1～7.
王荣，罗勇，杨艳，等 . 2014. 北京温榆河冲积扇地面沉降特征及机理研究 . 上海国土资源，35（4）：44～48.
王帅军，张先康，张成科，等 . 2007. 武清-北京-赤城二维地壳结构和构造，地球物理学报，50（6）：1769～1777.
王思敬，马凤山 . 1998. 论人类工程活动与地质环境的相互作用//孙枢，中国科学院、国家环保总局地质环境系统研究中心年报：地质环境系统研究，北京：海洋出版社：31～38.
王伟，等 . 2014. 辽西-冀东北地区前寒武纪地壳演化与壳幔作用过程. 北京大学 .
王五力 . 2004. 土城子阶、义县阶标准地层剖面及其地层古生物、构造-火山作用. 北京：地质出版社 .
王彦洁，武法东，张建平 . 2013. 北京延庆国家地质公园旅游资源类型与保护开发建议 . 旅游资源，29（1）：110～112.
王焰，张旗 . 2001. 八达岭花岗杂岩的组成、地球化学特征及其意义 . 岩石学报，17（4）：533～540.
王一峋，许焕林 . 1987. 北京平原西侧永定河古河道 . 北京师范学院报（自然科学版），8（1）：67～72.
王颖，等 . 2003. 潮滩沉积环境与岩相对比研究 . 沉积学报 . 21（4）：539～526.
王玉富，等 . 1988. 天津蓟县盘山石英二长岩体的 Rb—Sr 同位素年龄及其地质意义 . 地质找矿论丛 . 3（3）：94～97.
王泽九，黄枝高，姚建新，等 . 2014. 中国地层表及说明书的特点与主要进展 . 地球学报，35（3）：271～276.
王志宏，张兴辽，屠森 . 2008. 河南省地层古生物研究，第一分册，前寒武纪. 郑州：黄河水利出版社 .
王竹泉 . 1926. 中国地质图说明书：太原～榆林幅. 北京：商务印刷馆 .
王自强，王璞 . 1979. 北京西山坨里—大灰厂地区晚期中生代地层古生物简报 . 地层学杂志，(1)：42～52+85.
卫万顺 . 2008. 城市地质工作应走理念发展型之路 . 城市地质，3（4）：1～6.
魏春生 . 2000. A 型花岗岩成因模式及其地球动力学意义 . 地学前缘，7（1）：238.
魏兰英，彭贵，严富华，等 . 1997. 地区末次冰消期气候环境变化记录的初步研究 . 第四纪研究，5（2）：183～190.
魏亮，郭华明，谢振华，等 . 2010. 北京平原沉积物稀土元素地球化学特征及物源意义 . 地学前缘，17（6）：72～80.
魏文博，叶高峰，金胜，等 . 2007. 华北地区地壳 P 波三维速度结构 . 地球科学—中国地质大学学报，32（4）：441～452.
吴伯磊，沈淑敏 . 1982. 北京西郊第四纪砾石层的结构～构造特征及其成因的探讨 . 中国地质科学院地质力学研究所所刊，12：13～24.
吴超凡，杨为民，吴文金 . 2009. 北京门城镇煤炭采空区地面塌陷危险性分析 . 矿业安全与环保，36（5）：81～83.
吴春明，陈泓旭 . 2013. 变质作用温度与压力极限值的估算方法 . 岩石学报，029（05）：1499～1510.
吴福元，等 . 2008. 华北岩石圈减薄与克拉通破坏研究的主要学术争论 . 岩石学报 24（6）：1145～1174.
吴汉珍，等 . 2001. 中国大陆及邻区新生代构造-地貌演化过程与机理 . 北京：地质出版社 . 100～140.
吴立新，姜云，梁越，等 . 城市地下空间开发利用容量评估的基础研究 . 地理与地理信息科学，2004，20（4）：44-47.

吴梦源，等 . 1985. 北京地区雾迷山组碳酸盐岩潮汐沉积模式 . 河北地质学院学报 . 11（4）：37~50.
吴汝康，任美锷，朱显谟，等 . 1985：北京猿人遗址综合研究. 北京：科学出版社 .
吴锡浩，蒋复初，徐和聆，等 . 1997. 北京潭柘寺地区第四纪冰川遗迹的宏观和微观特征 . 地质力学学报，12（4）：3~12.
吴锡浩 . 1983. 北京地区第四纪冰期与构造~气候旋回，海洋地质与第四纪地质，3（2）：111~121.
吴珍汉，等 . 1999. 燕山南缘盘山岩体的热历史与构造-地貌演化过程 . 地质力学学报，5（3）：28~32.
伍家善，等，1991. 华北陆台早前寒武纪重大地质事件 . 北京：地质出版社 .
相振群，等 . 2014. 华北克拉通中元古代岩浆事件群与成矿作用 . 中国地质大学（北京）.
向宏发 . 2003. 隐伏活动构造探测研究的若干问题讨论 . 地震地质，25（3）：460~466.
肖庆辉，邱瑞照，邓晋福，等 . 2005. 中国花岗岩与大陆地壳生长方式初步研究，中国地质，32（3）：343~352.
肖庆辉，等 . 2002. 花岗岩研究思维与方法 . 北京：地质出版社 .
肖应凯，肖军，赵志琦，等 . 2008. 延庆杨户庄第四纪有孔虫的非海相生存环境：硼和锶同位素的证据 . 科学通报，53（7）：801~806.
萧宗正，关绍曾，何希贤，等 . 1996. 北京延庆第四纪盆地杨户庄剖面新信息-有孔虫与海相虫新发现 . 地质评论，42（3）：251~256.
萧宗正，杨鸿连，单青生 . 北京中生代地层及生物群研究. 北京：地质出版社，1993.
萧宗正 . 1981. 北京市南口、十三陵一带震旦亚界叠层石初步研究 . 北京地质（内刊）（1）：12~32.
萧宗正 . 1981. 燕山地区震旦亚界最早叠层石组合及其地层意义 . 地质论评，27（5）：14~20+103~104.
萧宗正 . 1985. 北京市南口、十三陵中上元古界叠层石. 第一届全国化石藻类学术会议论文选集，北京：地质出版社 .
萧宗正 . 1990. 北京西山宏观藻类化石的发现 . 地质科学，（4）：403~407.
萧宗正 . 1995. 北京地区第四纪洞穴地层学评介 . 北京地质，3：1~16.
谢广轰 . 2005. 大庙谢长岩和密云环斑花岗岩的岩石学与地球化学 . 北京：科学出版社 .
谢广轰，等 . 1980. 大庙斜长岩杂岩体的岩石学特征 . 地球化学 . 第 3 期：263~277.
谢广轰，等 . 1988. 北京及附近元古宙裂谷中球斑花岗岩套的地球化学 . 地质学报 . 第 4 期 .
谢广轰，等 . 2005. 大庙斜长岩和密云环斑花岗岩的岩石学和地球化学——兼论全球岩体型斜长岩和环斑花岗岩类的时空分布及其意义 . 北京：科学出版社 .
谢家荣，1936，北平西山地质构造概况 . 中国地质学会会志 . 16.
邢裕盛，段承华，梁玉左，等 . 1985. 中国晚前寒武纪古生物. 地质出版社 .
邢裕盛，刘桂芝，尹崇玉，等 . 1990. 中国末前寒武纪生物群演变及其地质意义// 中国地质科学院文集 .
邢裕盛，刘桂芝 . 1973. 燕辽地区震旦纪微古植物群及其地质意义 . 地质学报，（1）：3~53+140.
邢裕盛，刘桂芝 . 1982. 中国晚前寒武纪微古植物群及其地层意义 . 地球学报，4：57~69.
徐德斌，等 . 2002. 河北省兴隆地区大红峪组沉积古环境研究 . 地层学杂志 . 26（1）：73~79.
徐继山 . 2012. 华北陆缘盆地地裂缝成因机理研究 . 西安：长安大学 .
徐嘉炜 . 1995. 论走滑断层作用的几个主要问题，2（1−2）：125~136.
徐克勤 . 2007. 华南花岗岩类的成因系列和物质来源 . 北京：科学出版社 .
徐克勤，等 . 1982. 华南两个成因系列花岗岩及其成矿特征 . 矿床地质 .（2）：4~17.
徐克勤，等 . 1982. 华南两类不同成因花岗岩岩石学特征 . 岩矿测试 . 1（2）：3~14.
许继峰，王人镜，杨淑荣 . 1994. 北京云蒙山花岗岩基的岩石变形及其成因 . 地球科学 —中国地质大学学报，19（6）：806~814.
许靖华，孙枢，王清晨，等 . 1998. 中国大地构造相图（1：400 万）. 北京：科学出版社 . 1~155.
续海金，等 . 2013. 幔源岩浆底侵引发的壳幔相互作用以华北克拉通房山岩体为例 . 19 卷增刊 .
严富华，叶永英，刘粤霞 . 1986. 北京东斋堂黄圭的孢粉分析 . 中国第四纪研究，7（1）：39~43.
严富华，叶永英，麦学舜，等 . 1981. 从孢粉分析试论北京地区两泥炭矿时代和形成环境 . 地震地质，3（1）：53~67.
阎国翰，许保良，牟保磊 . 2001. 中国北方中生代富碱侵入岩钕、锶、铅同位素特征及其意义 . 矿物岩石地球化学通报 . 20（4）：234~237.
阎国翰，等 . 1995. 房山岩体中闪长岩质包体的矿物稀土元素地球化学及其包体成因 . 中国科学（B 辑）. 25（2）：219~224.
阎国瑜，等 . 2007. 华北克拉通板内拉张性岩浆作用与三个超大陆裂解及深部地球动力学 . 13（2）：161~174.

颜丹平，周美夫，宋鸿林，等．2005. 北京西山官地杂岩的形成时代及构造意义．地学前缘，12（2）：332~337.
颜丹平，等．1997. 变质核杂岩体．地质科技情报．16（3）：13~18.
颜丹平，等．2005. 北京西山官地杂岩的形成时代及构造意义．地学前缘．12（2）：332~337.
燕滨，等．2008. 北京西山房山岩体西北部强变形带的成因新解．大地构造与成矿学．32（4）：521~527.
杨曾威，等．1928. 北京西山周口店花岗岩．北京大学地质系学会志．
杨富全，等．2007. 天津蓟县盘山 I 型-A 型复合花岗岩体-区域构造环境转变的记录．岩石学报．23（03）529~546.
杨鸿连．北京中生代地层及生物群．地质出版社，1994.
杨建，陈庆寿．2004. 砂土液化影响因素及其判别方法．西部探矿工程，93（2）：1~2.
杨杰，等．1948. 北平西山之岩石性质与火山活动．地质论评．13（Z1）：28~54+167~175.
杨进辉，吴福元，邵济安，等．2005 燕山陆内造山带隆升的起始时间：北京西山中生代沉积岩中锆石 U-Pb 年代学和 Hf 同位素证据．全国岩石学与地球动力学研讨会．
杨进辉，等．2005. 北京密云环斑花岗岩锆石 U-Pb 年龄和 Hf 同位素及其地质意义．岩 石 学 报．21（6）：1633~1644.
杨进辉，等．2007. 辽东矿洞沟正长岩成因及其构造意义：锆石原位微区 U-Pb 年龄和 Hf 同位素制约．岩石学报．23（2）：263~276.
杨进辉，吴福元，邵济安，等．2006. 冀北张-宣地区后城组、张家口组火山岩锆石 U-Pb 年龄和 Hf 同位素．地球科学-中国地质大学学报．31（1）：71~80.
杨开辉．1990. 北京延庆大庄科花岗杂岩地球化学特征．现代地质，4（1）：78~89.
杨眉，等．1990. 北京地区元古代大红峪组火山岩岩石学特征．矿物学岩石学论丛，1995（73）：73~83.
杨明慧，等．2012. 华北克拉通晚三叠世沉积盆地原型与破坏早期构造变形格局．地质论评．2012（1）：1~18.
杨守仁，廖明和．北京西山门头沟群双壳纲化石新发现．地质论评，1983，29（5）：45.
杨涛，宫辉力，赵文吉，等．2010. 北京顺义区地裂缝分布特征及成因分析．自然灾害学报，19（6）：100~106.
杨晓勇，杨学明，等．1998. 碳酸岩是大陆岩石圈构造背景和地幔交代作用的指示岩石．地球物理学报．地球物理学报，41（S1）：228~235（S1）：228~235.
杨长秀．2008. 河南鲁山地区中-新太古代灰色片麻岩．地质论评，54（3）：327~334.
杨长秀，等．2008. 河南鲁山地区早前寒武纪变质岩系的锆石 SHRIMP U-Pb 年龄、地球化学特征及环境演化．地质通报，27（4）：517~533.
杨振升，刘午旭．1986. 华北克拉通太古宙上壳岩早前寒武纪构造变形与变质作用关系——以北京密云沙厂地区为例．长春地质学院学报，4：17~28.
杨钟健，1933，中国人类化石及新生代地质概论．地质专报乙种第 5 号．
杨主恩，应思淮，林传勇，俞理宝．1981. 北京密云北石城断裂带的断层岩特征及其地震事件的可能证据．地震地质，3（4）：1~14.
杨子庚，牟昀智．1981. 对周口店地区晚新生代地层的新认识．科学通报，13：807~810.
杨子赓，等，1985，周口店地区晚新生代地层研究．见：吴汝康，等，北京猿人遗址综合研究．北京：科学出版社，1~85.
姚丽景，颜丹平，胡玲．2007. 房山变质核杂岩基底拆离断层韧性剪切变形构造及环境分析．地球科学—中国地质大学学报，32（3）：357~365.
姚轶锋，叶超，寇香玉，等．2007. 北京天竺晚上新世以来植被演替与气候变迁．古地理学报，9（1）：45~57.
叶伯丹，等．1986. 全国同位素地质年龄数据汇编．北京：地质出版社．
叶浩．2014. 华北北缘东段早中生代构造变形与岩浆作用研究．中国地质大学（北京）．
叶浩，张栓宏，赵越，等．2014. 燕山褶断带晚三叠世都山复式岩基成因及侵位变形：华北北缘中生代早期构造背景的制约．地学前缘，21（4）：275~291.
叶良辅．1920. 北京西山地质志．地质专报．甲种，第 1 号．
叶盛基．1964. 北京密云地区超基性岩体特征与放马峪超基性岩体岩石学研究．北京地质学院科学研究论文集 3（岩石专辑）．
尹功明，卢演俦，陈杰，等．1997. 现代（零年龄）样品的红光释光（IRSL）测定及其 ED 值．核技术，20：489~491.
尹功明，卢演俦．1996. 北京延庆盆地 50 万年以来的主要构造事件及年代学的初步研究．华北地震科学，14（4）：19~30.
尹继才，吕林素．2005. 北京地区首次发现石炭纪石松化石．地球学报，26（5）：473~474.

尹金辉，计凤桔，郑荣章，等．2000. 永定河上游晚更新世晚期以来的堆积阶地年代测定．地震地质，22（2）：195~199.
尹赞勃，等．1963. 祁连山地质研究的新收获．科学通报：113~115.
油新华，何光尧，王强勋，张磊．我国城市地下空间利用现状及发展趋势．隧道建设（中英文），2019，39（02）：173-188.
于湘伟，陈运泰，王培德．2003. 京津唐地区中上地壳三维 P 波速度结构．地震学报，25（1）：1~14.
俞理宝．1983. 北京密云麻粒岩相中辉石的特征．地震地质，5（2）：25~30.
郁建华，付会芹．1996. 华北克拉通北部 1. 70Ga 非造山环斑花岗岩岩套．华北地质矿产杂志，11（3）：341~350.
郁建华，等．1990. 北京及附近元古宙裂谷槽中球斑花岗岩套的地球化学．地质学报．64（4）：322~336.
郁建华，等．1993. 国际地质对比计划 315 项美国密苏里会议简介．北京地质．
郁建华，等．1994. 环斑花岗岩套的赤城一古北口深成岩带及元古代裂谷作用．华北地质矿产杂志．9（1）：34~46.
郁建华，等．1994. 元古代北京裂谷槽中钾质碱性火山深成作用．中国区域地质．第 2 期．115~122.
郁建华，等．1996. 华北地台北部非造山环斑花岗岩及有关岩石．北京：中国科学技术出版社．1~182.
袁宝印，邓成龙，吕金波，等．2002. 北京平原晚第四纪堆积期与史前大洪水．第四纪研究，22（5）：474~482.
袁复礼，杜恒俭．1984. 中国新生代生物地层学. 北京：地质出版社．
袁洪林，柳小明，刘勇胜，等．2005. 北京西山晚中生代火山岩 U-Pb 锆石年代学及地球化学研究．中国科学：地球科学，35（9）：821~836.
苑东洋，等．2016. 周口店官地杂岩中斜长角闪岩的年代学、地球化学及其地质意义．西北地质．49（2）：149~164.
翟明国．2004. 华北克拉通 2. 1~1. 7Ga 地质时间群的分解和构造意义探讨．岩石学报，020（06）：1343~1354.
翟明国．2010. 华北克拉通的形成演化与成矿作用．矿床地质，029（01）：24~36.
张本仁．2001. 大陆造山带地球化学研究：岩石构造环境地球化学判别的改进，西北地质，34（3）：1~17.
张成科，张先康，盖玉杰，，等．1997. 文安-蔚县-察右中旗剖面地壳上地幔速度结构与构造研究．华北地震科学，15（3）：18~28.
张成科，张先康，赵金仁，等．2002. 渤海湾及其邻区壳幔速度结构研究与综述．地震学报，24（4）：428~435.
张福堂．1984. 北京断陷回升隆起．中国区域地质，2：98~100.
张福堂．1986. 永定河冲积扇及古河道在卫星图像上的显示特征．水文地质工程地质，1：52~53.
张国辉，等．1998. 下地壳及壳幔过渡带化学不均一性——河北汉诺坝地区深源捕虏体元素地球化学证据．地球化学. 27（2）：153~169.
张宏，王明新，柳小明．2008. LA-ICP-MS 测年对辽西-冀北地区髫髻山组火山岩上限年龄的限定．科学通报，（15）：1815~1824.
张宏，袁洪林，胡兆初，等．2005. 冀北滦平地区中生代火山岩地层锆石 U-Pb 测年及启示．地球科学-中国地质大学学报，30（6）：707~720.
张宏福，等．2004. 华北东南部中生代岩石圈地幔性质、组成、富集过程及其形成机理．岩石学报：21（4）：1271~1280.
张吉顺．1990. 北京房山花岗闪长岩体的侵位变形构造及气球膨胀式侵位机制，见：张吉顺，单文琅编北京西山地质研究武汉：中国地质大学出版社，48~63.
张佳华，孔昭宸，杜乃秋．1999. 北京房山 16000~7000 年以来的植被与环境变迁．微体古生物学报，16（4）：421~ 430.
张建平，邢立达，等．2012. 中国北京恐龙足迹的首次记录．科学通报，57（2-3）：144~152.
张建新，等．1997. 北京云蒙山地区花岗岩穹隆及伸展构造的探讨．地质论评，43（3）.
张建新，等．1997. 北祁连加里东造山带从挤压到伸展造山机制的转换．长春地质学院学报，27（3）：277~283.
张健，等．2015. 华北克拉通北缘 Columbia 超大陆裂解事件：来自燕辽裂陷槽中部长城系碱性火山岩的地球化学、锆石 U-Pb 年代学和 Hf 同位素证据．岩石学报．31（10）：3129~3146.
张静，徐建明，刘海坤，等．2009. 华北平原晚更新世以来岩石地层、生物地层及气候地层．现代地质，23（1）：64~74.
张磊，白凌燕，蔡向民，等．2014. 北京南口-孙河断裂北西段综合物探剖面定位及其活动性研究，现代地质，28（1）：234~342.
张磊，白凌燕，蔡向民，等．2014. 北京平原南口-孙河断裂带北西段活动性分析．中国地质，41（3）：902~911.

张磊，白凌燕，蔡向民，等．2014. 北京平原南口－孙河断裂南段第四纪活动性的磁性地层学研究．第四纪研究，34（2）：381～390.
张磊，何静，白凌燕，等．2016. 北京凹陷北缘第四纪凹陷盆地沉积速率变化特征与顺义断裂活动性的响应关系，中国地质，43（2），511～519.
张磊，张晓亮，白凌燕，等．2016. 北京平原沙河凹陷纪磁性地层学研究及其新构造动动的沉积响应．中国地质，43（3）：1076～1084.
张丽芬，曾夏生，张树明．2006. 北京周口店岩体地球化学特征及物源分析．岩石矿物学杂志，25（6）：480～486.
张连昌，翟明国，万渝生，等．2012. 华北克拉通前寒武纪 BIF 铁矿研究：进展与问题．岩石学报，028（11）：3431～3445.
张璐，等．2012. 全吉群碎屑锆石的 U-Pb 年龄谱和微量元素：基底热事件信息和早期演化启示．地球科学（中国地质大学学报）．37（增刊）：1～15.
张鹏远，谷淑琴．1986. 北京十三陵蓟县系雾迷山组微体化石群．地质学报，(4)：13～23+99～100.
张旗，翟明国．2012. 太古宙 TTG 岩石是什么含义．岩石学报，028（11）：3446～3456
张旗，等．2001a. 中国东部燕山期埃达克岩的特征及其构造－成矿意义．岩石学报．17（2）：236～244.
张旗，等．2001b. 燕山期的中国东部高原下地壳组成初探：埃达克质岩 Sr、Nd 同位素制约．岩石学报．17（4）：505～513.
张旗，等．2007. 花岗岩构造环境问题——关于花岗岩研究的思考之三．岩石学报．27（13）：2683～2689.
张清波，常英．1994. 北京地区大灰厂组孢粉组合及其地质意义．北京地质，(2)：1～7.
张清波，常英．1995. 北京地区芦尚坟组孢粉组合及地质时代．北京地质，(1)：1～9.
张清波，萧宗正．1996. 北京延庆盆地早更新世孢粉组合及其气候地层学意义．北京地质，1：1～9.
张瑞英，张成立，孙勇．2013. 燕山地区太古代麻粒岩相岩石退变质作用的研究．岩石学报，029（07）：2265～2280.
张树明，等．2010. 北京西山地区中生代岩浆作用、变质作用及其对华北东部岩石圈厚度变化的约束．北京：地质出版社．
张拴宏，赵越，叶浩，等．2013. 燕辽地区长城系串岭沟组及团山子组沉积时代的新制约．岩石学报，29（7）：2481～2490.
张拴宏，等．2004. 燕山构造带内蒙内古隆起东段晚古生代－早中生代构造活动及其地质意义．博士学位论文．北京：中国地质科学院．1～148.
张栓宏，等．2010. 华北地块北缘晚古生代－早中生代岩浆活动期次、特征及构造背景．岩石矿物学杂志．29（6）：824～842.
张双权，高星，陈福友，等．2016. 周口店第一地点西剖面 2009—2010 年发掘报告．人类学学报，35（1）：63～75.
张先康，王椿镛，刘国栋，等．1996. 延庆－怀来地区地壳细结构－利用深地震反射剖面，地球物理学报，39（3）：356～362.
张先康，赵金仁，等．2002. 三河－平谷 8.0 级大震区震源细结构的深地震反射探测研究．中国地震，18（4）：326～336.
张亚明，孟朴在，滕吉文．2008. 京津唐及邻区地壳结构的中新生代构造运动响应．地球物理学报，51（4）：1123～1134.
张英芳，姜均伟．2010. 北京西山中侏罗世植物群的古气候意义．中国地质，37（2）：515～524.
张英礼，黄兴根，焦振秀，等．1984. 北京平原地区晚更新世晚期地层．地层学杂志，8（1）：56～61.
张永栋，王均平，魏明建，等．2012. 北京市昌平区钻孔岩芯的磁化率与粒度祖坟变化的相关性研究．沉积学报，30（3）：572～579.
张玉泉，等．2000. 青藏高原东部钾玄岩系岩浆岩同位素特征：岩石成因及其构造意义．中国科学：地球科学，30（5）：493～67.
张岳桥，董树文，赵越，等．2007. 华北侏罗纪大地构造：综评与新认识．地质学报，81（11）：1462～1480.
张长厚，张勇，李海龙．2006. 燕山西段及北京西山晚中生代逆冲构造格局及其地质意义．地学前缘，13（2）：165～183.
张长敏．2009. 煤矿采空塌陷特征与危险性预测研究．北京：中国地震局地质研究所．
张志军．2002. 北京京西盆地早白垩世同翅目化石一新属（同翅目：蜡蝉总科）．Zoological Systematics（动物分类学报（英文）），27（1）：20～23.
张忠英，李正惠．1985. 燕山西段长城系高于庄组微体植物群．微体古生物学报，(03)：15～26+104～106.

张子斌，王丁，丁嘉贤 . 1981. 北京地区一万三千年来自然环境的演变 . 地质科学，3：259～268.
章雨旭，彭阳，高林志 . 1997. 北京西山寒武系—奥陶系主要副层序类型及其形成机制探讨 . 地质论评 ，43（2）：148～154.
赵崇贺，等 . 1964. 北京密云斜长环斑花岗岩地质学与岩石学研究 . 见中国地质学会编 . 第一届矿物岩石地球化学会议论文选集（岩石）. 124～143.
赵丰 . 2006. 北京市城市地质环境及其工程意义 . 北京：中国地质大学 .
赵刚，等 . 2008. 北京燕山地区燕山期火成岩分布及岩石圈减薄过程 . 中国地质大学（北京）.
赵国春，孙敏，S. A. Wilde. 2002. 早–中元古代 Columbia 超级大陆研究进展 . 科学通报，47（18）：1361～64.
赵国春 . 2009. 华北克拉通基底主要构造单元变质作用演化及其若干问题讨论 . 岩石学报，025（08）：1772～1792.
赵国春，等 . 2002. 燕辽地区燕山期火山活动节律与造山–深部过程 . 博士课题 .
赵华，卢演俦，陈杰，等 . 2000. 华北第四纪沉积物细颗粒组分红外释光与绿光释光测年 . 科学通报，45（21）：2332～2336.
赵金仁，张先康，等 . 2004. 利用宽角反射/折射和深反射探测剖面揭示三河–平谷大震区深部结构特征 . 地球物理学报，47（4）：646～652.
赵金仁，张先康，张成科，等 . 1999. 香河–北京–涿鹿及其相邻地区壳幔构造与速度结构特征 . 地震地质，21（1）：29～36.
赵金仁，张先康，张成科，等 . 2005. 北京西北地区壳幔结构的非均匀性特征与地震活动 . 地震学报，27（2）：119～127.
赵美涛，等 . 2014. 北京石城闪长岩和云蒙山花岗岩岩石地球化学特征及成因 . 矿物岩石 . 34（3）：60～69.
赵美涛，等 . 2014 北京长园带状岩体的岩石地球化学特征及成因探讨 . 矿物岩石地球化学通报 . 34（3）：820～829.
赵太平，等 . 2004. 河北大庙斜长岩杂岩体锆石 U–Pb 年龄及其地质意义 . 20（03）：685～690.
赵伟华，黄润秋，赵建军，巨能攀，等 . 2011. 强震条件下碎裂岩体崩塌机理及崩塌后壁对堆积体稳定性影响研究 . 工程地质学报，19（2）：205～212.
赵希涛，李容全 . 1981. 北京斋堂地区第四纪沉积与地层 . 中国科学院地质研究所集刊，第 2 号：112～122.
赵希涛，曲永新 . 1981. 北京斋堂雁翅地区的黄土 . 地质科学，1：47～54.
赵希涛，孙秀萍，张英礼，等 . 1984. 北京平原 30000 年来的古地理演变 . 中国科学 B 辑，6：544～554.
赵晓彦，胡厚田，时延兵 . 2005. 类土质边坡研究初探 . 工程地质学报，13（1）.
赵兴国，2009，北京地区燕山期火成岩与构造环境. 中国地质大学（北京）博士学位论文 .
赵勇，蔡向民，王继明，等 . 2013. 北京平原顺义 ZK12～2 钻孔剖面第四纪磁性地层学研究 . 地质学报，87（2）：288～294.
赵越，宋彪，张拴宏，刘健 . 2006. 北京西山侏罗纪南大岭组玄武岩的继承锆石年代学及其含义 . 地学前缘，13（2）：184～190.
赵越，徐刚，张拴宏，等 . 2004. 燕山运动与东亚构造体制的转变 . 地学前缘 ，11（3）：319～328.
赵越，等 . 1990. 燕山地区中生代造山运动及其构造演化 . 地质论评 . 36（1）：1～13.
赵越，等 . 2006. 北京西山侏罗纪南大岭组玄武岩的继承锆石年代学及其含义 . 地学前缘，13（2）：184～190.
赵忠海 . 2001. 北京地区东西向构造带对金矿化的有利条件分析 . 北京地质，13（3）：32～35.
赵忠海 . 2002. 北京延庆大海坨岩体的地球化学特征及其岩浆来源探讨 . 北京地质，14（2）：7～12.
赵忠海 . 2006. 北京地区地裂缝灾害的分布特征及其成因探讨 . 地质灾害与环境保护，17（3）：75～78.
郑公望，夏正楷，任秀生 . 2000. 北京王府井东方广场古人类遗迹热释光断代 . 北京大学学报（自然科学版），36（1）：128～130.
郑桂森，方景玲 . 1994. 北京西山青白口—下苇甸—带印支期侵入活动特征 . 北京地质，1：21～27
郑桂森，李安宁，1992，北京坨里—长辛店白垩纪新断陷沉积盆地的发展演化北京地质 . 4（3）. 11～19
郑桂森，刘振锋，方景玲 . 1994. 北京西山妙峰山地区中生界后城组地质特征及划分意义 . 北京地质，1：13～20.
郑桂森，王继明，何静，等 . 地下空间资源的属性特征 . 城市地质，2017，12（4）1-5.
郑桂森，徐吉祥，栾英波，等，2017，北京西山大石河山峡阶地发育特征 . 地质通报，36（7）：.
郑建平，等 . 1996. 古地幔流体及其在大陆克拉通地幔演化中的意义——以华北地台为例 . 地学前缘，3（3–4 期）：187～194.
郑建平，等 . 1999. 华北地台东部古生代与新生代岩石圈地幔特征及其演化 . 地质学报，73（1）：47～56.

郑建平，等.1999. 胶辽半岛金伯利岩中地幔捕虏体岩石学特征：古生代岩石圈地幔及其不均一性. 岩石学报，15（1）：65~74.
郑建平，等.1999. 中国东部地幔置换作用和中新生代岩石圈减薄. 武汉：中国地质大学出版社. 110~112.
郑建平，等.2000. 华北东部地幔改造作用和置换作用：单斜辉石激光探针研究. 中国科学：地球科学 30（4）：373~67.
郑斯成，等.1987. 房山花岗岩岩体氧氢碳的同位素研究. 岩石学报. 第3期，1987：13~22.
郑亚东.1999. 共轭伸展褶劈理夹角的定量解析. 地学前缘，6（4）：391~395.
郑亚东，等.2000. 华北克拉通北部古—中元古代富碱侵入岩同位素地球化学：岩石圈地幔富集的证据. 6（4）：391~395.
郑亚东，等.2000. 燕山带中生代主要构造事件与板块构造背景问题. 地质学报，74（4）：289~302.
中国地震局地质研究所，北京市地震局.2007. 北京市活动断层探测与地震危险性评价初勘总结报告.
中国地质学会城市地质研究会. 2005. 中国城市地质. 北京：中国大地出版社.
钟长汀，邓晋福，武永平，等.2006. 华北克拉通北缘中段古元古代强过铝质花岗岩地球化学特征及其构造意义. 地质通报，25（3）：389~397.
钟长汀，吴昌华，梅华林，等.1998. 晋冀蒙高级地体石榴石基性麻粒岩地质特征及其构造意义. 前寒武纪研究进展，21（1）：35~44.
周鸿黝，1983. 北京密云半截峪一带太古代变质铁矿的成因矿物学. 岩石矿物及测试，2（3）：197~205.
周昆叔，严富华，梁秀龙，等.1978. 北京平原第四纪晚期花粉分析及其意义. 地质科学，1：57~64.
周昆叔.1965. 对北京市附近两个埋藏泥炭沼的调查及其孢粉分析. 中国第四纪研究. 北京：科学出版社，1：118~134.
周昆叔.1984. 华北区第四纪植被演替与气候变化. 地质科学，2：166~171.
周丽清，邵德艳. 1994. 北京十三陵中元古界蓟县系雾迷山组原地风暴沉积的砾屑体. 沉积学报，(2)：72~76.
周绍林，等.1993. 北京早前寒武纪地质及铁矿. 北京：地质出版社.
周小平，张忠英. 1991. 北京十三陵雾迷山组微体植物群及沉积环境分析. 微体古生物学报，(3)：289~300.
周小平. 1990. 北京十三陵雾迷山组微体植物群落，地质学刊，(3)：27~32.
周小平. 1991. 北京十三陵雾迷山组几种特殊微化石的发现及其意义. 地质论评，37（5）：456~459.
周新华，等.2006. 中国东部中、新生代岩石圈转型与减薄研究若干问题. 地学前缘，13（2）：50~64.
周新华，等.2006. 中生代华北岩石圈地幔高度化学不均一性与大陆岩石圈转型. 地球科学，31（1）：8~13.
朱传庆，罗杨，杨帅，等. 2009. 北京西山寒武系层序地层. 中国地质，36（1）：120~130.
朱露培，曾融生，刘福田.1990. 京津唐张地区地壳上地幔三维P波速度结构. 地球物理学报，(03)：267~277.
朱日祥，邓成龙，潘永信.2007. 泥河湾盆地磁性地层定年与早期人类演化. 第四纪研究，27（6）：922~944.
朱士兴，黄学光，孙淑芬. 2005. 华北燕山中元古界长城系研究的新进展. 地层学杂志，29：437~449.
朱士兴.1994. 华北地台中、上元古界生物地层序列. 北京：地质出版社.
朱士兴.1995. 元古宙生物圈的研究意义、现状和关键. 地质调查与研究，(1)：61~72.
朱士兴. 1999. 关于开展元古宙重大生命演化事件研究的建议. 地质调查与研究，(2)：59~64.
朱士兴. 1999. 中国前寒武纪生物地层学研究的新进展. 现代地质，(2)：206~62.
祖金华，吴乾蕃，廉雨方.1997. 延庆-怀来盆地及其邻区地热研究. 地震学报，(04)：107~109.
Amp M K，Chen Y S，et al. 1935. Note on some thrusts in the western Hiuls of Peiping. Acta Geological Sinica. 4：112~144+194~195.
Anderson，et al. 1994. Isotopic constraints on age and sources of some minor rapakivi complexes，central Sweden. In：Lanphere M A，et al，（eds）Abstracts of the Eight International Conference on geochronology，cosmos－chronology and isotope geology. U. S. Geol. Surv. Circ. 1107. 8.（1995）.
Anderson J L and Smith D R. 1996. The effects of temperature and fO_2 on the Al-in-hornblende barometer. American Mineralogist. 80（5~6）：549~559.
Andersson，J G. 1923，Eassays on The Cenozic of Northern China：The Geological Surver of China，Memoirs Series A，No. 3.
Barbour，G B. 1967. Analysis of the Lushan glaciations problem：Bulletin of the geological Society of China，13：647~656.
Bromley A V. 1975，Tin mineralization of Western Europe：is it related to crustal subduction? Trans Instn Min Metall，84：B28~30.
Burchfiel B C. 1993，Tectonostratigraphic map of the Cordilleran Orogenic Belt conteminous United States；Published by：The Geol. Soc. Amer. Inc.

Castro A, et al. 1987. On granitoid emplacement and related structures. A review. Geologische Rundschau. 76 (1): 101~124.
Clayton R N, Stuart W, et al. 1989. Volcanic influences in a storm- and tide-dominated shallow marine depositional system: The late permian broughton formation, southern Sydney Basin, Kiama, NSW. Australian Journal of Earth Sciences. 36 (4): 569~584.
Condie & O'Neill, et al. 2007. Earth's changing thermal regime between 3 and 2.5 Ga: a primary control of planetary evolution?
Condie K C. 1998. Episodic continental growth and supercontinents: A mantle avalanche connection?. Earth and Planet science letters. 163 (1): 97~108.
Condie K C. Bickford M E, Aster R C, et al. 2011. Episodic Zircon ages-Hf isotopic composition, and the preservation rate of continental crust. Geological Society of America Bulletin. 123 (5~6): 951~957.
Davis, et al. 1996. The Huairou (Shuiyu) ductile shear zone, Yunmengshan Mts., Beijing. Field Trip Guide. 6 (T209) 1~25.
Davis G A, Wang C, Zheng Y D, et al. 1998. The enigmatic Yinshan fold- and- thrust belt of northern China: New views on its intraplate contractional styles. Geology, 26: 43~46.
Davis G A, Yadong Z, Cong W. 2002. Mesozoic tectonic evolution of the Yanshan fold and thrust belt with emphasis on Hebei and Liaoning Provinces, northern China. Beijing Geology, 10 (2): 171~197.
Davis G A, Zheng Y D, Wang C, et al. 2001. Mesozoic tectonic evolution of the Yanshan segment of the Yinshan fold-and thrust belt, northern China. In; Hendrix M S and Davis G A (eds). Paleozoic and Mesozoic tectonic evolution of central Asia: from continental assembly to intracontineental deformation, Boulder, Colonrado. Geological Society of America Memoir. 194: 171~197.
Davis G A, Zheng Y D, Wang C, Darby B J, Zhang CH and Gehrels G. 2001. Mesozoic tectonic evolution of the Yanshan fold and thrust belt, with emphasis on Hebei and Liaoning Provinees, northern China. In: Hendrix M S and Davis G A (eds.). Paleozoic and Mesozoic tectonic evolution Of central Asia: From continental assembly to intercontinental deformation. Geol. Soc. Of Am. Memoir. 194: 171~198.
Davis, 等. 2002. 变质核杂岩的定义、类型及构造背景. 21 (4~5期): 185~192.
Davis, 等. 2002. 中生代燕山褶皱冲断带的构造演化——以河北省和辽宁省为重点的研究. 北京地质. 14 (4): 1~40. 汪洋, 李凯明译自 Paleozoic and Mesozoic Tectonic Evolution of Central and Eastern Asia: From Continental Assembly to Intracontinental Deformation. Geological Society of America Memoir 194. 171~197.
Defant M J, Drummond M S, Mount St. Helens. et al. 1993. Potential example of the partial melting of the subducted lithosphere in a volcanic arc. Geology. 21 (6): 547~550.
Defant M J, Drummond M S. 1990. Derivation of some modern arc magmas by melting of young subducted lithosphere. Nature. 347: 662~665.
Deng F, Su S G, Mo X X, Zhao G C, Xiao Q H, Ji G Y, Qiu R Z, Zhao H L, Luo Z H, et al. 2004. The Sequence of Magmatic-Tectonic Events and Orogenic Processes of the Yanshan Belt, North China. Acta Geologica Sinica. 78 (1): 260~266.
Deway J F. 1972. Plate tectonics. Sci Amer, 226: 56~68.
Dewey J F, Kidd W S F. 1974. Continnental collisions in the Appalachian-Caledonian orogenic belt: Variations related to complete and incomplete suturing. Geology, 2: 543~546.
Emslie. R F, et al. 1991. Granitoids of rapakivi granite-anorthosite and related associations. Precamb. Res. 51: 173~192.
Friedman, Sanders et al. 1978. Chilingar G V. Principles of sedimentology. Wiley. 1978.
Gao et al. 2002. Re-Os evidence for replacement of ancient mantle lithosphere beneath the North China Craton. Earth and Planetary Science Letters, 198. 307~322.
Gao S, Rudnick R L, Carlson R W, et al. 2002. Re-Os evidence for replacement of ancient mantle lithosphere beneath the North China Craton. Earth Planet Sci Lett. 198: 307~322.
Goldfarb R J, Groves D I, Gardoll S. 2001. Orogenic gold and geological time: A synthesis. Ore Geology Reviews, 18: 1~75.
Groves D I, Goldfarb R J, Robert, et al. 2003. Golg deposits in metamorphic belts: Overview of understanding and outstanding problems, future research, and exploration significance. Economic Geology, 98: 1~29.
Shen G J, Gao X, Gao B, et al, 2009. Age of Zhoukoudian Homo erectus dedermined with 26Al/10Be burial dating. Nature. Vol. 458: 198~200.
Jahn, B M, Wu, F Y, Lo, C H, Tsai C H, et al. 1999. Crust-mantle interaction induced by deep subduction of the continental crust: geochemical and Sr-Nd isotopic evidence from post-collisional. 1998. Geochemical and isotopic characteristics of UHP

eclogites and ultramafic rocks of the Dabie orogen: implications for continental subduction and collisional tectonics. In: Hacker B R, Liou J G (Eds.), When Continental Collide: Geodynamics and Geochemistry of Ultrahigh—Pressure Rocks . Kluwer Academic Publishers, Wetherlands, pp. 203~239.

Lee J S, Chao Y T. 1924. Geology of the Gorges District of the Yangtze (from Ichang to Tzekuei) with special reference to development of the Gorges. Bull. Geol, Soc. (z1): 351~392.

Lee J S, et al. 1924. A Suggestion of a new method for geological survey of Igneous intrusions Bulletin of the Geological Society of China. 109~115.

Condie K, Allen P, Narayana B L. 1982. Geochemistry of the Archean low-to high-grade transition zone, Southern India. Contributions to Mineralogy and Petrology. 81 (3): 157~167.

Liu J L, Ji M, Guan H M, et al. 2008. Crustal detachment and destruction of the North China craton: constraints from Late Mesozoic extensional structures . Earth Science Frontiers, 15 (3): 72-81 (in Chinese with English abstract).

Lu F X, Zheng J P, Shao J A, et al. 2006. Asthenospheric upwelling and lithospheric thinning in late Cretaceous- Cenozoic in eastern North China . Earth Science Frontiers, 13 (2): 86~92 (in Chinese with English abstract).

Lü Z Y, Wu J P. 2010. 3-D P wave velocity structure of crust and upper mantle beneath North China. Acta Seismologica Sinica, 32 (1): 1~11.

Marsh J S. et al. 1982. Significance of $^{87}Sr/^{86}Sr$ ratios in the Merensky cyclic unit of the Bushveld Complex . Nature. 298 (5869): 53~55.

Matin H, et al. 1999. Adakitic magmas: modern analogues of Archaean granitoids. Lithos, 46 (3): 411~429.

Melinda A, Yang X G, Christoph T, et al. 2017. 40000-Year-old individual from Asia provides insight into early population structure in Eurasia . Current Biology, 27: 1-7.

Middlemost E A K, et al. 1994. Naming materials in the magma/igneous rock system . Earth-Science Reviews. 37 (3~4): 215~224.

Mitchell A H G. 1974. Southwest England granites: magmatism and tin mineralization in a post-collision tectonic setting . Trans Instn Min Metall, 83: B95~97.

Nutman A P, Wan Y S, Du L L, Clark F R L, Dong C Y, Xie H Q, Wang W, Sun H Y, Liu D Y. 2011. Multisitage late Neoarchaean crustal evolution of the North China Craton, eastern Hebei. Precambrian Reserch, 189: 43~65.

Pei W C, 1929. An account of the discovery of an adult Sinanthropus skull in the ChouKouTien deposit. Bulletin of Geological Society of China, 9, 203~205.

Pei W C, 1931a. An The age of ChouKoutien fossiliferous deposits-a tentative determination by comparison with other later Cenozoic (Psychozoic) deposits in China. A. Bulletin of Geological Society of China, 10, 165~178.

Pei W C, 1931b. Notice of the discovery of quartz and other stone artifacts in the Lower Pleistocene hominid-bearing sediments of the ChouKoutiencave deposit. Bulletin of Geological Society of China, 11, 109~146.

Pei W C, 1932. Preliminary note on some incised, cut and broken bones found in association with Sinanthropus remains and lithic artifacts from ChouKoutien. Bulletin of Geological Society of China, 12, 105~112.

Pei W C, 1939. The Upper Cave Industry of Choukoutien. Pal. Sin., S. D, 9.

Pei W C, 1940a. New fossil material and artifacts collected from the Choukoutien region during the years 1937 to 1939. Bulletin of Geological Society of China, 19, 207~234.

Pei W C, 1940b. The Upper Cave faunas of Choukoutien. Pal. Sin., S. C, 10.

Pei W C, 1934a. A preliminary report on the late palacolithic cave of ChouKoutien. Bulletin of Geological Society of China, 13, 337~358.

Pei W C, 1934b. On the Carnivora from Locality1 of ChouKoutien. Pal. Sin., S. C, 8 (1).

Peng P, Zhai M G, Li Q, et al. 2011. Neoproterozoic (~900Ma) Sariwon sills in North Korea: Geochronology, geochemistry and implications for the evolution of the south~eastern margin of the North China Craton. Gondwana Research, 20 (2011): 243~254.

Peng P, Jinghui Guo J H, Zhai M G, et al. 2010. Paleoproterozoic gabbronoritic and granitic magmatism in the northern margin of North China craton: Evidence of crust~mantle interaction. Precambrian Research, 183 (2010): 635~659.

Peng P, Guo J H. Zhai M G, Bleeker W, et a1. 2010. Paleoproterozoic gabbronoritic and granitic magmatism in the northern margin of the North China craton: evidence of crust-mantle interaction. Precambrian Research. 183: 635~659.

Peng P. Zhai M G, Guo J H, Zhang H F, Zhang Y B, et al. 2008. Petrogenesis of Triassic post-collisional syenite plutons in the Sino-Korean Craton: an example from North Korea. Geological Magazine. 145, 637~647.

Picher W S, et al. 1983. Granite type and tectonic environment. Mountain building processes London: academic press. 19~40.

Ramo, et al. 1995a. One hundred years of rapakivi granite. Mineral. Petrol. 52: 129~185.

Ramo, et al. 1995b. 1700Ma Shachang complex, northeast China: Proterozoic rapakivi granite not associated with Paleoproterozoic orogenic crust. geol, 23: 815~818.

Ramo, et al. 1995c. 1700Ma Shachang complex, northeast China: Proterozoic rapakivi granite not associated with Paleoproterozoic orogenic crust . Geology, 23 (9): 812~818.

Reading H G, 1978. Sedimentary Environments and Facies. Blackwell Scientific Publications.

Robertson A H F. 1994, Role of the tectonic facies concept in orogenic analysis and its application to Tethys in the Eastern Mediterranean region; Earth-Science Reviews, 37: 139~213.

Royse K R, Rutter H K, Entwisle D C, 2009. Property attribution of 3D geological models in the Thames Gateway, London: new ways of visualising geoscientific information. Bulletin of Engineering Geology and the Environment, 68: 1-16.

Schmidt M W. 1992. Amphibole composition in tonalite as a function of pressure: an experimental calibration of the Al-in-hornblende barometer. Contributions to Mineralogy and Petrology. 110 (2): 304~310.

Shen G J, Gao X, Gao B, et al. 2009. Age of Zhoukoudian Homo erectus dedermined with 26Al/10Be burial dating. Nature, 458: 198~200.

Song T R, et al. 1986. Are these the oldest metazoan trace fossils discovered from China? . Science Bulletin. 31 (12): 831~834.

Wang H Z, Yang S N, Liu B P, et al. 1990. Tectonopaleogeography and Paleobiogeography of China and Adjacent Regions. Wuhan: China University of Geosciences Press, : 35~88 (in Chinese).

Wang. Y. 2013. Yangshan fold-and-thrust belt of the North China craton: A Mesozoic analogue of Himalaya . Acta Geological silica (English Edition), 87 (supp): 325~327.

Wilde S A. Constraints on the timing of uplift of the Yanshan Fold and Thrust Belt, North China. 中国科学院地质与地球物理研究所 2006 年论文摘要集, 2007: 336-352.

Willis, Bailey, Blackwelder E, 1997. Research in China: Vol, 1. Washington.

Wintle A G, Murray A S. 1997, The relationship between quartz thermoluminescence.

Wong W H, 1927. Crustal movement and igneous activities in eastern China since Mesozoic time. Bulletin of Geological Society of China, 6 (1), 9~36.

Wong W H. 1926. Crustal movement in eastern China. Proceedings of 3th Pan Pacific science Congress, Tokyo, 642~685.

Wong W H 1929. , The Mesozoic orogenic movement in eastern China. Bulletin of Geological Society of China, 8: 33~34.

Xu, et al. 1995. Upper mantle geotherm for eastern China and geological implications. Science in China (Series B) 38: 1482~1492.

Xu Y G, et al . 2001. Thermo-tectonic destruction of the Archean lithospheric keel beneath the Sino-Korean craton in China: evidence, timing and mechanism. Phys. Chem. Earth, Part A Solid Earth Geod. 26: 747~757.

Xu K, Holland T and Blundy J, et al. 1994. Non-ideal interactions in calcicamhhiboles and their bearing on amphibole-plagioclase themometry. Contributions to Mineralogy and Petrology, 116: 433~447.

Yang X Y, Yongfei Z, Yang X M, et al. 2003. Mineralogical and geochemical studies on the different types of turquoise from Maanshan area. East China. Neues Jahrbuch für Mineralogie - Monatshefte. 2003 (3): 97~112.

Shi Y R, Simon A A, Wilde, Zhao X T, Ma Y S. 2012. Late Neoarchean magmatic and subsequent metamorphic events in the northern North China Craton: SHRIMP zircon dating and Hf isotopes of Archean rocks from Yunmengshan Geopark, Miyun, Beijing. Gondwana Research, 21 (2012) 785~800.

Zhang, et al. 2009. Contrasting Late Carboniferous and Late Permian-Middle Triassic intrusive suites from the northern margin of the North China craton: Geochronology, petro genesis, and tectonic implications. Geological Society of America Bulletin, January/February 2009: 181~200.

Zhang, et al. 2007a. Carboniferious granitic plutons from the northern margin of the North China Block: implications for a Late Paleozoic active continental margin. Journal of the Geological Society of London. 164: 451~463 .

Zhang, et al. 2007b. Petrogenesis of the Middle DevonianGushan diorite pluton on the northern margin of the North China Block and its tectonic implications. Geological Magazine. 144: 553~568.

Zhang S H, Zhao Y, Liu X C, et al. 2009. Late Paleozoic to Early Mesozoic mafic–ultramafic complexes from the northern North China Block: Constraints on the composition and evolution of the lithospheic mantle. Lithos. 110: 229~246.

Zhang S H, Zhao Y, Song B, et al. 2007. Carboniferous granitic plutons from northern margin of the North China Block: Implications for a late Paleozoic active continental margin. J Geol Soc. 164: 451~463.

Zhang S H, Zhao Y, Song B, et al. 2009. Contrasting Late Carboniferous and Late Permian–Middle Trassic intrusive suites from the northern margin of the North China craton: Geochronology, petrogenesis, and tectonic implications. Geol Soc Amer Bull. 121: 181~199.

Zhang Y. Stromatolitic microbiota from the middle Proterozoic Wumishan Formation (Jixian Group) of the Ming Tombs, Beijing, China. Precambrian Research, 1985, 30 (3): 277–302.

Zhang Y Q, Dong S W, Zhao Y, et al. 2007. Jurassic tectonics of North China: A synthetic view . Acta Geologica Sinica, 81 (11): 1462~1480 (in Chinese with English abstract).

Zhao T P, et al. 2009. Geochemical constraints on the tectonic setting of Paleoproterozoic A–type granites in the southern margin of the North China Craton.

Zheng T Y, Chen L, Zhao L, et al. 2007. Crustal structure across the Yanshan Belt at the northern margin of the North China Craton. Phys Earth Planet Inter. 161: 36~49.

摘　　要

本志以《北京市区域地质志》（1991）为基础，系统梳理北京地区近30年来的区域地质调查和科研成果，按照《中国区域地质志工作指南》编写。全志分区域地层、沉积岩及沉积作用、火山岩及火山作用、侵入岩及深成作用、变质地质、第四纪地质与地貌、岩石圈结构构造、地质构造、平原区三维地质结构、城市地质资源与环境、区域地质发展史11篇58章及绪言和结语等，约210万字。

附图：

1）北京市地质图（1∶25万）；

2）北京市地质构造图（1∶25万）；

3）北京市航磁 ΔT 等值线平面图（1∶25万）；

4）北京市第四纪地质与地貌图（1∶25万）；

5）北京市基岩地质图（1∶25万）。

一、区域地层

第一篇分为6章，分别对新太古代至新近纪上新世期间各类地层进行了系统总结。以岩石地层为主线，按照岩石地层序列属性，进行了年代地层、生物地层、层序地层、事件地层等多重地层划分和区域地层对比研究总结。

（1）综合原岩建造类型、变质程度、混合岩化作用、构造特征及矿物组成等，将新太古界表壳岩系划分为密云岩群和四合堂岩群2个非正式岩石地层单位，厘定了二者时代归属。密云岩群岩性组合为辉石麻粒岩、含辉石黑云斜长变粒岩、浅粒岩、石榴斜长辉石岩夹多层辉石磁铁石英岩，该岩群普遍遭受麻粒岩相区域变质，又叠加后期角闪岩相退变质作用。四合堂岩群岩性组合为斜长角闪岩、混合质黑云变粒岩、黑云角闪变粒岩、混合岩和磁铁石英岩，该岩群遭受角闪岩相区域变质，且产出位置因邻近中生代云蒙山花岗岩体而发生较强的剪切变形改造。密云岩群和四合堂岩群均难以划分岩群内部各岩性层层序，层间普遍发育小型褶皱、韧性剪切、片理化现象，及透镜状斜长角闪岩和磁铁石英岩透镜体。综合岩性组合、岩石化学与稀土元素特征、同位素年龄、接触关系等，认为密云岩群和四合堂岩群成岩时代分别为新太古代早期和晚期。

（2）中—新元古代地层分布广泛，出露面积约占山区面积的三分之一，共建立了12个组级岩石地层单位，即长城系常州沟组、串岭沟组、团山子组、大红峪组，蓟县系高于庄组、杨庄组、雾迷山组、洪水庄组、铁岭组，待建系下马岭组，青白口系龙山组和景儿峪组。本志根据最新的同位素年龄资料，将高于庄组置于蓟县系，将下马岭组划归于临时建立的待建系。地层岩性以一套未变质的（仅局部轻微变质）滨海-浅海相富镁碳酸盐岩为主，碎屑岩和黏土岩次之，局部发育碱玄岩系列火山岩。碳酸盐岩地层叠层石发育，并富含微体古生物化石。

（3）寒武纪—奥陶纪地层划分为7个组级岩石地层单位，即寒武系昌平组、馒头组、张夏组、炒米店组，奥陶系冶里组、亮甲山组、马家沟组。地层岩性主要为滨海-浅海碳酸盐岩，含有丰富的生物化石。寒武系以三叶虫为主，建立了中—上寒武统16个三叶虫化石带；奥陶系以头足类和牙形石为主，与华北地区有很好的对比性。

（4）石炭纪—三叠纪地层划分为4个组级岩石地层单位，即石炭系太原组、山西组，石炭系—二叠系石盒子组，二叠系—三叠系双泉组。该套地层总厚约700余米，下部为海陆交互相沉积，其中海相灰岩（或泥灰岩、钙质粉砂岩、钙质泥岩）夹层中富含海相动物化石䗴类、腕足类、双壳类等，上部为陆相含煤和杂色碎屑岩沉积，发育具有东方色彩的华夏植物群。

（5）晚三叠世—白垩纪地层划分为12个组级岩石地层单位，即上三叠统杏石口组，下侏罗统南大岭组、窑坡组，中侏罗统龙门组、九龙山组，中—上侏罗统髫髫山组，上侏罗统土城子组和下白垩统张家口组、东狼沟组、九佛堂组、坨里组、夏庄组。该套地层广泛分布于中生代新生盆地，总厚度达9700m，发育为陆相火山-沉积盆地内的巨厚多旋回复杂火山-沉积岩系，含有丰富的煤炭资源和门类众多的动植物化石群。

（6）古近纪—新近纪地层按照山区和平原区进行岩石地层单位划分，山区包括始新统长辛店组和上新统鱼岭组；北京凹陷包括始新统长辛店组、渐新统前门组、中新统—下上新统天坛组、上新统天竺组；大厂凹陷包括下始新统孔店组、上始新统—下渐新统沙河街组、上渐新统东营组、中新统馆陶组、上新统明化镇组。该套地层主要隐伏分布于本区平原区和延庆盆地，平原区西部山麓和延庆盆地边缘地带有零星出露，主要为河湖相沉积，也有部分洞穴沉积，含较为丰富的动植物化石。

二、沉积岩和沉积作用

第二篇分为7章，分别对中-新元古代、早古生代、晚古生代、中生代、古近纪和新近纪岩石地层岩石组合、沉积环境、沉积盆地、沉积建造和沉积古地理进行总结，并对各沉积作用与沉积矿产进行了概略总结。

（1）中—新元古代沉积岩性以富镁碳酸盐岩类为主，有少量的碎屑岩类和火山碎屑岩类。沉积作用则以海相为主，海陆过渡相次之，极少陆相沉积作用。海相沉积以障壁海沉积和内碎屑沉积为主，前者可划分障壁岛、潟湖、潮坪、滨岸、浅海等多类微相，后者可划分潮上（滨岸）、潮间（潮坪）、潟湖、潮下、浅海、深水等沉积微相。海陆过渡相沉积发育于常州沟组和龙山组，包括河口湾沉积和三角洲沉积，以前者为主。陆相沉积仅发育常州沟组山地河流相沉积作用。中—新元古代沉积作用先后发育于裂陷槽、坳陷槽和克拉通海盆中，沉积盆地隶属于燕山-辽西沉积盆地。伴随沉积盆地发育演化，沉积建造由次稳定型向稳定型转变，可划分出陆源碳酸盐岩建造、单陆屑（含磷）建造、单陆屑（铁质）建造、巨厚沥青藻礁碳酸盐岩建造、沥青碳酸盐岩建造、藻礁碳酸盐岩建造、火山复陆屑建造、灰色复陆屑建造等。

（2）早古生代沉积岩性为碳酸盐岩间夹碎屑岩组合，以石灰岩类为主，表现为克拉通陆表海沉积环境。平坦的浅水陆棚造成海水局限流通或半局限流通，波浪作用不强，以潮汐作用为主。可划分为蒸发台地沉积、局限台地沉积、开阔台地沉积及深水盆地沉积，间夹风暴沉积等多类沉积亚相。早古生代沉积作用发育于华北克拉通陆内陆表海盆地，属陆源碎屑-碳酸盐岩型及碳酸盐岩型陆表海盆地，盆地地势平坦、地形坡度很小，表现为以均匀缓慢沉降为主。沉积建造均为稳定型内源建造和陆源建造组合，沉积物横向岩性单一，沉积地层厚度大，表现出区域性整体同升同降特征。

（3）晚古生代沉积岩性主要为不同粒级的陆源碎屑岩，极少泥灰岩和火山碎屑岩，经历一个完整的海相-海陆过渡相-陆相复杂沉积演化过程。沉积环境由短暂陆表海障壁岛-潟湖沉积，过渡到海陆交互三角洲近海河流-湖泊沉积，最终演化为陆内河流-湖泊沉积环境。晚古生代沉积作用依次发育于稳定型陆表海盆地和次稳定陆内挤压型挠曲盆地。盆地内部沉积建造复杂，划分为稳定型海相单碎屑岩含煤建造、稳定型陆相碎岩含煤建造、稳定型河湖相碎屑岩建造、次稳定型陆相火山-碎屑岩建造和次稳定型河湖相屑碎屑岩建造。

（4）中生代沉积岩岩性以火山碎屑岩类、火山碎屑熔岩类和熔结火山碎屑岩类夹火山碎屑沉积岩、陆源碎屑岩占绝对，仅夏庄组和土城子组分别发育极少泥灰岩和粉晶灰岩。沉积环境为陆地河流-湖泊相，可划分出滨浅湖、辫状河、曲流河等亚相。中生代沉积作用发育于华北板内陆相盆地，沉积盆地受构造运动制约，形成三个世代盆地。每个世代由燕山构造旋回三幕次控制，每个构造幕次先伸展后挤压，分别演化出伸展机制下断陷盆地和挤压机制下坳陷（挠曲）盆地。各个世代的盆地在空间上有一定的分布规律，而且盆地的堆积物随着盆地的形成发展呈序列有规律地堆积产出。盆地内沉积建造为次稳定或不稳定陆相碎屑岩建造和陆相火山-沉积岩建造，可划分为陆相类磨拉石建造、陆相含煤建造、陆相河流相复成分砾岩建造、陆相湖泊相建造和陆相碎屑岩夹火山岩建造、陆相沉积岩-火山岩建造。

（5）古近纪—新近纪沉积岩性主要为不同粒级的陆源碎屑岩，局部地区夹有玄武岩，极少泥灰岩和火山碎屑岩。沉积作用主要发育于延庆盆地和北京平原区北京凹陷、大厂-廊固凹陷，分属于新生太行山山间断陷盆地和华北平原断陷盆地，主要为陆相河流-湖泊相沉积，局部地区由于海侵发育有海陆过渡相。盆地内沉积建造为陆相碎屑岩建造和陆相火山-沉积岩建造，以前者为主。

（6）与中元古代—新元古代沉积作用有关的沉积矿产有铁矿、锰矿、白云岩矿、硅石矿、钾矿、石灰岩矿、磷矿、重晶石矿、大理石矿、玉石矿和砖用页岩矿 11 种；与早古生代沉积作用有关的沉积矿产有铜矿、铝土矿、白云岩矿、石灰岩矿、大理石矿和页岩矿 6 种；与晚古生代沉积作用有关的沉积矿产有铁矿、耐火粘土矿、铁矾土矿、页岩矿和煤炭 5 种；与中生代沉积作用有关的沉积矿产有煤炭和页岩矿 2 种；与新生代沉积作用有关的沉积矿产有铸型用砂矿、泥炭、高岭土矿、建筑用砂或砂砾石、砖瓦黏土矿和建筑黄土 6 类。

三、火山岩和火山作用

第三篇分为 5 章，对各个时代的火山岩岩石类型、时空分布、岩石地球化学和稀土元素特征、火山岩相、火山岩建造、火山构造进行总结，并对火山活动与成矿作用进行概略总结。

（1）新太古代火山岩遭受强烈变质。密云岩群发育超镁铁质—镁铁质基性火山岩类和长英质酸性火山岩类，属拉斑玄武岩系列。四合堂岩群发育基性至中酸性火山岩类，构成玄武岩-安山岩-英安岩比较完整的火山喷发-沉积旋回。

（2）中元古代火山岩主要发育于长城纪大红峪组和团山子组，形成于大红峪期火山喷发活动。火山喷发中心位于平谷熊耳寨一带，大致组成 5 个喷发-沉积旋回，岩性组成为溢流相碱玄岩、响岩、粗安岩，火山爆发相火山角砾岩、火山角砾熔岩，属钾质碱性岩系列。岩浆来源于弱亏损型地幔源区，曾受地壳物质的混染，而分离结晶作用不显著。大红峪期火山活动形成于被动型裂谷大地构造背景环境，是燕山裂陷槽起始阶段岩浆活动重要表现，火山构造隶属于燕山-辽西火山喷发带（二级火山构造单元）之平谷-蓟县沉积-火山盆地（三级火山构造单元）内，是该期火山活动最为强烈的地段，境内发育将军关、黄松峪等 40 多个火山岩管。

（3）本区古生代火山岩仅见于西山地区双泉组，邻近河北省丰宁云雾山一带在二叠纪有强烈火山活动，并新建立云雾山组，推测涉及北京北部地区。本志参考《中国区域地质志·河北志》（2017），概略总结了古生代火山活动。古生代火山活动以溢流相为主，爆发相、碎屑流相次之，少许潜火山岩相。岩性组合为中性粗安质角砾凝灰岩、角砾熔岩、集块角砾岩、粗安岩、粗安质角砾熔结凝灰岩，酸性流纹质角砾熔结凝灰岩、角砾凝灰岩、流纹岩，属钙碱性系列，形成于大陆边缘造山带岩浆弧构造环境。

（4）本区中生代火山活动强烈。本志参考《北京燕山期火山地质及火山岩》（鲍亦冈等，1995）成果，对南大岭、髫髻山、张家口三期火山活动进行总结。

1）南大岭期火山活动可划分两个喷发周期，早期火山熔浆沿断层裂隙喷发带呈脉动式连续多次溢出，岩石类型主要为溢流相玄武岩。晚期火山熔浆呈中心式喷发为主，喷发中心位于门头沟桑峪法城一带，岩石类型主要为火山爆发相的凝灰岩、安山质角砾熔岩和火山角砾凝灰岩等。该期火山岩属钙碱性系列，形成时代 180~170Ma，属早侏罗世。岩浆来源于富集型地幔，早期岩浆表现出原始岩浆特征，晚期岩浆则受地壳混染，形成于板内前造山构造背景。该期火山构造隶属于承德-武安火山喷发带（Ⅱ级），区内发育西山和昌怀两个伸展型断陷火山-沉积盆地（Ⅲ级）。

2）髫髻山期火山活动可划分髫髻山期和土城子期两个喷发阶段，分别形成髫髻山组和土城子组火山岩。髫髻山火山喷发阶段又可划分为两个喷发周期和两个火山间歇周期，分别形成髫髻山组第一、第三和第二、第四岩性段。该期火山活动表现为先爆发后溢流特征，分别形成角砾熔岩或火山角砾岩和熔岩层。其中，第一喷发周期以粗安岩、粗安质角砾凝灰岩为主，少量玄武质、玄武安山质火山岩；第二喷发周期则岩石类型更加复杂，除发育第一喷发周期的岩类外，还发育少量粗面质、流纹质熔岩和火山碎屑岩等。土城子火山喷发阶段表现为髫髻山强烈火山喷发阶段之后出现的强弱不等的阵发性火山活动，总体火山活动强度较弱，也可划分为两个喷发周期。早期为流纹质角砾凝灰岩、流纹质凝灰岩及粗安质凝灰岩组合，晚期则为粗安岩、玄武岩、粗安质角砾凝灰岩、粗面质角砾凝灰岩组合。该期火山构造隶属于承德—武安火山喷发带（Ⅱ级），划分为百花山-新城子（南带）和燕羽山-四海（北带）两个亚带，内含髫髻山、新城子等多个火山-沉积盆地（Ⅲ级），发育黄安坨等众多火山机构（Ⅳ级）。该期火山岩跨碱性—钙碱性系列，具有埃达克岩特性，火山喷发过程中少有岩浆混染，但遭受强烈结晶分异作用。其中，髫髻山组火山岩岩浆来源于古老富集型EMⅠ型岩石圈地幔和俯冲交代地幔混合，形成于伸展拉张构造背景，形成时代170~150Ma，属中—晚侏罗世；而土城子组火山岩形成于晚侏罗世，岩浆来源于地壳，形成于挤压主压应力和左旋压扭构造背景。

3）张家口期火山活动可划分为张家口和东狼沟两个喷发阶段，构成张家口组和东狼沟组火山岩地层。张家口火山喷发阶段可划分为两个喷发周期和一个火山间歇周期，分别形成张家口组第一、第三岩性段和第二岩性段。其中，第一喷发周期岩性复杂，且各盆地中演化趋势也有所不同，主体以玄武安山岩、粗面岩、粗安岩、粗安质角砾凝灰岩等中性火山喷出岩为主。第二喷发周期则以中、酸性火山喷发占绝对优势，以流纹质熔结凝灰岩、流纹质角砾凝灰岩及流纹岩为主。东狼沟火山喷发阶段是继张家口强烈火山喷发阶段后的较弱火山活动阶段，可划分出两个橄榄玄武质熔岩-火山角砾岩-沉玄武质凝灰角砾岩喷发韵律。早期为玄武安山质、粗安质岩石，见少量酸性流纹质岩石，晚期火山活动略强，流纹质岩石占主导地位，各火山-沉积盆地略有所差异。该期火山构造同属于承德-武安火山喷发带（Ⅱ级），内含东岭等6个火山-沉积盆地（Ⅲ级），发育沿河城卫生院等众多火山机构（Ⅳ级）。该期火山岩跨碱性—高钾钙碱性系列，具有复杂的岩浆来源及岩浆演化特性，来源于幔源、壳幔混合源、壳源岩浆，且经历了岩浆混染和结晶分异作用。张家口组火山岩形成时代137.3~97.74Ma，属早白垩世，其火山喷发阶段主体仍处于燕山板内同造山阶段，表现为造山挤压构造环境和增厚地壳特性，形成的火山岩具有陆缘岛弧或陆缘火山弧岩石化学特征；而东郎沟火山喷发阶段则进入燕山板内造山后伸展阶段，表现出造山后伸展构造环境和逐渐减薄的岩石圈厚度，形成的火山岩具有类似陆内裂谷或陆缘裂谷岩石化学特征。

（5）新生代火山岩仅隐伏于平原区古近系前门组，火山活动时间为新生代始新世晚期—渐新世，推测火山喷发方式为裂隙式喷发，与南苑-通州、夏垫深断裂密切相关。火山岩以溢流相玄武质熔岩为主，并有爆发相玄武质角砾凝灰岩夹层，普遍发育3~5层，属拉斑玄武岩系列。

（6）与火山活动有关的矿产主要为与中元古代基性、偏碱性、碱性火山活动有关的金属矿产和中生代燕山期火山作用有关的矿产，多表现为与潜火山岩或隐爆角砾岩有关，如延庆大庄科隐爆角砾岩型钼矿。

四、侵入岩与深成作用

第四篇分为4章，分中基性—超基性和中性—酸性两类，划分新太古代、古元古代、中元古代、古生代、中生代、新生代6个期次，对侵入岩岩石类型、岩石构造、岩石地球化学和稀土元素特征、岩体分布、侵位时代、侵位方式等进行总结，建立岩浆演化序列，分析岩浆侵位构造背景，划分岩浆活动旋回和岩浆岩带，并对岩浆活动与地壳演化、侵入活动与成矿作用进行概略总结。

（1）基性—超基性侵入岩规模小，主要呈岩株、岩脉或岩枝状产出，形成时代有新太古代、古元古代、中元古代、古生代和中生代。综合本区所处的大地构造位置、构造演化历史和基性—超基性岩浆揭示的地幔源区特征，岩石圈经历古岩石圈（中太古代—古元古代）、新岩石圈（中元古代—晚古生代）和现今岩石圈（晚古生代—新生代）3个演化阶段。伴随岩浆侵位，形成岩浆岩型（岩浆结晶分异作用）和岩浆热液型两类矿产。典型矿床有古元古代平顶山铬铁矿床、放马峪铬铁矿床、红石湾铂钯矿床，中元古代新地钒钛磁铁矿床，中生代上庄钒钛磁铁矿床。

1）密云岩群和四合堂岩群中划分出黑卧、苇子峪、苍术会、镇罗营、西湾子、安州坝和官地7个新太古代片麻岩体。岩性普遍遭受麻粒岩相变质，部分发育后期角闪岩相退变质，原岩相当于辉长岩和辉长闪长岩，分属于碱性玄武岩和钙碱性岩序列，岩浆来源于地幔且受地壳物质混染，分别形成于大陆裂谷和洋-陆俯冲型活动大陆边缘安第斯型岛弧环境。

2）本志基于最新同位素年龄、岩石化学、岩体地质以及变质程度等，将《北京市区域地质志》（1991版）前长城期系列基性超基性侵入岩厘定为古元古代侵入岩，包含密云放马峪、平顶山、东老窝、四合、潮河主坝、白河涧、蔡家洼和延庆红石湾等岩体，岩性为橄榄岩类、辉石岩类和角闪石岩类三类岩性，属于碱性岩类，岩浆来源于原始地幔岩部分熔融析出榴辉岩成分后的地幔难熔残留体，且经历了强烈岩浆分异演化，向贫铁、富镁硅的方向演化。岩浆侵位于拉张构造背景下，表现为上地幔的一些碎块呈固体塑性状态沿深大断裂侵位。

3）中元古代基性侵入岩为新地岩体，岩性为斜长岩，属钙碱性系列，形成年代为中元古代早期。岩浆来源于上地幔，且经历了地壳物质混染和斜长石分离结晶作用。岩浆侵位于非造山构造背景环境，为燕山裂陷槽早期慢性扩展条件下岩浆活动产物。同时，本志还系统梳理了发育于新太古代密云岩群中的中元古代高岭岩墙群。

4）根据最新同位素年龄和岩石化学资料，结合岩体地质及变质地质特征，将原前长城期下窝铺、李家湾、老米沟和对角沟门4个基性岩体，厘定为古生代早二叠世侵入岩，岩性为辉长苏长岩、辉长岩、辉长闪长岩和闪长岩组合，普遍发生角闪岩相变质和绿片岩相退变质。岩石属钙碱性-高钾钙碱性系列，岩浆主要来源于古老下地壳，且早期富集地幔岩浆。岩浆侵位中经历一定分离结晶作用，形成于华北克拉通北缘活动大陆边缘大地构造背景，表现为此构造环境下远程（板块边缘）深大断裂岩浆侵位。

5）中生代基性—超基性侵入岩分别发育于晚侏罗世、早白垩世和晚白垩世。晚侏罗世为基性侵入岩，由分散零星小岩株组成，分布于密云抗峪、朱家湾、下庄头峪、太平山，延庆太平庄、五座山和昌平阳聂村。岩性为辉长岩及其演化出辉长闪长岩组合，均为强碱性岩系列，岩浆来源于下地壳，有少量上地幔熔体加入，形成时代相当于髫髻山晚期火山喷发阶段，具有和该阶段火山岩一致构造背景，表现出与之相似的岩石地球化学等特征。早白垩世侵入岩包括棋盘岩体和上庄岩体。岩性组成为橄榄辉长岩、橄长岩、苏长辉长岩、辉长岩及其演化的辉长闪长岩和二长闪长岩组合，属高钾钙碱性-钾玄岩系列，岩浆来源于原始地幔，岩浆侵位无分异，但遭受一定地壳混染，形成于板内伸展造山期构造背景环境。晚白垩世侵入岩包括高家窑和兴隆城岩体，岩性为辉绿岩，为高钾钙碱性-钾玄岩系列，岩浆来源于上地幔，形成于燕山板内造山后非造山构造环境。

（2）中—酸性侵入岩是区内主要侵入岩，岩体规模大、期序复杂，形成时代有新太古代、中元古代、古生代和中生代，形成了区内岩浆矿床、热液矿床、接触交代（矽卡岩）矿床、斑岩型矿床和非金属矿床五种具有工业价值的矿床。

1）新太古代系列中—酸性侵入岩体有沙厂英云闪长质片麻岩系列（包含大漕、沙厂、阳坡地、琉璃庙4个岩体）、宋营子英云闪长质片麻岩系列（包含大北湾、穆家峪等7个岩体）、荆子峪花岗闪长质片麻岩系列（包含荆子峪和王家窝铺2个岩体）、康各庄英云闪长质片麻岩系列（包括康各庄和得田沟2个岩体）和旗杆顶花岗质片麻岩系列（包括旗杆顶、和尚峪、土粉沟、车道沟4个岩体）。岩石遭受强烈变质变形和混合岩化，属钙碱性系列。沙厂英云闪长质片麻岩系列与表壳岩之间有成因联系，形成于不成熟的岛弧构造环境。而宋营子英云闪长质片麻岩系列、荆子峪花岗闪长质片麻岩系列和旗杆顶花岗质片麻岩系列组成TTG组合，岩浆来源于下地壳物质部分熔融，形成于成熟大陆边缘岩浆弧或与俯冲作用有关的岩浆弧构造背景环境。

2）中元古代中—酸性侵入岩体有涝洼、古洞沟、东辛店、兰营和沙厂5个，岩性为石英二长岩、石英正长岩、钾长花岗岩、环斑花岗岩、黑云母花岗岩组合，属钾玄岩系列，沿赤城-古北口深大断裂和密云-墙子路-兴隆深大断裂呈近东西向脊状分布。岩体侵入于新太古代变质岩中，侵入接触关系清晰，且沙厂岩体可见同期花岗斑岩脉被中元古界长城系不整合覆盖。岩浆来源于下地壳物质部分熔融，且侵位过程中经历强烈分离结晶作用。该套岩石同斜长岩构成 AMCC 组合，为燕山裂陷槽裂陷早期慢性扩展条件下岩浆活动产物。

3）晚古生代中—酸性侵入岩体有梨树沟、荞麦峪、栗树洼和老公营 4 个岩体，岩性为伟晶正长岩、正长岩、正长斑岩组合，难以划分岩相，但可见脉状、枝状穿插或过渡关系，表现出呈短期多幕侵位特征。岩体平行于赤城-古北口深大断裂带呈近 EW 向分布，侵入中元古界长城系和新太古宙片麻岩中，侵入关系清晰，又被中生界侏罗系髫髻山组地层不整合覆盖。岩石为强碱性岩，表现出火山弧火成岩特性，岩浆来源于下地壳变质岩，且有年轻碱性玄武质地幔物质加入，形成于华北克拉通北缘活动大陆边缘大地构造背景，表现为此构造环境下远程（板块边缘）深大断裂岩浆侵位。

4）中生代中性—酸性侵入岩分布面积大，发育有大小 210 余个侵入体，构成本区八达岭岩浆岩带。按形成时代可划分为早三叠世、中三叠世、晚三叠世、早侏罗世、中侏罗世、晚侏罗世、早白垩世早期、早白垩世中期、早白垩世晚期，分属印支构造岩浆旋回和燕山构造岩浆旋回。中生代侵入岩岩性复杂，几乎包含了中性至酸性全部岩类，跨碱性至高钾钙碱性系列，岩浆多数来源于下地壳，少部分为壳幔混合源，岩浆复杂演化，致使众多岩体表现为杂岩体。印支期岩浆活动与蒙古增生褶皱带碰撞造山（中亚造山）不断影响波及华北板块内部有关，为华北克拉通破坏早期提供重要动力。燕山期岩浆活动记录了燕山板内造山壳幔演化过程。

（3）岩浆活动揭示区域大陆地壳演化过程。新太古代晚期，发育二幕三期八类基性—酸性侵入岩，记录了本区古老地壳由不成熟的洋内岛弧（沙厂英云闪长质片麻岩侵位）-半成熟的大陆岛弧（黑卧辉长质片麻岩和苇子峪辉长闪长质片麻岩侵位）-成熟的大陆边缘岛弧（宋营子英云闪长质片麻岩、荆子峪花岗闪长质片麻岩、康各庄奥长花岗质片麻岩侵位）-大陆碰撞造山带岩浆弧（旗杆顶花岗质片麻岩侵位），反映了初始玄武质地壳转化为半成熟的大陆地壳的整个过程。古元古代本区仅发现存在于表壳岩及英云闪长质片麻岩中的为数不多的超镁铁质岩墙，侵入作用较弱，反映大陆地壳成熟，并产生一定分异作用。中元古代早期—中期发育二幕三期五类基性-酸性侵入岩研究，分别为早期高岭岩墙、双峰式斜长岩-环斑花岗岩和中期大红峪期火山活动，为燕山裂陷槽岩浆活动产物，反映了中元古代期间成熟古老大陆地壳和岩石圈一次伸展裂解，但未演化为大洋-岛弧等后期洋壳而夭折。中元古代晚期—晚古生代，本区没有岩浆活动，古老大陆地壳极为稳定。晚古生代—新生代，本区经历了中泥盆世、早二叠世、三叠纪、侏罗纪、早白垩世、晚白垩世及新生代多期次构造-岩浆热事件。特别是早侏罗世以来的岩浆活动，组合形成特征鲜明的岩浆岩带，记录了燕山造山作用岛弧-碰撞-碰撞后伸展垮塌壳幔作用过程。

五、变质地质

第五篇分 5 章，对各类变质岩地质特征、岩石学与矿物学、岩石化学与地球化学等进行总结，划分了变质相带和变质地质构造分区，并进行了原岩恢复与形成环境的探讨。

1. 区域变质岩系

本区变质岩主要有变质表壳岩（包括密云岩群和四合堂岩群）、变质深成侵入体和西山地区浅变质岩。

变质表壳岩包括密云岩群和四合堂岩群。密云岩群属麻粒岩相变质岩，主要岩性包括磁铁石英岩类、斜长角闪岩、斜长辉石岩、石榴二辉石岩、角闪二辉石岩、角闪石岩、变粒岩等，多以包体形式集中成片分布，重熔混合岩化作用普遍，混合程度不均匀，大部分属于混合质变质岩及条带状、条痕状混合岩；四合堂岩群属角闪岩相变质岩，岩石类型包括斜长片麻岩、变粒岩、斜长角闪岩等，混合岩化作用相对较弱，以钠质交代为主。

变质深成侵入体有英云闪长质-奥长花岗质-花岗闪长质片麻岩和石英二长质-奥长花岗质-花岗质

片麻岩两种岩性组合。前者包括沙厂英云闪长质片麻岩、黑卧辉长质片麻岩、苇子峪辉长闪长片麻岩、宋营子英云闪长质片麻岩，后者包括荆子峪花岗闪长质片麻岩、康各庄奥长花岗质片麻岩、旗杆顶花岗质片麻岩。

西山地区浅变质岩主要发育在中元古界上部至中生界侏罗系下统中，主要为一套变质程度较浅的片岩、板岩、千枚岩及变质砂岩、砾岩，硬绿泥石分布普遍，属中低压绿片岩相变质或低角闪岩相变质。受构造环境差异热流活动强弱影响，该类型变质岩在时空上表现为不同程度变质带产出。

2. 区域变质作用

基于不同时期和不同变质地质单元变质岩系中典型变质矿物共生组合、矿物时代和相互交代关系，主期变质与后期叠加变质矿物特征、变质相温压条件等，本区区域变质作用可划分为阜平期、五台期、吕梁期、华力西期、印支期、燕山期，依次形成麻粒岩相-角闪岩相、角闪岩相-绿帘角闪岩相、角闪岩相-绿帘角闪岩相、绿帘角闪岩相-绿片岩相、绿片岩相和绿片岩相变质作用。后期区域变质作用依次叠加于早期区域变质作用之上，形成相应共生矿物组合。

3. 其它变质岩系与变质作用

其他变质作用主要有接触变质作用、低温动力变质作用、气液变质作用和混合岩化作用。接触变质作用又包括热接触变质作用、低温交代变质作用及气—液变质或蚀变作用三类，均与岩浆活动有关。低温动力变质作用主要发生在较大规模的断裂带，诸如半城子断裂、石城断裂和崎峰茶断裂等。气液变质作用主要发育于中酸性侵入岩体内及边部（或发育于爆破岩浆角砾岩体中），表现为与岩浆期后热液蚀变有关。混合岩化作用又可分为注入式边缘混合岩化作用和区域性混合岩化作用，前者发育于四干顶、柏岔山、云蒙山等岩体与围岩接触地带，后者广泛发育于新太古代地层中。

4. 变质相带与变质地质构造分区

本区变质岩系可划分3个变质相带9个岩性矿物带，分别为密云岩群变质相带（包括单斜辉石-石榴石带、二辉石带）、四合堂岩群变质相带（包括低角闪岩带、高级角闪岩带）和晚古生代-早中生代变质相带（包括硬绿泥石-绿泥石带，黑云母带，红柱石带，铁铝榴石-十字石带，硅线石-白云母带）。综合地壳演化、构造旋回及地质作用（如深成侵入作用、变质作用、构造作用形式和变形等），本区变质岩系划分为2大3小变质地质构造小区，分别为北山（沙厂-大漕变质小区和四合堂变质小区）和西山变质岩区。

六、岩石圈结构构造

第六篇分4章，对区域地球物理场和岩石圈地球化学、结构构造进行全面总结，并初步探讨了深部地质作用及其对资源环境的制约。

（1）深地震剖面、电磁测深剖面和地震层析成像等资料，揭示了本区区域地球物理场特征，其表现为强烈的横向不均匀性。

1）在地壳和上地幔的速度结构有明显的层状结构，表现强烈的横向不均匀性，特别是浅层的速度结构与地质结构分布相关，深大断裂地带往往发育速度场梯度带，特别是延庆盆地边缘和太行山山前地带。

2）本区位于北北东向大兴安岭-太行山-武陵山重力梯度带上，布格重力异常等值线呈密集与舒缓相间分布，重力异常值由南东向北西逐渐降低，在深大活动断裂地带形成重力梯度带，在太行山山前地区水平梯度变化最大。区域重力场可划分为西山重力梯级带异常区、平原重力高异常区和北山重力低异常区。

3）本区航磁异常属冀北地区高强度正负磁异常复杂交变区，区域磁场从密（云）怀（柔）平（谷）的高异常区向四周低异常或平稳正负磁场区逐渐降低，正磁场强度一般在100~400nT之间，负磁场强度一般在-100~200nT之间。根据航磁异常类型、强度、走向和范围等特征，分为怀北磁场区、密怀磁场区、延昌西山磁场区和平原磁场区。

4）本区地壳及上地幔电性可划分为地壳表层高导层、壳上部高阻层、壳内高导层、下部高阻层和上地幔高导层计5层，对应电阻率值分别为1～10Ω·m、10^2～10^3Ω·m、1～10Ω·m、10^2～10^5Ω·m和1～10Ω·m。

5）本区西北部山区热流值较低，一般低于60mW/m^2，在背斜的轴部和活动断裂带热流值增高，东南部平原区热流值较高，一般高于60mW/m^2，大地热流值为（1.7～2.0）×41.686 mW/m^2，形成了大面积的地热异常区。

6）本区现今区域构造应力场特征表现为水平应力作用为主，最大水平压应力方向为北北东向，地应力场发育现状决定了现今地壳形变特征。

（2）地球物理测深和地幔包体研究表明，本区地幔由尖晶石/石榴子石二辉橄榄岩、各类辉石岩、二辉岩等组成；下、中地壳分别主要由麻粒岩相和角闪岩相岩石组成，而上地壳以绿片岩相岩石组成，部分为低角闪岩相；地壳内部局部发育有层状低速体，可能为盖层发育花岗岩体的深部岩浆房。

（3）本区莫霍界面等深线走向呈北东—北北东向及近东西向，埋深由东南向西北方向渐渐加大。综合莫霍面埋深、早前寒武纪结晶基底埋深、布格重力异常、深地震剖面及航磁等资料，可将本区深部构造划分为太行山-燕山幔坡带和平原幔隆区两个深部地质构造区。深部构造对区域岩浆活动、成矿作用及表生地质环境具有重要控制影响。

七、地质构造

第七篇分4章，对褶皱、断裂等各类构造变形系统总结，进行区域构造单元、构造旋回与构造层划分。

（1）本区发育2期结晶基底褶皱和4期盖层褶皱构造。其中，盖层褶皱分别形成于印支和燕山早、中、晚期，是区内主要褶皱变形。断裂构造尤其发育，不同方向、不同期次、不同类型的断裂交错切割，统计命名断裂达45条。重要区域断裂有赤城-长哨营-古北口-承德、南口-平谷-山海关等9条。本区还发育逆冲推覆构造、剥离断层、变质核杂岩、韧性剪切带、固态流变构造群落、穹隆和半穹隆构造等构造组合。

（2）本区一级构造单元属于柴达木-华北板块，二级构造单元属于华北陆块，三级构造单元为燕山-辽西裂陷带和华北盆地，在三级构造单元基础上划分出4个四级构造单元以及12个五级构造单元。

（3）本区地质构造可划分为4个阶段、9个构造旋回及相应的构造层，归纳为基底构造层（包括阜平、吕梁构造旋回）、盖层构造层（晋宁、扬子、加里东、华力西、印支构造旋回）和上叠构造层（燕山、喜马拉雅构造旋回）。其中，燕山运动对区内构造研究意义重大，按照陆内造山观点将燕山运动分前造山、造山期、后造山、造山后4个阶段，认为其起始于广大华北克拉通北缘的软流圈主动上涌和造陆运动背景下的，而这一过程就是华北克拉通东部去克拉通化。

八、第四纪地质与地貌

第八篇分7章，系统总结北京地区地貌、第四纪沉积物、新构造运动特征，开展地貌单元划分和沉积作用研究。

（1）山区更新统沿用泥河湾组、周口店组和马兰组3个组级岩石地层单位。泥河湾组主要分布于山区河谷的Ⅳ级阶地上，岩性以红棕色、灰/黄褐色的砂、亚砂土、亚黏土，夹砂砾石层，孢粉以松、桦栎属及草本蒿属菊科为主。周口店组主要分布于山区河谷的Ⅲ级阶地上，以坡积、残坡积和洪积为主，岩性以褐红色粘质砂土、砂质黏土为主，孢粉以灌木及草本花粉占优势，如榆（*Ulmus*）、椴（*Tilia*）、桦（*Betula*）和松（*Pinus*）等。马兰组主要分布于各河流Ⅱ级阶地和谷坡上，岩性以黄土状土（次生黄土）、黄色黏质砂土、砂质黏土夹数层透镜状砾石层为主，成因类型有洪积、冲积、坡积、残坡积等。

（2）洞穴堆积更新统沿用太平山组、周口店组和羊耳峪组3个组级岩石地层单位。太平山组主要分布于太平山北坡大砾岩山下、周口店第1地点猿人洞、军庄灰峪以及怀柔龙牙洞等地，各洞穴沉积物差异较大，颜色以棕黄色、黄色调为主，发育大量哺乳动物化石。周口店组主要分布于房山周口店猿人洞、房山上店洞和密云溪翁庄洞，堆积物以棕红色调为主的黏土、砂土和角砾，发育周口店动物群。羊耳峪组主要分布于房山羊耳峪东岭子洞、房山周口店猿人洞、昌平龙骨洞、房山云水洞等，堆积物以灰色、黄色、黄棕色角砾、粉砂、砂质黏土等，含大量哺乳动物化石。

（3）山区全新统主要分布在永定河、潮白河、大石河、清水河、错河及山区间歇性河流的河床、漫滩、Ⅰ级阶地以及山麓地带，沉积物以冲洪积、冲积、坡积为主，岩性主要为褐黄色、灰色、灰褐色等砂砾石、砂及黏土类。

（4）平原区和延庆盆地地质结构较复杂，主要发育马池口、后沙峪-俸伯、平谷等多个第四纪凹陷等。在同一时期不同地区由于所处地理、构造位置及沉积物源等差异，空间上岩性及沉积特征存在较大差异，地层横向对比具有较大困难，本志对各个凹陷第四系进行了总结，建立了构造-沉积单元岩性序列、磁性地层和生物地层。

（5）平原区第四纪沉积类型主要为河流、湖泊、沼泽、风积、冲洪积陆相沉积。早更新世主要发育以粗砾为主的冲洪积扇，形成了扇形堆积平原雏形；中、晚更新世继承早更新世格局，山前地带发育有洪积扇裙和扇缘洼地，平原低洼地区有河流-湖沼发育，以砂质黏土、黏质砂土沉积为主；全新世沉积主要为冲积扇上辫状河及废弃河道-湖沼沉积。

（6）本区地势总体呈现西北高、东南低的特征，山区剥蚀切割强烈、平原区堆积厚度大。本区地貌类型复杂多样，可划分为2个三级地貌单元和8个四级地貌单元，华北中、低山和丘陵区三级地貌单元包括燕山西段中低山和丘陵区、太行山北段中低山和丘陵区、延庆山间盆地，华北平原区三级地貌单元包括大清河流域、永定河流域、北运河流域、潮白河流域和蓟运河流域。

（7）本区新构造运动强烈，间歇性构造升降在山区形成多级夷平面和河流阶地，在平原区发育了沉积间断、不整合面以及沉积中心的多次迁移，断裂构造活动性明显，其中黄庄-高丽营、顺义、夏垫和南口-孙河断裂为区内最重要的全新世活动断裂。

九、平原区三维地质结构

第九篇分为3章，分基岩、新生界和工程层对平原区三维地质结构进行总结。

1. 平原区基岩地质

（1）平原区隐伏于新生界下的基岩地层与山区出露相应地层基本一致，除了缺少古元古界、新元古界的震旦系、古生界的上奥陶统至下石炭统外，从太古宙变质岩系到中生代地层均有分布。隐伏于新生界下的侵入体主要有辛相庄、阳坊、交道和窑上岩体等。

（2）平原区隐伏于新生界下的基岩褶皱构造主要为印支期和燕山期。印支期褶皱有小汤山背斜、板桥背斜、昆明湖背斜和凤河营向斜等；燕山期褶皱有黄村向斜、榆垡-礼贤背斜、通州向斜、张家湾背斜、北小营-昌平向斜、回龙观-平西府向斜、来广营-南法信向斜、大孙各庄向斜等。平原区基岩断裂构造发育，主要发育有东西向、北东向及北西向三组。桐柏和皮各庄断裂是主要东西向断裂，南口-孙河和永定河断裂是主要北西向断裂，黄庄-高丽营、南苑-通县、夏垫和礼贤等断裂是主要北东向断裂。这些断裂或断裂局段在新生代以来持续强烈活动。

2. 平原区新生界地质

（1）古近系、新近系主要分布于北京断陷和大厂断陷内。其中，北京断陷内沉积限定于黄庄-高丽营断裂和南苑-通县断裂间，呈北东向展布，古近纪沉积中心大致位于宛平、卢沟桥、丰台一带，最厚处沉积厚度超过400m；新近纪沉积中心位于北天堂—葆台—四合庄一线，厚度超过1700m。

（2）平原区总体上由多个巨大的冲洪积扇和沉积凹陷构成。最重要的沉积凹陷为马池口、后沙峪、顺义和平谷凹陷。由山前向沉积凹陷中心，岩性相变大，宏观表现为颗粒逐渐变细、厚度变薄、分层增多特征。

3. 平原区工程地质

（1）平原区浅表层（0~30m）是北京城市主要工程层，地层结构特征鲜明。自西部山前向东至通州，由均一厚层卵石填充砂砾结构过渡为多层不同厚度砂土和黏土复杂交互结构。

（2）综合区域地壳稳定性、地貌、浅表层承载力与压缩性、地下水条件、不良地质作用等，平原区浅表层工程地质可划分为6个工程地质区，即大清河水系、永定河水系、北运河水系、潮白河水系及泃河水系堆积工程地质区和台地剥蚀构造工程地质区。

4. 三维地质结构模型

“北京市多参数立体地质调查”项目利用各类地球物理和钻孔资料，搭建三维地层格架，基于ACCESS、MapGIS、三维V3DMain等软件，分层次构建了平原区基岩、新生界、工程层三维地质结构模型，可以实现三维地质信息输入、数据库管理和三维地质数据建模与可视化分析等多种功能。本志对该项目的上述成果进行了概要总结。

十、城市地质资源与环境

第十篇分为7章，基于“北京市多参数立体地质调查”等项目成果，对北京城市土壤、地下空间、地下水、矿产、能源、地质遗迹资源和崩滑流、地面沉降、地裂缝、采空塌陷等地质灾害等进行系统总结，进行城市地质环境系统划分，并以通州区为例，展示了城市多维数据模型构建进展。

1. 地质资源

（1）北京城市地下空间资源主要利用场所位于五环内，地下空间质量等级可划分为一级（适宜）、二级（较为适宜）和三级（不适宜）；地下0~50m范围内70%以上区域等级都为适建区，适宜进行地下空间资源开发利用。城市地下空间的开发利用，主要受地形地貌、人工填土等工程地质条件、水文地质条件，以及活动断裂、地裂缝、地面沉降、砂土液化、放射性气体等不良地质作用的影响。

（2）本区处于海河流域地下水系统之中，划分为潮白河-蓟运河-温榆河、永定河和大石河-拒马河3个相对独立的地下水系统。平原区第四系地下水是城市主要的供水水源，由山前向平原区，由单一含水层逐渐过渡为4个含水层组，地下水资源丰富。1991—2014年多年平均补给量为$23.59\times10^8m^3/a$，消耗总量为$28.46\times10^8m^3/a$。

（3）北京地区矿产种类繁多，截至2014年底，共发现各类矿产127种（含亚矿种），其中固体矿产121种，查明资源储量并已编入《北京市矿产资源储量表》的固体矿产有67种、354处矿产地。

（4）本区清洁能源资源主要为地热和浅层地热能资源。地热资源以中低温亚经济型为主，热储层主要为蓟县系，寒武系和奥陶系次之，热盖层主要为新生界；地热田热储总热量为$500.772\times10^{18}J$。浅层地热能资源主要赋存在地表以下15~200m内的岩土体和地下水中，受地貌、水文、地温、构造条件影响。浅层地热能资源利用方式以水源热泵为主，截至2012年底，已建浅层地热能项目约1073个，实现供暖面积$3348\times10^4m^2$。

（5）本区地质遗迹资源较丰富，共有重要地质遗迹点50处，分为基础地质类、地貌景观类、地质灾害类3大类，又可进一步细分为12类、24亚类。其中，世界级地质遗迹2处，分别为周口店古人类化石产地和千家店侏罗纪恐龙足迹化石产地；国家级地质遗迹7处，分别为下苇店寒武纪地质事件剖面、千家店侏罗纪木化石群、周口店动物群、十渡岩溶地貌、东关上-上方山洞穴群岩溶地貌、佛子庄洞穴群岩溶地貌和云蒙山花岗岩地貌；其余41处遗迹点均定级为省级。截止2014年，北京市共批建8个地质公园，包括2个世界级、5个国家级和1个市级。

2. 地质环境

（1）本区土壤主要为潮白河、永定河、温榆河及泃河等河流的冲积作用形成，土壤元素背景值与沉积物物源关系密切。土壤综合质量评价显示土壤质量总体较好，最好土壤发育地区主要位于山前平原地带，包括平谷地区、昌平—门头沟—房山一线平原边缘地区，及通州境内温榆河水系两侧、凉水河、凤港减河流域；平原区土壤也有少量污染，主要重金属污染元素为Hg，主要有机氯农药污染为

DDT 和 HCH。

（2）本区区地下水总体水质状况良好，平面上由南向北逐渐变好，垂向上由浅到深逐渐变好。伴随地下水资源开发利用，存在水位持续下降、局地水质污染、供水能力衰减等问题（截至 2015 年）。

3. 地质灾害

（1）地面沉降是北京平原区主要缓变性地质灾害之一，有南北 2 个沉降区 7 个沉降中心，沉降区累计沉降量呈上升趋势。自然因素包括地质构造运动、地层岩性及结构特征和土层次固结沉降等，是产生地面沉降的重要地质背景，而人为因素包括超量抽汲地下水、工程建设等。

（2）平原区地裂缝主要有顺义地裂缝、高丽营地裂缝、北小营地裂缝、羊房地裂缝、庙卷地裂缝、北彩地裂缝、北甸地裂缝和孙河地裂缝等。地裂缝成因类型多数属于构造型地裂缝，其次为液化型地裂缝，超量开采地下水对地裂缝活动具有诱发促进作用。

（3）砂土液化灾害主要分布在地势低洼、沉积物新、结构松散的潮白河、温榆河、小中河、泃河等河流的中下游沿岸地区。砂土液化的产生、发展和消散与动荷载条件、土的物理性质、埋藏条件、地下水位特征、受力条件和边界条件等有关。

（4）采空塌陷主要分布在京西产煤区，其影响因素较多，包括地质条件、地下采空区大小、煤层厚度及开采深度、重复采动以及地形地势等，自然重力或其他外界触发因素是其发展的动因。

（5）本区崩塌、泥石流地质灾害较为突出。崩塌主要分布在山区，尤以山区道路沿线分布最多，以滑移式、倾倒式和错落式崩塌最为常见。泥石流主要分布在密云北部、怀柔中部、延庆东部及房山、门头沟西部地区，以沟谷型高容重低黏性泥石流最为常见。崩塌和泥石流的形成与山地地貌、岩性、构造、地形、气候等条件有关。此外，北京地区还少量滑坡，一般多为古滑坡或由人类工程活动引起的边坡滑移。

4. 城市多维数据三维模型构建

以通州区为例展示城市地下资源环境三维模型，分为工程层、砂土液化、地下空间、地下水资源、地下水质量、地面沉降、浅表层（地表以下 20cm 范围）土壤地球化学及浅层地热能 8 个专业构建子模块，实现 8 个地质要素相关信息空间分布展示和属性信息查询。

十一、区域地质发展史

第十一篇分 4 章，分新太古代—古元古代、中元古代—中三叠世、晚三叠世—早白垩世、晚白垩世—第四纪 4 个构造阶段总结了区域地质发展演化过程。

1. 新太古代—古元古代阶段

新太古代早期，海底火山强烈而频繁的间歇性喷发，熔岩流和火山碎屑与侵位于上地壳的花岗岩类物质混合在一起形成了密云岩群和四合堂岩群表壳岩；晚期阜平期构造-热事件使密云岩群、四合堂岩群褶皱上隆，形成东西向构造线，伴有麻粒岩相区域变质，强烈韧性剪切，并伴随有大量基性（黑卧）、中性（苇子峪等）及酸性（旗杆顶等）岩浆侵入。密云、四合堂群表壳岩和后期变质侵入体构成了本区的大部分结晶基底。

古元古代，本区处于华北克拉通东部陆块，为剥蚀陆地区，缺失沉积地层记录。岩浆活动较弱，以橄榄岩、辉石岩、角闪岩、辉长岩及辉绿岩为代表的超基性—基性岩体和岩脉沿北东向张性断裂构造侵位，形成小型岩体和岩墙群，先期地质体发生浅层次韧性剪切和绿片岩相区域变质作用。

2. 中元古代—中三叠世阶段

中元古代早期，经华北克拉通地幔软流圈调整，多幕镁铁质岩墙群沿着古元古代拼合带（软弱带）或先期深断裂带侵位（高岭岩墙群等），并引起下地壳物质部分熔融，演化出中酸性岩和基性岩密切共生而又不同源非造山特性的岩浆侵位活动，开始了广大燕山裂陷槽形成演化阶段。地壳强烈伸展-裂陷构造背景下，尚义-隆化、康保-围场等大型断裂强烈活动，形成裂陷槽盆地，沉积长城系常州沟组至大红峪组地层。期间，地球外圈的含氧性大气圈基本形成，气圈和水圈中已开始拥有一定数量的游离氧，藻类生物开始繁殖。

中元古代中期，燕山裂陷槽逐渐消亡，伴随地幔失热和活化能量衰减，区域地壳应力的释放，岩石圈上部构造由裂陷扩展活动演化为坳陷槽不均匀沉降活动。浅部发育为具有过渡特性盆地，沉积高于庄组至雾迷山组地层。而深部能量衰减限制了该阶段岩浆活动，也没有经历显著的构造变形和变质作用。

中元古代晚期至中三叠世，本区主体处于稳定克拉通状态，地壳以升降运动为主，基本没有强烈的构造变形、岩浆活动和变质作用。构造抬升期区域普遍为陆，长期遭受剥蚀和夷平。构造沉降期演化为陆表海盆地，普遍发育属于陆块沉积的石英砂岩建造、碳酸盐岩建造和碎屑岩建造（含煤碎屑岩建造）。同时，区内怀柔、密云北部受蒙古-西伯利亚板块向南俯冲碰撞远程影响，先期尚义-隆化深断裂阶段性伸展正断活动，并伴有深源小规模多期次（中泥盆世、早—中二叠世、早三叠世）岩浆活动和接触变质作用。

3. 晚三叠世—早白垩世阶段

晚三叠世—早白垩世是燕山构造发展期，主要表现为陆内造山过程。境内先后形成3幕火山-沉积盆地，分别沉积杏石口组—九龙山组、髫髻山组—土城子组和张家口组—夏庄组地层。盆地空间格局经历了由北东东向→北东向→北北东向逆时针方向旋转。期间，构造变形较为复杂，以北东方向的显著褶皱和断裂为特征，岩浆多期喷发和侵入，是区内最重要的岩浆活动期。燕山运动中、晚期还是本区内生金属矿产的主要成矿期，直接控制了内生金属矿产的形成与赋存。

4. 晚白垩世—第四纪构造阶段

晚白垩世—第四纪是喜马拉雅构造发展期，本区主要表现为伸展的构造环境。其中晚白垩世至古新世处于燕山后造山稳定阶段，地壳处于隆升剥蚀状态，形成北台期夷平面。始新世—中新世，在伸展构造背景下，准平原沿断裂带开裂解体，以北东向黄庄-高丽营断裂、南苑-通县断裂和夏垫-礼贤断裂为界，在本区形成两隆两坳的盆-岭构造格局，即北京断陷、大厂断陷和京西隆起、大兴隆起。新近纪—第四纪，本区表现为弱伸展兼走滑构造环境，北东向断裂的继承并分段活动，北西向南口-孙河等断裂新生或再次活动，破坏了两隆两坳的盆-岭构造格局，演化出马池口凹陷、顺义凹陷等新的沉积中心。

Abstract

This publication, drawing upon the "*Regional Geological Annals of Beijing*" (1991), presents a comprehensive overview of regional geological surveys and research conducted in the Beijing area over the preceding three decades. Its structure adheres to the "*Work Guide for Regional Geology of China*" encompassing a wide range of geological topics. The book delves into Regional stratigraphy, Sedimentary rocks and deposition, Volcanic rocks and volcanism, Intrusive rocks and intrusion, Metamorphic geology, Quaternary geology and landform, Lithospheric structural tectonic, Geological structure, Three-dimensional geologic model of the Beijing plain, Urban geological resources and environment, and History of regional geology development in Beijing. Organized into eleven parts comprising 58 chapters, the text includes introductory and concluding remarks, totaling approximately 2. 1 million Chinese characters. The accompanying figures provide valuable visual representations of the region's geological features:

1) Geological Map of Beijing Municipality (1 : 250000).

2) Tectonic Map of Beijing Municipality (1 : 250000).

3) Aeromagnetic ΔT Contour Map of Beijing Municipality (1 : 250, 000).

4) Quaternary Geology and Geomorphological Map of Beijing Municipality (1 : 250000).

5) Bedrock Geological Map of Beijing Municipality (1 : 250000).

1. Regional Stratigraphy

This part comprises six chapters, systematically summarizing various strata from the Neoarchean to the uppermost Neogene. It focuses on lithostratigraphy, conducting multiple stratigraphic divisions and regional correlation by utilizing characteristics from lithostratigraphic sequences, including chronostratigraphy, biostratigraphy, sequence stratigraphy, and event stratigraphy.

A. Pre-Archean lithostratigraphy

The pre-Archean lithostratigraphy is classified into two informal lithostratigraphic units, the Miyun and Sihetang groups, based on the analysis of rock types, metamorphism, migatization processes, structural features, and mineral components. The Miyun Group, consisting of pyroxene leptynite, pyroxene-bearing biotite plagioclase granulite, leucoleptite, garnet pyriclasite interbedded with pyroxene magnetite quartzite, underwent metamorphism in the granulite facies, followed by retrograding superposition in the amphibolite facies. The Sihetang Group, composed of amphibolite, compound fabric biotite leptynite, biotite hornblende leptynite, migmatite, and magnetite quartzite, experienced regional metarmorphism in the amphibolite facies and strong ductile shear deformation due to its proximity to the Mesozoic Yunmengshan granite pluton. Both groups exhibit difficulties in identifying internal lithohorizon. Minor-scale folds, ductile shear, schistosity, and amphibolite and magnetite quartzite flaser structures are common. Based on lithological combination, lithochemistry and rare earth element geochemistry, isotope ages, and contact relationships, the Miyun Group and the Sihetang Group are interpreted to have formed during the early and late stages of the Neoarchean, respectively.

B. Meso to NeoProterozoic lithostratigraphy

Distributed across approximately one-third of the Beijing mountainous resion, twelve formations were established for this period. These include the Changcheng Systerm's Changzhougou, Chuanlinggou, Tuanshanzi, and Dahongyu formations; the Jixian Systerm's Gaoyuzhuang, Yangzhuang, Wumishan, Hongshuizhuang, and Tieling formations; the Undefined Systerm's Xiamaling Formation; and the Qingbaikou Systerm's Longshan and Jingeryu formations. Recent isotopic age data place the Gaoyuzhuang Formation within the Jixian Systerm and classify the Xiamaling Formation under the temporarily established Undefined Systerm and the Xiamaling Formation within the provisionally established Undefined Systerm. The lithology primarily consists of unmetamorphosed (with local slight metamorphism) littoral to neritic facies magnesium-bearing carbonate rocks. Clastic rocks and claystone are subordinate, with locally developed alkaline basaltic volcanic rocks. The carbonate formations exhibit stromatolith and are rich in microfossil remains.

C. Cambrian-Ordovician lithostratigraphy

Seven formations are distinguished for this period, including the Changping, Mantou, Zhangxia, and Chaomidian formations of the Cambrian System; the Yeli, Liangjiashan, and Majiagou Formations of the Ordovician System. These formations consist mainly of epicontinental carbonate deposits, rich in fossil organisms. The Cambrian strata are characterized by trilobites, with sixteen trilobite fossil zones established in the middle-upper Cambrian System. The Ordovician strata are dominated by cephalopods and conodontophoridia, demonstrating a good correlation with North China.

D. Carboniferous-Triassic lithostratigraphy

Four formations are identified for this period, including the Carboniferous System's Taiyuan and Shanxi formations; the Carboniferous-Permian System's Shihezi Formation; and the Permian-Triassic System's Shuangquan Formation. With a total thickness of approximately 700 meters, these formations represent deposition in transitional facies. The marine facies consist of limestone, marl, calcareous sandstone, and calcareous mudstone, intercalations rich in marine animal fossils, such as Brachiopods, Bivalves, and Cephalopods. The upper section, representing paralic deposition, contains coal seams and varicolored clastic rocks, characterized by the Cathaysian flora.

E. Late Triassic to Cretaceous lithostratigraphy

Twelve formations are distinguished for this period, including the Upper Triassic Xingshikou Formation; Lower Jurassic Nandaling and Yaopo formations; Middle Jurassic Longmen and Jiulongshan formations; Middle-Upper Jurassic Tiaojishan Formation; Upper Jurassic Tuchengzi Formation; and Lower Cretaceous Zhangjiakou, Donglanggou, Jiufotang, Tuoli, and Xiazhuang formations. Widely distributed in the Mesozoic Cenozoic continental basins, these formations collectively reach 9700 meters in thickness. They represent a thick, multicyclic, and complex volcanic-sedimentary rock series. These formations contain abundant coal resources and the association of plants and animals.

F. Paleogene-Neogene lithostratigraphy

listhostratigraphy between the mountainous resion and plain are differentiated for this period. In the torrent tract, the Paleocene Changxindian Formation and the Upper Eocene Yuling Formation are distinguished. In the Beijing Depression, the listhostratigraphy includes the Paleocene Changxindian Formation, the Oligocene Qianmen Formation, the Middle Miocene-Lower Pliocene Tiantan Formation, and the Upper Pliocene Tianzhu Formation. The Dachang Depression features the Lower Paleocene Kongdian Formation, the Upper Paleocene-Lower Oligocene Shahejie Formation, the Upper Oligocene Dongying Formation, the Middle Miocene Guantao Formation, and the Upper Pliocene Minghuazhen Formation. These strata are mainly buried in the plain and the Yanqing Basin, with sporadic outcrops in the western foothills of the plain and the marginal zone of the Yanqing Basin. They are primarily characterized by fluvio-lacustrine deposits, with some cave deposits, and contain abundant plant and animal fossils.

2. Sedimentary Rocks and Deposition

This part is divided into seven chapters and embarks on a chronological exploration of sedimentation patterns, basin development, and associated mineral resources in Beijing, spanning from the Mesoproterozoic to the

Neogene.

A. Meso to Nao Proterozoic: A Marine-Dominated Realm

Sedimentation during this period primarily yielded magnesian carbonate rocks, with subordinate clastic rocks and volcaniclastic rocks. Sedimentation occurred primarily in marine facies, followed by transitional facies, with limited continental facies. Barrier-lagoon complexes and intraclastic deposition dominated the marine realm, encompassing a variety of microfacies reflecting diverse sub-environments, from barrier islands and lagoons to tidal flats and deeper marine settings. Transitional (estuarine and deltaic) facies characterize the Changzhougou and Longshan Formations, while limited continental (fluvial) deposits are found within the Changzhougou Formation. Tectonically, this period witnessed a transition from rift to depression and finally to craton basins, all part of the broader Yanshan-Liaoxi Basin. Sedimentary structures reflect a shift from episodic to long-term stability, showcasing a variety of depositional systems, including terrigenous carbonatite formation, homogeneous terrigenous (phosphorus-bearing) formation, homogeneous terrigenous (iron-rich) formation, thick bituminous algal reef carbonatite formation, bituminous carbonatite formation, algal reef carbonatite formation, dust polygenic provenience formation, gray polygenic provenience formation.

B. Early Paleozoic: A Epicontinental Sea

The Early Paleozoic ushered in a regime of carbonate deposition, primarily limestone, interbedded with clastic rocks. A stable cratonic setting fostered the development of a shallow marine environment characterized by restricted circulation and dominance of tidal processes. Sedimentation occurred in a range of sub-environments, including evaporite platform facies, restricted platform facies, open marine platform facies, deep basin facies, and intermittent storm facies. These deposits accumulated within the epicontinental sea that blanketed the North China Craton, characterized by flat terrain, slow subsidence, and a predominance of terrigenous clastic-carbonate and carbonate sedimentation. Stable authigenic and terrigenous formation, marked by uniform lithologies and thick strata, attest to regional-scale uplift and subsidence cycles.

C. Late Paleozoic: From Marine to Continental Dominance

Terrigenous clastic rocks, spanning a wide range of grain sizes, dominated Late Paleozoic sedimentation, accompanied by subordinate mudstone, limestone, and volcaniclastic rocks. This period witnessed a dynamic transition from marine (barrier-lagoon) to transitional (deltaic-fluvial) and finally to continental (fluvial-lacustrine) environments. Tectonic evolution led to the development of stable epicontinental basins and, later, sub-stable intraplate foreland basins. Complex internal sedimentary formations reflect this dynamic setting, with five distinct formations types: stable marine homogeneous detrital (coal-bearing) formations, stable terrestrial detrital (coal-bearing) formations, stable fluviolacustrine detrital formations, sub-stable terrestrial volcaniclastic formations, and sub-stable fluviolacustrine terrestrial formations.

D. Mesozoic: Intraplate Basins and Volcanic Influence

Volcaniclastic rocks, volcaniclastic lavas, and welded pyroclastk rocks became prominent during the Mesozoic, interbedded volcaniclastic sedimentary rocks and terrigenous clastic rocks. The Xiazhuang Formation's marlstone and the Tuchengzi Formation's fine crystalline limestone reflect localized variations in depositional conditions. Terrestrial fluvial-lacustrine settings prevailed, encompassing shore shallow lakes, braided rivers, and meandering rivers. This period witnessed the development of intraplate basins within North China, driven by episodic tectonic movements associated with the Yanshanian orogeny. Three generations of basins, each shaped by three distinct tectonic pulses, evolved through time. Fault-bounded basins and sag basins (flexural) emerged, exhibiting distinct spatial distributions and sedimentary patterns. Sub-stable to unstable terrestrial clastic and volcanic-sedimentary rock formations dominate the stratigraphic record.

E. Cenozoic: Fault-Controlled Basin Formation

Cenozoic sedimentation, primarily from the Paleogene to Neogene, is characterized by terrigenous clastic rocks, with localized basalt, mudstone, and volcalastic rocks. The Yanqing Basin and the Beijing Plain, specifically the Beijing Sag and the Dachang–Langfang Sag, were the primary depositional centers. These sags formed within the broader context of the Neogene Taihang Mountains intermontane fault basin and the North China Plain fault basin. Fluvial–lacustrine environments dominate, with occasional marine transgressions evidenced by localized transitional facies. Terrestrial clastic rock constructions are predominant, with subordinate terrestrial volcan–sedimentary rock formations.

F. Mineral Resources: A Legacy of Diverse Sedimentation

The diverse and evolving sedimentary history of the Beijing region has endowed it with a wide array of mineral resources. The Mesoproterozoic–Neoproterozoic period yielded 11 mineral types, including iron, manganese, dolomite, silica, potassium, limestone, phosphate, barite, marble, jade, and brickkiln shale. The Early Paleozoic period yielded six mineral types, including copper, bauxite, dolomite, limestone, marble, and shale. The Late Paleozoic period yielded five mineral types: iron, apyrous clay, jarosite, shale, and coal. The Mesozoic period yielded coal and shale mineral types. Finally, the Cenozoic period yielded six mineral types: heap sand, peat, kaolin, building stone, brick clay, and building loess.

3. Volcanic Rocks and Volcanism

This part delves into the volcanic history in the Beijing region, encompassing Neoproterozoic to Cenozoic eruptions that shaped the landscape and left behind a legacy of diverse volcanic rocks and mineral resources.

A. Neoproterozoic Volcanism

Neoproterozoic volcanic rocks, heavily overprinted by metamorphism, offer glimpses into early volcanic activity. The Miyun Group comprises ultramafic–mafic and felsic acidic volcanic rocks, classified as tholeiitic basalt series. The Sihetang Group records a complete volcanic eruption–sedimentary cycle, characterized by basalt–andesite–dacite sequences.

B. Mesoproterozoic Volcanism

Concentrated in the Xiongerzhai area of the Pinggu district, Mesoproterozoic volcanism, primarily associated with the Dahongyu and Tuanshanzi Formations, documents five distinct eruption–deposition cycles. Effusive tephrite, ventrallite, phonolite, and andesitoid, along with explosive volcanic breccias and lavas, characterize this phase. These Shoshonite series rocks originated from a weakly depleted mantle source, influenced by crustal contamination and limited fractional crystallization.

This volcanic activity coincided with a passive rift tectonic setting, signifying a key stage in the early evolution of the Yanshan Rift Basin. Volcanic structures are preserved within the Pinggu–Jixian sedimentary–volcanic basin, part of the broader Yanshan–Liaoxi volcanic province. This region experienced the most intense Mesoproterozoic volcanism, marked by over 40 volcanic centers, including the prominent Jiangjunguan and Huangsongyu centers.

C. Paleozoic Volcanism

Paleozoic volcanic rocks are scarce in Beijing, represented only by the Shuangquan Formation. However, adjacent areas, the Yunwu Mountain of Fengning County, Hebei Province, experienced significant Permian volcanism, evidenced by the Yunwu Mountain Formation, which may extend into the Beijing region. This volcanic activity was characterized by plateau–forming eruptions, explosive events, pyroclastic flows, and limited subvolcanic intrusions. Intermediate trachyandesites and acidic rhyolites dominate the rock assemblages, suggesting a calc–alkaline magma series formed within a magmatic arc environment associated with a continental margin orogeny.

D. Mesozoic Volcanism

The Mesozoic witnessed a peak in volcanic activity across Beijing. This section draws upon the work of Bao et al. ("*Beijing Volcanic Geology and Volcanic Rocks during Yanshan Period*", 1995) to synthesize the volcanic history of the Nandaling, Tiaojishan, and Zhangjiakou periods.

a. Nandaling Period (180–170 Ma)

Early pulsating fissure eruptions of basalt transitioned to central vent eruptions centered at Sangyu and Facheng in the Mentougou district. These later eruptions produced volcanlastic deposits, including tuffs, agglomerates, and lavas, with a calc–alkaline affinity. Magmas originated from an enriched mantle source, showing evidence of crustal contamination during later stages. This activity occurred within an intraplate, pre–orogenic tectonic setting, associated with the Chengde–Wu'an volcanic province (Level II) and its extensional fault–bounded basins (e. g., Xishan and Changhuai basins, Level III).

b. Tiaojishan Period (170–150 Ma)

This period encompasses two sub–stages, the Tiaojishan and Tuchengzi, each is characterized by distinct eruption cycles and resulting formations. The Tiaojishan sub–stage involved both extrusive and effusive eruptions, producing a diverse suite of volcaniclastic rocks, including trachyandesites, basalts, and rhyolites. The Tuchengzi sub–stage witnessed less intense, intermittent eruptions, yielding rhyolitic and trachyandesitic tuffs and lavas.

Volcanic structures are associated with the Chengde–Wu'an volcanic province, specifically the Baihuashan – Xinchengzi and Yanyushan – Sihai subzones. These subzones host numerous volcanic sedimentary basins (e. g., Tiaojishan and Xinchengzi) and volcanic edifices (e. g., Huang'antuo). Tiaojishan volcanic rocks span the alkaline and calc–alkaline series, exhibiting adakitic characteristics and suggesting a mixed magma source involving enriched lithospheric mantle and subduction–modified mantle. Tuchengzi volcanic rocks, with a crustal origin, formed under a compression tectonic regime.

c. Zhangjiakou Period (137–98 Ma)

Two eruption stages, the Zhangjiakou and Donglanggou, characterize this period. The Zhangjiakou stage began with eruptions of intermediate composition (basaltic andesite, trachyte) followed by a shift to more acidic rhyolitic volcanism. The Donglanggou stage was marked by weaker, predominantly basaltic eruptions.

These volcanic rocks, spanning the alkaline to high–K calc–alkaline series, reflect a complex interplay of mantle–derived, crust–mantle, and crustal magmas, modified by contamination and fractional crystallization. The Zhangjiakou stage coincided with the syn–orogenic phase of the Yanshanian orogeny, characterized by compression and crustal thickening, while the Donglanggou stage transitioned into a post–orogenic extensional regime.

E. Cenozoic Volcanism

Recent discoveries have identified Cenozoic volcanic rocks within the Paleogene Qianmen Formation of the Beijing sub–plain. These late Eocene to Oligocene eruptions, potentially fissure–related and associated with regional faults, produced plateau basalts and basaltic tuffs, belonging to the tholeiitic series.

F. Volcanic–Hosted Mineralization

Volcanic activity throughout Beijing's history has contributed to the formation of valuable mineral deposits. Metal ores are associated with Mesoproterozoic and Mesozoic volcanic events, often concentrated within subvolcanic and cryptoexplosive settings, such as the Dazhuangke diatreme in Yanqing. This dynamic volcanic history, spanning over a billion years, has profoundly shaped the geology, landscape, and mineral potential of the Beijing region. Studying these diverse volcanic rocks provides critical insights into the region's tectonic evolution, magmatic processes, and metallogenic endowment.

4. Intrusive Rocks and Intrusion

This part unravels the intricate history of magmatism in Beijing, focusing on the diverse suite of intrusive rocks that record billions of years of crustal growth, tectonic reworking, and mineralization.

A. Ultrabasic–Basic Intrusions: Tracing Lithospheric Evolution

Ultrabasics–basite intrusions, primarily occurring as stocks, veins, and apophyses, punctuate Beijing's geologic record from the Neoarchean to the Mesozoic. These intrusions provide crucial insights into the evolution of the lithosphere, which can be broadly divided into three stages: the Paleolithosphere (Neoarchean–Paleoproterozoic, Characterized by the emplacement of primitive magmas derived from the mantle), Neolithosphere (Mesoproterozoic–Late Paleozoic, Marked by a shift towards more evolved magmas, reflecting increasing crustal influence), and the present lithosphere (Late Paleozoic–Cenozoic, Dominated by magmatism associated with plate interactions and intraplate processes). The emplacement of magmas has led to the formation of valuable mineral deposits, both through direct magmatic crystallization and magmatic–hydrothermal processes. Notable examples of these minerals include the Paleoproterozoic Pingdingshan chromite, the Fangmayu chromite, the Hongshiwan platinum palladian, the Mesoproterozoic Xindi schreyerite magnetite, and the Mesozoic Shangzhuang schreyerite magnetite.

a. Neoarchean

Seven gneissoid intrusions (Heiwo, Weiziyu, Cangshuhui, Zhenluoying, Xiwanzi, Anzhouba, and Guandi) within the Miyun and Sihetang groups showcase significant deformation and metamorphism. This gabbro and gabbro diorite, ranging from alkaline basalt and calc–alkaliic series, originated from the mantle–derived magmas that interacted with crustal material. Their emplacement likely transitioned from a continental rift to an active continental margin setting, potentially resembling an Andean–type arc.

b. Paleoproterozoic

Paleoproterozoic intrusions, including the Fangmayu, Pingdingshan, Donglaowo, Sihe, Chaohezhuba, Baihejian, and Caijiawa plutons in Miyun, and Hongshiwan pluton in Yanqing, are dominated by olivinfels, pyroxenite, and hornblendite, all exhibiting an alkaline series. These rocks originated from the partial melting of primitive mantle material, with subsequent differentiation leading to magnesium and silicon magnesium– and silicon–rich compositions. Emplacement occurred within an extensional regime, regime, possibly related to the emplacement of mantle fragments along discordogenic faults.

c. Mesoproterozoic

The Xindi pluton, composed primarily of anorthosite (the calc–alkaline series), exemplifies this period. Derived from mantle–derived magmas that experienced crustal contamination and plagioclase fractionation, this intrusion reflects the early extensional phase of the Yanshan Rift Basin. Mesoproterozoic gabbro dykes also provide insights into this extensional event.

d. Early Permian

Early Permian intrusions (the Xiawopu, Lijiawan, Laomigou, and Duijiaogoumen plutons) comprise gabbronorite, gabbro, gabbro diorite, and diorite, and commonly metamorphosed to amphibolite facies or greenschist facies. These calc–alkaline to High–K calc–alkaline series rocks originated from both the ancient lower crust and enriched mantle sources, and emplaced an active continental margin setting along discordogenic faults on the northern margin of the North China Craton.

e. Mesozoic

Mesozoic ultrabasics–basite intrusions developed during the Late Jurassic, Early Cretaceous, and Late Cretaceous periods in Beijing.

Late Jurassic plutons are represented by scattered stocks in Kangyu, Zhujiawan, Xiazhuangtouyu, Taipingshan in the Miyun district, Taipingzhuang in Yanqing district, Wuzuoshan, and Yangniecun in Changping district. These plutons consist of gabbro, transitioning to gabbro diorite, and belong to the alkaline series. Geochemical evidence suggests that their magmas originated from the lower crust with minor contributions from upper mantle melts. These plutons formed during the late volcanic eruption stage of the Tiaojishan Formation, exhibiting tectonic and geochemical characteristics similar to the coeval volcanic rocks.

Early Cretaceous intrusions include the Qipanyan and Shangzhuang plutons, composed of Aasbi-gabbro, allivalite, gabbronorite, gabbro, and their evolved gabbro diorite and monzodiorite. These rocks belong to the High-K calc-alkaline to Shoshonite series, suggesting that their magmas originated from the primitive mantle and were emplaced without differentiation but experienced some crustal contamination. These plutons were emplaced without significant differentiation but experienced some crustal contamination, and formed in an intraplate orogenic setting.

Late Cretaceous intrusions include the Gaojiayao and Xinglongcheng plutons, consisting of diabase and belonging to the High-K calc-alkaline to Shoshonite series. Their magmas originated from the upper mantle and were formed in a nonorogenic tectonic setting following the Yanshan intraplate orogenic movement.

B. Intermediate-Acidic Intrusions: Building Continental Crust

Intermediate-acidic intrusions dominate the plutonic record of Beijing, forming large, complex abyssolith/complexes with diverse lithology, lithofacies, and contact zones. Emplaced episodically from the Neoarchean to the Mesozoic, these intrusions are linked to five major economic mineral deposit types: magmatic deposits (Crystallization directly from the magma), hydrothermal deposits (Precipitation from hot, mineral-rich fluids associated with the intrusions), contact metamorphic (skarn) deposits (Formed by metasomatism at the contact between the intrusion and surrounding rocks), porphyry deposits (Associated with porphyritic intrusions and characterized by disseminated mineralization), and nonmetallic deposits (Industrial minerals such as feldspar and mica).

a. Neoarchean complexes

Several acidic intrusions, including the Shachang tonalite gneiss series (Dacao, Shachang, Yangpodi, and Liulimiao plutons), the Songyingzi tonalite gneiss series (Dabeiwan, Mujiayu plutons, and others), the Jingziyu granodioritic gneiss series (Jingziyu and Wangjiawopu plutons), the Kanggezhuang trondhjemite gneiss series (Kanggezhuang and Detiangou plutons), and the Qiganding granitic gneiss series (Qiganding, Heshangyu, Tufengou, and Chedaogou plutons), were formed during this period. These rocks have undergone significant metamorphism, deformation, and migmetization, and belong to the calc-alkaline series. The Shachang tonalite gneiss series exhibits a genetic relationship with the supracrustal rock type and was formed in an immature island arc tectonic setting. In contrast, the Songyingzi tonalite gneiss series, Jingziyu granodioritic gneiss series, the Kanggezhuang trondhjemite gneiss series, and Qiganding granitic gneiss series constitute the TTG (Tonalite-Trondhjemite-Granodiorite) assemblage. The magmas originated from the partial melting of lower crustal materials and were formed in a tectonic setting of a mature continental marginal arc or an island arc-trench system.

b. Mesoproterozoic complexes

Mesoproterozoic complexes in Beijing consist of five Intermediate-acidic plutons, Laowa, Gudonggou, Dongxindian, Lanying, and Shachang. These plutons were composed of adamellite, quartz syenite, moyite, rapakivi granite, and biotite granite, belonging to the Shoshonite series. They are distributed along the Chicheng-Gubeikou and the Miyun-Qiangzilu-Xinglong deep faults in a near latitudinal direction. These plutons intrude into the Neoarchean metamorphosed basement with well-defined intrusive contacts. The Shachang pluton exhibits synchronous granite-porphyry dikes that are overlain unconformity by the Mesoproterozoic Changzhougou Formation. The magmas responsible for these intrusions originated from partial melting of lower crustal material and experienced significant fractional crystallization during their emplacement. These rock assemblages, along with anorthosite and mangerite, form the AMCG (anorthosite-mangerite-charnockite-ganite) assemblage, representing a pristine slow extension of the Yanshan Rift Basin.

c. Late Paleozoic complexes

Late Paleozoic complexes in Beijing are represented by four Intermediate-acidic plutons: Lishugou, Qiaomaiyu, Lishuwa, and Laogongying plutons. These plutons were composed of pegmatoidal syenite, syenite, and syenitic porphyry. While intrusive facies plane determination is challenging, intrusive vein or apophysis, and gradational contact of lithology were observed, suggesting episodic multiphasic emplacement. These plutons are distributed in a near east-west orientation along the Chicheng-Gubeikou discordogenic faults. These plutons intruded into the Mesoproterozoic Changcheng system and Neoarchean metamorphosed basement and was unconformably overlain by the Jurassic Tiaojishan Formation. The rocks exhibit an alkalic series and are indicative of magmatic arc volcanite. Geochemical analysis suggests that the magmas originated from metamorphosed basement in the lower crust, with the addition of mantle material. Their emplacement occurred in a long-distance active continental margin tectonic setting on the northern margin of the North China Craton, representing the magma intrusion along discordogenic faults.

d. Mesozoic complexes

Mesozoic magmatism marks a period of intense plutonism in Beijing, responsible for over 210 intrusions that constitute the Badaling magmatite belt. They are classified into nine distinct periods based on their intrusive dating: Early Triassic, Middle Triassic, Late Triassic, Early Jurassic, Middle Jurassic, Late Jurassic, early stage of Early Cretaceous, Middle stage of Early Cretaceous, and Late stage of the Early Cretaceous, belonging to the Indosinian cycle and the Yanshan cycle. Mesozoic intrusions exhibit a diverse lithology, encompassing nearly all intermediate to acidic rock types, ranging from alkaline to High K calc-alkaline series. The majority of the magma originated from the lower crust, with a minor contribution from the upper crust and mantle. Their magmatic evolution was complex, showing an abyssolith of large-scale complexes. The Indosinian cycle magmatic activity, associated with the collision orogenesis of the Mongolian accretionary fold belt (Central Asian Orogenic Belt), had a long-distance impact on the interior of the North China Block. This activity played a crucial role in the early stages of the North China Craton's destruction. The Yanshan cycle magmatic activity provides evidence of the crust-mantle evolution process within the Yanshan intraplate orogeny.

C. Magmatism and Crustal Evolution

Magmatic activitiesin Beijing provide insights into the evolution of the continental crust. Eight types of basic to acidic magmatite were emplaced during the Neoarchean, representing the progression of ancient crust from the immature interoceanic arc (Shachang tonalite gneiss) to semimature continental arc (Heiwo gabbro gneiss and Weiziyu gabbro diorite gneiss) to mature continental marginal arc (Songyingzi tonalite gneiss, Jingziyu granodioritic gneiss, and Kanggezhuang trondhjemite gneiss) to continental collision tectogene magmatic arc (Qiganding granite gneiss) in the Beijing. This sequence reflects the transformation of the initial basaltic crust into a semi-mature continental crust. The presence of a few swarms of ultrabasite and basite in the Paleoproterozoic indicates weak intrusive activities and reflects the maturation of the differential continental crust. Five types of basic to acidic magmatite developed during the Early to Middle Mesoproterozoic, including Gaoling gabbro dike swarm, bimodal cosets between anorthosite and rapakiwi granite, and volcanic rock of the Dahongyu Formation. These rocks represent magmatic activities in the Yanshan Rift Basin, indicating the extensional rifting and crustal lithosphere separation of the mature ancient continental crust during this era. The rifting event did not evolve into later oceanic crust or island arcs and terminated prematurely. During the Late stage of the Mesoproterozoic to Late Paleozoic, magmatic activity in Beijing is lacking, suggesting a highly stable ancient continental crust. Since the Late Paleozoic, multiple tectonic-magmatic events occurred, such as the Middle Devonian, Early Permian, Triassic, Jurassic, Early Cretaceous, Late Cretaceous, and Palaeogene. Magmatic activity, particularly since the Early Jurassic, formed distinct magmatite belts, recording the processes of the Yanshan orogeny, such as island arc, collision, post-collision, and evolution of the crust and mantle.

5. Metamorphic Geology

Part five focuses on the diverse metamorphic rocks in Beijing, encompassing five chapters dedicated to geologic characteristics, petrology and mineralogy, lithochemistry and geochemistry of these rocks, metamorphic facies, and metastructure subdivisions. Additionally, the section reconstructs the protoliths of these rocks and explores their formation environments.

A. Metamorphic rocks

The metamorphite in Beijing are broadly classified into three types: supracrustal rock type, metamorphic intruded rock type, and epimetamorphic rock type.

Supracrustal rock type encompasses the Miyun and the Sihetang groups. The Miyun Group characterized by the granulite facies metamorphism, displays a variety of lithologies, including magnetite quartzites, amphibolite, pyriclasite, pirigarnite, piribolite, hornblendite, and leptynite, often exhibiting lenticular or banded structures. The Sihetang Group stands out as a distinct rock association composed of plagiogneiss, amphibolite, and leptynite, exhibiting amphibolitic facies and relatively weak migmatization dominated by Na metasomatism.

Metamorphic intruded rock encompasses two main rock associations: gabbro-gabbro diorite-tonalite gneiss and tonalite-trondhjemite-granitic gneiss. The former includes Shachang tonalite gneiss, Heiwo gabbro gneiss, Weiziyu gabbro diorite gneiss, and Songyingzi tonalite gneiss, while the latter encompasses Jingziyu tonalite gneiss, Kanggezhuang trondhjemite gneiss, and Qigandiding granitic gneiss.

Epimetamorphic rock types, primarily of Mesoproterozoic to Late Jurassic lithostratigraphy, are prominent in the Xishan area. These rocks consist of slightly metamorphosed lithologies, including schist, slates, phyllite, and metasandstones. The widespread occurrence of chloritoid points to greenschist facies or amphibolite facies metamorphism. Notably, variations in tectonic settings and the intensity of thermal activities have resulted in a spatially and temporally heterogeneous distribution of metamorphic zones within these rocks.

B. Regional metamorphism

Metamorphism records a complex metamorphic phase spanning multiple periods, as evidenced by the characteristic metamorphic assemblage, mineral generation and metasomatic reaction, and thermobaric condition. Six distinct Regional metamorphism phases have been identified: Fuping, Wutai, Lvliang, Haixi, Indosinian, and Yanshanian. Each phase corresponds to a specific metamorphic facies: granulite - amphibolite facies (Fuping), amphibolite - epidote amphibolite facies (Wutai), amphibolite - epidote amphibolite facies (Lvliang), epidote amphibolite - greenschist facies (Haixi), greenschist facies (Indosinian), and greenschist facies (Yanshan). The successive metamorphic events have overprinted earlier ones, resulting in the complex metamorphic assemblage observed today. Neogenic metamorphic processes are superimposed on older ones, creating a rich tapestry of metamorphic textures and mineral assemblage.

C. Other metamorphic rocks and metamorphism

In addition to the Regional metamorphism, local metamorphism has also shaped the rocks in Beijing, including contact metamorphism, dynamometamorphism, hydatopneumatic metamorphism, and migmatization. Contact Metamorphism, driven by magmatic intrusion, manifests in three forms: 1) Thermal Contact Metamorphism, characterized by recrystallization and the formation of new mineral assemblages due to the heat from intrusion; 2) Low-Temperature Metasomatism, involves the transfer of elements and fluids between the intrusion and surrounding rocks, leading to metasomatic alteration and the formation of skarns; 3) Fluid-Rock Interaction/Alteration, represents a broader category of alteration processes linked to magmatic fluids, encompassing both contact metasomatism and hydrothermal alteration. Dynamometamorphism is associated with deformation along major fault zones, such as the Banchengzi, Shicheng, and Qifengcha faults. Hydatopneumatic metamorphism is pervasive within and around intermediate to acidic intrusion or explosive volcanic breccias, and metasomatism reflects hydrothermal alteration processes active during the later stages of magmatism. Hybridization occurs in two main settings: 1) Intrusive Marginal Hybridization, localized at the contact between intrusion (e. g. , Siganding, Baichashan, and Yunmengshan) and their host rocks, reflecting heat transfer and fluid flux from the cooling magma; 2) Regional Hybridization, widespread within the Neoarchean strata, reflecting a more regionally extensive thermal event.

D. Metamorphic Facies Zones and Structural Zonation

The metamorphite in Beijing can be classified into three distinct metamorphic facies zones, further subdivided into nine litho-mineral zones, including the clinopvroxene-garnet faces zone and two-pyroxene facies zone within the Miyun Group, the low-grade amphibolite faces zone and the high-grade amphibolite facies zone within the Sihetang Group, the chloritoid-chlorite zone, biotite zone, andalusite zone, almandine-staurolite zone, and silimanite-rnuscovite zone within the Late Paleozoic to Early Mesozoic metamorphic rocks. Beyond these facies subdivisions, the metamorphism is also organized into two major and three minor metamorphic structural zonations based on crustal evolution, tectonic cycles, and relationship to magmatic events. Major structural zonations are the Beishan (Shachang-Dacao and Sihetang subzones) and Xishan structural zonation. These subdivisions reflect the intricate interplay of deep-seated intrusive activity, metamorphism, structural deformation, and tectonic uplift that shaped the region's metamorphic architecture.

6. Lithospheric Structure

This part synthesizes geophysical data, including deep seismic profiles, electromagnetic depth sounding, and tomographic inversion, to illuminate geophysical field and lithogeochemistry, and lithospheric structures, revealing strong lateral heterogeneity.

A. Geophysical field

Distinct layered structures are evident in the crust and upper mantle, with pronounced lateral heterogeneity in velocity. These variations correlate strongly with surface geological structures. Notably, prominent velocity gradients characterize deep fault zones, especially at particularly along the Yanqing Basin and the Taihang Mountains foreland region, highlighting active deformation.

Situated within the North-Northeast strike Daxing'anling-Taihang Mountains-Wuling Mountains gravity gradient zone, the region exhibits a complex gravity field with densely spaced, alternating high and low Bouguer gravity anomalies. Values decrease northwestwards, ranging from 5×10^{-5} m/s^2 to -120×10^{-5} m/s^2. Active fault zones coincide with pronounced gravity gradient, with the largest horizontal gradient variation (1×10^{-5} m/(s^2 · km)), observed in the Taihang Mountains piedmont. The regional gravity field can be subdivided into three zones: the Xishan Hills gravity gradient abnormaly zone, the Plain gravity high abnormaly zone, and the Northern Mountain gravity low abnormaly zone.

Aeromagnetic data reveal a complex pattern of alternating zones of high-intensity positive and negative magnetic anomalies. The magnetic field generally weakens outwards from the Miyun-Huairou-Pinggu high anomaly zone. Positive anomalies range from 100 to 400 nT, while negative anomalies vary from -100 to -200 nT. Based on the anomaly characteristics, the region can be divided into four magnetic zones: the Huaibei, the Miyun-Huairou, the Yanqing-Changping, and the Plain magnetic field zones.

Five distinct electrical layers are identified within the crust and upper mantle in Beijing: a high-conductivity crustal surface layer (1~10 Ω · m), a high-resistivity upper crust layer (10^2~10^3 Ω · m), a high-conductivity crust layer (1~10 Ω · m), a lower-resistivity lower crust layer (10^2~10^5 Ω · m), and s high-conductivity the upper mantle (1~10 Ω · m) layer.

Heat flow values are generally lower in the Beijing mountainous area (<60 mW/m^2), increasing within anticlinal cores and active fault zones. The Beijing Plain area exhibits higher heat flow values (>60 mW/m^2), with a geothermal anomaly characterized by $(1.7\sim2.0)\times 41.686$ mW/m^2.

The present-day stress field in Beijing is dominated by horizontal compression, with a maximum principal stress direction oriented north-northeast. This stress regime governs the ongoing crustal deformation patterns observed in the region.

B. Mantle and Crustal Composition

Geophysical surveying and mantle xenolith studies provide insights into the lithospheric composition. The mantle is primarily composed of akerite /garnet lherzolite, pyroxenite, and websterite. The crustal composition reflects the metamorphic zonation: the lower and middle crust consist mainly of granulite and amphibolite facies rocks, while the upper crust is dominated by greenschist facies rocks, locally transitioning to low-grade amphibolite facies. Notably, localized low-velocity zones within the crust likely represent magma chambers associated with granitic intrusions.

C. Deep Structure and Geological Controls

The Moho interface and isodepth contours generally trend northeast-southwest and east-west. Integrating data on Moho depth, Precambrian basement depth, Bouguer gravity anomalies, deep seismic profiles, and aeromagnetic patterns reveals two primary deep geological structure zones in Beijing: the Taihang Mountain-Yan Mountains Mantle Slope Zone and the Beijing Plain Mantle Uplift Zone. Structural intersections and zones of enhanced permeability within these deep structures focus fluid flow and contribute to ore formation. The deep structure influences regional topography, basin development, and the distribution of sedimentary facies.

7. Geological Structure

This part is divided into four chapters, providing a comprehensive overview of structural form, structural units, and structural cycle.

A. Structural form

This region exhibits complex structural forms characterized by two phases of basement folding and four phases of cover folding. Cover folding, primarily attributed to the Indosinian and early, middle, and late stages of the Yanshanian orogeny, represents the dominant structural form. Fault, another structural form, is well-developed, displaying intricate intersecting and cross-cutting relationships across various orientations, ages, and kinematics. Statistical analysis identified 45 named faults, including prominent regional faults such as the Chicheng-Changshaoying-Gubeikou-Chengde fault and the Nankou-Pinggu - Shanhaiguan fault. thrust belt, detachment faults, metamorphic core complexes, ductile shear zones, bedding rheid tectonic community, and dome and half-dome structures further contribute to the structural complexity.

B. Structural units

This region resides within the Qaidam-North China Block (Ⅰ level), specifically within the North China Craton (Ⅱ level). At a finer scale, it encompasses portions of the Yanshan-Liaoxi Rift and the North China Basin (Ⅲ level). These third-level tectonic units are further subdivided into four fourth-level and twelve fifth-level tectonic units, providing a detailed framework for understanding regional tectonics.

C. Structural cycle

The geological evolution is categorized into four stages, nine structural cycles, and corresponding structural layers. These include: ①basement structural layers (Fuping and Lvliang structural cycles), ②cover structural layers (Jinning, Yangtze, Kaledonian, Hualixi, and Indosinian structural cycles), ③superposed structural layers (Yanshan and Himalayan structural cycles). The Yanshan Movement plays a critical role in the region's tectonic development. An intraplate orogenic model divides the Yanshan Movement into four stages: pre-orogenic, syn-orogenic, late-orogenic, and post-orogenic. This model proposes that the Yanshan Movement resulted from asthenospheric upwelling and continental growth along the northern margin of the North China Craton, culminating in eastward delamination of the craton.

8. Quaternary Geology and Landform

This part presents a systematic analysis of the geomorphology, Quaternary deposits, and neotectonic movement in Beijing, emphasizing the interplay between landscape evolution and sedimentary processes.

A. Quaternary Stratigraphy

Distinct stratigraphic sequences characterize mountainous areas and cave deposits. In mountainous areas, the Quaternary is represented by the Nihewan, Zhoukoudian, and Malan formations, each defined by unique lithological and palynological characteristics. The Nihewan Formation, primarily found on fourth terraces, comprises reddish-brown, gray/yellow-brown sands, sand loam, and loam with interbedded sandy gravel, indicating a fluvial environment. The Zhoukoudian Formation, occupying third terraces, reflects diluvial, remain, and hillside processes, dominated by reddish-brown clay sands and sandy clays. The Malan Formation, primarily on the second terraces, consists of secondary loess, clay sand, sandy clays, and gravel lenses, indicative of diluvial, alluvion, hillside, and remain.

Cave deposits reveal a different stratigraphic succession: Taipingshan, Zhoukoudian, and Yangeryu formations. The Taipingshan Formation, characterized by variable brownish-yellow and yellowish tones and abundant mammal fossils, highlights the unique depositional environment within every cave. The Zhoukoudian Formation, primarily reddish-brown clay, sandy clay, and rubble with the characteristic Zhoukoudian fauna, suggests a shift towards more open conditions. The Yangeryu Formation, comprising gray, yellow, and yellow-brown rubble, silt, and sandy clay, with abundant mammalian fossils, indicates continued deposition within cave systems.

Holocene deposits, primarily found in river valleys and piedmont zones, such as the Yongding, Chaobai, Dashi, Qingshui, and Cuo rivers, consist of brownish-yellow to gray sands, gravels, and clays, representing alluvial, colluvial, and hillside processes.

B. Quaternary Basin

The Beijing Plain area and Yanqing Basin exhibit complex Quaternary sedimentation patterns within multiple depressions, including Machikou, Houshayu-Fengbo, and Pinggu. Varied tectonic settings and sediment sources result in significant lateral facies changes, posing challenges for stratigraphic correlation. This study establishes lithological, magnetic, and biostratigraphic frameworks for each depression to decipher the Quaternary evolution of these basins.

Sedimentary environments in the Beijing Plain transitioned from Early Pleistocene alluvial fan systems dominated by coarse gravels to Middle-Late Pleistocene fan-margin depressions and low-lying river-lake environments characterized by sandy clays and silty sands. Holocene sedimentation primarily involved braided rivers on alluvial fans and abandoned channel-lake systems.

C. Geomorphology and Neotectonics

The region's northwest high, and southeast low topography reflects the interplay between uplift and erosion in the mountainous areas and deposition in the plains. The Cenozoic basin and range landscape, encompassing the Yan Mountains, the Taihang Mountains, the Yanqing Basin, and the Daxing Uplift, Beijing Basin, provided the framework for Quaternary landscape evolution. Quaternary landforms, shaped by fluvial incision and deposition, are represented by multi-level planation surfaces, river terraces, and alluvial fans.

Intense neotectonic activity, characterized by intermittent uplift and subsidence, is evident in the region's geomorphology. Multi-level planation surfaces and river terraces in the mountainous areas, alongside sedimentary discontinuities, and shifting depositional centers in the Beijing Plain, attest to ongoing tectonic deformation. Active faults, including the Huangzhuang-Gaoliying, Shunyi, Xiadian, and Nankou-Sunhe faults, highlight the potential for seismic hazards in Beijing.

9. Three-dimensional Geologic Model of the Beijing Plain

This part includes three chapters, and summarises the three-dimensional geological structure of the Beijing Plain, focusing on its bedrock geology, Cenozoic Geology, applied geology, and three-dimensional geologic model.

A. Bedrock Geology

The lithostratigraphy, buried beneath the Cenozoic strata, generally correlates with those of adjacent mountainous areas in Beijing. A relatively complete stratigraphic sequence is present, encompassing strata from the Precambrian to the Mesozoic. However, Palaeoprotozoic, Upper Silurian to Lower Carboniferous (Paleozoic) stratas are notably absent. Several intrusions, the Zaixiangzhuang, Yangfang, Jiaodao, and Yaoshang, are also identified within the bedrock.

The buried bedrock exhibits fold structures primarily attributed to the Indosinian and Yanshanian movements. The Indosinian folds include the Xiaotangshan, Banqiao, Kunminghu anticlines, and Fengheying syncline. The Yanshanian folds include the Huangcun, Yufa-Lixian, Tongzhou, Zhangjiawan anticlines, and Beixiaoying-Changping, Huilongguan-Pingxifu, Laiguangying-Nanfaxin, Dasungezhuang synclines. The bedrock is dissected by a well-developed fault system, predominantly comprising three sets: East-West trending (e. g. the Tongbai and Pigezhuang faults, Northeast trending (e. g. the Huangzhuang-Gaoliying, Nanyuan-Tongzhou, Xiadian, and Lixian faults), and Northwest trending (e. g. the Nankou-Sunhe and Yongdinghe faults).

B. Cenozoic Geology

Paleogene and Neogene strata are primarily concentrated within the Beijing and Dachang sags. Within the Beijing Sag, the sedimentation is restricted to the zone between the Huangzhuang-Gaoliying and the Nanyuan-Tongzhou faults, exhibiting a northeastward trend. The Paleogene sedimentary center, characterized by a maximum thickness exceeding 400 meters, is located around the Wanping, Lugouqiao, and Fengtai. The Neogene center, with a thickness exceeding 1700 meters, lies along the Beitiantang-Baotai-Sihezhuang line. Multiple large alluvial fans and sedimentary basins dominate the Quaternary geology. Notable basins include the Machikou, Houshayu, Shunyi, and Pinggu basins. A significant transition in sediment characteristics is observed from the piedmont regions towards the basin centers, marked by a decrease in grain size, thinning of sediment thickness, and an increase in interbed.

C. Applied geology

The superficial part (0-30 meters depth), representing the primary engineering layer in Beijing, displays distinct lateral variations. A thick, homogeneous gravel layer in the western Piedmont transitions eastward towards Tongzhou into a complex sequence of interbedded sand and clay layers with variable thicknesses. Based on factors such as crustal stability, landforms, sola-bearing capacity and compressibility, groundwater conditions, and geological hazards, the superficial part of Beijing Plain can be divided into six engineering geological zones: the Daqing River system, the Yongding River system, the Beiyun River system, the Chaobai River system, the Ju River system, and the erosion and tectonic zone.

D. Three-dimensional Geologic Model

The "Beijing Multiparameter Stereoscopic Geological Survey" project utilizes a combination of geophysical and drilling data to develop a comprehensive three-dimensional geological framework. Utilizing software such as ACCESS, MAPGIS, and V3DMain, the project constructed hierarchical 3-D models for the bedrock, Quaternary deposits, and engineering layers. These models provide a powerful tool for visualizing strata structure, managing geological data, and supporting engineering and planning decisions.

10. Urban Geological Resources and Environment

This part is divided into seven chapters, and provides a systematic overview of Beijing's key urban geological resources and environment, highlighting their distribution, characteristics, potential for utilization, and associated challenges. Additionally, this section showcases the progress made in developing a multi-dimensional data model for urban geological and environmental management in the Tongzhou district.

A. Geological Resources

a. Underground cavity Resources

The primary areas suitable for underground cavity utilization in Beijing Plain are concentrated within the Fifth Ring Road. Based on geological and engineering factors, underground cavity quality is classified into three grades: Grade I (suitable), Grade II (relatively suitable), and Grade III (not suitable). Encouragingly, over 70% of the areas within a depth of 0-50 meters are deemed suitable for utilization. However, factors such as topography, geomorphology, hydrogeology, active faults, ground fissures, subsidence, soil liquefaction potential, and radioactive gases all pose challenges to underground cavity utilization and necessitate careful planning and mitigation measures.

b. Groundwater Resources

Situated within the Haihe River Basin groundwater system, Beijing encompasses three relatively independent groundwater systems: the Chaobai-Jiyun-Wenyu River system, the Yongding River system, and the Dashi-Juma River system. The Quaternary aquifer system underlying the Beijing Plain is the city's primary water source, transitioning from a single aquifer unit adjacent to the mountain to a complex of four aquifer groups across the plain. Although groundwater resources are generally abundant, with an average annual recharge of $23.59\times108m^3$ between 1991 to 2014, consumption ($28.46\times108m^3$) exceeds recharge, emphasizing the importance of sustainable management strategies.

c. Mineral Resources

Beijing possesses a diverse range of mineral resources, with 127 mineral commodities (including sub-mineral commodities) identified as of the end of 2014. 121 are classified as solid minerals. with 67 solid minerals and 354 mineral sites with confirmed reserves listed in the "Beijing Mineral Resource Reserves Table."

d. Energy Resources

Geothermal and shallow geothermal energy represents the most promising clean energy resources within Beijing. Geothermal resources are primarily of medium-low temperature and sub-economic viability. The Jixian System constitutes the principal heat storage layer, followed by the Cambrian and Ordovician, while the Neogene strata serve as the primary cap rock. The total estimated heat storage capacity of geothermal fields is $500.772\times10^{18}J$. Shallow geothermal energy resources, influenced by topography, hydrology, geotemperature, and geological structures, are mainly found within the sola and groundwater at depths of 15~200 meters. Water source heat pumps are the dominant utilization method, with approximately 1, 073 shallow geothermal energy projects, covering a heating area of 33.48 million square meters, operational by the end of 2012.

e. Geological Heritage Resources

Beijing boasts rich and diverse geological heritage resources, encompassing 50 sites of significance. These sites are categorized into three primary groups, which are further subdivided into twelve categories and twenty-four subcategories. Two world geoheritage sites, the Zhoukoudian Peking Man Site, and the Qianjiadian Jurassic Dinosaur Footprint Fossil Site, highlight the region's global geological significance. An additional seven sites hold the national designation, while the remaining 41 are classified as provincial. Reflecting the commitment to conservation and education, Beijing has established eight geological parks as of 2014, including two world, five national, and one municipal parks.

B. Geological Environment

a. Soil Quality

The soils in the Beijing Plain are predominantly alluvial in origin, shaped by the depositional processes of major rivers, including the Chaobai, Yongding, Wenyu, and Ju rivers. Consequently, sediment provenance plays a significant role in influencing background soil element concentrations. Overall, soil quality is generally good, particularly in areas characterized by well-developed soil profiles. These areas include the piedmont plains, such as Pinggu, Changping, and Fangshan, and the regions flanking the Wenyu, Liangshui, and Fenggang Jian rivers within Tongzhou. However, localized soil contamination is evident, primarily attributed to heavy metal pollution, such as Hg, and persistent organic pollutants like DDT and HCH.

b. Groundwater Quality

Groundwater quality in Beijing is generally favorable, exhibiting a spatial trend of improvement from south to north and with increasing depth. This pattern reflects the influence of recharge sources and natural attenuation processes. However, ongoing groundwater development and utilization have introduced several concerns, including declining water levels, localized pollution, and a reduction in the aquifer system's ability to meet demands. These challenges underscore the need for adaptive management strategies to ensure the long-term sustainability of this vital resource.

C. Geological Hazards

a. Ground subsidence

Ground subsidence isa major slow process hazard across the Beijing Plain, characterized by two main subsidence zones and seven subsidence centers. Subsidence rates show an increasing trend, posing significant risks to infrastructure and urban development. Both natural factors, including tectonic movements, and formation lithology, and anthropogenic factors, such as excessive groundwater withdrawal and engineering activities, contribute to subsidence.

b. Ground Fissures

Several prominent ground fissures, including Shunyi, Gaoliying, Beixiaoying, Yangfang, Miaojuan, Beicai, Beidian, and Sunhe fissures, transect the Beijing Plain. While most are classified as tectonic in origin, others are induced by soil liquefaction processes. Excessive groundwater extraction is a key factor exacerbating fissure development and activity.

c. Sand Liquefaction

Areas susceptible to liquefaction are concentrated in planitia along the middle and downstream reaches of rivers, such as the Chaobai, Wenyu, Xiaozhong, and Ju rivers, where young, unconsolidated sediments are prevalent. The complex interplay of dynamic loading, soil properties, burial conditions, groundwater fluctuations, stress regimes, and boundary conditions governs the occurrence, development, and dissipation of liquefaction hazards.

d. Mining-Induced Subsidence

Subsidence with coal mining activities is localized in western Beijing. Key contributing factors include geological conditions, the extent of underground voids, coal seam thickness, mining depth, repeated mining cycles, and topography. Gravity and other external triggers often initiate subsidence events.

e. Collapses and Debris Flows

Mountainous areas, particularly along road cuts, are highly susceptible to collapses, with sliding, toppling, and rockfalls being the most common failure modes. Debris flowspost significant threats in the northern

Miyun, central Huairou, eastern Yanqing, and western regions of Fangshan and Mentougou. Gully-confined, high-density, low-viscosity debris flows are prevalent. Steep topography, lithology, tectonics structures, terrain conditions, and climate all influence the formation of these hazards. Additionally, while less common, landslides, often triggered by human activities, pose localized risks.

11. Urban Multidimensional Data Model

A pilot project in Tongzhou District showcases the application of three-dimensional data modeling for urban geological and environmental management. This model integrates eight specialized sub-modules: Engineering Layers, Sand Liquefaction Potential, Underground Space Suitability, Groundwater Resources, Groundwater Quality, Ground Subsidence, Shallow Soil Geochemistry (0-20 cm), Shallow Geothermal Energy Potential. This integrated platform enables the visualization of spatial data layers, and querying of attribute information for each geological element, and provides a powerful tool for supporting urban planning and environmental decision-making.

12. History of Regional Geology Development

This part delineates the regional geological evolution of the Beijing area through four key tectonic stages: Neoarchean-Paleoproterozoic, Mesoproterozoic-Middle Triassic, Late Triassic-Early Cretaceous, and Late Cretaceous-Quaternary.

A. Neoarchean-Paleoproterozoic Stage

The Neoarchean witnessed intense seafloor volcanism, generating the Miyun and Sihetang groups through a combination of lava flows, volcanoclastic deposits, and granitic intrusions. Subsequent tectonic-thermal events during the late Fuping period resulted in east-west trending folding, uplifting, and regional metamorphism, characterized by granulite facies and ductile shearing. Contemporaneous mafic to felsic magmatism further modified the crustal architecture. These Neoarchean rocks constitute the crystalline basement of the region.

Paleoproterozoic evolution reflects a stable cratonic setting with limited sedimentation. Minor ultramafic-mafic intrusions, manifested as plutons (e. g. Hongshiwan) and dike swarms, intruded along pre-existing structures. Shallow ductile shearing and greenschist facies metamorphism indicate localized deformation.

B. Meso proterozoic to Middle Triassic Stage

The Early Mesoproterozoic witnessed significant crustal extension across the North China Craton, driven by mantle asthenosphere adjustments. This extension reactivated pre-existing Paleoproterozoic sutures and deep faults, resulting in the emplacement of mafic-ultramafic rocks (e. g. Gaoling dike swarm) and the eruption of the Dahongyu volcanics. These magmatic events, sourced from both the mantle and partially melting lower crust, marked a transition from a stable cratonic regime to an extensional rift tectonic setting. This shift facilitated the development of the extensive Yanshan Rift Basin, where the Changcheng System was deposited along major fault systems like the Shangyi-Longhua and Kangbao-Weichang. Concurrently, the Earth's atmosphere began to evolve, with the rise of an oxygenic atmosphere and the proliferation of early algal life.

By the Middle Mesoproterozoic, the Yanshan Rift Basin gradually subsided as mantle activity waned and regional crustal stress dissipated. The tectonic regime shifted from extension to uneven subsidence, resulting in the formation of shallow basins with transitional characteristics. Sedimentation continued, represented by strata from the Gaoyuzhuang Formation to the Wumishan Formation. However, reduced tectonic energy limited magmatic activity and precluded significant deformation or metamorphism during this period.

From the Late Mesoproterozoic to the Middle Triassic, the region remained largely stable, characterized by broad crustal uplift and subsidence cycles. Intense tectonic deformation, magmatic activity, or metamorphism were generally absent. Periods of uplift exposed the region to prolonged erosion and planation, while subsidence led to the development of an epicontinental basin. This basin became the site of extensive quartz sandstone, carbonatite, and conglomerate formations. Notably, the northern Beijing, including northern Huairou and Miyun areas experienced periodic extensional faulting along the Shangyi-Longhua discordogenic faults due to remote stresses from the Mongolian-Siberian Plate collision and subduction toward the Huabei Plate. This resulted in miniature, episodic magmatic activity and contact metamorphism during the Middle Devonian, Early-Middle Permian, and Early Triassic.

C. Late Triassic-Early Cretaceous Stage

The Late Triassic to Early Cretaceous marked a period of significant intraplate orogenesis in Beijing, manifested as the development of the Yanshanian Orogenic belt. This tectonic event resulted in the formation of three distinct volcanic-sedimentary basin episodes: the Xingshikou-Jiulongshan, Tiaojishan-Tuchengzi, and Zhangjiakou-Xiazhuang. These basins exhibit a counterclockwise spatial progression, with their orientations shifting from NEE to NE and finally to NNE. This period was characterized by complex tectonic deformation, including significant NNE trending folding and faulting. Multiple episodes of magmatic activity, both intrusive and extrusive, represent some of the most significant magmatism in the region's history. The late-orogenic and post-orogenic magmatism played a crucial role in the region's metallogeny, directly influencing the formation and distribution of endogenous metallic mineral deposits.

D. Late Cretaceous-Quaternary Stage

The Late Cretaceous to Quaternary witnessed the influence of the evolving Himalayan tectonic belt, ushering in a predominantly extensional tectonic regime across the region. The Late Cretaceous to Paleogene marked a post-orogenic phase for the Yanshanian movement, characterized by regional uplift and erosion that sculpted the Beitai planation surface. By the Eocene to Miocene, under an extensional tectonic setting, this peneplain underwent fragmentation and disintegration along pre-existing faults. Notably, the Huangzhuang-Gaoliying, Nanyuan-Tongzhou, and Xiadian-Lixian faults play a key role in shaping a basin-range structure, such as the Beijing Sag, Dachang sag, Jingxi Uplift, and Daxing Uplift. From the Neogene to the Quaternary, the region experienced weak extension and strike-slip tectonics. This reactivation impacted pre-existing northeast-trending faults and facilitated the development of new or reactivated northwest-trending faults, such as the Nankou-Sunhe fault. This tectonic reconfiguration further modified the basin-range structure, leading to the formation of new sedimentary depocenters, exemplified by the Machikou Depression and the Shunyi Depression.

图版 II

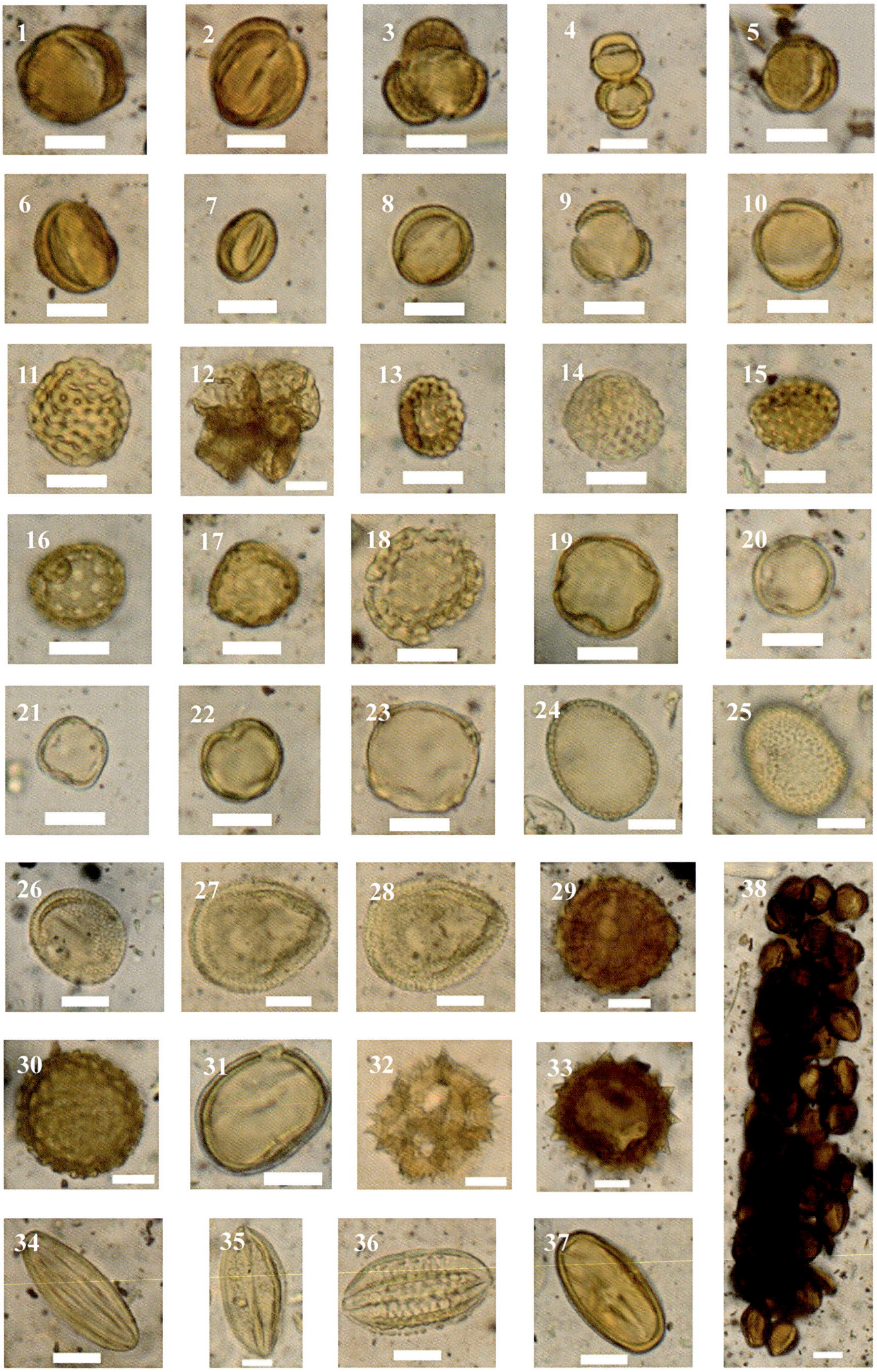
1
2
3
4
5
6
7
8
9
10
11
12
13
14
15
16
17
18
19
20
21
22
23
24
25
26
27
28
29
30
31
32
33
34
35
36
37
38

图版Ⅱ（续）

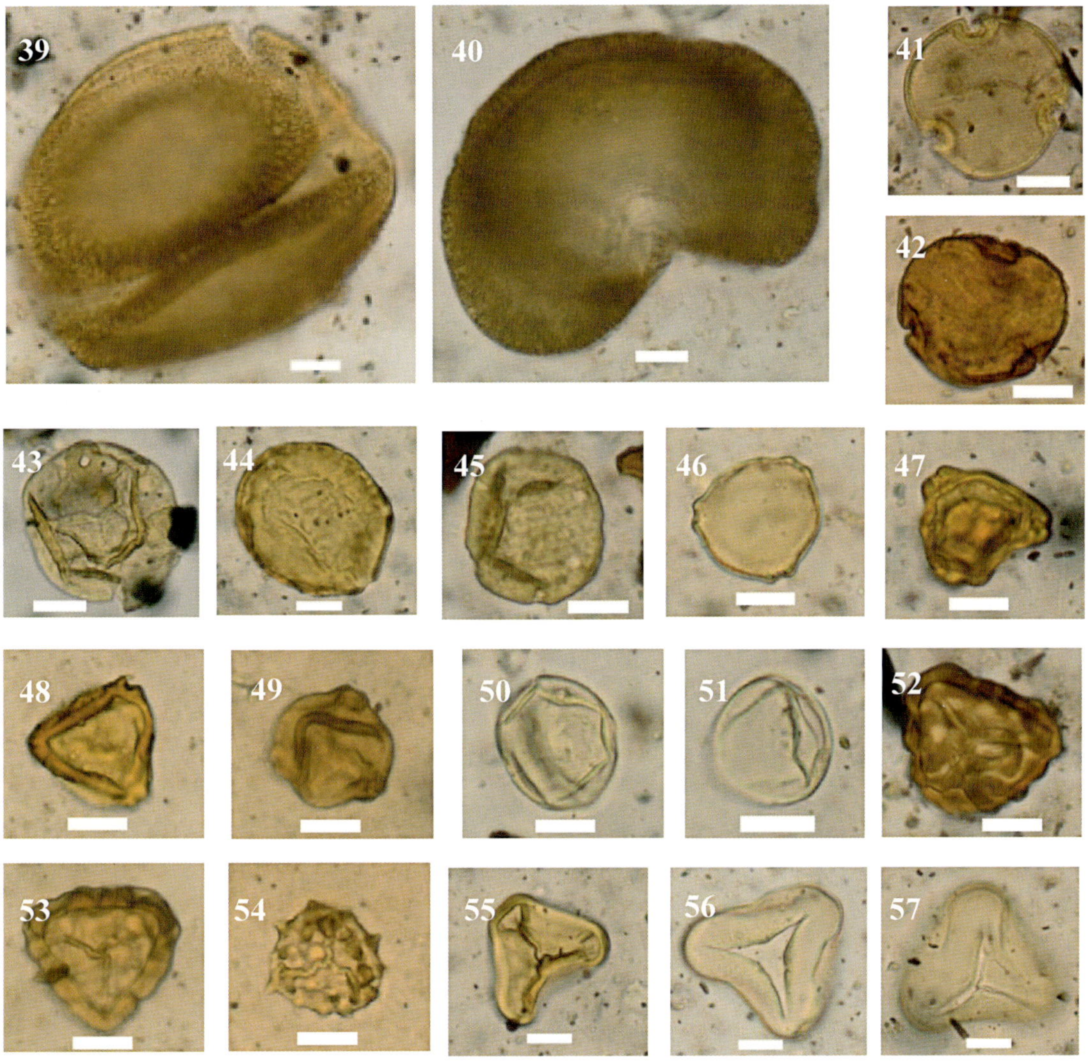

图版Ⅲ

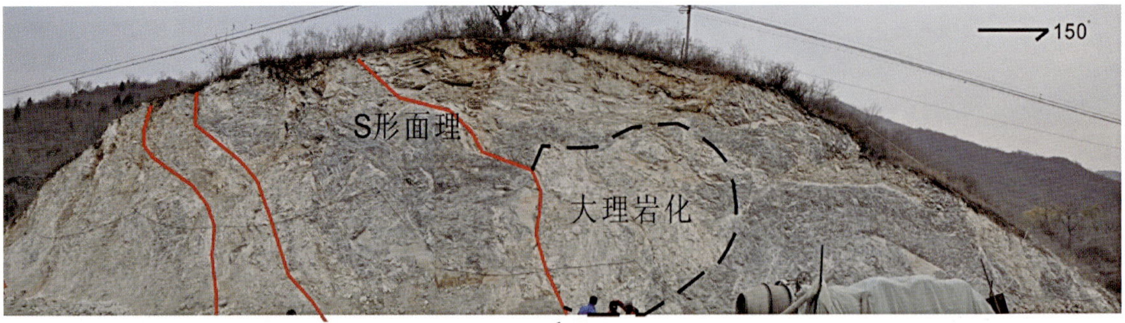

1

2

3

4

5

图版Ⅳ

1

2

图版Ⅳ（续）

3

4

图版Ⅳ（续）

5

6

图版Ⅳ（续）

7

8

图版 V

1

2

图版Ⅵ

1

2

3

图版Ⅵ（续）

4

5

图版Ⅶ（续）

1

2

图 版 说 明

图版 I

1. *Cervus* sp. （鹿属牙齿）
产地为东岭子洞第 6 层，时代为 Qp^3
2. *Cervus* sp. （马鹿牙齿）
产地为东岭子洞第 6 层，时代为 Qp^3
3. *Lepus wongi*（翁氏野兔牙齿）
产地为东岭子洞第 7 层，时代为 Qp^3
4. *Nyctereutes procyonides*（貉的牙齿）
产地为东岭子洞第 6 层，时代为 Qp^3
5. *Microtus brandtioides*（拟布氏田鼠牙齿）
产地为东岭子洞第 6 层，时代为 Qp^3
6. *Canis* sp. （犬属牙齿）
产地为北京太平山洞穴，时代为 Qp^1
7. *Allocricetus teilhardi*（德氏异仓鼠牙齿）
产地为东岭子洞穴，时代为 Qp^1
8. *Cricetulus* sp. （田鼠属牙齿）
产地为东岭子洞穴，时代为 Qp^1

图版 II

1-10. *Artemisia*（蒿属花粉）;
11-18. Chenopodiaceae（藜科花粉）;
19-22. *Thalictrum*（唐松草属花粉）;
23. *Myriophyllum*（狐尾藻属花粉）;
24-25. *Typha*（香蒲属花粉）;
26-28. *Potamogeton*（眼子菜属花粉）;
29. *Saussurea*（风毛菊属花粉）;
30-31. *Polygonum*（蓼属植物花粉）;
32. *Taraxacum*（蒲公英属花粉）;
33. *Aster*（紫菀属植物花粉）;
34-36. *Ephedra*（麻黄属植物花粉）;
37. *Nitraria*（白刺属花粉）;
38. *Artemisia* 花粉块（艾属花粉）;
39. *Picea*（云杉属花粉）;
40. *Cedrus*（雪松属花粉）;
41-42. *Tilia*（椴树属花粉）;
43-45. *Ulmus*（榆属花粉）;
46. *Corylus*（榛属花粉）;

47-49. *Betula*（桦木属花粉）；
50-51. Cupressaceae（柏科花粉）；
52-53. *Selaginella sinensis*（中华卷柏蕨类孢子）；
54. *Selaginella*（卷柏属蕨类孢子）；
55-57. Other ferns（其他蕨类植物孢子）；
（注：图版2中1-57标尺均为10μm，源于北京市地质调查研究院北京昌平区百善东沙屯村钻孔ZK3。）

图版Ⅲ

1. 十三陵推覆构造前锋带构造
2. 十三陵推覆构造带的膝折构造
3. 十三陵推覆构造前锋带倒转褶皱
4. 石景山潭峪村断裂擦痕与阶步
5. 石景山潭峪村断裂地表张裂隙

图版Ⅳ

1. 门头沟下苇店景儿峪组与昌平组接触关系（平行不整合），苏德辰供图
2. 门头沟九龙驾校窑坡组与龙门组接触关系（平行不整合），张宏仁供图
3. 门头沟雾迷山组白云岩层面发育的波痕，苏德辰供图
4. 门头沟下苇店炒米店组竹叶状灰岩
5. 门头沟发现的液化溢出丘，苏德辰供图
6. 门头沟大村斑脱岩
7. 顺义高丽营黄庄-高丽营断裂
8. 左：化石点上部的足迹群；右：鸟脚类恐龙的“三叶草”足迹

图版Ⅴ

1. 通州辛安屯附近被活动断裂错动的古墓（左）和涌入古墓的沙土液化脉（右）
2. 左：顺义天竺由顺义断裂引起的地表错动；右：延庆延矾断裂的破裂面

图版Ⅵ

1. 昌平南口双峰式脉岩-基性脉岩
2. 昌平南口双峰式脉岩
3. 房山十渡侵入蓟县系的煌斑岩脉
4. 平谷稻地村马兰黄土
5. 密云大城子常州沟组砾岩

图版Ⅶ

1-2. 朝阳区洼里砾岩